CELLULAR AND BIOCHEMICAL SCIENCES

CELLULAR AND BIOCHEMICAL SCIENCES

Editor

G. Tripathi

Associate Professor
Dept. of Zoology
J.N. Vyas University
Jodhpur, Rajasthan
India

I.K. International Publishing House Pvt. Ltd.

NEW DELHI • BANGALORE

Published by

I.K. International Publishing House Pvt. Ltd.
S-25, Green Park Extension
Uphaar Cinema Market
New Delhi – 110 016 (India)
E-mail: info@ikinternational.com
Website: www.ikbooks.com

ISBN 978-81-88237-85-2

10 9 8 7 6 5 4 3 2

Reprint 2012

Published by Krishan Makhijani for I.K. International Publishing House Pvt. Ltd., S-25, Green Park Extension, Uphaar Cinema Market, New Delhi – 110 016 and Printed by Rekha Printers Pvt. Ltd., Okhla Industrial Area, Phase II, New Delhi – 110 020.

Dedicated
to the Loving Memory of
My Mother

Preface

The fundamental aim of *Cellular and Biochemical Sciences* is to emphasize diversified topics of current interest to postgraduate students pursuing different courses in the area of Biological Sciences including Zoology, Botany, Biochemistry and Biotechnology. The text is also relevant to the students of Life Sciences, Biosciences, Cell Biology, Bioengineering and Pharmacology in all Universities, Colleges and other academic institutions. The selected topics are based on syllabi of various competitive examinations such as National Eligibility Test for Lecturership and Research Fellowship, Indian Civil and Forest Services, etc. Some of the topics are rarely found in other books of Biology. New information has been introduced which updates existing knowledge and enables the book to justify its claim as the most comprehensive text in the sphere of Cellular and Biochemical Sciences at the Postgraduate and Competitive Examination levels. Each and every chapter has been designed in a lucid and readable manner. There are references, suggested readings, long questions and objective questions at the end of chapters for revision of topics and helping students.

The book contains 58 chapters. It has been divided into 19 sections. Structure and function of cell organelles and regulation of cell cycle at molecular level have been described. Properties of catalytic molecules and their mode of action with enzyme kinetics have been discussed. Signal transduction and cell signaling have been explained. Synthesis of steroid hormones and identification of sex steroids have been elaborated. Structural polymorphism of nucleic acid, DNA replication and RNA synthesis have been advocated. Chapters on giant chromosomes, C-value paradox and heterochromatization have been included. Topics related to mutation, haploidy, cybride, transposon and genetic load are presented in the book. In addition, cell differentiation and homeotic genes have been discussed. There are some topics on immunology, hybridoma technology including applications of monoclonal antibodies. Genomics, proteomics and protein sequencing have been incorporated. Polymerase chain reaction, application of real time PCR and RNA mediated gene silencing have been placed in the book. Carcinogenesis, metabolic and genetic disorders, and drug detoxification have been explained. Chapters pertaining to recombinant DNA technology, cloning vectors, transgenic biology, gene targeting and application of genetic engineering in agriculture, health and industry have been given due consideration. Topics on protoplast culture, hybridization, somatic embryogenesis, tissue culture and liposome technology have been incorporated to make the book complete in its contents. Biochemistry and molecular biology of bioluminescence has been explained. There are descriptions about microbes, mycotoxins and antibiotics. A number of techniques currently used in cellular and biochemical sciences have been described. These techniques are the sequencing of nucleic acids, blotting, electrophoresis, cytometry, gel filtration, ion-exchange chromatography, high pressure liquid chromatography, hydrodynamics methods in biology and atomic absorption spectroscopy. A topic on concept and applications of bioinformatics has also been incorporated.

Overall the book covers many diverse aspects of modern biology and is a unique compilation to the students pursuing higher courses and appearing in various competitive examinations. Text of each chapter is the contribution from different authors who are solely responsible for the mistake, copyright and legal matter. Editor is no way responsible in these contexts.

This compilation would never have been accomplished without the encouragement and participation of many people. My personal appreciation goes to each of the contributor for accepting the challenge of preparing topics and sharing their ideas. I am grateful to all contributors who took pain and had patience till the appearance of compilation. My sincere thanks and gratitude are extended to those who supported and encouraged the preparation of the book.

In fact, the book presents materials that will be useful to a wide range of readers and provide an integrated understanding of cellular and biochemical sciences to the aspirants.

G. Tripathi

Contents

PART–1

Cell Organelles and Cell Cycle

CHAPTER 1

Plasma Membrane: Structure and Transport

S.B. Prasad
Dept. of Zoology, North-Eastern Hill University, Shillong – 793 022
E-mail: sbpnehu@hotmail.com

The basic unit of life, the cell, is enclosed by the plasma membrane or cell membrane which enables the cell to maintain the essential differences between the cytosol and the extracellular environment providing distinct biological properties to a cell. In eukaryotic cells, many sub-cellular organelles such as mitochondria, endoplasmic reticulum, Golgi apparatus, etc. are like wise membrane-bounded and maintain the characteristic differences between the contents of the organelles. Plasma membranes are semi-permeable and form selectively permeable barriers. Animal cells do not have cell walls but their activity is stabilized and specified by the interaction of plasma membrane with the extracellular matrix whose quantitative and qualitative chemical composition affects the properties of a tissue. Plasma membrane and membrane-derived specialized endomembrane cellular structures are involved in a variety of vital cellular functions like transmembrane transport of specific solutes, synthesis of ATP, receptors, signal transduction, etc. As the cells differ in plasma membrane dependent functions in different cells, it was an up hill task to find a common feature for all biological membranes on one hand and maintenance of differences in their functions at the other hand.

HISTORICAL PERSPECTIVES

Various suggestions and models were given by many scientists to unravel the molecular organization of biological membranes. Although, the *cell theory* or *cell doctrine* was proposed in 1839 by M.J. Schleiden and T. Schwann, the submission that the cells have boundary layer or plasma membrane regulating the passage of materials, came in 1890s. Sir Charles Overton (1899) found that water-soluble dyes entered cells less readily than fat (lipid)-soluble ones and coined the idea of the existence of the lipoidal membrane at the cell periphery. Subsequently, the observation that the rate at which different molecules pass into cells is directly related to their solubility in lipid, led Overton (1902) to postulate that the plasma membrane is composed of lipid molecules and speculated that it might contain fatty acids, cholesterol and lecithin. Irving Langmuir, a surface chemist, (1917) gave an insight on how lipid molecules might be arranged in cellular membranes. He discovered that purified phospholipids spread on a water surface spontaneously generate a film one molecule thick. In interpreting these results, Langmuir took into account the fact that these phospholipids contain both hydrophilic (polar) regions that are soluble in water, as well as hydrophobic (nonpolar) regions that are insoluble in water. Langmuir explained that when phospholipids are placed on a water surface, the molecule's hydrophilic head becomes aligned next to the water surface, while its hydrophobic tail is repelled by water and faces toward the air. Dutch scientists E. Gorter and F. Grendel (1925-26) adopted Langmuir's methods and studied the behavior of lipid molecules extracted from human red blood cells. Since mammalian red blood cells lack both, the nuclei and cytoplasmic organelles, all of the lipids extracted from the red blood cells can be assumed to have resided in the cells' plasma membrane. When spread on water, the lipid formed a monomolecular film that covered an area roughly twice, 1.8 and 2.2 to 1, as large as the surface area of red cells from which the lipids had been extracted. Hence, they suggested that the lipids in the plasma membranes are arranged in a bimolecular layer. They proposed that in the lipid bilayer the polar head group of each lipid was directed toward the out side/aqueous environment and the hydrophobic tail in the membrane interior away from water and this would be thermodynamically favoured arrangement. Thus, the polar head groups of lipids would face the cytoplasm on one edge and the blood plasma on the other.

The other indirect information came from the interfacial tension of different cells. Tension at a water-oil interface is about 10 to 15 dyn/cm, whereas, surface tension of cells is almost nil, a property that suggested the presence of a substance in the cell with the ability to lower the interfacial tension. The low tension was suggested to be due to the presence of proteins among the lipid components, as it was known that when a very small amount of protein is added to a lipid-water system, the surface tension is lowered. J.F. Danielli and H. Davson in 1935 proposed the membrane model which was supported by their surface tension studies of artificial lipid systems in which monolayer and bilayer films were formed. Danielli and Davson studied triglyceride lipid bilayers over a water surface and found that lipids arranged themselves with the polar heads facing outward and they always formed droplets (oil in water), and the surface tension was much higher than that of cells. However, if proteins were added to it, the surface tension reduced and the membranes flattened out. Thus, Danielli and Davson (1935) proposed that the plasma membrane contained a lipid bilayer, with protein adhering to both the lipid-aqueous interfaces. Later, the presence of protein lined pores were also suggested by them (Fig. 1a) but the functional aspects of the membrane could not be explained satisfactorily.

In 1959-60, J.D. Robertson proposed a modified version of the membrane model called as 'unit membrane model', which was primarily based on electron microscopic studies. These observations on many types of cells revealed a 'railway track like', three-layered structure with two outer dense layers of about 2 nm each and the middle one of about 3.5 nm. In this model, the electron microscopic images were interpreted along the lines of the Danielli-Davson model with the clear central layer corresponding to the hydrocarbon chains of the lipids and the dense surrounding layers to the proteins on both sides. However, one of the major weaknesses of Robertson's model was its failure to address satisfactorily the known permeability and transport properties of the membrane, because the model was supposed to be rigid much like electron microscopic observations.

The unit membrane model was eventually replaced in 1972, when S.J. Singer and G.L. Nicolson proposed the 'Fluid-mosaic model' which emerged as the universally accepted model for membrane structure (Fig. 1b). This was the first model where the structure of the plasma membrane successfully explained many of the dynamic and physiological functions of the membrane. The model retains the basic lipid bilayer structure first proposed by Gorter and Grendel. The proteins, however, were suggested to be globular and to float within the lipid bilayer rather than form the layers of the sandwich-type model. The proteins are either partially or wholly imbedded in the lipid bilayer. The lipid and proteins are disposed in a kind of mosaic arrangement, and are able to perform translational movement within the over all lipid bilayer. The hydrophobic tails of the phospholipids face inward, away from the water while its hydrophilic heads are on the outside where they interact with water molecules in the fluid environment of the cell. The concept of fluidity implies that the main components of the membrane are held in place only by means of non-covalent bonds. Later in 1974, they included the binding of oligosaccaharides at the outer surface (Fig. 1c).

CHEMICAL COMPONENTS OF PLASMA MEMBRANE

The basic chemical components of the plasma membrane are lipids, proteins and carbohydrates. The details about these chemical components are being described below.

(I) Membrane Lipids

The lipid bilayer has been established now as the universal basis for cell membrane structure. Membrane lipids constitute about 40-50% of the mass of most animal cell membranes. However, the ratio of percentage of lipids to proteins in the membranes of different cells is variable depending on the functional specificity. For example, in myelin sheath the lipid is about 80 per cent while in mitochondrial inner membrane it is almost reverse. Myelin is a stack of specialized plasma membrane sheets produced by a glial cell that wraps itself around the axon. In the peripheral nervous system, these glial cells are called Schwann cells, and in the central nervous system they are called oligodendrocytes. In both, vertebrates and some invertebrates, axons are accompanied along their length by glial cells, but specialization of these glial cells to form myelin occurs mainly in vertebrates. A myelin membrane, like all membranes, contains lipid bilayer with only few proteins. The predominant myelin protein in the peripheral nervous system and in the central nervous system, is P_0 and *proteolipid* respectively. Because of high percentage of lipids, myelin prevents transport of ions between the axonal cytosol and extracellular fluids, and the ions transport across the axonal membrane through the node of Ranvier only which is an unmyelinated area between two myelins on axonal membranes.

Membrane lipids are insoluble in water but dissolve readily in organic solvents, such as chloroform. Approximately 5×10^6 lipid molecules are packed in a 1μm x 1μm area of lipid bilayer. All of the lipid molecules in cell membranes are amphipathic, i.e. they have a

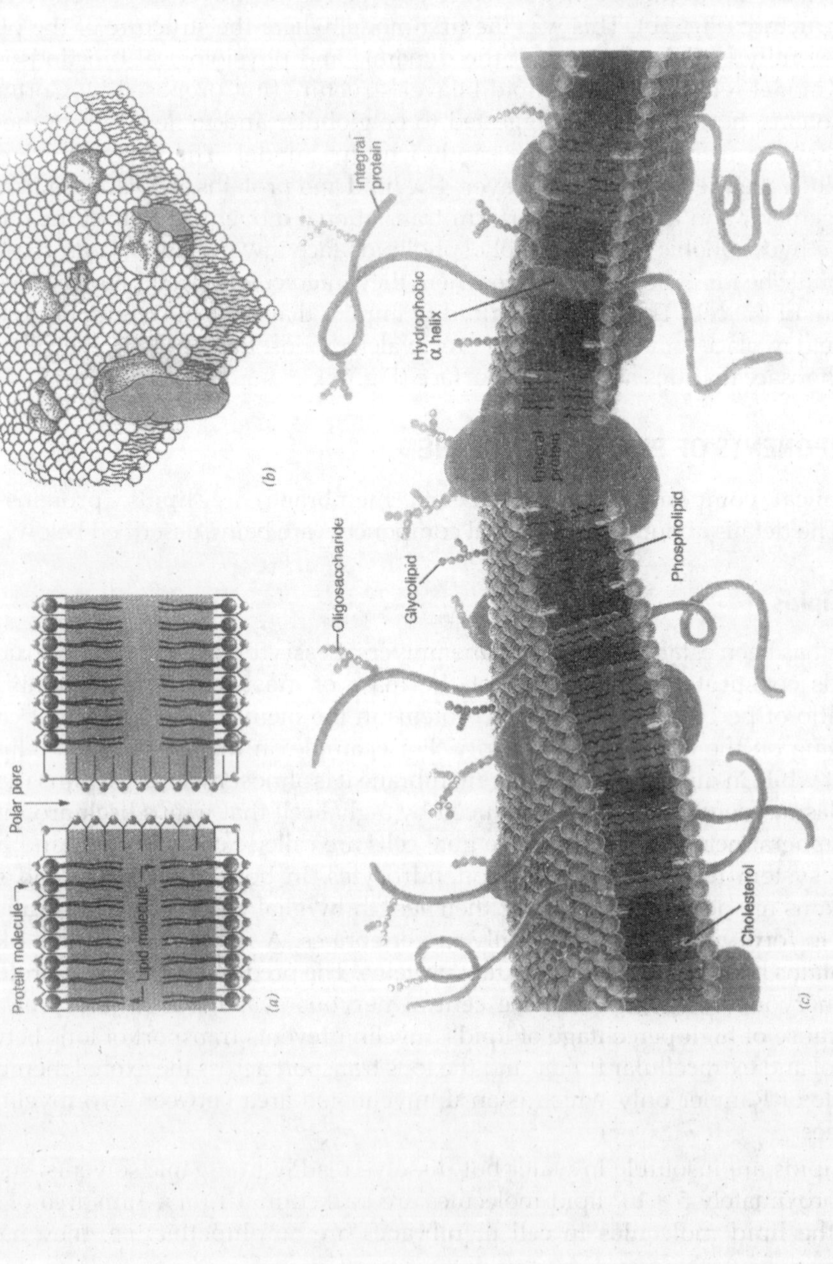

Fig. 1. Brief historical perspective of the structure of the plasma membrane. (a) Revised Davson-Danielli model (1954) showing the lipid bilayer which is covered on both surfaces by a layer of protein molecules that extend through the membrane to form protein-lined pores. (b) The fluid-mosaic model of membrane structure as proposed by Singer and Nicolson in 1972. Unlike previous models, proteins penetrate into and through lipid bilayer. (c) A current representation of the plasma membrane, showing the same basic organization as that proposed by Singer and Nicolson. The external surface of membranes contain short chains of sugars attached to membrane proteins forming glycoproteins, or to lipids forming glycolipids. (From Karp, G. (1999) *Cell and Molecular Biology : Concepts and Experiments*, John Wiley and Sons).

Table 1. Per cent ratio of protein and lipid in some membranes

Type of plasma membrane	Protein (%)	Lipid (%)	Ratio (Protein/ Lipid)
Human (myelin)	20	79	0.25
Rat (muscle)	65	35	1.86
Rat (liver)	60	40	1.50
Human (erythrocyte)	60	40	1.50
Rat (liver, mitochondrial inner membrane)	70	28	2.50

hydrophilic ("water loving") or polar end and a hydrophobic ("water fearing") or nonpolar end (Table 2). It is the shape and amphipathic nature of the lipid molecules that cause them to form bilayers spontaneously in aqueous environment. There are three types of lipids found in the membranes:

(a) Phospholipids
(b) Glycolipids and
(c) Cholesterol.

(a) Phospholipids: Phospholipids, the most abundant membrane lipids have a polar head group and two hydrophobic fatty acid hydrocarbon tails which can differ in length between 14 and 24 carbon atoms. The fatty acid tail may have one or more *cis*-double bonds i.e. it is unsaturated, or it may not have any double bond i.e. it is saturated. Each double bond creates a small kink in the tail (Fig. 5A). The differences in the length and saturation/unsaturation of the fatty acid tails influence the ability of phospholipids molecules to pack against one another, thereby affecting the fluidity of the membrane, which has been described later. Depending on the shape, i.e. wedge-shaped with single fatty acid tail or cylindrical-shaped with common two fatty acid tails, lipid molecules spontaneously aggregate to form spherical micelles (from wedge-shaped lipid) with the tails inward or they can form bimolecular sheets or *bilayers* as shown below in Fig. 2.

Being cylindrical, phospholipid molecules spontaneously form bilayers in aqueous environment which is energetically most-favoured arrangement with the hydrophilic heads facing water surface and the hydrophobic tails shielded from the water in the interior.

The main phospholipids in the plasma membrane of different cells are : phosphatidylcholine (PC), phosphatidylethanolamine (PE), phosphatidylserine (PS) and sphingomyelin (SM). The other phospholipids, such as phosphatidylinositol (PI) are present in smaller quantities, but have a crucial role in cell signaling. Mitochondrial inner membrane is very rich in cardiolipin or biphosphatidylglycerol constituting about 20% of total lipid. Cardiolipin is a double phospholipid joined by glycerol and contains four fatty acids. Sphingomyelin (SM), a phospholipid that lacks a glycerol backbone, contains sphingosine, an amino alcohol with a long unsaturated hydrocarbon chain. The terminal hydroxyl group of sphingosine is esterified to phosphocholine, so its hydrophilic head is similar to that of phosphatidylcholine. Structures of major phospholipids in a mammalian plasma membrane are shown in Fig. 3.

Based on the synthetic lipid bilayers, in 1970 it was recognized that individual lipid molecules are able to diffuse freely within lipid bilayers. Two types of preparations of bilayers were very useful in such studies: bilayers made in the form of spherical vesicles, called

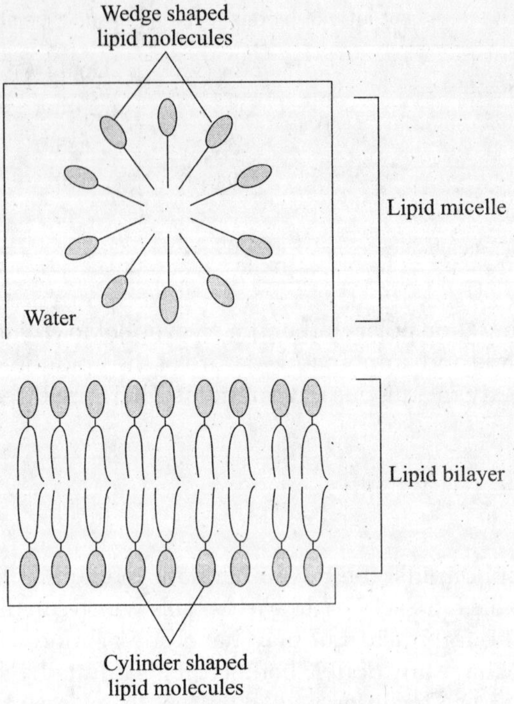

Fig. 2. A lipid micelle formed by wedge shaped lipid molecules, and a lipid bilayer formed by cylindrical phospholipids. Lipid molecules spontaneously form either of these two structures in water depending on their shape.

'liposomes', which can vary in size from about 25 nm to 1 μm in size depending on how they are produced, and planar bilayers, called 'black membranes', formed across a hole in a partition between two aqueous compartments. Based on the observations in synthetic bilayers, it is noted that phospholipids in synthetic bilayers or isolated biological membranes and in living cells, very rarely migrate from the monolayer on one side to that on the other, the process, known as 'flip-flop'. In contrast, lipid molecules readily exchange places with their neighbours within a monolayer, a process known as lateral diffusion (Fig. 4).

The fluidity of a lipid bilayer depends on both its composition and its temperature. Lowering of temperatures decreases the fluidity of membranes. If the temperature is slowly lowered, a point is reached where the lipid bilayer is converted from its normal fluid like state to a frozen crystalline gel in which the movement of the phospholipids is greatly restricted. The temperature at which this change occurs is called the *transition temperature* (Fig. 5A). The change of state of a lipid bilayer from a liquid state to a rigid crystalline or gel state at a characteristic freezing point is called a phase transition. The transition temperature of a particular bilayer depends on the particular lipids of which it is constructed. A shorter chain length reduces the tendency of the hydrocarbon tails to interact with one another. Thus, the shorter the fatty acyl chains of a phospholipids, the lower is its melting temperature. The presence of *cis*-double bonds in unsaturated fatty acids produce kinks in the hydrocarbon chains that make them more difficult to pack together, so that the membrane remains more

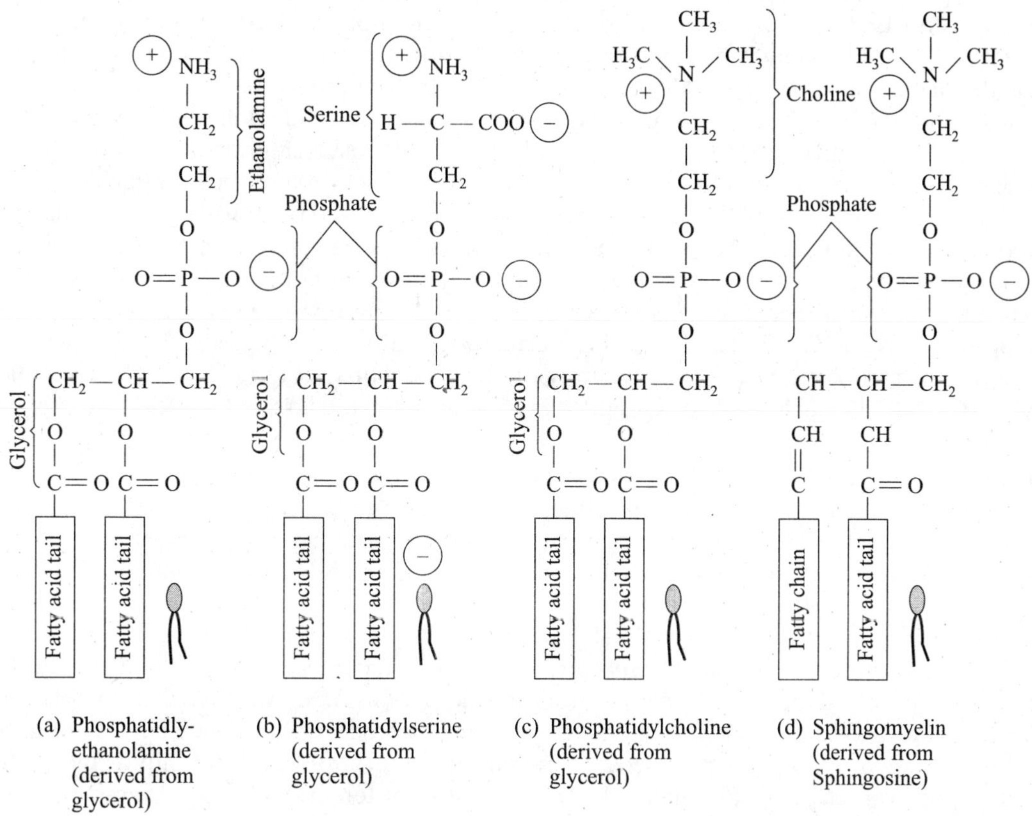

Fig. 3. Four representatives of phospholipids present in the typical plasma membrane.

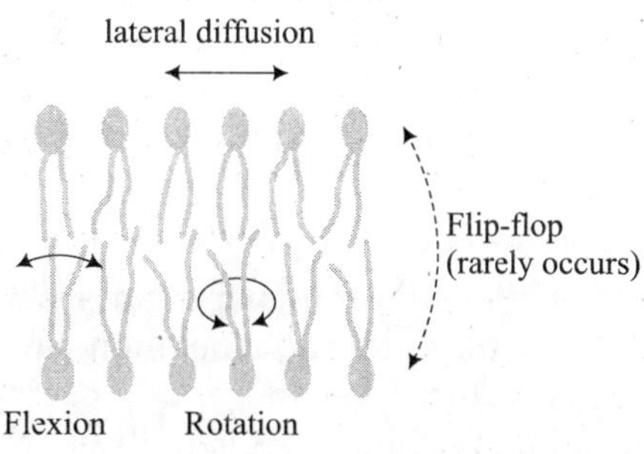

Fig. 4. The two types of phospholipids mobility in a lipid bilayer in the membranes.

fluid at lower temperatures (Fig. 5B). Bacteria, yeasts, and other organisms whose temperature fluctuates with that of their environment adjust the fatty acid composition of their membrane lipids to maintain a steady fluidity. For example, as the temperature falls the unsaturated fatty acids with more double bonds are synthesized, so the decrease in bilayer fluidity is avoided. Thus, another way for the cell to control the fluidity of its membrane is to regulate the ratio of saturated to unsaturated hydrocarbon chains of the phospholipids.

(b) Glycolipids: Carbohydrates are found in many membranes in the outer (exoplasmic) monolayer of lipid, as covalently bound proteins as constituent of glycoproteins or to lipids as constituent of glycolipids. The glycolipids are derived from sphingosine, so, it is also named as sphingolipids. Sphingosine in sphingolipids is linked to a fatty acid by its amino group and in this form it is called ceramide. The simplest glycolipid, glucosylcerebroside, contains a single glucose unit attached to ceramide. If the carbohydrate attached to ceramide is oligosaccharide, the glycolipid is called a ganglioside. Gangliosides contain one or more terminal sialic acid (N-acetyl neuraminic acid, NANA) residues, which give gangliosides a net negative charge. Since all the glycolipids have two long, hydrophobic hydrocarbon chains on one end and a hydrophilic region at the other, they are also amphipathic and similar in overall structure to phosphoglycerides. Glycolipids probably occur in all animal cell membranes. The nervous system is rich in glycolipids. The myelin membrane contains a high content of a particular glycolipid, called galactocerebroside which is formed when a galactose is added to ceramide (Fig. 6).

(c) Cholesterol: Another lipid component of certain membranes is the sterol, cholesterol. The basic structure of cholesterol is the four ring hydrocarbon and has a hydroxyl group at carbon-3 on ring. Thus, cholesterol is also amphipathic because its hydroxyl group can interact with water. Eukaryotic cell membranes, especially mammalian cells, contain large amounts of cholesterol. Cholesterol is absent from the plasma membranes of most plant and bacterial cells. Cholesterol breaks up the Van der Waals interactions and close packing of the phospholipids tails. This disruption makes the membrane more fluid. Therefore, variations in the level of cholesterol in the cell membrane may control the fluidity of its membrane. Thus, the cholesterol molecules enhance the permeability-barrier properties of the lipid bilayer. They orient in the bilayer with their hydroxyl groups close to the polar head groups of phospholipids molecules. In this position, their rigid, plate like steroid rings interact with those regions of the hydrocarbon chains closest to the polar head groups and partly immobilize. By decreasing mobility of the first few CH_2 groups of the hydrocarbon chains of the phospholipids molecules, cholesterol makes the lipid bilayer less deformable in this region and thereby decreases the permeability of the bilayer to small water-soluble molecules. At high concentrations, found in eukaryotic plasma membranes, cholesterol prevents the hydrocarbon chains from coming together and crystallizing, thus inhibits possible phase transitions (Fig. 7).

The lipid composition in different membranes is quite variable. A comparison of lipid composition in some membranes has been shown in Table 3.

A very important property of lipid bilayer in membranes is its asymmetrical nature i.e. the phospholipid composition differs in the outer and inner monolayer of lipid. For example, in the plasma membranes from human erythrocytes and certain canine kidney cells, almost all the sphingomyelin (SM) and phosphatidylcholine (PC), both of which have positively charged head groups, are found in exoplasmic leaflet of bilayer. Lipids with negative or neutral head groups, i.e., PE, PS, and PI are preferentially located in the cytoplasmic leaflet (Fig. 8).

The asymmetrical arrangement of membrane lipids may have a direct relationship to a particular membrane property. For example, glycolipids are restricted to the outer leaflet of plasma membrane because their carbohydrate groups are involved in signaling events that occur at the outer cell surface. PS and PI, on the other hand, are more concentrated in the inner leaflet of plasma membrane where they are involved in transmitting signals from the plasma membrane to the cell interior.

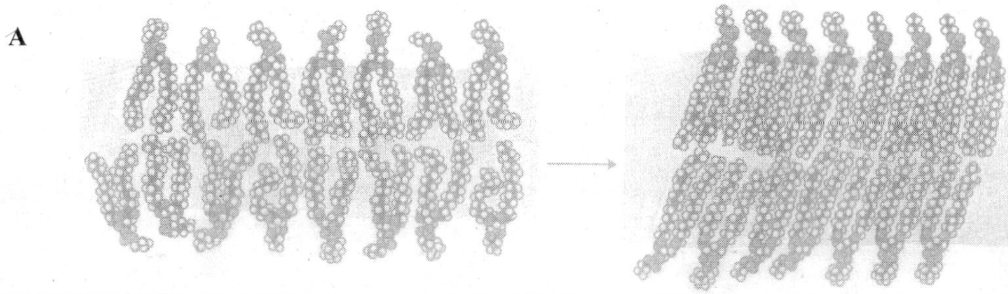

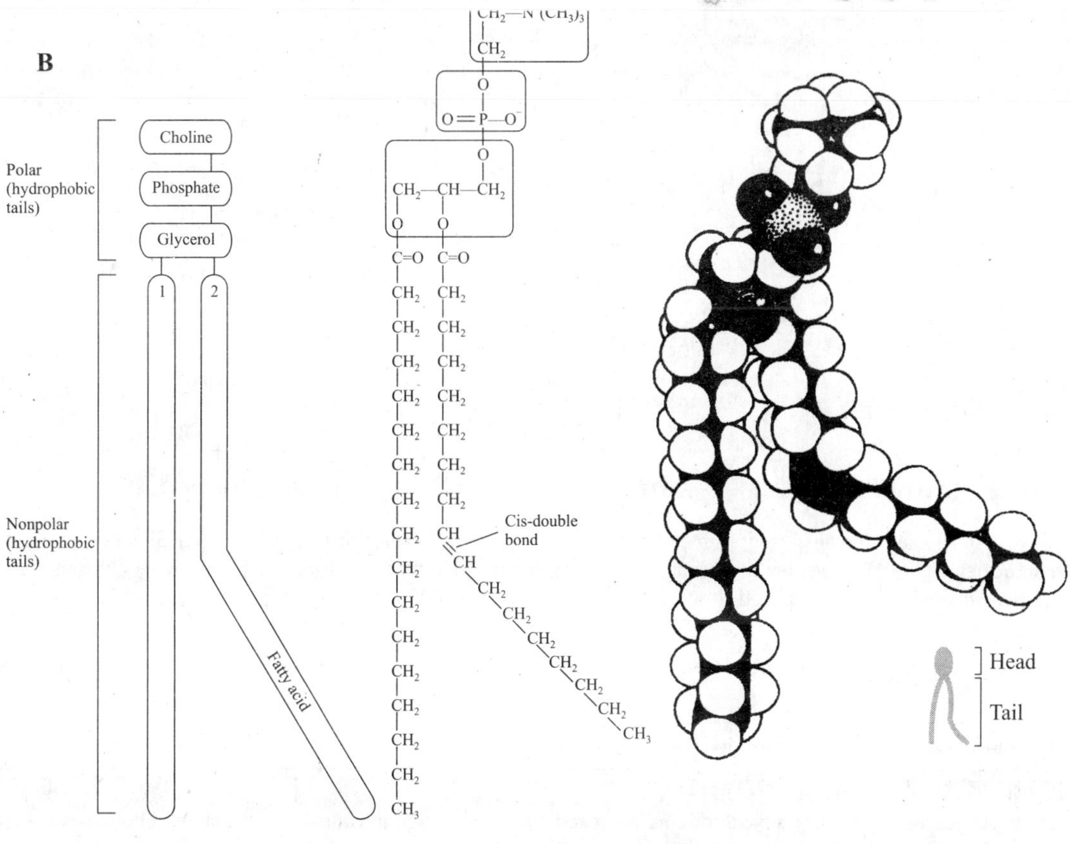

(a) Schematic representation (b) Structural formula (c) Space filling model (d) Symbolic representation

Fig. 5. (A) The fluidity of the lipid bilayer depends on the temperature. Above the transition temperature, the lipid molecules and their hydrophobic tails are free to move, increasing the fluidity. Below the transition temperature, the movement of the molecules is restricted and it is converted to crystalline gel, decreasing the fluidity. **(B)** The parts of a lipid molecule, phosphatidylcholine, showing a head and two tails. One of the fatty acid tails is shown to have a *cis*-double bond, thereby generating a kink in the fatty acid tail. Increase in the number of *cis*-double bonds in fatty acid tails increases the fluidity of membrane. (From Karp, G. (1999) *Cell and Molecular Biology : Concepts and Experiments*, John Wiley and Sons).

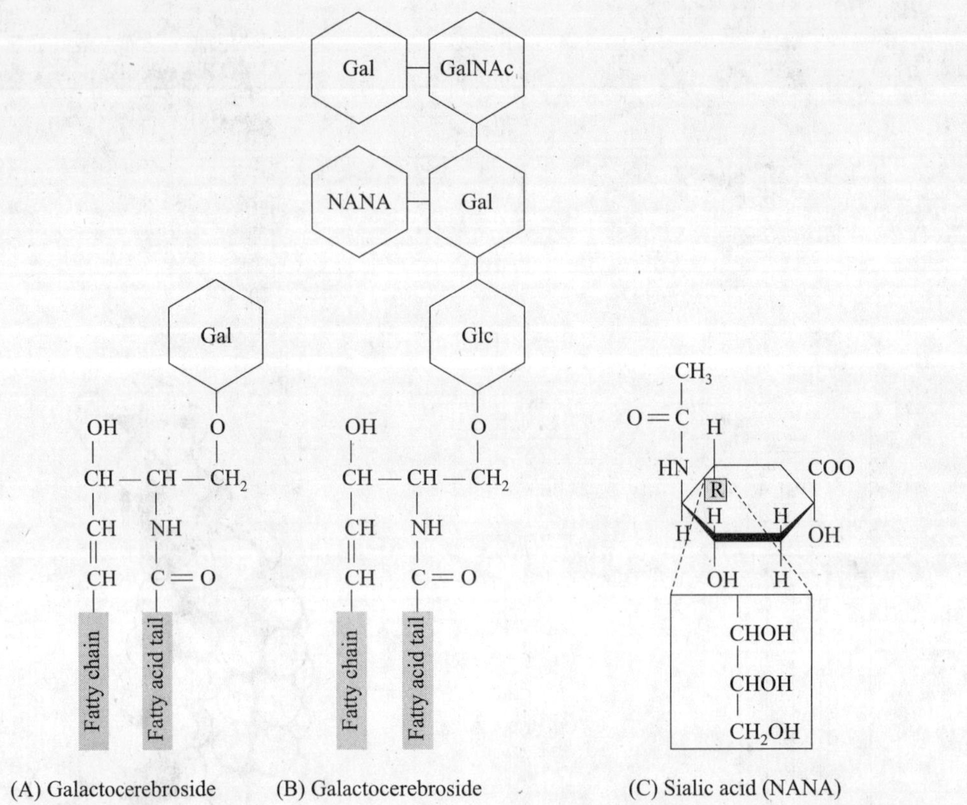

Fig. 6. Some examples of glycolipid molecules found in the lipid bilayer of membranes. **(A)** Simple glycolipid, galactocerebroside, **(B)** Complex glycolipid, a ganglioside always contains one or more negatively charged sialic acid residues (also called N-acetylneuraminic acid or NANA) whose structure is shown in **(C)**.

Table 2. Hydrophilic (polar) heads and hydrophobic (non polar) regions of different lipids in the plasma membrane

Type of Lipid	Hydrophobic region	Hydrophilic region
Phosphoglyceride	Fatty acid chain	Phosphorylated alcohol
Sphingomyelin	Fatty acid chain and hydrocarbon chain of sphingosine	Phosphoryl choline
Glycolipid	Fatty acid chain and hydrocarbon chain of sphingosine	One or more sugar residues
Cholesterol	Entire molecule except -OH group at C-3	-OH group at C-3

(II) Membrane Proteins

While the basic structure of biological membranes is provided by the lipid bilayer, membrane proteins perform most of the specific functions of cellular membranes. The amount and types of proteins in a membrane are highly variable and give each type of membrane in the cell its characteristic functional properties. In the myelin membrane, which serves as electrical insulation for nerve cell axons, less than 25% of the membrane mass is protein. By contrast, in

Table 3. Lipid composition (approximate percentage of total lipid by weight) of different cell membranes

Lipid	Liver cell plasma membrane	Red blood cell plasma membrane	Myelin	Mitochondrion (inner and outer membranes)	Endoplasmic reticulum	E. coli
Cholesterol	17	23	22	3	6	0
Phosphatidylethanolamine	7	18	15	25	17	70
Phosphatidylserine	4	7	9	2	5	Trace
Phosphatidylcholine	24	17	10	39	40	0
Sphingomyelin	19	18	8	0	5	0
Glycolipids	7	3	28	Trace	Trace	0
Others	22	13	8	21	27	30

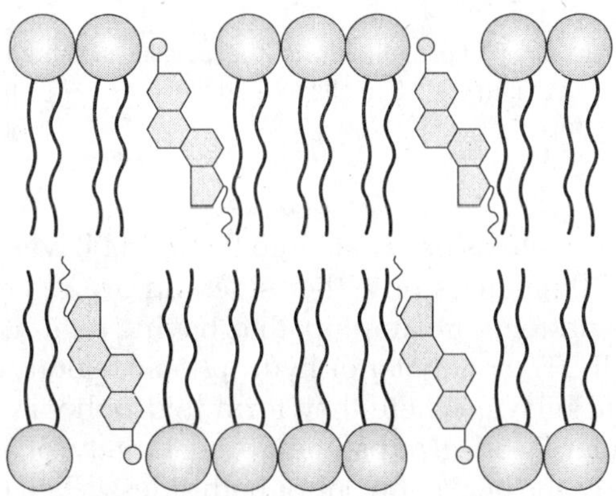

Fig. 7. The orientation of cholesterol molecules in the lipid bilayer of plasma membranes. The small hydrophilic –OH end faces the external surface of the bilayer while the bulk of their structure forming hydrophobic region is packed in among the fatty acid tails of the phospholipids. The placement of cholesterol molecules interferes with the tight packing of the phospholipids and tends to increase the fluidity of the bilayer. The cholesterol is often evenly distributed between the two layers.

the mitochondrial inner membranes involved in ATP production approximately 75% is protein. The membrane proteins are also asymmetrically situated as each protein is located and oriented in a particular position relative to the cytoplasm, so that the properties of the one surface of a membrane are very different from those of the other surface. This property is referred to as membrane 'sidedness'. For example, those parts of the proteins that interact with other cells or with extracellular ligands, such as hormones or growth factors, project outward into the extracellular space, whereas those parts of membrane proteins that interact with cytoplasmic molecules, such as G-proteins or protein kinases, project into the cytosol. Membrane proteins can be grouped into three distinct classes distinguished by the intimacy of their relationship to the lipid bilayer. These are integral proteins, peripheral proteins and lipid-anchored proteins.

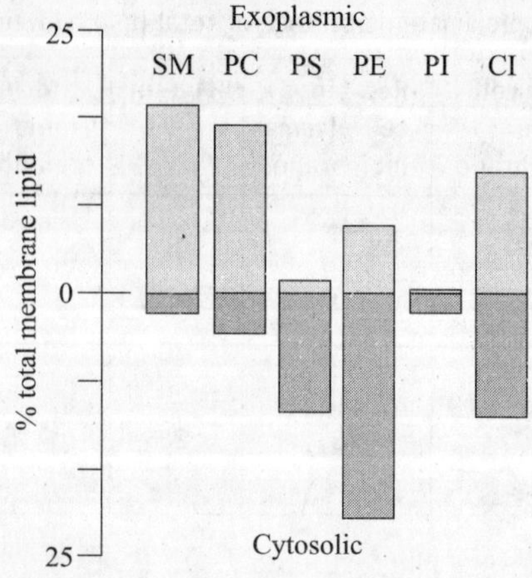

Fig. 8. The asymmetric distribution of phospholipids in the plasma membrane of human erythrocytes. (SM, sphingomyelin; PC, phosphatidylcholine; PS, phosphatidylserine; PE, phosphatidylethanolamine; PI, phosphatidylinositol; Cl, cholesterol).

(i) Integral proteins: These proteins penetrate into the lipid bilayer. Many extend through the lipid bilayer with part of their mass on either side and are referred to as transmembrane proteins. Transmembrane proteins are amphipathic, having regions that are hydrophobic and regions that are hydrophilic. The nonpolar or hydrophobic regions of the integral proteins are embedded within the lipid bilayer where they form hydrophobic interactions with the fatty acyl chains while the polar head groups composed largely of ionic amino amino acids, protrude beyond the edge of the bilayer on one or both sides. When the polypeptide crosses the membrane only once, it is described as 'single pass transmembrane protein' (e.g., glycophorin protein of R.B.C.) and when it crosses several times it is called a 'multipass transmembrane protein' (e.g., band 3 protein of human R.B.C.) or it may be present as barrels (e.g., porins). Most of the membrane receptors and transport proteins belong to the category of integral proteins. Because of their hydrophobic surfaces, integral membrane proteins are difficult to isolate in a soluble form. Removal of these proteins from membrane normally requires the use of detergent, such as SDS, Triton X-100 etc. Some multiple transmembrane polypeptide chains are arranged as a β-barrel, as seen in porin proteins. The porin barrel is formed from a 16-stranded antiparallel β-sheet which is sufficiently large to form a cylindrical structure. The β-barrel proteins are abundant in the outer membrane of mitochondria, chloroplasts, and many bacteria. The great majority of transmembrane proteins in animal cells are glycosylated. As in glycolipids, sugar residues are added in the lumen of endoplasmic reticulum and Golgi apparatus.

(ii) Peripheral or extrinsic proteins: They are located entirely outside of the lipid bilayer and are associated with the surface of the membrane by noncovalent bonds e.g., spectrin of R.B.C. membranes. These proteins can be easily released by extracting membranes with solutions of increased ionic strength.

(iii) Lipid-anchored proteins: These proteins are neither integral nor peripheral, but have characteristics of both. The polypeptide chains of this type of membrane protein are situated out side the bilayer and can protrude from either side of a membrane, but are covalently bound to lipid molecules in the bilayer. This lipid could be one or more fatty acid chains or prenyl groups. Other membrane proteins are entirely exposed at the external cell surface, being attached to the lipid bilayer only by a covalent linkage via a specific oligosaccharide to phosphatidylinositol in the outer lipid monolayer of plasma membrane (Fig. 9).

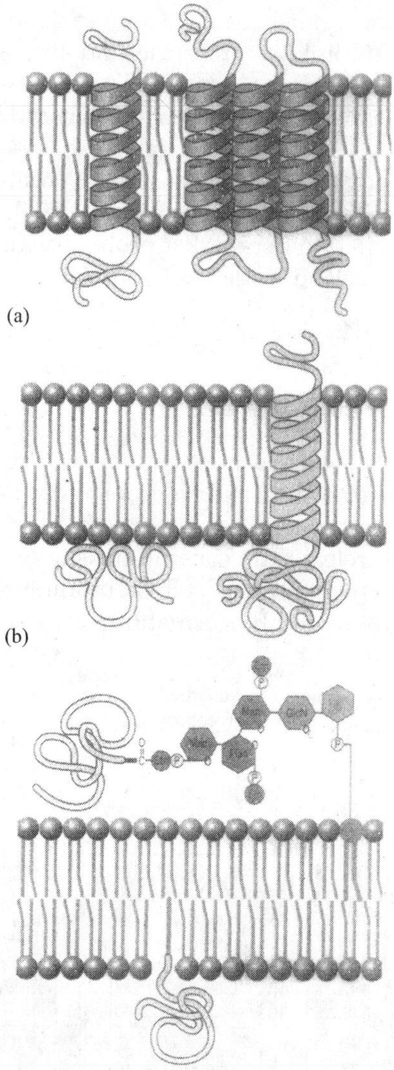

Fig. 9. Three classes of arrangement of proteins in plasma membranes. (a) Integral membrane proteins could be as single pass α-helix protein or as multipass protein. (b) Peripheral proteins are noncovalently bonded to the polar head groups of the lipid bilayer or to an integral membrane protein. (c) Lipid-anchored proteins are membrane proteins that are covalently bonded either to a phospholipids or a fatty acid that is embedded in one of the leaflets of the lipid bilayer. (From Karp, G. (1999) *Cell and Molecular Biology : Concepts and Experiments*, John Wiley and Sons).

(III) Membrane Carbohydrates

Plasma membrane proteins and most of the lipids in the outer monolayer of lipid bilayer are glycosylated and are present as glycoproteins and glycolipids respectively. Depending on the species and cell type, the carbohydrate content of the plasma membrane ranges between 2 and 10 per cent by weight, for example, human R.B.C. contains approximately 52% protein, 40% lipid and 8% carbohydrate, majority of which are present as glycoproteins. The carbohydrate of glycoproteins is present as short branched oligosaccharides having 10-14 sugars per chain, attached to either aspargine (N-linked) or serine and threonine (O-linked) amino acids (Fig. 10). The carbohydrates of red blood cell membranes determine whether a person's blood type is A, B, AB or O group. The monosaccharides which could be involved as the component of membrane oligosaccaharides are glucose, galactose, N-acetyl glucosamine, N-acetyl galactosamine, fucose, xylose, etc. The term cell coat or glycocalyx is often used to describe the carbohydrate rich zone on the cell surface and can be visualized by stains like ruthenium red or by lectins tagged with fluorescent or other marker probe. The function of the cell coat includes protection, cell recognition, cell-cell adhesion, inflammatory responses, etc. (Fig. 11).

MEMBRANE TRANSPORT

The selective transport of molecules across the cell membranes is one of the important functions of cells. When the bigger macomolecules like glycoproteins, etc. are secreted out or taken inside the cell, the term 'exocytosis' and 'endocytosis' respectively is used. In this case part of plasma membrane is detached and it forms transport vesicles during endocytosis, while during exocytosis similarly, secretory vesicles are generated from Golgi net work and it fuses with the plasma membrane to release the content out side the cell, it is called endocytosis-exocytosis cycle, helping to maintain the cell membrane size almost constant. As the endocytosis and exocytosis involves the formation of vesicles, this is also described as

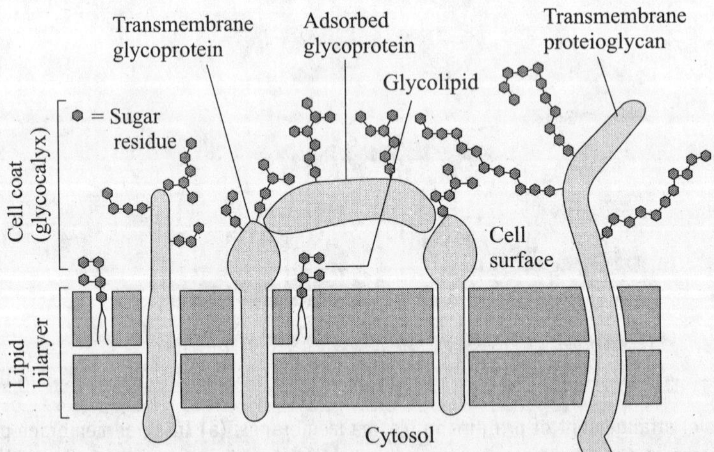

Fig. 10. The cell coat or glycocalyx is always on the external surface and made up of the oligosaccharide side chains of the glycolipids and glycoproteins.

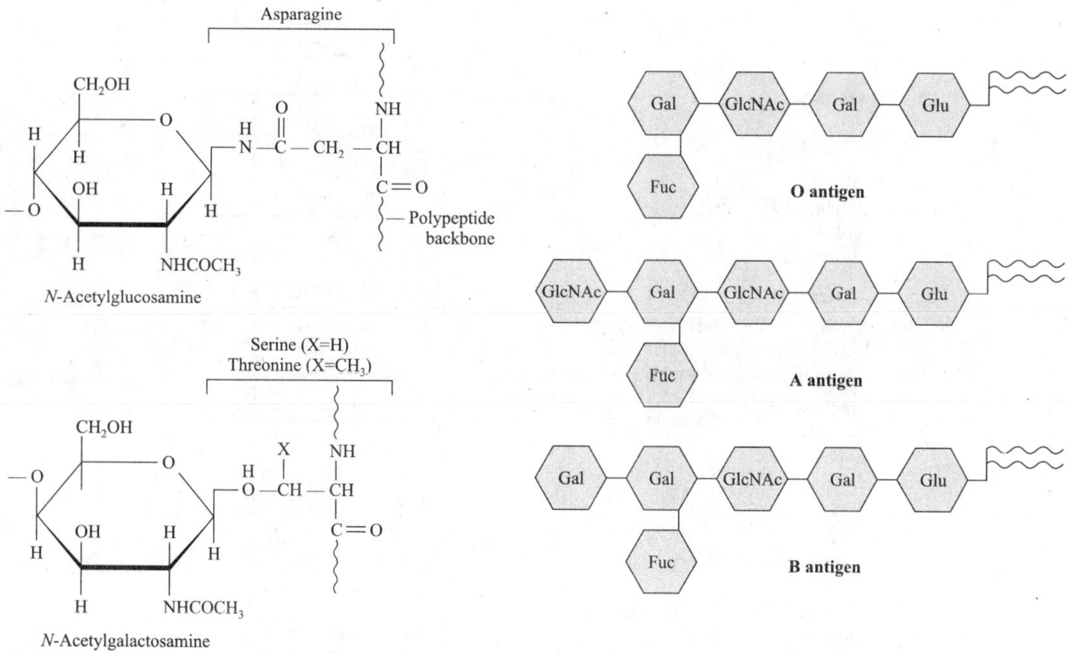

Fig. 11. (A) Two types of linkages that join sugars to a polypeptide chain. The N-glycosidic linkage formed between aspargine and N-acetylglucosamine is more common than the O-glycosidic linkage between serine or threonine and N-acetylgalactosamine. **(B)** Blood group A, B, AB or O is determined by a short chain of sugars covalently attached to membrane lipids and proteins of the red blood cell membrane. A person with type AB blood group has gangliosides with both the A and B structure.

'vesicular transport'. Here the discussion is mainly restricted on membrane transport which involves the transport of molecules across the membrane without detaching the membrane. This barrier function of membranes allows the cell to maintain specific concentrations of solutes in its cytosol and in each of the intracellular membrane-enclosed cell organelles. The substances transported may include a variety of molecules like nutrients, metabolites, and a variety of ions i.e. Na^+, Ca^+, Cl^-, H^+ etc. The rate at which molecules pass across the plasma membrane depends on the size of the molecule and its lipid solubility properties (Table 4). In general, the smaller the molecule and more hydrophobic (more soluble in oil/lipid), it will diffuse more rapidly across a lipid bilayer. Small nonpolar molecules such as O_2 and CO_2 gases readily diffuse across the lipid bilayer. Small uncharged polar molecules such as water or urea, also diffuse across the bilayer, but little slowly. By contrast, plasma membranes are highly impermeable to charged molecules including ions (Fig. 12).

The small hydrophobic molecules known as ionophores can be used to increase the permeability of membranes to specific ions. They shield the charge of the transported ion so that it can penetrate the hydrophobic interior of the lipid bilayer. The ionophores are not coupled to energy sources, thus, the movement of ions is only down their electrochemical gradient. Ionophores are mostly synthesized by microorganisms, and also used by scientists in various transport studies on cells or synthetic bilayers. Two classes of ionophores i.e. mobile

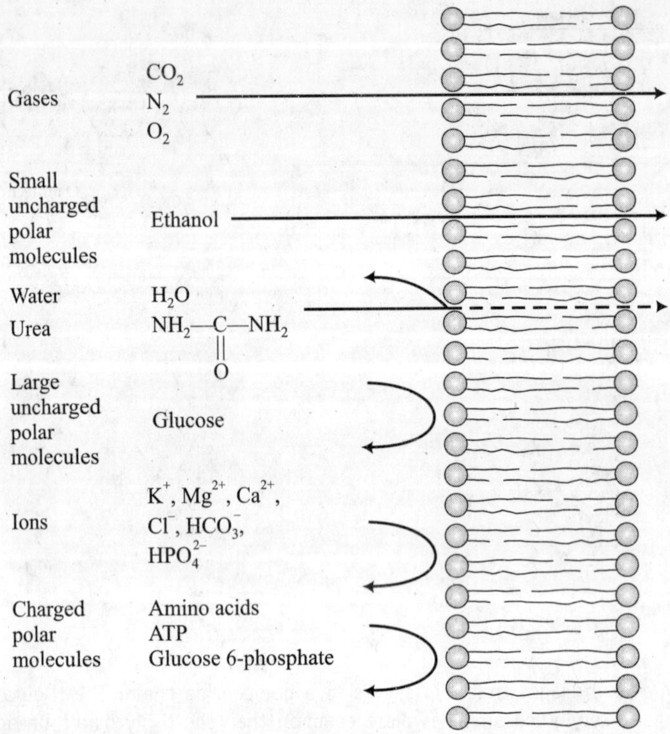

Fig. 12. Relative permeability a synthetic lipid bilayer to different molecules. The bilayer is permeable to small hydrophobic molecules and small uncharged polar molecules, but impermeable to charged molecules including ions.

ion carriers and channel formers, have been recognized. The examples of ionophores include valinomycin (a mobile ion carrier, which transports K^+), FCCP (a mobile ion carrier, that makes the membranes selective leaky to H^+, is often used to dissipate the H^+ electrochemical gradient across mitochondrial inner membrane, thereby blocking mitochondrial ATP production), A23187 (a mobile a mobile ion carrier, which transports bivalent cations such as Ca^{++} and Mg^{++}), and gramicidin A (a channel forming ionophore, made up of two linear 15 hydrophobic amino acid peptides, selectively allows monovalent cations to flow down their electrochemical gradients). The specific ionic composition of the cytosol as compared to its surrounding is

Table 4. Different types of molecules and their possible transport through plasma membrane

Type of molecules	Example of molecules	Transport through plasma membrane
Hydrophobic (nonpolar) molecules	O_2 (32), CO_2 (44), N_2 (28), Benzene (78)	Diffuse rapidly
Uncharged small polar molecules	H_2O(18), C_2H_5OH(46), urea (60), glycerol (92)	Diffuse rapidly
Uncharged large polar molecules	Glucose (180), sucrose (342)	Hardly diffuse
Charged molecules (ions, hydrated i.e. polar)	H^+, Na^+, K^+, Ca^{++}, Mg^{++}, Cl^-, HCO_3^-	Do not diffuse, transport using specific carrier/channel protein

Table 5. Typical ion concentrations inside and out side a typical mammalian cell and in squid axon (an invertebrate cell)

Ion	Intracellular concentration (mM)	Extracellular concentration (mM)
Mammalian cell		
K^+	140	5
Na^+	12	145
Cl^-	4	116
Mg^{++}	0.5 – 0.8	1.5
Ca^{++}	<0.0002	1.8
HCO_3^-	12	29
Squid axon		
K^+	400	20
Na^+	50	440
Cl^-	40-150	560
Ca^{++}	0.0003	10

necessary to be maintained to carry out various metabolic and biochemical reactions in the cells. The brief idea about the concentration of important ions inside and outside the cell is given in Table 5.

Membrane Transport Proteins

The transport of various polar molecules such as amino acids, sugars, nucleotides etc and ions is regulated by specialized membrane transport proteins. These proteins are generally multipass transmembrane proteins and categorized into two major classes of transport proteins: (i) Carrier proteins (also called as carriers, permeases or transporters) and (ii) Channel proteins. Carrier proteins bind the specific solute to be transported and undergo a series of conformational changes to transfer the bound solute across the membrane. Channel proteins interact with the solute to be transported much more weakly. They form aqueous pores that extend across the lipid bilayer. The transport through channel proteins occurs at a much faster rate than transport mediated by carrier proteins. All channel proteins and many carrier proteins allow solutes to cross the membrane only passively ("downhill"), a process called passive transport or facilitated diffusion. The facilitated diffusion may be exemplified by the transport of glucose by carrier protein, called glucose permease. This process should be considered differently than simple diffusion, for example, passage of gases across the plasma membrane is simple diffusion which does not involve any specific protein although it also occurs downhill (Table 6). On the other hand, if the transport of solute is actively pumped across the membrane against their electrochemical gradient ("uphill"), the process is known as active transport, that is why these carrier proteins are some time referred to as pumps and it is coupled with the metabolic energy, ATP hydrolysis. Thus, transport by carriers can be either active or passive, where as by channel proteins it is always passive (Fig. 13).

On the basis of number and direction of transport of solutes across the plasma membrane the membrane transport proteins are classified into three major classes, which are all integral

Table 6. Comparison of simple diffusion, facilitated diffusion and active transport

Property	Simple diffusion	Facilitated diffusion	Active transport
Rate of solute movement	Slow	Fast	Fast
Saturation kinetics	No	Yes	Yes
Solute specificity	Low	High	High
Direction relative to electrochemical gradient	Down	Down	Up
Inhibition by Competitive and noncompetitive inhibitors	No	Yes	Yes
Energy requirement	No	No	Yes
Type of membrane transport protein	None	Carrier and Channel proteins	Carrier (pumps) proteins

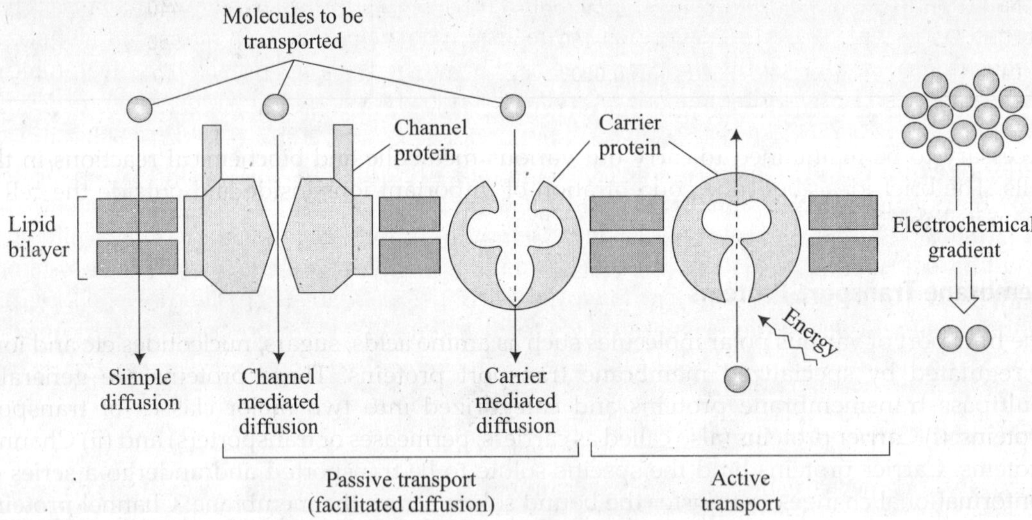

Fig. 13. Comparison of passive and active transport. Passive transport occurs down an electrochemical gradient either by simple diffusion through lipid bilayer or by facilitated diffusion through carrier proteins and channel proteins. Active transport is against electrochemical gradient, requires input of energy and always mediated by carrier proteins. (From: Alberts et al., (2002) *Molecular Biology of the Cell*, Garland Science).

proteins and exhibit a high degree of specificity for the substance transported. The carrier proteins which simply transport a single solute from one side of the membrane to the other, are called uniporters. Others function as coupled carriers in which the transfer of one solute strictly depends on the transport of a second. If the coupled transport involves the simultaneous transfer of a second solute in the same direction, it is called symporter, and if it occurs in opposite direction, it is called antiporter (Fig. 14). Different carrier proteins involved in specific ions and molecules have been given in Table 7 and 8.

Uniporter-catalyzed Transport

The plasma membrane of most cells contains several uniporters that enable amino acids, nucleosides, sugars and other small molecules to enter and leave cells down their

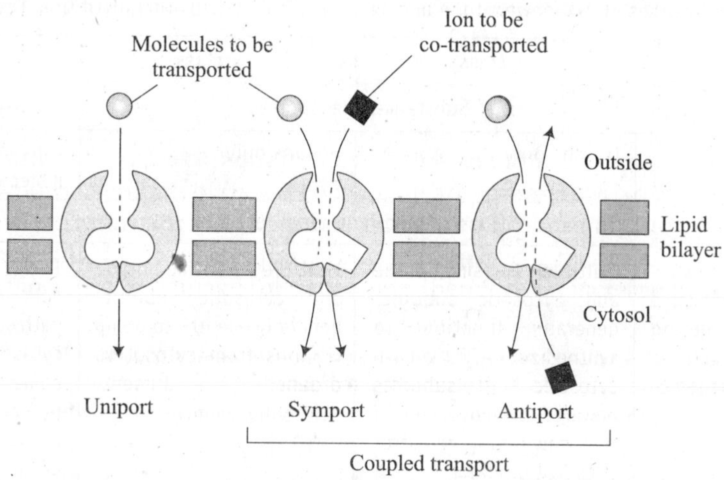

Fig. 14. Three types of carrier-mediated transport, i.e. uniport, symport and antiport.

Table 7. Different carrier proteins involved in transport

Family	Representative member	Transport type
Sugar transporter (Uniporter)	Glucose transporter GLUT1	Facilitated diffusion
Cation-transporting ATPases	(i) Na^+-K^+-ATPases (ii) Ca^{++} -ATPases (iii) H^+-K^+ -ATPases	Active transport
ABC (ATP-binding cassette) transporters	(i) Multidrug resistance (MDR) ATPases (Mammalian cells) (ii) Chloroquine resistance ATPase (*P.falciparum*) (iii) Cystic fibrosis transmembrane regulator (CFTR) protein	Active transport
Symporters (Na^+ driven)	(i) Na^+ - glucose symporter (intestine) (ii) Na^+ - proline symporter (bacteria) (iii) Na^+ - HCO_3^- symporter (glial cells)	Active transport
Cation antiporters	Na^+ - H^+ exchanger	Facilitated diffusion
Anion antiporters	Cl^- - HCO_3^- exchanger	Facilitated diffusion
Cation/anion antiporters	Na^+ - dependent Cl^- - HCO_3^- exchanger	Facilitated diffusion

concentration gradients. Each uniporter transports only a single species of molecule or a single group of closely related molecules. The details about some uniporters has been described below.

Glucose transport through GLUT1: Most mammalian cells express GLUT1, a plasma membrane uniporter that catalyzes movement of glucose down its concentration gradient and has been extensively studied in the mammalian erythrocytes. GLUT1 is an integral, transmembrane protein (molecular wt. 45,000) containing 12 α helices that span the lipid

Table 8. Different classes of ATP-powered ion and small-molecule pumps (modified from Lodish *et al.*, 2000).

P class	F class	V class	ABC class
		Substances transported	
H^+, Na^+, K^+, Ca^{++}	H^+ ions only	H^+ ions only	Ions and various small molecules
		Structural and functional features	
Large catalytic α subunits (often two) become phosphorylated during ion transport; smaller β subunits may regulate transport.	Multiple transmembrane and cytosolic subunits generally function to synthesize ATP on b cytosolic subunits powered by movement of H^+ down an electrochemical gradient	Multiple transmembrane and cytosolic subunits generally use ATP to pump H^+ ions from cytosol to organelle lumens, acidifying them.	Two transmembrane domains form the pathway for solute; two cytosolic ATP-binding domains couple ATP hydrolysis to solute movement. Domains may be in one or separate subunits.
		Location of specific pumps	
Plasma membrane of higher eukaryotes (Na^+-K^+ pump, Ca^{++} pump); Sarcoplasmic reticulum membrane in muscle cells (Ca^{++} pump)	Inner mitochondrial membrane	Endosomal and lysosomal membrane in animal cells	Mammalian endoplasmic reticulum (transporters of peptides associated with antigen presentation by MHC proteins)
Plasma membrane of plants, fungi, bacteria (H^+- pump)	Bacterial plasma membranes	Vacuolar membranes in plants, yeast	Bacterial plasma membranes (amino acid, sugar, and peptide transporters)
Apical plasma membrane of mammalian stomach cells (H^+-K^+ pump)	Thylakoid membrane of chloroplast	Plasma membrane of certain acid-secreting animal cells {e.g. osteoclasts (bone-resorbing macrophagelike cells) and some kidney tubule cells}	Mammalian plasma membranes (transporters of small molecules, phospholipids)

bilayers. The α helices in the transmembrane region contain mainly hydrophobic amino acids, but also bear other amino acids such as serine, threonine, aspargine and glutamine, whose side chains can form hydrogen bonds with the hydroxyl groups on glucose. These residues may be involved to form the inward-facing and outward facing glucose-binding sites in the interior of the protein. GLUT1 accounts for 2 per cent of the protein in the plasma membrane of erythrocytes. The glucose transporter, GLUT1 alternates between two conformational states; in one, a glucose-binding site faces the outside of the membrane and in the other, a glucose-binding site faces the inside. The details about the sequence of events has been depicted in Fig. 15.

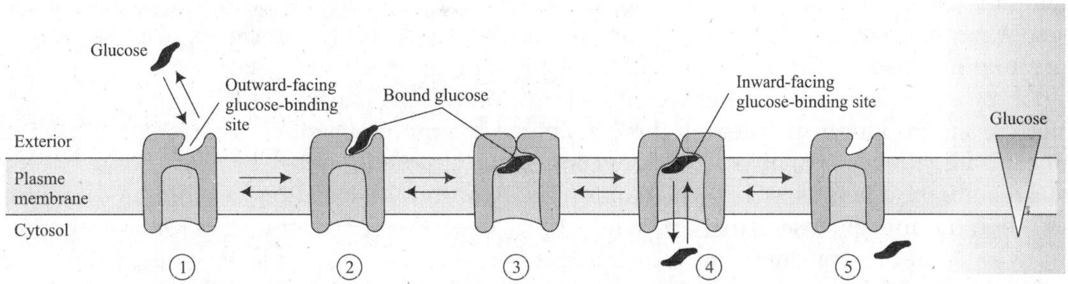

Fig. 15. Model of the mechanism of uniport transport of glucose by GLUT1 which is suggested to shuttle between two conformational states. In one conformation, as shown in (1), (2) and (5), the glucose-binding site faces outward while in the other, as shown in (3) and (4), the binding-site faces inward. Binding of glucose to to the outward-facing binding site triggers a conformational change in transporter, moving the bound glucose through the protein such that it is now bound to the inward-facing binding site. Release of glucose inside the cell brings back the original conformation. (From Lodish et al. (2000) *Molecular cell biology*, 4th edition, W.H. Freeman and Co.).

The GLUT1 has high affinity for normal substrate, D-glucose. For example, the K_m for D-glucose transport by GLUT1 is 1.5 mM and the K_m for nonbiological L-glucose is >3000 mM, thus the transport of L-glucose does not occur at a measurable rate. The K_m for D-mannose is 20 mM and that of D-galactose is 30 mM, so that considerably higher concentrations of these substrates than D-glucose are needed to half saturate the transport reaction. After glucose is transported into the erythrocyte, it is rapidly phosphorylated, forming glucose-6-phosphate, which cannot leave the cell and it helps to keep the intracellular concentration of free glucose low maintaining the concentration gradient.

Symporter-catalyzed Active Transport

Cells have another class of proteins that import or export ions and small molecules, such as glucose and amino acids, against a concentration gradient. These proteins use the energy stored in the electrochemical gradient of Na^+ or H^+ to power uphill movement of another substance, a process sometimes referred to as ***secondary active transport***. The Na^+ is the usual co-transported ion whose electrochemical gradient provides a large driving force for the active transport of a second molecule, for example, Na^+-driven glucose symport. The Na^+ that enters the cell during transport is pumped out by a separate ATP-driven pump in the plasma membrane. The ATP-driven carriers mediate ***primary active transport***. Intestinal and kidney epithelial cells contain a variety of secondary active transporters that are driven by the Na^+ gradient across the plasma membrane. Each system is specific for importing specific sugars or amino acid into the cell. The solute and Na^+ have different binding sites on the carrier protein. However, as Na^+ tends to move into the cell down its electrochemical gradient, the other solute i.e. sugar or amino acid is also 'dragged' into the cell with it.

Two-Na^+/one-glucose Symporter (Transport Across Intestinal Epithelium)

The plasma membrane of intestinal epithelial cells is organized into at least two discrete regions, each with different sets of transport proteins. The portion of the plasma membrane facing the intestine luman, the apical surface is specialized for absorption, and the rest of the membrane, referred to as basolateral surface, mediates transport of nutrients from the cell to surrounding fluid which lead to the blood. This type of specific transport from the specific region of a cell is called transcellular transport. Many cells, such as those, lining the small intestine and the kidney tubules use this Na^+/glucose symporter protein mediated transport to concentrate glucose against the concentration gradient. The protein contains 14 transmembrane a helices, out of which N-terminal portion, 1-9 α-helices, is required to couple Na^+ binding and glucose transport, while 10-14 α-helices function as glucose permeation pathway. It couples transmembrane movement of one glucose molecule with the transport of two Na^+ ions. The movement of Na^+ from the external medium into the cell is driven by two forces : (i) Na^+ concentration gradient and (ii) inside-negative membrane electric potential. In the steady state, all the Na^+ ions transported from the intestinal lumen into the cell, during Na^+/glucose symport, are pumped out by Na^+/K^+ -ATPase which is found exclusively in the basolateral membrane. The coordinated operation of these transporters allows uphill movement of glucose (and similarly amino acids) from the intestine into the cell. The glucose from inside the intestinal epithelial cells is exported down their concentration gradients into the blood via uniport protein, GLUT2, localized exclusively in the basolateral membranes of intestinal epithelial cells. The tight junctions between the epithelial cells prevent these molecules from diffusing back into the intestinal lumen. The model of transport mechanism by Na^+/glucose symporter is depicted in Figure 16, which has not yet been experimentally supported, but entails conformational changes analogous to those that occur in uniporter, such as GLUT1, described above. GLUT2 is a homolog of GLUT1, but GLUT1 generally functions to import glucose into many body cells. In the epithelial cells lining kidney tubules, there is a second type of Na^+/glucose symport protein, *one-Na^+/one-glucose symporter*. Both types are almost similar in amino acid sequence and mechanism of transport.

Na^+/Ca^{++} Antiporter

The plasma membrane of most cells contains one or more types of antiporters, which couple movement of one cotransported ion, usually Na^+, down its electrochemical gradient to movement of a different molecule in the opposite direction against a concentration gradient. In cardiac muscle cells, Na^+/Ca^{++} antiporter plays the principal role in maintaining a low concentration of Ca^{++} in the cytosol. The reaction of this cation antiporter can be written as :

$$3\,Na^+_{out} + Ca^{++}_{in} \leftrightarrow 3\,Na^+_{in} + Ca^{++}_{out}$$

The Na^+/K^+ ATPase in the plasma membrane of cells creates the Na^+ concentration gradient which is used to power export of Ca^{++} ions.

Cl^-/HCO_3^- Anion Antiporter

AE1 protein, earlier named as band 3 protein, (having about 930 amino acids) is the multipass integral protein of the mammalian erythrocytes membranes and it is folded into 12 transmembrane α helices. This anion antiporter catalyzes the one-for-one exchange of Cl^- and HCO_3^- across the plasma membrane. As there is no change in net movement of electric charge and the reaction is not affected by the membrane potential , the direction of the reaction is dependent only on the concentration gradients of transported ions. This anion exchange is

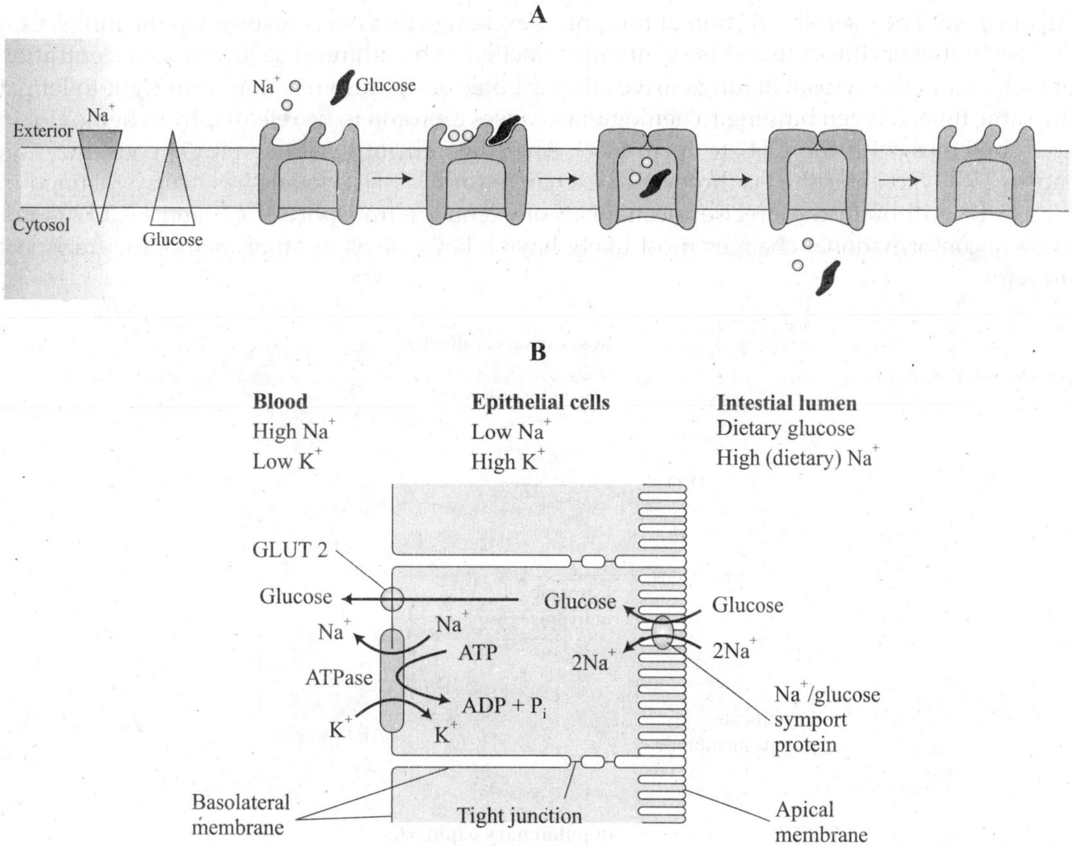

Fig. 16. **(A)** Model of the conformational change mechanism of the two-Na^+/one glucose symporter. The simultaneous binding of Na^+ and glucose to sites on exoplasmic surface induces a conformational change, generating a transmembrane pore or tunnel that allows both bound Na^+ and glucose to move through the protein to binding sites on the cytosolic domain and then to pass into the cytosol. **(B)** The involvement of this symporter in the apical region of the intestinal epithelial cells helping the absorption of glucose from the lumen of intestine. The glucose is then transported to blood by facilitated diffusion using glucose uniporter, GLUT2, present in the basolateral membrane of the intestinal epithelial cells. (From Lodish et al. (2000) *Molecular cell biology*, 4th edition, W.H. Freeman and Co.).

essential in the transport of CO_2 and O_2 by erythrocytes. The waste CO_2 released from cells in peripheral tissues into the capillary blood diffuses across the erythrocyte membrane. The potent enzyme carbonic anhydrase inside the erythrocyte converts CO_2 to water soluble bicarbonate (HCO_3^-) anion.

$$H_2O + CO_2 \leftrightarrow HCO_3^- + H^+$$

The release of oxygen from hemoglobin into the peripheral capillaries induces a conformational change in the globin polypeptide that enables histidine side chain to bind the proton produced by the carbonic anhydrase reaction. Meanwhile, the HCO_3^- formed by carbonic anhydrase is transported out of the erythrocyte in exchange for an entering Cl^- via

AE1 protein. The overall direction of this anion-exchange process is reversed in the lungs, CO_2 diffuses out of erythrocyte and is eventually expelled in breathing. The lowered concentration of CO_2 within the cytosol in lungs drives the carbonic anhydrase reaction from right to left, at the same time, oxygen binding to hemoglobin causes a proton to be released from hemoglobin and combines with the OH^- to form H_2O. The lowered intracellular HCO_3^- concentration causes HCO_3^- to enter the erythrocyte in exchange for Cl^-. The details has been explained in Figure 17. Although the precise mechanism of exchange transport of Cl^- and HCO_3^- is not known, conformational changes most likely have a key role as in other membrane transport proteins.

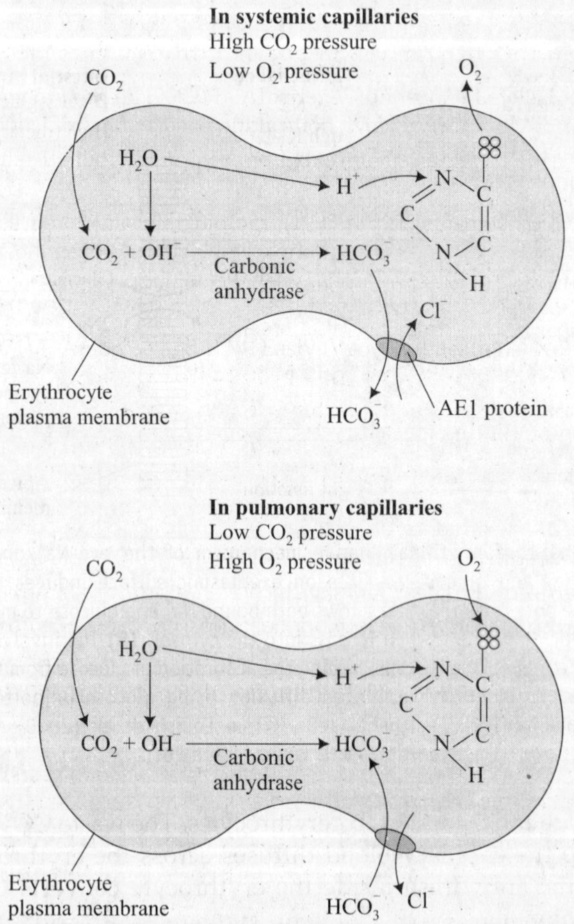

Fig. 17. Anion (Cl^- and HCO_3^-) transport across the erythrocyte membrane in systemic and pulmonary capillaries. AE1 protein, an anion antiporter, catalyzes the reversible exchange of Cl^- and HCO_3^- ions across the membrane and works in conjunction with carbonic anhydrase. In systemic capillaries, the overall reaction causes HCO_3^- to be released from the cell which is essential for CO_2 transport from the tissues to the lungs. In the lungs, the overall reaction is reversed.

Primary Active Transport by ATP-Powered Pumps

These have one or more binding sites for ATP on the cytosolic face of the membrane. The details about some of these pumps are described below.

Na^+-K^+ -ATPase: In 1957, Jens Skou, a Danish physiologist, discovered an ATP-hydrolyzing enzyme in the nerve cells of a crab that was only active in the presence of both Na^+ and K^+ ions (as well as Mg^{++}, which acts as a cofactor). Skou proposed that this enzyme was active in transporting Na^+ and K^+ ions. It is present in the plasma membrane of all animal cells and operates as antiporter, pumping out Na^+ ions and taking in K^+ ions. The Na^+ gradient produced by the pump drives the transport of most nutrients into animal cells and also has a crucial role in regulating cytosolic pH. This ion pump is a tetramer of two α and two β-subunits. The α subunit is a 120,000 molecular weight nonglycosylated polypeptide, contains Na^+ and ATP-binding sites on the cytosolic face, while the K^+ binding site faces the outer surface. The over all process of transport moves three Na^+ ions out of and two K^+ ions into the cell per ATP hydrolyzed and the movement of both types of ions is coupled. The inhibitor of this pump, drug ouabain binds to a specific region on the exoplasmic surface of the protein, competing the K^+ binding, and inhibiting its ATPase activity. The Na^+-K^+ - ATPase is an example of P-type transporter (Table 8), where 'P' stands for phosphorylation, indicating that, during pumping cycle, hydrolysis of ATP leads to the transfer of a phosphate group to an amino acid, aspartatic acid residue of the protein which causes an essential conformational change within the protein. The transport cycle depends on phosphorylation dephosphorylation cycle The structure and the mechanism of action of this ATPase is outlined in Figure 18.

In its E1 conformation, the protein has three high-affinity Na^+ -binding sites and two low-affinity K^+ -binding sites on the cytosolic-side of the protein. The K_m for binding of Na^+ to these cytosolic sites is 0.6 mM, much lower than the intracellular Na^+ concentration (12 mM), as a result Na^+ ions normally will fill these sites. When the protein binds three Na^+ ions on the inside of the cell and becomes phosphorylated after ATP hydrolysis, it shifts from E1 conformation to E2 conformation. In doing so, the Na^+ binding sites become exposed to the outside and protein loses its high affinity for Na^+ ions, generating three low affinity sites for Na^+ ions and they are released outside the cell. Transition to E2 conformation also generates two high affinity K^+ binding sites on the exoplasmic face thereby picking two K^+ ions, and the protein is dephosphorylated. Because the K_m for K^+ binding to these sites (0.2 mM) is very much lower than the extracellular K^+ concentration (5 mM), these two sites fill quickly with K^+ ions. Dephosphorylation of protein shifts back to original E1 conformation, releasing K^+ ions inside the cell. The cycle is then repeated.

Ca^{++} ATPase or Ca^{++} Pump

Ca^{++} ions take part in signal transduction pathways in all cells and control many functions including muscle contraction. For this, the free, unbound cytosolic Ca^{++} concentration should be kept below 0.1 – 0.2 μM. The low concentration of cytosolic Ca^{++} is maintained by Ca^{++} ATPases that transport Ca^{++} out of the cell against its electrochemical gradient. This is also a P-class ion pump and contains a catalytic α subunit with an ATP-binding site, as found in other P-class pumps. Plasma membrane Ca^{++} ATPase also contain the Ca^{++} -binding regulatory protein calmodulin. A rise in cytosolic Ca^{++} induces the binding of Ca^{++} ions to calmodulin, which

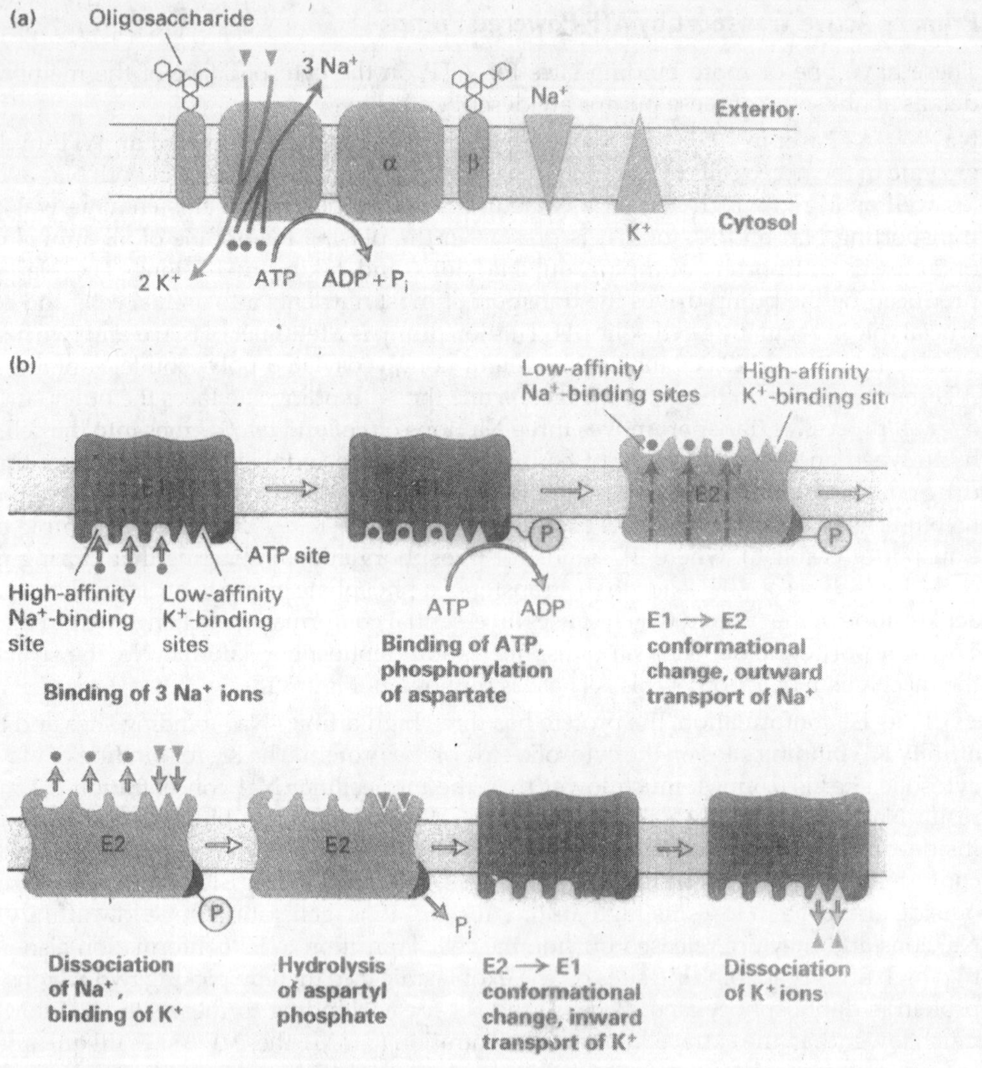

Fig. 18. Model of the structure and function of the Na^+/K^+ ATPase in the plasma membrane. **(a)** This P-class pump comprises two copies each of a small glycosylated β subunit and a large α subunit, which performs ions transport. Hydrolysis of one ATP to ADP and P_i is coupled to export of three Na^+ ions and import of two K^+ ions against their concentration gradients. It is not known whether only one α subunit, or both, in a single ATPase molecule transports ions. **(b)** Ion pumping involves a high-energy acyl phosphate intermediate (E1~P) and conformational changes. (From Lodish et al. (2000) *Molecular cell biology*, 4th edition, W.H. Freeman and Co.).

triggers an allosteric activation of the Ca^{++} ATPase, to accelerate the export of Ca^{++} ions from the cell and the original low cytosolic concentration of free Ca^{++} is restored rapidly. Besides the plasma membrane Ca^{++} ATPase, muscle cells contain a second, different Ca^{++} ATPase that transports Ca^{++} from the cytosol into the lumen of the muscle sarcoplasmic reticulum (SR) which stores Ca^{++} ions. It constitutes more than 80 per cent of the integral protein in the SR membranes and has been easily purified. Each muscle Ca^{++} ATPase catalytic α subunit (molecular weight, 100,000) has 10 transmembrane α helices and transports two Ca^{++} ions per ATP hydrolyzed. The release of Ca^{++} from the sarcoplasmic reticulum into the muscle cytosol causes contraction, and the removal of Ca^{++} ions from the cytosol by the muscle calcium pump induces muscle relaxation.

The mechanism of action of Ca^{++} ATPase involves ATP hydrolysis, conformational change and ion pumping. When the protein is in one conformation, termed E1, two Ca^{++} ions bind to high-affinity sites on the cytosolic surface (step 1, Figure 19). Then an ATP binds to its site on the cytosolic surface and hydrolyzed to ADP. The liberated phosphate is transferred to a specific aspartate residue in the protein. The protein then changes its conformation to E2 (step 2), generating two low-affinity Ca^{++}-binding sites on the exoplasmic surface which faces the SR lumen. This conformational change simultaneously propels the two Ca^{++} ions through the protein to these low affinity sites (step 3), and then Ca^{++} ions dissociate from the exoplasmic surface (step 4). Following this, the bound phosphate to aspartate residue is hydrolyzed causing E2 to revert to E1 (Fig. 19).

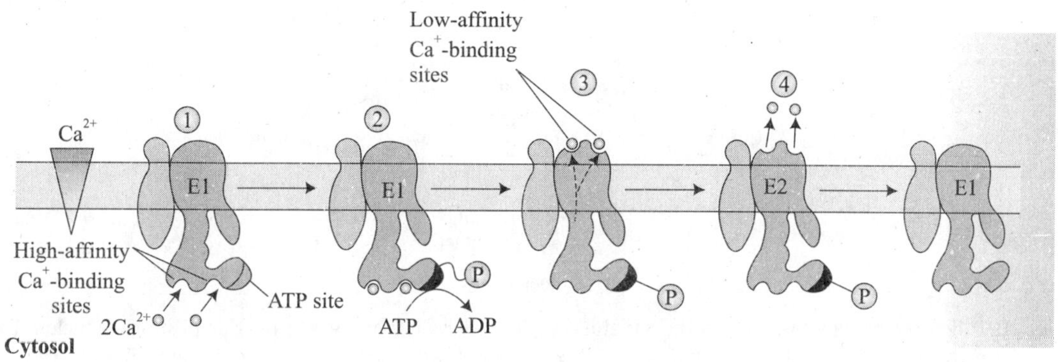

Fig. 19. Model of the mechanism of action of Ca^{++}ATPase. Only one of the two α subunits is depicted. E1 and E2 are alternate conformational forms in which Ca^{++} -binding sites are on the cytosolic and exoplasmic faces respectively. The conformational change from E1 to E2 is induced by ATP hydrolysis and phosphorylation of pump. (From Lodish et al. (2000) *Molecular cell biology*, 4th edition, W.H. Freeman and Co.).

The H^+/K^+ ATPase and Acidification of the Stomach Lumen

Mammalin stomach need strong acidic pH (~ pH 1-2) for the activity of various proteolytic enzymes and this is achieved by transcellular transport as described for intestinal epithelial cells. Hydrochloric acid (HCl) is secreted into the stomach by the parietal cells (also known as oxyntic cells) present in the gastric lining. These cells contain a H^+/K^+ ATPase in their apical membrane facing the stomach lumen. This enzyme is a P-class ATPase, similar in structure and function to the Na^+/K^+ ATPase, but it exports one H^+ and imports one K^+ for each ATP hydrolyzed, producing no net movement of charge. During acidification of the stomach lumen, H^+/K^+ ATPases, pump H^+ ions in the lumen, but the pH of the parietal-cell cytosol does not change. Parietal cells accomplish this feat by means of Cl^-/HCO_3^- antiporters present in the basolateral membrane. The excess cytosolic OH^-, generated by exporting protons, combines with CO_2 that diffuses into the cell from blood forming HCO_3^- in a reaction catalyzed by cytosolic carbonic anhydrase. The HCO_3^- is transported out across the basolateral membrane of the cell into the blood in exchange for an incoming Cl^- ion by means of an antiporter that is similar in structure and function to the erythrocyte AE1. The Cl^- ions, thus, imported into the cell exit through Cl^- channels in the apical membrane, entering the stomach lumen. The excess K^+ ions pumped inward by the H^+/K^+ ATPase, is returned across the apical membrane through a separate K^+ channel to the stomach lumen. The net result id the accumulation of both H^+ and Cl^- ions (HCl) in the stomach lumen (Fig. 20).

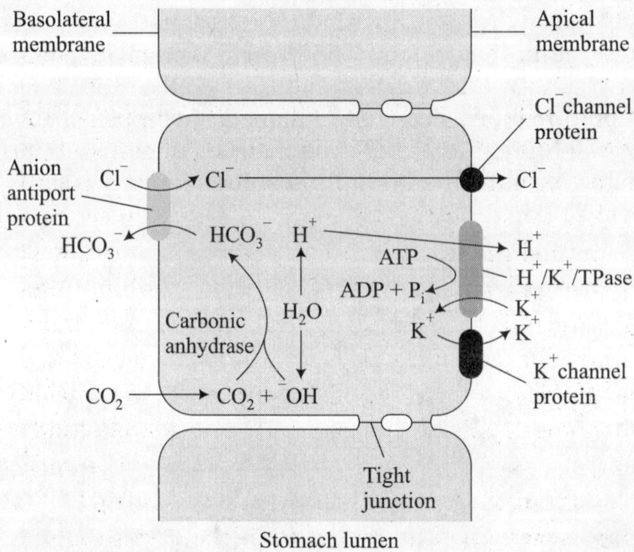

Fig. 20. The transcellular transport in the parietal cells (oxyntic cells) for the acidification of stomach lumen. The apical membrane contains a P-class pump, H^+/K^+ ATPase as well as Cl^- and K^+ channel proteins. The K^+ ions are pumped inward by the H^+/K^+ ATPase and exit via a K^+ channel. The basolateral membrane contains an anion antiporter that exchanges HCO_3^- and Cl^- ions. The combined operation of these four different transport proteins acidifies the stomach lumen while maintaining the neutral pH and electroneutrality of the cytosol.

V-Class H^+ ATPases

All V-class ATPases transport H^+ ions only, so they function as proton pumps. These are present in the membranes of lysosomes, endosomes and plant vacuoles, and function to acidify the lumen of these organelles. These V-class proton pumps contain two discrete domains : a cytosolic-facing hydrophilic domain (V_1) composed of five different polypeptides and a transmembrane domain (V_0) containing 9-12 copies of proteolipid **c**, one copy of protein **b**, and one copy of protein **a.** The subunit composition of the cytosolic domain is $\alpha_3\beta_3\gamma\delta\varepsilon$. The α and β subunits contain the sites where ATP binding and hydrolysis occur. The **c** and **a** subunits together form the proton conducting channel. Unlike P-class ATPases, V-class H^+ ATPases are not phosphorylated and dephosphorylated during proton pump.

ABC Superfamily Transporters

These transporters are a class of carrier protein ATPases that are called the ABC transporter superfamily because each member contains two highly conserved ATP-binding cassettes (ABC). They were characterized for the first time from bacteria. In *E.coli*, 78 genes, comprising about 5% of genes in this bacterium, encode ABC transporters. However, some ABC transporters are now being discovered from eukaryotes also. The transport proteins contain two transmembrane (T) domains and two cytosolic ATP-binding (A) domains. The T domains, each built of sis-membrane-spanning α helices, form the pathway through which the transported substance crosses the membrane and and determine the substrate specificity of each ABC protein. The sequence of A domains is about 30-40% homologous in all members of ABC superfamily, suggesting a common evolutionary origin. ABC superfamily transporters are considered to be of great clinical importance. Some examples of ABC superfamily transporters are as follows :

(i) **Bacterial Plasma-membrane permeases:** Bacterial plasma membranes contain numerous permeases of this family. They use the energy released by ATP hydrolysis to transport specific amino acid, sugars, vitamins, peptides into the cell. This helps for the bacteria to grow in soil or water where the concentration of nutrients is low. Bacterial permeases are inducible i.e. the quantity of the transport protein in the membrane is regulated by both the concentration of the nutrients in the medium and the metabolic need of the cell. In the gram negative bacteria such as *E. coli*, histidine permease is a typical example of ABC protein. In this permease, a soluble histidine-binding protein in the periplasmic space assists in transport. ATP binding leads to dimerization of two ATP-binding domains (A), and ATP hydrolysis leads to their dissociation and transmits structural changes in transmembrane domains (T) and transport molecules across the bilayer.

(ii) **Mammalian MDR transport proteins:** The first eukaryotic ABC transporter was discovered because of unexpected observation of their ability to pump hydrophobic drugs out of the cytosol. These cells develop resistance to several chemotherapeutic drugs, including anticancer drugs (adriamycin, etc.), and this resistance is due to enhanced expression of a multidrug-resistance (MDR) transport protein known as MDR1. In MDR transport proteins, the two T domains and two A domains are part of a single polypeptide, 170,000 MW protein. This protein uses the energy derived from ATP hydrolysis to export a large variety of drugs from the cytosol to the extracellular medium. The *MDR1* gene is highly amplified in multidrug resistant cells. The transporter pumps the drugs out of the cell, thereby reducing their toxicity and conferring resistance to a wide variety of therapeutic agents. MDR1 protein is expressed in abundance in the liver, intestine and kidney which are the sites to remove natural toxic products from the body. Thus, the natural function of MDR1 may be to transport a variety of natural and metabolic toxins into the bile, intestinal lumen, or forming urine. During the course of its evolution, MDR1 appears to have acquired the ability to transport drugs whose structures are similar to these toxins. Tumor developed from these cell tymes, such as liver cancers (hepatomas) are resistant to virtually all chemotherapeutic agents, because tumor cells exhibit increased expression of the MDR1 or MDR2. Similarly, malarial parasite, *Plasmodium falciparum* develops resistance to antimalarial drug, chloroquine and it hampers the control of malaria. The resistant *P. falciparum* have amplified a gene encoding an ABC transporter that increasingly pumps out the chloroquine.

(iii) **Cystic fibrosis transmembrane regulator (CFTR) protein:** In humans, the common lethal autosomal disease cystic fibrosis (CF) was shown to be due to a mutation in the CFTR gene which encodes a ABC transporter, chloride-channel protein, in the plasma membrane of epithelial cells in the lungs, sweat gands, pancreas and other tissues, and it is regulated by cAMP. An increase in cAMP stimulates Cl^- transport by such cells from normal individuals, but not from CF individuals. The structure of CFTR protein is very similar to those of MDR1 protein, except for the presence of an additional regulatory domain on the cytosolic face. The Cl^- channel activity of CFTR protein is enhanced by binding of ATP. Since CFTR protein is similar to MDR1 in structure, it may also function as an ATP-powered pump of some as-yet unidentified molecule.

ION CHANNELS

Unlike carrier proteins, channel proteins form hydrophilic pores across membranes. The rate of transport through ion channels is 1000 times greater than the rate of transport mediated by a carrier protein. Because these proteins are concerned specifically with inorganic ion transport, they are referred to as ion channels and they always mediate passive ('downhill') transport, showing ion selectivity (ion specific). The separate ion channels for Na^+, K^+, Ca^{++} and Cl^- allow the transport of specific ions down their electrochemical gradient across the plasma membrane. The ion channels are not continuously open, but are gated which allows them to open briefly and then close again. The main types of stimuli that are known to cause ion channels to open, are :

Voltage-gated channels: change in the voltage due to transport of Na^+, K^+ and Ca^{++} across the membrane.

Mechanically gated channels: mechanical stress.

Ligand gated channels: binding of a ligand (ligand could be a neurotransmitter i.e. acetylcholine, serotonin, glutamate and gamma aminobytyric acid (GABA), glycine-*transmitter gated channels*, or a nucleotide-*nucleotide gated channels*, or an intracellular mediator such as ion-*ion gated channels*).

More than 100 types of ion channels have been described so far. They are responsible for the electrical excitability of muscle cells and also in the electrical signaling in nervous system. But ion channels are not restricted to electrically excitable cells. They are present in all animal cells and are found in plant cells and microorganisms. Voltage-gated cation channels generate action potentials in neurons. Voltage-gated channels are formed either of a more or less cylindrically symmetric aggregate of multiple subunits or of tandem-repeated homologous domains within a single polypeptide. Voltage-gated ion channels have high degree of similarity in amino acid sequence suggesting their common evolutionary origin. K^+ channel, built of four identical subunits, equivalent to the four domains of the voltage-gated Na^+ channel, are arranged in the membrane around a central pore. Each subunit polypeptide contains six membrane-spanning a helixes, designated S1-S6. Transmitter-gated ion channels have 5-fold symmetry, with the subunit composition $\alpha_2\beta\gamma\delta$. The actual channel is a tapered central pore, with a maximum diameter of 2.5 nm, formed by segments from each of the five subunits. The channel opens when the receptor binds two acetylcholine molecules to sites located at the interfaces of the αä and αã subunits. GABA- and glycine-gated Cl^- channels are found at inhibitory synapses. These are also pentameric proteins like acetylcholine receptors, but are built of only one or two different types of subunits.

Water Channels

Plasma membrane is only slightly permeable to water, but small changes in extracellular osmotic strength causes most animal cells to swell or shrink rapidly. It has been now known that plasma membranes of erythrocytes and other cell types contain water-channel proteins that accelerate the osmotic flow of water. For the important discovery on the existence and significance of water channels in different cells, Peter Agre and Roderick MacKinnon, of USA were awarded the Nobel Prize in 2003 in chemistry. The absence of these water channels in frog oocytes and eggs protects them from osmotic lysis. The water channel protein, *aquaporin*, has been well characterized from erythrocytes. It is a tetramer of identical 28-kDa subunits, each of which contains six transmembrane α helices, two from each subunit. Aquaporin or homologous proteins are expressed in abundance in erythrocytes and other cells e.g. kidney cells that resorb water from urine, that exhibit high permeability for water.

SUMMARY

Cell membrane or plasma membrane provides a selectively permeable barrier to cells. Basically lipids, proteins and carbohydrates are the chemical constituents of the plasma membrane. The type and quantity of each constituent is variable from one cell membrane to other. Membranes are highly asymmetric structures whose two leaflets have different properties. The basic structure of plasma membrane is provided by lipid bilayer consisting of a bimolecular layer of amphipathic lipids which are held together by noncovalent bonds. The fluidity of a membrane is affected by variation in the composition of lipids, degree of saturation/unsaturation of fatty acids and temperature. Membrane proteins carry out most membrane related functions serving as receptors, enzymes, transporters etc. The proteins in the membranes are present as integral membrane proteins, peripheral membrane proteins and lipid-anchored proteins. The lipid molecules and many membrane proteins are able to diffuse in the plane of the membrane i.e. lateral mobility. Membrane carbohydrates are present as oligosaccharides on the outer monolayer of lipid bilayer, attached to proteins or lipids.

The regulated transport of various molecules across the membranes is one of the important membrane functions. Lipid soluble (nonpolar) molecules can pass the plasma membrane quite easily while charged molecules (polar) do not pass through the membrane. The transport of these specific molecules, ions is done by specific proteins in the membranes, which could be either as carrier proteins or channel proteins, which are specific for a specific solute. Carrier proteins bind specific solutes and transfer them across the lipid bilayer by undergoing conformational changes that expose the solute-binding site sequentially on one side of membrane and then on the other side. The transport by carriers can be either active or passive while solute flow through channel proteins is always passive. Water channels known as aquaporins, are found in plasma membranes of erythrocytes and other cell types e.g. kidney cells, that exhibit high permeability for water.

REFERENCES

Agre, P., Bonhivers, M. and Borgnia, J. (1998) The aquaporins, the blue prints for cellular plumbing systems. J. Biol. Chem., 273 : 14659-14662.

Bevers, E.M., Comfurius, P., Dekkers, D.W. and Zwaal, R.F. (1999) Lipid translocation across the plasma membrane of mammalian cells. Biochim Biophys. Acta, 1434 : 317-330.

Danielli, J.F. and Davson, H. (1935) A contribution to the theory of permeability of thin films. J. Cell Comp. Physiol., 5 : 495 - 508.

Eisenberg, D.(1984) Three dimensional structure of membrane and surface proteins. Annual Rev. Biochem., 53 : 595 - 623.

Langmuir, I. (1917) The constitution and fundamental properties of solids and liquids. II. Liquids. J Am. Chem. Soc., 39 : 1848-1906.

Overton, C. E. (1899) Über die allgemeinen osmotischen eigenschaften der zelle, ihre vermutlichen ursachen und ihre bedeutung für die physiologie. vierteljahrsschrift der naturforschenden gesellschaft in Zürich. 44 : 88-135.

Robertson, J.D. (1957) new observations on the ultrastructure of the membranes of frog peripheral nerve fibers. J

Singer, S.J. and Nicolson, G.L. (1972) The fluid mosaic model of the structure of cell membranes. Science, 175 : 720-731.

Suggested Readings

1. Alberts, B., Johnson, A., Lewis, J., Raff, M., Roberts, K. and Walter, P. (2002) Molecular Biology of the cell, 4th edition, Garland Science Publication, New York.
2. Karp, G. (1999) Cell and Molecular Biology: concepts and experiments, 2nd edition, John Wiley & Sons, New York.
3. Lodish, H., Berk, A., Zipursky, S.L., Matsudaira, Baltimore, D. and Darnell, J. (2000) 4th edition, W.H. Freeman and Company, New York.
4. Kleinsmith, L.J. and Kish, V.M. (1995) Principles of cell and molecular biology, 2nd edition, Harper Collins College Publishers, New York.
5. Evans, W.H. and Graham, J.M. (1989) Membrane structure and function. IRL press, Oxford, England.
6. Jain, M.K. (1988) Introduction to biological membranes, 2nd edition, John Wiley & Sons, New York.

REVIEW QUESTIONS

Long Questions

1. Discuss the salient features of different models for the structure of plasma membrane. How the fluid mosaic model could be acceptable to understand the structure of plasma membrane?
2. What are the chemical components of the plasma membrane? Discuss their salient features and molecular organization in the plasma membrane. Give an account on the factors which affect the fluidity of the plasma membrane.
3. Discuss the structural features of different phospholipids and glycolipids present in the plasma membrane. Add a note on the lipid asymmetry in plasma membrane.

4. Discuss the different types and ways of association of proteins in the plasma membrane. Illustrate with suitable examples. Give an account on the gycosylation of membrane proteins.
5. What do you understand by membrane asymmetry? Discuss the significance of asymmetry with reference to different chemical constituents of plasma membrane.
6. Write short notes on any four: (i) liposomes, (ii) Cholesterol, (iii) Glycocalyx, (iv) Single pass and multi pass transmembrane proteins, (v) Glycophorin, (vi) Membrane fluidity (vii) membrane asymmetry., (viii) Unit membrane model of plasma membrane
7. Distinguish between simple diffusion, facilitated diffusion and active transport. Discuss the significance of ion channels in membrane transport.
8. Describe the salient features of uniport, symport and antiport giving suitable examples. How does carrier proteins differ from channel proteins in the mechanism of transport.
9. Describe the structure and mechanism of transport through Na^+-K^+-ATPase.
10. Write short notes on any four: (i) ion channels, (ii) ionophores, (iii) Na^+-driven glucose symport, (iv) Ca^{++} pump, (v) Active transport, (vi) Water channels

Multiple Choice Questions

1. Plasma membrane's asymmetrical feature is associated with
   (a) lipids only (b) proteins only
   (c) carbohydrates only (d) all these three components.
2. Glycophorin of mammalian erythrocytes is a
   (a) peripheral protein (b) integral protein
   (c) cytosolic protein (d) lipid
3. Fluidity of plasma membrane will
   (a) increase with increase in unsaturated fatty acids in phospholipids of lipid bilayers
   (b) decrease with increase in unsaturated fatty acids in phospholipids of lipid bilayers
   (c) not change with increase in unsaturated fatty acids in phospholipids of lipid bilayers
   (d) decrease with increase in temperature.
4. Sphingomyelin has
   (a) glycerol and phosphoryl choline
   (b) sphingosine and phosphoryl choline
   (c) sphingosine and glycerol
   (d) sphingosine and serine
5. Membrane transport through carrier proteins could be
   (a) only active transport
   (b) only passive transport
   (c) active as well as passive transport
   (d) as endocytosis

6. Membrane transport through channel proteins could be
   (a) only active transport
   (b) only passive transport
   (c) active as well as passive transport
   (d) as endocytosis
7. Plasma membrane glycolipids are always derived from
   (a) sphingosine (b) glycerol
   (c) both sphingosine and glycerol (d) none of the above
8. Na^+-K^+ ATPase pump
   (a) passively transports three Na^+ ions out and two K^+ ions inside the cell
   (b) passively transports two Na^+ ions out and three K^+ ions inside the cell
   (c) actively transports three Na^+ ions out and two K^+ ions inside the cell
   (d) actively transports two Na^+ ions out and three K^+ ions inside the cell
9. Valinomycin is a
   (a) mobile ion carrier ionophore (b) channel-forming ionophore
   (c) antibiotic to kill the bacteria (d) hydrophilic moplecule
10. Water channels are
   (a) aquaporins, present in bacteria, plants and animals
   (b) aquaporins, present only in animals
   (c) porins, present in plants
   (d) porins, present in bacteria, plants ant animals

Answers

1. (d)	2. (b)	3. (a)	4. (b)	5. (c)	6. (b)
7. (a)	8. (c)	9. (a)	10. (a)		

CHAPTER 2

Endoplasmic Reticulum

Shelley Bhattacharya
Dept. of Zoology, School of Life Sciences, Visva-Bharati University, Santiniketan–731 235
E-mail: bhattacharyashelley@rediffmail.com

The term endoplasmic reticulum or ER has been derived from two simple words, endoplasmic meaning within the cytoplasm and reticulum meaning network. It is a member of the endomembrane organellar system and present in all eukaryotic cells. Prokaryotic organisms do not possess any organelle hence ER is also absent in these cells. The name endoplasmic reticulum was first used by Porter *et al.* (1945) when they noted the fibroblast-like cells in the cultures of chick embryonic tissues. The term endoplasmic reticulum is now applied to describe similar structures in all types of cells. However, the nomenclature differs in certain cases, such as ER in the muscle cells are known as Sarcoplasmic Reticulum (SR) and is specially adapted for calcium ion storage and release. The process is mediated by voltage gated calcium channels, and the calcium ATPase that pumps calcium ions into the SR. In striated muscle it is specially adapted to surround the myofibrils, forming triads with invaginations of the plasma membrane called T-tubules.

ORGANIZATION OF THE ER

The ER consists of an extensive membrane network of tubes and cisternae which are saclike structures enclosing a space, the cisternal space or internal lumen. Parts of the ER membrane are continuous with the outer membrane of the nuclear envelope, and the cisternal space of the ER is continuous with the space in between the two layers of the nuclear envelope. Parts of the ER covered with ribosomes which assemble amino acids into proteins based on instructions from the nucleus. Their rough appearance under electron microcopy led to their being called rough ER (rER), other parts are free of ribosomes and are called smooth ER (sER). The ribosomes on the surface of the rER insert the freshly produced proteins directly into the ER, which processes them and then passes them on to the Golgi apparatus. Rough and smooth ER differs not only in appearance, but also in function. The location of the ER corresponded to the basophilic regions of the cell seen in the light microscope. The earliest observations of this organelle in thin sections described it as a continuous three-dimensional reticulum made up of a network of membrane-enclosed spaces in the form of tubules that were in continuity with layers of flattened cisternae.In higher eukaryotes, the ER is a continuous, interconnected membrane system consisting of various subdomains, each associated with a particular function, In early studies, a fluorescent dye (dicarbocyanine) was used to study the continuity of the membrane organization of the ER (Jaffe and Terasaki, 1994). Fluorescence-loss-in-photobleaching (FLIP) experiments have been done to establish the continuity of the ER structure in sea urchin cells. GTP was repeatedly photobleached which resulted in loss of fluorescence throughout the cell (Terasaki, 2000). It has been shown that the interconnected dynamic ER network is structurally composed of tubules, 3-way junctions, cisternal sheets, and a polygonal lattice-like reticulum (Dreier and Rapoport, 2000). The basic movements undertaken by the structural components of the ER network in eukaryotic organisms include continuous tubule elongation and retraction, reorganization of polygons resulting in the addition or loss of a polygon from the network, branching of tubules, and sliding of the 3-way junctions (Fehrenbacher *et al.*, 2002).

THE ROLE OF CYTOSKELETON IN STRUCTURAL ORGANIZATION OF THE ER

ER tubules can arise from pre-existing tubules as evidenced by time lapse fluorescence imaging analysis. ER tubules are transported along the microtubule cytoskeleton along with a motor protein. This type of transportation has been termed the "sliding mechanism" model. ER tubules also extend by attaching themselves to the polymerizing end of a microtubule via a tip attachment complex (TAC). This type of motility has been referred to as the TAC mechanism model. The third model has been described as the "microtubule movement mechanism" which postulates that ER tubules form attachments to the side of a moving microtubule (Baumann and Walz, 2001). The ER in most eukaryotic cells is dispersed along the microtubule cytoskeleton. Microscopy studies using nerve cells have shown that kinesin drives transport of the ER towards the cell periphery in a plus-end directed manner (Feiguin *et al.*, 1994). The kinesin motor and the kinesin binding protein, kinectin, each associated with ER elements, and immunofluorescence experiments have demonstrated that the ER localizes with the kinesin motor and kinectin (Lane and Allan, 1999).

Regional Variations

The ER is best described as a complex branching network of membranes more concentrated in the core of the cytoplasm or the endoplasm than in the peripheral or ectoplasmic region. The structural appearance of this organelle varies from vesicles, tubules, large flattened sacs or cisternae and folded layers of membranes. The basic structure of ER is very similar to that of the plasma membrane and is an extension of the nuclear membrane. It also maintains communication with the Golgi complex. Considering the interlinking of all these organellar membranes in the cytoplasm, these structures are said to belong to the endomembrane system of the cell. The structure has been studied under video microscopy of cell cultures which reveals a complex branching network of membrane bound cavities or cisternae and is more concentrated in the endoplasmic region of the cell. Under electron microscopy the membrane appears as parallel pairs of the single or unit membranes. Traditionally on the basis of structure the ER has been classified into three regions, the rough ER or rER, the smooth ER or sER and the nuclear envelope. The rER is studded with ribosomes and is present in all rER appears more tubular and granular under the EM which is due to the attachment of ribosomes to its surface. The rough ER or granular ER manufactures and transports proteins destined for membranes and secretion. It synthesizes membrane, organellar, and secretory proteins. Immediately after proteins are synthesized most of them leave to the Golgi apparatus within vesicles. The rough ER also modifies, folds, and controls the quality of proteins.

On the other hand the sER appears rather smooth, more dilated and convoluted (Baumann and Walz, 2001; Voeltz *et al.*, 2002). The smooth ER or agranular ER has functions in several metabolic processes. It takes part in the synthesis of various lipids for building membranes such as phospholipids, fatty acids and steroids including hormones, and also plays an important role in carbohydrate metabolism, detoxification of the harmful chemicals by the enzymes in the smooth ER, and calcium storage. It is well developed in cells that produce steroid hormones and also in liver cells. Moreover it is known to be a large transporter of nutrient found in each cell. The physiological status of the cell determines the amount of rER and sER present in a cell at any time point. In the secretory cell type such as the cells of the pancreatic acini the rER is more abundant. While sER is abundant in the cell types involved in steroid synthesis such as adrenal and gonads, detoxification such as hepatocytes and storage of calcium as in the striated muscle cells. It has been found that both sER and rER contain most of the proteins synthesized by the cell, and those proteins which are involved in protein translocation or protein processing are found to be more abundant in the rER (Johnson and van Waes, 1999). Besides the divisions of sER and rER another type of ER is known as transitional ER (tER). This variety of tER can be classifed as a special type of sER involved in the packaging of proteins in the cell into ER-derived coat protein II (COPII) vesicles (Hobman *et al.*, 1998).

Biochemical Studies

The ER of the hepatocyte has been studied extensively and has been found to function in protein synthesis and transport, fatty acid synthesis, phospholipid and triglyceride synthesis, glycogen metabolism, including both deposition and depletion, and the metabolism of lipid-soluble xenobiotics as well as many endogenous substances. These same ER functions may be greater or diminished in other cell types, depending upon the cellular/organic activities of that cell and organ. Steroid-secreting endocrine cells possess an elaborate ER specialized for steroid hormone synthesis; the ER of intestinal absorptive cells can shift function toward lipid metabolism in response to a fatty meal. Membranes are involved also in the regulation of sequestering and/or transfer of ions. Ca^{2+}- ATPases bound to the ER membranes enable the movement of large quantities of Ca^{2+} from the cytosol to the lumen of the ER. Ca^{2+} sequestration is more highly developed in the muscle cell and occurs in the sarcoplasmic reticulum, a specialized form of ER.

FUNCTIONAL PROPERTIES OF THE ENDOPLASMIC RETICULUM

The endoplasmic reticulum serves many general functions, including the facilitation of protein folding, and the transport of proteins. Correct folding of newly made proteins is made possible by several ER proteins including Hsc70 family, calnexin, calreticulin, and the peptidylpropyl isomerase family. Only properly folded proteins are transported from the rER to the Golgi complex.

Transport of Proteins

Secretory proteins are moved across the ER membrane. Proteins that are transported by the ER and from there throughout the cell are marked with specific markers that are called signal sequences. The N-terminus of a polypeptide chain contains a few amino acids that work as the signal sequence, which are removed when the polypeptide reaches its destination. Proteins that are destined for places outside the ER are packed into transport vesicles and moved along the cytoskeleton towards their destination. The ER is also a part of a protein sorting pathway of the eukaryotic cell. The ER modifies proteins, makes macromolecules, and transfers substances throughout the cell. The ER is the site of translation, folding, and transport of proteins that are to become part of the cell membrane such as transmembrane receptors and other integral membrane proteins, proteins and that are to be secreted or "exocytosed" from the cell such as the digestive enzymes. Integral proteins need to be inserted into the ER membrane after they are synthesized which requires the correct topogenic sequences.

Adaptive Functions of the ER

The location and arrangement of ER within the cell also differ according to cell type and function. Some protein secretory cells have extensively developed arrays of rER arranged in parallel stacks, as seen in the pancreatic acinar cells. The rER is distributed throughout the cytoplasm in plasma cells which synthesize and secrete immunoglobin. Other cells may have rER mainly in an apical or basal orientation as in intestinal epithelial cells or thyroid follicular cells, respectively. While the amount and location of ER within various cell types are an aid to distinguishing one type from another, the biochemical explanation for the occurrence is not always clear. In nerve cells rough ER exists as large, flattened sacs lying on one another in lamellar fashion to form masses, Nissl bodies, identifiable by light microscopy. Hepatic parenchymal cells contain smaller blocks of rough ER. In plasma cells the rough ER is rather uniformly distributed throughout the cytoplasm except in the region of the cytocentrum. It may be tubular, vesicular, or flattened, depending on the phase of antibody secretion. Rough ER occupies the base of the pancreatic acinar cell. This rough ER, recognizable by light microscopy as basophilic material due to the affinity of its ribosomes for cationic dye, is also termed ergastoplasm.

The ER, as exemplified in hepatocytes responds to a wide range of stimuli to enable cellular adaptation to changing conditions in the extracellular milieu. In most cases there is a correlated structural-functional response. The liver's central role in the detoxification and metabolism of foreign as well as endogenous substances is underscored by the fact that 29% of the cardiac blood output is received by the liver. Therefore, any substance carried in the circulation can be rapidly acted on by the liver where the metabolism or "elimination" of lipophilic substances such as drugs, anesthetics, pesticides, toxins, and hormones takes place. A role of hepatic ER in carbohydrate metabolism has been suggested since Fawcett (1955) noted sER in close association with glycogen particles. Based largely on morphological observations it has been suggested that sER is associated with glycogen particles during all stages of feeding-mediated hepatic glycogen deposition and depletion. Apparently the development of sER follows a cyclic pattern. Proliferation of the organelle occurs during early stages of glycogen deposition and remains associated with glycogen particles throughout all stages of carbohydrate deposition until maximum quantities of glycogen are accumulated at which time the quantity of sER is not very much. The initiation of depletion again stimulates sER proliferation and hepatic sER is present throughout all stages of glycogen utilization. At the point where the cell is devoid of glycogen particles sER content decreases but nevertheless is always present.

Smooth ER occurs in a number of cell types and may have diverse functions. It has a role in the production of steroid hormones, and it is abundant in such cells as the Leydig cells of the testis, which produce the steroid testosterone. Smooth ER synthesizes complex lipids from fatty acids. It also detoxifies certain drugs and becomes very prominent in hepatocytes during the inactivation of phenobarbital. In striated muscle, smooth ER is distinctively organized as the sarcoplasmic reticulum whose functions include delivering high concentration of Ca^{++} and other ions to critical places in the sarcomere for muscular contraction and relaxation. Smooth ER in megakaryocytes delimits platelet zones in the cytoplasm and by fusing, freeze platelets from the megakaryocyte. Appropriately, this ER is termed demarcation membrane. Carbohydrate synthesis is associated with smooth ER and the Golgi apparatus. The reformation of the nuclear membrane in telophase is accomplished by smooth ER.

The membranes of the ER possess a self-healing capacity after disruption. When fractions rich in ER are recovered from disrupted cells, the ER is found as small vesicles known as microsomes. These microsomes could then be analyzed morphologically and characterized biochemically, and were shown to be rich in phospholipids and ribonucleic acid. Evidently the tubular system is fragmented, but the membranes reunite or "heal" to form small vesicles. All eukaryotic cells except enucleated red blood cells and sperm have some endoplasmic reticulum. In the pancreatic acinar cell, ER represents 60% of the total cell membrane compared to 15% in the hepatocyte (Garfield and Cardell, 1987). However, compared to the relative abundance of both rER and sER in the liver cell, greater than 99% of the ER in the pancreatic acinar cell is rER (Garfield and Cardell, 1987). In contrast, cells predominantly involved with steroid synthesis and lipid metabolism have a higher amount of sER than rER. The sER is greatly expanded in the intestinal epithelial cell during absorption of a fatty meal. The hepatocyte, therefore, lies between these two extremes of differing ER proportions.

RELATIONSHIP OF THE ER WITH OTHER CELLULAR ORGANELLES

The relationship of the ER with other cellular organelles is well established (Staehlin, 1997) demonstrating that the ER is closely associated with other organelles to facilitate the transfer of products synthesized in the ER. It has been further proposed that these sites of contact with different organelles constitute sub-compartments of the ER within the cell (Staehlin, 1997; Voeltz *et al.*, 2002). The best example of such an association is seen in the organization of the SR which is found to be juxtaposed to the plasma membrane. These sub compartments are known as T-tubules which extend to the interior of the muscle cell. In yeast cortical ER is found to be localized on the peripheral region adjacent to the plasma membrane while the ER around the nuclear envelope adjacent to the vacuole is known as perinuclear ER. It has been reported that a direct association between proteins localized in vacuoles and nuclear-envelope-specific proteins is observed at the nuclear-envelope-vacuole junction (Pan *et al.*, 2000).

ER-peroxisome linkages is observed in both mammalian and yeast cells under the electron microscope where peroxisomes are found adjacent to ER in rat liver and kidney epithelial cells, and in rabbit small intestinal cells (Zaar *et al.* 1987). More recently evidence has been provided demonstrating that peroxisomes are generated from domains of the ER (Hoepfner *et al.* 2005).

THE ER CONTACTS OTHER ORGANELLES

The ER appears to be associated with many other organelles in the cell, possibly to facilitate the transfer of products synthesized in the ER (Staehlin 1997). These sites of contact with different organelles are proposed to constitute subcompartments of the ER within eukaryotic cells (Staehilin 1997; Voeltz *et al.* 2002). For example, the ER in skeletal muscle, which is referred to as the sarcoplasmic reticulum, is juxtaposed to the plasma membrane and to specialized invaginations of the plasma a membrane, called T-tubules that extend into the muscle cell. In yeast a direct association between proteins localized to vacuoles and nuclear-envelope-specific proteins is observed at the nuclear envelope-vacuole junction. Membrane-bound peroxisomes are located adjacent to ER membranes in rat liver cells and kidney. The recent observation that peroxisomes are generated from domains of the ER provides an explanation for the intimate proximity of these two organelles (Hoepfner *et al.* 2005). The proximity of ER membranes and mitochondria was also visualized by fluorescence microscopy and 3-D reconstruction of electron micrographs in yeast cells. In addition to structural studies biochemical analyses have established a link between the ER and mitochondria A mitochondria-associated membrane (MAM) has been isolated from yeast extracts demonstrating ER enzyme activities as found in lysates from mammalian hepatocytes (Ardail *et al.* 1995) which led Prinz *et al.* 2000) to propose that the ER ultimately affects the mitochondrial ultrastructure. The tER or transitional ER is considered a subdomain of the ER. tER is also referred to as an ER exit site located adjacent to Gogi structures in various eukaryotic cells. The tER is believed to generate the ER-Golgi intermediate compartment, which then generates the Golgi (Hammond and Glick, 2000). Even in the parasites *Trypanosoma brucei* and *Toxoplasma gonadii*, a tight association between the Golgi and ER export sites has been observed. Juxtaposition of these two organelles is believed to reflect the extensive membrane flow between the tER and Golgi (Estrada de Martin *et al.*, 2005).

ER INHERITANCE IN ANIMAL CELLS

Inheritance of the ER is essential, since this organelle is required for protein assembly, lipid biosynthesis, and vesicular traffic in eukaryotic cells (Estrada de Martin *et al.*, 2005; Voeltz *et al.*, 2002). Dispersal of the ER in most eukaryotic cells is sensitive to the microtubule-destabilizing drug nocodazole. The motors kinesin and dynein have also been reported to colocalise with ER membranes suggesting that the microtubule cytoskeleton and its associated motors may be involved in ER partitioning during the cell cycle. Accordingly, two models for propagation of the ER in animal cells have been postulated. The first model states that the nuclear envelope and the ER network break down during mitosis, and ER elements are subsequently dispersed between the mother and daughter cells. Chromatin first condenses during prophase of the cell cycle. Then the nuclear envelope fragments at prometaphase generating vesicles with inner nuclear membrane proteins and components of the nuclear pore complex (Drummond *et al.*, 1999). Lamins subsequently disassemble and reorganize, resulting in the breakdown of the ER network throughout the cell (Voeltz *et al.*, 2002). The current model, also referred to as the selective retention model, proposes that the integrity of the ER network is maintained, and the ER network absorbs the fragmented nuclear envelope (Voeltz *et al.*, 2002). However, as reported earlier the structural integrity of the ER is maintained throughout the cell cycle, and it remains a continuous, interconnected network that is partitioned between the mother and daughter cell by cytokinesis (Ellenberg *et al.*, 1997).

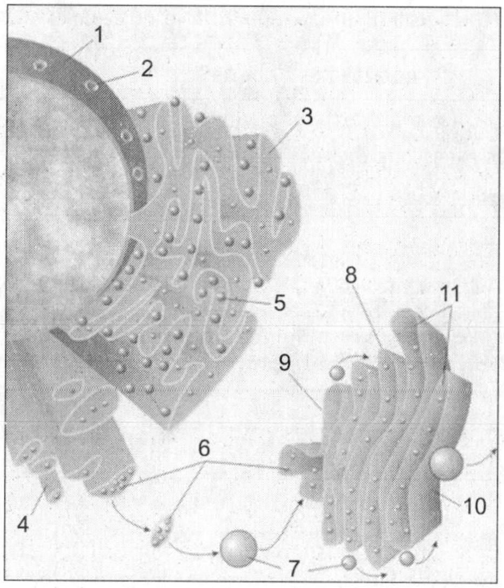

Fig. 1. Image of nucleus, endoplasmic reticulum and Golgi apparatus. (1) Nucleus. (2) Nuclear pore. (3) Rough endoplasmic reticulum (rER). (4) Smooth endoplasmic reticulum (sER). (5) Ribosome on the rough ER. (6) Proteins that are transported. (7) Transport vesicle. (8) Golgi apparatus. (9) Cis face of the Golgi apparatus. (10) Trans face of the Golgi apparatus. (11) Cisternae of the Golgi apparatus.

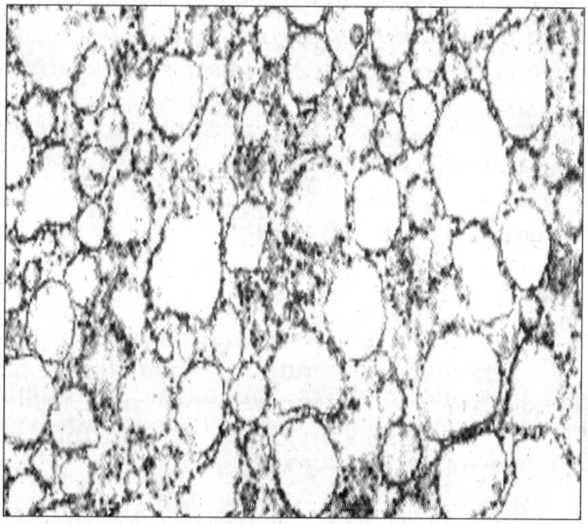

Fig. 2. Microsomes of rat liver. The liver was disrupted and various fractions recovered by ultracentrifugation. This figure shows the microsome fraction. It consists almost entirely of rough ER that had been disrupted and "healed"as vesicles. Ribosomes remain attached to the outer surface. x40,000. (From the work of Sabatini and Adelman.)

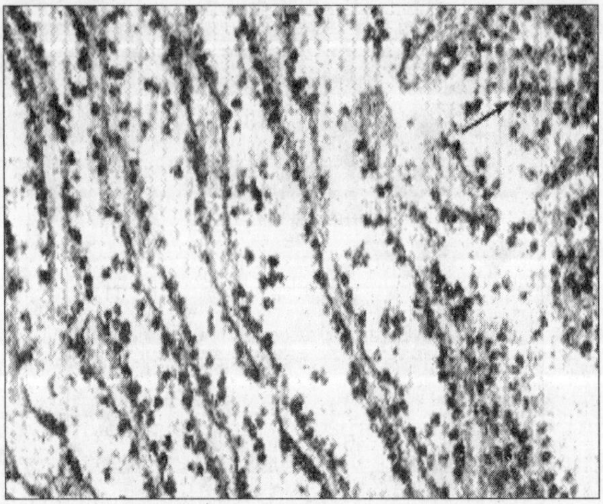

Fig. 3. Hepatocyte of a rat. In this portion of the cytoplasm most of the cisternae of the rough ER were cut transversely and others tangentially. In the latter (ARROW) the membrane of the ER and the attached polysomes are seen *en face*. A section of a mitochondrion is present. x 64,000. (From the work of G.E. Palade).

ER Tubule Fusion

Homotypic membrane fusion refers to the coalescence of similar types of membrane compartments (Wickner, 2002). ER tubule fusion is necessary for structural maintenance of the ER and the formation of the polygonal ER network. New polygon-like rings can form when two branched tubules fuse with a third, and formation of the rings continuously occurs throughout the ER network (Allan and Vale, 1991). The fusion events contribute to the dynamic nature of the ER and are responsible for reshaping and reorganizing the interconnected ER network. Various experimental results on fusion of ER vesicles and formation of the ER tubular network in kidney epithelial cells suggest that a GTPase is necessary to drive the ER fusion reaction (Allan and Vale, 1991; Watkins *et al.*, 1993).

SUMMARY

Endoplasmic reticulum (ER) is an endomembrane organellar system and present in all eukaryotic cells. It consists of an extensive membrane network of tubes and cisternae. ER are categorized as rough ER with ribosomes attached to the membrane, smooth ER without ribosomes and the transitional ER which is a special type of sER involved in packaging of proteins. ER tubules arise from pre existing tubules. These tubules along with some motor proteins are transported along the microtubule. Three models of mechanisms of transport have been described which are sliding mechanism, TAC mechanism and the microtubule movement mechanism. The structural appearance varies from vesicles, tubules and cisternae. The ER is closely associated with the nuclear envelope and the membranes of the Golgi. The regional variation of the ER is dependent upon the physiological status of the cell. In the secretory cell type the rER is preponderant while the cells engaged in steroid synthesis have more sER. Major biochemical functions of the ER are syntheses of proteins, fatty acid, phospholipids and triglycerides and metabolism of glycogen and lipid soluble xenobiotics. The ER serves various functions including the facilitation of protein folding and the transport of proteins to Golgi. Transport of proteins for secretion is also guided by the ER. The ER also has many adaptive functions in the different cell types and maintains intimate relationship with other cell organelles. The structural integrity of the ER is maintained throughout the cell cycle and it remains a continuous interconnected network partitioned between mother and daughter cells by cytokinesis.

REFERENCES

Allan, V. and Vale, R. (1991) Cell cycle control of microtubule based membrane transport and tubule formation in vitro. J. Cell Biol. 113 : 347-359.

Ardail, D. Gasnier, F. Lerme. F. Simonot, C. Louisot, P. and Gateau-Roesch, O. (1993) Involvement of mitochondrial contact sites in the subcellular compartmentalization of phospholipid biosynthetic enzymes. J. Biol. Chem. 268 : 25985-25992.

Baumann, O. and Walz, B. (2001) Endoplasmic reticulum of animal cells and its organization into structural and functional domains. Int. Rev. Cytol. 205 : 149-214.

Dreier, L. and Rapoport, T.A. (2000) In vitro formation of the endoplasmic reticulum occurs independently of microtubules by a controlled fusion reaction. J. Cell Biol. 148 : 883-898.

Drummond, S. Ferrigno, P. Lyon, C. Murphy, J. Goldberg, M. Allen, T *et al.* (1999) Temporal differences in the appearance of NEP-B78 and an LBR-like protein during *Xenopus* nuclear envelope reassembly reflect an ordered recruitment of functionally discrete vesicle types. J. Cell Biol. 144 : 225-240.

Ellenberg, J. Siggia, E.D. Moreira, J.E. Smith, C.L. Presely, J.F. Worman, H.J. and Lippincott-Schwartz, J. (1997) Nuclear membrane dynamics and reassembly in living cels : targeting of an inner nuclear membrane protein in interphase and mitosis. J. Cell Biol. 138 : 1193-1206.

Estrada de Martin, P. Novick, P. and Ferro-Novick, S. (2005) The organization, structure and inheritance of the ER in higher and lower eukaryotes. Biochem. Cell Biol. 83: 752-761.

Fawcett, D.W. (1955). Observations on the cytology and electron microscopy of hepatic cells. J. Nat. Cancer Inst. 15: 1475-1502.

Fehrenbacher, K.L. Davis, D., Wu. M., Boldogh, I. and Pon, L.A. (2002) Endoplasmic reticulum dynamics, inheritance, and cytoskeletal interactions in budding yeast. Mol. Biol. Cell 13 : 854-865.

Feiguin, F. Ferreira, A. Kosik, K. and Caceres, A. (1994) Kinesin mediated organelle translocation revealed by specific cellular manipulations. J. Cell Biol. 127 : 1021-1039.

Gaigg, B. Simbeni, R. Hrastnik, C. Paltauf, F. and Daum, G. (1995) Characterization of a microsomal subfraction associated with mitochondria of the yeast, *Saccharomyces cervesiae*. Involvement in synthesis and import of phospholipids into mitochondria. Biochim. Biophys. Acta, 1234 : 214-220.

Garfield, S. and Cardell, Jr., R.R. (1987) Endoplasmic Reticulum: Rough and Smooth. Int. Rev. Cytol. Suppl. 17: 255-272.

Hammond, A.T. and Glick, B.S. (2000). Dynamics of transitional endoplasmic reticulum sites in vertebrate cells. Mol. Biol. Cell, 11 : 3013-3030.

Hobman, T.C. Zhao, B. Chan, H. and Farquhar, M.G. (1998). Immunoisolation and characterization of a subdomain of the endoplasmic reticulum that concentrates proteins involved in COPII vescle biogenesis. Mol. Biol. Cell, 9 : 1265-1278.

Hoepfner, D. Schildknegt, D. Braakman, I. Philippsen, P. and Tabak, H.F. (2005). Contribution of the endoplasmic reticulum to peroxisome function. Cell, 122 : 85-95.

Jaffe, L.A. and Terasaki, M. (1994) Structural changes in the endoplasmic reticulum of starfish oocytes during meiotic maturation and fertilization. Dev. Biol. 164 : 579-87.

Johnson, A.E. and van Waes, M.A. (1999). The translocon : a dynamic gateway at the ER membrane. Annu. Rev. Cell Dev. Biol. 15 : 799-842.

Lane, J.D. and Allan, V.J. (1999). Microtubule-based endoplasmic reticulum motility in *Xenopus laevis* : activation of membrane associated kinesin during development. Mol. Biol. Cell, 10 : 1909-1922.

Novikoff, P.M. and Novikoff, A.B. (1972) Peroxisomes in absorptive cells of mammalian small intestine. J. Cell Biol. 53 : 532-560.

Pan, X. Roberts, P. Chen, Y. Kvam, E. Shulga, N. Hung, K. *et al.* (2000) Nucleus-vacuole junctions in *Saccharomyces cerevisiae* are formed through the direct interaction of Vac8p with Nvj1p. Mol. Biol. Cell, 11 : 2445-2457.

Porter K.A. Claude, A. and Fullam, E.F. (1945) A study of tissue culture cells by electron microscopy: Methods and preliminary observations. J. Exp. Med. 81: 233-246.

Prinz, W.A. Grzyb, L. Veenhuis, M. Kahana, J.A. Silver, P.A. and Rapoport, T.A. (2000). Mutants affecting the structure of the cortical endoplasmic reticulum in *Saccharomyces cerevisiae*. J. Cell Biol. 150 : 461-474.

Staehelin, L.A. (1997). The plant ER : a dynamic organelle composed of a large number of discrete functional domains. Plant J. 11 : 1161-1165.

Terasaki, M. (2000). Dynamics of the endoplasmic reticulum and Golgi apparatus during early sea urchin development. Mol. Biol. Cell, 11: 897-914.

Voeltz, G.K. Rolls, M.M. and Rapoport, T.A. (2002). Structural organization of the endoplasmic reticulum. EMBO Rep. 3 : 944-950.

Waterman-Storer, C.M., and Salmon, E.D. (1998). Endoplasmic reticulum membrane tubules and distributed by microtubules in living cells using three distinct mechanisms. Current Biology. 8 : 798-806.

Watkins, J. Hermanowski, A. and Balch, W. (1993). Oligomerization of immunoglobuin G heavy and light chains in vitro. A cell-free assay to study the assembly of the endoplasmic reticulum. J. Biol. Chem. 268 : 5182-5192.

Wickner, W. (2002). Yeast vacuoles and membrane fusion pathways. EMBO J. 21 : 1241-1247.

Zaar, K. Volkil, A. and Fahimi, H.D. (1987). Association of isolated bovine kidney cortex peroxisomes with endoplasmic reticulum. Biochim. Biophys, Acta, 897 : 135-142.

Suggested Readings

1. Molecular Biology of the Cell (2002) B. Alberts, A. Johnson, J. Lewis, M. Raff, K. Roberts, and P. Walter, 4th edition, Garland Science publication, New York.
2. Cell and Molecular Biology: Concepts and Experiments (1999) G. Karp, 2nd edition, John Wiley & Sons, New York.
3. Principles of Cell and Molecular Biology (1995) L.J. Kleinsmith and V.M. Kish, 2nd edition, Harper Collins College Publishers, New York.
4. Cell and Molecular Biology (1996) E.D.P. DeRobertis and E.M.F. DeRobertis Jr., B.I. Waverly Pvt. Ltd., New Delhi.
5. Cell Biology (1971) J. Ambrose and P.M. Easty, Vikas Publications, New Delhi.

REVIEW QUESTIONS

Long Questions

1. How would you categorize endoplasmic reticulum?
2. Can ER exist and function on its own? Give reasons for your answer.
3. Discuss the role of cytoskeleton in the structural organization of the ER.

4. Are ER in all cells similar? What are the adaptive functions of ER?
5. Why contact of ER with other organelles is advantageous to the cell?
6. Inheritance of ER is necessary – give your comments on the statement.

Multiple Choice Questions

1. Endoplasmic reticulum is found in
   (a) Prokaryotes
   (b) Erythrocytes
   (c) All cells
   (d) all cells except erythrocytes and prokaryotes
2. Which of the organelles mentioned below is not concerned with the electron transport system in a cell
   (a) Mitochondria (b) chloroplast
   (c) ER (d) none of the above
3. Which of the organelles mentioned below is concerned with protein synthesis
   (a) sER (b) rER
   (c) tER (d) nuclear envelope
4. Freshly produced proteins are inserted in the ER by
   (a) sER (b) tER
   (c) nuclear envelope (d) ribosomes
5. The ER in most eukaryotic cells is dispersed along
   (a) microtubule cytoskeleton (b) nuclear envelope
   (c) mitochondria (d) lysosomes
6. Continuity of the ER has been proved by
   (a) FRAP and GFP (b) FGAP and GFP
   (c) FLIP and GFP (d) none of these
7. Sarcoplasmic reticulum is the ER of
   (a) blood cells (b) muscle cells
   (c) liver cells (d) bacterial cells.
8. Under light microscope Ergastoplasm demonstrates basophilia in pancreatic acinar cells. This structure is
   (a) rER (b) sER
   (c) tER (d) nuclear envelope
9. sER has all the functions stated below except
   (a) lipid synthesis (b) steroid synthesis
   (c) detoxifies chemicals (d) protein synthesis.
10. ER is termed demarcation membrane in the following cell type
   (a) hepatocyte (b) megakaryocyte
   (c) oocyte (d) Leydig cells.

Answers

1. (d)	2. (c)	3. (b)	4. (d)	5. (a)	6. (c)
7. (b)	8. (a)	9. (d)	10. (b)		

CHAPTER 3

Golgi Complex

Shelley Bhattacharya

Dept. of Zoology, School of Life Sciences, Visva-Bharati University, Santiniketan–731 235

E-mail: bhattacharyashelley@rediffmail.com

The Golgi apparatus (also called a Gogli body, Golgi complex, or dictyosome) is an organelle found in most eukaryotic cells, including those of plants, animals, and fungi. The name was coined by an Italian anatomist Camillo Golgi, who identified it in 1889. Camillo Golgi (July 7, 1843 – January 21, 1926) was an Italian physician born in Corteno, Italy. His father was a physician and district medical officer. Golgi studied medicine at University of Pavia and spent much of his career studying the central nervous system. Tissue staining techniques in the later half of the nineteenth century were inadequate for studying nervous tissue. Therefore while working as chief medical officer in a psychiatric hospital he experimented with metal impregnation of nervous tissue. He discovered a method of staining nervous tissue which would stain a limited number of cells at random, in their entirety. This enabled him to view the paths of nerve cells in the brain for the first time. He called his discovery the black reaction. It is now known universally as the Golgi stain. This simple Black Reaction consisted in fixing silver chromate particles to the neurilemma by reacting silver nitrate with potassium bichromate. This resulted in a stark black deposit on the neural body or soma as well as on the axon and all dendrites, providing an exceedingly clear and well contrasted picture of neuron against a yellow background. The ability to visualize separate neurons led to the eventual acceptance of the neuron doctrine. Using his staining technique, Golgi identified the internal reticular apparatus in 1898 which bears his name, the Golgi apparatus. Golgi, together with Santiago Ramony Cajal, received the Nobel Prize in Physiology or Medicine in 1906 for his studies of the structure of the nervous system. Golgi died in Pavia, Italy, in January 1926.

The primary function of the Golgi apparatus is to process proteins targeted to the plasma membrane, lysosomes or endosomes, and those that will be formed from the cell and sort them within vesicles. Thus, Golgi functions as a central delivery system for the cell. Most of the transport vesicles that leave the endoplasmic reticulum specifically rough ER, are transported to the Golgi apparatus, where they are modified, sorted, and shipped towards their final destination. The Golgi apparatus is present in most eukaryotic cells, but tends to be more prominent where there are a lot of substances, such as proteins, being secreted. For example, plasma B cells, the antibody-secreting cells of the immune system, have prominent Golgi complex.

STRUCTURAL ORGANIZATION

The structure of the Golgi apparatus is quite complex and is the subject of scientific dispute. Like the ER, it consists of membranous structures. It is made up of a stack of flattened cisternae and similar vesicles. The cis face is the side facing the ER; the medial region is in the middle while the trans face is directed towards the plasma membrane. The following figures demonstrate the arrangement of the different stacks of cisternae.

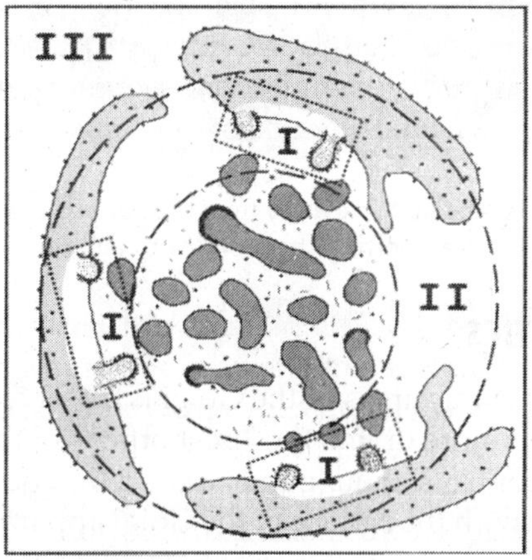

Fig. 1. Diagrammatic representation of the cisternal arrangements. Tier I shows budding from ER that is arranged facing a central zone at one end of the Golgi complex. These buds become vesicles and are coated with COPII protein coats. In Tier II the ER faces a central zone called a vesicular-tubular cluster (VTC). After they lose their COPII coat, they merge with the VTCs carrying the soluble and membrane proteins to the Golgi complex. Tier III illustrates the entire complex which is unique in the cytoplasm. It is termed the 'export complex' and contains unique proteins that suggest it is specialized for information flow to and from ER and the Golgi complex

The cisternal complex of the Golgi is found to be localized in definite cytoplasmic areas according to the special task it has to perform (Clermont and Tang, 1985; Lucocq and Montesano, 1985; Nemere *et al.*, 1985; Tassin *et al.*, 1985). It has been established that the location of the Golgi stacks of cisternae changes in response to alterations of the cell function and this is necessary for the smooth functioning of the several Golgi-connected processes in the cell (Rogalski *et al.*, 1984). The classic cytochemical Golgi reactions confirm the polarization of the stacks : after prolonged osmification osmium was found to be located specifically in the cis-region (Friend and Murray, 1965). On the other hand thiamine pyrophosphatase, nucleoside diphosphatase and acid phosphatase have been found to localize in the trans region (Hand, 1980; Novikoff and Novikoff, 1977) while medial cisternae stained positive to the nicotinamide adenine dinucleotide phospahatase reactions (Smith, 1980).

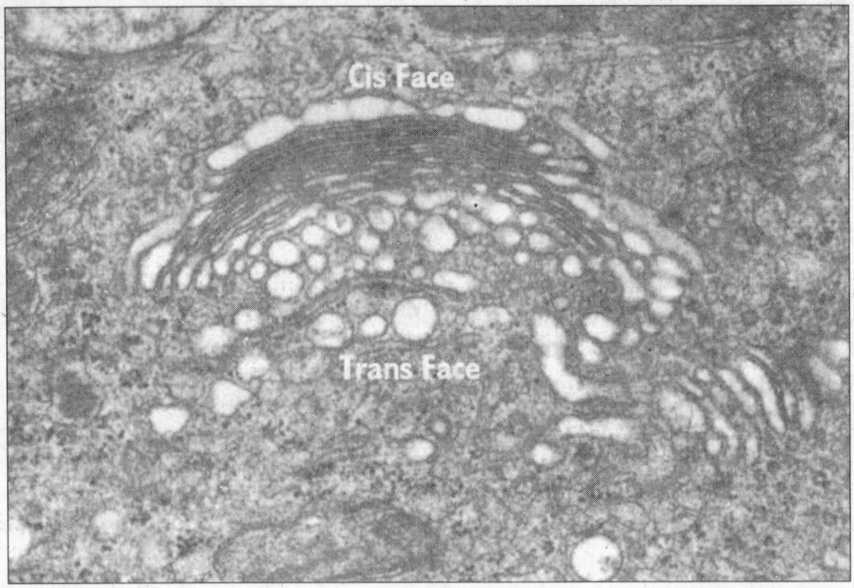

Fig. 2. Golgi Complex with its trans face pointing away from the nucleus toward the cell periphery. The numerous vesicles in the area are in the stage of transporting the proteins to and from cisternae (Fawcett, 1994).

FUNCTIONAL CHARACTERISTICS

The location of distinct cytoplasmic areas of the Golgi is closely related to its special tasks. The Golgi apparatus is considered more or less the "post office" of the cell. It handles all incoming lipids, proteins, etc., and controls their export as well. The transport vesicles from the Endoplasmic Reticulum fuse with the cis face of the Golgi apparatus to the cisternae and empty their protein content into the lumen. The proteins are then transported through the medial region towards the trans face and are modified on their way. Possible modifications include glycosylation or phosphorylation. The proteins are also labeled with a sequence of molecules according to their final destination. For example, the Golgi apparatus adds a mannose-6-phosphate label to proteins destined for lysosomes.

The transport mechanism itself is not yet clear, it could happen by cisternae progression (the movement of the apparatus itself, building new cisternae at the cis face and destroying them at the trans face) or by vesicular transport (small vesicles transport the proteins from one cisterna to the next, while the cisternae remain unchanged). It is also proposed that the cisternae are interconnected and the transport of cargo molecules within the Golgi is due to diffusion, while the localization of Golgi-resident proteins is achieved by an unknown mechanism. Once the proteins reach the trans face, the form of the vesicle is determined by the type of protein and the label it acquired.

Modification of Proteins in the Golgi

An example of the Golgi complex's functioning is the modification of glycoproteins. Vesicles from the ER contain simplified glycosylated proteins. In the Golgi complex, carbohydrates are attached and removed from these glycoproteins, creating a diversity of carbohydrate structures on the proteins. After they have been secreted into the cell the vesicles fuse to the cell membrane and release their contents. Besides protein modification, Golgi apparatus is involved in the transport of lipids around the cell and also creating lysosomes, the most important organelle involved in degradation of any cellular molecule.

Protein Processing in the Golgi

The Golgi is a cell structure mainly devoted to processing of proteins synthesized in the rER. Some of these proteins will end up as integral and constitutive membrane proteins and transported to the plasma membrane. There are plenty proteins ready for transport to the lysosomes via the Golgi. There are still other proteins which are in the line to be secreted or transported outside the cell by exocytosis. The major processing activity in the Golgi is addition of oligosaccharide moieties or glycosylation to form glycoproteins. Interestingly in some cells such as the mucus secreting epithelial cells, the proportion of carbohydrates added by glycosylation far exceeds the percentage of protein, where the product is given a different name, Mucopolysaccharides. In plant cells Golgi is known to secrete the cell plate and cell wall.

Ribosomes can synthesize proteins only upto a certain size. Thus very small peptides such as hormones and neurotransmitters are synthesized as large precursor proteins that are later chopped by the Golgi to the correct size. Proopiomelanocortin (POMC) is such a polypeptide of 265 amino acids from which adrenocorticotrophic hormone (ACTH), alpha and beta melanocyte stimulating hormone (MSH), beta endorphin and others are cut. How does the Golgi do it? The Golgi consists of a stack of membrane-bounded cisternae each of which is loaded with different enzymes to perform their designated functions. There are specific mechanisms to sort out the processed proteins and send them to their destinations while the processing enzymes such as the glycosylases are reclaimed for reuse.

Transport Pathways in Golgi

The transport mechanism in the Golgi is guided by two pathways, the outbound or forward and the inbound or backward. In the outbound pathway at first the transition vesicles or endoplasmic reticulum or tER are pinched off from the ER. The ER region is also specific as it carries integral membrane proteins, soluble proteins waiting their turn of processing and the processing enzymes. Pinching off of the tER requires that the vesicle be coated with COPII (coat protein II). Then the tER moves towards the cis Golgi network on microtubules. As they do so the COPII coat is removed and they may now fuse together forming larger vesicles and then with the cis Golgi. In the Golgi proteins are glycosylated in sequential steps each requiring a specific enzyme. These steps take place as shuttle vesicles carry the proteins from cis to medial to the trans Golgi compartments. At the outer face of the trans Golgi, vesicles are pinched off which carry the finished products to the specific destinations in the cell.

The inbound path of cisternal contents through the stack means that essential processing enzymes are also moving away from their proper site of action. Using a variety of signals, the Golgi separates the products from the processing enzymes that made them and returns the enzyme back to the endoplasmic reticulum. This transport is also done by pinching off vesicles but the inbound vesicles are coated with COPI (coat protein I). The movement of these special

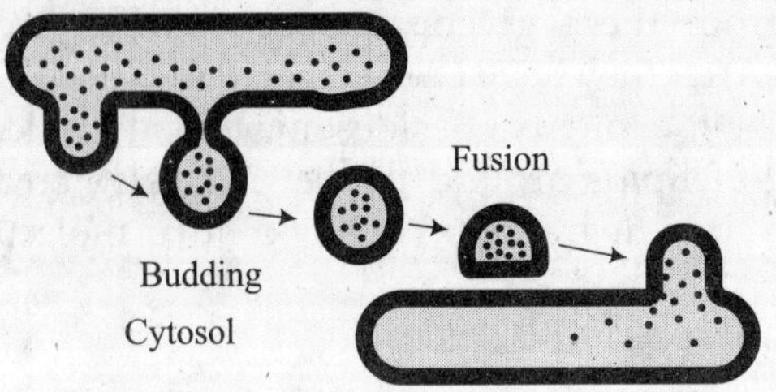

Fig. 3. Budding and fusion of vesicles.

transport vesicles is an energy requiring process. If one blocks production of ATP, the transport will not happen. The rough endoplasmic reticulum forms vesicles lose the attached ribosomes and carry the newly synthesized proteins to the Golgi complex. Thus the inside of the vesicle becomes continuous with the inside of the Golgi cisternae, so that protein groups pointing towards the inside, could eventually be directed to face the outside of the cell. Carbohydrate groups are attached and any subunits may be joined to the proteins in these cisternae. The protein is then passed to the final region of the Golgi called the "trans face". There it is placed in vacuoles that bud from this region of the Golgi complex. These may be of a certain size or density, characteristic of the cell itself. The vacuoles continue to condense the proteins and the final mature secretory granule is then moved to the membrane for secretion. Transport of material in and out of the Golgi complex essentially involves budding and fusion of vesicles. This (Fig. 3) shows that the membranes of each join and align themselves during the process so that the inside face remains in the lumen and the outside face remains towards the cytoplasm.

Regulatory and Constitutive Pathways

The Golgi complex controls trafficking of different types of proteins. Some are destined for secretion. Others are destined for the extracellular matrix. Finally, other proteins, such as lysosomal enzymes, may need to be sorted and sequestered from the remaining constituents because of their potential destructive effects. This (Fig. 4) shows the two types of secretory pathways. The regulated secretory pathway, as its name implies, is a pathway for proteins that requires a stimulus or trigger to elicit secretion. Some stimuli regulate synthesis of the protein as well as its release. The constitutive pathway allows for secretion of proteins that are needed outside the cell, like in the extracellular matrix. It does not require stimuli, although growth factors may enhance the process.

The (Fig. 5) demonstrates the compartmentalization of the Golgi complex. It has been found that phosphorylation occurs in the Cis region and in the other regions, different types of carbohydrates are added as a glycoprotein passes through the cisternae. The final sorting is done in the Trans Golgi complex. The functional differentiation of the Golgi complex can be studied with the electron microscope with specific techniques that detect enzymes. The cis region is rich in lipid-bearing membranes and can be delineated by osmium tetroxide labeling. The middle regions label for enzymes that add carbohydrates or other groups on the product. The inner, or trans region, is the area where the lysosomes are sorted. Therefore, it is heavily labeled for acid phosphatase.

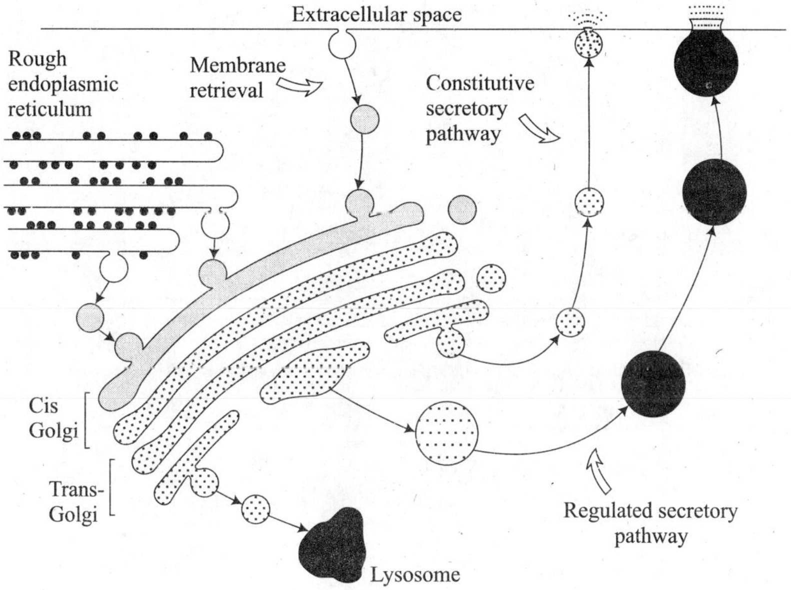

Fig. 4. Digrammatic representation of the regulatory and constitutive pathways operative in the Golgi . The important role of the Golgi Complex is to make certain the plasma membrane proteins reach their destination. This figure shows the route.

Differentiation of Golgi Stacks

There is much interest in understanding how the different Golgi cisternae are organized and differentiated. A number of models exist, however a favorite is called the "Maturational model" (Bannykh and Bakch, 1997).This model suggests that the new vesicles from the ER enter the cis Golgi network and retrograde vesicles bearing COPI coats move to merge with the cis region cisternae. These carry Golgi complex processing enzymes and their targeting to this region may be dependent on the low concentration of these processing enzymes. Then, as processing continues, the middle cisternae contain more mature product and lower amounts of the enzymes needed in the beginning. Finally, the trans region is specialized for sorting, containing receptors to sort and isolate lysosomal enzymes (Wooding and Pelham, 1998).

All proteins which reach the trans region are not meant for transport either by regulatory or constitutive pathways. It has further been studied that vital proteins needed in the rough endoplasmic reticulum are transported along with the other proteins in the Golgi complex. The Golgi complex has a mechanism for trapping them and sending them back to the rough endoplasmic reticulum (Fig.6). The rough endoplasmic reticulum has inserted a receptor protein on the membrane. The ER protein receptor captures all of the protein that carries the ER residency signal. . Vesicles then bud from the Golgi complex and move back to the rough endoplasmic reticulum. The receptor can circulate and continue to return the proteins needed by the endoplasmic reticulum. A drug called "brefeldin A" blocks the transfer of protein to the Golgi complex, however, the reverse transport is not blocked. (Cole, *et al.*, 1998).

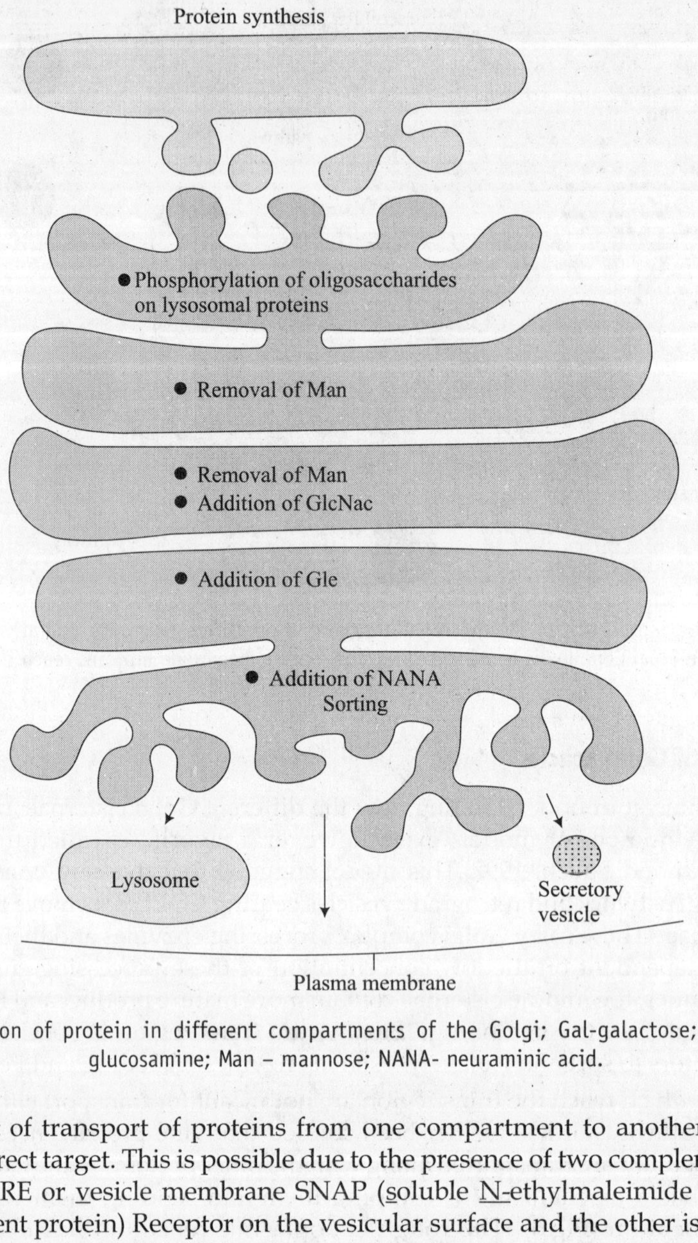

Fig. 5. Modification of protein in different compartments of the Golgi; Gal-galactose; Glc-NAC- n-Acetyl glucosamine; Man – mannose; NANA- neuraminic acid.

In the event of transport of proteins from one compartment to another a vesicle must recognize its correct target. This is possible due to the presence of two complementary integral proteins, v-SNARE or vesicle membrane SNAP (soluble N-ethylmaleimide sensitive fusion protein attachment protein) Receptor on the vesicular surface and the other is known as the t-SNARE or target membrane SNARE on the surface of the target membrane. Binding of the two complementary proteins, v-SNARE and t-SNARE is followed by the fusion of the two participating membranes. Thus, without t-SNARE and v-SNARE membrane fusions do not occur at all.

There is some evidence that in addition to the mechanism of Golgi traffic as described by pinching off and fusion of respective membranes, the cisternae of the Golgi actually migrate

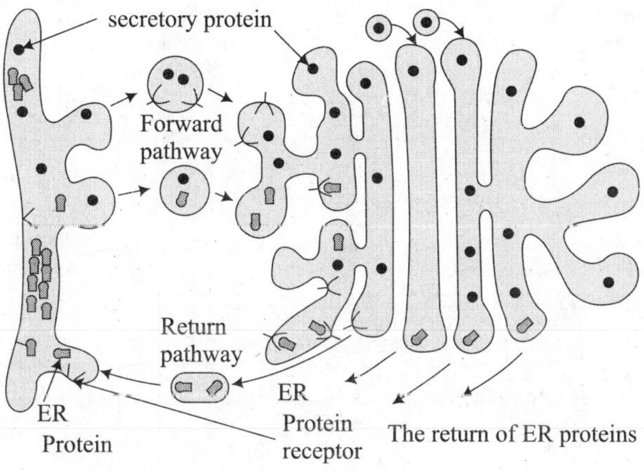

Fig. 6. Return and forward pathways in the Golgi stacks.

themselves; that is the cis Golgi gradually migrates up the stack becoming a medial and finally a trans Golgi vesicle. The Golgi is not a static organelle. It breaks up and disappears at the onset of mitosis. By the telophase of mitosis the Golgi reappears but it is still uncertain how the Golgi membranes are recreated.

The Role of Golgi in Alignment of Proteins in Plasma Membrane

It is known that plasma membrane proteins are inserted in the membrane at the level of the rough endoplasmic reticulum. The protein sequence is coded for membrane insert start and stop sites which directs the insertion and alignment points. Those that are multipass proteins have multiple start and stop sites. The important role of the Golgi Complex is to make certain the plasma membrane proteins reach their destination. (Fig. 7).

Mitosis Related Changes in Golgi Structure

It has been widely reported that structural disorganization takes place in the Golgi during mitosis (Schroeter *et al.*, 1985). In the late prophase the Golgi stacks are dispersed in the entire cytoplasm and during prometaphase to metaphase it becomes less and less distinct; the saccules are shortened and distended with a large number of 60–70 nm sized vesicles, which resembles the transport vesicles. In anaphase the stacked appearance of Golgi is further disturbed and the 60-70 nm vesicles are intermixed with the vacuolar remnants of the Golgi stacks. Despite the massive morphological changes the Golgi retains its cytochemical characteristics (Schroeter *et al.*, 1985). Furthermore during mitosis intracellular traffic of newly synthesized and internalized molecules is impaired (Warren, 1985). The disintegration of the Golgi has therefore been related to impaired traffic of ER membranes to Golgi (Feathersone *et al.*, 1985).

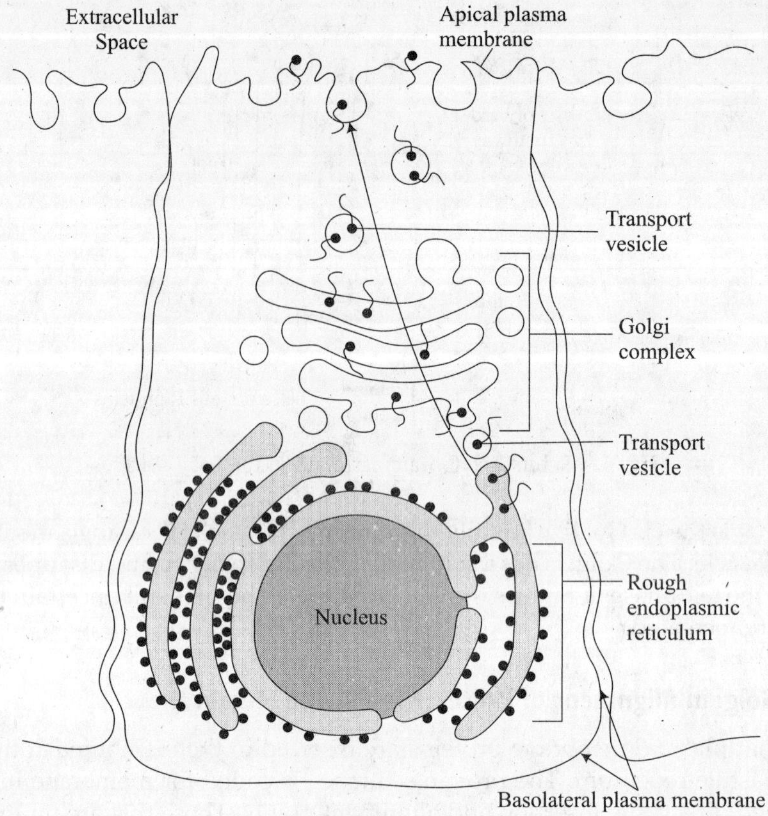

Fig. 7. The orientation of the protein is maintained in such a manner that the region destined to project outside the cell, is correctly aligned in that place. In order to do this, the plasma membrane proteins are placed facing inside the vesicle.

Association of Microtubules with Golgi Organization

Microtubules and the Golgi have been found to have a close relationship in the cytoplasm. Both elements are co localized (De Camelli *et al.*, 1986; Kupfer *et al.*, 1983) and sometimes the microtubules were found to form regular arrays in the vicinity of the Golgi stacks (Rogalski and Singer, 1984). It has been also demonstrated that disruption of microtubules specifically impairs the Golgi domains essential for (Banerjee *et al.*, 1976;) targeting plasma membrane glycoproteins.

SUMMARY

The Golgi is found in most eukaryotic cells including those of plants functioning as a central delivery system. The primary function of Golgi is to process proteins targeted to the plasma membrane, lysoosmes or endosomes. The structural organization is rather complex and divided in to cis and trans stacks. The cis stacks are situated near the ER and the trans are located near the plasma membrane. It handles all in coming lipids and proteins and controls their export as well. One of the major functions of the Golgi is the modification of the glycoproteins. The different stacks of the Golgi are loaded with different enzymes to perform their designated function in the processing of proteins. The transport pathways of the Golgi are guided by two pathways, the forward and the backward and there is constant budding and fusion of the membranes going on during the process. It is intimately related to both regulatory and constitutive secretory pathways. One of the important roles of Golgi is to ensure that the plasma membrane proteins reach their destination and also in their correct alignment. The Golgi shows mitosis related changes which can be identified at the different phases of mitotic cell division. The Golgi is also closely associated with the microtubule. It has been demonstrated that disruption of microtubules impairs the Golgi domains.

REFERENCES

Banerjee, D. Manning, C.P. Redman, C.P. 1976 The in vivo effect of colchicines on the addition of galactose and sialic acid to rat hepatic serum glycoproteins. J Biol Chem 251:3887-3892.

Bannykh S.I. and Bakch, W.E. 1997 Membrane Dynamics at the Endoplasmic Reticulum Golgi Interface J Cell Biol 138: 1-4

Clermont Y. and Tang, X. M. 1985 Glycoprotein synthesis in the Golgi apparatus of spermatids during spermiogenesis of the rat. Anat Rec 213:33-43

Cole, N.B., Ellenberg, J, Song, J, DiEuliis, D and Lippincott-Schwartz, J. 1998 Retrograde transport of Golgi-localized proteins to the ER. J Cell Biol 1-15,

D Fawcett._A 1994.Textbook of Histology Chapman and Hall, N.Y.

De Camelli, P. Moretti, M. Donini, S.D. Walter, U. and Lohmann, S.M. 1986 heterogeneous distribution of the cAMP receptor protein RII in the nervous system: Evidence for its intracellular accumulation on microtubules, microtubule-organizing centers, and in the area of the Golgi complex. J Cell Biol 103: 189-203.

Feathersone, C. Griffiths, G. and warren, G. 1985 Newly synthesized G protein of vesicular stomatitis virus is not transported to the Golgi complex in mitotic cells. J Cell Biol 101:2036-2046

Friend, D.S. and Murray, M.J. 1965 Osmium impregnation of the Golgi apparatus. Am J Anat 117:135-150

Hand, A.R. 1980 Cytochemical differentiation of the Golgi apparatus from GERL. J Histochem Cytochem 28:82-86.

Kupfer ,A. Dennert, G. and Singer, S.J. 1983 The reorientation of the Golgi apparatus and the microtubule-organizing center in the cytotoxic effecter cell is a prerequisite in the lysis of bound target cells. J Mol Cell Immunol 2:37-49.

Lucocq, J. and Montesano, R. 1985 Nonrandom positioning of Golgi apparatus in pancreatic B cells. Anat Rec 213:182-186.

Nemere, I. Kupfer, A. and Singer, S.J. 1985 Reorientation of the Golgi apparatus and the microtubule organizing center inside macrophages subjected toa chemotactic gradient. Cell Motility 5:17-29.

Novikoff, A.B. and Novikoff, P.M. 1977 Cytochemical contribution of differentiating GERL from the Golgi apparatus. Histochem J 9:525-551.

Rogalski, A.A. and Singer, S.J. 1984 Associations of elements of the Golgi apparatus with microtubules. J Cell Biol 99:1092-1100.

Rogalski, A.A. Bergman, J.E. and Singer, S.J. 1984 Effect of microtubule assembly on the intracellular processing and surface expression of an integral protein of the plasma membrane. J Cell Biol 99: 1101-1109.

Schroeter D. Eheman, V. and Pawaletz, N. 1985 Cellular compartments in mitotic cells : Ultrahistochemical identification of Golgi elements in PtK-1 cells. Biol cell 53 : 155-164.

Smith, C.E. 1980 Ultrastructural localization of nicotinamide adenine dinucleotide phosphatase (NADPase) activity to the intermediate saccules of the Golgi apparatus in rat incisor ameloblasts. J Histochem Cytochem 28: 16-26.

Tassin, A.M. Paintrand, M. Berger, E.G. and Bornens, M.1985 The Golgi apparatus remains associated with microtubule organizing centers during myogenesis. J Cell Biol 101: 630-638.

Warren, G. 1985 Membrane traffic and organelle division. Trends Biochem Sci 10: 439-443.

Wooding, S and H.R.B. Pelham, 1998. The dynamics of Golgi protein traffic visualized in living yeast cells. Molecular biology of the cell. 9: 2667-2680.

Suggested Readings

1. Molecular Biology of the Cell (2002) B. Alberts, A. Johnson, J. Lewis, M. Raff, K. Roberts, and P. Walter, 4th edition, Garland Science publication, New York.
2. Cell and Molecular Biology: Concepts and Experiments (1999) G. Karp, 2nd edition, John Wiley & Sons, New York.
3. Principles of Cell and Molecular Biology (1995) L.J. Kleinsmith and V.M. Kish, 2nd edition, Harper Collins College Publishers, New York.
4. Cell and Molecular Biology (1996) E.D.P. DeRobertis and E.M.F. DeRobertis Jr., B.I. Waverly Pvt. Ltd., New Delhi.
5. Cell Biology (1971) J. Ambrose and P.M. Easty, Vikas Publications, New Delhi.

REVIEW QUESTIONS

Long Questions

1. Golgi is the first organelle identified under the light microscope. Whose contribution is this? What was he by profession? What was his research interest? When did he share the Nobel Prize, and who was the second person who shared the Nobel prize with him?
2. Is it possible to discuss the structure and function of the Golgi apparatus as a separate entity? Give reasons for your answer.
3. The membrane complex of the Golgi is transient. Elaborate on the statement.
4. Describe the two transport pathways operative in the Golgi apparatus

Multiple Choice Questions

1. The staining reaction discovered by Camillo Golgi was named by him as:
   (a) Black Reaction (b) Silver impregnation
   (c) Golgi Reaction (d) Cajal reaction
2. The primary function of Golgi is to target proteins towards:
   (a) ER (b) rER
   (c) plasma membrane (d) nucleus

3. The Golgi apparatus was discovered in :
   (a) hepatocytes (b) nerve cells
   (c) oocytes (d) kidney cells
4. The Golgi is involved in :
   (a) formation of lysosomes (b) formation of sER
   (c) formation of nuclear envelope (d) formation of Rer
5. If Golgi apparatus is involved in secretion then we would expect most abundant Golgi in :
   (a) muscle cells (b) neurons
   (c) red blood cells (d) pancreatic acinar cells.
6. The cis face of the Golgi is nearer to :
   (a) plasma membrane (b) nucleus
   (c) ER (d) mitochondria
7. The trans face of the Golgi is nearer to :
   (a) plasma membrane (b) nucleus
   (c) ER (d) mitochondria
8. Acrosome of sperm is formed by :
   (a) ER (b) Golgi
   (c) lysosome (d) mitochondria
9. The golgi complex plays a major role :
   (a) as energy transducing organelle
   (b) in digesting proteins and carbohydrates
   (b) in trapping light quanta
   (c) in glycosylation of lipids and proteins
10. Secretory vesicle in the cell is formed by :
   (a) ER (b) trans Golgi network
   (c) cis Golgi network (d) medial Golgi cisternae.

Answers

1. (a)	2. (c)	3. (b)	4. (a)	5. (d)	6. (c)
7. (a)	8. (b)	9. (d	10. (b)		

CHAPTER 4

Mitochondrion and its Genomics

Ani Kurian
Dept. of Zoology, Lachoo Memorial College of Science and Technology, Jodhpur-342 003
E-mail: anikurianin@yahoo.co.in

Mitochondria are structures within cells that convert the energy from food into a form that cells can use. They are generally oblong organelles which range in size between 1 and 10 micrometers in length and occur in numbers that directly correlate with the cellular level of metabolic activity. They are especially abundant in cells and parts of cells that are associated with active processes. There are two specialized membranes encircling each mitochondrion (Fig. 1). These membranes divide it into a large inner matrix and a narrow inter membrane space. Inner membrane is highly convoluted to form a large number of infoldings or crista. This membrane is more selective than outer membrane to allow only those materials which are essential to the matrix. It utilizes a group of transport proteins that will only transport the correct molecules. The inter membrane space has an important role in the primary function of mitochondria, which is oxidative phosphorylation. The matrix contains highly specialized proteins. The outer membrane is a relatively simple phospholipid bilayer and has many channels formed by the protein porin. These channels act like a sieve to filter molecules that are big. Ions, nutrient molecules, ATP, ADP, etc. can pass through the outer membrane with ease. The matrix components can diffuse to inner membrane complexes and transport proteins within a relatively short time. The inner membrane is freely permeable only to oxygen, carbon dioxide, and water. Its structure is highly complex, including all of the complexes of the electron transport system, the ATP synthetase complex, and the transport proteins. Embedded in the inner membrane are proteins and complexes of molecules that are involved in the process called electron transport. The electron transport system (ETS), accepts energy from carriers in the matrix and stores it to a form that can be used to phosphorylate ADP.

The energy-conversion process takes place in mitochondria in presence of oxygen. The independent aerobic function of these organelles is thought to have evolved from bacteria that lived inside of other simple organisms in a mutually beneficial, or symbiotic, relationship,

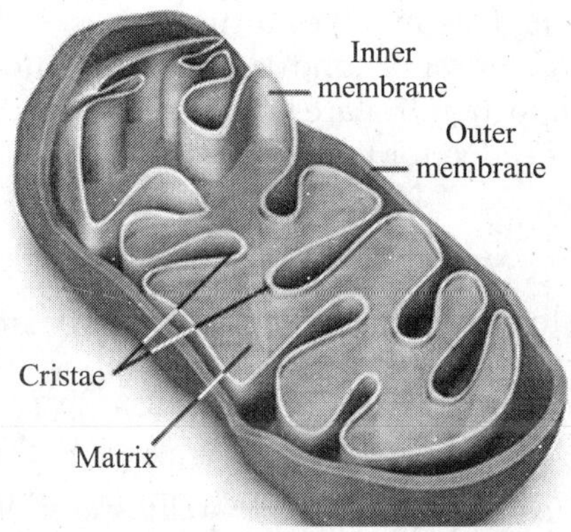

Fig. 1. Diagrammatic presentation of a mitochondrion.

providing them with aerobic capacity. The existence of mitochondrial DNA supports the endosymbiotic theory (Gray and Doolittle, 1982; Gray, 1989) which suggests that eukaryotic cells first appeared when a prokaryotic cell was absorbed into another cell without being digested. These two cells are thought to have then entered into a symbiotic relationship forming the first organ. Their genetic systems and cellular functions became integrated to form a single functioning cellular unit. Because mitochondria have their own DNA, RNA, and ribosomes, this scenario is quite possible. This theory is also supported by the existence of eukaryotic organism, called the amoeba, which lacks mitochondria. Therefore, amoeba must always have a symbiotic relationship with an aerobic bacterium

GENESIS

Mitochondria are found nearly in all eukaryotes, including plants, animals, fungi and protists. They are especially abundant in cells and parts of cells that are associated with active processes. For example, in flagellated protozoa or in mammalian sperm, mitochondria are concentrated around the base of the flagellum or flagella. In cardiac muscle, mitochondria surround the contractile elements. Hummingbird flight muscle is one of the richest sources of mitochondria. Thus, from their distribution alone one would suspect that they are involved in energy production. Multicellular organisms probably could not exist without mitochondria. Cells often have multiple mitochondria, according to its metabolic requirements.

Over the years it was known that the morphology of mitochondria is different in various cell types and organisms :small, bean shaped elongated tubules or highly branched reticula. Mitochondria are not considered to be distinct particles anymore but rather a dynamic network. They can change both shape (elongation , shortening , branching , buckling, swelling) and location inside living cells. (Bereiter-Hahn and voth, 1994). Mitochondrial fission and fusion are the two processes which control alterations in mitochondrial shape. (Chen and Chan, 2005). Mitochondrial fission and fusion occur during normal cell growth, cellular differentiation and development(Yaffe, 1999). During spermatogenesis in insects, fusion generates giant mitochondra that associate with the axoneme of sperm flagella (Dallai *et al.*, 2005; Hales and Fuller, 1997). Mitochondria replicate much like bacterial cells. When they get too large, they undergo fission. This involves a furrowing of the inner and then the outer membrane. Then the two daughter mitochondria split. The mitochondria must first replicate their DNA. Sometimes new mitochondria are synthesized in centers that are rich in proteins and polyribosomes which are needed for their synthesis.

F_1 PARTICLE

An ATP synthase is a general term for an enzyme that can synthesize adenosine triphosphate (ATP) from adenosine diphosphate (ADP) and inorganic phosphate by utilizing some form of energy. In mitochondria, the The F_1 portion of the F_1F_O ATP synthase (Fig. 2) is on the membrane, the F_O portion is within the membrane. It's easy to visualize the F_OF_1 particle with the head being the F_1 particle, the stalk being the gamma subunit of F1, and the base and "roots" being the F_O particle embedded in the membrane. The F_1 particle was first isolated by Ephraim Racker in 1961. The F_1 fraction derives it name from the term "Fraction 1" and F_O (written as a subscript "O", not "zero") derives it name from being the oligomycine binding fraction . The F1 particle is large and can be seen in the transmission electron microscope by negative staining (Fernandez-Moran et al., 1962) These are particles of 9 nm diameter and were originally called elementary particles

At the top of F_1 within F_OF_1, all six N-terminal regions derived from $\alpha + \beta$ subunits are shielded, indicating that one or more F_O subunits forms a "cap." At the bottom of F_1 within F_OF_1, the N-terminal region of the single δ subunit. The C-terminal regions of all three α subunits are shielded by F_O. Part of the γ subunit located at the bottom of F_1 is already shielded in F_1. The other F_1 subunits, are δ and ε. Specific regions at the top and bottom of F_1 are shielded by F_O and the F_O subunit called "F_6," unique to animal ATP synthases, is involved in shielding F_1.

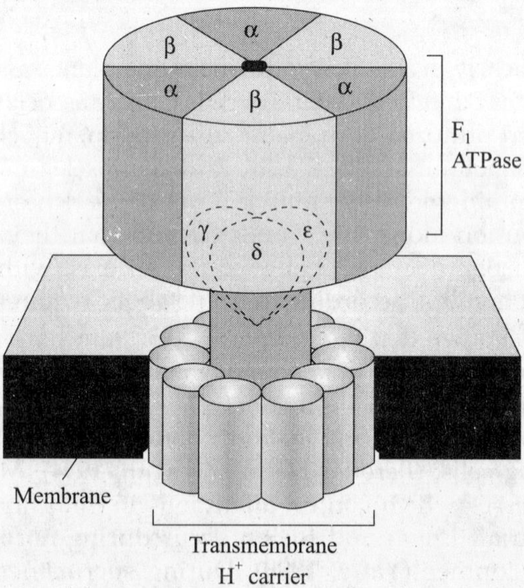

Fig. 2. ATP Synthase (http://grove.ufl.edu/~cleeuwen).

The F_1 component synthesizes ATP via chemiosmotic proton flux coming through Fo in the membrane. The function of F_1 is to use $\Delta\mu H^+$ set up by the electron transport chain to do chemiosmotic work, ie to synthesis ATP.The possible mechanism proposed for this are as follows:

1. A direct mechanism in which two protons pass through Fo to the active site of F1 where they act on one of the oxygen atoms of phosphate to form water.
2. An indirect mechanism where by ΔH^+ induces a conformational change in Fo and this change is transmitted to F1 via the interactions of many subunits. At this time nucleotides are bound to the active sites in F1.Once are the conformational change reaches these sites , ATP is released.

The complete breakdown of glucose to CO_2 occurs in two processes; glycolysis and citric acid cycle. As glycolysis and citric acid cycle produce a net total of only four ATP or GTP molecules per glucose molecules. This can not meet the physiological requirements of the eukaryotic cell. So for a higher yield of ATP cell has another process called oxidative phosphorylation which is carried out by a series of mitochondrial respiratory chain, a series of five enzyme complex embedded in the inner mitochondrial membrane.(Hatefi, 1985; Saraste, 1999). The Krebs' cycle reactions take place in the matrix of the mitochondria.

ELECTRON TRANSPORT SYSTEM (ETS)

Glycolysis and citric acid cycle generate NADH and $FADH_2$. NADH is an energy carrier. After release from glutamate dehydrogenase, it can be used to power other reactions. During oxidative phosohorylation electrons are transferred from NADH or to molecular oxygen via five enzyme complexes. It can also bind a specific enzyme on the mitochondria inner membrane, transferring some of its free energy to the *electron transport system*. In mitochondria, this is the principle role of NADH. The electron transfer system utilizes the energy released during electron transfer for the synthesis of ATP. Embedded in the inner membrane are proteins and complexes of molecules that are involved in electron transport (Table 1).

Oxidative phosphorylation is the process that couples the removal of hydrogen ions from one molecule and giving phosphate to another molecule. The ETS, accepts energy from carriers in the matrix and stores it to a form that can be used to phosphorylate ADP. Coupled with this transfer is the pumping of hydrogen ions. This pumping generates the gradient used by the ATP synthase complex to synthesize ATP.

Table 1. Some enzymes of mitochondrial matrix

S No	Enzyme	Type of reaction catalysed
1.	Citrate synthetase	condensation
2.	Isocitrate dehydrogenase	Oxydative decarboxylation
3.	α-ketoglutarate dehydrogenase complex	Oxydative decarboxylation
4.	Succinylnyl-COA synthetase	Substrate level phosphorylation
5.	Fumarase	Hydration
6.	Malate dehydrogenase	Oxidation

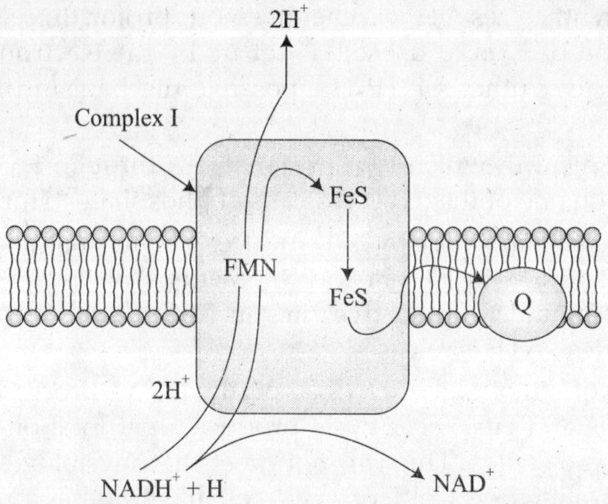

Fig. 3. Pumping of hydrogen ion across ETS.

At the start of the electron transport chain, two electrons are passed from NADH into the NADH dehydrogenase complex. Coupled with this transfer is the pumping of one hydrogen ion for each electron (Fig. 3).

The two electrons are transfered to ubiquinone. Ubiquinone is called a mobile transfer molecule because it moves the electrons to the cytochrome b complex. Each electron is then passed from the cytchrome b-c1 complex to cytochrome c. Cytochrome c accepts each electron one at a time. One hydrogen ion is pumped through the complex as each electron is transfered. The next major step occurs in the cytochrome oxidase complex. This step requires four electrons. These four electrons interact with oxygen and eight hydrogen ions. The four electrons, four of the hydrogen ions, and the molecular oxygen, are used to form two water molecules. The other four hydrogen ions are pumped across the membrane. This series of hydrogen pumping steps creates a gradient. The potential energy in this gradient is used by ATP synthase to form ATP from ADP and inorganic phosphate.

The five enzyme complexes of the respiratory chain consists of ~ 90 subunits , 13 of which are encoded in the mitochondrial genome (Table 2). Different observations suggest that these repiratory complexes are organized in supercomplexes. These super molecular structures are also called respirasomes.They have been observed in mitochondria of yeast (Schagger and Pfeiffer 2000), plants (Eubel et al., 2003; Krause et al., 2004) and mammals(Schafer et al., 2006; Schagger and Pfeiffer, 2001)

Electron transport is *driven* by the free energy that is available from the energy carriers, in turn obtained from substrates of Krebs intermediates. It is *restricted* by the chemiosmotic gradient. The only way electron transport can proceed is to the extent that the energy in the gradient is dissipated. In healthy mitochondria the gradient is maintained. That is, electron transport keeps up with the utilization of the energy stored in the gradient. Even in the presence of ADP, which allows ATP synthetase to exploit the gradient, the chemiosmotic gradient is maintained at a set energy level.

Table 2. composition of mitochondrial respiratory chain

Complex	I	II	III	IV	V
Enzyme	NADH-CoQ reductase	Succinate-CoQreductase	CoQ-cytochrome Creductase	Cytochrome C oxidase	ATP synthase
Inhibitor	Rotenoneamytal	TTFA malonate	Antimycin A	Cyanide, Carbon monoxide, Azide	Oligomycin
Mitochondrial DNA Subunits	ND1-6, ND4L	0	Cytochrome b	Cytochrome oxidaseI , II, III	ATPase 6 ATPase 8

OXIDATIVE PHOSPHORYLATION

Oxidative phosphorylation is the major energy conserver in the cell. Peter D. Mitchell(1960) proposed the chemiosmotic hypothesis for Electron transfer supported ATP synthesis (oxidative phosphorylation).

The theory suggests that most of the ATP synthesis in respiring cells comes from the electrochemical gradient across the inner membranes of mitochondria by using the energy of NADH and FADH2 formed from the breaking down of energy rich molecules such as glucose. Here electron transfer produces a proton (H^+)difference on either side of the inner mitochondrial membrane. The positive charges (H^+) accumulate on the outside of the mitochondrion and the negative charges (OH^-) on the inside.The carriers pass electrons to the ETC in the inner mitochondrial membrane, which in turn pass them to other proteins in the ETC. The energy available in the electrons is used to pump protons from the matrix across the inner mitochondrial membrane, storing energy in the form of a transmembrane electrochemical gradient. The protons move back across the inner membrane through the enzyme ATP synthase. The flow of protons back into the matrix of the mitochondrion via ATP synthase provides enough energy for ADP to combine with inorganic phosphate to form ATP. The electrons and protons at the last pump in the ETC are taken up by oxygen to form water (Fig. 4).

GENOMICS

Each mitochondrion has its own set of genes. (Wallace, 1982).The mitochondrial genome is contained in the matrix. Together with mitochondrial DNA (mt DNA) the matrix contains mitochondrial ribosomes, tRNAs and highly concentrated mixture of hundreds of enzymes, required for mitochondrial function. The mtDNA is not distributed homogenously within the mitochondrial compartment, but concentrates in structures called nucleoids, most often associated with mitochondrial inner membrane, (Hall et al., 1975, Kuroiwa, 1982, Echeverria et al., 1991, Dimmer et al., 2005).The functional basis for these nucleoid structures and their association with membrane is not yet clear, but they may have roles in inheritance , segregation , replication , transcription or perhaps recombination. The largest mitochondrial genomes and most complex mtDNA is of plants (Newton, 1988).Vertebrate mt DNA consists of structural genes which are joined to one another, with few or no spacer nucleotides between them(Attardi, 1985). In humans, mitochondrial DNA spans about 16, 500 bp. representing a fraction of the total DNA in cells. In contrast to the RNA components of the mitochondrial

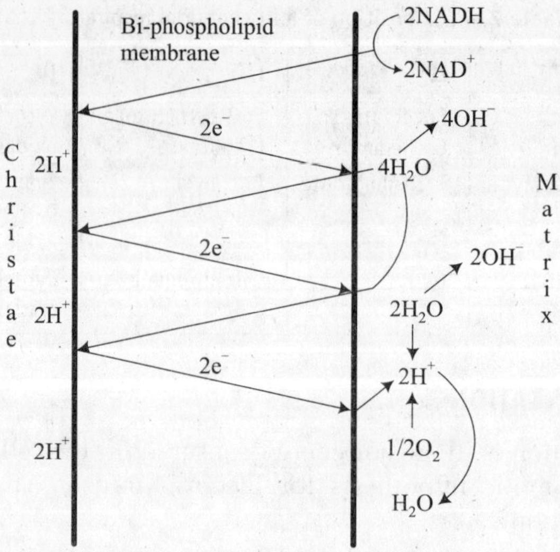

Fig. 4. *Chemiosmotic phosphorylation* (From Wikipedia).

translation apparatus (rRNAs and tRNAs), most mitochondrial genomes do not encode the proteins required for DNA replication, transcription, or translation. Instead, the genes that encode proteins required for the replication and expression of mtDNA are contained in the nucleus. In addition, the nucleus contains the genes that encode most of the mitochondrial proteins required for oxidative phosphorylation and all of the enzymes involved in mitochondrial metabolism (e.g., enzymes of the citric acid cycle). The proteins encoded by these genes (more than 95% of mitochondrial proteins) are synthesized on free cytosolic ribosomes and imported into mitochondria as completed polypeptide chains. (Shadel and Clayton, 1997)

Human mitochondrial DNA is a double stranded closed circular molecule of about 16.6 kb length . The double stranded DNA molecule can be seperated in to a light strand (L) and a heavy strand(H) by density gradients. The observed difference is due to uneven nucleotide content of the two strands. The H strand is guanine rich, whereas the L strand is poor in guanine.There are no introns in the genome and the 37 genes are highly compacted. They encode 13mRNAs , 22tRNAs and 2rRNAs (16Sand 12S) required for translation.(Anderson et al., 1981). Most of the genetic information of mtDNA is encoded on the heavy strand , including genes for both rRNAs 14tRNAs and 12 polypeptides.The light strand encodes 8 tRNAs and a single polypeptide, ND6 (Table 2). Some of the protein genes overlap and many termination codon are generated post transcriptionly by polyadenyllation of the processed mRNAs (Montoya et al., 1981; Temperley et al., 2003).

Nuclear and mtDNA are thought to be of separate evolutionary origin, with the mtDNA being derived from bacteria that were engulfed by early precursors of eukaryotic cells. Thus cells in present organisms, the vast majority of proteins found in the mitochondria (~1500 in mammals) are encoded by nuclear DNA. In mammals, 100% of the mtDNA contribution to a zygote is inherited from the mother and this is true for most, organisms. In rare cases mitochondria may also be inherited via a paternal route. Typically, a sperm carries mitochondria in its tail as an energy source for its long journey to the egg. When the sperm attaches to the egg during fertilization, the tail falls off. Consequently, the only mitochondria the new organism usually gets are from the egg its mother provided. Therefore, unlike nuclear DNA, mitochondrial DNA doesn't get shuffled every generation. So it is presumed to change at a slower rate. Currently, human mtDNA is present at 100-10,000 copies per cell, with each circular molecule consisting of 16,569 bp with 37 genes. Thirteen of these genes provide instructions for making enzymes involved in oxidative phosphorylation. The remaining genes provide instructions for making tRNAs and rRNAs. Mitochondria, accumulate mutations as organisms age and this results in reduced capacity for oxidative phosphorylation.

Mitochondrial DNA replication occurs independently of the cell cycle (Bogenhagen and Clayton, 1977). It is possible to reconstitute minimal mtDNA replisome *in vitro* with recombinant proteins (Korhonen, et al., 2004).The conventional model for mtDNA replication is called the strand -asymmetric model or Clayton model(Clayton, 1982). In this model two origins of replication exist, one for each mtDNA strand:the origin of heavy (O_H) strand and light (O_L) strand synthesis. DNA synthesis begins at the O_H utilizing a RNA primer transcribed from the light strand promoter (LSP). H strand DNA elongation continues around the entire length of the genome, with parental H strand displacement. Only a few initiation events at O_H actually result in the synthesis of a full length genome. Hence the site where H strand DNA synthesis is either prematurely terminated or allowed to to continue, functions as regulatory domain for mtDNA replication.(Madesen et al 1993). Synthesis of L strand DNA begins and proceeds in a direction opposite to that of H strand replication(Wong and Clayton, 1985). Another model ie. the symmetric model or Holt model has been proposed by Holt, et al., 2000. Replication of mtDNA starts at multiple sites (*ori*) inside the initiaton zone and subsequently proceeds bidirectional. O_H provides fork arrest and replication is then restricted to one direction only. A variety of double stranded replication intermediates were defined and they all pointed towards a conventional coupled leading and lagging strand replication.

MITOCHONDRIAL DISORDERS

The existence of different mitochondrial disorders are well established.(Wallace, 1992). Mitochondrial diseases are inherited either in a maternal or a mendelian way(Wallace, 1998). Mitochondria are semi-autonomous organelles as most mitochondrial proteins are encoded by nucleus. Thus functional mitochondria require polypeptides from both the nuclear and the mitochondrial genomes. Mutations in either genome may cause mitochondrial dysfunction (Larsson and clayton, 1995). Pathogenic mtDNA mutations were first reported by Holt et al.(1988) and Wallace, et al., (1988). It has been reported that nuclear genes could also be involved in mitochondrial disorders. (Zeviani et al., 1989). Many mitochondrial DNA mutations are linked to human diseases (Larsson and Clayton 1995). Because the mitochondria produce energy in cells, symptoms of mitochondrial diseases often involve degeneration or functional failure of tissue. For example, mtDNA mutations have been identified in some forms of diabetes, deafness, and certain inherited heart diseases. In addition, mutations in mtDNA are able to accumulate throughout an individual's lifetime. Mitochondrial DNA mutations can also concentrate in the mitochondria of specific tissues. There is even a theory, the 'mitochondrial theory of aging', which suggests that accumulation of mutations in mitochondria contributes to, or drives, the aging process. These defects are associated with

Parkinson's and Alzheimer's disease (Trifunovic et al., 2004) . The effects of ageing on matrix enzymes are not uniform, (Evelyne et al., 2004)

Somatic mutations in mitochondrial DNA have been reported in some forms of cancer including breast, colon, stomach, liver, and kidney tumors. These mutations also have been associated with cancer of blood-forming tissue (leukemia) and cancer of immune system cells (lymphoma). Mutations of mitochondrial DNA can lead to a number of illnesses including exercise intolerance and Kearns-Sayre syndrome (KSS), which causes a person to lose full function of their heart, eye, and muscle movements. The function of the heart is highly dependent on oxidative energy that is generated in mitochondria. Defects in mitochondrial structure and function can be found in association with cardiovascular diseases.

Mutations in four mitochondrial genes, MT-ND1, MT-ND4, MT-ND4L, and MT-ND6, have been identified in people with Leber hereditary optic neuropathy. Mutations in at least five mitochondrial genes, MT-ND1, MT-ND5, MT-TH, MT-TL1, and MT-TV, can cause the characteristic features of mitochondrial encephalopathy, lactic acidosis, and stroke-like episodes (MELAS). Mutations in the MT-RNR1 gene increase the risk of hearing loss. Mutations in two mitochondrial genes, MT-RNR1 and MT-TS1, are associated with non syndromic deafness (hearing loss without related signs and symptoms affecting other parts of the body). Genetic changes in the MT-TS1 gene cause a change in single building block (nucleotide) in the $tRNA^{Ser(UCN)}$ molecule. These mutations likely disrupt the normal production of the molecule or alter its structure. As a result, less $tRNA^{Ser(UCN)}$ is available to assemble proteins within mitochondria. These changes reduce the production of proteins needed for oxidative phosphorylation, which may impair the ability of mitochondria to make ATP. In addition to the critical cellular energy-related functions, mitochondrial genes are useful to evolutionary biologists because of their maternal inheritance and high rate of mutation. By studying patterns of mutations, scientists are able to reconstruct patterns of migration and evolution within and between species.

REFERENCES

Anderson, S., Bankier, aA.T., Barrell, B.G., deBruijn, M.H., Coulson, A.R, Drouin, J., Eperon. C, Nierlich, D.P., Roe, B.A., Sanger, F., et al. (1981). Sequence and organization of the human mitochondrial genome. Nature 290: 457-465

Attardi, G(1985).Animal mitochondrial DNA: an extreme example of genetic economy. Int.Rev. Cytol 93: 93-145

Bereiter- Hahn, J., and voth, M (1994) Dynamics of mitochondria in living cells:shape, changes, dislocations, fusion and fission of mitochondria. Microsc. Res. Tech 27: 198-219

Bogenhagen, D., and Clayton, D.A (1978) Mechanism of mitochondrial DNA replication in mouse L cells: Introduction of superhelical turns in to newly replicated molecules.J.Mol.Biol 119: 69-81

Chen, H., and Chan, D.C(2005)Emerging functions of mammalian mitochondrial fusion and fission.Hum.Mol.Genet.14 spec No.2: R283-289

Clayton, D.A., (1982)Replication of animal mitochondrial DNA.Cell 28: 693-705

Dallai, R., Lupetti, P., Osella, G., and Afzelius, B.A.(2005). Giant spermcells with acessory macrotubules in a neuropteran insect. Tissue cell 37: 359-366

Dimmer , K.S., Jakobs, S., Vogel, F., Altmann, K., and Westermann, B.(2005);Mdm31 and Mdm32 are inner membrane proteins required for maintenance of mitochondrial shape and stability of mitochondrial DNA nucleoids in yeast.J.Cell Biol. 168: 103-115

Evelyne Delaval, Martine Perichon and Bertrand Friguet(2004) Age-related impairment of mitochondrial matrix aconitaseand ATP-stimulated protease in rat liver and heart Eur. J. Biochem. 271: 4559–4564 _ FEBS

Eubel , H., Jansch, L and Braun, H.P.(2003).New insights in to the respiratory chain of plant mitochondria. Supercomplexes and a unique composition of complex II.Plant physiol. 133: 274-286

Echeverria, M., Robert, D., Carde, J.P., and Litvik, S.(1991)Isolation from wheat mitochondria of a membrane associated highmolecular weight complex involved in DNA synthesis Plant Mol.Biol. 16: 301-315

Fernandez-Moran et al., (1962)Journal of Molecular Biology, Vol 22, p 63.

Gray, M.W., and Doolittle, W.F (1982)Has the endosymbiont hypothesis been proven? Microbiol Rev. 46: 1- 42

Gray, M.W., (1989). The evolutionary origins of organelles.Trends Genet 5: 294-299.

Hales, K.G., and Fuller, M.T.(1997)Developmentally regulated mitochondrial fusion mediated by a conserved, novel, predicted GTPase. Cell 90: 121-129.

Hall , R.M., Mattick, J.S., Marzuki, S., and Linnane, A.W.(1975). Evidence for a functional association of DNA synthesis with the membrane in mitochondria of Saccharomyces cerevisiae.Mol.Biol Rep 2: 101-106

Hatefi, Y.(1985)The mitochondrial electron transport and oxidative phosphorylation system :Annu.Rev Biochem 54: 1015-1069

Holt, I .J., Lorimer, H.E., and Jacobs, H.T.(2000).Coupled leading –and lagging –strand synthesis of mammalian mitochondrial DNA.Cell 100: 515-524

Holt, I.J., Harding, A.E., and Morgan–Hughes, J.A.(1988).Deltions of muscle mitochondrial DNA in patients with mitochondrial myopathies.Nature 331: 717-719

Krause, F., Reifschneider, N.H., Vocke, D., Seelert, h., Rexorth, S. and Dencher, N.A.(2004)Respirasome - like supercomplexes in green leaf mitochondria of spinach .J Biol Chem. 279, 48369 - 48375

Kuroiwa, T.(1982)Mitochondrial nuclei.Int.Rev Cytol 75: 1-59

Larsson, N.G., and Clayton, D.A. (1995) Molecular genetic aspects of human mitochondrial disorders.Annu.Rev.Genet 29: 151-178

Madesen C.S., Ghivizzani, S.C., and Hauswirth, W.W. (1993). Protein binding to a single termination–associated sequence in the mitochondrial DNA D-loop region. Mol Cell Biol. 13: 2162-2171

Montoya, J., Ojala, D., and Attardi , G. (1981);Distinctive features of the 5′-terminal sequences of the human mitochondrial mRNAs.Nature 290: 465-470

Newton, K.J., (1988)Plant mitochondrial genomes:Organisation, Expression and Variation.Anu.Rev.Plant physiol.Plant mol boil. 39: 503-532

Saraste, M.(1999)Oxidative phosphorylation at the fin de siecle.Science 283: 1488-1493

Schafere, E., Seelert, H., Reifschneider, N.H Krause , F., Dencher, N.A and Vonck, J.(2006)Architecture of active mammalian respiratory chain supercomplexes.J.Biol Chem.281: 15370-15375

Schagger , H., and Pfeiffer, K.(2001).The ratio of oxidative phosphorylation complexes I-V in bovine heart mitochondria and the compositiiton of respiratory chain supercomplexes.J Biol.Chem 276: 37861-37867

Schagger, H., and Pfeiffer, K.(2000)Supercomplexes in the respiratory chains of yeast and mammalian mitochondria.Embo.J 19: 1777-1783

Shadel , G.S., and D.A. Clayton, (1997)Mitochondrial DNA maintenance in vertebrates. Annu.Rev. Biochem 66: 409-435

Temperley, R.J., Seneca, S.H., Tonska, K., Bartnik, E., Bindoff, L.A., Lightowlers, R.N., and Chrzanowska-Lightowlers, Z.M.(2003) Investigation of a pathogenic mtDNA microdeletion reveals a translation - dependent deadenylation decay pathway in human mitochondria.Hum.Mol. genet.12: 2341-2348

Trifunovic, A., Wredenberg, A., Falkenberg, M., Spelbrink, J.N., Rovio, A.T., Bruder, C.E., Bohlooly, Y.M., Gidlof, S., Oldfors, A., Wibom, R., *et al* (2004).Premature ageing in mice expressing defective mitochondrial DNA polymerase.Nature 429: 417-423

Wallace , DC., Singh, G., Lott, M.T., Hodge, J.A., Schurr, T.G., Lezza, A.M., Elsas, L.J 2nd and Nikoskelainen, E.K.(1988) Mitochondrial DNA mutation associated with Leber's hereditary optic neuropathy. Science 242: 1427-1430

Wallace, D.C., (1992) Diseases of the mitochondrial DNA. Annu. Rev. Biochem 61, 1175-1212

Wallace, D.C., (1998) Mitochondrial diseases in man and mouse.Science 283: 1482-1488

Wong T.W., and Clayton, D.A., (1985) In vitro replication of human mitochondrial DNA: accurate initiation at the origin of light strand synthesis. Cell 42: 951-958

Yaffe, M.P.(1999) Dynamic mitochondria. Nat.cell Biol I E 149-150

Zeviani , M., Servidei, S., Gellera, C., Bertini, E., DiMauro, S., and DiDonato, S.(1989) An autosomal dominantdisorder with multiple deletions of mitochondrial DNA starting at the D- loop region. Nature 339: 309-311

Suggested Readings

1. Alberts, B. *et al*. (2002) Molecular Biology of the cell, 4th edition, Garland Science Publication, New York.
2. Lehninger, A.L. (2005) Principles of Biochemistry 4th edition, W.H. Freeman and Company, New York.
3. Stryer, L. (1988) Biochemistry, W.H. Freeman and Company, New York.

REVIEW QUESTIONS

Long Questions

1. Describe the ultrastructure of a mitochondria.
2. Explain mitochondrion as a powerhouse of cell.
3. Discuss mitochondrial genomics and its DNA replication.
4. Discuss the structure of human mt DNA.
5. Describe the genetics of Mitochondrial Diseases.

Multiple Choice Questions

1. Which of the following statements about the electron transport chain is not correct
   (a) It is located in the inner mitochondrion membrane
   (b) Cytochrome C accepts electrons from complex II
   (c) Cytochrome oxidase (complex IV) accepts electrons from Cytochrome C
2. What happens when the energy needs of a cell are high,
   (a) Mitochondria grow and divide
   (b) Mitochondria are destroyed or become inactive
   (c) RBC increases

3. Cell of the body which lack mitochondria
   (a) Red blood cells (b) Muscle cells
   (c) Nerve cells
4. Aerobic respiration
   (a) is a part of glycolysis (b) is another name for lipid synthesis
   (c) is the process of energy conversion
5. The highly convoluted infoldings forms
   (a) Cristae (b) Matrix
   (c) Membrane
6. Primary function of mitochondria
   (a) Oxidative phosphorylation (b) Glycolysis
   (c) Kreb's cycle
7. The location of Krebs' cycle reactions in mitochondria
   (a) Matrix (b) Cristae
   (c) Membrane
8. The energy for electron transport is obtained from
   (a) Glutamate or Kreb's intermediates
   (b) ATP
   (c) Enzymes
9. The function of an electron in the electron transport chain is
   (a) To transfer energy from complex II to complex I
   (b) To pump hydrogen ions using complex II
   (c) To use its free energy to pump protons against their concentration gradient
10. The location of respiratory enzymes for electron transport system in mitochondrial
    (a) Matrix (b) Cristae
    (c) Membrane

Answers

1. (b) 2. (a) 3. (a) 4. (c) 5. (a) 6. (a)
7. (a) 8. (a) 9. (c) 10. (b)

CHAPTER 5

Cell Cycle and its Regulation at Molecular Level

A. Chatterjee
Dept. of Biotechnology, North-Eastern Hill University, Shillong–793 022
E-mail: chatterjeeanupam@hotmail.com

The ability to generate identical progeny is a fundamental task of living organisms. Faithful duplication of genetic material and proper distribution in the progeny cells is the essential task of the cell cycle. Under a microscope, a dividing eukaryotic cell shows a progressive increase in size in order to avoid a decrease in cell size with each division. After reaching a certain size, cells pass a point of no return (restriction point in mammalian cells, start point in yeast). The advent of phase-contrast microscopy and other techniques allowed the identification of a phase, termed mitosis (M), immediately before cell division during which cells condense their chromosomes, align them on a micritubular spindle, and segregate sister chromatids to opposite poles of the cell. Since little can be seen during the interval succeeding the M phase and cell division, it was termed interphase. The discovery that DNA is the genetic material stored in the chromosomes allowed Howard and Pelc (1953) to show that chromosome duplication occurs during a narrow and discrete window of time in interphase, termed DNA synthetic phase (S phase). The cell cycle was thus split into two major, non-overlapping phases, S and M phase, separated by phase without a clearly defined function, which were conveniently termed gap phases. The gap between M phase and S phase was named G1, whereas the gap between S and the M phase was named G2 (Fig. 1). This way cell cycle was operationally separated into four major phases. But how was progression from one phase to another regulated? How was the proper succession from one phase to the other achieved? Answers to these questions had to await new biochemical and genetic approaches.

The two most important events of the eukaryotic cell cycle are S phase and M phase, when chromosomes are replicated and segregated into the dividing cells. These events form an ordered developmental sequence that ensures that each newly divided cell receives a full

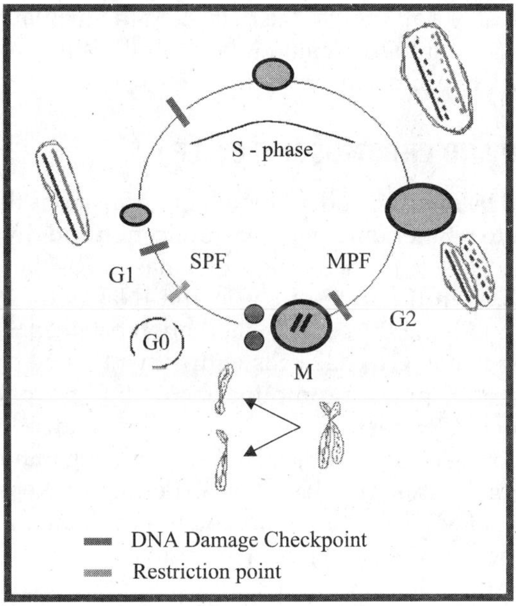

Fig. 1. The cell cycle phases showing restriction point and checkpoints.

complement of the hereditary material. In general eukaryotic cells do not reduplicate their chromosome before sister chromatids have been segregated in the previous mitosis, and do not initiate mitosis before DNA replication is completed. The regulatory mechanisms by which cells ensure the orderly succession through these phases have been the subject of a major scientific inquiry. Today the essential elements of the cellular machinery responsible for the execution of these functions have been defined, and a number of concepts governing its operation developed. The following sections outline our present state of knowledge in these topics.

Human cells share many cell cycle mechanisms with simpler organisms, such as yeast. It is also much easier to work with yeast organisms since they are readily genetically tractable and divide much faster. To define the circuit for cell cycle control, scientists need to identify the genes that regulate progression through the cycle. Therefore, they were looking for various mutants in yeast that block the cell cycle at a specific stage. However, blocking the cells cycle is a lethal mutation and therefore it was difficult to isolate cell cycle mutants in animal cells. Thus cell cycle mutants were obtained as conditional lethals, so that although they are unable to grow under the conditions of isolation, they can be maintained by growth in other conditions. Screens for such cell division cycle or cdc mutants identified about 80 genes in either fission (*S. pombe*) or in baker's (*S. cerevisiae*) yeast, in which the block to the cell cycle can be seen to affect the visible phenotype of the cell. Initial isolation relies upon the criterion that cells accumulate at a particular stage of the cell cycle at an elevated temperature (36°C), but continue normally through the cycle at 23°C. The mutant phenotype allows cells to continue growing larger while the cycle is blocked, causing an obvious aberration; in fission yeast the cells become highly elongated and in baker's yeast they fail to bud. Therefore, yeasts are ideal models for studying how cell cycle control may work in humans as well as how it can malfunction in cancer. Most of our knowledge regarding the regulation of the cell cycle comes from yeast. Thus, it is important to understand the central results from this model in order to apply it to the human cell cycle and cancer research.

CELL CYCLE TRANSITION AND CHROMOSOME CYCLE

The cell cycle is controlled by interdependent regulatory transitions that drive the chromosome cycle and bring the DNA to a state competent for duplication and division. This is achieved by 3 separate regulatory processes: (a) activation of kinases, (b) the operation of proteolytic pathways and (c) alteration in the state of chromatin (Nasmyth, 1996). To understand the regulatory mechanism underlying the orderly alternation between S and M phases cell fusion techniques were used. By fusing G1 / G2 cells with either S or M phases it is suggested that cells in S phase contain an S phase promoting factor (SPF) capable of inducing S-phase prematurely in G1 but not in G2 phase nuclei. On the other hand, cells in M-phase were found to contain M-phase promoting factor (MPF) capable of inducing mitotic events in cells in every phase of the cell cycle (Fig. 1). It is true that these experiments were informative but did not allow the characterization of MPF and SPF. Subsequently combination of biochemical studies in *Xenopus* eggs, genetic studies in yeast, and mammalian tissue culture led to the characterization of MPF and SPF as cyclin-dependent kinases (CDKs) (Nasmyth, 1993). This observation revolutionized the cell cycle field and laid the foundation to understand the mechanisms that regulate various transitions in cell cycle.

CYCLIN-DEPENDENT KINASES (CDKs)

Control of the cell cycle uses the similar types of components in animal cells and in yeasts, although there is more diversity in each component in animal cells. Transitions from G1 to S as well as from G2 to M phases are catalyzed by the phosphorylation of key substrates mediated by CDKs. This highly conserved group of enzymes is in its active form a complex having a catalytic subunit, the CDK, and a regulatory subunit, the cyclin (Miller and Cross, 2001). In mammalian cells the transition from :

$$\text{G1----S} \xrightarrow{\text{catalyzed by}} \text{SPF} \xrightarrow{\text{complex between}} \text{Cdk2/4 + Cyclin E or A}$$

$$\text{S--- M} \longrightarrow \text{MFF} \xrightarrow{\text{complex between}} \text{Cdk2/4 + Cyclin B}$$

Throughout the cell cycle the level of CDKs is fairly constant whereas cyclin plays an essential role in the regulation of the kinase activity of CDKs. Cyclins are proteins whose expression is periodic and highly regulated throughout the cell cycle. Table 1 shows different cyclins in mammals. Progression through the cell cycle is driven by an underlying biochemical cycle in which CDKs are sequentially activated by different cyclins. Thus, S phase cyclin-CDK complexes phosphorylates and activate proteins necessary for DNA replication, and M phase cyclin-CDK complexes phosphorylates different proteins necessary for mitosis and cytokinesis. Since the kinase activity of CDKs is absolutely dependent on the presence of the regulatory cyclin in the complex, the regulated expression of cyclins provides a first level of regulation of CDK activity.

In yeast, a single cyclin dependent kinase serves as a master controller of the cell cycle, assembling sequentially into active holoenzyme complexes with G1, S phase or mitotic cyclins

Table 1. Expression of different cyclins in mammalian cells

Cyclin	Period of expression
E	before and after S phase
A	begins in early S and peaks at the end of S-phase
B	begins during S phase and peaks in G2

Table 2. The components of the regulatory kinases in yeasts and animal cells

G2/M	Catalytic subunit	Regulatory subunit
Saccharomyces cerevisiae (budding yeast)	CDC28	CLB1-6 (B like)
Animal cells	Cdc2 or Cdk1	cyclins A, B1, B2
G1/S		
S. cerevisiae	CDC28	CLN1-3
Animal cells	Cdk2,4	cyclins A, D1, D2, D3, E

temporally to direct distinct transitions. Animal cells have more variation in subunits of the kinases. They use different catalytic subunits at each stage and also have a large number of cyclins.

Activation of CDK requires association of a cyclin and CDK that is both phosphorylated on a conserved threonine-residue and dephosphorylated on a tyrosine residue (Fig. 2). The activated CDK can be inactivated in several different ways as follows:

- Cyclin levels can be reduced by turning down transcription or by degradation of the protein itself.
- Dephosphorylation of threonine-residue.
- Inhibition of phosphorylation of threonine-residue.
- Interaction of CDK inhibitors (CKIs).

CELL CYCLE REGULATION IN NORMAL CELLS

Cell Progression in the Cycle

Work in yeast provides strong evidence that the order of SPF and MPF waves and their associated CDK activities cannot be alone responsible for ordering S and M phases. The state of

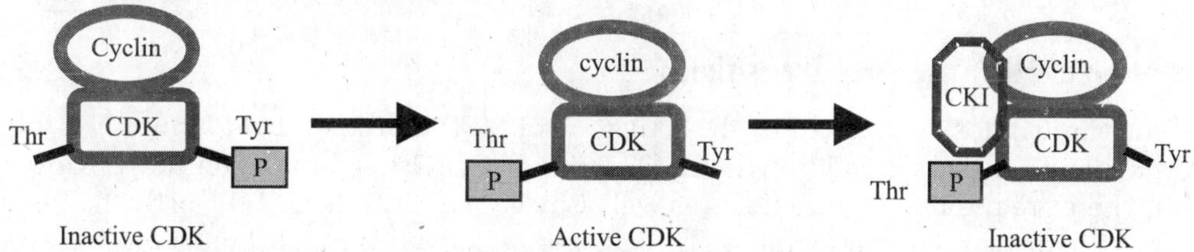

Fig. 2. About CDK.

their substrates must be equally important in determining whether a cell duplicates or segregates its chromosomes. In the budding yeast it is now clear that initiation of DNA replication requires the ordered assembly of several proteins at the DNA replication origin to form a competent prereplicative state. Once this state is formed, DNA replication can be triggered by the action of SPF. Similar processes are expected to be active in higher eukaryotes as well. To understand how CDKs regulate S phase, it is needed to describe the complex nucleoprotein structures that assemble at the origins of DNA replication initiation and the ways in which these complexes respond to signals that promote or prevent DNA replication. The discovery of budding yeast replication origins (autonomously replicating sequences; ARS) has led the groundwork for identifying replication initiators, proteins that connect DNA synthesis to the cell cycle. A multiprotein complex was identified that binds to the essential 11bp ARS consensus sequence in an ATP-dependent manner. This origin recognition complex (ORC) is a multisubunit protein that serves as a landing pad for protein-protein interactions at the origin in a cell cycle phase dependent manner. As a replication competent state, chromosomes have prereplication complexes (pre-RC) bound to the origins. Chromosomes inherit their pre-RCs as they exit mitosis, or form them in G1. Once DNA replication begins, the pre-RC is disrupted and the second state post-RC is established which is maintained until metaphase. Such alteration in the state of the chromosome from the pre-RC to post-RC state during cell cycle is important since it determine the result of activation of MPF and SPF-like kinases. It also explains why re-replication of the DNA cannot occur before the cell divides.

Regulation of Transitions in the Cell Cycle

Once a state of replication competence has been achieved in the chromosomes at the origins of replication, initiation of S phase requires activation of SPF. This is best known in the budding yeast and therefore is presented here (King et al., 1996) (Fig. 3). Cdc28, the catalytic subunit of SPF, is constitutively present in the cells binds to the regulatory cyclins (Clb5 and Clb6) after their expression. However, the kinase activity of this complex is also regulated by the association of Cki identified as the product of the Sic1 gene. In order to start DNA synthesis, Sic1 must be removed by proteolysis done by a cell cycle specific ubiquitin conjugation step that requires Cdc34. Thus, ubiquitin-dependent proteolysis by ubiquitin ligase, known as SCF complex (named after its three core components Skp1, Cdc53 and F-box protein), of a Cki is a crucial mechanism by which the onset of S phase is controlled.

Activation of SPF leads to inception of DNA replication in chromosomes at the pre-RC state. As replication progresses and origins one after the other fire, pre-RCs are destroyed and sister-chromatids with post-RCs are formed. During this period, as well as during G2, the formation of new pre-RCs on the chromosomes is blocked by the activity of Clb5-Cdc28 and Clb6-Cdc28, thus preventing replication.

Activation of MPF: The First Transition

The entry into mitosis is marked by the activation of MPF, a process that follows the synthesis of cyclin B and their association with Cdc28 (cdc2 or CDK1 in mammalian cells) (King et al., 1994). There are positive requirements as well as negative controls for MPF activation that are similar in both yeast and vertebrate cells. To generate active MPF, Cdc2 must bind to cyclin B and be phosphorylated on Thr-161. The kinase that phosphorylates this site is the Cdc2-

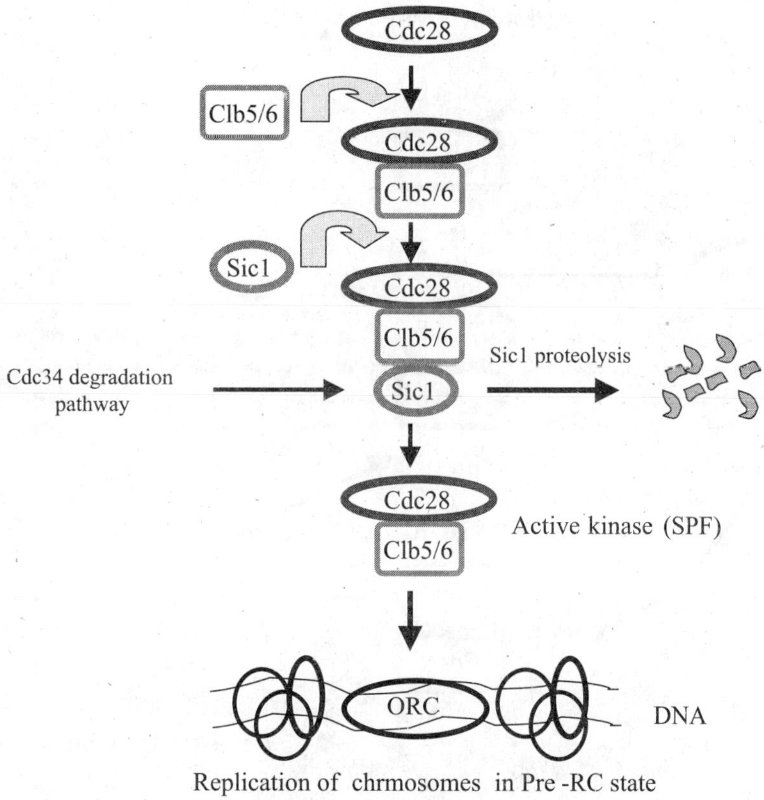

Fig. 3. The kinase cascade leading to the inception of DNA replication as known in budding yeast Similar processes are thought to be active in mammalian cells.

activating kinase (CAK), and its catalytic subunit is called CDK7. The regulatory subunit of CAK is homologous to other cyclins known as cyclin H. Since CAK activity does not appear to be tightly regulated in the cell cycle, it is unlikely that the absence of CAK activity restrains the activation of MPF, and therefore more emphasis has been placed on negative regulators of MPF.

In yeast and also in the mammalian cells, in late S phase and G2, Cdc2 is phosphorylated on Tyr-15 and Thr-14 and maintain Cdc2 in a catalytically inactive state. Such phosphorylation is performed by the gene Wee1, identified in yeast *S. pombe*. Dephosphorylation of these either in vivo or in vitro is correlated with activation of Cdc2. This activation is mediated by Cdc25 (tyrosine phosphatase), which removes inhibitory phosphorylation earlier introduced by Wee1. The importance of Cdc25 has been demonstrated genetically and biochemically in many systems. The phosphorylation and activity of Cdc25 dramatically increase as the cells enter mitosis and decrease upon exit from mitosis. Thus Cdc25 could be a major determinant of a positive feedback loop controlling MPF activation (Fig. 4). This active Cdc2-cyclinB can promote phosphorylation of a variety of mitotically active kinases, including forms of MAP kinases and the MPM-2 kinase in turn can activate and stabilize the mitotic state and probably MPF.

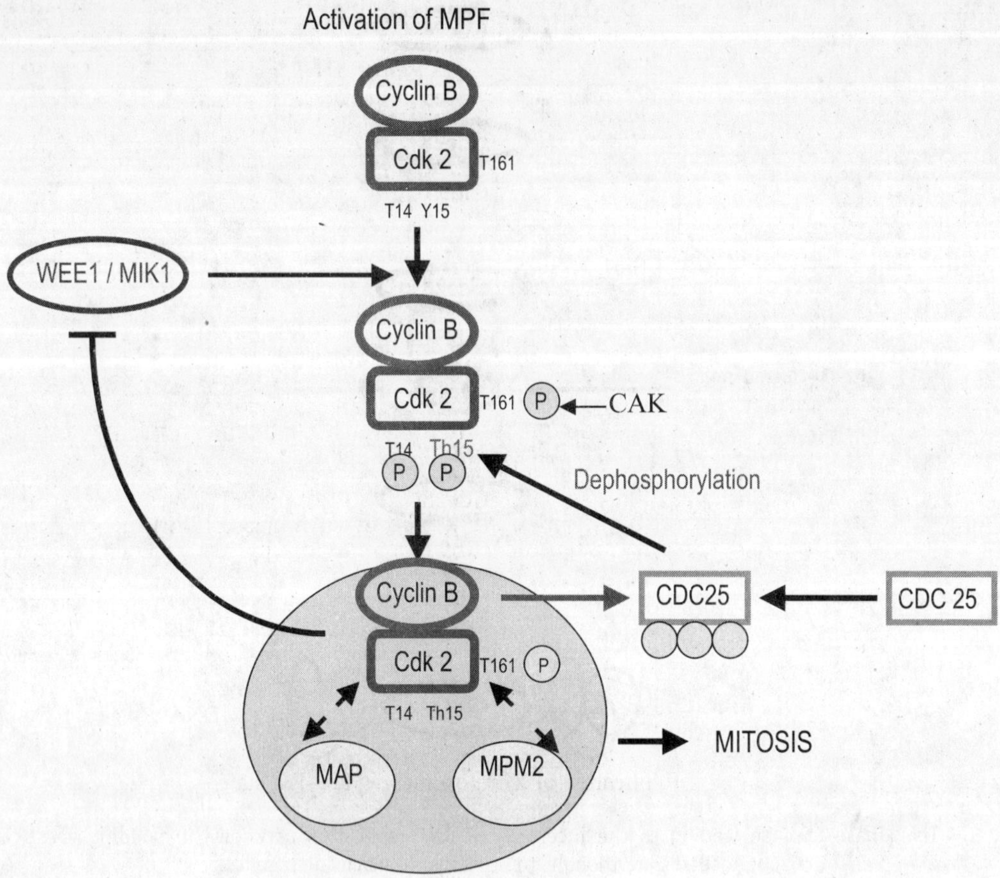

Fig. 4. The kinase cascade leading to mitosis.

Inactivation of MPF: The Second Transition

Active MPF initiates a phosphorylation cascade leads to the dissolution of the nuclear envelope via lamin phosphorylation and the condensation of the chromosomes via histone H1 phosphorylation. The spindle is formed and the chromosomes align at the metaphase plate. Once chromosome alignment at the metaphase plate is completed, MPF induces its own demise by activating the mitotic cyclin destruction system (King et al., 1996). The system is based on the ubiquitination of mitotic cyclins. The regulated component is the anaphase promoting complex (APC) (King et al., 1995), a multisubunit complex with diverse functions. The APC is important for sister chromatid separation and exit from mitosis and a stable subsequent G1 phase. The APC is activated by the regulatory subunits Cdc20 (APC^{Cdc20}) and Cdh1 (APC^{Cdh1}), which confer some substrate specificity. The APC recognizes two different amino-acid sequences. The destruction box (D box) with the conserved 9-amino acid motif in the amino terminal domain is recognized by APC^{Cdc20} and APC^{Cdh1}, and APC^{Cdh1} also recognizes the KEN box. While the D-box dependent proteolytic system for cyclins A and B may share some components, the degradation pathways may not be identical. For example cyclin A is degraded during metaphase, in advance of cyclin B. It seems that some components of the cyclin degradation system may be activated early in mitosis to degrade cyclin A, while sister chromatid separation and cyclin B degradation might require additional components activated later in mitosis.

To exit mitosis into a next G1 phase of the cell cycle, mitotic kinase inactivation is required. In the first phase cyclin B1/Clb2 proteolysis regulated by APC^{Cdc20} mediates spindle disassembly and cytokinesis. APC^{Cdc20} degrades cyclin A which may lead to decreased phosphorylation and activation of Cdh1. APC^{Cdc20} was found to be important for inactivating Clb2 during mitotic exit and the G1 phase of the cell cycle to regulate cell growth and the length of G1 allowing cell differentiation and correct assembly of pre-RCs at origins of replication and subsequent complete and accurate DNA replication.

APC^{Cdc20}-mediated degradation of securing at metaphase liberates separase, a protease that cleaves the Scc1 cohesin subunit responsible for sister chromatid cohesion. APC^{Cdc20} also initiates cyclin B degradation in mitosis, which is important for activation of separase. Securin degradation not only promotes sister chromatid separation but also releases the phosphatase Cdc14 which activates Cdh1 by dephosphorylation. Activated APC^{Cdh1} finally degrades mitotic cyclins further at the M to G1 transition and maintains their low state of activity during G1 (Fig. 5).

Resetting the Mitotic Cycle: The Third Transition

This final transition turns off cyclin degradation and depends on controls external to the mitotic circuitry, such as synthesis of G1 cyclins. The dominant ability of MPF to turn on cyclin degradation is supported by the fact that expression of G1 cyclins during mitosis in budding yeast does not inhibit cyclin B degradation. In somatic cells, where G1 cyclins are not

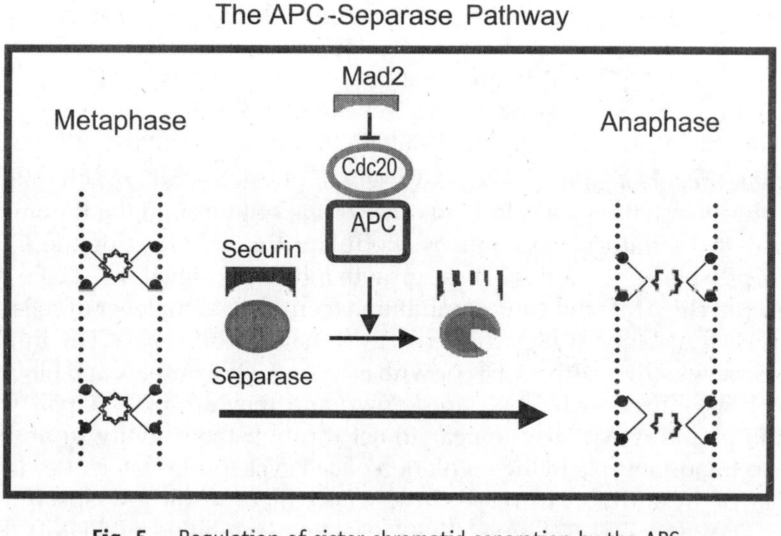

Fig. 5. Regulation of sister chromatid separation by the APC.

constitutively provided, the cyclin degradation system may continue to be active during early G1. In somatic cells, the mitotic cycle driven by cyclin B alone is not a closed cycle; G1 cyclins stimulated by exogenous growth signals are required to turn off cyclin degradation and restart the mitotic cycle, growth signaling and the cell cycle. The above described events that regulate the cell cycle build an autonomous circuitry that ensures DNA replication and segregation of sister chromatids without interruptions by extracellular signals. However, mammalian cells are only rarely cycling and spend a considerable fraction of their life span in a resting phase, termed Go, with a DNA content equal to that of a G1 cell. Cells entering G1 from mitosis or Go phase require serum mitogens continuously until several hours before the onset of S-phase; thereafter they become relatively serum-independent. This transition from a serum-dependent to serum-independent state is demarcated by a discrete point in time, which is known as R (restriction) point. By passing through the R point, the cell commits itself to traverse the remainder of its growth cycle through M.

In mammalian cells, passage of cells from Go to G1 is controlled by CDKs that are regulated by D-type cyclins. D-type cyclins act as growth factor sensors because their expression depends more on extracellular signals than on the position of the cell in the cycle. As cells enter the cycle from Go, one or more D-type cyclins (D1, D2, D3) are induced as part of the delayed early response to growth factor stimulation and assemble with CDK4 and CDK6 in a growth factor-dependent way. The kinase activity of CDK4/6-CyclinD1/D2/D3 is measurable by mid-G1, and increase to a maximum near the G1 to S transition. It persists at nearly maximum levels for as long as growth factor stimulation continues. D-type cyclins are labile proteins, and because their holoenzyme activity decays rapidly, cells deprived of mitogens exit the cycle (Sherr, 1994).

pRB as an integrator of positive signals: CyclinD-dependent kinases phosphorylate substrates whose modification is required for exit from G1. One such target is the Rb family of proteins (Rb, p130, p107). The Rb, product of the retinoblastoma tumour suppressor gene, operates in the midst of the cell cycle clock apparatus. The Rb family of proteins control gene expression mediated by a family of transcriptional regulators called E2Fs. E2Fs transactivate genes whose products are important for S phase entry. In its hypophosphorylated form, Rb binds to E2F and converts them to repressor that does not allow the expression of E2F target genes. Phosphorylation frees up E2F which can now transactivate target genes. Rb phosphorylation is accelerated late in G1 by the kinase activity of cyclinE-CDK2 complex, whereas cyclinA and cyclinB maintain the hyperphosphorylated state in the remainder of the cell cycle (Fig. 6).

pRB as an integrator of negative signals: A variety of physiological growth inhibitory signals prevent pRB phosphorylation and block advance of the cell through the R point and into late G1. These growth inhibitory mechanisms modulate the activities of the CDKs that are responsible for pRb phosphorylation. These growth inhibitory signals are TGFb (transforming growth factor b), cyclic AMP and contact inhibition (complete monolayer of cells stop growing in culture). In each instance, a negative growth signals mobilizes a CDK inhibitor (CDKI) protein that associates either with a CDK or with a cyclin-CDK complex and blocks its function (Koepp et al., 1999). Three such CDKI are known and they are $p27^{kip1}$, $p57^{kip2}$ and $p21^{cip1}$. Contact inhibition and cAMP also appear to act through their ability to mobilize $p27^{Kip1}$. $p21^{Cip1}$ plays an important role in the regulation of cell cycle progression after irradiation and is discussed later. In addition to these three CDKI there are a few specific polypeptide inhibitors of CDK4 and CDK6, termed INK4 proteins (Inhibitors of CDK4). Known INK4 proteins are $p16^{INK4a}$, $p15^{INK4b}$, $p18^{INK4c}$ and $p19^{INK4d}$.

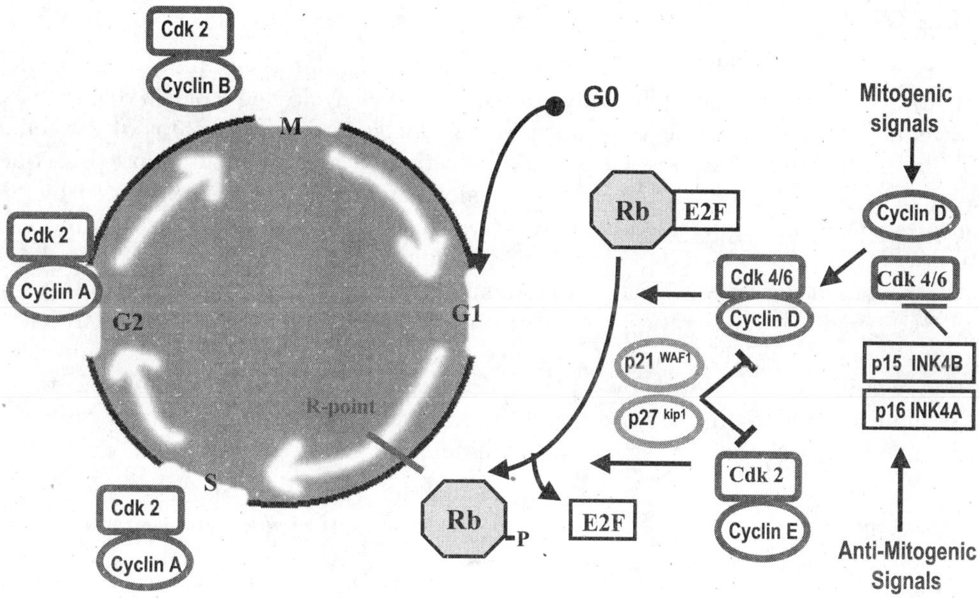

Fig. 6. Events leading to the inception of the cell cycle after stimulation to growth of noncycling cells.

pRB and regulation of the clock: The interplay of the cyclins, CDKs, CDKIs / INK4s and pRB enables the cell cycle clock to control its own forward progress. At least two types of regulatory circuits utilizing these clock components have been proposed. The first of these is CDKI $p27^{Kip1}$ which is expressed at constant levels throughout G1. Early in G1, any cyclinD-CDK4/6 that is present will associate with and be inactivated by $p27^{Kip1}$. The levels of cyclinD-CDK4/6 will increase gradually as cells progress through mid-G1. Eventually cyclinD-CDK4/6 levels will exceed a threshold beyond which they out-titer the limited amount of $p27^{Kip1}$ that is present. These excess cyclinD-CDK4/6 complexes can now be phosphorylated by CAK (CDK-activating kinases) and thus empowered, can initiate pRB phosphorylation and trigger the R-point transition.

The second of these proposed regulatory circuits involves $p16^{INK4a}$ that bind CDK4/6 and to preempt their association with D cyclins. The expression of $p16^{INK4a}$ increases strongly toward the end of G1 after pRB undergoes inactivation through phosphorylation. These observations, taken together, suggest the operation of a negative feedback loop which operates as follows:

First: CyclinD+CDK4/6 $\xrightarrow{\text{triggers}}$ pRB phosphorylation

Second: Inactive pRB releases E2F $\xrightarrow{\text{induces}}$ p16 expression

Third: p16 binds all CDK4/CDK6 $\xrightarrow{\text{disallow}}$ Association of CyclinD and CDK4/6

Fourth: CyclinD is degraded

After degradation of cyclin D, cyclin E binds to CDK2 and progress towards the S-phase.

THE ROLE OF CKIS

These proteins (CDK inhibitory proteins) are additional regulatory subunit for the CDKs. By binding to CDK-cyclin complex they put a brake on the cell cycle: some play a role in response to extracellular signals, whereas others appear to function in intrinsic steps of the cell cycle (Fig. 7). In all of these situations, CKIs must be activated at proper time and subsequently inactivated to allow further cell cycle progression. Synthesis of p21 is induced by p53 in response to DNA-damaging agents. Transcription of p27 gene (KIP1) is not induced in response to TGFb, and at least in 3T3 cells, p27 protein levels remain constant throughout the cell cycle. Several observations indicate that this inhibitor may bind to Cdk4-cyclinD complex, second, TGFb reduces the synthesis of Cdk4 and third cells over expressing of Cdk4 fail to arrest in response to TGFb. It is thus plausible that Cdk4-cyclinD complexes sequester p27: reduction of Cdk4-cyclin D levels by TGFb could release p27 and thereby allow p27 to inhibit Cdk2-cyclin E complexes to CKIs and arrest the cell cycle. In general, the proper ratio of Cdk-cycln complexes to CKIs might be a critical parameter for G1 progression.

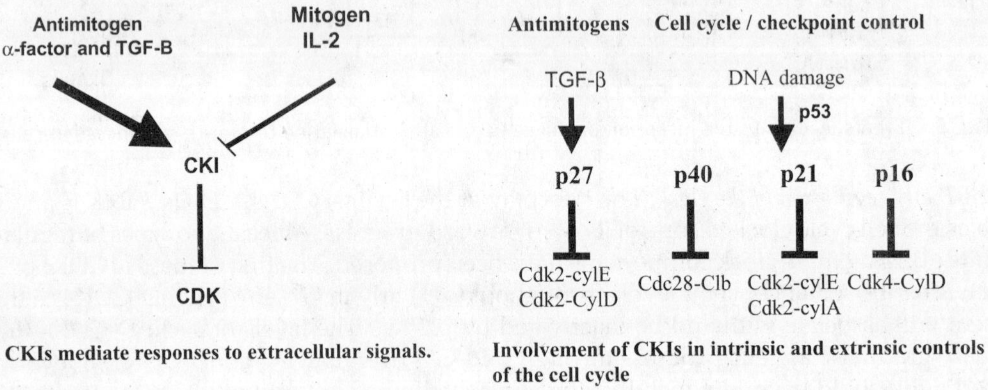

Figure 7 Role of CKIs.

Cell Cycle Regulation in Irradiated Cells

In general, cells do not allow progression into the next phase of the cell cycle before completing all events associated with the previous phase. Proper completion of the events of one phase before initiating the events associated with the following phase of the cell cycle assures the faithful duplication and transmission of the genome to the offsprings. Failure to solve the completion problem produces genetic damage. Furthermore, cells have developed ways to monitor genome integrity and to stall further progression if DNA damage is detected. These surveillance mechanisms are collectively known as checkpoints (Hartwell and Weinert, 1989) which are inhibitory pathways and are thought of as signal transduction pathways activated by a signal and resulting in an inhibition in cell cycle progression. Checkpoints assure that M follows S phase, and that anaphase does not start before complete alignment of the chromosomes at the metaphase plate.

The DNA Damage Checkpoint

Cell cycle checkpoints preserve genome integrity by monitoring the fidelity of DNA replication and segregation. In *Escherichia coli*, DNA damage or inhibition of DNA replication generates a signal that activates SOS response which is associated with the induction of an entire group of genes that either participate in DNA repair or block cell division. What is important in this type of response is that cell cycle arrest caused by DNA damage is not caused by the DNA damage *per se* but by specific surveillance mechanisms that detect mistakes and induce inhibitors of key cell cycle transitions (Fig. 8).

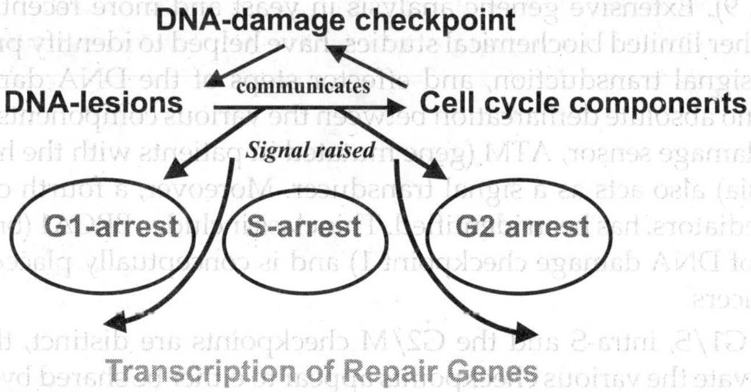

Fig. 8. The DNA damage checkpoint and its effect on various stages of the cell cycle.

In mammalian somatic cells, the best-characterized checkpoints are the DNA damage / replication checkpoints and the mitotic spindle checkpoint. The DNA damage/replication checkpoints result in cell cycle arrest if DNA is not fully replicated, or is damaged. The mitotic spindle checkpoint results in cell cycle arrest prior to anaphase if the spindle is not properly assembled. There is also evidence that defects in events relating to cell division itself can result in cell cycle arrest. With the use of fluorescence plus Giemsa (FPG) staining techniques for the investigation of cell cycle kinetics it has been established that X-ray induced cell cycle delay in human lymphocytes, irradiated in the Go-stage of the cell cycle, is of the order of a few hours after application of 4Gy (Lloyd et al., 1977) and that the cell cycle delay increases with dose (Vulpis et al., 1978), showing a linear relationship between the dose and mitotic delay of about 1 hour per Gray (Purrott et al., 1980). By employing FPG staining technique it has been reported that reduced glutathione pretreatment reduced the 2Gy induced delay in cell cycle and chromosome aberrations (CAs) significantly but was unable to do so consistently at higher doses of radiation in muntjac lymphocytes (Chatterjee and Jacob-Raman, 1986). The primary subcellular target responsible for radiation-induced delay in cell proliferation still remains undetermined. There are reports that indicate that damages to DNA, cell division related proteins, cellular and nuclear envelope and cAMP and cGMP metabolism are contributing to the induction of radiation-induced delay.

The DNA damage checkpoint communicates information between a DNA lesion and the components of the cell cycle. It detects damaged DNA and generates a signal that arrests cells in G1 phase, slow down cells in S phase and arrest cells in G2 phase. In addition to these effects on cell cycle progression, the DNA damage checkpoint also induces the transcription of repair genes. Lesions in the DNA are considered the initiating signal of the DNA damage checkpoint. How do checkpoints arrest cell cycle progression? The simplest model is that the checkpoints prevent the next CDK phase transition until repair is done. Each of the proteins that influence CDK activity are potential interfaces for signal transduction pathways that regulate cell cycle transitions and thus are potential targets for manipulations by checkpoint pathways. Evidence for this mechanism of action will be presented for the G1 and G2 phase arrest.

MOLECULAR COMPONENTS OF THE DNA DAMAGE CHECKPOINTS

The DNA damage checkpoint conceptually has three components: sensors, signal transducers and effectors (Fig. 9). Extensive genetic analysis in yeast and more recently in human cells, combined with rather limited biochemical studies, have helped to identify proteins involved in damage sensing, signal transduction, and effector steps of the DNA damage checkpoints. However, there is no absolute demarcation between the various components of the checkpoint. For example, the damage sensor, ATM (gene mutated in patients with the hereditary disorder ataxia telangiectasia) also acts as a signal transducer. Moreover, a fourth class of checkpoint proteins, called mediators, has been identified. This class includes BRCA1 (breast cancer gene 1), MDC1 (mediator of DNA damage checkpoint 1) and is conceptually placed between sensors and signal transducers.

Although the G1/S, intra-S and the G2/M checkpoints are distinct, the damage sensor molecules that activate the various checkpoints appear to either be shared by all the 3 pathways or to play a primary sensor role in one pathway and a back-up role in the others. Similarly, the signal-transducing molecules, which are protein-kinases and phosphatases, are shared by the different checkpoints to varying degrees. The effector components (proteins that inhibit phase transition) of the checkpoints are, naturally, what gives the checkpoints their unique identities.

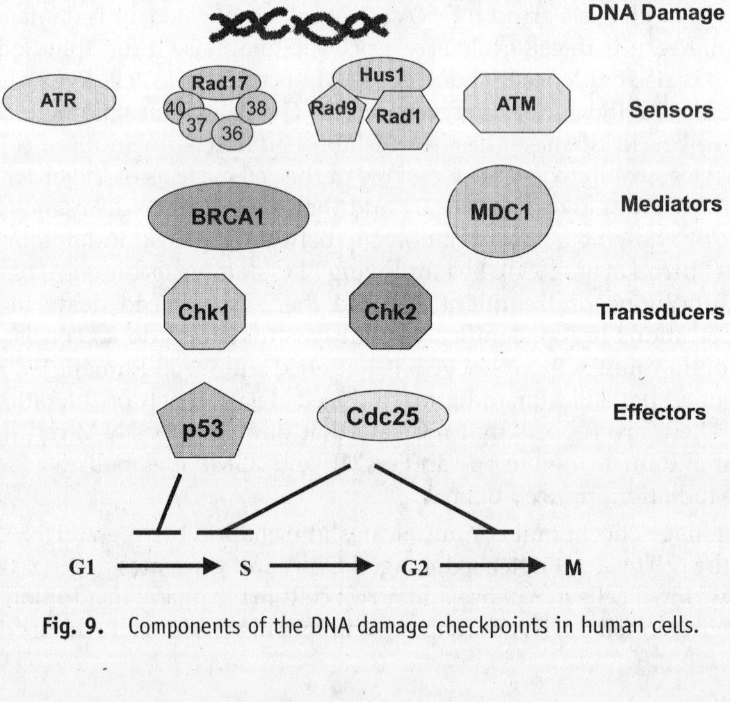

Fig. 9. Components of the DNA damage checkpoints in human cells.

However, various sensors, mediators, and signal transducers may play more prominent roles in one checkpoint than in others (Tanaka and Russell, 2001).

The GI Checkpoint

This is the best understood checkpoint in human cells in terms of signal transducers and cell cycle effectors. Known elements of the G1 checkpoint are summarized in Fig. 10. The presumed purpose of the G1 checkpoint is to regulate entry into S phase and to prevent cells from replicating damaged DNA. Its abrogation is expected to lead either to an increase in genomic instability, or an increase in damage tolerance, and therefore an increase in cell survival after radiation exposure. If the DNA damage is double-strand breaks (dsbs) induced by radiation or radiomimetic chemicals, ATM is activated and phosphorylates many target molecules, notably p53 and Chk2 (checkpoint kinase 2). These phosphorylation activate two signal transduction pathways, one to initiate and one to maintain the G1/S arrest. The reaction that initiates G1/S arrest is phosphorylation of Chk2, which in turn phosphorylates Cdc25A phosphatase, causing its inactivation by nuclear exclusion and ubiquitin-mediated proteolytic degradation. In absence of active Cdc25A inactive CDK2 (phosphorylated) will accumulate and unable to initiate replication from ORC-ORI complexes. If the DNA damage is caused by UV light or UV-mimetic chemicals, the signal is sensed by ATR, leading to phosphorylation of Chk1 which in turn phosphorylates Cdc25A, leading to G1 arrest.

This initial rapid response is followed by the p53-mediated maintenance of G1/S arrest, which becomes fully operational several hours after the detection of DNA damage. In this stage ATM or ATR phosphorylates Ser15 of p53 directly and Chk2 or Chk1 phosphorylates Ser20 of

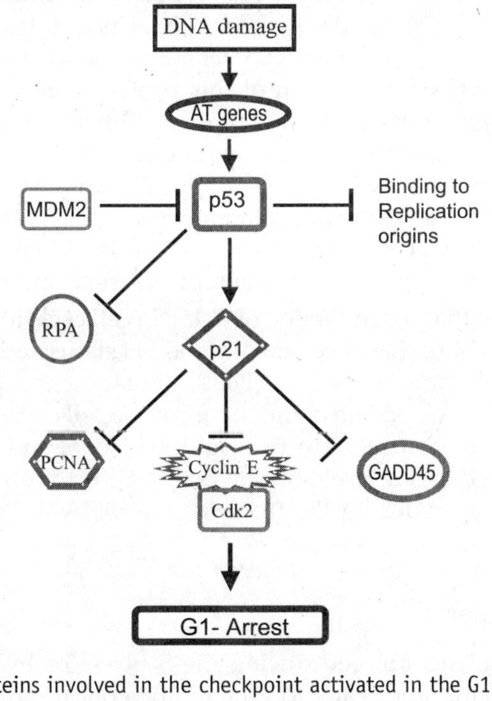

Fig. 10. Steps and proteins involved in the checkpoint activated in the G1 phase after DNA damage.

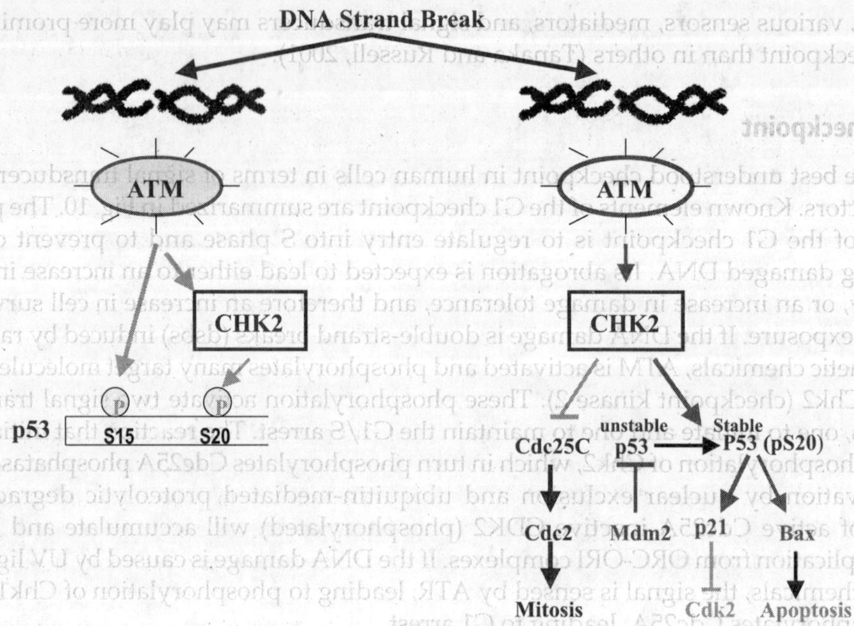

Fig. 11. Steps and proteins implicated in the checkpoint after DNA damage.

p53. Such phosphorylation inhibits its nuclear export and degradation and therefore the level of p53 rises (Levine 1997) (Fig. 11). The induction of p53 inhibits cell cycle progression by two different mechanisms. In the first mode of action, p53 acts as a transcriptional activator and induces the transcription of genes with p53 response elements. Such genes include the p21CIP1/WAF1, a known CKI. Interaction of this protein with CDK2/CyclinE complex is thought to inhibit the progression of cells into S phase. Other genes that are inducible by p53 include the growth arrest and DNA damage inducible gene GADD45 (growth arrest DNA damage-induced), which may contribute to overall inhibition in the progression to S-phase. The second mode of action involves the direct interaction of p53 with essential proteins of DNA replication, such as replication protein A (RPA) as well as the possible binding of the protein to replication origins, which may prevent initiation of DNA replication.

Recent studies showed that the induction of delay in cell cycle after radiation could be due to the involvement of two components-cell membrane-SH status and cytoplasmic proteins that regulate lymphocyte proliferation, and the DNA damage checkpoint that communicates information between a DNA lesion and components of the cell cycle, the latter trigger p53-p21 mediated cell cycle arrest. It seems that increase in the level of endogenous GSH reduces the radiation-induced cell cycle delay probably by acting directly on the first component and indirectly on the second, by reducing the frequency of DNA damages (Ray and Chatterjee 2007).

The Intra-S-phase Checkpoint

It is activated by damage encountered during the S phase or by unrepaired damage that escapes the G1/S checkpoint and leads to a block in replication. There are reports that in response to DNA damage sequestration of PCNA (proliferating cell nuclear antigen) by p21 or by ubiquitination of PCNA by scRad6 was happened and such modification slow down of replication fork progression. It seems that this checkpoint is a biochemical regulatory pathway which dictates the progression of cell cycle events (rather than phase transitions) in an orderly manner and that prevents the initiation of certain biochemical reactions before completion of the others within the cell. In response to a DNA dsb or dsb resulting from replication of a nicked or gaped DNA, ATM, the Mre11/Rad50/NBS1 complex, and the BRCA1 are all required for the activation of the checkpoint. All of these repair proteins function as sensors because biochemical data show that all bind to either dsbs or to special branched DNA structure (the Mre11/Rad50/NBS1 complex, BRCA1, BRCA2). It appears that the S-phase checkpoint initiated by dsbs proceeds through two pathways. The well understood ATM-Chk2-Cdc25A-Cdk2 pathway is strictly a checkpoint response. A second pathway depends on phosphorylation of SMC1 by ATM with the aid of BRCA1, FANCD2 and NBS1, likely plays a role in cell-cycle arrest.

The G2/M Checkpoint

It prevents cells from undergoing mitosis in the presence of DNA damage. As in other checkpoints, with certain types of DNA lesions induced by UV light, ATR-Chk1 signaling initiates cell cycle arrest, but the maintenance of the arrest is then performed by ATM-Chk2 signaling. With other types of lesions, such as ionizing radiation-induced DNA dsbs, the order of action is reversed. In any event the checkpoint kinases inhibit the entry into mitosis by downregulating Cdc25 and upregulating Wee1 which together control Cdc2/Cyclin B activity. Disruption of the Chk1-Cdc25A pathway abrogates ionizing radiation-induced S and G2 checkpoint, indicating that the Cdc25A phosphatase is the main effector of the G2/M checkpoint. Upon phosphorylation, the Cdc25 phosphatase binds to 14-3-3 proteins, becomes sequestered in the cytoplasm and is degraded by the ubiquitin-proteosome pathway. Inactivation of Cdc25 leads to accumulation of Y15-phosphorylated Cdc2 and mitotic arrest (Fig. 12).

DEREGULATED PROTEOLYSIS IN CANCER

Deregulation of cyclins and other cell cycle regulators can cause aberrant proliferation and genomic instability (Lengronne and Schwob, 2002). Aberrant expression of different cyclins can be detected in various malignant tumors (Yam et al., 2002), with important implications for clinical oncology. One recent example is the prognostic relevance of the elevated cyclin E expression in breast cancer (Sutherland and Musgrove, 2002). Various mechanisms, such as gene amplification or overexpression, missing inactivation by CKI and unregulated proteolysis, can result in deregulation of cyclins. In colon cancer, a decreased or absent activity of the CKI p27 could be found due to increased degradation. The APC substrate securin is important for sister chromatid separation and is involved in tumorigenesis (Zou et al., 1999). Securin is frequently overexpressed in hematopoietic neoplasms. Both cyclin A or cyclin B overexpression has also been reported in a variety of cancers and can stimulate genomic instability (Vaziri et al., 2003).

A specific mechanism to gurantee genomic integrity is supplied by the spindle assembly checkpoint (Cleveland et al., 2003). This cell cycle checkpoint is responsible for ensuring the

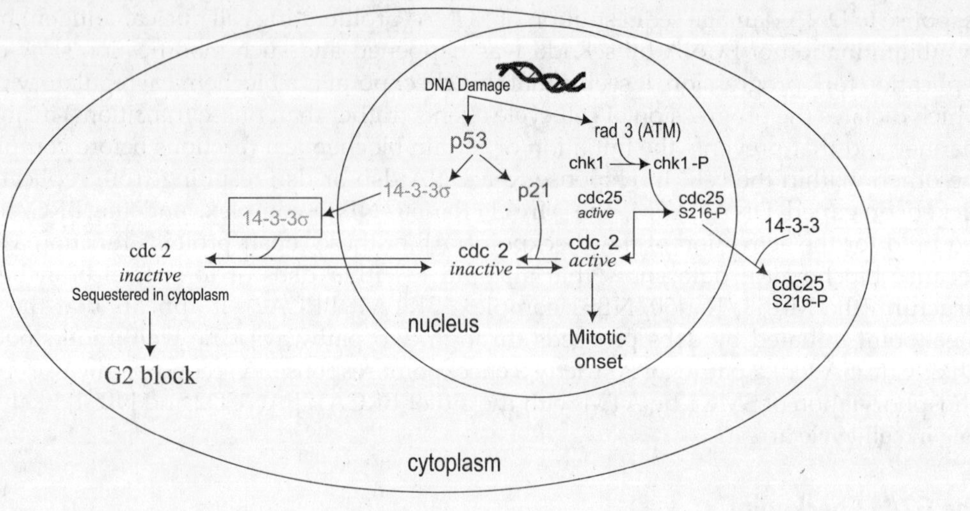

Fig. 12. Steps and proteins involved in the checkpoint activated in the G2 phase after DNA damage.

bilateral attachments of all pairs of sister chromatids to the mitotic spindle. Thereby, the precise spatial orientation of chromosomes is detected and the onset of anaphase is delayed until all the mitotic spindle microtubules are attached to the kinetochores at the centromeres of the chromosomes. The spindle checkpoint is regulated by signaling mechanisms, involving the Bub1, Mad1 and Mad2 gene products. The Bub and Mad genes are well conserved from yeast to humans and function as inhibitors of the APC, by preventing interaction with its activating phosphorylated subunit Cdc20. Recently it was shown that the deletion of one Mad2 allele results in a defective mitotic checkpoint in both human cancer cells and murine primary embryonic fibroblasts. Furthermore, $Mad2^{+/-}$ mice develop lung tumours at high rates after long latencies, implicating defects in the mitotic checkpoint in tumorigenesis (Michel *et al.*, 2001).

SUMMARY

Cell proliferation is under tight control to ensure accurate DNA replication and chromosome segregation. Cyclin-dependent kinasess (Cdks) and their activating subunits, the cyclins are the driving forces of the cell division cycle. Regulation of cyclin oscillation by ubiquitin-dependent proteolysis thereby has a central role in cell cycle regulation. DNA damage is a relatively common event in the life of a cell and may lead to mutation, cancer and cellular and organismic death. Cells have the ability to sense DNA damage, and to activate repair pathways that efficiently remove such damage and restore the integrity of the DNA. Highly sophisticated mechanisms further enable cells to actively stall growth and division after sensing DNA damage. Such inhibitory processes, termed checkpoints, may optimize repair and minimize the adverse consequences of DNA damage.

REFERENCES

Chatterjee A and Jacob-Raman M. (1986) Modifying effect of reduced glutathione on X-ray- induced chromosome aberrations and cell cycle delay. *Mutat. Res.* 175 : 73-82.

Cleveland DW, Mao Y and Sullivan KF (2003) Centromeres and Kinetochores: from epigenetics to mitotic checkpoint signaling. *Cell* 112: 407-421.

Hartwell LH and Weinert TA (1989) Checkpoints: controls that ensure the order of cell cycle events. *Science* 246: 629-634.

Howard A, and Pelc SR (1953) Synthesis of DNA in normal and irradiated cells and its relation to chromosome breakage. *Heredity* 6: 261-273.

King RW, Jackson PK, and Krischner MW (1994) Mitosis in transition. *Cell* 79 : 563-571.

King RW, Deshaies RJ, Peters JM. and Krischner MW (1996) How proteolysis drives the cell cycle. *Science* 274: 1652-1658.

Koepp DM, Harper JW and Elledge SJ (1999) How a cyclin became a cyclin: regulated proteolysis in the cell cycle. *Cell* 97: 431-434.

Lengronne A and Schwob E (2002) The yeast CDK inhibitor Sic1 prevents genomic instability by promoting replication origin licencing in late G1. *Mol. Cell*, 9:1067-1078.

Levine AJ (1997) p53, the cellular gatekeeper for growth and division. *Cell*, 88: 323-331.

Lloyd DC, Dolphin GW, Purrott RJ, Tipper PA. (1977) The effect of X-ray induced mitotic delay on chromosome aberrations yields in human lymphocytes. *Mutat. Res.* 42: 401-409.

Michel LS., Liberal V, Chatterjee A, Kirchwegger R, Pasche B, Gerald W, Dobles M, Sorger PK, Murty VVVS, and Benezra R. (2001) MAD2 haploinsufficiency causes premature anaphase and chromosome instability in mammalian cells. *Nature* 409: 355-359.

Miller ME. and Cross FR. (2001) Cyclin specificity: how many wheels do you need on a unicycle? *J Cell Science*, 114: 1811-1820.

Nasmyth K (1993) Control of the yeast cell cycle by the Cdc28 protein kinase. *Curr. Opin. Cell Biol.*, 274: 1643-1644.

Nasmyth K (1996) Viewpoint: Putting the cell cycle in order. *Science* 274: 1643-1644.

Purrott RJ, Vulpis N and Lloyd DC. (1980) The use of harlequin staining to measure delay in human lymphocyte cell cycle induced by in vitro irradiation, *Mutat. Res.* 69 : 275-284.

Ray S and Chatterjee A. (2007) Influence of endogenous glutathione level on X-ray induced cell cycle delay in human lymphocytes. *Cell Proliferation* 39: 37-47.

Sherr CJ (993) Mammalian G1 cyclins. Cell 73: 1059-1065.

Sutherland RL and Musgrove EA (2002) Cyclin E and prognosis in patients with breast cancer. *N. Engl. J. Med.*, 347: 546-1547.

Tanaka K. and Russell P. (2001) Mrc1 channels the DNA replication arrest signal to checkpoint kinase Cds1. *Nature Cell Biology* 3:958-965.

Vaziri C, Saxena S, Jeon Y, Lee C, Murata K, Machida Y, Wagle N, Hwang DS and Dutta A. (2003) A P53-dependent checkpoint pathway prevents rereplication. *Mol. Cell* 11: 997-1008.

Vulpis N., Purrott RJ and Lloyd DC. (1978) The use of harlequin staining to assess radiation effects on the in vitro cell cycle kinetics of human lymphocytes. *Mutat. Res.*, 51:145-155.

Yam CH, Fung TK, and Poon RY. (2002) Cyclin A in cell cycle control and cancer. *Cell Mol. Life Sci.* 59: 1317-1326.

Zou H, McGarry TJ, Bernal T and Krischner MW. (1999) Identification of a vertebrate sister-chromatid separation inhibitor involved in transformation and tumorigenesis. *Science* 285: 418-422.

Suggested Readings

1. The Cell Cycle : An Introduction (1993) A.W. Murray and T. Hunt, Oxford University Press, New York.
2. Cell Cycle Regulation (1997) R.R Ruffolo Jr, and B.W. Metcalf, George Poste.

3. Signaling Networks and Cell Cycle Control : The Molecular Basis of Cancer and Other Diseases (2000) J. Silvio Gutkind, Humana Press, INC, Totowa, NJ, USA.
4. Cell Cycle and Growth Control : Biomolecular Regulation and Cancer (2004) Gary S. Stein and Arthur B. Pardee, 2nd edition, John Wiley & Sons Inc., USA.

REVIEW QUESTIONS

Long Questions

1. What is a cyclin-CDK complex? What determines the substrate specificity of a cyclin-CDK complex? How do cyclin-CDK complexes exert their effects on cells? What counterpart class of enzymes reverses the effects of cyclin-CDK complexes?
2. How does the normal retinoblastoma protein function to hold mammalian cells at G1 restriction point ("Start")? Name two proteins that are targets of the anaphase promoting complex, and explain how this complex acts on these proteins.
3. What three cell cycle checkpoints were highlighted in this chapter? What event or events cause each checkpoint to be activated to stop the cell division cycle?
4. What protein is the major player in activating a DNA damage checkpoint? How is this protein normally kept from triggering the checkpoint? What happens to the protein when there is DNA damage?

Multiple Choice Questions

1. The time from the end of one mitosis to the start of the next is called
   (a) cell cycle (b) Interphase
   (c) M phase (d) none of the above
2. In which phage of a cell START in yeast and the restriction point in animal cells makes a commitment to divide.
   (a) G2-phase (b) Interphase
   (c) M-phase (d) G1-phase
3. The active component of the progesterone-treated egg extract could induce mitosis in oocyte is
   (a) SPF (b) CDK
   (c) MPF (d) none of the above
4. Phosphorylation and dephosphorylation of catalytic subunit are catalyzed by
   (a) Phosphatases and kinases (b) kinases and phosphatases
   (c) kinases and Invertases (d) none of the above
5. Cdc25 is a phosphatase that removes the inhibitory phosphate from Tyr-15 of
   (a) SPF (b) CDK
   (c) MPF (d) Cdc2
6. Weel is a kinase that antagonizes
   (a) MPF (b CDK
   (c) Cdc25 (d) none of the above

7. The important target for cdk-cyclinD is
   (a) Cdc25 (b) ATM
   (c) Rb (d) p53
8. Cyclin A and Cyclin B are degraded at two points during
   (a) metaphase and end of mitosis (b) end of mitosis and metaphase
   (c) S-phase and G1 (d) none of the above

Answers

1. (b)	2. (d)	3. (c)	4. (b)	5. (d)	6. (c)
7. (c)	8. (a)				

PART–2

Biocatalysis and Kinetics

CHAPTER 6

Catalytic Molecules in Cell

G. Tripathi
Dept. of Zoology, J.N.V. University, Jodhpur – 342 001
E-mail: drgst@rediffmail.com

Cell is a self perpetuating chemical system containing structurally and functionally different molecules. Some molecules speed up chemical reactions in cells. They are referred to as biological catalysts. They may be protein (enzyme and catalytic antibody) or nucleic acid (RNA). Enzyme is a Greek word, which means "in yeast". The term enzyme was first coined by Kühne in 1877. Enzyme is a protein catalyst which accelerates the rate of biochemical reaction. An enzyme molecule temporarily binds to a substrate during intermediate stages of reactions but is released in its original form at the end of the reaction. In 1897, Buchner extracted enzyme from yeast which catalyzes alcoholic fermentation. Later on, in 1926, Sumner isolated pure crystalline form of urease from extracts of jack bean and showed that enzyme is protein. However, the protein nature of enzyme was not accepted until the crystallization of pepsin, trypsin and chymotrypsin by Northrop during 1930 to 1936. Sumner and Northrop shared the Nobel Prize in 1946 for proving enzyme as protein. Enzymes interact with their substrates. The molecule acted upon by the enzyme to form product is the substrate. The substrate-binding site occurs on surface of the enzyme molecule. This binding site is a particular arrangement of amino acid side chains in the polypeptide of enzyme molecule that is specially designed to bind a specific substrate. Over 500 enzymes are found in a typical cell and many of these have been isolated in pure form. About 200 enzymes have been crystallized so far. The relation between structure and function of enzyme molecule has been established (Ferst, 1985). Some important properties of enzymes are as follows :

1. Enzymes are catalytic proteins present in small quantities in cells.
2. Enzymes increase the rate of reaction but do not change the equilibrium constant (Keq) and bring the equilibrium much earlier than would be possible without their presence. Thus they help in attaining biological rates of chemical reactions compatible with the life processes by reducing time frame of reactions from thousands of years to a minute.

3. Enzymes have high substrate specificity. For example, glucose oxidase will oxidize glucose but not galactose. The specificity resides in substrate-binding site. DNA and RNA polymerases have very high specificity and the rate of mistake in recognizing substrate is only 10^{-6}.
4. Absolute specificity is not seen in many enzymes. Some enzymes have a broad substrate specificity such as in case of hexokinase which phosphorylates glucose, mannose and fructose. The substrate binding site may contain catalytic site (active site). In some cases, the catalytic site may not be within the substrate binding site but may be contiguous to it in the primary sequence.
5. Each enzyme acts on a substrate which is stoichiometrically converted into product. Stoichiometric means in numerical proportion (1:1). Stoichiometric conversions are seen at most places in metabolic pathways.
6. Enzymes lower down the energy of activation and do not change the thermodynamic properties of the system with which it is interacting.

COFACTORS

Many enzymes contain a non-protein component which may be covalently or non-covalently linked with the enzyme molecule. Such enzymes are the conjugated proteins i.e., protein conjugated with non-protein part. This non-protein part is cofactor which is important for enzyme activity. Apoenzyme is the protein part of an enzyme without any cofactors or prosthetic groups.

Apoenzyme	+	Cofactor	=	Holoenzyme (Complete catalytic entity)
(Catalytically inactive)		(Catalytically not active)		(Catalytically active)

Cofactors are non-protein componant required for activity of some enzymes. Unlike enzymes, they are stable at relatively high temperatures. They may either remain unchanged at the end of a reaction or regenerated by a later process. The cofactor may be inorganic (metals) or organic (coenzyme) moiety. They may be defined as small organic or inorganic molecules that an apoenzyme requires for its activity. For an instance, in lysine oxidase copper is loosely bound but is required for the enzyme to be active. Similarly, carbonic anhydrase is a metallozyme which requires Zn^{2+} for catalyzing the conversion of carbon dioxide and water into carbonic acid in red blood cells. Metal acts as a catalytic center and bridge between the substrate and polypeptide chain of the enzyme. Hence it stabilizes catalytic conformation of enzyme molecule. Like metals, coenzymes (organic moiety) are also essential for enzyme activities and required in trace amount. They are derived from vitamin B e.g., nicotinic acid in NAD^+ and $NADP^+$, B_2 (riboflavin) in FAD and FMN etc. Coenzymes can be defined as small, dialyzable, heat-stable, organic molecules which may bind covalently or noncovalently to the apoenzyme (protein part). The term prosthetic group denotes a covalently bound coenzyme. It acts as an intermediate carrier of some atoms or groups during catalysis and is regenerated after catalysis is over. A prosthetic group is similar to a cofactor but is tightly bound to an apoenzyme. For example, in the cytochromes, the heme prosthetic group is very tightly bound and requires strong acids to dissociate it from the apocytochrome. Addition of a cofactor or prosthetic group to the apoprotein yields the holoenzyme (active enzyme). There are three recognized types of cofactors viz., metals, coenzymes and prosthetic groups.

(A) Metal Cofactors

Metals are required as cofactors for about two-third of all enzymes. They participate in enzyme reactions by acting as Lewis acids and by various modes of chelate formation. Lewis acids have empty d electron orbitals acting as electron sink. Chelates are organometallic coordination complexes e.g., complex between iron and protoporphyrin IX to form a heme. Metals acting as Lewis acid catalyst are Zn, Fe, Mn and Cu etc. In carbonic anhydrase, Zn acts as Lewis acid:

$$CO_2 + H_2O \xrightarrow[Zn^{2+}]{\text{Carbonic anhydrase}} H_2CO_3$$

Metals can also catalyze enzymatic reactions either by binding substrate and promoting electrophilic catalysis at the site of bond cleavage or by stabilizing intermediates in the reaction pathway.

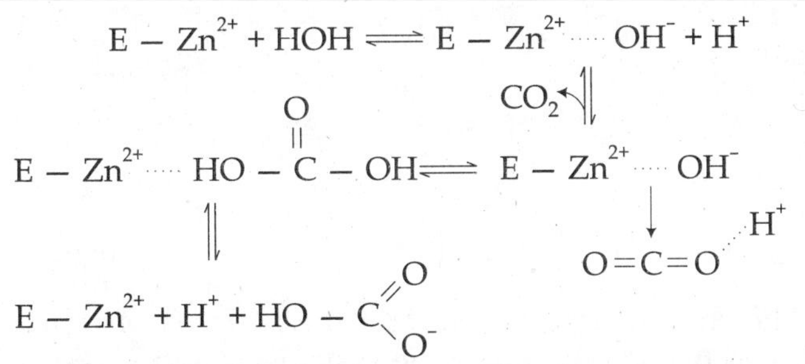

(Zinc, Zn^{2+}, functioning as Lewis acid in carbonic anhydrase)

In case of zinc proteases such as carboxypeptidase and thermolysin, zinc functions to generate a hydroxyl group from water and then to stabilize the transition state resulting from the attack of the hydroxyl on the peptide bond. The function of a metal as a Lewis acid in carbonic anhydrase and carboxypeptidase requires chelate formation. There are various modes of chelation between metal, enzyme and substrate where no acid catalysis takes place. For example, in case of kinases such as creatine kinase the true substrate is not ATP but Mg^{2+}–ATP. Here Mg^{2+} does not interact directly with the enzyme. It may serve to neutralize the negative charge density on ATP and facilitates binding to the enzyme. The ternary complexes of this conformation in E-S-M and referred to as "substrate-bridged complexes".

All kinases except muscle pyruvate kinase and phosphoenolpyruvate carboxykinase are substrate-bridged complexes. In pyruvate kinase, Mg^{2+} chelates ATP to the enzyme. Absence of metal cofactor results in failure of ATP to bind to the enzyme. Such enzymes are metal-bridged ternary complexes, E-M-S. They are known as metalloenzymes which contain a tightly bound transition metal such as Zn^{2+} or Fe^{2+}. Metals may also directly bind to enzyme to stabilize it in the active conformation. Mn^{2+}, Na^+, K^+ etc. may play role in this regard. The alkali earth metals such as Na^+ and K^+ (without empty d electron orbitals) stabilize the active conformator of the enzyme (pyruvate kinase) but are passive in catalysis.

Metals also play important role in oxidation and reduction. For example, iron-sulfur proteins are unique class of metalloenzymes in which the active center consists of one or more clusters of sulfur-bridged iron chelates. Cytochromes are heme iron proteins that function as cosubstrates for their respective reducatases. Copper is an active participant in several oxidases and hydroxylases. In case of dopamine β-hydroxylase, the enzyme contains one atom of cuprous ion that reacts with oxygen to form an activated oxygen-copper complex.

(B) Coenzymes

Coenzymes are small organic molecules that functions with the enzyme in the catalytic process. They are oftenly derived from vitamins and have affinity for enzyme similar to that of the substrate. Hence the coenzyme can be considered to be a second substrate. In some cases, the coenzyme is covalently bound to the apoenzyme and functions at or near the catalytic site. In other enzymes the role of the coenzyme falls between these two extremes. Coenzymes may exist in both a reduced and an oxidized form. In the oxidized state it functions in catalysis as a hydrogen acceptor –

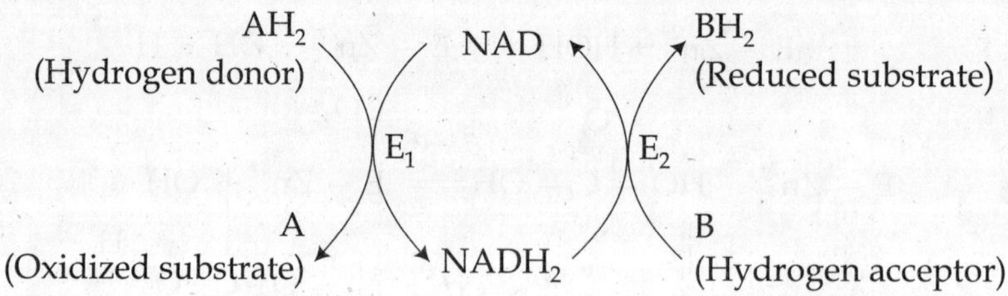

E_1 and E_2 are two different dehydrogenases. 2H are transferred from A to B. Here the coenzyme acts as a link between two different enzyme systems E_1 and E_2. Some of the examples of coenzymes are nicotinamide adenine dinucleotide (NAD), nicotinamide adenine dinucleotide phosphate (NADP), adenosine triphosphate (ATP), flavin mononucleotide (FMN), flavin adenine dinucleotide (FAD), coenzyme A (CoA) etc. NAD and NADP are the coenzyme forms of niacin. Similarly, FMN and FAD are coenzyme forms of riboflavin. Several coenzymes are derived from the B group vitamins. Vitamin B_6 requires little modification to form a coenzyme. ATP is synthesized *de novo* in all mammalian cells. It often functions as second substrate but can also serve as a cofactor in modulation of the activity of specific enzymes. The ebzymes requiring ATP as coenzyme have ATP binding site. The binding of ATP with enzyme changes the affinity or reactivity of the enzyme towards it substrates. In such cases, ATP acts as an allosteric effector. ATP is utilized as cosubstrate by the glucokinase for the transfer of the terminal phosphate to acceptors:

$$\text{Glucose} + \text{ATP} \xrightarrow{\text{Glucokinase}} \text{Glucose 6-phosphate} + \text{ADP}$$

Niacin (pyridine-3-carboxylic acid) is converted into two coenzymes (NAD and NADP). These coenzymes are involved in oxidoreductase reactions. Some dehyrogenases are specific for NAD and others for NADP. The reduced forms (NADH, NADPH) of these nucleotides act as a common 'pool' of electrons that arise from many oxidative reactions. They can be used for various reductive reactions. FAD and FMN coenzymes function in oxidoreduction reactions by accepting and donating $2e^-$ in the isoalloxazine ring (a heterocyclic ring). FAD participates in the following reaction:

$$\text{Succinic acid} + \text{FAD} \xrightleftharpoons{\text{Succinate dehydrogenase}} \text{Fumaric acid} + \text{FADH}_2$$

(C) Prosthetic Groups

Prosthetic groups are organic molecules tightly attached with apoenzyme in such a way they effectively assist the catalytic function of enzyme. In fact, tightly bound coenzymes are prosthetic groups. FAD, FMN, biotin and haem are prosthetic groups. Flavin coenzymes tend to be bound much tighter to their apoenzymes than the niacin coenzymes. Therefore, they often function as prosthetic groups rather than coenzymes. Haem is an iron-containing prosthetic group. It has a porphyrin ring with an iron atom at its centre. Haem performs basically three important biological functions :

(i) Electron carrier: Haem is the prosthetic group of cytochromes in respiratory chain where it acts as an electron carrier. In accepting electrons the iron is reduced to Fe (II), while in handing on electrons it is oxidized to Fe (III). Thus haem takes part in oxidation-reduction reactions by reversible changes in the valency of iron.

(ii) Oxygen carrier: Haemoglobin (Hb) and myoglobin (Mb) are oxygen-carrying proteins that contain haem groups. Here the iron remains in the reduced, Fe (II) form.

(iii) Other enzymes: Haem is found in catalases and peroxidases. These enzymes catalyze the decomposition of hydrogen peroxide into water and oxygen. Haem is also found in other enzymes.

Since prosthetic groups are tightly bound coenzymes, they may be described under coenzymes without creating third category of cofactors i.e., prosthetic groups.

ENZYME CLASSIFICATION

The Enzyme Commission (EC) of International Union of Biochemistry and Molecular Biology (IUBMB) has grouped enzymes into six classes (Table 1). Each class of enzyme is divided into subclasses that are further divided into sub-subclasses. Each enzyme is designated by four numbers. For example, lactate dehydrogenase (trivial name) is an oxidoreductase written as EC 1.1.1.27 (systematic name). The first 1 is for the class oxidoreductases, second 1 for the subclass dehydrogenases which catalyze removal of two atoms of hydrogen, third 1 for the sub-subclass acting on >CH-OH group and 27 for the enzyme acting on the substrate lactate.

Oxidoreductases catalyze oxidation-reduction reactions e.g., oxidation of an alcohol (ethanol) to an aldehyde by alcohol dehydrogenase (ADH) (alcohol : NAD^+ oxidoreductase):

$$\mathrm{R-\underset{\displaystyle H}{\overset{\displaystyle H}{\underset{|}{\overset{|}{C}}}}-O-H + NAD^+ \overset{ADH}{\rightleftharpoons} R-\overset{\displaystyle O}{\overset{\|}{C}}-H + NADH + H^+}$$

(Oxidation of ethanol by ADH)

Two electrons and two hydrogen atoms are removed from the alcohol to yield an aldehyde. The two electrons in the carbon-hydrogen bond of alcohol are transferred to the NAD^+, which is reduced.

Peroxidases utilize H_2O_2 rather than oxygen as the oxidant. NADH peroxidase catalyzes the reaction:

$$NADH + H^+ + H_2O \rightleftharpoons NAD^+ + 2H_2O$$

Table 1. Classes of enzymes, nature of reactions, and attack of subclass enzymes on functional groups of substrates.

Class	Nature of reaction	Subclass	Subclass acting on
1. Oxidoreductases	Oxidation and reduction	1. Dehydrogenases 2. Oxidases 3. Oxigenases 4. Oxidative deaminases 5. Hydroxylases 6. Peroxidases 7. Reductases 8. Catalases	> CH – OH > C = O > C = CH– > CH – NH_2– > CH – NH- > NADH, NADPH
2. Transferases	Exchange of groups between two substrates AB+CD = AC+BD	1. Aminotransferases 2. Transaldolase and transketolase 3. Acyltransferases 4. Glycosyltransferases 5. Kinases 6. Methyltransferases 7. Phosphoryltransferases 8. Phosphomutases	One carbon group Aldehylic or ketonic group Acyl group Glucosyl group Phosphate group Sulphur containing group

Table 1. Contd.

Table 1. Contd.

3. Hydrolases	Catalyze hydrolysis reaction $AB + H_2O = AOH + HB$	1. Esterases and phosphodiesterases 2. Glycosidases 3. Phosphatases 4. Peptidases 5. Deaminases 6. Amides and deamidases 7. Thiolases 8. Phospholipases 9. Ribonucleases	Ester Glycosidic bond Peptide bond Other C–N bond Acid anhydride
4. Lyases	Remove groups from substrates non-hydrolytically AB = A + B	1. Decarboxylases 2. Aldolases 3. Dehydratases 4. Hydratases 5. Synthases	$>C=C<$ $>C=O$ $>C=N-$
5. Isomerases	Catalyze isomerization of Substrates	1. Recemases 2. Epimerases 3. Mutases	Racemases Epimerases
6. Ligases	Catalyze joining together of two molecules coupled with the breakdown of a pyrophosphate bond in ATP	1. Synthetases 2. Carboxylases	C – O C – S C – N C – C

Catalase is unique in that H_2O_2 serves as both donor and acceptor. It functions in the cell to detoxify H_2O_2:

$$H_2O_2 + H_2O_2 \rightleftharpoons O_2 + 2H_2O$$

Transferases transfer functional groups between donors and acceptors e.g., aminotransferases (transaminases) transfer one amino group from one amino acid to an α-keto acid acceptor leading to the formation of a new amino acid and a new keto acid:

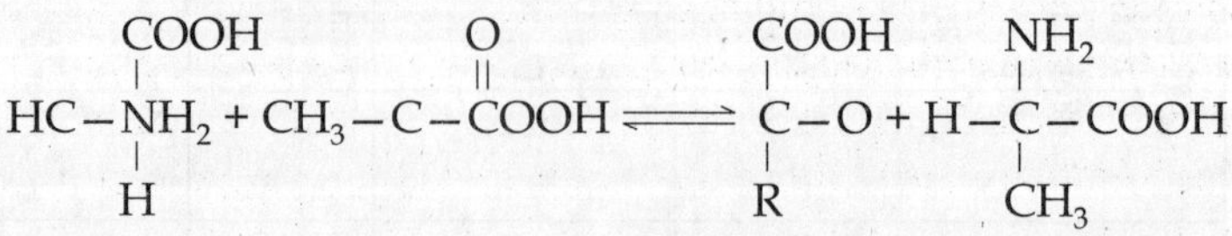

(Amino acid–1) (Keto acid-2) (Keto acid-1) (Amino acid–2)

Kinases transfer γ-phosphoryl group from ATP or other nucleoside triphosphate to alcohol or amino group acceptors:

$$ATP + R-OH \rightleftharpoons R-O-ADP + Pi$$

Hydrolases are a special class of transferases in which the donor group is transferred to water. The reaction involves the hydrolytic cleavage of C-O, C-N, O-P and C-S bonds e.g., cleavage of a peptide bond:

$$R_1-\overset{\overset{\displaystyle O}{||}}{C}-NH-R_2 + H_2O \rightarrow R_1-\overset{\overset{\displaystyle O}{||}}{C}-O + H_3N^+ + -R_2$$

Phosphatases replace a phosphate group with a hydroxyl group from water:

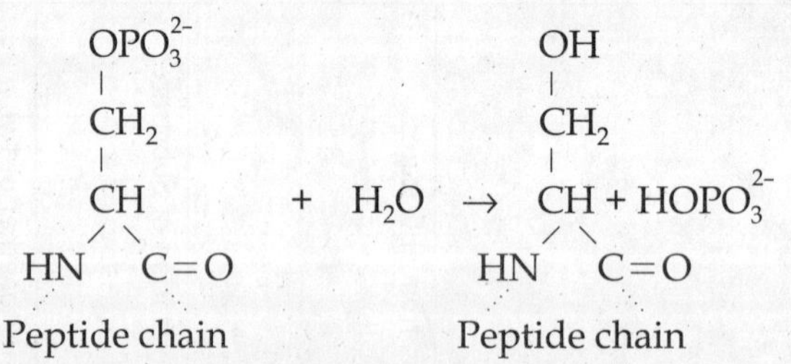

(Hydrolysis of a phosphorylated protein by a phosphate)

Lyases add or remove the elements of water, ammonia and carbon dioxide. Decarboxylases remove the element of CO_2 from α- or α- keto acids or amino acids:

$$R-\overset{\overset{\displaystyle O}{||}}{C}-\overset{\overset{\displaystyle O}{||}}{C}-O- + H^+ \rightarrow R-\overset{\overset{\displaystyle O}{||}}{C}-H + CO_2$$

Dehydratases remove H_2O in a dehydration reaction. Fumarase is a dehydratase which converts fumerate to malate:

$$\text{Fumarate} + H_2O \rightleftharpoons \text{L-Malate}$$

Isomerases catalyzing inversion at asymmetric carbon atoms are either epimerases or racemases:

$$\text{D-Xylulose 5-phosphate} \xrightleftharpoons{\text{Epimerase}} \text{D-Ribolose 5-phosphate}$$

$$\underset{\text{D-Lactic acid}}{\begin{array}{c}\text{COOH}\\ | \\ \text{HCOH}\\ | \\ \text{CH}_3\end{array}} \quad \xrightleftharpoons{\text{Racemase}} \quad \underset{\text{L-Lactic acid}}{\begin{array}{c}\text{COOH}\\ | \\ \text{HOCH}\\ | \\ \text{CH}_3\end{array}}$$

Ligases join two molecules at the expense of high energy phosphate bond of ATP. The term synthetase is reserved for this particular group of enzymes. Pyruvate carboxylase is a good example of a ligase enzyme:

$$\text{ATP} + \text{HCO}_3^- + \underset{\text{(Pyruvate)}}{\begin{array}{c}\text{COO}^-\\ | \\ \text{C}=\text{O}\\ | \\ \text{CH}_3\end{array}} \xrightarrow{\text{Enzyme + Biotin}} \underset{\text{(Oxaloacetate)}}{\begin{array}{c}\text{COO}^-\\ | \\ \text{C}=\text{O}\\ | \\ \text{CH}_3\\ | \\ \text{COOH}\end{array}} + \text{ADP} + \text{Pi}$$

MECHANISM OF ENZYME ACTION

A specific region of enzyme molecule involved in catalysis is called catalytic site. There may be another site on enzyme concerned with binding of effector and regulation of its activity. Though both the loci on enzyme molecule are active, it is better to refer the site involved in catalysis as 'catalytic site' in order to avoid overlap with effector binding or regulating site. During catalysis the shape of the enzyme molecule becomes such that it exposes the amino acids of the catalytic site so that the substrate molecule(s) can bind to it for catalysis. The specificity of enzyme resides in the substrate-binding site on the enzyme surface. The tertiary structure of the enzyme is folded in such a way as to create a region that has the correct molecular dimensions, the appropriate topology and the optimal alignment of counterionic groups and hydrophobic regions to accommodate a specific substrate. Most enzymes are far larger molecules than the substrates they act on. Only a very small portion of the enzyme, between 3-12 amino acids (catalytic site) comes into direct contact with substrate (Figs. 1 and 2). One can ask why an enzyme molecule is so large when only a portion of its structure is required for substrate binding and catalysis? The catalytic site is a place made up of few amino acids where binding of substrate takes place. The remaining amino acids of enzyme chain (i.e., bulk of enzyme part) function to maintain the correct globular shape of the molecule. The correct globular shape of enzyme molecule is necessary for function of catalytic site at the maximum rate. When an enzyme-substrate complex is formed, it is activated into forming the products of the reaction. Once formed, the products no longer fit into the active site and escape into the surrounding medium, leaving the active site free to receive further substrate molecules.

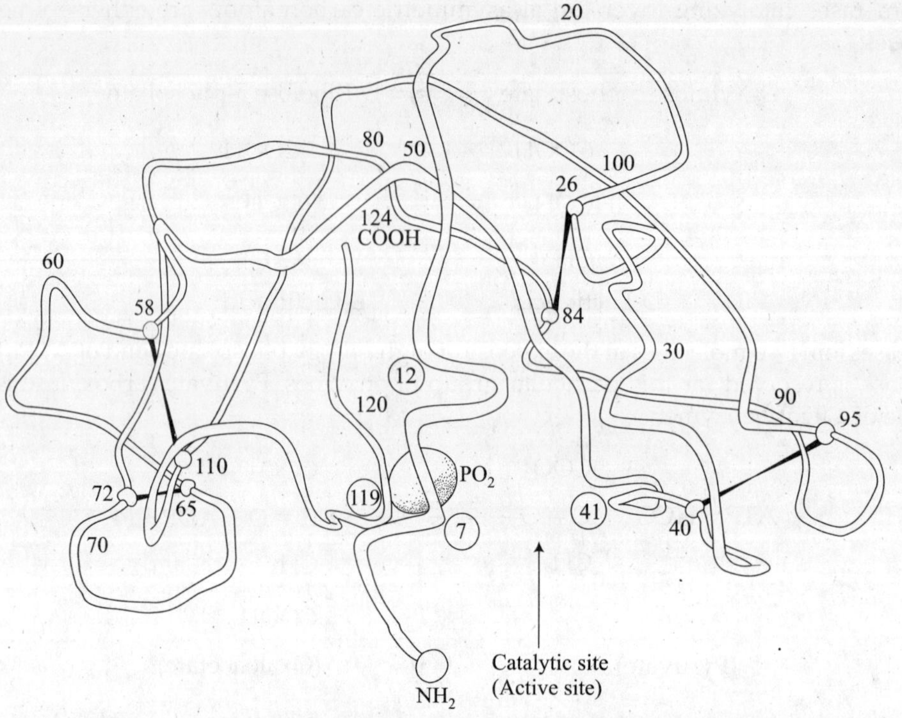

Fig. 1. Tertiary structure of an enzyme showing catalytic site (active site).

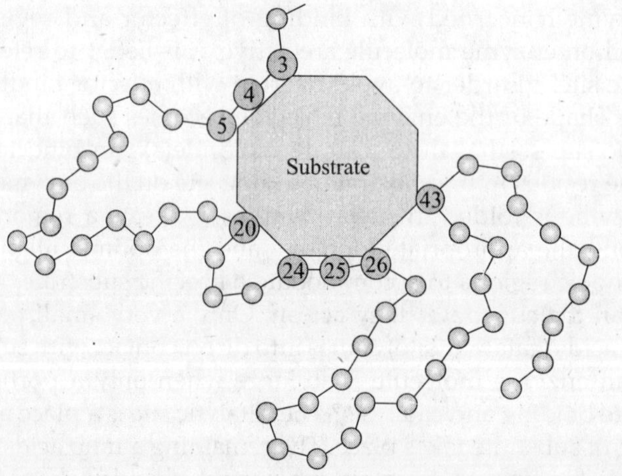

Fig. 2. Substrate attached with catalytic site (active site) of an enzyme.

Enzymes are the most studied yet enigmatic biomolecules. A lot of work has been done to elucidate the mechanism of enzyme action. However, it is not fully understood till date. Some models have been proposed to explain substrate specificity and mode of action of enzymes. They are lock-and-key model, induced-fit model, and substrate strain model.

(a) Lock-and-key Model

In 1890, Fisher proposed the rigid "lock-and-key" model, in which a negative impression of the substrate is considered to exist on the enzyme surface (Fig. 3). He explained interaction between substrate and enzyme in terms of a "lock-and-key" analogy. Here substrate is the key whose shape is complementary to the enzyme (i.e., like lock). Substrate fits in the binding site just as a key fits into the proper lock. In this model, the catalytic site is a rigid template and presumed to be preshaped to fit the substrate. It is also known as "Fisher template hypothesis" or "rigid template model". This hypothesis gives a rigid picture of the enzyme and can not account for the effects of allosteric ligands. Hence, it is not acceptable.

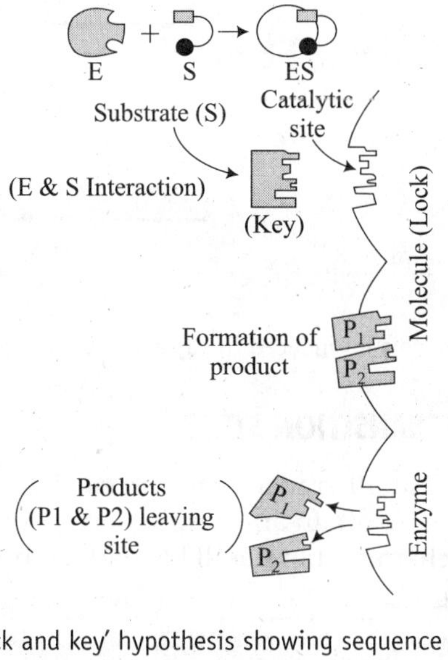

Fig. 3. Fischer's 'lock and key' hypothesis showing sequence of events in enzyme.

(b) Induced-fit Model

In 1959, Koshland, Jr. suggested a modification to the lock-and-key concept. According to Koshland hypothesis when a substrate comes in proximity of an enzyme, it induces conformational change in the enzyme molecule for a proper alignment (orientation) between enzyme and substrate. The amino acids present at the catalytic site are moulded into a precise formation which enables the enzyme to perform its catalytic function effectively (Fig. 4). This is referred to as the "induced-fit" hypothesis of enzyme action in which the catalytic site is a flexible structure than the rigid one (Koshland, 1973). It is widely acceptable model and have got experimental supports. In few cases, the substrate molecule also changes its shape slightly before binding. Hence the refinement to the induced-fit model has also been done.

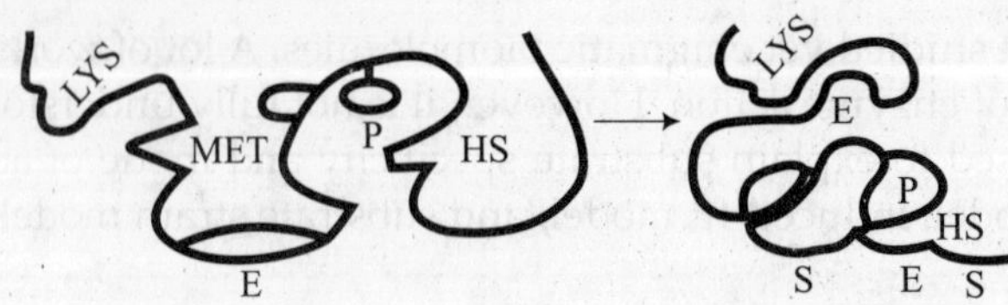

Fig. 4. Koshland's 'induced-fit' hypothesis showing conformational changes in protein structure. The relative positions of key residues before and after substrate binding are given.

(c) Induced-fit with Substrate-strain Model

A refined hypothesis to induced-fit model is a combination of induced fit with substrate strain. This has got more experimental supports for enzyme action than other hypotheses. In this model, substrate is "strained" troard product formation by an induced conformational transition of the enzyme (Fig. 5).

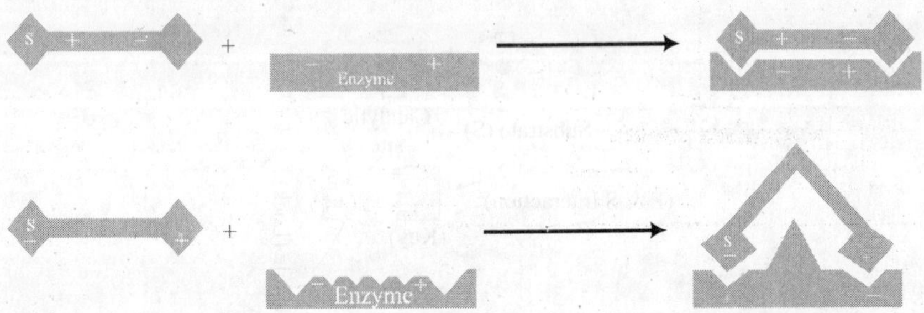

Fig. 5. Models for induced fit (a) and substrate strain (b).

ENERGY OF ACTIVATION AND TRANSITION STATE

Conversion of one molecule into another in a chemical reaction at a particular rate at any given instance requires enough energy to attain an 'activated state' or 'transition state' at which the probability is very high that a chemical bond will be made or broken down to form a product.

$$\underset{\text{(Substrate)}}{S} \rightarrow \underset{\text{(Transition state subtrate)}}{S^*} \rightarrow \underset{\text{(Product)}}{P}$$

$S^‡$ has more energy than S. As $S^‡$ is converted into P, the energy is lost. Thus the product (P) has less energy than the activated state substrate ($S^‡$). This transition state can be achieved by increasing temperature or using enzyme. Temperature and enzyme do not identically change the conformation of substrate to achieve its activated state. Transition state is not an intermediate because it can not be isolated. In case of enzyme catalyzed reaction, the energy of reactants and products is not different than in uncatalyzed reaction. Enzyme does not change the thermodynamics of the system but it changes the pathway for reaching the final state. The rate of reaction is actually not dependent on the concentration of substrate [S], rather it is dependent on the concentration of the energized state of the substrate molecule [$S^‡$]. The amount of energy needed to bring the S to $S^‡$ is called activation energy ($\Delta G^‡$) or free energy of activation. $\Delta G^‡$ is the amount of energy required to bring all the molecules of one mole of the substrate to their transition states at a given temperature. However, free energy (ΔG°) is a usable form of energy released in the process of reaction.

Enzyme combines transiently with reactants and generate transition state molecules requiring much lower energy of activation than those needed in activation of substrate in an uncatalyzed reaction (Fig. 6). In other words, a catalyst lowers down the requirement of energy of activation. For example, the estimated free energies of activation (ΔG°) for decomposition of H_2O_2 ($2H_2O_2 \rightarrow 2H_2O + O_2$) at 20°C in uncatalyzed, catalyzed by colloidal platinum and catatlyzed by the enzyme catalase are 18, 13 and 7 Kcal.mol^{-1}. This shows a remarkable decrease in the requirement of activation energy in an enzyme catalyzed reaction. The catalase can decompose 40,000 molecules of H_2O_2 (a toxic product of metabolism) within a second. On the average, an enzyme can enhance the rate of reaction by a factor of 10^9 –10^{12} times than that of the uncatalyzed reaction. Orotidine 5′-phosphate decarboxylase is the most efficient enzyme known with a rate enhancement of 10^{17}-fold.

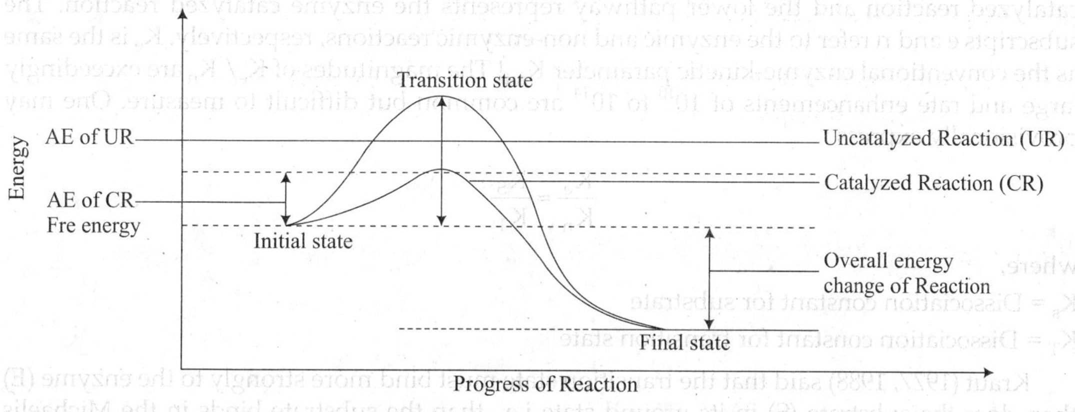

Fig. 6. Energy budget of a catalyzed and uncatalyzed reactions.

The transition-state theory of biocatalysis includes two assumptions. First shows that the reaction rate is controlled by decomposition of an activated transition-state-complex, and the second asserts that the system can be treated as the transition-state complex is in equilibrium with the reactants (Kreevoy and Truhlar, 1986). The resulting hypothesis is :

$$k' = k \, v \, K^{\ddagger}$$

where,

k' = Experimentally observable reaction-rate constant

k = Transmission coefficient

v = Frequency of the normal mode oscillation of the transition-state complex

$K^{\ddagger}$ = Equilibrium constant for formation of the transition-state complex from reactants

An elementary transition state theory may be applied to enzyme catalysis by using the conceptual device of the thermodynamic cycle (Kraut, 1988) :

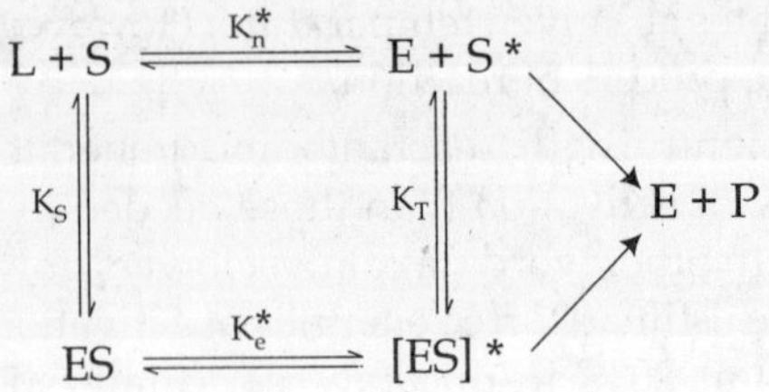

where,

E = Enzyme

S = Substrate

P = Product

The above thermodynamic cycle relates substrate binding and transition-state binding. Superscript double dagger (‡) denotes transition state. The upper pathway represents the non-catalyzed reaction and the lower pathway represents the enzyme catalyzed reaction. The subscripts e and n refer to the enzymic and non-enzymic reactions, respectively. K_e is the same as the conventional enzyme-kinetic parameter K_{cat}! The magnitudes of K_e / K_n are exceedingly large and rate enhancements of 10^{10} to 10^{14} are common but difficult to measure. One may provisionally express :

$$\frac{K_e}{K_n} \approx \frac{K_S}{K_T}$$

where,

K_s = Dissociation constant for substrate

K_T = Dissociation constant for transition state

Kraut (1977, 1988) said that the transition state must bind more strongly to the enzyme (E) than does the substrate (S) in its ground state i.e., than the substrate binds in the Michaelis complex ES– by a factor roughly equal to the enzymic rate acceleration. Internal molecular dynamics may also play a role in enzymic catalysis.

MECHANISM OF CATALYSIS

The mechanism of enzyme catalysis (enzymatic rate enhancement) can be explained with the help of acid-base catalysis, substrate strain (transition state stabilization), covalent catalysis, ground state destabilization, and entropy effects. A given enzyme may utilize one or more of these mechanisms to achieve the rate enhancement.

(a) Acid-base Catalysis

A general acid or base is a compound that is weakly ionisable. At the physiological pH, the protonated form of histidine is the most important general acid and its conjugate base is an important base. The acid and base forms of histidine are as follows :

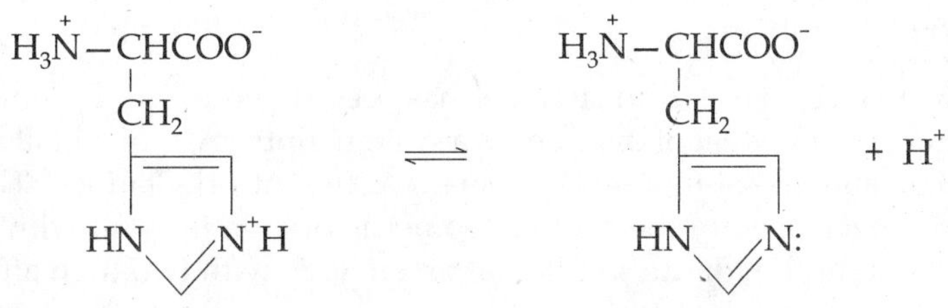

(Histidine as a general acid) (Histidine as a base)

The other acids are thiol-SH of cysteine, and Σ-amino group of lysine. Similarly, the other bases are carboxylic acid anions and the conjugated bases of the general acids. The enzyme ribonuclease (RNase) clearly shows the role of acid and base catalysis at the catalytic site. RNase cleaves an RNA chain at 32-phosphodiester linkage of pyrimidine nucleotides with an obligatory formation of a cyclic 22 32-phosphoribose on a pyrimidine nucleotide as intermediate. Here His 119 acts as a general acid to protonate phosphodiester bridge, whereas His 12 acts as a base in generating an alkoxide on the ribose-32-hydroxyl. The latter then attacks the phosphate group, forming a cyclic phosphate and breaking the RNA chain at this locus.

(b) Substrate Strain

Binding of substrate to a preformed site on the enzyme can induce strain in the substrate. The energy level of the substrate is raised, and the bond lengths and angles of the substrate more closely resemble those found in the transition state. A combination of substrate strain and acid-base catalysis has been described for the mechanism of action of lysozyme. In this case, the study showed X-ray that ring D of hexasaccharide substrate is strained to the half-chair conformation upon binding to isozyme.

(c) Covalent Catalysis

The attack of a nucleophilic (negatively charged) or electrophilic (positively charged) group in the enzyme catalytic site upon the substrate results in covalent binding of the substrate to the enzyme as an intermediate in the reaction sequence. Oftenly enzyme-bound coenzymes from covalent bonds with the substrate. In covalent catalysis, the enzyme or coenzyme-bound substrate is more labile than the original substrate. Enzymes such as transaminases and serine proteases are the best examples of covalent catalysis.

(d) Transition State Stabilization

The acid-base catalytic, substrate strain and covalent catalytic effects promote the substrate to enter the transition state. Since catalytic site (active site) binds the transition state with a much greater affinity than the substrate, small fraction of substrate molecules existing in a transition state will be converted quickly to products. Thus all substrate will be converted to products.

(e) Entropy Effect

Catalysis by entropic guidance from enzymes has been demonstrated (Young and Post, 1996). Entropy (ΔS) refers to the extent of disorder in a system. Entropy is maximal at equilibrium. For instance, two reactants in solution exist in many different orientations. The chances of two reactants coming together with the correct geometric orientation and with enough energy to react is small at 37°C and in dilute solution. If an enzyme with two high affinity binding sites for two reactants is poured into the diluted solutions of reactants, the two reactants will be bound to the enzyme molecule in the correct orientation for the reaction to occur. The substrates are strained toward the transition state when positioned and bound on the enzyme surface. Proper orientation and the nearness of the substrate with respect to catalytic groups enhance 10^3–10^5 fold reaction rate with enzyme.

(f) Ground State Destabilization

The destabilization of ground state results in lowering of the energy of activation. The charge-charge repulsion leads to destabilization of ground state. Thus energy barrier can be lowered by raising the energy of the ground state. The enzyme destabilizes ground state by selecting conformations of substrates and enzyme catalytic site reduces that approach the binding distances and angles found in transition state. The ground state can also be destabilized by bringing like charges between enzyme and substrate together.

FACTORS INFLUENCING CATALYTIC ACTIVITY

The rate of enzyme catalyzed reaction may be changed with the change in substrate concentration [S], temperature and pH. While studying the effect of a given factor on the rate of an enzyme catalyzed reaction, all other factors should be kept constant and at optimum levels.

(a) Effect of Enzyme Concentration

The rate of reaction is proportional to the enzyme concentration at high level substrate concentration, constant pH and temperature (Fig. 7). Invariably reactions are catalyzed by enzyme concentrations which are much lower than substrate concentrations. So the rate of the enzymatic reaction increases with the increase in enzyme concentration. The time required for optimum enzyme activity is also determined (Fig. 8).

(b) Effect of Substrate Concentration

The rate of an enzymatic reaction increases with the increasing substrate concentration for a given amount of enzyme (Fig. 9). There comes a point when any further increase in substrate concentration does not increase the reaction rate. It is due to the fact that at high substrate concentrations the catalytic (active) sites of the enzyme molecules at any given moment are virtually saturated with substrate. At this point, any extra substrate has to wait until the enzyme-substrate complex has dissociated into products and free the enzyme. Hence at high substrate concentration both the enzyme concentration and the time required for dissociation of the enzyme-substrate complex limit the rate of the enzymatic reaction.

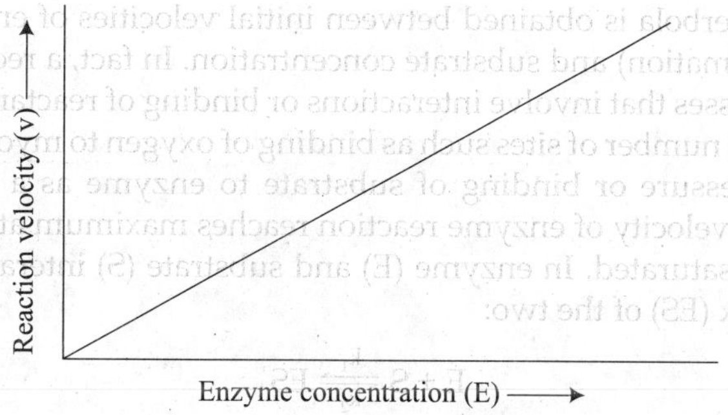

Fig. 7. Relationship between enzyme concentration and the rate of reaction.

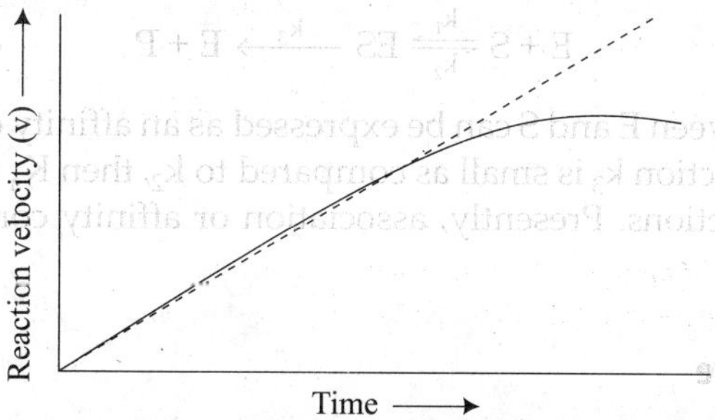

Fig. 8. Effect of time on enzyme catalyzed reaction.

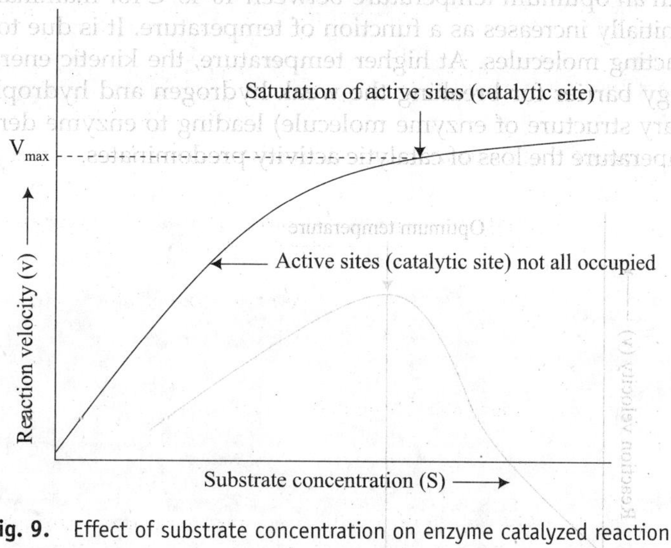

Fig. 9. Effect of substrate concentration on enzyme catalyzed reaction.

A rectangular hyperbola is obtained between initial velocities of enzyme reaction (initial velocity of product formation) and substrate concentration. In fact, a rectangular hyperbola is obtained for any processes that involve interactions or binding of reactants or other substances at a specific but limited number of sites such as binding of oxygen to myoblogin as a function of increasing oxygen pressure or binding of substrate to enzyme as a function of substrate concentration etc. The velocity of enzyme reaction reaches maximum at the point at which all the available sites are saturated. In enzyme (E) and substrate (S) interaction, initially there is formation of a complex (ES) of the two:

$$E + S \underset{k_2}{\overset{k_1}{\rightleftharpoons}} ES$$

The rate constant for formation of ES complex is k_1 and for dissociation of ES is k_2. The conversion of substrate to product (P) occurs from the ES complex with a rate constant k_3 :

$$E + S \underset{k_2}{\overset{k_1}{\rightleftharpoons}} ES \xrightarrow{k_3} E + P$$

The equilibrium between E and S can be expressed as an affinity constant (K_a). If the rate of the chemical phase of reaction k_3 is small as compared to k_2, then $K_a = k_1/k_2$. Earlier K_a is used to describe chemical reactions. Presently, association or affinity constant (K_a) is preferred in enzymological studies.

(c) Effect of Temperature

Over a limited range of temperatures, the velocity of enzyme-catalyzed reactions increases as temperature rises (Fig. 10). Plots of velocity versus temperature for most enzymes show a bell-shaped curve with an optimum temperature between 40-45°C for mammalian enzymes. The rate of reaction initially increases as a function of temperature. It is due to increased kinetic energy of the reacting molecules. At higher temperature, the kinetic energy of the enzyme exceeds the energy barrier for breaking the weak hydrogen and hydrophobic bonds (that maintain secondary structure of enzyme molecule) leading to enzyme denaturation. At this denaturation temperature the loss of catalytic activity predominates.

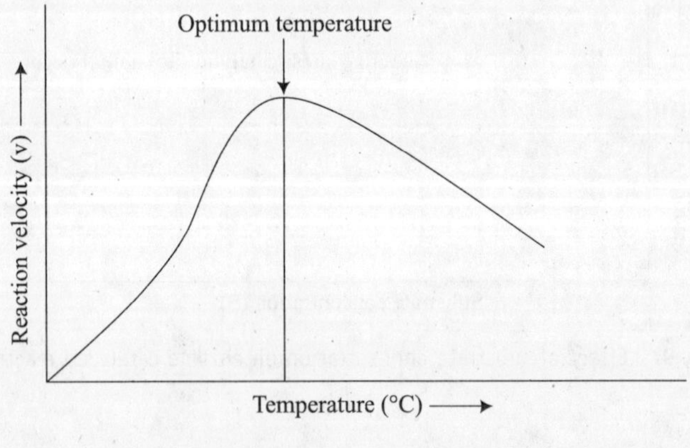

Fig. 10. Effect of temperature on the rate of enzyme catalyzed reaction.

Each and every enzyme has optimal temperature for its optimal activity. The optimal temperature depends upon the duration of the assay used to determine it. So the longer an enzyme is maintained at a temperature at which its structure is marginally stable, the more likely it is to be denatured. Over a range of 0-40°C, the temperature coefficient (Q_{10}) for an enzyme catalyzed reaction is 2 viz., the rate of an enzyme-controlled reaction is doubled for every rise of 10°C. Some microorganisms are adopted to grow in natural hot springs. The optimal temperatures of the enzymes in these organisms are close to boiling point of water.

The thermal inactivation (denaturation) of enzymes may be either reversible or irreversible, depending on whether return to ambient temperature results in the return of enzymatic activity within a reasonable period of time (Klibanov, 1983). The phenomenon of reversible thermal denaturation occurs due to temperature-induced conformational transitions in the molecules. However, the mechanism of reversible thermo-inactivation of enzymes is relatively less understood. Ahern and Klibanov (1985) proved that the processes causing irreversible inactivation of lysozyme at 100°C are deamidation of asparagine residue, hydrolysis of peptide bonds at aspartic acid residue, destruction of disulfide bond and formation of scrambled (incorrect) structures and their relative contributions depend on the pH. Our knowledge about irreversible thermo-inactivation of enzymes is essential for the development of rational approaches to enzyme stabilization and enzyme and protein engineering.

(d) Effect of pH

Each enzyme has an optimum pH for its most efficient function under the condition of a constant temperature. The optimum pH of an enzyme is that pH at which the maximal rate of reaction occurs (Table 2). The pH optimum varies greatly with different enzymes. The activity of enzyme declines on either side of the optimum pH and so the pH profile usually appears bell-shaped (Fig. 11). The bell-shaped optima of enzymes are because of acid-base behaviour (ionisable behaviour) of enzyme or substrate molecules. The pH effects enzyme activity by changing the charge on ionisable groups of the enzyme, and often of the substrate too. Since pH effects the ionization of enzyme molecules, the bell-shaped curve is obtained. If pH is high, NH_3^+ becomes NH_2 and COOH becomes COO^-, then no binding of substrate to active occurs.

Not all enzymes have bell-shaped curve for activity. The shape of pH activity curves is determined by enzyme denaturation at high or low pH, and alterations in the charged state of the enzyme and/ substrate molecule. Changes in pH alter the ionic charge of the acidic and

Table 2. pH values for optimal activity of some enzymes.

Enzyme	Optimum pH
Pepsin	2
Sucrase	4-5
Salivary amylase	6.8
Catalase	7.6
Chymotrypsin	7-8
Pancreatic lipase	9
Agrinase	9.7

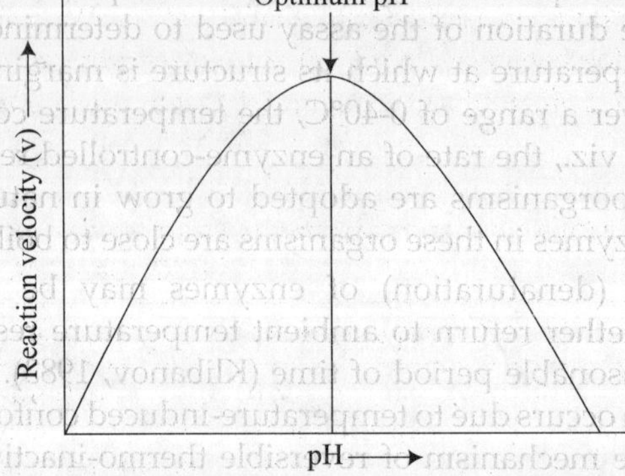

Fig. 11. Effect of pH on the rate of enzyme catalyzed reaction.

basic groups of amino acids that help to maintain the specific shape of an enzyme molecule. The pH change causes an alteration in enzyme shape, particularly at its catalytic site. If extremes of pH are encountered by an enzyme, then it will be denatured. Suppose a negatively charged enzyme (Enz^-) reacts with a positively charged substrate (SH^+):

$$Enz^- + SH^+ \rightarrow Enz.SH$$

Enz^- protonates and losses its negative charge at low pH in the following way :

$$Enz^- + H^+ \rightarrow EnzH$$

At high pH, SH^+ ionizes and losses its positive charge,

$$SH^+ \rightarrow S + H^+$$

The interacting forms are only SH^+ and Enz^- and extreme pH lowers the effective concentration of Enz^- and SH^+ thereby lowering the reaction velocity.

Enzymes may also undergo conformational changes as pH is varied. The pH alters the charges of catalytic site amino acid in an enzyme molecule. Ionization of amino acid is responsible for three dimensional (3D) conformation of the enzyme molecule. When 3D conformation changes, the catalytic site conformation also changes. A charged group distal to the region where substrate is bound may be essential to maintain an active tertiary or quaternary structure. As the charge on this group is changed, the enzyme may unravel, become more compact, or dissociate into protomers and causes loss of activity. Depending upon the severity of change, the enzyme activity may or may not be restored when enzyme is returned to its optimal pH. The optimum pH of an enzyme is not always the same as the pH of its immediate intracellular environment. The surrounding of enzyme may also be exerting some form of control over its activity.

ENZYME-ENZYME INTERACTIONS AND METABOLIC REGULATION

The properties of individual enzymes are influenced by interactions among enzymes, substrates and solvents. Hence the concentrations of cellular components affect the activity of the entire cellular milieu. Studies that focus on the catalytic activity of individual enzymes are almost always performed under conditions far from physiological one. Measurements of conversion of substrate to product at nanomolar concentrations of enzyme are possible over periods of several minutes. But the functional effects of enzyme-enzyme interactions are not evident at nanomolar concentrations except in case of some stable multienzyme complexes. However, at the high concentrations of many enzymes in the cell, enzyme-enzyme interactions are far more likely.

Srivastava and Bernhard (1986a,b) have done a lot of work and established metabolite transfer via enzyme-enzyme complexes. According to them the concentrations of enzyme sites in cells are usually higher than the concentrations of cognate intermediary metabolites. Therefore, metabolic pathways or a segment of pathways may proceed by direct transfer of metabolites from one enzyme site to the next by means of enzyme-enzyme complex formations. This mechanisms of metabolite transfer differs from that of the random diffusion of metabolite through the aqueous environment. These modes of transfers of metabolites are shown in Fig. 12.

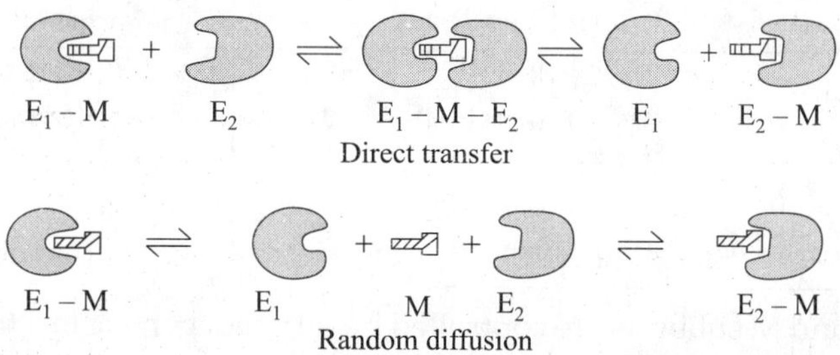

Fig. 12. Diagrammatic representation of the mode of transfer of metabolite (M) between two sequential enzyme sites (E_1 and E_2). in direct transfer, M does not dissociate from E_1 to reach at E_2 site. However, in random diffusion, M dissociates from E_1 into aqueous solvent before forming a complex with E_2 (After Srivastava and Bernhard, 1986).

The sequential enzyme-enzyme interaction is determined by specific protein surface interactions. Enzyme regulation of metabolism occurs at transcriptional stage, protein assembly state or allosteric level. The control mechanism involving changes in protein assembly state is the adaptive changes in the association states of subunits of multimeric enzymes and in the binding states of enzymes to other enzymes or to structural proteins (Somero and Hand, 1990). This process is called as compartmentalization. Enzymes that reversibly change their compartmentation in response to physiological signals have been referred to as ambiquitous enzymes (Wilson, 1988). Changes in subunit assembly state allow sharp 'on-off' regulation of enzymes. It is evident in case of regulation of phosphofructokinase (PFK) activity by transitions in pH. Changes in aggregation states of multiprotein complexes alter the proximity of several enzymes of a pathway which, in turn, change the efficiency of metabolite transfers among enzymes of a sequence.

ISOENZYMES

Isoenzymes (or isozymes) are structurally different forms of an enzyme catalyzing the same reaction. They differ in their physical properties (not necessarily) and migrate differentially on electrophoresis. For example, both rat liver and *E. coli* malate dehydrogenase catalyze interconversion of malate to oxaloacetate but their physical and chemical properties have many differences. Such physically different forms of the same enzyme may be present in different tissues of the same organism in different cell types, subcellular compartments or within a prokaryote (*E. coli*). Isozymes are commonly found in sera and tissues of vertebrates, invertebrates and other organisms. Dehydrogenases, oxidases, transaminases, phosphatases etc. exhibit their different isoenzymic forms. Even different tissues may contain different isoenzymes and these may differ in their affinity for substrates.

Isoenzymes are the products of two different genetic loci. The most commonest mechanism for the formation of isoenzymes involves the arrangement of subunits (protomers) arising from two different loci in different combinations. Therefore, oligomeric enzymes with dissimilar protomers can exist in several forms. For example, lactate dehydrogenase (LDH) is a tetrameric enzyme (MW 130,000) consisting of two types of subunits : H (for heart) and M (for muscle). Each subunit has about 34,000 MW. These two subunits (H and M) of LDH are combined in the following five ways:

Isoenzyme	Composition	Location
LDH_1	HHHH (H_4)	Heart and RBC
LDH_2	HHHM (H_3M)	Heart and RBC
LDH_3	HHMM (H_2M_2)	Brain and Kidney
LDH_4	HMMM (HM_3)	
LDH_5	MMMM (M_4)	Liver and Skeletal muscle

Synthesis of H and M subunits are controlled by different genetic loci that are differentially expressed in different tissues such as heart and skeletal muscle.

Creatine phosphokinase (CPK) exists as dimer with two types of subunits viz., M (muscle type) and B (Brain type). Both subunits in brain are electrophoretically the same, hence, referred to as B. Similarly, both subunits in skeletal muscle are M type. However, heart has CPK isoenzyme containing both M and B type subunits (MB). Other tissues contain variable amounts of the MM and BB isoenzymes. The three CPK isoenzymes (1,2,3) are numbered beginning with the isoenzyme species migrating the fastest to the anode on electrophoresis. They are as follows :

CPK_1 ------ BB
CPK_2 ------ MB
CPK_3 ------ MM

Isoenzymes have diagnostic use in medical science. Human sera contain various isoenzymes whose relative proportions varies significantly in some pathological conditions. For example, CPK_2 (MB) is released into the blood within the first 6-18hr after myocardial infarction. However, LDH release in blood occurs after 1-2 days of CPK_2 release. Under normal condition the activity of LDH_2 is higher than LDH_1 in blood. But, in case of myocardial infarction, the total LDH activity increases and LDH_1 becomes greater than LDH_2 between 12 and 24 hr. Virtually, LDH isoenzyme switch coupled with increased CPK_2 is indicative of myocardial infarction. Increase in LDH_5 is diagnostic of a secondary congestive liver involvement.

ENZYME REGULATION

Control of a metabolic pathway takes place through alteration in the activity of one or more key enzymes in the pathway. One of the key enzyme is the rate limiting enzyme. The rate limiting enzyme occurs early in the pathway and has lowest V_{max}. Another enzyme catalyzes the committed step in a metabolic pathway. Committed step is the first irreversible reaction that is unique to a pathway. The rate-limiting enzyme may not be necessarily the enzyme associated with the committed step. The activity of the enzyme associated with the rate-limiting step or committed step can be regulated mainly by the following ways :

(a) Synthesis and Degradation of Enzyme

The quantity of an enzyme present is determined by its rate of *de novo* synthesis (k_s) and rate of degradation (k_{deg}). The quantity of an enzyme in cell is increased either by an increase in its rate of synthesis or by a decrease in rate of degradation or by both. The combined processes of enzyme synthesis and degradation form enzyme turnover. Turnover of enzyme is a characteristic of cell which occur throughout life.

(b) Induction of Enzyme Synthesis

Inducers are low molecular weight substances which can induce the synthesis of specific enzymes e.g., lactose acts as inducer of β-galactosidase in *Escherichia coli*. Many inducers are substrates for the enzyme they induce. The compound structurally similar to the substrate may acts as inducer but not substrate. They are known as gratuitous inducers. In contrast, a compound may be a substrate but not inducer. Enzymes whose concentration in a cell is independent of added inducer are referred to as constitutive enzymes. Enzymes whose concentration in a cell is dependent on added inducer are called as facultative enzymes. Cells capable of being induced for a particular enzyme or other protein usually contain a small measurable basal level of that protein even when grow in absence of an added inducer. It is the genetic make up of a cell which determines the nature and magnitude of the response to an inducer. Therefore, the constitutive and inducible are relative terms like hot and cold that represent the extremes of a spectrum of responses.

(c) Repression of Enzyme Synthesis

Enzyme synthesis may be repressed by the presence of an end product. If the end product of a metabolic pathway begins to accumulate, it may act an allosteric or competitive inhibitor on the enzyme controlling the first step of the pathway. It is a short-term regulation in which the affinity of the enzyme for its substrate is lowered and further production of the end product is decreased or prevented. This is referred to as negative feedback inhibition or end-product inhibition. For example, sufficient cellular concentration of deoxyribonucleotides inhibits the key enzyme of the synthetic pathway. Sometimes the product of one pathway acts as an inhibitor or activator of an enzyme present early in another pathway. This is known as cross regulation. The cross-regulation of the production of four deoxyribonucleotides for DNA synthesis is one of the best examples because all four deoxyribonucleotides must be present in similar amounts for DNA synthesis.

(d) Enzyme Compartmentation

The activity of a metabolic pathway is also regulated to some extent by physical partitioning of the pathway from its initial substrate and by controlling access of the substrate to the enzymes of the pathway. This is known as enzyme compartmentation. Generally anabolic and catabolic pathways are segregated into different organelles for maximizing the cellular economy. For example, the oxidation of fatty acids does not occur at the same time and in the same compartment with the biosynthesis of fatty acids. If such situation occur, a futile cycle would exist. Thus by maintaining fatty acid synthesis in cytoplasm and oxidation in mitochondria, a control is done by regulating transport of common intermediates across the mitochondrial membrane.

RIBOZYME

Some small and stable RNA species function as ribonucleoprotein particles (RNPs) with one or more protein subunits attached. They are found in nucleus, cytoplasm and mitochondria. They are involved in RNA processing, transport, control of translation and recognition of proteins to be exported. In several cases, the RNA component of a ribonucleoprotein particle has catalytically active subunit. In other cases, the catalytic reaction has been demonstrated *in vitro* by RNA in the absence of any protein. The RNA subunit performing catalytic function is known as catalytic RNA or ribozyme (Cech, 1986). There are five classes of catalytic RNA. Three of these RNA species carry out self-processing reaction. However, the other two (ribonuclease P and rRNA) act on separate substrates. The first example of RNA catalysis was discovered in 1981 and 1982 from the studies in a ciliated protozoan *Tetrahymena thermophila*. Like other eukaryote, *T. thermophila* has four ribosomal RNAs. Three of these are transcribed as a single unit. The resulting primary transcript is then spliced and modified to produce finished RNA. The T. thermophila pre-rRNA intron is self-splicing, which is only one step in a cascade of reactions that the RNA caries out on its own. Its action is similar to the enzyme on the following aspects :

1. Catalytic RNA is highly specific and shows many folds acceleration of reaction. At splice site, folded structure of intron accelerates the rate of cleavage by a factor of 10 billion i.e., it reduces the time frame of the reaction from 19000 years to one minute.
2. Catalysis is dependent on 3D structure of the polynucleotide chain.
3. The RNA introns employs induced-fit strategy of action when binds guanosine and pyrimidines.
4. The folded structure of the intron activates the phosphate group at each reaction site, bringing it into a state favorable for cleavage.

The mechanism of self-splicing of the rRNA precursor of *Tetrahymena* has been shown in Fig. 13. A guanine nucleoside or nucleotide reacts with intron-exon phosphodiester linkage to displace the donor exon from the intron. This is called transesterification. The transesterification is promoted by the folded intron itself. Subsequently the free donor exon similarly attacks the exon-intron phosphodiester bond at the accepter end of the intron. This type of introns come under group I. Group I introns are found in a variety of genes in fungal mitochondria, bacteria and bacteriophage T4. An RNA from T4 has been shown to splice itself in *E. coli*. Virtually, Group I introns are not true ribozyme *in vivo* because they function for only

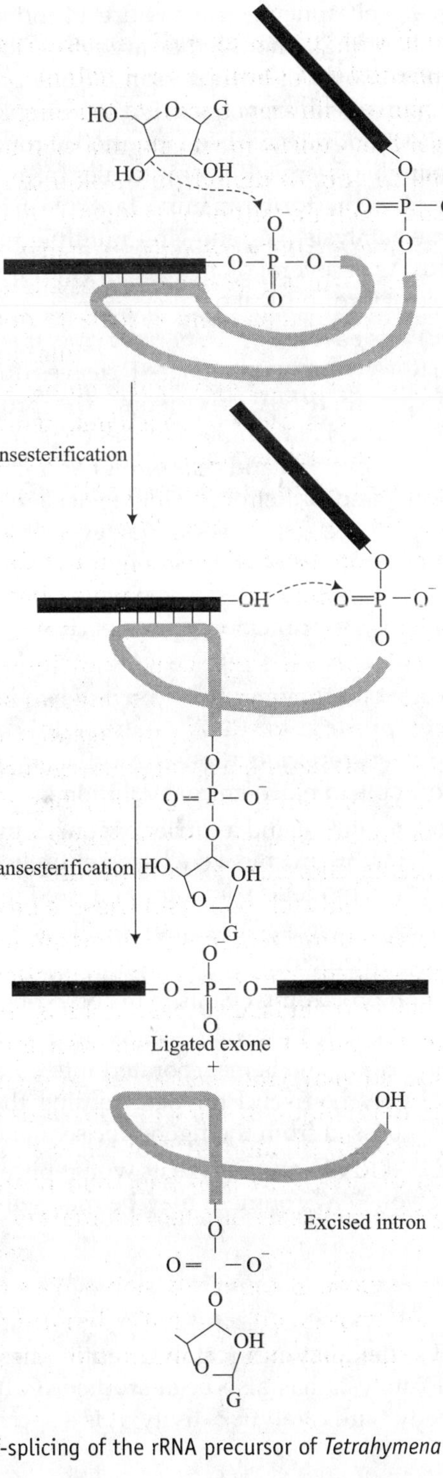

Fig. 13. Mechanism of self-splicing of the rRNA precursor of *Tetrahymena* (Woski and Schmidt, 2002).

one cycle. Whereas Group II self-splicing introns are found in the mitochondrial RNA precursors of the yeast and other fungi. Though they are self-splicing, there are many parallels between this reaction and the removal of introns form mRNA precursors during processing. An mRNA intron from the fungus *Neurospora crassa* is also self-splicing. The other self-splicing mRNA introns form yeast mitochondria have been reported. Group III self-cleaving RNAs self-cleave during generation of single genomic RNA molecules from large multimeric precursors. The amino group of a cytosine base performs an important role in cleavage reaction. These self-processing RNAs are the true catalysts (i.e., showing multiple turnover) *in vivo* and *in vitro.* They are found in genomic RNAs of several plant viruses. Hammer-head viral RNA (hammer-head ribozyme) is an example of Group III ribozyme.

Ribonuclease P (RNase P) acts as fourth type of ribozyme. It is a true ribozyme in cells and contains both a protein and a RNA component. RNase P cleaves tRNA precursors. It recognizes constant structure associated with tRNA precursors. The product of cleavage has a 52 phosphate in contrast to the products of hammer-head and similar RNAs. RNase P is a tRNA processing enzyme found both in bacteria and cells of higher organisms. It has been proved as a true catalyst and contain RNA and protein in a single package. But the RNA subunit alone can cut the pre-tRNA molecule at the correct spot. However, the protein does not have such capacity. Even some RNA subunits from a recombinant DNA template can catalyze the accurate maturation of tRNA precursors. Bacterial ribozome also acts as fifth type of ribozyme. The active site of ribosome is on 50S subunit. Some parts of rRNA chain serve as the catalyst for formation of peptide bond.

Fishing the best pool for novel ribozymes has been reviewed (Wedel, 1996). A number of other important papers have appeared on RNA catalysis. Ribozyme-catalyzed amino acid transfer reactions have also been reported (Lohse and Szostak, 1996). Discovery of RNA catalysis led the following concepts in biochemical evolution :

1. RNA can serve as both a catalyst and a carrier of genetic information. This reflects that possibly the earliest living organisms were based entirely on RNA and that DNA and protein evolved later. This model is sometimes referred to as the "RNA world".
2. RNAs of some pathogenic viruses are catalytic. This provides an opportunity for discovery of RNA-based medicines.
3. Many of the information processing events in mRNA splicing need RNA components. These RNAs may also act as biocatalyst.
4. When mRNA cleaving ribozyme is incorporated into cultured cells, it prevents gene expression. This tool may be useful for determining the function of genes whose metabolic activity are guessed from their genomic sequences.

If it can be demonstrated that RNA can catalyze its won replication, the explanation for the first step toward life in a prebiotic environment may be the "self-reflicating RNA molecule" itself.

ABZYME

Like enzymes and ribozymes, the abzymes (catalytic antibodies) have been characterized. Immunological evolution of catalysis has also been studied (Patten *et al.*, 1996). Abzyme is artificially synthesized antibody with catalytic activity. It is a specialized protein enzyme. The production of antibodies that can catalyze chemical reactions opens the way to make enzymes with any desired specificity. Antibodies and enzymes share a major point of similarity such as both bind with their target molecules with a high specificity and affinity. Essentially unlimited diversity of antibody molecule may be trapped to produce enzymes with whatever specificities are desired. Enzymes speed up biochemical reactions by stabilizing the transition state. It helps in lowering down the requirement of energy needed for conversion of reactants into products and, consequently, the rate of reaction is increased. The trick in producing catalytic antibodies lies in obtaining antibody molecules that stabilize the transition states of the selected chemical reactions. In abzyme, a transition state analog acts as the hapten. For lipase abzyme, a racemic phosphonate serve as a hapten. Abzyme generally enhances the rate of reaction by 10^3 to 10^6 fold which is close to the enzymatic rate.

The antibodies having catalytic activity can be synthesized on two principles :

1. The ability of immune system to recognize any arrangement of atoms in the foreign antigen and to make a binding site on the resulting antibody that is exquisitely to binding that antigen.
2. The strong binding of transition state-like substrates reduces the energy barrier along the reaction pathway.

The catalytic antibodies display a number of characteristics similar to enzymes :

1. They speed up chemical reaction.
2. They show substrate specificity.
3. They can be inhibited.
4. Their activity has a modest dependence on the pH of the reaction mixture.

Abzyme cleaving esters (Tramantano *et al.*, 1986) and the abzyme catalyzing carbonate hydrolysis (Pollack *et al.*, 1986) have been demonstrated. There may be following possible applications of catalytic antibodies :

1. Catalytic antibodies designed to cut proteins at any desired amino acid sequence could be used for selectively dissecting the proteins much as restriction enzymes used for cutting DNA.
2. Catalytic antibodies not only bind to protein but also cut them. This characteristic might be useful in medical science for dissolving blood clot and searching out and destroying tumor cells.
3. Antibodies genes can be mutated to produce catalytic antibodies with desired characteristics.
4. If it is possible to synthesize a semisynthetic catalytic antibodies by chemical modification of antigen-binding site, it would result in a combination of the specificity and high binding affinity of an antibody with the activity of a synthetic catalytic compound not normally found in antibodies (Max, 1986).

Virtually, the development of catalytic antibodies opens a way of tapping into vast repertoire of binding pockets on antibodies to do chemical work (Tramantano *et al.*, 1986).

SUMMARY

Biocatalysts accelerate the rate of chemical reactions in biological systems. They may be enzymes, catalytic antibodies and ribozymes. There are six classes of enzymes viz., oxidoreductases, transferases, hydrolases, lyases, isomerases and ligases. Some enzymes require coenzymes for catalytic function. 'Lock and key' and 'induced-fit' models have been proposed to explain the mode of action of enzymes. However, the mechanism of enzyme catalysis can be mainly described with the help of acid-base catalysis, substrate-strain and covalent catalysis. Enzyme concentration, substrate concentration, temperature and pH influence catalytic activity. Some enzymes have their isozymic forms. The enzyme regulation takes place at different levels. The RNA component of a ribonucleoprotein particle has catalytically active subunit and is known as catalytic RNA (ribozyme). Abzyme is artificially synthesized activiy with catalytic antibody. Different types of biocatalysts have paramount importance in biology and medicine. Biological catalysts are, in fact, partholes to our existence. They are the most studied but enigmatic biomolecules and many more things are yet to be discovered.

REFERENCES

Ahern, T.J. and Klibanov, A.M. (1985) The mechanism of irreversible enzyme inactivation at 100°C. *Science* 228 : 1280-1284.

Cech, T.R. (1986) RNA as an enzyme. *Scient. Am.* 255 : 64-71.

Lehninger, A.L. (1975) *Biochemistry*. Worth Publishers, Inc., New York.

Lohse, P.A. and Szostak, J.W. (1996) Ribozyme-catalysed amino acid transfer reactions. *Nature* 381 : 442-445.

Max, J.L. (1986) *Science* 234 : 1497-1498.

Michaelis, L. and Menten, M.L. (1913) Die kinetic der invertinwirkung. *Biochem. Zhurnal* 49 : 333-369.

Pollack, S.J., Jacobs, J.W. and Schultz, P.G. (1986) Selective chemical catalysis by an antibody. Science 234: 1570-1573.

Somero, G.N. and Hand, S.C. (1990) *Physiol. Zool.* 63 : 443-471.

Srivastava, D.K. and Bernhard, S.A. (1986a) Enzyme-enzyme interactions and the regulation of metabolic pathways. *Curr. Top. Cell. Regu.* 28 : 1-68.

Srivastava, D.K. and Bernhard, S.A. (1986b) Metabolic transfer via enzyme-enzyme complexes. *Science* 234 : 1081-1086.

Tramantano, A., Janda, K.D. and Lerner, R.A. (1986) Catalytic antibodies. Science 234 : 1566-1570.

Wedel, A.B. (1996) Fishing the best pool for novel ribozymes. *Trend. Biotechnol.* 14 : 459-464.

Wilson, J.E. (1988) In : *Microcompartmentation* (Ed., D.P. Jones), pp. 171-190, CRC Press, Boca Raton, Florida.

Suggested Readings

1. Enzyme Biotechnology (1999) G. Tripathi, Tecnoscience publications, Jaipur, India.
2. Text Book of Biochemistry (2002) Ed. T.M. Devlin, 5th Edition, John-Wiley & Sons, Inc., Publication, New York.
3. Biological Science (1995) Ed. R. Soper, 2nd Edition, Cambridge University Press, Cambridge, UK
4. Biochemistry (1975) Lehninger, A.L. Worth Publishers, Inc., New York.

REVIEW QUESTIONS

Long Questions

1. Define a biocatalyst and summarize characteristic features of enzymes.
2. Classify enzymes with suitable examples and comment upon enzyme nomenclature.
3. Discuss factors affecting the rate of enzyme reactions.
4. Give an account of coenzymes and prosthetic groups. What is enzyme-enzyme interaction?
5. What is catalytic (active) site? Discuss mechanism of catalysis.
6. Enumerate different theories of mode of action of enzyme. Comments upon transition state and activation energy.
7. Discuss ribozymes and abzymes and their importance.
8. What are isoenzymes? Discuss importance of isoenzymes.

Multiple Choice Questions

1. The catalytic (active) site in an enzyme contains
   (a) Substrate-binding site
   (b) A metal ion as a prosthetic group
   (c) Amino acid side chains involved in catalysis
   (d) All are possible but none is necessarily true
2. Malate dehydrogenase is
   (a) An oxidoreductase (b) A ligase
   (c) A hydrolase (d) A lyase
3. Which enzyme degrade H_2O_2
   (a) Peroxidases (b) Oxygenases
   (c) Oxidases (d) Dehydrogenases
4. Which of the following results in formation of enzyme-substrate intermediate?
   (a) Acid-base catalysis (b) Substrate strain
   (c) Covalent catalysis (d) Allosteric regulation
5. Metal cations can do all of the following except
   (a) Act as Lewis acids in enzymes.
   (b) Stabilize the active conformation of an enzyme
   (c) Participate in oxidation-reduction processes
   (d) Donate electron pairs to functional groups in enzyme molecule
6. Enzyme may be specific in connection to all of the following except
   (a) Type of reaction catalyzed
   (b) Optical activity of product
   (c) Chemical identity of the substrate
   (d) The atomic mass of the elements in the reactive group

7. All of the following can be chemically isolated except
   (a) Transition states (b) Enzymes
   (c) Enzyme-substrate complexes (d) Enzyme-inhibitor complexes
8. Lactate dehydrogenase isozymes have
   (a) Structural and functional similarities
   (b) Functional similarities
   (c) No functional similarities
   (D) No structural similarities
9. Abzymes are
   (a) Nucleic acid enzymes
   (b) Antibodies of all types
   (c) Hapten serving as transition state analog
   (d) Artificially synthesized antibodies with catalytic activity
10. Ribozymes
    (a) Function only in mRNA processing
    (b) Require a protein cofactor to form a peptide bond
    (c) Are enzymes whose catalytic function resides in RNA subunits
    (d) Are any ribonucleoprotein particles

Answers

1. (c)	2. (a)	3. (a)	4. (c)	5. (d)	6. (d)
7. (a)	8. (b)	9. (d)	10. (c)		

CHAPTER 7

Enzyme Kinetics and Inhibition

G. Tripathi
Dept. of Zoology, J.N.V. University, Jodhpur–342 001
E-mail: drgst@rediffmail.com

Kinetics is the study of the rate of change of reactants to products. Reaction velocity is the change in the concentration of substrate or product per unit time, whereas reaction rate refers to the change in total quantity (moles or grams) of starting material or product per unit time. Velocity and rate are used interchangeably. Initial velocity is the change in reactant or product concentration during the first few seconds of the reaction. The velocity of a reaction A → P is determined from its progress curve or velocity profile (Fig. 1). Velocity changes constantly as reaction proceeds to equilibrium, where it becomes zero. Velocity of a reaction can be expressed as :

$$\text{Velocity (v)} = -\frac{d[A]}{dt} = \frac{d[P]}{dt}$$

Rate equation is an equation that relates the experimentally determined initial velocity to the concentration of reactants. It can be expressed as :

$$\frac{-d[A]}{dt} = v = k[A]^n$$

Thus the observed initial velocity depends on the starting concentration of A to the nth power multiplied by a proportionality constant (k). The k is known as the rate constant. The enzyme catalyzed reactions may be classified either on the basis of number of molecules react to form products or on the basis of reaction order (Lehninger, 1975).

(a) **On the basis of reactant molecules:** The reactions may be monomoleculer (one reactant molecule), bimolecular (two reactant molecules), and trimolecular (three reactant molecules).

(b) **On the basis of reaction orders:** The reaction order depends on how the rate is influenced by the concentration(s) of the reactant(s) under a given set of conditions. They are as follows :

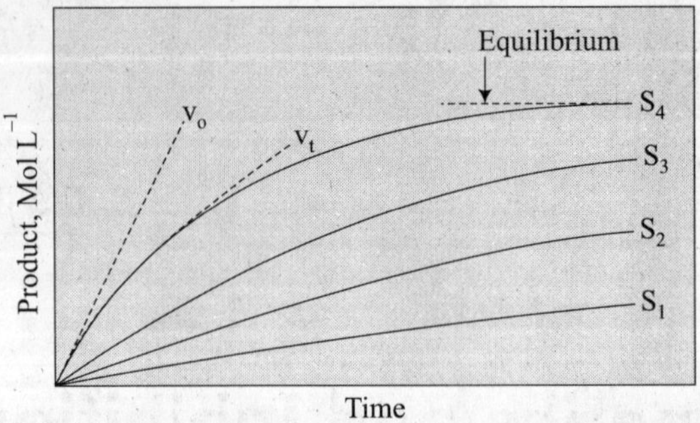

Fig. 1. Progress curves for an enzyme-catalyzed reaction (York, 2002).

(I) First Order Reaction

The reaction proceeds at a rate exactly proportional to the concentration of one reactant i.e., velocity (v) a [1 reactant] or v a [S]. The first order reaction rate at any given time 't' can be expressed as :

$$\frac{-dS}{dt} = K.[S]$$

(This is differential equation for first order reaction)

where,

[S] = Molar concentration of the reactant

$-dS/dt$ = Rate at which the reactant S disappears or is converted to product (P)

K = Proportionality or rate constant for 1st order reaction. K is reciprocal of time i.e., 1/t. Negative sign (–) is for disappearance of substance.

Integral form of 1st order equation will be,

$$\log \frac{[S_0]}{[S]} = \frac{Kt}{2.303}$$

where,

$[S_0]$ = Concentration of reactant at initial time

[S] = Concentration of reactant at time t.

$t_{1/2}$ is the time required for half of the initial reactants to be converted into product. Thus $t_{1/2}$ is not one-half the time required for the reaction to be completed rather it is half-life of the reaction. In first order reactions, the half time ($t_{1/2}$) of the reaction is:

$$t_{1/2} = \frac{0.693}{K}$$

Here, half-time ($t_{1/2}$) is independent of initial substrate concentration $[S_0]$.

(2) Second Order Reaction

The rate of reaction is directly proportional to the product of the concentration of two reactants or to the second power of a single reactant :

v α product of concentration of 2 reactants

or, v α [S^2]

Suppose the two reactants in equimolar concentrations are:

$$A + B \rightarrow P$$

The rate of reaction (–d[A]/dt, –d[B]/dt, or +d[P]/dt) is directly proportional to the product of concentrations of A and B i.e.,

$$= \frac{-d[A]}{dt} K\,[A]\,[B]$$

where, K = Second-order rate constant

Negative sign indicates the disappearance of reactants and positive sign shows the appearance or formation of product. If 2A → P, then second order rate equation becomes:

$$\frac{-d[A]}{dt} = K\,[A]\,[A] = K\,[A]^2$$

The rate constant of second order reaction has the following dimension:

$$K = \frac{1}{C_0 \times t}$$

where, C_0 = Initial concentration

Integration of second order rate equation will be:

$$t = \frac{2.303}{K([A]-[B])} \times \log \frac{[B_o][A]}{[A_o][B]}$$

(Initial concentrations) (Concentration at time t)

Under certain conditions, second order reaction appears as the first order. For example, if concentration of B is very high and that of A is very low, the reaction appears as first order because its rate will be nearly proportional to the concentration of one reactant. In other words, a second order reaction can be converted into pseudo or apparent first order reaction by keeping one of the reactant at a very high concentration of the other reactant. Hence such reaction is called as an apparent or pseudo first-order reaction.

$$\frac{-d[A]}{dt} = K \times A \times B$$ ...2nd order equation

If B is saturating (B at very high concentration),

$$\frac{-d[A]}{dt} = K \times A$$ ...Pseudo first-order reaction

(3) Third Order Reaction

Velocity of reaction is proportional to the product of three concentration terms, e.g., v a [S^3] :

$$\frac{-d[A]}{dt} = K \times A \times B \times C$$

(3 reactants A, B and C in equimolar concentrations)

This type of reaction is relatively rare.

(4) Mixed Order Reaction

The mixed-order reaction rates are neither the pure first order nor pure second order. Such reactions are oftenly referred to as mixed order reactions.

(5) Zero Order Reaction

Sometimes reactions are independent of the concentration of any reactant and this is called as zero-order reaction. Many catalyzed reactions are zero-order with respect to reactants. In this case, the reactants remain in saturating amount and are not limiting.

$$S_0 = \frac{d[A]}{dt} = K$$

Zero-order reaction can be made first-order by increasing the enzyme concentration ([E]).

BISUBSTRATE KINETICS

The enzyme catalyzed reactions with two or more reactants show a complex kinetics. This type of reaction is called as bimolecular or bisubstrate reactions:

$$A + B \rightarrow P$$

$$A + B \rightarrow P + Q$$

Bimolecular reactions may be of 2 types:

(a) Single-displacement Reaction

In this case, both reactants must be present at the active site simultaneously to give a ternary complex (3 component).

$$A + B \xrightarrow{E} E.A.B. \text{ (Ternary complex)}$$

$$E.A.B. \longrightarrow E + P + Q, \text{ or } E + P$$

The formation of ternary complex may be random or ordered.

$$E + A \longrightarrow EA$$

$$EA + B \longrightarrow EAB$$

or

$$E + B \longrightarrow EB$$

$$EB + A \longrightarrow EAB$$

Random formation of ternary complex occurs in case of phosphotransferases, creatine kinase etc.

$$\text{ATP} + \text{C (creatine)} \xrightarrow{\text{E (Creatine kinase)}} \text{C} \sim \text{P} + \text{ADP}$$

E.ATP (or) E.C binding shows the random formation of ternary complex. In case of ordered single displacement reaction, the reactants bind to the active site in a specific sequence. The reactant which binds first is called 'leading substrate' and it follows the binding of other substrate in sequence. For example, conversion of malate dehydrogenase (MDH) is an ordered single displacement reaction.

$$\text{NAD}^+ + \text{Malate} \xrightarrow{\text{E}} \text{Oxaloacetate} + \text{NADH}$$

If NADH is added before the two reactants (NAD^+ and malate), then reaction does not proceed. But if oxaloacetate is added before the malate + NAD^+, the reaction proceeds. Thus NADH can prevent only NAD^+ and not malate from binding to catalytic site.

Linear representation of ordered single displacement reaction may be as follows:

$$\begin{array}{c} \text{A}\ \ \text{B} \qquad \text{P}\ \ \text{Q} \\ \uparrow\ \ \uparrow \qquad \uparrow\ \ \uparrow \\ \hline \text{E} \rightarrow \text{EAB} \rightarrow \text{EPQ} \rightarrow \text{E} \end{array}$$

The above reaction is also referred to as 'bibireaction'.

(b) Double-displacement Reaction

In this case, one reactant binds first and one product goes out then another product goes out. It is also called as 'Ping-Pong' reaction. First reactant (A) binds to enzyme (E) to give a chemically modified E by transfer of a functional group and first product goes out. Then the second substrate (B) binds to which the functional group from the enzyme is transferred and thus the second product goes out. The double displacement occurs in case of transamination reaction e.g.,

$$\underset{\text{(Ala)}}{\text{Alanine}} + \underset{(\alpha\text{KG})}{\alpha\text{-Ketoglutarate}} \xrightarrow{\text{Alanine aminotransferase (AAT)}} \text{Pyruvate} + \text{Glutamate}$$

1st and 2nd displacements are:

(i) E + Ala-NH_2 ENH_2 + Pyruvate

(ii) E.NH_2 aKG E + Glutamate

It can be presented as follows:

$$\begin{array}{cccc} \text{Ala} & \text{Pyr} & \alpha\text{KG} & \text{Glu} \\ \downarrow & \downarrow & \uparrow & \uparrow \\ \hline \multicolumn{4}{c}{\text{E} \rightarrow \text{E.NH}_2 \rightarrow \text{ENH}_2 \rightarrow \text{E}} \end{array}$$

This is uni uni uni uni Ping-Pong catalysis.

Likewise, the other example may be:

$$\begin{array}{ccc} \text{F1, 6 dip} & \text{Gly3P} & \text{DHAP} \\ \downarrow & \uparrow & \uparrow \\ \hline \multicolumn{3}{c}{\text{E} \longrightarrow \text{E}} \end{array}$$

where,

F1,6 diP = Fructose 1,6 diphosphate

Gly3P = Glyceraldehyde 3 phosphate

DHAP = Dihydroxyacetone phosphate

The above reaction is called as 'unibireaction'.

There may also be 'biunireaction' as given below:

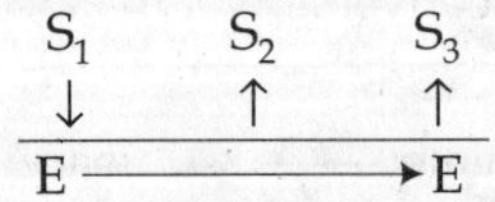

Enzymatic reactions are, in fact, the chemical reactions. So most of the laws of chemical reactions can be applied to the enzymatic reactions (Boyer, 1970; Dixon and Webb, 1979). Moreover, there are two basic differences between chemical and enzymatic reactions:

Enzyme catalyzed reactions	Chemical reactions
1. Rate of reaction increases up to optimum temperature. Afterwards, the rate begins to decrease rapidly as the enzyme molecules undergo denaturation. The native conformation of enzyme is altered so the catalytic site conformation is also changed. Hence, the enzyme molecule becomes incompatible to bind with substrate.	1. The rate of reaction goes on increasing with the increase in reaction temperature.
2. Enzymatic reactions show saturation with substrate beyond a certain concentration of substrate, because all the catalytic sites of enzyme molecules are occupied by substrate molecules.	2. The rate of chemical reaction goes on increasing with the increase in substrate obeying the law of mass action.

MICHAELIS-MENTEN THEORY OF ENZYME KINETICS

Enzyme and substrate react reversibly to form an enzyme-substrate complex (ES) before the formation of a product. The formation of ES is an essential step in enzymatic reactions. Michaelis and Menten (1913) derived a mathematical relationship between initial rate (v) of an enzyme catalyzed reaction, substrate concentration [S], affinity (Km) of enzyme (E) for substrate (S), and maximal velocity (V_{max}) at which the reaction can be catalyzed. This relation is expressed in the form of Michaelis-Menten equation (MM-equation). A hyperbola is drawn between substrate concentration and rate of reaction MM-equation is valid only for:

(a) One substrate reaction

(b) Steady state condition (Rate of ES formation is equal to the rate of (P formation)

(c) Initial period of reaction.

MM-equation is based on the following assumptions:

1. ES complex is in a steady state. It means that during the initial phases of reaction, the concentration of the ES complex remains constant even though many molecules of substrate are converted to products via ES complex.

2. All the enzyme molecules are converted to ES complexes and non is free under saturing conditions. This happens when substrate concentration is high.
3. If all enzyme molecules are in ES complex form, then the rate of formation of products will be maximal i.e.,

$$V_{max} = k_3\,[ES]$$

One can write the steady-state expression for formation and breakdown of ES complex as:

$$Km = \frac{k_2 + k_3}{k_1}$$

The events of a reaction can be shown as follows:

$$E + S \underset{k_2}{\overset{k_1}{\rightleftharpoons}} ES \xrightarrow{k_3} E + P$$

where,

k_1 = Velocity constant for ES formation

k_2 = Velocity constant for dissociation of ES

k_3 = Velocity constant for decomposition of ES with the formation of P.

The rate of formation of ES from E and S may be written as:

$$\frac{d[ES]}{dt} = k_1\,([E] - [ES])\,[S]$$

The rate of breakdown of ES may be written as:

$$\frac{-d[ES]}{dt} = k_2\,[ES] + k_3\,[ES]$$

At steady state,

Rate of formation of ES = Rate of decomposition of ES

$$k_1\,([E] - [ES])\,[S] = k_2\,[ES] + k_3\,[ES]$$

$$= [ES]\,(k_2 + k_3)$$

$$\frac{[S]([E] - [ES])}{[ES]} = \frac{k_2 + k_3}{k_1} = Km \text{ (Kinetic constant or Michaelis-Menten constant)}$$

$$[ES]\,Km = [S]\,([E] - [ES])$$

$$= [S]\,[E] - [S]\,[ES]$$

$$Km\,[ES] + [S]\,[ES] = [S]\,[E]$$

$$[ES]\,(Km + [S]) = [S]\,[E]$$

$$[ES] = \frac{[S]\,[E]}{Km + [S]} \qquad ...(1)$$

Initial rate (v) of the enzyme reaction is proportional to [ES] complex,

$$v \propto [ES]$$

$$v = k_3\,[ES] \qquad ...(2)$$

And when [S] is so high that all E is combined with S as ES complex, the maximal velocity is achieved. Then,

$$V_{max} \propto [E]$$

$$V_{max} = k_3 [E] \quad ...(3)$$

Substituting in equation (2) the value of [ES] from equation (1), one can obtain:

$$V = k_3 \frac{[S][E]}{Km + [S]} \quad ...(4)$$

Dividing the above mentioned equation (4) by equation (3):

$$v/V_{max} = k_3 ([S] [E] / Km + [S]) / K_3 [E]$$

$$v/V_{max} = k_3 ([S] [E] / Km + [S]) \times 1/ K_3 [E]$$

$$v/V_{max} = [S] / Km + [S]$$

$$v (Km + [S]) = V_{max} [S]$$

$$v = \frac{V_{max} [S]}{Km + [S]} \quad \text{...Michaelis-Menten equation}$$

When [S] is much less than Km, then the rate of reaction remains directly proportional to [S]:

$$v = \frac{[S] V_{max}}{Km}$$

When [S] is much greater than Km, then the rate of enzyme reaction remains independent of [S]:

$$v = V_{max}$$

An important numerical relationship emerges from MM-equation in a special case when the initial reaction rate remains exactly one-half of the V_{max} (i.e., when $v = ½ V_{max}$) then,

$$\frac{V_{max}}{2} = \frac{V_{max} [S]}{Km + [S]}$$

If divided by V_{max} one can get

$$\frac{1}{2} = \frac{[S]}{Km + [S]}$$

Solving for Km results:

$$Km + [S] = 2[S]$$

$$Km = [S], \text{ when v is exactly } ½ V_{max} \text{ (i.e., } v = V_{max}/2)$$

Km is equal to substrate concentration at which the reaction rate is half of its maximal value. In other words, Km is that substrate concentration at which half of the maximal velocity can be achieved. V_{max} reflects the amount of active enzyme present. MM-equation is basic to all aspects of enzyme kinetics. Even one can analyse bi and tri-substrate reaction quantitively by MM-theory. MM-model of enzyme reaction constitutes a strong background for the formation of ES complex. The MM-hypothesis explains the effects of substrate concentration on the rates of many enzyme-catalyzed reactions.

LINEAR EQUATION

One needs to know the values of V_{max} and Km to use the MM-equation. Measurement of initial rate with several different concentrations of substrate is required to determine the values of V_{max} and Km. However, it is not easy to draw a hyperbola for determination of V_{max} and Km through a set of experimental points specially if measurements extend only up to 80% of V_{max} due to substrate insolubility or expensiveness. Without an accurately extrapolated value of V_{max} one can not obtain ½ V_{max} in order to estimate Km. This practical problem has led the scientists to rearrange MM-equation in a linear form. Some of such equations are as follows.

Lineweaver-Burk Equation

In practice the determination of Km from substrate saturation curve is not very easy because V_{max} is approached asymptotically. Many enzymes give substrate saturation curves that do not readily permit evaluation of V_{max} (and hence of Km) when v is plotted versus [S]. A number of linear equations have been proposed to overcome the above problem. The linear plot given by Lineweaver and Burk (1934) is widely acceptable. This has been derived by rearranging MM equation and referred to as Lineweaver-Burk (LB) plot/ equation. Taking reciprocals of both sides of MM-equation, one can get:

$$\frac{1}{v} = \frac{Km + [S]}{V_{max}\,[S]}$$

Separating the components of numerator on right side:

$$\frac{1}{v} = \frac{Km}{V_{max}\,[S]} + \frac{[S]}{V_{max} + [S]}$$

It can be simplified as follows:

$$\frac{1}{v} = \frac{Km}{V_{max}\,[S]} + \frac{1}{V_{max}} \qquad \text{...Lineweaver-Burk}$$

One can easily draw LB plot (Fig. 2) with the help of LB-equation. This LB plot is also known as double reciprocal plot. This plot gives a link whose slop is Km/V_{max} and whose y-intercept is $1/V_{max}$. It is often easier to obtain Km from intercept which is $-1/Km$. In contrast to MM-plot, statistically significant values of Km and V_{max} can be obtained directly with six to eight data points.

Eadie-Hofstee Equation (Eadie, 1942; Hofstee, 1942)

Multiplying LB-equation by vV_{max} we get,

$$\frac{vV_{max}}{v} = \frac{Km\,v\,V_{max}}{V_{max}\,[S]} + \frac{vV_{max}}{V_{max}}$$

$$vV_{max} = \frac{Km\,v\,V_{max}}{V_{max}\,[S]} + v$$

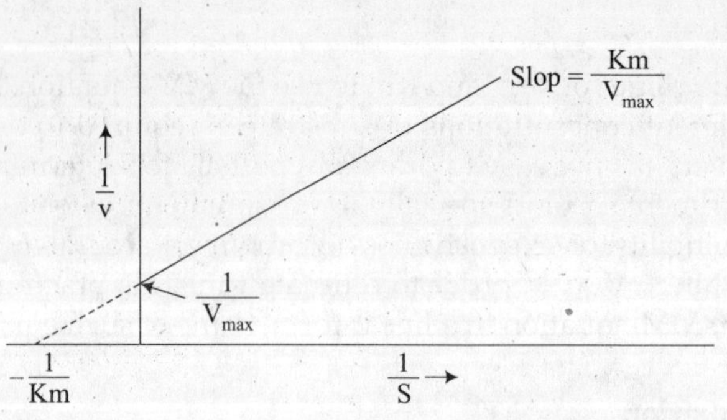

Fig. 2. Double reciprocal plot (LB-plot) for determining Km and V_{max} of an enzymatic reaction.

$$vV_{max} - \frac{v\,Km}{[S]} = v$$

or,

$$v = V_{max} - Km\,\frac{v}{[S]}$$

If we take v and v/[S] as two variables for plotting a graph, the equation again becomes in the form of a straight line (y = a + bx). This is known as Eadie-Hofstee plot.

Hanes Equation

Multiplying LB-equation by [S] we get,

$$\frac{[S]}{v} = \frac{Km\,v\,[S]}{V_{max}\,[S]} + \frac{[S]}{V_{max}}$$

$$= \frac{Km}{V_{max}} + \frac{[S]}{V_{max}}$$

$$= \frac{Km + [S]}{V_{max}}$$

$$\frac{[S]}{v} = \frac{Km + [S]}{V_{max}} \quad \text{...Hanes equation}$$

In Hanes equation, [S]/v versus [S] is plotted to get a linear graph.

Direct Linear Plot

The direct linear plot has been given by Eisenthal and Bowden in 1974. It is called as direct linear plot because each measurement is shown by a line rather than a point.

$$v = \frac{V_{max}\,[S]}{[S] + Km} \quad \text{...MM-equation}$$

After cross multiplication,

$$v[S] + vKm = V_{max} [S]$$

Dividing by both sides by v[S] we get,

$$1 + \frac{Km}{[S]} = \frac{V_{max}}{v}$$

Or, $$1 = \frac{V_{max}}{v} - \frac{Km}{[S]}$$

Or, $$\frac{V_{max}}{v} - \frac{Km}{[S]} = 1$$ ...Eisenthal and Bowden equation

Since the pioneering work of Michaelis and Menten (1913), most of the theoretical models of enzyme kinetics have been designed for enzymes studied in homogenous solutions *in vitro* (Segel, 1975). However, *in vivo* enzymes are often more or less tightly associated with the cellular structures such as membrane. Thus a question arises whether the reaction kinetics depend only on the molecular properties of the enzymes or whether they are affected by the structure of the enzyme environment. Vincent and Thellier (1983) have suggested that when structured enzyme or transport processes *in vivo* have sigmoidal kinetics, it is not the proof of allosteric protein. Likewise, when the apparent kinetics are dual-phasic, it is not the proof that the enzymes has also been developed to obtain the maximum velocity (V_{max}), kinetic constant (Km) and the parameters characterizing the inhibitors in an impure enzyme reactions (Kato and Shimotohno, 1984).

Unit of Enzyme Activity

Enzyme activity is generally expressed as μmol of substrate converted to product per minute under specified assay conditions. One unit (U) of enzyme activity is that activity which catalyzes transformation of one μmol min^{-1} of substrate. Specific activity of an enzyme is the number of enzyme units per milligram of protein (μmol min^{-1} mg of $protein^{-1}$ or U/mg of protein). However, specific activity does not show that whether the sample contains only the enzyme protein. The turnover number or catalytic constant of an enzyme is equal to the units of activity per mole of enzyme (μmol min^{-1} mol of $enzyme^{-1}$). Turnover number or catalytic constant gives direct comparison of relative catalytic ability between enzymes. For instance, the catalytic constants for catalase and alpha-amylase are 5×10^6 and 1.9×10^4, respectively, showing that catalase is about 2500 times more active than amylase. Km appears to have no physiological or clinical relevance. It is near the concentrations of substrate found in cells. Enzymes have evolved substrate-binding sites with affinities comparable to *in vivo* levels of their substrates.

ENZYME INHIBITION

There are various types of molecules which can deplete the rate of enzyme catalyzed reactions. Such molecules are called as enzyme inhibitors and the phenomenon is known as enzyme inhibition. Inhibition in enzyme activity may be reversible or irreversible (Murry *et al.*, 1990).

(a) Reversible Inhibition

Inhibitors reversibly bind with the enzyme molecule and affect the equilibrium constant of the reaction. It is of two types:

(i) Competitive inhibition: Inhibitor (structurally similar to substrate) binds with the catalytic (active) site of an enzyme. The inhibitor competes with the genuine substrate for a position in the catalytic site. Once bound, the enzyme can not convert the inhibitor to product. In this case, Km changes because it increases for the substrate in presence of inhibitor, but V_{max} remains constant (Fig. 3). The competitive inhibition can be reversed if substrate concentration is increased. Increasing the substrate concentrations will displace the reversibly bound inhibitor by the law of mass action. Thus the inhibition becomes reversible. Malonate competitively inhibits succinate dehydrogenase because it is structurally similar to the substrate succinate.

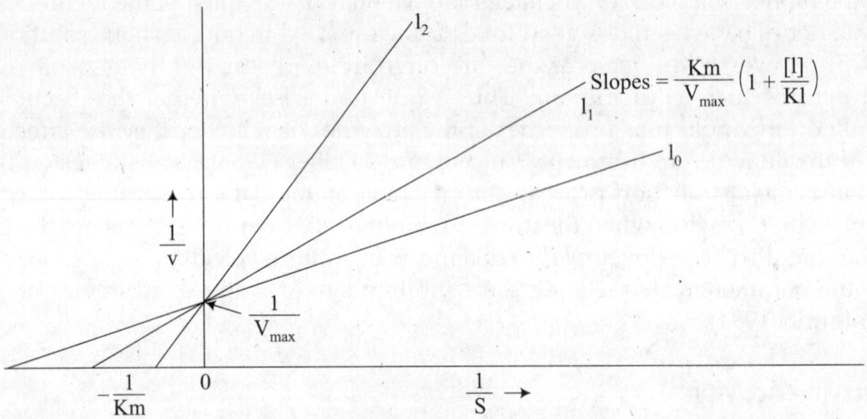

Fig. 3. Lineweaver-Burk (LB) plot showing competitive inhibition.

Competitive inhibitors are used to destroy infectious microorganisms without damaging host tissues. For example, sulphonamides were used extensively to prevent the spread of microbial infection during Second World War. Sulphonamides are similar in structure to paraaminobenzoate (PAB) which is essential for the growth of many pathogenic bacteria. The bacteria require PAB for production of an important enzyme cofactors– 'folic acid'. Sulphonamides interfere with the production of folic acid and kills microorganisms.

(ii) Non-competitive inhibition: Inhibitor molecule does not have real structural similarity to the substrate. It binds and forms an enzyme-inhibitor complex at a point other than catalytic site of enzyme molecule. It alters the globular structure of the enzyme in such a way that even the genuine substrate may be able to bind with enzyme but catalysis does not takes place. Non-competitive inhibitors do not prevent substrate from binding with enzyme and inhibition can not be reversed by increasing substrate concentration. Both binary (EI) and ternary (EIS) complexes are formed which are catalytically inactive and are dead-end complexes. Such inhibitors increase V_{max} of the enzyme. Inhibition may be reversed by exhaustive dialysis of the inhibited enzyme provided inhibitor has not reacted covalently with enzyme. Thus the value of V_{max} changes but Km remains constant (Fig. 4). The rate of reaction continues to decrease with the increasing concentration of inhibitor and it becomes almost nil when inhibitor saturation is

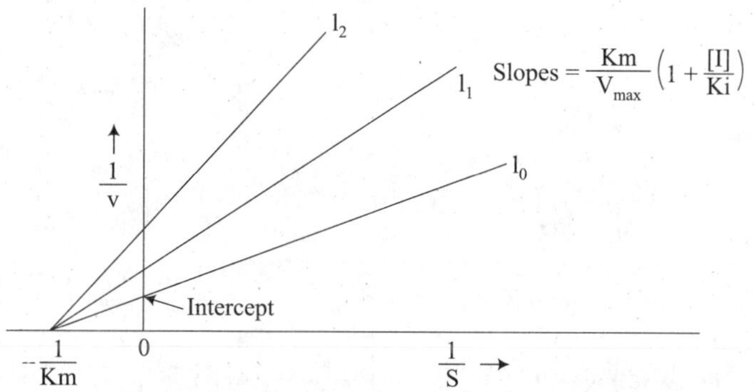

Fig. 4. LB-plot showing non-competitive inhibition.

reached. For example, cyanide combines with metallic ions (acting as prosthetic group) of certain enzymes (such as copper ion of cytochrome oxidase) and inhibits their activities.

An uncompetitive inhibitor binds only with the ES form of the enzyme in case of one-substrate enzyme. It causes apparent change in Km and V_{max} parallel to that of uninhibited enzyme.

(b) Irreversible Inhibition

Irreversible inhibition involves covalent modification of an enzyme site. When covalent modification occurs at the binding site or the active site, inhibition can not be reversed by dialysis. It occurs when an inhibitor covalently binds with the functional group of the enzyme at its catalytic site and physically block it. Such inhibition does not follow Michaelis-Menten Kinetics. Even very small concentrations of chemical agents such as mercury (Hg^{2+}), silver (Ag^{2+}) and arsenic (As^{+}), iodoacetic acid ($CH_2I.COOH$), diisopropyl flurophosphate (DEP) etc. completely inhibit the activities of some enzymes. They combine permanently with sulphydryl (-SH) group at catalytic site and thus the activity of enzyme is inhibited (Fig. 5). The active-site thiol group in glyceraldehyde 3-phosphate dehydrogenase reacts with p-chloromercuribenzoate to form a mercuribenzoate adduct of the enzyme. Such adducts are not reversed by dialysis or by addition of substrate. LB-plot of irreversible inhibition shows the characteristic pattern for non-competitive inhibition. DEP is an organophosphate compound and used as nerve gas in war. It forms an enzyme-inhibitor complex with the amino acid serine at the active site of the enzyme acetylcholinesterase. If acetylcholinesterase is inhibited, acetylcholine accumulates and nerve impulses are constantly propagated causing prolonged muscle contraction and finally paralysis or death takes place. Some organophosphorous insecticides such as parathion etc. are currently in use and have similar mode of action.

(c) Substrate or Product Inhibition

In some cases, the enzymatic reactions proceed very slowly if substrate concentration is very high. Similarly, product of a reaction inhibits the enzyme activity. Inhibition in enzyme activity at higher substrate or product concentration is because of attachment of several substrate or product molecules with the enzyme. The product becomes an inhibitor when it is the substrate

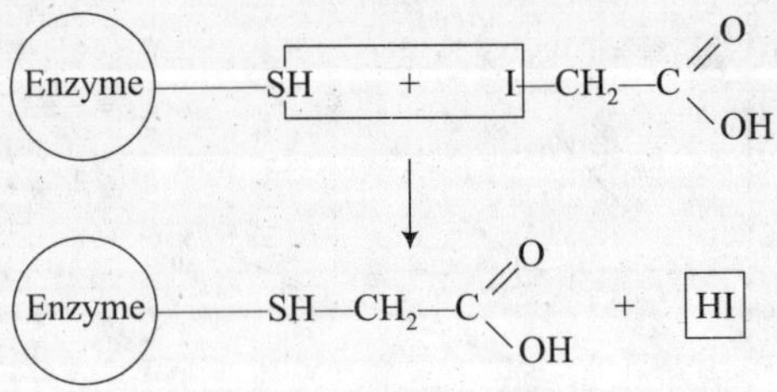

Fig. 5. Irreversible inhibition of an enzyme molecule by iodoacetic acid.

for the reverse reaction. In reversible reaction, a problem may arise if product has an affinity for the enzyme that is similar to that of substrate. In such case, the product can easily rebind to the catalytic (active) site of the enzyme and will complete with substrate for that site. This is referred to as product inhibition. A true estimation of inhibition potency of substrate, product, or coenzyme (acting as substrate/product) is difficult (Tripathi, 1998). Almost all the available methods for determination of substrate or product inhibition have been under some or other criticisms.

(d) Allosteric Inhibition

The activity of certain enzymes (i.e., allosteric enzymes) is regulated by compounds which are not their substrates and which bind to the enzyme at specific sites away from the catalytic site. They modify enzyme activity by producing a reversible change in the structure of the enzyme's catalytic site. They are called as allosteric inhibitors which slow down the reaction rate. There are also allosteric activators which speed up the enzyme reaction. Allosteric inhibition is exhibited by enzymes which possess at least two sites : catalytic site where the substrate binds, and the allosteric site (non-catalytic site) where metabolite of low molecular mass binds. The speeding up or slowing down the enzyme activity by allosteric effectors occurs by increasing or decreasing the affinity of the enzyme for its substrate. For example, ATP is an allosteric inhibitor of phosphofructokinase.

ENZYME INHIBITORS

Several drugs are enzyme inhibitors. Antimetabolic compounds (antimetabolites) are used as antiviral, antibacterial and antihumour drugs. They are enzyme inhibitors with some structural differences form the natural substrate. Sulfanilamide is a member of sulfa drugs (R-SO_2-NHR′). It acts as an antibacterial agent because of its competition with p-aminobenzoic acid (PABA), which is required for bacterial growth. Bacteria synthetize folic acid for their growth. Since sulfanilamide is a structural analog of p-aminobenzoate, the bacterial dihydropteroate synthetase is tricked into making an intermediate containing sulfanilamide. This intermediate can not be converted to folate. Synthesis of purine and pyrimidine bases require folic acid. Methotrexate (a structural analog of folate) is used to control childhood leukemia. It is a powerful competitive inhibitor of dihydrofolate reductase, which competes with dihydrofolate. Since methotrexate prevents the synthesis of thymidine monophosphate and other nucleotides, the leukemia cell can not divide. These drugs (sulfanilamide, methotrexate etc.) are classical antimetabolites.

Nonclassical antimetabolites are substrates for enzymes that upon action of enzyme produces a highly reactive species. Such highly reactive species make a covalent adduct with an amino acid side chain at the catalytic (active) site, causing irreversible inactivation of enzyme. Omeprazole is an example of this kind of enzyme inhibitor. It is used to treat excess acidity in stomach. Such nonclassical antimetabolites are very specific and known as sucide substrates. Another group of inhibitors have a reactive functional group that is situated outside the active site. For instance, dihydrotriazine is structurally similar to dihydrofolate. It binds specifically to dihydrofolate site on dihydrofolate reductase. Since triazine part of the inhibitor resembles the p-terine moiety, it binds to catalytic (active) site. Thus dihydrotriazine inhibits irreversibly dihydrofolate reductase by preventing access of substrate (dihydrofolate) to the catalytic site (Fig. 6).

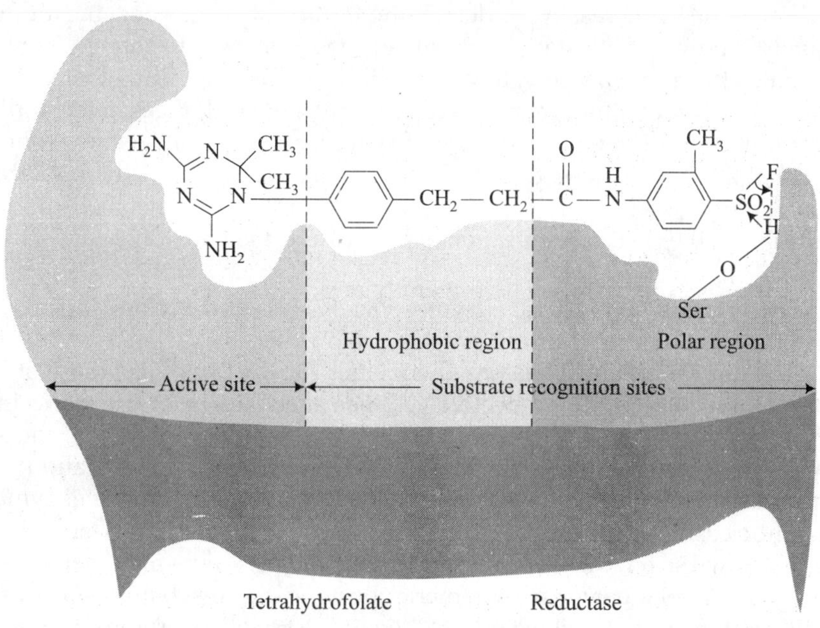

Fig. 6. Site-directed inactivation of tetrahydrofolate reductase (York, 2002).

Some of the other antimetabolites are fluorouracil (a thymine analog) and 6-mercaptopurine (analog of adenine, guanine). The deoxynucleotide of fluorouracil is an irreversible inhibitor of thymidylate synthetase. 6-mercaptopurine is a broad spectrum antimetabolite because it competes in reactions using adenine and guanine nucleotides. It is also very effective in treatment of childhood leukemias.

The therapeutic effect of some drugs depends on their ability to interact with enzyme molecule for which the drug acts as a specific inhibitor. The potency of an enzyme inhibitor is denoted as I_{50} which is defined as the inhibitor concentration required to produce 50% inhibition. However, the I_{50} is a relatively uninformative quantity since it depends on the type of inhibition (competitive/non-competitive), inhibition constant (K_i) and experimental conditions under which it is determined. Duggley (1988) has proposed a method for determining $I_{50.}$ His method is based on fitting the equation for non-competitive inhibition to data obtained by measuring the rate of reaction over a range of substrate and inhibitor concentrations.

ALLOSTERIC ENZYMES

Allosteric is derived from the Greek word allo means 'the other'. Allosteric site is additional site on enzyme molecule. Some enzymes have specific additional sites well away from the catalytic (active) site where small molecules bind and cause reversible change in the structure of the enzyme's catalytic (active) site. These enzymes are referred to as allosteric enzymes. The compounds producing allosteric effects are known as allosteric effectors. The allosteric effctors may either speed up (allosteric activators) or slow down (allosteric inhibitors) the reaction rate of an allosteric enzyme by increasing or decreasing the affinity of the enzyme for the substrate (Fig. 7). Actually binding of allosteric molecule at the allosteric site changes confirmation of enzyme molecule effecting enzyme affinity for substrate binding. Thus the catalytic (active) site of the enzyme becomes either more active or less active by changes in its binding affinity. An example of allosteric enzyme is phosphofructokinase. It catalyzes phosphorylation of fructose-6-phosphate to fructose-1, 6-diphosphate in glycolysis. When ATP is at high concentration, it allosterically inhibits phosphofructokinase. When metabolism increases and more ATP is utilized, the ATP concentration decreases and the pathway again starts operating (Fig. 8).

There are two classes of allosteric enzymes viz., K class and V class. In K class allosteric enzymes the effector alters the Km but not V_{max}. However, in V class enzyme, the effector changes V_{max} but not Km. K and V classes of enzymes show LB-plot like those of competitive and non-competitive inhibitors, respectively. Since mechanism of allosteric inhibition is different from that of simple (competitive and non-competitive) inhibition, the same terms (competitive and non-competitive) can not be used for allosteric modification. Most of the allosteric enzymes are oligomeric (several subunits). Only two monomeric enzymes allosteric enzymes (ribonucleoside) diphosphate reductase and pyruvate-UDP-N-acytylglucosamine transferase) are known so far. Monomeric (single unit) allosteric enzyme can undergo allosteric modification by effector. However, in oligomeric allosteric enzyme, binding of effector (ligand) to one subunit (protomer) affects the binding of ligand to the other protomer. This is known as homotropic interactions. Transmission of homotropic effects between protomers is an aspect of cooperativity. In contrast to homotropic interaction, heterotropic interaction is the effect of one

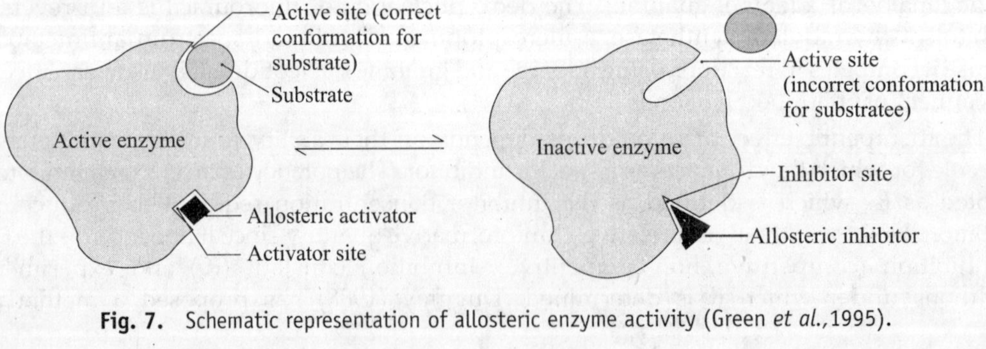

Fig. 7. Schematic representation of allosteric enzyme activity (Green *et al.*,1995).

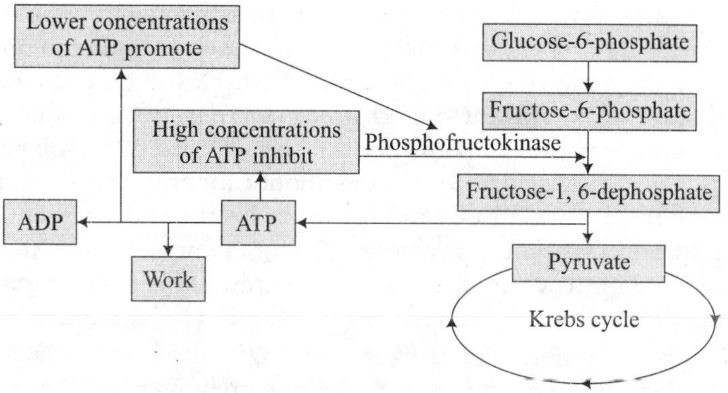

Fig. 8. A possible mechanism of the allosteric effect of ATP on phosphofructokinase.

ligand on the binding of a different ligand. The effect of a negative effector on the binding of substrate or on binding of an allosteric activator is heterotropic interactions. Homotropic interactions are always positive, while heterotropic interaction may be either positive or negative.

Allosteric enzymes show sigmoidal kinetics i.e., a sigmoid curve (S-shaped) is obtained between initial velocity versus substrate concentration (Fig. 9). Negative allosteric effector moves the curve towards higher substrate concentration and thus increases sigmoidicity. It means a higher concentration of substrate will be required to achieve ½ V_{max}. However, positive allosteric effector moves the curve towards a lower substrate concentration. It mean that ½ V_{max} is achieved at a lower substrate concentration. Positive effector or modulator shifts the curve towards hyperbolic type as observed in case of MM kinetics. The rates of allosteric enzymes are controlled by small fluctuations in the level of substrate. Large changes in enzyme activity are effected by small fluctuations in *in vivo* concentration of substrate.

COOPERATIVITY IN ENZYME

Cooperativity explains interaction between ligand sites in oligomeric enzyme. It is nothing but the influence that the binding of a ligand (effector) to one protomer (subunit) has on the

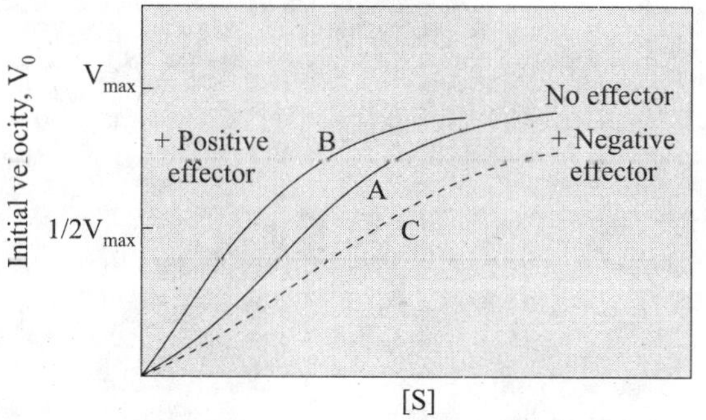

Fig. 9. Kinetic profile of K class allosteric enzyme (York, 2002).

binding of a ligand to an another protomer in an oligomeric protein molecule. Allosterism and cooperativity are different things. Conformational alteration in a given protomer in response to ligand binding at an allosteric site is an allosteric effect. However, cooperativity involves a change in conformation of an effector-bound protomer that in turn transforms an adjacent protomer into a new conformation with an altered affinity for the effector ligand or for a record ligand. An allosteric effect may also takes place without cooperativity. For example, in alcohol dehydrogenase (ADH), conformational modification occurs independently in each of the protomers upon addition of positive allosteric effectors. Since catalytic (active) site of each promoter is independent of the other, there is no cooperativity between promoters. Hence the conformational alteration in one protomer is not transmitted to adjacent protomers.

There are two important models of cooperativity viz., concerted model and the sequential induced-fit model. According to concerted model, the enzyme exists in only two states i.e., tense (T) and relaxed (R) states (Fig. 10). The T and R forms are in equilibrium. Activators and substrates favour the R form and shift the equilibrium towards R state. However, inhibitors favour the T form. A conformational alteration in one protomer produces a corresponding change in all protomers. Here no hybrid forms exist.

This model accounts for the kinetic behaviour of several enzymes but can not explain negative cooperativity. According to sequential model, ligand binding induces conformational modification in a protomer. It partially induces a conformational change in an adjacent protomer contiguous with the protomer containing the bound ligand. Thus the effect of ligand

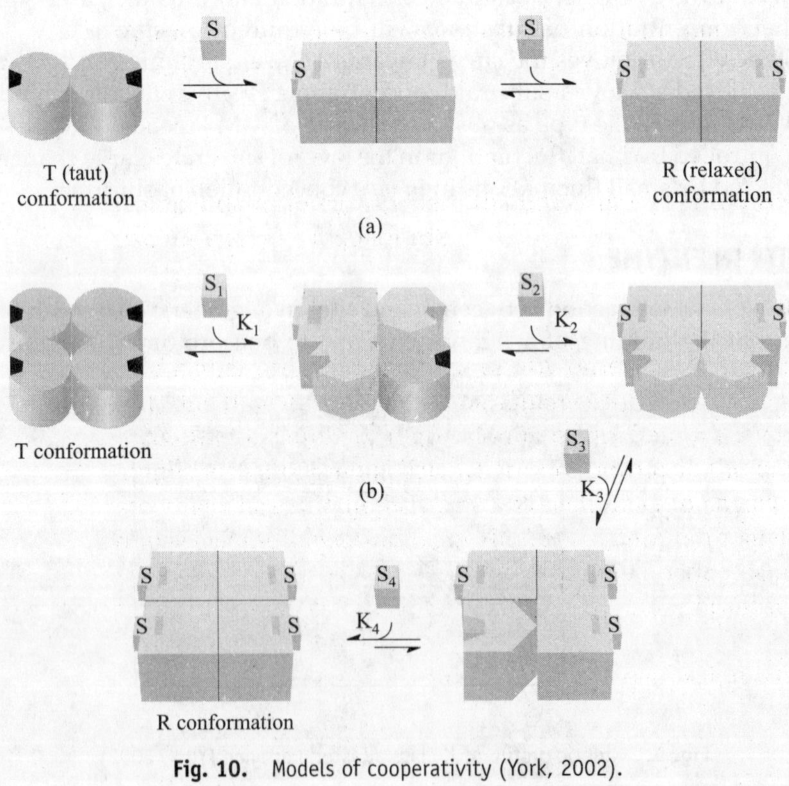

Fig. 10. Models of cooperativity (York, 2002).

binding is subsequentially transmitted through the oligomer causing increase or decrease in affinity for the substrate. In contrast to concerted model, the formation of several hybrid forms takes place (Fig. 10), giving rise to cooperativity and sigmoidal plot. The sequential model can explain both negative and positive cooperativity. For instance, a positive effector (modulator) induces a conformational change in the protomer, which has an enhanced affinity for the substrate. Similarly, a negative effector induces a different conformation in the protomer, which has a decreased affinity for substrate. Both of these effects (negative and positive) are cooperativity transmitted to adjacent protomers. In most allosteric enzymes, allosteric site resides on the same protomer as the catalytic site and all protomers are considered to be identical. However, in certain allosteric enzymes such as protein kinase A, there is a distinct regulatory subunit. This subunit has no catalytic function but its binding with catalytic protomers alters the activity of catalytic subunit through an induced conformational change.

SUMMARY

Enzymes affect the rate of chemical reactions. Therefore, it is important to understand basics of enzyme kinetics. Michaelis-Menten and Lineweaver plots are commonly used for determination of V_{max} and Km. Enzymes may exhibit reversible (competitive and non-competitive), irreversible, substrate and allosteric inhibitions. Allosterism and cooperativity occur in oligomeric enzymes.

REFERENCES

Boyer, P.D. (1970) *The Enzymes.* Academic Press, New York.

Dixon, M. and Webb, E.C. (1979) *Enzymes.* Longman, London.

Duggley, R.G. (1988) Determination of inhibition constants, I_{50} values and type of inhibition for enzyme-catalyzed reactions. *Biochem. Med. Metab. Biol.* 40 : 204-212.

Green, N.P.O., Stout, G.W. and Taylor, D.J. (1995), Enzymes. In : Biological Science (Ed. R. Soper), pp. 167-184, Cambridge University Press, Cambridge, U.K.

Kato, T. and Shimotohno, K. (1984) Estimation of kinetic parameters for substrate and inhibitor in a reaction with an enzyme sample containing different types of inhibitors. *Biochem. Biophys. Acta.* 801 : 157-162.

Lehninger, A.L. (1975) *Biochemistry.* Worth Publishers, Inc., New York.

Michaelis, L. and Menten, M.L. (1913) Die kinetic der invertinwirkung. *Biochem. Zhurnal* 49 : 333-369.

Murry, R.K., Granner, D.K., Mayes, P.A. and Rodwell, V.W. (1990) *Harper's Biochemistry.* Appleton and Lange, London.

Segel, I.H. (1975) *Enzyme Kinetics.* John-Wiley and Sons, New York. pp. 1-997.

Tripathi, G. (1998) True estimation of substrate inhibition in an enzyme catalyzed reversible reaction. *Sci. Cult.* (In Press).

Vincent, J.C. and Thellier, M. (1983) Theoritical analysis of the significance of whether or not enzymes or transport systems in structural media follow Michaelis Menten Kinetics. *Biophys. J.* 41 : 23-28.

Woski A.S. and Schmidt J.F. (2002), DNA and RNA : Composition and Structure. In : Text Book of Biochemistry (Ed. T.M. Devlin), pp. 27-92, John-Wiley & Sons, Inc., Publication, New York.

York, J. L. (2002) Enzyme : Classification, Kinetics and Control. In : Text Book of Biochemistry (Ed. T.M. Devlin), pp. 413-463, John-Wiley & Sons, Inc., Publication, New York.

Suggested Readings

1. Enzyme Biotechnology (1999) G. Tripathi, tecnoscience publications, Jaipur, India.
2. Text Book of Biochemistry (2002) Ed. T.M. Devlin, 5th Edition, John-Wiley & Sons, Inc., Publication, New York.
3. Biological Science (1995) Ed. R. Soper, 2nd Edition, Cambridge University Press, Cambridge, U.K.
4. Biochemistry (1975) Lehninger, A.L. Worth Publishers, Inc., New York.

REVIEW QUESTIONS

Long Questions

1. Describe Michaelis-Menten theory of enzyme kinetics.
2. Explain some equations for linear plots in enzyme kinetics. Why LB plot is better than MM plot?
3. Describe reversible and irreversible inhibition with suitable examples.
4. Explain cooperativity in enzyme kinetics.
5. Describe in detail the allosterism and allosteric inhibition.

Multiple Choice Questions

1. V_{max} is a point at which
   (a) All enzyme active sites are saturated
   (b) Three fourth of active sites are saturated
   (c) Half number of active sites are saturated
   (d) One- fourth of active sites are saturated
2. Temperature coefficient (Q_{10}) for an enzyme catalyzed reaction is
   (a) 1 (b) 2
   (c) 4 (d) 6
3. pH optimum curve of an enzyme is generally
   (a) Rectangular (b) Hyperbola
   (c) Bell- shaped (d) Sigmoidal
4. Rate of enzymatic reaction is directly proportional to the product of two reactants in
   (a) First order reaction (b) Second order reaction
   (c) Third order reaction (d) Mixed order reaction
5. When v is exactly ½ V_{max} then K_m is equal to
   (a) [S] at which ½ V_{max} is achieved (b) [S] at which 1/3 V_{max} is obtained
   (c) [S] at which ¼ V_{max} is obtained (d) [S] at which V_{max} is achieved
6. In case of competitive inhibition
   (a) V_{max} changes (b) K_m changes
   (c) Both K_m and V_{max} changes (d) Neither K_m and V_{max} changes

7. In case of non competitive inhibition
   (a) V_{max} changes
   (b) K_m changes
   (c) Both K_m and V_{max} changes
   (d) Neither K_m and V_{max} changes
8. Allosteric inhibition binds to the enzyme at
   (a) Catalytic site
   (b) Site away from the catalytic site
   (c) Both catalytic site and specific site away from the catalytic site
   (d) None
9. In K class allosteric enzymes the effector alters
   (a) The K_m but not V_{max}
   (b) The V_{max} but not K_m
   (c) V_{max} and K_m both
   (d) None
10. Cooperativety deals with
    (a) Ligand site in monomeric enzyme
    (b) Catalytic site in enzyme
    (c) Allosteric site in oligomeric enzyme
    (d) Interaction between ligand sites in oligomeric enzymes
11. Km reflects affinity between :
    (a) Enzyme-enzyme reaction
    (b) Substrate-substrate interaction
    (c) Enzyme-substrate interaction
    (d) Coenzyme-substrate interaction
12. Types of physiological regulation of enzyme activity deals with all of the following except
    (a) Competitive inhibition
    (b) Covalent modification
    (c) Suicide inhibition
    (d) Allosteric activation

Answers

1. (a)	2. (b)	3. (c)	4. (b)	5. (a)	6. (b)
7. (a)	8. (b)	9. (a)	10. (d)	11. (c)	12. (c)

PART–3

Hormone and Signal Transduction

CHAPTER 8

Second Messengers in Signal Transduction and Cell Signaling

N.J. Chinoy
Dept. of Zoology, School of Sciences, Gujarat University, Ahmedabad – 380 009
E-mail: zooldeptgu@satyam.net.in

In all multicellular organisms, the various cells have to communicate with one another in order to regulate their development, growth, and division, differentiation and metabolism. One of the principal method by means of which cells communicate with each other is by extracellular signaling molecules wherein, the cell is responding to a stimulus from its environment by relaying information to the internal compartment of the cell. The extracellular signaling molecules are specific substances synthesized and released into the extracellular space by another cell. These molecules elicit a specific response only in the target cells which have specific receptors (proteins in membrane, cytoplasm or nucleus). Cell signaling does not include the transfer of the stimulating agent across the membrane, but all that is transmitted is a signal that the stimulus has been received. The molecule that binds to the receptor is known as ligand. The signal so received will be passed through different substances in the membrane of the cell or its nucleus. This process is known as signal transduction. This indicates that the nature of the stimulus received by the cell-surface receptor is entirely different from the signal that is released to the interior of the cell. Signal transduction is a complex process, involving the passage of information along signal transduction pathways, normally consisting of a series of distinct proteins (Litwack, 1987). Cell-cell communication are of different types:

1. Endocrine signaling: The signaling molecules are called hormones which are released into the blood by one endocrine gland and it affects its specific target cell which is at a distance (Fig. 1a).

2. Paracrine signaling: The signaling molecule released by a cell affect target cells lying in close proximity to it. eg. conduction of an electric impulse from one nerve cell to another or from a nerve cell to a muscle cell by neurotransmitters and neurohormones. Other examples of paracrine signaling is by estrogen release from Sertoli cell of testis which affects the Leydig cells to produce androgens. Another example is secretion of somatostatin from delta cells of pancreatic Islet of Langerhans, which controls the secretion of α and β cells of the Islet to form glucagon and insulin respectively (Fig. 1b).

3. Autocrine signaling: Occurs when cells respond to substances which they themselves have released eg. Growth factors cause autocrine signaling (Fig. 1c).

4. Signaling by plasma membrane attached proteins: The proteins attached to the plasma membrane of one cell could interact directly with receptors on an adjacent cell (Fig. 1d).

5. Neuroendocrine signaling: The neurosecretory neurons of hypothalamus produce neurosecretory hormone the releasing factors, which act as signaling molecules for the pituitary which secretes its own hormones (Fig. 1e).

6. Signaling via gap junction : The gap junctions are specialized cell-cell junctions formed between closely apposed plasma membranes and which directly connect the cytoplasms of the cells via narrow channels. These channels allow the exchange of small intracellular signaling molecules or mediators such as calcium and C-AMP but not macromolecules. Hence these cells can communicate with each other via gap junctions (Fig. 1f).

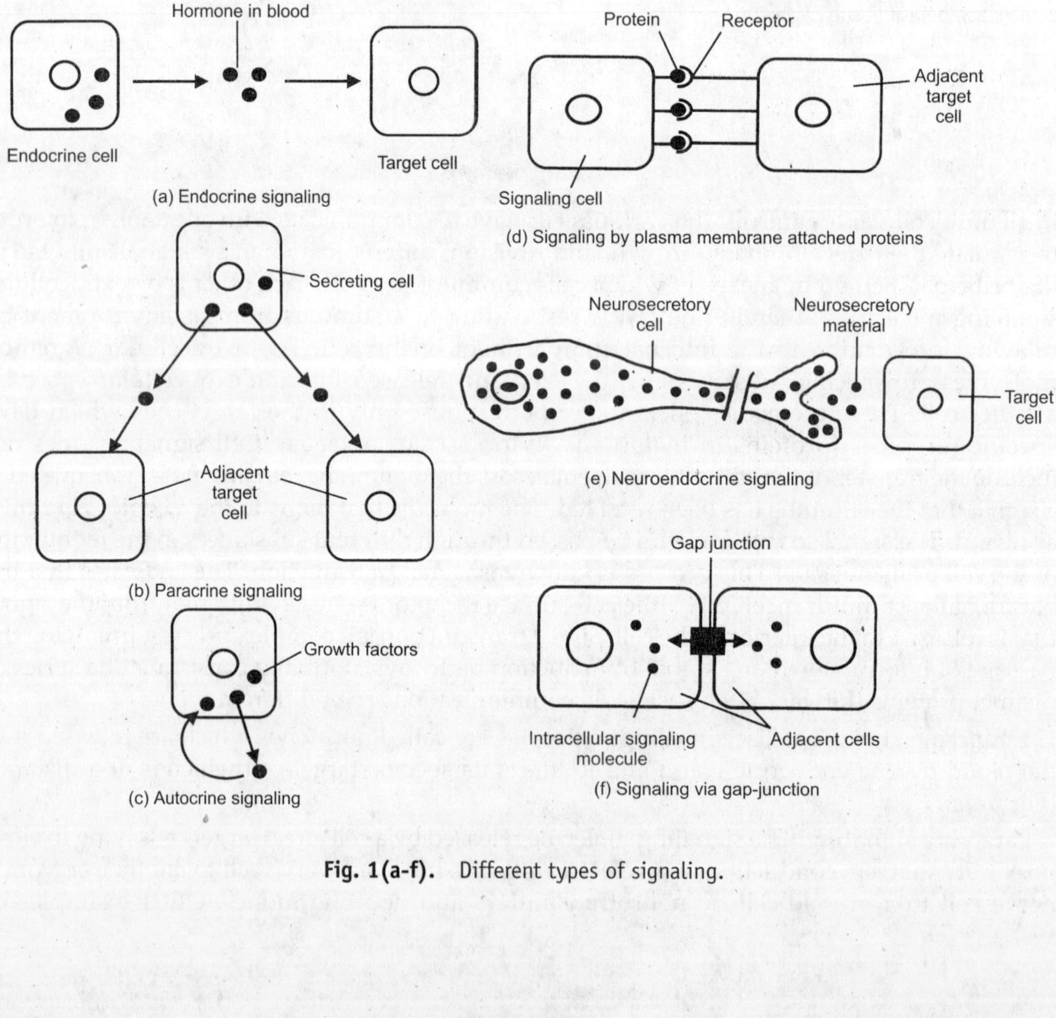

Fig. 1(a-f). Different types of signaling.

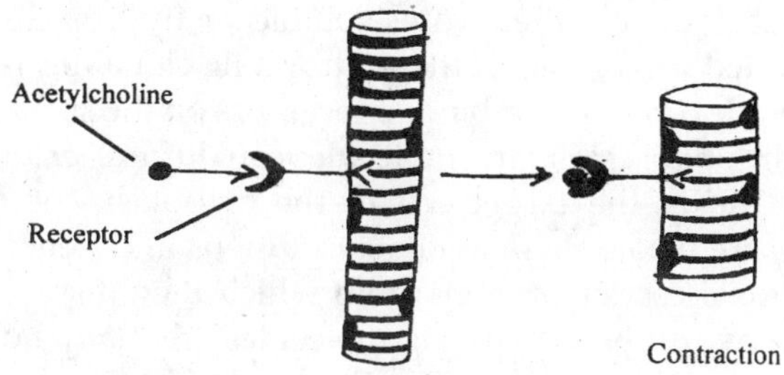

Fig. 2a. Skeletal muscle contraction by acetylcholine.

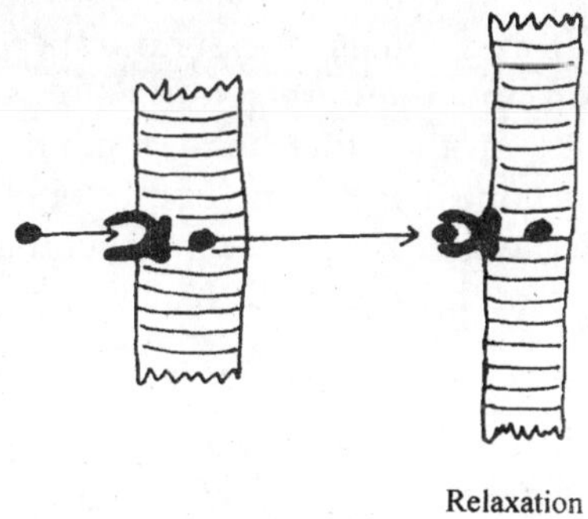

Fig. 2b. Heart muscle relaxation by acetylcholine.

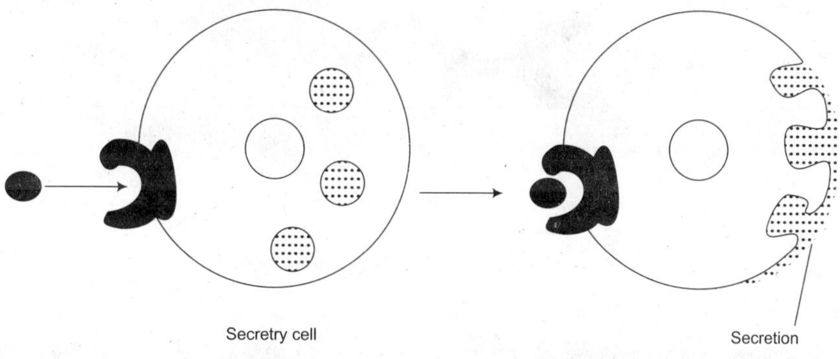

Fig. 2c. Secretion by secretory cell.

Different cells respond in different ways to the same chemical signal eg., a single signaling molecule has different effects in different target cells as illustrated by acetylcholine which stimulates the contraction of skeletal muscle cells but decreases the rate and force of contraction in heart muscle cells (Fig. 2a, b). This is due to the fact that the acetylcholine receptor proteins on skeletal muscle are different from the ones on heart muscle cells. In many cases the same signaling molecule binds to identical receptor proteins but still produces different responses in different types of target cells (Fig. 2a–c). In Fig. 2a and b, signaling molecule binds to different receptor proteins, whereas in Fig. 2b, and c, the signaling molecule binds to identical receptor proteins but they bring about different responses from the target cell (Fig. 2a–c).

The signaling molecules are either water-soluble or hydrophilic and lipid soluble or hydrophobic. As illustrated in Fig. 3a, b, the hydrophilic signaling molecules are unable to cross the plasma membrane directly and bind to receptors on the surface of the target cell (Fig. 3a). However, most hydrophobic signaling molecules can diffuse across the plasma membrane and bind to receptors inside the target cell in the cytoplasm or nucleus (Fig. 3b). The hydrophobic hormones are insoluble in aqueous solutions and hence are transported in the blood by binding to special carrier proteins from which they dissociate before entering the target cell (Fig. 3b). The hydrophilic signaling molecules are many hormones, local chemical signaling molecules and neurotransmitters. The hydrophobic signaling molecules are steroid and thyroid hormones. The length of time that these two classes of signaling molecules persist in the blood or tissue fluids is also different. The majority of water soluble signaling molecules are removed or broken down within minutes or lesser duration after entering the blood. Their responses are of short duration and they are referred to as quick acting molecules. The hydrophobic signaling molecules, on the other hand, remain in the blood for hours (steroid hormones) or days (thyroid hormones) and bring about longer lasting responses. These are hormones acting with a lag period where time is taken to synthesize new proteins which result in the ultimate response. They act at the level of DNA.

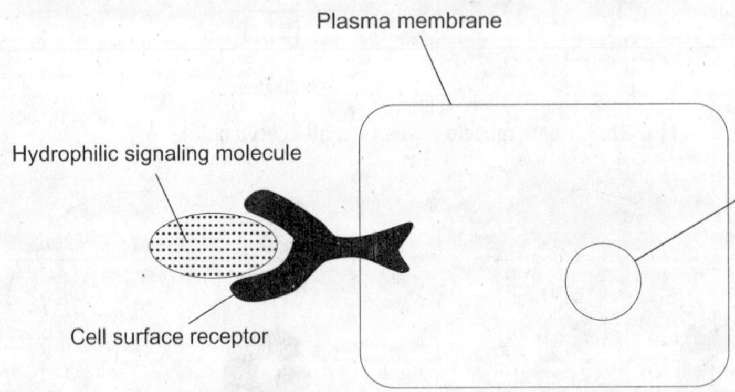

Fig. 3a. Signaling by hydrophilic signaling molecule.

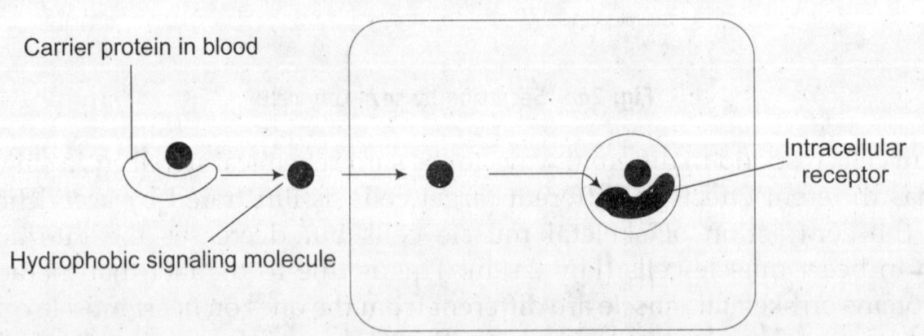

Fig. 3b. Signaling by hydrophobic signaling molecule.

MECHANISM OF ACTION OF QUICK ACTING HORMONES

In the 1950s, Earl Sutherland gave the mechanism of action of catecholamines, i.e. adrenalin (epinephrine) and nor-adrenalin (norepinephrine). These hormones bring about an elevation of blood glucose. The hormone interacts with the membrane bound receptor on target cell to form a hormone-receptor complex which inturn acts on the membrane bound adenylate cyclase and activates it. The activated enzyme acts on its substrate, ATP to form the cyclic form of AMP which is known as cyclic-AMP (C-AMP). C-AMP was called as the second messenger by Sutherland since once it is formed, it can takeover all the functions of the hormone or first messenger. C-AMP is a 150 kDa integral membrane protein (i.e. present in the membrane) that is regulated by the G proteins (described later). It can also be affected by calcium and calmodulin. C-AMP next activates several kinases. One such kinase is the inactive phosphorylase 'b' which is activated to phosphorylase 'a', the active form. The activated phosphorylase 'a' acts on its substrate, glycogen to from glucose-1-phosphate which is then converted to glucose-6-phosphate. The glucose-6-phosphatase enzyme now acts on glucose-6-phosphate and converts it to glucose Hence, its concentration rises in the blood.

The sequence of events are :

1. Hormone (H) + Receptor (R) on target cell $\longrightarrow$ HR Complex
2. HR Complex acts on membrane bound adenylate cyclase and activates it and it acts on ATP.
3. ATP $\xrightarrow{\text{Active adenylate cyclase}}$ C-AMP (Second messenger)
4. C - AMP activates several Kinases
5. Phosphorylase 'b' (inactive) $\longrightarrow$ Phosphorylase 'a' (active)
6. Glycogen $\xrightarrow[\text{Phosphorylase 'a'}]{}$ Glucose - 1 - Phosphate

$$\downarrow$$

Glucose - 6 - Phosphate

$$\downarrow \text{Glucose - 6 - Phosphatase}$$

Glucose

In 1971, Martin Rodbell and his coworkers working on rat liver cells found that a GTP binding regulatory protein called G-protein activates adenylate cyclase. Hence, Sutherland's cascade phenomenon as shown earlier is modified as:

ACTIVATION OF G-PROTEIN BY THE RECEPTOR

(1) Hormone (H) alters the cytoplasmic domain of the receptor (i.e. the part of the receptor protein which is towards the cytoplasm) and it binds to G-protein and activates it (Gilman, 1984). In other words, the hormone or ligand binding alters conformation of receptor exposing its binding site for G-protein (Fig. 4a, b). The G-protein is called Gs protein as 's' stands for stimulatory protein. The different types of G-proteins will be described later in the chapter. The association of ligand (hormone) –receptor complex with Gs-protein activates it for GTP-GDP exchange (Fig. 4c).

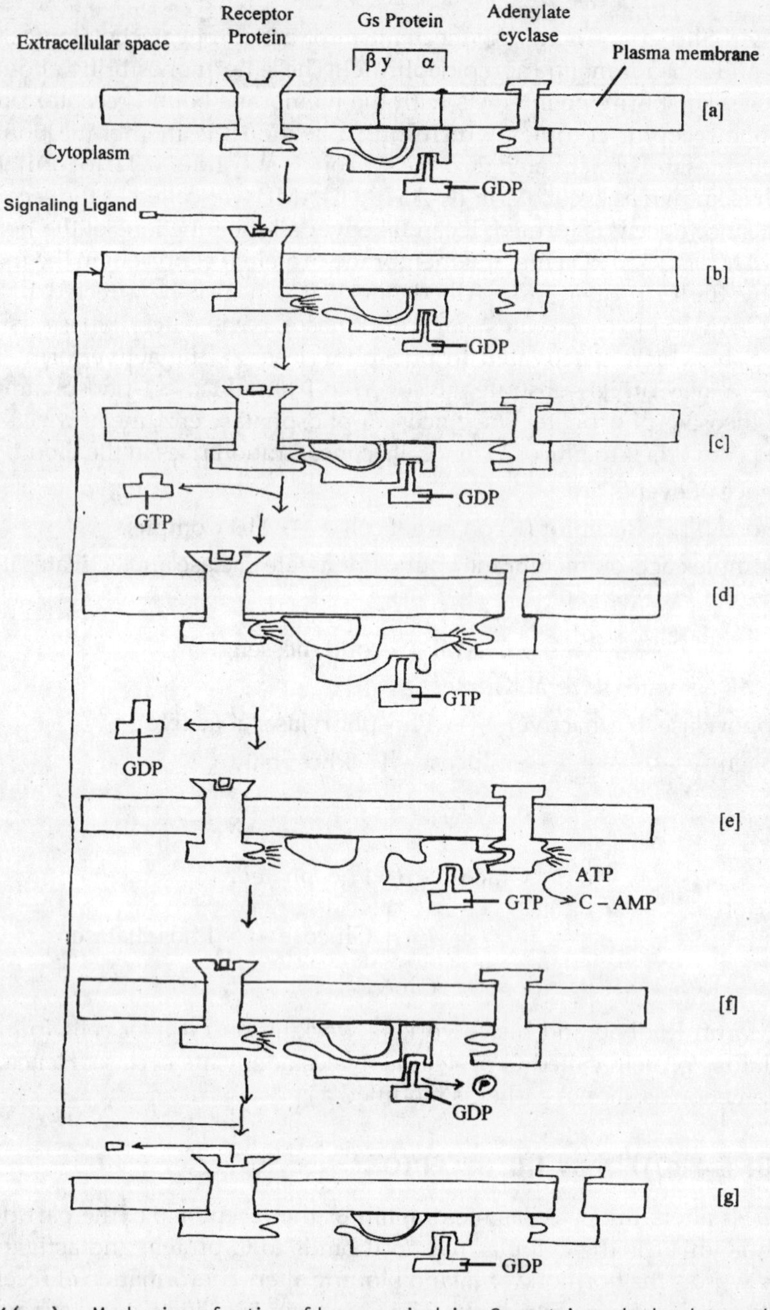

Fig. 4(a-g). Mechanism of action of hormone ivolving Gs protein and adenylate cyclase.

(2) The relay of signal from G-protein to effector occurs. The displacement of GDP by GTP causes the α-subunit of Gs-protein to dissociate from the Gs complex and thus exposes the binding site for adenylate cyclase (effector) on the α subunit (Fig. 4d). The G-proteins are made up of three subunits a, b and y and hence called trimeric proteins-(please see description in later part of chapter). The α-subunit binds to and activates adenylate cyclase to produce C-AMP (Fig. 4e). Next, the hydrolysis of the GTP by the α-subunit returns it to its original conformation, causing it to dissociate from the adenylate cyclase (which is inactivated) and reassociates with α, β, γ complex (Fig. 4f). The activation of adenylate cyclase is repeated until the dissociation of hormone (ligand) returns the receptor to its original conformation (Fig. 4g). It is clear from the above description that a family of proteins called G-protein (trimetric GTP binding proteins) are involved. They are attached to the cytoplasmic side of the plasma membrane and serve as relay molecules, coupling the receptor to enzymes or ion channels in the membranes. Thus, they have an important role in the steps between receptor activation and the release of the second messenger (Spiegel *et al.*, 1994).

The G-proteins are involved with a large variety of receptors and they themselves exist in various types, all having similar structure and same way of operation. The G-proteins are widely distributed and unique to eukaryotic cells and are involved in the responses produced by functionally different, numerous receptor types, such as ion channels in nerve and muscle cells, hormone receptors in all types of cells and photon receptors in the visual system. The name 'G' is given to these proteins since they are guanine nucleotide binding proteins. These G-proteins are anchored to the lipid of the cell membrane at its cytoplasmic or inner surface. Each G-protein consists of three subunits, which are different and hence, G-protein, are heterotrimers composed of an α chain which is loosely bound to a β γ dimer. The a subunit has a molecular weight of 40,000-50,000 daltons and are homologous. It has a GDP bound to it and is inactive. However, when stimulated by an activated receptor, the α-subunit releases the bound GDP allowing GTP to bind in its place which causes the trimer to dissociate into an α–subunit and a β γ complex. Several G-proteins share the same or very similar β subunit with molecular weight of 35,000 daltons and a γ subunit with 8,000 daltons molecular weight. There are about 27 genes which are involved in the encoding of these three polypeptide subunits. Atleast 20 different G_α subunits, 5 types of G_β subunits and seven different G_γ subunits are known to exist.

The G-protein Receptor

The G-protein receptor is a seven spanning structure in the plasma membrane with seven transmembrane loops. There are four extracellular loops which are present on the exterior side of the membrane, seven transmembrane α-helices which are in the membrane and four cytosolic loops which are towards the cytosol (inner side of the plasma membrane) (Fig. 5a). The receptor has sites of a number of mutations which are responsible for causing human diseases (Fig. 5b). The properties of mammalian G-proteins are shown in Table 1, whereas Table 2 lists the physiological processes mediated by G-proteins. Numerous inherited disorders have been attributed to either the defective G-protein itself (Table 3) or defects in its cell surface receptors (Table 4). The mutation in ADH receptor where the polypeptide is abnormally short in infants results in congenital nephrogenic diabetes insipidus wherein the kidney cells do not respond to the anti-diuretic hormone (ADH) and the patient suffers from

Table 1. Properties of mammalian G–proteins

Family	G proteins	Effect	Action mediatted by subunit	Associated effector protein	Second messenger
I	G_s	↑	α	Adenylate cyclase	C – AMP
	G_s	↑	α	Ca^{2+} channel	Ca^{2+} Change in
	G_s	↓	α	N^{a+} channel	membrane protential
	G_{olf}	↑	α	Adenylate cyclase in olfactory neurons	C–AMP
II	G_i	↓	α	Inhibits adenylate cyclase	C–AMP
	G_i	↑	βy	K^+ channel	Change in membrane potential
	G_i	↓	βy	Ca^{2+} channel	Ca^{2+}
	G_o	↑	βy	K^+ channel	Change in membrane potential
	G_o	↓	βy	Ca^{2+} channel	Ca^{2+}
	G_o	↑	α, βy	Phospholipase C–β	PIP_3, DAG
	G_t (transducin)	↑	α	Cyclic GMP Phosphodiesterase in rod cells	C–GMP
III	G_q or G_p	↑	α	Phospholipase C–β	PIP_3, DAG

Families are determined by amino acid sequence related to α-subunit; About 20α-subunits, 4 β and 7γ-subunits have been described in mammals.

↑ = stimulation: ↓ = inhibition;

PIP_3 = Phospho inositol 1, 4, 5 triphosphate; DAG – 1, 2 diacyl glycerol.

Table 2. Physiological processes mediated by G–proteins.

Stimulu	Receptor	Effector	Physiological process
Acetylcholine	Muscarinic receptor	Potassium channel	Slowing of pacemaket action
Serotonin	Serotonin receptor	Adenylate cyclase learning in Aplysia	Behavioural sensitization and
Epinephrine	β* – adrenergic receptor	Adenylate cyclase	Glycogen breakdown
Light	Rhodopsin	C – GMPVisual excitation phosphodiesterase (PDE)	
IgE antigens	Mase cell IgE receptor	Phospholipase C	Secretion
F – met peptide	Chemotactic receptor	Phospholipase C	Chemotaxis

Table 3. G-protein defect and human diseases

No.	Disease	Defective G-protein
1.	McCune – Albright Syndrome	$G_{s\alpha}$
2.	Albright's hereditary osteodystrophy and pseudohypoparathyroidism	$G_{s\alpha}$
3.	Pituitary, thyroid tumours	$G_{s\alpha}$
4.	Combined precocious puberty and pseudohypoparathyroidism	$G_{s\alpha}$
5.	Adrenocortical ovarian tomours	$G_{i\alpha}$

G_s is stimulatory while G_i is inhibitory

Table 4. Diseases due to defective G-protein receptor

No.	Disease	Defective G-protein receptor
1.	Hyperthyroidism (thyroid adenomas)	TSH receptor
2.	Familial male precocious puberty	LH receptor
3.	X-linked nephrogenic diabetes insipidus	ADH receptor
4.	Familial glucocorticoid deficiency	ACTH receptor
5.	Colour blindness	Cone opsin receptor
6.	Retinitis pigmentosa	Rhodopsin receptor

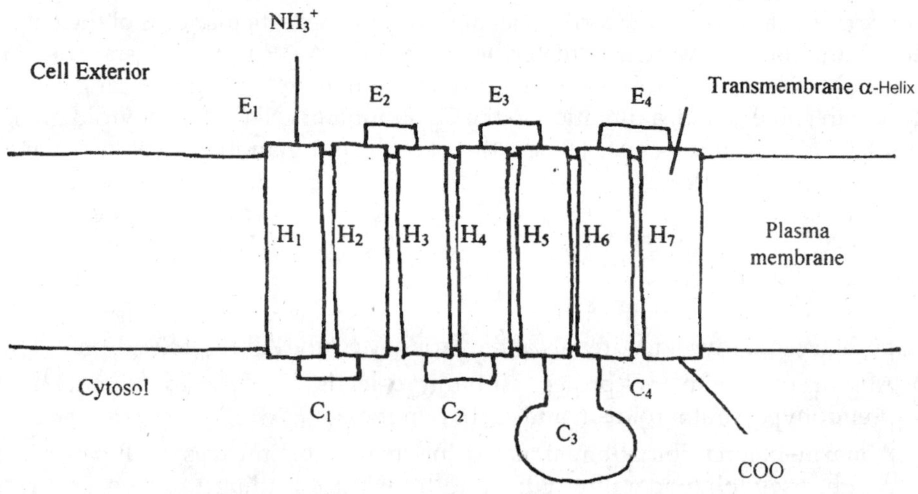

$E_1 - E_4$ = Extracellular loops
$H_1 - H_7$ = Transmembrane Helices
$C_1 - C_4$ = Cytosolic loops

Fig. 5a. Seven spanning G-protein receptor with seven transmembrane loops.

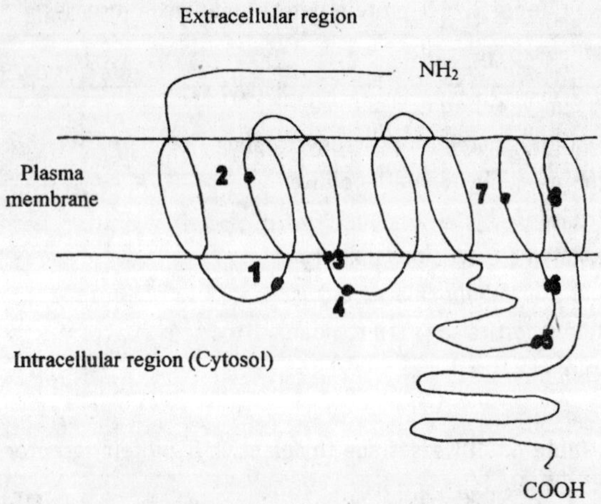

Fig. 5b. Seven-transmembrane G-protein receptor showing sites of a number of mutations causing Human diseases. Mutations No. 1, 2, 5, 6, 7, 8 cause stimulation of effector. 3, 4 block the receptor's ability to stimulate the effector. Mutations at sites 1, 2 are found in MSH (melanocyte Stimulating Hormone) receptor, 3 in ACTH receptor, 4 in ADH receptor, 5, 6 in TSH receptor, 7 in LH receptor and 8 in rhodopsin, the pigment of the retina.

serious dehydration. Similarly, a benign thyroid tumour an adenoma, is caused due to a mutation wherein the thyroid cells continuously secrete thyroid hormones without stimulation by TSH from pituitary unlike in the normal thyroid cell. The TSH receptor is mutated in these adenomas.

In male precocious puberty (early onset of puberty), the testicular cells of the patient had a G-protein mutation and were stimulated to produce C-AMP in the absence of LH ligand, which brought about premature synthesis of testosterone by the testis leading to precocious puberty. On the other hand, a mutation in the G_α subunit of cells of parathyroid gland caused the G-protein to remain inactive. Hence, the parathyroid gland cells could not respond to stimuli which in normal cells would make them secrete parathyroid hormone, resulting in hypoparathyroidism. The known G-protein diseases are rare endocrine disorders. A recent study suggests that mutations in G-protein can also lead to hypertension. The G-protein pathologies are of three types.

(1) Defective signal termination, i.e. excessive signals given leading to adenomas of pituitary and thyroid, adrenals and ovaries, McCune Albright Syndrome.

(2) Absent or inactive G-protein subunit α-leads to deficient signals and cause pseudohypoparathyroidism and night blindness.

(3) Abnormal signal initiation, signal is inadequate or in excess will cause whooping cough, pseudohypoparathyroidism, testotoxicosis leading to precocious puberty in boys, hypertension, etc.

The G-protein α subunit could also be modified by different bacterial toxins, e.g., the α_s subunit is altered by cholera toxin, the α_i by the bacterium, *Bordetella pertussis* (toxin) which causes whooping cough. Similarly, G_p α subunit and α_o are also affected by pertussis. The transducin α subunit is modified by both pertussis and cholera toxins.

The *ras* Gene and Ras Protein (A Small G-protein)

The *ras* gene is a viral oncogene, viz., a gene carried by certain tumor viruses which allows the virus to transform normal cells into a cancer cell. *ras* was also present as part of normal genome of animals, including humans. The product of *ras* gene is a protein called *Ras* protein which is a small G-protein located at the inner surface of the plasma membrane and is a member of several signaling pathways. The *Ras* protein consists of a single 21kDa subunit. Like the G-proteins, *Ras* cycles between an inactive GDP-bound form and an active GTP-bound form (Weinberg, 1985). *Ras* is one of the main components of signaling pathway extending from the outer surface of the plasma membrane to the nuclear DNA. The ligand is a growth factor like epidermal growth factor (EGF) or platelet derived growth factor (PDGF) which binds to extracellular domain of its receptor. The phosphotyrosines which are generated in the cytoplasmic domain of receptor tyrosine kinase (RTK) by autophosphorylation (described later), act as binding sites for a specific SH_2 protein called Grb_2.

The activation of *Ras* by RTKS occurs through some steps. The inactive receptors in the plasma membrane are monomers (Single subunits). The *Ras* is combined with GDP and is inactive (Fig. 6a). However, binding of the ligand such as growth factors EGF and PDGF triggers the receptors to dimerize (Fig. 6b), and autophosphorylation of the cytoplasmic domains of the receptor dimer occurs (Fig. 6b and 6d). The newly formed phosphotyrosine residues now act as sites for binding Grb_2-SOS which brings about the exchange of GTP for GDP on *Ras*. SOS is a guanine nucleotide exchange factor for Ras, i.e., Ras-GEF (Fig. 6a). It binds to Ras so that its GDP is replaced by GTP and *Ras* is activated. The GTP-activated *Ras* acts as a binding site for *Raf* (which is a protein) and localizes it in the plasma membrane where Raf is activated as a protein kinase (Fig. 6c). The kinase inturn triggers the mitogen activated protein kinase (MAP Kinase), so called as it is activated by mitosis stimulating growth factors (Fig. 6d, 30a).

SECOND MESSENGERS

As discussed earlier, Cyclic-AMP or C-AMP is a second messenger which is released into the interior of the cell due to the binding of a first messenger, which is a hormone or a ligand to a receptor at the outer surface of the cell. In eukaryotes, a number of other substances have been found to act as second messengers. The first messenger bind to a single receptor series, the second messengers formed within the cytoplasm are able to activate a variety of cellular activities. Hence, the cells can have a large–scale coordinated response by second messengers following stimulation by a single extracellular ligand.

Cyclic AMP

Cyclic AMP is formed by the action of the activated adenylate cyclase (an intergral membrane protein) by its action on ATP. Adenylate cyclase is called an effector as it brings about (effects) the cellular response by synthesizing C–AMP. When the hormone, epinephrine or glucagon bind to the outer surface of the target cell, namely, liver cell, it causes a change in the conformation of the specific receptor which is transmitted across the plasma membrane and activates adenylate cyclase at the inner surface of the membrane. The activated enzyme

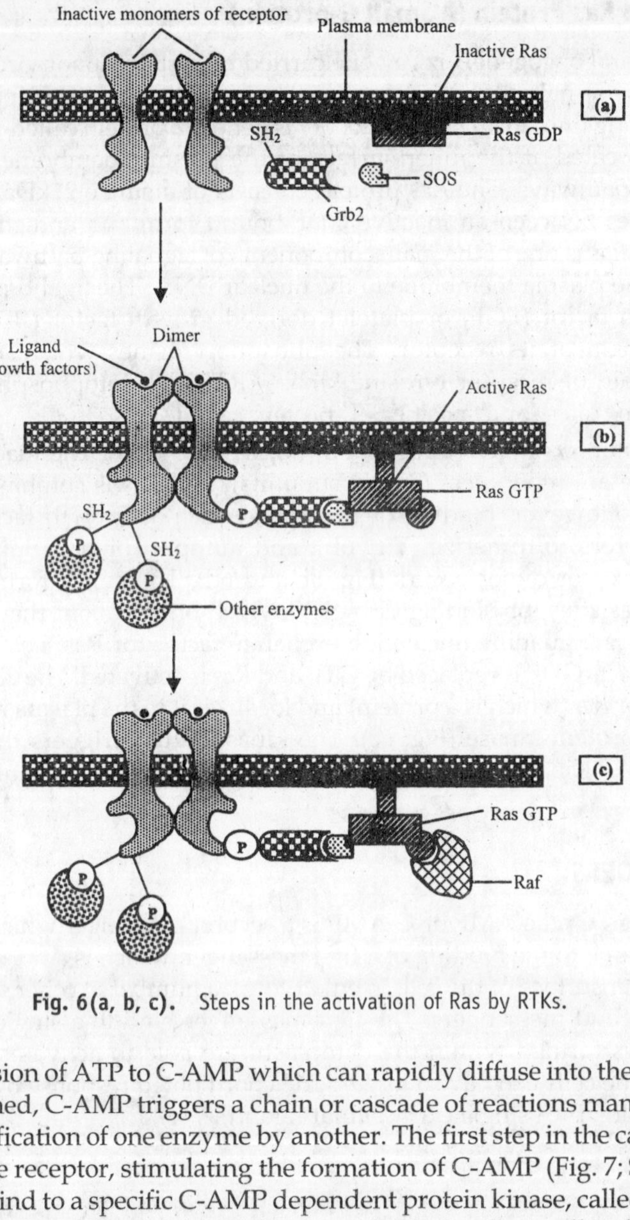

Fig. 6(a, b, c). Steps in the activation of Ras by RTKs.

catalyzes the conversion of ATP to C-AMP which can rapidly diffuse into the cytoplasm (Fig. 7; Steps 1,2). Once formed, C-AMP triggers a chain or cascade of reactions many of which consist of the covalent modification of one enzyme by another. The first step in the cascade is when the hormone binds to the receptor, stimulating the formation of C-AMP (Fig. 7; Steps 1,2). Each C-AMP molecule can bind to a specific C-AMP dependent protein kinase, called protein kinase A (PKA) and activates it (Fig. 7; Step 3). The substrates of PKA in a liver cell include two enzymes, glycogen synthase and phosphorylase kinase. (Fig. 7; Steps 4,5). If phosphorylation of glycogen synthase occurs, it inhibits its catalytic activity and hence prevents the conversion of glucose to glycogen. On the other hand, the phosphorylation of phosphorylase kinase activates the enzyme to catalyze the transfer of phosphate groups to phosphorylase molecules (Fig. 7; Step 6). The addition of a single phosphate group to a specific serine residue in the phosphorylase

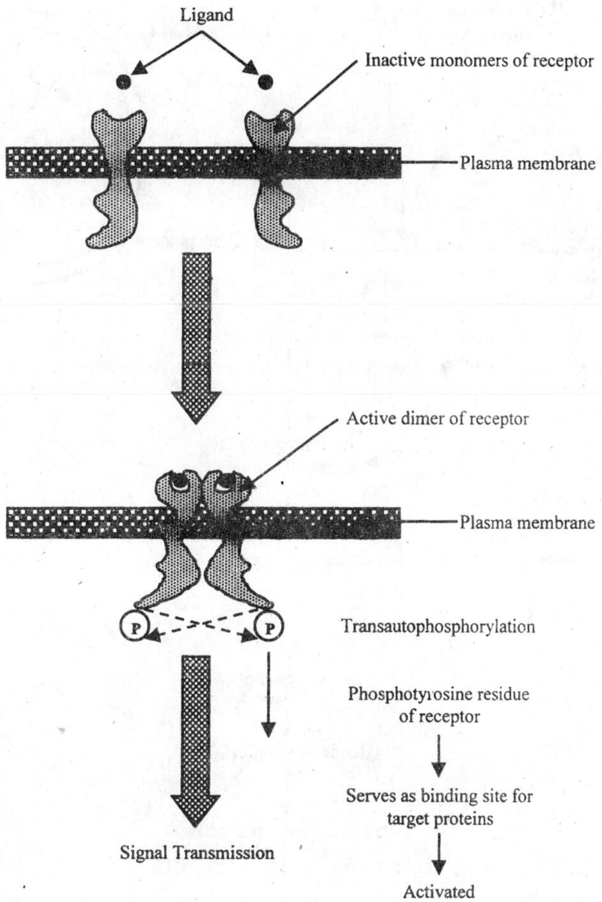

Fig. 6d. Autophosphorylation of receptor tyrosine Kinases (RTKs).

molecule activates this enzyme which acts on its substrate glycogen and breaks it down to glucose-1-phosphate (Fig. 7; Step 7). The Glucose-1-phosphate is converted to Glucose-6-phosphate and finally to glucose which diffuses in to the blood stream (Fig. 7; Step 8). Although most of the activated PKA molecules remain in the cytoplasm or cytosol, some of them translocate into the nucleus where they phosphorylate key nuclear proteins (Fig. 7; Step 9), a transcription factor called C-AMP response element binding protein (CREB). The phosphorylated CREB binds to sites on DNA (Fig. 7; Step 10) containing a particular nucleotide sequence, TGACGTCA, known as the C-AMP regulated enhancer, CRE. Thus m-RNA is transcribed.

In the cascade depicted in Fig. 7, a mechanism exists to reverse the steps, or else the cells would remain in an activated state. This is done by a number of phosphatases in liver cells which catalyze the removal of the phosphate groups added by the kinases. For example, Phosphatase-1-removes phosphates from phosphorylase kinase, glycogen synthase and phosphorylase. C-AMP is produced in several different cells in response to a wide variety of

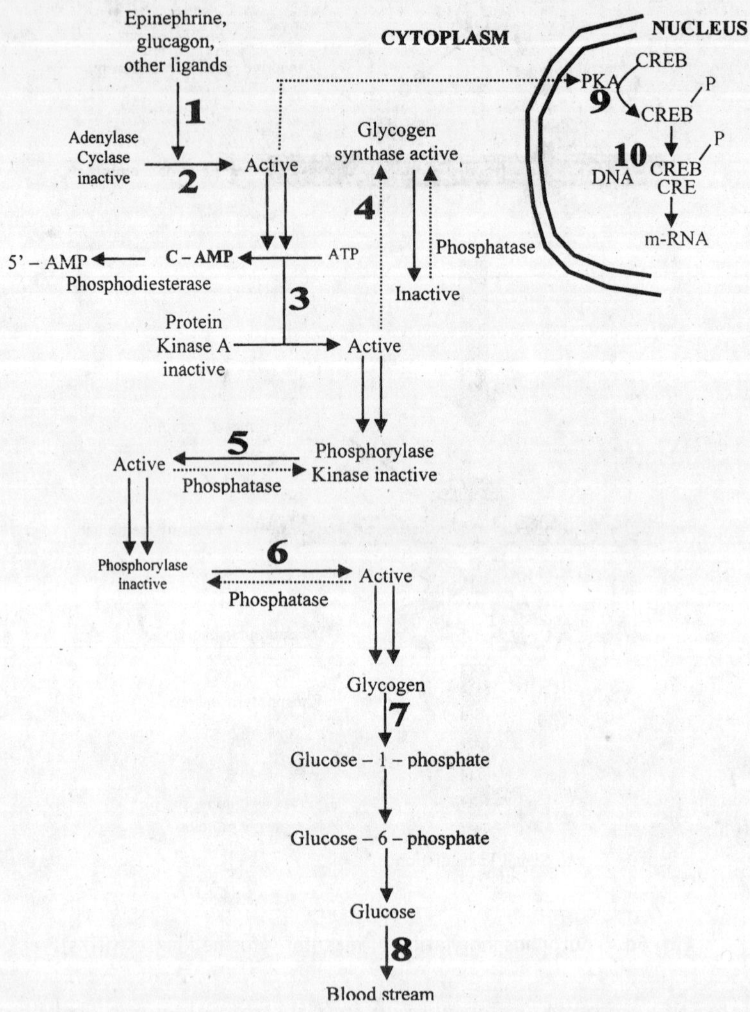

Fig. 7. The response of a lilver cell to epinephrine or glucagon, cascade reactions 1 – 8, 9, 10.

ligands or first messengers. Some responses mediated by C-AMP in mammalian cells are depicted in Table 5. C-AMP is inactivated and converted to 5′-AMP (which is not a second messenger) by the action of the enzyme, phosphodiesterase (PDE). PDE is inhibited by calcium and ascorbic acid and hence the levels of cyclic AMP will increase in the cell. A variety of processes which are affected by changes in C-AMP concentration are shown in Fig. 8. Figure 9 shows the two alternating pathways of signal transduction. First pathway is involving C-AMP and G-protein and in the second by activating a kinase of the receptor.

The cyclic nucleotide metabolism is depicted in Fig. 10 a and b. The cyclic nucleotides have been studied as second messengers of the intracellular events initiated by activation of the hormone and neurotransmitter receptor since the 1960s. The receptors that stimulate cyclic-AMP (C-AMP) production are linked to G-proteins. Receptor occupation by ligands such as

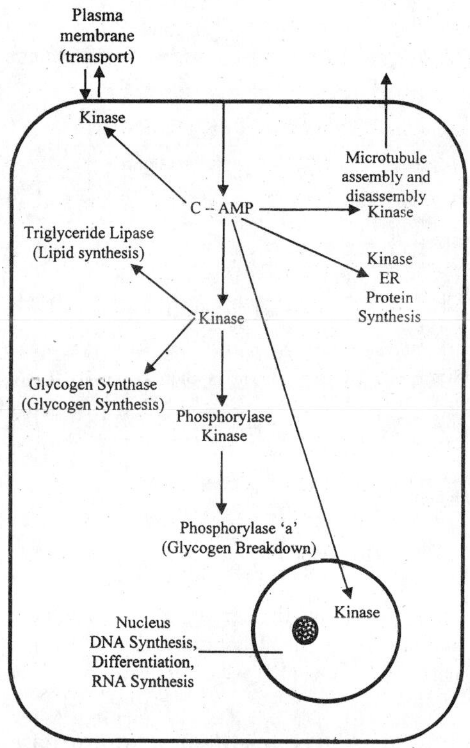

Fig. 8. Diagram showing a variety of processes which are affected by changes in C-AMP concentration. These events are mediated by different protein kinases.

Table 5. Hormone induced responses mediated by C-AMP

No.	Tissue	Hormone	Response
1.	Thyroid	TSH	Synthesis and secretion of thyroid hormones
2.	Ovary	LH	Increased secretion of steroid hormones
3.	Adrenal cortes	ACTH	Increased secretion of glucocorticoids
4.	Liver	Epinephrine, glucagon	Glycogen breakdown, glucose synthesis (gluconeogenesis) – Inhibition of glycogen synthesis
5.	Kidney	ADH	Increased permeability of epithelial cells to water or water resorption
6.	Bone	Parathyroid Hormone	Increased calcium resorption
7.	Adipose tissue	Epinephrine, ACTH and glucagon, TSH	Triacylglycerol catabolism
8.	Skeletal muscle	Epinephrine	Glycogen breakdown, inhibition of glycogen synthesis
9.	Cardiac muscle	Epinephrine	Increased contractions

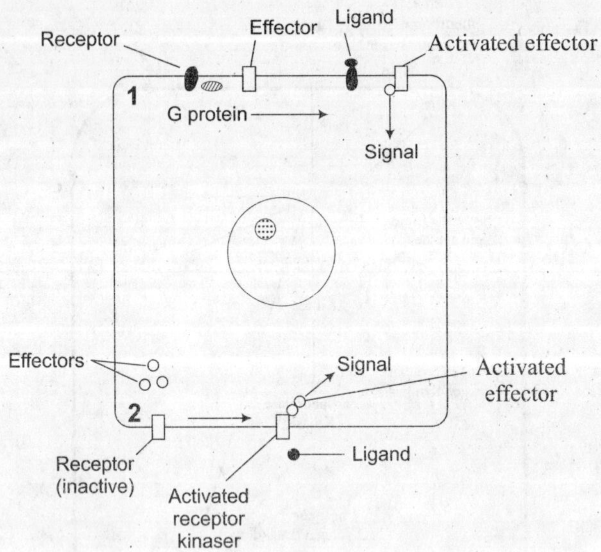

Fig. 9. Two alternate pathways of signal transduction. In the 1st pathway the bound ligand to receptor activates a G–protein which activates an effector and a signal is formed. In 2nd pathway the bound ligand activates an enzymatic activity (Kinase) of the receptor, which activates an effector, releasing a signal.

hormones (epinephrine, glucagon, ADH), results in a conformational change in the receptor that catalyzes the exchange of GTP for GDP on the α-subunit of the G-protein as described earlier. The activated α-subunit stimulates adenylate cyclase which promotes the conversion of ATP to C-AMP. In the neurons, C-AMP can activate some non ligand gated ion channels, inducing local depolarization. It should be noted that most of the actions of C-AMP are mediated through the activation of C-AMP dependent protein kinases, including protein kinase A (PKA). By catalyzing the phosphorylation (activation) of enzymes in the cell, C-AMP kinases help to regulate several metabolic and functional processes (Fig. 10a). Some peptide hormones such as atrial and brain natriuretic factors activate receptors (Struthers, 1990) which are linked to membrane bound guanylate cyclase so that receptor activation results in the conversion of GTP to cyclic GMP (C-GMP). Nitric oxide (NO) stimulates the synthesis of C-GMP by activating soluble guanylate cyclase. In the rod cells of retina C-GMP mediates the activation of ion channels in response to light. C-GMP could mediate many of its intracellular effects by the activation of specific C-GMP dependent protein kinases (PKG), similar to C-AMP which brings about the effect-activation of PKA (Fig. 10b).

Different types of phosphodiesterases (PDEs) (PDE I, II, III, IV, V) catalyze the degradation of C-AMP and C-GMP to adenosine-5′-monophosphate and guanosine-5′-monophosphate respectively. PDE I is affected by Ca^{2+} calmodulin, via phosphorylation by PKA. PDE II can cleave both C-AMP and C-GMP. PDE III is involved in the regulation of smooth muscle and cardiac muscle contraction. PDE IV is selective for C-AMP and is found in most cells. PDE V is selective for C-GMP and is localized in photoreceptors. C-CMP has been detected in tissues

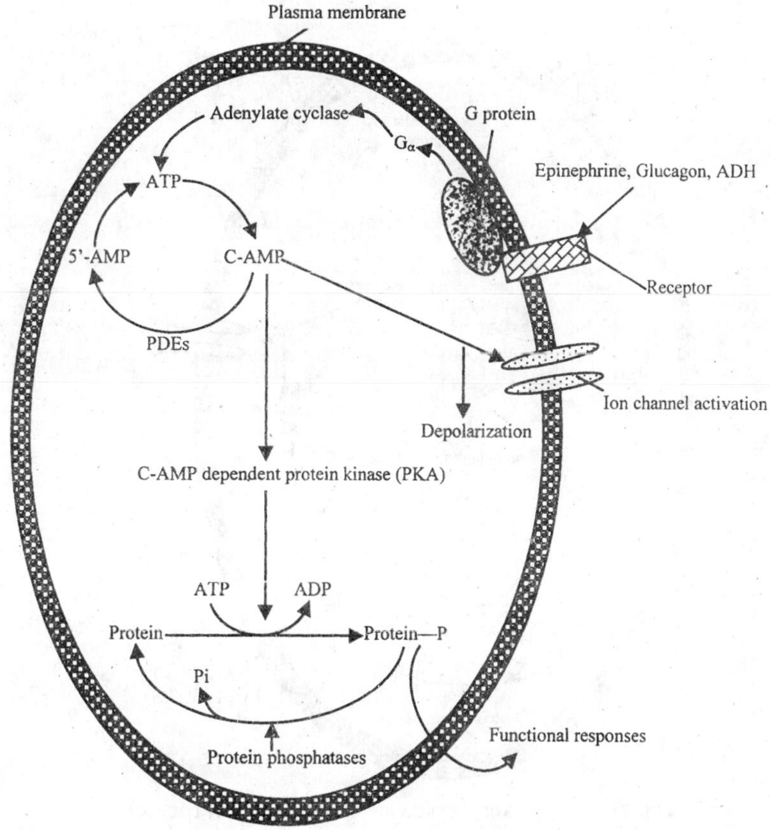

Fig. 10a. Cyclic nucleotide metabolism. PDEs = Phosphodiesterases. G_α = α subunit of G-protein.

and is elevated in rapidly proliferating tissues; e.g., levels in regenerating liver are 130 times those in normal liver. Furthermore, dibutyryl C-CMP is mutagenic in both leukemic cells and normal mammary epithelium. A cytidylate cyclase has also been identified and is similarly elevated in rapidly dividing cells. Finally, a very specific C-CMP PDE has been partially purified, it does not hydrolyze C-AMP or C-GMP and is not inhibited by methylxanthines such as theophylline (These compounds are competitive inhibitors and their net effect is to increase C-AMP concentration by blocking degradation. This PDE is low in rapidly proliferating tissues and can be inhibited by polyamines. Polyamines are associated with rapid growth. Therefore, their inhibition of this enzyme would allow C-CMP, a postulated mitogen, to accumulate and synergize with the polyamine and possibly other growth signals.

OTHER SECOND MESSENGERS

Cyclic AMP was one of the first second messenger to be discovered. However, there are other second messengers which will be discussed in the following pages.

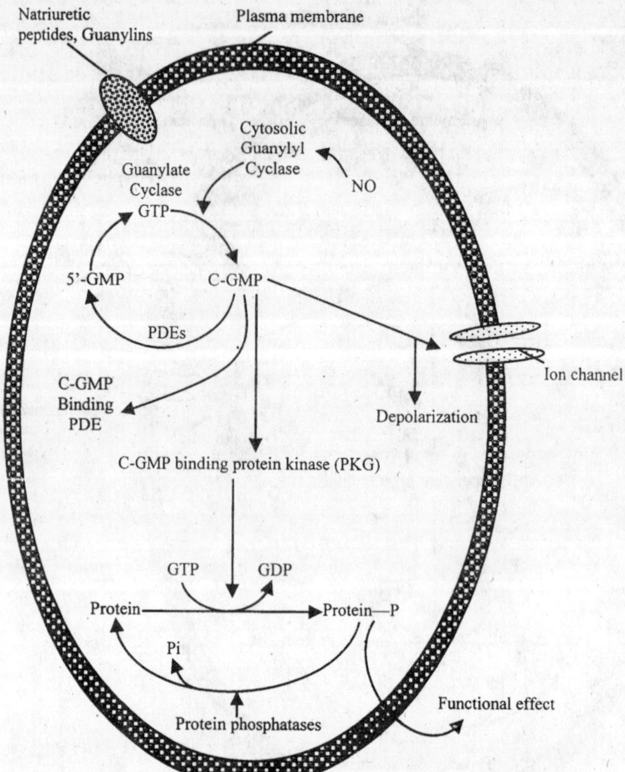

Fig. 10b. Cyclic nucleotide metabolism. NO = Nitric oxide.

Calcium

Calcium was found to function as a second messenger in 1947 when in an experiment the intracellular injection of a small amount of calcium caused a contraction of skeletal muscle cells. The calcium ion plays an important role in a variety of cellular activities as an intracellular messenger e.g. secretion, endocytosis, fertilization, synaptic transmission, metabolism, muscle movement or contraction, cell proliferation, blood clotting, sperm motility etc. Calcium can be used as a signal since its concentration in the cytosol is normally very low ($\sim 10^{-7}$ M), while its concentration in the extracellular fluid ($\sim 10^{-3}$ M) as well as in the lumen of endoplasmic reticulum (ER) is high. Hence, calcium tends to be driven into the cytosol across the plasma membrane and the ER membrane. When a signal transiently opens calcium channels in any of these membranes, then calcium goes into the cytoplasm (cytosol), increasing its concentration by 10-20 times and triggers calcium responsive proteins in the cell (Rasmussen, 1986a, b). Eukaryotic cells maintain a very low concentration of free calcium in their cytosol firstly, by active transport of calcium involving ATP and the calcium pump which removes calcium from cytosol to the exterior of the cell. Secondly, calcium is pumped into the ER, mitochondria and various molecules in the cell which bind free Ca^{2+} (Fig. 11a and b).

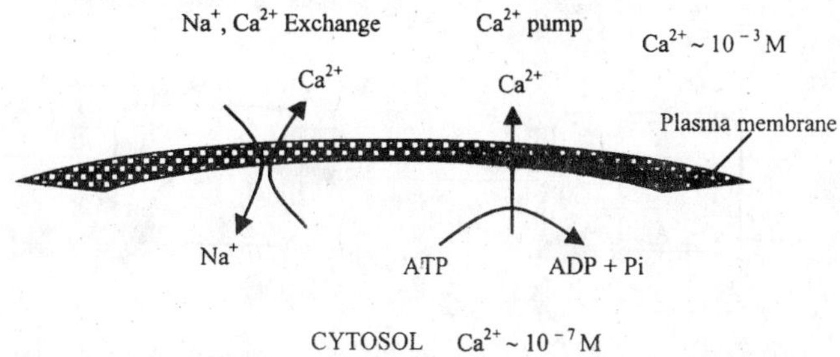

Fig. 11a Ca^{2+} is actively pumped out of cytosol to cell exterior.

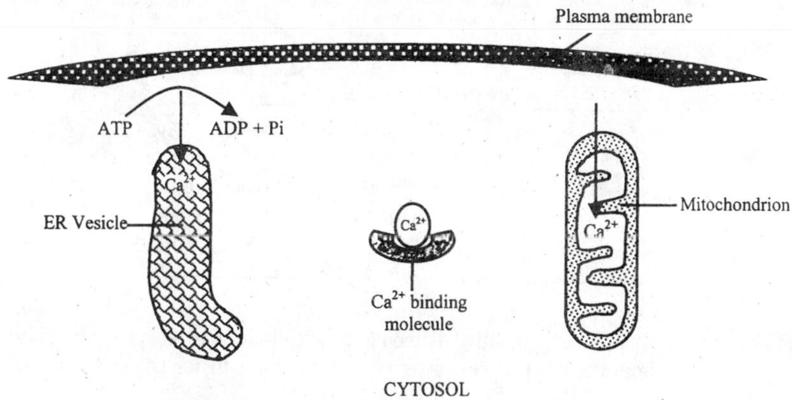

Fig. 11b. Ca^{2+} is pumped into ER, mitochondria and bound to Ca^{2+} binding molecule.

Three main types of calcium channels can bring about calcium signaling.

(1) Voltage dependent Ca^{2+} channels in the plasma membrane open in response to membrane depolarization and permit Ca^{2+} to enter the activated nerve terminals and trigger neurotransmitter secretion in nerve cells (Fig. 12 a).

(2) Inositol triphosphate (PIP_3) (an intracellular messenger molecule, described later) gated Ca^{2+} release channels allow Ca^{2+} to come out of ER when the inositol phospholipid signaling pathway is activated. In other words, the binding of extracellular signaling molecules to cell surface receptors causes the release of Ca^{2+} from the calcium-sequestering compartment such as the ER. The PIP_3 pathway is described later in the chapter (Fig. 12b).

(3) Ryanodine receptors (so named because they are sensitive to the plant alkaloid rhynodine) react to a change in plasma membrane potential to release Ca^{2+} from the ER which stimulate the contraction of muscle cells.

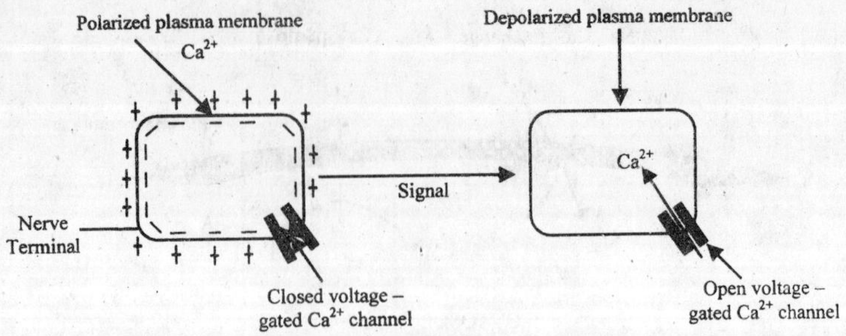

Fig. 12a. Calcium enters a nerve terminal from extracellular fluid through voltage – gated Ca^{2+} channels when the nerve terminal membrane is depolarized by action potential in nerve.

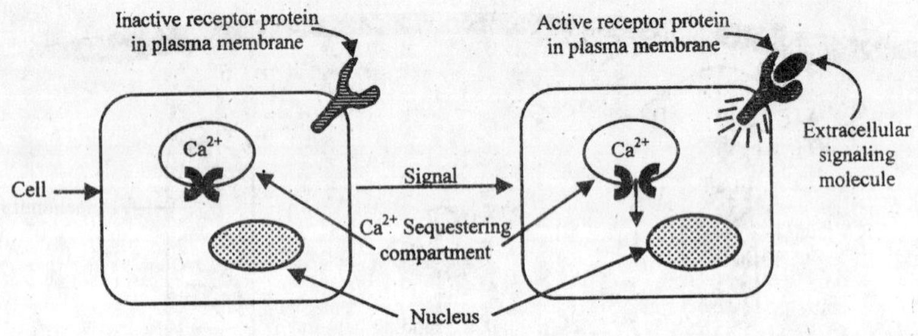

Fig. 12b. Binding of an extracellular signalling molecule to a cell-surface receptor stimulates the release of Ca^{2+} from the calcium-sequestering compartment inside the cell.

Calcium and Cyclic AMP Interaction

The second messengers calcium (Ca^{2+}) and cyclic AMP (C-AMP) are known to interact with each other and influence several metabolic processes (Blackmore and Exton, 1985). As depicted in Fig. 13, when: (1) A hormone acts on a specific target cell it influences the stored calcium and releases it into the cytosol as free calcium, (2) the hormone will cause the extracellular calcium entry into the cell and increase the free calcium pool, (3) The hormone will activate the receptor protein in plasma membrane of target cell which inturn will interact with Gs-protein to activate the adenylate cyclase and form C-AMP by acting on ATP as described earlier. C-AMP once formed will cause protein phosphorylation and activate several kinases. (4) Once free calcium is released in cytosol as by reaction (1) and (2) as described above, it itself i.e. calcium can cause protein phosphorylation and alter enzyme activity, (5) The C-AMP is metabolized by an enzyme phosphodiesterase (PDE) to 5′-AMP (which is not a second messenger). However, calcium inhibits PDE and does not allow it to act on C-AMP, hence, its levels increase in the cell and various metabolic reactions could continue unhampered (Fig. 13).

Calcium-C-AMP interactions involve :

(1) Calcium dependent regulation of C-AMP metabolism.

(2) C-AMP dependent regulation of calcium metabolism.

(3) Modulation of activity of one by the other.

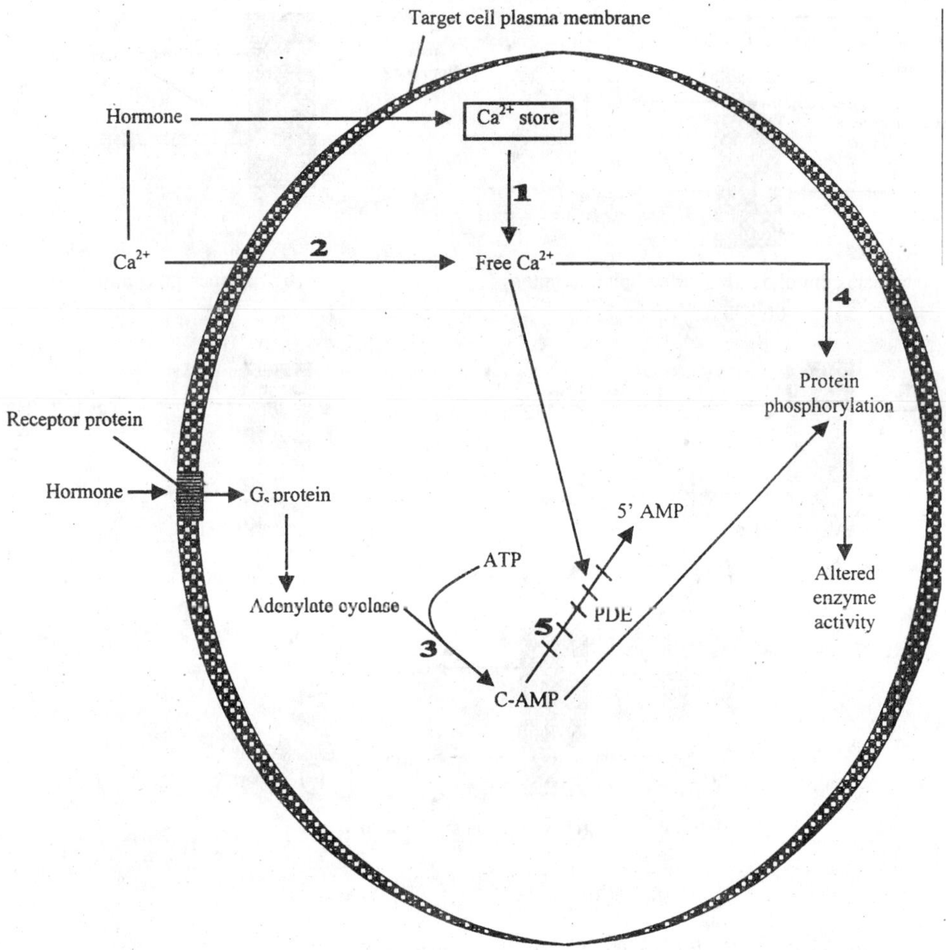

Fig. 13. Synergistic action of calcium and C-AMP on cell metabolic processes.

The interactions are of 5 different types :

(i) **Coordinate control:** For example, for salivary gland secretion and parathyroid hormone stimulation (Fig. 14a). In this type of control there is simultaneous increase in both messengers. The hormone acts on receptor R_1 in plasma membrane of target cell and causes an increase in calcium influx in the cytosol so that the levels of free calcium increase and bring about the end response or metabolic process. Simultaneously, the hormone acts on receptor R_2 to stimulate Gs-protein and finally form C-AMP.

(ii) **Sequential control:** For exmaple, for adrenal medullary secretion; cardiac muscle contraction (Fig. 14b). The reactions take place in sequential manner wherein calcium is released first in cytosol by the hormone acting on R_1 and calcium will now interact with receptor R_2 to form C-AMP and the final response. In some cases C-AMP is produced first and calcium next.

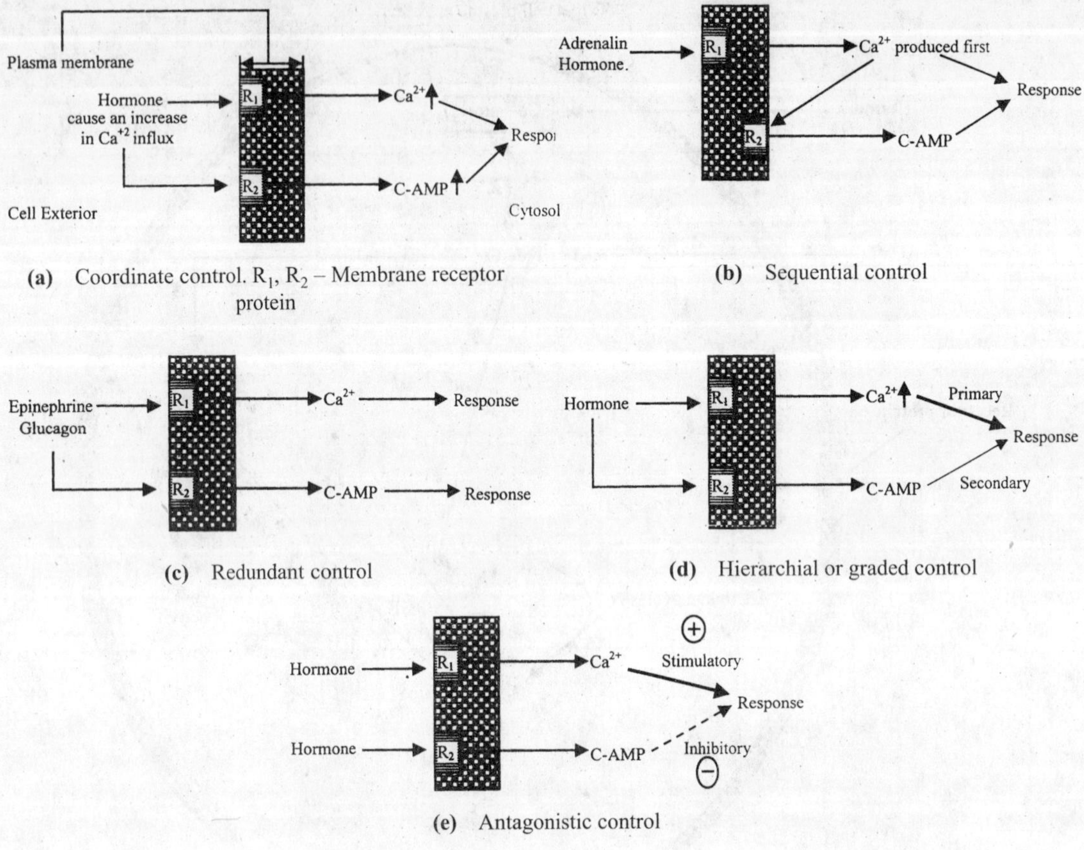

Fig. 14(a-e). Calcium, C-AMP interactions.

(iii) Redundant control: Redundant or superfluous control as one messenger is capable of bringing about the end response but both are produced e.g. for hepatic gluconeogenesis (formation of carbohydrate from non carbohydrate sources) and skeletal muscle glycogenolysis (glycogen breakdown). The hormone epinephrine from adrenal medulla or glucagon from alpha cell of islets of Langerhans of pancreas affect receptor R_1 and help in release of Ca^{2+} in cytosol which brings about the end response. Glucagon will interact with receptor R_2 and help in formation of C-AMP which will also bring about the end response (Fig. 14c).

(iv) Hierarchial or graded control: For exmaple, for exocrine secretion from parotid salivary gland, insulin secretion from pancreas, muscle contraction in Aplysia (a nudibranch mollusk), synaptic transmission etc. (Fig. 14d). The hormone insulin will interact with receptor (R_1) and cause increase in cytosolic calcium which is the primary response. Next the hormone will interact with receptor (R_2) and help in formation of C-AMP which is the secondary response.

(v) **Antagonistic control:** For example, Smooth muscle contraction, histamine release from mast cells, platelet release reaction etc. (Fig. 14e). The hormone interacts with the receptor (R_1) and brings about the release of Ca^{2+} in cytosol. Ca^{2+} will cause the stimulatory response. The hormone will interact with receptor (R_2) and C-AMP will be formed which causes the inhibitory response.

Role of Lipids as Second Messengers in Cell Signaling

Numerous lipids are involved as second messengers in cell signaling and they arise from arachidonic acid pathways (Cuatrecasas, 1986). Arachidonic acid (5, 8, 11, 14 eicosatetraenoic acid) is a 20 carbon unsaturated fatty acid and is a normal constituent of membrane phospholipids. It is released from these phospholipids by phospholipase A_2. Alternatively, cleavage of the glycerophosphate bond of membrane phospholipid by phospholipase C, results in formation of diglycerides from which arachidonic acid can be released by the action of diglyceride lipase (Fig. 15).

The action of cyclo-oxygenases on arachidonic acid leads to the formation of prostaglandin G_2 (Fig. 15), from which the biologically active prostaglandins PGD_2, PGE_2, $PGF_{2\alpha}$, PGH_2, Prostacyclin (PGI_2) and the thromboxanes (TxA_2 and TxB_2) are synthesized. Lipoxygenase (lipoxidase) acts on arachidonic acid to produce 5-hydroperoxyeicosatetraenoic acid (5-HPETE) that is converted to leukotriene A_4(LTA_4) which is the precursor of LTB_4, which induces inflammation by its chemotactic and degranulating actions on polymorphonuclear lymphocytes, and of LTC_4, LTD_4 and LTE_4, the aminoacid containing leukotrienes that induce vasoconstriction and bronchoconstriction and are involved in asthma and anaphylaxis. The synthesis of prostaglandins is widespread but thromboxanes and leukotrienes are synthesized in the platelets, neutrophils and the lung. Thromboxanes are also synthesized in the brain and leukotrienes in mast cells.

Hormones can also regulate cellular functions through the hydrolysis of phospholipids. These hormones bind to their receptors and activate a phospholipase C via a Gs like protein, other antagonistic hormones can inhibit this enzyme via Gi like protein. The phospholipase C then hydrolyses phosphatidylinositol 4,5-biphosphate (PIP_2) into phosphatidylinositol 1,4,5 triphosphate (PIP_3) and diacylglycerol (DAG). PIP_3 stimulates the release of calcium from internal sequestering compartments such as ER. PIP_3 is converted to PIP_4 (phosphatidyl inositol 1,3,4,5 tetrakisphosphate). PIP_3 is soluble and hence released in cytoplasm whereas, DAG remains embedded in the membrane. PIP_4 stimulates the influx of calcium from the extracellular fluid across the plasma membrane to replenish the internal stores (Figs. 16-18). The calcium can affect the cellular processes either directly or by combining with calmodulin. The other hydrolysis product, DAG, can either activate protein kinase C (PKC) or be further hydrolysed to release arachidonic acid for eicosanoid synthesis. The signal is terminated when calcium is pumped out of the cytoplasm and both PIP_3 and DAG are recycled to polyphosphoinositide (PPI or PIP_2).

In many systems PIP_3 and DAG act synergistically but in some cases DAG can be selectively produced by the activation of a phosphatidyl choline-specific phospholipase C; the DAG produced now stimulates PKC, but the phosphatidyl choline has no effect on calcium levels. There are some tissues with normal PKC content (Table 6) but unmeasurable PIP_3 receptors. In this system, both second messengers PIP_3 and DAG could be generated, but only

Table 6. Some responses mediated by protein kinase C.

No.	Tissue	Response
1.	Adrenal medulla	Secretion of adrenalin
2.	Pancreas	Secretion of insulin
3.	Thyroid	Secretion of calcitonin
4.	Pituitary	Secretion of GH and LH
5.	Testes	Testosterone synthesis
6.	Liver	Glycogen hydrolysis
7.	Adipose tissue	Fat synthesis
8.	Neurons	Dopamine release
9.	Smooth muscle	Increased contractility
10.	Mast cells	Histamine release
11.	Blood platelets	Serotonin release

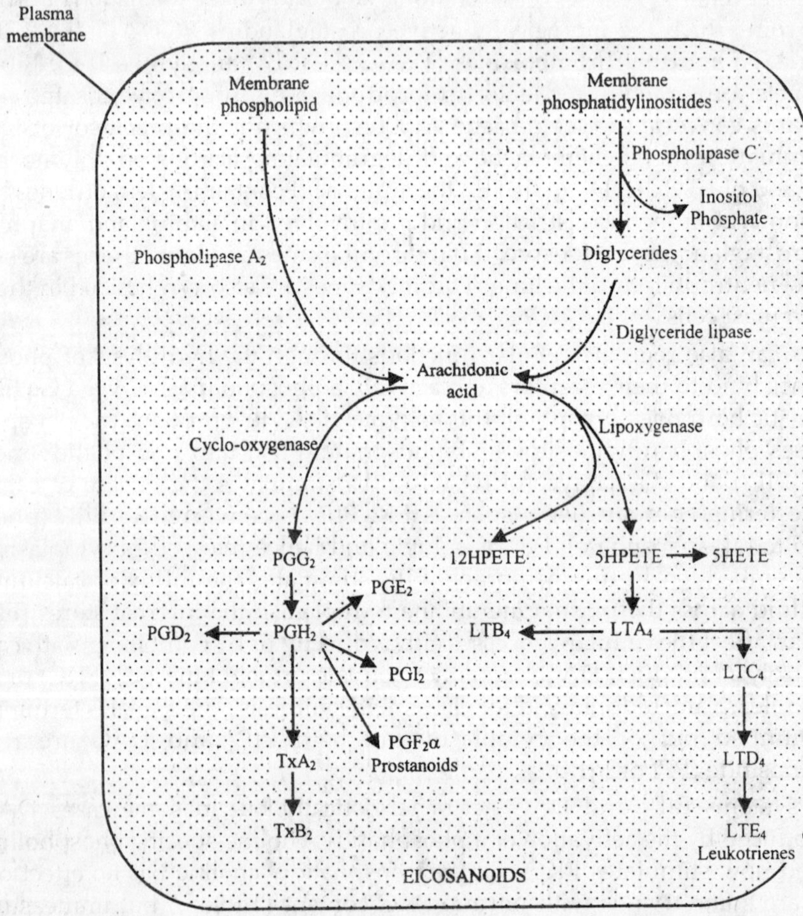

Fig. 15. Lipids in cell signaling. PG = Prostaglandin; LT = Leukotriene; Tx = Thromboxane; HETE = Hydroxyeicosatetraenoic acid; HPETE = Hydroperoxyeicosatetraenoic acid.

PKC activation would be observed since PIP_3 would not be able to act without its receptor. On the other hand, calcium levels can be selectively elevated by GTP or phosphokinase G (PKG). Phosphatidyl choline (PC) is another phospholipid that appears to be involved in hormone action. It can be a source for both arachidonic acid and DAG. It has also been suggested that formation of PC is involved with C-AMP production. The cellular responses to hormone induced rise in phosphatidylinositol 1, 4, 5 triphosphate (PIP_3) and rise in cytosolic calcium in various tissues are shown in Table 7. Table 8 depicts some cellular responses elicited by adding PIP_3 to cells. The Fig. 19a shows the general scheme of phosphoinositide pathway and Fig. 19b shows arachidonic acid cascade.

Table 7. Cellular responses to hormone induced rise in phosphatidylinositol 1, 4, 5-triphosphate (PIP_3) and rise in cytosolic calcium in various tissues.

No.	Tissue	Hormone induced rise in IP_3	Cell response
1.	Parotid salivary gland	Acetylcholine	Secretion of amylase
2.	Liver	Vasopressin	Formation of glucose from glycogen
3.	Pancreas (acinar cells)	Acetylcholine	Secretion of amylase and trypsinogen
4.	Pancreas (β cells of Islets)	Acetylcholine	Insulin secretion
5.	Smooth muscle of stomach	-do-	Contraction
6.	Blood platelets	Thrombin	Change in shape, aggregation
7.	Fibroblast	Peptide growth factors e.g., bombesin	Cell division, DNA synthesis
8.	Mast cells	Antigen	Histamine secretion

Table 8. Some cellular responses elicited by adding PIP_3 to cells.

Cell type	Response
Skeletal muscle	Contraction
Stomach smooth muscle	Contraction
Vascular smooth muscle	Contraction
Lacrimal gland	Increased potassium
Blood platelets	Shape change, aggregation
Sea urchin eggs	Membrane depolarization
Salamander rods	Modulation of light response
Xenopus oocytes	Calcium mobilization, membrane depolarization.

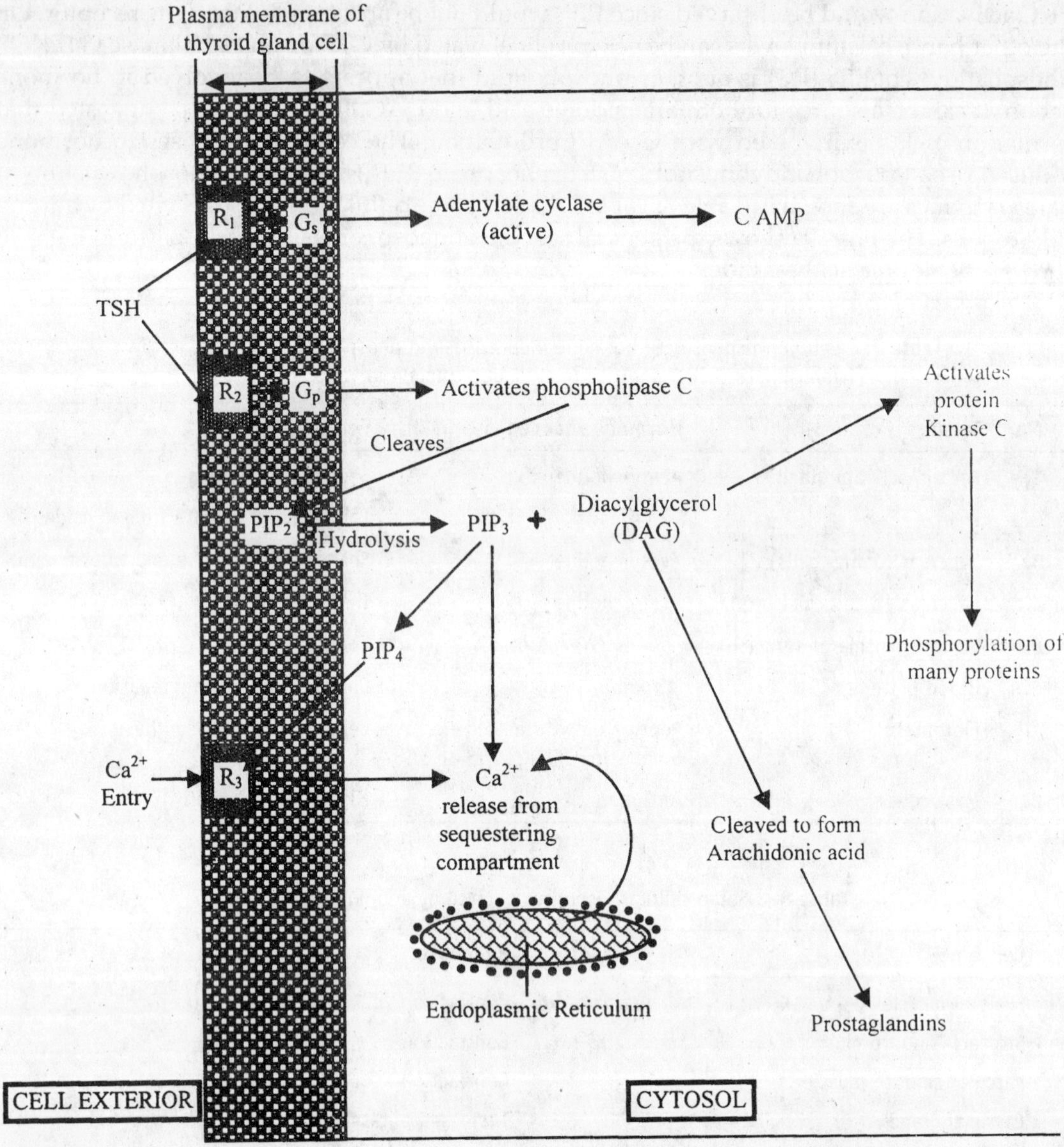

Fig. 16. Action of TSH on Thyroid. R_1, R_2, R_3 = membrane receptors; PIP_3 = Phosphatidylinositol 1, 4, 5 triphosphate; PIP_2 = Phosphatidylinositol 4, 5, biphospate; PIP_4 = Phosphatidyl inositol 1, 3, 4, 5 tetrakisphosphate.

MISCELLANEOUS SECOND MESSENGERS

In addition to the cyclic nucleotides and phospholipids, several other systems are known to affect signal transduction across the plasma membrane. They are polyamines, oligosaccharides and the cytoskeleton. Polyamines are required for DNA replication, transcription, translation

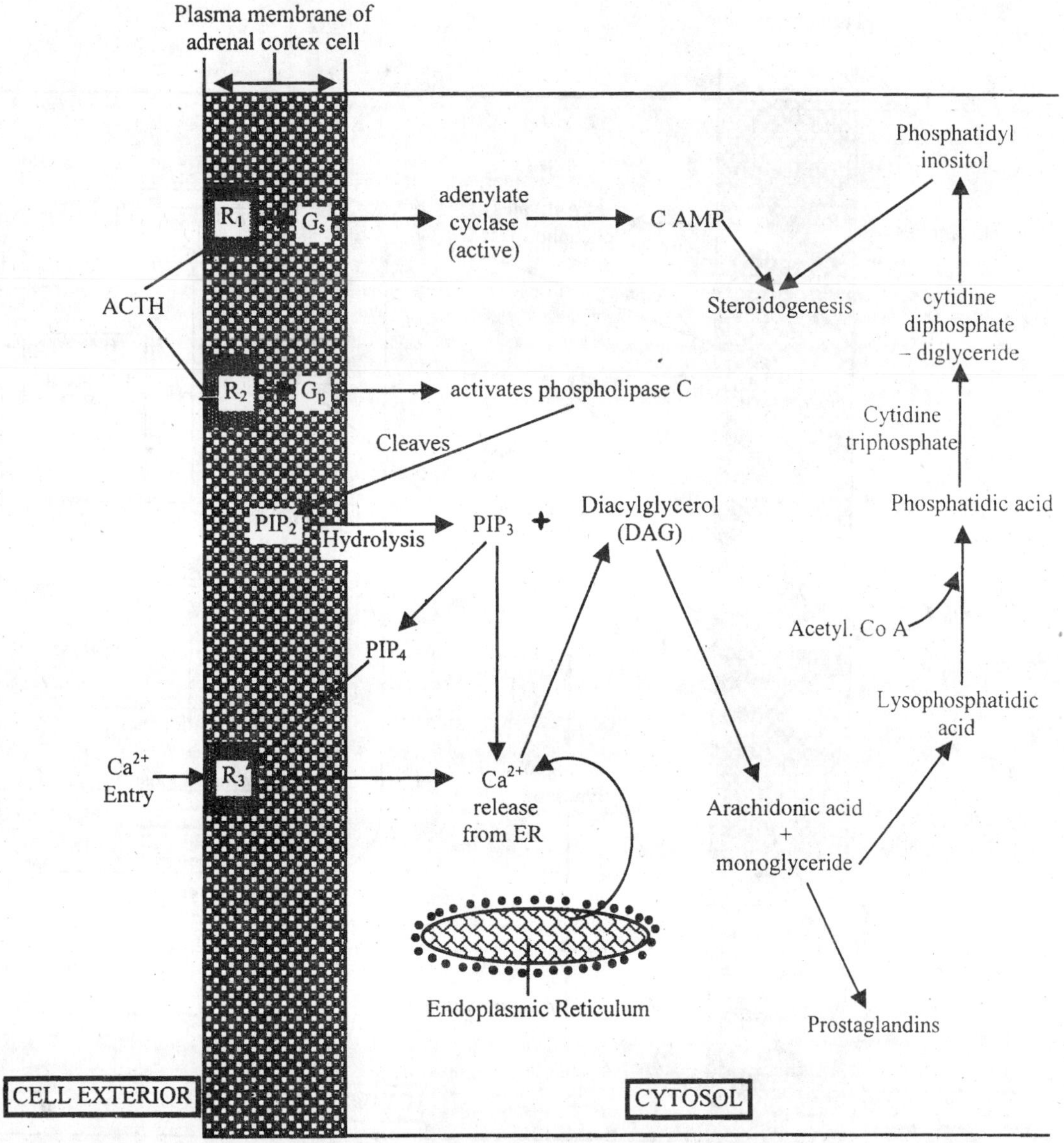

Fig. 17. Action of ACTH on Adrenal Cortex Cell. R_1, R_2, R_3 = membrane receptors; PIP_3 = Phosphatidylinositol 1, 4, 5 triphosphate; PIP_2 = Phosphatidylinositol 4, 5, biphospate; PIP_4 = Phosphatidyl inositol 1, 3, 4, 5 tetrakisphosphate.

and activation of a polyamine dependent protein kinase. The synthesis of polyamines and concentration of the kinase are under hormonal control. The oligosaccharides are released from a phospho-inositol glycan (PI-glycan) by a specific phospholipase C which releases the polysaccharide moiety – the sugar residues which in turn, affect the activities of several

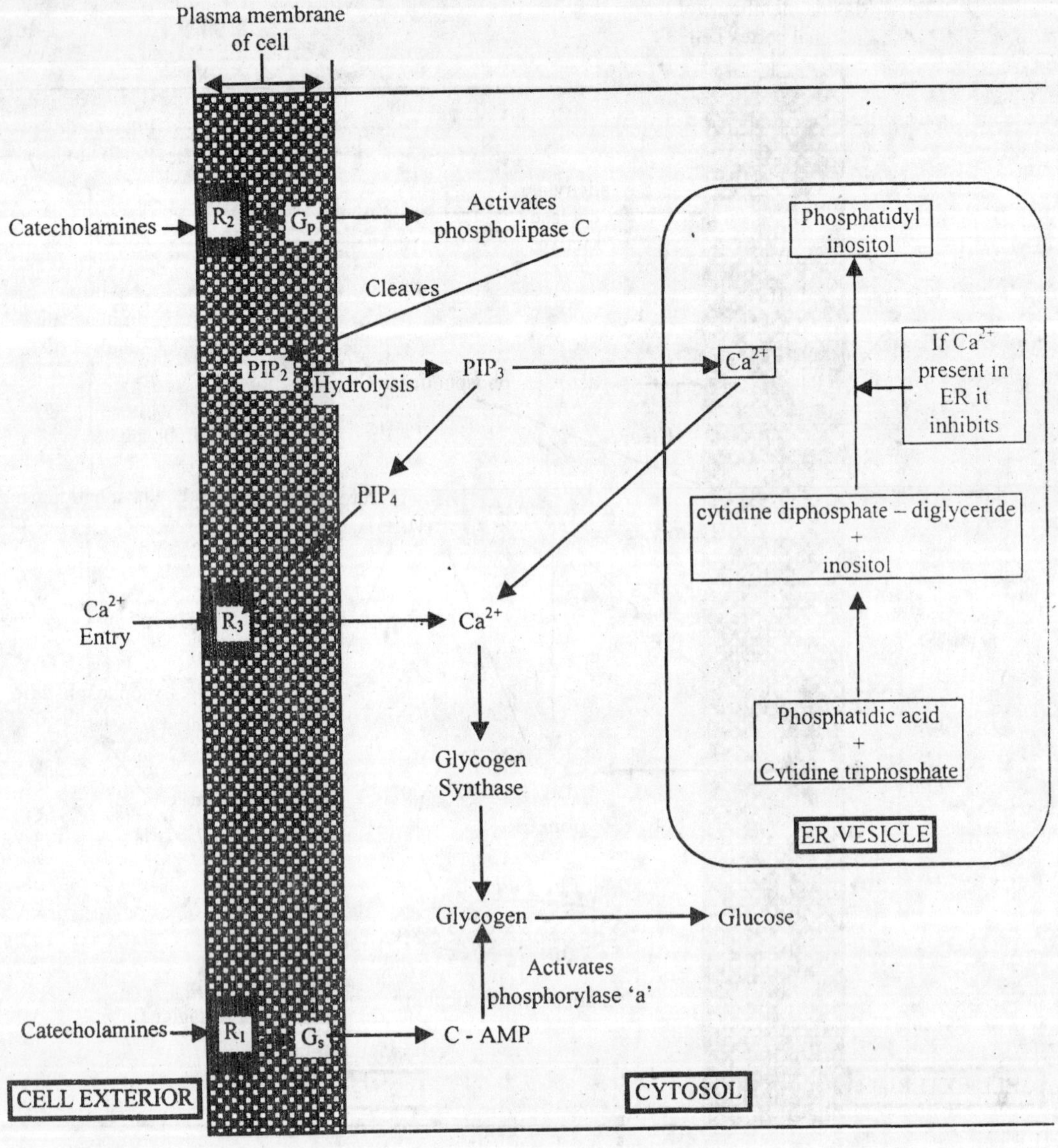

Fig. 18. Phosphatidylinositol turnover in liver cells and adipocytes. R_1, R_2, R_3 = mebrane receptors; PIP_3 = Phosphatidylinositol 1, 4, 5 triphosphate; PIP_2 = Phosphatidylinositol 4, 5, biphospate; PIP_4 = Phosphatidyl inositol 1, 3, 4, 5 tetrakisphosphate.

enzymes. The cytoskeleton (comprising of microtubules, microfilaments, intermediate filaments) is also involved in hormone action. The microfilaments are needed to transport cholesterol into the mitochondria for the human chorionic gonadotrophin (hCG) or adrenocorticotrophic hormone (ACTH) induced side chain cleavage. Cytoskeleton is also responsible in mediating hormone induced changes in cell morphology needed for specific

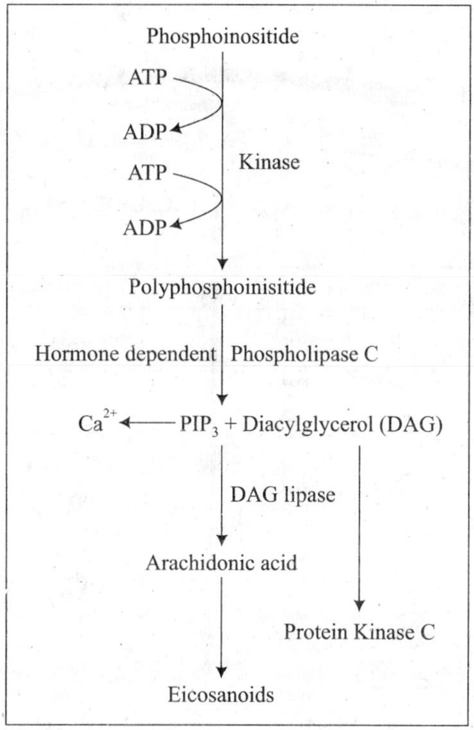

Fig. 19a. General scheme of polyphosphoinositide pathway.

functions. The elements of the cytoskeleton could be altered by concentration of calcium or phosphorylation. Some hormone receptors appear to be associated with these components and may directly affect their structure or function.

REACTIVE OXYGEN SPECIES/ FREE RADICALS AS SECOND MESSENGERS

Numerous factors such as a variety of drugs, chemicals, metal ions, some trace elements, environmental pollutants, etc., are responsible for generating a variety of free radicals (having an unpaired electron) which are highly reactive such as -OH° (hydroxyl radical), O_2^- (superoxide radical), ROO° (peroxyl radical), RO° (alkoxyl radical), RS° (thiyl radical), °NO, °NO_2 (oxides of nitrogen radicals), H_2O_2 (hydrogen peroxide) and HOCL (hypochloride) can cause severe damage to the cell structure especially the membranes. The O_2^-, -OH° and H_2O_2 are known as reactive oxygen species (ROS) and are highly toxic to the cell. They act as second messengers and cause oxidative stress. ROS are involved in cell proliferation and can lead to cancer, apoptosis, cell transformation, cellular injury, DNA damage (DNA strand breaks, deletion and point mutations causing oncogene activation, DNA base mispairing), lipid peroxidation of cell membranes resulting in formation of lipid peroxides. ROS can produce cytokines and growth factors whose role in signal transduction is described later.

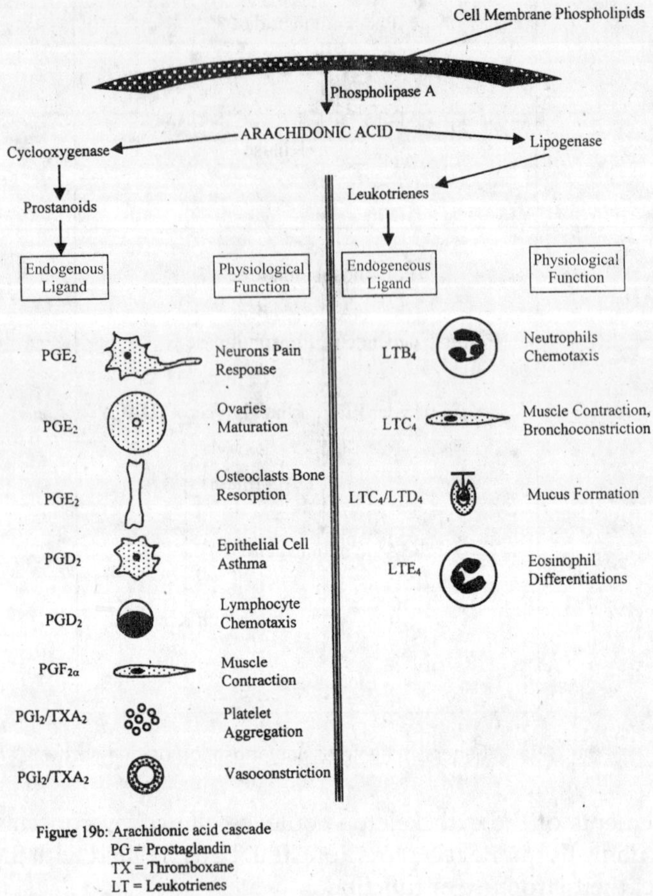

Fig. 19b. Arachidonic acid cascade. PG = Prostaglandin; TX = Thromboxane; LT = Leukotrienes.

PROTEIN PHOSPHORYLATION AND CELL SIGNALING

The phosphorylation of serine, threonine or tyrosine residues by intracellular protein kinases and their dephosphorylation by phosphatases serve as molecular switches which control enzyme activity and regulate most cellular processes (Saltiel, 1994). Many cell surface receptors are linked to enzymes such as phospholipases (PL), adenylate cyclase and guanylate cyclase which produce second messengers, such as phosphoinositol 1,4,5-triphosphate (PIP_3), diacylglycerol (DAG), C-AMP and C-GMP. DAG, C-AMP and C-GMP can stimulate protein kinases directly, while PIP_3 increases the level of intracellular calcium which bind to calmodulin (CAM) and activates protein kinase A and other calmodulin activated kinases. The activated protein kinase A catalyses the ATP dependent phosphorylation of enzymes and other cellular proteins, thus stimulating or inhibiting their activity (Fig. 20). The cell also contains

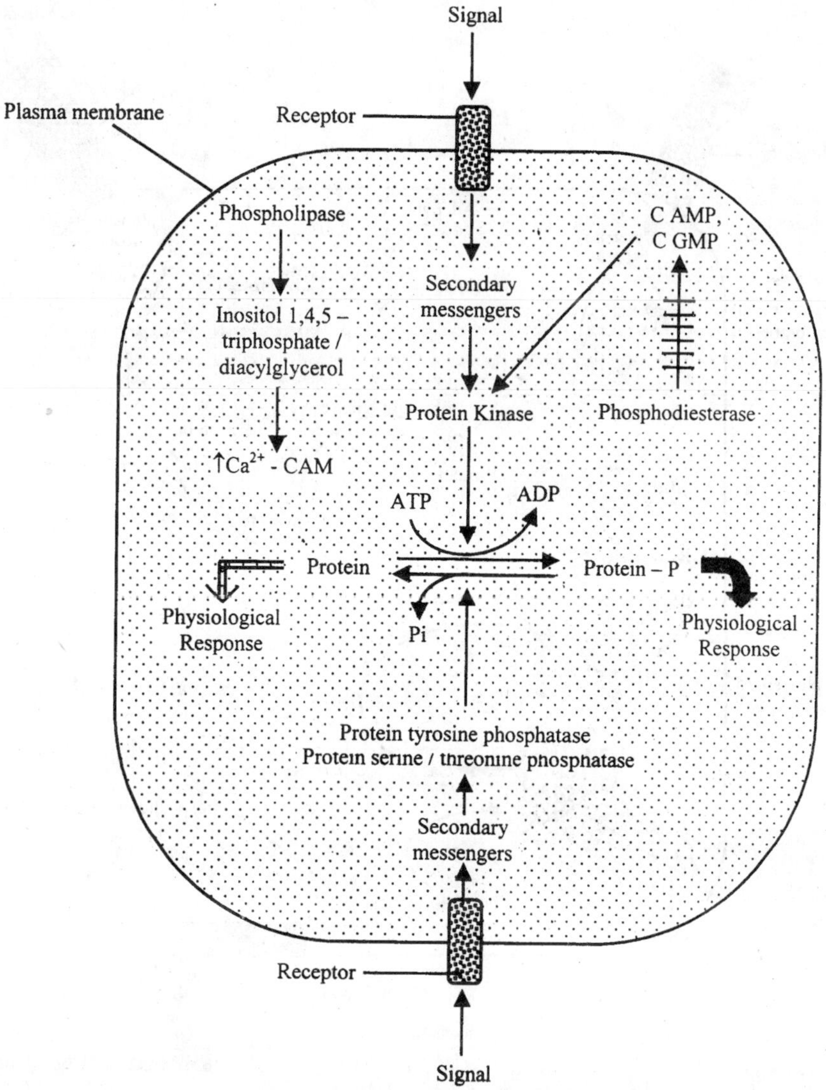

Fig. 20. Protein phosphorylation.

several tyrosine phosphatases and protein serine/threonine phosphatases that dephosphorylate cellular proteins, thus reversing the activity of protein kinases.

ACTION OF CYTOKINES, GROWTH FACTORS AND HORMONES ON TARGET CELL

The cytokines, growth factors and hormones mediate cellular communication. The modulation of cellular and nuclear functions by cytokines, growth factors and peptide/protein hormone is initiated by the activation of cell surface receptors which are linked to intracellular second

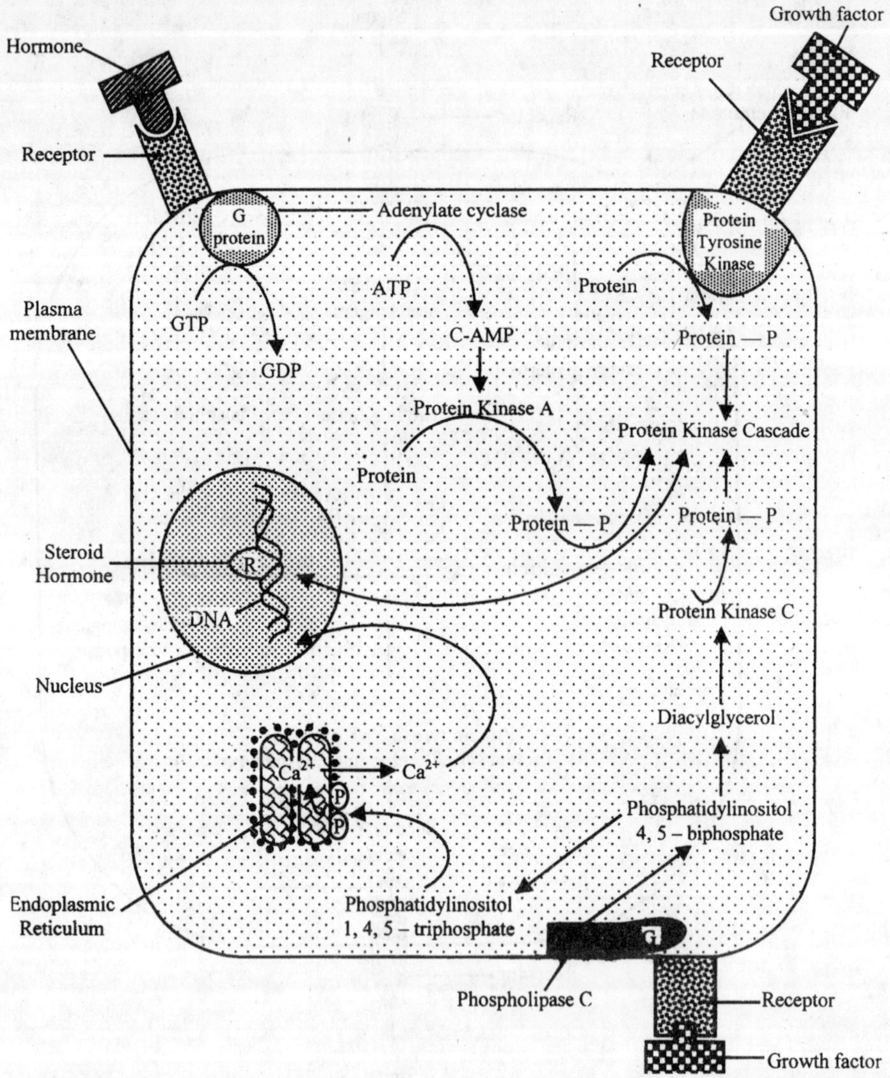

Fig. 21. Action of Cytokines, Growth factors and hormones. R = receptor; G = G protein.

messenger pathways (Fig. 21). Many growth factors bind to receptors that are linked through G-proteins to membrane bound phospholipase C (PLC). Activation of PLC cleaves phosphatidylinositol 4, 5-biphosphate (PIP_2) to form diacylglycerol and phosphatidylinositol – 1, 4, 5-triphosphate (PIP_3). PIP_3 once formed regulates intracellular Ca^{2+} by stimulating its release from the endoplasmic reticulum (ER). Ca^{2+} can bind to calmodulin, a small cytosolic protein, which mediates many cellular effects of Ca^{2+} ions. Each calmodulin molecule binds four Ca^{2+} ions. The binding of Ca^{2+} causes calmodulin to undergo a conformational change which enables the Ca^{2+}-calmodulin (Ca^{2+}-CAM) complex to bind to and activate many enzymes in the presence of cyclic–AMP. One such enzyme which is activated is protein kinase A (PKA) by binding of C-AMP to the regulatory subunit of the enzyme. Diacylglycerol (DAG) activates protein kinase C (PKC).

Other hormone receptors are linked through G-proteins to adenylate cyclase whose activation increases the cellular content of C-AMP and in the presence of Ca^{2+} - CAM complex will activate PKA. Some growth factors and cytokine receptors are protein tyrosine kinases (PTK) that are directly activated by receptor occupation. Activation of PKA, PKC or PTK catalyzes the phosphorylation of other proteins (mainly enzymes) within the cell. Enzymes activated (or inhibited) by phosphorylation may mediate functional processes within the cell while others may enter protein kinase cascades that regulate nuclear events. The steroid hormones, thyroid hormone, vitamin D_3, and retinoids are small lipophilic molecules that easily penetrate the cell and nuclear membrane and bind to intracellular receptors that are transcription factors. These ligand-receptor complexes bind to DNA and regulate gene expression (Fig. 21).

NITRIC OXIDE (NO) AS A SIGNALING MOLECULE

Nitric oxide (NO) is a major signaling molecule in neurons and immune system. The discovery of NO as a signaling molecule in cardiovascular system was done by three scientists, viz., Dr. R.F. Furchgott, State University of New York, Brooklyn, Dr. L.J. Ignarro of University of California, Los Angeles and Dr. Ferid Murad of Texas Medical School, Houston, USA. These three scientists were awarded the Noble prize in 1998 for this discovery.

RELATIONSHIP BETWEEN INFLAMMATION AND NITRATE EXCRETION

In 1985, a mutant strain of mice was isolated whose macrophages lacked the usual ability to phagocytise abnormal cells or bacteria. These mice also displayed an abnormality in nitrogen metabolism. Inflammation resulting from bacterial infection is accompanied by an increased excretion of nitrates in urine of normal mice, no such response was exhibited by the mutant strain. Earlier, the relationship between inflammation and nitrate excretion was totally obscure. However, it was discovered later that the presence of arginine in the medium controls phagocytic activity of normal macrophages in cell cultures. In the absence of arginine, the macrophages were unable to engulf foreign materials. Arginine was found to be a substrate for the enzyme nitric oxide synthase (NOS) that releases nitric oxide (NO) as one of the products of arginine oxidation (Fig. 22). The NO then acts as an intracellular messenger that stimulates the phagocytic activity of the cell. Thus, a substance, NO, that was once considered solely as a toxic pollutant in the atmosphere was found to be an essential physiological messenger. The metabolic deactivation of NO generates the nitrates which appear in the urine of animals whose macrophages are engaged in extensive phagocytic activity during inflammation.

NO is now found to be a signaling molecule in an increasing number of biological activities e.g., NO is produced by the endothelial cells that line the inner surface of blood vessels (Figs. 22a, 22b) in response to acetylcholine which is released by autonomic nerves that innervate the blood vessels. The binding of acetylcholine to the outer surface of the endothelial cell signals a rise in cytosolic calcium concentration which activates NO synthase which acts on arginine. The NO thus formed in the endothelial cell diffuses across its plasma membrane and into the

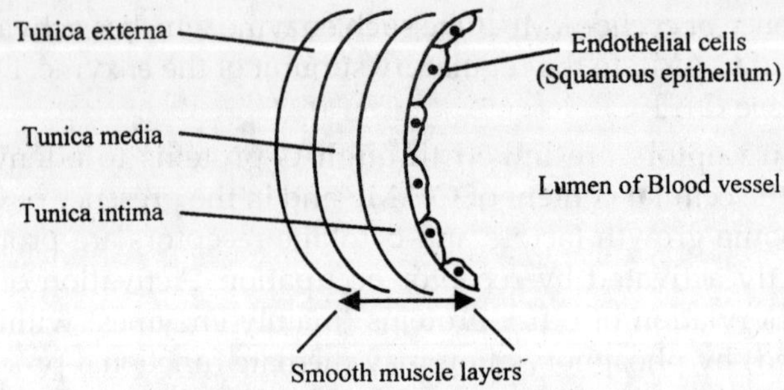

Fig. 22a. Histology of a blood vessel.

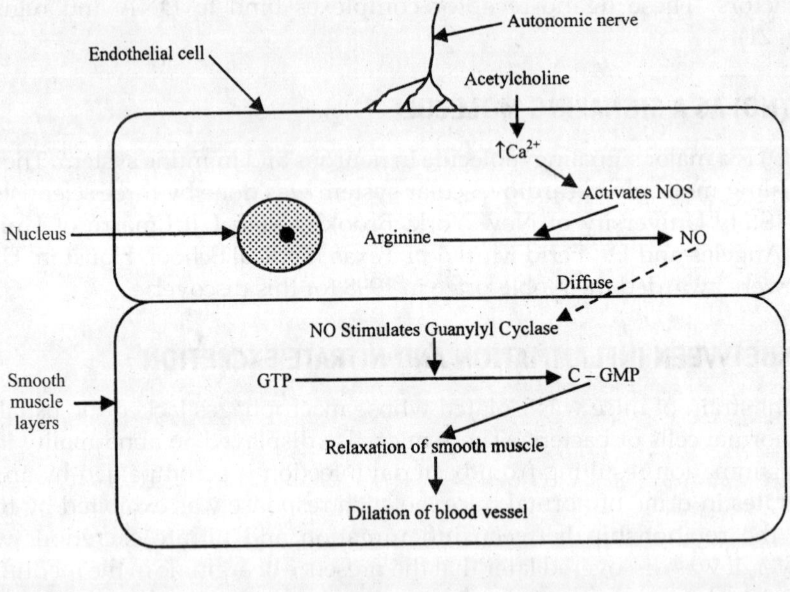

Fig. 22b. Action of NO on blood vessel.

adjacent smooth muscle cells where it stimulates guanylyl cyclase enzyme. The enzyme acts on GTP and produces C-GMP which is an important second messenger similar in structure to C-AMP. C-GMP now triggers the response leading to relaxation of the smooth muscle cells and subsequent dilation of the blood vessel (Fig. 22b).

NITRIC OXIDE (NO) METABOLISM

As described earlier, NO is a major signaling molecule in endothelium of blood vessels, neurons, immune system and other tissues. Within the cell, nitric oxide synthase (NOS) catalyses the conversion of arginine to citrulline and NO in the presence of molecular oxygen, NADPH and flavin cofactors. N-hydroxy-L-arginine is an intermediate in this reaction (Fig. 23). There are two major forms of NOS, constitutive NOS (cNOS) and inducible NOS

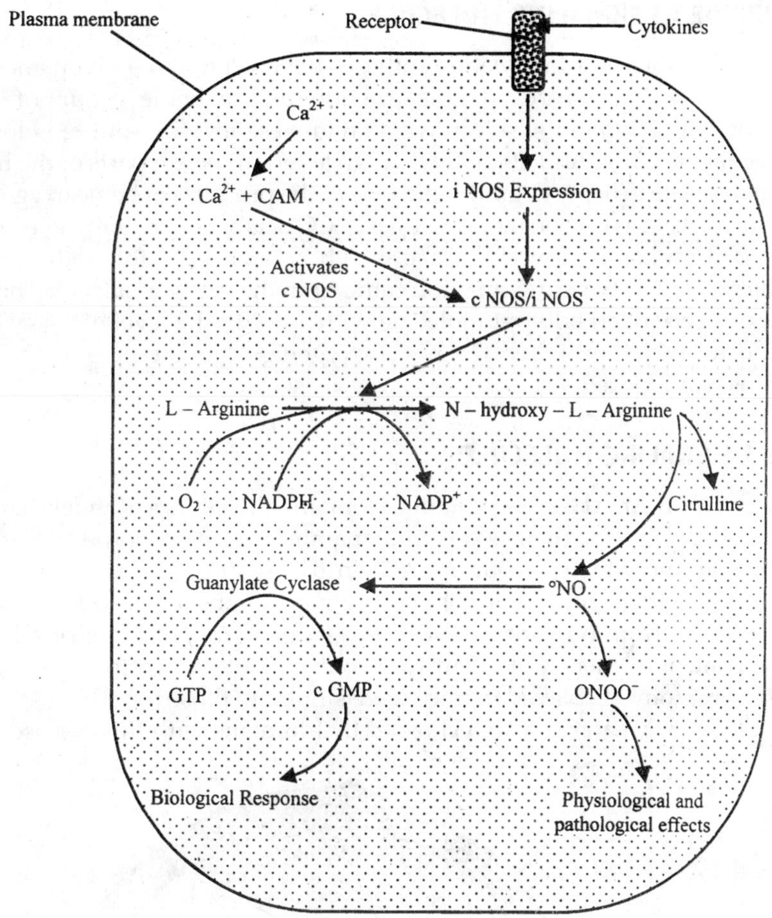

Fig. 23. Nitric Oxide Metabolism. cNOs = Constitutive nitric oxide synthase; iNOS = inducible NOS; °NO = Free radical; CAM = calmodulin; $ONOO^-$ = Peroxynitrite.

(iNOS). cNOS accounts for approximately 95% of the NOS found in neuronal tissue which is designated as nNOS or bNOS and in endothelial cells it is eNOS. cNOS is activated by calcium-calmodulin complex. On the other hand, iNOS or macNOS of macrophages is normally found only at low concentration in the cells, and its levels are induced by cytokines (interferon γ and tumor necrosis factor α) or by lipopolysaccharides.

Nitric oxide (NO) has a half life of only a few seconds *in vivo.* However, as it is soluble in both aqueous and lipid media, it readily diffuses through the cytoplasm and across cell membranes. NO has effects on neuronal transmission at synaptic region in the central nervous system. As described earlier, in the vasculature (blood vessels), NO activates guanylate cyclase, thus increasing the synthesis of C-GMP, which inturn, induces, smooth muscle relaxation and vasodilation. NO toxicity is related to its ability to combine with superoxide radical (O_2^o) to form peroxynitrite ($ONOO^-$), an oxidizing free radical which can damage DNA and other cell components (Fig. 23).

CARBON MONOXIDE AS SIGNALING MOLECULE

The discovery and studies on nitric oxide synthases have led to research on another enzyme, heme oxygenase, which catalyzes the destruction of the heme group of hemoglobin. Hemoglobin is broken down in the spleen which removes old, worn-out red blood cells from the blood. This enzyme was found at relatively high levels in certain parts of the brain where it is not likely to play a role in heme degradation. One of the products of heme oxygenase reaction on heme of old blood cells is carbon monoxide (CO). CO acts as a signaling molecule which acts via G-protein to stimulate guanylyl cyclase which acts on GTP to produce C-GMP. The evidence for the role of CO in neural function came from the study of olfactory neurons which respond to the presence of an odorant molecule with a rise in the level of C-GMP. In animals treated with a drug which inhibited the heme oxygenase, the increase in C-GMP in these olfactory neurons was blocked.

DOPAMINE AS A SIGNALING MOLECULE

Three scientists, viz., Dr. Arvid Carlsson, Paul Greengard and Dr. Eric Kandel shared the Nobel Prize for Medicine in 2000 for their work on Dopamine and its action. Dr. Carlsson's work revealed that dopamine which is concentrated in basal ganglia of the brain (Fig. 24) is a hormone and a transmitter which controls movement of limbs. He also found that blocking dopamine production (from tyrosine → L-DOPA → dopamine) in brain of mice caused the animals to lose control of their limbs, but that control could be restored by administering L-DOPA. These symptoms displayed in mice were similar to those found in patients suffering from Parkinson's disease. Later research showed that humans with this disease also had low

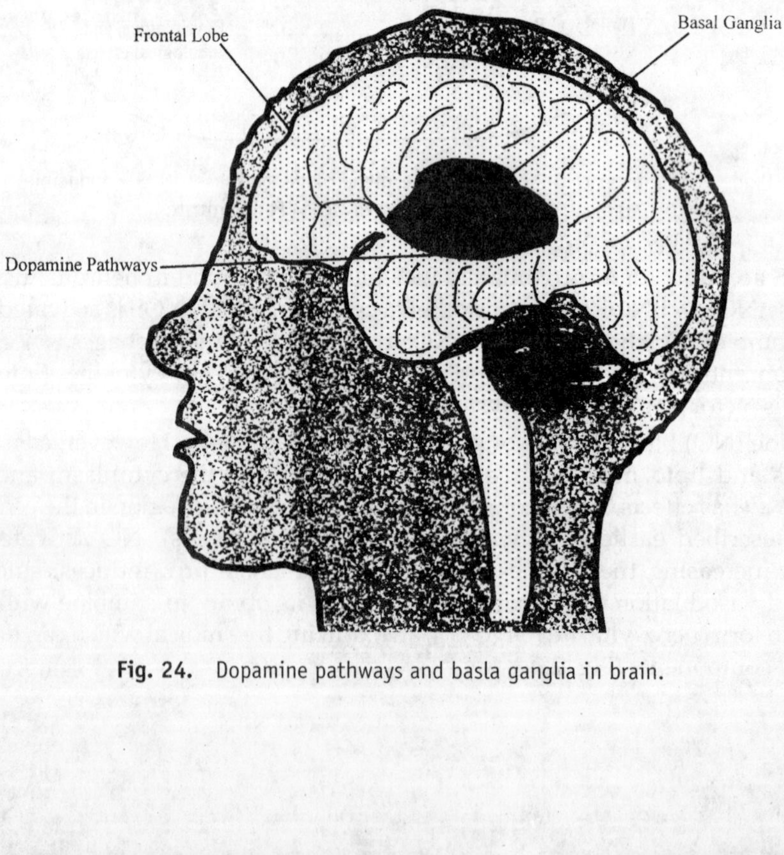

Fig. 24. Dopamine pathways and basla ganglia in brain.

levels of dopamine in their basal ganglia. Carlsson's researches have also shown that antipsychotic drugs, which are used against schizophrenia, also cause blockage of dopamine receptors. Paul Greengard's work has revealed how dopamine and other neuro-transmitters exert their effects at the synapses, the points of contact between individual brain cells. Dopamine is located in tiny vesicles at the end of the nerve axon called end bulb (Fig. 25). When the signal has to pass at the synapse, dopamine is released from the vesicles in the space in the synaptic region. It binds to a specific receptor in the nerve cell and helps in the formation of C-AMP which in turn activates protein kinase A. The activated enzyme affect certain key proteins which in turn regulate a variety of functions in the nerve cell. The proteins become modified as phosphate groups are added (phosphorylation) or removed (dephosphorylation) which causes a change in shape and function of the protein. It can open ion channels or enter the nucleus to produce new proteins (Fig. 26). By this mechanism the transmitter can carry their message from one nerve cell to another.

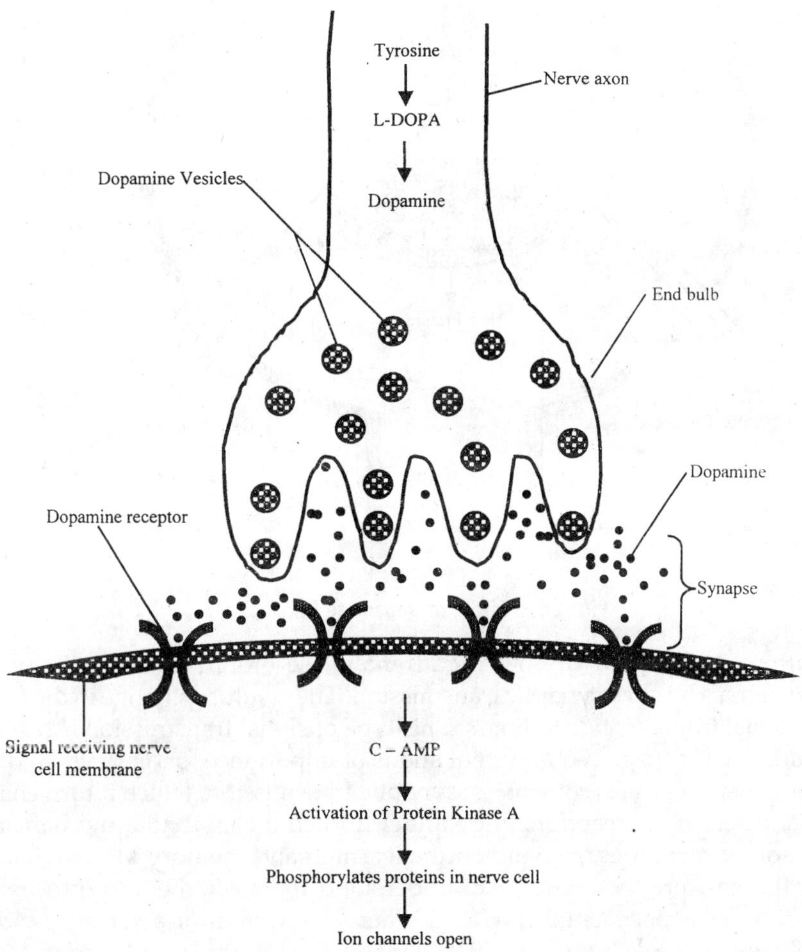

Fig. 25. Transmission of message in nerve cell.

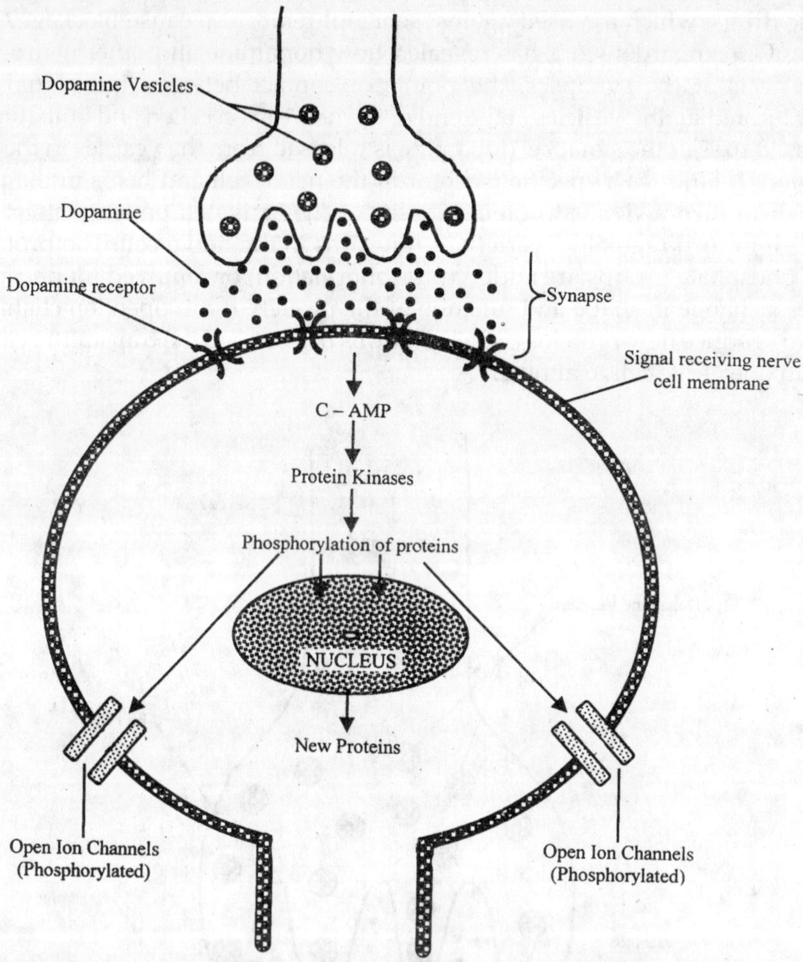

Fig. 26. Effect of Dopamine in nerve cell.

Transmitters such as dopamine, noradrenalin, serotonin and certain neuropeptides transmit their signals by slow synaptic transmission. The resulting change in the function of the nerve cell may last from seconds to hours. This type of signal transmission is responsible for a number of functions in the nervous system and is of importance for alertness and mood. Slow synaptic transmission can also control fast synaptic transmission, which inturn enables speech, movements and sensory perception. Eric Kandel studied the molecular mechanisms involved in the formation of memories. He studied the learning and memory in mammals, but found that the conditions were too complex to understand the basic memory processes. He then turned to a simpler experimental model, the nervous system of a sea slug, *Aplysia.* It has comparatively few nerve cells (around 20,000) many of which are large as compared to the 100 billion in the human brain. *Aplysia* has a simple protective reflex which protects the gills, which can be utilized to study basic learning mechanisms.

SHORT AND LONG TERM MEMORY

Eric Kandel showed that weaker stimuli gave rise to a form of 'short term memory which lasted from minutes to hours. The mechanism is that particular ion channels are opened in the cytosol so that more calcium ions will enter the nerve terminal. This leads to an increased amount of transmitter (dopamine) release at the synapse, which causes an amplification of the reflex (Fig. 27). A strong and long lasting stimulus will result in a form of long term memory which can remain for weeks. The strong stimulus will give rise to increased levels of the second messenger, C-AMP causing activation of protein kinase A. These will phosphorylate different proteins some of which will reach the nucleus (Fig. 27) and cause a change in a number of proteins in the synapse. The formation of certain proteins will increase, while others will decrease. The net result is that the shape of the synapse can increase and thereby create a long

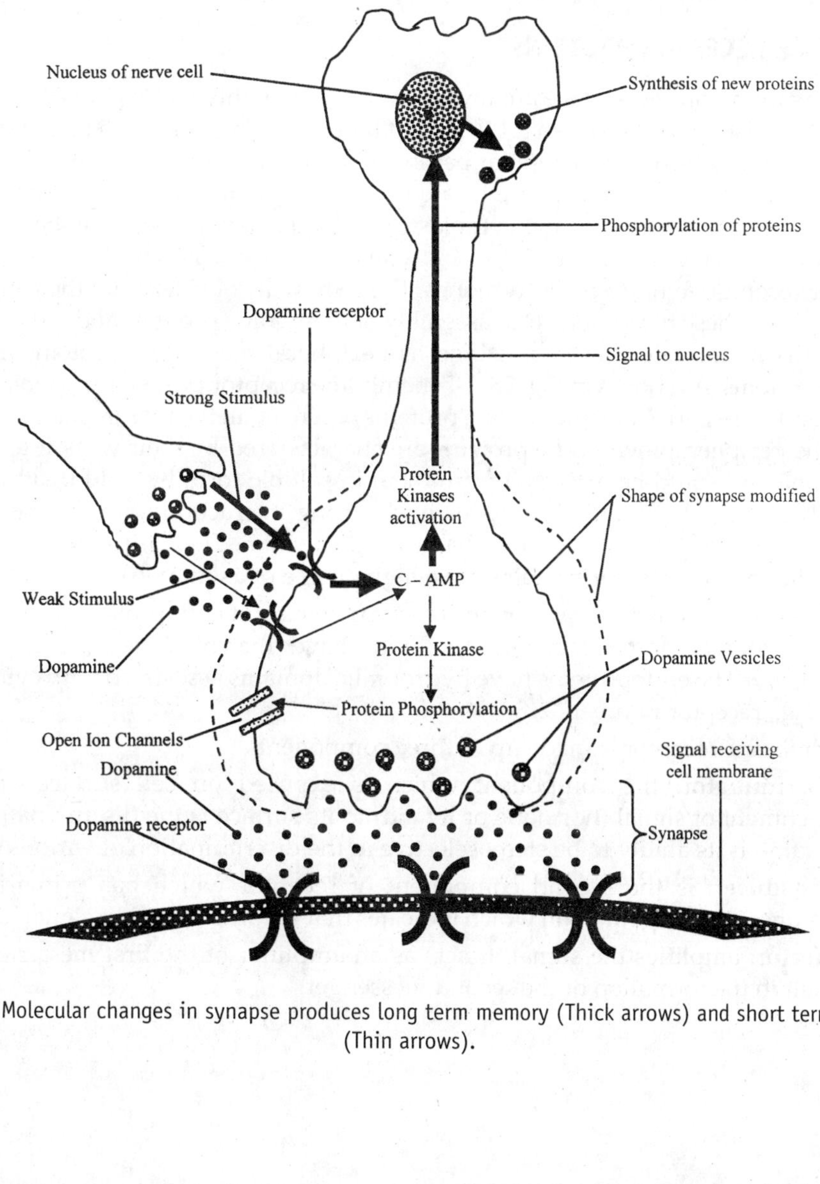

Fig. 27. Molecular changes in synapse produces long term memory (Thick arrows) and short term memory (Thin arrows).

lasting increase of synaptic function. In long term memory new proteins are synthesized. If this synthesis of new proteins is prevented, the long term memory will be blocked but not the short term memory.

It is evident from the work of Eric Kandel that short term memory and long term memory are located at the synapse in sea slug. Similar studies were carried out by him in mice during the 1990's wherein he demonstrated that the same type of long term changes of synaptic function which are seen during learning in the sea slug also applies to mammals and humans. Our memory is located in the synapse and changes in synaptic function are central, when different types of memories are formed. Since we now understand important aspects of the cellular and molecular mechanisms which make us remember, the possibilities to develop new types of medication to improve memory function in patients with different types of dementia might be increased.

CELL SURFACE RECEPTOR PROTEINS

The receptors found on cell surface are proteins. They are highly specific for each hormone or ligand and can distinguish between a 17α or 17β hormone. This is known as stereospecificity. They are synthesized through the regular pathway of protein synthesis. All hormone receptors (for protein and steroid hormones) are located in plasma membrane, cytosol or the nucleus. The cytosolic receptors found in the cytoplasm are larger having 8-10 Svedberg units while nuclear ones are smaller (4-6 S). The cytosolic could be transformed to the nuclear receptors. The nuclear receptors bind steroid hormones. Here we will consider only the cell membrane bound receptors. These receptors are an assembly of membrane proteins and according to their location in the membrane can be classified as peripheral membrane proteins and integral membrane proteins. As shown in Fig. 28, 1-7 membrane receptor proteins are depicted. Protein numbers 3 and 6 are peripheral membrane proteins whereas, all the rest are integral membrane proteins. The peripheral membrane proteins are characterized by their weak interaction with the membrane and could be extracted from them in soluble form by mild treatment and are lipid free. The integral membrane proteins have strong interaction with the membrane and have to be treated with strong organic solvents for extraction. They are associated with lipids and are insoluble. The receptor proteins have high degree of asymmetry in their structure.

Another classification of these proteins is ectoproteins and endoproteins. The ectoproteins have their hydrophilic domains beyond the lipid bilayer matrix of membrane. e.g., receptor numbers 1, 4, 5, 7. The endoproteins have hydrophilic domains associated with cytosolic site of membrane e.g., receptor numbers 2, 3.

The membrane receptor is made up of three components

(a) **Discriminator:** the component which is exposed on cell surface and initiates attachment of signal (hormone or ligand) to its surface is the discriminator. Its major function is its ability to be stereoselective in the discrimination of various signals.

(b) **Transducer:** is the second component of receptor which has primarily a signal response coupling function which indicates that binding of proper signal has occurred.

(c) **Effector:** amplifies the signal. It acts as an amplifier of the first messenger and will result in the formation of the second messenger.

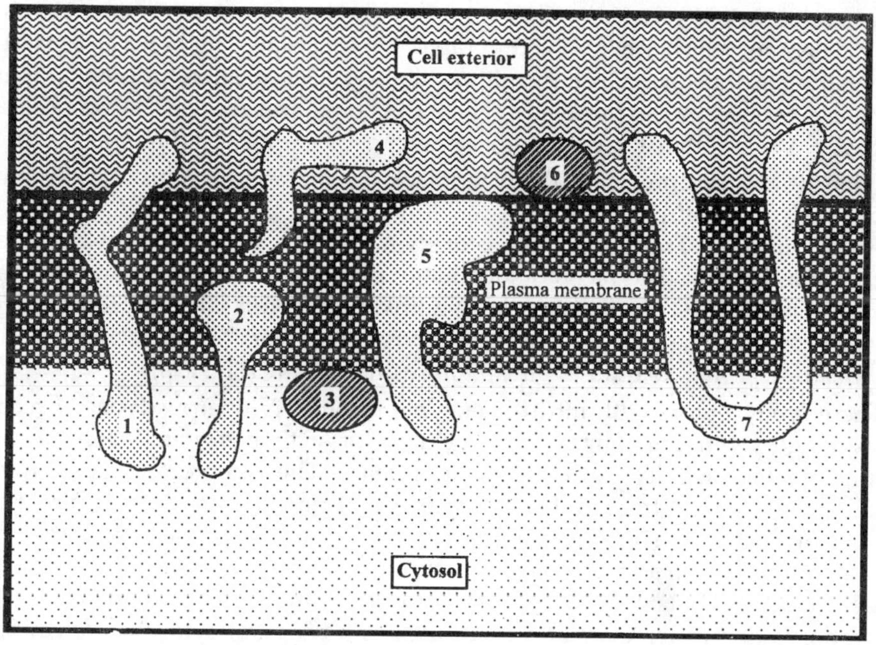

Fig. 28. 1–7 Membrane receptor proteins. 3, 6–Peripheral membrane proteins; 1, 2 4, 5, 7–Integral membrane proteins.

The carbohydrates in the membranes can be linked to each other in different ways yielding a large number of isomeric structures which have got the potential to provide a wide variety of recognition signals. These carbohydrates which are important in receptor recognition have regions known as glycoconjugates which could either attach to lipids or proteins. Some of the carbohydrates involved in receptor recognition are sialic acid, D-mannose, D-mannose-6-phosphate, D-galactose, N-acetyl-D-glucosamine etc.

Classification of cell surface receptors is based on the transduction mechanism they use. It has been mentioned earlier that all water soluble signal molecules (including neurotransmitters and all signal proteins) bind to specific receptor proteins on the surface of the target cells which they influence. These cell surface receptor proteins act as signal transducers. Hence, they convert an extracellular ligand-binding into intracellular signals that alter the behavior of the target cell. The different receptor proteins are:

G-protein linked receptors: They act indirectly to regulate the activity of a plasma membrane bound target protein, which can be either an enzyme or an ion channel. The interaction between the receptor and target protein is due to mediation of a G protein (trimeric GTP binding protein). The action of G-proteins has been described earlier in this chapter. The activation of target protein could change the concentration of the intracellular mediators-enzymes or ion channels. All of the G-protein linked receptors belong to a large family of homologous, seven-pass trans-membrane proteins (Fig. 5a). The hormones epinephrine and glucagon as well as serotonin act via G-protein linked receptors.

The ion-channel linked receptors: They are also called as transmitter-gated ion channels or ionotropic receptors which are involved in rapid synaptic signaling between nerve cells. This type of signaling is done by neurotransmitters like acetylcholine that transiently open or close an ion channel formed by the protein to which they bind and change the ion permeability of plasma membrane and thus the excitability of the post-synaptic cell (Fig. 10a, b, 26).

The enzyme linked receptors: (a) Receptors lacking intrinsic catalytic activity but directly associated with cytosolic protein tyrosine kinase. The ligand binding causes receptor monomers to dimerize. The dimeric receptor then interacts with and activates one or more cytosolic protein kinases. The receptors for many cytokines, interferons and human growth factor are of this type. These tyrosine kinase-linked receptors are referred to as cytokine-receptor superfamily. The activated tyrosine kinase phosphorylates tyrosine in receptor and help the substrate protein to bind to phosphotyrosine residue (Fig. 29a).

(b) There are receptors with intrinsic enzyme activity: Numerous types of receptors fall under this category. Binding of ligand to some receptors activates guanylate cyclase activity or protein phosphatase activity in the cytosolic domain. The receptors for insulin and many growth factors are ligand-triggered protein kinases. In most cases, ligand binding leads to dimerization and activation of either serine/threonine kinase or tyrosine kinase activity. These receptors are often known as receptor serine/threonine kinases and receptor tyrosine kinases (RTKs) which phosphorylate residues in the cytosolic domain of the receptor itself (Fig. 29b). Autophosphorylation of RTKs creates binding sites for several cytosolic enzymes, bringing them near their substrates in the plasma membrane (Fig. 6d). In some cases these enzymes

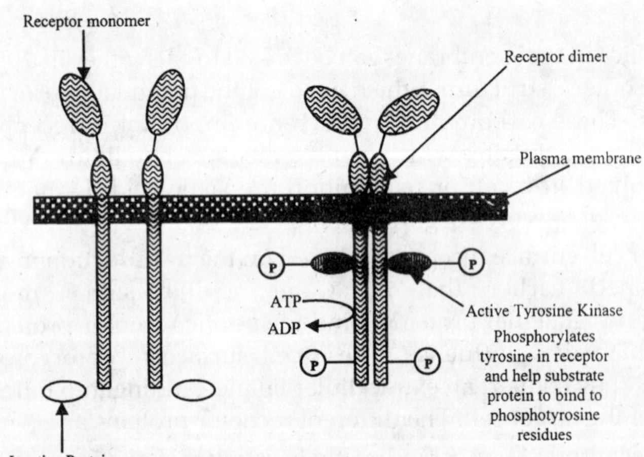

Fig. 29a. Tyrosine Kinase – Linked receptor.

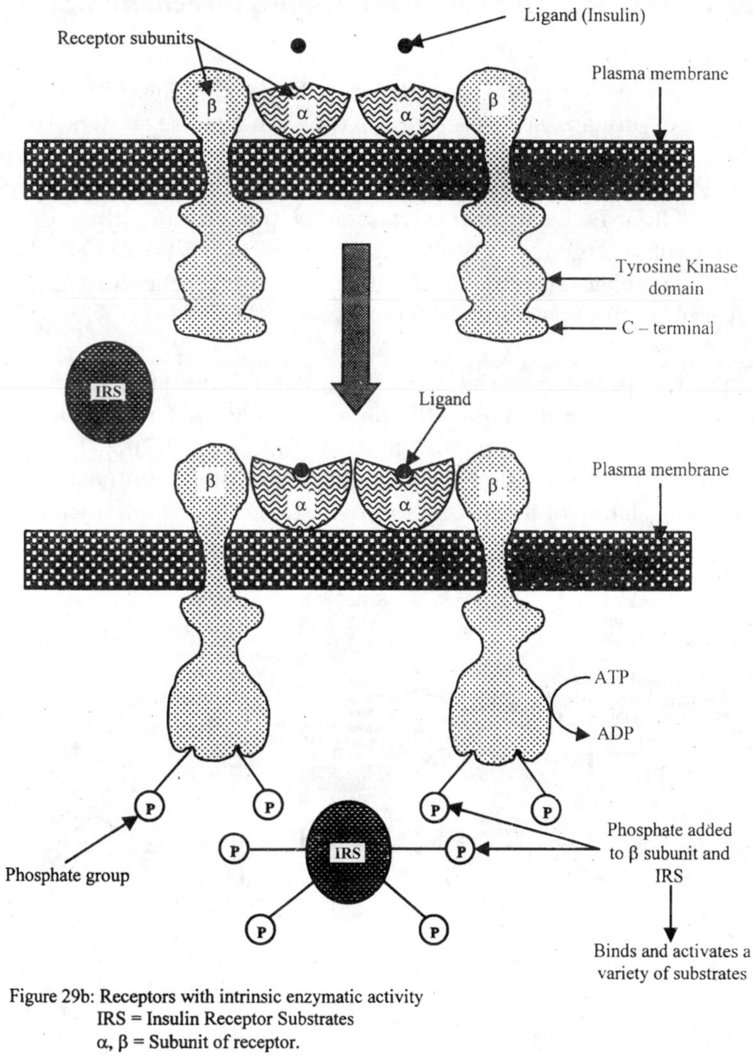

Fig. 29b. Receptors with intrinsic enzymatic activity. IRS = Insulin Receptor Substrates. α, β = Subunit of receptor.

generate second messengers. RTKs can phosphorylate various substrate proteins and alter their activity.

The insulin receptor: The binding of insulin to the insulin receptor occurs within seconds, then receptor autophosphorylation and activation of receptor protein tyrosine kinase occurs. The insulin receptor is a tetrameric protein composed of two α and two β-subunits. When insulin binds to the α-subunits it causes a conformational change in the β-subunit, which activates tyrosine kinase activity of the β-subunit. This in turn catalyze the transfer of phosphate groups to tyrosine residues located in the cytoplasmic domain of the receptor as well as to a variety of insulin receptor substrates (IRS). Next they will bind and activate numerous effectors (Fig. 29b).

CONVERGENCE, DIVERGENCE AND CROSSTALK AMONG DIFFERENT SIGNALING PATHWAYS

In Fig. 30a, b, the outlines of signal transduction pathways initiated by receptors that act via both heterotrimeric G proteins and receptor tyrosine kinases are shown. The two are seen to converge by the activation of different phospholipase C isoforms, both of which lead to the formation of the same second messenger, PIP_3 and DAG. The activation of receptor tyrosine kinase by either PDGF or EGF causes transmission of signals along three different pathways, which is an example of divergence. Crosstalk between the two types of pathways is depicted by calcium ions which are released from the smooth endoplasmic reticulum (SER) by PIP_3 and can then act on various proteins, including protein kinase C (PKC) whose activity is also stimulated by DAG.

Converging signaling pathways: The action of insulin provides the best example. Insulin binds to an insulin receptor and triggers the phosphorylation of an insulin receptor substrate (IRS) (Fig. 29b) and the activation of PI-3 hydroxyl kinase [PI(3)K] (Fig. 30b). Insulin also triggers other cellular responses such as activation of protein synthesis and stimulation of growth. The phosphorylation of the insulin receptor substrate forms a docking site for Grb_2, the

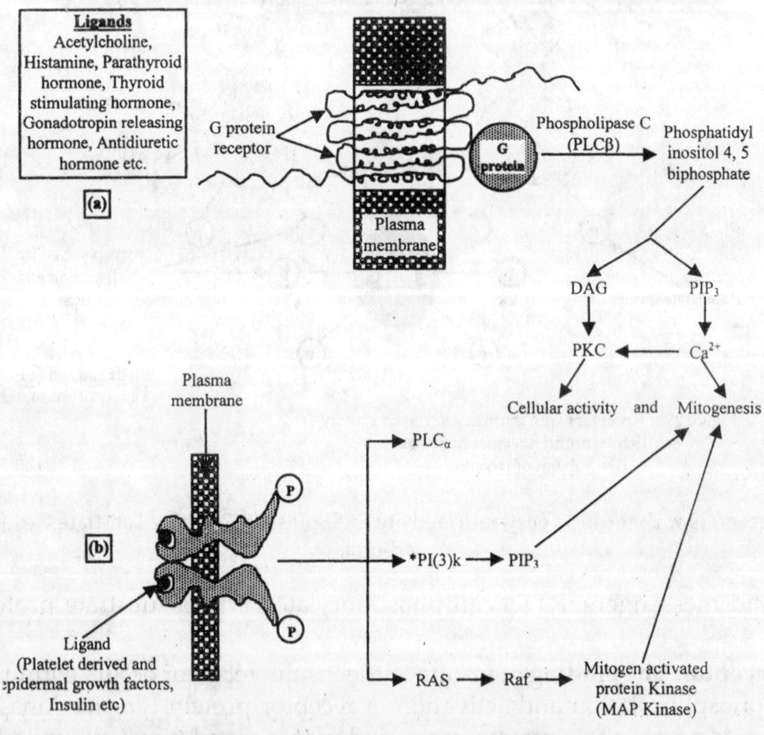

Fig. 30. (a) G-protein linked receptor. (b) Tyrsine Kinase linked receptor. PIP_3 = Phosphatidylinositol 1, 4, 5 triphosphate; PI = Phosphatidyl inositol; PKC = Protein Kinase C; DAG = Diacylglycerol; *PI(3)k = PI – 3 hydroxyl kinase which adds phosphate group to 3′ position of sugar of PI; RAS = Product of ras gene. It is a small G protein at the inner surface of plasma membrane; Raf = A protein bound by RAS-GTP to plasma membrane; Ⓟ = Phosphate group; PLCα, β = Phospholipase C.

same SH_2 protein that binds to the phosphorylated EGF or PDGF receptor. Hence, the signaling pathways initiated by the binding of the three ligands EGF, PDGF and insulin converge to activate Ras and MAP kinase cascade leading to transcription and translation of a similar set of growth promoting genes in the target cell (Fig. 6a, 30b).

Divergent signaling pathways: It is known that binding of growth factors (EGF or PDGF) to their respective receptor leads to autophosphorylation of the receptor (Fig. 6d) and recruitment of the SH_2 protein Grb_2 to the plasma membrane (Fig. 6a, b). Grb_2 is not the only protein containing a phosphotyrosine binding SH_2 domain which is capable of binding to a phosphorylated growth factor receptor. Two other SH_2 proteins, PI(3)K and phospholipase C, could also bind to these receptors (Fig. 30b), and get activated. PI(3)K is an enzyme which phosphorylates phosphatidylinositol (PI). The enzyme, phospholipase C hydrolyzes phosphorylated phosphatidylinositols into PIP_3 and DAG, which are both second messengers. Hence different (divergent) pathways are possible (Fig. 30b).

Crosstalk between signaling pathways: One example involves C-AMP. We have noted earlier in this chapter that C-AMP is an initiator of a cascade of reactions leading to release of glucose in the blood (Fig. 7). Recent studies have revealed that C-AMP can also inhibit the growth of numerous different cells such as fibroblasts and adipocytes (fat cells) by blocking the signals transmitted through the MAP kinase cascade. This is done by activating the C-AMP dependent kinase (PKA), which could phosphorylate and inhibit Raf.

The second example of crosstalk is the interaction between calcium and C-AMP, both of which function as second messengers causing distinct responses in target cells (Figs. 13, 14). Calcium-calmodulin complex could activate both adenylate cyclase (which synthesizes C-AMP) and C-AMP phosphodiesterase (which degrades C-AMP). On the contrary, C-AMP dependent protein kinase can phosphorylate calcium ion channels, thus altering their ability to release calcium into the cytosol. Moreover, a number of proteins such as transcription factor CREB (Fig. 7), glycogen synthase and phosphorylase kinase which play important roles in glucose metabolism, (Fig. 7), could be phosphorylated by both C-AMP dependent kinase and calcium-calmodulin dependent kinase. Thus it is clear that the signaling pathways of a cell are highly interdependent.

MEMBRANE RECEPTOR METABOLISM AND RECYCLING

Most of the protein hormones bind membrane receptors to elicit their actions. After binding the hormone receptor complex is metabolized. This process helps to terminate the signal leading to bring about hormone action. Cells have two functionally distinct classes of receptors; one for hormones and one for nutrients. Although some overlap exists in the way that each group is processed, there are few general differences that can be observed with respect to clustering, metabolism, internalization and recycling.

Clustering: After binding the hormone receptor complexes begin to aggregate into coated pits. These each containing 50-150 mm in diameter are formed by invagination of plasma membrane surrounded by a basket or cage of polygonal units. Each unit of this cage is triskelion or clathrin trimer having three 180 KDa clathrin molecules and three 33-36 KDa light chain or clathrin associated proteins. The former form the legs, while the latter act as a glue. Each leg is 445°A long with a bend 190°A from the vortex. The legs align with each other to form hexagon or

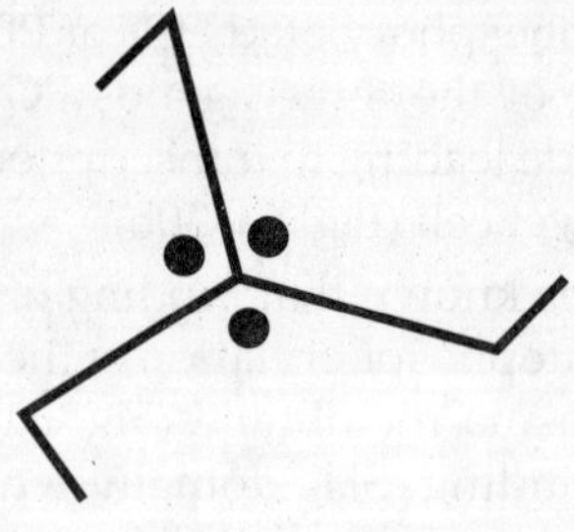

Fig. 31. Clathrin trimer of Triskelion.

pentagon as shown in Fig. 31. After aggregation, the pits bud off and quickly loose their clathrin coat. The coat is important for endocytosis. Injection of anticlathrin antibodies into cells inhibits this process. Clustering is highly variable and is essential for the biological activity of the hormone.

Internalisation and processing: In nutrient pathway the receptors are being continuously internalised and recycled even when unoccupied, as nutrients are present in the internal environment at enough concentrations to make such a system worthwhile. But in the hormone pathway, internalisation is ligand induced. For example, epidermal growth factor (EGF) increases internalisation 10 fold. The two pathways also differ in their processing of the internalised ligand-receptor complex. In the nutrient pathway, the endocytotic vesicle called as endosome or receptosome becomes acidified via an ATP dependent mechanism. At the same time, slender tubules arise from the surface of the endosome. This process is called compartment for uncoupling receptors and ligands (CURL). Due to acidification, ligand-receptor complex disrupts and unoccupied receptors migrate into the tubules while the free ligand remains in the vesicle (Fig. 32).

Actually segregation process involves simple diffusion, since receptors are membrane bound and ligands are soluble. However, some evidence suggests that the separation includes more active process. The tubules then detach and return to the plasma membrane, where receptors are recycled. Finally, the vesicle fuses with lysosomes and ligand is degraded. This pathway is called as the short loop. Similarly hormone receptor complex are treated excet that the segregated receptors first go to Golgi apparatus before returning to the plasma membrane. This is called long loop. Temporary storage, repair and/or purification from other membrane components of receptors have been postulated. For this reason in another less common variation, some hormone receptors are not recycled at all. For example, EGF receptors are totally degraded and is related to the tyrosine kinase activity of the receptor. Mutant receptors without such activity are internalized and also recycled. This phenomenon may also show tissue specificity of the hormone receptor. Insulin receptors in liver are recycled but those in adipose tissue are not. Thus tissue do not respond to the additional insulin hormone till they synthesize new receptors. This phenomenon is termed as refractory period of target tissue.

SUMMARY

The process of cell signaling and signal transduction has been described. The chapter describes the types of cell signaling found in different cells and their importance. This is followed by the

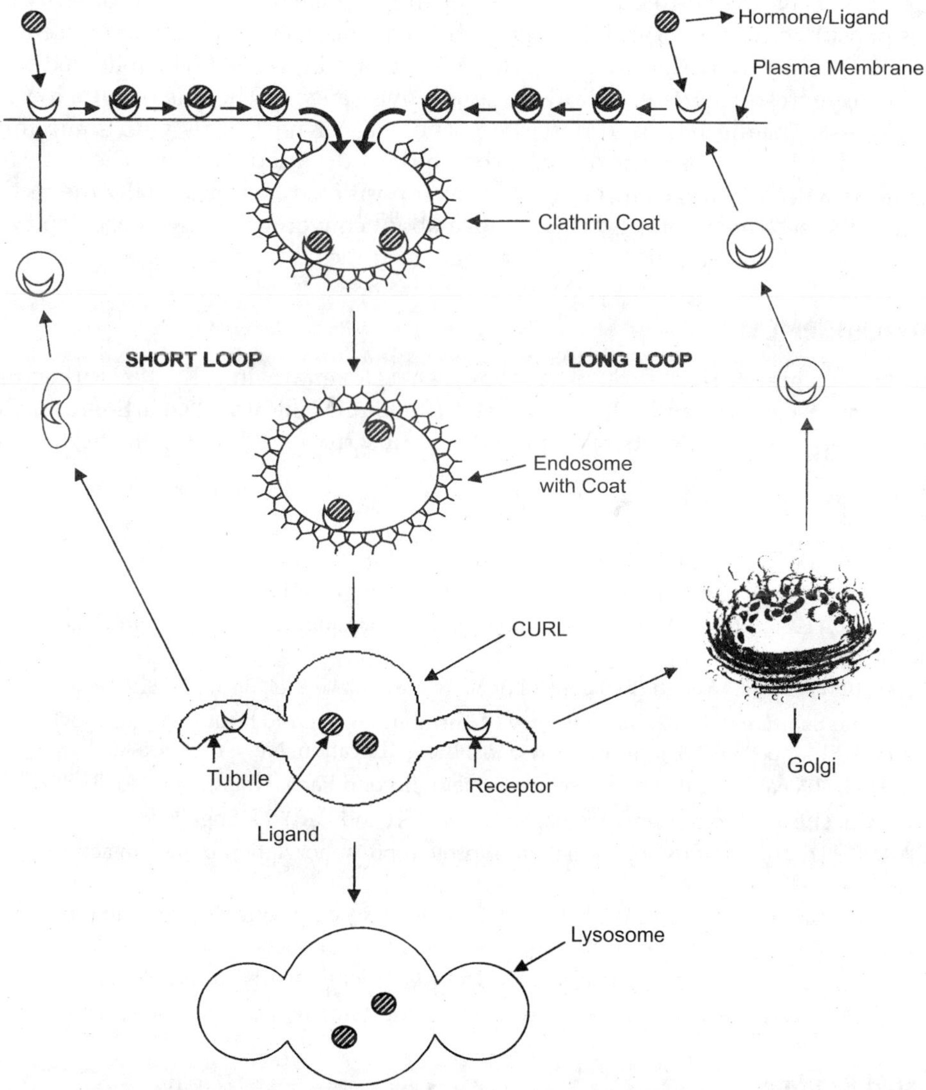

Fig. 32. Receptor Metabolism and Recycling.

work of Earl Sutherland and Martin Rodbell who have deciphered the mechanism of action of protein hormones. Rodbell's important discovery of G-proteins and cell signaling has paved the way in understanding the occurrence of several disorders associated with defects either in the G-protein itself or in its receptor. The next part deals with different second messengers such as Cyclic AMP, Cyclic GMP and Cyclic CMP which are known as cyclic nucleotides. Other second messengers such as calcium alone and its interaction with C-AMP in cell signaling has been described. Next, the role of lipids as second messengers has been described which includes phosphatidylinositol and its analogues. Some miscellaneous second messengers such as polyamines, oligosaccharides and the cytoskeleton have been dealt with followed by the role of reactive oxygen species/free radicals as second messengers. The role of cytokines, growth factors in cell signaling has been described. The chapter next deals with some important signaling molecules such as nitric oxide, carbon monoxide and dopamine. The last part of the chapter deals with cell surface receptors, their composition, types and finally the receptor for insulin and its mechanism of action. Some idea about convergence, divergence and crosstalk among different signaling pathways has been given at the end.

ACKNOWLEDGEMENTS

The author is grateful to Prof. M.V. Rao, Dr. Devendrasinh D. Jhala (Lecturer), R. Shivashankaran Chettiar and Alpesh R. Patel (Research Fellows), Department of Zoology, Gujarat University, Ahmedabad, for typing the manuscript and drawing the figures.

REFERENCES

Blackmore, P.F. and Exton, J.H. (1985) Mechanisms involved in the activations of calcium dependent hormones (Litwack, G., Ed.), Vol. 12, Academic Press, New York, USA.

Cuatrecasas, P. (1986) Hormone receptors, membrane phospholipids, and protein kinase. The Harve Lectures Ser. 80 : 89-93.

Gilman, A. (1984) G proteins and dual control of adnylate cyclase. Cell 36 : 577-582.

Litwack, G. and Schmidt, J. (2002) Biochemistry of hormones II : steroid hormones. In : Biochemistry (Ed., Devlin, T.M.), pp. 959-988, John Wiley & Sons, Inc. Publication, New York, USA.

Rasmussen, H. (1986a) The calcium messenger system (Second Part). N. Engl. J. Med. 314 : 1094-1099.

Rasmussen, H. (1986b) The calcium messenger system (Second Part). N. Engl. J. Med. 314 : 1164-1169.

Saltiel, A.R. (1994) The paradoxicol regulation of protein phosphorylation in insulin action. FASEB J. 8 : 1034-1038.

Spiegel, A.M., Shenker, A. and Weinstein, L.S. (1992) Receptor-effector coupling by G proteins: implication for normal and abnormal signal transduction. Endocr. Rev. 13 : 536-547.

Struthers, A.D. (1990) Atrial Natriuretic Factor. Blackwell Scientific Publication, Bostan, USA.

Weinberg, R.A. (1985) The action of oncogenes in the cytoplasm and nucleus. Science 230 : 770-773.

Suggested Readings

1. Molecular Endocrinology (1989) F. F. Bolander, Academic Press, New York.
2. Molecular Biology of the cell (2002) B. Alberts, A. Johnson, J. Lewis, M. Raff, K. Roberts and P. Walter, 4th Edition, Garland Science, New York.
3. Molecular Cell Biology (1995) H. Lodish, D. Baltimore, A. Berk, S.L. Zipursky, P. Matsudaira, J. Darnel, 3rd Edition, Scientific American Books, W. H. Freeman and Company, New York.
4. Cell and Molecular Biology (1999) Gerald Karp, 2nd Edition, John Wiley and Sons Inc., New York.
5. Sigma Catalog Supplement (1999).
6. Sigma RBI Catalog on Lipid signaling.

REVIEW QUESTIONS

Long Questions

1. Describe the mechanism of action of epinephrine on a target tissue and its role in increasing blood glucose.
2. What are G-proteins? Describe their types and the disorders caused by defects in G-protein or its receptor.
3. Describe the relationship between phosphatidylinositol, diacylglycerol, calcium ions and protein kinase C giving examples.
4. Explain the interaction of C-AMP and calcium for various metabolic processes.
5. Describe the steps in the signaling pathway by which nitric oxide mediates dilation of blood vessels.
6. How does dopamine act as a signaling molecule for the memory mechanism in the nerve cell?
7. Give an account of ion-channel receptors and enzyme linked receptors.
8. Describe the binding of an insulin molecule at the surface of a target cell and the activation of the effector PI(3)K.
9. What is the role of Ras in signal transduction pathways? How does it promote both convergence and divergence in these pathways.

Multiple Choice Questions

1. When the signaling molecule released by a cell affects target cells lying in close proximity, it is called
   (a) Neurocrine signaling
   (b) Autocrine signaling
   (c) Signaling by plasma membrane attached proteins
   (d) Paracrine signaling
2. Hydrophilic signaling molecules can
   (a) Diffuse across plasma membranes
   (b) Be carried by carrier proteins in blood
   (c) Interact with intracellular receptors
   (d) Interact with cell surface receptors
3. The G-protein receptor has
   (a) 2 transmembrane loops (b) 3 transmembrane loops
   (c) 5 transmembrane loops (d) 7 transmembrane loops
4. The action of C-AMP is mediated through the activation of
   (a) Phospholipase C (b) Protein Kinase G
   (c) Phosphodiesterase (d) Protein Kinase A

5. Which second messenger helps in the release of calcium from ER?
   (a) DAG (b) PIP_2
   (c) Polyamines (d) PIP_3
6. Nitric oxide acts a signaling molecule and causes
   (a) Striated muscle relaxation (b) Striated muscle contraction
   (c) Smooth muscle contraction (d) Smooth muscle relaxation
7. Which is the correct sequence?
   (a) C-AMP → L-DOPA → Dopamine → receptor binding → active protein Kinase A → key proteins stimulated
   (b) L-DOPA → C-AMP → Dopamine → active protein Kinase A → receptor binding → key proteins stimulated
   (c) L-DOPA → Dopamine → C-AMP → receptor binding → active protein Kinase A → key proteins stimulated
   (d) L-DOPA → Dopamine → receptor binding → C-AMP → active protein Kinase A → key proteins stimulated
8. Divergence or transmission of signals along three different pathways is caused by the signaling molecules
   (a) NO (b) Dopamine
   (c) CO (d) Growth factors, EGF and PDGF
9. Crosstalk between signaling pathways is caused by
   (a) Prostaglandins (b) Thromboxanes
   (c) NO (d) Calcium and C-AMP interaction
10. Which ligand binding causes dimerization and activation of either serine/threonine kinase or tyrosine kinase activity?
   (a) Glucagon (b) Epinephrine
   (c) Acetylcholine (d) Insulin

Answers

1. (d)	2. (d)	3. (d)	4. (d)	5. (d)	6. (d)
7. (d)	8. (d)	9. (d)	10. (d)		

CHAPTER 9

Signal Transduction by Steroid Hormones

M.V. Rao
Dept. of Zoology, School of Sciences, Gujarat University, Ahmedabad–380 009

Steroid receptors are intracellular in target tissue (Granner, 1990). It is known that most, if not all, steroid receptors are exclusively nuclear as demonstrated by the use of monoclonal antibodies to the receptors and by nuclear extrusion studies. With the former immunofluorescence is localized in the nucleus only. In the latter, cytochalasin is used to induce cells to expel the nuclei. After separation, steroid receptors are found in the nucleus and not the cytoplasm. Hence these receptors are known as nuclear receptors (NRs). The earlier studies were misleading. Dilution by the homogenization buffer and low affinity of unoccupied nuclear receptors for chromatin, the receptors are leached from the nucleus to cytoplasm. After binding and activation, the affinity increased and complex remained bound to the chromatin; it is translocated from cytoplasm to the nucleus. Nuclear localization of the receptors are now been accepted for various steroids including ecdysone in fruit fly. Hence parenthetically it is suggested that steroid receptors are present in both nucleus and cytoplasmic compartments and that steroid binding alters the distribution. The receptor is protein in nature and thermo labile. It is also sensitive to proteases like trypsin. The steroid receptor ranges in size from 65,000 (estrogen) to 105,000 (progesterone). In-activated form of the receptor is larger than the active form. The native cytosolic receptor complex of glucocorticoid usually is 8-9S whereas its active form is about 4S (Svedberg units). The complex constitutes two identical forms. Each has a homodimer in which the DNA binding site is encompassed, abutting a non- steroid binding 90,000 dalton heat shock protein (hsp), which like actual receptor is phosphorylated and may play a role in maintaining steroid binding capacity of the receptor (Fig. 1). The significance of phosphorylation of a specific receptor at different places is unclear.

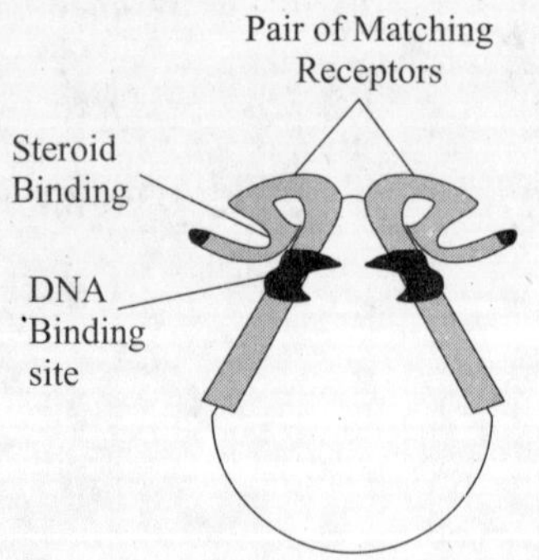

Fig. 1. Steroid receptors exist intracellularly as homodimers with a non-steroid binding Heat Shock Protein (hsp).

RECEPTOR STRUCTURE AND FUNCTION STUDIES

A lot of works have been done on receptors of steroid hormones (Evans, 1988); Schmidt and Meyer, 1994; Mester and Baulieu, 1995; Chen, 2000). The primary sequences of receptors for estrogen (ER), progesterone (PR), glucocorticoid (GR), aldosterone (ALR), T_3, retinoic acid, 1,25 dihydrocholecalciferol (DHCC) (Vitamin D_3) and androgens (AR) are known. These receptors fall into three major families: The glucocorticoid group includes glucocorticoids, progesterone, aldosterone and androgen receptor; the estrogen family contains only estradiol receptors and the non-steroid family contains the T_3, retinoic acid and vitamin D_3 nuclear receptors. Due to similarity in their structure they are studied together. Isoforms of a receptor also exist (ER, a,b). This prototypical receptor has three major domains (Fig. 2): an amino-terminal end; a carboxy terminal region or ligand binding domain (LBD); and a central DNA binding domain (DBD).

The central region is the most highly conserved and has about 70 amino acids. There is about 90% identity within glucocorticoid family and about 60% identity with other groups. Deletions, insertion and mutations in this domain all impair the ability of the receptor to bind DNA and induce transcription. This binding is very specific to turn on a specific gene(s) by the receptor. The arrangement of the basic amino acids in this domain is quite interesting. Each

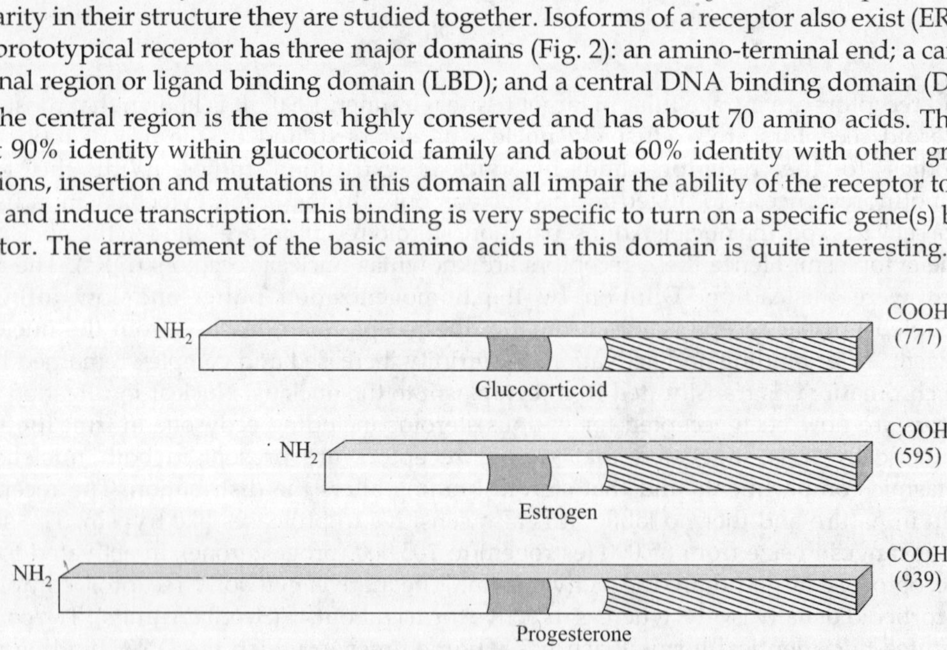

Fig. 2. Schematic representation of the nuclear receptors showing DNA binding domain (solid). The steroid binding domain (hatched) and the amino terminal end. Number in parenthesis indicates amino acids.

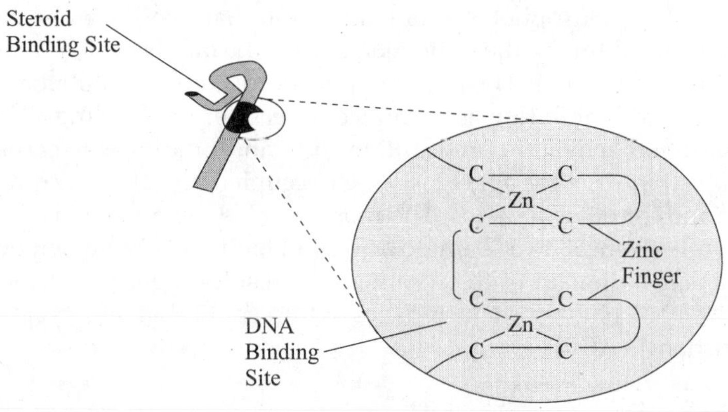

Fig. 3. DNA binding domain (DBD) of each hormone receptor.

DNA binding domain of a receptor has two elongated metal ion (M^{2+}) like zinc containing structure called zinc fingers or DNA binding fingers (Fig. 3).

These fingers enable the protein to position itself in the major grooves of DNA. Metal fingers have a cluster of 9-15 amino acids flanked by cysteine and histidine pairs capable of binding with metal ions (Fig. 4). A metal chelator can inhibit DNA binding region of receptor and this inhibition is attenuated by addition of zinc. Similar motifs are also reported for a number of transcription factors.

The carboxy terminus (LBD) of the receptor has about 250 amino acids and mutations in this region result in a loss of steroid binding. It has dimerization surfaces and transcription activation motifs. It is also called activation function–2(AF-2) domain. The sequence contains

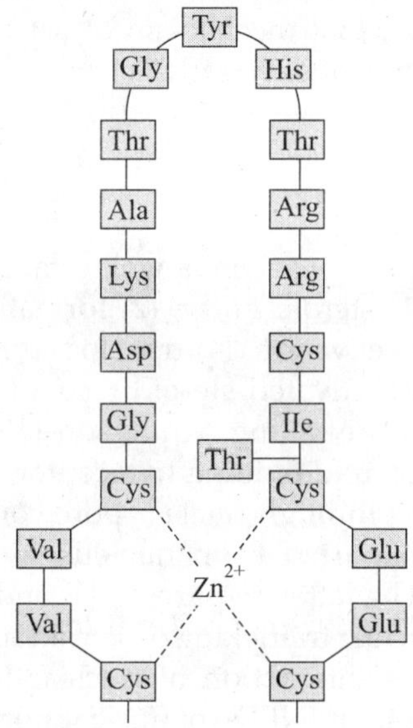

Fig. 4. A hypothetical DNA-Binding "Finger" or Zinc finger from the central domain of the steroid receptor.

loci that are very rich in hydrophobic amino acids and that may form the part of the steroid-binding site. The identity among the same members of the family is 50-60% and the homology between others is less (20-30%). This region appears to exert an inhibitory influence on the central region and this inhibition is removed after steroid binding. Thus, this domain influences transcription activation. In addition to ligand binding, the carboxy terminus (C-terminal) is involved in receptor aggregation. Unoccupied receptors tend to form complexes with themselves and/or other proteins. The most variable region of all is the amino terminus (N-terminal). It varies from 58 to 602 amino acids and has no similarity with other receptors. It is the immuno-reactive domain of the receptor and results suggest that the amino-terminus may modulate the DNA binding characteristics of the central domain. Its activity is referred to as activation function-1 (AF-1).

Activation-transformation

The molecular basis for activation of receptors, much is not known *in vivo.* But the process can be induced *in vitro* by variety of manipulations suggesting some possible mechanisms. Activation occurs when temperature, ionic strength or pH is increased. Such physiochemical effects disrupt subunit interactions. This indirect evidence is supported by direct measurements of the receptor in density gradients. Following transformation all inactivated receptor aggregates dissociate. The possible role of phosphorylation in receptor metabolism is complex. ATP stimulates hormone binding but phosphatases induce activation. These data suggest that activation may involve dephosphorylation, but recycling requires rephosphorylation in order to restore steroid binding. Another possible mechanism for activation is the removal of an inhibitor from the receptor. A major candidate is a 90 K heat shock protein (hsp 90) that appears to bind to the inactivated receptor, but dissociates upon hormone binding. Finally the binding stoichiometry appears to be variable causing "sticky nature" of the receptors. It is obvious that multiple mechanisms may operate during receptor activation. Depending upon the actual mechanism of activation, additional positive charge could have been enhanced by removal of negative charges before hormone receptor complex enters into the nucleus.

Steroid Effects on Gene Activity

Hormonal control target cell gene expression is well established. The signal transduction of steroid is deduced to involve: (1) steroid entry, (2) formation of steroid hormone complex, (3) activation of steroid receptor activation, (4) translocation of the complex into the nuclear compartment and binding of the activated steroid-receptor complex to chromatin or DNA, (5) changed gene expression and recycling. Anderson (1985) studied the effect of steroid hormones on trascription. After steroid binding to receptor, the receptor complex is activated and translocated into the nucleus through nuclear pore complex of nuclear membrane. The receptors are continuously transported from nucleus to cytosol and vice-versa through diffusion and active transport. The latter requires ATP and is temperature dependent. This process is called 'Cytoplasmic nuclear trafficking of steroid hormone receptors' or 'Shuttling of nuclear receptors'. This involves interaction of nuclear localization signals (NLs) of the receptors and nuclear binding proteins (NBPs) of nuclear pore complex. The NLs are specific to each receptor, which are small amino acid sequences of the receptors present at hinge region between central domain (DBD) and carboxy terminus (LBD) and others are present in the zinc finger of the receptor to facilitate their transport in and outside the nucleus.

Regulation of Gene Activity

The gene regulation by steroid hormones have been well documented (Gluzman *et al.*, 1983; Beato, 1989). All eukaryotic genes are transcribed from 5' to 3' beginning within 19 to 27 base pairs (bp) of a stretch of DNA containing the sequence of TATAAA and referred to as the 'hogness' or TATA box. Further 5' end of the TATA box is the CAAT box with the consensus sequence CGNCAATCT (N represents any of the four bases). These elements are also invariant and form the 'promoter' of the gene. Most genes also have elements further upstream that increase the activity or strength of the promoter and determine the tissue specificity of the gene expression. Similar elements that seem to affect transcription over long distances and are not dependent on orientation or position were described in viruses and formed enhancers. In addition to these elements found in the eukaryotic gene, the steroid-regulated gene has one or more hormone regulatory elements or hormone response elements (HREs) nearer to the 5'-end. When bound with the receptor these elements influence the transcriptional activity of the promoter (Fig. 5).

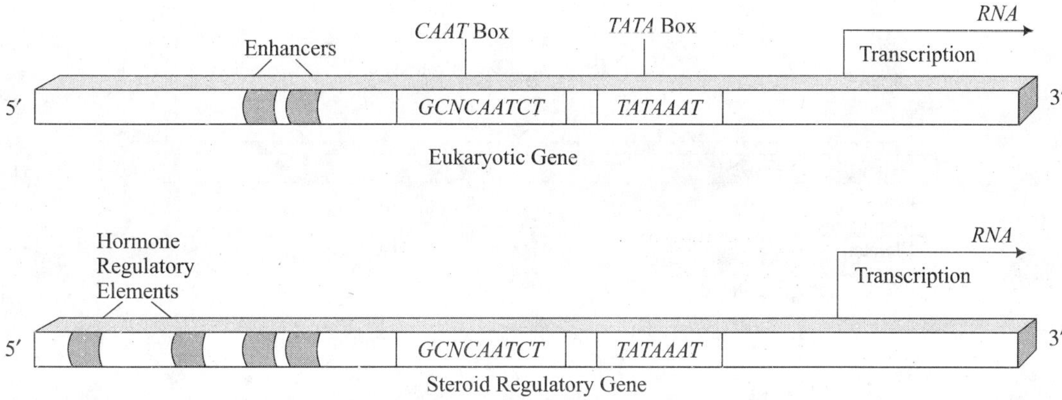

Fig. 5. The common organization of the eukaryotic gene(s) consists of the exon coding sequence which may be separated by one or more non coding introns.

The consensus sequences for the known HREs associated with upregulated genes fall into two classes (Table 1). HREs for glucocorticoid, androgen, progestin and mineralo-corticoid upregulated genes form one class of related sequences, while those for estrogen, thyroid, vitamin A, vitamin D_3 form another class. Thus, the minimum expected configuration for a gene, whose transcription is increased by steroid, is a promoter with one or more HREs usually localized in the 5' flanking region of a gene.

Early investigators found that the genomic response to a steroid could be divided into two broad categories. The first category consists of genes that responded within minutes of the steroid exposure and whose response was not blocked by inhibition of protein synthesis. The second category consisted of genes that might require hours of exposure before a response was measured and the response could be blocked by inhibition of protein synthesis. The gene in the

Table 1 Consensus sequences for hormone regulatory or response elements

HORMONE	HRE
CLASS 1	
GLUCOCORTICOID	GGTACANNNTGTTCT
PROGESTERONE	GGTACANNNTGTTCT
ANDROGEN	GGTACANNNTGTTCT
MINERALOCORTICOID	GGTACANNNTGTTCT
CLASS 2	
ESTROGEN	AGGTCANNNTGACCT
THYROID	TCAGGTCA—TGACCTGA
RETINOIC ACID	TCAGGTCA—TGAGCTGA
VITAMIN D	

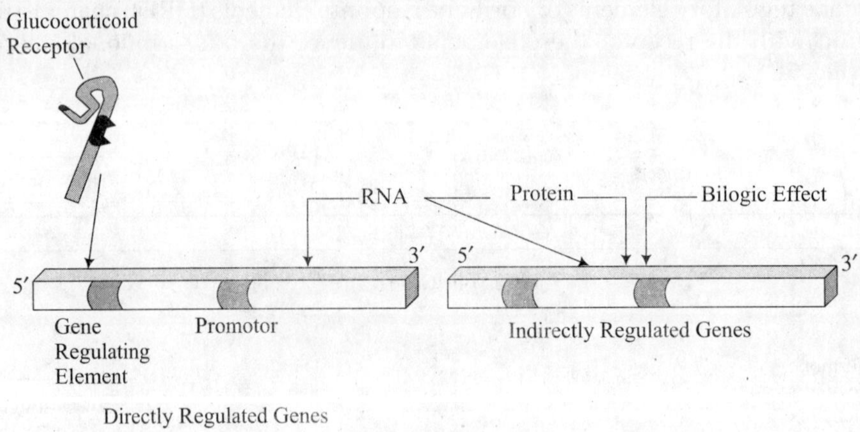

Fig. 6. Direct and indirect regulation of steroid hormone.

second category termed indirect responders that are regulated by proteins coded for genes in the first category, the directly regulated gene (Fig. 6).

MECHANISM

As soon as the steroid enters the cell and binds to the receptor component of the receptor-protein complex causing dissociation, release of hsp and activation of receptor. The activated receptor develops an affinity for DNA and translocated from cytosol to nucleus where it binds to DNA segment. The DNA binding results in a change in the transcriptional rate of specific gene(s). The resulting production of more or less mRNA under the control of transcription factors and RNA polymerase leads to changes in the contents of the specific proteins in the target cell (Fig. 7).

Recent studies have revealed the nuclear receptor (NR) superfamily can be divided into receptors with known ligands (Type I and II) and so called orphan receptors for which ligands

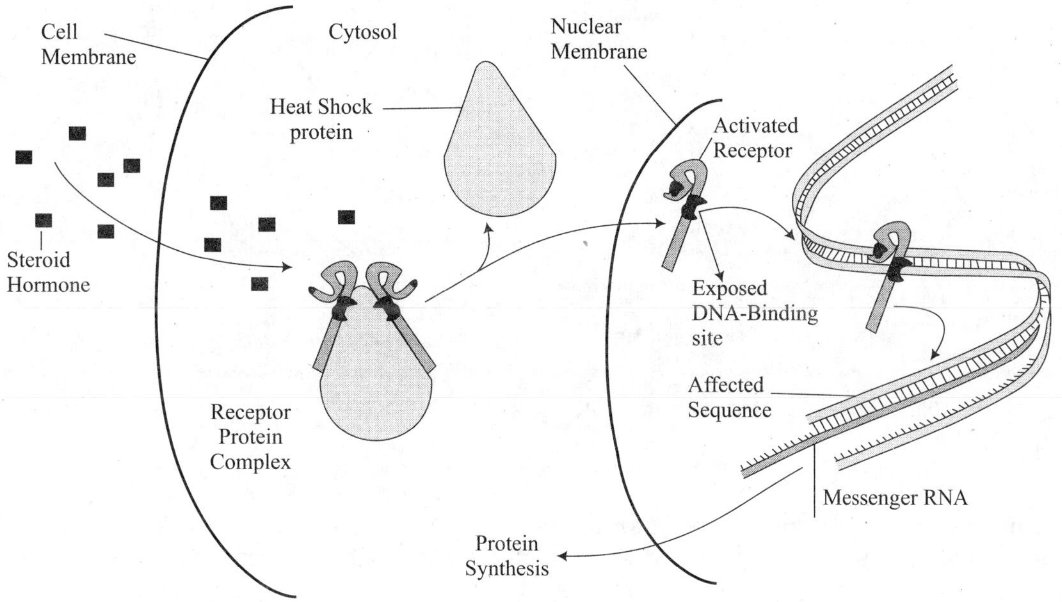

Fig. 7. Mechanism of steroid hormone action.

have not been identified or might not exist. Type I nuclear receptors include classic steroid receptors (glucocorticoids, estrogen, progesterone, androgen and mine-ralocorticoid that bind to DNA as homodimers. Type II nuclear receptors are of non-steroid group binding to DNA as heterodimers. These include receptors for thyroid hormones, vitamin D and retinoic acid. Many orphan receptors like steroidogenic factor-1 (SF-1) and DAX-1 bind DNA as monomers.

Nuclear receptors act primarily by altering rates of gene transcription, although there is evidence for interactions with other cellular signaling pathways such as the mitogen-activated protein kinase (MAPK) pathway. Hormone binding induces conformational changes that induce receptor interactions with transcriptional cofactors. For type I receptors, ligand binding creates a bend in the extreme carboxy-terminus (AF-2 domain) providing a docking site for co-activators (e.g. steroids receptor co-activators, SRCs). For Type II receptors, hormone binding induces similar conformational changes. However, in the absence of hormone, co-repressors (CoR) are bound and silence gene transcription. Many CoR complexes possess histone deacetylase (HDAC) which is thought to induce gene silencing. Hormone binding induces the dissociation of co-repressors and the recruitment of co-activators (CoA). CoA complexes possess histone acetyl transferase (HAT) which triggers changes in chromatin structure. Consequently gene expression can shift from being silenced in the absence of hormone to being stimulated in the presence of hormone.

RECYCLING

The steroid hormone binds to the inactivated receptor oligomer. After dissociation of inhibitor activation is completed as it undergoes dephosphorylation before it binds to DNA via newly exposed cationic sites as mentioned earlier. After gene activation removal and metabolism of

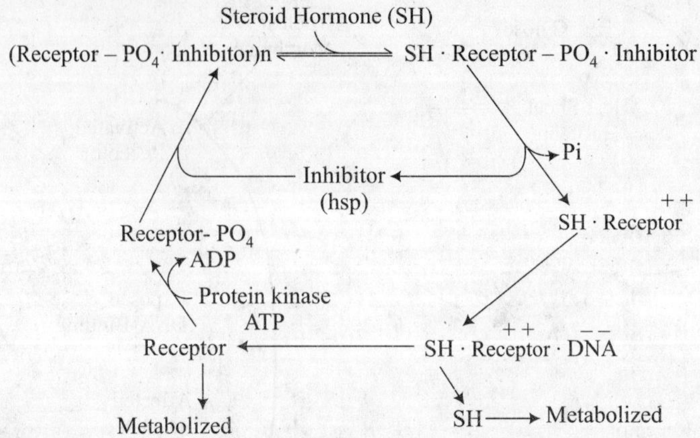

Fig. 8. A schematic representation of steroid hormone receptor metabolism.

hormone takes place. Then the receptor dissociates from the chromatin and is recycled after phosphorylation by use of ATP and re-aggregated. Although the evidence is strongest for recycling for glucocorticoid there are more data to support recycling for other steroid receptors as well (Fig. 8).

SUMMARY

The receptor recognizes the signal (steroid) and transduces its biological effects. Steroid hormones are smaller and hydrophobic molecules that bind intracellular receptors in the target tissue to elicit its action. The steroid hormones are C_{18}, C_{19}, and C_{21} compounds produced by the gonads, adrenal cortex and placenta. Their receptors belong to superfamily proteins, all of which are regulators of gene transcription. The superfamily consists of the receptors for steroid hormones, thyroid hormones, vitamin D_3 and retinoic acid and erb-A oncogene from the erythroblastocyst virus. Members of this family share three important properties. They each bind a hormonal ligand, they bind to specific DNA sequence and they act to modify the transcription activity of specific gene(s).

ACKNOWLEDGEMENTS

Author is grateful to Dr. Sunita Chawla, Mr. Alpesh Patel, Mr. Shiv Shankaran and Mr. Ashutosh Khamar for assistance in preparation of the manuscript.

REFERENCES

Anderson, J.E. (1985) The effects of steroid hormones on gene transcription. In : Biological Regulation and Development (Goldberger, R.F. and Yamamoto, K.R.), Vol. 3B, Plenum Press, New York.

Beato, M. (1989) Gene regulation by steroid hormones. Cell 56 : 335-339.

Chen, J.D. (2000) Steroid/ nuclear receptor coactivators. In : Vitamins and Hormones (Ed., Litwack, G.), Vol. 28, pp. 391, Academic Press, San Diego, USA.

Evans, R.M. (1988) The steroid and thyroid hormone receptor superfamily. Science 240 : 889-894.

Granner, D.K. (1990) Hormone action. In : Harper's Biochemistry (Eds., Murray, R.K., Granner, D.K., Mayes, P.A. and Rodwell, V.W.), pp. 446-477, Appleton & Lange, Connecticut, USA.

Mester, J. and Baulieu, E.E. (1985) Nuclear receptor superfamily. In : Endocrinology (Eds., DeGroot et al.), pp. 93, Saunders, Philadelphia, USA.

Schmidt, T.J. and Meyer, A.S. (1994) Autoregulation of corticosteroid receptors. How, when, where and why? Receptor 4 : 229-233.

Shenk, T. (1983) Current communications in Molecular Biology. Cold Spring Harbor Press, USA.

Suggested Readings

1. Molecular Endocrinology (1989) F. F. Bolander, Academic Press, New York.
2. Molecular Biology of the cell (2002) B. Alberts, A. Johnson, J. Lewis, M. Raff, K. Roberts and P. Walter, 4th Edition, Garland Science, New York.
3. Molecular Cell Biology (1995) H. Lodish, D. Baltimore, A. Berk, S.L. Zipursky, P. Matsudaira, J. Darnel, 3rd Edition, Scientific American Books, W. H. Freeman and Company, New York.
4. Cell and Molecular Biology (1999) Gerald Karp, 2nd Edition, John Wiley and Sons Inc., New York.

REVIEW QUESTIONS

Long Questions

1. Write an account on nuclear receptor superfamily?
2. Describe mechanism of action of steroid hormones?
3. Explain the gene regulation by steroids?
4. Classify steroid receptors and their properties?
5. Write notes on DBD, LBD and receptor recycling.

Multiple Choice Questions

1. Fill in the following blank
   (a) Steroid receptors are also called _______ receptors.
   (b) Steroid receptors belong to _______.
   (c) Estrogen receptors have _______ and _______ forms.
   (d) Role of Zn finger is _______.
   (e) Unoccupied receptor has _______ Svedberg units.
2. DNA finger consists of which metal ions
   (a) Zn^{2+} (b) Fe^{2+}
   (c) Ni^{2+} (d) Mg^{2+}
3. Chelation brings about loss of
   (a) Metal ions (b) DNA binding
   (c) Gene transcription (d) Steroid binding
4. Dephosphorylation of receptor is for
   (a) Activation (b) Inactivation
   (c) increasing - ve charge (d) none of the above

5. Example for Orphan receptor is
   (a) SF-1 (b) Estrogen receptor
   (c) SOX (d) Thyroid receptor
6. Highly conserved region of receptor is
   (a) DBD (b) LBD
   (c) N-terminus (d) Short chain

Answers

1. (a) Nuclear (b) Superfamily (c) α, β (d) To firmly bind with DNA (e) 9-10
2. (a) 3. (a) 4. (a) 5. (a) 6. (a)

CHAPTER 10

Biochemistry and Synthesis of Steroid Hormones

N.J. Chinoy
Dept. of Zoology, School of Sciences, Gujarat University, Ahmedabad – 380 009
E-mail: zooldeptgu@satyam.net.in

The lipids of the body are of different types such as simple, compound and derived lipids. Simple lipids are esters of fatty acids with different alcohols. They are fats which are esters of fatty acids with glycerol. In liquid state, it is known as an oil. Waxes are esters of fatty acids with higher molecular weight alcohols. Compound lipids are esters of fatty acids with groups in addition to an alcohol and a fatty acid, e.g., phospholipids have in addition to fatty acids and an alcohol, a phosphoric acid residue, nitrogen containing bases and other constituents. Glycolipids are compounds of fatty acids with carbohydrate, containing nitrogen but no phosphoric acid. Some other compound lipids are sulpholipids, aminolipids and lipoproteins. However, derived lipids are substances derived from simple and compound lipids by hydrolysis. These include fatty acids, (saturated and unsaturated), glycerol, steroids, alcohols in addition to glycerol and sterols, fatty aldehydes and ketone bodies. The glycerides, cholesterol and cholesteryl esters are termed neutral lipids as they are not charged.

The steroids are often associated with fat and are separated from it by saponification. All steroids in the mammals are derived from cholesterol and have a common steroid nucleus. The formation of steroid nucleus starts with a benzene ring. When two benzene rings are attached together it is called naphthalene and phenanthrene when a third benzene ring is attached. It gets fully saturated without any double bonds and it is now called as perhydrophenanthrene with rings A, B, C. Next, a cyclopentane, ring D is attached to it and the steroid nucleus is formed which is known as cyclopentanoperhydrophenanthrene, which has 17 carbon atoms and rings A, B, C and D (Fig. 1). All steroids in the mammalian body are of three types, C-21, C-

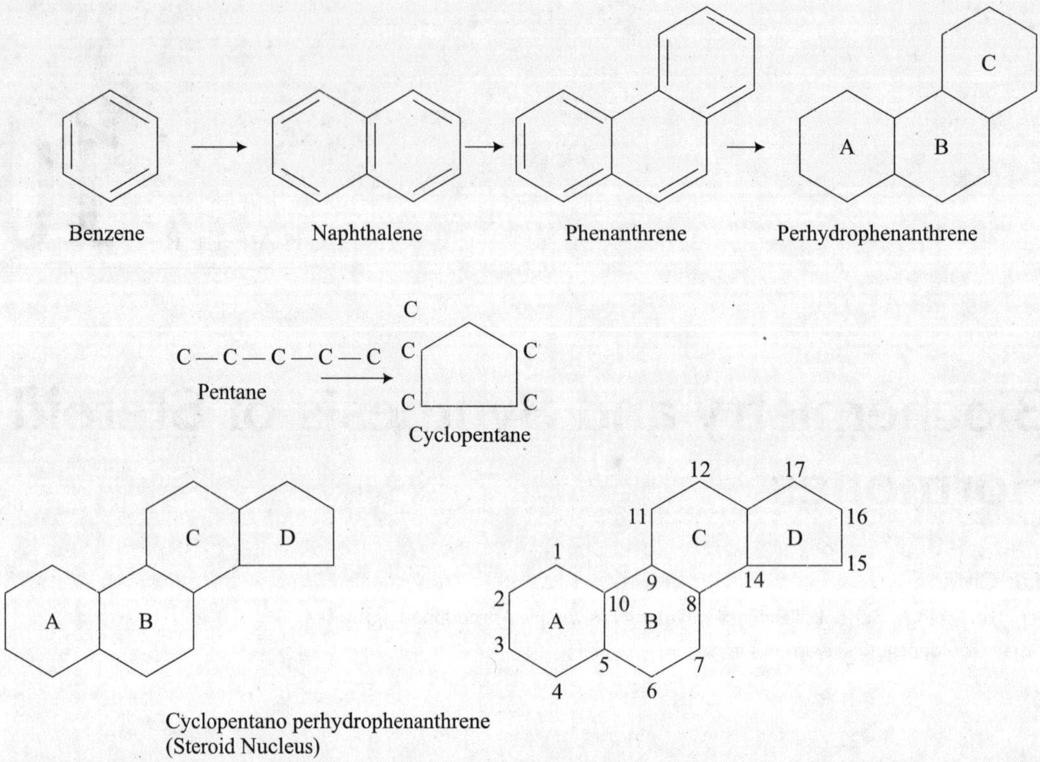

Fig. 1. Steroid nucleus.

19, and C-18. ie containing 21 carbon atoms, 19 carbon atoms or 18 carbon atoms. Hence the three basic nuclei are pregnane (C-21), Androstane (C-19) and Estrane (C-18) with side chains (Fig. 2).

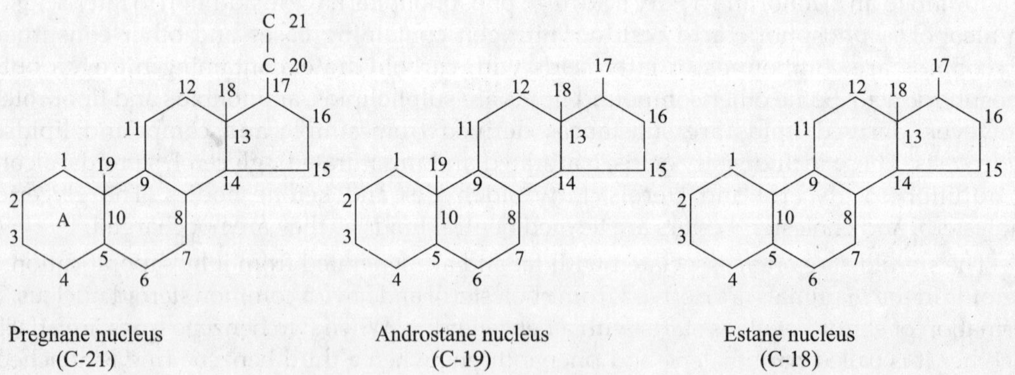

Fig. 2. Three basic nuclei.

Cholesterol (C-27) is the parent compound of all steroids synthesized in the body and has 27 carbon atoms. Cholesterol is derived from the diet. Acetyl-CoA is the source of all the carbon atoms in cholesterol (Mayes, 1990). It is widely distributed in all cells especially in nervous tissue. Many factors influence cholesterol balance in tissues (Rundey and Sexton, 1986). The steroids are synthesized mainly by the adrenals which produce C-21 steroids, the corticoids, and some C-19 and C-18 sex steroids, the testis which synthesizes C-19 steroids, the androgens and the ovary which produces C-18 steroids called estrogens and a C-21 steroid known as progesterone.

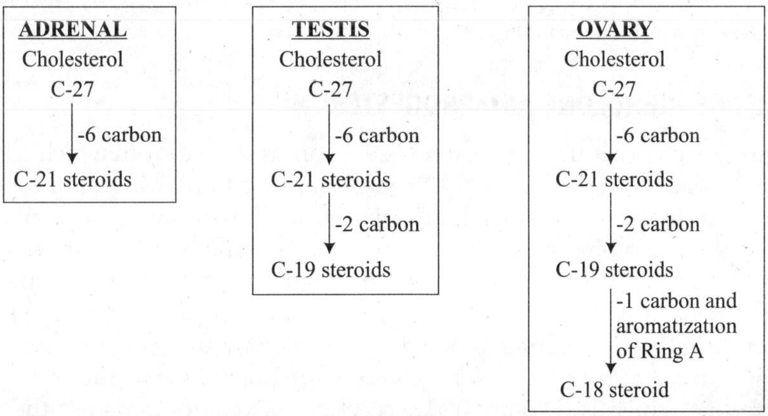

As shown above, the adrenal utilizes cholesterol which has a 6-carbon side chain which is cleaved off to yield a C-21 steroid. In the testis the same process is repeated, but further fission of two carbons (C–20, C–21) take place to yield C-19 steroid, the androgens. In the ovary, the above 2 steps are repeated and then further fission of C-19 takes place with the removal of carbon No. 19 and aromatization of ring A to yield the C-18 steroids, the estrogens. The pregnane and androstane nuclei have carbon atoms 18 and 19 which have a methyl group ($-CH_3$) attached to it and depicted by a straight line jutting out from C-13 and C-10 respectively. The estrane nucleus has only carbon 18 as in the previous two nuclei, but carbon 19 is not present.

BASIC RULES IN STEROID BIOCHEMISTRY

If a steroid has no double bonds in its structure its name ends with the suffix-ane as in pregnane, androstane and estrane. If one double bond is there in the ring structure, it is designated with the suffix, ene. If two double bonds then suffix is diene, if three, triene and so on. In case a keto group (=O) occurs the suffix is -one; if two keto groups, then it is designated as dione and if three double bonds, it is trione. The occurrence of a hydroxyl group (-OH) in a steroid, its name will end with the suffix -ol. If two -OH groups then it is -diol and if three, it is called -triol and so on.

In case the -OH group points down in a 3-Dimensional steroid model, it is designated as α-OH and written as – –OH, ie, with dotted lines. However, if the -OH group is pointing upwards away from the ground in a 3-D model, it is designated as β-OH and written with a solid line e.g., —OH. If the double bond occurs between carbons 4 and 5, in ring A, it is designated as delta $(\Delta)^4$ and has a keto (=O) group at carbon 3. In case the double bond occurs

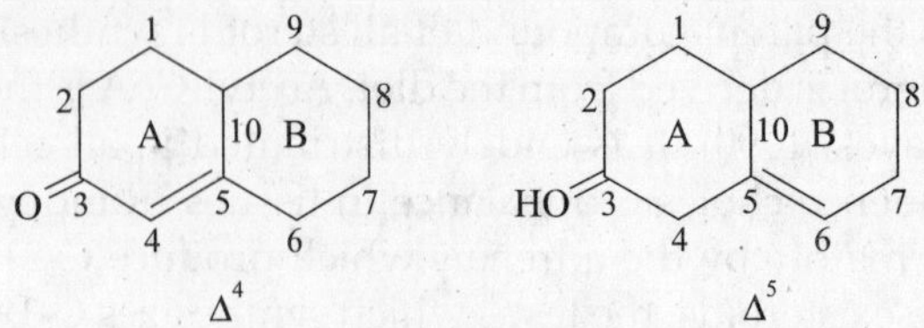

Fig. 3. The Δ^4 and Δ^5 positions.

between carbons 5 and 6 in Ring B, it is likewise designated as delta $(\Delta)^5$ and has a hydroxyl (-OH) group at position 3-carbon (Fig. 3).

SYNTHESIS OF PREGNENOLONE AND PROGESTERONE

It has been mentioned earlier that cholesterol is the parent steroid from which all other steroids are derived by fission of its side chain. The above reactions are common to all steroid synthesizing organs in the mammalian body, viz., adrenal, testis and ovary. In these reactions, first cholesterol is acted on by the 20, 22-hydroxylase enzyme which adds an alpha hydroxyl group (—OH) at carbons 20 and 22. In the next step, side chain fission occurs between carbons 20 and 22 and thereby, a six carbon chain is cleaved off forming isocaproic acid (a bye product) and a C-21 steroid called Δ^5- pregnenolone. It is acted on by two enzymes namely, the 3β-ol-dehydrogenase and $\Delta^{4\text{-}5}$ isomerase which convert Δ^5-pregnenolone to another C-21 compound called progesterone which is Δ^4 (Fig. 4). These two C-21 compounds are the main steroids which are involved in the steroidogenesis process forming different C-21, C-19 and C-18 compounds as described later. Steroid biosynthesis and control as well as release of steroid hormones have been described by Litwack and Schmidt (2002).

C-21 STEROIDOGENESIS IN ADRENAL CORTEX

The mammalian adrenal gland is situated above the kidney and is made up of the outer cortex region and the inner medulla. The adrenal cortex is divided into three zones, from the outside is zona glomerulosa which produces the C-21 aldosterone, followed by the zona fasciculata which synthesizes cortisol and the zona reticularis, the innermost region, next to the medulla which synthesizes small amounts of androgens (C-19) and estrogens (C-18) steroids.

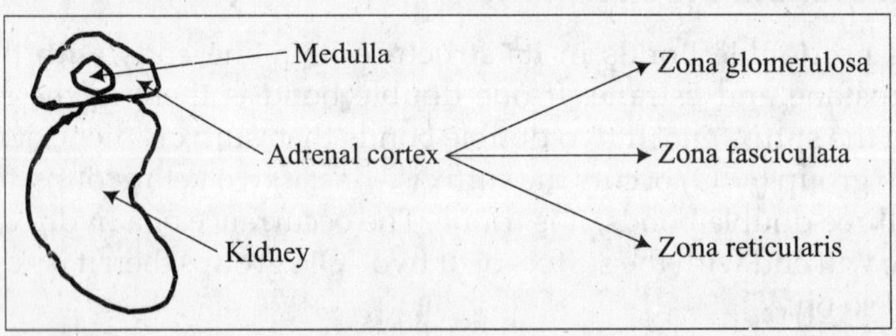

The C-21 steroidogenesis starts when the hypothalamus in the brain synthesizes adrenocorticotropic hormone-releasing hormone (ACTH-RH) which stimulates the pituitary gland to synthesize ACTH and release it in the circulation which inturn stimulates the adrenal

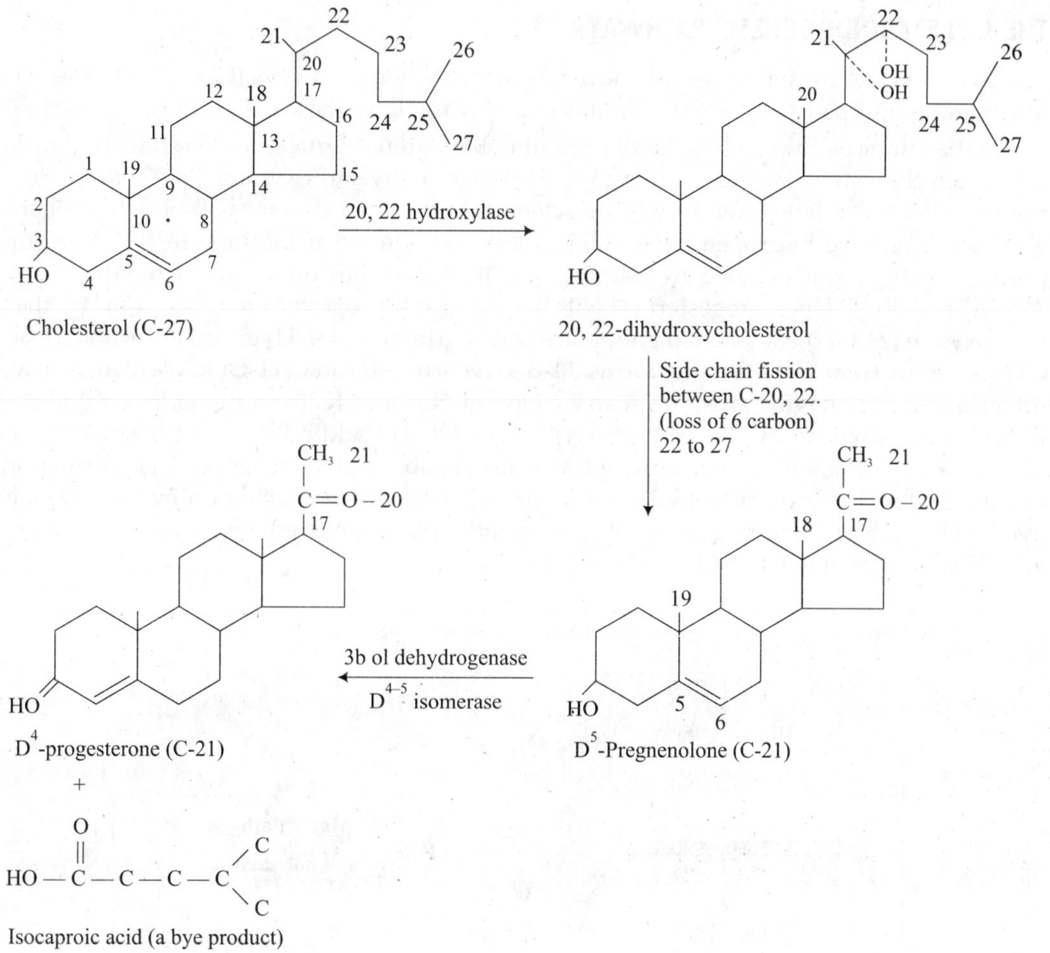

Fig. 4. Synthesis of pregnenolone and progesterone.

cortex zones-zona fasciculata and zona reticularis to synthesize various steroids. The zona glomerulosa is stimulated by the levels of sodium and potassium in the blood as well as renin from the kidney.

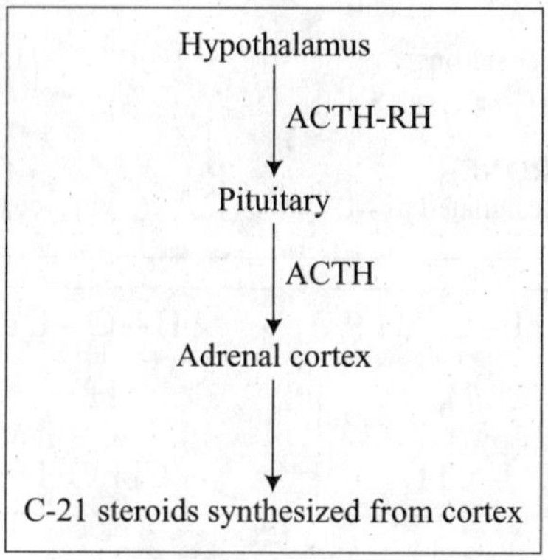

THE C-21 STEROIDOGENESIS PATHWAYS

The C-21 steroids are synthesized in adrenal cortex by two pathways (Granner, 1990a). These major and minor pathways are shown in the following flow chart.

In the minor pathway, cholesterol present in the adrenal cortical cells is converted to 20,22 dihydroxycholesterol which undergoes cleavage between carbons 20-22 to yield Δ^5-pregnenolone. The latter steroid will be acted on by the two enzymes 3β-ol dehydrogenase, which will remove hydrogen from -OH group and convert it into a keto (C=O) group at position 3. The second enzyme, Δ^{4-5} isomerase will transfer the double bond from ring B to ring A at Δ^4 position. Hence progesterone will be produced. Progesterone is acted on by the 21-hydroxylase (21-OH) enzyme which replaces one hydrogen of $-CH_3$ group at carbon 21 by a -OH group to from CH_2OH and forms 11-deoxycorticosterone (11-DOC) which is a weak mineralocorticoid having effect on reabsorption of Na^+ and K^+ from the kidney tubules. 11-DOC is next acted on by the 11–hydroxylase (11–OH) to add a β-OH at carbon 11 to form corticosterone which is a glucocorticoid as it brings about gluconeogenesis i.e. formation of carbohydrates from non-carbohydrate sources. Corticosterone is acted on by the enzyme 18 hydroxylase (18-OH) which adds a -OH group and replaces one hydrogen in the methyl group at C-18 to convert it to CH_2OH.

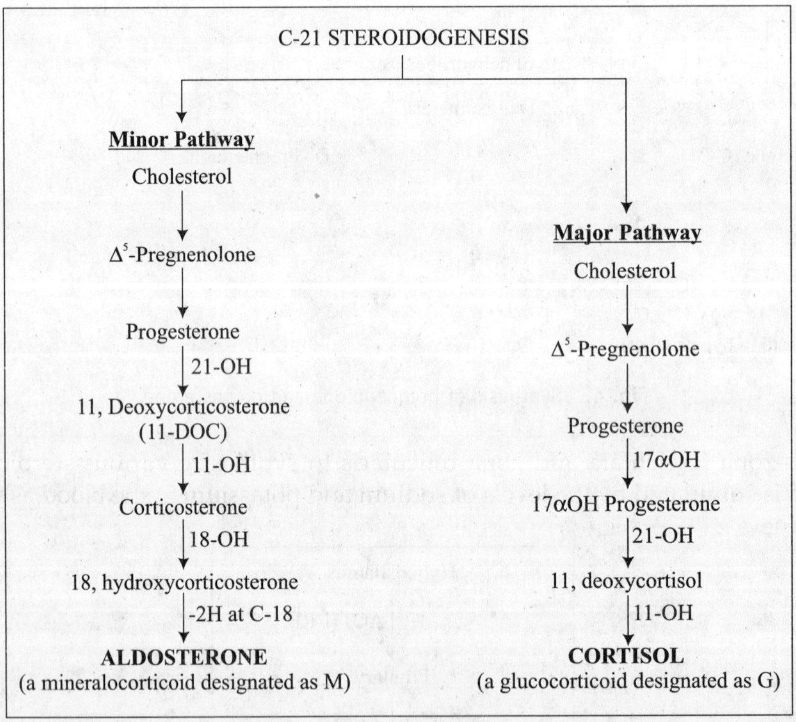


```
H–C–H    →    H–C–OH
  |             |
  H             H

 –CH3          CH2OH
```

The compound so formed is called 18-hydroxycorticosterone which is converted to aldosterone by removal of 2 hydrogens from $-CH_2OH$ at C-18 to convert it to an aldehyde (-CHO) group.

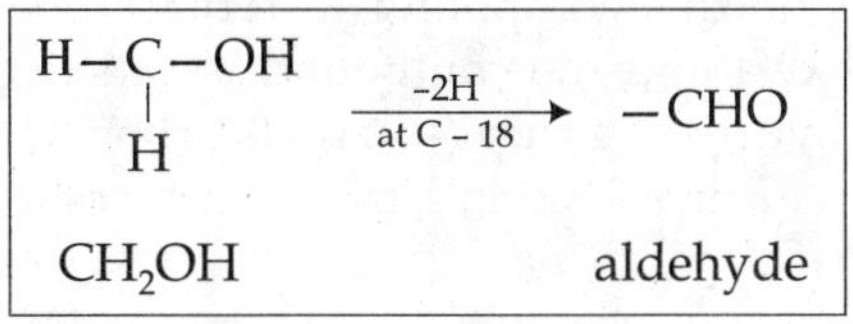

Aldosterone is named as such due to the presence of the -CHO group. Aldosterone is a potent mineralocorticoid (Fig. 5).

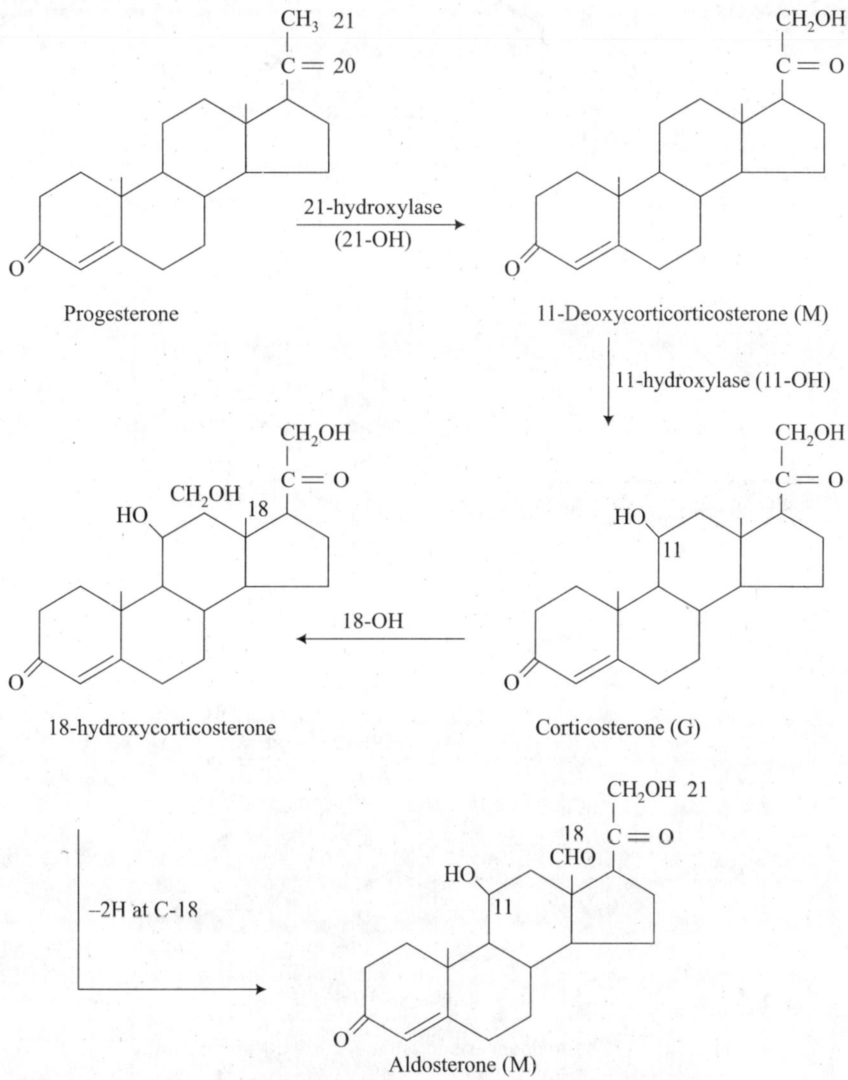

Fig. 5. C-21 steriodogenesis in adrenal cortex. Minor pathway.

The major pathway starts from cholesterol which is converted to Δ^5-pregnenolone and then progesterone which is acted on by 17 α-hydroxylase which adds an alpha -OH (–OH) group at position 17 carbon so that the compound formed is 17α–hydroxyprogesterone. The enzyme 21-hydroxylase (21-OH) acts on this compound to replace one hydrogen of methyl group ($-CH_3$) at C-21 to form $-CH_2OH$ and the compound is called 11-deoxycortisol. Finally, 11-hydroxylase enzyme adds a hydroxyl group (-OH) at carbon 11 to form cortisol which is the most potent glucocorticoid and is involved in gluconeogenesis (Fig. 6).

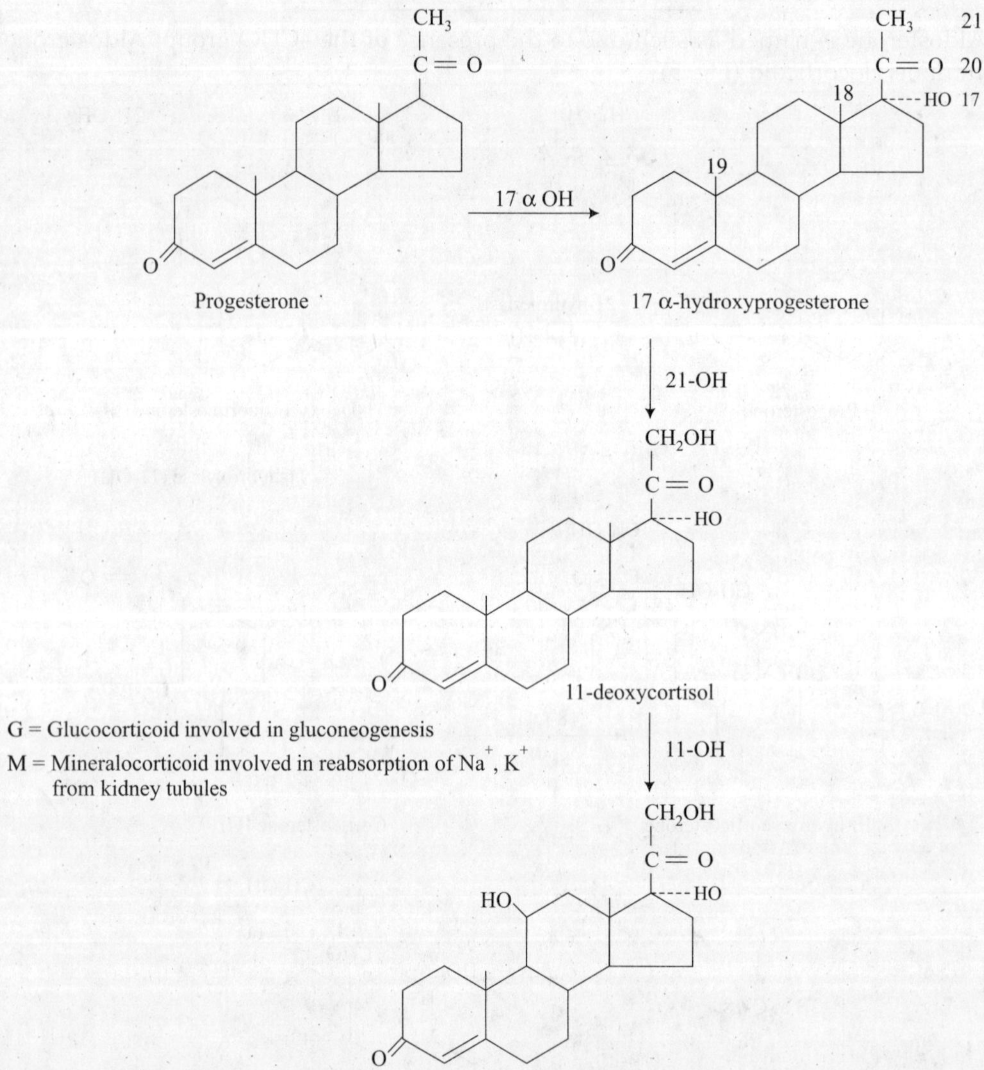

Fig. 6. C-21 steroidogenesis in adrenal cortex. Major pathway.

THE C–19 STEROIDOGENESIS

The C–19 steroidogenesis forming C–19 androgens, namely Δ^5-dehydroepiandro-sterone, Δ^4-androstenedione and testosterone takes place in the Leydig cells of the testis (Granner, 1990b). These cells are situated in the interstitium (connective tissue) region of the testis outside the seminiferous tubules of the testis.

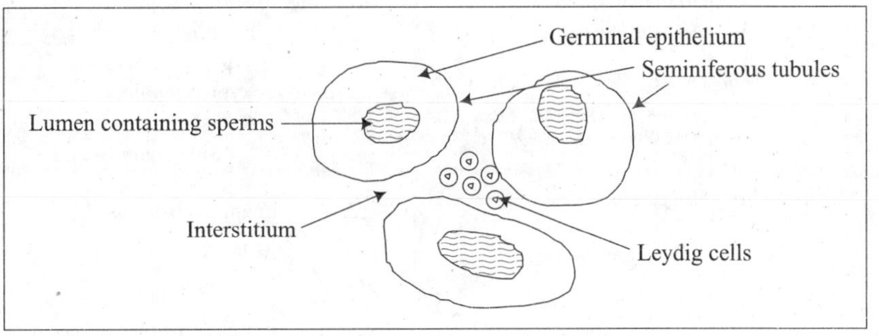

At the start of the C–19 steroidogenesis, the hypothalamus releases the gonadotropin - releasing hormones or Gn.RH which stimulates the pituitary gland to produce gonadotropins, FSH and LH. These are released in the circulation, and affect the testis. FSH stimulation causes growth of the testis cells and LH stimulation results in steroidogenesis by Leydig cells.

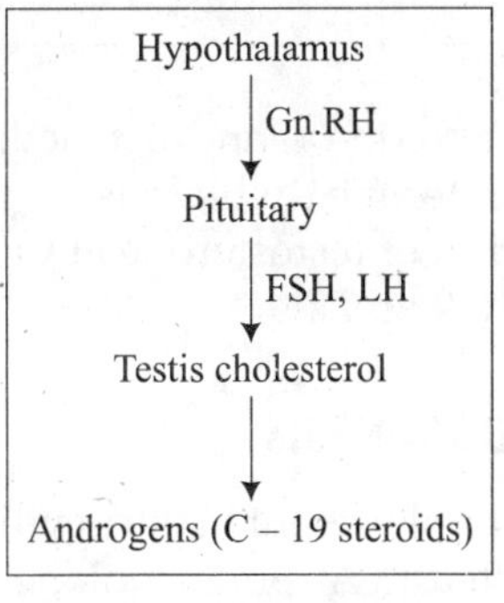

There are again two pathways, the Δ^4 and Δ^5 pathways for the synthesis of androgens. However, the conversion of cholesterol to first Δ^5–Pregnenolone and progesterone occur as described earlier. Progesterone is the starting steroid for the Δ^4–pathway and pregnenolone is the starting steroid for the Δ^5–pathway.

Δ^4 PATHWAY FOR C – 19 STEROIDOGENESIS

In this pathway most of the steroids formed are Δ^4 steroids. At first, cholesterol in Leydig cells is converted to Δ^5–pregnenolone by stimulation of LH-luteinizing hormone from the pituitary. Next pregnenolone is converted to progesterone by the enzymes, 3β–ol dehydrogenase and $\Delta^{4\text{-}5}$ isomerase. Progesterone is next acted on by the enzyme, 17α–hydroxylase (17α–OH) to convert it to 17α hydroxyprogesterone, which undergoes side chain fission so that carbons 20

C 19 Steroidogenesis

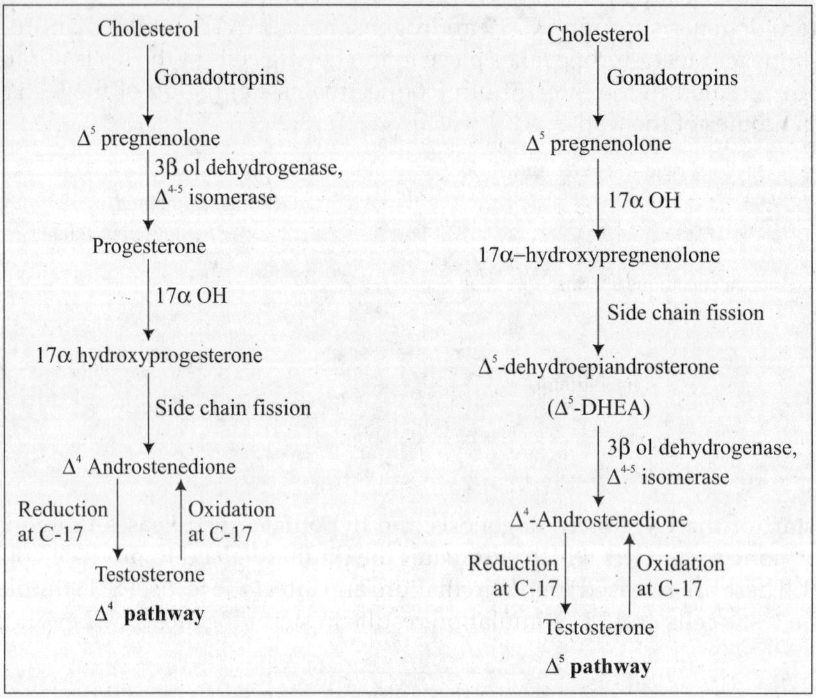

and 21 are cleaved off and a C-19 steroid, Δ^4 – androstenedione is formed. It can now undergo reduction at C-17 to form testosterone, which is a more potent androgen than Δ^4–androstenedione or else, by oxidation of testosterone at C-17, Δ^4 – androstenedione is formed again. This is the end of Δ^4 pathway (Fig. 7).

Δ^5 – PATHWAY FOR C–19 STEROIDOGENESIS

The biosynthesis starts from cholesterol present in the Leydig cells which under the stimulation of LH from the pituitary is converted to Δ^5-pregnenolone. It is acted on by the enzyme 17α-hydroxylase (17α-OH) and converted to 17α-hydroxy pregnenolone. It undergoes side chain fission to yield a weak androgen called Δ^5–dehydroepiandrosterone (Δ^5-DHEA). The Δ^5-pathway is extended to form the Δ^4 androstenedione and testosterone. Δ^5-DHEA is acted on by 3β-ol dehydrogenase and Δ^{4-5} isomerase to convert it to Δ^4-androstenedione which inturn is converted to testosterone by reduction at C-17 as in the Δ^4 pathway. Oxidation at C-17 in testosterone can again form Δ^4-androstenedione (Fig. 8). Hence the last two steps are common to both Δ^4 and Δ^5-pathways.

C–18, C–21 STEROIDOGENESIS IN OVARY

In order to understand the site of C–18 and C–21 steroidogenesis by the ovary, it is essential to know about its histology. The mammalian ovary is made up of the stroma or connective tissue which has large number of developing follicles. The majority are primary follicles which have

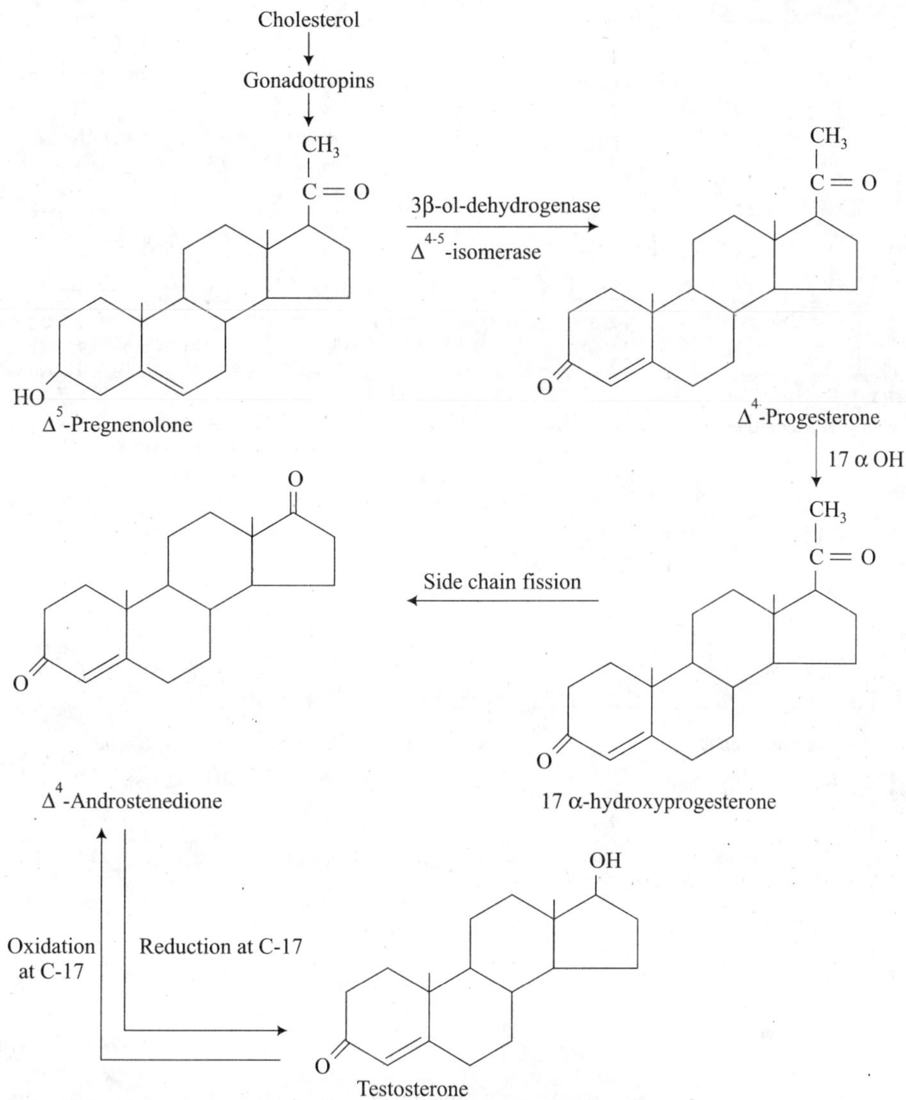

Fig. 7. Δ^4 pathway for androgenesis.

the ovum surrounded by a single layer of follicular cells. Under the action of FSH from the pituitary, the follicular cells multiply by mitosis and several layers now surround the ovum forming the secondary follicle. The tertiary follicle is formed by further action of FSH and multiplication of follicular cells which are now designated as granulosa cells since they have granular cytoplasm. The mature follicle or Graafian follicle is now formed by further action of FSH on tertiary follicle. The Graafian follicle is covered by two layers, the theca externa and theca interna. The granulosa cells multiply and increase in number pushing the ovum to one side and a cavity, the antrum appears in the Graafian follicle. The follicle is full of granulosa

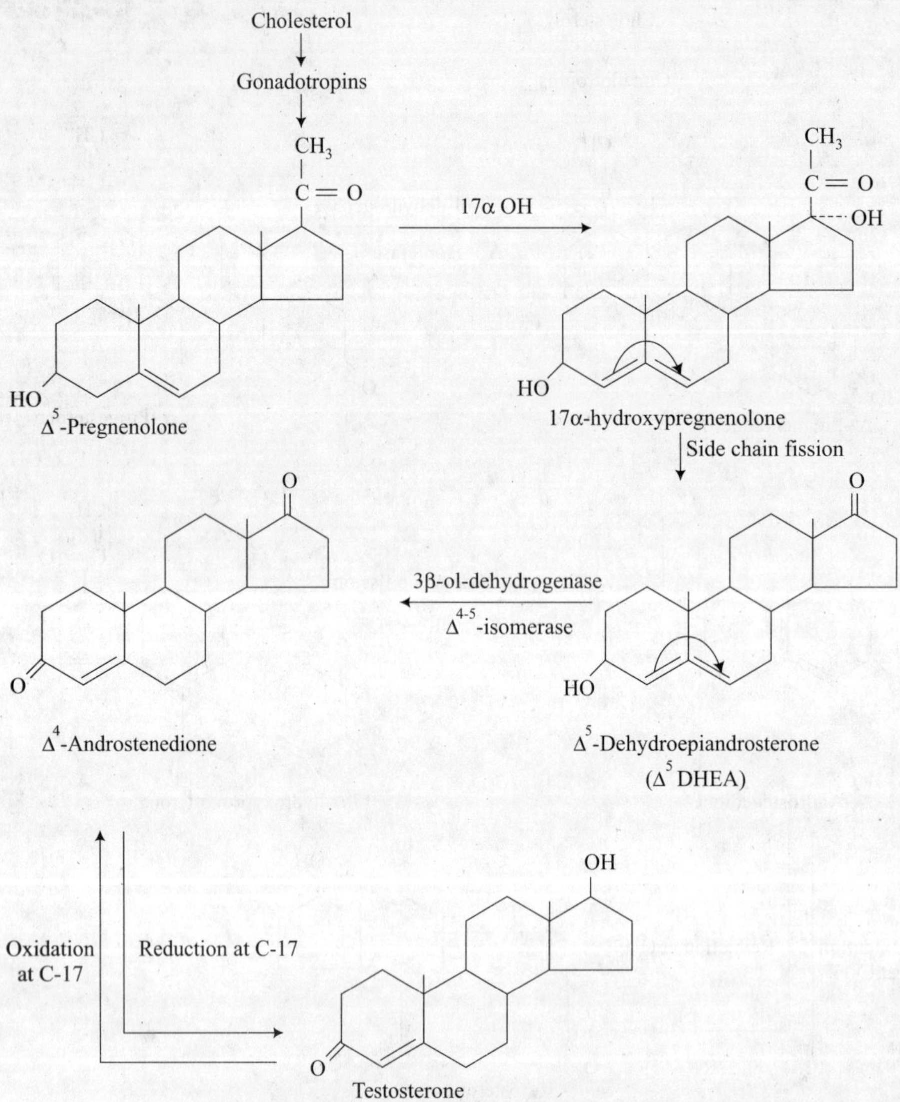

Fig. 8. Δ^5-pathway for androgenesis.

cells which occupy the entire follicle. The cells facing the antrum are called as corona radiata and those above the ovum as cumulus oophorus. The granulosa cells under the action of FSH produce estrogens. The antrum contains liquor folliculi, the secretion of follicular or granulosa cells. Now under the influence of LH, ovulation occurs i.e. the ovum is extruded out from the Graafian follicle. The remaining tissue, under the influence of LH forms the corpus luteum whose cells are luteinized by LH i.e., they synthesize and accumulate cholesterol. Hence, the cells are now called as granulosa lutein cells. These cells synthesize progesterone under the stimulation of LH (Fig. 9).

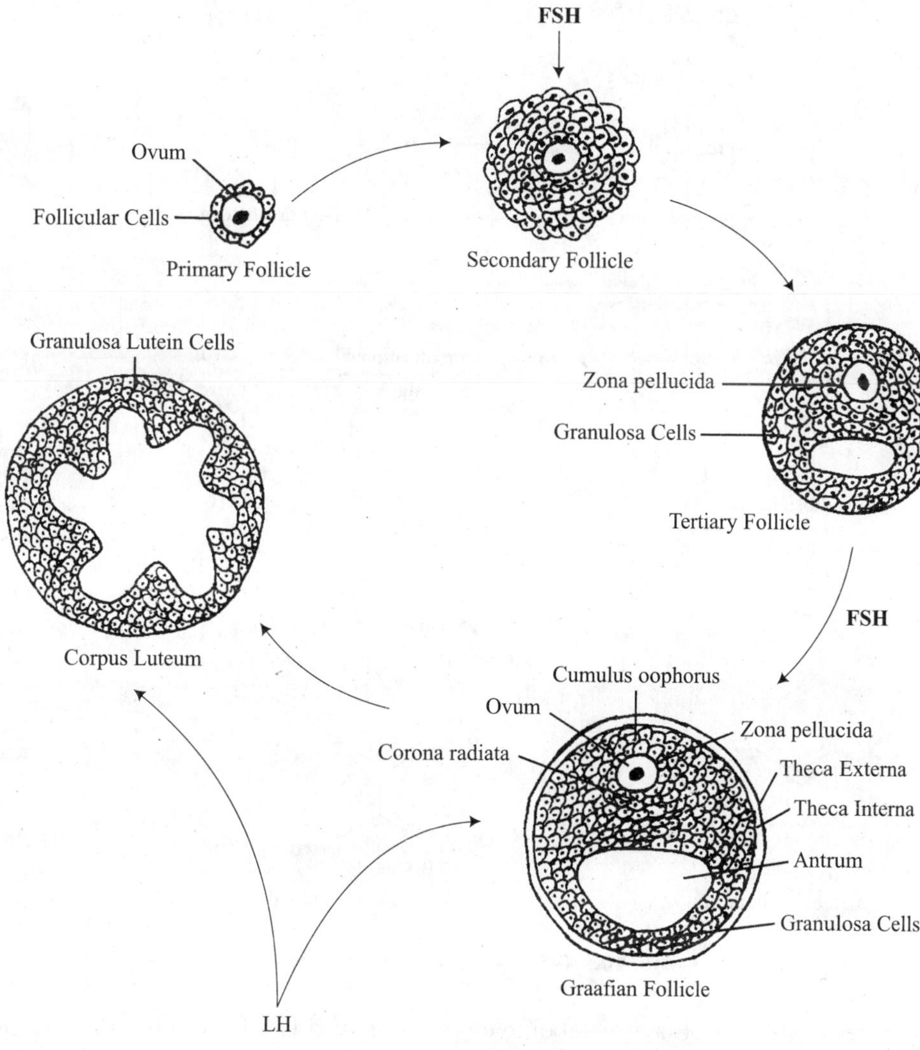

Fig. 9. The different follicles in the ovary.

The C-18 steroidogenic pathway occurs in the granulosa cells under the stimulation of FSH (Granner, 1990b). The cells contain cholesterol which is converted to Δ^5-pregnenolone which inturn forms Δ^5-DHEA and Δ^4-androstenedione as described earlier for the Δ^5-pathway of androgenesis. Δ^5-pregnenolone could also be converted to Δ^4-progesterone which inturn forms Δ^4-androstenedione as in the Δ^4-pathway of androgenesis. Δ^4-androstenedione undergoes aromatization of Ring A to form Estrone (E1) with a single –OH group at C-17. It is a C-18 steroid. Estrone undergoes reduction at C-17 to form the most potent estrogen, viz., 17 β estradiol with two-OH groups. Hence it is designated as E_2. Oxidation of 17β E_2 at C-17 could result in formation of estrone. However, reduction of ketone in estrone and 16α– hydroxylase action will form Estriol (E_3) which has three-OH groups, a β-OH at C–3, C–17 and an α–OH

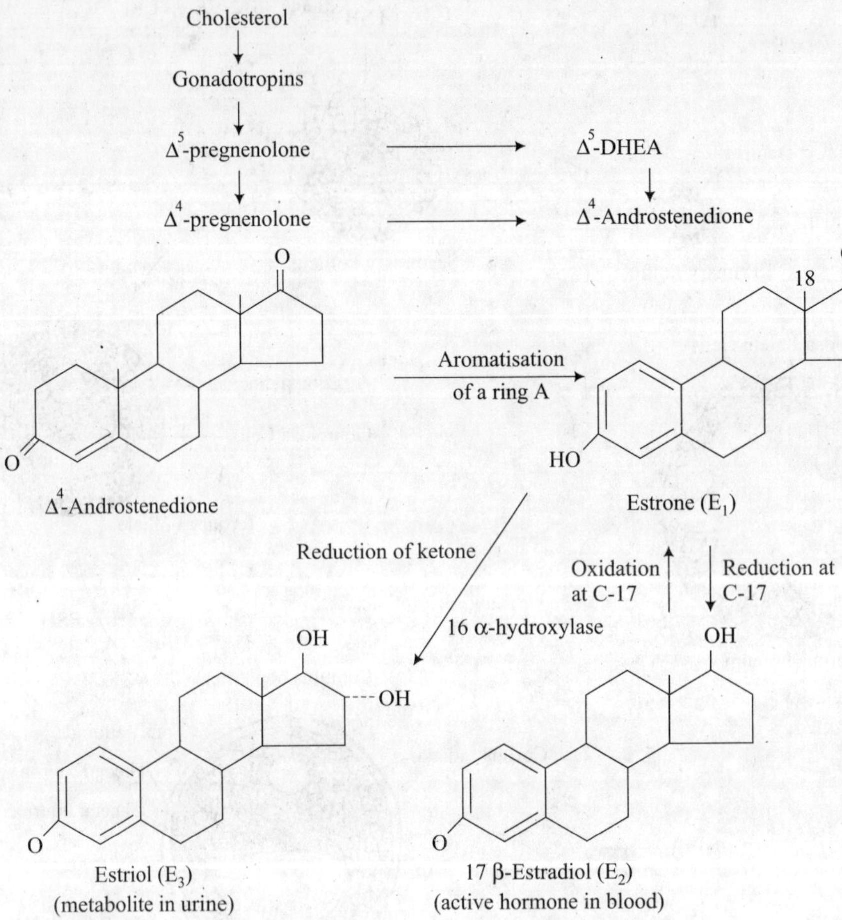

Fig. 10. C-18 steroidogenesis pathway.

(–OH) at carbon–16. Estriol is a metabolite of estradiol and is found in the urine, whereas, estradiol is the active estrogen found in the blood (Fig. 10).

After ovulation, which is brought about by the action of LH on Graafian follicle, the corpus luteum is formed whose granulosa lutein cells contain cholesterol and synthesize progesterone under the stimulation of LH. In the female reproductive system, estrogens act first to bring about growth of the organs and thereafter progesterone acts to bring about secretion from the estrogen primed organs.

SUMMARY

In the Introduction, steroids are described as derived lipids. The formation of the basic steroid nucleus, cyclopentanoperhydrophenanthrene has been described followed by the three basic nuclei of C-21 steroids, namely Pregnane, C-19 steroids, Androstane and of C-18 steroids, estrane have been described. Some rules in nomenclature of steroid molecules are given. The importance of cholesterol as the parent molecule is emphasized and the formation of pregnenolone and progesterone from cholesterol is described. The major and minor pathways for formation of C-21 corticosteroids in the adrenal have been shown with the formation of glucocorticoids and mineralocorticoids together with their site of synthesis in adrenal cortex. The C-19 steroidogenesis of androgens in testis from Leydig cells by the Δ^5 and Δ^4 pathways are described. Lastly, the site of C-18 steroidogenesis of estrogens in the ovary and the pathway from androstenedione is described.

ACKNOWLEDGEMENTS

Grateful thanks are due to Prof. M.V. Rao, Dr. Devendrasinh D. Jhala for typing the manuscript, to Dr. Nisha Shah, Professor, Department of Chemistry, Gujarat University, Ahmedabad, for all her help and Ms. Priya Santosh, Computer operator, Chemistry Department, Gujarat University, for painstakingly drawing the steroid structures.

REFERENCES

Granner, D.K. (1990a) Hormones of the adrenal cortex. In : Harper's Biochemistry (Eds., Murray, R.K., Granner, D.K., Mayes, P.A. and Rodwell, V.W.), pp. 499-510, Appleton & Lange, Connecticut, USA.

Granner, D.K. (1990b) Hormones of the gonads. In : Harper's Biochemistry (Eds., Murray, R.K., Granner, D.K., Mayes, P.A. and Rodwell, V.W.), pp. 516-529, Appleton & Lange, Connecticut, USA.

Hall, P.F. (1979) Testicular hormones : synthesis and control. In : Endocrinology (Ed., De Groot, L.J.), Vol. 3, pp. 1511-1520, Grune and Stratton, USA.

Litwack, G. and Schmidt, J. (2002) Biochemistry of hormones II : steroid hormones. In : Biochemistry (Ed., Devlin, T.M.), pp. 959-988, John Wiley & Sons, Inc. Publication, New York, USA.

Mayes, P.A. (1990) Cholesterol synthesis, transport and excretion. In : Harper's Biochemistry (Eds., Murray, R.K., Granner, D.K., Mayes, P.A. and Rodwell, V.W.), pp. 249-260, Appleton & Lange, Connecticut, USA.

Rudney, H. and Sexton, R.C. (1986) Regulation of cholesterol synthesis. Ann. Rev. Nutr. 6 : 245-260.

Wilson, J. (1975) Metabolism of testicular androgens. In : Handbook of Endocrinology (Eds., Hamilton, D.W. and Greep, R.P.), Vol. 5, pp. 491-508, American Physiological Society, Washington, DC, USA.

Suggested Readings

1. Introduction to Steroid Biochemistry and its chemical application (1970) M.A. Freedman and S.N. Freedman (Eds.), Thelma Clark and Harriet Lovejoy Harper and Row Publishers, New York.
2. Steroids (1954) L. F. Fieser, Scientific American, 192 : 52-60,.
3. Harpers' Review of Biochemistry (1981) D. W. Martin, P. A. Mayes and V. W. Rodwell, Lange Medical Publications Maruzen Asia PTE Ltd, Singapore.
4. Metabolic and Endocrine Physiology (1981) J. Tepperman, Year Book Medical Publishers, Inc., USA.

REVIEW QUESTIONS

Long Questions

1. Describe the formation of the basic steroid nucleus and the nucleus of C-21, C-19 and C-18 compounds. Add a note on their important features.
2. Explain Steroidogenesis of C-21 steroids in the adrenal cortex and indicate if they are gluco- or mineralocorticoids. Describe their function.
3. Give an account of C-19 steroidogenesis of androgens in the testis. Indicate the site of synthesis and its stimulation.
4. Describe formation of C-18 estrogens in the ovary. Indicate the site of synthesis. Does the ovary produce any C-21 steroid? If so, explain the site of synthesis and the pathway.
5. What are the basic rules in biochemistry for nomenclature of different steroids? Draw the structure of D^5-pregnenolone, progesterone, cholesterol, aldosterone, cortisol, D^5–dehydroepiandrosterone, D^4-androstenedione, estradiol, estriol, 17 a-hydroxy progesterone, testosterone.

Multiple Choice Questions

1. Cyclopentanoperhydrophenanthrene nucleus has.
   (a) one double bond and 17 carbon atoms
   (b) two double bonds and 17 carbon atoms
   (c) one double bond and 19 carbon atoms
   (d) no double bonds and 17 carbon atoms
2. Pregnane nucleus is present in.
   (a) all steroids (b) C-18 steroids
   (c) C-19 steroids (d) C-21 steroids
3. Estrane nucleus has.
   (a) 27 carbon (b) 21 carbon
   (c) 19 carbon (d) 18 carbon
4. Cholesterol has 27 carbons and – OH group at:
   (a) carbon 19 (b) carbon 18
   (c) carbon 6 (d) carbon 3
5. A methyl group is present in C -21 steroids on.
   (a) carbons 5,6 (b) carbons 4,5
   (c) carbons 3,4 (d) carbons 18, 19
6. A delta $(D)^5$ steroid has a double bond between.
   (a) carbons 1 and 2 (b) carbons 3 and 4
   (c) carbons 4 and 5 (d) carbons 5 and 6

7. The major pathway in C-21 steroidogenesis yields (in the last step).
   (a) aldosterone (b) corticosterone
   (c) 18 hydroxycorticosterone (d) Cortisol
8. The minor pathway in C-21 steroidogenesis yields (in the last step).
   (a) corticosterone (b) 11-deoxycortisol
   (c) Cortisol (d) Aldosterone
9. D^5-androstenedione undergoes C-17 reduction and yields.
   (a) 17 hydroxy androstenedione (b) Δ^5-dehydroepiandrosterone
   (c) Dihydrotestosterone (d) Testosterone
10. What is the metabolite of estrogens found in the urine?
   (a) Estradiol (b) Estrone
   (c) Estradiol and estrone (d) Estriol

Answers

1. (d)	2. (d)	3. (d)	4. (d)	5. (d)	6. (d)
7. (d)	8. (d)	9. (d)	10. (d)		

CHAPTER 11

Identification Techniques for Sex Steroid Hormones

Pratap B. Singh[*], Vandana Singh and S.S. Srivastava

Dept. of Zoology, Tilak Dhari College, Jaunpur – 222 002

E-mail: pratap_b_singh@rediffmail.com

Hormones play very important role in reproductive/pheromonal behavior which are species specific in several teleosts (Colombo *et al.*, 1973; Ueda *et al.*, 1984; Trant and Thomas, 1986; Kime, 1990; Asahina *et al.*, 1990; Ebrahimi *et al.*, 1995; Nagahama, 1997; Condeca *et al.*, 2001). Sulfates steroids are found in human ovarian tissue (Wallace and Silberman, 1964). Biosynthesis of 17,20αP from 17-P by sperms of *Cyprinus carpio* (Asahina *et al.*, 1990) and testicular fragments of *Silurus glanis* and *Carassius auratus* (Kime *et al.*, 1993) have been demonstrated *in vitro*. A shift in steroidogenesis from 11-KT (C19) to 17,20βP (C21) occurs during spermiating period (Joss *et al.*, 1996; Barry *et al.*, 1990a). Recently, Lower *et al.* (2004) reported electro-olfactogram recording of the olfactory epithelium of both male and female roach *Rutilus rutilus* and demonstrated that both sexes were able to detect free and glucuronidated 17α,20βP with high sensitivity. These authors did not find sulfated form of steroids. Scott *et al.* (2005) have demonstrated the rate of uptake of sex steroids from water by *Tinca tinca* and was influenced by their high affinity for sex steroids binding protein in plasma.

Most of the above information is restricted up to one or four steroids metabolites identification in fish. Our reports deal with the identification technique in roach spermiating and pre-spermiating testes. The object is to provide the complete microchemical reaction involved for identification which can be used for identification of steroid hormones by microchemical reactions and crystallization to constant specific activity.

[*]Corresponding author

EXTRACTION OF SEX STEROIDS (UNCONJUGATED AND CONJUGATED)

Spermiating and pre-spermiating roach fish (114-157 g wt range, 22-24 cm in size and 1.12-4.10, gonadosomatic index (GSI)) were obtained during breeding season in May. The fish were kept for 2-3 days in tanks of circulating fresh water and killed as required by decapitation for *in vitro* experiments. Chopped testes (100 mg) were incubated in 1 ml goldfish incubation medium (Ebrahimi *et al.*, 1995) for 3 hr at 20°C using 96.5 kBq 17[1, 2, 6, 7 - H^3]hydroxyprogesterone per incubation. After incubation the medium was frozen at -20°C until extraction of steroids.

The incubation medium was extracted twice with 5 ml distilled dichloromethane to give the unconjugated (free) steroid fraction and the aqueous residue (containing conjugated-glucuronide and sulphate steroids) was treated with 800 μg β-glucuronidase (Sigma G0251, from bovine liver, Type B-1 500,000 unit) to hydrolyse the glucuronide conjugate in 1 ml acetate buffer of 0.2M, 4.8pH in each extracting tube for 24 hr at 37°C. After incubation, steroid moieties of the glucuronides were extracted twice with 5 ml dichloromethane. The aqueous phase was extracted twice with 4 ml butanol-1 saturated with water, and extracts were evaporated. After evaporation for solvolysis, in each vial was added 20 μl distilled water and vortexed. Then the contents of each sample were treated with trifluroacetic acid (TFA) and ethyl acetate (1/100. v/v; 3 ml; 45°C for acid hydrolysis of sulfate conjugate steroids; 18 hr). After solvolysis steroid moieties of sulfates were extracted twice with 3 ml of ethyl acetate after adding 1 ml distilled water in each tube. Loss in residues/aqueous radioactivity in the medium were checked by counting the radioactivity in an aliquot of the medium in a BETAmatic Liquid Scintillation Counter, which was less than 1.2%. The schematic details of extraction, TLC and HPLC have been given in the text.

DIFFERENT TLC SYSTEM USED FOR SEPARATION OF RADIOACTIVE METABOLITE STEROIDS

TLC-I :Chloroform - Methanol (98 : 2) B_2, TLC-II :Chloroform. TLC - III: Chloroform - Hexane (1:1), TLC-IV: Chloroform-Methanol (9:1)B_{10}, TLC-V:2,4-Trimethylpentane–2-methylpropan-2-ol (7:3) A_{30}, TLC-VI:Chloroform-Methanol (95:5)B_5, TLC-VII:Chloroform-Methanol (80:20) $B_{20,}$ TLC-VIII:Chloroform-Methanol (94:6)B_6 were used in this study.

THIN–LAYER CHROMATOGRAPHY (TLC)

After extraction of unconjugated (free) and conjugated (glucuronide and sulfate) steroids, TLC was carried out on pre-coated plastic sheets (20x20cmx0.2mm, Silica gel 60 F254, E. Merck, Germany, Art.5735). Standards were detected by UV absorption at 254 nm, visualization in iodine vapor or by spraying with 10% molybdophosphoric acid in methanol and heating for 10-20 min at 100°C. The TLC System used were TLC-I, chloroform:methanol (98:2), TLC-II, chloroform; TLC-IV, chloroform:methanol (9:1), and TLC-VI, chloroform:methanol (95:5).

Before TLC in each sample mixture of authentic (AD, AT,17-P, T, 17,20βP, 17,20αP, 11-KT and 11β-T) standards (20 μg) were added in unconjugated and conjugated fractions for UV visualization. Then each sample was loaded separately in 2.5 cm lane of TLC pre-coated sheet, and plates were kept in TLC-VI solvent system. After solvent system plates were air dried and spots were encircled by pencil in UV. Now plates were cut off into 2.5 cm streak separately by scissors, and put colored isotopic markers 5 ml just above solvent front. Radioactive peaks were detected for each sample fraction on streak using a Packard Model 7220 TLC Scanner, were eluted with 4 ml of methanol, 1/20th counted for % yield. Radioactive peak in TLC-VI were recorded corresponding to added authentic standards which had Rf values: fraction 1- AD+AT, 0.60; fraction 2- 17-P, 0.44; fraction 3-T, 0.35; fraction 4- 11-KT+17,20βP+ 17,20αP, 0.26; fraction 5- 11β-OHT, 0.08. Further fraction 1 was re-chromatographed in TLC-I for further separation of AD and AT. TLC streak was scanned and Rf values for AD (0.47) and AT (0.41) were noted for this solvent system. For fraction 4, HPLC was done for further separation of steroids.

HIGH-PERFORMANCE LIQUID CHROMATOGRAPHY (HPLC)

HPLC used Philip PU4100 system with PU6030 data capture unit, a PU4110 detector set at 242 nm, and a Radiomatic Flo-one Beta detector using Packard Flo-Scent A Scintillion fluid (1.5 ml/ min). Metabolites and standards were eluted from a 25 cm × 4.6 mm column with 35% acetonitrile/65% water for 10 min at flow rate 1 ml/min. Retention times were :-11-KT/S = 7.50

Extraction of steroids from incubation medium

One ml incubation medium + 5 ml Dichloromethane [CH_2Cl_2] (double extraction)

↓

After double extraction with CH_2Cl_2 (solvent phase evaporated) *Free (unconjugated) steroids* TLC*/RIA**

Residue + 800 µg β-glucuronidase incubated in 1ml 0.2M acetate buffer at 37°C for 24 hr to hydrolyse glucuronide conjugates.

↓

Double extraction with 5 ml CH_2Cl_2 (solvent phase evaporated) *Glucuronide (conjugated) steroids* TLC*/RIA**

Aqueous phase extracted twice with water saturated Butanol-1 (extract evaporated). Residue counted for loss. For solvolysis, evaporated extract + 20 µl DW + 3 ml TFA in ethyl acetate (1 : 100, v/v) for 18 hr at 45°C to hydrolyse sulfate conjugates.

↓

After solvolysis + 1ml DW and double extraction with ethyl acetate (solvent phase evaporated). *Sulfate (conjugated) steroids.* TLC*/RIA**

Aqueous residue (waste) was counted for procedural loss

*TLC = Thin-layer chromatography for steroids produced from exogenous precursor (radiolabelled); **RIA = Radioimmunoassay for endogenous production of steroids. [Aliquots from the free, glucuronide, and sulfate fractions and the aqueous residues (wastes) after butanol and ethyl acetate extraction were counted for radioactivity. Less than 1.2% of the radioactivity remained in the residual medium] min, 17,20αP = 14.50 min, 17,20βP = 18.50 min. A standard mixture containing all the expected steroids [5 ml, mg/ml in ethanol and distilled water (DW) (1:1)] was injected with each extracted radioactive fraction [10 ml/100 kdpm in ethanol and DW (1:1)]. Full identification of the metabolites was followed as per method described (Ebrahimi *et al.*, 1995).

Thin-Layer Chromatography (TLC)

After extraction TLC for unconjugated (free) and conjugated (glucuronide + sulfate) steroids was done in the following solvent systems:

TLC – I Chloroform : Methanol (98 : 2)

TLC – VI Chloroform : Methanol (95 : 5)

Authentic standards (20 μg of each of AD, AT, 17-P, T, S, 17,20βP, 17,20αP, 11-KT and 11-βT was added for UV visualization at 254 nm.)

Radioactive peaks were detected using Packard Model 7200 TLC scanner, eluted with 4 ml of methanol which corresponded to added authentic steroids:

Fraction – 1 (F1) : AD + AT	Rf. 0.60
Fraction – 2 (F2) : 17-P	Rf. 0.44
Fraction – 3 (F3) : T	Rf. 0.35
Fraction – 4 (F4) : 11-KT/S+17,20βP+17,20αP	Rf. 0.26
Fraction – 5 (F5) : 11-βT/pregnentetrols	Rf. 0.08

F1 was rechromatographed on TLC–1 for AT and AD separation:

F1	:	AT	Rf. 0.41
		AD	Rf. 0.47

The components of F4 were further resolved by HPLC for separation of different steroids.

High-Performance Liquid Chromatography (HPLC)

HPLC used Philip PU4100 system with PU6030 data capture unit, a PU4110 detector set at 242nm, and a Radiomatic Flo-one Beta detector using Packard Flo-Scent A Scintillion fluid (1.5 ml/min)

Metabolites and standards were eluted from a 25 cm × 4.6 mm column with 35% acetonitrile/65% water for 10 min at flow rate 1 ml/min.

Retention times were:

11-KT/S = 7.50 min

17,20αP = 14.50 min

17,20βP = 18.50 min

A standard mixture containing all the expected steroids [5 ml, mg/ml in ethanol and distilled water (DW) (1 : 1)] was injected with each extracted radioactive fraction [10 μl/100 kdpm in ethanol and DW (1 : 1)].

Full identification of the metabolites was followed as per method described (Kime, 1990).

MICROCHEMICAL REACTIONS, IDENTIFICATION OF STEROIDS AND CRYSTALLIZATION OF STEROIDS

A steroid was considered positively identified when the peak of radioactivity was isopolar with authentic steroid in at least two chromatographic system, the products of at least one chemical reaction was isopolar with authentic derivative, and a constant (±5% of the mean) isotope ratio or specific activity was obtained through three successive crystallization with authentic steroid, using the following solvent combination: acetone-water, methanol-water, and ethanol-water. The various sex steroids interrelationship have been given in the Fig 1-3.

VARIOUS MICROCHEMICAL REACTION INVOLVED IN THE PRESENT STUDY

I. Acetylation

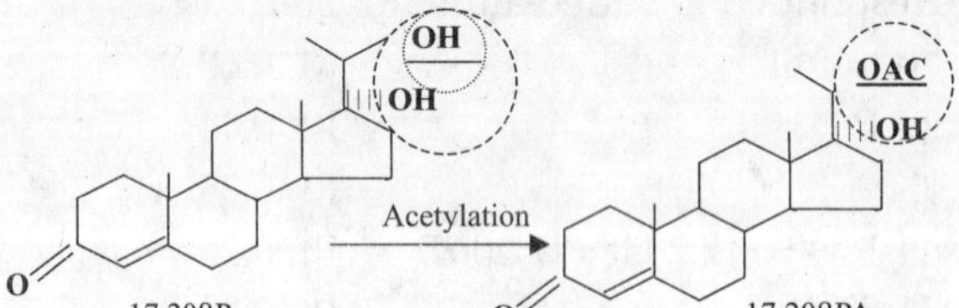

17,20βP 17,20βPAc

$$-\text{OH} \xrightarrow{\textbf{On Acetylation}} -\text{OAc}$$

On acetylation alcohol group (-OH) was converted to acetates (-OAc) which did not react with 17 or 11βOH. The steroid was dried into a stoppered test tube. Added 1 drop pyridine and 2 drops of acetic anhydride, stoppered and left overnight at room temperature. Added 1 ml M.HCl (to neutralize pyridine), shaken well and added 5 ml dichloromethane and shaken (to extract steroids). Allowed to settle or centrifuged and removed the top (aqueous) layer. Added 1 ml M. Sodium bicarbonate (to neutralize HCl), shaked and top layer was removed. Now, dichloromethane was washed twice with the distilled water (to remove remaining bicarbonate). Pipetted off dichoromethane into clean vial and evaporated.

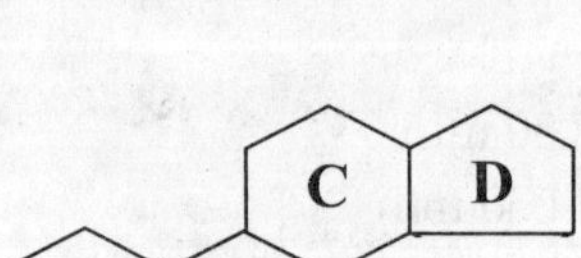

The Steroid skeleton – Cyclopentanoperhydrophenathrene

(Fig. 1 Contd.)

(Fig. 1 Contd.)

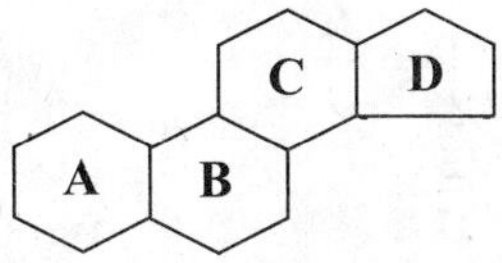

The Steroid skeleton Cyclopentanoperhydrophenathrene

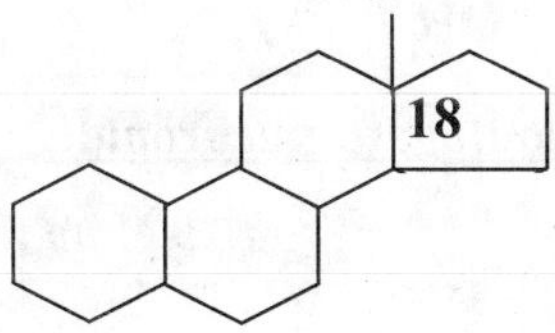

1. C18 (Estrogens)
   - **Estrone**
   - **Estradiol**

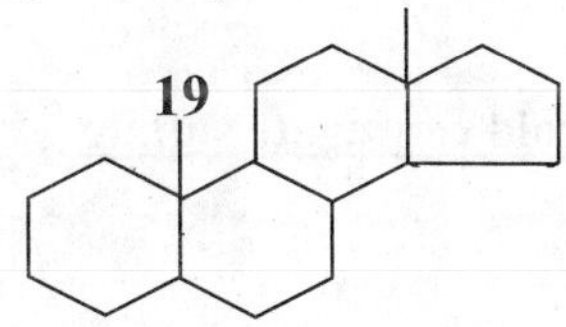

2. C19 (Androgens)
   - **Testosterone**
   - **11β-Hydroxytestosterone**
   - **11-Ketotestosterone**
   - **Androstenedione**
   - **Androstenetrione**

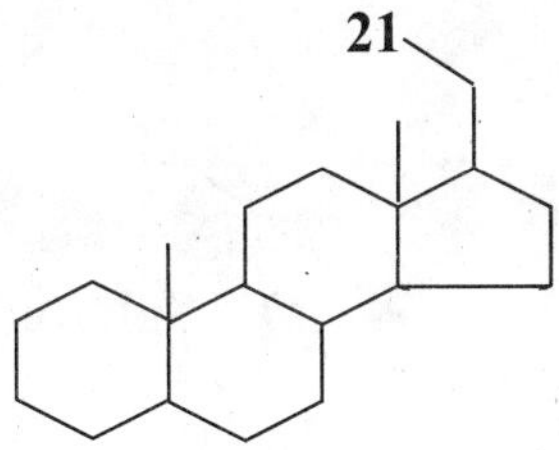

3. C21 (Progestogens)
   - **Pregnenolone**
   - **17-Hydroxyprogesterone**
   - **17α, 20αP**
   - **17α, 20βP**

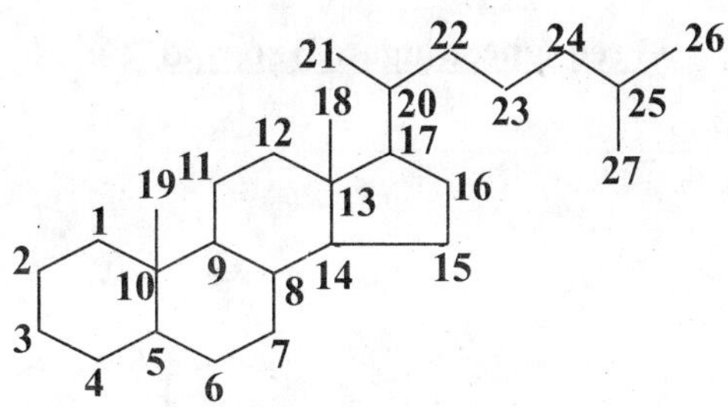

4. C21 (Corticosteroids)
   - **Cortisol**
   - **Cortisone**
   - **11-Deoxycortisol**
   - **Cholesterol**

Fig. 1. The Steroids classification.

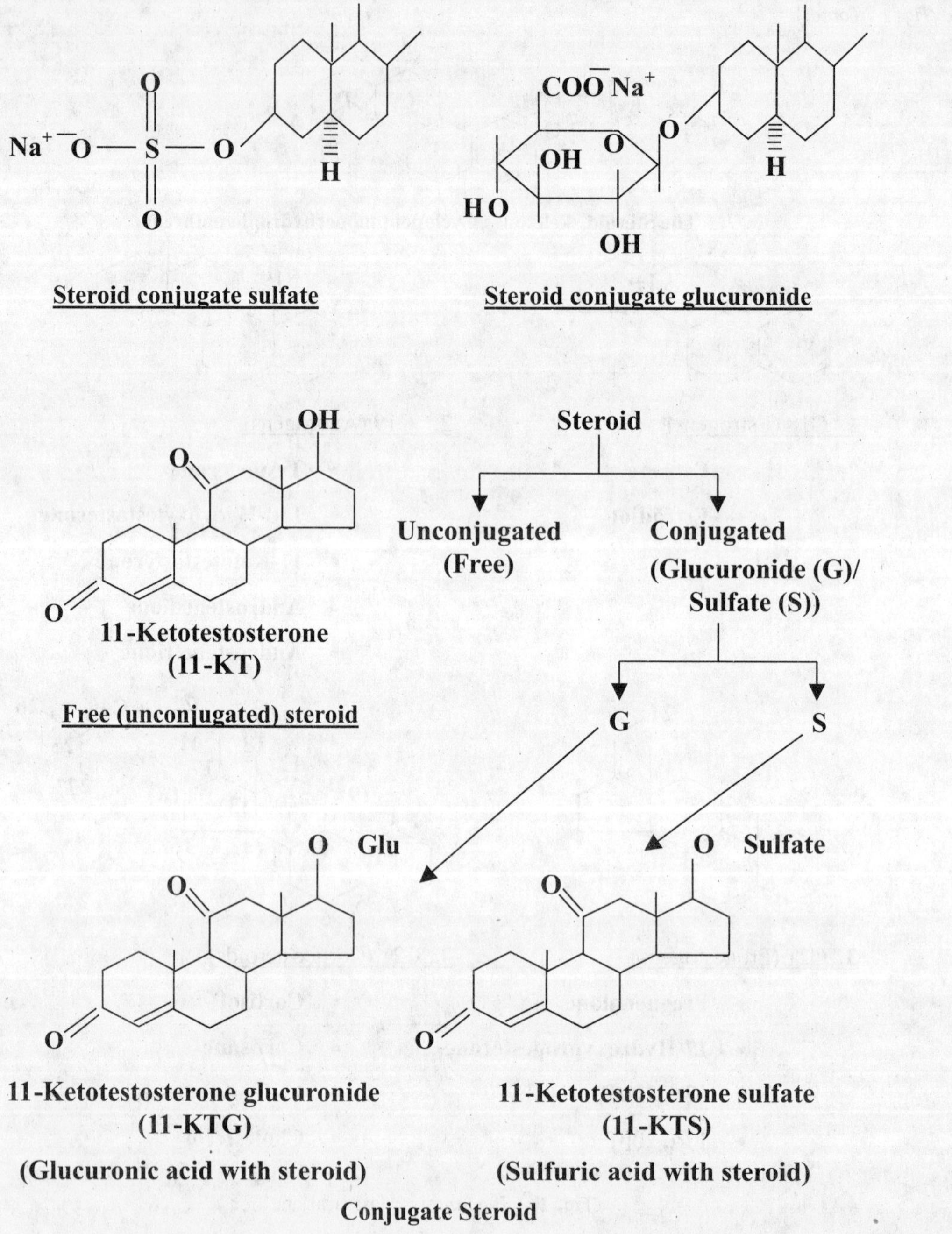

Fig. 2. Unconjugated/conjugated sex steroids.

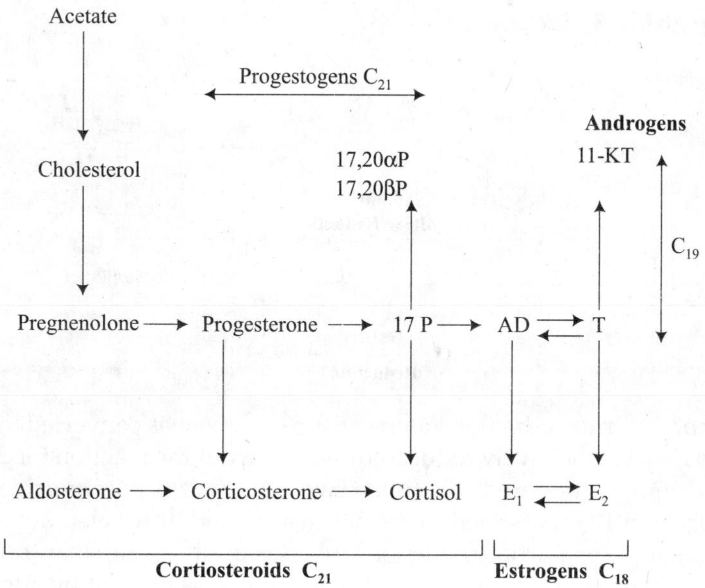

Fig. 3. Interrelationships of the steroid hormones in fish.

2. Chromium Trioxide Oxidation

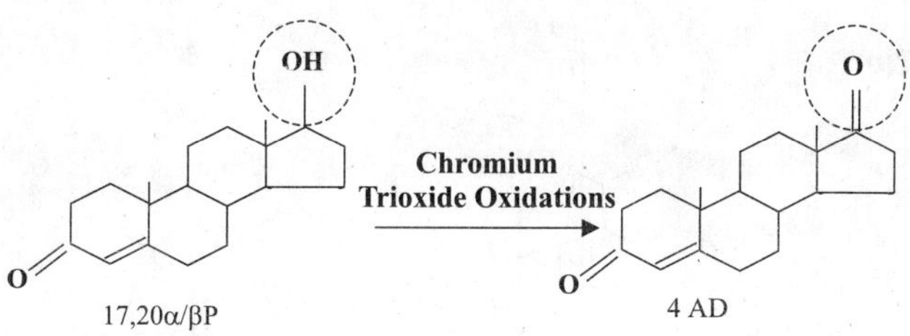

$$\text{-OH} \xrightarrow[\text{Oxidation}]{Cr_2O_3} \text{>C= O}$$

On oxidation, alcohol group (-OH) was converted to ketones (>C= O) which oxidized into 17-hydroxylated pregnenes with a C-20 ketone or alcohol to 17-ketoandrogens. The steroid was dried into a stoppered test tube. Five drops of glacial acetic acid and 3 drops of aqueous chromium trioxide (20 mg/ml chromium trioxide in water) were added. Stoppered and left overnight at room temperature. Added 1 ml 1% sodium metabisulphite solution to reduce excess reagent (turns green-blue). Five ml dichloromethane was added, shaked, allowed to settle and removed the top layer. It was washed successively with 1 ml M. sodium bicarbonate (to neutralize acetic acid). Transferred to clean vial and evaporated.

3. Sodium Borohydride Reduction

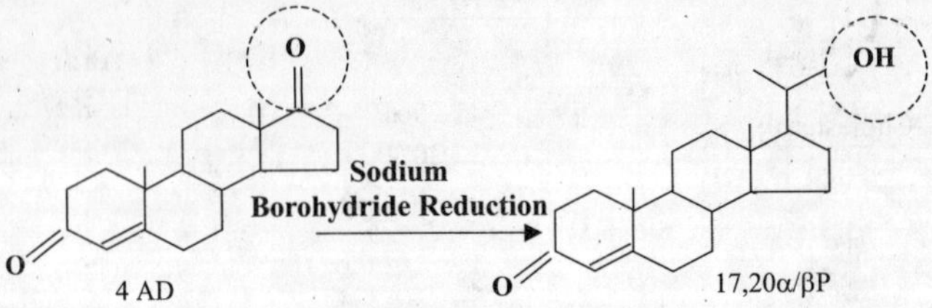

4 AD → 17,20α/βP

$$>C=O \xrightarrow[\text{Reduction}]{NaBH_4} -OH$$

On sodium borohydride reduction ketone group (>C=O) was converted to alcohols (–OH). The >C=O group may be selectively reduced by carrying out the reaction for 20 mins at –20°C. Dried the steroid into a stoppered test tube. Two hundred μl borohydride solution (4 mg sodium borohydride in 10 ml distilled methanol) was added. It was left at room temperature for 30 min (or longer). For partial reduction pre-cooled all solutions and tubes at -20°C in cardice/ethanol allowed to react for 20 min. Removed excess reagent by addition of 1 ml 1% aqueous acetic acid. Added 5 ml dichloromethane, shaken well, allowed to settle (or centrifuge) and removed the top layer. Washed successively with 1 ml sodium bicarbonate (to neutralize acetic acid) and again treated with 2 × 1 ml water. Transferred to clean vial and evaporated.

4. Periodate Oxidation

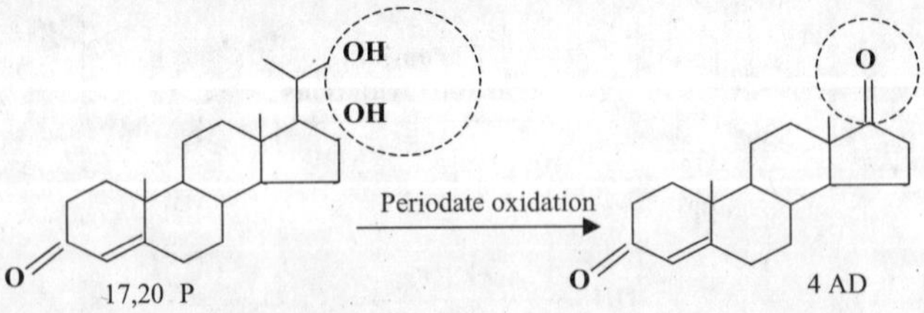

17,20 P → 4 AD

[Oxidises —C(OH)—C(OH)— to >C = O, eg. 17,20 diols to 17-keto (17,20βP to 4 AD).

On periodate oxidation —C(OH)—C(OH)— was oxidized to >C = O. As per example 17,20β-diols was oxidized to 17-keto (17,20βP to 4-androstenedione). The steroid was dried into a stoppered tube. Added 100 ml methanol and 100 ml periodate solution (40 mg periodic acid in 1 ml M sulphuric acid). It was left overnight in the dark at room temperature. Added 1 ml M sodium bicarbonate (to neutralize the acid) and added 5 ml dichloromethane. Shaked and allowed to settle or centrifuged and removed the top layer. Washed twice with water, transferred the dichloromethane to a clean vial and evaporated.

5. 17β-HSD Reduction

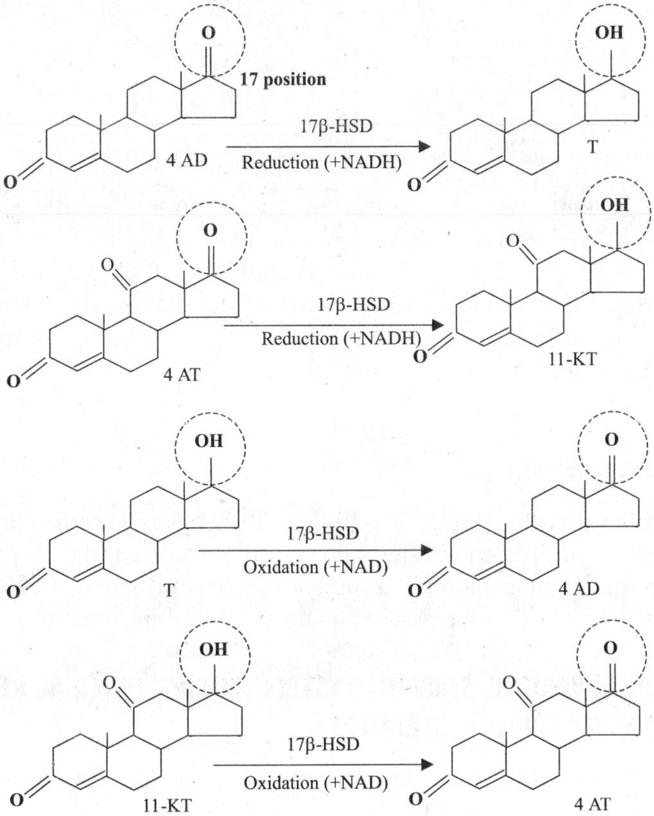

[17β-Hydroxysteroid dehydrogenase oxidized 17β-hydroxy to 17-keto, or reduces 17-keto to 17β-hydroxysteroid. Procedure was as for 20β-HSD]

6. 20β-HSD Reduction

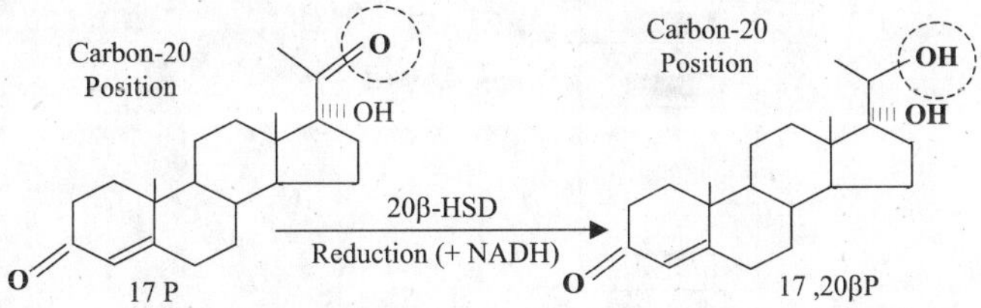

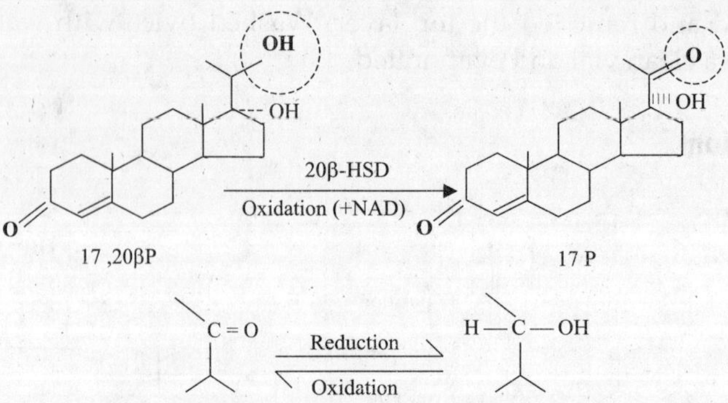

On reduction 20-keto was converted to 20β-hydroxysteroids. On oxidation, 20β-hydroxysteroids was converted to 20-keto. The steroid was dried into stoppered tube. Added 500 μl 0.05 M Tris buffer pH 7.6, 0.76 units 3α,20βHSD (40 μl; 19 units/ml) 500 μl buffer containing 2 mg/ml NAD (for oxidation) or NADH (for reduction). Contents were left for 2 hr at room temperature. Added 5 ml dichloromethane, shaked and removed the top layer. Washed twice with water and transfered organic phase was transferred to clean vial for evaporation.

7. Sodium Bismuthate Oxidation

On sodium bismuthate oxidation 21OH-20-keto-17OH was converted to 17-ketosteroids (cortisol to 11β-hydroxyandrostenedione). The steroid was dried into stoppered test tube and added 100 ml bismuthate suspension (2.5 mg sodium bismuthate in to 50 μl water and 50 μl acetic acid). Left at least 3 hr. Now extracted as the method followed for periodate oxidation.

AFTER SEPARATION IN VARIOUS SOLVENT SYSTEMS MICROCHEMICAL REACTION WAS DONE AS FOLLOWS FOR VARIOUS STEROIDS

Complete separations of metabolites were done on TLC Sheet in TLC-IV:

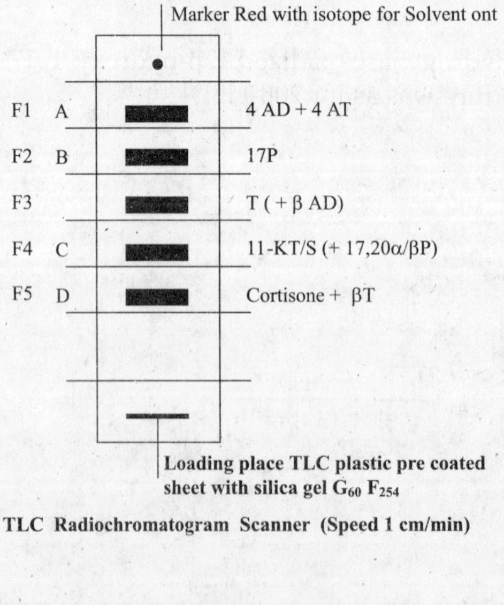

TLC Radiochromatogram Scanner (Speed 1 cm/min)

Chromatography in TLC-VI (TLC-VI, $CHCl_3 : CH_3OH$ (95 : 5) of the free fraction gave four peak (A, B, C and D) of the radioactivity and eluted which also corresponded to carrier androstenedione (AD) + androstenetrione (AT), 17-hydroxyprogesterone (17P), 11-KT + S + 17α,20αP + 17α,20βP + and 11β-hydroxytestosterone (11β-T). No distinct radioactivity peak was associated with carrier testosterone (T) and having very low % of yield.

1. Identification for Peak A(Fraction-1) :

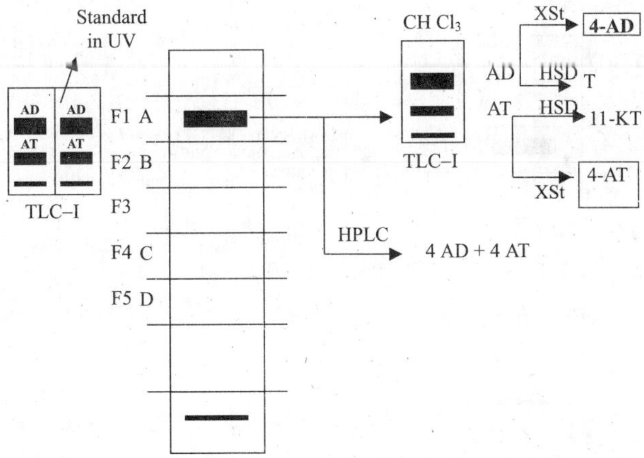

Peak A (Fraction-1) : Combined fraction of peak A when rechromatographed in TLC-I (TLC-VI, $CHCl_3 : CH_3OH$ (95 : 5) two distinct sub-peak (A1, and A2) were obtained, was isopolar with AD and AT which were converted by 17β-HSD to a product isopolar with T and 11-KT respectively in TLC-I. The two metabolites were identified as AD and AT by crystallization to constant specific activity.

Microchemical Reaction

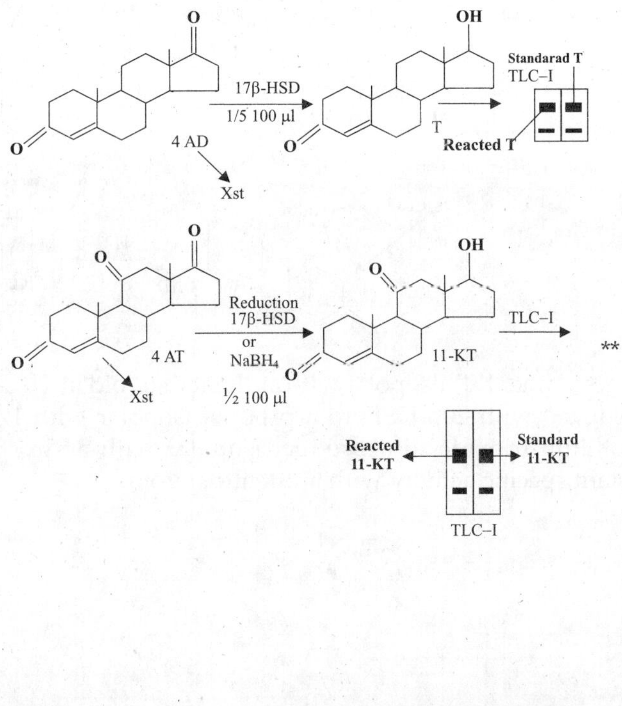

Androstenedione (AD) : Isopolar with authentic steroid in TLC-I. Reduced with 17β- HSD to a product isopolar with testosterone in TLC-I. Crystallized to constant specific activity with authentic steroid.

Androstenetrione (AT) : Isopolar with authentic steroid in TLC-I. Reduced with 17β-HSD to a product isopolar with 11-Ketotestosterone in TLC-I. Crystallized to constant specific activity with authentic steroid.

2. Identification for Peak B (Fraction-2):

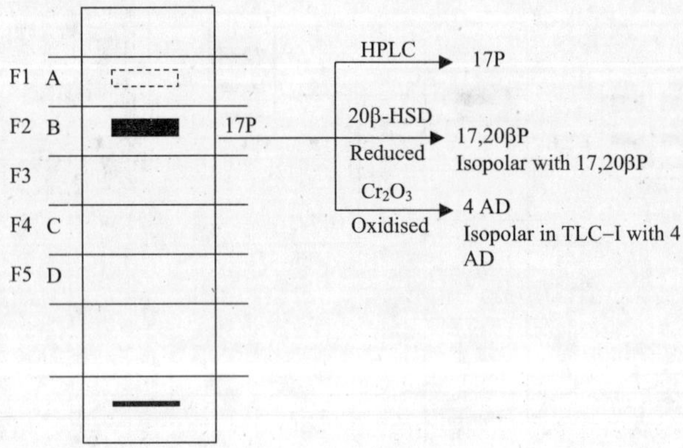

Peak B (Fraction 2) : Isopolar to authentic 17P in TLC-VI. Reduced with 20β-HSD to a product isopolar with 17α,20βP and oxidized to a product isopolar with AD in TLC-I. This metabolite was identified 17P by crystallization to constant specific activity.

Microchemical Reaction

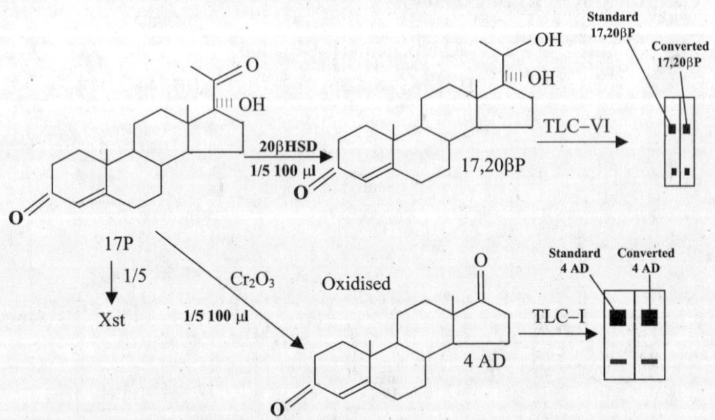

17-Hydroxyprogesterone (17P):Isopolar with authentic steroid in TLC- VI (TLC-VI, $CHCl_3$: CH_3OH (95 : 5). Reduced with 20β-HSD to a product isopolar with 17α, 20βP in TLC-VI. Oxidised with chromium trioxide to a product isopolar with androstenedione in TLC-I. Crystallized to constant specific activity with authentic steroid.

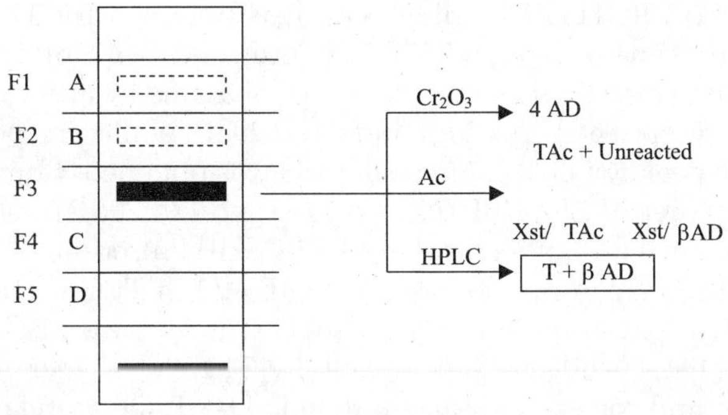

Fraction 3 : There was no distinct peak for T activity but was isopolar to authentic steroid in TLC-VI (TLC-VI, $CHCl_3$: CH_3OH (95 : 5),corresponding area was eluted (Fraction-3) but yield was very low (0.9 - 6.1%).

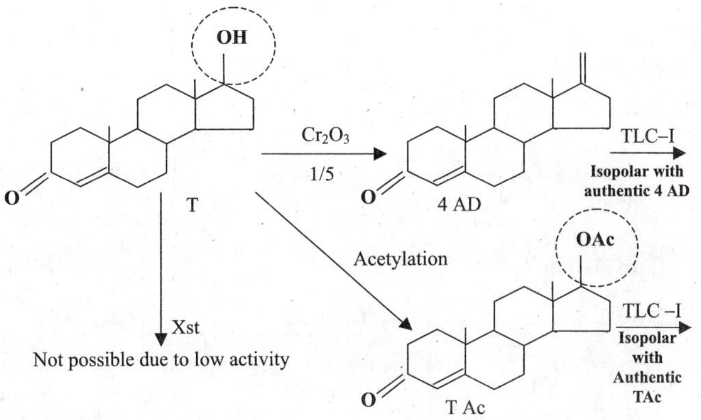

Testosterone (T) : Oxidized to AD and acetylated to its acetate and isopolar to its authentic standard. Crystallization was not possible for this steroid because of low activity.

3. Identification for Peak C (Fraction-4):

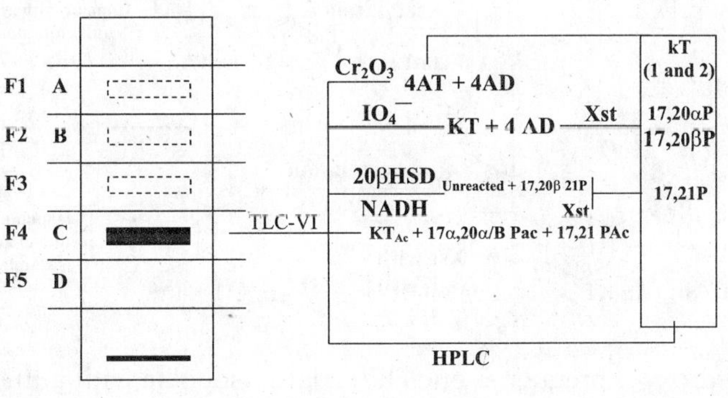

Peak C (Fraction 4) : In TLC-VI single peak was obtained for 11-KT, S 17α,20αP and 17α,20βP. But in radiochromatography 11-KT, 17α,20αP, and 17α,20βP gave distinct their peak separately corresponding to their authentic standard separated by HPLC. In spermiating testes incubation gave only the peak of 17α,20αP, and 17α,20βP, while prespermiating testes gave additionally one more peak for 11-KT. No peak for S was obtained. Combined fraction 4 after TLC-VI, corresponding area of 17α,20βP (Fraction 4A) and 17α,20αP (Fraction 4B) were eluted. Fraction 4A was isopolar with authentic steroid 17α,20βP but not its 20α-epimer in TLC-VI. Oxidized by periodate to a compound isopolar with AD in TLC-I, acetylated to a product isopolar with the acetate of authentic steroid in TLC-I. This metabolite was identified as 17α,20βP by crystallization to constant specific activity. There was other peak in TLC-I with AD which was more polar and corresponded to authentic 11-KT. On acetylation this product was isopolar with the acetate of authentic steroid in TLC-I. This metabolite was identified as 11-KT. Fraction 4B was isopolar with authentic steroid 17α,20αP but not its 20β-epimer in TLC-IV, oxidized by periodate, isopolar with AD in TLC-I, acetylated to a product isopolar with the acetate of authentic steroid in TLC-I. This metabolite was identified as 17α,20αP by crystallization to constant activity.

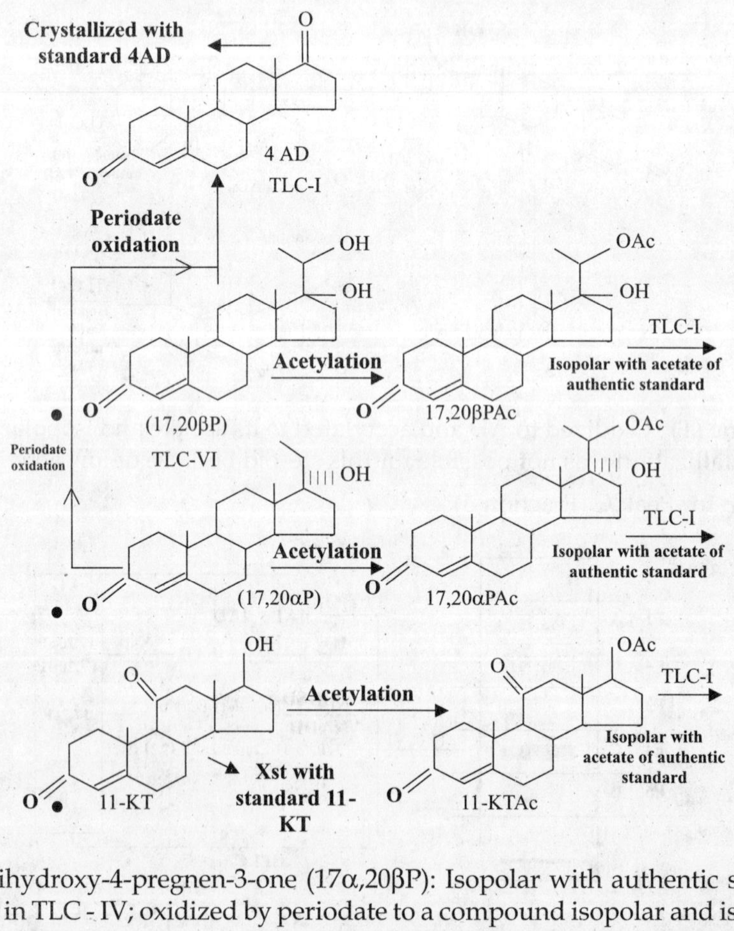

17α,20β-Dihydroxy-4-pregnen-3-one (17α,20βP): Isopolar with authentic steroid, but not its 20α-epimer in TLC - IV; oxidized by periodate to a compound isopolar and isomorphic with androstenedione in TLC-I, acetylated to a product isopolar with the acetate of the authentic steroid in TLC-I.

17α,20α-Dihydroxy-4-pregnen-3-one (17α,20αP): Isopolar with authentic steroid but not its 20β-epimer in TLC-VI, oxidized by periodate to a compound isopolar and isomorphic with androstenedione in TLC-I, acetylated to a product isopolar with acetate of authentic steroid in TLC-I.

17β-Dihydroxy-4-androstene-3,11-dione (11-Ketotestosterone): Isopolar with authentic steroid in TLC-VI, and I; acetylated to a product ispolar with acetate of authentic steroid in TLC -I.

4. Identification for Peak D (Fraction-5):

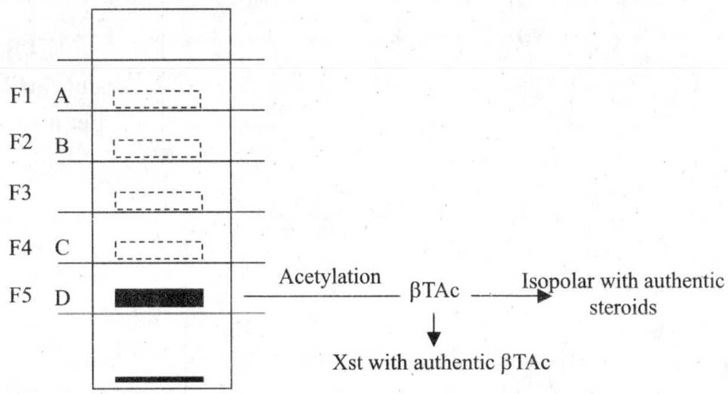

Peak D (Fraction 5) : The eluted metabolite of this fraction gave two peaks (major peak F5A, and minor F5B) in TLC-IV. Fraction 5A was isopolar with 11β-T in TLC-VI. Fraction 5B was isopolar to cortisol but more polar to 11β-T in TLC-IV. On acetylation gave two peak, major peak was isopolar to 11β-T acetate, second peak was not identified.Fraction 5A on periodate oxidation gave a product isopolar to epiandrosterone but fail to crystallize with authentic steroid.

Microchemical Reactions
F 5 A

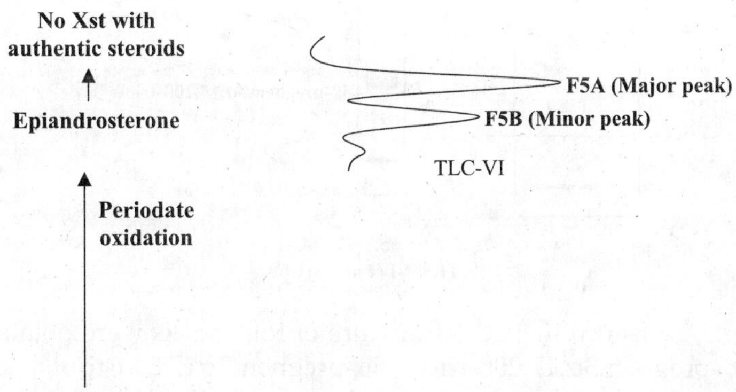

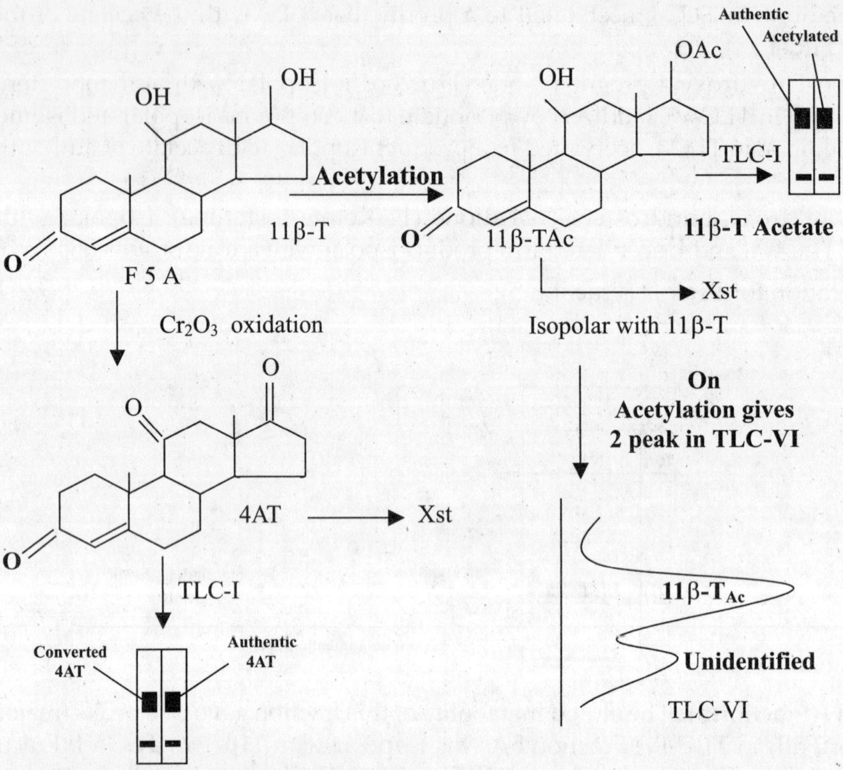

4AT was isopolar to authentic steroids Visible in UV

F 5 A

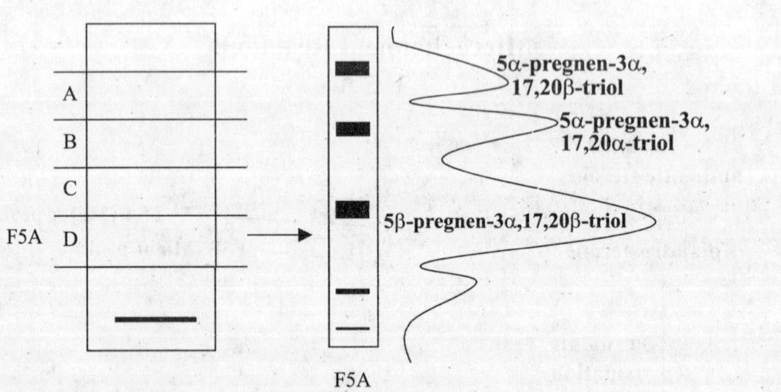

F5A
TLC-VI (4 peaks)

Again fraction 5A was run in TLC-VI mixture of four peaks were obtained, which were isopolar with 5α-pregnen-3α,17,20β-triol, 5α-pregnen-3α17,20α-triol, and 5β-pregnen-3α,1720β-triol. Due to insufficient activity, crystallization was not possible.

F 5 B

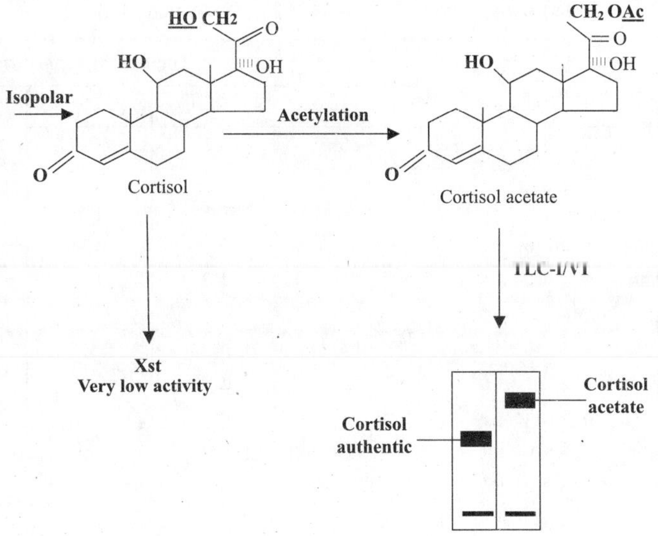

Fraction 5B on acetylation gave two peak in TLC-I, but was low activity.

Recovery of initial radioactivity was 99.0-100.0% in the media of all incubations. After extraction yields were: free (unconjugated) steroids-47.2-72.1%; glucuronides (conjugated) steroids-23.8-54.9%; and sulfates (conjugated) steroids-1.5-4.3%. Small amount of radioactivity as lost during the extraction/hydrolysis/residue and aqueous waste which was less than 1.2%. The details of % yield for individual fish for each unconjugated (free) and conjugated (glucuronide and sulfate) steroid fraction from the amount of radioactivity and the relative peak areas on HPLC are given in Table1 and their crystallization data in Table 2.

The pattern of androgens in roach testes is similar to other teleosts (Fostier *et al.*, 1983; Kime and Scott, 1993). There has been little work on testicular progestogens, although 17α,20βP peaks during spermiation in number of species (Baynes and Scott, 1985; Ueda *et al.*, 1984; Barry *et al.*, 1996), and both this steroid and its 20α-epimer have been found in plasma and incubation media of carp, *Cyprinus carpio* (Koldras *et al.*, 1990). 17α,20αP but not 17α,20βP, was noticed in male dab, *Limanda limanda*. (Canario and Scott, 1991). In incubation of goldfish testes Kime and Scott (1993) have recorded very low yields of a metabolite with the polarity of 17α,20βP (mainly as the sulfate), but a relatively high yield of 17α,20αP and its sulphate. In a similar cyprinid roach testes, yield of metabolite with isopolarity of 17α,20βP (mainly as the sulfate), but a relatively high yield of 17α,20αP and its sulphate. In a similar cyprinid roach testes, yield of metabolite with isopolarity 17α,20αP (mainly as free), but relatively very high yield of 17α,20βP and its glucuronides in spermiating as well as in prespermiating testes of roach. In pre-spermiating roach, 17α,20βP yield was low when compared to spermiating. The yield of 11-ketotestosterone (3.1%) was noticed only in spermiating roach. Since we have incubated one fish prespermiating testes, hence it requires more incubation for prespermiating testes to confirm shifting of steroidogenesis from androgen to progestogen.

Moore and Scott (1992) have reported the 17α,20β-dihydroxy-4-pregnen-3-one 20sulphate is a potent odorant in precocious male Atlantic salmon after pre-exposed urine of ovulated

Table 1. Percentage yield of metabolite of ^{3}[H]17α-hydroxyprogesterone by spermiating (s) and prespermiating (PS) roach testes (Fish No. 1S, 2S, 3S, 4S and 5PS)*

Recovered steroids		Incubation medium				
		1S	2S	2S	4S	5PS
• Free (F) - % recovered		65.2	72.1	40.9	69.3	47.2
Fraction-1.	Androstenedione					
	Androstenetrione	1.0	2.3	1.1	2.5	6.6
Fraction-2.	17α-hydroxyprogesterone	4.1	16.6	0.8	4.5	1.8
Fraction-3.	Testosterone	0.9	1.3	1.2	3.1	6.1
Fraction-4.	11-Ketotestosterone					
	17α,20βP	54.2	34.5	33.2	44.7	17.0
	17α,20αP	3.4	12.7	0.9	9.2	7.3
Fraction-5.	11β-hydroxytestosterone					
	5α-pregnen-3α,17,20β-triol	1.6	4.7	3.7	5.3	5.3
	5α-pregnen-3α,17,20β-triol					
	5β-pregnen-3α,17,20β-triol					
	Cortisol					
• Glucuronide(G) - % recovered						
Fraction-4.	17α,20βP					
Fraction-5.	11β-hydroxytestosterone					
	5α-pregnen-3α, 17,20β-triol	32.5	23.8	54.9	27.8	47.2
	5α-pregnen-3α, 17,20β-triol	30.7	21.1	51.6	26.0	43.6
	5β-pregnen-3α, 17,20β-triol	1.8	2.7	3.2	1.8	3.6
• Sulfate(S) - % recovered		1.5	3.0	3.3	2.0	4.3
Residue(R) waste - % loss		0.7	0.8	0.8	0.8	1.2
Aqueous(A) waste - % loss		1.0	0.3	0.2	0.1	0.2
Total F + G + S+ R + A %		100	99.0	99.0	99.1	99.0

* Individual gonadosomatic index - 1S = 1.57; 2S = 4.1; 3S = 1.3; 4S = 2.4 and 5PS = 1.6

females. Sorensen *et al.* (1991) have also reported that goldfish uses hormonal metabolite as sex pheromones. Becker *et al.* (1992) have measured 17α,20βP in the water in which the fish were maintained, and in the gonad of both sex. The present study also shows that the male roach may secretes glucuronides of 17α,20βP into water which may be as potent odorant/sex pheromones/chemical communicators during spermiation. Barry *et al.* (1989) have demonstrated that physiological levels 17α,20βP inhibits *in vitro* 11-ketotestosterone production by common carp, *Cyprinus carpio*, which supports our results obtained in another cyprinid, *Rutilus rutilus*. It is suggested that there is a switch in steroidogenic pathway from androgen to progestogen just before final sperm maturation and 17α,20βP is involved in the actual process of spermiation in this species. Our results also indicate that sperms in spermiating testicular fragments of roach contain enzyme 20βHSD and produces predominantly 17α,20βP as well as glucuronides by 20βHSD reduction of ^{3}[H]-17α-hydroxyprogesterone in *in vitro*. The yield of 17α,20βP was higher than 17α,20αP, which

Table 2 Identification of metabolite of ^{3}[H]-17α-hydroxyprogesterone in spermiating and prespermiating roach testes by crystallization to constant specific activity.

Metabolites	Specific activity (dpm/mg)				
	Original	1st	2nd	3rd	4th
Androstenedione	1203	1163	1125	1100	1064
Androstenetrione	2330	2112	2110	2307	2145
17α-hydroxyprogesterone	1910	1840	1845	1794	1908
17α,20βP[a]	16347	17548	17084	17146	16509
17α,20αP[a]	5054	4474	4640	4772	4499
11-KT	2801	2840	2841	2794	2805
11-βT	1011	1210	1294	1191	1166
Cortisol	100	110	160	200	205
17α,20β[a] gluc.	14336	15537	15173	15133	14476

Crystallization solvent : aqueous-acetone; aqueous-methanol; aqueous- ethanol. [a]as androstenedione after periodate oxidation. No crystallization data for sulfate steroids due to very low activity.

showed that 20βHSD activity was more than 20αHSD in spermiating roach. It is reasonable to say that 17α,20βP is the main regulating hormones during spermiation than 17α,20αP in this species.

The yield of 3[H]-17α-P in incubation media with low recovery of radiolabelled substrate indicates that this steroid is very rapidly metabolized. The yields of 11βT, AD and AT were also recorded. Asahina *et al.* (1990) have found more 17α,20αP during spermiation because of high 20αHSD rather than 20βHSD. Recently high 20αHSD activity has also been recorded in several of order-Pleuronectiformes and Perciformes by Canario (1992). Kime and Scott (1993) have found high yield (2.3-5.5 %) of 17α,20αP in free and sulphate steroid than 17α,20βP (1.1-1.6 %) after incubation with 3[H]17α-P with the testes of goldfish. It seems that there is diversity among cyprinid fish (common carp, goldfish and roach) for steroid production during spermiation responsible for pheromonal behavior. This also suggests that production is linked to gonadotropin surges caused by external behavior/pheromonal cues related to spermiation in addition to maturational changes of the testes.

SUMMARY

Thin layer Chromatography (TLC), high performance liquid chromatography (HPLC), microchemical reactions and crystallization to constant specific activity were used for identification of steroid hormones in *Rutilus rutilus*. For identification, testes of four spermiating and one pre-spermiating roach were incubated with 17-P and the major metabolites were identified. The 17α,20βP was the major product from all the incubations with its glucuronide (AD, AT, 11β-T). Polar metabolites were also recorded having no significant activity in sulfate fraction. The hormones after identification were found-free (F) % recovered AD, AT, 17-P, T, 11-KT, 17α,20βP, 17α,20αP and 11β-OHPg. In glucuronide (G) % recovered were 17α,20βP and 11β-OHPg. Sulfates (S) % recovered only 17α,20βP but its activity was low.

Abbreviation used for steroid nomenclature in the manuscript- AD= *Androstenedione*-4-androstene-3,17-dione;AT=*Androstenetrione*-4-androstene-3,11,17-trione;17-P=*17α-hydroxyprogesterone*-17α-hydroxy-4-pregnen-3,20-dione;T=*Testosterone*-17β-hydroxy-4-androsten-3-one; 11-KT=*11-ketotestosterone*-17β-hydroxy-4-androstene-3,17-dione; S(17,21P)=*11-Deoxycortisol*-17,21-dihydroxy-4-pregnen-3,20-dione;*17,20αP*=17α,20β-dihydroxy-4-pregnen-3-one;*17,20βP*=17α,20β-dihydroxy-4-pregnen-3-one;11OHT=*11β-hydroxytestosterone*-11β,17β-dihydroxy-4-androstene-3-one;*Epiandrosterone*-3β-hydroxy-5α-androstan-17-one; F=*Cortisol*-11β,17α,21-trihydroxy-4-pregnen-3,20-dione;E=*Cortisone*-17,21-dihydroxy-4-pregnen-3,11,20-trione; E_2= *Estradiol-17β*-1,3,5(10)-estratriene-3,17β-diol. (Italic words are approved by IUPAC-IUB).

REFERENCES

Asahina, K., Barry, T.P., Aida, K., Fusetan, N. and Hanyu, I. (1990) Biosynthesis of 17α,20α-dihydroxy-4-pregnen-3-one from 17α-hydroxyprogesterone by spermatozoa of the common carp, *Cyprinus carpio. Journal of Experimental Zoology* 255: 244-249.

Baynes, S.M. and Scott, A.P. (1985) Seasonal variations in parameters of milt production and in plasma concentrations of sex steroids of male rainbow trout (*Salmo gairdneri*). *General and Comparative Endocrinology* 57: 150-160.

Becker, D., Galili, N. and Degani, G. (1992) GCMS-identified steroids and steroid glucuronides in gonads and holding water of *Trichogaster trichopterus* (Anabantidae, Pallas 1770). *Comparative Biochemistry and Physiology* 10(1): 15-19.

Barry, T.P., Aida, K. and Hanyu, I. (1989) Effects of 17α, 20β-dihydroxy-4-pregnen-3-one on the *in vitro* production of 11-ketotestosterone by testicular fragments of the common carp, *Cyprinus carpio. Journal of Experimental Zoology* 251: 117-120.

Barry, T.P., Aida, K., Okumura, T. and Hanyu, I. (1990a) The shift from C-19 to C-21 steroid synthesis in spawning male common carp, *Cyprinus carpio*, is regulated by the inhibition of androgen production by progestogens produced by spermatozoa. *Biology of Reproduction* 43: 105-112.

Canario, A.V.M. and Scott, A.P. (1991) Levels of 17α, 20α-dihydroxy-4-pregnen-3-one, 3β17α,20α-trihydroxy-5β-pregnane and other sex steroids in blood plasma of male dab, *Limanda limanda* (marine flatfish) injected with human chorionic gonadotrophin. *General Comparative Endocrinolology* 83: 258-264.

Canario, A.V.M. and Scott, A.P. (1991) Levels of 17α, 20α-dihydroxy-4-pregnen-3-one, 3β17α,20α-trihydroxy-5β-pregnane and other sex steroids in blood plasma of male dab, *Limanda limanda* (marine flatfish) injected with human chorionic gonadotrophin. *General Comparative Endocrinolology* 83: 258-264.

Canario, A.V.M. (1992) 17α,20α-dihydroxy-4-pregnen-3-one, a male sex pheromone Abst. 2nd Intl. Symp on Fish Endocrinol held at St. Malo (France), June 1-4, p 81.

Condeca, J.B. and Canario, A.V.M. (2001) Gonadal Steroidogenesis in response to estradiol-17β administration in the sea bream (*Sparus aurata* L.). *General and Comparative Endocrinology* 124: 82-96.

Colombo, L., Bern, H.A., Pieprzyx, J. and Johnson, D.W. (1973) Biosynthesis of 11-deoxycorticosteroids by teleost ovaries and discussion of their possible role in oocyte maturation and ovulation. *General Comparative Endocrinology* 21: 168-178.

Ebrahimi, M. Singh, P.B. and Kime, D.E. (1995) Biosynthesis of 17,20α-dihydroxy-4-pregnen-3-one, 17,20β-dihydroxy-4-pregnen-3-one and 11-ketotestosterone by testicular fragments and sperm of the roach, *Rutilus rutilus. General and Comparative Endocrinology* 100:375-384.

Fostier, A., Jalabert, B., Billard, R., Breton, B. and Y. Zohar (1983) The gonadal steroids. In: *Fish Physiology* (Hoar, W.S., Randall, D.J. and Donaldson, E.M. ed.), Vol. IXA pp. 277-372, Academic Press, New York, USA.

Joss, J.M.P., Edwards, A. and Kime, D.E. (1996) *In vitro* biosynthesis of androgens in the Australian lungfish, *Neoceratodus forsteri. General and Comparative Endocrinology* 101: 256-263.

Kime, D.E. (1990) *In vitro* metabolism of progesterone, 17-hydroxyprogesterone and 17,20β-dihydroxy-4-pregnen-3-one by ovaries of the common carp *Cyprinus carpio*: production rates of polar metabolites. *General and Comparative Endocrinology* 79: 406-414.

Kime, D.E., Bhattacharya, S., Koldras, M. and Bieniarz, K. (1993) Steroidogenesis by ovaries and testes of the European catfish, the wells (*Silurus glanis*), *in vitro. Fish Physiology and Biochemistry* 10: 389-398.

Kime, D.E., and Scott, A.P. (1993) *In vitro* synthesis of 20α-reduced and of 11-and 21-oxygenated steroids and their sulfates by testes of the goldfish (*Carassius auratus*): testicular synthesis of corticosteroids. *Fish Physiology and Biochemistry* 11: 287-292.

Koldras, M., Bieniarz, K. and Kime, D.E. (1990) Sperm production and steroidogenesis in the testes of the common carp, *Cyprinus carpio* L., at different stages of maturation. *Journal of Fish Biology* 37: 365-645.

Lower, N. Scott, A.P. and Moore, A. (2004) Release of sex steroids into water by roach. *Journal of Fish Biology* 64(1): 16-33.

Moore, A. and Scott, A.P. (1992) 17α, 20α-dihydroxy-4-pregnen-3-one 20-sulphate is a potent odorant in precocious male Atlantic salmon (*Salmo salar* L.) parr which has been pre-exposed to the urine of ovulated females. *Proceedings of Royal Society London* B249: 205-209.

Nagahama, Y. (1997) 17α,20β-dihydroxy-4-pregnen-3-one, a maturation-inducing hormone in fish oocytes: mechanisms of synthesis and action. *Steroids* 62: 190-196.

Scott, A.P. Pinillos, M.L. and Huertas (2005) The rate of uptake of sex steroids from water by *Tinca tinca* is influences by their affinity for sex steroids binding protein in plasma. *Journal of Fish Biology* 67(1): 182-200.

Sorenson, P.W., Goetz, F.W., Scott, F.W. and Stacey, N.E. (1991) Recent studies indicate that goldfish use mixture of unmodified hormones and hormonal metabolites as sex pheromones. In Proc. 4th Intl. Symp. Reprod. Physiol. Fish, Norwich, U.K. 7-12 July 1991 p. 191-193. Edited by A.P. Scott., J.P. Sumpter, D.E. Kime and M.S. Rolf. FishSymp. 91, Sheffield.

Trant, J.M., Thomas, P. and Shackleton, C.H.L. (1986) Identification of 17α,20β,21-trihydroxy-4-pregnen-3-one as the major ovarian steroid produced by the teleost *Micropogonias undulates* during final oocyte maturation. *Steroids* 47(2-3): 89-99.

Ueda, H., Kambigawa, A. and Nagahama, Y. (1984) *In vitro* 11-ketotestosterone and 17α,20β-dihydroxy-4-pregnen-3-one production by testicular fragments and isolated sperm of rainbow trout, *Salmo gairdneri. Journal of Experimental Zoology.* 231: 435-439.

Wallace, E. and Silberman, N. (1964) Biosynthesis of steroid sulfates by human ovarian tissue. *Journal of Biology Chemistry* 203(9): 2809-2812.

Suggested Readings

1. Endocrine Disruption in Fish—David E. Kime
2. Comparative Vertebrate Endocrinology—P. J. Bentley
3. Fish Physiology Vol. IX.—W. S. Hoar and D. J. Randall
4. Fish Endocrinology—A. J. Matty
5. The Chemistry of Steroids—W. Klyne
6. Fish Reproduction—Rocha, Arukwe and Kapoor

7. Fish Spermatology – Alavi, Cosson and Rafiee
8. Fish Endocrinology – B.G. Kapoor

REVIEW QUESTIONS

Long Questions

1. Describe the categories of steroid hormones in fish.
2. Discuss in brief the work done on reproductive endocrinology of telost.
3. What is the need of identification of sex steroids?
4. Draw the figures of five sex steroids involved in fish reproduction.
5. What is crystallization of steroid hormones?

Multiple Choice Questions

1. Which is one of the main female sex steroid hormones?
   (a) Testosterone (b) Estradiol-17β
   (c) Estrone (d) Estriol
2. Maturational inducing steroid (MIS) in fishes is-
   (a) Testosterone (b) 17α, 20βP
   (c) Pregnenolone (d) Estrone
3. 11-Ketotestosterone is main sex steroid hormone in-
   (a) Male (b) Female
   (c) Male and Female (d) None
4. Androstenedione on 17β-HSD reduction gives-
   (a) 11-Ketotestosterone (b) Testosterone
   (c) Estrone (d) Estradiol-17β
5. 11-Ketotestosterone on oxidation gives-
   (a) Testosterone (b) Estrone
   (c) Androstenedione (d) Androstenetrione
6. 17-Hydroxyprogesterone on 20β-HSD reduction produces-
   (a) 17α, 20βP (b) 11-Ketotestosterone
   (c) Estrone (d) Testosterone
7. 17, 20βP on 20β-HSD Oxidation gives-
   (a) 17-Hydroxyprogesterone (b) 17,20αP
   (c) Estradiol-17β (d) None
8. C_{21} Progestogen is-
   (a) 17α, 20αP (b) Estradiol-17β
   (c) 11-Ketotesterone (d) None

9. Precursor for progesterone synthesis is-
   (a) Testosterone (b) Cholesterol
   (c) Pregnenolone (d) Estradiol-17β
10. The metabolites are confirmed by-
   (a) Oxidation (b) Reduction
   (c) Both oxidation and reduction (d) Crystallization to constant specific activity

Answers

1. (b)	2. (b)	3. (a)	4. (a)	5. (c)	6. (a)
7. (a)	8. (a)	9. (c)	10. (d)		

PART–4

Structure and Function of Nucleic Acids

CHAPTER 12

Nucleic Acids: Structure and Polymorphism

Y.K. Jaiswal* **and M.K. Jaiswal**

School of Studies in Biochemistry, Jiwaji University, Gwalior–474 011

E-mail: jaiswalyk@gmail.com

The hereditary nature of every living organism is defined by its genome, which consists of a long sequence of nucleic acid that gives the information needed to create the organism. The shape and activities of individual cells are, to a large extent, determined by the genetic instructions contained in DNA or RNA (in some viruses). There are two Kinds of nucleic acids—DNA and RNA, which can store the genetic information and make it available to cell. The structures of these molecules must be consistent because of following reasons:

(a) Genetic information must be stored in a form that is manageable in size and stable over a long time.

(b) Genetic information must be decoded—often many times—in order to be used. Transcription is the process by which nucleotide sequences in DNA are copied on to RNA so that they can direct protein synthesis, a process known as translation.

(c) The progeny of an organism must be equipped with the same set of instructions in the parent. Thus, DNA is replicated so that each daughter cell receives the same information.

Before any account is given of about the structure of the nucleic acids, it is necessary to describe the structure of their component parts. Nucleic acids are high-molecular-weight polymeric compounds, which on complete hydrolysis yield pyrimidine and purine bases, a sugar component and phosphoric acid. Partial hydrolysis yields compounds known as nucleosides and nucleotides.

*Corresponding author

COMPONENTS OF NUCLEIC ACIDS

Nitrogenous Bases

Two types of nitrogen-containing bases i.e., purines and pyrimidines are found in nucleic acids. Purines consist of a six-membered and a five-membered nitrogen-containing ring, which are fused together. Pyrimidines have only a six-membered nitrogen-containing ring (Fig. 1).

Purine Bases

Both types of nucleic acids contain the purine bases, adenine and guanine. They are derivatives of the parent compound purine, which is formed by fusion of a pyrimidine ring and an imidazole ring. It should be noted that the style of numbering of the pyrimidine ring in the purines differs from that used for pyrimidines. There are some other naturally occurring purine derivatives, which include hypoxanthine, xanthine and uric acid (Fig. 2-4).

Pyrimidine Bases

The pyrimidine bases are derivatives of the parent compound of pyrimidine. Nucleic acids contain pyrimidine bases, cytosine found in both RNA and DNA, uracil found in RNA and thymine and 5′methly-cytosine found in DNA. In certain of the bacterial virus cytosine is replaced by 5-hydroxymethycytosine or glucosylated derivative of 5-hydroxy methylcytosine.

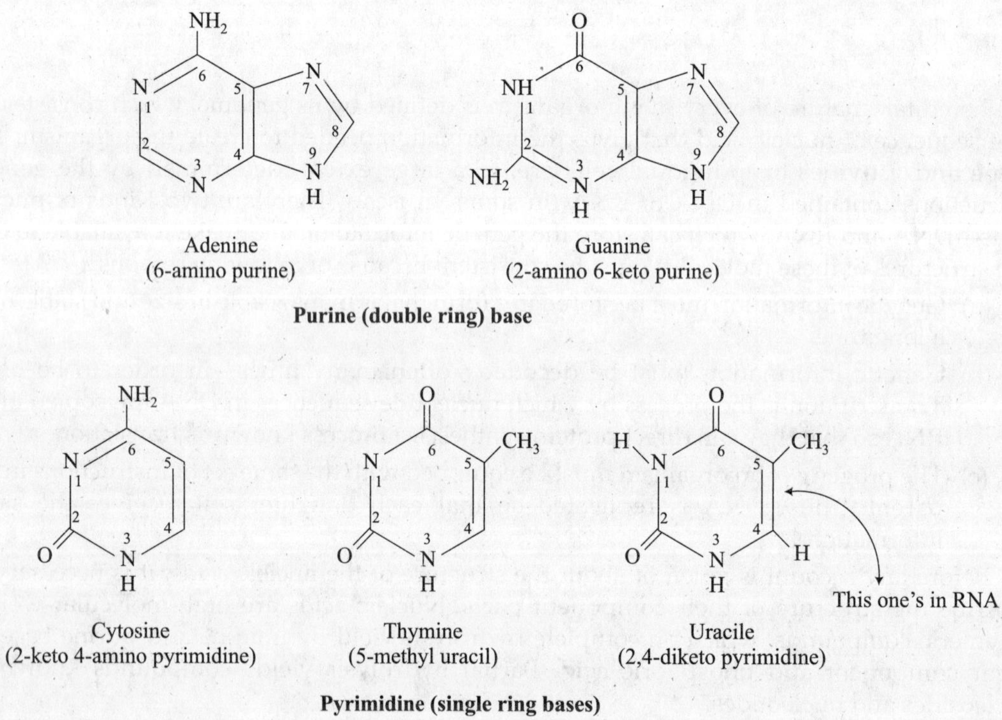

Fig. 1. Four bases of DNA (Plus one in RNA).

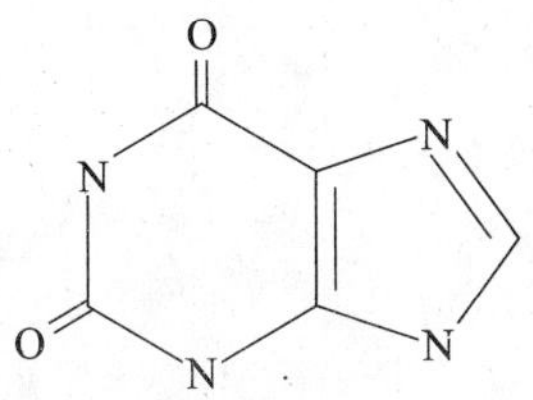

Fig. 2. Xanthine.

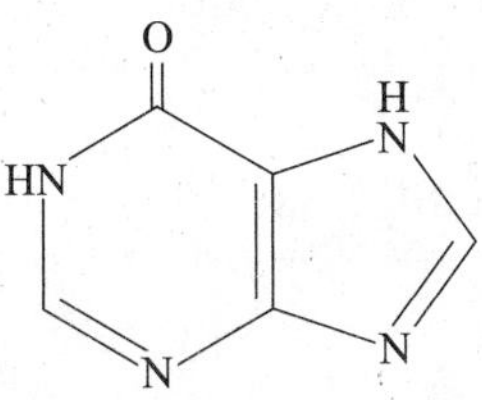

Fig. 3. Hypoxanthine.

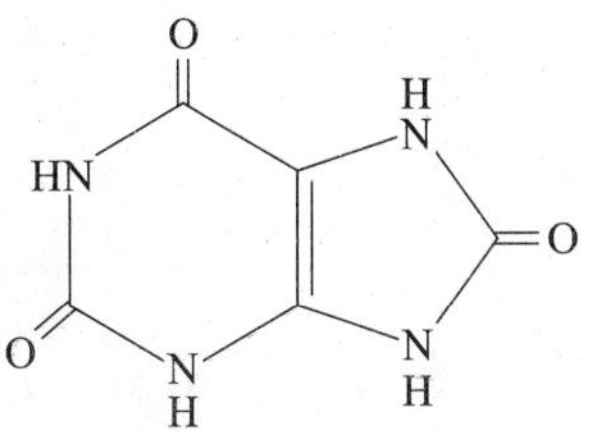

Fig. 4. Uric acid.

Minor Bases

Certain "minor bases" are also found in small amount in some nucleic acid. For example, tRNA contains a wide a variety of methylated bases including thymine. These unusual comprise less than five percent of the total base content of the tRNA and vary in relative amounts from species to species. Some of the minor bases found in RNA are given below (Table 1).

Table 1. Some of minor bases in RNA

1-methyladenine	5-carboxymethyl uracil
2-methyladenine	5-methyluracil
6-methyladenine	5-hydroxymethyluracil
6,6-dimethyladenine	2-thiouracil
6-isopentenyladenine	3-methyluracil
2-methyl thio 6-isopentenyladenine	3-methylcytosine
6-hydroxymethyl butenyladenine	4-methylcytosine
6-hydroxymethyl 2-methyladenine	5-methylcytosine
1-methylguanine	5-hydroxymethylcytosine
2-methylguanine	2-thiocytosine
2,2-dimethylguanine	4-acetylcytosine
7-methylguanine	6-aminoacyladenine
2,2,7-trimethylguanine	dihydrouracil
hypoxanthine	5-dihydrouracil
1-methyl hypoxanthine	
Xanthine	

The phenylalanine tRNA from yeast contains a most unusual base known as Wybutosine (Fig. 5).

Tautomerism in Nitrogen bases

Purine bases can exit in two tautomeric forms amino and imino (Fig. 6). The pyrimidines can undergo keto-enol tautomerism.

Pentose Sugar and Deoxy Sugars

The main sugar component of RNA is D-ribose, which is present in β-furanose form (Fig. 7). In DNA this sugar is replaced by 2-deoxyribose which is also in β-defuranose form. This small difference in the two types of nucleic acid has wide-ranging effects on both their chemistry and

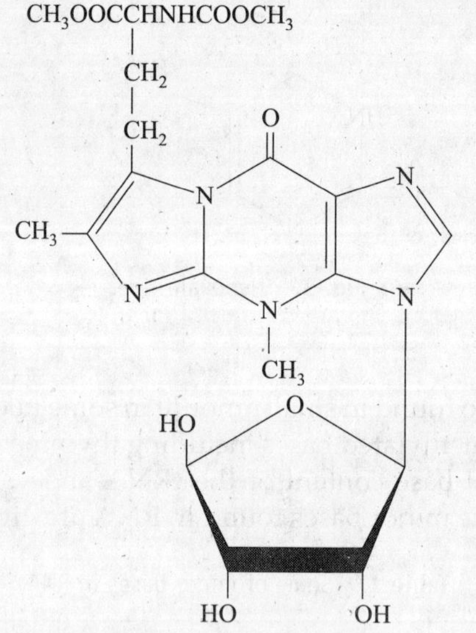

Fig. 5. Wybutosine.

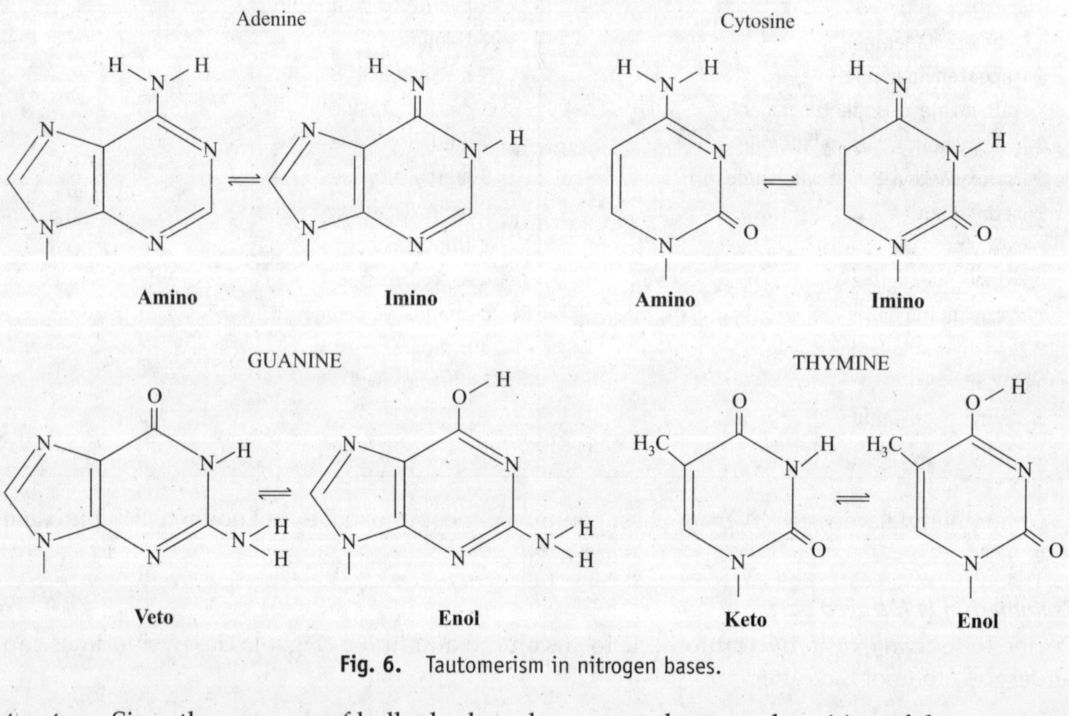

Fig. 6. Tautomerism in nitrogen bases.

structure. Since the presence of bulky hydroxyl group on the second position of the sugar not only limits the range of possible secondary structure available to the RNA molecules but also make it more susceptible for chemical and enzymic degradation.

Some RNAs like ribosomal RNAs contain very small amounts of 2-O-methylribose. When pentose sugar occurs in nucleic acid or nucleotides, carbon atom are numbered as 1′, 2′, 3′ etc to avoid the confusion with numbering of the ring atoms of the bases (Figs. 8 and 9).

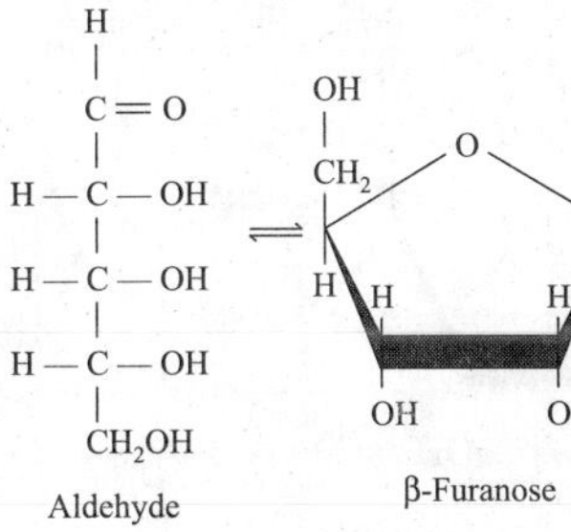

Fig. 7. β-furanose form.

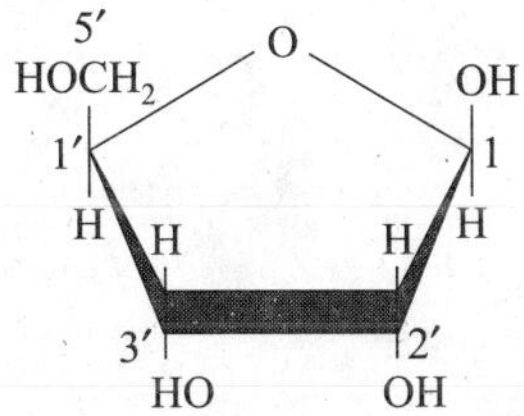

Fig. 8. Ribose sugar.

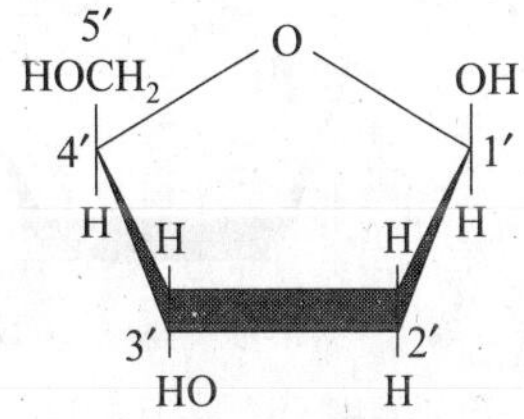

Fig. 9. Deoxyribose sugar.

Nucleosides

When a purine or a pyrimidine base is linked to ribose or de-oxyribose the resulting the compound is known as nucleoside. Purine base attached to the ribose or de-oxyribose moiety via a β-N_9-glycosidic bond while pyrimidine base moiety via a β-N_1-glycosidic bond. Adenine condenses with ribose to form nucleoside Adenosine, Guanine form Guanosine, Cytosine forms Cytidine, Uracil forms Uridine (Figs 10-14, Table 2). The Nucleoside derived from 2-deoxyribose are known as deoxyribonucleoside; deoxyadenosine, deoxyguanosine, deoxycytidine, deoxythymidine and so on (Figs 15-18).

Ribonucleosides can be form on partial hydrolysis on RNA. Ribonucleoside form hypoxanthine is named inosine. The nucleoside 5-ribouracil has been obtained in small amount from the digestion product of RNA particularly tRNA has been named as pseudouridine.

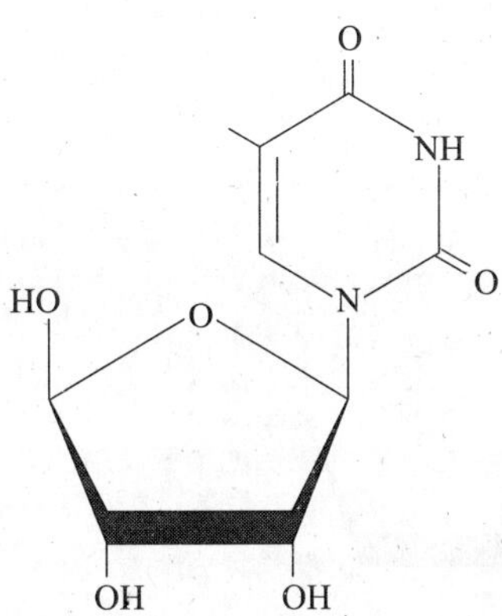

Fig. 10. Adenosine (A).

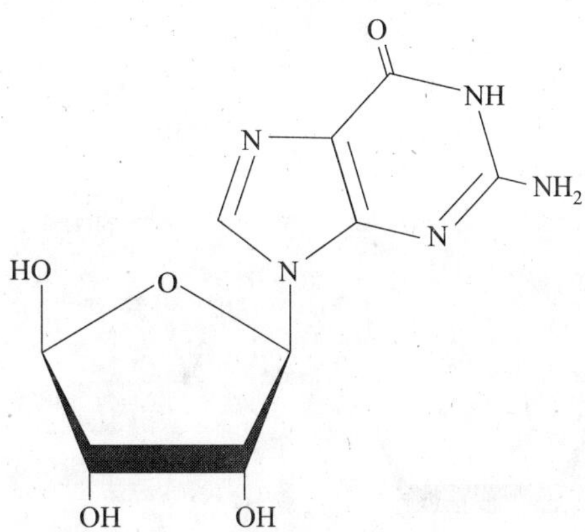

Fig. 11. Guanosine (G).

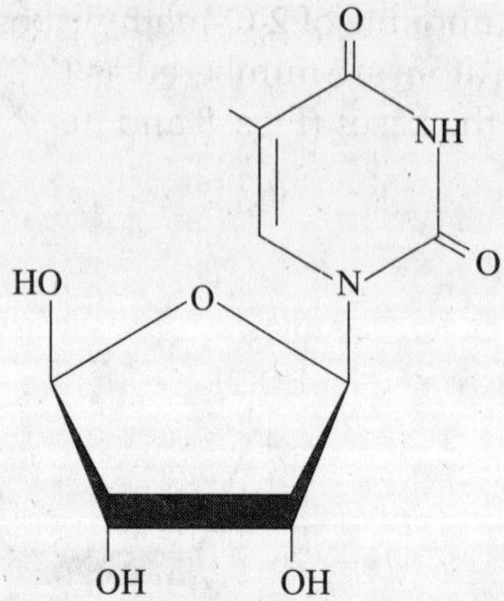

Fig. 12. 5-Methyluridine (m^5U).

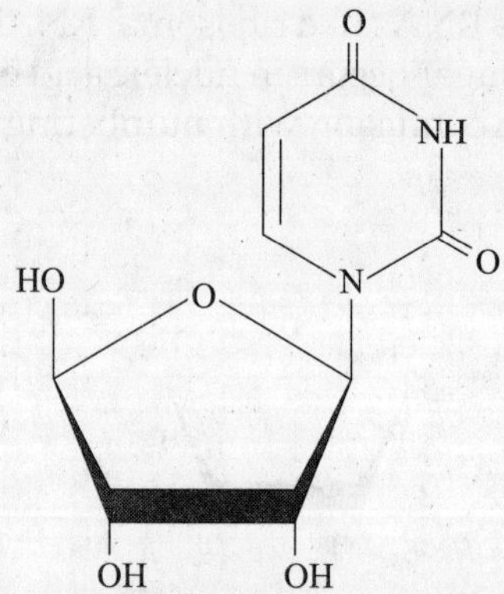

Fig. 13. Uridine (U).

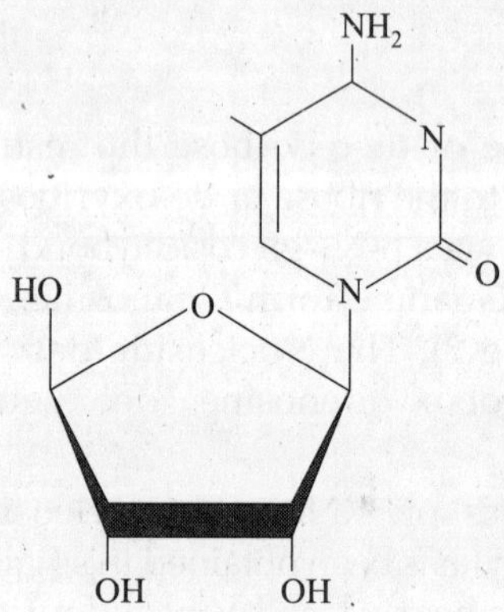

Fig. 14. Cytidine (C).

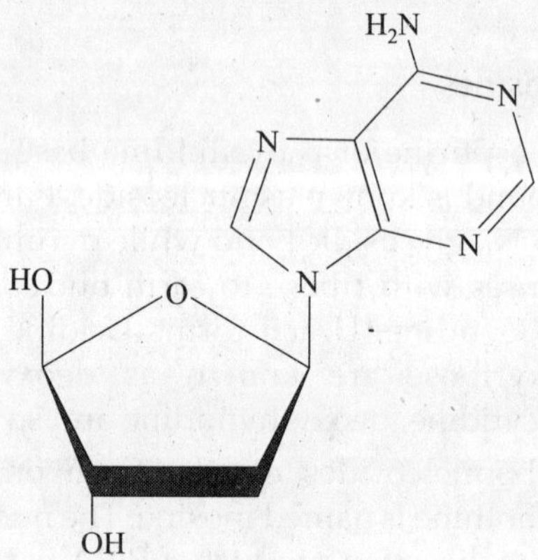

Fig. 15. Deoxyadenosine (dA).

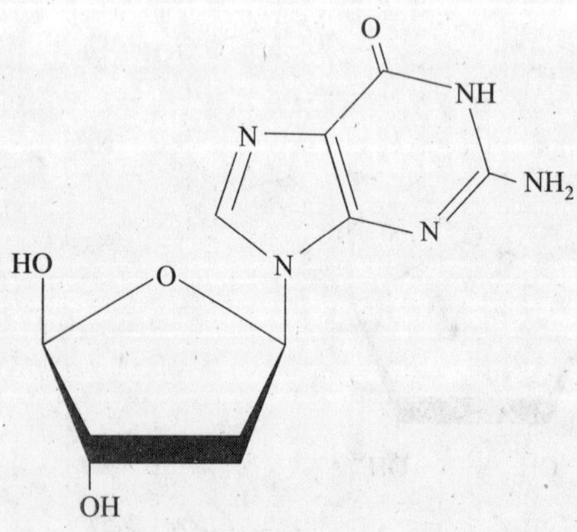

Fig. 16. Deoxyguanosine (dG).

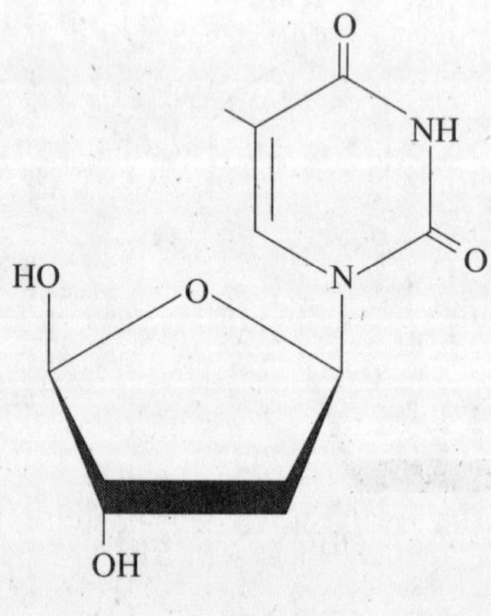

Fig. 17. Deoxythymidine (dT).

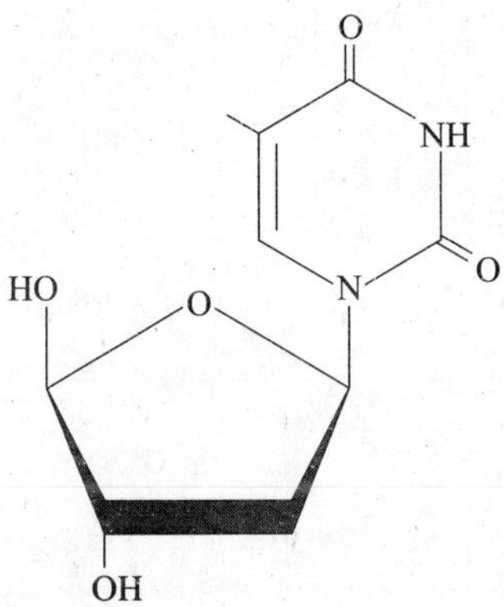

Fig. 18. Deoxyuridine (dU).

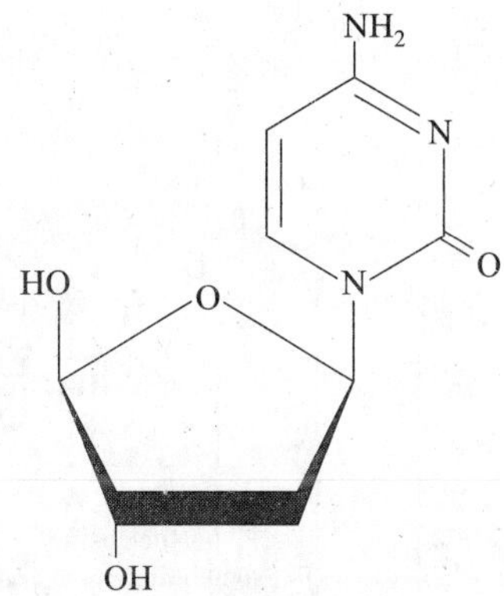

Fig. 19. Deoxycytidine (dC).

The base can exist in two distinct orientations about the N-glycosidic bond. These conformations are identified as, *syn* and *anti* (Figs. 20 and 21). It is the anti conformation that predominates in naturally occurring nucleotides.

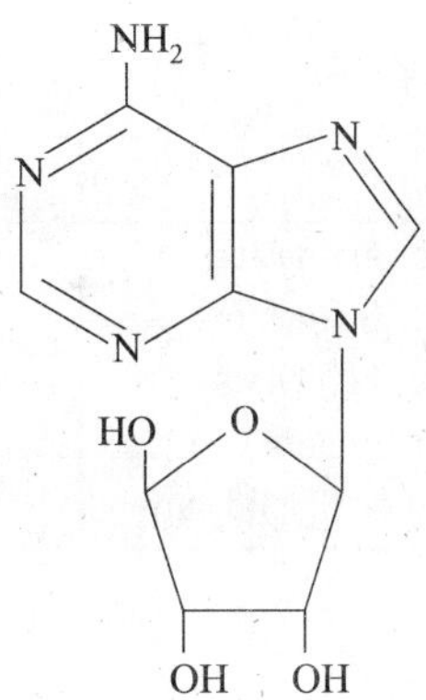

Fig. 20. syn-Adenosine.

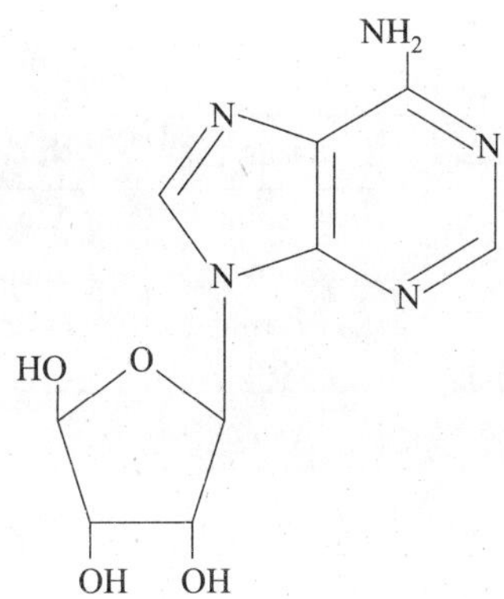

Fig. 21. anti-Adenosine.

Nucleotides

The structures of nucleotides are dealt with in more specialized term by Hutchinson and Michelson. They are all phosphoric acid ester of nucleoside. Those derived from the ribonucleosides are usually referred as ribonucleotides and those from deoxyribonucleosides as deoxyribonucleotides. These terms are sometimes abbreviated to riboside, ribotide, deoxyriboside and deoxyribotide (Fig. 22, Table 2).

THE BASES ABSORB UV LIGHT

All the nucleotides are present in DNA or RNA has λ_{max} near 260 nm, which is not affected by sugar phosphate configuration (Fig. 23).

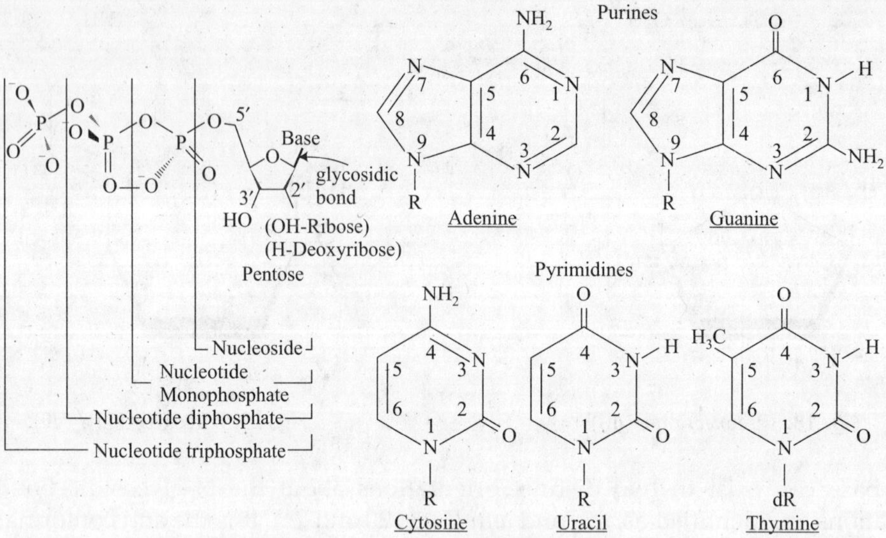

Fig. 22. Structure of nucleotide.

Table 2. Bases, nucleosides, nucleotides

Class of Bases	Bases	Nucleosides	Nucleotides	Abbreviations
1. Purine	Adenine	Adenosine	Adenylic acid or Adenylate	A
	Guanine	Guanosine	Guanylic acid or Guanylate	G
	Cytosine	Cytidine	Cytidylic acid or Cytidylate	C
2. Pyrimidine	Uracil	Uridine	Uridylic acid or Uirylate	U
	Thymine	Thymidine	Thymidylic acid or thymidylate	T

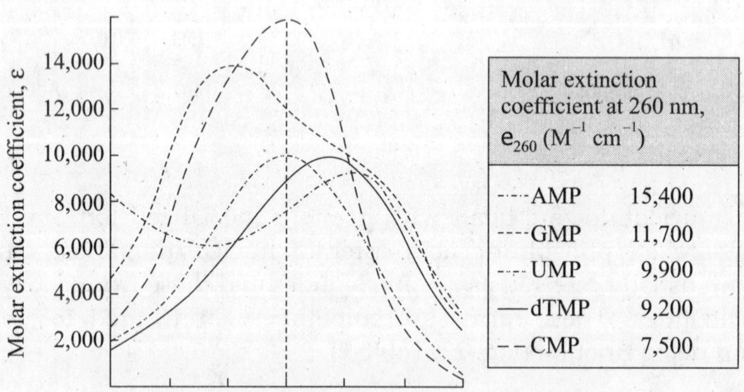

Fig. 23. Bases absorb UV light (Nelson and Cox, 2004)

NUCLEIC ACID IS POLYNUCLEOTIDE

Nucleotides join together through phosphodiester linkages between the 5' and 3' carbon atoms to form nucleic acids. The 3′–OH of the sugar group forms a bond with one of the negatively charged oxygen of the phosphate group attached to the 5' carbon of another sugar. When many of these nucleotide subunits combine, the result is the large single-stranded polynucleotide or nucleic acid (Fig. 24).

At one end of the large molecule, the carbon group is unbound and at the other end, the -OH is unbound. These different ends are called the 5'- and 3'-ends, respectively.

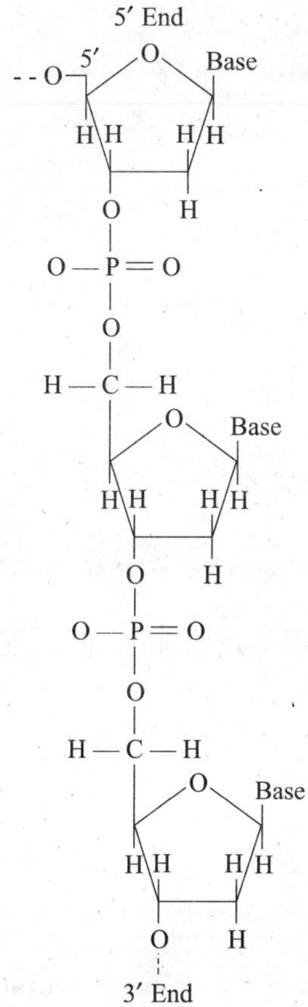

Fig. 24. Polynucleotide.

BASE PAIRING AND BASE STACKING

Base Pairing

The Watson-crick base pairs (Fig. 25) that are found in the usual double-stranded DNA and RNA are shown in fig 25. Their Geometry in such way that any sequence of base pairs can fit in to the nucleic acid helix without distortion. There are two H bonds in the AT base pair and three

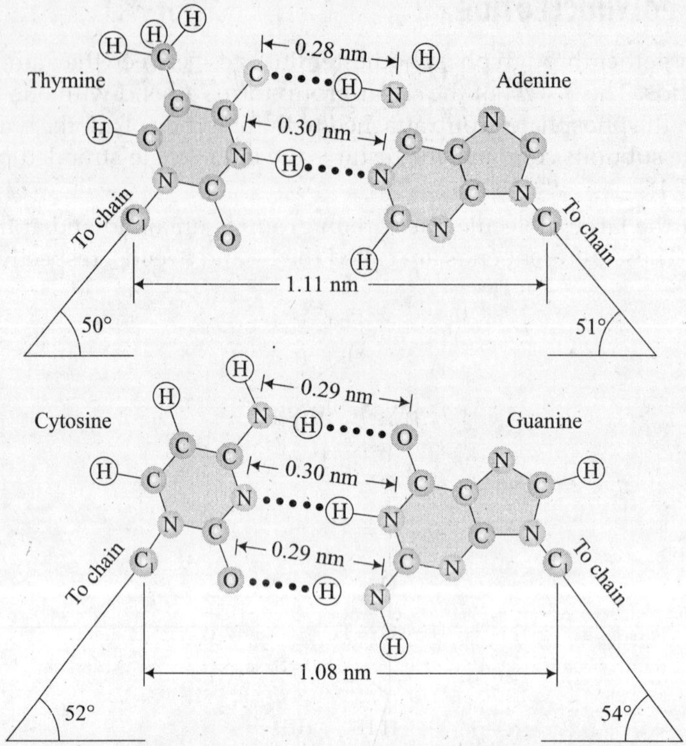

Fig. 25. Watson-crick base pairs (Nelson and Cox, 2004)

H bonds in the GC base pair. Therefore the stability of nucleic acid structure depend on the GC content, greater the GC content greater the stability.

The distance between C1′ atoms of sugars are opposite strands is essentially the same for A = T and G ≡ C base pairs.

Base-Stacking Interactions

From an energetic point of view, the most important contribution to the DNA helix is the stacking of the bases on the top of each other. The 'stacking energy' is the measure of how much energy is required to destack or melt a regain of double-stranded DNA. Table 3 lists the stacking energies for all the 16 different dinucleotide combinations. There is a strong sequence dependence on the amount of stabilizing energy from the base stacking. As a general trend, alternating pyrimidine-purine steps have less energy, and in particular T. A have the lowest (–3.82 kcal/mol). G.C steps have the largest value (–15 kcal/mol) and require the most energy to melt.

The propeller twist is measure of the angle between the planes of the two bases as shown in Figs. 26 and 27. Each bases is planar, but when two base pairs, they donot always line up perfect flat with each other; this angle is called propeller twist because the bases are twisted away from each other like an aeroplane propeller. This measure is related to the rigidity of the

Table 3. Dinucleotide Value for helical Parameter for B-DNA (Garrett and Grisham, 2007)

Dinucleotide step	Stacking Energy kcal Mol–1	Twist angle (°)	Propeller twist (°)
AA	–5.37	35.6	–18.66
AC	–10.51	34.4	–13.10
AG	–6.67	27.9	–13.10
AT	–6.57	32.1	–15.01
CA	–6.57	34.5	–9.45
CC	–8.26	33.7	–8.11
CG	–9.61	29.8	–10.03
CT	–6.78	27.9	–14.00
GA	–9.81	36.9	–13.48
GC	–14.59	40.0	–11.08
GG	–8.26	33.7	–8.11
GT	–1051	34.4	–13.10
TA	–3.82	36.0	–11.85
TC	–9.81	36.9	–13.48
TG	–6.57	34.5	–9.45
TT	–5.37	35.6	–18.66
Average	–7.2 ± 2.57	3.57 ± 8.0	–12.60 ± 3.2

Fig. 26. Propeller twist (Garrett and Grisham, 2007)

Fig. 27. Twist (Garrett and Grisham, 2007)

helix, such that a larger propeller twist reflects a more rigid helix. In B-DNA, the propeller twist angles are usually quite low.

Another important dinucleotide parameter related to base-stacking interactions is the twist angle of the two bases as shown in Fig 27. Table includes the values for twist angles for B-DNA solution. This range varies from 27.9° (which would correlate to a helix with 12.9 per turn) to 40° (corresponding to 9 bp per turn). Thus the pitch of the helix (i.e. the length of a full turn of the helix) can range considerably, based on the sequence. This has important biological consequences. If there are two protein-binding sites, each facing the same side of the helix, separated by a about 21 bp, or roughly two turns of the helix, it is possible to have different intervening sequences, resulting in quite different orientations of the sites (ranging over about 45° of the orientation relative to each other).

CHARGAFF'S RULES

In 1950, Erwin Chargaff (Fig. 28) analyzed the base compositions of DNA in a number of organisms. He reported that DNA composition varies from one species to another. Such

Fig. 28. Erwin Chargaff (http://post.queensu.ca/nforsdyke/bioinp2.htm).

evidences of molecular diversity, makes DNA a more credible candidate for the genetic material than protein.

Chargaff found that a peculiar regularity in the ratios of nucleotide bases. In the DNA of each species he studied, the number of adenines approximately equaled the number of thymine, and the number of guanines approximately equaled the number of cytosine. In human DNA, for example, the four bases are present in these percentages: A=30.9% and T=29.4%; G=19.9% and C=19.8%. The A=T and G≡C equalities, later known as Chargaff's rules, helped Watson and Crick to discover the structure of DNA. The following table shows some sample data that he collected (Table 4).

Table 4. * Plant DNA is an apparent exception to Chargaff's rules, however, plant DNA is often heavily methylated. As long as the amount of 5-methyl-cytosine is included in the total amount of cytosine, the rules are obeyed. Some bacteriophages DNA's (e.g. PhiX174) do not adhere to Chargaff's rules—these phages have single stranded genomes and are therefore not constrained by the requirements of a double-stranded structure (http://post.queensu.ca/nforsdyke/bioinp2.htm).

	mol % of bases				Ratios		
Source	A	G	C	T	A/T	G/C	%GC
PhiX-174	24.0	23.3	21.5	31.2	0.77	1.08	44.8
Maize	26.8	22.8	17.0 *	27.2	0.99	0.98	46.1
Octopus	33.2	17.6	17.6	31.6	1.05	1.00	35.2
Chicken	28.0	22.0	21.6	28.4	0.99	1.02	43.7
Rat	28.6	21.4	20.5	28.4	1.01	1.00	42.9
Human	29.3	20.7	20.0	30.0	0.98	1.04	40.7

THE PEOPLE RESPONSIBLE FOR THE DISCOVERY OF DNA

The people responsible for the discovery of the double-helical structure of DNA were Francis Crick, Rosalind Franklin, Linus Pauling, James Watson and Maurice Wilkins (Fig. 29). The story starts in 1951 when several things happened.

Francis Crick James Watson Maurice Wilkins Rosalind Franklin

Fig. 29. People responsible for the discovery of DNA (http://www.ba-education.demon.co.uk/for/science/dnamain.html).

Rosalind Franklin (1920-1958), a graduate in physical chemistry, had been working since 1947 on x-ray crystallographic methods at the Central State Chemistry Laboratory in Paris. As an expert on x-ray crystallography she was head hunted in 1951 by John Randall, of the Medical Research Council Biophysics Research Unit at King's college, London, to work on an x-ray picture of DNA taken by a graduate student Raymond Gosling. The working relationship at King's college did not get off to a good start because of a misunderstanding. New Zealander Maurice Wilkins (1916-2004), a physics graduate and member of the Medical Research Council Biophysics Research Unit, who had been working on nucleic acids and x-ray diffraction of DNA. Franklin was the kind of person that once they are given something to solve they keep it to themselves until it is complete; not the way that Wilkins liked to run the lab at King's college. When Wilkins returned he thought that Franklin was a high class technical assistant who was to supply the team with experimental data for it to analyse. This was not to be. Relationships were strained between Wilkins and Franklin, whereas Gosling got on with them both. Thus the seeds were sown for a rough ride for all concerned.

To solve the structure of DNA four ideas had to come together.

- The phosphate backbone was on the outside, bases on the inside.
- The molecule was a double helix.
- The strands were antiparallel.
- It had a specific base pairing.

Franklin soon found out that by bundling super thin strands of DNA and zapping them with a super fine x-ray beam there were two forms of hydration—the A form (easy to photograph) that is dry and the B form (hard to photograph) that is wet (Figs. 30-33). Her B form photographs showed a fuzzy cross which meant a helix. Since the water would be attracted to the phosphates in the backbone, and the DNA was easily hydrated and dehydrated, she guessed that the backbone was on the outside and the bases were on the inside. The first part of the problem was solved.

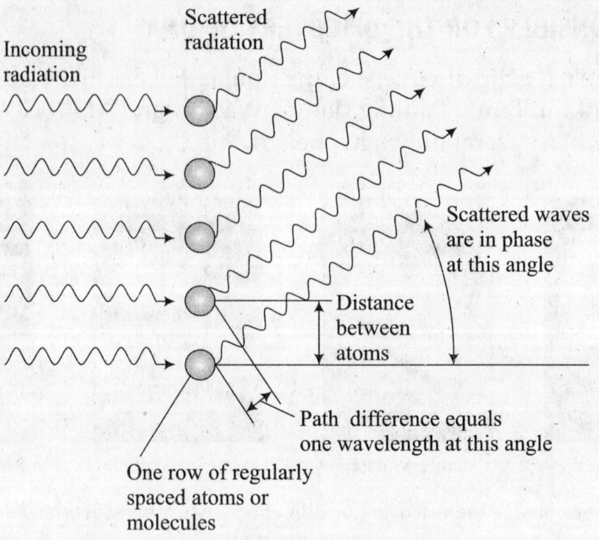

Fig. 30. Photo 51 x-ray diffraction of wet DNA showing the B form double helixtaken by Rosalind Franklin and Raymond Gosling on Friday, 2nd May, 1952 by long exposure (http://www.ba-education.demon.co.uk/for/science/dnamain.html; and Nelson and Cox, 2004).

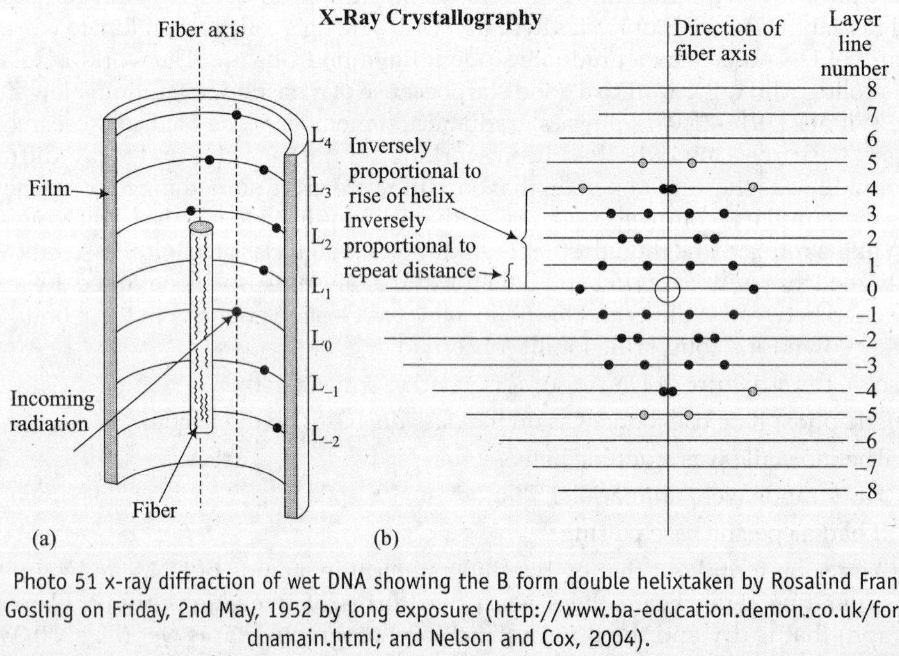

Fig. 31. Photo 51 x-ray diffraction of wet DNA showing the B form double helixtaken by Rosalind Franklin and Raymond Gosling on Friday, 2nd May, 1952 by long exposure (http://www.ba-education.demon.co.uk/for/science/dnamain.html; and Nelson and Cox, 2004).

In the November of 1951 Franklin gave a departmental seminar to bring the unit up to date on what she had achieved so far. In it she presented the A and B form data. In the audience was an American James Watson, a zoology graduate, who had become interested in molecular biology, with particular emphasis on DNA. He was on a National Foundation of Infantile

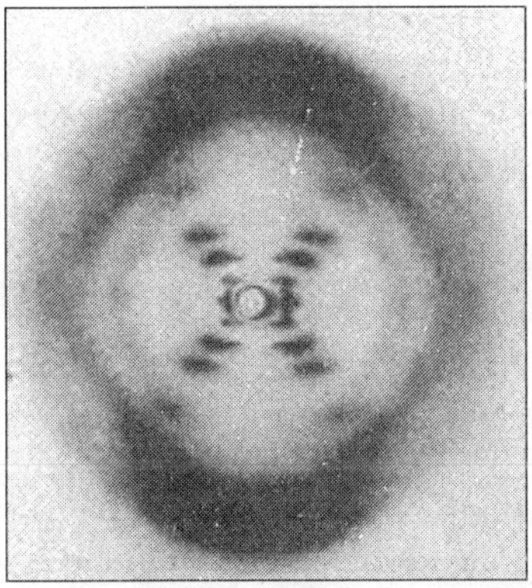

Fig. 32. Photo 51 x-ray diffraction of wet DNA showing the B form double helixtaken by Rosalind Franklin and Raymond Gosling on Friday, 2nd May, 1952 by long exposure (http://www.ba-education.demon.co.uk/for/science/dnamain.html; and Nelson and Cox, 2004).

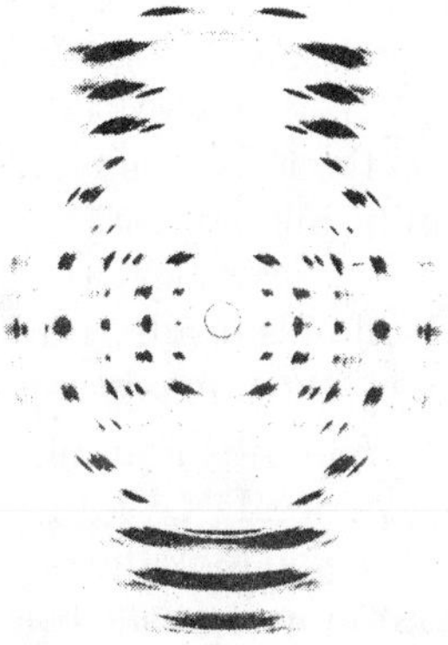

Fig. 33. X-ray diffraction of DNA showing the A.

Paralysis Fellowship at the University of Cambridge, England, and was working under Laurence Bragg and his x-ray team. After the seminar Watson caught the train back to Cambridge and, based on what he had heard and seen, he and Francis Crick (1916-2004) built their first model of DNA. This was slightly unethical as Bragg's Cambridge team had a gentleman's agreement that they would work on x-ray crystallography of protein and Randall's team in London would work on DNA. Watson had not taken notes, knew nothing about crystallography, and had only remembered bits of Franklin's talk. So, when Watson and Crick invited Franklin and Wilkins to view the model, Franklin tore its construction to shreds. The model was a triple helix with bases on the outside. Egos wounded, Bragg banned Crick and Watson from working on DNA.

In May, 1952 Franklin got the first good photograph of the B form of DNA. The photograph showed a double helix, which was the second part of the problem that had to be proved. Franklin, though, was a perfectionist when it came to research and would not release any data until she had more information on the A form which was giving the most data. Meanwhile, in America, Linus Pauling (1901-1994), was working on the structure of DNA and was very close to solving the puzzle. He was a physical chemist with an interest in biological chemistry, who in 1950 constructed the first satisfactory model of a protein molecule and was twice winner of the Nobel Prize (1954 chemistry and 1962 Peace). In May, 1952 he applied for a passport to visit England for a conference, but it was refused because of accusations of him being a communist. Had he attended the conference he would almost certainly have met Franklin. Had they collaborated, with her data and his knowledge, they would almost certainly have solved the structure of DNA.

Not releasing the information on the B form proved to be Franklin's downfall, for she got bogged down with calculations and obsessed in trying to determine whether the A form was helical. Gosling got so frustrated trying to visualise the geometry of the arcs that oranges were used to simulate the spatial relations of the several curves. Finally on Friday the 18th July, 1952 Rosalind took up her fountain pen and, in capital letters, wrote, on a 3 x 6 inch card with a hand inked black border, a death notice for the DNA helix (Crystalline) which she and Gosling signed; referring only to the A form "crystalline" DNA. Franklin continued to waste most of the winter of 1952 with work on the A form. She did meet with Crick who tried to offer advice, but, since his character was typically patronising, she rejected an opportunity to collaborate. An opportunity missed, for they would most certainly have solved the puzzle in months. Meanwhile, because of lack of communication at the King's College lab, Wilkins was trying to reproduce Franklin's results, but he failed to get the same quality. Back in Cambridge Watson and Crick (Fig. 34) were working with a lot of volatile enthusiasm, but no data.

Early in 1953 Linus Pauling sent his son Peter, who was studying at Cambridge, a draft copy of his paper on DNA for comment. Watson knew Peter and somehow got a sight of the DNA paper. Watson took the paper to Franklin for comment, but she dismissed it as rubbish, being further annoyed as she had written to Pauling for information, but none had been sent. Upset at his reception Watson went to talk to Wilkins, who showed Watson one of Franklin's B form photographs without her consent or knowledge. Then, in February, Max Perutz, head of the Medical Research Council Unit housed at the Cavendish Laboratory in Cambridge, where Crick was a research student, received a government report that contained the data presented by Franklin at her departmental seminar. The report was not confidential, but it was private. Perutz passed the report on to Crick without asking Randall or Franklin's permission.

Watson and Crick now had all Franklin's data, which showed that DNA was a multiple helix. Crick, who had worked on proteins, soon realized that Franklin's data implied an antiparallel double helix. According to Franklin's note book she had had the antiparallel idea for the A form, but had not applied it to the B form, had she done so she would have solved part three of the puzzle earlier.

The last part of the puzzle focused on the base pairing and required a thorough knowledge of Chargaff's rules. Watson and Crick had come to an elementary understanding of Chargaff's rules the year before, with Crick even arranging a meeting with Chargaff, where he had to

Fig. 34. Francis Crick shows James Watson the model of DNA that they started building on Wednesday, 4th March and finished in the evening of Saturday, 7th March, 1953 in their room number 103 of the Austin Wing at the Cavendish Laboratories, Cambridge (http://www.ba-education.demon.co.uk/for/science/dnamain.html).

admit that he did not even understand the basics, let alone the rules. Franklin was conversant with Chargaff's rules and so, with all the facts at her fingertips and her superior knowledge of chemistry, it was only a matter of time before she used logic to figure out the last part of the puzzle. In fact she had produced a draft paper, with Gosling, dated the 17th of March, 1953, which outlined that the molecule was a double helix, had specific base pairing and the antiparallel A form, which had not been applied to the B form. Franklin did not realize that Watson and Crick were racing to publish first, which they did on the 18th of March, 1953, so beating her because she had not published.

The Watson and Crick paper entitled "A Structure for Deoxyribose Nucleic Acid" written on the 2nd of April, 1953 and published in "Nature" on the 25th April, 1953 cites no authorities or historical record. It contains no experimental proofs. It contains only hypotheses. No acknowledgement is made to Franklin and Wilkins at King's College, London beyond the following statement. "We have also been stimulated by knowledge of the general nature of the unpublished results and ideas of Dr. M.H.F. Wilkins, Dr. R. E. Franklin and their co-workers at King's College London.

In 1962 Watson, Crick and Wilkins received the Nobel Prize in Physiology or Medicine. They proposed that the DNA molecule takes the shape of a double helix, an elegantly simple structure that resembles a gently twisted ladder. The rails of the ladder are made of alternating units of phosphate and the sugar deoxyribose; the rungs are each composed of a pair of nitrogen-containing nucleotides. In their Nobel lectures they cite 98 references none are Franklin's. Only Wilkins included her in his acknowledgements. Franklin died in 1958. The Nobel Prize is only awarded to living persons.

Deoxyribonucleic Acid (DNA) is a double-stranded, helical molecule consisting of two sugar-phosphate backbones on the outside, held together by hydrogen bonds between pairs of nitrogenous bases on the inside (Figs. 35 and 36). The bases are of four types (A, C, G, & T): pairing always occurs between A & T, and C & G. Watson & Crick realized that these pairing rules meant that either strand contained the information necessary to make a new copy of the entire molecule, and that the order of bases might provide information as a "genetic code". Watson, Crick, and Maurice Wilkins shared the Nobel Prize in 1962 for their discovery.

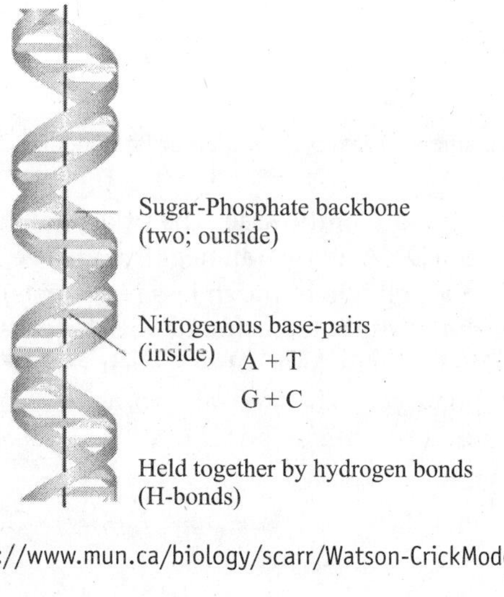

Fig. 35. Structure of DNA (http://www.mun.ca/biology/scarr/Watson-CrickModel.html; Voet and Noet, 2001).

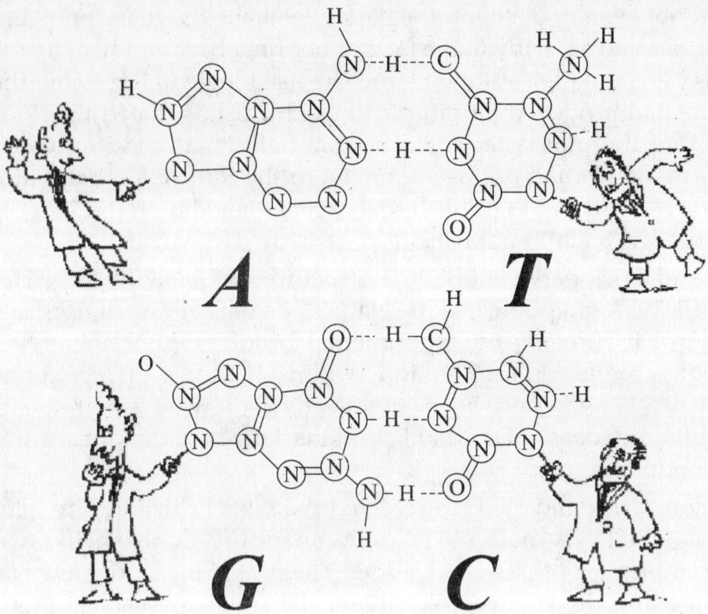

Fig. 36. The Watson-Crick model of DNA (1953) (http://www.mun.ca/biology/scarr/Watson-CrickModel.html; Voet and Noet, 2001).

DOUBLE HELICAL MODEL OF DNA

Phosphodiester Backbone

The Phosphate backbone consists of deoxyribose sugar molecules linked together by phosphate groups, as shown in the oligonucleotide. The backbone (Fig. 37) continues on in a simple repetitive pattern:

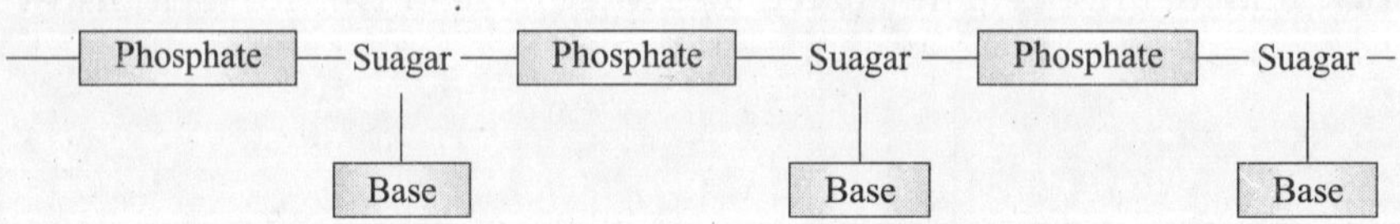

Fig. 37. Diagrammatic representation of Phosphodiester Backbone.

With the DNA bases sticking out at the side. The phosphate groups have been a negative charge, giving the outside of the DNA an overall negative charge. This charge is neutralized in solution by Na^+, although in the cell where much less Na^+ exists). The polyamines, spermine, and spermidine bind along the phosphate backbone and help neutralize some of the charge. (The charge is also neutralized by divalent cations such as $Mg2^+$, and many DNA-binding proteins often contain the positively charged amino acid lysine and arginine, which are attracted to the negative charged phosphate bones).

5' to 3' Direction of DNA Sequences

There is a biological reason for the convention of always writing DNA sequences in the 5′ to 3′direction. In part because of the structure of staring material (nucleotide 5′triphospahtes), the enzymes inside cells in which copy the DNA (e.g. DNA and RNA polymerase) always synthesize in the direction 5′ to 3′. The strands in DNA are anti-parallel i.e. they go in opposite directions (Figs. 38–40).

The double helix of DNA has the following features:

- It contains two polynucleotide strands wound around each other.
- The backbone of each consists of alternating deoxyribose and phosphate groups.
- The phosphate group bonded to the 5' carbon atom of one deoxyribose is covalently bonded to the 3' carbon of the next.
- The two strands are "anti-parallel"; that is, one strand runs 5′ to 3′ while the other runs 3′ to 5′.
- The DNA strands are assembled in the 5′ to 3′ direction and, by convention, we "read" them the same way.
- The purine or pyrimidine attached to each deoxyribose projects in toward the axis of the helix.
- Each base forms hydrogen bonds with the one directly opposite it, forming base pairs (also called nucleotide pairs). A pairs with T by two hydrogen bonds, while G pairs with C by three hydrogen bonds (Figs. 41, 42).
- Each base pair is rotated by 36 degrees from the adjacent base pair.
- The distance between adjacent base pair is 3.4 Å
- The double helix makes a complete turn in just over 10 nucleotide pairs, so each turn takes a 34 Å shown in the diagram.
- There is an average of 25 hydrogen bonds within each complete turn of the double helix providing a stability of binding about as strong as what a covalent bond would provide.
- The diameter of the helix is 20 Å.
- The path taken by the two backbones forms a major (wider) groove and a minor (narrower) groove.

Major and Minor Grooves

The terms 'major groove' and 'minor groove' are based on the two groove of the Watson-Crick B-DNA structure. The dimensions of the major and minor grooves are different (Table 5, Fig. 43) for the three different helix families. The definition of major groove and minor groove is as fallows:

For B-DNA helices, Protein binding in the major groove usually binds to specific sequences, often through the insertion of α helix in to major groove. In addition, the major groove of B-DNA is approximately the correct width to accommodate a third base pair (usually a pyrimidine), as happens in certain triplex DNA structures. Proteins that bind DNA nonspecifically (such as chromatin proteins) will often bind DNA in the minor groove, through interactions with a protein β-strand. In addition, water molecules and small ions can bind to,

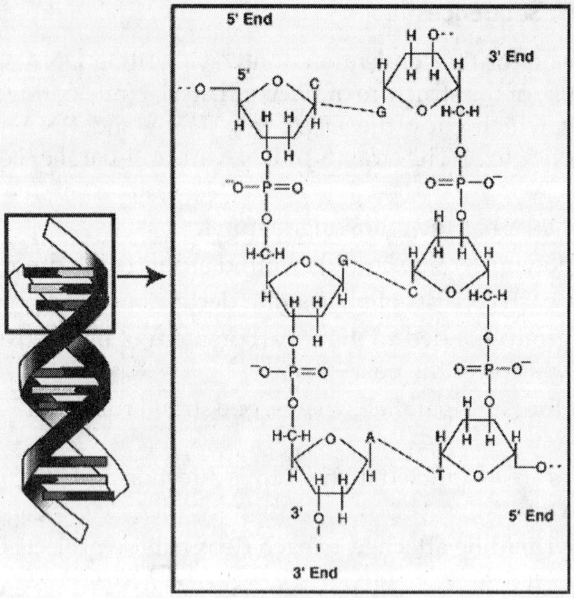

Fig. 38. DNA structure.

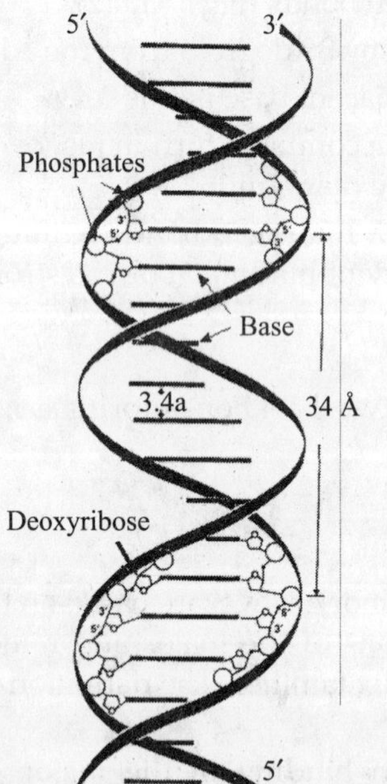

Fig. 39. Feature of DNA structure (Nelson and Cox, 2004).

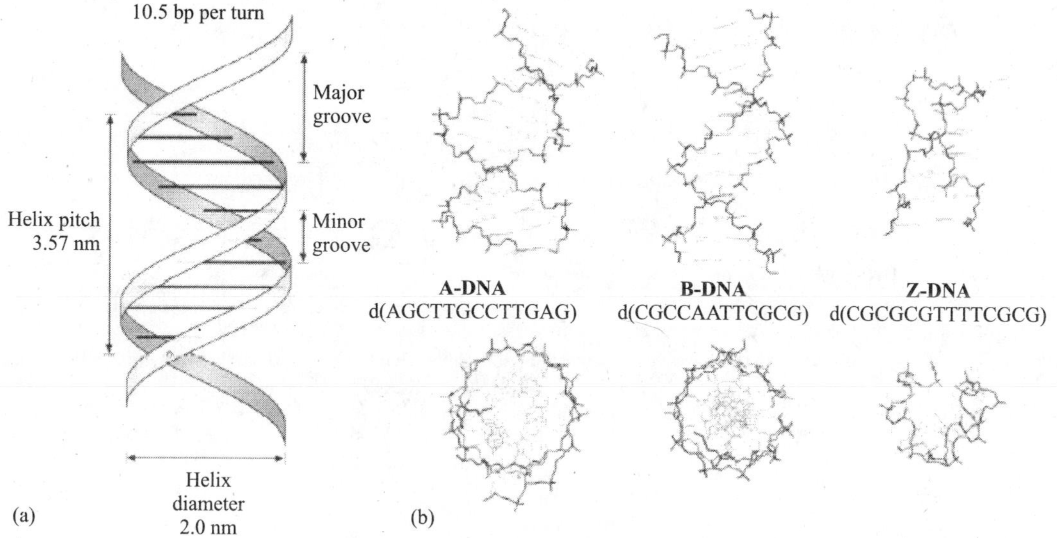

Fig. 40. Different views of the DNA helix (a) The structure of B-DNA as proposed by Watson and crick in 1953 based on the fibre diffraction studies. (b) A-, B- and Z-DNA, as seen from the side of the helix (above), and looking down the helix axis (below). The structures were drawn from the crystal structure using Cn3D program available on NCBI home page.

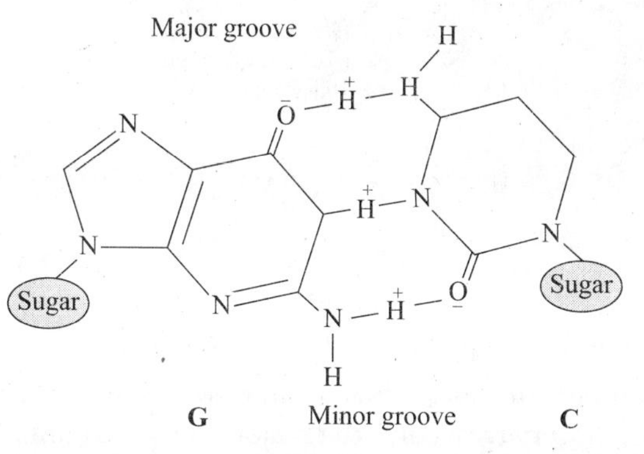

Fig. 41. Base Pairing between A and T.

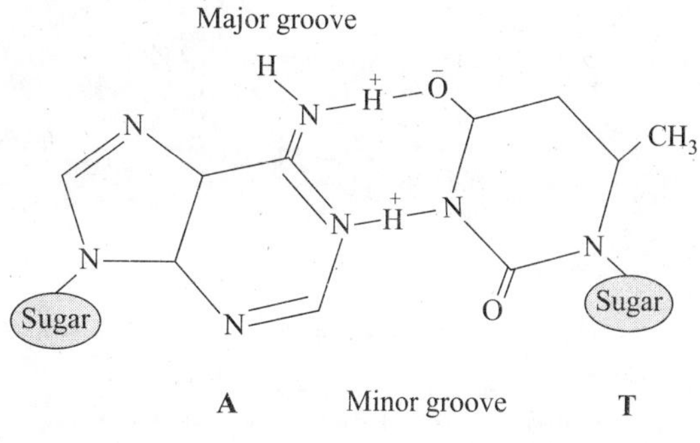

Fig. 42. Base Pairing between G and C.

Table 5. Major and minor groove

Major Groove		Minor Groove	
Adenine	N6 amino, N7	Adenine	N3
Cytosine	N4 amino	Cytosine	C2 carbonyl
Guanine	C6 carbonyl, N7	Guanine	N2 amino, N3
Thymine	C4 carbonyl, C5 methyl	Thymine	C2 carbonyl
Uracil	C4 carbonyl	Uracil	C2 carbonyl

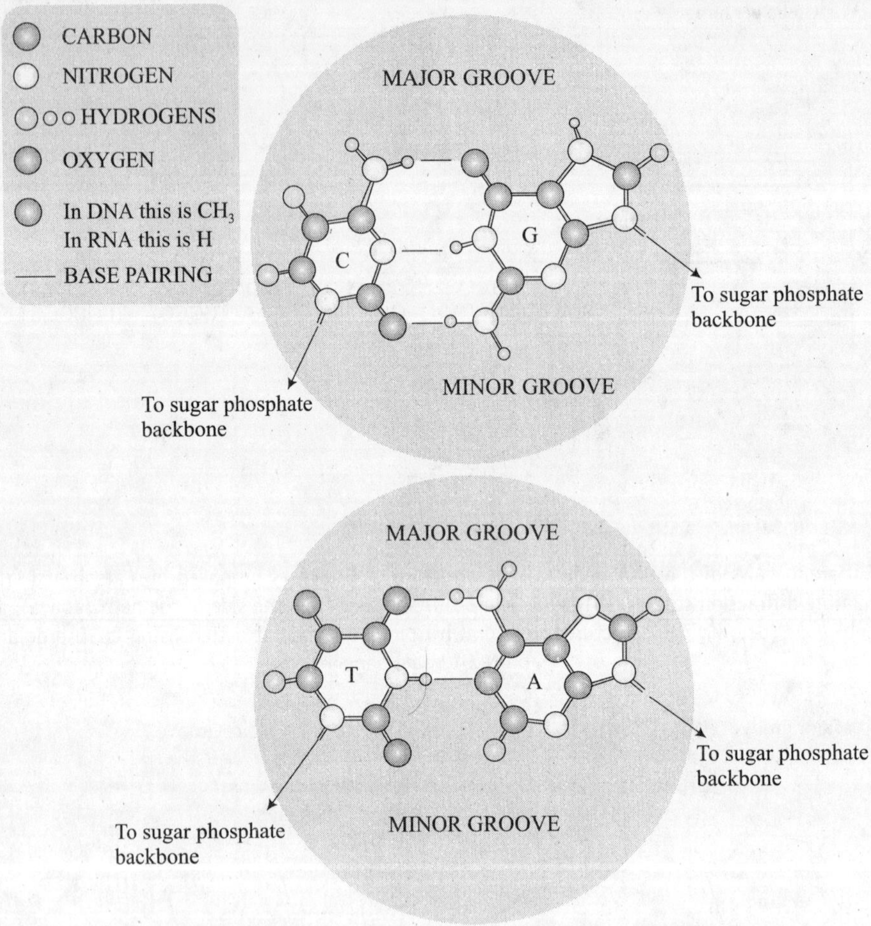

Fig. 43. Major and Minor Groove (Nelson and Cox, 2004).

and stabilize, the minor groove. In A-DNA, the minor groove is almost same size as the major groove, while in Z-DNA the minor groove is deep and narrow, and the major groove is almost non-existent.

OTHER FORMS OF DNA STRUCTURE

A-DNA

A-DNA was identified from the fibre-diffraction studies of DNA at 'low' (75%) relative humidity. When B-DNA is dehydrated, there is a reversible structural change to A-DNA, in which there is an increase in the tilt of the bases to about 20° with respect to the helical axis, which does not pass through the base pairs at all, but rather is shifted into the major groove. The result is a structure with a narrow and deep major groove and a wide and shallow minor groove, a diameter of about 26 Å, and 11 bases per helical turn. When looking at A-DNA end-on, there is a 6-Å diameter hole at the helical axis. An A-helix is the common form for DNA-RNA hybrids, as well as double-stranded RNA. This is due to the extra OH group on the ribose sugar, which cannot fit easily in to the tight space allotted to it in B-DNA.

Z-DNA

Unlike B-DNA and A-DNA, Z-DNA is a left-handed helix (Table 6, Fig. 44). It has a base tilt of about 7° and a diameter of about 18 Å. There are 12 bases per helical turn, a narrow and deep minor groove and a flat major groove. Sequences of DNA consisting of alternating purine and pyrimidine bases have been shown to adopt this conformation. Such sequences include poly d(GC), poly d(AC), poly d(GT) etc. in varying combinations and can be formed at high salt concentrations, under dehydrating conditions in the presence of Mn^{2+} or Co^{2+}, or in "supercoiled" DNA.The organization of the molecule is completely different from that which has been anticipated. Instead of all the residues having bases organized in the *ANTI*-CONFORMATION, in this molecule every other base rotated around the glycosyl bonds so that the bases alternated in *ANTI*- and *SYN*-CONFORMATIONS along the chain. Also, there was a zigzag arrangement of the backbone of the molecule (hence, the name Z-DNA) that looked different from the smooth continuous coil seen in models of B-DNA. Instead of having a helix with a major and minor groove, the base pairs are set off to the side, away from the axis, and only one groove is found that is analogous to the minor groove of B-DNA. The bases that form the major groove in B-DNA were reorganized in Z-DNA to form the convex outer surface.

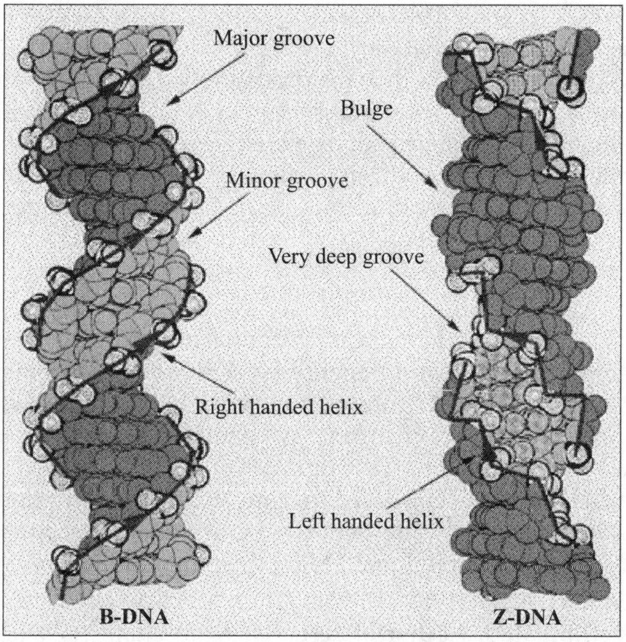

Fig. 44. Comparison of B-DNA and Z-DNA (Voet and Voet, 2001; Garrett and Grisham, 2007).

Table 6. Comparison of various forms of DNA

Geometry attribute	A-form	B-form	Z-form
Helix sense	Right-handed	Right-handed	Left-handed
Repeating unit	1 bp	1 bp	2 bp
Rotation/bp	33.6°	35.9°	60°/2
Mean bp/turn	10.7	10.0	12
Inclination of bp to axis	+19°	−1.2°	−9°
Rise/bp along axis	2.3Å	3.32Å	3.8Å
Pitch/turn of helix	24.6Å	33.2Å	45.6Å
Mean propeller twist	+18°	+16°	0°
Glycosyl angle	anti	anti	C: anti,G: syn
Sugar pucker	C3'-endo	C2'-endo	C: C2'-endo, G: C2'-exo
Diameter	26Å	20Å	18Å

CONFORMATION AND STABILITY

The relationship between Z-DNA and the more familiar B-DNA was indicated by the earlier work of Pohl and Jovin, who showed that the ultraviolet circular dichroism of poly (dG-dC) was nearly inverted in 4 M sodium chloride solution. The suspicion that this was the result of a conversion from B-DNA to Z-DNA was confirmed by examining the Raman spectra of these solutions and the Z-DNA crystals. The conversion of B-DNA to Z-DNA was associated with a 'flipping over' of the base pairs so that they had an upside down orientation relative to that of B-DNA. This flipping over resulted both in the production of a *syn*-conformation in every other base and a change in the deoxyribose-ring pucker in alternate bases. The net result of this reorganization was that the phosphate groups were closer together in Z-DNA than in B-DNA. Hence, under standard cellular conditions, the electrostatic repulsion of these charged phosphate groups would push the molecule into the B-DNA conformation. In the presence of a high-salt solution, the electrostatic repulsion of the phosphate residues is vastly decreased, and Z-DNA becomes the stable conformation.

Several studies quickly showed that chemical modification, including cytosine methylation and many other cations, such as spermine and spermidine, would stabilize the Z conformation. It emerged that the lowest energy-level ground state of DNA in a physiological solution was B-DNA, and that the Z conformation was a higher energy state. Because purines can form the *syn*-conformation without an energy penalty, it became apparent that the specific sequence of the base pairs was important in determining the energy that was required to change B-DNA to Z-DNA. The sequences that most readily converted had alterations of purines and pyrimidines, especially alternations of C and G. This change also occurred easily with alternations of CA on one strand and TG on the other. However, many other sequences were shown to be capable of forming Z-DNA.

This discovery stimulated a burst of research from chemists who were interested in studying DNA conformational changes. Although the ionic conditions that were suitable for stabilizing Z-DNA in most experiments were different from those present in a cell, spermine, spermidine and cytosine methylation, which also stabilize Z-DNA, are found *in vivo*.

Furthermore, the discovery that negative supercoiling would stabilize Z-DNA indicated biological involvement. Negative supercoiling requires energy and tends to unwind B-DNA. For example, in a plasmid with three negative supercoils, if one turn of the DNA helix changed from right-handed to left-handed, two negative supercoils would disappear and the energy of negative supercoiling would then stabilize a small segment of Z-DNA (Fig. 45). Supercoiling was known to be a part of biological systems, which indicated a connection between this alternative conformation and biological phenomena. Early experiments showed that the negative supercoiling of plasmids in prokaryotes would stabilize Z-DNA.

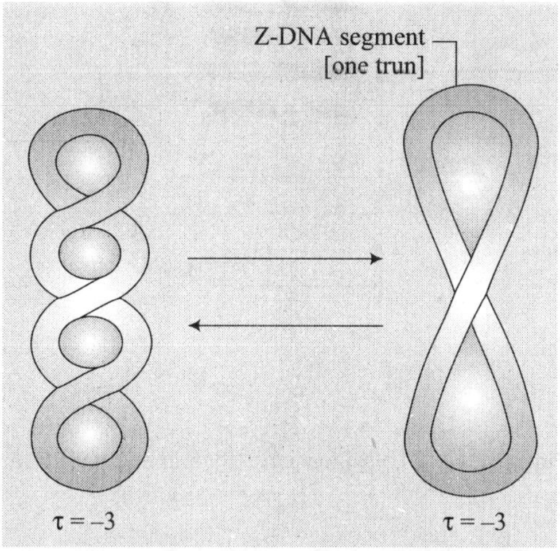

Fig. 45 Negative supercoiling stabilizes Z-DNA. A double helical plasmid with three negative supercoils is shown on the left. The energy of supercoiling is proportional to the square of the number of supercoils. On the right, one turn switches the DNA from right- to left-handed, and two negative supercoils are removed. The change of negative supercoiling energy stabilizes Z-DNA formation (Rich and Zhang, 2003).

The conformational change from B-DNA to Z-DNA is one mechanism for relief of the torsional strain found in B-DNA *in vivo*, and may serve as a switch mechanism to regulate gene expression. Formation of Z-DNA form B-DNA is a dynamic process where B-DNA is the relaxed state. When a Z-DNA segment is formed two B-Z junctions form. At each junction the hydrogen bonds between a Watson/Crick base-pair is broken and the bases are extruded. Extrusion of a base from the helix is a well-known reaction performed by enzymes (i.e. DNA glycosylase) that edit or repair DNA during Base Excision Repair (BER).

Structure of DNA with two B-Z DNA junctions (Fig. 46): It encompasses 1. breakage of a hydrogen-bond, where a Guanine rotates around its glycosyl-bond and the sugar thereby transforms into its syn-conformation. 2. Rotation of the corresponding second base (Cytosine) involving rotation of the sugar around the sugar-phosphate-bond. 3. At the B-Z junction hydrogen-bonds remain broken and bases are extruded.

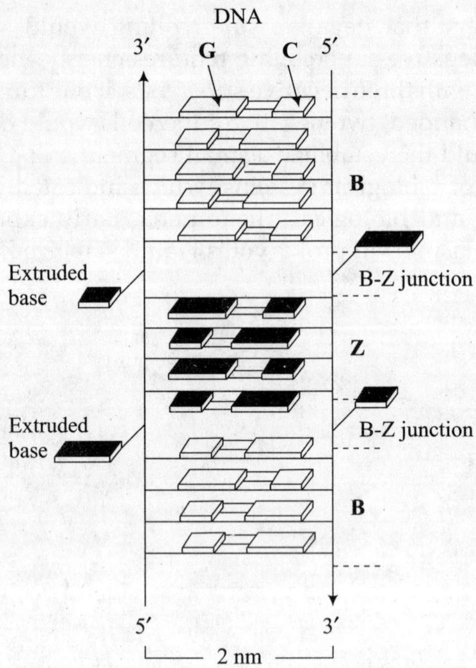

Fig. 46. B-Z DNA junctions (Rich and Zhang, 2003).

DENSITY OF DNA

When a concentrated solution of CsCl is centrifuged in the analytical ultracentrifuge at high speeds until the equilibrium is attained, the opposing processes of sedimentation and diffusion produce a stable concentration gradient of the CsCl i.e. there is a continuous increase in density along the direction of centrifugal force. The density gradient formed is proportion to centrifugal force according to the equation $d\rho/dr = \alpha\omega^2 r$, where ρ is the density as a function of the distance from the center of rotation r, ω is the speed, and α is a constant depending upon the nature of the salt. When the CsCl solution contains small amount of DNA, at equilibrium the molecules of DNA will be collected in bands at those zones of the centrifuge cell where their density and density of the medium are exactly equal (isopycnic) (Fig. 47). The buoyant density of DNA can be correlated with its G+C content by the empirical equation

$$P = 1.660 + 0.100(G + C \text{ frequency}) \text{ g/cm}^3$$

This property permits the fractionation of DNA molecules according to their G + C content.

DENATURATION OF DNA

In addition to hydrogen bonds, hydrophobic forces between the stacked purines and pyrimidines contribute to maintenance of the rigid two-stacked structure. Thus the reagent like urea and formamide, which increase the solubility of DNA in aqueous medium, also denature the DNA. Similar effect has been found in a variety of agents and conditions e.g., acid, alkali,

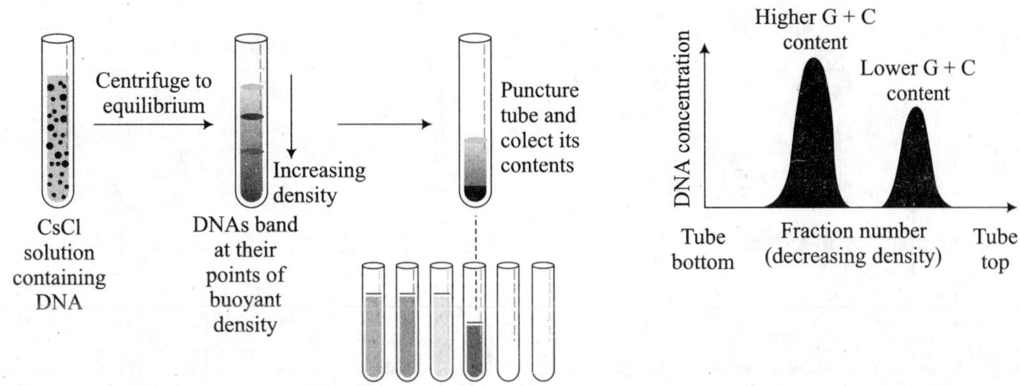

Fig. 47. Large DNA molecules separated by CsCl Density gradient centrifugation (Garrett and Grisham, 2007).

heat, and low ionic strength, on the DNA structure. Several methods are used to assess the transition from the native to a denatured state, as well as to determine other properties of DNA.

Ultraviolet absorption

All nucleic acid shows a strong absorption in the ultraviolet with a maximum near 260 nm (Fig. 48). When native DNA is altered, there is a marked hyper chromatic effect or increase in absorption. The change reflects a decrease in hydrogen bonding and observed not only with DNA but with RNA and with synthetic polynucleotides that have a hydrogen-bonded structure.

Optical rotation

Native DNA shows a strong positive rotation, which is markedly decreased by denaturation.

Viscosity

Solution of native DNA processes a high viscosity because of the relative rigid double–helical structure and long, rod like character of the DNA. Disruption of hydrogen bonds produces a marked decrease in viscosity.

Effect of Temperature

Heating of DNA sample in a given ionic condition produces an increase in ultraviolet absorption and a decrease on optical rotation and viscosity, the disruption of the hydrogen bonds between the two strands in the double helix.

Since the interaction between the bases in two strands are cooperative in a manner similar to the interactions between molecules in a crystal, the disruption of the ordered helical structure occurs over a small temperature interval, much like the melting of a crystal. For the reason, the heat denaturation of double-stranded is often referred to as a "meting" of the DNA and the temperature at which 50% of the DNA is denatured is termed the melting temperature Tm (Fig. 49).

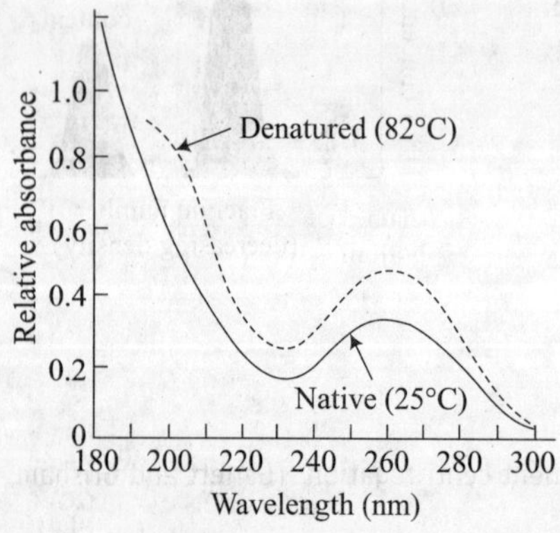

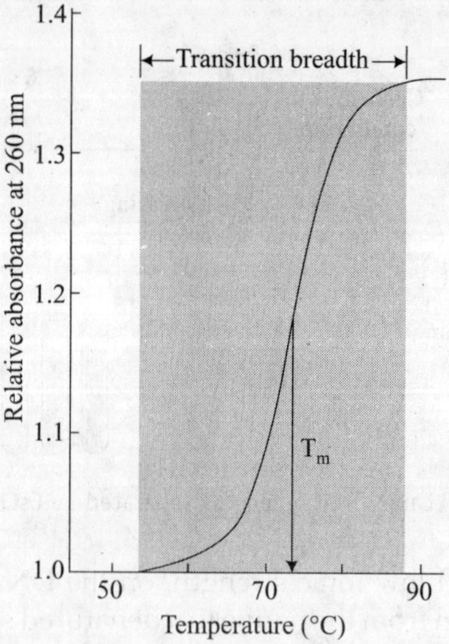

Fig. 48. Denaturation of DNA can be accessed by UV Absorption (Voet and Voet, 2001).

The nature of the melting transition is affected by several factors

The (G+C) content of the DNA

There are three hydrogen involve in the G-C base pair. Because of this, the higher the G+C content of DNA higher will be the stability of the molecule and melting of DNAs of different G+C content.

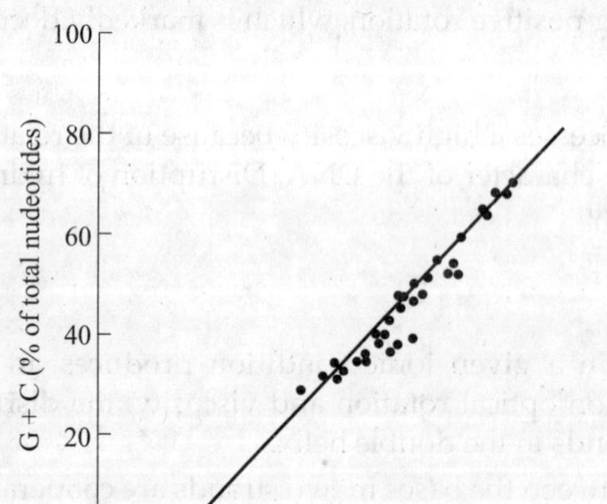

Fig. 49. Relationship between G+C and Tm value (Nelson and Cox, 2004).

Under standard ionic condition and pH, G+C amount is related to Tm as fallow:

$$\%G+C = (Tm-69.3) \times 2.44$$

The nature of the DNA

Most DNA molecules are mosaics of regions of varying G+C composition and the A+T rich regions will melt before the G+C rich regions. This can result in the two strands held together and hence in register by G+C-rich regions. On Cooling these strands regions will act as foci to allow the rapid reannealing of the two strands of DNA. Only when all the hydrogen bonds are broken does melting become irreversible on rapid cooling. Short viral DNAs, which may be more homogenous, exhibit sharper melting profiles.

The nature of solution

Tm value is lowered by the addition of urea, as agent known to disrupt hydrogen as well as hydrophobic bonds. For Example, in8M urea the Tm is decreased by nearly 20°C. In 95% formamide, DNA is completely separated in to single strands at room temperature.

Effect of pH

Disruption of the double-stranded structure of DNA also occurs at acid and alkaline pH values (Fig. 50) at which ionic charges of the substituents on the purine and pyrimidine bases can occur, near pH 12; ionization of enolic hydroxyl groups prevents the keto-amino group hydrogen bonding. Similarlyin acidic solution near pH 2-3, at which amino groups are protonated, the helix is disrupted.

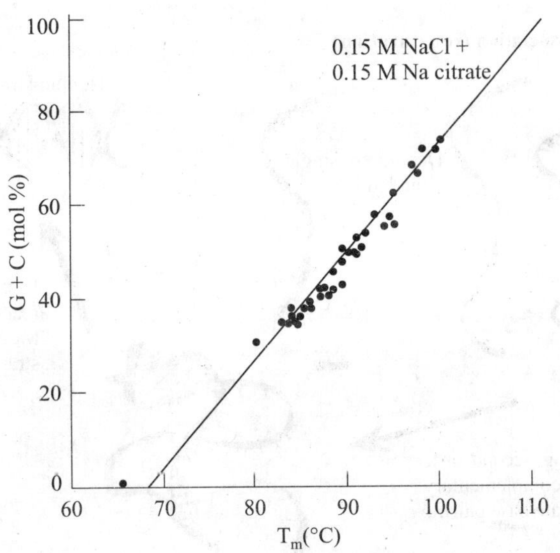

Fig. 50. Effect of pH and Tm (Garrett and Grisham, 2007).

DENATURATION OF DNA

Denaturation of a DNA double helix produces single-stranded DNA. The reverse reaction (the formation of double-stranded DNA) can be carried out if the complementary DNA strands are incubated under conditions that will promote renaturation (reassociation). This process involves two steps. The first step is a slower, rate-limiting reaction in which the complementary strands attempt to find each other. The single strands randomly interact until complementary base pairing can occur. This may occur over only a short region of each strand, but may be enough to "nucleate" the reaction. Because the two reacting strands are at the same concentration in solution, the reaction is second-order, with the rate of the reaction being defined by the second-order rate constant, k2. Once nucleation has occurred, the complementary strands rapidly zipper up. The complexity of a DNA is a function of how many base pairs it has. In other words, a low complexity DNA may consist of a few hundred or a few thousand base pairs, in contrast to a high complexity DNA that may contain millions of base pairs. Low complexity sequences are able to find each other much faster during a renaturation reaction (Fig. 51), than are high complexity sequences, since the low complexity sequences have to make fewer collisions to find a base-pairing partner. Thus, the rate of a renaturation reaction is a function of the complexity of the DNA. Or in other words, the complexity of a DNA can be determined by measuring its second-order reassociation rate constant.

Reasssociation kinetics

C = concentration of single-stranded DNA at time t (moles of nucleotide/liter)

C_0 = initial concentration of single-stranded DNA

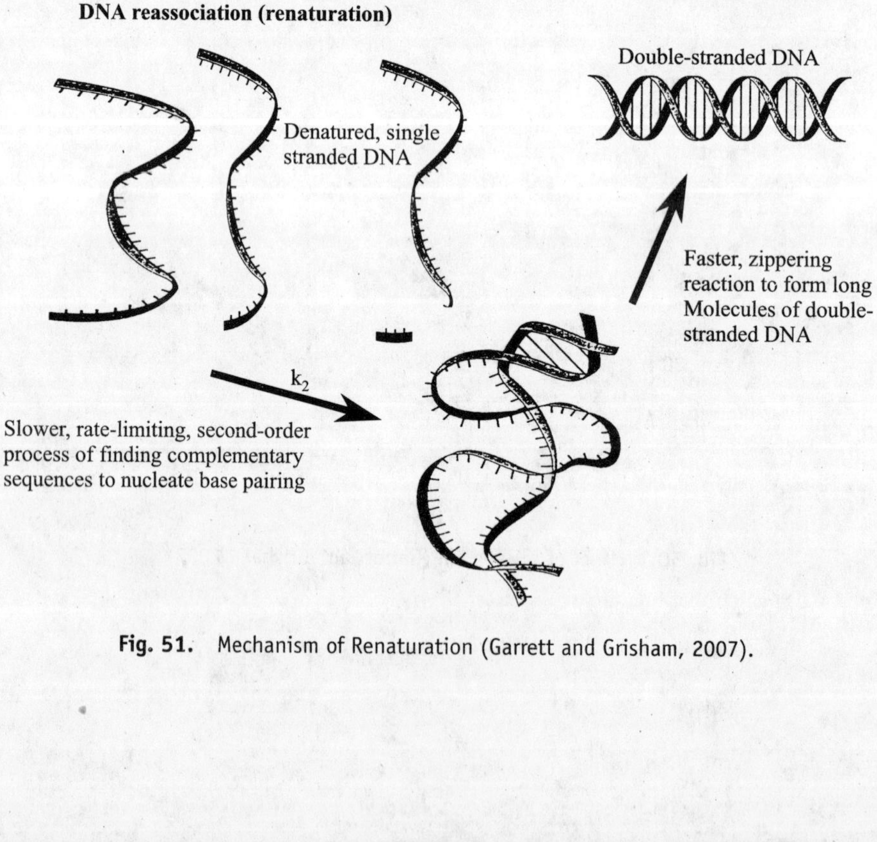

Fig. 51. Mechanism of Renaturation (Garrett and Grisham, 2007).

k = second order rate constant (depends on salt concentration, temperature, fragment size, viscosity, complexity of DNA, etc.)

t = time (seconds)

$$-dC/dt = kC^2 \quad ...(1)$$

$$-dC/C^2 = k\,dt \quad ...(2)$$

$$\int_{Co}^{C} -dC/C^2 = \int_{0}^{t} kdt \quad ...(3)$$

$$1/C\Big|_{C_o}^{C} = kt\Big|_{0}^{t} \quad ...(4)$$

$$(1/C) - 1/C_0 = kt \quad ...(5)$$

$$(C_0 - C)/CC_0 = kt \quad ...(6)$$

$$(C_0 - C) = CC_0kt \quad ...(7)$$

$$C_0 = C + CC_0kt \quad ...(8)$$

$$C_0 = C(1+C_0kt) \quad ...(9)$$

$$C/C_0 = 1/(1+C_0kt) = fs \text{ (fraction single stranded DNA)} \quad ...(10)$$

When t = 0 (start of the reaction), $C/C_0 = 1$ (all the DNA is single-stranded);

When t is very large (infinite time of reaction), $C/C_0 = 0$ (no single-stranded DNA);

When the reaction is half over ($C/C_0 = 1/2$ when $t = t_{1/2}$), then $C_0t_{1/2} = 1/k$ = kinetic complexity;

The $C_0t_{1/2}$ of a reaction is directly proportional to the amount of DNA in a genome i.e., the $C_0t_{1/2}$ is a molecular ruler (Fig. 52).

The Fig. 53 shows a DNA renaturation (reassociation) reaction or "Cot curve" for human genomic DNA. The reaction is graphed as a semi-log plot with "log Cot" on the x-axis and "% DNA reassociated" on the y-axis (0% at the top and 100% at the bottom). The reaction follows ideal second-order kinetics. Cot is the product of Co (initial DNA concentration) and t (time). DNA concentration is constant. The x-axis is simply the time course of the reaction. The reaction would proceed as follows: at time-zero the DNA is denatured into single strands and the conditions are adjusted to promote DNA base pairing to allow the renaturation of the DNA. During the reaction the various kinetic fractions find partners (complements) and reassociates until the reaction is complete (100% reassociation). This DNA shows complex reassociation kinetics because it is actually a mixture of three different species or populations of DNA: fast-reassociating, intermediate-reassociating, and slow-reassociating fractions. The fast and intermediate fractions are composed of repeated sequences in the genome, and the slow fraction is composed of single-copy (unique) sequences. Single-copy sequences are those that are present in one copy per haploid genome.

The $C_0t_{1/2}$ is equal to the inverse of the second-order reassociation rate constant, k2. In other words, the rate constant for a sample of DNA can be determined experimentally by measuring the $C_0t1/2$ of the reaction. Each reassociating component has its characteristic rate

Rate of ressociation is inversely proportional to the length of the reassociating DNA.

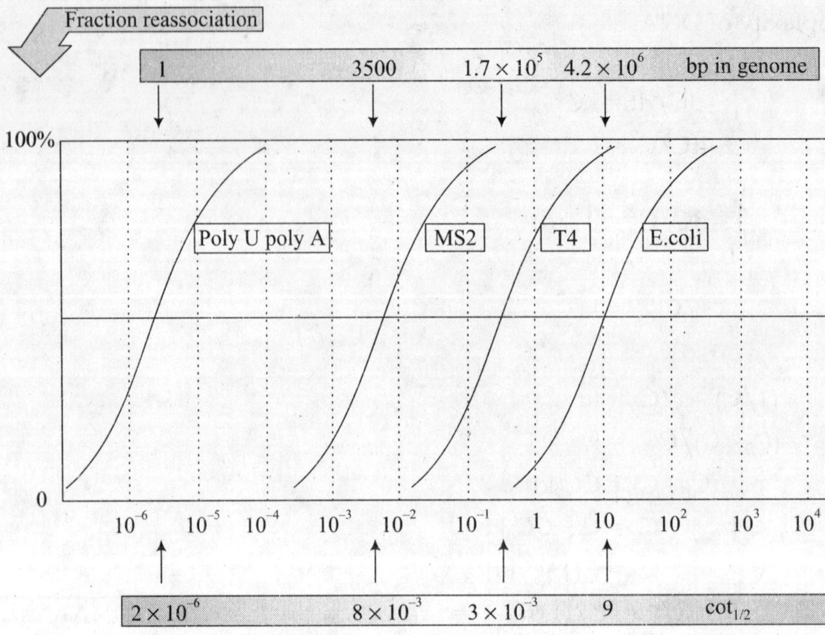

Fig. 52. Cot value and Genome complexity (Garrett and Grisham, 2007).

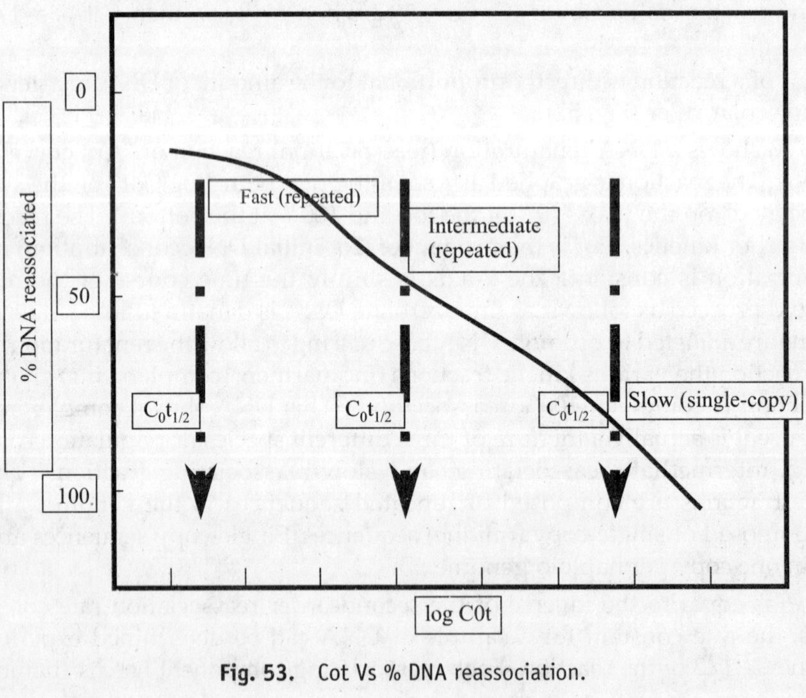

Fig. 53. Cot Vs % DNA reassociation.

constant and $C_0t1/2$. The rate constant provides the complexity of the DNA by comparison to the rate constants of other DNA samples of known complexity.

TOPOLOGY OF DNA

The double-stranded structure of DNA has many implications for biological function. Replication, at first appeared facilitated because genetic information is encoded twice in the DNA structure, once on each strand. However, the two parental strands must be separated and unwound to be copied.

Transcriptional enzymes must decide which of two complementary strands contains the correct information to copy. Transcription also involves transient unwinding of the DNA helix in a local region in order to be able to copy one strand.

Metabolic events involving unwinding impose great stress on the DNA because of the constraints inherent in the double helix.

In addition to the requirement to unwind DNA for replication and for transcription, there is an absolute requirement for the correct topological tension in the DNA (super-helical density) in order for genes to be regulated and expressed normally.

Supercoiled DNA

In 1965 Vinograd's group discovered that the circular DNA chromosomes isolated from small viruses like SV40 or polyoma were in a highly compact or folded conformation (Fig. 54).

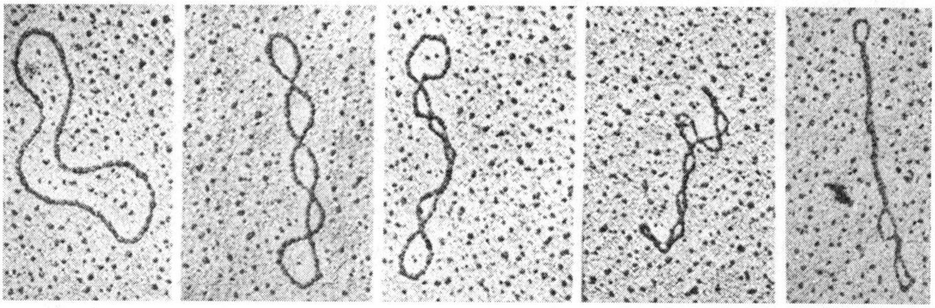

Fig. 54. Supercoiled DNA (Nelson and Cox, 2004).

This supercoiling or writhing of circular DNAs was a result of the DNAs being underwound with respect to the relaxed form of DNA. There are actually fewer turns in the DNA helix than one would expect given the natural pitch of DNA in solution (10.4 base pairs per turn). When a linear DNA is free in solution it assumes a pitch, which contains 10.4 base pairs per turn. This is less tightly wound than the 10.0 base pairs per turn in the Watson and Crick B-form DNA. In order to understand the origin of supercoiling, imagine a linear DNA 5200 base pairs in length. If the DNA were in the B-form one would expect the two strands of the helix to be wrapped around each other 500 times (5200 bp/10.4 bp/turn). Imagine a linear DNA in which the two ends become connected to form an open circle. This is referred to as a relaxed circular DNA. On the other hand, if the linear DNA were unwound 10%, say 50 turns, before its ends were joined, then the DNA molecule would be under stress. When the molecule is free in solution it will coil about itself in space as the two strands simultaneously twist about each other in order to return to equilibrium value of 10.4 base pairs per turn.

DNA that is underwound is referred to as negatively supercoiled. The helices wind about each other in a right handed path in space.

DNA that is overwound also will relax and assume a supercoiled conformation but this is referred to as a positively supercoiled DNA helix. Positively coiled DNA has its DNA helices wound around each other in a left-handed path in space.

Linking, Twisting and Writhing

The total number of times one strand of the DNA helix is linked with the other in a covalently closed circular molecule is known as the linking number Lk.

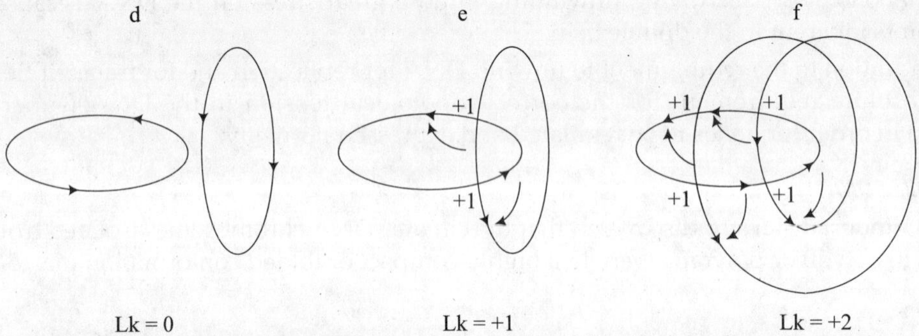

Fig. 55. Linking, twisting and writhing.

The *linking number* is only defined for covalently closed DNA and its value is fixed as long as the molecule remains covalently closed (Fig. 55).

The *linking number* does not change whether the covalently closed circle is forced to lie in a plane in a stressed conformation or whether it is allowed to supercoil about itself freely in space.

The *linking number* Lk of a circular DNA can only be changed by breaking a phosphodiester bond in one of the two strands, allowing the intact strand to pass through the broken strand and then rejoining the broken strand.

Lk is always an integer since two strands must always be wound about each other an integral number of times upon closure.

The *linking number* in relaxed DNA, Lk_0 is used as a reference. For the molecule has $Lk_0 = 200$. If two turns are removed from this molecule, Lk = 198. The change is described by the equation:

$$\Delta \text{Lk} = \text{Lk} - \text{Lk}_0 = 198 - 200 = -2$$

It is often convenient to express the change in the linking number in terms of a quantity that is independent of the length of the DNA molecule. This quantity, called the specific linking difference (σ), or superhelical density, is the measure of number of turns removed relative to the number present in relaxed DNA:

$$\sigma = {}'\Delta Lk / Lk_0$$

In above example, σ = –0.01, which means that 1% (2% of 200) of the helical turns present in the DNA has been removed. The degree of underwinding in cellular DNA falls in the range of 5% to 7%, that is, σ = –0.05 to –0.07. The negative sign indicates that the change in the linking number due to the underwinding of DNA. The supercoiling induced by underwinding is therefore defined by negative supercoiling.

In case of overwinding, a supercoiled conformation is referred to as a positively supercoiling.

The linking number of a covalently closed circular DNA can be resolved into two components called the twists Tw and the writhes Wr (Fig. 56).

$$Lk = Tw + Wr$$

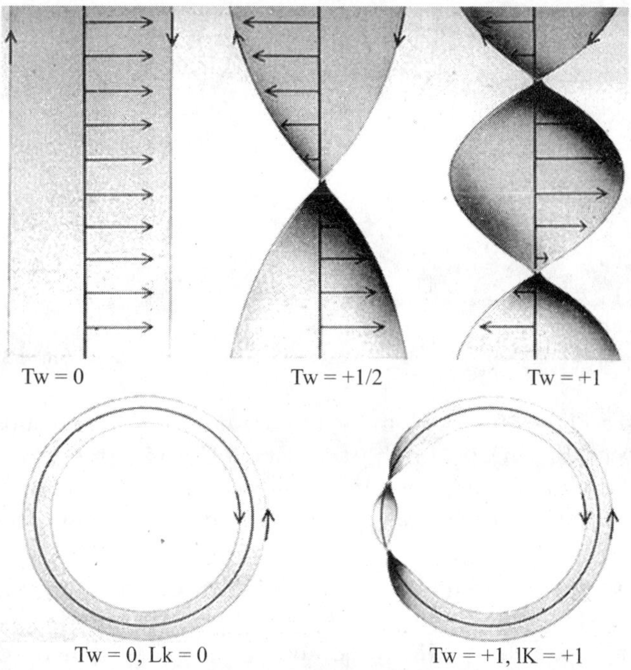

Fig. 56

The twists Tw are the number of times that the two strands are twisted about each other while the writhes Wr is the number of times that the DNA helix is coiled about itself in three-dimensional space.

The twist and the writhe are not necessarily integers. If SV40 DNA molecule for example, which is precisely 5243 base pairs in length, we would find that:

$$Lk = Tw + Wr$$

$$480 = 5243/10.4 - 24.13$$

$$480 = 504.13 - 24.13$$

Its length and its pitch in solution determine the twist of DNA. [Tw = Length (bp)/Pitch (bp/turn)]

The twist and the linking number, determine the value of the writhe that forces the DNA to assume a contorted path is space. [Wr = Lk – Tw] (Figs. 57 and 58).

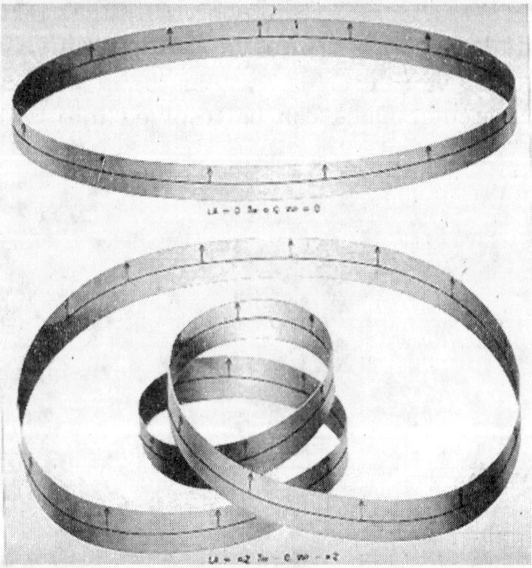

Fig. 57

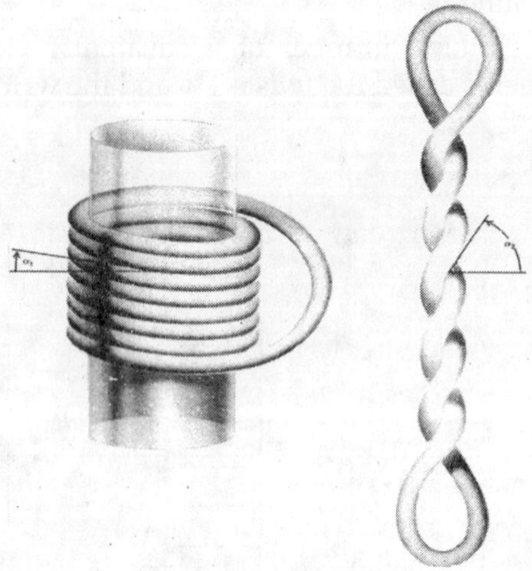

Fig. 58

Unlike the Twist and the Linking number, the writhe of DNA only depends on the path the helix axis takes in space (Fig. 59), not on the fact that the DNA has two strands. If the path of the DNA is in a plane, the Wr is always zero. Also if the path of the DNA helix were on the surface of a sphere (like the seams of a tennis ball or base ball) then the total Writhe can also be shown to be zero.

Writhes can also come in different forms. If a DNA molecule wraps around itself, then the writhes are known as supertwists. If a DNA molecule wraps around something else (another molecule for instance) then the writhes are known as solenoidial writhe. In solution, the writhes can isomerize between the two types.

Measuring Supercoils

The superhelical density of a circular DNA can be observed and measured in several ways by electron microscopy, sedimentation velocity, or by electrophoresis of DNA.

The first quantitative method for measuring supercoiling was the sedimentation procedure. Since supercoiled molecules are more compact they sediment faster in a centrifuge than when they are relaxed.

A more elegant and simpler modern method for determining the number of superhelical turns in DNA is by electrophoresis in an agarose gel. Supercoiled DNA migrates much more rapidly than does a relaxed molecule of the same length. DNA separates into discrete bands depending on the linking number.

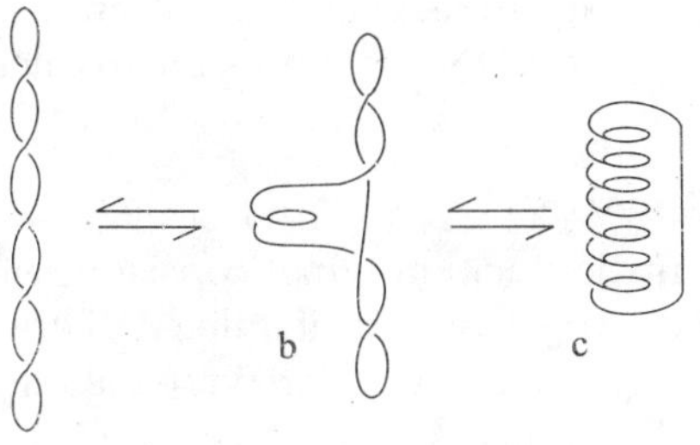

Fig. 59

Since DNAs resolved in this way differ from each other only in their topology, they are referred to as topological isomers, or topoisomers.

Molecules that differ by one unit in linking number can be separated by electrophoresis in agarose due to the difference in their writhe (i.e. due to difference in folding). The variation in linking number is reflected in a difference in the writhe. The variation in writhe is subsequently reflected in the state of compaction of the DNA molecule (Fig. 60).

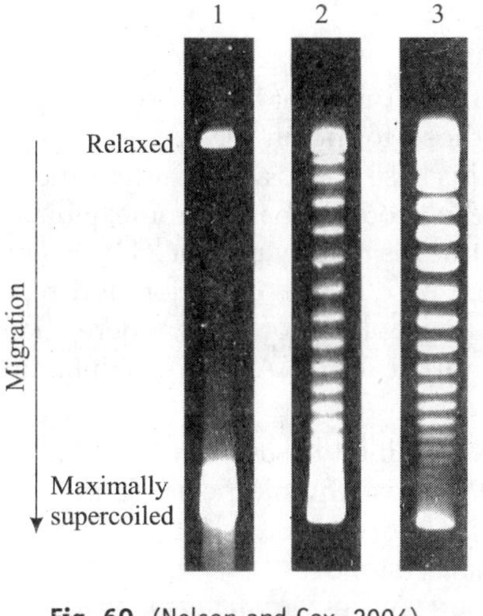

Fig. 60. (Nelson and Cox, 2004)

Neither of these methods by themselves can distinguish negatively and positively supercoiled DNA. Both forms sediment and electrophorese more rapidly than relaxed DNA. However by sedimenting or electrophoresing supercoiled DNA in the presence of an intercalating agent such as ethidium bromide or chloroquine, one can distinguish negatively supercoiled DNA from positively supercoiled DNA. When negatively supercoiled DNA binds an intercalating agent, the average pitch is reduced because the twist angle between adjacent base pairs on either side of the intercalating agent is reduced from 36° to as little as 10° of twist. This reduction of twist causes a compensatory increase in writhe in a covalently closed molecule. Hence a molecule that is initially negatively supercoiled will become more relaxed and a positively supercoiled molecule will become more twisted. By electrophoresing or sedimenting supercoiled DNA in increasing concentrations of an intercalating agent and knowing the affinity of the agent for DNA it is possible to calculate the writhe in the DNA.

RNA STRUCTURE

Ribonucleic acid (RNA) is a nucleic acid polymer consisting of covalently bound nucleotides. RNA nucleotides contain ribose rings and uracil unlike DNA, which contains deoxyribose and thymine. It is transcribed from DNA by an enzyme called RNA polymerase and further processed by other enzymes. RNA serves as the template for translation of genes into proteins, transferring amino acids to the ribosome to form proteins, and also translating the transcript into proteins.

History

The role of RNA in protein synthesis had been suspected since 1939, based on experiments carried out by Torbjörn Caspersson, Jean Brachet and Jack Schultz. The sequence of the 77 nucleotides of a yeast tRNA was found by Robert W. Holley in 1964, winning Holley the 1968 Nobel Prize for Medicine.

Chemical Structure

RNA is primarily made up of four different bases: adenine, guanine, cytosine, and uracil. The first three are the same as those found in DNA, but uracil replaces thymine as the base complementary to adenine. This base is also a pyrimidine and is very similar to thymine. Uracil is energetically less expensive to produce than thymine, which may account for its use in RNA (Fig. 61). In DNA, however, uracil is readily produced by chemical degradation of cytosine, so having thymine as the normal base makes detection and repair of such incipient mutations more efficient. Thus, uracil is appropriate for RNA, where quantity is important but lifespan is not, whereas thymine is appropriate for DNA where maintaining sequence with high fidelity is more crictical.

There are also numerous modified bases found in RNA that serve many different roles. Pseudouridine (ψ) and the DNA base thymidine are found in various places (most notably in the TψC loop of every tRNA). There are nearly 100 other naturally occurring modified bases, many of which are not fully understood.

Pseudouridine (abbreviated ψ) is the C-glycoside isomer of the nucleoside uridine, and it is the most prevalent of the over one hundred different modified nucleosides found in RNA. ψ is found in all species and in all classes of RNA except mRNA [1]. ψ is formed by enzymes called ψ synthases, which post-transcriptionally isomerize specific uridine residues in RNA.

Comparision from DNA

Unlike DNA, RNA is almost always a single-stranded molecule and has a much shorter chain of nucleotides. RNA contains ribose, rather than the deoxyribose found in DNA (there is a hydroxyl group attached to the pentose ring in the 2' position whereas DNA has a hydrogen atom rather than a hydroxyl group). This hydroxyl group makes RNA less stable than DNA because it is more prone to hydrolysis. Several types of RNA (tRNA, rRNA) contain a great deal of secondary structure, which help promote stabilty.

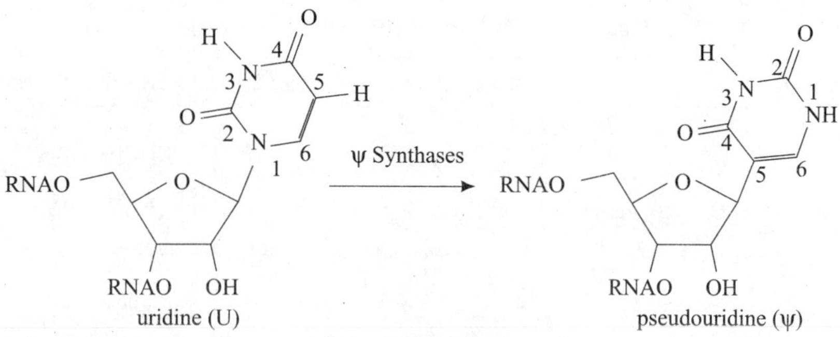

Fig. 61.

Synthesis

Synthesis of RNA is usually catalyzed by an enzyme, RNA polymerase, using DNA as a template. Initiation of synthesis begins with the binding of the enzyme to a promoter sequence in the DNA (usually found "upstream" of a gene). The DNA double helix is unwound by the helicase activity of the enzyme. The enzyme then progresses along the template strand in the 3′ -> 5′ direction, synthesizing a complementary RNA molecule with elongation occurring in the 5′ -> 3′ direction. The DNA sequence also dictates where termination of RNA synthesis will occur.

Types of RNA

The cells contain mainly three types of RNAs, i.e., messenger RNA (mRNA), transfer RNA (tRNA) and ribosomal RNA (rRNA). Cells also contain small RNA in cytoplam and nucleoplasm.

Abundancy of Different Types of RNA

A typical bacterium contains 0.05–0.10 pg of RNA, making up about 6% of its total weight (Table 7). A mammalian cell, being much larger, contains more RNA, 20–30 pg in all, but this represents only 1% of the cell as a whole. rRNAs are the most abundant RNAs in the cell, making up over 80% of the total in actively dividing bacteria. These molecules are components of ribosomes, the structures on which protein synthesis takes place. Messenger RNAs rarely make up more than 4% of the total RNA and are short-lived, being degraded soon after synthesis. Bacterial mRNAs have half-lives of no more than a few minutes and in eukaryotes most mRNAs are degraded a few hours after synthesis. This rapid turnover means that the composition of the transcriptome is not fixed and can quickly be restructured by changing the rate of synthesis of individual mRNAs. tRNAs are small molecules that are also involved in protein synthesis, carrying amino acids to the ribosome and ensuring that these are linked together in the order specified by the nucleotide sequence of the mRNA that is being translated. It make up about 10-12% of the total RNA. While other types of RNAs are about ~2-3%.

Table 7. Abundancy of different types of RNA

Types of RNA	Percentage of types of RNA in total cellular RNA
rRNA	~80-85%
mRNA	~1-5%
tRNA	~10-12%
Other types of RNA	~2-3%

Functions of RNA

Except for the case of certain viruses, RNA is not the genetic material and does not need to be capable of serving as a template for its own replication. Rather RNA functions as the intermediate, the mRNA, between the gene and the protein synthesizing machinery. Another function of RNA is as an adaptor, the tRNA, between the codon in the mRNA and amino acids. RNA can also play a structural role, as in the case of the RNA components of the ribosome. Yet another role for RNA is as a regulatory molecule, which through sequence complementarity binds to, an interfere with the translation of, certain mRNA. Finally some RNAs (including one of the structural RNAs of the ribosome) are enzymes that catalyze essential reactions in the cell.

Messenger RNA (mRNA)

Messenger RNA (mRNA) is RNA that encodes and carries information from DNA.

Life cycle of mRNA

The brief life of an mRNAs begins with transcription and ultimately ends in degradation. During their life, mRNAs may also be processed, edited, and transported prior to translation. Eukaryotic mRNAs often require extensive processing and transport, while prokaryotic mRNAs do not.

Transcription

During transcription, RNA polymerase makes a copy of a gene from the DNA to mRNA as needed. This process is similar in eukaryotes and prokaryotes. One notable difference, however, is that eukaryotic RNA polymerase associates with mRNA processing enzymes during transcription so that processing can proceed quickly after the start of transcription. The short-lived, unprocessed or partially processed, product is termed pre-mRNA; once completely processed, it is termed mature mRNA.

Eukaryotic pre-mRNA processing

Post transcriptional modification refers to the process by which, in eukaryotic cells pre-mRNA is converted into messenger RNA (mRNA). This process is vital for the efficient and correct translation of the genome by ribosomes and for the export of the mRNA from the nucleus for translation

There are three main forms of post transcriptional modification:

Splicing by the spliceosome;
5' capping and;
polyadenylation;

Splicing

Splicing is a modification of genetic information after transcription, in which introns are removed and exons are joined. Splicing is an essential process in eukaryotic pre-mRNA processing (Fig. 62) that must precede translation. Splicing occurs by a series of biochemical reactions between RNA nucleotides, which are catalyzed by proteins, RNA, or both.

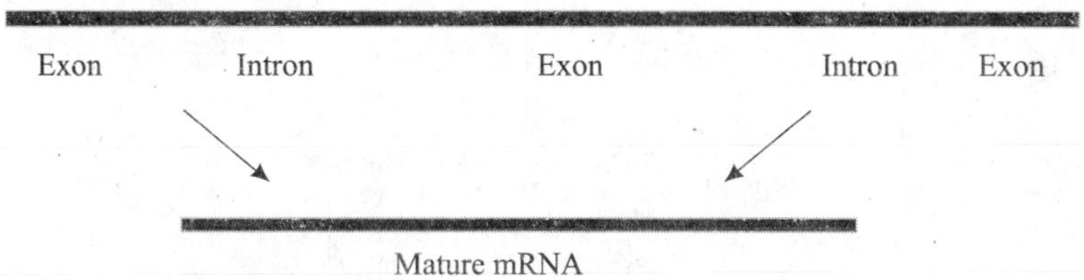

Fig. 62. Splicing.

Sometimes pre-mRNA messages may be spliced in several different ways, allowing a single gene to encode multiple proteins. This process is called alternative splicing. Splicing is usually performed by an RNA-protein complex called the spliceosome, but some RNA molecules are also capable of catalyzing their own splicing (*ribozymes*).

5' cap Formation

The 5' cap is a specially altered dinucleotide end to the 5' end of messenger RNA as found in eukaryotes. The process of 5' capping is vital to translation and stability of the messenger RNA, and is a highly regulated process which occurs in the nucleus. It is found on the 5' end of an mRNA molecule and consists of a guanosine nucleotide connected to the mRNA via an unusual 5' to 5' triphosphate linkage (Fig. 63). Further modifications include the methylation of the guanosine, and the possible methylation of the 2' carbons of the first 2 ribose sugars of the 5' end of the mRNA. The methylation of both 2' carbons is shown on the diagram. Functionally the 5' cap looks like the 3' end of an RNA molecule (the 5' carbon of the cap ribose is bonded, and the 3' unbonded). This provides significant resistance to 5' exonucleases.

The 5' cap has 4 main functions:

Regulation of nuclear export.

Prevention of degradation by exonucleases.

Promotion of translation.

Promotion of 5' proximal intron excision.

Polyadenylation

Polyadenylation is the covalent linkage of a polyadenylyl moiety to a messenger RNA molecule. In eukaryotic organisms, polyadenylation is the mechanism by which most messenger RNA (mRNA) molecules are terminated at their 3' ends. The poly(A) tail aids in mRNA stability by protecting it from exonucleases. Polyadenylation is also important for transcription termination, export of the mRNA from the nucleus, and translation. Some prokaryotic mRNAs also are polyadenylated, although the poly(A) tail's function is different from that in eukaryotes.

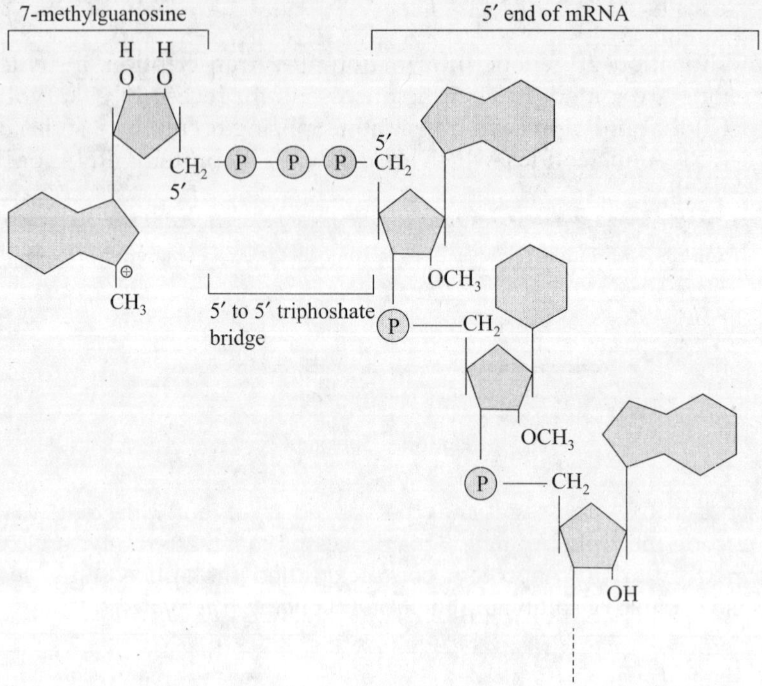

Fig. 63. 5' cap formation.

Polyadenylation occurs during and immediately after transcription of DNA into RNA. After transcription has been terminated, the mRNA chain is cleaved through the action of an endonuclease complex associated with RNA polymerase. The cleavage site is characterized by the presence of the base sequence AAUAAA near the cleavage site. After the mRNA has been cleaved, 80 to 250 adenosine residues are added to the free 3' end at the cleavage site. This reaction is catalyzed by polyadenylate polymerase.

Editing

In some instances, an mRNA will be edited, changing the nucleotide composition of that mRNA (Fig. 64). An example in humans is the apolipoprotein B mRNA, which is edited in some tissues, but not others. The editing creates an early stop codon, which upon translation, produces a shorter protein.

Transport

Another difference between eukaryotes and prokaryotes is mRNA transport. Because eukaryotic transcription and translation is compartmentally separated, eukaryotic mRNAs must be exported from the nucleus to the cytoplasm. Mature mRNAs are recognized by their processed modifications and then exported through the nuclear pore.

Degradation

After a certain amount of time the message degrades into its component nucleotides, usually with the assistance of RNases. Due to mRNA processing, eukaryotic mRNAs are generally more stable than prokaryotic mRNAs.

Mature mRNA Structure

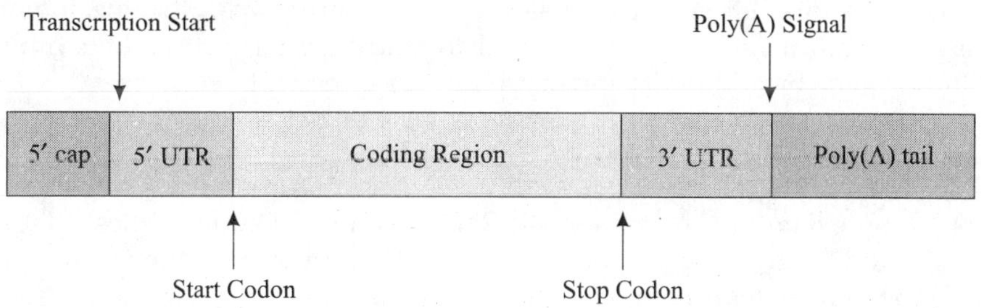

Fig. 64. Mature mRNA structure (Watson *et al.*, 2004).

Coding regions

Coding regions are composed of codons, which are decoded and translated into protein by the ribosome. Coding regions begin with the start codon and end with the one of three possible stop codons. In addition to protein-coding, portions of coding regions may also serve as regulatory sequences.

Untranslated regions

Untranslated regions (UTRs) are sections of the RNA before the start codon and after the stop codon that are not translated, termed the five-prime untranslated region (5' UTR) and three-prime untranslated region (3' UTR), respectively. These regions are transcribed as part of the same transcript as the coding region. Several roles in gene expression have been attributed to the untranslated regions, including mRNA stability, mRNA localization, and translational efficiency. The ability of a UTR to perform these functions depends on the sequence of the UTR and can differ between mRNAs. Stability of mRNAs may be mediated by the 5' UTR and 3' UTR due to varying affinity for certain RNA degrading enzymes (RNases), which can promote or inhibit the relative stability of the RNA molecule. The greater the stability of an mRNA, the more protein that may be produced from that transcript.

Cytoplasmic localization of mRNA is thought to be a function of the 3' UTR. Proteins that are needed in a particular region of the cell can actually be translated there; in such a case, the 3' UTR may contain sequences that allow the transcript to be localized to this region for translation. Translational efficiency, and even inhibition of translation altogether, can be mediated by UTRs. Proteins that bind to either the 3' or 5' UTR may affect translation by interfering with the ribosome's ability to bind to the mRNA.

Some of the elements contained in untranslated regions form a characteristic secondary structure when transcribed into RNA. These structural mRNA elements are involved in regulating the mRNA. Some, such as the SECIS element, are targets for proteins to bind. One class of mRNA element, the riboswitches, directly bind small molecules, changing their fold to modify levels of transcription or translation. In these cases, the mRNA regulates itself.

Anti-sense mRNA

During transcription, double stranded DNA produces mRNA from the sense strand; the other, complementary, strand of DNA is termed anti-sense. Anti-sense mRNA is an RNA complementary in sequence to one or more mRNAs. In some organisms, the presence of an anti-sense mRNA can inhibit gene expression by base-pairing with the specific mRNAs. In biochemical research, this effect has been used to study gene function, by simply shutting down the studied gene by adding its anti-sense mRNA transcript.

Transfer RNA

Transfer RNA (abbreviated tRNA) is a small RNA chain (74-93 nucleotides) that transfers a specific amino acid to a growing polypeptide chain at the ribosomal site of protein synthesis during translation. It has sites for amino-acid attachment and codon (a particular sequence of 3 bases) recognition. The codon recognition is different for each tRNA and is determined by the anticodon region, which contains the complementary bases to the ones encountered on the mRNA. Each tRNA molecule binds only one type of amino acid, but because the genetic code is degenerate, more than one codon exists for each amino acid. Transfer RNA is the "adaptor" molecule hypothesized by Francis Crick, which mediates recognition of the codon sequence in mRNA and allows its translation into the appropriate amino acid.

Structure of tRNA

tRNA has primary structure (the order of nucleotides from 5' to 3'), secondary structure (usually visualized as the *cloverleaf structure*) (Fig. 65), and tertiary structure (all tRNAs have a similar L-shaped 3D structure that allows them to fit into the P and A sites of the ribosome). The primary structure was reported in 1969 by Robert W. Holley. The secondary and tertiary structures were derived from X-ray crystallographic studies reported independently in 1974 by American and British research groups headed, respectively, by Alexander Rich and Aaron Klug.

Feature

- The 5'-terminal phosphate.
- The acceptor stem (also called the *amino acid stem*) is a 7-bp stem that incudes the 5'-terminal nucleotide and the 3'-terminal nucleotide with the 3'-terminal OH group (which can bind the amino acid). The acceptor stem may contain non-Watson-Crick base pairs.
- The CCA tail is a CCA sequence added to the 3' end of the tRNA molecule. This sequence is important for the recognition of tRNA by enzymes critical in translation.
- The D arm is a 4 bp stem ending in a loop that often contains dihydrouridine.
- The anticodon arm is a 5-bp stem containing the anticodon.
- The T arm is a 5 bp stem containing the sequence TψC.
- Modified bases are bases contained in tRNA that are not "canonical" bases, *i.e.* modified forms of the standard adenine, guanine, cytosine, and uracil bases.

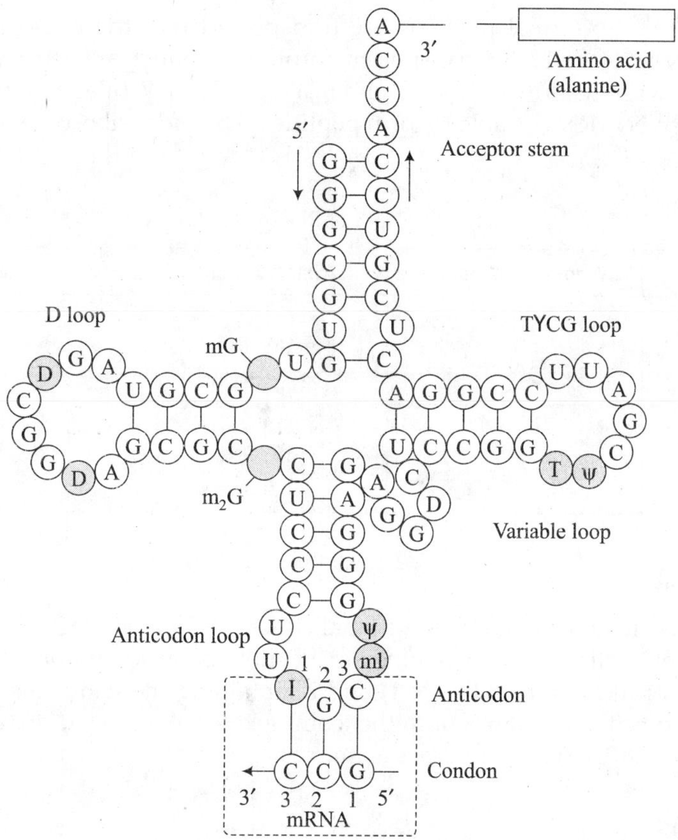

Fig. 65. Transfer RNA: *cloverleaf structure* (Voet and Voet, 2001).

Aminoacylation

Aminoacylation is the process of adding an aminoacyl group to a compound. Each tRNA is aminoacylated (or *charged*) with a specific amino acid by an aminoacyl tRNA synthetase. There is often just one aminoacyl tRNA synthetase for each amino acid, despite the fact that there can be more than one tRNA, and more than one anticodon, for an amino acid. Recognition of the appropriate tRNA by the synthetases in not mediated solely by the anticodon, and the acceptor stem often plays a prominent role.

Reaction:

amino acid + ATP → aminoacyl – AMP + Ppi

aminoacyl – AMP + tRNA → aminoacyl – tRNA + AMP

Ribosomal RNA

Ribosomal RNA (rRNA) is the primary constituent of ribosomes (Table 8). Ribosomes are the protein-manufacturing organelles of cells and exist in the cytoplasm. rRNA is transcribed from

DNA, like all RNA. Ribosomal proteins are transported into the nucleus and assembled together with rRNA before being transported through the nuclear membrane. This type of RNA makes up the vast majority of RNA found in a typical cell. While proteins are also present in the ribosomes, solely rRNA is able to form peptides. Therefore ribosome often is referred to as ribozyme.

Table 8. Types of Ribosomal RNA

Ribosome Source	Whole Ribosome	Small Subunit	Large Subunit
E. coli	70S	30S 16S RNA 21 proteins	50S 23S & 5S RNAs 31 proteins
Rat cytoplasm	80S	40S 18S RNA 33 proteins	60S 28S, 5.8S, & 5S RNAs 49 proteins

Small Nuclear RNA

Small nuclear RNA (snRNA) is a class of small RNA molecules that are found within the nucleus of eukaryotic cells. They are involved in a variety of important processes such as RNA splicing (removal of introns from hnRNA) and maintaining the telomeres. They are always associated with specific proteins, and the complexes are referred to as small nuclear ribonucleoproteins (snRNP) or sometimes as snurps.

Small Nucleolar RNA

Small nucleolar RNA (snoRNA) is a class of small RNA molecules that are involved in chemical modifications of ribosomal RNAs (rRNAs) and other RNA genes, for example by methylation. snoRNAs are a component in the small nucleolar ribonucleoprotein (snoRNP), which contains snoRNA and proteins. The snoRNA guides the snoRNP complex to the modification site of the target RNA gene via sequences in the snoRNA that hybridize to the target site. The proteins then catalyze modification of the RNA gene.

microRNA

microRNA (also miRNA) are RNA genes that are the reverse complement of another gene's mRNA transcript and inhibit the expression of the target gene.

gRNA

gRNAs (for guide RNA) are RNA genes that function in RNA editing. Thus far, RNA editing has been found only in the mitochondria of kinetoplastids, in which mRNAs are edited by inserting or deleting stretches of uridylates (Us). The gRNA forms part of the editosome and contains sequences that hybridize to matching sequences in the mRNA, to guide the mRNA modifications. The term "guide RNA" is also sometimes used generically to mean any RNA gene that guides an RNA/protein complex via hybridization of matching sequences.

eRNA

Efference RNA (eRNA) is derived from intron sequences of genes or from non-coding DNA. The function is assumed to be regulation of translational activity by interference with the transcription apparatus or target proteins of the translated peptide in question, or by providing a concentration-based measure of protein expression, basically introducing a fine-tuned analog element in gene regulation as opposed to the digital on-or-off regulation by promoters. Research into the role of eRNAs is in its infancy.

Signal recognition particle RNA

The signal recognition particle (SRP) is an RNA-protein complex present in the cytoplasm of cells that binds to the mRNA of proteins that are destined for secretion from the cell. The RNA component of the SRP in eukaryotes is called 4.5S RNA.

pRNA

At least one species of DNA-containing phages, phi-29, uses a complex of six identical short RNA sequences as mechanical components (utilizing ATP for energy) of its DNA packaging machinery. How common this phenomenon is has yet to be determined.

tmRNA

tmRNA has a complex structure with tRNA-like and mRNA-like regions. It has currently only been found in bacteria, but is ubiquitous in all bacteria. tmRNA recognizes ribosomes that have trouble translating or reading an mRNA and stall, leaving an unfinished protein that may be detrimental to the cell. tmRNA acts like a tRNA first, and then an mRNA that encodes a peptide tag. The ribosome translates this mRNA region of tmRNA and attaches the encoded peptide tag to the C-terminus of the unfinished protein. This attached tag targets the protein for destruction or proteolysis.

Ribozyme

Ribozymes are antisense RNA molecules that have catalytic activity. They function by binding to the target RNA moiety through Watson-Crick base pairing and inactivate it by cleaving the phosphodiester backbone at a specific cutting site.

Five classes of ribozymes have been described based on their unique characters in the sequences as well as three-dimensional structures (Bunnell,1997). They are denoted as: (1) the Tetrahymena group I intron, (2) RNase P, (3) the hammerhead ribozyme, (4) the hairpin ribozyme, and (5) the hepatitis delta virus ribozyme. They may catalyze self-cleavage (intramolecular or "in-cis" catalysis) as well as the cleavage of external substrates (intermolecular or "in-trans" catalysis) (Ohkawa, 1995). The leading study was conducted by Thomas Cech on ciliated protozoan *Tetrahymena thermophila* group I introns in 1980s, which showed that RNA can participate in the intramolecular catalysis of the self-splicing (Fig. 66) and acts as a protein enzyme. The definition of biological enzyme was broadened since then, by recognizing the enzymatic function of some RNA molecules.

The name of hammerhead ribozyme is given by the similarity between its secondary structure and the shape of a hammerhead (Fig. 67). They are the best understood subcategory of all ribozymes. As well as other ribozymes, the hammerhead ribozyme is an antisense RNA. Some of the ribonucleotides within the sequence selectively form Watson-Crick base pairs with

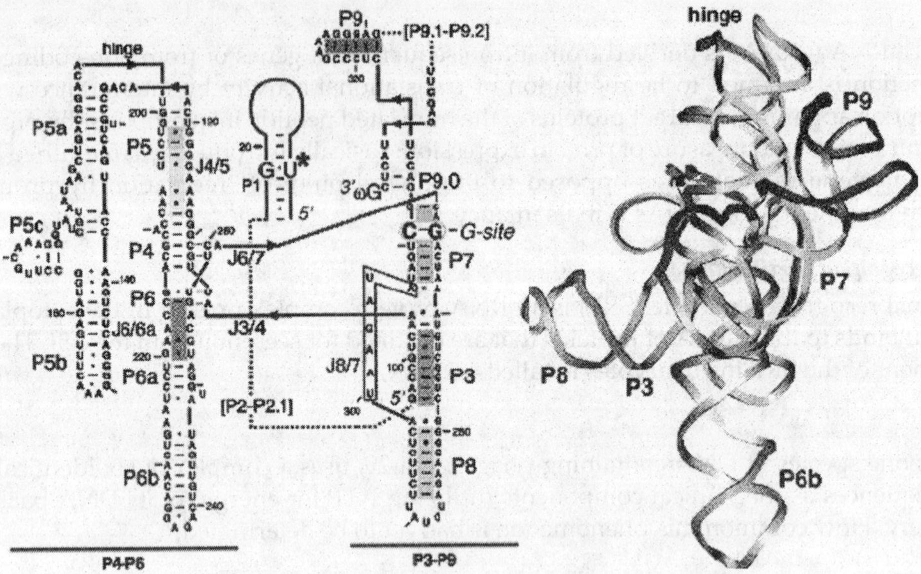

Fig. 66. Tetrahymena Group I intron secondary and tertiary structures (Golden B., Cech T.).

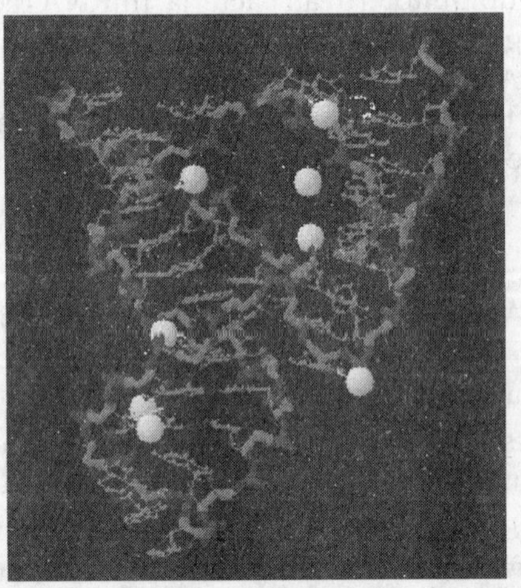

Fig. 67. A hammerhead ribozyme structure with magnesium cation binding sites (Murray, 1996).

others to form a stem, while the rest stay in single stranded state called loop. These loops and stems can be predicted at the secondary structure level using conformational energy analysis, such as RNAdraw and mfold; and three dimensional structures were obtained mainly by X-ray crystallography.

On the left of Fig. 68, the structure of the wild-type, cis-acting hammerhead ribozyme is shown. The three helices and the cleavage site are indicated. The self-splicing happens after the sequence G-U-C between helix I and III, and the G and U are conserved and crucial for the catalysis. Except for itself, a hammerhead ribozyme is not catalytic in cells since its nature of

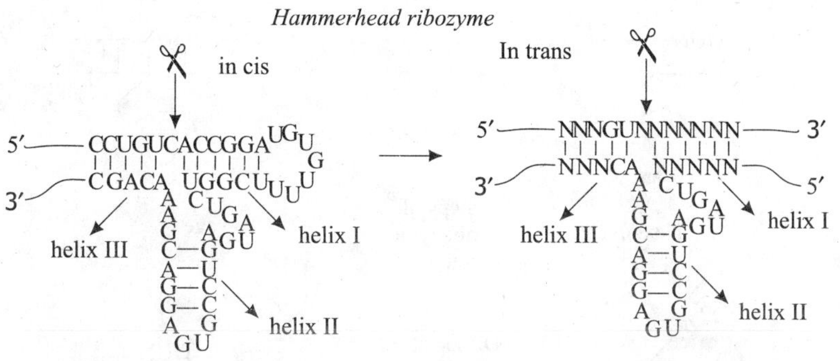

Fig. 68. Secondary structures of in cis and in trans hammerhead ribozymes.

intramolecular reaction (Symons, 1989). However, most ribozymes can be chemically engineered to cleave RNA in trans by separating the catalytic domain from ribozyme and attaching recognition (i.e., antisense) arms to the catalytic center, in order to target to a substrate. These trans-acting ribozymes can be very useful in the study of molecular biology and pharmaceutics.

In comparison to the conventional antisense RNAs, ribozymes provide the potential of turnover, with a single molecule being able to inactivate multiple target RNAs. Another important property of ribozyme is the specificity, which means the fidelity to cleave a unique target. Both of turnover and specificity are affected by binding arm length (helix I and III). If the length of binding arms is very short, the rate of dissociation of the target from the ribozyme may exceed the rate of cleavage, resulting in poor efficiency (Rossi, 1997). However, stable hybrids exhibit low catalytic activity because of slow dissociation of the cleaved substrate (Bertrand, 1994). Thus, the ideal situation is to have arm lengths that aid cleavage, yet provide for quick dissociation of the cleaved products (Fig. 69).

RNA World Hypothesis

The RNA world hypothesis proposes that RNA was actually the first life-form on earth, later developing a cell membrane around it and becoming the first prokaryotic cell.

Support of idea

The phrase "RNA World" was first used by Walter Gilbert in 1986. However, the theory of independent RNA life is much older and can be found in Carl Woese's book *The Genetic Code* (New York: Harper and Row, 1967). This hypothesis is supported by the RNA's ability to store, transmit, and duplicate genetic information, just like DNA does. RNA can also act as a ribozyme (an enzyme made of ribonucleic acid). Because it can reproduce on its own, performing the tasks of both DNA and proteins (enzymes), RNA is believed to have once been capable of independent life. Further, while nucleotides were not found in Miller-Urey's experiments, they were found by others' simulations, notably those of Joan Oro. Experiments with basic ribozymes, like the viral RNA Q-beta, have shown that simple self-replicating RNA structures can withstand even strong selective pressures (e.g., opposite-chirality chain terminators) (*The Basics of Selection* (London: Springer, 1997)).

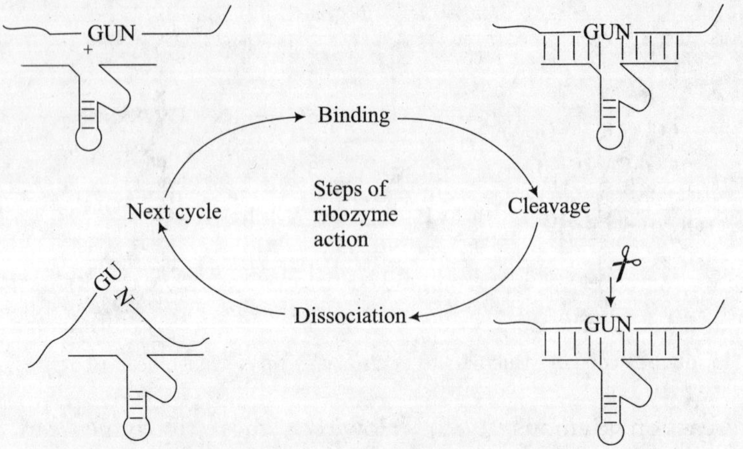

Fig. 69. Turnover cycle of RNA cleavage by a hammerhead ribozyme. Binding of the enzyme and substrate results in the formation of catalytically active structure. After cleavage, the two product strands dissociate and the ribozyme strand can go into the next cycle.

REFERENCES

Alexander, R. and Zhang, S. (2003) Nature Reviews Genetics 4: 566-572.

Garrett and Grisham (2007) Biochemistry (3rd ed.).

http://post.queensu.Ca/nforsdyke/hioinfo2.htm.

http://www.ha.education.demon.co.uk/for/science/dnamain.html.

http://www.mun.ca/biology/scarr/Watson-Crickmodel.html.

Nelson, D.L. and Cox, M.M. (2004) Lehminger Principles of Biochemistry (4th ed.), W.H. Freeman, USA.

The structure drawn from the crystal structure, using Cn3D program available on the NCBI home page.

Voet and Voet (2004) Biochemistry (3rd ed.) Wiley International Edition, USA.

Watson, Baker, Bell, Gann, Levine, Losick (2004) Molecular Biology of the Gene (5th ed.), Pearson Education Press, USA.

Suggested Readings

1. Franklin, R. and Gosling R.G. (1953) Molecular Configuration in Sodium Thymonucleate, Nature, 171:740-741.
2. Wilkins, M.H.F., Stokes A.R. and Wilson H.R. (1953) Molecular Structure of Deoxypentose Nucleic Acids, Nature, 171: 738-740.
3. Perutz, M.F., Randall, J.T., Thomson, L., Wilkins, M.H. Watson and J. D. (1969) DNA helixm, Science, 164:1537-1539.
4. Judson, H.F. (1979) The Eighth Day of Creation. Makers of the Revolution in Biology, Simon and Schuster.
5. Maddox, B. and Franklin, R. (2002) The Dark Lady of DNA.
6. Olby, R. (1974) The Path to the Double Helix: Discovery of DNA, published by MacMillan, with foreword by Francis Crick.
7. Watson, J.D. (1981) The Double Helix: A Personal Account of the Discovery of the Structure of DNA, Atheneum,Norton Critical Editions.

8. Wilkins, M. (2003) The Third Man of the Double Helix: The Autobiography of Maurice Wilkins.
9. Nelson, D. L. and Cox, M. M. (2004) Lehninger Principles of Biochemistry, 4th edition W. H. Freeman.
10. Lewin, B. (2004) Genes VIII, international edition, Pearson Prentice Hall.
11. Cook, P. R. (2001) Principles of Nuclear Structure and Function, first edition, Wiley-Liss publisher.
12. Voet and Voet (2004) Biochemistry, Third Edition, Wiley International Edition.
13. Berg, J. M., Tymoczko, J. L. and Stryer, L. (2006) Biochemistry, Sixth edition, Macmillan Publishers Ltd.
14. Simmons, S. (2006) Principles of GENETICS, fourth edition, Wiley Asia Student edition.
15. Watson, Baker, Bell, Gann, Levine, Losick (2004) Molecular Biology of the gene, fifth edition, Pearson Education Press.

REVIEW QUESTIONS

Long Questions

1. List three ways in which the mRNAs of eukaryotes differ from the mRNAs of prokaryotes?
2. Distinguish between DNA and RNA (a) chemically (b) functionally, and (c) by location in the cell.
3. What are differences between DNA and RNA?
4. Compare and contrast the structures of the A, B and Z forms of DNA.
5. Why did Watson and Crick choose a double helix for their model of DNA structure?
6. Why were hydrogen bonds placed in the double helix model to connect the bases?
7. If one strand of the DNA in the Watson-Crick double helix has a base sequence of 5′-GTCATGAC-3′, what is the base sequence of the complementary strand?
8. The temperature at which one-half of a double-stranded DNA molecule has been denatured is called the melting temperature, Tm. Why Tm does depend directly on the GC content of the DNA?

Multiple Choice Questions

1. A molecule of DNA is a polymer composed of
   (a) glucose (b) amino acids
   (c) fatty acids (d) nucleotides
2. The presence of DNA is important for cellular metabolic activities because DNA
   (a) directs the production of enzymes
   (b) is a structural component of cell walls
   (c) directly increases the solubility of nutrients
   (d) is the major component of cytoplasm
3. What type of ring structure is present in structures A and B?
   (a) Quinoline (b) purine
   (c) naphthalene (d) pyrimidine

4. To which of the following does thymine form hydrogen bonds in DNA?
   (a) cytosine (b) guanine
   (c) adenine (d) thymine
5. On which of the following molecules would you find anticodon?
   (a) messenger RNA (b) ribosomal RNA
   (c) transfer RNA (d) small nuclear RNA
6. The eukaryotic functional equivalent of the prokaryotic 16S rRNA is the
   (a) 5S rRNA
   (b) 16S rRNA (the same as the prokaryotic 16S rRNA)
   (c) 18S rRNA
   (d) 23S rRNA.
7. Zinc finger proteins and helix-turn-helix proteins are
   (a) types of DNA-binding proteins
   (b) involved in the control of translation
   (c) components of ribosomes
   (d) part of the hemoglobin in blood cells
   (e) bound to transfer RNA during replication
8. The sequence of one strand of DNA is 5′ TCGATC 3′. The sequence of the complementary strand would be
   (a) 5′ AGCTAG 3′ (b) 5′ TCGATC 3′
   (c) 5′ CTAGCT 3′ (d) 5′ GCTAGC 3′
   (e) 5′ GATCGA 3
9. The chemical bonds in DNA by which the sugar components of adjacent nucleotides are linked through the phosphate groups are called _______ bonds.
   (a) phosphodiester (b) hydrogen
   (c) hydrophobic (d) hydrophilic
   (e) ionic
10. The four ribonucleotide triphosphates incorporated into mRNA are
    (a) Inosine, Guanine, Uracil, Thymine
    (b) Adenine, Guanine, Cytosine, Thymine
    (c) Cytosine, Uracil, Adenine, Guanine
    (d) Thymine, Uracil, Guanine, Cytosine
    (e) Inosine, Guanine, Adenine, Cytosine
11. Which of the following is unique to eukaryotic mRNA synthesis?
    (a) Polycistronic mRNA (b) Coupled transcription-translation
    (c) Removal of introns (e) Polysomes

Answers

1. (d)	2. (a)	3. (b)	4. (c)	5. (a)	6. (c)
7. (a)	8. (a)	9. (a)	10. (c)	11. (a)	

CHAPTER 13

DNA Replication: The DNA Synthesis

Y. K. Jaiswal* and V. Agrawal
School of Studies in Biochemistry, Jiwaji University, Gwalior – 474 011
E-mail : jaiswalyk@gmail.com

DNA is the prime genetic molecule, carrying all the genetic information with in the chromosome. DNA is usually a double helix; it consists of two complementary polymeric chains twisted around each other in the form of a regular right handed double helix. Each chain is a polynucleotide, a polymeric collection of nucleotides in which the sugar of each nucleotides is linked via a phosphate group to the sugar of adjacent nucleotide. There are four main nucleotides each containing a deoxyribose residue, a phosphate group and nitrogenous bases; purine Adenine (A), Guanine (G); pyrimidine Thymidine (T), Cytosine (C). In order for cells to divide to give identical daughter cells, the DNA (genome) needs to be replicated. DNA replication is arguably the most fundamental process required for the proliferation of all living cells. During cell division, each daughter cell must receive essentially the same genetic information that was encoded in the DNA of the parent cell. This means that DNA replication must generate a perfect copy of the genomic DNA complement.

DNA replication only occurs at a specific step in the cell cycle i.e., S-phase. The table 1 describes the cell cycle for a hypothetical cell with a 24 hr cycle. Replication is the process of duplicating the DNA to make two identical copies, one of which will be passed along to each daughter cell when the cell divides. In other words the use of existing DNA as a template for the synthesis of new DNA strands. In humans and other eukaryotes, replication occurs in the cell nucleus. DNA replication is almost similar between prokaryotes (e.g. *E. coli*) and eukaryotes (e.g. humans). However, the *E. coli* genome is a single circular DNA molecule, whereas the human genome is on 46 linear molecules (chromosomes).

*Corresponding author

Table 1. Cell cycle

Stage	Activity	Duration
G1	Growth and increase in cell size	10 hr
S	DNA synthesis "DNA Replication"	8 hr
G2	Post-DNA synthesis	5 hr
M	Mitosis	1 hr

POSSIBLE MODES OF REPLICATION

Three Models have been Proposed for the Mode of Replication

Dispersive Replication

Dispersive replication suggested that each DNA molecule after replication might consist of segments of new and old DNA interspersed. This is the disproved model of DNA synthesis known to be incorrect (Fig. 1).

Conservative replication

Conservative replication suggested that parent strand not unwind and retains both the strands of the DNA duplex. But some how generate a new double stranded DNA copy of entirely new molecules. After replication one DNA molecule consists entirely of newly synthesized DNA whereas the other molecule is entirely original DNA (Fig. 1).

Semiconservative replication

Each DNA strand serves as a template for the synthesis of a new strand, producing two new DNA molecules, each with one new strand and one old. This is semiconservative replication. So that after replication each DNA molecule would consist of one of the original strands plus one newly synthesized strand (Fig. 1).

The hypothesis of semiconservative replication is proposed by James Watson and Francis Crick in 1953. This hypothesis was proved by ingeniously designed experiments carried out by Mathew Meselson and Franklin Stahl in 1957.

DNA REPLICATION: GENERAL FEATURES

Meselson and Stahl experiment: "The most beautiful experiment in the biology"

In 1957, a classic density labeling experiments was performed by Matthew Meselson and Franklin Stahl (Fig. 2) to determine which of the above models best represented DNA replication. In order to determine which of these models was true, the following experiment was performed:

A culture of bacteria was grown for many generations in a growth medium containing N^{15} (non-radioactive isotope of nitrogen) labeled NH_4Cl as a sole source of nitrogen (called heavy medium). In this way the parental DNA strands were labeled with the heavy isotope of nitrogen, N^{15} thereby increasing the density of DNA. The cells were transferred to a medium containing the common isotope of nitrogen N^{14} (called light medium).This DNA was allowed

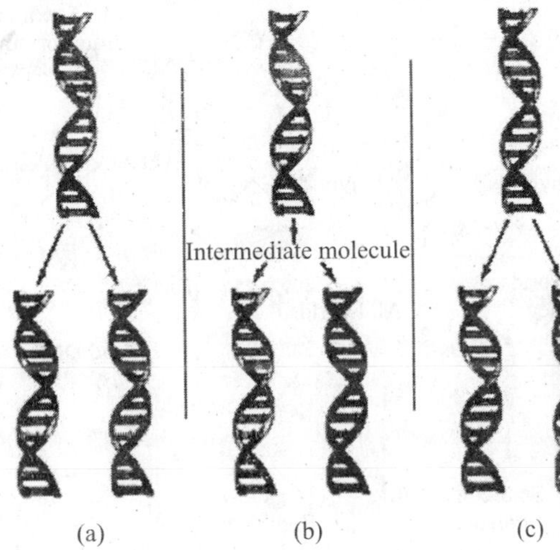

Fig. 1. Possible mode of DNA replication (a) Semiconservative replication (b) Conservative replication (c) Dispersive Replication (Watson, 2004).

Fig. 2. Photograph taken by F. L. Holmes of Matt Meselson and Frank Stahl in 1996, standing at the site where they met at Woods Hole 42 years earlier (Watson, 2004).

to go through one round of replication with N^{14}, and then samples of cells were collected and the DNA was isolated.

The isolated DNA was undergone to density gradient centrifugation in CsCl to measure the density of the molecules. These results show that all DNA has a hybrid density after one strand round of replication. After centrifugation the heavier DNA would form a band lower in the tube, and the intermediate (one N-15 strand and one N-14 strand) and light DNA (all N-14) would appear as a band higher in the tube. The expected results for each model were (Fig. 3). The actual results were as expected for the semiconservative model and thus Watson and Crick's suspicion was borne out.

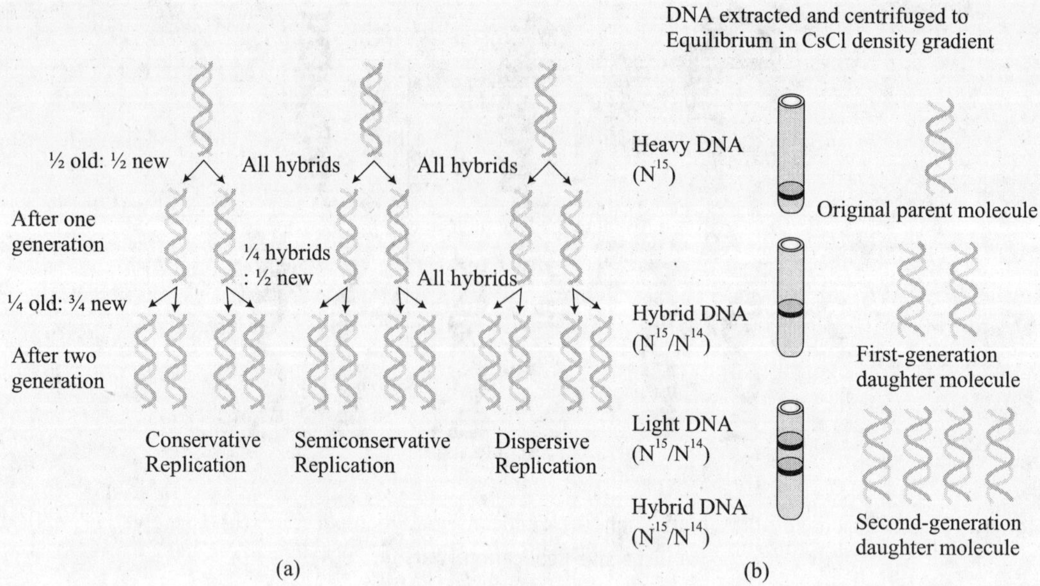

Fig. 3. "Meselson and Stahl" experiment (a) Three potential DNA replication models and their predicted outcomes (b) The actual data.

Replication begins at an origin and usually proceeds bidirectional

Although Meselson and Stahl showed that DNA replication was semiconservative, their results did not provide any information on chromosome replication – Where replication begins and what is the direction?

John Cairns (1963) was provided this idea by using autoradiography. He made *E.coli* DNA radioactive by growing cells in a medium containing tritium (^3H) labeled thymidine (Fig.4a). He then isolated the chromosomes by lysing the cells very-very gently and placed them on an electron micrograph (EM) grid which he exposed to X-ray film for two months. The radioactive thymidine residues generated "tracks" of silver grains in the emulsion, producing an image of the DNA molecule. These tracks revealed that the intact chromosome of *E. coli is* a single huge circle, 1.7mm long.

The replication intermediates first described by Cairns are known as theta forms due to their resemblance to the Greek letter theta θ. They are also known as Cairns forms (Fig. 4, 5). Cairns showed that:

The *E. coli* chromosome is circular. This had been suspected from genetic mapping evidence but direct physical confirmation was a nice demonstration that the genetic results were valid. There is a single origin of replication in *E. coli*. These results did not allow one to discriminate between unidirectional or bidirectional modes of replication and location of the origin of replication.

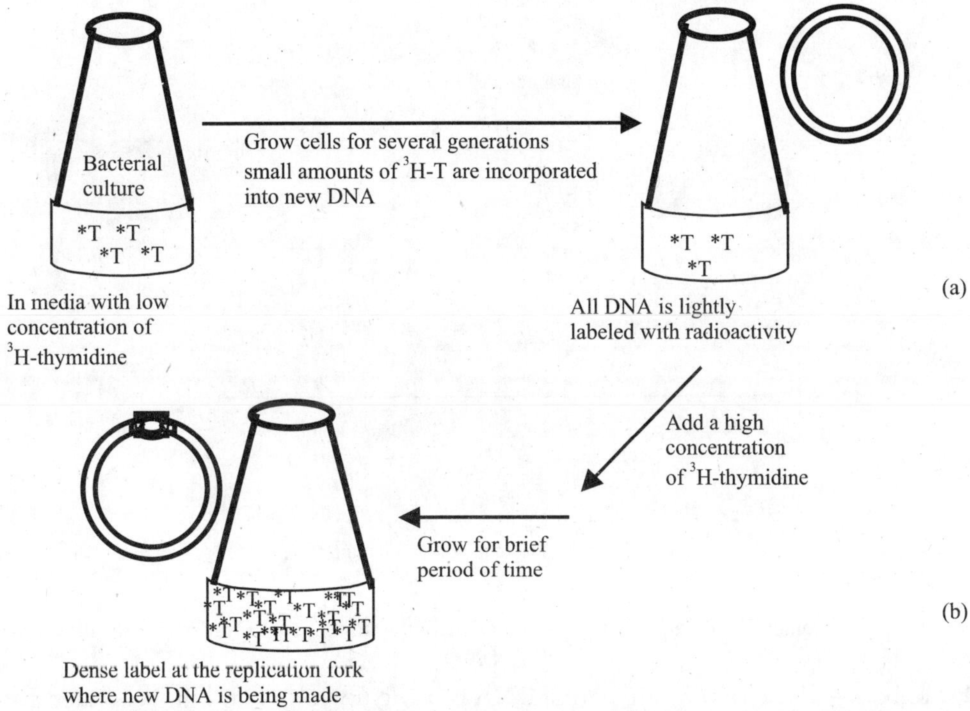

Fig. 4. Experiment performed by john cairns; He grows bacterial cells in a medium containing (a) low concentration of ^{3}H-thymidine (b) high concentration of ^{3}H-thymidine for several generations and for brief period of time to prove mode of replication. Then isolated DNA was undergone for autoradiography.

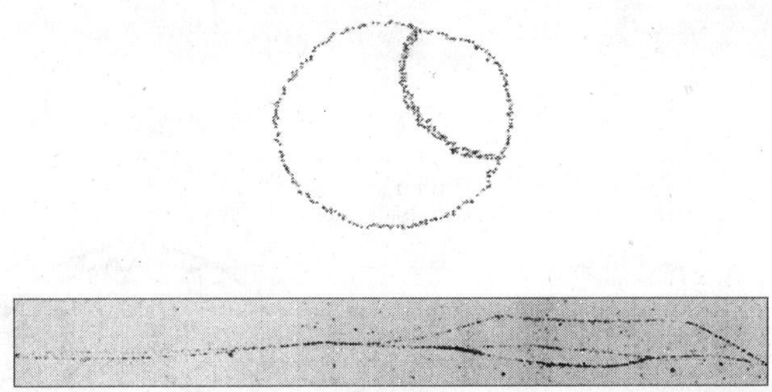

Fig. 5. An autoradiograph of replicating *E. coli* chromosome showing q form (Voet and Voet, 2005).

Replication can be Uni- or Bidirectional

The first uncertainty was resolved using a variation of the ^{3}H-thymidine labeling experiment (Fig. 4b). By increasing the concentration of ^{3}H-thymidine in the growth medium part way through replication, the most active regions of DNA synthesis can be seen in the background of a replicating chromosome. The result clearly showed that replication of the *E. coli* chromosome is bidirectional (Fig. 6).

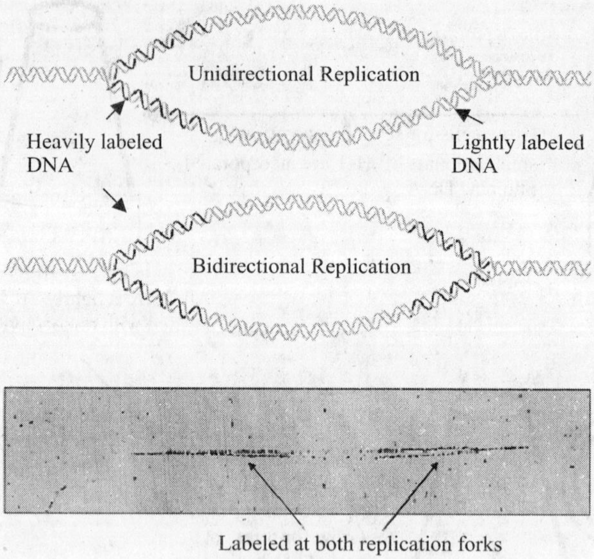

Fig. 6. The autoradiographic differentiation of unidirectional and bidirectional θ replication of DNA (Voet and Voet, 2004).

One or both ends of loop where parent DNA is being unwound and the separated strands quickly replicated termed replication forks. Labeling at both the replication forks shows that the replication usually bidirectional. For circular DNA molecules, the two replication forks meet at a point on the side of the circle opposite to the origin (Fig. 7). In eukaryotes DNA replication occurs at multiple origins along the DNA helix and replication bubbles get fused where they meet (Fig. 8).

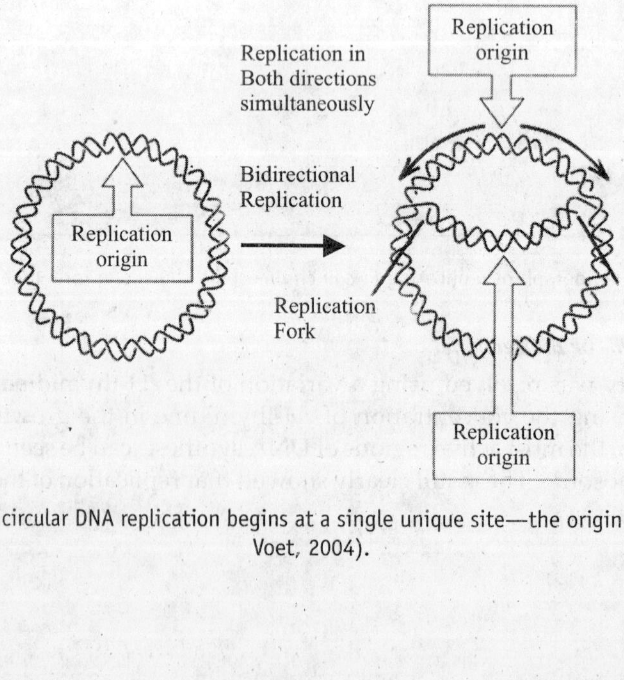

Fig. 7. In prokaryotic circular DNA replication begins at a single unique site—the origin of replication (Voet and Voet, 2004).

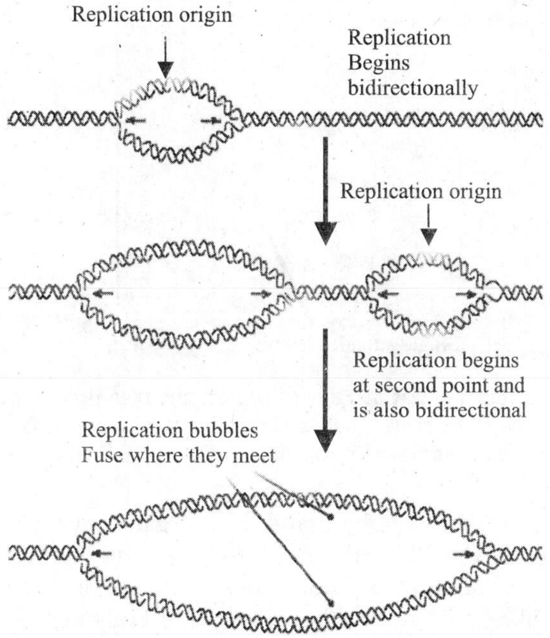

Fig. 8. In eukaryotes DNA replication occurs at multiple origins along the DNA helix. Initiation of replication commits the cell to continue the process until the entire genome has been replicated (Voet and Voet, 2004).

Does replication begin at random places or at a unique point?

Ross Inman and colleagues developed a technique called denaturation mapping. With this technique by using the 48,502 bp chromosome of bacteriophage λ, Inman showed that DNA could be selectively denatured at sequences unusually rich in A=T base pairs, generating a reproducible pattern single strand bubble (Fig. 9).

Most eukaryotic and prokaryotic DNAs replicate bidirectionally. Some extrachomosomal DNAs are replicated unidirectionally or using different replication modes. Isolated DNA containing replication loops can be partially denatured in the same way. This allows the position and progress of the replication forks to be measured and mapped, using the denatured regions as points of reference. The technique revealed that in this system the replication loops always initiate at a unique point, which was termed an origin. *E. coli* has a single origin of replication and the location of the origin was established. Eukaryotes have multiple origins of replication.

DNA synthesis proceeds in a 5'→3' Direction and is Semidiscontinuous

A new strand of DNA is always synthesized in the 5'→ 3' direction. The free 3'OH group serves as the point at which the DNA is elongated. Because the two DNA strands are antiparallel, the strand serving as the template is read from its 3' end toward its 5' end.

Why can DNA polymerase only act from 5'→3'? The reason is the relative stability of each end of DNA. A triphosphate is required to provide energy for the bond between a newly attached nucleotide and the growing DNA strand. However, this triphosphate is very unstable

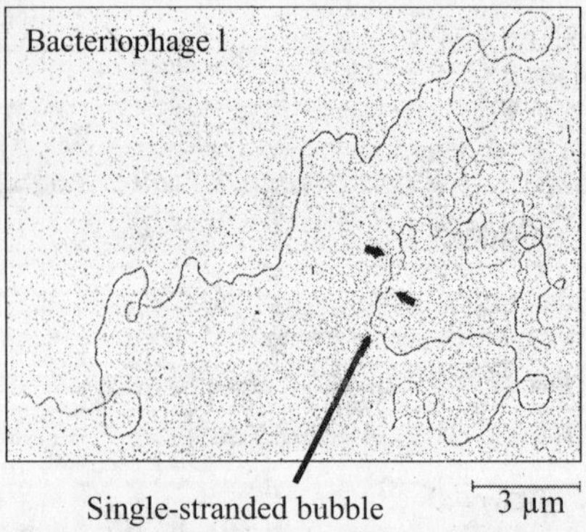

Fig. 9. Partially denatured DNA: This DNA was partially denatured, and then fixed to prevent renaturation during sample preparation. The arrows point to some single-stranded bubbles where denaturation has occurred. The regions that denature are highly reproducible and are rich in A=T base pairs (Voet and Voet, 2004).

and can easily break into a monophosphate and an inorganic pyrophosphate, which floats away into cell. At the 5' end of the DNA, this triphosphate can easily break, so if a strand has been sitting in the cell for a while, it would not be able to attach new nucleotides to the 5' end once the phosphate had broken off. On the other hand, the 3' end only has a hydroxyl group, so as long as new nucleotide triphosphate are always brought by DNA polymerase, synthesis of a new strand can continue no matter how long the 3' end has remained free.

This presents a problem, since one strand of the double helix is 5' → 3′, and the other one is 3'→5'. How can DNA polymerase synthesize new copies of the 5′ → 3′ strand, if it can only travel in one direction? This problem was resolved by Reiji and his wife Tsuneko Okazaki and their colleagues (1960s) by performing pulse and chase experiment (Fig. 10).

Replication of bacteriophage T4 in E. coli

Okazaki infected *E. coli* with bacteriophage T4 at a high multiplicity of infection (M.O.I. = 10). He then grew the infected cells for a total of 75 min. (5 min at 37°C followed by 70 min at 20°C). He knew that under these growth conditions, the bacteriophage was starting to replicate their DNA after this time.

He now added a pulse of ^{3}H-thymidine to the culture to label the newly synthesized DNA. He isolated the DNA from the cells soon—2 sec, 7sec, 15sec, 30sec, 60sec and 120sec—after adding the pulse and he analyzed it using an alkaline sucrose gradient. The alkaline conditions allowed him to monitor the incorporation of the radioactive label into ssDNA as opposed to dsDNA (Fig. 10). Okazaki saw the accumulation of label in a slowly sedimenting peak at early time points. At later time points, the label appeared in a faster sedimenting peak. This result demonstrated that DNA is synthesized as short precursor molecules that are subsequently joined to form larger molecules (Fig. 11).

The use of bacteriophage T4 mutants:

Okazaki studied DNA synthesis in two bacteriophage T4 mutants that carried temperature sensitive mutations in gene 30. However, 2 min before adding the pulse of ^{3}H-thymidine he raised the temperature to the restrictive temperature. Now when he analyzed the newly

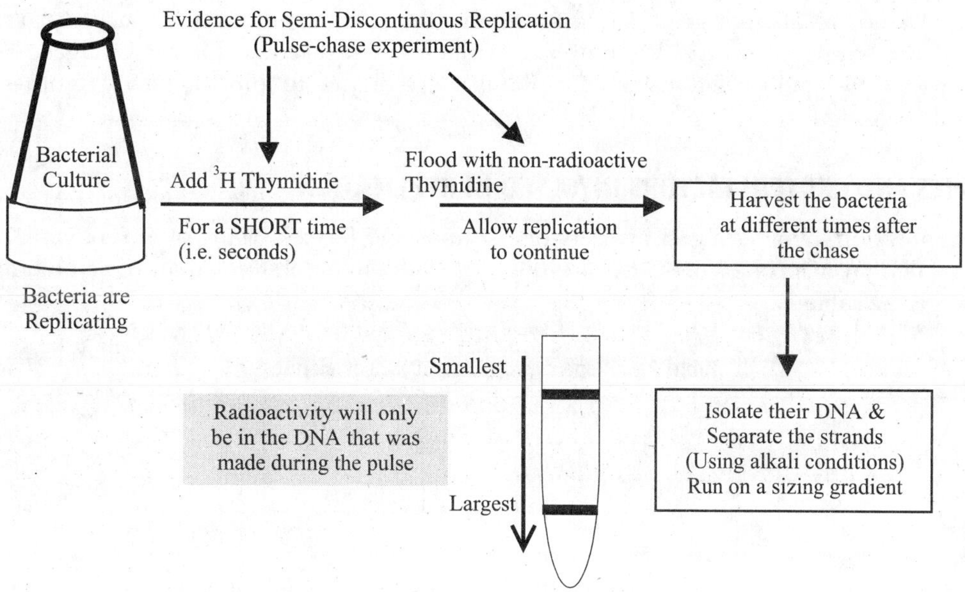

Fig. 10. Pulse and chase experiment performed by Reiji Okazaki to prove Semi-Discontinuous Replication He grows bacterial cells in a medium containing (a) radioactive ^{3}H-thymidine (b) non-radioactive thymidine for several generations to prove mode of replication. Then isolate their DNA and separate the strands by using a sizing gradient.

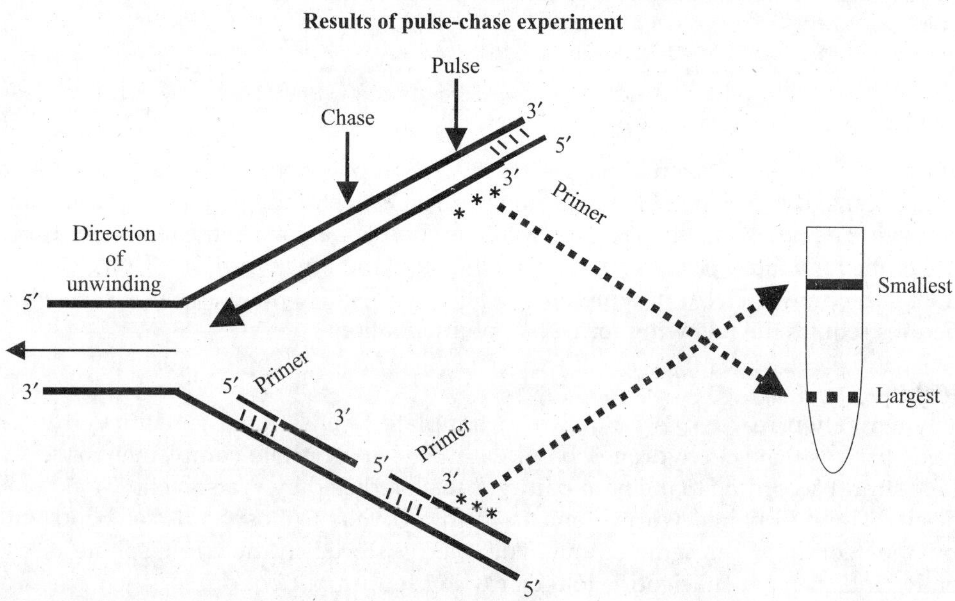

Fig. 11. Shows the results of pulse-chase experiment: larger DNA strands appear at the lower and smaller ragments appear at the upper portion of tube after doing separation on size gradient.

labeled DNA on an alkaline gradient, he found that the label accumulated exclusively in small slowly-sedimenting fragments. We now know that T4 gene 30 codes for a DNA ligase. So, at the restrictive temperature, Okazaki's mutants were unable to join the newly synthesized segments together.

ENZYMES AND PROTEIN FACTORS INVOLVED IN REPLICATION

DNA replication involves a great variety of enzymes and factors. Many of the enzymes were first isolated from prokaryotes and are therefore better understood than their eukaryotic counterparts (Table 2).

Table 2. Properties of E. coli DNA polymerases

	Pol I	Pol II	Pol III
Mass (kD)	103	90	130
Molecules/cell	400	?	10-20
Turnover number*	600	30	9000
Structural gene	*pol A*	*pol B*	*pol C*
Conditionally lethal mutant	+	-	+
Polymerization:5′ → 3′	+	+	+
Exonuclease: 3′ → 5′	+	+	+
Exonuclease:5′ → 3′	+	-	-
Processivity$	Upto 1,000	Upto 50	Upto 15,000

* Nucleotides polymerized molecules/ min at 37°C

$ The average number of nucleotides added before a polymerase dissociates.

DNA polymerase

A DNA polymerase is an enzyme that assists in DNA replication. Such enzymes catalyze the polymerization of deoxyribo nucleotides alongside a DNA strand, which they "read" and use as a template. The newly polymerized molecule is complementary to the template strand and identical to the template's partner strand. All DNA polymerases synthesize DNA in the 5′→3′ direction. No known DNA polymerase is able to begin a new chain (*de novo*). DNA polymerases require the following for DNA polymerization:

DNA template

The polymerization reaction is guided by a template DNA strand, recruiting free deoxy-nucleotide-triphosphates to hydrogen bond with their appropriate complementary dNTP on the single strand according to the base-pairing rules predicted by Watson and Crick: where A paired with T and G paired with C, and to form a covalent phosphodiester bond with the previous nucleotide of the same strand. The energy stored in the triphosphate is used to covalently bind each new nucleotide to the growing second nucleotide.

Divalent metal ion (Mg^{2+})

DNA polymerase is considered to be a holoenzyme since it requires a Magnesium ion as a co-factor to function properly. In the absence of the Magnesium ion, it is referred to as an apoenzyme.

Primer

A common feature of all DNA polymerases is that they cannot initiate synthesis of a chain of DNA *de novo*. Synthesis of new strand of DNA can only start from preexisting 3′-OH end and template strand must be converted to a single stranded condition. The 3′-OH end is called as primer. The primer can take various forms.

A sequence of RNA is synthesized on the template, so that free 3′-OH end of the RNA chain is extended by the DNA polymerase. This is commonly used in replication of cellular DNA and by some viruses.

A preformed RNA, pairs with the template to allow its 3′-OH end to be used to prime DNA synthesis. This mechanism is used by retroviruses to prime reverse transcription of RNA.

The primer terminus is generated with in duplex DNA. The most common mechanism is the introduction of a nick, as used to initiate rolling circle replication. In this case, the preexisting strand is displaced by new synthesis.

A protein primes the reaction directly by presenting a nucleotide to DNA polymerase. This type of reaction is used by certain viruses.

After adding a nucleotide to a growing DNA strand, DNA polymerase either dissociates or moves along the template and adds another nucleotide. Dissociation and reassociatetion of the polymerase can limit the overall polymerization rate-the process is generally faster if a polymerase adds nucleotide without dissociating from the template.

DNA polymerase can be processive or distributive

Processive polymerization is continuous synthesis by polymerase without dissociation from the template. A DNA polymerase that is distrubutive will dissociate from the template after each nucleotide addition i.e. Distributive polymerization. The average number of nucleotides added before a polymerase dissociates defines its processivity. Error correction is the property of some DNA polymerases. This process corrects mistakes in newly synthesized DNA.

DNA polymerases have a common structure

All the DNA polymerases share some common structural features. The enzyme structure can be divided into several independent domains, which are described by analogy with a human right hand. DNA binds in a large cleft composed of three domains. The "palm" domain has important conserved sequence motifs that provide the catalytic active site. The "fingers" are involved in positioning the template corectly at the active site. The "thumb" binds the DNA as it exits the enzyme, and is important in processivity. The most important conserved regions of each of these three domains converge to form a continous surface at the catalytic site. The N-terminal domain extends into the nuclease domain. DNA polymerases fall into five families based on sequence homologies; palm is well conserved among them, but the thumb and fingers provide analogous secondary structure elements from different sequences.

Based on sequence homology, DNA polymerases can be subdivided into seven different families A, B, C, D, X, Y, and reverse transcriptase.

Family A polymerases contain both replicative and repair polymerases. Replicative members from this family include the extensively studied T7 DNA polymerase as well as the eukaryotic mitochondrial DNA Polymerase γ. Among the repair polymerases are *E. coli* DNA pol I, Thermus aquaticus pol I and Bacillus stearothermophilus pol I. These repair polymerases are involved in excision repair and processing of Okazaki fragments generated during lagging strand synthesis.

Family B polymerases mostly contain replicative polymerases and include the major eukaryotic DNA polymerases α, δ, and ε. Family B also includes DNA polymerases encoded by some bacteria and bacteriophages, of which the best characterized are from T4 and RB69 bacteriophage. These enzymes are involved in both leading and lagging strand synthesis. A hallmark of the B family of polymerases is remarkable accuracy during replication and a strong 3′→5′ exonuclease activity.

Family C polymerases are the primary bacterial chrmosomal replicative enzymes and thus have polymerase and 3′→5′ exonuclease activity.

Family D polymerases are still not very well characterized. All known examples are found in the Euryarchaeota subdomain of Archaea. These are thought to be replicative polymerases.

Family X contains the well known eukaryotic polymerase pol β as well as other eukaryotic polymerases such as pol σ, pol λ and pol μ. In this family of polymerases pol β has been studied the most.

Family Y contains polymerases that repair damaged DNA, such as lesion caused by UV radiation and other mutagenic agents. Because of their functions, Family Y polymerases often have low fidelity and lack of exonuclease activity. Family Y polymerases from bacteria and eukaryotes include DNA polymerase IV and V in *E. coli* and pol μ in eukaryotes.

The reverse transcriptase family contains both retroviruses and eukaryotic polymerases. The eukaryotic polymerases are usually restricted to telomerases.

Prokaryotic DNA Polymerases

Bacteria have 5 known DNA polymerases (Table 2):

DNA Polymerase I is implicated in DNA repair and has both 5′→3′ polymerase activity, a 5′→3′ exonuclease, and a 3′→5′ exonuclease activity.

DNA Polymerase II is involved in replication of damaged DNA and has both 5′→3′ chain extension ability and 3′→5′ exonuclease activity.

DNA Polymerase III is the main polymerase in bacteria (elongates in DNA replication), as such it has 3′→5′ exonuclease proofreading ability.

DNA Polymerase IV is a Y-family DNA polymerase.

DNA Polymerase V is a Y-family DNA polymerase and participates in bypassing DNA damage.

DNA polymerase I

It is an enzyme that mediates the process of DNA replication in prokaryotes. This enzyme was discovered by Arthur Kornberg in 1958, it was the first known DNA polymerase (and, indeed, the first known polymerase of any kind), and was initially characterized in *E. coli*, although it is ubiquitous in prokaryotes.DNA polymerase I is a 928 amino-acid polypeptide (MW=103,118) encoded by the *polA* gene. It is often referred to as simply Pol I. DNA Polymerase I consists of a single polypeptide chain that has following enzymatic activities (Fig. 12):

A 5′→3′ DNA polymerase activity (Fig. 12a).

A 3′→5′ exonuclease activity that mediates proofreading. When an incorrect base pair is recognized, DNA polymerase reverses its direction by one base pair of DNA. This activity of the enzyme allows the incorrect base pair to be excised. Following base excision, the polymerase can re-insert the correct base and replication can continue (Fig. 12b).

A 5′→3′ exonuclease activity mediating nick translation with polymerase activity. Intact DNA polymeras I can replace a segment of DNA or RNA paired to the template strand. DNA polymerase I removes the RNA primer from the lagging strand and fills in the necessary nucleotides in 5′→3′ direction, proofreading for mistakes as it goes (Fig. 12c).

An inherent RNAase H activity which is essential for cell viability in *E.coli* but has not been used in molecular biology.

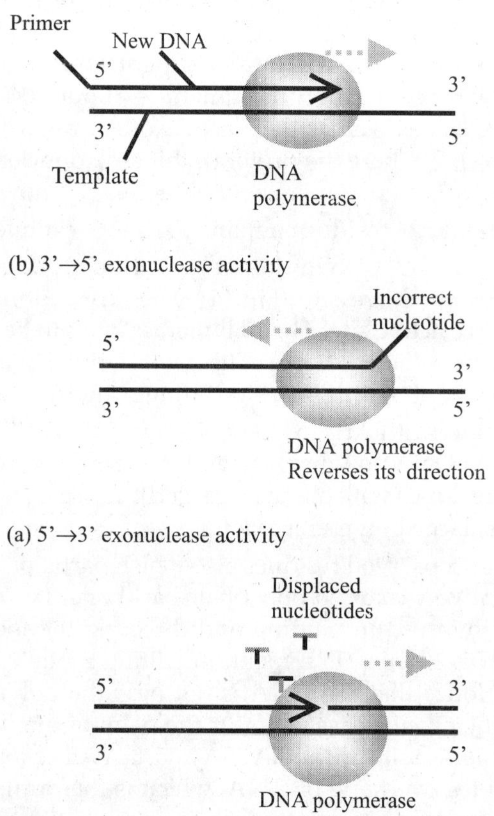

Fig. 12. DNA polymearse I enzymatic activities: (a) 5′→3′ polymerase activity (b) 3′→5′ exonuclease activity and (c) 5′→3′ exonuclease activity (Lewin, 2004).

DNA polymerase I is distributive, it does not remain associated with the lagging strand, but dissociates after each RNA primer is removed. Structurally, DNA polymerase I holoenzyme has two domains which can be separated by proteolysis with trypsin or subtilisin. Both enzymes cleave at the amino acid 324 (Fig. 13).

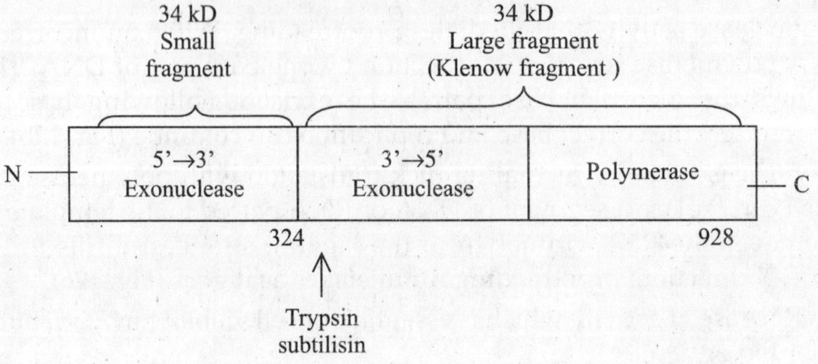

Fig. 13. DNA polymerase I.

The larger fragment is known as the Klenow Fragment after Hans Klenow, who first discovered this proteolytic cleavage. Klenow Fragment is about 604 amino acid residues contains the polymerase active site as well as the 3′→5′ exonuclease active site. This Klenow was crystallized in "editing mode", where the protein exhibits exonuclease activity. When the protein is in "polymerizing mode" the polymerase site is active (Klenow backbone).

The smaller fragment is about 324 residue contains the 5′→3′ exonuclease site.

The Klenow Fragment is consists of two domains. The small domain is about 324-517 residues contains the 3′→5′ exonuclease site (Fig.14). The larger domain is about 528-928 residues contains the polymerase active site. The polymerase site can be seen as being located at the bottom the verticle cleft on the molecule. This cleft is lined with positively charged molecules, is about 22 Å wide and 30 Å and forms the "finger and thumb" of the molecule.

The incoming dNTP's interact with Arg 754, Arg 682, Lys 758, His 734. These residues are highly conserved in the family of DNA polymerases, and interact with the phosphate backbone of incoming dNTPs (Positive residues which line large cleft). These conserved residues are all positively charged and line the large polymerase cleft.

There are a number of other conserved residues also which participate in the incorporation of incoming dNTP with a negatively charged phosphate backbone. Tyr 766, Asp 882, Asp 705, Glu 883, and His 881 are all involved in binding and directing the incoming dNTP (dNTP binders in polymerase site). This Klenow Fragment, in editing mode was cocyrstalized with and an 11 bp duplex DNA. Notice that the DNA is not near the polymerase site in editing complex; rather it located in the exonuclease site near the palm of the molecule (Fig. 14). The exonuclease site and the polymerase site are about 35Å apart. At the exonuclease site there is a single-stranded three nucleotides overhang of DNA, which is the primer strand. This crystal structure reveals that the primer strand enters the polymerase site through the exonuclease.

However when the klenow shifts between exonuclease mode and polymerase mode, it is the three nucleotide overhang on the primer that moves from the exonuclease site to the polymerase site (Fig. 14). However, because the two clefts are approximately perpendicular to

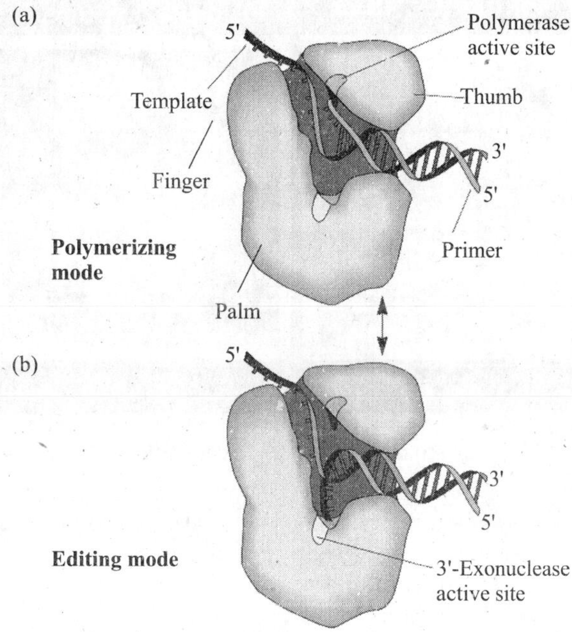

Fig. 14. Showing klenow fragment with (a) Polymerase active site (b) 3′→5′ exonuclease site (Snustad and Simmons, 2006).

each other it is clear that the duplex DNA must bend in order to go to polymerizing mode. Three very important residues in the polymerase cleft which facilitate this process are Asp 882, Glu 883, and Asp 705. These residues bind to divalent cations and enable the duplex DNA to bend toward the polymerase active site where incoming dNTPs can replicate the template strand. The primer and DNA enter through the exonuclease site. Previous mutation studies of exonuclease site have shown that Leu 361, Phe 473, His 660, and Glu 357 are significantly involved with primer interaction and proofreading activity through the help of divalent cations and hydrophobic interactions.

Since the DNA enters through the exonuclease site to reach the polymerase site, there are residues that are located on the thumb of the molecule which are directly involved in binding the DNA. These amino acids are highly conserved: Lys 635, Asn 678, and Asn 675 only interact with the backbone of the DNA primer strand (DNA binding residues on "thumb"). The "thumb" region (Fig. 15) goes through a conformational change and makes direct contact with duplex DNA by moving closer to the exonuclease site. Thus DNA causes a conformational change in the Klenow. When duplex DNA is present it resides in the new second cleft, which is perpendicular to the original top cleft. In addition to positively charged residues which contact the primer strand backbone, other conserved residues such as Arg 835, Asp 827, Ser 582, and Asn 579 also make interactions with the template strand backbone to stabilize the duplex inside the Klenow. Once duplex DNA is bound the protein makes contact only with the DNA backbone.

In summation the Klenow fragment exists in two different conformational forms when polymerase activity if required and when exonuclease activity if required. However it is clear from highly conserved amino acids that the DNA/primer strand causes a conformational

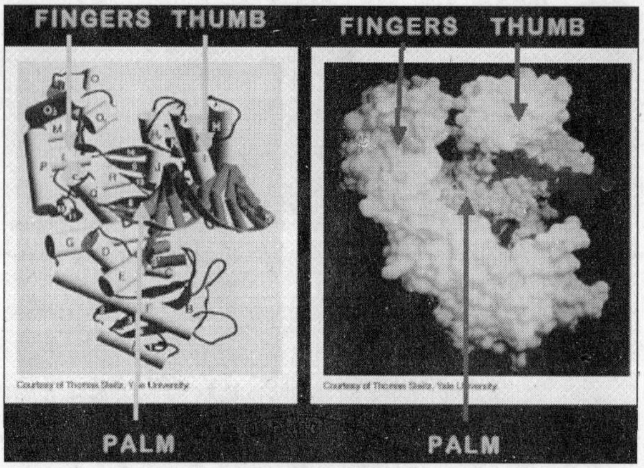

Fig. 15. X-ray structure of Klenow fragment (Beese *et al.*, 1993).

change which produces a new cleft near the exonuclease site. The DNA enters through this cleft and then the moving primer strand moves (along with a conformational change in the protein) to the polymerase site, which also contains positive residues. These positive residues interact with the duplex backbone. The presence of this intricate system increases the fidelity during replication by DNA polymerase I.

Despite the early characterisation of DNA polymerase I, it quickly became apparent that DNA polymerase I was not the enzyme responsible for most DNA synthesis. *E. coli* DNA replication proceeds at approximately 1,000 nucleotides/second, while the rate of synthesis by DNA polymerase I averages only 20 nucleotides/second. Moreover, its cellular abundance of approximately 400 molecules per cell did not correlate with the fact that there are typically only two replication forks in *E. coli.* Moreover, it is insufficiently processive to copy an entire genome, as it falls off after incorporating only 25-50 nucleotides.

DeLuca and John Cairns (1969) mutated the gene for DNA polymerase, *polA1*, and the bacteria grew just fine. This mutant of *E.coli* was viable even it lacked DNA polymerase I activity. Mutation was found to be an amber mutant. We now know from sequence information that the codon for tryptophan 342 is mutated to a stop codon in this mutant.

pol A I Mutant

- It multiplied at the same rate as the wild type.
  → DNA replication + normal
- It was much more easily killed by UV light.
  → DNA repair + impaired

Either the polymerase hypothesis was all wrong … or there were other DNA polymerases in *E. coli* that carried out DNA synthesis in the polA strains. **DeLuca** and **Cairns** assay the extracts of *E. coli* polA1 mutants and found the activities for two other types of DNA polymerases, i.e., for DNA polymerase II and for DNA polymerase III. After the discovery of DNA polymerase III, the main replicative DNA polymerase was finally identified.

DNA polymerase II

DNA polymerase II is a prokaryotic DNA polymerase most likely involved in DNA repair. The enzyme is 90 kDa in size and is coded by the *polB* gene. Synthesis of DNA polymerase II is induced during the stationary phase of cell growth. This is a phase in which little growth and DNA synthesis occurs. It is also a phase in which the DNA can accumulate damage such as short gaps, which act as a block to DNA polymerase III. Under these circumstances, DNA polymerase II helps to overcome the problem because it can reinitiate DNA synthesis downstream of gaps. DNA polymerase II has a low error rate but it is too slow to be of any use in normal DNA synthesis.

DNA polymerase III

DNA polymerase III **holoenzyme** is the primary enzyme complex involved in prokaryotic DNA replication. The complex has high **processivity**. The DNA polymerase III holoenzyme also has proofreading capabilities that correct replication mistakes by means of 3′→5′ exonuclease activity. DNA polymerase III is a component of the **replisome**, which is located at the replication fork.

The replisome is composed of the following (Table 3):

2 DNA Pol III enzymes, made up of α, ε, θ subunits; the a subunit has polymerization activity, the ε subunit has 3′→5′ proofreading activity, the θ subunit stimulates the ε subunit's proofreading.

2 β units which act as sliding DNA clamps, they keep the polymerase binded to the DNA. It has two subunits.

2 τ units which connect the 2 DNA Pol III enzymes.

ψ complex acts as a clamp loader for the lagging strand Okazaki fragments, helping the two β subunits to form a unit and bind to DNA. The γ complex is made up of 5 ψ subunits.

Table 3. Subunits of DNA polymerase III

Subunit	Mass(kD)	Gene	Subassembly	Activity
α	130	*dnaE*	Pol III core (there are 2 cores per DNA pol III holoenzyme)	5′→3′ polymerase
ε	27.5	*dnaQ*		3′→5′ exonuclease (editing exonuclease)
θ	10			Stimulates 3′→5′ exonuclease
τ	71	*dnaX*		dimerizes cores, activates DnaB helicase activity
γ	47.5	*dnaX*	γ complex, ATP dependent clamp loader (uses ATP energy when loading beta onto primed DNA)	binds ATP
δ	35			unknown
δ′	33			stimulates clamp loading
χ	15			interacts with SSB to allow removal of DnaG primase from primer
ψ	12			unknown
β	40.6	*dnaN*	β clamp	Processivity factor

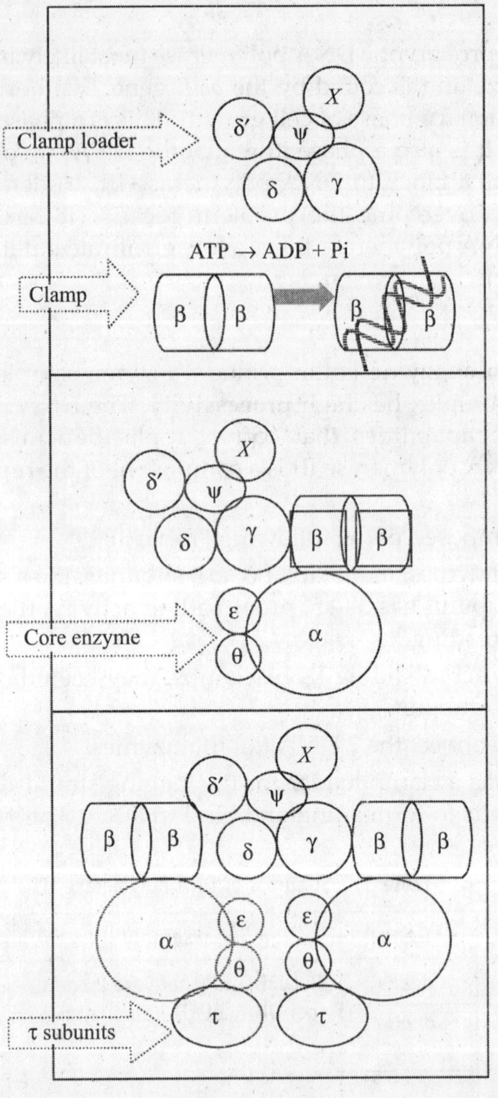

Fig. 16. DNA polymerase III holoenzyme assembles in above stages, generating an enzyme complex that synthesizes the DNA of both; lagging and leading strand, new strands. (a)Clamp loader cleaves ATP to load clamp on DNA (b) core enzyme joins (c) t and second core enzyme joins to give a symmetric dimer, t Subunits maintain dimeric structure (Lewin, 2006).

The holoenzyme assembles on DNA in three stages (Fig. 16):

Column I

Eukaryotic DNA polymerases

Animal cells contain at lease five distinct DNA polymerase designated in the order of their discovery (Table 4).

Table 4. Properties of Animal DNA polymerases

Enzymes	Pol α	Pol β	Pol γ	Pol δ	Pol ε
Location	Nucleus	Nucleus	Mitochondrion	Nucleus	Nucleus
Family	B	X	A	B	B
Mass(kD)	>250	36-38	160-300	170	256
Subunit composition	165,67,58,48	39	125,35	125,66,50,12	261,59,17,12
Structure	Tetramer	Monomer	dimer	Tetramer	Tetramer
Inhibitors: Aphidicolin	Strong	None	None	Strong	Strong
Dideoxy NTPs	None	Strong	Strong	Weak	Weak
N-ethyl Maleimide	Strong	None	Strong	Strong	Strong
Function	High fidelity replicase: Nuclear replication	High fidelity repair: Base excision repair	High fidelity replicase: Mitochondrial replication and repair	High fidelity replicase: Nuclear replication	High fidelity replicase: Nuclear replication
Associated activities	Primase	dRP lyase	$3' \rightarrow 5'$ Exonuclease	$3' \rightarrow 5'$ Exonuclease	$3' \rightarrow 5'$ Exonuclease
Exonuclease: $5' \rightarrow 3'$	-	-	-	-	-
Processivity	~100nt			∞	

DNA polymerase α

The Pol α which occurs only in nucleus was discovered as the first eukaryotic DNA polymerase more than 40 years ago. It is the DNA polymerase of family B and particularly sensitive to inhibition by butyl phenyl-dGTP but not ddNTPs. DNA polymerase α is inhibited by aphidicolin. Pol α has been isolated from a wide range of organisms and generally consists of a complex of four subunits.

The 180 kD polypeptide containing the catalytic activities, one polypeptide of about 86 kD with no detectable enzymatic activity but has regulatory functions, two polypeptides of 48 kD and 58 kDa that contain the DNA primase activity (Fig. 17). The essential catalytic primase activity is contained in the 48-kDa subunit, although full activity is only attained when both primase subunits come together. Pol α is devoid of proofreading ability.

DNA Polymerase α is the only eukaryotic enzyme capable of initiating DNA synthesis *de novo*, synthesizing short RNA primers of about 10 nucleotides first, which are then extended by Pol α to yield RNA-DNA primers which termed as initiator DNA, of approximately 30-40 nucleotides in length. The major function of Pol α is proposed to be the initiation of leading strand DNA replication as well as the continuous priming of Okazaki fragments on the lagging strand. It lacks exonulease activity.

Although expression of DNA polymerase α is almost constitutive, its activity is strictly regulated during the cell cycle. The p180 and p68 subunits of human Pol α are phosphorylated

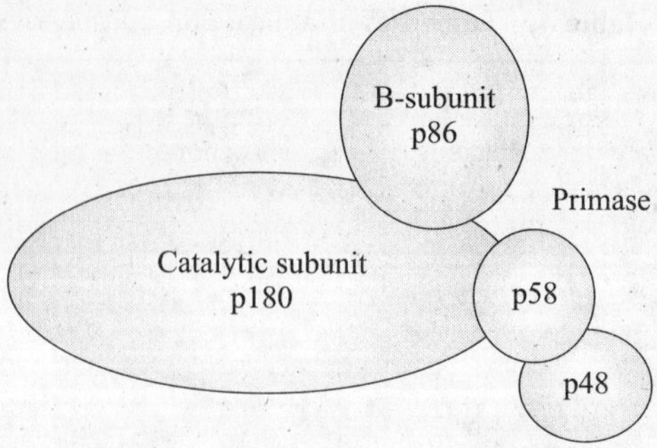

Fig. 17. DNA polymerase a subunits.

by cyclin A-dependent kinases during S phase, and the phosphorylated form of Pol α is dephosphorylated by protein phosphatases 2A and again phosphorylated by Cyclin E-Cdk2 during the G_1 phase. Phosphorylation by cyclin E-Cdk2 apparently renders Pol α initiation-competent, whereas further phosphorylation by the cyclin A-dependent kinase creates the form of the enzyme that synthesizes primers for the lagging strand.

All four subunits of Pol α interact with the large T antigen to form a ternary complex that also includes human single-stranded DNA binding protein, replication protein A (RPA) (Table 5). This complex formation is essential, and in the presence of topoisomerase activity, sufficient for initiation of SV40 replication. The interaction with RPA is thought to stabilize the Pol α -primer complex and to increase processivity and fidelity of the enzyme. Pol α interacts in a similar manner with several factors involved in the initiation of cellular DNA replication such as Mcm3 in a variety of organisms.

The essential role of Pol α in initiation makes it a principal target both for regulation and checkpoint control of DNA replication. Association of the growth suppressor $p12^{DOC-1}$ with Pol α inhibits initiation of DNA replication, probably by preventing phosphorylation of Pol α by Cdk 2. The primase subunits of Pol α are implicated in the replication checkpoint in different eukaryotes. In fact, both in fission yeast and *Xenopus*, it was shown that RNA synthesis by primase is required for the activation of this checkpoint.

DNA polymerase β

In the early 1970s, Pol β was isolated from several mammalian as well as invertebrate sources as a DNA polymerase of low molecular size. Pol β from mammalian sources is a single polypeptide of 39 kDa belongs to family X.

Table 5. DNA polymerase α associated units

Interacting protein	Pol α subunit(s)	Function
T-antigen	All subunits	Initiation of SV40 DNA Replication
RPA	Catalytic and primase subunits	Stimulation of Pol α activity and processivity
Cdc45	B-subunit	Recruitment of Pol α onto origins?
$p12^{DOC-1}$	All subunits	Regulation of Pol α activity and phosphorylation
PARP	Catalytic	

The small size and simple structure makes Pol β an ideal polymerase for studies. Pol β lacks any accessory proteins. Pol β consists of two domains, each of which has its own biochemical activity. An 8kDa N-terminal domain carries a deoxyribose phosphate lyase activity. This activity removes deoxyribose phosphate from DNA via a β-elimination mechanism, if the deoxyribose is not linked to a base and the DNA is cleaved at the 5′ position of this moiety. This is an intermediate step in the DNA base excision repair (BER) pathway. The same domain also binds very specifically to single stranded DNA gaps of five nucleotides with a 5′-phosphate at the margin of the gap.

The C-terminal 31 kDa domain carries the DNA polymerase activity of Pol β. The synthesis of DNA is highly distributive on normal template-primers, that is, the enzyme adds only one nucleotide at a time, and then dissociates from the template again. In contrast, on gapped substrates containing up to six nucleotides with a 5′-phosphate at the margin of the gap as described above, synthesis is processive, exhibiting also higher fidelity and efficiency. Pol β is insensitive to aphidicolin, but is strongly inhibited by ddTTP and other ddNTPs, which act as competitive inhibitors of the natural nucleotide triphosphates. This is in contrast to family A DNA polymerases like klenow or Taq Polymerase that readily incorporate dideoxynucleotides causing chain termination, and family B polymerases that discriminate against the binding of ddNTPs compared to the dNTP substrate.

Pol β is the only eukaryotic DNA polymerase for which there is a high resolution crystal structure and structural analysis of Pol β in different complexes with substrates and cofactors provide detailed insight into the molecular mechanisms of the DNA polymerase reaction. The overall structural commonality of Pol β with other DNA polymerases shows that it is not evolutionarily related to DNA polymerases of other families. Topology of the palm domain is different from the other DNA polymerases. Its architecture is mostly composed of β-sheets flanked on one site by two α-helices, as in other polymerases. The thumb and finger domains are positioned similarly in the structure, but when projected onto the primary structure; their position is reversed compared to DNA polymerases of other families. Binding of the primer-template and the correct dNTP substrate induces movement of the thumb domain to a closed conformation as in other DNA polymerases, effectively assembling the active site. This induced fit mechanism is thought to enhance the fidelity of the polymerase reaction.

DNA Polymerase γ

Mammalian Pol γ has been originally identified as an aphidicolin-insensitive DNA polymerase belongs to Family A. This occurs exclusively in mitochondria. DNA polymerase γ is a high fidelity replicase which involves in mitochondrial DNA replication and repair. DNA polymerase γ has 3′ → 5′ exonuclease activity.

DNA Polymerase δ

Mammalian DNA Polymerase δ has been originally identified as an aphidicolin-sensitive enzyme. DNA Polymerase δ (Fig. 18) is a nuclear enzyme, along with Pol ε; it is the main polymerase in eukaryotes. It exihibits a 3′→5′ exonuclease proofreading activity. Moreover, whereas DNA polymerase α exhibits only moderate processivity (~100nt), DNA Pol **δ** is essentially infinitely processive (it can replicate the entire length of a template) but only when it is in complex with a protein known as proliferating cell nuclear antigen (PCNA). It can also fill the gap after the primer has been removed but it doesn't have 5'→3' exonuclease activity (Table 6).

Table 6. DNA polymerase δ associated Subunits

Pol δ subunit	*S. cerevisiae*	*S. pombe*	*H. sapiens*	Function
Catalytic	Pol3	Pol3	p125	DNA polymerase activity
B-subunit	Hys2/Pol31	Cdc1	p50	DNA replication
Third subunit	Pol32	Cdc27	p58-p66	PCNA binding
Fourth subunit	-	Cdm1	p12	Stabilization of the holoenzyme?

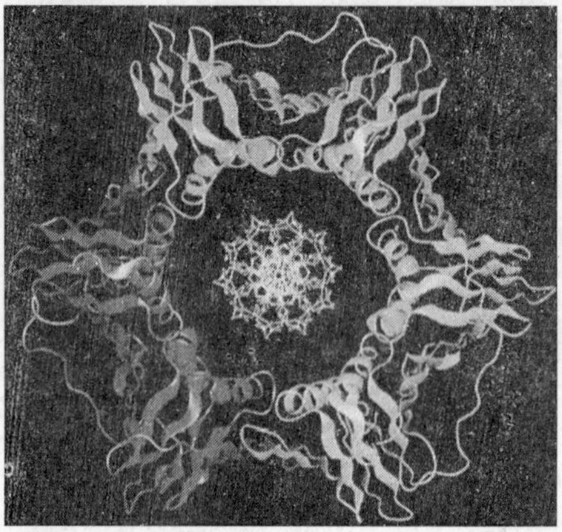

Fig. 18. DNA polymerase δ (Voet and Voet, 2004).

Pol δ has been isolated from numerous sources, and a large amount of comparative information on the primary structure is available. Although there appeared to be considerable heterogeneity in the subunit composition of enzymes from different sources, more recent work has shown that the structure and subunit composition of Pol δ is rather conserved among eukaryotes. The A subunit of about 125 kDa contains the catalytic activity for both the polymerase and the proofreading 3′→5′ exonuclease activity. The B subunit of about 35 kDa that is devoid of any catalytic activity is thought to be important for the enzyme stability as well as for protein-protein interactions. The C subunit of Pol Δ shows far less conservation than the A and B subunits. The C subunits have highly charged nature and extremely high lysine content. This subunit mediates the dimerisation of the enzyme.

Pol δ is implicated to be the major DNA polymerase required for DNA replication. Two factors RFC and PCNA provide the moving platform that mediates the dynamic functions of the enzyme. PCNA is a eukaryotic sliding clamp that is structurally and functionally related to the β subunit of *E. coli* Pol III holoenzyme. Three PCNA molecules form a closed ring able to encircle duplex DNA. Once loaded onto DNA by RFC in an ATP-dependent manner, PCNA cannot dissociate but can freely translocate on double-stranded DNA. PCNA then associates with Pol δ at the primer-template junction and supports processive DNA synthesis by the polymerase, and at the same time prevents nonproductive binding of Pol δ to single-stranded DNA. Pol δ in conjunction with PCNA and RFC is therefore generally referred to as the Pol δ holoenzyme.

Pol δ exhibits a complex interaction with PCNA. Two functional interaction sites with PCNA have been mapped in Pol δ, one residing in **the N-terminus** of the **catalytic subunit**, the other being a consensus PCNA interaction site in the **C-terminus** of the **C subunit**. The B subunit is required for the productive interaction of PCNA with the catalytic subunit, since stimulation of Pol δ activity by PCNA can only be observed in the presence of both A and B subunits.

DNA Polymerase ε

It was discovered earlier that Pol ε was originally a form of Pol δ that is processive in the absence of PCNA and RFC at low ionic strength. Pol ε was eventually proven to represent an independent DNA polymerase and the catalytic subunit was cloned from several species. Pol ε has a structure similar to that of other eukaryotic family B DNA polymerases (Fig. 19), but notably, the catalytic subunit possesses a large C-terminal domain of unclear function that accounts for half of the molecular weight of ~260 kDa. It is located between the DNA polymerase domain and the putative zinc-finger. The catalytic subunit is essential for viability in budding and fission yeast, and temperature-sensitive mutants fail to replicate chromosomal DNA at the restrictive temperature. It is rather surprising that in fact the carboxy-terminal domain is sufficient for cell viability, whereas the polymerase and exonuclease domains are dispensable.

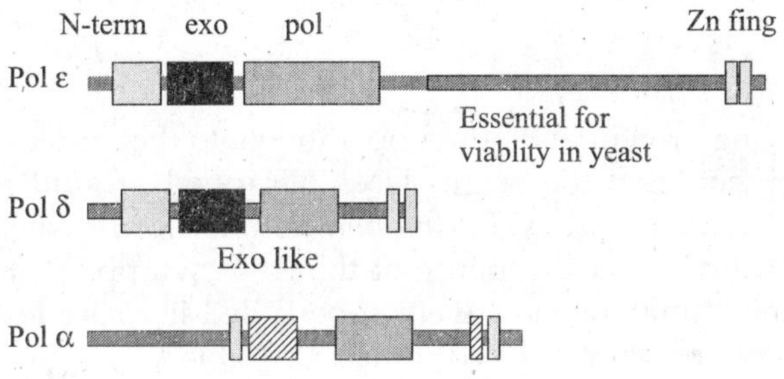

Fig. 19. The domain structure of human DNA polymerase e compared with the structure of other eukaryotic family B DNA polymerases. Although DNA polymerases a do not exhibit any proofreading exonuclease activity, regions of similarity to the exonuclease domain can be detected in their primary structure.

The putative zinc finger is particularly important, since several mutants in this domain or the adjacent C-terminus have growth defects. Furthermore, some of these mutants are defective in their response to DNA damaging agents during S-phase in budding yeast. It can be seen as consistent with the results from yeast that Pol ε, in contrast to Pol δ, is not required for DNA replication of SV40.

Apart from the catalytic subunit, Pol ε contains three more subunits that differ in size but are conserved in their primary structure from yeast to human. The B subunit of 60-86 kDa is also essential for viability in yeast and is implicated in homodimerisation of the enzyme. DNA polymerase ε is processive in the absence of PCNA. DNA Polymerase ε also has a 3'→5' exonuclease activity that degrades single-stranded DNA to 6-or 7-residue oligonucleotides rather than to mononucleotides. DNA Polymerase ε is required to repair UV-induced DNA lesions, but it may also be necessary for DNA replication *in vivo*.

None of the eukaryotic polymerases can remove primers (5'→3' exonuclease activity), that function is carried out by other enzymes. Only the polymerases that deal with the elongation (γ, δ and ε) have proofreading ability (3'→5' exonuclease).

Other eukaryotic DNA polymerases

There are also other eukaryotic polymerases known, which are not as well characterized: η, θ, τ, κ, λ, and μ (Table 7).

Table 7. Other eukaryotic DNA polymerases

DNA polymerase	Structure	Function
θ	Monomer	DNA repair of cross links
λ	Monomer	Meiosis-associated DNA repair
μ	Monomer	Somatic hyper mutation
Rev1	Monomer	Translesion synthesis
ξ	Heteromer	Low fidelity repair: Thymidine dimer bypass
τ	Monomer	Low fidelity repair: Required in meiosis
η	Monomer	Low fidelity repair: Base damage repair
κ	Monomer	Low fidelity repair: Deletion and base substitution

DnaA

DnaA is an origin-binding protein. It binds cooperatively to the four 9-bp repeats in oriC. The initial closed complex contains the origin DNA wrapped around an assembly of 10-20 monomers of DnaA complexed with ATP. An open complex forms when the three AT-rich 13-bp repeats in oriC unwind as a consequence of the DNA wrapping around the assembly of DnaA. The oriC plasmids must be negatively supercoiled in order to form open complexes (relaxed oriC plasmids can only form closed complexes). DnaA then guides the DnaB (helicase) hexameric protein from a DnaB-DnaC complex in solution to its places around each strand.

DnaB or DNA Helicase

An enzyme known as a helicase is required to unwind DNA from a double stranded structure to a single stranded structure to facilitate replication of each strand consistent with the semiconservative model of DNA replication.

Helicase unwinds DNA strands using ATP energy and moves processively (i.e., does not leave the DNA until replication is finished; it encircles the DNA strand) in the 5'-to-3' direction along DNA. DnaA together with the use of ATP energy is required to load DnaB (helicase) onto DNA in the form of a DnaB-DnaC complex. After loading DnaB onto the replication fork, DnaC is released from the DnaB-DnaC complex and leaves the DNA. DnaB helicase activity is stimulated more than 10-fold by making contact with DNA polymerase III holoenzyme.

DnaC

DnaC forms a complex with DnaB. It is required for loading DnaB onto DNA.

Single Strand Binding Protein

Single-strand binding protein, also known as SSB or SSBP, binds single stranded regions of DNA to prevent premature reannealing. The strands naturally want to coil back up so this protein attaches to the strands to keep the space made by the helicase open during DNA replication. Thus Single-stranded binding proteins are important to maintain the stability of the replication fork. Single-stranded DNA is very labile, or unstable, so these proteins bind to it while it remains single stranded and keep it from being degraded.

Primase

DNA primase is a form of RNA polymerase and a product of the dnaG gene. In bacteria, primase binds to the DNA helicase forming a complex called the primosome. Primase is activated by DNA helicase where it then synthesizes a short RNA primer approximately 11 ±1 nucleotides long, to which new nucleotides can be added by DNA polymerase III holoenzyme. Primase acts distributively (does not remain associated with DNA). It drops off DNA after primer synthesis, and then reloads onto DNA a second or so later by protein-protein interactions with DnaB to synthesize the next primer on the lagging strand. There are many places along the template where RNA primers can be made by primase.Primase is of key importance in DNA replication because no known DNA polymerases can initiate the synthesis of a DNA strand without initial RNA primers. A primer is a short RNA segment that is complementary to a DNA segment, and is necessary to begin DNA replication

Topoisomerase

DNA topoisomerases are involved in the regulation of DNA supercoiling.

Type I topoisomerases change the degree of supercoiling of DNA by causing single-strand breaks and religation, whereas

Type II topoisomerases (such as bacterial gyrase) causes double-strand breaks.

The different roles of DNA topoisomerase I and II may indicate an opposing pair of roles in the regulation of DNA supercoiling. Both activities are especially crucial during DNA transcription and replication, when the DNA helix must be unwound to allow proper function of large enzymatic machinery. Recently, new topoisomerases of both types I and II (classified as topoisomerase III and IV) have been discovered. Topoisomerase III enzymes implicated in regulation of recombination events, and topoisomerase IV implicated in the process of segregating newly replicated chromosomes. The *E. coli* DNA topoisomerase I is responsible primarily for the relaxation of negative supercoils. Topoisomerase I have also been implicated in knotting and unknotting DNA and in linking complementary rings of single-stranded DNA into double-stranded rings. The intact holoenzyme is a 97K protein with three Zn (II) atoms in tetracysteine motifs near its carboxy-terminus. Topoisomerase I appears to reverse supercoiling by transiently breaking a segment of single-stranded DNA, passing an intact single- or double-stranded strand of DNA through the gate, then rejoining the broken segment. To understand the function of this mechanism, a supercoil may be thought of as a knot on a string where the string is fixed at both ends, and so the knot may not simply be untied. Instead, the gating mechanism appears to open, untie, and reseal the "knotted" DNA precisely where it lies.

Topoisomerase I have following structural domains:

The 67K N-terminal fragment is capable of this gating/passage activity, but not of directly relaxing negative supercoils.

Zinc-binding domain implicated in the process of relaxing DNA supercoils and,

14 kD C-terminal DNA- binding domain.

Topoisomerase I has several unusual features. Unlike type II topoisomerases, topoisomerase I does not require ATP hydrolysis to catalyze the complex topological rearrangements of DNA for which it is responsible. Whereas most enzymes involved in complex rearrangements of DNA are oligomeric, topoisomerase I appears to be a fully functional monomer.

Structural features

The 67K N-terminal fragment of topoisomerase I is a single polypeptide with extensive secondary structure. There is a 27E hole in the center of the protein large enough to comfortably encircle either a single- or double-stranded piece of DNA with no steric hindrance between the DNA sugar-phosphate backbone and protein side chains within the torus. This hole is lined with basic (positively charged) residues that provide electrostatic attraction to the negatively charged phosphate oxygens of the DNA backbone. Thus, both the size and electrostatic character of the torus make it a likely candidate for temporarily holding a DNA segment.

DNA Cleavage

As mentioned earlier, the manipulation of DNA by topoisomerase involves transiently gating a segment of single-stranded DNA and passing either single- or double-stranded DNA through the gate.

One single-stranded and one double-stranded segment of DNA, both B-forms, are external to the protein. The single-stranded segment serves as the gating element and the double-stranded segment as the passing element. A key catalytic residue, Tyr 319, is shown.

A transesterfication occurs between Tyr 319 and a phosphodiester bond in the backbone of the single-stranded DNA segment. The 3' oxygen in the backbone is displaced by tyrosine oxygen, covalently linking the tyrosine oxygen to phosphorus at the 5' backbone nick. Thus, the 5' nick has been covalently linked to topoisomerase I in domain III. This covalent linkage causes a structural deformation in the protein which breaks the non-covalent interface between domain III and domains I and IV (recall that many of the active site residues are also implicated in this interface), and the rigid body formed by domains II and III swings away from the remainder of the molecule, forming the "open" conformation of the enzyme. The 3' nick remains non-covalently associated with domains I and IV (recall the DNA-binding character of domain I), while the 5' nick swings up with domains II and III, thus creating a gate in the single-stranded element.

The passing element travels through the transient gate and into the torus.

Domains II and III swing back in as the enzyme adopts a closed conformation, but one which is slightly gapped along the interface between domains I, III, and IV in order to accommodate the single-stranded DNA fragment. The single-stranded DNA segment now rejoins itself through nucleophilic attack of the 5' nick phosphorus by the 3' nick OH group.

The enzyme again adopts its open conformation. The single-stranded DNA segment remains non-covalently bound to domains I and IV.

The passing DNA element is released from the torus, completing the strand passage event. The single-stranded DNA element may remain non-covalently associated with the enzyme for repetition of the mechanism.

Since topoisomerase operates in the absence of ATP, this process must be driven solely by interactions within and between the enzyme and DNA. There is no defined directionality of this mechanism, so the overall process must involve the release of free energy. It seems likely that the supercoiled and deformed DNA conformations which topoisomerase manipulates are highly energetically unfavorable, and thus the lower energy state of the manipulated conformation may drive the mechanism. Although the model presented above involves a double-stranded passing element, it is important to recall that either single- or double-stranded DNA may serve as the passing element. Gating, however, has only been observed with single-stranded DNA.

The 14K C-terminal fragment of *E. coli* topoisomerase I enhance DNA binding activity and improve relaxation of negative supercoiling by the enzyme.

DNA ligase

DNA ligase is a particular type of ligase (EC 6.5.1.1) that can link together DNA strands that have double-strand breaks (a break in both complementary strands of DNA). The alternative, a single-strand break, is easily fixed by DNA polymerase using the complementary strand as a template. DNA ligase has applications in both DNA repair and DNA replication. In addition, DNA ligase has extensive use in molecular biology laboratories for recombination experiments.

Mammalian ligases

In mammals, there are four specific types of ligase.

DNA ligase I: ligates Okazaki fragments during lagging strand DNA replication and some recombinant fragments.

DNA ligase II: alternatively spliced form of DNA ligase III found in non-dividing cells.

DNA ligase III: complexes with DNA repair protein XRCC1 to aid in sealing base excision mutations and recombinant fragments.

DNA ligase IV: complexes with DNA protein XRCC4 and has similar function to DNA ligase III. It is more important in development.

DNA ligase can catalyze the formation of a phosphodiester bond given an unattached but adjacent 3'OH and 5'phosphate. This can fill in the unattached gap left when the RNA primer is removed and filled in. The DNA polymerase can organize the bond on the 5' end of the primer, but ligase is needed to make the bond on the 3' end (Fig. 20).

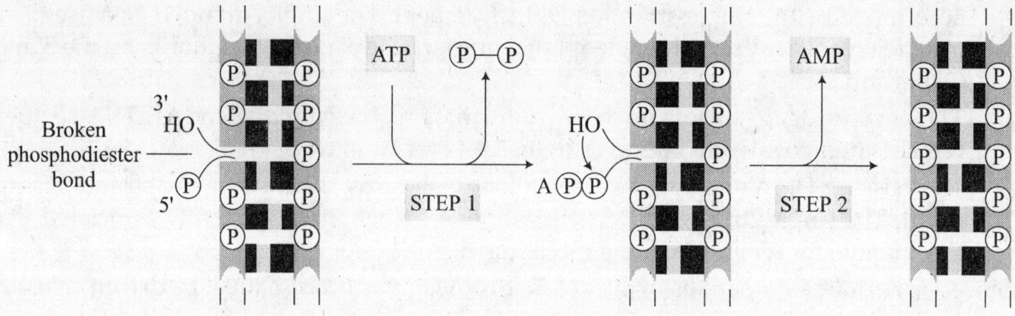

Fig. 20. DNA Ligase (Watson *et al.* 2004).

Enzymes Function without DNA Template

Reverse transcriptase

Reverse transcriptase was discovered by Howard Temin at the University of Wisconsin, and independently by David Baltimore at about the same time. These two scientist shared the 1975 Nobel Prize in Physiology or Medicine with Renato Dulbecco for this discovery.

Reverse transcriptase (EC 2.7.7.49), also known as RNA-directed DNA polymerase, is an enzyme that is able to transcribe RNA into DNA. That is, reverse transcriptase is able to copy genetic information from RNA to DNA, which is the reverse of the more typical direction i.e. DNA to RNA. Every retrovirus has a reverse transcriptase enzyme, which enables it to transcribe the genetic information from its RNA into DNA, which can then be integrated into the host genome. Similarly, retrotransposons copy themselves to RNA and then, via reverse transcriptase, back to DNA.

Reverse transcriptases also occur in group II introns, bacterial msDNAs, hepadnaviruses, and caulimoviruses. Reverse transcriptase has a high error rate (up to about 1 in 2,000 bases) when transcribing RNA into DNA. This high error rate allows retroviruses to mutate rapidly. Telomerase is a reverse transcriptase, with catalysis provided by a protein subunit that is similar to retroviral reverse transcriptases. The RNA template is contained in the telomerase itself. Reverse transcriptase enzymes include an RNA-dependent DNA polymerase and a DNA-dependent DNA polymerase, which work together to perform transcription. In addition to the transcription function, retroviral reverse transcriptases have an RNase domain, which belongs to the RNase H family.

Terminal transferase

Terminal transferase catalyzes the addition of nucleotides to the 3' terminus of DNA. It works on single-stranded DNA, including 3' overhangs of double-stranded DNA, and is thus an example of a DNA polymerase that does not require a primer. It can also add homopolymers of ribonucleotides to the 3' end of DNA. The much preferred substrate for this enzyme is protruding 3' ends, but it will also, less efficiently, add nucleotides to blunt and 3'-recessed ends of DNA fragments. Cobalt is a necessary cofactor for activity of this enzyme. Terminal transferase is useful for labeling the 3' ends of DNA, and adding complementary homopolymeric tails to DNA.

REPLICATION OF DOUBLE STRANDED CIRCULAR DNA

There are two modes of replication of circular DNA molecules:

θ-Replication

Visualization by electron microscopy of the replicating intermediates of many circular DNA molecules reveals that they retain a ring structure throughout replication. They always possess a theta-like shape that comes into existence by the initiation of replicating bubbles at the origin of replication. Replication can proceed either uni or bidirectionally. As long as each chain remains intact, even minor untwisting of a section of the circular double helix results in the creation of positive supercoils in the other direction. This supercoiling is relaxed by the action of topoisomerases, which creates single-stranded breaks (relaxed molecules) and then reseal them.

Replication of Bacteriophage λ

Bacteriophage lambda contains a linear dsDNA genome. However, the ends of the genomic DNA are single-stranded and are cohesive, i.e. they are complementary to one another. Two cohesive ends known as ***cos*** site, are 12 nucleotides in length. After adsorption of the phage to the bacterial cell surface and injection into the cell, the chromosome circularizes by means of these complementary cohesive ends. This helps to protect it from degradation by bacterial exonucleases. Bacteriophage lambda replicates in two stages.

Early replication

Bacteriophage lambda initially replicates by means of theta form intermediates. The origin of replication (*ori*) is located within the O gene, whose product is required for replication. The gene P product is also required for replication. The gene O product has a function analogous to that of DnaA. It binds to repeated sequences at the origin and initiates melting of the two strands nearby. The gene P product has a function analogous to that of DnaC. It helps DnaB to bind to the "melted" DNA. Thereafter, the other components of a bacterial replisome can bind and replication ensues. This mode of replication continues for 5 – 15 minutes after replication.

Late replication

After 15 minutes, bacteriophage lambda switches to replication by a rolling circle mechanism. It is not known what causes the switch from one mechanism to the other.

Rolling Circle Mechanism

An alternative way to replicate circular DNA is the rolling-circle mechanism though it generates linear copies of a genome rather than circular copies, or single-stranded multimers of the original sequence. This process named so, because the growing point can be foreseen as rolling around the circular template strand, or also known as ó replication mode, because of the resemblance of the greek lower-case letter sigma (Fig. 21). DNA synthesis starts with a cut in one strand at the origin of replication. The 5′ end of the cut strand is displaced from the duplex, permitting the addition of deoxyribonucleotides at the free 3′ end. As long as there is a free 3'

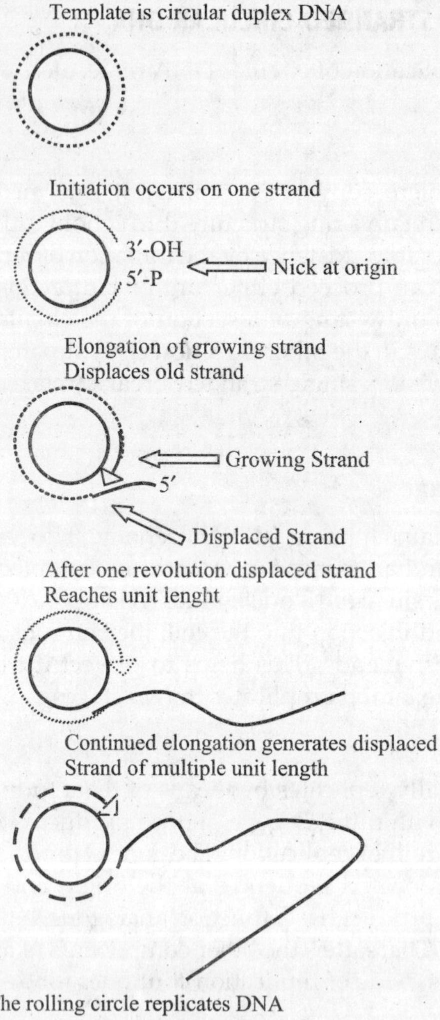

Fig. 21. The rolling circle generates a multimeric single-stranded tail (Lewin, 2004).

OH end, this can serve as a template for DNA polymerase. When the 3' OH end is extended, the 5' end can be displaced in a manner analogous to the strand displacement reaction. Synthesis on this strand is also analogous to leading strand synthesis. The displaced strand can, in turn, serve as a template for replication as long as a suitable primer is available. Synthesis on this strand is analogous to lagging strand synthesis.

As replication proceeds, the 5′ end of the cut strand is rolled out as a free tail of increasing length. When a full-length tail is produced, the replicating machinery cuts it off and ligates the two ends together. The double-stranded progeny can reinitiate replication, whereas the single-stranded progeny must first be converted to a double-stranded form.

If synthesis continues in this manner, the consequence of this mechanism of replication can be the production of concatemer copies of the circular molecule. As a result, multiple copies of a genome are produced. A rolling circle mode of replication is seen both during replication of bacteriophage lambda where rapid production of many copies of the genome is desired, and in the replication of bacteriophage M13 where only a single copy is produced each time.

Replication of Bacteriophage M13

Bacteriophage M13 (and other filamentous phage like it) has a circular ssDNA molecule in the capsid. When the phage attaches to an *E. coli* cell, this molecule is injected into the cell where most of it is coated with single-strand binding protein (SSB). Since bacteriophageM13 does not code for its own DNA polymerase, it must use the host cell machinery in order to replicate. It is, therefore, constrained by the requirements of the host cell replication machinery. Although its ssDNA genome is a perfect template for DNA synthesis, it is not such a suitable template either for RNA synthesis by either RNA polymerase or by primase.

However, although most of the genome is single-stranded, one part of it forms a double-stranded hairpin. This region somehow can serve as a promoter for the host cell RNA polymerase, which transcribes a short RNA primer. Transcription also disrupts the hairpin. DNA Pol III can then take over and synthesizes a dsDNA molecule. This dsDNA molecule is known as RFI-replicative form I.

Further replication of RFI does not proceed by means of theta intermediates but by a type of rolling circle replication. The gp2 endonuclease, which is encoded by the phage gene 2, nicks the RFI DNA at a specific site (+ strand origin). Rolling circle replication now occurs with displacement of a single strand. Concatemers are not formed; rather the gp2 endonuclease cleaves a second time after one complete copy has been synthesized. Thus the products of this one round of replication are a ssDNA circular molecule (the displaced strand-ligated into a circle) and a dsDNA RFI molecule. The circular ssDNA molecule can now be duplicated by repeating this entire sequence.

In order to synthesize the ssDNA strands that are to be packaged into the capsid, the displaced ssDNA molecules must be coated with a single strand binding protein-gp5-which is coded by the phage gene 5. These molecules are then packaged into new phage capsids.

Replication of ϕ X 174 phage genome

Phage $\phi \times 174$ consists of a single-stranded circular DNA, known as the plus (+) strand. A complimentary strand, called the minus (-) strand, is synthesized. This action generates the duplex circle. The duplex circle is converted to a covalently closed form, which becomes supercoiled. A protein coded by the phage genome, the A protein nicks the (+) strand of the duplex DNA at a specific site that defines the origin for replication (Fig. 22). After nicking the origin, the A protein remains connected to the 5′ end that it generates, while the 3′ end is extended by DNA polymerase.

The A protein is able to bind to a single-stranded decamer fragment of DNA that surrounds the site of the nick. This suggests that the supercoiling is needed to assist the formation of a single-stranded region that provides the A protein with its binding site. The nick generates a 5′- phosphate end i.e. covalently attached to the A protein, and a 3′-OH end, both

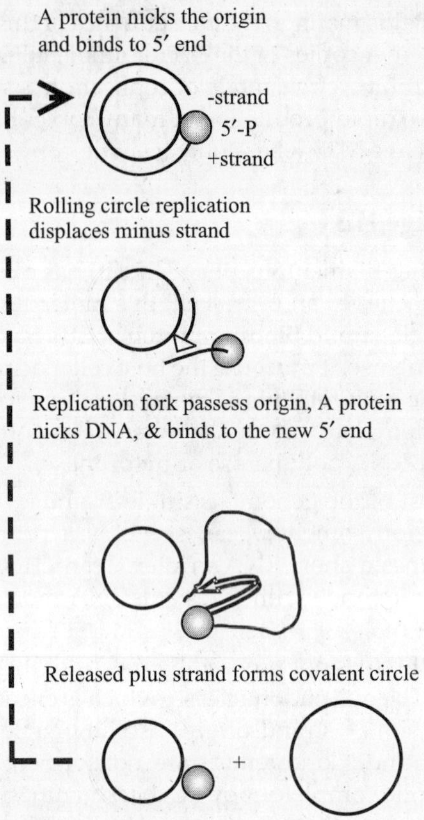

Fig. 22. ϕ X 174 RF DNA is a template for synthesizing single-stranded viral circles. The A protein remains attached to the same genome through indefinite revolutions, each time nicking the origin on the viral (+) strand and transferring to the new 5′ end.

of which have roles to play in ϕ × 174 replication. The 3′–OH end of the nick is extended into a new chain by rolling circle replication. The chain is elongated around the circular (–) strand template, until it reaches the starting point and displaces the origin. Now the A protein function again and remains connected with the rolling circle as well as to the 5′ end of the displaced tail, and it is therefore in the vicinity as the growing point returns past the origin. So the same A protein is available again to recognize the origin and nick it, now attaching to the end generated by the new nick. The cycle can be repeated indefinitely.

After this nicking event, the displacd single (+) strand is freed as a circle (Fig. 22). The A protein is involved in the circularization. The displaced (+) strand may follow either of two fates after circularization:

During the replication phase of viral infection, it may be used as a template to synthesize the complementary (–) strand. The duplex circle may then be used as a rolling circle to generate more progeny.

During phage morphogenesis, the displaced (+) strand is packaged into the phage virion.

D-loop Formation

Mitochondrial DNA is replicated by a process in which leading strand synthesis precedes lagging strand synthesis. The leading strand therefore displaces the lagging strand template to form a displacement or D loop (Fig. 23). The 15-kb circular mitochondrial chromosome of mammals normally contains a single 500-to 600-nt D loop that undergoes frequent cycles of degradation and resynthesis. During replication, the D-loop is extended. When the D loop has reached a point ~ 2/3 of the way around the chromosome, the lagging strand origin is exposed and its synthesis proceeds in the opposite direction around the chromosome. Lagging strand synthesis is therefore only ~ 1/3 complete when leading strand synthesis terminates (Fig. 23).

MECHANISM OF DNA REPLICATION

DNA replication proceeds sequentially and Semi-Discontinuously from one or more Unique, Heritable Origin(s) at one (Unidirectional) or two (Bidirectional) replication Fork(s) in an overall Semiconservative replication mode to a Terminus, followed by Daughter Chromosome separation and Segregation.

DNA replication is the process of duplication of the complete genome that has to precede each cell division. DNA replication is generally divided into three phases. During the initiation phase, the assembly of the prereplication complex leads to the local unwinding of the DNA double helix and the formation of the replication fork. During the elongation phase, the replication fork progresses to synthesize the bulk DNA. Finally, the termination phase, which is not been studied in detail yet, corresponds to the merging of two replication forks.

Genetic studies in the yeasts *S. cerevisiae* and *S. pombe* allowed the identification of factors involved in replication and its regulation. On the other hand, biochemical analysis of DNA replication has largely relied on model systems based on mammalian viruses like Simian virus 40 or other Papilloma viruses. Replication elongation of these viruses depends largely on the replication apparatus of their host cells. For the study of the initiation of DNA replication, *Xenopus* egg extracts have been widely used. In this embryonic model system, any DNA is replicated rapidly without particular sequence requirements.

Origin of Replication

The origin of replication (also called the replication origin) is a unique DNA sequence at which DNA replication is initiated. DNA replication may proceed from this point bidirectionally or unidirectionally. The specific structure of the origin of replication varies somewhat from species to species, but all share some common characteristics. The origin of replication binds a member of the pre-replication complex a protein complex that binds, unwinds, and begins to copy DNA (Fig. 24).

The two types of replication origin are—narrow or broad host range, and high- or low-copy number.

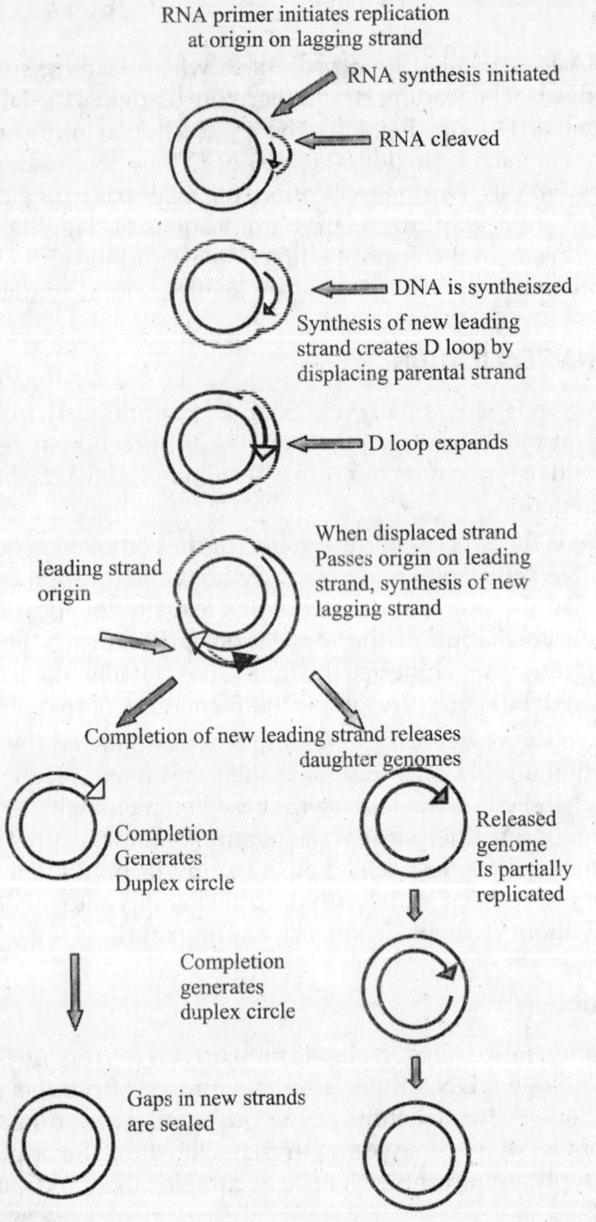

Fig. 23. The D loop maintains an opening in mammalian mitochondrial DNA, which has separates origins for the replication of each strand.

DNA Replication in Prokaryotes

Prokaryotes have a single circular molecule of DNA, and typically only a single origin of replication. Eukaryotes often have multiple origins of replication on each chromosome.

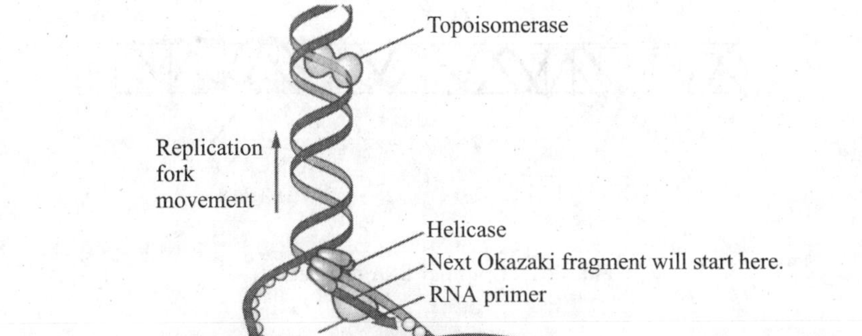

Fig. 24. Pre-replication complex (Snustad and Simmons, 2006).

Having many origins of replication helps to speed the duplication of their (usually) much larger store of genetic material. The segment of DNA that is copied starting from each unique replication origin is called a replicon. Origins of replication are typically assigned names containing "*ori*". For example, the *E. coli* replication origin is known as *ori*C.

Initiation of DNA replication

Initiation of replication at oriC in vitro starts with formation of a complex that requires six proteins: DnaA, DnaB, DnaC, HU, gyrase, and SSB. Of the six proteins involved in prepriming, DnaA is the only one protein uniquely involved in initiation vis-a-vis elongation. The first stage in complex formation is binding to oriC by DnaA protein.

The reaction involves action at two types of sequences: 9bp and 13bp repeats. Together the 9bp and 13bp repeats define the limits of the 245 bp minimal origin (Fig. 25). An origin is activated by the sequence of events (Fig. 25), in which binding of DnaA is succeeded by association with the other proteins.

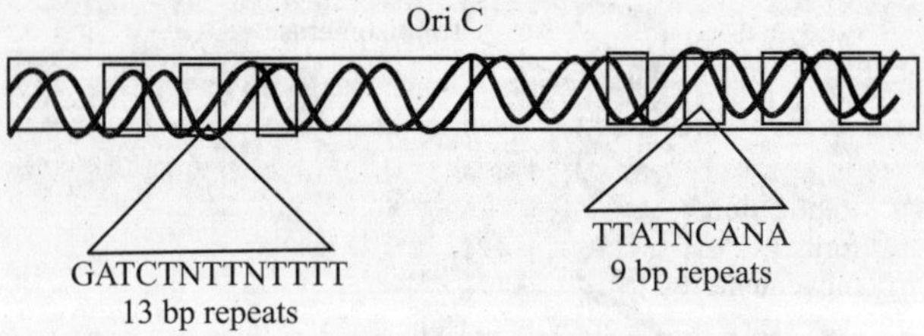

Fig. 25. Origin of DNA replication.

The four 9bp consensus sequences on the right side of oriC provide the initial binding sites for DnaA. It binds cooperatively to form a central core around which oriC DNA is wrapped. Then DnaA acts at three A-T rich 13 bp tandem repeats located in the left side of oriC. In the presence of ATP, DnaA melts the DNA strands at each of thee sites to form an open complex. All three 13bp repeats must be opened for the reaction to proceed to the next stage.

Altogether, 2-4 monomers of DnaA bind at the origin, and they recruit 2 "prepriming" complexes of DnaB-DnaC to bind, so that there is one for each of the two (bidirectional) replication forks. Each DnaB- DnaC complex consists of 6 DnaC monomers bound to a hexamer of DnaB. Each DnaB-DnaC complex transfers a hexamer of DnaB to an opposite strand of DNA. DnaC hydrolyzes ATP in order to release DnaB (Fig. 26). The region of strand separation in the open complex is large enough for both hexamers to bind, initiating the two replication forks. As DnaB binds, it displaces DnaA from the 13bp repeats, and extends the length of the open region. Then it uses its helicase activity to extend the region of unwinding. Each DnaB activates a DnaG primase, in one case to initiate the leading strand, and in the other to initiate the first okazaki fragment of the lagging strand.

To support the unwinding reaction two proteins are required: gyrase and SSB proteins. Gyrase provides a swivel that allows one strand to rotate around the other; without this reaction, unwinding would generate torsional strain in the DNA. This swivel action of gyrase requires ATP hydrolysis. The protein SSB stabilizes the single-stranded DNA as it is formed. The length of duplex DNA that usually is unwound to initiate replication is probably < 60 bp. The protein HU is a general DNA-binding protein in E.coli. Its presence is not absolutely required to initiate replication *in vitro*, but it stimulates the reaction. HU has the capacity to bend DNA, and is involved in building the structure that leads to formation of the open complex.

New DNA is Synthesised by DNA Polymerase

DNA polymerase requires a PRIMER

Priming activity is required to provide 3′-end to start off the DNA chains on the both leading and lagging strands. The leading strands requires only one such initiation event which occur at origin. The lagging strand requires a series of initiation events, since each okazaki fragment requires its own to start de novo. Each okazaki fragment start with a primer sequence of RNA ~ 10 bases long, which provides the 3′ end for the extension of DNA polymers.

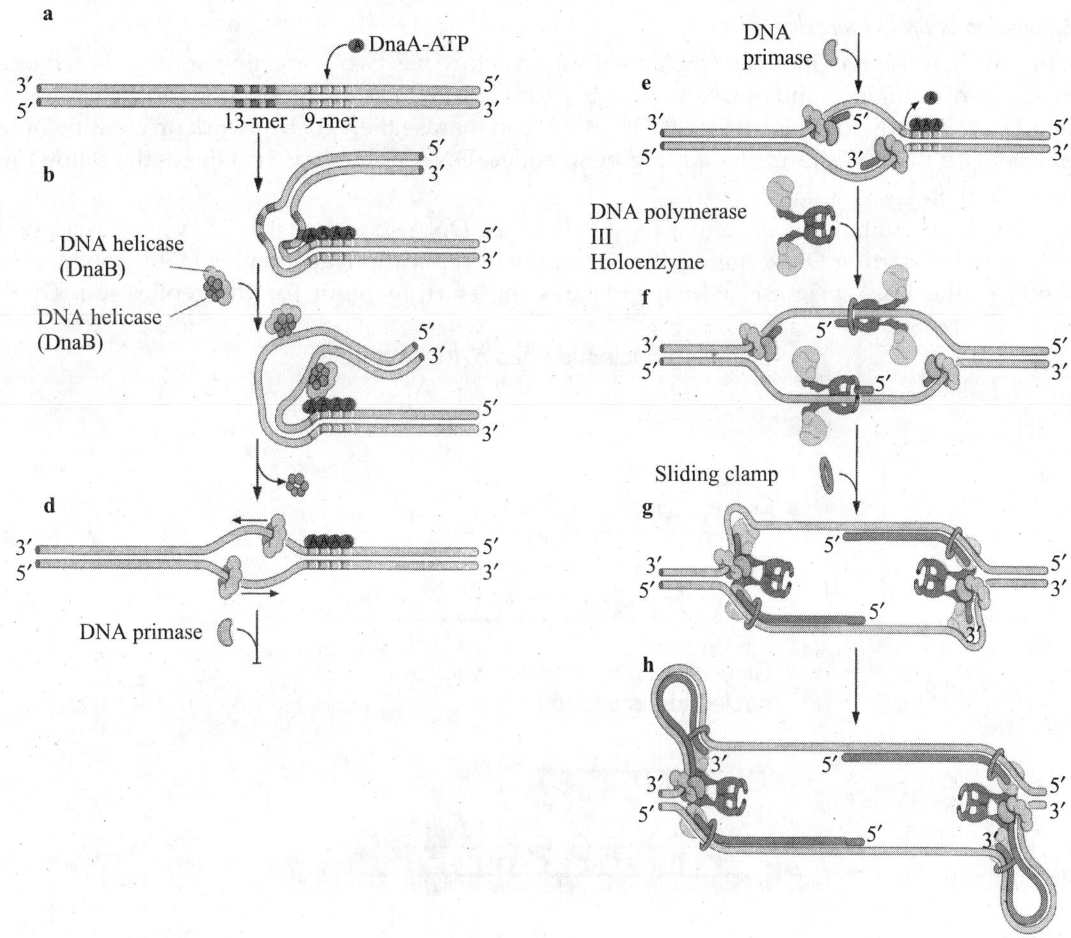

Fig. 26. Initiation of DNA replication (Watson *et al.*, 2004).

DnaG primase i.e. a special RNA polymerase associates with the replication complex, and typically synthesizes an 11-12 base primer. Primers start with the sequence pppAG, opposite the sequence 3′-GTC- 5′ in the template. There are two types of priming reaction in *E. coli*.

The oriC system, named for the bacterial origin, basically involves the association of the DnaG primase with the protein complex at the replication fork.

The $\phi \times$ system, named for phage $\phi \times 174$, requires an initiation complex consisting of additional components, called the primosome.

Priming of replication on double-stranded DNA always requires a DnaB helicase, Single-Strand Binding protein, and primase.

Elongation of DNA replication

After the helicase unwinds the DNA resulting structure has two branching "prongs" each one made up of a single strand of DNA. This strusture is known as the replication fork. This shows that DNA is ready to replicate itself. The DNA polymerase then goes to work on creating new partners for the two strands by adding nucleotides. DNA polymerase catalyzes the following reaction (Fig. 27).

Single-strand binding protein is used to hold the DNA strands apart. RNA primase is then bound to the starting DNA site. At the beginning of replication, an enzyme DNA polymerase binds to the RNA primase, which indicates the starting point for the replication. DNA

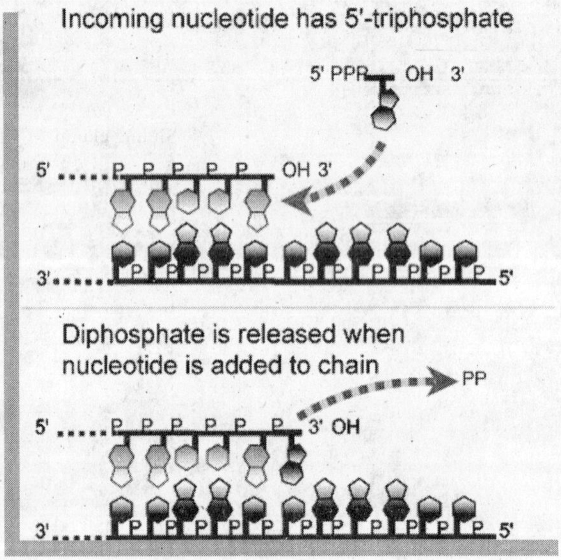

$$(dNMP)_n + dXTP \xrightarrow{Mg^{2+}} (dNMP)_{n+1} + PPi \rightarrow 2Pi$$

Fig. 27. DNA synthesis occurs by adding nucleotides to the 3′-OH end of the growing chain, so that the new chain grows in 5′→3′ direction. The precursor for DNA synthesis is a nucleoside triphosphate, which loses the terminal two phosphate groups in the reaction (Lewin, 2006).

polymerase can only synthesize new DNA from the 5′→3′ (of the new DNA). Because of this, the DNA polymerase can only travel on one side of the original strand without any interruption. This original strand, which goes from 3″→5′, is called the leading strand (Fig. 28). The complement of the leading strand, from 5″→3′, is the lagging strand. In prokaryotes, coupled leading strand and lagging strand synthesis is achieved by the action of the DNA polymerase III holoenzyme.

DNA replication occurs in both directions away from the origin. Both strands are synthesized at each replication fork by DNA polymerase III holoenzyme: one of the α subunit of the holoenzyme synthesizes the leading strand continuously (Fig. 29) in the direction of fork movement while the other α subunit of the holoenzyme synthesizes the lagging strand discontinuously in a piecemeal fashion as Okazaki fragments in the opposite direction of fork movement. Okazaki fragments have about 10 RNA nucleotides at the 5' end (RNA primer) and are about 1000-2000 DNA nucleotides long. The RNA is subsequently removed from Okazaki fragments by DNA polymerase I and the fragments are then joined together by DNA ligase into long, continuous DNA.

Each Okazaki fragment is initiated near the replication fork at a RNA primer created by primase, and extended by DNA polymerase III. The primer is later removed by nucleases such as Ribonuclease H (RNAse H) and flap endonucleases (FENs). In prokaryotes the FEN nuclease is a domain of DNA polymerase I while in eukaryotes FENs are separate enzymes. The excised RNA bases are replaced with DNA by DNA polymerase I in prokaryotes. Adjoining fragments are then linked together by DNA ligase, using phosphodiester bonds, to create a continuous strand of DNA.

Maturation of Okazaki Fragments

DNA polymerase III holoenzyme is released from the lagging strand when it collides into the previously synthesized downstream Okazaki fragment. Neighboring Okazaki fragments are thus separated by a nick. DNA polymerase I now bind to the nick and removes the RNA of the downstream okazaki fragment by the 5″→3′ exonuclease while simultaneously extending the DNA of the upstream okazaki fragment (Fig. 30). This is equivalent to the process of nick translation. The net result is that RNA primer is replaced by DNA. DNA polymerase I only synthesize about 10 to 20 nucleotides before it falls off of the DNA. This is enough to do the job. The remaining nick must be joined together with DNA ligase. *E. coli* DNA ligase uses NAD+ as the source of energy. The net result is that okazaki fragments are joined together as longer, continuous DNA without any RNA sequences.

Termination

Termination occurs when DNA replication forks meet one another or run to the end of a linear DNA molecule. Also, termination may occur when a replication fork is deliberately stopped by

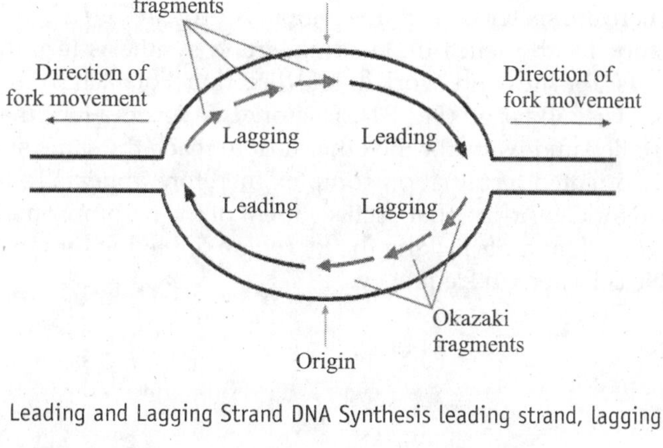

Fig. 28. Leading and Lagging Strand DNA Synthesis leading strand, lagging strand.

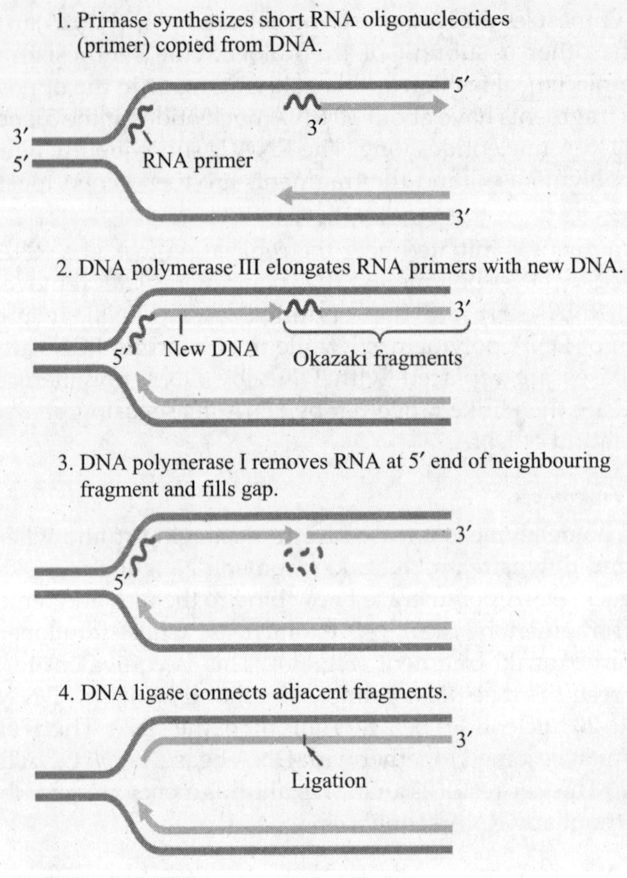

Fig. 29. Steps involves in the synthesis of the lagging strand.

a special protein, called a replication terminator protein, that binds to specific sites on a DNA molecule. The terminus of DNA replication is opposite the origin of replication on the circular *E. coli* chromosome. Ter is a "trap" replication forks enter, but do not leave, this region spanning 450 kb. There are six ter sites in this region. A protein called Tus binds to the ter sites, and this binding stops DnaB (helicase). Leading strand synthesis terminates one nucleotide away from bound Tus. Ter sites only work in one direction. For each round of replication, only one termination site is usually used (Fig. 31). The fork that meets a termination site in its active orientation first is stalled and waits for the other fork to reach the same site. A replication fork halted at a properly oriented termination complex therefore appears to form a barrier to the opposing replication fork, ensuring that each segment of the chromosome is replicated exactly once. Interestingly, *E. coli* mutants lacking the ter region or having mutated Tus are viable, but segregate non-viable cells at high frequency.

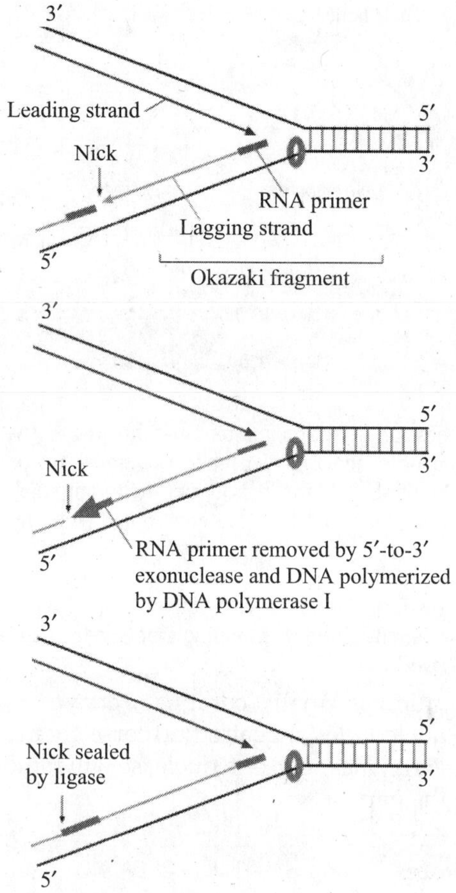

Fig. 30. Maturation of Okazaki Fragments (a) The RNA primer (red) portion of an okazaki fragment is about 10 nucleotides long and the DNA (green) portion is 1,000 to 2,000 nucleotides long (b)DNA polymerase I now bind to the nick and remove the RNA of the downstream okazaki fragment by the 5′→3′ exonuclease while simultaneously extending the DNA (brown) of the upstream Okazaki fragment (C) The remaining nick must be joined together with DNA ligase.

When the polymerase reaches the end of a length of DNA, there is a potential problem due to the antiparallel structure of DNA. Because an RNA primer must be regularly laid down on the lagging strand, the last section of the lagging-strand DNA cannot be replicated because there is no DNA template for the primer to be synthesied on. To solve this problem, the ends of most chromosomes consist of noncoding DNA that contains repeat sequences. The end of a linear chromosome is called the telomere.

The repeat DNA in the telomere is not essential for survival, because it does not contain genes, so cells can endure the shortening of the chromosome at the telomere. Many simple, single-celled organisms overcome the whole problem by having circular chromosomes such as a plasmid.

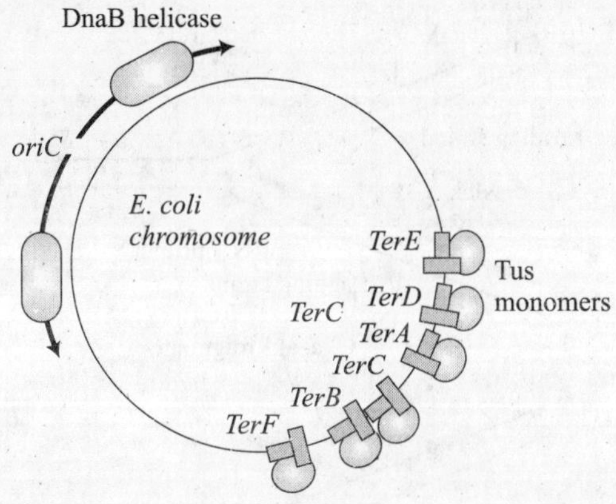

Fig. 31. Organization of the replication initiation and termination sites on the *E. coli* chromosome: The origin of bidirectional replication is labeled oriC, and the two replication forks are represented by the DnaB helicases. Termination sites are labeled TerA through TerF. The locations of the termination sites are not to scale: the sites are located between 23 and 29 min (TerA, TerD, and TerE), 33 and 36 min (TerB and TerC), and at 48 min (TerF); thus these sites are spread over a long distance (1 min is approximately equal to 50 kb). The T shape of the termination site denotes the polarity of the site; replication forks meeting the flat side (top of the T) are arrested (that is, the clockwise fork will pass through sites TerE, TerD, and TerA, but will stop at TerC, TerB, or TerF). A protein called Tus binds to the Ter sites, and this binding stops DnaB (helicase) action.

Before the DNA replication is finally complete, enzymes are used to proofread the sequences to make sure the nucleotides are paired up correctly in a process called DNA repair. If mistake or damage occurs, enzymes such as a nuclease will remove the incorrect DNA. DNA polymerase will then fill in the gap.

DNA Replication in Eukaryotes

Eukaryotic DNA replication is clearly a much more complex process than bacterial DNA replication. Eukaryotic DNA replication is slow compared to *E. coli* DNA replication i.e. only about 75 nucleotides/second. The deepest understanding of eukaryotic DNA replication comes from studying a model replication system using the small DNA tumor virus SV40 (monkey cells are the host for SV40).

Replication origins

Eukaryotic chromosomes have multiple origins of replication. In yeast, these are known as Autonomously Replicating Sequences (ARS). Yeast, which can be considered to be the eukaryotic equivalent of *E. coli*, has approximately 400 ARS elements in 12 chromosomes. Since the origins exist as short, well-defined sequences of about 150 bp. In the fission yeast *Schizosaccharomyces pombe*, however, the origins are more complex, lack a clear consensus sequence and consist of an AT-rich 500-1000 bp region. In mammalian cells the replication origin apparently consists of a large chromosomal region permissive for initiation, within which some regions are more frequently used than others.

All eukaryotes utilize similar proteins and a conserved mechanism for initiating and regulating initiation of DNA replication. DNA synthesis is initiated by the binding of initiator proteins to the origins of replication. In *S. cerevisiae* a multi-subunit protein called the origin recognition complex (ORC) binds specifically to ARS sequences. This complex consists of six proteins, Orc1p-Orc6p, which all are essential for viability and initiation of DNA replication. In *S. cerevisiae*, the ORC complex is bound to origins in all stages of the cell cycle and forms the core of the origin complex to which other components are loaded in a step-wise manner.

Organization of multiple replication sites

The average human chromosome contains 6 billion nucleotide pairs that are copied at about 50 base pairs per second by each replication fork. Yet, in a typical cell the entire replication process takes only about 8 hours. This is because there are many replication origin sites on a eukaryotic chromosome. Therefore, replication can begin at some origins earlier than at others. As replication nears completion, "bubbles" of newly replicated DNA meet and fuse, forming two new molecules.

There must be some form of regulation and organisation of these multiple replication sites to prevent conflict. To date, two replication control mechanisms have been identified: one positive and one negative. For DNA to be replicated, each replication origin site must be bound by a set of proteins called the origin recognition complex. These remain attached to the DNA throughout the replication process. Specific accessory proteins, called licensing factors, must also be present for initiation of replication. Destruction of these proteins after initiation of replication prevents further replication cycles from occurring. This is because licensing factors are only produced when the nuclear membrane of a cell breaks down during mitosis.

DNA replication and cell cycle

One of the safety mechanisms preventing re-replication is restriction of the pre-RC assembly to G_1-phase. In *S. cerevisiae*, Cdc6, the loading factor of MCMs is rapidly degraded at the onset of S-phase by CDK Cdc28. An additional mechanism is removal of the MCM proteins from chromatin during S-phase. In budding yeast this is at least partly accomplished by controlling the sub nuclear localization of MCM proteins. A recent report shows that MCM4, a key initiation factor, is transported into the nucleus only in late mitosis and G_1, the period of pre-RC assembly, and after S-phase, free MCM4, which is not bound to chromatin, is excluded from the nucleus. This process is strictly controlled by CDKs, as MCM4 is found in the nucleus only when CDKs are inactive. In human cells, however, the pre-RC assembly seems to be regulated not by Cdc6 proteolysis, but rather by controlling Cdc6 localization, as it is nuclear in S-phase and cytoplasmic in G_1-phase.

Assembly of replicative complexes onto origins

The binding of Cdc6 protein to the ORC is essential for the next step in initiation, which is loading of MCM (mini chromosome maintenance) proteins onto origins. A complex of ORC, Cdc6 and MCMs forms a pre-replicative complex called the pre-RC (Fig. 32), which is established at the end of mitosis, after separation of sister chromatids. Recently a novel loading factor for MCM proteins, Cdt1, was identified in both *S. pombe* and *Xenopus*, which appears to function coordinately with Cdc6. At the onset of S-phase, the pre-RC has to be converted into an initiation complex that leads to initiation of DNA synthesis. This activation of pre-RC is accomplished by the action of S-phase specific cyclin-dependent kinases (CDKs) and Dbf4-dependent kinases (DDKs), which activate the firing of replication origins.

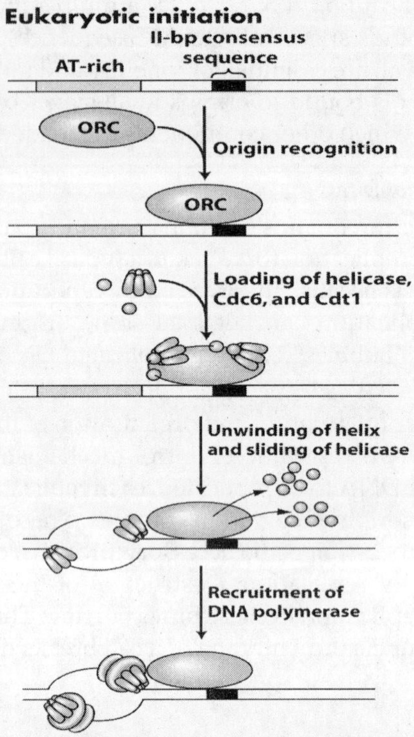

Fig. 32(a). Eukaryotic Initiation.

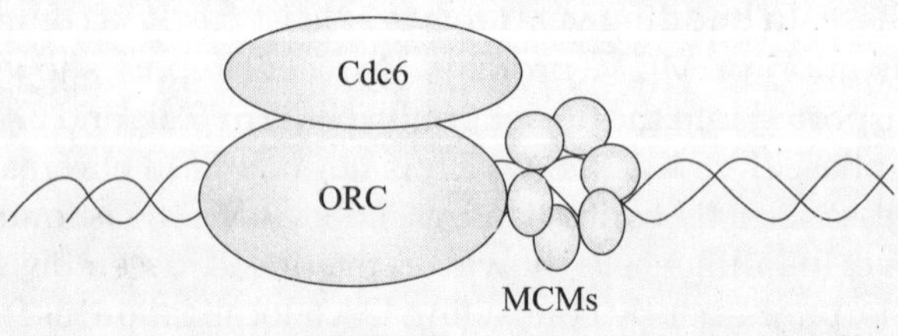

Fig. 32. Model for pre-RC assembly (Snustad and Simmons, 2006).

In the conversion process, Cdc6 and MCM proteins are released from the pre-RC complex and DNA polymerases are recruited onto origins to assemble a replicative complex. This assembly process takes place in a step-wise manner and the loading of Cdc45 is required for the loading of other replication proteins, such as the eukaryotic single-stranded DNA binding protein RPA. The ability of RPA to bind single-stranded DNA is essential for the first step of DNA replication i.e. unwinding of DNA at the origin of DNA replication.

RPA binding precedes loading of Pol α-primase which binds only, after the unwinding of origin. Recruitment of DNA polymerases onto origins is also controlled by the Dpb11 protein, which apparently associates with the origins after the pre-RC assembly, since loading of Dpb11 is dependent on MCM and RPA. Loading of Pol α and Pol ε onto replication origins is

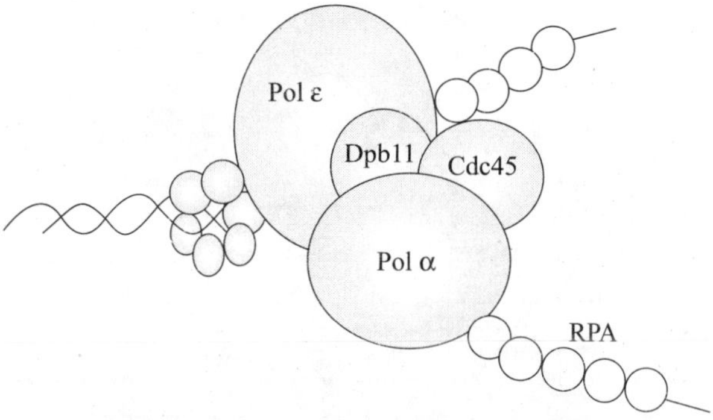

Fig. 33. Hypothetical model of the eukaryotic initiation complex. Origin unwinding by the putative MCM helicase allows recruitment of Cdc45, Dpb11 and DNA polymerases.

dependent on Dpb11, suggesting that the association of Dpb11 with origins is a prerequisite for the assembly of a replicative complex in order to initiate cellular DNA replication (Fig. 33).

The basic mechanism of eukaryotic DNA replication

Replication of the SV40 DNA has served as an excellent model system for eukaryotic DNA replication, because it requires only one viral protein, the large tumor antigen (T-antigen), while all other components must be of cellular origin. The T-antigen has a dual role and it functions both as an initiator protein that binds to the origin of replication, and as a helicase that unwinds duplex DNA. This unwinding is stimulated by the presence of the single-stranded DNA binding protein RPA and a single-stranded DNA substrate for the priming of DNA synthesis is produced.

In the following step, the T-antigen-RPA complex binds the DNA polymerase α-primase, which synthesizes a short RNA-DNA primer at the origin of replication. Further elongation of these initial primers by Pol α is inhibited by recruitment of RFC and PCNA at the primer terminus, which allows loading of Pol Δ, and the nascent DNA strands are extended to form the continuously synthesized leading strand. RPA plays a role in the polymerase switch, and a recent biochemical study shows that the clamp loader protein RFC also binds RPA and competes with Pol α-RPA interaction, thereby releasing Pol α and allowing PCNA assembly onto the primer terminus.

Pol α continues to prime and synthesize DNA discontinuously on the lagging strand template, but it does not synthesize full-length Okazaki fragments. In the current model based on the reconstituted SV40 DNA replication system, Pol α-primase is replaced by the processive RFC, PCNA and Pol δ complex, which completes the lagging strand synthesis. This polymerase switch is comprehensible, as Pol δ is a non-processive polymerase and does not possess a 3′ to 5′ exonuclease activity. The absence of processivity and proofreading activity would indeed make it a poor leading and/or lagging strand polymerase (Fig. 34). RNA primers are removed by RNase H and a 5′-3′exonuclease FEN1 and after gap-filling synthesis by Pol δ, DNA ligase I joins the ends of the Okazaki fragments.

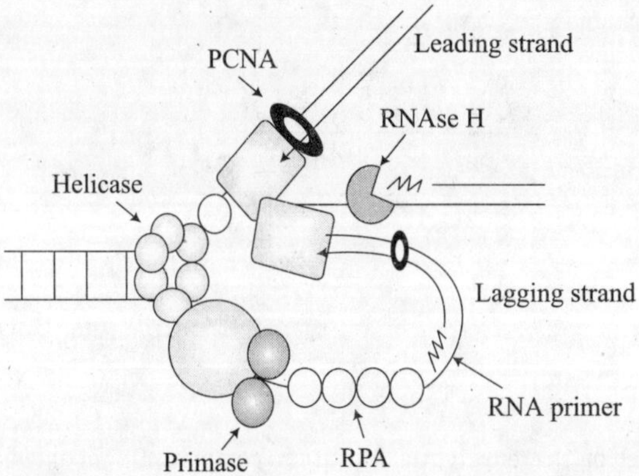

Fig. 34. Structural organization of the eukaryotic DNA replication fork based on the SV40 *in vitro* DNA replication model. Pol D is depicted as a dimer according to observations from budding yeast.

The Ends of Eukaryotic Chromosomes (Telomeres)

Eukaryotic chromosomes are linear DNA molecules. The ends (telomeres) are specialized sequences involved in chromosome stability. Telomeric DNA consists of tandem repeats of simple sequences. These sequences are usually short and rich in G residues on one strand. For example, the ciliate Tetrahymena contains tandem TTGGGG repeats, whereas humans and other mammals contain TTAGGG repeats. The orientation of the telomere sequences is also conserved; the G-rich strand runs 5' to 3' toward the end of the chromosome and thus makes up the molecular 3' end of the chromosome. In most organisms the number of repeats on any given end is not fixed. The tract length of these repeated sequences ranges from 38 base pairs in ciliates, like Oxytricha, up to tens of kilobase pairs in mammalian cells, each organism having a characteristic mean length. The length of the simple telomere repeat tracts is maintained by the enzyme telomerase. Telomerase is a specialized DNA polymerase that adds telomeric sequences onto chromosome ends. This de novo addition of sequences balances the loss of repeats during each round of replication of linear DNA molecules. DNA polymerases cannot synthesize DNA on template sequences at the very 3' end of linear chromosomes as shown in (Fig. 35).

Replication of telomers

Telomeres are extended by telomerases; Telomerase is an enzyme, discovered by Elizabeth Helen Blackburn and Carol Greider, that adds specific DNA sequence repeats, ("TTAGGG" in all vertebrates) to the 3' end of DNA strands, in the telomere regions at the ends of chromosomes which contain condensed DNA material during replication. The enzyme is a reverse transcriptase that carries its own RNA template; this RNA is used as a template for eukaryotic DNA replication (Fig. 36).

Telomer shortening

Telomer shorten because of the lagging strand phenomenon that is exhibited during DNA replication in eukaryotes only. Because DNA replication does not begin at either end of the DNA strand, but starts in the centre, and considering that all DNA polymerases that have been

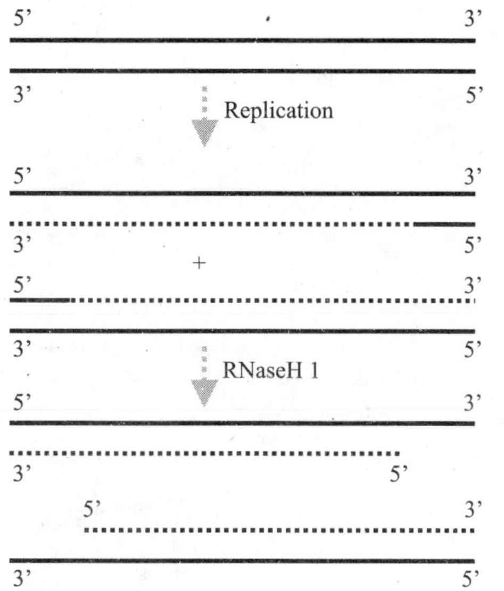

Fig. 35. Replication of a linear DNA molecule: After removal of the 5'-end-most RNA primers from newly synthesized strands, the new daughter strands will be shorter. Further rounds of DNA replication will further shorten daughter molecules.

discovered move from the 3' to 5' direction (polymerizing in the 5''→3' direction) one finds, on the DNA molecule being replicated, a leading and lagging strand.

On the leading strand, DNA polymerase can make a complementary DNA strand without any hurdles because it goes from 3' to 5'. On the other hand, there is a problem when they are expected to move from the 5' to 3' direction in the lagging strand. To counter this, short sequences of RNA acting as primers attach to the lagging strand a little way ahead of where the initiation site was. The DNA polymerase can start replication at that point and go to the end of the initiation site. This causes the formation of Okazaki Fragments. More RNA primers attach further on the DNA strand and DNA polymerase comes along and continues to make a new DNA strand.

Eventually, when the last RNA attaches, and DNA Polymerase and DNA ligase come along to convert the RNA (of the primers) to DNA, and seal the gaps in between the Okazaki fragments. But, in order to change RNA to DNA, they have to have another DNA strand in front of the RNA primer. This happens at all the sites of the lagging strand but, it doesn't happen at the end where the last RNA primer is attached. Ultimately, that RNA is destroyed by enzymes that degrade RNA on the DNA if it is left there. Thus, because they are the ones at the end, a section of telomeres is lost during each cycle of replication.

Structure of telomarase

Human telomerase is composed of at least two sub-units, human Telomerase Reverse Transcriptase (hTERT) and human Telomerase RNA (hTR or hTERC). These two subunits are coded for by two different genes in the genome. The coding region of the hTERT gene is 3396bp, and translates to a protein of 1131 amino acids. The polypeptide folds with the hTR

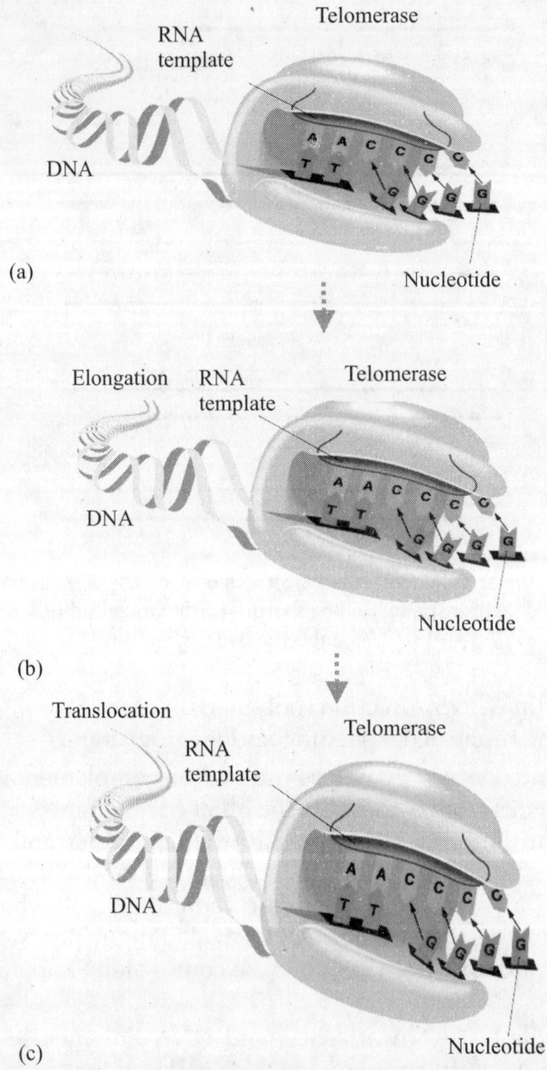

Fig. 36. Telomerase enzyme extending telomere (a) Telomerase enzyme (b) elongation (c) Translocation (Greider and Blackhum, 1996).

subunit (451 nucleotides long), which is not translated and remains as RNA. hTERT has a 'mitten' structure that allows it to wrap around the chromosome to add single-stranded telomere repeats.

hTERT is a reverse transcriptase, which is a class of enzymes that creates single stranded DNA using single stranded RNA as a template. Enzymes of this class (but not hTERT specifically) are utilized by scientists in the molecular biological process of Reverse Transcriptase PCR (RT-PCR), which allows the creation of several DNA copies of a target sequence using RNA as a template.

For example, the HIV virus uses a reverse transcriptase to make DNA copies of its RNA genome for insertion into human cell chromosomes. hTERT carries its own template around, hTR. Using this template, it can add a six nucleotide repeating sequence, 5'-TTAGGG (in humans and all vertebrates...the sequence differs in other organisms) to the 3' strand of chromosomes. This repeating TTAGGG sequence is called the telomere. The template region of hTR is 3'-CAAUCCCAAUC-5'. This way, telomerase can bind the first few nucleotides of the template to the last telomere sequence on the chromosome, add a new telomere repeat (5'-GGTTAG-3') sequence, let go, realign the new 3'-end of telomere to the template, and repeat the process.

Speed of Replication

Prokaryotes

The single molecule of DNA that is the *E. coli* genome contains 4.7 x 10^6 nucleotide pairs. DNA replication begins at a single, fixed location in this molecule, the **replication origin**, proceeds at about 1000 nucleotides per second, and thus is done in no more than 40 minutes. And thanks to the precision of the process (which includes a "proof-reading" function), the job is done with only about one incorrect nucleotide for every 10^9 nucleotides inserted. In other words, more often than not, the *E. coli* genome (4.7 x 10^6) is copied without error!

Eukaryotes

The average human chromosome contains 150 x 10^6 nucleotide pairs which are copied at about 50 base pairs per second. The process would take a month (rather than the hour it actually does) but for the fact that there are many places on the eukaryotic chromosome where replication can begin. Replication begins at some replication origins earlier in S phase than at others, but the process is completed for all by the end of S phase. As replication nears completion, "bubbles" of newly replicated DNA meet and fuse, finally forming two new molecules.

Endoreplication

This is the replication of DNA during the S phase of the cell cycle without the subsequent completion of mitosis and/or cytokinesis. (Endoreplication is also known as endoreduplication.) Endoreplication occurs in certain types of cells in both animals and plants. There are several variations e.g.,

(a) Replication of DNA with completion of mitosis but **no cytokinesis**

(b) Repeated replication of DNA without forming new nuclei in telophase, this can result in:

Polyploidy: the replicated chromosomes retain their individual identity.

Polyteny: the replicated chromosomes remain in precise alignment forming "giant" chromosomes.

Various intermediate conditions have also been observed between the variations as mentioned in a and b.

REFERENCES

Bell and Dutta (2002) Annu. Rev. Biochem, 71: 333–374.

Barry, E.R., and Bell, S.D. (2006) Microbiology and Molecular Biology Reviews : MMBR, DNA replication in the archaea, 70: 876-887.

Beese, L. et al. (1993) Biochemistry, Crystal Structure of the Klenow Fragment Bound to Duplex DNA, Science 260.

Beese, *L.* et al. (1993) Biochemistry, Crystal Structure of the Klenow Fragment of DNA Pol I, 32.

Berg, J. M., Tymoczko, J.L. and stryer, L. (2006) Biochemistry, (6th edition), Macmillan Publishers Ltd., USA.

Greider and Blackbum (1996) Scientific American, pp. 92-97.

Kelman, L.M., and Kelman, Z. (2003) Molecular Microbiology, Archaea: An archetype for replication initiation studies? 48: 605-615.

Lewin, B. (2004) Gene VIII, International Edition, Pearson Prentice Hall, UK.

Lewin, B. (2006) Gene IX, Jones and Barlett Poblishers, USA.

Russell, P.J. (2002) Genetics, Benjamin Cummings, San Francisco Press, USA.

Snyder and Champness (2003) Molecular Genetics of Bacteria, (2nd ed.) ASM Press, USA

Snustad and Simmons (2006) Principles of Genetics, (4th ed.), Wiley Asia Student Editions.

Voet and Voet (2004) Biochemistry, (3rd ed.), Wiley International Edition, USA.

Watson, Baker, Bell, Gann, Levine, Losick. (2004) Molecular Biology of the Gene, (5th ed.), Pearson/ Benjamin Cummings Publishers, UK.

Weem, Minka and Peeters (2001) International Baccalaureate Biology, (2nd ed.), IBID Press, Australia.

Suggested Readings

1. Barry, E. R., and Bell, S. D. (2006) Microbiology and molecular biology reviews : MMBR, DNA replication in the archaea, 70(4):876-887.
2. Beese L. et al. (1993) Biochemistry, Crystal Structure of the Klenow Fragment of DNA Pol I, 32.
3. Beese L. et al. (1993) The Structure of DNA Polymerase I Klenow Fragment Bound to Duplex DNA, Science 260.
4. Bell and Dutta (2002) Annu. Rev. Biochem, 71:333–74.
5. Berg, J. M., Tymoczko, J. L. and Stryer L. (2006) Biochemistry, 6th edition, Macmillan Publishers Ltd.
6. Kelman, L. M. and Kelman, Z. (2003) Molecular microbiology, Archaea: An archetype for replication initiation studies, 48(3), 605-6152.
7. Lewin B. (2004) Genes VIII, International Edition, Pearson Prentice Hall.
8. Peeters W.M. (2001) International Baccalaureate Biology, Second Edition, IBID Press, Australia.
9. Russell, P. J. (2002) iGenetics, Benjamin Cummings, San Francisco Press.
10. Simmons S. (2006) Principles of Genetics, Fourth Edition, Wiley Asia Student edition.
11. Snyder and Champness (2003) Molecular Genetics of Bacteria, Second Edition, ASM Press.
12. Voet and Voet (2004) Biochemistry, Third Edition, Wiley International Edition.
13. Watson, Baker, Bell, Gann, Levine, Losick. (2004) Molecular Biology of the Gene, Fifth Edition, Pearson/Benjamin Cummings Publishing.

REVIEW QUESTIONS

Long Questions

1. A culture of bacteria is grown for many generations in a medium in which the only available nitrogen in the heavy isotope (^{15}N). The culture is then switched to a medium containing only ^{14}N for one generation of growth; it is then returned to a ^{15}N-containing medium for one final generation of growth. If the DNA from these bacteria is isolated and centrifuged to equilibrium in a CsCl density gradient, how would you predict the DNA to band in the gradient?
2. Arrange the following enzymes in the order of their action during DNA replication in *E. coli* (1)DNA polymerase I, (2) DNA polymerase III, (3) DNA primase, (4) DNA gyrase, and (5) DNA helicase.
3. Seven distinct DNA polymerase have been characterized in mammals. What are the intracellular locations and functions of these polymerases?
4. Why must each of the giant DNA molecules in eukaryotic chromosomes contain multiple origins of replication?
5. (a) Why isn't DNA primase activity required to initiate rolling-circle replication? (b) DNA primase is required for the discontinuous synthesis of the lagging strand, which occurs on the single-stranded tail of the rolling circle. Why?
6. DNA polymerase I is needed to remove RNA primers during chromosome replication in *E. coli*. However, DNA polymerase III is the true replicase in *E. coli*. Why doesn't DNA polymerase III remove the RNA primers?
7. In *E. coli*, three different proteins are required to unwind the parental double helix and keep the unwound strands in an extended template form. What are these proteins, and what are their respective functions?
8. How similar are the structures of DNA polymerase I and DNA polymerase III in *E. coli?* What is the structure of the DNA polymerase III holoenzyme? What is the function of the *dnaN* gene product in *E. coli?*
9. The *dnaA* gene product of *E. coli* is required for the initiation of DNA synthesis at *oriC*. What is its function? How do we know that the DnaA protein is essential to the initiation process?
10. In what ways does chromosomal DNA replication in eukaryotes differ from DNA replication in prokaryotes?

Multiple Choice Questions

1. The number of replicons in a typical mammalian cell is
   (a) 40-200 (b) 400
   (c) 1000-2000 (d) 50000-100000
2. Which of the following activities is not needed for replication of the *E. coli* chromosome?
   (a) Telomerase (b) Ligase
   (c) Primase (d) Polymerase

3. All of the following statements about Type I topoisomerase (Topo I) are true except:
   (a) Topo I removes DNA supercoils in an ATP-dependent reaction
   (b) Topo I from eukaryotic cells can remove both positive and negative supercoils
   (c) Topo I from *E. coli* can remove only negative supercoils
   (d) Topo I is essential for viability in *E. coli*
4. Meselson-Stahl experiment established that:
   (a) DNA polymerase has a crucial role in DNA synthesis.
   (b) DNA synthesis in *E. coli* proceeds by a conservative mechanism.
   (c) DNA synthesis in *E. coli* proceeds by a semiconservative mechanism.
   (d) DNA synthesis requires dATP, dCTP, dGTP, and dTTP.
   (e) Newly synthesized DNA in *E. coli* has a different base composition than the preexisting DNA
5. An Okazaki fragment is a:
   (a) Fragment of DNA resulting from endonuclease action
   (b) Fragment of RNA that is a subunit of the 30S ribosome
   (c) Piece of DNA that is synthesized in the 3' '!5' direction
   (d) Segment of DNA that is an intermediate in the synthesis of the lagging strand.
   (e) Segment of mRNA synthesized by RNA polymerase
6. *E. coli* DNA polymerase III:
   (a) can initiate replication without a primer.
   (b) is efficient at nick translation.
   (c) is the principal DNA polymerase in chromosomal DNA replication.
   (d) represents over 90% of the DNA polymerase activity in *E. coli* cells.
   (e) requires a free 5'-hydroxyl group as a primer.
7. The 5'→3' exonuclease activity of *E. coli* DNA polymerase I is involved in:
   (a) formation of a nick at the DNA replication origin.
   (b) formation of Okazaki fragments.
   (c) proofreading of the replication process.
   (d) removal of RNA primers by nick translation.
   (e) sealing of nicks by ligase action.
8. In contrast to bacteria, eukaryotic chromosomes need multiple DNA replication origins because:
   (a) Eukaryotic chromosomes cannot usually replicate bidirectionally.
   (b) Eukaryotic genomes are not usually circular, like the bacterial chromosome is.
   (c) The processivity of the eukaryotic DNA polymerase is much less than the bacterial enzyme.
   (d) Their replication rate is much slower, and it would take too long with only a single origin per chromosome.
   (e) They have a variety of DNA polymerases for different purposes, and need a corresponding variety of replication origins.

9. In which process of information transfer would you find at work helicases, topoisomerases, ligases and a primase?
   (a) Replication (b) Transcription
   (c) Translation (d) Reverse transcription
10. Which of the following is used in rolling circle DNA replication but not in normal cellular DNA replication?
   (a) an endonuclease (b) an exonuclease
   (c) primase (d) DNA ligase
11. The two exonuclease activities of DNA polymerase I
   (a) are coupled during the formation of "gapped DNA".
   (b) degrade DNA in a 5'→3' direction.
   (c) degrade DNA in a 3'→5' direction.
   (d) occur at two different active sites.
12. A principal difference between prokaryotic and eukaryotic DNA replication is
   (a) Completely different proteins/enzymes in eukaryotes
   (b) multiple origins in eukaryotes
   (c) no requirement for topoisomerase activity in prokaryotes
   (d) the absence of a nucleus in prokaryotes

Answers

1. (d)	2. (a)	3. (a)	4. (c)	5. (d)	6. (c)
7. (c)	8. (b)	9. (a)	10. (a)	11. (a)	12. (b)

CHAPTER 14

Transcription: The RNA Synthesis

Y.K. Jaiswal* **and V. Agrawal**

School of Studies in Biochemistry, Jiwaji University, Gwalior–474 011

E-mail: jaiswalyk@gmail.com

Transcription is a process in which one DNA strand is used as a template to synthesize a complementary RNA. In other words, it is an enzymatic process that reads DNA and transcribes it into its complementary RNA sequence. In a prokaryotic cell, transcription and translation are coupled; that is, translation begins while the mRNA is still being synthesized. Because there is no nucleus to separate the processes of transcription and translation, when bacterial genes are transcribed, their transcripts can immediately be translated (Fig. 1 (a)).

In a eukaryotic cell, transcription occurs in the nucleus, and translation occurs in the cytoplasm. Transcription and translation are spatially and temporally separated in eukaryotic cells; i.e., transcription occurs in the nucleus to produce a pre-mRNA molecule. The pre-mRNA is typically processed to produce the mature mRNA, which leaves the nucleus and is translated in the cytoplasm (Fig. 1 (b)).

The DNA strand which serves as the template may be called "template strand, minus strand" or "antisense strand". The other DNA strand may be termed "non-template strand", "coding strand", "plus strand", or "sense strand". Since both DNA coding strand and RNA strand are complementary to the template strand, they have the same sequences except that T in the DNA coding strand is replaced by U in the RNA strand. The enzymes responsible for transcription in both prokaryotic and eukaryotic cells are called DNA-dependent RNA polymerase, or simply RNA polymerases. These enzymes are able to incorporate nucleotides, one at a time, into a strand of RNA (Fig. 2).

*Corresponding author

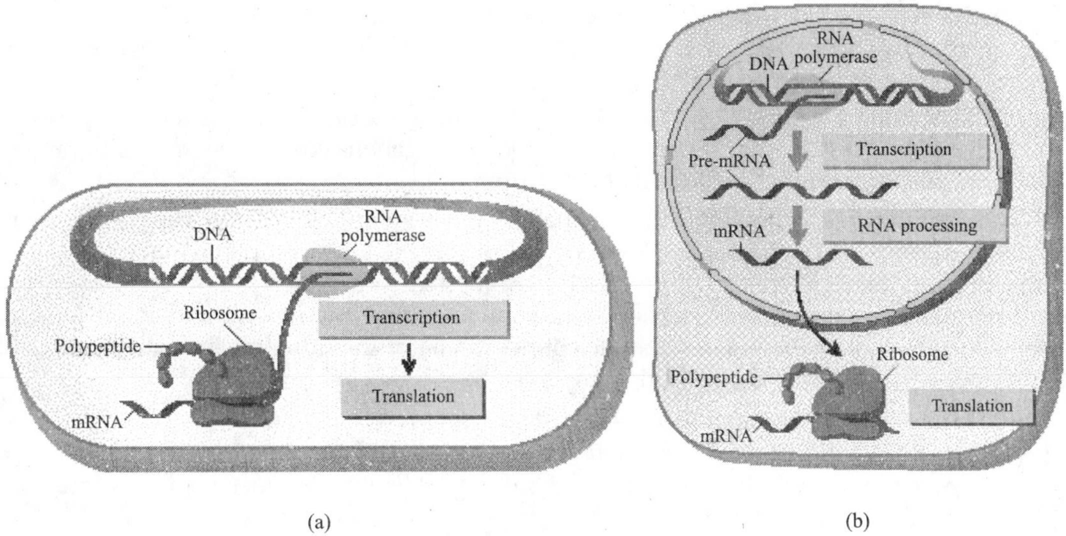

Fig. 1. Transcription in (a) Prokaryotic and (b) eukaryotic cell (Voet *et al.*, 2004).

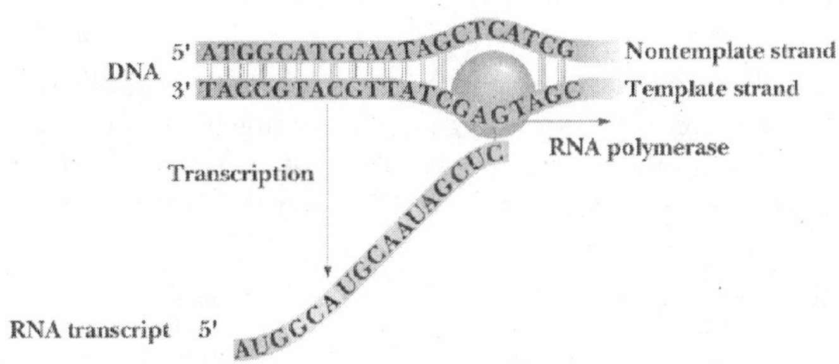

Fig. 2. Template and non template strands of DNA are defined by their function in transcription. The RNA transcript is synthesized on the template strand and is identical in sequence (with U in place of T) to the nontemplate strand, or coding strand.

The entire process of transcription involves the following essential steps:

(1) Binding of polymerases to the initiation site – The DNA sequence which signals the initiation of transcription is called the promoter. The prokaryotic polymerases can recognize the promoter and bind to it directly; however, the eukaryotic polymerases require the help of additional proteins called transcription factors. In addition to providing a binding site for the polymerase, the promoter contains the information that determines which of the two DNA strands is transcribed and the site at which transcription begins.

(2) Unwinding (melting) of the DNA double helix – The enzyme which can unwind the double helix is called helicase. Prokaryotic polymerases possess the helicase property, but eukaryotic polymerases do not. The unwinding of eukaryotic DNA is carried out by a specific transcription factor.

(3) Synthesis of RNA strand – It is dictated by the nucleotide sequence of the template strand of DNA. RNA polymerases use nucleoside triphosphates (NTPs) as the precursor to construct a RNA strand.

(4) Termination of synthesis – Prokaryotes and eukaryotes use different signals to terminate transcription. The "stop" codon in the genetic code is a signal for the end of peptide synthesis, not the end of transcription.

Transcription is chemically and enzymatically, very similar to DNA replication. Both involve enzymes that synthesize a new strand of nucleic acid complementary to a DNA template strand. However, some important differences exist between the two processes. The most notable difference in the case of transcription is that the new strand is made from ribonucleotides rather than deoxyribonucleotides. Some of the other mechanistic features of transcription that differ from replication are given below:

(a) RNA polymerase does not need a primer; rather it can initiate transcription de novo.

(b) The RNA product does not remain base paired to the template DNA strand - rather, the enzyme displaces the growing chain only a few nucleotides behind where each ribonucleotide is added (Fig. 2). This displacement is critical for the RNA to be (as is typically the case) translated to produce its protein product. Moreover, this displacement permits multiple RNA polymerase molecules to transcribe the same gene at a time, each following closely along behind another. Therefore, a cell can synthesize large numbers of transcripts from a single gene in a short time.

(c) Transcription, though very accurate, is less accurate than replication (i.e., one mistake occurs in 10,000 nucleotides added, compared to one in 10,000,000 for replication). This difference reflects the lack of extensive proofreading mechanisms for transcription, although two forms of proofreading for RNA synthesis do exist.

RNA POLYMERASES

In 1955, Marianne Grunberg-Manago and Severo Ochoa reported the isolation of an enzyme that catalyzed the synthesis of RNA. Later it was proved that Grunberg-Manago and Ochoa had isolated the enzyme polynucleotide phosphorylase which usually catalyzes the breakdown of RNA and not its synthesis.

The real E. coli RNA polymerase:

In 1960, the true enzyme was identified by Sam Weiss, Jerald Hurwitz, A. Stevens and J. Bonner. This enzyme requires a template, Mg^{++} and all four rNTPs as substrates to synthesize a product with a composition similar to that of the template. The principle enzyme responsible for RNA synthesis is RNA polymerases, i.e. the enzymes which are capable of copying the DNA so that a complimentary RNA copy is produced by the polymerization of ribonucleotides. In prokaryotes a single RNA polymerase enzyme is responsible for the transcription of DNA into RNA while in eukaryotes three RNA polymerase enzymes exist. RNA polymerase differs in their structure in prokaryotes and eukaryotes both. Transcription in eukaryotes is much more complicated than in prokaryotes, partly because eukaryotic DNA is associated with histones, which could hinder the access of polymerases to the promoter.

Bacterial RNA Polymerases

E. coli contains one RNA polymerase, which transcribes all three major types of RNA. It is made up of core enzyme and sigma subunit, which together forms the holoenzyme. This holoenzyme has many functions e.g., to scan DNA and identify promoters; to bind to promoters; to initiate transcription; to elongate the RNA chain; to terminate transcription; to be responsive to regulatory proteins (activators and repressors) etc.

Core Enzyme

Bacterial RNA core polymerases are multi-subunit complexes composed of five subunits: two a subunits and one for each β, β′, and ω, in the stoichiometry $a_2\beta\beta'\omega$ (390 kD) (Fig. 3). Subunits β (151 kD) and β′ (156 kD) are significantly larger than a (37 kD). Several different forms of a subunits have been identified, with molecular weights ranging from 28 kD to 70 kD.

The β and β′ subunits together make up the catalytic center, channel for DNA lies at the interface of these subunits. The β subunits can cross-linked to the template DNA, the product RNA, and the substrate ribonucleotides.

The α subunit is required for assembly of the core enzyme, promoter recognition and in the interaction of RNA polymerase with some regulatory factors. When phage T4 infects E. coli, the a subunit is modified by ADP-ribosylation of an arginine. The modification is associated with a reduced affinity for the promoters formerly recognized by the holoenzyme, so the a subunit plays a role in promoter recognition.

The ω subunit was for many years considered a curiosity since no function could be ascribed to it and it did not appear too necessary to function. However, it is now known that omega is necessary to restore denatured RNA polymerase *in vitro* to its fully functional form. It may function by binding simultaneously to the N-terminus and C-terminus of the β′ subunit. Core RNA polymerase can bind to DNA and catalyze the synthesis of RNA but it has no specificity. This form of the enzyme will bind quite well to non-specific DNA but it cannot recognize promoters.

Holoenzyme

The term "holoenzyme" refers to a complete and fully functional enzyme this includes the core polymerase and the Sigma (σ) subunit. Only the holoenzyme can initiate transcription. In *E. coli* holoenzyme has a molecular weight of ~ 465kD. The σ subunit performs two major functions i.e., (a) It reduces the affinity of the enzyme for non-specific DNA and (b) It greatly increases the affinity of the enzyme for promoters.

The following table summarizes the subunits of E. coli RNA polymerase and their properties (Table 1).

$$\underset{\text{Holoenzyme}}{\alpha_2\beta\beta'\omega\,\sigma^{70}} \;\underset{\text{Core Polymerase}}{\rightleftharpoons}\; \underset{\text{Sigma - 70}}{\alpha_2\beta\beta'\omega + \sigma^{70}}$$

The sigma subunit

The σ subunit is also known as the σ factor or specificity factor. The σ subunit with the core enzyme ($a_2\beta\beta'$) forms holoenzyme (Fig. 3). It plays an important role in recognizing the transcriptional initiation site, and also possesses the helicase activity to unwind the DNA

Table 1. Subunits of E. coli RNA polymerase and their Functions

Subunit	Size (kD)	Gene	Functions
alpha (α)	37	rpoA	required for assembly of the enzyme; interacts with some regulatory proteins; also involved in catalysis
beta (β)	151	rpoB	involved in catalysis: chain initiation and elongation
beta' (β')	156	rpoC	binds to the DNA template
sigma (σ)	70	rpoD	directs enzyme to the promoter
omega (ω)	11	rpoZ	required to restore denatured RNA polymerase in vitro to its fully functional form

double helix. It directs RNA polymerase to the promoter and ensures that transcription is initiated only where it is supposed to be initiated. The principal sigma factor in *E. coli* is σ^{70} – so called because the protein is 70 kD in size. The corresponding holoenzyme containing this sigma factor is sometimes abbreviated: $E{\cdot}\sigma^{70}$ (Table 2). σ^{32} is required for the expression of heat-shock genes, i.e. genes that are only expressed when the cell is exposed to a high temperature and it must make special proteins in order to survive. Some parts of the sequence of σ^{32} can be recognized in other bacterial sigma factors.

σ^{54} is present all the time and it is required for the expression of many genes that are involved for nitrogen metabolism. σ^{54} has a very different sequence and structure than the other sigma factors in *E. coli* and most bacteria; thus there are at least two different families of sigma factor proteins in bacteria. The importance of σ^{S} has been increasingly recognized in recent years. This sigma factor appears only as cells enter the stationary phase of growth. It is responsible for transcription of all of the genes whose products are required during stationary phase.

Anti-sigma factors

Anti-sigma factors form complexes with their cognate sigma factor, thereby inhibiting its function. One example is FlgM, which is an anti-sigma factor for the flagellar sigma factor σ^{F}. Another is Rsd, which is an anti-σ^{70} factor. It is not present in exponentially growing *E. coli* cells. However, when E. coli enters in stationary phase, Rsd is synthesized and acts to block the activity of σ^{70} thereby allowing σ^{S} to associate with the core RNA polymerase and direct expression of stationary phase genes.

Table 2. *E.coli* sigma factors and their functions

Sigma factor	Gene	Function
σ^{70}	rpoD	principal sigma factor
σ^{54}	rpoN (ntrA, glnF)	nitrogen-regulated gene transcription
σ^{32}	rpoH	heat-shock gene transcription
σ^{S}	rpoS	gene expression in stationary phase cells
σ^{F}	rpoF	expression of flagellar operons
σ^{E}	rpoE	involved in heat shock and oxidative stress responses; regulates expression of extracytoplasmic proteins
σ^{FecI}	fecI	regulates the fec genes for iron dicitrate transport

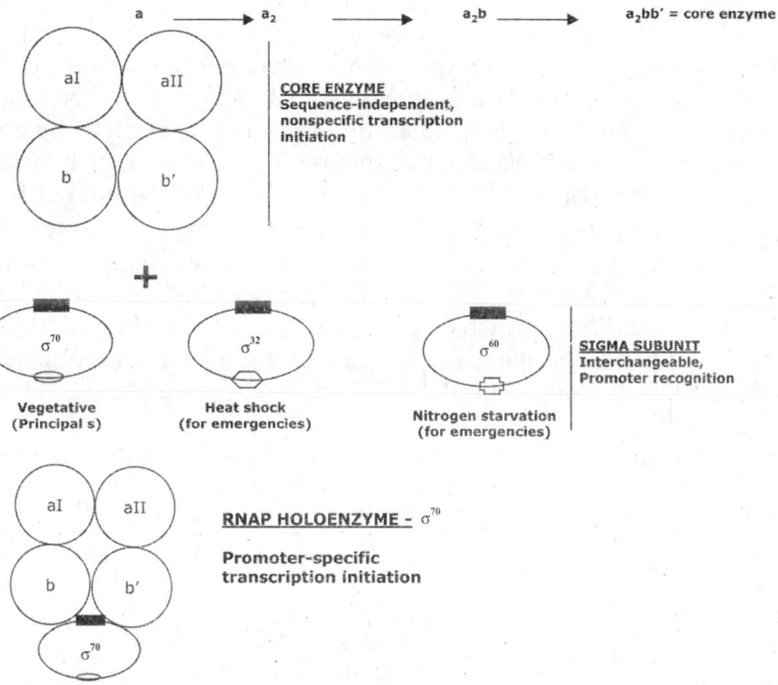

Fig. 3. The assembly pathway of the RNA polymerase holoenzyme.

Three Dimensional Structure

The three-dimensional structures of core enzyme (Fig. 4a) shows that the overall dimensions of a single core molecule in the crystal are 85 x 105 x 140 Å. The most striking feature of the core structure is a thumb-like projection surrounding a channel. The channel is 25 Å in diameter and can easily accommodate double stranded DNA.

The three-dimensional structure of RNA polymerase holoenzyme (Fig. 4b) shows that thumb in holoenzyme structure unlike in core RNA polymerase defines a deep open groove on the surface of the molecule. The irregularly shaped holoenzyme molecule is approximately 100 x 100 x 160 Å in size.

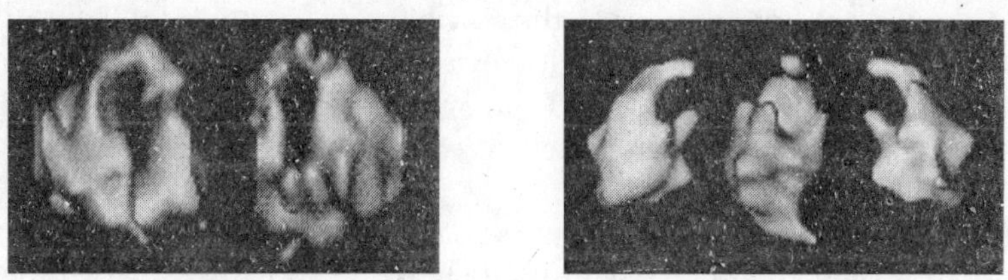

(a) (b)

Fig. 4. Shows three dimensional structures of RNA polymerase core enzyme and holoenzyme: (a) core enzyme (b) Holoenzyme (Voet et al., 2004).

Eukaryotes RNA Polymerases

There are three classes of nuclear eukaryotic RNA polymerases i.e., (a) RNA polymerases I, (b) RNA polymerases II and (c) RNA polymerases III. These RNA polymerases can be distinguished by their sensitivity to α-amanitin, a poisonous cyclic octapeptide from the poisonous mushroom *Amanita phalloides*. Polymerase I is insensitive, polymerase II is highly sensitive and polymerase III is only sensitive to the high concentration of α-amanitin. The mitochondrial and chloroplastidic RNA polymerases are insensitive to α-amanitin (Table 3).

Each eukaryotic RNA polymerase catalyzes transcription of genes encoding different classes of RNA. RNA polymerase I, located in the nucleolus, transcribe genes encoding precursor rRNA i.e. pre-RNA, which is processed into 28S, 5.8S, and 18S rRNAs. RNA polymerase II transcribes all protein-coding genes; that functions in production of mRNAs. RNA polymerase II also produces four of the five small nuclear RNAs U1, U2, U4 and U5 that

Table 3. Comparison of Eukaryotic DNA-dependent RNA polymerase

S. No.	Enzymes	Location	Products	Effect of alpha amanitin
1.	RNA Polymerase I	Nucleolus	Pre-rRNA (28S, 5.8S, 18S rRNAs)	Insensitive
2.	RNA Polymerase II	Nucleoplasm	Heterogeneous nuclear RNA (hnRNA i.e. precursors of mRNA) and U1, U2, U4, U5, snRNA	Highly sensitive
3.	RNA Polymerase III	Nucleoplasm	tRNAs , 5S rRNA and U6 snRNA	Sensitive only at high concentrations
4.	Mitochondrial	Mitochondria	Mitochondrial RNAs	Insensitive
5.	Chloroplast	Chloroplasts	Chloroplast RNAs	Insensitive

take part in RNA splicing. It is undoubtedly the most important among the three classes of RNA polymerases. RNA polymerase III transcribes genes encoding tRNAs and 5S rRNA. It also transcribes an array of small, stable RNAs, including U6 involved in RNA splicing and the RNA component of the signal-recognition particle (SRP) involved in directing nascent proteins to the endoplasmic reticulum (Table 3).

Each of the three eukaryotic RNA polymerase is more complex than *E. coli* RNA polymerase. All of the three enzymes comprise two large subunits and 12-15 smaller subunits. The two large subunits are homologous to the *E. coli* β and β′ subunits. Two smaller subunits are similar to the *E. coli* α subunit (Fig. 5). However, the eukaryotic RNA polymerase does not contain any subunit similar to the *E. coli* σ factor. Therefore, in eukaryotes, transcriptional initiation should be mediated by other proteins.

New RNA polymerase discovered in plants

A team headed by Craig Pikaard, Washington University Professor of Biology in Arts & Sciences, has discovered a fourth kind of RNA polymerase found only in plants and speculated to have been a plant feature for more than 200 million years. This entirely new cellular "machine" in plants plays a significant role in plant flowering and DNA methylation, a key chemical process essential for an organism's development.

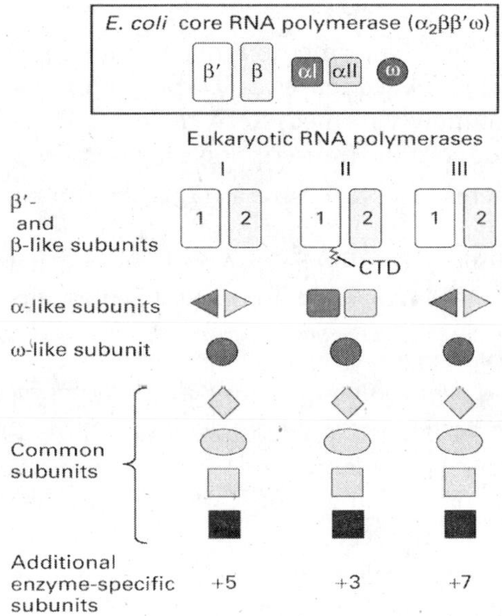

Fig. 5. Comparison of structures of eukaryotic and prokaryotic RNA polymerases subunits (Lodish, 1995).

Pikaard first noticed the evidence for a fourth polymerase when analyzing gene sequences after Arabidopsis thaliana, the "laboratory rat" of the plant world, was sequenced in 2001. It originally looked to him like an alternative form of either Polymerase I, which makes the largest of the ribosomal RNAs, Pol II which makes RNAs for protein-coding genes, or Pol III, a specialist in making the shortest of the ribosomal RNAs and tRNAs.

Pikaard and his collaborators discovered that Pol IV does not share in the duties of Pol I, II or III. But when the Pol IV subunits are knocked out, the most tightly packed DNA in the nucleus becomes less condensed, small RNAs called siRNAs corresponding to highly repeated 5S rRNA genes and retrotransposons (jumping genes) are completely eliminated and DNA methylation at 5S genes and retrotransposons is lost. Pikaard proposed that Pol IV helps make siRNAs that then direct DNA methylation to sequences matching the siRNAs.

TRANSCRIPTION EVENTS: DNA DEPENDENT SYNTHESIS OF RNA

Transcription in Prokaryotes

Initiation

Transcription initiation is the synthesis of the first nucleotide bond in RNA. RNA is synthesized by RNA polymerases. DNA-dependent RNA polymerase (Holoenzyme $\alpha_2\beta\beta'\omega$ σ^{70}) requires, in addition to a DNA template, all four ribonucleoside 5′- triphosphate (ATP, GTP, UTP and CTP) as precursors of the nucleotide units of RNA, and Mg^{2+}. The β′ subunit of enzyme also binds two Zn^{2+} which are thought to participate in the enzyme's catalytic function.

Template binding

RNA polymerase does not require a primer to initiate synthesis. Initiation occurs when RNA polymerase binds at specific DNA sequences called promoters. The 5′- triphosphate group of the first residue in a nascent (newly formed) RNA molecule is not cleaved to release PPi, but instead remains intact throughout the transcription process.

To enable RNA polymerase to synthesize an RNA strand complementary to one of the DNA strand, the DNA duplex must unwind over a short distance, forming a transcription "bubble". During transcription, the *E.coli* RNA polymerase generally keeps about 17 bp unwound. The 8 or 9 bp RNA-DNA hybrid occurs in this unwound region.

Transcription rate and accuracy

The *in vivo* rate of transcription is 20-50 nucleotides/sec at 37°C as indicated by the rate, by which E.coli incorporates ^{3}H-labeled nucleosides into RNA. Once an RNA polymerase molecule has initiated transcription and moved away from the promoter, another RNA polymerase can follow suit. The synthesis of RNAs that are needed in large quantities i.e. rRNAs, are initiated about once per second.

RNA polymerase lack a separate proofreading 3′ → 5′ exonuclease active site, and the error rate for transcription is higher than for chromosomal DNA replication—approximately one error for every 10^4 to 10^5 ribonucleotides incorporated into RNA. Many RNA polymerase, including bacterial RNA polymerase and the eukaryotic RNA polymerase II do pause when a mispaired base is added during transcription, and they can remove mismatched nucleotides from the 3′ end of a transcript by direct reversal of the polymerase reaction (pyrophosphorolytic editing). In second proofreading mechanism, the polymerase backtracks by one or more nucleotides and cleaves the RNA product, removing the error-containing sequence (hydrolytic editing). But we do not yet know whether this activity is a true proofreading function and to what extent it may contribute to the fidelity of transcription.

Sigma controls binding to DNA

In *E. coli*, a direct contact is formed between the promoter and RNA polymerase. The sequence specificity of the polymerase resides in its σ subunit: the 'core enzyme', which lacks σ subunit, can only make loose and non-specific attachments to DNA. Only 'holoenzyme' i.e. core enzyme and the σ subunit can initiate the transcription. The sigma factor released when the RNA chain reaches ~9 bases, after the stage of abortive initiation, leaving the core enzyme to undertake elongation. The core enzyme has the ability to synthesize RNA on a DNA template, but cannot initiate transcription at the proper site.

The release of sigma factor is indicated by the change in the shape of the RNA polymerase at the transcription from initiation to elongation. Once sigma has been released, the core enzyme begins to move along the template extending the RNA chain; the region of local unwinding moves with it. The unwound segment comprises only part of the total length of DNA bound by the polymerase. Transitions in shape and size identify three forms of the complex (Fig.6):

Initiation complex: When RNA polymerase holoenzyme initially binds to DNA, it covers some 77–80 bp, extending from –55 to +20. It remains in this location during the initiation phase, which involves RNA chain extension in the unwound region extending immediately beyond the start point.

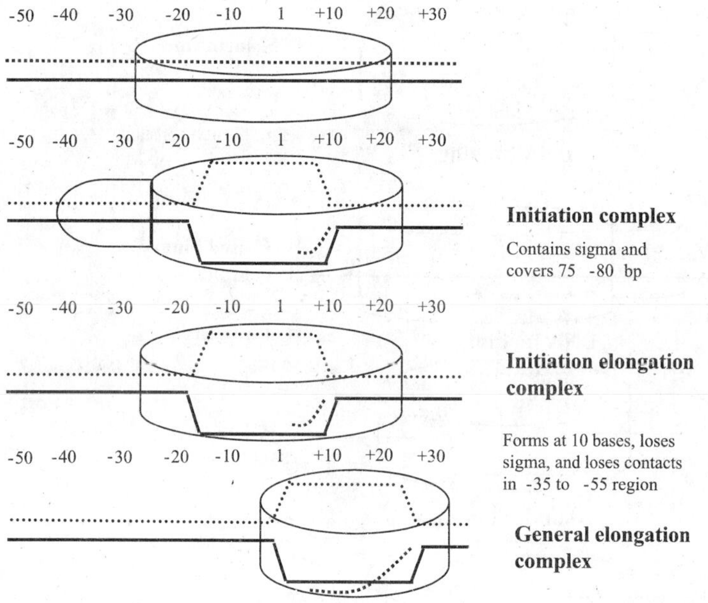

Fig. 6. RNA polymerase initially contacts the region from -55 to +20. When sigma dissociates, the core enzyme contracts beyond -30, and when the enzyme has moved a few base pairs, it becomes more compactly organized into the general elongation complex.

Initial elongation complex: As soon as the enzyme loses sigma factor (i.e. now core enzyme) and makes the transition to the elongation phase, it loses its contacts in the -55 to -35 region and covers only ~60 bp of DNA.

General elongation complex: When the RNA chain extends to 15-20 bases, the enzyme makes a further transitions, to form the complex that undertakes elongation; now it covers only 30-40 bp.

The function of sigma factor is to ensure that bacterial RNA polymerase binds stably to DNA only at promoters, not at other sites. The core enzyme has an affinity for DNA; any sequence of DNA that is bound by RNA polymerase in this general binding reaction is described as a loose binding site. The complex at such a site is stable; the half-life for dissociation of the enzyme from DNA is ~60 minutes. Core enzyme does not distinguish between promoters and other sequences of DNA. The holoenzyme (core enzyme and sigma factor) has a drastically reduced ability to recognize loose binding sites. The association constant for the reaction is reduced by a factor of $\sim 10^4$, and the half-life of the complex is < 1 second. Thus sigma factor destabilizes the general binding ability very considerably. The holoenzyme bind to promoter very tightly, with an association constant increased from that of core enzyme by 1000 times and a half-life of a several hours. The specificity of holoenzyme for promoters compared to other sequences is therefore $\sim 10^7$.

The stages of transcription in terms of the interactions between different forms of RNA polymerase and the DNA template are summarized Fig. 7.

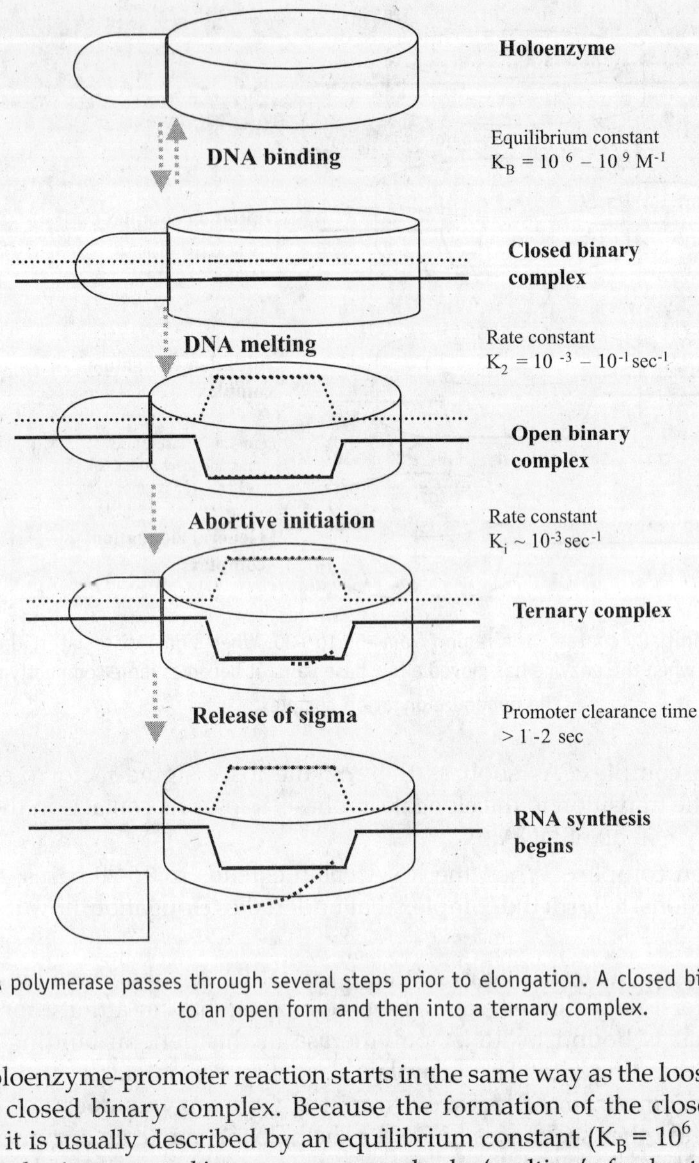

Fig. 7. RNA polymerase passes through several steps prior to elongation. A closed binary complex is converted to an open form and then into a ternary complex.

The holoenzyme-promoter reaction starts in the same way as the loose binding reaction, by forming a closed binary complex. Because the formation of the closed binary complex is reversible, it is usually described by an equilibrium constant ($K_B = 10^6 - 10^9\ M^{-1}$). The closed binary complex is converted into an open complex by 'melting' of a short region of DNA within the sequence bound by the enzyme. The series of events leading to formation of an open complex is called tight binding. Conversion into an open binary complex is irreversible, so this reaction is described by a rate constant (K_2). This reaction is fast. The next step is to incorporate the first two nucleotides; then a phosphodiester bond forms between them. This generates a ternary complex that contains RNA as well as DNA and enzyme. Formation of ternary complex is described by the rate constant K_i; this is even faster than the rate constant K_2. Further nucleotides can be added without any enzyme movement to generate an RNA chain of up to 9 bases. After each base is added, there is a certain probability that the enzyme will release the chain. This comprises an abortive initiation, after which the enzyme begins again with the first base. A cycle of abortive initiations usually occurs to generate a series of short (2-9 bases) oligonucleotides.

When initiation succeeds, the enzyme releases sigma and makes the transition to the elongation ternary complex of core polymerase-DNA-nascent RNA. The critical parameter here is how long it takes for the polymerase to leave the promoter so another polymerase can initiate. This is the promoter clearance time; its minimum value of 1-2 seconds establishes the maximum frequency of initiation as <1 event per second. The enzyme then moves along the template, and the RNA chain extends beyond 10 bases.

The Core enzyme and holoenzyme are distributed on DNA and very little RNA polymerase remain in the free form (Fig. 8). Excess core enzyme exists largely in the form of closed loose complexes, because the enzyme enters into them rapidly and leaves them slowly. There is very little, if any, free core enzyme.

There is enough sigma factor for about one third of the polymerase to exist as holoenzyme, and they are distributed between loose complexes at nonspecific sites and binary complexes (mostly closed) at promoters.

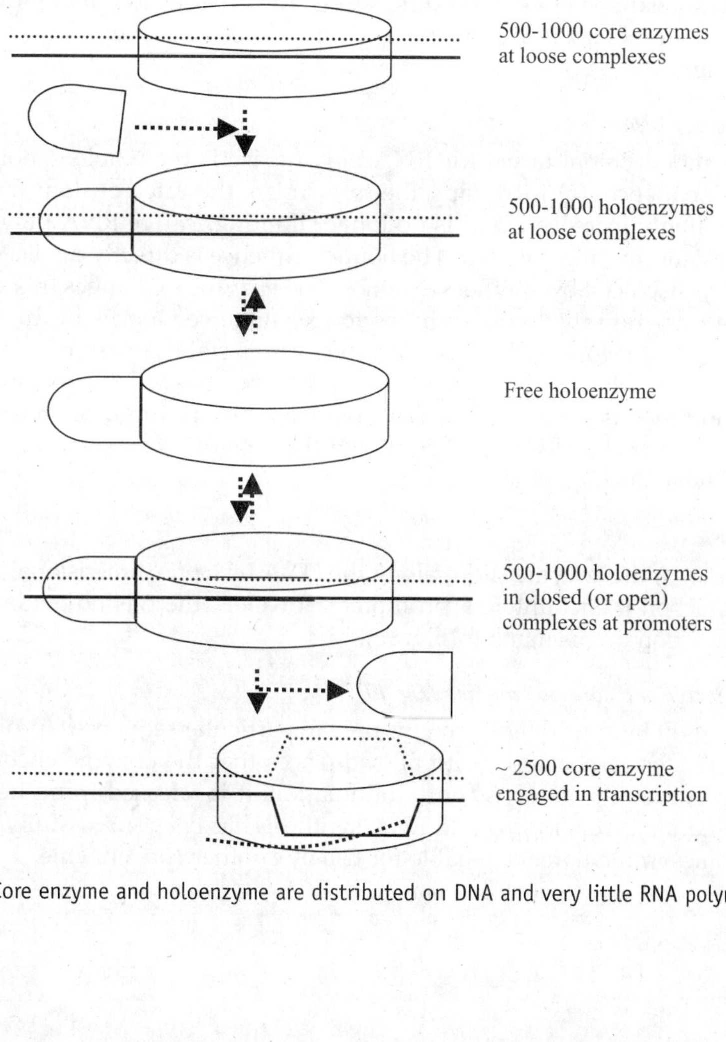

Fig. 8. Core enzyme and holoenzyme are distributed on DNA and very little RNA polymerase is free.

About half of the RNA polymerase consists of core enzymes engaged in transcription. How much holoenzyme is free? We do not know, but suspected that the amount is very small.

How RNA polymerase find out promoters from 4 X 10^6 bp; that comprise the E. coli genome (Fig. 9)

Three models are given to illustrate that how the RNA polymerase found promoters.

Random Diffusion to target (promoter)

According to this model RNA polymerase moves by random diffusion to find the promoter sequence. Holoenzyme very rapidly associates with, and dissociates from, loose binding sites. So it could continue to make and break a series of closed complexes in an agitated manner until it encounters a promoter. Then its recognition of the specific sequences would allow tight binding to occur by formation of an open complex. Movement of RNA polymerase from one site to another is limited by the speed of diffusion through the medium. Diffusion sets an upper limit for the rate constant for associating with a 60bp target of $< 10^8\ M^{-1}sec^{-1}$. But the actual forward rate constant for some promoters in vitro appears to be $\sim 10^8\ M^{-1}sec^{-1}$, at or above the diffusion limit. If this value applies in vivo, the time required for random cycles of successive associations and dissociations at loose binding sites is too great to accounts for the way RNA polymerase finds its promoters. RNA polymerase must therefore use some other means to seek its binding sites (Fig. 9a).

Random Diffusion to any DNA

According to this model initial target for RNA polymerase is the whole genome, not just a specific promoter sequence. By increasing the target size, the rate constant for diffusion to DNA is correspondingly increased, and is no longer limiting. A free RNA polymerase binds DNA and then remains in contact with it. The bound sequence is directly displaced by another sequence is directly displaced by another sequence. The enzyme exchanges this sequence with another sequence very rapidly, and continues to exchange sequences in this promiscuous manner until a promoter is found. Then the enzyme forms a stable, open complex, after which initiation occurs. The search process becomes much faster because association and dissociation are virtually simultaneous, and time is not spent communicating between sites. Direct displacement can give a 'directed walk', in which the enzyme moves preferentially from a weak site to a stronger site (Fig. 9b).

Sliding along the DNA

This idea supposes that the enzyme slides along the DNA by a one dimensional random walk, being halted only when it encounters a promoter. However, there is no evidence that RNA polymerase can function in this manner (Fig. 9c).

Recycling of sigma factor and core polymerase (Fig. 10)

Initiation requires tight binding of holoenzyme only to particular sequences (promoters), while elongation requires close association with all sequences that the enzyme encounters during transcription. Sigma factor is involved only in initiation. It is released from the core enzyme when abortive initiation is concluding and RNA synthesis has been successfully initiated. Free sigma factor becomes immediately available for use by another core enzyme.

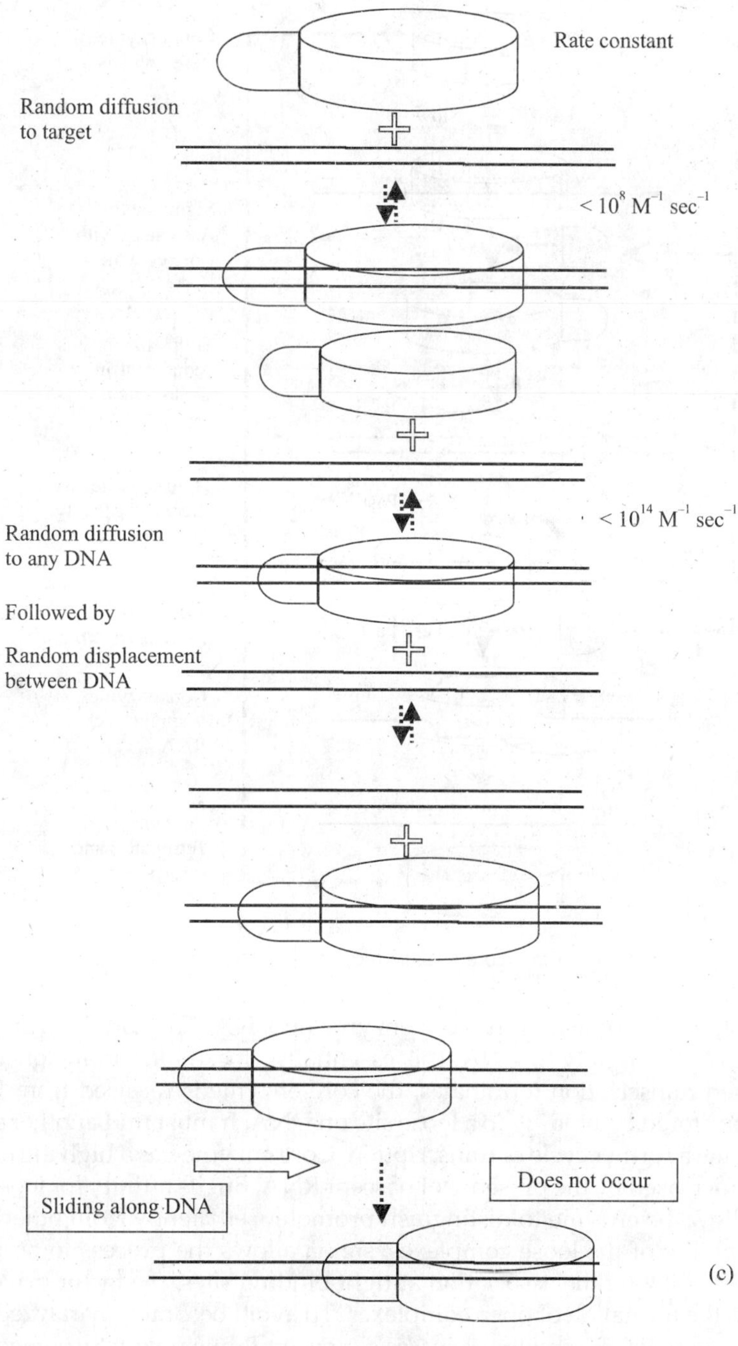

Fig. 9. How does RNA polymerase find promoters so rapidly on DNA? (a) Random diffusion to target DNA (b) Random diffusion to any DNA (c) Sliding along DNA.

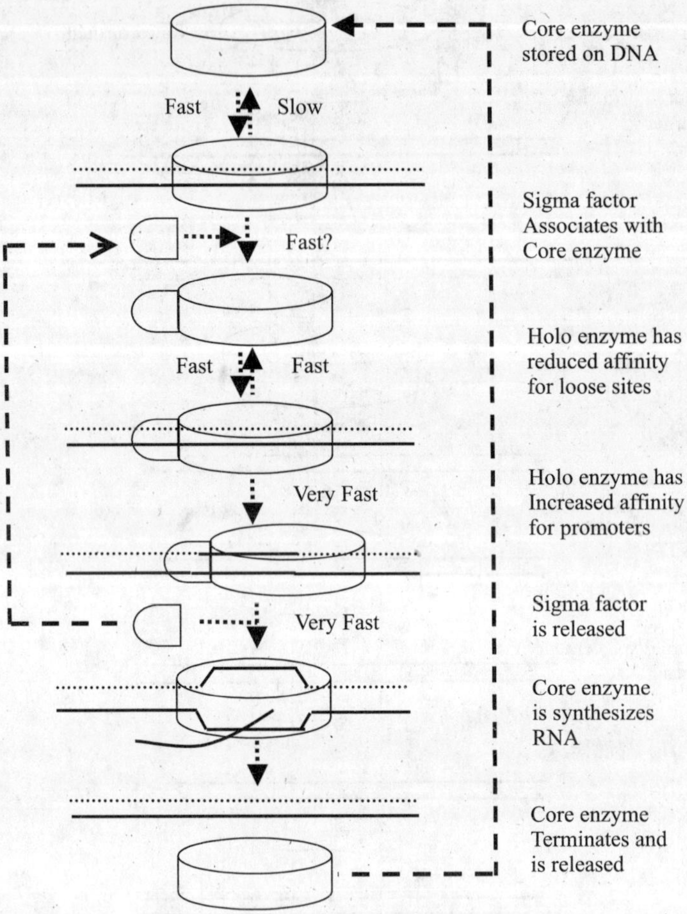

Fig. 10. Sigma factor and core enzyme recycle at different points in transcription.

Release of sigma factor leaves the core enzyme on DNA. The core enzyme in the ternary complex is bound very tightly to DNA. It is essentially "locked in' until elongation has been completed. When transcription terminates, the core enzyme is released from DNA as a free protein. It is then stored by binding to a loose site on DNA. It must find another sigma factor in order to undertake a further cycle of transcription. Core enzyme has a high intrinsic affinity for DNA, which is increased by the presence of nascent RNA. But its affinity for loose binding sites is too high to allow the enzyme to distinguish promoters efficiently from other sequences. By reducing the stability of the loose complexes, sigma allows the process to occur much more rapidly; and by stabilizing the association at tight binding sites, the factor drives the reaction irreversibly into the formation of open complexes. To avoid becoming paralyzed by its specific affinity for the promoter, the enzyme releases sigma, and thus reverts to a general affinity for all DNA, irrespective of sequence, that suits it to continue transcription.

RNA synthesis begins at promoters

The DNA sequence to which RNA polymerase binds to initiate transcription of a gene is called the promoter. Promoters are sequences of DNA that direct the RNA polymerase to the proper initiation site for transcription. The DNA sequences involved in promoter function were first identified by comparisons of the nucleotide sequences of a series of different genes isolated from *E. coli*. These comparisons revealed that the region upstream of the transcription initiation site contains two sets of sequences that are similar in a variety of genes. These common sequences encompass six nucleotides each, and are located approximately 10 and 35 base pairs upstream of the transcription initiation site. They are called the –10 and –35 elements, denoting their position relative to the transcription initiation site, which is defined as the +1 position. The sequences at the –10 and –35 positions in different promoters are not identical, but they are all similar enough to establish consensus sequences, i.e. the bases most frequently found at each position.

There are four conserved features in a bacterial promoter: the start point; the –10 sequence; the –35 sequence; and the separation between the –10 and –35 sequences:

The start point is usually a purine. It is common for the start point to be the central base in the sequence CAT, but the conservation of this triplet is not enough to regard it as an obligatory signal.

Just upstream of the start point, a 6 bp region is recognizable in almost all promoters. The center of the hexamer generally is close to 10 bp upstream of the start point; the distance varies in known promoters from position –18 to –9. Named for its location, the hexamer is often called the –10 sequence. Its consensus is (5′) TATAAT (3′) and known as TATA box or Goldberg-Hogness box after Michael Goldberg and David Hogness who deduced its existence in 1978.

Another conserved hexamer is centered ~ 35 bp upstream of the startpoint. This is called the –35 sequence. The consensus is (5′) TTGACA (3′).

The distance separating the –35 and –10 sites is between 16-18 bp in 90% of promoters; in the exceptions, it is little as 15 or as great as 20 bp. Although the actual sequence in the intervening region is unimportant, the distance is critical in holding the two sites at the appropriate separation for the geometry of RNA polymerase.

Some promoters have an A-T rich recognition element, located between positions -40 and -60 farther upstream. This is called as UP (upstream promoter) element. It interacts with the a subunit of the RNA polymerase. It is typically found in promoters that are highly expressed, such as the promoters for rRNA genes.

The optimal promoter is a sequence consisting of the –35 hexamer, separated by 17 bp from the –10 hexamer, lying 7 bp upstream of the start point. The sequence of a promoter is illustrated in (Fig. 11).

The upstream region is toward the 5′ side of the start site denoted by "–" sign. These are the nucleotide sequences preceding the start site and are denoted as –1, –2 and so on.

The downstream region is toward the 3′ side of the start site and denoted by "+" signs. These are the nucleotide sequences following the start site and are denoted as +1, +2 and so on.

+1 is the transcription start site. **There is no "zero".** (Fig. 12)

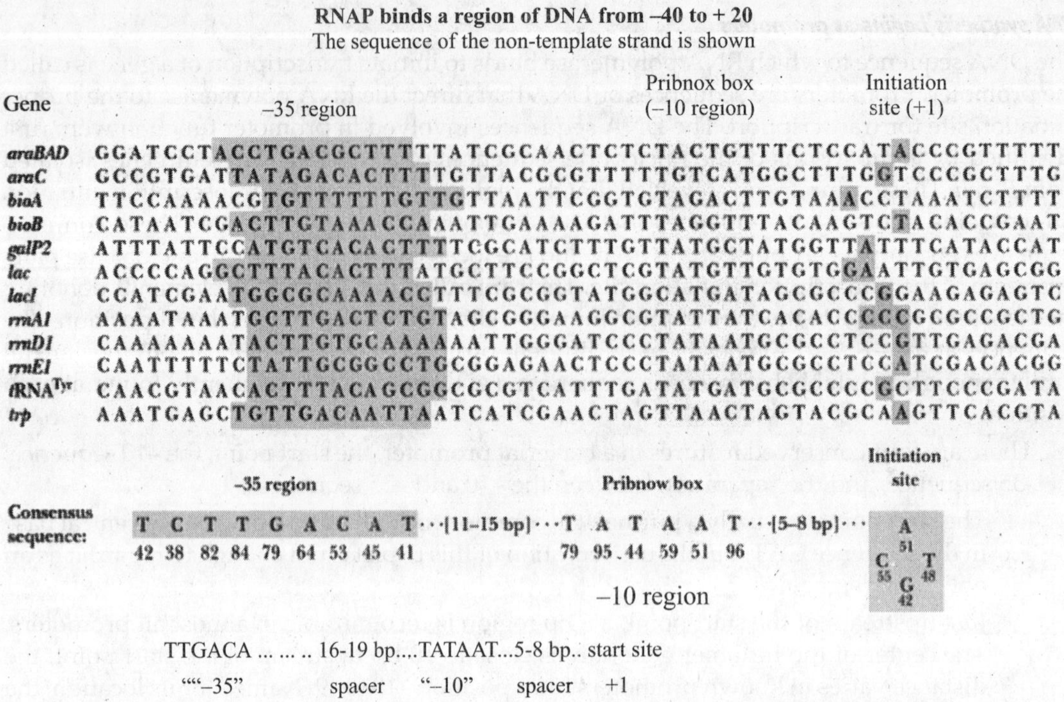

Fig. 11. The noncoding strand nucleotide sequences of selected E.coli promoters. The Pribnow box, a 6bp region centered around –10 positions, and an 8 to 12bp sequence around the –35 region are both conserved. The transcription initiation sites (+1), in most promoters occurs at a single purine nucleotide. The bottom row shows the consensus sequence of 112 promoters with the number below each base indicating its per cent occurrence (Voet *et al.*, 2004).

Experimental evidence supports the functional importance of the -10 and -35 promoter elements

Genes, with promoters that differ from the consensus sequences are transcribed less efficiently than genes whose promoters match the consensus sequences more closely. Mutations introduced in either the –35 or –10 consensus sequences have strong effects on promoter function.

The sites at which RNA polymerase binds to promoters have been directly identified by footprinting experiments, which are widely used to determine the sites at which protein bind to DNA (Fig.13).

RNA polymerase binds to one face of DNA

Foot Printing Experiment

The ability of RNA polymerase to recognize DNA can be characterized by footprinting. A sequence of labeled DNA bound to the protein of interest (i.e. RNA polymerase) is subjected to partial digestion with an endonuclease to attack individual phosphodiester bonds within the nucleic acid.

Principle of the method is that the regions of DNA to which the protein binds are protected from DNase digestion. These regions can therefore be identified by comparison of the digestion

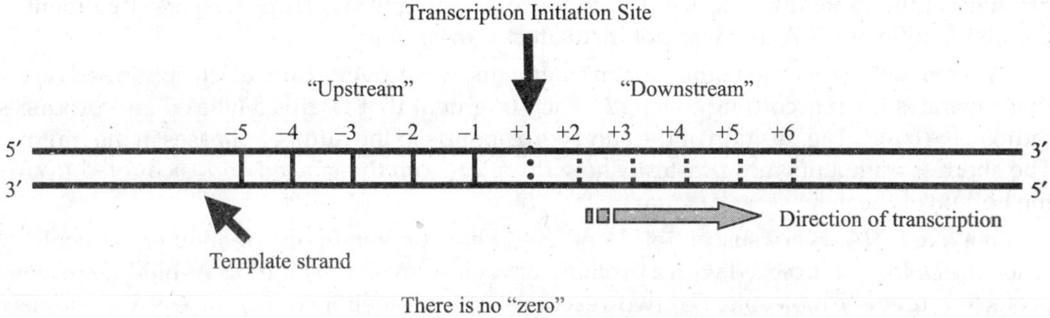

Fig. 12. The upstream and down stream region on DNA strand.

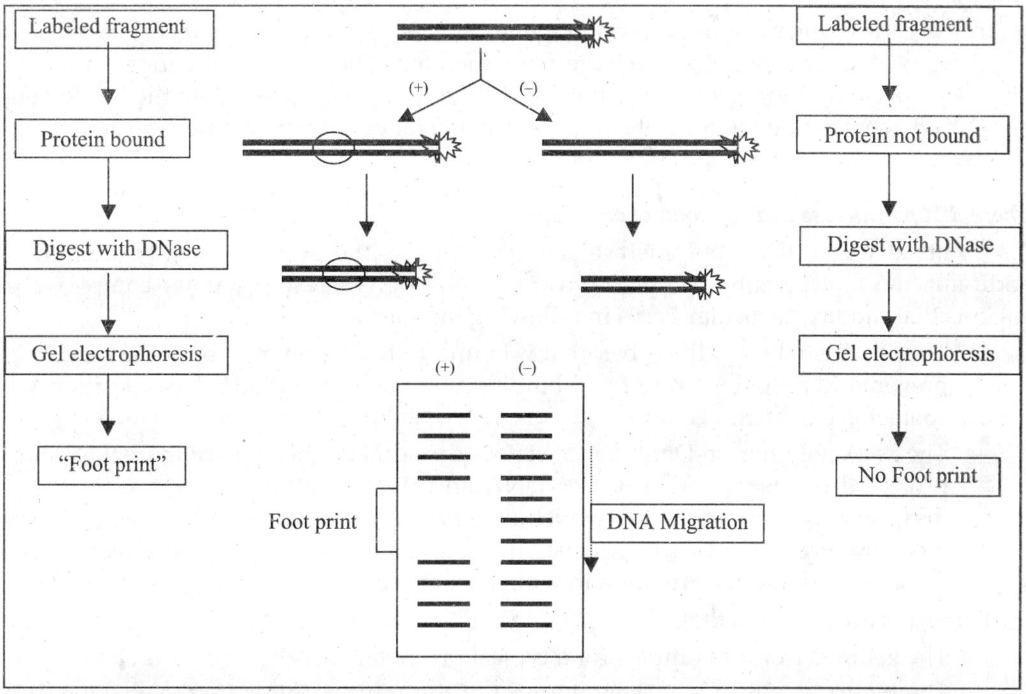

Fig. 13. DNA footprinting. A sample containing fragments of DNA radiolabeled at one end is divided into two, and one half of the sample is incubated with a protein that binds to a specific DNA sequence within the fragment. Both samples are then digested with DNase, under conditions such that the DNase introduces an average of one cut per molecule. The region of DNA bound to the protein is protected from DNase digestion. The DNA-protein complexes are then denatured, and the sizes of the radiolabeled DNA fragments produced by DNase digestion are analyzed by electrophoresis (as for DNA sequencing). Fragments of DNA resulting from DNase cleavage within the region protected by protein binding are missing from the sample of DNA that was incubated with protein. Hence, the DNA fragments corresponding to digestion in this region will not be formed and will appear as a "foot print" in the ladder of bands formed by DNase digestion at other, unprotected, parts of the DNA.

products of the protein-bound DNA with those resulting from identical DNase treatment of a parallel sample of DNA that was not incubated with protein.

After nuclease treatment, the broken fragments are recovered and electrophoresed on a gel that separates them according to length. Each fragment that retains a labeled end produces a radioactive band. The position of the band corresponds to the number of bases in the fragment. The shortest fragments move fastest, show distance from the labeled end is counted up from the bottom of the gel.

In a free DNA, every susceptible bond position is broken in one or another molecule. But when the DNA is complexed with a protein, the region covered by the DNA- binding protein is protected in every molecule. So, two reactions run parallel: a control of DNA alone; DNA molecule bound to the protein, an experimental mixture. When a bound protein blocks access of the nuclease to DNA, the bonds in the bound sequence fail to be broken in the experimental mixture.

In the control every bond is broken, generating a sequence of band, one representing fragments of the corresponding sizes are not generated. The absence of bands 9-18 (Fig. 13) identifies a protein-binding site covering the region located 9-18 bases from the labeled end of the DNA. By comparing the control and experimental lanes with a sequencing reaction we can identify the nucleotide sequence of the binding site.

Where RNA polymerase actually contact promoter

The points at which RNA polymerase actually contacts the promoter can be identified by modifying the foot printing technique to treat RNA polymerase-promoter complexes with reagents that modify particular bases in following manner:

The DNA can be modified before it is bound to RNA polymerase. If the modification prevents RNA polymerase from binding, we have identified a base position where contact is essential.

The RNA polymerase-DNA complex can be modified. Then we compare the pattern of protected bands with that of free DNA and the unmodified complex. Some bands disappear, identifying sites at which the enzyme has protected the promoter against modification. Other bands increase in intensity, identifying sites at which the DNA must be held in a confirmation in which it is more exposed.

This experiment shows that:

The geometry of the complex for a typical promoter.

The region at –35 and –10 contain most of the contact points for the enzyme. Within these regions, the same sets of positions tend both to prevent binding if previously modified, and to show increased or decreased susceptibility to modification after binding.

RNA polymerase binding does not depend on the presence of particular bases at some of the points of contact.

Some of the points of contact are not sites of mutation.

Where the initial melting starts

The region of DNA that is unwound in the binary complex can be identified directly by chemical changes in its availability. When the strands of DNA are separated, the unpaired bases become susceptible to reagents that can't reach them in the double helix. Such experiments implicate positions between -9 and +3 in the initial melting reaction. The region unwound during initiation therefore includes the right end of the -10 sequence and extends just past the start point.

The points of contacts lie upstream of the -10 sequence all on one face of DNA on the coding strand. These bases are probably recognized in the initial formation of a closed binary complex. This would make it possible for RNA polymerase to approach DNA from one side and recognize that face of the DNA. As DNA unwinding commences, further sites that originally lay on the other face of DNA can be recognized and bound.

Once the polymerase binds to the promoter and strand separation occurs, initiation usually proceeds rapidly (Fig. 14). The first, or initiating, NTP, which is usually ATP or GTP, binds to the enzyme. The binding is directed by the complementary base in the DNA template strand at the start site. A second NTP binds, and initiation occurs on formation of the first phosphodiester bond by a reaction involving the 3′-hydroxyl group of the initiating NTP with the 5′ phosphorus atom from the second NTP. Inorganic pyrophosphate derived from the second NTP is a product of the reaction.

The overall reaction is

$$\underset{\text{RNA}}{(\text{NMP})_n} + \text{NTP} \rightarrow \underset{\text{Lengthened RNA}}{(\text{NMP})_{n+1}} + \text{PPi}$$

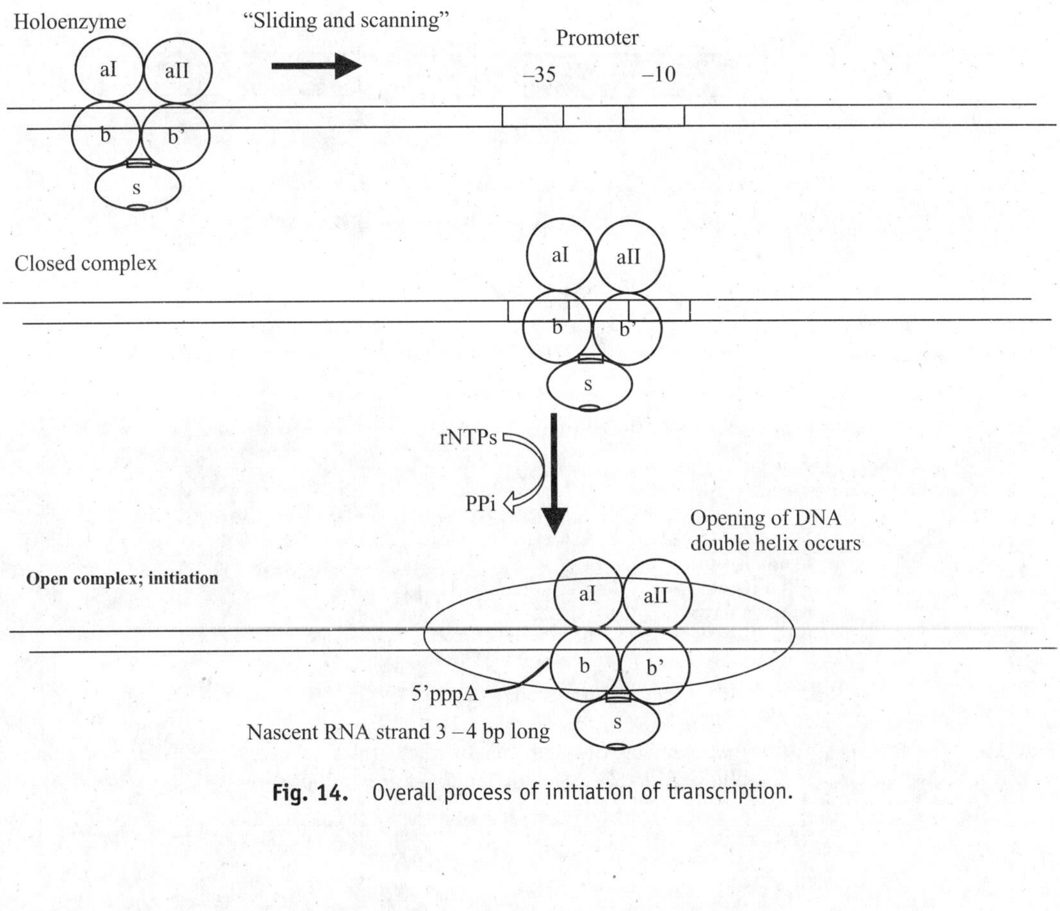

Fig. 14. Overall process of initiation of transcription.

Elongation

Transcription initiation is considered complete when the first two ribonucleotides of an RNA chain are linked by a phosphodiester bond. After 15-20 ribonucleotides have been polymerized, RNA polymerase dissociates from the promoter DNA and general transcription factors. Growth of a nucleic acid strand is always in the 5' to 3' direction. The direction of RNA chain elongation is determining by the rate that the radioactive label (γ-^{32}P) GTP is incorporated into RNA. The ratio of ^{32}P to total nucleotide was highest just after chain initiation and decreased with time thereby indicating that the 5′-terminal pppG was incorporated into the RNA chain first rather than last. Chain growth must therefore occur in the 5′→3′ direction.

This conclusion is corroborated by the observation that the antibiotic cordycepin, an adenosine analog that lacks a 3′-OH group, inhibits bacterial RNA synthesis. Its addition to the 3′ end of RNA, as is expected for 5′→3′ growth prevents the further elongation of RNA chain. Cordycepin would not have this effect if chain growth occurred in the opposite direction because it cannot be appended to RNA's 5′ end. The template DNA strand is copied in the 3′ → 5′ direction i.e. antiparallel to the new RNA strand which always grows in the 5' → 3′ direction (Fig.15). The 3′-hydroxyl group acts as a nucleophile, attacking the alpha phosphate of the incoming ribonucleoside triphosphate and releasing pyrophosphate (Fig.15).Each nucleotide in the newly formed RNA is selected by Watson-Crick base-pairing interaction; 'U' residue are inserted in the RNA to pair with 'A' residue in the DNA template, 'G' residue are inserted to pair with 'C' residue, and so on.

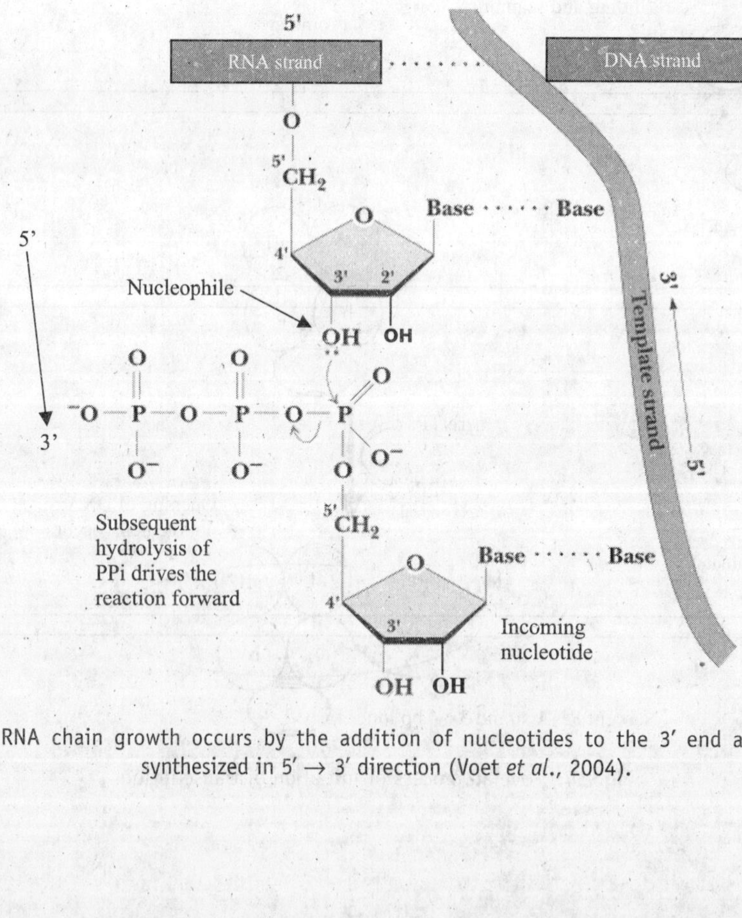

Fig. 15. The RNA chain growth occurs by the addition of nucleotides to the 3′ end and RNA chain get synthesized in 5′ → 3′ direction (Voet *et al.*, 2004).

One ribonucleotide at a time is added to the 3′ end of the growing (nascent) RNA chain during strand elongation by the polymerase. The enzyme maintains a melted region of approximately 17 base pairs, called transcription bubble (Fig. 16a). Approximately eight nucleotides at the template DNA strand in the transcription bubble. Elongation of a transcript by E.coli RNA polymerase proceeds at a rate of 50 to 90 nucleotides/second. Because DNA is a helix, movement of a transcription bubble requires considerable strand rotation of the nucleic acid molecules. DNA strand rotation is restricted in most DNAs by DNA-binding proteins and other structural barriers. As a result, a moving RNA polymerase generates waves of positive supercoils behind (Fig. 16b). In a cell topoisomerases rapidly eliminates the positive supercoils and regulate the level of negative supercoiling (Fig. 16b).

The elongation complex, comprising RNA polymerase (core polymerase- no sigma), template DNA (Fig. 17), and the growing (nascent) RNA strand, is extraordinarily stable. For example, RNA polymerase transcribes the longest known mammalian genes, containing H″2 × 10^6 base pairs, without dissociating from the DNA template or releasing the nascent RNA. The RNA synthesis occurs at a rate of about 1000 nucleotides per minute at 37°C.

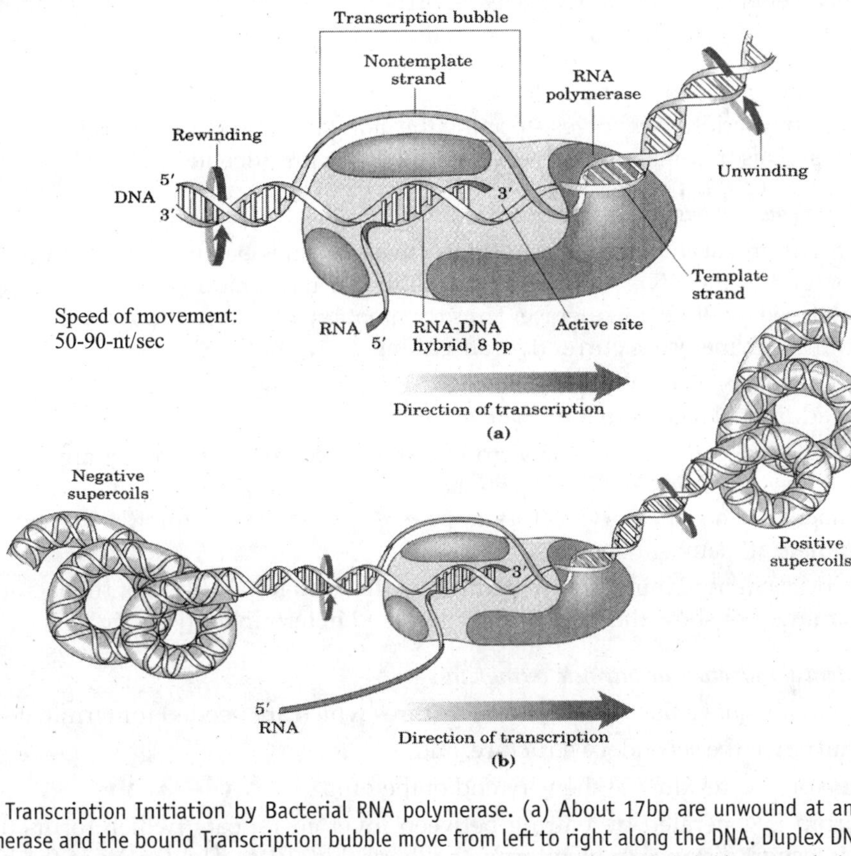

Fig. 16. Transcription Initiation by Bacterial RNA polymerase. (a) About 17bp are unwound at any given time. RNA polymerase and the bound Transcription bubble move from left to right along the DNA. Duplex DNA is unwound at the forward end of RNA polymerase and rewound at its rear end. The RNA-DNA hybrid rotates during elongation. (b) Supercoiling of DNA during transcription causes a requirement for topoisomerases (Snustad and Simmons, 2006).

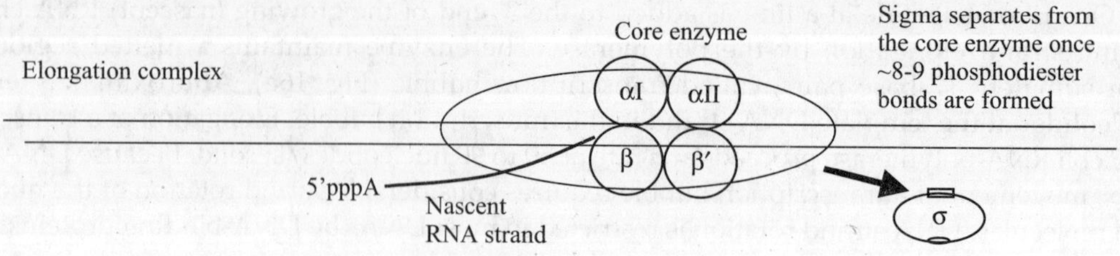

Fig. 17. Process of elongation, The elongation complex.

During the stage of strand elongation, the movement of enzyme is not monotonous process, but in fact the process is much less regular, and the forms of the enzyme represent only an average (Fig. 18). During elongation, the upstream boundary (left) boundary of the enzyme moves steadily as the RNA chain is extended, but the down stream (right) boundary remains stationary while several nucleotides are added; then suddenly it moves 7-8 bp along the DNA (Fig. 18). This mode of elongation involves steady compression and sudden expansion of the RNA polymerase on DNA; it contracts steadily from ~35 bp to ~28 bp, then expands back to ~35 bp.

Termination

Termination of transcription involves stopping the elongation process at a region on the DNA template that signals termination and release of the RNA product and the RNA polymerase.

Bacterial RNA polymerase has two modes of termination

The sequences at prokaryotic terminators show no similarities beyond the point at which the last base is added to the RNA. So the responsibility for termination lies with the sequences already transcribed by RNA polymerase. Thus termination relies on scrutiny of the template or product that the polymerase is currently transcribing.

Terminators have been distinguished in *E. coli* according to whether RNA polymerase requires any additional factors to terminate *in vitro*:

Core enzyme can terminate in vitro at certain sites in the absence of any other factor. These sites are called intrinsic terminators or Rho-independent terminators, reflecting the fact that no ancillary factors appear to be necessary for RNA polymerase to terminate at them.

Rho-dependent terminators are defined by the need for addition of rho factor in vitro; and mutations show that the factor is involved in termination *in vivo*.

Rho-independent terminators or Intrinsic terminators

Intrinsic terminators have the two structural features which are needed for termination:

A hairpin in the secondary structure; and

A run of ~6U residues at the very end of the unit

The hairpin is generated by pairing between inverted repeats (which forms the stem) separated by a short distance (which forms the loop) (Fig. 19 a). The length of the hairpin is variable: usually it contains a GC-rich region near the base of the stem (Fig. 19 b). All hairpins

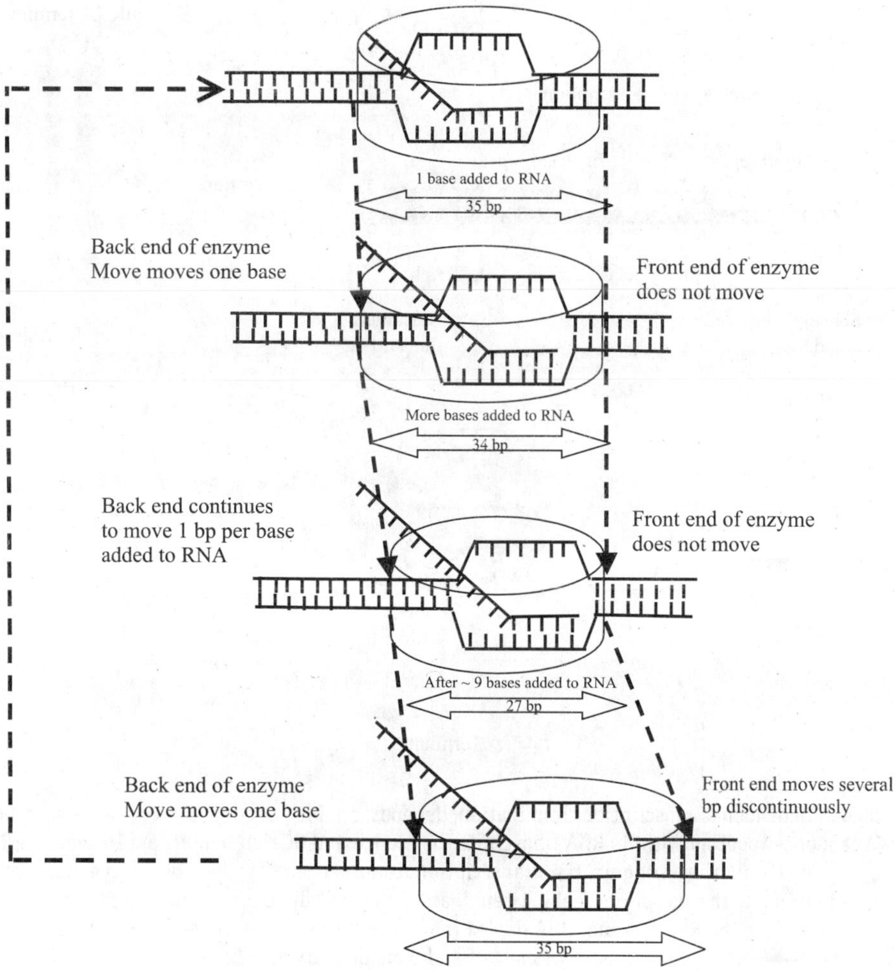

Fig. 18. RNA polymerase moves like a bellow during elongation, when it compresses from the back end and releases from the front end.

that form in the RNA product cause the polymerase to slow or pause in RNA synthesis. The length of the pause varies, but at a typical terminator lasts ~60 seconds.

Termination cannot depend simply on encountering a hairpin. (There is too much secondary structure in RNA.) Pausing creates an opportunity for termination to occur (Fig. 19c). But if no terminator sequences are present, usually the enzyme moves on again to continue transcription. The string of U residues probably provides the signal that allows RNA polymerase to dissociate from the template when it pauses at the hairpin. The rU-dA RNA-DNA hybrid has an unusually weak base-paired structure; it requires the least energy of RNA-DNA hybrid to break the association between the two strands. When the polymerase pauses, the RNA-DNA hybrid unravels from the weakly bonded rU-dA terminal region. Often the

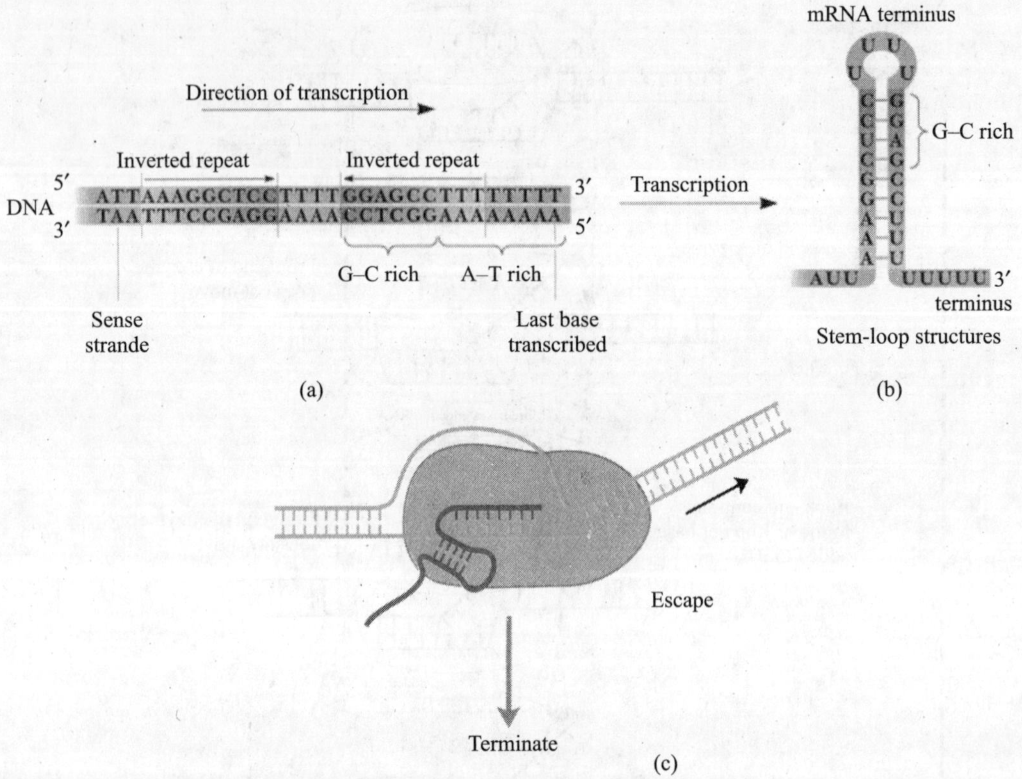

Fig. 19. Rho-Independent Transcription Termination depends on DNA sequence – NOT on the protein factor (a) The DNA sequence together with its RNA transcript. The AT rich and GC rich regions are shown. The RNA hairpin structure and poly U-tail that triggers the transcription termination (b) The core polymerase pauses after synthesizing a hairpin. If the hairpin is really a terminator, RNA will dissociate from the DNA strand as the A-U pairing is unstable. Once the RNA is gone, DNA duplex reforms and the core is driven off, as it has low affinity for dsDNA (Snustad and Simmons, 2006).

actual termination event takes place at any one of several positions toward or at the end of the U-run, as though the enzyme 'stutters' during termination.

Both, the sequence of the hairpin and the length of the U-run influence the efficiency of termination. However, termination efficiency in vitro varies from 2-90%, and does not correlate in any simple way with the constitution of the hairpin or the number of U residues in intrinsic terminators. Thus additional parameters which are undefined at present influence the interaction with RNA polymerase.

Rho-dependent terminators

Rho-dependent terminators use a termination factor called rho (ρ) factor to stop RNA synthesis at specific sites. When certain transcription units are used as templates for RNA synthesis in vitro, proper termination does not occur; the polymerase pauses at the terminator, but then resumes RNA synthesis. Rho factor was discovered as a protein whose addition to the in vitro system allows RNA polymerase to terminate at certain sites, generating RNA molecules with unique 3′ ends. This action is called rho-dependent termination. Rho is an essential protein in E. coli, although the genome has relatively few rho-dependent terminators; most of the known rho-dependent terminators are found in phage genomes. The sequences recognized by rho factor have taken a long time to identify. But as a general rule, the efficiency of rho-dependent terminator increases with the length of the C-rich/G-poor region.

How does rho factor work?

Rho factor functions solely at the stage of termination. It is a protein of ~46 kD, probably active as a hexamer (275 kD). It functions as ancillary factor for RNA polymerase; typically its maximum activity in vitro is displayed when it is present at ~10% concentration of the RNA polymerase. Rho factor acts via recognizing RNA. Rho has an ATPase activity, which is RNA-dependent; it requires the presence of a polyribonucleotide, >50 or so bases long. This suggest that rho binds RNA (Fig. 20). Rho factor moves along the transcript (nascent RNA) faster than RNA polymerase moves along the DNA. The enzyme pauses when it reaches a terminator, and termination occurs if rho catches it there.

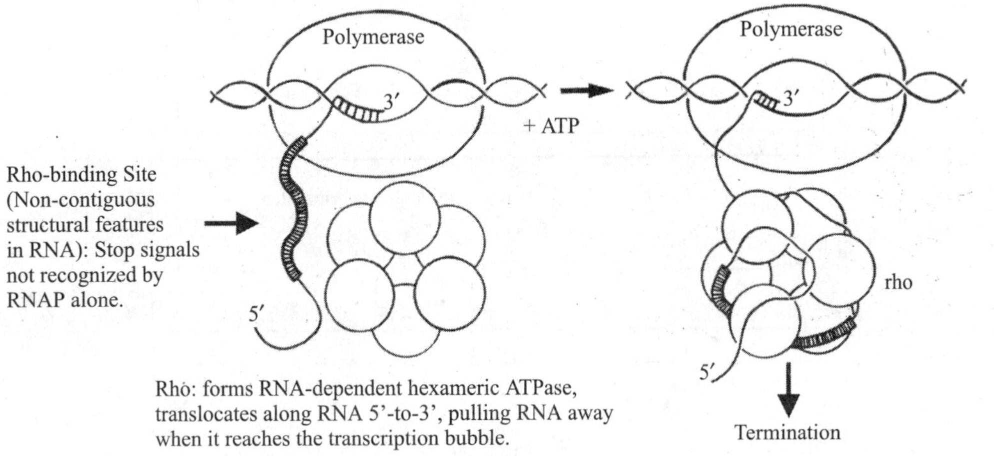

Fig. 20. Rho-Dependent Transcription Termination depends on a protein and a DNA sequence.

Rho factor release the transcript by acting directly on the DNA-RNA junction. In the absence of RNA polymerase, rho can cause an RNA-DNA hybrid to separate; hydrolysis of ATP is used to provide energy for the reaction. RNA can directly gain access to the stretch of RNA-DNA hybrid in the transcription bubble and cause it to unwind. Either as a result of this unwinding, or because of some interaction between rho and RNA polymerase, termination is complete by release of rho and RNA polymerase from the nucleic acid.

Antitermination

At some terminators, the transcription event can be prevented by specific ancillary factors that interact with RNA polymerase. Antitermination causes the enzyme to continue transcription past the terminator sequence, an event called read-through. Antitermination is used as a control mechanism in both phage regulatory circuits and bacterial operons.

During phage infection, different ancillary proteins (Antitermination factors) allow RNA polymerase to bypass specific terminator sequences. Such interactions control the ability of the enzyme to read past a terminator into genes lying beyond. In the example shown in the figure, the Antitermination factor positively regulates the expression of region 2. Terminators differ in their susceptibility to antitermination; in this example, the Antitermination factor can act at terminator 1, but does not prevent RNA polymerase from terminating at terminator 2. This mechanism of antitermination therefore depends upon the appearance of an anti-termination factor that is specific for a particular terminator (Fig. 21).

Antitermination can provide a link between translation and transcription, in which termination of transcription is prevented when the ribosome is unable to move along a leader segment of the mRNA. The basis of this effect rests with changes in RNA secondary structure that are determined by the positions of the ribosome on mRNA (Fig. 22).

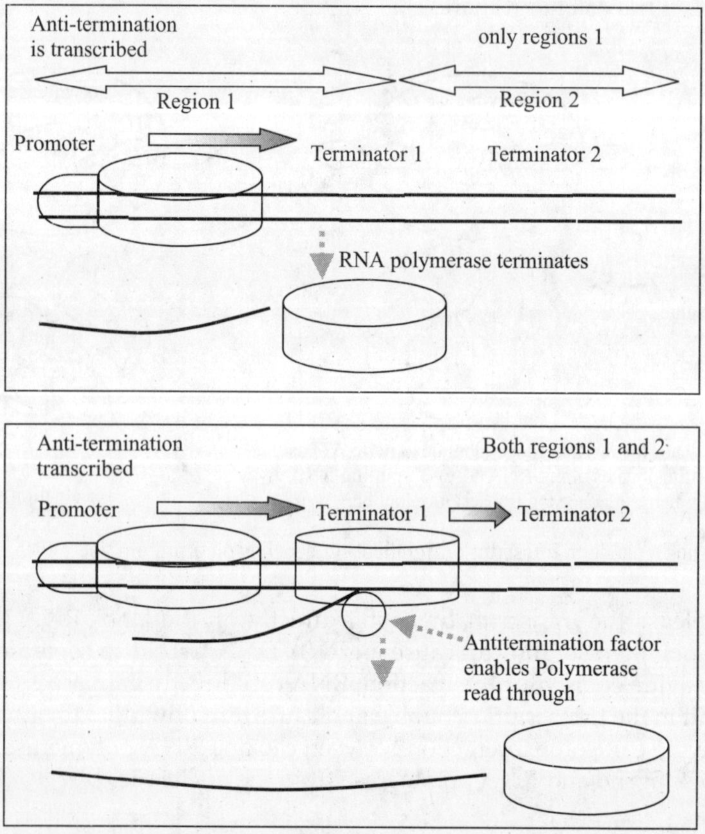

Fig. 21. Antitermination can be used to control transcription by determining whether RNA polymerase terminates or reads through a particular terminator into the following region.

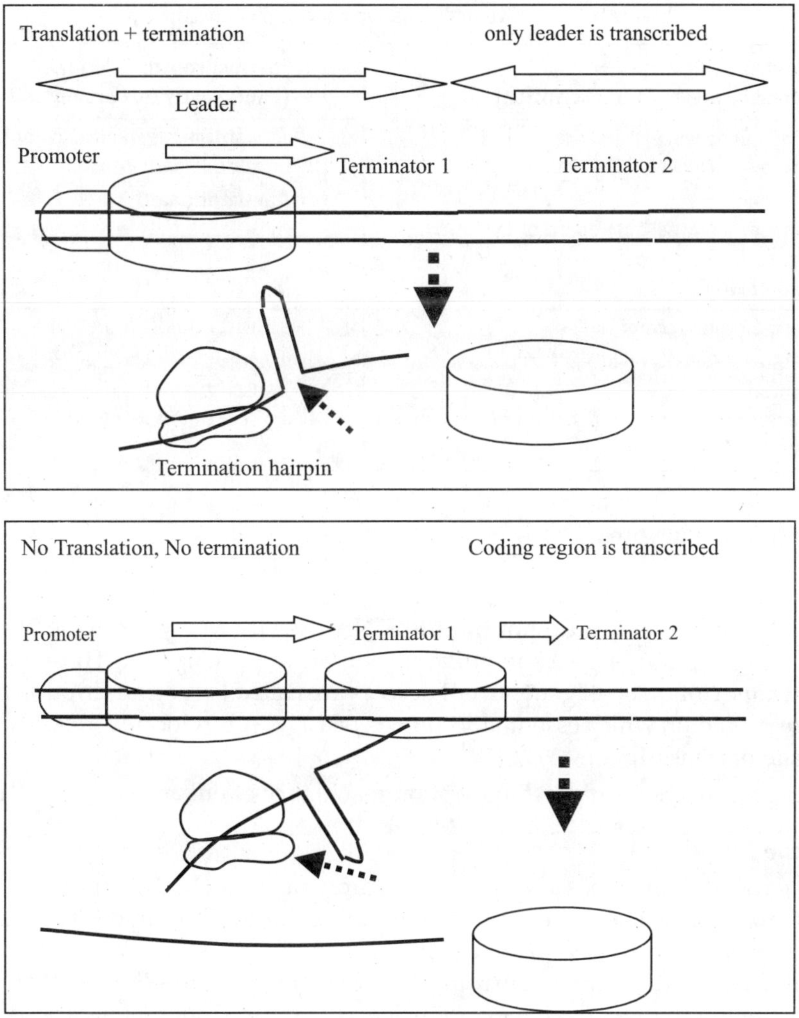

Fig. 22. Termination can be controlled via changes in RNA secondary structure that are determined by ribosome movement.

Regulatory factors of transcription

In prokaryotes transcription is regulated by repressor and activator proteins (sequence-specific). Vast number of prokaryotes regulatory, sequence-specific proteins known which are general and promoter-specific both. They can detect with gel-retardation assay and binding sites can be identifying by footprinting.

Binding of certain gene regulatory proteins to DNA depends on the presence or absence of a small regulatory molecule (ligand). They may increase or decrease its affinity for its DNA binding site, through induced conformational changes. The gene regulatory protein may be an activator or a repressor (Table 4).

Table 4. Positive and negative control of Transcription

Negative control (bound repressor prevents transcription)	Positive control (bound activator promotes transcription)
repressor binds in absence of ligand; increase ligand → removes repressor → Transcription on e.g. lac repressor removed from operator by inducer such as IPTG; RNA polymerase can now bind to the promoter.	activator binds in absence of ligand; increase ligand → removes activator → Transcription off
repressor binds in presence of ligand; decrease ligand → removes repressor → Transcription on e.g. Trp repressor binds in the presence of Trp and represses Trp biosynthesis.	activator binds in presence of ligand; decrease ligand → removes activator → Transcription off e.g. control of many genes by cAMP

Transcription in Eukaryotes

Initiation

Eukaryotic RNA polymerases do not bind to DNA directly Many more factors required for initiation of transcription of genes in eukaryotes than in prokaryotes. Those factors that are used for the transcription of all genes in eukaryotes are called basal transcription factors. These bind to a core promoter, which essentially consists of a TATA box located about 30 bp upstream of the start site of transcription.

Following figure shows basic transcription machinery of eukaryotes (Fig. 23).

RNA polymerase I

RNA polymerase I transcribes the genes encoding ribosomal RNA from a single type of promoter. Transcript includes the sequence of both enlarge ribosomal RNAs except 5S. The Promoter has been characterized in which it consist of two part sequence in a region preceding the start point. The Core promoter surrounds the start point from –45 to + 20 and is sufficient for the transcription to initiate. However efficiency of transcription is increased by upstream control element (UCE), which extends from –180 to –107. Both regions have an unusual composition foe a promoter, being rich in GC base pairs; they are ~85% identical. RNA polymerase I requires two ancillary factors:

UBF1

SL1

UBF1is a single polypeptide, that binds in a sequence-specific manner to a G-C rich element in the core promoter and UCE. Factor SL1 consist of four proteins one of them, called TBP, is a factor i.e. require also for imitation by RNA polymerase II and III. SL1 doesn't by itself have specificity for the promoter, but once UBF1 has bound, SL1 can bind cooperatively to extend the region of DNA that is cover. Once both factors are bound, RNA polymerase I can bind to core promoter to initiate transcription (Fig. 24).

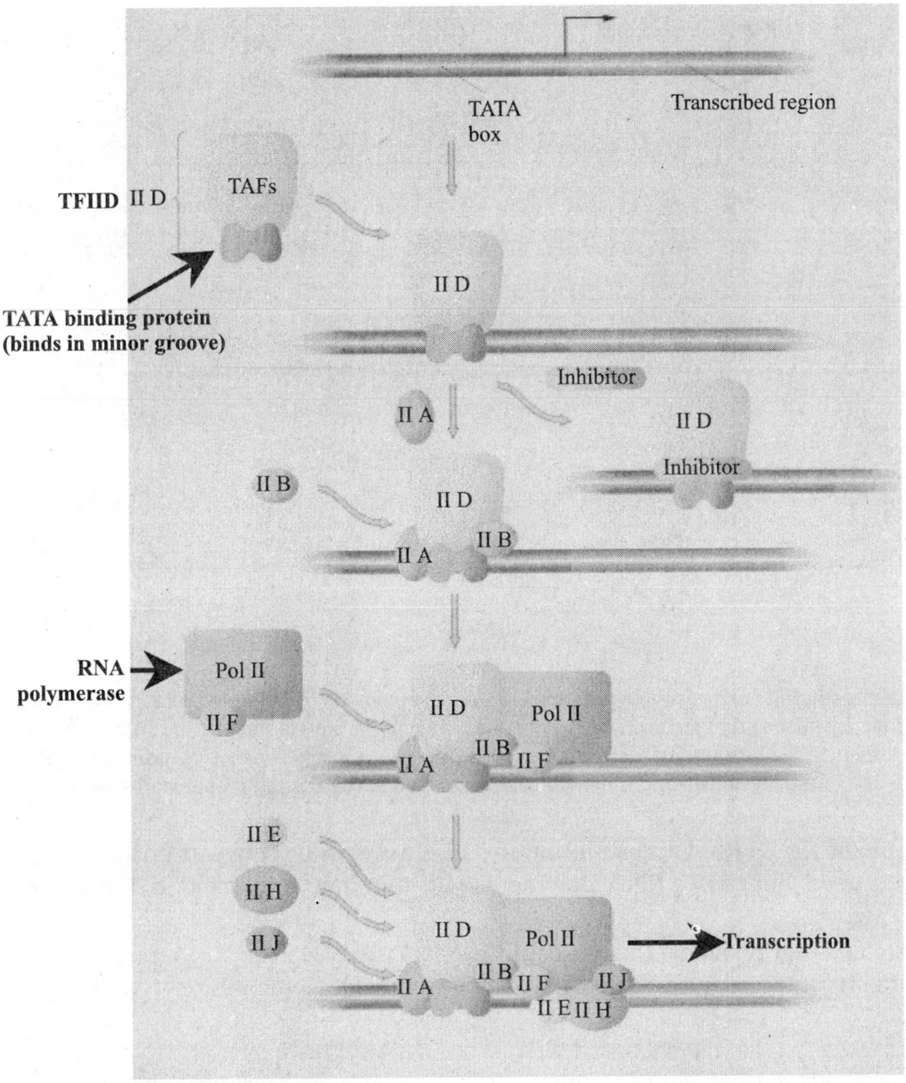

Fig. 23. Transcription in eukaryotes:-Basal transcription factors and TATA box (Lodish, 1995).

UBF1 and RNA polymerase I can function effectively with heterologous templates but factor SL1 has species specificity on the initiation reaction.

The behavior of SL1 resembles a bacterial sigma factor. It may have primary responsibility for ensuring RNA polymerase is properly localized at the start point.

RNA polymerase III

RNA polymerase III transcribes the sequences of t-RNA and 5S subunit of rRNA. Deletion analysis shows that the promoters for 5S and tRNA genes are internal; they lie downstream of startpoint. The promoters for snRNA (small nuclear RNA) genes lie upstream of the startpoint.

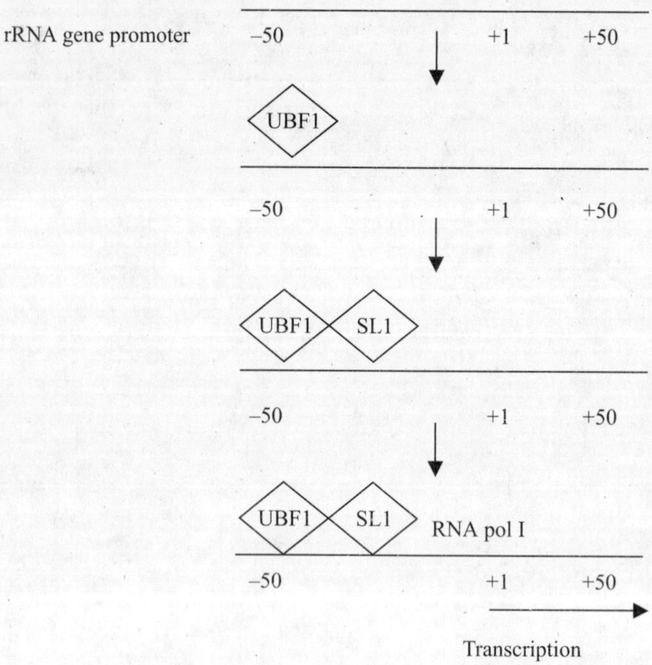

Fig. 24. Transcription initiation at the ribosomal RNA gene promoter. Binding of UBF is followed by the binding of the SL1 factor which in turn recruits RNA polymerase I via protein-protein interaction. In this and all subsequent figures, +1 refers to the first base transcribed into RNA, with other + numbers denoting bases within the transcribed region whilst – signs denote bases upstream of the transcriptional start site.

The promoter for 5S RNA transcription lies downstream of between positions +55 and +80 within the gene, but causes RNA polymerase III to initiate transcription a more or less fixed distance away.

There are two types of downstream, internal promoter for RNA polymerase III. Each contains a bipartite structure, in which two short sequence elements are separated by a variable sequence:

Type 1: It consists of a box A sequence separated from a box C sequence and

Type 2: It consists of a box A sequence separated from a box B sequence. The distance between box A and box B can vary quite extensively, but the boxes usually cannot be brought too close together without abolishing function (Fig. 25).

Initiation via the internal RNA polymerase III promoters involves three accessory factors TFIII A, TFIIIB and TFIIIC.TFIIIA is a member of zinc finer proteins. TFIIIB consists of TBP and two other proteins. TFIIIC is a large protein complex (500 kD), containing at least 5 subunits.

At type 1 promoters TFIIIA binds to a sequence that includes box C, and this enables TFIIIC to bind farther downstream (Fig.26). At type 2 promoters TFIIIC recognizes box B, but bind to a more extensive region including both boxes A and B. The binding of TFIIIC in turn enables TFIIIB i.e., function as 'positioning factor', to bind to a sequence surrounding the startpoint in either type of promoter. In the TFIIIB two proteins have been mapped on the DNA, binding

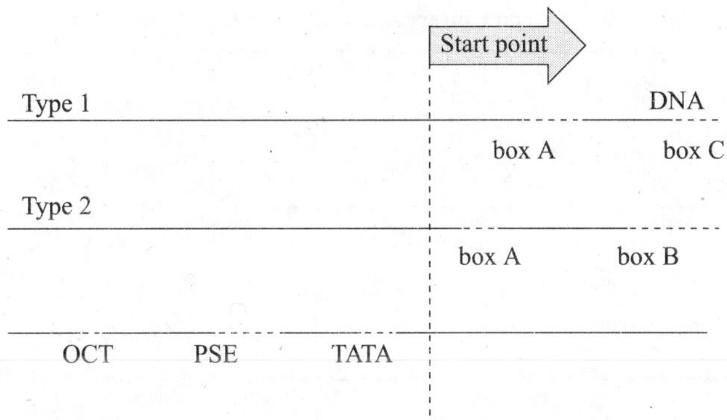

Fig. 25. Promoters for RNA polymerase II may consist of bipartite sequences downstream of the startpoint, with boxA separated from either boxC or boxB. Or they may consist of separated sequences upstream of the start point (OCT, PSE, TATA).

together to the region from – 40 to + 11.At this point, TFIIIA and TFIIIC can be removed from the promoter by high salt concentration in vitro without affecting the initiation reaction. TFIIIB remains bound in the vicinity of the startpoint and its presence is sufficient to allow RNA polymerase III to bind at the startpoint. Thus TFIIIB is the only true initiation factor required by RNA polymerase III; TFIIIA and TFIIIC are assembly factors, whose role is to assist the bonding of TFIIIB at the right location.

The upstream region has a more important role in the RNA polymerase III promoters. The upstream elements function in a similar manner in promoters for both polymerases II (for snRNA genes) and III. The TATA element appears to confer specificity for the type of RNA polymerase.

Initiation at an upstream promoter for RNA polymerase III can occur on a short region that immediately precedes the startpoint and contains only the TATA element. However, efficiency of transcription is much increased by the presence of proximal sequence element (PSE) and OCT elements. The factors that bind at these elements interact cooperatively. The PSE element may in fact be essential at promoters used by RNA polymerase II, whereas it is stimulatory in promoters used by RNA polymerase III.

The factor binds at the promoter before RNA polymerase itself can bind. They form a preinitiation complex that directs binding of the RNA polymerase III. The binding of RNA polymerase III is dependent on the presence of the stable transcription complex and not on the precise sequence of the DNA to which it binds.

RNA polymerase II

This eukaryotic RNA polymerase II catalyzes synthesis of mRNAs and a few small nuclear RNAs (snRNAs). Transcription of Protein-coding genes by RNA polymerase II can be initiated in vitro by sequential binding of the several initiation factors that are separated from the polymerase during purification. These initiation factors, which position polymerase molecules at transcription start sites and help to melt the DNA strands so that the template strand can enter the active site of the enzyme, are called **general transcription factors**.

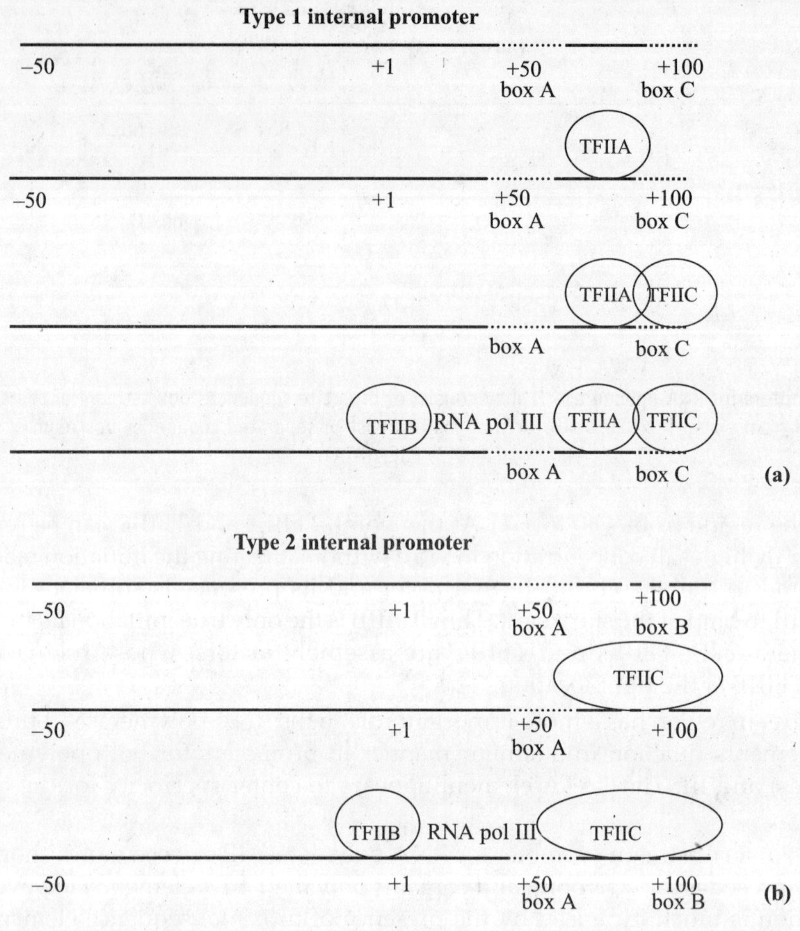

Fig. 26. Transcription initiation via the internal pol III at the 5S rRNA gene promoters involves the assembly factors TFIIA and TFIIC, the initiation factor TFIIB, and RNA polymerase III (a) Type I internal promoter (b) Type II internal promoter.

RNA polymerase II cannot initiate transcription itself, but is absolutely dependent on auxiliary transcription factors. The enzyme together with these factors transcription factors are required for synthesis of RNA from most genes.

Upstream region of the transcription start site which has different genes transcribed by RNA polymerase II contains an AT-rich sequence which is found approximately 30 bases upstream of the transcription start site. This TATA box plays a critical role in promoting transcriptional initiation and in positioning the start site of transcription for RNA polymerase II.

For the progressive assembly of the basal transcription complex for RNA polymerase II TATA box act as the initial DNA target site. Initially the TATA box is bound by the transcription factor TFIID whose binding is facilitated by the presence of another factor TFIIA (Fig. 27a).

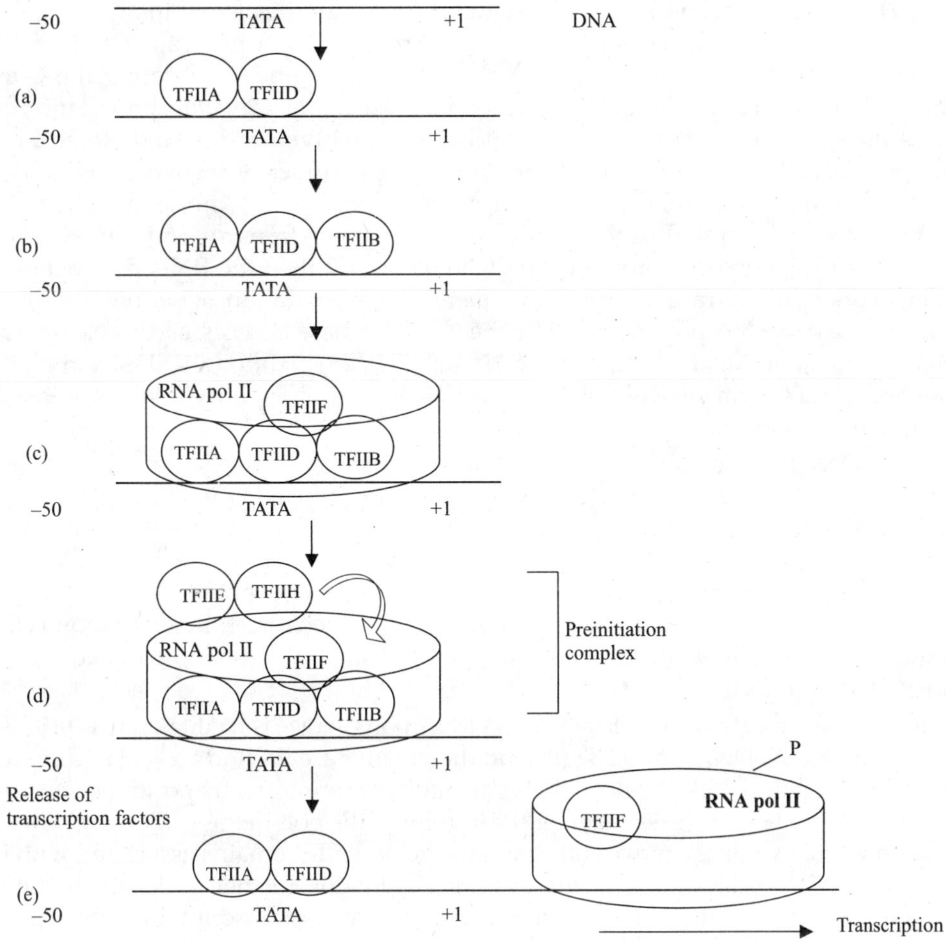

Fig. 27. Transcription initiation at an RNA polymerase II promoter. (a) Initially the TFIID factor binds to the TATA box together with another factor TFIIA. (b) TFIIB is recruited by interaction with TFIID (c) and TFIIB then recruits RNA polymerase II and its associated factor TFIIF. (d) TFIIE and TFIIH then bind and TFIIH phosphorylates the C-terminal domain of RNA polymerase II.(e)This convert RNA polymerase II into a form which is capable of initiating transcription and RNA polymerase II and TFIIF then move off down the gene producing the RNA transcript, leaving TFIIA and TFIID bound at the promoter.

TFIID It has a molecular clamp structure, which consists of four globular domains around an accessible groove. This groove can accommodate the DNA to which TFIID binds.

TFIIB The TFIID/DNA complex is recognized by another transcription factor TFIIB (Fig. 27b). TFIIB binds on the opposite side of TFIID which is bound by TFIIA.TFIIB binding is an essential step in the formation of the initiation complex.

TFIIF The binding of TFIIB allows the subsequent recruitment of RNA polymerase II to the initiation complex in association with another factor TFIIF (Fig. 27c). Subsequently, two other factors TFIIE and TFIIH associate with the complex (Fig. 27d).

TFIIH is a multi component complex which plays a critical role in allowing the RNA polymerase II to initiate transcription. It plays a key role in transcription and repair of damaged DNA both. TFIIH has kinase activity which is capable of phosphorylating the C-terminal domain (CTD) of the largest subunit of RNA polymerase II (Fig. 27e). Phosphorylation of the C-terminal domain plays a critical role in allowing the polymerase to initiate and continue transcription. This C-terminal domain contains multiple copies of sequence Tyr-Ser-Pro-Thr-Ser-Pro-Ser which is highly evolutionarily conserved. The kinase activity of TFIIH phosphorylates serine 5 within this repeat and this allows transcription initiation to occur. However, transcription only proceeds for 20-30 bases and then the RNA polymerase pauses and does not continue transcribing. Subsequent phosphorylation of serine 2 within the C-terminal domain of RNA polymerase II by the pTEF-b kinase is necessary to release this block and allow transcriptional elongation to occur (Fig. 28). Although SDNA recruited the dephosphorylated form of RNA polymerase II but its phosphorylation is necessary for transcription to produce the RNA product.

As the DNA is transcribed into RNA in the interior of the polymerase molecule, it encounters a wall of protein within the RNA polymerase (labeled W in Fig. 29). This forces it to make a right angled turn exposing the end of the nascent RNA and allowing ribonucleoside triphosphates to be added to it as transcription occurs. Subsequently, the newly-formed DNA-RNA hybrid produced as a consequence of transcription encounters another part of RNA polymerase, known as the rudder (labeled R in Fig. 29), which forces the separation of the RNA from the DNA. This allows the newly-formed part of the RNA molecule to exit and double-stranded DNA to reform.

This complex interaction of the DNA and RNA polymerase is facilitated by TFIIB. Thus by interacting with both DNA-bound TFIID and the polymerase itself in a very specific structure, TFIIB ensures that the DNA and the polymerase molecule are correctly positioned and oriented relative to one another for the DNA enter the interior of the polymerase, allowing transcription to occur. As the polymerase moves off down the gene, TFIIF remains associated with it while TFIIA and TFIID remain bound at the promoter, allowing further cycles of recruitment of TFIIB, RNA polymerase II etc., leading to repeated rounds of transcription (Table 5).

Table 5. RNA Pol II general transcription factors and coactivators

S. No.	Factor		Sub-units	Function
1.	TFIIA		2	Stabilizes TBP and TFIID-DNA binding. Blocks transcription inhibitors. Positive and negative gene
2.	TFIIB		1	Binds TBP, Pol II and promoter DNA. Helps fix transcription start site.
3.	TFIID	TBP	114	Binds TATA element and deforms promoter DNA.Platform for assembly of TFIIB, TFIIA and TAFs.
		TAFs*		Binds INR and DPE promoter elements. Target of regulatory factors.
Mediator *			24	Binds cooperatively with Pol II. Kinase and acetyltransferase activity. Stimulates basal and activated transcription. Target of regulatory factors.

Table Contd.

Table Contd.

4.	TFIIF		3	Binds Pol II and is involved in Pol II recruitment to PIC and in open complex formation.
5.	TFIIE		2	Binds promoter near transcription start. May help open or stabilize the transcription bubble in the open complex.
6.	TFIIH		10	Functions in transcription and DNA repair. Kinase and two helicase activities. Essential for open complex formation. Mutations in TFIIH can cause human disease.
SAGA		TAFs	5	Function unknown SAGA structure. Interact with TBP, TFIIA & Gcn5
		Spts,	9	
		Adas,	1	Histone acetyltransferase.
		Sgfs	1	Essential subunit, activator target. Third largest yeast protein. Component of the NuA4 HAT complex.
		Gcn5	1	Ubiquitin protease.
		Tra1		
		Ubp8		

Elongation and termination

TFIIF remains associated with RNA Polymerase II throughout the elongation process. During this stage, the activity of the polymerase is greatly enhanced by proteins called elongation factors. The elongation factors suppress pausing during transcription and also coordinate interactions between protein complexes involved in the posttranscriptional processing of mRNAs. Once the RNA transcript is completed, transcription is terminated. RNA Polymerase II is dephosphorylated and recycled, ready to initiate another transcript.

Fate of Nascent RNA

In prokaryotes RNA transcript undergo translation simultaneously with the process of transcription. It doesn't undergo any further post transcriptional processing or modification.

In eukaryotes RNA transcript undergo further post transcriptional modification or processing in the nucleus. These modified RNAs transport in cytoplasm through nuclear pores and perform their functions.

Processing of Eukaryotic mRNA

RNA processing achieves three things i.e., (a) RNA catalyzed Removal of introns; (b) Addition of a 5′ cap and (c) Addition of a 3′ tail. The processing signals that mRNA is ready to move out of the nucleus and may control its life span in the cytoplasm.

Processing of Eukaryotic tRNA

The tRNA acts as the adapter molecule between the genetic code on mRNA and the protein "language". These are 75-85 bases long. On each tRNA molecule a specific amino acid is covalently linked at the 3′ end. Elsewhere on the molecule is an anticodon complimentary to the specific amino acid codon on mRNA that codes for the amino acid carried by the tRNA.

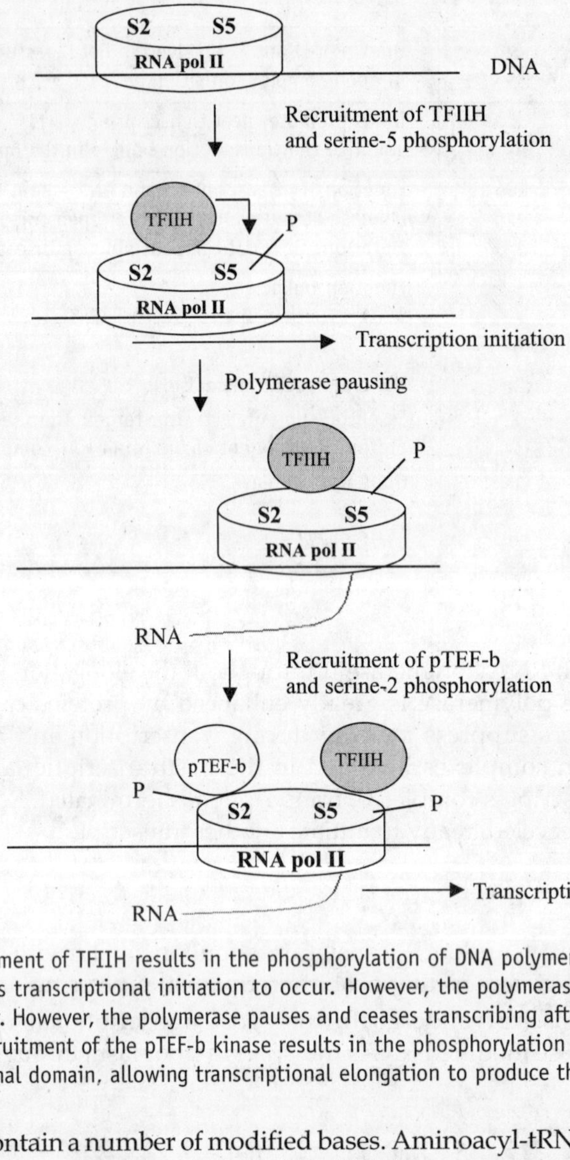

Fig. 28. Recruitment of TFIIH results in the phosphorylation of DNA polymerase Ii on serine-5 of its C-terminal domain and allows transcriptional initiation to occur. However, the polymerase pauses and allows transcriptional initiation to occur. However, the polymerase pauses and ceases transcribing after producing a short RNA transcript. Subsequently, recruitment of the pTEF-b kinase results in the phosphorylation of polymerase II on serine-2 of the C-terminal domain, allowing transcriptional elongation to produce the complete RNA product.

These RNAs contain a number of modified bases. Aminoacyl-tRNA Synthetase enzymes attach the correct amino acids to the correct tRNA. This is an energy consuming process. Aminoacyl-tRNA Synthetases recognize tRNAs on the basis of their looped structure, not by direct recognition of the anticodon.

Processing of Eukaryotic rRNA

The processing of pre-rRNA achieves two things: a) Methylation of pre-rRNA transcripts (i.e., methylation of 30S and 45S) and b) Cleavage of 30S pre-rRNA by the enzymes RNase III, RNase P and RNase E results 16S, 23S and 5S rRNA and that of 45S pre-rRNA results 18S, 5.8S and 28S rRNA.

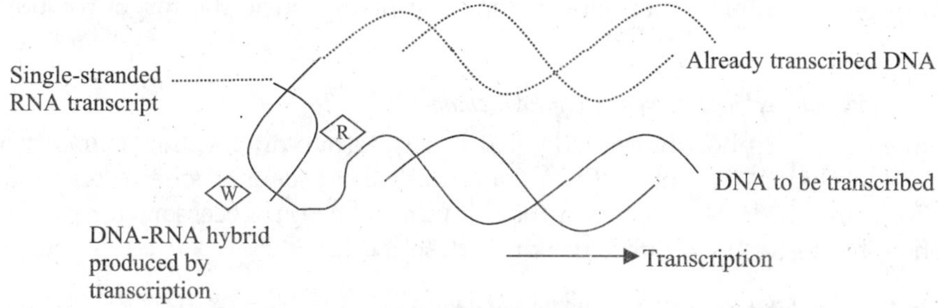

Fig. 29. Movement of the DNA being transcribed through the RNA polymerase molecule. "W" indicates the wall within the polymerase protein which forces the DNA-RNA hybrid to make a right-angled turn, thereby allowing transcription to occur by addition of ribonucleotide to the end of RNA chain. "R" indicates the rudder region of the polymerase which forces the DNA-RNA hybrid to melt releasing the newly-formed RNA and allowing double-stranded DNA to reform. The arrow indicates the direction of transcription.

Inhibitors of Transcription

Inhibitors act by binding to DNA

Actinomycin D, is a polypeptide-containing antibiotic produced by Streptomyces antibioticus, inhibits transcription. Actinomycin D binds tightly and specifically to double-helical DNA and thereby prevents it from being an effective template for RNA synthesis. It does not bind to single-stranded DNA or RNA, double-stranded RNA, or RNA-DNA hybrids. The results of spectroscopic and hydrodynamic studies of complexes of actinomycin D and DNA suggested that the phenoxazone ring of actinomycin slips in between neighboring base pairs in DNA. This mode of binding is called intercalation. At low concentrations, actinomycin D inhibits transcription without significantly affecting DNA replication or protein synthesis. Hence, actinomycin D is extensively used as a highly specific inhibitor of the formation of new RNA in both prokaryotic and eukaryotic cells.

Inhibitors bind to RNA polymerase

Rifamycins and rifampicin inhibit Prokaryotic transcription initiation:

The antibiotics, rifamycin B, which is produced by Streptomyces mediterranei, and its semi synthetic derivative rifampicin, specifically inhibit transcription by prokaryotic, but not eukaryotic, RNA polymerases. This selectivity and their high potency (bacterial RNA polymerase 50% inhibited by 2×10^{-8} M rifampicin) have made them medically useful bacteriocidal agents against gram-positive bacteria and tuberculosis. The isolation of rifamycin resistant mutants whose β subunits have altered electrophoretic mobilities indicated that this subunit contains the rifamycin-binding site. Rifamycins neither inhibit the binding of RNA polymerase to the promoter nor the formation of the first phosphodiester bond, but they prevent further chain elongation. The inactivated RNA polymerase remains bound to the promoter thereby blocking its initiation by uninhibited enzyme. Once RNA chain initiation has occurred, however, rifamycins have no effect on the subsequent elongation process. The rifamycins are useful research tools because they permit the transcription process to be dissected into its initiation and its elongation phases.

Streptolydigins inhibits chain elongation. The β-subunit is the target for both types of antibiotic.

Inhibitors are incorporated into the Growing RNA chain

Cordycepin in its 5′-triphosphorylated form is a substrate analog that is incorporated into growing RNA chains by most RNA polymerase. It causes chain termination after incorporation, since it does not contain the 3′-hydroxyl group necessary for the formation of the next phosphodiester bond (structure prev. described).

Intercalating Agents inhibit both DNA and RNA polymerases

Daunomycin and the closely related **Adriamycin**, which are chemotherapeutic agents in the treatment of certain human cancers, specifically bind to duplex DNA so as to inhibit its transcription and replication. These antibiotics presumably act by interfering with the passage of both polymerases. This antibiotic intercalated between the G-C pairs at both ends of the double helical fragment.

Ethidium bromide and proflavin are intercalating agents also inhibit nucleic synthesis, by intercalating between DNA and RNA strands.

MEASUREMENT OF TRANSCRIPTION RATES

Pulse-labelling to measure transcription:

The synthesis of RNA from DNA by the enzyme RNA polymerase involves the incorporation of ribonucleotides into an RNA chain. Therefore the synthesis of any particular RNA can be measured by adding a radioactive ribonucleotide which is usually uridine labeled with tritium, to the cells and measuring how much radioactivity is incorporated into RNA specific for the gene of interest.

The rate of transcription is measured by exposing cells briefly to he labeled uridine in a process referred to as pulse labeling. The uridine is incorporated into nascent RNA chains that are being made at this time and, even before a complete transcript has had time to form, the cells are lysed and total RNA is isolated from them. This RNA, which contains labeled partial transcript from genes active in the tissue used, is then hybridized to a piece of DNA derived from the gene of interest, the number of radioactive counts that bind providing a measure of the incorporation of labeled precursor in to its corresponding RNA (Fig. 30).

This method provides the most direct means of measuring transcription. For e.g., In Friend erythroleukemia induction of globin production occurs in repose to treatment with dimethylsulphoxide is mediated by increased transcription of the globin gene. This increased transcription produces the increased levels of globin-specific RNA in the nucleus and cytoplasm of the treated cells.

Although pulse labeling provides a very direct means of measuring transcription rates, the requirement to use very short labeling times to minimize any effects of RNA degradation limits its applicability. This method can be used only for those genes which are highly abundant or which have a very high rate of transcription e.g. globin. This technique has some limitations also. Primary limitation on the sensitivity of pulse labeling is the existence within the cells of a large pool of non-radioactive ribonucleotides, which are normally used by the cells to

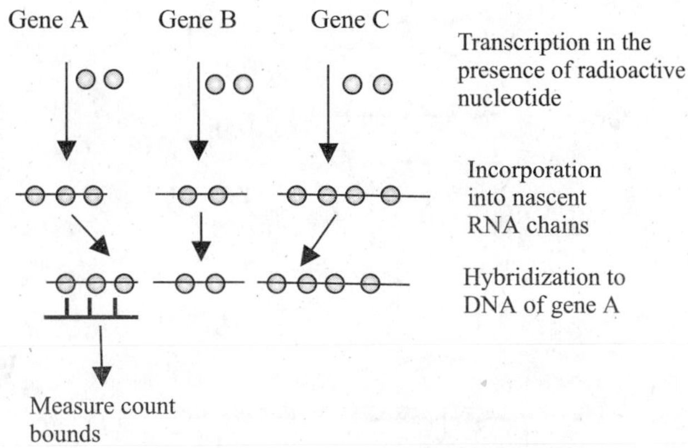

Fig. 30. Pulse-labeling assay to assess the transcription rate of a specific gene (gene A) by measuring the amount of radioactivity (dots) incorporated into nascent transcripts.

synthesize RNA. When labeled ribonucleotide is added to the cell, it is considerably diluted in this pool of unlabeled precursor. The amount of label incorporated into RNA in the labeling period is therefore very small, since the majority of ribonucleotides incorporated are unlabeled. The sensitivity of this method is thus severely reduced, resulting in its observed applicability only to genes with very high rates of transcription. However, although transcription takes place in the nucleus, most of the pool of precursor ribonucleotides is present in the cytoplasm. Hence it is possible, by removing the cytoplasm and isolating nuclei, to remove much of the pool of unlabeled ribonucleotide. This concept derived a new technique for measurement of transcription rate i.e. nuclear run-on assay.

Nuclear run-on assay to measure transcription

This method involves the removing of the cytoplasm and isolating nuclei, to remove much of the pool of unlabeled ribonucleotide. When this occurs the RNA polymerase ceases transcription due to lake of nucleotide but remains associated with the DNA (Fig. 31).Labeled nucleotides can then be added directly to the isolated nuclei in a test-tube and the RNA polymerase resumes transcribing and runs on to the end of the gene incorporating into any particular RNA transcript. Because the labeled ribonucleotides are not diluted in the unlabeled cytoplasmic pool, considerably more label is incorporated into any particular RNA transcript than is observed in pulse-labeling experiments. The label incorporated into any particular transcript is detected by hybridization to its corresponding DNA exactly as in pulse-labeling experiments.

This method, which is known as a nuclear run-on assay, is therefore much more widely applicable than pulse labeling, and can be used to quantify the transcription of genes that are never transcribed at levels detectable by pulse labeling. Many studies have now established that the RNA synthesized by isolated nuclei in the test-tube is similar to that made by intact whole cells. This method is therefore not only sensitive but also provides an accurate measure of transcription, free from artefact.

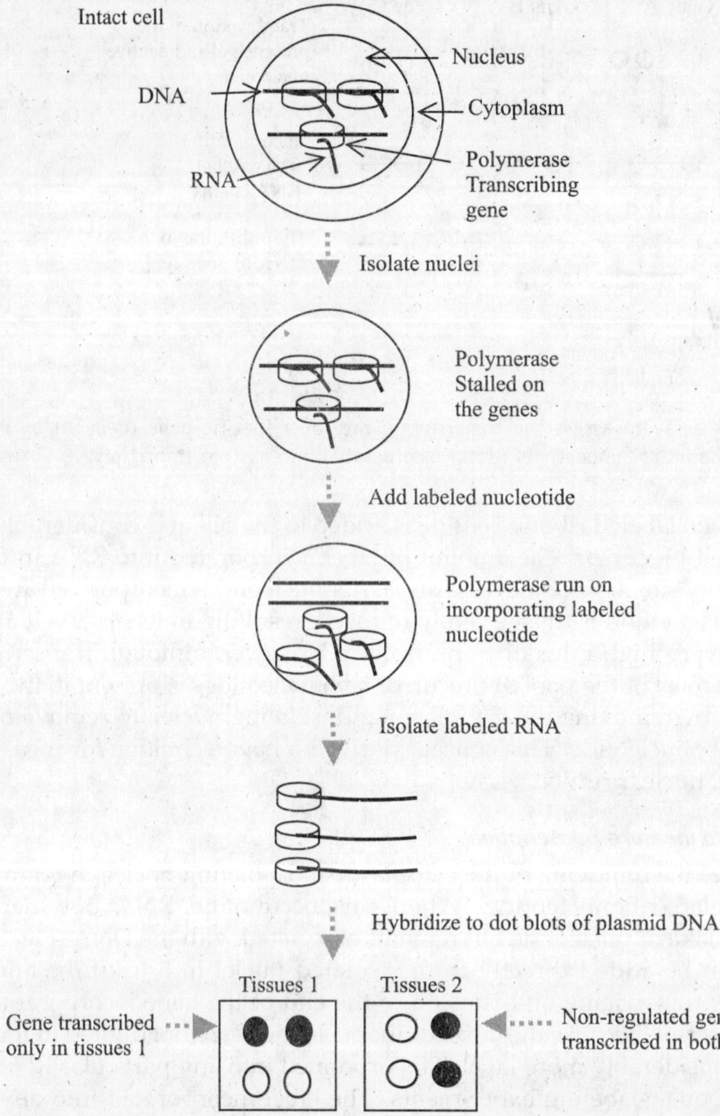

Fig. 31. Nuclear run-on assay. Following isolation of nuclei, RNA polymerase ceases transcribing due to lack of ribonucleotides but remains associated with the DNA. When radioactivity labeled ribonucleotide is added, the polymerase resumes transcription and runs on to the end of the gene incorporating the labeled ribonucleotide into the RNA transcript. The labeled RNA made in this ay in different tissues can then be used to probe filters containing DNA from genes whose transcription is to be measured in the different tissues.

REFERENCES

Berg, J. M., Tymoczko, J. L. and Stryer, L. (2006) Biochemistry, (6th ed.), Macmillan Publishers Ltd., USA.

Cook, P. R. (2001) Principles of Nuclear Structure and Function, Wiley-Liss Publishers, USA.

Hirota, Y.; Ryter, A., and Jacob, F. (1968) Quant. Biol, Thermosensitive mutants of *E. coli* affected in the processes of DNA synthesis and cellular division. Cold Spring Harbor Symp. 33: 677-693.

Lewin, B. (2004) Genes VIII, International Edition, Pearson Prentice Hall, UK.

Lodish, Baltimsu and Berk (1995) Molecular Cell Biology (3rd Edition), W.H. Freeman and Company, USA.

Nelson, D. L. and Cox, M. M. (2004), Lehninger Principles of Biochemistry, (4th ed.), W. H. Freeman, USA.

Russell, P. J. (2002) Essential of Genetics (1st ed.), Benjamin Cummings Publishers, UK.

Snustad and Simmons (2006) Principles of Genetics (4th ed.), Wiley Asia Student Edition, India.

Shamoo, Y. and Steiz, T.A. (1999) Building a replisome from interacting places: sliding clamp complexed to a peptide from DNA polymerase and a polymerase editing complex. Cell 99: 155-166.

Voet and Voet (2004) Biochemistry, (3rd edition), Wiley International Edition, USA.

Watson and Baker (2004) Molecular Biology of Gene (5th ed.), Pearson/Benjamin Cummings Publishers, USA.

Suggested Readings

1. Hirota, Y. Ryter, A., and Jacob, F. (1968) Quant. Biol, Thermosensitive mutants of E. coli affected in the processes of DNA synthesis and cellular division. Cold Spring Harbor Symp, 33:677-693.
2. Snustad and Simmons (2006) Principles of Genetics, Fourth edition, Wiley Asia Student Edition.
3. Shamoo, Y. and Steiz, T. A. (1999) Building a replisome from interacting places: sliding clamp complexed to a peptide from DNA polymerase and a polymerase editing complex, Cell 99, 155-166.
4. David, L. Nelson and Michael M. Cox (2004) Lehninger Principles of Biochemistry, 4th edition, W. H. Freeman.
5. Cook, P.R. (2001) Principles of Nuclear Structure and Function, wiley-Liss publisher.
6. Russell, P.J. (2002) Essential iGenetics, First edition, Benjamin Cummings publishers.
7. Lewin, B. (2004) Genes VIII, international edition, Pearson Prentice Hall.
8. Berg, J.M., Tymoczko, J. L. and Stryer, L. (2006) Biochemistry, 6th edition, Macmillan Publishers Ltd.
9. Voet and Voet (2004) Biochemistry, Third Edition, Wiley International Edition.

REVIEW QUESTIONS

Long Questions

1. If the template strand of a segment of a gene has the nucleotide sequence 3′-GCTAAGC-5′, what nucleotide will be present in the RNA transcript specified by this gene segment?
2. A scientist is studying the transcription of a gene in mouse cells growing in culture. How can she determine whether the gene is transcribed by RNA polymerase I, RNA polymerase II, or RNA polymerase III?
3. What bases in the mRNA transcript would represent the following DNA template sequence: 5′-TGCAGACA-3′?
4. Design an experiment to demonstrate that RNA transcripts are synthesized in the nucleus of eukaryotes and are subsequently transported to the cytoplasm.
5. What is the function of the introns in eukaryotic genes?
6. Many eukaryotic genes contain noncoding introns that separate the coding sequences or exons of these genes. At what stage during the expression of these split genes are the noncoding introns sequences removed?

7. What do the processes of DNA synthesis, RNA synthesis, and polypeptide synthesis have in common?
8. Compare the structures of primary transcripts with those of mRNAs in prokaryotes and eukaryotes. On average, in which group of organisms do they differ the most?
9. In what ways does RNA transcription in eukaryotes differ from RNA transcription in prokaryotes?
10. Why was the need for an RNA intermediary in protein synthesis most obvious in eukaryotes? How did researches first demonstrate that RNA synthesis occurred in the nucleus and that protein synthesis occurred in the cytoplasm?

Multiple Choice Questions

1. The sequence of bases located prior to the gene (along the DNA strand), to which a complex of RNA polymerase and sigma factors attaches itself to initiate transcription is called
   (a) Promoter (b) Terminator
   (c) Exon (d) Telomere
2. A segment of DNA that includes regions preceding and following the coding DNA (introns as well as exons) is called
   (a) Cistron (b) transposons
   (c) retroposon (d) operon
3. A G-C-rich stretch of nucleotides, followed by an A-T-rich stretch of nucleotides is the signal for
   (a) Rho dependent termination (b) Rho independent termination
   (c) Initiation (d) transcription
4. The bacterial promotor sequence usually found at the -10 position (ten base pairs upstream of the transcription start site) is
   (a) TATAAT (b) TAGACA
   (c) TTGACA (d) TATTAT
5. Sections of DNA within a gene that do not encode part of the protein that the gene produces are called
   (a) Transposons (b) intein
   (c) Introns (d) Exons
6. An RNA molecule that can catalyze either its own cleavage (self-splicing) or the cleavage of other RNA molecules is
   (a) An inducer (b) A chaperone enzyme
   (c) A spliceosome (d) A ribozyme
7. Which of the following is not true about introns
   (a) Intron sequences are found in both plant and animal genes
   (b) Intron sequences are not transcribed into RNA
   (c) Intron sequences are primarily found in eukaryotic genes
   (d) Introns sequences in RNA do not encode protein sequence.

8. In eukaryotic cells, the removal of introns occurs
   (a) During DNA replication
   (b) During transcription
   (c) Prior to mRNA leaving the nucleus
   (d) After mRNA leaves the nucleus
9. Termination of transcription in prokaryotes can be dependent on
   (a) Intron /exon boundary sequences
   (b) rho factor
   (c) the 5′UTR
   (d) all of the above
10. E. coli, attenuation and antitermination utilize which structure
    (a) Stem loop structures in RNA (b) stem loop structures in DNA
    (c) RNA/DNA hybrids (d) differential protein folding
11. Indicate the order in which the following steps occur in the production of a mature mRNA
    (a) Initiation of transcription, splicing, addition of 5' cap, addition of poly A tail, transport to cytoplasm
    (b) Initiation of transcription, addition of 5' cap, splicing, addition of poly A tail, transport to cytoplasm
    (c) Initiation of transcription, addition of poly A tail, addition of 5' cap, splicing, transport to cytoplasm
    (d) Initiation of transcription, addition of 5' cap, addition of poly A tail, splicing, transport to cytoplasm
12. Which event does not occur within the nucleus
    (a) RNA editing in mammals (b) RNA capping
    (c) polyadenylation (d) RNA editing in protozoans
13. Which type of RNA participates in nuclear export of mRNA
    (a) snRNA (b) hnRNA
    (c) tRNA (d) rRNA
14. Rho-dependent termination of transcription in E. coli
    (a) requiresATP
    (b) requires about 50 nucleotides of uncomplexed mRNA
    (c) removes mRNA and holoenzyme from the DNA
    (d) both a) and b) are correct.
15. RNA polymerase
    (a) Binds tightly to a region of DNA thousands of base pairs away from the DNA to be transcribed.
    (b) Can synthesize RNA chains de novo (without a primer).
    (c) Has a subunit called lambda, which acts as a proofreading ribonuclease.

(d) Separates DNA strands throughout a long region of DNA (up to thousands of base pairs), then copies one of them.

(e) Synthesizes RNA chains in the 3'→5' direction.

16. Binding by E. coli RNA polymerase, the correct order of events for transcription initiation is

(a) Closed complex formation, open complex formation, promoter clearance, start of RNA synthesis.

(b) Closed complex formation, open complex formation, start of RNA synthesis, promoter clearance.

(c) Open complex formation, closed complex formation, start of RNA synthesis, promoter clearance.

(d) Start of RNA synthesis, closed complex formation, open complex formation, promoter clearance

(e) Start of RNA synthesis, open complex formation, closed complex formation, promoter clearance.

17. "Footprinting" or DNase protection is a technique used to identify

(a) The specific binding site of a repressor, polymerase, or other protein on the DNA.

(b) E. coli cells that contain a desired, cloned piece of DNA.

(c) The position of a particular gene of a chromosome.

(d) The position of internally double-stranded regions in a single-stranded DNA molecule.

18. Which of the following is not known to be involved in initiation by eukaryotic RNA polymerase II

(a) Protein phosphorylation (b) DNA polymerase activity

(c) Formation of an open complex (d) Protein binding to specific DNA sequences

19. The synthesis of RNA and DNA has a number of features in common but there is one big difference. Choose the answer that gives a property of RNA synthesis that is not shared by DNA synthesis

(a) Synthesis proceeds in the 5' to 3' direction.

(b) Synthesis requires a cellular supply of nucleoside triphosphates.

(c) New chains are produced using an existing or parental strand as a template.

(d) RNA polymerases can begin a new RNA chain with the addition of the first nucleoside triphosphate but not DNA polymerase which requires a primer sequence.

20. When ó subunit dissociates from an initiated RNA polymerase

(a) It can bind a core enzyme to reform holoenzyme.

(b) It hydrolyzes ATP until rebound by core enzyme.

(c) It leaves behind an elongating species complexed with Rho factor.

(d) It remains bound to the promoter consensus sequence.

21. The RNA transcript which form after transcription is
    (a) Complimentary to coding strand of DNA
    (b) Identical to Non- Sense strand of DNA
    (c) Identical to template strand
    (d) Identical to minus strand
22. Which is the most likely prokaryotic rho-dependent transcription termination sequence?
    (a) 5′...UUAAAGGCUCCUUUGGAGCCUUUUUUUU...3′
    (b) 5′...UUAAAGGCUCCUUUGGAGCCUUGUCGAC...3′
    (c) 5′...UAGCUUCAGAUUCCAGGUUACUUUUUUU...3′
    (d) 5′...CAGCUCAGAUUGCCUUCAUUCAGAUAUG...3′
23. What mRNA sequence corresponds to this coding-strand DNA? 5′...AAGTGGAAGAAATGGAATGCTGGATGAGTTTCG...3′
    (a) 5′...CGAAACTCATCCAGCATTCCATTTCTTCCACTT...3′
    (b) 5′...CGAAACUCAUCCAGCAUUCCAUUUCUUCCACUU...3′
    (c) 5′...AAGUGGAAGAAAUGGAAUGCUGGAUGAGUUUCG...3′
    (d) 5′...GCUUUGAGUAGGUCGUAAGGUAAAGAAGGUGAA...3′
24. Which is not considered a transcriptional regulatory sequence?
    (a) TATA box
    (b) Enhancer sequence
    (c) Pribnow box
    (d) Shine-Dalgarno sequence
    (e) They are all transcriptional regulatory sequences.

Answers

1. (a)	2. (a)	3. (a)	4. (a)	5. (c)	6. (d)
7. (b)	8. (c)	9. (b)	10. (a)	11. (d)	12. (d)
13. (b)	14 (d)	15 (a)	16. (b)	17. (a)	18. (a)
19. (d)	20. (d)	21. (b)	22. ( )	23. (b)	24. (d)

PART–5

Chromosome and Heterochromatization

CHAPTER 15

Polytene and Lampbrush Chromosomes

Pragya Khanna
Govt. College for Women, Parade, Jammu–180 001
E-mail: pragyajmu2002@yahoo.co.in

(A) POLYTENE CHROMOSOMES

The Polytene chromosomes were observed for the first time by E.G. Balbiani in 1881 in the salivary gland nuclei of midges of the genus *Chironomus*. These structures were also studied separately by Korschelt and Carnoy in 1884, but the early workers did not attach any genetic significance to their observations. Later, various cytologists like Heitz and Bauer (1933), Painter (1933) and King and Beams (1934) worked out the cytological importance of these chromosomes. Polytene chromosomes although widespread in the order Diptera have also been observed in several species of protozoans and plants. In order Diptera besides occurring in salivary glands they are also present in several other tissues such as Malphigian tubules, gut, hepatic cecae and muscle cells. The term 'polytene chromosome' was coined by Kollar in 1935 which is suggestive of the presence of many chromonemal fibrils in them. The polytene chromosomes are larger in size in comparison to other somatic chromosomes. In *Drosophila melanogaster* these chromosomes are over 100 times the length of somatic metaphase chromosomes. The work of Heitz and Bauer showed that these chromosomes exist in haploid number, each being formed by intimate side-by-side pairing of two homologous chromosomes as a result of somatic pairing phenomenon.

STRUCTURE

From the structural point of view polytene chromosomes can be defined as special case of chromosomes, present during Interphase, characterized by (i) a *high content of DNA* and (ii) *particular chromatin structures* related to their high DNA contents. The large size of polytene chromosomes is attributed to the type of growth that occurs in the larval glandular tissues of dipterous insects. Salivaries and other glands grow by enlargement rather than by division. As

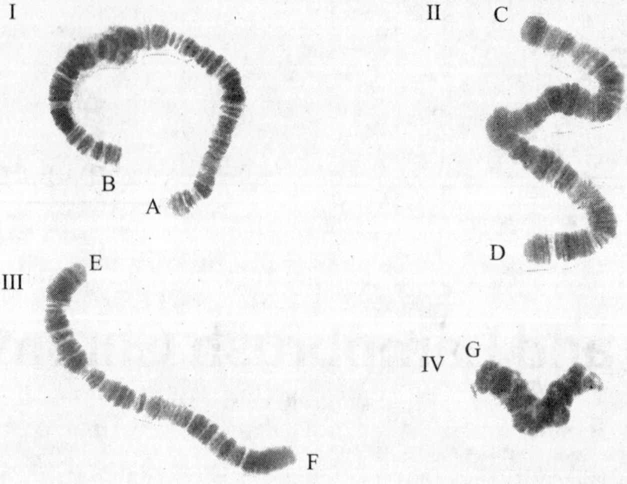

Fig. 1. A polytene chromosome complement of *Chironomus plumosus*.

a larva develops, the synapsed elements replicate nine to ten times giving rise to a polytene strand of about 1000 DNA double helix molecules. The process of recurrent duplication cycle without consequent mitosis is called Endo-reduplication. The probable function of polyteny is suggested to be gene amplification. Having multiple copies of genes permits a high level of gene expression; that is, abundant transcription and translation to produce the gene products in large, metabolically active cells (like salivary glands). For unknown reasons, the centromeric regions of the chromosomes do not endo-replicate very well. As a result, the centromeres of all the chromosomes bundle together in a mass called the Chromocenter (Fig.2).

The polytene chromosomes bear alternate transverse dark and light regions along their length. The dark bands observed in the polytene chromosomes are chromomeres, densely packed with chromatin, while the lighter interband regions are chromonemata made up by chromatin that is packed much more loosely. The bands are the regions of high DNA concentration, Feulgen positive, stain intensely with basic stains and absorb ultra-violet light at 2,600Å. The interbands on the other hand, are the regions of low DNA concentration, Feulgen negative, do not stain with basic dyes and absorb very little ultra-violet light. The polytene chromosomes of plants, however, do rarely display a clear banding pattern. A possible explanation is that chromomeres in this case are not lined up exactly in parallel and as a consequence the denser chromatin parts do not lie side-by-side to form a clear-cut banding pattern.

The work of Bridges (1936) showed that the bands of the polytene chromosomes are related to genes and the interbands are relatively inert linker regions. It is now believed that the interbands also contain many genes, though, chromomeres contain much folded or spiralized DNA while, in interband regions the DNA is in extended state. Du Praw and Rae (1961) suggest that a gene may overlap from band to interband or may even lie entirely in the interband. Polytene chromosomes, thus, correspond in linear structure with other chromosomes of the same species. The difference being that the polytene chromosomes are formed of about 1000 to 4000 unit chromatids which do not separate out to new cells through

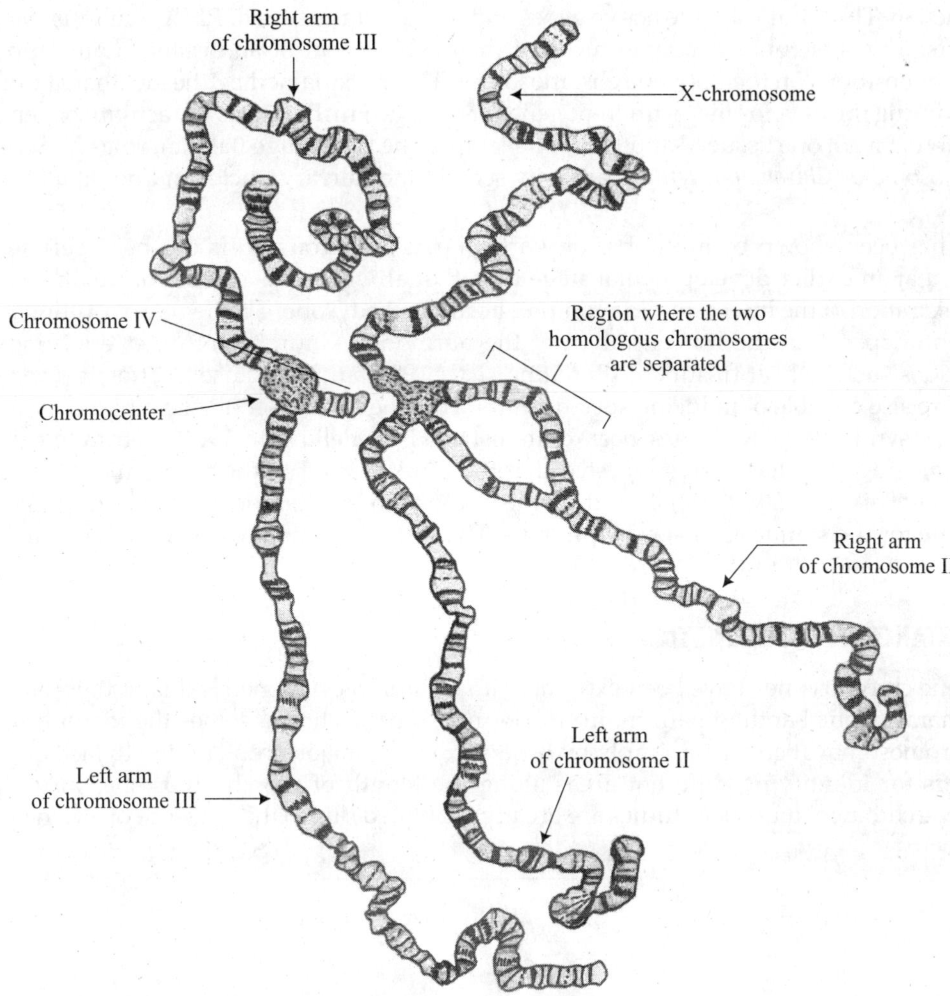

Fig. 2. A diagrammatic representation of polytene chromosomes of *Drosophila*.

cell division. About 5,000 bands have been observed on the four chromosomes of *Drosophila melanogaster*. The banding pattern is distinctive for each homologous pair of chromosomes in any given species.

PUFFS AND GENE ACTIVITY

At certain stages in development, some genetic loci in the polytene chromosomes become enlarged into swellings or puffs which have been called chromosome puffs or Balbiani rings. The process of puff formation has been studied in detail by Beerman and Bahr (1954) and it has been suggested that puffing involves unfolding or unraveling of DNA in the chromomeres of a particular band in the form of a loop. The process of puffing is due to the biosynthetic activity of the particular loci concerned and hence, plays an important role in the understanding of gene action. Thus, puffs are the active genes and represent the sites of RNA synthesis. Majority of puffs are visible only at certain times of the development of an organism and appear at different positions on the polytene chromosomes. This is explained by the fact that at different times during the development different genes are active. Furthermore, the activity of genes can be different from one tissue to another. For example, the three huge Balbiani rings in the fourth chromosome of *Chironomus tentans* are only seen in the salivary nuclei and not at all in other tissues.

It has been shown by a number of workers that numerous puffs can be experimentally induced at an earlier developmental stage than that at which they would normally occur by administration of the insect steroid moulting hormone Ecdysone. The degree of puffing and its duration depend on the concentration of the hormone. A number of other environmental conditions such as heat treatment (Barettino *et al.*, 1988) and galactose treatment (Diez *et al.*, 1990) are also capable of inducing specific puffs. That the phenomenon of puffing is associated with the synthesis of RNA has been demonstrated by Pelling in 1966 by tritiated uridine autoradiography. It has been suggested that the RNA formed in the puff is messenger RNA (mRNA). This mRNA is transferred to the cytoplasm as ribonucleoprotein and acts as a template for the synthesis of special proteins. The mRNA formed in one puff varies from the one made in another puff.

IMPORTANCE IN CYTOGENETICS

Polytene chromosomes have been extremely important in cytogenetics due to their large size and characteristic banding pattern. In the normal somatic chromosomes the identification of the chromosomal aberrations simply by observation is a major problem due to lack of visible markers for identifying different areas along the length of the chromosome. In the giant polytene chromosomes such studies are greatly facilitated due to the presence of alternate dark

Fig. 3. C.B. Bridges (1889-1938), an american cytogeneticist who worked on the chromosomes of *Drosophila melanogaster*.

and light bands. They provide a useful tool for comparing the individual chromosomes within a chromosome set or genome, as well as whole sets from different organisms. Comparison of banding sequences has been of great significance in the analysis of evolutionary cytogenetic processes. Also, the chromosomal rearrangements such as deletions, duplications, inversions, translocations can be studied in detail since any structural rearrangement is expressed in corresponding alteration in somatic synapsis.

Deletions: A single break near the end of the chromosome results in a terminal deficiency. In case two breaks occur a section may be deleted and an intercalary deficiency is created. In this case a loop like structure is observed in the polytene chromosomes due to alteration in the pairing of two homologues (Fig.4). By identifying the part of the polytene chromosome in which the formation of such a loop like structure is seen and then studying the phenotype of the organism carrying a recessive gene in the homologous chromosome opposite the deletion or deficiency, the gene can be spatially positioned on the chromosome. The physical location of many genes is now precisely known in *Drosophila melanogaster* and some other species of Diptera because of this technique.

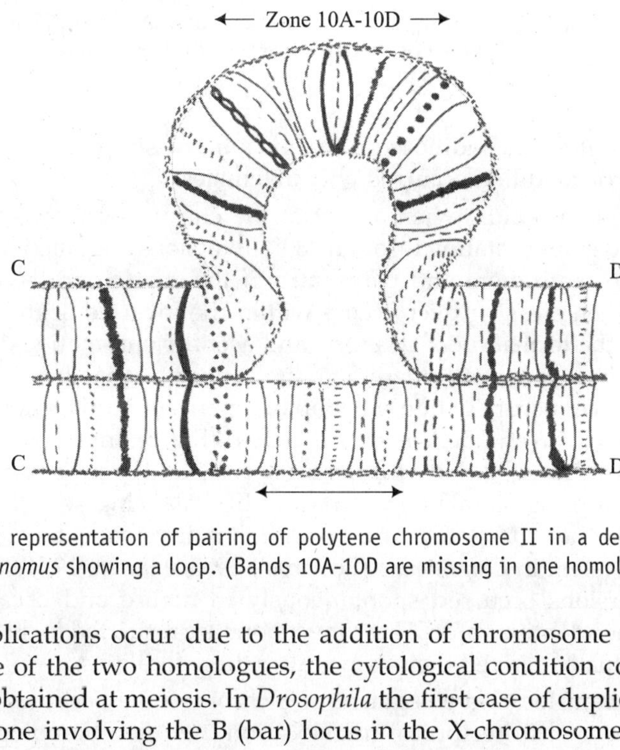

Fig. 4. Diagrammatic representation of pairing of polytene chromosome II in a deficiency heterozygote in *Chironomus* showing a loop. (Bands 10A-10D are missing in one homologue).

Duplications: Duplications occur due to the addition of chromosome parts. If duplication occurs only on one of the two homologues, the cytological condition comparable to that of deficiency will be obtained at meiosis. In *Drosophila* the first case of duplication to be critically analyzed was the one involving the B (bar) locus in the X-chromosome (Fig.5). Bar eye is a character where eyes are narrower compared to the normal eye shape and this condition appears due to the duplication in the region 16A in the X-chromosome. The wild type eye of *Drosophila* is large and has an average of 779 facets. It has two Bar regions on each X-chromosome. If this region undergoes duplication, the size of the eye is reduced and number of facets decreases. Though, there is no change in the amount of genetic material but the order of the genes gets altered. This gave rise to the concept of position effect. Barred individuals give

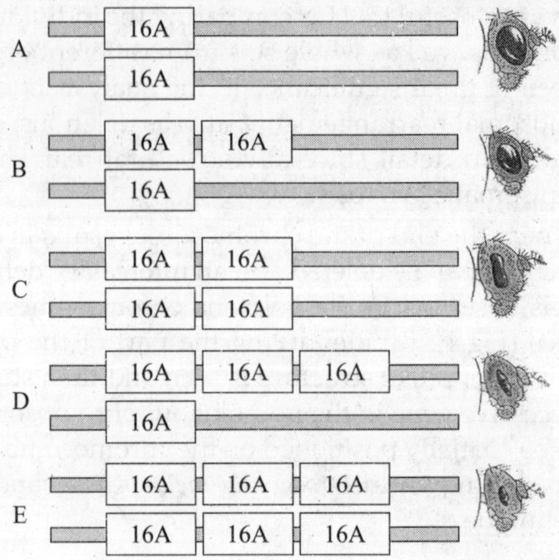

Fig. 5. Effects of different arrangements of duplicated sections of 16A in the X-chromosome of *Drosophila melanogaster* on size of the eye. (A) Wild type (B) Heterozygous Bar (C) Homozygous Bar (D) Heterozygous double Bar (E) Homozygous double bar.

rise to double bar and ultra bar conditions due to unequal crossing over. This was suggested by the fact that this type of condition occurred only in females.

Inversions: An inversion occurs when a part of the chromosome becomes detached and reunites in reverse order after rotating through 180°. Inversions are called pericentric when the segment includes the centromere and paracentric if the centromere is located outside the inverted segment. When crossing over occurs within the inverted segment of a paracentric inversion, it leads to the formation of dicentric and acentric chromatids during meiosis. The dicentric chromatids have two centromeres and are connected together by a bridge. If on the other hand, crossing over occurs within the loop of a pericentric inversion then the resulting chromatids with duplication and deficiency are formed (Fig. 6). Inversions are known to have played a significant role in evolution of different species and races of dipterans. It has been found that this condition is maintained in nature by balancing selection due to selective advantage of inversion heterozygotes. As a result, many inversions have become part of the stable chromosome system in *Chironomus* (Beerman, 1953; Martin, 1974). In *Drosophila* it has been seen that inversions occurred spontaneously in nature and became established in populations due to the adaptive value. These inversions are restricted to different localities and the different races actually derived their names after these localities.

Translocations: In a translocation a segment of a chromosome becomes attached to a non-homologous chromosome. In reciprocal translocation, two non-homologous chromosomes exchange segments. Translocations occur due to irregularities during crossing over. They may not involve a loss or an addition of chromosome material, but frequently they do become associated with deficiencies, duplications and unbalanced combinations of genetic units (Fig. 8).

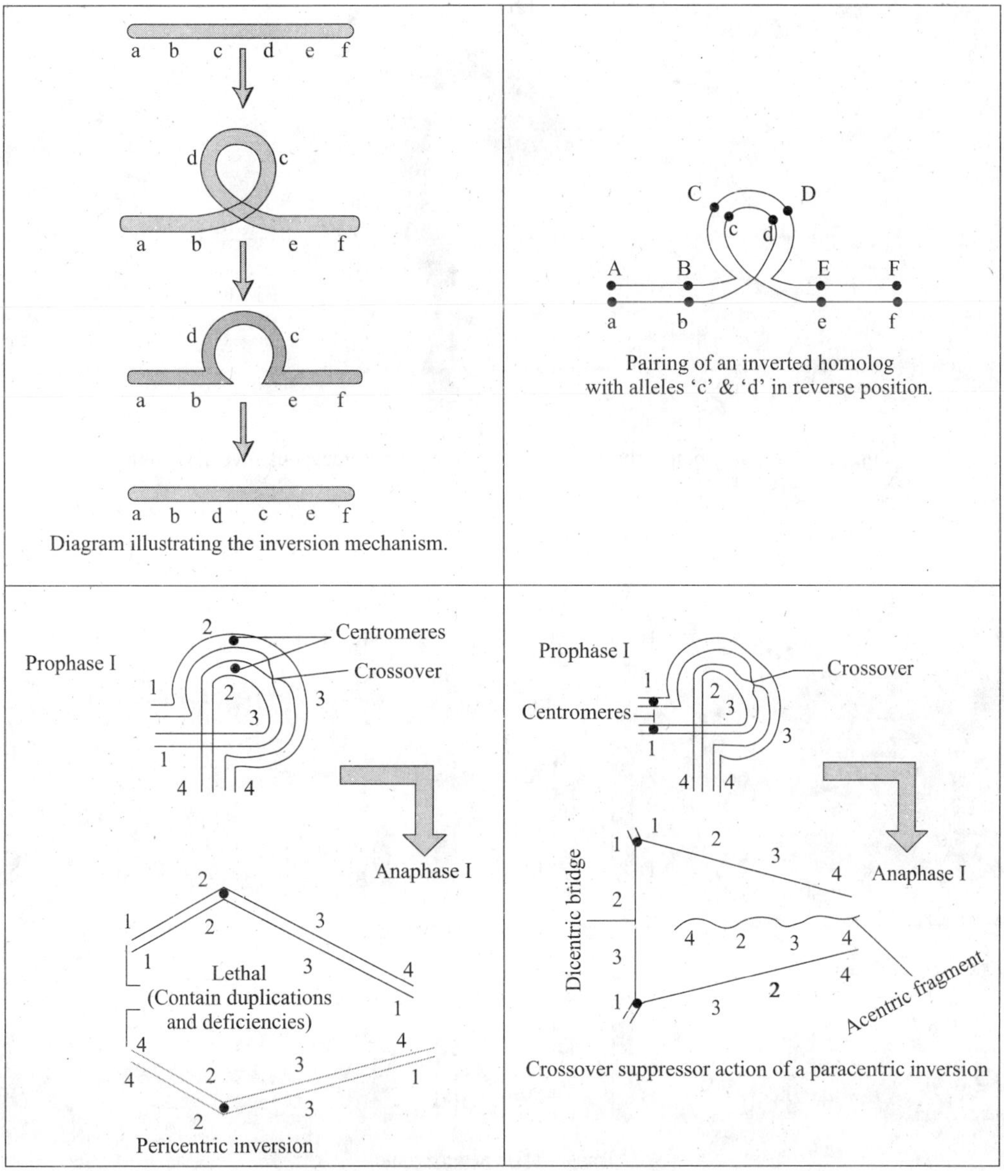

Fig. 6. Ozag. illustrating pericentric and paracentric inversions.

(B) LAMPBRUSH CHROMOSOMES

Lampbrush chromosomes are transitory structures that exist during an extended diplotene of the first meiotic division in female gametocytes of most animals, except mammals. They are found in the oocyte nuclei of vertebrates like sharks, amphibians, reptiles and birds and in

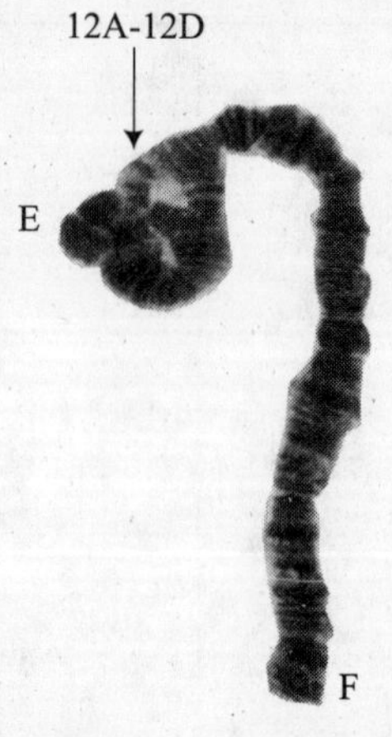

Fig. 7. A salivary gland chromosome illustrating a heterozygous inversion loop.

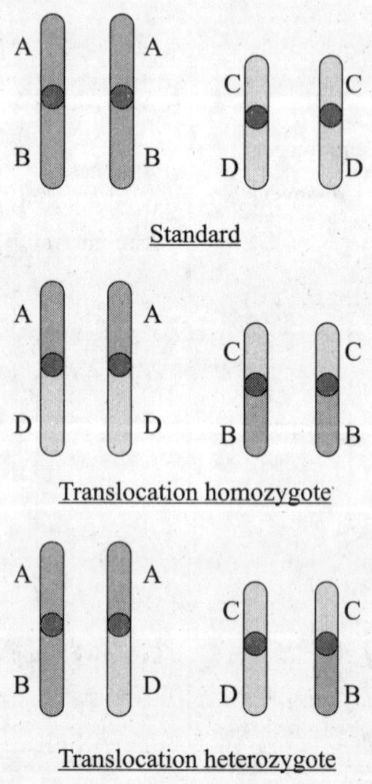

Fig. 8. Homozygous and Heterozygous translocation.

invertebrates like *Sagitta, Sepia, Echinaster* and in several species of insects. The name lampbrush is given to these giant chromosomes due to their resemblance to the brushes used to clean old-fashioned oil lamp chimneys. Lampbrush chromosomes were first seen in sections of salamander (*Ambystoma mexicanum*) oocytes by Flemming in 1882. Ten years later, in 1892, they were described in the oocytes of a dogfish by Ruckert. They have also been reported in plants (Grun, 1958).

STRUCTURE

Since lampbrush chromosomes exist in diplotene stage of meiosis, they are bivalents, i.e., they consist of two homologues (two pairs of sister chromatids) held together by chiasmata at one or more points (Fig. 9). They are about 800 μm long. At maximum development, they are even larger than the largest salivary gland chromosomes.

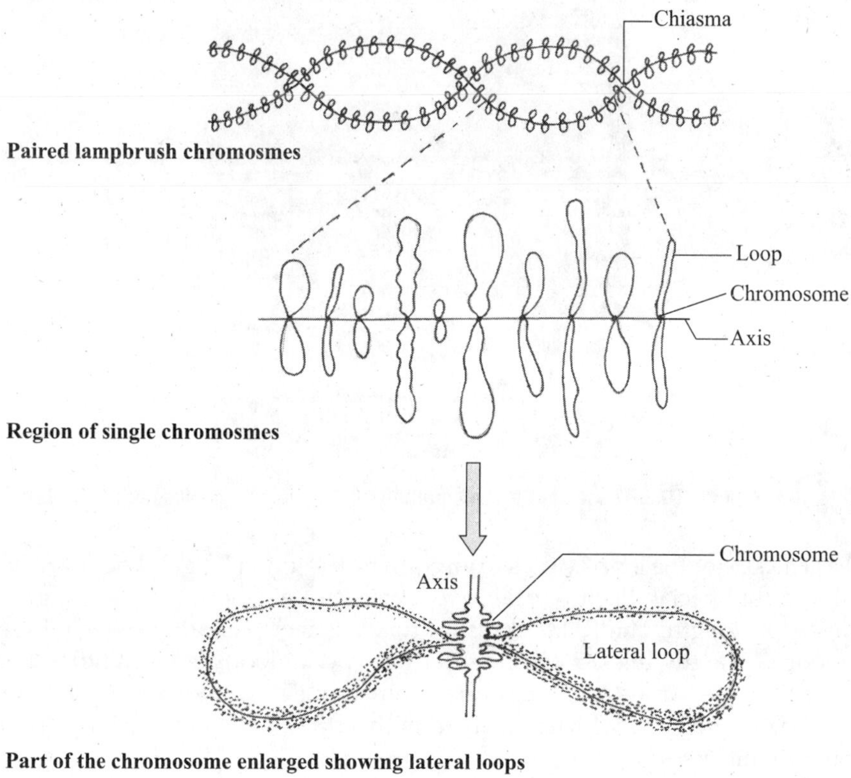

Fig. 9. Diagrammatic represenation of the structure of lampbrush chromosomes of amphibian oocytes.

Chromosomal axis: Lampbrush chromosomes are generally described as having a central chromosomal axis formed of four chromatids. The axis varies from 30-50 Å in diameter and is extensible and elastic, composed of DNA and protein.

Chromomeres: The chromosomal axis is dotted with a row of Feulgen-positive granules called chromomeres which are about 1 μm in diameter and 1-5 μm in length, connected by an extremely thin continuous chromonema. In addition to DNA, the chromomeres contain basic protein (Gall, 1954). These chromomeres are found in pairs, one for each homologue.

Loops: Most of the chromomeres bear pairs of lateral loops. It is the presence of hundreds of loops arising from each bivalent which give the chromosome the characteristic hairy 'lampbrush' appearance. Members of each pair of loops are similar in length and morphology, but, loops arising from different non-homologous chromomeres differ in length and

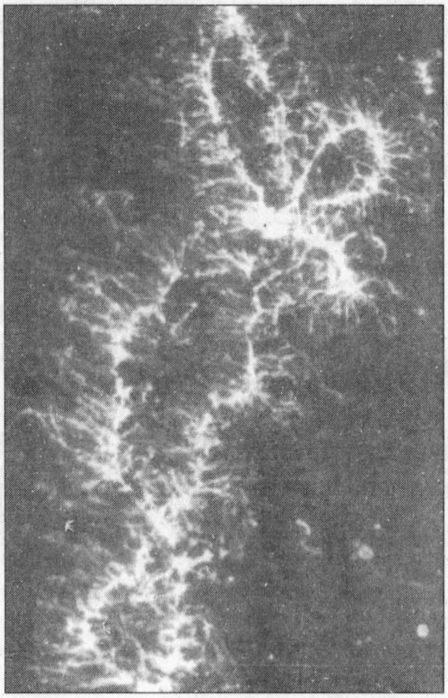

Fig. 10. Phase contrast micrograph of lampbrush chromosomes.

appearance. The size of the loops varies from 5 μm to 100 μm in length. Loops are of two types: (a) Typical and (b) Special. Each typical loop consists of a central axis which gives off RNA fibrils all along its length. The RNA fibrils on this axis increase progressively in length. This makes the loop markedly thicker on one side. The special loops bear granules at the ends of these RNA fibrils. One to nine loops may arise from a single chromomere. Each loop is thought to be composed of a loop axis which is covered with matrix. The loop axis is composed of DNA about 30-50 Å in thickness whereas, matrix has protein combined with RNA. The loops appear to be Feulgen-negative in stained preparations. The difference between the individual loops depends on the amount and kind of matrix. As meiosis proceeds towards metaphase I, the loops undergo regression and are eventually reabsorbed into the chromosomes which are finally converted into normal compact meiotic bivalents.

Master and Slave hypothesis: Callan and Lloyd in 1960 postulated the 'master and slave hypothesis'. According to this, each loop represents a long operon, consisting of a series of duplicate copies of the same structural gene (cistrons). Each gene locus produces an RNA molecule which interacts with protein to form ribonucleoprotein particles. There is a master copy of a particular gene in the chromomere which transfers information to several identical gene copies on the same loop (slave genes). Between the gene loci matrix free regions of 'spacer' DNA are present which do not produce RNA. It has been observed that only the slave genes take part in RNA synthesis. A high rate of synthesis is possible because of the presence of repeating similar genes.

FUNCTIONAL SIGNIFICANCE

Lampbrush chromosomes are thought to assist in fulfilling the high demand for transcripts during oogenesis. Studies on oogenesis in vertebrates reveal that extensive transcription occurs during prophase I (diplotene) of meiosis. The loops of the lampbrush chromosomes have been shown to be the regions of active transcription by means of pulse-labeling with [^{3}H] uridine and autoradiography. The transcribed genes are those whose products are required during early stages of embryogenesis.

SUMMARY

In certain organisms some cells at particular stages contain large nuclei with giant or large-sized chromosomes. Such giant chromosomes are of two types, viz., Polytene chromosomes and Lampbrush chromosomes. The polytene chromosomes are giant multistranded chromosomes commonly occurring in many dipteran flies. They are larger in size as compared to other somatic chromosomes. The enormous size of the polytene chromosomes is achieved by the process of endoreduplication in which the repeated rounds of DNA replication occur without any cell division. Another characteristic feature of these chromosomes is the occurrence of somatic synapsis as a result of which their number always appears to be half as compared to that in other normal somatic cells. They have a distinct pattern of transverse bands and interbands along their length. The bands become enlarged at certain points forming chromosome puffs or Balbiani rings which are associated with differential gene activation. The polytene chromosomes are used as a good model to examine the structural aberrations in chromosomes due to their large size and characteristic banding pattern. The lampbrush chromosomes are meiotic or 'germ-line' chromosomes, present during prophase I of oogenesis in many vertebrates particularly amphibians. Due to their large size they provide favourable material for cytological studies. Each lampbrush chromosome consists of a central axial region composed of two bivalent chromosomes and numerous pairs of lateral loops that project from the chromomeres present along the length of the axial filaments. The main axis is rich in DNA and protein whereas the loops mainly consist of RNA. The lampbrush chromosomes are associated with synthetic activity in the growing oocyte.

REFERENCES

Balbiani E.G. (1881) *Surla structure du noyau des celluler cellulaires chezles larvas de Chironomus.* Zool. Anz. 4 : 637-641.

Barettino, P., Morcillo, G. and Diez, J. (1988) Induction of the heat shock response by carbon dioxide in *Chironomus thummi*. Cell Differentiation 23: 27-36.

Beermann, W. (1953) Chromosomen polymorphismus und Bastardieurung Zwier *Chironomus*. Arten. Zool. Anz. Supplement. 17: 290-295.

Beermann, W. and Bahr, G.F. (1954) The submicroscopic structure of the Balbiani ring. Exp. Cell Res. 6:195-201.

Bridges, C.B. (1936) The Bar 'gene' a duplication. Science 83: 210-211.

Callan, H.G. and Lloyd, L. (1960) Lampbrush chromosomes. In: New approaches in cell biology (P.M.B. Walker, ed.) New York: Academic Press.

Carnoy J.B. (1884) *La Biologie Cellulaire*. Fasc. 1. Lierre.

Diez, J.L., Cortes, E., Merino, Y. and Santa-Cruz, M. (1990) Galactose induced puffing changes in *Chironomus thummi* Balbiani rings and their dependence on protein synthesis. Chromosoma 99: 61.

Gall, J.G. (1954) Lampbrush chromosomes from oocyte nuclei of the newt. J. Morph. 94: 283-352.

Heitz, E and Bauer, H. (1933) *Beweise fur die chromosomennatur der kernschleifen in den knavelkernen Von Bibio hotulanus*. L.Z. Zellforsch 17 : 67-82.

King, R.L. and Beams, H.W. (1934) Somatic synapsis in *Chironomus* with special refrence to the indivduality of chromosomes. J. Morph. 56 : 577-586.

Koller, P.C. (1935) The internal mechanics of the chromosome IV pairing and coiling in the salivary gland nuclei of *Drosophila*. Proc. Roy.Soc. Lond. Ser. B. 118 : 371-397.

Korschelt, E. (1884) *Uber die eigentumliche Bildung in den Zellkernen der speicheldruse von Chironomus plumosa*. Zool. Anz. 7 : 189-194.

Martin, J. (1974) A review of the genus *Chironomus* (Diptera, Chironomidae) IX. The cytology of *Chironomus tepperi* skuse. Chromosoma 45: 91-98.

Painter, T.S. (1933) A new method for the study of chromosome rearrangements and the plotting of chromosome maps. Science 78 : 585-586.

Pelling, C. (1966) A replicative and synthetic chromosomal unit-the modern concept of the chromomere. Proc. Roy. Soc. Lond. B. 164: 279-289.

Suggested Readings

1. Principles of Genetics (1991) E.J. Gardener, M.J. Simmons and D.P. Snustad, 8th edition, John Wiley and Sons Inc., USA.
2. Chromatin : Structure and Function (1998) A.P. Wolffe, Academic Press, UK.
3. Principles of Nuclear Structure and Function (2001) P.R. Cook, Wiley-Less, Inc., USA.
4. Chromosomes: Organization and Function (2003) T. Adrian, A. Summer and T. Sumner, Blackwell Publishing Co., UK.
5. Genetics (1996) Stickberger, Prentice-Hall of India Pvt. Ltd. India.

REVIEW QUESTIONS

Long Questions

1. Draw a Giant Chromosome. Label euchromatin, heterochromatin, and chromocenter.
2. How do polytene chromosomes form?
3. Describe the functional activity of dark and light bands.
4. Correlate formation of lampbrush chromosomes with the events of meiosis.
5. How do we know what the matrix of lampbrush chromosomes is composed of? How has transcription been demonstrated on lampbrush chromosomes?

Multiple Choice Questions

1. Polytene chromosomes were first of all reported by
   (a) Sutton and Boveri (b) C.B. Bridges
   (c) E.G. Balbiani (d) Fleming

2. Puffs are found in
   (a) Lampbrush chromosomes (b) Polytene chromosomes
   (c) B-chromosomes (d) All of the above
3. Polytene chromosomes occur at _______ stage of mitosis
   (a) Metaphase (b) Interphase
   (c) Anaphase (d) Telophase
4. A chromosome with large number of chromonemata is
   (a) Polytene chromosome (b) Lampbrush chromosome
   (c) Telocentric chromosome (d) Both a and b
5. Lampbrush chromosomes are found in
   (a) Oocytes of birds (b) Brain cells of reptiles
   (c) Salivary glands of diptera (d) Cancer patients
6. _______ first discovered lampbrush chromosomes
   (a) Ruckert (b) Kolliker
   (c) Balbiani (d) Bridges
7. A chromosome having a series of transversely disposed loops on either side is
   (a) Metacentric chromosome (b) Polytene chromosome
   (c) Lampbrush chromosome (d) Acrocentric chromosome
8. Lampbrush chromosomes are seen during
   (a) Diplotene of the first meiotic division
   (b) Prophase II
   (c) Leptotene of the first meiotic division
   (d) Metaphase I
9. Lampbrush chromosomes are not found in
   (a) Reptiles (b) Birds
   (c) Insects (d) Mammals
10. The part of the chromosome that shows greater degree of compaction and does not unfold for synthetic activity is called
    (a) Euchromatin (b) Heterochromatin
    (c) Chromonema (d) Chromocenter

Answers

1. (c)	2. (b)	3. (b)	4. (a)	5. (a)	6. (a)
7. (c)	8. (a)	9. (d)	10. (b)		

CHAPTER 16

C-Value Paradox

Pragya Khanna
Govt. College for Women, Parade, Jammu–180 001
E-mail: pragyajmu2002@yahoo.co.in

The C-value is the amount of DNA in the haploid genome of an organism. It varies over a very wide range with a general increase in C-value with complexity of organism from prokaryotes to eukaryotes. Vendrely and Vendrely (1948) pointed out a 'remarkable constancy in the nuclear DNA content of all the cells in all the individuals within a given animal species'. However, it was soon found that C-values, i.e., genome sizes vary enormously among species of the same genus as well as among genera of the same family and that this bears no relationship to the presumed number of genes as reflected by the complexity of the organism. For example, the cells of some salamanders may contain 40 times more DNA than those of humans. But, man is considered more complex in terms of genetic development. This lack of correlation between genome size and genetic complexity is considered paradoxical and is thus, referred to as C-value paradox. Fig. 1 summarizes the range of C-values found in different evolutionary phyla. There is an increase in the minimum genome size found in each group as the complexity increases (Fig. 2). But as absolute amounts of DNA increase in the higher eukaryotes, some wide variations in the genome sizes are seen within some phyla.

The term C-value paradox was coined by Thomas in 1971. There is yet another aspect of C-value paradox involving huge variation in C-values between species, whose apparent complexity does not vary correspondingly. This is especially seen in case of insects, amphibians and plants. The discovery of non-coding DNA resolved the C-value paradox, that is, why the genome size does not reflect gene number in eukaryotes (Table 1). Most eukaryotic (but not prokaryotic) DNA is non-coding and therefore does not consist of genes, and as such total DNA content is not determined by gene number in eukaryotes. The human genome, for example, is comprised of only about 1.5% protein-coding genes, with the other 98.5% being various types of non-coding DNA. This large proportion of non-coding DNA in most eukaryotes has been shown to be present in the form of multiple copies of identical DNA

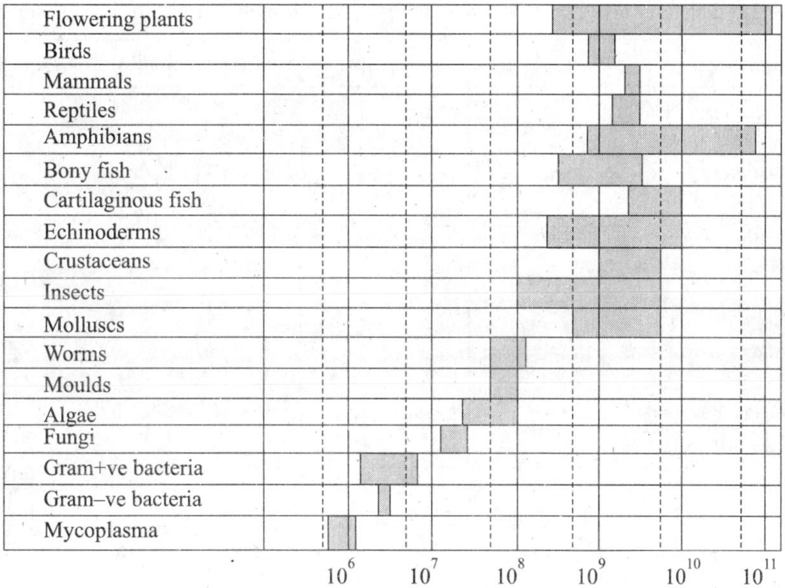

Fig. 1. Variation in genomic DNA in a variety of animals and plants.

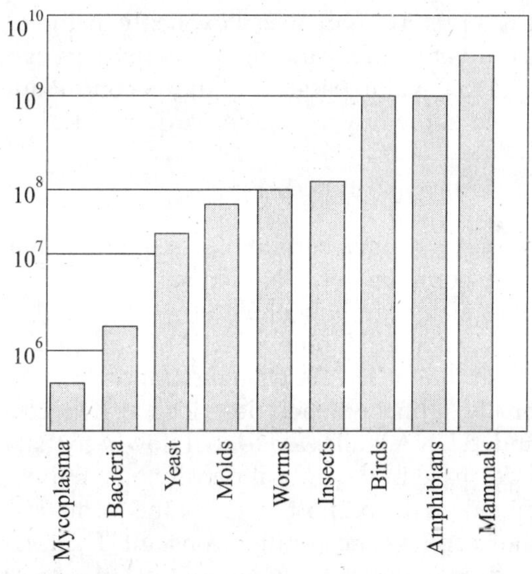

Fig. 2. Mininmum genome size increases with the phylum.

sequences; this multiplicity ranges from a few copies to millions of copies within the same cell. These DNA sequences are together described as repetitive DNA. The remaining DNA is described as Unique DNA and is found in the form of single copy DNA sequences. These sequences are dispersed throughout the genome and are responsible for coding proteins. The

Table 1. Range of DNA content of haploid genomes of representative groups of organisms

Group	Approximated size of haploid genome (bp)
Viruses & bacteria	$10^{-5} \times 10^6$
Algae & fungi	$5 \times 10^6 - 5 \times 10^7$
Worms	$5 \times 10^7 - 10^8$
Insects	$10^8 - 5 \times 10^9$
Echinoderms	$5 \times 10^8 - 5 \times 10^9$
Fish	$10^9 - 5 \times 10^{10}$
Amphibians	$5 \times 10^9 - 5 \times 10^{11}$
Birds	$5 \times 10^9 - 10^{10}$
Reptiles	$5 \times 10^9 - 5 \times 10^{10}$
Mammals	$5 \times 10^9 - 5 \times 10^{10}$
Flowering plants	$5 \times 10^9 - 10^{12}$

repetitive DNA, on the other hand, may be clustered at one place or interspersed with unique sequences: though they do not carry any genetic information, but they are believed to play some other structural or regulatory role.

Recently, it has been suggested that the term C-value paradox may be abandoned and the usage of a more appropriate term C-value enigma be considered. This term was coined by Gregory in 2001. This term is preferred because it explicitly includes all the questions that will need to be answered if a complete understanding of genome size evolution is to be achieved. It was endorsed in preference to C-value paradox at the 'Second Plant Genome Size Discussion Meeting and Workshop' held at the Royal Botanic Gardens, UK.

REPETITIVE DNA

The chromosomes of eukaryotes contain certain base sequences that are repeated many times in the haploid chromosome complement. DNA containing such repeated sequences, called repetitive DNA, represents a major component of the eukaryotic genome. In order to detect the presence of repetitive DNA in the genome, the DNA solution is heated resulting in the denaturation of the double stranded DNA into single stranded DNA. This is accompanied with increase in the optical density, a phenomenon described as hyperchromicity. In the next step, the solution of single stranded DNA is allowed to cool slowly so that the double stranded DNA forms again resulting in decrease in the optical density, i.e., hypochromicity. The amount of repetitive DNA is determined on the basis of extent of the formation of double stranded DNA in a definite period of time keeping concentration constant. The formation of double stranded DNA is actually measured over different values of *Cot*, that is, concentration × time. The patterns of reassociation in a number of organisms are shown by *Cot* curves (Fig. 3).

DNA RENATURATION KINETICS

The relationships among the rates of renaturation, complexity and repetition frequency in eukaryotes can be studied by DNA renaturation kinetics. If a long DNA molecule with no repeated sequences is broken into fragments of a particular length ad denatured and is then

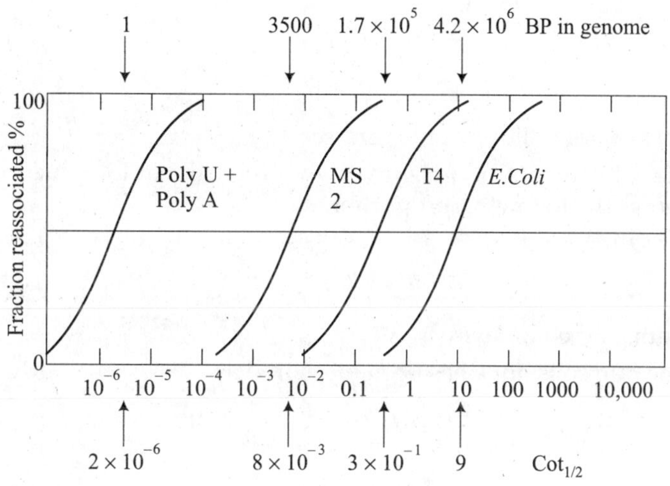

Fig. 3. Rate of reassocation in inversely proportional to the length of the reassociating DNA>

allowed to renature under appropriate conditions, the rate of renaturation will depend on, (1) the concentration of DNA in solution and (2) the complexity of the DNA, that is, the number of different fragments of a particular length. Let us consider a particular fragment composed of two complementary strands A and B. Reassociation of A and B will require a specific collision between these two strands and a collision of either of the two strands with any other single strand will not lead to hybridization. For any given concentration, the larger the DNA molecule, the more non-identical base-pair fragments of a particular length are seen and hence it results in the slower reassociation reaction because of a smaller proportion of the random collisions between the complementary single strands such as A and B. Hence, the rate of renaturation should be a function of the square of the concentration of single strands, which follows the second order or bimolecular reaction kinetics. The time required for half renaturation is inversely proportional to the rate constant. Let C = concentration of single stranded DNA at time *t* (expressed as moles of nucleotides per liter). The rate of loss of single-stranded DNA during renaturation is given by the following expression for a second-order rate process:

$$\frac{-dC}{dt} = kC^2 \quad \text{or} \quad \frac{dC}{C^2} = -kdt$$

That is, the decrease in concentration of single stranded DNA (-*dC*) with time (*dt*) is equal to the proportionality constant (*k*) times the square of the concentration of single stranded DNA. Integration from the initial conditions ($t = 0$ seconds and $C = C_0$, where C_0 equals the concentration of single stranded DNA at $t = 0$) shows:

$$\frac{C}{C_0} = \frac{1}{1 + kC_0{}^2}$$

Thus, at half renaturation when

$$\frac{C}{C_0} = 0.5 \quad \text{and} \quad t = t_{1/2}$$

One obtains:

$$Cot_{1/2} = \frac{1}{k}$$

where k is the rate constant in litres (mole nt)$^{-1}$ sec^{-1}

The rate constant for renaturation is inversely proportional to sequence complexity. The rate constant, k, shows the following proportionality:

$$k \, a \frac{\sqrt{L}}{N}$$

where L = length; N = complexity

Empirically, the rate constant k has been measured as

$$K = 3 \times 10^5 \frac{\sqrt{L}}{N}$$

in 1.0 M Na^+ at $T = T_m - 25°C$

The time required for half renaturation (and thus *Cot1/2*) is directly proportional to sequence complexity

Thus, $$Cot \times a \frac{N}{\sqrt{L}}$$

For a renaturation measurement, one usually shears DNA to a constant fragment length L (e.g. 400 bp). Then L is no longer a variable, and

$$Cot \times a \, N$$

The data for renaturation of genomic DNA are plotted as *Cot* curves.

DNA renaturation kinetics experiments not only show what proportion of a genome consists of a particular class of repetitive DNA, but also how many copies of that particular sequence are present.

SATELLITE DNA

There is remarkable variability in genome size among eukaryotes that has little correlation with organismal complexity, ploidy or number of coding genes. For example, a newt has six times the genome size of a human. Much of this variation is due to non-coding, tandemly repeated DNA. Indeed, a substantial fraction of the genomes of many eukaryotes is composed of repetitive DNA in which short sequences are tandemly repeated in small to huge arrays. Tandemly repetitive sequences, commonly known as "satellite DNAs" are classified into three major groups:

Satellites

They are highly repetitive with repeat lengths of one to several thousand base pairs. These sequences are typically organized as large (up to 100 million bp) clusters in the heterochromatic regions of chromosomes, near centrosomes and telomeres; these are also found abundantly on the Y chromosome.

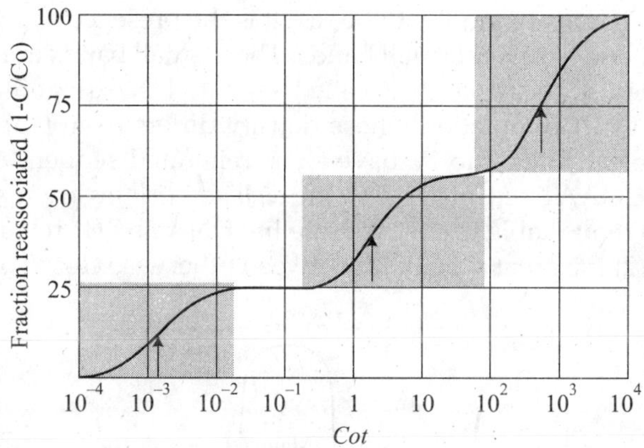

Components	Fast component	Intermediate component	Slow component
Percent of genome	25	30	45
$Cot_{1/2}$	0.0013	1.9	630
Complexity, bp	340	6.0×10^5	3.0×10^8
Repetition freq.	500,000	350	1

Fig. 4. The reassociation kinetics of eukaryotic DNA show three types of components (indicated by the shaded areas).

Minisatellites

They are moderately repetitive, tandemly repeated arrays of moderately sized (9 to 100 bp, but usually about 15 bp) repeats, generally involving mean array lengths of 0.5 to 30 kb. They are found in euchromatic regions of the genome of vertebrates, fungi and plants and are highly variable in array size.

Microsatellites

They are moderately repetitive, and composed of arrays of short (2-6 bp) repeats found in vertebrate, insect and plant genomes. The human genome contains at least 30,000 microsatellite loci located in euchromatin. Copy numbers are characteristically variable within a population, typically with mean array sizes on the order of 10 to 100. In general, satellite DNAs show exceptional variability among individuals, particularly with regard to the number of repeats at a given locus. Minisatellite loci are the most highly polymorphic sequence elements yet discovered in the human genome, and delineating the repeat lengths of these loci is the basis of most DNA typing systems used in forensic medicine.

Fast *reassociating* DNA often consists of tandem highly repeated DNA. The first evidence for repetitive DNA came from density-gradient analysis of eukaryotic DNA. In eukaryotes, since G-C base-pairs are denser that A-T base-pairs, the density of DNA increases as the %GC content increases. When such DNA is isolated, fragmented and centrifuged to equilibrium into a cesium chloride (CsCl) density gradient, the result is the presence of one large band of DNA (Mainband DNA) and one to several small bands. These small bands are called satellite bands and the DNA contained in these bands is called satellite DNA (Fig. 5). Therefore, 'satellite' actually refers to any DNA population whose density differs from that of the bulk of DNA. Human DNA has five satellites. I to IV have short reiterated sequences of less than 100 bp containing divergent GGAAT repeats and V, the *alphoid satellites* are made of 171 bp repeats and are A-T rich. Two remarkable features of satellite DNA are: (1) relative uniformity within the same species and (2) the great variability between otherwise closely related species.

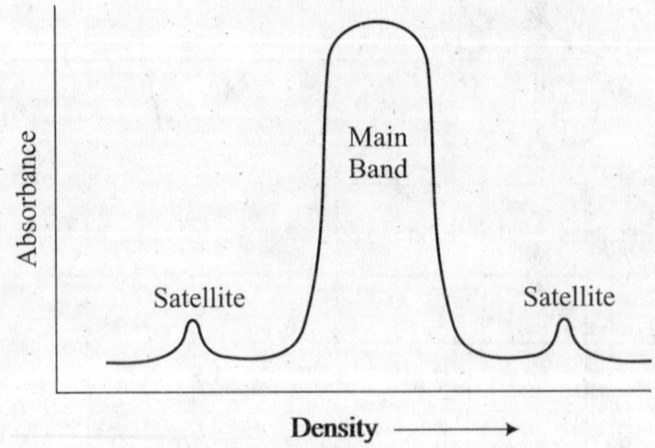

Fig. 5. Density centrifugation of embryonic DNA of *Drosophila melanogaster*, showing a main band and satellite bands.

The chromosomal locations of several satellite DNAs have been determined by a technique called *in situ* hybridization. It involves annealing single strands of isolated radioactive satellite DNA directly to denatured DNA in chromosome squash preparations. After washing out the non-hybridized radioactive material, the locations of the satellite DNA sequences in chromosomes are determined by autoradiography. Satellite DNAs are usually in heterochromatic regions of the chromosomes often around centromeres and near telomeres and are not transcribed into RNA. Other kinetically distinguished classes of DNA are the moderately repetitive and unique DNA.

SUMMARY

The C-value refers to the total amount of DNA in the haploid genome. The C-value paradox then refers to the fact that more complex organisms do not necessarily have a larger C-value. This inconsistent relationship between the C-value and the complexity of the organism has puzzled scientists for years until the realization that not all genomic DNA is transcribed and subsequently translated. Therefore, the primary reason for the C-value paradox lies with the non-coding and repetitious sequences of DNA that can vary tremendously between organisms due to a variety of reasons. DNA containing repeated sequences is called repetitive DNA. Such DNA is usually located in heterochromatin, while euchromatin consists of unique or single-copy sequences (protein coding).

REFERENCES

Britten, R.J. and Davidson, E.H. (1971) Repetitive and non-repetitive DNA sequences and a speculation on the origins of evolutionary novelty. Q. Rev. Biol. 46 : 111-133.

Britten, R.J. and Kohne, D. E. (1970) Repeated segments of DNA. Sci. Amer. 222 : 24-31.

Davidson, E.H., Hough, B.R., Amenson, C.S., Britten, R.J. (1973) General interspersion of repetitive with non-repetitive sequence elements in the DNA of Xenopus. J. Molecular Biology 77:1-23.

Gall, J.G. (1981) Chromosome structure and the C-value paradox. J. Cell Biol. 91 : 3-14.

Gregory, T.R. (2001) Coincidence, coevolution, or causation? DNA content, cell size, and the C-value enigma. Biological Reviews 76 : 65-101.

Gregory, T.R. (2005) Genome size evolution in animals. In: The Evolution of the Genome (T.R. Gregory, ed.), pp. 3-87. Elsevier, San Diego.

International Human Genome Sequencing Consortium. (2001) Initial sequencing and analysis of the human genome. Nature 409: 860-921.

Lewin, B. (2004) Genes VIII. Pearson prentice hall, Pearson Education Inc. NJ, USA.

Thomas, C.A. (1971) The genetic organization of chromosomes. Annual Review of Genetics 5: 237-256.

Vendrely, R. and Vendrely, C. (1948) La teneur du noyau cellulaire en acide désoxyribonucléique à travers les organes, les individus et les espèces animales: Techniques et premiers résultats. Experientia 4: 434-436.

Suggested Readings

1. Principles of Genetics (1991) E.J. Gardener, M.J. Simmons and D.P. Snustad, (8th edition), John Wiley and Sons, Inc., USA.
2. Genes VIII (2004) B. Lewin, Pearson Prentice Hall, Pearson Education Inc. NJ, USA.
3. Molecular Evolution (1997) W.H. Li, Sinauer Associates, Sunderland, MA, USA.

REVIEW QUESTIONS

Long Questions

1. Why do genome sizes vary by so many orders of magnitude among different organisms?
2. Why are members of some multigene families much more similar to one another within a species than they are between closely related species?
3. Write a short note on C-value paradox.
4. During DNA renaturation experiments, why is the rate of renaturation ($-dC/dt$) proportional to the square of the concentration of single strands (kC^2) rather than simply proportional to the concentration of single strands (kC)?
5. What is satellite DNA?

Multiple Choice Questions

1. Total amount of DNA in the haploid genome is called
   (a) Cot value (b) Junk DNA
   (c) C-value (d) Repetitive DNA

2. The term C-value paradox was given by
   (a) Beadle and Tatum (b) H.J. Muller
   (c) Barbara McClintok (d) C.A. Thomas
3. _______ is the term used to describe the discrepancy between nuclear genome size and the number of genes among eukaryotic species
   (a) C-value (b) C-value paradox
   (c) Cot value (d) All of the above
4. Repetitive DNA is found in the genome of
   (a) Prokaryotes (b) Eukaryotes
   (c) Both (d) None
5. C-values are usually larger in
   (a) Prokaryotes (b) Eukaryotes
6. Single copy DNA sequences are more prevalent in
   (a) Euchromatin (b) Heterochromatin
7. During DNA renaturation experiments the rate of renaturation ($-dC/dt$) is proportional to
   (a) kC (b) kC^2
   (c) kdt (d) $-kdt$
8. Repetitive DNA plays
   (a) Regulatory role
   (b) No role
   (c) Synthetic role
9. DNA sequences that are transcribed include
   (a) Moderately repetitive DNA
   (b) Highly repetitive DNA
   (c) Satellite DNA
   (d) Unique DNA
10. The C-value enigma was coined by
   (a) Vendrely (b) Gregory
   (c) Thomas (d) Mendel

Answers

1. (c)	2. (d)	3. (b)	4. (b)	5. (b)	6. (a)
7. (b)	8. (a)	9. (d)	10. (b)		

CHAPTER 17

Heterochromatization

Pragya Khanna
Govt. College for Women, Parade, Jammu–180 001
E-mail: pragyajmu2002@yahoo.co.in

It is a well known fact that gender is determined by the sex chromosomes. Humans usually have 46 chromosomes per diploid cell consisting of 22 sets of autosomes and a set of sex chromosomes—either XX or XY. In the usual course of events, individuals with the karyotype 46, XX are female and individuals with 46, XY, are male. The male being the heterogametic sex (XY) produces two types of spermatozoa in equal proportions, i.e., one carries the X chromosome and the other carries the Y chromosome. The female, on the other hand, being the homogametic sex (XX) produces only one kind of gametes. However, this is not true of all organisms, for example in birds and reptiles the female is heterogametic (ZW) and the male is homogametic (ZZ). But largely, the type of sex determination mechanism found to exist in all mammals and some insects is the common XY system wherein the individuals who possess XX chromosomes are female while male animals possess an X and a Y chromosome. But the question arises as to how the sex of an individual is determined. Is sex determined by the number of X chromosomes—with one X you are male or with 2 X's you are female? Or is sex determined by the presence or absence of the Y chromosome—the presence of a Y makes for a male or the absence of a Y produces a female? The answer was provided by individuals resulting from non-disjunction of the sex chromosomes. Some individuals have 45 chromosomes and have only one X chromosome, such individuals are phenotypically female; other individuals have 47 chromosomes and have two X chromosomes and a Y and are phenotypically male. If having two X chromosomes was the determining factor, then these individuals should be female. The table below indicates the sex of these individuals.

Chromosome Constitution	Name of Syndrome	Sex of Individual
46, XX	Normal	Female
46, XY	Normal	Male
45, XO	Turner Syndrome	Female
47, XXY	Klinefelter Syndrome	Male

Clearly, the presence or absence of a Y chromosome is more important to sex determination in humans than the number of X chromosomes present. Further research has shown that the gonads (which early in embryonic development can differentiate into either testes or ovaries) are directed to become testes if there is a Y chromosome present. If no Y chromosome is present, then the gonads differentiate as ovaries.

The region of the Y chromosome important for sex determination has been identified, and called SRY (for sex-determining region of Y). This region was identified in phenotypic males who were found to have two X chromosomes and no Y, although a small amount of Y chromosome DNA (the SRY) was found to have been transferred to one X. Phenotypic females who were XY were found to be missing this region of their Y chromosomes. Proof of the role of this region in sex determination came from work on mice, in which the SRY was inserted into the genome of XX embryos. These mice developed as perfectly normal males!

X-CHROMOSOME INACTIVATION

This strategy of sex determination creates a situation of inequity. Females, because of their two X chromosomes, have two copies of all X-linked genes, while males only have one. Therefore, females should produce twice as much of the proteins encoded by these genes as males. This would mean that females and males would have differences in color vision, among other things. (In fact, the overall effect would probably be lethal to female embryos.) To compensate for the extra X chromosome (the X chromosome 'dose'), one X chromosome (and all of its genes) is inactivated in every cell of early female mammalian embryos. The X chromosome is inactivated by converting it to heterochromatin. This highly condensed inactive X chromosome is called a Barr body and this phenomenon of bringing about equality in products synthesized under the control of genes carried on X chromosomes is termed dosage compensation (Fig. 1). Which chromosome is inactivated is randomly determined in every cell, and once inactivated the same X will continue to be inactivated in all the cells which are derived from it. This phenomenon of inactivation of X chromosomes was first described by Mary Lyon and was known as the Lyon Hypothesis, now considered as the Lyon Principle. So if a female is heterozygous for an X-linked trait, the dominant allele will be expressed in some cells, while the recessive allele will be expressed in others. Female mammals are therefore mosaics with regard to X-linked genes. This explains calico cats, females which have patches of different colored fur (one coat color gene for cats is found on the X chromosome). It turns out that during the first few days of embryonic life the X chromosome inactivation occurs. At that point in development there are hundreds or thousands of cells. In any one female cell the inactivated X chromosome may be either the paternal (X^p) or maternal (X^m). Which one is inactivated is a matter of chance but once that X has been inactivated in that cell, all that nucleus' descendants (clones through mitosis) will have the same inactive X chromosome. X inactivation is randomly determined but once it happens it's permanent. Each cell's descendants contribute to the

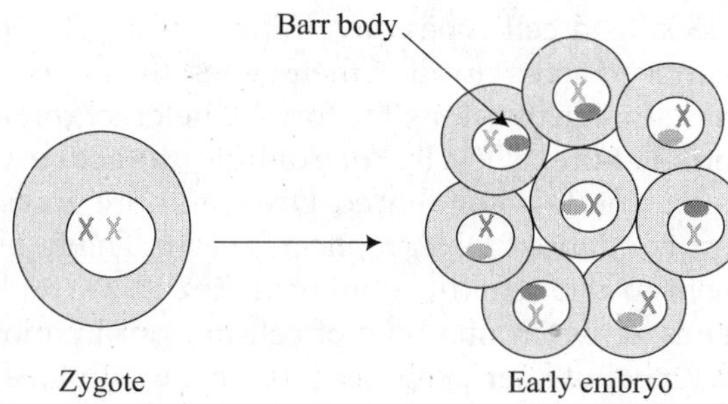

Fig. 1. Diagram illustrating the inactivation of X-chromosome and the formation of Barr body.

overall body of the organism. Because some cells have inactivated paternal Xs and others have inactivated maternal Xs, hence, the embryo and the organism become a mixture of two types of cells—expressing different alleles (Fig. 2). It has now been shown, that only parts of the inactivated X are in fact inactive, and that certain areas crucial to development remain active. This X is usually late replicating, highly coiled and condensed and can be visualized microscopically as a clear dense mass against the inner aspect of the nuclear membrane.

There are two classes of heterochromatin. Constitutive heterochromatin that is never transcriptionally active, located in the centromeric region of each chromosome and also found in the telomeres, the other class is called facultative heterochromatin. This type is sometimes active and somemtimes not, depending upon the cell type in an organism and on the organism's stage of development. There is little facultative heterochromatin in embryonic cells;

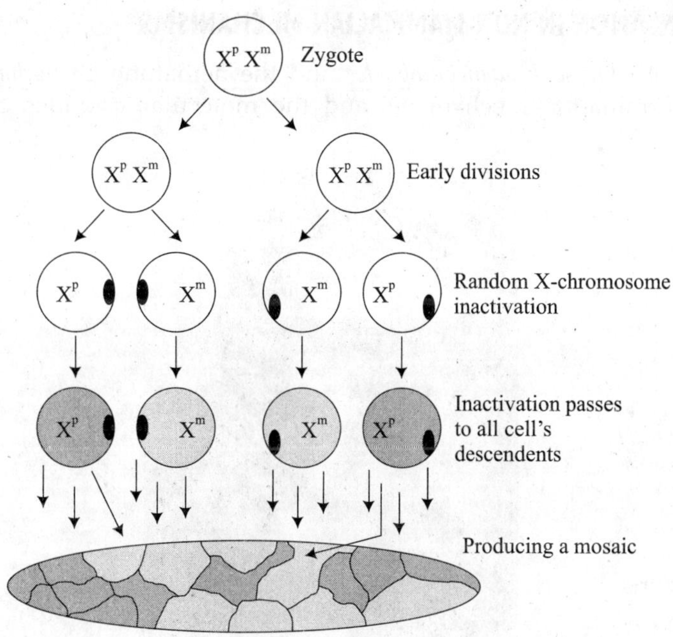

Fig. 2. X-inactivation allows for dosage compensation and explains the mosaicism seen in heterozygous female mice and cats, where it is called a "tortoise shell" or "calico".

whereas some highly specialized cell types have much more. This may mean that, as cells differentiate and no longer need access to all of their genes, the unnecessary genes are put into the equivalent of long-term storage by taking the form of heterochromatin. Heterochromatin is replicated late in the S period of the cell cycle. For example, the active X chromosome replicates before the inactive X does. The less condensed DNA is more accessible to the replication machinery. Therefore, less condensed DNA replicates early. The housekeeping genes, which are essential for the moment-to-moment functioning of the cell, are replicated first. If, in a given species, a particular gene is active in one type of cell but not in another type of cell, it will replicate early in the first type, but later in the second type. In humans the X chromosome that is inactivated is determined by chance. In marsupials, however, the paternal X chromosome is always inactivated. Inactivation is achieved by methylation of the X chromosome DNA, a common way in which the cell silences particular genes.

One might wonder, "If women with Turner syndrome (45, with XO) don't make a barr body, they don't inactivate their single X chromosome, so why do they have problems? Why aren't they as healthy as any male? On the other hand, why is Klinefelter syndrome (47, XXY) a problem since these men inactivate the extra X chromosome anyway (producing a barr body like a normal female)?"

Good questions! The fact is that a few of the thousands of genes on the X chromosome escape inactivation. Also, there are some delicate interactions between the X chromosome genes and other genes on the autosomes (and perhaps the Y chromosome too)! All together, these poorly understood effects contribute to the unusual medical problems caused by an "over dose" or "under dose" effect.

DOSAGE COMPENSATION IN NON-MAMMALIAN MECHANISMS

In both the fruit fly *Drosophila melanogaster* and the nematode *Caenorhabditis elegans*, the primary sex determination mechanisms and the molecular cascades controlling sexual

Fig. 3. Mary F. Lyon

differentiation have been studied extensively. Primary sex determination in these animals does not involve the Y chromosome but instead is determined by the ratio of the number of X chromosomes to autosomal (nonsex) chromosomes. By examining individuals with unusual numbers of various chromosomes it has been determined that in *Drosophila* those with one or fewer X chromosomes per diploid autosome set develop as males while those with two or more X chromosomes develop as females. Individuals with intermediate ratios such as those with two X chromosomes and a triploid set of autosomes develop as intersexes with both male and female characteristics. Although this ratio serves as the primary determinant of sex in both of these organisms, the specific gene products that influence this ratio assessment are different, demonstrating that different molecular mechanisms can be used for a similar purpose. The phenomenon of dosage compensation in *Drosophila* has been studied extensively by many workers and it has been shown that the X chromosome in male *Drosophila* shows hyperactivity rather than the X chromosome undergoing inactivation in female *Drosophila*.

REFERENCES

Hodgkin, J. (1992) Genetic Sex Determination Mechanisms and Evolution. *BioEssays* 4 : 253-261.

Lyon, M.F. (1962) Sex chromatin and gene action in the mammalian X-chromosome. Amer. J. Hum Genet. 14 : 135-148.

Mittwoch, U. (1973) Genetics and sex differentiation. Academic Press, New York.

Moore, K.I. (1966) The Sex Chromatin. Saunders, Philadelphia.

Suggested Readings

1. Principles of Genetics (1991) E.J. Gardener, M.J. Simmons and D.P. Snustad, 8th edition, John Wiley and Sons Inc., USA.
2. Genes VIII (2004) B. Lewin, Pearson Prentice Hall, Pearson Education Inc. NJ, USA.
3. Molecular Evolution (1997) W.H. Li, Sinauer Associates, Sunderland, MA, USA.

REVIEW QUESTIONS

Long Questions

1. How many sex chromatin bodies are expected to occur in cell nuclei with each of the following chromosome arrangements: (a) XY (b) XX (c) XXY (d) XXX (e) XXXX (f) XYY
2. What is dosage compensation? How is this achieved?
3. Discuss the different mechanisms of sex determination involved in humans, *Drosophila* and birds.

Multiple Choice Questions

1. The region of the Y chromosome important for sex determination is called

   (a) SRY (b) SDY

   (c) SDRY (d) SRX

2. This highly condensed inactive X chromosome is called
   (a) X-body (b) Barr body
   (c) X-chromatin body (d) Lyor body
3. This phenomenon of inactivation of X chromosomes was first described by
   (a) Mary Lyon (b) Mendel
   (c) Margaret Lyon (d) Martin
4. _______ is also found on much of the Y chromosome
   (a) Facultative heterochromatin (b) Constitutive heterochromatin
5. In humans the X chromosome that is inactivated is
   (a) Maternal (b) Paternal
   (c) determined by chance. (d) All of the above

Answers

1. (a) 2. (b) 3. (a) 4. (b) 5. (c)

PART–6

Mutation and Transposition

CHAPTER 18

Mutation

Pragya Khanna
Govt. College for Women, Parade, Jammu–180 001
E-mail: pragyajmu2002@yahoo.co.in

Mutations are sudden, permanent, transmissible (if the change is in a germ cell) changes in the genetic material. An organism exhibiting a novel phenotype as a result of the presence of a mutation is called a mutant. The term 'mutation' was used by Hugo De Vries (1901) to describe phenotypic changes which were heritable. Thus, he differentiated between environmental and heritable variations and called them somatic and germinal mutations. The heritable changes, in broad sense, refer to all those changes which result in altered pattern of heredity. These can be classified as : Chromosomal mutations (include the structural changes in chromosomes) and gene or point mutations (include changes that alter the chemical structure of the gene at molecular level).

The term mutation is now used in a restricted sense to cover only the changes in individual genes called gene mutations or point mutations and excluding the changes in chromosome structure or number. Hence, a mutation is any change in the sequence of the DNA encoding a gene. Our studies on mutations have lead to a better understanding about the nature of genes. Originally, genes were thought to have beads-on-string structure where each bead was a single entity responsible for a phenotype. This theory led to the concept that only single mutation was possible for a specific gene. Detailed genetic experiments proved that the gene actually consists of many individual units and specific changes in these units can lead to several mutant phenotypes. We now know that these units are nucleotides. Therefore, understanding the nature of mutations is important to our understanding of a gene.

Gene mutations or point mutations involve changes in single base-pairs, the substitution of one base-pair for another, or the duplication or deletion of single base-pairs. Without mutations, all genes would exist in only one form and the organisms would not be able to evolve and adapt to environmental changes. Hence, mutations are necessary to provide genetic variability to allow organisms to adapt to new environmental conditions.

There are different classifications of mutations depending on different aspects.

Classification of mutations on the basis of occurrence:

(i) **Spontaneous mutations:** These occur randomly under the natural environmental conditions and have a low frequency.

(ii) **Induced mutations:** These arise from exposure to mutagenic agents such as radiations and/or chemicals.

(iii) **Gametic/germinal mutations:** These occur in the germ-line cells and are heritable.

(iv) **Somatic mutations:** These occur in the non-sex cells and are non-heritable.

Classification of mutations on the basis of their effect on organism:

(i) **Morphological mutations:** These involve alterations in external form.

(ii) **Nutritional or biochemical mutations:** These result due to the deficiency of a nutrient or a chemical compound and can be overcome by supplying the same.

(iii) **Behavioral mutations:** These alter the behavior patterns of an organism.

(iv) **Regulatory mutations:** Such mutations may disrupt the regulatory mechanisms and permanently activate or inactivate a gene.

(v) **Lethal mutations:** These involve genotypic changes leading to death of an individual.

(vi) **Conditional mutations:** These allow the mutant phenotype to be expressed only under certain conditions called restrictive conditions.

HISTORICAL BACKGROUND

The earliest recorded point mutation was observed by Seth Wright in 1791 on his farm located on the banks of Charles River in Dover, Massachusetts. Wright noticed a peculiar male lamb with unusually short legs in his flock of sheep. He thought that it would be an advantage to have a whole flock of these short-legged sheep which could not jump over the low stone fences in his neighborhood. He used this new short legged ram for breeding his 15 ewes. Two of the 15 lambs produced had short legs, the short legged sheep were then bred together and a line was developed in which the new trait was expressed in all individuals. The short legged breed of sheep was known as Ancon breed (Fig. 1).

Morgan in 1910 began the scientific study on mutations. He worked on *Drosophila melanogaster* and reported white eyed male individuals among red eyed male individuals. It was found that the gene for this character is located on sex chromosome (X-chromosome) and expresses itself in a male individual. When these rare white eyed males were crossed to their sister red eyed females, white eyed females could also be obtained in some cases. This signified that the females involved were heterozygous (Fig. 2). After the discovery of white eyed mutant, Morgan and his coworkers studied different mutations in *Drosophila* and about 500 different mutations were observed by different scientists all over the world. Moreover, mutations were also studied in other organisms like rodents, fowl, pea, maize, snapdragon, *E. coli*, *Neurospora*, bacteriophage, man etc.

GERMINAL AND SOMATIC MUTATIONS

Eukaryotic organisms have two primary cell types — germ and somatic. Mutations can occur in either cell type. If a gene is altered in a germ cell, the mutation is termed a germinal mutation.

Fig. 1. Short-legged sheep of Ancon breed.

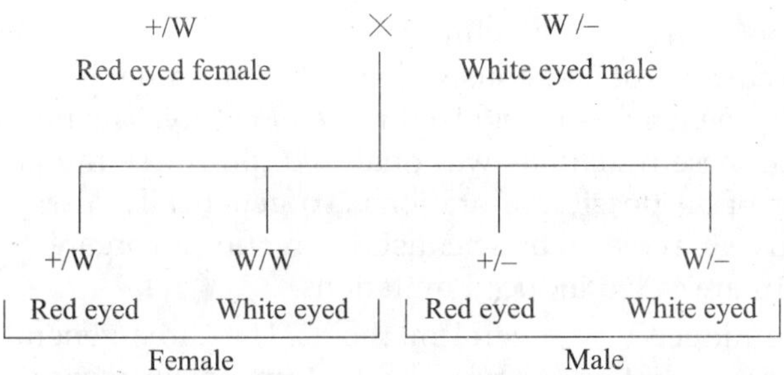

Fig. 2. Appearance of a white eyed male fly in a cross between red eyed femaled and white eyed male.

Because germ cells give rise to gametes, some gametes will carry the mutation and it will be passed on to the next generation when the individual successfully mates. Typically germinal mutations are not expressed in the individual containing the mutation. The only instance in which it would be expressed is if it affected gamete production.

Somatic cells give rise to all non-germ line tissues. Mutations in somatic cells are called somatic mutations. Because they do not occur in cells that give rise to gametes, the mutation is not passed along to the next generation by sexual means. To maintain this mutation, the individual containing the mutation must be cloned. Two examples of somatic clones are navel oranges and red delicious apples. Horticulturists first observed the mutants. They then grafted mutant branches onto the stocks of "normal" trees. After the graft was established, cuttings from that original graft were grafted onto tree stocks. In this way the mutation was maintained and proliferated.

Most tissues are derived from a cell or a few progenitor cells. If a mutation occurs in one of the progenitor cells, all of its daughter cells will also express the mutation. For this reason, somatic mutations generally appear as a sector on the mutated individual. Cancer tumors are a unique class of somatic mutations. The tumor arises when a gene involved in cell division, a protooncogene, is mutated. All of the daughter cells contain this mutation. The phenotype of all cells containing the mutation is uncontrolled cell division. This results in a tumor that is a collection of undifferentiated cells called tumor cells.

SPONTANEOUS VERSUS INDUCED MUTATIONS

Some mutations arise as natural errors in DNA replication (or as a result of unknown chemical reactions); these are known as spontaneous mutations. The rates of such mutations have been determined for many species. *E. coli* has a spontaneous mutation rate of $1/10^8$ (one error in every 10^8 nucleotides replicated). Humans have a higher spontaneous mutation rate: between $1/10^6$ and $1/10^5$ (probably as a result of the higher complexity of human replication). Mutations can also be caused by agents in the environment; these are induced mutations. Induced mutations increase the mutation rate over the spontaneous rate. Looking at a single mutation in an individual, one cannot tell if the mutation was spontaneous or induced. Induced mutations can only be discerned by looking at the mutation rate in a population, and comparing it to the spontaneous mutation rate for the species. If the observed mutation rate is higher, then induced mutations can be assumed. Agents in the environment that cause an increase in the mutation rate are called mutagens.

In general, the appearance of a new mutation is a rare event. Most mutations that were originally studied occurred spontaneously. This class of mutation is termed spontaneous mutations. Historically, geneticists recognized these in nature. The mutations were collected, and the inheritance of these mutations was analyzed. But these mutations clearly represent only a small number of all possible mutations. To genetically dissect a biological system further, new mutations were created by scientists by treating an organism with a mutagenizing agent. These mutations are called induced mutations.

Mutations can be induced by several methods. The three general approaches used to generate mutations are radiation, chemical and transposon insertion. The first induced mutations were created by treating *Drosophila* with X-rays. In addition to X-rays, other types of radiation treatments that have proven useful include gamma rays and fast neutron bombardment. These treatments can induce point mutations (changes in a single nucleotide) or deletions (loss of a chromosomal segment). Chemical mutagens work mostly by inducing point mutations. Point mutations occur when a single base pair of a gene is changed. These changes are classified as transitions or transversions. Transitions occur when a purine is converted to a purine (adenine to guanine or guanine to adenine) or a pyrimidire is converted to a pyrimidine (thymine to cytosine or cytosine to thymine). A transversion results when a purine is converted to a pyrimidine or a pyrimidine is converted to a purine.

Two major classes of chemical mutagens are routinely used. These are alkylating agents and base analogs. Each has a specific effect on DNA. Alkylating agents [such as ethyl methane sulphonate (EMS), ethyl ethane sulphonate (EES) and mustard gas] can mutate both replicating and non-replicating DNA. By contrast, a base analog (5-bromouracil and 2-aminopurine) only mutate DNA when the analog is incorporated into replicating DNA. Each class of chemical mutagen has specific effects that can lead to transitions, transversions or deletions.

Scientists are now using the power of transposable elements to create new mutations. Transposable elements are mobile pieces of DNA that can move from one location in a genome to another. Often when they move to a new location, the result is a new mutant. The mutant arises because the presence of a piece of DNA in a wild type gene disrupts the normal function of that gene. As more and more is being learned about genes and genomes, it is becoming apparent that transposable elements are a power source for creating insertional mutants.

The detailed knowledge of the structure and function of transposable elements is now being applied in the pursuit of new mutations. Stocks are created in which a specific type of element is present. This stock is then crossed to a genetic stock that does not contain the element. Once that element enters the virgin stock, it can begin to move around that genome. By carefully observing the offspring, new mutants can be discovered. This approach to developing mutants is termed insertional mutagenesis.

MUTATIONS OCCUR RANDOMLY IN NATURE

Most populations of pests are no longer affected by the anticoagulants and pesticides that have been traditionally used against them. Similarly, the pathogenic microorganisms are becoming resistant to antibiotics such as penicillin and streptomycin developed to control them. The development of these substances produced a new environment for the organisms involved. These organisms responded to the imposed environmental changes by evolving to forms that are resistant to these chemicals by the occurrence of mutations that produced resistance to these pesticides and antibiotics and the mutant organisms were at a large selective advantage in environments where these agents were present. The sensitive organisms were killed and the mutants multiplied to produce new resistant populations.

Lederberg developed a technique called replica-plating experiment to demonstrate the random nature of the mutation. Replica plating is a technique in which one or more secondary Petri plates containing different solid (agar-based) selective growth media (lacking nutrients or containing chemical growth inhibitors such as antibiotics) are inoculated with the same colonies of microorganisms from a primary plate (or master dish), reproducing the original spatial pattern of colonies. The technique involves pressing a velvet-covered disk to a primary plate, and then imprinting secondary plates with cells in colonies removed by the velvet from the original plate. Generally, large numbers of colonies (roughly 30-300) are replica plated due to the difficulty in streaking each out individually onto a separate plate (Fig. 3). The purpose of replica plating is to be able to compare the master plate and any secondary plates to screen for a selectable phenotype. For example, a colony which appeared on the master plate but failed to appear at the same location on a secondary plate shows that the colony was sensitive to a substance on that particular secondary plate. Common screenable phenotypes include auxotrophy and antibiotic resistance.

Replica-plating is especially useful for negative selection. For example, if one wanted to select colonies that were sensitive to ampicillin, the primary plate could be replica plated on a secondary Amp^+ agar plate . The sensitive colonies on the secondary plate would die but the colonies could still be deduced from the primary plate since the two have the same spatial patterns from ampicillin resistant colonies. The sensitive colonies could then be picked off from the primary plate. By increasing the variety of secondary plates with different selective growth media, it is possible to rapidly screen a large number of individual isolated colonies for as many phenotypes as there are secondary plates. This experiment suggests that the environmental stress does not direct or cause genetic changes rather, it simply selects rare pre-existing mutations that result in phenotypes that are better adapted to the new environment.

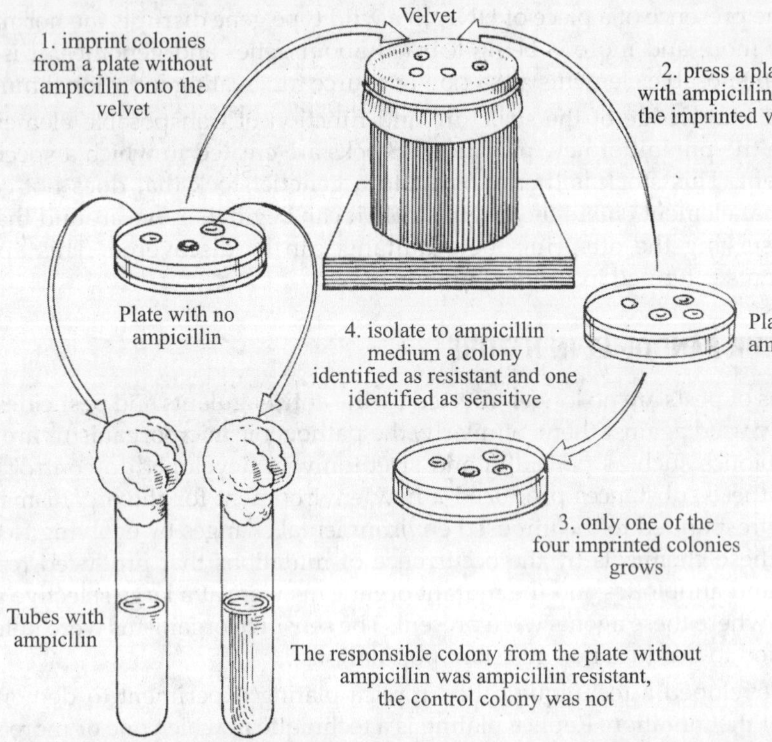

Fig. 3. Diagram illustrating the use of the replica-plating technique to demonstrate the random or non-directed nature of mutation.

THE MOLECULAR BASIS OF MUTATION

The double-helix structure of DNA proposed by Watson and Crick (1953), pointed out that the bases in DNA are not static. Hydrogen atoms can move from one position in a purine or pyrimidine to another position, for example, from an amino group to ring nitrogen. Such chemical fluctuations are called tautomeric shifts. In a DNA molecule, under normal conditions, the purine adenine (A) is linked to the pyrimidine thymine (T) by two hydrogen bonds, while the purine guanine (G) is linked to the pyrimidine cytosine (C) by three hydrogen bonds. Due to the tautomeric shifts, the more stable keto forms of thymine and guanine and amino forms of adenine and cytosine may shift to less stable enol and imino forms respectively (Fig. 4). When the bases are present in their rare imino or enol states, they can form adenine-cytosine and guanine-thymine base pairs and that results in two main types of point mutations: base substitutions and base additions or deletions.

Base substitutions are those mutations in which one base pair is replaced by another. Base substitutions again can be divided into two subtypes: transitions and transversions. A transition is the replacement of a base by the other base of the same chemical category (purine replaced by purine: either A to G or G to A; pyrimidine replaced by pyrimidine: either C to T or T to C). A transversion is the opposite – the replacement of a base of one chemical category by a base of the other (pyrimidine replaced by purine: C to A, C to G, T to A, T to G; purine

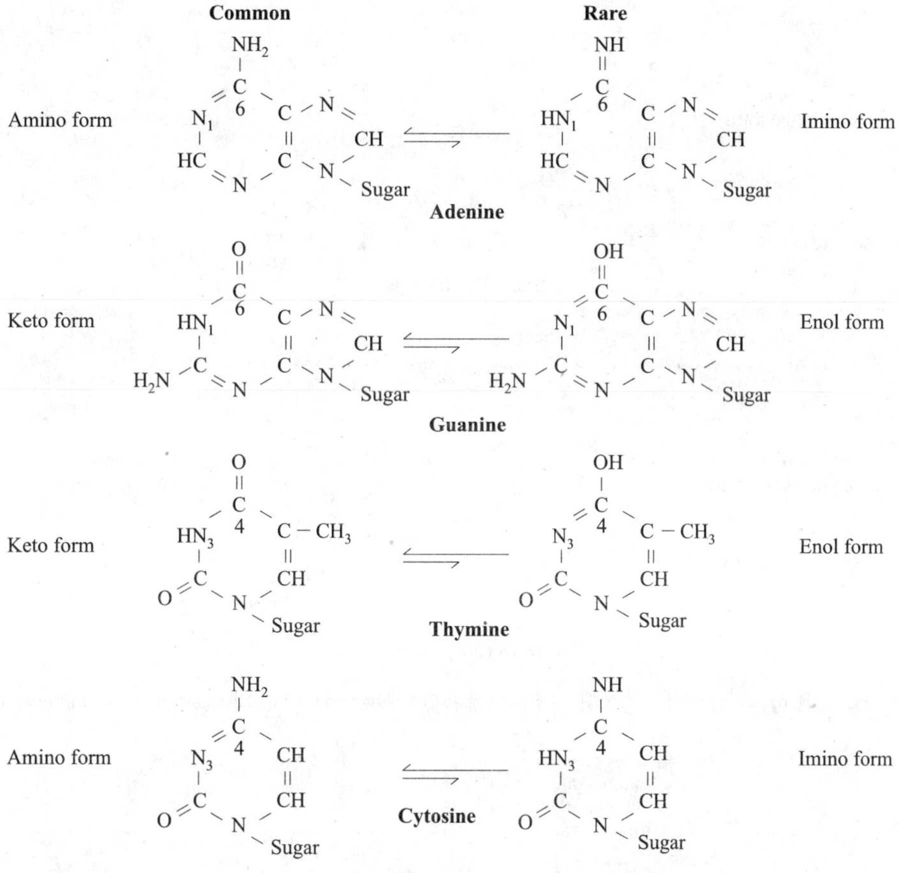

Fig. 4. Tautomeric forms of the four common bases of DNA.

replaced by pyrimidine: A to C, A to T, G to C, G to T) (Fig. 5). In describing the same changes at the double-stranded level of DNA, we must state both members of a base pair: an example of a transition would be G·C → A·T; that of a transversion would be G·C → T·A as the complementary change will take place on the other strand (Fig. 6).

Addition or deletion mutations are actually of *nucleotide* pairs; nevertheless, the convention is to call them *base*-pair additions or deletions. The simplest of these mutations are single-base-pair additions or single-base-pair deletions. There are examples in which mutations arise through simultaneous addition or deletion of multiple base pairs at once. For single-base substitutions, there are several possible outcomes, which are direct consequences of two aspects of the genetic code: degeneracy of the code and the existence of translation termination codons.

1. Silent substitutions: the mutation changes one codon for an amino acid into another codon for that same amino acid.
2. Missense mutations: the codon for one amino acid is replaced by a codon for another amino acid.

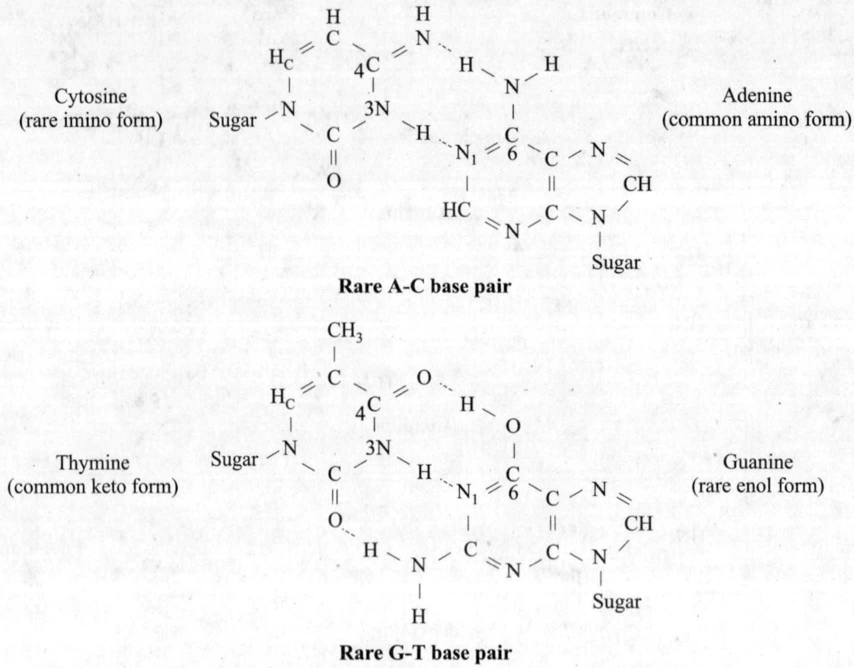

Fig. 5. Examples of mismatched AC and GT base pairs that form when one of the bases exists in rare tautomeric form.

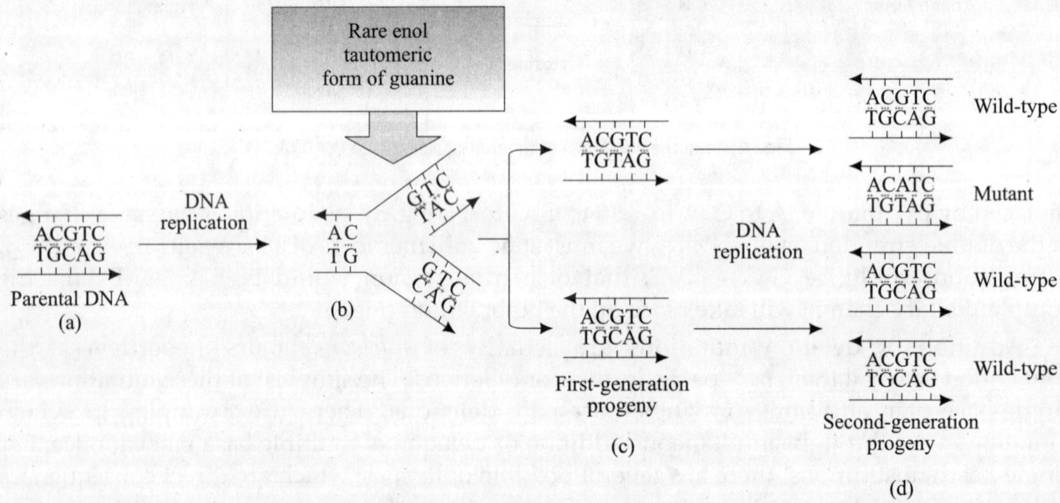

Fig. 6. Mutation via tautomeric shifts in the bases of DNA.

3. Nonsense mutations: the codon for one amino acid is replaced by a translation termination (stop) codon.

Silent substitutions never alter the amino acid sequence of the polypeptide chain. The severity of the effect of missense and nonsense mutations on the polypeptide will differ on a case-by-case basis. For example, if a missense mutation causes the substitution of a chemically similar amino acid, referred to as a synonymous substitution, then it is likely that the alteration will have a less-severe effect on the protein's structure and function. Alternatively, chemically different amino acid substitutions, called non-synonymous substitutions, are more likely to produce severe changes in protein structure and function. Nonsense mutations will lead to the premature termination of translation. Thus, they have a considerable effect on protein function. Typically, unless they occur very close to the 32 end of the open reading frame, so that only a partly functional truncated polypeptide is produced, nonsense mutations will produce completely inactive protein products.

Like nonsense mutations, single-base additions or deletions have consequences on polypeptide sequence that extend far beyond the site of the mutation itself. Because the sequence of mRNA is "read" by the translational apparatus in groups of three base pairs (codons), the addition or deletion of a single base pair of DNA will change the reading frame starting from the location of the addition or deletion and extending through to the carboxy terminal of the protein. Hence, these lesions are called frameshift mutations. These mutations cause the entire amino acid sequence to bear no relation to the original amino acid sequence. Thus, frameshift mutations typically exhibit complete loss of normal protein structure and function.

To illustrate the effects of these mutations, consider the following phrase, read as a triplet code (just like the genetic code):

The fat cat ate the hot dog.

A base substitution might have an effect like this:

The fat car ate the hot dog.

or perhaps:

The fat cat ate the hot hog.

In each case, the phrase still makes sense, but the meaning has been slightly changed.

An insertion, on the other hand, would have a more profound effect:

The fma tca tat eth eho tdo g.

Insertion of a single letter ('m' in this case) causes the phrase to become gibberish, because the reading frame has been changed. A deletion would have the same effect:

The atc ata tet heh otd og.

RADIATION INDUCED MUTATION

Radiations are capable of inducing mutations in DNA. Ionizing radiations, such as X-rays, gamma rays and cosmic rays are of high energy and can penetrate living tissues. In this process, these high energy rays collide with atoms and cause the release of electrons, forming ions. Ultraviolet rays dissipate their energy to the atoms they encounter, thereby raising the electrons in their outer orbitals to higher energy levels, a state referred to as excitation. Molecules containing atoms in either ionic forms or excited states are chemically more reactive and this forms the basis of mutagenic effects of ionizing radiation and ultraviolet light. Ionizing radiation produces a range of effects on DNA both through free radical effects and direct action. It can cause:

- breaks in one or both strands, this can lead to rearrangements, deletions, chromosome loss and death (if unrepaired)
- damage and/or loss of bases
- crosslinking of DNA to itself or proteins.

The genetic effects of radiation were reported by Muller in 1927 in *Drosophila* and by Stadler in 1928 in plants (barley); both showed that the frequency of induced mutations is a function of X-ray dose. Their experiments revealed that there was a linear relationship between X-ray dose and induced mutation level, that there was no threshold or "safe" dose of radiation and that all doses are significant, and finally, that "split dose" experiments showed that the genetic effects of radiation are cumulative (Fig. 7).

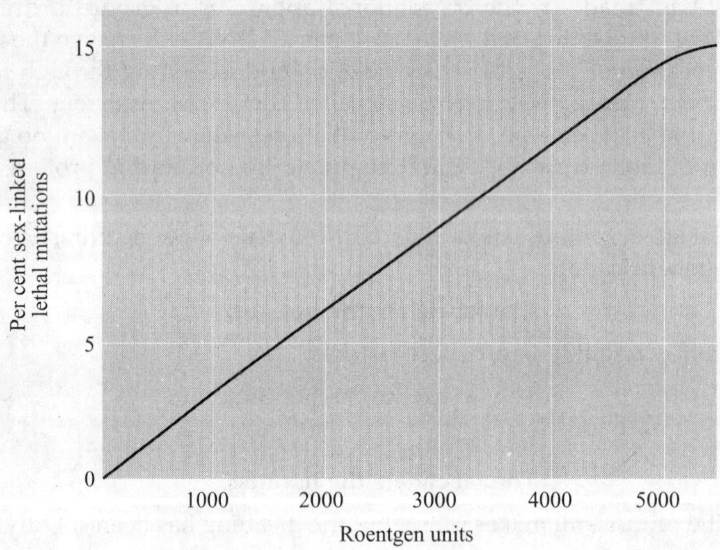

Fig. 7. Relationship between the frequency of sex-linked lethal mutations induced in *Drosophila* sperm and ionizing irradiation dosage.

In 1927, Muller demonstrated that the X-ray treatment increased the frequency of sex-linked recessive lethal mutations in *Drosophila melanogaster*. The different methods for the detection of sex-linked lethals are presented below:

The ClB Method

This method involves the use of a ClB stock which carries an inversion in the heterozygous state to work as crossover suppressor (C), a recessive lethal (l) on X-chromosome in heterozygous state and a dominant marker, *Barred* (B) for the barred eye. Male flies irradiated for induction of mutations were crossed to ClB females. The ClB female flies thus obtained as progeny are crossed to normal males while the male progeny receiving ClB X-chromosome die. In the next generation, in a cross between ClB females × normal males, 50% of males receiving ClB X-chromosome die and the other 50% receive the X-chromosome that may or may not carry

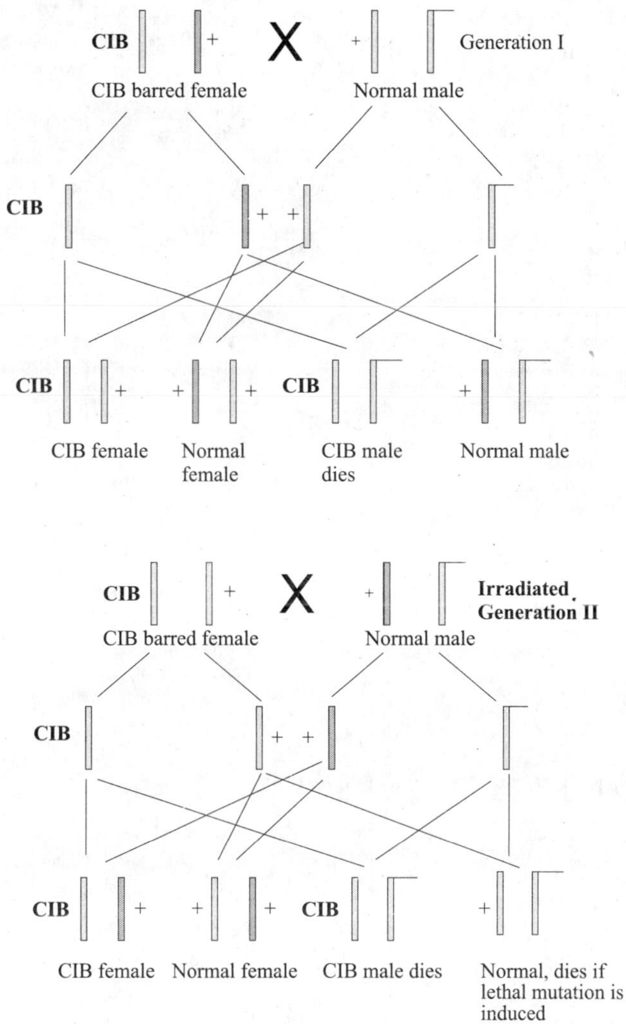

Fig. 8. Muller's CIB method for detection of sex -linked lethal mutations in *Drosophila*.

the induced mutation. In case lethal mutation was induced no males will be observed and if no lethal mutation was induced, 50% males will survive (Fig. 8). This method for the detection of sex-linked lethal mutations has been improved and is replaced by Muller-5 method.

Muller-5 Method

This method makes use of a *Muller-5 Drosophila* stock, which carries two marker genes, dominant 'Bar' (for barred eye) and recessive 'apricot' but does not carry a lethal as in ClB. In the F_2 generation, 50% males are Muller-5 in phenotype and remaining 50% are wild type. If a lethal mutation is induced in X-chromosome of irradiated male no wild type males would appear in F_2 generation. Hence, absence of wild type males in F_2 is an indication of an induced lethal mutation (Fig. 9).

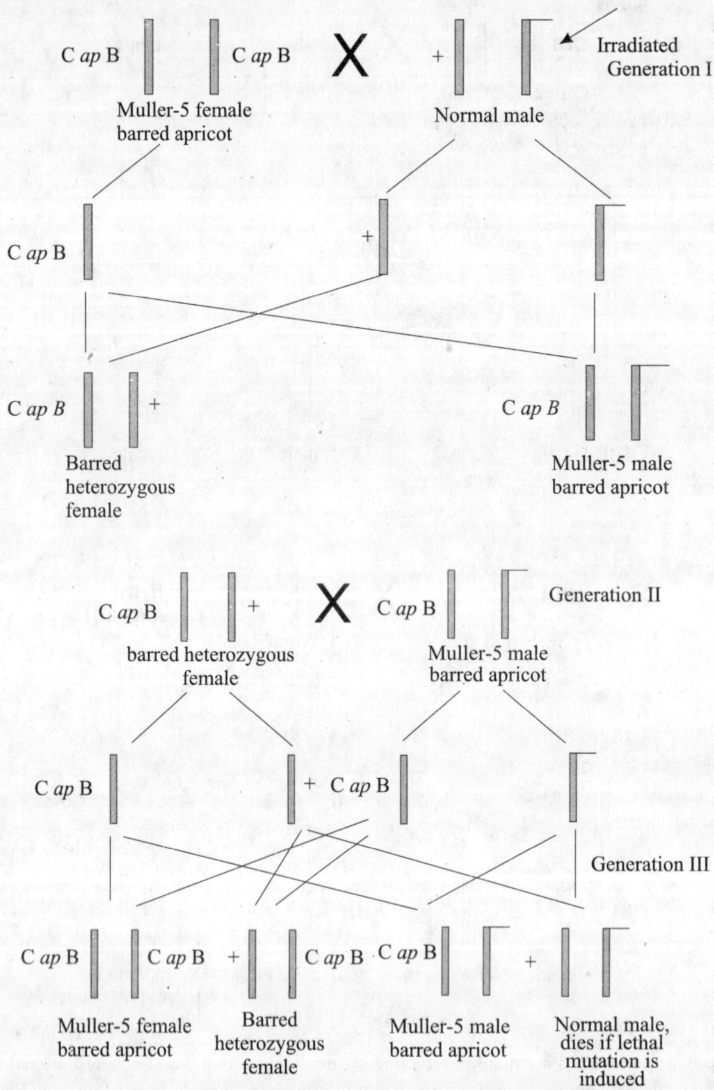

Fig. 9. Muller-5 method for detection of sex -linked mutations in *Drosophila*.

Ultraviolet (non-ionizing) radiation can cause mutations as well. UV radiation is less energetic, and therefore non-ionizing, but its wavelengths are preferentially absorbed by bases of DNA and by aromatic amino acids of proteins, so it, too, has important biological and genetic effects. UV is normally classified in terms of its wavelength: UV-C (180-290 nm)—"germicidal"—most energetic and lethal, it is not found in sunlight because it is absorbed by the ozone layer; UV-B (290-320 nm)—major lethal/mutagenic fraction of sunlight; UV-A (320 nm—visible)—"near UV"—also has deleterious effects (primarily because it creates oxygen radicals) but it produces very few pyrimidine dimers. The primary effect of UV on DNA is the creation of thymine dimers. Thymine dimers occur when two thymines are adjacent on a strand of DNA. UV radiation can cause the formation of a covalent bond between the two thymines, which prevents their participation in base pairing. Thymine dimers are very deleterious to a cell. They can completely interrupt replication, effectively causing a cell to die. These dimers, like bulky lesions from chemicals, block transcription and DNA replication and are lethal if unrepaired. They can stimulate mutation and chromosome rearrangement as well.

REPAIR OF THYMINE DIMERS

Several mechanisms are available for the removal or correction of thymine dimers from DNA. Which mechanism is used depends upon the circumstances of the cell.

Photoreactivation: It has been observed that a brief exposure to blue light following UV exposure can reverse the effects of the UV radiation. In other words, the blue light can cause a thymine dimer to be corrected. This is due to the function of an enzyme called photolyase or photoreactivation enzyme (PRE), which cleaves the covalent bonds linking the thymine dimers using the energy from a photon of blue light. This is essentially a reversal of the reaction that produced the thymine dimer in the first place.

Excision Repair: This is a repair system that doesn't require light. Instead of just breaking the bonds of the thymine dimer (as was done by photolyase), the excision repair system removes (excises) the region surrounding the offending nucleotides. Several proteins are involved in this process (in prokaryotes these are the products of the 'uvr' genes, for 'UV repair'). The steps of excision repair in prokaryotes are as follows (Fig. 10):

- The distortion in the DNA (caused by the thymine dimer) is recognized by a protein complex. A pair of endonucleases makes nicks in the DNA strand on either side of the thymine dimer (generally the nicks are 12 nucleotides apart).
- The 12-nucleotide piece of DNA between the nicks is removed, and DNA polymerase I fills in the gap left behind.
- DNA ligase seals the final nick in the DNA.

Recombination Repair: Sometimes, DNA replication will begin before a thymine dimer can be repaired by one of the other mechanisms. When the replication machinery hits the dimer, replication stops. Occasionally, the replication will reinitiate just beyond the dimer, leaving a gap in the DNA. Now if the cell tries to fix the dimer by excision repair, there is no template to use for resynthesis of the DNA. How, then, does the DNA get repaired? The answer is that the cell uses recombination to provide a template strand for repair synthesis.

First, the damaged region undergoes recombination with the complemenary strand on the other DNA molecule. One strand is exchanged between the two DNA molecules. This essentially transfers the gap to the DNA molecule that doesn't have the dimer. The gap can now be filled in by DNA polymerase I, and the dimer can be repaired by excision, since a template strand now exists (Fig. 11).

REPAIR OF MUTATIONS

Cells have mechanisms for minimizing the amount of mutation that takes place. As stated previously, these are not perfect, but they do reduce greatly the frequency of mutation. The two mechanisms we'll consider are proofreading and mismatch repair.

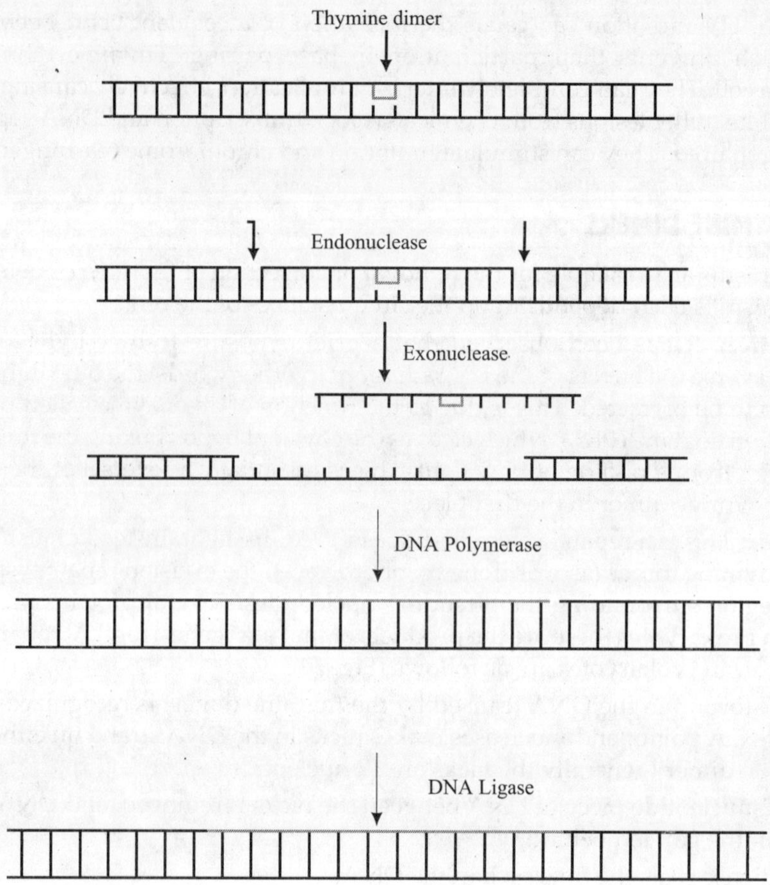

Fig. 10. Diagram of the excision repair pathway for the removal of thymine dimers from DNA.

Proofreading: This occurs during DNA replication. As DNA polymerase III adds nucleotides to the growing chain, it checks each one for correct base pairing. If the correct nucleotide has not been inserted, the polymerase uses its 3' to 5' exonuclease activity to remove the incorrect nucleotide. The polymerase can then carry on and insert the correct nucleotide. In bacteria, a wrong nucleotide gets inserted for every 10^5 nucleotides added during replication. Proofreading corrects most of these, so that the overall error rate in replication is one mistake for every 10^7 nucleotides added.

Mismatch Repair: This mechanism is used soon after replication, to correct errors that escaped proofreading. Because mismatched bases don't hydrogen bond, they create a distortion in the double helix, which can be recognized and repaired by excision repair. DNA under normal circumstances is methylated; these methyl groups do not interfere with the function of the DNA in any way. Newly replicated DNA is not methylated however; the methyl groups are added enzymatically after replication. If mismatch repair is done immediately after replication (before methylation occurs), the original DNA strand will be methylated, and the newly-synthesized strand (the one containing the error) will be unmethylated. The mismatch repair system therefore repairs the unmethylated strand.

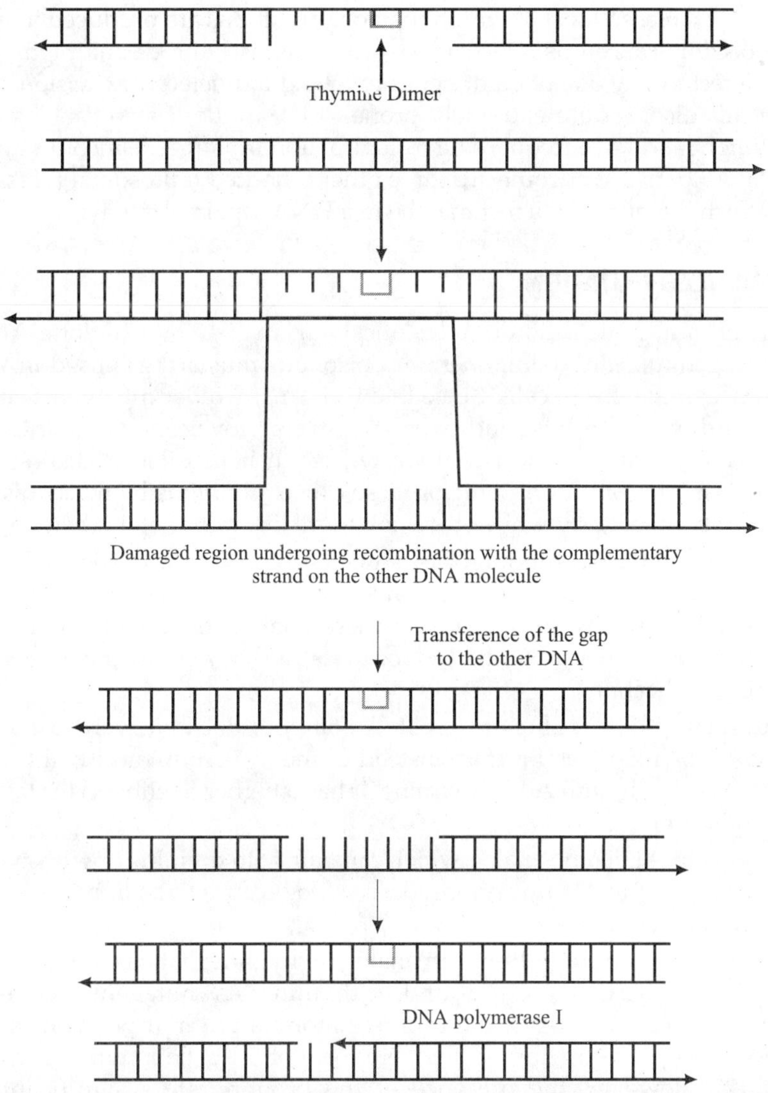

Fig. 11. Diagrammatic representation of recombination repair.

Excision repair can be used to correct other problems as well. For example, if a deoxynucleotide containing uracil is ever inserted into a DNA molecule, the base is detected and removed by an enzyme called uracil DNA glycosylase. This enzyme removes the base, but leaves the sugar and phosphate in the DNA molecule. This base-less site is then recognized by specific endonucleases, which initiate excision repair.

REPAIR DEFECTS AND HUMAN DISEASES

Several recessive human genetic diseases are known or suspected to be caused by defective genes in repair systems. These defects often lead to an increased incidence of cancer because, as part of the general increased level of mutation, more mutations are produced in genes that can cause a cell to become cancerous. One cancer-prone disease, xeroderma pigmentosum (XP), results from a defect in any one of eight genes involved in nucleotide excision repair. People suffering from this disorder are extremely prone to UV-induced skin cancers as a result of exposure to sunlight and have frequent neurological abnormalities. Fanconi's anemia, ataxia-telangiectasia and Bloom's syndrome are three other inherited (autosomal recessive) diseases in humans in which the primary defect may be in a DNA repair pathway.

CHEMICALLY INDUCED MUTATION

The first report of mutagenic action of a chemical was in 1942 by Charlotte Auerbach, who showed that nitrogen mustard (component of poisonous mustard gas used in World Wars I and II) could cause mutations in cells. Since that time, many other mutagenic chemicals have been identified and there is a huge industry and several government agencies dedicated to finding them in food additives, industrial wastes, etc. It is possible to distinguish chemical mutagens by their modes of action; some of these cause mutations by mechanisms similar to those which arise spontaneously while others are more like radiation in their effects.

Base analogs: Some chemical compounds are sufficiently similar to the normal nitrogen bases of DNA that they are occasionally incorporated into DNA in place of normal bases; such compounds are called base analogs. Many of these analogs have pairing properties unlike those of the normal bases; thus they can produce mutations by causing incorrect nucleotides to be inserted during replication.

- **Bromouracil** (BU)—artificially created compound extensively used in research. Resembles thymine (has Br atom instead of methyl group) and will be incorporated into DNA and pair with A like thymine. It has a higher likelihood for tautomerization to the enol form.
- **Aminopurine** —adenine analog which can pair with thymine or with cytosine; causes A:T to G:C or G:C to A:T transitions. Base analogs cause transitions, as do spontaneous tautomerization events.

The mutagen 5-bromouracil (5-BU) is an analog of thymine that has bromine at the carbon-5 position in place of the CH_3 group found in thymine. Its mutagenic action is based on enolization and ionization. In 5-BU, the bromine atom is not in a position in which it can hydrogen-bond during base pairing, so the keto form of 5-BU pairs with adenine, as would thymine (Fig. 12). However, the presence of the bromine atom significantly alters the distribution of electrons in the base ring, so 5-BU can *frequently* change to either the enol form or an ionized form, the latter of which pairs with guanine (Fig. 13). 5-BU causes G·C → A·T or A·T → G·C transitions in the course of replication, depending on whether 5-BU has been enolized or ionized *in situ* or as an incoming base. Hence the action of 5-BU as a mutagen is due to the fact that the molecule spends more of its time in the enol or ion form.

2-aminopurine (2-AP) is widely used in research. It is an analog of adenine that can pair with thymine but, when protonated, can also mispair with cytosine. Therefore, when 2-AP is incorporated into DNA by pairing with thymine, it can generate A·T → G·C transitions by mispairing with cytosine in subsequent replications. Or, if 2-AP is incorporated by mispairing with cytosine, then G·C → A·T transitions will result when it pairs with thymine. Genetic studies have shown that 2-AP, like 5-BU, is highly specific for transitions.

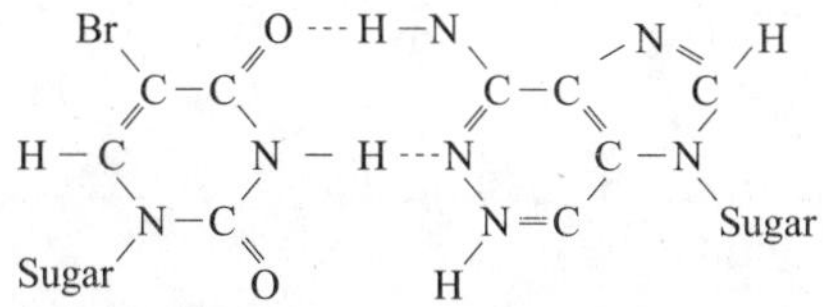

5- Bromouracil: (Keto from) Adenine

Fig. 12. 5-Bromouracil: Ademine base pair

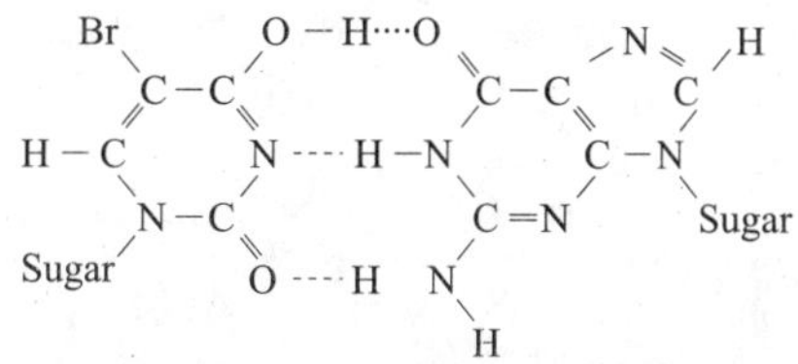

5-Bromouracil: (enol from) Guanine

Fig. 13. 5-Bromouracil: Guanine base pair

Other chemicals which alter structure and pairing properties of bases: There are many such mutagens; some well-known examples are:

- **Nitrous acid**—formed by digestion of nitrites (preservatives) in foods. It converts cytosine to uracil which base pairs with adenine insead of guanine, and adenine is deaminated to hypoxanthine which base pairs with cytosine instead of thymine. Thus, deamination by nitrous acid, like spontaneous deamination, causes transitions (Fig. 14).
- **Nitrosoguanidine, Methyl methanesulfonate, Ethyl methanesulfonate**—chemical mutagens that react with bases and add methyl or ethyl groups. Depending on the affected atom, the alkylated base may then degrade to yield a baseless site, which is mutagenic and recombinogenic, or mispair to result in mutations upon DNA replication.

Intercalating agents: Acridine orange, Proflavin, Ethidium bromide (used in labs as dyes and mutagens) all are flat, multiple ring molecules which interact with bases of DNA and insert between them. This insertion causes a "stretching" of the DNA duplex and the DNA polymerase is "fooled" into inserting an extra base opposite an intercalated molecule. The result is that intercalating agents cause frameshifts.

FREQUENCY OF MUTATIONS

The spontaneous mutation rate varies. Large gene provides a large target and tends to mutate more frequently. A study of the five coat color loci in mice showed that the rate of mutation ranged from 2×10^{-6} to 40×10^{-6} mutations per gamete per gene. Data from several studies on eukaryotic organisms shows that in general the spontaneous mutation rate is 2-12 $\times 10^{-6}$ mutations per gamete per gene. Given that the human genome contains 100,000 genes, we can conclude that 1-5 human gametes would contain a mutation in some gene.

PRACTICAL APPLICATIONS OF MUTATIONS

Most of the mutations are deleterious and make the organisms containing the mutations less efficient; yet, the possibility of developing new desirable traits through induced mutations has led to new avenues in plant breeding. Plant breeders through out the world have worked out such possibilities and have succeeded in developing mutant varieties of barley, wheat, oats,

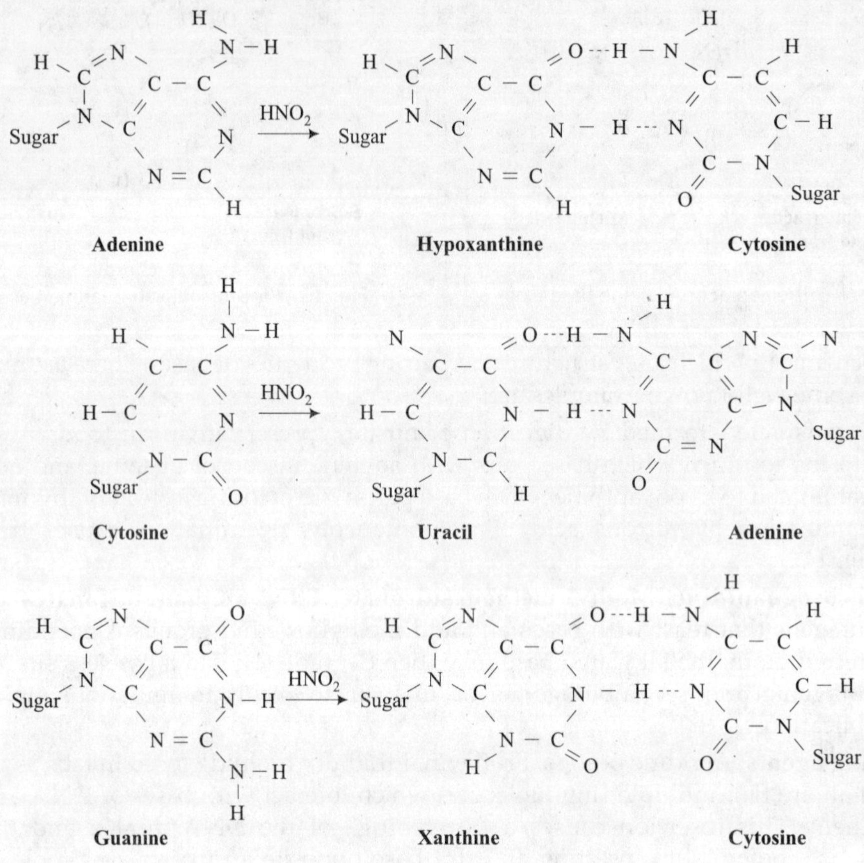

Fig. 14. Mutagenic action of nitrous acid

Fig. 15. H.J. Muller (1890-1967)

soyabeans, tomatoes and fruit trees etc. that provide increased yield, resistance to diseases, increased protein content etc.

SUMMARY

Mutations are the sudden, heritable changes in the genetic material. These may occur spontaneously or may be induced by different agents called mutagenic agents like radiations and/or chemicals. New mutations that arise provide the genetic variability that is necessary for evolution. Though, most mutations are considered to be harmful, yet, the possibility of developing new desirable traits through induced mutations has led to new avenues.

REFERENCES

Auerbach, C. (1962) Mutation, an introduction to research on mutagenesis. Oliver and Boyd, Edinburg, Great Britain.

Buratowski, S. (1993) DNA repair and transcription: the helicase connection. Science 260: 37-38.

Crow, J. F. (1993) How much do we know about spontaneous human mutation rates? Environmental and Molecular Mutagenesis 21: 122-129.

Lederberg, J. and Lederberg, E.M. (1952) Replica plating and indirect selection of bacterial mutants. J. Bacteriol. 63: 399-406.

Muller, H.J. (1927) Artificial transmutation of the gene. Science 66: 84-87.

Sancar, A. (1994) Mechanisms of DNA excision repair. Science 266: 1954-1956

Suggested Readings

1. Principles of Genetics (1991) E.J. Gardener, M.J. Simmons and D.P. Snustad, 8th edition, John Wiley and Sons Inc., USA.
2. Genes VIII (2004) B. Lewin, Pearson Prentice Hall, Pearson Education Inc. NJ, USA.
3. Cell and Molecular Biology (1997) E.D.P. De Robertis and E.M.F. De Robertis, Jr., 8th edition, B.I. Waverly Pvt. Ltd., New Delhi, India.

REVIEW QUESTIONS

Long Questions

1. Differentiate between a chromosomal mutation and a gene mutation.
2. Describe the occurrence of spontaneous mutations that arise naturally as a result of chemical rearrangements, tautomeric shifts or as the result of errors occurring during DNA replication.
3. Describe and illustrate a point mutation and its effects on a gene sequence.
4. Describe the effect of agents, such as radiation and acridine dyes, on a gene sequence with relation to their ability to cause frame shift mutations.
5. Explain why frameshift mutations are likely to be more detrimental than a point mutation.
6. Differentiate between excision repair and photoreactivation mechanisms of repair.
7. Compare the different types of DNA repair mechanisms known to counteract the effects of UV light and other DNA damage.

Multiple Choice Questions

1. _______ mutations have no specific agent other than natural forces
   (a) Spontaneous (b) Induced
   (c) Somatic (d) Germinal
2. _______ is an example of sex-linked recessive mutation
   (a) Haemophilia (b) Down syndrome
   (c) Polyploidy (d) All of the above
3. _______ mutation in single cell may not impare the organism even if mutation is detrimental
   (a) Germinal (b) Induced
   (c) Somatic (d) Chromosomal
4. Which mutation results due to the deficiency of a chemical compound and can be overcome by supplying the same
   (a) Morphological (b) Biochemical
   (c) Behavioural (d) Regulatory
5. Ancon breed was the name given by Seth Wright to mutant variety of
   (a) Horse (b) *Drosophila*
   (c) Sheep (d) Goat
6. The scientific study on mutations by T.H. Morgan began in the year
   (a) 1881 (b) 1927
   (c) 1910 (d) 1953
7. The changes induced when a purine is converted to a purine or a pyrimide is converted to a pyrimidine are known as
   (a) Transversions (b) Transitions
   (c) Frameshift mutations (d) Point mutations
8. When the codon for one amino acid is replaced by a codon for another amino acid the change is known as
   (a) Mis-sense mutation (b) Nonsense mutation
   (c) Silent substitution (d) None of the above
9. Biologically important tautomers involve
   (a) Keto-enol pairs for thymine and guanine
   (b) Keto-enol pairs for adenine and cytosine
   (c) Amino-imino pairs for thymine and guanine
   (d) None of the above
10. Xeroderma pigmentosum, a skin disorder that predisposes individuals to epidermal pigment abnormalities is
    (a) Autosomal recessive disorder (b) Sex-linked disorder
    (c) None of the above

Answers

1. (a) 2. (a) 3. (c) 4. (b) 5. (c) 6. (c)
7. (b) 8. (a) 9. (a) 10. (a)

CHAPTER 19

Haploidy and Cybrids

S. Rama Rao[1*], A. Kumar[2] and N. Gaur[2]

[1]Dept. of Biotechnology and Bioinformatics, North-Eastern Hill University, Shillong–793 022

[2]Dept. of Botany, J.N. Vyas University, Jodhpur–342 003

E-mail: srrao22@yahoo.com

HAPLOIDY

Speciation and evolution of living organisms has been influenced immensely by variations occurring at chromosomal level. Numerical as well as structural changes in chromosomes have played a significant role in evolution of many genera of plants as well as animals. The numerical changes of chromosomes may involve a complete basic set of chromosomes (x) (euploidy) or one or a few chromosomes of a chromosome complement (aneuploidy). Euploidy includes those numerical changes which are the exact multiples of complete basic chromosome set so that organisms comprise two (2x), three (3x), four (4x), etc. sets of chromosomes representing diploid, triploid and tetraploid individuals. Whereas aneuploidy includes those individuals whose somatic chromosome number deviate from exact multiplication of basic chromosome set and involve addition or deletion of one or few chromosomes only. Such organisms are designated as monosomics (2x–1), nullisomics (2x–2), trisomics (2x+1), tetrasomics (2x+2), etc.

Haploidy on the other hand represents one of the kinds of numerical changes with gametic number of chromosomes. Thus, haploidy can be defined as "sporophytic organisms with gametophytic chromosome number and genetic constitution." Blackslee *et al*. (1922) were the first to produce experimentally a haploid plant in *Datura*, which has opened many vistas for the plant breeder to experiment and produce genetically homozygous lines. Several genera and species have since been added to the list published by Kimber and Riley which appeared in 1963 including various crop plants like wheat, barley, cotton, coffee, pearl millet, tomato, rape

*Corresponding author

seed, coconut, phlox and *asparagus*. Schulz – schaeffer (1980) reported spontaneous occurrence of haploidy in 36 species of 26 genera belonging to 10 families. The list is still growing with each passing day (Singh, 1993).

CLASSIFICATION

There are two types of haploids viz.,euhaploids and aneuhaploids.

(i) Euhaploids

Haploid individuals belonging to diploid species are called monohaploids whereas those belonging to polyploid species are polyhaploids. Haploids belonging to autopolyploids (comprising multiple basic sets of the same genome) are known autopolyhaploids, whereas haploids belonging to allopolyploids (comprising multiple basic sets of different genomes) are allopolyhaploids.

(ii) Aneuhaploids

The haploid of a nullisomic plant (2n–2) is referred to as nullisomic haploid and comprised of n-1.Similarly, the haploid likewise tetrasomic (2n+2) are referred to as disomic haploid which comprises of n+1. Likewise a series of aneuhaploids can be developed. However, in some cases like substitution haploids, the haploids will have one chromosomes replaced by chromosome from another species e.g., the haploid of *Triticum aestivum* (2n = 6x = 42) comprising 21 chromosomes has one chromosome replaced by rye chromosome. Yet another type of haploids called misdivision haploids are produced as a result of misdivision of a chromosome leading to the presence of telocentric and isocentric chromosomes.

Haploids may originate at various stages of life cycle such as parthenogenetic development of gametes, chromosome elimination at zygote developmental stages or even by culture of meiospores. Fig. 1 explains various possible methods of haploid origin.

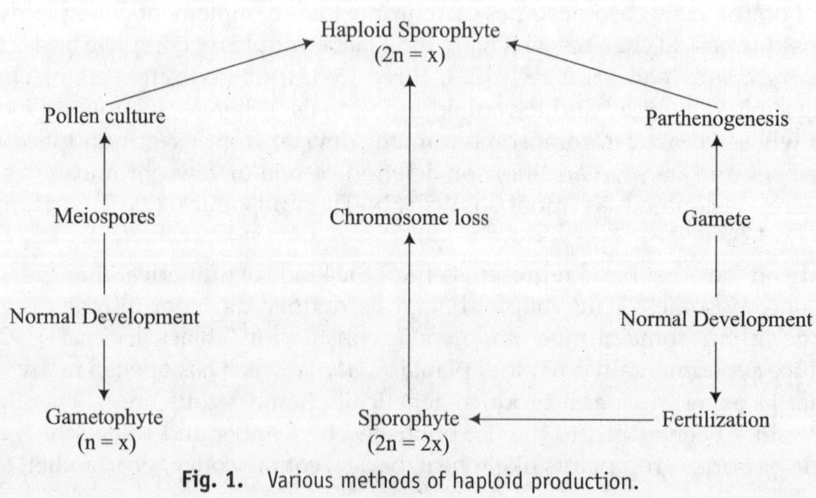

Fig. 1. Various methods of haploid production.

METHODS OF HAPLOID PRODUCTION

(i) Natural Occurrence

The development of an embryo directly from the egg cells without the participation of a male gamete is parthenogenesis. Whereas the development of an embryo without participation of any gametophyte cell is referred to as apogamy. In both the cases the resultant individuals develop into natural haploids.

(a) Spontaneous development: Spontaneous production of haploids is largely due to parthenogenetic development of the egg or any other cell of the embryo sac. Spontaneous haploid production has been reported in a number of plant species. For example, haploids of apogametic origin were obtained in flax (*Linum usitatissimum*) along with polyembryony.

(b) Delayed pollination: Depending on the viability of stigma and ovules, gynogenetic haploids can be induced to develop effecting delayed pollination. In some cases, as that of maize (*Zea mays*) many folds increase in the frequency of haploids has been obtained by Seany (1954).

(c) Genetic control: There are at least two important examples of specific genes that can induce haploidy in plants (Gupta, 1999). They are indeterminate gametophyte (*ig*) in *Zea mays*, which induces haploids of both androgenetic and gynogenetic origin. The other one is haploid initiator (*hap*) in *Hordeum vulgare* which induces haploids of gynogenetic origin only. Homozygous lines of *ig* and *hap* genes have been reportedly influence the frequency of haploids (30-40 %) in the progeny.

(ii) Synthetic Haploids

(a) Physical and chemical methods: In the case of a number of plants like *Nicotiana tabacum, N. glutinosa and N. rustica, Triticum aestivum, T. dicoccum, T. monococcum, Crepis tectorum, Antirrhinum majus* induction of haploids has been successfully achieved with irradiation treatment using X-rays and UV rays.

Besides the radiations, cold and temperature shocks immediately after the pollination, wounding of meristamatic tissue and certain types of chemical treatments given to pollen or to ovary or to both. They have been found to be very effective in synthesis of haploids.

(b) Wide hybridization: Distant and wide hybridization at interspecific and heteroploid level has been successfully utilized for stimulating the embryo to develop into haploid without undergoing fertilization process. The most successful example is *Solanum tuberosum* ($2n=4x=48$), with the stimulus provided by the pollinator *S. phureja* which results in formation of dihaploids.

(c) Alien cytoplasm: When the wheat nucleus is transferred to the cytoplasm of certain alien species by backcrossing method, haploid plants are produced in high frequency (Kihara and Tsunewaki, 1962).

(d) Semigamy: In semigamy, the sperm nucleus enters the egg cell but does not fuse with the nucleus. The nuclei of the egg and sperm may begin to divide synchronously leading to embryo development. The developing embryo often gives rise to chimeral plants for the two putative parental genotypes. One of them may take over to produce haploids (either female or the male). This method has been successfully used in cotton. In the progeny from crosses of a

Flower buds

↓

Surface sterilize with calcium hypochlorite or Bromine water

↓

Wash thoroughly with sterile water

↓

Dissect flower buds to remove anthers from the filaments

↓

Culture the anther on solid or liquid medium
(Potato II medium containing macro elements and Fe salt and thiamine)

↓

Haploid Plants

Fig. 2. Standard protocol of anther culture for production of haploids.

semigametic line as female, 0.7% of the germinated seeds are androgenetic haploids (Turcotte and Feaster, 1974).

(e) *In vitro* methods: Anther culture is one the popular methods of haploid production *in vitro*. First report of successful production of haploid from anther culture was reported by Guha and Maheshwari (1967) in *Datura inoxia*. In the following decades the number of plant species in which anther culture has been successfully utilized for production of haploids has increased to 153 species spanning over 23 angiospermic families (Maheshwari *et al.*, 1980). The standard protocol of anther culture for the production of haploids has been given in Fig. 2. Several factors including genotype of the donor plant, physiological state of the donor plant, development stage of microspores, composition of culture medium and shock pretreatment have a decisive role in production of haploid through anther culture.

Pollen or microspore culture is another way of *in vitro* haploid production. The anthers are used to prepare pollen suspension for culture of isolated pollen grains which subsequently give rise to haploid plants (Fig. 3). Pollen cultures have certain advantages over anther culture since they can be plated and manipulated like microorganisms especially for mutation and genetic transformation studies. But more often anthers rather than microspores are cultured since the extraction and culture methods of microspores differ and have been successful only in few species such as *Datura innoxia, Nicotiana sylvestris, Oryza sativa* etc.

Ovule culture is also done to produce haploids *in vitro*. Development of plants from unfertilized cells of female gametophyte (embryo sac) is becoming a viable alternative to anther culture for haploid production. It was first reported by San Noeun (1970). Since then gynogenetic haploids have been raised for 19 species belonging to 10 angiospermic families. Gynogenetic haploids mostly arise from unfertilized egg cell as reported in barley, sunflower and sugar beet. The gynogenetic plants may arise through direct embryogenesis, or the gametic cells may form a callus followed by plant regeneration on a different medium. However,

Anthers are pressed to release pollen grains

↓

Solution with pollen grain is filtered through a nylon sieve

↓

Centrifuge the solution at 500 – 800 rpm for 5 minute at room temperature

↓

Resuspend the pellet containing pollen in a fresh medium

↓ Repeat the centrifugation twice

Finally culture the pollen grain on a suitable medium at density of 10^3–10^4 grains ml^{-1} in petriplates

↓

Maintain the petriplates at 25°C and 16/8 hour photoperiod

↓

Callusing and embryogenesis

↓

Haploids

Fig. 3. Protocol of microspore culture for production of haploids.

certain aspects of haploidy like the events leading to induction of haploidy in female gametophytes and the differences in growth pattern of *in vitro* development of unfertilized female and the male gametophytes.

MECHANISMS OF HAPLOID PRODUCTION

(i) Somatic Reduction and Chromosome Elimination

Somatic reduction is a process where either spontaneously or due to specific treatments, the chromosome number is reduced to half in the somatic tissues. It has been found out that colchicines treatment can also lead to the reduction in chromosome number. This has been reported in *Sorghum vulgare* whose tetraploid cells respond to colchicine treatment and give rise to diploid cells which, in turn, develope into diploid individuals. They are haploids of the putative tetraploid progenitors. Haploidy has also been induced by a number of other chemicals such as parafluorophenyl alanine (PFPA), chloramphenicol etc.

(ii) Interspecific Hybridization

This method has been extensively used in the production of haploids in several crop plants viz., barley, wheat, bassicas etc. The chromosome elimination procedure for the production of haploids in barley involves interspecific hybridization between cultivated barley *Hordeum vulgare* (2n = 14) and *H. bulbosum* (2n = 14, 28), a wild relative which is a cross pollinated, self incompatible and perennial polyploid species. A cross between diploid *H. vulgare* and tetraploid *H. bulbosum* mainly produces triploid hybrids. The percentage of embryonic cells exhibiting gradual loss of chromosome increases from the day of pollination. After 10–12 days, 94% of the developing embryonic cells have become haploid due to chromosome elimination process. Elimination process is controlled by the genotype of both the parents. The crosses between *H. bulbosum* (4x) and *H. secalinum* (4x) also give rise to haploids of *H. bulbosum,* while *H. secalinum* (4x) x *H. vulgare* (2x) give haploids of *H. secalinum* along with the hybrids.

(iii) Intergeneric Hybridization

When an intergeneric cross is made between hexaploid wheat *Triticum aestivium* (var. chinese spring) and *Hordeum bulbosum* (both diploid and tetraploid species). The haploids were produced because of chromosome elimination in hybrids. It has been found that intergeneric hybridization involving many of the european varieties of wheat did not succeed. However, four varieties of Australian origin did produce haploids besides hybrid progeny. Further the chromosome substitution lines have provided valuable information. In case of the wheat variety"*Hope*", the cross-ability promoting genes were found on 1B, 1D, 3A, 3D, 4A, 4B and 6D, whereas genes that exhibit negative control were located on 5A and 5B. When the crosses are made between wheat and maize, the zygotes generally undergo elimination of maize chromosomes and thus lead to the production of wheat haploids. It was found that *H. bulbosum* is sensitive to wheat crossability genes Kr_1 and Kr_2, while maize is insensitive (Laurie and Reymondie, 1991).

DETECTION OF HAPLOIDS

Comparison of morphological features with that of diploid is one of the reliable methods used for detection of haploids. Significant reduction in overall dimensions of vegetative and floral parts showing reduction in cell size is often reported in haploids. The screening of haploids is often facilitated by observations such as reduction in stomata and pollen size, pollen stainability etc. Besides morphological features, genetic markers are also equally important in detection of haploids. The basic principle involved is that the hybrids and non-hybrids should differ, so that any failure of fertilization and parthenogenetic development of embryo into haploid individual can be recognized in seed/seedling/adult plant stage. If the recognition needs to be done at the seed stage itself then the marker should be such that it expresses at the late stage of embryo development, such as aleurone colour marker. Similarly, if the screening is to be done at the seedling stage, the marker should clearly expresses in the seedling (e.g. the seedling colour). This technique of genetic marker has been extensively used in developing *Zea mays* (maize) monoploids e.g., 'a_1' for brown colour aleurone. R-allele is another genetic marker which produces deep pigmentation of aleurone and thus allow screening at the dormant seed stage when the genetic markers are used. About 90–98% of the progeny can be discarded rather early and a very high frequency of haploids can be recovered in the remaining population.

(i) Morphological Characteristics of Haploids

The volume of the cell or nucleus as determined in epidermal cells, pollen or guard cells is reportedly proportionate to chromosome number, when compared with haploids (x), diploids (2x), triploids (3x) and tetraploids (4x) plants. Such a relationship was observed in several genera including *Datura, Lilium, Nicotiana* and *Gossypium* etc. Therefore, as expected the haploids exhibit significant reduction in height of the plant, leaf size, leaf index, size of the flowers, duration of flowering, pollen size, percentage of pollen stainability and seed size. All these characteristics make the haploids, especially monoploids, look much weaker than their putative disomic plants.

(ii) Cytological Characteristics of Haploids

The haploids of normal diploid plants are characterized by the presence of only one set of chromosomes in somatic cells. Since each chromosome is represented only once, it is not possible to resolve karyotype into distinct homogenous pairs. In case of polyhaploids arising out of autopolyploids we will find half of the somatic chromosome number and they may be resolved into pairs or groups depending upon ploidy level. On contrary the haploids of allopolyploid origin may or may not possess somatic chromosome complement which can be resolved into pairs or groups while constructing the karyotype.

Meiotically the monoploids behave quite differently than their putative diploids due to lack of homologous partners in their chromosome complements. Therefore, each chromosome in the absence of its homologous partner remain as univalent even at zygotene. In the subsequent course of meiosis these chromosomes neither under go recombination at pachytene nor they form chiasma at diplotene/diakinesis. At anaphase I these chromosomes may disjunct to opposite poles in quite erratic manner, often showing unequal distribution, which may eventually result in the formation of unbalanced gametes. At meiosis II the chromosomes may undergo equational division by separation of chromatids in more or less normal pattern. However, at the end of reduction division, the formation of gametes with abnormal chromosome numbers are inevitable. During meiosis the haploid plants of autopolyploid origin may behave normally since there will be more than one set of chromosomes making synapsis possible between homologous chromosomes. It is more or less true in case of haploids arising out of autopolyploids, comprising even number of chromosome sets. However, such haploids with odd number of chromosome sets may not undergo normal meiotic division, owing to unequal distribution of chromosome at anaphase I and II. The haploids of allopolyploid origin may also exhibit normal cause of meiotic cell division only if there are at least two sets of chromosomes belonging to one genome. However, if there are odd set of chromosomes belonging to one genome, the plants may behave like monoploids meiotically.

(iii) Breeding Behavior

Since monoploids or polyhaploids are largely self-sterile, progenies can be obtained only when they are used in crosses as male or female parent. Chase (1969) produced hybrids in maize by crossing diploid and hapoid plants with haploid as male parent. Pollen abortion was examined in resulting progeny and most of the progeny had low pollen abortion rate (0.10%). Alexander (1956, 1964) recorded translocations in haploid (x) X diploid (2x) crosses, involving six and seven chromosomes in maize but could not ascertain if same chromosomes were always independently involved in interchanges. The progenies obtained from crosses involving in most of the cases can be explained as a consequence of formation of restitution nuclei in the first division as observed in tomato by Lindstorm and Koos (1931). Although bivalents/multivalent chromosome associations are absent in monoploids, it is interesting to note that out of 34 species in which meiosis was studied in monoploids, occasional bivalents were observed in 18 species. Several factors may be responsible for such partial chromosome associations. They may be :

(a) The diploid might have had an archaic polyploid origin.

(b) The diploid might have originated due to hyperploid aneuploidy, so that the same chromosome originally became trisomic or tetrasomic. It is followed by differentiation and the polysomic nature may be represented as residual homology.

(c) The diploids must have resulted due to duplication of segments between the non-homologous chromosomes, originally arising through translocations. These duplications will then provide partial pairing ability among chromosomes in monoploids or polyhaploids.

APPLICATIONS

(i) Uses of Haploids in Plant Breeding Programmes

The most significant use of haploids is in crop improvement. The method used for utilizing haploids in plant breeding depends on the inheritance pattern observed in a crop, particularly when we use haploids derived from a polyploid. For an instance allopolyploids, particularly those which are self pollinated show disomic inheritance while autopolyploids especially cross-pollinated ones exhibit polysomic inheritance.

(a) Production of homozygous lines: Haploids are used for production of homozygous lines which can be used as cultivars or in breeding programmes. A double haploid 'Margloble tomato' has been commercially developed and is extensively used world over. In recent years a variety of techniques such as anther culture and the use of chromosome elimination method of *Hordeum bulbosum* in barley, have led to an increased seed production of haploids. A comparison of haploid method of breeding with conventional methods (pedigree method, bulk method, single seed discent) suggest that the time required in haploid breeding is only 4–5 years as against 10 years needed in pedigree and bulk method (Fig. 4).

(b) Hybrid sorting in haploid breeding: Hybrid sorting is the selection of superior plants among the haploids derived from F_1 hybrids through anther culture. The haploid method of breeding involving hybrid sorting is considered superior over pedigree and bulk methods because the frequency of superior gametes is higher than the frequency of corresponding superior plants in F_2 generation and the time required for development of new species is less in haploid breeding. The technique has been used for the development of new varieties in several crops. It has been very well exploited in China and Japan. Some examples are tobacco cultivar F_2 11 resistant to bacterial wilt, rice varieties Tanfong 1, Zonghua 10 having high yield and disease resistant, and the wheat varieties Huapei 1 and Yunhua 2 which are rust and cold resistant. Haploids have also been successfully utilized for breeding of barley, maize, bell peppers etc. In barley at least two varieties have been produced using haploids derived through *bulbosum* technique of chromosome elimination. These varieties are resistant against powdery mildew and barley yellow mosaic virus. In maize superior varieties were developed

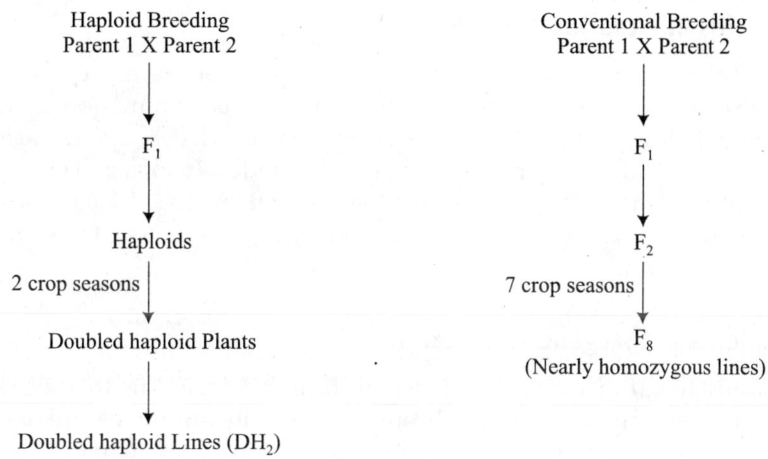

Fig. 4. Period required to develop homozygous line through haploid and conventional breeding.

through screening of haploids derived from anther culture. In sugarcane superior inbred lines were developed by anther culture technique. These plants have high sugar content. In *Brassica napus*, haploids were developed by anther culture technique. They have low erucic acid and glucosinolate content. Haploids derived from twin seedlings in *B. napus* led to the development of a superior cultivar 'Mairs Haplona'.

(c) Production of spontaneously doubled haploids for breeding: The utility of haploid in breeding is the production of doubled haploids through colchicines treatment. In a system where haploids are detected through genetic markers, we can also identify doubled haploids. This avoids colchicines treatment and the breeder can save time and plant material.

(d) Use of haploids in species with polysomic inheritance: Polyhaploids derived from polyploids exhibiting polysomic inheritance may be homozygous or heterozygous. For instance dihaploids of potato may be AA or Aa or aa depending upon the genotype of the putative tetraploid-parent, which may be heterozygous for this gene (AAAa, Aaaa, Aaaa). By doubling of haploids, homozygosity cannot be necessarily achieved in these cases unless dihaploids are again used for production of monoploids. The use of haploids in autotetraploid crops like potato differs from that of barley, rice etc.

(e) Analytical breeding: In a crop like potato, the genetic improvement through crosses and selection can be obtained at the dihaploid level. These improved dihaploids can be used for chromosome doubling to get increased tuber size. This process was described as analytical breeding. Advantages of analytical breeding are disomic inheritance in the dihaploids which provides simpler inheritance ratio, ease of selection and require smaller population size to recover recombinants. Dihaploids represent a level of homozygosity equivalent to three generations of self pollination in an autotetraploid. Analytical synthetic breeding scheme has been explained by Wenzel *et al.* (1979). This process takes the advantage of protoplast fusion and plant regeneration to bring together different genotypes into a completely heterozygous potato. In this procedure dihaploids are used for the production of monohaploids before the tetraploid is resynthesized. This technique can be easily used in potato because once the resynthesized tetraploids are regenerated they can be multiplied via tubers for commercial cultivation.

(ii) Use of Haploids in Mutation Research

Haploids possess only one set of chromosomes and, therefore, even recessive mutation is immediately expressed and used in application based on induced mutations. Mutant cell lines have been isolated which showed resistance to various drugs, amino acid analogues, pathotoxins, salts, herbicides, chilling, viruses and nematodes. In tobacco cell lines resistant to methionine-sulphoximine and wild fire disease (caused by *Pseudomonas tabaci*) has been developed. After differentiation and diplodization, such mutations can be incorporated in regenerated plants in homozygous condition.

(iii) Use of Haploids in Cytogenetic Research

Haploids are useful in various areas of cytogenetic research, including their use in production of aneuploids and determining the basic chromosome number or the nature of polyploidy. In wheat, initially monosomics were produced by Sears (1954), where haploids were pollinated by normal pollen. The monosomics are formed due to functioning of egg with restitution nuclei deficient for any individual chromosome. In tobacco several nullisomics were derived from haploids obtained from sterile monosomics. Haploids have been used for production of alien addition and alien substitution lines in a crop like wheat. A study of meiosis in haploids gives an evidence of the original base number of a genus. For example in *Pennisetum typhoides* occurrence of partial pairing (2 bivalents in some cells) suggested that the original base number (x) may be 5.

(iv) Use of Haploids in Evolutionary Studies

A comparison of dihaploids in potato with diploid wild species proved useful in drawing conclusion about the origin of cultivated potato. Since production of 2n gametes is common both in dihaploids and diploid species, the *andigenea*, (the progenitor of *tuberosum)* must have originated many times from different but related diploid (2x) populations in different locations through sexual polyploidization. Haploids in potato are crossed with diploid species to conduct genome analysis. Meiosis in hybrids and further generations revealed that wild species differs from cultivated potato few genes.

LIMITATIONS

Since haploids are highly sterile, they can not be maintained by sexual reproduction. Haploids produced from autotetraploid species are an exception since they may be normally fertile. Production of haploids requires special techniques, which demand skill, time and resources including a dependable green house. In many crops, large number of haploids can not be produced, owing to genomic constraints. Colchicine has to be applied for chromosome doubling of haploids. Colchicine is not effective in some cases, and it can sometimes induce mutations.

CYBRIDS

Plant growth and development are a result of the coordinated expression of genes located primarily in the nucleus, as well as in cytoplasmic organelles. Cytoplasmic hybrids or cybrids are the fusion products of nuclear genome of one parent and organelles of second parent, This process is known as cybridisation. Therefore a cybrid may be considered as a hybrid eukaryotic cell line which combines the nuclear genome from one source, with plastid and mitochondrial genomes from other source. In other words cybrids result from the fusion between cytoplasm and nuclear gene donors. Such genetic materials are an alternative means to broaden the genetic/breeding pool by non-sexual gene transfer.

MECHANISM AND METHODS OF CYBRID PRODUCTION

Cybrids can be produced naturally from anyone of the following methods:

(a) A nucleated cell fuses with an enucleated.

(b) Sudden loss of one nucleus from the heterokaryon having two types of nucleus.

(c) Loss of one complete set of chromosomes of one parent during cell division.

A general method to isolate and purify substantial number of viable cybrids from cultured mammalian cells immediately following cytoplast cell fusion letter, has been described by Kliot-Fields *et. al.* (1983). This method uses cytoplasts whose mitochondria are selectively sustained *in vivo* by the cationic fluorescent rhodamine and are fused to appropriate recipient cell lines, followed by the sorting of fusion mixture. Cybrids purified by cell sorting may be useful for detailed molecular studies of mitochondrial DNA gene expression and in the specific induction of new mitochondrial DNA mutants. Cybrids can be also produced artificially by fusing X-ray irradiated cell with a non- irradiated cell or fusion of an enucleated protoplast with nucleated protoplast.

PROGRESS OF CYBRID RESEARCH

Golmirzaie *et al.* (2003) produced cybrids through fusions between Y-245.7, (a potato clone with tetrad sterility) and Atzimba (a common clone). These synthetic cybrid plants exhibited significant differences for tuber yield, style length, and berry number among the hybrid off-springs. Bonnema *et al.* (1991) recovered a collection of tomato (*Lycopersicon esculentum* Mill.) mutants following cybridization. The cybrids were generated by fusing protoplasts of tomato cv. UC82 with irradiated protoplasts of *L. pennellii.* According to them the mutants displayed a five-fold reduction in reproductive growth and a two-fold reduction in vegetative growth. The mutant phenotype is further correlated with the presence of mitochondrial DNA from the wild species used in the protoplast fusion. Bonnema *et al.* (1992) compared the organization of the mitochondrial genome of the mutant cybrids, non-mutant cybrids, and the two parents. They reported that the number and organization of several mitochondrial genes in the cybrids related to tomato and opined have no large rearrangements around atpA, atp9, coxll, or rrn18 + rrn5 genes. Amplification of gene copy appears to have occurred for rrn26 in the mutant cybrids and the loss of gene copy appears to have occurred for atp6. All cybrids have the *L. pennellii* organization of rrnl8 + rrn5 gene cluster. All of the mutant cybrids have the *L. pennellii* organization of atp9. The mutant cybrids appear to have only one copy of atp6 derived from *L. pennellii*, while *L. pennellii* has two copies of this gene.

Vardi *et al*. (2000) have used nucellar calli of *Citrus* Its related wild genus *Microcitrus* acted as the sources of protoplasts, which upon appropriate conditions, are able to divide and result in the respective true-to-type trees. They used *Microcitrus* protoplasts as donors of cytoplasmic organelles and rough lemon or sour orange protoplasts as recipients using the donor-recipient protoplast fusion technique. Cybrid colonies and the subsequent cybrid plants were obtained having either rough lemon or sour orange morphological (nuclear coded) features. As expected restriction analyses for mitochondrial DNAs from cybrids revealed novel patterns which differ from those of either of the fusion partners. Southern blot hybridization indicated that the cybrid plants were homoplastomic in either type. Some plants contained *Microcitrus* chloroplasts while others contained the recipient's chloroplasts. This is an example of intergeneric cybridization

APPLICATION OF CYBRIDS

(1) Cybrids help in transfer of disease and herbicide resistance genes e.g., resistance to triazine herbicide has been discovered in *Solenum nigrum*. It is encoded by chloroplast DNA. It has been successfully transferred via somatic hybridization which expressed in *S. tuberosum* + *S. nigrum* somatic hybrid.

(2) Restoration of fertility in cytolasmic male sterile lines as reported in a CMS line of *Nicotiana sylvestris*.

SUMMARY

Haploidy refers to individuals/tissues that have somatic cells with gametic number of (n) chromosomes. Haploids of diploids (2x) are called monohaploids while that of polyploids (3x, 4x, etc.) are known as polyhaploids. Haploids may originate spontaneously through parthenogenetic development of unfertilized eggs or may be induced through heteroploid crosses at interspecific/intergenenic level. The irradiation of cells/tissues with ã-rays or culture of anther and/or pollen can successfully lead to the development of haploids in plants. Unlike plants, in case of animals such as insects and amphibians haploidy is prevalent where it plays an important role in sex-determination of the offsprings. Monohaploids contain one basic genome and, therefore meiosis is highly irregular due to lack of homologous chromosomes. The homologous chromosomes pair intragenomically as a result of either duplications or genetic redundancy in the haploid genome otherwise all the chromosomes appear as univalents starting from zygotene through metaphase I of meiosis. At anaphase I the disjunction of chromosomes to opposite poles occurs quite randomly. Contrary to this the polyhaploids may indulge in regular meiosis depending upon the ploidy level of progenitors. In case of plants, haploids are largely self-sterile; hence progenies can be obtained only when they are used in crosses as male or female parents. The most important use of haploids is their utility in production of homozygous lines, where these can be directly used as cultivars or may be used in appropriate breeding programs with plant improvement as a major objective. They require only 4-5 years for successful production of improved varieties. Haploids are also used in mutation research, evolutionary studies and in understanding genetic/cytogenetics mechanism underlying evolution of various plant/animal genera.

Cybrids are cytoplasmic hybrids produced by the fusion of two different cells and comprise only nuclear genome. The main advantage of developing cybrids is to understand the role and molecular regulation of cytoplasmic DNA, especially the mechanism of mitochondrial DNA gene expression and specific induction of new mitochondrial DNA mutants. Cybrids are of high value in mitochondrial research and provided several authentic clues for the role of mitochondria in various human aliments like Alizimer and Parkinson disease. Similarly, cybridization via protoplast fusion has also become an integral part of the breeding program of several horticultural species (e.g., *Citrus, Lycopersicon* etc.). Recently, a reliable methodology for isolation and purification of viable cybrids has been developed through somatic cybridization. It enhanced the utility of cybrids in disease control through drug development and production of desired horticultural plant varieties. Despite the multifacet advantage of cybrids, a lot of works are needed in the area of crop improvement and human disease control programs.

REFERENCES

Alexander, D.E. (1956) Further evidence for crossing over between non-homologus chromosomes during megasporogenesis of haploids. Maize Genetics coop News Letter 30: 37-38.

Alexander, D.E. (1964) Spontaneous reciprocal translocation during megasporogenesis of Maize Haploids. Nature, 201: 737-738.

Blackeslee, A.F., Bergner, A.D., and Avery, A.G. (1922) A haploid mutant in Jimson weed, *Datura stramonium*. Science, 55: 646-647.

Bonnema, C., Catherine, C., Stephan, S., Valerie and O'Connell, M. (1992) International plant Genome Conference, San Diego, California.

Bonnema, A., Castile, C., Corriz, S., Sisneros, V. and O'Connell, M. (1991) Organization of the mitochondrial genome in tomato cybrids Graduate Program of Molecular Biology, Department of Agronomy and Horticulture, New Mexico State University, Las Cruces, New Mexico 88003-0003.

Chase, S.S. (1969) Monoploids and monoploid derivatives of maize (*Zea mays* L.). Bot. Rev., 35: 117-167.

Golmirzaie, A. M., Tenorio, F., Serquén and Ortiz, R (2003) Cybrids and tetrad sterility for developing true potato seed hybrids : Ann. Appl. Biol. 143: 231-234.

Guha, S. and Maheshwari, S.C. (1967) Development of embroids from pollen grains of *Datura in vitro*. Phytomorphol., 17: 454-461.

Gupta, P. K. (1999) Cytogenetics. Rastogi Publications, Meerut, India.

Kihara, H. and Tsunewaki,K.(1962) Use of an alien cytoplasm as a new method of producing haploids. Jap. J. Genet, 37:310-313.

Kliot-Fields, T., Finney, D.A. and Wiseman, A. (1983) Purification of cybrids by fluorescence-activated cell sorting. Somatic Cell Genet. 9: 375-389.

Kimber, G. and Riley, R. (1963) Haploid angiosperms. Bot. Rev., 29: 480-531.

Laurie, D.A. and Reymondie, S. (1991) High frequencies of fertilization and haploid seedling production in crosses between commercial hexaploid wheat varieties and maize. Plant breeding. 106: 182-189.

Lindstorm, E.W. and Koos, K. (1931) Cytogenetic investigations of a Haploid Tomato and its diploid and tetraploid progeny. Am. J. Bot., 18: 398-410.

Maheshwari, S.C., Tyagi, A.K., Malhotra, K. and Sopory, S. (1980) Induction of haploidy from pollen grains in angiosperms. Theor. Appl. Genet., 58: 193-206.

Noeun, S. (1970) Haploids of *Hordeum vulgare* per culture. Nature, 225: 874-876

Schulz-Schaeffer, J. (1980) Cyatogenetics. Plants Animals Humans. Springer-Verlag New York, 15: 244.

Seanay, R. R. (1954) Monoploids in Maize. Maize Geney Cop. News Letter, 28: 22.

Sears, E.R. (1954) The aneuploids of common wheat. Missouri Agril. Exp. Stn. Res. Bull., 572: 1-58.

Singh, R.J. (1993) Plant Cytogenetics. CRC Press, Boca Raton, London.

Turcotte, E. and Feaster, C.V. (1974) Methods of producing haploids: Semigametic production of cotton haploids. In: Haploids in higher plants-Advances and Potential (Ed., K.J. Kashak), pp 53-64, University of Guelph, Guelph, Canada.

Vardi, A., Frydman, S., Galun, E., Gonen, P. and Bliechman, S. (2000) Citrus cybrids: Transfer of *Microcitrus* organelles into citrus cultivars ISHS Acta Horticulturae 280. Internatl. Symp. *In Vitro* Cult. Horticul. Breed.,

Wenzel, G., Schieder, O., Przewozny, T., Sopory, S.K. and Melchers, G. (1979) Comparison of single cell culture derived *Solanum tuberosum* L. plants and a model for their application in breeding programmes. Theor. Appl. Genet., 55: 449-55.

Suggested Readings

1. Principles and Procedures of Plant Breeding: Biotechnological and Conventional Approaches (2002) G. S.Chahal and S. S. Gosal, Narosa Publications, New Delhi, India.
2. Plant Breeding: Analysis and Exploitation of Variation (2000) R. Darbeshwar, Narosa Publishing House, New Delhi, India.
3. Cytogenetics (1999) P. K. Gupta, Rastogi and Company. Meerut, India.
4. Elements of Biotechnology (2004) P. K. Gupta, Rastogi Publications. Meerut, India.
5. Text Book of Plant Breeding (2000) B. D. Singh, Kalyani Publishers, New Delhi, India.

REVIEW QUESTIONS

Long Questions

1. What are haploids? Differentiate between the terms haploids, monoploids, dihaploids, doubled haploid and polyhaploids.
2. Discuss the origin and mechanism of haploid production.
3. Discuss is detail how haploids are detected.
4. Describe breeding behavior of haploids.
5. Explain the various uses of haploids.
6. What are cybrids? Discuss the mechanism of cybrid production.
7. Describe the recent progress of cybrid research and add a note on application of cybrids.

Multiple Choice Questions

1. Monosomics are designated as:
   (a) 2x-1 (b) 2x-2
   (c) 2x+2 (d) 2x+1
2. Trisomics are represented by:
   (a) 2x+2 (b) 2x+1
   (c) 2x-2 (d) 2x-1
3. Haploid individuals belonging to diploid species are called:
   (a) Polyhaploids (b) Allopolyhaploids
   (c) Monohaploids (d) None of the above

4. Experimentally first haploid plant was produced from:
   (a) Datura (b) Tobacco
   (c) Tomato (d) None of the above
5. The development of an embryo without participation of any gametophyte cell is referred to as:
   (a) Syngamy (b) Parthenogenesis
   (c) Apogamy (d) None of the above
6. Production of haploids through anther and/or microspore culture is known as:
   (a) Anther culture (b) Ovule culture
   (c) Embryo rescue (d) All of the above
7. Bivalents/multivalents chromosomes associations are absent in:
   (a) Diploids (b) Monoploids
   (c) Polyploids (d) All of the Above
8. The fusion products of nuclear genome of one parent and organelles of second parent are called:
   (a) Hybrids (b) Cybrids
   (c) Haploids (d) None of the above
9. The production of cybrids through fusion of X-ray irradiated cell with a non irradiated cell is known as:
   (a) Natural production method of cybrid
   (b) Artificial production method of cybrid
   (c) All of the above
   (d) None of the above
10. Which is the suitable example of cybrid:
    (a) Protoplast fusion of tomato cv.UC 82 with irradiated protoplasts of *L.pennellii*
    (b) Fusion between Y-245.7, a potato clone and a common clone
    (c) *S. tuberosum* + *S. nigrum*
    (d) Both A and B
11. Cybrids are most commonly useful in:
    (a) Transfer of disease resistance (b) Restoration of fertility in CMS lines
    (c) Transfer of herbicide resistance (d) All of the above

Answers

1. (a)	2. (b)	3. (c)	4. (a)	5. (c)	6. (a)
7. (b)	8. (b)	9. (b)	10. (d)	11. (d)	

CHAPTER 20

Transposable Elements

G. Tripathi[1*], B.M. Sharma[1], J. Shasmal[1] and U.K. Tomar[2]

[1]Depart. of Zoology, J.N.V. University, Jodhpur– 42 001

E-mail: drgst@rediffmail.com

[2]Arid Forest Research Institute, Jodhpur–342 005

E-mail: uktomar@afri.res.in

Deoxyribonuleic acid (DNA) was earlier believed to be a static entity. But now it is known that certain DNA segments can move from one location to another on the genome creating a genetic instability. They are known as transposable elements, transposons, translocatable elements, mobile sequence, movable sequence, insertion elements, jumping elements, parasitic elements or selfish DNA. In 1940s Barbara McClintock was first to identify transposable genetic elements, when DNA was not even considered to be the genetic material, while investigating the inheritance of pigmentation in the common corn plant, *Zea mays*. She reported that the variegation in pigments of kernel is due to action certain distinct genetic units which can move from one site to another on different chromosomes of maize. The placement of these genetic units into genes turns the expression of genes 'on' or 'off', regulating thereby their expression. She called them as 'controlling elements'. She was awarded the Noble prize in 1983 for her work. About 30 years after the discovery, a new class of mutation was observed in the genome of the common intestinal bacterium, *Escherichia coli*. It was found that the mutation was caused by insertion of certain DNA segments into the bacterial genome. These nucleotide sequences of DNA were called insertion sequences or IS elements. Initially they were recognized as small fragments of DNA inserted into genes (Jordan *et al.*, 1968; Shapiro, 1969), but later as performed movable elements (Malamy *et al.*, 1972).

Some genes are responsible for causing resistance to antibiotics in bacteria. These genes are able to move from one DNA molecule to another. In 1974, Hedges and Jacob showed that a DNA segment carrying antibiotic-resistance gene can be transposed from one DNA to another,

*Corresponding author

and called as transposable segment transposable element (Tn element) or transposons (Hedges and Jacob, 1974). After 1976 the research began to look into the mechanism of transposition of these mobile elements. It was soon found that during transposition the mobile elements were replicated, as a result of which one copy remained in the old position and the replicated one moved to the host genome.

Transposons encode the enzymes 'transposonases' or 'transposases', required for transposition of mobile genetic elements. Any alteration in the transposonase gene may abolish the transposition process. Another group of Tn element (e.g., Tn 3) express or encodes a repressor protein, which can bind to the transposonase and its own promoter region of the DNA. This repressor protein is considered as a site-specific recombination enzyme, acting as a part-time regulatory molecule.

There are two general groups of transposons. One that mobilizes via DNA and another has ability to make DNA copies of their RNA transcripts and these copies may integrate at new sites in the genome. The latter is related to retroviruses. Here we have mainly described the transposons that mobilize via DNA. The frequency of transcription varies from transposon to transposon. The overall rate of transcription is approximately 10^{-4} to 10^{-3} per element per generation.

INSERTION SEQUENCES

Insertion sequences (IS) are simple transposition modules. They are small elements (less than 2000 base pairs) encoding only transposition determinants. Like transposons, IS elements are capable of inserting themselves randomly into a DNA molecule and also have inverted terminal repeats of about 30 nucleotide base pairs (bp). It has been reported that the presence of an IS in a plasmid also acts as a transposon. Besides these, an IS has other functional similarities with the transposon. For instance, the incorporation of an IS into host DNA is associated with the replication of few bp in the recipient DNA. Integration of IS element with DNA molecule causes mutation which can be restored to normal after removing the insertion sequence. It also abolishes the function of the gene of an operon where it has been integrated because the IS element comprises nonsense and terminator sequences which cause rho- factor-dependent termination of transcription (Nevers and Saedlor, 1977). Insertion of an IS into a gene diminishes the activities of other genes downstream (5'-3') in the direction of transcription due to disturbances created in the continuity of nucleotide bp in the DNA.

Insertion sequences are categorized into non-homologous classes as IS1, IS2, IS3, IS4 and IS5. They appear to be normal components of chromosomes and plasmids of the gram-negative bacteria. For example, the genome of *Escherichia coli* possesses about eight copies of IS1, five copies of IS2 and three copies of IS3. The elements IS2 and IS3 have been reported to be present on various drug-resistance episomes. The smallest transposable elements (750-1500 bp) are known as 'transposition modules' or 'IS-like modules'. They encode only genetic determinants relevant to promoting and regulating their own transposition. They are IS1-IS5, IS10, IS102, etc. Insertions occur at a level comparable to spontaneous mutation rate, usually 10^{-5} to 10^{-7} per generation. Reversion (by excision of IS element) is infrequent with a range of rates of 10^{-6} to 10^{-10} per generation, which is approximately 10^{5} times less frequent than insertion (Lewin, 1994).

The composite transposons are elements, which encode not only transposition functions but other functions unrelated to transposition, e.g., antibiotic resistance. The drug-resistance causing transposable elements (Tn elements) have 4,500 nucleotide base pairs (Tn1, Tn2, Tn3) to 20,000 base pairs (Tn4). These Tn elements (Tn1 to Tn4) carry ampicillin-resistance and contain the inverted repeat sequences of 140 bp, while Tn5 (kanamycin-resistance) and Tn10 (tetracycline-resistance) have longer inverted repeat sequences (eg. Tn5, 2450 bp; and Tn10, 1400 bp). However, Tn9 (chloramphenicol-resistance) comprises a direct repeat sequences of about 750 bp. The repeat sequences of Tn9 and Tn10 are structurally more or less similar to the IS3 and IS1 respectively. Certain composite transposons are flanked by two copies of IS either in direct or in inverted orientation (Fig. 1A). There are other complex transposons not flanked by IS elements (Fig. 1B). A transposons having long, inverted repeat sequences can be visualized by electron microscope. When DNA strands of a transposon are separated and allowed to self-anneal, the complementary bases at both the ends join with one another, forming double-stranded circle 'lollipop' (Fig. 2). These lollipops are recognizable in electron micrograph.

BASIC EVENTS DURING TRANSPOSITION

The ends of transposons contain two copies of IS elements having terminal inverted-repeats. It has been documented that the incorporation of any gene between two copies of IS induces the transfer of gene into other DNA molecule through non-homologous recombination process. The transposition involves three steps: (i) Duplication of mobile elements; (ii) duplication at insertion site; and (iii) DNA rearrangements.

Duplication of Mobile Elements

When a mobile genetic element translocates from one plasmid (donor site) to another plasmid (recipient site), the formation of an intermediate (or byproduct?) takes place. The intermediate is composed of one copy of both the plasmids (donor and recipient) linked together by two copies of the mobile element. It infers that the mobile elements are replicated during transposition. Therefore, each element comprises exactly the same set of non-permuted DNA sequences as present in the donor DNA or on the same DNA molecule. The nucleotides present at both ends of a transposon join with the broken segments of DNA where it is going to be inserted. For example, there are three groups of transposons in gram-negative bacteria: IS modules, Tn3 group, Mu and its derivatives. It has been shown that transposition is a

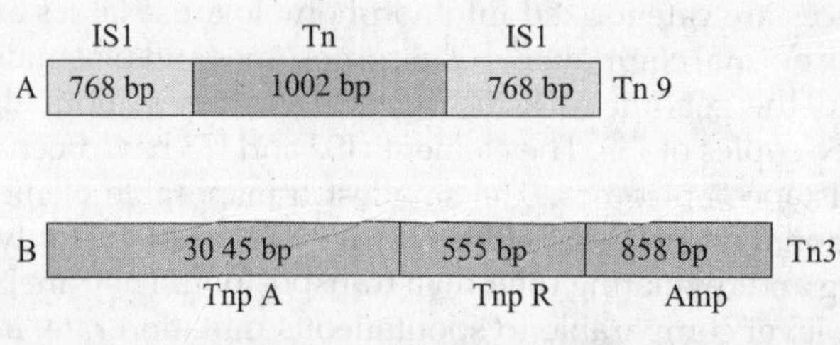

Fig. 1. Two types of composite transposons (Tn elements), one flanked by IS sequences (A) and the other in devoid of IS (B). Tn9 has a short inverted repeat of 23 bp but it is bound by direct repeat of IS1. Chloramphenicol (Cam)-resistance region is present on Tn element. Tn3 has three parts, two (TnpA and TnpR) of which an transposable and the third is ampicillin (Amp)-resistance region which shows gene amplification during resistance production.

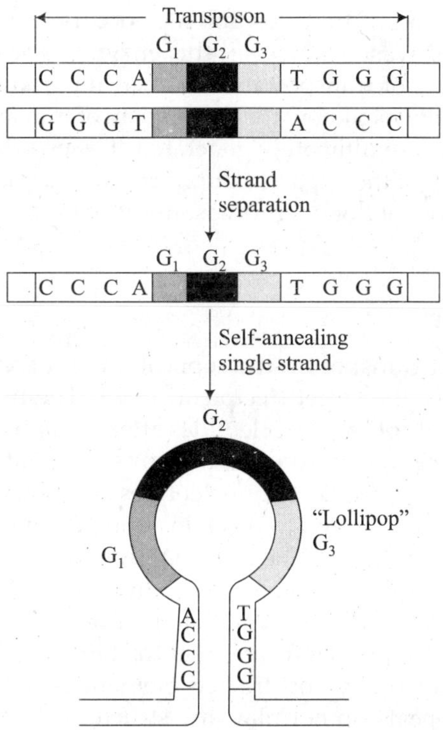

Fig. 2. Self-annealing of inverted repeat sequences produces a circle, the 'lollipop' which is visible under an electron microscope. Here, lollipop contains three, genes (G1, G2, G3) in the single stranded loop.

replicative process in the case of Tn3 family in which donor and target (recipient) replications fuse to produce a 'cointegrate' as an intermediate, which comprises two copies of Tn3, one at each junction between donor and target nucleotide sequences. The site-specific recombination between the two copies of a transposon can generate simple insertions from the cointegrate. Though the heterogenous collection of IS also promotes the fusion of replicons to form the cointegrate, the frequency is very low (1% of the frequency of simple insertions). Once the cointegrate is formed, it remains stable, i.e., no site-specific recombination occurs. If cointegrate is not broken down by recombination, how are simple insertions of an IS generated? Simple insertions are thought to be produced directly during incorporation of a single insertion sequence copy at a largest site of the DNA molecule. It is supported by the observation in the case of Tn5 elements that the cointegrates are produced by simple insertions from the dimers of donor replications rather than by replicons fusion between a monomer of donor and the target DNA.

It is thus evident that a fused structure or cointegrate is formed during transposition process. At the same time it remains unclear whether the cointegrate is a necessary intermediate or it is only a byproduct. In fact it is questionable because an enzyme should be present to resolve the fused structure if the cointegrate is a necessary intermediate. The function of this enzyme could be substituted by homologous recombination between two copies of the genetic element only in those cells which are proficient in recombination process.

Undoubtedly, the phenomenon of transposition also occurs in recombination-deficient cells mediated by an enzyme 'resolvase'. However, this enzyme is absent in several transposable elements except a single class of transposons. Another important fact reveals that a kind of molecular rearrangement is done to know the mechanism of transposition by DNA sequencing so that transposed sequences are ultimately generated. The possibility of error is not ignored, rather it has been suggested that the apparent duplication of the mobile genetic element during transfer may be due to the loss of donor chromosome after excision of a transposon-, 'suicidal transposition'.

Duplication at Insertion Site

In addition to replication of transposon, the replication of a short (3-12bp) DNA sequence occurs at the insertion site in the target (recipient) DNA. The nucleotide sequence of target DNA is found to have some duplicated nucleotides after transposition of the element, and the duplicated segments flank the transposon, i.e., the transposon integrated between duplicated DNA segments. The number of duplicated nucleotides is characteristic of particular element i.e., different nucleotides are duplicated for each insertion because the target DNA sequences are different at different insertion sites. The target DNA repeats in the case of IS1, IS2, IS3, IS4, and IS5 are 9,5,3 or 4, 11 or 12, and 4 bp respectively. Similarly, all TnA elements (Tn1, Tn2, Tn3, Tn501, Tn551, Tn801, Tn951, Tn1701, Tn2602 etc.) show target DNA repeat of 5 bp. The presence of short target DNA repeats indicates that the target positions on both the strands of DNA molecule. An experimentally constracted element is not flanked by direct repeats, indicating that the direct repeats do not play any structural role during transposition. Thus insertion of a transposon into a host DNA involves breakage of the target DNA not at random but in a staggered fashion, and a transposon gets inserted into and ligated with the cleaved or protruding chains of the target DNA molecule. It results into two gaps at both ends of target DNA, which are repaired by a small DNA duplication. The creation of flanking duplications is

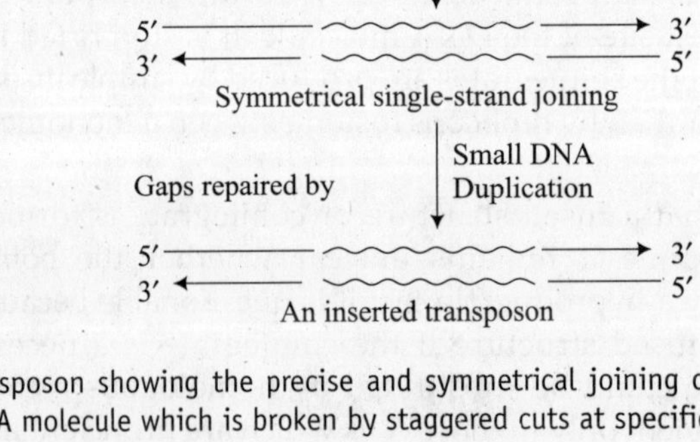

Fig. 3. Insertion of a transposon showing the precise and symmetrical joining of the transposon's ends to a target DNA molecule which is broken by staggered cuts at specific sites (0).

a characteristic of all prokaryotic transposons. The ligation of a transposon to the target DNA is always precise and symmetrical. The initial joining of the two is likely to be symmetrically-related single-stranded ligation at each end of the inserted element (Fig.3).

The duplication of a sequence of nine nucleotide pairs in host DNA by insertion of IS1 has been observed. Incorporation of such transposable element causes nine-nucleotide duplications. It is found that the staggered cleavage of recipient DNA occurs at positions 5, 9, or 11 nucleotide apart on opposite DNA stretches. After insertion of IS1 into the recipient DNA, the filling-in of gaps starts on both the strands of DNA. The process requires the synthesis of short single-strand fragments of complementary nucleotide sequences and in turn shows the nucleotide-sequence duplications (Fig. 4).

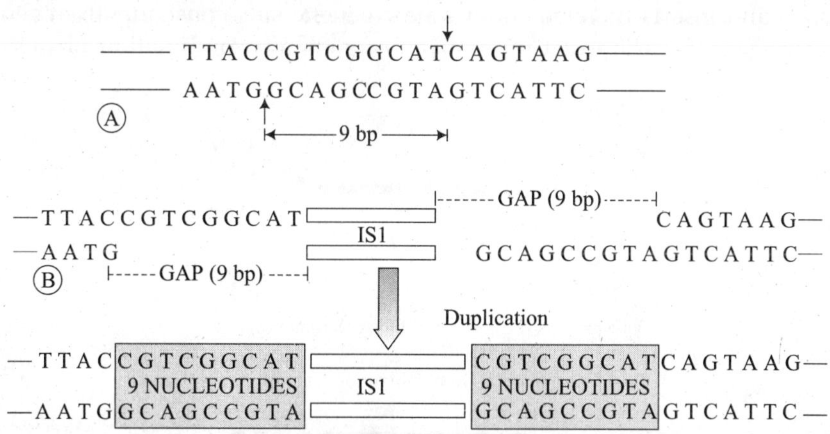

Fig. 4. Significance of nucleotide-sequence duplication in the recipient DNA molecule. Two strands of the DNA are cut at staggered sites (↑) which are 9 nucleotides apart from each other (A). Subsequently, an IS element (IS1) incorporates at the cut-sites (B) leaving 9 bases unpaired (gaps). Further, 9-nucleotide duplications of target DNA occur at both sides of the IS1 to fill-up the gaps. Thus, IS1 insertion generates a nine-nucleotide duplication.

Non-replicative Transposition

Transposition occurs by both replicative and nonreplicative ways. In case of non replicative transposition the transposon moves as a physical entity directly from one site to another and is conserved. Here the transposing element is lost from the donor site. After nonreplicative transposition the donor is either repaired or destroyed. Tn10 and Tn5 use the nonreplicative pattern of transposition. This type of transposition requires only transposase enzyme and not resolvase. Conservative transposition is a type of nonreplicative event in this case the mobile genetic element is excised from the donor site and inserted into a target site by a series of events in which every nucleotide bond is conserved. It resembles the mechanism of lambda integration. Transposes of such elements are related to λ integrate family. This way of transposition can mediate transfer of large elements so also of donor DNA from one bacterium to another. Actually, they are more appropriately regarded as episomes rather than transposons. Earlier they are classified as transposons.

Some transposons use only one type of pathway (replicative or nonreplicative) for transposition. However, other may utilized both pathways. The elements IS1 and IS903 use both pathways. Similarly, the ability of phase Mu to turn to either type of pathway from common intermediate has been established (Lewin, 1994). Replicative transposition is the only mode of mobility of the TnA family. TnA family includes several related transposons and consists of large transposons (45 kw) of which Tn3 and Tn1000 are well known.

DNA Rearrangements

Insertion of a mobile genetic element into the target DNA initiates replication in the transposed and the target nucleotide sequences leading to the formation of double-stranded DNA segments separately. Each of the double strands comprises one old and one new (i.e., replicated) DNA stretch. The reciprocal recombination then occurs between the two copies of the transposon and inserts the element at a new genetic site, consequently regenerating the donor molecule (Fig. 5). The exact mechanism of such a recombination process is still not

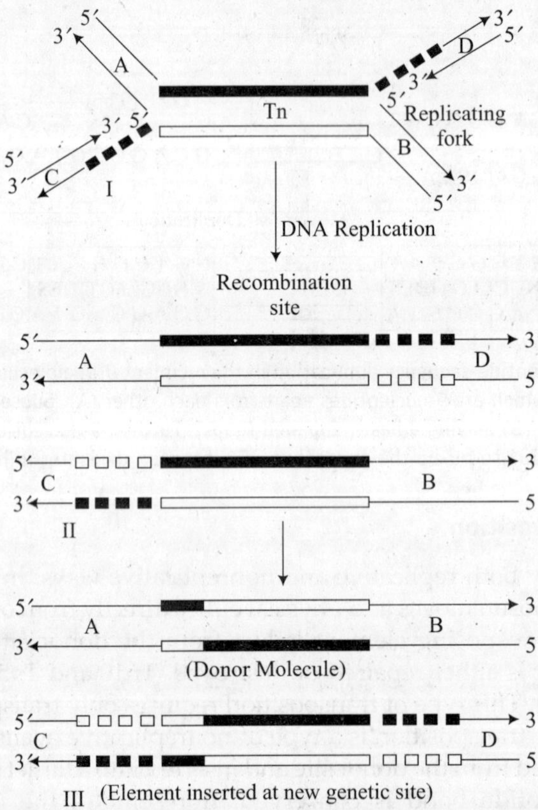

Fig. 5. Insertion of an element at a new genetic site by reciprocal recombination process. (1) : Replication of donor and recipient nucleotides after insertion. The donor segment originally comes from the A,B Genetic sites of a DNA molecule and is incorporated into recipient DNA between C,D sites; (II) : Two copies of a transposable element, each having one replicated strand, and (III) : Reciprocal recombination between these two copies of transposon incorporates the Tn element at a new genetic site and reproduces the donor DNA molecule (Cohen and Shapiro, 1980).

known. In case of Tn3, it is mediated through the nucleotide sequences present within the element. However, it is well known that such reciprocal recombination (non-homologous) does not require proteins which are needed in general homologous recombination process.

In addition to simple transposition process (Fig. 6A), the transposons produce a number of different types of DNA rearrangements involving the linkage of one or more transposon ends to a new target site. All types of DNA rearrangements can be explained on the basis of six types of events (1-6 in Fig. 6) mediated by transposons. Of these 1 and 3 are intermolecular events, while 2 and 4 are intramolecular in nature. Further in type (5), the incorporation of genetic element occurs without changing the intervening genes (e.g., ABC). It is not known whether this is due to inter- or intra-molecular transposition. The proper combination can lead to the formation of structures (products) similar to those generated by transposition alone. These structures are interconvertible. For instance, the homologous recombination can reduce the structure in type (5) to the structure in type (4), and can also interconvert the structures (5) and (2). Likewise, two circular DNA molecules each having a single copy of transposon from a cointegrate-type of structure by homologous recombination. It can be reverted through reverse reaction to generate two circular DNA molecules, each containing a copy of the element. Thus the homologous recombination process may be the most efficient way to produce some of the structures shown in Fig. 6 but under certain specific conditions.

From the above it is evident that IS are discrete DNA segments that can move from one genomic site to another and may cause genetic rearrangements. An interesting question arises whether these rearrangements are associated with duplication of IS that leads to rearrangements. In this connection, a strong evidence is provided by IS903, and it has been shown that any IS-mediated transcriptional recombination can occur by an efficient replicative mechanism (Weinert *et al.*, 1983). Based on several observations, Shapiro (1969) pointed out that the transposable elements bring together unrelated chromosomal segments to generate a variety of structural rearrangements including fusion, depletion, inversion and transposition. Hence the transposon play an important role in structural reshuffling of cellular DNA.

Tn rearrangements shown in Fig. 6 (e.g., 2, 3 and 5) generate products in which chromosomal (non-transposon) markers are bound by duplicate copies of an element. Such stretches are capable of their own transposition on the DNA molecules as discrete genetic units. These products come under the category of composite transposable elements which contain some accessory genes. In several cases the composite elements undergo the process of deletion or gene mutation to eliminate deleterious or unnecessary regions, thus refining the expression of accessory genes. Not only this rather composite elements can generate any of the structures shown in Fig. 6, but also the inner ends of IS modules can interact with a new target site to produce additional varieties of DNA rearrangements (Fig. 6:6). Several types of Tn elements have been identified which encode the accessory genetic determinants. For example, many transposons encoding genes for antibiotic resistance, heat-stable enterotoxin and lactose utilization have been shown in prokaryotes particularly in bacteria. The frequencies of many transposon-promoted genetic rearrangements range from 10^{-10} to 10^{-7} per generation.

Another example of DNA rearrangement has been reported from immunoglobulin (Ig) genes in higher organisms. Immunoglobulins (Igs) have two short or light (L) chains and two long or heavy (H) chains. Each chain possesses a N-terminal which constitutes the variable (V) region, and a C-terminal which has the constant (C) region. V and C regions are coded by two separate genes (V and C genes). Two types of L chain-kappa (*k*) and lambda (λ) are found

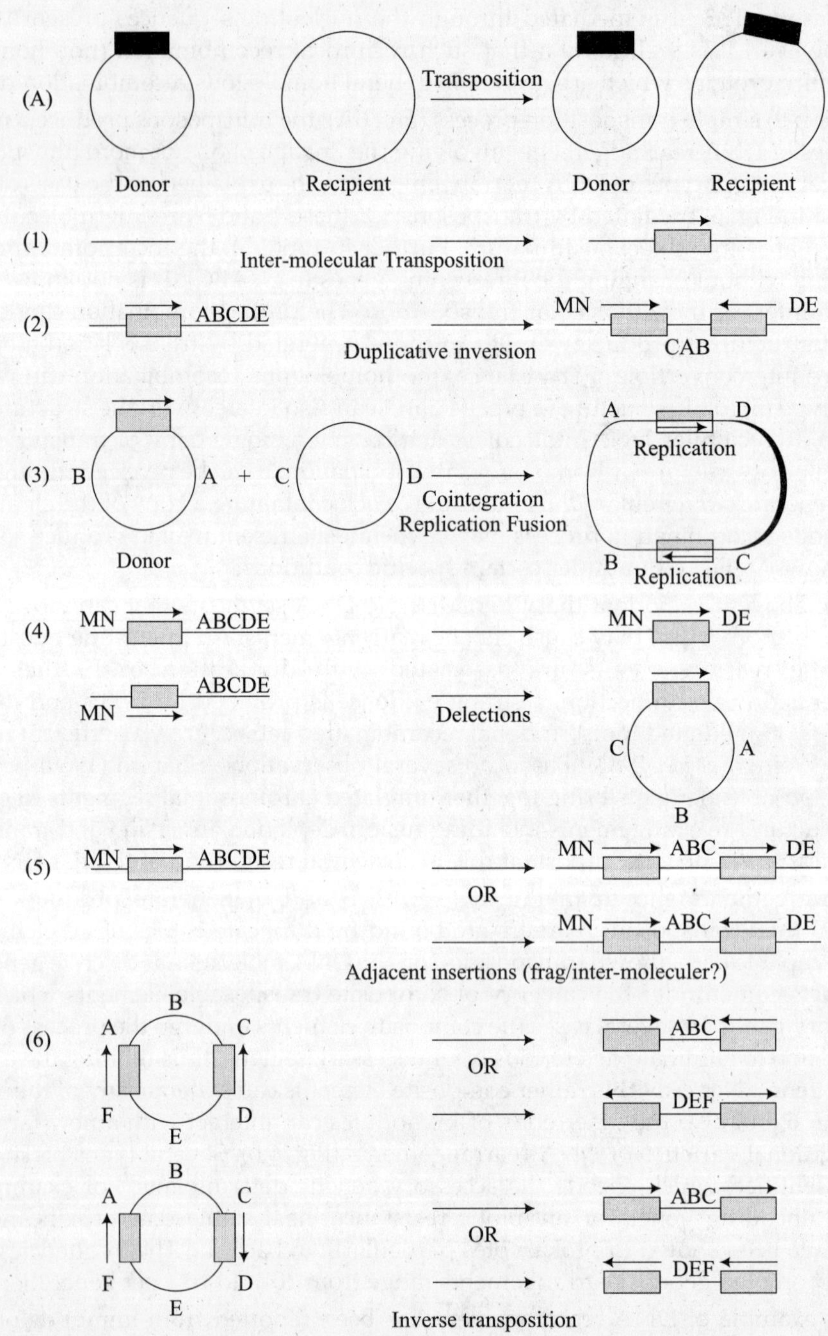

Fig. 6. Transposon-promoted genetic (DNA) rearrangements [(A) : Simple transposition of a genetic element between two different DNA molecules (donor and recipient). The numbers 1-6 indicate the varieties of transposon-induced events or rearrangements in which 1-5 are caused by a single transposon while 6 is due to a composite transposon.

which differ from each other in the amino acid (a. a) sequence of their C regions. Both the L chains of an Ig are always of one type; either *k* or λ. Likewise, H chains are of five types – alpha (α), delta (δ), epsilon (ε), gamma (γ) and mu (μ) based on their C region a.a. sequence. On the basis of their backbone, Igs are classified into five groups, IgA, IgD, IgE, IgG and IgM. The major immunoglobulins of the body are IgG, IgM and IgA. There are separate V and C genes for L and H chains. During differentiation of B-cells, transposition occurs in nucleotide sequences to bring a specific V gene into the proximity of a specific C gene (Fig. 7). This DNA rearrangement leads to transcription of a single mRNA from both V and C genes, which is translated into a L or a H chain from their respective mRNA molecules. In the body several thousand types of Igs are produced, which are necessary to combat millions of antigens (Ags). This is an important feature of the immune system to generate from a relatively small amount of genetic material a thousand types of antibody (Ab) molecules with different Ag binding specificities. Three families of Ig genes are recognized: (a) one consists of 20 to 30 genes coding for different V regions of both L and H chains; (b) nearby in the same chromosome is a second family of genes consisting of 2 to 4 genes coding for two C regions of L chain *k* and λ; and (c) the third family codes for the C-regions of various classes and subclasses of H chains (α, δ, ε, γ, μ).

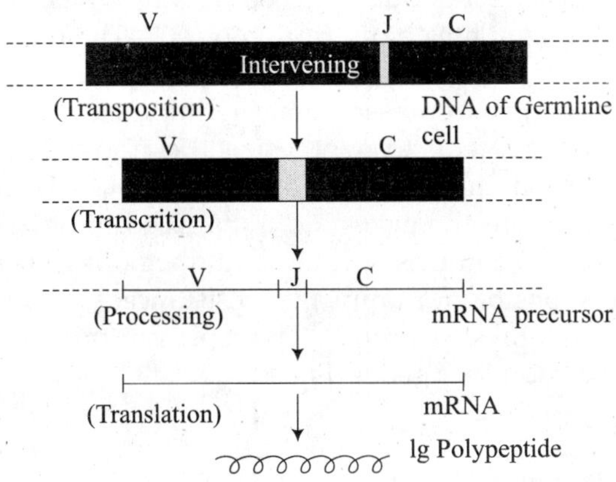

Fig. 7. Translocation of V and C genes through excision of intervening sequences and joining of the two by J segment. [Functional integration of V and C regions occurs by excision of the J, and forming an Ig polypeptide from V and C genes].

The genes within each family are linked but the families are not. On the basis of the above description it can be asked as to how an animal can produce about a million types of Igs? The mechanism for the expression of specific Ig genes must include selection of only one out of the two genes coding for *k* and λ; and the L chains should comprise a specific V gene alongwith one of the five classes of C genes for H chains (α, δ, ε, γ, μ). Thus antibody diversity arises by selection of one out of many V genes to combine with one out of some J genes to form the V region, which in turn, combinès with one out of three to four C genes. In case of H chain one out of five classes, and many subclasses of C genes are selected. Therefore, 'generation of diversity of Ig' (God Ig) is by simple permutation and combination of many V, J and C genes. Hence the transpositions and rearrangements of certain specific DNA sequences in the immunoglobulin genes lead to the production of millions of antibodies which have much evolutionary significance.

MECHANISM OF TRANSPOSITION

The process of excision and reintegration of the transposable element at another site is termed as transposition. The ends of a transposon is essential to conduct the transposition process and its deletion causes transposition defect. Almost all nucleotide bps present in the terminal inverted repeats at the ends of the elements Tn3 and IS10 are important to contribute the transpositional information and the latter element (IS10) also contains recognition sites for DNA or protein interactions. So there must be some specific enzymatic mechanism to identify or recognize the inverted-repeat ends and for the precise breakage of DNA segments. The gene responsible for the synthesis of these specific enzyme is located in the transposon itself. The mutation in this gene abolishes the transposition i.e., it changes the ability of the element to act as a transposable element. Mutations at another region of Tn3 lead to increased frequency of Tn3 transfer or movement into different plasmids. The mutated region may have a gene which alters the transposition capacity of Tn3 element! This gene of Tn3 element is known as 'transposonase gene', and the normal transposonase gene encodes an enzyme required for the transposition i.e., transposonase (a 100000 dalton peptide). Tn3 also contains a repressor gene encoding a repressor molecule (a peptide of 21000 dalton) which represses transposition of Tn3. The repressor molecule regulates transcription of Tn3. The repressor molecule regulates transcription of the transposonase gene as well as the synthesis of the repressor molecule (Gill *et al.*, 1979; Chou *et al.*, 1979). If repressor gene is mutated, there will be a constitutive transcription of transposonase gene to synthesize the transposonase enzyme. Thus the repressor molecule regulates the transcription at the promotor (P) sites of transposonase as well as repressor genes, and both the genes are transcribed in opposite direction (Gill *et al.*, 1979; Chou *et al.*, 1979). The Tn3 comprises three coding regions : (i) gene (bla) for the enzyme betalactamase which is responsible for ampicillin-resistance, (ii) gene coding transposonase, and (iii) gene encoding for a repressor protein. The repressor protein controls transcription of both the transposonase and repressor genes (Fig. 8).

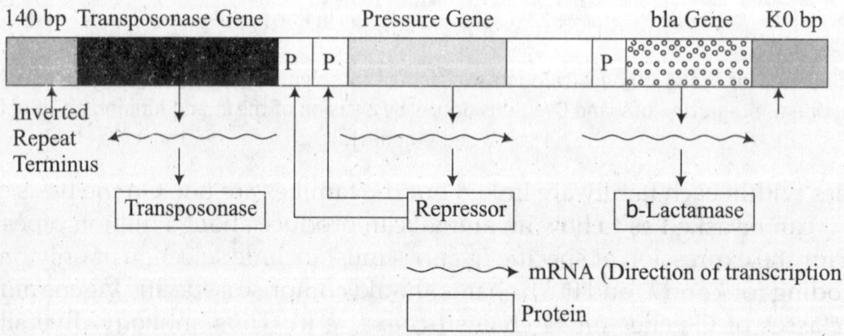

Fig. 8. Functional components of Tn3 transposon which is composed of three genes (transposonase, repressor and *bla* genes) [Transcription of transposonase and repressor genes is bidirectional to each other. The repressor molecule regulates transcription of both the transposonase and repressor genes through their promotor regions or promoters (P). β-lactamase is encoded from the *bla* gene and confers resistance to ampicillin as well as related antibiotics.

In some cases it has been suggested that the insertion site and orientation of transposon are decided by the ends of Tn element. For instance, if the ends of IS5 element is coupled to a promotor in an 'out-to-in' fashion, it directs the replication of recipient DNA. This is referred to as 'inceptor' activity. The inceptors are suggested to be the sites where transcription terminates to produce a primer for further DNA synthesis. In certain cases the ends of transposons possess active promoters, rho-termination sequences, Pribnow box like sequences or the sequences which can produce new promoters by DNA rearrangement. These indicate that the nucleotide sequences of the ends are A-T rich. Although the role of RNA polymerase in transposition is not known, it is speculated on the basis of above findings that the interaction of RNA polymerase and transposon ends could be important for transposition.

The insertion of transposon is not random. It can insert itself at multiple sites on a recipient DNA molecule. Certain parts of DNA are prone to multiple insertions. These specific regions of DNA are called as 'hot-spots'. Insertion, deletion and even inversion in Tn10 element occur regularly at hot-spots, and one such hot-spot has been shown per 1000 bp of the target DNA molecule. But it has also been documented that the Tn3 is inserted preferentially in the regions of recipient DNA, which are similar to the nucleotide sequences of the inverted-repeat ends of the transposon. Such insertions have been observed even in those bacterial cells which are unable to synthesize a protein (*recA* protein) needed for homologous recombination. Thus it appears that recognition of homologous DNA sequences may play some role in determining the frequency as well as site-specificity of transposon-associated recombinations (Cohen and Shapiro, 1980).

ROLE OF Tn ELEMENTS IN REGULATION OF GENE EXPRESSION

Insertion of transposable elements in DNA molecule controls the expression of adjacent genes. In some cases the genes are 'turned-off' while in others they are 'turned-on' by these elements. Hence the transposable elements act as switches to regulate the expression of genes.

Gene Turn-off

Many IS elements (IS-like modules) contain transcriptional terminators which turn-off the gene lying distal to the insertion sites. This causes polar mutation or polar effect (polarity) on the expression of genes present in an operon. Several such controls have been identified in classical operons (*gal* and *lac*) (Besemer and Herpers, 1977). Polarity of some IS elements induces the generation or synthesis of *rho*-termination factor for transcription, and the polar effect can be fully or partially relieved in *rho*-strains (Besermer and Herpers, 1977). Similarly, the polar mutations caused by insertions of the elements in lambda phage can also be suppressed by antitermination function (Kleckner *et al.*, 1978). It has also been suggested that the polar effects may be due to the stop codons present within the IS-element. So the multiplicity of stop signals and the absence of a nearby translation-reinitiation signals may lead to strong polarity.

Gene Turn-on

In some cases, insertion of the IS elements in DNA molecule activates the transcription of adjacent genes i.e., it turns-on the adjacent genes. Such regulation of genes by IS2 and IS3 interaction has been examined. Strong as well as weak turn-ons have been shown by IS2 insertions in both orientations. Tn5 insertion presents a weak turn-on of the gene. It has been reported (Saedler *et al.*, 1974) that IS2 activates the adjacent genes by the read-through transcription from a promotor region of the genes. However, it has been finally suggested that the transcriptional activation of adjacent gene is due to the formation of a new signal by juxtaposition of the element and target DNA sequences.

Several other evidences reveal that promoters present within IS elements are involved in the transcriptional activation of adjacent genes. For example, inversion of IS2 elements acts as a switch to control the expression of an adjacent gene (Fig. 9), and inversion of the loop is the result of a crossover between the two inverted repeat sequences (II) which can be reverted (I) by second crossover. The loop comprises a transcription terminator (T) and a promotor (P). Under normal condition (orientation in I), transcription of the promotor PG initiates the transcription of gene G. This transcription is terminated at T which ultimate inactivates the function of G gene. However, under inverted condition (orientation II) P starts transcription and in turn G is transcribed without initiation at the promotor PG. As shown in Fig. 10 the IS2 contains a strong transcriptional T and a P. So it may act as a switch for turning the genes on or off depending on its orientation.

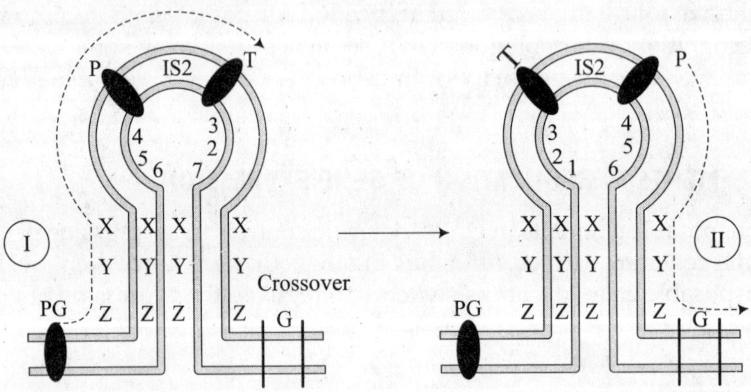

Fig. 9. Inversion of IS2 acting as a switch to control the transcription of an adjacent gene [Xx, Yy and Zz are the complementary base pairs of inverted repeat sequences. A crossover between these inverted repeat sequences forms an inversion of the loop which can be restored by another crossover. In orientation (I), the transcription of the promotor PG is terminated at T and in-turn inactivates the function of G gene. In case opposite orientation (II), P initiates transcription and G is transcribed without the transcriptional initiation at PG, thus turn-on the gene.

TRANSPOSITION OF ANTIBIOTIC RESISTANT GENES

It is well established that bacteria possess several antibiotic resistant genes in their genomes. They contain plasmids which are responsible to transmit different antibiotic resistances to a new bacterial cell. However, it was not clear until the discovery of transposition phenomenon as to how a number of antibiotic resistant genes were located on a single plasmid-DNA molecule? The answer to the above question is that the drug resistance-determinant segments of plasmids evolve as collections of transposons (Fig. 10), each carrying a gene conferring resistance to one or more antibiotics (Cohen and Shapiro, 1980). The plasmid, carrying several resistance genes, seems to be generated by the joining of a resistance-determinant segment and

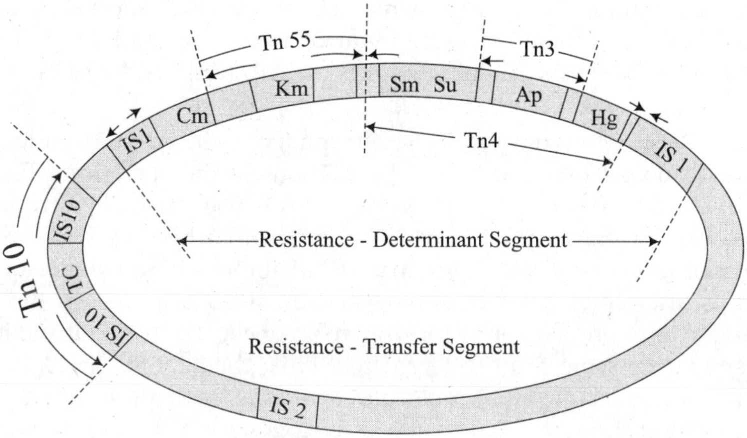

Fig. 10. A plasmid carrying several resistance genes represents the collection of transposons [Genes imparting resistance to chloramphenicol (Cm), kanamycin (Km), streptomycin (Sm), sulfonamide (Su), ampicillin (Ap, and mercury (Hg) are clustered on the resistance-determinant segment which is composed of many transposons (Tn55, Tn4, Tn3). Tn3 is present within the Tn4 element. While the transposon (Tn10) encoding resistance to tetracycline (Tc) is located on the resistance-transfer segment (Cohen and Shapiro, 1980).

resistance-transfer segment, the junction containing IS1 elements. Such junctions are the sites where the two segments (resistance-determinant and resistance-transfer segments) sometimes dissociate reversibly at a specific site. Each transposon present on the plasmid can be transferred at another site independently. IS elements are present at both the dissociable sites as well as at the sites where the interaction of plasmid and chromosomal DNA occurs to induce the transfer of genome between different bacterial cells. It has been shown that the antibiotic-resistant transposons can move even among the different bacterial species. For instance, certain DNA sequences similar to the Tn3 elements have been observed to be responsible for penicillin resistance in other bacterial species. The elements such as Tn1, Tn3, Tn501, and Tn1771 exhibit the phenomenon of transposition which is related to antibiotic resistance. Plasmid carrying one copy of an element does not allow the insertion of another copy of the same element in the plasmid. This inhibition of another or secondary insertion occurs to a lesser degree in cases of intramolecular transposition. The inhibition of secondary insertion is shown to be *cis*-acting. In Tn3 element, Tnp A end is essential for causing immunity and the immunity can occur even in the absence of Tnp R segment.

A GENERAL SURVEY OF TRANSPOSONS IN ORGANISMS

Transposons are present both in prokaryotic as well as eukaryotic cells. They are normal constituents of many bacterial genomes, plasmids and bacteriophages. Similarly, transposons have been identified in yeasts, maize, nematodes, fruitflies and sea urchins. IS2 to IS5 are present in *E. coli* K12 while IS1 in other strains of *E. coli* and many enterobacteria. ISR1 is located in the genome of *Rhizobium lupini* and IS8 in RP4 plasmid. Tn1 and Tn2 are present in plasmids of *Pseudomonas, Salmonella, N. gonorrhoea* and *H. influenzae*. Tn3, Tn4 and Tn2350 are located within the bacterial plasmids R1 drd 19, whereas Tn5 is present in *Klebsiella* plasmid JR67 and Tn6 in JR72. Tn10 is the resistance-determinant of R100, Tn9 and Tn981 (Tn204) are the products of deletion derived from Tn2671. Sources for TnD (TnD:TnA and TnD:Tn cam) are *H. influenzae.*

DNA of a temperate bacterial virus (bacteriophage) is inserted at multiple sites of the bacterial genome leading to several types of mutations in the bacterium. The phage Mu is refered to as 'mutator'. The transposition of Mu DNA causes several rearrangements of the host DNA molecule. Several such studies have shown that Mu is a transposable element. Another example of Tn element is the retrovirus (RNA tumour virus) which induces cancer in higher organisms after infection either by inserting itself adjacent to the chromosomal oncogene leading to its expression or by picking up an oncogene from cell and inserting it into a more active site of the same or of different cell. First, the genomic RNA of the retrovirus produces a double-stranded (ds) DNA molecule by reverse trascriptase, which gets integrated to host DNA and modulates the expression of oncogene which is located adjacent to the insertion site. The oncogene inactivates certain other genes after transposition to another site and causes DNA rearrangement. Retrovirus also contains short terminal inverted repeats of nucleotide sequences. Due to these properties sometimes retroviruses are also considered as transposons. This is an example virus Tn elements integrated into the eukaryotic DNA.

A transposable element (Tcl) has been studied in the genomes of two closely related strains of *Caenorhabditis elegans*. It is found that there are approximately 30 copies of Tcl in the Bristol strain whereas 200 to 300 copies are present in the Bergerac strain. Most of them are structurally highly conserved. This is an example of Tn element having strain-specific differences. The distribution of transposons appears to be ubiquitous, hence it may not be possible to describe them in all organisms. Here we will discuss about transposons in some eukaryotic organisms which are considered standard for the related genetic studies.

(i) Yeast

There is a family (Ty elements) of dispersed repetitive DNA sequences in *Saccharomyces cerevisiae.* Tyl and Ty917 are the two major classes of Ty elements and most of them fall into one of the two classes. Tyl is composed of 35 copies of certain DNA sequences having a total length of 5600bp. Many copies have related nucleotide sequences but not the identical sequences. Each copy has 250bp known as S. There are 100 copies of S in the genome, out of which 70 copies are associated with Tyl sequences and 30 copies are situated somewhere else. Direct repetitions of S sequences resemble the structure of some bacterial Tn elements. The function and mechanism of transposition of Tyl mobile element are not clearly known. It may be said that Ty element transcribes into two poly $(A)^+$ RNA species, but whether the Ty RNA is translated into protein or not, is not known.

(ii) Fruitfly

There are many types of transposable elements in fruitfly *Drosophila,* in which *Copia*, 297, 412, P-elements are found. *Copia* and 412 are dispersed in the genome and each is present in approximately 35 copies. These are bound by short terminal direct repeat sequences. The nucleotide stretch of *Copia* is 5000bp long having direct repeats of about 300bp. The length of 412 sequence is approximately 7300bp with direct repeats of 500bp. The sequence of 297 is not well known. The *White* mutation in *Drosophila* can be caused by insertion of *Copia* element. Another important example of transposons in *Drosophila* is confined to P-elements which are present in some lines and absent in others. In *Drosophila,* transposition of mobile elements generates a number of genetic disturbances known as 'hybrid dysgenesis' (Kidwell *et al.*, 1977) which comprises increased mutational rate, hybrid sterility and induction of chromosomal rearrangements. Hybrid dysgenesis is found only by a cross made between wild males and laboratory female flies, so the germ line of hybrids is affected. Hybrids produced in the first generation are morphologically normal but show reduced fertility and the progeny of these hybrids have other dysgenic traits or characters. P-elements of wild males cause hybrid dysgenesis in eggs of laboratory females lacking P-elements (M-female) and thus constitute a P-M system. Hence P-elements are found in P-stocks (30-50 elements/haploid genome) and absent in M-stocks. It has been suggested that hybrid dysgenesis involves interaction between paternal genome and maternal cytoplasm. P-element is believed to encode repressors of the transposition, and the P-element transposition is suppressed by the cytoplasm of females.

When certain genes cloned into P-element are injected in embryos of early stage the injected genes get inserted at multiple sites of the chromosomes of embryos and their expressions are in a regulated fashion. Through the expressions of these genes are different from that in mammalian cells, in which the clones are transplanted, it is not clear whether this is due to the differences in animals or transposons-guided regulation.

(iii) Maize

The transposition was first observed in the corn, *Zea mays*. At present much is known about the genetics and physiology of 'controlling elements' in maize. It was first discovered by McClintock that there are two genes in *Zea mays*, known as dissociator (Ds) and activator (Ac), which produce a drastic effect on colour-expression of the kernels. Ds affects the expression of genes adjacent to its location, while Ac acts on genes which are far apart from its position. Ac has 4300bp sequence and more than 30 copies of Ac derivatives are present in the genome which are defective in transposition. Simple internal deletion of Ac produces Ds hence, these are Ac derivatives. Complex structures having more than one Ds element can also be generated from these Ac.

An active maize Adhl-F gene, a Ds induced mutation of this gene and two independent Ac-induced revertants alleles have been described (Sutton *et al.*, 1984). Ds mutant has a 405bp insertion flanked by a direct repeat (8bp) which is a duplication of 8bp. The insertion sequence is A-T rich. It is suggested that the 405bp sequence is a Ds element. Low level of Adhl mRNA has been reported in Ds mutants. About 30 sequences are shown in the genome of maize, which are related to Ds element. It is significant to note that several elements have been found in various other plants having similar termini but different internal DNA sequences.

RETROPOSONS

Retroviruses have ability to insert DNA copies (proviruses) of an RNA viral genome into the chromosome of a host cell. Some eukaryotic transposons are related to retroviral proviruses in their organization. They transpose through RNA intermediates. Such mobile genetic elements are referred to as retro–transposons or retroposons. They range from retroviruses that freely infect host cell to the sequences that have transposed via RNA. The sequences do not themselves have ability to transpose that is why they are transposed via RNA. Life cycle of retroviruses shows transposon like events. An interesting part of life cycle in the occurrence of transducing viruses. The transducing viruses are variants that have acquired cellular sequences substituted for part of the viral sequence. The part of viral sequence is replaced by the *v-onc* gene, as a result *Gag-v-onc* whether speciation is due to transposable elements. It needs an protein is synthesized instead of usual Gag, Pol and Env proteins. The transformed virus is replication defective and cannot sustain and infective cycle by itself Ty elements of yeast are similar to retrovirses.

Retroposons can be divided into viral and non-viral superfamilies. Viral superfamilies retroposones have capacity to transpose because they code for reverse transcriptase and/or integrase activities. This retroposon differs from retrovirus in not passing through an independent infectious form. The non-viral superfamilies retroposons originated in RNA sequences. They do not code for protein that have transposition functions and the transposition occurs by an enzyme system coded elsewhere. A significant part of the moderately repetitive DNA of mammlian genomes consists of retroposons (Lewin, 1994).

ROLE OF TRANSPOSONS IN SPECIATION

The possible role of transposable elements has been ascribed to hybrid incompatibility and speciation (Ish-Horowicz, 1982). Many mutations caused by hybrid imcompatibility (hybrid dysgenesis) are genetically unstable. Two such separate dysgenic systems have been recognized. They are P-M and I-R systems which are induced by two different mobile elements. Much is known about the P-M system in *Drosophila* which is caused by insertion of P-elements. It is found that several mutations can be induced in the *White* locus by hybrid dysgenesis, in which five out of seven mutants are due to insertions of homologous sequences of differing length. It has been shown that dysgenesis leads to a high frequency transposition of P-elements into M chromosomes. Some of the P-elements behave like P chromosomes. For instance, they can induce sterility after crossing with M females. Thus there is a strong correlation between the number of acquired P-elements and the ability of M chromosome to produce dysgenic sterility. In most of the cases phenotypes accompanied with hybrid dysgenesis act specially to mobilize P-elements (Ish-Horowicz, 1982).

Few reports indicate that in the population of *D. melanogaster*, dysgenesis may be a step towards speciation (Schaefer *et al.*, 1979). It is a speculated that the different elements generated in isolated populations can lead to hybrid sterility and reproductive isolation which, in turn, lead to speciation (Ish-Horowicz, 1982). But it is not exactly clear extensive species-specific studies of mobile elements to explore a clear-cut concept about the role of transposable elements in speciation.

EVOLUTIONARY SIGNIFICANCE OF TRANSPOSONS

Evolution of transposon has occurred either into confer some selective advantage to organisms under varying conditions or they are just selfish sequences of DNA (selfish DNA) and survive because of a replicative advantage over the host genome. The first view indicates transposons as natural tools for genetic manipulations leading to variability or diversity among organisms for natural selection. In higher organisms, transposons play an important role for modifying gene activity during development, which shows a clear selective advantage. The second view holds that any advantage in hosts may be entirely incidental. Both the opinions regarding advantage conferred on the host can be explained. In a population of organisms residing in an antibiotic environment, transposon can confer adaptive advantage by causing antibiotic resistance. Transposition can also turn-on some silent metabolic pathways which may give an adaptive advantage to the population. This may be considered as incidental advantage.

Evans (1984) has thrown light on some evolutionary discussions about selfish or parasitic DNA. According to him an element, which imposes any cost on its own host, will be eliminated by natural selection (Evans, 1984). It creates the selection pressure which can be counteracted either by a mechanism for horizontal transfer (i.e., virus-like infection of cells) or by *de novo* production of new elements at a high rate. On the other hand, it has been suggested that during zygote formation, the parasitic elements capable of duplicative transposon are able to 'colonize' new chromosomes spreading them to new genomes by a process other than the horizontal transfer. Several studies incorporating negative selection into the model of selfish DNA suggests that the P-elements in *Drosophila* represents demonstrable negative effect on the fitness of the yet they have spread rapidly within the natural population. Therefore, the parasitic DNA is not only possible, rather highly probable, in eukaryotic organisms.

SUMMARY

Certain DNA sequences are capable of inserting themselves at specific sites on the chromosome(s) of the same or different cell(s). They are widely known as 'transposable elements' or 'transposons' because they transpose the blocks of genetic material back and forth. Transposition of these genetic elements causes multiple rearrangements in DNA leading to unstable mutations. Thus transposons control the expression of specific genes by turning them on or off. Transposons ignore the rules of general recombination by joining unrelated DNA segments. In addition to transposons that mobilize via DNA, there are also retroviruses and retroposons. Reverse trancription is the unifying mechanism for retroposons. Retroviruses are regarded from the perspective of the free viral (RNA) form, while retroposons are regarded from the stance of the genome (duplex DNA) form. Retroviruses have genomes of single-stranded RNA that are replicated through a double stranded DNA intermediate. An individual retrovirus contain two copies of its genome. The mobile genetic elements occur both in pro- and eukaryotes. They create genetic diversity (variation) and ultimately lead to evolution of newer traits. In addition to their medical importance in producing resistance to antibiotics, transposons are speculated to play an important role in speciation.

ACKNOWLEDGEMENTS

Authors are thankful to the Council of Scientific and Industrial Research, New Delhi for granting permission to reproduce major portion of an article published in Journal of Scientific and Industrial Research.

REFERENCES

Besermer, J. and Herpers, M. (1977) Suppression of polarity of insertion mutations within the gal operon of *Escherichia coli*. Mol. Gen. Genet., 151 : 295-304.

Chou, J., Lemaux, P.G. Casadaban, M.J. and Cohen, S.N. (1979) Transposition protein of Tn3: identification and characterization of an essential repressor-controlled gene product. Nature, 282 : 801-806.

Cohen, S.N., and Shapiro, J.A. (1980) Transposable genetic elements. Scient. Am., 242 : 36-45.

Evans, R.V. (1984) Limits on the selfishness of DNA. Nature, 310 : 456.

Gill, R.E., Heffron, F. and Falkow, S. (1979) Identification of the protein encoded by the transposable element Tn3 which is required for its transposition. Nature, 282 : 797-801.

Hedges, R.W. and Jacob, A.E. (1974) Transposition of ampicillin resistance from RP4 to other replicons. Mol. Gen. Genet., 132 : 31-40.

Isch-Horowicz, D. (1982) Transposable elements, hybrid incompatibility and speciation. Nature, 299 : 676-677.

Jordan, E., Saedler, H. and Stralinger, P. (1968) 0° and strong-polar mutations in the gal operon are insertions. Mol. Gen. Genet., 102 : 353-363.

Kidwell, M.G., Kidwell, J.F. and Sved, J.A. (1977) Hybrid dysgenesis in Drosophila melanogaster : A syndrome of aberrant traits including mutation, sterility and male recombination. Genet., 86 : 813-833.

Kleckner, N. Barker, D.F., Ross, D.G., Botstein, D., Swan, J.A. and Zubeau, M. (1978) Properties of the translocatable tetracycline-resistance element Tn10 in *Escherichia coli* and bacteriophage lambda. Genet., 90 : 427-450.

Lewin, B. (1994) Genes V. Oxford University Press, Oxford, U.K. pp 1272.

Malamy, M.H., Fiandt, M. and Szybalski, P. (1972) Electron microscopy of polar insertions in the lac operon of *Escherichia coli*. Mol. Gen. Genet., 102 : 353-363.

Nervers, P. and Saedlor, H. (1977) Transposable genetic elements as agents of gene instability and chromosomal rearrangements. Nature, 268 : 109-115.

Saedler, H., Reif, H.J., Hu, S. and Davidson, N. (1974) IS2, A genetic element for turn-off and turn-on of gene activity in *Escherichia coli*. Mol. Gen. Genet., 132 : 265-289.

Schaefer, B.E., Kidwell, M.G. and Fausto-Sterling, A. (1979) Hybrid dysgenesis in Drosophila melanogaster : Morphological and cytological studies of ovarian dysgenesis. Genet., 92 : 1141-1152.

Shapiro, J.A. (1969) Mutations caused by the insertion of genetic material into the galactose operon of *Escherichia coli*. J. Mol. Biol., 40 : 93-105.

Sutton, W.D., Gerlach, W.L., Schwartz, D. and Peacock, W.J. (1984) Molecular analysis of Ds controlling element mutations at the *adh*1 locus of Maize. Science, 223 : 1265-1268.

Weinert, T.A., Schaus, N.A. and Grindley, N.D.F. (1983) Insertion sequence duplication in transpositional recombination. Science, 222 : 755-765.

Suggested Readings

1. Genes V (1994) Lewin, B. Oxford University Press, Oxford, New York.
2. Tripathi, G. and Shukla, S.P. (1986) Transposons. J. Scient. Ind. Res. (India), 45 : 21-23.
3. Text Book of Biochemistry (2002) Ed. T.M. Devlin, 5th Edition, John-Wiley & Sons, Inc., Publication, New York.

REVIEW QUESTIONS

Long Questions

1. Define transposon and discuss insertion sequences in detail.
2. Explain basic events during transposition of DNA segment.
3. Distinguish between replicative and nonreplicative transposition.
4. Describe DNA rearrangement after insertion of a mobile genetic element
5. Elaborate mechanism of transposition in DNA.
6. Write an essay on general survey of transposon in organisms.
7. Discuss transposition in antibiotic resistance genes.
8. Write short note on the following :
   (a) Role of transposons in speciation.
   (b) Evolutionary significance of transposons.
   (c) Retroposons

Multiple Choice Questions

1. Transposon is
   (a) A static entity of genome
   (b) A loci at which no gene resides
   (c) An independent element like plasmid DNA
   (d) A discrete mobile sequence of the genome
2. There are two classes of transposons viz.,
   (a) DNA and mRNA
   (b) rRNA and DNA
   (c) Mobile DNA sequences and retroposons
   (d) Retroviruses and static genome
3. Transposons is a genetic restructuring which does not rely on
   (a) Donor sequences
   (b) Any relationship between donor and recipient sequences
   (c) Any relationship among additional sequences
   (d) Recipient sequences
4. Insertion sequences are
   (a) Simple transposition molecules
   (b) Direct terminal repeats
   (c) Inverted terminal repeats
   (d) Abnormal constituents of bacterial chromosomes
5. The rate of transposition per element per generation is
   (a) 10^{-7} to 10^{-5} (b) 10^{-10} to 10^{-8}
   (c) 10^{-4} to 10^{-3} (d) 10^{-8} to 10^{-6}

6. The enzyme which help in transposition is
   (a) Protein kinase (b) Transposase
   (c) Catalase (d) Superoxide dismutase
7. Transposition occurs by
   (a) Only replicative pathway
   (b) Only nonreplicative pathway
   (c) Both replicative and nonreplicative pathway
   (d) Only conservative pathway
8. TnA transposition requires
   (a) Transposase (b) Lamda integrates
   (c) Resolvase (d) Transposase and resolvase
9. Retroposon are
   (a) Only retroviruses
   (b) DNA sequences of eukaryotes
   (c) Only retrotransposons
   (d) Retroviruses infecting host cell to sequences that transpose via RNA
10. There are two superfamilies of retroposons viz
   (a) viral and nonviral (b) viral and bacterial
   (c) bacterial and fungal (d) non-viral and bacterial

Answers

1. (d)	2. (c)	3. (b)	4. (a)	5. (c)	6. (b)
7. (c)	8. (d)	9. (d)	10. (a)		

CHAPTER 21

Genetic Load and Counselling

S. Bhardwaj* and L. Choudhary
Dept. of Zoology, Govt. P.G. College, Banswara–327 001
E-mail: seema377@gmail.com

Lost of allels due to environmental pollution or selection in a population is known as Genetic load. It causes less fit individual, diseases and abnormalities or dysmorphogenesis in man. The term Genetic Counselling was coined by Sheldon Reed (1947). It is in fact a branch of Genetics which depends on diagnosis and medical genetics. In 1960, the fields of genetics like, clinical cytogenetics, biochemical genetics and dysmorphology developed. In 1970s techniques of diagnosis of prenatal genetical disturbance disorder and diseased were developed. After 1970, American Board of Medical Genetics was formed. At present American Board of Genetic Counselling (ABGC) is working in this field. According to American Society of Human Genetics (1945) Genetic Counselling is a communication process which deals with the human problems associated with the occurrence or risk of occurrence of a genetic disorder in a family.

GENETIC LOAD

In any population genetic burden (load) is a phenomenon of lost of alleles due to mutation, pollution or selection. The value of genetic load is ranged $0<L<1$. The concept was first formulated by the British population geneticist Haldane (1957). According to Wikipedia (the free encyclopedia) Genetic Load may be defined as, the extent to which the average individual in a population is inferior to the best possible kind of individual, which is equivalent to the relative chance that an average individual will die before reproducing because of deleterious genes.

(i) Type of Genetic Load

Genetic Load is of following types:

A. Pollutional Load: When Load due to Pollution.

*Corresponding author

B. Mutational Load: When Load due to Mutation.
C. Selectional Load: When load caused by selection.

Selection is generally occurs when fitness of two alleles are in equal so it positively create a load. The alleles frequencies will tend towards an equilibrium position in directional selection in this condition fittest allel reaching a frequency in Mutation - selection balance. If mutation is rare that it effectively fixation. In directional selection, heterozygote advantages exerts a load at equilibrium. In case the mean fitness is 0, the load is equal to 1, but the population goes extinct. According to Morris mutational load would increase over time and thus make populations inviable.

(ii) Formula of Genetic Load

$$L = \frac{(\text{Wopt} - v)}{\text{Wopt}}$$

where, Wopt = Fitness of genotype

v = Mean fitness

L = Load

Genetic load is number between zero and one (1). It can measures average individual in a population is inferior to the best kind of individual. For example there are so many types of genotypes in a population and one genotype has more fitness i.e. Wopt and average fitness of the whole population is v. It can calculate by the multiplication of frequency with fitness of each genotype. Suppose, in a population all individuals have optimal genotype, then v = Wopt and $L = 0$

$$L = \frac{(\text{Wopt} - v)}{\text{Wopt}}$$

if $v = 0$, then $L = 1$.

$$L = \frac{(\text{Wopt} - 0)}{\text{Wopt}} = 1$$

(iii) Causes of Genetic Load

Genetic load is due to several causes are such as,

1. Genetic load in any population is may be due to pollutional load; pollution may cause of mutation which can easily create the genetic load in this way pollution is indirect cause of Genetic load.
2. Inbreeding is also a cause of Genetic load. Population size of some wild animal species are become smaller due to inbreeding and that animal species became endangered because of small size of their population.
3. Selection is another main cause of Genetic load because alleles are lost by selection in population.
4. Mutation is also a major cause of Genetic load.
5. Genetic load may be due to some recessive genes. According to J.ARMSTRONG it is virtually impossible to avoid genetic disease and some species carry a Genetic load of

3-5 recessive lethal genes. At present their is lack of direct genetic test for recessive genes. When such types of tests do became available, they be used to identify the defective genes carried by some species. It will help in planning of crosses intelligently so that the defects (due to recessive gene) don't match upto create a homozygous affected individual.

GENETIC COUNSELLING

The term Counselling is lie the field of mental health and Psychotherapy. It based on medical genetic. It also needs accurate diagnosis. American Society of Human Genetics (1975) give following definition of genetic Counselling. Genetic Counselling is a communication process which deals with human problems associated which the occurrence or risk of occurrence of a genetic disorder in a family. This process involves an attempt by one or more appropriately trained person to help individual or family.

(1) Comprehenced the medical fact including the diagnosis, probable course of the disorder, and the available management.

(2) Appreciate the way heredity contributes to the disorders and risk of recurrence in specified relative.

(3) Understand the alternatives for dealing with the risk of recurrence.

(4) Choose a course of action which seems to them appropriate in view of their risk, their family goal, and their ethical and religious standards and act in accordance with that decision.

(5) Make the best possible adjustment to the disorder in an affected family member and / or to the risk of recurrence of that disorder.

By these views it is clear that Genetic Counselling is a complex task.

It is has some aspects. The first aspect is related with family history and diagnosis and relationship of both points. In this case generally treatment of genetic disease is same as found in other disease. Second aspects is related with risk determination. In unifactorial and multifactorial disorders, recurrence is estimated by empirical risk. For Mendelian disorders recurrence risks are predicated by inheritance patterns. However clinical issues are depend upon delayed age of on set, incomplete penetrance and variable expressions etc. Sometimes risk estimate can predicated by Baysian probability (Quantitative Genetics). The third and fourth aspects of genetic counselling are based on comparison between clinical biochemical approach and genetic model. These aspects are particularly related with Non directiveness, in this all decisions of future reproduction will depend upon family, because the main goal of genetic counselling is to help understanding of families about genetic diseases, not to reduce the incidence of genetic disease, so we can say that non directiveness is an important aspects of genetic counselling.

[A] Important Information for Genetic Counselling

Genetic Counselling practitioner must have following informations that are very useful in genetic counselling:

(i) Diagnosis and general idea about natural history.

(ii) Understanding of tenants of risk determination for some chromosomal and other diseases, and inheritance patterns of mendelian disorders.

(iii) Some statistical analysis such as Bayesian probability approach.

(iv) Knowledge of primary differences between traditional biochemical approach and genetic model.

(v) Discussion for reproductive options.

(vi) Counsellor should leave all decision about future reproduction up to family level (Non-directiveness) the main duty of Genetic Counselling is to help families understand and cope which genetic disease.

(vii) Knowledge of medical genetics.

(viii) Collection of family history data, and formation of pedigree.

Counsellor must have mutational and new mutational ideas. Idea about pathogenicity of genetic disorder. Dysmorphic syndromes require knowledge about mild malformation and some anomalies. Counsellor may have team of expertise from other disciplines. Knowledge of incomplete penetrance variable expressions. Accurate diagnosis is very important information by which he can made pedigree of patient's family. For pedigree formation standard pedigree symbol should be followed, some important symbols are as under:

S. No.	Symbols	Human Character
1.		Normal ♀
2.	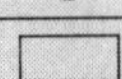	Normal ♂
3.		Unspecified Sex
4.	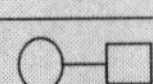	Mating
5.		Diseased ♀
6.		Diseased ♂
7.		Carrier Female
8.		Mating between relatives
9.	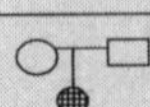	Parents with affected Female Child
10.	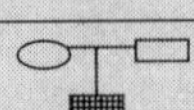	Parents with affected Male Child
11.	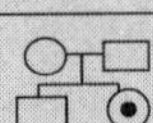	Parents and Children, Female-Carrier Male-Normal
12.	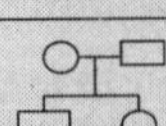	Parents with Normal Children
13.	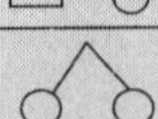	Identical twin (♀)
14.		Identical twin (♂)
15.		Fecternal twin (♂♀)

Fig. 1. Shows symbols of pedigree.

Though the usual way of pedigree formation is common but some variations can be there in different pedigree charts of different geneticists.

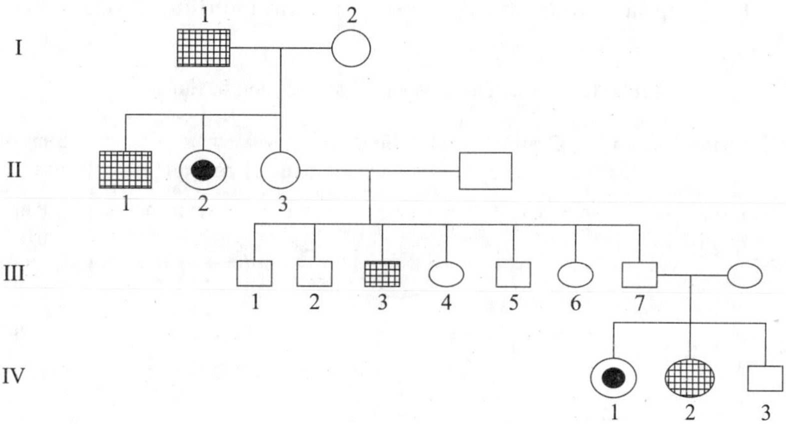

Fig. 2. Shows General pedigree.

Diagnosis of genetic disorder is a complex sequence. It depends upon biochemistry, dysmorphology, principles of medical genetics. In testing, test of adults are also important for genetic disorders because in HUNTINGTON'S disease symptom are not clear untill middle age. Some genetic conditions have symptoms that can be treated if they are detected very earlier. As we know in Phenylketonuria mental retardation, is generally found when phenylalanine is very low in diet. Some time diagnosis is a problem because the list of congenital malformational syndrome is very long. SMITH'S recognizable patterns of Human Malformation listed more than 400 conditions.

(ix) Age of Individuals – Age of parents and related individual is also very useful information. Because it may be related in some diseases such as breast cancer and ovarian cancer.

(x) Development of new disease – In some faculties, some family members develop newly diagnosed disease. Because such type of change can affect risk estimation.

(xi) Miscarriage in family and its percentage.

(xii) Ethnic Origin – Ethnic Origin of the family – It is also important because some diseases are directly related with different ethnic groups.

(xiii) Three generation family history should be known.

(xiv) Information of first degree relatives should be known.

(xv) Gender of individual and relation with other family members are also part of basic informations. Knowledge of some techniques of somatic hybridization because it will helpful in bio-chemical heritable, informations.

(xvi) Studies of Twins.

(xvii) Knowledge of human Karyotype and chromosomal classification. This arrangement of chromosomes explain relative morphology of structural informations such type of informations are helpful in chromosomal structural and numerical abnormally.

Table 1. Shows Chromosomal Classification of Human.

S. No.	Group	Length	Chromosomal No.	Type of chromosome on the bans of centromere	Sum of chromosome in cell
1.	A	Long	1, 2, 3	Metacentric/Sub-metacentric	3 Pair
2.	B	Smaller than A	4, 5	Sub-metacentric	2 Pair
3.	C	Medium	6, 7, 8, 9, 10, 11, 12	Sub-metacentric	7 Pairs + 1 in ♂ 7 Pairs + 1 in ♀
4.	D	Medium	13, 14, 15	Acrocentric/terminal	3 pairs
5.	E	Small	16, 17, 18	Metacentric/Submetacentric	2 pairs
6.	F	Smaller than E	19, 20	Metacentric	2 pairs
7.	G	Smallest	21, 22	Acrocentric/Terminal	7 Pairs + 1 in ♂ 7 Pairs + 1 in ♀

(Note :- 13, 14, 15, 21, 22 are sat chromosomes)

(xviii) Chromosomal banding techniques.

(xix) Lyon's Hypothesis. This hypothesis helps numerical allosomic position of man.

Table 2. Show Lyon's hypothesis for No. of Barrbodies related with No. of X Chromosome.

S NO.	Genotype	Bar bodies	Phenotypes	Result
1.	X	0	♀	Turner's Syndrome
2.	XY	0	♂	Normal
3.	XX	1	♀	Normal
4.	XXY	1	♂	Klinfelter's syndrome
5.	XXX	2	♀	3X – Syndrome
6.	XXXY	2	♂	3X-Y Syndrome
7.	XXXX	3	♀	4X-Syndrome
8.	XXXXY	3	♂	4X-Y Syndrome
9.	XXXXX	4	♀	5X –Syndrome
10.	XXXXXY	4	♂	5X-Y Syndrome

(xx) Genetic counsellor should have some knowledge of quantitative genetics. such as sib selection estimation of offspring's phenotype, etc.

Estimation of offspring's phenotypes :

$$t = a + i + e$$

where

t = Deviation between T, $\overline{T}$

a = Additive allelic effect

i = Interactions

e = Environmental deviation

T = Phenotypic value of an individual

$\overline{T}$ = Mean of the population.

by this formula we may get information about expected phenotype of an offspring. It is also known as 'Breeding value' of individual.

For father

$$a_f = t_f h^2$$

a_f = Additive allelic effect of father.

t_f = Deviation between T, $\overline{T}$ of father.

h^2 = Narrow sense heritability

For mother

$$a_m = t_m h^2$$

a_m = Additive allelic effect of mother.

t_m = Deviation between T, $\overline{T}$ of mother.

h^2 = Narrow sense heritability

Breeding value of offspring = a_O

$$a_O = \frac{(a_m + a_f)}{2}$$

$$= \frac{h^2(t_f + t_m)}{2}$$

Phenotypic deviation of offspring = t_O

$$t_O = a_O$$

(because e_O and i_O can't predicated)

Expected Phenotype is

$$T_O = \overline{T} + t_O$$

$$= \overline{T} + a_O$$

$$= \overline{T} + \frac{h^2(t_m - t_f)}{2}$$

$$\frac{(t_m + t_f)}{2} = \frac{(T_m + T_f)}{2} - \overline{T}$$

$$\frac{(T_m + T_f)}{2} = T^* \text{ --- } (T^* \text{ Mid parent value})$$

$$\frac{(t_m + t_f)}{2} = T^* - \overline{T}$$

$$T_O = \overline{T} + h^2 (T^* - \overline{T})$$

Example:

Father's Score of IQ = 110

Mother's Score of IQ = 120

Mean of population IQ = 100

Standard Deviation = 15

h^2 of IQ = 0.4

What is IQ of offspring of these parents?

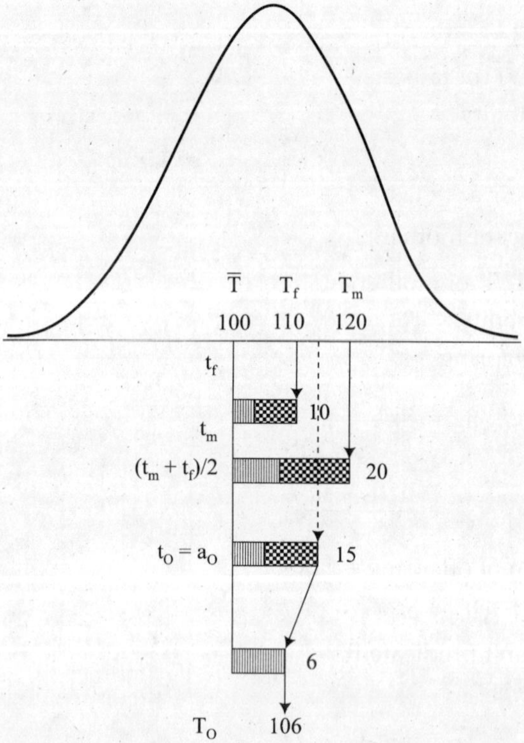

Fig. 3. Estimating an offspring's phenotype. The distribution of IQ scores from intelligence test is shown. The means score ($\overline{T}$) is 100 and the standard deviation is 15. The scores of a particular man (Tf) and a particular woman (Tm) are also shown, along with their phenotypic deviations from the mean (tf and tm respectively). The average of these deviation is $(t_m + t_f)/2 = 15$. By using the narrow sense heritability ($h^2 = 0.4$) this average allows the phenotypic deviation of a child born to this couple to be predicted; to = $h^2 (t_m + t_f)/2 = 6$. Consequently the IQ of the child is expected to be T + to = 106. [From principle of Genetics – Eldon john Gardner.]

Calculation for IQ of offspring

For Father

$$t_f = T_f - \overline{T} = 10$$

$$t_m = T_m - \overline{T} = 20$$

$$\frac{(t_f + t_m)}{2} = \frac{10+20}{2} = 15$$

Deviation of offspring

$$t_o = 0.4 \times 15$$

$$t_0 = 6$$

Expected IQ of offspring

$$IQ = T + t_o$$

$$IQ = 100 + 6$$

$$IQ = 106$$

$$T_o = \overline{T} + h^2 (T^* - \overline{T})$$

$$\overline{T}_o = ?$$

$$T_o = 100$$

$$h^2 = 0.4$$

$$T^* = \frac{110+120}{2} = 115$$

$$= 100 + 0.4 \times (115 - 100) = 106$$

(Such type of expecting results may change with variation, segregation and defective genes or alleles.

(xxi) Mendelian character and their inheritance in man.

[I] Autosomal Recessive Characters

Affected human character	Normal human character
Absense of sweat gland	Normal
Alkaptonuria	Normal
Attached ear lobes	Normal free ear lobed.
Blond hair	Dark hair
Blue or grey eyes	Normal blue eyes
Microcephilia	Normal
Micropnthalmus	Normal
Phenylketonuria	Normal
Red body hair	Normal hair
Retinal glioma	Normal
Schizopherenia	Normal
Sickle cell anemia	Normal
Xeroderma pigementosum	Normal

[II] Autosomal Dominant Traits

Affected human character	Normal human character
Absence of enamel of teeth	Normal teeth
Astigamatism (Cornea not spherical)	Normal eye
Brachydactyly	Normal
Cartilaginous growth of bones	Normal
Congential catract	Eye
Curly hair	Straight hair
Dark skin coulor	White skin colour
Ectrodactyly	Normal
Glucoma	Normal eye
Hair lip and clefted atrophy	Normal
Hairlessness	Normal hair
Ichthyosis (Scaly skin)	Normal skin
Long eye lashes	Short eye lashes
Mongoloid fold of skin around eye	Normal, no fold
Polydactyly	Normal
Progressive muscular atrophy	Normal
Syndactyly	Normal

[III] Sex-linked Dominant Characters

Xg. blood group (+)	Xg^3(–)
Hypophysphatemic rickets (vit D. resistance)	Normal

[IV] Sex-linked Dominant Characters (Recessive)

Colour Blindness	Normal Colour vision
Diabetes melletus	Normal
Haemophilia	Normal

[V] Characters Controlled by Multiple Alleles and Multiple Factors

Skin colour (Multiple Factor)

Intelligentia

Hepatoglobin (a blood serum protien)

A, B, AB and O blood group (Multiple alleles)

(xxii) Environmental causes of congenital malformation :

(A) Effect of German measles: During pregnancy if mother suffered from german measles positively infants will be effected.

(B) Effect of Thalidomide: If pregnant mother use thalidomide (Sleeping drug) their new born will affected and caused phocomelia. In this disorder arms of infents are reduced to flipper like stumps.

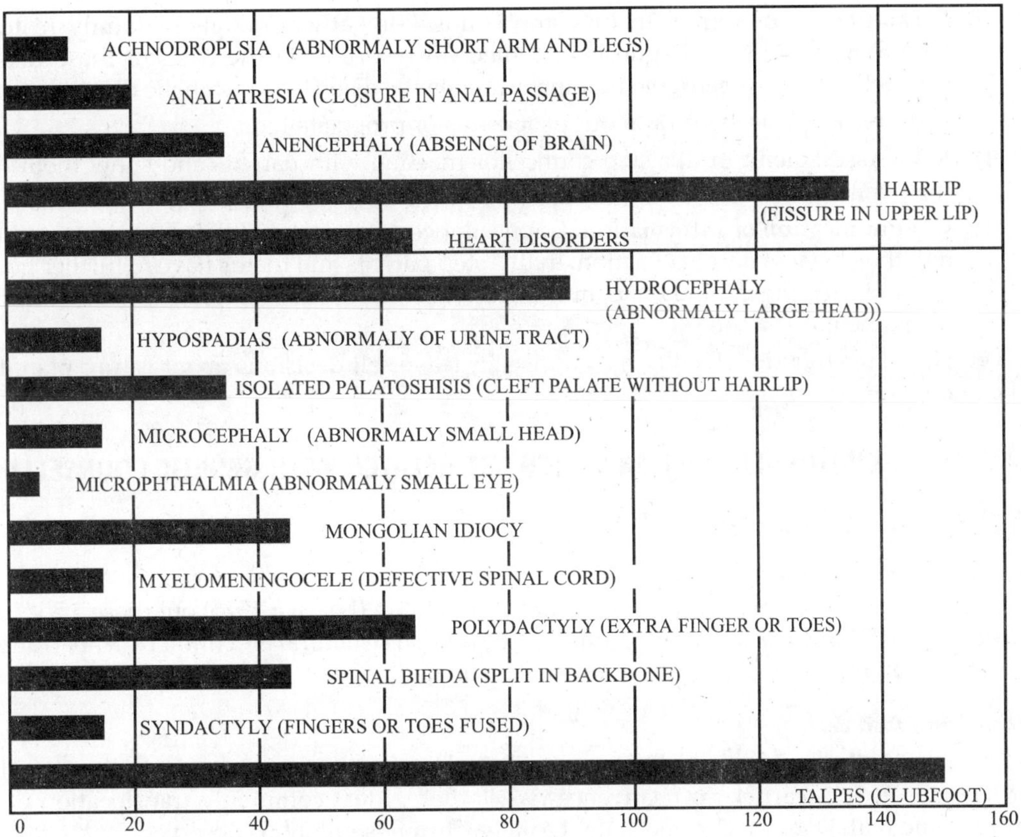

Fig. 4. Major Congenital deformities in man (Compiled by Dr. J.A. Book, Director of State Institute for Human Genetics)

(C) **Radiation:** Radiation produces chromosomal breakage in foetus.

(D) **Effects of drugs:** Some drugs are harmful during pregnancy, i.e. Salicylates, streptomycin, Actinomycin-D, Penicillin, Asperin, Quinine, Meprobromate, respirine etc.

(xxiii) Amniocentesis:

Amniocentesis is a process by which developmental defects, sex and heredity of foetus are tested. At fifteen week of pregnancy sample of amniotic fluid containing fetal cells is take out and cultured, then cultured cells are used for testing of wanting information.

[B] Process of Genetic Counselling

According Beth Lincoln Boyea and Robert Cefalo the process of Genetic Counselling has following steps :

I. **Formation of pedigree:** In this step counsellor gathers details of family history, laboratory results of diagnosis and medical records. On the basis of these details counsellor forms a pedigree that includes genetic disorders, metabolic disorders, etc. Pedigree helps us to know about propositus or proposita too.

II. **Risk Assessment:** In this step counsellor meeting with parents and show them that their children will express a genetic disorder that is present in pedigree.

III. **Communication of information:** Counsellor communicates all information to parents and they have to decide of action, In this step parents and foetus have to further go for genomic testing and then communicate second results corelate them with risk assessment.

IV. **Non-directiveness:** Finally the counsellor leaves all decisions about future planning on family.

SOME GENETIC DISORDERS OF MAN WHICH ARE RELATED WITH GENETIC COUNSELLING

(A) Chromosomal Disorders

First chromosomal abnormality in foetus was observed by Jacobson Barter in 1967. About one in 150 babies has a recognizable chromosomal abnormality (Nielsen 1975) out these 3.75/1000 autosomal and 2.06/1000 had a sex chromosome and mal structural and numerical aberrations are common in human out of these, numerical anamalies are easy for cytological diagnosis.

A_I Autosomal Disorders

(i) Chromosomal translocation –D/D group: Translocation is very common, If one parent is normal and other is carrier, so recurrence risk is there. Most commonly translocation of 21 chromosome with D group chromosome (Usually Chromosome 14) is observed.

Translocation of D group of 21 chromosome with 'G' group of 22 chromosome or the two '21' chromosomes are also present (G.C.R. Skinner). Balanced D/D group translocation is common and is not associated with unbalanced chromosome defects in live born offspring (de Grouchy 1976).

(a) Monosomic disorders : Main monosomic disorders are shown in Table 3.

Table 3. Monosomic disorders

S.No.	Monosomic	Syndrome	Diagnostic Feature
1.	Delation of ovum of chromosome – 21	Philadelphia	Granulocyte lukaemia
2.	Delation short arm of chromosome - 5	Cri du Chat	Mental retardation microcephaly, infants cry like a cat.

(b) Trisomics disorders: In trisomics disorders Down syndrome is most common. This disorder occurs approxi. 1 in 700 live-borns. Trisomy is mainly due to trisomy 21.95% Old mother and mother of previously affected child is show more risk for such type trisomy. R. Skinner (1983) shows age specific rates of Down syndrome these rates are shown in Table 4.

Table 4. Age specific rates of Down syndrome in live-birth and at amniocentesis

Material age in years	No, cases per 1000 live birth*	Risk at birth*	Risk of DS at amniocentesis	Risk of any aneuploidy at amntocentesis
35	3.09	1/324	1/222	1/133
36	1.96	1/510	1/204	1/102
37	2.94	1/340	1/130	1/75
38	3.10	1/322	1/110	1/68
39	7.57	1/132	1/76	1/54
40	10.50	1/95	1/83	1/43
41	14.50	1/69	1/43	1/33
42	12.59	1/80	1/30	1/16
43	15.72	1/64	1/56	1/25
44	36.91	1/27	1/18	1/13
45	33.64	1/30	1/30	1/20
46	28.11	1/36	1/12	1/07

*Data from Trimble and Bird 1978
Data from Ferguson - Smith 1979

From these data it can be clear that the risk of having a child with trisomy 21 has risen by a maternal age of 35 years. Except it some other trisomic syndromes are shown in Table 5.

Table 5. Some important trisomic syndromes

S. No.	Karyotype	Syndrome	Diagnostic Feature	Live Birth
	Trisomics			
1.	Trisomy – 13	D trisomy	So many defects, death at age of 1-3 month.	1/4500
2.	Trisomy – 15	-	Many clinical defects.	
3.	Trisomy – 18	E trisomy	Cardiac defects, ear deformed death by the age of 1 year.	1/4500
4.	Trisomy – 21	Down's Syndrome	Skin fold over eyes usually mangolian feature.	1/600
5.	Trisomy – 22	-	Skeletal deformation and other symptoms are same as found in Downs syndrom.	

(ii) Recessive Defective Autosomal Genes

(a) Thalassemia: This disorder is due by defective subunits of Hb.

α–Thalassemia: α–Thalassemia again of two types one type α–Thalassemia produces by absence of two genes (α - / α -, α α/- -) on 11th chromosome in such cases microcytic and less hypochromic erythrocyte appear. In other type of α–Thalassemia 3 genes are absent on 11th chromosome (—/α-) This condition develops microcytic hypochromic enythrocytes. This condition is much more harmful and risky.

β–Thalassemia: β–Thalassemia is also of two types:

1. β–Thalassemia Minor
2. β–Thalassemia Major

In former type one β–gene is defective and it cause microcytic RBC. RBC count is higher but 15% lower than normal percentage.

In later case both β–genes are defective. β–Thalassemia major develops microcytic hypochromic RBCs It is causes severe anaemia.

(b) **Phenylketonuria (P.K.U.):** In this case phenylalanine hydroxylase is lacking in liver. In the absence of this enzyme change of phenylalanine to tyrosine is blocked. This cause hyperphenylalaninemia and more phenylalanine in urine. Phenylalanine and its product are harmful. Mental retardation, eczema, hypopigmentation etc. are main symptoms of P.K.U.

(c) **Albunism:** In the absence of enzyme tryosinase, melanin formation from dihydrophenyl is stopped. One out of 20,000 individual shows albinism.

(d) **Alkaptonuria:** In this disorder patient lacking enzyme alkapton oxidase/ homogenetisase. This enzyme is necessary for metabolism of homogenetisic acid, due to absence of this enzyme alkopton accumulation in body producing artheritis, urine of affected person is of black brown in colour because oxidation of homogentisic acid occurs.

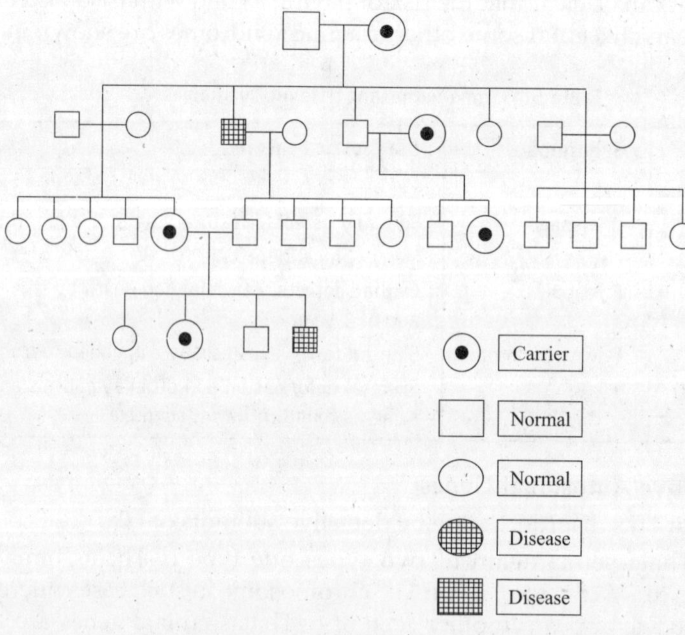

Fig. 5. Shows Pedigree of recessive gene inheritance.

(e) **Tay sach's disease:** Deficiency of enzym β-D-Nacetyl hexosaminidase, by this deficiency lipid Tay-sach's ganglioside or GM2 starts accumulation, this causes damage of nervous system (Brain, Spinal Cord etc.). It is also known as Infantile an auroticiliocy. This disease is more common in Jewish heritage and resulting in mental paralysis and mental retardation. At the time of birth child looks normal and healthy but with in 6 months signs of deterioration is exhibited. Chances of recessive disorders are more in related marriage because relatives share more same genes.

(iii) Dominant Defective Autosomal Genes

Some dominant defective characters in man are as:

Bracrydactyly - Palm and finger are short of this person.
Polydactyly - Presence of additional toe or fingers.
Huntington's Chorea - Muscular movement are affected show symptoms at the age of 25–50. In later stage mental problems develop.
Marfan's Syndrome - Long fingers, musculature weak, thin.
Dentinogenesis imperfecta - Crowns of teeth are affected in this disease.
Tylosis - Thick skin on sole and palms.
Achondroplasia - Diseased individual have short arms and legs. It is a kind of dwarfs but most of diseased individual die at the age of one year.
Anonychia - Absence of nail.

Lethal dominant defects generally not transmitte because affected one die before birth or not gain sexual maturity (except Huntington's chorea).

A_2-Allosomic Disorders

Allosomic disorders are also found in human populations some important disorders are as : Turner's syndrome, Klinefelter's syndrome and Triplofemale (Table 6).

Table 6. Allosomic disorders

S. No.	Karyotype	Syndrome	Diagnostic Feature	Live Birth
	Monosomic XO	Turner's	Webbed neek, breast poorly developed, ovaries degerative. Bars body absent slight mental retardation may or may not prenoticed.	1/2500 ♀
	Trisomic			
1.	XXY	Klinefelter	Testes degenerating Bars brodies present breast enlarged.	1/500 ♂
2.	XYY	-	-	-
3.	XXX	Triplo X	Female normal and fertile	1/1400 ♀

Sex Linked Inheritable Disorders

Haemophilia and colour blindness are two common human disease which transformed as recessive X-linked traits.

(a) Haemophilia: It is also called bleeders disease because persistent bleeding is main symptoms of this disease. Pedigree study of this disease in English families done by J. Otto (1803), Nasse (1820) Haldarse (1922) etc. A popular pedigree of Royal Families of Europe was discovered by Haldane.

(b) Colour blindness: This defect is also called red-green blindness. It is found 8% in white race 4% in black races. Horner (1876) discovered pedigree of this disorder.

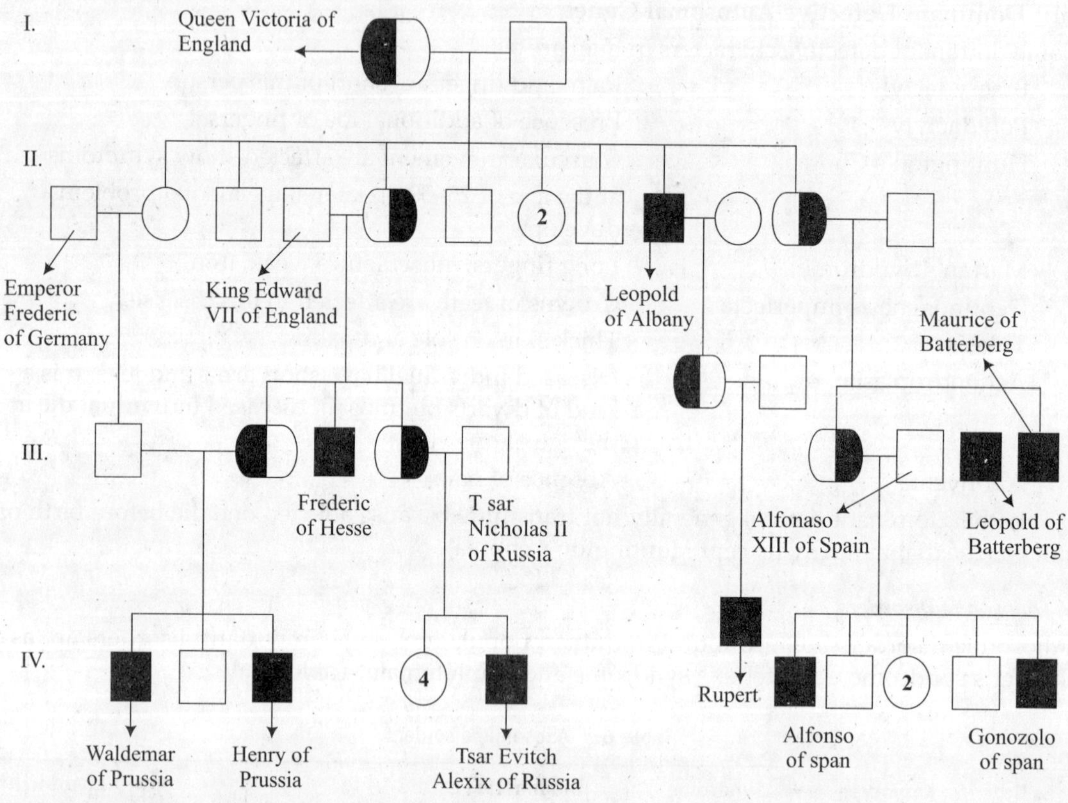

Fig. 6. Pedigree of Queen Victoria to illustrate the transmission of haemophilia.

GENETIC COUNSELLING OF UNIFACTORIAL DISORDERS

Unifactorial disorders (autosomal and allosomal disorder) can easily detected by pedigree data. These types of disorder can easily segregate in their affected families. In autosomal dominant disorders and risk of recurrence can be easily calculated by first principles, So genetic counselling in these case should be simple. If an individual has heterozygous autosomal dominant gene then there is a 50% risk of transfer of these gene to children. If date are available on age related occurrence rate then 'prior risk' can be made.

Some problems in counselling may be arises in case of new mutations when the fitness of affected person is lower then possibility of new mutation in new born will be higher.

Genetic Counselling in autosomal resessive disorder is comparatively difficult because mode of inheritance is not straight forward. Genetic Counselling of allosomic disorder in again straight forward like other unifactorial disorders. Prior risk of X-linked disorders can be easily estimated by first principles. In genetic counselling of X-linked disorders some problems are also present such as differentiation between normal and healthy carrier (heterozygous). In such problems prior risk can be modified by clinical or biochemical testings. According to Skineer (1983) some carrier detection tests in X-linked disorders can be detected by the Table 7.

Table 7. Carrier detection tests in X-linked disorders (From Skinner, 1983).

S. No.	Disorder	Manifestation in carrier
1.	Diabetes insipidus (nephrogenic)	↓ Urinary concentration
2.	G6 PD deficiency	↓ Erythrocytes G6PD
3.	Haemophilia A	Abnormal ratio of factor VIII to inactive antigen.
4.	Haemophilia B	↓ Factor IX
5.	Hunter syndrome	↓ Induronsulphate sulphatase in skin fibroblasts. Two cell population demonstrable.
6.	Hypogammaglobulinaemia (Bruton type)	↓ IgG
7.	Lowe (oculo-cerebral-renal) syndrome Ocular albinism	Lenticular opacities.
8.	Lesch-Nyhan syndrome	↓ HGPRT in skin fibroblasts or hair root bulbs. Two cell population demonstrable.
9.	Ocular albinism	Patchy depigmentation of the retina.
10.	Retinitis pigmentosa (x-linked form)	Peripheral pigmentary changes in retina. Abnormal ERG or flurescein angiography.
11.	Vit. D resistant rickets	↓ Serum phosphorus
12.	X-linked mental redardation	? Fragile X-chromosome (in positive families)
13.	X-linked ichthyosis	↓ Steroid sulphatase Corneal opacities
14.	Becker muscular dystrophy	↑ Serum creatine kinase.
15.	Fabry disease (angiokeratoma)	↓ α-galactosidase in skin fibroblasts. Two cell population demonstrable.
16.	Anhidrotic ectodermal dysplasia	Sweat pore counts reduced, dental anomalies.
17.	Choroideraemia	Pigmentary changes in retina.
18.	X-Linked ichthyosis	Corneal opacities steroid sulphatase.

GENETIC COUNSELLING OF MULTIFACTORIAL DISORDERS

Multifactorial disorders are due to combination of environmental and genetic factors. Therefore their Counselling and genetic make up is comparatively complicated. Recurrence risk from first principles is not possible. In these cases empiric risk of recurrence is generally used. Frequency of these disorders in family is comparatively low. Empiric risk of recurrence estimated by family data. Empiric risks are shown in Table 8.

Table 8. Empiric risks for some disorders (in per cent) (From Emery A.E.H.).

Disorder	Incidence	Sex ratio M/F	Normal parents having a second affected child	Affected parent having an affected child	Affected parent having a second affected child
Anencephaly	0.20	1:2	5*	–	–
Cleft palate only	0.04	2:3	2	7	15
Cleft lip ± cleft palate	0.10	3:2	4	4	10
Club foot	0.10	2:1	3	3	10
Cong. heart disease (all type)	0.50	2:1	1-4	1-4	–
Diabetes mellitus (Early on set)	0.20	1:1	8	8	10
Dislocation of hip	0.07	1:6	4	4	10
Epilepsy ('idiopathic')	0.50	1:1	5	5	10
Hirschsprung disease					
male proband	0.02	4:1	2	–	–
female proband	–	–	8	–	–
Hypospadias (inmales)	0.20	–	10	–	–
Manic depressive psychoses	0.40	2:3	–	10-15	–
Mental retardation ('idopathic')	0.30 0.50	1:1	3.5	–	–
Profound childhood deafness	0.10	1:1	10	8	–
Pyloric stenosis					
male proband	0.30	5:1	2	4	13
female proband	–	–	10	17	38
Renal agencies bilat					
male proband	0.01	3:1	3	–	–
female proband	–	–	7	–	–
Schzophrenia	1-2	1:1	–	16	–
Scoliosis (idiopathic, adolescent)	0.22	1:6	7	5	–
Spina bifida	0.30	2:3	5*	3*	–

*Anencephaly or spina bifida

Some factors, viz. sex, presence of affected individual in family environmental factor and parental consanguinity affect risks.

PROBLEMS OF GENETIC COUNSELLING

Genetic counsellor has to face some problems which may be sociological, clinical and diagnostic some common problem are as:

(i) **Expression of disorder:** Some genes shown mild express in parent phenotype.

(ii) **New mutation:** New mutation in some disorders create problems in diagnosis and expression of gene.

(iii) **Negative family history:** Negative family history is a basic problem for estimation of recurrence risk.

(iv) **Phenocopies:** Phenocopies are due to mimicry to genetic disorders. Phenocopies are caused by environmental factor. Microcephaly is common example of this problem. In general microcephaly is caused by autosomal recessive trait, but it may also result from intrauterine exposure of radiation: toxoplasmosis, amniotic bands, maternal disease etc.

(v) **False paternity:** False Paternity may be a problem for genetic counselling.

(vi) **Genetic heterogeneity:** Genetic Heterogeneity is a phenomenon in which some disorders superficially resemble with another disorders. Sometimes these disorders show same clinical results. Ehlers–Danlos syndrome is a good example of genetic heterogeneity. This disorder shows heterogeneity at both levels clinical and genetic. Genetical methaemoglobinaemia is another example of genetic heterogeneity.

(vii) **Small families:** The small size families makes a problem which is related with pedigree. Pedigree informations of such cases are not sufficient for Counselling.

(viii) **False diagnosis:** Genetic Counselling depended on right diagnosis. If diagnosis is not accurate may create problem in genetic counselling.

LINKAGE AND GENETIC COUNSELLING

Linkage studies are helpful in some genetic relationships of traits. Some important aspects of linkage are described here:

Linkage studies: Linkage disequilibrium of a gene with an mapped marker can be detected in a family. In-Situ-hybridization, may also be used to locate genes in specific chromosomal segment.

Pedigree analysis: By pedigree analysis we can estimate crossing-over frequencies of linked genes. Such types of studies are comparatively easy in allosomes than autosomes.

SUMMARY

Lost of alleles due to environmental pollution or selection in a population is known as genetic load. It causes less fit individual, disease and abnormalities or dismorphogenesis in man. According ASHG (1975) genetic Counselling is a communication process which deals with the human problems associated with the occurrence or risk of occurrence of a genetic disorder in a family. Genetic Counselling include some aspects like risk determination. Medical management, non-directiveness, Psychological aspects etc. Studies of spontaneous abortions of family history may be helpful in detection of chromosomal abnormalities. According to Boue and Boue; Jacobs et al. (1981) in spontaneous abortions before 12 weeks of gestation 60% cases show chromosomal abnormalities, So chromosomal abnormalities are common cause of abortion. Small sized modern families may create some problems in pedigree formation. Genetic Counselling is a partnership of clinical physician, psychologists and genetic counsellor.

REFERENCES

Antley, R.M., Seidenfeld M.J., (1978) Detailed description of mother's knowledge before Counselling for Down syndrome. American journal of Medical Genetics 2:357-364

Baker, Daine L., Jane Schuette, and Wendy Uhlmann (1998) A Guide to Genetic Counselling. Wiley-Liss : New York USA.

Bosk, Charles L. (1992) All God's Mistakes Genetic Counselling in a Pediatric Hospital. Chicago, 1992. University of Chocago, Chicago.

Boue, A., Boue J (1975) Chromosome abnormalities and abortions. In Coutinho BEM, Fuchs F (eds) Physiology and Genetic of Reproduction, Plenum Publishing Corporation, New York, USA.

Carter, C.O. (1977) Risk data : How good is empiric information In: Lubs HA, de la Cruz F (eds) Genetic Counselling, Raven Press, ch 5 p 407 New York, USA.

Emery, AEH (1979) Elements of Medical Genetics, 5th edn. Churchill Livingstone, Edinburghs.

Emery, AEH, Watt MS, Clack ER (1972) The effect of genetic Counselling in Duchenne muscular dystrophy. Clinical Genetics 3: 147-150.

Evers-Kiebooms G, van den Berghe H (1979) Impact of genetic Counselling : a review of published follow up studies. Clinical Genetics 15:465-474.

Ferguson-smith, M.A. (1979) Maternal age specific incidence of chromosome aberrations at amniocentesis In: Murken J D, Diagnosis, Stuttgart, Enke ch 1 P1.

Gardner, E.J. (1991) Principles of Genetics (8th Ed.) John wiley & sons Inc. New York USA.

Grouchy, J. de (1976) Human chromosomes and their anomalies In: Barltrop D (ed) Aspects of Genetics, Land ch1, p5.

Halden, J.B.S. (1957) the cost of natural selection. Jour of genetics 55:511-524.

Harper, P.S. (1998) Practical Genetic Counselling. Oxford : Butterworth-Heinemann.

Lippman-Hand, A. and Clarke, F. (1979) A Genetic Counselling : Provision and reception of information. American Journal of Medical Genetics 8: 113-127.

Lippman-Hand A, Clarke Fraser F (1979) A Genetic Counselling : the post Counselling period :11 making reproductive choices. American Journal of Medical Genetics 4:73-87.

Mikkelsen, M (1979) Previous child with Down syndrome and other chromosome aberration. In: Prenatal Diagnosis stuttagart Enke (Eds., Murken J D, Stengel Rutkowski S, Schwinger E.), pp. 22.

Muller, R.F. and young, ID (2001) Genetic Counselling, In : Emercy is elements of medical genetics (XIth ed.), pp. 241-248.

Nielsen J (1975) Chromosome examination of newborn children. Purpose and ethical aspects. Humangenetik 26:215.

Pearn, J. H., (1973) Patients subjective interpretation of risks offered in genetic Counselling. J. Med. Gen. 10 : 129-134.

Pearn, J. H., (1979) Decision-making and reproductive choice In : Counselling in Genetic (Eds., Hsia, Y.E., Hirchhorn, K., Silverberg, R.L. and Godmilow, L.), Alan R Liss Inc., New York, pp. 23.

Reynolds, B DeV, Puck M.H. Robinson A (1974) Genetic Counselling : an appraisal. Clinical Genetics 5 : 177-187.

Skinner, R., (1983) Genetic Counselling pp 1427-1436. Principles and practice of Medical Genetics. Churchill Livingstone.

Trimble, B.K., Baird, P.A., (1978) Maternal age and down syndrome: age specific incidence rates by single year intervals. Am. J. Med. Gen. 21 : 5.

Zallen, D.T. (1997) Does it run in the Family ? A Consumer's Guide to DNA Testing for Genetic Disorders. Ritgers University, New Brunswick.

Suggested Readings

1. Baker, Daine L., Jane Schuette, and Wendy Uhlmann. (1998) A Guide to Genetic Counselling. Wiley-Liss New York, USA.
2. Emery, A.E.H. and Rimoin D.L. (1983) Principle and practice of medical genetics . PP 1427-1436 Churchill Livingstone.
3. Gardner, E.J. (1991) Principles of Genetics (8th Ed.) John wiley & sons Inc. New York, USA.
4. Lincoln-Boyea, Beth and Robert Cefalo, (1996) Prenatal Diagnosis and Reproductive Genetic (Ed. Jeffery A. Kuller, Nancy C. Chescheir, and Robert C.), Cefalo St. Louis, Mosby.
5. Lippman, Abby (1994) Genetic Counselling: Practice and Principles (Ed., Angus Clarke). Routledge, New York.
6. Muller, R.F. and young, I.D. (2001) Emercy is elements of medical genetics (XIth Ed.), Churchill Livingstone.
7. Singh, B.D. (2001) Fundamentals of Genetics, Kalyani Publication, Ludhiyana, India.

REVIEW QUESTIONS

Long Questions

1. What is genetic load how we can calculate it ?
2. Describe the method of genetic counselling and mention its importance.
3. What are Basic informations, which are necessary for genetic counsellor ?
4. Give an account of unifactorial disorders and genetic counselling.
5. Describe some autosomal disorders which can reflect light on genetic counselling.
6. Write an essay on genetic load and genetic counselling.

Multiple Choice Questions

1. The sex chromosome complement of Turner's syndrome is
   (a) XXY (b) XYY
   (c) XO (d) YO
2. Turner syndrome has X chromosome
   (a) One (b) Zero
   (c) Three (d) Two
3. Phenylketonuria is
   (a) Sex linked recessive trait (b) Sex linked dominant trait
   (c) Autosomal dominant trait (d) Autosomal recessive trait
4. Amniocentosis is
   (a) With drawal of allantoic fluid.
   (b) Chemical analysis of fluid of pregnant woman.
   (c) Intraculative diagnosis
   (d) Culturing cells from amniatic fluid and studying chromosomal abnormalitics.

5. Word Genetic Counselling is given by
   (a) Pearn J.H. (1973)
   (b) Sheldon Reed (1947)
   (c) S. Read (1949)
   (d) J. Armstrong (1958)
6. Dysmorphology is
   (a) Study of morphometry
   (b) Study of Abnormal physical development
   (c) Study of normal morphology.
   (d) Study of mendelian characters
7. Clinical genetics is
   (a) Clinical care of genetic disorders.
   (b) Diagnosis
   (c) Biochemical tests of gene
   (d) Genetic Disorders
8. Non-directiveness
   (a) Decision of future reproduction upto family.
   (b) Suggestions to parrents.
   (c) Recommendation for treatment.
   (d) Choice for foetus sex.
9. Value of genetic load is ranged between
   (a) $0 < 1 < 1$
   (b) $0 < L < 1$
   (c) –1 and +1
   (d) 1 and –2
10. Microcephaly is an example of
   (a) False Paternity
   (b) Negative Family history
   (c) Genetic heterogeneity
   (d) Phenocopies

Answers

1. (c)	2. (a)	3. (d)	4. (d)	5. (b)	6. (b)
7. (a)	8. (a)	9. (b)	10. (d)		

PART-7

Differentiation and Development

CHAPTER **22**

Cell Differentiation and Commitment

Mangla Bhide
Dept. of Zoology, Dr. H.S. Gour Vishwavidyalaya, Sagar–470 003
E-mail: manglabhide@rediffmail.com

The term differentiation in the present context will be restricted to ontogenic development only. During development an egg undergoes many cell divisions to produce a large number of cells. The descendant cells that constitute an adult are not all alike but are of diverse types. The cells in adults differ from each other in their morphology, function, biochemistry and genetic make up. Therefore, although all the cells in an adult usually harbour the same genome, each has its own epigenotype i.e., the functional state of the genome of each cell type is different. The cells differ from each other and are specialized to perform different functions in an organism. Thus there are numerous cells differing from each another.

The process of the formation of a distinct adult cell type form a common progenitor cell type is referred to as differentiation. Development of entire organism from single fertilized zygote by differentiation of specialized cell types is a peculiar phenomenon in multicellular organism. Differentiation, specification and determination of a cell decide its fate. A cell may have single fate (in autonomous specification) or more than one fate (depending upon its interaction with neighboring cells as found in conditional specification). However, sometimes the fate of the cell is determined before the formation of cell membrane. So the development of specialized cell type from fertilized egg is called "differentiation". The cell differentiation involves change in biochemistry, function and commitment of cell. The cell commitment and differentiation are two different steps in the development of specialized cell type. The functions of cell change with the changes in biochemistry of cell as summarized in Table 1.

The differentiation of lens (Appleby and Modak, 1977; Clayton et al., 1986), erythrocytes (Clark and Kamen, 1987; Stamatoyannopoulos *et.al.* 1987), and skeleton muscle (Sutherland and Konigsherg, 1983; Bandman, 1985 a, b) has been described cell commitment includes two steps i.e., specification and determination. Commitment is of two types:

1. Irreversible commitment (differentiation is followed by determination)
2. Reversible differentiation (differentiation is not followed by determination).

Usually a cell become determined before it differentiates. In some cases these two processes occur simultaneously.

Table 1. Comparison of three differentiated cell types

S. No.		Cartilage	Muscle	RBC
1.	Morphology	Oval with a scalloped outline embedded in matrix	Elongated, syncytial	Smooth, rounded and float in blood
2.	Function	Resists compression and withstand tension	Contraction	O_2 and CO_2 transport
3.	Biochemistry	Synthesize substances mainly for extracellular use (collagen and chondroitin sulphate)	Synthesis mainly myosin and actin	Synthesis haemoglobin
4.	Epigenotype	Structural protein genes mainly collagen genes are active	Structural protein genes mainly myosin and actin genes are active	No DNA, no transcription

SPECIFICATION AND DETERMINATION

The cell is said to be specified when it differentiates autonomously only if placed in neutral environment. Whereas the cell and is said to be determinate if it becomes capable of differentiating autonomously even placed in another environment i.e., in another region of embryo by transplantation. There are following three major ways for cell commitment:

1. Autonomous specification and determination in mosaic development: In tunicate, molluscs (mosaic development), snails (spiral cleavage), and nematode (*Coenorhabditis elegans*).
2. Conditional specification and determination in regulative development: In deuterostome e.g., in sea urchin and higher vertebrates.
3. Syncytial specification and determination: In *Drosophila* and other insect's development.

I. Autonomous Specification and Determination

Autonomous specification and determination is the specificity due to cell's own cytoplasmic components. Here the cytoplasmic determinants play a major role in determining the fate of cell in early embryonic cleavage in mosaic development. The best example is of development of tunicates where cytoplasm of fertilized egg is differentiated into several regions giving rise to different kinds of cells (blastomeres). The fertilized egg having different presumptive regions develops into specific cells as follows:

(i) Clear animal cytoplasm: Develops into larval ectoderm.
(ii) Grey equatorial crescent: Develops into neural plate and notochord.
(iii) Yellow cytoplasmic crescent: Develops into muscle cells.
(iv) Yolky grey crescent: Develops into larval gut.

The presumptive cytoplasmic zone develops into organs where cytoplasmic components determine the differentiation of special cell types. This is known as "morphological determination" by cytoplasmic color differentiation. In Tunicate at eight cell stage the four pairs of blastomeres are separated and allowed to develop. Then they develop into separate specific structures showing autonomous cell specification. In mollusc embryo *(Patella coerulea)* the cytoplasmic components lead to mosaic development. Isolation of early blastomeres (trochoblast cells) develop cilia at the same time as they develop in intact trochoblast cells. In snails: spiral cleavage and development of polar lobe are the specific features. Polar lobe having mesodermal determinants are responsible for the development of mesodermal organs e.g., muscles, mouth, shell gland and foot etc. In nematode (*Coenorhabditis elegans)* asymmetric distribution of cytoplasm give rise to asymmetric division into specific blastomeres. Zygote of *Coenorhabditis elegans* divides to form founder cells, while stem cells form the cells containing germ line granules.

2. Conditional Specification and Determination or Regulative Development

It occurs due to ancestry of cell and depends on its neighboring cells. It means the specification of cell is dependant on neighboring cell interaction but at certain stage of development (as two celled stage or even four celled stage) each cell is able to develop into entire embryo. For example, in case of sea urchin at four blastomeres stage cells separate from each other and develop into four normal somewhat small sized plutei larva.

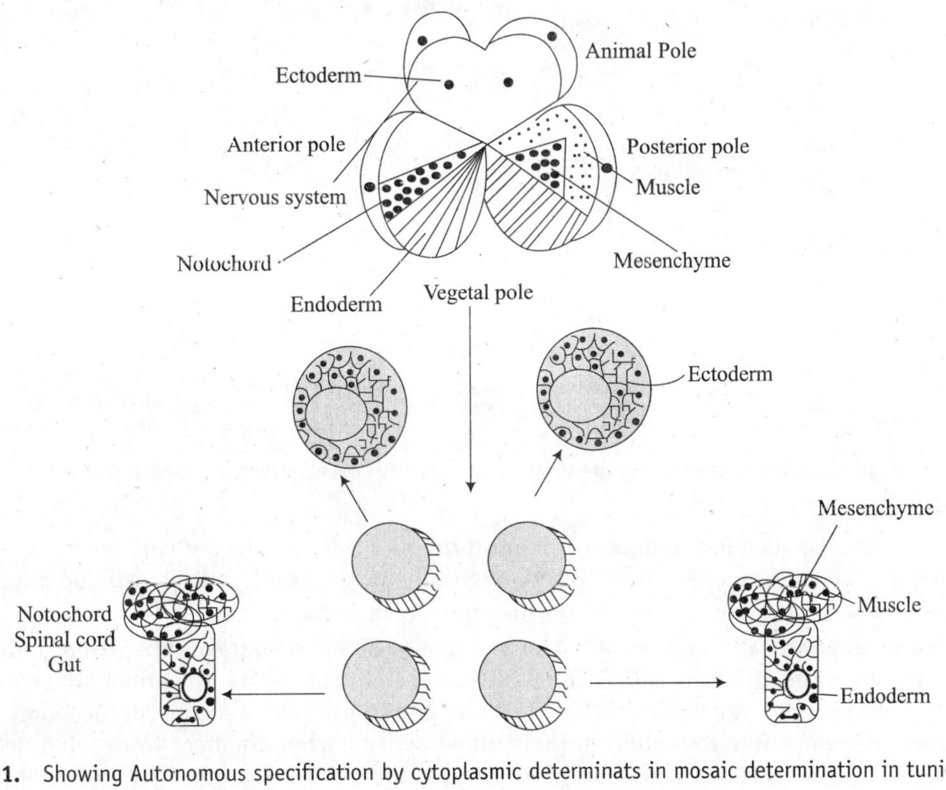

Fig. 1. Showing Autonomous specification by cytoplasmic determinats in mosaic determination in tunicate.

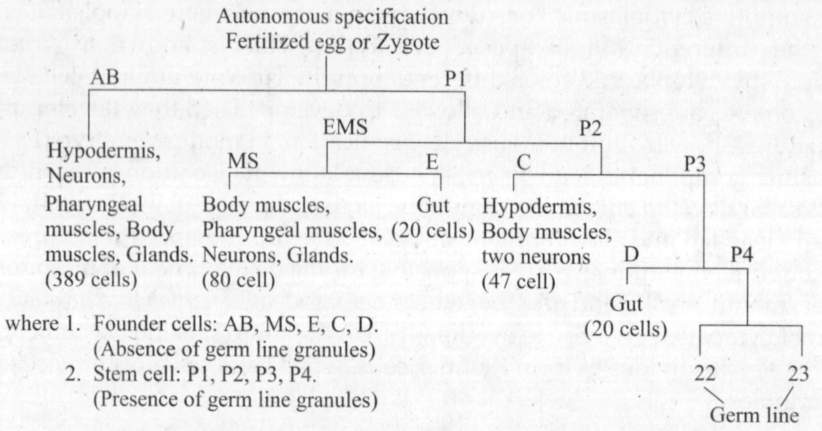

Fig. 2. Showing abbreviated cell lineage chart for *Coenorhabditis elegans.*

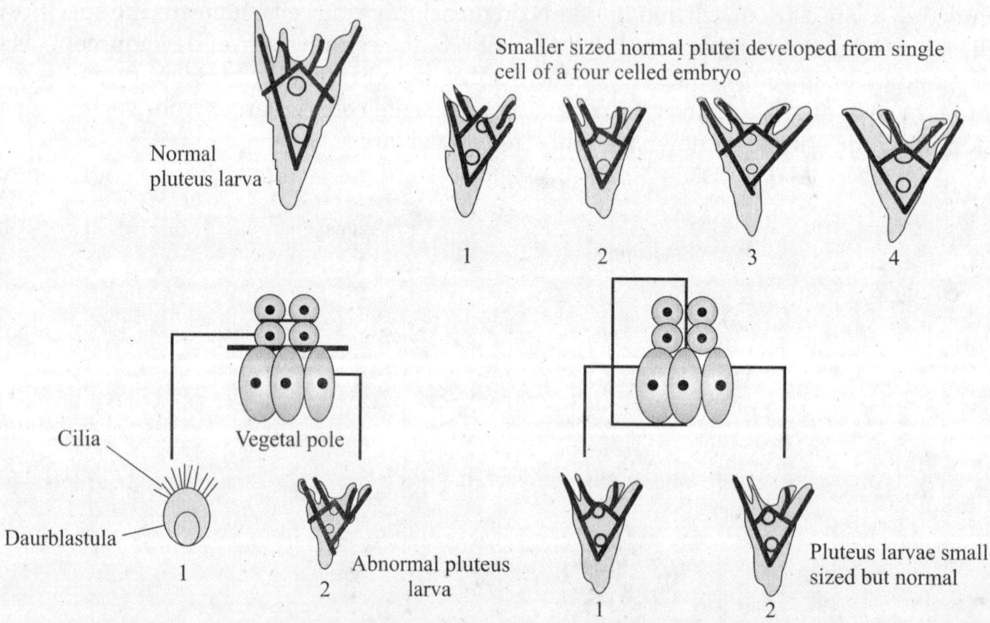

Fig. 3. Showing conditional specification or regulative development in sea urchin larva.

Condition specification and determination involve cells which are capable to opt more than one fate depending upon the interaction with the neighboring cells. It also depends upon the position of cell i.e., position of cell decides the fate of cell.

Experiment to show that in 64 celled embryo of sea urchin the fate of cells is determined by the interaction of neighboring cell. Animal layer and first vegetal layer normally develop into ectoderm. The second vegetal layer develops into endoderm and some larval mesoderm. The micromere develops into mesodermal skeleton. However, when the tiers were separated and recombined in various ways the original fate of these tires is altered by new neighbors suggesting a regular development.

Organizer and Primary Embryonic Induction

In sea urchin and vertebrate an individual cell has potential to develop into a whole organism till the embryo reaches a certain stage. Its fate is regulated by the neighboring cells. .For example in human being two identical twins is the result of two celled stage where two blastomeres separate and develop independently. In early gastrula the transplantation of dorsal blastopore lip to the embryo of same age (early gastrula with original blastopore lip) at same place (ventral side of the embryo) both the (original and transplanted) lips act as "*Organizer*". It decides the fate of neighboring cells to develop into specified tissue by initiating invagination and embryogenesis by induction (sending chemical massages to the neighboring tissue). Due to gastrulation the cells which are located at mid dorsal line (at roof) of archenteron (which act as organizer) specified the future notochord (presumptive). It sends chemical massages to the archenteron cells situated just above it to differentiate and form neural plate. The process by which fate of one tissue is determined by the other tissue is called as *induction*. The preliminary recognition or determination of tissue is called *primary embryonic induction*. The conditional specification in higher vertebrates depends on the following facts:

1. In early development (two or four celled stage) blastomeres separate and develop separately into complete organism (as in case of identical twins) due to the presence of all types of specific cytoplasmic components.
2. Organizer (dorsal blastopore lip or presumptive notochord) determines the fate of other neighboring cells by releasing chemical massages or primary embryonic induction.
3. Primary embryonic induction initiates a cascade of inductive events by which one tissue interacts with another to direct its fate. These inductive events are described as *secondary induction*.
4. Changes (transplantation) in the position of organizer or any other tissue in early gastrula change the fate of neighboring cells. This is due to specification of cells without determination. However, if transplantation of organizer of other tissue occur in late gastrula stage then it is not possible for organizer or other tissue to change the fate of cells.
5. Not only the development of specified cells determine the fate but also their orientation decide the fate of neighboring cells e.g., the development of wing and limb in chick.
6. Organizer produces stimulus and the specific tissue needs to respond. The ability to respond in a specific manner to a given stimulus is called *competence*. For example in early gastrula ectoderm in all regions of the embryo has the competence to form a new neural plate and embryonic axis but with increasing embryonic age (in late gastrula) the ectoderm loses its ability to respond. It can be said to be that the ectoderm loses its competence to respond to new blastopore lip at later gastrular stage.
7. The separation of somatic cells and germ cells is one of the events of cell differentiation. In animal system germ cells eventually migrate to the gonads where they differentiate into gametes by *gametogenesis*.

The chemical nature of evocators involved in induction due to cellular interaction has been identified. The active vegetalizing factor is a protein of about 14 KD molecular mass that

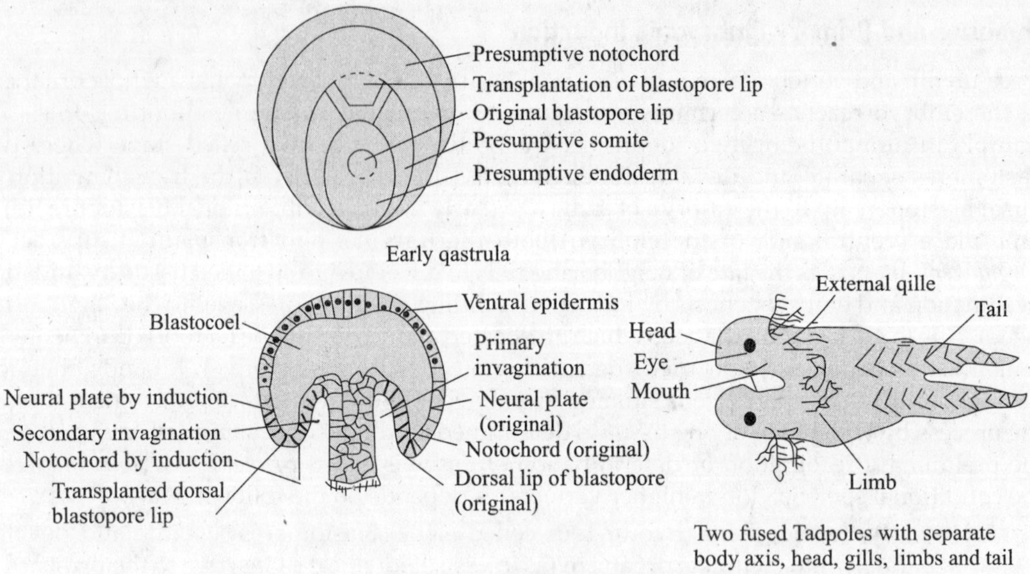

Fig. 4. Showing conditional specification in Amphibia (In newt).

induces the differentiation of a variety of mesodermal structures in insolated blastula animal pole cells (Tiedemann, 1967; Born *et al*; 1985). The other evocators are XTC cell line (Smith *et al.*, 1988), FGF (Rizzino, 1988) and activin (Sokol and Melton, 1991) in human and TGF- B in human and porcine (Knochel *et al.*, 1987).

3. Syncytial Specification and Determination

It involves cells whose fate is decided even before the cell membrane is formed due to segregation and interaction between cell components of a multinucleate cell or syncytium (before the nuclei are separated by the membrane). In *Drosophila* centro lecithal eggs are found, no cell membrane is formed up to thirteen nuclear divisions and the entire structure looks like a syncytium. This stage is known as Syncytioblastoderm.

At this stage antero-posterior and dorso-ventral axis are achieved by the interaction of cytoplasmic material within the single multinucleate cell. During ninth division about five nuclei reach the surface of posterior pole of embryo where poleplasm is already present. These nuclei become enclosed in cell membrane and generate the germ cells which give rise to gametes. Afterwards at tenth nuclear division most of the nuclei arrive at periphery and are surrounded by cell membrane to develop into *Cytoblastoderm* or 'cellular blastoderm' and afterwards conditional specification and determination occur. If the poleplasm of other blastoderm is transplanted to the anterior pole of the other blastoderm then this anterior region differentiates into germ cells suggesting that the presence of poleplasm is responsible for the development of germ cells in *Drosophila.*

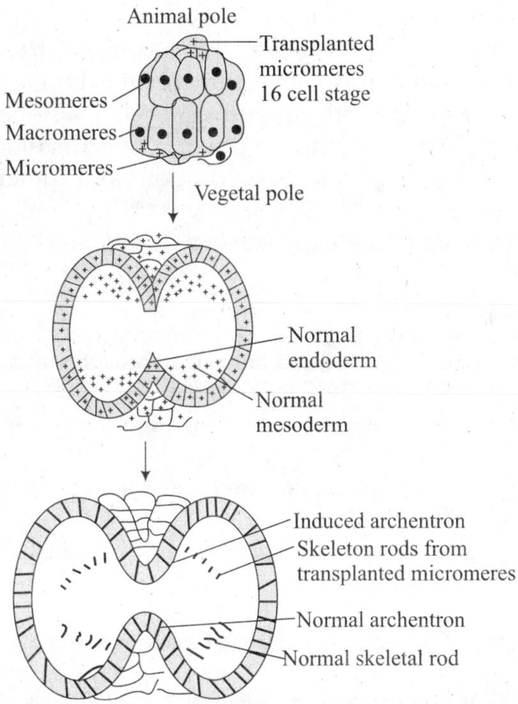

Fig. 5. Showing conditional specification in sea urchin.

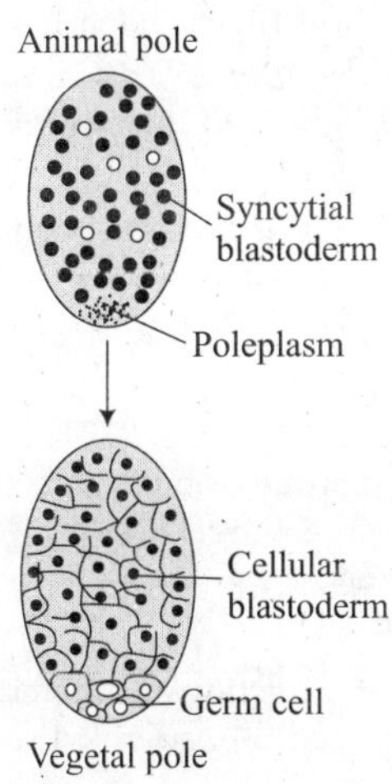

Fig. 6. Showing syncytial specification and germ cell determination by germinal poleplasm in *Drosophila*.

SUMMARY

Differentiation is the development of specialized cell type from fertilized egg. It involves change in biochemistry of cell, change in function of cell and commitment of the cell to a particular fate. Cell commitment is having two steps viz., specification and determination. Commitment is of two types e.g., irreversible and reversible. Irreversible differentiation is followed by determination whereas reversible differentiation is not followed by determination. There are three major ways for cell commitment viz., autonomous specification and determination in mosaic development, conditional specification and determination in regulative development and syncytial specification and determination. Without cell differentiation and cell commitment the development of any organism is not possible. These are the essential processes which decide the fate of a developing cell.

REFERENCES

Appleby, D.W. and Modak, S.P. 1977. DNA degradation in terminally differentiating lens fibre cells from chick embryos. *Proc. Nat. Acad. Sci.* USA, 74 : 5579-5583.

Bandman, E. 1985a. Continued expression of neonatal myosin heavy chain in adult dystrophic skeletal muscle. *Science,* 227: 780-782.

Bandman, E. 1985b. Myosin isozyme transition in muscle development , maturation and disease. *Internat. Rev. Cytol.* , 97 : 97-131.

Born, J., Hoppe, P., Schwarz, W., Tiedemann, H. and Wittmann- Liebold, B. 1985. An embryonic inducing factor: isolation by HPL chromatography and chemical properties. *Biol.Chem. Hoppe- Seyler,* 366: 729-735.

Clark, S.C. and Kamen, R. 1987. The human haematopoitic colony stimulationg factors. *Science,* 236: 1229-1237.

Clayton, R.M., Jeanny, J.C., Bower, J. D. and Errington, L.H.1986. The presence of extralenticular crystallins and its relationship with transdifferentiaton to lens. *Current Topics Dev. Biol.,* 20 : 137-151.

Knochel, W., Born, J., Hope, Loppnow- Blinde, H., Mckeehan, W.L. AND Grunz, H. 1987. Mesoderm inducing factor : Their possible relationship to heparin-binding growth factors and transforming growth factor â. *Naturwissenschaften,* 74 : 604-606.

Rizzino, A. 1988. Transforming growth factor â : Multiple effects on cell differentiation abd extracellular matrices. *Dev.Biol.,* 130 : 411-422.

Smithe, J.C., Yaqoob, M. and Symes, K. 1988. Purification, partial characterization and biological effects of the XTC mesoderm- inducing factor. *Development,* 103: 591-600.

Sokol, S. and Melton, D.A.1991. Pre- existent pattern in *Xenophus animal* pole cells revealed by insucton with activin. *Nature,* 351: 409-411.

Stamatoyannopoulous, G., Constantoulakis, P., Brice, M., Kuracchi, S. and Papayannopoulous, T. 1987. Coexpression of embryonic, fetal and adult globins in erythroid cells of human embryos : Relevance to the cell- lineage models of globin switching. *Dev. Biol.,* 123 : 191-197.

Sutherland, W.M. and Konigsberg, I.R. 1983. CPK accumulation in fusion- blocked quail myocytes. *Dev. Biol.,* 99 : 287-297.

Tiedmann, H. 1967. Biochemical aspect of induction and determination. In : The Biochemistry of animal development, vol.2 pp. 4-55. R.Weber (ed.). Academic Press, New York and London.

Suggested Readings

1. Development (1976) N. J. Berrill and G. Karp, McGraw Hill publishing Co. Ltd; New York.
2. Major Problems in Developmental Biology (1974) M. Locke, (Ed.) Academic Press, NewYork.
3. Epigenetics: A Treatise on Theoretical Biology. (1974): S. Lovtrup, John Wiley,

4. Towards a Theoretical Biology: 3 Drafts (1970) C.H. Waddington. (Ed.), Edinburg University Press, Edinberg.
5. Current topics in developmental biology. (1978) A.A. Moscona and A. Monroy, Academic Press, New York.
6. Developmental Biology : A Modern Synthesis (1994) K.V. Rao, Oxford and IBH Publishing Co. Pvt. Ltd; Delhi.

REVIEW QUESTIONS

Long Questions

1. Write an essay on cell differentiation and cell determination.
2. Explain major ways for cell commitment.
3. Discuss about organizer and primary embryonic induction in vertebrate development.

Multiple Choice Questions

1. The cell differentiation involves the following steps
   (a) Change in biochemistry of cell
   (b) Change in function of cell
   (c) Commitment of the cell to a particular fate
   (d) All the above
2. Cell commitment is
   (a) Reversible type (b) Irreversible type
   (c) Indefinite type (d) a and b type
3. Syncytial blastoderm is
   (a) Multinucleate blastula in insects
   (b) Multicellular blastoderm in insects
   (c) Blastula of vertebrate
   (d) Blastula of invertebrate
4. Competence means
   (a) Tissue making organizer produces stimulus, then the specific tissue needs to respond
   (b) The ability to respond in a specific manner to a given stimulus
   (c) Early gastrula ectoderm in all regions of the embryo has the ability to form a new neural plate and embryonic axis
   (d) All the above

Answers

1. (d) 2. (d) 3. (a) 4. (d)

CHAPTER 23

Homeotic Genes

Pragya Khanna
Govt. College for Women, Parade, Jammu–180 001
E-mail: pragyajmu2002@yahoo.co.in

Development is a process by which cells specialize and organize into a complex organism. It has four major components: growth, cell division, determination, differentiation and morphogenesis. A zygote is the single-celled result of fertilization. Because this single cell has the ability to give rise to every type of cell in the adult body we say it is totipotent. Totipotency is the condition of possessing all the genetic information and other capacities necessary to form an entire individual. Later in development cellular descendents of the zygote lose their totipotency and become determined. Cell differentiation is the process by which stable differences arise between cells. The zygote gives rise to the various tissues and organs. In molecular terms, cell differentiation means variable gene activity in different cells of the same organism. For example, red blood cells are highly specialized for the synthesis and storage of haemoglobin while nerve cells are specialized for synthesizing neurotransmitters. The mechanism by which different cells become specialized is explained by the fact that each eukaryotic cell expresses only a small percentage of the genes it contains, and cells of different tissues express different sets of genes. However, all the genes in a cell are not tissue specific. There are some genes called 'House keeping genes' that remain active in all types of cells and are required for building membranes, ribosomes, mitochondria, and the glycolytic enzymes, which are components common to all types of cells. The genes that are expressed differentially are the ones that dictate the bone cell progenitors to develop into bone cells, liver cell progenitors into liver cells, nerve cell progenitors into neurons and so on. Another important characteristic of development is that it is controlled by preprogrammed circuits of gene expression, in which each event triggers the expression of one set of genes and the product of one of these genes functions by turning off the transcription of first set of genes and/or turning on the transcription of a second set of genes.

CONTROL OF SEGMENTAL IDENTITY IN *DROSOPHILA*

Insects are composed of a series of separate segments that are more advanced and have become specialized for specific functions. For example, the thoracic segments are often specialized for locomotion and develop wings and legs. The development in *Drosophila* is controlled by a cascade of regulatory genes that act in a specific sequence to form different patterns of gene expression on different regions of fertilized egg. After the egg has been fertilized, the growth and development of any eukaryotic organism involves an amazingly complex yet marvelous choreography of gene expression. A great deal of research has been conducted in an effort to understand at the molecular level the sequence of events that direct and regulate development. The appropriate genes are expressed at the proper time in the correct sequence to produce the proteins that interact to produce a cell, a tissue, an organ or a complete fruit fly. For eg., the genes that are expressed in eye are not expressed in the leg or in the wing but these differences in the patterns of gene expression do not occur in response to a single stimulus. Instead, the different cell lineages that give rise to the cells of eye, leg or wing gradually swerve, wherein different programs of gene expression are turned on by a cascade of different regulatory genes. We can define 3 major classes of genes that control or regulate development in *Drosophila melanogaster*: Maternal genes, Segmentation genes and Homeotic genes (Fig. 2).

Maternal genes: These genes code for mRNA and proteins that are stored in the egg. They are responsible for defining the major axes of the developing embryo.

- The anterior (front): posterior (rear) axis is specified by approximately a dozen genes, the most important of which are *nanos, bicoid, hunchback., caudal* and *torso*.
- The dorsal (back): ventral (front) axis is also specified by approximately a dozen genes, the most important of which are ***toll*** and ***dorsal***.

Anterior/ Posterior Axis

In fact, there are eighteen different maternal-effect genes that influence the anterior-posterior polarity. For eg., the mRNAs for *nanos* and *bicoid* are laid down in localized deposits at the two opposite poles of the oocyte, thereby establishing the future poles of the developing embryo. Maternal mRNAs for *hunchback* and *caudal* are also deposited in the egg but are present in a uniform concentration throughout. The *bicoid* gene product activates expression of the zygote's

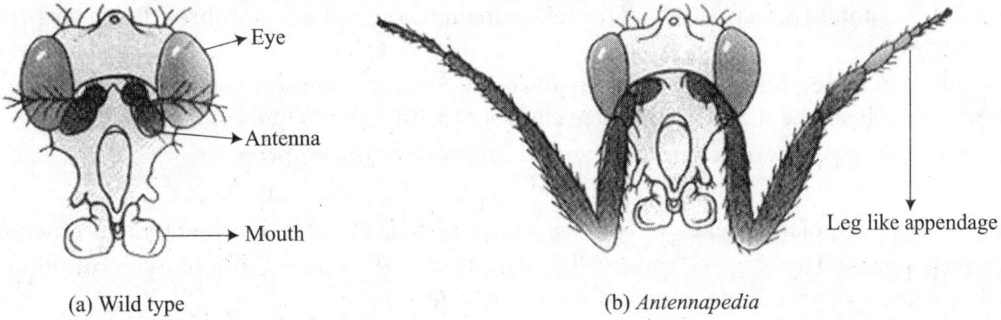

Fig. 1. Diagram illustrating the effects of the homeotic mutation *Antennapedia* on the phenotype of *Drosophila*.

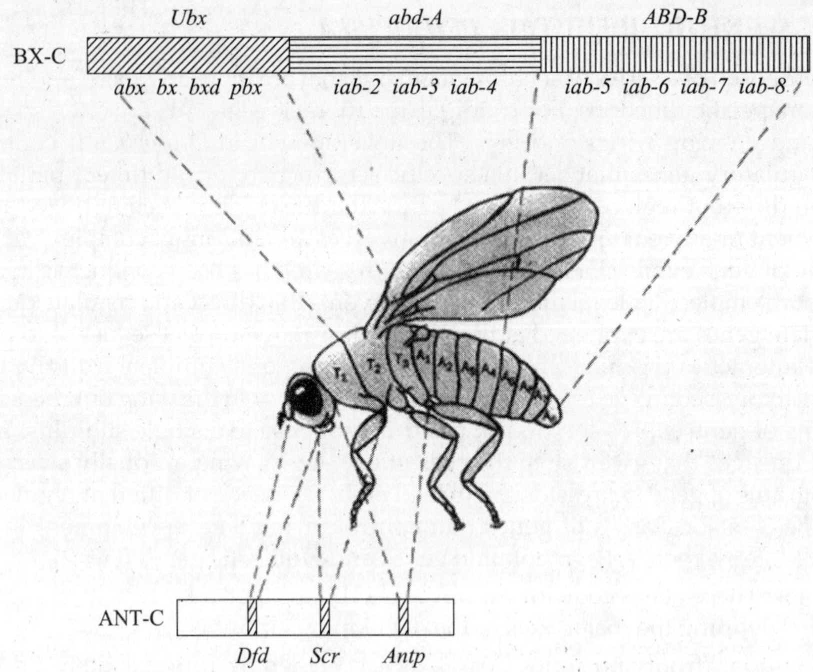

Fig. 2. Diagram showing the body segments of *Drosophila* affected by homeotic mutations in the two major clusters of homeotic genes on chromosome 3.

hunchback gene while *nanos* represses translation of *hunchback* mRNA (both maternal and zygote expressed). The net result is that the combined action of the *nanos* and *bicoid* proteins generates a concentration gradient of the *hunchback* protein between the two poles of the cell.

Dorsal/ Ventral Axis

The key maternal gene product in the establishment of the dorsal/ventral axis is the *toll* protein receptor. This protein is also deposited uniformly in the plasma membrane around the periphery of the entire oocyte. After fertilization, *toll* is activated by a fragment of the *spätzle* protein. This fragment is only generated in the perivitelline space opposite the ventral region of the oocyte/zygote. The activated *toll* protein transmits a signal which causes the *cactus* protein to be degraded.

Segmentation genes: These genes specify the correct number of body segments and the correct polarity of each segment. There are three classes of segmentation genes:

- Gap genes – which define the broad divisions of the embryo.
- Pair-rule genes – which specify the division of the gaps into segments.
- Segment polarity genes – which specify the polarity of individual body segments.

Homeotic genes: These genes specify the identity of each body segment by controlling the identity of the organs that develop within that segment.

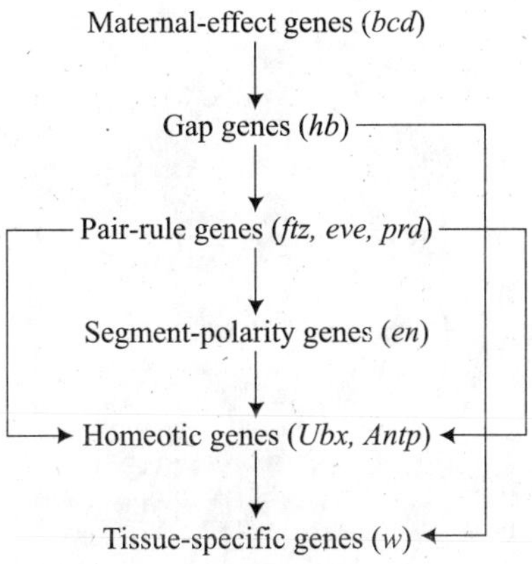

Fig. 3. Regulatory cascade in the control of development in *Drosophila*.

A mutation in a homeotic gene is usually characterized by the appearance of a body organ in the wrong place. Homeotic types of mutations were first noted in plants by Goethe in the late 18th century but the term was coined by William Bateson early in the 20th century.

The homeotic genes are expressed in different combinations along the length of the body (their expression is influenced by the products of the segment polarity genes) and their gene products tell each segment what to become. If homeotic gene products from one segment are put into another segment, the receiving segment shows characteristics of the donor segment. Mutations in these genes cause the transformation of body parts – called homeotic mutations. Homeotic genes of animals all contain a sequence of 180 base pairs of DNA called the homeobox. The homeobox nucleotides code for the homeodomain – 60 amino acids found in homeotic gene products. These products are usually transcription factors that control transcription of other genes. The homeobox sequence has been highly conserved during evolution and is very similar from yeasts to humans.

Thus, the homeotic genes can be thought of as genetic switches that turn different programs of cellular differentiation on or off. The activities of a number of homeotic genes are required to establish the identity of segments in the trunk region (posterior head, thorax, and abdomen) of the embryo. These genes are clustered in two major groups called the *Antennapedia* complex (ANT-C) and the *bithorax* complex (BX-C).

Antennapedia complex - anterior function :

labial

proboscidea

deformed

Sex combs reduced

Antennapedia

Ultrabithorax complex - posterior function :

Ultrabithorax

Abdominal-A

Abdominal-B

Fig. 4. William Bateson (1861-1926)

Homeotic genes code for proteins, all of which contain a conserved motif – the homeodomain – which is a DNA-binding motif in the helix-turn-helix family of DNA-binding proteins. The DNA segment that codes for the homeodomain is called a homeo box.

HOMEOTIC MUTANTS

Mutations in homeotic genes do not eliminate elements of the pattern but rather cause these elements to develop with inappropriate identities. Homeotic genes were first identified as dominant mutations that changed the identity of structures. Homeotic genes were discovered because of the fact that mutations in these genes can cause extrordinary phenotypes. In homeotic mutants, correctly formed body parts arise in ectopic places. Just as segmentation genes have the function to divide the body into repeated units, the homeotic genes are responsible for rendering these units different from each other. Every segment follows a similar building plan. Homeotic genes regulate, for example, whether the segment develops a haltere or a wing, by regulating a large number of downstream genes required for the development of segment specific structures. Accordingly, different homeotic genes are active in different segments. A large number of different homeotic mutants have been identified and studied in *Drosophila*. The majority of mutations map within one of the two clusters of homeotic genes on chromosome 3, viz., Antennapedia complex (ANT-C) and the bithorax complex (BX-C).

One of the best studied examples of these is the gene *Antennapedia (Antp)*, which as the name suggests causes the transformation of antennal structures to the corresponding (homologous) leg structures, resulting in a fly with a leg like appendage extending from its head (Fig. 1). A*ntennapedia* complex *(Antp*-C) spans over 100 kb and contains at least four genes: *Deformed (Dfd), Sex combs reduced (Scr), fushi tarazu (ftz) and Antennapedia (Antp).* Homeotic mutations in *Bithorax complex* alter the development of the thorax and the abdominal segments of the body. Studies of homeotics really began with a geneticist named Ed Lewis. He worked out the genetics of the bithorax complex and made the interesting observation that within the BX-C, the order of the genes on the chromosome is the same as the order of segments that they affect along the embryonic axis. The same turned out to be largely true for the ANT-C as well. However, it has been emphasized that homeotic mutations generally have such extreme effects on the phenotypes of the mutant fly that these mutants die before reaching the adult stage. Some mutations that have less extreme effects on phenotype produce flies that survive, but exhibit abnormal phenotypes such as flies with four wings which arise due to three mutations in the BX-C gene cluster that convert the halteres (balancing organs) on the third thoracic segment to extra pair of wings.

HOMEOBOX

A homeobox is a stretch of DNA sequence found in genes involved in the regulation of the development of animals and plants. Genes that have a homeobox are called homeobox genes and form the homeobox gene family. A homeobox is about 180 base pairs long and encodes a protein domain, the homeodomain which can bind DNA, which is about 60 amino acids long. A particular subgroup of homeobox genes are the Hox genes, which are found in a special gene cluster, the Hox cluster (also called Hox complex). Hox genes function in patterning the body axis. Thus, by providing the identity of particular body regions, Hox genes determine where limbs and other body segments will grow in a developing larva. The homeobox genes were first found in the fruit fly *Drosophila melanogaster* and have subsequently been identified in many other species, from insects to reptiles and mammals. The first genes found to encode homeodomain proteins were *Drosophila* developmental control genes, in particular homeotic genes, from which the name "homeo" box was derived. However, many homeobox genes are not homeotic genes; the homeobox is a sequence motif, while "homeotic" is a functional description for genes that cause homeotic transformations.

The term Hox cluster refers to a group of clustered homeobox genes, named Hox genes in vertebrates that play important roles in pattern formation along the anterior-posterior body axis. In fact, the first homeobox genes discovered were those of the *Drosophila* homeotic gene clusters, i.e. the "Antennapedia complex" and the "Bithorax complex", which summarily are referred to as HOM-C (homeotic complex). This HOM-C complex in *Drosophila* is the evolutionary homolog of the vertebrate Hox clusters and the evolutionarily related homeobox gene clusters in other animals (i.e. chordates, insects, nematodes, etc.) are now also called HOX clusters.

About a few decades back, men and flies were thought to share no more than a handful of genes. After the discovery of so-called 'homeotic genes' that controlled axial patterning, and the fact that their sequence and function was conserved between distant organisms, the unexpected realization that vertebrates and invertebrates were evolutionarily related through such an essential developmental process altered the course of developmental genetics — it turned flies into authorized models for animal development and kick started the field of vertebrate molecular embryology.

SUMMARY

The growth and development of any eukaryotic organism involves an amazingly complex yet marvelous choreography of gene expression soon after the egg is fertilized. A great deal of research has been conducted in an effort to understand at the molecular level the sequence of events that direct and regulate development. In *Drosophila*, the basic body plan unfolds through the establishment of anterior-posterior segmentation and dorsal-ventral polarity. These events are controlled by key regulator genes that function in intricately regulated pathways. Such genes that control pattern formation in *Drosophila* have been classified into five groups: (1) maternal-effect genes, (2) gap genes, (3) pair-rule genes, (4) segment-polarity genes, and (5) homeotic genes. The maternal-effect genes are expressed in the female during oogenesis and define the major axes of the developing embryo. The gap genes, pair-rule genes and segment polarity genes then act in a sequence to define divisions of the embryo. Later, the homeotic genes function and control the commitment of groups of cells to differentiate into specific tissues. Thus, homeotic genes determine the direction of developmental fates of groups of cells in a segment of the embryo. They include a DNA sequence called the homeobox that is similar in all homeotic genes. When altered the homeotic mutations such as Antennapedia, cause abnormal developments.

REFERENCES

Bürglin, T.R. (2005) Homeodomain Proteins. In : Encyclopedia of Molecular Cell Biology and Molecular Medicine (Meyers, R.A. ed.), Wiley-VCH Verlag GmbH and Co., Weinheim, pp. 179-222.

Gilbert, S.F. (2000) Specification of cell fate and the embryonic axis. In: Developmental Biology, 6th edition. Sinauer Associates, Inc., Sunderland, MA, pp. 569-577.

Kornberg, T. B., and Tabata, T. (1993) Segmentation of the Drosophila embryo. Curr. Opin Genet Dev 3 : 585-94.

McGinnis, W. and Krumlauf, R. (1992) Homeobox genes and axial patterning. Cell 68: 283-302.

Peifer, M.F.K. and Bender, W. (1987) The bithorax complex: control of segmental identity. Genes Dev. 1 : 891-898.

Suggested Readings

1. Principles of Genetics (1991) E.J. Gardener, M.J. Simmons and D.P. Snustad, 8th edition, John Wiley and Sons Inc., USA.
2. Genes VIII (2004) B. Lewin, Pearson Prentice Hall, Pearson Education Inc. NJ, USA.
3. Cell and Molecular Biology (1997) E.D.P. De Robertis and E.M.F. De Robertis, Jr. 8th edition, B.I. Waverly Pvt. Ltd., New Delhi, India.

REVIEW QUESTIONS

Long Questions

1. If all the cells in a given organism carry the same genes, how can gene expressions that are localized in time and space be explained?
2. What are homeotic genes?
3. How do the different classes of segmentation genes control the spatial patterns of homeotic gene expression?
4. How are the homeotic genes thought to control segmental identity?
5. Explain the homeotic mutants of *Drosophila*?

Multiple Choice Questions

1. The _______ genes can be thought of as genetic switches that turn different programs of cellular differentiation on or off.
   (a) Gap genes (b] Pair-rule genes
   (c) Segment-polarity genes (d) Homeotic genes
2. Studies of homeotics really began with a geneticist named _______.
   (a) Ed Lewis (b) Barbara Mc Clintock
   (c) William Bateson
3. Homeotic genes:
   (a) Determine the direction of developmental fates of groups of cells in a segment of the embryo.
   (b) Include a DNA sequence called the homeobox that is similar in all homeotic genes.
   (c) Are altered in homeotic mutations such as Antennapedia, causing abnormal development of the head.
   (d) Are similar beteween mice and flies for both structure and order on the chromosome.
   (e) All are true.
4. _______ genes determine the number, boundaries and polarity of segments.
   (a) Segmentation genes (b) Gap genes
   (c) Segment-polarity genes (d) Pair-rule genes
5. _______ genes organize large areas along the anterior-posterior axis.
   (a) Homeotic genes (b) Gap genes
   (c) Maternal-effect genes (d) Pair-rule genes
6. _______ genes divide the embryo into units of two segments each.
   (a) Pair-rule genes (b) Homeotic genes
   (c) Segment-polarity genes
7. Homeotic genes of animals contain a sequence of _______ of DNA called the homeobox.
   (a) 180 bp (b) 60bp
   (c) 240 bp (d) 180 kbp
8. *Nanos, bicoid, hunchback., caudal* and *torso* are examples of
   (a) Maternal genes (b) Segmentation genes
   (c) Homeotic genes
9. The DNA segment that codes for the homeodomain is called _______
   (a) Homeo box (b) Homeodomain
   (c) Hox cluster
10. Homeotic types of mutations were first noted in plants by _______
    (a) Goethe (b) William Bateson
    (c) Ed Lewis

Answers

1. (d)	2. (a)	3. (e)	4. (a)	5. (b)	6. (a)
7. (a)	8. (a)	9. (a)	10. (a)		

Multiple Choice Questions

PART–8

Immunity and Monoclonal Antibodies

CHAPTER 24

Antigen: Specificity and Variation

D. Gayathri[1]* and T.N. Devaraja[2]

[1]Dept. of Microbiology, Kuvempu University, Jnana Sahyatri, Shankaraghatta–577 451

[2]Taralabalu KVK, Anubhava Mantap Campus, Davanagere–577 004

E-mail: gayathridevaraja1@yahoo.com

An antigen is a substance that can elicit a specific immune response. It may be a substance from the environment such as chemicals, bacteria, virus, or pollen or may also be formed within the body, as with microbial toxins or tissue cells. Antigen has two specific functions - ability to induce antibody production and specific reaction with the antibody. Ability of an antigen to mediate immunological reaction is called immunogenicity. Typically, only a part of the antigen molecule is bound by an antibody. The part of the antigen that binds to the antibody is called antigenic determinant/epitope. The immune response is the first line of defense against infection and probably against cancer cells as well. Substances that elicit an immune response are also called immunogens, and a specific immunoglobulin that binds to this substance is called an antibody. If the invading particle is large, like a cell, a virus, or a protein, different classes of antibodies may be elicited, each type binding specifically to a given antigenic determinant or epitope, on the surface of an antigen. The types of immunoglobulin produced during immune response are:

(i) IgG: It is a gamma globulin, with gamma heavy chain. It is the most abundant extra and intravascular circulating antibodies.

(ii) IgM: An immunoglobulin produced during the early response to an invading microorganisms, and is the largest immunoglobulin. It contains 5 Y-shaped immunoglobulins of two light and two heavy chains each, connected to each unit by a J-chain.

(iii) IgA: An immunoglobulin found in body secretions, including saliva, sweat, tears, and along the walls of the intestine. It is the major antibody of colostrum, the initial secretion from a mother's breasts after birth, secretory IgA is usually dimeric.

*Corresponding author

(iv) IgD: An immunoglobulin found on the surface of B cells, resembles as that of IgG.

(v) IgE: An atopic, heat labile immunoglobulin, associated with some of the body's allergic responses.

Hapten, are the substances which when coupled with a carrier molecule, become antigenic. Because of their small size, they themselves can not be antigenic. An allergen, is a substance which mediates allergic manifestations. Tolerogen, is a substance that induces immunocompetent cells to non responsive which may be temporary and may result in disease and/or immunogenicity. An epitope or antigenic determinant is a smallest part of an antigen having specific configuration, recognized and bound by antibody. Corresponding part on antibody that bind an epitope is called paratope. To protect against infectious disease caused by a variety of microbial agent, antigen must recognize the foreign agent. Dendritic cells (DCs) are key to the maintenance of peripheral tolerance to self-antigens and the orchestration of an immune reaction to foreign antigens.

CHARACTERISTICS OF AN ANTIGEN

A number of factors contribute to an antigen to become highly specific in its reaction with antibody namely, size, chemical nature, foreignness and susceptibility to tissue enzymes.

(i) Size: The antigenicity of a molecule directly relates to its size. Too small (< 10,000 D) an antigen is less immunogenic and generally a size of >1, 00,000 D is considered as an immunogen. It was found that an antigen of 10 KD from the cellular extract of *Aphanomyces invadans*, a fungal pathogen was immunogenic in rabbit and fish while 43 KD was immunogenic in mice. This exemplifies the significance of antigen size causing variation in antibody production in vertebrates (Gayathri, 2002).

(ii) Chemical nature: Immune system recognizes infectious agent, either proteins or carbohydrates. Proteins are the most potent immunogens, followed by carbohydrate, lipid and nucleotides. However, latter ones are when combined with protein become potent immunogens (e.g., lipoprotein, lipoploysaccharide). For humoral immunity, proteins or polysaccharides are being used. However, for cell mediated immunity, only proteins, some lipids, few glycolipids serve as immunogens. Proteins are processed to smaller peptides and then recognized by MHC before recognized as antigen. Synthetic polymers lack immunogenicty although it (may) contain single amino acid or sugar. However, copolymers containing few aminoacids would be immunogenic. In addition to this, phenylalanine and tyrosine increase immunogenicity. Primary, secondary, tertiary and quaternary structure of protein enhances immunogenicity of an antigen with increase in chemical complexity. However, gelatin (protein) although a macromolecule, lack of an aromatic ring makes it non immunogenic.

(iii) Foreignness: Antibody response to an antigen takes place when any substance which is of non self. Generally, immune response will not mount to ones own antigens. This may be due to continuous exposure of tissue antigens during lymphocyte development starting from embryonic stage. The degree of immunogenicity increases with increase foreignness, that is, greater the phylogenetic distance between two species, the greater the structural dissimilarity and thus, the substance will be more antigenic. Bovine serum albumen is more immunogenic in rabbit than goat. Although, mounting of immune response to non self antigen, sometimes an indifferent mechanism could be observed, for example, upon disturbances to cornea or scrotal sacs, some of the sequestrated proteins (lens protein or sperms) release into circulation, and now acts as an antigen.

(iv) **Susceptibility to tissue enzymes:** Large insoluble macromolecules are more immunogenic than soluble small molecules in tissue fluids. The larger molecules are easily engulfed by phagocytes. D-amino acids are susceptible to tissue enzymes than that of L- amino acids and therefore D-amino acids are more immunogenic.

For effective development of immune defense, both humoral and cellular immune response are required. The interaction of T-cell with antigen processed cell and presented to MHC molecules. For the production of T-helper (T_H) cells antigen must be presented with MHC- II class molecules on an antigen presenting cell (APC). For the production of T-cytotoxic (T_C) cell, antigen must be presented with MHC-I class molecules.

SPECIFICITIES OF THE ANTIGEN

Antigen specificity is determined by single chemical group even single acid radical. The difference in antigenic specificity is also noted with the group of chemical compounds attached at the ortho, meta or para positions, in addition, dextro, levo or meso isomers. Antigen specificity can be further described with following heading. Specificities of antigen vary and they may be of species specificity, iso specificity, auto specificity, organ specificity and hetero specificity in nature.

- **Species specificity:** Individuals of a species possess distinct type of tissue antigens and a cross reaction can be noted among related species. Determination of immunological relationships thus helps in detecting evolutionary relatedness, in forensic application or in determining hypersensitivity reaction.
- **Iso specificity:** Some members of a species possess certain specific antigens, such antigens are isoantigens. For example, human blood groups – A, B, AB and O. Determination of blood group antigen helps in blood group matching for transfusion, isoimmunisation. It is also helpful in determining parents of disputed cases. Histocompatibility antigens specific to each individual of a species, can recognize genetically different individuals of the same species. These antigens are associated with the plasma membrane of tissue cells. Tissue compatibility determination helps in homograft transplantation. Non compatible tissue antigens reject the transplant.
- **Organ specificity:** Antigens of some organs of different species share same antigens, is referred as organ speicificity. Brain, kidney, lens protein of goat, sheep, human share similar antigens. However, a neuroparalytic complications develop following the administration of antirabies vaccine prepared from partially denatured sheep brain cells. The immune response against brain cells of sheep damages the recipient's brain tissues leading to convulsions.
- **Auto specificity:** Auto or self antigens are usually non antigenic. However, hidden or sequestrated antigens (those antigens that are not normally available in body fluids or circulation) are not recognized as self antigens. For example, upon damage to testis sperms are released to the circulation, for which immune response could be observed.

Similarly, damage to retina lead to the release of lens protein (which is usually embedded within the capsule), for which antibody production may take place. Immune response thus leads to autoimmune diseases.

- **Hetero specificity:** Similar or closely related antigens may occur in different species, like bacteria, erythrocytes, plant cells etc. The sharing of antigens among unrelated groups of species is having diagnostic importance. For e.g., Weil-Felix reaction, for detecting Rickettsial infection (typhus fever) using non motile strains of *Proteus vulgaris*. Paul Bunnel test, for detecting infectious mononucleosis caused by Epstein Barr virus, using sheep RBCs. A typical pneumonia caused by *Mycoplasma pneumoniae* can be detected using *Streptococcus* MG strain.

PLANTIGENS AND PLANTIBODIES

Plant Antigens or plantigens are those plant substances which induce human or animal production of antibodies. Plantibodies or plant antibodies are the human or animal antibodies made by and in Transgenic Plants.

Human Antibodies Produced by Field Crops Enter Clinical Trials

A Corn strain genetically engineered by Agracetus in Middleton, Wis., to secrete human antibodies. They have proposed a plan to begin injecting cancer patients with doses of up to 250 mg of antibodies purified from mutant corn seeds. If the treatment works as estimated, the antibodies will stick to tumor cells and deliver radioisotopes to kill them. Using antibodies as drugs is not new, but manufacturing them in plants is, and the technique could be a real boon to the many biotechnology firms promising medicines to market. So far most have failed, for two reasons, many early antibody drugs either did not work or provoked severe allergic reactions. They were not human but mouse antibodies produced in vats of cloned mouse cells. In recent years, immunogeneticists have bred cell lines that produce antibodies that are mostly or completely human.

Plantibodies might reduce another risk as well. This may include, no human disease is generally transmissible to plant cell, free from pesticides and other kinds of contaminants, it can forgo expensive screening for virus and bacterial toxins. Corn is not the only crop that can mimic human cells. Agracetus is also cultivating soybeans that contain human antibodies against herpes simplex virus 2, sexually disease causing agent, in the hope of producing a drug cheap enough to add to contraceptives. A biotechnology form is testing an anti-tooth-decay mouthwash made with antibodies extracted from transgenic tobacco plants and another group has modified tobacco to manufacture an enzyme called glucocerebrosidase in its leaves. People with Gaucher's disease pay up to $160,000 a year for a supply of this crucial protein, which their bodies cannot make. It's rather astonishing how accurately transgenic plants can translate the subtle signals that control human protein processing. However, there are some important differences as well. Human cells adorn some antibodies with special carbohydrate molecules. Plant cells can stick the wrong carbohydrates onto a human antibody. Which then the maladjusted antibodies cannot stimulate the body into producing its own immune response, and they are rapidly filtered from the bloodstream. Some of the important therapeutic and diagnostic plantibodies (Elsevier Science, 2001) are given in the Table 1.

Table 1. Therapeutic and diagnostic plantibodies

Target site	Application and specificity	Signal sequences	Antibody name or type	PLANT
Dental caries	streptococcal	Murine IgG signal peptides	Secretory IgA	*Nicotiana tobaccum*
Cancer treatment	carcinoembryonic antigen	Murine IgG signal peptide	KDEL - ScFvT84.66 (ScFv)	WHEAT
Cancer treatment	carcinoembryonic antigen	Murine IgG signal peptide	KDEL - ScFvT84.66 (ScFv)	RICE
Cancer treatment	carcinoembryonic antigen	TMV leader; murine IgG signal peptides	KDEL - T84.66 (IgG) -	*Nicotiana tobaccum* (transicntly with Agrobacterium infiltration)
B-cell lymphoma treatment	idiotype vaccine	- Rice a-amylase -) -	38C13 (scFv	*Nicotiana benthamiana*
Colon cancer; -;	surface antigen	Murine IgG signal peptide	KDEL - CO17-1A (IgG) - a	Nicotiana Benthamian
Herpes simplex virus 2		Tobacco extensin signal peptide	Anti-HSV-2 (IgG)	Soy bean
Diagnostic	anti-human IgG	Murine IgG signal peptides	C5-1 (IgG	Alfalfa

Fungal Antigens

Fungal antigens possess a variety of proteins or carbohydrates or lipids moieties as their virulence marker. In *Candida albicans*, cell wall of yeast and mycelial phase, contained a protein of 19 KD, a polysaccharide-rich, high-molecular-weight antigen linked through di-N-acetylchitobiose groups to proteins which is expressed on the surface of the germ tube (Ponton and Jones, 1986). Inhalation of fungal mycelium or spores often mount allergic manifestation in human and other vertebrates. Irritating effects in the respiratory tract and immune response is caused by fungal antigens, glucans and volatile metabolites. After subjecting an experimental animal to fungal antigen, the levels of specific antibodies (IgG, IgE, IgA) in serum and inflammatory mediators were higher. Conidiospores of *Aspergillus fumigatus* showed thirty-six antigens were identified using antisera from six rabbits by immunoblot. Denikus *et al.*, (2005) reported that *Aspergillus* antigens have value both in diagnostic tests for different forms of aspergillosis and as vaccine candidates for protection against invasive disease. *Aphanomyces invadans*, an important fungal pathogen in epizootic ulcerative syndrome, showed 5 antigens by immnublot (Gayathri, 2002).

Bacterial Antigen

Antibody-mediated immune (AMI) response and cell mediated immune (CMI) response are the protective defense observed by the host against against extracellular bacteria. On epithelial surfaces, the main antibacterial immune defense of the host is by the production of secretory

IgA. Once the epithelial surfaces have been passed by bacteria, the major immune defenses of AMI and CMI are encountered.

Immunological tolerance to a bacterial antigen: Tolerance is a property of the host in which there is an immunologically-specific reduction in the immune response to a given antigen tolerance to a bacterial antigen does not involve a general failure in the immune response but a particular deficiency in relation to the specific antigen(s) of a given bacterium. If there is a depressed immune response to relevant antigens of a parasite, the process of infection is facilitated. Tolerance can involve either AMI or CMI or both arms of the immunological response. Tolerance to an antigen man be due to,

1. Fetal exposure to antigen: If a fetus is infected at certain stages of immunological development, the microbial antigen may be seen as "self", thus inducing tolerance to the Ag which may persist even after birth.
2. High persistent doses of circulating antigen: Tolerance to a bacterium or one of its products might arise when large amounts of bacterial antigens are circulating in the blood.
3. Molecular mimicry: If a bacterial antigen is very similar to normal host antigens, the immune responses to this Ag may be weak giving a degree of tolerance.

Resemblance between bacterial antigen and host antigen is referred to as molecular mimicry. In this case the antigenic determinants of the bacterium are so closely related chemically to host "self" components that the immunological cells cannot distinguish between the two and an immune response cannot be raised. Some bacterial capsules are composed of polysaccharides (hyaluronic acid, sialic acid) so similar to host tissue polysaccharides that they are not so immunogenic.

Antigenic Disguise

Some pathogens can hide their unique antigens from opsonizing antibodies or complement. Bacteria may able to coat themselves with host proteins such as fibrin, fibronectin, or even Ig molecules. In this way they are able to hide their own antigenic surface components from the immunological system. For example, *Stapylococcus aureus* produces cell-bound coagulase and clumping factor that cause fibrin to clot and to deposit on the cell surface. It is possible that this disguises the bacteria immunologically so that they are not readily identified as antigens and targets for an immune response. Protein A produced by *S. aureus*, and the analogous Protein G produced by *Streptococcus pyogenes*, bind the Fc portion of immunoglobulins, thus coating the bacteria with antibodies and canceling their opsonizing ability. In addition, the fibronectin coat of *Treponema pallidum* provide an immunological disguise for the spirochaetes. In *E. coli* K1, causing meningitis in newborns, a capsule composed predominantly of sialic acid provides an antigenic mask, as hyaluronic acid capsule of *Streptococcus pyogenes*.

Persistence of bacterial antigens at bodily sites: Some pathogens can avoid exposing themselves to immune responses. Intracellular pathogens can escape host immune responses as long as they stay inside of infected cells and they do not allow microbial antigen to express on the cell surface. For example, *Brucella*, *Listeria* or *Mycobacterium leprae* that infects macrophages, the macrophages support the growth of these bacteria and at the same time give them protection from immune responses. *Yersinia*, *Shigella* (intracellular pathogens), are residents of cells that are neither phagocytes nor Antigen Presenting Cells and their antigens are not displayed on the infected cell's surface. Some pathogens persist in gastrointestinal tract, mouth cavity and urinary tract, or the lumen of the salivary gland, mammary gland or the kidney tubule. Some bacterial pathogens that grow at tissue sites generally inaccessible to the forces of AMI and CMI are, *Streptococcus mutans,* causes dental caries at any time after the eruption of the teeth, regardless of the immune status of the host. Either the host does not undergo an effective immune response or secretory IgA plays little role in preventing colonization and subsequent plaque development. *Vibrio cholerae,* multiplies in the gastrointestinal tract where the bacteria secrete a cytotoxin which causes loss of fluids and diarrhea in the host which is characteristic of the disease cholera. Although IgA antibodies act against cellular antigens of the cholera vibrios, are not completely effective in preventing infection. *Salmonella typhi,* the carrier state of typhoid fever results from a persistent infection by the typhoid bacillus. The organism is not eliminated during the initial infection and persists in the host for months, years or a life long. In the carrier state, *S. typhi* is able to colonize the biliary tract away from the immune forces, and be shed into urine and faeces. *Brucella abortus,* persistently infects mammary glands of cows and is shed in the milk. *Leptospira* multiplies persistently in the lumen of the kidney tubules of rats and is shed in the urine and remains infectious.

Soluble Bacterial Antigens

Some bacteria can liberate antigenic surface components in a soluble form into the tissue fluids. These soluble antigens are able to combine with antibodies before they reach the bacterial cells. For example, with lipopolysaccharide of Gram negative bacteria released into the body fluids combines with corresponding antibody such that the bacteria is free from interaction with antibody. Autolysis of Gram-negative or Gram-positive bacteria may release antigenic surface components in a soluble form. *Streptococcus pneumoniae* and *Neisseria meningitidis* are known to release capsular polysaccharides. These soluble bacterial cell wall components are powerful antigens and complement activators, so they contribute in a major way to the pathology observed in meningitis and pneumonia. Protein A, is produced by *Staphylococcus aureus* may remain bound to the cell surface or it may be released in a soluble form. Protein A will bind to the Fc region of IgG. On the cell surface, protein A binds IgG in the wrong orientation to exert its antibacterial activity, and soluble protein A agglutinates and partially inactivates IgG.

Antigenic Variation

Bacteria can avoid the immune response this is due to periodically changing antigens, i.e., they undergo antigenic variation. Some bacteria avoid the host antibody response by changing from one type of fimbriae to another, or by switching fimbrial tips. This makes the original AMI response outdated by using new fimbriae that do not bind the previous antibodies. Antigens may vary or change within the host during the course of an infection, or alternatively, antigens may vary among multiple strains (antigenic types) of a parasite in the population. Antigenic variation is an important mechanism used by pathogenic microorganisms for escaping the neutralizing activities of antibodies. Antigenic variation may be due to site-specific inversions or gene conversions or gene rearrangements in the DNA of bacteria. For example, *Borrelia recurrentis* is a spirochete that causes the human disease relapsing fever. The disease is characterized by episodes of fever which relapse for a period of weeks or months. After infection, the bacteria multiply in tissues and cause a febrile illness until the onset of an immune response a week or so later. Bacteria then disappear from the blood because of antibody mediated phagocytosis, lysis, agglutination, etc., and the fever falls. Then an antigenically distinct mutant arises in the infected individual, multiplies, and in 4-10 days reappears in the blood and there is another febrile attack. The immune system is stimulated and responds by conquering the new antigenic variant, but the cycle continues such that there may be up to 10 febrile episodes before final recovery. With each attack a new antigenic variant of the bacterium appears and a new set of antibodies is formed in the host. Thus, this bacterium can change its antigens during the course of an infection in a single host, and this variation in bacterial antigens contributes significantly to the course of the infection.

Neisseria gonorrhoeae can change fimbrial antigens during the course of an infection. During initial stages of an infection, adherence to epithelial cells of the cervix or urethra is mediated by pili (fimbriae). Equally efficient attachment to phagocytes have been observed. Rapid switching on and off of the genes controlling pili are therefore necessary at different stages of the infection, and *N. gonorrhoeae* is capable of undergoing this type of phase variation. Genetically controlled changes in outer membrane proteins also occur in the course of an infection. This finely controlled expression of the genes for pili and surface proteins changes the adherence pattern to different host cells, increases resistance to cervical proteases, increases resistance to phagocytosis and immune lysis, and is presumably necessary for successful infection.

There are number of multiple antigenic types or serotypes, i.e, they are variant strains of the same pathogenic species. For example, there are multiple serotypes of *Salmonella typhimurium* based on differences in cell wall (O) antigens or flagellar (H) antigens. There are 80 different antigenic types of *Streptococcus pyogenes* based on M-proteins on the cell surface. There are over one hundred strains of *Streptococcus pneumoniae* depending on their capsular polysaccharide antigens. Based on minor differences in surface structure there are multiple serotypes of *Vibrio cholerae, Staphylococcus aureus, Escherichia coli, Neisseria gonorrhoeae* and an assortment of other bacterial pathogens. As the immune response is the major defense against a pathogen, then being able to shed original antigens and present new ones to the immune system might allow infection or continued invasion by the pathogen. Furthermore, the infected host would be the ideal selective environment for the emergence of new antigenic strains of bacteria and may explain many bacteria exist in a great variety of antigenic types (Todar, 2002).

Viral Antigens

For many viral diseases a long lasting immune response has been observed for example, infection with polio virus, chicken pox or measles with residual effects. However, immunity against common cold (influenza virus) is short lived due to its antigenic variations.

Antigenic Variability

Four major antigenic sites have been located on the hemagglutinin monomer in avian flu virus. Evidence suggests that single amino acid substitutions within these four regions result in the ability of flu virus to escape immune surveillance and to spread worldwide every year. In addition to minor changes in the antigenic regions, major changes in the antigenic regions have produced the extremely virulent strains that caused the lethal flu pandemics of 1957 and 1968 (Bullough et al, 1994, Carr and Kim, 1994, Wiley *et al.*, 1981 and Wilson *et al.*, 1981).

HIV gp120

The spike protein of Human Immunodeficiency Virus 1 (HIV-1) responsible for binding to host cell receptor molecules is glycoprotein 120 (gp120), the core of which is shown at left in a complex with a portion of a host cell receptor protein (CD4) and an anti-gp120 antibody Fab fragment (Kwong, *et al.*, 1998). It has been observed that that CD4 binding induces significant conformational changes in gp120. Most of the envelope protein gp120 surface is hidden from circulating antibodies by glycosylation or by steric exclusion as gp120 (and gp41) form trimers (Wyatt, *et al.*, 1998). Conformational changes also provide means of escape. The conformation of gp120 prior to binding CD4 may display side-chain variability, employing the chameleon-like ability of HIV to change its molecular recognition profile. Also, since the CD4 binding pocket is recessed, antibodies may not see this important antigenic feature.

TYPES OF ANTIGEN

Based on antigenic stimulus for the production of antibody, antigens can be classified as,

T cell dependent antigens and T cell independent antigens.

(i) **T cell dependent antigens:** The antigens that could able to stimulate antibody production with the cooperation of T cells are called 'T cell dependent antigens'. Their molecular structure is complex, highly immunogenic, susceptible to phagocytosis and do not cause tolerance readily. They induce the production of IgM, IgG, IgA and IgE type of immunoglobulins. Before the production of antibody, antigens require processing and metabolism, this is followed by immunological memory.

(ii) **T cell independent antigens:** Some antigens that can directly stimulate antibody production by B cells without the cooperation of T cells are called T independent antigens. Their molecular structure is simple, with only few repeated epitopes (e.g., *Streptococcus pneumoniae*-capsular polysaccharide and LPS of Gram negative bacteria). Their immune response is dose dependent with the production of IgM and IgG3 type of antibody and such antigens induce immunological tolerance. They do not require antigen processing by macrophages and are metabolized very slowly. They do not produce immunological memory.

ANTIGEN RECEPTORS

B cell and T cell have have surface receptors for antigen. Each cell has thousands of receptors of a single specificity for an epitope. T cell receptors (TCR) enable the cell to bind along with additional signals, to be activated by and respond to an epitope presented by another cell called the Antigen-presenting cell (APC). The Antigen receptor on most T cells is a heterodimer made up of two transmembrane polypeptides designated as alpha and beta chains. B-cell receptors (BCR) enable the cell to bind along with additional signals, to be activated by and respond to an epitope on molecules of a soluble Antigen. This results in B cell secreting vast numbers of a soluble form of its receptors which are antibodies.

Antigen receptor genes: Each chain of a BCR or TCR is encoded by a gene that is assembled from separate gene segments during the differentiation of the cell. The resulting gene is transcribed into a mRNA to be translated into one chain of the receptor. The number of gene segments from which the variable regions are constructed is sufficiently large that both B cells and T cells can generate as many as 10^{10} different antigen-binding sites.

SUMMARY

Immunogenicity and immunoreactivity are two important attributes of an antigen. Specific antibody combining site is called epitope or antigenic determinant. Molecular weight, foreignness, chemical nature, susceptibility to tissue enzymes directly influence the antigenicity of a given substance. Plant Antigens or plantigens are plant substances which cause human or animal production of antibodies. Plantibodies or plant antibodies are the human or animal antibodies made by and in Transgenic Plants. Fungal antigens possess a variety of proteins or carbohydrates or lipids moieties as their virulence marker. Antibody-mediated immune (AMI) response and cell mediated immune (CMI) response are the protective defense observed by the host against against extracellular bacteria. For many viral diseases a long lasting immune response has been observed. Antigenic variations are common among virus and antigenic variability widely observed in avian flu virus and in HIV. T cell dependent antigens and T cell independent antigens have been classified based on antigenic stimulus for the production of antibody.

REFERENCES

Bullough, P.A., Hughson, F.M., Skehel, J.J. and Wiley, D.C. (1994) Structure of influenza haemagglutinin at the pH of membrane fusion. *Nature* 371, 37-43.

Carr, C.M. and Kim, P.S. (1994) Flu virus invasion: halfway there. *Science* 266 : 234-236.

Denikus, N., Orfaniotou, F. Wulf, G., Lehmann, P.F., Monod, M. and Reichard, U. (2005) Fungal antigens expressed during invasive aspergillosis. Infection and Immunity 73 : 4704-4713.

Elsevier Science (2001) Trends in Plant Science. Elsvier Science Publication 6 : 220.

http://www.agracetus.com/

http://www.antibodyresource.com/

http://www.croptech.com/

http://www.planetbiotechnology.com/

Kwong, P.D., Wyatt, R., Robinson, J., Sweet, R.W., Sodroski, J. and Hendrickson, W.A. (1998) Structure of an HIVgp120 envelope glycoprotein in complex with the CD4 receptor and a neutralizing human antibody. Nature 393 : 648-659.

Ponton, J. and Jones, J.M. (1986) Identification of two germ-tube-specific cell wall antigens of Candida albicans. Infect. Immuno. Dec. 54 : 864-868.

Richard, A. G., Thomas, J. K. and Barbara, A. O. (2000) Kuby immunology. 4th edition, W. H. Freeman and Co., New York.

Todar, K. (2002) Mechanisms of bacterial pathogenicity: Bacterial Defense against the Host Immune Responses. In : Text book of Bacteriology.

Wiley, D.C., Wilson, I.A., Skehel, J.J. (1981) Structural identification of the antibody-binding sites of Hong Kong influenza hemagglutinin and their involvement in antigenic variation. *Nature* 289 : 373-378.

Wilson, I.A., Skehel, J.J. and Wiley, D.C. (1981) Structure of the haemagglutinin membrane glycoprotein of influenza virus at 3Å resolution. *Nature* 289 : 366-373.

www.textbookof bacteriology.net/

Wyatt, R., Kwong, P., Desjardins, E., Sweet, R.W., Robinson, J., Hendrickson, W.A. and Sodroski, J. (1998) The antigenic structure of the HIV gp120 envelope glycoprotein. Nature 393, 705-711.

Suggested Readings

1. Gayathri Devaraja (2002) Antigen characterization, epitope analysis and detection of *Aphanomyces invadans*, the causative agent of epizootic ulcerative syndrome by monoclonal antibodies. Ph.D. Thesis. Mangalore University, India.
2. Richard, A. G., Thomas, J. K. and Barbara, A. O. (2000) Kuby immunology. 4th edition, W. H. Freeman and Co., New York

REVIEW QUESTIONS

Long Questions

1. Describe the specificities of an antigen.
2. Describe plantigens.
3. Describe types of antigen.
4. Describe antigen variability in bacteria.
5. Give an account on viral antigens.

Multiple Choice Questions

1. Hapten is a
   (a) Tolerogen (b) Antigen
   (c) Allergen (d) Antibody
2. Heat labile antibody is
   (a) IgG (b) IgE
   (c) IgD (d) IgM
3. An macrophage killing bacteria
   (a) Streptococcus (b) Staphylococcus
   (c) Gonococcus (d) Brucella
4. Avian flu virus antigen is in
   (a) Haemagglutinin (b) Neuraminidase
   (c) Glycolipid (d) Proteinase
5. The spike protein of Human Immunodeficiency Virus 1 (HIV-1) responsible for binding to host cell receptor molecules
   (a) gp120 (b) gp41
   (c) p17 (d) p24

6. Mention which of the following statements regarding hapten-carrier complex is true
   (a) Haptens are large protein molecules such as bovine serum albumen.
   (b) When this complex injected to an experimental animal, induce antibody production specific to hapten.
   (c) Carriers are required if CMI is required.
   (d) Carriers are smallest molecules such as streptomycin or dinitrosalicylic acid.

Answers

1. (b) 2 (b) 3 (d) 4. (a) 5. (a) 6. (b)

CHAPTER 25

Humoral Immunity

Janak Ahi
Dept. of Zoology, Dr. H.S. Gour Vishwavidyalaya, Sagar–470 003
E-mail : j_ahi@yahoo.co.in

Two types of immunity exist, innate and adaptive. Innate immunity , a first line of defense, is furnished by barriers such as skin, tears, saliva, and mucus, and the tissue inflammation that occurs after injury or infection. Adaptive immunity develops specific defenses against an invader that can be invoked whenever this particular intruder attacks again. The immune system responds to surface structures of the invading organism called antigens. There are two types of adaptive immune responses: humoral and cell mediated. In humoral immune responses antibodies appear in the body fluids and stick to and destroy antigens. The response is to toxic substances outside of the cell. In the cell-mediated immune response cells that can destroy other cells become active (T-cells). They destroy disease infected cells or cells making mutant forms of normal molecules. Humoral immunity is the aspect of immunity that is mediated by secreted antibodies, produced in the cells of the B lymphocyte lineage (B cell). Secreted antibodies bind to antigens on the surfaces of invading microbes (such as viruses or bacteria), which flags them for destruction.Thus the B-cells are responsible for humoral immunity. They arise from a separate population of stem cells of the bone marrow than that which gives rise to T-cells. These cells undergo multiplication and processing in lymphoid tissue elsewhere than in the thymus gland. In birds, the lymphoid tissue concerned has been located in the gut and called the bursa of Fabricius. In humans, the site is unknown although there is some evidence to suggest that such processing occurs in the bone marrow itself or in the fetal liver. Lymphocytes processed in this way are called B-lymphocytes after the bursa.

B-cells, like T-cells, have surface receptors which enable them to recognize the appropriate antigen, but do not themselves interact to neutralize or destroy the antigen. On recognition of the antigen they take up residence in secondary lymphoid tissue and proliferate to form daughter lymphocytes, processed in the same way as themselves. These B-cells then develop into short-lived plasma cells. The plasma cells produce antibodies and release them into the circulation at the lymphnodes. Some of the activated B-cells do not become plasma cells instead they turn into memory cells which continue to produce small amounts of the antibody long after the infection has been overcome. As with T-cells millions of B-cells are produced each with a different antigen binding specificity. If the B-cell comes into contact with the specific type of antigen to which it is targeted it divides rapidly to form a clone of identical cells. This antibody circulates as part of the gamma globulin fraction of the blood plasma.

Should the same antigen enter the body again this circulating antibody acts quickly to destroy it. At the same time memory cells quickly divide to produce new clones of the appropriate type of plasma cell.

ANTIBODY PRODUCTION

Humoral immunity refers to antibody production, and all the accessory processes that accompany it: Th2 activation and cytokine production, germinal center formation and isotype switching, affinity maturation and memory cell generation. It also refers to the effector functions of antibody, which include pathogen and toxin neutralization, classical complement activation, and opsonin promotion of phagocytosis and pathogen elimination.

B cells need two signals to initiate activation. Most antigens are T-dependent, meaning T cell help is required for maximal antibody production. With a T-dependent antigen, the first signal comes from antigen cross linking BCR and the second from the Th2 cell. T dependent antigens contain protein so that peptides can be presented on B cell Class II MHC to Th2 cells, which then provide co-stimulation to trigger B cell proliferation and differentiation into plasma cells. Isotype switching to IgG, IgA, and IgE and memory cell generation occur in response to T-dependent antigens. T-dependent responses require that B cells and their Th2 cells respond to epitopes on the same antigen. T and B cell epitopes are not necessarily identical; for example, T cells respond well to internal viral proteins while B cells produce neutralizing antibodies to viral coat proteins. (Once virus-infected cells have been killed and unassembled virus proteins released, B cells specific for internal proteins can also be activated to make opsonizing antibodies to those proteins.) Attaching a carbohydrate to a protein can convert the carbohydrate into a T-dependent antigen; the carbohydrate-specific B cell internalizes the complex and presents peptides to Th2 cells, which in turn activate the B cell to make antibodies specific for the carbohydrate. This linked recognition of T and B cells was first discovered with work on haptens. Some antigens are T-independent, meaning they can deliver both the antigen and the second signal to the B cell. Mice without a thymus (*nude* or athymic mice) can respond to T-independent antigens. Many bacteria have repeating carbohydrate epitopes that stimulate B cells to respond with IgM synthesis in the absence of T cell help.

B cell activation is initiated by the cross-linking of BCR. Antigen must have at least two identical epitopes situated so that they can cross-link B cell surface Ig to activate B cells. Aggregation of BCR-Igalgb complexes promotes phosphorylation of ITAMs by cytoplasmic tyrosine kinases Fyn, Blk, and Lyn, to initiate a second messenger cascade. Once antigen is bound, it is internalized with its BCR; antigen can then be degraded (processed), combined with Class II MHC, and presented to effector Th2 cells, which must bind specific antigen in order to provide help.

Effector Th2 cells bind B cells via CAMs and then specific peptide on Class II MHC with TCR. T cell CD40 ligand (CD40L) binds B cell CD40 and sends the co-stimulatory signal for B cell activation. The Th2 cell reorganizes its membrane molecules and golgi to concentrate CD40L binding and cytokine secretion towards the specific B cell to maintain the antigen-specificity of the response. Cytokines, predominantly IL-4, IL-5, and IL-6, signal the B cell to divide and differentiate into antibody-secreting plasma cells. In a primary immune response, IgM is secreted before isotype switching occurs and is the predominant isotype produced.

The cytokines direct isotype switching in activated B cells. Although Th1 cells promote cellular immunity and do not initiate antibody formation, they can induce isotype switching to certain isotypes. Depending on which lymphokines are in highest concentration near a given B cell, it can follow several differentiation pathways (these studies have been done in the mouse). Th1 cells secrete mainly IL-2 and IFNγ, while Th2 produce mainly IL-4, IL-5, and IL-6. IL-2 stimulates production of J chain. IFNg stimulates switching to IgG_{2a}. IL-4 stimulates switching to IgE or IgG_1, while IL-5 stimulates switching to IgA. IL-6 promotes antibody secretion. Th cells are thought to produce certain cytokines in response to their location and antigen density on the APC. For example, Th2 cells in mucosal tissues produce cytokines that promote isotype switching to α chain. Each antibody isotype has specific effector functions in humoral immunity.

Isotype switching requires DNA recombination. It increases the functional diversity of Ig molecules but does not affect their antigen-binding specificities. A few activated B cells differentiate into short-lived plasma cells that migrate directly to the medullary cords and begin secreting IgM and later IgG. Most activated B cells go to the B cell areas of the lymph node, called primary follicles, which are present in unstimulated peripheral lymphoid tissues, in the fetus, and in germ free animals. They contain mainly follicular dendritic cells (FDC) and IgM^+IgD^+ B cells (naïve resting B cells). B cell blasts (dividing cells) form secondary follicles (germinal centers) within a few days of antigen stimulation. Germinal centers are sites of intensive B cell proliferation and are surrounded by the Th2 cells that activated the B cells and migrated with them. By a week after initial antigen contact, many germinal centers are seen. Athymic mice have few germinal centers following immunization.

Isotype switching, affinity maturation, and generation of memory B cells occur late in a primary immune response. Antigen-binding specificity can be altered by somatic hypermutation, which occurs in rapidly dividing B cells in the germinal centers. Mutation occurs preferentially in the CDR (hypervariable regions) of H and L chains at a rate high enough that about half of B cells undergo mutation of their Ig antigen-binding regions. B cells which remain specific for the stimulating antigen are selected by binding antigen held on FcR and complement receptors on follicular dendritic cells (FDC), which do not express Class II MHC or act as APC for T cells. Only B cells which bind antigen with high enough affinity continue to secrete antibody and differentiate into plasma cells; low affinity B cells undergo apoptosis.

Plasma cells are antibody-producing cells that no longer divide or respond to antigen. They are larger than B cells and have more ribosomes, endoplasmic reticulum, and golgi, but no membrane Ig. Some survive only a few weeks, while others continue to produce antibody for longer periods, providing rapid protection against re-infection. Other B cells become memory cells, which can be rapidly re-stimulated by antigen and are present in higher frequency than the naïve resting B cells (more of the latter continue to be produced by the bone marrow). Memory B cells are functionally and physically distinguishable from naïve B cells.

They often have membrane IgG, IgA, or IgE and higher levels of ICAM-1 and CR then naïve B cells, and are thought to live longer. Low levels of antigen may remain on FDC for years, so that B cells may be continually activated at low levels to replenish memory cell populations.

ISOTYPE DISTRIBUTION AND FUNCTION

Microbes usually enter the body through the epithelial cells of the respiratory, digestive, or genital tracts or through skin broken by a scrape, cut, insect bite or hypodermic needle. Once inside, the microbe may begin replicating locally or get into the circulation and move throughout the body. Bacterial toxins also travel from the initial infection site. The Fc regions of the Ig isotypes allow them to bind Fc receptors and cross tissue barriers to reach pathogens throughout the body.

IgM is secreted first in a primary response. No somatic hypermutation has yet occurred, so it is low affinity antibody. IgM avidity is high, however, because it is a pentamer, and IgM fixes complement very efficiently to promote inflammation and pathogen lysis. Because IgM is so large, it cannot enter the tissues very efficiently; but it is effective in controlling pathogens in the circulation. Once isotype switching occurs, IgG predominates in serum and in tissues. IgG both neutralizes pathogens and their toxins and opsonizes them for phagocytosis by neutrophils and macrophages. IgG can also activate complement on the pathogen surface once concentrations are high enough for two IgG molecules to bind nearby epitopes.

FCR$^+$ ACCESSORY CELLS

In order to act as opsonins or to activate cells, antigen-bound antibodies bind to Fc receptors. Several different FcR have been identified on the membranes of granulocytes, dendritic cells, B cells, NK cells, and mast cells. Some are there constitutively, while expression of others is induced during an immune response. The isotype specificity of each FcR, its binding affinity, and its structure all influence its function. Free antibody (unbound to antigen) binds to FcR with a very low affinity except in the case of IgE. Antibody aggregated by antigen, with several nearby Fc regions, binds much more tightly to FcR. Binding of ligand to FcR acts through signal transduction complexes to alter gene expression.

SUMMARY

Development of B cells involves antigen-independent maturation in the bone marrow and antigen-dependent activation and differentiation of mature B cells in the periphery, resulting in antibody-secreting plasma cells and memory B cells. During B-cell maturation, sequential Ig-gene rearrangement transforms the pro-B cell, the earliest distinctly B-lineage cell, into an immature B cell, which expresses mIgM with a single antigenic specificity. Proliferation and development of pro-B cells into pre-B cells depends on the microenvironment provided by bone-marrow stromal cells. Pre-B cells express the heavy chain and a surrogate light chain associated with the Ig-a/Ig-b heterodimer. Although the function of this pre- B cell receptor is not entirely clears, it is required for maturation to continue to the immature B-cell stage. Further development, including changes in RNA processing, yields mature B cells, expressing both IgM and IgD. The antigen-induced activation and differentiation of mature B cells occur in the periphery, generating the humoral response. The B-cell response to soluble protein antigens requires involvement of TH cells; in contrast, the response to certain bacterial cell wall products (e.g., LPS) and highly polymeric molecules does not require T_H cells. The nature of the response to thymus-dependent and thymus-independent antigens differs considerably.

Activation of mature B cells involves signals generated by membrane events and, in some cases, cytokines. These signals are transduced by intracellular pathways, ultimately leading to changes in the expression of specific genes. B and T-cell activation share many parallels, including: compartmentalization of function within receptor sub-units; activation by membrane-associated protein tyrosine kinases; activating dephosphorylations by CD45; assembly of large signaling complexes with protein-tyrosine-kinase activity; and recruitment of several signal- transduction pathways. In both cell types, changes in gene expression are a consequence of these signal transduction processes. The B-cell coreceptor can intensify the activating signal resulting from crosslinkage of mIg. This may be particularly important during the primary response to low concentrations of antigen, because many naive B cells express mIg with low affinity for antigen. Activation induced by thymus-dependent antigens requires the delivery of contact-dependent help. The formation of T-B cell conjugates facilitates the interaction between CD40 on B cells and CD40L on activated T_H cells as well as the co-stimulation of T_H cells by interaction of B7 on the B cell with CD28 of its partnering T cell. The CD40/CD40L interaction is essential for B-cell survival the formation of germinal centers, the generation of memory cell populations and somatic mutation.

REFERENCES

Benschop, RJ and JC Cambier (1999) B cell development : signal transduction by antigen receptors and their surrogates.*Curr. Opin. Immunol.* 11 : 143.

Berek, C (1999) Affinity Maturation. In *Fundamental Immunology,* 4th ed., edited by WE Paul. Lippincott-Raven, Philadelphia and New York.

Clevers, HC (1996) Transcriptional control of lymphoid development: lessons from gene targeting. *Immunology Today* 17 : 336.

Kurosaki, T (1997) Molecular mechanisms in B cell antigen receptor signaling. *Curr. Opin. Immunol.* 9 : 309.

Reth, M (1994) B cell antigen receptors. *Curr. Opin. Immunol.* 6 : 3.

Stavenezer, J (1999) Immunoglobulin class switching. *Curr. Opin. Immunol.* 8 : 199.

Suggested Readings

Cooper, EL, Langlet, C, Bierne, J, eds. (1987) *Developmental and Commparative Immunology,* New York. AR Liss.

Manning, MJ, Tatner, RJ, *Comparative Immubobiology.,* Glasgow : Blackie.

Solomon, JB (1986) Invertebrate receptors and recognition molecules involved in immunity and determination of self and non-self. In : *Receptors in Cellular Recognition and Developmental Processes.* London : Academic Press.

Sissons, JG, Oldstone, MBA (1980) Antibody-mediated destruction of virus-infected cells. *Adv Immunol* 31 : 1.

REVIEW QUESTIONS

Long Questions

1. Discuss the origin of competence and progression signals require for the activation and proliferation of B cells induced by soluble protein antigens.
2. Describe the result of signal transduction in B cells during antigen induced activation and differentiation.

Multiple Choice Questions

1. T independent antigens do not
   (a) bind to BCR. (b) get presented on MHC Class II.
   (c) have repeating epitopes. (d) induce B cell proliferation.
2. The humoral immune response to T-independent antigens includes production of
   (a) IgA. (b) IgM.
   (c) memory B cells. (d) memory T cells.
3. T-independent antigens are often
   (a) components of self cell membranes.
   (b) polyclonal B cell activators.
   (c) repeating peptide epitopes
   (d) too small to be phagocytosed and presented.
4. Before a B cell can receive T cell help, the B cell must
   (a) express membrane B7.
   (b) express membrane CD40L.
   (c) express membrane IFNg receptors.
   (d) process and present peptide on Class II MHC.
5. Antigen-binding B cells entering the secondary lymphoid organs initially go to the
   (a) B cell areas where they can bind antigen presented by follicular dendritic cells.
   (b) B cell areas where they can process and present antigen to T cells.
   (c) plasma cell areas where they can secrete antibody.
   (d) T cell areas where they can find specific helper T cells.
6. Germinal centers are
   (a) areas of lymphocyte generation in the bone marrow.
   (b) common in unimmunized mice.
   (c) sites of rapid antigen-induced B cell division in the lymph nodes and spleen.
   (d) surrounded by naïve T cells waiting to be activated.
7. In order for T cells to provide help to B cells, T cell and B cell epitopes must be
   (a) covalently linked. (b) identical.
   (c) non-identical. (d) peptides.

8. B cell isotype switching is important for
   (a) increasing the avidity of the antibody.
   (b) providing antibodies of many different antigen specificities.
   (c) providing antibodies which can perform different effector functions.
   (d) signaling the B cells to become activated by follicular dendritic cells
9. Affinity maturation of the humoral immune response is due to
   (a) continued stimulation of B cells by high levels of antigen on the FDC.
   (b) DNA recombination by products of the RAG genes.
   (c) isotype switching.
   (d) positive selection of B cells with the highest affinity for antigen.
10. Plasma cells
    (a) are all very long-lived.
    (b) divide and differentiate into memory B cells.
    (c) produce most of their antibody at the site of infection.
    (d) None of the above are true.
11. Which of the following is NOT a similarity between the cellular and humoral immune responses?
    (a) Antigen-specific lymphocytes undergo clonal selection and expansion.
    (b) Cytokine signals promote effector cell differentiation.
    (c) Memory cells are generated.
    (d) Receptor isotype switching occurs.
12. Humoral immunity involves all of the following except
    (a) antibody-dependent cell-mediated cytotoxicity.
    (b) antibody secretion by plasma cells.
    (c) B cell activation by antigen plus cytokines.
    (d) macrophage activation by Th1 cells
13. Different Ig isotypes are found in different body locations because they
    (a) are secreted in different tissues.
    (b) bind to different FcR that allow them to cross tissue barriers.
    (c) have different addressins.
    (d) have different affinities for antigen.
14. An inactivated toxin used in a vaccine is called a(n)
    (a) adjuvant. (b) attenuated vaccine.
    (c) hemagglutinin. (d) toxoid.
15. Successful immune responses to bacteria which resist phagocytosis because of a polysaccharide capsule depends on the production of
    (a) active complement anaphylatoxins.
    (b) armed effector CTL.
    (c) neutralizing antibodies.
    (d) opsonizing antibodies.

16. Successful immune responses to bacterial toxins depend on the production of
    (a) active complement. (b) armed effector CTL.
    (c) neutralizing antibodies. (d) opsonizing antibodies.
17. In order for NK cells to do ADCC, they must bind
    (a) antibodies to virus proteins expressed on infected cell membranes
    (b) B7.
    (c) CD16.
    (d) toxic oxygen radicals.
18. NK cells kill their targets in ADCC using
    (a) complement-mediated lysis. (b) Fas and FasL.
    (c) oxidative burst. (d) perforins and granzymes.
19. Mast cells release their granule contents to stimulate inflammation in response to
    (a) antigen binding to IgE on mast cell FceR.
    (b) antigen binding to IgG on mast cell FcgR.
    (c) antigen binding to mast cell antigen receptors.
    (d) antigen binding to mast cell FceR.

Answers

1. (b)	2. (b)	3. (b)	4. (d)	5. (d)	6. (c)
7. (a)	8. (c)	9. (d)	10. (d)	11. (d)	12. (d)
13. (b)	14. (d)	15. (d)	16. (c)	17. (a)	18. (d)
19. (a)					

CHAPTER 26

Cell-mediated Immunity

Janak Ahi
Deptt. of Zoology, Dr. H.S. Gour Vishwavidyalaya, Sagar-470 003
E-mail: j_ahi@yahoo.co.in

Cell-mediated immunity is an immune response that does not involve antibodies but rather involves the activation of macrophages and natural killer cells, the production of antigen-specific cytotoxic T-lymphocytes, and the release of various cytokines in response to an antigen. Cellular immunity protects the body by activating antigen-specific cytotoxic T-lymphocytes that are able to lyse body cells displaying epitopes of foreign antigen on their surface, such as virus-infected cells, cells with intracellular bacteria, and cancer cells displaying tumor antigens; activating macrophages and natural killer cells, enabling them to destroy intracellular pathogens; and stimulating cells to secrete a variety of cytokines that influence the function of other cells involved in adaptive immune responses and innate immune responses.

T CELL ACTIVATION

Adaptive T cell-mediated immunity is driven by activation of T cells: cytotoxic T cells activated by endogenous antigen to kill infected cells, and helper T cells activated by exogenous antigen to stimulate macrophage killing of endosomal pathogens. The pathogens targeted by cellular immunity are protected from antibody and complement binding by their intracellular locations. Adaptive humoral immunity also usually involves T cell activation to produce cytokines that stimulate B cell antibody synthesis. Naïve resting T cells are stimulated by antigen peptide presented on Class I MHC to cytotoxic CD8 T cells or Class II MHC to helper CD4 T cells, along with co-stimulatory signals from APC, to proliferate and differentiate into clones of fully activated effector T cells. Only professional APC (dendritic cells, macrophages, and B cells) have both Class I and Class II MHC and can deliver co-stimulatory signals. Activated T cells perform their effector functions when they encounter MHC-presented peptide on their target cells.

Table 1. Effector Cells in Adaptive Immunity

	Effector T Cell	Pathogen Location	Antigen Presentation	Target Cell Action
Cellular Immunity	Tc CD8 cytotoxic	Cytoplasm	Infected cell MHC I	Infected cellapoptosis
	Th1 CD4 inflammatory	Macrophage vesicles	Macrophage MHC II	Macrophage activation to kill pathogen
Humoral Immunity	Th2 CD4 helper	Extracellular	APC MHC II	B cell antibody production

Naïve mature T cells constantly recirculate between secondary lymphoid organs and blood and lymphatic circulations. Antigen activation of T cells occurs in the secondary lymphoid organs: Peyer's patches or tonsils for pathogens infecting mucosal tissues (respiratory, digestive, and urogenital tracts), draining lymph nodes for pathogens in the peripheral tissues, and the spleen for pathogens which are in the blood. Traffic through these secondary lymphoid organs takes T cells past APC which are likely to be presenting foreign antigen if any is present, since both antigen and leukocytes collect in secondary lymphoid organs. T cell "sampling" of self peptide: MHC complexes on APC when no infection is present serves as a survival signal for mature T cells once they leave the thymus. Signaling between T cells and APC also depends on CAMs. Even if a T cell specific for a peptide were to encounter an APC with that peptide on the correct MHC, the probability of those molecules being close enough on both cell surfaces for binding to occur would be very small. Numerous CAMs on both cells allow them to adhere nonspecifically for a time so that if an antigen-specific interaction can occur, it will. As the T cell and APC adhere, movement of TCR and peptide:MHC complexes in the plane of the cell membranes brings them close enough to bind one another. Once TCR has bound peptide + MHC, T cell LFA-1 changes its conformation to bind even more tightly to the APC for the prolonged contact required for T cell activation. This contact may last several days, during which time the T cell divides and its progeny also adhere to the APC and receive activation signals.

To activate T cells, APC must deliver two distinct signals, one to TCR via peptide on MHC, and a second co-stimulatory signal. The professional APC that can provide co-stimulation are dendritic cells (DC), macrophages, and B cells. Of these cell types, DC deliver the best co-stimulation and are probably responsible for activating most naïve T cells. Signals received

Table 2. Key Adhesion Molecules in T Cell Activation

Ig Super family Adhesion Molecule	Tissue Distribution	Integrin Ligand
CD2 (LFA-2)	T cells	LFA-3 on APC
ICAM-1 (CD 54)	Activated endotheliumdendritic cells T cells	LFA-1 on T cellsMAC-1 on macrophages
ICAM-2 (CD102)	Resting endothelium	LFA-1 on T cells
ICAM-3 (CD50)	Naïve T cells	DC-SIGN on dendritic cells
LFA-3 (CD58)	LymphocytesAPC	CD2 on T cells
VCAM-1	Activated endothelium	VLA-4 on activated T cells

through CD28, when received with antigen signals, activate a T cell. Some antibodies to CD28 can provide co-stimulatory signals, while antibodies to B7 block the ability of the APC to activate T cells. When a T cell becomes activated it expresses CTLA-4 a molecule very similar to CD28 and also able to bind B7, alongside its CD28 molecules. CTLA-4 binds B7 about 20 times more avidly than does CD28, and CTLA-4 binding to B7 sends an inactivation signal to the T cell. The immune response, therefore, depends on the "push" of antigen stimulation to activate T cells into effectors and the "pull" of negative signals that curtail T cell responses. These combinations of activation and inactivation signaling mechanisms are common in the immune system and are there to control a very powerful response that can damage the body if not restricted. Mice born with a defective CTLA-4 gene suffer from over-proliferation of lymphocytes. Once immature DC have picked up antigen, they migrate to the peripheral lymphoid organs. There they enter the T cell areas and lose their ability to take up antigen, but express high levels of Class I and Class II MHC, B7, and the CAM DC-SIGN, making them excellent APC. They also secrete a chemokine (DC-CK) that attracts T cells. The signals that send DC to the lymphoid organs and change their phenotype are not known, although LPS signaling through Toll-like receptor TLR-4 (as described for macrophages in Innate Immunity) also occurs for DC. Langerhans cells in the skin are an example of immature dendritic cell which become active APC in the lymph nodes.

Macrophages are very active phagocytes, especially of bacterial pathogens; they do not take up soluble antigens (proteins) well. They are found throughout the body as resident macrophages in tissues, and circulating monocytes can enter tissues in response to infection and become macrophages. Resting macrophages express little Class II MHC and B7, but increase expression of these molecules once they have phagocytosed bacteria due to signaling through receptors such as TLR-4. Macrophages migrate to secondary lymphoid organs from infection sites to present antigen to T cells.

Macrophages in the red pulp of the spleen and Kupffer cells in the liver continuously phagocytose dead host cells and, if they were to present those cells to T cells, might induce autoimmunity. However these macrophages express little B7 and MHC and are not stimulated to do so by binding and phagocytosing host cells. They are also not in locations frequented by T cells.

B cells are unique among APC because they use their membrane Ig to specifically bind and internalize soluble antigens. Antigen cross-linking of BCR stimulates endocytosis and antigen presentation by B cells. B cells constitutively have high levels of Class II MHC. Co-stimulatory molecule expression increases following contact with some bacterial carbohydrates and LPS, making it unlikely that B cells will activate T cells to respond to self peptides in the absence of infection. B cells enter secondary lymphoid organs by passing through the T cell areas, where they are likely to contact T cells. B cells activate T cells at lower antigen concentrations than do macrophages. They present predominantly bacterial and insect toxins and allergens. The importance of B cells in activating naïve T cells is uncertain. Because specific B cells are not initially present in high frequencies, DC are probably responsible for more T cell activation during the early stages of a primary response. Once T cells are activated by peptide + MHC and receive co-stimulation, they begin to synthesize the a chain of IL-2 receptor (IL-2R, also called Tac after an early monoclonal antibody specific for it) and to secrete IL-2. Tac increases the affinity of the IL-2R, so that lower concentrations of IL-2 can signal the activated T cell to begin clonal proliferation and differentiation into an armed effector cell. Resting T cells express only

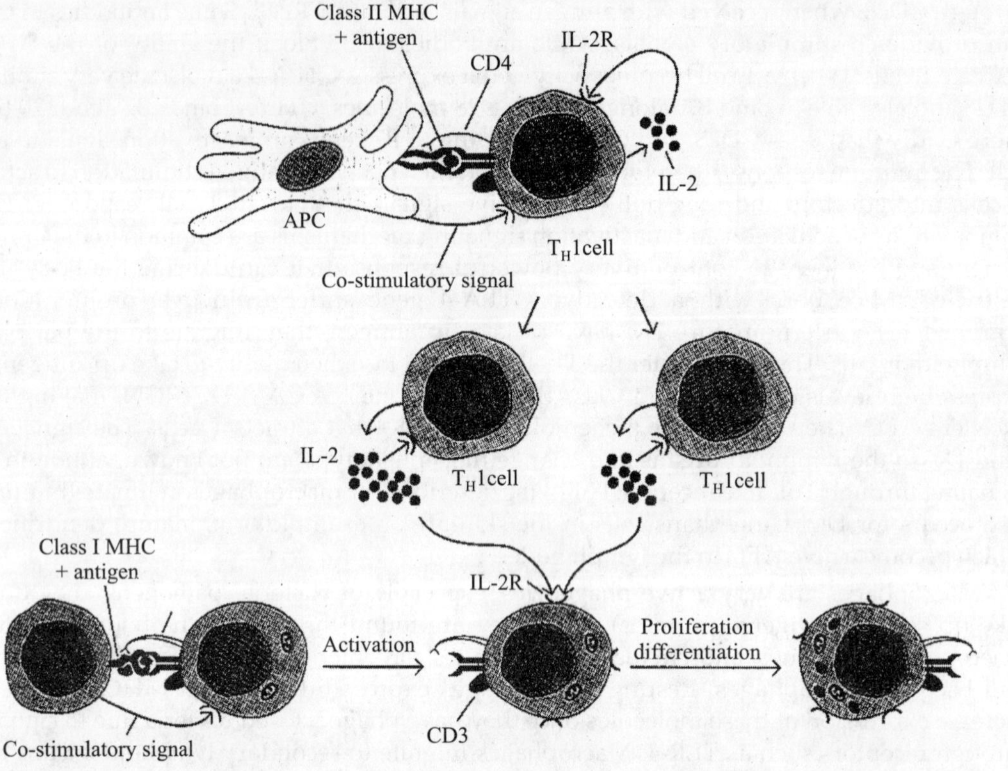

Fig. 1. Interaction of antigen-class 1 MHC complexes with appropriate target cell to generate effector CTL.

the β and γ chains of IL-2R and bind much less tightly to IL-2. The expression of more and higher affinity cytokine receptors by activated cells compared to resting cells keeps the immune response antigen-specific, even though the same cytokines are made in response to many different antigens. The response of T cells to a cytokine they secrete is called an autocrine response (Figure 1).

PROPERTIES OF ARMED EFFECTOR T CELLS

After 4-5 days of rapid clonal expansion, T cells differentiate into armed effector cells that no longer need co-stimulation to perform their effector functions: cytotoxicity or cytokine secretion. T cells must, however, continue to bind peptide + MHC in order to kill or activate their target cells. Three classes of armed effector cells are specialized to deal with three different kinds of pathogens. CD8 CTL kill infected cells displaying cytosolic pathogen peptides on Class I MHC. CD4 Th1 cells activate macrophages with persistent vesicular pathogens whose peptides are displayed on Class II MHC. They also activate B cells to produce opsonizing antibodies. CD4 Th2 cells activate B cells that have used their membrane Ig to internalize specific antigen and display peptides on Class II MHC.

CYTOTOXIC T CELLS

Cytotoxic T cells (CTL) mediate antigen-specific, MHC-restricted cytotoxicity and are important for killing intra-cytoplasmic parasites that are not accessible to secreted antibody or to phagocytes. Examples include all viruses, rickettsias (causes of Rocky Mountain spotted fever and typhus), some obligate intracellular bacteria (*Chlamydia*), and some protozoan parasites which export their proteins from macrophage vesicles to the cytoplasm (*Toxoplasma gondii*). The only way to eliminate these pathogens is to kill their host cells.

Effector CTL bind their targets first via nonspecific adhesion molecules such as LFA-1 (CD 11a/18), then with specific TCR. TCR-peptide-MHC-CD8 binding reorients the CTL cytoskeleton to focus release of effector molecules towards the target cell. CTL induce apoptosis in their targets. Cells undergoing apoptosis undergo chromatin condensation and membrane vesicle shedding. DNA is cut into pieces in multiples of 200bp, which look like steps on a ladder (and is called a "DNA ladder") when run on a gel that separates the DNA by size.

MACROPHAGE ACTIVATION BY THI (INFLAMMATORY T) CELLS

Macrophages phagocytose and kill many pathogens using lysosomal enzymes. Some pathogens, however, are phagocytosed but evade killing by preventing fusion of lysosomes with the phagocytic vesicle or activation of the lysosomal enzymes by acidification of the phagolysosome. Some pathogens can escape the phagolysosome and live in the macrophage cytoplasm. In all these circumstances, the pathogens are protected from complement and antibodies; where the pathogens remain in vesicles, their peptides are also not presented efficiently to CD8 T cells.

Macrophage activation by armed effector Th1 cells is required to eliminate vesicular pathogens. Macrophage activation requires T cell binding to antigen peptide on macrophage Class II MHC, co-stimulation of the macrophage via CD40, and activation of the macrophage by IFNγ. Th1 membrane CD40L binds to macrophage CD40 and signals the macrophage to express receptors for IFNγ, which is synthesized *de novo* by the Th1 cell. Synthesis takes several hours, so macrophage activation is slower than CTL-mediated cytotoxicity. CD40L and IFNγ secretion are directed towards the antigen-presenting macrophage by membrane polarization; activation is generally limited to infected macrophages. IFNγ produced by CTL can activate macrophages presenting cytosolic peptides on Class I MHC. Macrophages can be sensitized to IFNγ by low levels of LPS and membrane-bound TNFα or TNFβ can substitute for the CD40L signal in macrophage activation. Activated Th2 cells express CD40L but instead of IFNγ they secrete the macrophage-inhibitory cytokine IL-10.

IFNγ activates macrophages to upregulate Class II MHC, B7, CD40, and TNF receptor expression to recruit more effector Th1 cells and to become more sensitive to CD40L and TNFα. Activated macrophages also fuse their phagosomes and lysosomes more efficiently and increase their synthesis of nitric oxide, oxygen radicals, antimicrobial peptides, and IL-12, which pushes Th cells towards the Th1 pathway.

When pathogens are present which cannot be phagocytosed, such as large parasites, activated macrophages release oxygen radicals, nitric oxide, and proteases into the extracellular fluid to kill pathogen. Excretion of these compounds also damages surrounding host tissues. Chronically infected macrophages may become resistant to activation. Granulomas, giant cells consisting of fused macrophages surrounded by activated T cells, form when intracellular pathogens cannot be eliminated. As cells in the center of the granuloma die, the dead tissue resembles cheese and is called caseous necrosis. The local inflammatory response resulting from activated Th1 cells and macrophages is called Delayed Type or Type 4 Hypersensitivity (DTH). If macrophages express Fas, they can be killed by FasL-expressing Th1 cells.

In addition to activating macrophages, Th1 cells secrete a variety of cytokines that regulate cellular immunity. IL-2 stimulates clonal proliferation of T cells. IL-3 and GM-CSF stimulate macrophage differentiation in the bone marrow. Multiple cytokines, including TNFα, TNFβ, and Macrophage Chemotactic Protein (MCP), recruit macrophages to the site of infection.

MECHANISM OF NK-CELL KILLING

Natural killer cells appear to kill tumor cells and virus-infected cells *by* a process similar to that employed by CTLs. The cytoplasm of NK cells contains numerous granules containing perforin and granzymes. Unlike CTLs, which need to be activated before granules appear, NK cells are constitutively cytotoxic, always having large granules in their cytoplasm. After an NK cell adheres to a target cell, degranulation occurs with release of perforin and granzymes at the junction. of the interacting cells. The roles of perforin and granzymes in NK-mediated killing of target cells by apoptosis thought to be similar to their roles in the CTL-mediated process.

Despite these similarities, NK cells differ from CTLs in several significant ways. First, NK cells do not express antigen-specific T-cell receptors or CD3. In addition, recognition of target cells by NK cells is not MHC restricted; that is in many cases the same levels of NK-cell activity are observed with syngeneic and allogeneic tumor cells. Moreover, although prior priming enhances CTL activity, NK-cell activity does not increase after a second injection with the same tumor cells. In other words, the NL cell response generates no immunologic memory.

ANTIBODY DEPENDENT CELL MEDIATED CYTOTOXICITY (ADCC)

Cells with cytotoxic potential which also possess Fc receptor for IgG may bind to and lyse target cells coated with IgG when antobody is specifically bound to a target cell, these receptor-bearing cells can bind to the antibody Fc region, and thus to the target cells and subsequently cause lysis of the target cell. Although these cytobody cells are nonspecific for antigen, the specificity of the antibody directs them to specific target cells. This type of cytotoxicity is referred to as among the cells that can mediate ADCC are NK cells macrophages, monocytes, neutrophils and eosinophils.

The Mechanisms of Cell-Mediated Cytotoxicity

It appears that the mechanisms involved in killing are similar whether Tc cells, MK cells or lymphoid K cells are involved and whatever receptor-target interaction is responsible (Figure 2). Cells of the myeloid series use different pathways. One mode of killing, by lymphoid cells, involves three distinct phases.

1. The cells binds to the target.
2. A Ca^{2+} dependent phase, in which the vesicle contents of the cytotoxic cell are discharged these modify the target so that it is programmed for death.
3. A late phase, when the target cell is killed.

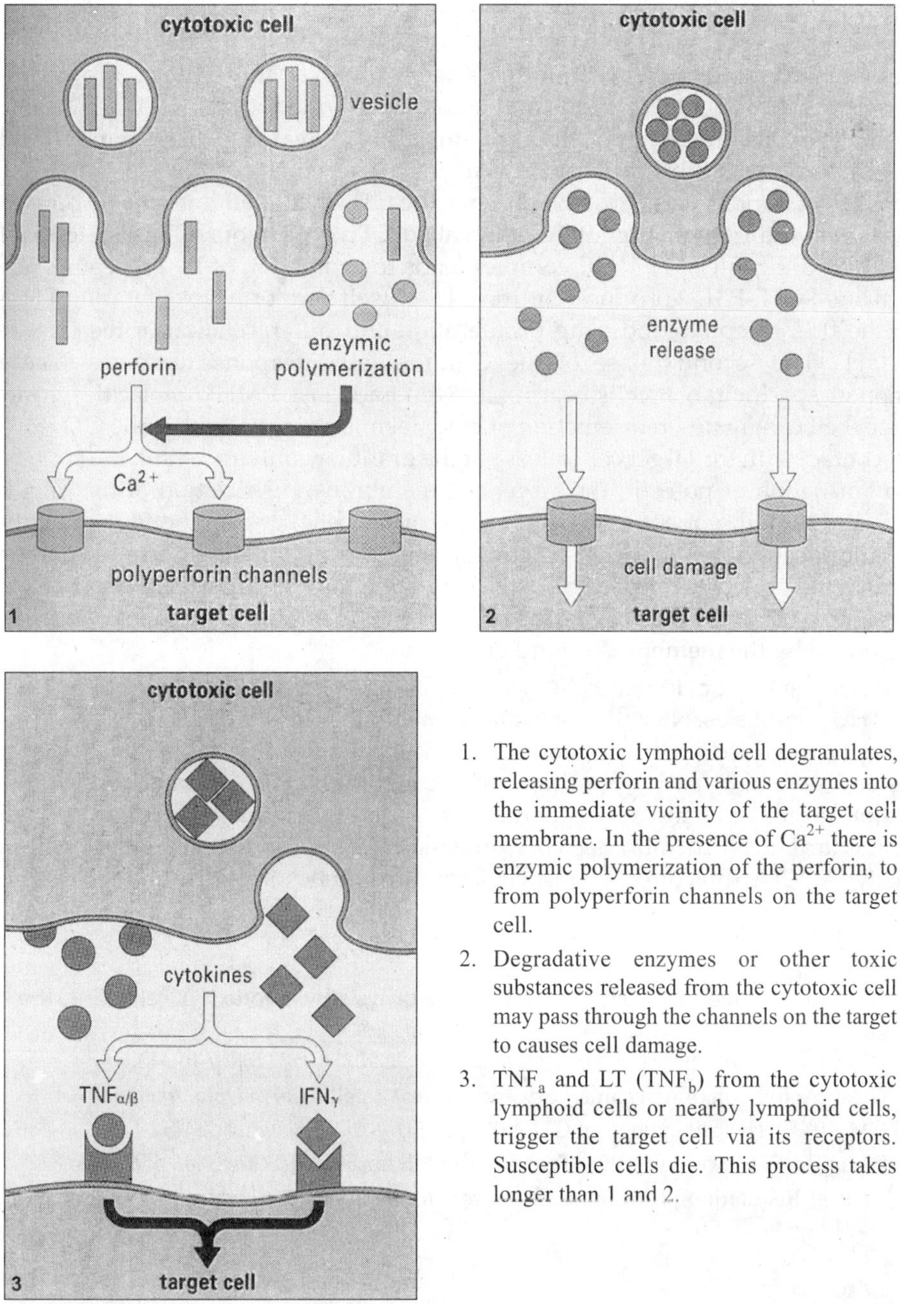

1. The cytotoxic lymphoid cell degranulates, releasing perforin and various enzymes into the immediate vicinity of the target cell membrane. In the presence of Ca^{2+} there is enzymic polymerization of the perforin, to from polyperforin channels on the target cell.
2. Degradative enzymes or other toxic substances released from the cytotoxic cell may pass through the channels on the target to causes cell damage.
3. TNF_a and LT (TNF_b) from the cytotoxic lymphoid cells or nearby lymphoid cells, trigger the target cell via its receptors. Susceptible cells die. This process takes longer than 1 and 2.

Fig. 2. Potential mechanisms for cytotoxic damage to target cells.

SUMMARY

The cell-mediated .branch of the immune system involves three types of antigen-specific effector cells: $CD4^+$ T_H1 and T_H2 cells and cytotoxic T lymphocytes (CTLs). Compared with naive T_H and Tc cells, the effector cells are more easily activated, express higher levels of cell adhesion molecules, exhibit different trafficking patterns, and produce both soluble and membrane effector molecules. The first phase of the CTL-mediated immune response involves the activation and differentiation of Tc cells, called CTL precursors (CTL-Ps). Interaction of a CTL-P with an antigen-class I MHC complex leads to expression of IL-2 receptors and of IL-2 on the activated CTL-P.IL-2 produced by the CTL-P itself and/or by proliferating TH cells then binds to the IL-2 receptors, inducing proliferation and differentiation of the CTL-P into an effector CTL. The second phase of the CTL-mediated response involves several steps: recognition of specific target cells bearing antigen and class I MHC molecules, formation of CTL/target-cell conjugates, reorientation of cytoplasmic granules within the CTL towards the region of contact with the target cell, release of the granular contents within the zone of cellular adhesion, formation of pores in the target-cell membrane, dissociation of the CTL from the target cell, and the subsequent destruction of the target cell. Perforin monomers released from the CTL granules insert into target-cell membrane and polymerize, forming pores through which granzyme proteases released from the CTL can enter the target cell. Target-cell killing by CTLs appears to occur through apoptosis induced by the soluble granzymes released from the CTL or induced by the membrane-bound Fas ligand.

Various nonspecific cytotoxic cells also can kill target cells without having to interact with antigcn-MHC complexes. NK cells. for example, mediate lysis of tumor cells and virus-infected cells by perforin-induced pore formation, by a mechanism similar to that employed by CTLs. NK cells and several other types of nonspecific cytotoxic cells can bind to the Fc region of antibody on target cells and subsequently release lytic enzymes, perforin, or TNF, which damage the target-cell membrane. This process, called antibody-dependent cell-mediated cytotoxicity (ADCC), thus directs nonspecific cytotoxic cells to specific target cells.

REFERENCES

Atkinson, EA, and RC Bleackley. (1995) Mechanisms of lysis by cytotoxic T cells. *Crit. Rev. Immunol.* 15:359.

Berke, G. (1995) Unlocking the secrets of CTL and NK cells. *Immunol. Today* 16:343.

Croft, M. (1994) Activation of naive, memory and effector T cells. *Curro Opin. Immunol.* 6:431.

Griffiths, GM. (1995) The cell biology of CTL killing. *Curro Opin. Immunol.* 7:343.

Henkart, PA, and MV Sitkovsky. (1994) Two ways to kill target cells. *Curr BioI.* 4:923.

Long, EO. (1999) Regulation of immune responses through inhibitory receptors. *Annu. Rev. Immunol.* 17:875-904.

Suggested Readings

1. Janeway CA. (1992) The imune system evolved to discriminate infectious self from noninfectious self. *Immunol Today*. 13(1):11-16.
2. McMichael AJ, Gotch F, Noble GR. (1983) Cytotoxic T cell immunity to influenza. *New Engl J Med.* 309:13.

3. Oppenheim JJ, Gery I. (1982) Interleukin-1 is more than an interleukin. *Immunol Today*. 3:113.
4. Ritz J, Schmidt RE, Michon J, Hercend T, Schlossman. (1988) Characterisation functional surface structures on human natural killer cells. *Adv Immunol*. 42:181.
5. Young JD, Liu C. (1988) Multiple mechanisms of lymphocyte mediated killing. *Immunol Today*. 9:140-44.

REVIEW QUESTIONS

Long Questions

1. Explain why NK cells from a given host will kill many types of virus infected cells but do not kill normal cells from that host.
2. What is an APC ? What type of cells can act as APCs ?
3. Compare and contrast the properties and functions of T_H and CTL.

Multiple Choice Questions

1. Adaptive cell-mediated immunity involves the generation of activated antigen-specific effector
   (a) antibodies. (b) B cells.
   (c) cytokines. (d) T cells.
2. T cells are activated to armed effector cells when they encounter antigen on
   (a) blood-borne pathogens. (b) bone marrow stromal cells.
   (c) professional APC. (d) all of the above.
3. T cells move from the circulation into the peripheral lymphoid tissues when they bind
   (a) antigen. (b) APC.
   (c) B7. (d) vascular addressins.
4. The antigen specificity of Th activation by macrophages is enhanced by the
   (a) ability of each macrophage to present only specific epitopes to T cells.
   (b) binding of T cell adhesion molecules only to macrophages presenting the antigen recognized by the T cell.
   (c) specificity of CD4 binding to Class II MHC on the macrophage.
   (d) upregulation of co-stimulatory molecules only on macrophages which have contacted foreign antigen
5. Membrane events required for Th activation include all of the following except binding of T cell
   (a) CD4 APC Class II MHC.
   (b) CTLA-4 to APC B7.
   (c) high affinity IL-2 receptor to IL-2.
   (d) LFA-1 to APC ICAM.

6. Cytoplasmic signals generated in response to antigen binding and co-stimulatory signals to Tc cells include all of the following except.
   (a) activation of transcription factors that upregulate synthesis of IL-2.
   (b) aggregation of TCR and CD8 in the T cell membrane.
   (c) increased free intracellular Ca^{++}.
   (d) induction of RAG synthesis.
7. Th1 cells secrete cytokines which
   (a) activate macrophages.
   (b) increase susceptibility to allergic reaction
   (c) inhibit macrophage secretion of inflammatory cytokines
   (d) stimulate B cells to secrete neutralizing antibodies
8. A T cell which binds antigen without undergoing a co-stimulatory signal from an APC
   (a) becomes inactivated.
   (b) can become a helper cell but not a killer cell.
   (c) can get a co-stimulatory signal from a different APC
   (d) dies
9. Dendritic cells are the best APCs for activating
   (a) CTL (b) helper T cells
   (c) naïve T cells (d) Th1 cells
10. Macrophages normally have low levels of B7 and MHC, but express more after they phagocytose
    (a) bacterial antigens (b) peptide antigens
    (c) Tc cytokines (d) Th cytokines
11. B cells differ from the other professional APC because B cells
    (a) activate T cells in the secondary lymphoid organs.
    (b) present exogenous antigen to T cells on Class II MHC.
    (c) present only peptides from antigens they bind with their membrane Ig.
    (d) use B7 as their co-stimulatory molecule.
12. Once T cells have received antigen plus co-stimulatory signals, they divide in response to IL-2 secreted by
    (a) B cells (b) dendritic cells
    (c) macrophages (d) themselves
13. Armed effector T cells differ from naïve T cells in that effector cells
    (a) are not antigen-specific.
    (b) do not need to bind co-stimulatory molecules.
    (c) do not need to bind peptide plus MHC.
    (d) do not use CAMs to bind other cells.

14. The effector molecules of cytotoxic T cells do not include
    (a) Fas (b) granzymes
    (c) IFNg (d) perforin
15. Cellular immunity includes all of the following except
    (a) activated macrophages killing phagocytosed bacteria.
    (b) CD8 T cells inducing apoptosis using Fas-FasL binding.
    (c) complement-mediated lysis of bacterial cells.
    (d) MHC-restricted cytotoxic T cells killing virus-infected cells.
16. CTL binding and destruction of target cells depends on
    (a) a cotion of cytokines by the infected cell to attract CTL.
    (b) antigen presentation on the surface of the target cell.
    (c) lack of a negative signal from MHC to prevent cytotoxicity.
    (d) presence of virus proteins in the membrane of the target cell.
17. T cells which activate macrophages do all of the following except
    (a) activate macrophages at the site of infection.
    (b) activate only macrophages presenting specific antigen on MHC.
    (c) become activated effector cells in response to antigen plus co-stimulation.
    (d) provide granzymes that the macrophages use to kill vesicular pathogens.
18. Th1 cells promote cellular immunity by secreting primarily
    (a) IL-1 and IL-6 (b) IL-2 and IFNg
    (c) IL-4 and IL-5 (d) IL-10 and IL-12
19. Macrophages kill pathogens using all of the following except
    (a) nitric oxide. (b) oxygen radicals.
    (c) perforins. (d) peroxides.

Answers

1. (d)	2. (d)	3. (d)	4. (d)	5. (b)	6. (d)
7. (a)	8. (a)	9. (c)	10. (a)	11. (c)	12. (d)
13. (b)	14. (a)	15. (c)	16. (b)	17. (d)	18. (b)
19. (c)					

CHAPTER 27

Major Histocompatibility Complex

Janak Ahi
Deptt. of Zoology, Dr. H.S. Gour Vishwavidyalaya, Sagar–470 003
E-mail: j_ahi@yahoo.co.in

During the first part of the twentieth century, clinical researchers discovered that blood could be transfused from one person to another, as long as the two individuals were compatible for the ABO blood group system. The success of blood transfusion led to the proposal that skin might also be grafted between individuals. This idea was tested during World War II when skin grafts were attempted on pilots and other military personnel who had received serious burns. The grafts were rapidly and completely rejected. After the war, researchers set out to determine the basis for tissue rejection. It was discovered that skin could be grafted successfully between mice of the same inbred strain, but that grafts between mice of different strain are were rapidly rejected. Mice of the same inbred strain are like identical twins; they are genetically identical. Subsequent studies revealed that the genes that governed tissue graft rejection were clustered in a region of the genome that was named the major histocompatibility complex (MHC).

The major histocompatibility complex (MHC) is a large cluster of genes found on the short arm of chromosome 6 in humans and on chromosome 17 in mice. The complex spans four million base pairs of DNA and contains 128 genes as well as 96 pseudogenes (non-functional gene remnants). Many, but by no means all of the genes in this complex play important roles in the immune system.

MHC Loci

The MHC is a region of multiple loci that play major roles in determining whether transplanted tissue is accepted as self (histocompatible) or rejected as foreign (histoincompatible). The MHC complex is divided into three subgroups called MHC class I, MHC class II, and MHC class III.

Class I MHC genes encode glycoproteins expressed on the surface of nearly all nucleated cells, the major function of the class I gene products is presentation of peptide antigens to T_C cells.

Class II MHC gene encode glycoproteins expressed primarily on antigen-presenting cells (macrophages, dendritic cells and B cells), where they present processed antigenic peptides to T_H cells.

Class III MHC genes encode several components of the complement system, a collection of soluble proteins found in the blood that targets foreign cells and breaks open their membranes. Adjacent to the class III region is a group of genes that control inflammation.

The class I and II MHC proteins perform the important function of antigen presentation. This is how the immune system finds out what is happening inside our cell's even though it can only survey them from the outside. Proteins inside the cell are broken into short fragments and displayed as peptide antigens by MHC proteins on the surface. This helps the immune system to discriminate between normal (self) antigens and those that are foreign and potentially dangerous. MHC antigen- presenting molecules are identical to cell surface molecules, which determine "self" or "non-self" for transplantation. They are called Human Leukocyte Antigens (HLA) in humans and H-2 antigens in the mouse and are encoded in the Major Histocompatibility Complex (MHC) genes. Class I MHC proteins are expressed on all nucleated cells however, leukocytes express the most class I MHC and neural cells the least. Class II MHC proteins are found constitutively (always) on B cells, dendritic cells, and thymic epithelial cells and can be induced on macrophages and human T cells.

Class I MHC molecules are encoded by the K and D regions in mice and by the A, B and C loci in humans. The two chains of the class II MHC molecules are encoded by the IA and IE regions in mice and by the DP, DQ and DR regions in humans. By contrast the class III MHC region, which is flanked by the class I and II regions, encodes molecules that are critical to immune function but have little in common with class I or II molecules. Class III products include the complement components C4, C2, BF and inflammatory cytokines, including tumor necrosis factor (TNF) and heatshock proteins. All the MHC genes map within a single region of

Table 1. Showing the organization of the major histocompatibility complex (MHC) in human and in mouse.

Human HLA complex								
Complex				**HLA**				
MHC class		II		III			I	
Region	DP	DQ	DR	C4,C2,BF		B	C	A
Gene products	DP αβ	DQ αβ	DR αβ	C′ proteins	TNF-α TNF-β	HLA-B	HLA-C	HLA-A
Mouse H-2 complex								
Complex				**H-2**				
MHC class	I	II		III		I		
Region	K	IA	IE	S		D		
Gene products	H-2K	IA αβ	E αβ	C′ proteins	TNF-α TNF-β	H-2D	H-2L	

the chromosome (hence the term *Complex)*; the physical map of the MHC of man and mouse has been determined by cloning, mapping and sequencing. All MHC loci are expressed co-dominantly that is to say both the set of alleles inherited from one's father and the set inherited from one's mother are expressed on each cell. In about 97% of cases the entire linked MHC alleles found on the same chromosome is called a haplotype.

Structure and functions of MHC molecules

Class I and class II MHC molecules are membrane-bound glycoproteins that are closely related both in structure and function. Both class I and class II MHC molecules have been isolated and purified and the three-dimensional structures of their extracellular domains have been determined by X-ray crystallography. Both types of membrane glycoproteins functions as highly specialized antigen-presenting molecules that from unusually stable complexes with antigenic peptides, displaying them on the cell surface for recognition by T cells. In contrast, class III MHC molecules are a group of unrelated properties that do not share the structural similarity and common function with class I and II molecules.

CLASS I MOLECULES

Class I molecules are composed of two polypeptide chains; one encoded by the BCA region and another (β2-microglobulin) that is encoded elsewhere. The MHC-encoded polypeptide is about 350 amino acids long and glycosylated, giving a total molecular weight of about 45 kDa. This polypeptide folds into three separate domains called alpha-1, alpha-2 and alpha-3. β2-microglobulin is a 12 kDa polypeptide that is non-covalently associated with the alpha-3

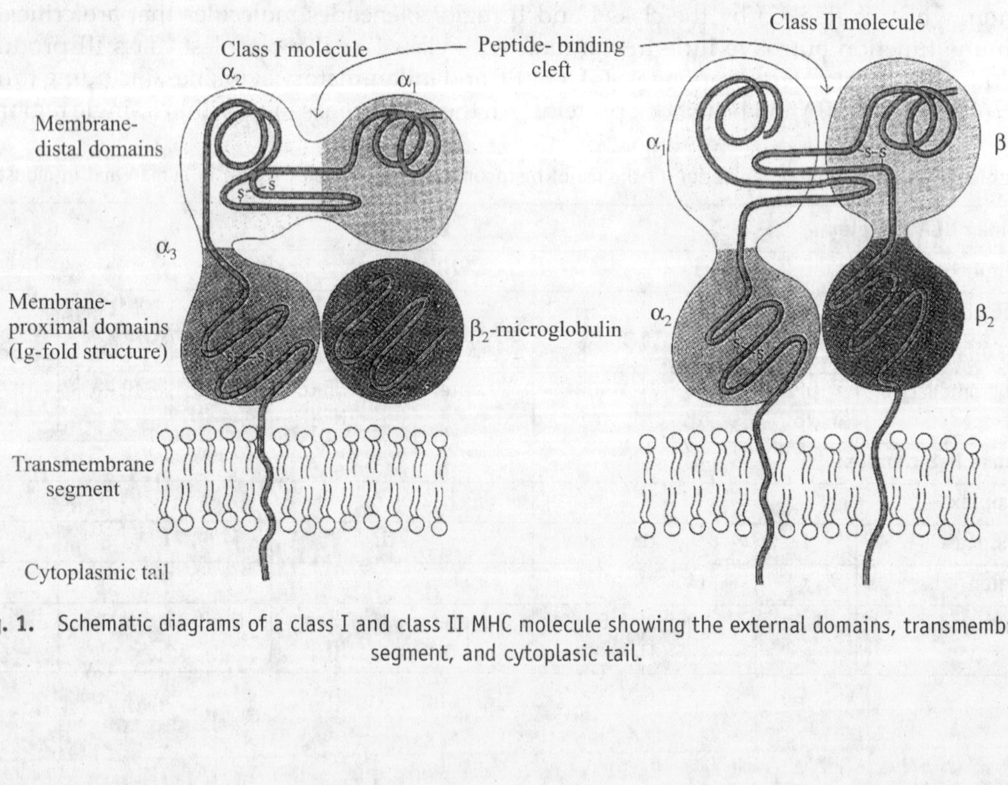

Fig. 1. Schematic diagrams of a class I and class II MHC molecule showing the external domains, transmembrane segment, and cytoplasic tail.

domain. Between the alpha-1 and alpha-2 domains lies a region bounded by a beta-pleated sheet on the bottom and two alpha helices on the sides. This region is capable of binding (via non-covalent interactions) a small peptide of about 10 amino acids. This small peptide is "presented" to a T-cell and defines the antigen "epitope" that the T-cell recognizes.

Class I MHC proteins are found on virtually all cell types and their job is to present fragments of proteins that are synthesised inside the cell. The peptide antigens presented in this manner are checked by killer T-cells, which have receptors for the class I MHC proteins. The purpose of this surveillance system is to identify abnormal body cells, such as those infected with viruses or those that have turned malignant. Such cells will display unfamiliar peptide antigens, e.g. fragments of viral proteins, and are attacked and destroyed. Because MHC class I molecules present peptides derived from cytosolic proteins, the pathway of MHC class I presentation is often called the *cytosolic* or *endogenous pathway*.

MHC class I molecules are loaded with proteins generated in the cytosol. As viruses infect a cell by entering its cytoplasm, this cytosolic, MHC class I-dependent pathway of antigen presentation is the primary way for a virus-infected cell to signal T cells. MHC class I molecules generally interact exclusively with $CD8^+$ ("cytotoxic") T cells (CTLs). The fate of the virus-infected cell is almost always apoptosis initiated by the CTL, effectively reducing the risk of infecting neighboring cells.

CLASS II MOLECULES

Class II molecules are composed of two polypeptide chains, both encoded by the D region. These polypeptides (alpha and beta) are about 230 and 240 amino acids long, respectively, and are glycosylated, giving molecular weights of about 33 kDa and 28 kDa. These polypeptides fold into two separate domains; alpha-1 and alpha-2 for the alpha polypeptide, and beta-1 and beta-2 for the beta polypeptide (Fig. 1). Between the alpha-1 and beta-1 domains lies a region very similar to that seen on the class I molecule. This region, bounded by a beta-pleated sheet on the bottom and two alpha helices on the sides, is capable of binding (via non-covalent interactions) a small peptide of about 10 amino acids. This small peptide is "presented" to a T-cell and defines the antigen "epitope" that the T-cell recognizes.

Class II MHC proteins are found only on immune cells such as phagocytes that engulf foreign particles such as bacteria. These cells are specially designed to present peptide antigens derived from such digested particles. The antigens are presented to helper T-cells, which have receptors for class II MHC proteins. The purpose of this surveillance system is to stop the immune system running out of control and attacking the body's own cells. Only if the presented antigen is recognised as foreign by the helper T-cell is the phagocyte allowed to survive. The peptides presented by class II molecules are derived from extracellular proteins (not cytosolic as in class I); hence, the MHC class II-dependent pathway of antigen presentation is called the *endocytic* or *exogenous* pathway.

Because class II MHC is loaded with extracellular proteins, it is mainly concerned with presentation of extracellular pathogens (for example, bacteria that might be infecting a wound or the blood). Class II molecules interact exclusively with $CD4^+$ ("helper") T cells (T_Hs). The helper T cells then help to trigger an appropriate immune response which may include localized inflammation and swelling due to recruitment of phagocytes or may lead to a full-force antibody immune response due to activation of B cells.

Major differences between MHC classes I and II

MHC class I	MHC class II
Comprised of an MHC-encoded α chain and a β2-microglobulin chain	Comprised of MHC-encoded α and β chains
Present on most cells	Present only on antigen-presenting cells
Bind endogenous antigens synthesized in a cell	Binds exogenous antigens
Present antigen to cytotoxic T cell lymphocytes	Present antigen to helper T cell lymphocytes
Bind CD8 adhesion molecules on cytotoxic T cells	Bind CD4 adhesion molecules on helper T cells
Presence of foreign or over-abundant antigens targets cell for destruction	Presence of foreign antigens induces antibody production, and attracts immune cells to area of infection

CLASS III MOLECULES

Several structurally and functionally diverse proteins are encoded within the third region of the MHC. These class III molecules include several complement components (C2, C4b, and factor B), two steriod 21-hydroxylase enzymes (21-OHA and 21-OHB), the inflammatory cytokines, tumor necrosis factors a and b TNF-a and TNF-b) and two heat-shock proteins.

Unlike class I and class II MHC molecules, the class III molecules are not membrane proteins, are not related structurally to class I and II molecules and have no role in antigen presentation, although most play some role in immune responses.

Expression of MHC Molecules

MHC class I molecules are widely expressed, though the level varies between different cell types. MHC class II molecules are constitutively expressed only by certain cells involved in immune responses, though they can be induced on a wider variety of cells.

Cell Type	MHC I	MHC II
T cells	+++	Varies, inducible in some species
B cells	+++	++
Macrophages	+++	+
Dendritic cells	+++ x10	+++ x10
Granulocytes	++	–
Endothelium	++	– (inducible)
Hepatocytes	+	–
Neurons	–	–

Two Pathways for Antigen Processing

The two classes of MHC molecule are specialised to present different sources of antigen. MHC class I molecules present endogenously synthesised antigens, e.g. viral proteins. MHC class II molecules present exogenously derived proteins, e.g. bacterial products or viral capsid proteins. The cell biology and expression patterns of each class of MHC are tailored to meet these distinct roles.

Processing of Class I MHC – Peptide Complexes

Antigens located in the cytosol of an APC are degraded into short peptides by proteases that are part of multimeric complexes called proteasomes. These proteases cleave the antigen into fragments approximately 8 to 10 residues long that are suitable for binding within a groove of an MHC class I molecule. The peptides are then transported across the membrane of the rough endoplasmic reticulum and into its lumen by a dimeric protein called TAP.

Once in the ER lumen, the peptide can bind to a newly synthesized MHC class I molecule, which is an integral protein of the ER membrane. The MHC-peptide complex moves through the biosynthetic pathway until it reaches the plasma membrane where the peptide is displayed.

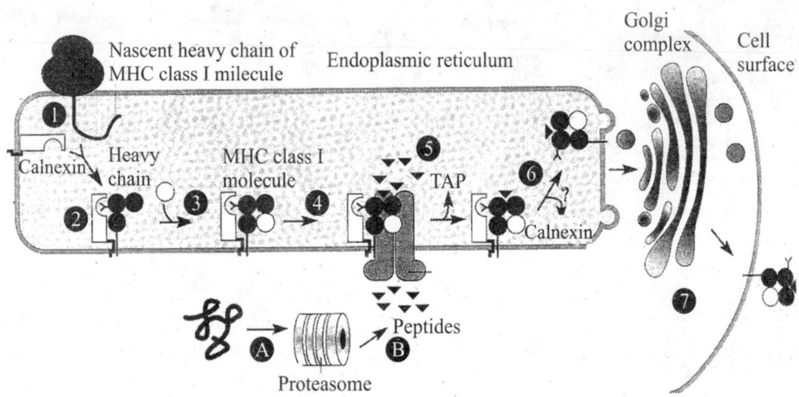

Fig. 2. The proposed for the assembly of an MHC class I-peptide complex.

The proposed pathway for the assembly of an MHC class I-peptide complex is as follows (Fig. 2). In step 1 the heavy chain to the MHC protein is synthesized by a membrane-bound ribosome and translocated into the ER membrane. The MHC heavy chain becomes associated with calnexin (step 2), a chaperone in the ER membrane, and the dimeric complex binds the invariant β_2m chain (step 3) The MHC complex then becomes associated with another ER membrane protein, TAP (step 4). Meanwhile, cytosolic antigens are taken into proteasomes (step A) and degraded into small peptides (step B). The peptides are transported into the ER lumen by the TAP protein, where they become bound within the groove of the MHC molecule (step 5). Calnexin and TAP dissociate form the MHC complex (step 6), which is transported along the biosynthetic/secretory pathway through the Golgi complex (step 7) to the plasma membrane, where it is the ready to interact with the TCR of a CTL.

Processing of Class II MHC – Peptide Complexes

MHC class II molecules are also synthesized as membrane proteins of the rough endoplasmic reticulum (RER) but they are joined noncovalently to a protein called invariant chain (Ii) which blocks the peptide binding site of the MHC molecule. Following its synthesis, the MHC class II-Ii complex moves out of the ER along the biosynthetic pathway, directed by targeting sequences located within the cytoplasmic domain to Ii. MHC class I and II molecules are thought to separate from one another in the trans Golgi network (TGN), which is the primary sorting compartment along the biosynthetic pathway. An MHC class I-peptide complex is directed toward the cell surface, whereas an MHC class II-Ii complex is directed into an endosome or lysosome, where the Ii protein is digested by acid proteases. An MHC class II molecule is then free to bind to peptides digested from antigens that were taken into the cell and directed along the endocytic pathway. Complex is then moved to the plasma membrane where it is displayed.Once on the surface of an APC, MHC molecules direct the cell's interaction with different types of T cells.

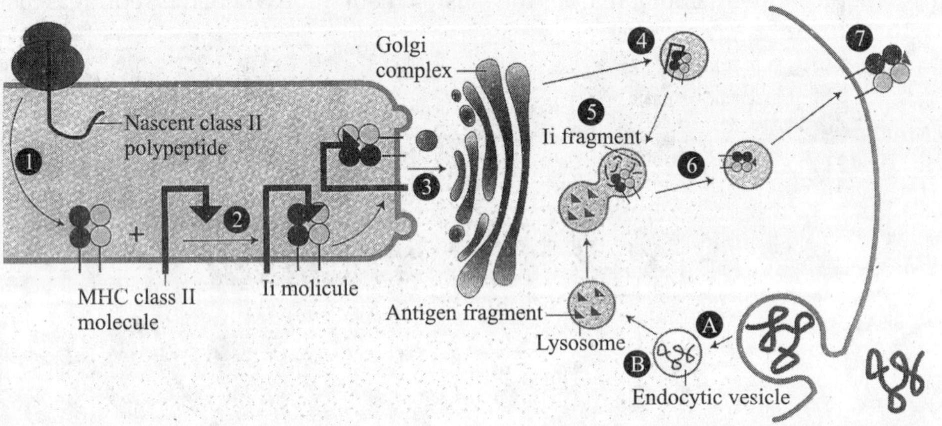

Fig. 3. The proposed for the assembly of an MHC class II-peptide complex.

The proposed pathway for the assembly of an MHC class II-peptide complex in as follows (Fig. 3). In (step 1), the MHC protein is synthesized by a membrane-bound ribosome and translocated into the ER membrane, where it becomes associated with Ii (step 2), a protein that blocks the MHC-peptide-binding site. The MHC–Io complex passes through the Golgi complex (step 3) and into a transport vesicle (step 4) meanwhile, an extracellular protein antigen is taken into the APC by endocytosis (step A) and delivered to a lysosome (step B), where the antigen is fragmented into peptides. The lysosome containing antigenic fragments fuses with the transport vesicle containing the MHC-Ii (step 5), leading to the degradation of Ii protein and association between the antigenic peptide fragment and the MHC class II moleules (step 6). The MHC peptide complex is transported to the plasma membrane (step 7) where it is ready to interact with the TCR of a T_H cell.

MHC Evolution Allelic Polymorphism

The MHC is highly polymorphic. There are a large number of genetic variants (alleles) at each genetic locus. Crucially many of these alleles are represented at significant frequency (> 1%) in the population, and in addition the alleles generally differ from one another by many (up to 30) amino acid substitutions. For example in humans there are more than 200 alleles described at some MHC loci. Such a remarkable degree of polymorphism implies a selective pressure to establish and maintain it.

The allelic diversity of MHC genes has created fertile grounds for evolutionary biologists. The most important task for theoreticians is to explain the evolutionary forces that have created and maintained such diversity. Most explanations invoke balancing selection, a broad term that identifies any kind of natural selection in which no single allele is absolutely most fit. Frequency-dependent selection and heterozygote advantage are two types of balancing selection that have been suggested to explain MHC allelic diversity.

Biological Role of the MHC

The products of the MHC play a fundamental role in regulating immune responses. T cells must recognise antigen as a complex with MHC molecules. This requires antigen to be processed by unfolding and proteolytic digestion before it complexes with the MHC molecule. Once formed the complex of antigenic peptide and MHC are generally very stable (half life ~ 24hrs). Thus the biological role of MHC proteins is to bind small peptides and to "present" these at the cell surface for the inspection of T cell antigen receptors. The allelic variation of MHC molecules is functionally reflected in the selection of peptides which can bind, each allelic product has a unique set of peptides which it can bind with high affinity (though rarely particular peptides may bind to more than one MHC allele).

In a normal animal or cell the majority of MHC molecules will be complexed with self peptides, "empty" MHC molecules are less stable especially in the case of class I products.

Association of HLA with Certain Diseases

A number of different diseases have been associated with particular MHC alleles. Among them are some autoimmune diseases, certain viral diseases, disorders of the complement system, some neurologic disorders, and several different allergies. Researchers have shown that some HLA alleles occur at a much higher frequency among those suffering from certain diseases than in the general population. This association between HLA alleles and a given disease may be quantified by typing the HLA alleles of the general population. By comparing the frequency of that allele in the patients with the allele frequency in the general population, the relative risk may be calculated (Table 2).

However, not all individuals with the at-risk allele will develop the disease and not all disease positive individuals possess the high-risk allele. In most cases diseases which show this HLA association have an obvious immune pathology.

SUMMARY

The major histocompatibility complex (MHC) refers to a cluster of tightly linked genes that encode proteins associated with intercellular recognition and antigen presentation to T lymphocytes. In mice the MHC is called the H-2 complex while in humans it is called the HLA complex. The MHC is organized into three regions based on the type of molecules encoded. Class I molecules are encoded by the K and D regions in mice and the A,B, and C regions in humans. Class II molecules are encoded by the IA and IE regions in mice and the DP, DQ and DR region in humans. Class III molecules in both humans and mice are encoded in the region between class I and II. Class I MHC molecules consist of a large glycoprotein α chain, encoded by the class I MHC genes, and a much smaller molecule of β_2-microglobulin, encoded by a gene

Table 2. Association of HLA Alleles with increased for various diseases

Disease	Associated HLA allele	Relative risk*
Insulin-dependent diabetes mellitus	DR4/DR3	20
Ankylosing spondylitis	B27	90
Multiple sclerosis	DR2	5
Myasthenia gravis	DR3	10
Narcolepsy	DR2	130
Reactive arthritis (Yersinia, Salmonella, Gonococcus)	B27	18
Reiter's syndrome	B27	37
Rheumatoid arthritis	DR4	10
Sjogrem's syndrome	Dw3	6
Systemic lupus erythematosus	DR3	5

*Relative risk is calculated by dividing the frequency of the HLA allele i the patient population by the frequency in the general population : $RR = \frac{(Ag^+/Ag^-)\text{disese}}{(Ag^+/Ag^-)\text{control}}$

outside of the MHC. The α-chain contains three external domains (α_1, α_2 and α_3) a hydrophobic transmembrane segment and a cytoplasmic tail. The latter two domains anchor the (α chain to the plasma membrane, and the α_1 and α_2 domains interact to from a cleft that binds antigenic peptides. Class II MHC molecules are heterodimers composed of two noncovalently associated glycoproteins, the α and β chain, which are encoded by separate class II genes. Each chain contains two external domains a transmembrane segment, and a cytoplasmic tail. The genes encoding the class I α chain and class II α and β chains are organized as a series of exons and introns, with each exon encoding a separate region of the polypeptide chain. Some portions of both class I and II MHC molecules exhibit the immunoglobulin-fold structure, these molecules are considered members of the immunoglobulin superfamily. Both class I and II MHC molecules present antigen to T cells. Class I molecules, which appear on the surface of nearly all nucleated cells, present processed endogeneous antigen to $CD8^+$ T cells. Class II molecules, which are expressed on a limited number of antigen-presenting cells (macrophages, dendritic cells, B cells), present processed exogeneous antigen to $CD4^+$ T cells. The class III region of the MHC encodes molecules that include a diverse group, nonmembrane proteins that play no role in antigen presentation. Disturbances in the expression of some class III molecules (e.g., certain complement components, tumor necrosis factors and heat shock proteins) may be associated with certain autoimmune diseses.

Complexes between antigenic peptides and MHC molecules are formed by degradation of a protein antigen in one of two different antigen-processing pathways. Those cells that process and present peptides associated with class II MHC molecules are called antigen-presenting cells (APCs); those that process and present peptides associated with class I MHC molecules are called target cells. Endogenous antigens are degraded into peptides within the cytosol by a large enzyme complex called the proteasome, which has several types of peptidase activity. The resulting antigenic peptides are transported into the lumen of the RER by TAP an ATP-dependent heterodimeric protein located in the RER membrane. Within the RER, the peptides assemble with class I MHC molecules; binding of peptide stabilizes the association between the class I α chain and β2-microglobulin. Peptide-class I complexes then are trans-ported from the RER through the Golgi complex to the plasmamembrane. Exogenous antigens, internalized by phagocytosis or endocytosis, are degraded by various hydrolytic enzymes within the acidic endocytic compartments Within the RER, newly formed class II MHC molecules associate with the invariant chain, which blocks binding of peptides. Class II molecules with the associ-ated invariant chain are routed to endocytic compartments, where the invariant chain is degraded by proteolytic enzymes, leaving the CLIP fragment in the peptide binding cleft. A reaction catalyzed by HLA DM allows antigenic peptides to displace CLIP.This reaction of HLA DM may be inhibited by HLA-DO.The peptide-class II complexes are then transported to the plasma membrane. Increased susceptibility to a number of diseases, predominantly but not exclusively of an autoimmune nature, has been linked to certain MHC alleles. There is a large range of risk carried by the MHC alleles; in all cases, other hereditary and environmental factors determine whether an individual will be afflicted.

REFERENCES

Borot, N, *et al.*, (1997). Mutations in the MHC class I like candidate gene for hemochromatosis in French patients. *Immunogenetics* 45 : 320.

Engelhard, VH, (1994). Structure of peptises associated with MHC class I molecules. *Curr. Opin. Immunol.* 6 : 13.

Gruen, JR and SM Weissman. (1997). Evolving views of the Major Histocompatibility Complex. *Blood* 90 : 4252.

Hughes, AL, M Yeager and M Carrington. (1996) Peptide binding function and the paradox of HLA disease associations. *Immunol. Cel Biol.* 74 : 444.

Nepon, GT, (1995). Class II antigens and disease susceptibility. *Annu. Rev. Med.* 46 : 17.

Stroynowski, I,and J Forman. (1995). Novel molecules related to MHC antigens. *Curr. Opin. Immunol.* 7 : 97.

Suggested Readings

1. Castellino F, (1997) Antigen presentation by MHC class II molecules : invariant chain function, protein trafficking and the molecular basis of diverse determinant capture. *Hum Immunol* 54(2), 159-169.
2. Krensky AM, (1997) The HLA system, antigen processing and presentation. *Kidney Int Suppl.* 58, S2-S7.
3. Li S, (1997) Cloning and functional characterization of a subunit of the transporter associated with antigen processing. *Proc. Natl Acad. Sci. USA* 94(16), 8708-8713.
4. Lundberg AS, (1992) Evolution of major histocompatibility complex class II allelic diversity : direct descent in mice and humans. *Proc. Natl. Acad. Sci. USA* 89(14), 65456549.
5. Maffei A, (1997) MHC class I antigen processing pathways. *Hum Immunol* 54(2), 91-103.
6. Ortmann B, (1997) A critical role for tapasin in the assembly and function of multimeric NHC class I Science 277(5330), 1306-1309.
7. Roitt, L, *et al.*, (1996). *Immunology*, 4th ed. Mosby.
8. Solheim JC, (1997) Prominence of beat 2-microglobulin, class I heavy chain conformation and tapasin in the teractions of class I heacy chain with calreticulin and the transporter associated with antigen processing. *J. Immunol* 158(5), 2236-2241.
9. Stumptner P, (1997) Interaction of MHC class II molecules with the invariant chain : role of the invariant chain (81-90) region. *EMBO J* 16(19), 5807-5818.

REVIEW QUESTIONS

Long Questions

1. Compare and contrast the structure of antigen receptors on B and T cells.
2. Described the steps in processing of a cytosolic antigen in an APC. What is the role of MHC proteins in this process?
3. Compare and contrast the roles of an MHC class I and class II protein molecule. What types of APCs utilize each class of MHC molecule, and what types of cells recognize them?

Multiple Choice Questions

1. Human Class I MHC a chain molecules are
   (a) β_2-microglobulin.
   (b) H-2 D, K, and L.
   (c) H-2 IA and IE
   (d) HLA-A. -B, and -C.
2. Class II MHC does not efficiently present endogenous antigen because
   (a) antigen synthesized inside the cell never makes it to the endosomal compartment.
   (b) endogenous antigen cannot be processed into peptides small enough.
   (c) HLA DM transports Class II to the surface before it can bind endogenous peptide.
   (d) invariant chain blocks binding of endogenous peptide in the ER.
3. Invariant chain (Ii)
   (a) inhibits binding of endogenous peptide to Class I MHC.
   (b) is degraded in the MIIC compartment to CLIP.
   (c) is released from Class II upon binding of $_2$-microglobulin.
   (d) is the constant region of Class I peptide binding site.
4. Antigen binding by Class I MHC molecules
   (a) accommodates many different peptides.
   (b) preferentially occurs for peptides 13-18 amino acids in length.
   (c) occurs at a site on Class I MHC formed by folding of a1 and?b_2-microglobulin domains.
   (d) occurs only on antigen presenting cells.
5. Both Class I and Class II MHC molecules are
   (a) composed of α and β chains with variable and constant regions.
   (b) expressed constitutively on all nucleated cells.
   (c) expressed on the B cell membrane.
   (d) part of the T cell receptor for antigen.
6. The major histocompatibility complex has
   (a) dozens of loci for Class I and Class II proteins.
   (b) genes that encode proteins associated with antigen processing.
   (c) only genes encoding Class I and Class II molecules.
   (d) single loci for Class I and Class II proteins.

7. MHC polymorphism
   (a) is generated by recombination of HLA A, B, and C gene segments.
   (b) is present primarily in the peptide-binding regions of MHC proteins.
   (c) is the result of random association of many alpha and beta genes.
   (d) restricts the ability of B cells to bind antigen.
8. T cells are MHC-restricted in their ability to respond to antigen because
   (a) all antigen must be processed and presented to activate lymphocytes.
   (b) during an infection, all cells in the body present antigen on MHC Class I.
   (c) MHC binds antigen more specifically than TCR does.
   (d) TCR must recognize both antigen and MHC molecules.
9. All of the following are associated with the expression of Class I MHC molecules except
   (a) antigen peptide presentation on membrane Class I MHC to Tc.
   (b) graft rejection.
   (c) increased risk of certain autoimmune diseases.
   (d) stimulation of antibody production.
10. Human Class II MHC molecules
   (a) are encoded by the genes HLA-A, B, and C.
   (b) are found on all nucleated cells.
   (c) have an antigen binding site formed from regions of two polypeptide chains.
   (d) must be associated with β_2-microglobulin molecules to bind peptide.
11. Which of the following statements is TRUE?
   (a) Each individual expresses all the diversity of MHC protein structure.
   (b) If a family has four children, no two of them will have the same MHC genotype.
   (c) Someone with bare lymphocyte syndrome who expressed no MHC proteins would die in infancy.
   (d) TCR on Tc cells binds $\alpha 1$ and $\beta 2$ domains of Class I MHC protein.

Answer

1. (d)	2. (d)	3. (b)	4. (a)	5. (c)	6. (b)
7. (b)	8. (d)	9. (d)	10. (c)	11. (c)	

CHAPTER 28

Hybridoma Technology and its Applications

D. Gayathri[1*] and T.N. Devaraja[2]

[1]Dept. of Microbiology, Kuvempu University, Jnana Sahyatri, Shankaraghatta–577 451

[2]Taralabalu KVK, Anubhava Mantap Campus, Davanagere–577 004

E-mail: gayathridevaraja1@yahoo.com

The technique of producing virtually unlimited quantities of a single antibody referred as monoclonal antibody, specific for particular antigenic determinant, has revolutionized immunology being a versatile tool in general research and clinical practices. Most antigen possess multiple epitopes, induce proliferation and differentiation of a variety of B-cell clones. The resulting serum antibodies are heterogeneous, containing mixture of antibodies, each specific for one epitope. Such a collection is referred as polyclonal antibody, which facilitate the localization, phagocytosis, and complement mediated lysis of target antigen. Thus, it has a clear advantage for the host system *in vivo*. However, this antibody heterogeneity reduces the efficacy for *in vitro* uses, therapeutic and diagnostic purposes. Hence, for such circumstances, antibody derived from a single clone specific for single epitope is desirable. Biochemical purification of monoclonal antibody from a polyclonal antibody preparation is not feasible as it results in damage of antigen binding sites of immunoglobulin and a very low amount is obtained which is not reproducible. In 1975, Georges Kohler and Cesar Milstein devised a method of preparing monoclonal antibody and each was awarded Nobel Prize in 1984 for the same. By fusing a primed B-cell with a myeloma cell, fused to generate hybridoma that possesses the immortal growth properties of the myeloma cell and secreted the antibody by the B-cell expression. The resulting clones of hybridoma cells, secreted large quantities of monoclonal antibody specific for the antigen.

PRODUCTION AND SELECTION OF HYBRIDOMA CELLS FOR MONOCLONAL ANTIBODY

The production of monoclonal antibody involves three stages:

1. Generation of B-cell hybridoma.
   (a) Immunization

*Corresponding author

   (b) Fusion of B-cell with its myeloma counter part
   (c) HAT (Hypoxanthine Aminopterin Thymidine) selection.
2. Screening the resulting clones for those clones that secrete desired antigenic specificity.
3. Propagation of desired hybridoma.

I. Generation of B- cell Hybridomas

(a) Immunization: Eight week old female Balb/C mice are immunized with desired purified/ crude antigen at 1:1 ratio of Freund's complete adjuvant. Further, booster doses are inject with 1:1 antigen and Freund's incomplete adjuvant on the 15th day. On the 25th day another booster dose with antigen in PBS intravenously. Blood (50 µl) from mice was collected from alternate orbital sinuses 4 days after final injection. Antibody titre in mice was estimated by an enzyme linked immunosorbant assay. Spleen cells from mice with serum titre more than 1:1000 was used for hybridoma production (Fig. 1).

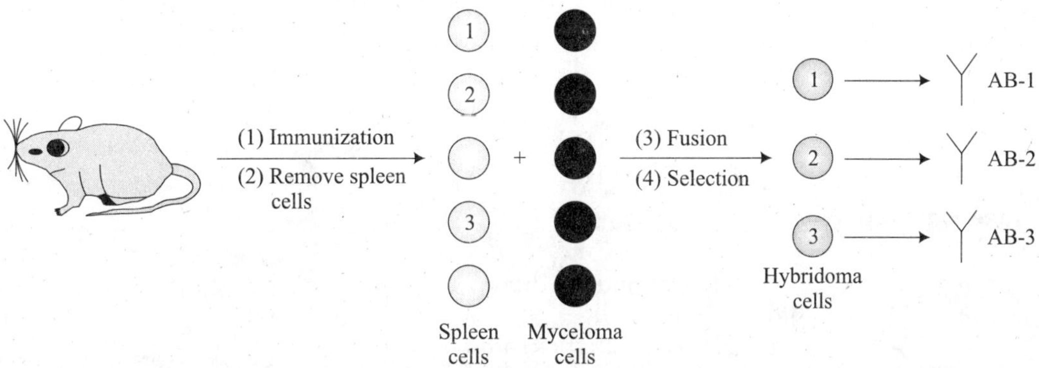

Fig. 1. An outline of the procedure used to produce monoclonal antibodies. In this scheme three different antigen-specific antibodies produced via immunization are shown. Hybridoma cells that produce antibodies that are not specific for the antigen are easily screened out with an ELISA (http://www.antibodyresource.com/).

Enzyme linked immunosorbant assay (ELISA): ELISA plate was cleaned with 70 % ethanol and air dried. Test antigen (100) µl in carbonate-bicarbonate buffer (pH 9.5) was coated to wells of ELISA plate and incubated at 4° C overnight. Unbound antigen was poured off and the plate was washed once with washing buffer (PBS/Tween-20). Free sites were blocked by incubating the wells with 200 µl blocking solution (5% skimmed milk powder in PBS) for 1 h at room temperature. The blocking solution was poured off and the plate was washed three times with washing buffer. Mouse antiserum (100 µl), with doubling dilution in PBS, were added and incubated for 1 h at room temperature. The antiserum was poured off and the plate was washed three times with washing buffer. Rabbit anti-mouse IgG peroxidase (100 ml) at 1:2000 dilution in washing buffer was added and incubated for 1h at room temperature. The conjugate was poured off and the plate was washed 3 times with washing buffer. Substrate (100 µl) [Tetramethylbenzidine (TMB)/H_2O_2 in distilled water 1:20] was added and incubated for 5–10 min. The reaction was stopped by adding 100 ml of 1M H_2SO_4. Optical density (OD) was measured at 450 nm using ELISA reader. Incubation with PBS instead of mouse antiserum was the negative control. Test OD (optical density) values equal to positive control and/ or twice the OD value of negative control were considered as positive.

(b) Fusion of B cell with myeloma cell: Immunized mice were killed by cervical dislocation and immersed in 70% ethanol. Spleen was dissected out aseptically and spleen cells were collected in serum free RPMI (Rosewell Park Memorial Institute) medium. Spleen cells from the immunized mice were fused with SP_2/O myeloma cells using polyethylene glycol (PEG). After fusion, the cells were resuspended in RPMI-HAT-20% FBS (Hypoxanthine Aminopterin Thymidine-Fetal Bovine Serum)/Dulbecco's modified Eagle medium and transferred to fourteen 96-well tissue culture plates containing 1 d old spleen cells feeder layer from unimmunized mouse. The plates were incubated in humidified CO_2 (5%) atmosphere in a CO_2. On the 8th day after fusion, 50% of medium in the wells was replaced with RPMI-HT-20% FBS. Hybridomas were screened from the 15th day onwards by an immunodot assay or ELISA.

(c) HAT selection: HAT selection is based on the principle that mammalian cells can synthesize nucleotides by two different pathways: *de novo* and the salvage pathways.

***De Novo* Pathway:**

Blocked by
Aminopterin

Phosphoribosyl pyrophosphate + Uridylate → → → // → → Nucleotides

Salvage Pathway:

TK$^+$ (thymidine kinase)
Thymidine → → → → → → → → → ↘
HGPRT$^+$ (hypoxanthine ↘
guanine phosphoribosyl → → → Nucleotides
transferase ↗
Hypoxanthine → → → → → → → → → ↗

In *de novo* pathway methyl or formyl group is transferred from an activated form of tetrahydrofolate, is blocked by aminopterin, a folic acid analogue. When de novo pathway is blocked, cells utilize the salvage pathway, which bypasses aminopterin block and directly utilizing hypoxanthine and thymidine available in HAT medium catalyzed by HGPRT and TK enzymes. Purine salvage usually involves phosphoribosyltransferase reactions, which generate ribonucleoside monophosphates (rNMP) from the purine bases and 5-phos phoribosyl-1-pyrophosphate (PRPP). These are phosphorylated to diphosphates (rNDPs) and then reduced to deoxyribonucleotides (dNDPs) with ribonucleotide reductase.

Thus, unfused myeloma cells do not survive as they are mutated and having HGPRT$^-$ and TK$^-$ genotype. B-cells cannot survive as they are mortal. Therefore, hybridoma cells only survive utilizing salvage pathway from B-cells and becomes immortal as that of myeloma cells. Hybridoma clones appear as cluster of cells attaching to the sides of the tissue culture plates which could be observed under inverted phase contrast microscope after 14 days of incubation at 37°C in 5% CO_2 incubator.

2. Screening the Resulting Clones for those Clones that Secrete Desired Antigenic Specificity

Once the hybridoma colonies observed, antibody secreting colonies must be selected, which is specific for target antigen. Although, hybridoma is produced against desired antigen, sometimes these hybridomas will produce antibody to unwanted antigens. Therefore, it is necessary to screen specific hybridoma producing antidody, for which following techniques are used:

(a) **ELISA** (as explained earlier)

(b) **Immunodot assay:** Hybrodomas were screened by immunodot assay according to Hawkes *et al.* (1982). Target antigen was dotted onto nitrocellulose (NC) paper which was then cut into squares and transferred into wells of a 96-well tissue culture plate. Free sites on NC were blocked with 3% BSA/1% normal rabbit serum in PBS for 30 min and washed 3 times with PBS/Tween-20. Hybridoma culture supernatant from a well was added to corresponding test well and incubated overnight at 4°C. After washing with PBS/Tween 20, NC was again blocked with 3% BSA/1% normal rabbit serum in PBS for 30 min. After washing with PBS/Tween-20, 100 ml rabbit anti-mouse IgG HRP in 3% BSA in PBS was added and incubated for 90 min. After washing with PBS/Tween, 100 ml of substrate (4-chloro 1 naphthol/ethanol/30% H_2O_2, in Tris buffer pH 7.6) was added. Clear purple blue dots developed on nitrocellulose paper were considered positive. Appropriate negative controls must be maintained. Hybridoma supernatant reacting with the antigen but not with negative controls was considered positive.

(c) **Radio immunoassay (RIA):** Less commonly used.

3. Propagation of Desired Hybridomas

Hybridomas of desired specificity are minicloned (subcultured) to highest dilution so as to obtain a viable clone at maximum dilution to ensure truly monoclonal. Then, the clones are propagated using one of the following techniques:

(a) **Tissue culture technique:** In tissue culture flask hybridoma cells were cultures, antibody is secreted into the medium (1-20 µg/ml), can be used.

(b) **In mice:** A hybridoma can be propagated in peritoneal cavity of histocompatible mice, in which antibody produced in ascites fluid is about 1-10 mg/ml.

APPLICATIONS OF MONOCLONAL ANTIBODIES

Monoclonal antibodies are widely used for diagnosing plant, animal, avian, fish and human pathogens. For the first time in India, we developed MAb against an aquatic fish pathogen *Aphanomyces invadans* for the detection of the pathogen (Gayathri *et al.*, 2004) MAbs are also used in testing blood sugar level, hormone level and histocompatibility matching. MAbs have wide applications in research and therapy particularly tumor imaging and detection. As an example for therapeutic antibody, a treatment for non-Hodgkin's lymphoma, will be only the third to go on sale in the US, although expensive. It is assumed by 2010, the World will experience 'molecular farming' as a huge industry of worth more than $100 billion which carry the potential of creating full human monoclonals.

For characterization of fungal antigen: MAbs to glycoprotein antigens of a fungal plant pathogen, *Phytophthora megasperma* f. sp. *glycinea* (Keith *et al.*, 1987) was developed. They produced MAbs against extracellular antigens of the fungal soybean pathogen *P. megasperma* f.sp. *Glycinea,* to identify antigenic determinants associated with particular avirulence genes.

For Characterization of viral antigen: MAbs were produced for identification and characterization of the virus-neutralizing epitopes of infectious bursal disease virus (IBDV). The results suggested that at least two non-overlapping epitopes recognized by the virus-neutralizing MAbs (Fahey *et al.*, 1991).

In tumor detection: For example, MAb labeled Iodine-131 against breast cancer cells, when introduced parenterally to detect tumor, the monoclonal imaging technique can detect breast cancer metastases, which would be undetectable by conventional scanning techniques.

Engineered MAbs with toxins: Immunotoxins composed of any tumor specific monoclonal antibody coupled to potentially lethal toxins, are valuable therapeutic reagents. For example, Shiga toxin (*Shigella)* or Diphtherin toxin (*Corynebacterium diphtheriae*), which inhibit protein synthesis. These toxins consist of two polypeptide chains, a toxin chain and a binding polypetide, which are required for interaction with receptors and harmless when binding polypeptide is absent. Thus, an immunotoxin is prepared by replacing binding polypeptide chain with a monoclonal antibody having specificity for a target tumor cell. The attached monoclonal antibody will target the tumor cell, where the toxin would inhibit protein synthesis of multiplying cells. This technology is more promising in clinical treatments against lymphoma, leukemia and other types of tumors.

Engineered MAbs: For therapeutic purposes, mouse monoclonal antibody in humans mounts immune response by allergic manifestation. To overcome this difficulty, either human MAbs are used or mouse MAb must be engineered. It is difficult to produce human MAb. To engineer monoclonal antibody, recombinant DNA is prepared containing promoter, leader and variable region sequences from a mouse antibody gene, which is combined with constant region exons from a human antibody gene. The antibody encoded by such a recombinant gene is 'chimeric mouse human DNA' commonly called 'humanised antibody'. The antigenic specificity of this is determined by the variable region which is derived from mouse DNA and its isotype which is determined by the constant region of human DNA. Since, constant regions are encoded by human gene, these chimeric antibodies will have fewer mouse epitopes and are less immunogenic than mouse MAb when it is administered to humans. In addition to this, chimeric antibody retains biologically effector functions of human antibody and may trigger complement activation and/ or Fc receptor binding reaction. Chimeric MAb that function as immunotoxins can also be prepared. Here, terminal constant region domain in tumor specific MAb is replaced with potential toxins. Thus, produced immunotoxins lack Fc terminal, unable to bind to cell bearing Fc receptors. These preparations would be more efficient as therapeutic agent in binding to tumor cells alone.

Recombinant single-chain variable fragment antibodies can be engineered: Antibodies are known to affect the morphology, growth, and metabolism of mollicutes and thus may serve as candidate molecules for a plantibody-based control strategy for plant-pathogenic spiroplasmas and phytoplasmas. Recombinant single-chain variable fragment (scFv) antibodies are easy to engineer and express in plants, but their inhibitory effects on mollicutes have been evaluated and compared with those of polyclonal and monoclonal antibodies (**Sylvie *et al.*, 2002**). They described the morphology, growth, and glucose metabolism of *Spiroplasma citri* in the presence of polyclonal, monoclonal, and recombinant antibodies directed against the immunodominant membrane protein spiralin. We showed that the scFv antibodies had no effect on *S. citri* glucose metabolism but were as efficient as polyclonal antibodies in inhibiting *S. citri* growth in liquid medium. Inhibition of motility was also observed.

Heteroconjugates: Heteroconjugates are hybrids of two different antibody molecules. One half of the antibody has the specificity for target tumor and another half has the specificity for a surface molecule on an immune effector cell, such as a natural killer cell, activated macrophages etc. Thus, heteroconjugates serve to link the immune effector cell to tumor cell by mediating lysis of tumor cell.

Abzymes: There is a similarity between the reactions of enzyme-substrate to that of antigen-antibody, in which, weak non-covalent interactions take place with high affinity and specificity. However, antibody does not alter the antigen, while enzyme converts substrate into product. Investigating such a similarity, Lerner and his co-workers, produced hapten–carrier complex in which the hapten structurally resembled the transition state of an ester undergoing hydrolysis. Then they generated anti-hapten MAb using this hapten-carrier complex. Thus, produced MAb when incubated with ester, accelerated the hydrolysis of ester by about 1000 folds. Therefore, such a preparation of MAb having both the characters of enzyme and antibody is called abzyme or catalytic antibody. The abzymes are very specific in their action, i.e., they hydrolyze esters whose transition state structure closely resembled that of the hapten in the immunizing conjugate. Abzymes may also be very useful in determining the structural and functional analysis of proteins, liquify blood clots, to dissolve viral glycoproteins at specific sites to block infectivity (HIV infection) and many more applications in various branches of science in future.

SUMMARY

Monoclonal antibody is a homogeneous preparation specific for an epitope. The production of monoclonal antibody involves: Generation of B-cell hybridoma, screening the resulting clones for those clones that secrete desired antigenic specificity, and propagation of desired hybridoma. ELISA Immuns dot assay and RIA are important methods for hybridoma screening. Monoclonal antibodies are widely used for diagnosing plant, animal, avian, fish and human pathogens. MAb are used for antigen characterization of fungal/viral antigens. MAb can be used for tumor detection, imaging and killing. Immunotoxins composed of any tumor specific monoclonal antibody coupled to potentially lethal toxins, are valuable therapeutic reagents. And, MAb can be engineered to avoid allergic reactions. Heteroconjugates are hybrids of two different antibody molecules. Abzymes have a dual role as enzyme and antibody.

REFERENCES

Gayathri D., Shankar, K. M. and Mohan, C. V. (2004) Monoclonal antibody- based immunodot test for epizootic ulcerative syndrome pathogen, *Aphanomyces invadans*. Current Science 87 : 289-291.

Fahey, K.J., McWaters, P., Brown, M.A., Erny, K., Murphy, V.J. and Hewish, D.R. (1991) Virus-neutralizing and passively protective monoclonal antibodies to infectious bursal disease virus of chickens. Avian Dis. 35 : 365-73.

Gayathri, D. and Devaraja, T. N. (2003) Hybridoma technology in aquaculture health management with an emphasis on *Aphanomyces invadans* of epizootic ulcerative syndrome in fish. In: Biotechnology in environmental management (Eds. T Chakrabarti, T K Ghosh, G Tripathi), pp. 361-373, APH Publishers, New Delhi.

Hawkes, R., Niday, N. and Gordon, J. (1982) A dot immunobinding assay for monoclonal and other antibodies. Anal. Biochem. 119 : 142-147.

http://www.antibodyresource.com/

Kohler, G. and Milstein, C. (1975) Continuous culture of fused cells secreting antibody of predefined specificity. Nature 256 : 495-497.

Lerner, R. A. and Tramontano, A. (1988) Catalytic antibodies. Scientific American. 258 : 58.

Malembic, S., Saillard, C., Bové, J.M. and Garnier, M. (2002) Effect of Polyclonal, Monoclonal, and Recombinant (Single-Chain Variable Fragment) Antibodies on *In Vitro* Morphology, Growth, and Metabolism of the Phytopathogenic Mollicute Spiroplasma citri, Appl. Environ. Microbiol. 68 : 2113-2119

Richard, A. G., Thomas, J. K. and Barbara, A. O. (2000) Kuby immunology. 4th edition, W. H. Freeman and Co., New York.

Wycoff, K.L., Jellison, J. and Ayers, A.R. (1987) Monoclonal Antibodies to Glycoprotein Antigens of a Fungal Plant Pathogen, *Phytophthora megasperma* f. sp. *glycinea*, Plant Physiol. 85 : 508–515.

Suggested Readings

1. Gayathri Devaraja, Shankar, K. M. and Mohan, C. V. (2004) Monoclonal antibody based immunodot test for epizootic ulcerative syndrome pathogen, Aphanomyces invadans. Current Science 87 : 289-291.
2. Kohler, G. and Milstein, C. (1975), Continuous culture of fused cells secreting antibody of predefined specificity. Nature 256 : 495-497.
3. Hawkes, R., N. Niday, and J. Gordon (1982): A dot immunobinding assay for monoclonal and other antibodies. Anal. Biochem. 119 : 142-147.

REVIEW QUESTIONS

Long Questions

1. Describe different stages of monoclonal antibody production.
2. Describe the mechanism of hybridoma selection.
3. Describe the applications of monoclonal antibody,
4. Describe the use of MAb in clinical fields.
5. Define Abzymes. Add a note on its application.
6. Write an account on enzyme linked immunosorbant assay.

Multiple Choice Questions

1. Hybridoma was produced by fusing B and myeloma cell by
   (a) Watson and Crick, (b) Kohler and Milstein,
   (c) Meselson and Stahl, (d) Harlow and Lane

2. Hybridoma cell uses _______ pathway in HGPRT/TK medium.
   (a) de-novo pathway, (b) HMP pathway,
   (c) salvage pathway (d) ED pathway
3. Function of aminopterin is,
   (a) Source of amino acid, (b) Blocks de-novo pathway,
   (c) blocks salvage pathway, (d) Allows myeloma cell growth
4. Genotype of hybridoma cell,
   (a) $HGPRT^+/TK^+$, (b) $HGPRT^-/TK^+$,
   (c) $HGPRT^+/TK^-$, (d) $HGPRT^-/TK^-$.
5. Heteroconjugate are useful in
   (a) Killing tumor cell, (b) In lysing macrophages,
   (c) Hydrolysing antibody (d) In ELISA test
6. Abzymes are useful in
   (a) Neutralising antibody, (b) Substrate to product conversion,
   (c) All chemical reactions, (d) Antigen-antibody reaction

Answers

1. (b) 2. (c) 3. (b) 4. (a) 5. (a) 6. (b)

PART–9

Genomics and Proteomics

CHAPTER 29

Genomics

Subodh Kumar Jain
Dept. of Biotechnology, Dr. H.S. Gour University, Sagar–470 003
E-mail: subjain@gmail.com

Thomas Roderick is credited with coining the term "genomics" in 1986 to describe the scientific discipline of mapping, sequencing and analyzing genomes. Genomics has the potential of offering new therapeutic methods for the treatment of some diseases, as well as new diagnostic methods e.g., for women newly diagnosed with breast cancer, a genomic test called Onco*type* DX assesses a patient's individual risk of breast cancer recurrence and likely benefit from chemotherapy, which can help doctors for more personalized treatment decisions. The major tools and methods related to genomics are Bioinformatics, Genetic analysis, Measurement of Gene expression, and Determination of gene function. Genomics appeared in the 1980s and took off in the 1990s with the initiation of genome projects for several species. The related field of genetics is the study of genes and their role in inheritance. The first genome to be sequenced in its entirety was that of bacteriophage $\phi \times 174$; (5,368 bp) in 1980. The first free-living organism to be sequenced was that of *Haemophilus influenzae* (1.8 Mb) in 1995, and since then genomes are being sequenced at a rapid pace. Recognition of DNA as the hereditary material, determination of its structure, elucidation of the genetic code, development of recombinant DNA technologies, and establishment of increasingly automatable methods for DNA sequencing set the stage for the Human Genome Project (HGP) began in 1990. The project's new research strategies and experimental technologies have generated a steady stream of ever-larger and more complex genomic data sets that have poured into public databases and have transformed the study of virtually all life processes. The genomic approach of technology development and large-scale generation of community resource data sets has introduced an important new dimension into biological and biomedical research. Interwoven advances in genetics, comparative genomics, high-throughput biochemistry and bioinformatics are providing biologists with a markedly improved repertoire of research tools that will allow the functioning of organisms in health and disease to be analysed and comprehended at an unprecedented level of molecular detail. Genome sequences, the bounded sets of information that guide biological development and function, lie at the heart of this revolution. In short, genomics has become a central and cohesive discipline of biomedical research.

The practical consequences of the emergence of this new field are widely apparent. Identification of the genes responsible for human diseases, once a herculean task requiring large research teams, many years of hard work, and an uncertain outcome, can now be routinely accomplished in a short period of time with access to DNA samples and associated phenotypes, an Internet access to the public genome databases, a thermal cycler and a DNA-sequencing machine. With the recent publication of a draft sequence of the mouse genome, identification of the mutations underlying a vast number of interesting mouse phenotypes has similarly been greatly simplified. Comparison of the human and mouse sequences shows that the proportion of the mammalian genome under evolutionary selection is more than twice that previously assumed.

Our ability to explore genome function is increasing in specificity as each subsequent genome is sequenced. Microarray technologies have catapulted many laboratories from studying the expression of one or two genes in a month to studying the expression of tens of thousands of genes in a single afternoon. Clinical opportunities for gene-based pre-symptomatic prediction of illness and adverse drug response are emerging at a rapid pace, and the therapeutic promise of genomics has ushered in an exciting phase of expansion and exploration in the commercial sector. The investment of the Human Genome Project (HGP) in studying the ethical, legal and social implications of these scientific advances has created a talented cohort of scholars in ethics, law, social science, clinical research, theology and public policy, and has already resulted in substantial increases in public awareness and the introduction of significant protections against misuses such as genetic discrimination.

GENOMICS CAN BE CLASSIFIED INTO FOUR MAJOR GROUPS

I. Comparative Genomics

Comparative genomics is the study of relationships between the genomes of different species or strains. Comparative genomics is an attempt to take advantage of the information provided by the signatures of selection to understand the function and evolutionary processes that act on genomes. While it is still a young field, it holds great promise to yield insights into many aspects of the evolution of modern species. The sheer amount of information contained in modern genomes (several gigabytes in the case of humans) necessitates that the methods of comparative genomics are mostly computational in nature. Gene finding is an important application of comparative genomics.

Comparative genomics exploits both similarities and differences in the proteins, RNA, and regulatory regions of different organisms to infer how selection has acted upon these elements. Those elements that are responsible for similarities between different species should be conserved through time (stabilizing selection, while those elements responsible for differences among species should be divergent (positive selection). Finally, those elements that are unimportant to the evolutionary success of the organism will be unconserved (selection is neutral).

Identifying the mechanisms of eukaryotic genome evolution by comparative genomics is one of the important goals of the field. It is however often complicated by the multiplicity of events that have taken place throughout the history of individual lineages, leaving only distorted and superimposed traces in the genome of each living organism. For this reason comparative genomics studies of small model organisms (e.g., yeast) are of great importance to advance our understanding of general mechanisms of evolution.

Having come a long way from its initial use of finding functional proteins, comparative genomics is now concentrating on finding regulatory regions and siRNA molecules. Recently, it has been discovered that distantly related species often share long conserved stretches of DNA that do not appear to code for any protein. It is unknown at this time what function such ultra-conserved regions serve.

2. Pharmacogenomics

Pharmacogenomics examines how your genetic makeup affects your response to drugs. Researchers in the field are working on applying human genome knowledge to pharmaceuticals by identifying genes that account for varying drug reactions in different people. Eventually, they hope to be able to customize drug therapies for specific patient populations or even individuals.

Pharmacogenomics combines traditional pharmaceutical sciences such as biochemistry with an understanding of common DNA variations in the human genome. The most common variations in the human genome are called single nucleotide polymorphisms (SNPs). There is estimated to be approximately 11 million SNPs in the human population, with an average of one every 1,300 base pairs.

An individual's response to a drug is often linked to these common DNA variations. In a similar manner, susceptibility to certain diseases is also influenced by common DNA variations. Currently, much of the research in the field of pharmacogenomics is focused on genes encoding either metabolic enzymes that can alter a drug's activity or defective structural proteins that result in increased susceptibility to disease.

Pharmacogenomics eventually will provide tailored drug therapy based on genetically determined variation in effectiveness and side effects. This will mean:

- ***More powerful medicines:*** Pharmaceutical companies will be able to produce therapies more targeted to specific diseases, maximizing therapeutic effects while decreasing damage to nearby healthy cells.
- ***Better, safer drugs the first time:*** Recovery time will go down and safety will go up as the likelihood of adverse reactions goes down or is eliminated altogether.
- ***More accurate methods of determining appropriate drug dosages:*** Current methods of basing dosages on weight and age will be replaced with dosages based on a person's genetics—how well the body processes the medicine and the time it takes to metabolize it.
- ***Better vaccines:*** Vaccines made of genetic material, either DNA or RNA, promise the benefits of existing vaccines without all the risks. They theoretically could activate the immune system but be unable to cause infections.

3. Functional Genomics

Functional genomics is a field of molecular biology that is dealing with dynamic aspects such as gene transcription and translation as opposed to the static aspects of the genomic information included in the nucleotide sequence. It makes use of the vast wealth of data produced by genome sequencing projects to describe genome function.

Functional genomics includes function-related aspects of the genome itself such as mutation and SNP analysis as well as all aspects of gene expression. The latter comprise a number of other "omics" such as transcriptomics, proteomics, or metabolomics. Together these fields of application describe the function and interactions of genes. The term "genome" refers to the full complement genes in an organism's cell. Similarly, "proteome" is the term used to describe all the proteins produced by the genome. Proteomics is the science of analyzing and cataloging all the proteins encoded by a genome. Two generalized categories of analysis are identified in proteomics: (1) expression proteomics (studying global changes in protein expression) and (2) cell map proteomics (systematic study of protein-protein interaction through the isolation of protein complexes). Proteomics tracks protein expression by cells, beginning with the functional protein and backtracking to the gene that encoded it.

Proteins are part of the dynamics of a living system, appearing and disappearing according to the needs of the cells. Proteomics is hence the study of global protein expression patterns that define cells in specific biological states. Furthermore, because proteins vary widely in their chemical behavior, it is difficult to prescribe one best technique to study all proteins. Unlike genomics, in which robotic gene sequencers were the workhorses of the large-scale projects, proteomics depend on a variety of key technologies, including the yeast two-hybrid system, mass spectrometry, two-dimensional gel electrophoresis, and DNA micro array hybridization.

Proteomics may also be described as protein expression profiling. The recent surge in large-scale protein study is attributable to the large amount of genomics data produced by genomics projects, especially the human genome project, coupled with advances in computing power and robotics that enable scientists to handle and manipulate huge databases. Most of all, scientists have not yet found out the cellular functions of the known and predicted proteins. Raw DNA and protein sequences operated by genomics projects provide little information. It is hoped that proteomics will be able to provide information about protein function to help scientists understand cellular metabolism more fully. The basic process in determining gene function is as follows:

(i) A protein of interest is first extracted from a tissue sample.
(ii) A fragment of the protein's amino acid sequence may be translated into its DNA base sequence.
(iii) With this information, a computer database is searched to identify the gene that encodes the protein as well as the complete protein.

Functional genomics uses mostly high-throughput techniques to characterize the abundance gene products such as mRNA and proteins. Typical technology platforms are DNA micro arrays and SAGE for mRNA or Two-dimensional Gel electrophoresis and Mass Spectrometry for Protein analysis. Because of the large quantity of data produced by these techniques and the desire to find biologically meaningful patterns, bioinformatics is crucial to this type of analysis. Examples of techniques in bioinformatics are data clustering or principal component analysis for unsupervised machine learning (class detection) as well as artificial neural networks or support vector machines for supervised machine learning (class prediction, classification).

4. Structural Genomics

Structural genomics includes determination of the three dimensional structure of all proteins of a given organism, by experimental methods such as X-ray crystallography, NMR spectroscopy or computational approaches such as homology modeling.

As opposed to traditional structural biology, the determination of a protein structure through a structural genomics effort often (but not always) comes before anything is known regarding the protein function. This raises new challenges in structural bioinformatics, i.e. determining protein function from its 3D structure.

One of the important aspects of structural genomics is the emphasis on high throughput determination of protein structures. This is performed in dedicated centers of structural genomics.

APPLICATIONS OF GENOMICS

I. Genomics to Biology

Structure and Function of Genomes

The broadly available genome sequences of human and some other organisms represent foundational information for biology and biomedicine. Embedded within this as-yet poorly understood code are the genetic instructions for the entire repertoire of cellular components, knowledge of which is needed to unravel the complexities of biological systems. Elucidating the structure of genomes and identifying the function of the myriad encoded elements will allow connections to be made between genomics and biology and will, in turn, accelerate the exploration of all realms of the biological sciences.

For this, new conceptual and technological approaches will be needed to:

- Develop a comprehensive and comprehensible catalogue of all of the components encoded in the human genome.
- Determine how the genome-encoded components function in an integrated manner to perform cellular and organismal functions.
- Understand how genomes change and take on new functional roles.

I. Comprehensively Identify the Structural and Functional Components Encoded in the Human Genome

Although DNA is relatively simple and well understood chemically, the human genome's structure is extraordinarily complex and its function is poorly understood. Only 12% of its bases encodes proteins[1], and the full complement of protein-coding sequences still remains to be established. A roughly equivalent amount of the non-coding portion of the genome is under active selection[2], suggesting that it is also functionally important, yet vanishingly little is known about it. It probably contains the bulk of the regulatory information controlling the expression of the approximately 30,000 protein-coding genes, and myriad other functional elements, such as non-protein-coding genes and the sequence determinants of chromosome dynamics. Even less is known about the function of the roughly half of the genome that consists of highly repetitive sequences or of the remaining non-coding, non-repetitive DNA.

The next phase of genomics is to catalogue, characterize and comprehend the entire set of functional elements encoded in the human and other genomes. Compiling this genome 'parts list' will be an immense challenge. Well-known classes of functional elements, such as protein-coding sequences, still cannot be accurately predicted from sequence information alone. Other types of known functional sequences, such as genetic regulatory elements, are even less well understood; undoubtedly new types remain to be defined, so one must be ready to investigate novel, perhaps unexpected, ways in which DNA sequence can confer function.

Comparison of genome sequences from evolutionarily diverse species has emerged as a powerful tool for identifying functionally important genomic elements. Initial analyses of available vertebrate genome sequences1, 2, (and Aparicio *et.al.* 2002) have revealed many previously undiscovered protein-coding sequences. Mammal-to-mammal sequence comparisons have revealed large numbers of homologies in non-coding regions[2], few of which can be defined in functional terms. Further comparisons of sequences derived from multiple species, especially those occupying distinct evolutionary positions, will lead to significant refinements in our understanding of the functional importance of conserved sequences (Sidow, 2002). Effective identification and analysis of functional genomic elements will require increasingly powerful computational capabilities, including new approaches for tackling ever-growing and increasingly complex data sets and a suitably robust computational infrastructure for housing, accessing and analysing those data sets. Simultaneously, investigators will need to become increasingly adept in dealing with this new information. As a better understanding of genome function is gained, refined computational tools for *de novo* prediction of the identity and behaviour of functional elements will emerge (Zhang, 2002).

Complementing the computational detection of functional elements will be the generation of additional experimental data by high-throughput methodologies. One example is the production of full-length complementary DNA (cDNA) sequences e.g. http://mgc.nci.nih.gov and http://www.fruitfly.org/EST/full.shtml. Major challenges inherent in programmes to discover genes are the experimental identification and validation of alternate splice forms and messenger RNAs expressed in a highly restricted fashion. Even more challenging is the experimental validation of functional elements that do not encode protein e.g. regulatory regions and non-coding RNA sequences. High-throughput approaches to identify them will be needed to generate the experimental data that will be necessary to develop, confirm and enhance computational methods for detecting functional elements in genomes.

2. Elucidate the Organization of Genetic Networks and Protein Pathways and Establish How They Contribute to Cellular and Organismal Phenotypes

Genes and gene products do not function independently, but participate in complex, interconnected pathways, networks and molecular systems that, taken together, give rise to the workings of cells, tissues, organs and organisms. Defining these systems and determining their properties and interactions is crucial to understanding how biological systems function.

Moreover these systems are far more complex than any problem that molecular biology, genetics or genomics has yet approached. On the basis of previous experience, one effective path will begin with the study of relatively simple model organisms (Banerjee and Zhang, 2002), such as bacteria and yeast, and then extend the early findings to more complex organisms, such as mouse and human.

Understanding biological pathways, networks and molecular systems will require information from several levels. At the genetic level, the architecture of regulatory interactions will need to be identified in different cell types, requiring, among other things, methods for simultaneously monitoring the expression of all genes in a cell. At the gene-product level, similar techniques that allow *in vivo*, real-time measurement of protein expression, localization, modification and activity/kinetics will be needed. It will be important to develop, refine and scale up techniques that modulate gene expression, such as conventional gene-knockout methods (Van der *et al*, 2002), newer knock-down approaches (Hannon, 2002) and small-molecule inhibitors (Stokwell, 2000) to establish the temporal and cellular expression pattern of individual proteins and to determine the functions of those proteins. This is a key first step towards assigning all genes and their products to functional pathways.

Establishing a true understanding of how organized molecular pathways and networks give rise to normal and pathological cellular and organismal phenotypes will require more than large, experimentally derived data sets. Once again, computational investigation will be essential and there will be a greatly increased need for the collection, storage and display of the data in robust databases. By modelling specific pathways and networks, predicting how they affect phenotype, testing hypotheses derived from these models and refining the models based on new experimental data, it should be possible to understand more completely the difference between a 'bag of molecules' and a functioning biological system.

3. Develop a Detailed Understanding of the Heritable Variation in the Human Genome

Genetics seeks to correlate variation in DNA sequence with phenotypic differences (traits). The greatest advances in human genetics have been made for traits associated with variation in a single gene. But most phenotypes, including common diseases and variable responses to pharmacological agents, have a more complex origin, involving the interplay between multiple genetic factors (genes and their products) and non-genetic factors (environmental influences). Unravelling such complexity will require both a complete description of the genetic variation in the human genome and the development of analytical tools for using that information to understand the genetic basis of disease.

Establishing a catalogue of all common variants in the human population, including single-nucleotide polymorphisms (SNPs), small deletions and insertions, and other structural differences, began in earnest several years ago. Many SNPs have been identified, and most are publicly available.

A comprehensive understanding of genetic variation, both in humans and in model organisms, would facilitate studies to establish relationships between genotype and biological function. The study of particular variants and how they affect the functioning of specific proteins and protein pathways will yield important new insights about physiological processes in normal and disease states. An enhanced ability to incorporate information about genetic variation into human genetic studies would usher in a new era for investigating the genetic bases of human disease and drug response.

4. Mechanism to Understand Evolutionary Variation Across Species

The genome is a dynamic structure, continually subjected to modification by the forces of evolution. The genomic variation seen in humans represents only a small glimpse through the larger window of evolution, where hundreds of millions of years of trial-and-error efforts have created today's biosphere of animal, plant and microbial species. A complete elucidation of genome function requires a parallel understanding of the sequence differences across species and the fundamental processes that have sculpted their genomes into the modern-day forms.

The study of inter-species sequence comparisons is important for identifying functional elements in the genome. Beyond this illuminating role, determining the sequence differences between species will provide insight into the distinct anatomical, physiological and developmental features of different organisms, will help to define the genetic basis for speciation and will facilitate the characterization of mutational processes. The recent finding that mutation rates vary widely across the mammalian genome raises numerous questions about the molecular basis for these evolutionary changes. At present, our understanding of DNA mutation and repair, including the important role of environmental factors, is limited.

Genomics will provide the ability to substantively advance insight into evolutionary variation, which will, in turn, yield new insights into the dynamic nature of genomes in a broader evolutionary framework.

II. Genomics to Health

I. Translating Genome-based Knowledge into Health Benefits

The sequencing of the human genome, along with other recent and expected achievements in genomics, provides an unparalleled opportunity to advance our understanding of the role of genetic factors in human health and disease, to allow more precise definition of the non-genetic factors involved, and to apply this insight rapidly to the prevention, diagnosis and treatment of disease. The report by the US National Research Council that originally envisioned the HGP was explicit in its expectation that the human genome sequence would lead to improvements in human health, and subsequent five-year plans reaffirmed this view. But how this will happen has been less clearly articulated. With the completion of the original goals of the HGP, the time is right to develop and apply large-scale genomic strategies to empower improvements in human health, while anticipating and avoiding potential harm.

Such strategies should enable the research community to achieve the following:

- Identify genes and pathways with a role in health and disease, and determine how they interact with environmental factors.
- Develop, evaluate and apply genome-based diagnostic methods for the prediction of susceptibility to disease, the prediction of drug response, the early detection of illness and the accurate molecular classification of disease.
- Develop and deploy methods that catalyse the translation of genomic information into therapeutic advances.

2. Develop Robust Strategies for Identifying the Genetic Contributions to Disease and Drug Response

For common diseases, the interplay of multiple genes and multiple non-genetic factors, not a single allele, usually dictates disease susceptibility and response to treatments. Deciphering the role of genes in human health and disease is a formidable problem for many reasons, including impediments to defining biologically valid phenotypes, challenges in identifying and quantifying environmental exposures, technological obstacles to generating sufficient and useful genotypic information, and the difficulties of studying humans. Vigorous development of crosscutting genomic tools to catalyse advances in understanding the genetics of common disease and in pharmacogenomics is needed. Prominent among these will be a detailed haplotype map of the human genome that can be used for whole-genome association studies of all diseases in all populations, as well as further advances in sequencing and genotyping technology to make such studies feasible.

More efficient strategies for detecting rare alleles involved in common disease are also needed, as the hypothesis that alleles that increase risk for common diseases are themselves common will probably not be universally true. Computational and experimental methods to detect gene"gene and gene"environment interactions, as well as methods allowing interfacing of a variety of relevant databases, are also required. By obtaining unbiased assessments of the relative disease risk that particular gene variants contribute, a large longitudinal population-based cohort study, with collection of extensive clinical information and ongoing follow-up, would be profoundly valuable to the study of all common diseases.

3. Develop Strategies to Identify Gene Variants that Contribute to Good Health and Resistance to Disease

Most human genetic research has traditionally focused on identifying genes that predispose to illness. A relatively unexplored, but important, area of research focuses on the role of genetic factors in maintaining good health. Genomics will facilitate further understanding of this aspect of human biology and allow the identification of gene variants that are important for the maintenance of health, particularly in the presence of known environmental risk factors. One useful research resource would be a 'healthy cohort', a large epidemiologically robust group of individuals with unusually good health, who could be compared with cohorts of individuals with diseases and who could also be intensively studied to reveal alleles protective for conditions such as diabetes, cancer, heart disease and Alzheimer's disease. Another promising approach would be rigorous examination of genetic variants in individuals at high risk for specific diseases who do not develop them, such as sedentary, obese smokers without heart disease or individuals with *HNPCC* mutations who do not develop colon cancer.

4. To Develop Genome-based Approaches to Prediction of Disease Susceptibility and Drug Response

The discovery of variants that affect risk for disease could potentially be used in individualized preventive medicine including diet, exercise, lifestyle and pharmaceutical intervention to maximize the likelihood of staying well. For example, the discovery of variants that correlate with successful outcomes of drug therapy, or with unfortunate side effects, could potentially be rapidly translated into clinical practice. Turning this vision into reality will require the following: (1) unbiased determination of the risk associated with a particular gene variant, often overestimated in initial studies; (2) technological advances to reduce the cost of genotyping; (3) research on whether this kind of personalized genomic information will actually alter health behaviours; (4) oversight of the implementation of genetic tests to ensure that only those with demonstrated clinical validity are applied outside of the research setting; and (5) education of healthcare professionals and the public to be well-informed participants in this new form of preventive medicine.

Systematic analyses of somatic mutations, epigenetic modifications, gene expression, protein expression and protein modification should allow the definition of a new molecular taxonomy of illness, which would replace our present, largely empirical, classification schemes and advance both disease prevention and treatment. The reclassification of neuromuscular diseases and certain types of cancer provides striking initial examples, but many more such applications are possible.

Such a molecular taxonomy would be the basis for the development of better methods for the early detection of disease, which often allows more effective and less costly treatments. Genomics and other large-scale approaches to biology offer the potential for developing new tools to detect many diseases earlier than is currently feasible. Such 'sentinel' methods might include analysis of gene expression in circulating leukocytes, proteomic analysis of body fluids, and advanced molecular analysis of tissue biopsies. An example would be the analysis of gene expression in peripheral blood leukocytes to predict drug response. A focused effort to use a genomic approach to characterize serum proteins exhaustively in health and disease might also be highly rewarding.

5. Use New Understanding of Genes and Pathways to Develop Powerful New Therapeutic Approaches to Disease

Pharmaceuticals on the market target fewer than 500 human gene products. Even though not all of the 30,000 or so human protein-coding genes will have products targetable for drug development, this suggests that there is an enormous untapped pool of human gene-based targets for therapeutic intervention. In addition, the new understanding of biological pathways provided by genomics should contribute even more fundamentally to therapeutic design.

The information needed to determine the therapeutic potential of a gene generally overlaps heavily with the information that reveals its function. The success of imatinib mesylate (Gleevec), an inhibitor of the BCR-ABL tyrosine kinase, in treating chronic myelogenous leukaemia relied on a detailed molecular understanding of the disease's genetic cause. This example offers promise that therapies based on genomic information will be particularly effective. A particularly promising example of the gene-based approach to therapeutics is the application of chemical genomics. This strategy uses libraries of small molecules (natural compounds, aptamers or the products of combinatorial chemistry) and high-throughput screening to advance understanding of biological pathways and to identify compounds that act as positive or negative regulators of individual gene products, pathways or cellular phenotypes. Although the pharmaceutical industry applies this approach widely as the first step in drug development, few academic investigators have access to this methodology or are familiar with its use.

Providing such access more broadly, through one or more centralized facilities, could lead to the discovery of a host of useful probes for biological pathways that would serve as new reagents for basic research and/or starting points for the development of new therapeutic agents (the 'hits' from such library screens will generally require medicinal chemistry modifications to yield therapeutically usable compounds).

Also needed are new, more powerful technologies for generating deep molecular libraries, especially ones tagged to allow the ready determination of precise molecular targets. A centralized database of screening results should lead to further important biological insights. Generating molecular probes for exploring the basic biology of health and disease in academic laboratories would not supplant the major role of biopharmaceutical companies in drug development, but could contribute to the start of the drug development pipeline. Scientists would doubtless find many of these molecular probes of interest for further exploration through optimization by medicinal chemistry, target validation, lead compound identification, toxicological studies and, ultimately, clinical trials.

6. Genetic Risk Information

Investigate how genetic risk information is conveyed in clinical settings, how that information influences health strategies and behaviours, and how these affect health outcomes and costs.

Understanding how genetic factors affect health is often cited as a major goal of genomics, on the assumption that applying such understanding in the clinical setting will improve health. But this assumption actually rests on relatively few examples and data, and more research is needed to provide sufficient guidance about how to use genomic information optimally for improving individual or public health.

Theoretically, the steps by which genetic risk information would lead to improved health are: (1) an individual obtains genome-based information about his/her own health risks; (2) the individual uses this information to develop an individualized prevention or treatment plan; (3) the individual implements that plan; (4) this leads to improved health; and (5) healthcare costs are reduced. Scrutiny of these assumptions is needed, both to test them and to determine how each step could best be accomplished in different clinical settings.

Research is also required that critically evaluates new genetic tests and interventions in terms of parameters such as benefits, access and cost. Such research should be done by using the tools and expertise of many fields, including genomics, health education, health behaviour research, health outcomes research, healthcare delivery analysis, and healthcare economics. Some of these fields have historically paid little attention to genomics, but high-quality research of this sort could provide important guidance in clinical decision-making – as the work of several disciplines has already been helpful in caring for people with an increased risk of colon cancer as a result of mutations in *FAP* or *HNPCC* .

7. Development of Genome-based Tools to Improve the Health

Disparities in health status constitute a significant global issue, but can genome-based approaches to health and disease help to reduce this problem? Social and other environmental factors are major contributors to health disparities; indeed, some would question whether heritable factors have any significant role. But population differences in allele frequencies for some disease-associated variants could be a contributing factor to certain disparities in health status, so incorporating this information into preventive and/or public-health strategies would be beneficial. Research is needed to understand the relationship between genomics and health disparities by rigorously evaluating the diverse contributions of socioeconomic status, culture, discrimination, health behaviours, diet, environmental exposures and genetics.

It is also important to explore applications of genomics in the improvement of health in the developing world where both human and non-human genomics will play significant roles. If we take malaria as an example, a better understanding of human genetic factors that influence susceptibility and response to the disease, and to the drugs used to treat it, could have a significant global impact. So too could a better understanding of the malarial parasite itself and of its mosquito vector, which the recently reported genome sequences should provide. It will be necessary to determine the appropriate roles of governmental and non-governmental organizations, academic institutions, industry and individuals to ensure that genomics produces clinical benefits for resource-poor nations, and is used to produce robust local research expertise.

To ensure that genomics research benefits all, it will be critical to examine how genomics-based health care is accessed and used. What are the barriers to equitable access, and how can they be removed? This is relevant not only in resource-poor nations, but also in wealthier countries where segments of society, such as indigenous populations, the uninsured, or rural and inner city communities, have traditionally not received adequate health care.

III. Genomics to Society

1. Promoting the Use of Genomics to Maximize Benefits and Minimize Harms

Genomics has been at the forefront of giving serious attention, through scholarly research and policy discussions, to the impact of science and technology on society. Although the major benefits to be realized from genomics are in the area of health, as described above, genomics can also contribute to other aspects of society. Just as the HGP and related developments have spawned new areas of research in basic biology and in health, they have also created opportunities for research on social issues, even to the extent of understanding more fully how we define ourselves and each other.

In future, society must not only continue to grapple with numerous questions raised by genomics, but must also formulate and implement policies to address many of them. Unless research provides reliable data and rigorous approaches on which to base such decisions, those policies will be ill-informed and could potentially compromise us all. To be successful, this research must encompass both 'basic' investigations that develop conceptual tools and shared vocabularies, and more 'applied', 'translational' projects that use these tools to explore and define appropriate public-policy options that incorporate diverse points of view.

As it has in the past, such research will continue to have important ramifications for all three major themes of the vision presented here. Following are the research that focuses on society itself, more than on biology or health. Such efforts should enable the research community to:

- Analyse the impact of genomics on concepts of race, ethnicity, kinship, individual and group identity, health, disease and 'normality' for traits and behaviours.
- Define policy options, and their potential consequences, for the use of genomic information and for the ethical boundaries around genomics research.

2. Develop Policy Options for the Uses of Genomics in Medical and Non-medical Settings

Surveys have repeatedly shown that the public is highly interested in the concept that personal genetic information might guide them to better health, but is deeply concerned about potential misuses of that information. Topping the list of concerns is the potential for discrimination in health insurance and employment. A significant amount of research on this issue has been done, policy options have been published, and many US states have now passed anti-discrimination legislation. The US Equal Employment Opportunity Commission has ruled that the Americans with Disabilities Act should apply to discrimination based on predictive genetic information, but the legal status of that construct remains in some doubt. Although an executive order protects US government employees against genetic discrimination, this does not apply to other workers. Thus, many observers have concluded that effective federal legislation is needed, and the US Congress is currently considering such a law.

Making certain that genetic tests offered to the public have established clinical validity and usefulness must be a priority for future research and policy making. In the United States, the Secretary's Advisory Committee on Genetic Testing extensively reviewed this area and concluded that further oversight is needed, asking the Food and Drug Administration to review new predictive genetic tests prior to marketing. That recommendation has not yet been acted on; meanwhile, numerous websites offering unvalidated genetic tests directly to the public, often combined with the sale of 'nutraceuticals' and other products of highly questionable value, are proliferating.

The use of genomic information is not limited to the arenas of biology and of health, and further research and development of policy options is also needed for many other applications of such information. The array of additional users is likely to include the life, disability and long-term care insurance industries, the legal system, the military, educational institutions and adoption agencies. Although some of the research informing the medical uses of genomics will be useful in broader settings, dedicated research outside the healthcare sphere is needed to explore the public values that apply to uses of genomics other than for health care and their relationship to specific contextual applications e.g., should genetic information on predisposition to hyperactivity be available in the future to school officials? Or should genetic information about behavioural traits be admissible in criminal or civil proceedings? Genomics also provides greater opportunity to understand ancestral origins of populations and individuals, which raises issues such as whether genetic information should be used for defining membership in a minority group.

3. Understand the Relationships between Genomics, Race and Ethnicity, and the Consequences of Uncovering These Relationships

Race is a largely non-biological concept confounded by misunderstanding and a long history of prejudice. The relationship of genomics to the concepts of race and ethnicity has to be considered within complex historical and social contexts.

Most variation in the genome is shared between all populations, but certain alleles are more frequent in some populations than in others, largely as a result of history and geography. Use of genetic data to define racial groups, or of racial categories to classify biological traits, is prone to misinterpretation. To minimize such misinterpretation, the biological and sociocultural factors that inter relate genetics with constructs of race and ethnicity need to be better understood and communicated.

This will require research on how different individuals and cultures conceive of race, ethnicity, group identity and self-identity, and what role they believe genes or other biological factors have. It will also require a critical examination of how the scientific community understands and uses these concepts in designing research and presenting findings, and of how the media report these. Also necessary is widespread education about the biological meaning and limitations of research findings in this area, and the formulation and adoption of public-policy options that protect against genomics-based discrimination or maltreatment.

4. Understand the Consequences of Uncovering the Genomic Contributions to Human Traits and Behaviours

Genes influence not only health and disease, but also human traits and behaviours. Science is only beginning to unravel the complicated pathways that underlie such attributes as handedness, cognition, diurnal rhythms and various behavioural characteristics. Too often, research in behavioural genetics, such as that regarding sexual orientation or intelligence, has been poorly designed and its findings have been communicated in a way that oversimplifies and overstates the role of genetic factors. This has caused serious problems for those who have been stigmatized by the suggestion that alleles associated with what some people perceive as 'negative' physiological or behavioural traits are more frequent in certain populations. Given this history and the real potential for recurrence, it is particularly important to gather sufficient scientifically valid information about genetic and environmental factors to provide a sound understanding of the contributions and interactions between genes and environment in these complex phenotypes.

It is also important that there be robust research to investigate the implications, for both individuals and society, of uncovering any genomic contributions that there may be to traits and behaviours. The field of genomics has a responsibility to consider the social implications of research into the genetic contributions to traits and behaviours, perhaps an even greater responsibility than in other areas where there is less of a history of misunderstanding and stigmatization. Decisions about research in this area are often best made with input from a diverse group of individuals and organizations.

5. Assess how to Define the Ethical Boundaries for Uses of Genomics

Genetics and genomics can contribute understanding to many areas of biology, health and life. Some of these human applications are controversial, with some members of the public questioning the propriety of their scientific exploration. Although freedom of scientific inquiry has been a cardinal feature of human progress, it is not unbounded. It is important for society to define the appropriate and inappropriate uses of genomics. Conversations between diverse parties based on an accurate and detailed understanding of the relevant science and ethical, legal and social factors will promote the formulation and implementation of effective policies. For instance, in reproductive genetic testing, it is crucial to include perspectives from the disability community. Research should explore how different individuals, cultures and religious traditions view the ethical boundaries for the uses of genomics — for instance, which sets of values determine attitudes towards the appropriateness of applying genomics to such areas as reproductive genetic testing, 'genetic enhancement' and germline gene transfer.

6. Technology Development

Many areas of critical importance to the realization of the genomics-based vision for biomedical research require new technological and methodological developments before pilots and then large-scale approaches can be attempted. Recognizing that technology development is an expensive and high-risk undertaking, the NHGRI is nevertheless committed to supporting and fostering technology development in many of these crucial areas, including the following:

DNA sequencing: There is still great opportunity to reduce the cost and increase the throughput of DNA sequencing, and to make rapid, cheap sequencing available more broadly. Radical reduction of sequencing costs would lead to very different approaches to biomedical research.

Genetic variation: Improved genotyping methods and better mathematical methods are necessary to make effective use of information about the structure of variation in the human genome for identifying the genetic contributions to human diseases and other complex traits.

The genome 'parts list': Beyond coding sequences and transcriptional units, new computational and experimental approaches are needed to allow the comprehensive determination of all sequence-encoded functional elements in genomes.

Proteomic: To focus on the development of appropriate, scalable technologies for the comprehensive analysis of proteins and protein machines in human health and in both rare and complex diseases.

Pathways and networks: As a complement to the development of the genome 'parts list' and increasingly effective approaches to proteome analysis, it is possible to develop new technologies that generate a synthetic view of genetic regulatory networks and interacting protein pathways.

Genetic contributions to health, disease and drug response: There is a high priority on creating and applying new crosscutting genomics tools, technologies and strategies needed to identify the genetic bases of medically relevant phenotypes. Research on the genetic contributions to rare and common diseases, and to drug response, will typically involve biological systems and diseases of primary interest.

Molecular probes, including small molecules and RNA-mediated interference, for exploring basic biology and disease: Exploration of the feasibility of expanding chemical genomics in the academic and public sectors, particularly with regard to the establishment of one or more centralized facilities, may be established.

Databases

Another type of community resource for the biological and biomedical research communities is represented by databases. But their support represents a potentially significant problem. Funding agencies, reflecting the interest of the research community, tend to prefer to use their research funds to support the generation of new data, and the ongoing need for continued and increasing support for the data archives and robust access to them is often given less attention. Both the scientific community and the funding agencies must recognize that investment in the creation and maintenance of effective databases is as important a component of research funding as data generation.

The major source of support for several major genetics/genomics-oriented databases, including the Mouse Genome Database (http://www.informatics.jax.org/mgihome/MGD/aboutMGD.shtml), the *Saccharomyces* Genome Database (http://genome-www.stanford.edu/Saccharomyces), FlyBase (http://flybase.bio.indiana.edu), WormBase (http://www.wormbase.org) and Online Mendelian Inheritance in Man (http://www.ncbi.nlm.nih.gov/omim).

REFERENCES

Aparicio, S. *et al.* Whole-genome shotgun assembly and analysis of the genome of *Fugu rubripes*. *Science* 297, 1301-1310 (2002).

Banerjee, N. and Zhang, M.X. Functional genomics as applied to mapping transcription regulatory networks. *Curr. Opin. Microbiol.* 5, 313-317 (2002).

Francis S. Collins, Eric D. Green, Alan E. Guttmacher and Mark S. Guyer . A vision for the future of genomics research. *Nature* 422, 835-847 (24 April, 2003)

Hannon, G. J. RNA interference. *Nature* 418, 244-251 (2002).

Sidow, A. Sequence first. Ask questions later. *Cell* 111, 13-16 (2002).

Stockwell, B. R. Chemical genetics: Ligand-based discovery of gene function. *Nature Rev. Genet.* **1**, 116-125 (2000).

The International Human Genome Sequencing Consortium. Initial sequencing and analysis of the human genome. *Nature* 409, 860-921 (2001).

The Mouse Genome Sequencing Consortium. Initial sequencing and comparative analysis of the mouse genome. *Nature* 420, 520-562 (2002).

The Chipping Forecast II. *Nature Genet.* 32, 461-552 (2002).

Van der Weyden, L., Adams, D. J. & Bradley, A. Tools for targeted manipulation of the mouse genome. *Physiol. Genomics* 11, 133-164 (2002).

Zhang, M. Q. Computational prediction of eukaryotic protein-coding genes. *Nature Rev. Genet.* 3, 698-709 (2002).

Suggested Readings

1. Apley, L.: DNA sequencing: From experimental methods to bioinformatics. New York: Springer-Verlag.(1997).
2. Schena, M., and R.W.Davis.: Genes, genomics and chips. In DNA Microarrays: A practical approach (ed. M. Schena). Oxford, UK: Oxford University Press (1999).
3. Lipman, D.J. and W.B. Pearson. Rapid and sensitive protein similarity searches, Science, 227:1435-1441.
4. Southern, E.M. Detection of specific sequences among DNA fragments separated by gel electrophoresis. J.Mol.Biol.,98: 503-517 (1975).
5. Fields, S. and O. Song. A novel genetic system to detect protein-protein interactions. Nature, 340:245-246. (1989).

Internet Resources

1. Intro to proteomics: http:*//www.e-proteomics.net/tech.html*
2. *Proteomics journal: http://www.mcponline.org/*
3. *Proteomics glossary: http://www.genomicglossaries.com/content/proteomics.asp*

REVIEW QUESTIONS

Long Questions

1. What is genomics? Describe its major groups.
2. Describe the techniques used in the field of genomics.

3. What is the purpose and expected outcome of proteomics?
4. Describe major source of support for genomics-oriented databases.
5. Describe various applications of genomics in various fields.

Multiple Choice Questions

1. All of the genes and other DNA of an organism is called its
   (a) Open reading frame (b) Genome
   (c) Intron (d) Multigene family.
2. Why the first genome sequence was considered a milestone?
   (a) It proved that viruses contained nucleic acids
   (b) It proved that DNA could be isolated from a virus
   (c) It showed the coding sequences for all the proteins produced by a virus
   (d) It proved that the coding sequences within a genome could be located and identified.
3. How do we recognize where a gene begins within a long sequence of DNA nucleotides?
   (a) Most genes begin with the nucleotide adenine
   (b) The process of trial and error eventually leads investigators to locate where a gene begins.
   (c) A gene starts with a three-nucleotide "start" codon
   (d) Genes begin at either end of a DNA sequence, never in the middle.
4. When the first free-living organism, the bacterium Haemophilus influenzae, was sequenced in the mid-1990s, about what percentage of its genes were shown to have a known function?
   (a) About half (b) Around 10%
   (c) Nearly all (d) Less than 1%
5. Why did the first organisms that had their genomes sequenced have comparatively small genome sizes?
   (a) Organisms that could be kept easily in the lab were the focus of early research.
   (b) Until the advent of automated DNA sequencers, researchers had to use organisms with small genomes because the procedure was very time-consuming.
   (c) Only organisms that reproduced rapidly were selected and these had small genomes.
   (d) It was only coincidental that the first organisms to be sequenced had small genome sizes
6. How many genes constitute the human genome?
   (a) More than 1 million (b) 4 Mb
   (c) About 30,000 (d) 46

7. Groups of distinctly different genes that often occur together in a cluster are called
   (a) Single-copy genes (b) Segmental duplications
   (c) Multigene families (d) Tandem clusters
8. Whole blocks of genes that have been copied from one chromosome to another are referred to as
   (a) Single-copy genes. (b) Segmental duplications
   (c) Multigene families (d) Tandem clusters
9. Bits of DNA that literally jump from one chromosome to another are
   (a) Introns (b) Transposable elements
   (c) Pseudogenes. (d) Tandem clusters.
10. A discrete collection of gene fragments on a stamp-sized chip is called a
    (a) Reference sequence (b) SNP profile
    (c) Gene microarray (d) Semiconductor
11. Which type of genomics studies the physical nature of genomes?
    (a) Comparative genomics (b) Structural genomics
    (c) Functional genomics (d) None of the above
12. Which type of genomics studies the transcripts and proteins expressed by a genome?
    (a) Comparative genomics (b) Structural genomics
    (c) Functional genomics (d) None of the above
13. Which type of genomics studies similarities and differences among the genomes of multiple organisms?
    (a) Comparative genomics (b) Structural genomics
    (c) Functional genomics (d) None of the above
14. An open reading frame (ORF) is
    (a) The sequence of a complete genome
    (b) A plasmid vector used in genome sequencing
    (c) A possible gene predicted by DNA sequencing
    (d) None of the above
15. Which of these organisms has the smallest genome?
    (a) Haemophilus influenzae (b) Mycoplasma genitalium
    (c) Mycobacterium tuberculosis (d) *E. coli*

Answers

1. (b)	2. (d)	3. (c)	4. (a)	5. (d)	6. (c)
7. (c)	8. (b)	9. (b)	10. (c)	11. (b)	12. (c)
13. (a)	14. (c)	15. (b)			

CHAPTER 30

Proteomics and Protein Sequencing

B. Bajpai and V. Sharma*

Dept. of Bioscience and Biotechnology, Banasthali Vidyapith, Banasthali–304 022

E-mail: vinaysharma30@yahoo.co.uk

There have been a number of revolutions in molecular biology in the past few decades- most recently the completion of the sequencing of human genome. Apart from the information inherent in the sequence itself, the genome provides researcher with an inventory of genes with which to pursue comprehensive global inquiries into the working of cell and organism. As a genome describes the genetic content of an organism, a proteome defines the protein complement of the genome. However, the protein complement (proteome) generated from a given genome is much more complex than the corresponding genome itself. In human for example, about half a million proteins (~500,000) are generated from some 30,000 genes. After processing and modifications a single gene may express between one and a few dozen different protein products. Most importantly while the genome is a rather constant entity, the proteome is constantly changing through the biochemical interactions within the genome. One organism will have radically different protein expression in different parts of the body and in different stages of its life cycle. Thus, the proteome is highly complex and dynamic network generated from an essentially static genomic blueprint.

The term 'proteomics' was coined in 1995 and was defined as the large-scale characterization of the entire protein complement of a cell line, tissue or organism (Wilkins *et al.*, 1995). Today, proteomics has been defined variously by different workers and the commonly encountered definitions of proteomics may be divided into two categories. The first is the more classical definition of proteomics restricting the study to large-scale analysis of proteins, particularly their structure and functions. The second and more inclusive definition combines protein studies with analyses that have a genetic readout such as mRNA analysis, genomics and the yeast two hybrid analyses. Some of the definitions of Proteomics are as below:

*Corresponding author

1. The study of entire protein systems (Proteomes): What are the complement proteins, how they interact with each other, what kinds of metabolic network or signaling network they form, etc.
2. The study of the set of proteins produced (expressed) by an organism, tissue or cell and the changes in the protein expression patterns in different environments and conditions.
3. The study of the full set of proteins encoded by genome.
4. The study of all the proteins expressed in a cell.
5. Systematic analysis of the protein expression of healthy and diseased tissues.

Proteomics attempts to study the structure, functions and control of biological systems and processes by the systematic and quantitative analysis of many properties of proteins. These include the sequence, abundance, activity and structure of the proteins expressed in a cell as well as the modifications, interactions and translocation each protein might experience.

To catalogue all human proteins and ascertain the functions and interactions presents a daunting challenge for scientists. However, there has been a great progress owing to novel instrumentation, improved experimental strategies and bioinformatics methods An international collaboration to achieve these goals is being coordinated by Human Proteome Organization (HUPO) (http://www.hupo.org)). The HUPO was formed on 9th February 2001 to promote a 'broader understanding of the importance of proteomics and the opportunities it offers in the diagnosis, prognosis, and therapy of disease'. However, the project has yet to take on the apparent global focus of its predecessor, the Human Genome Project, the reason for this is both financial and technical.

CLASSIFICATION OF PROTEOMICS

Proteomics is mainly concerned with determining the structure, expression, localization, biochemical activity, interactions, and cellular roles of as many proteins as possible. It can be broadly classified into three categories (Graves *et al.*, 2002):

1. Protein Expression Profiling
2. Functional Proteomics
3. Structural Proteomics.

I. Protein Expression Profiling

The quantitative study of protein profiling between samples that differ by some variable is known as Expression Proteomics. In this approach protein expression of the entire proteome or of sub-proteome between samples can be compared. This is done by the analysis of mRNA expression using various methods such as, Serial Analysis of Gene Expression (SAGE) and DNA microarray technology. However, the analysis of mRNA is not a direct reflection of the protein content in the cell. The formation of mRNA is only the first step in a long sequence of events resulting in the synthesis of proteins. The average number of protein isoforms that result from differential splicing, co-translation modifications (PMT) was predicted to be one or two in bacteria, three in yeasts and five to seven in humans per open reading frame. In addition some bodily fluids such as serum or urine have no mRNA source and therefore cannot be studied by mRNA analysis.

The comparative profiling of proteins regarding their presence or absence under different cellular or physiological conditions is often performed after separating the proteins via 2D-gel electrophoresis and identifying them by mass spectrometry. However, the main constraint here is of low abundance cellular protein constituents and the absence of a PCR equivalent for proteins. To substitute for this, affinity ligands (protein microarrays) were developed for signal enrichment. These affinity ligands may take the form of either: (a) Polyclonal antibody (b) monoclonal antibody (c) single chain antibody (d) domain antibody (e) Receptins (e.g. affibodies, antibody mimics, anticalins (f) aptamers/photoaptamers or (g) other molecules such as peptides, RNA's or combinatorial libraries with desirable affinities. Once Ab/Ag complex is formed, they can be detected by different methods namely, light and color conjugation, isotopic labeling, enhanced chemiluminescence, rolling circles, resonance light scattering, etc.

2. Functional Proteomics

Functional proteomics is a broad term for many specific, directed proteomics approaches. Functional proteomics is the systematic perturbation or functional inactivation of protein within a given physiological environment to elucidate the potential role of target proteins in a cellular process. In some cases, specific sub-proteomes are isolated by affinity chromatography. This approach allows a specific group of proteins to be studied, characterized and provide information about protein signaling, disease mechanism or protein-drug interactions. It is becoming clear that proteins can perform multitude of functions depending on their cellular location, cell types where they are to be expressed, multimeric state and the substrate. Approaches in functional proteomics include specific blocking molecules such as neutralizing antibodies, pharmacological inhibitors, aptamers, trans-dominant mutations and chromophore associated laser inactivation.

3. Structural Proteomics

Structural proteomics is large-scale determination of the 3D-structure of protein complexes or the proteins present in a specific cellular organelle. The structural proteomics or cell map attempts to identify all the proteins within a protein complex or organelle, determine where they are located and characterize all protein-protein interactions. In the cells, proteins do not act in isolation but usually form transitory or stable complexes in order to participate in pathways and act in networks. Protein-protein interactions thus constitute an essential aspect of the normal workings of the living cells and the knowledge about the various interactions in which individual proteins are involved will help in understanding protein functions. Identification of all potential protein folds (domains) and their corresponding functional properties will prove useful in deducing all possible functions from protein 3D-structure. This information will help piece together the overall architecture of cells and explain how expression of certain proteins gives a cell its unique characteristics.

TOOLS AND TECHNIQUES IN PROTEOMICS

To characterize functional protein networks and their dynamic alterations during physiological and pathological processes proteins have to be identified, sequenced,

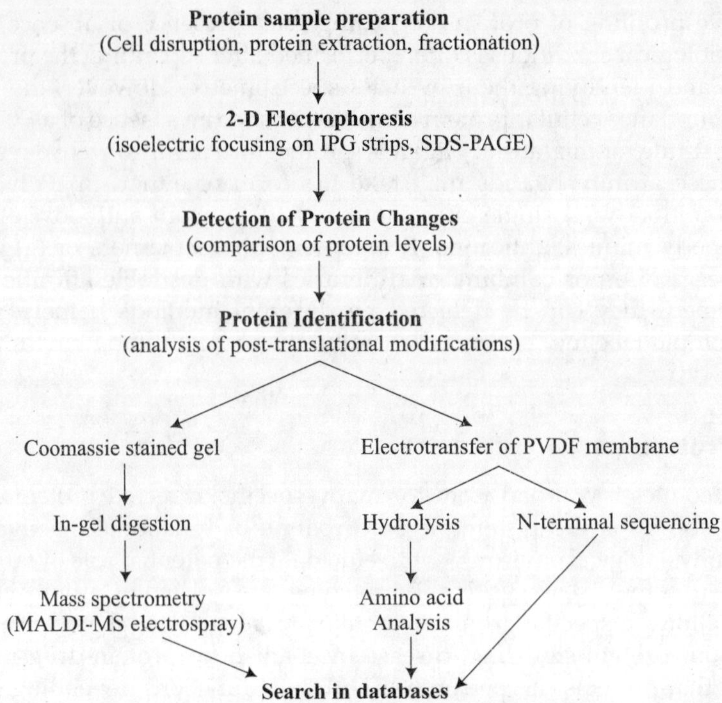

Fig. 1. Workflow in proteomics.

categorized and classified with respect to their functions and interaction patterns in a protein network (Fig. 1). Quantitative protein profiling is an essential part of proteomics and requires new technologies that accurately, reproducibly and comprehensively identify and quantify proteins contained in biological samples. This is achieved by the combination of high-resolution protein separation techniques and modern sequence database mining tools. However, there is currently no single experiment or platform that permits the systematic and quantitative analysis of various properties of proteins on proteome wide scale (Table 1).

KEY TECHNOLOGIES FOR PROTEOMICS

1. 1-D and 2-D electrophoresis are used for separation and visualization of proteins
2. To identify and characterize proteins Mass Spectrometry, X-ray Crystallography, Electron Tomography and NMR are used.
3. To characterize protein-protein interactions, a number of chromatography techniques are used especially affinity chromatography. Protein expression systems like Yeast two hybrid system (Y2HS) and Fluorescence Resonance Energy Transfer (FRET), Isotope Coded Affinity Tag (ICAT), Protein microarrays, etc. can also be used to characterize protein-protein interactions.

Table 1. Key technologies of proteomics

Method	Description	Applications	Practicality
Mass Spectrometry	Digest proteins and fragment peptides to identify proteins.	Protein Identification, sequence, post-translational modifications, quantitation.	Need access to instrument/expertise to run samples
Yeast 2- Hybrid	Genetic approach based on transcription factor reconstitution that identifies protein interactions.	Protein-protein interaction: protein-DNA interaction	No specialized instruments required but high false positive rate
Chips	Synthesize proteins, peptides, antigens or antibodies into an array format and spot on to slides.	Protein interactions with proteins, lipids and small molecules; drug discovery; post-translational modifications, clinical diagnosis.	Need access to facility that can provide chips and analysis capabilities.
2-D gels	Separation of protein samples on the basis of change and molecular weight	Post-translational modification studies, differential protein expression.	Reproducibility issues, dynamic range problems.
Bio-informatics	*In silico* proteomics	Mining databases and/or large-scale data sets, predicting protein interactions.	Relies on experimental data for prediction and confirmation
Structure-based	X-ray crystallography, NMR spectroscopy, electron tomography, 2D electron microscopy	Structure of macromolecular assemblies, determination of protein sub-unit contacts and proximity	Need access to equipment used for structure determination
Small molecules screen	Screen a library of small molecules, looking for compounds that affect a specific cellular function.	Protein-small molecule interactions, drug discovery, enzyme specificity studies.	
Activity based profiling	Active site directed chemical modification, directed chemical modification, directed towards a specific class of biological activity (e.g. Serine hydrolases)	Detection, isolation and identification of active enzymes from complex protein mixtures.	
Imaging	Fluorescence energy resonance transfer between fluorophores that are in close proximity (<100A°): transfected cell arrays. GFP fluorescence	Protein-protein interactions in live cells: protein localization	Clones required relatively simple
Reagents	Antibodies, cDNAs, recombinant protein sets, siRNA constructs	Protein-protein interaction, localization, affinity pulldowns, protein validation	Significant outlay of resources to construct clones, purify proteins, generate antibodies.

2-D GEL ELECTROPHORESIS

In most of the proteomics studies protein is resolved from a mixture by using 1-D or 2-D polyacrylamide gel electrophoresis. In 2-DE, isoelectro focussing (IEF) is applied in the first dimension to resolve proteins by their net charge and then in the second dimension polyacrylamide gel electrophoresis (SDS-PAGE) sieves the components according to their molecular weight. The combination of these two techniques produces resolution far exceeding than that obtained in 1-DE. An overview gel is first run in the pI range of pH 3-10 and the additional gels help zoom in the crowded ranges, for example pH 4-5 or 5.5-6.5. The total number of detected spots increases with higher resolution. The protein spots on 2-DE are visualized by staining with Coomassie blue, Silver nitrate or fluorescent dyes. A protein profile map is then drawn up by image analysis, displaying and matching the spot pattern on a computer screen. In order, to study a new, yet unknown proteome the spots have to be determined and a 2-DE database is built-up as a reference map. In comparison to this reference, related proteomes are analysed and the interpretation will highlight differentially expressed proteins. The first step in the basic analysis of 2-DE images involves the digitization of the two dimensional gels. In most of the cases it is achieved by the use of a laser densitometer, a CCD camera or a Phosphor imager. Once loaded onto the software, spot detection and quantitation is performed automatically, producing a repository of all protein spots contained in the 2-D images, as well as the related quantitative data, such as the spots optical densities, area and volume etc.

One of the greatest strength of 2-DE is the ability to resolve proteins that have undergone some form of post-translational modification. This resolution is possible in 2-DE because many types of protein modifications confer a difference in charge as well as change in mass of protein. The primary application of 2-DE is in protein expression profiling of any two samples: the appearance or disappearance of spots can provide information about differential protein expression, while the intensity of those spots provides quantitative information about protein expression levels. Another application of this technique is in cell map proteomics. 2-DE is used to map proteins from microorganisms, cellular organelles and protein complexes. It can also be used to resolve and characterize proteins in sub-proteomes that have been created by some form of purification of a proteome. 2-DE is a sensitive tool that enables the simultaneous separation and visualization of many unknown proteins, no other method can currently do that, hence it remains a powerful tool for the cataloging of proteins. Many 2-DE databases have been constructed and are available on the world wide web. 2-DE also improves the quality of the results but is a time consuming and labor intensive method with limitations of reproducibility. The number and the types of proteins that can be resolved by 2-DE is also limited, for example, the protein mixture obtained from a eukaryotic cell lysate is too complex to be completely resolved on a single gel. Many large or hydrophobic proteins will not enter the gel during the first dimension and proteins of extreme acidity or basicity (proteins with pIs below pH 3 and above pH 10) are not well represented. Another limitation of 2-DE is the inability to detect low-copy proteins when a total-cell lysate is analysed.

However, the development of sensitive fluorescence staining methods with a linear signal response over a wide dynamic range and the introduction of a two color fluorescent labeling system allowing the concurrent electrophoresis of two differentially labeled protein samples in the same 2-DE gel and thus the determination of the ratio of abundance by spectral analysis of each protein spot have eliminated or at least alleviated some of the challenges related to 2-DE gel imaging. Further, the speed of running 2-DE has been improved and 2-DE gel can now be run in mini gel format. The availability of strips with immobilized pH gradients (IPG) has been an important step for the accessibility and reproducibility of the technique. In addition, there have been efforts to automate the process and the use of computers has aided in the analysis of complex 2-DE gels.

MULTIDIMENSIONAL PROTEIN IDENTIFICATION TECHNOLOGY (MUDPIT) OR ORTHOGONAL SEPARATIONS

MudPIT is a non-gel approach for the identification of proteins from complex mixtures. The method, developed by Link *et al.*, 1999 was initially called as *Direct Analysis of Large Protein Complexes (DALPC)* and involved the integration of two dimension chromatography of peptides rather than proteins and tandem mass spectrometer (MS/MS). In tandem mass spectrometer peptides are fragmented in a predictable fashion, which allows for the computational determination of the peptide sequence and therefore the identification of the protein from which the peptide was derived. DALPC is now commonly referred as *Multidimensional Protein Identification Technology (MudPIT).*

MudPIT couples a biphasic microcapillary column packed with reversed phase (RP) and strong cation exchange (SCX) high-performance liquid chromatography (HPLC) grade materials to a tandem mass spectrometer. By exploiting a peptide's unique physical property of charge and hydrophobicity, a complex mixture can be separated prior to sequencing by tandem MS. The first dimension is normally a strong cation exchange (SCX) column, as these have high loading capacities. The second dimension is reverse phase chromatography (RP), which complements the SCX as it is efficient in removing salts and has the added advantage of being compatible with electrospray mass spectrometry.

Fig. 2 depicts the experimental process of MudPIT. Sample preparation is relatively simple, the samples are denatured, the cysteines reduced and alkylated and the proteins digested with a protease such as trypsin. The samples are then acidified and loaded onto the SCX column. Charged peptides bind to the SCX column, whereas, any uncharged peptides pass through and bind to a reverse phase trap column. The peptides are then eluted from the trap column onto an analytical RP column, using a reverse phase gradient, separated and eluted into a tandem mass spectrometer. Peptide fragmentation data is then obtained to identify the peptides and hence the proteins from which they are derived. In the next step, salt at a particular concentration is injected onto the SCX column, displacing further peptides from it onto the RP trap column. Salt is removed by washing, again an analytical RP separation is performed and the eluting peptides analysed by mass spectrometry. Incremental increases of salt are used (salt step gradient from around 0 – 200 mM). The end result is multiple protein identifications from each salt step. The data are searched using the SEQUEST algorithm, which interprets the MS/MS generated data and identifies the peptide sequence from which it was generated resulting in the determination of the protein content of the original sample.

The procedure is intended to maximize the probability that every protein is represented by at least one peptide in the subsequent analysis. The number of peptides identified by this method is 1500 as compared to only a few hundred for 2-DE based methods. However, the MudPIT derived proteome cannot accurately quantify proteins, only determine whether they are present or absent.

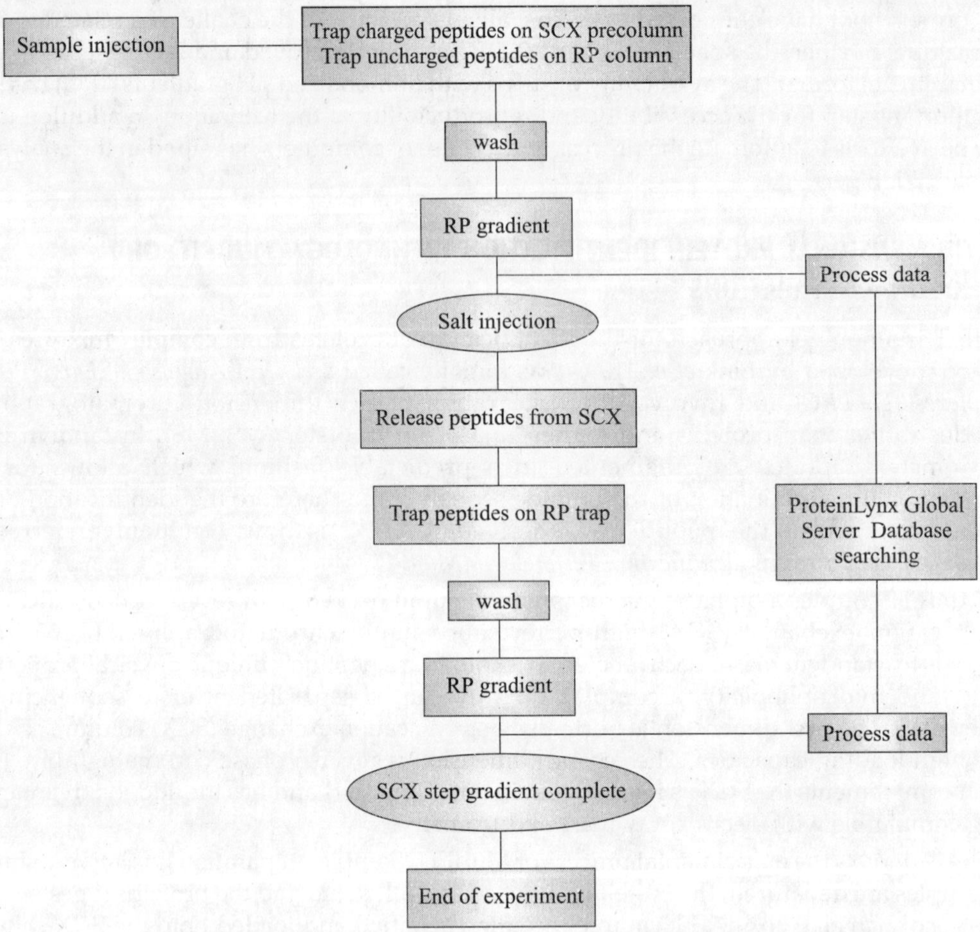

Fig. 2. The experimental process of multi dimensional protein identification technology.

ISOTOPE CODED AFFINITY TAG

In order to study the protein expression, quantitative methods based on isotope tagging of peptides have recently been developed, in particular ICAT method by Gygi *et al.*, 1999; N-terminal labeling of peptides by Munchbach *et al.*, 2000 and using ^{18}O / ^{16}O water by Stewart *et al.*, 2001 and Mirgorodskaya *et al.*, 2000). The ICAT method developed by Gyagi et al, 1999 has been by far the most popular method that has emerged as an alternative to protein electrophoresis, mainly because the kits that provide all the chemicals for protein labeling are commercially available. This method allows the quantitative protein profiling between different samples and relies on the labeling of protein samples from two different sources with two chemically identical reagents (ICAT) that differ only in mass as a result of isotope composition (Fig. 3). The ICAT reagent consists of: (i) a biotin affinity group (ii) a linker region that can incorporate heavy (Deuterium) or light (Hydrogen) atom and (iii) thiol reactive end group.

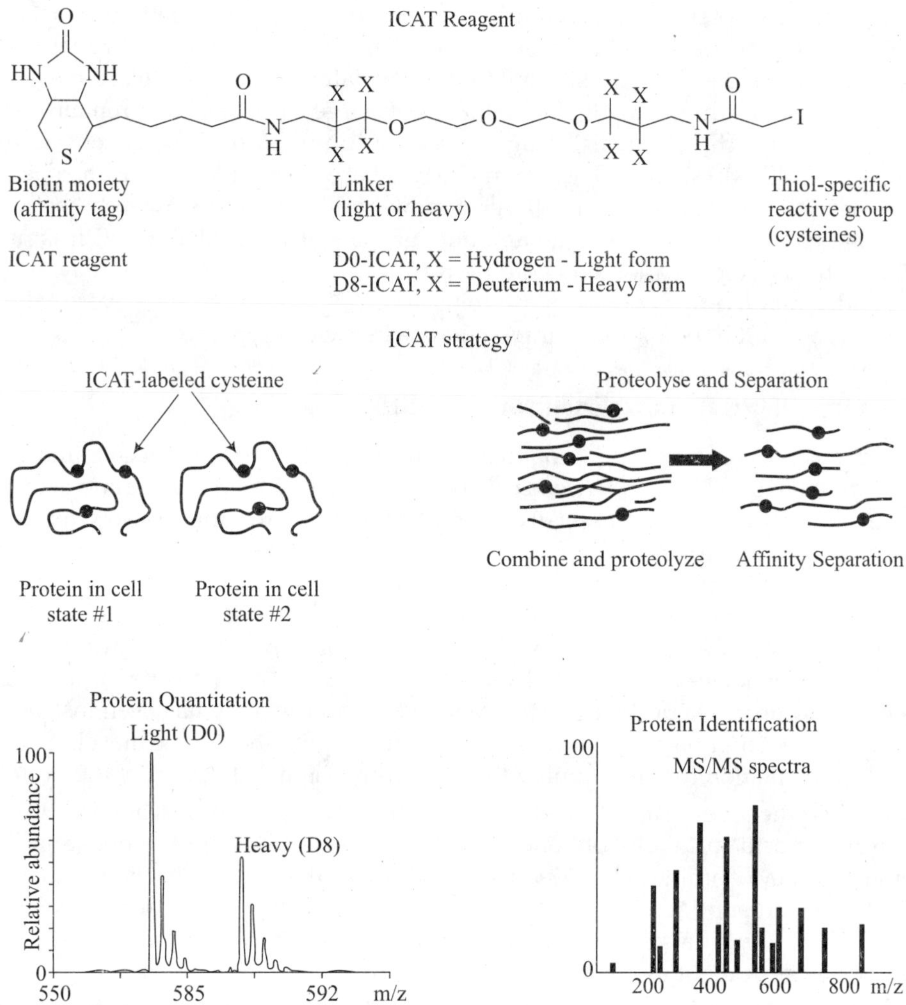

Fig. 3. ICAT strategy.

ICAT REAGENT

The ICAT reagent reacts with cysteine residues in proteins via a thiol reactive group and biotin moiety facilitates its purification. Cell extract from two different samples is reacted with one of the two forms of ICAT reagent, an isotopically light form in which the linker region contains light hydrogen atom or a heavy form in which linker contains deuterium atom. Peptides are recovered on the basis of the biotin tag by avidin affinity chromatography and are then analysed by MS. The difference in peak height between heavy and light peptide ions directly correlate with the difference in protein abundance in the cells. Following quantitation of the peptides, they can be fragmented by MS/MS and the amino acid sequence can be obtained. Thus, using this approach protein can be identified and then expression level can be compared in the same analysis. The major advantage of this method is that it is a gel free approach and even low abundance proteins can be quantified. The bottle- neck of the method is that currently this works only for proteins containing cysteine, even though this includes majority of the proteins. In addition peptides must contain appropriately spaced protease cleavage sites flanking the cystein residues. Finally the ICAT label is large (~500kDa) and remains with each peptide throughout the analysis, fragments resulting from the reagent complicate interpretation of MS/MS spectra. This can make database search more difficult especially for small peptides with limited sequence. ICAT is powerful technique; however, the reagent is expensive, the technique requires large amounts of the starting material and it is not easily compatible with cell fractionation and column purification steps. Sensitivity may also be low tagged peptides derived from low copy proteins are likely to be poorly recovered during the affinity step as a result of non-specific interactions with avidin sepharose.

PROTEIN SEQUENCING BY CONVENTIONAL METHODS

The most important properties of a protein are determined by the sequence of amino acids in the polypeptide chains. This sequence is called the primary sequence of amino acid. In 1953, Frederick Sanger of Cambridge University in England reported the amino acid sequence of two polypeptide chains composing the protein insulin. Since then we know the sequence of thousands of peptide and proteins through the use of methods developed by Sanger' laboratory. Today the amino acid sequences of almost 20,000 different proteins are known. Knowledge of amino acid sequence is extremely useful in a number of ways: (1) it permits comparisons to be made between normal and mutant proteins (2) it permits comparisons to be made between comparable proteins in different species and thereby has been instrumental in positioning different organisms on the evolutionary tree (3) finally and the most important, it is vital piece of information for determining the three dimensional structure of the protein.

Determining the order of amino acids involves the sequential removal and identification of successive amino acid residues from one or the other free terminal of the polypeptide chain. The overall accuracy of amino acid sequencing declines as the length of polypeptide increases. The very large polypeptides in proteins must be broken down into smaller pieces to be sequenced efficiently. The strategy for determining the amino acid sequence of a protein usually involves basic eight steps:

1. If the protein contains more than one polypeptide chain, the chains are separated and purified.
2. Interchain S-S (disulfide) cross-bridges between cysteine residues in the polypeptide chain are cleaved . (If these bonds are interchain linkages then step 2. precedes step1.)
3. The amino acid composition of each polypeptide chain is determined.
4. The N-terminal and C-terminal residues are identified.
5. Each polypeptide chain is cleaved into smaller fragments and the amino acid sequence of each fragment is determined.
6. Step 5. is repeated, using a different cleavage procedure to generate a different and therefore overlapping fragments.
7. The overall amino acid sequence of protein is reconstructed from the sequences in the overlapping fragments.
8. The positions of S-S cross-bridges formed between cysteine residues are located.

Each of these steps is discussed in greater detail in the following sections.

Step I. Separation of Polypeptide Chains

If the protein of interest is heterodimer (composed of more than one polypeptide chain) then the protein must be dissociated and its component polypeptide subunits after separation must be sequenced individually. The non-covalent forces of most of the multimeric proteins can be easily dissociated by exposure to pH extremes, 8M urea, 6M guanidine hydrochloride or a high salt concentration. The individual polypeptide chains then can be separated from one another on the basis of size and/or charge.

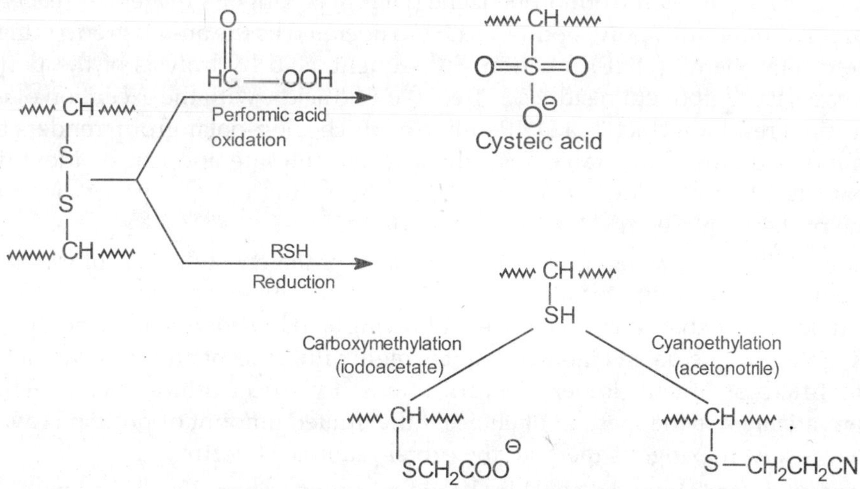

Fig. 4. Methods for cleavage of disulphide bond in proteins.

Step 2. Cleavage of Disulfide Bridges

1. **Performic acid oxidation:** Performic acid oxidation results in the formation of two equivalents of cysteic acid from a disulfide. The electrostatic repulsion between the ionized SO_3^- groups on cysteic acid residues prevents S-S recombination.
2. **Reduction and alkylation of the disulfide** : Sulfhydryl compounds such as 2-mercaptoethanol readily reduce S-S bridges to regenerate two cysteine -SH side chains. This must be followed by treatment with alkylating agent such as iodoacetate or acrylnitrile, which modifies the SH groups and block disulfide bridge formation.

Step 3. Analysis of Total Amino Acid Composition

The complex amino acid mixture in the hydrolysate obtained after digestion of a protein for one to three days in 6N HCl is resolved by ion-exchange chromatography which separates the component amino acids. They can then be individually quantified by spectrophotometric assay. The chromatographic separation and analysis is completely automated in instruments called amino acid analyzer, which can perform the determination on as little as 1 nmol. of protein in less than 1 hour. If the molecular weight and the exact amount of the protein analysed are known (or the number of amino acid residues per molecule is known), the molar ratios of amino acid in the proteins can be calculated.

Step 4. Identification of N- and C-Terminal

End-group analysis identifies the N-and C-terminal residues in the polypeptide chain and the number of ends in the protein. That is, if the protein consists of two or more different polypeptide chains, then more than one end group will be discovered.

N-terminal analysis

The amino acid residue at the N-terminal can be identified by a number of ways, some of which are destructive to protein sample (Fig. 5).

(i) ***Sanger's reagent***: Dinitrofluorobenzene (DFNB) is Sanger's reagent, it reacts with free amino terminus of a polypeptide and HF is liberated as the amino group is tagged with the dinitrophenyl (DNP) function. Subsequent, acid hydrolysis of the polypeptides releases its component residues as free α-amino acids, with the exception of the amino terminal residue, which is a DNP-amino acid. The non-polar group renders this DNP-amino acid ether-extractable from the acid hydrolysate and can be identified by its characteristic migration rate on thin-layer chromatography or paper electrophoresis. The protein sample is destroyed during the analysis.

(ii) ***Dansyl chloride***: Dansyl (or dimethylaminonaphthalenesulfonyl) chloride also reacts to label the free amino terminus of proteins. Like Sanger's reagent, it yields an amino acid derivative that can be recovered following acid hydrolysis of the polypeptide. The advantage of the dansyl label is that it is highly fluorescent and thus can detected and identified at much lower concentrations by spectrofluorometry. The dansyl derivatization is the method of choice if the limited amount of protein is available for analysis. As in Sanger's method, the protein sample is destroyed.

(iii) ***Edman degradation***: In weakly basic solutions, phenylisothiocyanate (Edman's reagent) will combine with the free amino terminus of the protein, which can be excised from the end of the polypeptide chain and recovered as a phenylthiohydantoin (PTH) derivative. This PTH derivative can be identified by chromatographic methods. In this method, the rest of the chain remains intact and can be subjected to further rounds of Edman degradation to identify successive amino acid residues in the chain. Often, the carboxyl terminus of the polypeptide under analysis is coupled to an insoluble matrix, allowing the polypeptide to be easily recovered by filtration following each round of Edman reaction. Thus, Edman reaction not only identifies the N-terminus of the proteins but can also reveal information regarding sequence. In the instrument automated to carry out the reaction cycles of the procedure, so called Edman sequenators, can yield the sequence of the first 30-60 amino acid residues in a polypeptide chain under appropriate conditions.

(iv) ***Leucine Amino peptidase***: An enzymatic method of N-terminal analysis uses leucine aminopeptidase, an enzyme from hog kidney, to hydrolyse the peptide bond linking the N- terminal to the polypeptide chain. It works best when non-polar residues such as Leu are N-terminal. Lecine aminopeptidase will not attack peptide bonds in which the N is contributed by proline.

C-Terminal Analysis

Fewer methods of analysis are available for the identification of C-terminal residue of polypepetides. Chemical methods for carboxyl end-group determination are less satisfactory. The commonly employed methods are as given below:

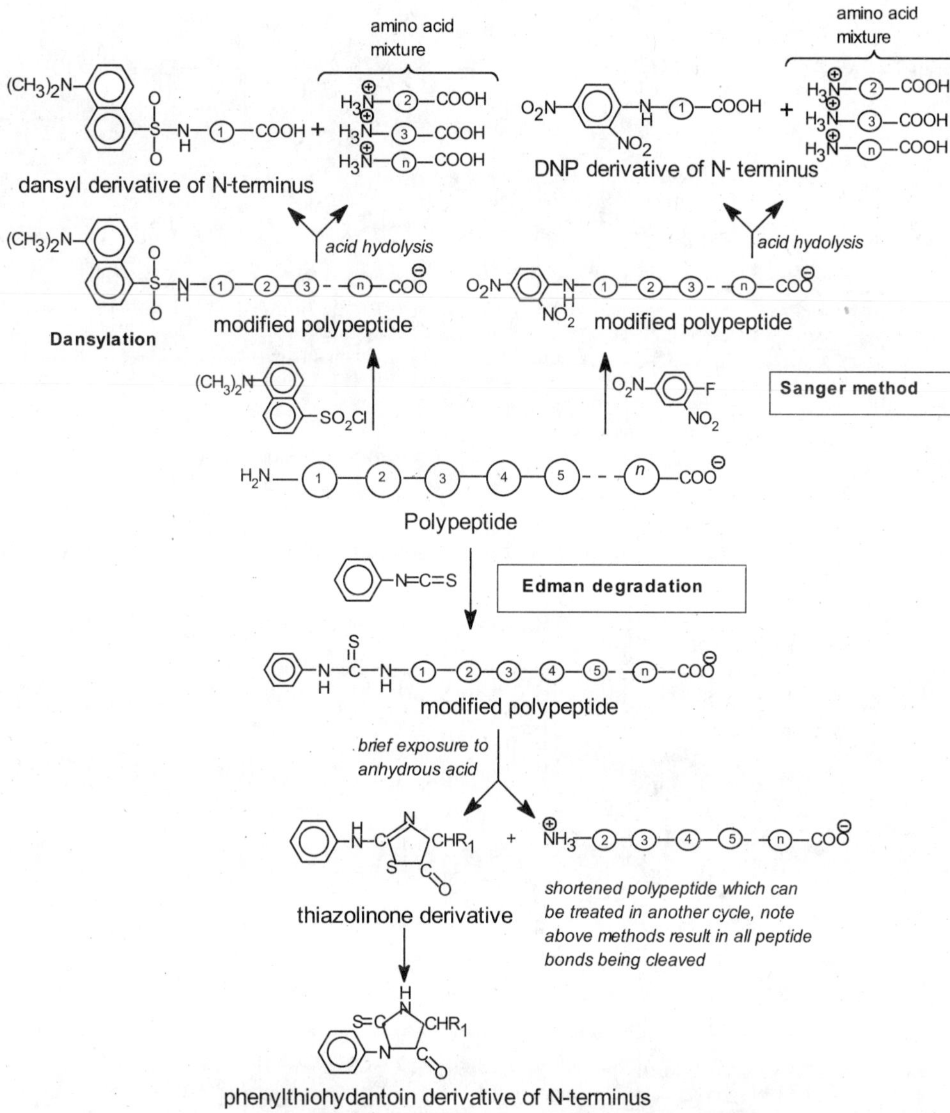

Fig. 5. Methods for determination of N-terminal residues.

1. Hydrazinolysis: Also called as *Akabori* reaction. Treatment of polypeptides with anhydrous hydrazine (H_2N-NH_2) at 100°C results in the conversion of all amino acid residues to amino acid hydrazides except for the carboxyl terminal residue, which remains as the free amino acid which can be isolated and identified chromatographically (Fig. 6). Despite complicating side reactions this is the most satisfactory chemical method presently available.

2. Reduction of C-terminus by $LiAlH_4$: The free α-amino group at the end of a peptide chain can be reduced to an alcohol by the strong reducing agent lithium aluminum hydrazide ($LiAlH_4$). Subsequent acid hydrolysis of the protein yields a mixture of free amino acids and a single free amino alcohol representing the C-terminal residue (Fig. 7).

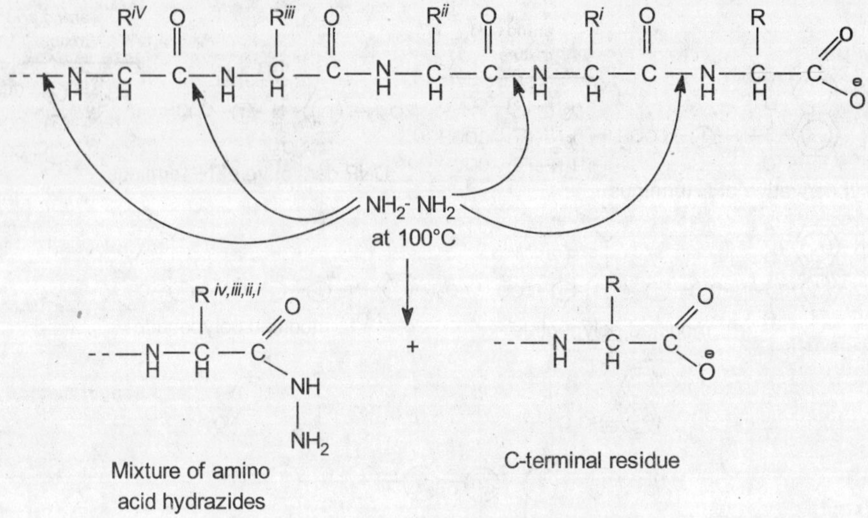

Fig. 6. C-terminal analysis by hydrazinolysis.

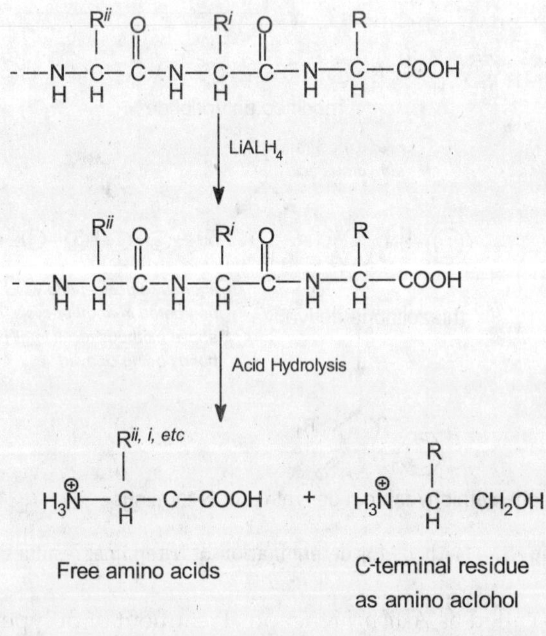

Fig. 7. C-terminal analysis by lithium aluminium hydrazide.

3. Enzymatic analysis with carboxypeptidases : Carboxypeptidases are enzymes that cleave amino acid residues from the C-termini of polypeptides in a successive fashion. The free C-terminal amino acid can then be isolated and identified chromatographically. Four carboxypeptidases are in general use: A, B, C and Y. Carboxypeptidase A (from bovine pancreas) works well in hydrolyzing all the C-terminal peptide bond of all residues except proline, arginine and lysine.The analogous enzyme from hog pancreas, carboxypeptidase B, is effective only when Arg or lys are the C-terminal residues. Thus, a mixture of carboxypeptidases A and B will liberate any C-terminal residue amino acid except proline. Carboxypeptidase C from citrus leaves and carboxypeptidase Y from yeast act on any C-terminal residue (Fig. 8).

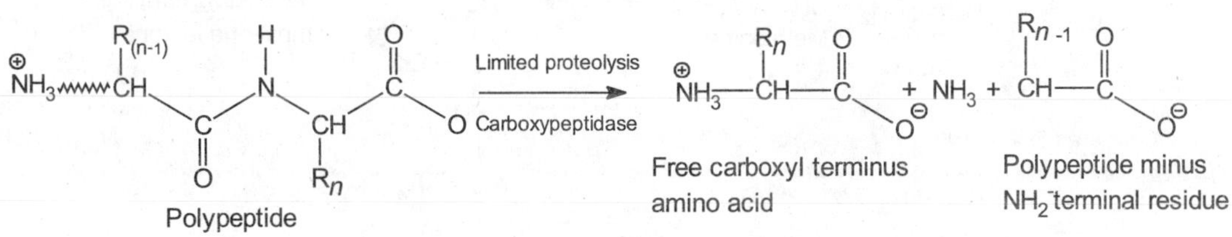

Fig. 8. Enzymatic hydrolysis of peptides with carboxypeptidases.

Step 5 & 6. Fragmentation of the Polypeptide Chain

The aim at this step is to break down the polypeptide chain into shorter, well-defined fragments for subsequent sequence analysis. Fragmentation can be achieved by using endopeptidases, which are enzymes that catalyzed polypeptide-chain cleavage at specific sites in the protein. Table 2 shows the specificity of some of the enzymes commonly used for this purpose. Another specific chemical method for polypeptide chain cleavage involves reaction with cyanogen bromide (CNBr). This reaction cleaves specifically at methionine residues, resulting in the conversion of free carboxyl-terminal methionine to homoserine lactone.

A number of other chemical methods give specific fragmentation of polypeptides, including cleavage at asparagines–glycine bonds by hydroxylamine (NH_2OH) at pH 9 and selective hydrolysis at aspartyl–prolyl bonds under mildly acidic conditions.

The cleavage products generated from the enzymatic/chemical methods must be isolated and individually analyzed with respect to amino acid composition, end group identity and amino acid sequence to accumulate the information necessary to reconstruct the protein's total amino acid sequence. In the past, sequence was often deduced from exhaustive study of the amino acid composition and end-group analysis of small, overlapping peptides. Peptide

Table 2. The specificity of some common methods for fragmenting polypeptide chains

Reagent	Cleavage points
Trypsin	Lys, Arg (C)
Submaxillarus Protease	Arg (C)
Chymotrypsin	Phe, Trp, Tyr (C)
Staphyloccus aureus V8 Protease	Asp, Glu (C)
Asp-N-protease	Asp, Glu (N)
Pepsin	Phe, Trp, Tyr (N)
Endoproteinase Lys C	Lys (C)
Cyanogen bromide	Met (C)

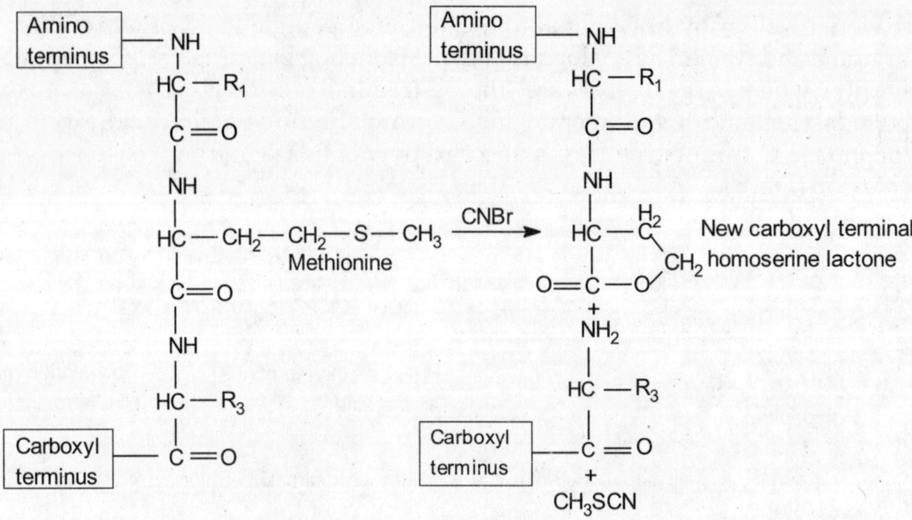

Fig. 9. Cleavage of peptides at methionine by cyanogen bromide.

sequencing today is done most commonly by Edman degradation of relatively large peptides. The Edman degradation is carried out on an automated machine, called a sequenator, that mixes reagents in the right proportions, separates the products, identifies them and records the results. These methods are extremely sensitive. Often, the entire amino acid sequence can be determined starting with only a few micrograms of protein.

Step 7. Reconstruction of Overall Amino Acid Sequence

The sequences obtained from the sets of fragments derived from two or more cleavage procedures are now compared, with the objective being to find overlaps and establish how they are connected together in the intact protein. The sequence analysis of peptide fragments obtained from different cleavage methods makes it possible to piece together the overall sequence. That is, the free amino and carboxyl residues of peptides originally interconnected in the intact protein and liberated by one specific cleavage method will recur in the internal sequence of peptides liberated by a second specific method.

Step 8. Location of Disulfide Cross-bridges

Although disulfide bonds formed between cysteine residues in a protein are not a part of its primary structure, nevertheless, information about their location can be obtained by procedures used in sequencing. Once the primary sequence has been determined, the location of disulphide bonds in the intact protein can be established by repeating a specific enzyme cleavage on another sample of protein with intact disulphide bonds. Separation of the resulting peptides will show the appearance of one new peptide and the disappearance of two other peptides, when compared with the enzymatic digestion product of the material whose disulphide bonds have first been chemically cleaved.

MASS SPECTROMETRY

Mass spectrometry has been a significant driving technology in the development of the field of proteomics. The invention of instruments capable of relatively soft ionization of proteins/ peptides and other biological molecules with very fine sensitivity, have made mass spectrometry an indispensable tool for proteomics studies.The analysis of proteins or peptides by MS can be divided in two general categories:

(i) Peptide mass fingerprinting

(ii) Amino acid sequencing.

Most mass spectrometers consist of four basic elements: (i) an ionization source (ii) one or more mass analysers (iii) an ion mirror and (iv) a detector (Fig. 10).

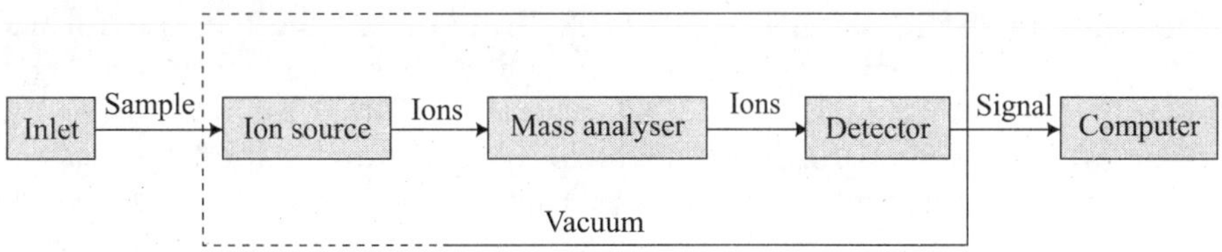

Fig. 10. Basic diagram for a mass spectrometer.

Sample Introduction

The sample can be inserted directly into the ionization source, or can undergo some kind of chromatography *en route* to the ionization source. The latter method of sample introduction commonly involves the mass spectrometer with being coupled directly to a high pressure chromatography (HPLC), gas chromatography (GC) or capillary electrophoresis (CE) separation, hence the sample is separated into a series of components which then enter the mass spectrometer sequentially for individual analysis.

Methods of Sample Ionization

Many ionization methods are available and each has its own advantages and disadvantages. The ionization method to be used should depend on the type of sample under investigation and the mass spectrometer available.

Ionization method includes the following types:

1. Atmospheric Pressure Chemical Ionization (APCI)
2. Chemical Ionization (CI)
3. Electron Impact (EI)
4. Electrospray Ionization (ESI)
5. Fast Atom Bombardment (FAB)
6. Field Desorption / Field Ionization (FD / FI)
7. Matrix Assisted Laser Desorption Ionization (MALDI)
8. Thermospray Ionization (TSP)

The ionization methods used for the majority of the biochemical analysis are Electrospray Ionization (ESI) and Matrix Assisted Laser Desorption Ionization (MALDI). With most ionization methods there is possibility of creating both positively and negatively charged sample ions, depending on the proton affinity of the sample. If the sample has functional groups that readily accept a proton (H^+) then positive ion detection is used e.g. amines R-NH_2 + H^+ to R-$NH_3{}^+$ as in proteins, peptides.

If the sample has functional groups that readily lose a proton then negative ion detection is used e.g. carboxylic acids R-CO_2H to R-$CO_2{}^-$ and alcohols R-OH R-O^- as in saccharides, oligonucleotides.

Analysis and Separation of Sample Ions

The main function of the mass analyzer is to separate or resolve the ions formed in the ionization source of the mass spectrometer according to their mass-to-charge (m/z) ratios. The most commonly used mass analyzers are Quadrupoles, Time-of-Flight (TOF) analyzers, Magnetic Sectors, Fourier Transform and Quadrupole Ion traps. All of the mass analyzers listed above can be coupled with ESI, except that MALDI is not usually coupled to a Quadrupole analyzer.

Detection and Recording of Sample Ions

The detector monitors the ion current, amplifies it and the signal is then transmitted to the data system where it is recorded in the form of mass spectra. The m/z values are plotted against their intensities to show the number of components in the sample, the molecular weight of each component, and the relative abundance of various components in the sample. The type of the detector depends on the analyzer, the more common ones are the photomultiplier, the electron multiplier and the micro-channel plate detectors.

Electrospray Ionization (ESI)

In ESI, a liquid sample flows through a microcapillary tube into the orifice of the mass spectrometer, where a potential difference between the capillary and the inlet to the mass spectrometer results in the generation of the mist of charged droplets. As the solvent evaporates, the size of the droplet decreases, resulting in the formation of desolvated ions. As the droplet dry, the peptide within the drop concentrate until electrostatic repulsion forces the droplet to explode into finer droplets. Ions generated by ESI usually bear multiple charges (M+nH)n+ (Fig. 11). A quadrupole analyzer is frequently used with ESI. Only ions of certain m/z will have the correct oscillation path that enables it to reach the detector in the presence of an electric field.

A significant improvement in the ESI technology occurred with the development of nanospray ionization. In nanospray ionization, the microcapillary tube has a spraying orifice of 1 to 2 μm and flow rates as low as 5 to 10 nl/min. With such low flow rates not only far less sample is consumed but the sample lasts for several minutes, enabling multiple experiments to be performed. A common application of this technique is for a protein digest mixture to be analyzed to generate a list of molecular weights for the components present and then each component is analyzed further by tandem mass spectrometric amino acid sequencing

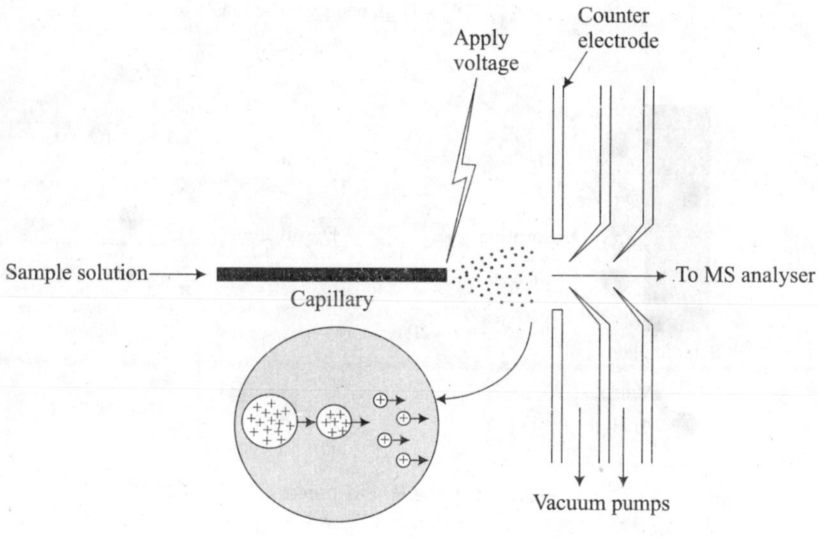

Fig. 11. Schematic diagram of an ESI interface and the ESI process.

techniques. ESI and nanospray ionization are very sensitive analytical techniques but the sensitivity deteriorates with the presence of non-volatile buffers and other additives, which should be avoided as far as possible.

In positive ionization mode a trace of formic acid is often added to aid protonation of the sample molecules, in negative ionization mode a trace of ammonia solution or a volatile amine is added to aid deprotonation of sample molecule. Proteins and peptides are usually analysed under positive ionization conditions and saccharides and oligonucleotides under negative ionization conditions. In all cases, the m/z scale must be calibrated by analyzing a standard sample of a similar type to the sample being analysed (e.g. a protein standard for a protein sample) and then applying a mass correction.

MATRIX ASSISTED LASER DESORPTION IONIZATION (MALDI)

MALDI is based on the bombardment of sample molecules with a laser light to bring about sample ionization. In MALDI, sample is incorporated into matrix molecules and then subjected to irradiation by a laser. The laser promotes the formation of molecular ions. The matrix is typically a small energy absorbing molecule such as 2,5-dihydroxybenzoic acid or alpha-cyano-4hydroxycinnamic acid. The analyte is spotted along with the matrix on metal plate and allowed to evaporate, resulting in the formation of crystals. The plate, which can be 96-well format, is then placed in the mass spectrometer and the laser is automatically targeted to specific places on the plate. Since the sample application can be performed by a robot the entire process can be automated, this is the single biggest advantage of MALDI. Another advantage of MALDI over ESI is that samples can often be used directly without any purification after in-gel digestion. MALDI is a soft ionization method and so result predominantly in the generation of singly charged molecular-related ions regardless of the molecular weight, hence the spectra are relatively easy to interpret. Fragmentation of the sample ions does not usually occur.

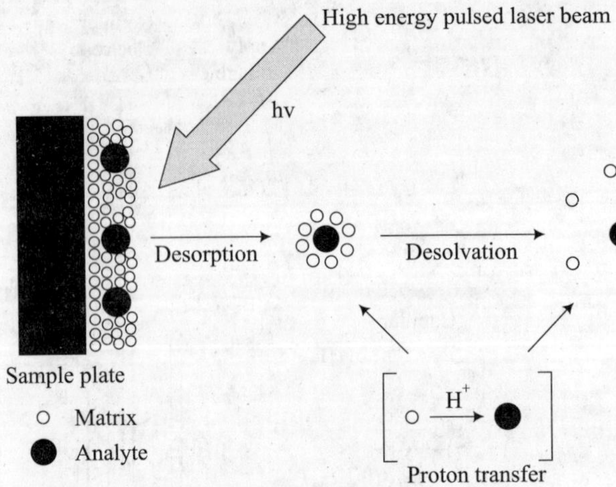

Fig. 12. The MALDI process.

Mass Analysis

Mass analysis follows the conversion of proteins or peptides to molecular ions. This is accomplished by mass analyzer in a mass spectrometer, which resolves the molecular ions on the basis of their mass to charge ratio in a vacuum. Some of the different types of mass analyzers are:

(i) Quadrupole: One of the most commonly used mass analysers is the quadrupole mass analyser. In this ions are transferred through an electric field created by an array of four metal rods, the quadrupole. A quadrupole can act to transmit all ions or as a mass filter to allow the transmission of ions of a certain mass to charge (m/z) ratio. If multiple quadrupoles are combined they can be used to obtain information about the amino acid sequence of a peptide.

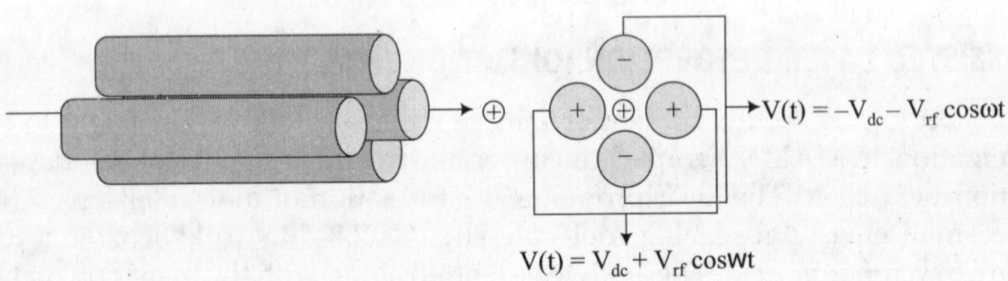

Fig. 13. Schematic diagram of a quadrupole mass analyzer, showing the direction of ion travel and the equations for the potentials applied to the rods. V(t)+voltage at time t, V.

(ii) **Time of flight:** A time–of–flight (TOF) instrument is one of the simplest mass analysers. It measures the m/z ratio of an ion by determining the time required for it to traverse the length of a flight tube. Some TOF mass analysers include an ion mirror at the end of the flight tube, which reflects ions back through the flight tube to a detector. In this way, the ion mirrors serves to increase the length of the flight tube. The ion mirror also corrects for the small energy differences among the ions. Both of these factors contribute to an increase in mass resolution.

(iii) Ion trap: Ion trap mass analysers function to trap molecular ions in a 3-D electricfield. In contrast to quadrupole mass analyser where ions are discarded before the analysis begins, the main advantage of ion trap mass analyser is the ability to allow ions to be stored and then selectively ejected from the trap increasing sensitivity.

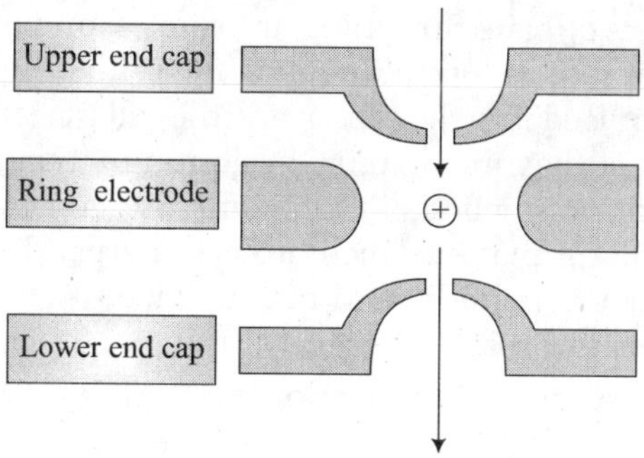

Fig. 14. Schematic diagram of an ion trap mass analyser.

TYPES OF MASS SPECTROMETER

The names of the various instruments are derived from the name of their ionization source and the mass anlayser. Tandem mass spectrometers (MS/MS) are instruments that have more than one analyzer and therefore can be used for structural and sequencing studies. Two, three and four analyzers have all been incorporated into commercially available tandem instruments and the analyzers do not necessarily have to be of the same type, in which case the instrument is a hybrid one. Most popular tandem mass spectrometer includes those of the quadrupole-quadrupole, magnetic sector-quadrupole and more recently, the quadrupole-time-of-flight geometries.

(i) Triple quadrupole: Triple quadruple mass spectrometer is most commonly used to obtain amino acid sequences. In the first stage of analysis, the machine is operated in MS scan mode and all ions above certain m/z ratio are transmitted to the third quadruple for mass analysis. In the second stage, the mass spectrometer is operated in the MS/MS mode and a particular peptide ion is selectively passed on to collision chamber. Inside the collision chamber the peptide ions are fragmented by interactions with an inert gas (argon) by a process known as collision induced dissociation (CID) or collision activated dissociation (CAD). The peptide ion fragments are then resolved on the basis of their m/z ratio by the third quadruple. Since two mass spectra are obtained in this analysis, it is referred to as tandem mass spectrometry (MS/MS). MS/MS is used to obtain the amino acid sequence of peptides by generating a series of peptides that differ in mass by a single amino acid. ESI typically uses a triple-quadruple mass spectrometer with collision induced dissociation to perform these analyses.

(ii) Quadruple - TOF (QqTOF): In recent years several "hybrid" mass spectrometers have emerged from the combination of different ionization sources with mass analyser. In this machine, the first quadruple (Q1) and the quadruple collision cell (q) of a triple-quadruple machine have been combined with a time-of-flight analyser. The main application of a QqTOF mass spectrometer is protein identification by amino acid sequencing and characterization of protein modifications.

(iii) Fourier Transform Mass Spectrometer (FTMS): A Fourier transform mass spectrometer is (FTMS), also known as an ion-cyclotron resonance mass spectrometer (ICR-MS), is a type of ion-trap consisting of a cubic cell inside a strong magnetic field. The cell has three sets of plates: trapping, transmitter and receiver. For mass analysis, a continuous beam of ions is formed outside the cell and guided up to the trap and a portion of beam is pulsed into the cell. Inside the cell the ions are constrained by strong magnetic field so that they move in the cyclotron motion (circular motion) in a plane perpendicular to the magnetic field; they are also constrained by electric potentials applied to the trapping plates, which are also perpendicular to the magnetic field. Cyclotron movement is periodic and characterized by its cyclotron frequency with which an ion repeats its orbit.

(iv) MALDI-TOF: The principal application of MALDI-TOF mass spectrometer is in peptide mass fingerprinting, it can be completely automated. A MALDI-TOF instrument can also be used to obtain amino-acid sequence of peptides by a method known as post-source decay, however, it is not a very reliable method.

PEPTIDE MASS FINGERPRINTING (PMF)

PMF is a method of identifying proteins that relies on the use of a protease to digest the protein in question into a number of smaller peptides. Not all proteases are suitable and the most commonly used one is trypsin. Trypsin cuts at C-terminal of the two basic amino acids lysine (K) and arginine (R), which are fairly common residues. As different proteases cleave at different amino acid residues the PMF of the protein will depend on the protease used, but will always be the same for a particular enzyme used. This means that as long as the digestion is complete, that is the molecule is cleaved at all the possible sites it will produce a set of peptides, of varying masses, which are characteristic of that protein. The mass of each peptide will be the sum of the amino acids present including any modifications that those amino acids might have undergone. Peptide mass fingerprinting is limited to the identification of proteins for which sequences are already known, it is not a method of structural elucidation.

PMF MALDI-TOF is extremely effective for the analysis of proteins and peptides. Measurement of peptide masses are most easily performed by Matrix Assisted Laser Desorption Ionization-Time of Flight (MALDI-TOF) mass spectrometers. The masses of about 4-5 peptides are sufficient to identify a protein of known amino acid sequence unambiguously. However, as databases of known protein sequences have become larger, the amount of data required to identify a specific protein has increased

Amino Acid Sequencing

Protein identification by amino acid sequencing is less ambiguous than those from PMF because the peak pattern in MS/MS provides information additional to the measured mass of the peptides. However, the amount of data generated is huge. The commonly used search algorithms for the identification of proteins are SEQUEST, MASCOT and MS-Tag44.

Precursor or parent ion scanning: The first analyser allows the transmission of all sample ions, whilst the second analyser is set to monitor specific fragment ions, which are generated by bombardment of the sample ions with the collision gas in the collision cell. This type of instrument is particularly useful for monitoring groups of compounds contained within a mixture which fragment to produce common fragment ions, e.g. glycosylated peptides in tryptic digest mixture, aliphatic hydrocarbons in an oil sample, or glucuronide conjugates in urine.

Product or daughter ion scanning: The first analyser is used to select user-specified sample ions arising from a particular component; usually the molecular related (i.e. M+H+) or (M-H-) ions. These chosen ions pass into the collision cell, are bombarded by the gas molecules which cause fragment ions to be formed, and these fragments are analysed i.e. separated according to their mass to charge ratios, by the second analyser. All the fragment ions arise directly from the precursor ions specified in the experiment and thus produce a fingerprint pattern specific to the compound under investigation. This type of experiment is particularly useful for providing structural information concerning small organic molecules and for generation of peptide sequence information.

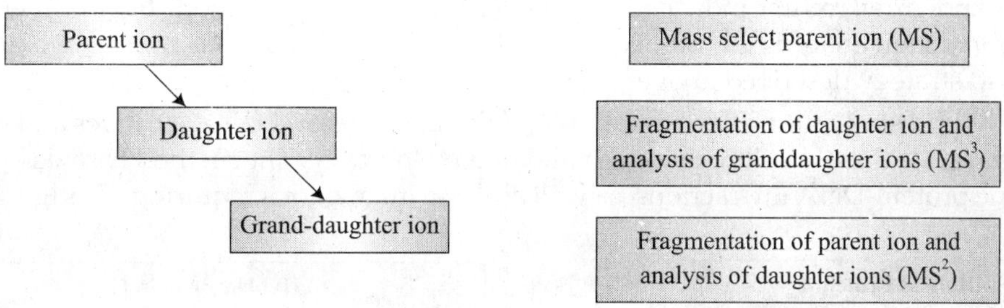

Fig. 15. Generation of fragment ions via collision-induced dissociation (CID) and the mass analysis (MS^n) of the progeny fragment ions.

Protein-protein Interactions

Protein interactions fundamentally can be characterized as stable or transient, both stable and transient interactions can be either strong or weak. Stable interactions are those associated with proteins that are purified as multi-subunit complexes. The subunits of the complex can be identical or different. Hemoglobin and core RNA polymerase are two examples of stable multi-subunit complex interactions. Stable interactions are best studied by co-immunoprecipitation, pull-down or far-Western methods. Transient interactions are expected to control the majority of cellular processes. As the name implies, transient interactions are on/off or temporary in nature and typically require a set of conditions that promote the interaction. Transient interactions can be strong or weak, fast or slow. While in contact with their binding partners, transiently interacting proteins are expected to be involved in the whole range of cellular processes including protein modification, transport, folding, signaling, cell cycling, etc. Transient interactions can be captured by cross-linking or label transfer methods.

THE "BAIT-PREY" MODEL – THE YEAST TWO-HYBRID SYSTEM

The yeast two hybrid system was used in 1989 as a tool to elucidate protein-protein interaction (Fileds and Song, 1989). It is now a mature technology and has been utilized extensively by many laboratories in a low through-put manner. It has also been automated and applied to various large-scale projects. The yeast two-hybrid system uses the transcription process to make predictions about protein interaction. This method is based on the ability of an interacting protein pair to bring together the DNA binding domain and activation domain of a transcription factor *in vivo* to produce a functional activator of transcription. The interaction can be detected by expression of the linked reporter genes. The system requires that two yeast hybrids be prepared, one in which the bait protein is fused with a transcription factor and one in which the prey protein is fused with a related transcription factor. If the bait and prey proteins indeed interact then the two factors fused to these two proteins are also brought into proximity with each other, creating a functional transcription factor. As a result a specific signal is produced (for example, transcription of the intended reporter *ß-galactosidase* gene), indicating an interaction has taken place (Fig. 16). The "bait-prey" nomenclature has been adopted and applied to *in vitro* methods used to study protein interactions. Most interactions currently known were first indicated by the yeast two-hybrid method. *In vitro* methods for protein interaction analysis are often employed to confirm interactions indicated by the yeast two-hybrid strategy described above.

Two-hybrid studies are often fraught with false positives and false negatives and on a high throughput scale these results cannot be validated biologically. The method has been modified to examine protein-DNA interactions, as well as those interactions requiring a bridging protein

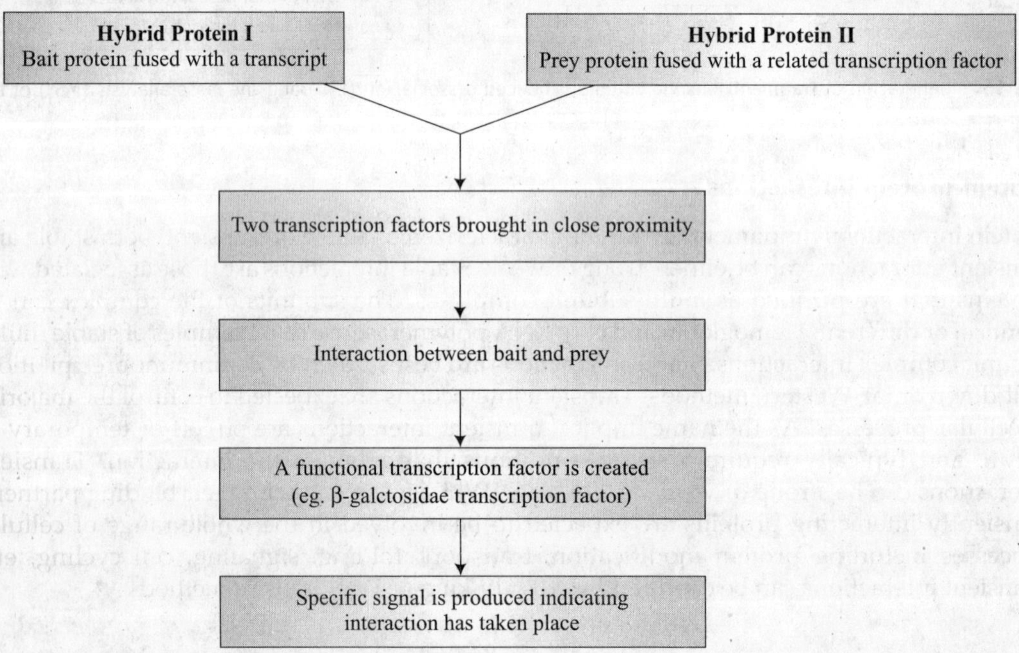

Fig. 16. Bait-prey model to study protein-protein interaction.

or a post-translational modification; there have been other modifications to the technique that reconstitute a protein activity other than that of a transcription factor, making the method more suitable for those proteins which cannot be studied by the traditional yeast two-hybrid .

PROTEIN MICROARRAYS (PROTEIN CHIPS)

Protein microarrays are another approach for analyzing large sets of proteins. In principle protein microarrays are extremely versatile and could be adapted for use in many areas of biology (e.g. lipids, carbohydrates, proteins, nucleic acids). Microarrays were first developed as a tool for high-throughput genotyping and gene expression analysis. Gene expression arrays operate under the pretense that changes in mRNA levels ultimately correlate to changes in encoded protein levels; often this assumption does not hold true. Additionally, gene expression arrays provide no information on protein post-translational modifications (phosphorylation, glycosylation, etc.) that affect cell function. To examine expression at the protein level and otherwise acquire quantitative and qualitative information on proteins of interest, the protein microarray (chips) was developed. Such chips would require two key components: (i) spatially addressed molecules that recognize individual protein moieties and (ii) a method to detect the interaction of the individual proteins in the mixture with their corresponding recognition molecules. In this technique, individually purified ligands or recognition molecules such as proteins, peptides, antibodies, antigens, aptamers, carbohydrates or small molecules are spotted on to a derivatized surface and are generally used for examining the protein expression level for protein profiling and clinical diagnostics. The production of synthetic phage-displayed antibodies provides considerable scope for the generation of protein recognition molecules to support the development of protein chips. Apart from antibodies, any group of molecules that can specifically recognize individual protein with appropriate affinity and avidity and can do so in immobilized form would be suitable. There are disadvantages to the use of antibody proteins as protein recognition molecules: (i) the chips might suffer from poor storage properties and (ii) the presence of any protease activity within the protein mixtures to be analyzed may degrade the chip. The large array of molecules derived from combinatorial chemistry could also provide a good source of protein recognition molecules. Aptamers are single-stranded oligonucleotides possessing high affinity for conformational biomolecules, such as proteins. Libraries of up to a few hundred nucleotides in length may be immobilized to an appropriate surface for screening and may provide information on conformation, position of hydrogen bonds and other data which help to build up a three-dimensional model of the target protein. This methodology has the potential for the development. Artificial receptors or antigen-binding sites have also recently been generated using molecularly imprinted polymers. Arrays containing recombinant proteins can also be used to monitor protein activities such as binding properties to small molecules, proteins, antibodies and drugs and for post-translational modifications. In addition, the ability to assay small molecules and drug-binding properties makes this technique attractive to pharmaceutical companies for use in drug discovery and validation. Apart from the difficulty in expressing large numbers of human proteins in soluble form, a main challenge with protein arrays is presenting the proteins in native state such that they can still participate in interactions. Additionally, these are in vitro assays and need to be followed in vivo. An interesting application of protein microarrays is to check the specificity of antibodies by incubating recombinant protein arrays with them. A brief overview of some of the protein array is as follows:

Antibody-pair Protein Arrays

Antibody microarrays fall into one of two subtypes: those using matched antibody pairs for sandwich-type assays and those utilizing single antibodies and a sample labeling methodology. This type of microarray consists of arrayed capture antibodies and appropriate control and orientation elements. Assays are performed by adding an antigen standard or test sample followed by a detector antibody (Fig. 17). The detector antibody is either modified with a directly detectable label (enzyme, fluorescent molecule, isotope, etc.), as illustrated in Figure, or it is biotinylated for detection after subsequent probing with labeled streptavidin. Antibody-pair microarrays essentially are multiplexed ELISAs. Sandwich-style antibody pair microarrays can be used for qualitative or comparative (i.e., treated versus non-treated) detection of protein analytes or for protein quantification when appropriate standards are used for the preparation of standard curves. One of the limitations of ELISA-type sandwich assay is that it is difficult to produce specific antibodies for large number of human proteins, and most applications so far have been limited to cytokines where these antibodies exist.

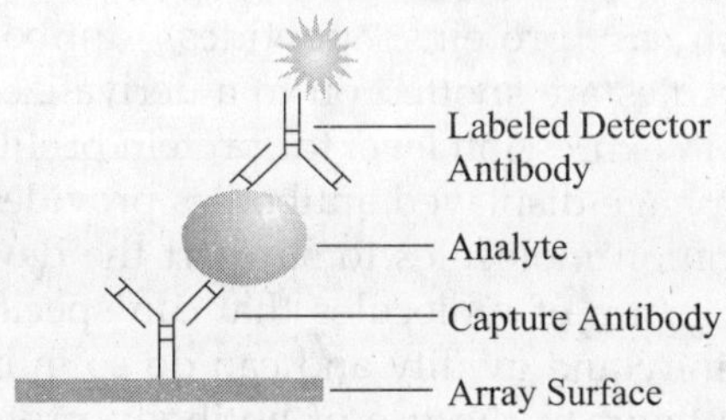

Fig. 17. Architecture of sandwich-style antibody-pair microarray assays.

Single Antibody/ Labeled Sample Protein Assay

When matched antibody pairs are not available, single-antibody protein microarray protocols involving labeled samples can be used. As in sandwich antibody pair arrays, the array platform consists of arrayed antibody (or antibodies). In this assay format, however, the captured protein analytes are themselves labeled for direct detection, obviating use of a detector antibody (Fig. 18). The method requires that protein samples be labeled beforehand (e.g., with fluorescent molecule, isotope, or biotin). The label enables detection of any proteins in the sample that interact with the microarrayed antibody and associated elements.

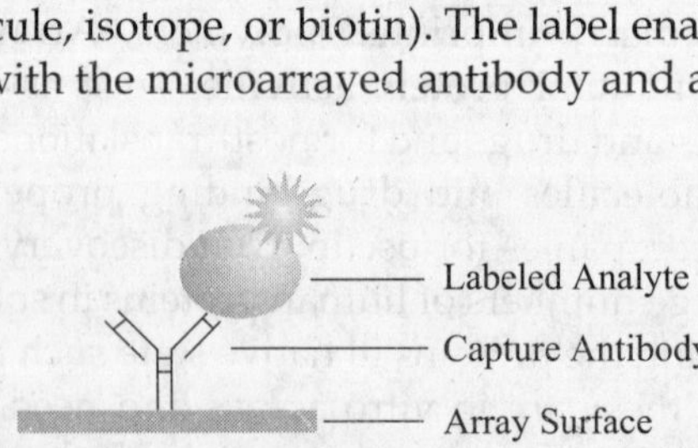

Fig. 18. Architecture of single antibody/labeled sample microarray assays.

As stated previously, this technique is useful for examining protein targets, such as poorly characterized cell signaling proteins, for which paired antibodies do not yet exist. The main drawback to this method is its lack of antibody redundancy, which helps to ensure specific antigen recognition. Additionally, since all sample constituents are labeled (i.e., the target as well as other proteins in the sample), nonspecific background signal increases. This technique is primarily used for comparative and qualitative studies.

Cellular Lysate Protein Arrays

Microarrays of cellular proteins are either arrayed as complex protein mixtures, or arrayed as purified or over-expressed proteins. Complex protein mixture arrays are essentially dot blots of cellular lysates. Creating microarrays consisting of libraries of purified or over-expressed proteins allows for screens for protein-protein interactions and kinase activities. A representation of this type of array is the yeast kinase microarray produced by Zhu *et al.*; this array consisted of 119 of the 122 known yeast kinases.

This diverse set of applications clearly demonstrates that the protein microarray is a powerful tool. Unfortunately, costs for equipment such as arraying robots and slide scanners limit the number of researchers able to take advantage of this technology.

FRET (FLUORESCENCE RESONANCE ENERGY TRANSFER)

Optical imaging with fluorescent fusion proteins allows dynamic measurements of several tagged proteins in live cells. The method describes an energy transfer mechanism between two fluorescent molecules. A fluorescent donor is excited at its specific fluorescence excitation wavelength. By a long dipole-dipole coupling mechanism, this excited state is then non-radiatively transferred to a second molecule, the acceptor. The donor returns to the electronic ground state. The described energy transfer mechanism is termed "Foster Resonance Energy transfer " after the name of German scientist Theodar Foster. When both molecules are fluorescent, the term "Fluorescence resonance energy transfer (FRET) is often used. The FRET efficiency is determined by three parameters:

1. The distance between the donor and the acceptor.
2. The spectral overlap of the donor emission spectrum and the acceptor absorption spectrum.
3. The relative orientation of the donor emission dipole moment and the acceptor absorption dipole moment.

FRET is a useful biophysical tool to quantify molecular dynamics, such as protein-protein interactions, protein-DNA interactions and protein conformational changes. For monitoring the complex formation between two molecules, one of them is labeled with a donor and the other with an acceptor and these fluorophore-labeled molecules are mixed. When they are dissociated, the donor emission is detected upon the donor excitation. On the other hand, when the donor and acceptor are in close proximity (1–10 nm) due to the interaction of the two molecules, the acceptor emission is predominantly observed because of the intermolecular FRET from the donor to the acceptor. For monitoring protein conformational changes, the target protein is labeled with a donor and an acceptor at two loci. When a twist or bend of the protein brings the change in the distance or relative orientation of the donor and acceptor,

FRET change is observed. If a molecular interaction or a protein conformational change is dependent on ligand binding, this FRET technique is applicable to fluorescent indicators for the ligand detection.

The most popular FRET pair for biological use is a cyan fluorescent protein (CFP)-yellow fluorescent protein (YFP) pair. Both are color variants of green fluorescent protein (GFP). While labeling with organic fluorescent dyes requires troublesome processes of purification, chemical modification, and intracellular injection of a host protein, GFP variants can be easily attached to a host protein by genetic engineering. By virtue of GFP variants, the use of FRET techniques for biological research is becoming more and more popular. One advantage of FRET is that it can be used to provide information on proteins in physiological state and in their correct cellular location and to perform temporal experiments to follow transient interactions.

Multiplexed Proteomics

The Multiplexed Proteomics (MP) approach is a new technology that permits quantitative, multicolor fluorescence detection of proteins in two-dimensional (2-D) gels and on Western blots. This methodology allows the parallel determination of both post-translational modification patterns and protein expression level changes within a single 2-D gel experiment. The linear responses of the fluorescent dyes utilized allow rigorous quantitation of changes in protein expression over a broad 3-log linear dynamic range. The Multiplexed Proteomics technology relies on a set of protein stains that enable the global detection of changes both in specific protein expression (phosphoproteins, glycoproteins or membrane proteins) and in total-protein expression in 1-D and 2-D polyacrylamide gels or on blots. Some of the fluorescent and luminescent protein gel stains include:

Pro-Q Diamond phosphoprotein gel stain

Pro-Q Emerald glycoprotein gel stains

Pro-Q Amber transmembrane protein gel stain

SYPRO Ruby protein gel stain

The Pro-Q Diamond phosphoprotein gel stain, Pro-Q Emerald glycoprotein gel stains and SYPRO Ruby protein gel stain are used in serial detection of phosphoproteins, glycoproteins and total proteins on a single protein sample separated by 1-D or 2-D gel electrophoresis. After each staining step, an image of the gel is collected. Once collected, the three images can be overlaid in any combination for analysis of phosphorylation, glycosylation and total-protein expression between samples. Because all three stains are used on the same gel, unambiguous spot matching of phosphoproteins and glycoproteins is made simple by direct comparison with the total-protein profile provided by the SYPRO Ruby protein gel stain. This simultaneous measurement of several variables ensures perfect spatial registration of signals and increases the amount of data that can be collected in a single experiment, leading to more controlled experiments, more accurate data comparisons and fewer ambiguities. When used in combination with the SYPRO Ruby protein gel stain, the Pro-Q Amber transmembrane protein gel stain allows selective identification of transmembrane proteins. The staining procedure is rapid, simple to perform, readily reversible and fully compatible with modern microchemical analysis procedures, such as matrix-assisted laser desorption/ionization-time of flight (MALDI-TOF) mass spectrometry.

Table 3. *In vitro* methods to study protein-protein interactions.

***In vitro* Methods**	**Description**
Co-Immunoprecipitation (Co-IP)	An immunoprecipitation (IP) experiment designed to affinity purify a bait protein antigen together with its binding partner using a specific antibody against the bait.
Cross-linking Reagents	Strategies involve homo- or heterobifunctional reagents whose chemical cross-links may or may not be reversed. Nearest neighbors (suspected to interact) *in vivo* or *in vitro* can be trapped in their complexes for further study.
Far-Western Analysis	Similar strategy to Western blotting with one key difference. The antibody probe in a typical Western blot detection is substituted with an appropriately labeled bait protein as the probe. Detection can be radioisotopic, chemiluminescent or colorimetric, depending on the probe label.
Label Transfer	Involves a specialized cross-linking agent with several important features. These include heterobifunctionality for stepwise cross-linking, a detectable label and reversibility of the cross-link between binding partners. Upon reduction of the cross-linked complex a binding partner (prey protein) acquires the label from a bait protein that was first modified with the reagent. The label is typically used in the detection process to isolate or identify the unknown prey protein.
Protein Arrays/ Protein Chips	Antibody-based or bait-based arrays that allow for screening and detection of specific interactions of proteins from complex mixtures. Primary applications include high-throughput assays of protein expression profiling, protein:protein interaction and enzyme activity. Most of the current protein chips are based on the binding between the capture proteins immobilized on a surface and the target proteins in the sample solution.
Protein Interaction Mapping	Utilizes an "artificial protease" on a bait protein to initiate contact-dependent cleavages in the prey protein in the presence of specific reactants. The nonspecific cleavage fragments produced by the artificial protease can be analyzed to map the contact sites or interface of a known protein:protein interaction.
Pull-Down Assays	An affinity chromatography method that involves using a tagged or labeled bait to create a specific affinity matrix that will enable binding and purification of a prey protein from a lysate sample or other protein-containing mixture.
Surface Plasmon Resonance	Relates binding information to small changes in refractive indices of laser light reflected from gold surfaces to which a bait protein has been attached. Changes are proportional to the extent of binding. Special labels and sample purification are not necessary, and analysis occurs in real time.

X-ray Diffraction

X-ray crystallography currently is a powerful technique for structural proteomics, because if provided with a well diffracting crystal it is possible to determine a 3D structure in hours. X-rays are diffracted at a crystal of protein or a derivative of protein containing a heavy metal atom in order to determine secondary and tertiary structures. The rays are scattered in a pattern dependent on the electron densities in different portions of a protein. The crystal must remain super cool during data collection. This minimizes radiation damage, and scatter, increases resolution and allow for long-term storage and reuse of crystals. Images are translated into electron density maps which superimposed on one another either manually or by specialized computer programs. The bottleneck of the process is the production of well diffracting crystals, which is unpredictable and may take between hours and months. Crystallography is time consuming, expensive and requires very specialized training and equipment. However, it reveals very precise and critical structural data about amino acid orientation that is then used to understand protein interactions and design drugs in structure based drug designing.

NUCLEAR MAGNETIC RESONANCE (NMR)

NMR spectroscopy investigates transition between spin states of magnetically active nuclei in a magnetic field. The most important magnetically active nuclei for the proteins are proton (^1H), (^{13}C) carbon, nitrogen –15 (^{15}N) and phosphorous-31 (^{31}P). All are nuclei of spin one-half, which in an external magnetic field will have two spin states, one at lower energy with the magnetic spin opposing the external field. The magnetic moment of each nucleus precesses about the external field and is influenced by other fields. Influences on a given spin are generated by neighboring spins in the molecule, giving rise to intrinsic NMR parameters, or by radio-frequency pulses and/or pulsed field gradients as programmed by the NMR spectroscopist. In an NMR experiment, a sequence of radio-frequency pulses and pulsed field gradients are applied to the spins present in the molecule studied. The excited spins are allowed to interact with one another and with the external magnetic field. Then the state of the system is read out by detection of the current induced by the nuclear spins of the sample in the receiver coil of the NMR spectrometer.

The NMR spectrometer can be programmed to operate on all nuclei of a given atom type, or selectively on nuclei with particular special features (chemical- shift range, spin-spin coupling range, etc.). The classical approach to structure determination is to first use multidimensional, multinuclear NMR methods to determine 'sequence specific assignments', that is, resolve sequence from the 1H, 15N and 13C nuclei of a protein and assign them to specific nuclei in the covalent structure of the molecule.

NMR methods can be used routinely to determine small to medium sized proteins domain structures. Key milestones in the development of this technique include NOE-based (Nuclear Overhauser Enhancement) resonance assignment and structure refinement, efficient stable isotope labeling methods, detection of multinuclear correlations coupled with indirect detection and multinuclear NMR spectroscopy. Protein domains of up to 25KDa are generally amenable to NMR structure determination, provided that soluble, stable U-13C and U-15N-labeled samples can be obtained. This isotope enrichment enables the use of sensitive heteronuclear correlation experiments for the task of matching all chemical shifts in the protein with specific amino acid residues and atoms. Structure of small proteins can be determined with samples at natural abundance. Selected labeling approaches make it possible to obtain information about selected regions of proteins even as large as 150 kDa.

A Protocol for Protein Structure Determination by NMR Spectroscopy

Protein emerging from the protein pipeline are divided by molecular weight: those >60 kDa are diverted to the X-ray cryatallization screens; those <60kDa are labeled with 15N (from $^{15}NH_3Cl$ in the minimal growth medium as the sole source of nitrogen) and subjected to automated NMR screening to determine their state of folding and aggregation under a variety of solution conditions. Proteins that pass these screens are double labeled (M1 <20,000) from $^{15}NH_3Cl$ and (U–^{13}C) glucose in the growth medium or triple labeled (m1> 20,000); high throughput methods for triple labeled are still under development.

NMR structure determination is limited currently by size constraints, lengthy data collection and analysis times (often months) and the method is best applied to proteins smaller than 250 amino acids. On the other hand, NMR experiments do not require crystals, and samples appropriate for structure determination can be identified within minutes of the protein being purified. In summary, X-ray crystallography and NMR spectroscopy seem to have complementary deficiencies, and the relative success of these methods in structural proteomics remains to be determined.

Comparison of NMR Spectroscopy and X-ray Diffraction

Currently, NMR spectroscopy and X-ray diffraction are the only methods considered for high-throughput protein structure determination. The two methods have complementary features. X-ray crystallography represents a mature and rapid approach for proteins that form suitable crystals. NMR has advantages for structural studies of small proteins that are partially disordered, exist in multiple stable conformations in solution, show weak interactions with important cofactors or ligands, or do not crystallize readily. Low-resolution structures derived from NMR data can be used in phasing X-ray data. NMR spectroscopy is an incremental method that can rapidly provide useful information concerning overall protein folding, local dynamics, existence of multiple-folded conformations or protein-ligand or protein-protein interactions. This information can be useful in designing strategies for structure determination by either NMR or X-ray crystallography. Several ongoing structural genomics pilot projects are employing a combination of X-ray crystallography and NMR spectroscopy. Solid-state NMR strategies shows great promise for structural studies of proteins that may not be amenable to investigation in solution, such as membrane proteins and for functional investigation of proteins in the solid state.

Electron Microscopy

With the advent of computer-controlled electron microscopes and the automation of data acquisition it became possible to obtain molecular-resolution (2-4 nm) tomograms of structures as large as organelles or whole cells. Cryoelectron tomography involves non-invasive three-dimensional (3-D) imaging of vitrified cells. Tomograms of cells at molecular resolution are essentially 3-D images of the cell's entire proteome and they reveal the spatial relationships of macromolecules in the cytoplasm, the 'interactome'. Many of the restrictions of X-ray crystallography or NMR spectroscopy do not apply to electron microscopy. Crystalline order is helpful but not necessary; there is no upper limit for the structures studied, the quantities of sample needed is very small and cryomethods enable the observation of the molecules in their native aqueous environment. Imaging of large and multicomponent cellular machinery close to physiological conditions is possible using electron microscopy and imaging analysis.

Three-dimensional Reconstruction

The determination of 3D structure by electron cryomicroscopy follows a common scheme for all macromolecules. Briefly, each sample must be prepared in a relatively homogenous, aqueous form. This specimen is then rapidly frozen (vitrified) as a thin film, transferred to the electron microscope, and imaged under low-dose conditions (less than 5000 electrons/nm2). Before image analysis, the best micrographs are selected in which the electron exposure is correct, there is no specimen movement, minimal astigmatism and a minimal amount of defocus. The exact steps of image analysis and image acquisition varies according to the symmetry and the nature of the specimen. Three basic tasks are common to all samples. First, images of the object must be obtained in a sufficient number of orientations. This task can be achieved using the natural design of the object (helical filaments or icosahedral viruses), using experimental designs (by tilting the sample holder in the microscope to specified angles) or through a random distribution of orientations (single particles). Second, the orientations and the center of the object need to be determined. Iterative refinement of theses parameters is usually carried out by cross-comparison between different images or by projection images of preliminary models. Third, image shifts must be applied computationally either real space or Fourier space to bring all views of the object to a common origin.

Electron Tomography

The most general method of obtaining 3D information by electron microscopy is tomography. The method is applicable to isolated particles as well as the pleomorphic structures such as mitochondria, other organelles and even whole cells. In this technique a series of images is taken of a single specimen as the specimen is tilted over a wide range of angles. Sometimes for better angular coverage, another tilt series is taken with the specimen rotated by 90°. Tomography is the only method available for the reconstruction of specimens with unique structure. The main disadvantage of tomographic approach is that radiation damage builds up during the multiple exposures as the specimen is being tilted. Although data collection with extremely low doses of radiations is under development, the experimental realization and image processing under such conditions still poses great challenges. Similar to electron crystallography, there is a missing wedge of data because of the maximum tilt angle, resulting in anisotropic resolution. Features perpendicular to the electron beam are better resolved than the features parallel to the beam.

POST TRANSLATIONAL MODIFICATIONS

Post-translational modifications (PTMs) are covalent processing that change the properties of a protein by proteolytic cleavage or by addition of modifying groups to one or more amino acids.

Various covalent modifications often occur, either during or after assembly of the polypeptide chain. Most proteins undergo co- and / or post-translational modifications. Knowledge of these modifications is extremely important because they may alter physical and chemical properties, folding, conformation distribution, stability, activity, and consequently, function of the proteins. Examples of the biological effects of protein modifications include phosphorylation for signal transduction, ubiquitination for proteolysis, attachment of fatty acids for membrane anchoring and association, glycosylation for protein half-life, targeting, cell:cell and cell:matrix interactions. Consequently, the analysis of proteins and their post-translational modifications is particularly important for the study of heart disease, cancer, neurodegenerative diseases and diabetes.

Because a modification results in a change in the molecular mass of the affected amino acid, MS is the method of choice for characterizing PTMs. FTMS coupled with ECD (electron capture dissociation), a relatively new technology for peptide fragmentation, has great potential for PTMs mapping especially labile modifications. There have been more that 200 modifications described, some of the important ones are shown in table 4.

DATA MINING IN PROTEOMICS

The analysis of proteome requires a combination of number of procedures, both experimental (wet-lab experiments) and Bioinformatics (dry–lab experiments). Comprehensive databases

Table 4. Types of post-translational modification

PTM Type	Function and notes
Phosphorylation	Reversible, activation/inactivation of enzyme activity,
pTyr, pSer, pThr	modulation of molecular interactions, signaling.
Acetylation	Protein stability, protection of N-terminus. Regulation of protein-DNA interactions (histones)
Methylation	Regulation of gene expression
Acylation, fatty acid modification	Cellular localization and targeting signals, membrane tethering,
Farnesyl	mediator of protein-protein interactions
Myristoyl	
Palmitoyl	
Glycosylation	Excreted proteins, cell-cell recognition/signaling, *O*-GlcNAc
N-linked	reversible, regulatory functions
O-linked	
GPI anchor	Glycosylphosphatidylinositol (GPI) anchor. Membrane tethering enzymes and receptors, mainly to outer leaflet of plasma.
Hydroxyproline	Protein stability and protein-ligand interactions
Sulfation(sTyr)	Modulator of protein-protein and receptor-ligand interaction
Disulphide bond formation	Intra- and inter-molecular crosslink, protein stability
Deamidation	Possible regulator of protein-ligand and protein-protein interactions, also a common chemical artifact
Ubiquitination	Destruction signal, after tryptic digestion, ubiquitination site is modified with the Gly-Gly dipeptide.

are a pre-requisite for successful protein identification, and identified proteins are then in turn used to populate specific secondary databases. The success of database searching depends on the quality of the data obtained from analysis (e.g. MS or Edman degrdation), the quality of database searched and the method used to search the database. Existing database searching programs have become both more sophisticated and (Web) accessible. Bioinformatics methods can predict modification sites *in silico*, with certain limitations and are no substitute for direct experimentation data. Some of the general and specific databases used in proteome analysis are listed in Table 15.

Table 5. Software tools for proteome data exploration available on internet

Tool	Description	URL
MOWSE	Peptide mass mapping and sequencing	http://srs.hgmp.mrc.ac.uk/cgi-in/mowse
PROWL	Peptide mass mapping and sequencing	http://prowl.rockfeller.edu/
PeptIdent	Peptide mass mapping and sequencing	http://www.expasy.ch/tools/petident
MASCOT	Peptide mass mapping and sequencing	http://www.matrixscience.com/
Prospector	Peptide mass mapping and sequencing	http://prospector.ucsf.edu/
PepMAPPER	Peptide mass mapping and sequencing	http://wolf.bms.umist.ac.uk/mapper/
PeptideSearch	Peptide mass mapping and sequencing	http://mann.embl-heidelberg.de/
FindMod	Posttranslational modification	http://www.expasy.ch/tools/findmod/
PSORT	Posttranslational modification	http://psort.nibb.ac.jp/
SignalP	Posttranslational modification	http://cbs.dtu/services/SignalP/
DIPTM	Protein-protein interactions	http://dip.doe-mbi.ucla.edu/
SEAQUEST	Uninterpreted MS/MS searching	http://fields.scripps.edu/sequest/
MSDB	MS protein database	http://csc.fserve.hh.med.ic.ac.uk/msdb.html
SWISS-2DPAGE	2D-PAGE database	http://us.expasy.org/ch2d/
Melanie 3	2-D gel analysis tool	http://us.expasy.org/melanie/

SIGNIFICANCE OF PROTEOMICS RESEARCH

The genome sequences of important model systems are available and the focus is now shifting to large-scale experiments enabled by this data. Following in the footsteps of genomics, we have functional genomics, proteomics and even metabolomics, roughly paralleling the biological hierarchy of the transcription, translation and production of small molecules. After the completion of human genome project, now the emphasis is to study all the proteins produced by cell types and organisms. Researchers are now looking at how genes and proteins interact to form other proteins. Many types of information cannot be obtained from the study of the genes alone. For example, proteins, not genes are responsible for the phenotype of cells. It is impossible to elucidate mechanisms of disease, aging and the effects of environment signaling by studying the genome alone. Only through the study of proteins can protein modification be characterized and targets of drugs identified. Proteome analysis will play an increasingly critical role in drug discovery and evaluation, particularly because proteomics can measure many pieces of vital information that genomics cannot predict. These are:

1. Annotation of genome
2. Protein expression studies:
   (i) If and when gene products are expressed
   (ii) The abundance of these gene products and their modifications
   (iii) The regulation of expression of these proteins
3. The extent of post-translational modifications
4. Protein structure and functions
5. Protein localization and compartmentalization
6. Protein–protein interactions
7. The phenotype of a cell experiencing multi-enzyme phenomenon resulting from modulation of protein expressions due to drug administration, cell cycle, ontogeny, aging, stress and disease. Proteome analysis will be particularly relevant for those disease states where multiple genes are involved (polygenic diseases) and therefore make excellent target for proteome analysis.

SUMMARY

A complete understanding of protein function and structure is the primary goal of post-genomic era. Proteome is the protein complement of a given cell at a specific time. Proteomics includes the identification of proteins in biological tissues, characterization of their physico-chemical properties (complete sequence, post-translational modifications) and the description of their behavior (function, expression level). The discipline of proteomics can be broadly divided into three classes: structural proteomics, protein expression profiling, and functional proteomics. Proteomics is a technology driven field, currently there is no single experiment or platform that permits the analysis of all protein properties on a proteome-wide scale. Three main analytical tools of proteomics that can be distinguished are: (i) mass spectrometry, coupled with 2-Dimensional gel electrophoresis/ICAT (Isotope Coded Affinity tag)/2-Dimensional Liquid Chromatography. Conventional protein sequencing methods (Sanger's method, Edman's N-terminal degradation method etc.) can also be used for determining amino-acid sequence. (ii) to characterize protein-protein interactions affinity chromatography, yeast two hybrid system (Y2S), protein arrays and Fluorescence Energy Transfer (FRET) can be used. Array based methods are generally used for examining the protein expression levels for protein profiling and clinical diagnostic and (iii) the structural and imaging techniques, where, the 3-D shape, localization and dynamics of individual protein or protein complexes are investigated on large scale, using X-ray crystallography, NMR spectroscopy, electron tomography and 2-D electron microscopy. Imaging techniques used for protein localization are FRET and GFP fluorescence.). The most important applications of proteomics are believed to be in the discovery of new drug markers and drug targets. Proteomics analysis will be particularly relevant to disease states where multiple genes are involved (polygenic diseases), notably in the development of cardiovascular dysfunction, cancer, diabetes type II, asthma, hypertension and neurological diseases.

REFERENCES

Binz, P, Gasteiger, E. and Sanchez, J., Bairoch, A., Hochsrasser, D. F. and Apple, R. D. Proteome analysis

Garret, R. H. and Grisham, C. M. (1995) Proteins: Their biological functions and primary strucutre. In: Biochemistry (Garret, R.H. and Grisham, R.H. eds.) pp.81-131, Saunders College Publishing Harcourt Brace College Publishers.

Graves, P. R. and Haystead, A. J. (2002) Molecular Biologist's guide to proteomics Microbiol. Mol. Biol. Rev. 66(1): 39-63.

Gygi. S. P., Rist, B., Gerber, S. A., Frantisek, T., Gelb, M. H. and Abersold, R. (1999) Quantitative analysis of complex protein mixtures using isotope-coded affinity tags Nat. Biotechnol. 17: 994-999.

Hoog, C. amd Mann, M. (2004) Proteomics Ann. Rev. Genomics Hum. Genet. 5: 267-293.

Ito, T., Ota, K., Kubota, H., Yamguchi, Y., Chiba, T., Sakuraba, K. and Yoshida, M (2002) Roles for the two-hybrid system in exploration of the yeast protein interactome Mol. Cell. Proteomics 1: 561-566.

Kellner, R. (2000) Proteomics. Concepts and perspectives Freseneus J. Anal. Chem. 366: 517-524.

Lane, C. S. (2005) Mass spectrometry-based proteomics in the life sciences Cell. Mol. Life Sci. 62: 848-869.

Link J, Eng J, Schieltz DM, Carmack E, Mize GJ, Morris DR, Garvik BM, Yates JR (1999) Direct analysis of large protein complexes using mass spectrometry Nat. Biotechnol. 17: 676-682.

Mann, M. and Jensen, O. (2003) Proteome analysis of post-translational modifications Nat. Biotechnol. 21: 255-261.

Markley, J. L., Ulrich, E. L., Westler, W. M. and Volkman, B. F. (2003) Macromolecular structure determination by NMR spectroscopy. In: Structural Bioinformatics (Bourne, P. E. and Weissig, H. eds.) pp. 89-108, Wiley-Liss Inc.

Mirogoroddskaya, O. A., Kozmgin, Y. P., Titov, M. I., Korner, R.,Sonksen, C. P., Roepstoff, P. (2000) Quantitation of peptides and proteins by matrix-assisted laser desorption/ ionization mass spectrometry using 18O labeled internal standards. Mass Spectrom 14: 1226-1232.

Munchbach, M., Quadroni, M., Miotto, G., James, P. (2000) Quantitation and facilitated de novo sequencing of proteins by isotopic N-terminal labeling of peptides with a fragmentation directing moiety Anal. Chem. 72: 4047-4057.

Rastogi, S. C., Mendiratta and Rastogi, P. (2004) Proteomics. In: Bioinformatics Methods and Applications (Rastogi, S. C., Mendiratta and Rastogi, P. eds.) pp. 269-290, Prentice Hall of India Pvt. Ltd., New Delhi.

Siuzdak, G. (1994) The emergence of mass spectrometry in biochemical research Proc. Natl. Acad. Sci. 91: 11290-11297.

Smolka, M., Zhou, H. and Abersold, R. (2002) Quantitative protein profiling using two-dimensional gel electrophoresis, isotope-coded affinity tag labelling and mass spectrometry Mol. Cell. Proteomics 1: 19-29.

Stewart, I. I., Thompson, T., Figeys, D. (2001) ^{18}O labelling: a tool for proteomics Mass Spectrom. 15: 2456-2465.

Sydor, J. R. and Nock, S. (2003) Protein expression profiling arrays: tools for the multiplexed high-throughput analysis of proteins Proteome Sci. 1: 3

Volkmann, N. and Hanein, D. (2003) Electron micorscopy. In: Structural Bioinformatics (Bourne, P. E. and Weissig, H. eds.) pp. 115-130, Wiley-Liss Inc.

Wilkins, M. R., Sanchez, J. C., Gooley, A., Appel, R. D., Humphery, I., Hochstrasser, D. F. and Wiliams, K. L. (1995) Progress with proteome projects: why all proteins expressed by a genome should be identified and how to do it Biotechnol. Genet. Eng. Rev. 13: 19-50.

Zubay, G. L. (1998) Protein structure and function. In: Biochemistry (Zubay, G. L. ed.) pp. 43-140, 4th Edition, W.M.C. Brown Publishers.

Suggested Readings

1. Biochemistry and Molecular Biology (1997) W. H. Elliott and Elliott, Oxford University Press, UK.
2. Biochemistry (1995) R. H. Garret and C. M. Grisham, Saunders College Publishing and Harcourt Brace College Publishers.
3. Lehninger's Principles of Biochemistry (2005) D. L. Nelson and M. M. Cox, 4th Edition, WH Freeman and Company, NY, USA.
4. Pennington, S. R. and Dunn, M. J. (2002) Proteomics: From protein sequence to function. BIOS Scientific Publishers Ltd., Oxford, UK.
5. Biochemistry (1998) G. L. Zubay, 4th Edition, W.M.C. Brown Publishers.

REVIEW QUESTIONS

Long Questions

1. Define proteomics and write about significance of proteomics research.
2. Describe different methods used for characterizing protein-protein interactions.
3. Discuss protein sequence determination using mass spectrometry.
4. What are methods for determination of N and C terminus of a polypeptide?
5. Compare and contrast the applications of X-ray diffraction and NMR spectroscopy in structural proteomics.

Multiple Choice Questions

1. The term proteomics was coined in
   (a) 1990 (b) 1991
   (c) 1995 (d) 1994
2. In 2-DE, proteins are resolved on the basis of
   (a) Conformation and molecular weight
   (b) Net charge and molecular weight
   (c) net charge and density
   (d) hydrophobicity and molecular weight
3. The full form of ICAT is:
   (a) Isotope coded affinity test (b) Isobar coded affinity tag
   (c) Isomer coded affinity test (d) Isotope coded affinity tag
4. MudPIT couples a biphasic capillary column packed with
   (a) Reversed phase and strong cation exchanger
   (b) Reversed phase and strong anion exchanger
   (c) Strong cation and anion exchanger
   (d) Reverse phase and ligand

5. The automatic protein sequencing device works on the principle of
   (a) Sanger's method (b) Maxam–Gilbert method
   (c) Edman's Method (d) none of the above
6. Electron tomography is _______ electron microscopy
   (a) 1-Dimensional (b) 2-Dimensional
   (c) 3-Dimensional (d) none of the above
7. SAGE is a profiling technique for
   (a) m-RNA (b) t-RNA
   (c) r-RNA (d) cDNA
8. Quadrupole is an example
   (a) Detector (b) Ionizer
   (c) Mass analyzer (d) Sample injector
9. Post-translational modifications results in alteration of the following property of a protein molecule
   (a) Physical and chemical (b) Conformation
   (c) Activity (d) all the above
10. GPI-anchor is a type of
    (a) Post-translational modification (b) ICAT reagent
    (c) GFP variant (d) A protein gel stain

Answers

1. (c)	2. (b)	3. (d)	4. (a)	5. (a)	6. (c)
7. (a)	8. (c)	9. (d)	10. (a)		

PART–10

Polymerase Chain Reaction and Gene Silencing

CHAPTER 31

Polymerase Chain Reaction: *In vitro* Amplification of DNA

Y. K. Jaiswal* and M. K. Jaiswal
School of Studies in Biochemistry, Jiwaji University, Gwalior–474 011
E-mail: jaiswalyk@gmail.com

Polymerase chain reaction (PCR) is a technique which is used to amplify the number of copies of a specific region of DNA, in order to produce enough DNA to be adequately tested. It is a rapid, inexpensive and simple means of producing relatively large numbers of copies of DNA molecules from minuscule quantities of source DNA material even when the source DNA is of relatively poor quality (Fig. 1). The PCR is the oldest in theory and more versatile in practice. The method was proposed by in the early 1970 by H. Govind Khorana and his colleagues as a strategy to lessen the labor involved in the chemical synthesis of genes. Their idea however did not seem practicable at a time when gene is not sequenced, thermostable DNA polymerase had not described, and synthesis of oligonucleotide primers was more of an art than a science. The technique was independently conceived in 1985, given its present name, and put in practice by Kay Mullis and his coworker by at Cetus Corporation (a biotechnology firm in Emeryville, California) who described the *in vitro* amplification of mammalian genes using the Klenow fragment of E. coli DNA polymerase I. The use of a thermostable polymerase from *Thermus aquaticus* greatly increase the efficiency of PCR and opened the door to automation of the method. By the end of 1980s, cloning was no longer the way to isolate genes: DNA sequencing had been revolutionized and PCR had become a fundamental cornerstone of genetic and molecular analyses.

Kary Mullis (Fig. 2) won the Nobel Prize in 1993 in Chemistry for his invention of the polymerase chain reaction, a chemical procedure that allows scientists to "see" the structures of the molecules of genes.

*Corresponding author

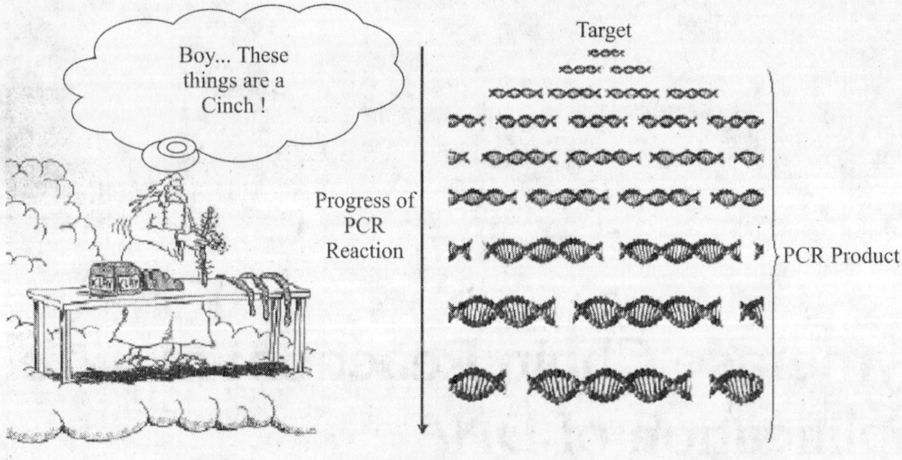

Fig. 1. How Does PCR WORK?

Fig. 2. Kary Mullis (http://archieve.salon.com/health/future/2000/03/29/mullis/).

Polymerase Chain Reaction—The Early Years

Back in the early days of PCR work, the polymerase would need to be added fresh to the reaction at each temperature cycle, because thermostable enzymes were not commercially available. Of course, there were also no thermocycler then, and so moving the tubes from one temperature bath to another (Fig. 3) every 30 seconds to a minute for several hours was a mind numbing job that fell to graduate students.

PARAMETERS THAT AFFECT POLYMERASE CHAIN REACTION

Essential Component of Polymerase Chain Reaction

PCR contain seven essential components e.g., (a) A thermostable DNA polymerase to catalyze template-dependent synthesis of DNA, (b) A pair of oligonucleotides to prime DNA synthesis,

Fig. 3. Polymerase Chain Reaction – In early years.

(c) Deoxynucleoside triphosphates, (d) Mg^{++} as divalent cation, (e) Buffer to maintain, pH, (f) Monovalent cations, (g) Template DNA etc.

Thermostable DNA Polymerases

Fig. 4. Yellowstone National Park (Wyoming) (Sambrook and Russel, 2001).

Thermophilic Archaea, whose chief DNA polymerase belongs to the polymerase α family. *Thermus aquaticus* an organism from thermophilic Archaea family. Taq (*T. aquaticus)* DNA polymerase, is first purified from the hot springs bacterium *Thermus aquaticus,* was published in 1976. The first commercially-available thermostable DNA polymerase for PCR came from the thermophilic eubacterium *Thermus aquaticus*. Several other thermostable DNA polymerase have been isolated and characterized which offer several advantages for specialized PCR assays.

In the 1960s, microbiologist Dr. Thomas Brock and Hudson Freeze of Indiana University traveled to Yellowstone National Park (Wyoming) (Fig. 4) to research heat-loving microorganisms called thermophiles. He discovered a bacterial species that thrived in high temperatures. Twenty years later, scientists at Cetus Corporation realized that the DNA polymerase from Thermus aquaticus would solve their problem of denatured enzyme after each cycle (Mullis added fresh E. coli DNA polymerase enzyme after each cycle). A number of other thermophilic organisms have since been discovered and exploited for PCR applications.

The primary requirements for a polymerase are used in the Polymerase chain Reaction are good activity at temperatures around 75°C and ability to retain that activity after prolonged incubation at higher temperature (95°C).

Taq DNA polymerase

Taq, a thermostable DNA-dependent DNA polymerase, was first isolated from the thermophilic eukaryote Thermus aquaticus in 1976 (Chien et al., 1976; Kaledin 1980). Some year later, the enzyme became famous for its use in PCR (Saki et al., 1988) and was designated in 1989 as Molecule of the year (Koshland 1989), for whatever that may mean.

Source: *Thermus aquaticus* is a species of bacterium that is the source of the enzyme Taq DNA Polymerase, one of the most important enzymes in molecular biology. It is one of several thermophilic bacteria that belong to the Deinococcus-Thermus group.

The gene for Taq (Lawyer et al., 1989) encodes an 832-amino acid protein consiting of two domain.

General properties

Size and activity Taq DNA polymerase has a single polypeptide chain with a molecular weight of approx. 95 kD (Fig. 5). It has 5′→3′ DNA polymerase and 5′→3′ exonuclease activity but lacks 3′→5′ exonuclease activity. The enzyme is most active at a pH of 9 (adjusted at 20°C) and temperatures around 75°C.

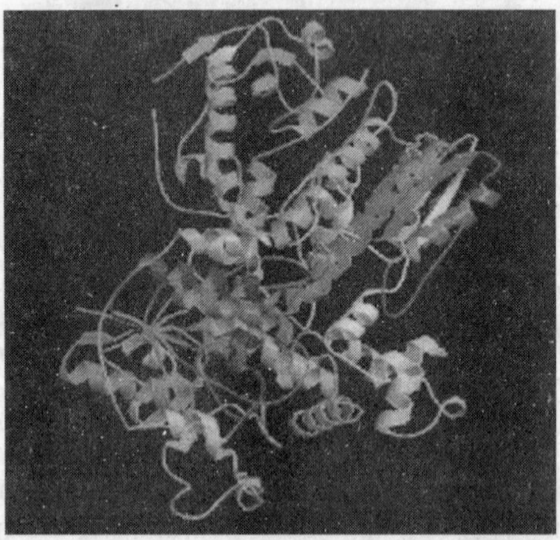

Fig. 5. Taq DNA polymerase (http://en.wikipedia.org/wiki/Taq polymerse).

Thermal Stability Taq DNA polymerase has a half life of less than 5 minutes at 100°C. However, the enzyme retains its activity for a long times (half life up to 40 minutes) at 95°C. Its optimal temperature is 75–80°C.

Magnesium Concentration Taq DNA polymerase prefers $MgCl_2$ as a metallic cofactor. The standard concentration for PCR with the polymerase is 1.5 mM.

dNTP Concentration The nucleotide concentration during PCR should be 200μM of each dNTP

Detergent and other additives The non-ionic detergent Tween 20 (0.5–1.0%) has been used to enhance the efficiency of Taq DNA Polymerase in certain PCRs. Other additives reported to have enhancing effects include DMSO, Gelatin, Glycerol and Ammonium Sulfate.

Prevention of carry-over contamination with dUTP/UNG, Taq DNA polymerase readily accepts dUTP (a substitute for dTTP) as a substrate. Therefore, you may use Taq DNA polymerase with uracil DNA glycosylase (UNG) to prevent carry over contamination.

Km for dNTP is 10-16μM

Km for DNA is 1.5 nM

Extension Rates (dNTPs/sec/enzyme molecule):

75°C	150
70°C	<60
55°C	24
37°C	1.25
22°C	0.25

Processivity (dNTP/sec/enzyme molecule): 42

Error Rates: Error rates ranging from 8.9×10^{-5} to 1.1×10^{-4} have been reported by various groups over the years.

Special Properties

Modified Nucleotide as substrate Taq DNA polymerase accepts modify deoxyribonucleoside triphosphate as substrate, and can be used to label DNA fragments either with radionucleotides digoxigenin, biotin, fluorescein.

DNA sequencing The processivity, absence of exonuclease and temperature optima of Taq DNA polymerase makes the enzyme useful for DNA sequencing.

rTth polymerase

Source: *Thermus thermophilus*

General properties

It's fidelity is low. Km for dNTP is 115μM.

Activites It has 5′→3′ DNA polymerase activity and 5′→3′ exonuclease activity.

Thermal Stability rTth polymerase is stable at 95°C for 20 min. Its optimal temperature is 75–80°C.

Special properties

rTth, the recombination form of Tth DNA polymerase, can use Mg^{2+} or Mn^{2+} as a cofactor. In the presence of Mn^{2+}, the reverse transcriptase activity of rTth is enhanced.

Tfl polymerase

Source: *Thermus flavus*

General properties

It's fidelity is low. Km for dNTP is 63μM.

Activites It has 5′→3′ DNA polymerase activity and lacks 3′→5′ and 5′→3′ exonuclease activity.

Thermal Stability Tfl polymerase is stable at 70°C for 120 min. Its optimal temperature is 75°C.

Special properties

This polymerase is useful for the amplification of large segments of DNA.

Hot tub polymerase

Source: *Thermus ubiqitus*

General properties

It's fidelity is low.

Activites It has 5′→3′ DNA polymerase activity and lacks 3′→5′ and 5′→3′ exonuclease activity.

Tbr polymerase

Source: *Thermus brockianus*

General properties

It's fidelity is low.

Activites It has 5′→3′ DNA polymerase activity and 5′→3′ exonuclease activity.

Thermal Stability Tbr polymerase is stable at 96°C for 150 min.

UlTma polymerase

Source: *Thermotoga maritima*

General properties

It's fidelity is high.

Activites It has 5′→3′ DNA polymerase and 3′→5′ exonuclease activity but lacks 5′→3′ exonuclease activity.

Thermal Stability Ultma polymerase is stable at 95°C for 50 min.

rBst polymerase

Source: *Bacillus sterothermophilus*

Activites It has 5′→3′ DNA polymerase 3′→5′ exonuclease (suspected) and 5′→3′ exonuclease activity.

Thermal Stability Its optimal temperature is 60–65°C.

Pwo Polymerase

Source: *Pyrococcus woesei*

General properties

It's fidelity is high.

Size and activites Pwo DNA polymerase has a molecular weight of about 90 kDa. It has 5′→3′ DNA polymerase and 3′→5′ exonuclease but lacks 5′→3′ exonuclease activity.

Thermal Stability Pwo polymerase is stable at 100°C for more 120 min. Its optimal temperature is 60–65°C.

Special properties

Pwo generates blunt-ended DNA fragments and is recommended for cloning of PCR products.

Tli polymerase

Source: *Thermococcus litoralis*

General properties

It's fidelity is low. Km for dNTP is 66μM.

Activites It has 5′→3′ DNA polymerase 3′→5′ exonuclease but lacks 5′→3′ exonuclease activity.

Thermal Stability Pwo polymerase is stable at 100°C for 100 min. Its optimal temperature is 70–80°C.

Special properties

Tli is recommended of DNA sequencing.

Deep vent polymerase

Source: *Pyrococcus* strain GB-D

General properties

It's fidelity is high. Km for dNTP is 50μM.

Activites It has 5′→3′ DNA polymerase 3′→5′ exonuclease but lacks 5′→3′ exonuclease activity.

Thermal Stability Deep vent polymerase is stable at 100°C for 480 min. Its optimal temperature is 70-80°C.

Special properties

Deep Vent is suitable for primer extension and PCRs requiring a high fidelity highly stable DNA polymerase.

Pfu polymerase

Source: *Pyrococcus furiosus*

General properties

It's fidelity is high.

Activites It has 5′→3′ DNA polymerase 3′→5′ exonuclease but lacks 5′→3′ exonuclease activity.

Thermal Stability Pwo polymerase is stable at 95°C for 240 min. Its optimal temperature is 72–78°C.

Special properties

Pfu is suitable for primer-extension reaction and PCRs requiring a high-fidelity high stable DNA polymerase.

For routine PCRs, Taq polymerase (0.5–2.5 units per standard 25–50 µl reaction) remains the enzyme choice.

A pair of oligonucleotides to prime the DNA Synthesis

A pair of oligonucleotides, which prime the synthesis of DNA, is design carefully to obtain the desired product in high yield, to suppress amplification of unwanted sequences, and facilitate subsequent manipulation of amplified product. Primers are heavily influence the success or failure of PCR product. It is ironic that the guidelines for their design are large qualitative and are based more on common sense than on well-understood thermodynamic or structural principles. In certain situations, it may desirable to amplify several segment of DNA simultaneously. In these cases, an amplification reaction termed "Multiplex PCR" is used that includes more than one pair of primers in the reaction mix. Oligonucleotide primers synthesized on an automated DNA synthesizer can generally be used in standard PCRs without further purification.

Design of primer for basic PCR

The chief goal the primer design is specificity which is achieved when a pair of primer anneal in a steady way to its target sequence in the template DNA According to the rule of thumb, the longer primer has a higher specificity for a picky target DNA. In 1979 Nei and Li gave the equation for calculating the probability that a sequence exactly complementary to a string of nucleotides within the DNA sequence space that consists of a random sequence of nucleotides.

$$K = [g/2]^{G+C} \times [(1-g)/2]^{A+T}$$

K = Expected frequency of occurrence within the sequence space

g = relative concentration of G+C content of the sequence space

G, C, A and T are the number of specific nucleotides in the oligonucleotide.

For a double-stranded genome of size N (in nucleotides), the expected number (n) of sites complementary to the oligonucleotide is n = 2NK.

This equation predict that an oligonucleotide of 15 nucleotides would be represented only once in a mammalian genome where N is ~ 3.0×10^9. In the case 16-mer, there is only once chance in the ten that a typical mammalian cDNA library (complexity ~10^7) will fortuitously contain a sequence that exactly matches that of the oligonucleotide. However this is based on the postulation that the distribution of nucleotides in mammalian genome is random. But this is not a case because of codon usage and a significant fraction of genome is consisting of repetitive DNA sequences and the gene families. To minimize this problem of nonspecific annealing, it is advisable to use primers longer than the statistically indicated minimum. Because of the presence of repetitive elements, not more than 85% of the mammalian genome can be targeted precisely, even by primer 20 or more nucleotides in length.

During the designing of primer various following properties of oligonucleotides should be kept in mind that influences the efficiency of amplification:

- Base composition: G+C content should between 40% and 60%.
- Length should be 18-25 long. Member of primer pair should not differ in length more 3 bp.

- Primer should not have inverted repeat sequences or self complementary sequences >3 bp in length. Otherwise its form hairpin structures, which, if stable under PCR conditions, can effectively, prevent the primer from annealing to its target DNA.
- The 3′ sequence of primer should not able to bind to any sites of any other primer. Because primers are present in high concentration in PCR, even weak complementarity between them leads to hybrid formation and consequent synthesis and amplification of primer dimers. If primer dimers form early in PCR, they can compete for DNA polymerase, primer, nucleotides and so can suppress amplification of the target DNA.
- Melting temperatures Tm of primer pair should not differ by >5°C. The Tm of the amplified product should not differ from the Tm values of the primer pairs by > 10°C. This property ensures that the amplified product will be efficiently denatured during each cycle of PCR.
- The 3′ of base of primer should be G or C. However, the primers with a ….NNCG or ….NNGC sequence at their 3′termini are not recommended because GC bases promotes the formation of hairpin structures and may generates primer dimer.
- The GC content of both the primers pair and amplified product should be similar and lie between 40% and 60%.

Calculation of melting temperatures between hybrids of primers and their target sequences

Several equations are available to calculate the melting temperature of hybrids formed between an oligonucleotide primer and its complementary target sequence.

According to the Wallace in 1986, the Tm for perfect duplex is 15-20 nucleotides in length in solvents of high ionic strength (e.g., 1 M NaCl):

$$\text{Tm (in °C)} = 2(\text{A+T}) + 4(\text{G+C})$$

Where (A+T) is the sum of A and T residue of oligonucleotide and (G+C) is the sum of G and C residue in the oligonucleotide.

The equation derived originally by Bolton and McCarthy (1962) and later modified by Baldion et al. (1989) predict reasonably well the melting temperature of oligonucleotides, 14-70 nucleotides in length, in cation concentration of 0.4M or less:

$$\text{Tm (in °C)} = 81.5\text{°C} + 16.6\,(\log_{10}[\text{K}^+]) + 0.41(\%\ [\text{G+C}]) - (675/n)$$

Where n is the number of bases in the oligonucleotide.

The equation can also be used to calculate the melting temperature of an amplified product whose sequence and size are both known. When PCR amplification is carried out under standard condition, the calculated melting temperature of the amplified product should not exceed ~85°C, which will ensure complete separate of its strands during denaturation step.

Computer-assisted design of primers

To avoid boredom, save time, and minimize problems, use computer programs to optimize the design, selection, and placement of primers.

Most of the programs use graphics tools, employ user-friendly interfaces, rank potential primers and primer pair according to the weight assigned to various parameters. Some of the programs contain, for example, facile searching of databases for unintentional matches to the primer, optimization of conditions for the amplification reaction, translation of amino acid sequences in to populations of degenerate oligonucleotides, and the elimination of primer capable of forming stable secondary structures.

Deoxynucleoside triphosphates (dNTP)

Standard PCRs contain equimolar amounts of dATP, dTTP, dCTP and dGTP. Concentrations of 200-250 µM of each dNTP are recommended for Taq polymerase in reactions containing 1.5 mM $MgCl_2$ In a 50µl reaction, these amounts should allow synthesis of ~6-6.5 µg of DNA. High concentrations of dNTPs (> 4 mM) are inhibitory, perhaps because of sequestering of $MgCl_2$. A satisfactory amount of amplified product can be produced with dNTP concentration as low as 20 µM– 0.5-1.0 pmole of amplified fragment ~1kb in length.

Divalent cations

All thermostable DNA polymerase require free divalent cations – usually Mg^{2+}. Some polymerase will also work, although less efficiently with buffers containing Mn^{2+}. In most of the cases a concentration of 1.5mmol/l will produce satisfactory PCR results with Taq polymerase. In general the higher Mg^{2+} concentrations increase PCR yield, but decrease the specificity of the reaction and increase the incidence of primer dimers. Lower the Mg^{2+} concentration increase the specificity but decrease the yield. So it is impossible to determine a concentration of Mg^{2+} that is optimal in all circumstances.

The optimal concentration of Mg^{2+} must be determined empirically for each combination of primers and template.

If possible, preparation of DNA template should not contains significant amounts of chelating agents such as EDTA or negatively charged ions, such as phosphate ions, which can sequester Mg^{2+}.

Buffer to maintain pH

Tris-Cl, is used to adjust the pH between 8.3 and 8.8 at the room temperature, is included in standard PCRs at a concentration of 10 mM. At 72°C which is usually for the extension of phase of PCR, pH of the reaction mixture is decrease by one unit ~ 7.2.

Monovalent cations

Standard PCR buffer contains 50 mM KCl and works well for the amplification of segment of DNA > 500 bp on length. Rising of KCl concentration to ~ 70-100 mM often improves the yield of shorter DNA segments.

Template DNA

Template DNA containing target sequences can be added to PCR in single or double-standard form. Closed circular DNA templates are amplified slightly less efficiently than the linear DNAs. Although the size of the template DNA is not critical, amplification of sequences embedded in high-molecular-weight DNA (>10 kb) can be improved by digesting the template with the restriction enzyme that doesn't cleave within the target sequence.

PCR require only a single copy of a target sequence as template. More typically, however, several thousand copies of the target sequence as template are seeded in to per reaction. In the case of mammalian genomic DNA, up to 1.0 µg of DNA is utilized per reaction, an amount that contains ~ 3 x 10^5 copies of a single-copy autosomal gene. The typical amounts of yeast, bacteria, and plasmid DNAs used per reaction are 10 ng, 1ng, and 1pg.

PCR ADDITIVES

A variety of PCR additives and enhancing agents have been used to increase the yield, specificity and consistency of PCR reactions. Whilst these additives may have beneficial effects on some amplification it is impossible to predict which agents will be useful in a particular context and therefore they must be empirically tested for each combination of template and primers. Some of the more popular of these additives are:

DMSO

2–10% DMSO may be necessary for amplification of some templates, however 10% DMSO can reduce *Taq* polymerase activity by up to 50% (Gelfand 1989) so it should not be used routinely. DMSO is thought to reduce secondary structure and is particularly useful for GC rich templates.

Formamide

1-5% and 10% Formamide is reported (Gelfand 1989) to have no effect on the activity of *Taq* polymerase, however, Sarkar et al. (1990) found that 1.25% formamide worked as well as 2.5% and 5%, and no amplification was seen at 10% so it seems prudent not to use concentrations of formamide greater than strictly necessary for optimal amplification.

Non-ionic detergents

It stabilizes *Taq* polymerase and may also suppress the formation of secondary structure. 0.1–1% Triton X-100, Tween 20 or NP-40 may increase yield but may also increase non-specific amplification. As little as 0.01% SDS contamination of the template DNA (left-over from the extraction procedure) can inhibit PCR by reducing *Taq* polymerase activity to as low as 10%, however, inclusion of 0.5% Tween-20 or -40 will effectively neutralize this effect (Gelfand 1989).

Others

TMAC (Tetramethylammonium Chloride) is generally used at a final concentration of 15-100 mM to eliminate non-specific priming. TMAC has is also used to reduce potential DNA-RNA mismatch. The base analogue 7-deaza-2'-deoxyguanosine may facilitate amplification of templates with stable secondary structures when used in place of dGTP in a ratio of 3: 1, 7-deaza-2'-deoxyguanosine: dGTP. BSA may also be added to the reaction mixture when PCR inhibitors e.g., melanin appear as contaminants with the ancient DNA or templates used for amplification.

BASIC PROGRAM OF POLYMERASE CHAIN REACTION

PCR is the interactive process, consisting of three elements which are repeated for 30 or 40 cycles:

(a) Denaturation of template DNA, (b) Annealing of primer to template DNA and (c) Extension of primers by thermostable of DNA polymerase. The steps of the process are carried out in an automated thermal cycler, which can heat and cool the tubes with the reaction mixture in a very short time.

Denaturation

During the denaturation, the double strand melts open to single stranded DNA at 94–95°C (Fig. 6).

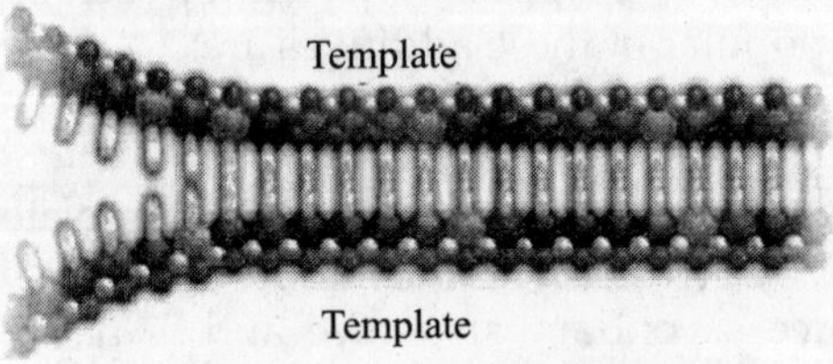

Fig. 6. Denaturation.

Hydrogen bond denaturation (94°C)

Double-stranded DNA templates denature at a temperature that is determined in part by their G+C contents. The higher the proportion of G+C, the higher the temperature required separating the stand of template DNA. The longer the DNA molecules, the greater the time required at chosen denaturation temperature to separate the two strand completely. If the temperature for denaturation is too low or if the time is too short, only AT-rich regions of the template DNA will the denaturted.

Annealing

The primers are jiggling around, caused by the Brownian motion. Ionic bonds are constantly formed and broken between the single stranded primer and the single stranded template. The more stable bonds last a little bit longer (primers that fit exactly) and on that little piece of double stranded DNA (template and primer), the polymerase can attach and starts copying the template. Once there are a few bases built in, the ionic bond is so strong between the template and the primer, that it does not break anymore.

Annealing of primers (58-68°C)

Annealing is generally carried out at 54°C (Fig. 7). The temperature used for the annealing step (T_a) is critical. If annealing temperature is too high, the oligonucleotide primers anneal poorly, if at all, to the template and the yield of amplified product is very low. If annealing temperature is too low, nonspecific annealing of primers may occur, results the amplification of unwanted segments of DNA. Annealing is usually carried out 3–5°C lower than the calculated melting temperature at which the oligonucleotide primers dissociate from their templates.

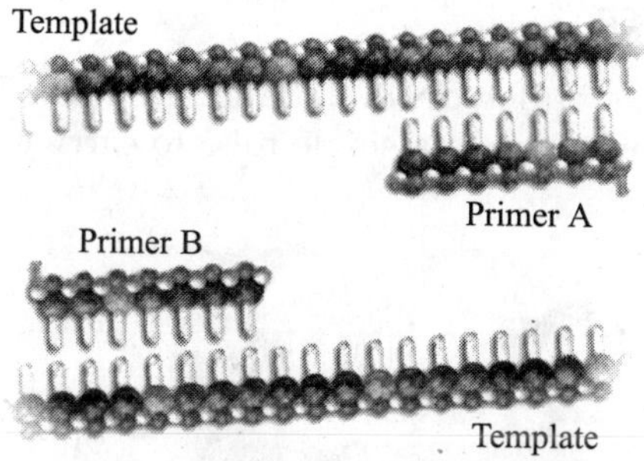

Fig. 7. Annealing.

Extension

This is the ideal working temperature for the polymerase. The primers, where there are a few bases built in, already have a stronger ionic attraction to the template than the forces breaking these attractions. Primers that are on positions with no exact match get loose again (because of the higher temperature) and don't give an extension of the fragment. The bases (complementary to the template) are coupled to the primer on the 3' side (the polymerase adds dNTP's from 5' to 3', reading the template from 3' to 5' side). Bases are added complementary to the template (Fig. 8).

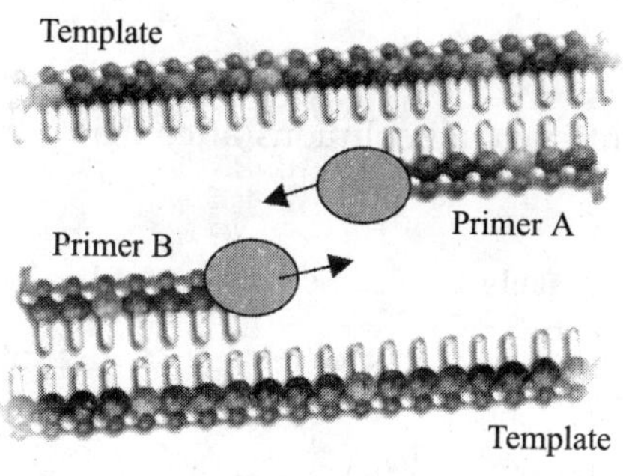

Fig. 8. Extension.

Extension of the target sequence copy (72°C)

The polymerization rate of Taq Polymerase is ~2000 nucleotides/min at the optimal temperature 72–78°C. According to the thumb rule, extension is carried out for 1min for every 1000 bp of product.

Number of Cycle

The number of cycle is depending upon the number of copies of template DNA present at the beginning of the reaction and the efficiency of primer extension and amplification.

METHODOLOGY OF BASIC PCR

A master mix containing water, buffer, dNTPs, primers, and Taq DNA polymerase, $MgCl_2$ and template DNA solutions is prepared in a single tube to carry out polymerase chain reaction (Fig. 9).

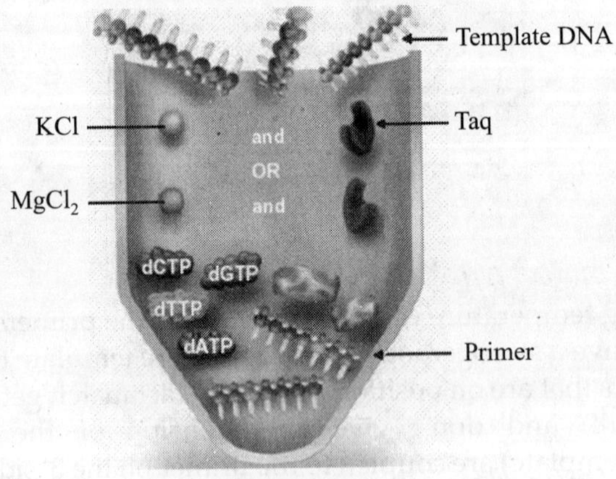

Fig. 9. Master Mix Components.

Master Mix Components

Reaction Mixture Set Up

Gently vortex and briefly centrifuge all solutions after thawing. Add, in a thin-walled PCR tube, on ice (Table 1).

Table 1. Reaction Mixture Set Up

Reagent	Final concentration	Quantity, for 50µlof reaction mixture
Sterile deionized water	-	variable
10X *Taq* buffer	1X	5 µl
2 mM dNTP mix	0.2 mM of each	5 µl
Primer I	0.1-1 µM	variable
Primer II	0.1-1 µM	variable
Taq DNA Polymerase	1.25u/50 µl	variable
25 mM $MgCl_2$	1-4 mM	variable*
Template DNA	10 pg-1 µg	variable

Mix the sample gently and briefly centrifuge to collect all drops from walls of tube. Overlay the sample with half volume of mineral oil or add an appropriate amount of wax. This step may be omitted if the thermal cycler is equipped with a heated lid. Place the tubes containing samples in a thermal cycler and start PCR (Tables 2 and 3).

Table 2. Selection of 25 mM $MgCl_2$ solution

Final concentration of $MgCl_2$ in 50µl reaction mix, mM	**1.0**	**1.25**	**1.5**	**1.75**	**2.0**	**2.5**	**3.0**	**4.0**
Volume of 25 mM $MgCl_2$, µl	2	2.5	3	3.5	4	5	6	8

Table 3. Programming for PCR

	Temperature	**Time**	**No. of cycle**
Initial Denaturation	94°C	2 min	
Denaturation	94°C	15-30sec	
Annealing	50-65°C	30-60sec	25-30x
Elongation	72°C	45sec-3min	
Final Elongation	72°C	7min	
Store	4°C		

PCR IN THEORY

When amplifying a small fragment of larger double-stranded DNA fragment, the desired blunt ended fragment first appear in the third cycle of PCR. The amplified product then accumulate exponentially, as described in the following equation

$$N_f = N_o (1 + Y)^n$$

where,

N_f = copy number of amplified sequence after n cycle of amplification

N_o is the initial number of the target sequence in the template DNA

Y is the efficiency of amplification per cycle. For Taq polymerase Y is ~ 0.7

This equation is only applies to the exponential phase of amplification, which continues until one of the component in the reaction becomes limiting. In a standard PCR containing ~1 unit of thermostable DNA polymerase, the enzyme becomes limiting when the copy number of the target sequences approaches 10^{12}. At this point, the efficiency of the reaction drops precipitously and the product no longer accumulates in a geometric fashion. Knowing the final yield of amplified product at the end of the exponential phase, the overall efficiency of the exponential reaction can be calculated from the following equation 2

$$Y = [N_f / N_o]^{1/n} - 1$$

where,

Y is the efficiency of amplification per cycle,

N_f/N_0 is the number of amplified molecules produced in n cycle of exponential amplification (Fig. 10).

The efficiency of PCR is determined by the quality of the thermostable polymerase. Because of the geometric nature of PCR, the penalty for using an inefficient enzyme is severe.

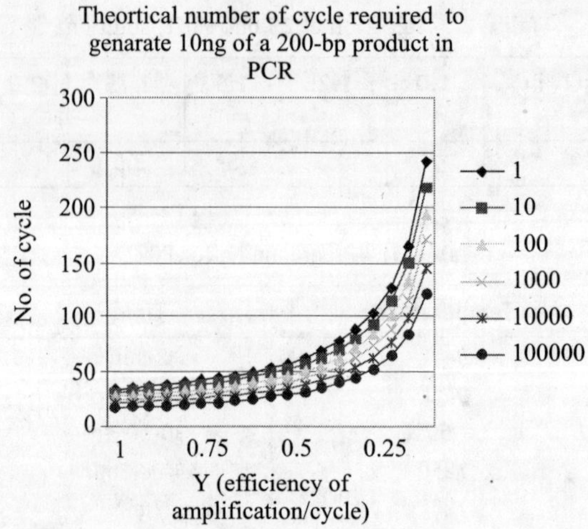

Fig. 10. Theortical number of cycle required to genarate 10ng of a 200-bp product in PCR.

PCR CYCLE

Step I

In step of PCR, Target sequences are denature at 94–95°C and open in to single strand.

Step 2

Primer pair anneals to the target sequence at 58–68°C (Fig. 11)

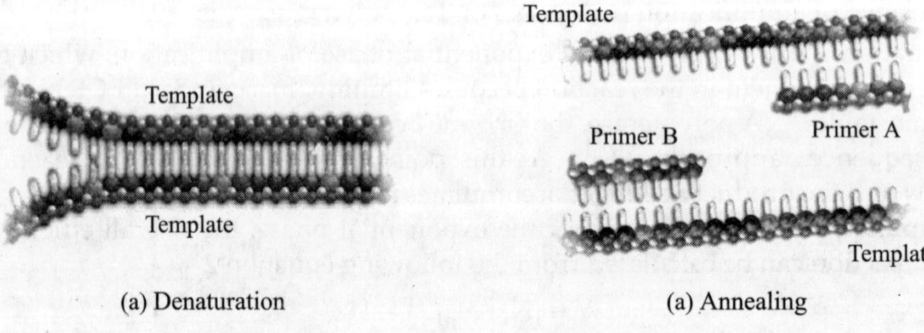

Fig. 11

Step 3

Taq DNA polymerase catalyses DNA primer extension as incorporation of complementary nucleotides to the template (Fig. 12). The ideal working temperature for the polymerase is 72°C.

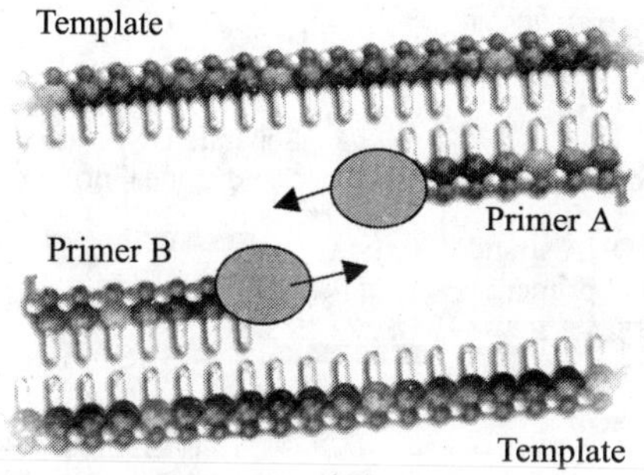

Fig. 12. Extension.

End of the Ist PCR Cycle results in two copies of the target sequence (Fig. 13).

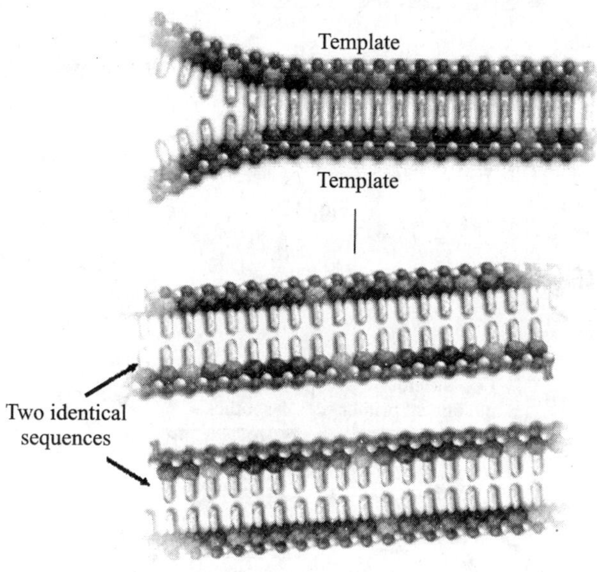

Fig. 13. Two copies of the target sequence.

PCR: The First Cycle (Fig. 14)

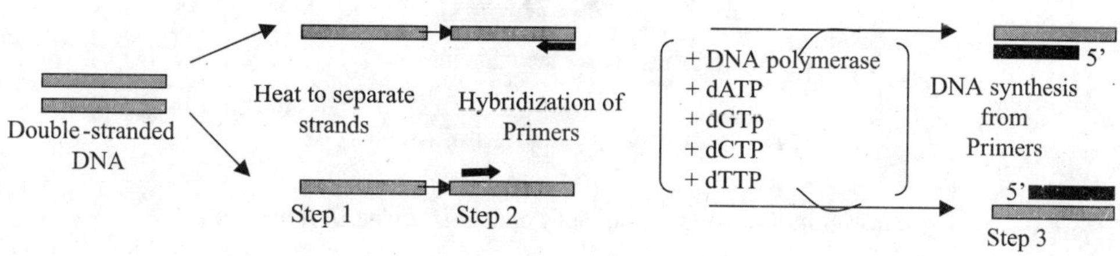

Fig. 14

PCR: The Second Cycle (Fig. 15)

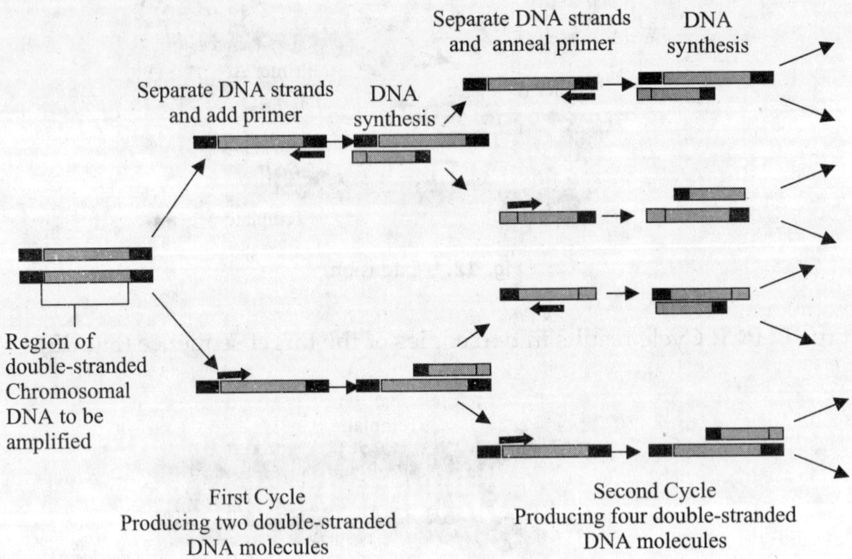

Fig. 15

PCR: The Third Cycle (Fig. 16)

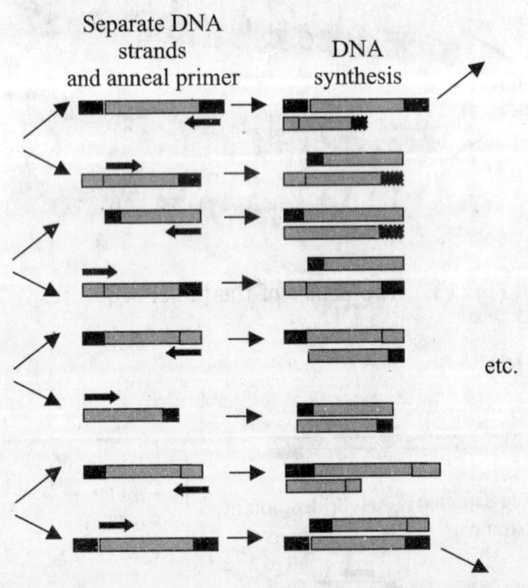

Fig. 16. Third Cycle Producing eight double-stranded DNA molecules.

PCR VARIANTS

AFLP (Amplified Fragment Length Polymorphism)

AFLP is a DNA fingerprinting technique which detects multiple DNA restriction fragments by means of PCR amplification. The AFLP technology usually comprises of the following steps (Fig. 17):

(a) The restriction of the DNA with two restriction enzymes, preferably a hexa-cutter and a tetra-cutter.

(b) The ligation of double-stranded adapters to the ends of the restriction fragments.

(c) The amplification of a subset of the restriction fragments using two primers complementary to the adapter and restriction site sequences, and extended at their 3' ends by "selective" nucleotides.

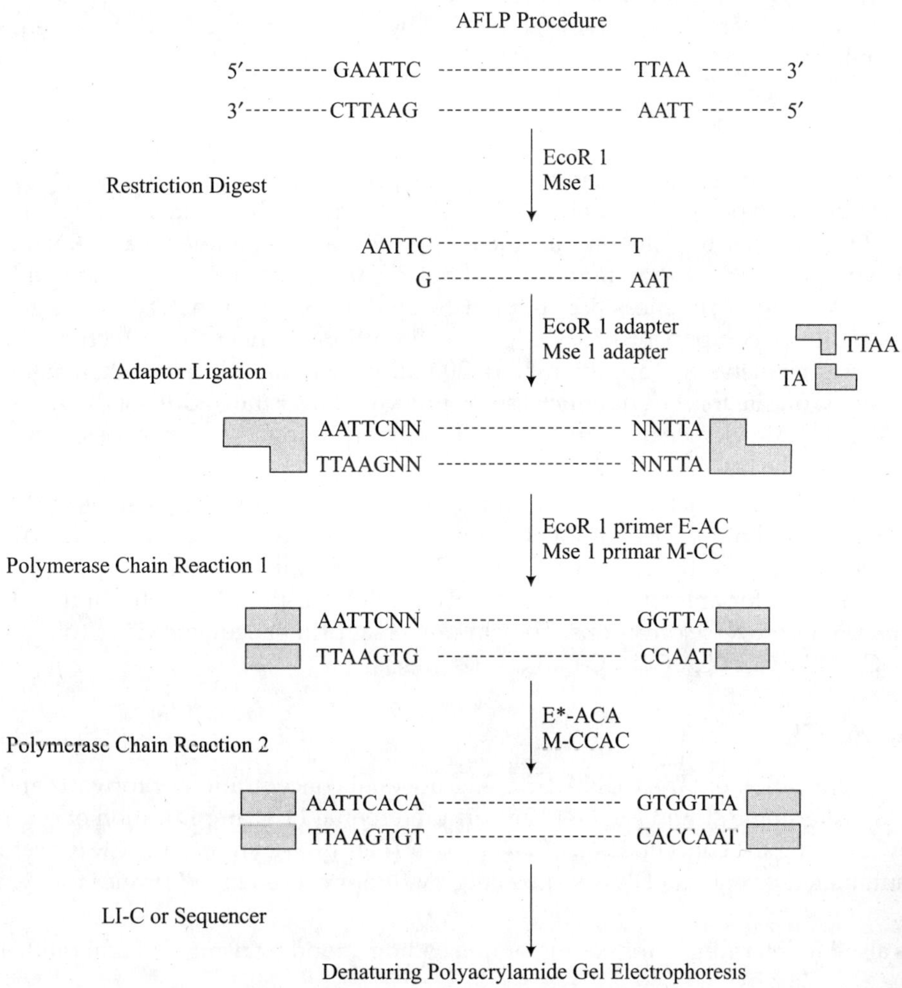

Fig. 17. Amplified Fragment Length Polymorphism (Ulrich *et al.*, 1999).

(d) Gel electrophoresis of the amplified restriction fragments on denaturing slab gels or capillaries.

(e) The visualization of the DNA fingerprints by means of autoradiography, phospho-imaging, or other methods.

A major advantage of the technology is the high marker density that can be obtained. A typical AFLP fingerprint contains between 50 and 100 amplified fragments. The frequency with which AFLP markers are detected depends on the level of sequence polymorphism between the tested DNA samples. The molecular basis of AFLP polymorphisms will usually be single nucleotide polymorphisms in the restriction sites, or in the selective nucleotides adjacent to the restriction sites (causing the AFLP primers to mispair at the 3' end). In addition, deletions, insertions and rearrangements affecting the presence or size of restriction fragments will result in detectable polymorphisms.

The AFLP technology can be applied in wide scope of restriction enzymes and all feasible combinations of selective nucleotides. Depending on the degree of polymorphism between samples, individual samples can be fingerprinted by using more and/or different enzyme and primer combinations.

Alu-PCR

It is the amplification of Alu fragments using PCR primers specific to Alu sequences. The Alu family is short interspersed repeated DNA elements that are distributed throughout primate genomes. Over the past 65 million years, the Alu sequence has amplified via an RNA-mediated transposition process to a copy number of about 500,000-comprising an estimated 5% of the human genome. Alu sequences are thought to be derived from the 7SL RNA gene which encodes the RNA component of the signal recognition particle that functions in protein synthesis. Alu elements are approximately 300-bp in length and derive their name from a single recognition site for the endonuclease Alu 1 located near the middle of the Alu sequence. About 500-2,000 Alu elements are present to the human genome. A few of these have inserted recently, within the last one million years.

Animal cell lines are also containing "alu gene". There are approximately 900,000 copies of the alu gene distributed throughout the human genome, and multiple copies distributed through the genome of other animal cells, as well. Thus, the alu gene provides the sequence for a universal primer for animal cell lines. The alu primer is especially useful in that it binds in both forward and reverse directions. The alu universal primer sequence is as follows: 5' GTG GAT CAC CTG AGG TCA GGA GTT TC 3' (26 mer).

Asymmetric PCR

It is the amplification of target DNA by using unequal concentration of forward and reverse primer to make single-strand (ss) DNA or it is preferential PCR amplification of one strand of DNA by lowering the concentration of one primer (Fig. 18). Asymmetric PCR is useful for the direct automated or manual DNA sequencing by fluorescence tagged primer for synthesis of ssDNA.

It is also for determining the coding or non coding strand by using unequal ratio of primer.

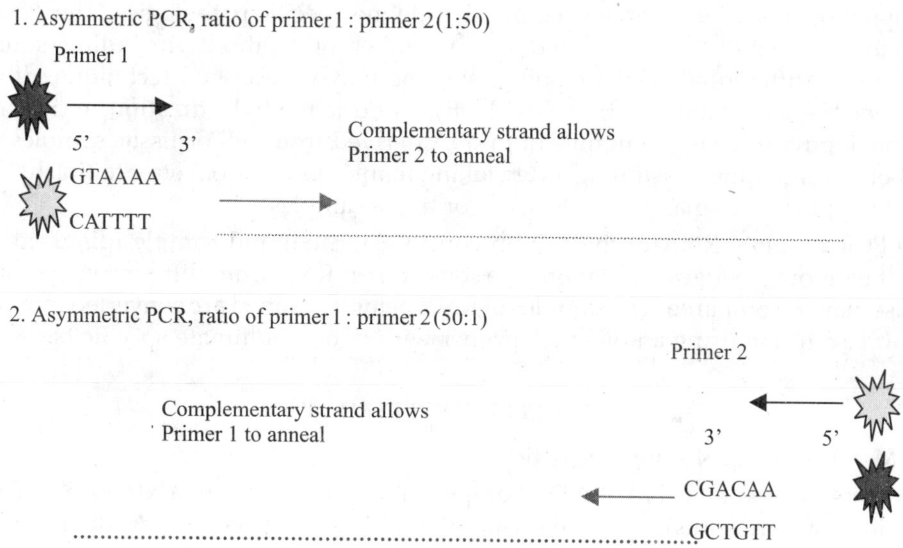

Fig. 18. Overview of Asymmetric PCR.

Colony PCR

Colony PCR is designed to quickly screen for transformation success and to determine the length and orientation of a cloned insert directly from colonies. A typical colony PCR reaction contains following reaction mixture:

38 µl sterile distilled water
5 µl 10 × PCR buffer
1.0% Triton × 100)
3 µl 25 mM $MgCl_2$
1 µl 10 mM dNTPs
1 µl 20 µM forward primer
1 µl 20 µl reverse primer
0.2-1 µl *Taq* polymerase
50 µl total volume

To each cold PCR tube containing the PCR reaction, add a small amount of colony. Sufficient mixing will result in complete cell lysis and high yields.

Differential Display PCR

The human genome contains over 100,000 genes, each one coding for specific RNA and/or a protein product. At any given time there are in all cells only a limited number of these genes expressed. Different tissues at different stages of growth and development express genes in a highly specific manner. In addition, during disease progression various genes are expressed inappropriately, resulting in abnormalities that define specific pathologies. These expressed genes can be identified and isolated by a number of methods, including subtractive-hybridization, differential cDNA cloning, and the newly described technique, differential display-polymerase chain reaction (DD-PCR). Subtractive-hybridization and differential cDNA cloning require large amounts of mRNA derived from cell or tissue samples, and are very laborious and time consuming, often taking many months from start to finish. However DD-PCR requires only small amounts of cell or tissue samples.

DD-PCR requires RNA derived from control (normal) and sample (diseased) cells or tissues; if temporal changes/alterations are sought then RNA from different stages of normal or disease may be compared. The samples of RNA being examined are converted into cDNA by reverse transcription using an oligo-dT primer with two penultimate specific bases to the 3′ end:

5′ TTTTTTTTTTTTMN 3′

where, M = A, G, or C; N = any nucleotide

Upon conversion to cDNA the PCR amplification is done with arbitrary 5′ primer (8-10 mers) using random sequences, and the same 3′ primer (shown above). In order to define most of the genes expressed in these cells, different sets of random primers may be used. After ~30 amplification cycles using radioactive S-35 labeled nucleotides the expressed cDNA fragments are displayed using DNA sequencing gels. Differentially expressed cDNAs are easily visualized following autoradiography. The X-ray films can be used directly as a template to identify bands in the gels that are differentially expressed. The DNA from these gels slices is recovered by boiling, and re-amplified by PCR using the original primers. The amplified DNAs are characterized by sub-cloning, sequencing and data-base searches. To verify the differential expression status of each of these sub-clones, Northern-Blot assays are performed using RNA isolated from the original sample sources.

Degenerate PCR

Degenerate PCR is in most respect identical to ordinary PCR, but with one major difference instead of using specific PCR primers with a given sequence, use mixed PCR primers. That is, if sequence of the gene not known insert "wobbles" in the PCR primers where there is more than one possibility (degenerate primer). **Degenerate Primers** which have a number of options at several positions in the sequence so as to allow annealing to and amplification of a variety of related sequences e.g.

5′-TCG AAT TCI CCY AAY TGR CCN T-3′

Y = pyrimidines = C/T (degeneracy = 2X)

R = purines = A/G (degeneracy = 2X)

I = Inosine = C/G/A/T

N = Nucleotide = C/G/A/T (degeneracy = 4X)

Why degenerate primers are use?

(a) To amplify (find out) conserved sequences of a gene or genes from the genome of an organism.

(b) To get the nucleotide sequence after having sequenced some amino acids from a protein of interest.

Degenerate Primer Design

I. Sequence alignment

The Amino acid sequences of similar or homologous proteins can be retrieved from a database such as GeneBank or it can be your own. These sequences are then aligned. At least 2 blocks of conserved amino acids should be present to enable the design of PCR primers. A further alignment can be done at nucleotide level if desired. There should be conserved base throughout the alignment then it can be 'guessed' that this particular base will also be the same in the case of interest. This can decrease degeneracy but can also be straight out 'wrong'. The primers should be 20-30 mer in length (minimum 20 mer). The sequence alignment will also give the expected size of the PCR product.

2. Terminal a.a. sequence information

Should this data be available then it can also be used as a starting point for primer design other than using the alignment method above.

3. Primer degeneracy

Degeneracy of the primers depends on a multitude of factors including the template. 1000-10,000 fold degeneracy has been done. To get the degeneracy, simply multiply all the degeneracy values incorporated into the primer sequence. Another factor that must also be taken into consideration is that as the degeneracy of the primers increase, the concentration of a specific primer will decrease.

Hot Start PCR

The activity of PCR polymerases is optimal at elevated temperatures (70°–80°C); most of these polymerases (e.g. Taq DNA Polymerase) have significant lower activity at much lower temperatures. Since the components of the reaction are all mixed together at 4°C or at room temperature, the PCR primers may anneal non-specifically to template and the polymerase may elongate primer from these non-specific annealing sites prior to the initial cycle. The result is a series of non-specific amplification products, which contaminate the desired product. To prevent this contamination, a number of strategies, known as "Hot Start" techniques, have been developed. Hot start is a method to optimize the yield of the desired amplified product in PCR and simultaneous, to suppress nonspecific amplification. This is done by withholding an essential component of the PCR- the DNA polymerase or Mg^{2+} for example – until the reaction mixture has been heated to a temperature that inhibits nonspecific priming and primer oligimerization.

Overview of "Hot Start" Techniques

In a "Hot Start" PCR, some component essential for polymerase activity is "separated" from the reaction mixture until the temperature in the tubes has exceeded the optimal primer annealing temperature (usually 55°–65°C). Since the polymerase cannot function in the incomplete reaction mixture, there is no primer elongation during the period when primers can bind non-specifically. "Hot Start" PCR has been shown to significantly improve the specificity, sensitivity, and yield of PCR.

The separation of the essential component can be accomplished in a number of ways:

Manually

A component such as polymerase or Mg^{2+} is omitted from the original reaction mixture, and is only added to the tubes after the temperature is >70°C.

Comment

Manual addition of components to the heated tube is generally inconvenient and may lead to problems such as lack of reproducibility and carry-over contamination.

Physical separation

Reaction components are divided into two mixes, which are separated by a barrier, such as a wax plug. Alternatively, wax beads containing Mg^{2+} ions are added to the PCR mix. During the initial denaturation step, the wax melts (at 75°–80°C) and allows all the reaction components to mix.

Polymerase antibodies

In a pre-incubation step, a heat-sensitive antibody is added to the polymerase (usually Taq or Tth DNA Polymerase) to keep it inactive in the reaction mixture. As the temperature in the tube rises, the antibody is inactivated, setting the active polymerase free.

Chemical modification of polymerase

Addition of heat-labile blocking groups to some of the amino acid residues on Taq DNA Poly-merase makes the modified enzyme (e.g. Fast Start Taq DNA Polymerase) inactive at room temper-ature. It is "activated" (by removal of the blocking groups) at high temperatures.

Other inhibitors

Other polymerase inhibitors (such as polymerase-binding oligonucleotides) are added to the reaction mixture, keeping the enzyme inactive until increasing temperature dissociates the inhibitor from the enzyme, setting it free to begin polymerization.

However Hot start is not essential in optimized simple PCRs that contains a single pair of well-designed primer and generate high yields of a specific amplified product. It is require when non-specific amplification is a problem, for example, when fewer than 104 copies of template DNA are present in the reaction, when the template DNA is highly complex or PCR contain several pairs of primer (multiplex PCR).

In situ PCR

In situ hybridization (ISH) has proven to be a very important molecular tool in diagnostic and research and has significantly advanced the study of gene structure and expression at the level of individual cells. However, the usefulness of ISH is occasionally limited by a detection sensitivity of about 20 copies of mRNA per cell, an obstacle that may now be overcome by exciting new technologies.

In recent years, strategies to improve threshold levels for detection in ISH studies have included protocols which either amplify the detection signals e.g., using several antibody steps or increase the absolute amount of hybridized probes by using e.g., cocktails of oligonucleotide or multiple cRNA probes (Fig. 19). A third strategy employing a PCR based amplification of the target nucleic acid sequences prior to ISH was developed by several laboratories independently in the late 1980's. This approach has variously been termed "*in situ* PCR", "PCR *in situ*", "PCR *in situ* hybridization", "in-cell PCR", "PCR-driven ISH" or "cycling PAINS". More recently, an intracellular reverse transcription (RT) step to generate complementary DNA from mRNA templates prior to in situ PCR has been used for the detection of low copy mRNA sequences. This modification of *in situ* PCR has been termed "*in situ* RT-PCR" or less precisely "RT *in situ* PCR" or "*in situ* cDNA PCR".

Prior to *in situ* PCR, cells or tissue samples are fixed and permeabilized to preserve morphology and permit access of the PCR reagents to the intracellular se-quences to be amplified (Figs. 20 and 21). PCR amplification of target sequences is next performed either in intact cells held in suspension in micro-Eppendorf tubes or directly in cytocentrifuge preparations or tissue sections on glass slides. In the former approach, fixed cells suspended in the PCR reaction mixture are thermal cycled in micro-Eppendorf tubes using conventional block cyclers. After PCR the cells are cytocentrifugated onto glass slides with visualization of intracellular PCR products by ISH or immunohistochemistry. *In situ* PCR on glass slides is performed by overlaying the samples with the PCR mixture under a coverslip which is then sealed with nail polish, rubber cement or mineral oil to prevent evaporation of the reaction mixture. Thermal cycling is achieved by placing the glass slides either directly on top of the heating block of a conventional or specially designed thermal cycler or by using thermal cycling ovens. Detection of intracellular PCR-products is achieved by one of two ways (Fig. 19).

Applications

In situ PCR has a number of potential research and diagnostic applications. To date several different groups have reported successful *in situ* PCR detection of single copy nucleic acid sequences in cell preparations and also of low copy DNA or RNA sequences in tissue sections. Most of the studies to date have focused on the detection of viral or proviral DNA, but *in situ* PCR has also been applied to the study of endogenous DNA sequences including human single copy genes, rearranged cellular genes and chromosomal translocations and to map low copy number genomic sequences in metaphase chromosomes. Recently, successful amplification

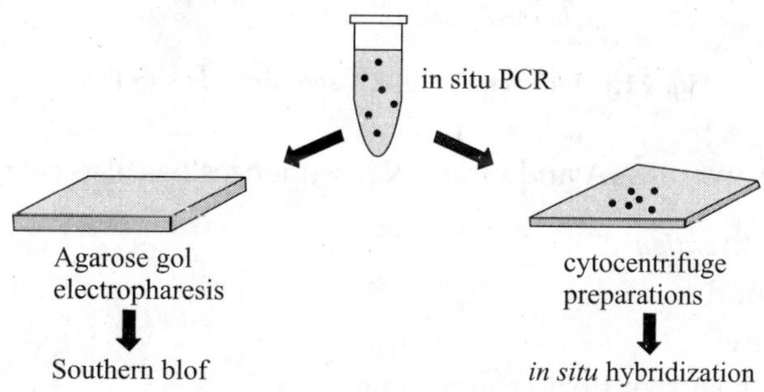

Fig. 19. Detection of intracellular PCR-products. Southern Blot and *in situ* Polymerase Chain Reaction.

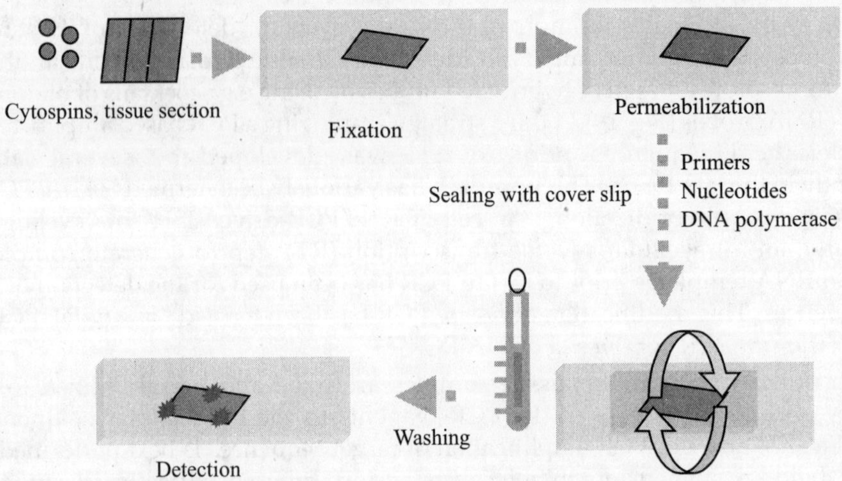

Fig. 20. Overview of In Situ Polymerase Chain Reaction.

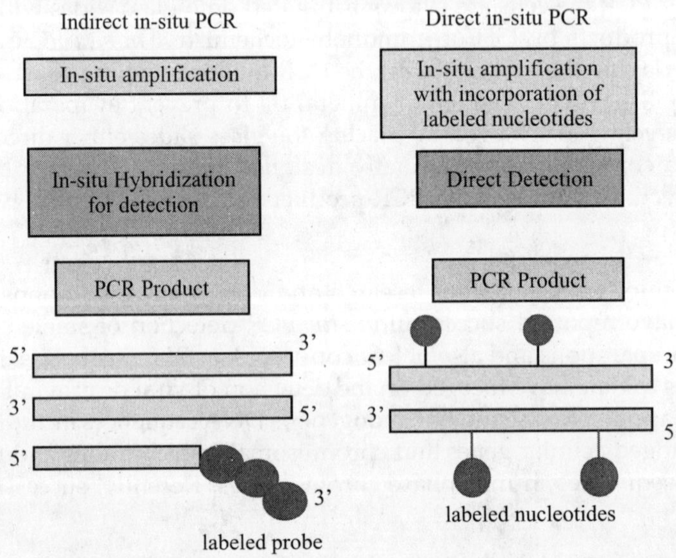

Fig. 21. Indirect *in situ* PCR and Direct *in situ* PCR.

and detection of low copy mRNA and viral RNA sequences has also been reported.

Main Advantages of this Invention

(a) Low background
(b) High specificity
(c) Fast assay with shorter turn-around time
(d) No radioactive chemicals needed

Inverse PCR

Standard PCR amplifies segments of DNA that lie between two inward-pointing primers. By contrast, inverse (also known as inverted or inside-out) PCR is used to amplify and clone unknown DNA that flanks one end of a known DNA sequence and which no primers are available. The technique, developed independently by several groups (Ochman et al., 1988; Trigila et al., 1988; silver and Keerikatte 1989), involve digestion by restriction enzyme of a preparation of DNA containing the known sequence and its flanking region (Fig. 22).

The individual restriction fragments are converted into circles by intramolecular ligation, and the circularized DNA is then used as a template in PCR. The unknown sequence is amplified by two primers that bind specifically to the known sequence and point in opposite directions. The product of the amplification reaction is a linear DNA fragment containing a single sites for the restriction enzyme originally used to digest the DNA.

Inverse primer

There is nothing unusual about by primers used in inverse PCR except the placement on the template molecule.

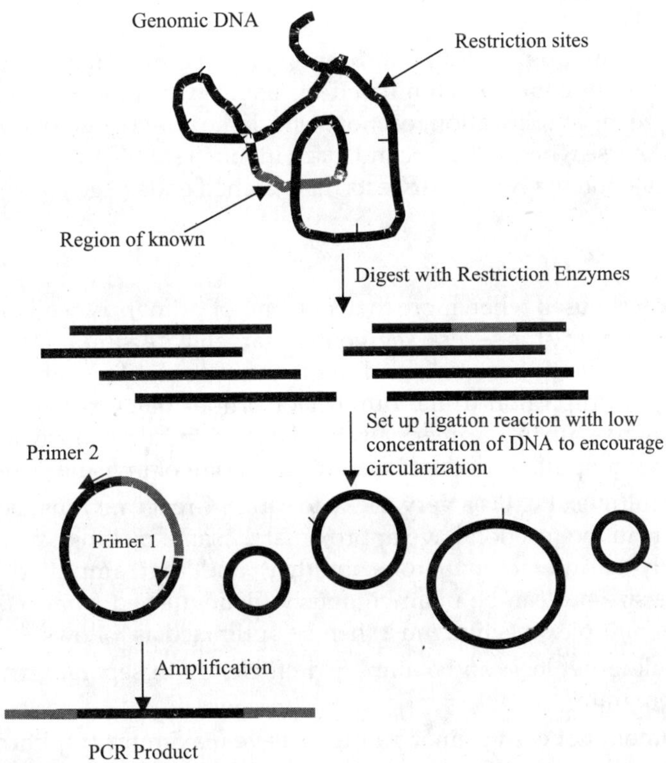

Fig. 22. Overview of Inverse PCR.

Applications

Inverse PCR has been used chiefly to generate end-specific probes for chromosome walking. It is also useful to clone unknown cDNA sequence from total RNA and to recover integration sites used by viruses, transgenes, and transposons.

Long PCR

Standard PCR is capable of amplifying segment of DNA 1-2 Kb in length. This capacity is sufficient for many routine manipulation of DNA. But it is not enough to an entire mammalian gene even one of modest size nor a cDNA of average dimensions. This inability is because of the following reasons:

(a) Damage to the template and product DNAs during exposure to high temperature in buffers that may not adequately maintain control over pH
(b) The presence of stray divalent cations
(c) Difficulties in denaturing very long DNA molecules
(d) The inability of DNA polymerases to attached
(e) The high rate of incorporation of incorrect bases.

Four of these problems were fixed by simple adjustments to the reaction conditions. The fifth, however, required an original thought. Barnes (1994) realized that the obstacles created by mismatched 3′ termini could be eliminated by using two different thermostable DNA polymerase to catalyze the amplification reaction. One of the polymerase, would be an efficient but error-prone workhorse whereas the second, used in much small amounts, would provide a 3′→5′ exonuclease function that would dissects mismatched ends (e.g., *pfu*).

Multiplex PCR

Multiplex PCR is the term used when more than one pair of primers is used in a PCR. The goal of Multiplex PCR is to amplify the several segments of target DNA simultaneously and thereby to conserve the template DNA, save time and minimize expense. The yield of each amplified product is reducing in proportion to the number of primer pairs included in the reaction. However, as a general rule, up to 8 primer pairs can be used simultaneously in a standard PCR before the yield of each amplified product is reduce to a point of invisibility on an agarose gel.

Unfortunately Multiplex PCR is a very tricky to setup. Great care must be taken to ensure that all of the primers in the reaction have approximately same melting temperature, that the primers are unlikely to interact with one another, and that amplified product are of approximately same size but can be unambiguously distinguished from one another by gel electrophoresis. The multiplex reaction must then be optimized as follows:

(a) Check that all target loci can be amplified efficiently in separate reactions using the same PCR program
(b) Titrate the amount of each primer pair to achieve maximum amplification in separate reactions using the same PCR program and reaction condition
(c) Balance the amount of each primer pair in the multiplex reaction to achieve acceptable amplification of all target regions

Nested PCR

Nested PCR means that two pairs of PCR primers were used for a single locus. The first pair amplified the locus as seen in any PCR experiment. The second pair of primers (nested primers) bind within the first PCR product and produce a second PCR product that will be shorter than the first one. The logic behind this strategy is that if the wrong locus were amplified by mistake, the probability is very low that it would also be amplified a second time by a second pair of primers (Figs. 23–27).

```
5' ..GCAT...TTGG....//....GCGC...ATAT.. 3'
3' ..CGTA...AACC....//....CGCG...TATA.. 5'
```

Fig. 23. Template DNA (http://www.bio.davidson.edu/courses/genomics/method.NestedPCR.html).

Segments of DNA with dots representing non-descript DNA sequence of unspecified length. The double lines represent a large distance between the portions of DNA illustrated in this figure. The portions of DNA shown with four bases in a row represent PCR primer binding sites, though real primers would be longer.

```
5' ..GCAT...TTGG....    ....GCGC...ATAT.. 3'
                                 ←tata

     gcat→
3' ..CGTA...AACC....    ....CGCG...TATA.. 5'
```

Fig. 24. First round of amplification (http://www.bio.davidson.edu/courses/genomics/method.NestedPCR.html).

The first pair of PCR primers binds to the outer pair of primer binding sites and amplify all the DNA in between these two sites.

```
5' GCAT...TTGG....//....GCGC...ATAT 3'
3' CGTA...AACC....//....CGCG...TATA 5'
```

Fig. 25. PCR product after first round of PCR (http://www.bio.davidson.edu/courses/genomics/method.NestedPCR.html).

PCR products after the first round of amplification

```
5' GCAT...TTGG....    ....GCGC...ATAT 3'
                         ←cgcg

          ttgg→
3' CGTA...AACC....    ....CGCG...TATA 5'
```

Fig. 26. Second round of amplification (http://www.bio.davidson.edu/courses/genomics/method.NestedPCR.html).

Second pair of **nested primers** binds to the first PCR product. The binding sites for the second pair of primers are a few bases "internal" to the first primer binding sites.

```
5' TTGG..../ /....GCGC 3'
3' AACC..../ /....CGCG 5'
```

Fig. 27. Final PCR product after second round of PCR (http://www.bio.davidson.edu/courses/genomics/method.NestedPCR.html).

The length of the product is defined by the location of the internal primer binding sites. Nested PCR protocols is use to detect rare deletions in DNA from complex populations.

PCR-ELSIA

PCR-ELISAs (also known as PCR ELISA) are a capture assay for nucleic acids that mimic enzyme linked immunosorbant assays. In this assay, PCR products hybridized to an immobilized capture probe. The assay thus measures sequences internal to the PCR product and is a less expensive assay and an alternative to real time PCR.

PCR-ELISA has been in use since the late 1980s and has developed into an assay for detecting specific sequences within polymerase chain reaction products. Though many methods are available for detecting specific sequences, PCR ELISA is useful for detecting and differentiating between multiple targets. The PCR ELISA is also useful for screening multiple samples.

PCR-ELISA has following steps (Figs. 28–33).

Production of PCR product

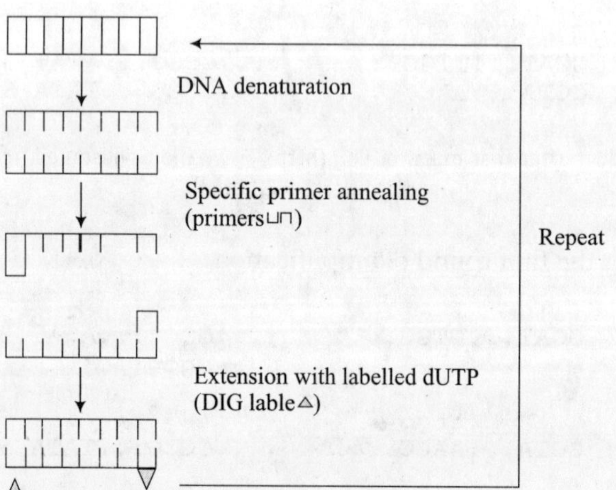

Fig. 28. Production of PCR product.

Hybridization of probe and PCR product

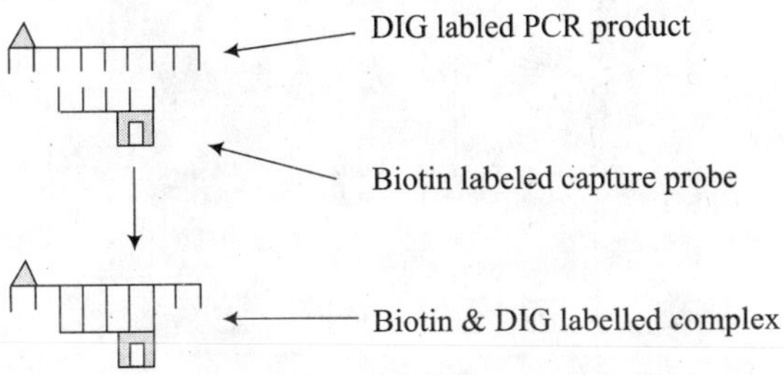

Fig. 29. Hybridization of probe and PCR product.

Immobilization of the probe/DNA hybrid on a microtiter plate

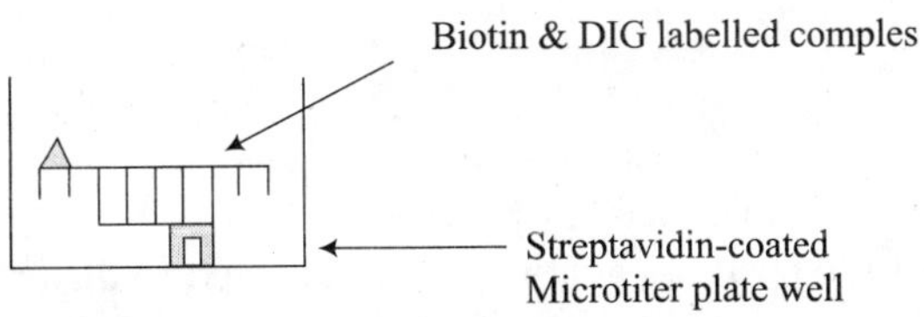

Fig. 30. Immobilization of the probe/DNA hybrid on a microtiter plate.

Binding of the antibody/enzyme complex to the PCR Product

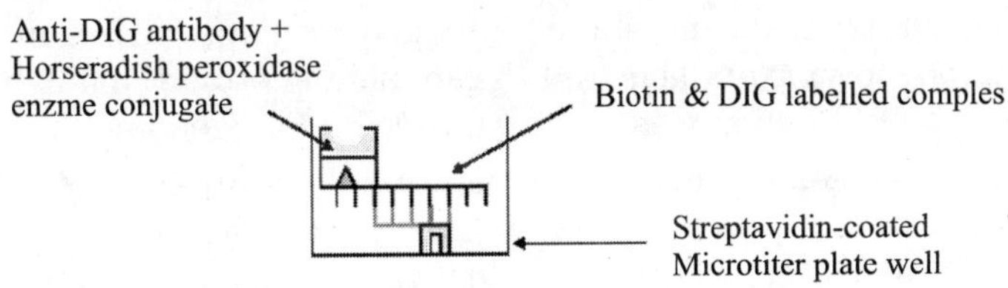

Fig. 31. Binding of the antibody/enzyme complex to the PCR Product.

Binding of enzyme substrate and colorimetric detection

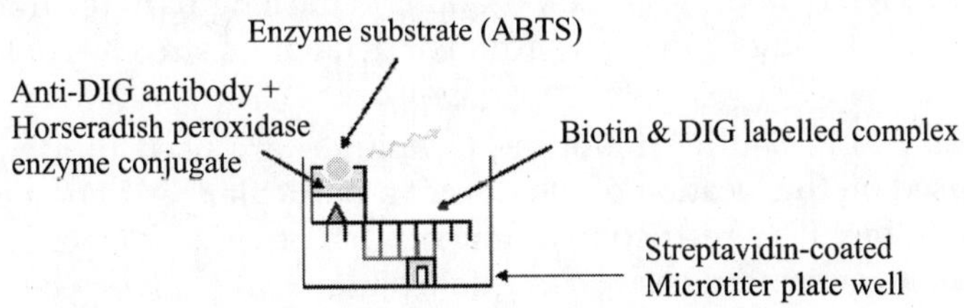

Fig. 32. Binding of enzyme substrate and colorimetric detection.

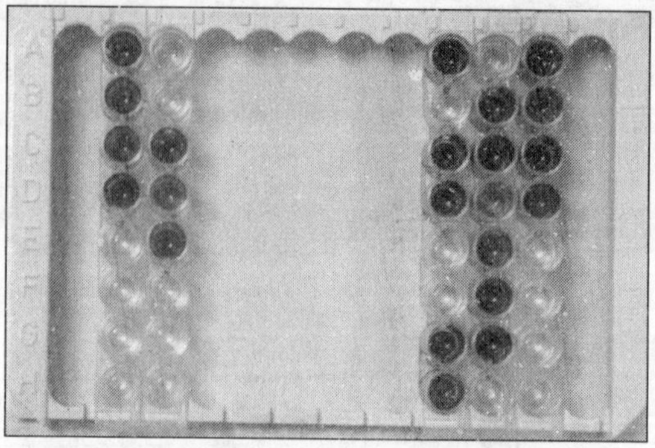

Fig. 33. PCR ELISA Plate (http://nl.wikipedia.org/wiki/ELISA).

Restriction Fragment Length Polymorphism-PCR (RFLP-PCR)

Restriction Fragment Length Polymorphism (RFLP) is a technique in which organisms may be differentiated by analysis of patterns derived from cleavage of their DNA. If two organisms differ in the distance between sites of cleavage of a particular restriction endonuclease, the length of the fragments produced will differ when the DNA is digested with a restriction enzyme. The similarity of the patterns generated can be used to differentiate species (and even strains) from one another. It is useful for detection of point mutations by RFLP of PCR amplified DNA sequences.

Restriction fragment length polymorphism (RFLP) is a powerful technique for the characterization of DNA at the molecular level. These markers are inherited in the same manner as genes that code for visible phenotypes. The recombination frequency between RFLP and a detectable phenotype can be measured. Thus, genetic maps can be constructed to include both genotypic and phenotypic markers. RFLPs can thereby provide a link between genes that lie far apart. In 1980, an RFLP map was created for the human genome.

Restriction maps that result from different patterns of distribution of restriction sites in the DNA of individuals within a population of organisms are called restriction fragment length polymorphisms (RFLPs). Differences in individual base pairs between comparable sequences of any two individual chromosomes occur at a frequency of greater than 1 change per kilobase. Highly polymorphic regions are usually located between genes, where small variations in sequence do not affect gene expression. The polymorphisms can be identified by the digestion of genomic DNA with a restriction enzyme (endonuclease). Differences in sequences that result in the gain or loss of a restriction site cause variations in the lengths of the fragments produced. Polymorphisms can also result from the insertion or deletion of stretches of DNA between two restriction sites.

To visualize an RFLP, Southern blotting techniques are used to identify fragments of various lengths based on the location of the sites of a particular restriction enzyme. Probes of known sequence that highlight restriction fragments that often vary in length among different individuals are used to generate clearly discernible patterns. A probe specific for a portion of a particular chromosome will reveal differences within that region of the genome from one individual to another.

Because restriction polymorphisms should occur near any particular target gene, RFLPs can be identified that show tight linkage with a mutant (or disease) phenotype. Comparison of restriction maps of patients suffering from a particular disease with those of unaffected individuals can reveal specific bands that are always present or absent in affected individuals.

After the identification of an RFLP that is tightly linked with a disease, it may be used as a molecular procedure to detect the disease, either in a prenatal screen or after birth. For instance, an RFLP has been identified that is associated with the genetic disease sickle cell anemia. Sickle cell anemia is caused by a mutation in the alpha-globin gene and results in an abnormal form of hemoglobin. Digestion of DNA with the restriction enzyme Hpa1 and Southern analysis using a probe specific for the alpha-globin gene results in the production of a 7-kb fragment in normal individuals. In contrast, patients with sickle cell anemia display a fragment of 13-kb. Carriers of the disease, those who do not have sickle-cell anemia, but who have inherited one copy of the mutant gene from a parent, can also be identified by RFLP analysis. They will produce both the 7- and 13-kb fragments. It is important to note that the change in DNA sequence that causes this RFLP is not the same change that causes sickle cell anemia itself. Instead, it is tightly linked to the alpha-globin gene. Other human genetic diseases that can be detected via RFLP analysis include Huntington's Chorea, phenylketonuria, and cystic fibrosis.

RFLP analysis is particularly useful for diagnosis of disease because it assays directly for a genotype (DNA sequence) and does not depend on expression of a gene or even phenotypic expression of the disease itself. Thus, a disease can be identified in an individual before symptoms of the disease are apparent. Additionally, a fetus can be monitored for diseases before birth.

Additionally, RFLPs provide a beginning point for the isolation of the gene responsible for a disease. If the mutation that causes an RFLP in fact lies within the gene responsible, an RFLP at this gene must occur in all cases of the disease. Therefore, it is difficult to prove that a defect in a particular gene is in fact responsible for causing a disease. However, mapping and RFLP analysis can exclude certain genes as candidates. They may also provide a point from which researchers may proceed along the DNA to identify the causal gene itself.

The technique termed DNA fingerprinting utilizes RFLPs and other polymorphic markers to identify individuals based on their particular patterns. Through this technique, hair, blood or other bodily fluids found at a crime scene and blood obtained from a suspect can be used to compare DNA fingerprint patterns. A match between the two patterns provides strong evidence against the suspect. RFLP patterns are also used to establish parent-child relationships by comparison of the map of a suitable region of the chromosome between potential parents and the child.

Reverse Transcriptase-PCR (RT-PCR)

Reverse Transcriptase-PCR is a method used to amplify cDNA copies of RNA. Sensitivity and versatile, RT-PCR is used to amplify to retrieve and clone the 5′ and 3′ termini of mRNAs and to generate large cDNA libraries from very small amounts of mRNA. Depending on the purpose of the experiment, the primer for the first-strand cDNA synthesis can be specifically designed to hybridize to a particular target gene or it can bind generally to all mRNAs. Oligo (dT), which binds to the endogenous poly $(A)^+$ tails of mammalian mRNAs, can be used as a universal primer for convention first-strand cDNA synthesis. Subsequent amplification by

PCR generally utilizes one or more internal, gene-specific primers and oligo (dT) to generate copies of the 3′-treminal sequences of a particular mRNA.

cDNA synthesis can be primed by a synthetic antisense oligonucleotide that hybridizes specifically to a chosen region in a particular target RNA or family of RNAs. Amplification of the desired portion of cDNA can be achieved in PCRs primed, for e.g., by sense and antisense oligonucleotide primers corresponding to specific sequences in a particular cDNAs. Random hexanucleotides, which are capable to priming cDNA synthesis at many points along the RNA templates, generate fragmentary copies of entire population of RNA molecules. They are useful when target DNA is extremely long or contains so much secondary structure that cDNA synthesis can't be efficiently primed by oligo (dT) or synthetic oligonucleotides.

Reverse transcriptase used in RT-PCR

The discovery of DNA-dependent RNA polymerase in 1970 solved the long-standing puzzle of how the RNA genomes of certain tumor viruses could be copied into DNA in infected and/or transformed cells the name reverse transcriptase was coined as a joke by John Tooze, who first used it in an anonymous News and Views article in nature. Much to the annoyance of purists and the delight of John Tooze, the name stuck and quickly became iconic.

Three form of RNA dependent RNA polymerase used *in vitro* to catalyze the synthesis of DNA complementary to an RNA template (Fig. 34).

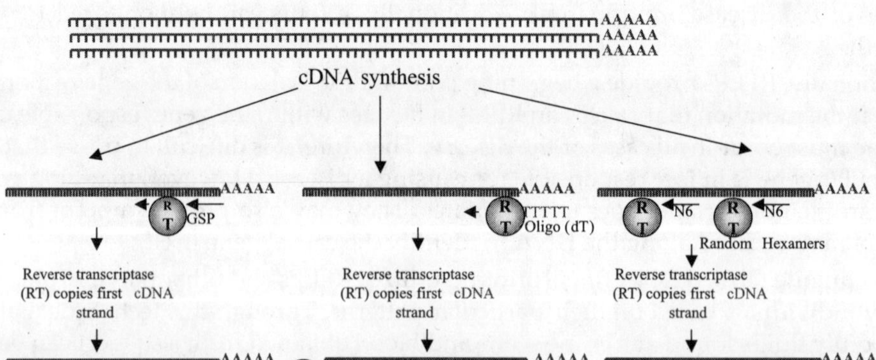

Fig. 34. Three form of RNA dependent RNA polymerase used *in vitro* to catalyze the synthesis of DNA complementary to an RNA template (Sambrook and Russel, 2001).

Mesophilic enzymes encoded by avian myoblastosis virus (AMV) and Molony strain of murine leukemia virus (Mo-MLV). Both enzymes require an RNA or DNA template with an RNA or DNA primer bearing a 3′-OH group. Lacking an editing 3′ → 5′ exonuclease activity, both enzymes are prone to error. The K_m values for their substrate dNTP are very high, in a millimolar range. The enzymes encoded by AMV has a powerful RNase H activity hat can digest RNA moiety of RNA-DNA hybrids and can cleave the template near the 3′-terminals of the growing DNA strand if reverse transcriptase pauses during synthesis.

The murine enzyme is far better suited for RT-PCR because its RNase H activity is comparatively weak. Forever Mo-MLV enzyme reaches maximum activity at lower temperature (37°C) than the avian enzyme (42°C) which may be slightly disadvantage if the RNA template has a high degree of secondary structure.

Variants of Molony strain of murine leukemia virus (Mo-MLV) reverse transcriptase that lack RNase H activity. The modified reverse transcriptase transcribes a greater proportion of template molecules and synthesizes longer cDNA molecules than the wild-type enzyme. In addition, they are capable of cDNA synthesis at very high temperature (up to 50 °C) which is an advantage when the template RNA is rucked into secondary structures.

Thermostable Tth DNA polymerase, which is encoded by thermophilic eubacterium *Thermus thermophilus* and exhibits reverse transcriptase activity in the presence of Mn^{2+}. The chief advantage of using Tth DNA polymerase in RT-PCR is that both phase of reaction (reverse transcription and amplification) are carried out in a same reaction tube.

Real Time PCR

Real Time PCR (sometimes called kinetic PCR) is used in research laboratories to quantify gene expression and confirm the differential expression of genes detected by array technology.

The Evolution of PCR to Real-Time

Traditional PCR has advanced from detection at the end-point of the reaction to detection while the reaction is occurring. Real-Time chemistry allows the detection of PCR amplification during the early phases of the reaction. Measuring the kinetics of the reaction in the early phases of PCR provides a distinct advantage over traditional PCR detection. Traditional methods use Agarose gels for detection of PCR amplification at the final phase or end-point of the PCR reaction.

Limitations of End-Point PCR

Agarose gel results are obtained from the end point of the reaction. Endpoint detection is very time consuming. Results may not be obtained for days. Results are based on size discrimination, which may not be very precise. The end point is variable from sample to sample. While gels may not be able to resolve these variabilities in yield, real-time PCR is sensitive enough to detect these changes. Agarose Gel resolution is very poor, about 10 fold changes. Real-Time PCR can detect as little as a two-fold change. Some of the problems with End-Point Detection are poor precision, low sensitivity, short dynamic range < 2 logs, low resolution, non-automated, size-based discrimination only, results are not expressed as numbers, ethidium bromide for staining is not very quantitative, post PCR processing etc.The samples in the gel contain 10 copies and 50 copies, respectively. It is hard to differentiate between the 5-fold changes on the Agarose gel. Real-Time PCR is able detect a two-fold change (i.e. 10 Vs. 20 copies).

PCR Phases

To understand why end-point PCR is limiting, it is important to understand what happens during a PCR reaction. A basic PCR run can be broken up into three phases (Fig. 35):

Exponential: Exact doubling of product is accumulating at every cycle (assuming 100% reaction efficiency). The reaction is very specific and precise.

Linear (High Variability): The reaction components are being consumed, the reaction is slowing, and products are starting to degrade.

Plateau (End-Point: Gel detection for traditional methods): The reaction has stopped, no more products are being made and if left long enough, the PCR products will begin to degrade.

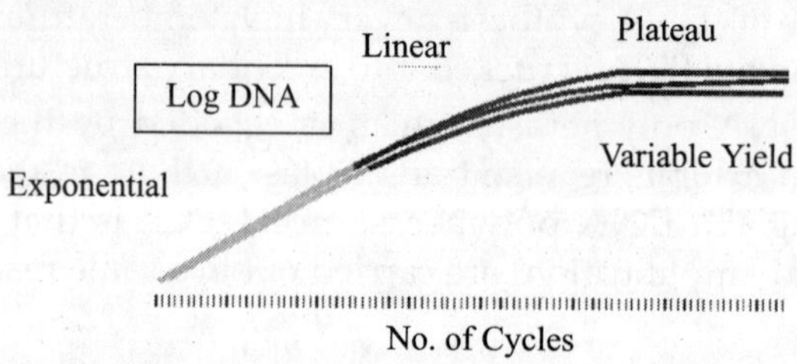

Fig. 35. Three replicates of a sample. The replicates have the same starting quantity. As the PCR reaction progresses, the samples begin to amplify in a very precise manner.

The replicates have the same starting quantity. As the PCR reaction progresses, the samples begin to amplify in a very precise manner. Amplification occurs exponentially, that is a doubling of product (amplicon) occurs every cycle. This type of amplification occurs in the exponential phase. Exponential amplification occurs because all of the reagents are fresh and available, the kinetics of the reaction push the reaction to favor doubling of amplicon.

However, as the reaction progresses, some of the reagents are being consumed as a result of amplification. This depletion will occur at different rates for each replicate. The reactions start to slow down and the PCR product is no longer being doubled at each cycle. This linear amplification can be seen in the linear phase of the reaction. The three samples begin to diverge in their quantities during the linear phase. Eventually the reactions begin to slow down and stop all together or plateau. Each tube or reaction will plateau at a different point, due to the different reaction kinetics for each sample. These differences can be seen in the plateau phase. The plateau phase is where traditional PCR takes its measurement, also known as end-point detection. Figure also shows that the three replicate samples, which started out at the same quantity in the beginning of the reaction; reflect different quantities at the plateau phase. Since the samples are replicates they should have identical quantities. Therefore, it will be more precise to take measurements during the exponential phase, where the replicate samples are amplifying exponentially. The amplification phases can be viewed differently to assess the PCR phases. The figures that follow show the phases of PCR in a Logarithmic scale view and a Linear scale view (Figs. 36 and 37).

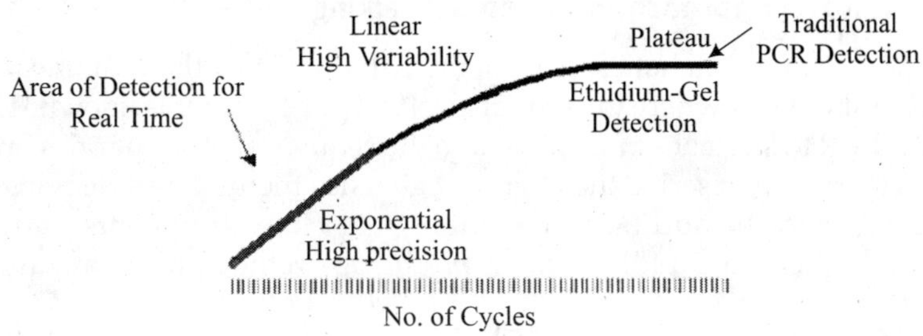

Fig. 36. Phases of PCR in a Logarithmic scale view.

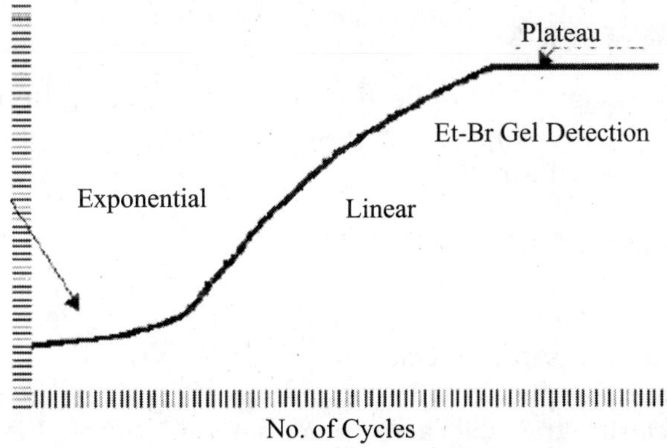

Fig. 37. PCR Phase in Linear scale view (http://www.rt-per.com/).

Touchdown PCR

Touchdown PCR is a simple method to optimize yields of amplified DNA when the melting temperature of hybrids between the oligonucleotide primers and their target sequence is not known with certainty e.g., (a) when the primers have designed from amino acid sequence, (b) when amplifying members if a multigene family, (c) when the template DNA is extremely complex and occupies a large sequence space, (d) when amplifying DNA form a species other than that for which the primers were designed, (e) when using primers containing "universal" bases such as inosine.

The earliest steps of a Touchdown PCR cycle have high annealing temperatures. For every subsequent cycle, the annealing temperature is decreased by 1 degree Celsius. The primer will anneal at the highest (and therefore, least-permissive of nonspecific binding) temperature at which it is able. Thus, the first sequence amplified is the one between the regions of greatest primer specificity; it is most likely that this is the sequence of interest. These fragments will be further amplified during subsequent rounds at lower temperatures, and will swamp out the nonspecific sequences to which the primers will bind at those lower temperatures. If the primer initially binds to the sequence of interest at a low temperature, subsequent rounds of PCR can be performed upon the product to further amplify those fragments.

Vectorette PCR: A Novel Approach to Genomic Walking

Vectorette PCR (or chemical genetics) is a method that enables the amplification of specific DNA fragments in situations where the sequence of only one primer is known. Thus, it extends the application of PCR to stretches of DNA where the sequence information is only available at one end. In this report, we describe the chemical genetics method and demonstrate its use in one specific application. In addition, it demonstrates how fragments generated by this procedure can be sequenced rapidly and simply using methods that have the potential for automation.

APPLICATIONS OF PCR

Diagnostic Applications of PCR

There are three primary diagnostic applications of PCR i.e., (a) Detecting pathogens using genome-specific primer pairs, (b) Screening specific genes for unknown mutations and (c) Genotyping using known STS markers.

Detecting pathogens using genome- specific primer pairs

Multiplex PCR can be used to simultaneously screen for pathogenic and nonpathogenic strains of E. coli using strain-specific primer pairs. By sing at least two strain-specific primer pairs for each organism and replicate samples reactions. In this example, two food samples were analyzed for the presence of a specific pathogenic strain of *E. coli*. By using appropriate control samples for the two bacterial strains, it can be seen that food sample 1 is contaminated.

The use of quantitative Real-Time RT-PCR to detect viral genomes such as HIV (AIDS) or HPV (human papilloma virus in association with cervical cancer) is another diagnostic application of PCR (Fig. 38).

Identifying genetic mutations

The PCR technique of single strand conformational polymorphism (SSCP) is one of the most widely used methods for detecting single base pair changes in genomic DNA (Fig. 39). SSCP is based on the principle that two single strand DNA molecules of identical length, but unlike sequence, will migrate differentially in a non-denaturing acrylamide gel. This is due to sequence-dependent intramolecular folding reactions which contribute to the overall structure of the molecule. In the following example, AMG primer pairs were used to detect nucleotide alterations in a specific region of the AMG gene known to co-segregate as dominant mutation. Using DNA from three individuals (two disease heterozygote and a homozygote normal), PCR amplification was performed with AMG-specific primers that flank a region where multiple dominant alleles have previously been identified (Figs. 40 and 41).

A variation on this assay system is the WAVE DNA Fragment Analysis System by Trangenomic. In this method the DNA sample is heated and cooled, but by cooling slowly under optimal hybridization conditions, both homoduplex and heteroduplex PCR products will form. The sample is then run through a high permformance liquid chromatography (HPLC) system, and OD_{260} peaks are resolved as they are eluted from the column under gradient conditions. This Denaturing HPLC (DHPLC) system is used in the detection of

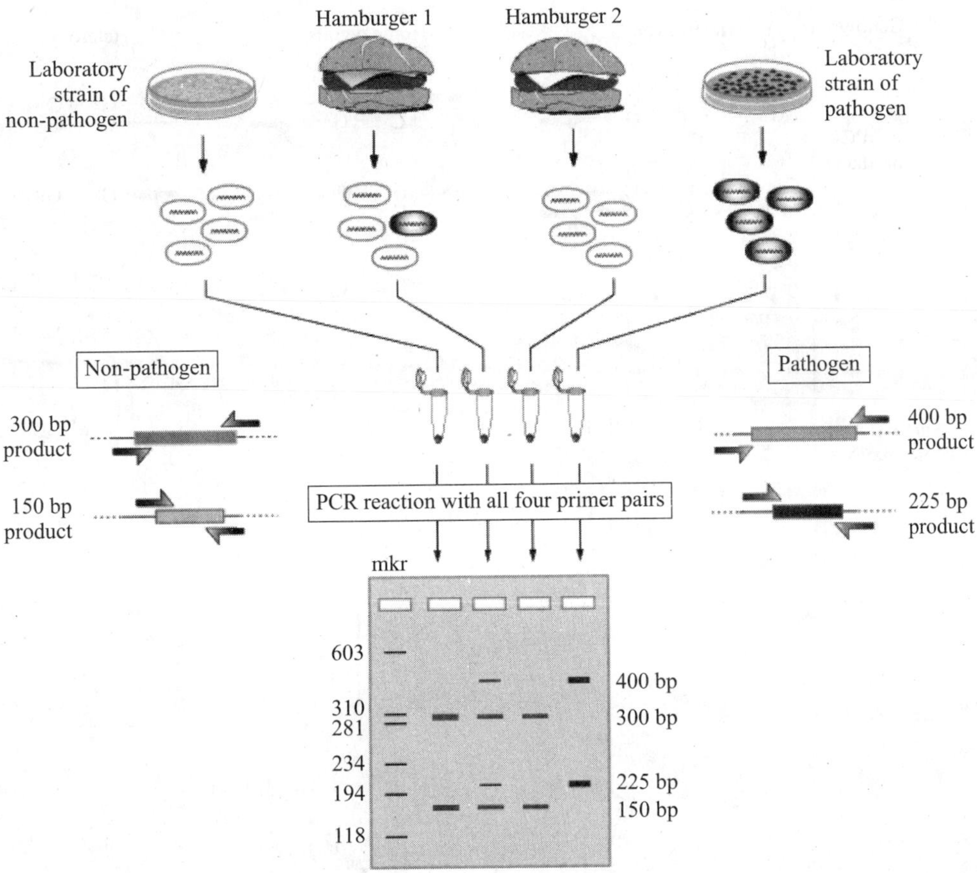

Fig. 38. Detecting pathogens using genome- specific primer pairs (http://www.biochem.arizona.edu/classes/bioc471/pages/Lecture14/Lecture14.html).

mutation screening, polymorphic marker mapping, linkage analysis, cancer epidemiology, population studies, forensic and paternity analysis (STR), loss of heterozygosity (LOH) etc.

PCR genotyping using sequence tagged sites

STSs (Sequence Tagged Sites) are defined as genomic DNA segments for which sequence information and physical mapping data are available. These markers can be defined by PCR primer pairs which can be used to generate distinct PCR products. Many of the STS markers that have been used for pedigree and forensic genotyping are VNTRs (Variable Nucleotide Tandem Repeats), STRPs (Short Tandem Repeat Polymorphisms) and SNPs (Single Nucleotide Polymorphisms) (Figs. 42 and 43). PCR primers that specifically amplify across polymorphic repeat sequences can be used to determine genotype based on the type of PCR products generated.

Laboratory Applications of PCR

PCR is a valuable laboratory tool that is used routinely for tasks that can be done other ways, but which are easier and quicker when using PCR. Two general applications have been chosen

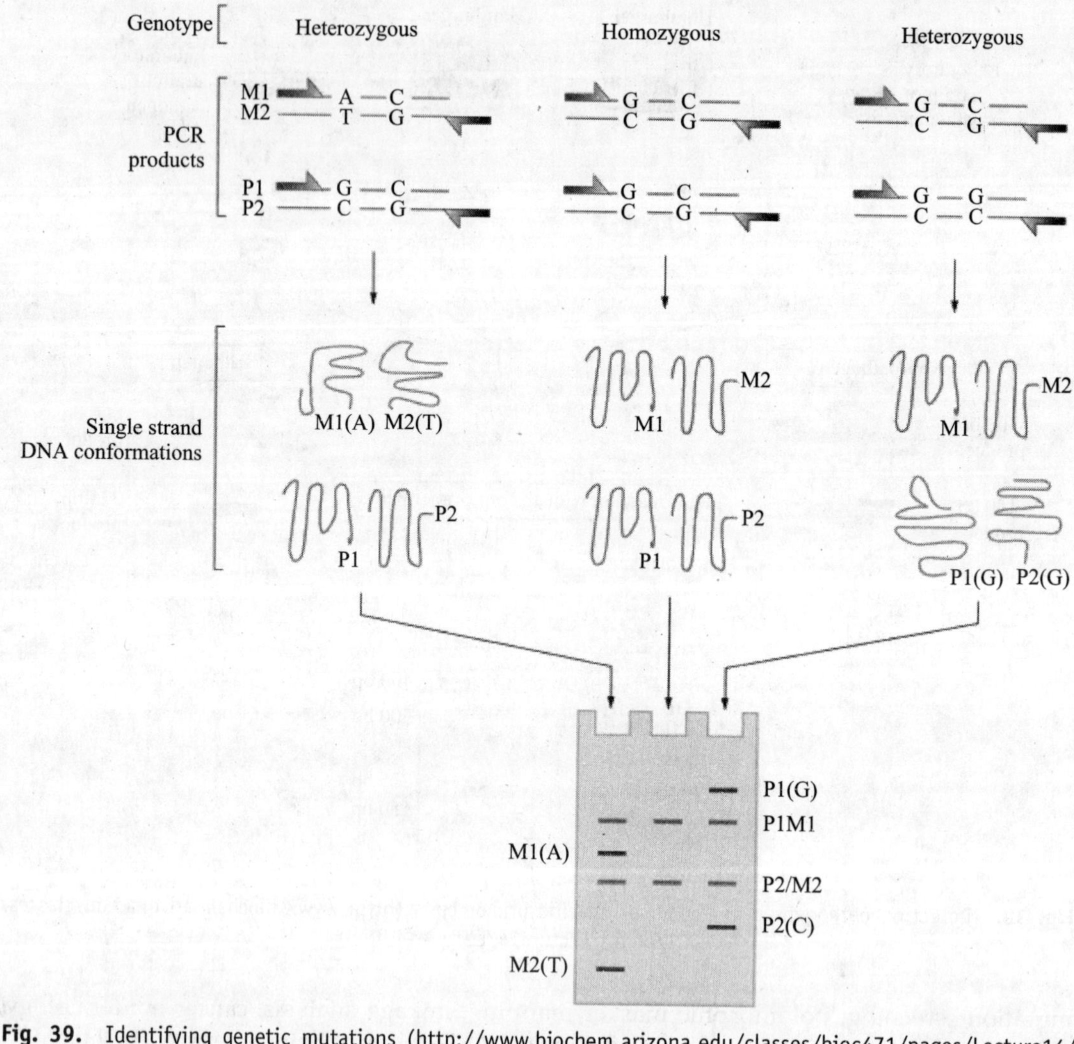

Fig. 39. Identifying genetic mutations (http://www.biochem.arizona.edu/classes/bioc471/pages/Lecture14/Lecture14.html).

as representative examples; (1) subcloning DNA targets using PCR, and (2) PCR-mediated in vitro mutagenesis.

Subcloning DNA targets using PCR

There are numerous situations in which a specific DNA segment needs to be cloned into a plasmid vector, but there are no convenient restriction enzymes sites to facilitate the design of a subcloning strategy. Three strategies are commonly used to subclone PCR products i.e., T/A cloning, restriction site addition, blunt-end ligation. The most common method is restriction site addition which utilizes PCR primer pairs that incorporate restriction enzyme recognition sites into the 5′ end of the oligonucleotide (Fig. 44). Another method is to take advantage of the propensity of Taq polymerase to spuriously add a dATP residue to the 3′ terminus of extended chains. A specially-prepared dT cloning vector is often used that includes lacZ coding

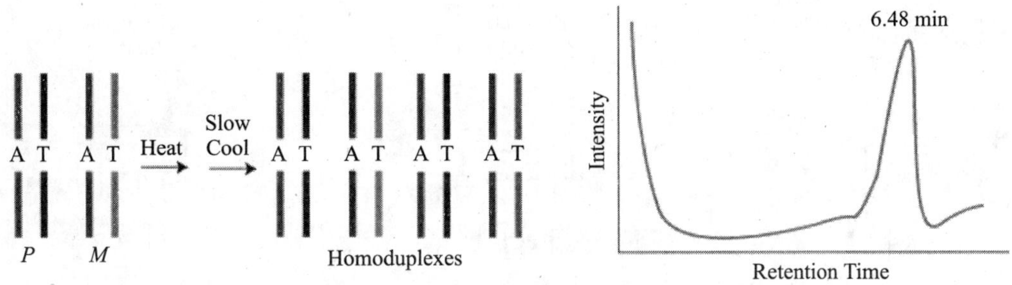

Fig. 40. Detection of Homozygous genotype through PCR (http://www.biochem.arizona.edu/classes/bioc471/pages/Lecture14/Lecture14.html).

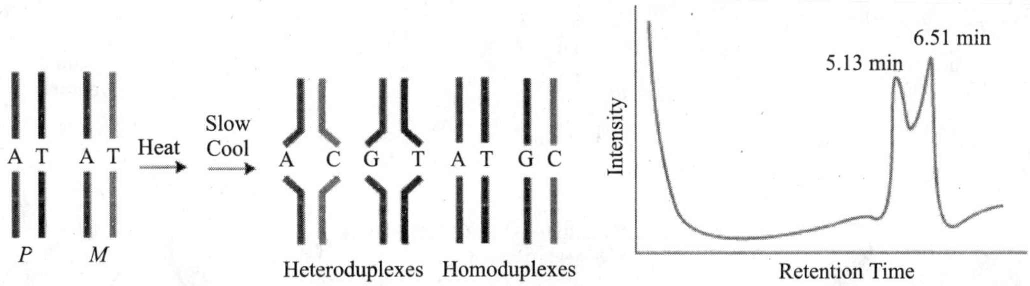

Fig. 41. Detection of Heterozygous genotype through PCR (http://www.biochem.arizona.edu/classes/bioc471/pages/Lecture14/Lecture14.html).

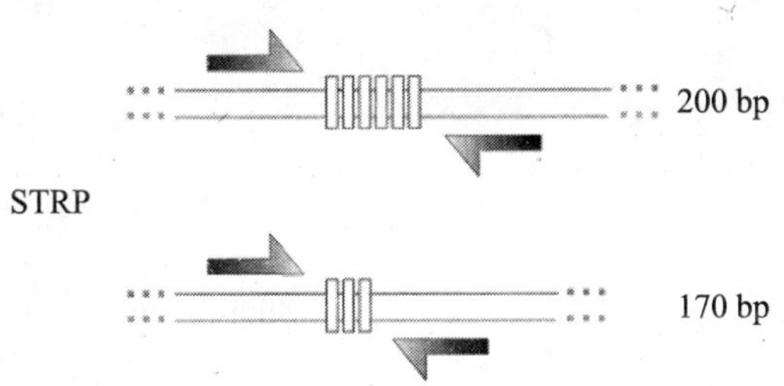

Fig. 42. Detection of Short Tandem Repeat Polymorphisms through PCR (http://www.biochem.arizona.edu/classes/bioc471/pages/Lecture14/Lecture 14.html).

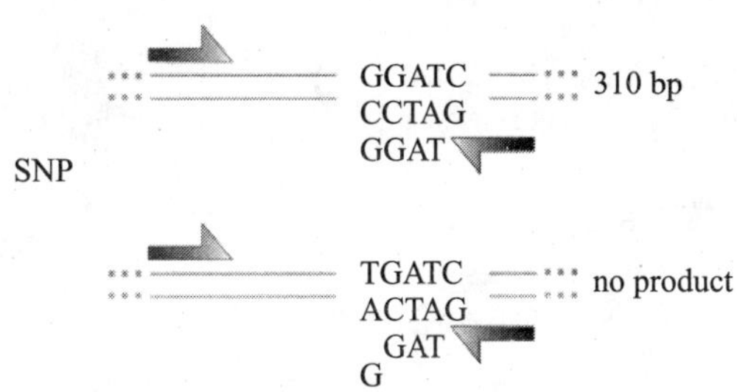

Fig. 43. Detection of Single Nucleotide Polymorphisms through PCR (http://www.biochem.arizona.edu/classes/bioc471/pages/Lecture14/Lecture14.html).

sequences for colony screening. Blunt-end cloning of PCR products can be done using a lacZ plasmid vector that has been digested with an appropriate enzyme such as SmaI.

PCR-mediated in vitro mutagenesis

Mutagenic PCR is a random mutagenesis technique that exploits the elevated error rate of Taq polymerase in the presence of $MnCl_2$ and high $MgCl_2$. Mutagenic PCR is similar to random chemical mutagenesis with the added advantage that the chosen primer pair provides a

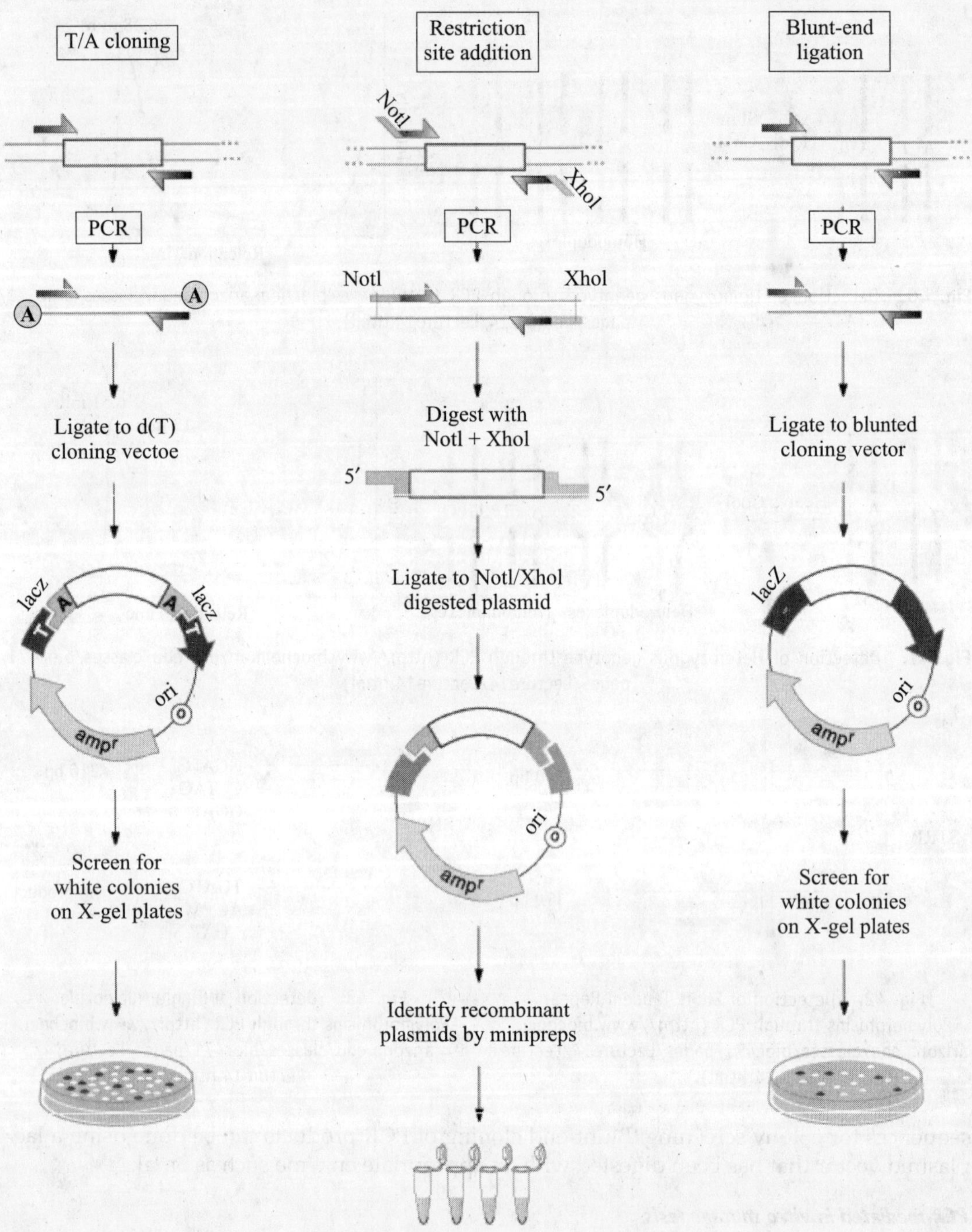

Fig. 44. Subcloning DNA targets using PCR (http://www.biochem.arizona.edu/classes/bioc471/pages/Lecture14/Lecture14.html).

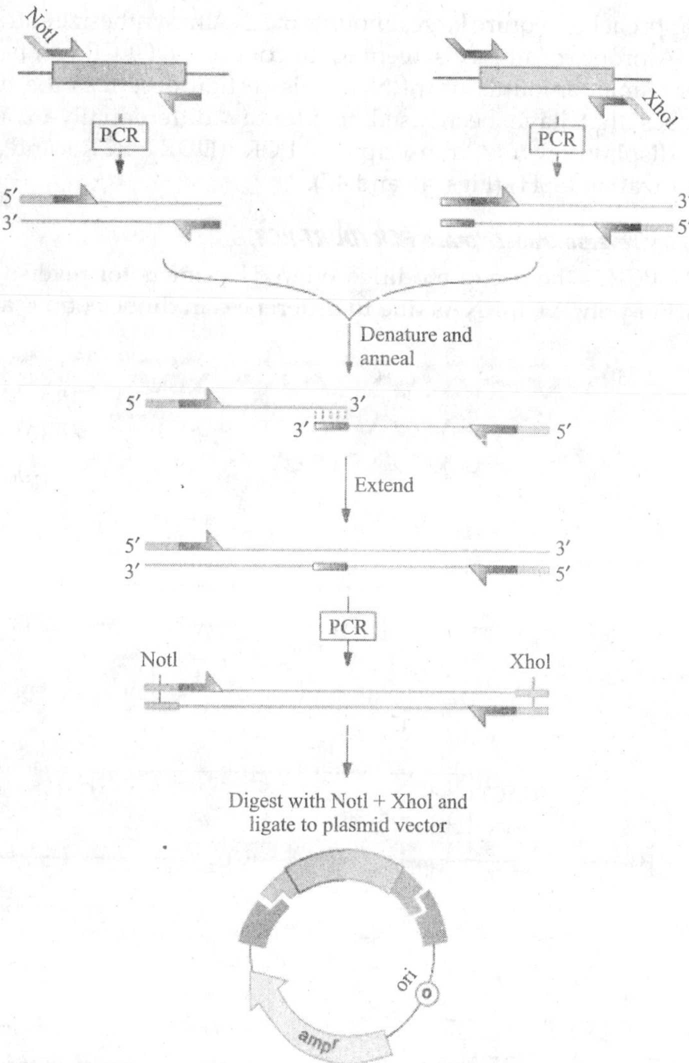

Fig. 45. Overview of PCR-mediated in vitro mutagenesis (http://www.biochem.arizona.edu/classes/bioc471/pages/Lecture14/Lecture14.html).

convenient way to target the random bp mutations to a defined segment of DNA. Strategies have also been developed that permit the construction of unique gene fusions using consecutive PCR reactions that utilized the PCR products of one reaction to prime DNA synthesis during the second amplification (Fig. 45). This technique is sometimes called PCR-mediated gene SOEing (splicing by overlap extension).

Amplification of differentially-expressed gene sequences

Differential screening and subtraction hybridization are two techniques that can be used to isolate gene transcripts which are present at different levels in two RNA populations.

However, these approaches require large amounts of RNA to synthesize sufficient quantities of an enriched cDNA probe for library screening. In contrast, RT-PCR can be used to generate cDNAs from very small amounts of mRNA. This section describes the two basic RT-PCR cloning approaches that have been used to identify differentially-expressed genes i.e., (a) Differential display reverse transcriptase PCR (DDRT-PCR) and (b) Suppression subtraction hybridization (SSH) (Figs. 46 and 47).

(a) Differential display reverse transcriptase PCR (DDRT-PCR)

The key to DDRT-PCR is the use of modified oligo dT primers for reverse transcription that anneal to a subset of poly A+ mRNAs due to differences in dinucleotides at the 3′ end of the

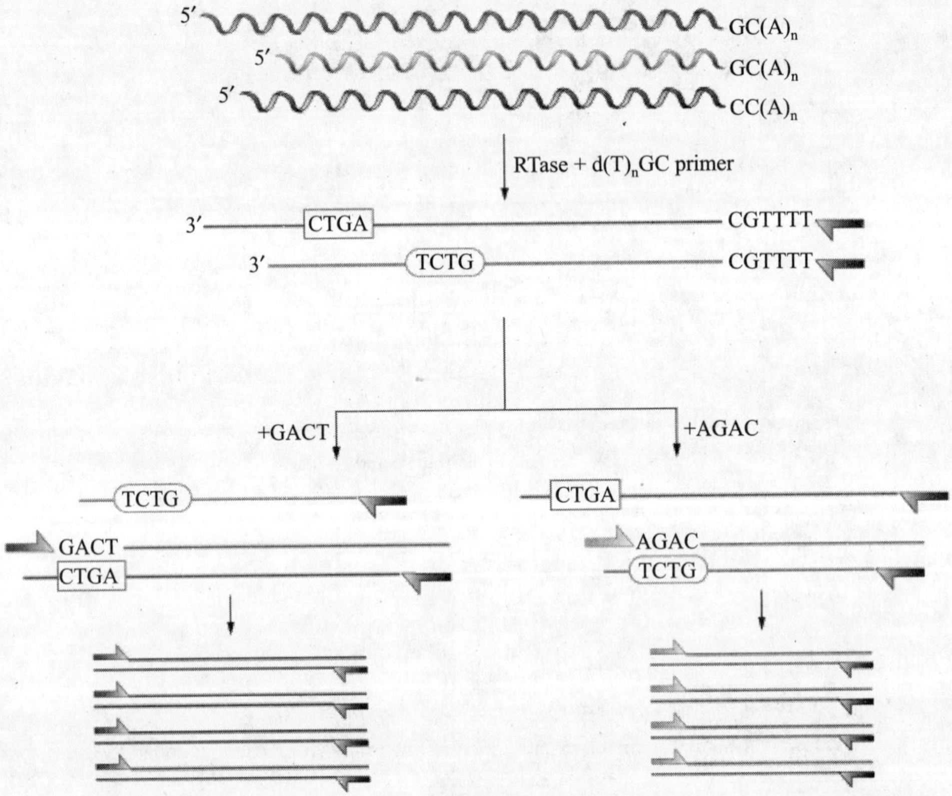

Fig. 46. Amplification of differentially-expressed gene sequences (http://www.biochem.arizona.edu/classes/bioc471/pages/Lecture14/Lecture14.html).

primer. DDRT-PCR generates subsets of cDNA products based on differences in 3′ dinucleotides and the location of 5′ arbitrary priming sites for PCR. Schematic diagram showing how a 5′-dT-GC-3′ primer is used to synthesize cDNA from a subset of RNAs containing the appropriate 3′ dinucleotide adjacent to the poly A tail. Of these first strand cDNAs, only a fraction will serve as appropriate templates for separate PCR reactions containing a known arbitrary primer.

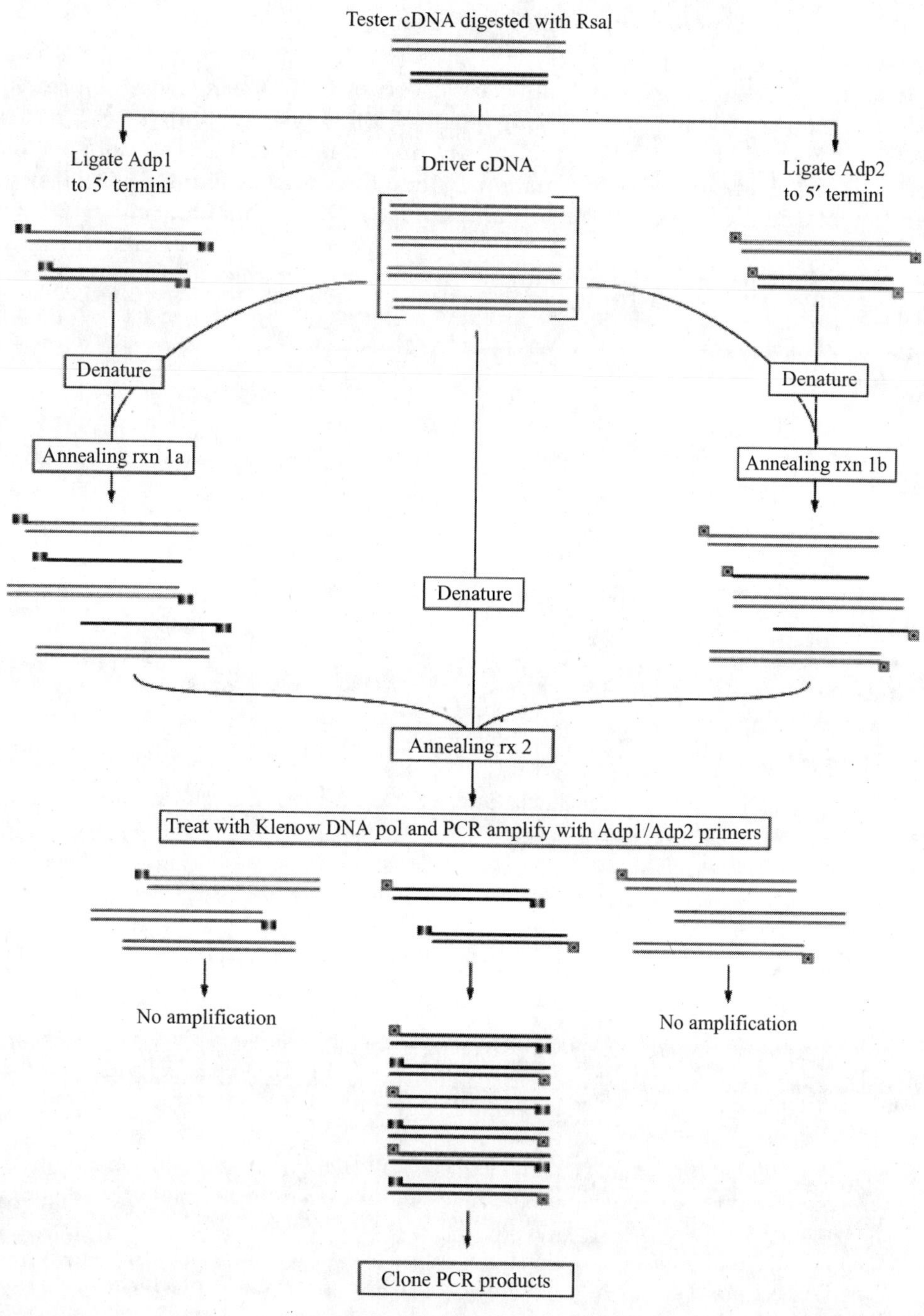

Fig. 47. Suppression subtraction hybridization SSH (http://www.biochem.arizona.edu/classes/bioc471/pages/Lecture14/Lecture14.html).

(b) Suppression subtraction hybridization (SSH)

The basis for differential amplification in SSH is two-fold. First, conditions are used that promote rapid reassociation kinetics with excess "driver" cDNA sequences to normalize two "tester" cDNA pools. Following this driver-tester hybridization reaction, the two tester pools are mixed and the unhybridized single strand low abundance molecules are allowed to anneal under conditions that favor duplex formation. In the subsequent PCR amplification step, these unique "tester" duplexes are exponentially amplified by a pair of adaptor primers. Second, the PCR amplification of non-differential cDNAs is suppressed because these undesirable sequences contain inappropriate priming sites, or are capable of forming intrastrand "panhandle" structures that are poor substrates for PCR.

REFERENCES

http://archive.salon.com/health/feature/2000/03/29/mullies/

Sambrook, J. and Russel, D.W. (2001) Molecular Cloning: A Laboratory Manual, 3rd ed., Cold Spring Harbor, New York, USA, Chapter 8: *In vitro* amplification of DNA by the polymerase chain reaction.

http://en.wikipedia.org/wiki/Taq polymerase

Ulrich, 6. *et al.* (1999) AFLP genotyping and fingerprinting. Review Tree Vol. 14

http://www.bio.davidson.edu/courses/genomics/method/NestedPCR.html

http://nl.wikipedia.org/wiki/ELISA

http://www.rt-per.com/

http://www.biochem.arizona.edu/classes/bioc471/pages/Lecture14/Lecture14.html

Suggested Readings

1. The history of PCR: Smithsonian Institution Archives, (2006) Institutional History Division. Retrieved 24.
2. Cheng S, Fockler C, Barnes WM, Higuchi R (1994) Effective amplification of long targets from cloned inserts and human genomic DNA, Proc Natl Acad Sci. 91: 5695-5699.
3. Sambrook J. and Russel D.W. (2001) Molecular Cloning: A Laboratory Manual, 3rd ed., Cold Spring Harbor, *N.Y*, Chapter 8: In vitro Amplification of DNA by the Polymerase Chain Reaction.
4. Pavlov AR, Pavlova NV, Kozyavkin SA, Slesarev AI (2004) Recent developments in the optimization of thermostable DNA polymerases for efficient applications, Trends Biotechnol.
5. Sharkey D.J., Scalice E.R., Christy Jr., K.G., Atwood S.M. and Daiss J.L (1994) Antibodies as Thermolabile Switches: High Temperature Triggering for the Polymerase Chain Reaction, Bio/ Technology 12: 506-509.
6. Chou Q., Russell M., Birch D.E., Raymond J. and Bloch W. (1992) Prevention of pre-PCR mis-priming and primer dimerization improves low-copy-number amplifications, Nucleic Acids Research 20: 1717-1723.
7. Pavlov AR, Pavlova NV, Kozyavkin SA, Slesarev AI (2006) Thermostable DNA Polymerases for a Wide Spectrum of Applications: Comparison of a Robust Hybrid TopoTaq to other enzymes, in Kieleczawa J: DNA Sequencing II: Optimizing Preparation and Cleanup. Jones and Bartlett, 241-257.
8. Paul R. (1996) Making PCR: A Story of Biotechnology, University of Chicago Press.
9. Mullis K.B. (1990) The unusual origin of the polymerase chain reaction. Sci Am 262(4):56-61, 64-55.

REVIEW QUESTIONS

Long Questions

1. For perfectly matched PCR primers, it generally isn't considered of any benefit to make them longer than about 28 bases. Why?
2. Does the amount of PCR product at the end of the amplification reflect the amount of original template in the sample? What conditions must be met for the amount of PCR product to reflect the amount of the original template?
3. Overly stable PCR primers often raise a background of nonspecific products, but overly stable fluorogenic probes don't cause any particular problem. Why?
4. A real time reverse transcriptase/PCR assay for a message is designed with one gene-specific PCR primer paired with oligo dT as the other PCR primer, and detection by SYBR Green. What is wrong with this design?
5. One PCR primer in a particular set is diagnosed as having a hairpin of marginal significance. What symptoms might you expect if this hairpin causes a problem? How might conditions be adjusted to overcome the problem?
6. What are the various properties require designing primer?
7. What are various applications of Polymerase chain Reaction?

Multiple Choice Questions

1. The polymerase chain reaction or PCR is a technique that
   (a) was used to demonstrate DNA as the genetic material
   (b) is used to determine the content of minerals in a soil sample
   (c) uses short DNA primers and a thermostable DNA polymerase to replicate specific DNA sequences in vitro.
   (d) measures the ribosome transfer rate during translation
   (e) detects the level of polymerases involved in replication
2. The presence of a ______ with a free 3'-OH group is essential for DNA polymerase to synthesize DNA since no known DNA polymerase is able to initiate chains.
   (a) origin of replication (b) restriction endonuclease
   (c) palindrome (d) primer
   (e) promoter

Answers

1. (c) 2. (d)

CHAPTER 32

Applications of Real-time PCR in Gene Quantification and Disease Diagnostics

S.A. Deepak[1*], R. Rakwal[2], G.K. Agrawal[3], Junko Shibato[2] and G. Oros[4]

[1]Dept. of Applied Botany and Biotechnology, University of Mysore, Manasagangotri, Mysore–570 006

E-mail: deepak@affrc.go.jp

[2]National Institute of Advanced Industrial Science and Technology, Tsukuba, Japan

[3]Research Laboratory for Agricultural Biotechnology and Biochemistry, Kathmandu, Nepal

[4]Plant Protection Institute, Hungarian Academy of Sciences, Budapest, Hungary

The RNA quantification can be performed by northern blotting, ribonuclease protection assay (RPA), *and in situ* hybridization or by PCR. Among these methods PCR is the most sensitive method which can discriminate closely related mRNAs and the amplification process involved makes it possible to detect low concentrations of the target. In recent years, a revolution has been made for detecting even a single copy of DNA or RNA by PCR. In conventional PCR, the theoretical calculations on starting material and the end product of PCR is often not accurate among the replicate reactions which is actually due to consumption of reagents or inaccuracy in starting materials (Fig. 1). This makes the conventional PCR not very accurate for quantification. Advancements, such as development of competitive PCR, sequence analysis gene expression (SAGE) and high-throughput microarray have been made in order to address this issue. However, these techniques are time-consuming, cumbersome and require several manipulations of the samples.

The disadvantages of traditional PCR with end point quantification includes poor precision, low sensitivity, short dynamic range, low resolution, non-automatic, size-based discrimination, results not expressed in specific units, ethidium bromide (EtBr) staining does not ensure accurate quantity and post PCR processing. During 1992, Higuchi and co-workers developed a simultaneous amplification and detection of specific DNA sequences in "real

*Corresponding author

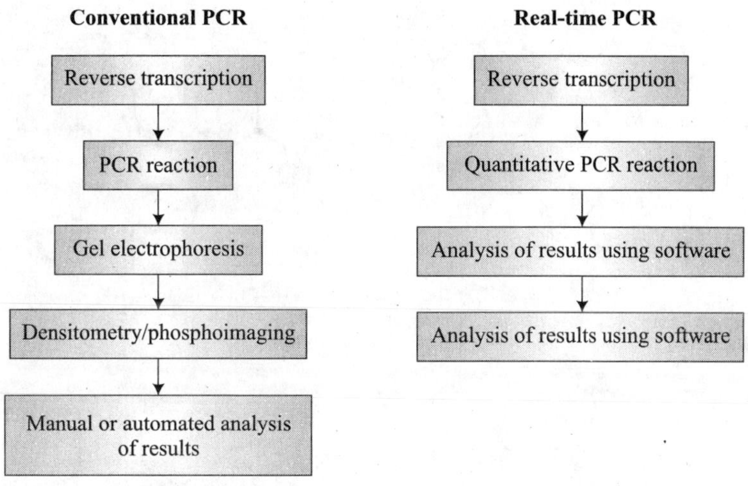

Fig. 1. Conventional PCR vs. real-time PCR.

time" using EtBr as intercalator, which binds to the increasing amounts of the amplified double stranded DNA and resulting in increased fluorescence. Since then, several fluorescent dyes have been developed (Fig. 2).

Idaho technologies light cycler developed a typical machine during 1996 and was licensed to Roche in 1998. However, the first high-throughput real-time cycler, ABI Prism 7700 was launched by Perkin-Elmer during 1997. To date, Applied Biosystems is the largest player in the market. A low budget-end cycler iCycler iQ with 96 well 4 color cycler was launched by Bio-Rad and another by Idaho Technologies, namely, Ruggedized Advanced Pathogen Identification Device (RAPID) that was commercialized in 1999. Since, 2000 real-time cyclers that provide high-throughput solutions and are less expensive have been introduced (Table 1).

The tendency of vigorous development is slowed down due to consumption of reaction components in the plateau phase (also varies among replicates) and start to degrade in the linear phase while, in plateau phase, the reaction has almost stopped accompanied by degradation of the products. The latter phase is chosen in the end point PCR for comparison. In fact, the acquisition of data will progress from the exponential phase itself (Fig. 3).

Real-time PCR is mostly used for gene expression as primary validation or to confirm the microarray data followed by pathogen detection, allelic discrimination/SNP analysis, transgene detection/quantification, biological diversity/contamination, zygosity testing etc. Real-time PCR assays are 10000- to 100000-fold more sensitive than Rnase protection assays, 1000-fold times more sensitive than dot blot hybridization and can even detect a single copy of a specific transcript. The quantification of gene expression by real-time PCR requires template in the form of genomic DNA, plasmid DNA or cDNA generated from RNA by first running a separate reverse transcriptase reaction. 10-100 ng of template RNA is required for the reaction (Fig. 4). The cDNA levels in samples are then analyzed by real-time PCR. Apart from conventional methods for RNA extraction (phenol-chloroform method), commercial kits can be effectively used for obtaining high quality RNA. Most of the commercial kits for instance, Qiagen kit provides high quality total RNA and aid quick extraction.

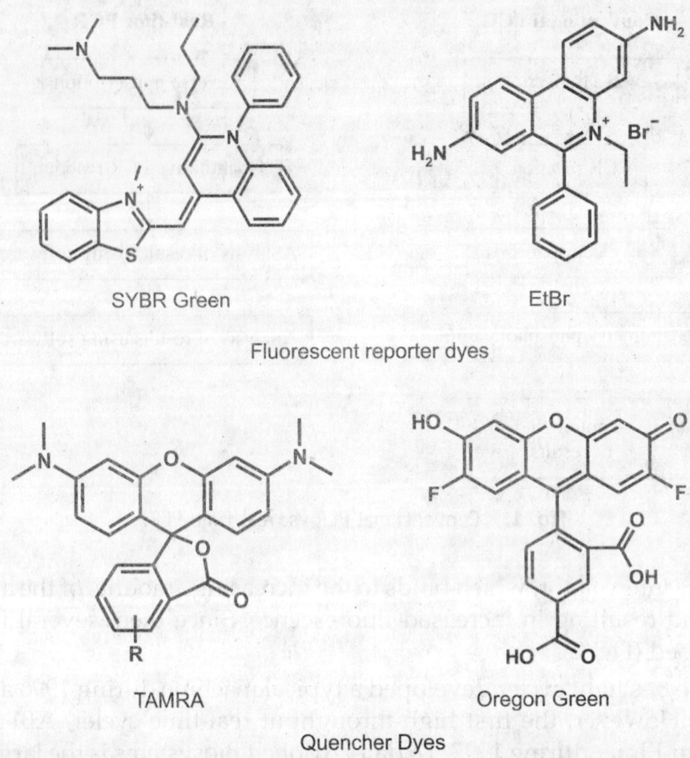

Fig. 2. Fluorescence dyes used in various PCR techniques.

RNAse protection should be ensured in all stages to avoid denaturing of RNA. Although the kits effectively separate DNA from RNA extracted, the DNase treatment is required which digests both single and double stranded DNA. Long incubations at elevated temperatures can cause Mg^{++}-dependent hydrolysis of the RNA. Later, RNA is transcribed to cDNA by reverse transcriptase.

Instrumentation

The equipments such as ABI 7700 consists of a 96 well thermal cycler connected to a laser and charge-coupled device (CCD) optics system. An optical fiber inserted through a lens is positioned over each well and laser light is directed through the fiber to excite the fluorochrome in the PCR solution. The emissions are sent through the fiber to the CCD camera, where they are analyzed by the software's algorithms. This data is subsequently sent to the computer. Emissions are measured at every seven seconds (Grove, 1999).

Detection chemistries

The two types of principle probes are the double strand DNA binding dyes (SYBR Green etc.) and the hybridization probes such as 5′ nuclease (Taqman), molecular beacon, hybridization probes etc. The sequence detection systems (SDS) instruments from Applied Biosystems (ABI)

Table 1. Real-time cyclers available in the market and their characteristics

Cycler	Source	Detector	Applications	No. of samples	Footprint
ABI Prism 7000	Tungsten-halogen	CCD camera	SYBR, FAM, HEX, TET, TAMRA, VIC	96	39 × 51 cm
Bio-Rad iCycler iQ	Tungsten-halogen	CCD camera	SYBR, FAM, HEX, TE, TA, VIC	96	33 × 62 cm
Cepheid Smartcycler	LED	Silicon detectors	SYBR, FAM, TET, ROX, Cy3, Cy5	16	30 × 25 cm
Corbett Research Rotor-Gene 3000	LED	PMT	SYBR, FAM, HEX, TET, TAMRA, VIC	72	38 × 48
MJ Research DNA Engine Opticon 2	LED	2 PMTs	SYBR, FAM, HEX, TET, TAMRA, VIC	96	34 × 47 cm
Roche Light Cycler 2	LED	Fluorimeters	SYBR, FAM, HEX, VIC, LightCycler Red Stains	32	30 × 45 cm
Stratagene Mx3000P	Tungsten-halogen	1 PMT scanner	SYBR, FAM, HEX, TET, TAMTA, VIC	96	33 × 46 cm
Techne Quantica	Halogen	PMT	SYBR, FAM, HEX, TET, TAMRA, VIC	96	45 × 50 cm

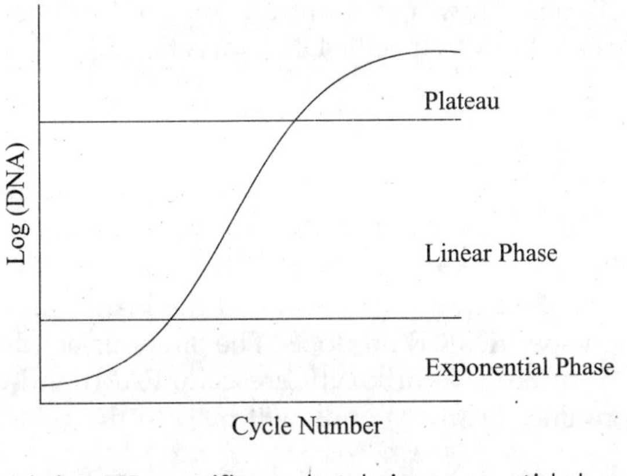

Fig. 3. Phases of PCR. The real-time PCR quantifies product during exponential phase wherein the doubling of the product accumulating is very specific and precise.

can perform two types of chemistries namely TaqMan® chemistry (also known as flurogenic 5′ nuclease chemistry) and SYBR® Green I dye chemistry (Table 2). The characteristics of an individual experiment determine the type of detection chemistry, DNA binding dyes to be used for validation of results from microarray data etc.

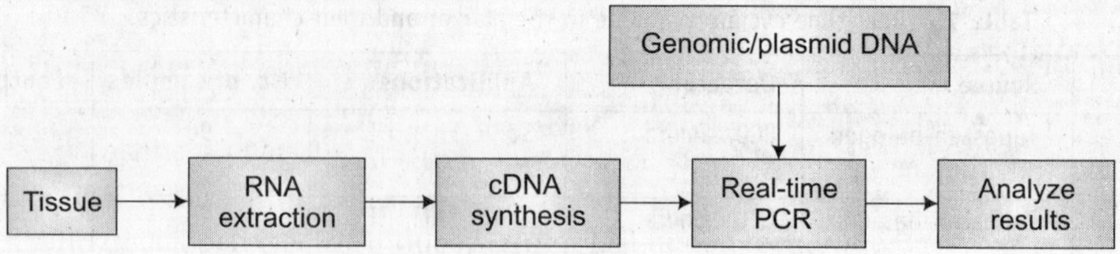

Fig. 4. Subsequent steps of RT-PCR.

Table 2. Applications of TaqMan and SYBR Green I dye chemistries

TaqMan Chemistry	SYBR Green Chemistry
1. One-step RT-PCR for RNA quantification 2. Two-step RT-PCR for RNA quantification 3. DNA/cDNA quantitation 4. Allelic discrimination 5. Plus/Minus assay	1. One-step RT-PCR for RNA quantification 2. Two-step RT-PCR for RNA quantification 3. DNA/cDNA quantitation

SYBR Green I dye chemistry involves small molecules that can bind to double-stranded DNA, namely, intercalators and minor-groove binders (Fig. 5).

These bind to the double-stranded DNA, but not single-stranded DNA in the sample as well as the PCR produced amplicons resulting in increase in fluorescence which is read during each PCR cycle. This is a direct method of measuring the amplicons. The fluorescence is much brighter than EtBr. The cost of experiments with this dye is the most inexpensive compared to all other detection chemistries. However, multiplexing and allelic discrimination is not possible since the dye does not bind to specific DNA targets.

Hybridization Probes

TaqMan chemistry is an indirect method of measurement which uses a flurogenic probe (TaqMan® probe with TAMRA™ as the quencher dye and TaqMan® MGB probe) to detect specific PCR products (Fig. 6) accumulated during PCR cycles (Livak *et al.*, 1995).

The intercalator dyes are used in this protocol and the major disadvantage is that it will detect both specific and non-specific PCR products. The oligonucleotide probe which contains both the reporter dye at 5′ (usually 6-carboxyfluorescein, FAM) and quencher dye at 3′ (6-carboxy-tetramethyl-rhodamine, TAMRA) ends will bind to the target sequence during the PCR .

The fluorescence of the reporter dye from intact probe is masked or not detected due to the presence of quencher dye in close proximity because the quencher dye reduces reporter dye fluorescence by fluorescence resonance energy transfer (FRET). However, as the primer amplifies the target DNA, and reaches the probe location, the Taq DNA polymerase with 5′ nuclease activity will remove the reporter dye from the quencher, and the fluorescence from the reporter dye is detected by the instrument. As the number of PCR cycles increases, the amount of reporter dye increase and makes it possible to detect the increase in amounts. The

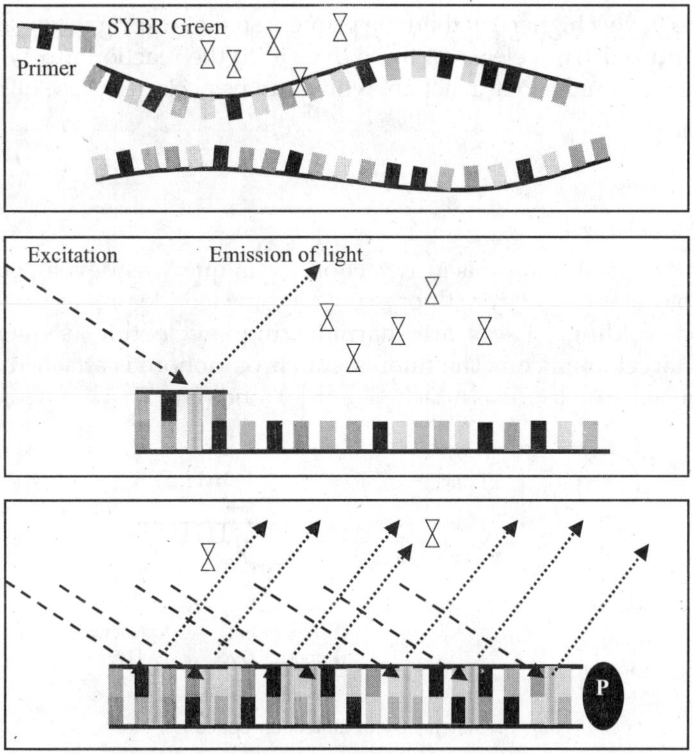

Fig. 5. SYBR Green chemistry.

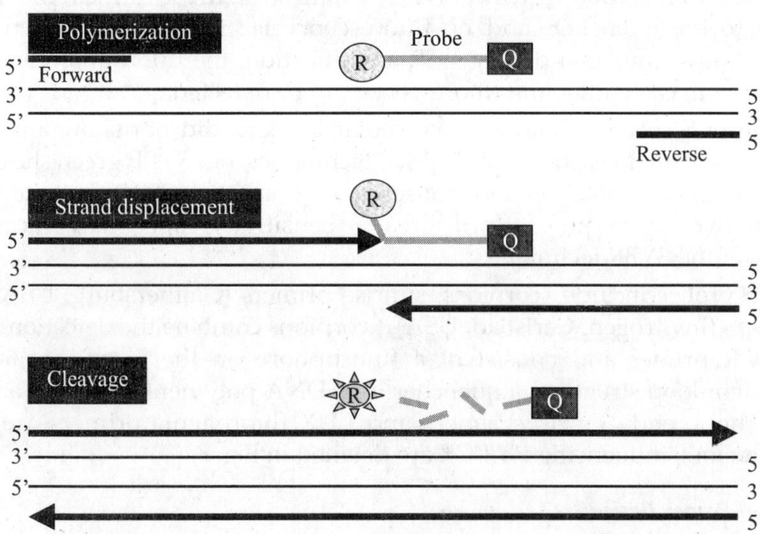

Fig. 6. Process of TaqMan sequence detection chemistry.

probe is designed to have a higher *Tm* than the primers so that during the extension phase the probe remains hybridized. The release of these dyes in to the reaction mix will not have any kind of inhibitory action on the PCR reaction, which can be noticed in case of EtBr staining.

HAIRPIN PROBES

Molecular Beacons

A novel fluorescence based nucleic acid detection technique was developed by Tyagi and Kramer in 1996. Molecular beacons are fluorescent oligonucleotide probes that are designed to include stem-loop folding. They are harmonizing nucleotide sequences that are complementary to target amplicon. The fluorescent chromophore is attached at the 5′ end of the probe and a quencher molecule is attached at the 3′ end (Fig. 7).

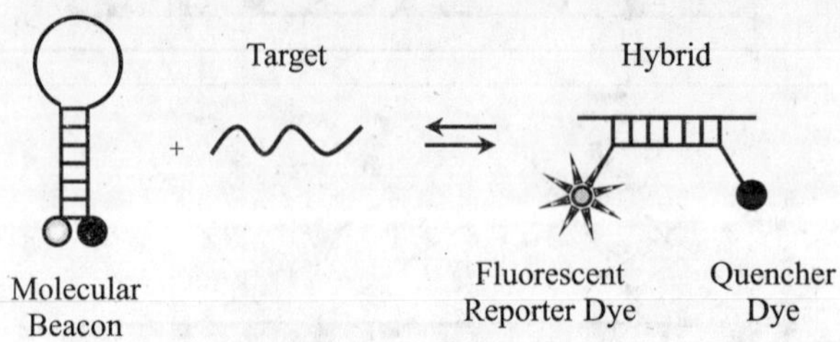

Fig. 7. Chemistry of molecular beacons.

A stem structure is formed by annealing of the complementary arm sequences that are added on both sides of the probe sequence. When a stem structure is formed, the fluorophore transfers energy to the quencher, and no fluorescence is emitted. But, when the probe hybridizes to the target amplicon during PCR amplification, the fluorophore and quencher become separated from each other and fluorescence can be detected.

A comparison of SYBR, TaqMan and molecular beacons did not show any significant variation between molecular beacons and TaqMan techniques, but SYBR green showed 25-fold higher copy number of beta actin genes in mouse tissues (Tian *et al.*, 2004). This study suggests that the former two techniques afford greater sensitivity, specificity, accuracy and reproducibility over the SYBR technique.

Other hairpin probes include scorpions, sunrise primers (Gaithersburg, USA) and LUX fluorogenic primers (Invitrogen, Carlsbad, USA). Scorpions combine the detection probe with the upstream PCR primer and consist of a fluorophore on the 5′ end, followed by a complementary stem-loop structure, a quencher dye, DNA polymerase blocker and finally a PCR primer on the 3′ end. Sunrise Primers and LUX fluorogenic primers are similar to scorpions. Some factors, influencing RT PCR are detailed in Fig. 8.

Target Sequence and Primer Design

The specificity of PCR (classical or real-time) ultimately depends upon designing proper PCR primers that are unique to the target organism gene. The specific primers for real-time PCR are designed for target DNA, cDNA, total RNA or plasmid and are designed using the available

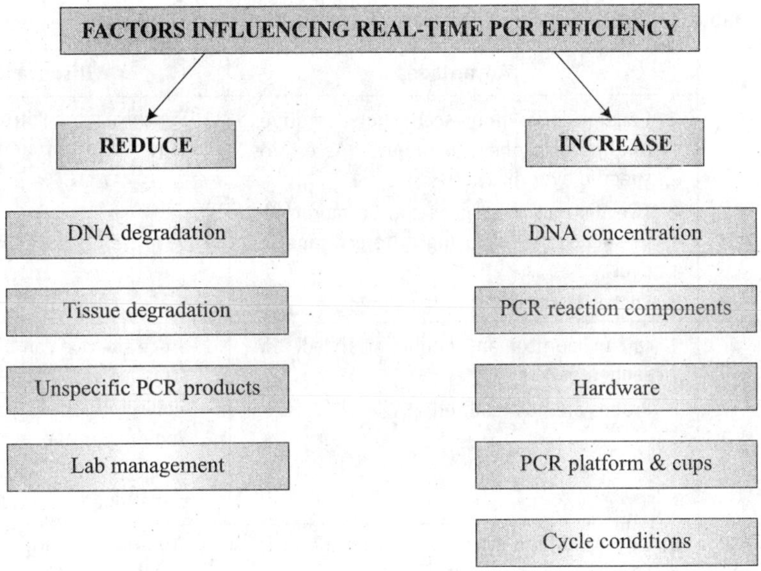

Fig. 8. Factors influencing RT-PCR.

software. Most popular softwares used are Primer Express (ABI), Primer3 (MIT-free on the web) or Beacon Designer (Primer Biosoft), Oligo (MBI) etc. Most consistent results are obtained when shorter amplicons of 50-150 bp range are used within the target sequence. The GC content range recommended is 20-80 % and runs of over four guanine nucleotides are avoided and the last five nucleotides should not contain more that two guanine/cytosine. The *Tm* optimum range is 58-60 °C. In case the primers are designed using genomic DNA, the amplicon should span one or more introns to avoid amplification of the target gene. The concentration of primer for PCR should be optimized so as to get maximum ΔRn and lowest threshold cycle (C_T) when minimum primer concentrations are used. In our experiments, the primer volume of 2 ml at strength of 1.4 to 1.6 μM in a final reaction of 25 μl provided best results in several cases. Primers are either designed as original or adopted from literature or commercially (in case of commercial protocols). The primers should have high efficiency, specific and should not form primer-dimers. Primer-dimers can be detected from the dissociation curves wherein they characteristically show lower *Tm* than the specific product (Fig. 9). They can be more prevalent in No Template Control (NTC) and in sample wells containing low concentrations of template.

The target nucleotide sequence of interest such as DNA, cDNA, total RNA or plasmid of short segments is identified and primers are designed within that segment. In some cases, SYBR lacks specificity and can detect other products as well. However, molecular beacons and TaqMan probes are highly specific. On the other hand, with respect to their designing, specific primers for SYBR technique and TaqMan probes are easily designed using the Primer Express software (Applied Biosystems, Foster City, CA), while design of molecular beacon probes are more complicated and require more caution, experience and time. A typical example of primers designed for cucumber beta actin gene using the nucleotide sequence is retrieved from the data base (Genbank) is given below (Fig. 10).

Table 3. Advantages and disadvantages of TaqMan and SYBR Green Chemistries

Chemistry	Advantages	Disadvantages
TaqMan	1. Generation of fluorescent signal requires binding of probe and target – Ascertains specific hybridization 2. Two distinct sequences can be monitored in single tube by using different reporter dyes 3. Post-PCR process is not required	1. Synthesis of different probes is required for different sequences
SYBR Green	1. It can monitor any double stranded DNA sequence 2. Probes are not required	1. Since the dye can bind to any double stranded DNA, amplification of non-specific target cannot be avoided 2. Requires optimization of assay 3. Longer amplifications makes stronger signal due to CDC camera saturation
Molecular Beacons	1. Extremely high signal to background ratio 2. Detection of amplicons as they are synthesized during real-time PCR 3. Discrimination between DNA sequences that differ form one another by as little as single nucleotide substitution (selective genotypic assay) 4. Rapid and cost effective high throughput screening on the level of individual genotyping (multiplex detection of single nucleotide variations)	1. Studies detailing interlaboratory variability are lacking 2. The identification of problems by eye may not always be possible due to large amount of data generated 3. Using multiple fluoropheres may lead to temperature dependent spectral overlapping

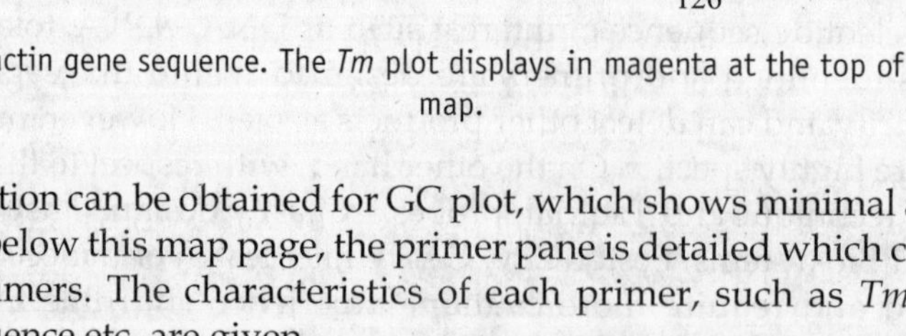

Fig. 9. *Tm* plot for actin gene sequence. The *Tm* plot displays in magenta at the top of the Map page Feature map.

Similar description can be obtained for GC plot, which shows minimal and maximal GC of the sequence. Just below this map page, the primer pane is detailed which comprises of several sets of suitable primers. The characteristics of each primer, such as *Tm*, amplicon length, location in the sequence etc. are given.

The software provides the best primers suited for this gene amplification, which is always on top, and several others are also shown in the result output page.

```
GAAAAGATGACGCAGATAATGTTTGAGACATTCAATGTGCCTGCTATGTA    50
TGTTGCCATCCAGGCCGTTCTGTCCCTCTACGCTAGTGGACGTACCACTG    100
GTATCGTGCTGGATTCTGGTGATGGTGTGAGTCACACTGTTCCCATCTAT    150
GAGGGTTACGCCCTCCCTCATGCCATTCTCCGTTTGGACCTTGCTGGTCG    200
TGACCTTACTGATGCTCTCATGAAAATTCTCACTGAAAGAGGTTACATGT    250
TCACCACCACTGCCGAACGGGAAATTGTCCGTGACATGAAAGAGAAGCTG    300
GCATATGTTGCTCTTGACTACGAACAAGAACTCGAGACTGCAAAGAGCAG    350
TTCCTCTATTGAGAAGAACTATGAACTTCCTGATGGACAAGTCATCACCA    400
TCGGAGCTGAGAGATTCCGTTGCCCAGAAGTTCTATTCCAGCCATCTCTC    450
ATCGGTATGGAAG                                         463
```

Fig. 10. Primers designed using Primer Express Software for beta actin gene of cucumber. Forward and reverse arrows indicate forward and reverse primers. Green box is TaqMan Probe.

Quantification Assays

Quantification of amplicons from DNA, cDNA or RNA that are generated during PCR process is monitored during each cycle. Absolute quantification refers to knowing the exact copy number of RNA in a biological sample while relative quantification provides a comparative picture on the differences viz., increase or decrease of the target gene. For absolute quantification, serially diluted standards with known concentrations are required to produce a standard curve. This curve can provide a linear relationship between C_T and initial amounts of the RNA or cDNA that allows the determination of unknown based on their C_T values. Majority of researchers do not need absolute data such as copy number of transcripts or nanograms of DNA. Relative quantification can be made by several mathematical models such as standard curve, comparative C_T, Pfaffl, Gentle, Q-gene, Liu and saint and DART-PCR. These methods use an external standard and the results expressed are based on the target / reference gene ratio. Absolute quantification is considered to be more suited for time-course experiments. Furthermore, when fold changes of a gene expressed is considered both methods produce comparable results (Peirson *et al.*, 2003).

Standards

The standards used are most commonly the purified genomic DNA, PCR product, *in vitro* transcribed RNA or plasmid, and other options can be oligonucleotide, etc. The absolute quantities of the standards are known by measuring the concentrations at A260 and converted to the number of copies based on the molecular weight of DNA or RNA. The range of dilutions should be broader so as to cover the copy number of the unknown samples.

Assay

Quantification of mRNA can be performed either by one-step or two-step real-time PCR. One-step method can minimize experimental variation, but RNA used as starting material could be degraded if not handled properly and is less sensitive than two-step method. Moreover, DNA standards cannot be used at the same time as doing one-step process of absolute quantification. In the two-step protocol, reverse transcription and real-time PCR are done separately resulting higher reproducibility. RNA should not be degraded and should be free from protein, DNA and PCR inhibitors. Minus RT controls are essential. This controls for the effectiveness of DNAse treatment as positive amplification in real-time PCR of a minus RT sample indicates genomic contamination of that sample. Template concentration range of 10-100 ng of genomic DNA and 1-10 ng of cDNA is generally used. Both the template and primers concentrations should be optimized by conducting preliminary experiments. Sample loading needs certain pattern, one such pattern is given below (Fig. 11).

Each sample should be either duplicated or triplicated. The total assay volume is 20, 25 or 50 ml. In general either 20 or 25 ml reaction mixture is preferred to minimize the costs. Most real-time PCR users do not use robotics for setting up of reactions, but prefer to use single or multichannel pipettes. However, it is recommended to use at least an electronic pipettor to avoid pipetting errors. It is recommended to load the samples to 96 well plates as quickly as possible to avoid denaturing of cDNA. Air bubbles should be removed. The instrument will take about 2 hours to complete the process. Pre-developed assays for real-time PCR are provided by ABI, Qiagen etc. In order to detect contamination (like genomic DNA) in the sample, minus RT controls are added.

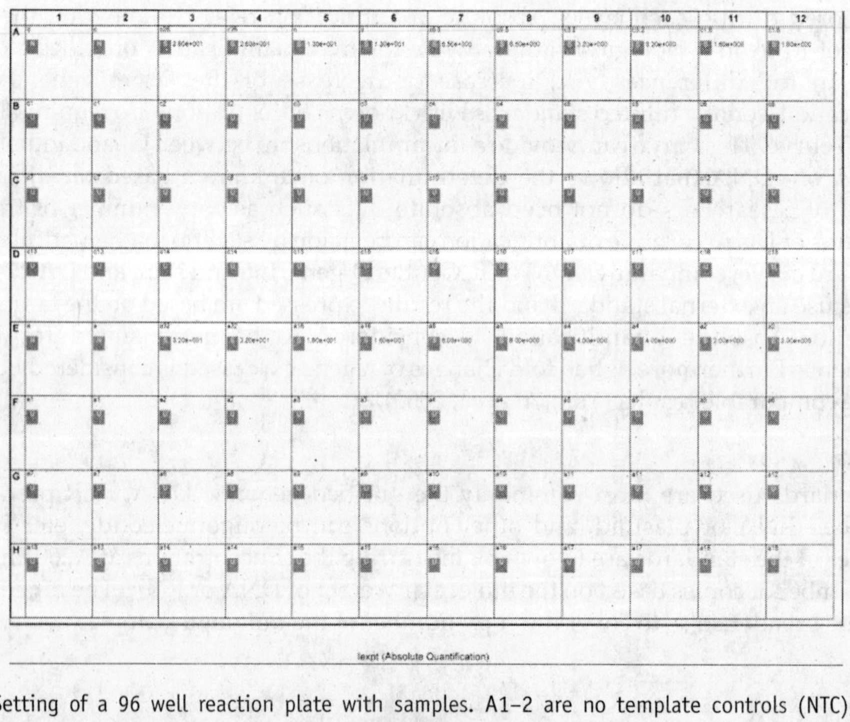

Fig. 11. Setting of a 96 well reaction plate with samples. A1–2 are no template controls (NTC); A3–12 are different concentrations of standards; B to D1–12 are unknown samples of target gene. Similar pattern is followed from E to H for reference gene.

Viewing Results

The results can be viewed as shown in Fig. 12. A typical amplification curve has plateau phase, linear phase, geometric phase, background and a baseline. Baseline fluorescence emission is calculated during the linear ground phase (during the first 10-15 cycles of PCR) wherein the emission at each cycle is not raised above the background.

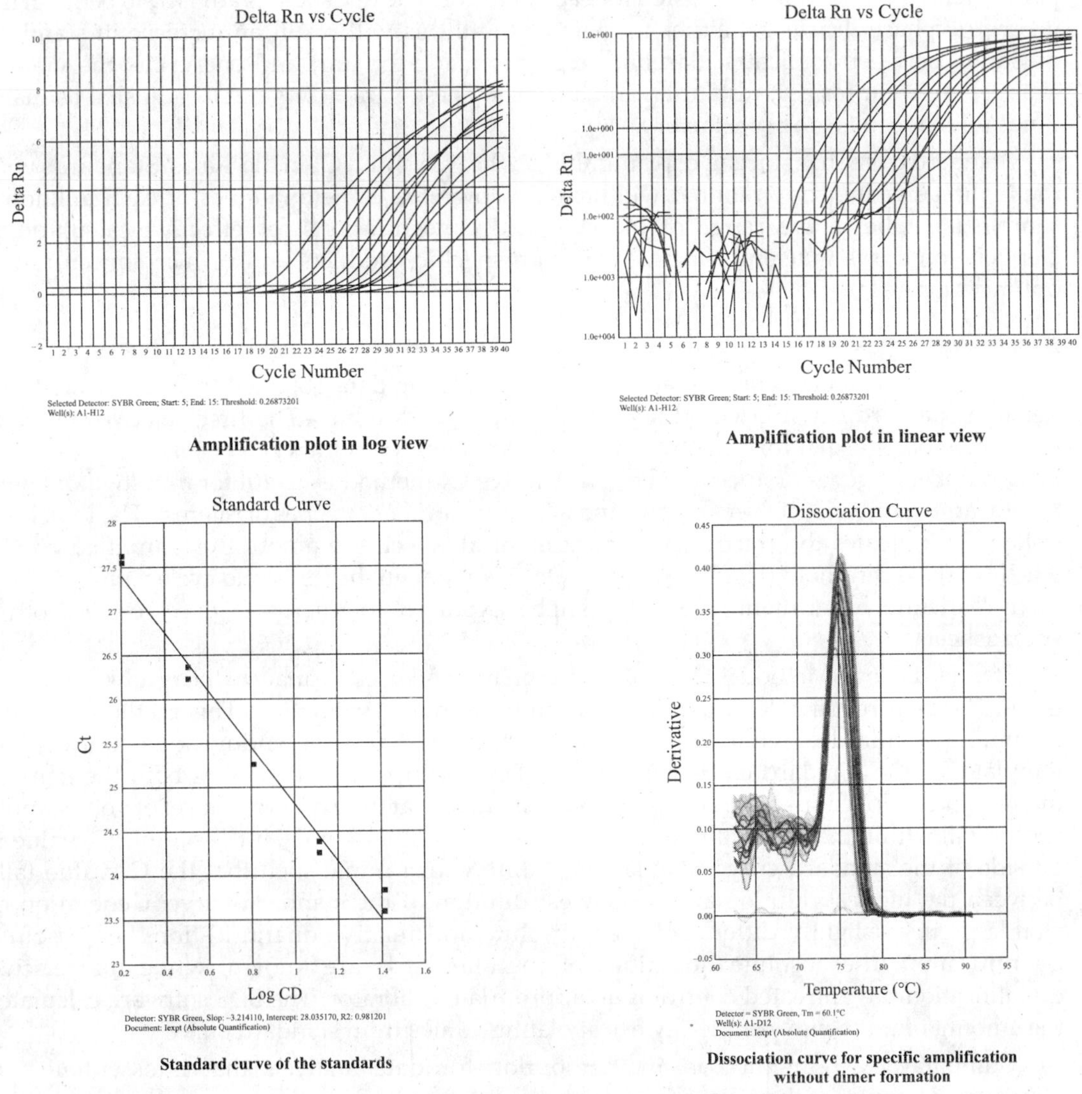

Amplification plot in log view

Amplification plot in linear view

Standard curve of the standards

Dissociation curve for specific amplification without dimer formation

Fig. 12. Typical results of RT PCR.

The baseline, representing the starting gene copy number, used for calculation of experimental results is referred as C_T that generally 10 times over the standard deviation of the process measured during this interval of the emission of fluorescence reaching the threshold level. At this phase, the doubling of the products occurs and finally the process reaches the plateau phase where the intensity of fluorescence gets saturated. The SDS software algorithm generates baseline and threshold values for the associated detector. However, these parameters can be manually adjusted for each detector. The threshold is advised to be set in the geometric phase. If it is set either too low or high, this will result an increase in standard deviation between replicates. Baseline setting is made for start and end cycle, the view is changed to linear for best resolution and analysis. These parameters can be adjusted as many times till the results look satisfactory.

Standard curve has starting copy number on the X axis and C_T in the Y axis. Standards with the highest starting copy number will have the lowest C_T. The quantities of each unknown sample must lie in the range of the standard concentrations for effective calculations. Dissociation curve is useful for detecting non-specific products and primer concentration optimization.

Data Analysis

There are two methods, both equally valid, for analyzing data obtained from real time PCR: Relative Standard Curve Method and Comparative C_T Method. The first, relative standard curve method is useful for investigators that have a limited number of cDNA samples and a large number of genes of interest. The comparative C_T method is useful for investigators who have a large number of cDNA samples and a limited number of genes of interest. The threshold cycle (C_T) indicates the fractional cycle number at which the amount of amplified copies reaches a fixed threshold (Heid *et al.*, 1996). Once you obtain the threshold cycle values (C_T) for your real time experiment, the data can be exported to Microsoft Excel or any other spreadsheet.

Standard Curve Method is the most common method used to analyze the results from real-time PCR, as it requires least amount of optimization and validation. This method requires running a standard curve for both the gene of interest and the endogenous control. The template for the standard curve can be any cDNA sample that expresses both the gene of interest and endogenous control. It is important to note that the expression level of your sample should fall within the limits of your standard curve. In other words, if the sample C_T value is outside of the standard curve, you should dilute your sample such that the C_T value falls between the highest dilution and the lowest dilution of the standard curve. Generation of standard curve slightly differs between absolute and relative quantification. For absolute quantification, the absolute quantities of the standards are known, while for relative quantification the untreated control is an appropriate calibrator. The SDS software calculates the amount of unknown samples by extrapolating values from standard curve.

Comparative C_T method is another option for data analysis that is essentially an abbreviated version of the relative standard curve method. Using the comparative C_T method utilizes the assumption that the primer efficiencies are relatively similar, thus the standard curve can be omitted from the assay.

In 5′ nuclease assay or using TaqMan probes, the software computes a standard curve from the C_T values of the diluted standards and extrapolates absolute quantities for the unknown samples based on their C_T values. The results from this assay approach 100% for quantitative assays using TaqMan probes when latest assay design guidelines by Applied Biosystems is coupled. In SYBR Green assay, absolute quantification calculation is made based on standard curve method. Absolute quantification can be obtained if known standards used are of known concentrations. Plasmid DNA or *in vitro* transcribed RNA are commonly used to prepare absolute standards. The concentration is measured by A260 and converted to the number of copies using the molecular weight of the DNA or RNA.

Normalization

It is quite often noticed that sampling of RNA results in different quantities depending on the cell number in the amount of tissues although weighed accurately. In general a reference gene which does not change its expression irrespective of the experimental conditions is chosen for this purpose. Nowadays, standard reference genes have been well documented for numerous experimental systems. However, if a new system is being experimented, then search and standardization are required for such constitutively expressing genes according to the experimental conditions (Fig. 13).

Reference gene used are glyceraldehyde-3-phosphate dehydrogenase mRNA, major histocompatability complex I (MHC I) mRNA, cyclophilin mRNA, 28S or 18S rRNA, beta actin mRNA, HPRT, beta 2 microglobulin, mRNA of certain ribosomal proteins, etc. These endogenous reference genes can correct the variation (if any) in the efficiency of reverse transcriptase. A recent study by Nicot *et al.* (2005) examined the expression of seven house keeping genes, namely, beta tubulin, cyclophilin, actin, elongation factor, 18S rRNA, adenine phosphoribosyl transferase and cytoplasmic ribosomal protein, during biotic and abiotic stress and revealed that elongation factor was the most stable gene while others varied depending on

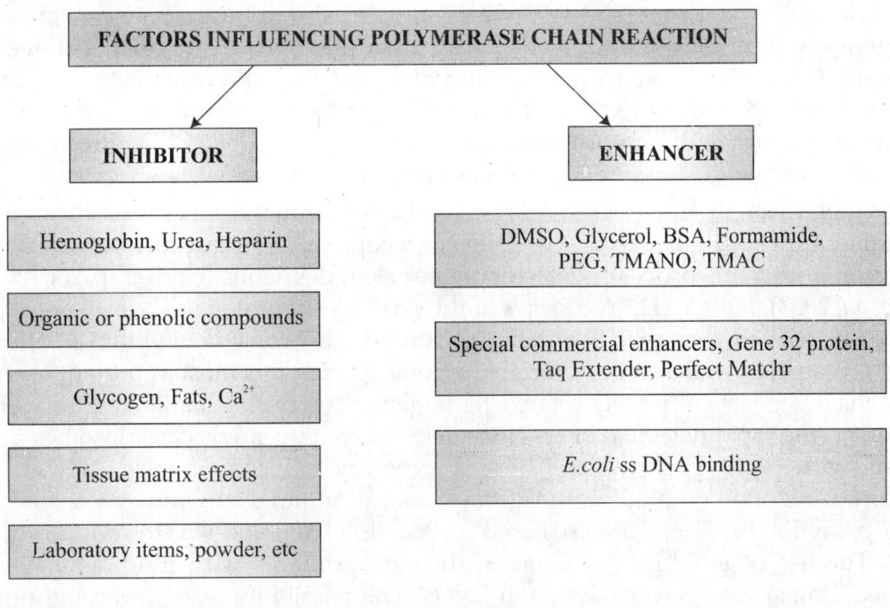

Fig. 13. Factors influencing exactness of PCR.

the type of stress. Further, ten times increase for some of these reference genes have been observed in the works reporting the use of real-time PCR. This indicates that for each type of experiment, there is a need to search for stable genes. Moreover, a passive reference dye like ROX is often included in the master mix to account for subtle differences in PCR master mix volumes and non-PCR related variation resulting in fluorescence signal. The PCR efficiency [E=10^(-1/slope)] is considered to be acceptable as >90 % (slope less than –3.60).

A general formula, which provides fold increase of gene expression in samples, is given below. Data can be imported to excel program for easy calculations.

$$\text{Fold gene expression} = \frac{\text{Quantity of target gene/Quantity of reference gene}}{\text{Quantity of target gene in DW treatment at 0h/Quantity of reference gene}}$$

Applications in Disease Diagnostics

The control of any existing or newly occurring diseases can be made by detection and proper identification of the casual organism. Biological specimens not showing any symptoms are difficult to evaluate for the presence of any pathogens if not by doing laborious processes. Quick and sensitive techniques for detection are needed. Selective detection of pathogens is a vital aspect. Closely related organisms can be specifically detected by molecular-based techniques that are not possible by conventional diagnostic techniques. Gene targets can be screened randomly in different parts of genome or can be done with a reference gene that is prone to specificity to that organism. Random amplified polymorphic DNA (RAPD) and amplified fragment length polymorphism (AFLP) are the random screening approaches (Vos *et al.*, 1995) indicated that extensive screening is required to ensure specificity of the potential marker candidate. Currently a vigorous diversification of the use of RT-PCR technique can be observed for various diagnostic purposes (Table 3). Basing on the occurrence of single nucleotide polymorphism (SNPs) there is easy to differentiate between closely related pathogens. A multiplex real-time PCR can quantitatively distinguish between races of pathogens found. For instance, methicillin resistant *Staphylococcus aureus* (MRSA) found associated in nosocomial infections, but coagulase negative staphylococci as well. Multiplex PCR represented a rapid and powerful method for detection of MRSA directly from clinical specimens containing a mixture of staphylococci.

The high-throughput DNA sequencing technology with the aid of sophisticated software's has further improved the detection of SNPs. Detection of SNPs can be achieved by using PCR primers, hybridization probes or detector oligonucleotides (Consolandi *et al.*, 2001; Livak, 1999; Papp *et al.*, 2003). Fungal rDNA consists of three rRNA subunit genes, which are separated by internal transcribed spacers (ITS) and this region is highly conserved within a pathogenic race (in certain cases) as well as highly variable among species. It is most important to consider the pathogenicity aspects of closely related pathogens. One of the challenging tasks for genetic diagnosis is the rapid detection of specific alleles located on single chromosomes in single cell analysis with high speed, reliability and experimental convenience. New technologies introduced are often employ more rigorous criteria to judge the same. The first single cell genetic assay for Tay-Sachs disease based on real-time PCR and was introduced by Rice *et al.* (2002). The use of molecular beacons in their experiments with individual lymphoblasts provided almost cent percent detection levels. The possibility of high-throughput real-time

Table 4. Time course of developments in application of real-time PCR used for diagnosis.

Implications of RT PCR	Reference
Molecular diagnosis of chronic myeloid leukaemia	Mensink *et al.*, 1998
Molecular diagnosis of haematological malignancies	Morgan and Pratt, 1998
Molecular diagnosis of follicular lymphoma	Luthra *et al.*, 1998
Molecular diagnosis of non-Hodgkins lymphoma	Johnson *et al.*, 1999
Quantitatification of HTLV-1	Kamihira *et al.*, 2000
Dignostics of acute lymphoblastic leukemia	Eckert *et al.*, 2000
Molecular diagnostics, Epstein Barr virus, Hepatitis C virus, Human cytomegalovirus, Varicella zoster virus	Niesters, 2001
Real-time quantitation of E2A-Pbx1 fusion gene; leukemia	Pennings *et al.*, 2001
Prostate-specific antigen detection	Straub *et al.*, 2001
Diagnostics of Epstein-Barr virus	Kudo *et al.*, 2001, Stevens *et al.*, 2002, Ruiz *et al.*, 2005
Diagnostic of recessive juvenile Parkinsonism	Hedrich *et al.*, 2001
Identification of *Helminthosporium solani*	Cullen *et al.*, 2001
Minituarized genotyping of thrombosis genetic risk factors	Cabeda *et al.*, 2002
Clinical diagnosis of Epstein-Barr virus, Hepatitis B virus, Human cytomegolovirus, Varicella zoster viruset *al.*, 2002, 2004	Niesters, 2002, Mackay
Detection of *Legionella spp*.	Raggam *et al.*, 2002
Diagnosis of plant pathogens, *Clavibacter michiganensis subsp. sepedonicus, pseudomonas syringae pv phaseolicola, Xylella fastidiosa, Pseudomonas solanacearum*	Schaad and Frederick, 2002
Quantification of human herpesvirus 8, Kaposi's sarcoma; multicentric Castleman's disease	Boivin *et al.*, 2002
Detection of *Colletotrichum coccodes*	Cullen *et al.*, 2002
Quantitative use of real-time PCR technology	Klein, 2002
Discrimination of single nucleotide sequence variants	Tewsa *et al.*, 2003
Diagnostics of influenza A virus	Perdue, 2003
Diagnostics of respiratory virus infection	van Elden *et al.*, 2003
Quantification of plant nucleic acid molecules	Liu, 2003
Analysis of low abundant point mutations in K-ras oncogenes	Wabuyele *et al.*, 2003
Hematologic neoplasia, human cytomegalovirus	Ohyashiki *et al.*, 2003
Detection of *Chlamydia pneumoniae*	Reischl et al, 2003
Diagnosis of plant pathogens, *Gaumannomyces-Phialophora* complex, *Pyrenopeziza brassicae; Leptosphaeri maculans*	McCartney *et al.*, 2003
Microarray multiplex diagnostics; *Escherichia coli, Francisella tularensis, Bacillus anthracis*	Cirino *et al.*, 2004
Detection and differentiation of North American swine influenza viruses	Richt *et al.*, 2004
Bluetongue virus differentiation	Orru *et al.*, 2004
Quantitative analysis of methylated alleles, retinoblastoma	Zeschnigk *et al.*, 2004
Identification of swine fever virus	Sandor *et al.*, 2004

Table 4 Contd.

Table 4 Contd.

Plant pathogen diagnostics, *Pyrenopeziza brassicae, Clavibacter michiganensis subsp. sepedonicus; Gaeumannomyces graminis*	Ward *et al.*, 2004
Prostate cancer identification	Stolk *et al.*, 2004
Quatification of plant pathogen, *Alternaria brassicicola, Botrytis cinerea*	Gachon *et al.*, 2004
Diagnostics of minimal residual disease; chronic myeloid leukemia, acute lymphoblastic leukemia	Muller *et al.*, 2004, Thiede *et al.*, 2004
Molecular diagnostics of infectious diseases; *Staphylococcus aureus*	Peterson and Dalhoff, 2004
Rapid detection of filoviruses; Ebola virus; Marburg virus	Weidmann *et al.*, 2004
Advanced method of asymmetric PCR	Sanches *et al.*, 2004
Quantification of *Campylobacter jejuni*	Rudi *et al.*, 2004
Human brucellosis, *Brucella spp., Borrelia burgdorferi*	Navarro *et al.*, 2004
Rapid diagnosis of pulmonary tubercolosis, *Mycobacterium tubercolosis*	Cheng *et al.*, 2005
Potato diseases, Mop-top virus, *Colletotrichum coccodes, Spongospora subterranea, Phytophthora infestans*	Lees *et al.*, 2005
pathogenicity of *Fusarium sambucinum*	Cullen *et al.*, 2005
Chronic myeloid leukemia	Khalil, 2005
Lung cancer, oncogen mutations	Schmiemann *et al.*, 2005
Acut respiratoty syndrome, chronic myeloid leukemia colorectal cancer	Bustin and Mueller, 2005
Cutaneous melanoma	Lewis *et al.*, 2005
ATP-binding cassette transporters; cystic fibrosis; familial HDL deficiency; recessive retinitis pigmentosa, acute myeloid leukemia	Schuierer and Langmann, 2005
Thyroid cancer	Hesse *et al.*, 2005
Apolipoprotein E polymorphism, type-2 diabetes	Camsari *et al.*, 2005
Musculoskeletal infections, *Staphylococcus aureus, Borrelia burgdorferi, Neisseria gonorrheae*	Tarkin *et al.*, 2005
Ulcer disease, *Helicobacyter pylori*	Huynh, 2005
Allelic discrimination in prenatal diagnosis, single nucleotide polymorphism, cytikine gene expression	Arya *et al.*, 2005
Cancer diagnostics, non-Hodgkin lymphomas, B-cell lymphoma, follicular lymphoma	Stahlberg *et al.*, 2005
Microbiological diagnosis of periopathogens; *Porphyromonas gingivalis, Fusobacterium*	Jervoe-Storm *et al.*, 2005
Bovine viral diarrhoea	Balint *et al.*, 2005
Acut respiratory syndrome, rapid diagnosis of SARS coronavirus	Tang *et al.*, 2005, Yam *et al.*, 2005
Human papillomavirus	Molijn *et al.*, 2005
Prenatal clinical diagnosis of hereditary optic neuropathy and external ophtalmoplegia	Wong et Boles., 2005
Escherichia coli	Liu *et al.*, 2005
Molecular diagnostics and genotyping of *Toxoplasma gondii*	Switah *et al.*, 2005
Rapid detection of Angelman syndrome and Prader-Willi syndrome	Raca *et al.*, 2005
Behcet's disease, N-acetyltransferase 2 polymorphisms	Tamer *et al.*, 2005
Rapid detection of Hippel-Lindau disease	Hoebeeck *et al.*, 2005
Normalization of gene expression measurements in tumor tissues	de Kok *et al.*, 2005
Appliaction of RT-PCR to intraoperative cancer diagnostics	Raja *et al.*, 2005

PCR assays of the human and numerous other genomes provides an immense source for the development of new assays (Klein, 2002).

Forensic labs use this technique for identifying DNA samples isolated from crime scene samples. In most of the diagnostic laboratories, PCR techniques have almost completely replaced southern and slot blot methods. Advancement in PCR technique such as real-time PCR is an important progression. Markers, such as Alu sequences are primate specific and are nearly 800000 copies in human genome.

Severe acute respiratory syndrome coronavirus (SARS-CoV) which caused 812 deaths over 27 countries of the world during 2003 are inconsistently detected by both immunofluorescence and ELISA. However, real-time PCR using cloned version of the 3′ terminal portion of the genome showed a detection limits <5 copies per assay and was consistent with serological results up to 20 days after the onset of symptoms. Furthermore, the viruses are characterized by a high mutation rate, which may influence on the exact quantification.

In a comparative study using real-time PCR and ELISA, positive findings for 90 and 100 % respectively bronchoalar lavage (BAL) fluid specimens from patients with probable confirmed invasive pulmonary aspergillosis (IPA) suggested that combined use of these methods could improve the diagnosis (Sanguinetti *et al.*, 2003). Similarly, detection of the induction of cytokine and chemokine expression in sputum of patients following viral excrebation of asthma has been documented. In another study, group B streptococci were detected by both conventional and real-time PCR in terms of both sensitivity and specificity, but, real-time PCR was quick (about 30 min) for amplification (Ke *et al.*, 2000). For the sake of improved diagnostic procedures, embedded biopsies are recommended. The methods described by Lehmann and Kreipe (2001) are helpful for quantitative analysis of nucleic acids extracted from processed formalin-fixed and paraffin embedded pathological specimens by real-time PCR. Using this technique, Christopher *et al.* (2002) could quantify human DNA in about 2 hr with high accuracy. The amount of sample required for human DNA quantification was minimal compared to the dot blot approach.

Detection of the presence of *Camplyobacter* on chicken carcasses which was traditionally done by standard culturing and characterization methods; this method used to take at least 5 days. A protocol of real-time PCR method with detection limits of 600 cfu/ml approximately representing 25 gene copies is also available. However, lower concentrations revealed a below threshold detection.

Crops can be infected by several pathogens during their different stages of growth; simultaneous detection of several pathogens can be achieved by multiplex PCR. Analysis of results by viewing bands on agarose gels is one option and another being the real-time PCR. However, the availability of limited number of dyes emitting fluorescence at different wavelengths and monochromatic character of the energizing light source makes this task limited to detection of only a few pathogens in a single reaction.

Serological methods can efficiently identify fungal and bacterial pathogens (Hampton, 1990), however, there are not available reliable techniques for viruses other than host indexing or other than DNA based techniques. Rassmusen and Wulff (1991) first developed PCR-based detection assay for plant pathogens. In the same year, Prosen *et al.* (1991) published a protocol relating to the above in infected seeds. Since, then PCR-based detection systems have been well developed for several plant pathogens and are being routinely utilized for identification and disease diagnosis purposes. However, some plants such as strawberry leaves are known to contain PCR inhibitors, modifications in protocols for DNA extraction or additional purification step may be required (H. Ishii, personal communication). Another important aspect is that DNA fragments amplified by PCR, evidenced on gels may not guarantee the fact of presence of the pathogen, since samples may not essentially harbor a particular type of pathogen. Non-specific products can mislead on several occasions. Taq DNA polymerase errors as well as chimeras and heteroduplex molecules are often experienced which lead to sequence artifacts and bias (Acinas *et al.*, 2005). Under such circumstances, southern blotting or ELISA may be essential; on the other hand, isolation of pathogens on agar media could provide a cheaper option. Such concerns are not experienced while dealing with real-time PCR (Livak *et al.*, 1995). Several instances have been documented for detection of plant pathogens and are tabulated in Table 4. This approach is undertaken only if the marker gene is unknown for that pathogen. However, fungal pathogen detection can be made by choosing ubiquitously conserved genes, rDNA and house keeping genes such as beta tubulin, actin and elongation factor.

The identification and quantification of virus, biotrophic organisms as well as some bacteria such as ammonia-oxidizing bacteria are difficult to quantify since they are notoriously difficult to culture. In these cases the molecular techniques play an important role. Literature survey reveals several attempts for quantification of the ammonia oxidizing bacteria in soil (Okano *et al.*, 2004), wheat yellow dwarf virus, ectomycorrhizal mycelium in soil (Landeweert *et al.*, 2003). Eun and Wong (2000) designed two pairs of molecular beacons specific to the RNA dependent RNA polymerase and coat protein genes for the detection of the two most prevalent orchid viruses namely Cymbidium mosaic virus (CymMV) belonging to genus Potexvirus and Odontoglossum ringspot virus (ORSV) belonging to Tobamovirus in infected orchid leaf tissues. This technology was successfully applied to detect 0.5 ng of viral RNA of both orchid viruses in 100 mg co-infected *Oncidium* orchid leaves.

Future Prospects

The development in biological sciences during the last century helps in comprehensive understanding of information about interacting network of various gene modules that coordinately carry the molecular mechanism of phenotypic expression of genotype. Nevertheless, the function of a major part of the genome is still unknown and the relationship between enzymes, hormones, signalling substances and various small molecules is still rather limited. In order to fully understand the genetic mechanisms and to regulate the phenotypic expression successfully more information of recognition of DNA by proteins, transcription factors, drugs and other small molecules is required. With the advent of sophisticated PCR machines that incorporate software-driven fluorescence analysis systems, real-time PCR is rapidly becoming a more common method for the analysis and quantification of gene ex-pression.

Development of robust and efficient methods for analyzing and interpreting high dimension gene expression profiles continues to be a focus in computational biology. The accumulated experiment evidence supports the assumption that genes express and perform their functions in modular fashions in cells. Therefore, there is an open space for development of the timely and relevant computational algorithms that use robust functional expression

Table 5. Plant pathogens determined by quantitative real-time PCR.

Pathogen	Host	Reference
Clavibacter sepedonicus	Potato tubers	Schaad *et al.*, 1999
Ralstonia solanacearum	Potato tubers	Weller *et al.*, 2000
Acidovorax avenae subsp. citrulli	Watermelon	Schaad *et al.*, 2001
Agrobacterium strains	Several plants	Weller and Stead, 2002
Xylella fastidiosa	Grape vine	Schaad *et al.*, 2002; Schaad and Fredrick; 2002
Erwinia amylovora	Apple	Salm and Geider, 2004
Spongospora subterranea	Potato	van de Graaf *et al.*, 2003
Synchytrium endobioticum	Potato	vanden Boogert *et al.*, 2005
Fusarium solani f. sp. phaseoli	soil-french beans	Filion *et al.*, 2003
Ophiosphaerella narmari	Bermuda grass	McMaugh and Lyon, 2003
Phytophthora infestans	Potato	Avrova *et al.*, 2003
Verticillium dahliae	oliva tree	Mercado-Blanco *et al.*, 2003
Alternaria brassicicola	*Arabidopsis*	Gachon and Saindrenan, 2004
Botrytis cinerea	*Arabidopsis*	Gachon and Saindrenan, 2004
Fusarium solani f. sp. glycines	Soybean	Gao *et al.*, 2004
Phytophthora ramorum	Sudden oak	Hayden *et al.*, 2004; Tomlison *et al.*, 2005
Tilletia spp.	Wheat	MacNeil *et al.*, 2004; Eibel *et al.*, 2005
Biscogniauxia mediterranea	Oak	Luchi *et al.*, 2005
Fusarium oxysporum f. sp niveum	Melon and soil	Zhang *et al.*, 2005
Mycosphaerella melonis	Melon and soil	Zhang *et al.*, 2005
Oculimacula sps.	Wheat	Walsh *et al.*, 2005

profiles towards precise classification of complex biological associations at the modular level. Although much progress in applying microarray technology to versatile biological kingdoms has been witnessed in recent time, further advancing its efficiency and power in elucidating complex biological mechanisms would very likely rely on our ability to handle the high dimension genetic information mixed with measurement noises, intrinsic biological variance, and a large number of irrelevant genes.

Real-time PCR technologies open increasing opportunities to detect and study phytopathogenic and antagonistic fungi. Accurate disease diagnosis and precise identification of any pathogens involved is an essential prerequisite for understanding plant diseases and controlling them effectively. Rapid-cycle real-time polymerase chain reaction (PCR) methods may revolutionize the manner in which plant pathogens are identified and diseases are diagnosed. As the genomics age progresses and more and more DNA sequence data become available, highly specific primers and fluorescent probe sequences can be designed to yield target amplicons to unique regions of a pathogen's genome. Portable real-time PCR instruments are now allowing for diagnostic assays to be run directly in the field or at remote locations other than the standard diagnostic laboratory. Rapid real-time PCR diagnosis can result in appropriate control measures and (or) eradication procedures more quickly and accurately than traditional methods of pathogen isolation. Disease losses are minimized and control costs reduced.

PCR based detection technologies utilizing species-specific primers are proving indispensable as research tools providing enhanced information on the ecology, aetiology and epidemiology of plant pathogenic micro-organisms. However, successful and routine application of PCR diagnostics to lignified plant tissues is often limited due to the lack of quality template due to inefficient DNA extraction methodologies, but also the presence of high levels of unidentified, co-precipitated PCR inhibitory compounds, presumably plant polyphenolics and polysaccharides. Development of new, easy-to-handle protocols fro the extraction of DNA from plants and filamentous fungi is required

Real-time PCR has become very useful in the quest towards achieving quantitative PCR methodologies for determination of GMO content and prior knowledge of the type of genetic modification in cultivated plants, such as resistance traits and controlling elements such as promoters and terminators. The use of multiplex PCR systems using combined probes and primes targeted to sequences specific to var-ious GMOs is becoming more important than standard PCR, which is proving to be insufficient for GMO-products. Against, sampling procedures are of great importance towards the validation of analytical methods for GMO analysis as largest single source of error in the analysis is the sampling procedure.

Several new methods are currently being used or are being developed to produce rapid detection methods. These include the use of biosensors, micro arrays, chip-based systems and the use electrophoresis combined with laser-induced fluorescence. Either conventional or real-time PCR based detection technologies utilizing species-specific primers are proving indispensable as research tools providing enhanced information on the ecology, aetiology and epidemiology of plant pathogenic microorganisms.

SUMMARY

The advancements in bioscience during the last century help in comprehensive understanding of information about interacting network of various gene modules that coordinately carry out integrated cellular function in somewhat isolated fashion, i.e., the molecular mechanism of phenotypic expression of genotype. The function of a major part of the genome is still unknown and the relationship between enzymes, signaling substances and various small molecules is rather limited. In order to understand the regulation of metabolism and to alter that successfully more information of gene expression, recognition of DNA by proteins, transcription factors, drugs and other small molecules is required. The invention of polymerase chain reaction (PCR) by Kary Mullis in 1984 was considered as a revolution in science (Mullis, 1987). Later, scientists revolutionized the basic biology as well as the diagnostics and forensics. Real-time PCR is becoming a common tool for detecting and quantifying expression profiles of selected genes. The technology to detect PCR products in real-time, i.e., during the reaction, has been available for the past 10 years, but has seen a dramatic increase in use over the past 2 years. A search using the key word real-time and PCR yielded 7 publications in 1995, 357 in 2000, and 2291 and 3021 publications in 2003 and 2004, respectively. At the time of this writing, there were 4398 works in 2005. In this compilation, we concentrate on the technique and applications of real-time PCR on gene expression analysis as well as disease diagnostics.

REFERENCES

Acinas, S.G., Rupavtarm, R.S., Ceraj, V.K. and Polz, M.F. (2005) PCR induced sequence artifacts and bias: Insights from comparison of two 16S rRNA libraries constructed from the same sample. Applied and Environmental Microbiology 71:8966-8969.

Christopher, S.Y.N., Jessi, S.I.M. and Tan, W.F. (2002) Quantification of human DNA using real time PCR. 16th Meeting of the International Association of Forensic Sciences, 2-7th September, Montpellier, France.

Consolandi, C., Castiglioni, B., Bordoni, R., Busti, E., Battaglia, C. and De Bellis, R. (2001) Development of oligonucleotide arrays to detect mutations and polymorphisms. Minerva Biotechnol. 13:261-268.

Eun, A.J.C. and Wong, S.M. (2000). Molecular beacons: A new approach to plant virus detection. Phytopathology 90:269-275.

Hampton, R.O., Ball, E.M. and De Boer, S.H. (1990) Serological tests for detection of viral and bacterial pathogens. American Phytopathological Society, St. Paul, Minn.

Heid, C.A.J. Stevens, K.J., Livak and Williams, P.M. (1996) Real time quantitative PCR. Genome Research 6:986-994.

Higuchi, R., Dollinger, G. and Walsh, P. (1992) Simultaneous amplification and detection of specific DNA sequences. Biotechnology 10:413-417.

Ke, D., Menard, C., Picard, F.J. Boissinot, M., Ouellette, M., Roy, P.H. and Bergeron, M.G. (2000) Development of conventional and real time PCR assays for the rapid detection of group B streptococci. Clinical Chemistry 46:324-331.

Klein, D. (2002) Quantification using real-time PCR technology: applications and limitations. Trends in Molecular Medicine 8:257-260.

Landeweert, R., Veenman, C., Kuyper, T.W., Fritze, H., Wernars, K. and Smit, E. (2003) Quantification of ectomycorrhizal mycelium in soil by real time PCR compared to conventional quantification techniques. FEMS Microbiology Ecology 45:283-292.

Livak, K.H., Flood, S.J. A., Marmaro, J., Giusti, W. and Deetz, K. (1995) Oligonucleotides with fluorescent dyes at opposite ends provide a quenched probe system useful for detecting PCR product and nucleic acid hybridization. PCR Methods Ap. 4:357-362.

Livak., K.J. (1999) Allelic discrimination using fluorogenic probes and the 5′ nuclease assay. Genet. Anal. 14:43-149.

Mullis, K. (1987) Process for amplifying nucleic acid sequences. United States Patent No. 4 683 202.

Nicot, N., Hasuman, J.F., Hoffmann, L. and Evers, D. (2005) Housekeeping gene selection for real-time PCR normalization in potato during biotic and abiotic stress. Journal of Experimental Botany 56:2907-2914.

Okano, Y., Hirstova, K.R., Leutenegger, C.M., Jackson, L.E., Denison, R.F., Gebreyesus, B., Lebauer, D. and Scow, K.M. (2004) Application of real time PCR to study effects of ammonia on population size of ammonia oxidizing bacteria in soil. Applied and Environmental Microbiology 70:1008-1016.

Papp, A.C., Pinsonneault, J.K., Cooke, G. and Sadee, W. (2003) Single nucleotide polymorphism genotyping using allele-specific PCR and fluorescence melting curves. BioTechniques 34:1068-1072.

Peirson, S.N., Butler, J.N. and Foster, R.G. (2003) Experimental validation of novel and conventional approaches to quantitative real time PCR data analysis. Nucleic Acids Res. 31:e73.

Prosen, D., Hatziloukas, E., Panopoulos and Schaad, N.W. (1991) Direct detection of the halo blight pathogen *Pseudomonas syringae pv. phaseolicola* in bean seed by DNA amplification. Phytopathology 81:1159 (Abstr.).

Rasmussen, O.F. and Wulff, B.S. (1991) Detection of *Pseudomonas syringae pv. pisi* using PCR. In Proceedings 4th International Working Group on Pseudomonas syringae Pathovars. Kluwer Adacemic Publishers, Dordrecht, The Netherlands, pp. 369-376.

Rice, J.E., Sanchez, J.A., Pierce, K.E and Wangh, L.J. (2002) Real time PCR with molecular beacons provides a highly accurate assay for detection of Tay-Sachs alleles in single cells. Prenatal Diagnosis 22:1130-1134.

Sanguinetti, M., Posteraro, B., Pagano, L., Fianchi, L., Mele, L., Sorda, M.L., Franco, A. and Fadda, G. Comparison of real time PCR, conventional PCR and galactomannan antigen detection by enzyme linked immunosorbent assay using bronchoalveolar lavage fluid samples from hematology patients for diagnosis of invasive pulmonary aspergillosis. Journal of Clinical Microbiology 41:3922-3925.

Tian, B., Rentz, S.S., Gorman, G.S, Rogers, T. and Page, J.G. (2004) Comparison of real time PCR assay methods in detection and quantification of beta actin genes in mouse tissues. Preclinica 2:214-220.

Tyagi, S. and Kramer, F.R. (1996) Molecular beacons: Probes that fluoresce upon hybridization. Nature Biotechnology 14:303-308.

Vos, P., Hogers, R., Bleeker, M., Reijans, M., van de Lee, T., Hornes, M., Frijters, A., Pot, J., Peleman, J., Kuiper, M. and Zabeau, M. (1995) AFLP: A new technique for DNA finger printing. Nucleic Acids Res. 23:4407-4414.

Suggested Readings

1. Arya, M Shergill, IS Williamson, M Gommersall, L Arya, N. and Patel, HRH. 2005. Basic principles of real-time quantitative PCR. Expert Review of Molecular Diagnostics, 5(2):209-219.
2. Bustin, SA. and Mueller, R. Real-time reverse transcription PCR (qRT-PCR) and its potential use in clinical diagnosis. 2005. Clinical Science, 109(4):365-379.
3. Eun, AJC. and Wong, SM. 2000. Molecular beacons: A new approach to plant virus detection. Phytopathology 90:269-275.
4. Heid, C. A., J. Stevens, K. J., Livak. and Williams, PM. 1996. Real time quantitative PCR. Genome Research 6:986-994.
5. Schaad, N. W. and Frederick R. D. 2002. Real-time PCR and its application for rapid plant disease diagnostics. Can. J. Plant Pathol. 24:250-258.

REVIEW QUESTIONS

Long Questions

1. Write a detailed note on different chemistries available for real-time reverse transcriptase polymerase chain reaction.
2. Explain the differences between conventional and real-time PCR.
3. Highlight the events involved in conducting real-time PCR.
4. Write a note on applications of real-time PCR in various fields with examples.
5. Discuss on the types of real-time PCR equipments available and explain their features.

Multiple Choice Questions

1. An no treatment control (NTC) is used in real-time PCR for detecting
   (a) Dyes (b) Primer dimer
   (c) cDNA (d) Genomic DNA

2. Which of the following is a dye used in PCR techniques
   (a) SYBR (b) TAMRA
   (c) EtBr (d) All
3. ABI Prism 7000 real-time PCR machine uses which of the following as detector
   (a) CCD camera (b) PMT
   (c) Silicon detector d) Fluorimeters
4. The real-time PCR quantifies product during
   (a) Plateau phase (b) Linear phase
   (c) Exponential phase (d) All
5. Which of the following assays can quantify DNA with highest specificity comparatively?
   (a) RNA protection assay (b) Dot blot assay
   (c) Real-time PCR d) RT-PCR
6. Which of the following is unique to TaqMan chemistry and not to SYBR chemistry?
   (a) DNA quantification (b) Two step RT-PCR for RNA quantification
   (c) cDNA quantification (d) Allelic discrimination
7. A novel fluorescence based nucleic acid detection technique, Molecular beacons was developed by
   (a) Tyagi and Kramer (b) Mullis
   (c) Higuchi *et al.* (d) Malinen *et al*
8. Which of the following order PCR techniques is from more sensitive to less sensitive
   (a) TaqMan, Molecular beacons, RT-PCR, SYBR
   (b) SYBR, TaqMan, Molecular beacons, RT-PCR
   (c) TaqMan, Molecular beacons, SYBR, RT-PCR
   (d) Molecular beacons, RT-PCR, SYBR, TaqMan
9. Which of the following factors reduce the efficiency of real-time PCR?
   (a) DNA concentration (b) Cycle conditions
   (c) Reaction components (d) Non-specific products
10. The standards used in real-time PCR is
   (a) Genomic DNA (b) PCR product
   (c) Plasmid (d) any of the above

Answers

1. (b)	2. (d)	3. (a)	4. (d)	5. (c)	6. (d)
7. (a)	8. (c)	9. (d)	10. (d)		

CHAPTER 33

RNA-mediated Gene Silencing: Mechanism and Possible Therapeutic Applications

R. Mishra[1]* and A.S. Srivastava

[1]Dept. of Medicine, Division of Nephrology, Case Western, Reserve University, Cleveland OH 44106, USA

E-mail: Stangnath.mishra@case.edu

[2]Cancer Center, School of Medicine, University of California San Diego, La Jolla, CA-92093, USA

The discovery of RNA-mediated inhibition of gene expression (RNAi), considering all the steps involved in process and their novelties, can arguably be stated as most impressive discovery in biological sciences in recent times and has taken molecular biologists by surprise. For years RNAs were known to be involved in protein translation machinery either by transmitting information from DNA to protein via mRNAs or by participating directly in protein translation machinery, like ribosomal RNAs or transfer RNAs. The central dogma of molecular biology that DNA makes RNA makes protein came under scrutiny with the discovery that in some cases RNA makes DNA. The scenario has become quite complicated with the discovery that RNAs not only can make DNA but also can inhibit expression of genes in certain conditions which is now called as RNAi. Though the phenomenon has been recognized for more than a decade, it is currently receiving tremendous research focus. Numerous studies demonstrate that the process of RNA-mediated inhibition of genes plays a very critical role in maintaining genetic stability, regulation of expression of a number of genes, their translation, and sometimes even in secretion of proteins. There has been an explosion in RNAi research recently which demonstrates that RNAi is another regulatory safety valve evolved in maintaining normal cellular physiology.

RNAi is a process by which dsRNAs silence gene expression, either by inducing the sequence-specific degradation of mRNAs having complementary sequences or by inhibiting

*Corresponding author

translation. It is noteworthy that introduction of long dsRNAs in cells provokes an interferon like process leading to the cell apoptosis. In principle, RNAi can target any gene transcript very specifically; therefore, RNAi provides a rapid means for assessing the effects of the loss of gene function. Even genome-wide RNAi screens have been successfully carried out in plants and invertebrates which have helped in discovering novel genes. An advantage of these RNAi screens is that a loss-of-function phenotype can be readily linked to a specific gene which compares favorably with mutagenesis screens, which require the laborious mapping of the induced mutation. The four major classes of small RNAs, which may play an active role in RNAi, are listed in Table 1. Out of these four classes of small RNAs, two classes, small/short interfering RNAs (siRNAs) and microRNAs (miRNAs), are studied in greater detail and will be discussed.

Table 1. Short RNAs — story in a nutshell

Short RNA	**Properties and possible functions**
Short/ small interfering RNAs (siRNAs)	This class of 21-22 nucleotides long double stranded RNAs is generated by cleaving dsRNAs. These siRNAs silence genes by promoting the cleavage of mRNAs having exactly complementary sequences or some times by recruiting inhibitory proteins to, or directing the modification of DNAs with exactly complementary sequences.
Micro RNAs(miRNAs)	miRNAs constitute a class of 19-25 nucleotides long single stranded RNAs and are encoded in the genome of most of the multicellular organisms studied. These RNAs are evolutionarily conserved and are regulated during development and in normal physiological conditions. They can act at the level of protein synthesis by impeding translation by ribosomes in addition to affecting stability of mRNAs. miRNAs regulate expression of genes having varying degrees, from perfect to partial, of complementarities.
Tiny noncoding RNAs (tncRNAs)	A recently discovered class of 20-22 nucleotides long RNAs which are encoded in the genome of *C. elegance*. These are conserved evolutionarily and few are developmentally regulated. Their function is yet to be worked out.
Small modulatory RNAs (smRNAs)	A short dsRNA identified in mice allows expression of neuron specific genes only in adult neurons.

SHORT/ SMALL INTERFERING RNAS (siRNAs)

Small interfering RNAs (siRNAs) are one of the most recent additions to the wide repertoire of nucleic acid molecules used to silence gene expression. siRNAs are the effector molecules of the RNAi pathway, discussed in mechanisms. siRNA constitutes a class of RNAi in which short RNA fragments hybridize with complementary mRNA and guide them to RNA degradation machinery. Mechanism of action of siRNA has been studied in greater detail and is far more understood than that of any other RNA-mediated inhibition. It is noteworthy that the discovery of RNA-mediated gene silencing is rooted in a paradoxical observation recorded few years earlier. In late 1980s and early 1990s, plant biologists were trying to enhance brightness of color in Petunias by expressing multiple copies of gene Chalcone Synthase, a critical enzyme in synthesis of deep purple flower color (Napoli, Lemieux and Jorgensen, 1990; van der Krol *et al.*, 1990). To their surprise, the color of flowers in the plants overexpressing extra copies of the gene, varied from absent to patchy. The only explanation for the observation at the time was that by some unknown mechanism extra copies of the gene led to degradation of both endogenous as well as the artificially overexpressed copies of the gene, since the investigators recorded a lower level of the transcripts by Northern blotting. These findings were further confirmed by another additional technique such as by infecting plant cells with genetically engineered viruses containing plant genes. Plants infected with genetically modified viruses had that particular gene down-regulated which was incorporated in the viral genome (Wassenegger *et al.*, 1994).

An explanation for such a puzzling observation was conceived only after the publication of Fire and colleague's work (Fire *et al.*, 1998), who demonstrated that double stranded RNAs (dsRNAs) help degrade the mRNA in the lysates of Drosophila eggs. It also helped explain the observations recorded in plant cells infected with genetically modified virus. It is well known that viruses make multiple copies of their genome through double stranded RNA (dsRNA) intermediates in order to multiply. During the process of multiplication, they also make multiple copies of the gene artificially inserted in their genome. Multiple copies of the artificially introduced gene in the viral genome are cleaved quickly to produce siRNA and thus inhibited the gene expression *in vivo*.

Mechanism of Gene Silencing by siRNAs

Indeed, it was a very interesting as well as intriguing observation that RNAs can inhibit gene expression at the post transcriptional level. However, it was still an unanswered question how RNA-mediated gene silencing worked. An early breakthrough in determining the mechanism was revealed with the observation that plants in which RNA-mediated gene silencing was observed, had unusually short RNAs. These RNAs were 20-25 nucleotides (nt) long and, interestingly, their sequences matched with the genes being down-regulated (Hamilton and Baulcombe, 1999). However, their unusually small size (smaller than any typical RNAs like mRNA, tRNA or rRNA) posed a conceptual problem in interpreting their importance until recently. Biochemical reactions reconstituted in the extracts of eggs of fruit flies revealed that long dsRNAs were quickly diced into smaller pieces with a very specific pattern (Fig. 1) (Tuschl *et al.*, 1999; Zamore *et al.*, 2000). The processing of long dsRNAs, hairpin miRNAs (miRNAs) or plasmid-synthesized short hairpin RNAs (shRNAs) by Dicer (an RNAse III family member) leads to the formation of small interfering RNAs (siRNAs)- 21-23 nt RNA duplexes with symmetric 2-3-nt 32 overhangs and 52-phosphate groups. Exogenously provided synthetic siRNAs are converted into active functional siRNAs by an endogenous kinase that provides 52-phosphate groups in the presence of adenosine triphosphate (ATP). siRNAs associate with cellular proteins to form an RNA-induced silencing complex (RISC), which contains a helicase that unwinds the duplex siRNA by an ATP-dependent reaction. In an ideal situation the antisense strand guides RISC to the target mRNA for endonucleolytic cleavage. In theory, each of the siRNA strands can be incorporated into RISC and direct RNA interference. The antisense strand of a siRNA can direct the cleavage of a corresponding sense RNA target, whereas the sense strand of a siRNA can direct the cleavage of an antisense target. An RNA having a perfect match with the target mRNA behaves like a siRNA and leads to mRNA degradation, whereas an RNA with a partial match functions as a miRNA and causes translational repression. Interestingly, recent data show that miRNAs also can induce the degradation of fully complementary mRNA targets. The finding that artificially synthesized dsRNAs also silenced the genes having complementary sequences was very important and provided a great impetus in the field (Elbashir, Lendeckel and Tuschl, 2001). Thus, artificially synthesized siRNAs enter directly at RISC level and suppress the expression of the target gene.

Pioneering studies in plants regarding RNA silencing, revealed that RNAi can affect transcription or translation of genes by impeding the process via mechanical hindrance as well (Matzke *et al.*, 2004). Follow up studies indicated that both of the strands of a gene are transcribed in order to make dsRNA which participate in gene silencing. This observation was in stark contradiction to the mainstream thought that is only sense strand of the gene is transcribed to make mRNA. Transcription of both the strands yields complementary RNAs which anneal to make dsRNA. The cleavage of the dsRNA by Dicer yields small siRNAs, which, in turn, participate in gene silencing by binding to transcription start site. In simple terms, it can be safely stated that modification of chromatin organization to inhibit gene transcription is a process of adding few more chemical groups to the DNA in addition to many chromatin proteins to render it too compact to be transcribed as dictated by the sequence of siRNA (Fig. 1).

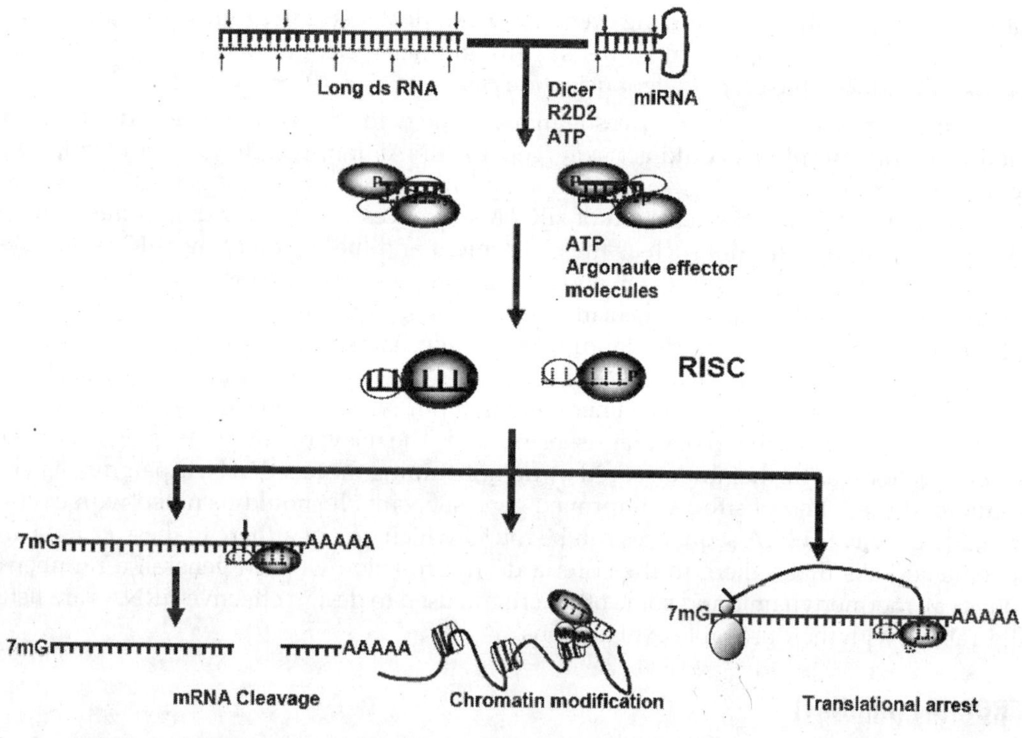

Fig. 1. Hairpin or dsRNA introduced in cells are cleaved into small dsRNAs by the enzyme dicer. RNA induced silencing complex (RISC) binds with antisense strand executes process of gene silencing. Once the siRNA binds to the mRNA having complementary sequences, it guides the mRNA degradation machinery to degrade mRNA and thus down-regulating the gene. RNAi can affect expression of genes by affecting chromatin organization and inhibiting translation also.

In some instances, amplification of siRNA mediated RNAi also takes place. The amplification of RNA-mediated gene silencing requires its own setup which varies in different organisms. For example, in some organisms, gene silencing by RNA requires RNA dependent RNA polymerase (RdRP) which uses the antisense strand of a siRNA as a template and then makes multiple copies of antisense RNAs to amplify the effect of the process. In other cases like plants, cell to cell transfer of the RNA may spread through out a plant except the reproductive tissues. Such a phenomenon is observed in generating resistance to viral infections. A similar incidence is not thought to take place in mammals or fruit flies as a gene similar to RdRP has not been discovered yet.

Designing siRNAs for Gene Silencing in Mammalian Systems

Tuschl and colleagues demonstrated that chemically synthesized short 21-mer dsRNAs (siRNAs) were also able to silence genes in a sequence-specific manner when introduced into mammalian cells (Tuschl *et al.*, 1999). Their findings were of immense importance as introduction of long dsRNAs in cells imitates viral infection and initiates an interferon-mediated stress response pathway which leads to apoptosis/cell death. This exciting observation was quickly reproduced by other groups, and subsequent work rapidly defined the basic structure of siRNAs that were able to induce gene silencing, including the need for a 19-nt RNA duplex with 2-nt overhang on the 3' ends. Most research groups substituted the 2 nucleotides at 3′ overhangs with (2'-deoxy) thymidine nucleotides, which did not affect silencing but reduced the cost of oligonucleotides.

With the increasing number of investigations in the field, it was soon realized that not all sequences within an mRNA could act as targets for RNAi and, therefore, a set of rules was necessary to for designing RNA sequences aimed for siRNA-mediated gene silencing in a laboratory set up. Initially the selection of siRNA sequences was largely determined using a limited set of empirical factors such as the GC-content and the region of the mRNA targeted and two base overhangs on both the 3′ ends. Based on these assumptions, most investigators were able to achieve efficient gene silencing with about one quarter of all siRNAs. However, a breakthrough came from the understanding of the mechanisms by which siRNAs associate with the RISC complex. Studies revealed that only the antisense strand of the siRNA is incorporated into this enzymatic machinery and that miRNAs and effective siRNAs exhibited decreased stability of the 5' end of the antisense strand. The relevance of this finding to enhance siRNA efficiency was further underscored by demonstrating the fact that base pair mismatches introduced at the 5' end of siRNAs improved gene silencing. It should be noted with caution that highly effective siRNA sequences can be found which do not adhere to these criteria and that some siRNAs that adhere to the criteria do not function well. In general, a number of biophysical, thermodynamic and some other criteria used to design effective siRNAs are listed in the Table 2 with their probable explanations.

MICRORNAs (miRNAs)

The existence of siRNAs procured in response to experimentally introduced long dsRNAs and transgenes prompted a search for similar short RNAs encoded by the genome. Efforts to purify small RNAs of 10-25 long nucleotides resulted in a wonderful surprise and that was discovery

Table 2. Biophysical criteria considered for the designing a siRNA with probable explanations

Criteria	Explanation
Biophysical, thermodynamic and structural considerations	
Low to medium GandC content between 30–50%	Facilitates interaction with RISC and unwinding
Low internal stability at the 52 antisense strand	Promotes antisense-strand selection by RISC
High internal stability at the 52 sense strand	Blocks sense-strand selection by RISC
Absence of internal repeats or palindromes	Increases the concentration of functional, stable hairpins
A-form helix between siRNAs and target mRNA	Enhances RNA–RNA interactions and promotes cleavage
Base preferences at specific positions in the sense strand	
Presence of an A at position 3 and 19 of sense strand	Promotes antisense-strand selection by RISC
Absence of a G or C at position 19 of sense strand	Promotes antisense-strand selection by RISC
Presence of a U at position 10 of sense strand	Promotes RISC mediated cleavage of mRNA and dissociation of the RISC–siRNA complex
Absence of a G at position 13 of sense strand	Promotes efficient unwinding
Enhancing specificity of siRNA-mediated gene silencing	
Perform stringent homology searches	Minimizes potential nonspecific gene silencing
Avoid low-stringency sequence interactions between siRNA and 32 UTR	Minimizes potential nonspecific gene silencing

of all together a totally new class of short RNAs in place of siRNAs and it was named as miRNAs. These miRNAs are numerous and single stranded. Their copy number may vary from few to 40,000 per cell (Lim *et al.*, 2003b). miRNAs are encoded in the genome (Fig. 2) and do not participate at any step in protein synthesis rather they regulate expression of a gene by affecting the stabilities of target mRNAs or sometimes by affecting their translation (Fig. 1). The miRNAs have been found in all species studied so far and many studies demonstrate that knocking out the miRNA genes either affects development detrimentally or is lethal. The miRNA encoding genes seem to make 0.5%-1% of the predicted genes in various organisms. It is predicted that there are almost 350 genes encoding miRNAs in human (Lim *et al.*, 2003a). About 330 genes have been already confirmed by employing various biochemical analyses in human by time this article was prepared (Griffiths-Jones, 2004). An updated but still growing list of miRNAs can be found at the website http://www.sanger.ac.uk/cgi-bin/Rfam/mirna/browse.pl.

On the basis of data available, it is suggested that miRNA caused inhibition of gene expression is mediated by base-pairing with fully to partially complementary sequences in mRNAs and leading to inhibition of their translation either by leading to degradation of mRNA or by mechanical interference with translation process. Therefore, unlike siRNAs, miRNAs can inhibit a group of transcripts whose complementarities can vary from partial to perfect. The importance of this process can be appreciated in the light of the finding that expression of miR-124 miRNA shifts global gene expression profile in HeLa cells, a tumor cell line, towards that of brain where as expression of miR-1 shifts expression profile toward muscle (Lim *et al.*, 2005). miR124 and miR-1 miRNAs are expressed in brain and muscle respectively. Degree of inhibition by miRNAs depends on the number of copies and degree of complementarities to the target gene. As stated above, one species of miRNAs can inhibit expression of a number of mRNAs having even partial complementarities. Therefore, inhibitory effects of any one particular species of miRNA on global expression profile are difficult to predict. However on basis of several experimentally substantiated assumptions it is thought that each miRNAs can inhibit up to 5 mRNAs. Keeping the fact in view the fact that at least 330 miRNAs have been recognized in human, there are more than 1,600 possible target genes which constitute a significant fraction of all known genes (~20-25,000 genes (2004). Keeping the possible number of genes those could possibly be regulated by miRNAs in mind these tiny miRNAs appear to have mega importance in maintaining normal cellular and physiological processes.

Biology of miRNAs

Significant strides have been made in our understanding of the biochemical mechanisms by which endogenous miRNAs silence gene function. Initially, miRNAs are transcribed as single-stranded precursors up to 2 kb in length that exhibit significant secondary structure owing to the presence of stretches of bases that can undergo extensive base pairing followed by stretches that adopt loop structures. Importantly, the pairing regions, or stems, present in miRNAs often contain a small number of mismatched bases that create "bubbles" in the miRNA structure. Together with loop structures, these are important for the recognition of miRNAs by cellular enzymes and for the ability of these RNAs to silence genes.

Primary miRNA (pri-miRNA) transcripts are identified and processed in the nucleus by the RNase III enzyme, Drosha, into approximately 70-nt-long precursors, known as pre-miRNAs. These are exported to the cytoplasm, where they are cleaved by a second RNase III enzyme, Dicer. Dicer converts miRNAs into double-stranded 21 to 23-nt-long mature miRNAs. Mature miRNAs associate with an enzymatic machine known as the RNA-induced silencing complex (RISC). The composition of the RISC is not completely defined but includes Argonaute family of proteins. The RISC unwinds miRNAs and associates stably with the (antisense) strand that is complementary to target mRNA. The complete enzymatic machinery required to process miRNAs appears to be expressed in most eukaryotic cells and is essential because mice and zebra fish that lack *Dicer1* fail to complete development.

Depending on the degree of homology between a miRNA and its target mRNAs, miRNA-RISC complexes inhibit gene function by two distinct pathways. Most miRNAs pair imperfectly with their targets and silence gene expression by translational repression. This RNAi mechanism appears to operate most efficiently when multiple miRNA-binding sites are present in the 3'UTR of the target mRNAs. In some cases, miRNAs exhibit perfect sequence identity with the target mRNA and inhibit gene function by triggering mRNA degradation. As discussed below, this appears to be the dominant mechanism by which synthetic siRNAs and plasmid-expressed shRNAs silence gene expression.

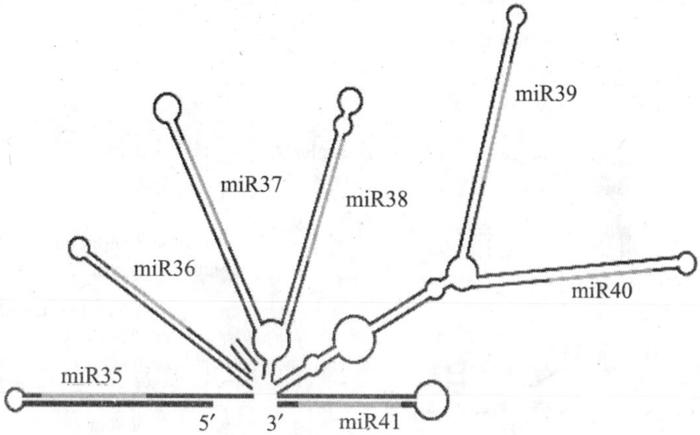

Fig. 2. Gene organization and transcription of miRNAs: Approximately 60% of miRNAs are expressed independently, 15% of miRNAs expressed in clusters and 25% in introns.

Functional significance of RNAi

Considering involvement of RNAi at different critical steps, it is reasonable to ponder about their physiological and evolutionary significances. From an evolutionary point of view, RNAi is an ancient phenomenon which is considered to precede even the divergence of worms and plants. The evolutionary significance of this process is in impeding propagation of rogue and parasitic genetic elements and viruses which multiply via a step involving double stranded RNA and thus conferring genetic stability to the genome which would be detrimentally affected otherwise. However, efforts to characterize the functional significance of RNA-mediated silencing have led to the fact that this is a very important regulatory element at many steps during various developmental stages and also in normal physiological conditions.

The answer to this question can be appreciated only in the light of understanding the cost and profit of allowing transposons and other unwanted genetic elements to multiply in the genome. For example, about half of the human genome was generated by duplication and insertion of the transposable/parasitic elements. In general, siRNAs play an important role in silencing transposable elements. If not regulated stringently and effectively, the whole genome would be loaded with transposons which would exert an extreme load on the replication system in addition to inflicting a significant instability to the genome. Therefore, an additional safety valve evolved to inhibit propagation of transposons and other parasitic genetic elements in early evolutionary environment considering the fact that RNA played a primal role in the evolution of life. Experimental proofs for such a hypothesis have been demonstrated in worms and plants. Instability caused in the genome of reproductive cells of *C. elegans* has been demonstrated after mutating a gene in the siRNA-mediated silencing pathway (Ketting *et al.*, 1999; Tabara *et al.*, 1999). This observation was further strengthened by a similar observation in plants also (Xie *et al.*, 2004).

The same process of RNAi mediated gene silencing is employed as an additional safety measure in some cases to control protein level only beyond a critical number of gene transcripts. A typical example of such a phenomenon is observed in fruit flies where gene silencing occurs only beyond the critical number of gene transcripts. In this particular case,

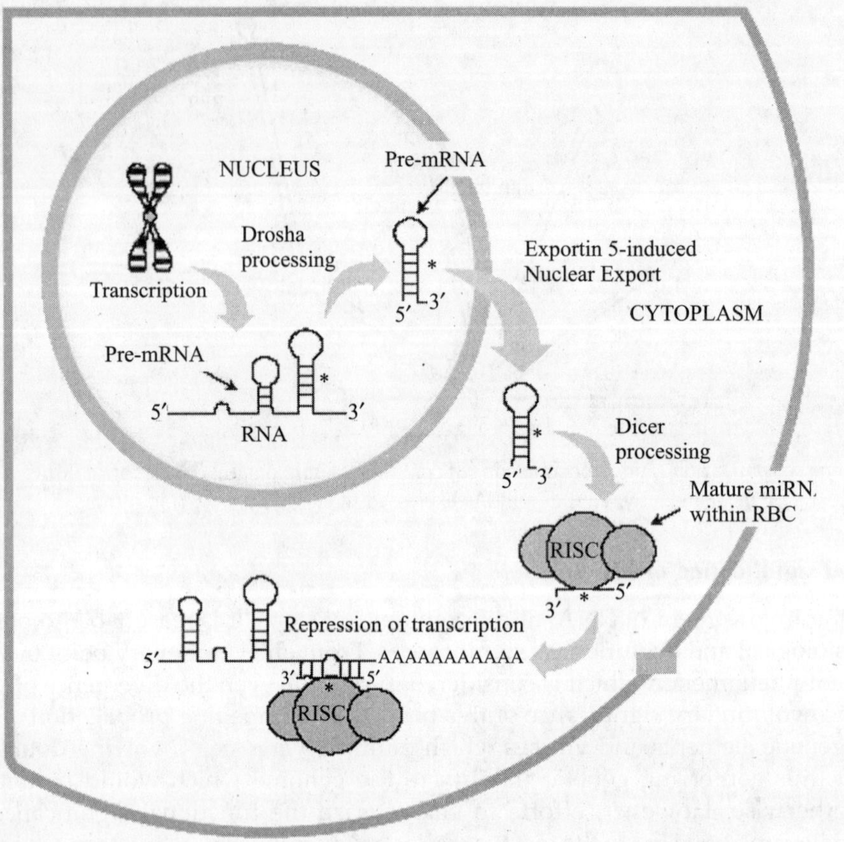

Fig. 3. Transcription of cluster of miRNA gene yields pri-miRNA which is cleaved by DROSHA in nuclei itself. Pri-miRNAs thus yield hairpin pre-miRNA, which after being exported to cytoplasm, are cleaved by the enzyme DICER. Mature miRNAs, with help of RISC, inhibit translation of the mRNA having perfect to partial homology.

increase in the level of protein corresponds to increasing copies of the RNA but only up to certain critical limit only. Beyond this critical limit, gene expression is inhibited by RNAi mediated gene silencing (Pal-Bhadra *et al.*, 2004). In this and similar other cases the set of DNA binding proteins are altered to inhibit transcription and thus gene expression. Such a phenomenon is observed in yeast also with an additional significant step in inhibiting the expression of transposons (Schramke and Allshire, 2003). In yeasts, the gene silencing spreads to neighboring regions in addition to affecting expression of transoposons by some intriguing process which is not well understood yet. Recent studies have indicated that miRNAs play a regulatory role for a number of genes like ras, a regulatory signaling molecule, is regulated by miRNA lin7. Insulin secretion is regulated by miR-375 as its overexpression in cultured cells suppressed glucose-induced insulin secretion, and conversely, inhibition of endogenous miR-375 function enhanced insulin secretion (Poy *et al.*, 2004). Recent surge in miRNA literature demonstrates their active role in carcinogenesis. Thus, with increasing understanding of the mechanisms and functions, miRNAs are getting increasingly more importance in regulation of biological functions. Table 3 lists a number of miRNAs whose function has been deciphered by using loss of function or gain of function techniques.

Table 3. Function of animal miRNAs *in vivo*

Organism/ Process	miRNA	Targets	Function	Evidence
C. elegans				
Developmental timing	*lin-4* miRNA	*lin-14, lin-28*	Stem cell differentiation	LOF
Developmental timing	*let-7* miRNA	*lin-41, hbl-1, daf-12, pha-4*	Stem cell differentiation	LOF
Developmental timing	miR-48, miR-84, miR-241	*hbl-1*	Stem cell differentiation	LOF
Developmental timing	miR-48, miR-84	Unknown	Cessation of molting	LOF
Organogenesis	miR-84	*let-60*	Differentiation/proliferation	GOF
Differentiation	*lsy-6* miRNA	*cog-1*	Left-right asymmetry	LOF
Differentiation	miR-273	*die-1*	Left-right asymmetry	GOF
D. melanogaster				
Growth control and programmed cell death	*bantam* miRNA	*Hid*	Proliferation/programmed cell death	LOF
Programmed cell death	miR-14	Unknown	Programmed cell death	LOF
Patterning and embryogenesis	miR-2a, -2b, -6, -7	*E(spl)/bHLH, bearded* families	Notch signaling	GOF
Embryogenesis and programmed cell death	miR-2, -6, -11, -13, -308	Unknown	Programmed cell death	2-O-Me-RNA
D. rerio				
Differentiation and organogenesis	miR-430	Unknown	Neurogenesis	Dicer rescue
M. musculus				
Differentiation and organogenesis	miR-1	Hand2	Angiogenesis	GOF
Differentiation and organogenesis	miR-181	Unknown	Hematopoiesis	GOF
H. sapiens				
Insulin secretion	miR-375	*Myotrophin (Mtpn)*	Exocytosis	2-O-Me-RNA
Human disease	miR-17, -18, -19a, -20, -19b-1, -92-1	Unknown	Tumorigenesis	GOF
Human disease	miR-32	PFV-1	Viral defense	LNA

Abbreviations used in Table 3 : PFV-1, primate foamy virus type 1; E(spl)/bHLH, Enhancer of Split/basic helix loop helix transcription factor cluster; LOF, miRNA has been studied using loss-of-function mutations; GOF, miRNA has been studied using gain-of-function approaches only; 2-O-Me-RNA, miRNA has been studied by depletion using 2' O-Methyl antisense oligoribonucleotides; LNA, miRNA has been studied by depletion using locked oligonucleotides (LNA); Dicer rescue, miRNA has been studied using rescue experiments in a Dicer mutant background.

Table 3 includes miRNAs that have been analyzed *in vivo* using loss-of-function studies, and miRNAs for which a likely function has been demonstrated by using an indirect approach, e.g. mis-expression experiments. It does not contain miRNAs for which a target mRNA has been predicted and validated using overexpression experiments, but for which no further functional characterization at the cellular or organismal level has been carried out (Alvarez-Garcia and Miska, 2005).

The findings in this rapidly evolving field have established that miRNAs are involved in numerous cellular processes including development, differentiation, proliferation, apoptosis, and the stress response and pathological biology like carcinogenesis. Studies within the last three years have demonstrated that there is altered expression of miRNA genes in several human malignancies, including chronic lymphocytic leukemia, pediatric Burkitt's lymphoma, gastric cancer, lung cancer, and large cell lymphoma. Many other studies confirm that altered expression of specific miRNAs is involved in tumorigenesis, while a fourth reveals that miRNAs may be important for enabling stem cells to overcome the G1/S cell cycle checkpoint (Croce and Calin, 2005). Moreover, Hariharan et al (2005) have demonstrated that many human miRNA have target HIV transcripts suggesting a possibility that repressed expression of miRNA could lead to some pathological progression (Hariharan *et al.*, 2005).

RNAi IN FUNCTIONAL GENOMICS AND THERAPEUTIC CONSIDERATIONS

Opting the Ubiquitous RNAi Setup for Experimental and Therapeutic Purposes

Success of RNAi as a research and potential therapeutic tool is because of the following facts: (a) their extreme specificity to target a particular gene or allele or even mutant form of a transcript, and (b) the enzymatic setup required to process RNAi is ubiquitously expressed which can be co-opted by exogenous RNAs to direct sequence-specific gene silencing. Sequential activity of Drosha and Dicer converts viral, transgene-encoded, and synthetic dsRNAs into siRNAs that trigger gene silencing in plants. As stated above long dsRNA are known to induce an interferon mediated stress response pathway in animal cells. To avoid such a stress responses induced by long dsRNAs in addition to achieve gene silencing in mammalian cells, small exogenous RNAs that enter the RNAi pathway further downstream are used. Synthetic siRNAs are designed to mimic Dicer products which can directly associate with the RISC and target homologous mRNAs for degradation. Plasmid-expressed shRNAs are also thought to enter the RNAi pathway because they are recognized and cleaved by Dicer into products that bind the RISC. A recent study suggests that it may even be possible to trigger RNAi in mammalian cells by providing a long dsRNA substrate for Drosha, as long as the dsRNA is confined to the nucleus by removing its 5' cap structure.

Stable Induction of RNAi in Mammalian Cells Through Expression of shRNAs

In some organisms, notably *C. elegans*, RNA-dependent RNA polymerases exist that are able to amplify siRNAs and even pass them on through the germ line. As a consequence, introduction of dsRNA triggers long-lived, stable gene silencing in this organism. These polymerases do not exist in mammalian cells, and consequently, gene silencing induced by siRNAs is limited by the number of RNA molecules introduced into a cell. The number of siRNAs decreases with time by dilution as cells divide and probably also as a consequence of degradation by cellular enzymes. Accordingly, in many cell culture systems, gene silencing is seen for only a few days after siRNAs are administered. On the other hand, siRNA-induced gene silencing has been observed for weeks in slowly proliferating or non-dividing cell types such as macrophages and hepatocytes.

The solution for obtaining more universal stable gene silencing through RNAi has been to develop expression systems that stably produce siRNAs using the promoter of either the U6 or H1 splice factor to express a shRNA which can be processed by Dicer to produce siRNAs. The duplex stems in these shRNAs has varied length from 19 to 29 nt connected by a short loop sequence. The sequence of the loop is critical for effective target silencing and several different sequences have been shown to be functional. The U6 and H1 promoters recruit RNA polymerase III, a specialized polymerase responsible for transcription of most of the cell's small RNAs (e.g., splicing RNAs and tRNAs). Use of pol III promoters to create shRNAs confers following benefits: (i) well defined and highly conserved transcription initiation and termination sites, (ii) highly active promoters functional in most, if not all, mammalian cells, and, (iii) very flexible pol III promoter shRNA expression cassettes which could be introduced in a number of different expression systems, including virus-based and those used to create transgenic animals. Several pol III promoters have been extensively characterized and co-opted to express shRNAs.

In contrast to promoters controlling transcription of small RNAs, those for most cellular genes recruit RNA polymerase II and usually generate transcripts that initiate in a less conserved manner and require a long termination poly (A) sequence. With a few exceptions both features have largely precluded the use of pol II promoters to generate shRNAs. This is not the case for endogenous miRNAs, which are expressed from pol II promoters in a developmentally regulated or tissue-specific manner. Furthermore, miRNAs and, possibly, also slightly longer shRNAs may enter the RNAi pathway more efficiently than typical 19-mer shRNAs, potentially leading to more potent gene silencing. For these reasons, new stable expression systems for siRNAs that are based on miRNA structures have been developed by a number of groups. Expression of a long nuclear-restricted dsRNA in cells and mice has also been shown to silence gene expression effectively, suggesting that further improvements and modifications will be made to stable siRNA delivery systems as our understanding of the RNAi machinery improves.

Regulated and Tissue-Specific Gene Silencing by RNAi

Constitutive expression of siRNAs in cells, tissues, and animals has provided important insights into biological processes on one hand, the ability to control gene silencing more tightly is likely to significantly extend the applications of experimental RNAi on the other. For these reasons alone, a significant focus has been on the creation of regulated and tissue-specific siRNA delivery systems. To date, most of these systems are based on engineered pol III promoters that are controlled by small molecules or the Cre recombinase. A number of groups have demonstrated that gene silencing can be initiated or inhibited by Cre-driven recombination of modified pol III promoters in cells and in mice and by introduction of DNA elements that bind tetracycline- or ecdysone-regulated transcriptional activators or repressors (see Table 4). More improvements are likely to result in more effective and flexible RNAi systems to interfere with gene function in mammals. An alternative approach to obtain

Table 4. Vector-based versus synthetic siRNA-based RNAi in mammalian cells

Approach	Advantages	Disadvantages
Vector-based	Delivery to non-transfectable cells. Stable silencing for non-essential genes. Inducible expression. Enzymatic preparation of hairpin libraries using cDNAs and cDNA libraries. Flexibility of shuttling of hairpin insert between different vectors (for example, between lenti-, retro- or adenoviral expression vectors; variation of promoters). Stable positive-readout screening using complex polyclonal libraries.	Prone to nonspecific interferon- response-related effects caused by high expression of hairpin RNA. Difficult to select and construct highly effective hairpin RNAs. Decreased potential for systemic delivery in therapeutic applications.
siRNA-based	Less prone to induce nonspecific side effects due to greater control over amount of transfected reagent. Ease of chemical synthetic production and quality control. Small size and chemical modifications hold best potential for therapeutic applications. Useful for structural functional studies of RNAi machinery.	Duration of silencing depends on the rate of cell division.

regulated expression of siRNAs might be to use pol II promoters. With the exception of the constitutive cytomegalovirus (CMV) promoter, this approach has not yet been shown to be successful, although it is likely that by using miRNA-based structures to deliver siRNAs that these promoters will be used to trigger RNAi in a regulated manner in the future.

Off-Target Effects and Interferon Responses : Possible Limitations to the Use of RNAi

Though a key feature of siRNAs is that they inhibit genes in a highly specific manner and miRNAs can affect expression of a number of genes. Indeed, the initial description of these siRNAs also demonstrated that alteration of a single base pair was sufficient to disrupt gene silencing. For many subsequent studies with siRNAs, this specificity was shown to hold true. However a number of studies have suggested that there are situations where mismatches between the siRNA and target sequence can be tolerated. These observations raised the concern that siRNAs may have effects on genes that are not considered targets, so-called off-target effects. This concern has been addressed by a number of groups that have examined genome-wide changes in gene expression following the introduction of siRNAs. Some of these studies found that a number of genes unrelated to the target are changed in expression, mostly by a factor of two. These off-target effects have been correlated with the concentration of siRNAs, as well as similarities between the off-target transcripts and the 5' ends of siRNAs. It seems plausible on the basis of recent work that the decreased off-target mRNA levels are the consequence of siRNAs adopting miRNA-like properties, resulting in slightly decreased levels of mRNAs, possibly through alterations in mRNA stability. It remains to be determined whether translational inhibition is seen on these off-target mRNAs. Currently it is not possible to predict whether a particular mammalian siRNA will induce off-target effects. It is widely assumed that the most informative parameter will be the degree of homology between siRNAs and other gene products, and therefore most experimental siRNAs are designed to have no known perfect matches with mRNAs other than the intended target. It is interesting that off-target effects are not observed when dsRNAs are used in primitive organisms. This may be because the dominant species of siRNAs generated from these dsRNAs are selected by Drosha and Dicer and other components of the endogenous RNAi machinery, which might have a proofreading activity that guards endogenous genes from silencing. Thus it is possible that mammalian siRNAs generated from dsRNAs precursors through the action of Drosha and Dicer may be less prone to induce off-target effects.

A second major concern among researchers using RNAi in mammals is the possibility that introducing exogenous dsRNAs may trigger an antiviral interferon response mediated by the PKR. Indeed, many early attempts at silencing gene expression using dsRNAs strategies analogous to those developed for primitive organisms failed because they triggered the production of interferon, nonspecific gene silencing, and apoptosis in mammalian cells. Early work by Tuschl and colleagues suggested that dsRNAs that were less than 30 bases in length were able to silence gene expression in a specific manner, while eluding the molecular machinery responsible for triggering the interferon response. This finding has been corroborated by the successful use of siRNAs as reagents to interfere specifically with gene function in a wider variety of different mammalian systems. However, a number of recent studies suggest that even short dsRNAs can trigger the expression of some of the target genes of the interferon response and, in some cases, can induce the cellular changes associated with this process. As of now, it is not clear how often siRNAs and shRNAs trigger the interferon pathway and which conditions favor this response to these RNAs. Chemical features of dsRNAs, as well as their expression levels and delivery routes, may determine whether they become visible to the interferon response machinery.

Similar to the interferon response, evidence exists that siRNAs and shRNAs can activate dendritic cells and other cells of the immune system through a much more specific and restricted class of receptors, the Toll-like receptors (TLRs), that can recognize foreign nucleic acids including dsRNAs. While the consequence of this remains to be determined, these findings do raise the possibility that RNAi reagents may trigger adverse immune responses *in vivo*.

In vivo Delivery of siRNAs to Induce RNAi

Due to efficient and cost-effective RNA-mediated gene silencing, RNAi provides a significant potential to study the genetic basis of normal tissue physiology, as well as disease processes in animal models. These incentives lure many groups to work on diverse strategies to deliver siRNAs or shRNAs to cells and tissues of experimental animals. Early efforts focused on direct administration of synthetic siRNAs, and three major delivery methods have been shown to be promising. In first approach, intravenous injection of siRNAs in a large volume (1 ml) of saline solution, works by creating a back-flow in the venal system that forces the siRNA solution into several organs (mainly the liver, but also kidneys and lung with lesser efficiency). In second approach, gene silencing also has been achieved *in vivo* by injecting smaller volumes of siRNAs packaged in cationic liposomes. siRNAs, administered intravenously using this strategy, display silencing primarily in highly perfused tissues, such as the lung, liver, and spleen. Local delivery of siRNAs has been shown to be successful in the central nervous system. Finally, gene silencing has also been achieved by electroporation of siRNA duplexes directly into target tissues and organs, including muscle, retina, and the brain.

Although successful, it is likely that these strategies to silence genes are limited by the stability of siRNAs molecules *in vivo* and the efficiency with which they are taken up by target cells and tissues. Much effort has been directed to increasing the half-life of the siRNAs by modifying the chemistry of the RNAs used. A number of groups have also used plasmid-based shRNAs, instead of siRNAs, to obtain relatively longer-life gene silencing *in vivo*. A number of approaches have also been shown to improve cell and tissue delivery of siRNAs and shRNAs, including conjugating RNAs to membrane-permeating peptides and by incorporating specific binding reagents such as monoclonal antibodies into liposomes used to encapsulate siRNAs (Fig. 4).

Use of Viral Vectors to Induce RNAi in Primary Cells, Tissues and Experimental Animals

In order to obtain efficient and long-lived gene silencing using RNAi in cells and tissues, many researchers have developed a variety of viral vectors to deliver siRNAs both *in vitro* and *in vivo*. Retrovirus-based vectors that facilitate stable insertion of genetic material into cycling cells have been engineered to express shRNAs and to trigger RNAi in transformed as well as in primary cells. Since they can infect and express in certain adult stem cells, notably hematopoietic stem cells, retrovirus-based vectors have also been used to create "knockdown" tissues in mice.

Since lentiviral vector can infect noncycling and postmitotic cells such as neurons, more wide-ranging applications of RNAi have been reported using these vectors. Transgenic knockdown animals generated by infecting embryonic stem cells or single-cell embryos using lenitviral vectors have displayed loss-of-function phenotypes and transmit the RNAi vector to their offspring, suggesting that this technique represents an efficient, low-cost alternative to knockout technologies to study normal tissue physiology and disease processes in a variety of experimental animal systems.

Highly effective siRNA delivery systems have also been created that are based on adenoviruses and adenovirus-associated viruses (AAV). Adenoviruses can infect a wide range of cells and have been shown to silence gene expression *in vivo*. Since they do not integrate into the genome and tend to induce strong immune responses, their use in some circumstances may be limited. In contrast, AAV does not cause disease in humans and can integrate into the genome of infected cells. Unlike retroviruses and lentiviruses, AAV tends to integrate at a defined location in the genome, thus minimizing the chance of a mutagenic effect of the integrated virus. Effective gene silencing mediated by AAV-based vectors has been demonstrated following systemic or tissue-specific injection of viral particles.

Creation of RNAi Transgenic Animals

In addition to the use of lentiviral vectors, more traditional transgenesis strategies have been employed successfully to create loss-of-function models to study gene function in rodents using RNAi. This provides another strategy by which RNAi might provide an alternative to create gene-knockout animals without disrupting the indigenous gene and minimally altering the genome. Inheritable RNAi transgenesis has been achieved both through expression of shRNAs and long dsRNAs whose expression is restricted to the nucleus or oocyte. On the basis of this small number of pioneering studies, it appears that RNAi is effective at silencing gene expression in many, if not all, tissues. In some instances, attempts to create RNAi transgenic animals by injection of plasmids encoding shRNAs into single cell embryos have been unsuccessful, whereas injection of DNA into blastocysts has succeeded. This may reflect a toxic effect of overexpression of siRNAs during early development, possibly due to competition with miRNA pathways. How significant this impediment could be for the generation of RNAi transgenic animal remains to be determined.

Whereas RNAi in transgenic animals has been shown to recapitulate some loss-of-function phenotypes established in knockout animals, there is mounting evidence that the RNAi phenotype will often appear more variegated than the knockout phenotype. This is probably due to the fact that RNAi does not abrogate gene expression but rather reduces it to varying levels. Although this may in some cases limit the use of RNAi *in vivo*, it is also likely to provide important new insights into the genetic basis of normal tissue physiology, disease processes, and therapeutic strategies by demonstrating the effects of altering gene expression to varying degrees. In particular, RNAi may prove especially important in the creation of animal models of human diseases in which susceptibility and resistance are encoded by alleles that show relative, rather than absolute, differences in expression levels.

RNAi in Genetic Screening

The ability of RNAi to efficiently suppress target genes in mammals has been heralded as the most powerful reverse-genetics approach in recent times. Most researchers use RNAi as a simple reverse-genetics tool to understand the function of one or a few genes. On the other hand, RNAi represents an ideal strategy to decipher the role of hundreds or thousands of genes simultaneously in screens for specific phenotypic changes. These forward genetic approaches may help to gain insight into complex physiological and pathological processes. However, for RNAi to be widely accepted as a gene-silencing tool in mammals, significant technical challenges still need to be overcome. These challenges include developing efficient ways of designing, identifying and delivering effective siRNAs: specifically; improving the accuracy with which susceptible sites in the target RNA molecules can be identified, and obtaining inducible tissue- and cell-specific regulation of siRNA both *in vitro* and *in vivo*.

RNAi Screening in Lower Organisms

The discovery of RNAi and sequencing of the genome of popular model organisms such as *C. elegans* and *D. melanogaster* provided the impetus for functional genetic screens (Table 3). The first such screens were performed in *C. elegans* and focused on easily detectable phenotypic changes, such as viability and sterility, and identified the biological role of a few hundred genes located on chromosomes I and III. These studies were significantly facilitated by the finding that RNAi can be induced simply by feeding this worm with bacteria overexpressing dsRNA molecules of interest. More recently, genome-wide RNAi screens targeting ~90% of the predicted transcripts have led to the functional annotation of an additional ~2000 genes. Importantly, these and other RNAi-based screens have identified the role of genes that have conserved orthologs in mice and humans and are involved in processes as diverse as genomic stability, fat metabolism, mitochondrial function, and embryogenesis. They have also provided evidence that RNAi pathways are under genetic control in this organism, a discovery that may have significant impact on the use of RNAi both in basic research and in clinical applications.

RNAi can also be induced with ease in cells derived from fruit flies by adding dsRNAs to the culture medium. RNAi-based genetic screens have been performed to systematically explore important signal transduction cascades, such as the Wnt-, Hh-, PI3K- and MAPK-pathways, and more broadly to examine the function of most known kinases and phosphatases. Other high throughput screens identified genes involved in the regulation of cell shape, cytokinesis, and phagocytosis, as well as in heart development and innate immunity. Recently, Boutros and colleagues targeted almost all of the predicted *Drosophila* mRNAs to analyze their possible roles in cell growth and viability.

RNAi Screening in Mammalian Systems

Soon after the basic requirements to create successful RNAi reagents for mammalian systems were established, multiple groups started to assemble si/shRNA libraries that targeted increasingly larger numbers of genes. These have been successfully applied to study a variety of cellular processes. Initial studies examined the function of a relatively small set of genes in specific signaling or metabolic pathways.

Very recently, much larger RNAi libraries attempting genome-wide coverage have been created (Table 3). Generating such libraries and delivering them into mammalian cells has presented real challenges that have been solved in part by using strategies borrowed from technical breakthroughs in the fields of gene sequencing and chemical genetics. Many recent reports document the results of such large-scale RNAi-based genetic screens in mammalian cells using retrovirus-based vectors to express shRNAs (see (Cullen and Arndt, 2005) for a review). A notable detail of the strategy used is the inclusion of short unique DNA sequences or the use of the ~ 60-nt-long shRNA coding sequences as "molecular bar codes" in RNAi vectors. Both strategies have produced encouraging results that these sequences could be detected using high-density oligonucleotide arrays and thus could circumvent the need to identify shRNAs introduced into cells by sequencing. Zheng and colleagues have performed a large-scale RNAi-based screen in tissue culture cells using a PCR-based approach to generate siRNAs in cells (Zheng *et al.*, 2004).

The large-scale genetic screens have successfully identified a number of novel genetic elements of the interrogated signaling or metabolic pathways. On the other hand, certain known components of these pathways were not identified even if the respective targeting constructs were present in the applied RNAi libraries. This discrepancy may be explained by the fact that sometimes not all pathway components can be assessed by the same phenotypic screen. Moreover, despite careful design and use of multiple targeting sequences, some genes may not be effectively silenced to produce the assayed phenotype. However, functional validation of each si/shRNA molecule represents an even bigger challenge than the construction of the library. The recently reported high-throughput methods are based on overexpression of the target gene in fusion with a reporter construct (typically green fluorescent protein; GFP), which require in-frame cloning of the coding sequences of the targeted genes and data handling of hundreds of gigabytes of visual images. Therefore, these methods are laborious and expensive and do not allow for the selection of si/shRNA molecules that target untranslated regions of the gene.

RNAi also promises to provide an efficient approach to systematically investigate the genetic basis of normal and pathological physiology in animal models. Although RNAi screenings performed *in vitro* have identified a number of physiologically or pathologically relevant genes, these models can often be biased by specific genetic background and/or environmental conditions. Furthermore, many complex physiological or disease processes have no *in vitro* correlates, such as organ development or metastasis. These limitations will likely be overcome by performing RNAi screens *in vivo* once appropriate strategies have been developed.

RNAi — A Therapeutic Strategy

The accurate potency and flexibility exhibited by RNAi in experimental systems has stimulated efforts to use RNAi-based reagents in the clinic as "molecular targeting" therapeutics to shut down genes associated with a disease in humans. In principle, therapeutic RNAi has ability to alter the expression and function of genes in a wide range of disease settings. Indeed, RNAi has been used to silence the expression of exogenous disease-causing genes, such as those of pathogens, as well as endogenous genes that play an essential role in the disease process. The high degree of specificity of RNAi has even been able to distinguish between alleles of genes exhibiting spontaneous or inherited polymorphisms and alternative splicing events that underlie the development of cancer and other diseases.

These findings suggest that the expression of disease-associated genes could be inhibited by RNAi-based reagents. These reagents have been shown to be effective as potential therapeutic agents in a variety of tissue culture and animal preclinical model systems, including those for cancerous disorders, microbial infections, autoimmune and inflammatory disease, and neurological disorders.

Different siRNAs have recently been described that effectively silence mutated genes related to cancer such as Ras, Bcr-Abl, vascular endothelial growth factor (VEGF), focal adhesion kinase (FAK), Bcl-2, multiple drug resistance-1 (MDR-1), human papillomavirus (HPV) E6 and E7 proteins, CDK-2, MDM-2, PKC-alpha and -beta; and TGF-beta1. siRNAs alone or in combination with anticancer drugs have been shown to promote apoptosis and reduce tumor burden in mice, demonstrating that RNAi for cancer-related genes is feasible *in vivo* and may prove to be helpful in cancer therapy.

The possibility of using siRNAs to combat infections, especially viral infections, is being extensively explored. Many genes from important human viral pathogens, including HIV, HBV, HCV, influenza virus, and SARS corna virus, have been shown to be targets for RNAi. Inhibiting these viral genes has been shown to interfere with viral replication *in vitro* and in mouse models of viral infection (McCaffrey *et al.*, 2003). Because many viruses, notably HIV, exhibit high mutation rates, RNAi-based therapeutic strategies have also been explored that target host genes that are required for viral entry into cells or that contribute to the pathogenic sequelae of virus infection. Similar approaches have also been shown to be successful at modulating inflammatory gene expression in experimental models of immune-mediated diseases.

Practically, RNAi-based therapeutics can also selectively target any RNA transcript including mutant forms of genes that underlie the development of neurodegenerative disorders. Inhibition by RNAi of genes encoding proteins involved in polyglutamine-induced neurological disorders (spinocerebellar ataxia type 1 and Huntington's disease), Alzheimer's disease, amyotrophic lateral sclerosis, and prion-based diseases has been shown and may represent a promising therapeutic strategy.

RNAi-Based Therapeutics: Delivery Routes and Concerns

The most significant barrier to molecular therapy including for RNAi-based therapy is an efficient and precise delivery of the reagents to the affected organs/cells of patients. Though, a number of strategies have been developed that allow siRNAs and shRNAs to be delivered effectively in animals, hydrodynamic delivery of siRNAs which involves the intravascular injection of large fluid volumes in order to locally increase intravascular pressure can be adapted for local administration of siRNAs by arterial or venous catheterization in organs, such as liver, kidney, heart or lungs, only and cannot be performed for systemic treatment in clinics. In the mouse, effective silencing of genes in tumor tissues has been reported following intravenous, intraperitoneal, and subcutaneous injections of siRNAs, suggesting that effective delivery of RNAi reagents may be achieved by different parenteral routes.

A number of carrier systems and chemical modifications have been explored to enhance the efficiency and specificity of RNAi-based therapeutics. The use of cationic lipids has been shown to significantly enhance gene silencing. More sophisticated strategies, in which receptor-specific monoclonal antibodies or other targeting proteins are incorporated into pegylated immunoliposomes (PILs), have been shown to direct gene silencing in a number of tissues, including the brain in laboratory animals (Fig. 4).

Virus-based RNAi delivery systems have also been shown to achieve effective gene silencing *in vivo*. Systemic or tissue-directed injection of adenoviruses encoding shRNAs has been shown to be effective at inhibiting gene expression in the liver, as well as the central nervous system. Retroviruses for RNAi could potentially be applied for *ex vivo* cellular manipulations, including those of dendritic cells for the modulation of immune responses. However, the use of these vectors may be associated with a risk of insertional mutagenesis and should be carefully evaluated. Other virus-based systems for RNAi, such as those based on AAVs, are also being considered as delivery vehicles for therapeutic RNAi.

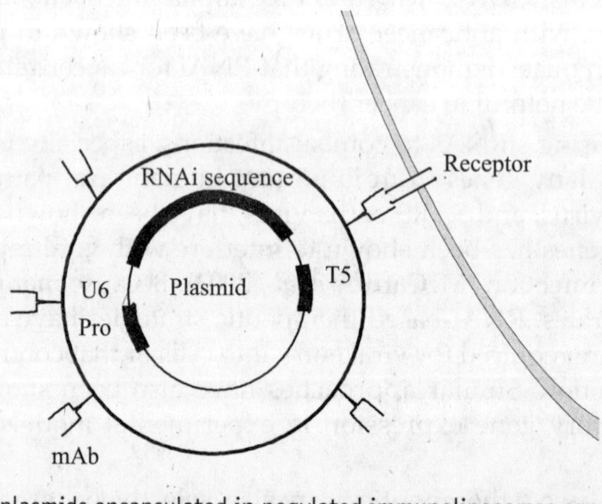

Fig. 4. miRNA expressing plasmids encapsulated in pegylated immunoliposomes can be directed to a specific cell type. Monoclonal antibodies directed against a cell type specific biomolecule incorporated in the liposomes help direct the delivery of plasmids to desired cell types. (Zhang *et al.*, 2002)

SUMMARY

In last decade and a half itself, RNAi has been recognized as an extremely essential biological process which regulates gene expression at multiple levels. It also has emerged as an extremely powerful experimental tool with potential therapeutic values. New discoveries coupled with technological breakthroughs in the field have prompted the creation of effective RNAi reagents with the potential to study normal as well as pathological physiology in a wide range of *in vitro* and *in vivo* settings. However, further significant improvements in the field of understanding of RNAi biology and also in delivery technologies for RNAi reagents are required before these strategies can become more effective with manageable or no undesired side effects and easily available. Though several questions related to the safety and efficacy of using RNAi as a therapeutic strategy have to be addressed before these RNAi reagents could be tried in clinics, RNAi is emerging as a very promising field with unfathomable research and therapeutic application potentials without any doubt. It is anticipated that current and future studies in animal models will help optimize therapeutic applications of RNAi in humans.

REFERENCES

Alvarez-garcia, I. and Miska, E. A. (2005) MicroRNA functions in animal development and human disease. Development 132, 4653-62.

Comment. (2004). Finishing the euchromatic sequence of the human genome. Nature 431, 931-45.

Croce, C. M. and Calin, G. A. (2005) miRNAs, cancer, and stem cell division. Cell 122, 6-7.

Cullen, L. M. and Arndt, G. M. (2005) Genome-wide screening for gene function using RNAi in mammalian cells. Immunol Cell Biol 83, 217-23.

Elbashir, S. M., Lendeckel, W. and Tuschl, T. (2001) RNA interference is mediated by 21- and 22-nucleotide RNAs. Genes Dev 15, 188-200.

Fire, A., Xu, S., Montgomery, M. K., Kostas, S. A., Driver, S. E. and Mello, C. C. (1998). Potent and specific genetic interference by double-stranded RNA in Caenorhabditis elegans. Nature 391, 806-11.

Griffiths-Jones, S. (2004). The microRNA Registry. Nucleic Acids Res 32, D109-11.

Hamilton, A. J. and Baulcombe, D. C. (1999). A species of small antisense RNA in posttranscriptional gene silencing in plants. Science 286, 950-2.

Hariharan, M., Scaria, V., Pillai, B. and Brahmachari, S. K. (2005). Targets for human encoded microRNAs in HIV genes. Biochem Biophys Res Commun 337, 1214-1218.

Ketting, R. F., Haverkamp, T. H., van Luenen, H. G. and Plasterk, R. H. (1999). Mut-7 of C. elegans, required for transposon silencing and RNA interference, is a homolog of Werner syndrome helicase and RNaseD. Cell 99, 133-41.

Lim, L. P., Glasner, M. E., Yekta, S., Burge, C. B. and Bartel, D. P. (2003a). Vertebrate microRNA genes. Science 299, 1540.

Lim, L. P., Lau, N. C., Garrett-Engele, P., Grimson, A., Schelter, J. M., Castle, J., Bartel, D. P., Linsley, P. S. and Johnson, J. M. (2005). Microarray analysis shows that some microRNAs downregulate large numbers of target mRNAs. Nature 433, 769-73.

Lim, L. P., Lau, N. C., Weinstein, E. G., Abdelhakim, A., Yekta, S., Rhoades, M. W., Burge, C. B. and Bartel, D. P. (2003b). The microRNAs of Caenorhabditis elegans. Genes Dev 17, 991-1008.

Matzke, M., Aufsatz, W., Kanno, T., Daxinger, L., Papp, I., Mette, M. F. and Matzke, A. J. (2004). Genetic analysis of RNA-mediated transcriptional gene silencing. Biochim Biophys Acta 1677, 129-41.

McCaffrey, A. P., Nakai, H., Pandey, K., Huang, Z., Salazar, F. H., Xu, H., Wieland, S. F., Marion, P. L. and Kay, M. A. (2003). Inhibition of hepatitis B virus in mice by RNA interference. Nat Biotechnol 21, 639-44.

Napoli, C., Lemieux, C. and Jorgensen, R. (1990). Introduction of a Chimeric Chalcone Synthase Gene into Petunia Results in Reversible Co-Suppression of Homologous Genes in trans. Plant Cell 2, 279-289.

Pal-Bhadra, M., Leibovitch, B. A., Gandhi, S. G., Rao, M., Bhadra, U., Birchler, J. A. and Elgin, S. C. (2004). Heterochromatic silencing and HP1 localization in Drosophila are dependent on the RNAi machinery. Science 303, 669-72.

Poy, M. N., Eliasson, L., Krutzfeldt, J., Kuwajima, S., Ma, X., Macdonald, P. E., Pfeffer, S., Tuschl, T., Rajewsky, N., Rorsman, P. and Stoffel, M. (2004). A pancreatic islet-specific microRNA regulates insulin secretion. Nature 432, 226-30.

Schramke, V. and Allshire, R. (2003). Hairpin RNAs and retrotransposon LTRs effect RNAi and chromatin-based gene silencing. Science 301, 1069-74.

Tabara, H., Sarkissian, M., Kelly, W. G., Fleenor, J., Grishok, A., Timmons, L., Fire, A. and Mello, C. C. (1999). The rde-1 gene, RNA interference, and transposon silencing in C. elegans. Cell 99, 123-32.

Tuschl, T., Zamore, P. D., Lehmann, R., Bartel, D. P. and Sharp, P. A. (1999). Targeted mRNA degradation by double-stranded RNA in vitro. Genes Dev 13, 3191-7.

van der Krol, A. R., Mur, L. A., Beld, M., Mol, J. N. and Stuitje, A. R. (1990). Flavonoid genes in petunia: addition of a limited number of gene copies may lead to a suppression of gene expression. Plant Cell 2, 291-9.

Wassenegger, M., Heimes, S., Riedel, L. and Sanger, H. L. (1994). RNA-directed de novo methylation of genomic sequences in plants. Cell 76, 567-76.

Xie, Z., Johansen, L. K., Gustafson, A. M., Kasschau, K. D., Lellis, A. D., Zilberman, D., Jacobsen, S. E. and Carrington, J. C. (2004). Genetic and functional diversification of small RNA pathways in plants. PLoS Biol 2, E104.

Zamore, P. D., Tuschl, T., Sharp, P. A. and Bartel, D. P. (2000). RNAi: double-stranded RNA directs the ATP-dependent cleavage of mRNA at 21 to 23 nucleotide intervals. Cell 101, 25-33.

Zhang, Y., Jeong Lee, H., Boado, R. J. and Pardridge, W. M. (2002). Receptor-mediated delivery of an antisense gene to human brain cancer cells. J Gene Med 4, 183-94.

Zheng, L., Liu, J., Batalov, S., Zhou, D., Orth, A., Ding, S. and Schultz, P. G. (2004). An approach to genomewide screens of expressed small interfering RNAs in mammalian cells. Proc Natl Acad Sci U S A 101, 135-40.

Suggested Readings

1. Gene Silencing by RNA Interference: Technology and Application, Editor and Author: Muhammad Sohail, CRC press. First edition (July 26, 2004).
2. RNAi: A Guide to Gene Silencing by Gregory J. Hannon (Editor). Cold Spring Harbor Laboratory Press; first edition (September 2003).
3. RNA Interference in Practice: Principles, Basics, and Methods for Gene Silencing in C. elegans, Drosophila, and Mammels by Ute Schepers (Author). Wiley-VCH; First edition (January 31, 2005).
4. RNA Interference Technology: From Basic Science to Drug Development by Andrew Fire, Marshall Nirenberg, Krishnarao Appasani (Editor). Cambridge University Press (January 17, 2005)
5. RNA Interference (RNAi): The Nuts and Bolts of siRNA Technology (Nuts and Bolts Series) by David Engelke (Author). Dna Press; First edition (February 6, 2004).

6. Elbashir S.M., Harborth J., Lendeckel W., et al. Duplexes of 21-nucleotide RNAs mediate RNA interference in cultured mammalian cells. *Nature*. 2001; 411: 494-498.
7. Harborth J., Elbashir S.M., Bechert K., et al. Identification of essential genes in cultured mammalian cells using small interfering RNAs. *J. Cell Sci.* 2001;114(pt 24): 4557-4565.
8. Starr D.A., Hermann G.J., Malone C.J., et al. unc-83 encodes s novel component of the nuclear envelope and is essential for proper nuclear migration. Development. 2001;128: 5039-5050.
9. Xia H., Mao Q. Eliason S.L., et al. RNAi suppresses polyglutamine-induced neurodegeneration in a model of spinocerebellar ataxia. *Nat Med.* 2004;10: 816-820.

REVIEW QUESTIONS

Long Questions

1. What are the four major classes of small RNAs, which may play an active role in RNAi?
2. Describes the mechanism of gene silencing by siRNAs?
3. Give the details of biophysical criteria considered for the designing a siRNA with probable explanations?
4. What is the functional significance of RNAi?
5. Describes about the functional genomics and therapeutic consideration of RNAi?
6. Write down the uses of viral vectors to induce RNAi in primary cells, tissues and experimental animals?

Multiple Choice Questions

1. The siRNAs silence genes by
   (a) promoting the cleavage of complementary mRNAs
   (b) recruiting inhibitory proteins
   (c) directing the modification of complementary DNAs sequences
   (d) All of the above
2. Length of siRNAs is
   (a) 1-12 nucleotides (b) 21-22 nucleotides
   (c) 40-50 nucleotides (d) 5-10 nucleotides
3. miRNAs constitute of
   (a) 19-25 nucleotides (b) 28-32 nucleotides
   (c) 40-50 nucleotides (d) 5-10 nucleotides
4. Dicer is a member of
   (a) RNAse III family (b) RNAse II family
   (c) RNAse I family (d) None of the above
5. Gene silencing by RNA requires
   (a) RNA dependent RNA polymerase
   (b) DNA polymerase

(c) DNA dependent RNA polymerase
(d) None of the above

6. To design the siRNA G&C content in medium should be
(a) 10-20% (b) 30-50%
(c) 5-6% (d) 50-60%

7. Low to medium G&C content between 30-50% facilitates the
(a) Block sense-strand selection by RISC
(b) Promotes antisense-strand selection by RISC
(c) Interaction with RISC and unwinding
(d) all of the above

8. To design the siRNA presence of an A at position 3 and 19 of sense strand
(a) Promotes antisense-strand selection by RISC
(b) Block sense-strand selection by RISC
(c) Interaction with RISC and unwinding
(d) None of the above

9. Total number of miRNAs have been recognized in human
(a) 330 miRNAs (b) 440 miRNAs
(c) 100 miRNAs (d) 200 miRNAs

10. To enhance the gene silencing by RNAi the efficient carrier system is
(a) Carbohydrates (b) Anionic lipids
(c) Proteins (d) Cationic lipids

11. When miRNAs are encoded in the genome in which step it not participate
(a) Protein synthesis (b) Gene expression
(c) Translation (d) Transcription

12. Which factors have been affected by miRNAs
(a) expression of proteins (b) expression of number of genes
(c) transcription factors (d) none of the above

Answers

1 (d)	2 (b)	3 (a)	4 (a)	5 (a)	6 (b)
7 (c)	8 (a)	9 (a)	10 (d)	11 (a)	12 (b)

PART–11

Carcinogensis, Disorders and Detoxification

CHAPTER 34

Molecular and Biochemical Aspects of Cancer, Oncogenes and Chemical Carcinogenesis

S. Tripathi[1], U. Ghoshal[1] and U.C. Ghoshal[2*]

[1]Dept. of Gastroenterology and [2]Dept. of Microbiology, Sanjay Gandhi Postgraduate Institute of Medical Sciences, Lucknow–226 014

E-mail: ghoshal@sgpgi.ac.in

WHAT IS CANCER?

Cancer cells stop responding to signals that control normal cell growth and differentiation leading to uncontrolled proliferation. These cells invade normal tissues and subsequently spread throughout the body. This property of cancer cells is called as metastasis and it accounts for lethality of cancers because it disrupts the tissue and organ functions needed for the survival of an organism as a whole. Thus, the term cancer refers to any malignant growth. The word cancer is derived from Latin word meaning "crab" as the general outline of many malignancies resembles a crab with body being the main tumor mass and the claws being the invasive tumor margins. Tumors are defined as "masses of tissue whose growth exceeds and is uncoordinated with that of normal tissues" (Willis, 1948). In Latin the word tumor literally means swelling. Thus, tumor, which can be benign or malignant, is a swelling caused by excessive and continued growth of cells in a tissue. The term neoplasm means a new and diseased form of tissue growth. In clinical practice the words neoplasm and tumors are used interchangeably. The branch of science, which deals with cancers, is called as oncology (Greek: ogkos – a swelling; logos – science). Cancer occurs in numerous tissues and result from multiple causes. Therefore, cancer exhibits enormous diversity and heterogeneity fuelled by various combinations of genetic and epigenetic alterations. It is represented by more than 100 forms of the disease. In fact, subtypes of cancer can be found even within a specific organ.

*Corresponding author

However, it is a well accepted fact that most of the human cancers share similar cellular, molecular and biochemical characteristics, which will be discussed later in this chapter. How cancer develops is no longer a mystery. The extensive research done in the field of cancer in the last decades has generated extensive knowledge in this disease. But still there are many aspects, which need to be explored in order to develop a complete understanding of this complex disease.

Cancer: A Genetic Disease

Cancer is a genetic disease. It can be called as "disease of genes" because all the characteristics of cancer tissue are result of inappropriate or altered expression of various genes, which regulate the cell phenotype especially, cell growth and differentiation. Malignant transformation of a normal cell involves series of dynamic changes in its genome. There are sufficient evidences, which show that cancer is a genetic disease. However, there are two key differences between cancer and other genetic diseases. First, cancer is generally caused by somatic mutations whereas other genetic diseases arise as a result of germ-line mutations with the exception of a minority (~5%) of cancers are caused by germ-line mutations. Secondly, cancers originate not from a single mutation but from several sequential mutations whereas other genetic diseases are usually associated with a single specific mutation.

Multistep Nature of Cancer Development and Heterogeneity of Cancers

With a few exceptions, cancers are monoclonal in origin that is, cancers are derived from a single somatic cell that has undergone mutation and all cells of a tumor possess the genetic characteristics of the original transformed cell. However, the clonal origin of tumors does not mean that the original cell that gives rise to tumor has initially acquired all the characteristics of cancer cells. This cell though altered at gene level is phenotypically normal. This genetic abnormality of the cell provides some additional growth advantage leading to its abnormal proliferation. This marks the first step of carcinogenesis ("tumor initiation"). This cell divides to produce a genetically homogeneous clone (clonal expansion). Tumor initiation is followed by "tumor progression" during which additional mutations occur in cells, which further increase growth potential of the cells. This results in formation of sub clones within the tumor, each with different characteristics. Therefore, most tumors are heterogeneous. Some of these mutations confer a selective advantage to the cells over other cells such as increased growth rate, survival and invasion allowing a part of clones of cells to overgrow its neighbor. This is called as "clonal selection". The process of tumorigenesis involves various phases of clonal expansion and clonal selection. Thus, tumors acquire more aggressive characteristics as they develop, invade neighboring tissue and finally metastasize. Most importantly, cancers do not involve a single mutation but several sequential mutations such that each event of carcinogenesis is determined by activation, mutation or loss of specific genes (Fig. 1). This concept is called as multi-hit or multistage theory of carcinogenesis.

As described earlier, though cancer development is a multistep process, each of the steps involved in the process may not be clearly defined in every cancer. However, colorectal cancer is a classic example of tumor progression in which each step is characterized by specific genetic events (Fearon *et al.* 1990) (Fig. 2). Development of colorectal cancer involves increased proliferation of colon epithelial cells. One of the proliferative cells gives rise to a small benign neoplasm (an adenoma or poly p). Further phases of clonal expansion and selection lead to

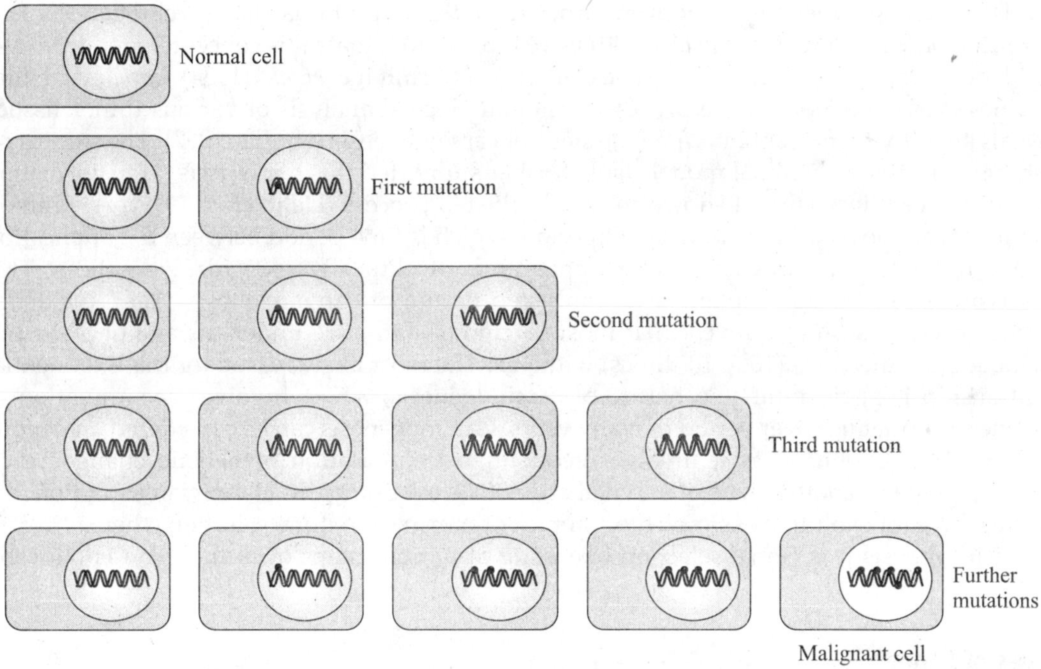

Fig. 1. Accumulation of mutations during carcinogenesis.

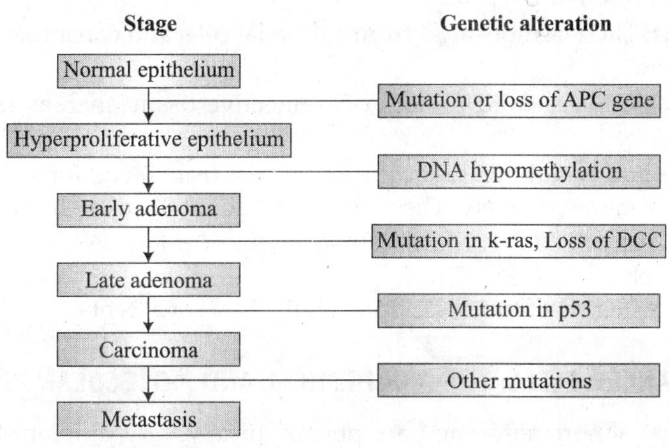

Fig. 2. A model for colorectal tumorigeneis.

increase in size and growth potential of adenomas. Benign adenomas then invade neighboring connective tissue through basal lamina, that is, acquire malignant characteristics. The cancerous cells continue to proliferate and penetrate the wall of colon and invade other abdominal organs. These may even invade blood and lymphatic vessels and thus metastasize throughout the body.

There are sufficient evidences in support of the hypothesis that carcinogenesis is a multistep process. Berenblum *et al.* (1949) were the first to initiate this concept. Later, from age incidence curves of 17 common types of cancer Armitage *et al.* (1954) concluded that carcinogenesis is a 6 or 7 stage process. Histopathological analysis of various tumor tissues reveals areas that represent intermediate steps of carcinogenesis (Foulds, 1957). The studies on various cell lines and animal models have further supported this hypothesis. Transformation of cultured cell lines is itself shown to be a multistep process (Hahn *et al.* 1999). A common feature of carcinogenesis is a long latent period, which is time period between the application of a carcinogenic stimulus and first appearance of clinically recognizable cancer. This observation can be best explained by multistep nature of carcinogenesis. Age/ incidence curves of various cancers reveal that most carcinomas are rare under the age of 30-yr but incidence rate increases (10^3 – 10^4 times) with age. The most likely reason for this exponential relationship is again multistep nature of carcinogenesis, which involves accumulation of genetic abnormalities over period of many years. Carcinogenesis can be considered analogous to Darwinian evolution as it involves sequential accumulation of genetic changes each conferring one or another type of survival advantage over other. In Darwinian evolution, the driving force for evolution of simpler to complex organisms is natural selection whereas in case of carcinogenesis it is clonal selection operating at various steps of tumorigenesis (Nowell, 1976).

Types of Cancer

On the basis of the type of tissue from which tumors originate (histogenesis), malignant tumors can be classified into following types:

1. **Carcinomas:** These tumors arise from epithelial cells and constitute 90% of all human malignancies.
2. **Sarcomas:** These are solid tumors of connective tissue such as muscle, bone, and cartilage.
3. **Leukemias and lymphomas:** These tumors arise from blood forming cells and cells of immune system respectively. These constitute about 8% of all human malignancies.

Tumors are also classified according to their organ of origin for example, breast cancer, lung cancer, stomach cancer etc. Grading of tumors is done on the basis of extent of differentiation while staging of tumors is done on the basis of extent of local invasion.

PROPERTIES OF CANCER CELLS: THE BIOCHEMICAL AND MOLECULAR ASPECTS

Cancers cells display abnormalities and are phenotypically very different from normal cells (Table 1). Recently, Hanahan *et al.* (2000) have identified six "hallmarks features" of cancer cell phenotype. According to them these alterations in cell phenotype are acquired during conversion of a normal cell to a malignant one and are essential for successful establishment of cancer in the host, surpassing all anticancer defense mechanisms. These "hallmarks of cancer" are not associated with any specific type of cancer but are shared by all types of human cancers. These "hallmarks of cancer" are shown in figure 3.

Table 1. Differences between normal cell and cancer cell

No.	Normal cell	Cancer cell
1.	Display density dependent inhibition of cell proliferation in culture	Insensitive to density dependent inhibition of cell proliferation and proliferate in an uncontrolled manner to high cell densities
2	Proliferate only in presence of mitogenic growth signals	Exhibit extremely reduced dependence on exogenous growth stimulation and autocrine growth stimulation
3.	Stringently regulated by cell- cell and cell matrix interactions	Less stringently regulated by cell- cell and cell matrix interactions due to reduced cell adhesion molecules
4.	Movement and proliferation inhibited by cell- cell contact in culture (contact inhibition) forming an orderly array of cells	Insensitive to such contact inhibition forming disordered multilayered pattern
5.	Do not invade adjacent tissues	Secrete certain proteases that digest extra cellular matrix components and allows tumor cells to invade normal tissues
6.	Strictly regulated angiogenesis (neovascularization)	Secrete growth signals that promote angiogenesis in order to cope with extra requirement of O_2 and nutrients.
7.	Differentiate normally	Differentiation is usually blocked at an early stage
8.	Undergo programmed cell death or apoptosis	Increased life span because of failure to undergo apoptosis

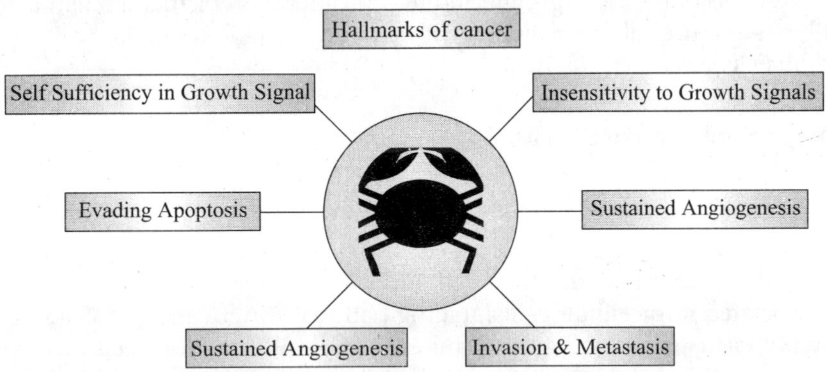

Fig. 3. The six "hallmark features" of cancer as proposed by Hanahan *et al.* (2000).

I. Self Sufficiency in Growth Signals

Normal cells proliferate only in the presence of certain growth signals or growth factors (GF). These growth factors bind to and activate specific receptors. In contrast, cancer cells have significantly reduced requirements for extra cellular GF. When grown in culture, normal cells exhibit density dependent inhibition of cell proliferation, that is, they proliferate only until a finite cell density is attained. This final density is proportional to the concentration of available

GF in the culture medium. However, cancer cells are not sensitive to such density dependent inhibition and proliferate up to high cell densities irrespective of concentration of GF. In some cases, cancer cells themselves secrete GF, which stimulate their own proliferation. This is called as autocrine growth stimulation. Example: Production of platelet derived growth factor (PDGF) and tumor growth factor (TGFα). At the molecular level, cancer cells exhibit self-sufficiency in growth signals by alterations in the growth signaling pathways. These alterations may include any of the followings:

(a) Alteration in extra cellular GF: GF can be either immobilized extra cellular components and cell-to-cell adhesion molecules (CAMs) or diffusible molecules that bind to growth factor receptors expressed on cell surface.

(b) Alteration in transmembrane receptors that bind growth signals and transmit to the cell: GF receptors are usually over expressed in cancers so that cancer cells are stimulated by GF even at concentrations, which are insufficient for stimulating proliferation of normal cells. For example, over expression of epidermal growth factor receptor in stomach, brain and breast cancers (Yarden Y. *et al.*, 1998). In some cases over expression of growth factor receptors can even lead to GF independent stimulation. GF independent stimulation can also be caused by abnormalities in structure of such receptors.

(c) Alteration in intracellular signaling systems that convert these signals into action: The most important intracellular signaling pathway that transmits signals from growth factor activated receptors and controls various cellular processes is SOS-Ras-Raf-MAPK pathway (mitogen activated protein kinases). *Ras* (Rous Sarcoma) proteins were first identified as oncogenic proteins of tumor viruses that cause sarcomas of rats. Later, mutations in Ras encoding gene (ras gene) were found to be associated with various tumors. Experiments have shown that microinjection of Ras protein induces proliferation of mammalian cells. The Ras protein exhibits structural abnormalities in tumors and gets stimulated even in absence of external stimuli (Medema *et al.* 1993).

2. Insensitivity to Antigrowth Signals

Antigrowth signals maintain tissue homeostasis by allowing cell proliferation only when required. These signals can be soluble growth inhibitors, components of extra cellular matrix (ECM) or components of surface of adjacent cells. Like growth promoting signals, antigrowth signals also bind to and thus stimulate cell surface receptors, which in turn transmit the stimulus to associated intracellular cell signaling pathways. Activation of these cell signaling pathways finally elicits such biological changes in the cell that block cell proliferation. The signaling pathways stimulated by antigrowth signals are usually associated with the cell cycle. We all know, the cell cycle includes various phases that finally lead to DNA duplication and cell division. In eukaryotes, the cell cycle consists of four distinct phases as follows:

(a) **M Phase:** During this phase division of nucleus or mitosis occurs and it is followed by cytokinesis.

(b) **G1 Phase (Gap 1):** The phase between mitosis and initiation of DNA replication is called as G1 phase.

(c) **S phase:** It is the synthesis phase during which DNA replication occurs.

(d) **G2 Phase (Gap 2):** During this phase proteins are synthesized in preparation for mitosis.

In some cases such as skin fibroblasts cells, cells of organs like liver, kidney or lung divide only occasionally in presence of appropriate extra cellular signals. These cells are metabolically active but do not proliferate unless stimulated. These cells exit G1 and enter a quiescent stage of the cell cycle called as G0 phase. The progression of cell from one phase to other phase of cell cycle is strictly regulated by various mechanisms. These regulatory mechanisms not only control progression of cell cycle but also link the cell cycle with extra cellular signals that control cell proliferation. In animal cells there are following two check points in cell cycle control, through which extra cellular signals can control cell proliferation:

(a) **G1 Checkpoint (Restriction point):** At this checkpoint G1 to S phase progression is controlled.

(b) **G2 checkpoint:** At this checkpoint G2 to M phase progression is controlled.

G1 checkpoint plays a pivotal role in controlling cell cycle progression. The antigrowth factors associated cell signaling pathways usually control the progression of cell cycle from G1 to S phase, that is, they act at G1 check point. Retinoblastoma protein (Rb) plays the pivotal role in transmission of antigrowth signal and it acts at G1 checkpoint. Rb was first identified as a protein coded by a gene causing retinoblastoma (a rare childhood cancer). However, subsequent studies revealed that mutations in Rb gene resulting in loss of functional Rb protein are not only associated with retinoblastoma but also with many other human cancers. In unphosphorylated state Rb protein binds to E2F family of transcription factors. The E2F transcription factors are responsible for expression of genes required for progression from G1 to S phase including genes encoding cyclin E. Thus unphosphorylated Rb protein acts as repressor of E2F and blocks progression of G1 to S phase which in turn blocks cell proliferation. When phosphorylated by cyclin:CDK complexes it does not bind E2F. E2F is released and transcription of genes responsible for G1 to S transition is resumed. Any disruption of Rb pathway thus causes failure of regulation of G1 to S transition by antigrowth factors, which leads to uncontrolled proliferation of cells. The best reported antigrowth factor till date is TGF β, which prevents phosphorylation of Rb by many ways, most still elusive. In some cases TGF β causes synthesis of $p15^{INK4B}$ and p21 proteins, which block the cyclin:CDK complexes responsible for Rb phosphorylation. Rb pathway can be disrupted by alteration in any of the associated proteins such as, TGF- β, TGF- βR, Smad, P15, CDK, Cyclin, or Rb resulting in insensitivity to antigrowth factors and uncontrolled cell proliferation. In fact, the cell signaling pathway associated with Rb is found to be disrupted in majority of human cancers (Fynan *et al.* 1993; Markowitz *et al.* 1995; Chin *et al.* 1998).

3. Evading Apoptosis

Apoptosis or programmed cell death plays an important role in maintaining tissue homeostasis by maintaining a balance between cell proliferation and cell death. Apoptosis involves a series of genetically controlled events that leads to death of aging, damaged or other unwanted cells. Apoptotic cells exhibit cytoplasmic blebbing, chromatin condensation, cell shrinkage, nuclear fragmentation. Finally cell is fragmented into membrane bound apoptotic bodies that are ingested by phagocytes. Development of tumor is not only determined by uncontrolled proliferation of cells but also by rate of cell death. Apoptosis is mainly responsible for cell death and evidences suggest that cancer cells are resistant to apoptosis.

There are two pathways that lead to apoptosis as follows:

(a) Extrinsic or receptor linked pathway: This pathway is triggered by binding of a ligand (death or survival factor) to cell surface receptors. e.g. - IGF-1, IGF-2 binds to IGF1R, IL-3 binds to IL-3R, FAS binds to FAS receptor, TNF α binds to TNF-R1. The ligand receptor binding triggers downstream molecules that finally end in activation of "effectors of apoptosis" called caspases (cysteine aspartyl proteases). The caspase activated in this case is caspase 8 (one of the two "gatekeeper' caspases"). Caspases are proteases that act as effectors of apoptosis. They execute the process of cell death by destruction of subcellular organelles and genome.

(b) Intrinsic apoptotic pathway or mitochondria mediated pathway: Most of the apoptotic signals (DNA damage, hypoxia, oncogene activation, cell—extra cellular matrix (ECM) or cell—cell interaction abnormalities) induce apoptosis by mitochondria mediated pathway. Mitochondria in response to such signals release cytochrome c which acts as "catalysts of apoptosis". Bcl-2 family of proteins play important role in cytochrome c release by mitochondria. Bax, a proapoptotic bcl-2 family of proteins oligomerizes and forms pores in mitochondrial membrane. This results in either decrease in the inner mitochondrial transmembrane potential or opening of the voltage-dependent anion channel releasing cytochrome c from the space between the inner and outer mitochondrial membranes. p53 may act as stimulator of Bax expression and thus increase apoptosis in response to DNA damage. Cytochrome c induces formation of apoptosome, which consists of apoptotic protease activating factor - (Apaf-1), caspase-9 and either ATP or dATP. Caspase-9 acts as second "gatekeeper caspase" by stimulating many downstream caspases.

The most important molecular strategy that cancer cells adapt for evasion of apoptosis is inactivation of p53 protein. More than 50% of human cancers are associated with inactive p53 protein. The p13kinase-AKT/PKB pathway, which acts as an anti-apoptotic pathway is also considered to play an important role in evasion of apoptosis by cancer cells. AKT prevents cytochrome c release, inactivates caspase-9 and prevent transcriptional activation of cell death promoting genes by phosphorylation of transcription factors. Loss of pTEN is also associated with various cancers. pTEN acts as a tumor suppressor gene and inhibits AKT pathway. Reduced expression of IGF-1 (insulin like growth factor) and IGF-2, loss of Apaf-1, over expression of bcl-2 and bax are also important mechanisms that lead to reduced apoptosis in cancer cells. MYC along with MAX protein exhibits dual role in cell. Beside growth promoting tendencies, it can also induce apoptosis under conditions of stress, genotoxic damage or reduced growth factors. Myc stimulates release of cytochrome c and activated capase-9. Thus, over expression of myc in cancer cells also plays important role in controlling apoptosis.

4. Limitless Replicative Potential

Self-sufficiency in growth signals, insensitivity to antigrowth signals and evasion of apoptosis makes a tumor cell deaf to regulatory mechanisms of cell growth. However this is not sufficient for tumor expansion. For a tumor to attain a size that is life threatening the tumor cells should have limitless replicative potential. Normal cells in culture have finite replicative potential. After certain number of divisions, the cells stop dividing. Thus, there is a limit (Hayflick limit) to the number of cell divisions, which they can undergo. This phenomenon is called as cell senescence. Cells that have acquired the ability to multiply without a defined limit are said to be immortalized. Most of the tumor cells exhibit this immortalization trait, that is, they can divide much more times than a normal healthy cell.

Tumor cells acquire the limitless replicative potential by a phenomenon called as Telomere maintenance. Telomeres are ends of chromosome that contain several thousand repeats of short 6 base pairs sequence. Telomeres are essential for stability of chromosomes. During replication DNA polymerase fails to completely replicate the 3′ end of DNA thus during each cell cycle 50-100bp of DNA is lost; thus, successive cell cycle leads to progressive erosion of telomeres. Loss of telomeres is followed by end-to-end chromosomal fusion and finally death of effective cells. In germ cells and few specialized cells telomere length is maintained by a specialized reverse transcriptase, called telomerase, which adds 6bp repeats to the ends of chromosomes using a tightly associated RNA template. (McEacher *et al.* 2000). However, telomerase activity is absent or very low in most of the normal human cells (Shay *et al.* 1997). In tumor cells telomeres are maintained at the length which is more them that of normal cells. Thus tumor cell can replicate unlimited number of times. In most tumors (80-90%) telomere maintenance is achieved by over expression of telomerase enzyme (McEacher *et al* 2000).

5. Sustained Angiogenesis

The process of formation of new blood vessels is called as angiogenesis or neovascularization. During tissue growth, it plays an important role in supplying oxygen and nutrients to the growing cells and also removes catabolites from the cells. The process of angiogenesis is strictly controlled by various stimulating and inhibitory signals. These signals can be both soluble and immobilized factors. Soluble factors that promote angiogenesis include vascular endothelial growth factor (VEGF) and fibroblast growth factor (FGF1/2), which binds to their respective receptors expressed on endothelial cells. The immobilized factors include integrins, which connect the cells to ECM and cell-to-cell adhesion molecules. Thrombospondin-1, which binds to CD 36, a trans membrane receptor present on endothelial cells is an important inhibitor of angiogenesis. Endothelial cells also release GF such as basic fibroblast growth factor (bFGF), granulocyte colony stimulating factors, PDGF, Insulin like growth factor (IGF-1) and heparin binding epidermal growth factor like growth factor. Theses growth factors act as paracrine stimulators of cells and also maintain endothelial activation.

Various experimental evidences point toward essential role of angiogenesis in rapid clonal expansion required for tumor growth. In the absence of angiogenesis tumor growth cannot proceed beyond a size of few millimeters. Tumor cells exhibit the tendency to induce angiogenesis by stimulating angiogenesis promoters and inhibiting the inhibitors. A common strategy used to achieve this goal is to alter gene transcription leading to altered expression of promoters and inhibitors. Tumors are usually associated with increased expression of VEGF and FGF or decrease expression of thrombospondin-1. p53 protein acts as activator of thrombospondin-1; thus, loss of p53 in tumors is also associated with decrease in thrombosondin-1 and hence increased angiogenesis. Secondly there are certain protease enzyme, which controls the bioavailability of angiogenic activator and inhibitors. Alteration of these proteases is an important factor in increased angiogenesis and tumors.

6. Tumor Invasion and Metastasis

As discussed earlier cancer cells possess a unique property of invading adjacent healthy tissues and colonization at distant sites by circulating through blood and lymphatic vessels. This is called as metastasis and it is the most life threatening property of cancers resulting in 90% of all cancer deaths. Invasion and metastasis are very complex processes and their molecular and biochemical aspects are still elusive. Both processes are closely related and are caused due to changes in interaction of cell to its surrounding tissues and activation of extra cellular proteases. The extra cellular matrix (ECM) acts as a barrier for metastasis and invasion of tumor cells. ECM is a dense network of collagen and elastin proteins embedded in stroma composed of glycoproteins and proteoglycans. During metastatic tumor invasion the tumor cells have to overcome this barrier. A three-step hypothesis has been proposed (Liotta LA.1986) to describe the sequence of events involved in the invasion of ECM by cancer cells. This hypothesis is summarized in the Fig. 4.

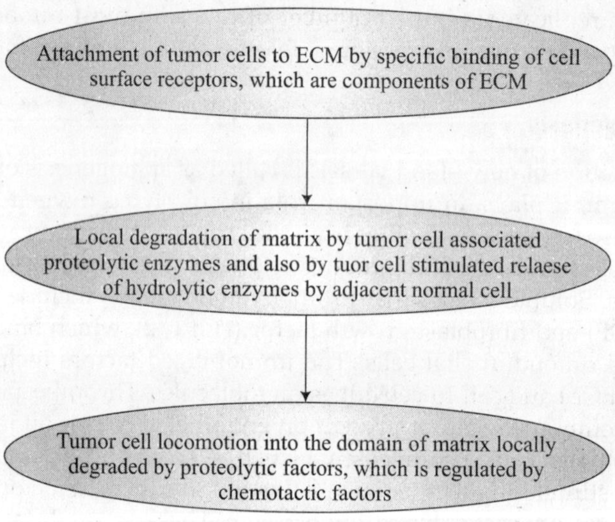

Fig. 4. Hypothetical sequence of events involved in invasion of ECM by cancer cells

Normal cells interact with adjacent cells and ECM by specific proteins. Cell-cell adhesion molecules (CAMs) such as cadherin family of proteins and immunoglobulin family of CAMs act as mediators of cell-cell interactions. Integrins act as mediators of cell to ECM interaction. E-cadherin interacts with cytoplasmic protein β-catenin and transmits antigrowth signals to intracellular signaling pathway involving Lef/Tcf transcription factors. Cancers exhibit loss of functional E-cadherin protein by various mechanisms such as mutations in E-cadherin gene or by repression of transcription or proteolysis of extra cellular cadherin domain. In some cases β-catenin gene is inactivated by mutation. N-CAM, an immunoglobulin family of CAM, is expressed in low levels in cancers. In some cancers such as Wilm's tumor, neuroblastoma and small cell lung cancer, a poorly adhesive form of N-CAM is expressed, instead of normal highly adhesive form. Interaction of α and β subunits of integrins with ECM defines the "area code" of a cell in its tissue microenvironment. Cancer cells need to evade this coding system in order to invade or metastasize in environments to which they do not belong. This is achieved by changes in expression of integrins. Cancer cells usually express integrins such as α 3β1 and αVβ3, which bind to, degraded ECM produced by extra cellular proteases, instead of integrins, which bind to normal epithelium. Matrix-degrading extra cellular proteases genes are up regulated in cancers, which enhances their capability of invading surrounding tissues.

GENETIC AND EPIGENETIC ALTERATIONS IN CANCER

Transformation of a normal cell to cancerous cell progresses through a series of genetic and epigenetic changes. Similar types of cancers may be associated with different combinations of genetic and epigenetic changes. This is the basis of predicting clinical nature of tumors by "molecular profiling" (Liotta *et al.* 2000).

Genetic Alterations in Cancers

Genomic instability is described as a characteristic feature of cancers. It is caused by DNA damage or genetic alterations. Evidences suggest that genetic instability occurs at two levels-nucleotides and chromosomes (Lengauer *et al.* 1998). At the nucleotide level genetic instability involves subtle sequence changes such as base substitutions, deletions or insertions of few nucleotides while at chromosome level it involves gross chromosomal abnormalities. The genetic alteration can range from point mutations to chromosomal abnormalities. The genetic alterations found in tumors are of following four types (Lengauer *et al.* 1998):

1. **Subtle sequence changes:** These include base substitutions, deletions, or insertions of few nucleotides. These types of instabilities are usually rare in cancers.
2. **Alterations in chromosome number (Aneuploidy):** It includes loss or gain of whole chromosomes. Such genetic alterations are associated with almost all types of human cancers. In colon, breast, pancreas and prostate cancers about 25% of alleles are lost (Lengauer *et al.* 1998). The molecular basis of chromosomal instability is still being explored. However, evidences point towards following factors as cause of chromosomal instability:
   (a) Disruption of genes encoding spindle checkpoint: Spindle checkpoint proteins ensure that the chromatid separation occurs only after chromosomes are aligned properly along the mitotic spindle during cell division.
   (b) Failure of DNA damage checkpoints: DNA damage checkpoints prevent replication of damaged DNA during cell cycle.
   (c) Abnormal centrosomes: Multipolar spindles and abnormal number of centrosomes are observed frequently in cancers and may be an important cause of aneuploidy.
3. **Chromosomal translocations:** Transfer of genetic material between two or more chromosomes resulting in net gain and loss of genetic material is also associated with cancers. Translocations may also lead to production of a novel fusion gene usually not seen in nature.
4. **Gene amplification:** Presence of multiple copies of a gene or gene amplification also plays an important role in carcinogenesis. Genomic segment that is amplified is called as amplicon. It usually occurs late in tumorogenesis and affects single or few genes only. Amplification of genes involved in metabolism or inactivation of drugs is an important mechanism of drug resistance in cancer cells (Brodeur *et al.* 1998).

Epigenetic Alterations in Cancers

Epigenetic alterations are those that donot change the coding sequence of the DNA. DNA methylation is the most important epigenetic alteration. In humans, DNA methylation involves the covalent addition of a methyl group (CH3) to the 5′ position of cytosine that precedes a guanosine in the DNA sequence (the CpG dinucleotide). These nucleotides are usually clustered in small stretches of DNA called "CpG islands". "CpG islands" are often associated with the promoter regions of genes. The methylation of CpG islands in the promoter region silences gene expression. Thus, DNA methylation regulates transcription and hence plays important role in maintaining cellular function. Therefore, changes in methylation patterns are important contributors of carcinogenesis. Aberrant DNA methylation has been documented in cancers for more than 20 years (Brown *et al.* 2002) and is found to be frequently associated with cancers. Epigenetic silencing of genes is associated with the inactivation of almost all the pathways involved in carcinogenesis, such as DNA repair (hMLH1, BRCA1, MGMT), cell cycle regulation (p16, p14, p15), apoptosis (DAPK, APAF-1), carcinogen metabolism (Glutathione-S-transferase P1), hormonal response (RARb2), and cell adherence (CDH1, CDH3). Despite the well-documented presence of abnormal DNA methylation in cancer, the mechanism of these events is incompletely understood.

Epigenetic alteration is also an important mechanism by which dietary components can regulate gene expression. Various studies suggest that the cancer-protective effects of many dietary components are associated with DNA methylation patterns. Dietary factors that are involved in one-carbon metabolism provide the most compelling data for the interaction of nutrients and DNA methylation because they influence the supply of methyl groups, and therefore the biochemical pathways of methylation processes. These nutrients include folate, vitamin B12, vitamin B6, methionine, and choline (Davis *et al.* 2004).

MOLECULAR BIOLOGY OF CANCERS

The progression of a normal cell to a tumor cell occurs via accumulation of a series of genetic and epigenetic events. Similar types of cancers can have different combinations of these events. The genetic events involved in malignant transformation usually involve alterations in following three categories of genes:

1. Oncogenes
2. Tumor suppressor genes
3. DNA repair genes.

If we compare a dividing cell to a running vehicle, then oncogene is analogous to its accelerator, tumor suppressor gene is analogous to brake and DNA repair genes are analogous to mechanic which senses and repairs defects in the cell.

I. Oncogenes

Proto-oncogenes are genes, which code for various cytoplasmic and nuclear proteins, which act as positive regulators of cell growth and differentiation. These become permanently activated or "turned on" by various mechanisms, such that the proteins encoded by these genes become overactive and the affected cells proliferate excessively. The activated counterparts of proto-oncogenes are called as oncogenes. Oncogenes are thus cancer-promoting genes and activation of oncogenes represents a "gain-of-function" genetic event. Oncogenes are dominant in nature. Alteration of one allele is sufficient to drive the malignant transformation. Oncogenes work in collaboration with other carcinogens. Oncogenes, by themselves cannot produce cancers

More than 100 oncogenes are recognized till date, and undoubtedly more will be discovered in the future. Oncogenes function at different cellular levels (Fig. 5).

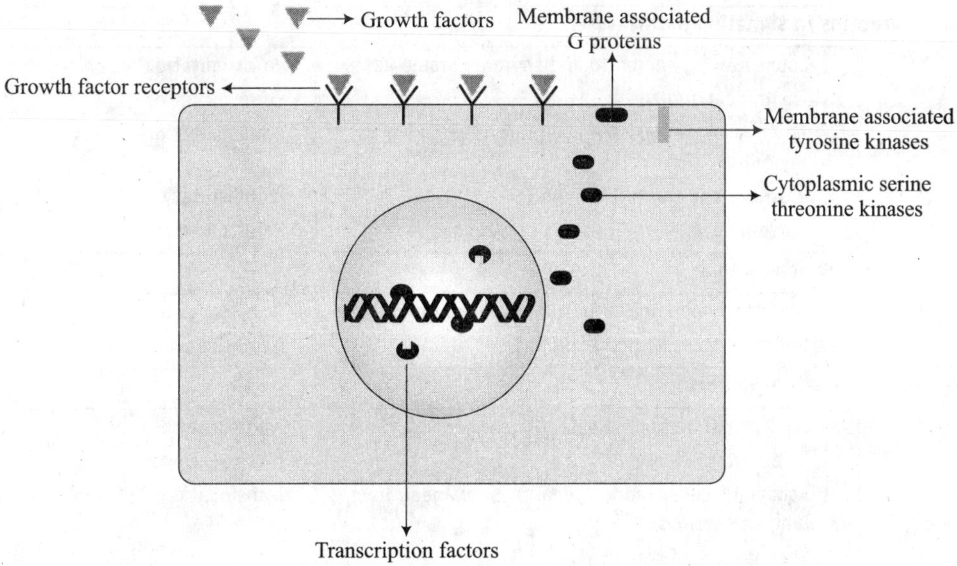

Fig. 5. Different cellular levels of action of oncogenic proteins.

Accordingly, scientists have divided oncogenes into different classes as described below (Table 2)

- Growth factors
- Membrane and cytoplasmic growth factor receptors
- Signal transducers: Membrane associated guanine nucleotide binding proteins, Membrane associated/ cytoplasmic protein tyrosine kinases, cytoplasmic serine-threonine kinases
- Transcription factors
- Nuclear proteins

  Cell division stimulants (Cyclins)

  Tumor suppressor antagonists

  Apoptosis antagonists

Table 2. Representative oncogenes of human cancer

Oncogene	Type of cancer	Mechanism of activation
Genes for growth factors and their receptors		
PDGF	Glioma (Brain cancer)	
PDGFR	Chronic myelomonocytic leukemia	Translocation
erb-B2	Breast and ovarian carcinomas	Amplification
ret	Multiple endocrine neoplasia, thyroid carcinoma	Point mutation, DNA rearrangement
Genes for proteins in signaling pathways		
rasK	Colon, lung, pancreatic and thyroid carcinomas	Point mutation
rasH	Thyroid carcinoma	Point mutation
rasN	Acute myelogenous and lymphocytes leukemia's, thyroid carcinoma	Point mutation
gsp	Pituitary and thyroid tumors	Point mutation
gip	Adrenal cortical and ovarian carcinomas	Point mutation
Genes for transcription factors		
c-myc	Burkitt's lymphoma, Breast cancer, lung cancer	Translocation, amplification
Nmyc	Neuroblastoma, lung carcinoma	Amplification
Genes for other kinds of molecules		
Bcl-2	Follicular B-cell lymphoma	Translocation
D1	Parathyroid adenoma	Translocation
D1	Squamous cell, bladder, breast, esophageal, liver, lung carcinomas	Translocation
MDM2	Wide variety of sarcomas	

Mechanisms of Oncogene Activation

A proto-oncogene can be activated to an oncogene by various mechanisms (Fig. 6). However, it should be noted that no mechanism is specific for any oncogene. Thus one oncogene can be activated by different mechanisms in different cancers or even same type of cancer. These mechanisms can be broadly classified into following categories:

(a) Structural alterations: Mutations in oncogenes can lead to production of an abnormal protein. Point mutations have been described in several oncogenes. Balanced translocations involve transfer of genetic material between two or more chromosomes. These may be very complex involving large-scale transfer of genes between chromosomes but without gains or losses. In an unbalanced translocation, there is both a qualitative and a quantitative defect or defects. There is transfer of genetic material but there is also a net gain or loss of genetic material. Translocations may lead to production of a novel fusion gene not seen in nature, or they may lead to a functional gene coming under the influence of an inappropriate control gene. An example of chromosomal translocation in an oncogene leading to production of a fusion protein is seen in chronic myeloid leukaemia. In chronic myeloid leukaemia, the ABL gene (Abelson oncogene) on chromosome 9 is fused with the gene BCR (Breakpoint Cluster Region) on chromosome 22. This produces a tyrosine kinase which is constantly switched on

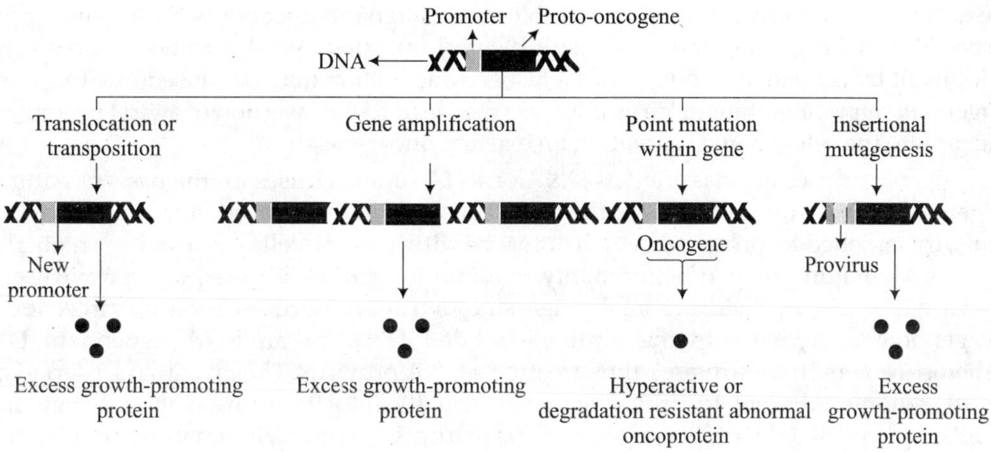

Fig. 6. Mechanisms of oncogene activation.

and which drives cell division and other aspects of malignant transformation. In some cases an oncogenic mutation can lead to deletions in the resulting protein.

(b) Amplification of proto-oncogene: Amplification of proto-oncogene means presence of multiple copies of the same gene. An amplified proto-oncogene functions as oncogene because it leads to overproduction of the normal protein.

(c) Loss of transcriptional control: Loss of transcriptional control can also lead to oncogene activation. This can occur by insertional mutagenesis, chromosomal translocation etc. Like amplification, loss of transcriptional control of a proto-oncogene is also manifested as overproduction of the gene product.

(d) Activation by viruses: Cancer causing viruses are called as oncogenic viruses (Table 3). The characteristic feature of the genome of oncogenic viruses is that they have lost their replicative genes. Instead they carry nucleotide sequences termed as *v-onc*, which is the basis of their cancer causing properties. Each retroviral oncogene has a three-letter notation specific to the virus in which it was first identified. Oncogenic sequences of viruses are conserved throughout the eukaryotic evolution. These are present in many normal hosts that are tumor free. The normal cellular counterparts of viral oncogenes are termed as *c-onc*. The presence of oncogenes

Table 3. Oncogenic viruses

Virus family	Cancer
DNA tumor viruses	
Hepatitis B viruses	Liver cancer
SV40 and polyomavirus	None
Papillomaviruses	Cervical carcinoma
Adenoviruses	None
Herpesviruses	Burkitt's lymphoma, nasopharynegeal carcinoma, Kaposi's sarcoma
RNA tumor viruses	
Retroviruses	Adult T-cell leukemia

in the normal genomes suggests either possible non-malignant functions of these genes or their incorporation during evolution as a result of viral infection. Viral infection causes tumor development by activation of oncogenes. Oncogene activation may be caused due to gene re-arrangement or point mutation. Virus infection may lead to the insertion of a viral promoter or enhancer into the cell genome, activating an existing oncogene.

Oncogenic viruses are classified as DNA or RNA tumor viruses on the basis of nature of their genome. DNA tumor viruses are the major cause of virally induced human cancers. DNA tumor viruses encode proteins, which interact with critical cellular growth –regulatory molecules sabotaging their function. Only one family of RNA viruses, the retroviruses, is known to cause tumors. These are small single stranded animal viruses. Soon after infection the RNA genome is copied into the double stranded DNA by an RNA dependent DNA polymerase (reverse transcriptase), thus creating a DNA provirus. The provirus is inserted into the host genome adjacent to proto-oncogene, usually into the transcriptional regulatory elements, upstream of the proto-oncogene. This disrupts the transcriptional control leading to over expression of the oncoprotein. Proviruses thus act as muatgens and this process is called as insertional muatgenesis. Retroviruses can also cause oncogene activation by mechanisms such as transactivation and oncogene transduction.

(e) Role of bacteria in carcinogenesis: Besides viruses, other infectious agents such as bacteria are also known to be associated with cancers. For example, *Helicobacter pylori,* a Gram negative, spiral shaped bacterium, is found to be associated with various gastric pathologies including gastric adenocarcinoma and gastric lymphoma and is classified as type I carcinogen by WHO.

2. Tumor Suppressor Genes

Tumor suppressor genes control cell growth and differentiation by encoding proteins that inhibit cell division. In contrast to oncogene, loss or inactivation ("turned off") of tumor suppressor genes is associated with tumorigeneis. Thus, inactivation of tumor suppressor genes represent "loss-of-function" genetic event. Another major difference is that while the majority of oncogenes develop from mutations in normal genes (proto-oncogenes) during the life of the individual (*acquired* mutations), abnormalities of tumor suppressor genes can be inherited as well as acquired. They act in a recessive fashion. One normal allele is adequate to protect against cancer or in other words loss of tumor suppressor function of these genes requires inactivation of both alleles. Our knowledge of tumor suppressor genes is based on studies of hereditary cancers. Inactivation of tumor suppressor genes occurs usually by chromosomal deletion, point mutation or both. About 30 tumor suppressor genes have been identified, including p53, *BRCA1, BRCA2, APC,* and *Rb.* Mutations of p53 are associated with up to 50% of human cancers. Therefore, p53 is the most common target of genetic alterations in human malignancies. A list of important tumor suppressor genes is given in Table 4.

3. DNA Repair Genes

DNA repair genes are responsible for maintaining genomic integrity of a cell by repairing the damaged DNA. DNA damage is much more harmful for a cell as compared to RNA or proteins as DNA sequence information drives the synthesis of RNA and RNA to proteins – the ultimate effector molecules of a cell. The DNA repair genes are highly conserved across species

Table 4. Representative tumor suppressor genes of human cancer

Tumor suppressor gene	Type of cancer
APC	Colon cancer, stomach cancer
DPC4	Pancreatic cancer
NF-1	Neurofibroma, Myeloid leukemia
NF-2	Meningioma (brain cancer)
MTS1	A wide range of cancers
P53	A wide range of cancers
Rb	Retinoblastoma, bone cancer, bladder cancer
BRCA1	Breast cancer, ovarian cancer
BRCA2	Breast cancer
VHL	Renal cell cancer
PTEN	Brain tumors, melanoma, prostate, endometrial, kidney and lung carcinomas
WT1	Wilms' tumor
PTC	Basal cell carcinoma
INK4	Melanoma, lung carcinoma, brain tumors, leukemias, lymphomas
MADR2	Colon/Rectal carcinoma

implying that they emerged very early in eukaryotic evolution and are of great importance to the cell. DNA damage can be caused due to various external and endogenous factors, which are summarized below:

Endogenous DNA damaging factors:

(a) Spontaneous mutations: Certain chemical changes occur in DNA spontaneously under physiological conditions. These chemical changes are mainly of two types—Deamination (of adenine, cytosine and guanine) and depurination (loss of purine bases). Depurination creates an apurinic (AP) site in DNA. About ten thousand AP sites are generated in a cell per day (Lindahl, 1993). The frequency of spontaneous mutations in normal cells is 1×10^{-7} per cell division (Loeb, 2001). This frequency of spontaneous mutation is inadequate to account for genomic instability in cancers.

(b) Errors made by DNA polymerase during replication.

(c) Bye-products of normal cellular metabolism such as reactive oxygen (superoxide ions, hydroxyl radicals and hydrogen peroxide) and nitrogen species. Over 100 oxidative modifications have been identified in DNA (Cadet *et al.*, 1997).

Exogenous DNA damaging factors: These include environmental factors such as sunlight, radiations, chemicals, oncogenic viruses etc. As soon as DNA damage occurs, the cell cycle regulatory mechanisms sense the DNA damage and arrest cell cycle at specific checkpoints and allow DNA repair mechanisms to restore the damaged DNA (Zhou *et al.*, 2000). When the damage is widespread in a cell and beyond the capacity of repair mechanisms then an alternative method is adapted to get rid of it. In such cases the whole cell is subjected to apoptosis. The repair of damaged DNA is necessary because it can have deleterious consequences on the cell and organism as a whole. Immediate effects include inhibition of replication, transcription, abnormal chromosome segregation during mitosis and cell cycle arrest. Long-term consequences include generation of mutations, which can eventually lead to cancer (Hoeijmakers, 2001).

There are various pathways of DNA repair each suited for a particular type of DNA defect. There are four main repair pathways functional in mammals-nucleotide-excision repair, base-excision repair, mismatch repair, homologous recombination and end joining (Hoeijmakers, 2001). In some cases there is direct reversal of DNA damage particularly of damage caused due to UV light and also of damage due to interaction of alkylating agents with DNA. More than 130 DNA repair gene products have been identified (Wood *et al.*, 2001). All the DNA repair mechanisms are genetically controlled. Inactivation of DNA repair genes also represents "loss-of-function" genetic event as in case of tumor suppressor genes. However, inactivation of these genes is not essential for acquisition of hallmark features of cancer. Instead loss of function of these genes accelerates the process of carcinogenesis. DNA repair genes are therefore appropriately called as "caretakers" (Kinzler *et al.*, 1997). In contrast, tumor suppressor genes play an essential role in carcinogenesis and are called as "gatekeepers" (Kinzler *et al.*, 1997). Individuals having mutations or defects in DNA repair machinery are more susceptible to cancer as compared to normal counterparts.

Chemical Carcinogenesis

Cancer is a very complex disease. It is the end result of interaction of various endogenous and exogenous factors, which act either simultaneously, or sequentially leading to disruption of normal cell growth and proliferation. Besides the genetic factors described earlier in the chapter there are variety of other exogenous factors (environmental factors) also that have significant association with cancer development. Association between cancer and occupational exposures dates back to 16th century (Luch, 2005). These exogenous factors are called as carcinogens (cancer causing substances). A carcinogen can be defined as a discrete physical, chemical, or biological entity, which is capable of inducing cancer or that significantly increases the incidence of cancer. In brief, substances that cause cancer are called as carcinogens. These can be broadly classified into following three categories:

1. Chemicals
2. Physical (Radiation)
3. Biological (Viruses, bacteria).

The transformation of a normal cell to a malignant one under the influence of a chemical carcinogen is called as chemical carcinogenesis. However, chemical carcinogenicity should not be considered as an inherent property of a chemical but rather as an outcome of the interaction of a chemical with a complex biological system and is influenced by many other factors. The International Agency for Research in Cancer (IARC) has defined chemical carcinogenesis as "the induction by chemicals of neoplasms that are not usually observed, the earlier induction by chemicals or neoplasms that are usually observed, and the induction by chemicals of more neoplasms than are usually found." (IARC, 1982).

Chemical carcinogenesis is a multistep process. Three main steps are tumor initiation tumor promotion and tumor progression. Each of these two steps itself consist of several steps. During tumor initiation the chemical carcinogen reacts with DNA to produce a mutation. This step is irreversible. However, an initiated cell is not a cancer cell and does not necessarily become one. Under certain conditions or certain carcinogens can also initiate tumorogenesis indirectly. This can be due to chronic cytotoxicity (resulting in cell turnover and natural errors in cell replication), activation of cellular oncogenes, or other mechanisms. Initiation can be modulated by factors that change the efficiency of DNA repair or immune surveillance, and, in the case of chemicals that require metabolic activation; initiation will be affected by factors that modify their metabolism. Promotion refers to the subsequent progression and proliferation of the "transformed" cell through a variety of pathological states (e.g., hyperplasia, neoplasia) leading eventually to a malignant tumor (Office of Science and Technology Policy, 1985). Most chemical carcinogens or their metabolites form covalent bonds with DNA and induce mutations. Thus, their mechanism of action is "genotoxic". However, there are certain chemicals that contribute to carcinogenesis by non-genotoxic mechanisms. They stimulate cell proliferation rather than inducing mutations. These are called as tumor promoters (e.g., phorbol esters such as Tetardecanoyl phorbol acetate (TPA) and Phenobarbital). Thus, chemical carcinogens may or may not be mutagenic and also, mutagens may not be carcinogenic. Promoters generally are not mutagenic. Chemical carcinogens that act by genotoxic mechanisms can be of following two types:

(a) Activation Independent or direct carcinogens: These compounds bind directly to DNA without being metabolized. For example, Ethylene oxide, bis (chloromethyl) ether and some aziridine or nitrogen-mustard derivatives. Among known human carcinogens only a few belong to this category.

(b) Activation Dependent (procarcinogens): Most of known carcinogens are actually procarcinogens. These are nucleophilic or chemically inert compounds that require cellular enzymatic metabolism to be converted into an ultimate carcinogen in order to exert their carcinogenic action (Fig. 7). For example, aromatic and heterocyclic amines, aminoazodyes, PAHs (polycyclic aromatic hydrocarbons), n-nitrosamines, halogenated olefines etc. The activation independent carcinogens and activated intermediates of procarcinogens are electrophilic in nature. Nucleophilic sites are abundant on DNA, RNA and proteins. These carcinogenic electrophiles react with nucleophilic sites on these biological macromolecules and bind covalently. Nucleophilic sites on DNA bases can react with one or more electrophilic carcinogens or carcinogen metabolites. Sites involved in hydrogen binding between purine-pyrimidines bases which may lead to miscoding are considered to be most prone to carcinogenic action. However, role of other types of interactions leading to DNA recombinations, gaps, translocations, etc can not be ignored. A chemically induced mutation does not always result in cancer owing to presence of highly efficient host repair mechanisms. The frequency with which mutations are not corrected by host repair systems is a key host determinant in the carcinogenic effect of a chemical.

Identification of Chemical Carcinogens

Till date about 200 different chemicals are anticipated to be carcinogens (Luch, 2005) (Table 5) There are large numbers of physiological (e.g., age, sex, hormone balance, nutritional status) and exogenous (e.g., stress, dietary fiber, and fat) factors, which are often, key determinants in the development of cancer. Should they be defined as carcinogens? Identification of carcinogenic potential of a chemical is of academic importance as well as it has conceptual and practical implications in relation to the development of a national policy for the regulation of chemical carcinogens. It is also of importance to human society that is confused due to an

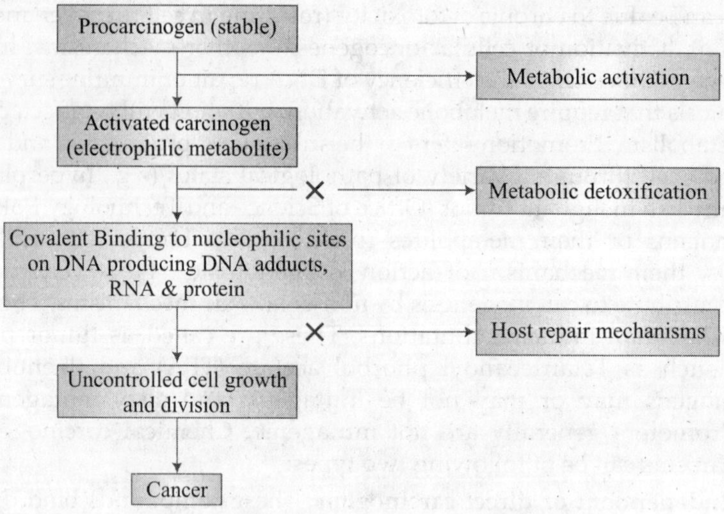

Fig. 7. Mechanism of action of procarcinogens.

increasing list of "carcinogens" especially those that are associated with their day to day life. Therefore, what chemicals may be considered as a carcinogen is a critical question and it needs to be answered appropriately.

Identification of chemicals as carcinogens is based mainly on following types of studies:

1. Chronic animal bioassays
2. Short-term *in vitro* tests for genotoxicity
3. Epidemiological studies in human populations.

Chronic bioassay with laboratory animals, mainly rats and mice, remains the major and most practical experimental procedure for identifying potential carcinogens. Major limitations of chronic animal bioassays are that they are insensitive and highly variable in nature. These tests also require the extrapolation of data obtained under one set of conditions (i.e., with rodents exposed to very high doses in the laboratory) to predict those likely to occur under an entirely different set of conditions (i.e., with humans exposed to very low doses in the real world). Extrapolation of data is also not acceptable due to significant interspecies differences in the metabolism of carcinogens and drugs. During the past two decades, a variety of short-term tests for genotoxicity have been developed to augment chronic animal bioassays for carcinogenesis. These include the well-known Ames Salmonella test for mutagenicity and several other *in vitro* and *in vivo* assays based on a number of genotoxic end points such as sister chromatid exchange, unscheduled DNA synthesis, and chromosome aberrations. Short-term tests are widely used, but there is little correlation between the results of these tests and those of the chronic bioassays. This may be due to the fact that the short-term tests usually fail to replicate the overall metabolic and pharmacokinetic conditions that exist in animal models used in bioassays. Moreover, there is inconsistency among the results of different short-term tests themselves. This reflects involvement of different biological systems in every test. However, the short-term tests are shown to correlate better with each other than with bioassays (Tennant *et al.* 1987). The IARC has developed a qualitative classification scheme that divides

Table 5. Important human carcinogens

Compounds	Main sources/uses	Associated cancers
Aminoazo dyes		
o-Aminoazotoluene	Pigments; colouring oils; immunosuppressant	Liver, lung, bladder
N, N-dimethyl-4-aminoazobenzene	Colour polishes; waxes (no longer in use)	Lung, liver
Anticancer drugs		
Melphalan	Chemotherapy	Leukaemia
Thiotepa	Chemotherapy (no longer in use)	Leukaemia
Aromatic amines/amides		
Naphthylamine Dyes	antioxidant (no longer in use)	Bladder
4-Aminobiphenyl Dyes	antioxidant (no longer in use), research tool	Bladder
2-Acetylaminofluorene	Model compound; tested as a pesticide	Liver, bladder
Natural carcinogens		
Aflatoxin B 1	A mycotoxin produced by the mold *Aspergillus flavus*. (found in contaminated food, damp corn and grain stores, especially in tropical regions)	Liver
Asbestos (fibrous silicates)	Thermal insulation; gaskets (declining usage), Brake-lining	Lung, mesothelioma
Heterocyclic amines (E.g. IQ (2-amino-3-methylimidazo [4,5-*f*] quinoline)	Derived from amino acids during high temperature cooking (>180°C) especially of meat products	Colon, Breast and stomach
Acrylamide	Starch based foods such as potato chips and French fries cooked using high temperature deep-frying and baking methods	
Polycyclic aromatic hydrocarbons (PAH)		
Benzo[a]pyrene BaP	Tobacco smoke, unburned tobacco, petroleum combustion bye-product, coal tar	Lung, larynx, oral cavity, cervix, stomach
Dibenzene[a,h]anthracene	Tobacco smoke, unburned tobacco, petroleum combustion bye-product	Lung, larynx, oral cavity, cervix, stomach
Polychlorinated biphenyls	Flame retardants; hydraulic fluids	Liver, skin
Metals		
Arsenic	Natural ores, alloys, insecticide/herbicide, sprayers, tanners, oil refiners, pharmaceutical agent	Skin, lung, liver
Cadmium	Natural ores; pigments; batteries; ceramics, unburned tobacco	Lung, prostate, kidney
Nickel	Natural ores; alloys; electrodes; catalysts	Lung, nasal cavity

Table 5 Contd.

Table 5 Contd.

Nitrosamines		
4-(methylnitosamino)-1-(3-pyridyl) -1-butanone (NNK)	Tobacco smoke, unburned tobacco	Lung, nasal, oral cavity, esophagus, liver, pancreas, cervix
N—nitosocorticone (NNN)	Tobacco smoke, unburned tobacco	Nasal, oral cavity, esophagus
Aromatic amines		
4-aminobiphenyl	Tobacco smoke, chemicals, dyes	Bladder
2-naphthylamine	Tobacco smoke	Bladder
Benzidine	Chemical, dyes, rubber	Bladder

chemicals into groups with varying degrees of evidence for carcinogenicity in humans (International Agency for Research on Cancer, 1982). These groupings are based on bioassays, short-term tests, and human epidemiology.

Ames test: The Ames test is a standard used for predicting the potential mutagenic and carcinogenic property of chemicals. A mutant strain of *Salmonella* that has a mutation in its histidine coding gene is used. This mutation makes *Salmonella* inefficient in growing on histidine containing growth medium. Mutated *Salmonella* is exposed to suspected carcinogen and cultured on a histidine-containing medium. Only those *Salmonella* that have undergone backward mutations (a mutation that corrects the histidine-coding gene), due to mutagenic effect of carcinogen, grow successfully on histidine plate. The numbers of mutant colonies seen are counted and it gives a measure of muatgenicity of the chemical in question. The more mutagenic the chemical, the greater is the number of mutant colonies that grow on histidine containing growth medium. Rat liver extract is also added as it contains enzymes that convert procarcinogens to carcinogens (Fig. 8). This enables the testing of procarcinogens that needs metabolic activation for exhibiting its mutagenic effect. Carcinogenicity of non-genotoxic chemicals (e.g., promotors) cannot be assessed with the Ame's assays. A positive result in the Ames test carries a high probability (70%) that a chemical will be associated with carcinogenic activity (Tennant *et al.*, 1987).

Metabolism of Carcinogens

Metabolism plays an important role in the activity of many carcinogens and is the primary determinate of their level of activity, the site of action (organotropism), sex specificity and species specificity. The entire set of physical and chemical changes that occur when foreign substances are introduced in body, from its uptake to excretion is called as Xenobiotic metabolism. It is an important defense mechanism of our body against environmental toxicants and chemical carcinogenesis. There are sufficient experimental evidences, which show that xenobiotic detoxification pathway plays an important role in chemical carcinogenesis. There are considerable inter-individual genotypic variations in these pathways, which could be one of the reasons for differences in cancer susceptibility (Gonzalez, 1997). Appropriate bodily response to environmental toxicants including chemical carcinogens requires proper functioning of the xenobiotic metabolism. Xenobiotic metabolism involves two phases –

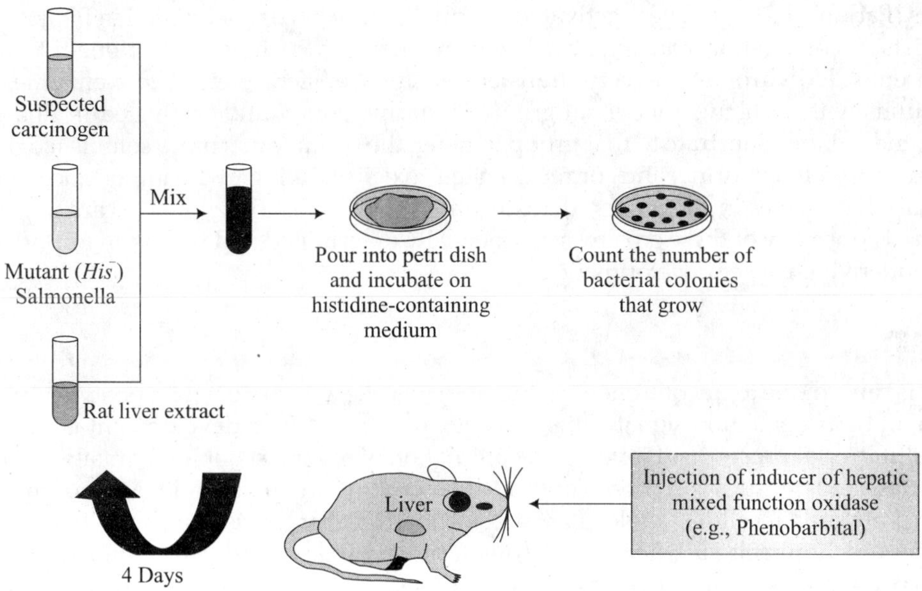

Fig. 8. Ames test.

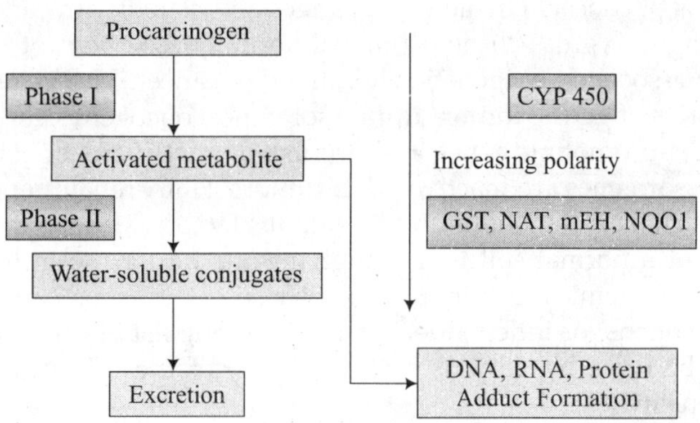

Fig. 9. Carcinogen metabolism: xenobiotic metabolism.

Phase I (activation phase) and Phase II (detoxification phase) (Fig. 9). Each phase is characterized by specific enzyme activities. The normal function of these enzymes is to protect our body against chemical carcinogens by inactivating and excreting them from the body. The competing detoxifying and activating pathways determine the ultimate effect of carcinogen exposure.

Phase I consists of metabolic activation of procarcinogens by enzymes such as Cytochrome P450. Non-polar fat compounds are activated to electrophilic intermediates by oxidation/ reduction, hydrolysis reactions catalyzed by Phase I enzymes. These electrophilic intermediates are usually unstable epoxides. Epoxides are organic compounds with 3 oxygen atoms in a strained configuration and hence are highly reactive and unstable. Epoxides cause cellular damage by inducing mutations in the genetic material and by forming RNA and protein adducts (Parkinson, 1999). These activated metabolites are in part detoxified in Phase II. Phase II consists of conjugation reactions catalyzed by enzymes such as glutathione-S-transferses (GSTs), epoxide hydrolase, N-acetyl transferase, sulfotransferase etc. These enzymes act on compounds with an active functional group, including compounds activated in Phase I and add an endogenous substrate to that group to make the compounds more soluble (so that they are readily excreted in urine) and/or reduce their toxicity. Owing to the importance of Phase I and Phase II enzymes in carcinogen detoxification, role of these enzymes in carcinogenesis is obvious. The activity of these enzymes is genetically determined and is thus an important host factor underlying cancer development.

SUMMARY

Cancer is caused due to uncontrolled proliferation of cells, which do not respond to normal cell growth and differentiation signals. It is a genetic disease. Cancer development is a multistep and multifactorial process and involves accumulation of several sequential mutations. Tumors can be classified on the basis of type of tissue involved or organ of origin. Most of the human cancers share similar cellular, molecular and biochemical characteristics. The hallmark features of cancers include self-sufficiency in growth signals, insensitivity to antigrowth signals, evasion of apoptosis, sustained angiogenesis, limitless replicative potential and tissue invasion and metastasis. Each of these features is achieved by various genetic and epigenetic alterations in cancer cells, which is phenotypically expressed at protein level too. This forms the molecular and biochemical basis of cancer. The genetic changes associated with cancers can range from subtle sequence changes to gross chromosomal abnormalities. Aberrant DNA methylation is the most commonly associated epigenetic alterations in cancer. The genetic changes usually involve three categories of genes, tumor suppressor genes, oncogenes and DNA repair genes. Oncogenes are cancer-promoting genes and are usually activated in cancers. On the other hand, tumor suppressor genes are inactivated in tumors. DNA repair genes, which repair the DNA damaged by various endogenous and exogenous factors, are also inactivated in tumors. The transformation of a normal cell to a malignant one under the influence of a chemical carcinogen is called as chemical carcinogenesis. These carcinogens can be either activation dependent (procarcinogens) or independent. There is a long list of carcinogens. Identification of carcinogens can be done by various methods. One of these is Ames test, which is an important carcinogenicity test.

REFERENCES

Armitage, P. and Doll, R. (1954) The age distribution of cancer and a multi stage theory of carcinogenesis. Br. J. Cancer 8 : 1-12.

Berenblum, I. and Shubick, P. (1949) The persistence of latent tumor cells induced in the mouse's skin by a single application of 9:10-dimethyl-1: 2-benzanthracene Br J Cancer, 3 : 384-386.

Brodeur, G.M. and Hogarty, M.D. (1998) The Genetic Basis of human cancer 1st edn Vol1 (eds., Kinzler, K.W. and Vogelstein. B.) 161-179 (McGraw-Hill, New York.

Brown, R. and Strathdee, G. (2002) Epigenomics and epigenetic therapy of cancer. Trends Mol. Med. 8 : S43–S48.

Cadet, J., Berger, M.; Douki, T. and Ravanat, J.L. (1997) Oxidative damage to DNA : formation measurement, and biological significance. Rev. Physiol. Biochem. Pharmacol. 131: 1-87.

Chin, L., Pomerantz, J. and DePinko, R.A. (1998) The INK 4a/ARF tumor suppressor: one gene - two products - two pathways. Trends Biochem Sci 23: 291-296.

Davis, C.D. and Uthus, E.O. (2004). DNA Methylation, Cancer Susceptibility, and Nutrient Interactions. Exp Biol Med 229: 988–995.

Fearon, E.R and Vogelstein, B. (1990) A genetic model for colorectal tumorigenesis. Cell, 61: 759- 767.

Foulds L (1957) Tumor progression. Cancer research 17: 355-6.

Fynan, T.M and Reiss M. (1993) Resistance to inhibition of cell growth by TGF β and its role in oncogeneis. Crit Rev Oncog, 4: 493-540.

Gonzalez F.Z. (1997) The role of carcinogen metabolizing enzyme polymorphism in cancer susceptibility. Reprod. Toxicol.11: 397 - 412.

Hahn W.C., Counter C.M., Lundberg A.S., Beijersbgern R.L., Brooks W.M., Weinberg R.A. (1999) Creation of human tumor cells with defined genetic elements. Nature 400: 464-468.

Hanahan, D. and Weinberg, R.A. (2000) The hallmarks of cancer. Cell 100: 67-70.

Hoeijmakers, J. H. J. (2001) Genome maintenance mechanisms for preventing cancer. 411: 366-374.

International Agency for Research on Cancer (IARC) (1982) Evaluation of the Carcinogenic Risk of Chemicals to Humans. IARC Monographs, Lyons, France, Supp. 4, pp. 7-23.

International Agency for Research on Cancer (IARC). IARC Monographs on the Evaluation of the Carcinogenic Risk of Chemicals to Humans"; IARC: Lyons, France, 1982; Vol. 29, p. 16.

Kinzler, K. W. and Vogelstein, B. (1997). Gatekeepers and caretakers 386: 761-763.

Lengauer, C., Kinzler. K.W. and Vogelstein B. (1998) Genetic instabilities in human cancers 396: 643-649.

Lindahl, T. (1993) Instability and decay of primary structure of DNA.362: 709-715.

Liotta LA. (1986) Tumor invasion and metastasis - role of the ECM: Roads Memorial Award Lecture. Cancer Res 4: 1-7.

Liotta, L. and Petricoin, E. (2000) Molecular profiling of human cancers. Nature Rev Genet. 1 : 48-56.

Loeb, L.A. (2001) A mutator phenotype in cancer. Cancer Res 61: 3230-3239.

Luch, A. (2005) Nature and nurture: lessons from chemical carcinogenesis. Nature 5: 113-125.

Markowitz S, Wang J, Meyeroff L, Parsons R, Sun L and Lutterbaugh J. (1995) Inactivation of the type II TGF b receptor in colon cancer cells with microsatellite instability. Science 268: 1336-1338.

McEacher, M.J., Krauskopf, A. and Blackburn, E.H. (2000) Telomeres and their control. Ann Rev.Genet. 34: 331-358.

Medema RH, Bos JL. (1993) The role of p21-ras in receptor tyrosine kinase signaling. Crit Rev Oncog, 4: 615-661.

Nowell, P.C., (1976) The clonal evolution of tumor cell populations. Science 194: 23-28.

Office of Science and Technology Policy. (1985) Chemical Carcinogens: A Review of the Science and its Associated Principles. Fed. Regist 50: 10371-442.

Parkinson A (1999) in Biotransformation of Xenobiotics: Casarett and Doulle's Toxicology (ed., Klasen, C.D.), pp. 113-186, McGraw-Hill, Inc, New York.

Shay, J.W. and Bacchetti, S. (1997) A survey of telomerase activity in human cancer. Eur. J. cancer 33: 787-791.

Tennant, R.W. *et al.* (1987) Science 236: 933-941.

Willis R.A., (1948) Pathology of tumors. St Louis, MO, Mosby.

Wood R.D., Mitchell M., Sgouros J. and Lindahl. T. (2001) Human DNA repair genes. Science 29: 1284-1289.

Yarden Y., Ullrich, A. (1998) EGF and erbB2 receptor over expression in human tumors, growth factor receptor tyrosine kinases. Ann. Rev. Biochem. 57: 443-478.

Zhou, B.B. and Elledge, S.J. (2000) The DNA damage response : putting checkpoints in perspective. Nature 408 : 433-439.

Suggested Readings

1. Macdonald, F. and Ford, C.H.J. (1997) Molecular biology of cancer, bios Scientific publishers.
2. Mendelsohn, J., Howley P.M, Israel, M.A and Liotta, L.A. (2001) The Molecular Basis of cancer, 2nd edition, W.B Saunders company.
3. Alison, M. R. and Sarraf C.E. (1997) Understanding cancer: From Basic science to clinical practice, Cambridge University press.
4. Mihich, E. and Croce, C. (1998) The Biology of tumors. Pezcoller Foundation Symposia Volume 9, Plenum Press.
5. Friedberg, E.C., G.C. Walker and W. Siede (1995) Washington, D.C. WSM Press.

REVIEW QUESTIONS

Long Questions

1. What is cancer? Describe the multistage theory of carcinogenesis.
2. Differentiate between a normal cell and cancer cell.
3. Describe the various types of cancer. What is the difference between a benign and a malignant tumor?
4. What is the difference between genetic and epigenetic alterations in cancer? Describe the various types of genetic alterations in cancer.
5. What are "hallmark features" of cancer? Describe the various mechanisms by which a cancerous cell can evade apoptosis.
6. Differentiate between oncogenes and tumor suppressor genes with examples. Describe the various mechanisms of oncogene activation.
7. What do you understand by the terms chemical carcinogenesis and xenobiotic metabolism? List the various methods used for carcinogen identification. Describe Ames' test.
8. Describe the role of DNA repair genes and genes encoding drug metabolizing enzymes in carcinogenesis.

Multiple Choice Questions

1. Which of the following property is not common to both malignant tumor and benign tumor?
   (a) Evasion of apoptosis (b) Limitless replicative potential
   (c) Tissue invasion and metastasis (d) Multistage nature
2. Overexpression of telomerase enzyme enables the tumor cells
   (a) To evade apoptosis (b) To metastasize to distant organs
   (c) To exhibit sustained angiogenesis (d) To undergo limitless replication

3. Which of the following statements is correct?
   (a) Grading of tumors is done on the basis of extent of local invasion while staging of tumors is done on the basis of extent of differentiation
   (b) Both grading and staging of tumors is done on the basis of extent of differentiation
   (c) Grading of tumors is done on the basis of extent of differentiation while staging of tumors is done on the basis of extent of local invasion
   (d) Both grading and staging of tumors is done on the basis of extent of local invasion
4. Majority of human cancers are
   (a) Sarcomas (b) Carcinomas
   (c) Lymphomas (d) Leukemias
5. Which of the following genetic changes is most commonly associated with cancers?
   (a) Aneuploidy (b) Base substitutions
   (c) Gene amplification (d) Chromosomal translocations
6. Which of the following is most common genetic mechanism of drug resistance in cancers?
   (a) Point mutations (b) Aneuploidy
   (c) G ene amplification (d) Insertional mutagenesis
7. Which of following represents a "loss-of-function" genetic event in cancer?
   (a) Activation of oncogenes
   (b) Inactivation of tumor suppressor genes
   (c) Inactivation of DNA repair enzymes
   (d) Both (b) and (c)
8. The "effector" molecules of apoptosis are
   (a) Caspases (b) cytochrome c
   (c) p53 protein (d) None of the above
9. The catalysts of apoptosis are
   (a) Caspases (b) cytochrome c
   (c) p53 protein (d) None of the above
10. Which of the following statements is correct?
   (a) Both carcinogens and pro-carcinogens are electrophiles
   (b) Carcinogens are electrophiles but pro-carcinogens are nucleophiles
   (c) Both carcinogens and pro-carcinogens are nucleophiles
   (d) carcinogens are nucleophiles pro-carcinogens are electrophiles
11. Liver extract is used in Ames' test
   (a) as a nutrient supplement to the *Salmonella*
   (b) as a source of enzymes required for conversion of pro-carcinogens to carcinogens
   (c) for maintaining pH in the culture media
   (d) none of the above

12. Aflatoxin B1, a natural carcinogen, is obtained from
    (a) Bacteria
    (b) Algae
    (c) Virus
    (d) Fungus

Answers

1. (c)	2. (d)	3. (c)	4. (b)	5. (a)	6. (c)
7. (d)	8. (a)	9. (b)	10. (b)	11. (b)	12. (d)

CHAPTER 35

Metabolic and Genetic Disorders

K. Ramalingam
Dept. of Advanced Zoology and Biotechnology, Govt. Arts College, Nandanam, Chennai–600 035
E-mail: drkrimmunotoxicol@yahoo.co.in

Current advances that are made in genetics at molecular level, after the discovery of double helix DNA have given insights to the various intricacies of phenotype expression and manifestation of biochemical characters and/or diseases. During embryonic development of an organism, differentiation of a quite a lot of cell types occur. The phenotypes of all these cell types are conspicuously different, although they share the same genotype. Each cell/cell type manifests only a part of its genome. According to earlier contention, different cell types behave as if different sets of genes (determinants) were allotted to them during embryonic development. Later on, it was realized that it is not the genes that are distributed unequally, rather the chemical environments which are expressed through differential regulation of the gene function. Thus, distinctions among the differentiated cells/cell types are largely attributed if not entirely to the different proteins viz., enzymes which they contain (Stanbury *et al.*, 1960; Wagner and Mitchell 1964).

DIFFERENTIAL GENE FUNCTION

Cells within single individuals show conspicuous differences in protein content as well as in gene function (Gupta, 1996). Each cell is characterized by its repertoire of proteins/enzymes which are specific for each stage of its differentiation. Some proteins appear early and then disappear (HbF), while others appear late and persist throughout the remaining life of the cell. To put it another way, the assortment of genes that is identifiable by function is specific for each cell at each stage of its differentiation. As a matter of fact, differential gene function is intrinsic and fundamental aspect of cell differentiation and whatever is responsible for regulating gene function is also responsible for cellular differentiation (Watson, 1987).

According to the central dogma of protein synthesis the DNA of the chromosome acts as a template for the synthesis of messenger RNA molecules (mRNA) which in collaboration with transfer RNA (tRNA) and with the assistance of ribosomal RNA (rRNA), brings about the synthesis of specific proteins. Several enzymes are involved in this process of protein synthesis. It also requires the integration of a cascade events and interference at any of the points in the cascade chain of events from gene to protein, results in the absence of the protein per se. So absence of a protein means to imply that the gene for that protein is not functioning in that cell and contrarily the appearance of a specific protein can be construed as a satisfactory indicator of gene function (Ridley, 2002).

LACTATE DEHYDROGENASE

One of the best understood proteins in terms of ontogenic behaviour, physiological function, macromolecular structure and genetic control is the enzyme lactate dehydrogenase (LDH). LDH is an ubiquitous enzyme in vertebrate tissues and is found in some invertebrates as well. It is an oxidoreductase that catalyzes the interconversion of pyruvate and lactate using the coenzyme NAD. Figure 1 shows the location of the above reaction in the glycolytic metabolic pathway viz., glycolysis. In vertebrate tissues/organs lactic acid/lactate appears to be at the end of a metabolic blind ally. In muscular tissues, at times of oxygen debt or during anaerobic periods, the pyruvate is converted to lactate. This side reaction occurs as a metabolic mechanism that provides a temporary storage reservoir for hydrogen during anaerobic tissue state. When oxygen is in short supply all available NAD is reduced to $NADH_2$ which accumulates. The accumulated $NADH_2$ will depress the rate of oxidation of glyceraldehydes phosphate which requires NAD and thus reduce the supply of chemical energy from glucose metabolism (Graham and Wareing, 1984).

However, the oxidized coenzyme NAD is regenerated by the transfer of hydrogen from $NADH_2$ to pyruvate to yield lactate. The lactate accumulates until the replenishment of oxygen supply to oxidize the lactate back to pyruvate becomes available. The lactate is referred to as a fatigue poison and it may also enhance metabolic (respiratory) acidosis. However, lactate is readily transported from one cell to another so that excess production in one tissue can be compensated by vigorous oxidation of lactate in another, where oxygen is relatively more available (Fig. 1).

LDH exists in five principal isozymic forms and is present in nearly all cells of all vertebrates. The relative proportions of the five isozymes is characteristic for each tissue and for each stage of its differentiation. LDH is a tetramer and the monomers are of two varieties viz., A and B. LDH_1 to LDH_5 are synthesized by the combinations of the above AB subunits. The isozyme formula for the different LDHs (1-5) and their polypeptide composition are given in Table 1. Two non-allelic genes are responsible for the basic five isozymes of LDH. Each of these genes is subject to mutation and can manifest pathological symptoms. Some coronary diseases in human are the manifest functions of the LDH genes and their position.

GENETIC ERRORS

It is an established fact that any property of a protein can virtually be changed as a result of gene mutation. A mutation can result in the replacement of a single amino acid in a polypeptide chain. So the linear sequence of amino acids in proteins is strictly related to the linearly arranged triplet code of a gene segment. Thus each genetic locus is responsible for the

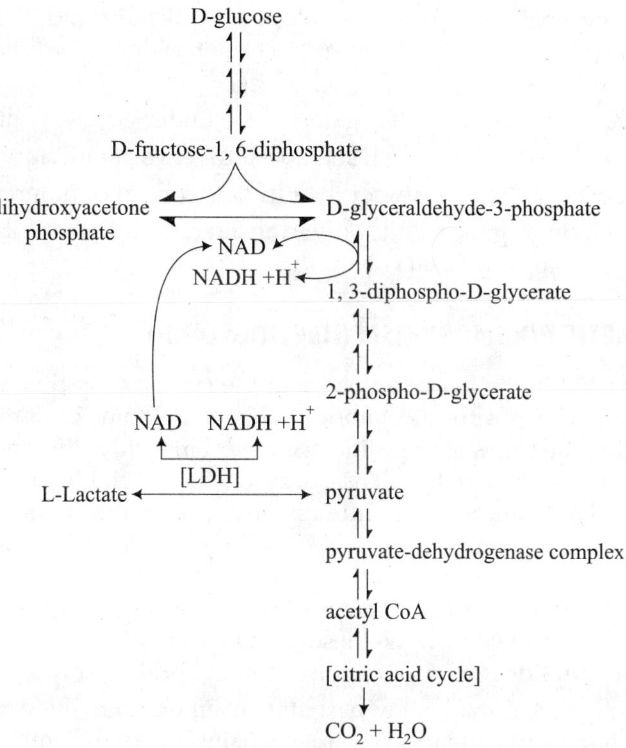

Fig. 1. Glycolytic metabolic pathway.

Source: Ideas in Modern Biology by John Moore (Synthesized) Page no: 234. Published by National Academy of Science, Washington (1965).

Table 1. LDH isozymes and their polypeptide composition.

LDH-1	A_0B_4
LDH-2	A_1B_3
LDH-3	A_2B_2
LDH-4	A_3B_1
LDH-5	A_4B_0

synthesis of a specific protein. Though, the above idea was expressed by Beadle and Tatum (1941) as "one gene one enzyme" theory, recent developments in genetics in understanding the structures of certain proteins like haemoglobin and also certain enzymes revealed that a protein may be comprised of two or more different polypeptide chains and that each chain is controlled by a separate genetic locus and that atleast two loci are involved in the synthesis of a single "complete" molecule. The later developments gave a new twist to the original concept of 'one gene one enzyme' and introduced the term 'one gene one polypeptide'. Modern genetics has rephrased it as "one cistron one polypeptide".

The one gene-one enzyme concept developed by Beadle and Tatum (1945) by their experiments of induced mutations in *Neurospora crassa* has delineated the following features: viz.,

1. All biochemical processes in all organisms are under generic control.
2. These biochemical processes are resolvable into series of individual stepwise reactions.
3. Each biochemical reaction is under the ultimate control of different single gene.
4. Mutations of a single gene results only in an alterations in the ability of the cell to carry out a single primary chemical reaction.

BIOCHEMICAL/GENETIC POLYMORPHISM (HAEMOGLOBIN)

It is elegantly illustrated in the genes involved in the synthesis of human haemoglobin. The haemoglobin genes and the synthesis of abnormal haemoglobins in human reveal how these result due to mutations and how the specific haemoglobins with differences in the amino acid sequence become an inherited defect. The study on haemoglobins is also a revealation of genetic or biochemical polymorphism. In human the haemoglobin exists in different forms viz., foetal haemoglobin (HbF) adult haemoglobin (HbA), Sickle cell haemoglobin (Hbs) and abnormal haemoglobins C and D. Ingram studied the chemical composition of these haemoglobins and elucidated the differences in the amino acid composition among them. He revealed that the only difference between these haemoglobins involves a single peptide (in beta chain) while all other peptides are indistinguishable and presumably identical (Ingram 1957).

Crystallographic studies had shown that a haemoglobin molecule consists of four polypeptide chains having two identical α (alpha) chains and two identical β (beta) chains. The foetal haemoglobin exhibits a clearly different amino acid composition from adult and it is produced by a different synthetic and presumably genetic mechanism. The genotype is HbFHbF. The normal adults have haemoglobin A with the genotype Hb^AHb^A. The abnormal haemoglobin S with the homozygous genotype Hb^SHb^S causes sickle cell anaemia disease (Edelstein 1986). The heterozygous genotype Hb^SHb^A produces the sickle cell trait (Figs. 2 and 3). The abnormal haemoglobins C and D are due to gene Hb^C and Hb^D respectively.

Hunt and Ingram finger printed the peptides derived from different haemoglobins viz., A,S, C and revealed that their β chain exhibits a change at the 6th amino acid position, due to a missense mutation, which is illustrated below.

	1 2 3 4 5 6 7 8
Haemoglobin A	Val-his-leu-thr-pro-<u>glu</u>-glu-lys-
Haemoglobin S	Val-his-leu-thr-pro-<u>Val</u>-glu-lys-
Haemoglobin C	Val-his-leu-thr-pro-<u>Lys</u>-glu-lys-

CONCEPT OF INBORN ERROR OF METABOLISM AND METABOLIC BLOCK

Sir Archibald Garrod developed a concept by his studies on certain human diseases such as alcaptonuria, albinism, cystinuria, pentosuria etc, that these diseases and the like of which of life long duration, arise due to a genetic error and the consequent metabolic block in the biosynthetic pathways involving several enzymes. The functionally deficient gene makes the enzyme governing a single metabolic step reduced in activity or missing altogether (Garrod, 1909).

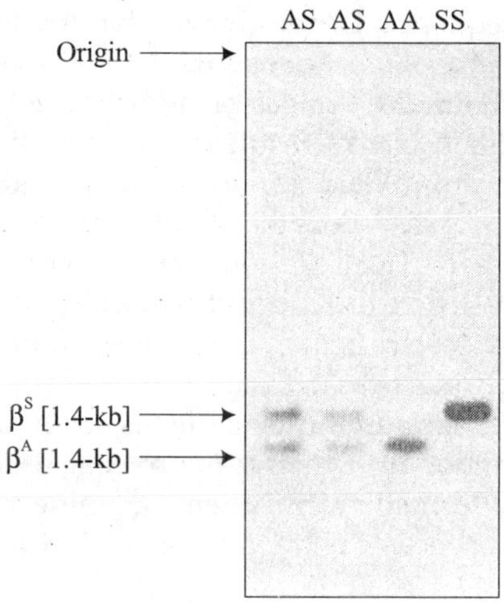

Fig. 2. Restriction enzyme analysis of DNA for pre-natal diagnosis of haemoglobin genes. A – normal genes; S – Sickel gene.

Source: The sickled Cell by Edelstein, S.J. Page number 143, Published by Harvard University Press (1986).

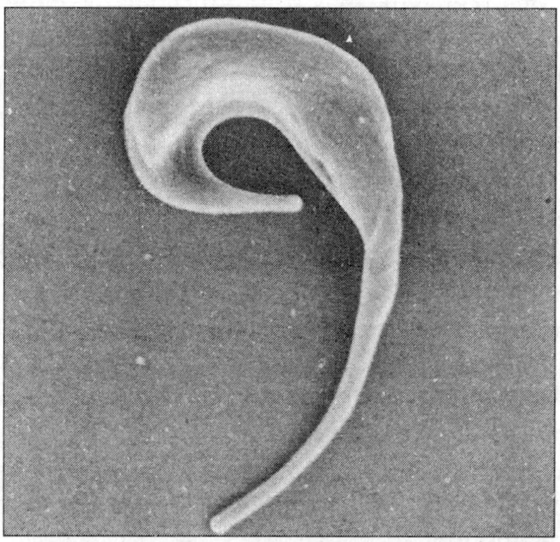

Fig. 3. A highly sickled RBC in Sickle cell anemia.

Source: The sickled cell by Edelstein Page number: 133, Published by Oxford University Press (1986).

A biosynthetic pathway involves the collective actions of several enzymes wherein a specific substrate undergoes sequential conversion into several intermediate products before the final product is reached to manifest the specific biochemical phenotype. As genes are responsible for the synthesis of enzymes, mutations of genes will affect the synthesis of enzymes or result in defective enzymes, which are inactive. Absence of an enzyme or the defective/inactive enzyme may block a particular reaction in the biosynthetic pathway causing thereby a blocked metabolic sequence. Depending on the step where the enzyme involved in the reaction sequence, the metabolic sequence may be blocked/interrupted either in the beginning itself or in any of the middle stages or prior to the last stage. Ultimately the final product is not synthesized. Such blocks result in accumulation of either the substrate or intermediate products. The accumulated abnormal metabolites or their intermediates are excreted through urine. In several disorders the abnormal products may be found excessively in body fluids or stored in tissues and cause lesions. In Von Gierke's disease absence of glucose-6-phosphatase leads to the gross accumulation of hepatic glycogen. In phenyl ketonuria the block in phenylalanine hydroxylase results in an accumulation of phenylalanine and a gross over production of phenylketone products. In maple syrup urine disease the block in decarboxylation of branched chain amino acids leads to the accumulation and abnormal excretion of branched chain amino, α-keto and α-hydroxy acids. In alkaptonuria, the alkapton, viz., homogentisic acid an intermediate in phenyl alanine metabolism, accumulate and excreted in urine due to the absence of homogentisic acid oxidase activity in the liver of the patient (Fig. 4).

Several metabolic defects in human beings are due to the absence or reduced activity of single specific enzyme. Table 2 illustrates some important hereditary disorders due to lack of enzymes/proteins or their modifications. Most of these defects are phenotypically patent only in the homozygous condition of the alleles. In heterozygotes with one dominant and one mutant gene, sufficient protein/enzyme may be produced to meet the physiological needs. For example persons with the heterozygotes for sickle cell trait, Hb C type possess one normal gene A and an abnormal mutant gene for the aberrant β chain. Similarly the heterozygotic parents of galactosemic infants have approximately 50 per cent of the normal amount of galactose-1-phosphate uridyl transferase.

Thus the metabolic blocks due to genetic mutations result into consequences in different ways, as mentioned below.

1. There may be complete absence of the final product of the pathway eg., the lack of melanin in albinism (Fig. 5).
2. There may be an accumulation of an intermediate metabolite for which there is no significant alternative pathway. For example, Glycogen storage disorders in which glycogen is synthesized but cannot be utilized or alkaptonuria in which homogentisic acid accumulates.
3. There may be an accumulation of products of an alternative pathway of metabolism as in phenyl ketonuria leading to the excretion of large amounts of compounds ordinarily produced in trace amounts.
4. A degradative enzyme may be lacking leading to excessive accumulation of normal components that continue to be synthesized as in various glycosphingo lipo dystrophies, or
5. the presence of an abnormally large amount of an enzyme resulting in the diversion of an excessive quantity of substrate into one pathway at the expense of another eg., hereditary hemolytic anemia in which 50 to 70 times as much adenosine deaminase (ADA) as in normal erythrocytes are being routed into a different pathway thus leading to a reduction in the levels of various adenosine derivatives including ATP and NADP etc.

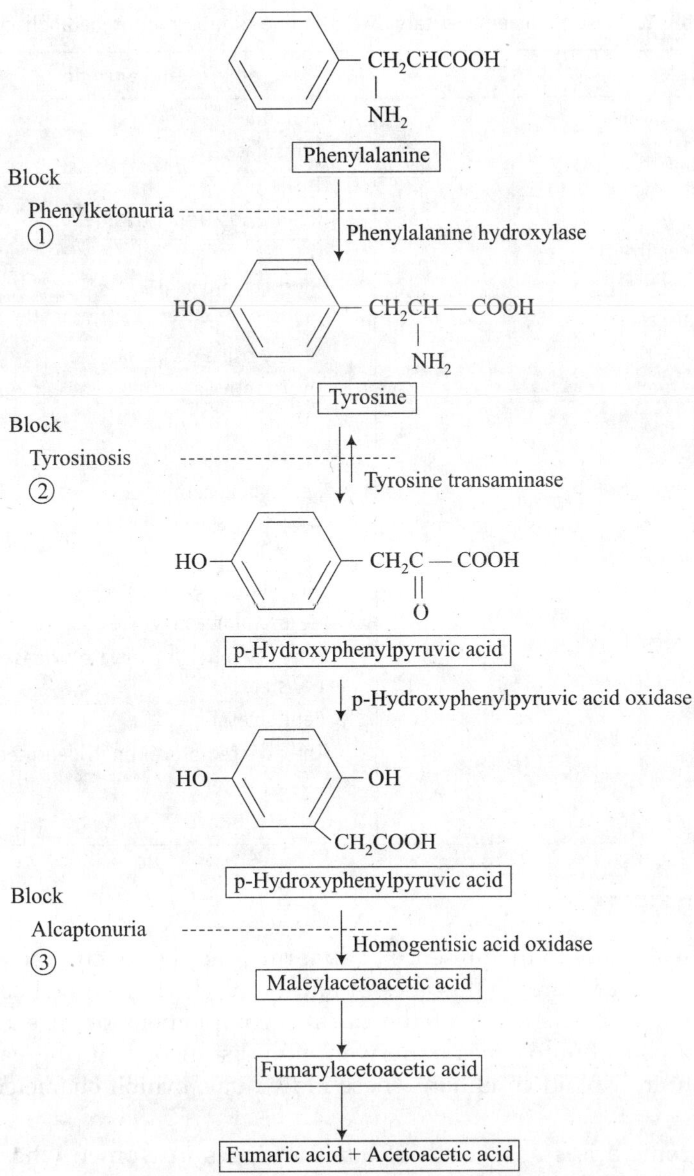

Fig. 4. Phenyl alanine metabolic pathway.

Source: The metabolic basis of inherited disease by John B. Stanbury *et al.*, (synthesized) Page No: 450, Published by McGraw Hill Book Co., Inc, New York (1960).

In several hereditary disorders, the affected protein has not yet been identified. Such protein aberration related disorders include congenital steatorrhea, cystic fibrosis, cystinuria, cystinosis, Fanconi syndrome, Hypomagnesemia, congenital lactate acidosis, etc.

Table 2. List of some hereditary disorders due to inborn error metabolism.

Name of the disorder	Affected protein/ enzyme
Agammaglobulinemia	γ-Globulin
Albinism	Tyrosinase
Infantile Diabetes	Insulin
Galactosemia	Gal-1-Phos-Uridyl trasnferase
Gangliosidosis (generalized)	β-galactosidase
Gaucher's disease	Glucocerebrosidase
Glycogen storage disease	,,
Familial goiter	Iodotyrosine dehalogenase
Gout, Primary (one form)	Hypoxanthine-guanine phosphoribosyl transferase
Haemophilia	,,
Lesch-Nyhan Syndrome	,,
Maple syrup Urine disease	Ketoacid decarboxylase
Myeloperoxidase deficiency with disseminated candidiasis	Myeloperoxidase
Phenylketonuria	Phenylalanine hydroxylase
Tay-Sach's disease	Hexosaminidase A
Tyrosinemia	p-hydroxyphenyl pyruvate oxidaseTyrosinase amino transferase
Wilson's disease	Ceruloplasmin
Cystinuria	The affected protein not identified
Fanconi's syndrome	,,
Lactate acidosis	,,

CHROMOSOMAL DEFECTS

Genetic defects are also due to the presence of abnormal number of chromosomes in human. Due to non-disjunction of paired autosomes or allosomes, the zygote will be produced with one or two chromosomes deficient of the normal complement or in excess. In Down's syndrome the 21st pair of autosomes underwent non-disfunction during gamete formation and so it is present in triploid condition. These individuals exhibit characteristic symptoms viz.,

Down's syndrome arises as a result of late marriages in women who happen to bear children beyond age 40. It is also called Mongolism or Mongaloid idiocy. Their characteristic appearance-short stature, plum bodies, narrow eyes, happy faces is immediately familiar. They are mentally retarded, gentle and destined to age rapidly, often developing a form of Alzheimer's disease, and die before they reach the age of forty. Down syndrome babies are generally born to older mothers. The chances of having down syndrome babies is 1% at the age of forty (Gardner 1975).

Sex chromosome involved defects are klinefelters syndrome and Turner syndrome. These arise due to non-disfunction of the paired XX chromosomes during gametogenesis (oogenesis). As a result an ovum with 22 autosomes and XX which upon fusion with a normal Y carrying sperm result into a genotype of 44A + XXY (47) who happen to be Klinefilters. These

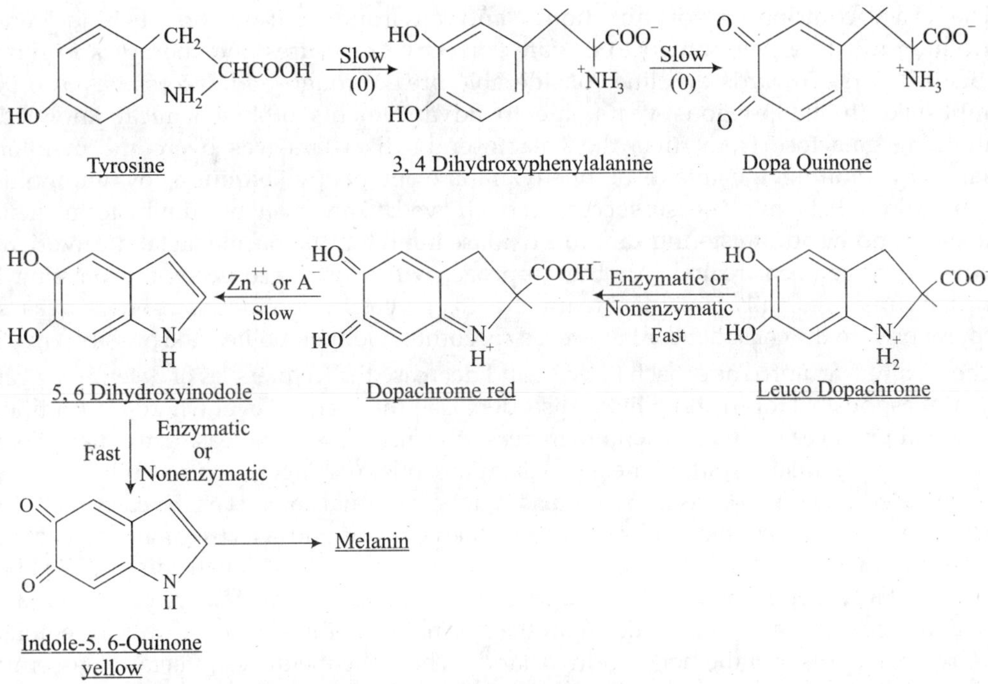

Fig. 5. Tyrosine metabolic pathway.

Source: The metabolic basis of inherited disease by John B. Stanbury *et al.*, (synthesized) Page No: 450, Published by McGraw Hill Book Co., Inc, New York (1960).

individuals appear to be phenotypically male but do not undergo sexual maturity and are frequently mentally retarded. Individuals with the genotype 44A + X (45) are known as Turner's. They are phenotypically females but sexually they are infertile and exhibit features like webbed neck, short stature bloated face, absence of menstrual cycle etc. They also exhibit associated defects such as mental retardation, cardiac deformities and deafness.

OTHER FACTORS

Genetic abnormalities due to mutations resulting in enzymic abnormalities and the consequent metabolic blocks have been successfully identified for most of the hereditary disorders. However, not all physical, psychological and metabolic disorders could be solely ascribed to inherited genetic differences. Other factors such as social, environmental, may also be attributed for the hereditary or congenital abnormalities.

In the case of rickets, a disorder due to vitamin D deficiency or to the inadequate exposure of skin to ultraviolet radiation, the reason could be ascribed more to environmental and nutritional factors than genetic. Similarly, the endemic goiter disorder in many 'goitre belt' regions was believed to be an inherited one. But it is later recognized succinctly that an environmental factor viz., deficiency of iodine in the water is the main etiological factor responsible for the enlarged thyroid gland. Nevertheless some forms of goiter are inherited.

The exact contribution of nutritional and environmental factors, behind genetic contribution must be determined to initiate corrective measures and therapies to human genetic disorders. Towards this line considerable breakthrough and success has also been accomplished in the previous years due to advancements of biochemical knowledge. Withholding galactose from diets the galactosemia disorders was overcome in infants. Similarly, by limiting the intake of phenylalanine, in phenylketonuria; by withholding phytamic acid in Rafsom's disease, success was achieved. Conversely by administering insulin in diabetics; and by administering xanthine oxidase inhibitor, the purine metabolism oriented gout disease have been treated. Another approach which is yet to become promising but attempted unsuccessfully in the treatment to glycosphingolipodystrophies was the incorporation into the cells that contain stored/accumulated metabolites, the missing enzyme.

The mostly warranted approach to avert and decrease the frequencies of deleterious genes which are responsible for the hereditary disorders is at the societal level where the traditional customs and practices knowingly tend to increase the disorders in increasing frequency of the genes per se, in human population e.g., consanguineous marriages, and marriages between disease afflicted and disease free individuals with their mutual consent as the case in HIV^{+}ves and HIV^{-}ves. Nevertheless the potential of prenatal diagnosis and selective abortion to reduce the population with a genetic disease can be illustrated in case of a genetic disease viz., beta-thalassemia. Homozygous individuals for thalassemic genes are much more certain of facing severe and life threatening difficulties than the normal individuals. As a result in regions of Italy, Greece, Cyprus and the neighboring islands where the incidence of beta-thalassemia is high, prenatal diagnosis and abortion have become routine and the number of births of thalassemic babies has been greatly reduced since 1970s.

EXAMPLES OF GENETIC DISORDERS-DIAGNOSTIC CHARACTERISTICS

I. Amaurotic Family Idiocy (AFI)

An inheritable defect in the metabolism of sphingolipid, manifests clinically in three diseases viz., the infantile amaurotic family idiocy (Tay-Sach's disease), Neimann Pick disease and Gaucher's disease. These are differentiated according to the nature of lipids accumulated intracellularly. The accumulation of gangliosides either in nerves or reticuloendothelial characterizes Tay-Sachs disease. Similarly the accumulation of sphingomyelin and cerebroside characterizes Neimann Picke disease and gauchers disease respectively. The abnormalities in Tay-Sachs are confined to the nervous system. Neimann-Pick and gaucher's disease may involve the viscera and also show extensive brain damage. Amauratic family idiocy (AFI) is characterized by dementia, paralysis and blindness. Tay-Sachs, the infertile form of it (AFI) begins within the first year of life and is fatal by the third. The juvenile form becomes manifest at the age 5 to 10. In all forms of AFI, the pathological symptoms are restricted to the nervous system. In the infantile form the ganglion cells are overloaded with gangliosides,which are hydrophilic lipids containing sphingosine, fatty acid, hexose, hexosamine and sialic acid. The eventual death of ganglion cells is accompanied by demyelineation or failure of myelination and an increase in lipid laden glacial cells. Both infantile and juvenile AFI appear to be inherited by autosomal recessive traits. The metabolic sequence of gangliosides synthesis and the physical state of gangliosides in brain are not yet understood completely. Therefore, what enzymic defects manifest AFI remain to be still obscure and enigmatic (Stanbury *et al.*, 1960).

2. Thalassemia

Thalassemia is an erythoblastic anaemia beginning in early life of the individuals. Its severe form is associated with enlargement of spleen and liver and characteristic bone changes. The severe form is called Thalassemia major while its milder form, the thalassemia minor. However disease symptoms clinically expressed in between the major and minor is termed as thalassemia inter media. When the signs and symptoms are less than in thalassemia minor it is described as thalassemia minima. The thalassemic gene in homozygous state manifests severe thalassemia major while its heterozygous state expresses the disease in mild form viz., thalassemia minor. But all heterozygotes cannot be considered to belong the minor type (Fig. 6).

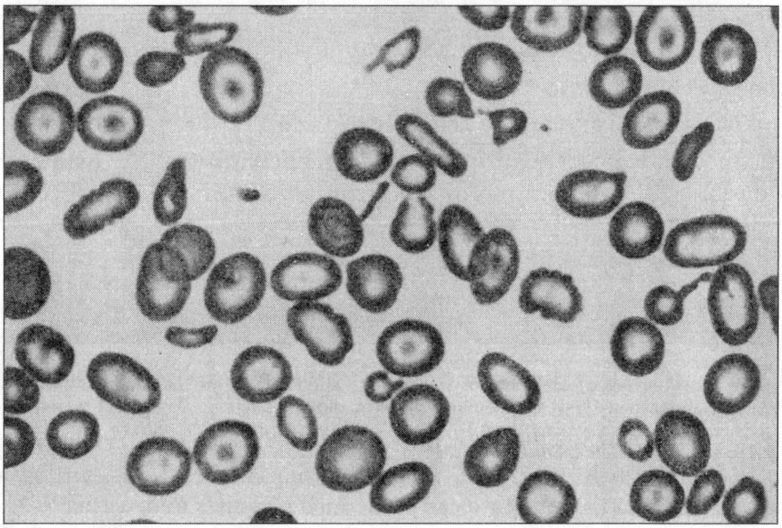

Fig. 6. Red cells picture in thalassemia major showing thin, unequal and irregular cells.
Source: The metabolic basis of inherited disease by John B. Stanbury *et al.*, (synthesized) Page No: 450, Published by McGraw Hill Book Co., Inc, New York (1960).

Thalassemia is characterized by the principle disability to produce red cells with normal haemoglobin content. The haemoglobin concentration is much below normal. Red cell production or erythropoiesies per unit bone marrow is decreased. There is also increased haemolysis in the peripheral blood. Splenomegaly, hepatomegaly are noted early. The disease was widely prevalent in Mediterranean regions and Asia and was also found among Turks, Persians, Kurdistan Jews and seen even in India and China (Carey *et al.*, 1998).

3. Fanconi Syndrome

The Fanconi syndrome is characterized by the features of osteomalacia, renal glycosuria, aminoaciduria and hyper phosphaturia. In infants the disease begins after a normal post natal period of 4 to 6 months. Death is usually due to progressive renal failure or intercurrent infection. In the adult the syndrome shows the same pathological and clinical features as in the infant viz., osteomalacia with multiple fractures, hypophosphatemia with hyperphosphaturia, elevated alkaline phosphatase, aminoaciduria and renal glucosuria. Crippling bone disease with deformities dominates the clinical picture. The hereditary nature of this syndrome was first noted by Abderhalden and has been referred to repeatedly since by many workers. All of them concluded that it is a recessive character. They have also considered the defect to be one of seven possible combinations of defects deriving from abnormalities of tubular reabsorption of phosphate, glucose and amino acids (Stanbury *et al.*, 1960):

1. Glucose alone (glucosuria)
2. Phosphate alone (Vit-D resistant hypophosphatemic rickets)
3. Amino acids alone (cystinuria or hepatolenticular degeneration)
4. Glucose with phosphate (combination of renal glucosuria and hypophosphatemic rickets)
5. Phosphate and amino acids (amino aciduria with Vit-D deficiency)
6. Glucose and amino acids
7. Phosphate, glucose and amino acids (syndrome).

Thus Fanconi syndrome constitutes no clear cut entity with distinct, functional or etiological boundaries but merges into the spectrum of disturbances arising from all possible renal tubular defects.

4. Gout

Among the various metabolic diseases, gout is a notable one rich in historical lore. Its descriptions can be traced to the dawn of medical literature. It is a disease of purine and pyrimidine metabolism. In 1776, the chemical history of gout was traced to the compound uric acid. Primary gout is a genetically determined disorder of purine metabolism in which the cardinal feature is hyperuricemia. The hyperuricemia is initially asymptomatic, and may remain so, but in a small percentage of affected individuals the disorder is complicated by episodes of acute arthritis of unknown cause, by chronic arthritis due to depositions of urate in cartilage and other tissues, by formation of renal calculi composed of uric acid, and possibly by an accelerated development of vascular and renal disease (Sarada Subrahmanyam and Madhavankutty 2004).

Though recent studies attributed gout to metabolic derangement, the precise biochemical defect or defects leading to hyperuricemia in subjects with this disease is still unknown. Though several factors are attributed, for the disease, they cannot adequately explain the mechanisms operating to initiate the acute attack of gouty arthritis in a person genetically endowed with the gouty trait. The most common complication of hyperuricemia is acute gouty arthritis. Primary gout may occur at any age but this is predominantly a disease of adult males. In a large series of gouty patients, only about 5% are females. The chronic gouty arthritis begins incidiously and progresses with or without the superimposition of further attacks in the involved or other joints. The duration of time from the initial attack to the beginning of chronic symptoms is highly variable ranging from 3 to 42 years. In the chronic case, the accumulations may be enormous and the deformities of the hands or feet grotesque. The deposition of solid urate crystals are not limited to joint regions alone but also in regions such as cartilaginous tissues of the body (helix and antihelix of the ear, olecranon or Patellar bursa, etc). In the chronic gouty arthritis, the characteristic tophaceous nodule (tophus) consisting of multicentric deposits of urate crystals and intercrystaline matrix together with the inflammatory reaction is found. The free purine bases that result from the nucleoside cleavage are adenine, guanine, hypoxanthine and xanthine. Since purine nucleosides phosphyrylase acts most readily upon inosine and guanosine the major bases generated are very likely hypoxanthine and guanine. In mammalian tissue free adenine is not deaminated as it is in certain bacterial systems, so that adenine does not give rise directly to hypoxanthine. It is not reconverted to its nucleoside or nucleotide, it may be excreted unchanged, and normal human subjects excrete adenine in small quantities in urine. In contrast, the other purine bases are readily converted to uric acid. Guanine is deaminated by guanase to yield xanthine. Hypoxanthine is oxidized by xanthine oxidase to yield xanthine, which in turn is further oxidized by the same enzyme to yield uric acid.

The heredity of hyperuricemia of gout is well established. Studies conducted in different generations within the gouty families revealed the genetic mechanism or mechanisms underlying the gouty patients. The trait for hyperuricemia behaves as a dominant factor in most gouty families. They have also revealed that it is due to autosomal dominance as far as males are concerned. However the above hyperuricemia and the consequent gout is determined by an autosomal dominant gene with incomplete penetrance (monomerism) or by a cumulative gene action (polymerism) remains to be elucidated (Stanbury *et al.*, 1960).

5. Albinism

It is an inherited disorder of melanin metabolism characterized by the absence of melanin in the skin, hairs and eyes (Smith *et al.*, 1983). Melanin is a dark brown or black insoluble pigment contained on melanin granules in the cytoplasm of melanocytes and formed from tyrosine by tyrosinase. The clinical characteristics of this disease are the milk-white coloured skin and marked photophobia. In albinism, complete lack of melanin pigment (a dark brown pigment) causes the albino to suffer. The incidence of albinism in human has been reported to be from 1:5000 to 1:25000. The albinism may be generalized albinism, localized albinism of the eye (ocular albinism) or partial albinism (skin and hair) (Table 3). The recessive genes 'aa' do not produce the tyrosinase enzyme, which converts DOPA (3,4-dihydroxy phenyl alanine) into melanin in the melanocytes.

SUMMARY

Since metabolic disorders in homozygous allelic condition are lethal while heterozygous condition entitle the person to be carrier of the disease, diagnosis of such diseases in foetus stage itself through techniques like amniocentesis could reduce the incidence of individuals in the population by opting abortion. Although this seems to be a useful tool for reducing the incidence and frequency of diseases, in principle, such prenatal diagnosis and selective abortion may be opposed and the implementation and practice of such medical technique legally, may be opposed by such factors as political, social, religious and economic factors, etc. in both developed as well as developing countries. The current advancements in genetic engineering, biotechnology, and the pharmacological approach of drug designing may show a promising feature for the discovery of "made to order drugs" for many hereditary disorders.

REFERENCES

Beadle, G.W. and Tatum, E.L. (1941) Genetic control of biochemical reactions in Neurospora. Proc. Natl. Acad. Sci., USA, 27 : 499-506.

Beadle, G.W. and Tatum, E.L. (1945) Neurospora II. Methods of producing and detecting mutations concerned with nutritional requirements. Amer. J. Bot., 32 : 724-740.

Carey, C.F., Lee, H.H. and Woeltje, K.F. (1998) The Washington Manual of medical therapeutics. 29th edition, Lippincott-Raven publications, New York.

Edelstein, S.J. (1986) The Sickled Cell, Harvard University Press, England.

Gardner, E.J. (1975) Principles of Genetics, 5th Edition John Wiley, New York.

Garrod, A.E. (1909) Inborn Errors of Metabolism, Oxford University Press, Oxford.

Graham, C.F. and Wareing, P.F. (1984) Developmental Control in Animals and Plants, Blackwell Scientific Publications, Oxford.

Gupta, P.K. (1996) Genetics, 3rd Edition, Rastogi Publications, Meerut, India.

Ingram, V.M. (1957) Gene mutations in human haemoglobin. Nature, 180 : 326-328.

Ridley, M. (2002) Genome. Books Today, New Delhi, India.

Sarada Subramanyam and Madhavankutty, K. (2004) Textbook of human physiology, 6th edition, S. Chand & Company, New Delhi, India.

Smith, E.L., Hill, R.L., Lehman, I.R., Handler, P. and White, A. (1983) Principles of biochemistry – General aspects, McGraw-Hill Book Co., Singapore.

Stanbury, J.B., Wyngaarden, J.B. and Fredrickson, D.S. (1960) The Metabolic Basis of Inherited Disease, McGraw-Hill Book Company, Inc., London.

Wagner, R.P. and Mitchell, H.K. (1964) Genetics and Metabolism, John Wiley, New York.

Watson, J.D. (1987) Molecular Biology of Gene, 4th edition, Benjamin, New York.

Suggested Readings

1. The Sickled Cell (1986) S.J. Edelstein, Harvard University Press, England.
2. Developmental Control in Animals and Plants (1984) C.F. Graham and P.F. Wareing, Blackwell Scientific Publications, Oxford, U.K.
3. Genome (2002) M. Ridley, Books Today, New Delhi, India.
4. Principles of Biochemistry-General Aspects (1983) E.L. Smith, R.L. Hill, I.R. Lehman, P. Handler and A. White, McGraw-Hill Book Co., Singapore.
5. The Metabolic Basis of Inherited Disease (1960) J.B. Stanbury, J.B. Wyngaarden and D.S. Fredrickson, McGraw-Hill Book Company, Inc., London.

REVIEW QUESTIONS

Multiple Choice Questions

1. Who propounded the famous 'one gene one enzyme theory'
   (a) Gregor Johann Mendel
   (b) Thomas Hunt Morgan
   (c) Jacob and Monad
   (d) Tatum and Beedle

2. In which of the following haemoglobin chains the genetic mutation occurs for sickle cell anemia
   (a) α-chain (b) β-chain
   (c) In both α and β (d) None of these
3. In the haemoglobin polymorphism, which amino acid is variable
   (a) 6th position (b) 7th position
   (c) 8th position (d) none of these
4. When homogentisic acid accumulates in the urine the metabolic disease is called
   (a) Alkaptonuria (b) Phenylketonurai
   (c) Alzhemeir's (d) Korsakov's
5. When the genotype is 44A+XXY, the person is said to be
   (a) Down's syndrome (b) A Klinefilter
   (c) A Turner (d) A Fanconi
6. Thalassemia is prevalent in the following people
   (a) Turks and Persians (b) Kurdistan Jews
   (c) Both a and b (d) Negros
7. Gout, the metabolic disease is caused due to the accumulation of
   (a) Uric acid (b) Lactic acid
   (c) Pyruvic acid (d) Hippuric acid
8. In albinism, the following is not produced
   (a) Melanin (b) haemozoin
   (c) Zymogen (d) Keratin
9. Sickle cell anemia is due to the following genotype
   (a) Hb^AHb^A (b) Hb^SHb^S
   (c) Hb^AHb^S (d) Hb^AHb^C
10. Metabolic disorders and diseases are due to
    (a) Recessive genes (b) Mutator genes
    (c) Lethal genes (d) Silent genes

Answers

1. (d) 2. (b) 3. (a) 4. (a) 5. (b) 6. (c)
7. (a) 8. (a) 9. (c) 10. (a)

CHAPTER 36

Drug Metabolism and Detoxification

K. Ramalingam
Dept. of Advanced Zoology and Biotechnology, Govt. Arts College, Nandanam, Chennai–600 035
E-mail: drkrimmunotoxicol@yahoo.co.in

Several bio-active substances or chemical compounds operate the cellular metabolism and are responsible for the health, reproduction and survival of organisms. These molecules/ compounds, which are common to both animals and human, include vitamins, co-enzymes, enzymes, hormones, neurotransmitters, inorganic ions, nutritional fragments, respiratory pigments, adenosine triphosphate (ATP) and the purine and pyrimidine components of nucleic acids (RNA, DNA), etc. The interactions between various biomolecules and their complementary biopolymers viz., the enzymes are responsible for the cellular and organismal integrity (Smith *et al.*, 1985; Stryer, 1981).

DRUGS AND ANTAGONISTS

Pharmokokinetics is the derivation of information and data by animal model studies about the degree of absorption of the drug, its routes, and rate of elimination, the nature of chemical changes to the drug and the temporal distribution of the drug and its metabolites in the body tissues especially those involving the targets of the drug action. These metabolic studies serve to validate the choice of species selection as a model for human. Drugs represent the biologically active agents which are used for the treatment of diseases in man and his domesticated economic animals. Besides their usage for curative purposes, some drugs or chemical compounds have been deployed to eradicate the pests, attacking agricultural crops as well as the vectors involved in the transmission of epidemics. On the contrary antagonists are drugs, which are devised to counteract an ill effect *in vivo*. Such drugs may be designed to overpower an invading pathogenic microbe or help to reinstate normalcy from a disordered metabolism by suppressing abnormal or unwanted cycles or rhythms. The former is designated as chemotherapy while the latter, pharmacodynamic therapy. Various categories of antagonist drugs include the general anaesthetics, local anaesthetics, hypnotics, antihistamines, analgesics, antiepileptics, neurotransmitters antagonistics and selective toxicants against pathogens such as bacteria, viruses, fungi, protozoans, parasitic worms, etc.

Drugs mostly are organic chemicals of low molecular weight and are synthesized in the chemical industries. Some drugs are natural plant products and derivatives of animals. In this context, the alkaloids represent the nitrogenous excreta of plants, and the antibiotics represent the secretory products of fungi or bacteria. The hormone cortisone and vitamins belong to latter mentioned category. In addition to these, antigens and antibodies of vaccines and sera are high molecular weight immunochemicals.

Community Diseases (Infectious Diseases)

The various diseases of human that warranted the drugs discovery throughout the world include parasitic diseases, bacterial epidemics, viral diseases, fungal (mycotic) diseases, etc. Among the parasitic diseases malaria, trypanosomiasis, leishmaniasis and amoebioasis are the most serious ones caused by protozoa. Besides the above helminthic diseases caused by parasitic worms are also a serious threat to adults, children in human and to various domesticated and wild animal species. Among bacterial epidemics some of them such as cholera, TB, trachoma, leprosy, brucellosis, gonorrhoea and syphilis were found to be inadequately controlled in world in general and more specifically in underdeveloped and developing countries. In spite of effective drugs being available, various social factors are hindering their application. Improvement in nutrition, sanitation and water supply, need for prophylactic measures, quarantine measures become the determinants. to supplement the beneficial effects of the antibiotic drugs. Of virus caused diseases, immunization by vaccines have steadily decreased the incidence of such diseases as small pox, chicken pox, measles, german measles, polio, etc. But in recent years, the emergence and onslought of certain viral diseases in human population which broke out as transdemic/pandemic diseases and also as zoonotic (SARS and Bird flu) and sexually transmitted ones (AIDS) have become a cause for concern. The current recourse to immunotherapy with sera and vaccines seemed to be successful only to a few viral diseases. Turning to fungal diseases of man (mycoses) they are being treated by internal medication but there is still scope for better drugs. As far as the parasitic diseases are concerned, the containment of them, need, not only the modern anthelmintic drugs but several other factors such as the control of vectors, improvement of water supply to population, disinfection and bioremediation of contaminated water bodies, new irrigation schemes are necessary to delink the infection chain viz., environment-vectors-parasites – human/animal populations. Many farm animals have suffered from severe parasitic diseases which sapped their vitality and immunity and decreased their market value. The yields from such farm animals are also drastically reduced.

Non-infectious Diseases

While infectious diseases take a great toll in underdeveloped countries, the more prosperous developed countries suffer due to non-infectious diseases. These diseases, which arise as a consequence of life style changes, stress, metabolic derangement, require pharmacodynamic rather than chemotherapeutic agents. Heart attacks, stroke, cancer, auto-immune diseases, rheumatoid diseases, mental ill health, Alzheimers disease and other psychosomatic illness, etc. account for seventy percent of all deaths, besides bringing a high percentage of incapacitation. For all these modern diseases, novel therapeutic treatment and psychological treatment are urgently required. Still some non-infectious diseases like cancer which includes about 100 different types of malignant tumours remains a challenge to our scientific research arena and some leads are required to replace the existing toxic drugs (chemotherapy) and radio therapy.

BIOLOGICAL RESPONSE AND RECEPTORS

The cellular biological responses to drugs depend on the action and kinetics of various metabolites, enzymes and hormones. The latter in turn is highly dependent on fine details of their chemical structure. For instance, the activity of thiamine (B_1) drops to 5%, if the methyl group is removed from the pyrimidine ring and its activity completely disappears if an extra-methyl group is inserted. Similarly, in drugs, the structure and activity are very much correlated to their manifestation of pharmacological properties. Understanding of enzyme and protein chemistry and cellular properties are pertinent to the discovery of drug formulations and their biological/therapeutic actions.

The action of drugs reveals very strongly that they are concentrated by cells on small, specific areas called receptors. Drug-receptor combination which is of reversible nature has been established as the principle, behind drug action, drug specificity, drug efficacy and for biological effect. The drug receptors have been identified as enzymes proteins, nucleic acids, coenzymes, small molecules etc. The action of a drug on a drug receptor is often correlated to the desired physiological result through a cascade of chemical and biological reactions (O'Brien, 1979).

DRUG DISTRIBUTION

Drug distribution is an important aspect in disease therapeutics. Differential distribution is attributed to the selective action of several drugs, which would otherwise be toxic to all kinds of cells. Selective partitioning and selective distribution by cells and tissues enable the drugs to undergo sequestration/storage, inactivation and excretion and concurrently effect their pharmacological action. Sequestration and storage of drugs take place in several ways. Lipophilic drugs like barbiturate-thiopentone are accumulated in deposits of fats; anionic drugs such as aspirin, sulphonamides and anticoagulants are complexed with serum albumin; cationic drugs like chloroquine are stored by nucleic acids. The above binding and storage depends on the fine details of the drug's molecular structure and plays an essential part in replenishing the plasma level as and when it begins to fall.

The inactivation of drugs *in vivo* involves the breaking of covalent bonds which is not freely reversible. The excretion of drugs is effected mainly through kidneys, bile ducts and to certain extent through other organs like lungs and skin. In several situations, the drugs after being subjected to storage and inactivation and the consequent loss, cannot be excreted unless and otherwise it penetrated back into the blood stream. This seems to be a limitation which tends to maintain the drug concentration in vivo. The drugs after surviving all these hazards of storage, inactivation and excretion only, enter the compartment (foci) that contains the receptors with which they bind and effect their characteristic pharmacological response.

The receptor-drug binding is covalent in only a few cases (Mercurials, Pencillin organophosphate compounds) and in majority it takes place by easily reversible secondary bonds. As a result the concentration of the drug may fall steadily leading to insufficiency and inefficacy of drug action. Hence a steady course of medication must be practiced in order to maintain an effective concentration of the drug near the receptor.

The concept of apparent volume of distribution (V_D) is often used in bioassays to enable drug action. It is defined as :

$$V_D = \frac{\text{Weight of drug in the body}}{\text{Plasma concentration of the free drug}}$$

V_D refers here to the fluid volume in which the drug seems to be dissolved. Value of V_D is an inference of drug confinement in tissue compartment. Values of V_D greater than the total body volume indicate that the drug is deposited in a tissue. Some relevant volumes of body compartments are (in litres) given below:

Circulating plasma of blood	: 3 L
Erythrocytes	: 3 L
Extra cellular water	
Other than blood	: 11 L
Intracellular water	: 24 L.

The total volume may constitute about 41 litres or 58% of body weight.

Biomembranes and Drug Distribution

The distribution of a drug is highly dependent on its ability to penetrate the semipermeable and selectively permeable membranes. Studies on the permeability behaviour of biomembranes have categorized four main types of membranes (Chappell, 1966; Phillips *et al.*, 1965).

Type 1 membranes: These seem to be the most common membrane type in which majority of drugs in common usage penetrate through. These molecules possessing some lipophilic groups, few hydrophilic groups and no ionic charge, diffuse fast.

Type 2 membranes: These membranes incorporate a carrier, which provides facilitated diffusion. The transported molecule combines reversibly with the carrier, which then makes 180° turn towards the inner surface of the membrane in order to release the molecule into the cytoplasm. Permanently ionized substances, which cannot enter by simple diffusion through 'type 1' membranes are transported by facilitated diffusion. The transport of glucose into human erythrocytes is by facilitated diffusion. The carrier in this, esterifies the a-hydroxy group. Carrier facilitated diffusion occurs in mitochondria for several of its metabolites viz, succinate, malate, malonate, tartrate mesotartrate, maleate, fumarate, citrate, cis-aconitate, phosphates and adenosine nucleotides. By designing drugs with groups, which will facilitate carrier-mediated uptake, the increased selectivity of the drugs could be accomplished.

Type 3 membranes: These are the most complex of all types, which concentrates substances against a gradient and operates by expenditure of energy. The penetrating molecule is thought to combine with a carrier as in Type 2 but this carrier is subjected to chemical modification in the membrane. The transport of organic cations in mammalian cells, iodide anions into the thyroid gland cells occur in this way. The absorption and excretion of a wide range of ionized and non-ionized substances by the kidney tubules and to a less extent by the gastrointestinal membrane are effected in this fashion (Misra Hunger and Keberle, 1966).

Type 4 membranes: These are membranes provided with pores, which enable the transport of small anions such as chloride and larger molecules like insulin (mol wt. 5000) e.g., glomerulus. These pore channels exclude cations by coloumbic repulsion. New drugs are designed by stepwise increases in lipophilicity suited for Type 1 membranes. For type 2 and 3 membranes, the drugs are designed to resemble in the relevant part, a natural substance for which a specific transport mechanism exists in particular cells. The designing of anticancer drugs like nitrogen mustard is based upon the above principle. To obtain excellent results amino acids and natural sugars are attached to enable efficient penetration and drug action (Faust and Shearin 1974).

Mammalian Tissues

The absorption and distribution of foreign substances into tissues follows a much simpler pattern than those of natural cell substrates and active constituents. Among mammalian tissues, the gastro-intestinal epithelium, the renal tubule epithelium, the blood brain barrier and the blood-cerebrospinal fluid (CSF) offer a Type 1 lipid barrier to the passage of many foreign molecules.

Digestive tract: Drugs are well absorbed from the stomach by the change in pH. When the pH of the stomach contents is increased, basic drugs could be better absorbed because they are less ionized. On the contrary this pH change decreased the absorption of acidic drugs because less of the non-ionised form remains. Weakly acidic drugs such as salicylic acid, aspirin and the more lipophilic barbiturate are readily absorbed because they are not ionized at the pH of the stomach (pH1). On the contrary, basic substances such as quinine, ephedrine and amidopyrine are not absorbed because they are totally ionized at this pH. The epithelial lining of small intestine permits the penetration of non-ionized drugs and impede the passage of the corresponding ions. The foreign substances that closely resemble the natural substrates like amino acids, glucose are transported from small intestine to plasma by the same active transport mechanism, which does for the natural substrates. The absorption of drugs from the colon is found to be similar to that from the small intestine. The distribution of drugs between blood plasma and the tissues is also the same as between the gastrointestinal tract and the blood plasma. However, the above distribution holds good only to those fractions not bound to plasma proteins.

Blood: Drugs strongly bound to blood albumin are excreted with much difficulty by the glomerulus. On the contrary, if the drug constitution makes it suitable and eligible for the active transport mechanisms of the renal tubules, the protein binding is of little avail against excretion. These principles are of value for designing and redesigning a drug with a view to reducing the effective dose. Molecules much larger than those of the common antibiotics and synthetic drugs pass readily from blood to bile. The biliary pathway of absorption and excretion of drugs is however insufficiently understood.

Brain: The brain is protected by a membrane called the blood- brain barrier, which lines the capillaries of the cerebral vessels. Substances with a very high lipid/water partition coefficient can penetrate this barrier. Thus, highly lipophilic drugs can penetrate this barrier. Next to the barrier the drugs penetrate the meninges. Sulphanamides and barbiturates being more lipophilic, readily penetrate the blood brain barrier. But penicillin and salicylates could not. When the barrier and brain meninges are inflamed, a wider range of drugs pass through into the brain tissue. Lipophilic drugs are also differentially distributed within the brain tissue. Generally the white matter being rich in lipids, drugs concentrate more in it than the grey matter.

Permeability of drugs within cells depends upon the type of membranes, the cytoplasmic organelle belong to. The membranes of mitochondria and endoplasmic reticulum appear to be type 1 with patches of types 2 and 3. The nuclear membrane appears to be the highly porous type 4. Specific binding of drugs within cells also seem to play an important role in the selective toxicity of the drugs. Much remains to be understood still towards this end.

ACTIVATION OF PRODRUGS

Though most drugs appear to act in the form in which they are introduced into the body system, some drugs are administered as prodrugs to be converted to the true active forms by the enzymes inside the body. The prodrug concept (Valdivieso *et* al., 1976; Foster *et* al., 1981; Bodor, 1982) has some advantages in drug therapy. Certain chemical compounds are ineffective in their form but they undergo some conversions to become true drugs. For instance, in treating trypanosomiasis (sleeping sickness) phenylarsonic acids are given, which diffuse into the site of infection in the central nervous system and there they become reduced to the true drug form viz., arsenoxides, which do not penetrate the blood brain barrier as such. In this example arsenoxide administered in its corresponding prodrug form phenylarsonic acids is advantageous to bring the efficacy. In some cases the administration of true drug forms confer efficacy. For instance the antibacterial sulphonamids were previously thought to be true drug form. But these showed infact, no intrinsic antibacterial properties. The true drug form was later identified as the sulphanilamides which formed by the reductive fission in the patients bowels. Accordingly, the sulphamilamides replaced the sulphonamides in clinical therapeutics since the former acted more promptly and directly. In cancer therapy, 6-azauridine is converted into triacetyl derivative to be absorbed well in the intestine.

In the drug manufacture the contemporary trend is to introduce substituents into a drug which deactivate the functional group of the drug. Inside the body the enzymes remove the substituents and the drug becomes active. The deactivating groups are used as substituents to deactivate the chlorine atoms in the nitrogen mustard class of anti-cancer drugs. The enzymes in cancer cells remove the deactivating substituents to make the drugs more potent. Whenever no correlation could be found between plasma concentration and therapeutic effect, the drug may probably be construed a prodrug (Elion *et al.*, 1977).

Table 1. Prodrug-True drug nomenclature

Predrug	True Drug	Disease
Succinyl Sulphathiazole		Sulphathiazole
Bacterial dysentery		
Proguanil	Cycloguanil	Malaria
Arsonic acid	Arsenoxide	Trypanosomiasis
Sulphonamides	Sulphanilamides	Antibacterial

METABOLISM AND DRUG INACTIVATION

Three principal mechanisms by which an active drug can be lost before it reaches the effective receptor are the binding and storage, (sequestration) excretion and chemical inactivation. Veldstra referred to the loci where these three major mechanisms of drug loss occur as "sites of loss". He has also suggested that blocking of these sites of loci will allow a higher concentration of the drug to reach the receptor. In the above mechanisms inactivation of drug may be effected by metabolic changes, contrary to the mechanism where metabolic enzymes activate the prodrugs to true drugs. These metabolic alteration or inactivation of drugs involves making or breaking of covalent bonds. It is also less reversible than the process of storage. Several hydrophilic drugs are excreted as such in mammalian body. But most of the drugs undergo conjugation with some small metabolites before being excreted. Thus, organic acids are conjugated with glycine; A few amines undergo acetylation but most of them are conjugated with glucoronic acid or sulphuric acid. Glucoronides are formed by alcohols, amides and carboxylic acids in the celis endoplasmic reticulum. Such conjugations occur almost in organs such as gut, liver, spleen, lungs, kidney, etc. The endoplasmic reticulum is endowed with many kinds of scavenging enzymes and serves as a versatile organellae, to accumulate more lipophilic drugs/compounds and chemically alters them into hydrophilic substances to be excreted readily. For example, toluene is oxidized by the endoplasmic reticulum to benzyl alcohol, and its further oxidation in cytoplasm gives rise to benzoic acid which then conjugates with glycine in the mitochondria. The resultant benzoyl glycine (hippuric acid) is readily eliminated in the urine. The microsomes formed as spherules from endoplasmic reticulum often contain more potent oxidases and also other reductases, hydrolases. These enzymes perform the degradative processes viz., aliphatic-c-hydroxylation, aromatic-c-hydroxylation, N-oxidation, S-oxidation, O- and S- dealkylations, N-dealkylation and deaminations etc. These metabolic alterations of drugs/foreign substances are collectively referred to as "Detoxification" (Adrien, 1985).

SYNERGISM OF DRUGS

Synergism of drugs refers to the action of a substance, the diphenylprophyl acetic ester of diethyl amino ethanol (SKF-525A), which synergizes a wide variety of drugs and prevents their metabolic alteration in the endoplasmic reticulum. The synergetic action of this substance is by the non-competitive inhibition of all hydroxylation reactions and by the competitive inhibition of hydrolytic reactions and does not exert any effect on the endoplasmic reticulum membrane (e.r). Thus synergism is a mechanism of blocking the sites of loss. But in the pharmacodynamics there is no place for synergism. Antagonism refers to the inhibition of drug loss by metablic inactivation more often accidentally by other drugs. These instances are examples of unfortunate synergism by antagonistic drugs. But favourable examples also exist. For instance pencillins, which are moderately liposoluble acids may be eliminated through the proximal tubules of the kidney by facilitated transport. But several inert substances can physiologically block the above excretion of pencillins. Hence probenecid [Benwenaid-P(dipropylsulphamoyl) benzoic acid] is used clinically to enhance the action of pencillins when highly resistant strain of bacterium infects a person. The drug loss by enzymic degradation can also be overcome through use of synergists.

PHARMACOGENETICS

Generally, in animals as well as human, detoxification of foreign substances require detoxicating enzymes viz., (MFO) multiple function oxidase eg., cytochrome P-450. Sometimes in some persons or in a race, the genes for the above detoxifying enzymes may not be present. This genetic deficiency causes a patient to react even to a small dose of drug as though, it were a large one. This genetic cum immune response is termed as pharmacogenetics.

The foregoing account of drug metabolism elucidates the mechanism of drug action, selectivity and the factors that interact with a drug's selective/specific action and distribution and also its loss. The complete understanding of the above drug metabolism still prompts more information and knowledge on various factors, and agents both good and bad, and will go a long way to add more insight to the drug biochemistry and its implication in medicine.

In animal species differences exist in the ways in which foreign substances are broken down. Comparative biochemistry has revealed such differences between species (interspecific) as well as within species (intraspecific). Such differences in the different tissues and in the same tissue of different regions of a single organism are also not uncommon. For example, aconitase and oxaloacetate transacetase are much more abundant in heart than skeletal muscle and the reverse is true for aldolase. The enzymes xanthine oxidase, catalase and D-amino acids oxidase are abundant in liver and kidney but absent in the muscle tissue of mammal. Similarly, glutamine synthetase, which condenses glutamic acid with ammonia is abundant in brain and liver but almost absent from all other human tissues.

ANALOGOUS ENZYMES

Enzymes carrying out apparently identical functions in two dissimilar organisms are themselves dissimilar. These are known as analogous enzymes. Analogous enzymes have been demonstrated by their differences in kinetics, electrophoresis, specificity of co-enzymes, substrates or inhibitors. The selective action of analogous enzyme, one localized in the parasite (Schistosome) and the other in the host (man) provide the best example. These worms are blood flukes belonging to the class trematoda (Helminthes) parasitic in man and cause a chronic debilitating illness wherein the infected man's urinary bladder becomes inflamed by the eggs of the parasite and the patient passes bloody urine (Menstruating men disease). The phosphofructokinase of these blood flukes is much more sensitive to inhibition by antimonial drugs than is mammalian phosphofructokinase. The therapeutic success of antimonial drugs lies in the above differential sensitivity of the analogous enzymes of the host and parasite. These drugs are also successful in treating the bilharziasis in Human. The current biochemical investigations and studies have revealed that sequentrial blocking of two or more enzymes in the parasites which are responsible for their growth within the host may become a successful therapy since the drugs arrest the parasites sustainability on one hand while keep the patients welfare agile on the other.

ENZYMES AS DRUGS

In malignant tumours it has been revealed that striking differences exist between the normal tissue and tumour tissue. For instance, in some strains of human leukemia cells the enzymes for synthesis of L-asparagine, is necessary for their metabolism and growth. By injecting asparaginase into a patient, his peripheral body part of this amino acid can be depleted without causing him great distress. In the cancer wards, drug resistant lymphoblastic leukemia of children respond to this treatment, but for a short time only. Similarly the brain tumour where in the guanase enzyme is lacking as compared to the normal counterpart, the use of purine analogue 8-azaguanine showed remissions after all other treatment modalities have been exhausted.

METABOLITE ANALOGUES AS DRUGS

Metabolite analogues are drugs or other foreign substances especially designed to antagonize metabolite. The molecule of each such analogue has a region similar to that region of the metabolite which makes contact with the enzyme. The molecule of analogue must be similar not only in dimensions but also in electron distribution, because most of the active sites on enzymes are highly polar. The metabolite analogue exerts its antagonism by occupying and blocking the enzyme reactive site used by the substrate metabolite. This relationship between metabolite and analogue and the inhibition of the enzyme action is known as competitive inhibition. If x molecules of metabolites are antagonized by y molecules of analogue then 10x molecules of metabolite require 10y molecules of analogue to give the same biological endpoint. Since competitive reactions are freely reversible, the antagonism of x molecules of metabolite by y molecules of analogue can be abolished by another x molecules of metabolite.

Each metabolite + antagonist pair has an unique index of inhibition defined as the ratio of the number of molecules of analogue (to those of metabolite) required to give 50% inhibition. This ratio, which varies with the biological species, expresses the relative affinity of analogue and metabolite for the appropriate receptor, but it also includes a hidden term for differences in the penetrative ability of the two substrates when the site of action is not exposed in the test. The dissociation constant (K_i) for an inhibitor is expressed:

$$K_i = [E]\,[I]/[EI],$$

Where [E] is the concentration of the enzyme, [I] is that of the inhibitory, and [EI] is that of the complex which they form. Hence the index of inhibition (or inhibitory index) is K_i/K_m, in which K_m is the Michaelis-Menton constant:

$$K_m = [E]\,[S] / [ES]$$

Where [S] is the concentration of the substrate. The smaller the index of inhibition, the more efficient the inhibitor. That for the antagonism by sulphanilamide of *p*-aminobenzoic acid in streptococci is 300, a rather large but still quite advantageous figure. The index 0.001 for the antagonism by methotrexate of dihydrofolic acid (on its reductase) is outstandingly good.

Most inhibitory indices have been found to be high because enzymes have evolved, slowly and under the strong selection pressure of nature, to do their function efficiently. They are usually undersaturated with substrate and although adsorption of the antagonist may initially inhibit the enzyme, accumulation of extra substrate can overcome this inhibition, thus leading to a new series of steady states, which may be advantageous for the organisms greatly. In view of this, formulating metabolic analogues that are selective in action proved to be hard. Nevertheless, metabolite analogues that have surmounted all hurdles in the more complex systems of living cells have been discovered and are in commercial production and regular use.

ANALOGUES OF VITAMINS

Amprol – an analogue of thiamine, which has proved highly successful in coccidiosis, a protozoan infection of poultry.
Oxamycin – An analogue of D-alanine used in bacterial infection.

The antibiotic azaserine (O-diazoacetyl-L-Serine) – A structural analogue of glutamine used as a sequential blocking agent in the chemotherapy of cancer, interferes with one of the earliest stages of purine biosynthesis. Similarly some successful analogues of purine and pyrimidines are currently the treatment of choice for cancers in human.

DRUG METABOLIC PATHWAYS

Pathway I

Pathway 1 involves oxidation of the drug and is catalyzed by the mixed function oxidase enzyme system. The mixed-function oxidase is considered to be the principal drug-metabolizing enzyme system in the body because it has a broad substrate specificity (i.e., it catalyzes the oxidation of a wide variety of substances). Multiple forms of the mixed-function oxidase, each of which possessing unique but overlapping substrate specificity exist. The reactions catalyzed by this enzyme system increase the hydrophilicity (water solubility) of drugs and facilitate their easy elimination by the kidney.

Pathway 2

Pathway 2 of drug metabolism consists of a number enzymes that catalyze the formation of conjugates with the oxidized drug. The enzymes catalyzing these reactions are classified as transferases. The conjugates produced are ionized at physiological pH. A variety of transferases exist, including the glucuronyl transferases, sulfotransferases and glutathione trasnferases, as well as several specific amino acid trasnferases. Conjugation of drugs leads ultimately to the generation of highly water soluble, ionic compounds, which are readily excreted in the urine. The net effect of Pathway 1 and 2 is the conversion of lipid-soluble drugs to ionically charged water-soluble agents that are removed efficiently from the body.

Mixed-Function Oxidase Enzyme System

The mixed-function oxidase system is considered to be the major enzyme system responsible for the metabolism of drugs and other xenobiotics. High levels of the mixed functions oxidase are present in the liver, which is generally considered to be major site of drug metabolism. However, this enzyme system is later on found in almost all organs. Thus, drug metabolism can occur throughout the body. The mixed-function oxidase is present predominantly on the smooth endoplasmic reticulum of liver cells, although activity is associated also with the rough endoplasmic reticulum. The enzyme system has been purified, and three distinct components have been isolated: viz., (1) cytochrome c reductase (cytochrome P-450 reductase), (2) cytochrome P-450, and 3) a phospholipid.

Cytochrome c Reductase

The first component, cytochrome c reductase, catalyses the initial transfer of an electron nicotinamide-adenine dinucleotide phosphate (NADPH) to the cytochrome. The enzyme has an absolute requirement for NADPH and requires Mg^{2+} for activity. The enzyme contains one molecule each of flavin adenine dinucleotide (FAD) and flavin mononucleotide (FMN) as cofactors, This reductase is capable of donating two electrons to the cytochrome, the first of which reduced the heme iron to the ferrous form and the second to oxidized oxygen to the superoxide anion.

Cytochrome P-450

The second component, cytochrome P-450, is the terminal electron acceptor and is the binding site of drugs. The designation cytochrome P-450 is actually a generic term that denotes a family of cytochromes that are immunologically and biochemically distinct, although all accept an electron from NADPH cytochrome c reductase. Specific antibodies have been raised to the different forms of cytochrome P-450, demonstrating specific structural differences between the different species of this cytochrome. At least 16 different species of the cytochromes have been identified to date. The unique substrate specificities of the different forms of cytochrome P-450 are imparted by the structure of the apoprotein. The cytochromes contain iron that is chelated to a porphyrin ring system that is identical to that of hemoglobin and protoporphyrin IX. Because sex steroid hormones have been shown to induce different forms of cytochrome P-450 species, selective induction of specific forms of cytochrome P-450 with various drugs and other exogenous agents has been used in the purification and characterization of the different species of this cytochrome.

Phospholipid

The last essential component of the mixed-function oxidase system is a phospholipid that is required to act as a membrane surface to allow for the exchange of electrons from the reductase to the iron of the cytochrome. The second electron that oxidizes oxygen to the superoxide anion can come also from cytochrome b_5 and cytochrome b_5 reductase, a nicotinamide-adenine dinucleotide (NADH)-specific reductase. Under conditions in which NADPH is limiting, it is believed that the second electron can be donated by the cytochrome b_5 system. This latter enzyme system cannot take the place of NADPH cytochrome reductase as a donar of the first electron.

As noted in the reaction, for every molecule of oxygen consumed in the reaction, one atom is utilized to oxidize the substrate, whereas the other atom of oxygen results in the production of one molecule of water. NADPH is capable of donating two electrons in the reaction: the first is utilized to oxidize iron to the ferrous state after substrate binds, whereas the second electron from a second molecule of NADPH reacts with oxygen to produce a highly reactive superoxide anion, which reacts with substrate to produce product. The reduced cytochrome is capable of complexing with carbon monoxide, resulting in a complex that absorbs light at approximately 450 nm. The wavelength for the absorbance of light observed is specific for the different forms of P-450 and normally varies between 447 and 553 nm. Because carbon monoxide is capable of binding to the reduced cytochrome, it can act as a potent inhibitor of the mixed-function oxidase, and this interaction is used often to determine experimentally whether a drug is oxidized by this enzyme system.

REACTIONS CATALYSED BY THE MIXED-FUNCTION OXIDASE

The mixed-function oxidase is a relatively unique enzyme system which catalyzes the oxidation of a variety of different types of reactions, mentioned here under:

Aromatic Ring Hydroxylation: The oxidation of benzene to phenol. Almost all drugs or xenobiotics that contain an aromatic ring system will undergo ring hydroxylation. In some cases, hydroxylation can occur on a given aromatic ring more than once. Ring hydroxylation may or may not lead to the total inactivation of the drug.

Side-chain hydroxylation: Side chain hydroxylation is not as common an occurrence as ring hydroxylation, although when it does occur, it often results in the inactivation of the parent drug.

N-dealkylation: In this case, the tertiary amine is demethylated to form the secondary amine along with production of one molecule of formaldehyde. N-dealkylation reactions are relatively common in drugs that contain a tertiary or secondary amine group. N-dealkylation does not necessarily result in the inactivation of the parent drug because the product may have either more or less pharmacological activity than the parent drug.

O-dealkylation: This reaction is similar to that of N-dealkylation but in this case involves the oxidation of an ether.

S-alkylation: This reaction is involved in the activation of the drug to the pharmacologically active form.

Sulfoxidation: Many drugs that contain a thioether linkage can undergo oxidation to form the sulfoxide. The phenothiazines undergo sulfoxidation to form the inactive catabolite. Formation of the sulfoxide of the vitamin biotin can also be catalysed by the mixed-function oxidase enzyme system.

Desulfuration: Desulfuration does not result in inactivation of the parent drug but leads to production of the oxy-derivative, which possesses lower pharmacological activity. Desulfuration of parathion leads to formation of the acetylcholineesterase inhibitor paraoxon.

Deamination: Several enzymes catalyze the deamination of endogenous and exogenous amines. These include monoamine oxidase (MAO), diamine oxidase (DAO), plasma amine oxidase, and the mixed-function oxidase. MAO, as well as plasma amine oxidase, catalyzes the deamination of the biogenic amines and structurally related monoamine drugs, whereas DAO is specific for compounds containing diamines such as cadaverine and spermine. The mixed-function oxidease is responsible primarily for the deamination of sympathomimetic amines that contain an α-methyl group on the carbon atom adjacent to the primary amine as the case of amphetamine.

Amphetamine is a structural analog of phenylethylamine, a compound that is readily deaminated by MAO. The presence of the α-methyl group on phenylethylamine causes this compound to no longer be a substrate for MAO, and thus its deamination occurs exclusively through the mixed-function oxidase. This latter reaction is a slow event compared with the deamination of phenylethylamine by MAO.

Dehalogenation: Several enzymes are responsible for the dehalogenation of drugs *in vivo*. The mixed-function oxidase reaction is responsible for the dehalogenation of several aromatic ring and aliphatic halogens. For example, anesthetic agent halothane. Halothane undergoes a dehalogenation reaction that leads to the production of an unstable free radical intermediate. The majority of this intermediate is converted into trifluroacetaldehyde and subsequently trifluroacetic acid. However, a small percentage of the free radical may bind to components within the liver and act as the hapten, which ultimately leads to hepatitis.

PHASE II (CONJUGATION REACTIONS)

Phase II of the drug-metabolism enzyme system consists of a variety of enzymatic reactions all of which involve conjugation of either the parent of the oxidized drug. These reactions are catalyzed by several different transferases, all of which generate an end product that is ionized at physiological pH. With few exceptions, conjugation results in inactivation of the parent drug or its active catabolites and also facilitates its excretion within the kidney. Thus, conjugation reactions represent an important process involved in inactivation of drugs *in vivo*. The following examples describe some of the important conjugation reactions.

Glucuronide Conjugation

Formation of the glucuronide conjugate represents one of the major degradatiive process involved in drug detoxification and inactivation. The reactions involved in synthesis of the glucuronide donor uridine diphosphate (UDP)-glucuronic acid are presented. UDP-glucuronic acid transferase represents a family of enzymes that is located in the endoplasmic reticulum of the cell. Like the mixed-function oxidase, this transferase is inducible upon the exposure to certain drugs. These enzymes can catalyze the conjugation of hydroxyl groups, sulfhydryl groups, and primary amines as well as carboxylic acids. The end product of these reactions is the formation of the ionized glucuronide drug conjugate.

Sulfate Conjugation

Sulfate conjugation and glucuronide conjugation are the two major phase II conjugation reactions. Some enzymatic steps are involved in sulfate conjugation. The first step in the reaction catalyzed by the enzyme sulfurylase thermodynamically favors the reverse reaction. However, this reaction proceeds in the forward direction because the enzyme is coupled to APS-kinase, which subsequently converts adenosine-5′-phosphosulfate (APS) to the sulfate donor-3′-phsophoadenosine-5′-phosphosulfate (PAPS). In addition, a number of pyrophosphatases rapidly break down inorganic pyrophosphate formed by the sulfurylase reaction and thus prevent the reverse reaction from occurring.

Similar to the glucuronide transferases, the sulfotransferases, also represent a family of enzymes. More recent studies have idenfitied two functionally distinct forms of phenolsulfo transferases that possess distinct but overlapping substrate specificity. These enzymes are involved in the conjugation of a wide variety of drugs and other xenobiotics. Sulfoconjugation can occur on both phenolic compounds and aromatic amines. The end product of this reaction is the ionized sulfate ester of the drug. A variety of other sulfotransferases have been identified that appear to selectively sulfate steroids. These steroid sulfotransferases also utilize PAPS as the sulfate donors.

In general, both glucuronide and sulfate conjugation result in the formation of a biologically inert catabolite. Although this is usually the case for most drugs, in several instances conjugation has been reported to promote the biological response of drugs and other agents. For example, the sulfate ester of minoxidil produced *in vivo* appears to be the active form of the parent drug, because minoxidil sulfate is both an antihypersensitive agent and a hair growth promoting factor. Similarly, the sulfate conjugates of N-hydroxy-2-acetylaminofluorene and structurally related compounds have been reported to be the potent carcinogens. In the absence of formation of the sulfate ester within the cell, these unconjugated compounds have been shown not to be carcinogenic. A variety of compounds are also capable of serving as esterifying agents for a variety of drugs. These conjugating substrates include glutathione and a variety of amino acids. In all cases, the conjugate formed is ionized at physiological pH, thus assuring that the esterified drug of other exogenous agent will be excreted rapidly by the kidney.

OTHER PATHWAYS INVOLVED IN DRUG METABOLISM

Aromatic amines and hydrazine undergo acetylation reactions. The reaction is catalyzed by the enzyme N-acetyltransferase, which proceeds by a double displacement (ping-pong) reaction mechanism as illustrated for isoniozid.

Enzyme + Ac Co A ↔ Ac-enzyme _ CoA

Ac-enzyme + isoniazid ↔ Ac-isoniozid + enzyme

AcCoA + isoniozid↔Ac-isoniozid + CoA

Transsulfuration Reactions (Detoxification of CN^-)

There are two reactions important in the detoxification of CN^-. The first reaction is catalyzed by a mitochondrial sulfotransferase (rhodanase; transfuralase) that utilizes thiosulfate as the donor molecule. The second reaction utilizes β-mercaptopyruvic acid as the sulfur donor and is catalyzed by a cytosolic sulfotransferase. Thiosulfate of β-mercaptopyruvic acid can be used clinically to treat individuals exposed to toxic levels of cyanide.

O, N and S Methylation Reactions

A variety of methyltrasnferases exist that are capable of methylating hydroxyl, amine, or free sulfhydryl groups on drugs and other xenobiotics. These methyltransferases utilize S-adenomethionine (SAM) as the methyl donor.

Alcohol and Aldehyde Oxidations

Alcohols and aldehydes are normally oxidized *in vivo* by alcohol dehydrogenase and aldehyde dehydrogenase. Both enzymes require nicotinamide adenine dinucleotide (NAD) as the cofactor and have a broad substrate specificity. For example, The oxidation of chloral hydrate to trichloroacetic acid (TLA).

Oxidation reactions are also important in the metabolisms of aldehydes produced from the deamination of sympathomimetic amines by MAO as well as in the detoxification of ethanol.

Reduction Reactions

Aldehydes can be reduced also to the alcohol *in vivo* by the enzyme aldehyde reductase, which requires NADH. Although in general reduction reactions are not as common as oxidation, they do occur in specific cases. For example, the aldehyde product of norepinephrine (NE), 3-methoxy-4-hydroxyphenyl-β-hydroxyacetaldehyde, is preferentially reduced to form 3-methoxy-4-hydroxyphenylglycol (MHPG) in brain, whereas in the peripheral tissues the major metabolite of this aldehyde is the oxidized acidic product.

Hydrolysis

A variety of specific substrate and non specific exterases and amidases involved in hydrolysis are present *in vivo*.

Deamination

Several enzymes catalyzes the deamination of amines. These are MAO, plasma amine oxidase (benzylamine oxidase), DAO, and the mixed-function oxidase. A brief description of the properties and specificities of these enzymes follows:

Monoamine Oxidase

Two forms of MAO, viz., type A and type B are found on the outer mitochondrial membrane. These enzymes have different substrate specificities and are responsible for the deamination of catechol and phenolic amines *in vivo*. The substrate specificities of the two enzymes are presented below.

Type A MAO(Substrates)	Type B MAO(Substrates)
Norepinephrine	Norepinephrine
Dopamine	Dopamine
Tyramine	Tyramine
5-hydreoxytryptamine	Phenyletyhylamine

In the brain the A form is localized to neurons, whereas the B MAO in some serotonergic neurons.

Plasma amine oxidase (benzylamine oxidase): The function of plasma amine oxidase is unknown. It has a substrate specificity similar to that of the B form MAO. It is a soluble enzyme and is found predominantly in the plasma and blood vessel walls. The rate of deamination by plasma amine oxidase is extremely slow compared with that with MAO, and thus its role *in vivo* is uncertain.

Diamine oxidase: As the name implies, DAO is responsible for the deamination of diamines, such as putrescine and cadaverine. The enzyme has an absolute specificity for diamines and will not deaminate monoamines. The enzyme is soluble and contains FAD as a cofactor.

DRUG RESISTANCE BY PATHOGENS AND CELLS (TRANSFORMED)

Drug resistance in organisms arises in at least five distinct mechanisms.

Type I Resistance

This takes place by exclusion of the agent. The bacteria Staphylococcus aurous becomes resistant to tetracycline by abolishing the drug accumulating mechanism. The drug's target remains as sensitive as ever but the drug no longer reaches it.

Type 2 Resistance

This is accomplished by the organism bringing a destructive enzyme into play. This arises by the selection of a favuorable mutant or by induction. Penicillin resistant strains of S. aureus isolated from patients contain pencilinase. The enzyme hydrolyses the drug to useless pencilloic acid. Similar to the bacterial species developing resistance consequent to the consistent use of a specific drug over two hundred species of insects have also developed resistance against pesticides which were once effective.

Type 3 Resistance

This sort of resistance is brought about by the organisms or by the tissues, with drawing a necessary drug modifying enzyme. Many purine and pyrimidine analogues do not act against cancer until intracellular enzymes metabolize them to ribonucleotides. E.g., Mouse leukemia cells become resistant to 6-mercaptopurine in this way.

Type 4 Resistance

In this the organism secretes an excessive amount of substance to which the drug is a metabolite antagonist. For example, *Pneumococci* becomes resistant to antibacterial sulphonamides by forming extra quantities of P-amino benzoic acid.

Type 5 Resistance

This develops due to the exchange of genetic materials between bacteria through conjugation or transduction. Many gram negative bacteria in the intensities carry a transferable particle of DNA which can invade other bacteria and thereby transfer the drug resistance eventually in the recipient. In this type, the drug is chemically altered to a substance. For example, Bacterial resistance to tetracycline and chlorampenicol.

REFERENCES

Adrien, A. (1985) Selective toxicity - The physico-chemical basis of therapy, Chapman and Hall Ltd., London.

Bodor, N. (1982) Trends Pharmacol. Sci., 3: 53.

Chappell, J. (1966) Biochem. J., 100: 43.

Elion, Furman, P., Fyfe, J., de Miranda, P., Beauchamp, E. and Schaeffer, H. (1977) Proc. Natl. Acad. Sci., U.S.A., 74: 5716.

Faust, R. and Shearin, S. (1974) Nature, Lond., 248: 60.

Foster, A., Jarman, M., Kinas, R., Van Maanen, J. and Taylor, G. (1981) J. Med. Chem., 24: 1399.

Misra, A., Hunger, A. and Keberle, H. (1966) J. Pharm. Pharmacol., 18: 246,531.

O'Brien, R. (ed.) (1979) The Receptors: A comprehensive treatise, New York: Plenum Press.

Phillips, R., Love, A., Mitchell, T. and Neptune, E. (1965) Nature, 206: 1367.

Smith, E.L., Robert, L. Hill, I. Robert Lehman, Robert J. Lefkowitz., Philip Handler, Abraham White, (1985) Principles of Biochemistry – General Aspects. McGraw Hill Book Company, Singapore.

Stryer, L. (1981) Biochemistry, W.H. Freeman and Company, New York.

Valdivieso, M., Bodey, G., Gottlieb, J. and Freireich, E. (1976) Cancer Res., 36: 1821.

Suggested Readings

1. Microbiology (1995) Michael. J. Pelczar, Jr., E.C.S. Chan and Noel. R. Krieg, Tata McGraw Hill Publishing Company Limited, New Delhi.
2. Toxicity of Pesticides to Fish (1986) A.S. Murthy, Vol. I & II, CRC Press. Inc. Florida.
3. Quality Control in Toxicology (1977) G.E. Paget, MTP Press Limited, England.
4. Clinical Significance of Drug-Drug Interactions (1985) Philip. D. Hansten, Lea and Febriger, Philadelphia.
5. Selective Toxicity- The Physicochemical Basis of Therapy (1985) A. Adrien, Chapman and Hall Ltd., London.
6. Biochemistry (1981) L. Stryer, W.H. Freeman and Company, New York.

REVIEW QUESTIONS

Multiple Choice Questions

1. Steroid hormones belong to the category of drug
   (a) Agonist type (b) Antagonist type
   (c) Narcotic type (d) None of these
2. Which of the following are antagonist type of drugs?
   (a) Anesthetics b) Anti-histamines
   (c) Analgesics d) all the three
3. Which of the following are called bacterial epidemics?
   (a) Cholera (b) TB
   (c) Trachoma (d) All the three
4. Bird flu and SARS belong to
   (a) Zoonotic disease (b) Pandemic disease
   (c) Endemic disease (d) Both a and b
5. Which of the following are non-infectious diseases
   (a) Leprosy (b) TB
   (c) Cancer (d) AIDS

6. What is the membrane type in which majority of drugs can penetrate?
   (a) Type I (b) Type II
   (c) Type III (d) Type IV
7. Which of the following is a prodrug?
   (a) Proguanil (b) Cycloguanil
   (c) Arsenoxide (d) Sulphanilamides
8. The cell organelles involved in drug inactivation is
   (a) Golgi complex (b) Microsomes
   (c) Vacuoles (d) None of these
9. Which of the following are called detoxicating enzymes?
   (a) Acid phosphatase (b) Multiple function oxidase
   (c) cytochrome a (d) cytochrome c
10. MFO catalyses the following reaction
   (a) Aromatic ring hydroxylation (b) N-dealkylation
   (c) Sulfoxidation (d) All the three

Answers

1. (a)	2. (d)	3. (d)	4. (d)	5. (c)	6. (a)
7. (a)	8. (b)	9. (b)	10. (d)		

PART-12

Gene Technology and its Application

CHAPTER 37

Recombinant DNA Technology and Cloning Vectors

P. K. Agrawal[1*] and A. Srivastava[2]

[1]Biotechnology Unit, Vivekananda Parvatiya Krishi Anusandhan Sansthan, Almora–263 601

[2]Indian Institute of Pulses Research, Kanpur-208 024

E-mail: pawancrri@yahoo.co.in

In early 1970s, the recombinant DNA technology (rDNA) was recognized as powerful tool in biotechnological research. During the same period the laboratory techniques like use of restriction enzymes, cutting and joining of DNA molecules, cloning of DNA fragments in *E. coli* and their improvements strengthened molecular biology. In a gene cloning experiment, individual gene(s) sequences are inserted into a bacterium. Cloning of DNA fragments allows its large scale multiplication. The requirements for gene cloning experiment are :

1. Different types of enzymes for cutting and joining the DNA fragments into vector molecules.
2. Cloning vehicles or vectors.
3. Desired DNA fragments and their transformation into the host bacterial cells.
4. Selection of transformed cells containing the desirable recombinant DNA molecules.

CUTTING AND JOINING OF DNA MOLECULES

Earlier it was apparent that the phenomenon of host-controlled restriction and modification can help for cutting and ligation of DNA molecule. The bacteria identify the incoming DNA into its cell as foreign DNA and destroy them by cleaving into small pieces. It also uses the process of methylation for the protection of DNA from cleavage. The endonucleases are those which cut the DNA is called restriction endonucleases or restriction enzymes. The nucleases are in general the enzymes those degrade the DNA molecule by breaking the phosphodiester

*Corresponding author

bonds present between two nucleotides in a DNA strand. In any experiment related to recombinant DNA technology production of a recombinant or chimeric DNA, cutting and joining of vector DNA as well as DNA to be cloned constitute vital steps.

There are basically two kinds of nucleases :

(A) **Exonucleases:** cut the DNA by removing bases one at a time from the end of DNA molecule.

(B) **Endonucleases:** break the DNA molecules internally i.e., by breaking the internal phosphodiester bonds between two bases within the molecule.

Both of these nucleases are not specific in there action i.e., they do not have any specific site for digestion (cutting) of DNA molecule. However, cloning requires a specific fragment of DNA to be excised. Thus, for cloning experiment a unique group of endonucleases is needed, that unlike other nucleases can cleave the DNA molecule from a defined area out of whole nuclear DNA molecule. Such 'unique' endonucleases are known as, 'Restriction endo nucleases.'

'Restriction endonucleases' or 'restriction enzymes' are part of restriction-modification system, or R-M system. The R-M system functions to protect the bacterium genome from getting degraded by potentially lethal incoming genome of foreign origin chiefly from phages. Restriction endonucleases do this by recognizing specific sequences in the incoming DNA and digesting the DNA into fragments, either at specific or at more randomly chosen sites.

During the construction of a recombinant DNA molecule, the circular DNA (vector) is cut at a point and the target fragment of DNA is inserted. The restriction and ligation during the process need to be very precise and the cuts have to be site specific. The first enzyme was isolated in 1970s from *Haemophilus influenza*. Smith and Nathens (1973) proposed a three-letter code in italics for each restriction enzymes. For example, *Haemophilus aegypticus* is represented by Hae and *Haemophillus influenza* by Hin. When the enzymes are isolated from different serotypes of a species, generally four-letter code is used. For example, Hinf for *Haemophilus influenzae* serotype f. When more than one enzyme is isolated from a single origin, they are detonated by Roman numericals (I, II, III etc.). For example, Eco RI is the first enzyme isolated from *Escherichia coli* serotype R.

TYPES OF RESTRICTION ENZYMES

The restriction enzymes are divided into 3 types (Yual, 1981) viz., Type I endonucleases, Type II endonucleases and Type III endonuclease which are being discussed later on in this chapter.

Type I: The characteristics of this group of endonucleases are:

1. The type I restriction enzyme is one of the first restriction enzyme to be characterized. Messelson and Yuan purified this enzyme in 1968 from *Escherichia coli* K-12.
2. They are multifunctional enzymes having three different subunits i.e., restrictive subunit, modification subunit and specificity subunit. The specificity subunit determines the recognition of the specificity.
3. They require ATP, Mg^{++}, and S-adenosyl methionine as cofactors.
4. The recognition site of Class I enzyme is 15 kb in length. It can be methylated at the adenine position. This type of restriction endonucleases cuts the DNA approximately 1000 bp away from the 5′ end of the 15 bp recognition site. The cleavage is possible only when the restriction site is unmethylated.

5. This enzyme consists of several subunits each type for different purposes, it has two modification or more precisely two methylation subunits, one recognition subunit and two restriction subunits. The recognizable regions are usually very long and do not have any symmetry in sequences *per se*.
6. These enzymes exhibit high ATPase activity. For each phophodiester cleavage, more than 10,000 molecules of ATP are hydrolysed.
7. This group of enzymes have specificity for recognition site but not for cleavage site.
8. They produce a heterologous population of DNA fragments. Hence they are not desirable for gene cloning purpose. It is this property of the enzyme that makes unfit for use in DNA manipulation or cloning experiments.

Type II: The characteristics of this group of endonucleases are:

1. This consists of two separate enzymes, each performing their individual functions of cleavage and modification of the recognized sequence. The cleavage of the DNA is done in absence of modification.
2. They consists of single polypeptide and have separate endonuclease and methylase activities.
3. They do not require any cofactor like ATP or S-adenosylmethionine (as required by type-I) making them amenable to work with.
4. They recognize a specific nucleotide and cut the DNA at this site.
5. Most of these enzymes recognize a hexa-nucleotide sequence but some other recognize tetra-, penta-, and octanucleotide sequences on the DNA.
6. They require Mg^{++} as cofactor.
7. They are very stable.
8. The majority of recognition sequences have palindromic sequences. The recognition sequence of Eco RI is as follows :

   5′-GAATTC-3′
   3′-CTTAAG-5′

9. Many restriction endonucleases make a simple double-stranded cut resulting in blunt or flush end.

   Hae III
   ↓
   5′-GGCC-3′
   3′-CCGG-5′
   ↑

10. Many other restriction endonucleases do not cut exactly at the same position and the cleavage is staggered. The resulting DNA fragment have short, single-stranded overhangs at each ends called sticky or cohesive ends.

   Eco RI
   ↓
   5′-GAATTC-3′
   3′-CTTAAG-5′
   ↑

11. The DNA ends of one molecule generated by an enzyme may have complementary base pairs with any other DNA molecules.
12. Some enzymes posses the same recognition site but differ in the cleavage site. SmaI and Xma I have same recognition site but different cleavage sites.
13. Some restriction endonucleases posses identical recognition and cleavage sites. Ex. Hind III and Hsu I. They are called isoschimers.
14. Some restriction endonucleases differ in their recognition sequences, but produce same sticky ends. Ex. Bam H1, Bgl II and Sau 3A produce same sticky ends.
15. Star activity is another property of Type II restriction endonucleases. Eco RI normally recognizes and cleaves the sequence 5′-GAATTC-3′ and upon change of reaction conditions, it recognizes a tetra-nucleotide sequence 5′-AATT-3′.

Because of the above mentioned characteristics the type II system of restriction enzymes enjoy the most preferred status by molecular biologists for DNA cloning and many other manipulation experiments.

Type III: The characteristics of this group of endonucleases are:

1. Each restriction enzyme is made up of two different subunits, one sub-unit does the recognition and modification of target site while the other one does the function of cleavage.
2. The cleavage occurs approximately 24-26 base pairs away from the site of recognition.
3. The recognition sites are asymmetric. The nuclease cleave the DNA at sequences which are nonpalindromic.
4. The cleavage takes place by nicking of the recognition site, usually 24-26 bp down stream of the recognition site. Hence two sites in opposite orientation are required to break the DNA duplex.
5. Require ATP and Mg^{++} as cofactors.
6. Lack ATPase activity.

They are not used for cloning because cleaved products are not uniform.

Other Restriction Enzymes

Besides the above mentioned endonucleases, there are some other restriction enzymes those have different characteristics. Few of them are discussed here. Eco 571 consists of single polypeptide that has both modification and nuclease activities. However it has many properties similar to Type II endonuclease. The characteristics of few more endonucleases are given below:

Enzyme	Description
Eco 571 (Pertusyte *et al.* 1988)	Single polypeptide having both restriction and modification activities.
Mcr	Can cause cleavage of DNA that is methylated at specific positions. (Modified by CpC methylase) two types viz, Mcr A and Mcr BC are found in *E. coli*.
Mrr	Attacks and cleaves the DNA with methylated specific sequences like methylated adenine.
Doubling homing endonucleases	These are double stranded DNases derived from introns and inteins. Have a large asymmetric recognition sequence.

DNA ligase: When restriction endonucleases are engaged in cutting or cleavage of DNA molecules, there should be a group of enzyme whose function should be to join the cleaved DNA molecules in order to complete the process of DNA cloning. This function is performed with the help of DNA ligases, by forming a phosphodiester bond between two nucleotide. This process is known as DNA ligation (Enger *et al.*, 1982).

The phenomenon of ligation may proceed in either of the following three ways depending on the type of restriction enzymes used to produce fragments:

(a) Using DNA ligase to form phosphodiester bonds between two cohesive or sticky ends produced by the restriction enzyme.

(b) Using DNA ligase to anneal the blunt ends by forming phosphodiester bonds directly between the two successive nucleotides.

(c) The enzyme terminal deoxynucleotidyltransferase is used to synthesize homopolymeric 3′ single stranded tails at each blunt ends of the cleaved DNA fragments. It is followed by annealing with the help of DNA ligases.

The DNA ligases are encoded by *E. coli* and T-4 bacteriophage, with a basic function of sealing the nick between adjacent nucleotides in a duplex DNA chain (Lobban and Kaiser, 1973). While the reaction mechanism between the ligases (of *E. coli* and T-4 phases) is almost same but they differ in their requirement for cofactors. The ligases from *E. coli* needs ATP as co factor and ligase from T-4 phage requires NAD^+ as cofactors.

Mechanism of Action of DNA Ligase

With the living cell DNA ligases perform a very important function of repairing any discontinuous strands (Fig. 1a).

The action of ligase mainly depends on the type of ends of the DNA molecules to be joined, that is where ligation is going to occur. If it is a sticky end then ligation is very efficient. This is because of the fact that the compatible sticky ends can easily base pair with one another by hydrogen bond forming a relatively stable structure for the enzyme to work on (Lehman, 1974). Ligation in a blunt end is not so efficient. This is because the ligases on their own are unable to hold the two different molecules to be ligated.

Like other enzymes ligases are very sensitive to temperature. The optimum temperature for their action is 37°C, but this temperature is not appropriate for stabilizing the hydrogen bonded sticky ends. Thus the optimum temperature for ligation process should lie in the range of 4-15°C. Therefore, a compromise between the rate of enzyme action and association of the termini is done so as to make this process viable. For joining the blunt ends by DNA ligases efficiently, strategy of formation or addition of linkers and adapters is followed. Such adapters and linkers are given in detail in the following section.

Linkers and Adapters

As stated in the preceding section, compatible sticky ends are desirable for the DNA molecule to be ligated together in a gene cloning experiment. The sticky ends can be generated by digesting both the vector and the DNA fragment containing gene of interest with the same restriction endonuclease or, with different enzymes that produce the same sticky end.

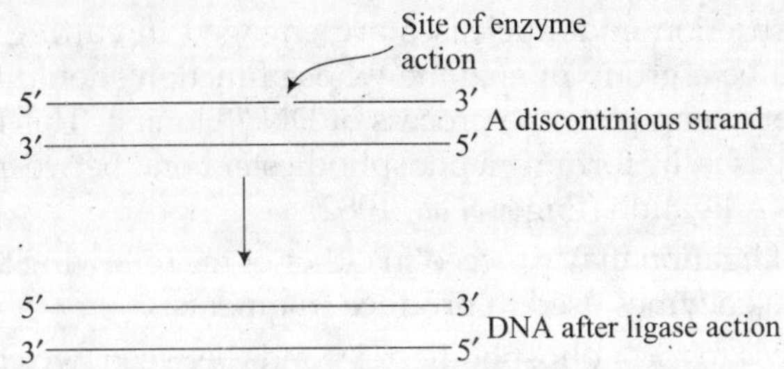

Fig. 1a. Mechanism of action of DNA ligase. Here discontinuity is quite simply because of absence of a phosphodiester bond between adjacent nucleotides.

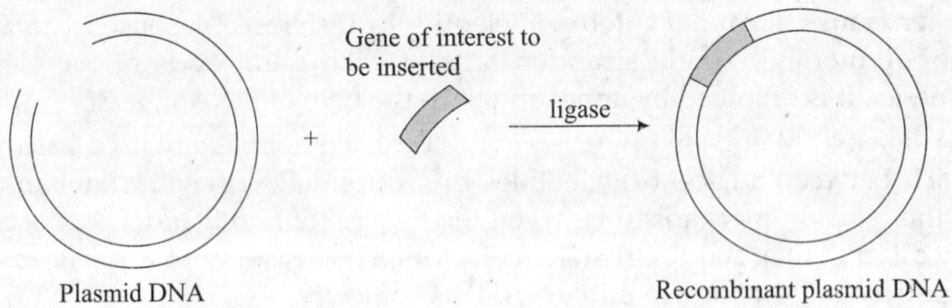

Fig. 1b. Mechanism of action of DNA ligase. Here in the final step of DNA cloning i.e., during the construction of recombinant DNA molecules, DNA ligases are used in joining together of the vector molecule and the DNA to be cloned.

However, this is not the situation that occurs every time. Many times condition is encountered like the vectors have sticky ends while the DNA fragment to be cloned is having the blunt ends and *vice-versa.* Under such condition following methods are employed to install correct sticky ends to the DNA fragments.

I. Linkers: Linkers are short pieces of double-stranded DNA of known nucleotide sequence, synthesized *in vitro.* Such linkers are blunt ended, but encompass restriction site being very small in size. The DNA ligases attach these linkers to a much larger blunt ended DNA molecules. Without much difficulty many such linkers are added producing a chain and are then subsequently digested with the restriction enzyme producing a large number of sticky ends with the restriction enzyme recognition site.

II. Adaptors: The second method of attaching a sticky end to blunt end is the attachment of adaptors. Adapters are also short synthetic oligonucleotides but unlike linkers, adaptors are synthesized in such a manner that its one end is blunt, while the other end has over hangs or sticky ends. The blunt ends of large DNA molecule and sticky ends are thus made available for ligation. There is one hindrance in doing this is that the sticky ends of adaptors will have the tendency to join with another to form dimmers so that the new DNA molecule will still remain blunt ended. To overcome this problem, the adaptors are synthesized in a little bit different way. Here the blunt bends of adaptor molecule is like 'normal' DNA strands but the sticky ends are different. The 3′-OH terminus is the same as usual, but the 5′-P terminus is modified. It lacks the phosphate group and is in fact a 5′-OH terminus (Hung and Wensink, 1984). Now the DNA ligase is unable to form the phosphodiester bond. As a result the base pairing always occurs between the sticky ends of adapters but the association is never stabilized by ligation. After the adaptors, have been attached to the abnormal 5′-OH terminus, OH is converted to 5′-P by action of enzyme called polynucleotide kinase.

DNA Modifying Enzymes

There are a few more enzymes that are directly or indirectly involved in the process of DNA cloning.

1. **Kinase:** The enzyme kinase can phosphorylate a DNA molecule by transferring the phosphate from ATP to 5′ terminus of DNA. In cloning experiments bacteriophage T4 polynucleotide kinase is used to transfer γ^{32P} of ATP to 5′ terminus of DNA. It is also used in phosphorylating the linkers and other fragments of DNA that lacks 5′ phosphate ends.
2. **Alkaline phosphatase:** This enzyme can catalyze the removal of 5′ phosphate groups from DNA modifying the terminal ends. These enzymes are essential because they prevent re-circularization and plasmid dimerisation, as ligase cannot join the ends.
3. **DNA polymerase:** Polymerase is an important component in cloning experiments. It converts the single stranded DNA molecule into double stranded one like the restriction fragments having blunt ends. A detail of polymerase enzyme is beyond the scope of this section and will be describe in detail in other chapter.
4. **Terminal transferase:** This enzyme can add oligodeoxynucleotides. Because of this property this enzyme is regularly used in synthesizing homo-polymeric tails during the creation of sticky ends.
5. **S-1 Nuclease :** It converts cohesive ends of the DNA to blunt ends by digesting the short single stranded overhangs. It is required when two incompatible ends anneal and the overlapping ends are required to be removed.
6. **λ-Exonuclease :** This enzyme is used for the 3′ end modification. It removes the nucleotides from 5′end duplex DNA to create a substrate for terminal transferase.
7. **Bal 31 Nuclease :** This enzyme can simultaneously modify both 3′ and 5′ ends of a DNA molecule, by degrading both ends resulting in creation of blunt ends at the termini of DNA molecule.

CLONING VECTORS

In a gene cloning experiment, a vehicle is needed to carry the fragment of DNA containing gene of interest into the host cell and carry out its replication. Such vehicles are termed as vectors. Thus vector is an element of gene cloning experiment that functions as a vehicle to transport the gene of interest containing DNA fragment into the host cell. For a molecule, in order to perform the function of vectors, few essential qualities are needed to be fulfilled. It should have :

1. The vector should have independent existence and identity.
2. The vector should have ability to enter and replicate within the host cell.
3. Small genome size (8-10 kb ideally) is desirable.
4. Can exist/ grow inside a bacterial cell.

5. The vector should have several marker genes those help to differentiate the transformed cells from the non-transformed cells. For example, genes for ampicillin and tetracycline resistance.
6. The vector DNA should contain suitable control elements like promoter and terminator for the expression of cloned DNA.

There are two naturally existing cloning vectors available with that satisfies all the above laid conditions. They are plasmids and bacteriophages. Both of these prospective vectors are described in detail in the following sections.

I. Plasmids

Plasmids are extra chromosomal replicons that live and divide independently within the bacterial cell. Most plasmids are double stranded covalently closed circular (CCC) DNA. Few exceptions are: linear plasmids found in bacteria, like *Streptomyces* sp. and *Borrelia burgdorferi* B31.

Table 1. Plasmids and their sizes

Species	DNA molecule	Size (Mb)
Borrelia burgdorferi B31	Circular Plasmid cp9	0.009
	Circular Plasmid cp26	0.026
	Circular Plasmid cp32	0.032
	Linear Plasmid Ip17	0.017
	Linear Plasmid Ip25	0.024
	Linear Plasmid Ip28-1	0.027
	Linear Plasmid Ip28-2	0.027
	Linear Plasmid Ip28-3	0.029
	Linear Plasmid Ip28-4	0.027
	Linear Plasmid Ip36	0.037
	Linear Plasmid Ip38	0.039
	Linear Plasmid Ip54	0.054
	Linear Plasmid Ip56	0.056

Characteristics of Plasmids

Genome of plasmids: The genome of a plasmid is very small with size ranging from 1.0 kb to 250 kb. Plasmids carry only one or few genes often responsible for some useful characters. The expression of such genes is required only under some peculiar circumstances (see classification table) otherwise a bacterium can live comfortably well in their (plasmids) absence under normal or amenable conditions. Thus, plasmid is useful but not essential for bacterial growth, making them a dispensable entity, that can be manipulated without disturbing the genomic integrity of bacteria.

Copy number: 'Copy Number' term refers to the number of plasmids molecules that are normally present in a single bacterial cell. A bacterial cell may have multiple copies of different types of plasmids. The characteristic copy number values generally ranges from one to more than 50. From cloning point of view, large copy number is beneficial, as it will allow, to form large number of chimeric DNA molecules.

Plasmid Classification

A plasmid can be classified in the following ways :

1. On the basis of characteristics coded by the plasmid genes such as given in table 2.
2. There are certain plasmids, to which phenotypic characteristics have not yet been ascribed, such plasmids are known as 'cryptic plasmids'.
3. The other model of classification is based on the presence or absence of 'tra' or transfer gene, and is therefore classified as conjugative and non-conjugative plasmids, respectively.

Table 2. Classification of plasmids

S. No.	Plasmid Type	Gene Coded by Plasmid	Gene Function
1.	'R' Plasmids	RP 4 for *Pseudomonas,* Rbk for *E. coli* and all other bacteria	Antibiotic resistance
2.	'F' Plasmid	F for *E. coli*	Promotes Conjugal transfer of genetic material
3.	Col / Killer Plasmids	Gene for colicins	Synthesis of toxins that can kill bacteria.
4.	Degradative Plasmids	TOL by *Pseudomonas putida*	Synthesis of toluene a unusual molecule for bacterium
5.	Virulence	Ti for *Agrobacterium tumefaciens*	Induction of tumour (Crown Gall) by bacterium.

Many types and numbers of plasmids may be present in the same bacterial cell, including more than one different conjugative plasmids at any one time (Barnes, 1977). For e.g., *E. coli* can have upto seven plasmids each differing from one another in any of the above defined characteristics. To be able to coexist in a single cell such different plasmids must be compatible to each other. If two plasmids are incompatible than one of the two will be eliminated as the bacterium divides. Based on their different incompatibility factors plasmids are assigned different groups.

Preparation and Purification of Plasmids

The most pivotal step in using a plasmid as a vector in a gene cloning experiment is its isolation or preparation. The preparation of plasmid DNA from bacteria is a little bit complex issue, in comparison to the total cell DNA preparation. In isolation of plasmid DNA it is always required to separate the small sized plasmid DNA from the large amount of bacterial chromosomal DNA present in the cells. Following steps are performed for isolation of plasmid from a bacterial culture.

Growth: A culture of bacterial cell containing plasmid is grown in a liquid (broth) medium. The culture medium is taken such that it provides all the nutrients at optimum concentration so that the bacterium grows in its maximum capacity. Two kinds of media are taken: Defined Medium, in which all the components are known along with their concentration etc. e.g., M9 medium for bacterial culture (Table 3). Such defined medium is used when there is need to grow some specific set of bacteria. However, when culture of bacteria is done only for isolation of DNA a more lineaments – Undefined Medium is used. The commons example of this type is the LB or Luria Bertani medium (Table 3). In this media the precise information of the constituent is unavailable because most of its ingredients are complicated mixtures of unknown chemical composition.

Table 3. Media composition of the bacterial cell

S. No.	Media Name	Component	Concentration (g/l)
1.	M 9	$Na_2 HPO_4$	6.0
		KH_2PO_4	3.0
		NaCl	0.5
		NH_4Cl	1.0
		$MgSO_4$	0.5
		$C_6H_{12}O_6$	2.0
		$CaCl_2$	0.01
2.	LB(Luria-Bertani)	Tryptone	10
		Yeast Extract	5
		NaCl	10

LB media is generally taken for culturing bacteria for DNA isolation purpose because it does not require any further supplement to optimize its capacity and more over, this media can support growth of large species of bacteria. Bacteria grown on this media are provided with all necessary environmental conditions like 37°C temperature and proper aeration by shaking it at 150-250 revolution per minute (RPM) on rotatory platform. The growth of culture can be monitored by reading optical density (OD) at 600nm. As a standard 1OD unit at 600 nm corresponds to 0.8×10^9 cells/ml (Barnes, 1977).

Harvesting: A small amount of bacteria is needed for forming cell extract. Bacterial culture is subjected to centrifuge for 5000 RPM for 10 minutes to pellet down the bacteria. In general one liter of bacterial cell culture yields 10ml of bacterial cell.

Preparation of cell extract: Bacterial cells are enclosed in cell wall, followed by a cytoplasmic cell membrane. There are certain bacteria like *E. coli*, where the cell wall may itself is enveloped by a second outer membrane. For obtaining a cell extract, these barriers are required to be broken.

The disruption of bacterial cell (Fig. 2) can be done by physical or chemical method. For plasmid DNA isolation generally chemical method of cell disruption is adopted. Chemical method involves breaking of cell wall by EDTA (ethylenediamine tetracetate) and lysozyme. EDTA removes Mg^{+2} ions that are essential for preserving the overall structure of the cell

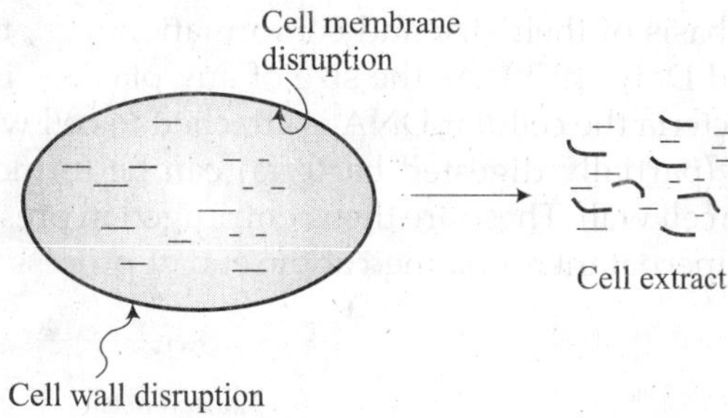

Fig. 2. Cell disruption.

envelope and also inhibits cellular enzymes that can degrade DNA. Lysozyme are enzymes that digests many polymeric compounds present in the cell membrane which give it rigidity. A small amount of detergent like SDS is also added to ease out the process of cell disruption.

After cell disruption, the cell extract is centrifuged. As a result of centrifugation cell wall fractions are pelleted, leaving DNA, RNA and protein etc. as supernatant, which is subsequently pipetted out (Fig. 3).

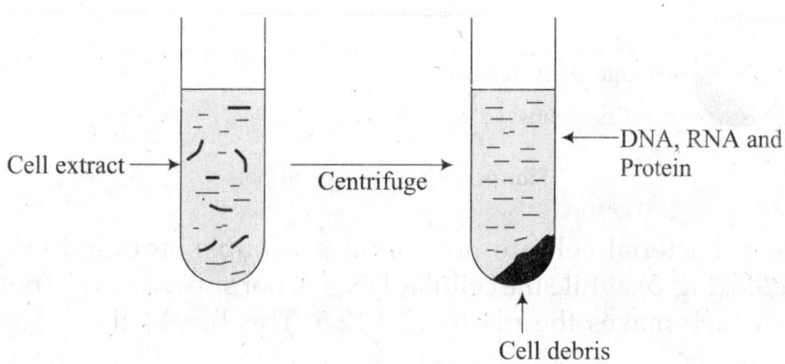

Fig. 3. Preparation of cell extract.

Purification of DNA: The cell extract contains DNA, RNA and proteins. Methods can be employed to remove these contaminants, leaving DNA in its pure form. The general method of deprotinization is the addition of phenol and chloroform. These organic solvents precipitate proteins but leaves nucleic acids (DNA/RNA) in aqueous solution. The nucleic acid solution is then purified for DNA by treating the solution with ribonuclease, which digests the RNA molecule into ribonucleotides.

Measurement of purity of DNA concentration: Ultra violet spectroscopy is used for checking whether the extracted is free from contamination or not. The sample of DNA is subjected to UV spectroscopy and measurement of absorbance is taken at 260 nm and 280 nm. The ratio of absorbance at 260 nm and 280 nm (A260/A280) should be 1.8. Ratio of less than 1.8 indicates that the solution is contaminated, either with protein or phenol.

Separation of plasmid from cellular DNA: The separation of plasmid DNA from cellular DNA is most important step in gene cloning experiment. A slight contamination from cellular DNA may produce undesirable results. There are several methods available for plasmid separation, chiefly on the basis of their size and conformation (e.g., rapid alkaline procedure described by Birnboim and Doly, 1979). As the size of any plasmid is only 8% of the cellular DNA and more over in bacteria the cellular DNA is attached to cell wall. Even in a careful cell disruption, sepheroplasts (partially digested bacteria) can be formed in which the cellular DNA is still attached to the cell wall. These are then centrifuged in presence of sucrose and non-ionic detergents. A clear supernatant is obtained at the end of process which is largely made of plasmids (Fig. 4).

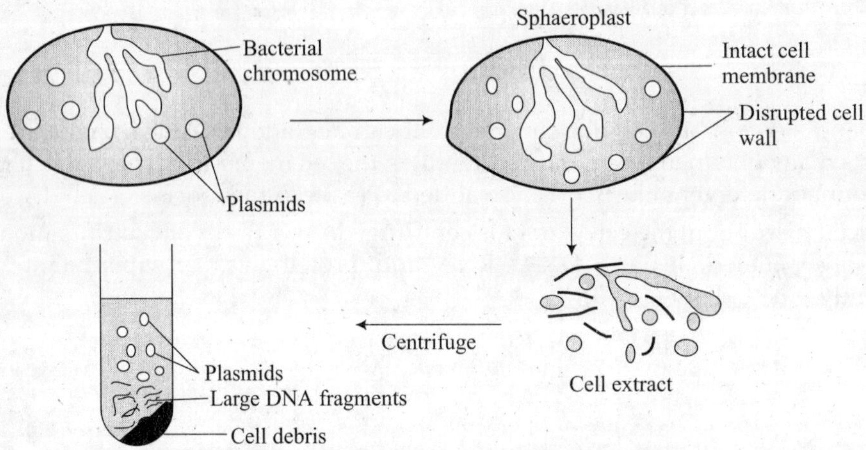

Fig. 4. Plasmid separation.

The plasmids in bacterial cell are not found as double stranded circle, but exist in a supercoiled molecule (Fig. 5), while the cellular DNA is not supercoiled. If now NaOH is added to the cell extract which makes the pH to 12.0-12.5. This breaks the weak hydrogen bonds present in non-supercoiled DNA molecules, causing the helix to unwind. Further, centrifugation will again separate the denatured cellular DNA from intact plasmid DNA. Thus these are the some of ways by which plasmid DNA is isolated/seperated from the bacterial cellular or chromosomal DNA. The plasmid DNA is then subjected for their amplification so that gene cloning experiment can be initiated.

Types of Plasmid Based Vectors

pBR 322: The conventional plasmids like Col E1 and PSC01 have many limitations like single site for the restriction endonuclease, limited genetic markers and small fraction of plasmid attachment foreign gene etc. As a result of these factors the conventional plasmids were capable of incorporating only small sized DNA fragments, and that too at very limited locations on the plasmid's genome. In order to overcome these limitations, superior, need based cloning vectors were designed *in vitro* and subsequently synthesized. One of the most popular and widely used artificially synthesized plasmid is pBR322 (Fig. 6) (Boliver *et al.*, 1977).

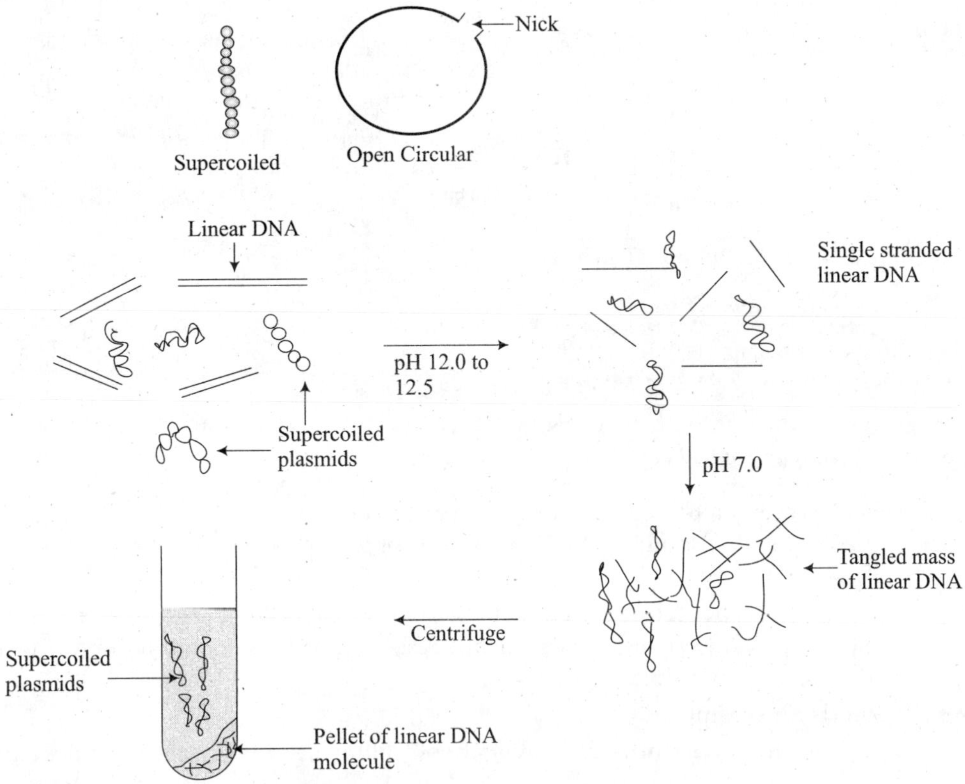

Fig. 5. Plasmid isolation from bacterial cells.

The nomenclature of 'pBR322' is done according to the standard rules of plasmid nomenclature where,

- 'p' stands for plasmids.
- 'BR' identifies the laboratory of Boliver and Rodriguez where this plasmid was synthesized.
- '322' is used to differentiate this plasmid with other plasmids developed in the same laboratory.

The plasmid pBR322 is superior to conventional vectors in many ways. The most useful feature of pBR322 is its size, which is 4363 bp. Such size helps this plasmid in its easy purification and also approximately 6kb of additional alien DNA fragment can be incorporated. The second feature is that it contains two sets of antibiotic resistance genes namely ampicillin resistant gene (Ap^R) and tetracycline resistant gene (Tc^R). Either of Ap^R or Tc^R can act as a selective marker in cloning experiment. The marker genes are created in such a way that each of them contains a unique restriction site.

The third feature of pBR322 is that it has a reasonably high copy number. Generally there are 15 or more molecules present in *E. coli* cell. Their numbers may increase up to 1000-3000 by plasmid amplification, in presence of protein synthesis inhibitor such as chloromphenicol.

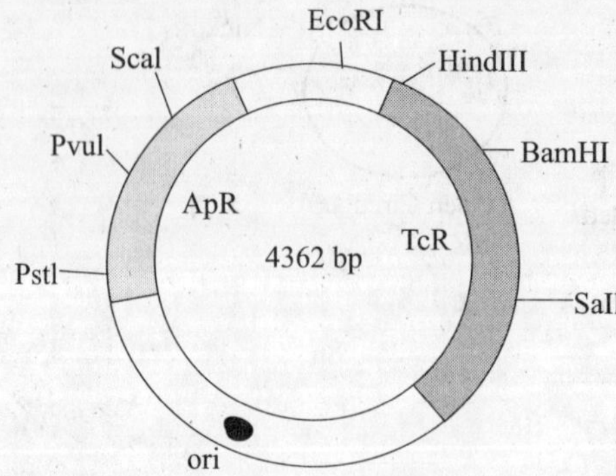

Fig. 6. pBR 322.

Some Other Improved Vectors

pBR325: This is basically a pBR322 plasmid with an additional segment of DNA (Fig. 7). This new additional segment contains gene for chloramphenicol resistance. This gene contains restriction site for EcoR1 only.

pUC vectors: pUC series of vectors are basically derivatives of pBR322. Although these vectors have taken only Ap^R and *ori* regions from pBR322, rest of their genomic organization is appreciably different from pBR vectors where,

- 'P' stands for plasmid
- 'UC' stands for laboratories of University of California where, they were developed.

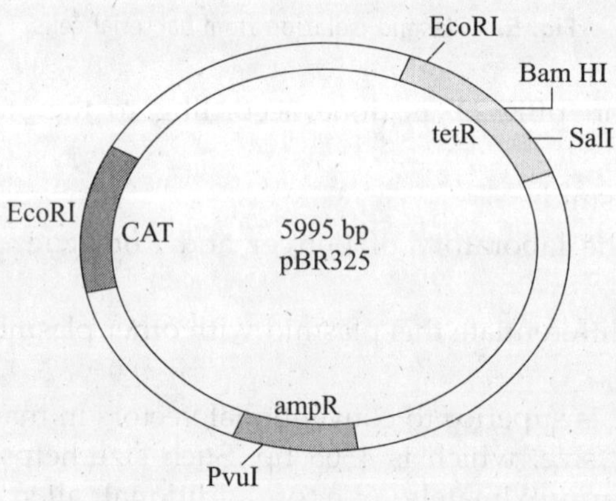

Fig. 7. pBR 325 plasmid.

The pUC is a small sized plasmid (2.7 kb) containing an *ori* region, Ap^R and a *lacZ′* gene (Fig. 8). Almost all of the unique restriction sites are clustered in the segment containing *lacZ′* gene segment. Because of this, a new marker has been developed that works on the phenomenon of insertional inactivation, making the identification of chimeric cells formed after cloning a very larg task.

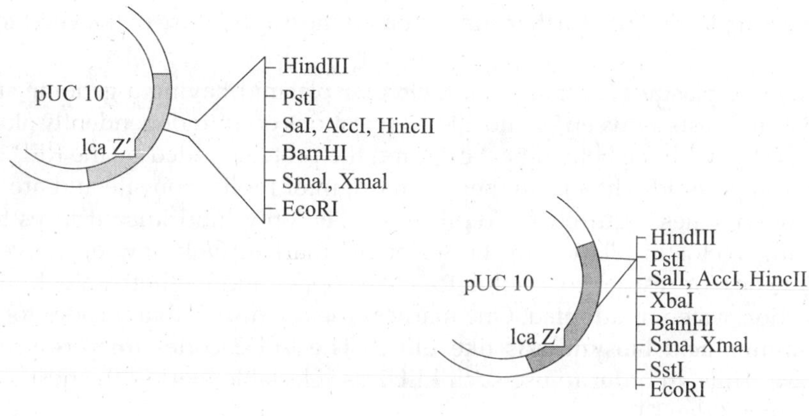

Fig. 8. pUC 10.

The *lacZ'* gene is responsible for synthesis of β-galactosidase, an important component of lac operon. β-galactosidase is one in the series of three metabolic enzymes involved in the breakdown of lactose to glucose and galactose. It is normally coded by gene *lacZ* present in *E. coli* genome. However, some strains of *E. coli* have a modified form of *lacZ* gene referred to as *lacZ'*. In *lacZ'* gene, an important component of β-galactosidase is missing. This missing part is α- peptide portion of enzyme. Thus, such *E. coli's* with *lacZ'* gene can synthesize the whole *lacZ* enzyme only when they have plasmids like pUC that carry the missing *lacZ'* segment of gene in their genomes. Screening for β-galactosidase activation/ absence or inactivation/ absence can differentiates between a normal pUC and recombinant pUC respectively. The pUC has two important advantages that makes them important component of gene cloning experiments. The first one, being that the cloned or chimeric *E. coli* cells can be identified in a single step by plating on agar (a media containing agar, antibiotic ampicillin and X-gal). The X-gal (C5-bromo-4-chloro-3-indolyl-β-d-galactose) is an analogue of lactose which is broken down by β-galactosidase to a blue coloured product. The appearance of blue colour indicates presence of a normal, non-recombinant pUC plasmid, whereas in a disrupted *lacZ'* gene β-galactosidase is inactivated and gives a white coloured colony indicating a successful recombination. The second, advantage of pUC is the clustering of restriction site which allows cloning of more than one different DNA strands such as addition of a polylinker segments etc.

pACYC 184: This is another *in vitro* generated plasmid developed from plasmid pl5A. In structure it is very close to pBR322. It has a genome size of 4.0 kb (pBR322 = 4.36 kb). Unlike pBR322, the presence of two antibiotic resistance genes equip these plasmids and their host cell with resistance to antibiotics like chloramphenicol and tetracycline. Insertional inactivation helps in identifying the recombinant cells as described in previous section.

Yeast Plasmid Vectors

Yeast (*Sacchromyces cervisiae*) is a unicellular eukaryote. It is one of the most important organisms for biotechnological research as it has multiple applications in biotechnology. Apart from its major role in industry like brewing and bread making, it has also been exploited as a host organism for production of many important genes in molecular biology. The discovery of presence of a 2 μm plasmid has further substantiated the candidature of the yeast to be used as vectors.

Most strains of plasmid contain a 2 μm circular plasmid having a genome size of 6.0kb (Fig. 9). It exists in yeasts between 70 and 200. Replications occur independently along with the chromosomal DNA with the help of host cells and the proteins coded by the REP 1 and REP 2 genes present on plasmid. They have their own origin of replication site and are also able to produce many enzymes required for replication. The only hindrance it faces to act as an efficient cloning vector in relation to the selectable marker. Only few of the yeast cloning vectors carry genes for resistance to inhibitors like copper and methotrexate. In most cases a different selection system is adopted. One marker is selected such that it codes for an enzyme involved in amino acid biosynthesis like LEU2. The LEU2 codes for β-isopropyl malate dehydrogenase. Thus, in order to use such LEU2 as selectable marker the host should be an auxotrophic mutant (leu2⁻).

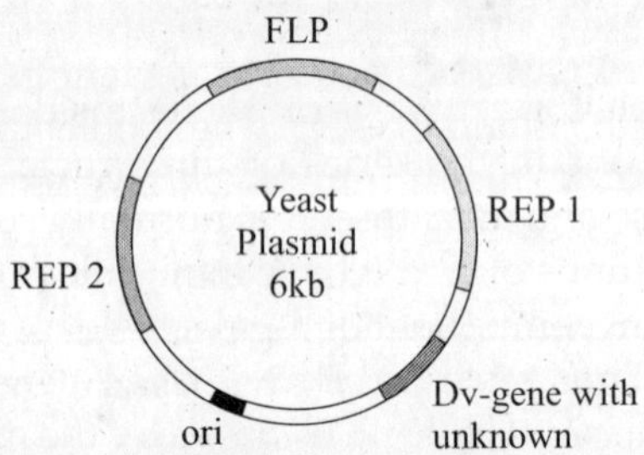

Fig. 9. Yeast plasmid.

Vectors Based on Yeast Plasmids

YEps : This vector is derived from 2 μm circular plasmid (Fig. 10). Although it can replicate independently of yeast's chromosome, still it is found attached to the main chromosome. Hence, it is called yeast episomal plasmid or YEps. The most prominent example of YEps is pJDB219.

pJDB219 is a shuttle vector-a vector that can perform cloning in two different species. The genome of pJDB219 includes the entire sequence of pBR322 and hence can replicate and can be selected for both yeast and *E. coli*.

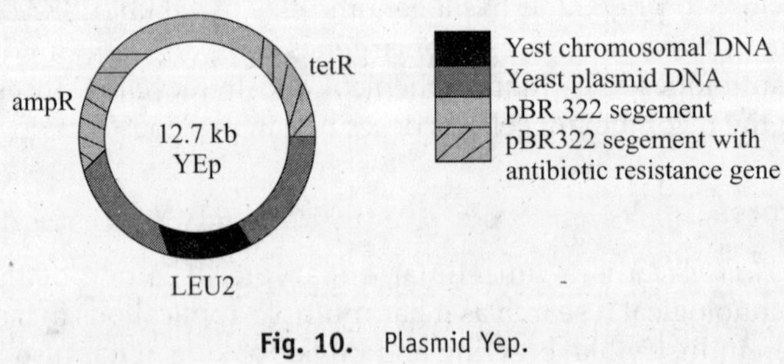

Fig. 10. Plasmid Yep.

YIps: these are Yeast integrative plasmids, which are basically bacterial plasmids in structure carrying a yeast gene. The most commonly used plasmid is YIp 5, which is pBR322 with an URA 3 gene (Fig. 11). This gene codes for orotidine 5′-phosphate decarboxylase-an enzyme involved in pyrimidine biosynthesis. The URA3 gene is used as selectable marker similarly as LEU2 gene in YEps. A YIp cannot replicate on itself and is therefore dependent upon its integration with yeast chromosomal DNA for its survival.

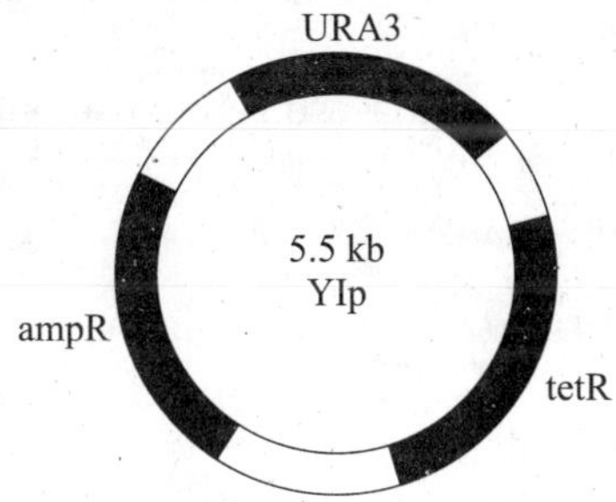

Fig. 11. Plasmid YIp.

YRps: The YRps are yeast replicative plasmids. These can multiply as independent plasmids because they carry a chromosomal DNA sequence that includes an origin of replication region. YRp7 is the commonest example of this type (Fig. 12). It is made up of bacteria based plasmid pBR322 and yeast gene TRP1. This gene is involved in tryptophan biosynthesis and is located adjacent to the origin of replication. Out of these vectors the need of experiment decides, which type of vector is to be taken. In general two factors are taken into consideration. The first being 'frequency of transformation. YEps have found to have maximum frequency of transformation that lies between 1000-10,000 transformed cell per μgm. The second factor, is the stability of transformants. YEp transformants are not stable, while the most stable one is Yip. In general many experiments of transformation involves YEp and YRp but if in any experimental stability is of cardinal importance then YIp becomes more suitable.

Ti plasmids: These are bacterial plasmids found in *Agrobacterium tumefaciens*—bacteria responsible for causing crown gall disease in many species of dicotyledonous plants. The manifestation of bacterial infection occurs in form of a cancerous proliferation of the stem tissue. The *A. tumefaciens* gains entry into the plant body through an existing wound or lesion in the stem of plant.

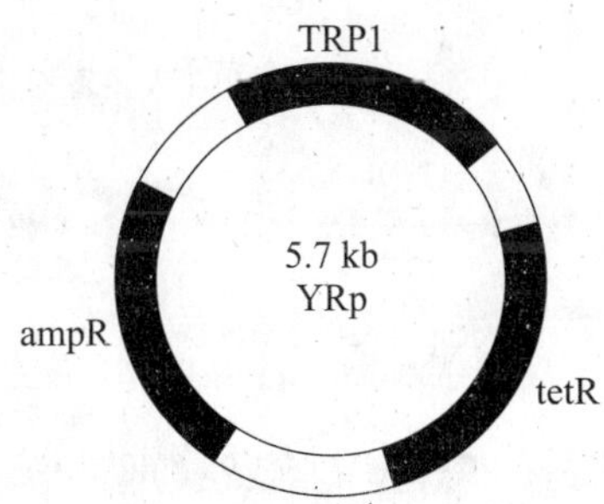

Fig. 12. Plasmid YRp.

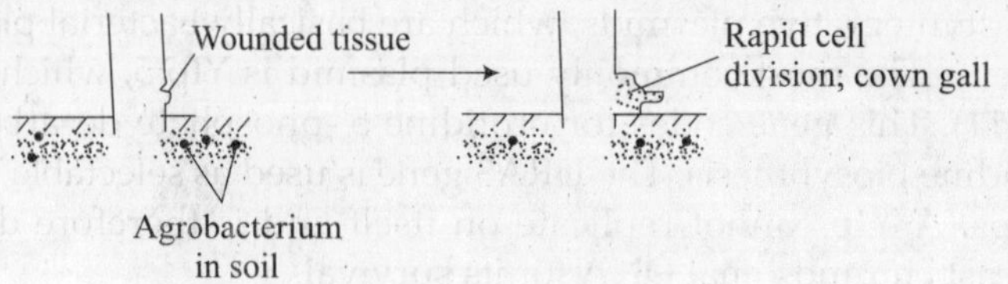

Fig. 13. *Agrobacterium* infecting the plant.

The ability of a bacterium causing a disease is because of presence of Ti (tumor inducing) plasmid within the bacterial cell (Fig. 14). A Ti plasmid has genomic size of 200 kb and carries numerous genes involved in causing the disease.

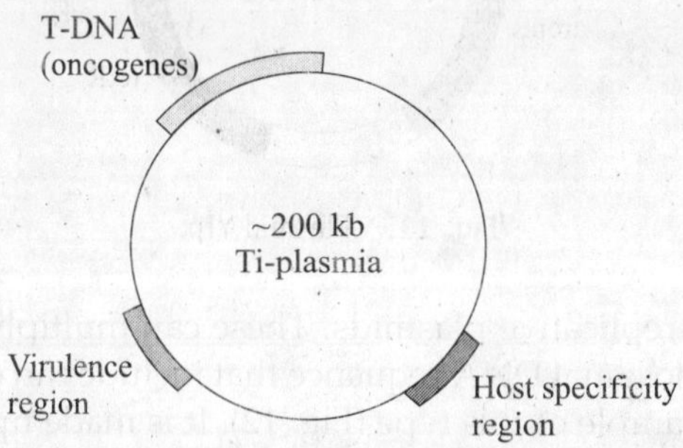

Fig. 14. Ti-plasmid.

The remarkable feature of Ti plasmid is that the plasmid integrates with the chromosomal DNA of the host (eukaryotic host). The segment that does this function known as T-DNA or transfer DNA. It is between 15-30 kb in size. Such properties of plasmid makes it an ideal candidate for transfer of gene in plant cell (Fig. 15).

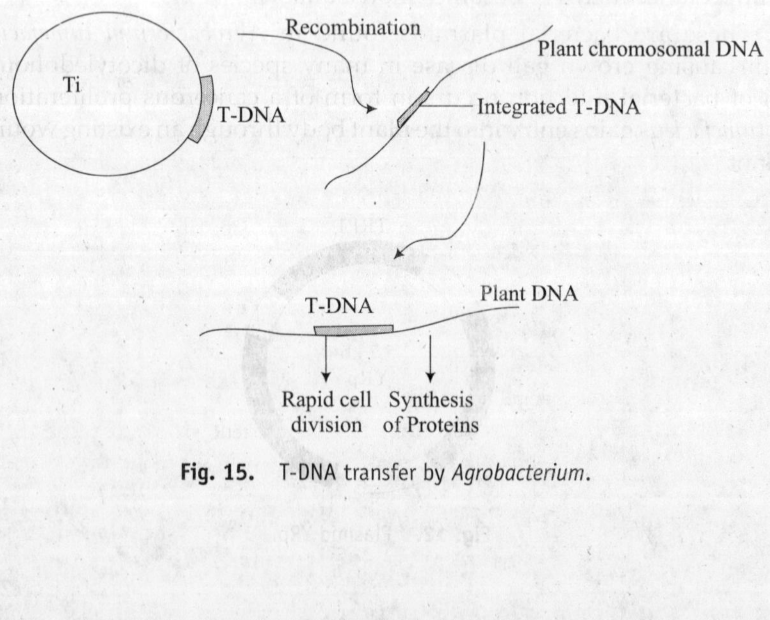

Fig. 15. T-DNA transfer by *Agrobacterium*.

The large genome size of Ti plasmid makes the manipulation of hosts genome a difficult task. It is nearly impossible to find a unique restriction site in a genome of 200kb. Two novel strategies are adopted to overcome this problem :

1. The binary vector strategy: It was observed that T-DNA does not need to be physically attached to the whole of the remaining Ti-plasmid. Hence, a binary plasmid system (Fig. 16) was developed with T-DNA on a relatively small molecule. The rest of plasmid is in normal form. The T-DNA plasmid is small enough to have a unique restriction site and can easily be manipulated by using standard techniques. e.g., pGV3850.

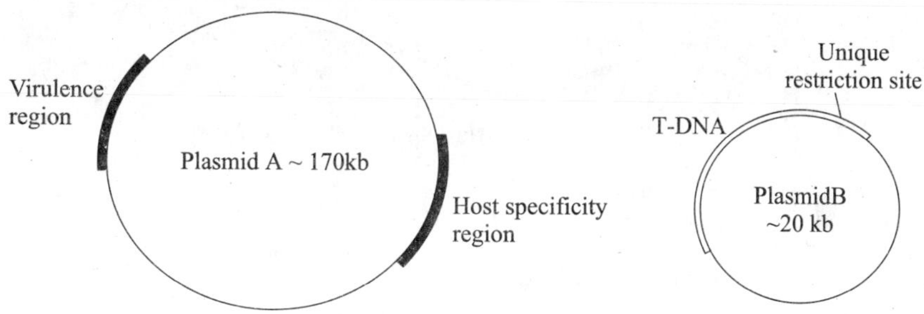

Fig. 16. Binary vectors.

The cointegration strategy: This strategy uses an entire new plasmid which is based on pBR322 or any other *E. coli* vector plus a small portion of T-DNA fragment. The presence of a small fragment of T-DNA on a plasmid like pBR322 creates a homologous site for recombination with the main Ti plasmid (Fi.g 17). A gene that is to be cloned is inserted to the unique restriction sites of pBR322. They are subsequently passed to the Ti plasmid by integration of small T-DNA fragment present on pBR322 plasmid and T-DNA of Ti-plasmid. e.g., Brin 19.

The *A. tumefaciens* bacterium with the reorganized Ti plasmid is only capable of bringing transformation to cells of crown gall tumor and not to the whole plant. Thus in order to get a whole plant regenerated with new genome in place of mature plant, new plantlets are needed to be cultured in liquid medium. It is followed by transformation by using Ti-plasmid based method. Such plant cells are cultured in the same way as microorganisms. They are placed in selected media and provided nutrients externally. All other physical factors such as temperature, humidity, light, etc. are maintained as per requirement for a proper growth of plant. A mature plant regenerated from transformed cell will contain the transgene in every cell that will be passed to all subsequent new cells ultimately forming a whole new transgenic plant.

Bacteriophages as Vectors

Bacteriophages or 'phages' are viruses that infect the bacterium. Like any other viruses they are very simple in structure (Fig. 18). They are made up of two parts, a protective protein coat called 'capsid' that encloses very small genome which can be either double stranded or single stranded DNA or RNA but never both. The phage particles attacks or infects bacteria by

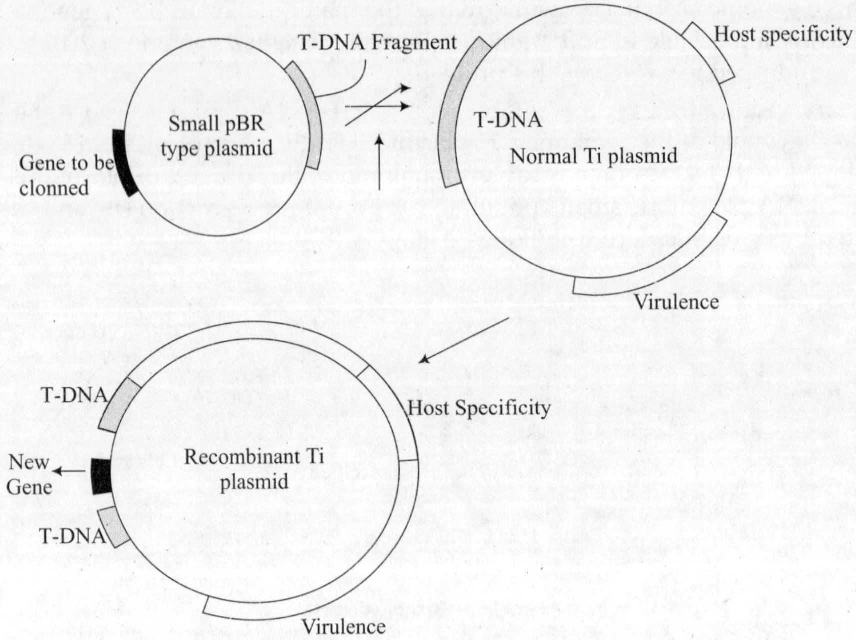

Fig. 17. Cointegrate vector.

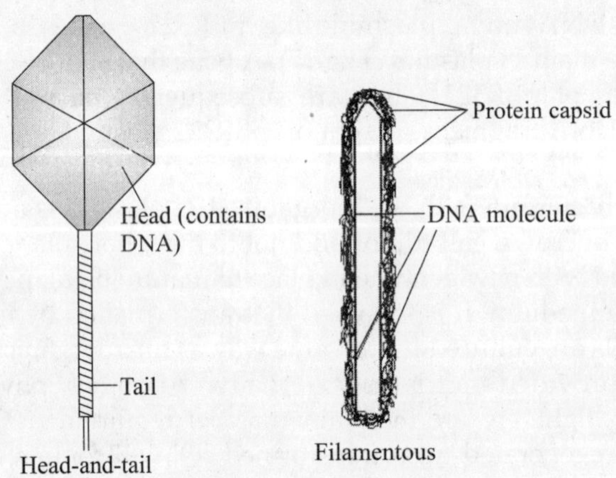

Fig. 18. Bacteriophage.

injecting their to them. The phage genome then uses the host (bacteria) machinery to synthesize its own genome enzymes and various other proteins. They are finally are released from the host's cell by its lysis after infection. In some phage types, the entire infection cycle is completed in very short period of time. This type of cycle is called lytic cycle. In contrast to lytic cycle, there is lysogenic infection cycle, which is characterized by the retention of the phage DNA molecule in the host bacterium. In many lysogenic phages, the phage DNA is inserted into the bacterial genome in a manner similar to episomal insertion in bacteria by plasmids.

This integrated form of the phage is called prophage. This property in forming of prophages is exploited in using it as another potential vector for gene cloning (Young and Davis, 1983).

Gene organization in the λ DNA molecule: The λ DNA is 49kb in size and the genome of λ has been fully sequenced. An important feature of λ genome is that the functionally related genes are clustered together. A second feature of λ is such that, at each 5′end of linear DNA there is a 12 base single stranded regions or overhang. This linear molecule can get paired forming a circular, (completely double stranded) molecule due to presence of such complementary strands.

Complementary single strands are often referred as 'sticky ends' or 'cohesive ends'. The λ cohesive ends are also called the *cos* sites and they play two very important role during λ infection cycle. Firstly, they allow the linear DNA molecule to get themselves injected into the cell by making them circularized—a necessary feature for their insertion to bacterial genome. The second role of the *cos* site comes after prophage has excised from the host genome. At this stage a large number of new λ DNA is being produced by rolling circle mechanism of replication. As a result continuous new DNA molecules are being rolled off from the template molecule. The series of linear λ DNA formed in such ways that they are joined at *cos* site. The *cos* site acts as recognition site for restriction endonucleases which cleaves the newly synthesized continuous DNA strand producing discrete single λ genome. Thus, endonuclease creates the single stranded sticky ends, and also aids in their packaging into λ head.

It has been observed that the internal structure of the λ genome can be changed by insertion of new genes. Such reorganization of λ genome causes no major effect on the phage molecule so long as these changes do not bring much alteration in the overall length of λ genome. This makes the λ phage as an ideal choice for using them as vector in gene cloning experiments as new genes can now be inserted in phage genome which will be successfully transported and incorporated into the bacterial genome.

λ Phage as vectors: The λ phage vector is 48 kb in size and can incorporate another 25 kb of DNA. This put the vector in an advantageous position than the other plasmids which can carry only up to 10 kb of DNA fragments with them. By partial digestion of λ genome average 15-20 kb sized DNA fragments are obtained. The gene of interest containing fragments is added to them. Usually, when the phage genome is subjected to endonuclease digestion, three major fragments are obtained (as already established that functionally related genes remain together in λ genome). One carrying genes for λ head and tail, second one (or central) fragment containing red and gamma gene that promotes lysogeny and third or last fragment containing genes for lysis (Rackwitz et al., 1984).

The foreign DNA or the DNA molecule to be cloned is generally inserted into the central region and are subsequently packaged into the λ phage. Absence of central region causes the lytic cycle to occur. Thus a collection of chimeric phage is formed by amplifying the phages through several generations of growth on a strain of bacterium. Thus, at the end, a large number of sets of chimeric phage carrying the gene of interest are isolated. There are various phage based cloning vectors presently available like λ gt 11, EMBL3, etc. They are used in development of full length cDNA library (Okayama and Berg, 1982).

The isolation of phage is a big limitation in using them as vector. However, there are certain unique advantages that makes them a very important vector, like they can carry relatively large fragment of DNA (upto 25 kb), their multiplication is autonomic hence, phage molecules containing the foreign DNA also multiplies very rapidly with a very little assistance from either host or from elsewhere, producing large amount of cloned DNA very quickly. The selection and maintenance of clone is relatively much easier in comparison to plasmids.

M-13 phages: The bacteriophage M-13 is an *E. coli* phage and one of the most popular vectors. It is especially used in those DNA cloning experiments that finally lead to sequencing experiments (Messing, 1993). The phage M-13 has a completely different structure than λ phage. The genome size of M-13 is only 6.4kb. It is circular but consists of only a single stranded DNA molecule. Due to small size large amount of foreign DNA can be taken up by the M-13 phage genome for cloning. Only three genes are required by the M-13 for its sustainability, whereas in case of λ approximately 15 genes are required for synthesis of head-and-tail structure making the manipulation of its genome easier.

The infection cycle of M-13 (Fig. 19) is also simple than that of λ phage cycle. The M-13 DNA molecule is injected into *E. coli* via a conjugation pilus. The single stranded DNA undergoes replication to form double stranded DNA inside the cell. The molecule is not inserted into bacterial genome but undergoes replication to form more than 100 copies of double stranded DNA. When the bacterium divides, each daughter receives copies of the phage genome. It continues to replicate to maintain its overall number per cell. Thus new phage particles are continuously being produced during each generation of an infected cell.

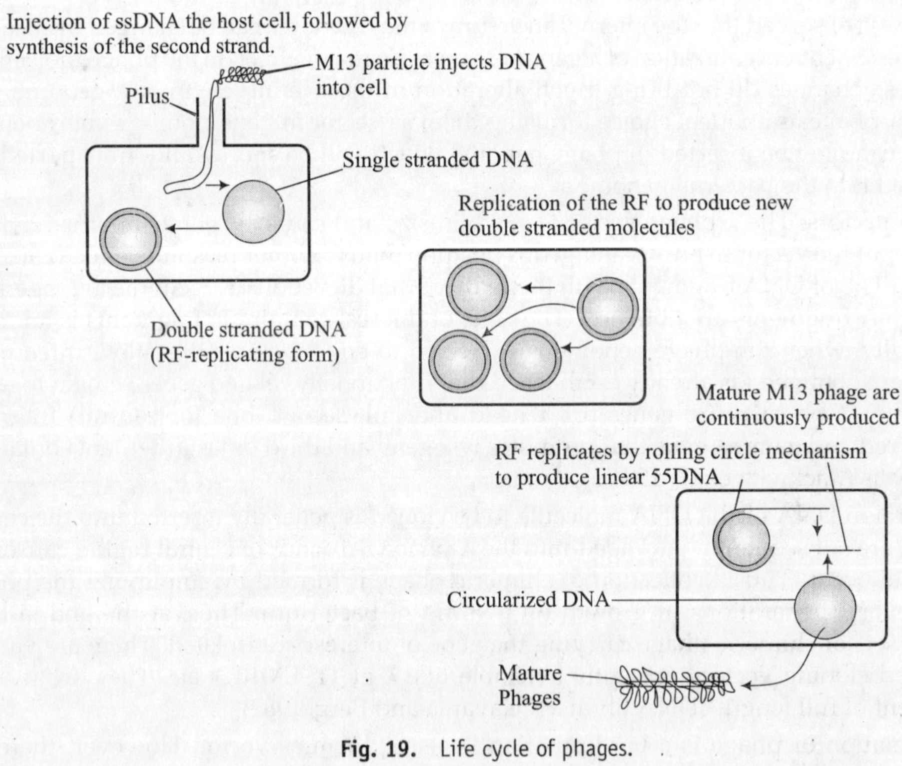

Fig. 19. Life cycle of phages.

There are several features that make M-13 phage an amenable cloning vector. The genome size is less than 10kb. It can easily take up large sized fragment of DNA. The M-13 behaves more like plasmid than like phage. Hence, it can easily be prepared from *E. coli* and can also be easily reintroduced by transfection. The most important property that M-13 bears is that the genes cloned with M-13 based vector can be obtained in the form of a single stranded DNA. Such single stranded cloned genes are pre-requisites of several experiments like DNA sequencing, *in vivo* mutagenesis, etc.

Cosmids

There are many purpose built vectors that are available to molecular biologists. Cosmids are one of them. Cosmids are plasmids that are constructed in such a manner that they contain a fragment of λ DNA including the *cos* (cohesive) site (Collins and Hohn, 1978). Cosmid is evolved by combining *cos* and plasmid (Fig. 20). A cosmid contains all the properties of plasmid including genes for antibiotic resistance. Cosmids are proved to be an efficient vector because of their small genomic size (5kb). They can take upto 32-47 kb of foreign DNA which is much more than a phage vector or plasmids (Rackwitz *et al.*, 1984). The Cosmids are particularly used in construction of eukaryotic clones because plasmids cannot clone a large size eukaryotic gene *per se* and phage (λ) can take up very small size fragments. Hence, cosmids are an attractive alternative for eukaryotic gene cloning.

The cloning experiment with cosmid is carried out by inserting the foreign DNA fragment of interest at its unique restriction site. During ligation step, due to presence of *cos* sites, the linear chains or concatamers of DNA are packaged into the λ head. After packaging *in vitro*, the 'λ' particle is used to infect a suitable host that may be a eukaryotic cell. The recombinant cosmid DNA remains as circularized phage DNA but replicates like bacterial plasmids. Further the transformed cell is selected on the basis of drug resistance markers.

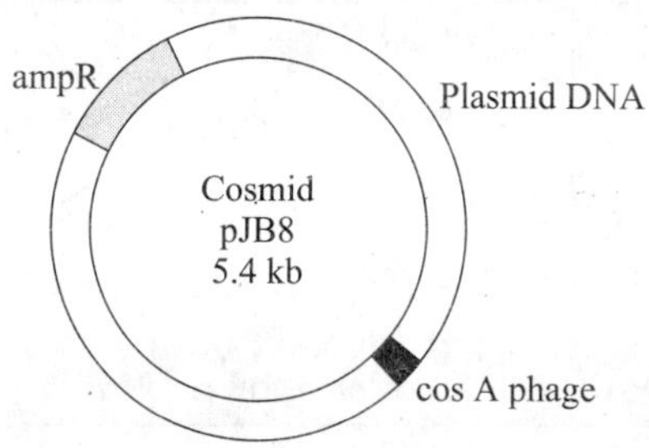

Fig. 20. Cosmid pJB8.

Phagemids

Phagemids is one more purpose-built cloning vector. This vector is made by combining elements from both plasmids and bacteriophage M-13. The elements that is usually combined is '*ori*' region of both plasmids and M-13 phages. Phagemids are basically a specific type of "phasmids" which is formed by combination of elements from plasmids and bacteriophages. Most common example of phagemids is pBluescript11 KS^{TM} (Fig. 21). The pBluescript11 KS^{TM}

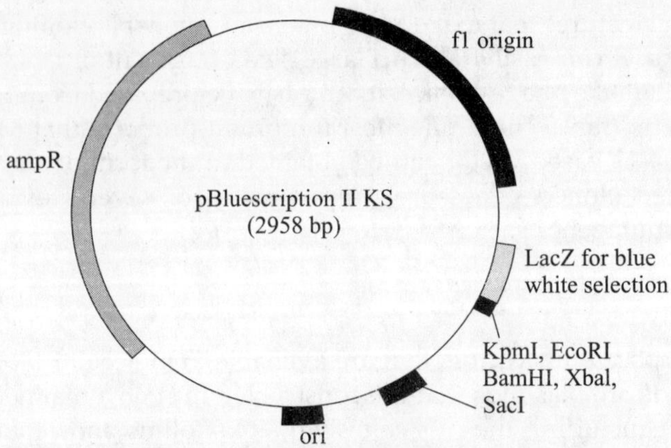

Fig. 21. pBluescript II KS.

is a pUC based vector of size 2.9 kb. Thus the selection of recombinant cells are also based on similar pattern with which pUC mediated recombinants are selected i.e., through blue-white colonies on X-gal plates.

Yeast Artificial Chromosome (YAC)

For the purpose of construction of gene libraries and genome sequencing of large eukaryotes, the various vectors described so far are not effective (Table 4). The reason for their ineffectiveness is that they can take up and transport very small size of alien DNA.

Table 4. Different vector and their insert size

S. No.	Vector Name	Host	Insert Size
1.	λ phage	*E. coli*	5-25 kb
2.	λ cosmids	*E. coli*	35-45 kb
3.	P1 phge	*E. coli*	70-100 kb
4.	PACs	*E. coli*	100-300 kb
5.	BACs	*E. coli*	50-300 kb
6.	YACs	*Saccharomyces cervisae*	200-2000 kb

(Monaco and Larin, 1994)

To increase the capacity of DNA incorporation into the vector, several artificial chromosomes have been designed. One such artificial chromosome built for the purpose of DNA cloning is YAC or Yeast Artificial Chromosome (Burke and Olson, 1987) (Ramsay, 1994). This vector can take up a DNA molecule of 1000-2000kb in size. YAC is basically a minichromosome, that is made by centromere (CEN 4 elements), two telomere sequences (TEL), a site of origin for replication, variable polylinker site guarded by two BamHI sites and one or more genes that may act as a selectable marker.

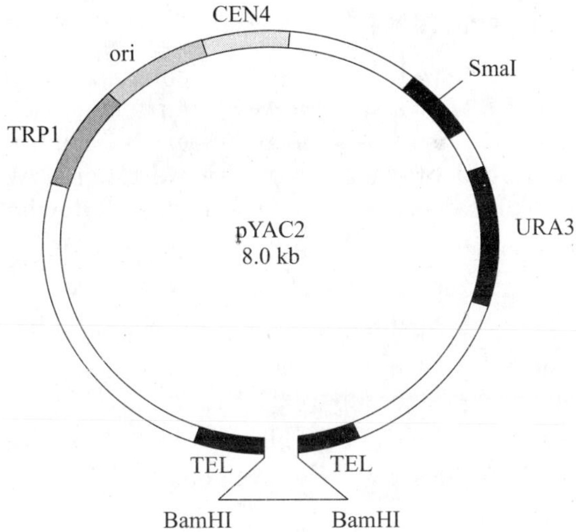

Fig. 22. Plasmid pYAC2.

For construction of YAC the various elements of chromosome are isolated by recombinant DNA technology and are then joined together *in vitro* creating an artificial chromosome. Szostak and Blackburn (1982) developed the first YAC based vector by cloning yeast telomere in YRp. Several YAC vectors have now been developed on the same line. pYAC2 (Fig. 21) is a typical example of YAC. pYAC is essentially a pBR322 plasmid into which a number of yeast genes have been inserted like URA3 and TRP1. Although URA3 and TRP1 genes are also present in YIp5 (yeast integrative plasmids) and YRp7 (yeast replicative plasmids), respectively. However, in pYAC2, URA3 and TRP1 are present along with the sequence CEN4 (a sequence for centromere from chromosome 4 of yeast).

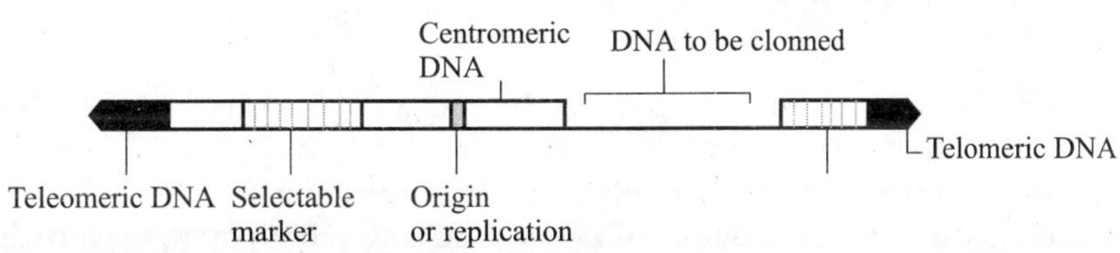

Fig. 23. Representative diagram of a YAC.

The third component for telomeres is provided by the two sequences called TEL. The last part of pYAC2 is the SUP4 which is in fact a selectable marker into which the new DNA is inserted during experiment. During cloning, the pYAC2 is treated with Bam H1 and Sma1 for cutting the molecules. Then Bam H1 fragments are removed and the DNA to be cloned which is having a blunt end is ligated between the two arms in place of Bam H1. By protoplast fusion method the chimeric chromosome is then introduced inside the yeast cell. The transformants are then plated on the minimal media to identify and isolate the correctly constructed artificial chromosome.

Bacterial Artificial Chromosome (BAC)

Like YAC, BAC is a cloning vector which is more efficient than YAC. It can carry a genome of 50-300 kb or more. The first BAC was pBeloBAC11. Shizuya and Gibbins (1992) developed mapping system based on BAC. As the pBeloBAC11 lacks selectable marker, the recombinant clones constructed by them are identified by colony hybridization. BAC vectors are grown in *E. coli* as large single molecule. Ideally a BAC vector is made of following components (Fig. 24) :

- A single copy of sex factor 'F' of *E. coli*.
- *E. coli* plasmid gene for replication e.g., oriS repE
- parA and parB genes to maintain the copy number at a level of one or two phage genome.
- Cm^R chloramphenicol resistance marker.
- Two cloning sites as HindIII and Bam HI.
- Several G+C restriction sites. e.g., Sfi I, Not I etc.

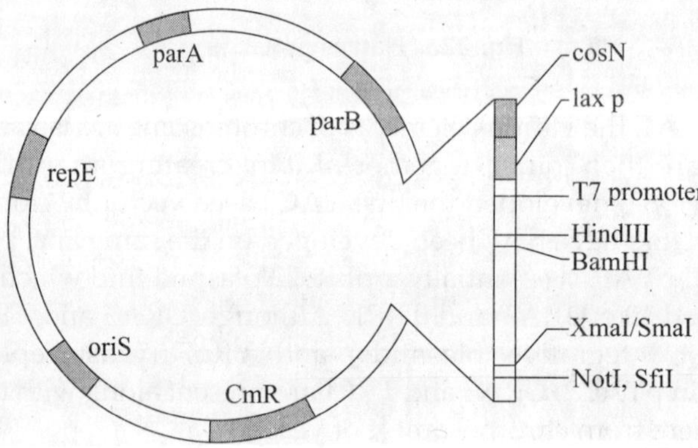

Fig. 24. Bacterial artificial chromosome.

BACs are capable of maintaining large sized human and plant genome for over a hundred generations with a high degree of stability. Therefore, BACs have been extensively, used in construction of genome libraries and in many genome sequencing projects. Presently, two BAC vectors are widely used. They are pBeloBAC11 and pECBAC1. Both of these were derived from pBAC108L where the cloning site is replaced by *lacZ* gene carrying the polylinker DNA. Frengen and Monaco (1999) have presented an improved version of BAC, where he managed to include a site for insertion of transposons. This enables the chimeric genome to be modified after there transfer to bacteria. The procedure of post cloning modification is termed as 'reterofitting'. The method of reterofitting helps in introduction of reporter gene.

PAC (P-I derived artificial chromosome)

It is a vector formed by combination of P-1 and F-factor. The P-1 is an element of vector P-1 phage, which is responsible of packaging of P-1 phage DNA into the viral head. F-factor is the sex factor present in some *E. coli* plasmids that are responsible for formation of sex pilli (Pierce and Sternberg, 1992). A PAC vector can be transfected to *E. coli* by means of electroporation and biolistic methods. Once inside the cell, a PAC can take up a fragment of DNA of 100-300kb. This exhaustive capacity of gene cloning and relative ease of handling of this vector makes it an important alternative to other complex vectors as cosmids, YACs, etc.

Expression Vectors

These are unique kind of functional vectors where the cloned gene can not only be transcribed into mRNA but in many cases can be translated into proteins. This is why these vectors are termed as expression vectors. By using the property of expression, the expression vector can be employed to form a collection of various cDNA. The various proteins encoded by the expression vector can be synthesized in the host cells itself. Such proteins and the cDNA clones can, therefore, be easily isolated and assayed. Depending on requirement and need of experiment the expression vector of either RNA expressing or protein expressing type or both can be selected.

SUMMARY

The recombinant DNA technology involves insertion of a gene sequences into a suitable vector followed by their replication in the host. The selection of transformed cells containing the gene of interest requires various enzymes for cutting-and-joining of DNA molecules and suitable vectors for transferring the DNA fragments. The nucleases are used for cutting the DNA molecules. Endonucleases are preferred because they break the DNA molecules internally. There are three types of restriction endonucleases. Among them Type II endonucleases are routinely used in cloning experiments becasue their recognition and restriction sites are very specific. Upon restriction, the endonucleases produce two kinds of ends viz., sticky or cohesive ends, and blumt ends. DNA ligases are used to join the DNA fragments. Oftenly specialized small DNA fragments like linkers and adapters are used during the cloning experiments.

Vectors are the cloning vehicles to carry the DNA fragments of interest. The vectors may be plasmids, cosmids, phasemids, BACs, YACs and PACs. Some of the popular plasmids routinely used are pBR 322, pBR 325, pUC (all plasmid bases vectors) and certain yeast based vectors like YEps, YRps, etc. Ti-plasmids obtained from *Agrobacterium tumefaciens* are specifically used for engineering plant genomes. Plasmids can carry small DNA fragments. However, YACs (Yeast artificial chromosome), BACs (Bacterial artificial chromosome) and PACs or P-1 derived artificial chromosome are some artificially designed vectors used for efficiently transferring large size DNA fragments.

REFERENCES

Barnes, W.M. (1977) Plasmid detection and sizing in single colony lysates. Science 195 : 393-394.

Birnboim, H.C., and Doly J. (1979) A rapid alkaline procedure for screening recombinant plasmid DNA. Nucle. Acid. Res. 7 : 1513-1523.

Boliver, F., Rodriguez, R.L., Betlach, M.C., Heyneker, H.L., Boyer, H.W., Crosa, J.H., and Fallow S. (1977) Construction and characterization of new cloning vehicles. A multipurpose cloning system. Gene 2 : 95-113.

Burke, D.T. and Olson, M.V. (1987) Cloning of large segments of exogenous DNA into yeast by means of artificial chromosome vectors. Science 236 : 806-812.

Collins, J. and Hohn, B. (1978) Cosmids: A type of plasmid gene cloning vector that is packageable in vitro in bacteriophage lambda heads. Proc. Natl. Acad. Sci. 75 : 4242-4246.

Engler, M.J. and Richardson, C.C. (1982) DNA Ligases in The Enzymes, Academic Press, New York.

Frengen, J.E. and Monaco A.P. (1999) Retrofitting vectors for E.coli based artificial chromosome (PACs and BACs) with markers for transfection studies. Genome Res. 7 : 179-186.

Hung, M.C. and Wensink, P.C. (1984) Different restriction enzyme- generated sticky ends can be joined in vitro. Nucle. Acid. Res. 12: 1863-1874.

Lehman, I.R. (1974) DNA ligase: structure, mechanism and function. Science 186 : 790-787.

Lobban, P.E. and Kaiser, A.D. (1973). Enzymatic end to end joining of DNA molecules. J. Mol. Bio. 78 : 453-471.

Messing, J. (1993) M13 cloning vehicles. Their contribution to DNA sequencing. Meth. Mol. Bio. 23 : 9-22.

Monaco A.P. and Larin Z. (1994) YACs, BACs, PACs, and MACs: Artificial chromosomes as research tools. Trend. Biotechnol. 12 : 280-286.

Okayama, H. and Berg, P. (1982) High effiency cloning of full-length cDNA. Mol. Cell Biol. 2 : 161-170.

Pierce, J.C. and Sternberg, N.L. (1992) Using the bacteriophage P1 system to clone high molecular weight genomic DNA. Meth. Enzymol. 216 : 549-574.

Rackwitz, H.R., Zehetner, G., Frischauf, A.M., and Lehrach, H. (1984) Rapid restriction mapping of DNA clonned in lambda phage vectors. Gene 30 : 195-200.

Ramsay, M. (1994) Yeast artificial chromosome cloning. Mol. Biotechnol. 1 : 584-603.

Shizuya R. and Verrinder Gibbins, A.M. (1992) Construction and characterization of a large-fragment chicken bacterial artificial chromosome library. Genomics. 42:217-226

Young R.A. and Davis R.W. (1983) bacteriophage lysis: Mechanism and regulation. Microbiol. Rev. 56 : 430-481.

Suggested Readings

1. Chawala, H.S. Introduction to Plant Biotechnology, Oxford & IBH Publishing Co. Pvt. Ltd; New Delhi, India.
2. Brown, T.A. Gene cloning, Chapman and Hall, London, England.
3. Old, R.W. and Primrose, S.B. Principles of gene analysis: An introduction to genetic engineering, Blackwell Science, Oxford, London.
4. Gupta, P.K. Elements of Biotechnology, Rastogi Publishers, Meerut, India.
5. Maniatis, T., Fritch E.F. and Sambrock, J. Molecular Cloning : A Laboratory Manual, Cold Spring Harbor Laboratory Press, New York, USA.
6. Brown, T.A. Genetics a molecular approach, Chapman and Hall, London, England.
7. Lewin, B. Gene VIII. Pearson-Prentice Hall, New York, USA.

REVIEW QUESTIONS

Long Questions

1. Describe the steps of a cloning experiment.
2. Give outline of the roles of restriction endonucleases and DNA ligases in DNA cloning.
3. Describe the types of vectors that are available for cloning genes in plants and animals.

4. Discuss on similarities and dissimilarities between PCR and gene cloning.
5. Describe different types of restriction enzymes.
6. What are linkers and adaptors. Discuss their role in DNA cloning?
7. Describe in detail how plasmid DNA is prepared.
8. '*Agrobacterium tumefaciens*' are usually referred as natural genetic engineers, comment. Also describe why can't any other bacteria be used as vector in plant transformation experiments?
9. Explain why gene cloning is important?
10. 'Bacteriophages as vector in a cloning experiment', comment upon this statement.
11. Define recombinant DNA technology, cloning vector, clone, genomic library, and Ti-plasmid.
12. What are the main advantages of an M-13 cloning vector?
13. What are the key features of a YAC vector? How is the *lacZ'* is used in gene cloning experiment?
14. What are key features of a BAC vector? Outline the useful properties of pBR322?
15. Describe briefly BAC and YAC? What are expression vectors?
16. Differentiate between the following :
    (a) Endonuclease and Restriction endonuclease
    (b) Plasmid, Cosmid and Phagemids
    (c) cDNA library and Genomic library
    (d) Blunt ended DNA and Sticky ended molecule
    (e) YEp and YIp

Multiple Choice Questions

1. An enzyme that induces double stranded breaks in DNA and re-joins them is
   (a) Restriction endonuclease (b) DNA gyrase
   (c) DNA ligase (d) DNA polymerase
2. Which of the following enzymes is used in recombinant DNA technology
   (a) Exonuclease (b) Endonuclease
   (c) Restriction endonuclease I (d) Restriction endonuclease II
3. The genome of plasmid ranges from
   (a) 1 bp – 250bp (b) 100bp – 2500bp
   (c) 1000bp – 25000bp (d) 1000bp – 250000bp
4. *Rbk* gene codes for
   (a) Recombination (b) Antibiotic resistance
   (c) Restriction enzyme (d) Environmental toxicity resistance
5. Linear plasmids are found in
   (a) Eukaryotes like sacchromyces (b) Phages
   (c) *Borrelia burgdorferi* B31 (d) Ti-Plasmids

6. Which of the followings is a artificially synthesized plasmid
   (a) pBR322 (b) Ti- plasmid
   (c) Both (d) None
7. For DNA sequencing experiment, if cloning is done then which will be the most suitable vector
   (a) pBR 322 (b) l Phage
   (c) M-13 phages (d) Cosmids
8. Which of the following vectors is having the largest insert size
   (a) P1 phage (b) PACs
   (c) BACs (d) YACs
9. R-M system is associated with
   (a) Plasmids of bacteria (b) Recombinant DNA technology
   (c) Bacterial immune system (d) Bacterial reproductive system
10. For pure DNA the ratio of absorbance at 260nm and 280nm (A260 nnm/ A280 nm) should be
    (a) Less than 1.8 (b) Equal to 1.8
    (c) More than 1.8 (d) In the range of 1.0–2.0

Answers

1. (b)	2. (d)	3. (d)	4. (b)	5. (c)	6. (a)
7. (c)	8. (d)	9. (c)	10. (b)		

CHAPTER 38

Development of Transgenics

P. K. Agrawal

[1]Biotechnology Unit, Vivekananda Parvatiya Krishi Anusandhan Sansthan, Almora–263 601

E-mail: pawancrri@yahoo.co.in

Development of transgenics essentially involves transfer of foreign gene into the genome, its stable integration and development of a whole organism from the transformed cell. In this process, the host organism is transformed with the targeted gene (transgene). The reproductive component of the transgenic organism in the first generation (T0) needs to have stable integration of the transgenes so that the offsprings carry the transgene stably. In case of plants, the explants for the gene transfer can be a cell, protoplast, organized tissues, callus or even a whole plant. In most of the transformation methods, help of tissue culture technique is taken because reliable and repeatable methods are still not available for most of the plant system to bypass the tissue culture steps. Even if regeneration systems are reported in many of the crop plants, the tissue culture and whole plant regeneration is still influenced by environmental and genetic components (genotypes). There is a wide genotypic variability for their responses to tissue culture conditions and regeneration. This chapter will focus on the available transformations methods, selectable markers, agronomically useful genes available, biosafety issues, recent advances in transformation methods and the future prospects of gene transfer in crop plants.

GENE CONSTRUCTS

A gene construct is a stretch of DNA that essentially has a promoter, coding region and terminator. In addition to the gene of interest, it may have selectable or reporter gene(s), regulatory sequences, upstream or down stream of the gene cassette (promoter + coding region + terminator) for controlling the level of expression (Fig. 1). It has also an untranslated region or leader sequence (Fig. 1). The promoter is the region where the RNA polymerase gets bound and initiates transcription. The enhancers/silencer sequences help in the regulation of extent of

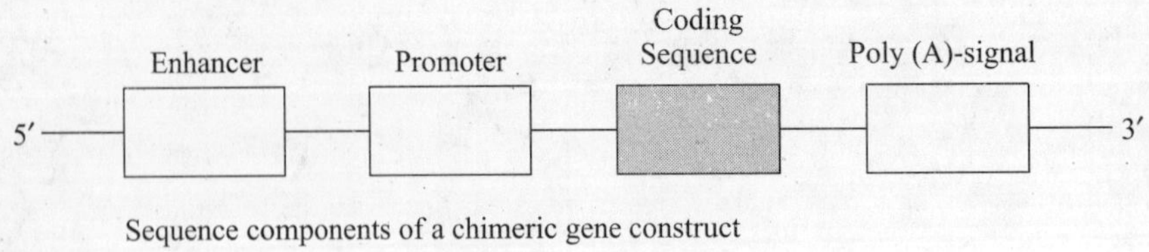

Fig. 1. Gene construct.

gene expression. The end of the gene coding sequence is with the chain termination codon TAG (it is UAG in mRNA) or one of the two non-sense codons (TAA and TGA). The presence of polyadenylation site denotes the end of transcription.

REPORTER GENES

Reporter gene is a test gene whose phenotypic expression can be quantified. This kind of reporter genes are useful to study and standardize the parameters for successful gene transfer in crop plants. The transgenic plants can also be screened on the basis of the expression of the reporter genes. The assay of the reported gene is usually carried out at the level of protein quantification.

Characteristics of a Novel Reporter Gene

1. It should have low endogenous background.
2. The detection should be with high sensitivity.
3. Quantitative assay for the protein expression should be available.
4. The assay should be preferably non-destructive.
5. The cost for the assay should be minimum and the time requirement should be less.

Among many reporter genes available, few most commonly used reporter genes are β-glucuronidase (*gus*), green fluorescent protein (*gfp*), bacterial luciferase (*Lux* F2), anthocyanin regulatory genes, octopine synthase (*ocs*), nopaline synthase (*nos*), chloramphenicol acetyl transferase (*cat*) (Table 1).

β-glucuronidase (GUS): In recent past *gus* has become the most widely used reporter gene in plant transformation experiments. GUS is encoded by the bacterial gene *uid* A. GUS is a soluble enzyme, monotetramer of approximately 68 Kda. It breakdowns glucuronide substrate to give a coloured reaction. GUS has high reaction using fluorogenic substrate 4-Mu gluc and histochemical assay using X-gluc. Using this reporter gene, it is possible to study the cell and tissue-specific expression of the transgenes. Expression studies of this reporter gene do not require methods like DNA extraction, electrophoresis or autoradiography.

Bacterial luciferase (Lux F2): In bacteria like *Vibrio harvey* and *V. fischerii*, luciferase is expressed as heterodimeric flavin monooxygenase, responsible for catalyzing a light-emitting reaction. The luciferase from *V. harvey* (Lux) is encoded by two cistrons *viz*., Lux A and Lux B, part of a longer operon. The fused protein in plant has been designated as Lux F2.The activity of luciferase is measured in terms of light intensity after mixing the enzyme with its aldehyde and reduced flavin mononucleotide substrates (FMNH2) in the presence of oxygen. The light

Table 1. Reporter genes used in plant transformation

Reporter gene	Substrate	Identification
β-glucuronidase (gus)	X-gluc, 4-Mu gluc	Fluorescence detection, colorimetric, histochemical,
Firefly luciferase (Luc)	ATP + Luciferin	Bioluminiscence
Bacterial lucuferase(Lux F2)	Decanal FMNH2	Bioluminiscence
Green fluorescent protein (gfp)	UV light irradiation	Flurescence detection
Anthocyanin regulatory genes	No substrate	Visual
Octopine synthase (*ocs*)	Arginine + Pyruvate + NADH	Electrophoresis, Chromatography

intensity is measured as photons or quanta per second (qu/s). Normally a luminometer is used to detect and quantify the light intensity and thereby the luciferase activity.

The firefly luciferase (Luc): The firefly luciferase is isolted from the North American firefly *Photinus pyralis.* It catalyzes the oxidative decarboxylation of luciferin to oxyluciferin in the presence of ATP, oxygen and magnesium ions. The luminescence (light production) is recorded by a luminometer.

Green fluorescent protein (GFP): This gene has been obtained from jellyfish (*Aequorea victoria*). It is a small protein of 238 aminoacids. The b-hydroxy benzylidine imidazolinone chromophore is responsible for the visible absorbance and fluorescence. The emitted green fluorescence is detected under a fluorescent microscope.

Anthocyanins: Anthocyanins are prominent coloured pigments of the flowering plants. They are often used as visible markers in a transformation experiment. Many regulatory genes (c1) and structural genes of the biosynthetic pathways (B/R) coordinately regulate the activity of the structural genes. The chimeric gene constructs with C1 and B/R when transferred into plant cells produce purple red colouration which can be visually or microscopically observed. The advantage of this system is that the analysis is non-destructive and no additional substrate is added for the transgene expression.

Chloramphenicol acetyl transferase (CAT): The cat gene is commonly used as a reporter gene in any transformation experiment. This structural gene codes for the enzyme, chloramphenicol acetyl transferse (CAT). It carries out acetylation of chloramphenicol. This enzymatic assay is very sensitive and simple. The labeled chloramphenicol is incubated with the crude extract of the plant tissue and acetyl CoA is used. The acetyl chloramphenicol is detected through autoradiography.

Opine synthase: The opine synthase genes are present in the T-DNA of the Ti or Ri plasmids of *Agrobacterium.* The screening using opines is very simple and the enzymatic assay is inexpensive. This is also very easy to perform. Two simple methods are followed for detection of the enzymatic activity and the synthesis of the opines. In one method, the protein extract of the plant is incubated with appropriate opines precursor like arginine, pyruvate and NADH for *ocs* and the reaction products are separated by paper chromatography. The second method involves extraction of the plant metabolites from the tissue. The separation of these compounds by paper electrophoresis is followed by appropriate staining which leads to the detection of the expression of the transgene.

SELECTABLE MARKERS

In many gene transfer experiments, the transformation events happen in a small percentage of the cells of the explants. Hence it is essential to find out or select the transgenic cells/tissues over the non-transgenic cells/tissues. The selectable markers are used to help select the transgenic cells/tissues or to help selective growth of the transgenic cells/tissues. The gene of interest is generally put along with the selectable markers for the indirect selection of the targeted transgenic plants (plants with the target genes). When a vector is used for the transformation, the selectable markers are placed in the vectors along with the gene of interest. A large number of selectable markers are available *viz.*, antibiotic and herbicide resistance genes, genes conferring resistance to toxic levels of aminoacids or aminoacid analogs, and genes related to hormone biosynthetic pathways.

Antibiotic Resistance Markers

NPT II enzyme is coded by the gene npt II, derived from the Tn5 transposon. It inactivates a number of aminoglycoside antibiotics such as kanamycin, neomycin, geneticin (G418), and paromycin. Kanamycin is used as a selective agent with a concentration of 50 to 500 mg/litre. Kanamycin can also be sprayed to the plant or applied to soil to select the kanamycin resistant plants.

Hygromycin phosphotransferase (*hpt*): This gene was derived from *E. Coli*. The hygromycin B is more toxic than kanamycin and fastly kills the sensitive cells. The screening of the plants can also be done by using hygromycin B in the potted soil or media containing the antibiotic.

Herbicide Resistance Markers

Bar gene: Glutamin synthase (GS) is an essential enzyme for the assimilation of NH_3 and helps in the regulation of nitrogen metabolism in plants. Inhibition of glutamin synthase results in accumulation of NH_3 which causes death of plant cells. The *bar* gene codes for the enzyme phosphinothricin acetyl transferase (PAT). The enzyme converts bialaphos/phosphinothricin (PPT) into acetylated form by transferring the acetyl group from acetyl CoA on the free amino group of PPT and induces resistance to the cells. The *bar* gene has been derived from two different strains, *Streptomyces hygroscopicus* and *S. Viridochromogenes*.

Enolpyruvylshikimate phosphate (EPSP) synthase (*aro* A) gene: Glyphosate (N-phosphonomethyl-glycine) is a broad spectrum herbicide which inhibits the essential photosynthetic process. This inhibits the enzyme ESPS synthase. Genes (Aro) resistant to glyphosate has been derived from *Salmonella typhimurium* which is resistant to glyphosate. Transgenic plants with ESPS Synthase escapes the effects of glyphosate and can grow normally on application of glyphosate. This gene has been used as a selectable marker in many transformation experiments.

Acetolactate synthase (*als*) gene: Sulphonyl ureas and imidazolinones have ability to inhibit the enzyme acetolactate synthase (Als) in the pathway leading to the production of aminoacids like valine, leucine, and isoleucine. Genes have been cloned from *E. Coli* and yeast and have been used as selectable markers in plant transformation experiments.

Bromoxinil nitrilase (*bxn*): Bromoxynil, an herbicide inhibits the photosystem II. The bromoxynil nitrilase gene, isolated from *Klebsiella ozaene*, can convert bromoxynil into 3,5-

dibromo-4-hydroxybenzoic acid. This gene has also been used as selectable marker. Of course this gene is not used routinely as a selectable marker.

GENE TRANSFER METHODS

The various gene transformation methods can be groupd under two categories viz., vector mediated gene transfer, and direct DNA transfer. The vector mediated gene transfer essentially needs a vector like plasmid, cosmid, phasmid, and uses biological systems like *Agrobacterium* and *Vir*uses to transfer the gene of interest into the host (plant) organism. Among the vector mediated transformation systems, *Agrobacterium tumefaciens* is the most used system. The inherent quality of this bacteria is to transfer the TDNA component to the host for its survival and growth. It has been exploited to transfer the gene of interest into the targeted host plants.

Agrobacterium-mediated Gene Transfer

The first gene transfer in plant was done in 1983 using the *A. rhizogenes*. It is considered as the most effective natural plant genetic engineer. *Agrobacterium* are gram-negetive rods that belong to the family Rhizobiaceae. They are found in the soil near to the junction of plant stem and root. *Agrobacterium tumefaciens* induces crown gall disease. *A. rhizogenes* induces hairy root disease and *A. radiobacter* is an virulent strain. Recently, the strains of *Agrobacterium* have been reclassified as biotype 1 and 2 based on their growth pattern.

Ti Plasmids and Tumour-inducing Principle

Bacterium is the causative agent of crown gall tumours. It was showed that the continued presence of viable bacteria is not required for tumour maintenance in plants. It was also observed that the bacteria do not penetrate into the plant cells to convert them to tumour cells, rather they stay in the intercellular spaces and penetrate into the injured cells. However, they also stay attached to the wall of the healthy cells. This necessitated to find the tumour inducing principle in the *Agrobacterium* system. It was seen that the virulence trait is plasmid borne and carried in a large plasmid. These large plasmids in *A. tumefaciens* is called Ti plasmid and in *A. rhizogenes* as Ri plasmids. Later on it was observed that a small fragment of the Ti plasmid, called T-DNA get integrated into the host genome.

The naturally occurring crown gall differs from the non-transformed tissue in the sense that it has ability to grow autonomously on synthetic media in the absence of hormones. It shows phytohormone independent growth and contains one or more of unusual aminoacid derivatives known as opines. Opines are useful to the bacterium as a source of carbon and nitrogen. Some opines also inducc conjugation between bacteria. Two different families of bacteria responsible for crown gall formation produces two different aminioacid derivatives called octopine and nopaline. The octopine family has carboxyl-ethyl derivatives of arginine where as nopaline family has related compound that are dicarboxypropyl derivatives of arginine. For the synthesis of octopine and nopaline, the corresponding enzyme octopine synthase (*ocs*) and nopaline synthase (*nos*) are coded by T-DNA.

Organization of Ti Plasmid

The Ti plasmid of *Agrobacterium* are large circular molecules. They are upto 200kb in length and covers about 3 to 8% of the *Agrobacterium* chromosome. They are independent replicating genetic units. The genetic organization of Ti plasmid (octopine) has been shown in Fig. 2. The major regions are for virulence, origin of replication, conjugation, oncogenicity and catabolism of opines. The essential regions of Ti plasmid required for T-DNA transfer to plants are discussed here.

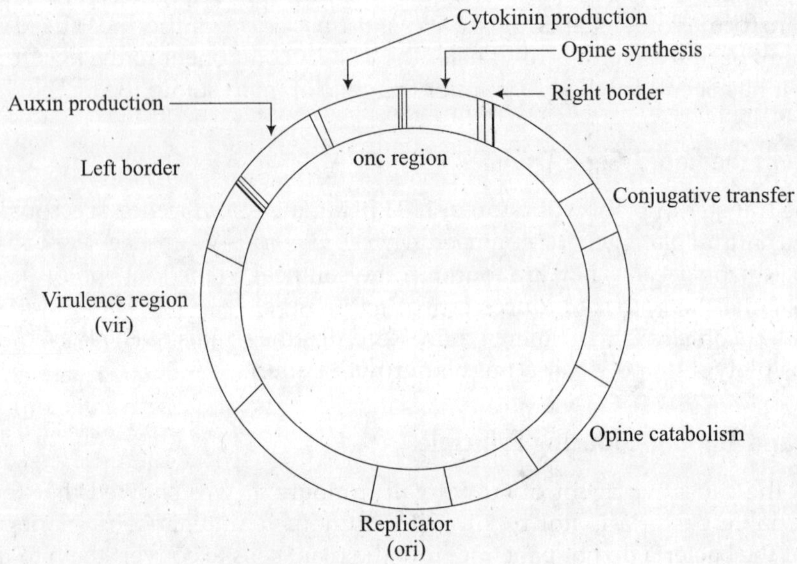

Fig. 2. Genetic organization of Ti-plasmid (octopine).

T-DNA : T-DNA is the transferred DNA or the DNA that is transferred from the Ti plasmid to the target tissue of the host plant. The length of the T-DNA varies in Ti and Ri plasmids. The T-DNA region of the Ti plasmid in pTiC58 is 23kb long whereas the octopine type plasmid (i. e., pTiAch5) has two closely adjoining regions of 13kb left and 8kb right hand piece. The T-DNA carries genes for the auxin production, cytokinin production and opine synthesis (*nos* or *ocs*). The oncogenic regions (*onc*) contain genes for root inhibition (*roi*) or tumour morphology root (*tmr*) developing abnormal roots and a shooty locus (*shi*) or tumour morphology shoot (*tms*) developing abnormal shoots. The tms (*tms* 1 and *tms* 2) genes are responsible for the synthesis of the phytohormone, indole 3-acetic acid from tryptophan. The *tmr* gene codes for isopentyl transferase, which catalyses the formation of cytokinin. The phytohormones alter the developmental gene regulation in the plant and induces the formation of crown gall. The opine synthesis (*os*) region is responsible for the synthesis of unusual aminoacid or sugar derivatives. The octopine synthesis is controlled by octopine synthase (gene: *ocs*) and nopaline synthase is controlled by nopaline synthase (gene: *nos*).

The T-DNA region has 25bp highly conserved direct repeated DNA sequence flanking the T-DNA and referred to as left and right borders. These borders are very important in the transfer process. It is interesting that the DNA sequences located between these borders are transferred to the plant and it is this significant observation that has been exploited for use in gene transfer experiments. The other interesting information is that there is no gene within the T-DNA of naturally occurring Ti plasmids that is necessary for the T-DNA transfer. The 'oncogenes' harboured within the T-DNA segment in the naturally occurring plasmids are necessary for the tumorigenesis and opine production. Thus the T-DNA can be disarmed by deleting the oncogenes and disabling them to cause tumours. Novel genes (genes of interest) can be placed within the T-DNA borders in these plasmids and be used for gene transfer experiments.

***Vir*ulence Genes:** The *Vir*ulence (*Vir*) genes are essential for the transfer of T-DNA. However, they are outside the T-DNA. In octopine Ti plasmid, the *Vir* region consists of 40 kb-region, encompassing 24 genes, present in eight operons i.e. *Vir* A to *Vir* H. Except *Vir* A, *Vir* F and *Vir* G, most of the operons contain several genes (polysystronic). Except *Vir* A and *Vir* G genes, other *Vir* genes do not transcribe during the normal vegetative growth. The *Vir* A and *Vir* G genes have a regulatory role for the transcriptional activation of the remaining *Vir* genes. Those remaining *vir* genes have a role in processing of the T-DNA from the Ti-plasmid, T-DNA transfer from the bacterium to the plant cell, protection during the transfer process, targeting the T-DNA into the plant nucleus and integration of the T-DNA into the plant DNA. The manipulation of the *vir* gene help in increased T-DNA transfer to the target plant cell.

Bacterial chromosome genes: Many of the genes present on the chromosome of the bacteria have a role in the transfer of the T-DNA. Some of these genes are *chv* A, *chv* B and *psc* A (exo C). Genes *chv* A and *chv* B code for exopolysaccharide production which is helpful for the attachment of the bacterium to the plant cells. These two genes are constitutively expressed. The genes *chv* E encodes a glucose/galactose transfer and is helpful in binding specific sugars that act as a coinducer to the *vir* genes. Other chromosomal genes like *chv* D, *ilv* and *att* help in virulence, their exact functions are to be understood still. Two more loci *noc* and *occ* control the fully catabolism of nopaline and octopine respectively. These two genes do not form part of T-DNA.

Transfer of T-DNA: Wounding is a necessary requirement for the introduction of T-DNA into the plant genome. It also induces the expression of the *vir* genes. Acetosyringone and B-hydroxy acetosyringone are the two phenolic compounds produced due to wounding of plants. *Vir* A gene product is a periplasmic membrane protein which senses phenolic compounds. *Vir* A protein also interacts with another protein encoded by *chv* E and is useful in binding of sugar coinducers. The phenolics and sugar induction helps in autophosphorylation of *vir* A which in turn activate the *vir* G gene by transferring the phosphate group to the *vir* G protein.

Nicking the Ti plasmid is an important step for the T-DNA transfer. The nicking of the DNA is done by the *vir* D genes i.e., *vir* D1 and *vir* D2 which have a endonuclease activity. The nick is made at the right border 25bp repeat and it provides a priming point. Synthesis of a new DNA strand starts from this point in $5' \rightarrow 3'$ direction. The *vir* D2 protein stay attached covalently to the 5′ terminus and protects the 5′ end from the exonucleases. Besides it also acts as a pilot to direct the T-strand to the nucleus of the transformed cell.

A small stretch of DNA, immediately adjacent to the right border is called overdrive. This stretch of DNA enhances the transfer process. *Vir* C1 and *vir* C2 also act as enhancers. The *vir* E2 protein, 69-k Da in size binds the single-stranded T-DNA and is involved in transporting T-DNA into the plant nucleus. It protects the single stranded T-DNA. The *vir* B operon is essential for the virulence. *Vir* B11 gene has an ATP-binding site and the protein has an ATPase activity. Similarly, *vir* H operon consists of two genes and has a role in detoxification of certain plant compounds that might otherwise adversely affect the growth of *Agrobacterium*. Presence of *Vir* H gene has been associated with enhanced tumorogenised. *Vir* F plays a role in T-DNA delivery. It is still not very clear how the transferred DNA is integrated into the plant genome. At some stage, the single stranded DNA is converted into a double stranded DNA.

Types of Agrobacterium Vectors

Two major types of *Agrobacterium* vectors are cointegrate and binary vectors.

Cointegrated vectors: In the cointegrate vectors, the vector gets recombined into a resident Ti plasmid because of the DNA homology between them. The gene of interest lies with the donor vector and this vector is covalently linked to the disarmed Ti vector to form a cointegrate vector. In this type of vectors, a part of wild type of T-DNA is replaced with a segment of DNA which is common between the modified T-DNA and the donor vector with identical sequences. These identical sequences provide a site for recombination resulting in a hybrid or cointegrate Ti plasmid. The cointegrate vector is produced by a single recombination event, thus placing the gene of interest and the entire donor vector and its marker gene within the border of the T-DNA of the disarmed Ti-plasmid (Fig. 3a,b).

A standard cointegrate vector contain an antibiotic selectable marker, a site for the insertion of the gene of interest, a selectable marker gene functional in the plant system, and origin of replication from *E. coli*.

Binary vectors: Binary vector systems consist of two autonomous replicating plasmids. It is a shuttle vector that contains the gene of interest within the T-DNA borders and a helper Ti plasmid that provides the *Vir* gene products which helps the transfer process. The oncogenes are removed from the helper Ti plasmids whereas the *vir* genes required for the transfer of desired genes are retained within them (Fig. 4).

Characteristics of binary vectors:

1. They contain multiple cloning site,
2. The origin of replication is functional in both *E. Coli* and *A. tumefaciens*.
3. They contain selectable markers for both bacteria and plants.
4. The shuttle vector contains the T-DNA border sequence.

Advantages of using a binary vector over cointegrate vector:

1. Binary vectors do not need *in vivo* recombination for maintenance in *A. tumefaciens*.
2. Transfer of binary vector into the target bacterium is very quick.
3. The binary plasmid exists as separate replicons. Hence the copy number is not limited by the Ti plasmid and the transformation event is accomplished via *Agrobacterium* mini preps.
4. The binary vectors are easy for *in vitro* and *in vivo* DNA manipulations and higher transformation efficiencies.

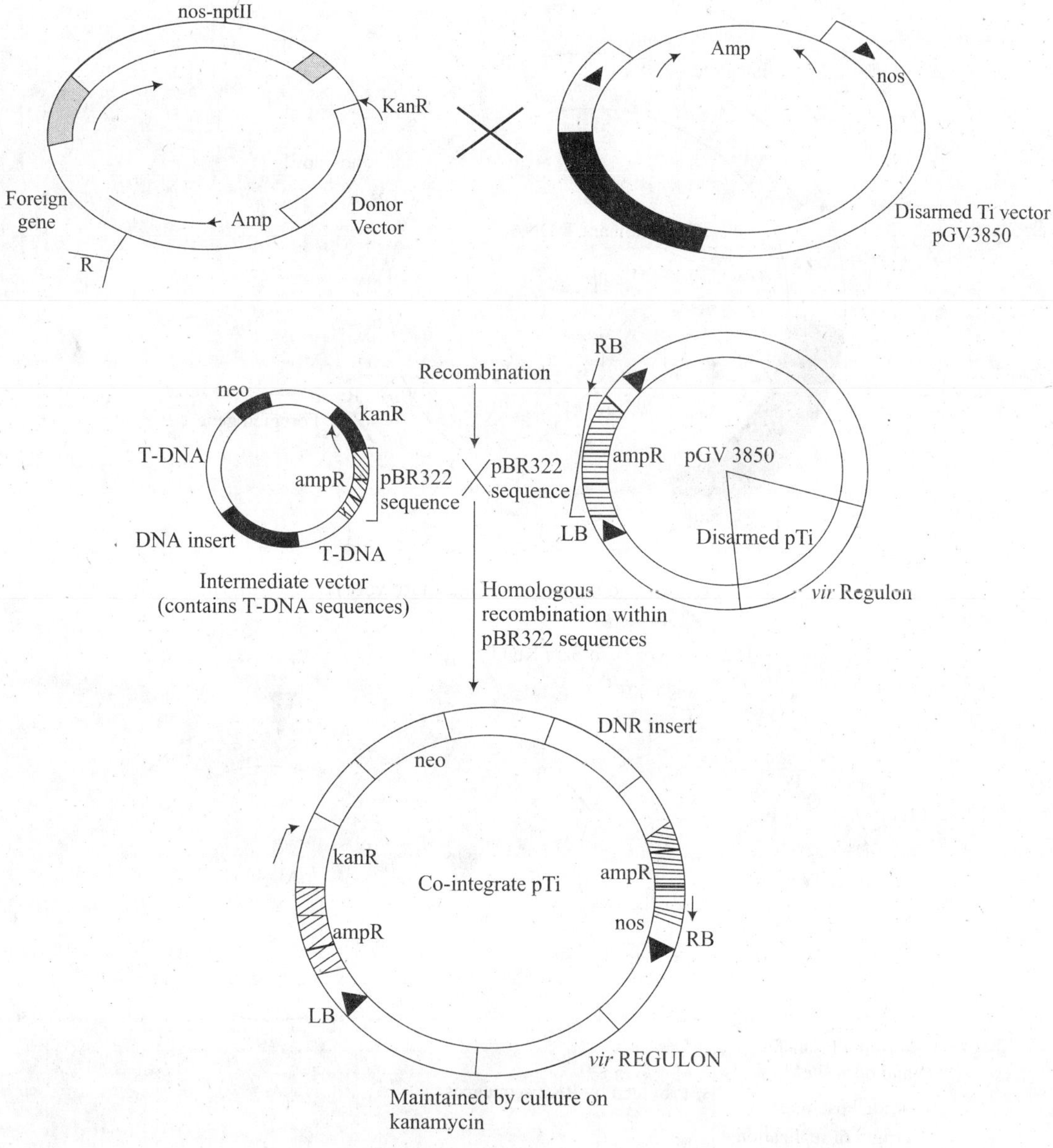

Fig. 3a. Cointegrate vectors.

Requirements for *Agrobacterium*-mediated gene transfer in higher plants :

1. The plant cells of the target tissue should be competent for the transformation.
2. The explants must produce acetosyringone and related compounds for the inactivation of the *vir* genes or acetosyringone should be applied additionally to the explants.

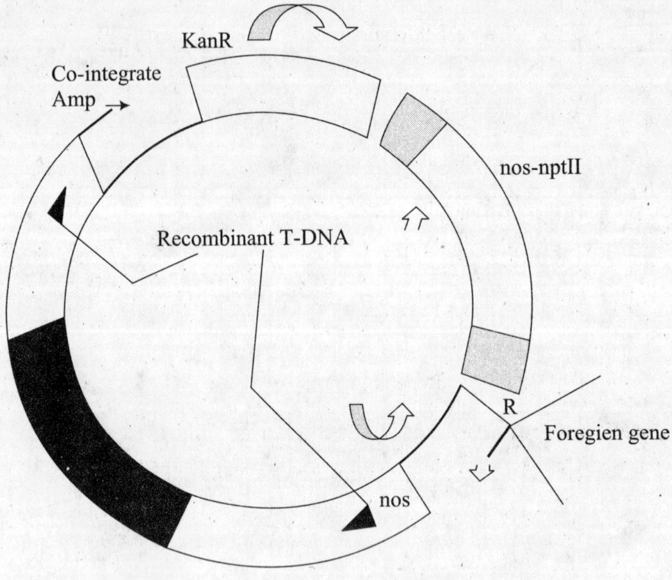

Fig. 3b. Recombinant cointegrate vector.

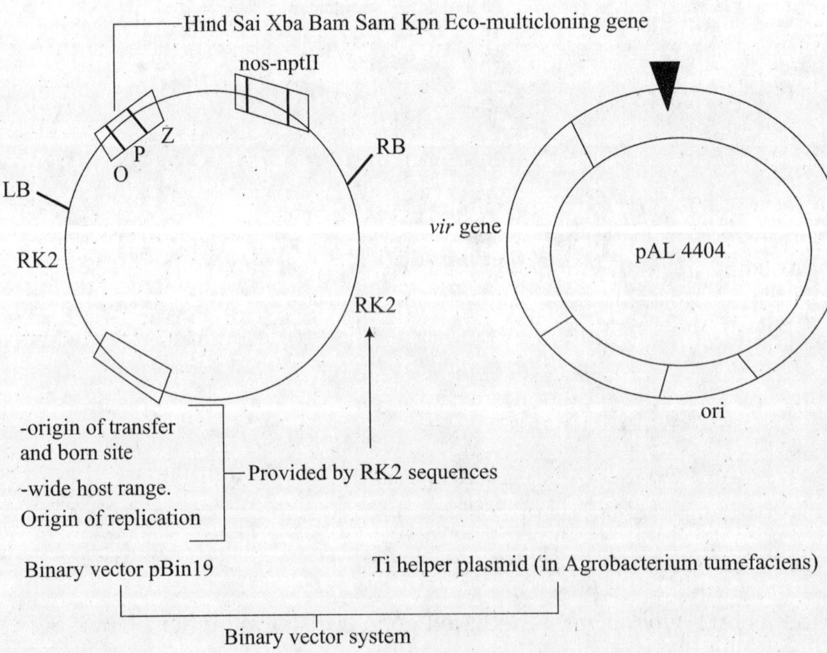

Fig. 4. Details of a binary vector.

3. The induced agrobacteria should have access to the competent explant cells.
4. The transformed plants should have capability to regenerate into whole plants.

The explants for the *Agrobacterium* mediated gene transfer include protoplasts, cells, callus, organized tissues, whole organs, or even whole plants.

The general approaches for *Agrobacterium* mediated gene transfer are:

1. Infection of wounded plants
2. Cocultivation
3. Leaf disc method.

Agrobacterium-mediated gene transfer has been very successfully reported in plant transformation (Herrera-Estrella *et al.*, 1983; Hiei *et al.*, 1994) and is being routinely used now-a-days.

Agroinfection (*Agrobacterium*-mediated Virus Infection)

The transfer of viral or viroidal sequences to the plant system using a bacterium is called 'agroinfection'. It has been shown that when the genome of a plant *Virus* is placed within the Ti plasmids, the viral genome can be transferred from the Ti-plasmid into the plant genome. On their transfer, they initiate the typical infection in the plants. This method of *Agrobacterium* mediated viral infection is termed as agroinfection. Using this method, Grimsley *et al.* (1987) constructed a vector containing the double stranded maize streak virus (MSV, a Gemini virus) and successfully transformed the maize plants which developed all the symptoms of MSV infection.

Advantages of *Agrobacterium*-mediated transformation :

(i) *Agrobacterium* is capable of infecting intact plant cells, tissues and organs. Even gene transformation can be made into the reproductive organ of a plant.
(ii) *Agrobacterium* is capable of transferring large fragments of DNA very efficiently.
(iii) The copy number of the transgenes in the host genome is low.
(iv) The integration of the T-DNA is relatively precise.
(v) The stability of the transgenes within a generation and over generations is excellent.
(vi) Since *Agrobacterium* uses natural system, the cost of gene transfer is relatively very low.

Disadvantages of *Agrobacterium* mediated gene transfer:

(i) The host range for the *Agrobacterium* mediated gene transfer is limited. Even if many monocots have been successfully transformed using *Agrobacterium* system, the genotype and species of the explants have a significant effect on the successful gene transfer using *Agrobacterium*.
(ii) Leaky expression of plant selection marker genes such as *npt* II and *hpt* (under CaMV35S or *nos* promoter) in *Agrobacterium* contributes to bacterial overgrowth during selection and to the emergence of false-positive. Often during the histo-pathological analysis of expression of *gus* as a reporter gene, the *gus* (reporter gene) expresses in *Agrobacterium* despite the use of plant promoters. This problem was overcome by inserting introns which are in plants but not in *Agrobacterium*. The intron 2 of potato STT-LS1 gene was being sucessfully placed in the coding sequence of *gus* gene.

(iii) Multiple gene transfer using *Agrobacterium* is still very difficult and often it is highly affected by the crop plants, genotype of the crop plant and the stage of the explants.

(iv) There are reports that in low frequencies, the T-DNA transfer initiates from the left border and the non T-DNA portions may be transferred to the plant genome. The other way is that, often the T-DNA transfers starting from the right border skips the left border and the entire binary plasmid is thus transferred to the plant genome. This process is termed as 'Long transfer'.

(v) Transgene silencing has been observed in plants derived from both *Agrobacterium* and biolistic mediated transformation.

Virus-mediated gene transfer: Virus-based vectors are highly efficient for gene transfer and amplification of the transgenes. Viral infection is also systemic. So when one cell of the plant is infected, all the other cells of the plant also get transformed. The vectors for the viral gene transfer are based on DNA or RNA molecules. However the viruses are non-integrative vectors compared with those vectors based on the T-DNA of *A. tumefaciens* which are integrative gene vectors. In most cases, the viral genomes have been modified to mimic their wild counterpart.

Characteristics of a viral vector:

1. It should have a broad host range, virulence and ease of mechanical transmission.
2. For the virus to be suitable vector, the targeted genetic materials should be able to be manipulated and infectious.
3. The virus-vectors should have the potential to carry additional genetic materials.

Types of Viruses used for Vector Development and Transmission

The three groups of viruses used now-a-days for vector development and transmission purpose are,

1. Caulimo viruses
2. Gemini viruses
3. RNA viruses.

I. Caulimo Viruses

In this group of viruses, the cauliflower mosaic virus (CaMV) is the most cited and potential vector for gene introduction into crop plants. It has double stranded DNA which is easy to be manipulated for the purpose of vector development. The group of caulimo viruses consists of 6-19 viruses. Some of the commonly observed viruses are cauliflower mosaic virus (CaMV), carnation etched virus (CERV) and dahlia mosaic virus (DaMV). The viruses are transmitted by aphids. However, they are also easily transmitted mechanically. These virus particles are found in high copy number and spread to the entire plant in a systemic way in 3-4 weeks. The virus particles are spherical, isometric and about 50 nm in diameter. The DNA exists in linear, open circular and twisted or knotted forms.

The genome size of CaMV is 8kb. The CaMV replicate via reverse transcription. The infecting CaMV DNA enters the plant nucleus. Here the single stranded overlaps are digested and the gaps are ligated to form a supercoiled minichromosome. This minichromosome act as a template for the plant nuclear RNA polymerase II. The transcript thus formed is transformed, transported to the cytoplasm and is replicated by reverse transcription or is translated. The RNA transcript is then copied into the minus-strand DNA. Synthesis of the plus strand DNA starts at two primer binding sites near gap 2 and 3. The DNA molecule gets packaged into the *Virus* particles or re-enter the nucleus and undergoes another round of transcription and/or translation/replication. Foreign protein encoded by the bacterial dihydrofolate reductase gene in turnip plant cells by a CaMV vector has been seen to be expressed.

Advantages of Caulimo viruses as vector:

(i) The naked DNA is infective when rubbed to a leaf mechanically.

(ii) It is a DNA virus and transcribed by RNA polymerase II, hence suitable as an experimental material.

Disadvantages of Caulimo viruses as vector:

(i) Maximum of 1000 nucleotides of foreign DNA can be inserted into this vector. Usually it is less than 500bp.

(ii) This group of viruses infect the Cruciferae those can be infected by the viral DNA.

(iii) CaMV DNA has multiple cleavage sites for most of the commonly used restriction enzymes.

(iv) The virus spread throughout the whole plant upon infection. However, it lacks inheritance through the germline.

2. Gemini Viruses

Gemini viruses cause disease in many plant species. Cury top virus (CTV) and maize streak virus (MSV) (Grimsley *et al.*, 1987) are some of the important examples.

Characteristics of Gemini viruses:

(i) They posses single-stranded (ss) DNA.

(ii) They have a small capsid size, 18-20 nm x 30 nm.

(iii) They have covalently closed circular (ccc) topography of the ss DNA, with a molecular weight of 7×10^5 to 9×10^5 daltons.

(iv) The single coat protein subunit is in the range of $2.7\text{-}3.4 \times 10^4$ Daltons.

(v) The genome consists of either one or two circular, single stranded DNA molecules.

(vi) The genomic ss DNA replicates in the nucleus through double-stranded DNA (ds DNA) intermediates, most likely by a rolling circle mechanism.

(vii) Most of the Gemini *Virus* vectors are relatively primitive and simply permit the investigator to replace the coat protein coding sequence with a reporter or other gene of interest.

Disadvantages of Gemini virus as a cloning vector:

(i) They are not readily transferred by mechanical means.

(ii) The small particle size present packaging problems for modified DNA molecules.

RNA Viruses

Advantages of RNA viruses in gene transfer:

(i) It is very easy to use the RNA viruses.

(ii) The level of expression of genes is very high.

(iii) The *Viruses* can infect and spread in differentiated tissue.

The examples of RNA viruses are tobacco mosaic virus (TMV) (Powell-Abel *et al.*, 2002) and brome mosaic virus (BMV) (French *et al.*, 1986).

Direct DNA Transfer

The direct DNA transfer methods can be done in the following ways:

(i) Physical methods
(ii) Chemical methods
(iii) Others (includes imbibition by cell, tissue and organs).

In the direct DNA transfer method, no parental vector is involved. Here a physical force or non-biological transfer system is used for the transfer process. However, as for integration of the transgenes are concerned, the host machinery is used here also like in any other methods of gene transfer.

Electroporation: In this process protoplasts are used as explant for transformation. The cells/ tissues are treated for the digestion of the cell walls and release of protoplasts. The electroporation method uses a relatively highly initial field strength (1-1.5 kv) and a low capacitance, resulting in a short decaying time. The pulse is applied by discharge of a capacitor across the cell. Use of polyethylene glycol (PEG) can increase the transformation frequency. PEG helps in the association of DNA with the membrane. Following this method, intact cells or tissues can also be transformed after a pretreatment of mechanical wounding or by treating the cells or tissues with hypertonic or enzyme-containing (e.g., 0.3% macerozyme) solutions. The advantage with the system is that this is independent of the genotype. However, regeneration of whole plant from the protoplasts is difficult, time consuming and needs special skill in handling the protoplasts (Krens *et al.*, 1982; Zhang *et al.*, 1988; Abdullah *et al.*, 1989; Shimamoto *et al.*, 1989; Datta *et al.*, 1990).

Particle bombardment (Microprojectile, biolistic): This technique uses a tungsten or gold particle of 1-3 micrometer. The DNA (gene construct) is wrapped around the particle. The DNA-bearing gold or tungsten particle referred to as microprojectile (equivalent to bullet for a gun) is bombarded to the target cells/ tissues. The DNA gets integrated into the host genome upon its entry into the cell. The microprojectile is carried by a macrocarrier (macroprojectile) and accelerated by use of helium or any inert gas with a pressure of 500 to 1700 psi. In order to make the microprojectile at proper speed and direction, the chamber is partially made vacuum (Klein *et al.*, 1987; Christou *et al.*, 1988; Sanford *et al.*, 1991). The following parameters are important for successful particle bombardment.

Physical components of biolistic: The particle (gold or tungsten) should be of high mass in order to gather adequate momentum so that it can penetrate the cells/tissues. The other metals like platinum, iridium and rhodium can also be used. The most commonly used particle size is around 1 micrometer. The chemical used for the particle purpose need to be chemically inert. Some additives such as spermidine and calcium chloride appear to be useful for the DNA binding to the gold or tungsten particle. The target cells/tissues for particle bombardment should be capable for cell division and transformation. Penetration of the particle DNA to a specific layer is done by controlling the speed of the particle (Fig. 5).

Environmental component: The factors like temperature, photoperiod and humidity of the donor plants, explants and the tissue are important. They influence the receptiveness of the target tissue and the capacity to recover the injury caused by the gold/tungsten particles.

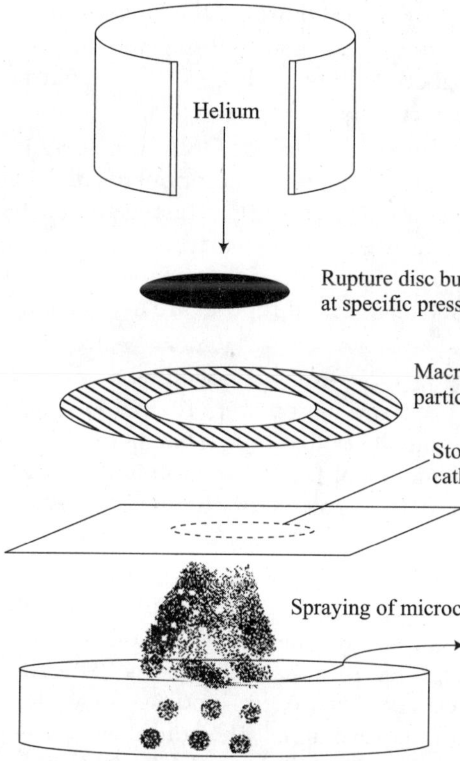

Fig. 5. A typical figure showing the biolistic movement for transformation.

Biological component: The explants collected from plants grown under stress are not ideal for transformation. Osmotic pre- and post treatment of explants with mannitol has been shown to be important in transformation.

Advantages of particle bombardment:

(i) The transformation process is independent of genotype and stage of the explant.

(ii) The transformation can be made into potentially regenerable tissue.

(iii) Transformation into both monocots and dicots, including most of the agronomically important crops is possible.

(iv) Co-transformation of many genes is possible.

(v) It is clean and safe.

Disadvantages of particle bombardment:

(i) The cost of instrument and consumables is very high.

(ii) Most of the reports indicate multiple copies and gene silencing resulting from the biolistic method. However, Agrawal *et al.* (2000, 2005) reported multiple gene transformation in rice using biolistic method. It shows simple integration pattern and higher transgene expression.

Injecting DNA into the developing floral parts: The development of two kanamycin resistance rye plants by injecting the transgenes using syringe into the developing flower of the tillers of the plants has been described. In this method, 5–10 microlitres of DNA solution is injected. This method is hardly reproducible, hence, not used.

Microinjection: By this method of transformation, the DNA (transgene) is introduced into the targeted tissue/embryo, cells, protoplasts or a component of a single cell by glass micropipettes of 0.5 to 10.0 micrometer diameter of tip. The recipient cells are immobilized by using agarose embedding, polylysine treated glass surface or suction holding pipettes. The cells/tissues after the gene transfer are regenerated into a whole plant. Lot of chimaeric plants are developed following this method. Successful reports are available in tobacco and *Brassica napus*.

DNA transfer via pollen: The use of DNA-treated pollen as a vector for pollinating fertile plants of maize have been made. There are various reports that DNA can be taken up by the germinating pollen or reach the zygote via pollen tube. This method is simple, fast and cheap. However, the repeatability and frequency of transformation is very low. The presence of cell wall and nucleases of the DNA create problem for the gene transfer. This method is not frequently used.

Liposome mediated gene transfer: Successful gene transformation has been reported in tobacco and carrot by using liposome containing the DNA molecules and fusion of liposome with protoplasts. The liposomes (containing the DNA) are taken up by the protoplasts by endocytosis and the DNA gets integrated into the host genome. This method of gene transfer has many advantages like reduced level of toxicity and use of wide variety of cell types. However, because of the stage of requirements of protoplasts as explants (followed by regeneration) and poor efficiency of the system, this method could not be very popular.

Ultrasound-mediated gene transformation: Ultrasound has been reported to stimulate the uptake of DNA by plant protoplasts and leaf segments in tobacco. Generally sonication of 0.5 w/cm^2 for 30 minutes is given to the explants. Even if the technique is simple and inexpensive, very little success has been achieved using this technique.

PEG-mediated gene transfer: The DNA uptake of Ti-plasmid DNA by protoplasts in the presence of polyethylene glycol (PEG) or poly L-ornithine has been demonstrated in Petunia and tobacco. The isolated protoplasts in suspension and plasmid DNA are taken in a tube and 40% PEG is added slowly and incubated for few minutes. The concentration of magnesium or calcium ions plays an important role for the efficiency of the gene transfer. The requirement of protoplasts as an explant and their regeneration limit the use of this technique. After the advent of biolistic technique, all the methods involving protoplasts as an explant for gene transfer have not been emphasized.

GENE SILENCING

Inactivation or down regulation of transgene expression within or over generations is known as gene silencing. Gene silencing can be transcriptional gene silencing or post-transcriptional gene silencing. The ability of a cell to recognize 'self' from 'non-self' is a characteristics feature that occur at cellular level.

Causes of Gene Silencing

(i) **DNA methylation:** When foreign DNA enters into a cell, the restriction and modification system of cell causes methylation of foreign sequences to make it non-functional. The cytosine molecule within symmetrical target sequences get methylated. Methylation is known to prevent transcription by affecting the trans-acting factors. The methylated cytosine residue also provides basis for the methylation of the newly synthesized complementary strand with the help of the enzyme methyl transferase. Methylation results in both kind of silencing viz., i) transcriptional gene silencing, and ii) post-transcriptional gene silencing. DNA methylation can occur because of DNA-DNA pairing when multiple transgene copies integrate as a concatamer or due to effect of stresses and other factors or when the transgenes get integrated into a hypermethylated region which upon spreading of hypermethylation pattern leads to inactivation of the transgenes.

(ii) **Homology-dependent gene silencing:** Homologous sequences affect the stability of the transgenes and affect the activity of the endogenous genes also. It may be by inactivation of homologous transgenes, cosupression of the transgenes or by paramutation.

(iii) **Position effect:** When a transgene is integrated in a region which are hypermethylated or telemetric or heterochromatic, the gene gets inactivated by the effects of the adjacent regions.

(iv) **Increased Copy number:** It is broadly agreed that high copy number results in more gene silencing. This is more true for the genes in *cis* position. The multiplicity also include either duplication of promoter or addition of truncated transgene-coding regions that cause decreased expression. However, it is difficult to say always high copy is related to gene silencing. Lot of examples with high copy number and high transgene expression are available. It is also true that transgenic plants with complex transgene integration need to be avoided.

(v) **Antisense genes:** When an antisense transgene is present in the genome, it affects the expression of the wild gene and generate mutant phenotype. Commercial transgenic tomato 'Flavr-Savr' is a good example. This kind of transgene silencing/inactivation results from post-transcription inhibition. The antisence RNA affects the wild RNA, in many occasions sense and antisense RNAs form a double stranded RNA intermediate that is rapidly degraded by ds RNA specific ribonucleases.

Ways to Avoid/ Reduce Gene Silencing

(i) Avoid integration of complex inserts. It can be achieved by vector mediated gene transfer or modified biolistic system (clean DNA transformation).

(ii) Homology at the DNA and RNA level need to be avoided.

(iii) The stability and transgene expression need to be observed at different stages of plant development within a generation and between the segregating generations.

Many modified methods of gene transfer (Cheng *et al.*, 2001; Ebinuma and Komamine, 2001; Lucca *et al.*, 2001; Reddy *et al.*, 2002; Agrawal *et al.*, 2005) have been reported. They help in better expression, eliminating the selectable antibiotic marker or non-use of antibiotic selectable marker.

TRANSGENICS IN CROP PLANTS

Transgenic plants have been developed for rice, wheat, tomato, pulses, sugarcane, horticultural and ornamental plants etc. for resistance to biotic stresses, abiotic stresses, specialized products like edible vaccines, and biodegradable plastics. The biotic stress includes resistance against pests, diseases and viruses. Few most important genes deployed for this purpose are, *bt*, proteinase inhibitors and lectins for insect resistance, coat-protein-mediated cross protection and antisense and sense-mediated resistance for virus and Xa-21, chitinase etc. for disease resistance (Table 2) (Vaeck *et al.*, 1987; Tu *et al.*, 2000). Similarly, genes like g-pyrroline-5-carboxylate synthetase (P5CS) for salinity and drought tolerance, dehydration responsive transcription factors (DREB) for drought tolerance, heat shock protein transcription factor (AtHSF) for high-temperature stress-tolerance have been described (Table 3) (Cheng *et al.*, 2002). The *bar* genes are classical example of herbicide resistance. For quality characters like shelf life of fruits, 'flavr-Savr' tomato is a classical example where antisense RNA technology has been used (Table 4) (De Block *et al.*, 1984; Horsch *et al.*, 1984; Kasuga *et al.*, 1999).

In recent years, trangenics have become very important in global agriculture. Transgenic soyean, cotton, canola and many other importamt transgenic crops occupy more than 70 million ha globally. However, there is a need to refine the transformation systems and the gene constructs to have temporal and spatial specificity of the transgene expression. It is always desirable to have transgenic plants free from antibiotic markers. The level of expression of the transgenes, cotransformation and coexpression of multiple genes, manipulation of metabolic pathways are few desirable components that the future transgenic systems are looking for. In

Table 2. Transgenic plants for resistance to diseases, insect-pests and viruses

Crop	Resistance gene	Pathogen
Cotton	Bt	Insect resistance
Potato	Bt	Insect resistance
Maize	Bt	Insect resistance
Rice	Bt	Resistance against rice stem borer
Potato	PR 5	*Phytopthora infestans*
Rice	Chitinase	*Rhizoctonia solani*
Tobacco	Chitinase	*Rhizoctonia solani*
Potato	H_2O_2 gene	*Phytopthora*

Table 3. Transgenic plants for tolerance to abiotic stresses

Crop	Genes	Traits
Tobacco	Mannitol 1-phosphate dehydrogenase	
Tobacco	g-pyrroline-5-carboxylase synthetase	Tolerance to osmotic stress
Rice	Cod A	Tolerance to salinity
Arabidopsis	Cod A	Tolerance to salinity and cold
Tobacco	Cu/ Zn super oxide dismutase from rice	Oxidative and chilling stress
Tobacco	Bet A	Tolerance to salinity

Table 4. Transgenic plants for quality characters

Crop	Gene	Traits
Tobacco	Rat-desaturase	Mono-unsaturated fatty acids
Canola	Desaturase	γ-Linolenic acid
Potato	Bacillus subtilis- Fructosyl transferase	Fructans
Tomato	Maize sucrose phosphate synthase (SPS)	Sucrose
Oilseed rape	Thioesterase	Medium chain fatty acids

coming years, we may have novel genes/gene construct(s) and marker free technology to develop desirable transgenic crop plants.

SUMMARY

Development of transgenic organism essentially involves transfer of foreign gene(s) into the genome of the organism. The stable integration and expression of the foreign gene is advantageous. A desirable gene construct with promoter, coding region and the terminator are required. The gene construct is made on the basis of the purpose and the expression capacity of the gene in the host and the source gene. Along with the gene of interest, one or more reporter and/or selectable markers are used in order to identify the transgenic organism. The commonly used reporter genes are GUS, LUC and GFP. Similarly, the commonly used selectable markers are NPT II, hpt, bar and ESPS Synthase. *Agrobacterium* mediated gene transfer is the most commonly used gene transfer method in dicot plants. This method is also used monocots. Among the type of vectors used for *Agrobacterium*-mediated gene transfer, binary vectors are very common. Besides viruses can also be used for gene transfer in plants. Particle bombardment (biolistic) is the most accepted direct gene transfer method in crop plants. One of the most commonly encountered problems with transgenics is the silencing or down regulation of the transgenes. Recently, many modifications have been made to reduce the gene silencing. One of the major demand of the policy makers and environmentalists is the development of transgenic plants free from antibiotic markers. Many options for the development of marker free plants are available. In coming years, we hope to get newer novel genes/gene construct(s) and marker free technology to develop desirable transgenic crop plants.

REFERENCES

Abdullah, R., Thomson, J.A., Khush, G.S., Kaushik, R.P., Cocking, E.C. (1989) Protoclonal variation in the seed progeny of plants regenerated from rice protoplasts. Plant Sci. 65 : 97-101.

Agrawal, P.K., Christou, P., Kohli, A. (2000) Clean DNA Transformation: Co-integration and expression analysis of minimal transgene cassette in rice (*Oryza sativa* L.), 4th International Rice Genetics Symposium, IRRI, Los Banos, Philippines 202.

Agrawal, P.K., Kohli, A., Twyman, R., Christou, P. (2005) Transformation of plants with multiple cassettes generates simple transgene integration patterns and high expression level. Mol. Bree. 16 : 247-260.

Cheng, Z.Q., Targolli, J., Huang, X.Q., Wu, R. (2002) Wheat LEA genes, PMA80 and PMA1959, enhance dehydration tolerance of transgenic rice (*Oryza sativa* L.). Mol. Bree. 10 : 71-82.

Cheng, Z.Q., Targolli, J., Wu, R. (2001) Tobacco matrix attachment region sequence increased transgene expression levels in rice plants. Mol. Bree. 7 : 317-327.

Christou, P., McCabe, D.E., Swain, W.F. (1988) Stable transformation of soybean callus by DNA-coated gold particles. Plant Physiol. 87 : 671-674.

Datta, S.K., Peterhans, A., Datta, K., Potrykus, I. (1990) Genetically engineered fertile indica-rice recovered from protoplasts. Biotechnol. 8 : 736-746.

De Block, M., Herrera-Estrella, L., Van Montagu, M., Schell, J. and Zambryski P. (1984) Expression of foreign genes in regenerated plants and their progeny. Embo. J. 3 :1681-89.

Ebinuma, H. and Komamine, A. (2001) MAT (Multi-Auto-Transformation) vector system, the oncogenes of Agrobacterium as positive markers for regeneration and selection of marker-free transgenic plants. *In Vitro* Cell Dev. Biol. 37 : 103-113.

French, R., Janda, M. and Ahiquist, P. (1986) Bacterial gene inserted in an engineered RNA Virus: Efficient expression in monocotyledous plant cells. Science 23 : 1294-1297.

Grimsley, N., Hohn, T., Davis, J.W. and Hohn, B. (1987) Agrobacterium-mediated delivery of infectious maize streak Virus into maize plants. Nature 325 : 177-179.

Herrera-Estrella, L., Depicker, A., Van Montagu, M. and Schell, J. (1983) Expression of chimaeric genes transferred into plant cells using a Ti plasmid derived vector. Nature 303: 209-213.

Hiei, Y., Ohta, S., Komari, T. and Kumashiro, T. (1994) Efficient transformation of rice (O. sativa L.) mediated by Agrobacterium and sequence analysis of the boundaries of the T-DNA. The Plant J. 6 : 271-282.

Horsch, R.B., Fraley, R.T., Rogers, S.G., Sanders, P.R., Lloyd, A. and Hoffman, N. (1984) Inheritance of functional foreign gene in plants, Science 223 : 496-498.

Kasuga, M., Liu, Q., Miura, S., Yamaguchi-Schinozaki, K. and Shinozaki, K. (1999) Improving plant drought, salt, and freezing tolerance by gene transfer of a single stress inducible transcription factor. Natl. Biotechnol. 17 : 287-291.

Klein, T.M., Wolf, E.D., Wu, R. and Sanford, J.C. (1987) High velocity microprojectiles for delivering nucleic acids into living cells. Nature 327 : 70.

Krens, F.A., Molendijk, L., Wullems, G.J. and Schilperoot, R.A. (1982) In vitro transformation of plant protoplasts with Ti-plasmid DNA. Nature 296 : 72-74.

Lucca, P., Ye, XuDong., Potrykus, I. And Ye, X.D. (2001) Effective selection and regeneration of transgenic rice plants with mannose as selective agent. Mol. Bree. 7 : 43-49.

Powell-Abel, P.A., Nelson, R.S., De, B., Hoffman, N., Rogers, S.G., Fraley, R.T. and Beachy RN. (1986) Delay of disease development in transgenic plants that express tobacco mosaic Virus coat protein. Science 232 : 738-743.

Reddy, V.S., Leelavathi, S., Selvapandiyan, A., Raman, R., Giovanni, F., Shukla, V., Bhatnagar, R.K., Leelavathi, S., Selvapandiyan, A., Raman, R., Giovanni, F. and Shukla, V. (2002) Analysis of chloroplast transformed tobacco plants with cry1Ia5 under rice psbA transcriptional elements reveal high level expression of Bt toxin without imposing yield penalty and stable inheritance of transplastome. Mol. Bree. 9 : 259-269.

Sanford, J.C., Devit, M.J., Russel, J.A., Smith, F.D., Harpending, P.R., Roy, M.K. and Johnston, S.A. (1991) An improved helium driven biolistic device. Techniq. 3 : 3-16.

Shimamoto, K., Terada, R., Izawa, T. and Fujimoto, H. (1989) Fertile rice plants regenerated from transformed protoplasts. Nature 338:274-276.

Tu, J., Zhang, G., Datta, K., Xu, C., He, Y., Zhang, Q., Khush, G.S., Datta, S.K., Tu, J.M., Zhang, G.A., Datta, K., Xu, C.G., He, Y.Q. and Zhang, Q.F. (2000) Field performance of transgenic elite commercial hybrid rice expressing Bacillus thuringiensis delta-endotoxin, Nature Biotechnol. 18 : 1101-1104.

Vaeck, N., Reynaerts, A., Hofte, H., Jansens, S., Beukeleer, M.D., Dean, C., Zabeau, M., Montagu, M.C. and Leemans, J. (1987) Transgenic plants protected from insect attack. Nature 328 : 33-37.

Zhang, H.M., Yang, H., Rech, E.L., Golds, T.J., Davis, A.S., Mulligan, B.J., Cocking, E.C. and Davey, M.R. (1988) Transgenic rice plants produced by electroporation-mediated plasmid uptake into protoplasts. Plant Cell Rep. 7 : 379-384.

Suggested Readings

1. Chawala, H.S. Introduction to Plant Biotechnology, Oxford & IBH Publishing Co. Pvt. Ltd; New Delhi, India.
2. Brown, T.A. Gene Cloning, Chapman and Hall, London, England.
3. Old, R.W. and Primrose, S.B. Principles of Gene A nalysis: An introduction to genetic engineering, Blackwell Science, Oxford, London, England.
4. Gupta, P.K. Elements of Biotechnology, Rastogi Publishers, Meerut, India.
5. Bhojwani, S.S. and Razdan, M.K. Plant Tissue Culture : Application and Limitation, Elsevier Science Publishers, Amsterdam, The Netherlands.

REVIEW QUESTIONS

Long Questions

1. Explain *Agrobacterium*-mediated gene transfer as an ideal transformation vector.
2. Discuss various methods of direct gene transfer in crop plants.
3. What are the cointegrate and binary vectors. Discuss their advantages and disadvantages?
4. What is gene silencing? What are the types of gene silencing? Write about different measures to reduce gene silencing in crop plants.
5. Bt gene has been widely used in crop improvement programme. Discuss their merits, demerits and successful utilization in crop improvement programme.

Multiple Choice Questions

1. For formation of adventitive embryos the culture frequently used is
   (a) Callus culture (b) Nucellus culture
   (c) Organ culture (d) Seed culture
2. Which of the following is not used in plant tissue culture technique?
   (a) Bud culture (b) Flower culture
   (c) Seed culture (d) Meristem culture
3. Which of the following is not significant for haploids
   (a) Induction of mutants
   (b) Hybrid development
   (c) Generation of exclusively male plants
   (d) All of the above
4. Dilpoidization of haploids can be done by
   (a) Endomitosis (b) Enomeiosis
   (c) Using DNA synthesis inhibitors (d) Cold treatment

5. Cybrids are
   (a) Nuclei and cytoplasm from a single parent
   (b) Nuclei from one and cytoplasm from both parents
   (c) Nuclei from one and cytoplasm from another parent
   (d) Nuclei from any one parent, cytoplasm from any other source.
6. Ploidy statuses cannot be estimated by
   (a) Counting of plastids and stomata
   (b) Number of nucleoli
   (c) Flow cytomerty
   (d) Optical denisity measurements
7. Protoplast fusion routinely involves
   (a) PEG treatment (b) Treatment with sodium nitrate
   (c) Electrofusion (d) All of the above
8. Gynogenic haploids are made from
   (a) Flower cult (b) Ovary culture
   (c) Seed culture (d) Fruit culture
9. Which of the following is a anti metabolite marker gene
   (a) aroA (b) Luc
   (c) LuxF2 (d) GUS
10. Gene silencing does not occur due to
   (a) DNA methylation (b) Paramutation
   (c) Cosuppression (d) None

Answers

1. (b) 2. (b) 3. (d) 4. (a) 5. (d) 6. (d)
7. (d) 8. (b) 9. (a) 10. (d)

CHAPTER 39

Gene Targeting

Rajnikant Mishra
Dept. of Zoology, Banaras Hindu University, Varanasi-221 005
E-mail: rmishraa@hotmail.com

Nature has provided some animal models of disease and some have been generated by random mutagenesis. The use of cultured cells and cell extracts are extremely valuable in studying gene expression and function. However, the ability to insert genes into whole animals or to selectively delete or alter single predetermined genes (gene-targeting) in an animal provides power in studying gene function. The new technologies are promising in altering at will even single gene within a living animal. Such animals having alteration in an analogous gene of humans provide a higher chance of resembling human disease phenotypes (Sedivy and Dutriaux, 1999). When a foreign DNA molecule is artificially introduced into the cells of an animal, a transgenic animal is produced. The foreign DNA molecule is called a transgene and may contain one or many genes. By inserting a transgene into a fertilized oocyte or cells from the early embryo, the resulting transgenic animal may be able to transmit the foreign DNA stably in its germ line. Many different types of transgenic animals have been created including transgenic *Drosophila,* transgenic frogs, transgenic fish and a variety of transgenic mammals including mice, rats and various livestock animals (McClenaghan *et al.*, 1991; Carver *et. al.*, 1993; Ebert *et al.*, 1994; Prokop and Kivirikko, 1995; Stacey *et al.*, 1998).

Although transgenes often integrate into the host chromosomes without affecting the expression of any endogenous genes, occasionally the integration event alters endogenous gene expression (insertional mutation), producing a recognizable phenotype. This constitutes a form of *in vivo* mutagenesis, at an unselected target gene. Gene-targeting was developed as a method of *in vivo* mutagenesis in which the mutation is introduced into a preselected endogenous gene. This can be achieved in somatic cells, but gene targeting in cultured ES cells is particularly powerful because it can lead to the construction of an animal in which all nucleated cells contain a mutation at the desired locus. In mammals, gene targeting has been possible only in mice but research on ES cells from other species may extend the capacity for gene targeting in the near future. It is a point to be considered that in some cases gene targeting is used to produce a skilful mutation and as a result the ES cells used for blastocyst implantation do not contain any foreign sequences. The resulting mice may be described as genetically modified but not as transgenic.

It is over a decade since the first demonstration that mouse embryonic stem cells could be used to transfer a predetermined genetic modification to a whole animal (Thomson *et al.*, 1989). The process of gene-targeting involves engineering of mutation in a pre-selected endogenous gene that replaces wild-type alleles with mutant ones. It is also referred as gene knockout. The mutation leads to either inactivation of specific gene expression or corrects pathogenic mutations to restore normal phenotype. It would be beneficial to recapitulate following concepts before understanding the methods in detail.

ES CELL AND DNA INSERTION

During early mouse development, there is a stage when only two cell types are present. The outer trophoblast cells that forms the portion of the placenta, and the inner cell mass, whose cells give rise to the embryo. The inner cell mass is isolated from the embryo and cultured *in vitro*. The cultured cells are called embryonic stem cells (ES cells). The ES cells retain their totipotency, and each of them can contribute to all organs if injected into a host embryo (McWhir *et al.*, 1996). Cloned pieces of DNA are isolated, modified if required, and placed into cells by following means. The microinjection is very direct technique, in which a solution containing the cloned gene is injected very carefully into the nucleus of a cell. This is an especially useful technique for injecting genes into newly fertilized eggs, since the haploid nuclei of the sperm and egg are relatively large.

The DNA is incorporated directly into cells by incubating them in a solution that makes them "drink" it in. This process is referred as transfection. The chances of a DNA fragment being incorporated into the chromosomes in this way are relatively small. The DNA of interest is usually mixed with another gene, such as a gene encoding resistance to a particular antibiotic. This enables cells to survive under culture conditions that incorporate the DNA and kills all the other cells. Another technique is electroporation in which pulses of high voltage are delivered to cells. It causes temporary relaxation of the selective permeability properties of the plasma membranes and pushes the DNA into the cells.

The transposable element or retroviral vector provides more "natural" way of getting genes into cells. These are naturally occurring mobile regions of DNA that can integrate themselves into the genome of an organism. Retroviruses are RNA-containing viruses and make a DNA copy themselves within a host cell using their own virally encoded reverse transcriptase. The DNA copy then becomes double-stranded and integrates itself into a host chromosome. The integration is facilitated by two identical sequences, long terminal repeats, at the ends of the retroviral DNA. Retroviral vectors are made by removing the viral packaging genes from the center of a mouse retrovirus. The packaging genes are needed for the exit of viruses from the cell. The extraction creates a vacant site where other genes can be replaced. These retroviral vectors infect mouse cells with high efficiency. Similarly, in *Drosophila,* new genes can be carried into a fly via P elements. The P elements are naturally occurring DNA sequences that can integrate like viruses into any region of the *Drosophila* genome. The genes can be inserted into the center of the P element. When the recombined P element is injected into a *Drosophila* oocyte, it can integrate into the DNA and provide the embryo with the new gene.

CHIMERIC MICE

The ES cells are treated to incorporate genes of interest for targeting. The treated ES cell with targeting vector and/or cloned DNA can then be injected into another early-stage mouse embryo. It usually gets integrated into the host embryo. The developing embryo results a chimeric mouse that contains mixture of cells derived from the host embryonic stem cells and treated embryonic stem cell. It is very critical that the treated cells become part of the germ line of the mouse and some of its gametes are derived from the donor cell. The chimeric mouse is mated with wild-type mouse and some of its progeny may carry one copy of the inserted gene. When these heterozygous progeny are mated to one another, about 25% of the resulting offsprings carry two copies of the inserted gene in every cell of their bodies. Thus, in three generations the chimeric mouse may give progeny of different genetic background like heterozygous and homozygous mouse for a gene that was cloned from other organism. Strains of such transgenic mice have been particularly useful in determining how genes are regulated during development.

METHODS OF GENE-TARGETING

Gene targeting is a more powerful method of genetic manipulation and requires essentially the same procedures of transfection and drug selection of cultured cells. The experimental gene targeting was first carried out in somatic cell lines (Lin *et al.*, 1985; Smithies *et al.*, 1985) but the use of embryonic stem (ES) cells now predominates. Currently, there has been no definitive comparison of gene targeting efficiency in primary somatic cells and ES cells (Yanez and Porter, 1998). The gene-targeting utilizes imaginative uses of ES cells and application of homologous recombination. The ES cells are remarkable cell lines to be cultured and manipulated *in vitro* without losing their potential when returned to the embryo. The unique property of homologous recombination introduces alteration in specific endogenous gene when a cloned gene or gene segment is transfected into the appropriate cells. The selected genes are closely related in sequence to endogenous target gene. In some of the cells, homologous recombination occurs between the introduced gene and its chromosomal homolog.

The gene-targeting by homologous recombination has been achieved in some somatic mammalian cells. The myoblasts is one of the example, however, the most important application involves mouse ES cells. The desired mutation is created into a specific mouse gene within the ES cells. It is a laborious process and need high background of molecular biology techniques. It requires use of replacement and/or insertion vectors with recombinant alleles and critical screening for targeted events with appropriate selection markers. The identification of the desired homologous recombination events of the targeting vector, often a plasmid vector, is assisted through a marker gene (e.g., *neomycin* gene), which permits selection for cells that have taken up the introduced DNA. The targeting construct is transferred into cultured mouse ES cells by electroporation.

The single colony for homologous targeting is screened through PCR or Southern-blot analysis. The reactions is monitored by using a marker-specific primer plus a primer derived from a sequence present in the target gene but absent from the introduced homologous gene segment. The modified ES cells can then be injected into the blastocyst of a foster-mother and eventually a mouse can be produced with the mutation in the desired gene in all nucleated cells.

STEPS OF GENE-TARGETING

The general plan of gene targeting is depicted through the Fig. 1. In brief, a targeting vector construct with antibiotic resistant gene is prepared. The ES cells are cultured from a mouse blastocyst using standard protocols. The mutant genes are electroporated into ES cells. In some of the cells, homologous recombination exchanges a wild-type gene for the mutant copy. These cells are selected by their gene for antibiotic (e.g., neomycin) resistance. The selected heterozygous ES cells are inserted into the inner cell mass of a wild-type blastocyst, and the blastocyst is returned to the uterus of mouse. The progeny of this mouse is a chimera composed of heterozygous tissues for mutated-gene and wild-type tissues. The mating of chimeric mouse to a wild-type mouse produces heterozygous offspring if the ES cells have contributed to the germ line. These heterozygous mice can be bred together for their progeny having homozygous for the mutant allele.

The efficiency of gene-targeting through homologous recombination is very low (i.e. about once in a thousand times) but it replaces one of the two copies of the normal gene into the chromosomes at random. The technique is exploited for these rare events to target any specific gene in a mouse cell by a direct gene replacement. It is completed into following two-steps.

The first step: The DNA fragment containing a desired mutant gene or part of a gene is inserted into a vector and then introduced into ES cells that grow in cell culture. These cells are capable of producing cells of any tissue and are allowed to grow for cell proliferation. The rare colonies of cells in which a homologous recombination event is likely to have caused a gene replacement to occur are isolated by a double drug selection. The *correct colonies* that contain recombinant DNA sequences in which the inserted fragment has replaced all or part of one copy of the normal gene are identified by PCR or by Southern blotting.

The second step: The individual cells from the identified *correct colonies* are taken up into a fine micropipette and injected into an early mouse embryo. The transfected embryo-derived stem cells collaborate with the cells of the host embryo to produce a normal-looking mouse. The large parts of this chimeric animal, including cells of the germ line, often derives from the artificially altered stem cells. It is critical to note here the difference between a chimera *and* a hybrid. The hybrid *results from the union of two different genomes within the same cell.* The offspring of an AA genotype parent and an aa genotype parent is an Aa hybrid. The chimera *results when cells of different genetic constitution appear in the same organism.* These mice are bred to produce both a male and a female animal, each heterozygous for the gene replacement (i.e. they have one normal and one mutant copy of the gene). The heterozygous mice are mated for their progeny with homozygous for the altered gene. The studies of these homozygotes allow the function of the altered gene to be examined in the absence of the corresponding normal gene.

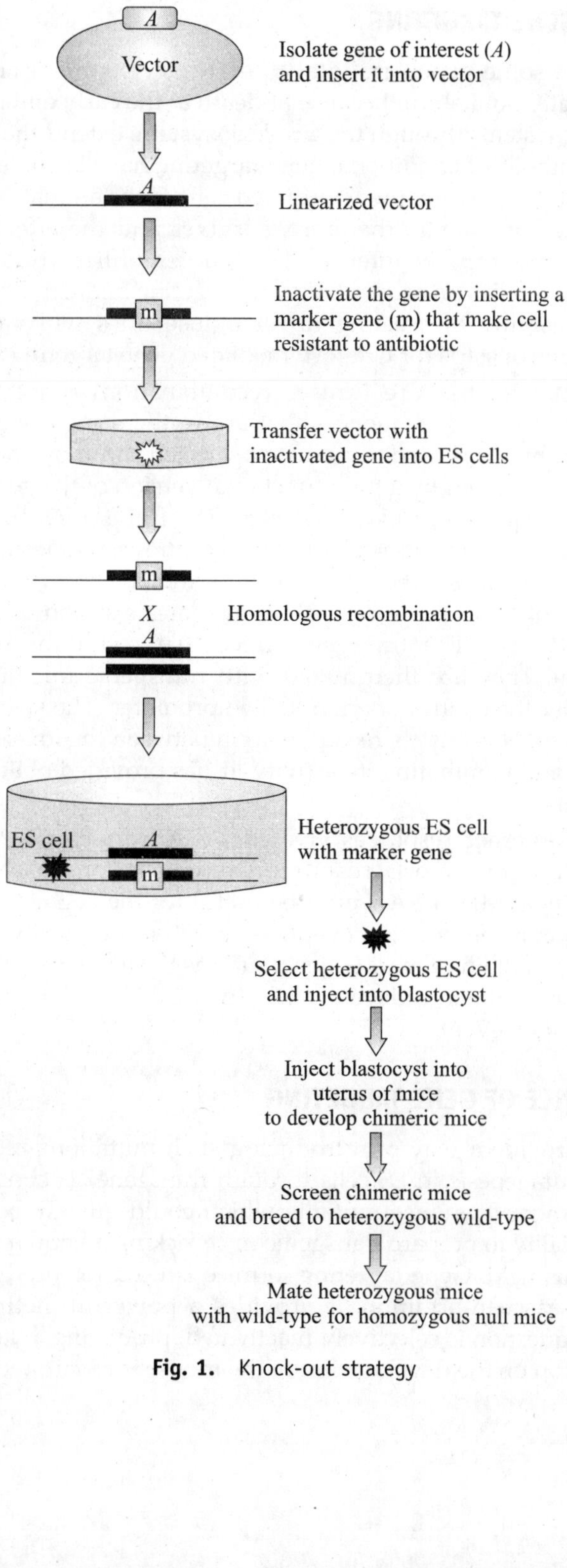

Fig. 1. Knock-out strategy

Cre-*lox*P MEDIATED GENE-TARGETING

It has been found that some genes are vital to early development and simple knock-out experiments are generally not helpful because of death at the early embryonic stage. The site-specific recombination systems through the Cre-*lox*P system, extend the power of conditional gene targeting. The methods of conditional gene-targeting have been developed to inactivate expression of the target gene in only predetermined cells of the animal. This is also referred as conditional knock-out. The animal, therefore, survives and the effect of the knock-out is studied in a tissue or cell type of interest. The studies utilize strategy for chromosome engineering in ES cells. It relies on sequential gene targeting and Cre-*lox*P recombination that offer the exciting possibility of creating novel mouse lines with specific chromosomal abnormalities. An outline of steps for Cre-*lox*P mediated gene targeting is illustrated in Fig. 2.

The natural function of the Cre (causes recombination) recombinase is to mediate recombination between two *lox*P sequences that are in the same orientation. The transient expression of Cre-recombinase is used to mediate a selected chromosomal rearrangement. The target gene in ES cells is replaced by a fully functional version of the gene that is flanked by a pair of the short DNA sequences, called *lox*P-sites. The *lox*P sequence consists of 34bp and comprises two inverted 13-bp repeats separated by a central asymmetric 8bp spacer. The *lox*P site is used to integrate at the desired chromosomal locations. The *lox*P-sites are recognized by the Cre-recombinase protein. The recombination mediates excision of intervening sequence between the two *lox*P sites. The transgenic mice that result through this process are phenotypically normal. They are then mated with transgenic mice that express the Cre-recombinase gene under the control of an inducible promoter. The specific cells or tissues in which Cre is switched on, it catalyzes recombination between the lox-sequences. It facilitates excising a target gene and eliminating its activity. It has provided ability for tissue- and cell type-specific knock-outs.

An user friendly internet resources created by Aurora Burds Connor from Rippel Transgenic www.cellmigration.org/resource/komouse/komouse_resources_targeting 4beginners.shtml facility at MIT, USA would be useful for the beginers. The identification of the desired homologous recombination events is monitored critically. The targeting vector (often a plasmid vector) contains a marker gene, such as the *neo* gene, which permits selection for cells that have taken up the introduced DNA. The PCR is used to screen for evidence of a homologous recombination event.

TECHNICAL IMPORTANCE OF GENE TARGETING

The mouse ES cells provide a way of introducing such mutations into the germ line. The random insertional mutagenesis in ES cells through transgenes is considered useful way of producing a large number of mouse mutants which could quickly be characterized at the molecular level. The ability to prepare transgenic mice lacking a known normal gene has been a major advance in this field. Gene targeting in mice is used for producing artificial mouse models of human disease. In addition, it provides a powerful method of studying gene function. The gene in question is selectively inactivated, producing a 'knock-out' mouse, and the effect of the mutation on the development of the mouse is monitored carefully.

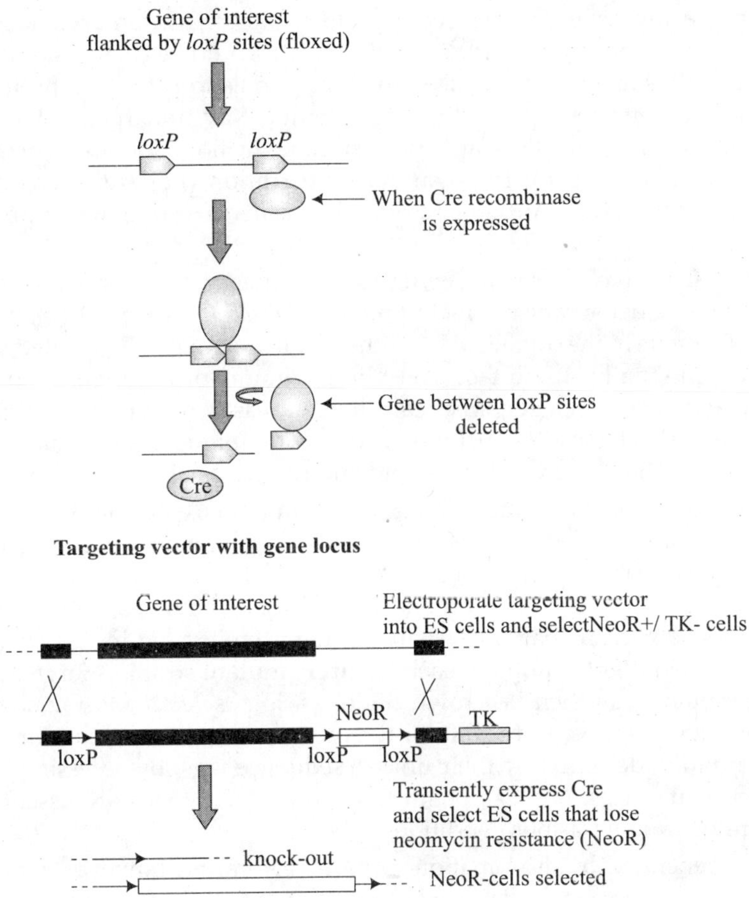

Fig. 2. General mechanism of Cre-*loxP* mediated gene targeting.

The extension of this technique to other mammalian species, particularly livestock, might bring numerous biomedical benefits, for example, ablation of xenoreactive transplantation antigens, inactivation of genes responsible for neuropathogenic disease and precise placement of transgenes designed to produce proteins for human therapy. Nuclear transfer from cultured somatic cells provides an alternative means of cell-mediated transgenesis (Campbell *et al.*,1996; Schnieke *et al.*, 1997).

The mutation that is induced by inserting a known foreign DNA sequence (transgene) into the mouse genome has a major advantage over one induced by chemical mutagens or X-rays. The technique is now being used to dissect the functions of a large number of mammalian genes. The similar techniques are used to produce conditional mutants, in which a selected gene becomes disrupted in a specific tissue at a certain time in development. If the cell used for this gene transfer is a fertilized egg (for an animal) or a totipotent plant cell in culture, transgenic organisms can be produced that express the mutant gene and will pass it on to their progeny.

The use of nuclear transfer does not require embryonic stem or embryonic germ cells, and circumvents the generation of chimaeric animals, which would be costly and time consuming in livestock. Fibroblasts are also being used in clinical trials to provide a protein production system after *ex vivo* gene therapy in human patients (Kay and High, 1999). It has been suggested that the introduction of therapeutic transgenes by homologous recombination could avoid undesirable effects arising from random integration (Lai and Lien, 1999). Nuclear transfer in animals such as sheep provides a rigorous means of testing the suitability of specific loci for transgene placement.

Extensive collaborative efforts are being made to generate comprehensive libraries of mutations in several model organisms, including *S. cerevisiae, C. elegans, Drosophila, Arabidopsis,* and the mouse. The gene in the organism has either been systematically deleted or altered such that it can be conditionally disrupted. The library would provide an invaluable tool for investigating gene function on a genomic scale. In some cases, each of the individual mutants within the collection will have a distinct molecular tag (a unique DNA sequence) designed to make identification of the altered gene for rapid and routine analysis.

In *S. cerevisiae*, the task of generating a set of 6000 mutants, each missing only one gene, is made simpler by yeast's propensity for homologous recombination. For each gene, a "deletion cassette" is prepared. The cassette consists of a special DNA molecule that contains 50 nucleotides identical in sequence to each end of the targeted gene, surrounding a selectable marker. In addition, a special "barcode" sequence tag is embedded in this DNA molecule to facilitate the later rapid identification of each resulting mutant strain. A large mixture of such gene knockout mutants can then be grown under various selective test conditions, such as nutritional deprivation, temperature shift, or the presence of various drugs and the cells that survive can be rapidly identified by their unique sequence tags. By assessing how well each mutant in the mixture fares, one can begin to assess which genes are essential, useful, or irrelevant for growth under various conditions.

The insights generated by examination of mutant libraries, however, will be great. For example, studies of an extensive collection of mutants in *Mycoplasma genitalium*, the organism with the smallest known genome have identified the minimum complement of genes essential for cellular life. Analysis of the mutant pool suggests that 265-350 of the 480 protein-coding genes in *M. genitalium* are required for growth under laboratory conditions. Approximately 100 of these essential genes are of unknown function, which suggests that a surprising number of the basic molecular mechanisms that underlie cellular life have yet to be discovered.

LIMITATIONS OF GENE-TARGETING

There is also possibility of genetic redundancy when another gene is able to carry out the function of the gene that has been knocked out. The phenotype in such cases is often lacking after inactivating a gene that would be expected to be crucially important. There are some examples of *Hox* genes or genes which encode a transcription factor known to be expressed in early embryonic development. The studies on such gene function have come through double or even triple gene knock-outs.

The loss of function of a single gene in the mouse can affect many different tissue types at different stages of development whereas the loss of other genes is found to have no obvious effect. Adequately characterizing mutant phenotypes in mice often requires a thorough examination, along with extensive knowledge of mouse anatomy, histology, pathology, physiology, and complex behavior.

Unlike gene targeting, the transgene should not be related to endogenous sequences, but should integrate essentially at random by non-homologous recombination. However, because large sections of the mouse genome are non-coding, many random insertion events may not result in gene disruption.

The challenge in deriving information from the study of yeast mutants lies in deducing a gene's activity or biological role based on a mutant phenotype. Some defects like inability to live without histidine, for example point directly to the function of the wild-type gene. Other connections may not be so obvious.

REFERENCES

Campbell, K. H. S., McWhir, J., Ritchie, W. A. and Wilmut, I. (1996) Sheep cloned by nuclear transfer from a cultured cell line. Nature 389 : 64-66.

Carver, A. S. et al. (1999) Transgenic livestock as bioreactors: stable expression of human alpha-1-antitrypsin by a flock of sheep. Bio-Technology 11 : 1263-1270.

Ebert, K. D. et al. (1994) Induction of human tissue plasminogen activator in the mammary gland of transgenic goats. Bio-Technology 12 : 699-702.

Kay, M. A. and High, K. (1999) Gene therapy for the hemophilias. Proc. Natl Acad. Sci. USA 96 : 9973-9975.

Lai, L.W. and Lien, Y.H. (1999) Homologous recombination based gene therapy. Exp. Nephrol. 7 : 11-14.

Lin, F. L., Sperle, K. and Sternberg, N. (1985) Recombination in mouse L cells between DNA introduced into cells and homologous chromosomal sequences. Proc. Natl Acad. Sci. USA 82 : 1391-1395.

McClenaghan, M. et al. (1991) Production of human a1-antitrypsin in the milk of transgenic sheep and mice targeting expression of cDNA sequences to the mammary gland. Anim. Biotechnol., 2 : 161-176.

McWhir, J. et al. (1996) Selective ablation of differentiated cells permits isolation of embryonic stem cell lines from murine embryos with a non-permissive genetic background. Nat. Genet. 14 : 223-226.

Prockop, D. J. and Kivirikko, K. I. (1995) Collagens: molecular biology, diseases, and potentials for therapy. Annu. Rev. Biochem. 64 : 403-434.

Schnieke, A. S. et al. (1997) Human Factor IX transgenic sheep produced by transfer of nuclei from transfected fetal Fibroblasts. Science 278 : 2130-2133.

Sedivy, J. M. and Dutriaux, A. (1993) Gene targeting and somatic cell genetics: a rebirth or coming of age? Trends Genet. 15 : 88-90.

Smithies, O., Gregg, R. G., Boggs, S. S., Koralewski, M. A. and Kucherlapati, R. S. (1985) Insertion of DNA sequences into the human chromosomal b-globin locus by homologous recombination. Nature 317 : 230-234.

Stacey, A. et al. (1988) Perinatal lethal osteogenesis imperfecta in transgenic mice bearing an engineered mutant pro-alpha 1(I) collagen gene. Nature 332 : 131-136.

Thompson, S., Clarke, A. R., Pow, A. M., Hooper,M. L. andMelton, D.W. (1989) Germ line transmission and expression of a corrected HPRT gene produced by gene targeting in embryonic stem cells. Cell 56: 313-321.

Yanez, R. J. and Porter, A. C. (1998) Therapeutic gene targeting. Gene Ther. 5 : 149-159.

Suggested Readings

1. Genes IX (2008) B. Lewin, Pearson Prentice-Hall, UK.
2. Genes VIII (2004) B. Lewin, Pearson Prentice-Hall, UK.
3. Cancer Medicine (2003) D.W. Kufe, R.E. Pollock, R.R. Weichselbaum, R.C. Bast Jr., T.S. Gansler, J.F. Holland and E. Frei III (Eds.) 6th ed., BC Decker Inc., Hamilton, Canada.
4. Molecular Biology of the Cell (2002) B. Alberts, A. Johnson, J. Lewis, M. Raff, K. Roberts and P. Walter, 4th ed., Garland Publishing, New York.
5. Molecular Cell Biology (2000) H. Lodish, A. Berk, S. L. Zipursky, P. Matsudaira, D. Baltimore and J. E. Darnell, 4th ed., W. H. Freeman and Co., New York.
6. Developmental Biology (2000) S.F. Gilbert, 6th ed., Sinauer Associates, Inc., Sunderland, Myrialand.
7. Basic Neurochemistry, Molecular, Cellular, and Medical Aspects (1999) G.J. Siegal, B. W. Agranoff, R.W. Albers, S.K. Fisher and M.D. Uhler (Eds.), 6th ed., Williams and Wilkins, Lippincott, Philadelphia.
8. Human Molecular Genetics 2 (1999) T. Strachan and A.P. Read, 2nd ed., Taylor and Francis, London.
9. Gene Targeting : A practical approach (1993) A.L. Joyner, Oxford University Press, New York.

REVIEW QUESTIONS

Long Questions

1. Draw a flow-chart to demonstrate gene-targeting strategies.
2. Describe general mechanism for Cre-loxP mediated gene targeting.
3. Discuss limitations of gene targeting technology
4. Enumerate the role of ES cells in gene targeting
5. Write short-notes on following :
   (A) Knock-out animals (B) Genetic redundancy
   (C) Homologous recombination

Multiple Choice Questions

1. Gene targeting is a type of mutagenesis of endogenous gene
   (a) un-selected (b) pre-selected
   (c) random (d) none of the above
2. Embryonic Stem (ES) cells are cultured cells from early developing embryo
   (a) Outer trphoblast (b) inner cell mass
   (c) differentiated cells (d) none of the above
3 Screening of targeting vector is facilitated through
   (a) Marker-specific primer (b) antibiotic selection markers
   (c) Southern-blotting (d) All

4. Which of the following is a false statement
   (a) loxP is recognized by Cre-recombinase protein
   (b) A chimera and a hybrid animal is similar
   (c) Random insertion events may not result in gene disruption
   (d) Recombinant vectors are isolated by double drug selection

Answers

1. (b) 2. (b) 3. (d) 4 (b)

CHAPTER **40**

Applications of Genetic Engineering in Agriculture, Health and Industry

R. Prakashkumar[1] and P. E. Rajasekharan[2*]

[1]Kerala State Council for Science Technology and Environment, Pattom, Thiruvananthapuram–695 004

[2]Indian Institute of Horticultural Research, Hessaraghatta Lake P.O. Bangalore–560 089

E-mail: rprakum@yahoo.co.in

Ever since the latter part of the 19th century, when Gregor Mendel discovered that characteristics in pea plants could be inherited, scientists have been improving plants by changing their genetic makeup. Typically, this was done through hybridization in which two related plants were cross-fertilized and the resulting offspring had characteristics of both parent plants. Breeders then selected and reproduced the offspring that had the desired traits. Today, to change a plant's traits, scientists are able to use the tools of modern genetic engineering to insert a single gene—or, often, two or three genes-into the crop to give it new, advantageous characteristics. Most genetic modifications make it easier to grow the crop. About half of the American soybean crop planted in 1999, for example, carries a gene that makes it resistant to an herbicide used to control weeds. About a quarter of US corn planted in 1999 contains a gene that produces a certain conventional pesticides.

Since Mendel, plant breeders have modified the genetic material of crops by selecting plants that arise through natural or, sometimes, induced changes. Gardeners and farmers and, at times, industrial plant breeders have crossbred plants with the intention of creating a prettier flower, a hardier or more productive crop. These conventional techniques are often imprecise because they shuffle thousands of genes in the offspring, causing them to have some of the characteristics of each parent plant. Gardeners or breeders then look for the plants with the most desirable new trait. With the tools developed from genetic engineering, a gene can be inserted into a plant to give it a specific new characteristic instead of mixing all of the genes

*Corresponding author

from two plants and seeing what comes out. Once in the plant, the new gene does what all genes do: It directs the production of a specific protein that makes the plant uniquely different. This technology provides much more control over, and precision to, what characteristic breeders give to a new plant. It also allows the changes to be made much faster than ever before. No matter how a new crop is created-using traditional methods or biotechnology tools-breeders must conduct field-testing for several seasons to make sure only desirable changes have been made. They must check to make sure the plant looks right, grows right, and produces food that tastes right. They also must perform analytical tests to see whether the levels of nutrients have changed and whether the food is still safe to eat.

Genetic engineering refers to "any technique that uses living organisms (or parts of organisms) to make or modify products, to improve plants or animals, or to develop microorganisms for specific uses." Given this definition, genetic engineering has been used for centuries in making cheese, yogurt, bread, and wine and in cross-breeding plants and animals to enhance desired qualities. However, when most people talk about biotechnology, they are referring to newer uses developed since the 1970s when isolation and duplication of foreign DNA in bacteria was accomplished and recombinant DNA (rDNA) technology was made possible. With rDNA technology, a copy of a piece of DNA containing one or more genes is transferred between organisms or recombined within an organism. Given this context, genetic engineering may be defined as "the use of biotechnical methods to modify the genetic material of living cells so they will produce new substances or perform new functions." the potential genetic engineering has for reducing the need for agricultural chemicals, improving the productivity, efficiency, and profitability of food production, and improving the quality, safety, cost, and convenience of consumer food products. Because people with this philosophy do not see genetic engineering as being that much different in basic principle than cross-breeding, they tend to oppose special labeling of food products made using biotechnology. They see such labeling as being unnecessary and potentially misleading. The advent of recombinant DNA technology and advancement in knowledge relating to nucleic acid, biochemistry, molecular biology and biotechnology has led to an era marked by dramatic changes in the methodologies and altered perceptions about living systems. One such development is the utilization of the genetic engineering for crop improvement through the production of transgenic crops. The greatest advantage of this technology is that gene transfer can take place transcending taxonomic barriers. Virtually genes from bacteria viruses, or other microbes, animals or other plants can be transferred to plants.

APPLICATION OF GENOMIC TECHNOLOGIES TO CROP PLANTS: OPPORTUNITIES AND CHALLENGES

Enormous genomic resources have been developed for model plants such as *Arabidopsis thaliana* (L.) Heynh. and rice (*Oryza sativa* L.), including detailed genetic maps, huge numbers of expressed sequence tags (ESTs), deep-coverage large-insert [such as bacterial artificial chromosome (BAC)] libraries with extensive contig assemblies, and both targeted and complete genome sequencing and annotation. These resources, coupled with the development of mutant stocks by knock-outs or targeted induced local lesions in genomes, will allow for the efficient identification of gene(s) controlling phenotypes in model systems. However, it is not clear how broadly applicable genetic associations revealed in model systems will be to economically important plants. Unquestionably, some genes identified in model plants will also condition economically important phenotypes in a crop plant. An example is the FLC locus that controls flowering in *Arabidopsis* and in the closely related brassicas. Similar successes are readily communicated to the scientific community through publications; however, it will be difficult to publish, and therefore assess, the opposite scenario in which genes identified in model systems are not associated with similar traits in economically important plants.

These challenges not withstanding, the genomic resources developed for model plants will reveal a plethora of candidate genes and provide great insights into gene expression. In some cases genes identified in model systems will condition economically important phenotypes in crop plants. However, it remains imperative that we identify, clone, and understand specific gene(s) conditioning economically important phenotypes in our major crops. In many cases, large genomic clones of specific crop plants will be imperative for the isolation of genomic regions, either up or down stream from the structural gene, interacting with important trans-acting factors controlling gene expression. Deep coverage genomic libraries and targeted sequencing will be required for the identification, cloning, and manipulation of specific genes affecting economically important phenotypes in major crop plants.

The alleles that are incorporated by MAS (Marker Aided Selection) are generally present within the gene pool of a particular crop and are transferred by meiotic chromosome recombination. One of the positive aspects of this approach is that these genes reside at their natural chromosomal locations, thereby minimizing the possibility of gene silencing. Another important aspect of cultivars developed by MAS is that they are not transgenic and therefore, do not face the public resistance against transgenic crops. The MAS strategy is a way to capitalize on available markers and to incorporate valuable traits into elite lines that are suitable for cultivar release. In addition, release of these MAS-improved cultivars is an efficient way of demonstrating the power of these technologies to the public.

GENETIC ENGINEERING IS ONLY THE NEWEST TOOL TO IMPROVE PLANTS

Conventional breeding for a long time was limited to taking genes from closely related relatives because in essence it moves genes by pollination. Over the last 80 years mutation breeding has been widely used. By means of radiation or chemical mutagens, breeders are stipulating random genetic changes. Cell culture or Embryo rescue, some more recent methods, have allowed so called conventional breeding to overcome the sexual barriers between non-compatible plants and is the basis for the development of hybridized crops.

Conventional breeding with its uncertainties has come to rely on very careful testing, analysis and selections, to eliminate plants that have undesirable characters. In many respects Plant Biotechnology is more precise than conventional plant breeding. Much more targeted questions about the effect of those particular genes can be asked. The experience accumulated with conventional breeding is a very important source of know how for the evaluation of the GM crops. With GMO assisted breeding advances will be much more targeted, require fewer resources and offer plant improvements that seemed impossible through accessing a wide available gene pool.

The application of Plant Biotechnology can be used to produce a wide array of new products. On the basis of the value added 3 areas of applications can be distinguished:

(i) Reduced cost of production through improved agronomic characteristics. Objective is to improve the agronomic characteristics of plants. Widely commercialized are Herbicide Resistance and the BT-Technology. Other major areas in research focus on projects to improve the plants ability to fight pests and diseases as well as the tolerance for abiotic stress-like tolerance to cold, heat or salty soils.

(ii) Improved value of crop production through optimized plant ingredients. Advances in plant biotechnology will help provide solutions to nutrition and health challenges, especially those facing developing countries. For example, a worldwide team of scientists, has developed a strain of rice (Golden Rice) that contains increased levels of beta-carotene, a precursor to vitamin A and increased iron, which will potentially prevent blindness, anemia and early death through the food source billion of people eat every day. Several projects aim at the development of plants with increased vitamin content or healthier oils with fewer saturated fats. The prevention of cardiovascular diseases is only one application of this new technology.

(iii) Production of specialty chemicals through plants as factories. Examples are vitamins, industrial fatty acids and enzymes. Biotechnology can make it possible to grow more food on the same land, especially under tough growing conditions.

APPLICATIONS OF GENETIC ENGINEERING

There are three main applications of genetic engineering. First, it is an important laboratory tool. Genetic engineering allows scientists to examine the role of individual genes in great detail, and to dissect and analyse the regulatory systems that control the expression of genes. Second, techniques of genetic engineering may be used in conventional programmes of plant and animal breeding to accelerate genetic improvement. This allows breeders to introduce specific isolated genes in a highly controlled way, in contrast to traditional breeding in which large numbers of genes are combined during crosses. There are about 160,000 genes in the sets of chromosomes of a plant. This means that 80,000 from each parent are swapped in conventional crossing, compared with the two or three which might be moved in genetic engineering. Genetic engineering overcomes the problem, which can arise if a desirable gene is closely linked, and therefore co-inherited, with a gene or genes coding for undesirable traits. Using recombinant DNA technology, just the desirable gene can be transferred.

The third application of genetic engineering is that which offers the greatest opportunities for innovative improvements in the way we use biological materials industrially. It is also the most controversial because it can involve moving genes from one species to another; and in extreme cases from, for example, animals and microorganisms into plants.

Genetically Engineered Organisms

Some technologies based on genetic engineering are already used commercially. Organisms derived from genetic engineering are often termed transgenic or transformed. One example is the enzyme used in vegetarian cheese. This is made in bacterial cells, which have been genetically engineered to contain copies of the calf enzyme rennin, which is used in conventional cheese making. In 1994 saw the launch in the USA of the first commercial transgenic tomato: the Flavr-Savr tomato, developed by Calgene Inc. to stay ripe longer. This tomato is engineered using an "antisense" gene to suppress the activity of a gene involved in the softening processes associated with ripening. Several other biotechnological manipulations are close to commercialization. There are prospects of genetically engineering crops so that they contain genes, which confer resistance to pests and disease. This would reduce reliance on agrochemicals. Longer term there is the potential to engineer crops for enhanced nutritional characteristics, for example higher amounts of beneficial antioxidants in fruits; or for greater stress tolerance that would allow plants to be grown in environments that are hostile in terms of temperature or soil salinity.

It should even be possible to engineer plants to produce high-value therapeutics and fine chemicals. One possibility is the manipulation of oilseed crops such as rape to produce industrial raw materials for the plastics, cosmetic and chemical industries. This would provide an environmentally friendly renewable source of materials currently made by chemical synthesis.

The applications of genetic engineering in agriculture leading to the development of transgenic crops have transformed the landscape of agriculture globally. The acreage of transgenic crops has expanded dramatically since 1996 and is estimated to be 167 million acres in 2003. The first generation of transgenic plants were developed during the last ten years with emphasis largely on incorporation of input traits like resistance to insects, viruses, herbicides as well as abiotic stresses. This decade is witness to the development of second generation transgenics with incorporation of output traits like improved nutritional quality, shelf life, flavours and fragrances. During the past five years, there has been rapid progress in the sequencing of plant genomes leading to better understanding of the structure, function and regulation of genes. This knowledge could be utilized to alter plant architecture, manipulate flowering time, fruit size and colour, improve photosynthetic efficiency, improve nutrient assimilation and exploit and manipulate heterosis and apomixis, resulting in a third generation of transgenics. To incorporate these traits and to continually meet future demands, the repertoire of genes has to be increased and new genes have to be identified.

Most of the transgenics in the field today have been developed with genes from distant organisms, mostly prokaryotes and rarely from crop plants. With the advances made in the field of genomics, it is now possible to mine genes of high agronomic value from secondary gene pool of species and genera related to the crop species.

Several molecular and computational techniques are now available to unlock the genetic diversity present in crop plants and harness the potential of these genes in crop improvement programmes.

Genetic Engineering in Agriculture

The most recent research and development advances in modern biotechnology have introduced an ever-widening range of genetically engineered products to agriculture. While traditional biotechnology improves the quality and yields of plants and animals through, for example, selective breeding, genetic engineering is a new biotechnology that enables direct manipulation of genetic material (inserting, removing or altering genes). Protagonists argue that genetic engineering entails a more-controlled transfer of genes because the transfer is limited to a single gene, or just a few selected genes, whereas traditional breeding risks transferring unwanted genes together with the desired ones. Against that advantage, antagonists argue that the side effects in terms of potentially adverse impacts on the environment and human health are unknown. Genetic engineering techniques and their applications have developed rapidly since the introduction of the first genetically modified plants in the 1980s. In 1999, genetically modified crops occupied 40 million hectares of land – making up 3.4% of the world's total agricultural area and representing a considerable expansion from less than 3 million hectares in 1996. Cultivation of transgenic crops has so far been most widespread in the production of soybeans and maize, accounting for 54% and 28% of total commercial transgenic crop production in 1999, respectively. Cotton and rapeseed each made up 9% of transgenic crop production in 1999, with the remaining GM crops being tobacco, tomato, and potato.

To date, genetic engineering in agriculture has mainly been used to modify crops so that they have improved agronomic traits such as tolerance to specific chemical herbicides and resistance to pests and diseases. Development of plants with enhanced agronomic traits aims at increasing farmer profitability, typically by reducing input requirements and hence costs. Genetic modification can also be used to improve the final quality characteristics of a product for the benefit of the consumer, food processing industry, or livestock producer. Such traits may include enhanced nutritional content, improved durability, and better processing characteristics. The United States holds almost three-fourths of the total crop area devoted to genetically modified crops. Other major GM-producers are Argentina, Canada, and China. At the national level, the largest shares of genetically engineered crops in 1999 were found in Argentina (approximately 90% of the soybean crop), Canada (62% of the rapeseed crop), and the United States (55% of cotton, 50% of soybean and 33% of maize). The USDA (2000) figures for the United States are similar in magnitude: it is estimated that 40% of maize and 60% of soybean areas harvested in 1999 were genetically modified. Continued expansion in the use of transgenic crops will depend in part on the benefits obtained by farmers cultivating transgenic instead of conventional crops relative to the higher cost for transgenic seeds. So far the improvements have been not so much in increased yields per hectare of the crops, but rather by reducing costs of production. Empirical data on the economic benefits of transgenic crops are still very limited, however. The effects vary from year to year and depend on a range of factors such as crop type, location, magnitude of pest attacks, disease occurrence, and weed intensity. In developing countries one of the main reasons for low crop yields is the prevalence of biotic stresses caused by weeds, pests, and diseases. The first generation of improved transgenic crops, into which a single trait such as herbicide tolerance or pesticide resistance has been introduced, can provide protection against several of these. The development of more complex traits such as drought resistance, which is a trait controlled by several genes, is underway and highly relevant for tropical crops that are often growing under harsh weather conditions and on poor-quality soils. There are not many estimates of the potential productivity impact that widespread cultivation of transgenic crops may have in developing countries, World Bank panel has estimated that transgenic technology can increase rice production in Asia by 10 to 25 percent in the next decade (Christian, 2000).

Biotic Stresses

Disease and insect resistance: Varieties resistant to viral, fungal and bacterial diseases as well as insect pests have to be continuously evolved since the disease scenario is also changing.

Pathogen-derived resistance would continue to be the strategy and the quest for genes from these sources would have to go on. However, recent advances made in our understanding of plant-pathogen interactions have led to the discovery of several resistance genes (R) from plant sources. Hence, there is a need to look for host plant resistant genes for fungal and bacterial resistance. One of the strategies suggested is to balance R gene polymorphism.

Antimicrobial peptides from plants can also be candidate genes for fungal and bacterial resistance. However, synthetic peptides have to be generated since natural peptides are degraded in the plant system. Genes can be mined from microbial sources from organisms that are effective biocontrol agents for fungal and insect resistance. For insect resistance, genes for serine, cysteine proteinase inhibitors can be used along the vast arsenal of *Bacillus thuringiensis* endotoxin genes.

Abiotic Stresses

Abiotic stresses, such as drought, salinity, extreme temperatures, chemical toxicity and oxidative stress are serious threats to agriculture and result in the deterioration of the environment. Abiotic stress is the primary cause of crop loss worldwide, reducing average yields for most major crop plants by more than 50%. Drought and salinity are becoming particularly widespread in many regions, and may cause serious salinization of more than 50% of all arable lands by the year 2050. Abiotic stress leads to a series of morphological, physiological, biochemical and molecular changes that adversely affect plant growth and productivity. Drought, salinity, extreme temperatures and oxidative stress are often interconnected, and may induce similar cellular damage.

A variety of genes that are expressed by plants in response to abiotic stresses like drought, salinity and cold have been described. Functional analysis of some of these has been demonstrated by overexpression in model systems. Such genes could be candidate genes for conferring tolerance to abiotic stresses. These are: (a) Dehydration response element and C-repeat for induction of *rd*29A (b) Galactinol synthase (c) Glyoxalase I & II (d) *sp*1 (e) 9-cis-epoxy carotenoid dioxygenase.

Increased shelf life: Several genes have already been isolated for the manipulation of ripening and shelf-life. These include genes for enzymes involved in ethylene biosynthetic pathway (ACC synthase, ACC oxidase, SAM decarboxylase and SAM deaminase) and cell wall hydrolysis (polygalacturonase, cellulase, beta-galactosidase). The strategy currently being suggested is to manipulate of ethylene responses through modification of ethylene receptors. However, as ethylene is involved in several aspects of plant physiology and development, a number of negative effects are likely to be manifested. The key to effective manipulation of ethylene sensitivity will be the use of tissue-specific or development–specific promoters thereby controlling spatial and temporal expression of the transgenes. In ornamental crops, there is also potential to increase cytokinin synthesis which has been shown to delay leaf senescence resulting in "stay green" phenotypes.

Improvement of tomato quality: Genetic engineering can also be employed to manipulate the levels of endogenous enzymes in raw material. The classic example of this is the removal of polygalacturonase activity in tomato fruit, which has been achieved by the application of gene silencing technology by Calgene in the USA and by Zeneca/Nottingham University in Europe. The resultant fruit have extended shelf lives and produce pastes with increased solid content and viscosity.

The softening of tomato fruit results primarily from the degradation of the walls surrounding the pericarp cells. In particular the pectic components of these cell walls are degraded becoming depolymerized and more soluble during ripening. In addition, compositional changes occur to the pectic components, the major alterations being a reduction in the percentage of uronic acid residues, which are esterified with methyl groups—i.e., the degree of esterification drops — and a loss of neutral sugars such as galactose and arabinose. These changes are brought about by the action of hydrolytic enzymes such as polygalacturonase, pectinesterase and galactanase.

A cDNA clone encoding the tomato fruit polygalacturonase has been isolated and used to reduce polygalacturonase in transgenic plants by the application of gene silencing techniques. One approach was to use antisense RNA technology. In this case the DNA coding sequence, or part thereof, for polygalacturonase is used to construct an artificial antisense gene in which the promoter drives the synthesis of an antisense RNA complimentary to the naturally occuring mRNA for polygalacturonase. The interaction of this antisense gene with the endogenous polygalacturonase gene is such that expression of the latter is prevented. The precise mechanism for antisense inhibition of gene expression is unknown but may involve the formation of RNA hybrids between the antisense RNA and natural mRNA. An alternative gene silencing technique is calledco-suppression. In this case gene silencing is brought about by the interaction of the natural gene sequence with that of an extra copy of part of this sequence introduced as a transgene. Again the precise mechanism of silencing in this instance is not clear. The down regulation of polygalacturonase has been achieved using both these gene silencing techniques. Transgenic plants containing either an antisense polygalacturonase gene or co-suppressed with a partial sense polygalacturonase gene both exhibit levels of polygalacturonase activity around 0.5 per cent that found in normal fruit.

Floral colour: Several ornamental plants have been engineered for modified flower colour. Research has been focused on manipulation of either anthocyanins (red and blue colours) or carotenoids (yellow and orange colours). The crops modified are carnation, rose and gerbera. In carnation, a commercial variety is already available with mauve to violet shades.

Floral scent: Attempts are being made to isolate genes involved in scent production with the objective of "putting the scent back" in flowers that have lost their scent or are without scent. In spite of the vast number of compounds involved, surprisingly only a small number of metabolic pathways are involved, which have been shown to be amenable to manipulation.

Two genes responsible for the unique aroma of roses, two O-methyl transferases have been isolated.

Nutritional quality: There are several opportunities to enhance the nutritional value of horticultural products. These include vitamins, minerals and nutraceuticals. Levels of organic nutrients like tocopherols have been altered through manipulation of gamma-tocopherol methyl transferase activity. Iron content can also be manipulated through overexpression. Carotenoid content can be improved in fruit crops like banana as in the case of golden rice. However, holistic understanding of relevant transport and partitioning mechanism is required before attempting genetic manipulation for increasing nutritive quality. This is because a number of whole plant factors like mechanism of acquisition, regulation of deficiency responses and the plant organ that assimilates these nutrients would influence the expression of these nutrients.

Tree architecture and plant size: Growth habit of trees is an important consideration especially with regard to improving productivity per unit area. The horticultural practices currently being used may be enhanced through the introduction of transgenes that play a role in growth and development. The best candidate genes at present are the *rol* genes A, B, C, D from *Agrobacterium rhizogenes* which have been shown to influence internodal distance, adventitious rooting, apical dominance and seed set. Manipulation of plant size of ornamental plants is also receiving attention in recent years. Overexpression of GA-20 oxidase gene in *Arabidopsis* resulted in stem elongation while antisense expression led to dwarfing. It has also been shown that overexpression of *Arabidopsis* DWF 4 gene in tomato resulted in phenotypes with increased branching (Vasil, 2003).

To meet the challenges of crop production, the list of desirable traits will continue to grow and more and more novel genes will have to be identified. To ensure proper expression of the traits without negative side effects, targeted gene expression would also have to be resorted to. This would entail isolation of specific promoters that could control temporal and spatial expression of transgenes. The rapid advances being made in molecular biology would facilitate more versatile techniques for gene discovery and the years to come would definitely witness the expansion of the repertoire of genes available with us today.

TRANSGENIC CROPS — CURRENT SCENARIO IN THE COUNTRY

Several programmes on developing transgenic crops are going on in government-funded research laboratories in India and abroad. The traits of interest have been insect and disease resistance to a great extent as well as extended shelf life and better nutritional qualities. These include: (a) development of insect-resistant brinjal, tomato, cauliflower, cabbage and potato through transfer of *Bt* gene (b) development of transgenic tomato resistant to Tomato Leaf Curl Virus (TLCV), Cucumber Mosaic Virus, Tomato Spotted Wilt Virus and fungal diseases like early blight and development of transgenic chillies fungal resistance (c) development of transgenic watermelon resistant to watermelon bud necrosis virus (d) genetic manipulation of tomato, banana and papaya for extended shelf life (e) development of transgenic papaya resistant to papaya ringspot virus and (f) development of transgenic potato with increased protein content.

Molecular Markers and DNA Fingerprinting

Another important area of biotechnology that has a lot of important applications is the area of molecular markers and DNA fingerprinting. DNA fingerprinting is an important application, which is all the more relevant in the present scenario of plant variety protection rights. India is endowed with a rich wealth of horticultural plant resources comprising 66 genera and 899 species, of which 190 (109 species in fruits, 54 in vegetables, 27 in spices and condiments) are of economic importance. The country falls within the primary centre of diversity in several important fruits (mango, banana, jackfruit, some species of citrus and plums), vegetables (brinjal, snake gourd, pointed gourd, cucumber and yams and spices (pepper, ginger and turmeric). Besides, rich diversity exists in even some introduced crops like tomato, chillies and bean. We need to protect this diversity using DNA fingerprinting. For example, a wealth of diversity exists in our country for mango — the king of tropical fruits. This diversity is manifested in the form of several unique characters and molecular marker techniques can help in documenting these. India is also the treasure – house of medicinal and aromatic plants and these have to be characterized. Realising the need and potential of molecular characterization in the present scenario of intellectual property, breeders' and farmers' rights, the ICAR(Indian Council of Agricultural Research) has been supporting projects on DNA fingerprinting in a number of its institutes. A National Research Centre on DNA fingerprinting has also been set up with a mandate to develop and interlink programmes on DNA fingerprinting of agricultural bioresources including crops, animals and fish.

The process of developing new crop varieties involves many steps and takes several years. In recent times, biotechnological advances have helped in reducing these considerably. Apart from the techniques of haploid culture and genetic engineering, one of the tools which is making it easier for scientists to select plant traits and develop new varieties is the use of molecular markers. Polygenic characters which are very difficult to analyze using traditional plant breeding methods can now be easily tagged using molecular markers. Molecular markers can be used for: (a) characterization of germplasm (b) varietal identification and clonal fidelity testing (c) assessment of genetic diversity (d) validation of genetic relationships and (e) marker-assisted selection. These have widespread applications in management of genetic resources as well as in breeding programmes. Marker-assisted selection has practically revolutionized breeding programmes. Traditionally, plant breeders have selected plants based on their visible or measurable traits called the phenotype. But this process can be difficult especially in the case of multigenic or quantitative traits. Molecular markers can be used to trace linkages with traits of importance. In the area of horticulture, this is a very useful tool in tree breeding, where the breeder has to wait for several years to see the expression of the desired trait. Several molecular markers are available in the world today in horticultural crops to detect traits of interest. More markers need to be developed for tree species and other perennial horticultural crops.

Techniques of Gene Transfer in Plants

Enthusiasm about the potential benefits of genetic engineering can tend to obscure the technical complexities. In principle, the recombining of DNA molecules is not difficult: different enzymes can be used to "cut out" segments of DNA and to "paste" them into a (hybrid) recombinant molecule. But, in practice, highly specialized techniques are required to ensure that genes are transferred in such a way that they are targeted to the right location, are properly regulated and are "read" correctly to produce active gene products. And before any of this can happen the genes have to be correctly identified and isolated.

Several methods can be deployed to transfer genes, although not all are suitable for all species. Three important techniques are: the use of microbial vectors to carry genes into host cells; the uptake of DNA from a bathing solution into protoplasts (cells from which the tough outer wall has been removed); and a ballistic method for firing genes into cells by coating them on to tiny gold or tungsten particles.

Microbial vectors: In nature, there are several routes through which "foreign" genes become inserted into plant cells. These have obvious potential in genetic manipulation. One mechanism of particular interest is that used by the natural soil bacterium Agrobacterium. *Agrobacterium tumefaciens* causes "crown gall" infection. It attacks wounds in plants and induces the cells there to multiply, creating a tumour. The Agrobacterium does this by inserting genes into the plant chromosome from a plasmid (a small loop of DNA, separate from the bacterium's own chromosome) which it carries. The plasmid contains genes which enable it to insert into the plant DNA, and others which trigger the growth of the tumour. Before it can be used as a vector for gene transfer, the plasmid must be "disarmed" by removing its tumour-producing genes. These can then be substituted with "useful genes" for transfer to the plant.

Viruses are potentially very useful for ferrying genetic information into plants because they enable the introduction of genes by simple inoculation. Since viruses can multiply, viral vectors may allow more copies of the gene product to be made. At first, research efforts concentrated on using viruses which contain DNA. However, unlike other organisms, many viruses have RNA as their only genetic material. Recently, systems have been developed that allow the use of RNA viruses as vectors. The production of an RNA viral vector is started by preparing a suitable length of DNA. This is then transcribed to RNA, which is put into the virus, and through it into the plant. The process requires the construction of a piece of DNA coupled to a promoter sequence so that it can be transcribed. Then any non-essential regions of the DNA are deleted to make space for the genes that are to be transferred. Besides their use in transformation studies, viral vectors can be useful tools for plant molecular biologists and pathologists. It is also likely that antisense viral constructs will be used to "turn off" virus genes, providing a form of resistance to virus diseases.

Gene transfer into protoplasts: Protoplasts are "naked" cells from which the outer wall has been removed by treatment with enzymes, and from which whole plants can be regenerated. They can be produced in a range of crop plants including potato, oilseed rape, sugar beet, Italian ryegrass and clover. But in order to be capable of regenerating into normal fertile plants, protoplasts must be isolated from very young tissues or cell suspensions.

Several techniques have been developed to facilitate the uptake of "foreign" DNA by protoplasts. For example, DNA molecules are taken up more readily from a solution in the presence of polyethylene glycol. Scientists at the Institute of Grassland and Environmental Research to introduce genes into forage grass – the first time this transformation has been achieved in the UK, have recently used this approach.

Protoplasts can also be made to take up DNA by a process known as electroporation. Rapid pulses of electricity (typically 2 kilo-volts per cm at 20-50 microseconds) produce temporary pores in the plasma membranes of the protoplasts through which pieces of DNA from the surrounding medium can enter. Electroporation has been used to introduce complete packages of genes ("gene constructs") into potato protoplasts. This has allowed scientists to study how well the same genes are expressed in protoplasts made from different parts of the potato plant.

Gene gun: A recent development has been the use of a "particle gun" to shoot DNA-coated tungsten or gold particles directly into plant cells. This has proved to be effective for intractable species such as wheat and other major cereals, which are not readily infected by Agrobacteriumor suitable for protoplast technology. The particles are about 1 [mu] m in diameter; they are fired from the barrel by a plastic bullet propelled by a 0.22 cartridge. (A gun licence is required to operate this machine.) The bullet strikes a plastic screen with which it fuses, and which stops it dead. The DNA-coated particles carry on, flying through a tiny hole in the centre of the screen. They drive through the walls of plant cells placed in their path and lodge within them. Scientists at IACR-Rothamsted to introduce genes into cereal plants have used this approach. In this case a helium particle gun has been used to fire gold pellets coated with the DNA.

A key element in the design of effective gene transfer techniques for wheat has been the identification of cell types which will tolerate particle bombardment and be able to proliferate and regenerate at high efficiency. Researchers at Rothamsted used highly regenerable cultures of immature inflorescence tissue, which gave much higher rates of transformation than the use of standard immature embryo plant tissues.

In all types of gene transfer technology; scientists need quick and easy methods for identifying and selecting those cells that have taken up the new gene(s) and for eliminating those that have not. This is usually achieved by co-transferring an easy-to-find "marker gene". For example, this might be a gene conferring resistance to an antibiotic; plants cells that have received the marker gene can readily be selected by culturing in the presence of the antibiotic. Advancement in cell biology to generate plants from single cells and the discovery of techniques to transfer and expression or foreign genes into plant cells provided the pre requisite for the practical use of genetic engineering in crop improvement. The essential requirements for the production of transgenic plants are the availability of a target tissue having cells competent for both plant generation and transformation, a procedure to select transformed cells and system to regenerate plants from the transformed cells at a satisfactory rate. Many techniques available for the introduction of transgene into the host plant cell involve physical chemical and biological method. Physical method involves transformation by bombardment of plant cell with micro projectiles, microinjection electroporation etc. The transformation-mediated through polyethylene glycol (PEG) and $CaCl_2$ etc. constitute the chemical way to transfer foreign genetic information. Biological method utilizes the natural ability of the soil microbe viz Agrobacterium species the gram-negative rod shaped aerobic bacterium to transform plant cell.

In 1984, the group of Jeff Schell and Marc van Montagu in Cologne and Gent and the group of Robert Horsch and collaborators of Monsanto Company in US simultaneously published procedure for the transfer of foreign gene in to the genome of plant utilising Ti plasmids of *Agrobacterium tumefaceins*. This method has made it possible to alter the protein compliment of a plant specifically to meet special requirements for example to render plant resistant to pest or herbicides, to achieve a qualitative or quantitative improvement in the productivity of crop plants and to adapt plants so that they can produce defined sustainable raw materials for the chemical industry. In the decade after the first publication of this transformation procedure and immense number of transformed plants have been licensed by 1995 fourteen transgenic cultivars had been licensed to be grown commercially.

Agrobacterium tumefaciens, the gram negative bacterium first reported as a causative agent of a wild spread neoplastic plant disease crown gall by Smith and Townsend (1907). Agrobacterium received extra ordinary attention after the discovery that the bacteria is capable of modifying genetic material of host cells by transferring part of its Ti plasmid which gets incorporated into the host cell DNA and express itself with the normal host DNA. Other than *Agrobacterium tumefaciens, A.rubi, A.rhizogenes, A.radiobacter* are used as a genetic architect.Thus Agrobacterium gene transfer system has now become one of the most significant discoveries triggering a series of existing developments in molecular biology and plant genetic engineering. Using this technique transgenic plants for model species such as tobacco to agronomical valuable species have been generated.However practical transformation of many plant species is still limited by the combination of low frequency of transformation and high frequency of undesired genetic change or unpredictable transgene expression.

GENE TRANSFER PROTOCOLS

Direct Gene Transfer

Different gene transfer methods are available:

1. Direct gene transfer method.
2. Vector mediated gene transfer method.

In Direct gene transfer methods foreign DNA is introduced into the cytoplasm or the nucleus of a plant cell, which becomes incorporated into the plant chromosome. These are number of innovative means by which the DNA has been physically delivered into the plant cell. The primary advantage of direct gene transfer system is that they are not subject to host range restriction. However there are two main draw backs, the recovery of transgenic plants requires regeneration from protoplasts and frequency of incomplete sequence arrangement or concatemerised DNA insert observed in transgenic plants is relatively high.

Electroporation: Introduction of DNA into cells by exposing them for very brief periods to high voltage electrical pulses, which is thought to induce transient pores in the plasmalemalla, is called electroporation. Electroporation with high voltage is currently the most efficient method for transforming *E.coli*. This technique was based upon the original observation by Zimmevan *et.al* (1983), that high voltage electric pulses can induce the cell plasma membrane to fuse. Subsequently it was found that when subjected to electric shock, the cells take up exogenous DNA from the suspending solution, apparently through pores momentarily created in plasma membrane. The procedure described may be used to transfer freshly prepared cells to transform cells that have been frozen.

Particle Bombardment: In this method 1-2 micro meter tungsten or gold particles coated with the DNA to be used for transformation, are accelerated to velocities, which enable their entry into plant cells/nuclei. Particle acceleration is achieved by using a device, which varies considerably in design and function. The most successful devices accelerate particles in one of the two ways. By using pressurized helium gas by electrostatic energy released by a droplet of water exposed to a high voltage. This method is also called biolistic or ballistic method of DNA delivery. Using the alternative approach of microprojectile bombardment-mediated transformation, carnation transgenics have been generated from stem explants.

Microinjection: In this method, the DNA solution is injected directly inside the cells using capillary glass micropipettes with the help of micromanipulators of a microinjection assembly. It is easier to use protoplasts than cells since cell wall interfere with the process of microinjection.

Pollen mediated transformation: Mature pollen grains are collected from anther and incubated at conditions with foreign DNA fragments. Pollinations are done artificially, recently transgenic rice plants have been produced by using this technique.

PEG mediated: The method came about when Dr. Kao (1974) added a strong solution of Polyethylene glycol (PEG) to a population of protoplasts. The protoplasts became totally transformed, but apparently did not break. Then he gradually diluted the PEG. The protoplasts then began to reform when that procedure was repeated with a mixture of protoplasts from a cell culture and from leaves (green), it became clear that protoplast fusion had occurred. The PEG method has become adapted and is in use for fusion, also of yeast as well as mammalian cells.

Vector-mediated Gene Transfer

The development of plant gene transfer systems was one of the key breakthroughs enabling such rapid progress. In 1983 the first foreign gene was introduced and expressed in tobacco cells using *Agrobacterium tumefaciens* as a vector (Bevan *et al.*, 1983; Herrera -E.Strella *et al.*, 1983 and Fraley *et al.*, 1983). Since then many plant species have been transformed and regenerated and several new gene transfer techniques have been developed. One of the most significant achievements in the field of plant biology is the elucidation of the mechanism by which *A.tumefaciens* deliver the transfer DNA (T-DNA) to the plant cells. This natural gene transfer system is highly efficient, frequently yielding transformants containing single copies of T DNA which have relatively, uncomplicated integration pattern compared with other transformation procedure. The key aspect with regard to gene transfer is that none of the T DNA genes are involved in the transfer process and therefore any or all of these genes can be removed, mutated or replaced.

Application

Using genetic engineering, scientists can, for the first time, insert genes from different species, families, or even kingdoms, something inconceivable in traditional breeding. Despite all of the unknowns, proponents of genetic engineering continue to push forward with previously unheard of combinations. Some of these combinations that have been field tested in the U.S. include:

- Chicken genes in corn, wheat, and creeping bentgrass;
- Human genes in barley, corn, tobacco, rice, and sugarcane;
- Mouse genes in corn, along with human genes;
- Cow genes in tobacco;
- Carp genes in safflower;
- Pig genes in corn;
- Simian immunodeficiency virus (SIV) and Hepatitis B genes in corn;
- Jellyfish genes in corn, rhododendrons, Bermuda grass, pink bollworms, and rice;
- Fruit fly genes in potatoes; and
- Rat genes in soybeans.

To protect public health and the environment, genetically engineered food ingredients or crops should not be allowed on the market unless:

- Independent safety testing demonstrates they have no harmful effects on human health or the environment;
- They are labeled to ensure the consumer's right-to-know; and
- The biotechnology corporations that manufacture them are held responsible for any harm.

PLANT HEALTH

Crop Protection: Pest and Weed Control

Biopesticides, biofungicides and bioherbicides: These innovative crop protection products do not require an existing agrochemicals business, or even a seed business; however, these biologically based products add-on nicely to the product lines of those firms that do already market agrochemicals, and new biological products are often less expensive to develop than new chemicals. Thus, both small biotechnology firms and large agrochemical producers can be found developing and indeed already marketing such products to farmers. Among the biotechnology firms, biosys, Ecogen, and Mycogen have put a raft of biologically based crop protection products onto the market.

Resistance to insects and other pests: Often, the very same compounds, such as Bt toxin, that act as effective biopesticide applications can be engineered directly into crop plants, providing an innate resistance to pests. Agrochemical- and pesticide-producing firms are motivated to incorporate insect resistance into plant because again they are less costly to develop and to get regulatory approval for than are chemical pesticides. It can costs four times as much to develop a new pesticide chemical than a new pest-resistant seed. This set of technology is also avidly pursued by smaller biotech research firms because it does not require an existing pesticide chemicals business; however, a seed business outlet for the traits developed is necessary. Pest resistance has furthermore been a focus of public research organizations, with an environmental aim at reducing the overall use of chemicals in agriculture.

Resistance to infection and disease: This again is an improved agronomic feature of crop plants that has been traditionally pursued by seed firms to improve the vigor and health of their product lines. Biotechnological research to these ends continues in those organizations and constitutes a significant focus of public research organizations.

Plant diagnostics: Increasing the speed and ability of farmers to detect and diagnose infestations and crop disease can greatly save in costs. Many biological assay and diagnostic test kits have been developed through the application of biotechnologies. These products are easy for small research firms to develop and market independently, and it is an area of plant biotechnology that is close to medical (diagnostics) biotechnology. Furthermore, the market is naturally rather small. Thus, plant diagnostics tends to be the purview of small biotech firms.

WHAT IS "BIOPHARMING"?

"Biopharming" is an experimental application of biotechnology in which organisms are genetically engineered to produce pharmaceutical proteins and chemicals they do not produce naturally. While most of these substances are kept secret as confidential business information, a few known examples include a contraceptive, potent growth hormones, a blood clotter, blood thinners, industrial enzymes, and vaccines (Schillberg *et al.*, 2002). Corn is by far the most popular biopharm plant, followed by soybeans, tobacco and rice. Some 400 biopharm products are reportedly in the pipeline, and over 300 open-air field trials have already been conducted in unidentified locations across the country.

'PHARM' FACTORIES OF THE FUTURE

Plants may become miniature "factories" for pharmaceuticals through genetic engineering. Scientists are growing plants that produce antibodies to help fight cancer, heart disease and tooth decay. And researchers are experimenting with growing fruits and vegetables that contain vaccines for measles, hepatitis B, Norwalk virus, diarrhea, cholera and more. These edible vaccines could be pureed into an applesauce-like consistency and fed to children and adults alike. They could be produced in larger quantities and at less expense than current vaccines, although they will have to be grown and handled under strict conditions imposed by the U.S. Department of Agriculture (USDA) aimed at ensuring that they do not inadvertently enter the food supply. The Food and Drug Administration regulates pharmaceuticals, whether they are manufactured in traditional factories or in crops in the field, to ensure their safety, purity and potency. The FDA and the USDA have jointly published draft guidance specifically addressing the use of bioengineered plants to produce pharmaceutical products (Schillberg *et al.*, 2002).

Metabolic Engineering — Regulation of the Production

The knowledge on the various biosynthetic pathways has been used already for altering biosynthetic pathways. The goals one can envisage for metabolic engineering are many. Some important goals are:

- Improving the production of a desired compound/enzyme for subsequent extraction and isolation
- Improving resistance against pests and diseases
- Lowering level of undesired compounds in food plants
- Increasing the level of a desired compound in food (e.g., vitamins)
- Giving a new trait (colour, taste, smell) to food, flowers or ornamental plants.

MOLECULAR FARMING

Molecular Farming' is a novel approach to the production of pharmaceuticals, where valuable recombinant proteins can be produced in transgenic organisms on an agricultural scale. Plants have been traditionally used as a source of medicines, but the use of transgenic plants in molecular farming represents a novel source of molecular medicines that include plasma proteins, enzymes, growth factors, vaccines and recombinant antibodies. Until recently, the wide use of these molecular medicines was limited because of the difficulty in producing these proteins outside animals or animal cell cultures. The application of molecular biology and plant biotechnology in the 1990s showed that many molecular medicines could be synthesised in plants. The goal of this Molecular Farming technology is to produce pharmaceuticals that are safer, easier to produce and less expensive than those produced in animals or microbial cultures. Molecular Farming in plants has become possible because several molecular technologies have come together. Plant transformation has advanced to the point where many commercially interesting crops can be routinely genetically modified (Verpoorte *et al.*, 2002).

HEALTH

Edible Vaccines

Plants are ideal production systems for diagnostic or therapeutic human proteins needed in large amounts. Mammalian genes are readily expressed in plants, and plants are likely to become the production systems of choice for mammalian proteins that are needed as diagnostics for detecting human diseases or therapeutics for treating human and animal diseases. The term biopharming is used for producing monoclonal antibodies, blood plasma proteins, peptide hormones and cytokines in plants. Why plant system is used for this the answer is simple purity and cost.

Plants can be used to deliver edible vaccines for serious diseases of humans and domestic animals. Humans and other warm blooded animals the primary task of the immune system is to defend against invading organisms-particularly against pathogenic bacteria and viruses. If the plant make the antigenic determinant, then eating the plant (under the right conditions) will amount to vaccination.

Genetic engineering techniques are also being used to put vaccines in foods. Potatoes have been developed to contain vaccines against a variety of diseases, including cholera, Norwalk virus and hepatitis B. One draw back from the use of potatoes as the vehicle is that they must be eaten raw. Dr. Charles Arntzen, a pioneer in this field, is now inserting a vaccine against hepatitis in bananas, a more palatable vehicle when eaten raw and a food that can be readily eaten by small children. The benefits of food-borne vaccines are that they could be provided for a fraction of what it costs to develop and administer vaccines as injections. However, there are many questions about how these products would be controlled, so again new policies and procedures will need to be developed. Issues such as proper distribution, quality control and control of access will need to be resolved.

Much attention has been given recently to the potential of using plants as a production system for high value compounds of medical importance. Examples include the expression of antibodies that may reduce the growth of bacteria associated with dental caries or that can be used in the treatment of human tambours. Other medically related projects include the production of recombinant blood factors, human milk proteins, interferon mammalian growth factor (Taihan Sugar Industry Company, 1999), gonadotrophin and the human secretory

Table 1. Production of vaccines in plants.

Potential application	Protein expressed in plants
Dental caries	*Streptococcus mutans* surface protein
Cholera and E.coli diarrhea	*E.coli* heat labile enterotoxin
Hepatitis B	Hepatitis B antigen
Hoof and mouth disease	Hoof and mouth viral antigen
Malaria	Malarial B cell epitope
Influenza	Hemagglutinin
Rabies	Rabies virus glycoprotein
HIV	HIV epitopes gp41 and gp120

Table 2. Examples of Indian health biotechnology products

Sector	Type	Product name	Application	Producer
Vaccines	Recombinant hepatitis B surface antigen	Shanvac-B	Hepatitis B	Shantha Biotechnics
	Recombinant hepatitis B surface antigen	Revac-B	Hepatitis B	Bharat Biotech
	Recombinant hepatitis B surface antigen	Gene Vac-B	Hepatitis B	Serum Institute of India
	Purified capsular polysaccharide Vi of *Salmonella typhi*	Typbar Vi	Typhoid	Bharat Biotech
Therapeutics	Recombinant human insulin	wosulin	Diabetes	Wockhardt (Mumbai)
	Recombinant human erythropoietin α	Epox	Anemia	? Wockhardt
	Recombinant human interferon α2b	Shanferon	Cancer	Shantha Biotechnics
	Recombinant streptokinase	Shankinase	Cardiovascular	Shantha Biotechnics
	Bharat Biotech	Indikinase	Cardiovascular	Shantha Biotechnics
	Liposomal amphotericin B injection	Fungisome	Visceral leishmaniasis	Lifecare Innovations (New Delhi
	Recombinant human granulocyte colony-stimulating factor	Gramstim	Neutropenia	Gramstim Neutropenia Dr. Reddy's Laboratories
Diagnostics	Immunoblot assays using recombinant HIV-1 antigens gp41 and C-terminus of gp-120 and HIV-2 antigen gp-36	HIV TRI-DOT	HIV-1 and HIV-2	J. Mitra (New Delhi)
	Immunoblot assay using recombinant HIV-1 antigens gp-41 and gp-120, HIV-2 antigen gp-36, and HCV antigens NS-3, NC-4 and NC-5	HIV-HCV Combo	HIV and hepatitis C	Bhat Biotech India (Bangalore)
	Enzyme-linked immunosorbent assay for recombinant HCV core antigens 1b & 3a, together with peptides for HCV antigens NS-3, NS-4 1, NS-4 2 and NS-5	HEP-Chex C	Hepatitis C	xCyton Diagnostics (Bangalore)
	Enzyme-linked immunosorbent assay for recombinant version of *Taenia solium* excretory/ secretory antigens	Cysti-Chex	Neurocysticercosis	xCyton Diagnostics

Some products have more than one Indian manufacturer; only selected examples are shown (Nandini *et al.*, 2004).

protein somatotrophin. In a recent study of this type, the bovine follicle stimulating hormone (bFSH), a pituitary heterodimeric glycoprotein hormone, was chosen as a model to assess the ability of plant cells to express a recombinant protein that requires extensive N-glycosylation for subunit foldingand assembly, intracellular trafficking, signal transduction and circulatory stability. Stimulation of cAMP production in a CHO cell line expressing the porcine FSH receptor confirmed the native-like structure of the recombinant singlechain (sc) bFSH particularly in terms of the sufficient degree of N-glycosylation required for signal transduction. Furthermore, in superovulatory treatments of mice, sc-bFSH displayed significant *in vivo* bioactivity, although much lower than that of pregnant mare serum gonadotrophin. It was concluded that plants may have a broad utility as hosts for the recombinant expression of proteins even where glycosylation is essential for function. An interesting related example in this area is the study of the processing and subcellular localization of a chimeric gene consisting of the bovine milk protein, beta-casein, under the control of a soybean seed lectin promoter in the seeds of transgenic soybean plants. The casein proteins were purified from the soybean seeds by immunoaffinity chromatography and were analyzed by two-dimensional gel electrophoresis, blotting, and amino terminal sequencing. Results from these analyses suggest that the lectin expression cassette could be used successfully for production of pharmaceutical or other recombinant proteins of added value in the developing soybean seed. Recent advances in the use of plants to produce vaccines for veterinary or medical purposes have extended greatly. Examples of the former category include the expression of antigens against respiratory syncytial virus, foot and mouth virus, a virus causing enteric disease, and the bovine pneumonic pasteurellosis agent *Mannheimia haemolytica*. Perhaps the best known example for human disease is hepatitis B. Oral immunogenicity of recombinant hepatitis B surface antigen (HBsAg) derived from yeast (purified product) or in transgenic potatoes (uncooked unprocessed sample) was compared, with an oral adjuvant, cholera toxin, used to increase immune responses. Transgenic plant material containing HBsAg was found to be the superior means of both inducing a primary immune response and priming the mice to respond to a subsequent parenteral injection of HBsAg.

Electron microscopy of transgenic plant samples revealed evidence that the HbsAg accumulated intracellularly and it was concluded that natural bioencapsulation of the antigen may provide protection from degradation in the digestive tract until plant cell degradation occurs near an immune effector site in the gut. Mice fed HBsAg-transgenic potatoes produced HBsAg-specific serum antibodies that exceeded the protective level and, on parenteral boosting, generated a strong long-lasting secondary antibody response. These results also confirmed the effectiveness of oral delivery by using a parenteral prime-oral boost immunization schedule. This demonstrated success of oral immunization for hepatitis B virus with an 'edible vaccine' provides a strategy for contributing a means to achieve global immunization for hepatitis B prevention and eradication. Other medically important examples include a cholera toxin rotavirus fusion protein, measles virus, Puumala virus (the rodent-born hantavirus which is a causative agent of haemorrhagic fever in humans), and human papillomavirus (implicated in cervical cancer). Similarly, glycoprotein C from human cytomegalovirus has been produced successfully in the protein storage vesicles of transgenic tobacco. The potential advantages of using plants as a production system include the reduced risk of mammalian viral contaminants, the ability to scale up at low cost and the low maintenance requirements. Issues that remain to be confirmed include the glycosylation status of human proteins expressed in plants (though see data on sc-bFSH above. Examples of plant-produced recombinant enzymes used in medical therapies include canine lipase for treatment of pancreatic dysfunction, and acetylcholinesterase used in the treatment of organophosphate poisoning. Possibly the most commercially advanced use of plants to produce high value, medically-related enzymes is the example of transgenic maize expressing recombinant egg white avidin and bacterial glucuronidase; both products are now marketed by Sigma. An example of an animal protein useful in a great diversity of food and medical purposes is collagen, the constituent of gelatin. At present large amounts of this protein are purified from animal sources and there is some concern, of unquantifiable proportion, about the possibility of developing the prion disease, new variant Creutzfeldt-Jakob Disease (nvCJD) and possibly other diseases, through injection of animal-derived gelatin or, possibly, from ingesting collagen or gelatin from infected animal sources. The human version of this product has already bee n produced in plants (Ruggiero *et al.*, 2000), and there is commercial progress reported concerning the co-expression of the collagen gene along with prolyl hydroxylase, the intracellular enzyme required for the synthesis and formation of all 21 known types of collagen.

Prolyl hydroxylase acts by modifying the individual polypeptide alpha chains of procollagen to allow them to form stable, triple helical structures. This structure gives the collagen molecule its stiffness, strength and insolubility. In addition to the production of proteins of direct therapeutic value, transgenic plants are also being investigated as sources of other pharmaceutical compounds such as alkaloids. These compounds represent one of the largest groups of natural products, and many of them are pharmacologically active. One such study of genes in the biosynthetic pathways for scopolamine, nicotine, and berberine examined the expression of two branching-point enzymes, putrescine N-methyltransferase (PMT) in transgenic plants of *Atropa belladonna* and *Nicotiana sylvestris* and (S)-scoulerine 9-O-methyltransferase (SMT) in cultured cells of *Coptis japonica* and *Eschscholzia californica*. It was found that overexpression of PMT increased the nicotine content in *N. sylvestris*, whereas suppression of endogenous PMT activity severely decreased the nicotine content. Ectopic expression of SMT caused the accumulation of benzylisoquinoline alkaloids in *E. californica*.

Antibodies

Functional antibodies produced in tobacco plants were first reported over a decade ago. The basic protocol used to generate these 'plantibodies' involved the independent cloning of H and L chain antibody genes in transformation vectors, the regeneration of transgenic plants expressing individual chains, and combining them in sexual crosses to allow the production of fully assembled and functional antibody. In mammalian cells, the antibodyH and L chains are produced as precursor proteins that are translocated into the endoplasmic reticulum (ER), under the guidance of signal sequences. Within the ER, the signal peptides are cleaved, and several stress proteins act as chaperonins to bind the unassembled antibody chains, and direct their subsequent folding and tetramer formation. A similar process occurs in plant cells, and expression can be directed via signal sequences (even of foreign origin) into the aqueous environment of the apoplast, or to be accumulated in other specific plant tissues, including tubers, fruit, or seed. Plants can also efficiently assemble secretory IgA, which is comprised of four chains, H and L chains, J chain and secretory component. It is expected that plant 'bioreactors' might yield over 10 kg of therapeutic antibody/acre in tobacco, maize, soybean, and alfalfa. Compared with conventional steel tank bioreactors using mammalian cells, or microorganisms, the costs of 'plantibodies' are expected to be perhaps one tenth. Although there are known differences (see below) in glycosylation patterns between plants and animals, (N-linked glycans of plants differ from human by having fucose-linked alpha 1,3 and the sugar xylose), no adverse effects of human antimouse antibodies have been observed in >40 patients receiving topical oral application of a plant produced secretory IgA specific to *Streptococcus mutans*, one of the bacteria responsible for the onset of dental caries. [An alternative approach to the prevention of this disease is provided by the production of non-cariogenic sugars in transgenic plants. The progressive improvement of expression vectors for plantibodies, and purification strategies, as well as the increase in transformable crop species, is expected to lead to almost limitless availability of inexpensive (even edible forms of) recombinant immunoglobulins free of human pathogens for human and animal therapy, and for novel industrial applications (e.g. catalytic antibodies). Other recent advances in this area of antibody production include the crossing of plants expressing antibodies with ones expressing beta 1,4-galactosyltransferase (BGT), the enzyme that converts typical plant N-glycans into mammalian-like N-glycans.

Genetically Engineered Foods

Genetically engineered foods are produced from crops whose genetic makeup has been altered through a process called recombinant DNA, or gene splicing, to give the plant a desirable trait. Genetically engineered foods are also known as biotech, bioengineered, and genetically modified, although "genetically modified" can also refer to foods from plants altered through other breeding methods. Hybrid corn and tangelos (hybrid of a tangerine and grapefruit), for example, are the result of genetic modification through traditional methods of plant breeding. And the many varieties of apples we eat today were produced through genetic modification. Using traditional genetic modification methods, such as cross-fertilization, scientists can produce a desired trait, such as a hardier plant. But in doing so, they mix thousands of genes from several plants, requiring many attempts over many years to weed out the unwanted traits that occur. Newer methods of genetic modification, in the form of genetic engineering, are more precise and predictable-and faster. By controlling the insertion of one or two genes into a plant, scientists can give it a specific new characteristic without transferring undesirable traits (Fresco, 2001).

Production of a Combined Colourant/ Antioxidant Food Additive

Genetic engineering is not the only aspect of biotechnology, which has, or could, enhance the use of enzymes in the food industry. Enzyme systems can be employed in the production of a wide range of food additives. This aspect of enzyme biotechnology can be exemplified by the in vitro production of colourants and antioxidants from apple tissue. This technology represents developmental work carried out at Nottingham in collaboration with the department of Nutrition and Dietetics, Kings College, London and has not as yet been exploited commercially.

Apple contains a high percentage of a phenolic compound called phloridzin. This can be converted enzymatically by apple polyphenoloxidase into a coloured pigment. Indeed this pigment is largely responsible for the natural colour of apple juice. The pigment has stability and tinctorial properties which would render it capable of acting as a natural food colourant with applications similar to those currently served by the artificial dye tartrazine. The bioconversion of phloridzin to the coloured pigment can be manipulated to produce a range of other products which include phloretin, 3-hydroxyphloretin and 3-hydroxyphloridzin. These compounds all demonstrate significant antioxidant activity as measured by their ability to prevent liposomal lipid peroxidation. In fact 3-hydroxyphloretin proved to be as effective an antioxidant as quercetin, the most potent naturally occuring dietary flavonoid so far tested with this system. Dietary antioxidants are thought to protect against a range of diseases such as atherosclerosis and cancer. This potent antioxidant activity could mean that these phloridzin derivatives may be useful additives in functional foods.

As a result of genetic engineering for new traits that has been applied to plants, the technology is available for plants to provide a host system for new products. Expanded use of genetic engineering in plants will allow the development of new products that would have been limited or unattainable by alternative host production systems. Currently, the most economically and technically viable class of products are obtained from engineering genes to produce protein products. This has resulted in a $40-billion industry of new therapeutics and industrial enzymes. While this has been extremely useful and lucrative, the host systems used today can still be limiting. Microbial systems are the most popular and cost effective but are limiting because of downstream processing. Yeast can cause hyperglycosylation, and bacteria do not glycosylate. Animal cell cultures are better hosts for downstream processing. However, the cost of animal cell culture systems is orders of magnitude higher than their microbial counterparts. There is the potential to use transgenic animals in the future as a host. These will undoubtedly prove useful; however, one disadvantage is that the protein product will be derived from an animal source. This is significant because there is a mandate from many pharmaceutical companies to remove all animal source materials from their products. Therefore, an opportunity exists for plant systems to increase the safety margin of these products since plants do not harbor human pathogens. Plants may also provide a lower cost of goods and a greater convenience. Increased use of plants for protein production will result in new products that can help improve our quality of life as well as provide safe alternatives to environmental risks.

Proteins: There are many, varied applications of transgenic technology in modifying food quality The improvement of either the quantity or quality of seed protein is at the centre of many plant breeding programmes and there is a wide diversity of approaches to the modification of specific traits. A recent example of expression in transgenic plants of an animal enzyme used widely in the food industry is that involving production of chymosin.

Carbohydrates: As well as the projects designed to modify protein quality, there are many concerned specifically with changing the profile of starch produced either in the seed or other storage organ such as the tuber of potatoes. Examples include the use of a beta-amylase gene, a rice starch granule-bound protein, a debranching enzyme, or starch synthase. It is also claimed that an overall increase in the level of stored carbohydrate can be achieved by over expressing dihydro-orotase, a gene of pyrimidine metabolism. In a similar study designed to improve the storage capacity of potato tubers, the amount of the two enzymes alpha glucan L-type and alpha glucan H-type phosphorylases has been reduced. The transgenic tubers demonstrate a reduced conversion of starch to sugar during storage, thus increasing the storage life. by prolonging dormancy and reducing the incidence of disease.

Oil: There are a large numbers of studies and field trials in which either the total amount or the specific oil profile has been altered. Metabolic engineering can regulate essential oil quantity and quality. Using new transgenic techniques, it is now possible to introduce new qualities into crops that it would be very difficult or costly to obtain through traditional selection methods. Notwithstanding, these modifications do need to offer some benefit to society. There are at present numerous on-going projects involving oilseeds and protein crops, with a first generation of modified plants – notably crops resistant to the herbicides glufosinate, glyphosate, bromoxynil – already being cultivated on an experimental basis by several companies (AgrEvo, Monsanto, Rh ne-Poulenc, etc.) and by INRA, France's National Institute for Agronomic Research. However, none of these experimental varieties will be authorised for at least two years, since time is needed to measure their impact on the environment and their economic potential for producers.

A second generation of oilseeds and protein crops will mainly consist of plants whose intrinsic qualities have been modified for industry, animal feed and pharmaceuticals. These crops will, for example, be able to resist fungal and viral attack and may in the longer term, be made less sensitive to the environment in which they grow, improving their ability to withstand drought and salt, etc. (Murphy, 2002) .

- Lauric acid-rich rapeseed (40%). Uses: food industry, detergents, cosmetics. This variety is already being cultivated on experimental fields.
- Erucic acid-rich rapeseed (more than 60%). Uses: lubricants and plastics industry.
- Rapeseed producing phytase enzymes for animal feed. Phytase enables animals to more easily digest the phosphorus in oil cake, so reducing the discharge of the element into the environment.
- Rapeseed producing gastric lipase in dogs for the treatment of mucoviscidosis.
- Fungus-resistant rapeseed and sunflower.

Other food constituents: In additions to the modifications of proteins, starches and oils there are several studies concerned with minor, though important constituents of food (Tucker, 1996). Perhaps the best known is the production of golden rice, a GM rice variety that has increased amounts of vitamin A in the endosperm, thus the golden epithet induced by vitamin A deficiency and found in many developing countries where rice is the staple crop. Another good example concerns the modification of tomatoes, fruit that are an excellent source of the carotenoid lycopene, a compound that is thought to be protective against prostate cancer. They also contain small amounts of flavonoids in their peel (approximately 5–10 mg/kg fresh weight), mainly naringenin chalcone and the flavonol rutin, a quercetin glycoside. In a study designed to generate fruit with increased antioxidant capacity and a wider range of potential health benefit properties, flavonol biosynthesis was upregulated by transformation of tomato with the Petunia *chi-a* gene encoding chalcone isomerase. Other recently described GM food products with possible beneficial health effects include those with increased levels of vitamin C, the antioxidant tocopherol, and those expressing an oxysterol-binding protein a compound with increased ability to act as a cholesterol lowering agent.

The risks and uncertainties surrounding the process of genetic modification have prompted many countries to regulate the development and use of GMOs. Mandatory labeling is one such response. GMO labeling allows consumers to choose products according to their ethical, religious, cultural and dietary preferences. Many consumers, citing religious and ethical convictions, prefer not to use products created by transferring genetic material across species boundaries. The consumer's right to know provides a compelling justification for labeling of all GMO products.

Genomics

Genomics is a rapidly emerging area of research, which came into existence in the closing years of the last century and promises to become a dominant in the coming decades, revolutionizing our understanding of biology as never before. Plant genomics is a part of this larger field and embraces the study of whole genomes of plants, their physical and molecular organization, evolution and functions of the myriads of constituent genes. Included in it are also studies on genome-wide or global patterns of gene expression, that have recently become possible by the new DNA chip technology, allowing visualization of activity of hundreds and thousands of genes simultaneously. The sequencing of entire genomes is in fact propelling the development of many other novel technologies and the shaping of a new discipline of bioinformatics. Genomes of eukaryotes comprise thousands of genes and millions of base pairs – storing the output of massive data and information not only on genes themselves, but relating also to their expression, and later retrieving and analysing this information requires new computational tools, even new mathematical techniques, not envisioned before. The advent of genomics is a consequence of the discovery of the elegant and simple procedure of sequencing of nucleic acids pioneered by Sanger and coworkers in UK and by Maxam and Gilbert in USA in 1977. But it is not so much this technical ability *per se* which has aroused world-wide interest in genomics rather the discipline has assumed tremendous importance today because deciphering the sequence of bases of DNA would unlock the whole blueprint of the development of an organism. Appropriately, therefore, the impact of research on genomics, of which the Human Genome Project (HGP) is a prime example, has been compared to the discovery and consolidation of the Periodic Table in chemistry1. The study of various genomes also holds the key to understanding the origin and evolution of plants and animals – as put so aptly by Lander and Weinberg2, it is a journey to the centre of biology. But genomics is not merely of fundamental or academic interest; it indeed will be of vital importance for the agriculture of tomorrow as it is proving already for medicine and human welfare. Only a few plant species stand between prosperity, on the one hand, and hunger and starvation on the other. Clearly, if we understand the genomes of crops like rice, wheat, maize, beans and potato, we can ensure a better future with the capability of surer, more perfect and precision genetic manipulation for enhanced yield and survival under adverse conditions.

Regulation of GM Foods

The question is whether there should be any governmental oversight of transgenic crops needs to be addressed. One must look at the reasons why the government regulates products or technologies and whether there are facts to justify government regulation of transgenic crops.

There are many reasons why governments regulate and oversee processes and products. One reason for regulation is to ensure human safety. Governments regulate food and drug products to ensure that what humans ingest or put in their bodies is safe. For example, many countries have mandatory regulation of food additives to ensure that those man-made substances are safe when we eat them. Another reason for regulation is to protect the environment. Many countries have established environmental laws, which limit pollution that might harm our air, water, or land. Whether regulation is established to 'protect' or 'safeguard' human health or the environment, the government's goal is to minimize or eliminate any real or potential hazards. The fact that there is government oversight does not mean that the activities, products, or technologies regulated are inherently unsafe, only that there is a potential that some specific applications could prove harmful.

The European Union requires mandatory food labeling of GM foods and has adopted a 1% threshold for affirmative GMO labeling. If any ingredient in a finished food product has GMO content above 1%, the product must state it clearly. The purpose of the 1% threshold is to ensure that food ingredients obtained from non-GM sources do not need to be labeled if they contain low levels of GM material, as a result of accidental cross-contamination. These values are being re-assessed in the light of new knowledge. Some experts are proposing a 0.1% threshold, which would prevent accidental GMO contamination. In Japan, GMO testing has been mandatory from April 2001. In the US, the regulatory process is complicated because there are three different government bodies that have jurisdiction over GM foods. The FDA's stance has been that GM foods are substantially equivalent to unmodified, natural foods, and therefore not subject to FDA regulation (Dunwell, 2002).

Public Perception of Transgenic Plants

(1) Creation of distrust of public regulation and expertise. This distrust is key to the observed difference in public attitudes between the U.S. and Europe.
(2) Following these crises, the media have been receptive to information on public health or environmental problems.
(3) Because of the first two effects, GM crops have become a very risky issue for policy makers This approach to decision making has led to unclear, even incoherent, decision processes.

There is a need for a rational approach to the benefits of genetically modified foods, as well as consideration of the rights of consumers to know what is in their food. Genetically modified foods have the potential to solve many of the world's hunger and malnutrition problems, and to help protect and preserve the environment by increasing yield and reducing reliance upon chemical pesticides and herbicides. However, there are many challenges ahead for governments, especially in the areas of safety testing, regulation, international policy and food labeling. Demand for testing GMOs and for certifying non-GMO foodstuffs has increased dramatically and an expanding range of techniques are available for identification and quantification (Dunwell, 2002).

GENETIC ENGINEERING AND INDUSTRY

Structure and Evolution of the Biotechnology Industry

The structure of the biotechnology industry remains a complex network of corporate players, dominated by large firms with strong marketing capabilities and start-up firms that focus on research and development. Many technological advances in industrial and environmental biotechnology originate from leading firms with strong backgrounds in chemical engineering. Frontier technologies are concentrated in a few firms in industrialized countries, many of which are the same firms that are involved in other fields of biotechnology. The biotechnology industry in recent years has undergone major changes associated with waves of mergers, acquisitions, and spinoffs. For a while, firms experimented with the model of "life science" industries—seeking to bring together various applications in the fields of health, agriculture, and industry under one enterprise and common technology platform. This corporate model, represented by firms such as Monsanto, has yet to demonstrate its viability because of differences in approaches to research and development, regulatory pathways, and markets in the health and agricultural sectors. As a result, firms such as Monsanto, Novartis, and Astra Zeneca have had to scale back their expectations.

The formation of alliances still remains a central feature in the development of industrial and environmental biotechnology. Cargill, specializing in polymer research, has formed an alliance with Dow Chemical, a respected name in plastics. In turn, Cargill Dow Polymers has formed alliances with polymer users and support industries (plastic molding firms). There are similar alliances between enzyme research and development centers, producers, and users. These alliances, which have been concentrated in industrialized countries, share the risks and benefits of research, development, commercialization of innovations, and information. Extending alliances and interactions to developing countries on a selective basis through appropriate institutional arrangements is an important element in promoting the wider application of industrial and environmental biotechnology. Biotechnology applications in agriculture are well known, but its industrial applications are now beginning to emerge. The chemical industry would be transformed through biotechnology innovations. Progress in various areas such as industrial enzyme development, progress in bioplastics production and biofuels, applications in the paper industry, biomining and environmental applications.

Purifying and selling bacterial enzymes for use in cheese making, food manufacturing, cotton weaving, washing powders, etc., has existed for decades, but in 1988 the Danish company Novozymes produced the first transgenic enzyme, a fat-digester for detergents. The company supplies, among others, enzymes for the scouring process, which consists of removing the brown, non-cellulose parts of cotton that in the conventional process involves the use of relatively harsh chemical solutions. Thanks to advanced genetic techniques, multiple enzyme variants can be created at high speed, which are then screened to fit the desired applications. Novozymes has become the world's biggest enzyme manufacturer, and also renowned for its transparency procedures, for involving the consumers in the discussion of its activity report, and its commitment to a healthier environment and to sustainable development. India too is a player in biotechnology with the bulk of India's industrial biotechnology sector focuses on producing enzymes for the textile and paper industries. The sector was the second-largest exporter ($37 million in 2002-2003) to other Asian and European countries. In 2002-2003, this sector produced revenues of $72 million and was rapidly growing.

Bioplastics

Plants could produce the monomeric units of the plastic polymers, as some monomers are difficult to elaborate from oil. C. Chapple's work consisted of trying to stabilize some monomers and to produce them in high concentrations in plants. The viability of the whole process depended on the storage of sufficient amounts of the plant-made plastics in the cell vacuoles from which they could be extracted. The gene was eventually patented in favor of the University and the US corporation. Currently, the two most advanced plastics projects are those of E.I. DuPont de Nemours and Co., Inc., one of the world's biggest chemical corporations (which invented nylon), and of Cargill-Dow Chemical Company, a joint venture between an agricultural and a chemical firm (Sasson, 1991).

Feedstocks and Biofuels

As a chemical feedstock glucose syrup is a refined product, made out of maize starch, i.e. maize grains. The latter costs about $80 a ton; that is cheaper than oil, weight for weight, but this could be improved if the plant waste is also used, as it fetches only about $30 a ton for silage. But it consists mainly of cellulose, also a polymer of glucose, but more difficult to break down; therefore, the search began for efficient cellulases (Sasson, 1991).

Ethanol Production

Biofuel production is part of 'white' biotechnology. Ethanol, a biofuel, is produced from the fermentation of cane sugar. In Brazil, they have mastered the technology as well as that of building motor engines that use this rather corrosive fuel. The decision was made to decrease the amount of the oil bill and to seek energy independence. Ethanol served to power the first car of Henry Ford, while nowadays in Brazil an important proportion of the car fleet uses a mixture of gasoline and 20 per cent ethanol. Even in the USA, nearly a tenth of all motor fuel sold is a blend of 90 per cent petrol and 10 per cent ethanol. The maximum biofuel content of current fuels is 5 per cent; it can be increased to 30 per cent in the case of collective transport means or of vehicles under public administration. In France, the average rate of incorporation of biofuels is only 1 per cent.

The Paper Industry

The paper industry's basic raw materials are trees. Separating cellulose from lignin in wood is costly and uses non-environment-friendly chemical processes. Researchers at the State University of North Carolina have bred aspens with only half the lignin of ordinary ones and, it turned out, they had the additional advantage of growing faster. However, the commercial use of transgenic trees is at least ten years off but is on its way.

Biominining and Bleaching of Ores

Bioleaching had been used only for the recovery of copper from wastes until the mid 1980's. The process was upgraded thanks to multidisciplinary and multi-institutional research efforts by several universities, research institutes and the productive sector, with the support of the

Chilean government, the United Nations Development Programme (UNDP) and the United Nations Industrial Development Organization (UNIDO).

ENVIRONMENTAL BIOTECHNOLOGY: BIOREMEDIATION

Bioremediation is a key area of 'white' biotechnology, because the elimination of a wide range of pollutants from water and soils is an absolute requirement for sustainable development. There are numerous processes of cleaning water, industrial effluents and solid wastes, using microorganisms aerobically and anaerobically. Some of them are quite sophisticated, while others are simple and adapted to the conditions of developing countries. For instance, using microalgae (and in particular blue-green algae or *cyanobacteria*) in ponds to eliminate nitrogen and phosphorous, after organic matter has been degraded by bacteria, leads to water that can be recycled for irrigating non-food crops (e.g. cotton) or for industrial purposes; in addition, microalgal biomass can be used as feed (Sasson, 1991).

Bioremediation and Genetic Engineering

The molecular basis of heavy metal accumulation is being studied with a view to transferring the relevant genes to plant species having a wider geographic and ecological distribution. Transgenesis applied to phytoremediation is certainly incipient. Its application on a large scale is confronted with the evaluation of risks relating to the transfer of the bacterial transgenes to plants consumed by herbivorous animals that might acquire the property of hyper-accumulating toxic metals or compounds. Genetic transformation of the microorganisms involved in bioremediation could enhance the process through the introduction of genes controlling specific degradation pathways; it also aims at degrading recalcitrant compounds such as pesticides and other xenosubstances.

Metagenomics

Metagenomics could help us understand environmental problems such as carbon dioxide take up by bacteria to reduce its concentration, the breakdown of pollutants by micro-organisms, It could also help to detect environmental damage, because once we know what it is that is out there, it is a step to be able to determine any changes to the environment.

Fermentation Technology

For thousands of years humankind has been taking advantage of the activities of micro-organisms to produce foodstuffs and drinks without understanding the microbial processes (fermentation). It was not until the end of the nineteenth century that the first attempts were made to standardize the fermentation processes in order to obtain products of uniform quality. Fermentation is used in a number of different ways: (1) for the production of cell matter or biomass (e.g., bakers yeast, single cell protein); (2) for the production of cell components (e.g., enzymes, nucleic acids); (3) for the production of metabolites, including both primary metabolites (e.g., ethanol and lactic acid, and secondary metabolites (e.g., antibiotics); (4) for catalysis of specific single-substrate conversions (e.g., glucose to fructose, penicillin to 6-amino-penicillanic acid and stereospecific synthesis of fine chemicals (e.g., pharmaceuticals); and (5) for catalysis of multiple-substrate conversions (e.g., biological waste treatment).

Table 3. Top ten sub-categories of the biotechnology industry

Category	Number of companies with involvement	Percent of total
1. Therapeutics	330	5 %
2. Clinical diagnostics	201	19.8
3. Plant agriculture	92	9.1
4. Reagents	83	8.2
5. Specialty chemicals	68	6.7
6. Environmental testing/treatment	49	4.8
7. Animal agriculture	42	4.1
8. Testing/ analytical services	38	3.7
9. Immunological products	36	3.5
10. Equipment	28	2.8
Top Ten Total:	967	Top 10 Total: 95.2 %

Enzymes

Enzymes are important in food processing for a variety of reasons but particularly because of their high substrate specificity, rapid reaction rates and ability to work effectively at relatively low temperatures and under mild conditions. Enzymes are found in all raw food material and their presence may often represent an advantage to the food processor. A classic example of this is the action of amylases, found in malt, to produce substrates for fermentation during the brewing of beer. However, in other instances the continued action of enzymes during food processing may be detrimental to the quality of the final product. A classic example of this is the continued action of pectinases, such as polygalacturonase and pectinesterase, in fruit, which ultimately can result in oversoftening of the tissue. These problem enzymes can often be eliminated by the processing treatment, especially if this involves high temperatures or extremes of pH. However, this treatment is costly and may result in a reduction in product quality by, for example, a reduction in flavour volatiles (Biesgen *et al.*, 2002).

Biotechnology, and in particular genetic engineering, has been used to enhance the use of enzymes in the food industry. This is exemplified by the production of chymosin in recombinant microorganisms as a replacement for the traditional source of this enzyme, namely calf stomach. This technology has also been used to manipulate levels of endogenous enzymes in the raw material so as to enhance the quality of the end product. Again this approach can be exemplified by the use of gene silencing techniques to reduce the level of polygalacturonase activity in tomato fruit and, hence, improve transportation, flavour and processing qualities. This technology is only in its infancy at the present time and hopefully more applications will be forthcoming in the future. However, it must not be forgotten that other aspects of biotechnology, such as the further development of immobilized enzyme systems will also play a significant role in the future.

Genetic engineering enables the transfer of DNA from one organism to another. Thus, given suitable DNA sequences, the ability to synthesize any enzyme can be transferred between organisms. This technology allows enzymes to be synthesized in large amounts and in systems, which are commercially viable. This technology has been employed in the production of chymosin and has been reviewed by Teuber. Chymosin is used in the production of cheese, the industry as a whole requiring an estimated 56,000 kg per annum. The traditional source of this enzyme is calf stomach, from which extraction can be difficult and for which the supply of raw material is declining.

These micro-organisms are easy to grow on an industrial scale. The end product is indistinguishable from calf stomach derived chymosin and the cheese produced using these recombinant enzymes is also identical to that manufactured using the calf enzyme. The use of this micro-organism derived chymosin has been approved for use in over 20 countries where it is marketed under several trade names Maxiren,Chymogen and Chymax.

Biofertilizers

Many Soil micro-organisms may enhance the nutrient uptake of plants. Those microorganisms which have a direct beneficial effect on the plant may have considerable potential as biofertilizers. Three groups of plant-beneficial micro-organisms can be distinguished: nitrogen-fixing microorganisms, mycorrhiza fungi and plant growth-promoting rhizobacteria.

Embryo Technology

Over the last 30–40 years, the techniques for recovery, storage and implantation of animal embryos have been perfected. This technique is already applied in cattle, sheep and goat breeding. The principal benefit of embryo transfer is the ability to produce more offspring from a female animal than would be possible with normal reproduction. For example, each cow would normally give birth to about four calves in an average lifetime. With embryo transfer, this could be increased to at least 25 calves. This could be used to enhance the rapid expansion of rare genetic stock (e.g., a new breed). Embryo transfer can also reduce costs of imported cattle because importing an animal as an embryo is much cheaper. Moreover, by raising these animals in their new home country they are better able to adapt to local environmental conditions.

Table 4. History of Genetic Engineering and Biotechnology

8-10,000 BC - domestication of animals and plants
8-10,000 BC - selective breeding of companion animals and poultry
17th Century - selective breeding of agronomic animals and plants
1946 - DNA as hereditary material
1953 - DNA structure
1961 - DNA mRNA protein
1970 - RNA DNA
1972 - DNA splicing concept
1973 - First reports of gene cloning
1974 - Moratorium on certain types of gene cloning experiments
1975 - Asilomar Conference

Table 5. Applications of genetic modification

Agriculture
To delay ripening to allow time to transport foods
To delay rotting after ripening to prolong retail shelf life
To add color to natural fibers before harvesting
To reduce the need for fertilizers
To confer resistance to pests and fungi
To facilitate the use of herbicides on harmful weeds
Medicine
Production of human growth hormone
Production of blood clotting factors VIII and IX
Production of insulin
Production of edible vaccines
Treatment of cystic fibrosis
Research into human and animal diseases

Table 6. Implications of genetic manipulation and human health

Inserted gene may have adverse effects
May code for proteins that are toxic for human
May produce an allergenic reaction
May cause activation of previously silent genes
May be transferred to other organisms
Consumption of GMO may alter balance of existing micro-organisms in the gut
Javed Akhter (2001)

FUTURE TECHNOLOGIES

As biotechnological processes become cheaper, the economic contribution of industrial biotechnology will increase. Discovering new materials for the industry while drawing on natural resources also speaks well for the prospects of industrial biotechnology. A number of different approaches have been taken to combine or 'pyramid' transgenes in one plant and have met with varying degrees of success. These approaches include sexual crossing, re-transformation, co-transformation and the use of linked transgenes (Dunwell, 2002). These methods have been utilised to generate the 'stacked trait' varieties now increasingly being grown in the USA. For example, in the year 2000 the area of cotton with combined herbicide resistance and insect tolerance was 1.7 million ha, and there were 1.3 million ha of corn with both traits. In the future it is likely that novel, alternative technologies will be developed that use single transgenes to coordinate the expression of multiple genes. Amongst the constructs being tested at present are some modelled on viral polyproteins can be used to simultaneously introduce multiple protein-coding genes into plants.

SUMMARY

Advances in biotechnology-related fields such as genomics, genetic engineering, chemical engineering, and cell technology are transforming the industrial and environmental biotechnology landscape. Genetic engineering is one of the most promising technologies of the twenty-first century and as such is seen as the key technology for managing the challenge of health, abundance of food, and sustainable environments for a continuously growing world population. In this age of globalization and interdependent economies, genetic engineering is unquestionably assuming an outstanding position. This chapter reviews the contributions of genetic engineering in agriculture, health and industry with special emphasis on practical applications. In agriculture the production of transgenic is discussed in detail. Some of the technical complexities of gene transfer in plants are discussed in detail. Genetic engineering in food technology is likely to be perceptions of benefit resulting from application of the technology also discussed. Genetic engineering (GE) refers to techniques used to transfer a single, specific gene from a donor to a host organism. Applied to foods, GE is used to enhance taste, nutrient content, shelf life, processing, yield, and production economy. the use of renewable raw materials (biomass) to replace raw materials derived from fossil fuels, and the use of biological systems such as cells or enzymes (used as reagents or catalysts) to replace conventional, non-biological methods.

REFERENCES

Akhter (2001) Genetically modified foods: health and safety issues. Annl. Saudi Med. 21 : 3-4.

Bevan, J. M. (1984) Binary *Agrobacterium* vectors for plant transformation. nuclic acids Res. 12: 8711-8721.

Biesgen, C., Hillebrandl, H. and Herbers, K. (2002) Technical enzymes produced in transgenic plants Phytochemistry Reviews 1: 79–85, 2002.

Christian, J. P. (2000) Genetic engineering in Agriculture who stands to benefit? J. Agril. Environ. Ethics 13 : 313–327.

Denis, J. M (2002) Biotechnology and the improvement of oil crops-genes, dreams and realities. Phytochem. Rev. 1: 67–77.

Dunwell, J.M. (2002) Future prospects for transgenic crops Phytochem. Rev. 1: 1–12.

Fralely, R.T., Grasser, C.S. (1984) Genetically engineering plants for crop improvement. Science 244 : 1293-1309.

Fresco L. O. (2001) Genetically Modified Organisms in Food and Agriculture: Where are we? Where are we going? Food and Agriculture Organization.

Herrera-Estrella L, De Block, M, Van Montagu, M.D. Schell, J. (1983) Chimeric genes as dominant selectable marker in plant cells. EMBO J. 2 : 987-995.

John, A.H. (2005) Commercialization of Biopharmaceutical and Bioindustrial Proteins from Plants. Crop Sci. 45 : 468-473.

Kao, K.N., Constable, F, Michayluk, M.R., Gamberg, O.L. (1974) Plant protoplast fusion and growth of intergenic hybrid cells. Planta 120 : 215-227.

Murphy, D. J. (2002) Biotechnology and the improvement of oil crops – genes, dreams and realities *Phytochemistry* Reviews 1: 67-77, 2002.

Nandini, K.K., Uyen, Q.H., Thorsteinsdóttir, H., Somsekhar, A., Daar, S. and Singer, P.A. (2004) Indian biotechnology-rapidly evolving and industry led. Nat. Biotechnol. 22 : 31-36.

Sasson, A. 1991. Production of useful biochemicals by higher-plant cell cultures: biotechnological and economic aspects *Options Méditerranéennes* - Série Séminaires - no 14: 59-74.

Stefan, S., Neil, E. and Rainer, F. (2002) Antibody molecular farming in plants and plant cells Phytochem. Rev. 1 : 45–54.

Tucker, G. (1996) Biotechnology and enzymes in the food industry. British Food J. 98 : 14.

Vasil, I.K (ED.) (2003). Plant Biotechnology 2002 and beyond, Kluwer Academic.

Verpoorte, R., Continl, A. and Memelink, J. (2002) Biotechnology for the production of plant secondary metabolites. Phytochem. Rev. 1: 13–25.

Suggested Readings

1. Balasubramanian, D.C.F.A., Dharmalingam, B.K., Green, J. and Jayaraman K. (2002) Concepts in Biotechnology, University Press, Hyderabad.
2. Dunwell J.M. (2002) Future prospects for transgenic crops Phytochemistry Reviews 1: 1–12.
3. Gregory Jaffe (2004) Regulating transgenic crops: a comparative analysis of different regulatory processes Transgenic Research 13: 5–19.
4. Stefan Schillberg, Neil Emans and Rainer Fischer (2002) Antibody molecular farming in plants and plant cells Phytochemistry Reviews 1: 45–54.
5. Vasil I.K. (Ed.) (2003) Plant Biotechnology 2002 and beyond, Kluwer Academic Publishers, 2003 pp 337-342.

REVIEW QUESTIONS

Long Questions

1. Why to regulate the transgenic crops?
2. What are the impacts of genetic engineering in agriculture, health and industry?
3. What in you view the greatest contribution of genetic engineering?
4. What is the future of genetic engineering?
5. How the application of genetic engineering transformed the life on earth?

Multiple Choice Questions

1. What is a DNA fingerprint?
   (a) Characteristic pattern of DNA deposited by human fingers on glasses, door knobs, pistols, etc.
   (b) Pattem of DNA fragments generated by a restriction endonuclease ("DNA Scissors").
   (c) Analysis procedure that can be used to identify the person who left behind a blood stain.
   (d) Analysis procedure based on subtle differences in DNA sequences among individuals.
2. What are some purposes of genetic engineering as practiced today?
   (a) To produce many copies of specific genes.
   (b) To change specific characteristics of plants and animals.
   (c) To make bacterial cells that synthesize large amounts of particularly useful proteins.
   (d) To facilitate the analysis of how genes work.

3. Which of these are true of the "DNA scissors" used by molecular biologists?
   (a) Most of them clip the DNA at random.
   (b) Their main use is for generating many copies of DNA from one or a few starting DNA molecules.
   (c) Their real function is for defense of bacteria against antibiotics.
   (d) They recognize short sequences of nucleotides that are palindromes (same sequence fonvard and backward).
4. What is a plasmid?
   (a) The contents of a eukaryotic cell.
   (b) A circular DNA molecule.
   (c) A "minichromosome" in plants.
   (d) A good "vector" for preparation of"designer genes".
5. Which must be done successfully to introduce a gene into a cell so that it can be copied indefinitely?
   (a) Somehow we have to obtain the specific DNA that we wish to introduce.
   (b) The gene of interest must be spliced together with the vector (i.e. carrier) DNA.
   (c) A DNA "fingerprint" must be prepared to prove that you have DNA and not RNA.
   (d) The hybrid DNA molecuie must be treated with a teratogen before it can be inserted into the host cell.
6. Proteomics is the study of
   (a) all proteins in an organism (b) gene-transcribing proteins
   (c) structural proteins of a cell (d) regulating proteins in humans
7. A plant useful in the study of plant genomes because of its small genome size is
   (a) Avena sativa (b) *Petunia porodii*
   (c) *Tulipa* spp. (d) *Arabidopsis thaliana*
8. The genetic engineering of crop plants can make them healthier and more beneficial to humans.
   (a) True (b) False
9. Using special enzymes, scientists have successfully removed the gene that controls the production of interferon and have inserted this gene into the DNA of certain bacteria. These bacteria can now produce interferon. This technique is known as
   (a) amniocentesis (b) Genetic engineering
   (c) Differentiation (d) karyotyping
10. In the synthesis of proteins, what is the function of messenger RNA molecules?
    (a) They act as a template for the synthesis of DNA.
    (b) They remove amino acids from the nucleus.
    (c) They carry information that determines the sequence of amino acids.
    (d) They carry specific enzymes for dehydration synthesis.

Answers

1. (b)	2. (b)	3. (a)	4. (b)	5. (b)	6. (b)
7. (d)	8. (b)	9. (c)	10. (c)		

PART–13

Tissue Culture and Liposome Technology

CHAPTER 41

Protoplast Culture and Somatic Hybridization

U.K. Tomar[1*] and P.K. Dantu[2]

[1]Arid Forest Research Institute, New Pali Road, Jodhpur–342 005

[2]Dept. of Botany, Dayalbagh Educational Institute, Dayalbagh, Agra–282 005

E-mail: uktomar@afri.res.in

In plants, where fairly distant species could be crossed, it has always not been possible to obtain full hybrids between desired individuals because of sexual incompatibility barriers. This has often proved to be a serious handicap in crop improvement programs through hybridization. With the ability of isolating protoplasts from plant cells using enzymes in the early 1960 by E. C. Cocking, the interest in genetic modification of somatic cells in higher plants has developed. Within a decade, Takebe *et al*. (1971) regenerated completes plants from leaf protoplasts of tabacco increased the potential of prtoplast culture techniques.

Not only isolated protoplasts but their fusion product, a somatic hybrid, can also be regenerated into whole plants. Plant protoplasts can also take up foreign DNA, through their naked plasma membrane, under specific chemical and physical treatments. Protoplasts also provide an experimental system for a wide range of biochemical and molecular studies ranging from investigations into the growth properties of individual cells to membrane transport.

Present chapter describes the methods of protoplast isolation, purification, culture techniques as also protoplast fusion techniques and somatic hybrid selection procedures.

PROTOPLAST ISOLATION

The term protoplast refers to the spherical plasmolysed content of plant cell enclosed by plasma membranes or naked cell without cell wall. Before going into the details about

*Corresponding author

protoplast culture, it is important to isolate as gently and as quickly as possible viable and uninjured protoplasts. The two methods for isolation of protoplasts are: mechanical and enzymatic method.

Mechanical Isolation

Klercker (1892) was the first to isolate protoplasts from plasmolysed cell of *Stratiates aloides*. These studies were later extended for protoplasts isolation from tissues of onion bulbs. Scales were immersed in 1.0 M sucrose until the protoplast shrunk away from their enclosing wall and then the plasmolysed tissue was cut with a sharp knife at such a thickness that only the cell walls are cut without damaging the protoplasts in strips. The protoplasts were released by osmotic swelling when these strips of the tissue are placed in low concentration sucrose solution (Fig. 1).

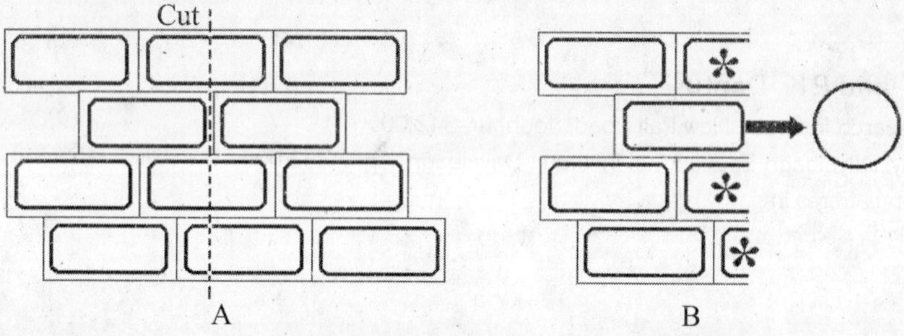

Fig. 1. When tissue is cut at the dotted lines (A) with a sharp razor blade, some cells release uncut complete protoplast and rest of the cells produced broken dead protoplasts as shown in Figure 1B marked with stars (*).

Enzymatic Isolation

This method involves the use of enzymes to dissolve the cell wall for releasing protoplasts. Initial studies were carried out on protoplasts isolation of yeast cell by digesting the cell wall using gastric juices obtained from the snail, *Halix poneatia*. However, credit of developing high yield protoplast isolation technique from higher plant protoplasts goes to Cocking (1960). He employed a crude cellulase preparation from the fungus *Myrothecium verrucaria* to dissolve cell wall and release the protoplasts from tomato roots. Later this method with suitable modifications and using purified enzymes has been extensively used by other group of workers. The enzymatic method could be used as a one step method (direct method), or as a two-step method (sequential method).

In the one step method, protoplasts are isolated directly from the tissue by using two enzymes, cellulase and pectinase, simultaneously. While, in the two-step method, cells are first isolated from callus or tissue by using pectinase and to this cell suspension cellulase is added to digest the cell wall and release protoplasts.

An enzymatic solution used for protoplast isolation contains the enzymes and a sugar as an osmoticum to prevent the plasma membrane from rupturing. Some salts and nutrients are also used as osmoticum. Generally 50 mM $CaCl_2$ is added to increase the stability of released protoplasts (Rose, 1980). The enzyme solution is filter sterilized through a 0.45µ filter membrane and sterilized tissue cut into small pieces is added. Incubation for some time releases protoplasts from the tissue.

PROTOPLAST PURIFICATION

The successful culture of protoplasts requires a pure population of intact and viable protoplasts at a high yield. So the protoplasts require to be purified by removing the undigested material (debris), burst protoplasts and enzymes.

Removal of Debris and Enzymes

Debris can be removed from the protoplast suspension by filtering the preparation through a steel or nylone mesh of 100µ pore size. Enzyme is removed by centrifuging the protoplast suspension at 600 rpm for 5 minutes. The protoplasts settle to the bottom of the centrifuge tube while the supernatant is removed with the help of a pipette. The protoplasts are then re-suspended in a washing medium containing an osmoticum only or osmoticum with nutrient medium or hydrated calcium chloride. The suspension is centrifuged again to settle the protoplasts and the washing medium is decanted. Traces of enzyme are removed by washing the protoplasts twice or thrice with the medium.

Removal of Broken Protoplasts

Intact protoplasts are separated from the broken debries by suspending the protoplast preparation in 20-40% sucrose solution and centrifugation at 350 rpm for three minutes. Intact protoplasts collect at the top of the sucrose solution and are carefully removed with a pipette (Gregory and Cocking, 1965; Power and Cocking, 1970; Evans *et al.*, 1972).

Schenk and Hilderbrandt (1971) used ficoll solution while Larkin (1976) used density buffer for purification of protoplast. In this method 0.5-3.0 volumes of crude protoplast preparation after filteration through sterile muslin cloth is layered on lymphoprep in the centrifuge tube and spun at 50-200 g for about 10 minutes. The protoplasts collect as a ring between the enzyme solution and lymphoprep and debris settle to the bottom (Fig. 2A). The protoplasts are removed from the interphase with a pasture pipette (Fig. 2B).

Protoplast Viability and Cell Wall Formation Tests

To a small volume of the protoplast suspension add equal volume of 0.1% calcoflour solution, incubate for 5 minutes and then observe under fluorescent microscope. The cell wall will fluoresce and protoplasts remain dark.

Protoplast Viability

Fluorescein diacetate (FDA) solution in acetone (5 mg/l) is added to protoplasts suspension to give a final concentration of 0.01%. After 5 minutes at room temperature the protoplasts are examined using fluorescent microscope. Only viable protoplasts can be seen.

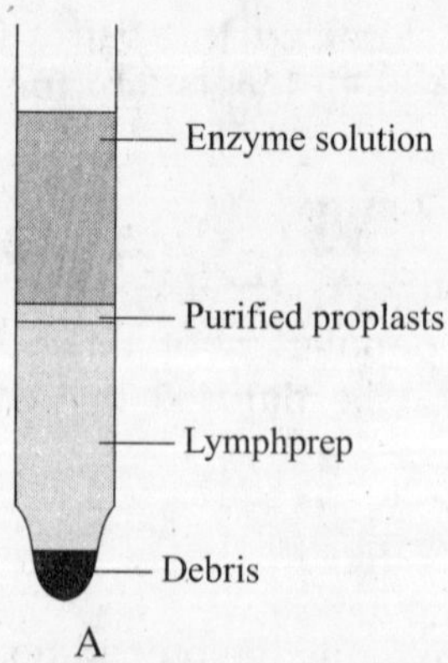

Fig. 2A. Purified Protoplast ring formation and debris get settled at the bottom

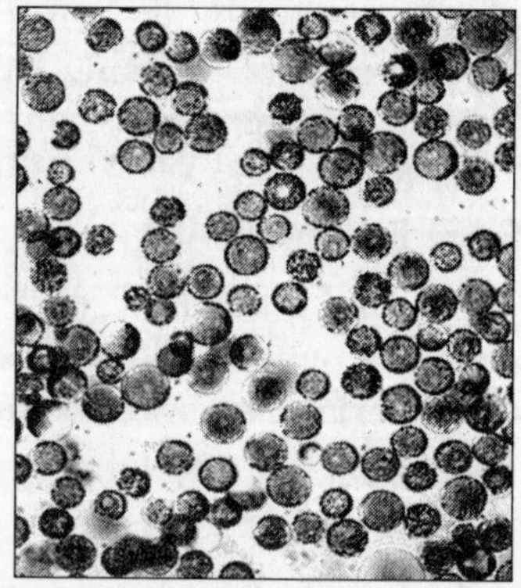

Fig. 2B. Purified protoplast population

FACTORS AFFECTING THE PROTOPLASTS ISOLATION AND ITS VIABILITY

To achieve a high yield of viable protoplasts, it is necessary to standardize the type and concentration of osmoticum (e.g., sucrose, mannitol, sorbitol, glucose and other sugars). Mannitol is the most commonly used osmoticum. The other condition to be standardized is that of enzymes (cellulase, pectinase, hemicellulase, meicellase, pectinol, or dresilase) whether they are to be used singly, in combination or sequentially with the osmoticum.As a thumb rule, low enzyme concentration at low temperature and high pH (5-8) for short incubation period proved to be better than longer incubation periods with high enzyme concentration, high temperature and low pH value.Though the ionic salts when used with osmoticum degrade the enzymes but increase the stability of protoplasts. Protoplasts isolated in presence of Ca^{++} or Mg^{++} ions showed a greater capacity for cell wall regeneration as compared to protoplasts isolated in the absence of these ions (Rose, 1980).

PROTOPLAST CULTURE

After viability test the protoplasts are cultured at a known density. Different methods have been used for culturing the isolated protoplasts.

Suspension Culture

In this method protoplasts are suspended in a liquid medium with a suitable concentration of osmoticum. Protoplasts at a density of 10^5/ml is generally plated on 25-50 ml of medium in an Erlenmeyer flask. The cultures are shaken slowly to provide sufficient aeration for growth. Sometimes shaking can cause bursting of protoplasts, so rpm of shaking required for a given species should be standardized. Evans and Cocking (1975) suggested that 2 ml suspension of protoplasts could be cultured in 25 ml flasks to facilitate aeration. Vasil (1976) suggested the addition of ficoll to the medium to keep protoplasts floating and thereby allowing better growth.

Hanging Drop Method

Kao and his group developed the hanging drop method in 1970 which was subsequently used by others (Bawa and Torrey, 1971). In this method, a suspension of protoplasts at a density of 10^4/ml to 10^5/ml is placed as 50 μl drops in plastic Petri plates, sealed with parafilm and incubated in an inverted position at 25-30°C under lw light intensity (100-500 lux) or even in dark. After cell wall regeneration and the initiation of cell division, fresh medium is added to make cell suspension. The small size of drop helps in providing enough aeration to protoplasts (Vasil, 1976).

Agar Plating Method

The method is almost similar to the one developed by Bergmann (1960) to grow callus cells of tobacco and beans and was first employed for the protoplast culture by Nagata and Takabe (1971). The major advantages of this technique are: (i) a large number of protocols can be handled simultaneously, and (ii) plating efficiency can be determined easily. At the same time the major drawback of this technique is that after regeneration of cell wall and induction of cell division, osmotic potential of the medium can not be altered by addition of fresh medium lacking osmotic stabilizer, since the initial medium is semi-solid. However, this problem can be solved by transferring blocks of agar (in which protoplasts are cultured initially) to a fresh medium with lesser concentration of osmoticum or altogether in its absence. This technique has been modified in different ways viz., as feeder technique to support division of cells plated at low density (Raveh et al., 1973), as micro vessel (Button, 1978), and as multiple drop arrays (Potrykus et al., 1979).

Micro Culture Technique

This technique developed by Jones et al. (1960) for culturing isolated cells was used by Vasil and Hilderbrandt (1965) to raise tobacco plants from isolated cells. Micro culture technique has been successfully used for culturing protoplasts of tobacco and *Petunia* (Durand et al., 1973). A drop of culture medium containing one or more protoplasts is put on a microscopic slide. On either sides of this microscopic slide are kept two cover slips. A third one is put over these two cover slips to shield the protoplasts suspensions.

Multidrop Array (MDA) Technique

This is a further improvement of hanging drop method and allows screening of large number of hormonal and nutritional factors (up to 4900) using small amount of plant material (Potrykus et al., 1979)

FACTORS AFFECTING PROTOPLASTS CULTURE

Successful growth and regeneration of protoplasts is dependent on a number of factors ranging from the status of the donor plant to the culture conditions.

Plant Species and Varieties

It is a well documented fact that even very small genetic difference leads to varying protoplast responses to culture conditions.

Plant Age and Organ

A second important factor for the successful culture is the age of the donor plant and developmental stage of the donor part used for isolation of the protoplasts. The stages with respect to their responses are germinating embryos, followed by seedling of one week, plantlets, leaves, juvenile tissue from mature plant and mature tissue. Though having the same genetic information protoplasts isolated from these parts behave differently in culture conditions. Plant regeneration has been most successful from leaf protoplasts of herbaceous species.

Pre Culture Conditions

Besides the factors mentioned above the behaviour of protoplasts in culture is highly influenced by climatic factors and pre culture conditions. Different culture conditions of seedling yields protoplasts having different responses when cultured.

Pretreatment to the Tissue, Before Isolation of Protoplasts

Pretreatments such as cold treatment, plasmolysis, and hormone treatment to the tissue increases the chances of recovery of viable protoplasts and their plating efficiency.

Nutritional Requirement for Protoplasts

Nutritional requirement for the growth of protoplasts differ from the nutritional requirement for tissues and cell culture. Protoplasts leak in cultures as they are devoid of cell walls, and moreover, they have a greater surface area for the diffusion of metabolites than the tissues. So the concentration of different metabolites in protoplasts is less than that in the tissues and cells. To compensate for these loses from protoplasts the number of metabolites provided in their culture media is more (Kao and Meichyaluk 1975).

Protoplasts Density

Protoplasts density is very crucial as it influences the plating efficiency and better surviving of protoplasts. When plated at higher density protoplasts compete with one another while at lower density losses of metabolites from protoplasts is more. The later situation of protoplast leakiness can be circumvented by addition of required metabolites to make medium isotonic. Alternatively, one can use nurse callus or feeder layer technique to provide essential organic molecules to protoplast for survival and their subsequent growth (Menczel et al., 1978).

PROTOPLAST FUSION OR SOMATIC HYBRIDIZATION

As plant cells have an inhibiting cell wall it is very difficult to fuse them. But isolated protoplasts were observed to fuse spontaneously because now the only barrier between the cytoplasm of two cells is the plasma membrane. After lot of refinements the techniques for protoplast fusion became important to produce hybrids from sexually incompatible species. Protoplast fusion is possible even between a plant cell and an animal cell and thus can have wider application. The process of somatic hybridization involves: (i) protoplast isolation, (ii) protoplast fusion, (iii) selection of somatic hybrids, and (iv) culture of somatic hybrids to regenerate complete plants. Different methods of protoplast fusion are described by Bengochea and Dodds, (1986). Three important methods are generally used for protoplast fusion: (a) high Ca^{++} and high pH, (b) polyethylene glycol (PEG), and (c) electric field. Literature survey indicates that the second and third methods are most commonly used due to better results. The three methods are discussed below.Sexual hybridization occurs when haploid cells generated in a previous meiosis fuse. The fusion of somatic diploid cells should generate a tetraploid fusion product provided that the nuclei fuse, too. If this is the case, it is spoken of a synkaryon. A fusion product where the nuclei stay separate is called a heterokaryon.

High Ca^{++} and High pH Induced Protoplast Fusion

Physical contact of two protoplasts is essential for their fusion. However, protoplasts do not fuse easily due to two main reasons: (i) they have a net negative charge on their membrane surfaces and force of repulsion works between protoplasts, and (ii) it is difficult to remove water from hydrophilic surfaces of protoplasts which also create a repulsive force between two protoplasts. It was observed that positively charged ions reduce the net negative charges of membranes reducing the repulsive force considerably. Keller and Melcher (1973) found calcium ions were suitable for such purposes and developed this method of protoplast fusion using high Ca^{++} ions in a high pH solution. Later the method was improved by Melcher and Labib (1974). The technique involves the following steps: Freshly isolated protoplasts of selected parents are mixed in a ratio of 1:1 with a final density of 2.5×10^5 protoplasts per ml. The protoplasts are collected as a pellet by centrifuging at 50 g for 3 to 5 minutes. The supernatant is removed and 2 ml of fusion mixture containing 50 mM $CaCl_2.2H_2O$, 50 mM Glycine-NaOH buffer and 400 mM Mannitol is added. pH of the mixture is adjusted to 10.5. Protoplasts are re-suspended in solution by gently shaking the centrifuge tubes.The protoplasts are collected by centrifuging at 50 g for 3 to 5 minutes. The centrifuge tubes with protoplasts are incubated in a water bath at 37°C for 10 to 30 minutes. The fusion mixture is replaced by washing medium (600 mM Mannitol and 50 mM $CaCl_2.2H_2O$) and protoplasts are left for 30 minutes.The protoplasts are washed twice with the washing medium.The protoplasts are re-suspended in culture medium. Protoplast fusion products can be observed under microscope. These protoplasts can be cultured on selection medium, which allows only somatic hybrids to grow.

PEG Induced Fusion

Polyethylene glycol (PEG) induced protoplast fusion was developed by Kao and Michayluk (1974). This technique is relatively more efficient than the previous one. Polyethylene glycol molecules have polarity like membrane phoshopholipid molecules and get attached with membrane proteins. When the attached PEG between two protoplasts is removed it results in breakdown of membranes at the contact points causing protoplasts to fuse with subsequent rejoining of plasmamembarnes of the adjacent protoplasts. The following steps are involved for protoplast fusion through polyethylene glycol. Mix the freshly isolated protoplasts (while still in enzyme solution) of selected parents in a ratio of 1:1. Pass the suspension through a 62 µm pore size filter and collect the filtrate in a centrifuge tube. Seal the mouth of the tube with screw cap. Centrifuge the filtrate at 50 g for 60 minutes to sediment the protoplasts. Remove the supernatant with pasture pipette. Wash the protoplasts with 10 ml of solution I (solution I: 500 mM Glucose, 0.7 mM $KH_2PO_4.H_2O$ and 3.5 mM $CaCl_2.2H_2O$ and pH 5.5). Remove and wash the protoplasts in solution I and make a suspension with 4-5% v/v protoplasts per ml. Put a 2-3 ml drop of silicon 200 fluid (100 cs) in a 60 x 15 mm sterile Petri dishes.Place a 22x22 mm cover slip on the drop. Pipette 150 µl of protoplast suspension onto the cover slip with a pasture pipette. Allow about 5 minutes for the protoplasts to settle on the cover slip forming thin layer. Add drop-by-drop 450 µl of PEG solution (50% PEG-1540, 10.5 mM $CaCl_2.2H_2O$, 0.7 mM $KH_2PO_4.H_2O$) to the protoplast suspension and observe the protoplast adhesion under inverted microscope. Inoculate the protoplast in PEG solution and keep at room temperature (24°C) for 10 to 20 minutes. Gently add 0.5 ml aliquots of solution II (50 mM Glycine, 50 mM $CaCl_2.2H_2O$, 300 mM Glucose, pH 9-10.5) at 10 minutes intervals. After another 10 minutes add 1 ml of protoplasts culture medium. Wash the protoplasts five times at five minutes intervals with 10 ml of the fresh protoplasts culture medium. At the end of each washing do not remove entire medium from the cover slip but leave behind a thin layer of the old medium adding fresh medium to it. If parent protoplasts are distinguishable visually, it may be possible to assess the frequency of heterokaryon formation at this stage. Culture the fused protoplasts together with the unfused protoplasts on the same cover slip in a thin layer of 500 µl of culture medium. Put additional 500-1000 µl medium in the front of droplets around the cover slip to maintain humidity inside the Petri plates.

Electric Field Induced Fusion

In this method, developed by U. Zimmerman, protoplasts are placed in an electric field and are exposed to high intensity electric pulse for a short duration (nano to micro second). This exposure to electric field reversibly increases permeability of cell membrane. Local electrical charge break down occurs in the plasma membranes resulting in fusion of adjacent protoplasts. The original properties of membrane are restored within micro second to minutes depending on the experimental conditions and membrane properties (Zimmerman and Scheurich, 1981; Zimmermann and Vienken, 1982). An instrument designed by U. Zimmerman is used for protoplast fusion and for genetic transformation through electroporation. The process of electric field induced protoplast fusion can be explained in the following five steps:

Dielectrophoresis
Mutual dielectrophorasis
Membrane contact
Electric break down of membranes, and
Protoplast Fusion

(i) Dielectrophoresis: Close membrane contact of two protoplasts is one of the prerequisite for protoplast fusion. This step is achieved by dielectrophoresis. Dielectrophoresis is concerned with motion of neutral particles in a non uniform electric field. Neutral particles become polarized in the presence of electric field with negative side towards anode and positive charge of the same magnitude on the opposite side, i.e. towards cathode.In uniform electric field the field strength is equal on both the sides. And in such a field these induced bipolar particles will

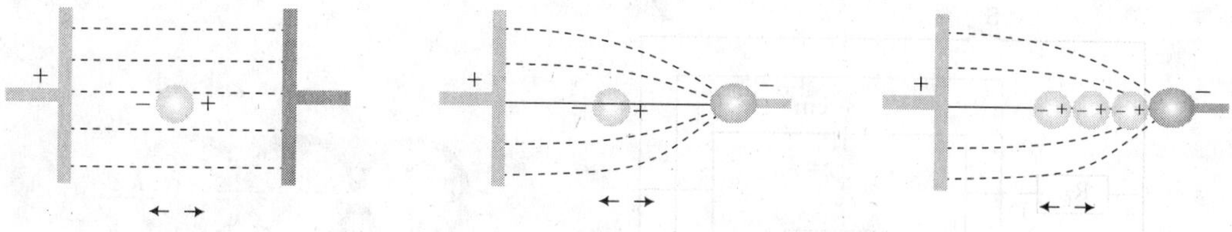

A. Uniform field developed equal force on both side of charged protoplasts B. Unequal electrical field developed net force towards higher field strengthC. Non uniform electric field thus forms pearl chain of protoplasts

Fig. 3. Movement of protoplast in non uniform electric field towards higher strength electric field and forms protoplast chain.

not move to either side due to equal electric force on both sides (Fig. 3A). However, under non uniform field (field strength on the two sides of particles is unequal) a net force acting upon the particle results in linear motion towards the region of highest field intensity (Fig. 3B). This is termed as dielectrophoresis.

(ii) Mutual Dielectrophoresis: One protoplast approaching another polarized protoplast during the movement towards the region of highest field intensity, will encounter an enhancement of local field divergence and will tend to move towards that protoplast since the field strength is higher near that cell. As a result protoplasts in non uniform field form chain like aggregates (so called pearl chain) with point to point membrane contact (Fig. 3C). This effect is termed as mutual dielectrophoresis.

The attraction forces arising from dipole generation within protoplasts overcome both the electrostatic repulsive forces between membrane surfaces of neighboring protoplasts bearing net negative charges and repulsive hydration forces. Repulsive hydration force is assumed to be a consequence of work required to remove water from hydrophilic surface as the protoplasts approach one another. Dielectrophoresis and pearl chain formation usually have to be performed in virtually non conductive solution (conductivity less than 10^{-4} cm^{-1}).

(iii) Electric Break Down: Reversible electric breakdown in the zone of membrane contact is the primary process responsible for the initiation of fusion (Fig. 4). Phospholipids are arranged in a planer bi-layer into which peripheral or integral structural carrier proteins are embedded in mosaic like fashion. The lateral fluidity of phospholipids is very high, while the movement vertical to the membrane surface is severely limited. A flip-flop movement of lipids in this direction is, therefore, highly difficult.

In an equivalent electrical circuit the membranes can be represented by a plate capacitor of specific capacitance (Cm) and resistance (Rm) connected in parallel (Fig. 4A). The aqueous external solution and polar heads of liquids represent the plate of capacitor, while membrane interior forms a dielectric with relative dielectric constant of 2 to 3. The resistance of aqueous external solution R_E is in series with the membrane. The specific resistance of cell membrane is in the order of 10^2 to 10^4 Ωcm^{-2}, while that of specific membrane bilayer is three to four orders of magnitude higher. The specific capacitances of artificial and biological membranes on the other hand are comparable of having values of 0.3 to 0.7 μFcm^{-2} and 1 μFcm^{-2}, respectively.

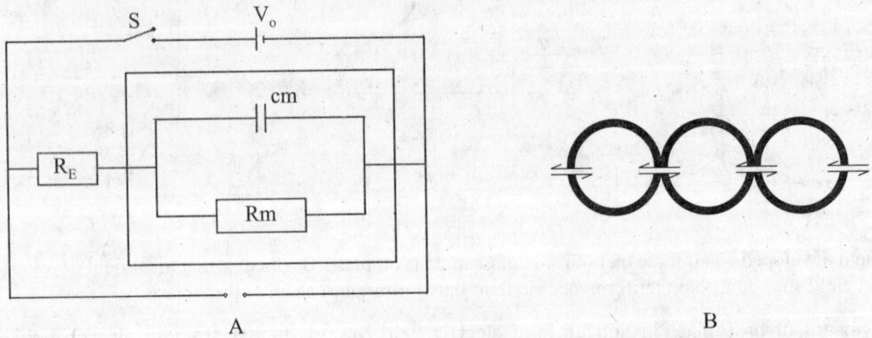

Fig. 4. (A) Diagrammatic representation of Charge pulse technique, (B) Electric break down occurs at the poles of the cells and the zone of contact between two protoplasts.

It is known that capacitors can only be charged to a certain maximum voltage. Above this voltage level (which is dependent on separation of plates, common area between plates and dielectric material of plates) electrical (dielectric) breakdown is observed in the capacitor. Electrical breakdown is associated with an extreme increase in the electrical conductivity of the capacitor, which is usually irreversible i.e. the capacitor is destroyed. In the self-regenerating capacitors on the other hand the original resistance and capacitance is restored under certain experimental conditions. Biological membranes and artificial lipid bilayer behave in an analogous way to the electric breakdown of the self-regenerating capacitors described above. The reversible electrical breakdown of protoplast membranes leads to perturbation of membrane structure, which permits an exchange of materials between the protoplast and its environments. In figure 4B it is shown that the electric field effects the protoplast adhering to each other in the orientation to the field direction only. In this case, electrical breakdown occurs at the poles of the cells and zone of contact between two protoplasts.

Merits of Electrically Induced Protoplast Fusion

1. The fusion process between any two protoplasts of different species in a mixed population can be followed under microscope. This would be of particular interest when producing hybridoma cells.
2. Fusion process is synchronous and extends over a short duration. So the hybrids do not lose the viability.
3. Viability can also be affected by fusogenic compounds, which are thus able to interact with total membrane surface in an uncontrolled manner. In this method there is no use of such fusogenic compounds.
4. Yield of somatic hybrids is very high.
5. During the fusion, loss of intercellular substances is generally very low.

7. SELECTION OF SOMATIC HYBRIDS

Successful somatic hybridization requires the induction of protoplast fusion resulting in the formation of heterokaryocytes and somatic hybrid, their survival, selection and recovery of hybrids, subsequent division of fusion products and morphogenesis giving rise to hybrid plants.

Different methods have been developed for the selection of somatic hybrids. These methods can broadly be divided as follows:

Group 1 Selection of somatic hybrids by culturing them on such a medium on which only somatic hybrid protoplast can grow (Power et al., 1977).

Group 2 Complementary selection of somatic hybrids on specific culture medium.

Group 3 Mechanical isolation by visual means and knowledge of identification of somatic hybrids.

Group 4 Morphology of the plant after regeneration.

Group 1 - Selection Medium: Smith et al. (1976) used a model system involving the knowledge from sexual interspecific *Nicotiana* hybrids that can be cultured and selected on media that fails to support parental species. Chupeau et al. (1978) were the first to demonstrate the successful use of this technique for selecting somatic hybrids. Fusion of protoplasts from *Nicotiana longsdorfii* and *N. gluca* were induced using PEG. Parasexual hybrid colonies were selected for their ability to grow without growth substances (Fig. 5).

Similarly protoplasts of sexual hybrid plants of *Petunia parodii* and *P. hybrida* can grow and regenerate complete plants on specific medium containing Actinomycin D, where as protoplasts of parents can grow and form colonies on this medium but can not regenerate complete plants. Power et al. (1977) successfully used this medium for selecting somatic hybrids.

Group 2 - Complementary selection: This method developed by Melcher and Labib (1974) utilizes biochemical markers such as albino and chlorophyll deficient mutants (Fig. 6). Somatic hybrids can be selected on the basis of complementation of characters expressed by the hybrids, which are not present in any of the parents. Sometimes, hybrid is selected on the basis of characters of parent protoplasts present in somatic hybrid.

Melcher and Labib (1974) took two mutants of *Nicotiana tabacum*, one of which was sublethal, that did not grow more than few inches (2-3 inches) in low light intensity (800 lux), but at higher light intensities (10,000 lux) was able to grow into complete plants. The other

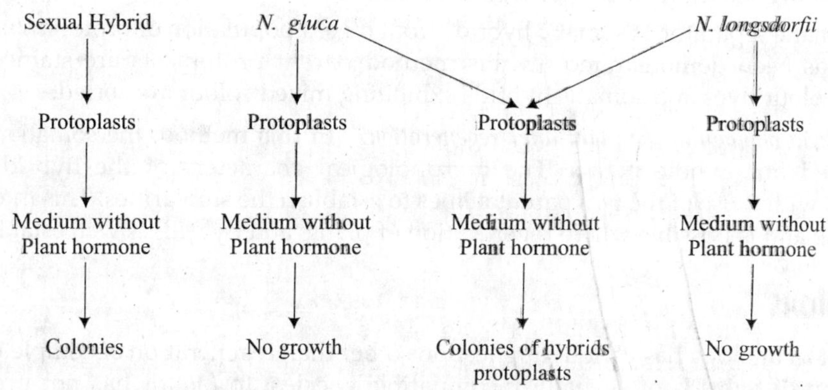

Fig. 5. Flow chart showing selection of hybrids using selection medium on which only hybrid can grow.

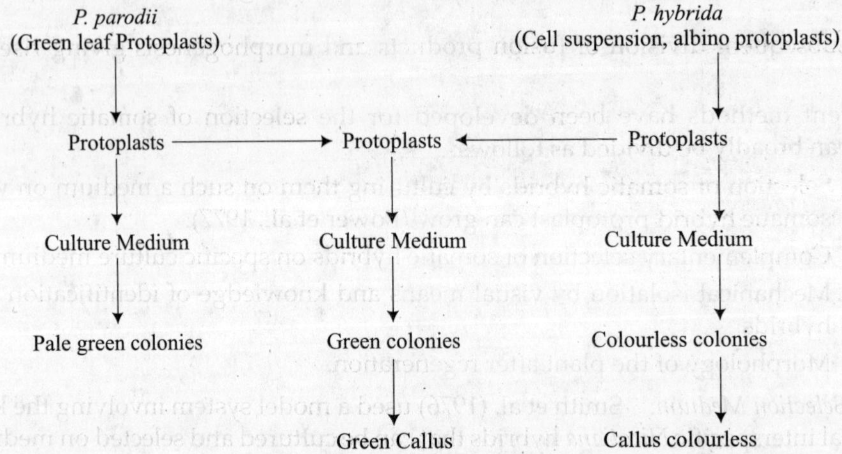

Fig. 6. Flow Chart showing complementary selection of somatic hybrids on specific culture medium.

mutant was chlorophyll deficient that could not grow more than few inches. When protoplasts of these two mutants were fused the hybrid produced green colonies in culture and later on regenerated into full plants in low light intensity (800 lux).

Gleba et al. (1975) developed a similar complementation selection procedure that involves complementation between plastome and genome chlorophyll mutant and does not utilize light sensitivity as a selection marker. Cocking et al. (1977) used albino protoplasts of *P. hybrida* and green leaf mesophyll protoplasts of *P. parodii* (Fig. 6). These somatic hybrids included strict amphidiploids.

Group 3 - Mechanical Isolation: In this method heterokaryons are selected and isolated visually under microscope. This method is particularly useful for those interspecies fusions where the post zygotic incompatibility prevents sexual hybridization. In such cases of interspecies fusion chromosome elimination may result in failure to select the somatic hybrids, while other interspecies fusion may result in somatic hybrid protoplasts with no evidence of specific chromosome elimination. In these instances the stable conditions may fail to be maintained on organogenesis. Through this selection approach such barriers to whole somatic hybrid plant formation can be identified and perhaps overcome.

Automatic separator of somatic hybrids from mixed population of protoplasts after fusion process has been demonstrated. In this method parent protoplasts are stained with two different colour dyes and somatic hybrids exhibiting mixed colour are sorted.

Group 4 - Morphology of the plant after regeneration: In this method, the somatic hybrids are regenerated into whole plants. The morphological characters of the hybrid plants are compared with those of the two parental lines to establish the similarities. This method is time consuming and is possible where regeneration of plants and hybrids is well established.

CONCLUSIONS

Protoplast technology has various applications other than regeneration of complete plants and production of hybrids of sexually incompatible species. In fact, it has not produced any wonderful research results so far as it has promised in early 1960s. There is not even a single successful production of agronomically useful hybrid plant through this technology. Nevertheless, these techniques have been instrumental in generating basic scientific information on cell biology, plant incompatibility, membrane functions, cell organelle studies and cell wall regeneration. Ultrastructure and molecular architecture of plant cells. These techniques are now being used for transfer of cytoplasmic male sterility. Protoplast can take up macromolecules (nucleic acids and proteins), viruses, cell components like chromosomes and chloroplasts by phagocytosis.

REFERENCES

Bawa, S.B. and Torrey, J.G. (1971) "Budding" and Nuclear Division in Cultured Protoplasts of Corn, Convolvulus, and Onion. *The Bot. Gaz.* 132: 240-245.

Bengochea, T. and Dodds, J.H. (1986). Plant Protoplasts: A biotechnological Tool for Crop Improvement.

Bergmann, L. (1960) Growth and division of single cells of higher plants in vitro. *Journal of General Physiology* 43: 841–851.

Button, J. (1978) The effect of some corbohydrate on growth and organization of Citrus ovular callus. *Z. Pflanzenphysiol.* 88: 61-68.

Carlson, P. S., Smith, H. H. and Dearing, R. D. (1972) Parasexual interspecific plant hybridization. *Proc. Nat. Acad. Sci.* US 69: 2292-2294.

Chupeau, Y.; Missonier, C.; Hommel, M-C; Goujaud, J. (1978): Somatic hybrids of plants by fusion of protoplasts. *Molecular and General Genetics* 105 (3): 239-45.

Cocking E.C. (1960) A method for the isolation of plant protoplasts and vacuoles. *Nature* 187: 962-963.

Cocking, E. C ; George, D.; Price-Jones, J. J.; Power, J. B. (1977) Selection procedures for the production of inter-species somatic hybrids of *Petunia hybrida* and *P. parodii.* II Albino complementation selection. *Plant Science Letters* 10:7-12.

Durand, J. Potrykaus, I. and Donn G. (1973) Plantes isuues protoplastes de Petunia. *Z. Pflanzenphysiol.* 69: 26-34.

Evans P. K. and Cocking E. C. (1975) The techniques of plant cell culture and somatic cell hybridisation.. In: *New Techniques in Biophysics and Cell Biology* (R. H. Pain and B. J. Smith, Eds.), pp 127-158, John Wiley and Sons Ltd.

Evans, P. K., Keates, A. G. and Cocking, E. C. (1972) Isolation of protoplasts from cereal leaves. *Planta* 104: 178-191.

Gleba, Y. Y. (1979) Nonchromosomal Inheritance in Higher Plants as Studied by Somatic Hybridisation. In: *Proceedings Fourth Annual College of Biological Sciences Colloquium, "Plant Cell and Tissue Culture-Principles and Applications"*, Pp.775-788. Ohio State Univ. Press, Columbus.

Gregory, D. W., and Cocking, E. C. (1965) The large-scale isolation of protoplasts from immature tomato fruits. *J. Cell Biol.* 24:143-146.

Jones, L. Hildebrandt, A. Riker, A. Wu, J. (1960) Growth of somatic tobacco cells in microculture. *Am. J. Bot.* 47: 468-475.

Kao, K. N. (1976) A method for fusion of plant protoplasts with polyethylene glycol. In: *Cell Genetics in Higher Plants* (D. Duddits et al. Eds.). pp.233-237.Akad. Kiado, Budapest.

Kao, K.N. Keller, W.A. and Miller, R.A. (1970). Cell divisions in newly formed cells from protoplasts of soybean. *Exp. Cell Res.* 62: 338-340.

Kao, K. N. and Michayluk, M. R. (1974) A method for high-frequency intergeneric fusion of plant protoplasts. *Planta* 115:355-67.

Kao, K.N., Constabel, F., Michayluk, M.K. and Gamborg, O.L. (1974) Plant protoplast fusion and growth of intergeneric hybrid cells. *Planta* 120:215-27, 1974.

Keller, W.A. and Melchers, G. (1973) The effect of high pH and calcium on tobacco leaf protoplast fusion. *Z. Naturforsch.* 28, 737-741.

Klercker, J. von (1892) Eine Methode zur isolier lebender protoplasten. *Ofvers Vetensk Akad Forhandl* 49: 463-474.

Larkin, P.J. (1976). Purification and viability determinations of plant protoplasts. *Planta* 128, 213–216.

Menczel, L., Lazar, G. and Maliga, P. (1978) Isolation of somatic hybrids by cloning *Nicotiana* heterokaryons in nurse cultures. *Planta* 143: 29-32.

Melchers, G. and Labia, G. (1975) Somatic hybridization of plants by fusion of protoplasts. Selection of light resistant hybrids of "haploid" light sensitive varieties of tobacco. Molec. Gen. Genet. 135: 277-294.

Nagata, T. and Takebe, I. (1971) Plating of isolated tabacco mesophyll protoplasts on agar. Planta Vol. 99: 12-20.

Potrykus, I. Harms, C.T. and Lorz, H. (1979) Multiple-drop-array (MDA) technique for the large scale testing of culture media variations in hanging microdrop cultures of single cell systems. I. The technique. Plant Sci. Lett. Vol. 14, 231-235.

Power, J. B. and Cocking, E. C. (1971) Fusion of plant protoplasts. *Science Progress (Oxford) 59:* 181-98.

Power, J. B., Berry, S. F., Frearson, E. M. and Cocking, E. C. (1977) Selection procedures for the production of inter-species somatic hybrids of *Petunia hybrida* and *P. parodii.* I Nutrient media and drug sensitivitycomplementation selection. *Plant Science Letters* 10: 1-6.

Raveh, D., Huberman, E. and Galun, E. (1973) *In vitro* culture of tobacco protoplasts. Use of feeder techniques to support division of cells plated at low densities. *In Vitro* 9: 216-22.

Raveh, D. Huberman, E. and Galun, E. (1973) *In vitro* culture of tobacco protoplasts. Use of feeder techniques to support divison of cells plated at low densities. *in vitro* 9: 216-22.

Rose, R. J. (1980). Factors that influence the yield, stability in culture and cell wall regeneration of. spinach mesophyll protoplasts. *Aust. J. Plant Physiol.* 7: 71-725.

Schenk, R. U., and Hildebrandt, A. C. (1969) Production of protoplasts from plant cells in liquid culture using purified commercial cellulases. *Crop. Sci.* 9:629-631.

Smith, H.H. Kao, K.N. and Combatti, N.C. (1976) Interspecific Hybridization by Protoplast Fusion in *Nicotiana.* The J. Heredity 67: 123-128. 1976.

Takebe, I., Labib, G. and Melchers, G. (1971) Regeneration of whole plants from isolated mesophyll protoplasts of tobacco. *Naturwissenschaften* 58: 318–320.

Vasil, I. K. (1976) The progress, problem, and prospects of plant protoplast research. *Adv. Agron.,* 28: 119-160.

Vasil, V. and Hildebrandt, A. C. (1965) Differentiation of Tobacco Plants from Single, Isolated Cells in Microcultures. Science Vol. 150: 889-892.

Zimmerman, U. (1982) Electric field-mediated fusion and related electrical phenomena. *Biochimica et Biophysica Acta* 694: 227–277.

Zimmerman, U. and Scheurich, P. (1981) High frequency fusion of plant protoplasts by electric fields. *Planta*151: 26-32.

Zimmermann, U. and Vienken, J. (1982) Electric field-induced cell-to-cell fusion. *J Membr Biol* 67: 165-182.

Suggested Readings

1. Bengochea, T. and Dodds, J.H. (1986). Plant Protoplasts: A biotechnological Tool for Crop Improvement. Chapman & Hall.

2. Dodds, D.J. and Bengochea, T. (1987) Plant Protoplasts: A Biotechnological Tool for Plant Improvement, Chapman & Hall.
3. Davey, M.R. and Kumar, A. (1983) Higher plant protoplasts: retrospect and prospect. In: *Plant Protoplasts. International Review of Cytology,* (Giles, K.L. Ed.) Supp. 16, 219-263. Academic Press, New York.

REVIEW QUESTIONS

Long Questions

1. Write a note on five important applications of protoplast techniques.
2. Write short note on protoplast isolation techniques and which one is more efficient?
3. Write a note on cell wall and protoplast viability test.
4. What are the methods of protoplast fusion? Describe any one method in detail.
5. Describe three methods of somatic hybrid selection.

Multiple Choice Questions

1. Who discovered enzymatic method of protoplast isolation?
   (a) Klercker (b) E.C. Cocking
   (c) I. K. Vasii (d) U. Zimmermann
2. Which enzyme is used to dissolve middle lamella in plant cells?
   (a) Cellulase (b) Pactinase
   (c) Both of above (d) Non of above
3. Crude cellulase enzyme was isolated from which organism?
   a) Bacteria (b) Snail
   (c) Fungus (d) Algae
4 Which chemical is used in cell wall test?
   (a) Saffranin (b) Calcoflour
   (c) Fluorescence diacetate (d) Non of the above
5 Which chemical is used in protoplast viability test?
   (a) Analine blue (b) Calcoflour
   (c) Fluorescence diacetate (d) Saffranin
6 High Ca^+ ions and high pH is used for protoplast fusion to
   (a) Increase positive charge (b) Increase membrane fluidity
   (c) Make protoplast neutral (d) All of above
7. What effect on protoplast can be observed under an electric field
   (a) Bursting (b) Shrinking
   (c) Dipole formation (d) Non of the above
8. Regeneration of completed plants through protoplasts was first reported in
   (a) Datura (b) Petunia
   (c) *Nicociana tabcum* (d) *Daccus carrota*

9. First successful regeneration of complete plants from somatic hybrid have been reported by
   (a) Takebe, I. et al
   (b) E.C. Cocking,
   (c) I.K. Vasil
   (d) J.B. Power
10. Who developed electric induced protoplast fusion technique?
   (a) P. S. Carlson
   (b) E.C. Cocking
   (c) I. K. Vasil
   (d) U. Zimmermann

Answers

1. (b) 2. (b) 3. (c) 4. (b) 5. (c) 6. (d)
7. (c) 8. (c) 9. (a) 10. (d)

CHAPTER 42

Somatic Embryogenesis

P. K. Dantu[1] and U. K. Tomar[2*]

[1]Dept. of Botany, Dayalbagh Educational Institute, Dayalbagh, Agra–282 005

[2]Arid Forest Research Institute, New Pali Road, Jodhpur–342 005

E-mail: uktomar@afri.res.in

The process of development of an embryo from zygote is termed as embryogenesis and fertilization of the egg cell with sperm is a prerequisite for this process. Additionally, in some plants such as, *Citrus* and *Mangifera* embryos may be formed by nucellar cells outside the embryo sac. Such embryos are termed adventive embryos. However, no instance of ex-ovulo embryo development is known in nature (Bhojwani and Bhatnagar, 1990). Plants are unique in that each of its somatic cell can exhibit totipotency when given proper in vitro conditions. The capability of a somatic cell to reinitiate an entire ontogenic program to form embryos is called somatic embryogenesis. It, thus, leads to the formation of a bipolar structure with a root/shoot axis and a closed vascular system. These somatic embryos formed whether in dicots or monocots trace all the stages of their respective zygotic embryogenesis.

It was the German Botanist, Haberlandt, who prophesied in the early 1900s that vegetative cells of plants can be induced to form embryos. The earliest report of Somatic Embryo (SE) formation in cultured cells of *Daucus carota* is generally credited to Reinert (1958) and Steward *et al.* (1958). Since then, there has been an explosion in the number of species in which somatic embryo formation is reported. Somatic embryogenesis offers a great opportunity for clonal propagation of a given genotype on a large scale. This becomes all the more important for transgenics. The whole process of somatic embryo formation can give us better insights into the cellular levels of regulation underlying the flexible developmental strategy of plants, which rely on coordinated responses to complex environmental and endogenous signals. It also provides an excellent *in vitro* experimental system to study molecular, biochemical and physiological aspects of plant embryo development.

*Corresponding author

Ever since the first report, carrot has been widely used to investigate various aspects of somatic embryogenesis. Other plants in which this phenomenon has been studied in some details are *Citrus* (Kochba and Spiegel-Roy, 1977; Tisserat and Murashige, 1977; Gavish et al., 1991, 1992), *Coffea* sp. (Nakamura et al., 1992), *Macleaya cordata* (Kohlenbach, 1977), *Medicago* sp. (McKersie and Bowley 1993), *Ranunculus sceleratus* (Konar and Nataraja, 1969; Konar et al., 1972), and *Zea mays* (Emons and Kieft, 1991). Embryos formed in cultures have been variously designated as accessory embryos, adventive embryos, embryoids, and supernumerary embryos. In this chapter, the embryos formed in cultures have been referred to as somatic embryos (SE).

FACTORS AFFECTING SOMATIC EMBRYOGENESIS

In vitro somatic embryogenesis can be successfully achieved through judicious selection or manipulation of the explant, medium, growth regulators, and physical culture environment. Successful protocols of somatic embryogenesis need to fulfill the following four steps: induction of embryogenesis, embryo development, embryo maturation, and conversion. The first two steps are concerned with inducing changes in the cells to orient them for forming and development of embryos. The third step of embryo maturation prepares the embryo for germination and the fourth step of conversion refers to the ex-vitro survival of somatic seedlings and their continued growth.

Explant

The choice of explant for the production of SE is generally limited to immature or less differentiated parts, although success has been achieved with fully differentiated explants as well (mesophyll cells of *Macleaya cordata*). In majority of cases zygotic embryo in the early stages of development has been used to initiate cultures, as its cells already possess embryogenic competence and are termed PEDCs (pre-embryogenic determined cells). Quite often hypocotyls and leaf explants have also been used as explants. Other more differentiated explants have cells that must be induced to become embryogenic and are termed IEDCs (induced embryogenic determined cells). Success in obtaining somatic embryos from previously recalcitrant species has largely been due to a shift in emphasis from media manipulation to explant selection.

Genotype

It is now clear that even though the explant and media used are same genotypes of a species respond differently. Striking examples are found in rice, alfalfa, groundnut, soya bean and maize. Genotypic variations could be due to varying endogenous levels of growth regulators. The endogenous levels of auxin were 16-20% less and that of cytokinin was 10-15% less in the ovules of highly embryogenic lines of maize than in the non-embryogenic and poorly embryogenic lines.

It has been shown that somatic embryogenesis/caulogenesis in alfalfa is genetically controlled. Inheritance studies have clearly demonstrated that the capacity to regenerate plants in alfalfa is controlled by two dominant genes. Crossing and recurrent selection from two poorly responding parents 'Du Puits' with 10% regeneration and 'Sarnac' with 14%

regeneration produced a highly regenerating tetraploid line 'Regan-s' with 67% regeneration. And, through chromosome manipulation and breeding it has been possible to select a diploid line of alfalfa 'HG2' having a regeneration ability of 96%.

Growth Regulators

Auxin. Most of the studies on somatic embryogenesis show a similar pattern in the induction and development of somatic embryos. For induction of embryogenic clumps presence of an auxin is necessary in the initiation medium. These clumps on transfer to an auxin-free or a very low auxin medium will develop somatic embryos. The most commonly used auxin for the induction of somatic embryogenesis is the synthetic auxin 2, 4-D, also used as a herbicide. However, instances are known where embryogenesis was obtained without the aid of growth regulators. In one of the cases high sucrose concentration was used and in another instance pH of the medium was manipulated to get embryogenic cultures.

Generally, the embryogenic cultures of carrot are initiated and multiplied in a medium, termed as proliferation medium, containing 2,4-D in the range of 0.5-1.0 mg l^{-1}. On this medium callus differentiates localized groups of meristematic cells called 'proembryogenic masses' (PEMs). In repeated subcultures on the proliferation medium the embryogenic callus continues to multiply without the appearance of embryos. These PEMs develop into embryos only if transferred to a medium, referred to as embryo development medium, from which either auxin has been removed or very low level of auxin (0.01-0.1 mg 1^{-1}) is present. The presence of an auxin in the proliferation medium seems essential to orient the callus to develop embryos in the later embryo development medium. Therefore, the proliferation medium is regarded as the 'induction medium' for somatic embryogenesis and each PEM an unorganized embryo (Kohlenbach, 1978). Tissues maintained continuously either in auxin-containing or auxin-free medium would not form embryos.

Among the large number of auxin tested, 2, 4-D is by far the most effective for producing embryogenic cultures even in major crop species such as cereals and legumes, although exceptions are there. The other auxin, though used less frequently, that have been shown to promote somatic embryogenesis are dicamba (3,6-dichloro-o-anisic acid) in orchardgrass (Gray et al., 1993), 2-(2,4- Dichlorophenoxy) propionic acid in alfalfa, picloram in *Arachis hypogaea,* and IAA and NAA in *Podophylum hexandrum*.

A balance of endogenous and exogenous levels of auxin is required for the development of SEs. This was clearly demonstrated in the habituated cultures of *Citrus* which after continuous growth for two years in the induction medium containing IAA failed to produce somatic embryos in the presence of even very low concentration of IAA (0.00 1 mgl^{-1}). But when auxin synthesis inhibitors such as 2-hydroxy-5-nitro-benzyl bromide or 7-aza-indole were used in the medium, somatic embryogenesis could again be observed (Kochba and Spiegel-Roy, 1977).

It has been suggested that the 2, 4-D induced suppression of embryo development may be because of endogenous ethylene production. In suspension cultures of carrot 2-Chlorophosphonic acid (Ethephon), which releases ethylene in plant tissues, also suppresses the development of mature SEs without an appreciable reduction in the growth and multiplication of the PEMs. Moreover, it has been shown that the auxin-grown tissue cultures of carrot produce more ethylene than auxin-free cultures. High ethylene content would result in enhanced activity of cellulase or pectinase or both, causing breakdown of the clumps before polarity is established in the proembryos for further organized development. Thus, in 2,4-D medium tissue multiplication goes on but mature embryos do not appear.

Cytokinin. Cytokinin, such as BAP is reported to inhibit embryogenic potential but promote cell division in carrot cultures. This may be due to selective multiplication of non-embryogenic components of the cultures. However, in carrot itself zeatin at a concentration of 0.1 μM promoted the process, especially when added after 3-4 days of transfer from the proliferation medium. Similarly, for coffee and alfalfa, a cytokinin was found to be necessary. Zygotic embryos of *Podophyllum hexandrum* could be induced to form globular embryos in a medium containing 2 μM BAP and 0.5 μM IAA.

Others. Other growth regulators may be needed during embryo development. In caraway a balance between ABA, zeatin and gibberellic was required (Ammirato, 1974) and 2ip and GA_3 was needed for grapevine. In carrot exogenous GA_3 decreased the number of heart and torpedo stage embryos, thus reducing the number of somatic embryos formed (Fujimura and Komanine, 1975).

Medium

In 70% of the reports of successful cases of somatic embryogenesis the basal medium used was that of Murashige and Skoog's (1962) salt formulation or its modification. Very rarely White's or SH basal medium have been used albeit with suitable supplements. The most commonly used carbon source is although sucrose, in scarlet runner bean glucose was found to be superior for embryogenic cell induction. Similarly, for citrus nucellus cultures galactose and lactose were found to be superior (Kohlenbach, 1977). It has been found that sucrose and auxin concentrations interact with each other, with the optimal concentration of one being dependent on the concentration of the other.

The form of nitrogen in the medium significantly affects in vitro embryogenesis. Carrot cultures initiated in a medium containing KNO_3 as the sole source of nitrogen failed to form somatic embryos when transferred to auxin-free medium. However, addition of a reduced nitrogen source as NH_4Cl (5 mM) in the presence of 55 mM KNO_3 allowed somatic embryo development. Similarly, for alfalfa requirement of ammonium was absolute for induction and differentiation of somatic embryos. In general, it appears that at least 5-12.5 mM NH_4^+ in the medium is required for the induction of somatic embryogenesis.

Other nitrogen sources such as casein hydrolysate (CH) and amino acids have also been found to increase somatic embryo formation. Suspension cultures of orchardgrass showed rapid development of somatic embryos only when CH ($3gl^{-1}$) was added to the established medium and in the absence of which the cultures differentiated only root primordia. Substitution of CH with a combination of proline and serine/threonine was superior with respect to quality of SEs formed. These somatic embryos showed higher percentage of conversion (98% versus 69%), considerably less incidence of precocious germination (14% versus 46%) and had a smoother epidermis as against the irregular surface of the embryos developed in the presence of CH. In alfalfa also, somatic embryo production dramatically increased by 100 times in the presence of amino acids such as proline, alanine, arginine and glutamine. The importance of nitrogen requirement during embryogenesis may be a reflection of continued synthesis of protein, nucleic acids and reserve substances and may also be in maintenance of an appropriate pH.

Selective Subculture

Multicellular explants are generally heterogeneous in terms of the morphogenic potential of its constituent cells. Only a small proportion of these cells are able to express their cellular totipotency under a set of culture conditions. Therefore, the calli derived from such explants are also heterogeneous. Sometimes the embryogenic portions of the callus are distinct from the non-embryogenic tissue on the basis of their morphological appearance and it is essential to make artistic subcultures to establish regenerating tissue cultures.

Callus formed in cereals could be distinguished as: (i) white, off white or pale yellow, compact and often nodular, and (ii) soft, granular and translucent. Of these only the first type of calli exhibited embryogenic differentiation (Vasil and Vasil, 1991). Similarly, callus of *Gossypium hirusitum* could be green, yellow, white, brown and red but only yellow callus yielded embryogenic cultures (Finer, 1988). Selective or artistic subculture has also been practiced in producing embryogenic cultures of *Daucus carota, Zea mays* (Duncan et al., 1985), *Sorghum bicolor* (Bhaskaran and Smith, 1988), and *Cucumis melo* (Kathal *et al.*, 1986).

Other Factors

Some miscellaneous factors have been found to be either essential or could induce better embryogenesis in certain plants. Some of these are discussed as follows.

The gaseous phytohormone ethylene plays an important role in plant morphogenesis. The gas phase above the culture includes acetaldehyde, O_2, CO_2, and ethanol. Ethylene applied exogenously inhibited somatic embryogenesis in carrot. On the other hand, ethylene biosynthetic inhibitors such as, cobalt, nickel and salicylic acid, and action inhibitor, silver nitrate, increased embryogenesis in carrot (Tisserat and Murashige, 1977) and rubber.

Oxygen tension has been shown to promote embryogenic development in cultures. Reducing dissolved oxygen in wheat cultures improved somatic embryogenesis by almost 6-fold (Carman, 1988; 1990). Precocious germination of somatic embryos and frequency of abnormal scutellar enlargement significantly decreased in low levels of O_2. The need for reduced dissolved O_2 could be substituted by the addition of ATP to the medium. However, there are contradictory reports of the affectivity of reducing oxygen on embryogenesis in carrot (Kassel *et al.*, 1977).

Interestingly, compounds of bacterial origin have been shown to enhance somatic embryogenesis. Soybean when co-cultured with *Pseudomonas maltophilia* differentiated somatic embryos, which otherwise could be obtained only from immature cotyledons. Extracts from blue green algae such as *Nostoc, Anabaena, Syriechococcus,* and *Xenococcus* have also been shown to increase somatic embryo formation in carrot cultures following 2, 4-D treatment. It is not yet clear whether compounds from these organisms act as signal molecules or have growth promoting properties.

Light is a major factor of the culture environment and has been shown to have definite effect on embryogenesis of some plants. Somatic embryo induction in *Solanum melongena* has an absolute requirement for light. In contrast, poplar has an absolute requirement for darkness. For induction of somatic embryos in carrot green light, red light, or darkness was equally effective, while high intensity of white light and blue light were inhibitory. It has been suggested that in vitro cultured immature embryos of wheat respond to 2,4-D and to light signals by modifying expression of G-proteins, NDP Kinase and arrestin proteins.

Humidity stress and starvation of cultures has recently been shown to improve embryogenesis in carrot. When the embryogenic callus was cultured on half-strength MS medium or MS medium without sucrose or cultured under conditions of reduced humidity (69.3%) the production of somatic embryos increased 3.4-4.5 fold compared to controls. Embryogenic callus cultured on MS medium after 5 days of starvation by placing in empty culture plates showed a 20-fold increase in somatic embryo production.

Often, cultures for somatic embryogenesis are maintained on semi-solid medium as growth is slower and subculture is less frequent. In liquid cultures some embryogenic tissues will not form good quality somatic embryos capable of converting to plantlets with high frequency. Cultures of alfalfa, poinsettia and yellow poplar must be plated on a semi-solid medium for proper somatic embryo formation.

Induction and Development

Acquisition of embryogenic competence largely relies on dedifferentiation when the existing developmental information must be stopped or altered in order to make the cells responsive to new signals. Reactivation of cell division in somatic plant cells is required for dedifferentiation and establishment of embryogenic competence by increasing receptivity to developmental signals due to general reorganization of chromatin structure, reprogramming of gene expression and altering cellular metabolism. The different events leading to embryogenesis though, intrinsically coupled to one another, are often referred to as dedifferentiation, competence acquisition, induction, and determination.

Depending on the nature of the explant and culture conditions, somatic embryos may arise either directly from cells of the explant or indirectly via a phase of unorganized growth (callus formation) after several cycles of cell divisions. As stated earlier in the former case explants carry competent or PEDCs while the latter have to acquire competence and so are the IEDCs. Both can be maintained and multiplied in the embryonic state under appropriate conditions.

Induction

To acquire embryogenic competence the prerequisite is dedifferentiation of cells. In carrot, which is the most studied in this respect, commonly used inductive signals are plant growth regulators, such as auxin. Changes in pH and application of heavy metals also elicit embryogenic response.

Auxin is considered to be the main plant hormone required for the activation of cell division in differentiated plant cells both *in vivo* and *in vitro*. One possible target of auxin is the induction of expression of the cdc2 gene coding for the key regulatory protein kinase of the cell cycle. In alfalfa leaf protoplast-derived cells it was shown that auxin alone can result in the accumulation of this protein in high amounts but for activation of the kinase presence of cytokinin is required. These protoplast-derived cells enter the cell cycle faster under embryogenic rather than non-embryogenic conditions although it could not be correlated with increased activity of the cdc2 kinase. Under embryogenic conditions the pH of cytoplasm and vacuole increase at a high degree during protoplast reactivation as compared to non-embryogenic conditions.

An auxin, particularly 2,4-D, is generally necessary to induce embryogenesis in plants such as carrot, alfalfa and cereals. However, the requirement of exogenous auxin for the induction of somatic embryogenesis depends on the nature of the explant used. For example, petiole explants (Ammirato, 1985), hypocotyls explants and single cells isolated from established suspension cultures of carrot required exposure to 2,4-D for 1, 2 or 7 days, respectively, to acquire competence to form embryos on embryo development (ED) medium devoid of 2,4-D. Microcalli of alfalfa required even a shorter pulse (a few minutes to a few hours) of relatively high concentration (100 μM) of 2,4-D to produce embryos in ED medium. An important phenomenon associated with the induction of somatic embryogenesis is the change of cellular polarity.

Polarity. Several observations support the hypothesis that plant growth regulators and other treatments employed for the induction of embryogenesis do this by altering the cell polarity and promotion of subsequent asymmetric division (de Jong *et al.*, 1993). On a cytokinin-containing medium the immature embryos of white clover develop several adventive embryos directly from the hypocotyl epidermis. The first microscopically observable sign of induction of embryogeny is a shift from the normal anticlinal divisions in the epidermal cells to periclinal or oblique divisions. Asymmetric cell divisions following stimulation by auxin have been frequently observed in leaf protoplast cultures of an embryogenic cultivar of alfalfa; the protoplasts from non-embryogenic cultivar divided symmetrically (Bogre *et al.*, 1990). And also, in single cell suspensions of carrot, the first division was asymmetric (Backs-Husemann and Reinert, 1970; Komamine *et al.*, 1990) and only the smaller daughter cell ultimately developed into somatic embryo. Since the root pole of the SE is always oriented towards the larger cell the polarity of the entire SE is determined much before the first division of the embryogenic cell.

Isolated small carrot cells become embryogenic in the presence of low concentrations of auxin and cytokinin, go through unequal division and exhibit polarized synthesis of macromolecules. Video cell tracking demonstrated that different cell-types of carrot with equal and unequal first divisions are able to form somatic embryos. However, differential cell fate is determined not by morphological but by physiological asymmetries, with respect to unequal distribution of molecules, such as arabinogalactans, and cell constituents (Feher *et al.*, 1999).

In very young spherical SEs of carrot electric gradient exists along the future longitudinal axis (Brawley *et al.*, 1984). Exposure of freshly isolated mesophyll protoplasts to an electric field (0.02 V DC current for 20 h) considerably promoted the embryogenic response in alfalfa (Dijak *et al.*, 1986). The electric stimulus seems to promote the differentiation of organized structures (shoots/embryos) by affecting cell polarity through changes in the organization of microtubules (Dijak and Simmonds, 1988; de Jong *et al.*, 1993). Two days after culture the treated alfalfa protoplasts contained more microtubules, which were thinner and more branched than the untreated control cells. In the untreated cells the microtubules were arranged in parallel strands whereas in the treated cells they occurred as a disordered network. Another striking effect of electric treatment was the induction of an asymmetric first division in 47% of cells coupled with a relatively short period of cell expansion, resulting in spherical structures composed of many small irregularly shaped cells and a few large ones. In contrast, the first division in untreated cells was symmetrical, and this and subsequent divisions were followed by a period of cell expansion, such that the colonies formed consisted of cells of similar size.

Development

The usual strategy to start an embryogenic suspension culture of carrot is to expose explants to a high concentration of 2, 4-D. After re-initiation of cell division and a period of cell proliferation in the presence of the auxin embryogenic cells are released into the medium (de Vries et al., 1988). These cells are in the form of clusters of cytoplasmic cells (PEMs). The suspension cultures also contain single cells of two types: (i) small, cytoplasmic, and (ii) long, vacuolated. The former type of cells may give rise to PEMs (Komamine et al., 1990) but most of the PEMs are derived from pre-existing PEMs (de Vries *et al.*, 1988; Emons *et al.*, 1992). The PEMs comprise embryogenic cells, which are small (400-800 μm^3), angular, connected with the adjacent cells by many plasmodesmatas (2-4 per μm^2), have several small vacuoles (30% of cell volume), many starch grains (5-25 per cell), high density of ribosomes, numerous profiles of rough endoplasmic reticulum, spherosome-like vesicles, high dehydrogenase activity, many exocytosis configurations on the plasmalemma (0.6 per μm^2) and polylamellate walls at all wall facets (McWilliam *et al.*, 1974; Emons *et al.*, 1992; Emons, 1994). The embryogenic cells are held together by non-embryogenic cells which are larger (1000-3000 μm^3), rounded, have intercellular spaces, fewer plasmodesmata (0.1-1.0 per μm^2), larger vacuoles (80% of cell volume), few starch grains (1-2 per cell), low population of ribosomes, very few endoplasmic reticulum profiles, little or no spherosome-like vesicles, low dehydrogenase activity and sometimes fewer exocytosis configurations.

Besides inducing embryogenesis, auxin also causes cell elongation and disruption of formerly adhering cells. Consequently, with the continued presence of the auxin non-embryogenic cells of the PEMs elongate leading to breakdown of the PEMs. The released embryogenic cell clusters, which are also referred to as 'pre-globular embryos' or 'globules' (Emons, 1994), develop into new PEMs and the non-embryogenic cells of the disintegrated PEMs further elongate to form the non-embryogenic component of the suspension. Thus, in the presence of 2,4-D this cycle of PEM or pro-embryo proliferation continues as long as cytoplasmic cells remain present and their daughter cells adhere. However, if auxin concentration is too high or subculturing is too often the population of small embryogenic cells drop because of their disruption and elongation, and the embryogenic potential of the culture is eventually lost. The embryogenic cells secrete certain proteins into the culture medium which not only help in maintaining the embryogenic potential of the cultures by restricting cell elongation in the presence of 2,4-D but also induce the appearance of small embryogenic cells in previously non-embryogenic cultures (Kreuger and Van Holst, 1993).

When embryogenic cultures of carrot are transferred to auxin-free medium, the disruption of cells from each other stops and the globules develop into globular embryos. In this process, the first differentiation step is the formation of a protoderm outside the globule. The globular embryos then continue further development and form typical embryos. In maize, globular and later stage embryos develop in the presence of 2,4-D but the differentiation of protoderm occurs only after transfer to 2,4-D-free medium (Emons, 1994). ABA stimulates epidermis development. The developing embryos now synthesize their own auxin and further morphogenesis beyond globular stage is dependent on proper polar transport of auxin.

As the development proceeds, the globular embryo attains an oblong shape. The elongation of the inner cells of the carrot somatic embryos mark the differentiation of the procambium, which becomes distinguishable from the adjacent ground tissue, characterized by large, vacuolated cells (Schiavone and Cooke, 1985). In somatic embryos, cotyledons originate by local mitotic activity of cells in the periphery of the embryo development. Cotyledons might be important for the regeneration capacity of the embryos, possibly by regulating the auxin flow, thus fixing a physiological gradient within the embryo, and for a prompt conversion at the inception of germination. Apical meristem differentiation usually begins at the late globular stage of embryo development, and structurally organized meristems are visible at the cotyledonary stage.

SEs formed on the surface of the callus have intrinsic polarity because these are attached to the callus cells at one end. The future root of such SEs is always oriented towards the callus center. In maize, attachment of SEs to callus is very important. Only such embryos that are attached to the callus until the formation of proper shoot meristem develop into complete embryos. Those embryos grown from cell clusters in liquid medium do not form a proper shoot meristem (Emons an Kieft, 1991). In carrot, which easily forms complete embryos in liquid medium, the globular embryos exhibit electric polarity in the form of ionic current; there is influx of potassium ions at the future plumular end and efflux of hydrogen ions at the radicular end (Brawley et al., 1984). The PEMs of carrot also show polarity in the distribution of calcium-calmodulin complex before morphological polarity is visible. The concentration of activated calmodulin was higher in the region of the root pole. The production of embryos in liquid cultures of sweet potato was improved by anchoring PEMs to alginate beads to maintain a physiological polarity.

Somatic embryogenesis can be considered as an extreme response of plant cells towards specific stress conditions. During this process plant cells try to establish a new homeostasis through changes of pH gradients of all cell compartments, arresting different functions, starting the endogenous synthesis of growth regulators like auxin, reactivating the cell cycle and reorganizing gene expression as well as metabolism. 2,4-D as a specific, metabolically stable molecule with strong auxin effects acts as an efficient inducer of these changes as it can reactivate the somatic cells, and also act as a stressor inhibiting their further division and directing them towards an embryogenic pathway (Feher et al., 1999).

SYNCHRONIZATION OF SOMATIC EMBRYO DEVELOPMENT

The population of cells undergoing embryogenesis whether on a semi-solid medium or in a liquid suspension is highly heterogenous, because of asynchronous differentiation of different stages of embryogeny. It is desirable to have all the differentiating cells and the somatic embryos at the same developmental stage for the system to be used effectively for artificial seed technology or for studying physiological, biochemical and molecular changes associated with the process of somatic embryogenesis. Various synchronic developments observed in nature are because of signal transmissions through plasmodesmatcl connections, which are either absent in or not functional under cultural conditions or the right signals are absent or are insufficient. In cultures synchronization has been achieved by, (i) physical separation of different embryogenic stages, and/or (ii) using growth regulators.

Heterogenous culture suspensions, as in carrot, are passed through nylon sieves to separate cell clusters of different sizes. These clusters are centrifuged in a Ficol gradient for further fractionation. Single cells and clusters measuring 50-100 μm obtained from such fractionation are highly homogenous and on culturing result in synchronized embryogenesis

(Fujimura and Komamine, 1985). Recent development of the image analysis technique for counting somatic embryos could make way for computer assisted sorting methods.

In the second method of physiological synchronization of somatic embryogenesis exogenous application of ABA induces reversible cessation of embryo development at a particular stage. In carrot ABA inhibits root growth and enriches suspension with torpedo shaped embryos. ABA has been successfully used to synchronize conifer suspensions as well.

Another well-characterized synchronous system is based on leaf explants of *Chicorium intibus.* In this, somatic embryos are induced directly from the leaf explants by growth regulators or by heat stress (35°C) and embryo development can be synchronized by the addition of glycerol to the incubation medium. The disadvantage of the system is that embryogenic cells are embedded in a mass of unresponsive cells.

Maturation and Conversion of Somatic Embryos

Although somatic embryogenesis has been reported for several crop species, the quality of SEs with regard to their germinability or 'conversion' into plants has been generally very poor. As poor as 3-5% conversion has been observed in many cases. This is because the apparently normal looking SEs are actually incomplete in their development. Unlike seed embryos, the SEs normally do not go through the final phase of embryogenesis, called 'embryo maturation', which is characterized by the accumulation of embryo specific reserve food materials and proteins which impart desiccation tolerance to the embryos; embryo size does not increase during this phase.

ABA, which prevents precocious germination and promotes normal development of embryos by suppression of secondary embryogenesis and pluricotyledony (Ammirato, 1983) is reported to promote embryo maturation in several species. A number of other factors, such as temperature shock, osmotic stress, nutrient deprivation and high density inoculum, can substitute for ABA, presumably by inducing the embryos to synthesize the hormone. ABA is known to trigger the expression of genes which normally express during the drying-down phase of seeds. Probably the products of these genes impart desiccation tolerance to the embryos (Florin et al., 1993). ABA has been shown to increase desiccation tolerance in SEs of carrot (Kitto and Janick, 1985) and celery (Kim and Janick, 1989).

Aminoacids, such as glutamine enhanced protein synthesis and dry weight accumulation during maturation of somatic embryos of alfalfa and proline and serine reduced precocious germination in orchardgrass (Trigiano and Conger, 1987). During maturation supply of carbohydrates appears to be important for embryo quality and number. In soybean somatic embryo conversion increases from 50% to 96% when matured in the presence of 10% sucrose (Buccheim et al., 1989). High sucrose has been used for maturation of somatic embryos of maize as well (Emons and Kieft, 1993). Other carbohydrate, such as maltose has been used for maturation of somatic embryos of alfalfa, soybean and white clover (Kielly and Bowley, 1992).

Somatic embryos accumulate storage proteins that are similar in size and biochemical characteristics as those of zygotic-embryos. Storage lipids have also been analyzed in developing somatic embryos.

Desiccation and/or starvation of somatic embryos have been found to enhance their germination and conversion. Soybean SEs desiccated in empty plates until they shriveled to 40-50% of their volume rapidly imbibed water following transfer to medium and germinated with at least seven times the frequency of non-desiccated embryos (Hammatt and Davey, 1987). Gradual drying of the alfalfa SEs with progressive and linear loss of water gave better response and improved the embryo quality as compared to uncontrolled drying. A series of relative humidity (RH) generated in desiccators over saturated salt solutions of NaCl (78% RH), NH_4NO_3 (63% RH), $Ca(NO_3)_2$ (51% RH) and K_2CO_3 (43% RH) are used for desiccation. The embryos in a petri plate, without a nutrient medium, are equilibrated at each humidity for a day and then air-dried to final moisture of 10-15% (McKersie and Bowley, 1993). Starving somatic embryos of carrot in empty culture plates enhanced maturation and germination of the embryos to 92% compared to 23% in controls.

The embryos removed from maturation conditions often display poor or aberrant germination, growth and/or vigour and poor conversion without some post-maturation treatment. Conversion refers to the ex-vitro survival of somatic seedlings and their continued growth. Sometimes, the residual ABA, used during maturation, may interfere with further development. Culture regime during maturation may also induce dormancy. Thus dormancy breaking treatments such as cold or gibberellic acid application enhanced somatic embryo germination and plantlet growth in barley, grape and rapeseed.

Synthetic Seeds

Somatic embryogenesis is expected to be the only clonal propagation system economically viable for crops currently propagated by seeds. However, it would require mechanical planting of SEs. Although suggestions have been made to use naked somatic embryos for large scale planting, it would be desirable to convert them into synthetic seeds or 'synseeds' by encapsulating in a protective covering. The coating of the synseeds need to have the following qualities: (i) it must be non-damaging to the embryo, (ii) the coating material should be sufficiently durable to withstand rough handling during manufacture, storage, transportation and planting at the same time is mild and protective to the somatic embryo, (iii) it should be possible to incorporate nutrients, growth regulators and other components necessary for germination, and (iv) the artificial seeds should be amenable to existing farm machinery. The success of synthetic seed technology would also depend on the quality of somatic embryos, uniform development stage matured with reversible arrested growth and high rates of conversion on planting. Currently synthetic seeds being developed are of two types: (a) desiccated, and (b) hydrated.

Desiccated Synthetic Seeds

The first synthetic seeds developed are the desiccated type. The encapsulaton is achieved by suspending somatic embryos in a solution of polyoxyethylene, which is then dispensed as drops on a teflon sheet and dried to thin wafers. Polyoxyethylene has been chosen because of its ready solubility in water with an ability to form thin films on drying, and also it does not support growth of microorganisms and is non-toxic to the somatic embryos. For germination the wafers are redissolved in embryogenic medium and the rehydrated somatic embryos are cultured on a medium. The germination and survival of seedlings has been very poor so far with this type of encapsulation.

Hydrated synthetic seeds

Currently the most popular method for encapsulation of single, hydrated somatic embryos is coating with Ca-alginate (Redenbaugh and Walker, 1990). In practice, somatic embryos are mixed with 2% solution of Na-alginate and dropped from a plastic pipette into a 100mM solution of $Ca(NO_3)_2$. An ion exchange reaction occurs to form Ca-alginate. In one of the modifications the gel beads are impregnated with microcapsules containing 3% sucrose coated with a mixture of 8% Elvax 4260 and beeswax, and 0.1% Topsin M, a fungicide.

Despite considerable research several major problems remain with regard to commercialization of artificial seed production. Lack of high quality somatic embryo production is a major limiting factor. Additionally, germination of synthetic seeds is poor because of lack of nutrients and oxygen, microbe invasion and mechanical damage of somatic embryos. Occurrence of high levels of somaclonal variation in tissue cultures is another aspect to be considered seriously for scaling up and production of synthetic seeds.

PROTOCOL FOR INDUCING SOMATIC EMBRYOGENESIS IN *DAUCUS CAROTA* (After Smith and Street, 1974)

(i) Surface sterilize the seeds in 10% calcium hypochlorite for 15 min and, after washing three times in sterile distilled water, germinate them on sterilized moistened filer paper in a petri dish, in dark at 25°C.

(ii) Cut 1 cm long segments of roots from 7-day-old seedlings and culture them individually on a semi-solid medium containing the inorganic salts of Murashige and Skoog's medium, organic constituents of White's medium, 100 mg l^{-1} myo-inositol, 0.2 mg l^{-1} 2, 4-D, 2% sucrose, and 1% agar (Difco or a any other brand). Incubate the cultures in dark.

(iii) After 6-8 weeks transfer pieces of root calli (0.2 g fresh weight) to fresh medium of the original composition and maintain the cultures in light at 25°C. The tissues may be multiplied by sub-culturing every 4 weeks in a similar manner.

(iv) After the first passage initiate suspension cultures by transferring ca. 0.2 g of callus tissue to a 200 ml Erlenmeyer flask containing 20-25 ml of liquid medium of the same composition as used for callus growth (without agar). Incubate the flasks on a horizontal rotary shaker at 100 rev. min^{-1} in light at 25°C.

(v) Sub-culture the suspensions every 4 weeks by transferring 5 ml to 65 ml of fresh medium (1: 13).

(vi) To induce embryo development, transfer callus pieces or portions of suspension to 2,4-D-free medium of otherwise the same composition as used before.

(vii) After 3-4 weeks the cultures contain numerous embryos in different stages of development.

PROTOCOL FOR INDUCING SOMATIC EMBRYOGENESIS IN *CITRUS* SP. (After Tisserat and Murashige, 1977)

(i) Take a 6-8-week-old fruitlet of a local cultivar and surface-sterilize it with 1% sodium hypochlorite for 15-20 min. follow the subsequent steps under aseptic conditions.

(ii) Bisect the fruit and transfer the ovules to a sterile petri dish.

(iii) Excise nucellus tissue from the ovules using a dissecting microscope. Hold the chalazal region of the ovules with pointed forceps. Give a shallow incision longitudinally through the integuments from the chalazal region to the micropylar tip. Cut the ovule transversely into two halves. From the micropylar half, remove the integuments, endosperm and any embryo (especially zygotic), if present. Transfer the nucellar section to the culture vessel in a manner that its cut end is in contact with the medium.

(iv) Use semi-solid medium containing inorganic salts of Murashige and Skoog's medium, 100 mg l^{-1} myo-inositol, 0.2 mg l^{-1} thiamine-HCl, 1 mg l^{-1} nicotinic acid, 4 mg l^{-1} glycine, 500 mg l^{-1} malt extract, 5% sucrose, and 1% agar.

(v) Incubate the cultures in light (1000 lx) at 27± 1°C.

(vi) Within 4-6 weeks multiple embryos would develop from the callused nucellar tissue. To stimulate full plantlet development transfer the embryos to another medium that differs from the previous medium in having 1 mg^{-1} GA_3, in place of malt extract.

REFERENCES

Ammirato, P. V. (1974) The effects of abscisic acid on the development of somatic embryos fromcells of caraway (Carum carvy L.). *Bot. Gaz.* 135: 328 337.

Backs-Husemann, D. and Reinert, J., (1970) Embryo formation by isolated single cells from tissue cultures of Daucus carrota. *Protoplasma*, 70:49-60.

Bhaskaran, S. and Smith, R. H., (1988) Enhanced somatic embryogenesis in Sorghum bicolor from shoot tip culture. *In Vitro Cell Dev. Biol.* 24: 65-70.

Bogre, L. Stefanor, I., Abraham, M., Somogyi, I. and Duddits, D., (1990) Differences in responses to 2, 4-dichlorophenoxy acetic acid (2,4-D) treatment between embryogenic and nonembryogenic lines of alfalfa. In: *Progress in Plant Cellular and Molecular Biology* (H. J. J. Nijkamp et al., Eds.) pp.427-436, Kluwer, Dordrecht.

Brawley, S. H., Wethrel, D. F. and Robinson, K. R., (1984) Electrical polarity in wild carrot embryos precedes cotyledon differentiation. *Pro. Nat. Acad. Sci.* USA, 81: 6064-6067.

Buccheim, J. A., Colburn, S. M. and Ranch, J. P., (1989) Maturation of soybean somatic embryos and the transition to plantlet growth. *Plant Physiol.*, 89: 768-775.

Carman, J. G., (1988) Improved somatic embryogenesis in wheat by partial simulation of the in-ovulo oxygen, growth-regulator and desiccation environments. *Planta*, 175: 417-424.

Carman, J. G., (1990) Embryogenic cells in plant tissue cultures: occurrence and behavior. *In Vitro Cell Dev. Biol.*, 26: 746-753.

De Jong, A. J., Schmidt, E. D. L. and de Vries, S.C., (1993) Early events in higher-plant embryogenesis. *Plant Mol. Biol.*, 27: 367-377.

De Vries, S. C., Boolj, H, Meyerink, P., Huisman, G., Wilde, H. D., Thomas, T. L., and Van Kammen, A., (1988). Acquisition of embryogenic potential in carrot cell-suspension cultures. *Planta*, 176: 196-204.

Dijak, M., and Simmonds, D. H., (1988) Microtubule organization during early direct embryogenesis from mesophyll prtoplasts of Medicago sativa L. *Plant Sci.* 58:183-191.

Dijak, M., Smith, D.L., Wilson, T.J. and Brown, D.C.W., (1986) Stimulation of direct embryogenesis from mesophyll protoplasts of Medicago sativa. *Plant Cell Rep.*, 5: 468-470.

Duncan, D.R., Williams, M.E., Zehr, RE. and Widholm., J.M:, (1985) The production of callus capable of plant regeneration from immature embryos of numerous Zea mays genotypes. *Planta*, 165: 322-332.

Emons, A.M.C. and Kieft, H., (1991) Histological comparison of single somatic embryos of maize from suspension culture with somatic embryos attached to callus cells. *Plant Cell Rep.*, 10: 485-488.

Emons, A.M.C., Vos, J.W. and Kieft, H., (1992) A freeze fracture analysis of the surface of embryogenic and non-embryogenic suspension cells of Daucus carota. *Plant Sci.*, 87: 85-97.

Emons, AM.C. (1994) Somatic embryogenesis: cell biological aspects. *Acta Bot. Neerl.*, 43: 1-14.

Finer,J.J., (1988) Plant regeneration from somatic embryogenic suspension cultures of cotton, Gossypium hirsutum L. *Plant Cell Rep.*, 7: 399-402.

Florin, B., Tessereau, H., Lecouteux, C., Didier, C. and Petiard, V., (1993) Long-term preservation of somatic embryos. In: *Synseeds: Applications of Synthetic Seeds to Crop Improvement* (K Redenbaugh, Ed.), pp. 133-161, CRC Press, Boca Raton, FL

Fujimura, T. and Komamine, A., (1975) Effects of various growth regulators, on the embryogenesis in a carrot cell suspension culture. *Plant Sci. Lett.*, 5: 359-364.

Gavish, H., Vardi, A. and Fluhr, R., (1991). Extracellular proteins and early embryo development in Citrus nucellar cell cultures. *Physiol. Plant.*, 82: 606-616.

Gavish, H., Vardi, A. and Fluhr, R., (1992) Suppression of somatic embryogenesis in Citrus cell cultures by extracellular proteins. *Planta*, 186: 511-517.

Gray, D.J., Trigiano, R.N. and Conger, B.V., (1993) Liquid suspension, culture production of orchardgrass somatic embryos and their potential for the breeding of improved varieties. In: *Synseeds: Applications of Synthetic Seeds to Crop Improvement* (K Redenbaugh, Ed.), pp. 351-366, CRC Press, Boca Raton, FL

Hammatt, N. and Davey, M. R., (1987) Somatic embryogenesis and plant regeneration from cultured zygotic embryos of soybean (Glycine max L. Merr.) *J. Plant. Physiol.*, 128: 219-226.

Kathal, R., Bhatnagar, S. P., and Bhojwani, S. S. (1986) Regeneration of shoots from hypocotyls callus of *Cucumis melo* cv. Pusa Sharbati. *J. Plant Physiol.*, 126: 59-62.

Kessell, R.H.J., Goodwin, C. and Philip, J., (1977) The relationship between dissolved oxygen concentration, ATP and embryogenesis in carrot (Daucus carota) tissue cultures. *Plant Sci. Lett.*, 10: 265-274.

Kielly, G.A. and Bowley, S.R., (1992) Genetic control of somatic embryogenesis in alfalfa. *Genome*, 35: 474-477.

Kim, Y.H. and Janick, J. (1989) ABA and Polyox-encapsulation or high humidity increases survival of desiccated somatic embryos of celery. *HortScience*, 24: 674-676.

Kitto, S. K. and Janick, J. (1985) Production of synthetic seeds by encapsulating asexual embryos of carrot. *J. Am. Soc. Hortic. Sci.*, 110: 277-282.

Kochba, J. and Spiegel-Roy, P., (1977) The effects of auxins, cytokinins and inhibitors on embryogenesis inhabituated ovular callus of the 'Shamouti' orange (Citrus sinensis). *Z. Pflanzenphysiol.*, 81: 283-288.

Kohlenbach, H.W., (1977) Basic aspects of differentiation and plant regeneration from cell and tissue cultures. In: Plant Tissue Culture and its Bio-technological Application (W. Barz Ed.), pp. 355-366, Springer, Berlin.

Kohlenbach, H.W., (1978) Comparative somatic embryogenesis. In: Frontiers of Plant Tissue Culture (T.A. Thorpe Ed.), pp.59-66, University of Calgary Press, Canada.

Komamine, A., Matsumoto, M., Tsukahara, M., Fujiwara, A., Kawahara, R., lto, M., Smith, J. Nomura, K and Fujimura, T., (1990) Mechanisms of somatic embryogenesis in cell-cultures, physiology, bjochemistry and molecular biology. In: *Progress in Plant Cellular and Molecular Biology* (H.J.J. Nijkamp et al. Eds.),. pp 307-313, Kluwer, Dordrecht.

Konar, R.N. and Nataraja, K., (1969) Morphogenesis of isolated floral buds of Ranunculus sceleratus L. in vitro. *Acta Bot. Neerl.*, 18: 680-699.

Konar, R.N., Thomas, E. and Street, H.E., (1972) Origin and structure of embryoids arising from epidermal cells of the stem of Ranunculus sceleratus L. *J. Cell Sci.* 11: 77-93.

Kreuger, M. and Von Holst, G. J. (1993) Arabinogalactan proteins are essential in somatic embryogenesis in *Daucus carota* L. *Planta*, 189: 243-248.

McKersie, B.D. and Bowley, S.R., (1993) Synthetic seeds of alfalfa. In: *Synseeds: Applications of Synthetic Seeds to Crop Improvement* (K. Redenbaugh, Ed.), pp. 231-255, CRC Press, Boca Raton, FL.

McWilliam, A.A, Smith, S.M. and Street, H.E., (1974) The origin and development of embryoids in suspension cultures of carrot (*Daucus carota*). *Ann. Bot.*, 38: 243-250.

Murashige, T. and Skoog, F., (1962) A revised medium for rapid growth and bioassays with tobacco tissue cultures. *Physiol. Plant.*, 15: 473-493.

Nakamura, T., Taniguchi, T., and Maeda, E., (1992) Studies on somatic embryogenesis in coffee by scanning electron microscope. *Jpn. J. Crop sci.* 61: 476-486.

Redenbaugh, K. and Walker, K., (1990). Role of artificial seeds in alfalafa breeding. In: Plant Tissue Culture: Applications and Limitations,. (S. S. Bhojwani, Ed.), pp. 102-135, Elsevier, Amsterdam.

Reinert, J., (1958) Morphogenese und ihre kontrolle an Gewebekulturen avs karotten. Naturwissen Schaften 45: 344-345.

Schiavone, F. M. and Cooke, T. J., (1987). Unusual patterns of somatic embryogenesis in the domesticated carrot: developmental effects of exogenous auxins and auxin transport inhibitors. *Cell Differ.*, 21: 53-62.

Smith, S. M. and Street, H. E., (1974) The decline in embryogenic potential as callus and suspension cultures of carrot (Daucus carota L.) are serially subcultured. *Ann. Bot.*, 38: 223-241.

Steward, F.C., Mapes, M. O., and Mears, K., (1958) Growth and organized development of cultured cells. II. Organization in cultures grown from freely suspended cells. *Am. J. Bot.*, 45: 705-708.

Tisserat, B., and Murahige, T., (1977) Probable identity of substances in citrus that repress asexual embryogenesis. *In Vitro*, 13: 785-789.

Trigiano, R. N. and Conger, B. V., (1987) Regulation of growth and somatic embryogenesis by proline and serine in suspension cultures of Nicotiana tabacum L. *Planta*, 119:149-159.

Vasil, I. K. and Vasil, V., (1991) Embryogenic callus, cell suspension and protoplast cultures of cereals. In: Plant Tissue Culture Manual, (K. Lindsey, Ed.),. pp.B1 1-16, Kluwer, Dordrecht.

Suggested Readings

1. Bhojwani, S. S. and Bhatnagar, S. P. (1990) The Embryology of Angiosperms.Vikas, Delhi, 370 pp.
2. Bhojwani, S.S. and Razdan, M. K. (1996) Plant Tissue Culture: Theory and Practice, A Revised Edition. Elsevier, Amsterdam.
3. Emons, AM.C., (1994) Somatic embryogenesis: cell biological aspects. *Acta Bot. Neerl.*, 43: 1-14.
4. Redenbaugh, K. (ed.) (1992) Synseeds - Application of Synthetic Seeds to Crop Improvement. CRC Press, Boca Raton.

REVIEW QUESTIONS

Long Questions

1. Write a short note on applications of somatic embryogenesis.
2. Describe four important factors that influence the process of SE.

3. Write a short note on "synthetic seeds".
4. Write step wise protocol for developing somatic embryos in *Daucus carota.*
5. Write advantages of somatic embryogenesis over organogetic pathway of plant regeneration.

Multiple Choice Questions

1. Development of embryos from somatic cell under *in vitro* condition is termed as
   (a) Somatic Embryogenesis (b) Organogenesis
   (c) Caulogenesis (d) Totipotency
2. In which year the first report on Somatic Embryogenesis in cultured cells of *Daucus carota* appear
   (a) 1956 (b) 1958
   (c) 1960 (d) 1962
3. Who reported *in vitro* somatic embryogenesis for the first time?
   (a) K. Redenbaugh (b) E.C. Cocking
   (c) J. Reinert (d) O.L. Gamborg
4. The ability of a cell to proceed through all the stages of development and thus produce a normal adult is termed as
   (a) Somatic Embryogenesis (b) Organogenesis
   (c) Caulogenesis (d) Totipotency
5. Advantage of regenerating plants through somatic embryogenesis over organogenesis
   (a) High frequency regeneration
   (b) Shoot and root development simultaneously
   (c) Synthetic Seed
   (d) All of above
6. Somatic embryogenesis was first time reported in which plant species?
   (a) *Nicotiana tabacum* (b) *Daucus carota*
   (c) *Ranunculus sceleratus* (d) *Albizia lebbeck*
7. Which hormone plays an important role in somatic embryo maturation
   (a) IAA (b) BAP
   (c) GA_3 (d) ABA
8. Following process triggers somatic embryogenesis in embrogenic callus cultures of *Daucus carota*
   (a) Addition of Auxin (2,4-D) in the medium
   (b) Addition of cytokinin (BA) in the medium
   (c) Removal of Auxin (2,4-D) in the medium
   (d) Removal of cytokinin (BA) in the medium

9. Synthetic seeds are
   (a) Quality SEs
   (b) Encapsulated SEs in jelly like chemicals
   (c) Mature and desiccated SEs
   (d) All of the above
10. Which part of somatic embryos are generally poorly developed
   (a) Plumule (b) Radicle
   (c) Cotyledon (d) All of above

Answers

1. (a)	2. (b)	3. (c)	4. (d)	5. (d)	6. (b)
7. (d)	8. (c)	9. (d)	10. (c)		

CHAPTER 43

Plant Cell and Tissue Culture

P. E. Rajasekharan[1] and R. Prakashkumar[2*]

[1]Indian Institute of Horticultural Research, Hessaraghatta Lake, P.O. Bangalore–560 089

[2]Kerala State Council for Science Technology and Environment, Pattom, Thiruvananthapuram–695 004

E-mail: rprakum@yahoo.co.in

The main advantage of tissue culture technology lies in the production of high quality and uniform planting material that can be multiplied on a year-round basis under disease-free conditions anywhere irrespective of the season and weather. Plant cell and tissue culture is fundamental to most aspects of plant biotechnology and is great interest to molecular biologist, plant breeders and industrialists. One of the significant contributions to the manipulative powers of modern biologists has been the development of cell and tissue culture technique. Tissue culture is generally used as a broad term to describe all types of in vitro plant cultures. Strictly speaking it should refer only to those of unorganized aggregates of cells. *In vitro* propagation is also termed 'Micropropagation', as miniature shoots and plantlets are initially obtained. Traditionally, plants have been propagated vegetatively by skills acquired for many centuries. However, plant propagation by tissue culture is a relatively new method, only around 64 years old, which utilizes growth of very small plant organs or pieces of tissue under aseptic conditions. Tissue culture techniques are termed *in vitro* (literally meaning in a glass) because cultures are grown within the glass or clean plastic vessels. The tissues or cells obtained from parts of the plants like stem, root, leaf, shoot apex, etc. are induced to proliferate and produce a new plant. Plant tissue culture falls under 2 categories – organized growth and unorganized growth.

Organised Growth

It occurs when plant organs such as apical meristems of shoots or roots, leaf, initials, young flower buds, etc. are cultured, where they grow with their structure preserved or when

*Corresponding author

structures are formed during cultures of previously undiffertiated tissues. This process is called organogenesis or morphogesis.

Unorganised Growth

In this case, the tissue formed is an undifferentiated mass of cells called callus, which is no recognizable structure. They can be maintained as such for long periods on media and can be increased in volumes by subculturing. Tissue culture (Fig. 1) is based on the principle of totipotency, that is, the concept that every living cell has he genetic potential to produce the entire organisms. A second major principle is the hormonal regulation of the shoot and root generation made possible by the discovery of auxin and cytokinin. Development of cell suspension culture and concept of competency and determination have leaded to further development of tissue culture. The various kinds of in vitro plant tissue cultures generally recognized are:

1. Callus (or) tissue culture : These are cell aggregates that arise from unorganized growth of small plant parts.
2. Suspension (or) cell culture : Clumps of plant cells are dispersed in an agitated liquid media.
3. Protoplast cell culture : Culture of plant cell isolated without cell wall.
4. Anther cultures : The culture of complete anthers containing immature microspores.
5. Organ cultures : Culture leaf primordial, immature flower or fruit. The most important being
    - Meristem or shoot tip cultures
    - Single node cultures of separate lateral buds
    - Embryo cultures (fertilized or un fertilized).

Fig. 1

HISTORY

In 1838 Schwann and Schleiden put forward the totiotency theory, which states that cells are automatic and are capable of regenerating to give a complete plant. Their theory was in fact the foundation of tissue culture. The first attempt at tissue culture technique was made by the German botanist Haberlandt (1854-1945). He was the first person to culture the isolated and differentiated cells as early 1898. He is known as 'father of plant tissue culture'. In 1902 he chose single isolated cells from leaves, but failed in achieving the desired result as none of the cells divided. He, however, made numerous predictions in his paper, which he unfortunately could not test experimentally. Between 1907 and 1909 Harrison, Burrows and Carrel succeeded in culturing animal and human tissue culture *in vitro*. In 1939 Nobecourt, Gautheret and white succeeded in obtaining the first real plant tissue culture.After the Second World War (1939-1970) development in this field was especially rapid and numerous. Results of importance to agriculture, forestry and horticulture have been published. In 1941 Van overbeek and his co-workers demonstrated for the first time the stimulatory effect of coconut milk. Since then, further detailed work by Raghavan and Tourey (1963), Norstog (1965) and others led to the development of synthetic media for the culture of younger embryos.

The first growth regulators to be discovered was the auxin namely; indole acetic acid (IAA) which created great opportunities for *in vitro* culture of plant tissues. The discovery of another growth regulator Kinetin(a cytoknin) by Miller et al in 1955 was further stimulus. It was in 1957 that Skoog and Miller put forth the concept of hormonal control of organ formation. This concept of hormonal regulation of organogenesisis now applicable to most plant species. Murashige put forth the techniques of in vitro cultures as a practical approach to the propagation of horticultural plants. In 1962 Murashige and Skoog developed their famous medium, which is extensively used in plant tissue culture. By this time methods of in vitro culture had become reasonably well developed and had gained some recognition. During this time the botany school at the university of Delhi guided by Maheshwari became actively involved with the *in vitro* culture of reproductive organs of flowering plants. In 1964 Maheshwari and Guha produced the first haploid plants from pollen grains. Kantha in 1962 was promoted by Maheshwari's success and came up with the technique of test tube fertilization. As a result, India slowly became aware of this new and interesting field of Tissue culture.

APPLICATIONS OF PLANT TISSUE CULTURE

I. Micropropagation

The regeneration of the whole plant by plant tissue culture is called Micropropagation. It is essentially a technique where callus mass has been initiated from a single explant taken from any living part of a plant and within a short period of time and space a large number of plantlets can be produced from such a callus tissue. The number of plantlets produced depends on the number of shoot obtained from callus tissue. The number of plantlets produced depends on the number of shoot primordial that can be induced to form within these cell aggregates. As a result a large number of plant species can be propagated all year round so that the plant cultivator or breeder no longer is confined by season in the production of a large number of plants.

2. Clonal Propagation

It is a very important procedure in the fields of horticulture and silviculture. In vitro clonal propagation is a kind of micro propagation. The cultured plants cultivated from tissue culture

History of Plant Tissue Culture

Schleiden and Schwann	1838 and 1839	Cell theory, cell is the functional unit of living organisms
Habertland (Father of tissue culture)	1839	
Nobecourt	1939	Cultivating plant tissues for unlimited period
Gautheret	1955	**Totipotency** Establishment of habituated cultures, tumour cells, importance of light and temperature for the root growth. Two important functions: 1. The meristem of plant tissue cultures. 2. Plant tissue culture
Georges Morel		First to culture monocotyledonous tissue, meristem tip culture for elimination of virus disease of orchids, discovered two unique opines of crown gall tissues, Established about protocom
White	1934	First reported for the first time successful continuous cultures of tomato root tips in liquid culture and obtained indefinite growth., in vitro cultivation of viruses on excised roots and growth of tumour tisues, Important publication1.the cultivation of animal and plant cells
Caplin and Steward	1948	Use of coconut milk for the first time, cell cultures of single cells, various types of vessels for culture work like rotating nipple flasks.
Steward and Ammirto	1969	Somatic embryogenesis in carrot for the first time, , completely established the totipotency, Important publication Volumes in "Plant Physiology"
Reinert	1959	Bipolar embryo formation in carrot and showed embryogenesis, cryopreservation of cells, regeneration of plantlets, Important publication. First state of art in plant tissue culture "Plant cell, tissue and organ culture"
Vasil and Vasil	1965	Land mark work on single isolated cells of tobacco which paved way for cells cuttures in genetics, morphogenesis and vegetative propagation and proved totipotency in cells, has contributed significantly for cells culture, protoplast, somatic embryogenesis of cerals
Maheswari	1964-1967	Experimental embryology, important publication, an introduction to the embryogenesis, phytomorphology.
Guha and Meheswari		Development of haploids through anther and pollen culture for the first time

are obtained asexually and also proliferate within the culture vessel by asexual means. Asexual reproduction, on the other hand, gives rise to plants, which are genetically identical to the original (parent) plant. Multiplication of genetically identical copies of asexual reproduction is called clonalpropagation. A plant population derived from a single donor plant in tissue culture makes up a clone. Therefore, the variations that normally arise from sexual production are omitted. For orchids, *in vitro* clonal propagation is the only commercially possible method of micro propagation.

3. Production of Genetically Variable Plant

Larkin and scowcroft have proposed a general term "somaclones" for plant variants obtained from tissue culture, irrespective of their origin. There are several useful characteristics got from various plants like stress tolerance, resistance to diseases and herbicides, hardiness, etc., Such changes are valuable for crops which are normally propagated by vegetative methods.

4. Plant Pathology

Tissue culture knowledge has been used in trying to solve problems in plant pathology particularly in viral eradication by apical meristem culture and also in application in treating plant tumor and especially crown gall.

Viral Eradication

The distribution of viruses in an infected plant body is uneven and the apical meristems are generally with very low concentrations or are free of viruses. The only way to obtain a clone of virus free plants, which can be multiplied vegetatively under controlled conditions, are that would protect them from the chance of reinfection is by apical meristem culture. The virus eradication by apical meristem culture is of significant agricultural and horticultural value. Techniques of tissue culture have a capacity of restoring the original proprieties of the variety by eliminating infection. These virus free stocks could provide ideal material for national and international distribution of plants, either as breeding material or for further propagation.

Improvement in Plant Breeding by Plant Tissue Culture

Usually, conventional breeding methods are most commonly used for crop improvement. However, sometimes tissue culture techniques have been used to supplement these methods to increase efficiency or reach a goal which, cannot be attained, by other conventional methods. Embryo culture is useful in propagation of orchids, shortening the breeding cycle and overcoming seed dormancy. All year round rapid clonal propagation using plant tissue culture techniques have shown possibilities for new plant improvement techniques.

Production of Useful Secondary Metabolites

Useful Metabolites can also be produced using this method. The compounds obtained from plants can be used as medicines, pigments, hormones, latex, flavouring agents and tannins. Sometimes somatic cells can naturally produce metabolites. Certain cell structures can produce specialized metabolites found in the intact parent, like alkaloids such as nicotine, atropine, caffeine, epinephrine and codeine. In 1856 the patent for producing metabolites commercially by plant cell cultures were issued (Verpoorte *et al.*, 2002).

TISSUE CULTURE LABORATORY

Laboratory space: The organization of a tissue culture laboratory depends mainly on the nature and the scale of activity. In general, space for the following is needed: (1) washing, drying and storage of vessels, (2) preparation, sterilization and storage of media, (3) aseptic handling of explants and cultures, (4) maintenance of cultures and (5) observation of cultures. In a modest laboratory, provision for activities 1 and 2 can be made in a single room (media room), while the remain work can be done in other room (culture room). In such situation, the following precautions should be taken.

1. The washing area should be physically separated, even by a temporary partition, from that used for medium preparation.
2. The weighing balances should be kept in a separate enclosed space.
3. Autoclave, refrigerator, deep freeze and incubators may be located in the corridor.
4. For aseptic manipulation laminar flow hoods are commonly used which can be housed in the culture room. A small table having a stereoscopic microscope may be adequate for culture observations.

Culture vessels and their washing: Generally, glass culture vessels are used, as they are cheaper, reusable and autoclavalble. It is desirable to use only borosilicate or Pyrex glassware's as ordinary soda glass may be toxic to some tissues. Culture vessels made of clear plastic are available for a variety of purposes; these vessels are generally presterilized and disposable but certain types are autoclavable and, therefore, reusable. In general, plastic vessels, in the long run, much costlier than glass vessels. Tissues are generally cultured in culture tubes, (rimless tubes of 25x150mm or larger), flasks (long neck or even ordinary conical flasks) and petri plates; but many especially designed dishes are also used. Wide mouth bottles, including milk bottle are often employed especially for micro propagation work. Suspension cultures, of necessity, are maintained in long neck culture flasks. Culture tubes and flasks are usually stoppered with cotton plugs which are often wrapped in these cloth, but preparing such plugs on a large scale may be time consuming and inconvenient. Many workers use caps made of their polypropylene or a metal (aluminum/stainless steel) for this purpose; these caps have proved quite effective but their design may considerably affect their performance. Culture vessels and other lab wares are generally soaked in a suitable nontoxic detergent solution overnight, washed with a suitable brush, thoroughly rinsed clean with tap water followed by a rinse with distilled water and dried in a hot air oven (75-80 degree Celsius). Culture vessels having contamination or some dried agar sticking to the wall are first autoclaved. Washed culture vessels should be stored preferably in a dust-proof cupboard or cabinet.

Requirements

1. **Washing area:** The washing area of the tissue culture lab should contain a large sink with both running cold and hot water brushes detergents and buckets for soaking the glassware.

2. **Hot air oven:** After having washed the bottles thoroughly, it is necessary to store the bottles in a hot air oven to dry them and keep them microbe-free.
3. **Refrigerator:** A refrigerator is needed to store the chemicals like hormones, amino acids, coconut water etc. Stock solutions may also be kept.
4. **Balances:** Essentially two types of balances are required namely; electronic balance and the rough balance for weighing agar and sucrose, etc.
5. **pH meter or pH paper:** After the media has been prepared it is vital that one checks that using a pH paper or meter. It should be within the range of 5.6 –5.8.
6. **Distillation unit:** A distillation unit is essential to prepare the distilled water. An all glass distillation unit is preferable.
7. **Autoclave:** The autoclave is essential for the sterilization of the media, glassware, instruments etc. the temperature used for the sterilization of the media is 121°C, 15 pounds pressure for 25 minutes.
8. **Heater:** This is used for heating purposes for e.g., heating of the media so that the agar melts and distributes uniformly throughout the media.
9. **Microtome:** Is an instrument, which takes thin sections of the tissue growing in the media.
10. **Shaker:** Required while preparing a liquid media. The shaker helps in thorough distribution of chemicals in the media. This shaker is fitted with a regulator of 50-100 RPM.
11. **Laminar airflow cabinet:** This instrument is very important in tissue culture since all the inoculation procedures are carried out here. The laminar airflow cabinet is used for aseptic work. This instrument is placed in a separate room, which is kept sterile and dust free. The room is kept pressurized so that no air from outside can enter. It makes use of filters to prevent the contaminated air from entering the room. The cabinet is also installed with fluorescent tubes for lighting purposes, which is fixed at the top of the cabinet. The cabinet also has a number of small blower motors to blow air, which passes through a number of high efficient particulate air (HEPA) filters. These filters remove particles larger than 0.3 microns. The filters of laminar airflow cabinet should be regularly vacuumed and replaced annually. The cabinet should be cleaned with 70% alcohol prior to inoculation and UV light should be switched on approximately for 15-20 minutes to attain complete sterility.
12. **Other instrument:**
    (a) Scalpel: Used for cutting the plant material. Generally, they should be 10 inches in length.
    (b) Forceps: It should be of 10 inches in length with corrugated marking at the pointed end for holding the plant material.
    (c) Glass: Tissue culture bottles or test tubes with plastic caps, Klinwrap, Glass rods, pipettes, reagent bottles, conical flasks, glass marker etc.,

Culture Room or Growth Room

This room is primarily used for the storage of the cultures and it is also under controlled conditions. Again, shoes and other footwear should be left outside the room to maintain high standards of sterility. The room should be provided with conditioner or air coolers. The temperature should be maintained around 25 degree Celsius (+or–2). The room should be provided with specially designed shelves or racks, which are illuminated to place the cultured bottles. A thermometer and a hygrometer are fixed to the wall to check temperatures and humidity respectively. The humidity being 64%. Fluorescent light is a must to maintain light for the growth of the callus.

Sterility

Sterility is one of the most important requirements in tissue culture. Sterility means a microbe-free environment that has to be provided for the growth of the plants *in vitro*. These contaminants compete with the plant for nutrients and they grow at a faster rate compared to the plant, cause disease in the plant material eventually killing it. Therefore, much of the efforts made in tissue culture go in the exclusion of microorganisms. Media that is prepared has to be sterilized before it can be used. Heat labile compounds are sterilized by filtration or irradiation. The laminar airflow in which all culture work is done is used to provide a sterile environment, so that air borne microorganisms do not contaminate the culture vessel. Tissue culture laboratories are fumigated regularly and cleaned thoroughly every day with the intention of keeping the microbial level in the lab as low as possible. Hence, sterility is the key factor to the success of in vitro growth of any plant.

INITIATION OF PLANT MATERIAL INTO TISSUE CULTURE

Process of Micropropagation

The process of plant micro-propagation aims to produce clones (true copies of a plant in large numbers). The process is usually divided into the following stages (Fig. 2)

Stage 0 – pre-propagation step or selection and pre-treatment of suitable plants.

Stage I – initiation of explants — surface sterilization, establishment of mother explants.

Stage II – subculture for multiplication/proliferation of explants.

Stage III – shooting and rooting of the explants.

Stage IV – weaning/hardening.

This process requires consideration of several factors. First and foremost what are you trying to achieve by tissue culturing a certain plant, and what type of growth you want. If you are establishing a germplasm collection then you are concerned with clonal propagation, therefore all the procedures that you use must reduce any possibility of somaclonal variation. This has implications for the type of explants that you use and the culture medium.

Explant Selection

Therefore the first consideration is the explant. The term explant is used to describe the initial piece of plant introduced in *vitro*. If the main aim of establishing tissue culture is clonal propagation then the explant should be of a differentiated or organized nature, i.e. a shoot-tip axillary bud. As these explants are composed of pre-formed meristematic tissue the clonal integrity of their 'off-spring' is guaranteed. The use of explants, such as stem internodes and

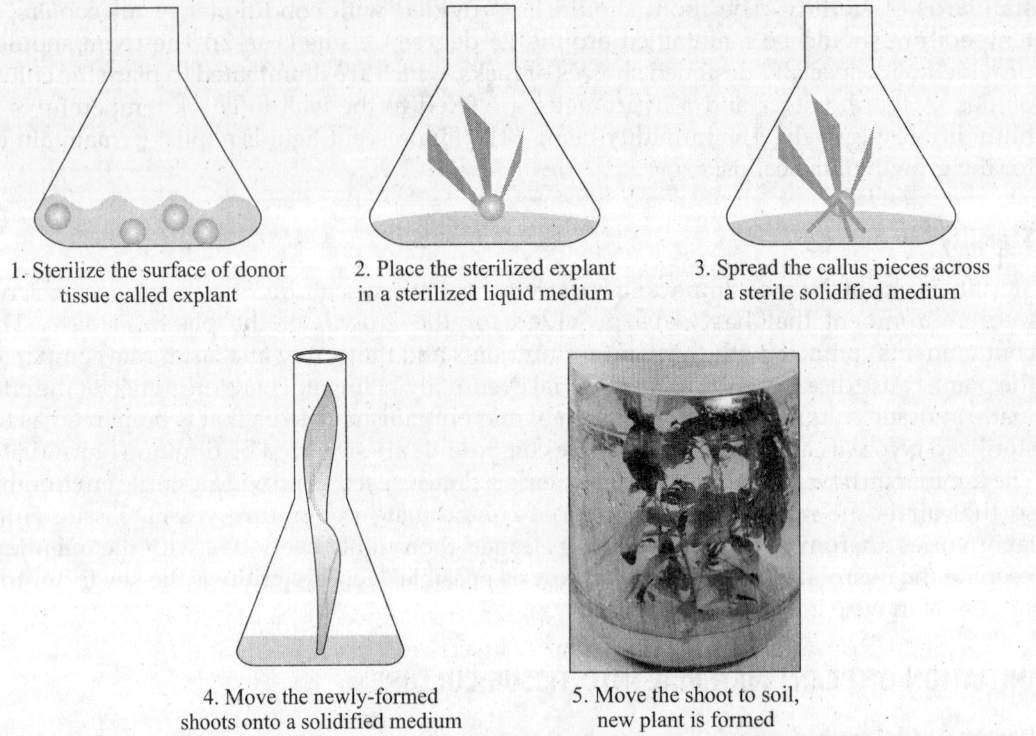

Fig. 2

the subsequent induction of adventitious shoots do not ensure that the offspring are clones of the parent plant. Explant selection plays a paramount part in successful plant regeneration studies, and there are several factors, which have to be considered when selecting the explants.

1. **Genotype:** Within the plant kingdom there is a wide range of regenerative capacity, with the dicotyledons regenerating better than the monocotyledons. Regenerative capacity can also be vary within a species, example taro.
2. **Age of plant:** As a plant becomes older its regenerative capacity often decreases, and parts of juvenile plants are preferred to those from adult, especially in the case of trees and shrubs.
3. **Age of tissue or organ:** Young, soft (non-woody) tissues are generally more amenable for culture than older woody tissue.
4. **Physiological state:** Bud taken from plants which are still in a resting state are more difficult to culture than buds from plants which are no longer dormant.
5. **Health:** Explants must be obtained from healthy, vigorous plants.
6. **Effect of different years:** If material is taken from the field, then successful establishment in vitro can be influenced by the climatic conditions encountered that year.
7. **Growth conditions:** In genera material from a greenhouse regenerates more readily than materials taken from outside.

8. **Position of the explant on/within the plant:** Topophysis is the phenomenon whereby the position of the explant in the plant influences the in vitro growth and development after isolation. Regeneration gradients are observed with nodal bud explants. Also, the higher up a shoot is isolated from a tree then the lower the probability that roots will be formed.
9. **Size of explant:** As a general rule the larger the explant the easier it will grow and develop. This is due to a grater reserve of food and hormones than what would be found in a smaller explant. There is also the effect of wounding, in that the %of wounded to non-wounded surface must be considered. Increasing the wound surface area increases the uptake of nutrients and regulators, while at the same.

Pre-treatment of the Parent Plant Material

Pre-culture plant conditions may greatly influence the subsequent growth of explants in culture, for example, rate of nature of chemical fertilizer treatment, season and environmental conditions. It is also possible to treat parent plant so that in vitro growth is optimized. Some of these treatments are spraying the mother plants with growth regulators, e.g. cytokinins, injection of the starting material with regulators, putting explant sources in a forcing solution, spraying with mixtures of antibiotics and fungicious to reduce contamination. Rejuvenation treatments are often used with mature trees. These severe pruning of trees to stimulate the production of lateral juvenile meristems and the spraying of plants with cytokinin before culture to name a few. Forcing solutions using sugars, BAP, GA, have been effective in influencing softwood growth from several woody species. A mixture of sucrose and 9 hydroxyquinoline has also been shown to induce softwood growth from quiescent buds of cut decreduous stems of woody species.

Sterilization of the Explant

In principle there are four sources of infection in tissue culture, the plant, the nutrient medium, the air and the worker. The most important of these is the plant itself, and the plant material should be well sterilized before it is inoculated onto any culture medium. As the source of the explant is derived from field or screen house material there will obviously be soil, dead parts. Excess of leaves etc, to be removed. This must be carried out in an area outside of tissue culture lab, and should be followed by washing in water. Once this is done the material is ready for chemical sterilization, which will aim to remove all associated microorganisms.

Chemical Sterilization

There are several ways in which plant material can be sterilized, and selecting the optimum method is a case of trial and error. As a general rule the first step is to look at sterilizing material with bleach or sodium hypochlorite, this necessitates investigating a range of concentrations over different times. Usually more delicate the tissue, lower the concentration and the Chlorox, (0.525% sodium hypochlorite) for 10 minutes are effective at eliminating contaminants. To all solutions of bleach a few drops of Tween 20 or 80, (polyoxyethylene sorbitan monolaureate) should be added. If treatment with Chorox is not successful, then an initial dip with 70% ethanol can be useful. The ethanol merely removes air bubbles and so allows for better penetration of sterilizing agent into the tissue. As alcohol is very dehydrating, any treatment should only last for a short time as it takes up water. If sterilization is a problem, i.e. contaminants are difficult to remove then the effectiveness of process can be improved by:

1. Washing the plant very intensively with very clean water before beginning sterilization, and allowing the water to drain off.
2. Shaking/stirring during the sterilization with blench stage.
3. Carrying out sterilization with blenches under vacuum: this results in loss of air bubbles.

If all attempts to sterilize using hypochlorite fail then mercuric chloride can be used. This is an extremely toxic substance for plants as well as animals. The concentration used in the order of 0.01-0.05%for 2-12 minutes, and thorough rinsing is crucial. Because of its extreme toxicity it should not be washed down the sink.

Pre-treatment of Parent Plant Material

It is possible to treat the parent plant to try and reduce the level of contaminants present. If material can be taken from the field and grown in screen house conditions, then this material tends to be cleaner. Once in the screen house the plants can be sprayed with antibiotic/ fungicide mixtures, which can assist in the clean-up process. Spraying the stock plants of black pepper with fungicide prior to isolation of the explant reduced contamination level. In addition, the avoiding of overhead watering can also help reduce contamination. In tropical countries, there can be very significant difference in contamination levels depending on which season it is, and therefore with some species it might be necessary to only initiate cultures in the dry season. In some cases, exposure of the stock plants to dry conditions for 3-4 weeks prior to the start of the tissue culture work can assist in the decontamination process. This was shown to be effective with tomatoes, reducing contamination from 100%in the field to 30% in a growth cabinet.

Tissue Culture Medium

It is the substance on which the piece of plant excised from the mother plant is placed. It is the equivalent of soil in the field or the pot. Their function is to provide the plant with all the substances necessary for its healthy growth and development. When the culture medium is in a semi-solid state then the culture medium is also providing support. The culture medium is just one of factors that is required for in vitro growth and development. Nutrients are essential for the growth and development of a plant: without water and mineral nutrients a plant cannot live in vitro or in vivo. Sugars have also got to be added to the culture medium, since plants in this condition are not completely autotrophic. Organic substances, such as vitamins growth regulators are often required in a culture medium. Growth regulators, such as auxins and cytokinins, regulate the development of organs on parts of plants grown in vitro. There is no one specific medium, nutritional requirements for optimum growth is very different for most species. The medium may be chemically defined, i.e., only known nutrients are included in the medium, or it can be undefined meaning that it contains some complex additives such as coconut milk or orange juice, which contain unknown growth factors. Gelling agents include agar and agar substitutes such as Phytagel or Gel rite. Agar is used at a concentration of 0.7-

1.0%, whereas Gel rite and Phytagel are used at 1.75 g/I.As a general rule, water, macronutrients, micronutrients and sugars are essential to growth and development in vitro. The best approach is to consult the literature and to see if any related plants, (same family or genus) have been successfully cultured. If this does not offer any information then as Murashige and Skoog medium have been successful for many plants you can start with this and investigate a range of say 4 concentrations of 2 growth regulators. Alternatively you can look at the broad-spectrum method of de Fossard (1976), where 4 different categories, minerals, auxins, organics and cytokines are examined at each of 3 concentrations. The drawback with this approach is that the experiment requires 81 different treatments. With woody plants a useful approach is to try ½ MS with no hormones or McCowan's Woody plant Medium without hormones (George and Sherrinton, 1993). 95% of the nutrient medium is composed of water therefore it has to be of very good quality. It should be glass distilled and for meristems, protoplasts, etc., it should be double distilled. Ion exchange columns can be used but these can result in contamination of the water with organic residues. For storage of distilled water, polythene should be used as glass can contain traces of lead, sodium, arsenic, which can be released into the water

Sugar: There needs to be a carbon source in the medium as photosynthesis is insufficient. Usually sucrose is used at a concentration of 1.5%. With autoclaving some of the sucrose is converted to glucose and fructose, which may be used more efficiently by the plants in culture.

Mineral nutrition: The Table gives the combinations of macro-and micronutrients used in some of the more important mixtures. Murashige and Skoog medium(1962) is generally considered to be a high salt medium.

Agar: Agar is a seaweed derivative, used as a gelling agent. It is purified and washed so there are no toxic materials present. Agar is generally used at a concentration of 0.7% to 1.0%. Agar is a polysaccharide with a high molecular mass that has the capability of getting medium. There are alternatives to agar, for e.g., gelrite, a polysaccharide. It has the advantage that it forms a very clear, transparent gel, and so allows for better observation of bacterial contamination. Gelrite requires both a heating cycle and the presence of cations for gelation to take place. The concentration advised is 1.75g/l. The setting of gelrite is more easily affected by the concentration of the salts in the medium than agar/other substances are available now which are cheaper, for e.g., Carrageenan, which is a gelling agent prepared from Irish Moss, and is being used by a commercial lab in NZ, the latter can be purchased for 50% of the price of the agar. Obviously, a gelling agent need only be used if solid medium is required.

Growth regulators: Hormones are organic compounds, which are synthesized in the tissues of plants, and required, in low concentration to influence plant growth and development. Many synthetic organic molecules have been found to have a biological activity similar to the hormones, as a group these compounds plus the hormones are known as growth regulators. As tissue culture requires the manipoulation of plant tissue, growth regulators can have a very significant role to play (Tables 1 and 2). Although most explants are capable of endogenous production of auxins and cytokinins, an exogenous supply has to be provided to stimulate growth and development.

Auxins: These include the naturally occurring IAA is used the concentration is in the order of 0.01–10.0 mg/l. The use of IAA is avoided because of the ease with which a break down on exposure to light, plus it is not heat stabile. With the synthetic auxins, the concentration is in the order of 0.001–10.09 mg/l. Auxins generally cause cell elongation and swelling of the tissue, cell division, the formation of adventitious roots, and the inhibition of adventitious and axillary shoot formation. With low auxin levels root Formation predominates whereas with high auxin levels there is tendency for inhibition of photosynthesis.

Cytokinins: These include the naturally occurring 2iP and zeatin, and the synthetic cytokinins, BAP and kinetin. The latter have a very high biological activity, are not expensive and so are widely used in tissue culture. They promote cell division, inducing adventitious shoot formation but generally inhibiting root formation. They promote axillary shoot formation by reducing special dominance.

Gibberellins: This group includes about 60 compounds, which are not generally used in tissue culture. The most commonly used one is gibberellic acid, GA. The substance is very heat sensitive with 95% of its activity being lost on autoclaving. It can be used in the culture medium to encourage elongation of the internodes.

Table 1. Commercially available plant hormones used for *in vitro* propagation (formula weights are given in parentheses)

Name	Abbreviation	Comments
I. Auxins		
Indole-3-acetic acid	IAA	A natural auxin,
Indole-3-butyric acid	IBA	unstable
Alpha-napthaleneacetic acid	NAA	Stable
2,4 dichlorophenoxyacetic acid	2,4D	Strong, stable
II. Cytokinins, adenine	BA	
Benzyladenine (6-denzylaminopurine)	BAP	A natural cytokinin
Kinetin (6-furfurylaminopurine)	DMAA	Weak cytokinin activity
Isopentyladenine (dimethylallyl- Aminopurine)	Ad SO_4	
III. Gibberellins	GA_3	May promote growth, sometimes inhibits bud initiation.
Gibberellic acid A_3		

Other Substance with Growth Regulating Activity

Abscisic acid: ABA is regarded as a growth inhibitor, and so can be used I slow-growth media. It has shown potential with sweet potato, though its effects subject to variental influence. Plants in nature can synthesize their own food material. In contrast, plants growing *in vitro* hetrotrophic that is they cannot synthesize their own food material. Plant tissue culture therefore require all essential minerals plus a carbohydrates source usually added in the form of sucrose and also other growth hormones (regulators and vitamins). Growth and morphogenesis of plant tissue *in vitro* are largely governed by the composition of culture media. Although the basic requirements of cultured plant tissues are similar to those of whole plants, in practice nutritional components promoting optimal growth of a tissue under laboratory conditions may vary with respect to the particular species. Media compositions are thus formulated considering specific requirements of particular culture system. For example, some tissues show better response on a solid medium. While others prefer liquid medium.

Table 2. Storage of Auxins, Cytokinins, Vitamins and Enzymes

Compound	Storage requirements (Dry powder/stock solution
Auxins	
Indole acetic acid (IAA)	Desiccate 20°C
Nephthylene acetic acid (NAA)	4°C
2,4-Dichlorophenoxy acetic acid (2,4-D)	4°C
Indole butyric acid (IBA)	4°C
Cytokinins	
6-Furfurylaminopyurine (Kinetin)	Desiccate - 20°C
6-Benzyladenine (BA)	Desiccate -20°C
6-6 7-7-Dimethylallyamino)	Desiccate -20°C
6-(4-hydroxy-3-methyl-trans-2-butenylamino) purine (Zeatin)	Desiccate -20°C
Vitamins	
Thiamin Hcl	4°C in dark bottle
Nicotinic acid	4°C or room temp.
Folic acid	4°C or room temp.
Pyridoxine Hcl	4°C or room temp.
Myso-Insoitol	4°C or room temp.
Enzymes	
Cellulase	Desiccate - 20°C
Macerase	Desiccate - 20°C
Pectinase	4°C in the Dark

Significant contributions to formulation of define growth media suitable for a wide range of application was made by Murashige and Skoog (1962) made wide range of applications. In their work to adapt tobacco callus cultures for use as a hormone bioassay system, they evaluated many medium constituents to achieve optimal growth of calluses. In so doing, they improved upon existing types of plant tissue culture media to such an extent that their medium (MS-medium) has since proved to be one of the most widely used in plant tissue culture

Media Composition

Components of media for the growth of plant callus and suspension cultures can be classified into five groups. These groups are :

1. Inorganic nutrients (macro nutrients and micro nutrients)
2. Organic suppliants
3. Carbon source
4. Growth regulators
5. Gelling agent.

Some tissues are grown on simple media containing only inorganic salts and utilizable carbon (sugar) source, but for most other it is essential to supplement the medium with the vitamins, amino acids and growth substances. Very often, complex nutritive substances are added to the medium. Such a medium, composed of "chemically defined " compounds is referred to as a "synthetic" medium.

Macronutrients: The macronutrients six major elements nitrogen (N), phosphorus (P), potassium (K), calcium (Ca), magnesium (Mg) and sulphur (S) present as salts that constitute varies media. All are essential plant cell and tissue growth.

Micronutrients: Inorganic elements required in small quantities but essential for plant cell and tissue growth constitutes microelements. These are iron (Fe), manganese (Mn), zinc (Zn), boron (B), copper (Cu), molybdenum (Mo.). Of these, iron seems more critical since chelated forms of iron and zinc are commonly used in preparing culture media. The essential micronutrients are coponents of plant cell proteins of metabolic and physiological significance. Several micronutrients are involved with the activity of growth.

Carbon and energy sources: The most preferred carbon source in plant tissue culture is sucrose. Glucose supports equally good growth while fructose is less efficient. Sucrose, while autoclaving the medium, is converted into glucose and fructose. In the process, first glucose is used then fructose.

Organic supplements: Vitamins; plant synthesize vitamins endogenously and these are used as catalysts in various metabolic processes. When plant cells and tissues are grown in in vitro, some essential vitamins are synthesized but only in sub-optimal quantities. Hence, it is necessary to supplement the medium with required vitamins and amino acids to achieve the best growth of tissue. Thiamine (B1), nicotinic acid (B3), pyridoxine (B6), calcium pentothenate (B5), and myoinositol are used more often; of these, Thiamine is the basic vitamin required by all tissues. Nicotinic acid and pyridoxine are usually added to culture media but may not be essential for cell growth in many species.

Uses and Effects of Micronutrients Present in Media

Iron: it is important for the formation of chlorophyll. Low iron content in the media causes deficiency of chlorophyll (Table 3). It is required for the formation of the precursors of chlorophyll such as aminolaevulinic acid. It is a component of the ferrodoxin proteins, which are the electron carriers in the photosynthesis.

Manganese: It is the one of most important micronutrient element of tissue culture media. Number of initiated decreases with omission of Mn. Where as a high level of Mn compensates for the lack of Mo.

Zinc: Deficient plants show decrease in chlorophyll content and poorly developed chloroplast. Zn appears in oxidase enzymes. It is essential for tryptophan synthesis, which is a precursor for auxins. It was also found to be essential for normal root development.

Boron: It has stimulatory effect on callus cultures and on root growth. An excess of boron however is toxic to plants. A continuous supply of boron is essential for merista matic activity.

Copper: Found in oxidase enzymes and a component of super oxide dismutase enzyme responsible for removing super oxides formed from molecular oxygen.

Table 3. Some of the elements important for plant nutrition and their physiological function. These elements have to supplied by the culture medium in order to support the growth of healthy cultures *in vitro*

Element	Function
Nitrogen	Component of proteins, nucleic acids and some coenzymes Element required in greatest amount
Potassium	Regulates osmotic potential, principal inorganic cation
Calcium	Cell wall synthesis, membrane function, cell signalling
Magnesium	Enzyme cofactor, component of chlorophyll
Phosphorus	Component of nucleic acids, energy transfer, component of intermediates in respiration and photosynthesis
Sulphur	Component of some amino acids (methionine, cysteine) and some cofactors
Chlorine	Required for photosynthesis
Iron	Electron transfer as a component of cytochromes
Manganese	Enzyme cofactor
Cobalt	Component of some vitamins
Copper	Enzyme cofactor, electron-transfer reactions
Zinc	Enzyme cofactor, chlorophyll biosynthesis
Molybdenum	Enzyme cofactor, component of nitrate reductase

Cobalt: Required for activation of DNA and RNA polymerase enzyme.

Molybdenum: Very important part of the chlorophyll.

Iodine: Though not considered an essential micronutrient it improves the growth of roots in tissue culture.

Preparation: The above ingredients are weighed separately and dissolved one after the other in about 500 ml of distilled water. Vitamins and hormones are added in the required quantity. This depends upon the plant being cultured and effect desired. The pH is adjusted to the required range (5.6-5.8). The media is now heated and sucrose and agar are added. It is further heated till the agar melts, following which it is dispensed into the culture vessels. These are sterilized by autoclaving.

Stock solution: The preparation and manufacture of a medium is a very tedious and time-consuming process. It is not practical and possible to weight and mix all the chemicals each time a medium has to be prepared. For convenience it is advisable to prepare the stock solutions of macronutrients, micronutrients, KI, Chelating agent and other additives. These are stored in amber colored bottles at 4°C and must be checked periodically to make certain there is no contamination with microorganisms or precipitation of components. (Note – they cannot be stored for a very long time).

Growth regulator :

BAP 0.25-20mg/l (varied according to requirements)
NAA 0.1-5mg/l (varied according to requirements)

HARDENING: THE TRANSFER FROM NUTRIENT MEDIUM TO THE SOIL

A plant, which has originated *in vitro*, differs in many respects from one that has been produced in vivo, and these differences can affect the efficiency of the transfer process.

These differences are as follows:

(1) A poorly developed cuticle, (wax layer) often occurs because of the relative humidity (90-100%) *in vitro*. Therefore when the plant is transferred to soil, extra water loss occurs because the humidity of the air *in vivo* is much lower.

(2) Leaves of an in vitro plant are often thin, soft and photo synthetically not very active they have fewer palisade cells to use light effectively and have larger mesophyll air spaces.

(3) Stomata do not operate properly in tissue culture plants; they tend to stay open for too long resulting in significant water stress.

(4) Poor vascular connections can exist between roots and shoots, therefore reducing water conduction.

(5) *In vitro* plants have been raised as heterotrophs, whereas in vivo sugar must be replaced by photosynthesis.

(6) Roots developed in vitro have few or no root hairs and so do not function properly *in vivo*.

(7) Plants, which under normal growth conditions live symbiotically with fungi or bacteria, lack these symbiotic when they are transferred from in vitro to *in vivo*.

It is obvious from the above that great care has to be taken in the transfer from in vitro to *in vivo* to prevent the waste of valuable resources. Acclimatization can take place by allowing the *in vitro* plants to gradually get used to the lower humidity tents, automatic mist systems, etc. can be used be encouraged by lowering the relative humidity *in vitro*. Better development of roots can occur in a liquid medium. And a so liquid phase can be introduced in the final stage of micropropagation. Alternatively the rooting phase can take place in vivo, with the plant let being transferred to the soil or other substrate. It is possible for *in vitro* grown plants to be inoculated with fungi or bacteria and this has been shown to stimulate growth and development. Other considerations in this final process of producing plants by tissue culture are as follows:

(1) All gelling agent should be well rinsed of.

(2) Sterilized soil should be used.

(3) Any immediate attack by pathogens of any type should be dealt with instantly because the *in vitro* plant is weak.

(4) To avoid damaging the roots plants can be transferred to finely sieved soil.

(5) Development in vivo is often optimized by the use of sand, perlite, vermiculite, with the soil to improve drainage and aeration.

Artificial Seeds or Synthetic Seeds

Development of micropropagation techniques is preferred to ensure abundant supply of desired plant species. In some crop species seed propagation has not been successful. This is mainly due to heterozygosity of seed, minute seed size, presence of reduced endosperm and the requirement of association of seed with mycorrhizal fungi for germination (e.g., orchids), and also in some seedless varieties of crop plants like grapes, watermelon, etc.,. Some of these species can be propagated by vegetative means. However, *in vivo* vegetative propagation techniques are time consuming and expensive. Development of artificial seed production technology is currently considered as an effective and efficient alternate method of propagation in several commercially important agronomic and horticultural crops. It has been suggested as a powerful tool for mass propagation of elite plant species with high commercial value.Artificial seed technology involves the production of tissue culture derived somatic embryos encased in a protective coating. Artificial seeds have also been often referred to as synthetic seeds. However, the term synthetic seed should not be confused with commercial seeds of a synthetic cultivar, which is defined as an advanced generation of an open pollinated population, composed of a group of selected inbred clones or hybrids. These synthetic seeds would also be a channel for new plant lines produced through biotechnological advances to be delivered directly to the green house or field.. This synthetic seed production technology is a high volume low-cost production technology. High volume propagation potential of somatic embryos combined with formation of synthetic seeds for low-cost delivery would open new vistas for clonal propagation in several commercially important crop species(Ara *et al.*, 2000).When somatic embryoids derived from plant tissue culture are encapsulated by a hydrogel and such encapsulated embryoids behave like true seeds if grown in soil and can be used as substitutes of natural seeds.

Basic Requirement for Production of Artificial Seeds

Recently, production of synthetic seeds by encapsulating somatic embryos has been reported in few species. One prerequisite for the application of synthetic seed technology in micropropagation is the production of high quality, vigorous somatic embryos that can produce plants with frequencies comparable to natural seeds. Inability to recover such embryos is often a major limitation in the development of synthetic seeds. Synthetic seed technology requires the inexpensive production of large numbers of high quality somatic embryos with synchronous maturation. The overall quality of the somatic embryos is critical for achieving high conversion frequencies and, encapsulation and coating systems, though important for delivery of somatic embryos, are not the limiting factors for development of synthetic seeds. The characteristic lack of developmental synchrony in embryogenic systems stymies multi-step procedures for guiding somatic embryos through maturation for the present time. The lack of synchrony of somatic embryos is, arguably, the single most important hurdle to be overcome before advances leading to wide spread commercialization of synthetic seeds can occur. Synchronized embryoid development is required for the efficient production of synthetic seeds.

Procedure for Production of Artificial Seeds

Establish somatic embryogenesis
↓
Mature somatic embryos
↓
Synchronize and singulate somatic embryos
↓
Mass production of somatic embryos
↓
Standardization of encapsulation
↓
Standardization of artificial endosperm
↓
Mass production of synthetic seeds
↓
Green house and field planting

Types of Gelling Agents Used for Encapsulation

Several gels like agar, alginate, polyco 2133 (Bordon Co.), carboxy methyl cellulose, carrageenan, gelrite (Kelko. Co.), guargum, sodium pectate, tragacanth gum, etc., were tested for synthetic seed production, out of which alginate encapsulation was found to be more suitable and practicable for synthetic seed production. Alginate hydrogel is frequently selected as a matrix for synthetic seed because of its features including moderate viscosity and low spinnability of solution, low toxicity for somatic embryos and quick gellation, low cost and bio-compatibility characteristics. The use of agar as gel matrix was deliberately avoided because, it has been described as inferior to alginate with respect to long term storage. Alginate was chosen for ease of capsule formation as well as, the rigidity of alginate beads provided better protection to the encased somatic embryos against mechanical injury than did agar.

Principle and Conditions for Encapsulation with Alginate Matrix

Alginate is a straight chain, hydrophilic colloidal, polyuronic acid composed primarily of an hydro-b-D-mannuronic acid residues with 1-4 linkages. The major principle involved in the alginate encapsulation process is that the sodium alginate droplets containing the somatic embryos when dropped into the $CaCl_2.2H_2O$ solution will result in the formation of round and firm beads as a result of an ion exchange between the Na^+ in sodium alginate with that of Ca^{2+} in the $CaCl_2.2H_2O$ solution. The hardness or rigidity of the capsule mainly depends upon the number of sodium ions exchanged with calcium ions. Hence, the concentration of both the gelling agents i.e., sodium alginate and $CaCl_2.2H_2O$ and the complexing time should be optimized for the formation of the capsule with optimum bead hardness and rigidity. In general, sodium alginate 3% upon complexation with $CaCl_2.2H_2O$ 75 mM for half an hour gives optimum bead hardness and rigidity for the production of viable synthetic seeds.

Utilization of Artificial Seeds

The artificial seeds can be used for specific purposes, notably multiplication of non-seed producing plants, ornamental hybrids which are currently done by cuttings or the propagation of polyploid plants with elite traits. The artificial seed system can also be employed in the propagation of male or female sterile plants for hybrid seed production. Cryo-preserved artificial seed may also be used for germplasm preservation, particularly in recalcitrant species (such as mango, cocoa, coconut), where seeds do not undergo desiccation. Furthermore, transgenic plants which require separate growth facilities to maintain original genotypes may also be preserved using somatic embryos. Somatic embryogenesis is a potential tool in the genetic engineering of plants. Potentially, a single gene can be inserted into a somatic cell and if the plants are regenerated by somatic embryos from single transgenic cell the progeny will not be chimeric. Multiplication of elite plants selected in plant breeding programs via somatic embryos avoids the genetic recombination, and therefore does not warrant continued selection inherent in conventional plant breeding, saving considerable amount of time and other resources. Since artificial seeds are produced in tissue culture they are free of pathogens. Thus, another advantage is the transport of pathogen free propagules across international borders avoiding bulk transportation of plants, quarantine and spread of diseases.

Potential Uses of Artificial Seeds

Delivery systems:

- Reduced costs of transplants
- Direct green house and field delivery of:
    - elite, select genotypes
    - hand-pollinated hybrids
    - genetically engineered plants
    - sterile and unstable genotypes
- Large-scale mono cultures
- Mixed-genotype plantations
- Carrier for adjuvants such as micro organisms, plant growth regulators pesticides, fungicides, nutrients and antibiotics
- Protection of meiotically-unstable, elite genotypes
- Conceivably be handled as seed using conventional planting equipment

Sl. No.	Gel	Complexing agent
1.	Sodium alginate	Calcium salts
2.	Sodium alginate with gelatin	Calcium chloride
3.	Carnagenan with locust beam gun	Potassium or ammonium chloride
4.	Gelrite	Temperature lowered

Requirements for Encapsulation

Plant material: Apical and axillary shoot tips or nodes, rhizome from *in vitro* plantlets.

Equipments / glasswares: Binocular dissecting microscope, capillary tubes, pasteur pipettes Cryovials, petri plates, conical flasks, filter papers, scalpel and forceps.

Media /reagents/chemicals:

- Sodium alginate dispensed into small bottles/vials
- Calcium chloride solution dispensed in 30 ml aliquots into conical flasks/beakers
- Ms basal media /sterile water
- Agar solidified recovery growth medium in culture tubes
- 5ml gilson 'pippetman' fitted with a tip.

PROTOCOL

Shoot Tip Dissection

1. With the aid of a dissecting microscope and 2 fine needles, remove the larger expanded leaves from the shoot tips, while retaining several non or partly expanded leaf primordial and small amount of the adjacent tissue.
2. Cut the shoot tip to a size of 0.5 to 1.5 mm.
3. Transfer shoot tips to a filter paper soaked with pre-growth medium immediately after dissection.

Encapsulation

1. Transfer freshly dissected shoot tips to a 3% alginate solution; swirl gently to submerge the shoot tips. Avoid agitation which will form air bubbles.
2. Using a 5ml pippetteman withdraw 3 ml of alginate solution at the same time capture 2 to 4 shoot tips.
3. Hold the pipette vertically and wipe away any alginate solution from, the exterior of the tip using sterile filter paper.
4. Place the pipette vertically over the flask having high calcium containing medium and with slow and controlled pressure dispense droplets of the alginate into the liquid. On contact with ca^{2+} the droplets will solidify and form a bead encapsulating the shoot tips.
5. When all the shoot tips are encapsulated, leave the flask for 30 minutes, in order to allow the beads to polymerize. Remove the excess liquid and dispense the beads in to large petridish. Sort out the beads, which contain the shoot tips. Wash 3-4 times in sterile distil water or MS basal medium.
6. Inoculate these beds into media and incubate them at 26°C and observe for the plantlet regeneration or transfer the beeds into a cryovial and plunge directly into liquid nitrogen for storage.

PRODUCTION OF SECONDARY METABOLITES

What is Secondary Metabolite?

The secondary metabolites are those cell constituents which are not essential for their survival and not involved in primary metabolic activities like photosynthesis, respiration, protein synthesis etc. These secondary metabolites include alkaloids, glycosides (steroids, phenolics), terpenoids and a variety of flavours, perfumes, agrochemicals, etc. Plants are the chief source of a wide range of biochemicals which are of medicinal importance. It has been found that the cells of the respective plants when cultured in specific medium *in vitro,* they produce the secondary metabolites (Fig. 3) which can be conveniently harvested and further processed.

Why Synthesis of Secondary Metabolites through Cell Culture?

1. The yield and quality of the product is more consistent in cell cultures because it is not influenced by the environment
2. The production schedule can be predicted and controlled in the laboratory or industry.

A list of some groups of biochemicals obtained from plants:

Group	Examples
Alkaloids	Morphine, codeine, quinine, nicotine, cocaine, hyoscyamine, lysergic acid etc
Terpenoids	Menthol, camphor, carotenoid, polyterpenes
Phenylpropanoids	Anthocyanin, coumarins, flavonoids, isoflavonoids
Quinines	Anthraquinone, nebzoquinones
Steroid	sterols

A list of some pharmaceutically valuable biochemicals obtained from plants (Verpoorte *et al.*, 2002).

Compound	Plant species	Medicinal value
Shikonin	*Lithospermum erythrorhizon*	Antiseptic
Berberine	*Coptis japonica*	Antibacterial
Quinine	*Cinchona* sp.	Antimalarial
Taxol	*Taxus* sp.	Breast and ovarian cancer treatment

What is a Bioreactor?

Bioreactors are vessels designed for large-scale cell, tissue or organ culture in liquid media. For an increasing number of plants, they have demonstrated a number of important advantages over conventional semi-solid micropropagation, including several fold increase in multiplication rates, and reduction in space, energy and labor. These cost-saving advantages have been the driving force for increased attention to the use of liquid systems generally. There are, however, many disadvantages because of a number of problems associated with the use of bioreactors in micropropagation. These include contamination, lack of protocols and production procedures, increased hyperhydricity, and problems of foaming, shear stress and release of growth inhibiting compounds by the cultures. Because of the inherently difficult

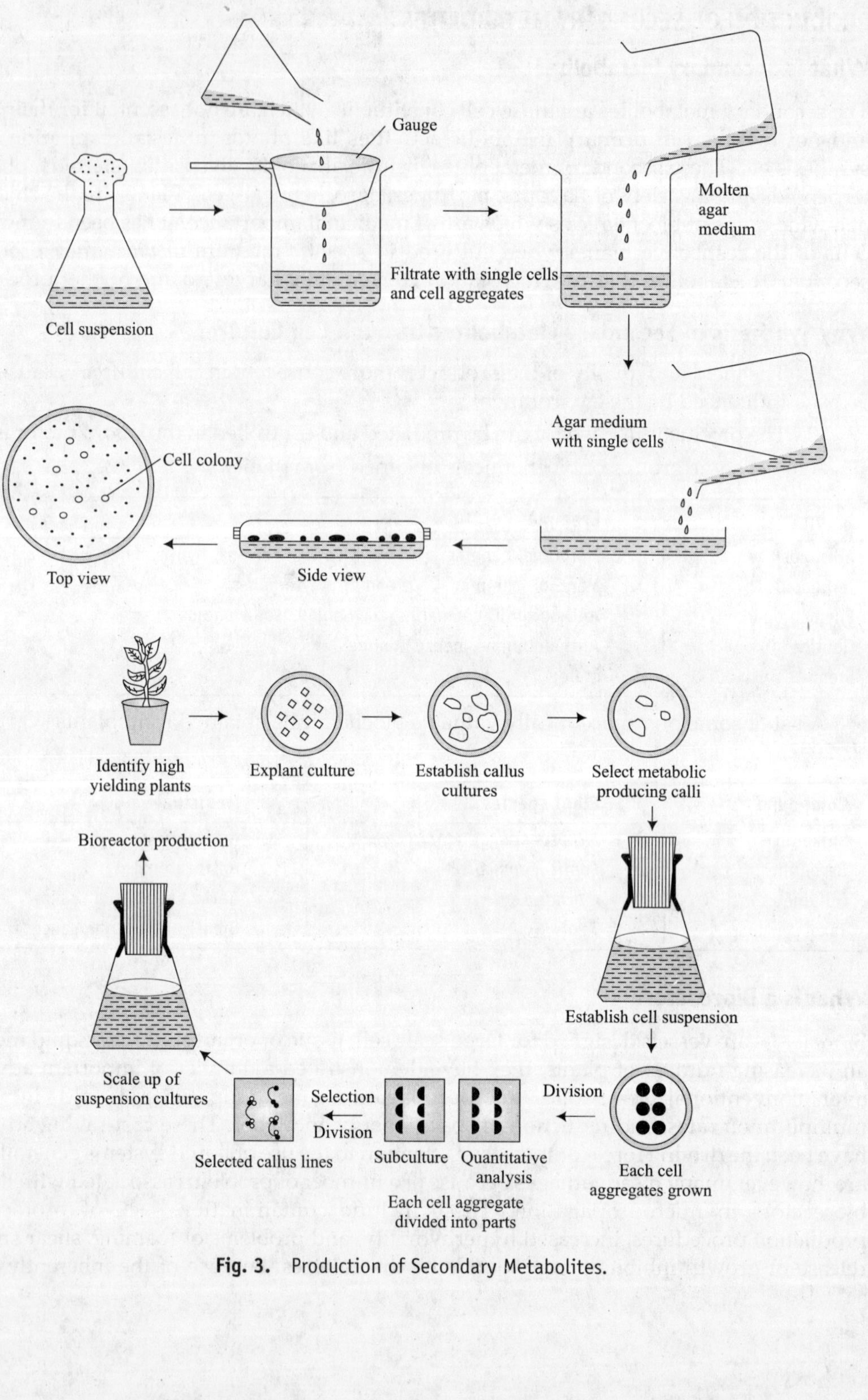

Fig. 3. Production of Secondary Metabolites.

problems of contamination in liquid systems, and the lack of problem solving information compared to semi-solid growth media, such systems involve more risk and require greater skill (Verpoorte, 1999). It is a culture vessel generally of a large volume which has provisions for aeration, stirring to achieve medium and cell mixing , contamination control, replacement of used medium and or cells.

Four types of bioreactors are :

(1) batch bioreactors
(2) continuous bioreactors
(3) multistage bioreactors
(4) immobilized cell bioreactors

Commercial production of shikonin:

Shikonin was the first commercial product from cell culture of *Lithospermum erythrorhizon*. Steps involved in the production of shikonin:

1. the high producing clone cells to be used as inoculum i.e., stock or seed clutres and maintained in jar fermentors,

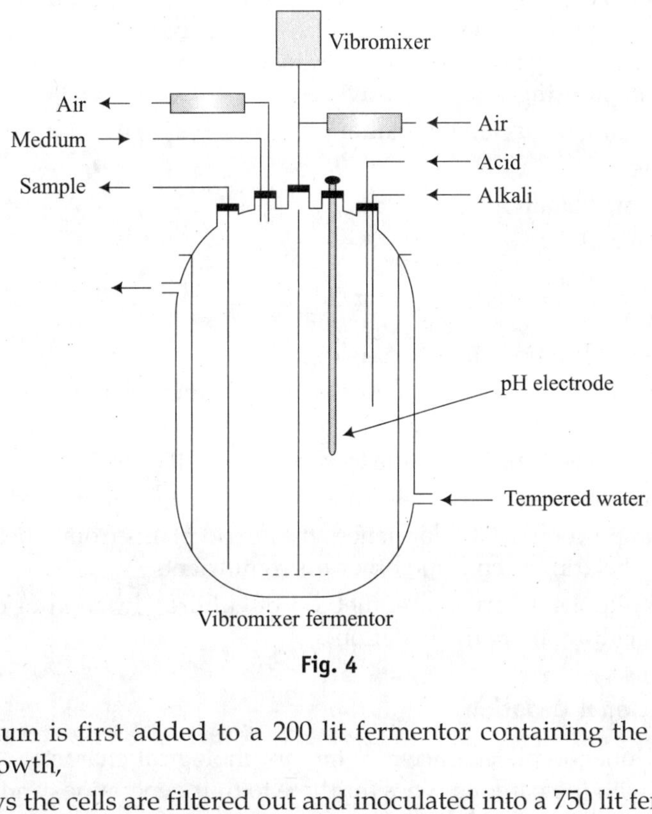

Fig. 4

2. the inoculum is first added to a 200 lit fermentor containing the MG5 medium for culture growth,
3. after 9 days the cells are filtered out and inoculated into a 750 lit fermentors with M 9 medium and incubate for 14 days,
4. the cells are harvested by simple filteration techniques and shikonin derivatives are extracted from the cells.

A 750 lit bioreactor with 600 lit medium would yield 1.2 kg Shikonin in 2 weeks. In contrast *Lithospermum erythrorhizon* roots from one hectare land would yield about 9 kg Shikonin after 4 years. Shikonin is used for making cosmetics, lotion and soap in Japan.

Advantages of secondary metabolite synthesis:

1. threat on endangered or extinct species could be eliminated (e.g. *Lithospermum erythrorhizon*),
2. when the target plant is limited in supply (*Trichosanthes* sp.) or slow in growth (L.erythrorhizon) cell culture is highly useful for faster and continuous supply,
3. cell cultures produce more secondary metabolites than the plant tissues in the field (20% more production of shikonin in cell culture),
4. cells can be stored.

Somaclonal Variation

Somaclonal variation is the genetic variability which is regenerated during tissue culture . Such variations manifests themselves as heritable mutation and present in the plant population even after transplantation to the field. Larkin and Scowcroft (1981) proposed the term somaclone to describe the plants originating from any type of tissue culture. The somaclonal variation may be attributed to either (i) pre existing variation in the somatic cells of the explants, or (ii) variation generated during tissue culture.

Mechanisms causing somaclonal variation

1. endomitosis
2. chromosomal abberation
3. DNA amplification
4. transposable elements
5. point mutation.

Why or how these chromosomal changes occur?

1. Alterations in DNA methylation: a decrease in methylation causes the genes more active
2. lack of nucleic acid precursors due to very rapid cell division
3. role of growth regulators like 2,4-D and sometimes kinetin in inducing variation
4. nutrient composition: KNO_3 influences the albino plants from wheat cultures, level of organic N_2 chelating agents and other micro nutrients
5. Culture conditions: temperature, method of culture, duration of culture (longer the duration of culture more the variation).

Selection of Somaclonal Variation

1. analysis of phenotypic characters : the morphological characters of somaclones are compared with the parental types for three to four generations and finally the genetic variability is assessed
2. cytological study of variants
3. mesurement of DNA content of the variants

4. gel electrophoresis of proteins and enzyme
5. by exerting selection pressure using particular chemicals in the media:
   (a) selection for disease resistance – the toxin produced by the pathogen can be used in the media (phytophthora resistance in potato)
   (b) herbicide resistance – by adding interested herbicide in the media (tobacco)
   (c) salt tolerance – by adding NaCl (tobacco)
   (d) drought/water logging tolerance – by adding PEG (tomato)
   (e) mineral toxicity/deficiency tolerance – adding and reducing the quantity of interested nutrient in the media (tomato, sorghum)
   (f) temperature stress / chilling response – exposing the culture to very high or low temperature (tobacco for chilling response, pear for temperature stress).

Somaclonal variation among regenerated plants from callus and protoplast culture has been presented to a lot of significant contributions to plant science. Somaclonal variation among regenerated plants from callus and protoplasts culture has been suggested as a useful source of potentially valuable germplasm for plant breeding and improvement. The major benefit of somaclonal variation is to create variation in adapted genotype.

List of crop species where desirable and heritable somaclonal variation has been reported

Species	Characters which were modified
Allium sativa	Bulb size, shape, clove no. aerial bulbil
Avena sativa	Plant height, heading date, awns
Hordeum spp	Plant ht, tillering
Oryza sativa	Plant ht, heading date , seed fertility, grain no., wt
Saccharam officinarum	Disease (eye spot, fiji disease, downey mildew)
Triticum aestivum	Plant and ear morphology, awns, gliadins, amylase, grain wt., yield
Zea mays	T toxin resistance, male fertility , Mt DNA
Lycopersicon esculentum	Leaf morphology, branching habit, fruit colour, pedicel, male fertility, growth
Medicago sativa	Multifoliate leaves, elongated petioles, growth branch number plant height, dry matter yield
Solanum tuberosum	Tuber shape, maturity date, plant morphology, resistance for early andlate blight, photoperiod, leaf color, vigour, height, skin colour

Problems with somoclonal variation:

- variation is cultivar dependent
- frequencies of change vary
- many changes are undesirable
- some changes are unstable
- many changes are not novel
- characters of interest may not change.

Methods of assessment of somaclonal variation:

- phenotypic parameters(quantitative (leaf size, plant height, qualitative, branching pattern, flowercolour)

- physiological parameters (protein patterns by electrophoresis, secondary products formation)
- genetic parameters (chromosome number and structure, banding pattern).

Anther Cutlure and Microspore Culture

By carefully selection of developing anthers at a precise and critical stage it is possible to establish the anther culture that will give rise to haploid plants. Guha and Maheswari (1964) first reported the direct development of embryos form pollen grains of *Datura* by anther culture (Fig. 5).

What is androgenesis?: Androgenesis is the *in vitro* development of haploid plants originating from totipotent pollen grains through a series of cell division and differentiation.

Direct androgenesis: Microspores behave like a zygote and undergoes change to form embryoid which ultimately gives rise to a plantlet.

Indirect androgenesis: In contrast to the direct androgenesis the microspore instead of undergoing embryogenesis divide repeatedly to form a callus tissue which differentiates into haploid plantlet.

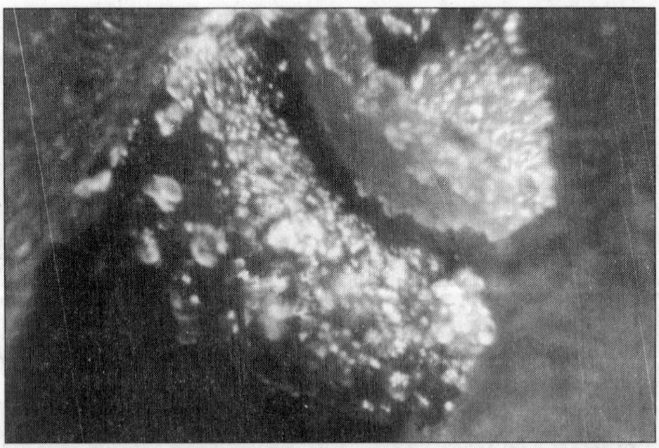

Fig. 5. Embryogenic response from cultured anthers.

Principle of anther culture: The basic principle of anther and pollen culture is the production of haploid plants exploiting the totipotency of microspore and the occurrence of single set of chromosome (n) in microspore. The principle behind the anther culture is that without disturbing the natural habitat and environment of the enclosed anther, pollen can be grown by culturing the intact anther. In culture condition, the diploid tissue of anther will remain living without proliferation at the selective medium and at the same time it will encourage the development of pollen by nursing and providing nutrient. In anther culture the gametophytic pathway of pollen is broken and it undergoes sporophytic pathway. Flower buds of the appropriate developmental stage are collected, surface sterilized and their anthers are excised and placed horizontally on culure medium. Flower buds with small anthers may themselves be cultured and in some case, the entire inflorescence has been cultured. Care should be taken to avoid injury to anthers since it may induce callus formation from anther walls. Alternatively pollen grain may be separated from anthers and cultured on a suitable medium.

Microspore Culture or Pollen Culture

Isolated pollen grains when cultured *in vitro* gives rise to haploid embryos or callus and this approach is called pollen culture. Pollen may be isolated either by squeezing or float culture of anthers.

Squeeze culture: About 50 anthers may be placed in 20 ml of medium and squeezed with a glass rod; the solution is filtered through a nylon mesh of suitable pore size and centrifuged at 500 - 800 RPM for 5 minutes. The pollen pellet is collected washed twice and suspended at a final density of 10^3 – 10^4 pollen / ml.

Float culture: Excised anthers are floated on a shallow liquid medium in petridishes; the anthers dehisce in a few days releasing their pollen grains into the medium. Initially isolated pollen grains were cultured either in hanging drops or on a filter paper raft placed on cultured anthers. Subsequently Nitsch and Co Workers first replaced the nurse tissue by an extract of cultured anthers and finally devised a completely synthetic medium for pollen culture the crucial ingredients of which were glutamine, L-Serine, and inositol.

Slit technique: Cutting the anther wall to release the microspore calluses/embryos rather than relying on natural dehiscence but this is a time consuming process eg. Tobacco, *Pennisetum*.

Advantages of haploids / anther culture / pollen culture:

(1) Haploid are useful in cytogenetic studies

(2) Production of homozygous inbreds / isogenic diploids within a year as compared to the long inbreeding method which takes 4-6 years

(3) To trace the parents of the hybrids

(4) Haploids are valuable for mutation studies

(5) By comparing the heterozygous diploid with haploid or homozygous diploid population recessive phenotypic characters can be identified

(6) Development of pure lines and 100% male plants in Asparagus

(7) Recovery of sexual inter specific hybrids between wild and domestic species (tomato)

(8) For understanding the phylogenetic relationship between species the study of meiotic behaviour of haploid provide clues for chromosome duplication within a species

(9) To study the inheritance pattern

(10) Use of haploids in the production of monosomics, nullisomics, and other anueploids

(11) Double haploid, that are homozygous and fertile are readily obtained enabling the selection of desirable gene combination.

How to double the chromosomes in haploid?

Spontaneous duplication: Homozygous diploid callus or embryoids may be formed by the spontaneous fusion of two similar nuclei of cultured pollen after first division. Haploid cells are unstable in culture and have a tendency to undergo endomitosis.

Induced duplication: Colchicines treatment: the young plantlets regenerated through anther culture are treated with 0.5% colchicines solution for 24–48 hours. Treated plantlets are replanted in the medium after through washing. In case of mature haploid plantlets , 4% colchicines paste may be applied to the axil of the leaves.

Ovary Culture

Gynogenesis: Culture of unfertilized ovaries to obtain haploid plants from egg cell or other haploid cells of embryo sac is called ovary culture and the process is known as Gynogeneis. The first report on gynogenesis was by San Noem in 1976 in case of barley. During induction of ovaries are floated on a liquid medium having low auxin and kept in dark while for regeneration they are transferred to an agar medium with higher auxin concentration and incubated in light. *In vitro* parthenogenesis, haploid plants generally originate from egg cell in most of species (rice). *In vitro* apogamy, when haploid plants originate from other than egg cell called synergid (rice) or antipodal (Allium tuberosum).

Limitation:

- The frequency of responding ovaries (1 – 5%) and the number of plants / ovary (1-2) is low
- It has been successful only in lesser number of plant species.

Advantages of ovary culture:

(1) Ovary/ovule culture may be useful when there is male sterility in crop species
(2) For *in vitro* pollination and fertilization or for embryo rescue
(3) To produce parthenogenetic haploids (wheat, barley,)
(4) To understand the physiology of fruit development
(5) Reduction in the frequency of albino plants.

Following pollination, whole flower buds are excised (2-15 days pollination) Calyx, corolla and stamen are removed. Ovaries are then surface sterilized and inoculated. To obtain unpollinated ovaries, flower buds are removed 24-48 hours prior to anthesis.

Ovule Culture

Flowers are sterilized inoculated on pre-culture medium. After 10 to 14 days ovules are removed from flowers and plated on medium. Ovule requires exclusive skill . Ovule culture is mainly tried only in those cases where embryo aborts very early, and embryo culture is not possible due to difficulty of its excision at a very early stage.

Embryo Culture

Embryo culture is the aseptic isolation and growth of sexually produced embryos *in vitro* with the objective of obtaining viable plants. In 1904, Hanning published the first report on embryo culture in Cruciferae.

Types of embryo culture:

(1) Culture of immature embryos (hybrid seed)
(2) Culture of mature embryos (to avoid inhibition in germination)

Isolation of embryo: The embryo excision operation is performed aseptically in a laminar flow hood. A stereomicroscope equipped with a coolary fluorescent lamp is required for excision of small embryos. Soaking a hard coat seed in water facilitates easy excision. The excision of smaller or immature embryos require careful dissection.

Culture strategies: Sometimes young embryos may prove difficult to culture directly on the medium. In such cases, embryos may be placed onto or implanted into developing endosperm that are cultured in vitro. This technique is known as embryo-nurse endosperm technique.

Precocious germination: When excised immature plant embryos are grown *in vitro*, they do not proceed further for its maturation and start germinate. This phenomenon without completing normal embryogenic development is known as precocious germination.

Practical applications of embryo culture:

1. Rescuing embryos from incompatible crosses in interspecific and intergeneric hybridization programmes, incompatability barriers often prevent normal seed development and production of hybrids due to post zygotic barriers and causes embryo abortion. In such cases isolating the embryo and culturing in vitro may give useful results.
2. Overcoming dormancy and shortening breeding cycle:
   (a) the life cycle of Iris was reduced from 2 – 3 years to less than one year. It is possible to obtain two generations of flowering against one in Rosa spp.
   (b) germination of the immature embryos of orchid seeds
   (c) dry storage wild oat (*Avena fatua*).
3. Overcoming seed sterility in early ripening fruit cultivars seeds do not germinate because their embryos are still immature, e.g. Peach, apricot, plum, coconut.
4. Production of monoploids by bulbosum technique in interspecific hybridization between Hordeum vulgare and Hordeum bulbosum the zygote is formed but during the development of embryo subsequently the chromosomes of H. bulbosum are eliminated and monoploid embryos of H.vulgare are formed. However the monoploid embryos are weak and show slow growth and on the other hand when these monoploids are dissected and cultured in vitro they develop into monoploid plants are produced.

Plant Protoplasts: Status and Biotechnological Perspective

The 1980s witnessed many protoplast-based articles, particularly those reporting novel protoplast-to-plant systems for genetic manipulation(Davey *et al*,2005) . During the 1990s, protoplast-based technologies for gene transfer were overshadowed by Agrobacterium and Biolistics R-mediated gene delivery to plants. However, public antagonism (especially in Europe) to recombinant DNA technologies renewed interest in exploiting protoplasts in somatic hybridisation, cybridisation, protoclonal variation studies, proteomics, and metabolomics. Cells of primary plant tissues possess cellulosic walls with a pectin-rich matrix, the middle lamella, joining adjacent cells. The living cytoplasm of each cell, bounded by the plasma membrane, constitutes the protoplast. Normally, intimate contact is maintained between the plasma membrane and the wall, since this membrane is involved in wall synthesis. However, in hypertonic solutions, the plasma membranes of cells contract from their walls.

Subsequent removal of the latter structures releases large populations of spherical, osmotically fragile protoplasts (dnakedT cells), where the plasma membrane is the only barrier between the cytoplasm and its immediate external environment. Protoplast isolation is now routine from a wide range of species; viable protoplasts are potentially totipotent. Therefore, when given the correct chemical and physical stimuli, each protoplast is capable, theoretically, of regenerating a new wall and undergoing repeated mitotic division to produce daughter cells from which fertile plants may be regenerated via the tissue culture process. Protoplast-to-plant systems are available for many species, with an extensive literature relating to their exploitation. It is noteworthy that the basic procedures for protoplast isolation have undergone little change since first reported. However, remarkable progress has been made in the number of species for which protoplast-to-plant systems exist. Furthermore, the later decades of the 20th century witnessed dramatic developments in the genetic manipulation of plants through protoplast fusion and transformation The physiological status of the source tissue influences the release of viable protoplasts. Furthermore, seasonal variation, which affects the reproducibility of protoplast isolation from glasshouse-grown plants, can be effectively eliminated using in vitro grown (axenic) shoots, seedlings, and embryogenic cell suspensions. Nevertheless, Keskitalo reported that protoplast isolation from cultured shoots of *Tanacetum vulgare* and *Tanacetum cinerariifolium* was most successful during winter and spring (December to April), suggesting the persistence of a seasonal dclockT even in vitro. In contrast, other workers have not observed seasonal variation *in vitro*.

Mliki *et al.* (2003) isolated protoplasts from Tunisian varieties of grape (*Vitis vinifera*) and concluded that the highest yields were from leaves of cultured shoots 4–5 weeks after transfer of the shoots to new medium. An advantage of seedlings is that protoplasts can be isolated from radicles, hypocotyls, cotyledon tissues, roots, and root hairs within a few days of seed germination. Procedures for protoplast isolation Mechanical procedures, involving slicing of plasmolysed tissues, are now rarely employed for protoplast isolation, but are useful with large cells and when limited (small) numbers of protoplasts are required. Recently, this approach has been used successfully to isolate protoplasts of the giant marine alga, Valonia utricularis, for patch clamp analyses of their electrical properties, including physiological changes of the plasma membrane induced by exposure of isolated protoplasts to enzymes normally used to digest cell walls (Davey *et al.*, 2005).

CULTURE TECHNIQUES FOR ISOLATED PLANT PROTOPLASTS

Culture Media

The necessity to develop protoplast-to-plant systems, particularly for economically important species, has demanded a major investment of resources. Typically, isolated protoplasts commenced cell wall regeneration within a short time (often minutes) following introduction into culture. However, they require osmotic protection until their new primary walls can counteract the turgor pressure exerted by the cytoplasm. In some cases, gradual reduction of the osmotic pressure by diluting the culture medium with a solution of similar composition, but of reduced osmotic pressure, is essential for sustaining mitotic division, leading to the formation of daughter cells and tissues. Innovative approaches for protoplast culture The prime objective in developing effective protoplast-to-plant systems is to maximize cell growth and differentiation, particularly for those systems exploited to generate novel plants, often allied to some form of genetic manipulation. Whilst guidance can be obtained from the literature, a universal protocol does not exist in terms of medium composition and physical parameters to maximize protoplast growth. Consequently, such parameters must be determined empirically. Several novel approaches have been described, some of which have been successful earlier for animal systems. These include electrical treatment of cells (Lynch and Davey, 1996) and manipulation of the gaseous environment (Lowe *et al.*, 2003) Conclusions Protoplasts are naked cells that are the equivalent, in general terms, to cultured animal cells. However, unlike the latter, protoplasts exhibit the unique property of totipotency. Consequently, protoplasts provide a cell system that can be manipulated readily, using physiological and pharmacological perturbations, and such experimentation can be followed through differentiation pathways to the whole (plant) organism and subsequent generations. It is clear that future research using protoplasts will be informed, to some extent, by developments in animal cell culture and that closer liaison between animal and plant biologists will increasingly become a feature of 21st century biology. One relevant example is the exciting developments in the field of molecular farming in which human therapeutic proteins are synthesised in plant cells. Protoplasts will increasingly provide an underpinning technology for these developments to be more fully realized (Davey *et al.*, 2005).

Sources of protoplast:

(a) mesophyll cells of leaves
(b) cultured suspension cells
(c) callus culture
(d) preconditioned plant material.

Isolation of protoplast:

Methods of protoplast isolation can be classifed into two main groups

(1) mechanical method (non enzymatic)
(2) enzymatic
    (a) sequential method (two step method)
    (b) mixed method (single step method).

Mechanical method: It is done by cutting plasmolysed cells with a sharp edged micro scalpel or knife after keeping the material under microscope. The protoplasts are released and and the cells are deplamolysed. This method is useful for isolation of protoplasts from vacuolated cells (eg. Onion bulbs, scales, radish roots). This method gives poor yield of protoplasts and is not suitable for isolating protoplasts from meristematic and less vacuolated cells. The mechanical method though was used as early as 1982, in now only rarely used for isolation of protoplasts

Enzymatic method: Commercially available enzymes:

(a) pectolyase Y 23
(b) cellulase = onozyka R10
(c) meicelase
(d) rhozyme
(e) macerozyme R 10
(f) hemicellulase

(g) pectinase
(h) drieselase.

The role of enzymes is to dissolve middle lamella and dissolving cell wall. A combination of these enzymes in a concentration of 0.5 to 20% is used . In many cases, the macerozyme and cellulase are sufficient to obtain protoplast in significant umber. The enzyme solution is prepared in 10-15% sorbitol or mannitol containing small amount of $CaCl_2$ (7mM) for membrance stability.

Sequential method: This involves initial incubation of macerated plant tissue with pectinase (macroenzyme) which inturn are then converted into protoplasts by cellulase treatment.

Mixed enzymatic: Plant tissue are plasmolysed in the presence of a mixture of pectinase and cellulases thus inducing simultaneous separation of cells and degradation of cell wall to release protoplast directly. After enzyme treatment protoplast suspensions are collected by centrifugation (60 –100 rpm) for 2 – 5 minutes. Then washed in medium without enzyme. Cell debries are removed and the protoplasts are placed in a medium with appropriate concentration of sucrose or mannitol. This method is widely used since:

(1) large quantities of protoplasts can be obtained
(2) the cells are not broken as in case of mechanical isolation
(3) Osmotic shrinkage is less
(4) Takes reduced time
(5) Lesser contamination.

Osmoticum: Osmoticum is a solution causing changes in osmatic pressure. During isolation and culture , protoplasts require osmotic protection until they regenerate a strong wall. Inclusion of an osmoticum in both isolation and culture media prevents rupture of protoplasts. The most widely used osmotica are sorbitol, mannitol, glucose, or sucrose.

Subprotoplasts: Subprotoplast do not contain the entire contents of plant cells and include the following :

Cytoplast: The protoplasts lacking a nucleus and each contains entire cytoplasm of a cell

Miniprotoplast / karyoplast: Isolated protoplasts contain a nucleus surrounded by some cytoplasm and the original outer plasma membrane

Microplast: The protoplast contain only a fraction of cytoplasm and outer membrane

Microprotoplast: The protoplats containing only a few of all chromosomes and a fraction of the cytoplasm

Viability of protoplast: The viability of protoplast is tested by

(1) Observing the presence of cytoplasmic streaming
(2) Exclusion of Evans Blue dye
(3) Change in protoplast size due to change in the level of osmoticum
(4) Presence of photosynthetic and respiratory activity
(5) FDA (Fluorescein DiAcetate) or CFW Calcofluor white (CFW) test.

Minimum plating density: A minimum plating density of protoplasts is requied for growth to begin eg. 5×10^3.to 1×10^5protoplast / cm^3 for tobacco.

Culture of Protoplast

The first step in the protoplast culture is the development of a cell wall around the membrane of isolated protoplast. This is followed by induction of division in the protoplast derived new cell giving rise to a small cell colony. By manipulation of the nutritional and physiological conditions in the nutrient media, cell colonies may be induced to grow callus continuously or to regenerate whole plants.

Protoplast fusion: Protoplasts fusion or somatic hybridization is one of the most important uses of protoplast culture. This is particularly significant for hybridization between species or genera, which can not be made to cross by conventional method of sexual hybridization.

Somatic hybridization: *In vitro* fusion of plant protoplasts derived either from somatic cell of somatic plant or from two genetically different plant is called somatic hybridization.

Methods of Protoplasts Fusion

(1) Spontaneous fusion
(2) Induced fusion.

Spontaneous fusion: Protoplasts during isolation often fuse spontaneously and this phenomenon is called spontaneous fusion. Simply physical contact is sufficient to bring about the spontaneous fusion among similar parental protoplasts. The occurrence of multinucleate fusion bodies is common when cells are prepared from actively dividing cells. Spontaneous fusion usually gives rise to homokaryon usually intraspecific. Protoplasts of young leaves undergo spontaneous fusion frequently. This type of fusion can be done with the help of micromanipulators or micropipettes.

Induced fusion: For achieving interspecific and intergeneric fusion the following methods are followed:

(1) Chemical method
(2) Electric method
(3) Mechanical method.

The induced fusion requires a suitable agent called fusogen. The inducing agents first brings the protoplasts together and then causes them to adhere to one another for bringing about fusion (Fig. 6).

Chemofusion: The following chemicals are used as fusogens

(1) $NaNO_3$ (isolated protoplasts are suspended in an aggregation mixture of 5.5% $NaNO_3$ in 10% sucrose solution.

(2) Poly Ethylene Glycol (PEG)(1 ml of protoplasts suspended in a culture medium with 1ml of 56% of PEG and tube shaken for 5 seconds. The protoplasts are allowed to sediment for 10 minutes washed with growth medium and examined for successful agglutination and fusion).

When the quantity of protoplast is less drop cultures can be used. Protoplasts are placed as microdrops in petriplates to which PEG (50ml each)are added to each drop and kept for 5-10 minutes.

(3) Calcium ions (Ca^{++})(spinning the protoplasts in a fusion inducing solution (0.05M $CaCl_2 2H_2O$ in 0.4M Mannitol at high pH of 10.5).

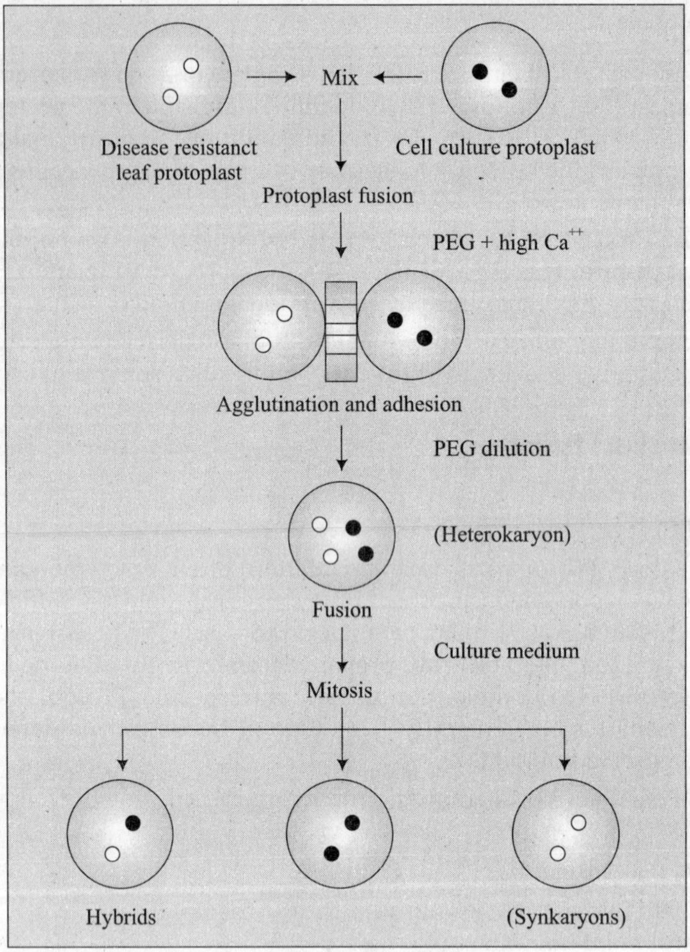

Fig. 6

(4) Proteins: (Gelatin and early products of its degradation at a concentration of 2.5% induced aggregation at high frequency within one hour (e.g. Vicia, glycine, allium).

Electric fusion: If protoplasts are placed into a small vessel containing electrodes and a potential difference is applied then the protoplasts will line up between the electrodes. Afterwards an extremely short square wave electric shock is applied protoplasts can be induced to fuse.

Mechanical fusion: The isolated protoplasts are brought into intimate physical contact mechanically under microscope using micromanipulatorand perfusion micropipette.

Selection of somatic hybrids: The protoplasts suspension recovered after a treatment with fusogens consists of following types:

1. unfused prototplasts of two species/strains
2. products of fusion between two or more protoplasts of the same (homokaryon)

3. hybrid protoplast produced by fusion between protoplasts of two species (heterokaryon) effective strategy has to be employed for the identification somatic hybrids of heterokaryon in nature and they should be isolated. This step is called selection of hybrid cells.

Methods of Selection of Somatic Hybrids

1. Somatic visual markers

   Hybrids of Petunia parodii (green) with P.hybrida (white) could be isolated in the form of green callus which represented only hybrid cells.
2. biochemical basis for complementation

   This selection strategy exploits thse natural properties of the two parental species which show complementation in the hybrid cells and at same time permit their selection.
   (a) auxin autotrophy: the parental protoplasts of Nicotina glauca and Nicotiana longsdorffi requires an auxin compound in order to proliferate, whereas hybrid callus tissue needs no such requirement since the hybrid cells are auxin autotrophic.
   (b) protoplast of Petunia hybrida form calli on the MS medium while those of P.parodii produce only small cell colonies. Further actinomycin (1mg/ml) inhibits cell division of P.hybrida protoplasts, but it has no effect on those of P.parodii, Thus protoplasts of both these Petunia species fail to produce macroscopic colonies on MS medium supplemented with 1mg/ml actinomycin D whereas hybrid cells divide normally and produce macroscopic calli.
3. to culture the entire protoplast population
4. labeLling: protoplasts of two parents may be labeled by different fluorescent agents, which will then enable the selection of hybrids
   (i) octadeconyl amino fluorescent
   (ii) octadecyl palamine
5. Fluorescent Assorted Cell Sorter (FACS).

Symmetric hybrids: Some somatic hybrid plants retain the full or nearly full somatic complements of the two parental species ; they are called symmetric hybrids.

Asymmetric hybrids: Many somatic hybrids exhibit the full somatic complement of one parental species are lost during the preceding mitotic divisions; such hybrids are referred to as asymmetric hybrids.

CYBRIDS

Cybrids or cytoplasmic hybrids are cells or plants containing nucleus of one species but cytoplasm from both the parental species. Cybrids can be produced in the following methods:

1. fusion of normal protoplasts from one parent with enucleated protoplasts from the other parent. Enucleated protoplasts can be obtained by high speed centrifugation (20,000 – 40,000g for 45 – 90 minutes) of protoplasts or by irradiation treatment
2. fusion of normal protoplasts form one parent and protoplasts containing non viable nuclei from the other

3. selective elimination of one of the nuclei from the heterokaryon
4. selective elimination of chromosomes of one parent at a later stage after fusion of the nuclei.

Major Application of Cybrids

- to transfer the cytoplasmic male sterility (tobacco, tomato)
- to transfer antibiotic resistance character(tobacco)
- to transfer herbicide resistance (brassica)

Production of cytoplsmically male sterile lines in tomato

1. Mesophyll protoplasts of tomato (Lycopersicon esculentum) were treated with iodoacetamide (IOA) to inactivate mitochondria and
2. Mesophyll protoplast of Solanum acaule (*S. tuberosum*) were irradiated with ϒ or X-rays to inactivate nuclei. The protoplast were mixed in 1:1 ratio and induced to fuse using $Ca^{++}$and PEG, leading to the production of heterologous hybrids. Among the fusion from the original cultivars with respect to morphology, physiology and chromosome number (2n = 24), but exhibited various degrees of male sterility
3. The nuclear genotype of cultivar remains unaffected 100% probability of getting somatic hybrids

Applications of somatic hybridisation and cybridisation

(i) For effecting wide hybridization
Interspecific hybrids
in genus Daucus and its relatives
in genus Nicotiana and Brasscia
Intergeneric
Rice + Echnichloa = oryzochloa
Raphanus sativus + B.oleracea = raphanobrassica
N.tabacum + Lycopersicon esculentum = solanopersicon
Solanum tuberosum + L.esculentum = solanopersicon

(ii) To create recombinants in asexually propagated crops or sterile plants, e.g. Potato

(iii) Overcoming the barriers of self incompatability, e.g. Nicotiana tabacum x N.nesophila (disease resistant)

(iv) Production of cms lines

(v) Production of herbicide tolerant plants (brassica).

Conservation of Tissues

Six major steps defined in the conservation use cycle are collection, quarantine, propagation, characterization, evaluation, monitoring, storage and distribution. The role of *in vitro* conservation techniques in the overall conservation strategies should be indicative of the fact that it should complement other conservation strategies within the total programme of a given species or population. The methods chosen should be carefully considered taking into account the feasibility, practicality, economy and security (Grout, 1995 and Ashmore, 1997 and Rajasekharan and Ganeshan, 2002).

Generally, field conservation of medicinal plants requires more space, labour intensive and expensive. They also run the risk of being damaged by natural calamities and biotic stress factors. Techniques to conserve such species *in vitro* have recently been developed. For some species, while *in situ* conservation is the only option available, tissue culture systems offer advantages, which are listed below:

(1) Very high multiplication rates.
(2) Aseptic System
  - free from fungi, bacteria, viruses and insect pests.
  - Production of pathogen free stocks
(3) Reduction of space requirements
(4) Genetic erosion reduced to zero under optimal storage conditions.
(5) Reduction of the expenses in labour costs.

In vitro collections of species could be maintained at the same or separate site, but should have clear linkages with field gene banks. The properties required for a successful *in vitro* conservation system as defined by Grout (1995) are:

The ability of the biological system to:

1. Minimize growth and development *in vitro.*
2. Maintain viability of stored material at the highest possible level along with minimum risk of genetic stability.
3. Maintain full developmental and functional potential of the stored material when it is returned to physiological temperatures.
4. Make significant savings in labour input, materials and commitment of specialized facilities.

Some of the advantages favouring this conservation strategy are:

1. Collection may occur at anytime, independent of flowering periods for each species.
2. There is in the potential for virus elimination from contaminated tissue through meristem culture.
3. Clonal material may be produced
4. Rapid multiplication.
5. Germination of 'difficult' immature seed/embryo rescue may be facilitated for breeding
6. Distribution across borders may be safer.

Tissue culture techniques are of great interest for the collecting, multiplication and storage of plant germplasm (Fig. 7). Tissue culture systems allow propagation plant material with high multiplication rates in an aseptic environment. Virus-free plants can be obtained through meristem culture in combination with thermo therapy, ensuring the production of disease-free stocks and simplifying quarantine procedures for the international exchange of germplasm. The miniaturization of explants allows reducing space requirements and consequently labouring costs for the maintenance of germplasm collections (Ashmore, 1997).

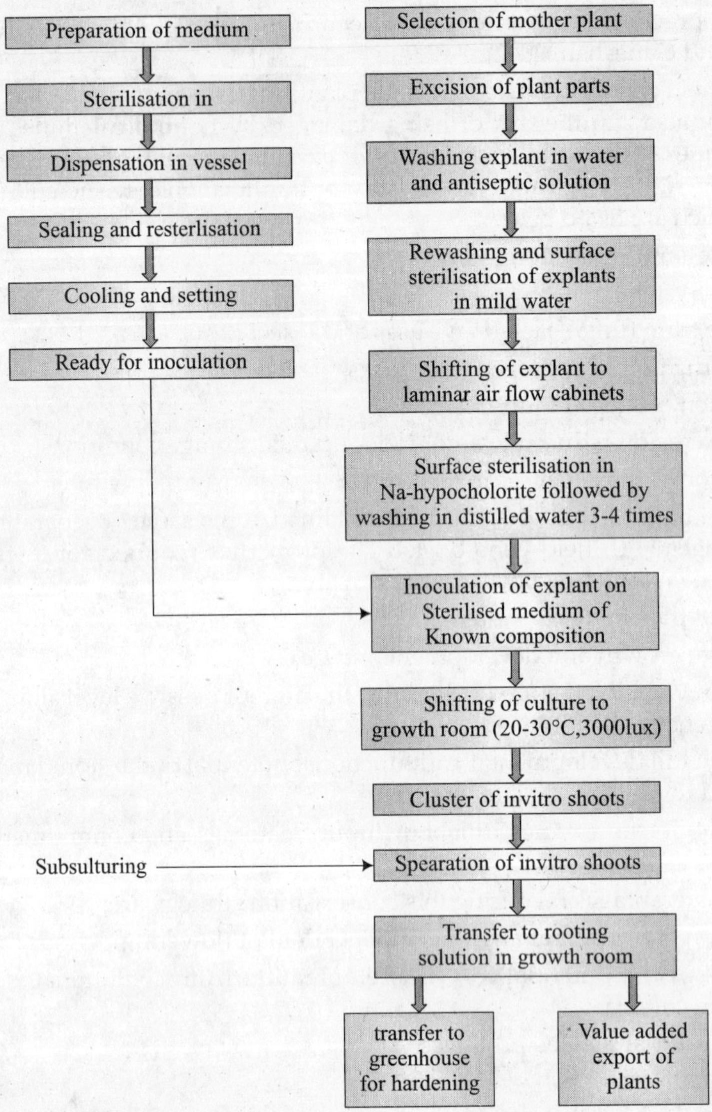

Fig. 7. Flow chart for plant propagation by tissue culture

The *in vitro* methods employed vary according to the storage duration requested. For medium-term conservation, the aim is to reduce growth, thus increasing intervals between subcultures. For long-term conservation, cryopreservation, i.e. conservation at ultra-low temperature, usually that of liquid nitrogen (–196°C), is the only method currently available. Indeed, all metabolic processes as well as cell divisions are stopped at this temperature. Plant material can thus be conserved without modification or alteration for extended periods, under a reduced volume and maintenance. Standard culture conditions can only be used for medium-term storage of slow growing species. However, in most cases, environmental conditions and/

or culture medium have to be modified to induce growth reduction. This is most frequently achieved by temperature reduction, often in association with a decrease in light intensity or even its complete suppression.

Temperatures in the range of 0–5°C are employed to cold-tolerant species but higher temperatures are requested in the case of tropical species, which are generally cold sensitive. It is also possible to limit growth by modifying the culture medium, mainly by reducing the sugar and/or concentration of mineral elements. Finally, the type of culture vessel, its volume as well as the type of enclosures influence survival of cultures. Notably, it is important to limit the evaporation of the culture medium by using a type of enclosure, which is airtight.

Medium-term conservation techniques have been developed for a wide range of plant species, but they are used routinely for *in vitro* conservation for few. New medium-term conservation techniques are presently being researched. They include reduction of the oxygen level available to cultures (achieved by covering explants with a layer of liquid medium or mineral oil, or by placing them in controlled atmosphere), desiccation and encapsulation of explants in alginate beads. However, these techniques are still at the experimental stage. Dramatic progress has been made during the last 10 years in the development of in vitro culture techniques for the conservation of plant germplasm. If slow growth techniques are immediately applicable in most cases, cryopreservation techniques still require fine tuning by more research inputs, before they can be applied routinely to plant species (Engelmann, 1997).

In most laboratories and genebanks of developing countries the use of *in vitro* conservation techniques offer main advantages. Even though basic infrastructures already exist in many cases, they are not necessarily adaptable to the often-complex conservation procedures described in literature. It is therefore indispensable to assess the flexibility of conservation protocols, their limits of applicability and simplify them to achieve efficient conservation. Finally, it is essential to underline the complementarities of *in vitro* techniques with other strategies of plant diversity conservation. *In vitro* conservation does not intend to replace conventional approaches for *in situ* and *ex situ* conservation. *In vitro* conservation offers researchers and genebank curators a set of additional tools to allow improvement in conservation of indigenous germplasm collections used in plant systems, placed under their responsibility.

CRYOPRESERVATION

The most promising approach to secure, long-term conservation lies in cryopreservation; that is, storage at ultra-low temperature in liquid nitrogen. Cryopreservation effectively suspends metabolism and therefore should eliminate time related biochemical phenomena Thus, the challenge in cryopreservation relates to the transition to and from the exposure to that temperature for a particular period of time. There is a growing awareness of the need to conserve plant genetic resources, not just to maintain biodiversity, but also to support plant breeding and biotechnology programmes. Cryopreservation, the storage of viable biological material at ultra-low temperatures, provides a means for the long-term stable storage of plant germplasm. The techniques could be applied to a range of plant species belonging to the tropical sub tropical and temperate category. It has important applications in plant genetic resources management and plant biotechnology. Cryopreservation of *in vitro* cultures is especially important for long-term storage and international exchange of germplasm of vegetatively propagated crops and recalcitrant seed species. Similarly in plant biotechnology, cryopreservation is the method available for the production of secondary metabolites or of cell lines derived from genetic transformation somatic hybridization etc (Engelmann, 1997).

Preservation of the *germ*plasm at –196°C, the temperature of liquid nitrogen (L.N) has been actively utilized for the base collections of gene banks of animal sperm, embryos, cells and microorganism. The first report of successful cryopreservation by Quatrano (1986) of plant cell suspension and production of somatic embryos from cryopreserved cells led to numerous studies on cryopreservation of plant systems. Cryogenic storage is a new emerging technology; long-term conservation of germplasm at ultra low cryogenic temperature of liquid nitrogen is based on the reduction and subsequent arrest of metabolic functions of all biological material.

Cryogenic storage in liquid nitrogen (–196°C) provides the most promising method of in vitro germplasm conservation, where in all or most of the metabolic activities are immobilized. Cryopreservation has been successfully applied to seed, pollen, anthers, cell suspension culture, calli, protoplast culture, shoot apical and lateral meristems, shoot tips and embryos (both somatic and zygotic).

However there is a small risk of injury to macromolecules, including the genetic material, in cryopreserved subjects resulting from free radical activity during processing and from external ionizing radiation during storage. Any such injury will be preserved in the stored tissues and accumulate, due to the nonfunctioning of repair mechanism at the cryogen temperature.

In practice cryopreservation provides a genuinely long-term storage technique but, as with all storage systems, there is a finite point, probably measured in decades, at which conserved material should be recovered and re-grown *in vivo* before being re-evaluated and re-entered into storage .One of the major advantages of cryopreservation is that a reduced frequency of viability testing of individual samples especially for seeds. Materials stored are not subjected to loss from electrical and mechanical failure of refrigeration units since LN is the source for temperature maintenance and its stock can be easily maintained. A good LN storage tank should be self contained, vacuum insulated vassel with LN reservoir and samples in the same cavity LN in spite of being an inert gas can cause cold burns. Cryogloves and safety goggles are thus essential. In conclusion the various equipment required for a cryopreservation laboratory are:

1. Reliable source of liquid nitrogen.
2. Safety equipment (gloves, apron, face shield or goggle) pumps for dispensing liquid nitrogen, from and to large storage tanks, trolleys for transport of tanks.
3. Small (11–35 lit) liquid nitrogen tanks
4. Cryotanks for routine storage of LN in large capacities (1000 lits etc)
5. Tanks for long term storage of specimens., These may be equipped with an alarm system which is activated if the liquid nitrogen level falls below a critical level.
6. Crovials, cryogloves, boxes, canes and racks.
7. Programmable freezer.
8. Water bath for thawing at 35–50°C.

The techniques of cryopreservation basically involves several important steps such as follows:

1. Pretreatment
2. Dessication
3. Freezing
4. Cryopreservation in liquid nitrogen
5. Rewarming/thawing
6. Post treatment/recovery
7. Testing of viability and reculture.

Cryoprotection

1. Pre treatment: In general, the plant tissues show a natural resistance to freezing, most plant materials require special manipulations and/or culture conditions in order to present a specimen with maximal freeze-tolerance, that is the specimen can be made dehydration tolerant and freezing tolerant by pretreatment with cryoprotectants. Discovery of chemicals with cryoprotective properties has contributed significantly towards the development of cryopreservation procedures for biological material, these refered to as cryoprotectants. Most commonly used are Dimethylsulfoxide (DMS) and glycerol, glycerol and sucrose in combination.

The application of certain chemicals to the culture for approximately 1hour before freezing will bring about changes in cell permeability, freezing point and response to the stresses of freezing and thawing essential to survival. The cryoprotectant used must be relatively non - hazardous to the plant tissue. The most commonly used compounds are DMSO and glycerol. The former is some times applied alone where as glycerol is rarely effective and is used in combination with DMSO and/or other compounds such as sorbitol, sucrose, glucose, mannitol, proline or polyethylene glycol. Cryoprotectants are usually prepared in culture medium at double strength, to be diluted at the time of application, adjusted to the standard medium pH, and filter sterilized to avoid the risk of caramellisation or other breakdown processes during autoclaving.

2. Dessication: The plant material to be cryopreserved should be devoid of moisture. Seeds can be dessicated by simple sun drying or by using some dessicants, which are able to absorb excess moisture content from the plant material. Pollen can be dried using dessicators.

3. Freezing: Between the initial sub-zero temperatures and approximately –100°C, significant changes take place as ice formation ensues extra cellular and intracellularly. The nature of changes depends upon the specimen, cryoprotectant treatment and upon the cooling rate. Essentially 2 types of freezing, either by rapid or slow cooling is employed.

A. Freezing by rapid cooling causes relatively early intracellular freezing and relatively little dehydration. Unless cooling is very rapid indeed, the ice crystals forming within the cell will be damaging.

B. Freezing by slow cooling: Slow cooling permits the flow of water from the cell to the outside thereby promoting extracellular ice formation instead of the lethal intracellular freezing. It is found that slow freezing is preferable to rapid freezing in order to retain maximum viability.

4. Storage in liquid nitrogen: To avoid damaging changes involving ive re-crystallization, it is necessary to store frozen cultures at an adequately low temperature of at least –100°C. For practical convenience, storage in liquid nitrogen itself at –196°C or liquid nitrogen vapor at approximately –150°C is adopted.

5. Thawing: Damaging ice re-crystallization will occur in cells, which are thawed too slowly. The extent of damage will be in proportion to the amount of intracellular ice. Rapid thawing is therefore a sensible general precaution. This is carried out by dropping the specimen's into warm or culture medium as appropriate. In few cases, slow thawing may be advantageous.

6. Recovery: Major biophysical changes may have occurred in the cell during freezing and thawing, the freshly thawed cell is strongly predisposed to further damage and requires appropriate 'nursing'. This is done by direct transfer to recovery medium without washing, although special washing conditions may be used to overcome damaging effects, while permitting removal of cryoprotectants.

NEW CRYORESERVATION TECHNIQUES

Vetrification

The term "vitrification" refers to the amorphous solidification of water without the formation of damaging ice crystals. Hydrated cells, tissues and organs do not survive storage of LN because of the detrimental intracellular freezing that occurs during cooling. In order to avoid intracellular freezing, cells should be sufficiently dehydrated so that remaining water vitrifies during the rapid cooling when the cells are immersed in liquid nitrogen. The term "vitrification" refers to the amorphous solidification of water without the formation of damaging ice crystals. This involves the avoidance of ice crystallization at low and ultra low temperature and exploits the ability of water to form an amorphous lass in high viscosity solution and thus, a glassy "vitrified" state results. Thew tissue can be directly plunged into liquid nitrogen without the need of a controlled cooling step. Because ice formation is avoided, theoretically this technique has considerable potential in cryogenic storage. However, there remain in practice several disadvantages. Vitrification solutions (30% <w/v> glycerol + 15% <w/v> ethylene glycol + 15% <w/v> DMSO) can be toxic and temperature and Tate of their application is critical in order to minimisre damaging osmotic effects. Glasses are unstable and the return of vitrified solution to "normal" growth temperatures can result in ice nucleation.

Encapsulation/ Dehydration

First developed by Fabre and Dereuddre (1990) this procedure involves the encapsulation of meristamatic tissues in calcium alginate beads. Subsequent bead dehydration using osmotic (sucrose) and air drying treatments can be followed by direct freezing using liquid nitrogen or controlled cooling to an intermediate temperature (using a programmable freezer) before transfer to liquid nitrogen. After storage, the beads may be warmed at room temperature and returned to standard culture conditions. In most species it is possible to regenerate whole plants as the developing shoot emerges from the beads. Cryopreservation of organized plant material is, after twenty years of research, still in the experimental phase. It is important that empirical research be extended and the more fundamental aspects of cryopreservation

Media for Plant Tissue and Cell Cultures (mg/L)

Components	Murashige-Skoog (1962)	White (1963	Gamborg (1968)	Nitsch (1951)	Heller (1953)	Schenk - Hildebrandt (1972)	Nitsch - Nitsch (1967)	Kohlenbach - Schmidt (1975)	Knop (1865)
$(NH_4)_2SO_4$	-	-	134	-	-	-	-	-	-
$MgSO_4 \times 7H_2O$	370	720	500	250	250	400	125	185	250
Na_2SO_4	-	200	-	-	-	-	-	-	-
KCl	-	65	-	1,500	750	-	-	-	-
$CaCl_2 \times 2H_2O$	440	-	150	25	75	200	-	166	-
$NaNO_3$	-	-	-	-	600	-	-	-	-
KNO_3	1,900	80	3,000	2,000	-	2,500	125	950	250
$Ca(NO_3)_2 \times 4H_2O$	-	300	-	-	-	-	500	-	1,000
NH_4NO_3	1,650	-	-	-	-	-	-	720	-
$NaH_2PO_4 \times H_2O$	-	16.5	150	250	125	-	-	-	-
$NH_4H_2PO_4$	-	-	-	-	-	300	-	-	-
KH_2PO_4	170	-	-	-	—	-	125	68	250
$FeSO_4 \times 7H_2)$	27.8	-	27.8	-	-	15	27.85	27.85	-
Na_2EDTA	37.3	-	37.3	-	-	20	37.25	37.25	-
$MnSO_4 \times 4H_2O$	22.3	7	10 (1 H_2O)	3	0.1	10	25	25	-
$ZnSO_4 \times 7H_2O$	8.6	3	2	0.5	1	0.1	10	10	-
$CuSO_4 \times 5H_2O$	0.025	-	0.025	0.025	0.03	0.2	0.025	0.025	-
H_2SO_4	-	-	-	0.5	-	-	-	-	-

Contd.

Contd.

$Fe_2(SO_4)_3$	-	2.5	-	-	-	-	-	-	-
$NiCl_2 \times 6H_2O$	-	-	-	-	0.03	-	-	-	-
$CoCl_2 \times 6H_2O$	0.025	-	0.025	-	-	0.1	0.025	-	-
$A1Cl_3$	-	-	-	-	0.03	-	-	-	-
$FeCl_3 \times 6H_2O$	-	-	-	-	1	-	-	-	-
$FeC_6O_5H_7 \times 5H_2O$	-	-	-	10	-	-	-	-	-
Kl	0.83	0.75	0.75	0.5	0.01	1.0	-	-	-
H_3BO_3	6.2	1.5	3	0.5	1	5	10	10	-
$Na_2M_0O_4 \times 2H_2O$	0.25	-	0.25	0.25	-	0.1	0.25	0.25	-
Sucrose	30,000	20,000	20,000	50,000 or 36,000	20,000	30,000	20,000~30,000	10,000	-
Glucose	-	-	-		-	-	-	-	-
Myo-Inositol	100	-	100	-	-	1,000	100	100	-
Nicotinic Acid	0.5	0.5	1.0	-	-	0.5	5	5	-
Pyridoxine HCl	0.5	0.1	1.0	-	-	0.5	0.5	0.5	-
Thiamine HCl	0.1-1	0.1	10	1	1	5	0.5	0.5	-
Ca-Pantothenate	-	1	-	-	-	-	-	-	-
Biotin	-	-	-	-	-	-	0.05	0.05	-
Glycine	2	3	-	-	-	-	2	2	-
Cysteine HCl	-	1	-	10	-	-	-	-	-
Folic Acid	-	-	-	-	-	-	0.5	0.5	-
Glutamine	-	-	-	-	-	-	-	14.7	-

investigated. With the new cryopreservation techniques, the number of species or cultivars that can be cryopreserved has greatly increased over the last few years. These new techniques are the operational simplicity, since they will be applied mainly in developing tropical countries where the largest part of genetic resources of problem species is located. Cryopreservation is now practical for base collections of Clonal crops formerly preserved only as field collections. For further development of cryopreservation of tropical plants, studies on preconditioning for the induction of dehydration tolerance appear to be most important (Engelmann and Takagi, 2000).

Technique for Single Cell Culture

Bergman (1960) first used this method. This technique is particularly useful where attempts are made to obtain single cells clones. The objective is to establish evenly distributed suspension in a thin layer of culture medium with a low cellular density as is compatible with the growth of colonies from high proportion of the cellular units of single cells.

SUMMARY

Tissue culture technology is used for the production of doubled haploids, cryopreservation, propagating new plant varieties, conserving rare and endangered plants, difficult-to-propagate plants, and to produce secondary metabolites and transgenic plants. Various facets of tissue culture are dealt with examples in this chapter. History of tissue culture along with other the developments are also described. Application of tissue culture, requirements for a tissue culture laboratory its requirements, synthetic seeds and encapsulated embryos, production of secondary metabolites, bioreactor technology, somaclonal variation, anther and microspore culture, ovary culture, embryo culture, protoplast culure, fusion and somatic hybridization are some of the aspects which finds its place in this chapter. Plant tissue culture techniques are now being used globally for the multiplication of important plant species. *In vitro* techniques offer a rapid means for multiplication besides serving as alternate method for *ex situ* conservation. Due to limitations with conventional methods of propagation establishment of aseptic cultures and development of an efficient protocol for regeneration and multiplication of plants are required for developing *in vitro* strategies. Tissue and cell culture and cryopreservation provide excellent avenues and opportunities for *ex situ* conservation of these genetic resources. Successful applications have been accomplished in the science of conservation biology with intervention of tissue culture.

REFERENCES

Ara, H., Jaiswal, U. and Jaiswasl V. S. (2000) Synthetic seed. Prospects and limitations Curr. Sci., 78 : 1439 . 1444.

Ashmore, S. E. (1997). Status report on the development and application of in vitro techniques for the conservation and use of plant genetic resources. IPGRI,. Rome., Italy.

Davey, M.R., Anthony, P., Power, J.B. and Lowe, K.C. (2005) Plant protoplasts : status and biotechnological perspectives. Biotechnology Advances 23 : 131–171

De Fossard (1993) Plant Tissue Culture Propagation.

Engelmann, F. (1997) *In vitro* conservation methods : in Biotechnology and Plant Genetic Resources Conservation and use. (Eds., B. V. Ford-Lloyd, H.J. Newbury and J. A. Callow), CABI, Oxon. UK.

Engelmann, F. and Takagi, H. (2000) Cryopreservation of tropical plant germplasm. current research and application. IRCAS, Tsukuba, and IPGRI, Rome, Italy.

Fabre, J. and Dereuddre, J. (1990) Encapsulation-dehydration: a new approach to cryopreservation of *solanum* shoot tips. Cryo-Letters. 21: 43-426.

George, E.F., Sherrington, P.D., 1984. Plant Propagation by Tissue Culture. Exegetics, UK.

George, E. F. (1996) Plant propagation by tissue culture. Exegetics Edinborough UK.

Grout, B. (1995) Genetic Presevation of Plant Cells *in vitro*. Springer Verlag Laboratory Manual, Springer Verlag, Berlin.

Larkin and Scowcroft (1981) Somaclonal variation – a novel source of variability from cell cultures for plant improvement. Theor. Appl. Genet. 60 : 197-214.

Murashige T, Skoog F.1962. A revised medium for rapid growth and bioassay with tobacco tissue culture. Physiol. Plant15 : 473-497.

Rajasekharan PE, Ganeshan S .(2002). Conservation of medicinal plant biodiversity-an Indian perspective. J Med Aroma Plant Sci 24 : 132-147.

Verpoorte, R. A. Contin1 and J. Memelink (2002) Biotechnology for the production of plant secondary metabolites Phytochemistry Reviews 1: 13–25.

Verpoorte R. R. van der Heijden1, H.J.G. ten Hoopen and J. Memelink (1999).Metabolic engineering of plant secondary metabolite pathways for the production of fine chemicals Biotechnology Letters 21: 467-479.

Suggested Readings

1. Evans, D., Sharp, W.R., Ammirato, P.V. and Yamada, Y. (Eds.) Handbook of Plant Cell Culture. MacMillan, New York.
2. Bajaj.Y.P.S. Biotechnology in Agriculture and Forestry. Vol. 30. Springer-Verlag, Berlin.
3. Bhojwani, S.S., and Razdan, M.K. (1983) Plant Tissue Culture: Theory and Practice. Elsevier Science Pub. Amsterdam.
4. George, E.F. (1993) Plant Propagation by Tissue Culture. Part 1. The Technology. Exegetics Ltd., Edington.
5. Jain, S.M., Brar, D.S. and Ahloowalia, B.S. (1998) Somaclonal Variation and Induced Mutations in Crop Improvement. Kluwer Acad. Pub. Dordrecht.
6. Pierik, R.L.M. (1989) *In Vitro* Culture of Higher Plants. Martinus Nijhoff, Dordrecht.
7. Trever. A.Thorpe (1981) Plant tissue culture-methods and applications in agriculture.

REVIEW QUESTIONS

Long Questions

1. What are the advantages and disadvantages of using tissue culture to produce seedlings as opposed to seeds?
2. Which enzymes would be needed to degrade the walls of plant cells, and how could you tell when this had been achieved?
3. How to conserved germplasm using tissue culture and cryopreservation
4. How to produce haploids
5. What are the Indian contributions to the development of tissue culture?

Multiple Choice Questions

1. Increased genetic diversity following extended time in a tissue culture is a problem called
   (a) gene alteration (b) temporal modification
   C) somaclonal variation (d) culture shock
2. A mass of dividing, undifferentiated cells in a tissue culture is called
   (a) a shield (b) a callus
   (c) an embryoid (d) an aggregate
3. Electroporation is a technique used with
   (a) calluses (b) pollen
   (c) protoplasts (d) organs
4. To produce plants that are homozygous for all traits, the best choice is
   (a) cell suspension culture (b) callus culture
   (c) anther/pollen culture (d) plant organ culture
5. Cell suspension cultures require
   (a) organogenesis (b) electroporation
   (c) differentiation (d) disaggregation
6. If you want to use a plant tissue culture as a chemical factory for vitamins, choose
   (a) suspension cultures (b) callus cultures
   (c) organ cultures (d) anther pollen cultures
7. Most plant tissue cultures are initiated from
   (a) calluses (b) explants
   (c) plantlets (d) protoplasts
8. Unlike the somatic cells of animals, plant cells can be grown in tissue culture and regenerate new plants
   (a) because each cell contains the entire genome
   (b) because plant cells can dedifferentiate and revert back to seeds
   (c) because plant cells are able to express genes that weren't previously expressed
   (d) both a and b
9. Which of the following are mismatched?
   (a) cell suspension culture-somatic cell embryogenesis
   (b) protoplast isolation-hybrids
   (c) homozygous plants-anther/pollen culture
   (d) callus culture-differentiated tissues regenerate
10. Plant transformation is
(a) When a plant grown in culture generates increased genetic variation
(b) When plant cells in suspension cultures form individual embryos that can grow into plants

(c) The incorporation of foreign DNA into the plant genome
(d) When dedifferentiated callus cells develop into tissue that is different from the original source tissue

Answers

1. (c)	2. (b)	3. (c)	4. (d)	5. (d)	6. (a)
7. (b)	8. (b)	9. (a)	10. (c)		

CHAPTER 44

Liposome Technology and its Applications

K. Pandima Devi*, N. Suganthy, P. V. Kiruthiga and S. Karutha Pandian
Dept. of Biotechnology, Alagappa University, Karaikudi–630 003
E-mail: devikasi@yahoo.com

INTRODUCTION

Lipids are a group of chemical compounds (such as oils and waxes), which occur in living organisms, which are sparingly soluble in water and readily soluble in organic solvents. Lipids were once the primary sources of aliphatic carbon compounds used by industry. With the arrival of petroleum, the consumption of lipids declined in most industrial applications, although obviously they still retain their strong position in foods and feeds. However importance of lipids in pharmaceutical industry increased with the discovery of liposome. Phospholipids are a special group of lipids containing phosphate. Phospholipids are the building blocks of liposomes and cell membranes. Lipids in general are hydrophobic, also called non-polar (not able to be mixed in water). But the phosphate group in phospholipids is hydrophilic, also called polar (able to be mixed in water). When phospholipids are mixed in water they arrange themselves so that their hydrophilic regions point towards the water and their hydrophobic regions point away from the water. This unique simultaneously hydrophilic/hydrophobic structure of phospholipids is the key to their ability to organize as a double layer (bilayer formation) when mixed in water. The interaction and rejection forces between phospholipids and water cause phospholipids to organize themselves as bilayers and phospholipid bilayers are the core structure of liposomes.

Liposomes are microscopic, manmade vehicles used as sustained-action delivery vehicle for a wide variety of drugs, vaccines, enzymes, non-enzyme proteins, and genetic material—and now, for some nutritional supplements as well. They can encapsulate water-soluble

*Corresponding author

ingredients in their inner aqueous core, and oil-soluble ingredients in their phospholipid membranes. Liposomes are made up of one or more concentric lipid bilayers, and range in size from 50 nanometers to several micrometers in diameter.

Professor Alec Bangham of Babraham Institute, Cambridge of England in 1965, first discovered liposome during his study on phospholipids and blood clotting. Since then, liposomes have been employed as excellent model of biological membranes and very useful carriers of both lipid-soluble and water-soluble drugs due to their biodegradability, ease of preparation, non-toxicity and non-immunogenicity. Liposome encapsulated amphoterian B was approved by Ireland in 1990 and in 1995, FDA approved the liposome encapsulated Doxorubicin for cancer therapy. In recent years, the applications of these lipid spheres have been extended to enhance the drug efficacy and potency, reduce toxicities of encapsulated drugs and target specific tissue sites.

CLASSIFICATION

Liposomes may be classified based on various characteristics such as size, lamellarity, lipid composition and applications. All of them consist of a spherical cell membrane (lipid bilayer) that encapsulates a payload of therapeutic molecules. All of them bypass the digestive tract, thus increasing the bioavailability of their payloads, which remain biologically inert until the cell membrane ruptures. When and where that occurs, and why and how it occurs, is what makes the difference. It depends on many different physical and chemical properties of the liposomes and on the physiological environment in which they find themselves. Common types of liposomes are:

(a) Classification based on Circulation *in vivo*

(i) Conventional liposomes (classical liposomes): Classical liposomes are "naked," i.e., they're ordinary liposomes made from simple phospholipids, with nothing to protect them from phagocytes. Conventional liposomes are made of neutral phospholipids like phosphotidylcholine along with varying concentration of cholesterol and sometimes-small percentage of acidic phospholipids. Cholesterol is added to increase the stability of liposome in the presence of plasma, and the negative charge is added to avoid aggregation and to increase encapsulation efficiency. Their surfaces (both exterior and interior) are either electrically neutral or negatively charged depending on the specific nature of the lipid molecules used in making the bilayer. Their half-life decreases with increasing diameter, negative surface charge and fluidity. The problem with conventional liposomes is that they're seen as alien invaders by phagocytes of liver and spleen. These liposomes are used mainly to deliver drugs to the phagocytes that eat them, so as to treat disorders of the phagocytes themselves or to treat diseases of the liver or spleen, where most phagocytes are found. The latter form of treatment is made possible by drugs that are released prolonged and gradually by macrophage-mediated drug release. Liposome uptake by the macrophage of liver and spleen (the reticuloendothelial system - RES) hampered the development of liposomes as drug delivery systems for over 20 years. Discovery of stealth and polymeric liposomes (cationic/fusogenic) helped liposomes transfer from the bench to the clinic (Bangham, A. D. 1972).

(ii) Sterically stabilised liposomes (Stealth or long circulating liposome): They are conventional liposomes stabilised by addition of specific compounds, which enhances stability in biological fluids and reduces reactivity towards plasma protein and cell surface receptors. These liposomes used either GM1 ganglioside or phosphatidyl inositol (PI) as a stabilising component. Recently Polyethylene glycol (PEG) has superseded the use of GM1 or PI for achieving liposomes with long circulation. PEG's are attached, at one end of the polymer chain, to the surface of the liposome. The long, slender, highly flexible PEG molecules slosh around the liposome like spaghetti boiling in a pot. Because of their chemical affinity for water molecules, they are heavily hydrated. To phagocytes, this molecular "cloak" of water of hydration makes the PEGylated liposome (Fig. 1) looks like little watery blobs rather than something edible, so they tend to remain non-immunogenic (Papahadjopoulos, D. *et al.*, 1991).

- More over it prevents liposomes from sticking to each other and to blood cells or vascular wall and also prevents association and binding of opsonins in plama thereby making it invisible to the immune system. So the circulation half-life period of stealth liposomes increases to days.
- It can be directed to specific organs by incorporating targeting ligands on the surface of the liposomes.
- The polymer layer also exerts a normal and lateral surface pressure that opposes close approach by macromolecules preventing dissolution of liposomes by lipoprotein particles in the bloodstream. This is called exotropic exclusion.
- Increases the bioavailability of drugs or supplements by bypassing the digestive tract.
- Minimizes any potential toxicity or side effects of these agents by remaining in the circulation for a long time and releasing their payloads slowly. As a bonus, they're passively targeted to tumors and to inflamed tissues, where they're preferentially absorbed because of the increased permeability of the capillaries that nourish these tissues (Rosenecker, J. *et al.*, 1996).

Other methods for extending the circulation time of liposome in blood include the incorporation of cholesterol, polyvinyl-pyrrolidone polyacrylamide lipids, glucuronic acid lipids or the high phase transition temperature phospholipid distearoyl phosphatidylcholine into liposomes. Liposome size also affects biodistribution and a size of between 70-200 nm is necessary to achieve prolonged circulation times with stealth liposomes.

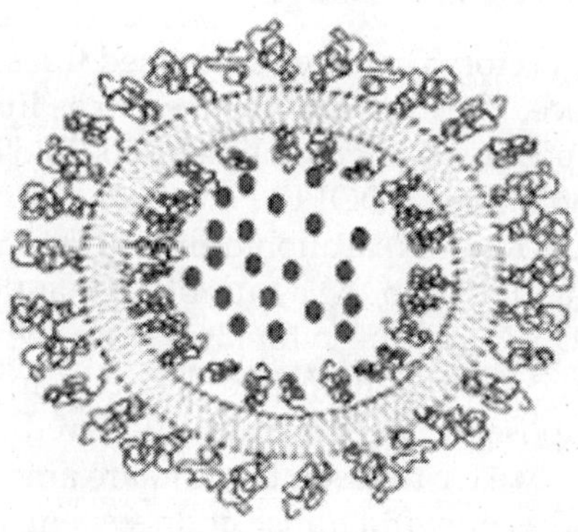

Fig. 1. Hydrophilic polyoxyethylene lipids incorporated into liposome.

(b) Classification According to Size and Lamellarity

(i) Small Unilamellar Vesicles (SUV) are "small, unilamellar vesicles" or "sonicated, unilamellar vesicles" are usually prepared by sonication using a cuphorn, bath, or probe tip sonicator. SUV's comprise of a single internal aqueous space enclosed in a single lipid bilayer. They are usually spherical with a minimum surface/volume ratio and diameter ranges from 20 to 100 nm. SUV will spontaneously fuse on storage when the temperature drops below the phase transition temperature of the lipid forming the large unilamellar vesicle.

(ii) Large Unilamellar Vesicles (LUV): LUV vesicles range in size from a few hundred nm to several microns. The thickness of the membrane (phospholipid bilayer) measures approximately 5 to 6 nm. LUV can be prepared by a variety of methods including extrusion (LUVET or "Large, Unilamellar Vesicles prepared by Extrusion Technique"), detergent dialysis (DOV or Dialyzed Octylglucoside Vesicles), reverse evaporation (REV or Reverse Evaporation Vesicles), and ethanol injection. LUV is stable on storage. LUVs can also be produced by calcium-induced fusion of sonicated SUV to form large (0.2 to 2 microns in diameter) vesicles with a single limiting membrane.

(iii) Large Multilamellar Liposomes (LML): Multilamellar liposomes have many bilayers of lipid enclosing the aqueous volume. Although the total internal aqueous space of multilamellar liposomes is relatively large, the individual compartments between each lipid lamella are too small to accommodate large macromolecules. Multilamellar liposomes may be used as informative model systems for studying entrapment of materials and monitoring the properties of liposomal systems e.g., susceptibility to hydrolysis by phospholipases. However, it is relatively more difficult to produce a homogenous size distribution in multilamellar liposomes. The trapping of material varies in multilamellar and unilamellar liposomes. Multilamellar liposomes are oblate cylinders and/or spheroids.

(c) Classification According to Application

(i) Diagnostic
(ii) Therapeutic.

(d) Classification According to Surface Charge

(i) Cationic/Liposomal DNA vector: They are specialized liposomes with net positive charge on their inner and outer surface. They are also referred to as liposomal DNA delivery vector. Cationic liposomes consist of positively charged lipid, a co-lipid and a cationic amphiphile, e.g. dioleyl phosphatidyalethanolamine (DOPE) or N-[1-(2,3- dioleyloxy)propyl]-N,N,N,-trimethylammonium chloride (DOTMA). Cationic amphiphile acts as transfectant- membrane destabilizing agent. DOPE is added with the assumption that the fusogenic properties at acid pH will allow the release of DNA into the cytoplasm from the endosomal compartment and after transfection they undergo degradation there by reducing the possibility of toxicity of these compounds. Cationic liposome are generally preferred over viruses because they are non-immunogenic, are relatively easy to assemble, and are amenable to scale-up for industrial production. So it is also termed as nonviral transfection system. In addition, liposomal delivery vehicles offer versatility with a diverse range of morphologies, size, and release characteristics;

can be used for tissue targeting; and provide protection of the plasmid DNA from attack by degradative nucleases. Since their introduction as potential delivery systems in 1987, DNA-cationic lipid complexes, also known as cationic lipoplexes, have been used in numerous research protocols for DNA delivery in a range of cell types and are currently being investigated in several gene therapy clinical trials (Farhood, H. *et al.*, 1995).

Disadvantages: Inactivated by the presence of serum in the transfecting medium. It exhibits high degree of cellular toxicity especially to phagocytic macrophages. The commercially available cationic lipid-based transfection agent lipofectAmine has been reported to induce toxicity in various tissues such as pulmonary cells, and in arterial cell walls in a porcine restenosis model for localized gene therapy. Need for efficient and well-tolerated delivery system in gene transfer has led to the discovery of Anionic liposomes.

(ii) Anionic liposomes: Anionic liposomes (ALs) occur naturally in eukaryotic cell membranes, and it has been used as DNA delivery systems recently as an alternative to cationic liposomes because of their low cytotoxicity. They are classified as intrinsically pH-sensitive liposomes or external non-lipid trigger liposomes. pH sensitivity of these liposomes is dependent on the combination of phosphoditylethanolamine (PE) either with one of a number of acidic amhiphiles and titrable polymer. Liposomes are prepared from mixtures of endogenously occurring anionic and zwitterionic lipids, 1,2-dioleoyl-*sn*-glycero-3-[phospho-*rac*-(1-glycerol)] (sodium salt) (DOPG) and 1,2 dioleoyl-*sn*-glycero-3-phosphoethanolamine (DOPE). Anionic lipoplexes were formed by complexation between anionic liposomes and plasmid DNA molecules using Ca^{2+} ions. Toxicity of naturally occurring anionic lipoplexes is demonstrated to be lower than the cationic liposome lipofectAmine and Ca^{2+}-DNA complexes. A mechanism of anionic lipoplex transfection is proposed on the basis of the formation of nonbilayer hexagonal phases in the lipoplex that facilitate cellular internalization. Anionic lipoplexes might be useful in *ex vivo* and *in vitro* applications for DNA transfer particularly in cell lines sensitive to the toxic effects of cationic lipids.

(e) Specialised Liposomes

(i) Targeted liposome: These are liposomes (Fig. 2) that are directed to recognize and bind to specific cells *in vivo*. This can be achieved by incorporating a surface ligand if there is no decrease in the half-life. When the surface ligand is an antibody or antibody fragments, it is referred to as an immunoliposome.

(ii) Immunoliposome: Immunoliposomes are lipid vesicles coated with antibodies that have been attached with hydrophobic anchors. These surface antibodies direct the specific interaction of liposomes with antigen-expressing cells. These cells take up liposomes by receptor-mediated endocytosis. Attachment of whole antibodies to the surface of conventional liposomes increases their clearance from circulation, resulting in shorter half-life of few minutes. Immunoliposomes formed from sterically stabilized liposomes have circulation half-lives of several hours. Sterically stabilised liposomes with PEG-PE have indicated that such liposomes may be targeted to specific cells without losing their property of a long half-life in blood. They are used mostly for tumor therapy

(iii) pH Sensitive Liposomes: Liposomes that exhibit triggered release properties have potentially important applications in drug delivery. It has been shown that liposomes can be constructed sensitive to a variety of physical and chemical stimuli, including temperature,

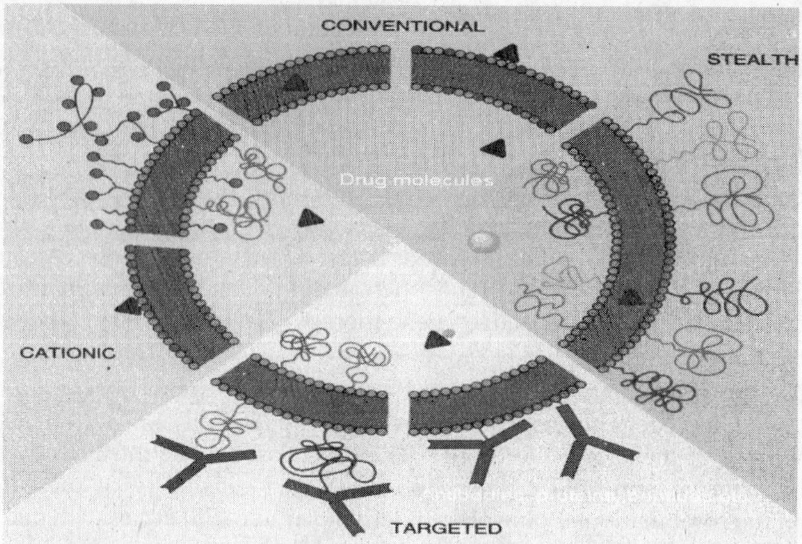

Fig. 2

light, or pH. Liposomes that can be triggered to release their contents or fuse in response to pH stimuli are of particular interest, as they can potentially respond to acidic environments *in vivo*. Such environments include those encountered in tumor and primary endocytic vesicles.

Liposomes dependent upon the pH of the environment are referred to as pH-sensitive liposomes. It is composed of unsaturated phosphatidylethanolamines and mildly acidic amphiphiles such as cholesteryl hemisuccinate. Lipids such as phosphatidylserine, palmitoylhomocysteine, and α-tocopherol hemisuccinate have also been used in concert with DOPE to prepare pH-sensitive liposomes. The pH sensitive liposomes release a larger proportion of their contents than do non-pH sensitive liposomes, when the pH of the surrounding environment decreases from physiological pH (pH 7.4) to a pH between about 3.5 and 6.5. This characteristic can be exploited as advantage for *in vitro* and *in vivo* uses. Within the cells, majority of the liposomes is internalized through an endocytotic pathway and are exposed in the endosome to a decreasing pH. Since liposomes can be targeted to cells using appropriate ligands such as antibodies directed to epitopes on cells of interest, a liposome stable at pH 7.4 but unstable at a pH encountered in the endosome can be used to deliver its contents preferentially to the targeted cell population. Since many tumor cells tend to have a pH range from 5.8 to 6.5, liposomes stable at pH 7.4, but destabilized at a pH of 5.8-6.5, can be used to deliver cytotoxic agents, imaging agents, or other desired agents preferentially to tumor cells. The use of pH-sensitive liposomal systems requires targeting of liposomes to specific cells that are capable of internalizing substance-filled liposomes by means of endocytosis. Liposomes internalised via endocytosis will experience a gradual pH decrease and this environmental change constitutes the basic idea for triggered release from pH-sensitive liposomes. Although it is well established that pH-sensitive liposomes do collapse and release their contents upon acidification, one problem still remains, that is, the active substance must also be able to cross the endosomal membrane. A number of *in vitro* studies indicate that internalised DOPE-based pH-sensitive liposomes are indeed able to deliver hydrophilic substances to the cytosol of target cells. A cytosolic delivery depends primarily on the chemical properties of the drug. A drug with suitable hydrophilic/hydrophobic properties will be able to cross the endosome membrane by simple diffusion. This is not the case, however, for most liposome-encapsulated drugs and transportation into the cytosol thus requires other mechanisms. One possibility is that the liposomes, upon acidification, fuse with the endosomal membrane. This would lead to a "microinjection" of the drug into the cytosol (Fig. 3). A second alternative is that the liposomes, as a result of a lowered pH, first collapse and releases their contents into the endosomal compartment. In a second step the DOPE molecules, initially situated in the liposomes, may interact with the endosome membrane, which could lead to a higher permeability and perhaps also major structural rearrangements of the endosome membrane.

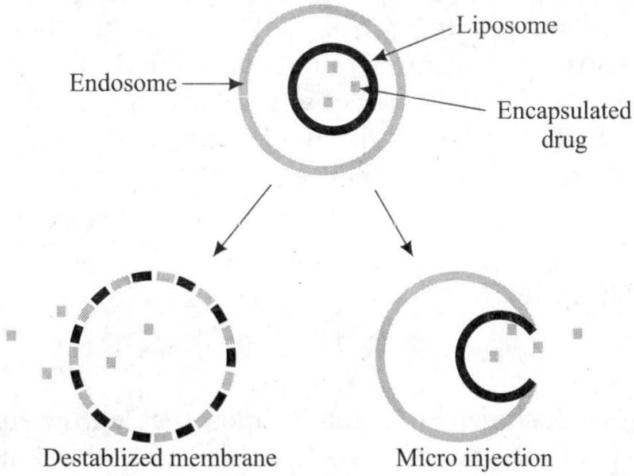

Fig. 3. Two possible mechanisms for a cytosolic delivery. (A) The destabilization mechanism and (B) The fusion or the "microinjection" mechanism.

Second major problem of pH sensitive liposomes is that liposomes are stabilized by certain plasma components, which inserts into the membrane bilayer and reduce the liposome's sensitivity to pH (Liu and Huang, 1989, 1990). Recently, two approaches have been attempted to induce acid-mediated leakage of water-soluble content markers. First attempt is to incorporate lipid anchored copolymers into egg PC/Cholesterol liposomes (Meyer, O. *et al.*, 1998b). Incorporation induced the release of content when the pH was lowered below 5.5. Unlike other pH-sensitive liposomes, release of the marker is not due to fusion but rather to a collapse of the polymer at the phase transition and subsequent collapse of the liposomes or resulting local defects in the membrane that allow for contents leakage. The second approach is the design of cleavable PEG-DSPE conjugates. PEG-DSPE is known to stabilize 1,2-dideoyl-3-*sn*-phosphatidylethanolamine containing membranes and prevent fusion of liposomes. Release of PEG from the surface with a sulfhydryl- or an acid-sensitive trigger gives a fusion-competent liposome, capable of releasing its contents. pH-sensitive liposomes have been suggested as a promising alternative to the conventional PC-liposomes.

(iv) Cationic Liposome-Entrapped Polycation Condensed DNA (LPDI): Polycations such as poly-L-lysine are known to efficiently condense DNA. The binding of DNA to a polycation is stoichiometric, and the condensation process is highly cooperative. The LPDI particles have a higher performance than the corresponding liposome/DNA complexes. Unlike the liposome/DNA complexes, which are large and heterogeneous in size, LPDI particles are highly compact with a size d"100 nm. These particles would be more favorable for entering cells via an endocytosis pathway. Also, LPDI particles offer better protection of DNA from enzymatic digestion (Gao, X. and Huang, L. 1995).

(v) Anionic Liposome-Entrapped Polycation Condensed DNA (LPDII): Structurally, these particles are similar to LPDI. Both contain a highly condensed core of polylysine and DNA and a lipid shell. DNA is first condensed with polylysine with the positive charge in excess. The resulting cationic complex is then mixed with anionic liposomes carrying a targeting ligand. Folate is commonly chosen as a targeting ligand because many tumors, especially ovarian carcinoma, over-express folate receptors (Lasic, D. D. 1998). The advantages of LPDII are that besides being highly compact, LPDII preparations do not require purification and are single vial formulations. Compared with the traditional anionic and neutral liposomal vectors, DNA is highly condensed in LPDII and is quantitatively encapsulated without the requirement of using excess amounts of lipids. Finally, because anionic liposomes are relatively compatible with biological fluids, LPDII can be potentially used for tissue-specific gene delivery.

CHARACTERISTICS OF LIPOSOMES

(a) Size

Large, multiple-layer liposomes are actually liposomes within liposomes (Fig. 4). They have a limited ability to penetrate narrow blood vessels or into the skin. The materials that are entrapped in the inner layers of these liposomes are practically neither accessible nor releasable. Large liposome can be prepared by shaking phospholipids with water. These liposomes have very limited functions and are usually made of commercial lecithin, commonly found in food products. Commercial lecithin's main function is as an emulsifying agent, improving the ability of oil and water to remain mixed. Very small, single-bilayer liposomes are called nanosomes which differ from large liposomes in many ways. They contain high percentage of ingredients like phosphatidylcholine (PC), one of the essential components of cell membranes. Nanosomes are very difficult to make, which requires subjecting large, multiple-layer liposomes to ultrasonic energy. This process is very long, extremely delicate, and is done in small batches. Nanosomes can easily penetrate into small blood vessels by intravenous injection; they can also penetrate into the skin by topical application. The contents entrapped inside them, are efficiently transported and can be delivered to desired targets such as cells.

There are several techniques suitable for determining the size of liposome preparation; the most straightforward one is analysis by quasi-elastic or dynamic light scattering. This provides the mean diameter and distribution of the particles. It can also distinguish whether the particle population is uniformly distributed around one or more particle sizes (unimodal vs. bimodal). Particle size can also be determined by electron microscopy. This technique does not allow for good characterization of distribution of the particle sizes and may suffer from changes to

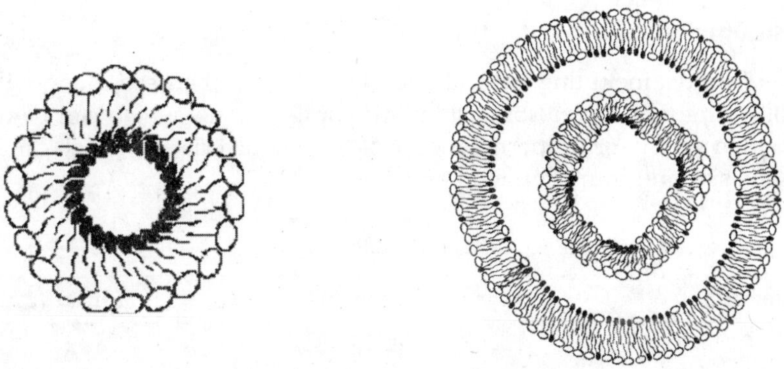

Very Small, Single Bilayer Liposome Large, Multiple-Layer Liposome

Fig. 4

particle size induced by sample fixation and staining. Finally, size can be determined by measuring the volume of trapped internal contents using a fluorescent probe. In this technique a fluorophore is trapped in the internal compartment of the liposome and the liposomes are separated from non-trapped fluorophore. The trapped fluorphore can be released and concentration is measured. This information is used to calculate the internal volume of the liposome and thus the size of the particle that matches its volume. This technique assumes the liposome is unilamellar and therefore is not suitable for multilamellar liposomes. The size of liposomes plays an important role in the clearance of liposomes by reticuloendothelial system—the extent of its importance differing for classical and sterically stabilised liposomes. In classical liposomes increase in size increases the rate of clearance from blood circulation by reticuloendothelial cells. In case of sterically stabilised liposomes, for liposomes up to 250 nm diameter, uptake into the liver occurs slowly, and liposomes are found primarily in liver Kupffer cells. Above 300 nm, increased uptake of sterically stabilised liposomes into the spleen occurs. The ability to produce liposomes of controlled size distribution was initially obtained by sonication and more recently by extrusion.

(b) Surface Charge

Positively or negatively charged liposomes are cleared more rapidly than neutral classical liposomes. In case of large liposomes negative liposomes are readily taken up by spleen while brain and lung tissues take up positive liposomes. In contrast, in the case of small-sonicated liposomes, positively charged liposomes concentrate in bonemarrow, brain and spleen while negatively charged liposomes concentrate in lung tissues. In multilamellar liposomes entrapment of enzymes or solutes in the aqueous spaces between the lamellae depends upon the capacity of the phospholipids to swell and form hydrated liquid crystals consisting of a series of concentric bilayers that alternate with aqueous compartments. The volume of these compartments is determined by the net charge of the lipids and the ionic strength of the swelling solution (Bangham, A. D. 1968, Papahadjopoulos, D. and Miller, N. 1967). Increase in the net anionic surface charge on the lamellae increases the capture of enzyme as the like-sign repulsion of adjacent lipid layers causes an increase in the volume of the aqueous compartments.

(c) Composition

Many types of lipids, including natural and synthetic, can be used along with sterols to aid stability. Following table gives some standard compositions, which can be used to produce liposomes with a variety of properties. Some of the characteristics, which can be modified are surface charge, size and transition temperatures.

Table 1

Neutral liposomes	Negative liposomes	Positive (Cationic) liposomes
High transition composition	High transition composition	
L-a-Phosphatidylcholine, distearoyl	L-a-Phosphatidylcholine, distearoyl, L-a -Phosphatidyl-DL-glycerol, distearoyl	1,2-Dioleoyl-sn-Glycero-3-Phosphoethanolamine (DOPE)
L-a -Phosphatidylcholine, distearoyl/cholesterol	L-a-Phosphatidylcholine, distearoyl, L-a-Phosphatidyl-DL-glycerol, distearoyl /Cholesterol	L-a–Phosphatidylcholine / stearylamine/cholesterol
Low transition composition	Low transition composition	
L-a-Phosphatidylcholine, cholesterol	L-a –Phosphatidylcholine/ Dicetyl phosphate/cholesterol	
L-a-Phosphatidylcholine, b-oleoyl-g –palmitoyl	L-a -Phosphatidylcholine, b -oleoyl-g -palmitoyl, L-a -Phosphatidyl-DL-glycerol, dioleoyl	
L-a-Phosphatidylcholine, b -oleoyl-g -palmitoyl/cholesterol	L-a -Phosphatidylcholine, b -oleoyl-g -palmitoyl, L-a -Phosphatidyl-DL-glycerol, dioleoyl/cholesterol	

(d) Storage

Once lipid particles have been formed, maintaining the physical properties of the particles can be difficult. Size distribution can change on storage due to degradation of the components. Permeabilization of the membrane can lead to leakage of encapsulated material. Stability issues due to hydrolytic degradation are a general problem with lipid products. Aqueous formulations of drug products tend to be less stable since the presence of excess or bulk water leads to rapid hydrolytic degradation in lipid preparations.

After the sizing process is complete, lipid suspensions should be stored at close to pH 7 as possible. Lipids containing ester-linked hydrocarbon chains are susceptible to acid and base hydrolysis. Hydrolysis rate is dramatically affected by temperature; therefore lipid suspensions should be kept refrigerated during storage. Lipid suspensions should not be frozen as the freezing process could fracture or rupture the vesicles leading to a change in size distribution and loss of internal contents. Change in particle size can have a dramatic effect on functionality. The use of cryoprotectants such as dextrose, sucrose, and trehalose may increase stability from hydrolysis. Also, samples may experience oxidation upon storage. The addition of small amounts of antioxidants during processing may stabilize the suspension and limit oxidation of the product.

Storage time depends on a number of factors including temperature, pH, medium, etc. Liposomes stored in a buffer at pH 7.4 and at ~4°C do not usually display membrane structural changes for 5-7 days. Beyond that time the entrapped components tend to leak out of the liposome indicating the presence of membrane destabilizing components, presumably lyso lipid and free fatty acid generated by hydrolysis of the lipid.

(e) Concentrating a liposome

There is no good general procedure for concentration of liposomes. Large liposomal particles can be concentrated using centrifugation. The lowest speed possible to achieve pelleting is best since higher speeds could induce deformation and/or fusion of particles. Centrifugation is only possible with larger particles (>100 nm). SUV particles will not readily pellet at lowest speed due to their small size. Other methods of concentration may be possible but could lead to particle shape deformation depending on the sample and technique.

(f) Permeability of Liposome Membrane

In diagnostic applications, liposomes with membranes impermeable to an entrapped marker are advantageous, whereas in therapeutic applications, it is mandatory that the liposomes deliver the drug at the target site by release through its membranes. The rate of elution of an entrapped marker from liposomes depends both on the nature of the enclosed molecule and the lipid bilayer of the liposome. Molecular size and charge are important characteristics of the entrapped substance in this regard. Small molecules cross lipid bilayers more rapidly than large ones, and anions cross more rapidly than cations.

(g) Captured Volume

The volume of a liposome is the sum of the volume occupied by the hydrated lipid and the volume of the internal aqueous phase. However, captured volume refers only to the amount of aqueous solution entrapped per lipid. Although hydrating dried lipid films or powders produce large multilamellar liposomes, the high number of lamellae limits the aqueous solution capturing efficiency of these liposomes. Another problem is the phenomenon of solute exclusion (Perkins, W.R. *et al.*, 1993). Aqueous addition allows hydration of surface lipid that organizes into bilayers and because water is more membrane permeable than most solutes, it permeates through these outer layers to hydrate interior lipid with solute partially excluded. These problems are overcome by solvent removal techniques like solvent injection method, in which lipid is first dissolved in solvent that is then mixed with the aqueous phase in which the liposomes are to be suspended at the time solvent is removed. Liposomes of high captured volume have advantages in applications as slow releasing drug depositories, carriers of diagnostic imaging agents and as model systems to study bilayer properties.

ISSUES TO CONSIDER WHEN SELECTING LIPIDS FOR LIPOSOME PREPARATION

(a) Phase Transition Temperature

The phase transition temperature is defined as the temperature required to induce a change in the lipid physical state from the ordered gel phase, where the hydrocarbon chains are fully extended and closely packed, to the disordered liquid crystalline phase, where the hydrocarbon chains are randomly oriented and fluid (Small, D.M. 1986). Several factors directly affect the phase transition temperature including hydrocarbon length, unsaturation, charge, and head group species. As the hydrocarbon length is increased, vanders Waals interactions become stronger requiring more energy to disrupt the ordered packing, thus the phase transition temperature increases. Likewise, introducing a double bond into the acyl group puts a kink in the chain that requires much lower temperatures to induce an ordered packing arrangement.

When developing a new product, procedure, or method, controlling the transition temperature of the lipid could be useful. Choosing a high transition lipid where the lipid vesicle would always be in the gel phase would provide a non-leaky packaging system. Alternatively, a lipid with a transition temperature between the starting temperature and the ending temperature of the system would provide a means of releasing packaged material as the lipid passes through its phase transition temperature and the vesicle becomes leaky. Liposomes with high phase transition phospholipids (tighter bilayers) are cleared more slowly than liposomes with fluid bilayers.

(b) Stability

The lipid species used in the formulation can dramatically affect the long-term stability or shelf life of a drug product containing lipids. Generally, the more unsaturated a compound, the easier the product is oxidized, and thus the shorter the shelf life of the product. Lipids from biological sources (e.g., egg, bovine, or soybean) typically contain significant levels of polyunsaturated fatty acids and therefore are inherently less stable than their synthetic counterparts . While saturated lipids offer the greatest stability in terms of oxidation, they also have much higher transition temperatures and thus present other difficulties in formulation. If unsaturation is a requirement, the degree of unsaturation should be kept as low as possible. In most cases, compounds containing oleic acid (18:1, cis$\ddot{A}^9$) are sufficient to satisfy the need for unsaturation, and since they are monounsaturated, oleoyl-containing products are much more stable than polyunsaturated compounds (Frrkjaer, S. *et al.*, 1982).

Stability issues due to hydrolytic degradation are a general problem with lipid products. Aqueous formulations of drug products tend to be less stable since the presence of excess or bulk water leads to rapid hydrolytic degradation in lipid preparations. This hydrolysis is dependant on several factors including pH, temperature, buffer species, ionic strength, acyl chain length and head group, and the state of aggregation. The cause of this hydrolysis is possibly due to the penetration of water to near the deeper carbonyl group into the membrane. In membranes containing cholesterol, water only penetrates to the glycerol backbone, which indicates that cholesterol acts as a stabilizing agent protecting the membrane from hydrolysis. Recently, lipid preparations have been stabilized using carbohydrates. The possible reason for the stabilizing effect carbohydrates have on lipid membranes is that the carbohydrate could intercalate into the head group region near the membrane/water interface and displace water from that region. In dry lipid formulations, this would serve to maintain a "hydrated" lipid membrane and keep the liposome structure intact. If this is true, it stands to reason that in an aqueous environment, the carbohydrate could still enter this region and displace water. This would tend to stabilize the membrane to hydrolysis from the bulk water phase.

(c) Lipid Mixtures

In many cases, a single lipid species does not yield the exact physical properties needed for a particular system, or does not adequately mimic the natural system for which it is intended to replace or reproduce. For these issues, a complex lipid mixture composed of two or more individual lipid species, the composition designed to create or reproduce a particular charge ratio, unsaturation ratio, phase transition temperature, or biological function is considered.

(d) Cholesterol

Cholesterol is a membrane constituent widely found in biological systems which serves a unique purpose of modulating membrane fluidity, elasticity, and permeability. It literally fills in the gaps created by imperfect packing of other lipid species when proteins are embedded in the membrane. Cholesterol serves much the same purpose in model membranes. But disadvantage is cholesterol of high purity is not available. Most cholesterol commercially available is derived from egg or wool grease (sheep derived). These animal sources are potentially not suitable for human pharmaceuticals due to the potential viral contamination. Also, cholesterol is readily oxidized creating a stability problem for lipid based drug products. Some of these oxidation by-products tend to be rather toxic in biological systems. The oxidation products 25-hydroxy cholesterol, 7-keto-cholesterol, 7a- and 7b-hydroxycholesterol, cholestane-3b,5a,6b-triol and the 5- and 7-hydroperoxides, were found in a concentrate which had activity causing aortic smooth muscle cells to die.

(e) Source

The two basic sources of phospholipids are synthetic and tissue-derived. Tissue-derived lipids are generally either egg-derived or bovine-derived. For clinical applications, either of these sources is not suitable due to stability problems and the possibility of viral or protein contamination. Regardless of the regulatory issues, animal-derived products do not offer any advantage to synthetic lipids. They are inherently less stable due to the polyunsaturated fatty acids, and in most cases, the synthetic counterpart cost the same or less than the tissue-derived product. Synthetic lipids can be prepared from glycerol or glycero-3-phosphocholine (GPC) derived from a plant or animal source. The latter is sometimes referred to as semi-synthetic lipids because a portion of the molecule is derived from a natural source. Lipids derived from glycerol require the chiral center be synthetically prepared which may lead to stereochemical impurities present in the final product. Lipids prepared using GPC obtained from an animal source may suffer from the same viral and protein contamination issues outlined above. The typical plant source for GPC is soybean lecithin. Other factors to be considered are nature (size and charge) of the drug and bilayer configuration of the lipid.

METHOD OF LIPOSOME PREPARATION

Properties of lipid formulations can vary depending on the composition (cationic, anionic, neutral lipid species) however; the same preparation method can be used for all lipid vesicles regardless of composition. The general element of the procedure involves preparation of the lipid for hydration, hydration with agitation, and sizing to a homogeneous distribution of vesicles (Danilo D. Lasic 1997).

(a) Preparation of Lipid for Hydration

When preparing liposome with mixed lipid composition, the lipids must first be dissolved and mixed in an organic solvent to assure a homogeneous mixture of lipids. Usually this process is carried out using chloroform or chloroform: methanol mixtures. The intent is to obtain a clear lipid solution for complete mixing of lipids. Typically, lipid solutions are prepared at 10-20 mg lipid/ml organic solvent, although higher concentrations may be used if the lipid solubility and mixing are acceptable.

Once the lipids are thoroughly mixed in the organic solvent, the solvent is removed to yield a lipid film. For small volumes of organic solvent (<1mL), the solvent may be evaporated using dry nitrogen or argon stream in a fume hood. For larger volumes, the organic solvent is removed by rotary evaporation yielding a thin lipid film on the sides of a round bottom flask. The lipid film is thoroughly dried to remove residual organic solvent by placing the vial or flask on a vacuum pump overnight. If the use of chloroform is objectionable, an alternative is to dissolve the lipid(s) in tertiary butanol or cyclohexane. The lipid solution is transferred to containers and frozen by placing the containers on a block of dry ice or swirling the container in a dry ice-acetone or alcohol (ethanol or methanol) bath.

After freezing completely, the frozen lipid cake is placed on a vacuum pump and lyophilized until dry (1-3 days depending on volume). The thickness of the lipid cake should be no more than the diameter of the container being used for lyophilization. Dry lipid films or cakes can be removed from the vacuum pump, the container closed tightly, taped and stored frozen until ready to hydrate.

(b) Hydration of Lipid film/Cake

Hydration of the dry lipid film/cake is accomplished by simply adding an aqueous medium to the container of dry lipid and agitating. The temperature of the hydrating medium should be above the gel-liquid crystal transition temperature (Tc or Tm) of the lipid with the highest Tc before adding to the dry lipid. After addition of the hydrating medium, the lipid suspension should be maintained above the Tc during the hydration period. For high transition lipids, hydration is accomplished by transferring the lipid suspension to a round bottom flask and the flask is placed on a rotary evaporation system without vacuum. Spinning the round bottom flask in the warm water bath maintained at a temperature above the Tc of the lipid suspension allows the lipid to hydrate in its fluid phase with adequate agitation. Hydration time may differ slightly among lipid species and structure, however, a hydration time of 1 hour with vigorous shaking, mixing, or stirring is highly recommended. It is also believed that allowing the vesicle suspension to stand overnight (aging) prior to downsizing makes the sizing process easier and improves the homogeneity of the size distribution. Aging is not recommended for high transition lipids as lipid hydrolysis increases with elevated temperatures. The hydration medium is generally determined by the application of the lipid vesicles.

Suitable hydration media include distilled water, buffer solutions, saline, and non-electrolytes such as sugar solutions. Physiological osmolality (290 mOsm/kg) is recommended for *in vivo* applications. Generally used solutions in hydration media are 0.9% saline, 5% dextrose, and 10% sucrose. During hydration some lipids form complexes unique to their structure. Highly charged lipids have been observed to form a viscous gel when hydrated with low ionic strength solutions. The problem can be alleviated by addition of salt or by downsizing the lipid suspension. Poorly hydrating lipids such as phosphatidylethanolamine have a tendency to self-aggregate upon hydration. Lipid vesicles containing more than 60 mol% phosphatidylethanolamine form particles having a small hydration layer surrounding the vesicle. As particles approach one another, there is no hydration repulsion to repel the approaching particle and the two membranes fall into an energy well where they adhere and form aggregates.

The aggregates settle out of solution as large flocculates, which will disperse on agitation but reform upon sitting. The product of hydration is a large, multilamellar vesicle (LMV) analogous in structure to an onion, with each lipid bilayer separated by a water layer. The spacing between lipid layers depends on composition, with poly hydrating layers being closer together than highly charged layers, which separate based on electrostatic repulsion. Once a stable hydrated LMV suspension is produced, the particles can be downsized by a variety of techniques, including sonication or extrusion.

(c) Sizing of Lipid Suspension

(i) Sonication: Disruption of LMV suspensions using sonic energy (sonication) typically produce SUV with diameters in the range of 15–50 nm. The most common instrumentation for preparation of sonicated particles are bath and probe tip sonicators.

(a) Probe tip sonicators: The tip of a sonicator is directly immersed into the liposome dispersion. The energy input into lipid dispersion is very high in this method. The dissipation of energy at the tip results in local overheating and therefore the vessel must be immersed into an ice/water bath. During the sonication up to one hour more than 5% of the lipids can de-esterify. Also, with the probe sonicator, titanium will slough off and contaminate the solution, which must be removed by centrifugation prior to use. For these reasons, bath sonicators are widely used for the preparation of SUV.

(b) Bath sonicator: Sonication of LMV dispersion is accomplished by placing a test tube containing the suspension in a bath sonicator and sonicating for 5–10 minutes above the Tc of the lipid. The lipid suspension should begin to clarify to yield a slightly hazy transparent solution. The haze is due to light scattering induced by residual large particles remaining in the suspension. These particles can be removed by centrifugation to yield a clear suspension of SUV.

Composition, concentration, temperature, sonication time, power, volume, and sonicator tuning influence mean size and distribution. Since it is nearly impossible to reproduce the conditions of sonication, size variation between batches produced at different times is common. Also, due to the high degree of curvature of these membranes, SUV are inherently unstable and will spontaneously fuse to form larger vesicles when stored below their phase transition temperature.

(ii) Extrusion *via* polycarbonate membranes: Lipid extrusion is a technique in which a lipid suspension is forced through a polycarbonate filter with a defined pore size to yield particles having a diameter near the pore size of the filter used. Prior to extrusion through the final pore size, LMV suspensions are disrupted either by several freeze-thaw cycles or by prefiltering the suspension through a larger pore size (typically 0.2–1.0 μm). This method helps prevent the membranes from fouling and improves the homogeneity of the size distribution of the final suspension.

As with all procedures for downsizing LMV dispersions, the extrusion should be done at a temperature above the Tc of the lipid. Attempts to extrude below the Tc will be unsuccessful as the membrane has a tendency to foul with rigid membranes, which cannot pass through the pores. Extrusion through filters with 100 nm pores typically yields large, unilamellar vesicles (LUV) with a mean diameter of 120–140 nm. Mean particle size also depends on lipid composition and is quite reproducible from batch to batch.

Box 1. Preparation of Multilamellar Liposomes (LMV)

A brief outline of the protocol for lyophilizing (freeze-drying) lipid mixtures for preparation of liposomes (Szoka and Papahadjopoulos, 1980)

- Dissolve the lipids in chloroform and combine the lipids in the appropriate ratio.
- Carefully evaporate the organic solvent using a dry nitrogen stream.
- Resuspend the lipid mixture in cyclohexane. If the mixture is not completely soluble in the cyclohexane, add a small amount of ethanol (1-2% of the cyclohexane volume). Do not use too much ethanol as the solution will not freeze with excessive ethanol present.
- Freeze the cyclohexane solution using dry ice.
- Quickly place the frozen mixture on a high vacuum system (lyophilization system).
- Leave the sample on the vacuum system for 3-5 hours or until the sample is completely dry
- Suspend the lipid mixture in the aqueous buffer (buffer temperature should be above the phase transition of the lipid) and allow the mixture to hydrate above the transition temperature of the lipid for 30-60 minutes (vortexing occasionally).

This procedure yields large, multilamellar vesicles (LMV or MLV). For water-soluble compounds to be entrapped, the same protocol is followed except the compound is dissolved in the aqueous buffer used to reconstitute the dry lipid and the external compound (not encapsulated) is remove by gel filtration.

Box 2. Other Liposome preparation methods

(1) **Hand-shaken method:** In order to produce liposomes, lipid molecules must be introduced into an aqueous environment. When dry lipid film is hydrated the lamellae swell and grow into myelin fibers. Only mechanical agitation provided by vortexing, shaking and swirling or pipetting causes myelin fibers (thin lipid tubules) to break and reseal the exposed hydrophobic edges resulting in the formation of liposomes. Large multilamellar liposomes can be made by hand-shaken method.

(2) **Reverse-phase evaporation method:** Historically this method provides a breakthrough in liposome technology, since it allowed for the first time the preparation of liposomes with a high aqueous space-to-lipid ratio and able to entrap a large percentage of the aqueous material presented. Reverse-phase evaporation is based on the formation of inverted micelles. These inverted micelles are formed upon sonication of a mixture of a buffered aqueous phase, which contains the water soluble molecules to be encapsulated into the liposomes and an organic phase in which the amphiphilic molecules are solubilized. The slow removal of the organic solvent leads to transformation of these inverted micelles into a gel-like and viscous state. At a critical point in this procedure, the gel state collapses and some of the inverted micelles disintegrate. The

excess of phospholipids in the environment contributes to the formation of a complete bilayer around the remaining micelles, which results in formation of liposomes. Liposomes made by reverse phase evaporation method can be made from various lipid formulations and have aqueous volume-to-lipid ratios that are four times higher than multilamellar liposomes or hand-shaken liposomes.

(3) **Freeze-dried rehydration method:** Freeze-dried liposomes are formed from preformed liposomes. Very high encapsulation efficiencies even for macromolecules can be achieved using freeze-dried hydration method. During dehydration, the lipid bilayers and the materials to be encapsulated into the liposomes are brought into close contact. Upon reswelling the chances for encapsulation of the adhered molecules are much higher. The rehydration is a very important step and is to be performed very carefully. The aqueous phase should be added in very small portions with a micropipette to the dried materials. After each addition, the tube should be vortexed thoroughly. Generally, the total volume used for rehydration must be smaller than the starting volume of the liposome dispersion.

(4) **Detergent depletion method:** The detergent depletion method is used for preparation of a variety of liposomes and proteoliposome (liposome with lipophilic protein inserted into the membrane) formulations. Detergents can be depleted from mixed detergent-lipid micelles by various techniques, which lead to the formation of very homogeneous liposomes. In practice, all lipids below their phase transition temperature can be used with this preparation method. Not all detergents are suited for this method and only a few detergents can be used for detergent depletion method. The most popular detergents used are sodium cholate, alkyl(thio)glucoside and alkyloxypolyethylenes. Mixed micelles are prepared by adding the concentrated detergent solution to multilamellar liposomes (the final concentration of the detergent should be well above the critical micelle concentration (CMC) of the detergent). Equilibrium of the mixed micelles in the aqueous phase takes quite some time and the equilibrium does not happen during a short period of time. The use of different detergents results in different size distributions of the vesicles formed. Faster depletion rates produce smaller size liposomes. The use of different detergents also results in different ratios of large unilamellar vesicles/oligolamellar vesicles/multilamellar vesicles.

Detergent depletion is achieved by following four approaches:

(a) Dialysis: The dialysis can be preformed in dialysis bags immersed in large detergent free buffers (equilibrium dialysis) or by using continuous flow cells, diafiltration and cross filtration.

(b) Gel filtration: In this method, the detergent is depleted by size exclusive chromatography. Sephadex G-50, Sephadex G-100, Sepharose 2B-6B and Sephacryl S200-S1000 can be used for gel filtration. The liposomes do not penetrate into the pores of the beads packed in a column. They percolate through the inter-bead spaces. At slow flow rates, the separation of liposomes from detergent monomers is very good. The swollen polysaccharide beads adsorb substantial amounts of amphiphilic lipids, therefore pre-treatment is necessary. The pre-treatment is done by pre-saturation of the gel filtration column by lipids using empty liposome suspensions.

(c) Absorption: Detergent absorption is achieved by shaking of mixed micelle solution with beaded organic polystyrene adsorbers. The great advantage of using detergent adsorbers is that they can remove detergents with a very low critical micelle concentration (CMC), which are not completely depleted by dialysis or gel filtration methods.

(d) Dilution: Upon dilution of aqueous mixed micellar solution of detergent and phospholipids with buffer the micellar size and the polydispersity increases dramatically, and, as the system is diluted beyond the mixed micellar phase boundary, a spontaneous transition from polydisperse micelles to monodisperse vesicles occurs.

Box 3. Method for incorporating proteins into liposomes

REAGENTS:

1. Dioleoyl Phosphatidyl Choline
2. Buffer of choice / distilled water

METHOD:

1. Dry 0.5 mmole of dioleoyl phosphatidyl choline under nitrogen in a disposable glass tube.
2. Evacuate in dessicator under vacuum for 30 minutes.
3. Add buffer / dH20 to required volume and scrape the sides of the glass tube to dislodge the lipid.
4. Add protein at 1 mg/ml of lipid used.
5. Vortex for 30 seconds.
6. Sonicate twice in a bath sonicator at 7 degree for 15 sec.

This makes multilamellar vesicles that become small unilamellar vesicles (SUV) with prolonged sonication time. To make large unilamellar vesicles extruder is used.

PHARMACOKINETICS OF LIPOSOMES

Clearance of classical liposomes from the circulation has a biphasic character- a rapid initial phase leading to 75% disappearance from the blood in the first five minutes for large unsonicated liposomes and about 50% elimination for the smaller sonicated ones; this is followed by a considerably slower rate of clearance after fifteen minutes for both types of liposomes. The rapid initial uptake by the liver may be ascribed to endocytic activity of both Kupffer and endothelial cells of the liver. The slow second phase of clearance from the blood probably reflects the uptake of lecithin by hepatocytes mediated by lipoproteins (Juliano, R. L. and Stamp, D. 1975).

ROUTES OF ADMINISTRATION OF LIPOSOMAL DELIVERY SYSTEM

The *in vivo* applicability of liposomal delivery system depends on its routes of administration- oral, intravenous, subcutaneous, dermal, transdermal, intraperitoneal, intramuscular or inhalation through the bronchial track. All these pathways have specific characteristics and limitations. The liposomes, when administered orally, can survive stomach digestion, but are lysed by the lipolytic enzymes in the intestine. It has been shown that they can protect the entrapped materials from such degradation and hence, this route is used for liposomes in oral vaccination. When injected intravenously, liposomes are rapidly cleared from the blood and absorbed mainly by the phagocyte cells of the reticuloendothelial system. Liposomes injected through subcutaneous, transdermal or intramuscular routes may remain in the circulation longer. Thus, it may act as a depot of drugs and facilitate the slow release of the entrapped materials from the vesicles. One report indicated the supremacy of the subcutaneous route over other routes (Allen, T.M. *et al.*, 1993).

GENE THERAPY

Gene therapy has emerged to be one of the most potent and promising strategies for the development of cures for inherited and acquired disorders with underlying genetic defects or malfunction. More than 500 gene therapy protocols have been approved for a wide range of diseases that include cancer, cystic fibrosis, hemophilia, and neuromuscular disorders. In addition, genes encoding for various antigens may be administered as vaccines. Successful gene therapy necessitates targeted transfection of plasmids containing transgenes of interest into cells. The administration of the gene of interest should theoretically result in the expression of the therapeutic protein. However, the delivery of the large anionic bioactive DNA across cell membranes is by no means a simple feat. DNA is easily degraded by circulating and intracellular deoxyribonucleases and also must be delivered intact across the cell and nucleolar membranes to the nucleus. Hence DNA transfection has largely been achieved using viral or nonviral delivery vectors. Such delivery systems can protect the DNA from degradation and have been able to successfully transfer it both *in vitro* as well as *in vivo*.

(a) Cationic Liposomal DNA Delivery

Nonviral transfection systems such as cationic liposomes are generally preferred over viruses because they are nonimmunogenic, are relatively easy to assemble, and are amenable to scale-up for industrial production. Both membrane fusion and endocytosis have been implicated as mechanisms for DNA transfer using liposomes (Fig. 5). Genes present in the transferred DNA can be expressed transiently or the transferred DNA may also integrate into chromosomes. Liposomes for DNA delivery are prepared from phospholipids with an amine hydrophilic head group. The amines may be either quaternary ammonium, tertiary, secondary or primary and the liposomes prepared in this way are commonly referred to as cationic liposomes, because they possess a positive surface charge at physiological pH, both exterior and interior [most liposomal membrane surfaces are negatively charged (anionic) or neutral]. In cationic liposome-mediated transfection, the cationic lipid is usually mixed with a "helper lipid" like dioleoylphosphatidylcholine (DOPC) or phosphatidylethanolamine (dioleoyl PE) to increase its transfection potency. Cationic liposomes interact with DNA through charge interaction, and there is an extensive lipid rearrangement during the complexation of liposomes with DNA.

There are several advantages of using cationic liposomes for gene delivery.

(a) First, unlike neutral or anionic liposomes that require entrapment of DNA inside vesicles, cationic liposomes form complexes with negatively charged DNA via charge interaction.

(b) Second, liposome/DNA complexes are normally prepared in such a way that the complexes are in slight excess in positive charge. This allows an efficient interaction of the positively charged complexes with the negatively charged cell membrane.

(c) Third, complexation of cationic liposomes with DNA may help protect the DNA against physical forces and enzymatic digestion.

The knowledge gained in the development of anticancer liposomes, such as the prolongation of liposome circulation using polyoxyethylene lipids, may be applied in the development of gene delivery liposomes in order to produce efficient carriers. This strategy should aid liposome targeting in gene delivery. Thus far, targeting efforts with cationic liposomal gene formulations have been rare. A rare example is the targeting to mouse hepatocytes by the portal vein administration of cationic liposome DNA complexes coated with chylomicron remnants. Such liposomes are specifically taken up by the apolipoprotein receptor. However, these chylomicron remnant coated liposomes were not targeted to the hepatocytes on tail vein administration. Although clinical trials with liposomal gene delivery

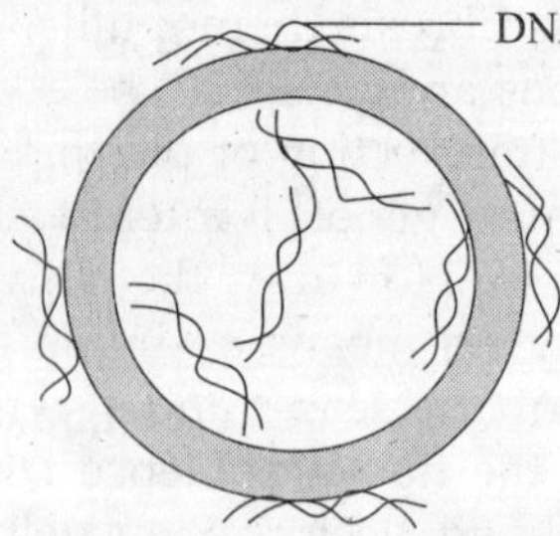

Fig. 5

in cystic fibrosis patients have concentrated on administration via non-parenteral routes, for the obvious reason that the disease manifests itself in the alveolar epithelium, it may be possible to treat this condition by the intravenous administration of the replacement gene-liposome complex. This is because such a strategy results in gene expression in distal regions of the alveolar epithelium.

Liposomal DNA is superior to free DNA in the treatment of experimental tumours. A cationic liposome formulation of the human interleukin 2 (IL-2) gene improves the tumoricidal activity of IL-2 in experimental xenografts. Cell kill in this case is mediated by expression of the IL-2 cytokine, which stimulates natural killer (NK) cell activity. Efficient arterial gene transfer will be beneficial in the treatment of arteriosclerosis and gene transfer into porcine arteries, via catheter, is enhanced by complexing the gene with liposomes.

Although the experimental data indicate that cationic liposomes are able to facilitate the transfer of DNA into live mammalian cells, there are still major problems that need to be overcome, which include a reduction in the rapid clearance of cationic liposomes and the production of efficiently targeted liposomes. At the cellular level, the problems may be overcome by improving receptor mediated uptake using appropriate ligands, the endowment of liposomes with endosomal escape mechanisms, a more efficient translocation of DNA to the nucleus and the efficient dissociation of the liposome complex just before the entry of free DNA into the nucleus.

Although cationic liposomes are relatively efficient in delivering DNA into cells, they can be inactivated in the presence of serum, and there have been reports of instability upon storage. Inactivation in serum is due to interaction between liposome/DNA complexes and anionic molecules in the serum, which would neutralise the positive charges and decrease the performance of liposome/DNA complex.

In addition, cytotoxicity of cationic liposomes remains a concern irrespective of the preferred route of DNA transfer, and such effects have been shown to occur both *in vitro* and *in vivo*. Under *in vitro* and *ex vivo* conditions, massive cell death caused by toxic delivery systems can significantly reduce their overall potential for DNA delivery. Dose-dependent pulmonary toxicity of cationic lipids due to the production of reactive oxygen intermediates has been demonstrated in mice upon intratracheal instillation of a cationic liposome preparation. Cationic lipids were demonstrated to be cytotoxic to phagocytic macrophages. The commercially available cationic lipid-based transfection agent lipofectAmine has been reported to induce toxicity in various tissues such as pulmonary cells, and in arterial cell walls in a porcine restenosis model for localized gene therapy. It is evident that there is a need for efficient and well-tolerated delivery systems to exploit the benefits of gene therapy medicine.

(b) Anionic Liposomal DNA Delivery

A few studies that use anionic liposomal DNA delivery vectors as an alternative to cationic liposomes have been reported in recent years. Anionic liposomes have been previously used as models to simulate the cell surface to study non–receptor-mediated transport of DNA across cells. For DNA delivery purposes, however, anionic liposomal formulations have predominantly been used to transfer oligonucleotides into hippocampal neurons. However, these investigations are limited owing to: (1) inefficient entrapment of DNA molecules within anionic liposomes, and (2) lack of toxicity data. Lack of further development of these systems may be attributed to the inefficient association between anionic lipids and DNA molecules resulting from the repulsive electrostatic interaction between these negatively charged species. It was demonstrated that anionic liposomes could successfully entrap high levels of DNA by complexation with divalent Ca^{2+} ions . Since anionic lipoplexes form by interaction of anionic residues on the surface of liposomes with the phosphates in the plasmid DNA backbone, mediated by divalent Ca^{2+} bridges, increase in Ca^{2+} can increase the amount of DNA complexed to the liposomes and is predicted to increase transfection efficiencies. But increase in Ca^{2+} concentration increases the cytotoxicity in the cells.

APPLICATIONS OF LIPOSOMES IN MEDICINE

The idea of using liposomes for drug delivery has been around since the early 1970s. The goals are to protect the body from unwanted side effects of various drugs and, to be targeted to specific tissue, to achieve desired concentrations of these drugs at a target site. The liposomes, smaller than a red blood cell, can be used to carry drugs inside them, where they do not come into contact with the cells. The lipids can be broken down by metabolism inside the body and the drug is then released. Lipid/drug aggregates are easy to form and vulnerable to structural manipulations, allowing for the adjustment of their properties for particular purposes. The major advantages of the liposome application are the protection of active compounds from degradation.

As shown in the following Fig. 6, the amphiphilic phospholipid molecules form a closed bilayer sphere in an attempt to shield their hydrophobic groups from the aqueous environment while still maintaining contact with the aqueous phase via the hydrophilic head group. The resulting closed sphere may encapsulate lipid soluble drugs within the bilayer membrane or aqueous soluble drugs within the central aqueous compartment. The encapsulation within/ association of drugs with liposomes alters drug pharmacokinetics, and this may be exploited to achieve targeted therapies. Alteration of the liposome surface is necessary in order to optimize liposomal drug targeting.

Though liposomal drug technology was developed before 30 years, it was in 1990 in Ireland the first liposomal drug was approved and it was used to treat mycosis. The obstacle associated with liposomal drug delivery includes lack of a general strategy for tissue targeting, drug concentrations, which are sometimes, lower than desired, difficulty of large-scale

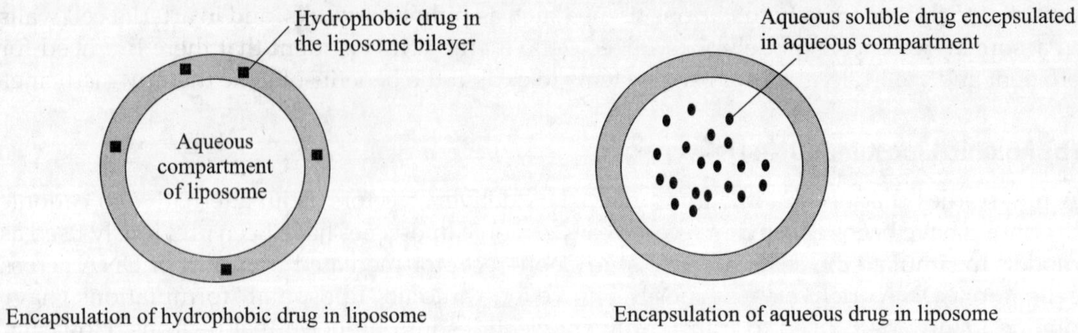

Fig. 6

production, and unpredictable shelf life. One of the first products that came out was a doxorubicin-containing liposome to treat cancer. It was not as toxic to the heart–but instead of heading for the tumor site, it stayed in circulation for only half an hour or so, and then lodged in the liver and spleen. This is because the liver and spleen are two major organs of the reticuloendothelial system (RES), which are known to accumulate and metabolize liposomes (Juliano and Stamp, 1978; Derksen *et al.*, 1988). Hence long-circulating liposomes are used nowadays for drug delivery.

(a) Localised Application

Topical application of liposomes has great potential in dermatology. Wolf *et al.*, (1995) studied the delivery of a liposome-encapsulated DNA repair enzyme, T4 endonuclease V, in pH-sensitive liposomes (phosphatidylcholine, phosphatidylethanolamine, oleic acid and cholesteryl hemisuccinate (2:2:1:5 molar ratio)) and found that liposome encapsulation enhanced the delivery of T4 endonuclease across mouse skin *in vivo* leading to increased protection against UV radiation. This reduced the incidence of skin cancer in a UV-induced mouse model and prevented systemic immunosuppression of contact and delayed type of hypersensitivity.

- Topical formulation of interferon encapsulated in liposomes for the treatment of viral infection like Herpes simplex virus and warts was effective than *in vivo* treatment (Vonka, V. *et al.*, 1995).
- *Transferosomes* are readily deformable liposomes and can penetrate across the stratum corneum in response to osmotic gradients.
- Topical delivery of plasmid DNA to keratinocytes would raise the possibility of obtaining controlled gene expression in localized tissues. This may be very beneficial for the design of vaccines and the treatment of cancer.
- Liposomes can passively deliver melanins, proteins and other bioactive molecules to hair follicles and hair shafts. Liposomal delivery of these molecules is time dependent. Negligible amounts of delivered molecules enter the dermis, epidermis or bloodstream, thereby demonstrating selective follicle delivery. Naked molecules are trapped in the stratum corneum and are unable to enter the follicle. Liposomes thus have high potential in selective hair follicle targeting. Gene delivery of liposomes,

opened the field of gene therapy and other molecular therapy to restore hair growth, physiologically restore or alter hair pigment and to prevent hair loss.

(b) Systemic Applications

(i) In drug delivery for cancer

Several potential barriers interrupt effective delivery of chemotherapeutic drug in its active form to solid tumors. Most small-molecule chemotherapeutic agents have a large volume of distribution on intravenous administration, which results in high level of toxicity in healthy tissues e.g., anthracycline antibiotics such as daunomycin, induces a dose-dependent, heart toxicity. One way to overcome the problem of toxicity and to enhance the circulation time of drug is to encapsulate the drug in a biocompatible material that can be injected into the blood stream with the intention of delivering drug to a diseased site, (e.g., solid tumor tissue). Over the last 30 years, lipid-based drug carrier systems have been developed that can retain drug, evade the body's defenses, and so target (passively and specifically) the interstitial tissue of tumors. Liposomal encapsulation of drug significantly decreases the volume of distribution, which reduces non -specific toxicities and increases the concentration of drug that can be effectively delivered to the tumor (Papahadjopoulos and Gabizon, 1995; Gabizon and Martin, 1997; Martin, 1998). During circulation the drug is carried within the liposomal aqueous space protecting it from metabolism and inactivation in the plasma. Moreover the continuous endothelium of the healthy tissues prevents leakage of liposome in to the tissues. In the case of tumour vasculature discontinuous endothelium with gaps ranging from 100 to 780 nm, allows liposomes to extravasate and due to impaired lymphatics leads to increased accumulation of drug in the tumour tissues. This ability to control and produce a burst release would be extremely advantageous and may prove to be an essential step in providing efficacious levels of drug in the tumour. Conventional liposomes have been clinically evaluated and approved in a variety of diseases e.g., Doxorubicin encapsulated in polyethylene glycol coated liposome (Doxil) has been used as standard drug for the treatment of Kaposi's sarcoma, breast cancer and ovarian cancer. Circulation half-life of doxil is 55 hours, which makes the drug very effective. The *in vivo* use of classical liposomes, however, is hampered by their very rapid clearance from the circulation by the reticuloendothelial system. The liposome half-life in the circulation can be considerably prolonged by steric stabilization of the liposomes with PEG, which is referred as stealth liposomes. Stealth liposomes exhibit sustained release of the drug and show tendency to accumulate in tumor tissues. Both conventional and stealth liposomes are composed of lipid membranes that are impermeable to both amphipathic and highly water-soluble molecules at physiological temperatures (37°C) which is an important feature for the maintenance of stable liposome drug formulations, both during storage and in plasma. Non-targeted liposomes accumulate in the intersitium surrounding the tumor and not within the tumor cells. The limited distribution of liposomes within the tumor interstitium results in high interstitial pressure, which makes it less effective for the treatment of large tumors as high interstitial pressure prevents the access of the drug in to the necrotic core. Discovery of immunoliposome helped to overcome this obstacle. The single biggest challenge, therefore, now facing drug delivery (for liposomes and other carriers) is to initiate the release of the encapsulated drug only at the diseased site and at controllable rates. For this to happen, the carrier/drug relationship must be triggered to change from the stable or kinetic trap requirement of the delivery phase to one of gross instability at the site. Hyperthermia increases the permeability of tumor microvasculature, which in turn increases the bioavailability of liposomal drugs in the tumor area and provides uniform distribution of the drug. In 1978, Yatvin, M.B. *et al.*, suggested using a temperature-sensitive liposome Doxorubicin to achieve this type of effect, and their groundbreaking efforts paved the way for many additional discoveries. Liposomes can be rendered thermosensitive by replacing some of the DSPC lipid component with DPPC, resulting in an increased leakage of the encapsulated material when heated to 42°C. This effect was found to be dependent on the presence of plasma proteins. At 37°C these liposomes are stable and do not release DOX. However, heating to 42°C for 30 min results in a release of >60% of the encapsulated DOX. The combination of hyperthermia and L-DOX appears to be a very promising strategy for the treatment of cancer due to its ability to enhance three important characteristics of liposomal drug delivery: tumor accumulation, intratumoral distribution, and bioavailability. Liposomes are suggested as long awaiting "magnetic bullet" cancer therapy due to their ability to accumulate in the tumor tissues. However not all cancers and not all patients respond to the bullet equivalently. The drug being delivered by the liposomes play an important role in response to the tumor. Some of the anticancer drugs encapsulated in liposomes used for therapy are doxorubicin, mitoxantrone and others.

(ii) Liposomal Anti-microbial agents

Although a large number of antimicrobial agents are available, many of these drugs are unable to penetrate the cell interior so there was a need for efficient delivery vehicle. Liposome encapsulated antimicrobial drug effectively gained access with microbes and exhibited the drug action. Liposomes have been used to administer drugs in infective conditions like malaria and leishmaniasis. The use of amphotericin B, a polyene antibiotic, in the treatment of systemic fungal infections is associated with extensive renal toxicity. Amphotericin B acts mainly by binding to sterols such as ergosterol in membranes of sensitive fungi, thus increasing membrane permeability. The toxicity associated with this compound is probably the result of its interaction with cholesterol in mammalian cells. Liposomal amphotericin B (Ambisome), the first liposomal preparation to be licensed for clinical use, is used for the treatment of systemic fungal infections. Liposomal amphotericin B, by passively targeting the liver and spleen, reduces the renal and general toxicity of the drug at normal doses. Ambisome are also used to treat drug-resistant leishmaniasis, a parasitic infection of the reticuloendothelial system. The ability of liposomes to be taken up by macrophages and to concentrate in the liver and spleen undoubtedly makes them ideal for the treatment of diseases of the liver and spleen, such as leishmaniasis. Another polyene antibiotic with a similar mechanism of action to amphotericin B, hamycin, improves the survival of mice infected with systemic Candida albicans when encapsulated in liposomes. Liposome encapsulation of the aminoglycosides—amikacin and gentamicin has significantly enhanced the activity of these antibiotics in the treatment of intracellular mycobacterial infections. Liposomes may be targeted to the lung by coating vesicles with O-stearoyl amylopectin and polyoxyethylene or monosialoganglioside. The encapsulation of the anti-tuberculosis agents rifampicin or isoniazid in lung targeted liposomes modulates the toxicity and improves the efficacy of these drugs. Intra cellular transport of drugs for the removal of unwanted materials (metabolic products, metals like iron, mercury, gold) stored in the lysosomes is also facilitated by liposomes (Rahman, Y.E. 1980). Thus, the liposomes by passively targeting the liver and spleen and also by actively targeting the lung have been shown to improve the efficacy and modulate the toxicity of certain anti-infectives.

(iii) Treatment of respiratory disorders

Aerosolised liposomes can be used for inhalation therapy and is effective in relieving bronchial constriction in asthma. Treatment of adult respiratory distress syndrome with liposomal prostaglandin E1 (Lip-PGE1) injected intravenously decreased lung leak and lung lavage neutrophil accumulation. Intratracheal administration of a liposomal alpha-tochopherol decreased acute lung injury (Suntres, Z. E. *et al.*, 1995).

(iv) As an oxygen-carrying fluid

Liposome-encapsulated hemoglobin (LEH) is encapsulating hemoglobin in an inert shell. Inert outer shell can be derived from synthetic non-lipid materials, egg-lecithin-based lipid mixtures and distearoyl-phosphatidylcholine-based blends. LEH acts as best oxygen-carrying fluid whose circulation time ranges from 16–20 hours. Improvement of the LEH to increase its biocompatibility includes carbohydrate moiety addition as carbohydrates are expressed on the majority of biological membranes including the red blood cell. Addition of gangliosides to the lipid bilayer of the liposomes increases the circulation time of liposomes (Rabinovici, R. 1990).

(v) As vaccine adjuvants

The use of liposomes as immunological adjuvants in vaccines is another promising development with important implications in medicine (Gregoriadis, 1981b). It can act as powerful adjuvants (without the presence of side effects seen with other adjuvants) to induce humoral and cellular immunity against bacterial and viral antigens relevant to disease such as diphtheria toxoid, *plasmodium falsiparum* antigens, hepatitis B surface antigens, rubella haemaglutinin, cholera toxin and influenza virus subunits. Liposomal immunoadjuvants act by slowly releasing encapsulated antigen on intramuscular injection and also by passively accumulating within regional lymph nodes. Strategies that enhance the targeting of liposomes to regional lymph nodes, for example the use of phosphatidylserine liposomes, may thus improve adjuvanticity. Liposomal vaccines can be made by associating microbes, soluble antigens, cytokines or deoxyribonucleic acid (DNA) with liposomes, the latter stimulating an immune response on expression of the antigenic protein. Liposomes encapsulating antigens, which are subsequently encapsulated within alginate lysine microcapsules, to control antigen release, improve the antibody response. Liposomal vaccines may also be stored dried at refrigeration temperatures for up to 12 months and still retain their adjuvanticity. Liposomes are unique in their flexibility in composition and structure, which makes them an effective immunological adjuvant. Manipulation of size, surface charge, phospholipid composition influences the rate of clearance from blood and their uptake by tissues. These modifications in conjugation with control of antigen diffusion from liposomes *in situ* led to pattern of antigen distribution and fate that would lead to agumented immunogenicity, and thus helps in producing "single shot vaccine".

(vi) Immunoliposomes for active targeting

Immunoliposomes (Fig. 7) are produced by conjugating various antibodies to the surface of liposomes, which is used for active targeting.

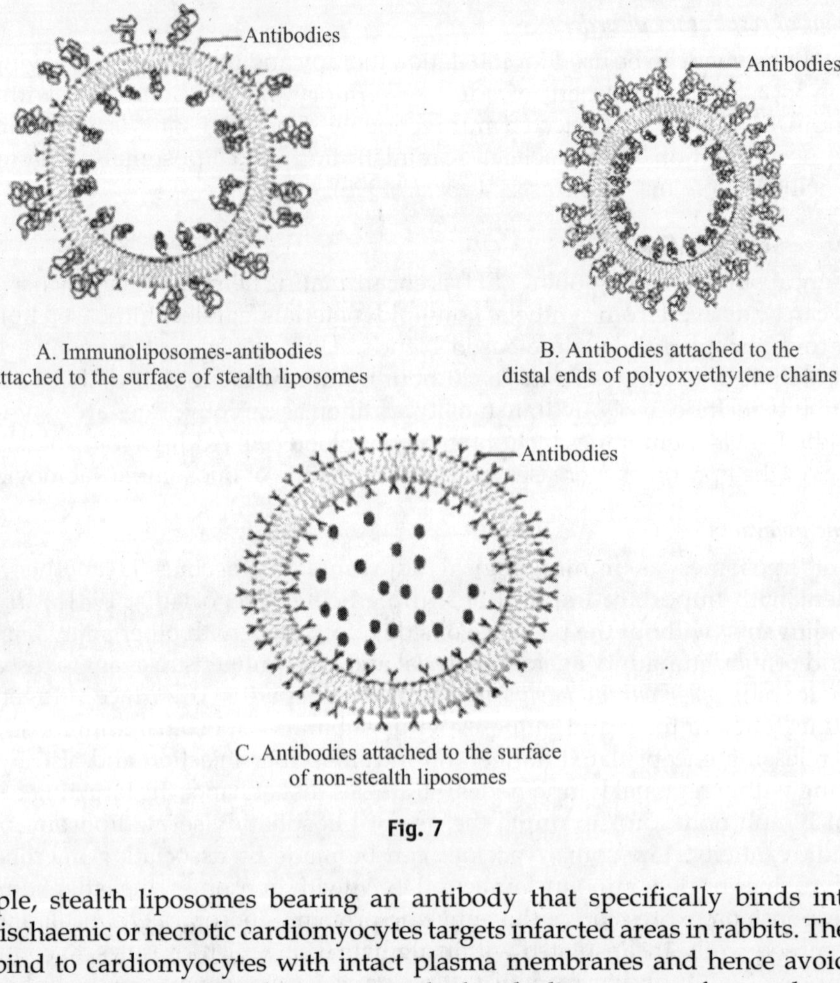

A. Immunoliposomes-antibodies attached to the surface of stealth liposomes

B. Antibodies attached to the distal ends of polyoxyethylene chains

C. Antibodies attached to the surface of non-stealth liposomes

Fig. 7

Example, stealth liposomes bearing an antibody that specifically binds intracellular myosin in ischaemic or necrotic cardiomyocytes targets infarcted areas in rabbits. The antibody does not bind to cardiomyocytes with intact plasma membranes and hence avoids healthy tissue. The infarcted heart antibodies were attached to the liposome surface as shown in Fig.A and the effectiveness of the antibody in homing the liposome was found to be directly related to the level and molecular weight of polyoxyethylene on the surface. A high polyoxyethylene density and polyoxyethylene with a molecular weight in excess of 2000 hampers the directing ability of the antibody by shielding the antibody from its target. To overcome this masking effect, surface antibodies may be attached to the distal ends of polyoxyethylene chains (Fig. 7B). This attachment of antibodies improves the targeting ability when compared with the attachment of antibodies to the surface of non-stealth liposomes (Fig. 7C). This indicates that the extended circulation time achieved by stealth liposomes improves targeting simply because the prolonged circulation allows an extended period of time for liposomes to make contact with their target.

Several molecularly targeted drugs are designed for directing the standard chemotherapy drugs specifically to cancer cells for better treatment of cancer and to cut back their side effects. To accomplish this goal, tens of thousands of chemotherapy drug molecules are loaded into synthetic nanoparticles called liposomes (Antibody-Directed Nanoparticles), which have a knack for penetrating cells. Researchers then take multiple copies of portions of antibodies that specifically recognize receptors highly expressed by cancer cells and attach them to the liposome's' perimeter. The end product is a novel targeted immunoliposome, which can target chemotherapeutic agents more specifically to tumors. Example of receptors targeted are: HER2, which is often overexpressed in breast cancer, and EGFR, which is overexpressed in many cancers including lung, colorectal, brain, pancreatic, ovarian, breast, and prostate tumors (Park, J.W. 1997). Both receptors are involved in cell growth and development. Several chemotherapy drugs, including doxorubicin, vinorelbine, epirubicin, topotecan, and irinotecan, were loaded successfully into the immunoliposomes. In studies with cancer cells in culture and tumor models in mice, the immunoliposomes demonstrated a remarkable affinity for cancer cells, with minimal release of their contents into the bloodstream. In essence, instead of just dumping the chemotherapy drugs into the bloodstream for dissemination, the immunoliposome serves as a protective package with a way to target the delivery of its contents.

While most immunoliposomes have been developed to target potentially toxic cancer chemotherapeutic agents, they may also prove beneficial in the treatment of other pathological conditions. The conjugation of interleukin 2 (IL-2) to the surface of non-stealth liposomes allows the targeting of toxic immunosuppressants to areas of the immune system participating in graft rejection, such as T-cells expressing the IL-2 receptor, without affecting other parts of the immune system.

Limitations to the clinical use of immunoliposomes include a failure to demonstrate an advantage in experimental cancer chemotherapy plus the fact that these systems are rather complex and active targeting with immunoliposomes still faces tremendous obstacles in terms of the effective delivery of drugs to specified tissues or organs. Although the technology has been under development for several decades, its first clinical application using antibodies as the targeting molecules was a phase I trial published in 2004.

In this trial, the anticancer drug doxorubicin (which, like many such drugs, is highly toxic to various tissues) was administered intravenously to 23 patients with stomach cancer. The targeting molecules used were certain cancer-reactive human monoclonal antibodies. The results clearly demonstrated the technical feasibility of active targeting in clinical practice. The treatment itself, however, was unsuccessful, despite the known clinical efficacy of doxorubicin (much research still needs to be done on immunoliposomal delivery).

(vii) Drug targeting using ligand bearing liposomes

An alternative targeting strategy that may ultimately have a wider application than antibodies described above utilizes surface moieties that are up regulated in certain diseases as targets. These include the folate receptor, over expressed in ovarian carcinoma, and the cell adhesion molecules (eg, selectins and integrins), which are implicated in metastatic events. Liposomes bearing specific ligands such as folate may be used to target ovarian carcinomas, while specific peptides or carbohydrates may be used to target integrins and selectins. Targeting with small ligands appears more likely to succeed than the use of antibodies since these ligands are easier to handle and manufacture. Specifically, hamycin liposomes bearing mannosyl human serum albumin, which directs them to macrophages *in vivo*, are superior at suppressing leishmanial parasites than non-targeted liposomes. Hence, although liposomes are rapidly cleared by macrophages *in vivo*, an enhancement of this macrophage uptake by the use of targeting ligands results in an enhancement of the antileishmanial activity. Liposomes may also be targeted to hepatocytes by conjugating with asialofetuin. Such liposomes may be useful in the delivery of genes to hepatocytes for the treatment of specific diseases.

Conventional liposomes can also be rendered mitochondria-specific via the attachment of known mitochondriotropic residues to the liposomal surface, as molecular mitochondrial drug targets for selectively delivering biologically active molecules to the site of mitochondria, which will eventually launch a multitude of new therapies for the treatment of mitochondria-related diseases (Sarathi V. Boddapati *et al.*, 2005).

(viii) Liposomal based DNA vaccines

DNA vaccines administered as liposomal complexes also improve the antibody response over that seen with free DNA. Cell and humoral mediated immunity after the intramuscular administration of plasmids encoding for the S region of hepatitis B was found to be enhanced by liposomal complexation. Cationic liposomes, by virtue of being able to transfer DNA more efficiently into cells than can be achieved with naked DNA alone, lead to an increase in the level of antigen expression and hence an increase in the immune response to the antigen. For e.g. Delivery of human leukocyte antigen gene, HLA B7, aided by liposomes has been to correct melanoma. In cancer gene therapy, the degree of growth inhibition of malignant cells depends upon lipid to DNA ratio and total DNA complex administered to the cells (Hofland, H. and Huang, L. 1995).

OTHER APPLICATIONS OF LIPOSOMES

(a) Liposomes in Diagnostic Imaging

Radiolabeled liposomes have been successfully used to monitor pharmacodynamic changes of liposomes and image tumors, abscesses, ischemic and infarcted regions. Gamma-emitting radionuclides, such as ^{99m}Tc, ^{111}In and ^{67}Ga, have been encapsulated in liposomes for use as imaging agents. Liposomes made of radio-opaque iodinated or brominated phospholipids have been used for contrast image enhancement of liver and spleen so as to improve the diagnostic accuracy of computed tomography (Gregoriadis, 1983). Similar success has been obtained in lymph node imaging or treatment, after subcutaneous or intramuscular injection of liposomes containing appropriate agents (Segal *et al.*, 1975). Some clinical trials for tumor imaging have already been started using labeled liposomes. Generally, two methods have been used to label liposomes for diagnostic imaging purposes. One involves the entrapment of radionuclides during the hydration of the lipids with the aqueous phase. However, low encapsulation efficiency and batch production variability are two disadvantages of this approach. The second method is the labeling of preformed liposomes. Labeling of preformed liposomes with radionuclides is ideally suited for clinical imaging applications wherein the liposomes can be formulated and distributed to the clinician as an unlabelled kit. This preparation is subsequently labeled on site with a short half-life radionuclide prior to clinical use.

(b) Liposomes in Delivery of Cosmetic Materials

Topical application of liposomes has great potential in dermatology. There is an array of effective agents, including proteins, amino acids, peptides, vitamins and bioactives, for skin care applications. Most conventional methods release the active with no delivery or minute value. Therefore, there clearly exists a need for an effective system that delivers these bioactive compounds at the site of action, while minimizing peak-trough fluctuations. Ideally, such a system would eliminate undesirable side effects and reduce dosage and frequency of administration while improving visible effects. Many technologies are already in place, including multiple emulsions, microemulsions, liposomes etc. Among all these technologies, liposomes are most suitable for transferring cosmetic actives into the sub-epidermal level.

Liposomes frequently favour the disposition of encapsulated active ingredients in the epidermis and dermis, while the permeation rate is decreased. This helps to fix active ingredients to the outermost skin layers as desired for cosmetic products. Simultaneously, the washing out may be delayed so that, for example, aqueous sun care products containing liposome-encapsulated UV filters are water-resistant. Liposomes has the ability to hold more cosmetic active with high efficiencies while maintaining the integrity of the liposome structure and also have high moisturizing qualities. However, these positive effects mentioned above depend on the composition, size, and the amount of liposomes.

The cosmetic active loading can be achieved either passively (the active is encapsulated during liposome formation) or actively (after liposome formation). A hydrophobic cosmetic active can be directly incorporated into liposomes during vesicle formation, and the extent of uptake and retention is governed by active-lipid interactions. Trapping efficiencies of 100% are often achievable, but this is dependent on the solubility of the cosmetic active in the liposome membrane. Passive encapsulation of a water-soluble cosmetic active relies on the ability of liposomes to entrap an aqueous buffer containing a dissolved active during vesicle formation. Trapping efficiencies (generally less than 30%) are limited by the trapped volume contained in the liposomes and the solubility parameters of the cosmetic active.

The risk factors concerning the use of phospholipids in cosmetics is that high doses of phospholipids applied topically over a longer period may lead to irritations on dry and normal skin. Likewise, it has been mentioned that due to a biochemical feedback mechanism, a long-term application of phospholipids may have an impact on the dermal lipid metabolism.

(c) Liposomes in Enzyme Immobilisation

Liposome plays an important role in enzyme immobilisation or encapsulation.

Steps involved in encapsulation are:

- Phospholipid is dissolved in suitable organic solvent.
- Solvent is evaporated under vacuum in desiccator for 30 min.
- Dry phospholipid film is dispersed with enzyme solution in suitable buffer, vortexed for 30 min and sonicated in bath sonicator for 15 seconds.
- Small unilamellar liposomes encapsulating enzyme is obtained from multilamellar vesicles.

Enzymes are delivered to the target organs in two different ways. One way is by interaction of phospholipid bilayer of appropriate charges with liver and spleen cells, penetration of the liposomal membrane in to the cell and subsequently fuse with cell lysosome and the contents are liberated or the smaller substrate a penetrate in to the phospholipid bilayer and react with enzyme in aqueous phase and the product are liberated through the bilayer. Enzyme encapsulation finds important application in treatment of enzyme deficiencies or malfunction in liver and spleen cells related to metabolism of macromolecules.

SUMMARY

Liposomes are artificial lipoid systems consisting of one or more phospholipid bilayers arranged concentrically with aqueous compartments and a core and range in size from 50 nanometers to several micrometers in diameter. The sources of phospholipids for making the liposomes are usually synthetic (derived from glycerol or glycero-3-phosphocholine) and tissue-derived (either from egg or bovine). Liposomes are often distinguished according to their number of lamellae and size as small unilamellar vesicles (SUV), large unilamellar vesicles (LUV) and large multilamellar vesicles (MLV) or multivesicular vesicles (MVV). Liposomes are usually prepared by methods like Hand-shaken method, Sonication method, Reverse-phase evaporation method, Freeze-dried rehydration method, Detergent depletion method etc. The general elements of the procedure involve preparation of the lipid for hydration, hydration with agitation, and sizing to a homogeneous distribution of vesicles. Liposomes are used in gene therapy to treat a wide range of genetic disorders, where in the DNA is administered to patients along with liposomes by either cationic or anionic liposomal DNA delivery methods. Liposomes are made to enclose a variety of drugs and the liposomal drug delivery system is used for delivering drugs like cancer drugs, anti-infectives, immunoadjuvants, antitumour drugs etc. Liposome's sensitive to a variety of physical and chemical stimuli (temperature, light, or pH) can be constructed so that they can be triggered to release their contents in response to the stimuli of particular interest, the triggered release properties having potentially important applications in drug delivery. Liposomes used in drug delivery systems can also be attached to hydrophilic polymers like polyethylene glycols (PEG) to make long circulating liposomes or stealth liposomes so that they circulate in the body for long time and can deliver the drug for more time. Recently the discovery of immunoliposomes, pH sensitive liposomes and thermosensitive liposomes increased the therapeutic index of liposomes in medicine. Apart from the medical applications, liposomes are also used in other industries like cosmetic industries.

ACKNOWLEDGEMENTS

One of the authors (Dr. K. Pandima Devi) gratefully acknowledges Department of Science and Technology, India for the SERC Fast Track Proposals for Young Scientist scheme

REFERENCES

Allen, T.M., Hansen, C.B. and Guo, L.S. (1993) Subcutaneous administration of liposomes: a comparison with the intravenous and intraperitoneal routes of injection. *Biochim. Biophys. Acta* 1150: 9-16.

Bangham, A. D. (1968) Membrane Models with Phospholipids. *Prog Biophys. Mol. Biol.* 18: 29–95.

Bangham, A. D. (1972) Lipid Bilayers and Biomembranes. Annu. Rev. Biochem. 41: 753–776.

Boddapati, S.V., Tongcharoensirikul, P., Hanson, R.N., D'Souza, G.G.M., Torchilin, V.P. and Weissig, V. (2005) Mitochondriotropic Liposomes. Journal of Liposome Research.15: 49 – 58.

Danilo, D. Lasic (1997) Liposomes in Gene Delivery. CRC Press LLC.

Farhood, H., Serbina, N. and Huang, L. (1995). The Role of Dioleoyl Phosphatidylethanolamine in Cationic Liposome Mediated Gene Transfer. *Biochim. Biophys. Acta* 1235: 289–295.

Frank Szoka, Jr. and Papahadjopoulos, D. (1980) Comparative Properties and Methods of Preparation of Lipid Vesicles (Liposomes). Ann. Rev. Biophys. Bioeng. 9:467-508.

Gao, X. and Huang, L. (1995) Cationic Liposome-Mediated Gene Transfer. *Gene Ther.* 2: 710–722.

Gregoriadis, G. (1981b) Liposomes: a role in vaccines? *Clin.Immunol. Newslett.* 2:33-36.

Hofland, H. and Huang, L. (1995) Inhibition of ovarian-carcinoma cell proliferation by liposome-plasmid DNA complex. *Biochem. Biophys.Res.Commun.* 207: 492-496.

Ijeoma, F. Uchegbu (1999) Parenteral drug delivery: 1. *Pharmaceutical Journal* 263: 309-318.

Juliano, R. L. and Stamp, D. (1975) The Effect of Particle Size and Charge on Clearance Rates of Liposomes and Liposome Encapsulated Drugs. *Biochem. Biophys. Res. Commun.* 63: 651–658.

Juliano, R. L. and Stamp, D. (1978) Pharmacokinetics of liposome-encapsulated antitumor drugs. Studies with vinblastine, actinomycin D, cytosine arabinoside, and daunomycin. *Bioch. Pharmacol.* 27: 21–27.

Lasic, D.D. (1998) Novel Applications of Liposomes. *Trends Biotechnol.* 16: 307–321.

Li, L. and Hoffman, R. M. (1995) Model of Selective Gene Therapy of Hair Growth: Liposome Targeting of the Active Lac-Z Gene to Hair Follicles of Histocultured Skin. *In vitro Cell Dev. Biol. Anim.* 31: 11–13.

Liu, D. and Huang, L. (1989) Role of cholesterol in the stability of the pH-sensitive, large unilamellar liposomes prepared by the detergent dialysis method. *Biochim. Biophys. Acta* 981: 254–260.

Liu, D. and Huang, L. (1990) pH-sensitive, plasma-stable liposomes with relatively prolonged residence in circulation. *Biochim. Biophys. Acta* 1022: 348–354.

Meyer, O., Papahadjopoulos, D. and Leroux, J.C. (1998b) Copolymers of *N*isopropylacrylamide can trigger pH sensitivity to stable liposomes. *FEBS Lett.* 421: 61–64.

Papahadjopoulos, D. and Miller, N. (1967) Phospholipid Model Membranes. I. Structural Characteristics of Hydrated Liquid Crystals. *Biochim. Biophys. Acta* 135: 624–638.

Lee, K. D., Woodle, M. C., Lasic, D. D., Redemann, C. and Martin, F. J. (1991) Sterically stabilized liposomes: improvements in pharmacokinetics and antitumor therapeutic efficacy. *Proc. Natl. Acad. Sci. U.S.A.* 88:11460–11464.

Park, J. W., Hong, K., Kirpotin, D. B., Papahadjopoulos, D. and Benz, C. C. (1997) Immunoliposomes for cancer treatment. *Adv. Pharmacol.* 40: 399–435.

Patil, S., Rhodes, D. and Burgess, D. (2004) Anionic Liposomal Delivery System for DNA Transfection. AAPS J. 6 (4): article 29.

Perkins, W. R., Minchey, S. R., Ahl, P. L. and Janoff, A. S. (1993) The Determination of Liposome Captured Volume. *Chem. Phys. Lipids* 64: 197–217.

Rabinovici, R., Rudolph, A. S., Ligler, F. S., Yue, T. L. and Feuerstein, G. (1990) Liposome-Encapsulated Hemoglobin: An Oxygen-Carrying Fluid. *Circ Shock* 32: 1–17.

Rosenecker, J., Zhang, W., Hong, K., Lausier, J., Geppetti, P., Yoshihara, S., Papahadjopoulos, D. and Nadel, J. A. (1996) Increased liposome extravasion in selected tissues: Effect of substance P. *Proc. Natl. Acad. Sci. U.S.A.* 93: 7236–7241.

Sarathi, V. Boddapati and Pakamas Tongcharoensirikul (2005) Mitochondriotropic Liposomes. *Journal of liposomal research* 15: 49-58.

Segal, A.W., Gregoriadis, G. and Black, C.D.V. (1975) Liposomes as vehicles for the local release of drugs. *Clin. Sci. Mol. Med.* 49: 99-106.

Suntres, Z. E. and Shek, P. N. (1995) Liposomal Alpha-Tocopherol Alleviates the Progression of Paraquat-Induced Lung Damage. *J. Drug Target* 2: 493–500.

Vonka, V., Petrovska, P., Borecky, L. and Roth, Z. (1995) Increased Effects of Topically Applied Interferon on Herpes Simplex Virus-Induced Lesions by Caffeine. *Acta Virol.* 39: 125–130.

Wolf, P., Cox, P., Yarosh, D. B. and Kripke, M. L. (1995) Sunscreens and T4N5 Liposomes differ in their ability to protect against ultraviolet-induced sunburn cell formation, Alterations of dendritic epidermal cells, and local suppression of Contact Hypersensitivity. *J Invest Dermatol.* 104: 287–292.

Yatvin, M.B., Weinstein, J.N., Dennis, W.H. and Blumenthal, R. (1978) Design of liposomes for enhanced local release of drugs by hyperthermia. *Science (Wash. DC)* 202: 1290–1293.

Suggested Readings

1. Gregoriadis, G. (1977) Liposomes as carrier of enzyme and proteins in medicine. In: *Biomedical application of immobilized enzymes and proteins* (T.M.S Chang, ed.), Vol. I, pp.191-218, Plenum press, New york.
2. *Handbook of Nonmedical Applications of Liposomes* (1996) Vol. I, II, III, & IV (Danilo D. Lasic and Yechezkel Barenholz eds.) CRC Press, Inc.
3. Lasic, D.D. (1992) Liposomes: Synthetic lipid microspheres serve as multipurpose vehicles for the delivery of drugs, genetic material and cosmetics. *American Scientist,* Vol. 80, pp. 20-31.
4. Nicolau, C. and Cudd, A. (1989) Liposomes as carriers of DNA. *Critical Reviews in Therapeutic Drug Carrier Systems,* Vol. 6, pp. 239-271.
5. Papahadjopoulos, D. and Gabizon, A.A. (1995) Sterically stabilized (Stealth®) liposomes: Pharmacological properties and drug carrying potential in cancer. In: *Liposomes as Tools in Basic Research and Industry* (J.R. Philippot and F. Schuber, ed.), pp. 177–188, CRC Press, Boca Raton, FL.
6. Rahman, Y.-E. (1980) Liposomes and chelating agents. In: *Liposomes in Biological systems* (G. Gregoriadis and C. Allison, eds.), pp.265-298, Wiley, New york.
7. Small, D.M. (1986) The Physical Chemistry of Lipids, From Alkanes to Phospholipids. In: *Handbook of Lipid Research,* Vol. 4, Plenum Press, New York.

REVIEW QUESTIONS

Long Questions

1. Liposomes are characterized on what basis? Describe in detail about the different kinds of liposomes
2. What is gene therapy? How are liposomes used in gene therapy?
3. Write in detail about the applications of liposomes in treatment of cancer
4. Describe the different methods applied in the preparation of liposomes
5. How are drugs actively targeted using liposomes?

Multiple Choice Questions

1. Liposomes were first demonstrated by
   (a) A.D. Burg (b) A.D. Bangham
   (c) A.D. Clark (d) A.D. Daniel
2. SUVs are
   (a) Single Unit Vesicle (b) Small Unit Vesicle
   (c) Single Unilamellar Vesicles (d) Small Unilamellar Vesicles

3. The hydrated molecule present in PEGylated liposomes are
   (a) polyethoxy glucose (b) polyethoxy glycol
   (c) polyethylene glycol (d) Polyethylene glucose
4. Liposomes attached to antigen/antibody fragment for active targeting is called as
   (a) PEGylated liposomes (b) Immunoliposomes
   (c) ationic liposomes (d) Conventional liposomes
5. An example of cationic liposome is
   (a) L-a-Phosphatidylcholine, cholesterol
   (b) L-a –Phosphatidylcholine /Dicetyl phosphate/cholesterol
   (c) 1,2-Dioleoyl-sn-Glycero-3-Phosphoethanolamine (DOPE)
   (d) L-a-Phosphatidylcholine, distearoyl
6. Optimal storage pH for liposome is around
   (a) pH 5 (b) pH 8
   (c) pH 6 (d) pH 7
7. The method of producing liposomes by vortexing is called as
   (a) Sonication Method (b) Hand-Shaken Method
   (c) Freeze-dried rehydration method (d) Detergent depletion method
8. pH sensitive liposomes responding to _______ environment is suitable for drug delivery to tumour
   (a) Basic (b) Neutral
   (c) Acidic (d) None
9. Liposomes for gene delivery, prepared using phospholipids with an amine hydrophilic head group are called as
   (a) Cationic liposomes (b) Anionic Liposomes
   (c) Neutral Liposomes (d) None
10. The first liposomal preparation to be licensed for clinical use for the treatment of systemic fungal infections is
    (a) Liposomal hamycin (b) Liposomal amphotericin B
    (c) Liposomal rifampicin (d) Liposomal isoniazid

Answers

1. (b)	2. (d)	3. (c)	4. (b)	5. (c)	6. (d)
7. (b)	8. (c)	9. (a)	10. (b)		

PART–14

Microbes, Toxins and Antibiotics

CHAPTER 45

Biochemistry and Molecular Biology of Bacterial Bioluminescence and Applications of Lux Genes

P. Bramha Chari and S.K. Dubey*

Dept. of Microbiology, Division of Bacterial Genetics and Environmental Biotechnology, Goa University, Goa–403 206

E-mail: dubey_santosh1@rediffmail.com

Bioluminescence refers to the process of visible light emission in living organisms mediated by an enzyme catalyst. The phenomenon of bioluminescence has been observed in many different organisms such as bacteria, fungi, fish, insects, algae, and squids. The enzymes that catalyze the bioluminescence reaction are called luciferases and the substrates are often referred to as luciferins. Bioluminescent organisms comprise a diverse set of species that are widely distributed, inhabiting terrestrial, freshwater and marine ecosystems. The discoveries of physiology, biochemistry, molecular biology and genetic control of bacterial bioluminescence have revolutionized the area of environmental microbiology and biotechnology, ecology, industrial and medical significance. The elucidation of luciferase genes regulation permitted the discovery of intercellular communication among bacteria, which in turn, has led to a better understanding of bacterial pathogenesis and the associations of microorganisms in the environment. With the advent of molecular biology, it has been possible to construct bioluminescent bacteria that are naturally dark by insertion of *lux* genes. The present communication illustrates the biochemistry, molecular biology and genetic control of bacterial bioluminescence and its impact on new areas of research and biotechnological applications using luminescent systems as environmental biosensors.

*Corresponding author

DISTRIBUTION

Bioluminescent bacteria are the most widely distributed and abundant luminescent organisms in nature (Wilson and Hastings 1998). Although their primary habitat is the ocean and they are as free-living or associated with other marine organisms often in a symbiotic relationship. Some luminescent bacteria are also found in terrestrial habitats as well as in freshwater. Luciferase and other enzymes involved in the luminescent *(lux)* system as well as the corresponding *lux* genes have been isolated from marine bacteria *Vibrio* and *Photobacterium* and also from terrestrial bacteria *Xenorhabdus sp.*. The most widely studied bioluminescent bacteria are *Vibrio harveyi, Vibrio fischeri, Photobacterium phosphoreum, Photobacterium leiognathi,* and *Xenorhabdus luminescens*. However, other species of luminescent bacteria which are of interest include light-emitting *Vibrio cholerae* found in freshwater and the aerobic *Shewanella hanedai* and the obligate symbionts of the flashlight fishes, the deep-sea anglerfishes, and the pyrosomes. Some of the major advantages of bioluminescence are defense, schooling of fish, luminous lure feeding, navigation or communication, mating and camouflage.

BIOCHEMISTRY

The light-emitting reaction in bacteria involves the oxidation of reduced riboflavin phosphate ($FMNH_2$) and a long chain fatty aldehyde decanal with the emission of blue-green light. Because the structure of the substrate is relatively simple ($FMNH_2$ and RCHO) is part of or closely related to the normal metabolites in the cell, the term luciferin has generally not been used to refer to the substrates for the reaction catalyzed by bacterial luciferases (Hastings *et al.*, 1985). The luciferase mediated light reaction is as follows:

$$FMNH_2 + RCHO + O_2 \rightarrow FMN + H_2O + RCOOH + h\nu \text{ (490 nm) (blue green light)}$$

The reaction is highly specific for $FMNH_2$. Modification of the flavin ring or removal of the phosphate group decreases the activity significantly (Fig 1). The natural aldehyde for the bioluminescence reaction is believed to be tetradecanal on the basis of identification of this compound in lipid extracts, the preference for tetradecanal by luciferases at low (nonsaturating) substrate concentrations, and the specificity of the *lux*-specific fatty acid reductase system, which catalyzes the synthesis of the fatty aldehyde substrate (Baldwin et al 1995;Tu and Mager, 1995). However, differences in aldehyde specificity do exist among different bacterial luciferases. Particularly noteworthy are the high luminescence responses of *V. harveyi* and *X. luminescens* luciferases to decanal at saturating substrate concentrations, whereas higher light intensity can be obtained with dodecanal for *P. phosphoreum, P. leiognathi, and V. fischeri* luciferases. This property may be very important in terms of expression of light emission *in vivo* in prokaryotic and eukaryotic cells missing the aldehyde substrate, because decanal appears to cross the cell membrane much more readily than longer-chain aldehydes do (Baldwin *et al.*, 1995).

The mechanism of the bioluminescence reaction catalyzed by luciferase and the intermediates in the reaction have been studied extensively, primarily because of the very slow turnover of the enzyme (Lee *et al.*, 1985). The reduced flavin, $FMNH_2$, bound to the enzyme, reacts with 0_2 to form a 4a-peroxyflavin.This complex interacts with aldehyde to form a highly stable intermediate, which decays slowly, resulting in the emission of light along with the oxidation of the substrates (Fig. 1). Luciferase undergoes only a single catalytic cycle in most

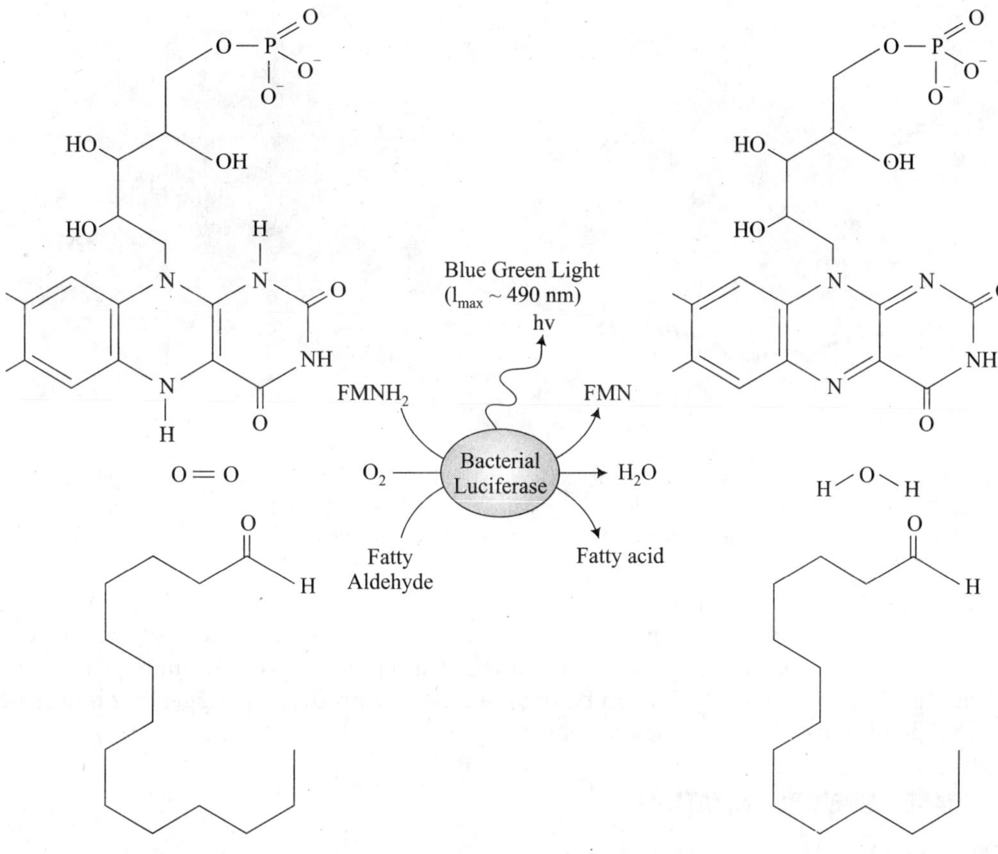

Fig. 1. The net chemical equation of the bacterial luciferase catalyzed light reaction.

assays, because the rate of chemical oxidation of $FMNH_2$ is higher than the turnover rate of luciferase in the bioluminescence reaction. Consequently, the decay of luminescence with time is a first-order process and reflects the turnover number of the enzyme under the assay conditions. Half times for turnover are dependent on experimental conditions, including the particular luciferase and aldehyde. For example *Xenorhabdus luminescens* luciferase has the lowest turnover rates, with half times of 3 and 20 s with decanal and dodecanal respectively.

Luciferases from all bioluminescent bacteria studied are heterodimers of α (40 kDa) and β (35 kDa) subunits. The *lux*A and *lux*B genes encoding the α and β subunits are adjacent in the *lux* operon (Baldwin *et al.*, 1995), which also contains, among others, three genes (*lux* C, D, and E) encoding proteins that make up the fatty acid reductase complex (for aldehyde synthesis) (Fig. 2). *Lux* A and B have been cloned and expressed heterologously and extensively used as molecular reporters. There is a higher degree of similarity among α subunits of different strains than between the β subunits (Baldwin *et al.*, 1995); the α and β subunit of each strain are themselves quite similar (32% identity); β subunit probably has evolved by gene duplication. The organization of the complete *lux* regulon is of special interest.

Structurally, the luciferase enzyme is simple and has no metals, nor prosthetic groups or amino acid residues. The α subunit appears to dictate the primary kinetic properties, such as

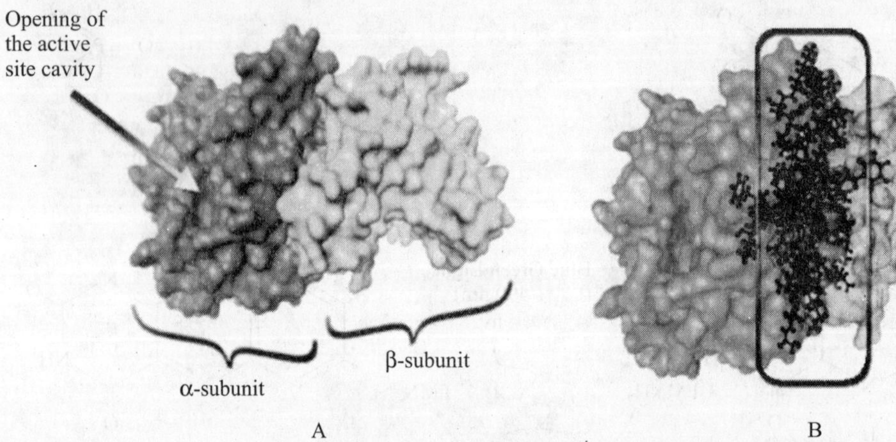

Fig. 2. (A) Bacterial Luciferase and (B) The network of extensive inter-subunit interactions, highlighted in the rectangular box, and involving ionic attractions, hydrogen bonds, and hydrophobic contacts, is responsible for the assembly of the functional bacterial luciferase.

substrate binding ($FMNH_2$ and aldehyde), light emission and turnover rate, while the β subunit appears to influence the interaction with flavin and to contribute to an active center. However, it has now been demonstrated that purified and immobilized subunits do not exhibit luciferase activity and that activity can be regained if renatured, thus suggesting that neither the α- and β-subunits possess an active center (Figs. 2A and B).

LUCIFERASE GENE ORGANIZATION

Common *Lux* Genes

The α and β subunits of luciferase are encoded by the *lux*A and *lux*B genes, respectively. In *Vibrio fischeri,* these genes are adjacent in the *lux* operon, which also contains the genes encoding for the proteins that make up the fatty acid reductase complex for aldehyde synthesis (*lux*C, D, and E), which divert tetradecanoic acid away from the fatty acid biosynthesis pathway into the long chain fatty aldehyde substrate for the luminescent reaction (Meighen, 1994). *Lux*D, the first thioesterase whose crystal structure has been elucidated, cleaves tetradecanoyl-ACP. The fatty acid product is activated by *luxE* with ATP to form fatty acyl-AMP. In the presence of *lux*C, the acyl group is transferred from acyl-AMP to *lux*E and then to *lux*C before being reduced with NADPH.

In most luminescent bacteria, the luciferase genes *lux* AB are flanked upstream by *lux* CD and downstream by *lux* E with transcription from left to right (Meighen, 1993). However, Mancini and coworkers (1989) found that in *P. phosphoreum,* the gene *lux* F is located between *lux* B and *lux* E (Fig. 3). Also, the nucleotide sequences of *lux* A and *lux* B genes in *P. phosphoreum* are considerably different from other bacterial luciferases (Ferri et al 1991). The regulation of autoinduction in the *lux* system of *V. fischeri* has been studied in detail (Engebrecht *et al.,* 1983) It is regulated by two genes (*lux*R and *lux*I), which are present in two divergent operons.

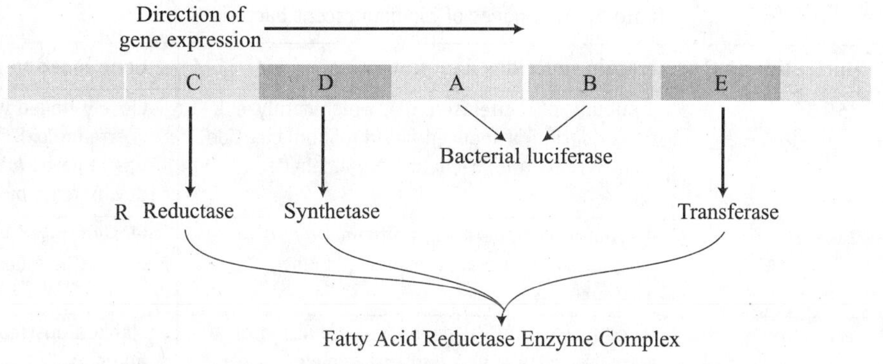

Fig. 3. The arrangement of the *Lux*CDABE open reading frame (genes in *lux* operon of *V. fischeri*).

$$\text{RCO-X} \xrightarrow{LuxD} \text{RCOOH} \xrightarrow[\text{ATP}]{LuxE} \text{RCO-AMP} \xrightarrow{LuxC} \text{RCO-}Lux\text{E-} \rightarrow \text{RCO-}Lux\text{C} \rightarrow \text{R-CHO}$$

$$\text{NADPH}$$

The three proteins form a weakly associated fatty acid reductase complex in *P. phosphoreum* that generates tetradecanal. Shorter chain aldehydes, particular decanal and dodecanal, are routinely used in the in vitro assays for luciferase activity, although tetradecanal is thought to be the natural substrate for the luminescent reaction, consistent with the specificity of the fatty acid reductase complex.

Species-specific *Lux* Genes

Although the luciferase and fatty acid reductase genes are the only common *lux* genes in luminescent bacteria, new *lux* genes are continually being discovered in specific luminescent bacteria, adding to the complexity and diversity. Their discovery and the identification of their roles help in understanding how the luminescent system is connected with other metabolic pathways and related to the overall physiology of different bacteria. The *lux* operons containing the common *lux* structural genes (*lux*ABCDE) of different species show their organization and their relationship to other genes that are part of, or closely linked to, the primary *lux* operon (Fig. 3) (Engebrecht, 1984). In all *lux* systems, the luciferase genes (*lux*AB) are flanked by the *lux*CDE genes in the order of *lux* CDABE (Meighen, 1993).

***Lux* F gene:** A gene *lux* F encoding a flavoprotein related in sequence to the luciferase subunits is located between *lux* B and *lux* E in certain *Photobacterium* species. The flavin prosthetic group is composed of the products of the luminescent reaction, flavin and tetradecanoic acid, covalently linked to each other (Kita *et al.*, 1991). The *lux* F protein (23 kDa) has almost 40% sequence identity with the carboxyl half of the *lux* B subunit (36 kDa) and appears to have arisen by gene duplication followed by internal deletion of 100 codons in the amino terminal region (Meighen and Dunlap, 1993). The recent crystallization of the *lux* F protein (Kita et al 1991) and elucidation of its structure should be of value in modeling the structure of luciferase. Although the *lux* F protein is apparently not necessary for luminescence, it may play a specific role in the luminescence of *Photobacterium* species.

***Lux* G gene:** The *lux G* gene is very closely linked to the *lux* E gene in marine luminescent species with its ribosomal binding site located in some species in the *lux E* coding region

Table 1. *Lux* genes of bioluminescent bacteria

Genes	Nucleotide (bp)*	Protein functions and properties	Gene location
*lux*A	354-360	α subunit of luciferase (αβ), which catalyze the oxidation of reduced flavin mononucleotide ($FMNH_2$) and a long-chain fatty aldehyde.	Closely linked with upstream *lux*B flanked upstream by *lux*CD and downstream by *lux*E.
*lux*B	324-328	β subunit of luciferase. Approximately 30% identical in amino acid residue sequence with *lux*A	Closely linked with downstream *lux*A
*lux*C	477-480	r (reductase) NADPH-dependent reduction of activated fatty acyl group to aldehyde and/or transferred back to the synthetase.	Flanked upstream *luxAB* and closely linked with upstream *luxD*
*lux*D	305-314	t (transferase). Generation of fatty acids for the luminescence reaction	Flanked upstream *luxAB*
*lux*E	370-378	s (synthetase). ATP-dependent activation of fatty acyl group to form acyl-AMP	Flanked downstream *luxAB*
*lux*F	228-231	Non-fluorescent flavoprotein (23 kDa) with homology (40%) in sequence with luciferase β subunits	In most *Photobacterium* the strains between *luxB* and *luxE*, excluded *Ph. leiognathi* species.
*lux*G	233-236	A 27 kDa protein related to that of a flavin reductase and other enzymes in electron transport, possibly implicating it in producing $FMNH_2$ for luminescence reaction.	Found after *luxE* in marine but not terrestrial luminescent bacteria, *X. luminescence*
*lux*H	230	A 27kDa protein related in sequence (64% identity) to a new protein (HtrP) of *E. coli* which catalyzes the synthesis of riboflavin precursor, implicating the *luxH* gene in riboflavin synthesis	Found in *V. harveyi* after *luxG*
*lux*I	193	Regulatory protein (22 kDa) required for synthesis of autoinducer, implicating *Lux*I protein as an autoinducer synthetase	Found in *V. fischeri* upstream of *luxC*
*lux*R	250-252	Regulatory protein (29 kDa), a receptor for the autoinducer and activates the expression of the *lux* operon in *V. fischeri* an operon containing them	Found in *V. fischeri* in a divergent operon immediately upstream of *lux* structural genes
*lux*R	205	Regulatory gene locus controlling expression of *lux* genes in *V. harveyi*. No homology in sequence to *luxR* in *V. fischeri*	Gene locus in *V. harveyi* not linked to the *lux* structural genes
*lux*L	189	*Lux*L protein tightly binds lumazine or riboflavin and can affect the wavelength, intensity, and decay of light produced by bacterial luciferase	Found 600 bp upstream of *luxC* in *Ph. phosphoreum* transcribed in the opposite direction

Table 1 Contd.

Table 1 Contd.

***lux*Y**	194	A yellow fluorescence protein (YFP, 22 kDa) that cause a shift in the luminescence from 490 nm to a higher wavelength resulting in the emission of a yellow rather than a blue-green light.	Found in one strain of *V. fischeri* (Y1). Not linked to the *lux* structural genes

The number does not include a stop codon. The range is indicated for different bacterial species. The nucleotide sequences of common *lux* genes (*lux* CDABE) were reported for three genera (i.e. *Vibrio*, *Photobacterium*, and *Photorhabdus*) and *lux* genes sequences of other species were identified in only specific luminescent strains (see reviewed by Meighen and Dunlap, 1993).

(Table 1). In Spite of this linkage, evidence connecting its expression with luminescence has not been obtained with dark mutants arising from *lux* regulatory genes (Engebrecht, *et al.*, 1983). Moreover, *lux* G is not present after *lux* E in the terrestrial bacteria, *X. luminescens*. The amino acid sequence of the *lux* G protein has recently been shown to be related to that of a flavin reductase and other enzymes in electron transport, possibly implicating it in producing $FMNH_2$ for the luminescent reaction (Meighen 1994).

GENES RELATED IN SEQUENCE TO THE RIBOFLAVIN SYNTHESIS GENES

The *lux* H gene has only been found as part of the *V. harveyi lux* operon (Swartzman, 1990). The close relationship of its amino acid sequence (60% identity) and molecular weight with an *E.coli* protein catalyzing the synthesis of a riboflavin precursor, 3,4-dihydroxy-2-butanone 4-phosphate, implicates the *lux*H gene in riboflavin synthesis. Downstream of the *lux*G gene in *P. leiognathi* and *P. phosphoreum* are genes with protein products that have sequences closely related to enzymes implicated in riboflavin synthesis in *Bacillus subtilis*. Denoted as ribI, ribII, and rib III in Fig. 4 are genes with sequences corresponding to riboflavin genes involved in the conversion of 6, 7-dimethyl-8-ribityl lumazine (lumazine) to riboflavin and the biosynthesis of 3,4-dihydroxy- 2-butanone 4-phosphate and lumazine, respectively (Lee, *et al.*, 1992). An *in vivo* transcription termination site could not be detected between the *lux* and putative *rib* genes in *P.leiognathi*. These results along with reduced ribofiavin-5-phosphate being a substrate in the luminescent reaction suggest a possible linkage between the *lux* and *rib* genes in luminescent bacteria (Fig. 3). Two other genes, *lux* L and *lux* Y, code for proteins with 30% identity in sequence to an enzyme catalyzing the conversion of lumazine to riboflavin. The *lux* L protein tightly binds lumazine or riboflavin and can affect the wavelength, intensity, and decay of light produced by bacterial luciferase (Lee *et al.*, 1990). The lumazine protein causes a small shift in the intensity of the light toward lower wavelengths. This protein is found in large amounts in some *Photobacterium* species, with the *lux*L gene being located upstream and transcribed in the opposite direction, to the *P. phosphoreum lux* CDABEG operon (Fig. 4).

The *lux* Y gene codes for a yellow fluorescence protein (YFP) in a few *V.fischeri* strains. This gene is apparently not linked to the *lux* operon. The YFP causes a shift in the luminescence from 490 nm to a higher wavelength resulting in the emission of a yellow rather than a blue-green light has so far been detected in X *luminescens* strains isolated from insects.

Regulatory *lux* genes: Regulatory genes controlling the expression of luminescence in bacteria have been cloned and sequenced from *V. fischeri* and *harveyi*. Two regulatory genes, *lux* I and *lux* R, are located immediately upstream of the *lux* CDABEG genes in the *V. fischeri lux* operon

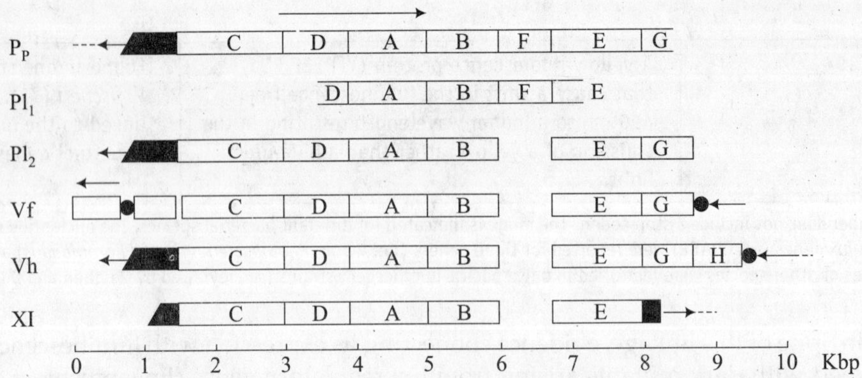

Fig. 4. *Lux* gene organization for *P. phosphoreum* (Pp), *P. leiognathi* (P1), *V. fischeri* (Vf), *V. harveyi* (Vh), *and X. luminescens* (Xl). The nucleotide sequences have been determined for all regions represented.

(Engebrecht, *et al.*, 1983) (Fig. 4). *Lux* I is cotranscribed in the same direction as the *lux* structural genes, is required for the synthesis of an autoinducer that controls expression of luminescence in *V. fischeri.* The *lux* R gene product is believed to function as a receptor for the autoinducer and to activate the expression of the *lux* operon. A regulatory gene controlling expression of the *V. harveyi lux* operon has also been cloned and sequenced (Martin, *et. al.*, 1991). Although designated as *lux* R, it has no sequence identity with *V.fischeri lux* R (Fig. 5). In the *lux* systems from other species, regulatory genes have not yet been isolated; however, the DNA immediately upstream of the *lux* operon is comparable to that of the *V. harveyi lux* operon, being highly AT rich without any extended open reading frames (Miyamoto, *et al.*, 1989).

REGULATION OF BIOLUMINESCENCE BY AUTOINDUCTION

Bioluminescence in bacteria can also be regulated through a phenomenon known as autoinduction (Nealson, 1977). Autoinduction or quorum sensing was first discovered in *V. fischeri,* it is cell-to-cell communication that ties gene expression to bacterial cell density. Quorum sensing involves the self-production of a diffusible pheromone called an autoinducer (AI), which serves as an extra cellular signal molecule that accumulates in the medium and evokes a characteristic response from cells. In bioluminescence, once the concentration of the AI reaches a specific threshold (above 10^7 cells mL^{-1}). It triggers the energetically costly synthesis of luciferase and other enzymes involved in luminescence. Thus, by sensing the level of AI, the cells are able to estimate their density and ensure that the luminescent product will be sufficiently high to cause an impact in the environment, (Wilson and Hastings 1998), making the process cost effective. The AI for *V. fischeri*, N-acyl homoserine lactone (AHL), was once thought to be species-specific, however recent studies have established that AHL can serve as a signaling molecule for more than 16 genera of gram-negative bacteria. This suggests that the AI protein can facilitate interspecies communication, allowing quorum-sensing bacteria to monitor the population of other species as well as their own. Quorum sensing is now known to be a widespread regulatory mechanism in bacteria, particularly among a number of pathogens, influencing their ecology and associations with eukaryotic organisms.

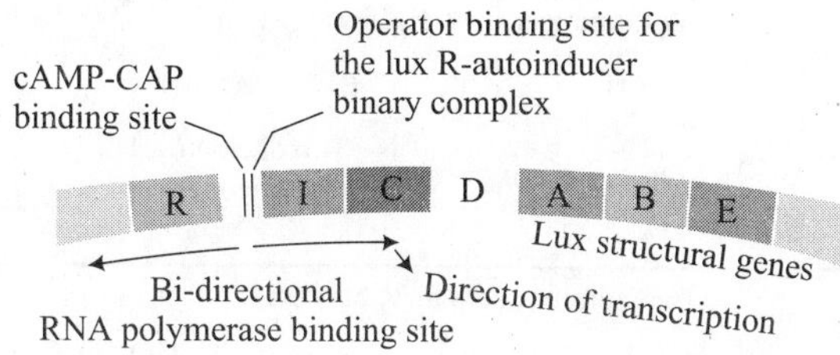

Fig. 5. The arrangement of the regulatory elements in the upstream region of the *lux* structural genes in *V. fischeri*.

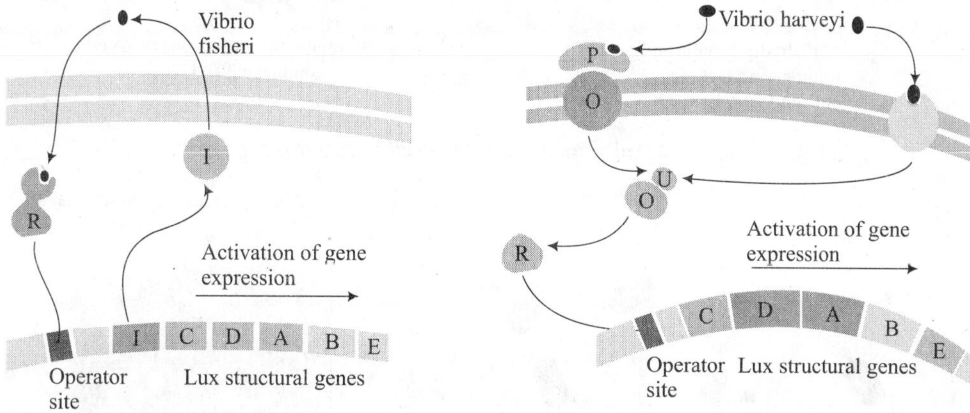

Fig. 6. Comparison of the different mechanisms involved in the regulation of expression of the *lux* structural genes in *V. fischeri* and *V. harveyi*.

The *lux*I regulatory gene is in the rightward operon together with the *lux* CDABE structural genes while the *lux*R is in the leftward operon. The *lux*I codes for an autoinducer synthase that is responsible for the production of the AI. The *lux*R gene encodes the *lux*R protein, which serves as both a receptor for the AI and a transcriptional activator of the *lux* operon. Binding of the AI to the *lux*R protein forms a complex that acts as a transcriptional regulator, activating transcription from the *lux* operon promoter. Once induction begins, the level of AI increases rapidly because the gene for AI synthase is part of the *lux* operon. In this way, the AI controls its own synthesis through a positive feedback circuit. Advances in molecular biology have made possible the cloning and expression of the *lux*AB genes in bacteria that are normally non-luminescent, transforming them into bioluminescent bacteria. The resulting transformations provide enormous applications in industry, medicine, microbial ecology, and environmental fields.

APPLICATIONS OF *LUX* GENES

I. Clinical and Diagnostic Applications

Luminescence immuno and gene probe assays, detection techniques, in particular, have proven to be very sensitive, allow obtaining analytical performance similar to or better than

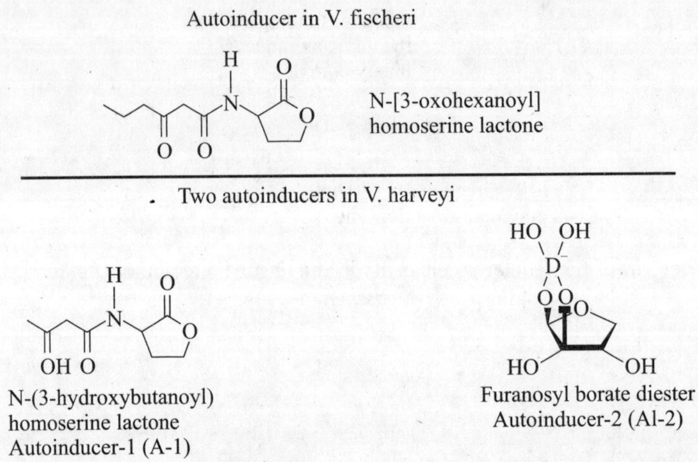

Fig. 7. Chemical structure of the autoinducers of luminous bacteria.

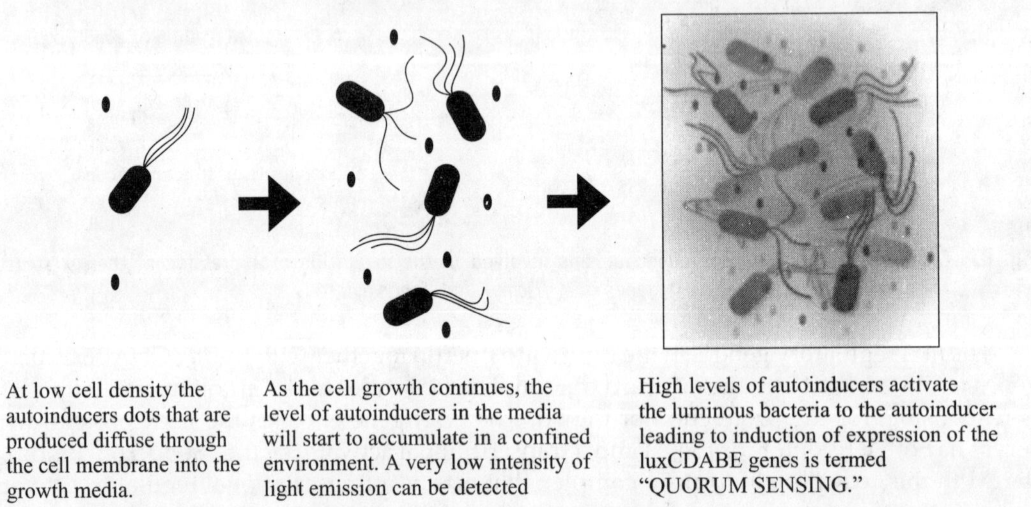

Fig. 8. The confined environment, in which the produced autoinducers can accumulate, is the key to the activation of the quorum sensing mechanism.

those achievable using radioisotopes. Measurement of light from a chemical reaction is very useful because the light output (intensity) is directly related to the analyte concentration, thus allowing precise and accurate quantitative analysis.

2. Bioluminescent Imaging: Applications to Cancer Biology and Endocrinology

Numerous cell lines expressing bioluminescent reporter genes can study mechanism of action of nuclear receptors. Cells expressing the luciferase gene under the control of a constitutive promoter were used to follow *in vivo* proliferation of cancer cells. The bioluminescent models

Table 2. Quorum-sensing systems of various bacteria

Species	Function	System	N-AHL acting as AI
Photobacterium fischeri	Bioluminescence	*Iux*I-*lux*R	N-(3-oxohexanoyl)-L-homoserine lactone (OHHL), AI-1
		*ain*S-*ain*R	N-Octanoyl-Lhomoserine lactone (OHL), AI-2
Vibrio harveyi	Bioluminescence	*lux*L, *lux*M	N-(3-hydroxybutanoyl)-L-homoserine lactone
		*lux*R, *lux*N	(HBHL), AI-1
		*lux*S-*lux*Q	AI-2 (unidentified)
Erwinia carotovora	Production of antibioic carbapenem	*car*I-*car*R	N-(3-oxohexanoyl)-L-homoserine lactone (OHHL)
	Production of hydrolases (pectinases, cellulases, etc.)	*eca*I(exp)-*eca*R(*exp*R)	
Agrobacterium tumefaciens	Production of factors for *conjugal transfer of Ti plasmids*	*tra*R-*tra*I	N-(3-oxooctanoyl)-L-homoserine lactone (OOHL)
Pseudomonas aeruginosa	Production of virulence factors (toxin A, elastase)	*las*I-*las*R	N-(3-oxododecanoyl)-L-homoserine lactone (OdDHL, 3OC12-HSL)
	Production of virulence factors and secondary metabolites (rhamnolipids)	*rhl*I-*rhl*R	N-(3-butanoyl-L-homoserine lactone) (BHL, C4-HSL)

are new tools for evaluating cancer treatment efficiencies and the role of hormone receptors in invasion. Cells expressing the luciferase gene under the control of hormones are used as *in vivo* biosensors for studying analog bioavailabilities and *in vivo* response kinetics.

3. Heat Shock Gene-bioluminescence Encoding Gene Fusions

Rapid and sensitive pollutant detection is possible by induction of heat shock gene-bioluminescence gene fusions. The utility of this response for pollutant detection using two *Escherichia coli* heat shock promoters, *dnaK* and *grpE*, were fused to the *lux* genes of *Vibrio fischeri*. Metals, solvents, crop protection chemicals, and other organic molecules rapidly induced light production from *E. coli* strains containing these plasmid-borne heat shock gene-bioluminescence gene fusions.

4. Bioluminescent Sensors

Bioluminescent sensors for detection of bioavailable Hg (II) in the environment. By using fusions of the well understood Tn21 mercury resistance operon (*mer*) with promoterless *lux*CDABE from *Vibrio fischeri*, three biosensors for Hg (II) have been constructed and tested. Bioluminescence specified by pRB28, carrying merRo/pT, by pOS14, mediating active transport of Hg (II), and by pOS15, containing an intact *mer* operon, was measured in rich and minimal media. Selifonova *et al.*, 1993 constructed a biosensor using fusions of the Tn21 mercury resistance operon (*mer*) with promoterless *lux CDABE* from *V. fischeri*. This *mer-lux* biosensor demonstrated the semi quantitative detection of inorganic Hg (II) in natural water in the 0.1 to 200 ppb range and was a good system for distinguishing bioavailable from unavailable forms of mercury. Recombinant luminescent bacteria have been constructed and used for general toxicity testing (including heavy metals) by Rychert and Mortimer (1991), Virta *et al.* (1994), Lampinen *et al.,* (1995). Metal-specific recombinant bacterial sensors have been constructed and used for the determination, e.g., inorganic mercury (Selifonova *et al.*, 1993; Virta *et al.*, 1995), organomercurials (Ivask *et al.*, 2001), zinc, cadmium, cobalt and lead, (Corbisier *et al.*, 1996), cadmium, lead, cobalt and nickel (Tauriainen *et al.*, 1998). In a metal-specific bacterial sensor the expression of a reporter gene is controlled by a genetic regulatory unit (receptor), which responds to the given heavy metal, i.e., receptor–reporter concept (Lewis *et al.*, 1998) is used. Most of the regulatory units used in the construction of metal-specific sensor bacteria originate from bacteria that possess natural precisely regulated resistance systems towards heavy metals. Heitzer *et al.*, 1992 developed a bioassay to assess the bioavailability of naphthalene and salicylate in contaminated soils. They used the genetically engineered bacterium *Pseudomonas fluorescens* HK44 carrying the *nah-lux* reporter plasmid capable of degrading both salicylate and naphthalene. Applegate et al 1998, has constructed a *tol-lux* fusion and introduced it into *Pseudomonas putida* F1. This new strain (TVA8) was used as a whole-cell reporter for benzene, toluene, ethylbenzene, and xylene (BTEX) sensing and bioavailability determination.

5. Quantitative Assessment of Hydrocarbons

A bioassay was developed and standardized for the rapid, specific, and quantitative assessment of naphthalene and salicylate bioavailability by use of bioluminescence monitoring of catabolic gene expression. The bioluminescent reporter strain *Pseudomonas fluorescens* HK44, which carries a transcriptional *nahG-lux CDABE* fusion for naphthalene and salicylate catabolism, was used. 2,4-D, DCP, Phenols, nitrates, mercury, heavy metals, explosives, and organic compounds have all been detected in this manner.

6. Monitoring Virulence Factors

Use of bacterial luciferase for monitoring the environmental regulation of expression of genes encoding virulence factors in *Listeria monocytogenes* is an important example. A promoter probe vector, which utilized the *lux AB genes* from *Vibrio fischeri* as reporter of gene expression, was constructed for use in *Listeria monocytogenes.* Using this system gene expression can be monitored non-destructively and in real-time, simply by measuring cellular bioluminescence. Derivatives of the promoter probe were also constructed that contained the cloned promoters from the *hlyA* and *plcA genes* of *L. monocytogenes.*

7. Tumor Research

Tumor growth and metastases as well as drug efficiency have been monitored in living animals by injecting a mouse with bioluminescent recombinant tumor cells and imaging the produced light.

8. *In vivo* Gene Expression

Bioluminescent reporter gene has been introduced in transgenic mice or rats, having a modified endogenous gene as well as BL reporter gene integrated into their genome which are used as models of human disease, in drug development and discovery and for target validation.

9. Infectious Disease Progression Monitoring

Disease progression and the efficiency of therapeutic agents have been assessed by injecting a mouse with BL marker such as pathogenic microorganisms like *Salmonella enteridis*, engineered to emit light and then imaging the light transmitted through the layers of animal tissues.

10. Bioluminescence Reporter Genes

Reporter genes are DNA sequences that encode an easily detectable protein or enzyme. These can be put easily into a cell to monitor the gene expression and to attain the whole cell biosensors for a cell localization purpose. In bacteria, the genes responsible for the light-emitting reaction (the *lux* genes) have been isolated and used extensively in the construction of bioreporters that emit a blue-green light with a maximum intensity at 490 nm (Fig. 2). Three variants of *lux* are available, one that functions at < 30°C, another at < 37°C, and a third at < 45°C. The *lux* genetic system consists of five genes, *luxA*, *lux*B, *lux*C, *lux*D and *lux*E. Depending on the combination of these genes used, several different types of bioluminescent bioreporters can be constructed.

11. *Lux* AB Bioreporters

Lux AB bioreporters contain only the *lux* A and *lux* B genes, which together are responsible for generating the light signal. However, to fully complete the light-emitting reaction, a substrate must be supplied to the cell. Typically, this occurs through the addition of the chemical decanal at some point during the bioassay procedure Numerous *lux* AB bioreporters have been constructed within bacterial, yeast, insect, nematode, plant and mammalian cell systems. Some of the chemical and biological agents are capable of detection by *lux* AB-based bioreporters.

12. *Lux* CDABE Bioreporters

Instead of containing only the *lux*A and *lux*B genes, bioreporters can contain all five genes of the *lux* cassette, thereby allowing for a completely independent light generating system that requires neither extraneous additions of substrate nor any excitation by an external light source. So in this bioassay, the bioreporter is simply exposed to a target analyte and a quantitative increase in bioluminescence results, often within less than one hour. Due to their rapidity and ease of use, along with the ability to perform the bioassay repetitively in real- time and on-line, makes *lux* CDABE bioreporters extremely attractive (Fig. 9). Consequently, they have been incorporated into a diverse array of detection methodologies ranging from the sensing of environmental contaminants to the real-time monitoring of pathogen infections in living mice.

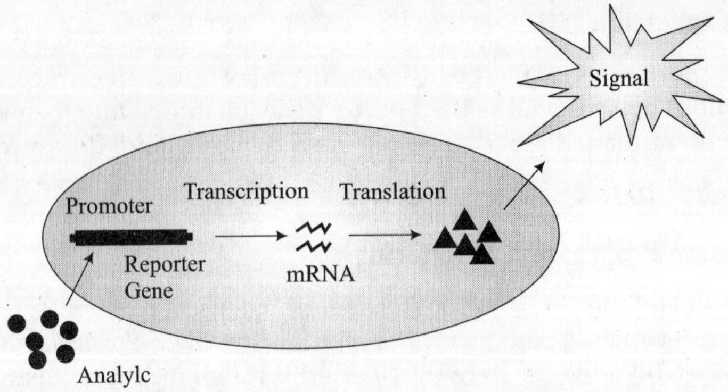

Fig. 9. Bioreporter organism exposure to a specific analyte, the promoter/reporter gene complex is transcribed into messenger RNA (mRNA) and then translated into a reporter protein that is ultimately responsible for generating a signal.

Table 3. Chemical and biological agents detectable by *lux*AB-based bioreporters

Antibiotic effectiveness	Gene expression/regulation
Antimicrobial agents	Growth phase regulation
Bacterial biofilms	Immunoassays
Bacterial biomass	*In vivo* expression technology (IVET)
Bacterial stress response	Industrial waste runoff
Bacterial transport mechanisms	Metabolic regulation
Bioremediation process monitoring	Mutagenicity tests
Cell viable counts	Plant pathogens
Circadian rhythms	Toxicity assays
DNA damaging agents	Tumor burden
Environmental contaminants	Viral infection
Food borne pathogens	Xenobiotic detection

Research is being conducted using bioluminescent bacteria to detect NO_2 gas, one of the main chemicals released when TNT decomposes underground and in detection of landmines. Rapid and effective diagnostic tests based on bioluminescence are the 'Microtox test' for water quality/toxicity testing which uses the luminous marine bacteria *Photobacterium phosphoreum*. When a toxin challenges the organism or pollutant the luminescent reaction pathway is disrupted, resulting in decrease in luminescence intensity.

SUMMARY

Bioluminescence refers to the visible light emission in living organisms that accompanies the oxidation of organic compounds, luciferins or decanals mediated by an enzyme luciferase. Significant advances have been made in characterization of bacterial bioluminescence with reference to *Lux* A, B encoded luciferase enzymes and other *lux* gene encoded proteins as well as their molecular organization and regulation in several marine bacteria viz. *Vibrio fischeri,*

Vibrio harveyi, Photobacterium phosphoreum, Photobacterium leiognathi and one terrestrial bacterium, *Xenorhabdus luminescens*. These microbes exhibit a fantastic cell to cell signaling (Quorum sensing) pathway.Infact the quorum sensing mechanism was first discovered in *Vibrio fischeri*. Bacterial quorum sensing using acyl-homoserine lactones (acyl-HSLs) as cell-density dependent signalling molecules is important for the transcriptional regulation of many genes essential in the establishment and the maintenance of bacteria-host associations. This bacterial *lux* system provides an ideal system for investigating developmental changes in bacteria, and has also served as an excellent bioreporter gene to develop several microbial luminescent sensors for monitoring environmental pollutants viz. Hg, As, Cd, Naphthalene, Xylene, Salicylate and several other poly-aromatic hydrocarbons.Additionally, the ease in cloning and transferring the *lux* genes into almost any biological organism has greatly increased the number and diversity of applications utilizing light emission for the detection of almost any metabolic function or compound, involving luciferase enzyme.Ease of light detection has also revolutionized the application of *lux* genes in detection of cancer, DNA probe development, pathogen detection and other biotechnological applications. This chapter elaborately deals with the biochemistry and molecular biology of bacterial bioluminescence along with several applications of bacterial *lux* genes in biology, medicine, agriculture and environmental sciences.

REFERENCES

Applegate, B.M., Kehrmeyer, S.R. and Sayler, G.S. (1998) A chromosomally based tol-*lux*CDABE whole-cell reporter for benzene, toluene, ethyebenzene, and xylene (BTEX) sensing. Appl. Environ. Microbiol. 64: 2730-2735.

Baldwin, T.O., Christopher, J.A., Raushel, F.M., Sinclair, J.F., Ziegler, M.M., Fisher, A.J. and Rayment, I. (1995) Structure of bacterial luciferase. Curr. Opin. Struct. Biol. 5: 798-809.

Chatterjee, J. and Meighen, E.A. (1995) Biotechnical applications of bacterial bioluminescence (*lux*) genes. Photochem. Photobiol. 62 : 641-650.

Corbisier, P., Nuyts, G., Mergeay, G., and Silver, S. (1993) *lux*AB gene fusions with the arsenic and cadmium resistance operons of *Staphylococcus aureus* plasmid pI258. FEMS Microbiol. Lett. 110 : 231–238.

Eberhard, A., Widrig, C. A., McBath, P. and Schineller, J. B. (1986) Analogs of the autoinducer of bioluminescence in *Vibrio fischeri*. Arch. Microbiol. 146 : 35-40.

Engebrecht, J., Nealson, K. H. and Silverman, M. (1983) Bacterial bioluminescence: isolation and genetic analysis of functions from *Vibrio fischeri*. Cell 32 : 773-781.

Engebrecht, J. and Silverman, M. (1984) Identification of genes and gene products necessary for bacterial bioluminescence. Proc. Nat. Acad. Sci. (USA) 81 : 4154-4158.

Ferri, S.R., Soly, R.R.; Szittner, R.B. and Meighen, E.A. (1991) Structure and properties of luciferase from *Photobacterium phosphoreum*. Biochem. Bioph. Res. Com. 176 :541-548.

Hastings, J. W., Potrikas, C. J., Gupta, S. C., Kurfurst, M. and Makemson, J.C. (1985) Biochemistry and physiology of bioluminescent bacteria. Adv. Microbiol. Physiol. 26 : 235-29.

Heitzer, A., Webb, O.F., Thonnard, J.E. and Sayler, G.S. (1992) Specific and quantitative assessment of naphthalene and salicylate bioavailability by using a bioluminescent catabolic reporter bacterium. Appl. Environ. Microbiol. 58 : 1839-1846.

Ivask, A., Hakkila, K. and Virta, M. (2001) Detection of organomercurials with whole-cell bacterial sensors. Anal. Chem. 73 : 5168–5171.

Kricka, L. J. (1988) Clinical and biochemical applications of luciferases and luciferins. Anal. Biochem. 175 : 14-21.

Lee, J. (1985) The mechanism of bacterial bioluminescence. In : Chemiluminescence and Bioluminescence (Ed., Burr, J. G.), Marcel Dekker, Inc., New York, pp. 401-437.

Lee, J. (1989) Bioluminescence. In : The Science of Photobiology (Ed., Smith, K. C.), Plenum Publishing Corp., New York, pp. 391-417.

Lee, J., Matheson, I. B. C., Muller, F. C. Y., Kane, D. J., Vervoort, J. and Visser, A. J. W. G. (1990) The mechanism of bacterial bioluminescence. In: Chemistry and Biochemistry of Flavoenzymes (Ed., F. Muller), Vol. II, CRC Press, Boca Raton, Florida, pp. 109-151.

Lee, C. Y. and Meighen, E. A. (1992) The *lux* genes in *Photobacterium leiognathi* are closely linked with genes corresponding in sequence to riboflavin synthesis genes. Biochem. Biophys. Res. Commun. 186 : 690-697.

Lewis, J.C., Feltus, A., Ensor, C.M., Ramanathan, S. and Daunert, S. (1998) Applications of reporter genes. Anal. Chem. 70 : 579–585

Martin, M., Showalter, R., and Silverman, M. (1989) Identification of a locus controlling expression of luminescence genes in *Vibrio harveyi*. J. Bacteriol 171 : 2406-2414.

Mancini, J., Boylan, M., Soly, R., Ferri, S., Szittner, R. and Meighen, E. A. (1989) Organization of the *lux* genes of *Photobacterium phosphoreum*. J. Biolum. Chemilum. 3 : 201-205.

Meighen, E.A. (1991) Molecular biology of bacterial bioluminescence. Microbiol. Rev. 55 : 123-42.

Meighen, E. A. and Szittner, R. B. (1992) Multiple repetitive elements and organization of the *lux* operons of luminescent terrestrial bacteria. J. Bacteriol. 174 : 5371-5381

Meighen, E.A. (1993) Bacterial bioluminescence: organization, regulation, and application of the *lux* genes. FASEB J. 7 : 1016-1022.

Meighen, E. A. and Dunlap, P. (1993) Physiological, biochemical, and genetic control of bacterial bioluminescence. Adv. Microb. Physiol. 34 : 1-67

Meighen, E. A. (1994) Genetics of bacterial luminescence. Annu. Rev. Genet. 28 : 117–139.

Miyamoto, C., Boylan, M., Cragg, L. and Meighen, E. A. (1989) Comparison of the *lux* systems in *Vibrio harveyi* and *Vibrio fischeri*. J. Biolumin. Chemilumin. 3 : 193-199.

Nealson, K.H. (1977) Autoinduction of bacterial luciferase : Ocurrence, Mechanism and Significance. Arch. Microbiol. 112 : 73-79.

Nealson, K.H. (1993) Bacterial bioluminescence: three decades of enlightenment. Naval. Res. Rev. 45 : 13–20.

Olsson, O., Koncz C. and Szalay, A.A. (1988) The use of the *lux*A gene of the bacterial luciferase operon as a reporter gene. Mol. Gen. Genet. 215 : 1–9.

Rychert, R. and Mortimer, M. (1991) Inhibition of bioluminescence in a recombinant *Escherichia coli*. Environ. Toxicol. & Water Qual. 6 : 415– 421.

Selifonova, O., Burlage, R. and Barkay, T. (1993) Bioluminescent sensors for detection of bioavailable Hg (II) in the environment. Appl. Environ. Microbiol. 59 : 3083-3090.

Stewart GSAB, Williams P. (1992) *Lux* gene and the applications of bacterial bioluminescence. J. Genet. Microbiol. 138 : 1289-1298.

Tu, S.C. and Mager, H.I. (1995) Biochemistry of bacterial bioluminescence. Photochem. Photobiol. 62 : 615-624.

Turpeinen, R., Salminen, J. and Kairesalo, T. (2000) Mobility and bioavailability of lead in contaminated boreal forest soil. Environ. Sci. Technol. 34, 5152–5156.

Virta, M., Karp, M. and Vuorinen, P. (1994) Nitric oxide donor-mediated killing of bioluminescent *Escherichia coli*. Antimicrob. Agents Chemoth. 38 : 2775-2779.

Virta, M., Lampinen, J. and Karp, M. (1995) A luminescence-based mercury biosensor. Anal. Chem. 67 : 667–669.

Wilson, T. and Hastings, J.W. (1998) Bioluminescence. Ann. Rev. Cell Dev. Biol. 14: 197–230.

Suggested Readings

1. Bioluminescence Methods and Protocols (1998) (Ed.) R.A. Larossa, Humana Press, USA.
2. Microbial Signalling and Communication (1999) (Eds.) R. England, G. Hobbs, N. Bainton and D. McL. Roberts, Cambridge University Press, Cambridge, UK.
3. Proceedings of XIth International Symposium on Bioluminescence and Chemiluminescence (2001) (Eds.) J.F. Case, P.J. Herring, B.H. Robinson, S.H.D. Haddock, L.J. Kricka and P.E. Stanley, World Scientific, USA.
4. Bioluminescence and Chemiluminescence (2002) (Eds.) P.E. Stanley, L.J. Kricka, World Scientific, USA.

REVIEW QUESTIONS

Long Questions

1. Define bioluminescence with reference to bioluminescent bacteria?
2. What is autoinduction? Explain the role of autoinduction during bioluminescence?
3. What is quorum sensing? Explain the quorum sensing mechanism in bacteria?
4. What is cell density dependent induction of luminescence?
5. Describe the *lux* gene organization and regulation of *lux* genes?
6. Describe the enzymology and biochemistry of luciferases?
7. Write a brief note on *lux* genes and its applications?
8. What is bioreporter? Explain the use of bioreporters in environmental biotechnology?

Multiple Choice Questions

1. The active site of the bacterial luciferase is located within the subunit
   (a) α (b) β
   (c) γ (d) δ
2. Quorum sensing mechanism is observed in
   (a) Actinomycetes (b) gram positive bacteria
   (c) Fungi (d) Gram negative bacteria.
3. Fatty acid reductase complex codes for the genes
   (a) *Lux*CDE (b) *Lux*CDABE
   (c) *Lux*CDABEG (d) *Lux* AB
4. Bacterial quorum-sensing signal molecule (AHL) contains
   (a) Manganese (b) Magnesium
   (c) Boron (d) Molybdenum

5. The synthesis of _______ for the bioluminescence reaction is catalyzed by a multienzyme fatty acid reductase complex
   (a) Ketones (b) Aldehydes
   (c) Carboxylic acids (d) Esters
6. Autoinducer signal molecules are chemically termed as
   (a) Homoserine butyrates (b) Homoserine lactones
   (c) Homoserine esters (d) Homoserine pyruvates.
7. Five genes in the *lux* operon _______ are needed to produce light emission
   (a) *Lux*CDABH (b) *Lux*CDABE
   (c) *Lux*CDABG (d) *Lux* CDABI
8. *Lux* AB reporter genes are used as an application in
   (a) Gene expression (b) Conjugation
   (c) Transduction (d) Replication
9. The regulatory genes in the *lux* operon are
   (a) *Lux* I and R (b) *Lux* A and B,
   (c) *Lux* C and D, (d) *Lux* E and G.
10. The transcriptional activator gene in the *lux* operon is
   (a) *Lux*A (b) *Lux* B
   (c) *Lux*R (d) *Lux* I

Answers

1. (a)	2. (d)	3. (a)	4. (c)	5. (b)	6. (b)
7. (b)	8. (a)	9. (a)	10. (c)		

CHAPTER 46

Microbes as Pathological Agents

Sharad Bissa[1], Dimple Songara[2], Avinash Bohra[1] and A. Bohra[2*]
[1]Mahila PG Mahavidyalaya, Jodhpur
[2]Dept. of Botany, J.N.V. University, Jodhpur–342 001
E-mail: profabohra@yahoo.co.in

A microbe is any living thing that spends its life at a size visible only with a microscope. It is too tiny to be seen with the naked eye. Microbes are the oldest form of life on Earth. Some types have existed for billions of years. They may live as individuals or cluster together in communities. Microbes live in the water you drink, the food you eat, and the air you breathe. Microbes include bacteria, viruses, fungi, algae, and protozoa. These single-cell organisms are invisible to the eye, but they can be seen with microscopes. The term microbe is short for microorganism, which means small organism. To help people understand the different types of microbes, they are grouped or classified in various ways. Microbes are the part of our lives in more ways than most understand. They have shaped the present environment and their activities will greatly influence our future. Microbiology is one of the most applied branch of all the biological sciences.

EVOLUTION OF MICROORGANISMS

Oparin and Haldane (1928) independently proposed the origin of life by chemical evolution. Earth was formed about 4600 million years ago. The atoms of nitrogen, hydrogen, oxygen, argon, carbon etc. formed the primitive atmosphere. The atmosphere was reducing, not oxidizing as it is today. As the earth began to cool, large quantities of H_2, N_2, water vapour, CO_2, CH_4 and NH_3 were present. Further fall in temperature allowed water to remain in liquid form so that oceans and water bodies were formed containing large amounts of dissolved NH_3, CH_4, HCN, nitrides, carbides and various gases. Reacting with water, simple saturated hydrocarbons like CH_4 formed unsaturated hydrocarbons like ethylene and acetylene. From

*Corresponding author

the latter aldehydes, Ketones, alcohols and organic acids were formed. Using abundant energy available, the organic molecules of ocean water formed complex compounds like amino acids, sugars, glycerol, fatty acids, nitrogenous organic bases. These molecules further formed macromolecules like proteins, carbohydrates and fats. The oceanic water became a rich mixture of organic compounds called Prebiotic soup. Stanely Miller and H.C. Urey (1953) verified the Oparin-Haldane theory and showed that amino acids and diverse other molecules can be made under atmospheric conditions thought representative of those on the early earth. Due to zwitterionic nature, protein molecules formed colloidal hydrophilic complexes which got surrounded by water molecules.

Coalescence of such structures produces a separation of colloids from their aqueous phase (coacervation). The substances within the coacervates, due to chemical reactions and by absorbing metal ions into the coacervates, formed enzymes. The alignment of lipid molecules along the boundary between the coacervates and the external medium conferred stability. A primitive type of cell was thus produced. The first organism were heterotrophs. Some of the molecules were able to use solar energy for synthesizing new cellular materials. In cooperation of such molecules into the cells lead to evolution of autotrophs. It is believed that oxygen evolving photosynthetic organisms similar to present day cyanobacteria developed. The increase in atmospheric oxygen and its ionization to form ozone layer reduced the ultra violet rays striking the earth. It provided stability to successful forms of life. The first eukaryote evolved at least 1.5 billion years ago. One interesting hypothesis is that the more complex cells evolved partly as a result of certain prokaryotes taking up residence inside other cells. Those prokaryotes that were able to convert to the use of oxygen in ATP increase. Some of them evolved into modern forms of aerobic bacteria.

MICROBIAL TAXONOMY

Taxonomy is described as Classification, nomenclature and identification of microorganisms. On the basis of common characters or properties , a set of organism is considered into groups (taxa). Nomenclature is the name given for defined taxa as also governed by bacteriological code of nomenclature. For identification, serological tests (agglutination, fluorescent antibody techniques), have enormous values. Genetic tools are the modern ways of identification of bacteria.

There are some general methods of classifying bacteria. In *Intutive method*, different taxonomist classify microorganism in very different groupings, based on all characters because a character may be important to a particular microbiologist may not be that important to another. In *Numerical Taxonomy*, the classification is based on several characteristics for each strain and each character is given equal weightage. An another classification is based on *genetic relatedness* between organisms. The % G + C determines the organism whether it is the same or different species. Further, characterization is based on the *DNA-DNA hybridization*. In this method after annealing i.e. separation of two strands and converting them into single strand, the later one is mixed with those obtained from other organism. If the two organism are similar the pairing will occur in strands of both the organism and form heteroduplexes otherwise not. Another characterization is based on *16S rRNA sequencing* , due to the highly conserve nature of rRNA genes, the 16S rRNA nucleotide sequence give better informations than DNA molecule. It is important to mention that if there is no DNA-DNA relatedness between two organisms, still nucleotide sequence of their rRNA cistrons may be studied so as to observe the similarity.

OCCURRENCE OF MICROBES

Microbes are present in both natural and man-made worlds. Microbes has special qualities which enable it to survive in such unique places as the soil, oceans, rivers, and streams, ice, water pipes, concrete, hot springs, the human intestine, root of plants, and even oil wells. They are omnipresent and are characterized by very high degree of adaptability. Soil is rich in microbial flora like bacteria, algae, and fungi. Microbes are responsible for the decomposition and decay of dead organic matter (plants and animals). If it were not for microbial decomposition, the earth would be covered with various types of dead organic matter. They are responsible for recycling the dead organic matter and make it available for reuse. Sewage treatment plants are, in part, gain microbial cultures for speeding decay of waste products. Water contains microbes and cause diseases of various kinds. Drinking water and industrial water are pumped from lakes, rivers, and wells which should be cleared from harmful bacteria, viruses and other microbes to control diseases and prevent of fouling of industrial equipments. Water transfers cholera and typhoid fever in humans. Therefore, it is essential to purify water before use. Air is the greatest reservoir of microorganisms. Generally all the pathological agents are transmitted from an infected person to other persons usually by inhaling bacteria-carrying air droplets. When a person coughs, sneezes, or speaks, small particles that carry bacteria surrounded by a layer of moisture are released in the air. When another person inhales these particles, the bacteria may lodge in that person's lungs and multiply. Foods are transported and stored for a long period, during these process food may be contaminated by microbes. The role of microbes in food poisoning, spoilage, and preservation is studied in the branch of food and diary microbiology. *Pasteurization* is one of the best-known method to prevent the food from spoilage and the transfer of disease. Human beings, like other animals, harbour a wide array of microorganism either on or in their bodies. A knowledge of the normal microflora of the body is essential to understand the interaction of human beings and their pathogen laden environment. The human skin is constantly bombarded by organisms present in the environment. Cultures from skin have frequently demonstrated diptheroids, staphylococci; Gram positive aerobic spore bearing bacilli: *Streptococcus viridans*; Gram negative bacilli such as *E.coli, Proteus* and other intestinal organisms; *Candida albicans*. The predominant organisms of the eye are *Corynebacterium xerosis, Moraxella* species, *Staphylococci* and nonhemolytic *Streptococci*. The nasopharynx can be considered the natural habitat of the common pathogenic bacteria which cause infection of nose, throat, bronchi and lungs. The mouth contains a plethora of organisms pigmented and nonpigmented *micrococci*, Gram positive aerobic spore bearing *Bacilli, Coliforms, Proteus and Lactobacilli*. Due to the low pH of the stomach it is virtually sterile except soon after eating. The number of bacteria increases progressively beyond the duodenum to the colon, being comparatively low in the small intestine. *Mycobacterium smegmatis*, is found in the smegma of the genitalia of both men and women.

TYPES OF MICROBES

Microbes are very diverse and represent all the great kingdoms of life. In fact, in terms of numbers, most of the diversity of life on Earth is represented by microbes. Here is an outline of the major groups of microorganisms.

Bacteria consist of only one cell, but they're a very complex group of living things. Some bacteria can live in temperatures above the boiling point and in cold below the freezing point. Some bacteria cells exist as individuals while others cluster together to form pairs, chains, squares or other groupings. Some bacteria can make their own food from sunlight, just like plants. Also like plants, they give off oxygen. Other bacteria absorb food from the material they live on or in. Some of these bacteria can live off iron or sulfur! The bacteria that live in your stomach absorb nutrients from the digested food you've eaten. Some bacteria move about their environment by means of long, whip-like structures called flagella. Other bacteria secrete a slime layer and ooze over surfaces like slugs. Others stay almost in the same spot. Bacteria live on or in just about every material and environment on Earth from soil to water to air, and from your body to the Arctic ice to the Sahara deserts.

Viruses were originally distinguished from other infectious agents because they are especially small (filterable) and they are obligatory intracellular parasites- that is, they require living host cells in order to multiply. Viruses are the entities that contain a single type of nucleic acid, either DNA or RNA and contain a protein coat (sometimes itself enclosed by an envelope of lipids, proteins and carbohydrates) that surrounds the nucleic acid. They multiply inside living cells using the synthesizing machinery of the cell. Viruses cause the synthesis of specialized structures that can transfer the viral nucleic acid to other cells. Viruses can infect every living thing. However, viruses tend to be somewhat picky about what type of cells they infect. Sometimes, a virus may infect one animal and do no harm, but cause a great deal of damage when it gets into a different but closely related animal.

Prions are proteinaceous infectious particles, which are composed of only protein. No nucleic acid from this infectious agent can be isolated. For the outstanding work on prions, Stanely B. Prusiner was awarded Nobel Prize in Medicine in 1997. Prions are known to be causal agents for neurological disorders such as mad cow disease in cattle and Kuru disease in humans.

Fungi are organisms that scientists once confused with plants. However, scientists have found that, at the cell level, the fungi are more like animals than they are like plants. For one thing, fungi cannot synthesize their own food like plants do, but instead they eat other organisms as do animals. All fungi are chemoheterotrophs, requiring organic compound for energy and carbon. The majority of fungi are saprophytes in soil and water; there they primarily decompose plant material. Fungi come in a variety of shapes and sizes and different types. They can range from single cells to enormous chains of cells that can stretch for miles. Fungi include single-celled living things that exist individually, such as yeast, and multicellular clusters, such as molds or mushrooms. However, thousands of fungi are pathogenic to plants and animals.

Protozoa are one celled, eukaryotic organisms that belong to the kingdom Protista. Protozoans inhabit soil and water and feed upon bacteria and small particulate nutrients. All protozoa digest their food in stomach-like compartments. As they digest, they make and give off nitrogen, which is an element that plants and other higher creatures can use. Of the nearly 20,000 species of protozoans, relatively few cause disease. Protozoans are classified into phyla on the basis of their means of motility. Four phyla include disease causing species. One such phylum , the Sarcodina consists of amoebas. Another phylum, the Mastigophora, or flagellates, posses flagella. Species in the phylum Ciliata have cilia that are similar to but shorter than flagella. The species in the fourth phylum, the Sporozoa are incapable of independent movement.

SOURCES AND RESERVIORS OF MICROBIAL AGENTS

Living agents or inanimate objects that serve as the bridge for contact between the host and the microorganism are referred to as sources of infection. Animate or inanimate objects that are necessary to support growth of microorganism are termed reservoirs of infection.

Animate reservoirs: Humans represent the most important reservoirs of infectious agents. The infected persons may be a carrier of the infectious agent. Some carriers harbor the infectious agents in the bloodstream, and bloodsucking insects may become infected by biting the host. The infected insect in turn can support the growth of the infectious agent and transmit it to susceptible host. Insects that transfer infectious disease in a passive way from one host to another are called Mechanical vectors. Flies, for example, can act as passive vectors because, whereas many of them do not parasitize humans or animals. Insects that serve as a host and reservoir for infectious agents are called Biological vectors, such as *Anopheles* mosquito, which carries agent of malaria.

Inanimate reservoirs: The primary inanimate reservoirs for infectious agents are soil, water and food. Most of the microbes found in the soil that cause disease in humans are inhabitants of plants and animals and are shed or excreted into the soil. Many of these organisms remain in the soil as spores and await transfer to a viable host before germinating. Foods are important reservoirs of microorganisms because they contain nutrients that support the growth and maintenance of microorganism. Food-borne disease results from the ingestion of foods contaminated by microorganisms that either directly or indirectly cause gastrointestinal symptoms. This is not a true infection but an intoxication. Some microbes such as *Shigella*, that contaminate food are ingested, replicate, and the intestine to cause disease. Improper cooking or preservation of foods is the cause of infection or intoxication in nearly all food-borne diseases.

Water, which is seldom distilled, usually contains inorganic ions and is often contaminated with organic material derived from soil, air, animals or plants. Waterborne diseases results from consumption of contaminated water that has not been properly prepared for drinking. Most outbreaks involve semipublic and municipal water systems, including recreational areas.

TRANSMISSION OF MICROBES

Microorganism are transmitted to susceptible hosts by four routes: contact, air, common vehicle and vector borne. Vector borne disease was discussed previously.

Contact: Direct contact implies actual physical interaction with the source of microbes. Human-to-human transmission occurs mostly through touching, kissing or sexual intercourse. Sexually transmitted diseases (STD's) are transmitted primarily by sexual contact, but infected nursing mothers may also transmit the microbial agent to their offspring by direct contact. The agent of syphilis, as well as some other microbes, can be transferred across the placental barrier to the developing fetus.

Sexually Transmitted Diseases (STDs), formerly known as venereal diseases, more than 25 infectious diseases passed from one person to another primarily during sexual contact. STDs are among the most common infections known. Despite the prevalence of STDs, studies show that many people are unaware of their risks for STD infection or the serious, and sometimes deadly, health consequences that may result from an untreated infection. Some STDs, such as gonorrhea or chlamydia, may cause no symptoms. People who do not know they are infected risk infecting their sexual partners and, in some cases, their unborn children. If left untreated, these diseases may cause debilitating pain or may destroy a woman's ability to have children. Some STDs can be cured with a single dose of antibiotics, but many, such as acquired immunodeficiency syndrome (AIDS), are incurable. People with these diseases remain infectious to others for their entire lives. Those most at risk for contacting STDs are people who have unprotected sex; those who have multiple partners; and those whose sex partners include intravenous drug users who share needles. STDs are transmitted by infectious agents—microscopic bacteria, viruses, parasites, fungi, and single-celled organisms called protozoa—that thrive in warm, moist environments in the body, such as the genital area, mouth, and throat. Some STDs are transmitted in ways other than by sexual contact. Certain viral STDs, such as AIDS and some types of hepatitis, may be transmitted by contact with infected blood. For instance, viral STDs may pass between people who share infected needles, and a person can become infected from a transfusion of infected blood. Some STDs may pass from an infected mother to her child. Infection may occur before birth, when the infectious agent crosses the placenta (organ in a pregnant woman's uterus that links the blood supplies of mother and baby) and enters the baby's bloodstream. Infection also may occur during childbirth, as the baby passes through the birth canal, or after birth, when the baby consumes infected breast milk. STDs cannot be transmitted through shaking hands or other casual contact, or through contact with inanimate objects such as clothing or toilet seats. The most common STDs include chlamydia, gonorrhea, syphilis, herpes, AIDS, hepatitis, genital warts, and trichomoniasis.

Indirect Contact: Indirect contact implies that the infectious agent persists on an object from which there is transfer to susceptible host. A variety of objects such as clothes, eating utensils, Hospital thermometers, drinking glasses and bedding may be the intermediary.

Air: Infectious microbes can reach the air in a number of different ways. They may be dispersed from the respiratory tract by sneezing or coughing or other related activities.These air borne secretions may remain suspended in air for prolonged periods of time and are called droplet nuclei. Smallpox, measels, tuberculosis are spread by droplet nuclei. The very large secretions expelled from respiratory tracts can contaminate clothing and various inanimate surfaces. Many soil microbes produce spores, some of which are easily disseminated in the air. During the hot, dry summers violent dust storms spread the spores in the air. Inhalation of large numbers of spores can cause serious disease or allergy, even in healthy host.

Common vehicles: This type of transmission refers to the acquisition of microorganism by a large group of individuals from a common inanimate vehicle or source. This type of transmission is frequently associated with hospital- associated infections.

CLINICAL STAGES OF INFECTIOUS DISEASES

The clinical response of the host to an infectious disease agent may be characterized as acute or chronic.

- Acute diseases are those in which symptoms are manifested for a short period but subside when the host's immune system has responded.
- Chronic diseases are characterized by the appearance of symptoms over an extended period of time due to an inefficient immune response.

Acute diseases may be divided into four distinct periods : The incubation period is that time between contact with the infectious agent and the appearance of the first symptoms. It is during this period that the infectious agent begins to multiply at the initial site of infection. Diseases in which organisms are spread throughout the body to infect organs or tissues are referred to as ***systemic***. The incubation period may be as short as few hours such as in food poisoning, or as long as several years, such as in leprosy. The first sign of symptoms is called the prodromal period and usually lasts 1 or 2 days. The symptoms might include a headache or upset stomach , but usually a general feeling of not being well. The acute period represents the peak of microbial activity in the host and therefore the peak expression of the host symptoms. Fever, which is characteristic of the acute stage in most infectious diseases is often an indicator of the disease. Fluctuations in fever during the course of a disease are often associated with an increase of microorganisms in the blood stream. The convalescent period is a stage of recovery from the disease and is characterized by a rapid decline in fever and other symptoms because of an effective response to the infection. Not all infections by a microorganism result in acute disease manifestations. The immune system of the host, because of previous encounters with the microbe, responds rapidly and the infection ceases. Such diseases are called subclinical or inapparent.

TYPES OF INFECTIONS

Epidemiological investigations may be concerned with diseases that are divided into three types based on their occurrence in the population :

- Epidemics are defined as either small or large numbers of cases of disease that are always in excess of what is expected .
- Endemic refers to disease that is continuously present in the population and the frequency of disease remains at predictable levels (endemic levels) within the community.
- Pandemic implies overwhelming outbreaks of disease that involves countries or are worldwide. Influenza, for example, has been implicated in several pandemics.

PATHOLOGICAL AGENTS

BACTERIA

Bacillus: *Bacillus* is a genus of rod-shaped bacteria. Likewise, Bacilli refers to the particular class *Bacillus* belongs to, while bacilli are any rod-shaped bacteria.The genus *Bacillus* belongs to the Firmicutes, and like most other members has a Gram-positive stain. They are either obligate or facultative aerobes, and test positive for catalase. *Bacillus* are ubiquitous in nature, including both free-living and pathogenic species. The important disease caused by *Bacillus* are :

Anthrax

Anthrax, disease of warm-blooded animals, including humans, caused by the bacterium *Bacillus anthracis*. Anthrax most commonly occurs in cattle and other plant-eating mammals, but it can also affect humans who come in contact with infected animals. Anthrax is a sporeforming bacterium. The spores have protective coats and can withstand extreme heat, drought, and other harsh conditions. They can live for centuries in soil. Anthrax spores also have the potential for use in biological warfare because of their ability to survive and because they spread easily in air and can be inhaled. Once the spores are inside the lungs, the bacteria develop and begin to multiply.

In humans, the disease can appear in two forms: cutaneous or by inhalation,. The cutaneous, or external form, primarily involves the skin and is contracted mainly by those who handle contaminated hides, wool, or carcasses. The bacteria enter through a cut or other opening in the skin, and a dark, itchy bump that resembles an insect bite appears. The bump then develops into an open sore with a black area in the center. The cutaneous form of anthrax can be treated with antimicrobial drugs. Humans contract internal forms of anthrax by inhaling anthrax spores or by eating contaminated meat. People who work with animal hair and wool are most likely to inhale the spores, especially in areas where anthrax occurs in animals. Inhaled anthrax is fatal unless the infected person has been vaccinated or antibiotics are administered soon after inhalation. Ciprofloxacin, penicillin, and other antibiotics are effective treatments, except in rapidly progressing cases.

Botulism

Botulism, poisoning caused by eating food containing *Clostridium botulinum,* a poisonous bacterium. The organism, which is derived from the soil, grows in many meats and vegetables. Its spores are killed by boiling for 30 minutes, but the toxins may be destroyed by moist heat at 80°C (176°F) for the same period. Because the spores grow best in the absence of air, improperly processed foods in sealed containers offer a perfect environment for their development. Commercially prepared foods are now a rare source of the disease, but improperly sterilized home-canned nonacidic foods cause many cases. Symptoms appear usually 18 to 36 hours after ingestion. The toxin, which is not destroyed by the enzymes of the gastrointestinal tract, affects the central nervous system and interrupts nerve impulses, but the mind continues to function normally. Disability progresses from difficulty in walking and swallowing and impaired vision and speech to occasional convulsions and ultimately to paralysis of the respiratory muscles, suffocation, and death, all within a few hours or days, depending on the amount of toxin ingested. Two-thirds of those attacked may die.

Tetanus

Tetanus, serious infectious disease of the nervous system, in which a bacterial toxin causes severe muscle spasm. Also called lockjaw, the bacterial infection occurs when wounds are contaminated with a bacillus, *Clostridium tetani,* which is found everywhere but which is especially likely to be present in street dirt. It is an anaerobic bacillus—that is, one that lives without oxygen, and thus when introduced into human tissue, usually at the time of an injury, it can multiply where the oxygen is deficient. This occurs both in deep, penetrating wounds and in crushing wounds with extensive tissue destruction. As the bacillus multiplies, large amounts of exotoxin are released into the surrounding tissues.

Treatment, besides cleaning the wound, involves eliminating the bacilli both surgically and with antibiotics, neutralizing the exotoxin with antitoxin, reducing muscle spasm, and supporting or aiding respiration. Although about 60 percent of all cases are fatal, active immunization with tetanus toxoid before an injury, with periodic reimmunization, or so-called booster injections, is extremely effective; tetanus rarely occurs in people so immunized. A series of three injections is recommended if the tetanus toxoid is combined with both diphtheria and toxoid pertussis (whooping cough) vaccine in the commonly administered DTP (diphtheria, tetanus, pertussis) injections. The supplementary booster injections should be given one year after completion of the primary immunization series and every five to ten years thereafter.

Diphtheria

Diphtheria, an acute and highly infectious disease, affecting children particularly, characterized by the formation of a false membrane in the passages of the upper respiratory system. The cause of the disease is *Corynebacterium diphtheria,* a bacillus discovered in 1883. The diphtheria bacilli enter the body through the mouth and nose and attack the mucous membranes, where they multiply and secrete a powerful toxin. The toxin damages the heart and central nervous system, and can lead to death. Before the development of diphtheria antitoxin in 1894 by French bacteriologist Pierre Roux and German physician Emil von Behring, the mortality rate from the disease averaged 35 per cent and was as high as 90 per cent in cases of diphtheria of the larynx. The universal use of antitoxin in treatment has cut the mortality rate to approximately 5 per cent. Even more effective has been the development of harmless forms of diphtheria toxin, called *toxoids*. Given to infants during the first year of life as part of a combined injection, these toxoids immunize the children against serious infection and have drastically reduced the incidence of diphtheria in most parts of the world. The Schick test, used to determine whether an individual is immune or susceptible to diphtheria, was developed by the Hungarian-American pediatrician Béla Schick in 1913.

Leprosy

Leprosy or Hansen's Disease, chronic infectious disease caused by the bacterium *Mycobacterium leprae*. Leprosy can be treated effectively with several drugs, but if left untreated, the disease can result in severe disfigurement, especially of the feet, hands, and face. It is rarely fatal. Leprosy has long been one of the most feared diseases worldwide. The stigma attached to leprosy has often caused those who contracted the disease to be shunned by family, friends, and society. Although leprosy was once widespread throughout the world, today it is primarily a tropical disease. More than 95 per cent of leprosy cases occur in just 11 nations, and India and Brazil have the largest numbers of cases. Due to effective treatment and leprosy eradication programs, the prevalence of the disease has declined dramatically in recent years.

Leprosy has two main forms, known as tuberculoid and lepromatous disease. In tuberculoid leprosy the skin lesions are few and small, with only a few bacteria present in each. In lepromatous leprosy, the more severe form of the disease, the lesions may be much more widespread and contain many leprosy bacteria. As lepromatous leprosy progresses, hard nodules and folds of skin may form on the face and the nose may collapse, giving a person a characteristic lionlike appearance. In both forms of leprosy, there is usually some degree of irreversible nerve damage resulting from either of these two processes. Because of the lack of sensation in affected areas of the skin, people with leprosy often do not notice burns and injuries to their fingers and toes and fail to treat them. These injuries can then become infected with other types of bacteria that cause tissue damage. Gradually, damaged tissue and bone are resorbed by the body, causing the digits to become shorter. However, leprosy does not, as myth would have it, cause parts of the body to fall off.

Three antibiotics, Dapsone, Rifampin, and Clofazimine, are currently used to treat leprosy. The drugs must be taken for a long period, typically six months in cases of tuberculoid leprosy and two years for lepromatous leprosy. Treating leprosy using multidrug therapy is much more effective than using any one drug alone. These drugs cannot reverse the nerve damage and deformities of the hands, feet, and face that are characteristic of the disease. However, they can often halt the progression of the disease and help prevent it from being passed on to anyone else. Currently the only strategy for preventing the spread of leprosy is treatment of existing cases.

In 1991 WHO launched a leprosy elimination program to provide multidrug therapy to leprosy patients all over the world. The disease has proven difficult to eliminate for several reasons. One reason is that scientists do not understand how the bacterium is transmitted from person to person. Also, scientists do not know how to identify people who are infected but have not yet developed the disease. Efforts to eliminate leprosy are also hampered by the stigma still attached to the disease: Many patients are reluctant to seek treatment for fear of being abandoned by their family and ostracized by society in general. Scientific understanding and effective drug therapy have debunked many of the myths surrounding leprosy, and life for most patients today is quite different than it was in Medieval Europe. Only a better-informed public can eliminate the stigma of the disease.

Tuberculosis

Tuberculosis (TB), chronic or acute bacterial infection that primarily attacks the lungs, but which may also affect the kidneys, bones, lymph nodes, and brain. The disease is caused by *Mycobacterium tuberculosis,* a rod-shaped bacterium. Symptoms of TB include coughing, chest pain, shortness of breath, loss of appetite, weight loss, fever, chills, and fatigue. Children and people with weakened immune systems are the most susceptible to TB. Half of all untreated TB cases are fatal. In 1993, the World Health Organization (WHO) declared TB to be a global emergency, the first such designation ever made by that organization.

TB is transmitted from person to person, usually by inhaling bacteria-carrying air droplets. When a person sick with TB coughs, sneezes, or speaks, small particles that carry two to three bacteria surrounded by a layer of moisture are released in the air. When another person inhales these particles, the bacteria may lodge in that person's lungs and multiply. A less common route of transmission is through the skin. Pathologists and laboratory technicians who handle TB specimens may contract the disease through skin wounds. TB has also been reported in people who have received tattoos and people who have been circumcised with unsterilized instruments. If a person contract the infection, disease can develop in two stages: primary and secondary.

In primary TB, a person has become infected with the TB bacteria but often is not aware of it, since this stage of the disease does not produce noticeable symptoms. If the bacteria multiply, active primary tuberculosis will develop. Symptoms include coughing, night sweats, weight loss, and fever. If the bacteria are inhibited, rather than destroyed, the immune cells and the bacteria form a mass known as a granuloma or tubercle. In effect, the immune cells form a wall around inactive bacteria. As long as the immune system remains strong, the TB bacteria remain walled off and inactive. If the immune system becomes weakened, the tubercle opens, releasing the bacteria, and the infection may develop into secondary TB. In secondary TB, the formerly dormant bacteria multiply and destroy tissue in the lungs. They also may spread to the rest of the body via the bloodstream. Fluid or air may collect between the lungs and the lining of the lungs, while tubercles continue to develop in the lung, progressively destroying lung tissue. At this secondary stage, carriers of TB can infect others.

General preventive measures can be taken to reduce the spread of TB in public places. Ventilation systems lessen the chance of infection by dispersing the bacteria. Vaccines, such as the bacillus Calmette Guerin (BCG) vaccine, prepared from bacteria that have been weakened, are another preventive measure. The BCG vaccine is most effective in preventing childhood cases of TB. With the advent of effective antibiotics for TB, drug therapy has become the cornerstone of treatment. Single-drug treatment often causes bacterial resistance to drugs. Therefore, all recommended therapies include multiple drugs given for at least 6 months, often for as long as 9 to 12 months. A combination of antibiotics, including Isoniazid, Rifampin, Streptomycin, Pyrazinamide, and Ethambutol, is usually prescribed. In 1998, scientists successfully decoded the entire gene sequence, or genome, of the tuberculosis bacteria. This advance is likely to lead to the development of new methods for treatment and prevention of TB.

People with HIV have weakened immune systems that increase their susceptibility to TB, and in these people, TB often progresses rapidly from the primary to the secondary stage. A second factor contributing to TB resurgence is the failure of patients to complete the full six to nine months of antibiotic therapy required to cure the disease. When patients fail to follow the prescribed treatment, they may become actively infectious, spreading the disease to others. Migration, international air travel, and tourism also have contributed to the global spread of TB. The emergence of strains of bacteria that are resistant to multiple drug therapy is a serious problem, particularly because no ready drug treatment is available to combat newly emerging strains. To improve compliance, the WHO strongly recommends that all countries, especially those in Africa and Asia, adopt a program called directly observed treatment short-course (DOTS). DOTS requires health workers to monitor patients to make sure that they follow the complete course of treatment.

Typhoid Fever

Typhoid Fever, acute infectious disease caused by the typhoid bacillus *Salmonella typhi.* The bacillus is transmitted by milk, water, or solid food contaminated by feces of typhoid victims or of carriers, that is, healthy persons who harbor typhoid bacilli without presenting symptoms. The incubation period of typhoid fever lasts one to three weeks. The bacteria collect in the small intestine, from which they enter the bloodstream. This induces the first symptoms, chills followed by high fever and prostration. Victims may also experience headache, cough, vomiting, and diarrhea. The disease spontaneously subsides after several weeks in most instances, but in about 20 per cent of untreated cases the disease progresses to pneumonia, intestinal hemorrhage, and even death. Deaths from typhoid fever were greatly reduced by the isolation of the first antibiotic effective against the typhoid bacillus, Chloromycetin, or Chloramphenicol. This drug is still the preferred treatment in most cases. For infection with typhoid bacilli resistant to Chloramphenicol or for treatment of carriers, Ampicillin is recommended.

Typhoid is slowly disappearing. Compulsory inspection of milk and water supplies, and the pasteurization of milk in particular, have greatly reduced the incidence of the typhoid bacilli. Of equal importance in the control of typhoid fever has been the recognition of carriers, who can then be prevented from handling food, and improvement of sewage facilities.

Dysentery

Dysentery, acute or chronic disease of the large intestine of humans, characterized by frequent passage of small, watery stools, often containing blood and mucus, accompanied by severe abdominal cramps. Ulceration of the walls of the intestine may occur. Although many severe cases of diarrhea have been called dysentery, the word properly refers to a disease caused by either a specific amoeba, *Entamoeba histolytica,* or a bacillus that infects the colon.

Bacillary Dysentry : Bacillary dysentery is caused by certain non-motile bacteria of the genus *Shigella*. Bacillary dysentery is spread by contaminated water, milk, and food. Feces from active cases and those from healthy carriers as well contain immense numbers of the disease-producing bacteria. Flies carry the bacteria on their feet or in their saliva and feces and deposit them on food; ants are also believed to spread the disease. In the treatment of bacillary dysentery, proper replacement of fluid is important. Sulfonamides, tetracycline, and streptomycin were effective in curing acute cases until drug-resistant strains emerged.

Plague

Plague is an acute, infectious, contagious disease of rodents and humans, caused by a short, thick, gram-negative bacillus, *Yersinia pestis.* In humans, plague occurs in three forms: bubonic plague, pneumonic plague, and septicemic plague. Bubonic plague is the best-known form and is so called because it is characterized by the appearance of buboes, or enlarged, inflamed lymph nodes, in the groin or armpit or on the neck. Bubonic plague is transmitted by the bite of any of numerous insects that are normally parasitic on rodents, and that seek new hosts when the original host dies. The most important of these insects is the rat flea *Xenopsylla cheopis,* which is parasitic on the brown rat. Pneumonic plague, so called because the lung is the site of infection, is most often transmitted by droplets sprayed from the lungs and mouth of infected persons; the infection may spread from the lungs to other parts of the body, resulting in septicemic plague, which is infection of the blood. Septicemic plague may also be initiated by direct contact of contaminated hands, food, or objects with the mucous membranes of the nose or throat. In bubonic plague, the first symptoms are headache, nausea, vomiting, aching joints, and a general feeling of ill health. The lymph nodes of the groin or, less commonly, of the armpit or neck, suddenly become painful and swollen. The temperature, accompanied by shivering, rises to between 38.3° and 40.5°C (101° and 105°F).

Many preventive measures, such as sanitation, killing of rats, and prevention of the transport of rats in ships arriving from ports in which the disease is endemic, are effective in reducing the incidence of plague. Famine, which reduces resistance to the disease, results in spread of plague. Individuals who have contracted the disease are isolated, put to bed, and fed fluids and easily digestible foods. Sedatives are used to reduce pain and to quiet delirium. Antibiotics, such as streptomycin and tetracyclines, were found to be more effective in controlling the disease.

***Chlamydia*:** Chlamydia are a large group of obligate perasites closely related to gram-negative bacteria. The two diseases caused are Urethritis and Conjuctivitis.

Urethritis

Chlamydia trachomatis causes sexually transmitted diseases (STDs) in both men and women, including nongonococcal urethritis (NGU), an infection of the urinary tract. In both men and women symptoms of sexually transmitted *Chlamydia trachomatis* may include a watery discharge and pain when urinating; in women *Chlamydia trachomatis* also causes inflammation of the vagina, cervix, uterus, fallopian tubes, and ovaries. Women may also experience vaginal discharge, fever, abdominal pain, and pain in the genital area. In as many as 50 to 70 percent of *Chlamydia trachomatis* infections, however, women experience no symptoms. As a result, the infection remains untreated and may develop into more serious conditions, including pelvic inflammatory disease, ectopic pregnancy, and infertility. Pregnant women with chlamydia infections can also pass it to their babies during birth.

Conjunctivitis

Conjunctivitis, inflammation of the conjunctiva, a mucous membrane that lines the inner surface of the eyelids and joins with the cornea of the eyeball. Conjunctivitis may be caused by infection, allergy, or injury and is marked by redness and swelling, accompanied by a feeling of roughness as the membrane passes over the exposed surface of the eyeball. The eye may be sensitive to light; in more severe cases a mucoid sticky fluid, or even pus, may be discharged, depending on the cause of the infection. The acute form of conjunctivitis is commonly called pinkeye. In newborn babies it may result from several kinds of cocci, especially the gonococcus (gonorrheal conjunctivitis), or from a strain of the parasitic bacterium *Chlamydia trachomatis* (inclusive conjunctivitis). A strain of fly-borne *Chlamydia trachomatis* causes trachoma, a severe form of conjunctivitis found mostly in Africa and Asia. As in gonorrheal and inclusive conjunctivitis, trachoma often results in blindness because of the involvement of the cornea. Conjunctivitis caused by exposure of the eye to a welding arc or other source of ultraviolet rays is called flash kerato-conjunctivitis. Most cases of bacterial conjunctivitis are treated successfully with antibiotics and sulfonamides.

***Coccobacillus*:** Coccobacillus is the singular of coccobacilli, which are bacteria of a slightly elongated shape. Coccobacillus is derived from the words cocci (spherical shaped) and bacilli (elongated) bacteria. The disease caused are :

Pertussis (Whooping Cough)

Whooping Cough, also known as pertussis, common name applied to an acute, infectious disease of the respiratory tract, caused by the bacillus *Bordetella pertussis.* The disease is characterized in its late stages by a deep cough ending in a peculiar high-pitched whooping sound. Infection with pertussis usually occurs by inhaling droplets sprayed into the air during coughing or sneezing spells by an infected person. Once inside the respiratory tract, pertussis bacteria produce toxins (poisonous chemicals) that interfere with the respiratory tract's ability to eliminate microbial infections. Pertussis bacteria also produce chemicals that cause inflammation of the lining of the respiratory tract. Physicians treat whooping cough with antibiotics. Other treatments include controlling coughing with sedatives and replacing fluids lost by vomiting. Use of pertussis vaccine during the first year of life provides immunity to the infection in most children. Pertussis vaccine is usually given together with diphtheria and tetanus vaccines in a shot called DTaP.

Meningitis

Meningitis, inflammation of the meninges, the membranes that surround the brain and spinal cord. Meningitis may be caused by a physical injury, a reaction to certain drugs, or more commonly, infection by certain viruses, bacteria, fungi, or parasites. Most cases of bacterial meningitis are caused by one of three species of bacteria—*Haemophilus influenzae*, *Streptococcus pneumoniae*, and *Neisseria meningitidis*. Bacterial meningitis can be fatal if not treated promptly. Some children who survive the infection are left with permanent neurological impairments, such as hearing loss or learning disabilities. Many of the microorganisms that cause meningitis are quite common in the environment and are usually harmless. The microorganisms typically enter the body through the respiratory system or, sometimes, through the middle ear or nasal sinuses. Nearly all patients with meningitis experience vomiting, high fever, and a stiff neck. Meningitis may also cause severe headache, back pain, muscle aches, sensitivity of the eyes to light, drowsiness, confusion, and even loss of consciousness.

Any delays in treating bacterial meningitis can lead to stroke, severe brain damage, and even death. Patients with bacterial meningitis are usually hospitalized and given large doses of intravenous antibiotics. The specific antibiotic used depends on the bacterium responsible for the infection. To prevent the spread of bacterial meningitis, antibiotics are sometimes given to family members and other people who have had close contact with patients who develop the disease.

Coccus: Cocci (singular-coccus) are any spherical or near spherical bacteria. When they are formed in pairs they are known as *diplococcus*, in beadlike chains as *Streptococcus*, in grapelike clusters as *Staphylococcus*, and in groups of eight as *sarcina*.

Bacterial Pneumonia

Infection with the *Streptococcus pneumoniae* bacterium, also called pneumococcus, is the most common cause of bacterial pneumonia. Pneumococcus usually causes lobar pneumonia, attacking an entire lobe or portion of a lobe of the lung; in double pneumonia, pneumococcus attacks both lungs. Pneumococcal lobar pneumonia often occurs in winter after an acute, viral upper respiratory infection. Usual symptoms include a shaking chill followed by a fever of about 40°C (104°F), pain in the chest while breathing, a cough, and blood-streaked sputum. *Staphylococcus*, genus of round, parasitic bacteria, commonly found in air and water and on the skin and upper part of the human pharynx. These bacteria are known to cause pneumonia and septicemia as well as boils and kidney and wound infections (see Abscess; Carbuncle; Infection). The antibiotic drug penicillin was once effective for the treatment and control of staphylococci, but the increase of resistant strains requires use of other antiobiotic agents such as semi-synthetic Penicillins, Cephalosporins or Vancomycin. Two common species of Staphylococcus include *Staphylococcus aureus,* which is commonly responsible for skin infections, and *Staphylococcus epidermis,* which does not normally cause infection. However, either of these bacteria can cause serious infections under the right conditions. Other bacteria that cause pneumonia include *Klebsiella pneumoniae, Haemophilus influenzae, Legionella pneumophilia* (the bacterium that causes Legionnaires' disease), and various *Staphylococci* and *Streptococci* bacteria. Infections with these organisms primarily cause bronchopneumonia.

Gonorrhea

Gonorrhea, infectious sexually transmitted disease of humans, which primarily involves the mucous membranes of the urogenital tract. Characterized by a discharge of pus, it is caused by the gonococcus bacterium, *Neisseria gonorrhoeae.* Gonorrhea is much more obvious in males, who develop an acute discharge of pus from the urethra. Scant at the start, it becomes progressively thicker and heavier and causes frequent urination, often with a burning sensation. Should the prostate become infected, the passage of urine is partly obstructed. In females the infection occurs in the urethra, the vagina, or the cervix. Gonorrhea can be diagnosed by examining discharge from the penis or vagina for the presence of bacteria. A urine test can also detect the presence of bacteria and it is a noninvasive way to check people who are not exhibiting symptoms. Treatment in the early stages is usually effective. If the disease is untreated in the male, the early symptoms may subside but the infection may spread to the testicles, causing sterility. In the untreated female the infection usually spreads from the cervix into the uterus and fallopian tubes, causing pelvic inflammatory disease. Antibiotics are commonly used against gonorrhea, although over the years an increasing number of antibiotic-resistant strains of the gonorrhea bacteria have developed, causing a rise in the incidence of the disease in the 1990s. Still effective are antibiotics that are administered in a single dose, including Ceftriaxone, Cefixime, Ciprofloxacin, and Ofloxacin.

Listeria*: *Listeria monocytogenes is a Gram-positive bacterium, in the division Firmicutes, named for Joseph Lister. It is motile by means of flagella. Some studies suggest that 1 to 10% of humans may carry *L. monocytogenes* in their intestines. *L. monocytogenes* are pathogenic to some degree. Infection by *L. monocytogenes* causes the disease listeriosis. The manifestations of listeriosis include septicemia, meningitis (or meningoencephalitis), encephalitis. Influenza-like symptoms, including persistent fever, usually precede the onset of the aforementioned disorders. Gastrointestinal symptoms such as nausea, vomiting, and diarrhea may precede more serious forms of listeriosis or may be the only symptoms expressed. When listeric meningitis occurs, the overall mortality may reach 70%; from septicemia 50%, from perinatal/neonatal infections greater than 80%. In infections during pregnancy, the mother usually survives. Reports of successful treatment with parenteral penicillin or ampicillin exist. Trimethoprim-sulfamethoxazole has been shown effective in patients allergic to penicillin.

Mycoplasma*: *Mycoplasma is a genus of small bacteria which lack cell walls. Several species are pathogenic in humans, including *M. pneumoniae,* which is an important cause of pneumonia and other respiratory disorders, and *M. genitalium,* which is believed to be involved in pelvic inflammatory diseases. *Mycoplasma* are members of the class Mollicutes. Mycoplasmas are most often parasites or commensals of humans, other animals including insects, and plants.

Pneumonia

Pneumonia, inflammation of one or both lungs. In people with pneumonia, air sacs in the lungs fill with fluid, preventing oxygen from reaching blood cells and nourishing the other cells of the body. Sometimes the inflammation occurs in scattered patches in the tissue around the ends of the bronchioles, the smallest air tubes in the lungs. This is known as bronchopneumonia. In other cases the inflammation is widespread and involves an entire lobe of the lung. This condition is called lobar pneumonia.

Pneumonia has more than 30 different causes. Most cases of pneumonia result from infection with microorganisms, primarily viruses, bacteria, mycoplasmas (small, free-living particles with characteristics of both bacteria and viruses), and fungi. Pneumonia may also result from certain kinds of allergic reactions, inhalation of fluids or some gases, and the inhalation of ingested foods. About 50 per cent of pneumonia cases are caused by viruses, particularly those viruses that cause upper respiratory infections, such as the viruses that cause influenza, adenoviruses, and rhinoviruses. Most cases of viral pneumonia are mild and resolve spontaneously without specific treatment. One common type of pneumonia, formerly called primary atypical pneumonia, is caused by *Mycoplasma pneumoniae*, a mycoplasma. Epidemics of mycoplasma pneumonia occur in schools and in the military. The most prominent symptom of mycoplasma pneumonia is a violent dry cough. Some patients experience nausea or vomiting. *Pneumocystis carinii* pneumonia (PCP) is caused by a normally harmless fungus that may become deadly in people with impaired immune systems. PCP is the most common cause of death in people with acquired immunodeficiency syndrome (AIDS).

The incubation period varies from two weeks to several months, but generally is about two weeks. The first symptoms are usually headache and depression, followed by difficulty in swallowing and in opening the jaws. Stiffness of the neck develops and gradually a spasm of the cheek muscles sets the face in a peculiar, sardonic grin. Eventually, the spasms spread to other muscles of the body. Antibiotics can cure bacterial pneumonia and speed recovery from mycoplasma pneumonia and PCP. Antibiotics rarely have an effect on pneumonia caused by viruses. However, patients with viral pneumonia often receive antibiotics to prevent bacterial pneumonia from developing during the course of their illness. In addition to drug treatment, a patient with pneumonia should stay in bed, eat healthy meals, and drink large amounts of liquids.

Rickettsia**:** *Rickettsia* are non-motile, gram-negative, non-sporeforming highly pleomorphic organisms that could be shaped as cocci , rods or thread-like. Being obligate intracellular pathogens, the *Rickettsia* depend on entry, growth, and replication within the cytoplasm of a eukaryotic host cell (typically endothelial cells). In the past they were regarded as microorganisms positioned somewhere between viruses and real bacteria. The majority of *Rickettsia* bacteria are susceptible to antibiotics of the tetracycline group. Bacteria of the genus ***Rickettsia*** are carried as parasites by many ticks, fleas, and lice, and cause diseases such as Epidemic typhus, Brill-Zinsser disease (spread by lice) caused by *Rickettsia prowazekii* ; Rocky Mountain spotted fever (spread by ticks) caused by *Rickettsia rickettsii*; Endemic typhus (murine typhus, spread by rat fleas) caused by *Rickettsia typhi*.

***Spirillum*:** These are spiral bacteria having a distinctive helical shape, like a corkscrew and fairly rigid body. They are motile due to presence of flagella. Campylobacteriosis (bacterial diarrheal illness) is caused by *Campylobacter fetus jejuni* and Rat-bite fever is caused by *Spirillum minor*.

***Spirochete*:** The spirochaetes (or spirochetes) are a phylum of distinctive bacteria, which have long, helically coiled cells. They are distinguished by the presence of flagella running lengthwise between the cell membrane and cell wall, called *axial filaments*. These cause a twisting motion which allows the spirochaete to move about. Most spirochaetes are free-living and anaerobic, but there are numerous exceptions. It has been suggested by Lynn Margulis that eukaryotic flagella were derived from symbiotic spirochaetes, but few biologists accept this, as there is no close structural similarity between the two. Important members of this phylum include *Leptospira* species, which causes leptospirosis; *Borrelia burgdorferi* which causes Lyme disease, and *Treponema pallidum*, which causes syphilis.

***Vibrio*:** Rod- or comma-shaped, gram-negative, aerobic bacteria; commonly with a single flagellum; include Vibrio cholerae, cause of cholera, and luminescent forms symbiotic with deep-water fishes and squids.

Cholera

Cholera, severe infectious disease endemic in India and some other tropical countries and occasionally spreading to temperate climates. The symptoms of cholera are diarrhea and the loss of water and salts in the stool. In severe cholera, the patient develops violent diarrhea with characteristic "rice-water stools," vomiting, thirst, muscle cramps, and sometimes circulatory collapse. Death can occur as quickly as a few hours after the onset of symptoms. The causative agent of cholera is the bacterium *Vibrio cholerae,* which was discovered in 1883 by the German physician and bacteriologist Robert Koch. Virtually the only means by which a person can be infected is from food or water contaminated by bacteria from the stools of cholera patients. Prevention of the disease is therefore a matter of sanitation. Control of the disease is still a major medical problem worldwide. Treatment consists mainly of intravenous or oral replacement of fluids and salts. Packets for dilution containing the correct mixture of sodium, potassium, chloride, bicarbonate, and glucose have been made widely available by the WHO. Most patients recover in three to six days. Antibiotics such as tetracyclines, ampicillin, chloramphenicol, and trimethoprim-sulfamethoxazole can shorten the duration of the disease.

New Bacterial Diseases

Legionnaires' Disease: Legionnaires' Disease, severe form of pneumonia, characterized by headache, chest pain, lung congestion, and high fever. The name is derived from an outbreak at an American Legion convention in a Philadelphia hotel in July 1976. The causative agent as a previously unknown rod-shaped bacterium (*Legionella pneumophila*). The bacterium is most commonly transmitted through contaminated water sources; it is not readily transmitted from person to person. Treatment with the antibiotic erythromycin is effective.

Lyme Disease: Lyme Disease, infectious bacterial disease transmitted to people by deer ticks. Lyme disease was first recognized in Lyme, Connecticut, in 1975 and is now the most common tick-borne illness in the United States. The disease is characterized by flulike symptoms, skin rash, or joint inflammation in the early stages. If the disease is left untreated, it may lead to chronic problems affecting the nervous system and heart. Lyme disease is caused by *Borrelia burgdorferi,* a spiral-shaped bacterium that is frequently carried by deer ticks of the genus *Ixodes.* When the deer tick bites a human being, the bacteria are transmitted to the human bloodstream, where it travels throughout the body. Treatment of Lyme disease is most effective in the early stages of the illness. Oral antibiotics may be necessary. The majority of patients respond after two to four weeks of medication. Patients may have persistent fatigue and achiness after appropriate treatment, but these symptoms generally resolve within six months. In more advanced cases, 75 to 80 per cent of people with.

Protection against Bacteria

Our immune system is designed to protect us against harmful bacteria. It works to keep our normal microflora in check and also to eliminate invaders from outside the body. Some immune-system defenses are built in: The skin acts as a barrier to bacterial invaders, and antimicrobial substances in body secretions such as saliva and mucus can kill or stop the growth of some disease-causing bacteria. We acquire another immune-system defense through exposure to disease-causing bacteria. After recovering from many bacterial infections, people have the ability to resist a second attack by the same bacteria. They can do so because their immune system forms disease-fighting proteins called antibodies designed to recognize specific bacteria. When next exposed to those bacteria, the antibodies bind to the surface of the bacteria and either kill them, prevent them from multiplying, or neutralize their toxin. Vaccines also can stimulate the immune system to form disease-fighting antibodies. Some vaccines contain strains of the bacterium that lack the ability to cause infection; others contain only parts of bacterial cells.

VIRUS

Adenovirus*:** Adenoviruses are viruses of the family ***Adenoviridae. They infect both humans and animals. Adenoviruses were first isolated in human adenoids, from which the name is derived. Adenoviruses are medium-sized (90-100 nm), nonenveloped icosahedral viruses containing double-stranded DNA. There are 51 immunologically distinct types (6 subgenera: A through F) that can cause human infections. Adenoviruses are unusually stable to chemical or physical agents and adverse pH conditions, allowing for prolonged survival outside of the body. These are group of medium sized, non enveloped DNA viruses, which infect human beings, animals and birds, showing strict host specificity. These are relatively stable, remaining viable for about a week at 37°C. They are inactivated at 50°C. They resist ether and bile salts.

Adenoviruses cause infections of respiratory tract, eye, bladder and intestine. Common syndromes associated with Adenoviruses are respiratory disease in children, Sore throat, febrile cold, pneumonia, ARD in military recruits, follicular conjunctivitis, epidemic keratoconjunctivitis, Diarrhea.

Bunyavirus*: *Bunyaviridae is a family of viruses. Their morphology is envelope, spherical, and helical with the diameter of 80-120 nm. Bunyaviridae are single-stranded RNA viruses with negative polarity, the genome is circular with the size around 11-19 kbp.

The genome consists of 3 segments which are L, M, and S, called tripartite. For the Genus of *Phlebovirus*, the S partite is ambisense. Ambisense means that one segment of RNA strand consist of half positive sense and half negative sense. The replication for this RNA virus is still in cytoplasm. *Bunyaviridae* can be also called vector-borne virus, which is the Biosafety level 4 viruses.

Calicivirus: A virus is picked up by and sticks to the legs of a fly. The fly lands on a rabbit and the virus is transferred to the new host. Within a few hours the rabbit feels lethargic. Within 24 hours a major organ of the rabbit hemorrhages, the rabbit dies.

In the 18 years since its' emergence it has killed 90% of all rabbits in China and Europe. The virus, RHD, (Rabbit Hemorrhagic Disease), is one of the constantly mutating Caliciviruses and no safeguards are in place to diagnose diseases these viruses are currently causing or preventing them from mutating into a form that creates human fatalities. The Calicivirus family, in many of its forms, though exhibiting no symptoms, is endemic in the livestock of the US. It has been found to create long-term clinical disease in primates. Evidence shows that 5%, a conservative figure, of the human population is now infected with a Calicivirus and as a result many are now carrying an undiagnosed hepatitis or elevated liver enzymes. In addition, of women tested who suffered spontaneous abortions from undiagnosed causes, 40% showed evidence that they were infected with the virus. Caliciviruses are known to be an abortive agent in many other species.

Coronavirus: It is a genus of animal virus belonging to the family *Coronaviridae* Coronaviruses are enveloped viruses with a positive-sense single-stranded RNA genome and a helical symmetry. The name "coronavirus" draws reference to the "corona" – the "ring-like radiating structure" formed by the outermost part of the atmosphere of the sun. When observed under electron microscopy (E.M.), coronavirus particles exhibit a characteristic corona-like morphology. The corona-like structures are actually formed by the viral spike (S) peplomers, which are proteins that populate the surface of the virus and determine host tropism.

Coronaviruses primarily infect the upper respiratory and gastrointestinal tract. Four to five of total isolated strains of coronaviruses infect humans. The most publized human coronavirus, SARS-CoV, has a unique pathogenesis because it causes both upper and lower respiratory tract infections and can also cause gastroenteritis.

In 2003, following the outbreak of Severe Acute Respiratory Syndrome which had begun the prior year (SARS) in Asia, and secondary cases elsewhere in the world, the World Health Organization issued a press release stating that a novel coronavirus identified by a number of laboratories was the causative agent for SARS.

Filovirus: Filovirii are viruses belonging to the family *Filoviridae.* These viruses are single stranded negative sense RNA viruses that target primates. There are two general viruses, the Ebola virus (*Ebolavirus*, with four species) and the Marburg virus (*Marburgvirus*).

These viruses cause viral hemorrhagic fevers, characterized by massive bleeding from every orifice of the body. These diseases are shockingly lethal, with very high mortality rates and possibility of transfer. They are among the most lethal and destructive viruses known to man.

Flavivirus: *Flavivirus* (also Japanese Encephalitis Antigenic Complex) is a genus of the family *Flaviviridae*. This complex viruses which cause encephalitis. *Flavivirus* share a common size (40-60 nanometres), symmetry (enveloped, icosahedral nucleocapsid), nucleic acid (positive-sense, single stranded RNA approximately 10,000-11,000 bases), and appearance in the electron microscope.

Hepadnavirus: These viruses belong to the family *Hepadnaviridae*. Hepadnaviruses cause liver infections in humans and animals. Hepadnaviruses have very small genomes of partially double-stranded, partially single stranded non-circular DNA. They have a peculiar mode of replication, because they replicate through an RNA intermediate (which they transcribe back into cDNA using reverse transcriptase). Because of this mode of replication, Hepadnaviruses are classified as RNA viruses.

Hepatitis

Hepatitis, inflammation of the liver caused by viruses, bacterial infections, or continuous exposure to alcohol, drugs, or toxic chemicals, such as those found in aerosol sprays and paint thinners. Hepatitis can also result from an autoimmune disorder, in which the body mistakenly sends disease-fighting cells to attack its own healthy tissue, in this case the liver. No matter what its cause, hepatitis reduces the liver's ability to perform life-preserving functions, including filtering harmful infectious agents from the blood, storing blood sugar and converting it to usable energy forms, and producing many proteins necessary for life.

The Hepadnavirus Family derives its name from its hepatotropic nature and double-stranded DNA genome. While hepatitis B virus (HBV) is the only causative agent for a human disease within the family, a profile on hepatitis D virus or delta agent has also been included due to its specific dependence upon HBV for its propagation. The human hepatitis B virion consists of an icosahedral nucleocapsid (core) enclosing a circular, double-stranded DNA genome. The hepadnaviral genome is unique in that the plus strand is incomplete, leaving 15-50% of the molecule single-stranded and the minus strand while being complete, possesses a discontinuity at a unique site. The core is surrounded by an envelope containing HBsAg (hepatitis B surface antigens) derived from budding through the lipid bilayer of the endoplasmic reticulum (ER).

Hepatitis B remains a major global concern as some 300 million people are chronic carriers of the virus and of these, a significiant minority suffer severe pathologic consequences including cirrhosis and hepatocellular carcinoma (HCC). Hepatitis B virus is second only to tobacco among the known human carcinogens and studies indicate that HCC may develop with 100 times greater frequency among HBV carriers than controls. Over 1 million of such affected individuals die each year. While a majority of affected adults and older children suffer from acute infection, neonates and infants primarily develop chronic infection. Therefore, patients who develop HCC are individuals who typically became infected with HBV early in life and developed liver cancer 20 to 50 years following primary infection.

Safe and effective vaccines are available to prevent hepatitis B infection. Vaccination against HBV, which will also prevent HDV, for all newborn babies, infants, adolescents, and people in jobs that put them at risk for hepatitis, such as health. The best protection against these viruses is to avoid high-risk activities, including preventing exposure to body fluids of infected individuals, and always washing hands after using the toilet or changing an infant's diapers.

Herpesvirus: All herpesviruses are composed of relatively large double stranded linear DNA genomes encoding 100-200 genes encased within an icosahedral protein cage called the capsid which is itself wrapped in a lipid bilayer membrane called the envelope. This particle is known as the virion.

Chickenpox

Chickenpox, also called varicella, contagious viral disease that affects mainly children. Typically, chickenpox begins with a low fever, headache, rash, and a general feeling of sickness, or malaise. The rash, which usually covers the face, scalp, and trunk of the body, starts as red bumps but quickly develops into small blisters. The rash and the blisters are extremely itchy. As the disease progresses, the blisters break open and form scabs, which fall off after about one to two weeks. Chickenpox is caused by *varicella-zoster* virus, a type of herpes virus. The virus spreads through the air via infected droplets emitted from the nose or mouth while coughing or sneezing. Touching the fluid from a chicken pox blister can also spread the disease. Chickenpox is contagious for approximately seven days during a person's period of infection. Contagiousness begins about two days before symptoms appear and continues until all blisters have formed scabs. Doctors recommend keeping the infected person isolated from others during those seven days.

Contracting chickenpox provides immunity, or lifelong resistance, against the disease. However, after the symptoms disappear, the virus remains in the body's nerve cells and occasionally reactivates later in life, causing a disease known as shingles, an infection of the nerve fibers. Shingles usually occurs in people over 50, due to an age-related weakening of the immune system, and causes pain, burning, itching, inflammation, and blisters.

Treatment of chickenpox is usually limited to bed rest, acetaminophen for relief of fever and discomfort, and measures that soothe the itching, including lukewarm baths and application of topical medicines such as calamine lotion. Excessive scratching can cause infection of blisters, which can lead to scarring. Acyclovir, an antiviral drug, is used to treat severe cases of chickenpox, particularly in patients with a weakened immune system.

Orthomyxovirus: The ***Orthomyxoviridae*** are a family of RNA viruses which infect vertebrates. It includes those viruses which cause influenza.

Influenza

Influenza, also known as flu, contagious infection primarily of the respiratory tract. It is characterized by cold like symptoms plus chills, fever, headaches, muscle aches, and fatigue. The virus consists of an inner core of the genetic material ribonucleic acid (RNA) surrounded by a protein coat and an outer lipid (fatty) envelope from which project spikes of proteins called hemagglutinin and neuraminidase. These proteins act as antigens – that is, they elicit an immune response in the human or other host organism that the virus invades. In addition to their role as antigens, hemagglutinin enables the virus to bind to and invade cells and neuraminidase allows the virus to move among cells.

Influenza viruses pass from person to person mainly in droplets expelled during sneezes and coughs. When a person breathes in virus-laden droplets, the hemagglutinin on the surface of the virus binds to enzymes in the mucous membranes that line the respiratory tract. The enzymes, known as proteases, cut the hemagglutinin in two, which enables the virus to gain entry into cells and begin to multiply. Influenza is an acute disease with a rapid onset and pronounced symptoms. Classic symptoms include sore throat, dry cough, stuffed or runny nose, chills, fever with temperatures as high as 39°C (103°F), aching muscles and joints, headache, loss of appetite, occasional nausea and vomiting, and fatigue. For most people flu symptoms begin to subside after two to three days and disappear in seven to ten days. However, coughing and fatigue may persist for two or more weeks.

There is no specific cure for influenza. Recommended treatment usually consists of bed rest and increased intake of nonalcoholic fluids until fever and other symptoms lessen in severity.

No drugs can cure influenza, but certain antiviral medicines can relieve flu symptoms. The drugs amantadine (sold under the brand name *Symmetrel*) and rimantadine (*Flumadine),* both in pill form, work against hemagglutinin and are effective in treating influenza. A flu vaccine consists of greatly weakened or killed flu viruses, or fragments of dead viruses. Antigens in the vaccine stimulate a person's immune system to produce antibodies against the viruses. If the flu viruses invade a vaccinated person at a later time, the sensitized immune system recognizes the antigens and quickly responds to help destroy the viruses.

***Papovavirus*:** A papovavirus is a member of the Papovaviridae family of viruses. Papovaviridae includes two genus: papillomavirus and polyomavirus.Papovaviruses are viruses containing double-stranded DNA, are icosahedral in shape, and do not have a lipoprotein envelope. They are commonly found in humans and other species, mostly mammals. The one that most often causes disease in humans is the human papillomavirus.

***Picornavirus*:** These viruses that belong to the family *Picornaviridae*. The name *picornavirus* means *small RNA virus*. Picornaviruses are non-enveloped, positive-stranded RNA viruses with an icosahedral capsid.

Poliomyelitis

Poliomyelitis, infectious viral disease that sometimes results in paralysis. The infection chiefly affects children and young adults and is caused by any one of three related viruses called polioviruses. In more than 95 per cent of cases, the infection may pass without creating severe or even noticeable illness because the body's immune system neutralizes the invading virus and protects against future infection. In a small percentage of cases the virus penetrates to the central nervous system, infecting cells that control muscle function. Such infection can result in permanent paralysis of limbs. In its most perilous form, the infection attacks the brain, creating complications that sometimes result in death. The term *poliomyelitis* derives from Greek words referring to inflammation (*itis*) of the gray (*polios*) matter of the spinal cord (*myelos*). The shortened term *polio* is commonly used by the public and medical professionals, and the disease has also been referred to as *infantile paralysis*. While still a health threat in many parts of the world, polio also represents one of the greatest medical success stories of the 20th century. The number of polio cases found worldwide today constitutes a 90 per cent reduction from the number of cases reported in the late 1980s. Most health experts credit this achievement to a mass vaccination program called the Global Polio Eradication Initiative that WHO initiated in 1988 to eradicate polio worldwide by the year 2005.

Three types of poliovirus have been identified: Type 1 (also known as Brunhilde), Type 2 (Lansing), and Type 3 (Leon). Type 1 is the most common form and the one most closely associated with polio's more severe, paralytic progression. A person who develops immunity from exposure to one poliovirus type will not have immunity against the other polioviruses.

Poliovirus typically enters the body through the mouth and multiplies in the tonsils and lymph nodes of the upper respiratory tract. Infection proceeds from the mouth into the gastrointestinal tract through the stomach to the intestines. The virus multiplies in the intestines and is shed from the body in feces, often resulting in further infections. Large quantities of poliovirus multiply in intestinal tissue known as Peyer's patches, where cells of the body's lymphatic system are concentrated. Passage of the virus into the body's lymphatic system stimulates the production of antibodies. These specialized immune-system defenders, in time, will destroy the viral intruder. From the lymphatic system, the virus typically invades the bloodstream. Once the virus enters the bloodstream, the virus may cause one of four types of illnesses marked by varying severity. Some infections result in *abortive poliomyelitis*, a mild form of the disease characterized by fever, headache, sore throat, fatigue, nausea, and vomiting. Some poliovirus infections of nerve cells, or neurons, result in *nonparalytic poliomyelitis*. In addition to the fever and other symptoms seen in abortive poliomyelitis, nonparalytic poliomyelitis causes pain and stiffness in the neck and back. In perhaps 1 or 2 per cent of cases a more disabling form of the disease occurs, called *paralytic poliomyelitis*. In this form viral infection of neurons in the spinal cord may cause temporary damage to these cells—or permanent destruction. The muscles activated by the involved neurons become painful, and muscular weakness in the arms and legs may develop, sometimes followed by paralysis. In the most serious cases of polio infection the virus attacks the brain, causing *bulbar poliomyelitis*. Various nerves in the head and face, including those that send signals to the ears, eyes, and the muscles controlling chewing and swallowing, may be affected. Sometimes the virus affects the part of the brain that controls breathing and heartbeat, resulting in death.

In 1908 Austrian pathologist Karl Landsteiner first isolated and identified poliovirus from the spinal fluid of a young polio victim. At the same time, American pathologist Simon Flexner confirmed the viral origins of polio. Once polio was recognized as a viral disease, scientists focused on developing a vaccine to prevent the disease. American physician Jonas Salk developed the first successful vaccine in the early 1950s. The Salk vaccine was a killed-virus type that contained all three poliovirus types inactivated by formaldehyde. Around the same time American virologist Albert Sabin performed large-scale tests of his oral vaccine in Mexico and the Soviet Union, also with success. In 1961 the U.S. government approved the Sabin oral vaccine for widespread use. By the 1970s the Sabin oral vaccine had replaced the Salk killed-virus vaccine as the most commonly used vaccine worldwide.

Paramyxovirus: These are viruses of the *Paramyxoviridae* family. They are negative-sense single-stranded RNA viruses responsible for a number of human and animal diseases. Virions are enveloped and can be spherical, filamentous or pleomorphic. Fusion proteins and attachment proteins appear as spikes on the virion surface. Matrix proteins inside the envelope stabilise virus structure. The nucleocapsid core is composed of the genomic RNA, nucleocapsid proteins, phosphoproteins and polymerase proteins.

Measles

Measles, also rubeola, acute, highly contagious, fever-producing disease caused by a filterable virus. Measles is characterized by small red dots appearing on the surface of the skin, irritation of the eyes (especially on exposure to light), coughing, and a runny nose. About 12 days after first exposure, the fever, sneezing, and runny nose appear. Coughing and swelling of the neck glands often follow. Four days later, red spots appear on the face or neck and then on the trunk and limbs. In 2 or 3 days the rash subsides and the fever falls; some peeling of the involved skin areas may take place. Infection of the middle ear may also occur. Measles was formerly one of the most common childhood diseases. Since the development of an effective vaccine in 1963, it has become much less frequent. In other parts of the world measles is still a common childhood disease and according to the World Health Organization (WHO), about 1 million children die from measles each year. No specific treatment for measles exists.

Mumps

Mumps, acute infectious disease caused by a virus that mainly attacks glandular and nervous tissues, frequently characterized by swelling of the salivary glands. The disease is worldwide in distribution and can occur in epidemics. Mumps is spread from person to person by droplets sprayed from the respiratory tract of infected persons, and it is highly contagious. The incubation period of the disease varies from 15 to 21 days. Few fatalities result from mumps, and one attack usually confers complete immunity, because only one antigenic type of virus causes this disease. In children, the first symptoms are usually a mild fever, a feeling of illness and chilliness, loss of appetite, and dryness of the throat. This is followed by soreness and swelling around the ears, and a higher fever. These symptoms are usually gone by 12 days. In adult males, inflammation of the testes occurs in up to 20 per cent of the cases, but resultant sterility is rare. In children, infection of the auditory nerve can cause deafness, but this is also a rare result.

Parvovirus*: *Parvovirus, commonly called *parvo*, is a genus of the *Parvoviridae* family of DNA viruses. Parvoviruses are some of the smallest viruses found in nature. Like all members of the parvoviridae family, they infect only mammals. Parvoviruses can cause disease in some animals. For example, *Canine parvovirus* is a particularly deadly disease among young puppies, causing gastrointestinal tract damage and dehydration as well as a cardiac syndrome in very young pups. The most accurate diagnosis of parvovirus is by ELISA. Dogs and cats can be vaccinated against parvovirus. Many types of mammalian species have a strain of parvovirus associated with them. A parvovirus tends to be specific about the taxon of animal it will infect.

***Poxvirus*:** The name of the family, *Poxviridae*, is a legacy of the original grouping of viruses associated with diseases that produced poxs in the skin. Modern viral classification is based on the shape and molecular features of viruses, and the smallpox virus remains as the most notable member of the family.

Smallpox

Smallpox, highly contagious viral disease that is often fatal. The disease is chiefly characterized by a skin rash that develops on the face, chest, back, and limbs. Over the course of a week the rash develops into pustular (pus-filled) pimples resembling boils. In extreme cases the pustular pimples run together – usually an indication of a fatal infection. Death may result from a secondary bacterial infection of the pustules, from cell damage caused by the viral infection, or from heart attack or shock. In the latter stages of nonfatal cases, smallpox pustules become crusted, often leaving the survivor with permanent, pitted scars.

The scientific name for smallpox is *variola,* from a Latin word meaning "spotted." Scientists recognize two main forms of smallpox caused by related subspecies of viruses. *Variola major* is the more serious form of the disease, believed to kill anywhere from 20 to 40 percent of those infected. *Variola minor,* on the other hand, produces milder symptoms and is fatal in around 1 percent of cases. Smallpox viruses are shaped somewhat like bricks, measuring in size from 250 to 300 nanometers across—almost big enough to be visible with an ordinary light microscope. Inside a variola virus, within an outer shell made up of scores of proteins, lies a coiled strand of deoxyribonucleic acid (DNA), the virus's genetic material.

Smallpox is primarily spread via airborne, virus-laden droplets exhaled by an infected person. Prolonged contact is not required. The smallpox virus typically enters the body through the mouth or nose, ultimately invading the cells of the lymph nodes and the lungs. Once inside a cell, the smallpox virus employs the same stealthy act of hijacking employed by all viruses, commandeering the infected cell's own genetic machinery to make as many as 100,000 copies of itself. Some of these viruses instruct the host cell to send out fingerlike tendrils that make contact with other cells. It is through these extensions that new cells become infected with virus particles and the viral replication process repeats. After the virus has used a cell to replicate, it abandons the cell, leaving it to die. Eventually smallpox virus particles circulate in the bloodstream—a condition known as viremia. The virus may then invade the liver, spleen and other internal organs. This viral invasion of the skin, in turn, leads to the painful, pus-filled lesions that are the hallmark of the disease. Meanwhile, throughout the body, infected and dying cells release toxins of their own that also contribute to illness.

After a person is infected with the smallpox virus, about 12 days pass before the person begins to feel sick, a time known as the incubation period. By the 12th day the virus moves into the bloodstream, producing sudden and dramatic symptoms that include high fever of 39° to 41°C (102° to 106°F), headache, muscular and abdominal pain, and vomiting. Within two to three days the virus spreads to the skin and the rash appears. At first the rash takes the form of skin spots known as macules. By the second day of the rash, as infection worsens within the skin cells, the spots become raised lesions called papules; by the seventh day these papules fill with pus and are referred to as pustules. In the most extreme forms of smallpox, the virus causes bleeding underneath the skin, giving the skin a charred appearance, as if burned. In these cases the virus causes massive tissue damage to skin and to internal organs, and victims bleed heavily from the mouth and other body openings. Such cases are referred to as black pox and they are almost invariably fatal. There is no cure, or even a specific treatment, for a smallpox infection. In modern outbreaks, medical care usually consisted of bed rest and the treatment of symptoms: the use of intravenous fluids to prevent dehydration, sedatives to keep patients calm, and aspirin or codeine to relieve pain in the head and muscles.

A more effective medical blow against smallpox, and one of history's landmark biomedical achievements, took place in 1796 with an experiment performed by the British physician Edward Jenner. He had observed that young women who milked cows for a living often contracted a minor skin infection known as cowpox—and that these milkmaids subsequently seemed to be protected from smallpox. Jenner arranged to perform an experiment on eight-year-old James Phipps, scratching his arm with pus taken from the cowpox lesion of a milkmaid. Six weeks later, when Jenner scratched Phipps's arm with pus from a smallpox lesion, the boy failed to show any reaction or illness. The cowpox virus had created natural immunity against smallpox. Thus, with this new procedure, which was later dubbed vaccination (from the Latin word *vacca* for "cow"), Jenner set the course that ultimately led to victory over smallpox. In the battle against smallpox in the 20th century, two weapons were key: (1) an effective vaccine and (2) the fact that the smallpox virus exists only in humans. Smallpox requires a chain of human hosts to keep its life cycle going.

In 1967 the World Health Organization (WHO) launched a worldwide vaccination campaign against smallpox. At the time, about 10 to 15 million cases of the disease occurred each year, with more than 2 million deaths. Initially, the effort consisted of mass vaccinations, with the aim of inoculating as many people as possible. Two technological innovations aided this campaign. One was a process that freeze-dried the vaccine so that it could be transported easily and would remain potent even in the most extreme climates. The other was a special bifurcated (two-pronged) needle. Unlike the syringes and hollow needles used in other vaccinations, the bifurcated needles could be used in any setting by personnel who required no advanced medical training to perform vaccinations. Within just a few years, WHO's effort began recording victories over smallpox. In 1977 a young man in Somalia had the last naturally occurring case of smallpox recorded in the world. In 1979 WHO declared victory when it announced that naturally occurring smallpox had been wiped from the face of the Earth. Today the smallpox virus exists officially in only two laboratories, at the United States Centers for Disease Control and Prevention (CDC) in Atlanta, Georgia, and at a virology laboratory located outside Novosibirsk, Russia.

***Reovirus*:** The *Reoviridae* are a family of viruses that includes some viruses that affect the gastrointestinal system and some that cause respiratory infections. The genetic material of viruses in this family is double-stranded RNA.

Diarrhea

Diarrhea, frequent passage of abnormally loose, watery stool. Diarrhea usually develops suddenly and may last from several hours to a few days. It is often accompanied by abdominal pains, low fever, nausea, and vomiting. If the attacks are severe or increasingly frequent, exhaustion and dehydration can result. In normal digestion the large intestine absorbs excess water from liquid food residues produced by earlier phases of the digestive process before excreting semisolid stools. When the mucous membrane lining the large intestine is irritated or inflamed, food residues move through the large intestine too quickly and the resulting stool is watery because the large intestine cannot absorb the excess water. Diarrhea is not a disease. It is a symptom of numerous disorders, such as food poisoning from contaminated foods or beverages, infections by viruses and bacteria, or anxiety. Chronic diarrhea, which lasts weeks or months, may be caused by amoebic dysentery (intestinal infection), tumors, and other serious intestinal disorders such as Crohn's disease, ulcerative colitis, or irritable bowel syndrome. The usual treatment for diarrhea consists of bed rest, drinking liquids to replace fluids and salts lost from the body, and eating soft foods. Dehydration is a serious concern in infants and the elderly. If the condition lasts more than a few days, a physician should be consulted.

***Retrovirus*:** A Retrovirus is a virus which has a genome consisting of two RNA molecules, which may or may not be identical. It relies on the enzyme revers transcriptase to perform the reverse transcription of its genome from RNA into DNA, which can then be integrated into the host's genome with an integrase enzyme. The virus itself stores its nucleic acid genome and serves as a means of delivery of that genome into targeted cells, which constitute the infection. Once in the host's cell, the RNA strands undergo reverse transcription in the cytosol and are integrated into the host's genome, at which point the retroviral DNA is referred to as a provirus.

AIDS

Acquired Immunodeficiency Syndrome (AIDS)

AIDS, human viral disease that ravages the immune system, undermining the body's ability to defend itself from infection and disease. Caused by the human immunodeficiency virus (HIV), AIDS leaves an infected person vulnerable to opportunistic infections. Infection with HIV does not necessarily mean that a person has AIDS. Physicians prefer to use the term *AIDS* for cases where a person has reached the final, life-threatening stage of HIV infection. While cases of AIDS have been reported in every nation of the world, the disease affects some countries more than others. More than 95 percent of all HIV-infected people live in the developing world.

Causes of Aids

AIDS is the final stage of a chronic infection with the human immunodeficiency virus. There are two types of this virus: HIV-1, which is the primary cause of AIDS worldwide, and HIV-2, found mostly in West Africa. On its surface, HIV carries a protein structure that recognizes and binds only with a specific structure found on the outer surface of certain cells. HIV attacks any cell that has this binding structure. However, white blood cells of the immune system known as T cells, which orchestrate a wide variety of disease-fighting mechanisms, are especially vulnerable to HIV attack. Particularly vulnerable are certain T cells known as CD4 cells (Fig. 1). When HIV infects a CD4 cell, it commandeers the genetic tools within the cell to manufacture new HIV virus. The newly formed HIV virus then leaves the cell, destroying the CD4 cell in the process. No existing medical treatment can completely eradicate HIV from the body once it has integrated into human cells. The loss of CD4 cells endangers health because these immune cells help other types of immune cells respond to invading organisms. The average healthy person has over 1,000 CD4 cells per microliter of blood. In a person infected with HIV, the virus steadily destroys CD4 cells over a period of years, diminishing the cells' protective ability and weakening the immune system.

PROCESS OF SPREAD OF HIV INFECTION

Sex with an Infected Person

HIV transmission occurs most commonly during intimate sexual contact with an infected person. During sexual intercourse, the virus gains access to the bloodstream of the uninfected person by passing through openings in the mucous membrane – the protective tissue layer that lines the mouth, vagina, and rectum.

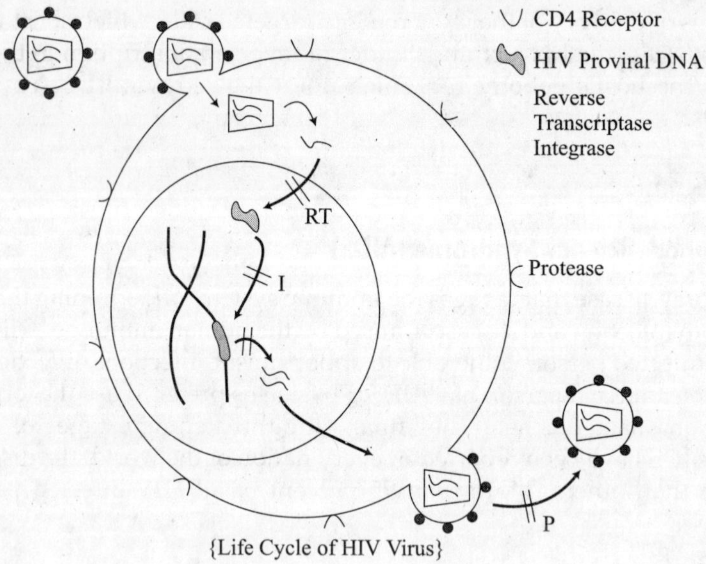

Fig. 1. Life Cycle of HIV Virus.

Contact with Infected Blood

Direct contact with HIV-infected blood occurs when people who use heroin or other injected drugs share hypodermic needles or syringes contaminated with infected blood. Sharing of contaminated needles among intravenous drug users is the primary cause of HIV infection. Less frequently, HIV infection results when health professionals accidentally stick themselves with needles containing HIV-infected blood or expose an open cut to contaminated blood.

Mother-to-Child Transmission

HIV can be transmitted from an infected mother to her baby while the baby is still in the woman's uterus or, more commonly, during childbirth. The virus can also be transmitted through the mother's breast milk during breastfeeding. Mother-to-child transmission accounts for 90 per cent of all cases of AIDS in children.

DETECTION OF HIV INFECTION

Since HIV was first identified as the cause of AIDS in 1983, a variety of tests have been developed that help diagnose HIV infection as well as determine how far the infection has progressed.

The standard test to detect HIV antibodies in the blood is the enzyme-linked immunosorbent assay (ELISA). In this test, a blood sample is mixed with proteins from HIV. If the blood contains HIV antibodies, they attach to the HIV proteins, producing a telltale color change in the mixture. This test is highly reliable when performed two to three months after infection with HIV. Doctors routinely confirm a positive result from an ELISA test by using the Western Blot test, which can detect lower levels of HIV antibodies. In this test a blood sample is applied to a paper strip containing HIV proteins. If HIV antibodies are present in the blood, they bind to the HIV proteins, producing a color change on the paper. The combination of the ELISA and the Western Blot test is more than 99.9 per cent accurate in detecting HIV infection within 12 weeks following exposure. Once tests confirm an HIV infection, doctors monitor the health of the infected person's immune system by periodically measuring CD4 cell counts in the blood. The progressive loss of CD4 cells corresponds to a worsening of the disease as the immune system becomes increasingly impaired. A modified ELISA test that detects p24 antigen, a protein produced by HIV, can determine if specific drug treatments are having a positive effect on a patient. Blood banks, plasma centers, clinical laboratories, private clinics, and public health departments also use this p24 antigen test to screen for the presence of HIV in blood, blood components, and organs before they are used in medical procedures.

TREATMENT

While no medical treatment cures AIDS, in the relatively short time since the disease was first recognized, new methods to treat the disease have developed rapidly. Health-care professionals focus on three areas of therapy for people living with HIV infection or AIDS

Antiretroviral Therapies

Understanding the specific steps in the HIV replication cycle is critical in order for scientists to develop drugs that attack vulnerable stages within the cycle. HIV belongs to a unique group of viruses known as retroviruses, so named because these viruses reverse the usual flow of genetic information within an infected cell. Most viruses store their genetic material in deoxyribonucleic acid (DNA), the double-helix structure that makes up genes. When a virus infects a cell, the viral DNA forms the template for the creation of messenger RNA, a type of ribonucleic acid. This messenger RNA directs the formation of specific proteins, and these proteins, in turn, build new virus particles. In HIV, however, genetic material is stored in two single-stranded RNA molecules. When HIV infects a cell, an enzyme called reverse transcriptase copies the genetic instructions in the virus's RNA and moves it into the DNA. This movement of genetic information from RNA to DNA is the opposite of that which occurs in most cells during protein synthesis. Another HIV enzyme, called integrase, helps the newly formed viral DNA to become part of the structure of the infected cell's DNA. The viral DNA then forces the infected cell to manufacture HIV particles. A third HIV enzyme, called protease, packages these HIV particles into a complete and functional HIV virus. Over the last decade researchers have created a variety of drugs that block the action of some of the enzymes used in HIV replication. The three main classes of drugs used against HIV are nucleoside analogues, non-nucleoside reverse transcriptase inhibitors, and protease inhibitors.

Drug Resistance

Clinical studies of treatment with antiretroviral drugs immediately showed that their benefits are short-lived when a single drug is used alone. This short-term effectiveness results when HIV mutates, or changes its genetic structure, becoming resistant to the drug. Genes mutate during the course of viral replication, so the best way to prevent mutation is to halt replication. Studies have shown that the most effective treatment to halt HIV replication employs a combination of three drugs taken together—for instance, a combination of two nucleoside analogues with a protease inhibitor. This regimen, called triple therapy, maximizes drug potency while reducing the chance for drug resistance. Despite phenomenal success, triple therapy has some drawbacks. This multidrug therapy is quite complicated, requiring patients to take anywhere from 5 to 20 pills a day on a specific schedule. Some drugs must be taken with food, while others cannot be taken at the same time as certain other pills. Even the most organized people find it difficult to take pills correctly. Yet, just one or two lapses in treatment may cause the virus to develop resistance to the drug regimen.

Development of New Drugs

Scientists continue to develop more powerful HIV treatments that have fewer side effects and fewer resistance problems. Some drugs under investigation block the HIV enzyme integrase from inserting viral DNA into the infected cell. Other drugs prevent HIV from binding with a CD4 cell in the first place, thereby barring HIV entry into cells. Some scientists focus on ways to fortify the immune system. A biological molecule called interleukin-2 shows promise in boosting the immune system's arsenal of infection-fighting cells. Interleukin-2 stimulates the production of CD4 cells. If enough CD4 cells can be created, they may trigger other immune cell responses that can overpower HIV infection. In other research, doctors hope to bolster the immune system with a vaccine. Most vaccines available today, including those that prevent measles or poliomyelitis, work by helping the body to create antibodies. Such vaccines mark specific infectious agents, such as the measles and polio viruses, for destruction. But many experts believe that an effective HIV vaccine will need to do more than just stimulate anti-HIV antibodies. Studies are underway to develop vaccines that also elevate the production of T cells in the immune system. Scientists hope that this dual approach will prime the immune system to attack HIV as soon as it appears in the body, perhaps containing the virus before it spreads through the body in a way that natural immune defenses cannot. The genetic variability of HIV frustrates efforts to develop a vaccine: A vaccine effective against one type of HIV may not work on a virus that has undergone genetic mutation.

Prevention

With a vaccine for AIDS years away and no cure on the horizon, experts believe that the most effective treatment for AIDS is to prevent the occurrence of HIV infection. Health officials focus public education programs on altering risky behaviors linked to HIV transmission, particularly unsafe sexual practices and needle-sharing by intravenous drug users. Education programs instruct about the proper way to use condoms to provide a protective barrier against transmission of HIV during sexual intercourse. Public education about AIDS has also proven effective in other countries.

Origin of the Virus

Using computer technology to study the structure of HIV, scientists have determined that HIV originated around 1930 in rural areas of Central Africa, where the virus may have been present for many years in isolated communities. The virus probably did not spread because members of these rural communities had limited contact with people from other areas. But in the 1960s and 1970s, political upheaval, wars, drought, and famine forced many people from these rural areas to migrate to cities to find jobs. During this time, the incidence of sexually transmitted infections, including HIV infection, accelerated and quickly spread throughout Africa. As world travel became more prevalent, HIV infection developed into a worldwide epidemic. In 1970, at about the same time that the HIV epidemic was taking hold in Africa, American molecular biologist David Baltimore and American virologist Howard Temin independently discovered the enzyme reverse transcriptase, which could be used to identify retroviruses. Over the next ten years, many retroviruses were identified in animals. But not until 1980, shortly before the first AIDS cases were recognized in the United States, did American virologist Robert Gallo identify the first human retroviruses, HTLV-I and HTLV-II (HTLV stands for human T cell lymphotropic virus).

Rhabdovirus: These are viruses belonging to the family *Rhabdoviridae*. Rhabdoviruses infect a broad range of hosts throughout the animal and plant kingdom. Animal rhabdoviruses infect insects, fish, and mammals, including humans. Rhabdoviruses carry their genetic material in the form of negative-sense single-stranded RNA. They typically carry genes for five proteins: large protein (L), glycoprotein (G), nucleoprotein (N), non-structural protein (NS), and matrix protein (M). Rhabdoviruses that infect vertebrates are bullet-shaped. The main disease caused is Rabies.

Rabies

Rabies, acute, contagious infection of the central nervous system, caused by a specific virus that enters the body through the bite of an animal. All warm-blooded animals are susceptible, but the disease is most common in skunks, foxes, bats, raccoons, dogs, and cats. Most of the cases of rabies in humans are caused by the bite of one of these animals. The incubation period in humans varies from three weeks to 120 days, with an average of about four to six weeks. Rabies is virtually always fatal when vaccine is not administered.

In humans, at the end of the incubation period the site of the now healed wound becomes irritated and painful, and the local tissues may become numb. Depression and anxiety are common. This initial stage lasts for about two days. In the next stage, the period of excitation, the patient becomes irritable and hypersensitive; the general attitude is one of terror, intensified by the onset of difficult breathing and swallowing and a feeling of strangulation, caused by spasmodic contractions of the diaphragm and larynx. The patient is extremely thirsty but experiences spasms of the larynx when water is presented or even mentioned, whence the original name of the disease, hydrophobia (Greek *hydor,* "water"; *phobos,* "fear"). Vomiting, pallor, and fever of about 39°C (102°F) are common during this stage. A thick secretion of mucus collects in the mouth and throat, and the individual expectorates frequently or attempts to cough. This stage lasts three to five days and usually terminates in death from a convulsive seizure or from cardiac or respiratory failure.

In animals rabies takes two forms, furious or irritable rabies and a dumb or paralytic rabies. The stages of furious rabies are similar to those through which an infected human passes; during the stage of excitement the animal usually runs amuck, biting and snapping at any living thing in its path. In dumb rabies, which is not as common as furious rabies, the stage of excitement is of very short duration or is absent, and the paralytic stage sets in early in the disease, first attacking the muscles of the jaws and larynx. Domestic animals such as dogs and cats are generally immunized against rabies; in many areas this is required by law.

In 1884 the French bacteriologist Louis Pasteur developed a preventive vaccine against rabies, and modifications of Pasteur's methods are still used in rabies therapy today. The Pasteur program, or variations of it, has greatly reduced the fatalities in humans from rabies. Modern treatment, following a bite by a rabid or presumed rabid animal, consists of immediate and thorough cleansing of the bite wound and injection into the wound and elsewhere of hyperimmune antirabies serum. A 14- to 30-day course of daily injections of rabies vaccine is then given; booster doses are given 10 days after this course and again 20 days later.The traditional vaccine contains inactivated rabies virus grown in duck eggs. A newer vaccine, which contains virus prepared from human cells grown in the laboratory, is safer and requires a shorter course of injections.

Togavirus: The genome of the *Togaviridae* viruses is a linear, single-stranded RNA of positive sense that is 10,000-12,000 nucleotides long. The 5'-terminus carries a methylated nucleotide cap and the 3'-terminus has a polyadenylated tail. Viruses in this family have an icosahedral capsid and a membrane envelope. The disease caused :

German Measles

German Measles, also rubella, contagious disease of short duration, caused by virus infection. The disease is characterized by a rose-colored rash and frequently by other mild symptoms, such as a slight fever, sore throat, and swelling of the lymph glands behind the ears. German measles is most common among teenagers and young adults and rarely occurs in infants or in adults over the age of 40. Although far less severe than measles, German measles can have severe consequences for women in the first three months of pregnancy. The newborn child may be afflicted with various congenital abnormalities, including heart defects, mental retardation, deafness, and cataracts. An attack of rubella after the fourth month of pregnancy rarely causes birth defects. Pregnant women who are exposed to German measles are given gamma globulin in an effort to prevent contraction of the disease. Women of childbearing age are advised to be immunized with attenuated live virus vaccines several months before anticipated pregnancy.The rubella virus was isolated and identified in 1961, and by the early 1970s the vaccine was readily available.

Natural Defence against Viral Infection

Viral infections are encountered by the host immune system. The immune system consists of a wide range of morphologically and antigenically distinct white cells with different functional responses to viral invasion. The major type of white cells are lymphocytes, these being T cells and B cells. T cells (cytotoxic(CD8) or helper(CD4) cells) pass through the thymus during their maturation. B cells give rise to antibody producing plasma cells. T4 helper cells help B cells to react to highly specific antigenic stimuli. In addition to these specific responses the host mounts a series of non-specific responses, for example, production of cytokines (e.g. interferon), natural killer cells, mucocilliary responses and macrophages. It is the outcome of the competition between the immune system and the progress of viral replication that determines the outcome of the infection. In self-limiting infections, viruses are destroyed by the immune system. Good humoral immunity requires the induction of high levels of IgG antibodies.

PRIONS

Prions are proteins found on the plasma membrane (the membrane that surrounds a cell and defines its physical boundary). In mammals, prions are found in the highest concentration in cells of the central nervous system. If the prion is an infectious agent, it is the first infectious agent identified that does not contain the nucleic acids-deoxyribonucleic acid (DNA) or ribonucleic acid (RNA). While the prion's role in causing disease is still controversial, its discovery has opened the door to alternative ideas about how infectious diseases may be transmitted. Discovered by American neurologist Stanley B. Prusiner, prions are a mutated form of a normal protein. The normal protein is found on the surface of nerve cells in the brain, white blood cells, muscle cells, and cells of many other tissues. In mammals and yeast, there are several genes coding for different prions (Fig. 2). The function of normal prions (denoted PrP^C) is unknown. Aberrant or mutant prions (denoted PrP^{Sc}) are thought to be the causative agents of a set of neurological disorders, among which are BSE in cows and Creuzfeldt-Jacob Disease (CJD) in humans. The term prion comes from "proteinaceous infectious particles". The term was coined during early research of the sheep disease called scrapie. During that time, all that was known about the particle that caused scrapie was that it was a protein and that it was infectious.

Prions have the same system of organization as other proteins. They may have up to four levels of organization. The primary structure of a protein is the amino acid sequence that makes up the protein. The secondary structure of a protein is the local organization of the primary sequence of the protein -helical or -sheet formations. The tertiary structure of a protein is the overall shape of the protein, caused by interactions between the various local structures. Some proteins have a quaternary level of organization, which is defined by the interactions between the tertiary structures of two or more protein subunits.

Infection of normal cells may occur when an aberrant prion acts as a template for the refolding of a normal prion into a new aberrant prion. It is thought that there is at least one

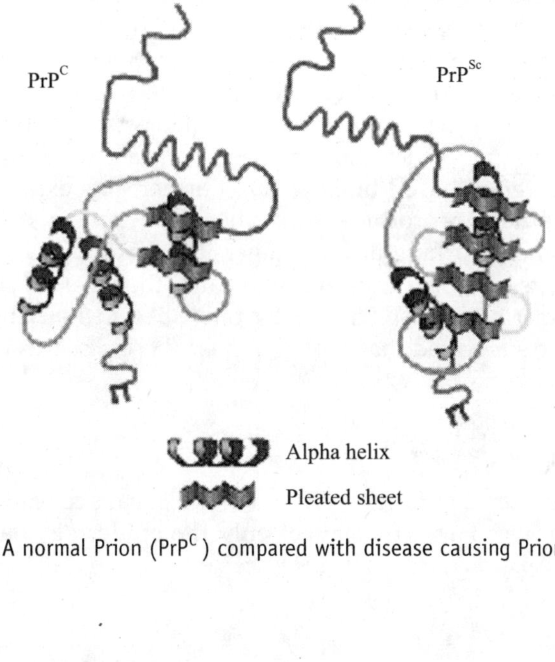

Fig. 2. A normal Prion (PrP^C) compared with disease causing Prion (PrP^{Sc}).

more protein involved: the as-yet-unidentified Protein X. This protein is believed to mediate the folding from a normal into an abnormal prion. When proteins are synthesized inside of a cell there are other special proteins (known as chaperones) that help in this process. Chaperones are proteins that bind to the newly synthesized protein or protein subunit, in order to ensure that the protein is properly folded into its secondary or tertiary structure. It has been hypothesized that Protein X is a type of chaperone.

The role of the normal protein is not yet understood, but the mystery of its structure has been solved. A hundred times smaller than the smallest virus, the normal protein is composed of 208 amino acids twisted into three long telephone cord-like coils known as helices. A floppy tail of 97 amino acids extends from the end of one of the helices. The mutated form of the protein is built of the same amino acids. However, instead of the coil shape, the mutated protein is folded like the flat pleats of a partly opened accordion. Some scientists think that the mutated protein causes disease when it contacts the normal protein and triggers part of it to switch from the coiled to the pleated form. A chain reaction follows, resulting in a cluster of tangled, nonfunctional proteins called plaques. These plaques are found in the brains of animals that die from prion-related diseases. The plaques destroy the brain cells, resulting in one of the diseases collectively known as transmissible spongiform encephalopathies (TSEs).

TSEs cause inflammation and characteristic spongelike holes in the delicate membranes surrounding brain cells. This physical damage results in loss of coordination, dementia, and, eventually, death. Perhaps the best-known TSE is bovine spongiform encephalopathy (BSE), more popularly known as mad cow disease. BSE made headlines in 1996 when about a million cattle in the United Kingdom became infected with the disease. The cows contracted BSE when they ate feed made from the processed animal parts of sheep, pigs, and chickens that were infected with the disease. Ingestion of the infected cow meat caused about 20 people in Britain to develop an unusual form of Creutzfeldt-Jakob disease. Other TSEs include kuru, a rare disease contracted by natives of New Guinea who ate the infected brains of their dead relatives during ritual cannibalism; and scrapie, which affects sheep and goats and is so named because diseased sheep sometimes scrape off their own wool. The road to the discovery of the prion began with investigations into the cause of TSEs. In 1967, British scientist Tikvah Alper and her colleagues at the Hammersmith Hospital in London extracted brain tissue from scrapie-infected sheep. They chemically treated the tissue to isolate a virus, bacteria, or other agent that might cause the disease. This processed tissue was then injected into healthy sheep to see if the disease would be transmitted. The healthy sheep contracted scrapie, indicating that the infectious agent was in the diseased brain tissue. Further, this experiment showed that the infectious agent could reproduce in healthy animals to cause disease.
In the early 1980s, Prusiner, working at the University of California at San Francisco, found that similar tissue extracts no longer caused disease after they were exposed to treatments that destroy proteins. Prusiner concluded in a study published in 1982 that proteins alone were responsible for TSEs. He suggested that proteins cause these diseases by replicating in tissues of the nervous system.

FUNGAL DISEASES

Mycoses: The term *myco* refers to a fungus and *osis* or *iasis* means condition such as *phycomycosis* and *candidiasis* respectively. Fungi that infect only the epidermis, hair and nails are called dermatophytes and their infections are called cutaneous mycoses. Subcutaneous mycoses are fungal infections beneath the skin and they are caused by saprophytic fungi that live in soil and on vegetation. Infection occurs by direct implantation of spores or mycelial fragments into a puncture wound in the skin.

Dermatomycoses: It is a type of cutaneous mycoses. Dermatophytes secretes keratinase, an enzyme that degrades keratin. Keratin is a protein found in hair, skin and nails. Infection is transmitted from human to human or from animal to human by direct contact with infected hairs and epidermal cells. Dermatophytes , on the basis of clinical aspects of the disease, are grouped into three genera as *Trichophyton, Microsporum* and *Epidermophyton.* Therefore, the fungi responsible for infection of tineas (ringworm) are grouped into above three genera. The scaly annular skin lesions are generally called as tinea. Tinea are further classified according to the affected parts such as :

- *Tinea pedis* (athletes foot) : Lesions begins in the web between thr fourth and fifth toes. The fissures are bordered by narrow zone of peeling epidermis. The causal agent are *T.rubrum, T.mentagrophytes* and *Epidermophyton floccosum.*
- *Tinea capitis* (ringworm of scalp) : It has dry diffuse and scaly lesions of the scalp on which are found black dots characterized by broken off hairs at follicle. *T.capitis* is caused largely by the species of *Microsporum viz., M.audouinii* and *M.canis.*
- *Tinea carporis* (ringworm of the non-hairy skin of body) : These are characterized by circular lesions with inflammation of various degree. Primary lesions develop on the face, shoulders, arms due to exposure to the infected pets or cattle. Secondary infection may be caused by autoinoculation or exposure to to these infected animals. The etiologic agents in cats and dogs are *Microsporum canis* and *T.mentagrophytes* which are often associated with dogs.

Mycetomas: A chronic infection involving the feet and characterised by the formation of localized lesions with tumefactions and multiple draining sinuses. The exudate contains granules that may be yellow, white, red, brown, or black, depending upon the causative agent. Mycetoma is caused by two principal groups of microorganisms:

- Actinomycotic mycetoma is caused by actinomycetes, including species of Streptomyces, Actinomadurae, and Nocardia.
- Eumycotic mycetoma is caused by true fungi, including species of *Madurella, Exophiala, Pseudallescheria, Curvularia, Neotestudina, Pyrenochaeta, Aspergillus, Leptosphaeria, Plemodomus, Polycytella, Fusarium, Phialophora, Corynespora, Cylindrocarpon, Pseudochaetosphaeronema, Bipolaris, and Acremonium.* Any tumour with draining sinuses is produced by filamentous fungi.

Sporotrichosis: Sporotrichosis is a fungal infection that usually affects the skin. It is mainly an occupational disease of farmers, gardeners, and horticulturists. The fungus that carries sporotrichosis is found in the soil and in sphagnum moss, hay, and other plant materials. The fungus enters the skin through small cuts or punctures from thorns, barbs, pine needles, or wires. Sporotrichosis is usually treatable with fungus-killing medicine. To prevent sporotrichosis: (1) wear gloves and long sleeves when handling wires, rose bushes, hay bales, pine seedlings, or other materials that can prick the skin; and (2) avoid skin contact with sphagnum moss. Until recently, sporotrichosis was usually treated with potassium iodide, taken by mouth in droplet form. A new drug, itraconazole (Sporanox), is now the first choice for treatment because it causes fewer side effects than potassium iodide. Treatment often takes several weeks, until the skin sores are completely healed.

Chromomycosis: Chromomycosis is a cutaneous infection affecting normal, immuno competent persons mostly in tropical or subtropical areas, characterized by formation of papillomatous nodules that tend to ulcerate. Most infections begin on the foot or leg, but other exposed body parts may be infected, especially where the skin is broken. Dematiaceous fungi also may cause other patterns of infection in normal hosts. Lymphatic obstruction may occur, itching may persist, and secondary bacterial superinfections may cause ulcerations and, occasionally, septicemia. Itraconazole is the most effective antifungal drug, although not all patients respond.

Candidiasis: Mycosis caused by *Candida albicans* is called candidiasis. It is an acute or chronic mycosis. *C. albicans* grows in the sufficiency of diets high in fruits and carbohydrates. Mostly oral candidiasis is observed in new borns, aged and debilitated patients of both the sex. Infection by *C. albicans* is endogenous, therefore, it is found in mouth and intestinal tract of a large number of humans. *C. albicans* causes most frequently the meningeal candidiasis. It may also occur as vulvovaginal candidiasis during pregnancy.

C.albicans also colonizes the mucosal surface of respiratory tract. Bronchocandidiasis is characterized by cough, varying amount of sputum and appearance of peribronchial thickening of hazy linear fibrosis.

Nystatin in the form of ointment or candida cream is recommended as tropical therapy of candida infection forming skin lesions, whereas powder, suspension or tablets for oral or vaginal uses.

Actinomycosis: It is a chronic actinomycetous disease. It spreads to contiguous tissues without limitation by anatomical barriers. Sinus tracts are formed which drain suppurative lesions. In pus firm, lobulated or angular colonies of the etiological agent are present. *Actinomyces israelli* is the etiological agent in man and *A. bovis* in cattle or swine. In addition *Bifidobacterium eriksonii* has been isolated from pleural fluid and lung abscess.Usually the actinomycosis contain bacteria that help infection. Actinomycosis is sporadic in occurrence but not contagious. Mainly there are three main sites of primary lesions of actinomycosis e.g., facial, thoracic and abdominal.It produces sulphur granules in tissues which may be sparse.

Traditionally, solution of KI has been used to cure actinomycosis. Thereafter, oral benzylpenicillin is also given to control the disease. In addition chloramphenicol is also recommended. Moreover, for cure of actinomycosis, surgical drainage of abcess is necessary.

Aspergillosis: It is caused by the species of *Aspergillus* especially *A. fumigatus.* Aspergillosis is an opportunistic mycosis and occurs in persons who have debilitating lung diseases or cancer and have inhaled *Aspergillus* spores. Occasionally aspergillosis occurs in sinuses, bronchi, lungs and other parts of body as well. The disease is sporadic and worldwide in distribution. The contaminated moldy cereal grains are also one of the sources. Many clinical types of aspergillosis have been reported. Pulmonary aspergillosis is characterized by granulomatus lesions in lung parenchyma, which may spread through hyphae to kidney and other organs. Bronchopulmonary aspergillosis is marked by frequent pyrexial attacks associated with severe cough and blood tinged sputum with flecks of whitish material containing mycelium and a high eosinophilia.

Surgery is suggested in certain pulmonary aspergilosis. Treatment of aspergillosis with sodium iodide intravenously and potassium iodide orally has cured disease effectively. The only known useful antibiotic for aspergillosis is amphotericin B.

Otomycosis: It is the mycosis of ear which is characterized by inflammation, scaling, pruritis and pain. It develops as a result of superficial chronic or subacute infection on the outer ear canal. Otomycosis is caused by fungi especially *Aspergillus niger* and *A. fumigatus*.

In addition species of *Mucor, Rhizopus, Candida* and dermatophytes are also occasional etiological agents. The disease is characterized by inflammation, pruritis and exfoliation of epithelium of the ear. When the ear canal is occluded by mycelial plug, cerumen and epithelial debris, it results in partial deafness. Otomycosis is chronic and may occur even after cure. Therefore, ear canal should be cleaned by using 5% aluminum acetate solution. This will reduce edema and remove epithelial debris present in ear canal.

Penicillinosis: The species of *Penicillium* is ubiquitous in distribution and contaminates the lesions, open ulcers, urine, sputum etc. through air-borne conidia and grow rapidly in culture from pathological specimens. It has been accepted that pulmonary mycosis is caused by *Penicillium. P. commune* caused disseminated pulmonary and cerebral penicillinosis in a patient with acute leukemia. This fungus grows profusely in lungs and brain resulting in vascular invasion, thrombosis and infection of lungs.

Histoplasmosis: This disease superficially resembles tuberculosis. Although the lungs are the organs most likely to be initially infected, the organism cause lesions in almost all organs of the body. Normally the symptoms are rather poorly defined and mostly subclinical and the disease passes for a minor respiratory infection. The causative organism, *Histoplasma capsulatum* , is a dimorphic fungus; that is, it has yeast like morphology in tissue growth and in soil or artificial media it forms a filamentous mycelium carrying reproductive conidia. Humans acquire disease from air-borne conidia produced under conditions of particular moisture and pH levels. The organism is found in the spleen and bone marrow as well as in phagocytic cells. Currently the most effective chemotherapy is with amphotericin B and ketoconazole. New research might lead to a vaccine for dimorphic pathogenic fungal infections such as histoplasmosis.

Coccidioidomycosis: Another fungal pulmonary disease is coccidioidomycosis. The causative organism is *Coccidioides immitis,* a dimorphic fungus. The spores are found in dry, highly alkaline soils. In tissues, the organism forms a thick walled body filled with spores. The symptoms of the disease includes chest pain, fever, coughing and loss of weight. Most infections are inapparent and almost all patients recover in a few weeks. Amphotericin B has been used to treat serious cases but less toxic imidazole drugs, such as ketocanazole, might replace it.

Blastomycosis: The disease is caused by Blastomyces dermatidis, a dimorphic fungus found most often in moist soils. Most of the infections are asymptomatic. The infection begins in the lungs and can spread rapidly. Cutaneous ulcers commonly appear and there is extensive abscess formation and tissue destruction. The organism can be isolated from pus and biopsy specimens. Amphotericin B is usually an effective treatment.

Mycotoxicoses: It refers to poisoning by ingestion to fungal toxins i.e. mycotoxins in food which have been deteriorated by the growth of toxin-producing fungi. The examples of mycotoxins are aflatoxins (by *Aspergillus flavus*), patulin (by *Penicillum claviforme*), ochratoxin (by *A. ochraceus*), zeralenone (by *Fusarium graminearum*) and one important example is of ergot production by *Claviceps purpurea* in rye seeds. The mushroom poisoning (Mycetismus) has its effects on gastrointestinal and nervous system.

Allergy: A number of studies have suggested that exposure to indoor fungi can trigger allergy symptoms. Fungal exposure is also associated with exacerbations of asthma, although it remains unclear that this exposure causes the disease. Adverse health symptoms associated with fungal contact also include fungal infection, irritant reaction to spores or fungal metabolites, and toxic reaction to mycotoxins, described as follows:

- Infection: Most common indoor fungi grow near room temperature and do not easily grow inside the human body. *Aspergillus* is the most likely to cause respiratory infections, where most other fungi tend to infect the skin, nails and other areas of the body that function a lower temperatures.
- Irritant: An irritant reaction to fungi exposure causes inflammation and can happen without previous hypersensitivity. Substances produced by fungi that are suspected to cause irritant reactions include microbially derived volatile organic compounds (MVOCs), ergosterols and glucans. MVOCs are produced by fungi as a byproduct of their metabolism, and their health effects are undetermined. Indoor levels of airborne fungi are generally below outdoor levels of similar species in a well-constructed home without water damage. There can be some contribution from indoor sources, such as plants, pets and mold contaminates brought in on footwear and clothing.

Regular inspection for sources of moisture and their elimination is the most important strategy to control indoor fungal growth. Since fungal growth depends on moisture and a carbon source, the most important strategy for reducing or eliminating its growth is controlling the amount of moisture present.

HUMAN PATHOGENIC PROTOZOA

Entamoeba histolytica: It is an anaerobic parasitic protozoan, classified as an entamoebid. It infects predominantly humans and other primates. The active (trophozoite) stage exists only in the host and in fresh feces; cysts(the environmental survival form of the organism) survive outside the host in water and soils and on foods, especially under moist conditions on the latter. When swallowed they cause infections by excysting (to the trophozoite stage) in the digestive tract. The disease caused by *E. histolytica* are amoebiatic dysentery i.e. amoebiasis throughout the world. Most infections occur in the digestive tract but other tissues may be invaded. Complications include ulcerative and abscess pain, usually involving the liver, and, rarely, intestinal blockage. This parasite is transmitted through contamination of water and food.

Disease occurs when amoeba comes in contact with the cells lining the intestine. It then secretes toxic substances, including enzymes that destroy cell membranes and allow it to penetrate and digest human tissues, resulting in flask-shaped ulcers in the intestine. These are the same enzymes that are normally used to digest the bacteria it usually feeds on. *E. histolytica* also feeds on the destroyed cells by phagocytosis and is often seen with red blood cells inside.Theoretically, the ingestion of one viable cyst can cause an infection. *E. histolytica* infections occur in both the intestine and (in people with symptoms) in tissue of the intestine and/or liver. As a result two different sorts of drugs are needed to rid the body of the infection, one for each location. Both types of drug must be used to treat infections, with Metronidazole usually being given first, followed by Paromomycin.

Entamoeba coli: It is a non-pathogenic species of entamoebid that is important clinically in humans only because it can be confused with *Entamoeba histolytica,* which *is* pathogenic, on microscopic examination of stained stool specimens. A simple finding of *Entamoeba coli* trophozoites or cysts in a stool specimen requires no treatment. There is no clear evidence of the pathogenic role of *E.coli* in man. Proper disposal of faecal matter preventing contamination of water and food is an important measure to control this pathogen.

Balantidium coli: It is a species of ciliate protozoan, and is the only one that is a pathogen of humans. It is carried by pigs, rats, primates (including humans), rats, horses, cattle and guinea pigs. It is transmitted within or between these species mostly via fecal transmission. Pigs are the most significant reservoir hosts, though they show few if any symptoms. Cysts are the parasite stage responsible for transmission of balantidiasis. The host most often acquires the cyst through ingestion of contaminated food or water. Following ingestion, excystation occurs in the small intestine, and the trophozoites colonize the large intestine. Both cysts and trophozoites are identifiable by a large, "sausage shaped" macronucleus. The trophozoites reside in the lumen of the large intestine of humans and animals, where they replicate by binary fission, during which conjugation may occur . Trophozoites undergo encystation to produce infective cysts . Some trophozoites invade the wall of the colon and multiply. Some return to lumen and disintegrate. Mature cysts are passed with feces. Symptoms can be local due to involvement of the intestinal mucosa, or systemic in nature and include diarrhea. Balantidiasis can be treated by Tetracycline and Diiodohydroxyquin.

Plasmodium: ***"Plasmodium"*** is a genus of parasitic protozoa, four species of which cause malaria in humans. Other species infect other animals, including birds, reptiles and rodents. In 1898 Ronald Ross demonstrated the existence of *Plasmodium* in the stomach of the *Anopheles* mosquito. For this discovery he won the Nobel Prize in 1902. However some credit must also be given to the Italian professor, Giovanni Battista Grassi, who showed that human malaria could only be transmitted by *Anopheles* mosquitoes.

The four species of *Plasmodium* that attack humans are

- *Plasmodium falciparum* (the cause of malignant tertian malaria),
- *Plasmodium vivax* and the much rarer
- *Plasmodium ovale* (the causes of "benign" tertian malaria), and
- *Plasmodium malariae* (the cause of "benign" quartan malaria.)

The life cycle of *Plasmodium* is very complex. The infective stage is sporozoites which are released from the salivary glands of mosquito into the blood of man and migrate to the liver, where the parasite invades hepatocytes. The parasite replicates into thousands of merozoites, which then invade red blood cells. The infected RBC's are disintegrated which release the toxic substances. The toxic substances include the symptoms of malaria i.e., chills and fever. After reaching the body temperature upto 40°C fever subsides. Due to breakdown of RBCs the patient suffers from anaemia, Liver and spleen enlargement.

However, no effective immunity against Plasmodium develops besides some resistance in endemic areas. Quinine, chloroquinine, primaquine are effective drugs against the parasite. Several time malaria eradication movement was launched by the government organizations to eradicate mosquitoes by spraying DDT, but the mosquitoes developed resistance. Unfortunately, no significant control measure has been reported so far.

***Leishmania*:** It is a genus of Trypanosome protozoa, and is the parasite responsible for the disease leishmaniasis. It is spread through sandflies of the genus *Phlebotomus* in the Old World, and of the genus *Lutzomyia* in the New World. They primary hosts is a vertebrate; *Leishmania* commonly infects hyraxes, canids, rodents, and humans. *Leishmania* cells have two morphological forms: promastigote (with an anterior flagellum) in the insect host, and amastigote (without flagella) in the vertebrate host. Infections are regarded as cutaneous, mucocutaneous, or visceral:

- Cutaneous (localized and diffuse) infections appear as obvious skin reactions. The most common is the Oriental Sore (caused by Old World species *L. major*, *L. tropica*, and *L. aethiopica*). In the New World, the most common culprits are *L. mexicana* and *L. (Viannia) braziliensis*..
- Mucocutaneous (espundia) infections will start off as a reaction at the bite, and can go metastasis into the mucous membrane and become fatal.
- Visceral infections are often recognized by fever, weight loss, swelling of the liver and spleen, and anemia. They are known by many local names, of which the most common is probably *Kala azar* and are caused exclusively by species of the *L. donovani* complex (*L. donovani*, *L. infantum* syn. *L. chagasi*). Visceral infections are most common in Bangladesh, Brazil, India, Nepal and Sudan.

An important aspect of the *Leishmania* protozoan is its glyconjugate layer of liphosphoglycan (LPG). This is held together with a phosphoinosite membrane anchor, and has a tripartite structure consisting of a lipid domain, a neutral hexasaccharide, and a phosphrorylated galactose-mannose, with a termination in a neutral cap. Not only do the little unfriendlies not develop post-phlebotomus digestion but, it is thought to be essential to oxidative bursts, thus allowing passage for infection. Chararacteristics of digestion include an endosome fusing with a lysosome, releasing acid hydrolases which degrade DNA, RNA, proteins and carbohydrates. Treatments of the disease generally involve a solution containing antimony, which is toxic. A solution which breaks down the LPG is also a form of treatment being looked at. Those who feel they are infected should seek immediate attention. Transmission is mainly through the bite of sand flies but also through needle-sharing by drugs users. Most at risk are humans infected with HIV, canines and rodents.

***Trypanosoma*:** It is a notable genus of trypanosomes, a group of parasitic protozoa. The name is derived from the Greek *trypaô* (boring) and *soma* (body) because of the way the organisms move. Different species infect a variety of different vertebrates, including humans, causing the trypanosomiasis diseases. Most species are transmitted by invertebrates such as biting insects and leeches.Trypanosoma undergo a complex lifecycle which may include several different morphological forms especially in the species which are transmitted by invertebrates. They may go through a variety of different forms in the invertebrate host, but in the vertebrate host the cells take a characteristic form called a trypomastigote, where the flagellum is runs from the posterior to the anterior of the cell and is connected by an undulating membrane.

Sleeping sickness or African trypanosomiasis is a parasitic disease in humans. Caused by protozoa of genus *Trypanosoma* and transmitted by the tsetse fly, the disease is endemic in certain regions of Sub-Saharan Africa, covering about 36 countries and 60 million people. There are two forms of African sleeping sickness, caused by two different parasites: *Trypanosomiasoma brucei gambiense* and *Trypanosomiasoma brucei rhodesiense*. It multiplies by binary fission in blood stream of man and reaches lymph system. Later the parasite penetrates into the cerebro-spinal fluid and causes damage to the brain which makes the patient lethargic which is characteristic of sleeping sickness.

Chagas disease (also called ***American trypanosomiasis***) is a human tropical parasitic disease which occurs in the Americas, particularly in South America. Its pathogenic agent is a flagellate protozoan named *Trypanosoma cruzi*, which is transmitted to humans and other mammals mostly by hematophagous insects of the subfamily Triatominae. Those insects are known by numerous common names varying by country, including assassin bug, benchuca, vinchuca, kissing bug, chipo, barbeiro, et cetera. Primarily it infects skin or mucous but later obstructs lymph nodes. In response to the parasite when inflammation occurs, it damages the macrophages and central nervous system, leading to weakness, loss of appetite and patient finally dies.

Toxoplasma gondii: It is a species of parasitic protozoa, belonging to the Apicomplexa, that can cause the disease toxoplasmosis in humans. The life cycle of *Toxoplasma gondii* has two phases. The sexual part of the lifecycle takes place only in cats, the definitive host of *Toxoplasma gondii*. The asexual life cycle can take place in any warm-blooded animal, like other mammals and birds. In these intermediate hosts, the parasite invades cells forming a so-called parasitophorous vacuole. Within this vacuole *T. gondii* propagates by a series of binary fissions until the infected cell bursts. This fast replicating form of the asexually reproducing *T. gondii* is called tachyzoite. Additionally, a slow replicating form exists, the bradyzoite. Bradyzoite vacuoles can form tissue cysts in the infected animals/humans (mainly within the muscles and brain). Resistance to antibiotics varies, but the cysts are very difficult to eradicate entirely. The immune system of the host does not detect these cysts, while free tachyzoites are efficiently cleared by the immune response. Tissue cysts can be ingested by a cat, e.g. by feeding on an infected mouse. The cysts survive passage through the digestive system of the cat and the parasites infect epithelial cells of the small intestine where they undergo sexual reproduction and oocyst formation. Oocysts are shed with the feces. Animals and humans that ingest oocysts (e.g. by eating unwashed vegetables etc.) become infected. The parasite enters macrophages in the intestinal lining and is distributed via the blood stream throughout the body. Latent infection is normally asymptomatic; but often gives flu-like symptoms in the early acute stages. However, in the case of immunocompromised patients (such as those infected with HIV, or transplant recipients on immunosuppressive therapy) toxoplasmosis can develop. The most notable manifestation of toxoplasmosis in immunocompromised patients is toxoplasmic encephalitis, which can be deadly. To control infection antiprotozoan drugs e.g. pyrimethamine should be given in combination with sulfonamides.

Giardia: ***Giardia lamblia*** (formerly also ***Lamblia intestinalis*** and also known as ***Giardia duodenalis***) is a flagellated protozoan parasite that infects the gastrointestinal tract of humans. Infection causes giardiasis, a type of gastroenteritis that manifests itself with severe diarrhea and abdominal cramps. Other symptoms can include bloating, flatulence, fatigue, nausea, vomiting and weight loss. Giardia is a major cause of intestinal disease worldwide and the most frequent non-bacterial cause of diarrhea in North America. Infection from giardia can occur from consuming contaminated food or water. It can also be transferred from animal or human feces.One of Giardia's techniques is to alter the proteins on its surface, which confounds the ability of the infected animal's immune system to detect and combat the parasite. This work reveals why *Giardia* infections are extremely persistent and prone to reoccur. Treatment of drinking water for Giardia typically involved some form of high efficiency filtration and/or chemical disinfection such as chlorination or ozonation.

SUMMARY

Microbe is any living thing, visible only with a microscope. It is too tiny to be seen with the naked eye. Microbes live in the water you drink, the food you eat, and the air you breathe. Microbiology is one of the most applied of all the biological sciences. Microbes are omnipresent and are characterized by very high degree of adaptability. Living agents or inanimate objects that serve as the bridge for contact between the host and the microorganism are referred to as sources of infection. Microorganism are transmitted to susceptible hosts by four routes: contact, air, common vehicle and vector borne. Any microbe which cause diseases is known as pathological agent. These include bacteria, viruses, fungi, algae, and protozoa. The main diseases caused by bacteria are Anthrax, Botulism, Tetanus, Diptheria, Leprosy, Tuberculosis, Typhoid fever, Dysentery, Plague, Pertussis, Meningitis, Pneumonia, Cholera and even sexually transmitted disease like Gonorrhea. Our immune system is designed to protect us against harmful bacteria. It works to keep our normal microflora in check and also to eliminate invaders from outside the body. Viruses also cause many diseases like Hepatitis, Chicken pox, Flu, Polio, Measels, Mumps, Small pox (eradicated), Diarrhea, Rabies, Rubella and AIDS, which is a threat to whole world and no proper cure for this disease is available. Another pathological agents are Prions, proteinaceous infectious particles, which cause *Kuru* disease. Fungi also cause a wide range of infections namely, Mycoses, which may be Dermatomycoses, Actinomycosis, Histoplasmosis, Blastomycosis, Aspergillosis, Candidiasis, Penicillinosis. Fungal infection may also give rise to Mycotoxicoses or Allergies. Protozoa are one celled, eukaryotic organisms that belong to the kingdom Protista. Different human pathogenic protozoa causes Amoebiasis, Malaria, leishmaniasis, trypanosomiasis, toxoplasmosis, giardiasis.

REFERENCES

Ajellow, L. (1968). A Taxonomic review of the dermatophytes and related species. Sabouradia 6: 147-159.

Arunachalam, S. (1987). AIDS: An update. Sci. Reporter : 24: 459-463.

Bellet, A.J.D. (1967). Classification of Viruses. Virology. 37: 117-123.

Blake, P.A.; Weaver, R.e. and Hollis, D.G.(1980). Diseases of humans caused by vibrios. Ann.Rev. Microbiol.34: 341-367.

Bloom, B. 1994. Tuberculosis: Pathogenesis, Protection and control: Am. Soc. Microbiol.

Bull, A.T. and Hardman, D.J. (1991). Microbial diversity. Curr. Opin. Biotechnol. 2: 421-28.

Emmons, C.H., Binford C.H. Utz J.P. and Kwon-Chung, K.J. (1997). Medical Mycology. Lea and Febiger, Philadephia pp.592.

Fauci, A.S. (1987). AIDS-pathogenic mechanisms and research strategies. ASM News 53(5): 263.

Huang, S.N. and Harris, A.s. (1963). Acute disseminated Penicillosis. Am.J.Clin.Pathol.39: 167-174.

Kaplan, M.M. and Koprowski, H. (1980). Rabies.Sci. Am. 242: 120-134.

Lord F.T. (1993). The etiology, pathogenesis and diagonosis of actinomycosis. Med. Clin.N.Amer. 16: 829-844.

Mackowiak, P.A. (1984) Microbial latency. Rev. Infect. Dis. 6(5): 649.

Mayer, K.H., and Zinner, S.H. (1985). Bacterial pathogens of increasing significance in hospital -acquired infections. Rev. Infect. Dis.7: 371.

MMWR (1993). Revised classification system for HIV infection and expanded surveillance case: definition for AIDS among adolescents and adults. No. RR17.

Prusiner, S.B. (1982). Novel proteinaceous infectious particles cause scrapie. Science 216: 136-144.

Prusiner, S.B. (1984). Prions. Sci. Amer. (Oct.) 15: 45-57.

Prusiner, S.B. (1994). Biology and genetics of prion diseases. Ann. Rev. Microbiol. 48: 655-686.

Quagliarell, O.V. and Scheld, W.(1992). Bacterial meningitis: pathogenesis, physiology and progress. N. Engl. J. Med. 327 (12): 864-871.

Segretain, G. (1962). Pulmonary aspergillosis. Lab. Investigation. 11: 1046-1052.

Stevens, J.G.1989. Human herpesviruses: a consideration of latent stage. Microbiol Rev. 53: 318-332.

Tiollasis,P. and Buendia, M.A.(1991). Hepatitis B virus. Sci. Am. 264: 116-124.

Suggested Readings

1. Tortora, G.J., Funke, B.R., and Case, C.L. MICROBIOLOGY: An Introduction. The Benjamin/ Cummings Publishing Company, Inc., California, 1989.
2. Boyd, R.F. General Microbiology. Times Mirror/ Mosby College Publishing, Missouri, 1988.
3. Dubey, R.C., and Maheshwari, D.K. A textbook of Microbiology. S. Chand and Company Ltd, 1999.
4. Purohit, S.S. Microbiology: Fundamentals and Applications. Agrobios, (India).2003.
5. Ketchum, Paul. Microbiology: Concepts and applications. John Wiley, 1994.
6. Krivoshein, Y.S. Hand Book of Microbiology. Mir Publisher, Moscow, 1989.

REVIEW QUESTIONS

Long Questions

1. What are Microbes? Give a detail account of types of microbes and their occurrence in nature. Briefly describe the sources and reservoirs of Microbial agents.
2. Give a detailed account of tetanus and anthrax.
3. Discuss in brief the causal organism, types and control measures of Leprosy.
4. What is polio? Discuss in detail the hazards of polio and its control measures.
5. What do you mean by AIDS? Write in detail about HIV, working of immune system and control measures.

Multiple Choice Questions

1. In an acute disease the time between contact with the infectious agent and the appearance of the first symptoms is known as

   (a) Incubation period (b) Prodromal period

   (c) Acute period (d) Convalescent period

2. The outbreaks of disease that involves countries or are worldwide are called
   (a) Endemic (b) Latency
   (c) Pandemic (d) Epidemic
3. Which of the following is a disease caused by bacteria
   (a) Chicken pox (b) Flu
   (c) Mycetomas (d) Anthrax
4. Which of the following disease is a sexually transmitted disease
   (a) Tuberculosis (b) Gonorrhea
   (c) Botulism (d) leprosy
5. The WHO strongly recommends that all countries should adopt a program called directly observed treatment short-course (DOTS), whish is related with cure of
   (a) Diptheria (b) Tuberculosis
   (c) Tetanus (d) Malaria
6. HIV infection is not spread by
   (a) Sex with an Infected Person
   (b) Contact with Infected Blood
   (c) Eating food with an infected person
   (d) Mother-to-Child Transmission
7. Bovine Spongiform Encephalopathy (BSE) is caused by
   (a) Bacteria (b) Prions
   (c) Virus (d) Fungus
8. *inea pedis* (athlete's foot) is a kind of
   (a) Fungal disease (b) Viral disease
   (c) Bacterial disease (d) Protozoan disease
9. Which of the following disease have been completely eradicated from the World
   (a) Chicken pox (b) Rubella
   (c) Plague (d) Small pox
10. The malignant form of malaria is caused by
   (a) *Plasmodium vivax* (b) *Plasmodium ovale*
   (c) *Plasmodium falciparum* (d) *Plasmodium malariae*

Answers

1. (a)	2. (c)	3. (d)	4. (b)	5. (b)	6. (c)
7. (b)	8. (a)	9. (d)	10. (c)		

CHAPTER 47

Microbes as Biofertilisers

N. Mathur, J. Singh, S. Bohra, A. Bohra and A. Vyas*
Dept. of Botany, Jai Narain Vyas University, Jodhpur (Rajasthan)
E-mail: dranilvyas@scientist.com

For normal growth and development, plants need sufficient nutrients in proper balance. Sixteen chemical elements are known to be essential for plants. Except carbon, hydrogen and oxygen, rest 13 elements are taken up by plant from the soil has increased by over four times since 1950-51 putting four-fold pressure on soil. Although annual fertiliser ($N + P_2O_5 + K_2O$) consumption has also increased by over 12 million tonnes during the period, it was much below the annual nutrient removal by crops resulting in a yearly negative balance of about 10 million tonnes between nutrient removal and supply. Depletion of soil nutrients poses a major threat to sustainability of crop production and underlines the need for maintaining it by tapping other plant nutrient sources.

Some soil microorganisms play an important role in improving soil fertility and crop productivity due to their capability to fix atmospheric nitrogen, solubilise insoluble phosphate and decompose farm wastes resulting in the release of plant nutrients. The extent of benefit from these microorganisms depends upon their number and efficiency, which, however, is governed by a large number of soil and environmental factors. When the number and activity of specific microorganism is suboptimal, artificially multiplied culture of that microorganism called microbial inoculant or biofertiliser is used to hasten biological activity to improve availability of plant nutrient.

RESEARCH LANDMARKS

Although the beneficial effect of legumes in improving soil fertility was known since ancient times, the field of biological nitrogen fixation opened up with the discoveries by Boussingault and Hellreigel in 1886. Nobbe and Hiltner produced for the first time a laboratory culture of

*Corresponding author

Rhizobia under the name "Nitragin" in 1895. Starting with Rhizobia, a vigorous search for other N fixing microorganisms began and soon it was found that there were other non-symbiotic bacteria such as *Azotobacter, which* could fix atmospheric nitrogen. It was followed by a third group called blue green algae. Subsequently, microorganisms capable of solubilising phosphate were discovered. In 1970, a new group of bacteria called *Azospirillum* was identified. Vesicular-Arbuscular Mycorrhizae (V AM) are the latest introduction in the list of biofertilisers which mobilize phosphorus.

In India, systematic research on biofertilisers started with the first study of N.V. Joshi in 1920. This was followed by extensive research by Gangulee, Sarkaria and Madhok on the physiology of the nodule bacteria and its inoculation for increasing crop yields. Important milestones in production, development and promotion of biofertilisers in India are given in Table 1.

Table 1. Some milestones in research, production and promotion of biofertilisers in India.

Year	Events
1920	First study on Legume-*Rhizobium* symbiosis by N. V. Joshi
1934	Earliest documented production of *Rhizobium* inoculant by M.R. Madhok
1956	First commercial production of Biofertiliser
1957	Study on solubilisation of phosphate by microorganisms by Sen and Pal
1964	Spurt in demand of Biofertiliser for soybean particularly in Madhya Pradesh
1970	Scope for use of charcoal, lignite and FYM as alternate carriers to peat reported by V. Iswaran
1977	Use of ISI mark for *Rhizobium*
1983	Setting up of National Project on Development and Use of Biofertilisers Ministry try of Agriculture, Government of India
1986	Setting up of National facility for BGA Collection at IARI. New Delhi by Department of Biotechnology, Government of India
1990	Setting up of National Facility for *Rhizobium* Germplasm Collection at the Division of Microbiology, IARI, New Delhi by Department of Biotechnology, Government of India.

EMERGING NEED

One of the essential plant nutrients on which successful agriculture depends to a great extent is nitrogen. High crop requirement, susceptibility to gaseous and leaching losses has made nitrogen the most in demand fertiliser material followed by phosphorus. Although air has about 80% of nitrogen, plants cannot make use of atmospheric N. The discovery of bacteria which had the ability to reduce non usable (atmospheric nitrogen) into usable from (ammonical) had been a major breakthrough in the field of agricultural research. Of late, increasing attention is being paid to harness the potential benefits from these renewable sources of plant nutrients because of the following reasons:

- Depleting soil fertility due to widening gap between nutrient removal and supplies.
- Depleting feedstock/fossil fuels and increasing cost of fertilisers.
- Growing concern about environmental hazards.
- Increasing threat to sustainable agriculture.

CLASSIFICATION

A number of biofertilizers arc now available in India. Depending upon the nutrients provided, these can be broadly classified as follows:

Nitrogen Fixing Biofertilisers

After the advent of the acetylene test for assessing nitrogen fixing potential, rapid progress has been made in identifying nitrogen fixing organisms. Out of a large number of microorganisms possessing the property of nitrogen fixation, only a few such as *Rhizobium, Azotobacter, Azospirillum,* BGA, *Azolla,* etc. have been commercially exploited. The reaction in biological nitrogen fixation is essentially the same as in production of chemical fertilisers (Haber-Bosch process) i.e. the catalytic reduction of dinitrogen (N_2) to ammonia (NH_3).

Rhizobium Inoculant

Among biofertilisers, *Rhizobium* inoculants are of greater importance and are being gainfully exploited. Rhizobia have the ability to fix atmospheric nitrogen in symbiotic association with legumes and certain non-legumes like *Parasponia.* Rhizobia colonize the roots of specific legume, enter through root hairs, multiply and form tumor like growth called root nodule. A mature nodule consists of a central 'bacteroid zone' surrounded by several layers of cortical cells. The process of N-fixation is wholly dependent on the activity of the enzyme nitrogenase which is located within the bacteroids. The volume and number of bacteroids have a direct positive relationship with the amount of nitrogen fixed.

Rhizobium Specificity

Both leguminous plants and species of *Rhizobium* show a degree of specificity, only a particular species of rhizobia infects a particular group of legumes. Seven cross-inoculation groups are recognised on the basis of ability of rhizobia to form nodules with legume plants. The cross-inoculation grouping system is not perfect, since rhizobia have often been found to cross-infect between groups. The newly proposed classification of root nodule bacteria recognizes two genera- *Rhizobium* (fast growers) and *Bradyrhizobium* (slow growers). The detailed classification is given in Table 2.

Isolation and Identification

Rhizobium Inoculant

Leguminous plants are carefully uprooted and their roots are washed in running water to remove soil particles. Healthy intact and pink nodules preferably on the tap root are selected and cut out with a portion of the root. The nodules are surface sterilised by immersion in acidified 0.1 % $HgCl_2$ for 5 minutes and repeated washing with sterile water. The nodules are then dipped in 70% ethanol for 30-60 seconds followed by several washings with sterile water. Surface sterilised nodules are individually crushed in sterile water with a sterile glass rod and the resulting suspension is streaked on yeast extract mannitol agar (YEMA) plates containing congo-red, poured in 24 hours earlier to ascertain the sterility of plates and culture media.

Table 2. New system of classification of root nodule bacteria

Genus	Species	Biovars	Host legumes
Rhizobium	*leguminosarum*	a. Viceae	*Vicia*
		b. Trifolii	*Trifolium*
		c. Phaseoli	*Phaseolus*
		meliloti	Medicago
		loti	Lotus
		fredii	*Glycine*
BradyRhizobium	*Japonicum*	-	*Glycine*
BradyRhizobium	species	-	*Cicer, Cajanus, Vigna* etc.

Upon incubation at 28-30°C for 3 to 10 days depending on the nature of *Rhizobium* (slow or fast growing), the colonies of bacteria normally appear as white, translucent, glistening and elevated with entire margin. *Agrobacterium*, the most common contaminant of *Rhizobium* takes up the colour of Congored and appears red in colour.

There are several presumptive tests which are carried out to identify *Rhizobium* isolate. Rhizobia grow very poorly on glucose peptone agar medium, does not grow on Hofers alkaline broth (pH 11.0), and does not utilise lactose. But the best way of identifying rhizobia is to conduct plant infectivity tests and see if effective nodules are produced. A plant test is done under controlled and sterile conditions with assemblies ranging from small tubes to pots. Small test tubes are used for small seeded legumes like clover, berseem, moong and urd, whereas Leonard jar assembly or small pots are used for testing bold seeded legumes like chickpea, arhar and french bean. Agar-agar, vermiculite or sand can be used as the substrate for growing plants. Surface sterilised seeds of specific host are sown in appropriate assembly and inoculated with the bacterial isolate under test. The plants are watered at regular intervals with any N-free nutrient solution. After 5-6 weeks the plants are examined for nodulation, nitrogenase activity and dry biomass. Thus, an efficient strain for a particular crop can be selected. Such strains are then subjected to pot and field trials to assess their potential as inoculants.

After selecting efficient strains, it is necessary to lyophilize them to maintain their efficiency since sub-culturing at monthly intervals on artificial medium may lead to deterioration of the activity of organisms.

Mass Production

The Rhizobial Inoculant Industry has come a long way since its inception in 1895 when Nobbe and Hiltner introduced a laboratory culture of Rhizobia under the name 'Nitragin'. The last 100 years have witnessed number of developments towards the better exploitation of this technique. Carrier based rhizobial inoculants have almost completely replaced liquid cultures and agar slope cultures which were commonly used a few years ago.

For preparation of carrier-based inoculants the rhizobial broth culture must attain a minimum population of 100 cells/ml. It is then blended with the finely powdered (100-200 mesh sieves) and sterilized carrier. The sterilization of carrier is carried out in autoclaves at 15 lb psi for 3-4 hrs continuously or 1 hr each on three successive days. Blending is done by mixing of broth culture with sterilized carrier in trays either manually (with hands washed and smeared with alcohol) or by using a mechanical mixer. The broth is added equivalent to one-third of water holding capacity of the carrier.

After mixing the carrier with the broth culture, raw-blended carrier is kept for 24 hours for curing. During this time the rhizobia get acclimatized with carrier. After curing, the inoculant is packed in the polyethylene bags. The polyethylene bags are sealed by an electric sealer leaving about two-third vacant space to give proper aeration to the inoculant. Directions for use of inoculant are printed on the packet.

Carriers for *Rhizobium* Inoculants

The medium in which rhizobia are allowed to multiply is an important factor in rhizobial culture preparation. The carrier is used for this purpose. The carrier is a crucial ingredient in the formulation of inoculant because the shelf life of the products largely depends on it. A variety of carriers are in use at present while some are under testing. Some materials whose usefulness as carrier has been established and used as carriers are given in Table 3. Even though typical peat is not available in large quantities in our country, the peat like material available in Nilgiris has been found to be a good carrier. Lignite available from Neyveli Lignite mines in South India is another carrier material widely used. The characteristics of good carrier are:

- It should be indigenously available and cheap
- Should be non-toxic to the biofertiliser bacteria
- Should have around neutral pH
- Should have an Organic matter content around 40%
- Should have high water holding capacity and surface area.

Method of Inoculation

A 10% sugar solution or 10% solution of pharmaceutical grade of gum arabic is prepared to serve as sticker for *Rhizobium* cells seeds. This solution is sprinkled on the seeds required for one hectare and the seeds arc spread on a polyethylene sheet and mixed uniformly. The content of the inoculant packet are sprinkled over sticker coated seeds and mixed thoroughly. If required more water is added but only in sufficient amount to coat the seeds fully with the

Table 3. Carriers used for *Rhizobium* inoculants in India

Carrier	Availability	Location	Water holding capacity (%)	Organic matter (%)
Peat	5.5 milliontonnes (total)	Ootacamund (Nilgiri Hills)	52-120	21-56
Lignite	3 milliontones (annually)	Neyveli	92	28
Farmyard manure	Plenty	Everywhere	40-50	30
Charcoal	Plenty	Everywhere	90	77

inoculant. The seeds are spread uniformly for drying on gunny bag or cement floor in shade avoiding direct sunlight. After drying the sowing is done immediately.

In order to protect the rhizobia from adverse situation of soil conditions (acidity, alkalinity, hazard of pesticides, etc.), pelleting method of seed inoculation is recommended. Finely ground calcium carbonate (300 mesh sieve) is added to freshly inoculated wet seeds in a container and mixed rapidly for 2 minutes and until seeds are uniformly coated. Coated seeds are allowed to harden for an hour by spreading on a clean surface. Besides lime, other pelleting agents like gypsum, bentonite, rock phosphate, charcoal etc. have also been recommended.

Crop Response

The largest contribution of BNF to agriculture is derived from the symbiosis between legumes and species of *Rhizobium*. The *Rhizobium*-legume association can fix upto 50-100 kg N/ha in one crop season and in certain situations leave substantial N for the following crop. The quantity of N fixed varies with the kind of Legumes and agro-climatic conditions. Crops like groundnut, chickpea, green gram and soybean can leave behind 20-30 kg N/ ha for succeeding crops, while crops like Lucerne, pasture legumes and green manuring crops like sunnhemp, *dhaincha* add about 50 kgN/ ha.

The field response to *Rhizobium* inoculation is quite variable. The average response to *Rhizobium* inoculation ranged from 66 kg/ha at Hisar to 152 kg/ha at Badanpur in chickpea. In general, most types of legumes are reported to respond to *Rhizobium* inoculation but not at all locations. Studies carried out in All India Coördinated Agronomic Research Project taking 5 legumes (moongbean, urdbean, soybean, cowpea, groundnut) indicated significant response at 7 out of 17 locations. Residual effect of *Rhizobium* inoculation on succeeding cereal crop was noticed in 4 cases only. Average response to *Rhizobium* inoculation ranged from 6 % in soybean to 18 % in cowpea. A residual response of 1 to 3 % was noticed on succeeding wheat crop.

Factors Affecting Crop Response

Although legume inoculation is a long established practice to maintain soil-health, it has not become popular largely because of its erratic response. Some of the common reasons of poor or no response to inoculation are:

(a) Failure of germinate: This is due to the use of poor quality seed and lack of sufficient moisture for germination.

(b) Failure of seedlings to become nodulated: This may be the result of poor quality inoculant, adverse soil conditions, inoculation with wrong inoculant, compatibility of rhizobia with pesticides, etc.

(c) Ineffective nodulation: This situation may result from the presence of a large native population of rhizobia which may compete with rhizobia applied for nodule formation and makes it ineffective in fixing N with introduced legume.

(d) Nutritional deficiencies: Nodule formation is severely affected by nutritional deficiencies particularly phosphorus, calcium and molybdenum because of their role in protein synthesis, cell wall development and in nutrition of rhizobia, respectively.

(e) Plant competition: Failure of the legume plant to promptly nodulate and establish is frequently due to competition for moisture, nutrients, light and space.

Azotobacter Inoculant

Azotobacter is one of the most important non-symbiotic N-fixing micro-organism and considered to be very important for fixation of nitrogen in non-leguminous plants. Based on morphological and physiological features, the genus *Azotobacter* has been classified in 3 species i.e. *Azotobacter chroococcum, A. beijerinckii* and *A. vinelandii*. The first two species are deemed to be the most commonly occurring species in India. *Azotobacter chroococcum* appeared in acid soil while *A. beijerinckii* in neutral and alkali soils.

Isolation

Azotobacter spp. are isolated by soil dilution plating method. Soil sample of 10 g is transferred in 100 ml of sterilised water and serial dilution of the suspension are prepared by further dilutions using sterile distilled water. Petriplates and pipettes are sterilised. Anyone of the nitrogen free agar media specific for *Azotobacter* is prepared and poured into petriplates. One ml aliquot of appropriate dilution are spread evenly over cooled and set agar medium in petriplates. The plates are incubated at 30° C in an incubator. After 3-4 days of incubation, soft, milky and mucoid colonies of *Azotobacter* are formed on agar plates. Direct isolation of *Azotobacter* has been practiced by Russian Scientists. Lumps of soils are spread out on a nitrogen-free agar medium and the plates are incubated at 28°C. After three days of inoculation *Azotobacter* colonies develop on the agar plates.

Another method is to prepare a paste of soil in nitrogen free liquid medium. This paste is incorporated in a layer of washed charcoal placed in a petriplate. Soil is aerated by burying a glass tube in it. After incubation of the plates at 28° C for 3-4 days, small *Azotobacter* colonies, if present, develop on the soil surface. This method is popularly known as 'soil plate' method.

Mass Production

Mass production of *Azotobacter* for preparing inoculants in India is done either by growing the culture in large flasks on a rotary shaker flasks on a rotary shaker or in batch fermentors depending on the quantity needed. All precautions and sterilisation production in the case of *Rhizobium* are also observed in case of *Azotobacter*.

Method of Application

A slurry of carrier based culture is made with minimum amount of water and the seeds are inoculated by mixing with slurry. The seeds are dried in the shade and then sown. For transplanted crops, seedlings are dipped in the slurry for certain time (20-30 minutes) and planted immediately. For sugarcane, many packets of culture are recommended at regular intervals in the early stages of growth. Second and subsequent inoculations are made by pouring slurry near root zone. The inoculant is also mixed in FYM and broadcasted for the root zone.

Crop Response

The results from a large number of experiments conducted in last four decades, have shown a positive response to *Azotobacter* application on a wide range of crops like cereals, millets,

Table 4. Effect of *Azotobacter* Inoculation on crop yields

Crops Increase	Fertiliser supplementation (kg/ha)			% yield over control
	N	P	K	
Wheat	120	-	-	11
	-	-	-	10-30
Rice	120	60	60	23
Maize	-	-	-	34
Sorghum	-	-	-	15-20
Pearl millet	-	-	-	0-27
Onion	-	50	100	22
Tomato	-	-	-	2-24
Cotton (fibre)				
Irrigated	63	30	-	10-20
Non-irrigated	-	-	-	11-16
Sugarcane	-	-	-	24

vegetables, cotton, sugarcane etc. *Azotobacter* inocu-lation increased the crop yields by 0-30% (Table 4). The beneficial effect of *Azotobacter* is attributed to its N_2 fixing capacity and also the ability to produce growth promoting substances. The actual contribution under field conditions has been doubled due to its low population in soil. The intensive research efforts are required to develop the *Azotobacter* strains, which can multiply in soil and establish high populations.

Azospirillum Inoculant

The ability to fix nitrogen by certain spirilla (*Spirillum lipoferum*) was first recorded by Beijerinck in 1925 and later on confirmed by several workers using not only conventional micro-kjeldahl assay but also by the methods such as acetylene reduction technique and the isotopic enrichment method involving ^{15}N. Tarrand and his associates (1978) renamed this organism as *Azospirillum* (N_2 fixing *Spirillum*).

Major Species

There are four major species of *Azospirillum* viz. *A. lipoferum, A. brasilense, A. amazonense* and *A. seropedicae.* Of the Four species, first two species are commonly found. The first species is *A. lipoferum* commonly found in the roots of C^4 plants while the second one *A. brasilense* is found in C^3 plants in tropical and subtropical regions.

Isolation

Half centimeter long root pieces are surface sterilized in 70% alcohol for 3-5 seconds and washed repeatedly in phosphate buffer (pH 7.0). These root pieces are then placed in semi-solid nitrogen-free sodium malate medium. The tubes are incubated at 35°C for 48 hours.

Characteristic growth of *Azospirillum* is indicated by formation of white pellicle 2-4 mm below the surface of the medium. The isolates thus obtained are purified in two steps. Firstly, the isolates are enriched by repeated transfer into a semi-solid sodium malate medium in small screw cap tubes. These cultures are then subjected to dilution upto 10^{10}. From each dilution 0.1 ml is inoculation in to a fresh lube of sodium malate medium. The least dilution which gives good growth is found to be pure and used for further studies.

Mass Multiplication

For mass multiplication of *Azospirillum,* the organism is allowed to grow flasks containing malic-acid medium augmented with 1 % ammonium chloride and incubated at 35°C for 3 days. When there is good growth, the broth culture is mixed with carrier and the carrier-based culture is then packed in small polyethylene packets. Soil and FYM in the ratio of 1:1 sterilized for 3 hours consecutively for 3 days were found to be the best carrier wherein bacterium was able to survive upto 6 months with counts of 10^6 cells/g of carrier material.

Method of Inoculation

The most suitable method used for carrier based culture is the seed inoculation. The technique adopted for inoculating the seed with *Azospirillum* culture is same as that described in the case of *Rhizobium*. In case of transplanted rice, the dipping of roots of rice seedlings in the slurry of inoculant is practiced.

Conditions Needed for Successful Inoculation

Seed inoculation with effective strains should be done in shade and subsequently the sowing should be done on the same day. The field should be well saturated during rice transplanting and after transplanting, the flooding should be done only after 5-6 days. For upland crops, moisture at field capacity should be maintained at the time of seeding so that added inoculum may get congenial condition for its growth and development.

Crop Response

The *Azospirillum* inoculation has been found beneficial in a variety of crops including cereals, forage and other crops. The crop response, however, varies greatly with types of crop and variety, location, season, soil fertility level, native microorganism interaction, etc. Crop response to *Azospirillum* inoculation is attributed to its ability to fix atmospheric nitrogen and to produce phytohormones.

A number of field experiments conducted on various crops to find out crop response to *Azospirillum* inoculation observed a wide variation in increase in yield (0-64%) over control (Table 5). Significant increase in yield was found at about, 50% of the locations. On an average, an increase of about 2-3 q/ha in crop yields and an economy of 20-30 kg N/ha equivalent is possible by *Azospirillum* inoculation. Among various field crops, sorghum and millets respond better than others under rainfed conditions. In Tamil Nadu, rice responds very well to *Azospirillum* inoculation.

Table 5. Effect of *Azospirillum* inoculation on crop yields

Crops	N-fertiliser (kg/ha) supplementation	% yield increase over control
Wheat	80	16-20
Rice	40	3-17
Maize	120	14
Barley	40	17
Sorghum	-	23-64
Pearl millet	-	0-37

Blue-green Algae Inoculant

As early as 1939, De suggested that fertility of tropical rice fields was largely due to activity of the N-fixing BGA. Blue-green algae are photosynthetic nitrogen fixers i.e. they use energy derived from photosynthesis to fix atmospheric nitrogen. They are also called cynobacteria and are free living organisms. In general BGA show dominance over the other classes of algae in India.

Distribution

The principal genera consisting of maximum number of nitrogen fixing species are *Aulosira, Nostoc, Anabaena, Tolypothrix, Sytonema, Plectonema, and Cylindrospermum.* Contrary to general belief that BGA are present invariably in tropical rice soils, the distribution of soils harbouring N fixing BGA varies from as low as 4% to as high as 80%. Species of *Nostoc* and *Anabaena* were ubiquitous in all regions in varying proportion and formed the major constituent of the algal biotypes in 9 out of 16 states. An all India survey has indicated that out of 2213 soil samples examined, only 33% were found to contain useful N fixing BGA. This, indicate the need for inoculation of efficient strain of BGA.

Isolation

BGA are generally covered with mucilage and it is easy to locate them as colonies on rice fields. Isolation of BGA involves the transfer of an algae from its natural habitat to medium where it can be grown without any contamination.

Mass Production

Currently, the technology for mass production of soil-based algal inoculants is well developed and in active use in India. Production involves a simple process with a few modifications to enable the production of algae both at small and large scale. The four common methods of algal production are:

(a) Trough method
(b) Pit method
(c) Field method
(d) Nursery cum algae production method.

Based on the natural ecology of these algae; a simple rural oriented open air method of producing them in bulk has been developed at the IARI, New Delhi. The basic principle is to grow them in natural sunlight under conditions simulating those in the rice field. The starter culture – a soil based mixture of *Aulosira, Tolypothrix, Nostoc, Anabaena* and *Plectonema* is supplied by the Indian Agricultural Research Institute, New Delhi. The details of multiplication from the starter culture are as follows:

Shallow tray (15 cm × 7.5 cm × 22.5 cm) of galvanized iron sheet, or brick and mortar, or pits lined with polythene sheets are prepared. The size can be increased if more material is to be produced. About 10 kg soil is placed and mixed with 200 g superphosphate. The trays are then filled with water (5-15 cm) depending upon the local conditions and rate of evaporation; the pH of the soil should be around neutral. After soil settles down saw-dust and the starter culture are sprinkled on the surface of the standing water. The whole assembly is allowed lo expose to sun. In hot summer months, the growth of the algae will be rapid and in about a week, a thick algal mat will be formed on the surface of the soil and sometimes even float up. If the daily rate of evaporation is high, more water is added intermittently. When the algal growth becomes sufficiently thick, addition of water should be discontinued and the water is allowed to dry up in the sun. The dried algal flakes are collected from the surface and stored in bags for future use in the fields. The trays are again filled with water and a small amount of the dry algal flask is added, as further inoculum. The process is continued as above. Once the soil in the tray is exhausted (usually 3-4 harvests), fresh soil is put and mixed with superphosphate and process is repeated as before. To prevent the breeding of insects, application of Folidol (0.001 ppm) or Malathion (0.00075 ppm) or Cabofuran (3% granules) is recommended.

Method of Field Application

Algae are applied @ 10 kg/ha over the standing water in the field one week after transplantation. The field is kept water-logged at least for a couple of days immediately after algal application.

Recommended pest control measures and other management practices generally do not interfere with the establishment activity of these algae in the fields.

The sun-dried algal material can be stored for long and used in the field. The algal material should not be stored in direct contact with chemical fertilisers or other agricultural chemicals.

Crop Response

The results of the field trials conducted in India have indicated about 10-15 per cent increase in rice yields with BGA application. The response of BGA application is better in dry season than in wet season and long duration rice varieties show greater response compared to short duration ones. Use of mixed BGA inoculation is superior to unialgal inoculation. Among different strains, agronomic potentiality of *Aulosira* is well established in Indian condition. Grain yield measureme-nts suggest that algalization produces a cumulative residual effect. Comparison of N fertiliser and BGA indicates that BGA inoculation is equivalent to the application of 20-30 kg- N/ha. Besides nitrogen, these algae are known to provide growth promoting substances.

Table 6. Performance of BGA on farmers' fields

Treatments	Grain yield (kg/ha)		
	Vellore (11 trials)	Trichy (8 trials)	Cuddalore (10 trials)
$N_{50}P_{25}K_{25}$	4235	4557	3789
$N_{50}P_{25}K_{25}$ + BGA	4455	4735	3884
$N_{75}P_{37.5}K_{37.5}$	4400	4485	3953
$N_{75}P_{37.5}K_{37.5}$ + BGA	4565	5117	4156

Investigations have shown that HYVs of rice respond to BGA inoculation even in presence of moderate level of N application (Table 6). The observed BGA effect in presence of fertiliser may be due to the supply of growth promoting sub-stances to the crop as well as to the N-fixed by the organism.

Factors Affecting Crop Response

The factors like the shortage of P in soil, the presence of combined nitrogen in flood water, presence of predators of BGA and use of herbicides in rice often limit the growth and activity of cyanobacteria. The non-availability of BGA inoculants at large scale due to scarcity of water and high temperature during June in North West India, and cubmbersome method of multiplication are the two major hinderances in their wider adoptability. The development of BGA which multiply fast and could be inoculated directly in the field may go a long way in the popularization of its use.

Azolla-Anabaena Symbiosis

Azolla is a aquatic fern, commonly found floating in ponds, pools, tanks, shallow ditches and channels. It is also found in rice fields. Nitrogen fixing Blue-green Algae (BGA) *Anabaena azollae* is found in the cavities on its dorsal leaves and fixes atmospheric nitrogen efficiently which is available to *Azolla* plants. The recognition of *Azolla* as an organic manure in rice cultivation was first shown in North Vietnam in 1957. *Azolla* contains 0.2–0.3% nitrogen on fresh weight basis and 3–5% nitrogen on dry weight basis.

Classification

About 90 strains belonging to seven species are being maintained at CRRI, Cuttack, which is the premier centre for *Azolla* research. Among the species, *Azolla pinnata* is most widely distributed in India.

Mass Production and Application

Azolla can be conveniently grown round the year. However, the multiplication of *Azolla* is rapid in winter season than in summer. The occurrence of standing water and application of phosphorus are essential requirement for rapid spread of *Azolla*. In India fresh weight of *A. pinnata* increased 2-6 fold in a week.

Two methods of *Azolla* application have been advocated. One is as a green manure by amending it in fields prior to rice cultivation and the other is dual cropping with rice when the fern is grown along with principal crop for sometime. The rate of inoculation is usually higher for multiplication plots than for dual cropping. The inoculum of 0.1-0.3 kg/square metre is recommended for multiplication plots and 0.5-1.0 kg/ square metre (0.5-1.0 t/ha) for dual cropping.

In green manure crop, the fields are prepared for rice cultivation, flooded with water followed by seeding with *Azolla*. After 5-10 days, the water of fields drained off and *Azolla* mat is ploughed into the soil. The process may be repeated for two crops of *Azolla*. In dual cropping *Azolla* is grown simultaneously with rice seedlings after transplanting. Normally, it is inoculated after one or two weeks of transplanting and field is flooded with water. When *Azolla* forms a mat on water surface, water is drained off and *Azolla* mat is incorporated into the soil. The *Azolla* mat required repeated incorporation to avoid chocking of rice plants due to *Azolla* mat.

Crop Response

Studies carried out in India have shown that *Azolla* green manuring produces 10-20 t/ha of green matter containing 20-40 kg N, whereas its dual cropping provides 20-30 kg N/ha. The supply of N can be further increased by raising more than one crop of *Azolla* , first as green manure and then as a dual crop along with rice. *Azolla* green manuring followed by dual cropping is equivalent to that from 60 kg N /ha as fertiliser. Combined use of *Azolla* and N fertiliser give better results. Besides other management practices grain yield response varies greatly with *Azolla* species and strains. On an average, the increase in grain yield due to *Azolla* green manuring ranges from 0.5-2.0 t/ha, whereas dual cropping produces additional yield of 0.5-1.5 t/ha over control.

Factors Affecting Response

The growth of *Azolla* and crop yield are significantly influenced by water availability, fertiliser application, rate and time of inoculation, method of rice planting and insects, pests and diseases. Water is prerequisite for multiplication of *Azolla*. *Azolla* cannot withstand desiccation and cannot be used in upland rice. Water depth of 5-10 cm is maintained during *Azolla* multiplication. *Azolla* requires at least an application of 20-30 kg P_2O_5 /ha. Split application of fertiliser P helps in producing higher *Azolla* biomass. Soluble P Sources are found to be more effective than rock phosphate.

The amount of inoculum significantly affects the growth rate of *Azolla*. Under poor growing conditions and in areas of excessive algal growth, higher inoculum rate should be used. *Azolla* inoculation in early stages of rice growth is preferred because a delay reduces the fern growth due to increased shading of rice. Temperature higher than 34°C + 1°C during day time is detrimental for growth. *Azolla* is often attacked by insects and the entire crop is damaged within a few days, if they are not controlled timely. Pests like *Pyralia* and snails (*limnae acuminata*) and diseases particularly black rot caused by *Rhizoctonia solani* are serious ones affecting *Azolla* multiplication. To control pest damage, insecticide application like Carbofuran is usually recommended.

PHOSPHATE MOBILISING BIOFERTILISERS

The efficiency of phosphatic fertilisers is very low (15-20%) due to its fixation in soil. Besides, native soil phosphorus is mostly unavailable to crops because of its solubility. The introduction of efficient P solubilizers in the rhizosphere has been found to increase the availability of phosphorus from both applied and native soil P (Fig. 1). The phosphorus which gets fixed in soil is made available to crops by the action of microorganisms which solubilize insoluble/ fixed forms of phosphates into forms which are readily taken up by the plants. Another factor which affects the effective utilisation of P is its low mobility in soil. Fortunately some filamentous organisms like *vesicular arbuscular mycorrhizae* (V AM) which form symbiotic association with plant roots have the ability to mobilise phosphorus from soil and thereby helping in absorption of P by plant roots. Commercial exploitation of phosphatic biofertilisers can play an important role particularly in making the direct use of abundantly available low grade phosphate possible.

The first Posphatic biofertiliser under the name of 'phosphobacterin' incorporating *Bacillus megaterium* var. *phosphaticum* was prepared at the then USSR.

Phosphate Solubilisers

Phosphate solubilising microorganisms (PSM) include various bacterial, fungal and actinomycetes forms, which help to convert insoluble/phosphate into simple and soluble forms. Species of *Pseudomonas, Micrococcus, Bacillus, Flavobacterium, Penicillum, Fusarium, Sclerotium* and *Aspergillus* are some of the PSM. Among these, soil bacteria belonging to the genera *Pseudomonas* and *Bacillus* and fungi belonging to the genera *Penicillium* and *Aspergillus* are more common. The solubilization effect of PSM is generally due to the production of

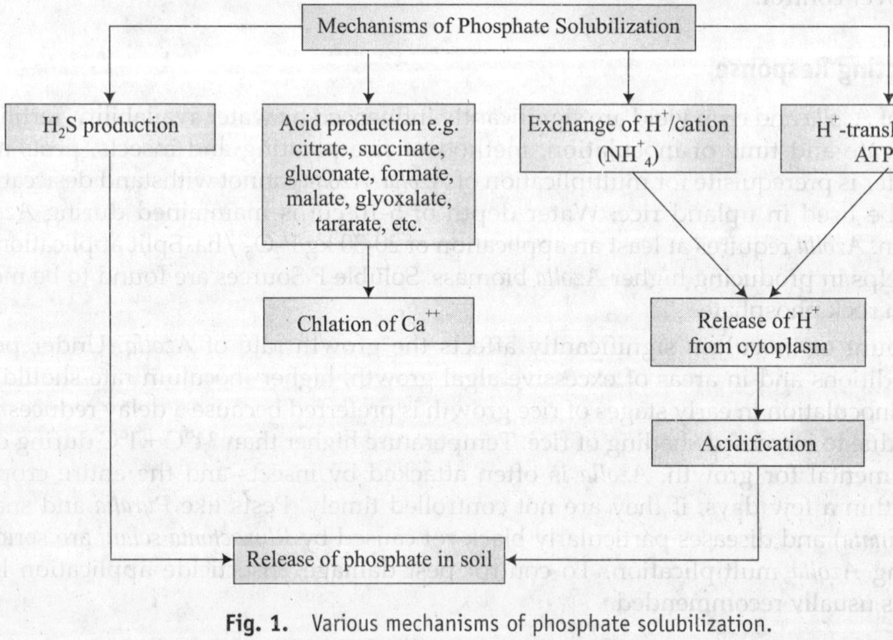

Fig. 1. Various mechanisms of phosphate solubilization.

organic acids by these organisms. They are also known to produce amino acids, vitamins, growth promoting substances like Indole Acetic Acid (IAA) and Gibberalic acid (GA) which help in better growth of plants.

Isolation

PSM can be isolated from rhizosphere and non-rhizosphere soil and also from other environments such as phosphate deposits and marine environments. PSM vary in their efficiency in solubilizing insoluble phosphates. The initial isolation of phosphate solubilizers is made by using a medium suspended with insoluble phosphates such as tri-calcium phosphate. The production of clearing zone around the colonies of the organism is indication of the presence of P-solubilising organisms. Such cultures are isolated, identified and the extent of solubilization determined quantitatively. It has been observed that the capacity of bacteria isolated to solubilise phosphate depends on the Zone of their origin. Those derived from rhizosphere have the highest, the rhizosphere one has intermediate and non-rhizosphere has the least capacity.

Several rock phosphate dissolving bacteria, fungi, yeast and actinomycetes were isolated from soil samples collected from rock phosphate deposits and rhizosphere soils of different leguminous crops. Many isolates solubilize a very high quantity of rock phosphate. The most efficient bacterial isolates were identified as *Pseudomonas striata, Pseudomonas rathonis* and *Bacillus polymyxa* and fungal isolates as *Aspergillus awamori, Penicillium digitatum, Aspergillus niger* and *Sacchromyces occidentalis*. These efficient micro organisms have shown consistently their capability to solubilize chemically-fixed soil phosphorus and rock phosphate from different sources – Mussorie, Udaipur, Matoon, Singhbhum, Morocco, Gafsa and Jordan. In addition, these microorganisms were found to mineralize organic phosphorus to soluble form due to enzymatic activity. By developing improved techniques for isolation of rock phosphate solubilizing microorganisms and by systematic investigations new efficient bacteria such as *Pseudomonas striata, Bacillus polymyxa* and fungi like *Aspergillus awamori* have been selected for preparation of carrier-based inoculant in addition to several basic studies. This preparation is known as 'IARI Microphos Culture' is now being mass produced and made avail-able to farmers.

Crop Response

Inoculation of seedlings with phosphate solubilizing organisms resulting in increased uptake of P and crop yields was first shown in 1948 by Gerrtsen. Since then beneficial influence of artificial inoculation with phosphate dissolving organisms has been reported for different crops under diverse agroclimatic conditions. In India, the microphos culture containing *Pseudomonas striata* and *Bacillus polymyxa* was tested in multilocational field trials With wheat, rice, chick pea, soybean, lentil and potato (Table 7). Results showed that inoculation of seed and seedlings increased grain yields of crops. The use of cultures increased the efficiency of rock phosphate and superphosphate applied to neutral to alkaline soils.

The role of phosphate solubilising organisms in increasing crop yield is subjected to controversy due to inconsistent results. Of the total inoculation experiments conducted so far, a significant increase in yield of 5 to 10% has been observed only in about 25% experiments.

Table 7. Effect of microphos Inoculant (phosphobacteria) on grain yield of crops in multi-location trials

Crop	Number of trials	Per cent Increase due to Inoculant	
		P_o	RP_{60}
Paddy	7	10-20	5-15
Wheat	8	7-42'	5-50
Gram	3	10-30	15-25
Lentil	2	-	13-15
Potato	2	30-50	

P_o - Without phosphorus; RP_{60} - 60 kg P_2O_5 /ha as rock phosphate

Several factors such as carbon and nitrogen sources, rock phosphate concentration, particle size of rock phosphate, pH, temperature, aeration, incubation period, etc. affect the efficiency of phosphate solubilizers. In general, the response to phospho-bacterins was found to be positive only in soils with high organic matter content and low in available phosphorus.

Vesicular-Arbuscular Mycorrhizae (V AM)

The role of fungi particularly those selectively colonising the roots surface of growing plants roots in increasing the availability of phosphorus has received scanty attention. These organisms are known not only to mobilise phosphates but also perform the function of transferring phosphorus from soil into roots as observed in the case of typical ectotrophic mycorrhizae. The term mycorrhiza refers to the association of fungi with roots of higher plants. Mycorrhizae are indeed the chief organisms involved in nutrient uptake of most land plants. V AM occur in most plant species in the two families of major economic importance, the leguminoseae and Graminae. They are virtually ubiquitous, being present in tropical, temperate and arctic regions. Their impact in tropical agriculture will be greater because in the tropics phosphate-deficient and phosphate-fixing soils are wide- spread, phosphatic fertilisers are in short supply and the temperature is more congenial for their activity.

V AM fungi inoculated to crop plants colonize the plant root system and increase the growth and yield of crop plants. The improved plant growth is due to increased nutrient uptake particularly P, Zn and other micronutrients, production of growth promoting substances, tolerance to drought and salinity and resistance to plant pathogens (Mathur and Vyas, 2000a).

Trials conducted with the locally isolated *Glomus macrocarpum* on chilli showed the possibility of saving 50 % phosphatic fertiliser without affecting the yield levels. Increase in the nutrient uptake and yield of various crops in India and elsewhere has been common with V AM inoculation (Mathur and Vyas, 2000b). Table 8, 9 and 10 shows the effect of mycorrhizal inoculation on some crop yields (Singh et al., 2005; Bohra et al., 2005). However, it has not been possible to exploit this organism on commercial scale due to difficulties in culturing the organism on laboratory media.

Table 8. Effect of mycorrhizal (VAM) inoculation on crop yields

Crop	P-fertiliser supplementation (Kg P_2O_5 / ha)	% yield increase over control
Finger millet	19	18
Soybean	25-50	19
Chilli	37.5	55
Chickpea	40	25
Groundnut	-	10-20

Table 9. Biomass production and nutrient uptake in *Trigonella foenum-graceum* by different arbuscular mycorrhizae.

S. No.	Treatment	Plant biomass (Fr. wt. g. plant^{-1})	Phosphorus (mg^{-1} dry wt.)	Nitrogen (mg g^{-1} dry wt.)
1.	Control	7.2	0.355	3.20
2.	*Gigaspora gigantea*	8.0	0.372	3.35
3.	*Glomus ambisporum*	8.5	0.384	3.46
4.	*Glomus fasciculatum*	9.2	0.398	3.59
5.	*Scutellospora nigra*	9.7	0.424	3.82
6.	*Sclerocystis coremioides*	10.4	0.441	3.97
7.	*Acaulospora laevis*	10.8	0.453	4.08
8.	*Glomus macrocarpum*	11.2	0.484	4.36
	CD at 5 % level	0.5	0.02	1.2

Table 10. Nutrient uptake and productivity of *Dolichos biflorus* as effected by different arbuscular mycorrhizae.

S. No.	Treatment	Phosphorus (mg^{-1} dry w.)	Nitrogen (mg g^{-1} dry wt.)	Plant dry weight (g)	No. of pods per plant
1.	Control	0.282	2.54	12.5	16
2.	*Acaulospora morrawae*	0.296	2.66	15.4	18
3.	*Scutellospora calospora*	0.392	3.52	17.6	22
4.	*Glomus constrictum*	0.400	3.60	18.7	25
5.	*Sclerocystis rubiformis*	0.483	4.35	19.6	27
6.	*Glomus aggregatum*	0.500	4.50	21.0	29
7.	*Gigaspora margarita*	0.519	4.67	22.4	31
8.	*Glomus desrticola*	0.562	5.06	25.2	36
	CD at 5 % level	0.02	1.2	2.2	-

BIOFERTILISERS: PRESENT POSITION AND FUTURE PROSPECTIVES

Nitrogen, Phosphorus and potassium are the three major plant nutrients. Fertilisers are chemicals supplying nutrients to plants. Fertiliser industry greatly depends on petroleum reserves and it is expected that in the next few decades, these reserves will be almost exhausted.

Further use of chemicals is not only deleterious to environment, but injurious to human and soil health. Under these circumstances, alternate sources for fertilisers are looked for. The current population of India is nearly 100 million. The current gross cropped are is 190.6 million hectares and the food grain production is 190.3 million tonnes.

With the current rate of population increase it is projected that India will have a population of 1500 million by 2025 A.D. To feed this population we must produce 310 MT of food grains. The scope for increasing the gross cultivated area (horizontal expansion of production) is going to be marginalised the most. It may increase to 200 m ha (addition of 10 m ha from now) through reclamation of ravines, saline alkaline, waterlogged and other categories of cultivable wasteland. That means with limited available gross cultivable area we have to produce more food per hectare. This points out that our food production per hectare (currently is 1.3 t/ha), in the next 30 years must be doubled to feed the bugling population i.e. we should have vertical expansion of production. For this purpose use of external inputs, especially optimum use of fertiliser (off farm,) supplement with farm manures, composts, city wastes, green manures and biofertilisers (on farm) is indispensable.

It is estimated that annually 25.1 MT of nutrients are removed from the soil. Whereas, only 15.0 MT are supplied from soil sources including organics. The fertiliser (NPK) production and requirement, in million metric tones of our country is given in Table 11.

Table 11. Fertiliser requirement and production in India- Projections (Million tonnes)

Year	1996	2000	2011	2031	2051
Requirement	16.0	19.0	20.2	27.3	31.3
Production	12.8	14.9	15.8	20.9	23.9
Gap	3.2	4.1	4.4	6.4	7.2

Table 1 brings out that the fertiliser production in out country is less than the required amount. To fill this gap alternate sources of nutrients have to be looked for. Organic wastes and biofertilisers are the alternate sources to meet the nutrient requirement of crops and to bridge the future gaps. Further, knowing the deleterious effects of using only the chemical fertilisers on soil health, use of chemical fertilisers supplemented with organic wastes and biofertilisers will be an environmentally benign approach to nutrient management and ecosystem function. Such integrated approach will help to maintain soil health and productivity and thus should be advocated and popularized among the farming community (Bagyaraj and Suvarna, 1999).

Biofertilisers are preparations containing micro-organisms insufficient numbers helping plant growth and nutrition. Biofertilisers are widely accepted as low cost supplements to chemical fel1ilisers and have no deleterious effect either on soil health or environment.

Biofertilisers are known make a number of positive contributes in agriculture (Motsara et., al., 1995).

For example,

- They supplement fertiliser supplies for meeting the nutrient needs of crops.
- They can add 20-200 N / ha (by fixation) under optimum conditions and Solubilize/ mobilize, 30-50 kg P_2O_5 / ha.

- They liberate growth promoting substances and vitamins and help to maintain soil fertility.
- They suppress the incidence of pathogens and control diseases.
- They increase crop yield by 10-50%, N-fixers reduce depletion of soil nutrients and provide sustainability to the farming system.
- They are cheaper, pollution free and based on renewable energy sources.
- They improve soil physical properties, tilth, and soil health in general.

Biofertilisers, which can supply nitrogen and phosphorus, two major plant nutrients have been developed (Verma, 1993). Organisms capable of fixing atmospheric nitrogen like *Rhizobium, Azotobacter* and *Azospirillum* are commercially produced as biofertilisers to supply nitrogen and phosphate solubilizing organisms are mass multiplied as biofertilisers aiding phosphorus nutrition (Tilak and Saxena, 1996).

Biofertiliser demand and availability

The projected production of different biofertilisers in our country is given in Table-12 (Singh, 1993).

Table 12. Target of production of different biofertilisers by 2005 A.D.

Biofertilisers	Quantity (tonnes)
Rhizobium	14.460
Azotobacter	7 ,230
Azospirillum	7 ,230
PSM	7,230
BGA	3.015
Total	39.165

Taking the cropped area in our country to be around 200 mha, the demand for bio-fertilisers at present (2005 A.D.) should be :

- For 1 m ha, biofertiliser required is = 500 T
- For 200 m ha, biofertiliser required = 200 x 500 = 1,00,000 tonnes.

It would be oversimplification to multiply the entire cropping area by 0.5 kg biofertiliser/ ha to compute demand for biofertilisers of the country. From a realistic point of view, it is better to accept 25% of the theoretical figure (i.e.1,00,000 tonnes) as the present annual biofertiliser requirement in India i.e. 25,000 tonnes/annum. The present annual production of biofertiliser in the country is around 15,000 tonnes (Bhattacharya and Kumar, 2002). Thus the annual requirement and present production capacity for different biofertilisers shows not only a tremendous gap, but also indicates the potential for organized production (Venkataraman and Tilak, 1990).

Constraints in Biofertiliser Technology

At present there are several constraints in biofertiliser production and its commercialization. They may be physical, chemical, Biological, technological, infrastructural, financial, market related or concerned with human resource development.

Physical and Environmental Constraints

In summer months in some parts of India temperature is very high. Care must be taken to avoid transportation of biofertiliser packets in direct sunlight. Storage of biofertilisers should not be done in containers/sheds where the temperature becomes very high. While making packets care must be taken to see that the carrier material has enough moisture (about 40%). If the carrier material is very dry, population of the organisms will decrease rapidly.

Chemical Constraints

Highly acidic or saline soils will adversely affect the population of introduced biofertiliser in acidic soil, pelleting of seeds (after treating with biofertiliser) with lime and in saline soils with gypsum can be recommended. In soils P availability is less than 10 ppm, and application of P fertiliser is essential without which the nitrogen fixing biofertilisers will not function effectively.

Biological Constraints

Soil flora and fauna can parasitize or devour biofertiliser added to soil. Bacteriophages can destroy rhizoba added to soil. Some protozoa like *vorticella* and nematodes like *Meloidogyne* can act as predators of biofertiliser added to soil. *Azolla* is subjected to attack of pests, especially larvae of various Lepidopterous and Dipterous insects and Aphids. Cynobacteria can be attacked by phages, myxobacteria and grazers like snails and mosquito larvae.

Technical Constraints

Raw material: Peat and lignite which are good sources of carrier materials are available only in very limited places like Nilgiris and Neyveli, thus resulting in high cost of transportation to biofertiliser manufacturing units.

Strains: suitable strains of biofertilisers for different crops have been identified through research. Many produces do not pay attention to this aspect and use whatever strains they get, resulting in poor quality biofertilisers.

Suitable technology: In India biofertiliser production is generally done with non sterile carrier, which encourages considerable contamination. Suitable technology to minimize contamination should be developed, ex. need for gamma irradiation of the carrier material, injection of liquid cultures into polythene bags containing sterile carrier material etc.

Recently, there is some interest to produce liquid biofertilisers. They are special liquid formulation containing not only the desired microorganisms and their nutrients but also special cell-protectants or chemical that promotes formation of resting spores or cysts for longer shelf life and tolerance to adverse conditions (Bhattacharya and Kumar, 2002).

The liquid biofertiliser of good quality holds great promise for following benefits over the carrier material, saving on carrier material, transport, pulverization and sterilization, convenience in handling, storage and transportation and better performance (Hegde, 2002). The use of liquid biofertiliser is the still at research and development stage and more experiments are needed before accepting this technology.

Lack of quality assurance: Substandard quality of inoculants is one of the most important factors resulting in failure at the field and lack of farmer's confidence in the product. The bureau of Indian standards prepared 151 specifications for *Rhizobium* and *Azotobacter* inoculants in 1977. Revised specification for *Rhizobium* and *Azotobacter* and specification for *Azospirillum* and Phosphobacteria have been formulated recently.

Quality control consists of strain testing and selection service, and mixing of broth before incorporation into the carrier materials like lignite, charcoal, peat, vermiculite etc., for the carrier-based culture after maturation and control of culture during the period of storage up to final broth stage, so each individual manufacturer has to do his own quality control.

The BIS has no network to test biofertilisers samples. They send samples to Agricultural Universities or Research Institutes in order to make ISI mark. Once a manufacturer gets in ISI mark, they continue it without routine checking. This amounts to sub-standard product reaching the farmer. It is suggested that there should be separate testing laboratories with adequate infrastructure and man power to check the quality of biofertilisers at various stages of production, marketing and application. Perhaps this can be done through strengthening National Biofertiliser Development Center and its Regional Centers. More than anything, there should be honesty in the production units to produce quality product, especially as it is going to because mostly by marginal farmers.

Infrastructural Constraints

Many production units do not have adequate space and other infrastructure facilities including electric supply needed for quality biofertilisers.

Human Constraints

Some biofertiliser production units do not have technically well-qualified (Microbiologists) and trained persons who can produce high quality biofertilisers. It has been experienced that some organizations depend more of non-skilled laborers working on contract basis, which leads to sub-standard biofertilisers.

Most of the farmers in the country do not have sufficient and clear knowledge on the use of biofertilisers. The farmers need to be educated. In order to educate farmers, educating and training extension workers is utmost important. This can be done through demonstration trials on the cultivator's fields. The impact is greater and long lasting when farmers see the beneficial results for themselves on their lands.

Marketing Constraints

This includes weakness in the marketing network, retail outlet, storage facility etc. There is no standardization in packing, labeling and prices of biofertilisers. The size of the packet varies

Table 13. Some suggestions for distribution agencies.

Do	Don't
• Keep biofertiliser packets away from direct heat and sunlight. • Store it in cool and dry place. • Sell only biofertiliser packets which contain batch number. • If the expiry period of the packet is over, discard it. It will not be effective.	• Don't store biofertiliser packets under heat and sunlight. • Don't sell biofertiliser packets after expiry period is over. • Don't holes into the packets or puncture them· Don't store heavy items above biofertiliser packets.

from 75 to 500 g/packet. Some manufacturers engage private dealers with lucrative trade discount. A uniform capacity of 200 g packet is recommended so as to facilitate the designing of packing machine that can be manufactured and marketed allover the country .Some suggestions for distribution agencies are given in Table 13 :

Incentives/Support/encouragement from the Government

Government of India has sanctioned the National project on Biofertilisers. One of the important component of this project is to encourage people to set up biofertiliser production units. For this, there is a provision to provide financial assistance to the state government, fertiliser or seed companies, state institutional agencies, well-established private companies, NGOs etc.

The Government of India also gives national productivity award to the best organization for encouraging biofertilisers production in the country. "Mission Mode Project" was initiated in 1993 with and aim to demonstrate the use of biofertilisers under integrated nutrient management system.

Establishment of National Bureau of Agriculturally Important Microorganisms for collection, documentation, cataloguing, conservation and domestication of microorganisms important in agriculture including biofertilisers at New Delhi.

FUTURE RESEARCHABLE AND POLICY ISSUES

Researchable issues

- Development of efficient strains of different biofertilisers through natural selection and genetic engineering.
- Enhancement of shelf life of the produce.
- Development of easy techniques (could be some intrinsic or introduced tag) for quality assurance.
- Cost effective methods for mass- producing arbuscular mycorrhizal fungi.

Policy issues

- Mass awareness about the use of biofertilisers.
- Provision for subsidy on biofertilisers.

- Strengthening financial support to public or private sectors for production of biofertilisers.
- Training of farmers on the use of biofertilisers.
- Standardization of quality control aspects on biofertilisers.

SUMMARY

Biofertilisers are biologically active (living or temporarily inert) inputs and contain one or more types of beneficial microorganisms such as, bacteria, algae or fungi. Every microorganism and hence each type of biofertiliser has a specific capability and function. Nitrogen fixing biofertilisers include—rhizobia, azotobacter, azospirillum, blue-green algae and azolla. Phosphate mobilising biofertilisers include phosphate-solublising microorganisms and mycorrhiza. Biofertilisers on application remain in soil, multiply and benefit crops. They do not get depleted as chemical fertilisers. They are cost effective, environment friendly and improve soil health. Their effect on Plant growth may not be spectacular as that of chemical fertilisers, being biological material. Their use is of great significance in developing countries because even the marginal farmers can afford to buy and use them. Hence it is essential that the manufacturer of biofertilisers adhere to the quality control protocols in the production units ensuring a quality. Care should also be taken to overcome other constraints coming in the way of biofertiliser technology. We should recognize that in adopting a rational approach to the use of management of biofertilisers in sustainable agriculture, the microbial labour force holds a vast potential for the future.

REFERENCES

Bagyaraj, D.J and Suvarna, C. V. (1999) Biofertilisers and their application. *Agrieast*, 1: 16-21.

Bhattacharya, P. and Kumar, R. (2002) Liquid biofertiliser—current knowledge and further prospect. Proceeding National Seminar on Development and Use of Biofertilisers, Biopesticides and Organic Manures, Bidan Chandra Krishi Viswavidyalaya, Mohanpur, pp.10-21.

Bohra, S., Singh, J., Mathur, N. and Vyas, A. (2005) Increased nutrient uptake and productivity of *Dolichos biflorus* by mycorrhizal fungi. *J. Arid legumes* 2(1): 187-189.

Hegde, S.V. (2002) Liquid biofertilisers in Indian agriculture. *Biofertiliser newsletter* 10: 17-25

Mathur, N. and Vyas, A. (2000a). Mycorrhizal dependency of *Tamarix aphylla* in Indian Thar Desert. *Naturalia*, 25: 105-110.

Mathur, N. and Vyas, A. (2000b). Influence of arbuscular mycorrhizae on biomass production, nutrient uptake and physiological changes in *Ziziphus mauritiana* Lam. under water stress. *J. of Arid Environ.*, 45: 191-195.

Singh, J., Mathur, N. Bohra, S., Singh, R. and Vyas, A. (2005) Mycorrhizal dependency of *Trigonella foenum-gracecum* in Indian Thar desert. *J. Arid legumes* 2(1): 190-192.

Tilak. KV.B.R and Saxena, A.K. (1996) Biofertilisers - their role in crop production. *Fertl. industry*, 19; 187-194.

Venkataraman, G.S. and Tilak, K.V.B.R. (1990) Biofertilisers in sustainable agriculture in: soil Fertility and Fertilisers, Vol. IV, Nutrient Management and supply System for Sustaining Agriculture in 1995, Kumar, V. Shortiya, G.C. and Kavre, S.V. (EDS), India Farmers Fertilisers Co-operative Ltd., New Delhi, pp. 132- 148.

Verma. L.N. (1993) Biofertilisers in agriculture. In: Organics in soil health and crop production, Thampan, P.K. (ed). Peekay Tree Crops Development foundation, cochin, pp. 151- 184.

Suggested Readings

1. Motsars, M.R., Bhattacharyya, P. and Srivastava, B. (1995) Biofertiliser Technology, Marketing and usage, Fertiliser Development and consultation organization, New Delhi.
2. Prospects and potential of Biofertilisers: A mission mode approach. Proceedings of National Conference on Biofertilisers and Organic Farming, Madras (1996).

REVIEW QUESTIONS

Long Questions

1. What is a carrier (for bacterial inoculants) ? Write the characteristics of a good carrier and types of carrier for *Rhizobium* inoculant in India.
2. Define the process of isolation and mass production of *Azotobacter* ?
3. What are the conditions required for successful inoculation of *Azospirillum* ? What is the response of *Azospirillum* inoculation ?
4. Write about the four common methods of mass production of BGA. How BGA is multiplied from the starter culture (mass production) ?
5. What are the factors affecting the crop response of *Azolla* ?
6. Write a short note on V AM.
7. Write a note on the future prospects of biofertilisers.
8. Write a detail note on – "Constraints in biofertiliser technology".

Multiple Choice Questions

1. Nobbe and Hiltner were the first to produce a laboratory culture of *Rhizobium* under the name.
   (a) Rhizin (b) Nitragin
   (c) Rhizobactin (d) Nutrien
2. It is the most common contaminant of *Rhizobium,* it takes the colour of congored & appear red in colour.
   (a) *Brady Rhizobium* (b) *Azotobacter*
   (c) *Agrobacterium* (d) *Azospirillum*
3. *Azotobacter* _______ appears in acid soil while *A.* _______ in neutral and alkaline soil.
   (a) *beijerinckii* and *vinelandii* (b) *chroococcum* and *beijerinckii*
   (c) *vinelandii* and *chroococcum* (d) *beijerinckii* and *chroococcum*
4. *Azospirillum* inoculation is attributed to ability of.
   (a) N_2 – fixation and production of phytohormones
   (b) N_2 – fixation and inhibition of other algal/fungal growth.
   (c) P – Solubilization and production of growth promoting substances.
   (d) P – Solubilization and inhibition of other algal/fungal growth.

5. In India the premier centre for *Azolla* research is
   (a) NBPGR, Delhi (b) CPRI, Shimla
   (c) CCRI, Cuttack (d) IARI, New Delhi
6. First phosphatic biofertiliser was named as.
   (a) Phosphobacterin (b) Rhizo-phosphin
   (c) Bactino fertilizer (d) Bacterio-phosphate
7. V AM occur mostly in two plant / crop families of major economic importance.
   (a) Leguminoceae and Brassicaceae (b) Brassicaceae and Graminae
   (c) Brassicaceae and Fabaceae (d) Leguminoceae and Graminae
8. In soil P-availability is less than ______ and application of P-fertilizer is essential without which N_2 - fixing biofertilizer will not function effectively.
   (a) 10 ppm (b) 20 ppm
   (c) 25 ppm (d) 5 ppm
9. Project initiated with the aim to demonstrate the use of biofertilizer under integrated nutrient management system.
   (a) "Mission Fertilizer" (b) "Bio-organic Land"
   (c) "Mission-Mode Project" (d) "Project-Organic Fertilizer"
10. *Azospirillum* ______ commonly found in roots of C^4 plants while *A.* ______ is found in roots of C^3 plants.
    (a) *lipoferum* and *brasilense* (b) *brasilense* and *lipoferum*
    (c) *amazonense* and *seropedicae* (d) *seropedicae* and *amazonense*

Answers

1. (b) 2. (c) 3. (b) 4. (b) 5. (c) 6. (a)
7. (d) 8. (d) 9. (c) 10. (a)

CHAPTER 48

Mycotoxins and their Effects

N.K. Bohra[1*], P. Gehlot[2] and D.K. Purohit[2]

[1]Arid Forest Research Institute, New Pali Road, Jodhpur–342 005

[2]Dept. of Botany, J.N.V. University, Jodhpur–342 001

E-mail: nkbohra@rediffmail.com

What we eat actually depends on what we get and not what we want. Approximately 20% food becomes unsafe before reaching the consumers. Moulds are known to produce a wide variety of biologically active compounds on Agricultural and other consumable commodities. Human poisoning by fungal metabolites caused due to consumption of contaminated foodstuffs has been a matter of serious concern in different parts of the world. The deleterious effects of eating poisonous mushrooms are known since the down of civilization but it was only in the 18th century when the people could realize the effect of certain filamentous fungi (moulds) as serious health hazards to human beings. Such fungi may be saprophytic or pathogenic. Fungi produce a vast and bewildering array of toxic metabolites, which abound in organic environments (Turner, 1971). The magnitudes of their effects depend upon the nature of the producing fungus and the situation under which it interacts with other organisms. Toxins play a pivotal role in determining inter and intra species relationships among the fungi. Some of the fungal toxins, which are mostly secondary metabolites, are highly toxic to the vertebrates and termed as mycotoxins.

Human poisoning by fungal metabolites caused due to consumption of contaminated foodstuffs has been a matter of serious concern in different parts of the world. Natural contamination of agricultural produce with mycotoxins is comparatively higher in underdeveloped and developed countries because of socioeconomic backwardness and faulty, unhygienic, outdated agricultural and storage practices. In India, high temperature, fairly high relative humidity, unseasonable rains and floods prevails in different eco-regions of the country at the time of harvesting of major crops. This situations leads to excessive moistening of the grains resulting in intense mould growth and mycotoxins elaboration. Mycotoxins are fungal secondary metabolites formed by consecutive series of catalytic reactions from a few

*Corresponding author

simple intermediates of primary metabolism. These mycotoxins can enter the human and animal food chain by direct or indirect contamination in mycotoxins is equally harmful to animal as well as human beings. Realizing the importance of mycotoxins to the health of man and animals there have been concerted efforts to develop highly sensitive analytical methods to determine mycotoxins in food, mixed feed and feed ingredients, animal tissue, blood, urine and milk. Most of the mycotoxins can now readily be identified qualitatively and quantitatively. Most of the researches are concentrated on increasing sensitivity, accuracy, reproducibility and reduction of the time of determination.

The term mycotoxins is derived from Greek words *Mykes* meaning fungus and Latin word *toxicum*- meaning poison. Mycotoxins represent a wide variety of chemical compounds produced as secondary metabolites by some toxigenic strain of fungi during their development on food and feed materials. The acute disease or toxic syndrome that results from the ingestion of mycotoxins contaminated foods and feeds are called mycotoxicosis. The term toxin stands for a poisonous substance elaborated from microbe, bacterium, plant or animal. However, the concept of toxin differs in different disciplines. Toxin in human pathology is a protein of microbial origin and has got antigenic properties. In plant pathological test, the term toxin is commonly used for low molecular weight microbial products which cause distinct damage to plant tissue and are involved in disease development.

Knowledge on mycotoxins is very important particularly to developing countries because of the health problems they cause. Several human and animal diseases have been shown in the past to be associated with the consumption of mycotoxins contaminated foods and feed. Mycotoxins are produced by fungi that grow on grain and other seeds in preharvesting or in storage condition. About 160 mould species are known to produce 182 mycotoxins and other metabolites (Frisvad *et. al.*, 1991).

International society for human and animal mycology (ISHAM) at its meeting held in Newzealand in 1982 defined mycotoxins as a metabolite of micro- fungus which when ingested by a natural group in sufficient concentration will cause ill effects in animals including birds and man. This definition includes both primary and secondary metabolites as the distribution between the primary and secondary metabolites is not always clear. This definition excludes the compounds, produced by mushrooms and other higher fungi, compounds toxic only to plants, micro-organisms and insects and compounds produced *in vivo* by a fungal pathogen in an animal. Therefore, the term "Mycotoxins" may be defined as "A group of chemically unrelated secondary fungal metabolites which are detrimental to living organisms and cause illness and deaths in humans and animals and in certain cases are harmful to green plants also. These chemical compounds are produced by certain toxigenic moulds during their development on food and feed materials. The toxic syndromes induced by the ingestion of mycotoxins contaminated foods and feeds are referred to as primary "Mycotoxicoses". Mycotoxins may also pass through the food chain into animal products such as milk or meat and illness caused as a result of the consumption of these products is termed as secondary mycotoxicosis.

HISTORY

The first clearly recognized episode involving mycotoxins in human toxicosis was that of ergot poisoning though the relationship between ergotism and casual fungus (*Claviceps purpurea*) which infected the rye ears but was not conclusively established till the 18th century. The ergot alkaloids are derivatives of lysergic acid. Toxicologically these compounds are oxytoxics i.e., they stimulate the smooth muscles of the uterus. They are also weak vasoconstrictors. The last outbreak of ergot poisoning is described by Gabbai *et al.* (1951) in France where about 200 people got ill due to consumption of moulded rye which was used as flour and bakery product. Authentic reports about mould produced toxins in cattle (1924), Stachybotryotoxicosis in horse (1931), *Fusarium* toxicity in swine (1936) and mouldy corn poisoning in horses (1940) started appearing in scientific literature since the early part of the last century. Maximum concern was felt in the post Second World War period when the disease outbroke in human being in Russia, alimentary toxic aleukia (ATA). This disease was first recorded in eastern Siberia in 1913 but its deleterious effects were not fully realized at that time. Russians eating mouldy (due to infestation of *Fusarium/Cladosporium*) with overwintered and Snow covered grains suffered from several dermal necrosis, leucopenia, haemorrhage, rashes and destruction of bone marrow. In acute cases nasal, gastric and intestinal haemorrhages occured and necrotic lesions developed in throat, on the lips and skin of the nose, jaws and the eyes. Subsequently, some interesting observations were made and it was found that over wintered sample of cereals retained their toxicity even after seven years of mould infestation. The culture filtrates of *Fusarium* and *Cladosporium* grown at low temperature were more toxic than those maintained at room temperature. Highest toxicity was associated with materials collected at the time of maximum spore production (Joffe, 1983). In certain areas, mortality was up to 60%.

Historically the year 1960 was very important because of the scientific concern generated about the mycotoxins. During that year, about one lakh beautiful game birds Turkey perished in England within a short time without leaving any trace about the cause of death. This mysterious disease greatly baffled the veterinary scientists who were unable to find out the causes or cure for the disease. Since it was difficult to give any scientific name to the disease, it was designated as "Turkey × disease". It was however soon realized that the sudden doom of turkey poults was due to consumption of meal served to them. Systematic analysis of the meals was, therefore, initiated but it was found to be free of bacteria, viruses or any other known toxic chemicals. The affected birds expressed loss of appetite; letharginess, weakening of the wings and drying of legs etc. It was found that the meal served to the `turkey birds' was prepared of peanuts, which were actually imported from Brazil by a ship named Rosette. The meal was thus nick named as "Rosette meal".

A vigorous search by British microbiologists led by Sergeant concludes that the meal which had killed the birds was heavily infested by the fungus *Aspergillus flavus link ex fries* (Sergeant *et. al.*, 1961). This fungus was very well distributed in nature and its conidia are widespread in air and soil. Unable to chemically characterize the nature of toxin, they named the toxic factor as aflatoxin (Sergeant *et. al.,* 1961). Several mycotoxins are elaborated on cereals, pulses, oil seeds, dry fruits, spices, green fruits, vegetables, dried fish and shrimps, milk and mild products, different types of meat as well as wide variety of other consumable articles. Most of these toxic substances find their way to human system through different routes. These harmful chemical are in one food chain but unfortunately their long range effects have not been properly evaluated.

MYCOTOXIN PRODUCTION AND ITS EFFECTS

Mycotoxins are produced by a wide range of fungi but the main producers belong to the genera viz. *Aspergillus, Fusarium* and *Penicillium*. Some of the important mycotoxin producing fungi and their toxic metabolites implicated in the outbreaks of mycotoxicosis are given in Table 1. The mycotoxins producing fungi are strongly aerobic and require optimum quantity of oxygen, the seed moisture, humidity, temperature, length of storage incubation etc. The other factors are also responsible for mycotoxins productions. On biological effects of mycotoxins are as follows:

Table 1. Various toxins produced from fungi and their biological effects.

S. No.	Toxins	Fungi	Biological effects
1.	Aflatoxins	*Aspergillus flavus*	Hepatotoxic
		A. parasitious	carcinogen
2.	Ochratoxin	*A. ochraceus*	Nephorotoxin
3.	Citrinin	*Pencillium viridicatium*	Nephratoxin
		P. Citrinum	
4.	Patulin	*P. Patulum*	Death of Cattle
5.	Trichthecenes	*Fuxarium tricinctum*	Dermal necrosis
		F. solani	Haemorrhage
6.	Zearalenone	*F. graminearum*	Vulvovaginitis
		F. roseum	Abortion

(a) On plant system: It was found that mycotoxins can inhibit plant growth by restricting seed germination and seedling formation, minimizing chlorophyll synthesis and loosing down the protein and starchy contents of seeds. Mycotoxins also suppress the synthesis of RNA, DNA and also induce several abnormalities such as increase in number of chromosomes etc. In this way it decreases the nutrient content of plants and makes the plants part useless.

(b) On animal system: Several studies have been done in the past on human and animals. The food and feed stuffs contaminated by mycotoxins producing fungi and can cause several harms such as hemorrhagic rash, bleeding from nose, throat and gums, extreme leucopenia, exhaustion of the bone marrow etc. Among mycotoxins "aflatoxin" is most carcinogenic. The statistical analysis of human epidemiological studies reveled that the intake of potential aflatoxin (AFB) in higher dosage by susceptible human can result in liver cancer and cirrhosis. It is believed that aflatoxin being hepato carcinogenic may induce similar tumors, with variations in the type of cells involved. European Economic Community and Protein Advisory group have listed the tolerance limits of Aflatoxin B in some of the countries (Table 2). It is quite clear that the mycotoxins problem is a problem now a day all over the world.

MYCETISMUS VERSUS MYCOTOXICOSIS

Toxicity of fungi to man and animals can arise by mycetismus (mushroom poisoning) or by mycotoxicosis (Emmons *et al.,* 1977). Mycetismus occurs by ingestion of certain mushroom type fungi which can cause a toxic response as a result of: (a) protoplasmic poisoning (by

Table 2. Tolerance limit of aflatoxin in some countries

S. No.	Country	Commodity	Tolerance limit (Aflatoxin B1PPb)
1.	Japan	All foods	10
2.	Malaysia	All foods	0
3.	Poland	All foods and feeds	5
4.	Sweden	All foods and feeds	5
5.	USA	Consumer peanut products All other foods and feeds	1 (Total Aflatoxin) 20
6.	Canada	Nuts and nut products	15–40 (Total)
7.	Denmark Peanut and Brazil	nuts	5–10
8.	Denmark	Feeds	20–50 (E.E.C.I.Record)*
9.	India	Peanut Meal (Feed)	30 (Based on P.A.G. Record)**
10.	Italy	Peanut (Imports)	50
11.	Norway	Oil seed meal (Import)	600
12.	Rhodesia	Feed	50–400
13.	UK	Peanut meal	0–500
14.	Belgium	Peanut meal (export)	50
15.	France	Feed	20–50 (E.C.E. Record)
16.	Israel	All feeds	20
17.	Brazil	Peanut	50

*European Economic Community (EEC)

**Protein Advisory group. (PAG)

Amanita Phalloides, A. verne, A. tennifolia and *Galerina venenata*) (b) neurologic effects (by *Amanita pantherina, Inocybe, Clitocybe, Lepiota, Hebeloma*) and (c) gastrointestinal irritation (by *Lectarius terminosis, Tricholoma pardinum, Agaricus, Boletus, Clavaria, Russula, Clitocybe*). In contrast to mycetismus, mycotoxicosis is poisoning by ingestion of toxins of fungal origin in foods which have been altered as damaged by the growth of toxic producing mould fungi. Mycotoxins are in general low molecular weight, non antigenic fungal metabolite capable of eliciting a toxic response is man and animals. Mycotoxins are known to produce elicit effect on a variety of food commodities including agricultural crops. They enter into host by ingestion, inhalation or contact. Quite small amounts of the compounds can cause significant health hazards (WHO, 1979; Hayes, 1980). At high concentration many mycotoxins can produce acute disease syndrome, while at lower levels they can be carcinogenic, mutagenic, teratogenic or estrogenic and may reduce the growth rate of young animals. They can even interfere with the mechanism of resistance which impairs immunologic responsiveness making the animal more susceptible to infections (Pier, 1980).

The diversity of symptoms of mycotoxicoses association differs with the different fungal species. The main genera of fungi associated with mycotoxins include species of *Aspergillus, Penicillium, Fusarium, Claviceps, Alternaria, Pithomyces, Stachybotrys, Phoma* and *Diploidia*. Mycotoxins are fungal secondary metabolites formed by consecutive series of enzyme catalysed reactions from a few simple intermediates of primary metabolism e.g., acetate mevalonale, malonate and certain amino acids (Turner, 1971; Steyn 1980). The main biosynthetic reaction includes condensation, oxidation/reduction, alkylation and halogenation steps which create a remarkable range of secondary compounds. The principal pathways involved in the formation of mycotoxins are through polyketide (e.g., aflatoxins), terpene, (e.g., trichothecene), amino acids (gliotoxin) and tricarboxylic acid routes (rubrotoxin). Some mycotoxins (e.g., cyclopiazonic acid) are formed from a combination of two or more of the principal pathways (Turner, 1971).

SOURCES OF MYCOTOXINS

Mycotoxins can enter the human and animal food chain by direct or indirect contamination. In direct contamination the food material support the toxigenic mould growth. Possible routes for mycotoxin contamination of human and animal food may be as follows (Jarvis 1976).

(i) Agricultural produce e.g., cereals oil seeds (groundnut) fruits, vegetables.

(ii) Consumer foods (secondary infections) e.g., animal feeds (secondary infections), residues in animal tissue and animal products, milk and dairy produce meat.

(iv) Mould ripened foods e.g., cheese fermented meat products, oriental fermentations.

(v) Fermented products e.g., microbial proteins, enzymes, food additives, and vitamins.

Almost all foods are susceptible to mould growth at some stage during their production, processing, transportation and storage. In indirect contamination the food ingredient is contaminated with mycotoxins.

(A) Mycotoxins in Agricultural Crop

Cereals are highly susceptible to fungal growth when in the field, during processing or storage (Stoloff, 1976). Small grains (wheat sorghum, oats, rye, barley and rice), unless abused in storage or after preparation, appear to be less susceptible to mycotoxin formation than the larger grains such as maize. The main mycotoxins that have been detected in cereals in regular occurrence are the aflatoxins, sterigmatocystin, ochratoxin A,zearalenone, T2 toxin and vomitoxin (Scott, 1973; Stoloff, 1976; Bennett and Shotwell 1979; Pathre and Mirocha, 1979; WHO, 1979; Anon., 1980; Vensonder and Hesseltine, 1981. The incidence of occurrence varies with climatic conditions prevailing at the time of harvest, transportation and later storage conditions. During harvesting of the grains moisture plays an important role for mould colonization which may further leads to production of mycotoxins by secondary metabolites, e.g., Maize which is harvested at a high moisture level and therefore it is more susceptible to mycotoxins producing fungi. A list of mycotoxins in agricultural product is given in Table 3.

Realizing the importance of seed health, a lot of work has been done on the seed mycoflora of various agricultural commodities including cereals, pulses, millets, oil seeds spices and dry fruits. Seed mycoflora in medicinal plants as well as in food and feed were also reported by various workers (Bilgrami *et al.*, 1980; Chandra *et al.*, 1981; Srivastava and Gupta, 1984; Esteves, 1984; Abate and Abegaz, 1985; Basak and Mirdha, 1985; Kingsland, 1986; Hegde and Hiremath, 1987; Rudrignes and Thorne, 1990; Podder and Purohit, 1990; Frisvad and Samson, 1991; Kanchanlata, 1991; Kumar and Sinha, Handry and Mugene 1993; 1992; Ursula Bosch and Mirocha, 1992; Kumar and Roy, 1993; Weidenborner and Kunz, 1993; Dodan *et al.*, 1994; Kolhe *et al.*, 1994; Ranganathaiah, 1994; Rajkumari and Reddy, 1993; Nelson *et al.*, 1994; Sinha, 1994 Thirumaladevi, K. *et al.*, 2001; Reddy, S.V.2001; Moss, 2002, Szabo *et al.*, 2003).

Table 3. Natural occurrence of mycotoxins in food

S. No.	Mycotoxin	Commodity	Reference
1.	Aflatoxin	Maize	Krishnamachari *et al.*, 1975a Mishra 1977 Bilgrami *et al.*, 1981; Mall *et al.*, 1983
		Paddy	Sreenivasamurthy, 1965 Nusrath and Ravi, 1983
		Wheat	Mishra, 1977
		Sorghum	Bhadraiah and Ramarao, 1983
		Groundnut	Sreenivasmurthy *et al.*, 1965 Mehan and Mc Donald, 1983
		Cotton	Singh, Anjana, 1983
		Dry Fruits	Singh Anjana, 1983
		Spices	Singh Anjana, 1983
2.	Zearalenone	Maize	Bilgrami *et al.*, 1980
		Paddy	Nusrath and Ravi, 1983
		Dry fruits and spices	Singh, Anjana, 1983.
3.	Ochratoxin	Paddy	Nusrath and Ravi, 1983
		Dry fruits and spices	Singh, Anjana, 1983
4.	Citrinin	Groundnut	Subrahmanyam and Rao, 1974
		Dry fruits and spices	Singh, Anjana, 1983

(B) Mycotoxins Aflatoxicosis in Farm Animals

Toxicological studies on test animals indicate that farm animals will vary considerably in their susceptibility to aflatoxicosis. Some animals, notably rabbits, ducklings, trout, pigs, calves and turkeys are susceptible to the acute toxicity of aflatoxin, and the aflatoxin intake of these animals must be controlled very closely if heavy losses are to be prevented Other animals are more resistant, but can suffer chronic effects which manifest themselves in poor productivity and general unthrift ness. Some animals, whilst not showing symptoms of aflatoxicosis, produce toxic metabolites, which can enter the human food chain, so aflatoxin intake must again be controlled. Although many farm animals are susceptible to the carcinogenic effects of aflatoxin, this is rarely of importance as they are usually sacrificed before such effects would manifest themselves.

Only a relatively few in-field cases of aflatoxicosis are no record, and this is almost certainly because few farmers and veterinary officers are sufficiently aware of the aflatoxin problem, especially in the tropics where animal losses can be caused by a greater variety of factors. Even when aflatoxicosis is suspected, it is often impossible to prove the presence of aflatoxin in the feed, because the suspect feed has either been consumed completely, or sufficient has been consumed to make good sampling impossible.

There is evidence that aflatoxicosis in farm animals may be prevalent in areas where locally grown animal feeds are contaminated with aflatoxin. A survey of aflatoxicosis in farm animals was carried out over a 22-month period in North Carolina, USA, where locally grown maize is susceptible to aflatoxin contamination. Over the survey period 278 cases of suspected

Table 4. Mycotoxicosis associated with acute primary disease of live starle of poultry (Pier *et al.,* 1980)

S. No.	Mycotoxicosis	Animal Species	Primary Syndrome
1.	Aflatoxicosis	Poultry, swine, cattle, dog	Acute hepatitis, haemorrhagic
2.	Ergotism	Cattle, sheep, chicken	Gangenous, necrosis; nervous seizures; reproductive failure
3.	Facial eczema	Sheep, cattle	Cholangio hepatitis; photosensitivity
4.	Fusariotoxicoses		
	(a) vervitoxicosis	Swine	Emesis
	(b) T-2 toxicosis	Swine, cattle, poultry	dermonecrosis; gastro-enteritis
5.	Diacetoxy scirpenol	Swine	Oral, gastroenteric ecrosisharmorrhages
6.	Leukoencephalomalacia	Horses	Nervous depression, in coordination
	F-2 toxicosis (Zearalenone)	Swine	strogenism
	Ochratosicosis	Swine, turkeys	Nephropathy
		Cattle, sheep, horse	Ataxia,
	Paspalum staggers	cattle, sheep	salivation,diarrhoea, polyurea
	Slaframine toxicosis	Horses	Dermonecrosis,gastro-enteritis,
	Stachybotryo toxicosis	Cattle, sheep, dog	aematopoietic depression
	Tremorgen intoxication		Fasciculation; atoxia, prostration.

aflatoxicosis were studied, and of these 94 cases were confirmed on the basis of finding aflatoxin in the feed and also finding bile duct proliferation of the liver of affected animals. The mean concentration of aflatoxin in feed samples from the confirmed cases was 3890 ppb, and the mean value for the corn used in making the feed was 5180 ppb. There were 83 cases of aflatoxicosis in swine, 7 cases in cattle, and 4 cases in poultry. The associated symptoms were varied and included: gross liver damage, hemorrhage or oedema of various organs, elevated body temperature, mortality of up to 4-0 per cent, anoresis and unthrift ness.

Mycotoxins contaminated feeds fed to some animals, e.g. aflatoxin to dairy cattle, ochratoxin to pigs, can pose a threat to health because of the possible carry over of toxic residues into the animal products and thus into the food chain. Mycotoxicoses associated with acute primary disease of live starle of poultry is given in Table 4. Mycotoxicosis and epidimidogic data with human involvent is given in Table 5.

CHEMICAL PROPERTIES OF MYCOTOXINS

Mycotoxins comprise of a group of chemically unrelated fungal metabolites. These may be polypeptide, alkaloid, benzoquinone, anthraquinone, xanthone, coumarin, terpene and their various derivatives. Aflatoxins are a group of highly oxygenated heterocyclic bisfusarocoumarin metabolites. The main member of this group is designated as B_1, B_2, G_1 and G_2. The aflatoxin fluoresces strongly in ultra violet light (Ca, 365 nm). B_1 and B_2 produce blue fluorescence where as G_1 and G_2 Produces green fluorescence. B_2 and G_2 are dihydro derivatives of parent compounds B_1 and G_1.

Ochratoxins are a group of seven isocoumarin amides of 1 phenyl analine, among which ochratoxins A is most toxic and a dihydro isocoumarin carboxylic compound. Citrinin is a

Table 5. Mycotoxicosis in which analytic and/or epidemiologic data suggest human involvement (Hayes, 1980)

S. No.	Disease	Species	Substrate	Etolofic Agent
1.	Alimentary toxic Aleukic (ATA) (on Septic Angina)	Man	Cereal grains and Toxic Bread	*Fusarium sporotrichoides*
2.	Dendroclochiotoxicosis	Horse	Fodder (Skin-contact/inhaled)	*Fusarium sporotrichoides*
	Dendioduchium	Man	Fodder particles	*Fusarium sporotrichoides*
3.	Kashin Beck Disease 'Urov Disease'	Man	Cereal grains	*Fusarium sporotrichiella*
4.	Stachybotryotoxicosis	Man Horse and other livestock	Hay, cereal grains Fodder (Skin contact inhaled haydust)	*Stachybotrys atra*
5.	Cardiac beriberi	Man	Rice	*Fusarium species*
6.	Ergot	Man	Rye, cereal grains	*Claviceps purpurea*
7.	Balkan enphropathy	Man	Cereal grains	*Penicillium sp.*
8.	Rey's syndrome	Man	Cereal grains, groundnuts	*Aspergillus sp.*
9.	Hepatocarcinoma	Man	Cereal grains, groundnuts	*Aspergillus sp.*
10.	Pink rot	Man	Celery	*Sclerotinia sp.*
11.	Onyalai	Man	Millet	*Phoma Sorghina*

yellow crystalline chemical compound having p quinone methide structure. The presence of hydrogen bonding between the carboxyl group and both orthohydroxyl groups in citrinin molecule has been confirmed by x-ray diffractor. Zearalenone is an enone derivative of 6 B resorcyclic acid lactone.

DETECTION OF MYCOTOXINS AND CONTROL

Most of the main mycotoxins can now readily be identified quantitatively and qualitatively. Most of the recent research is related to increasing sensitivity, accuracy and reproducibility and to decrease the time of determination (Romer, 1976; Pohland *et. al.*, 1979) Accurate and reliable analytical methods are the corner stone for a sound understanding of all aspects of mycotoxicology. The major Steps involved in mycotoxins determination are given in Table 6. The time between sampling and analysis is critical and samples held in environmental conditions are fit for mould growth and can easily develop mycotoxins within one or two days. Among the analytic technique generally used HPLC (high performance liquid chromatography) is the most widely accepted analytic technique for detections of mycotoxin.

There is no doubt that the exclusion of mycotoxins from human and animal food chain would result in significant avoidance of human illness, reduced productivity, loss of food or feed, livestock mortality, reduced growth rate and feed efficiency in live stock, livestock infertility and reduced market price for commodities (FAO, 1977). Three basic strategies i.e. prevention, inactivation and detoxification have been adopted to control the adverse effects of

Table 6. Basic steps in chemical analysis of mycotoxins (after Romer, 1976)

S. No.	Step	Description	Purpose
1.	Sampling	Probe of automatic sampler	Representative sample
2.	Sample preparation	grinding, mixing, subsampling	Representative sample
3.	Extraction	Shaker or blender	Separate the toxins from Compounds insoluble in the extraction solution.
4.	Clean up	Liquid liquid partitioningSeparate the toxin from(separatory funnel) column chromatography Divalent metal clean up (Pb^{+2},Fe^{++},Cu^{+2})	groups of compounds in the *Fusarium* sample extract.
5.	Final separation	Thin layer chromatography (TLC) Gas liquid chromatography (GLC) Liquid liquid chromatography (LC) Mini column chromatography	Separate the toxin from remaining compounds in the sample extracts that might inter-fare with the toxins.
6.	Detection and quantization	Fluorescence on TLC Plate Fluorescence in solution U.V. absorption in solution GLC Flame detector	Detection and measurement of response
7.	Confirmation	TLC Separation and detection of derivative of mycotoxins Biological test Mass Spectrometry	Identification of chemical compound

mycotoxins. However low temperature, storage of grains treatment of grains with propionic acid and sodium salt during storage, radiation and UV treatment, treatment with aqueous plant extracts have been proved to be responsible for reducing or slowing down the growth of mycotoxins producing fungi and production of mycotoxins.

RECENT ADVANCES

Several workers all over the world have investigated the problem of mycotoxins in different food commodities. The world health organization (WHO) has fixed certain limits for mycotoxins in different food and feed stuffs. In India, scientists gave attention on mycotoxins problem and showed considerable awareness. The problem of mycotoxicosis has been reported from different parts of the country. The fungal seeds responsible for mycotoxicosis are available in air. They deteriorate the seed content and produce toxins too especially during storage. In Rajasthan people normally use alternative foods for their survival during famine and store them in traditional storage systems. These foods were found toxicated during a study and produce toxins (Bohra, 1997). In groundnuts both biotic (soil insects and soil pathogens) and abiotic (seasonal drought and soil temperature) factors during crop growth can cause aflatoxin contamination. The propagation of plants is accomplished by seeds and the fungal microbes may be systematically transferred from one generation to another through seeds (Campbell and White, 1995).

Study of aflatoxin structure revealed the relationship of toxicity with the presence of a double bond on the furan ring of aflatoxin. Aflatoxin B1 and G1 have this double bond and they are carcinogenic, whereas B2 and G2 lack the bond and are only weak toxic. These aflatoxins are absorbed from gut and then pass to liver where they become highly reactive. The epoxide bridge formed at the double bond on furan ring enables the toxin to bind to DNA and depurinate it causing genomic lesion that can lead to hepatomas.

Mycotoxin is a serious problem. All foods and feeds must be examined carefully for mycotoxigenic fungi. Mouldy or infected foods and feeds should be destroyed. The prevention of mouldy food is the best and only answer to prevent of mycotoxin production in food and feed. Consumption of these mycotoxins is bound to produce ill effects. To avoid mycotoxins, the simplest way is to check the growth of fungi responsible for toxin production. Education of farmers to improve the method of storage ,surveillance of food stuffs and animal feeds for the presence of mycotoxins and application of appropriate food processing technology to separate the contaminated food components from the clean ones are some of the appropriate measure to overcome the problem mycotoxin related health hazards.

SUMMARY

Human poisoning by fungal metabolites is caused due to consumption of contaminated food stuffs. It is a matter of serious concern in different parts of world. Natural contamination of food with fungal toxins is higher in underdeveloped and developing countries due to poor agricultural and storage practices. Mycotoxins are poisonous secondary metabolites of fungi that are produced on grains and seeds, which are detrimental to living organisms and cause illness and death in humans, animals and plants. The toxic syndromes induced by ingestion of mycotoxins contaminated foods are referred to as mycotoxicoses. Seed moisture, humidity, temperature and other factors are responsible for growth of mycotoxin producing fungi. The mainly occurring mycotoxins are aflatoxin, ochratoxin, zearalenore, T-2 toxin and vomitoxin. Most of the mycotoxin can be identified quantitatively as well as qualitatively. Recently detoxification of mycotoxin from food chain is a big challenge.

REFERENCES

Abate, D. and Abegaz, B. (1985) Prevalence of *Aspergillus flavus* in Ethiopian cereals grains a preliminary survey. Ethiopian Med. J., 23: 143-148.

Anon (1980) Survey of mycotoxins in the United Kingdom. Food surveillance paper no. 4 Ministry of agriculture, fisheries and food, London, Her Majesty's stationary office.

Basak, A. B. and Mdridha, A. U. (1985) Mycoflora associated with seeds of different varieties of rice collected from chittagong and chittogong hill tracts districts of Bangladesh. Seed Res., 13: 78-84.

Bennett, G. A. and Shotwell, O. L. (1979) Zearalenone in cereal grains. J. Am. Oil Chem. Soc., 56: 12-819.

Bilgrami, K. S., Prasad, T. S., Mishra, R. S., Mishra, K.K. and Sinha, K. K. (1980) Survey and study of mycotoxin producing fungi associated with grains in the standing maize crop. ICAR Final Report. Bhagalpur University, Bhagalpur.

Bohra, N. K. (1997) Survey and study of Mycotoxin producing fungi associated with alternative food commodities in Rajasthan Ph.D. thesis. J.N.V. University, Jodhpur, Rajasthan.

Campbell, K. W. and White, D. G. (1995) Inheritance of resistance to *Aspergillus* ear rot and aflatoxins production in corn genotypes. Phytopathol., 85: 886-896.

Chandra, S., Narang, M. and Srivastava, R. K. (1981) Studies on seed Myco flora of oil seed in India part-I, qualitative and quantitative estimation. Int. Bio. Bull., 17: 71-75.

Doden, D.S., Singh, R. and Sunder, S. (1994) Survival of Fusarium moniliforme in infected rice grains and its chemical control. Indian J. Myco. Pl. Path., 24: 135-138.

Emmons, C. W., Binford, C. H., Utz, J.P. and Kwon-Chung, J. (1977) Medical mycology, 3rd edition. Lea and Febiger, Philadellphia.

Esteves, A. (1984) Incidence of diseases under field conditions on maize (*Zea mays*) cultivars in Dourados Fitopatologia. Brasileira, 9:155-16.

F.A.O. (1977). Mycotoxins. Report of the joint FAO/ WHO/ UNEP conference on mycotoxins. Nairobi, 19-27 Sept, 1977, Rome. pp. 95-128.

Frisvad, J. C. and Samson, R. A. (1991) Filamentous fungi in food and feeds; ecology, spoilage and mycotoxin production. In: Handbook of applied mycology Vol. 3 (Eds., D.K. Arora et. al.), pp. 121-177, Marcel Dekker, New York.

Gabbai, Lisbonne and Poourquir (1951) Ergot poisoning at Pont st. Esprit, Brit. Med. J., 2: 650- 656.

Hayes, A. W. (1980) Mycotoxins : A review of biological effects and their role in human diseases. Clin. Toxicol., 17: 45-83.

Hedge, D. G. and Hiremeth, R. V. (1987) Seed mycoflora of cowpea and its control by fungicide. Seed Res., 15: 60-65.

Hendry, M. and Hugene, C. C. (1993) A. review of mycotoxins in Indoor Air. J. Toxicol. Environ. Hlth., 38: 183-198.

Jarvis, B. (1976) Mycotoxin in food. In: Microbiology in Agriculture, Fisheries and Food (Eds., F.A. Skinner and J.G. Carr), pp. 251-267, Academic Press, London.

Joffe, A.Z. (1983) Food borne disease-Alimentary toxic Aleukia. In : Handbook of foodborne disease of biological origin (Ed., M. Recheigle), pp. 353-495, CRC Press, Florida.

Kanchanlata, (1991) Seed mycoflora of some cereals grains from Gorakhpur. Proc. Natl. Acad. Sci., 61: 433-437.

Kingsland, G. C. (1986) Relationship between temperature and survival, *Aspergillus flavus* link and Fr. on naturally contaminated grain. J. Stored Prof. Res., 22: 29-32.

Kolhe, A. S., Verma, R. J. and Dube, H. C. (1994) Aflatoxin contamination in oil cakes. Indian phytopath., 47: 270-274.

Kumar, S. and Roy, A. K. (1993) Occurrence of aflatoxins in some liver. Curative herbal medicines. Lett. Appl. Microbiol. 17: 112-114.

Moss,O. Maurice(2002). Mycotoxin Review-2. Fusarium , Mycologist 16: 158-161.

Nelson Paul, E., Jean, H., Juba, P., Ross Frank and Rice Larry, G. (1994) Fumonisin production by *Fusarium* species on solid substrates. J. AOAC, 77: 522-525.

Kumar, N. and Sinha, K.K. (1992) Natural occurrence of mycotoxins in some stored food grains. J. Indian Bot. Soc., 71: 191-194.

Pathre, S.V. and Mirocha, C.J. (1979) Trichothecene : Natural occurrence and potential hazard. J. Am. Oil Chem. Soc., 86: 820-823.

Pier, A.C., Richard, J.L. and Cysewski, S.J. (1980) Implication of mycotoxin in animal disease. J. Amer. Vet. Med. Assoc., 176: 719-724.

Podder, S.R. and Purohit, D.K. (1994) Aflotoxion contamination of mustard and taramira seeds under storage in Rajasthan. Ann. Arid Zone, 33: 67-68.

Pohland, A.E., Thorpe, C.W. and Nesheim, S. (1979) Newer development in mycotoxin methodlogy. Pure Appl. Chem., 52: 213-223.

Rajkumari, D. and Reddy, S. M. (1993) Influence of water activity on seeds mycoflora of fennel. Ind. Phytopath., 46: 389-392.

Ranganathiah, K. G. (1994) Seed borne disease of minor millets. Vistas Seed Biol., 1: 251-261.

Reddy, S.V., Mayi, D., Kiran D., Reddy, Uma M., Thirumala Devi, K. and Reddy, D.V.R. (2001) Aflatoxin B1 in different grades of chillies (Capsicum annum L.) in India as determined by indirect competitive –ELISA. Food Add. Contam., 18: 553-558

Romer, T. (1976) Methods of detecting mycotoxins in mixed feeds and feeds ingredients. Feedstuffs, 12: 18-21

Rudrignes, B. F. and Thorne, S. G. (1990) Seed mycoflora of the Canavalia species in Goa. Ind. J. Mycol Pl. Pathol., 20:164-165.

Sargeant, K., Sheridan, A., Kelly, J.O. and Carnagham, R.B.A. (1961) Toxicity associated with certain samples of groundnuts. Nature, 192: 1096-1097.

Scott, P.M. (1973) Mycotoxin in stored grains. Feed and otehr cereal products. In : Grain storage-Part of a system (Eds., R.N. Sinha and W.E. Muir). Avi Publishing Co., WestPoint, USA.

Sinha, K.K. (1994) Mycoflora of sugarcane seed (Fuzz) and disease problems. Vistas Seed Biol., 1: 346-352.

Srivastava, S and Gupta, J.S. (1984) Seed mycoflora from Indian seed lots of cosmos bipinnatus and their control. Ind. Phytopath., 37: 83-87.

Steyn, P. (1980) The biosynthesis of mycotoxins, Academic Press, London.

Stoloff, L. (1976) Occurrence of mycotoxins in foods and feeds. In: Mycotoxins and other fungal related food problems.

Szabo,B.,Tyihak,E., Szabo G. and Botz, L. (2003) Acta Botanica Hungarica, 45: 409-417.

Thirumala devi K., Mayo, M.A., Reddy, Gopal, Emmanuel, K.E., Larondello, Yvan and D.V.R., Reddy (2001) Food Add. Contam.,18: 830-835

Turner, W. B. (1971) Fungal metabolites. Academic Press, London.

Ursula, Bosch and Mirocha, C. J. (1992) Toxin production by *Fusarium* species from sugar bets and natural occurrence of zearalenone in beets and beet fibres. Appl. Environ, Microbiol., 58:3233-3239.

Vensonder, R. F. and Hesseltine, C.W. (1981) Vomitoxin, natural occurrence on cereal grains and significance as a refusal and emetic factor to swine. Proc. Biochem. J., 16: 12-14.

Weidenborner, M. and Kunz, B. (1993) The mycoflora of stored cereal grains. Med. Facullteit Land Bauwweten schappen Rijks Universitea, Gent. 8.

WHO (1979) Environmental health criteria 11. Mycotoxins. World Health Organization, Geneva.

Suggested Readings

1. Medical Mycology 3rd edition. Lea and Febiger, Philadellphia, 1977.
2. The biosynthesis of Mycotoxins, Academic Press, London, 1980.
3. Fungal Metabolities. Academic Press, London, 1971.
4. Mycotoxins In Food and Feed, the Allied press Bhagalpur, 1983.
5. Mycotoxic Fungi, Mytoxins, Mycotoxicoses, Marcel Deker. New York, 1978.

REVIEW QUESTIONS

Long Questions

1. What do you mean by mycotoxins? Give a detailed list of mycotoxins derieved from different fungi?
2. Discuss about the mycotoxin effects of on plant and animal systems.
3. Give a brief account of mycotoxicosis.

4. Discuss the detection and biocontrol of mycotoxins in naturally occurring products.
5. Explain the sources and biological effects of aflatoxin and citrinin.

Multiple Choice Questions

1. Aflatoxin is produced from
   (a) *A. flavus* (b) *A. niger*
   (c) *F. solani* (d) *F. roseum*
2. The ergot alkaloids are derived from
   (a) Acetic acid (b) Lysergic acid
   (c) Aspartic acid (d) Glutamic acid
3. Which toxin is produced from *Fusarium roseum*
   (a) Ochratoxin (b) Citrinine
   (c) Patuline (d) Zearalenone
4. Etolofic agent for ergot disease is
   (a) *Penicillium* (b) *Aspergillus*
   (c) *Claviceps* (d) *Phoma*

Answers

1. (a) 2. (b) 3. (d) 4. (c)

CHAPTER 49

Antibiotics: Chemotherapeutic Agent

Dimple Songara[1], Sharad Bissa[2], Avinash Bohra[2] and A. Bohra[1*]

[1]Dept. of Botany, J.N.V. University, Jodhpur–342 001

[2]Mahila PG Mahavidyalaya, Jodhpur.

E-mail: profabohra@yahoo.co.in

Antibiotics (Greek *anti,* "against"; *bios,* "life") are chemical compounds used to kill or inhibit the growth of infectious organisms. Originally the term *antibiotic* referred only to organic compounds, produced by *bacteria* or *molds that* are toxic to other microorganisms. The term is now used to include synthetic and semisynthetic organic compounds. Drugs such as sulfonamides are called antibiotics, although they are synthetic chemotherapeutic agents, not microbially synthesized. *Antibiotic* refers generally to antibacterials; however, because the term is not well defined, it is preferable to specify compounds as being antimalarials, antivirals, or antiprotozoals. All antibiotics share the property of selective toxicity: They are more toxic to an invading organism than they are to an animal or human host. Some of these drugs (chemicals) that were obtained naturally were put to chemical modifications (By removing some chemical groups and adding other) in attempts to enhance beneficial effects while minimizing the toxic effects. The resultant modified product is termed as Semisynthetic antibiotics. Most antibiotic now used are semisynthetic. The chemist has synthesized many drugs that have got the antibacterial property and less toxicity. These drugs are called synthetic antibiotic drugs. Penicillin is the most well known antibiotic and has been used to fight many infectious diseases, including syphilis, gonorrhea, tetanus, and scarlet fever. Another antibiotic, Streptomycin, has been used to combat tuberculosis.

HISTORY

The principle of using organic compounds to fight infection has been known since ancient times. Crude plant extracts were used medicinally for centuries, and there is anecdotal evidence for the use of cheese molds for topical treatment of infection. The first observation of

[*]Corresponding author

what would now be called an antibiotic effect was made in the 19th century by French chemist Louis Pasteur, who discovered that certain saprophytic bacteria can kill *Anthrax bacilli*. In the first decade of the 20th century, German physician and chemist Paul Ehrlich began experimenting with the synthesis of organic compounds that would selectively attack an infecting organism without harming the host organism. Ehrlich put forth fundamental theories explaining the chemistry of immunity and developed a number of chemical treatments for diseases. Paul Ehrlich won the 1908 Nobel Prize in medicine. His experiments led to the development, in 1909, of *salvarsan*, a synthetic compound containing arsenic, which exhibited selective action against spirochetes, the bacteria that cause syphilis. Salvarsan remained the only effective treatment for syphilis until the purification of penicillin in the 1940s. In the 1920s British bacteriologist Sir Alexander Fleming, who later discovered penicillin, found a substance called lysozyme in many bodily secretions, such as tears and sweat, and in certain other plant and animal substances. Lysozyme has some antimicrobial activity, but it is not clinically useful. Penicillin, the archetype of antibiotics, is a derivative of the mold *Penicillium notatum*. Penicillin was discovered accidentally in 1928 by Fleming, who showed its effectiveness in laboratory cultures against many disease-producing bacteria.

(Paul Ehrlich)

(Sir Alexander Fleming)

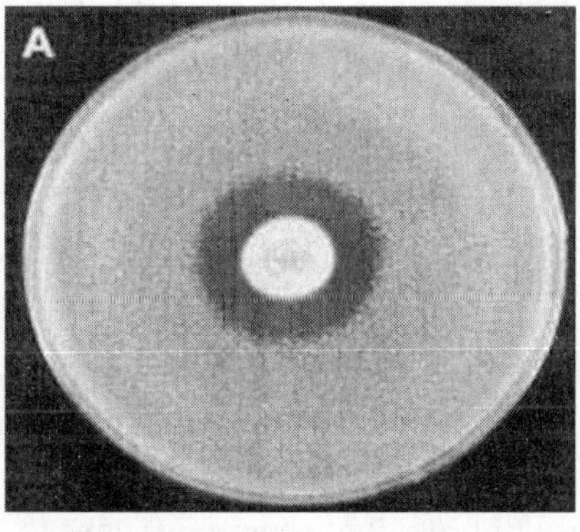

(*Penicillium notatum*)

Fig. 1

This discovery marked the beginning of the development of antibacterial compounds produced by living organisms. Penicillin in its original form could not be given by mouth because it was destroyed in the digestive tract and the preparations had too many impurities for injection. No progress was made until the outbreak of World War II stimulated renewed research and the Australian pathologist Sir Howard Florey and German-British biochemist Ernst Chain purified enough of the drug to show that it would protect mice from infection. Florey and Chain then used the purified penicillin on a human patient who had staphylococcal and streptococcal septicemia with multiple abscesses and osteomyelitis. The patient, gravely ill and near death, was given intravenous injections of a partly purified preparation of penicillin every three hours. Because so little was available, the patient's urine was collected each day; the penicillin was extracted from the urine and used again. After five days the patient's condition improved vastly. However, with each passage through the body, some penicillin was lost. Eventually the supply ran out and the patient died.

The first antibiotic to be used successfully in the treatment of human disease was Tyrothricin, isolated from certain soil bacteria by American bacteriologist Rene Dubos in 1939. This substance is too toxic for general use, but it is employed in the external treatment of certain infections. Other antibiotics produced by a group of soil bacteria called Actinomycetes have proved more successful. One of these, Streptomycin, discovered in 1944 by American biologist Selman Waksman and his associates, was, in its time, the major treatment for tuberculosis. Since antibiotics came into general use in the 1950s, they have transformed the patterns of disease and death. Many diseases that once headed the mortality tables—such as tuberculosis, pneumonia, and septicemia—now hold lower positions. Surgical procedures, too, have been improved enormously, because lengthy and complex operations can now be carried out without a prohibitively high risk of infection. Chemotherapy has also been used in the treatment or prevention of protozoal and fungal diseases, especially malaria, a major killer in economically developing nations. Slow progress is being made in the chemotherapeutic treatment of viral diseases. New drugs have been developed and used to treat shingles (a disorder causing blisters on skin) and chicken pox. There is also a continuing effort to find a cure for acquired immunodeficiency syndrome (AIDS), caused by the human immunodeficiency virus (HIV). The broad-spectrum antibiotic Chloramphenicol is isolated from a mulched field in Venezuela. The drug is approved for use two years later and manufactured under its trade name, Chloromycetin, by the Parke-Davis pharmaceutical company.

CRITERIA FOR EVALUATING ANTIMICROBIAL DRUGS

Antimicrobial drugs can be evaluated by a number of criteria, including the following:

1. The drug should demonstrate selective toxicity. This means that, at optimum concentration, the drug should be toxic for the microorganism, but not for the host.
2. The drug should not produce hypersensitivity (allergy) in most hosts.
3. A drug must have solubility in body fluids so that it can rapidly penetrate body tissues.
4. Microorganisms should not readily become resistant to drug.
    - The therapeutic index is the ratio of the toxic dose to therapeutic dose. The larger the therapeutic index, the better the chemotherapeutic agent.

Table 1. Table shows major discoveries of Antibiotics with their representative sources

Discovery	Year	Antibiotics	Source
Sir Alexander Fleming	1928	Penicillin	*Penicillium notatum* *P. chrysogenum*
Dubos	1939	Tyrothrycin	*Bacillus bravis*
Waksman	1943	Streptomycin	*Streptomyces griseus*
Ehrlich *et al.,*	1947	Chloramphenicol	*S.venezuelae*
Duggar	1947	(Tetracyclines) Chlorotetracycline Oxytetracycline	*S. aureofaciens* *S. rimosus*
Clerk	1953	(Macrolides) Erythromycin Carbomycin Ravomycin	*S. erythraeus* *S. halstedii* *S. ambofaciens*
Hazen and Brown	1954	(Antifungal) Nystatin Griseofulvin	*S. noursei* *Penicillium gryiseofulvum*
Moore *et al.,*	1954	Azaserine	*S.fragilis*

CLASSIFICATION

Presently over 6000 antibiotic substances have been described and it is very difficult to handle these great varieties of antibiotics unless they are summarized properly. There are several approaches to classify the antibiotics. Antibiotics may broadly be classified according to types of organisms against which they act. The most common method classifies them according to their action against the infecting organism. Some antibiotics attack the cell wall; some disrupt the cell membrane; and the majority inhibits the synthesis of nucleic acids and proteins, the polymers that make up the bacterial cell. Another method classifies antibiotics according to which bacterial strains they affect: *Staphylococcus, Streptococcus,* or *Escherichia coli,* for example. Antibiotics are also classified on the basis of chemical structure, as Penicillins, Cephalosporins, Aminoglycosides, Tetracyclines, Macrolides, or Sulfonamides, among others. Various types of classification of antibiotics:

Classification According to Type of Organism or Biological Origin

Antibacterial drugs

(A) Antibiotics produced by microorganisms of order Eubacterials.

(a) Antibiotics form by representatives of genus *Pseudomonas*

Pyocyanin – *Pseudomonas aeruginosa*

Viscocin – *Pseudomonas viscocin*

(b) Antibiotics form by representatives of genera *Micrococcus, Streptococcus, Diplococcus, E.coli, Proteus.*

Nisin – *Streptococcus lactis*

Diplomycin – *Diplococcus*

Coliformin – *E.coli*

Protaptins – *Proteus vulgaris*

(c) Antibiotics form by genus *Bacillus*
Gramicidins – *Bacillus brevis*
Subtilin – *Bacillus subtilis*
Polymixins – *Bacillus polymyxa*

(d) Antibiotics produced by microorganisms belonging to the order streptomycetales
Streptomycin – *Streptomyces griseus*
Tetracyclins – *Streptomyces aureofaciens*
Actinomycins – *Streptomyces antibioticus*

Antifungal drugs

(a) Antibiotics formed by imperfect fungi
Penicillin – *Penicillium chrysogenum*
Griseofulvin – *Penicillium griseofulvum*
Trichothecin – *Trichothecium roseum*

(b) Antibiotics formed by fungi of the Basidiomycetes and Ascomycetes class
Thermophillin – Basidiomycetes
Chetomin – Ascomycetes

Antibiotics formed by Lichen, Algae and lower plants

Usnic acid – Lichen
Chlorellin – *Chlorella vulgaris*

Antibiotics formed by higher plants

Allicin – *Allium sativum*
Raphanin – *Raphanus sativum*

Antibiotics of animal origin

Lysozyme
Interferon
Ecmolin

Antiviral drugs

Amantidine
Ribavirin
Acyclovir
Interferon

Antiprotozoal drugs

Chloroquine
Diidohydroxyquin
Metronidazole

Antihelminthic drugs

Praziquantel
Niclosamide

Classification According to Mechanism of Action

Most antibiotics act by selectively interfering with the synthesis of one of the large-molecule constituents of the cell like the cell wall or proteins or nucleic acids. Some, however, act by disrupting the cell membrane.

Some important and clinically useful drugs interfere with the synthesis of peptidoglycan, the most important component of the cell wall. These drugs include the β-lactam antibiotics, which are classified according to chemical structure into Penicillins, Cephalosporins, and Carbapenens. All these antibiotics contain a β-lactam ring as a critical part of their chemical structure, and they inhibit synthesis of peptidoglycan, an essential part of the cell wall. They do not interfere with the synthesis of other intracellular components. The continuing buildup of materials inside the cell exerts even greater pressure on the membrane, which is no longer properly supported by peptidoglycan. The membrane gives way, the cell contents leak out, and the bacterium dies. These antibiotics do not affect human cells because human cells do not have cell walls.

Many antibiotics operate by inhibiting the synthesis of various intracellular bacterial molecules, including DNA, RNA, ribosomes, and proteins. The synthetic sulfonamides are among the antibiotics that indirectly interfere with nucleic acid synthesis. Nucleic-acid synthesis can also be stopped by antibiotics that inhibit the enzymes that assemble these polymers—for example, DNA polymerase or RNA polymerase. Examples of such antibiotics are Actinomycin, Rifamicin, and Rifampicin, the last two being particularly valuable in the treatment of tuberculosis.

The quinolone antibiotics inhibit synthesis of an enzyme responsible for the coiling and uncoiling of the chromosome, a process necessary for DNA replication and for transcription to messenger RNA. Some antibacterials affect the assembly of messenger RNA, thus causing its genetic message to be garbled. When these faulty messages are translated, the protein products are nonfunctional. There are also other mechanisms: The Tetracyclines compete with incoming transfer-RNA molecules; the aminoglycosides cause the genetic message to be misread and a defective protein to be produced; Chloramphenicol prevents the linking of amino acids to the growing protein; and Puromycin causes the protein chain to terminate prematurely, releasing an incomplete protein.

Antibiotics inhibiting synthesis of the cell wall

Penicillin
Bacitracin
Vancomycin
Cephalosporin

Antibiotics inhibiting cell membrane

Gramicidins
Candicidins
Nystatins
Trichomycin
Endomycin

Antibiotics inhibiting selectively the synthesis of nucleic acids

(a) Inhibiting synthesis of RNA
Actinomycin
Griseofulvin
Neomycin

(b) Inhibiting synthesis of DNA
Mitomycin
Novobiocin
Actidion
Sarcomycin

(c) Antibiotics inhibiting synthesis of Purines and Pyrimidins
Decoinin
Sarcomycin

Antibiotics inhibiting synthesis of Protein

Bacitracin
Metimycin
Tetracyclins
Chloramaphenicol
Erythromycin

Antibiotics inhibiting Respiration

Antimycin
Oligomycin
Pyocyanin
Usnic acid

Antibiotics inhibiting Oxidative phosphorylation

Gramicidins
Colicins
Oligomycins

Antibiotics having Antimetabolite properties

Antibiotic substances formed by some Streptomycetes and Molds. These antibiotics act as antimetabolites of amino acids, vitamins and nucleic acids. Antibiotic antimetabolites are Furanomycin. Antibiotic antagonist of metabolism of Arginine and Ornithine formed by *Streptomycetes griesovariabillis.* Antibiotic antagonist of methionine and thiamine isolated from the culture of *Streptomycetes globisporus.*

Classification According to Range of Effectiveness

In some species of bacteria the cell wall consists primarily of a thick layer of peptidoglycan. Other species have a much thinner layer of peptidoglycan and an outer as well as an inner membrane. When bacteria are subjected to Gram's stain, these differences in structure affect the differential staining of the bacteria with a dye called gentian violet. The differences in staining coloration (gram-positive bacteria appear purple and gram-negative bacteria appear colorless or reddish, depending on the process used) are the basis of the classification of bacteria into gram-positive (those with thick peptidoglycan) and gram-negative (those with thin peptidoglycan and an outer membrane), because the staining properties correlate with many other bacterial properties. The range of microbes affected by an antimicrobial is called spectrum. Antibacterials can be further subdivided into narrow-spectrum and broad-spectrum agents.

Narrow-spectrum antibiotics acts against only a few groups of bacteria for example the narrow-spectrum Penicillins act against many gram-positive cocci, gram-negative cocci, and spirochetes and ineffective against gram-negative rods. Aminoglycosides, also narrow-spectrum, act against many gram-negative as well as some gram-positive bacteria.

Broad-spectrum antibiotics inhibit gram-negative and gram-positive bacteria. The tetracyclines and chloramphenicols are both broad-spectrum drugs because they are effective against both gram-positive and gram-negative, aerobic, anaerobic bacteria, spirochetes, mycoplasmas, rickettsias, chlamydias, and some protozoa.

All-important antibiotics can conventionally be divided in to the following groups:

Antibacterial antibiotics of Narrow spectrum, active against gram- positive organisms

Penicillin group

(a) Biosynthetic Penicillins:
  Benzyl Penicillin and its salt
  Bicillin
  Phenoxymethyl Penicillin

(b) Semisynthetic Penicillins
  Acid resistant- Inactive against penicillase
  Proricillin
  Pheneticillin
  Oxacillin
  Cloxacillin

(c) Semisynthetic cephalosporins
  Cephaloridin
  Cephalotin
  Cephaloxin
  Albomycin
  Bacitracin

(d) Macrolides
  Erythromycin
  Leucomycin
  Pyostatine
  Tylosine

Broad-spectrum antibacterial antibiotics

Tetracyclins
Chloramaphenicol
(a) Aminoglycosides:
Streptomycin
Neomycin
Gentamycin
(b) Polymyxins
Gramicidin S
Semisynthetic penicillins
Ampicillin

Antituberculosis antibiotics

Streptomycin
Canamycin
Viomycin
Cycloserin

Antifungal antibiotics

Nystatine
Griseofulvin
Candicin
Trichothecin

Antitumour antibiotics

Actinomycin C
Mitomycin C
Olivomycin

Antiamoebic antibiotic

Fumagillin

Classification of Antibiotics by Their Chemical Structure

According to Shemyakin *et al.*, 1961 and also in compliance with recent development, the following groups of antibiotics can be distinguished.

Antibiotics of Acyclic structure

Allicin
Bioformin
Raphanin
Nystatine
Trichomycin
Fumagillin

Depending on the structure this group comprise the following main sub groups:

Fatty acids, acetylenes (CH=CC=CCH$_2$–), Polyenes, Sulphur and nitrogen containing compounds.

Polyens occupy an important position among these compounds. They are characterized by presence of a system containing from 3 to 8 conjugated double bond (–CH=CH–). Many of these contain amino sugar.

Mycosamine

Perosamine

Some substances have second nitrogen compounds.

P-aminoacetophelon: This group of antibiotics comprises cyclopentane derivatives.

Sarcomycin

Actidione

Tetracycline: This group includes compounds similar in their structure.

Aromatic antibiotics: antibiotics in this group are benzene derivatives:

Chloramphenicol

Quinine antibiotics: Antibiotics substances in this group are as a little practical use. These are

Benzoquinones(Raphanon)

Naphthoquinones(Plumbagin)

Anthraquinones(Endocricin)

The group of quinone antibiotics comprises also Celicomycin whose molecules contain chromophoric carbohydrate and peptide groupings. Antibiotics of Celicomycin group are blue pigments of Actinomycetes.

Antibiotics-heterocyclic oxy compounds: This group includes many antibiotics among which are:

Antimycins

Penicillic acid

Griseofulvin

Usnic acid

Oxygen containing heterocyclic antibiotics can be divided in to the following major groups:

(a) Antibiotics with one five membered O–heterocycle. Carlina oxide contained essential oil of carlina ocaulis plant.

(b) Antibiotics with six membered O–heterocycle. Kojic acid is a simple natural derivative of V–pyrone.

Macrolides: The characteristic feature of these antibiotics is presence in their molecule of macrocyclic lactone ring connected with one or more carbohydrate residues.

Methymycin

Erythromycin

Nitrogen containing hetrocyclic compounds: This group includes many antibiotics whose molecules contain various often very complicated cyclic systems.

Azomycin
Nocardiomycin
Penicillins

Actinomycins: This group includes antibiotics that contain noxasine chromophore group.

Actinomycin–C

Metal containing compounds: These are iron and copper containing antibiotics.

Iron containing–Sideromycin
Copper containing–Fleomycin

Classification According to Cell Death and Growth Suppression

Antibiotics may also be classed as bactericidal (killing bacteria) or bacteriostatic (stopping bacterial growth and multiplication).

Bacteriostatic drugs are nonetheless effective because bacteria that are prevented from growing will die off after a time or be killed by the defense mechanisms of the host. The tetracyclines and the sulfonamides are among the bacteriostatic antiobiotics.

Bacteriocidal drugs that damage the cell membrane cause the cell's metabolites to leak out, thus killing the organism. Such compounds, including Penicillins and Cephalosporins, are therefore classed as bactericidal.

MODE OF ACTION OF ANTIBIOTICS

Antibiotics interfere with growth and metabolism of microorganisms, which suppress the growth and other activities of cells. Although bacteria are orthodestroyed by host immune system. According to their mode of action the antibiotics are classified as follows:

- Inhibitors of Bacterial CELL WALL synthesis.
- Inhibitors of Bacterial PROTIEN synthesis.
- Inhibitors of Bacterial CELL MEMBRANE synthesis.
- Inhibitors of Bacterial NUCLEIC ACID synthesis.

Inhibitors of Bacterial Cell Wall Synthesis

Substances which interfere with formation of bacterial cell wall either inhibit the enzyme involved in cell wall synthesis. If cell wall is absent these antibiotics are not active. These are:

Penicillin

Penicillins are bactericidal, inhibiting formation of the cell wall. There are four types of penicillins: The narrow-spectrum penicillin-G types, ampicillin and its relatives, the penicillinase-resistants, and the extended spectrum penicillins that are active against *pseudomonas.* Penicillin-G types are effective against gram-positive strains of *streptococci, staphylococci,* and some gram-negative bacteria such as *meningococcus.* Penicillin-G is used to treat such diseases as syphilis, gonorrhea, meningitis, anthrax, and yaws.

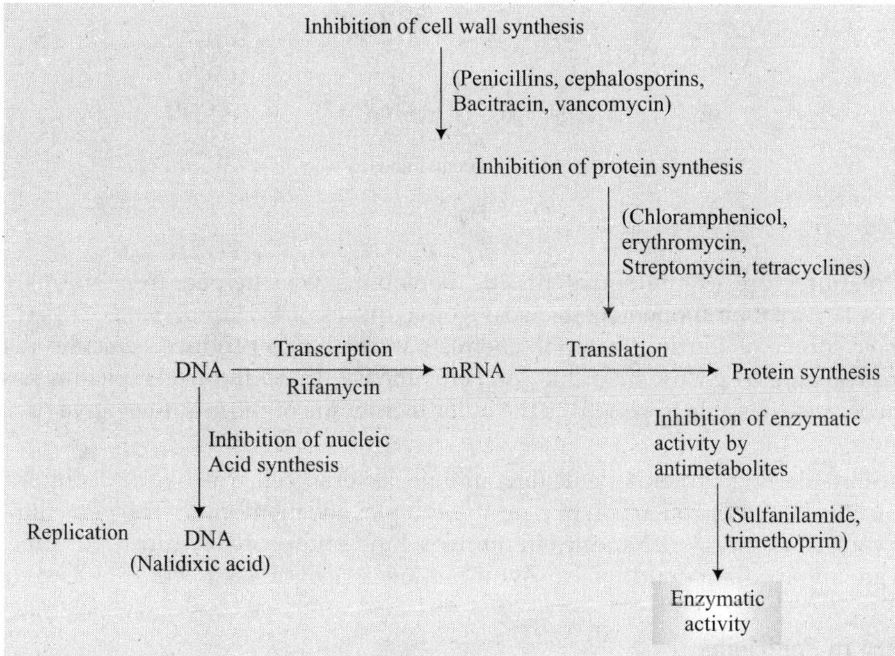

Fig. 2. The Figure shows summary of action of Antimicrobial Drugs.

The related penicillin V has a similar range of action but is less effective. Ampicillin and amoxicillin have a range of effectiveness similar to that of penicillin-G, with a slightly broader spectrum, including some gram-negative bacteria. The penicillinase-resistants are penicillins that combat bacteria that have developed resistance to penicillin-G. The antipseudomonal penicillins are used against infections caused by gram-negative *Pseudomonas* bacteria, a particular problem in hospitals. They may be administered as a prophylactic in patients with compromised immune systems, who are at risk from gram-negative infections.

Side effects of the penicillins, while relatively rare, can include immediate and delayed allergic reactions specifically, skin rashes, fever, and anaphylactic shock, which can be fatal. Penicillin G, the antibiotic naturally produced by *Penicillium notatum* and *P. chrysogenum*, inhibits the synthesis of the peptidoglycan of the bacterial cell wall. Most penicillins are narrow-spectrum antibiotics effective against gram-positive bacteria, spirochetes, and few gram-negative cocci. They are largely ineffective against gram- negative bacilli because they cannot penetrate the outer membrane of cell wall. The high therapeutic indices of the penicillins make them the drugs of choice for treating the majority of cases of gonococcal, treponemal, and streptococcal infections.

The Penicillin Family, Penicillium *chrysogenum* produces different penicillins depending on its growth medium. Although penicillin G (Benzylpenicillin) is a very effective antibacterial agent, it has the disadvantage of being inactivated by penicillase (β-lactamase) and slowly decomposes in the acid environment of the stomach. To surmount this problem, chemists make semisynthetic penicillins that possess improved chemotherapeutic properties.

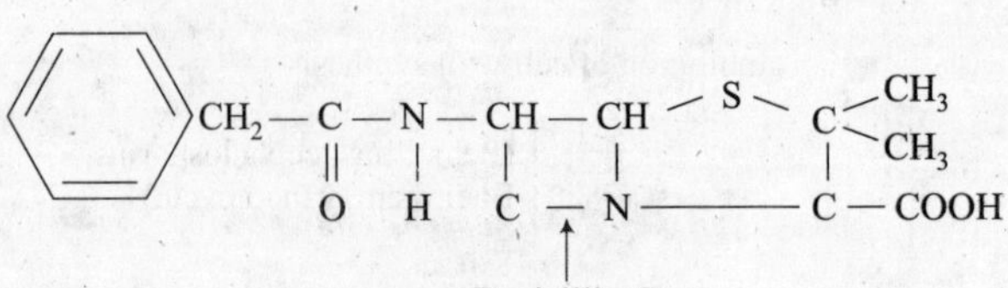

Penicillin G

Fig. 3

The manufacture of the semisynthetic penicillins was helped tremendously by the isolation in 1959 of 6-aminopenicillanic acid from cultures of *P. chrysogenum*. Armed with this inexpensive source of starting material, chemists were able to produce penicillin derivatives that were resistant to gastric acid and/or penicillinase. In addition, Ampicillin has a wide spectrum because it is able to penetrate the outer membrane of the gram-negative bacterial cell wall.

Penicillin–Binding Proteins Penicillins inhibit bacterial cell wall synthesis by interacting with penicillin-binding proteins on or near the cytoplasmic membrane. The penicillin-binding proteins vary among bacterial species in number, kind and probably function. Some of these proteins are involved in peptidoglycan synthesis and in cell division.

Resistance to Penicillins

Resistance to Penicillins comes about when bacteria produce penicillinase: (or β- lactamase), which hydrolyzes the β-lactam ring of the penicillin. This inactivates the an-tibiotic before it is able to inhibit cell wall synthesis. Penicillin-resistance strains of staphylococci produce an extracellular penicillinase, produced by gram-negative bacteria occur in the peri-plasmic space. Penicillinase-producing bacteria can be killed with other antibiotics (for example, spectinomycin for penicillin-resistant gonococcus) or with the penicillin derivatives that are modified to be penicillinase resistant (such as methicillin). Bacteria can become resistant to penicillins by mutational events. These mutations probably affect the penicillin-binding proteins or the autolytic enzymes that cause the cell to lyse.

Pharmacology of Penicillin: The pharmacological advantage of the penicillin is their high therapeutic index. They have a low toxicity except in the case of a few people who are allergic to them and experience an anaphylactic reaction or serum sickness. A person who is allergic to one penicillin should be considered allergic.

Cephalosporins

The *Cephalosporium* fungus produces the antibiotic cephalosporin C as an extra cellular product. Cephalosporin C is chemically similar to the penicillin by virtue of its β-lactam ring and its action as a cell-wall inhibitor. The family of cephalosporin drugs includes the naturally produced cephalosporin C and the semisynthetic cephalosporins that are grouped as "generations" based on their antibacterial spectra.

The "first-generation" cephalosporins are active against gram-positive cocci including penicillinase-producing *Staphylococcus aureus* and some of the enterics. The "second generation" cephalosporins have an expanded spectrum that includes species of *Proteus, Enterobacter,* and *Citrobacter.*The "third-generation" cephalosporins are particularly active against gram-negative bacilli and are very important in the control of nosochemical infections.

Cephalosporins are similar in their mode of action to the penicillins. Although the biochemical reactions they inhibit are not entirely understood, it is known that they bind to enzymes and membrane proteins that appear to be involved in cell wall synthesis. Cephalosporins are inactivated by cephalosporinase, which is produced by resistant strains. The first generation cephlosporins are sensitive to the β-lactamases produced by gram-negative bacilli; however, this problem has been overcome with the third generation cephalosporins. The cephalosporins are widely used and relatively safe. Some patients are allergic to the cephalosporins; however, of these only a few (5 to 10 per cent) are also allergic to penicillin. The third-generation cephalosporins reach the cerebrospinal fluid and are effective in treating meningitis caused by gram-negative bacteria.

Cycloserine

It is a relatively simple compound, is related in structure to alanine it was originally discovered as an antibiotic produced by streptomyces and now manufactured through chemical synthesis. The main use of this antibiotic is in tuberculosis therapy. However because of potential undesirable effects, its utilization is limited. Cycloserine manifests its inhibitory effect on peptidoglycan synthesis by interference with synthesis of the peptide moiety of peptidoglycan.

Thienamycins: Imipenem

A class of β-lactam antibiotics called thienamycins (thi-en' ah-my'cin) was discovered in the 1970s as natural products of *Streptomyces cattleya.* The prototype drug is Imipenem (im-I' pen-em), which possesses novel stereochemical features that differentiate it from the penicillins and cephalosporins. Imipenem inhibits bacterial cell-wall synthesis by interacting with penicillin-binding proteins, and is effective against a broad spectrum of bacteria, including the gram-positive cocci, the enteric gram-negative bacilli, most anaerobic species, and *Pseudomonas aeruginosa.* Another advantage of this drug is its resistance to hydrolysis by most of the known β-lactamases, penicillinases, and cephalosporinases. Imipenem is given by intravenous infusion because gastric acids hydrolyze it.

β-Lactamase Inhibitors

Specific inhibitors of the β-lactamases produced by bacteria are now used in combination with penicillins in chemotherapy. *Streptomyces clavuligerus* produces clavulanic acid, which inhibits β-lactamases from a number of gram-positive and gram-negative organ-isms. This β-lactam analog is strongly bound by β-lactamases, thus initiating the destruction of the enzyme. Clavulanic acid is combined with amoxicillin as an effective treatment against certain penicillin-resistant bacteria.

OTHER ANTIBIOTICS THAT INHIBIT CELL WALL SYNTHESIS

Vancomycin

It is a glycopolypeptide antibiotic produced by *Streptomyces* spp. that is chemically unrelated to other known antibiotics. It inhibits the transfer of the peptide to the peptidoglycan and therefore inhibits cell wall synthesis. Vancomycin is a narrow-spectrum antibiotic with activity against the gram-positive cocci. It must be injected because it is poorly absorbed from the gastrointestinal tract. Vancomycin is used to treat staphylococcal infections caused by methicillin-resistance *staphylococci.*

Bacitracin

It is a peptide antibiotic produced by *Bacillus subtilis* and *B. licheniformis.* The first antibiotic-producing strain of *B. subtilis* was isolated from the dirt in a wound suffered by a girl named Tracy and was thus named bacitracin This antibiotic prevents peptidoglycan synthesis and is effective against gram-positive bacteria. It is not used to treat systemic infections because of its high toxicity, but is applied to wounds and in skin infections, often in combination with polymyxin and neomycin, as a tropical ointment.

Inhibitors of bacterial protein synthesis

Bacterial ribosomes are composed of 50S and 30S subunits, which join with messenger RNA to form a 70S mRNA complex. Protein synthesis begins with the binding of tRNA to this complex. At the termination of translation, the ribosomes revert to their subunit state and remain as 50S and 30S subunits subunits until they form another complex with mRNA. Some antibiotics inhibit translation on the 70S ribosomes. These antibiotics also inhibit the functioning of the 70S ribosomes found in mitochondria and chloroplast, but this inhibition is not lethal to eukaryotic cells.

Chloramphenicol, Erythromycin and Clindamycin

These three antibiotics are produced by different species of *Streptomyces*. They are structurally diverse antibiotics that inhibit protein synthesis by binding to the 50S bacterial ribosomal subunit. The probable mechanism of inhibition is to prevent peptide-bond formation by binding to the peptidyl transferase center on the ribosome.

Chloramphenicol is a broad-spectrum antibiotic originally derived from a culture of *Streptomyces;* it is now produced by chemical synthesis. This bacteriostatic antibiotic binds to the 50S ribosomal subunit and blocks the peptidyl transferase reaction necessary for protein synthesis. Chloramphenicol penetrates the blood-brain barrier and quickly enters the cerebrospinal fluid, a characteristic that makes it useful in the treatment of bacterial meningitis.

Chloramphenicol carries certain risks. It disrupts normal development of red blood cells in one of every 50,000 persons causing an extremely severe and some times fatal condition called aplastic anemia. Aplastic anemia also occurs in persons who are not taking chloramphenicol, so it is clear that other factors are involved in causing this disorder.

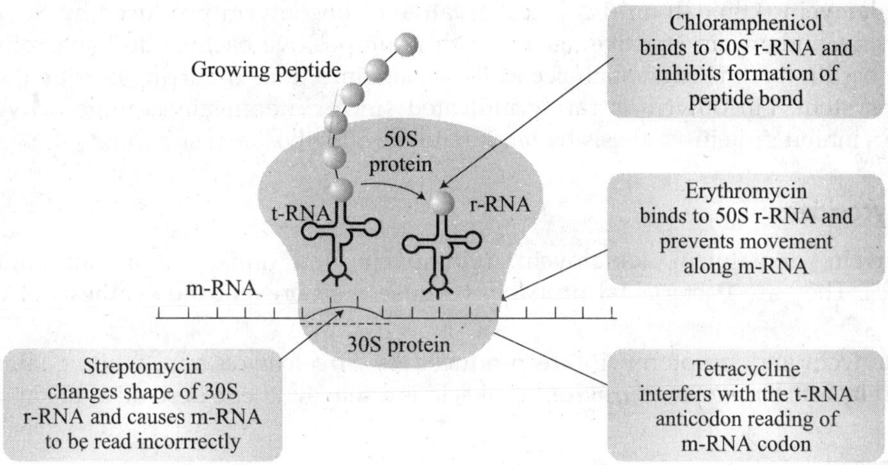

Fig. 4. Inhibition of protein synthesis by antibiotics.

Chloramphenicol is the drug of choice in typhoid fever and serious infections such as meningitis and epiglottitis when no alternative antibiotic is appropriate Long-term therapy with the drug is, obviously, contraindicated.

Erythromycin is the most commonly prescribed member of the macrolide family of antibiotics. It is produced by *Streptomyces erythraeus* and it inhibits protein synthesis by binding to the 50S ribosomal subunit and blocking translocation.

It is effective against gram-positive bacteria *Legionella, mycoplasma, and chlamydias*. It is the drug of choice in *mycoplasmal pneumonia, pertussis, chlamydia trachomatis, Legionella pneumophila,* and *streptococcal* infections of the respiratory tract in patients allergic to penicillin. Erythromycin is administered orally, is readily absorbed, and has a high therapeutic index.

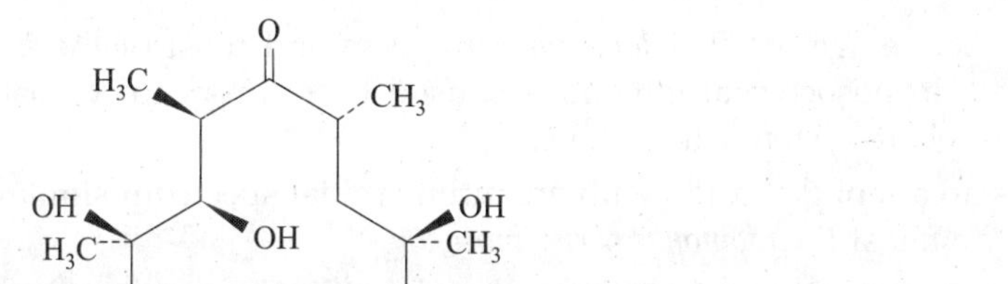

Fig. 5. Erythromycin.

Clindamycin (klin'dah-mi'sin) is a derivative of lincomycin produced by *Streptomyces lincolnensis.* Clindamycin is active against both gram-positive bacteria and anaerobic gram-negative bacilli. It is the drug of choice in *Bacteroides* infections occurring outside the central nervous system. Lincomycin is rarely indicated since clindamycin is more active. These antibiotics inhibit protein synthesis by binding to the 50S ribosomal subunit.

Aminoglycosides

Streptomycin, gentamicin, kanamycin, tobramycin and amikacin are aminoglycoside antibiotics. They are Bactericidal probably because they prevent the synthesis of essential proteins.

Kanamycin and streptomycin are produced by *Streptomyces* spp. while gentamicin is produced by *Micromonospora purpurea.* Amikacin is a semisynthetic derivative of kanamycin.

Streptomycin

It is rarely prescribed except to treat tuberculosis and, in combination with penicillin to treat bacterial endocarditis. Streptomycin was the first effective aminoglycoside antibiotic to be discovered and is important historically for its effectiveness against *Mycobacterium tuberculosis.* In therapy it is combined with other drugs because bacteria readily develop resistance to it. Combinations of streptomycin and isoniazid are credited with significantly reducing the incidence of tuberculosis during the 1950s. Streptomycin must be injected and is toxic to the auditory and vestibular branches of the eight cranial nerve, so it can cause loss of hearing and balance.

Streptomycin binds to one protein on the 30S ribosomal subunit thereby interferes with protein synthesis. Cells become resistant to streptomycin when the chromosomal gene for that protein is mutated such that the streptomycin no longer binds to the ribosome. Resistant cells can also produce enzymes encoded by plasmid-borne genes that enzymatically inactivate streptomycin.

Gentamicin is active against *Pseudomonas aeruginosa* and most of the *Enterobacteriaceae.* Gentamicin is given in nosocomial infections caused by gram-negative bacilli and with a second antibiotic in severe systemic infections.

Tobramycin is an aminoglycoside with an antimicrobial spectrum similar to gentamicin, but it is more active against *Pseudomonas aerug inosa*

Amikacin is a semisynthetic antibiotic that has the broadest spectrum of all the aminoglycosides. It is not inactivated by many of the aminoglycoside-inactivating enzymes and therefore can be used against gentamycin-or to-bramycin-resistant bacteria.

Spectinomycin

This antibiotic is an aminocyclitol that was isolated in 1960 from *Streptomyces spectabilis.* Spectinomycin prevents protein synthesis by binding to the 30s ribosomal subunit. Its primary use is to treat gonorrhea caused by penicillin-resistant *Neisseria gonorrhoeae*; other bacteria readily develop resistance to the spectinomycin. It is not effective in sexually transmitted diseases other than gonrrhea; *Treponema pallidum*(syphilis) is insensitive to it, and most strains of *Chlamydia trachomatis* are resistant.

Tetracycline

The tetracyclines are a family of broad-spectrum antibiotics that inhibit the initiation of protein synthesis. They are effective against many bacteria and are indicated in brucellosis, chancroid, chlamydial and rickettsial infections, cholera, relapsing fever and *mycoplasmal pneumonia.* Tetracyclines have a low therapeutic index in large part because they bind to proteins in food and to serum Proteins in blood.

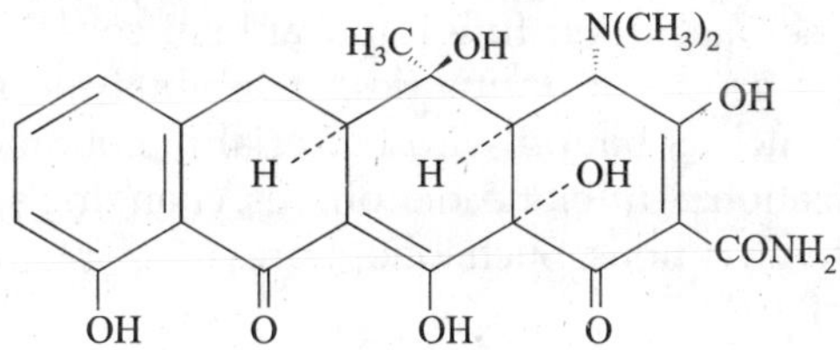

Fig. 6. Tetracycline.

The tetracyclines have bothersome side effects. Gastrointestinal disturbances including diarrhea is one. They cause staining and deformity of developing teeth, which is especially pronounced in young children whose permanent teeth are still developing. For this reason, children under 8 years of age and pregnant women are not usually treated with tetracyclines.

Resistance to a tetracycline occurs in some bacteria, which apparently prevent the antibiotics entry into their cytoplasm.

Inhibitors of Bacterial cell membrane synthesis

Several polypeptide antibiotics produced by Bacillus species have the ability to damage cell-membrane structure. They adversely affect the normal permeability characteristics of the cell membrane.

Included in this category are:

Polymyxins

Gramicidins

Tyrocidines

These agents are bactericidal; they cause a leakage from the cytoplasmic content of the cell. Because of their toxicity to tissue they have limited application in chemotherapy.

Polymyxins

The polymyxins are cyclic polypeptide antibiotics produced by *Bacillus polymyxa* that disrupt the function of the cytoplasmic membrane. Polymyxin B and polymyxin E are active against gram-negative bacteria and may be indicated for certain, *Pseudomonas aeurginosa* in-fections. They are rarely given parenterally because of their toxicity. Topical ointments containing polymyxin B are used in treating skin, ear, and eye infections. The polymyxins bind to phospholipidsin the cell membrane and prevent membrane transport. They kill resting cells without causing lysis.

Gramicidin S

The antibiotic disorganizes lipoprotein system and permeability of cells. This antibiotics changes permeability of membrane and causes a quick loss of vital compounds like nucleotide, phosphorus, and amino acids and also inhibit the energy metabolism, especially at its initial stage that is dehydration. Another category referred to, as Polyenes are large ring structures with many double bonds. viz: Nystatin produced by *Streptomyces noursei* and amphotericin, produced *Streptomyces nodusus*. Polyene antibiotics act upon cells, which have sterols in their cytoplasmic membrane. They act upon fungi and animal cells but do not affect bacteria. Oligomycin belongs to this group of antibiotics. It inhibits the mitochondrial adenosine triphosphatase (ATPase), which is bound with bacterial membranes and is involved in the synthesis of ATP or its utilization. Antibiotic ionophores, changing specifically permeability of biological membranes for some cations. Such ionophores are:

Valinomycin

Gramicidins (except for gramicidin S)

Macrotetralides (actins)

Under the action of these antibiotics, cells sensitive to them lose the potassium ion (K+) and absorb ions of sodium, lithium or ammonium instead. Potassium deficit in the cell stops the synthesis of protein and thus kills cell. In the presence of valinomycin, which is atypical ionophore, the membranes of sensitive bacterial cells become selectively permeable to potassium ions. Valinomycin binds to potassium ion to transfer it through the membrane thousands of times more actively than it can transport the sodium ion.

Inhibitors of Bacterial Nucleic Acid Synthesis

Rifampin selectively binds to the bacterial RNA polymerase and inhibits the initiation of mRNA synthesis. It does not interfere with the activity of RNA polymerases of eucaryotes. Rifampin is a semisynthetic derivative of rifamycin, and is active against the gram-positive cocci as well as *Mycobacterium tuberculosis*.

Antibiotics Inhibiting Synthesis of Purines and Pyrimidines

Azaserine is an antitumour antibiotic derived from *Streptomyces fragilis*. It inhibits involvement of glycine in biosynthesis of purine. Azaserine is structural analogue of glutamine. The antibiotic presumably competes with glutamine in binding with the enzyme, which irreversibly inactivated by it. The –N=N– group of azaserine molecule reacted with the group of the enzyme which is responsible for the transport of =NH from the amide group of glutamine to the substrate.

Azaserine inhibits biosynthesis of purine by combining with the enzyme responsible for the conversion of formylglycinamide ribonucleotide in to formylglycinamidine ribonuclotide. The antibiotics act mainly on cysteine.

Sarcomycin has a multiple effects on the cell metabolism in the ascites cell tumor. It inhibits inclusion of glycine in protein, blocks the synthesis of pyridine nucleotides and the inclusion of phosphorus in nucleotides and the inclusion of phosphorus in nucleotides and nucleic acid.

Antibiotcs Inhibiting Synthesis of Nucleic Acid

The power of actinomycins to inhibit the synthesis of protein and RNA without affecting the formation of DNA. According to Sazykin and Borisova (1963), Aurentin (an actinomycin) immediately stops formation of RNA by *Staphylococcus aureus.* Discontinued synthesis of RNA and hence of messenger RNA, bears a direct effect on the protein synthesis.

Mitomycins are broad-spectrum antibiotics. They inhibit the growth of many bacteria, protzoa and slow down the growth of tumour cells. The mechanism of biochemical action of these biological potent substances is unlike that of actinomycins: Mitomycins block the synthesis of DNA without practically affecting the synthesis of RNA and protein. But the synthesis of viral DNA is very stable against Mitomycin. According to some authors, Mitomycin C is responsible for cross-linking of molecules of DNA.

Novobiocin there is evidence that like penicillin, novobiocin upsets the normal synthesis of cell walls. This is confirmed by accumulation of the mucopeptide fragment of the cell wall in sensitive bacteria exposed to action of novobiocin. But there are data showing that it stimulate accumulation of uridine nucleotides inside the cell and inhibits synthesis of RNA. This antibiotic also inhibits cell division and deceases the DNA content of the growing cell.

Daunomycin and Rubomycins, these antibiotics has antitumour action. It has been shown that the mechanism of biological effect of Daunomycin is due to its ability to set between pairs of DNA bases. The antibiotic is associated with those parts of DNA where G-C pairs are available.

ANTIFUNGAL ANTIBIOTICS

Nystatin: Nystatin is an antifungal agent useful in the therapy of nonsystemic fungi infections. It is produced during fermentation by a strain of *Streptomyces iloursei.* The antimicrobial activity is restricted to yeasts and other fungi viz., *Candida ,Aspergillus, Penicillium* and *Botrytis.*

Griseofulvin: Griseofulvin is obtained from *Penicillium griseofulvin.* It is used in treatment of many superficial fungus infections of skin and body surfaces and also effective in treatment of some systemic mycoses.

Polyenes

Commonly used polyene antibiotics, which are products of *Streptomyces* spp., are *Nystatin* and *Amphotericin B.* Both drugs are fungicidal and exert their action by damaging fungal plasma membranes; the drugs combine with the sterols in the plasma membranes. Nystatin is used to treat local infections of the vagina and the skin and can be taken orally for *Candida* infections of the intestinal tract. It is so poorly soluble that it does not cross the intestinal lining, so it does not enter the body tissue in toxic amounts. For many years, amphotericin B has been a mainstay of clinical treatment for systemic fungal diseases, such as histoplasmosis, coccidioidomycosis and blastomycosis. The drug's toxicity, particularly to the kidneys, is a strongly limiting factor in these uses.

Imidazoles

Imidazole antifungals, such as Clotrimazole, Miconazole, and Ketoconazole, primarily interfere with sterol synthesis in fungi, although they probably have other antimetabolic effects. Clotrimazole and Miconazole are generally used in treatment of cutaneous mycoses such as athlete's foot. When taken orally, ketoconazole is effective and is a less toxic alternative to amphotericin B for many systemic fungal infections, although occasional liver damage has been reported. Ketoconazole also has the advantage of an unusual wide spectrum of activity.

Other Antifungal Drugs

Tolnaftate is a common alternative to miconazole as topical agent for the treatment of athlete's foot. Its mechanism of action is not known. Undecylenic acid is a fatty acid that has antifungal activity against athlete's foot, although it is not as effective as tolnaftate or the imidazoles.

ANTIVIRAL CHEMOTHERAPEUTIC AGENTS

In the developed parts of the world, it is estimated that at least 60% of infecting illness are caused by viruses and only about 15% are caused by bacteria. The first antiviral drug licensed for systemic use in USA was Amantadine. It prevents the virus from entering the cell or from uncoating after it enters. The drug has limited usefulness in the prevention of influenza A. it has limited effect on the course of the disease once it has been contracted, but in confined situations, such as institutions or nursing homes, it can limit the spread of an outbreak.

Another drug with activity against influenza is Ribavirin, an analog of guanine, a purine in DNA and RNA. Ribavirin has also been tested on infants suffering from a viral pneumonia. Interest in ribavirin remains high because it is effective in vitro against a wide range of viruses.

Acyclovir is an analog of a guanine containing nucleoside. It is effective only if acted on by thymidine kinase, an enzyme present in cells infected by herpes virus.

Vidarabine has also been approved for intravenous use in the treatment of encephalitis caused by herpes virus.

Zidovudine is an analog of thymidine that has been promising in chemotherapy of AIDS patients.

Interferon: Among the more promising of the chemotherapeutic agents for treating viral disease is interferon. Interferons are small glycoprotein substances of which 2 types are leukocytic interferon and fibroblast interferon.

Acycloguanosine is a nucleoside analog, which is active against the herpes virus in animals.

ANTIPROTOZOAN DRUGS

Quinine still has very limited use in the control of the protozoan disease malaria, but synthetic derivatives, such as chloroquine, have largely replaced it. Quinacrine is the drug of choice for treating giardiasis. Diiodohydroxyquine(iodoquinol) is an important drug prescribed for several intestinal amoebic diseases. Metronidazole is one of the most widely used antiprotozoan drugs. It is unique in having activity not only against parasitic protozoans but also against obligate anaerobic bacteria. Pentamidine isothionate is used in the early stages of African trypanosomiasis. The drug's mode of action is unknown, but it appears to bind to DNA. Nifurtimox is a member of the synthetic nitrofuran drug s and is effective against the trypanosome caused Chagas' disease.

ANTIHELMINTHIC DRUGS

Niclosamide is the first choice of treatment, in case of tapeworm infestations. The drug is effective because it inhibits ATP production under anaerobic conditions. Praziquantel is about equally effective for treatment of tapeworms. It is highly recommended for treatment of several fluke- caused diseases, especially schistosomiasis. Mebendazole is an antihelminthic drug frequently used in the treatment of several of the most common intestinal helminthes-ascariasis. Piperazine is used in the treatment of pinworms and ascariasis. It paralyzes the organisms, which soon pass out of the victim's body.

SYNTHETIC ANTIMICROBIAL DRUGS

Isoniazid (INH)

It is a very effective synthetic antimicrobial drug against *Mycobacterium tuberculosis.* It has little effect on nonmycobacteria. It is a structural analog of vitamine B_6 (pyridoxine), which serves as coenzyme in metabolic reactions. The primary effect of INH might be inhibit synthesis of mycolic acid, which are components of cell walls only of the *mycobacteria,* but the specific mode of action is not entirely understood. When used to treat tuberculosis, INH is usually administered simultaneously with other drugs, such as streptomycin, rifampin or ethambutol.

Ethambutol

It is effective only against *mycobacteria.* The drug apparently kills the microbes by inhibiting the synthesis of certain cellular metabolites. It is comparatively weak Antitubercular drug; its principal use is as the secondary drug to avoid resistance problems. For this purpose, it has essentially replaced para-aminosalicylic acid (PAS).

Sulfonamides

Sulfa drugs were among the first synthetic antimicrobial drugs used to treat microbial diseases. Unfortunately many pathogens have developed resistance to them. Moreover, sulfonamides cause allergic reactions in many people. For these reasons, antibiotics have tended to diminish the importance of sulfonamides in chemotherapy. The sulfonamides continue to be used to treat certain urinary tract infections and have other specialized uses, as in combination drug **silver sulfadizine,** used in control of infections in burn patients. Sulfonamides are bacteriostatic.

Nitrofuranes

It constitutes a family of drugs that are active against a large number of gram- positive and gram-negative bacteria, Protozoans and fungi. Nitrofurantoin is widely administered orally for urinary tract infections. Nalidixic acid stops bacterial replication of DNA and is especially effective against gram–negative enterics. Because it accumulates in the urine, this drug is used to treat urinary tract infections in which *Escherichia, Proteus, or Klebsiella* species are implicated. The quinolones are an emerging group of synthetic antimicrobials. They are related to Nalidixic acid and may be taken orally. Their mode of action is interfering with an aspect of DNA replication that is unique to prokaryotes.

SENSITIVITY OF ANTIBIOTICS

Besides isolating and identifying the agents, clinical microbiologist are responsible for advising physicians of their isolates to test susceptibility to anti microbial. Sensitivity is determined by in vitro testing of isolates against chemotherapeutic agents species and strains of microorganisms have varying degrees of susceptibility to different antibiotics. Furthermore the susceptibility of an organism to a given antibiotic may change especially during treatment. It is therefore important for the clinician to know the identity of the microbe and the specific antibiotic, which may be expected to give the most satisfactory results in the treatment.

For this information the microbiologist will be called upon, to make an accurate microbiological diagnosis and to determine the susceptibility of the organism to various antibiotics. From time to time during the course of therapy the microbiologist may be required to make estimates of any change in the susceptibility of the pathogen to the drug and possibly even to assay the antibiotic concentration in the body fluids. The basic measure of this susceptibility is of two types:

1. Minimum lethal concentration that is MLC:

   The least concentration of anti microbial agent to kill a portion usually 99% within a specified Period.

2. Minimum inhibitory concentration that is MIC:

   It is the least amount of anti microbial agent that prevents visible growth of microbes under standard conditions.

The two basic techniques for determining MIC:

(a) Tube dilution technique

(b) Diffusion sensitivity disc method.

[A] Tube Dilution Technique

Susceptibility test effort the most direct method of measuring susceptibility of antimicrobials. In a typical micro dilution test it is called as 2, 4 dilution of antimicrobials are prepared in growth media that concentration of drug covered a clinical significant range. The equal volume of broth containing 10^5 -10^6 ml of drug added to each tube and to control tube that contain no antimicrobials. The tubes are read to determine visible change after overnight incubation. The micro dilution is the method for determine anti microbial susceptibility in liquid media. All are use to determine MIC upon which an appropriate drug therapy is based. In most clinical situation the MIC determined in vitro. Thus, MIC serves as guide to organism relative susceptibility. The increasing amounts of the antibiotics under examination are placed in a series of culture tubes containing a suitable broth medium inoculated with the test organism. The dosage of the drugs used in the test illustrated is indicated by the labels on the tubes. The tubes at left is the control tube and contain no antibiotic. After incubation the concentration of drug required to inhibit the growth of the organism used in determined by observing the absence of growth.

[B] Diffusion Sensitivity Disc Method

In this method which is a simple and economical and a substitute for micro dilution method. In this method known concentration of culture spread over an appropriate Agar media. Filter paper disc containing predetermined concentration of antimicrobials are placed and put in Agar plates in set pattern with equal spacing in Agar. Normally 8 or more antimicrobials can be used. Each disc is identified by anti microbial and its concentration. During incubation the antibiotic diffuses out creating a concentration gradient that decreases according to distance away from the disc. The plate is incubated to 16–18 hrs at 35°C. After which sensitivity is measured on the basis of size of zone of inhibition around disc. This measure is recorded in millimeter by comparing chart. The most common method of Disc Diffusion is **Kirby Bauer Method**.

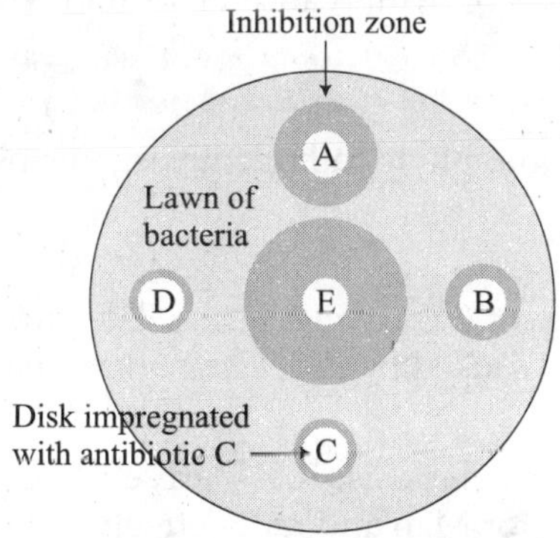

Fig. 7. Photograph showing that Bacteria are more susceptible to antibiotics A & E; have Intermediate effect to B and resistant to antibiotic C & D.

NON-MEDICAL USES OF ANTIBIOTICS

Antibiotics are now widely used not only in medicines but also in agriculture and in food industry.

[A] Antibiotics in Plant Growing

Antibiotics are used to control various agents and to prevent propagation of infection. Here follow the requirements of antibiotics used for plant disease control:

1. The antibiotic should be active against the phytopathogenic agent.
2. The antibiotic should easily penetrate the plant tissue.
3. The effective dose should be harmless to the plant.
4. The antibiotics should not be inactivated inside the plant, but should easily be inactivated by soil micro flora.
5. The antibiotic should produce its biological effect inside the plant. The techniques are based on the inhibition of growth or destruction of phytopathogenic organism on the plant surface or added to soil.
   (a) Adding antibiotics to soil: Antibiotic solutions added to soil are absorbed by the roots of plants and are soon found in the other organs of the plant. When antibiotics are added to the soil, serious obstacles arise, first of al the antibiotics are quickly destroyed by various microbial metabolites and other soil organisms consume the antibiotics as the source of nutrients.

(b) Treatment of affected plants by spraying antibiotic solution: It is effective during the entire vegetation period. Spraying is one of the most efficient methods to control plant diseases. Dusting and Dipping are also used. Various antibiotics or their combinations are now used to control phytopathogenic organisms.

1. Streptomycin:

   It is used for fight haricot and soybean wilt and some other diseases of agriculture plants.

   2.Griseofulvin:

   It is used against diseases of plants caused by fungi. The antibiotic active against Rust and Mildews.

3. Trichothecin:

   It is effective against some phytopathogenic organisms including *Botrytis cinerea* and Helminthosporium

4. Blasticidin S:

   Derived from *Streptomyces griseochromogenus.* The antibiotic was successfully used in Japan to fight Rice blast caused by *Piricularia oryzae.* It is however toxic to human.

5. Kasugemycin:

   It is used to control Piriculariosis. It is derived from *Streptomyces kasugaensis.* The preparation is less toxic for Man and more effective in treatment of rice blast.

6. Polyoxins:

   This group includes nine components. The antibiotics have a specific chemical structure and are peptide nucleotide compounds. They inhibit the growth of phytopathogenic *Alternaria, Piricularia* and *Cochliobalus* fungi.

7. Tetranastin A:

   It is derived from *Streptomyces aureus.* It is accumulated in the mycelium. It is effective against tick and spiders, parasitizing on fruit tree.

[B] Antibiotics in Nutrition and Veterinary

First of all antibiotics were used to treat many diseases of domestic animals.

1. Griseofulvin: is very effective in the treatment of mastitis of cattle.
2. Metimycin: Only inhibit the growth of Brucellosis microbe but does not cure the animal completely.
3. Streptomycin, Chloramphenicol, Chlorotetracycline: are used to treat turkeys with sinusitis.
4. Oxytetracycline: is used to treat bee families with European or American rot.
5. Antibiotics are used as growth stimulatory preparation. When small quantities of antibiotics are added to the animal feeds, survival of young animals increases.
6. Wine production wastes and boiled wheat meal are used to produce antibiotics for animal Nutrition. They contain chlorotetracycline.

Although the stimulatory effect of antibiotics on the growth of animals is studied thoroughly by many investigators, it mechanism is not yet clear. It can probably be explained by:

(a) Effects of antibiotics on intestinal micro flora:

1. Antibiotics stimulate the growth of useful microorganisms producing vitamins.
2. Antibiotics decrease the number of harmful microbes that consume vitamins and from toxins.
3. Antibiotics affect microorganisms present in the intestine of the animal to stimulate formation of stable strains that less harmful to animal.
4. Antibiotics stimulate migration of microorganism in the animal intestine.
5. Antibiotics reduce the occurrence of sub clinical infection in the young animals which otherwise slow down their growth.
6. An antibiotic decrease the pH of the intestinal contents and decreases the surface tension of cells to accelerate their division.

(b) Direct effect of antibiotics on animals:

The antibiotics effect on the growth of animals is in the first instance connected with the changes occurring in the micro flora of the intestine. The forms of this effect are varied.

1. The absorption rate increases and metabolite consumption is stimulated in animal tissues along with reduction in the energy of assimilation of nutritive.
2. Synergism of hormones occurs in the body the quantity of growth hormones increases nutrient consumption is intensified and the adaptability to disadvantageous condition increases.
3. Antibiotics decrease the vitamin demand; the synthesis of sugars and vitamin A from carotene is stimulated. The animal body forms smaller amounts of side product. The rate of enzyme synthesis increases.

[C] Antibiotics in Food Preservation

Many of the methods for food preservation that we use how, such as fermentation, heat processing, salting, smoking and even freezing were used by properties of natural food changes during high temp processing fermentation and other preservation techniques spoilage of food during storage may be caused by a variety of factors. These include bacterial spoilage and the action of enzymes together with oxidation by atmospheric oxygen. The strongest deteriorating effect is produced by microorganisms that liberate various metabolites to spoil the quality of food because of formation of strong poison.

(a) Antibiotics in canning:

Antibiotics were first reported to be used in canning industry in1943. Subtilin, Nisin and some other antibiotics are now used. Phytoncides of higher plants are also often used for food canning. Antibiotics are used to can meat, fish, poultry, cheese and dairy products, fruits and vegetables. Most microorganisms are killed at higher temperature but some valuable properties of food are also lost vitamins are destroyed, flavor and consistency of foods changes. If antibiotics are used for food canning, the time of thermal processing may be reduced significantly. Subtilin kills clostridial and thermophillic bacteria that stand high temperature. Nisin the antibiotic formed by a lactic acid *streptococcus* is also effective as a food preservative. It is used in medicine as well as in food industry. This antibiotic is used for canning tomatoes, green peas, cauliflower etc. Nisin inhibits the growth of some thermophilic bacteria and is harmless to man. If nisin is used, the time of heat processing can be halved.

(b) Using antibiotics to preserve fresh foods:

The meat of slaughtered animals soon becomes spoiled because of the micro flora contained in the alimentary tract of the animal, which contaminates meat during carcass cutting or during the slaughter. There are 2 methods by which quick spoilage of meat can be prevented.

1. Ante mortem feeding of animal with food containing an antibiotic.
2. Postmortem injection of the antibiotic into blood system, this prolongs the time of sage storage of meat to three days and improves its quality. Fish are among most perishable foods and its preservation is very important problem, especially so that fishing boats may be far from shore. Antibiotics are very effective in preservation of fish spoilage various methods are used for the purpose.
   [a] Fish may be dipped for 1-5 minutes in seawater containing chlortetracycline (5-100 mg/l) this methods prolongs that storage time significantly.
   [b] Fish may be dipped in seawater containing only 2 mg/l of chlortetracycline but the temperature of water should be 1-1.5°C.
   [c] Fish can be kept fresh for more than 5 days on ice containing 1-2 mg/l chlortetracycline.

(c) Preservation of milk and dairy product:

Nisin is used in the manufacture and storage of cheese. The antibiotic is derived from various strains of laticiferous *streptococci.* Its anti microbial spectrum is narrow and antibiotic can only inhibit the growth of *Clostridial* and other bacteria responsible for decay of cheese. During pasteurization of dairy products conditions are created that stimulate the growth of *Clostridia,* which produces toxins and spoil cheese, frequently with gas formation. Nisin prevent cheese from premature spoilage. Despite the apparent advantages of using antibiotics for nutrition of animals and preservation of foods, they should be used with great precautions. Getting inside man with food, antibiotics in small concentration may stimulate formation of antibiotic resistant microorganisms. This will evolve new problem in therapy of human diseases in this connection, antibiotics used for treatment of various diseases in humans may not be used for nutrition of animals or for preservation of food.

PRODUCTION

The production of a new antibiotic is lengthy and costly. First, the organism that makes the antibiotic must be identified and the antibiotic tested against a wide variety of bacterial species. Then the organism must be grown on a scale large enough to allow the purification and chemical analysis of the antibiotic and to demonstrate that it is unique. This is a complex procedure because there are several thousand compounds with antibiotic activity that have already been discovered, and these compounds are repeatedly rediscovered. After the antibiotic has been shown to be useful in the treatment of infections in animals, larger-scale preparation can be undertaken.

Commercial development requires a high yield and an economic method of purification. Extensive research may be needed to increase the yield by selecting improved strains of the organism or by changing the growth medium. The organism is then grown in large steel vats, in submerged cultures with forced aeration. The naturally fermented product may be modified chemically to produce a semisynthetic antibiotic. After purification, the effect of the antibiotic on the normal function of host tissues and organs (its pharmacology), as well as its possible toxic actions (toxicology), must be tested on a large number of animals of several species. In addition, the effective forms of administration must be determined. Antibiotics may be topical, applied to the surface of the skin, eye, or ear in the form of ointments or creams. They may be oral, or given by mouth, and either allowed to dissolve in the mouth or swallowed, in which case they are absorbed into the bloodstream through the intestines. Antibiotics may also be parenteral, or injected intramuscularly, intravenously, or subcutaneously; antibiotics are administered parenterally when fast absorption is required.

In the United States, once these steps have been completed, the manufacturer may file an Investigational New Drug Application with the Food and Drug Administration (FDA). If approved, the antibiotic can be tested on volunteers for toxicity, tolerance, absorption, and excretion. If subsequent tests on small numbers of patients are successful, the drug can be used on a larger group, usually in the hundred.

Finally a New Drug Application can be filed with the FDA, and, if this application is approved, the drug can be used generally in clinical medicine. These procedures, from the time the antibiotic is discovered in the laboratory until it undergoes clinical trial, usually extend over several years.

EFFECTIVENESS OF CHEMOTHERAPEUTIC AGENTS

Drug Resistance

Bacteria become resistant to antimicrobials in a number of different ways. Some resistance is due to the ability of a particular mutant to destroy a given antibiotic. For example, as we noted earlier, some *staphylococci* produce the enzyme penicillase, which destroy natural penicillins. Resistance to tetracycline is usually related to cellular changes that decrease the cell's in take of the drug. Bacteria become resistant to sulfa drugs and antibiotics when receptor sites for the antibiotic develop less affinity for it. Resistance to trimethoprim can arise when the microbe begins synthesizing very large amounts of the enzyme against which the drug is targeted. Conversely, polyene antibiotics can become less effective when resistant organisms produce smaller amounts of the sterols with which the drug produces its effects.

All these forms of resistance are transmitted genetically by the bacterium to its progeny. Genes that carry resistance can also be transmitted from one bacterium to another by means of plasmids, chromosomal fragments that contain only a few genes, including the resistance gene. Some bacteria conjugate with others of the same species, forming temporary links during which the plasmids are passed from one to another. If two plasmids carrying resistance genes to different antibiotics are transferred to the same bacterium, their resistance genes can be assembled onto a single plasmid. The combined resistances can then be transmitted to another bacterium, where they may be combined with yet another type of resistance. In this way, plasmids are generated that carry resistance to several different classes of antibiotic. In addition, plasmids have evolved that can be transmitted from one species of bacteria to another, and these can transfer multiple antibiotic resistance between very dissimilar species of bacteria.

Strains of bacteria that are resistant to antibiotics are particularly common among persons who work in hospitals, where antibiotics are in constant use. *Staphylococcus aureus*, a common opportunistic pathogen that is carried in nasal passage, develops resistance to antibiotics very frequently. Everyday in hospitals, surprisingly large volumes of antibiotics are dispensed as aerosols by the universal practice of clearing syringe of air before an injection is made. As these aerosols are generally inhaled in small amounts, the staphylococci harmlessly residing in a nurse's nose, for example, quickly develop resistance. If these resistant microbes are transferred to surgical wounds and other exposed sites, they can cause nosochemical infections.

The problem of resistance has been exacerbated by the use of antibiotics as prophylactics, intended to prevent infection before it occurs. Indiscriminate and inappropriate use of antibiotics for the treatment of the common cold and other common viral infections, against which they have no effect, removes antibiotic-sensitive bacteria and allows the development of antibiotic-resistant bacteria. Similarly, the use of antibiotics in poultry and livestock feed has promoted the spread of drug resistance and has led to the widespread contamination of meat and poultry by drug-resistant bacteria such as *Salmonella*.

In the 1970s, tuberculosis seemed to have been nearly eradicated in the developed countries, although it was still prevalent in developing countries. Now its incidence is increasing, partly due to resistance of the tubercle bacillus to antibiotics. Some bacteria, particularly strains of staphylococci, are resistant to so many classes of antibiotics that the infections they cause are almost untreatable. When such a strain invades a surgical ward in a hospital, it is sometimes necessary to close the ward altogether for a time. Similarly, plasmodia, the causative organisms of malaria, have developed resistance to antibiotics, while, at the same time, the mosquitoes that carry plasmodia have become resistant to the insecticides that were once used to control them. Consequently, although malaria had been almost entirely eliminated, it is now again rampant in Africa, the Middle-east, South-east Asia, and parts of Latin America. Furthermore, the discovery of new antibiotics is now much less common than in the past.

Effects of Combinations of Drugs

It sometimes happens that the chemotherapeutic effect of two drugs given simultaneously is greater than the effect of either given alone, a phenomenon called synergism. For example, in the treatment of bacterial endocarditis, penicillin and streptomycin are much more effective when taken together than when either drug is taken alone. Damage to bacterial cell wall by penicillin makes it easier for streptomycin to enter. The most widely used sulfa today is a combination of trimethoprim and sulfamethoxazole (TMP-SMZ). This combination is an excellent example of drug synergism. When the drugs are used in combination, only about one-tenth of the concentration is needed as when each is used alone. Other combinations of drug can be antagonistic.For example, the simultaneous use of penicillin and tetracycline is often less effective than when either drug is used alone. By sopping the growth of the bacteria, the bacteriostatic drug tetracycline interferes with the action of penicillin, which requires bacterial growth. Combination of antimicrobial drugs should be used only for the following purpose:

1. To prevent or minimize the emergence of resistant strains.
2. To take advantage of the synergistic effect.
3. To provide optimal therapy in life–threatening situation before a diagnosis can be established with certainty.
4. To lessen the toxicity of individual drugs by reducing the dosage of each in combination.

THE FUTURE OF CHEMOTHERAPEUTIC AGENTS

Antibiotics are clearly one of the greatest triumphs of medical science, although they are not without their problems and disadvantages. Our most pressing concern is with the misuse of antibiotics, which increases the likelihood of the emergence of resistant strains of pathogens. Antiviral drugs are urgently needed. As research discovers additional aspects of the internal working of viruses and cells, more chemicals that are effective against viruses will probably be developed. There is considerable similarity between the requirements of cancer therapy and requirements for treatments for viral disease. It will be interesting to see whether research for one will benefit the other.

SUMMARY

Antibiotics (Greek *anti,* "against"; *bios,* "life") are chemical compounds used to kill or inhibit the growth of infectious organisms. Originally the term *antibiotic* referred only to organic compounds, produced by bacteria or molds that are toxic to other microorganisms. All antibiotics share the property of selective toxicity: They are more toxic to an invading organism than they are to an animal or human host. Penicillin is the most well known antibiotic and has been used to fight many infectious diseases, including syphilis, gonorrhea, tetanus, and scarlet fever. Another antibiotic, streptomycin, has been used to combat tuberculosis. Penicillin, the archetype of antibiotics, is a derivative of the mold ***Penicillium notatum***. Penicillin was discovered accidentally in 1928 by Fleming, who showed its effectiveness in laboratory cultures against many disease-producing bacteria. Antibiotics may broadly be classified according to types of organisms against which they act. The most common method classifies them according to their action against the infecting organism. Antibiotics may also be classed as bactericidal (killing bacteria) or bacteriostatic (stopping bacterial growth and multiplication). Penicillin-G types are effective against gram-positive strains of *streptococci, staphylococci,* and some gram-negative bacteria such as *meningococcus.* Penicillin-G is used to treat such diseases as syphilis, gonorrhea, meningitis, anthrax, and yaws. Cephalosporin C is chemically similar to the penicillins by virtue of its β-lactam ring and its action as a cell-wall inhibitor. Chloramphenicol is a broad-spectrum antibiotic originally derived from a culture of *Streptomyces.* Chloramphenicol is the drug of choice in typhoid fever and serious infections such as meningitis and epiglottitis. Streptomycin was the first effective aminoglycoside antibiotic to be discovered and is important historically for its effectiveness against *Mycobacterium tuberculosis.* The tetracyclines are a family of broad-spectrum antibiotics that inhibit the initiation of protein synthesis. The polymyxins are cyclic polypeptide antibiotics produced by *Bacillus polymyxa* that disrupt the function of the cytoplasmic membrane. Nystatin is an antifungal agent useful in the therapy of nonsystemic fungi infections. It is produced during fermentation by a strain of *Streptomyces Ilourse*i . The antimicrobial activity is restricted to yeasts and other fungi. Sulfa drugs were among the first synthetic antimicrobial drugs used to treat microbial diseases. Nisin the antibiotic formed by a lactic acid *streptococcus* is also effective as a food preservative. It is used in medicine as well as in food industry. Antibiotics are clearly one of the greatest triumphs of medical science, although they are not without their problems and disadvantages.

REFERENCES

Abrahm, E. P. (1981). The beta- lactam antibiotics. A discussion of semisynthetic penicillins and cephalosporins. Scientific American 244(6): 76-86.

Abraham, E.P., 1974" Some Aspect of the Development of the Penicillin and Cephalosporin E.Don Murray and A. W. Bourquin.

Baldry, P. (1976). The battle against Bacteria.The history of fight against diseases and the development of antibacterial drugs. Cambridge: Cambridge University Press.

Bryan, L. E., ed (1984). Antimicrobial drug resistance. Discusses mechanism of resistance to antibacterial, antifungal, antiviral, and antimalarial drugs. New York: Academic press.

Bull, A.T., Ellwood, D.C. and ratledge, C. (1979). Microbial technology: Current state, future prospects.Cambridge: Cambridge university Press.

Casida, L. E. (1986). Industrial microbiology. Published by Wiley Eastern Limited, New.

Cruegar, W and crueger 1984 Biotechnology : A textbook on industrial Microbiology.

Egorov, N. S. (1985). A scientific approach. Published by MIR publishers, Moscow. Egorov N.S.1965: Microbe-antagonistc and Biological Assessment of their Antibiotics Activity, Moscow Vysshaya Shkola Publishers.

Franklin, D.J. and G.A. Snow 1981:Biotechnology of Antimicrobial Action 3 ed. Chapman and Hallm, Ltd. London.

Garrod, L.P., H.P. Lambert and F.O.'Grady 1981 Antibiotic and Chemotherapy, Churchill Livingstone New York.

Gauri, K. K. (1981). Antiviral chemotherapy: Design of inhibitors of viral functions. Describes inhibition of viral entry, protein synthesis, and nucleic acid replication as well as the toxicity of the drugs. New York: Academic press.

Hirsch, M. S. and J.C. Kaplan. (1987). Antiviral therapy. A review of mechanism by which antiviral drug work. Scientific American. 246 (4): 76-85.

Hockenhull, D. J. D. (1963). Antibiotics IV. Streptomycin. Biochemistry of industrial microorganisms. New York: Academic press. 249-267. Pratt, W. B., and R. Fekety (1986). The Antimicrobial drugs. Avery good summary of most of the commonly used antibiotics, with uses, mode of action, pharmacology, toxicity and other information. New York: Oxford University Press.

Ketchum , Paul A (Paul Abbtt)1942- Microbiology : Concept and application.

Khokhlov A.S. and Tovarova I.I., Post et al Z. Allg. 1973 Microbiol. 13, 647.

Korzybski, Tandeusz, et al. 1978 : Antibiotics Origin, Nature and Properties, American Society for Microbiology Washington, D.C.

Sanders, C. C. (1987). Chromosomal cephalosporinases responsible for multiple resistance to newer b-lactam antibiotics. Describes methods of action of b-lactamases and methods to detect production of these enzymes by bacteria. Annual review of microbiology 41: 573-593.

Stewart, G.T.,1965 The Penicillin Group of Drug Elsevier, Amsterdam.

Waksman S.A.,1959: The Actinomyctes,Vol. I, Balimore.

Waksman S.A., Katz E., and Vining L.1958, Proc. Natl. Acad. Sci. USA, 44 602.

Wishnow, R. M. and J. L. Steinfeld (1976). The conquest of the major infectious diseases in the United states: a bicentennial retrospective. A concise summary of control of diseases, including tuberculosis, cholera malaria and yellow fever.

Suggested Readings

1. Tortora, G.J., Funke, B.R., and Case, C.L. MICROBIOLOGY: An Introduction. The Benjamin/ Cummings Publishing Company, Inc., California, 1989.
2. Ketchum, Paul A. Microbiology: Concepts and applications. John Wiley.1994.
3. Egorov, N. S. Antibiotics: A scientific Approach. Mir publications.
4. Murray, P.P., Drew, L.W., Kobayashi, G. And Thompson, J. Medical Microbiology. The CV Mosby Company, Baltimore, 1990.
5. Salle, A.J. Fundamental principles of Bacteriology. Tata-McGraw Hill Publishers, New Delhi, 1984.
6. Boyd, R.F. General Microbiology. Times Mirror/ Mosby College Publishing, Missouri, 1988.
7. Pelzer, Chain and Kring. A text book of Microbiology. McGrawHill, International Edition, 1986.

REVIEW QUESTIONS

Long Questions

1. Define a chemotherapeutic agent? Distinguish between a synthetic chemotherapeutic agent and an antibiotic.
2. List and explain five criteria used to identify an effective antimicrobial agent.
3. What similar problems are encountered with antiviral, antifungal, antiprotozoan and antihelminthic drugs?
4. Identify three methods of action of antiviral drugs. Give an example of a currently used antiviral drug for each method of action.
5. Compare and contrast the tube dilution and agar dilution test. Identify at least one advantage of it.
6. Describe the disk - diffusion test for microbial susceptibility. What information can you obtain from this test?
7. Define drug resistance? How it is produced? What measures can be taken to minimize drug resistance.

Multiple Choice Questions

1. The first antibiotic Penicillin was discovered by
   (a) Sir Alexander Fleming (b) Louis Pasteur
   (c) Paul ehrlich (d) Sir Howard Florey
2. Antibiotic Chloramphenicol produced by which microorganism
   (a) Bacillus subtilis (b) Bacillus brevis
   (c) Streptomyces venezuelae (d) Streptomyces aureofacien

3. An antimocrobial drug, kill microorganisms directly, termed as
   (a) Bactericidal (b) Bacteriostatic
   (c) Broad-spectrum (d) Narrow-spectrum
4. Chloramphenicol and Erythromycin is an antibiotic which inhibit
   (a) Bacterial cell wall synthesis
   (b) Bacterial Protein synthesis
   (c) Bacterial cell membrane synthesis
   (d) Bacterial nucleic acid synthesis
5. Which antibiotic inhibit mRNA synthesis by binding with bacterial RNA polymerase
   (a) Gramicidin S (b) Rifampin
   (c) Polymixins (d) Aminoglycosides
6. Amantidine is a type of drug
   (a) Antifungal (b) Antihelminthic
   (c) Antiprotozoan (d) Antiviral
7. The lowest concentration of chemotherapeutic agent capable of preventing microbial growth, is called as
   (a) MIC (b) MBC
   (c) MLC (d) MOC
8. Griseofulvin interferes with eukaryotic cell division and is used primarily to treat skin infection caused by
   (a) Virus (b) Bacteria
   (c) Fungi (d) Protozoa
9. The antibiotic Blasticidin S a plant antibiotic used to treat against
   (a) Botrytis cinerea (b) Piricularia oryzae
   (c) Helminthosporium (d) Alternaria
10. The antibiotic Nisin is used in the manufacture and storage of cheese, is derived from
    (a) Lactic acid staphlococcus (b) Lactic acid micrococcus
    (c) Lactic acid streptococci (d) Streptococcus

Answers

1. (a)	2. (c)	3. (a)	4. (b)	5. (b)	6. (d)
7. (a)	8. (c)	9. (b)	10. (c)		

PART–15

Nucleic Acid Sequencing and Hybridization

CHAPTER 50

Sequencing of Nucleic Acids

B. Bajpai and V. Sharma*

Dept. of Bioscience and Biotechnology, Banasthali Vidyapith, Banasthali–304 022

E-mail: vinaysharma30@yahoo.co.uk

In genetics terminology, sequencing is the process of determining the nucleotide order of a given DNA/RNA fragment. The sequence of DNA encodes the necessary information for living things to survive and reproduce. The complete set of instructions to make an organism is contained within its genome. This contains all the information required to perform all cellular activities over the course of an organism's lifetime. The precise nature of the genome differs from species to species but is always based upon the chemical deoxyribonucleic acid (DNA). In humans and other higher organisms this DNA is bound with protein molecules in structures known as chromosomes. DNA consists of two strands that wrap around each other resembling a twisted ladder. The backbone of this ladder is comprised of a set of repeating sugar and phosphate molecules. The identity of this sugar is a distinguishing feature between DNA and a similar molecule ribonucleic acid (RNA) used for other processes within cells. The rungs of the ladder are made of nitrogenous bases. There are four different bases in DNA: adenine (A), thymine (T), cytosine (C) and guanine (G). These are also present in RNA except that thymine is replaced by uracil (U). The order of the bases along the backbone is known as the sequence. Determining the sequence is therefore useful in 'pure' research into why and how organisms live, as well as in applied subjects.

Prior to the mid-1970s no method existed by which DNA could be directly sequenced. Knowledge about gene and genome organization was based upon studies of prokaryotic organisms and the primary means of obtaining DNA sequence was so-called reverse genetics in which the amino acid sequence of the gene product of interest is back-translated into a nucleotide sequence based upon the appropriate codons. Given the degeneracy of the genetic code, this process can be tricky at best. In the mid-1970s two methods were developed for directly sequencing DNA. These were the Maxam-Gilbert chemical cleavage method and the

*Corresponding author

Sanger chain-termination method. In 1975, Frederick Sanger and his colleagues, and Alan Maxam and Walter Gilbert developed two different methods for determining the exact base sequence of a cloned piece of DNA. These spectacular breakthroughs revolutionized molecular biology. This has allowed biologists to determine the sequence of thousands of genes, and they have even made it possible to sequence all 3 billion base pairs of the human genome.

MAXAM – GILBERT METHOD

In the late 1970s, A. M. Maxam and W. Gilbert devised the first method for sequencing DNA fragments containing up to ~500 nucleotides. In this method, samples of full-length 5′end labeled DNA fragments are chemically cleaved with base specific reagents. This is a two-step catalytic process involving piperidine and two chemicals that selectively attack purines and pyrimidines (Maxam and Gilbert, 1977). Purines react with dimethyl sulfate and pyrimidines react with hydrazine in such a way as to break the glycoside bond between the ribose sugar and the base displacing the base (Step 1). Piperidine then catalyzes phosphodiester bond cleavage where the base has been displaced (Step 2). Moreover, dimethyl sulfate and piperidine alone will selectively cleave guanine nucleotides but dimethyl sulfate and piperidine in formic acid will cleave both guanine and adenine nucleotides. Similarly, hydrazine and piperidine will cleave both thymine and cytosine nucleotides whereas hydrazine and piperidine in 1.5M NaCl will only cleave cytosine nucleotides (Fig. 1). The reactions are controlled so that each labeled chain is likely to broken only once. The labeled sub-fragments created by all four reactions have the label at one end and the chemical cleavage point at the other. This labeled substrate is subjected to four separate cleavage reactions, each of which creates a population of labeled cleavage products ending in known nucleotides. The reactions are loaded on high percentage polyacrylamide gels and the fragments are resolved by electrophoresis. The gel is then transferred to a light-proof X-ray film cassette, a piece of X-ray film placed over the gel, and the cassette placed in a freezer for several days. Wherever a labeled fragment stopped on the gel the radioactive tag would expose the film due to particle decay (autoradiography). Since electrophoresis, whether in an acrylamide or an agarose matrix, will resolve nucleic acid fragments in the inverse order of length, that is, smaller fragments will run faster in the gel matrix than larger fragments, the dark autoradiographic bands on the film will represent the 5′-3′ DNA sequence when read from bottom to top (Fig. 2).

The process of base calling involves interpreting the banding pattern relative to the four chemical reactions. For example, a band in the lanes corresponding to the C only and the C + T reactions would be called a C. If the band was present in the C + T reaction lane but not in the C only reaction lane it would be called a T. Similarly, bands appearing in the G and C lanes can be read directly, bands in lane A+G lane that are not duplicated in G lane are read as A and vice-versa. The disadvantages of this method are that it requires large amounts of radioactive material, either 35S or 32P, large, paper thin acrylamide gels and hydrazine is a neurotoxin.

SANGER METHOD (DIDEOXY METHOD)

At about the same time as Maxam-Gilbert DNA sequencing was being developed, Sanger was developing an alternative method, rather than using chemical cleavage reactions, Sanger opted for a method involving a third form of the ribose sugars. As shown in Fig. 3, Ribose has a hydroxyl group on both the 2′ and the 3′ carbons whereas deoxyribose has only the one

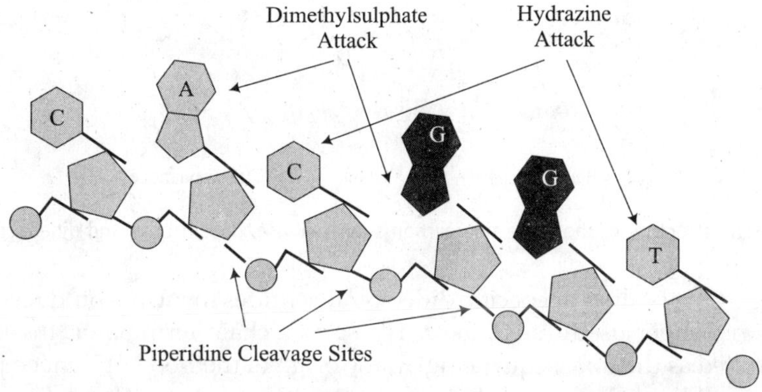

Fig. 1. Chemical targets in the Maxam-Gilbert DNA sequencing strategy. Dimethylsulphate or hydrazine will attack the purine or pyrimidine rings respectively and piperidine will cleave the phosphate bond at the 3′ carbon.

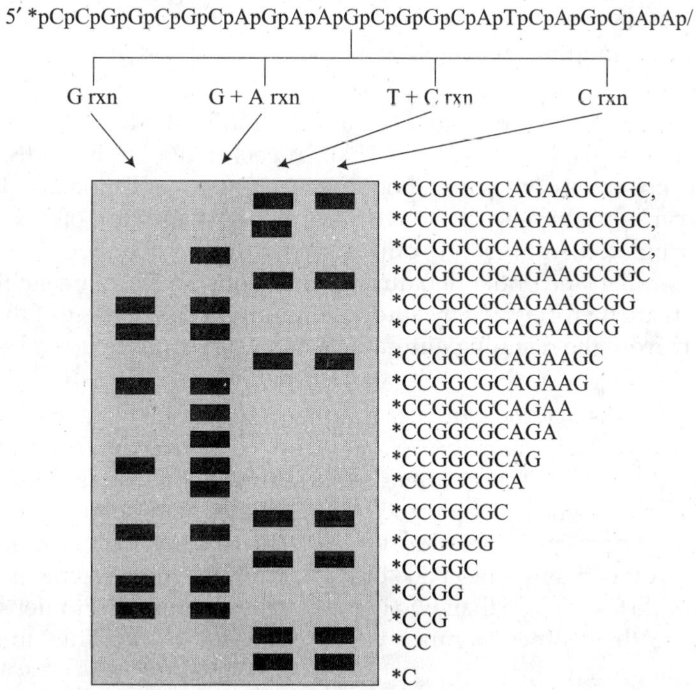

Fig. 2. The Maxam-Gilbert manual sequencing scheme. The target DNA is radiolabeled and then split into the four chemical cleavage reactions. Each reaction is loaded onto a polyacrylamide gel and run. Finally, the gel is autoradiographed and base calling proceeds from bottom to top.

hydroxyl group on the 3′ carbon and there is a third form of ribose in which the hydroxyl group is missing from both the 2′ and the 3′ carbons, this is dideoxy ribose.

Whenever a dideoxynucleotide is incorporated into a polynucleotide, the chain would irreversibly stop or terminate since coupling in DNA and RNA occurs through the 3′ hydroxyl

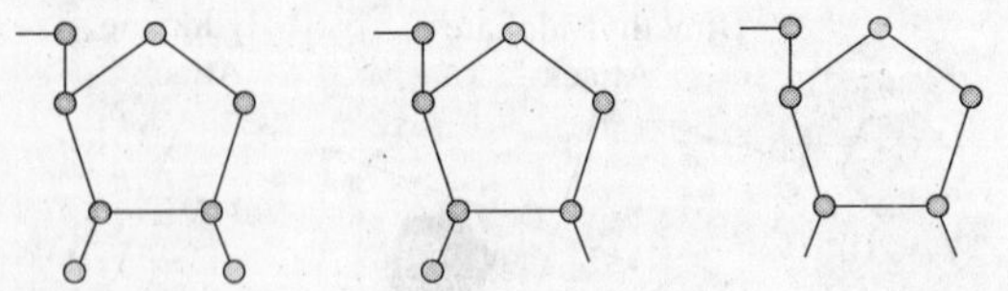

Fig. 3. The structure of the three five carbon sugars ribose, deoxyribose and dideoxyribose.

group. Thus, the incorporation of specific dideoxynucleotides in vitro would result in selective chain termination. Sanger used this dideoxy ribose as a chain terminator, therefore Sanger's method is also called as dideoxy sequencing. In the original method M13 phage particles were used for the generation of single stranded DNA (details in section). These days, one can start with double-stranded DNA and simply heat it to create single stranded DNAs for sequencing. This single-stranded DNA to be sequenced serves as the template strand for *in vivo* DNA synthesis, a synthetic 5'-labeled oligonucleotide primer is designed to hybridize to a sequence adjacent to the multiple cloning region of the vector and oriented with its 3'-end pointing toward the insert in the multiple cloning region.

As shown in the Fig. 4, four separate polymerization reactions are performed, each with a low concentration of one of the four ddNTPs in addition to the higher concentration of the normal deoxynucleotide triphosphates (dNTPs). In each reaction, the ddNTP is randomly incorporated at the positions of the corresponding dNTP; such addition of ddNTP terminates polymerization because the absence of 3'-hydroxyl prevents addition of next nucleotide. The mixture of the terminated fragments from each of the four reactions is subjected to gel electrophoresis in parallel lane under denaturing conditions, so DNA are single stranded. The separated fragments are then detected by autoradiography. The sequence of the original strand can be read directly from the resulting autoradiogram. The sequence is read starting from the bottom.

The one new twist was that the sequence fragments on the gel were the complement of the actual template (Fig. 4). The major improvements of Sanger sequencing method over the Maxam-Gilbert method were the elimination of some of the dangerous chemicals, like hydrazine and better efficiency of the process. All things being equal, when dealing with nucleic acids, enzymatic processes are more efficient than chemical processes, *Taq* polymerase makes DNA strands off of a template at 500 bases per minute whereas chemical synthesis of a 25-mer oligonucleotide takes more than two hours. Under optimum conditions Sanger method could deliver two to three times as much confirmed data in the same amount of time as Maxam-Gilbert sequencing.

The advent of Sanger sequencing gave a boost to DNA sequencing in general and led to an even more rapid accumulation of sequence data for various genes and organisms. This increase in sequence data in the scientific literature also resulted in the establishment of the first DNA sequence repository by Walter Goad at Los Alamos National Laboratories in 1979. This repository has since become GenBank.

AUTOMATED FLUORESCENCE SEQUENCING

The manual sequencing as described above is powerful, but it is relatively slow. If a large amount of the DNA is to be sequenced, such as the 3 billion base pairs found in the human

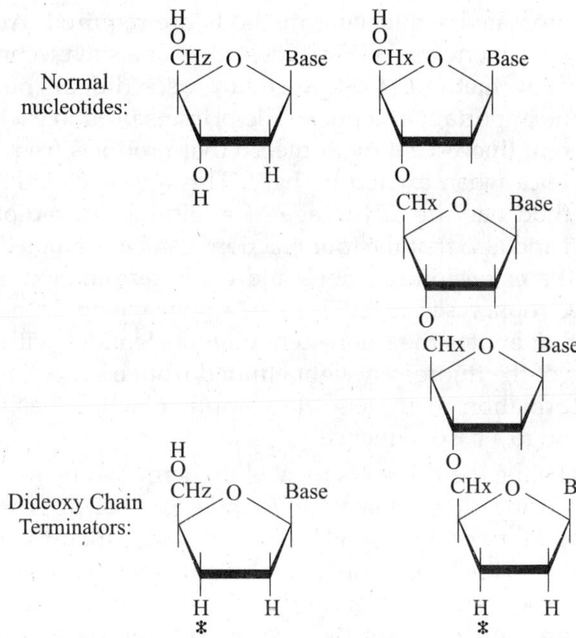

Fig. 4. The structure of normal and dideoxy nucleotides used in Sanger's chain termination method.

5′ pCpCpGpGpCpGpCpApGpApApGpCpGpGpCpApTpCpApGpCpApApA 3′

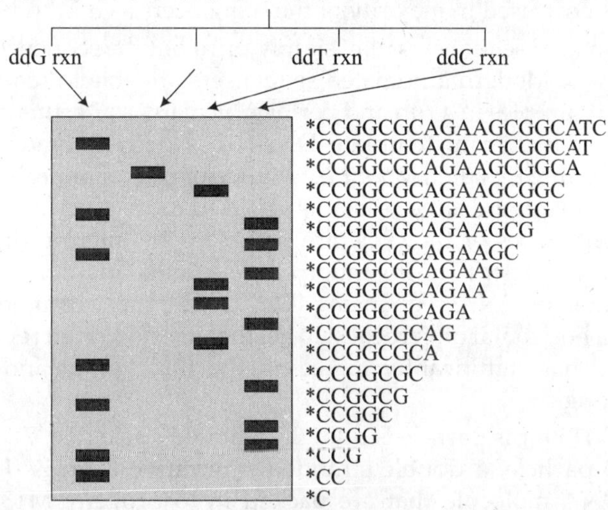

Fig. 5. A Sanger sequencing scheme. Here, a sequencing primer is radiolabeled and the reaction involves generation of sequence fragments that are the complement of the template DNA. The sequence fragments are resolved on a polyacrylamide gel and the gel is autoradiographed. Base calling for the template is the complement of the gel band.

genome, then rapid, automated sequencing methods are required. Automated sequencing method has been in use for many years, Fig. 6 describes one such technique, again based on Sanger's chain termination method. This procedure uses dideoxynucleotides, just as the manual method with one important exception. The primers used in each of the four reactions are tagged with a different fluorescent molecule, so that products from each tube will emit a different color fluorescence when excited by light. This approach is known as "dye primer sequencing". This method has the advantage of avoiding the radiolabelling of primers, increasing safety, speed and also that the four reactions can be combined and run in the single lane of the gel. After the extension reactions and chain termination are complete, all four reactions are mixed electrophoresed together in the same lane on the gel. Near the bottom of the gel is an analyzer that excites the fluorescent oligonucleotides with a laser beam as they pass by. Then the color of the fluorescent light emitted from each oligonucleotide is detected electronically. This information then passes to computer, which has been programmed to convert color information to a base sequence.

An alternative to labeling the primer is to label the terminators instead, commonly called "dye terminator sequencing". The major advantage of this approach is that complete sequencing set can be performed in a single reaction, rather than the four needed with the primer labeling approach. This is accomplished by labeling each of the dideoxynucleotide chain-terminators with a separate fluorescent dye, which fluoresce at different wavelengths. This method is easier and quicker than the dye primer approach, but may produce more uneven data peaks (different heights), due to varying incorporation of the larger, bulkier chain terminators interacting with the previously transcribed sequence of the template. To some extent this drawback has been reduced with modifications of the chemicals and the enzymes used. This method is now used in majority of the sequencers as it is both easier and cheaper.

Nowadays, automated sequencers may simply print out the sequence or send it directly to a computer for analysis. Modern automated sequencers are able to sequence as many as 396 fluoresecently labeled samples in a run and perform as many as 24 runs in a day. To produce a detectable labeled product from template DNA, it is current practice to perform cyclic sequencing. By performing repeated (about 26) rounds of primer annealing, product extension and dissociation of DNA strands, the reactants are used more efficiently, this helps to reduce the cost. The different stages of the cycle are produced by altering the temperature of the reaction using a 'thermal cycler'. The invention of the automated fluorescence DNA sequencer has advanced the level of DNA sequencing. Earlier, the most commonly used format for electrophoresis was a horizontal or vertical slab gel that has now been replaced by the capillary electrophoresis. This has significantly increased the throughput and decreased the time required for sequencing.

The DNA of an M13 virus particle is a single stranded circle (Fig. 7). When an *E.coli* cell is infected by an M13 particle, a double stranded replicative form is first produced. Single stranded circular DNA molecule that are packed in to progeny M13 phage particles are synthesized from this double stranded molecule. Cells that are infected with M13 do not lyse: instead they continually secrete infective virus particles into the surrounding medium. By removing a non-essential segment of M13 DNA, it became possible to add cloned DNA without destroying the infectivity of the new phage particles. Bacteriophage M13 has been used as combined cloning and DNA sequencing vector system. Usually, target DNA that is not much larger than 500 bp is cloned into the multiple cloning sites that forms part of a cloned

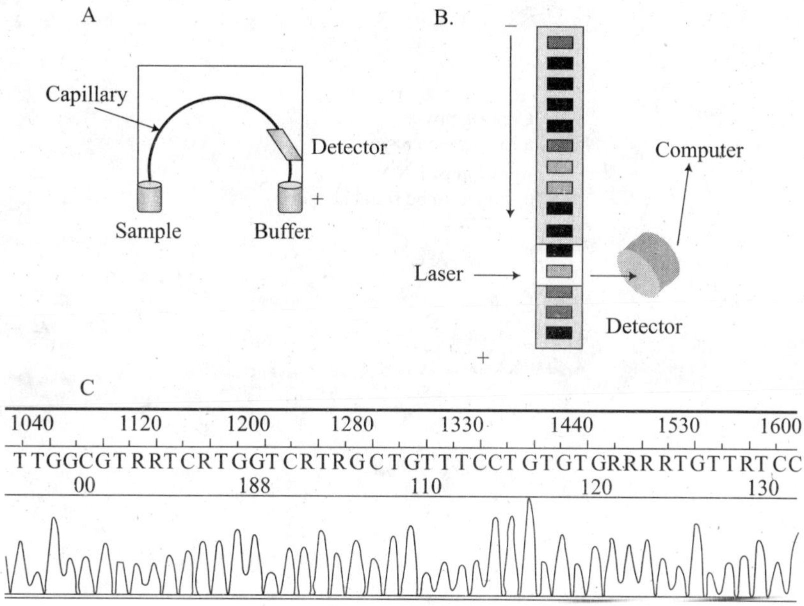

Fig. 6. Schematic representation of a capillary-based DNA sequencing system. A. The basic set up of a capillary system. The sequencing reaction is placed in the sample holder, the electrophoresis buffer is held in the second holder, the capillary has been filled with polymer, and the electrophoresis current is applied through the capillary. B. False color representations of the resolving sequence fragments running past the optical window, being excited by the laser, the detector reading emission wavelengths and sending that information to the computer where the software has been trained for base calling. C. The final sequence electropherogram output.

modified *lac* Z gene on the double-stranded replicative form of M13. *E. coli.* Cells containing the insert are picked up following the blue-white colony selection on X-gal containing medium. The bacteriophage particles from the white colony are isolated and used as a source of single stranded DNA for DNA sequencing. A primer oligonucleotide that hybridizes to the vector DNA near the insertion site of the cloned DNA is added to the single-stranded M-13 cloned DNA construct, the dideoxy reactions are carried out, and the sequence of the insert is read from the autoradiograph.

When a large piece of DNA (< 2,000 bp) needs to be sequenced, there are a number of strategies that may be employed. One approach is to clone the DNA fragment into a conventional plasmid cloning vector and create a detailed restriction map of the insert. With this information, overlapping segments of 100 to 500 bp of the insert DNA can be identified, and each segment can be subcloned into a different M13 vector and sequenced. The compete sequence of the cloned DNA is reconstructed by determining the sequences of all of the subcloned pieces. To ensure that correct sequence is obtained and there is no ambiguity regarding the identity of a single nucleotide, both strands must be sequenced several times. The sequencing of both the strands is facilitated by constructing the M13 system so that each fragment of the original DNA insert can be subcloned in two orientations. In one of the orientations the primer initiates the synthesis of one of the strands, and in the other orientation the same primer directs the synthesis of the other strand.

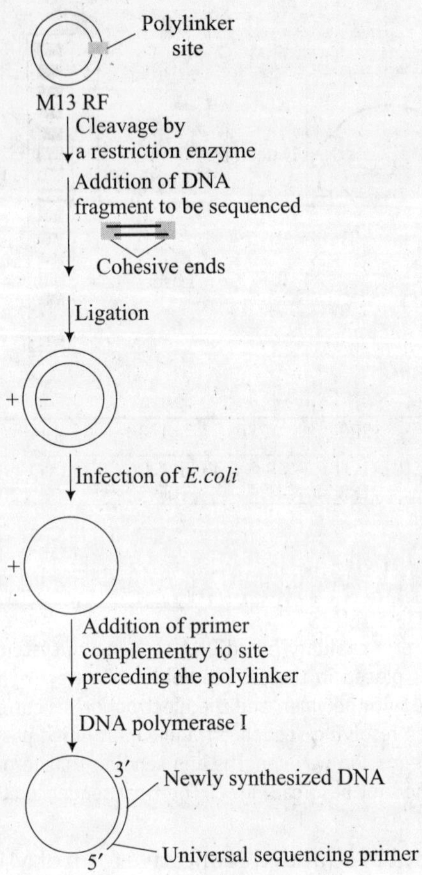

Fig. 7. Replication in M13 phage particle.

Primer Walking

For very long pieces of DNA (>5,000 bp), the M13 sequencing system, although feasible, becomes time consuming because the number of M13 vectors containing overlapping subsequences that have to be created and maintained is very large. To overcome this inconvenience, double stranded plasmid DNA sequencing protocols that do not require subcloning steps have been developed. In these protocols, plasmid DNA containing the target DNA is purified and annealed with a synthetic oligonucleotide that base pairs with one strand of the vector near the insertion site of the cloned DNA. The dideoxyribonucleotide sequencing reactions are carried out to determine the identity and order of the first 250 to 350 nucleotides of the insert DNA. On the basis of this analysis, a second oligonucleotide primer designed to hybridized to a region about 300 nucleotides away from the binding site of the initial primer is chemically synthesized and then used to determine the sequence of the next 250 to 350 nucleotide. In a similar manner, a third binding site is selected, another oligonucleotide is synthesized and the sequence of the next 250 to 350 bases is determined. This "primer walking" strategy proceeds until the complete cloned DNA is sequenced.

The complementary strand may be sequenced in the same way, starting with a oligonucleotide primer that hybridizes to this strand near the insertion site of the cloned DNA. To prevent the non-specific amplification, the primers used in this case are at least 24 nucleotides long. In addition, stringent annealing conditions that do not permit the spurious binding of the primer are maintained. Pieces of DNA that have been cloned into bacteriophage λ (~20 kb) or a cosmid vector (~40 kb) have been sequenced by this technique.

Polymerase Chain Reaction

The polymerase chain reaction is an effective procedure for generating large quantities of a specific DNA sequence in vitro. The amplification, which can be more than a million fold, is achieved by a three step cycling process. The essential requirements for PCR are: (1) two synthetic oligonucleotide primers (~20 nucleotides each) that are complementary to regions on opposite strands that flank the target DNA sequence and that, after annealing to the source DNA, have their 3′-hydroxyl ends oriented toward each other; (2) a target sequence in a DNA sample that lies between the pair of primers which can be from 100 to 5000 bp in length; (3) a thermostable DNA polymerase that can withstand heating to 95°C or higher and (4) the four deoxyribonucleotides.

A typical PCR entails a number of cycles for amplifying a specific DNA sequence; each cycle has three successive steps.

1. Denaturation: The first step in the PCR amplification system is the thermal denaturation of the DNA sample by raising the temperature within a reaction tube to 95°C. In addition to the source DNA, this tube contains a vast molar excess of the two oligonucleotide primers, a thermostable DNA polymerase and four deoxyribonucleotides. The temperature is maintained for about one minute.
2. Renaturation: For the second step, the tempeature of the mixture is slowly cooled to ~55°C. During this step, the primers base pair with their complementary sequences in the source DNA.
3. Synthesis: In the third step, the temperature is raised to ~75°C, which is optimum for the catalytic functioning of *Taq* DNA polymerase. DNA synthesis is initiated at the 3′-hydroxyl end of each primer.

The duration of each step and the temperature changes required during a PCR cycle are usually carried out in an automated, programmable block heater in which the reaction tubes are immersed. One cycle generally lasts from 3 to 5 minutes.

Chromosome Walking

Chromosome walking is a method in genetics for identifying and sequencing long parts of a DNA strand, e.g., a chromosome. As long DNA strands cannot be sequenced, this method works by dividing the long sequence into several short ones. Basically, chromosome walking works as follows:

A primer that matches the beginning of the DNA to sequence is used to synthesize a short DNA strand complementary to the unknown sequence, starting with the primer (see PCR). The new short DNA strand is sequenced. The end of the sequenced strand is used as a primer for the next part of the long DNA sequence. In this way, the short part of the long DNA that is sequenced keeps "walking" along the sequence. The method can be used to sequence entire chromosomes (thus, *chromosome walking*). A different method with the same purpose that became more popular for large-scale sequencing (e.g., the Human Genome Project) is shotgun sequencing.

Shotgun Sequencing

Shotgun sequencing is a method used in genetics for sequencing long DNA strands. Small segments of DNA are cloned randomly, and then these clones are sequenced. By doing this for several copies of the same long DNA strand, overlapping fragments are created. Finally, computer programs align these overlapping sequences and determine the original (long) sequence.

An extremely simplified example with only two sequences is as follows:

Original strand: XXXAGCATGCTGCAGTCATGCTTAGGCTAXXXX
First shotgun sequence: XXXAGCATGCTGCAG
TCATGCTTAGGCTAXXXX
Second shotgun sequence: TTAGGCTAXXXX
XXXAGCATGCTGCAGTCATGC
Reconstructed strand: XXXAGCATGCTGCAGTCATGCTTAGGCTAXXXX

In real-world applications, where thousands or millions of sequences are to be aligned, the computational power of computers is used to realign the sequences. For the shotgun sequencing of the human genome in the Human Genome Project run by Celera Genomics in 2000, several supercomputers were running some months nonstop to align all human DNA correctly.

FUTURE TECHNOLOGY

With the completion of human genome sequence map, there is a desire to sequence many individual human genomes to provide high-resolution genotypes that can be used to determine the complex relationship among disease, pharmaceutical efficacy and genetic variability. Further, mRNA sequencing is valuable to determine exon splicing patterns and as a tool to discover gene function from cotext-specific expression data, many areas that are highly polymorphic in both exons and introns of the genome will be known. Resequencing of the polymorphic areas that are linked to disease development will greatly contribute to the understanding of the disease and therapeutic development. Thus, high-throughput accurate methods for resequencing the highly variable intron/exon regions in the genome are needed in order to explore the potential of the complete human genome sequence map and move onto the next logical step: massive comparative genomic studies, disease gene discovery and personalized medicines. The current high-throughput DNA sequencing methods have inherent limitations especially in cost and read length and are not suitable for large scale DNA sequencing projects, also the accuracy required for mutation detection needs to be improved for a wide variety of applications ranging from disease gene discovery to forensic identification. In addition, the first few bases after the priming site are often masked by the high fluorescence intensity from the excess dye-dye labeled primers or terminators, and are therefore difficult to identify. The limitations of electrophoretic approach have prompted researchers to work on alternative methods viz. sequencing by hybridization, mass spectrometry, microfluidic system, pyrosequencing, polymerase colonies, sequencing using nanopores and massively parallel single molecule sequencing. While some of these methods are promising, none has yet bested electrophoretic separation. However, they may find important niche applications in areas such as SNP determination, mini-sequencing and gene expression analysis.

Sequencing by Hybridization (SBH)

Sequencing by hybridization (SBH) is a novel DNA sequencing technique in which an array (SBH chip) of short sequences of nucleotides (probes) is brought in contact with a solution of the target DNA sequence. In general, arrays are described as *macroarrays* or *microarrays*, the difference being the size of the sample spots. Macroarrays contain sample spot sizes of about 300 microns or larger and can be easily imaged by existing gel and blot scanners. The sample spot sizes in microarray are typically less than 200 microns in diameter and these arrays usually contain thousands of spots. Microarrays require specialized robotics and imaging equipment. DNA microarray or DNA chips are fabricated by high-speed robotics, generally on glass but sometimes on nylon substrates, for which probes with known identity are used to determine complementary binding, thus allowing massively parallel gene expression and gene discovery studies. An experiment with a single DNA chip can provide researchers information on thousands of genes simultaneously – a dramatic increase in throughput. In this technique a "probe" is the tethered nucleic acid with known sequence, whereas a "target" is the free nucleic acid sample whose identity/abundance is being detected.

There are two variants of the DNA microarray technology, in terms of the property of arrayed DNA sequence with known identity:

Format I: probe cDNA (500~5,000 bases long) is immobilized to a solid surface such as glass using robot spotting and exposed to a set of targets either separately or in a mixture and is traditionally called as DNA microarray. **Format II:** an array of oligonucleotide (20~80-mer oligos) or peptide nucleic acid (PNA) probes is synthesized either *in situ* (on-chip) or by conventional synthesis followed by on-chip immobilization. The array is exposed to labeled sample DNA, hybridized, and the identity/abundance of complementary sequences are determined. This method, "historically" called DNA chips, was developed at Affymetrix, Inc., which sells its photolithographically fabricated products under the *GeneChip®* trademark.

Arrays of tens of thousands of clones are interrogated with short oligomer probes to determine sets of oligomers that are present in individual clones. SBH requires highly discriminative hybridizations with a large number of probes. A DNA probe chip is a biochemical chip that contains all possible combinations of sub-sequences or probes (of a specified length N) of nucleotide bases. Hence total number of probes in a DNA probe chip will be 4 N. For example, if N equals 4, then the probe chip will contain all possible probes of length 4 starting from AAAA to CCCC. The probes on the chips are built by combinatorial chemical synthesis. In hybridization technique, a solution containing the DNA fragment that is to be sequenced is washed over the DNA probe chip. The sub-sequences of the DNA strand will hybridize the complementary probes (A-T and G-C are complementary to each other) in the DNA chip. For, example if the original DNA fragment is GACCATCGGA, then the hybridized probes will be GTAG(CATC), TAGC(ATCG), TGGT(ACCA), CTGG(GACC), GGTA(CCAT), GCCT(CGGA), AGCC(TCGG). The strings within the bracket represent the sub-sequences present in the original string. In a way, the oligonucleotide hybridization technique involves single biochemical experiment in contrast with the conventional methods of DNA sequencing. However, the main problem in hybridization technique is to reconstruct the target DNA sequence from the information available from the hybridized probes. Both the precision (inter-spot distance and roundness of spots) and density of microarrays (Table 1) are increasing rapidly. There are two major applications of the DNA microarray technology: Identification of sequence (gene/gene mutation) and determination of expression level (abundance) of genes.

Table 1. Steps in the design and implementation of a DNA microarray experiment

(1) Probe (cDNA/oligo with known identity)	(2) Chip fabrication (Putting probes on the chip)	(3) Target (fluorescently labeled sample)	(4) Assay	(5) Readout	(6) Informatics
Small oligos, cDNAs, chromosome, ... (whole organism on a chip?)	Photolithography, pin-based fluid transfer, piezoelectric (ink-jet dispensers),	RNA, (mRNA==>) cDNA	Hybridization, long, short, ligase, base addition, electric, MS, electrophoresis, fluocytometry, PCR-DIRECT, TaqMan, ...	Fluorescence, probeless (conductance, MS, electrophoresis), electronic, ...	Robotics control, Image processing, DBMS, WWW, bioinformatics, data mining and visualization

Microfluidic System

Microfluidic devices take advantage of the chemical properties of liquids and gases and the electrical properties of semiconductors and combine them on a single chip. The minuscule systems, whose dimensions are measured in nanometers, perform tasks that range from detecting airborne toxins to analyzing DNA sequences (Fig. 8). The devices offer several advantages over conventionally sized systems, including compact size, disposability, increased functionality, and they require smaller volumes of reagents and samples. Microfluidic technology has made its most significant mark in biological chips designed to facilitate identification of DNA sequences. The two main schemes used in miniaturized fluidic systems are "array based systems where sample or reagent is immobilized in large arrays on a plate or chip fluids are flushed over the surface" or "microfluidic channels that form complex manifolds for fluid manipulation and controlled delivery of samples and reagents". Advances in microfluidic chip technology are expected to play a major role in the development of cost-effective and rapid DNA analysis methods.

The first lab-on-a-chip device available for sale was the Agilent 2100 Bioanalyzer, the only desktop device that can analyze DNA, RNA, proteins, and cells. Biotechnology researchers using the device simply load multiple microliter-sized samples on the chip, press a button, and minutes later view the resulting data on a laptop computer. A sequencing by synthesis (SBS) method that uses microfluidic plumbing (μSBS), involves exposing a primed DNA template to a mixture of known type of standard nucleotides, its fluorescently tagged analog and DNA

Fig. 8. Microfluidic device.

polymerase. If the tagged nucleotide is complementary to the template base next to the primer's end, the polymerase extends the primer with it and fluorescence signal is detected after a washing step, as the tag would then be attached to the DNA template. Since the type of the tagged nucleotide is known in advance, the presence of signal reveals the template base. Iteration with each type of nucleotide reveals the DNA sequence. This approach to DNA sequencing has advantages in economy of material and integrability under the lab-on-chip paradigm. The microfluidic and surface chemistry technologies can be applied to many other systems involving DNA studies, pyrosequencing being one example.

Pyrosequencing

Pyrosequencing is a DNA sequencing technique that is based on the detection of released pyrophosphate during (Ppi) during DNA synthesis (Fig. 9). Pyrosequencing is a user-friendly, real-time and non-electrophoretic technique, which avoids the need for labeled nucleotides and labeled primers. This technique employs enzymatic reactions, catalyzed by ATP sulfrylase and luciferase, to monitor the inorganic pyrophosphate released during nucleotide incorporation (Fig. 10). Unreacted nucleotides are degraded by apyrase, allowing the iterative addition of nucleotides.

The kinetics of the pyrosequencing enzymes can be studied in real time by following the pyrosequencing signals (a pyrogram). The slope of the ascending line in a pyrogram is determined mainly by the activities of the polymerase and ATP sulfurylase; the height of the signal is determined by the activity of the luciferase, and the slope of the descending curve by the efficiency of the nucleotide removal. In solid-phase system using microfluidics, which employs the three enzyme system, the descending curve is determined by the washing efficiency. In the four-enzyme system of liquid phase pyrosequencing, the accumulation of inhibitory substances decreases the efficiency of luciferase and apyrase (Figs. 11 and 12). In both the systems, the activity of ATP sufurylase is relatively constant during the sequencing reaction. In pyrosequencing, the most critical reactions are DNA polymerization and nucleotide removal by either washing or enzymatic degradation. Nucleotide removal (descending curve) competes with the polymerization reaction (ascending curve) (Fig. 13). Therefore, slight changes in the kinetics of these reactions directly influence the performance of the sequencing reaction.

$$(NA)_n + \text{Nucleotide} \xrightarrow{\text{Polymerase}} (NA)_{n+1} + PPi$$

$$PPi + APS \xrightarrow{\text{ATP sulfurylase}} ATP + SO_4^{2-}$$

$$ATP + \text{Luciferin} + O \xrightarrow{\text{Luciferase}} AMP + PPi + \text{Oxyluciferin} + CO_2 + \text{light}$$

Fig. 9. The general principle behind different pyrosequencing reaction systems. A polymerase catalyzes incorporation of nucleotide(s) into a nucleic acid chain. As a result of the incorporation, a pyrophosphate (PPi) molecule(s) is released and subsequently converted to ATP, by ATP sulfurylase. Light is produced in the luciferase reaction during which a luciferin molecule is oxidized.

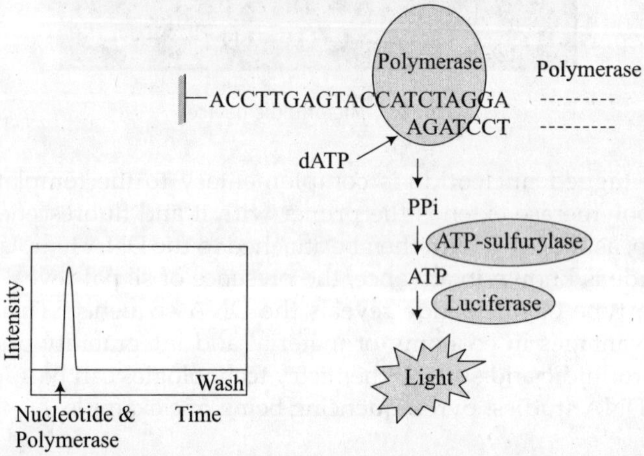

Fig. 10. Schematic representation of the progress of the enzyme reaction in solid-phase pyrosequencing. The four different nucleotides are added stepwise to the immobilized primed DNA template and the incorporation event is followed using the enzyme ATP sulfurylase and luciferase. After each nucleotide addition, a washing step is performed to allow iterative addition.

There are several advantages with genotyping by pyrosequencing compared with the available methods: (i) pyrosequencing determines the exact sequence thereby providing the same accuracy as conventional sequencing methods (ii) the technique dispenses with the need for labeled nucleotide, labeled primers and electrophoresis and (iii) large number of samples can be analysed in a short time. This technique has now been multiplexed which enables multiple analyses of one or several target DNA thereby decreasing the cost for PCR, template preparation and genotyping. Furthermore, the analysis time can be shortened allowing processing of a large number samples. In addition several peaks are obtained in a pyrogram of multiplex pyrosequencing and their heights can be compared in order to confirm the number of incorporated nucleotides, thereby making the typing very accurate.

One inherent challenge with standard pyrosequencing is proportionality of the signal intensity in a pyrogram when >5 nt in a row are incorporated. This problem is due mainly to the longer time needed for polymerization by DNA polymerase and rapid degradation of nucleotides by apyrase. Two fully automated machines using this system are now available that analyse 96 or 384 samples simultaneously, which enables analysis of 5000 samples per 8h or 50,000 per day, respectively.

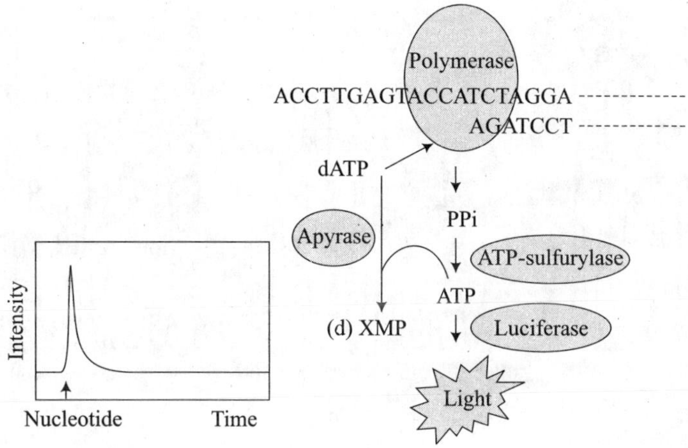

Fig. 11. Schematic representation of the progress of the enzyme reaction in liquid-phase pyrosequencing. Primed DNA template and four enzymes involved in liquid-phase pyrosequencing are placed in a well of a microtiter plate. The four different nucleotides are added stepwise and incorporation is followed using the enzyme ATP sulfurylase and luciferase. The nucleotides are continuously degraded by nucleotide-degrading enzyme allowing addition of subsequent nucleotide. dXTP indicates one of the four nucleotides.

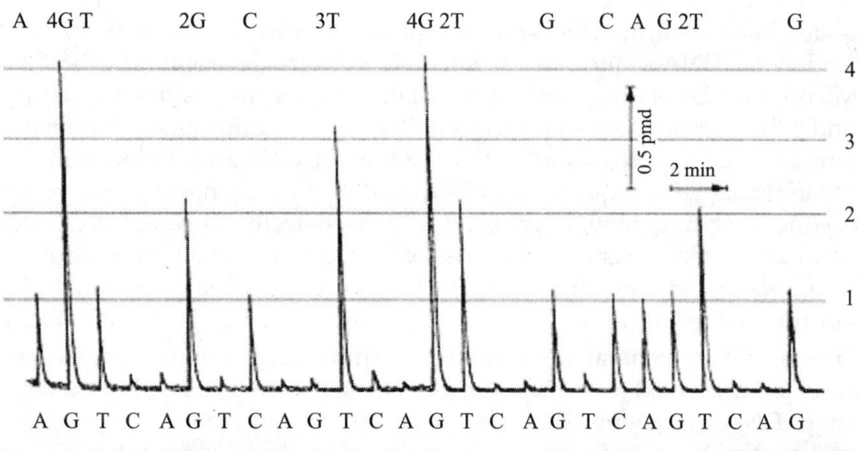

Fig. 12. Pyrogram of the raw data obtained from liquid-phase pyrosequencing. Proportional signals are obtained for one, two, three, and four base incorporations. Nucleotide addition, according to the order of nucleotides, is indicated below the pyrogram and the obtained sequence is indicated above the pyrogram.

Applications of Pyrosequencing

Pyrosequencing has opened up new possibilities for performing sequence-based DNA analysis. The availability of an automated system for liquid-phase pyrosequencing has allowed the technique to be adapted for high-throughput analysis. Some of the potential applications of pyrosequencing are: Genotyping of Single-Nucleotide Polymorphisms, Microbial typing, Resequencing, Tag sequencing and analysis of difficult secondary structures.

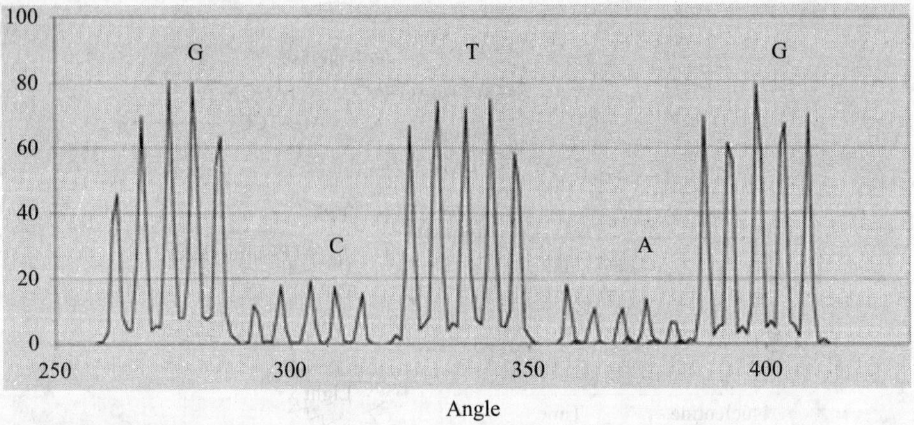

Fig. 13. Pyrogram obtained from five different chambers using microfluidics in a spinning CD-format device developed at Gyros AB (Sweden). All chambers contained the same DNA template. The addition order of nucleotides was GCTA. Nucleotides G, T, and G were correctly incorporated. Background signals for nucleotides C and A were due to pyrophosphate (PPi) contamination in the pyrosequencing reaction mixture.

MASS SPECTROMETRY

Matrix assisted laser desorption ionization time-of-flight mass spectrometry (MS) has been explored widely for DNA sequencing. Compared with gel-electrophoresis based sequencing systems, MS produces a very high resolution of DNA-sequencing fragments, fast separation on microsecond time scales and completely eliminates the compression associated with gel electrophoresis. The high resolution of MS allows accurate mutation and heterozygote detection. One challenge to sequencing DNA with MS is the stringent purity requirement for the DNA sequencing fragments introduced to mass detector, because DNA sequences are determined by accurately measuring the mass of the DNA-sequencing fragments, which must be free from alkaline and alkaline earth salts as well as other contaminants. Biotin and streptavidin are used for purifying the DNA fragments from MS sequencing (Fig. 14).

DNA fragments are generated enzymatically in Sanger's sequencing reactions, with the order of the peaks generated in MALDI corresponding to the sequence. For the separation and the detection DNA fragments that are produced in enzymatic termination reactions, conventional approaches rely on denaturing polyacrylmide gel electrophoresis coupled with radioactive or fluorescence based detection schemes.

Capillary Electrophoresis

Automated DNA sequencers using laser-based fluorescence detection in junction with conventional slab gel electrophoresis have theoretical capacities of about 10,000 bases of raw sequence data per day. The efficiency of this system is limited by the number of the samples that can be analysed at one time and the speed of the separation process. The separation speed is limited in slab gel by the magnitude of the electric field which can be applied to the gel without excessive heat production. Recently, it has been shown that if electrophoretic separation is brought about in very small diameter capillaries (typically 50-100 μm i.d.), a much

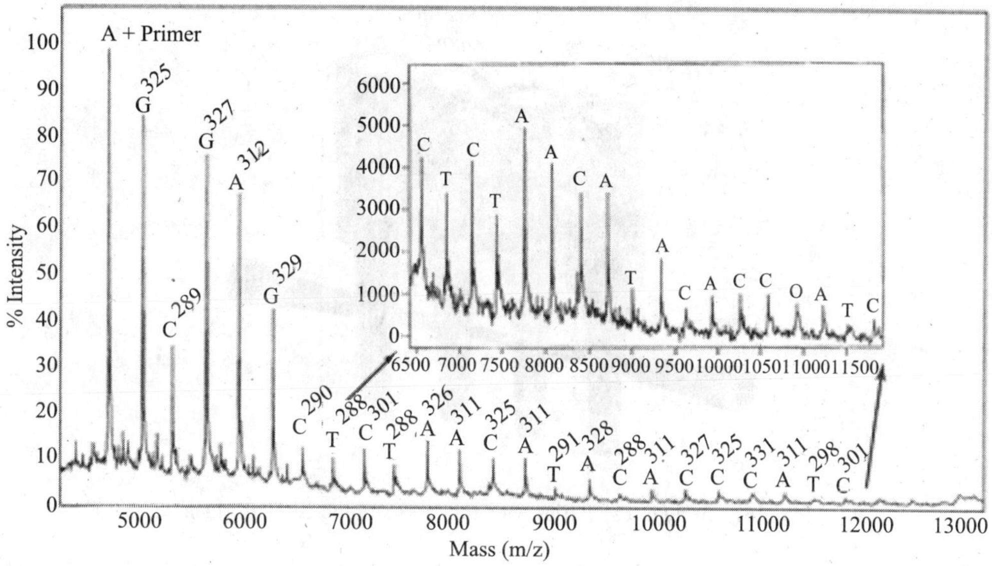

Fig. 14. Mass-sequencing spectrum generated using Biotin-11-ddNTPs. The inset shows a magnification of the lower intensity region. The mass difference (indicated above each base) between a peak and the previous one is used to determine the identity of the nucleotide.

greater field may be applied due to reduced Joule heating (Fig. 15). RIKEN (Tsukuba and Wako, Japan) has developed RISA (RIKEN Integrated Sequence Analyser), a 384 capillary sequencer with disposable capillaries and cross-linked acrylamide. In RISA 384 samples are simultaneously electro-kinetically injected and detected by laser fluorescence.

Capillary-based Sequencers

The separation speed of the slab gel electrophoresis is limited by the magnitude of the electric field which can be applied to the gel without excessive heat production. However, if electrophoresis is performed in very small diameter capillaries (typically 50-100 μm i.d.), a much greater field may be applied due to the reduce Joule heating. Separation performed by this method, known as capillary electrophoresis (CE), are generally characterized by very high speed and resolutions. It is possible to use liquid polymers in CE, which provides additional advantages over traditionally used agarose and polyacrylamide gels, it allows gel matrices to be replaced during runs thus avoiding cross-contamination. Some of the common polymers used are: cross-linked polymers, polyvinyl alcohol, dextran etc.

Online DNA Sequencers

Online systems have been developed to automate and integrate thermal cycling, purification, loading and capillary electrophoresis steps. Samples are placed in the loop of an injection valve, amplified inside of a rapid air thermal cycler, separated by liquid chromatography, loaded in a continuous, flow through manner onto a polymer-filled separation capillary and detected by laser-induced fluorescence.

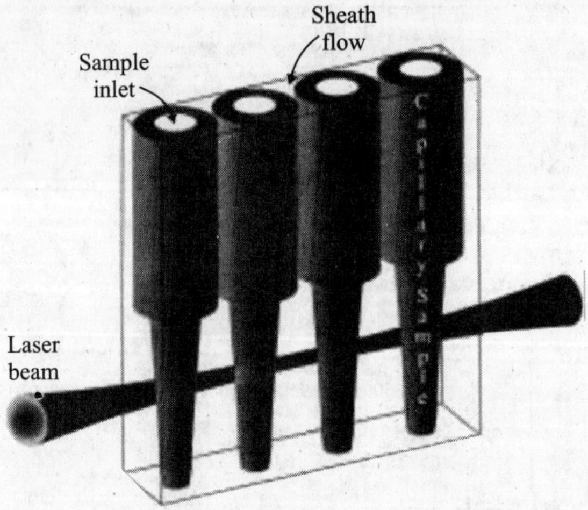

Fig. 15. Array of capillaries with the laser shining through them.

Single Molecule Analysis

Sequence information can be obtained from a single DNA molecule and research is going on to develop new sequencing technologies either by observing the interaction of particular proteins with DNA or by using ultra-high resolution probe-microscopy. Although none of these methods have been demonstrated experimentally, but unlike conventional technology, their speed and read length would not be inherently limited by the resolving power of the electrophoretic separation. Single molecule sensitivity might permit direct sequencing of mRNA from rare cell populations or perhaps even individual cell. A major obstacle in the development of singe molecule sequencing scheme is that DNA has extraordinarily high linear data density, with a pitch of only 3.4Å between successive bases.

RNA Sequencing

RNA is less stable in cell and also more prone to nuclease attack experimentally. As RNA is generated by transcription from DNA, the information is already present in the cell's DNA. However, it is sometimes desirable to sequence RNA molecules. In particular, in eukaryotes RNA molecules are not necessarily co-linear with their DNA templates, as introns are excised. To sequence RNA, the usual method is first to reverse transcribe the sample to generate DNA fragments, this can then be sequenced using any of the DNA sequencing method.

SUMMARY

DNA sequencing is the determination of the order of nucleotides or bases in a sample of DNA to obtain definitive information about it. The function of a gene can be deduced from its nucleotide sequence, which also provides a starting point for research into the thousands of diseases that have a genetic basis. In addition, knowledge of DNA sequence is a pre-requisite for planning any substantial manipulation of the DNA. Essentially, there are two conventional

DNA sequencing methods: (i) a chemical procedure, originally developed by Alan Maxam and Walter Gilbert (1977), which makes use of the chemical reagents to bring about specific cleavage of DNA and (ii) an enzymatic procedure developed by Fred Sanger (1977), often called as the dideoxy nucleotide method or chain termination method. The refinement to the dideoxy chain termination method and its easy automation has made it a preferred method. M13 bacteriophage was used to obtain the complete sequence of a cloned piece of DNA, the limitation is that large DNA inserts are unstable and though sequencing is feasible, it is time consuming. To overcome this inconvenience, a new strategy called "primer walking" (primer extension) was developed, however, sometimes false priming can give erroneous and ambiguous results. Moreover, this also involves repeated synthesis of primers which is expensive. Pieces of DNA cloned in λ phage (~20 kb) and cosmid vector (~40 kb) have been sequenced by this method. Baker's yeast genome was sequenced by "Clone by Clone" technique, which is reliable but slow method. Another method based on the alignment of overlapping sequences is "whole-genome shotgun", which is potentially very fast but it can be extremely difficult to put together so many tiny pieces of sequences all at once. Constant efforts by scientists have led to the development of high-throughput sequencing techniques with enhanced speed, reliability and cost. Automated DNA sequencers, which are based on Sanger's method but involve attachment of different fluorescent tags to each of the dideoxy analogue, offer a number of advantages over the manual sequencing. In a new gel based sequencer (capillary array electrophoresis), high resolution fragment separation is obtained in much lesser time because the tube dissipate heat well and allow the use of high electric field. The limitations of electrophoresis based methods can be alleviated by the use of Mass spectroscopy (MS), which measures the m/z ratio of particles. MS has advantage of producing very high resolution of DNA sequencing fragments, fast separation on microsecond time scale and completely eliminates the compressions associated with gel electrophoresis. DNA microarrays or gene chips, based on the principle of base-pairing or hybridization, can identify the target sequence by complementary binding of target sequence to the oligonucleotide probes immobilized on a matrix called chip. Recently, pyrosequencing has emerged as a new real time sequencing methodology. Pyrosequencing has the potential advantages of accuracy, flexibility, parallel processing and can be easily automated. In RNA sequencing, RNA molecule is first reverse-transcribed into cDNA by RT-PCR (reverse transcription polymerase chain reaction) followed by sequencing of cDNA using any one of the DNA sequencing method.

REFERENCES

Braslavsky, I., Hebert, B., Kartlov, E. and Quake, S. R. (2003) Sequence information can be obtained from single DNA molecules *Proc. Nat. Acad. Sci.* 100 : 3960 3964.

Edwards, J. R., Itagaki, Y. and Ju, J. (2001) DNA sequencing using biotinylated dideoxynucleotides and mass spectrometry Nucleic Acid Res. 29 : 104.

Kartalov, E. P. and Quake, S. R. (2004) Microfluidic devices read upto four consecutive base pairs in DNA sequencing-by-synthesis Nucleic Acid Res. 32 : 2873-2879.

Lodish, H., Berk, A., Matsudaira, P., Kauser, C. A., Krieger, M., Scott, M. P., Zipursky, S. L., Darnell, J. (2000) Molecular Cell Biology 4th edition, WH Freeman and Company, New York.

Luckey, J. A., Drossman, H., Kostichka, A. J., Mead, D. A., Cunha, J. D., Norris, T. B. and Smith, L. M. (1990) High speed DNA sequencing by capillary electrophoresis *Nucleic Acid Res.* 18(5): 4417-4421.

Meldrum, D. (2000) Automation for genomics, Part two: Sequencers, Microarrays and future trends *Genome Res.* 10 : 1288-1303.

Pourmand, N., Elahi, E., Davis, R. W. and Ronaghi, M. (2002) Multiplex pyrosequencing *Nucleic Acid Res.* 30 : 1-31.

Ronaghi, M. (2001) Pyrosequencing sheds light on DNA sequencing *Genome Res.* 11 : 3-11.

Weaver, R. F. (2003) Molecular Biology 2nd edition, McGraw Hill.

Suggested Readings

1. Molecular Biology of Gene (2004) T. A. Baker, S. P. Bell, A. Gann, M. Levine and R. Losick, 5th edition, Pearson Education.
2. Molecular Biology and Biotechnology : A comprehensive desk reference (1995) A. M. Robert, VCH Publishers Inc.
3. Short protocols in molecular biology (1992) F. M. Ausubel, R. Brent, J. G. Seidman, R. E. Kingston, J. A. Smith, 2nd edition, Green Publishing Associated and John Wiley and Sons Inc., USA.

REVIEW QUESTIONS

Long Questions

1. What is a dideoxynucleotide? What is the precise function of these in Sanger's technique of sequencing?
2. Compare and contrast the features of sequencing techniques given by Sanger and Maxam-Gilbert.
3. What is the principle of automated sequencing machine? How it works?
4. Discuss the principle and applications of pyrosequencing.
5. Describe the gel-free methods of DNA sequencing.

Multiple Choice Questions

1. Chain termination method was developed by
   (a) Sanger (b) Maxam and Gilbert
   (c) Edman (d) Muller
2. Dideoxy nucleotide lacks two
   (a) -COOH (b) $-PO_4$
   (c) -OH (d) $-SO_4$
3. Human genome has approximately
   (a) 35,000 genes (b) 40,000 genes
   (c) 30,000 genes (d) 25,000 genes
4. In a sequencing gel base calling id done from
   (a) Top to bottom (b) Bottom to top
   (c) Left to right (d) Right to left

5. Following method of sequencing was used in the human genome project
   (a) Primer walking (b) Chromosome walking
   (c) Shot gun (d) Single molecule analysis
6. Macroarrays contain spot size of more than
   (a) 300μ (b) 200μ
   (c) 250μ (d) 400μ
7. Microarrays contain spot size of less than
   (a) 100μ (b) 200μ
   (c) 150μ (d) 300μ
8. Pyrosequencing is based on the detection of ______ released during the sequencing reaction
   (a) Pyrophosphate (b) Inorganic phosphate
   (c) Diphosphate (d) Triphosphate
9. The dideoxynucleotides used in sequencing by mass spectrometry are
   (a) Biotinylated (b) Acetylated
   (c) Phosphorylated (d) Glycosylated
10. Nanopore sequencing is a method of sequencing for
    (a) Protein (b) DNA
    (c) RNA (d) Polysaccharide

Answers

1. (a)	2. (c)	3. (c)	4. (b)	5. (c)	6. (a)
7. (b)	8. (a)	9. (a)	10. (b).		

CHAPTER 51

Blotting Techniques

Mangla Bhide
Dept. of Zoology, Dr. H. S. Gour University, Sagar–470 003
E-mail: manglabhide@rediffmail.com

Southern, Northern and Western blottings are used to search out the desired DNA, RNA fragments or protein molecules by hybridization with complimentary probe or antibodies, respectively. In this process a mixture of DNA, RNA or protein fragments are separated by gel electrophoresis. The separated bands are stained and visualized directly in the gel. Molecular probes are available to confirm the identity of these bands or to find out similarities between one or more bands. It is possible to hybridize these bands with a labeled probe. Although hybridization of separated DNA fragments with a probe of small size (10-20 base) can be conducted within the gel (in gel hybridization), but to facilitate hybridization with relatively bigger probes, the bands are often transferred to a nitrocellulose membrane through blotting technique. When DNA bands are blotted, it is called Southern blotting (after the name of E. M. Southern) and when RNA bands are transferred it is described as Northern blotting. Similarly, when protein bands are transferred, the technique is known as Western blotting.

SOUTHERN BLOTTING

A technique to transfer DNA species separated by agrose gel electrophoresis to a filter for analysis was developed in the 1978. Later on Southern (1975) developed a technique for transferring single stranded DNA and denatured double stranded DNA from an agrose gel to nylon membrane or nitrocellulose filter. The blotted DNA is tightly bound to the membrane but it can still hybridize to labeled nucleic acid probe. For Southern blotting DNA sample is digested with a restriction enzyme and these fragments are electrophoresesd on agrose gel (Fig. 1). The DNA bands are denatured into single strand in the gel with the help of an alkali solution. Afterwards the gel is laid on the top of a buffer saturated filter paper, placed on a solid support (e.g., glass plate), with its two edges immersed in the buffer. A sheet of nitrocellulose membrane is placed on top of the gel. A stack of many papers (paper towels) are kept on top of this membrane. A weight of about 0.5 kg. is placed on top of the paper towels. The buffer solution is drawn up by filter paper wick, and passes through the gel to the nitrocellulose membrane and finally to the paper towels. While passing through the gel, the buffer carries with it single strand DNA. The DNA strand binds on to the nitrocellulose membrane, when the buffer passes through gel to the paper towels. After leaving this arrangement for a few hours. or overnight, paper towels are removed and discarded with labeled DNA or RNA probes. The nitrocellulose membrane with blotted single stranded DNA bands is baked at 80°C for 2-3 hours to fix the DNA permanently on the membrane. Now it has a replica of DNA bands from agrose gel, and thus can be used for hybridization with labeled DNA (Fig. 2) or RNA probes. Then the membrane is washed to remove any unbound DNA. X-ray film is exposed to the hybridized membrane to get autoradiograph. The steps involved in Southern blotting are shown in Fig. 1.

1. Preparation of DNA samples for electrophoresis by digesting it with appropriate restriction enzymes to get the desired sequence and desired length of DNA fragment.
2. Electrophoresis of DNA fragments on agrose gel for blotting.
3. Blotting can be done in different ways such as capillary blotting, electrophoretic transfer or electro blotting, vacuum blotting or positive pressure blotting.

All these methods are suitable for blotting agrose gels. However, capillary blotting is the most popular method for blotting the DNA fragment from agrose gel.

Southern blotting has many applications in molecular biological research. It is used in important steps of gene cloning project, Genomic DNA having gene of interest to be cloned is blotted and hybridized to identify one or more restriction fragments containing the desired

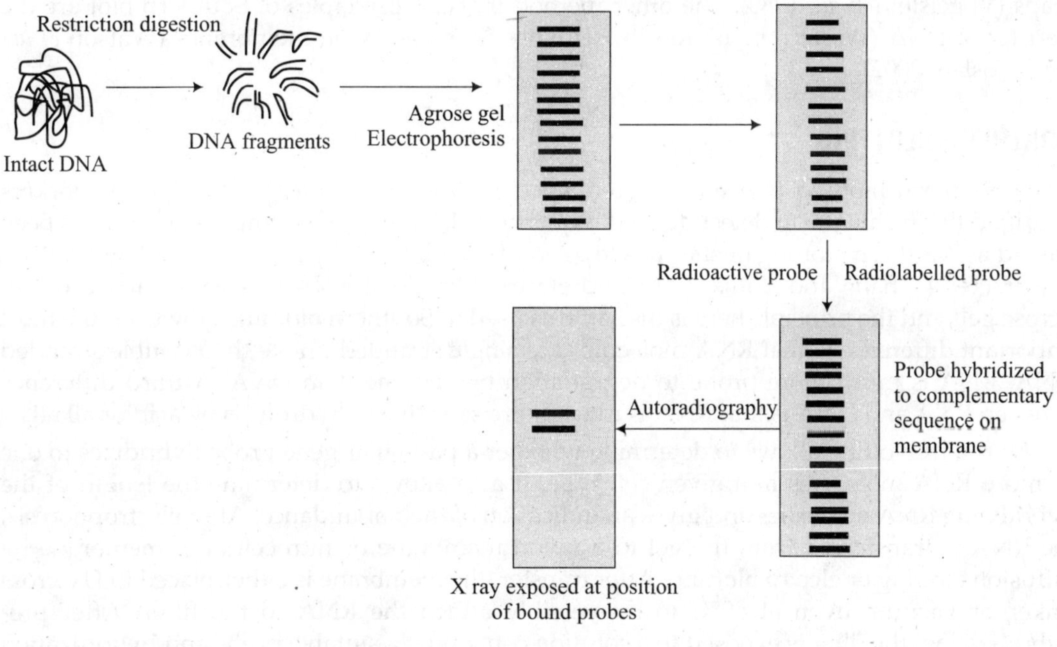

Fig. 1. Showing southern blotting.

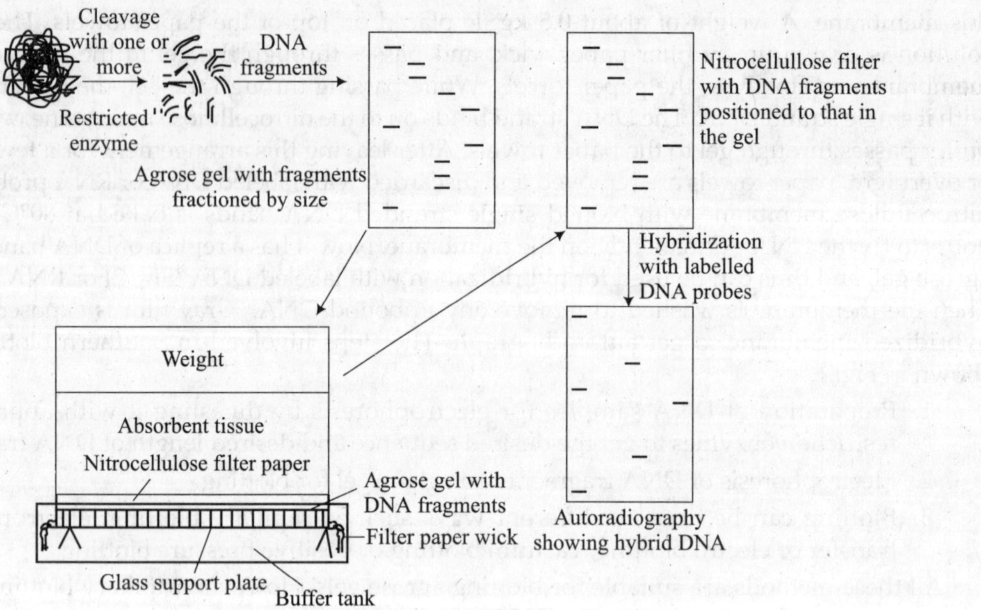

Fig. 2. Showing hybridization of DNA with labelled probes.

gene. Another application of Southern blotting is in restriction fragment length polymorphism (RFLP) analysis, which is important in several contexts including construction of genomic maps (Vogelstein *et al.*, 1985). The other method that use principles of Southern blot are the transfer of RNA (Western blots) to nitrocellulose filters or nylon membranes (Watson *et al.*, 1983; Soslau, 2002).

NORTHERN BLOTTING

Since Southern blotting is used to detect specific DNA molecules, a number of workers modified this technique to detect desired sequence of RNA. The resulting procedure has been named as Northern blotting (just opposite to Southern blotting) by molecular biologists. RNA is extracted (Zhang and Yunis, 1998) and electrophoresis of RNA is usually carried out in agrose gels and the principle here is the same as used in Southern blotting. However, the most important difference is that RNA molecules are single stranded unlike than double stranded DNA. RNA is much more prone to degradation by enzyme than DNA. A third difference between RNA and DNA is that RNA is much more sensitive to hydrolysis by acid or alkali.

Northern blotting allows to determine whether a particular gene probe hybridizes to one or more RNA molecules in a given cell type. It also allows to determine the length of the hybridizing RNA molecules and gives an indication of their abundance. After electrophoresis, the RNA is transferred from the gel to a nylon membrane or nitrocellulose membrane by diffusion blotting or electro blotting. After transfer, the membrane is either placed in UV cross linker or vacuum oven at 80°C to irreversibly attach the RNA to the filter. After pre-hybridization, the filter is exposed to a solution containing a suitable probe and hybridization is started. Either DNA or antisense RNA can be used as probes. The probe is usually radioactively labeled so that a hybridization signal can be visualized on a film or photo imager system. Northern blotting has a sensitivity for detection of approximately 1-5 mg of a particular species of mRNA The term "reverse Northern" refers to the procedure in which DNAs are attached to a nylon membrane followed by hybridization with labeled RNA probes. The procedure can provide quantitative data regarding mRNA abundance but not on mRNA size.

WESTERN BLOTTING

The Western blotting technique has been developed for transferring proteins from SDS polyacrylamide gel to nitrocellulose filters so that they can be detected with labeled antibodies, lectins or other protein binding molecules (Kerr, 1995). The transfer of proteins from the gel to nitrocellulose filters can be achieved in one or two ways i.e., by capillary blotting or electro blotting. The nitrocellulose filters with transferred proteins is called 'blots'. The blots can be further examined to detect a specific protein (Fig. 3). The detection process is just like a sandwitch reaction in which the binding of transferred protein with specific antibodies or lectins is followed by binding with radioactive molecule so it could be possible to search the desired protein by UV light or X-ray diffraction. Application of Western blotting may be done for the following purposes:

1. To detect the desired protein one can synthesize the mRNA of that protein and also synthesize the DNA segment from this specific mRNA by using reverse transcriptase method.
2. To detect the susceptible antigens so it could be possible to decide the therapy.
3. To detect the quantity of the desired protein or antigens in the sample.

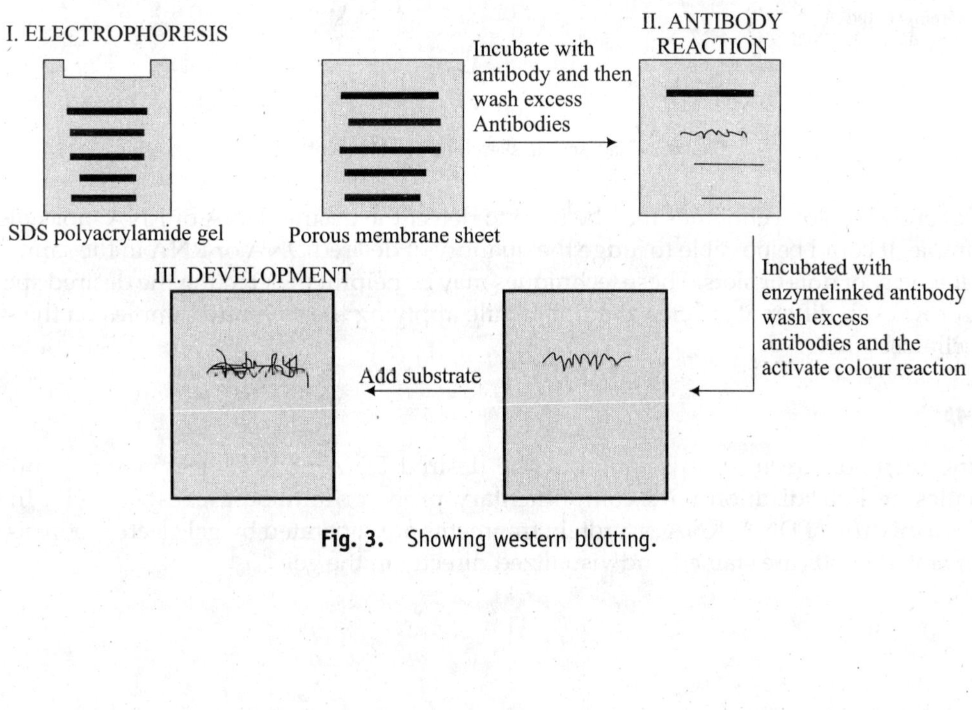

Fig. 3. Showing western blotting.

DOT BLOT AND SLOT BLOT TECHNIQUES

These are simple techniques for the detection of desired DNA molecule or mRNA molecules in the sample. They neither need electrophoresis of sample on agrose gel nor any blotting on nitrocellulose filter paper. Nitrocellulose filters are required only for application of samples. The samples (extracted DNA) applied on nitrocellulose filters in the form of dots. Then they are denatured and baked at 80°C to attach the single stranded DNA onto the filters. These filters are pretreated and dipped in the solution containing desired single stranded radioactive probes. These probes are hybridized with those DNA molecules which are complimentary to the specific probes. These probes are radioactive and could be detected by autoradiography. This technique is known as *dot blot technique* (Fig. 4). If these dots are applied on nitrocellulose filers by the help of slots by machine then technique is known as *slot blot technique*. These slots are oblong in appearance. This is the only difference otherwise the entire technique is the same as used in dot blot technique.

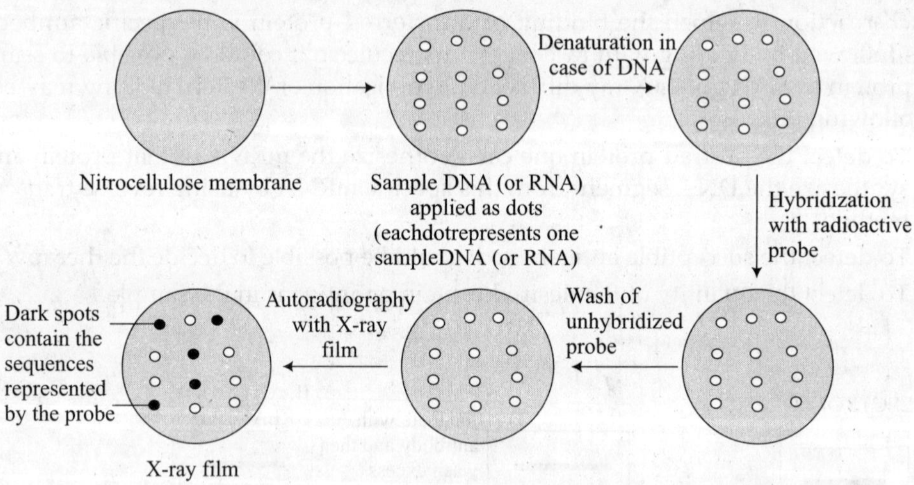

Fig. 4. Diagramatic sketch of dot blot method.

Dot and slot blot techniques may be used to detect the desired DNA or RNA molecules in the sample. It could be possible to judge the quantity of desired DNA or RNA in the sample by color intensity of dots or slots. These techniques may be helpful in detecting the desired specific DNA or RNA in different species at a time while applying the so many samples on the same nitrocellulose filters.

SUMMARY

Blotting methods are used to search out the desired DNA or RNA fragments or protein molecules by hybridization with complimentary probe or antibodies, respectively. In this process a mixture of DNA, RNA or protein fragments are separated by gel electrophoresis and the separated bands are stained and visualized directly in the gel.

Southern blotting has many applications in molecular biology research. It is used in important steps of gene cloning project. Genomic DNA having desired gene to be cloned is blotted and hybridized to identify one or more restriction fragments containing the gene in question. Since Southern blotting is used to detect specific DNA molecules, a number of workers modified this technique to detect desired sequence of RNA. The resulting procedure is known as Northern blotting. Western blotting is used to detect the desired protein, susceptible antigens and the quantity of the desired protein or antigens in the sample.

REFERENCES

Kerr, L. D. (1995) Electrophoratic mobility shift assay. Methods in Enzyme. 254 : 619-627.

Soslau, G. (2002) Recombinant DNA and biotechnology. In : Text Book of Biochemistry (Ed. Devlin, T.M.) Willy-Liss, New York, USA. 279-327

Southern, E.M. (1975) Detection of specific sequences among DNA fragments separated by gel electrophoresis. *J.Mol. Biol.* 98 : 503-507.

Vogelstein, B., Fearon, E.R., Hamilton, S.R. and Feinberg, A.P. (1985) Use of restriction fragment length polymorphism to determine the clonal origin of human tumor. *Science.* 227 : 642-644.

Watson, J.D., Tooze, J. and Kurtz, d.t. (1983) Recombinant DNA: A short Course. Scientific American Books Freeman, San Francis, USA.

Zhang, Y. and Yunis, J.J. (1998) Improved blood RNA extraction microtechniques for RT-PCR. Bio Techniques 18 : 783-788.

Suggested Readings

1. Battey, J.F., Davis, L.G. and Kuchl, W.M. (1994). Basic methods in molecular Biology, Application, New York, USA.
2. Benjamin, L. (1994). Genes V, Oxford University Press, Oxford, UK.
3. Maniatis, T., Fritsch, E.F. and Sambrook, J. (1989) Molecular Cloning. Cold Spring Harbor Laboratory Press, New York, USA.
4. Sambrook, J. and Russell, D.W. (2001) Molecular Cloning. Cold Spring Harbor Laboratory Press, New York, USA.

REVIEW QUESTIONS

Long Questions

1. Describe Southern blotting technique in detail.
2. Discuss Northern blotting technique with its importance.
3. Explain Western blotting with its possible applications.
4. What are Dot blot and Slot blot techniques and how they differ from Southern, Northern and Western blotting.

Multiple Choice Questions

1. Southern blotting is
   (a) In important steps of gene cloning.
   (b) In hybridization of one or more restriction fragments containing the desired gene.

(c) In restriction fragment length polymorphism (RFLP) analysis
(d) In all the above mentioned cases.

2. Northern blotting is used for
   (a) Detection of desired sequence of RNA.
   (b) Determination of whether a particular gene probe hybridize to one or more RNA molecules in a given cell type.
   (c) Determining the length of the hybridizing RNA molecules and gives an indication of their abundance.
   (d) All the above.
3. Western blotting is used for
   (a) Detection of desired protein by synthesizing the mRNA of that protein.
   (b) Detection of susceptible antigens so it could be possible to decide the therapy.
   (c) Detection of quantity of the desired protein or antigens in the sample.
   (d) All the above.
4. Dot blot and slot blot techniques are
   (a) Simple techniques to detect the desired DNA or RNA molecules in the sample.
   (b) Useful to judge the quantity of desired DNA or RNA in the sample.
   (c) Useful to detect the desired DNA or RNA in different species at a time while applying the so many samples on the same nitrocellulose filters.
   (d) All the above.

Answers

1. (d) 2. (d) 3. (d) 4. (d)

Part–16

Electrophoresis and Cytometry

CHAPTER 52

Electrophoresis

Mangla Bhide[1*] and G. Tripathi[2]

[1]Dept. of Zoology, Dr. H. S. Gour University, Sagar-470 003

[2]Dept. of Zoology, J. N. V. University, Jodhpur-342 001

E-mail: manglabhide@rediffmail.com

Arne Tiselius firstly described electrophoresis in 1933. It is a technique for the separation of charged molecules. If a mixture of substances (applied as sample on an adsorbent supporting media) in buffer solution is introduced to electric field then the charged particles (sample molecules) move towards the opposite poles. Cation (+ ion) moves towards cathod (-pole) and anion (-ion) moves towards anode (+ pole).The rate of migration depends on the size, shape, molecular weight, total or net charge, pH of buffer solution, current, pH, voltage and other factors (Andrews, 1983). Many biological molecules e.g., amino acids, peptides, proteins and nucleic acids possess ionisable groups and, therefore at any given pH they exist in solution as electrically charged molecules either as cation (+) or as anion (-). When an electrolyte is subjected to an electric field the differential migration of ions takes place. Sample materials must be suspended or dissolved in buffer for electrophoresis and the supporting medium must also be saturated with buffer for conduction of current.

Cathod (– pole)	Anode (+ pole)
$2e + H_2O = 2\,OH^- + H_2$	$H_2O = 2H^+ + O_2\,e^-$

The components of the mixture may be separated according to their electrophoretic mobility. Separation of charged substances ranging from small inorganic ion to large macromolecules may be done by electrophoresis. There are many types of supporting media available e.g., sheet or absorbent paper or cellulose acetate, a thin layer of silica or alumina or gel of starch, agar or polyacrylamides. A common feature of the use of all supporting media is that the substances migrate as distinct zones, which at the end of the analysis can be readily detected by suitable analytic techniques. The term zone electrophoresis has been applied to this

---

[*]Corresponding author

method, which has widespread applications in both preparative and analytic work. Separation of protein based on molecule mass or size has been well documented (Schultz and Liebman, 2003).

Electrophoresis

Means (Transport by electricity. It helps in migration of charged particles)

Biological molecules such as amino acids, proteins and nucleic acids remain as electrically charged species in suitable solution at a given pH.

Velocity of the charged particle (V) :

$$V = Eq / f$$

where,

E = Electric field

q = Net charge of molecule

f = Frictional coefficient

Zwitter ions possess anionic and cationic groups and net charge depends on the pH

Migration is affected by ionic strength and heat.

Types of gel electrophoresis

Based on buffer system

(A) Continuous type gel electrophoresis system : Same buffer is used for the gel medium and electrode. Sample is loaded directly on to gel.

(B) Discontinuous type gel electrophoresis system : Buffer systems differ in composition and ÞH. Sample is loaded on to stacking gel.

Based on the supporting media

(A) Cellulose acetate : Migration takes place in the buffer film on the gel surface.

(B) Agarose, starch and polyacrlamide :Migration occurs through the network of pores of relatively constant size.

INSTRUMENTATION

Power pack (Fig. 1) and an electrophoresis unit (Figs. 2, 3 and 4) are required for electrophoresis. The equipment described here is for low voltage work. High voltage electrophoresis is a more specialized technique. The power pack provides a stabilized direct current and has control for both voltage and current output. For low voltage, power packs are available with an output of 0 – 500 V and 0 – 150 mA. It can be operated either on constant voltage or constant current. The electrophoresis unit contains the electrodes, buffer reservoirs, a support for the electrophoresis medium and a transparent insulating cover or lid (Fig. 2). The two buffer reservoirs are normally partitioned into two sections, the electrode and the wick compartments. Electrical contact between the buffers in the two compartments is maintained by small holes or slots in the partition between compartments or by means of porous wicks. Contact between the supporting medium (always saturated in buffer prior to electrophoresis) and the buffer in the reservoirs is normally maintained by wicks of several layers filter paper.

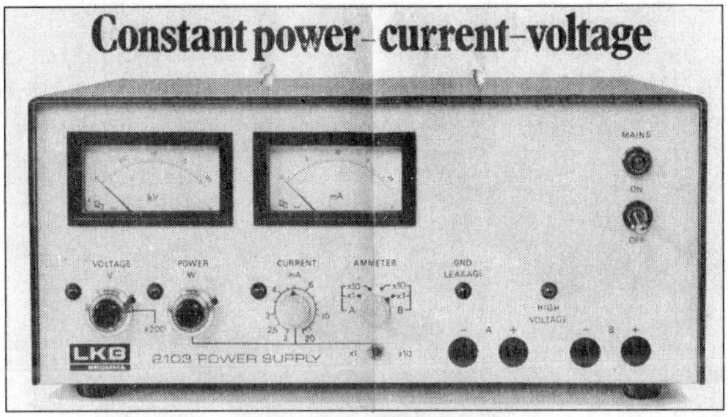

Fig. 1. Power supply apparatus.

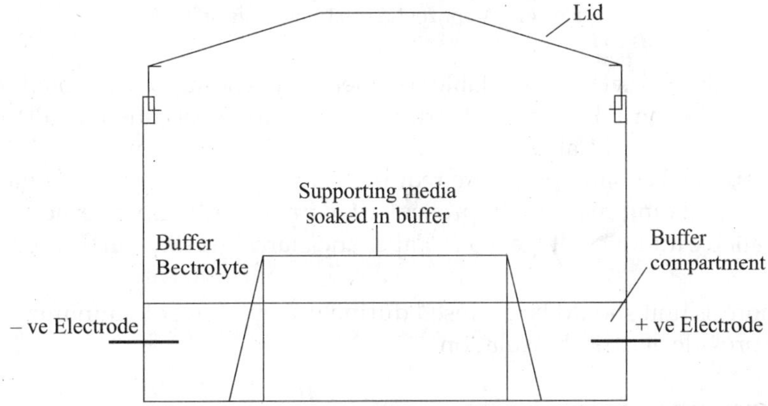

Fig. 2. An electrophoresis tank.

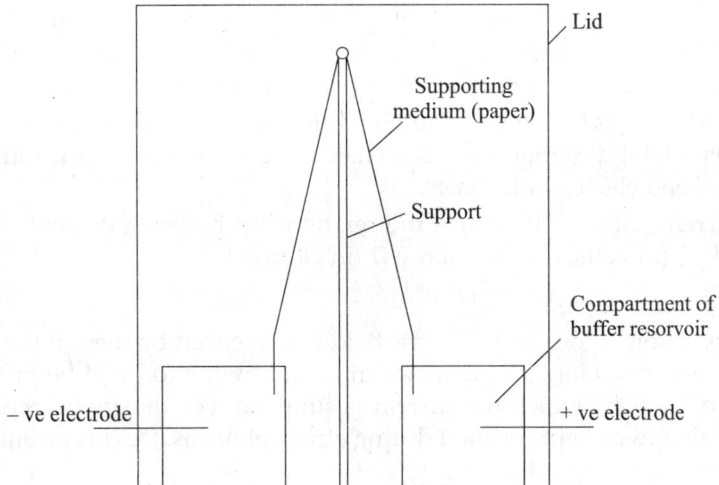

Fig. 3. Vertical electrophoresis unit.

Multiphor Electrophoresis Equipment

Fig. 4. A horizontal electrophoresis unit.

Reusable encapsulated wicks are available commercially. The supporting medium is normally arranged horizontally on a flat surface. Horizontal electrophoresis unit is available for use with paper, cellulose acetate, gel slab and thin layer (Fig. 4). Although for the higher voltage work with thin layers, metal cooling plates are required. Gel and thin layers are sometimes covered with a plate of insulating material to prevent evaporation. A simple vertical electrophoresis unit is also available for low voltage work. Paper and starch gel slab can be used vertically (Fig. 3).

Electrophoresis unit should be enclosed during use by a cover to minimize evaporation of buffer and to provide electrical insulation.

FACTORS AFFECTING MIGRATION RATE

Nature of charged compounds affects migration rates in the following ways:

1. Charge: The rate of migration increases with the magnitude of the charge.
2. Size: The rate of migration decreases for large molecules due to the increase frictional and electrostatic force, which are exerted by the surrounding media.
3. Shape: Molecules of similar size but different in shapes e.g., fibrous and globular proteins exhibit different migration characteristics because of the differential effect of frictional and electrostatic forces.
4. The current: Ohm's law states the relationship between current 'A' (in amperes), voltage 'V' (in volts) and resistance Ω (in Ohms).

   $$A = V/\Omega$$

   The separation of ions in an electric field is influenced by these three factors (current, voltage and resistant). The distance migrated by the ion will be proportional to the length of time for which the current is supplied. For maximum reproducibility, the current should be kept constant during electrophoresis. Direct current must always be used.
5. Voltage: This governs the current. The rate of migration is proportional to the potential difference across the supporting medium. It is voltage gradient. High voltage is used

mainly for the separation of low molecular weight compounds. It is also used for rapid separation of substances within a short time.

6. Resistance: The rate of migration is inversely proportional to the resistance, which depends on the type and size of the supporting medium and on the ionic strength of the buffer. Resistance increases with the length of the supporting medium and decreases with both width of the medium and buffer ionic concentration.
7. Buffers: Formate, acetate, citrate, barbiton, phosphate, Tris, EDTA, pyridine and borate buffers are commonly used. Buffer acts as solvent for the samples
8. pH: It has little effect on fully ionized compounds e.g., inorganic salts but for organic compounds pH determines the extent of ionization. The ionization of organic acids increases as the pH increases and the ionization of organic base increases as the pH decreases. The pH for separation of acid compounds must be 7.1 to 14 (alkaline buffer), while for basic compounds it must be 6.9 to 1 (acidic buffer). The degree of migration will be pH dependent e.g., amino acids have both basic and acidic properties (ampholytes) as follows:

$$H_2N - \overset{\displaystyle R}{\underset{\displaystyle H}{\overset{|}{\underset{|}{C}}}} - COOH \underset{H^+}{\overset{OH^-}{\rightleftharpoons}} H_2N^+ - \overset{\displaystyle R}{\underset{\displaystyle H}{\overset{|}{\underset{|}{C}}}} - COO^- \underset{H^+}{\overset{OH^-}{\rightleftharpoons}} H_2N - \overset{\displaystyle R}{\underset{\displaystyle H}{\overset{|}{\underset{|}{C}}}} - COO^-$$

The separated samples are located by staining with appropriate dye or by using autoradiography. On the basis of supporting media the electrophoresis is of two types:

(1) Paper electrophoresis

(2) Gel electrophoresis.

SUPPORTING MEDIA FOR ELECTROPHORESIS

The main supporting media for electrophoresis are filter paper, cellulose acetate membrane and gel (starch, polyacrylamide and agar) (Bhide, 2000). They are described as follows:

I. Filter Paper

Filter paper is an extremely useful and convenient medium. It is appropriate medium for high voltage electrophoresis. However, there are certain shortcomings. Firstly, the random structure of the paper causes irregularities in the separation pattern. Secondly, the relatively polar nature of cellulose shows some absorptive effects which binders the appearance of zones. Thirdly, diffusion of sample results into the broadening of zones but Whatman No. 1 filter paper gives the fine resolution within 3 – 4 hrs. while largely supplanted by more sophisticated techniques, paper electrophoresis it will find application in amino acid sepatation (Rodwell, 1990; Hawcroft, 1996).

2. Cellulose Acetate Membrane (CAM)

CAM was introduced as a better supporting medium as it contained more regularly defined pore structure. Separation could be made in 1-2 hrs. Absorption effects were minimum. Small amount of sample is required. Soaking in an oil or organic solvent could also make membrane transparent for densitometry. However, CAM still showed electroendosmotic effects.

3. Gel

A variety of gels have been used in order to develop preparative techniques and to eliminate some of the disadvantages of the solid support media (Dunn,1993). Gels of agar, starch, polyacrylamide etc. may be prepared in buffer shortly before use. This is often a time-consuming process. The composition of gels is such that there is very little adsorption, electro-osmosis or zone spreading due to diffusion. Gel electrophoresis has a preparative use, but is more widely used analytically and particularly useful for resolving mixtures of substances which have similar charges slightly different masses.

(i) Starch gel: All the media described above have a large pore size, which does not interfere with the separation process. However, starch and polyacrylamide gels have a restricted pore size and show a molecular sieving effect. The starch gel is also prepared by heating starch in the appropriate buffer in a boiling water bath until a mixture becomes translucent. A vacuum is applied to the flask to remove air bubbles and the solution is quickly poured into a heated tray and allowed to cool to form a gel of about 5 mm thick.

Gel electrophoresis

Starch gel electrophoresis

Supporting media	- starch or starch amylose
Advantages	- Easy to handle and to optimize.
	- Cathodal and anodal bands are distinguishable
	- Economical as many enzymes can be stained from a single gel
Disadvantages	- Time consuming

Polyacrylamide gelelectrophoresis

Gel is formed by vinyl polymerization

Acrylamide monomer (CH_2=CHCONH$_2$)

Polymerised with

N,N' methylene-bis-acrylamide {CH_2=(CHCONH=CH_2)}

Oxygen inhibits the polymerization as it removes the free radicals

(ii) Polyacrylamide gel: Polyacrylamide gel shows many advantages over starch gel because of its synthetic nature and pore size. The gel is formed by the polymerization of two monomers, acrylamide and a cross-linking agent N, N-methylene–bis–acrylamide. Acrylamide gels are defined in terms of the total percentage of the acrylamide present. Changing the concentration of both acrylamide and bis–acrylamide, one can vary pore size to some extent in the gel to suit particular analyses. Acrylamide concentration may vary from 3 to 30%. Low percentage of acrylamide is used for separation of DNA. Gel between 10 to 20% is used in SDS gel electrophoresis.

A diagrammatic representation of the gel network is as follows:

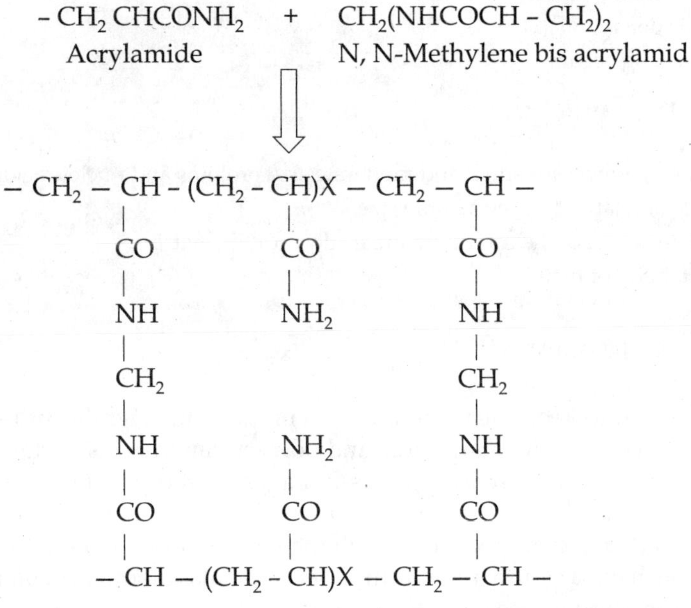

Polyacrylamide gel electrophoresis (PAGE) is performed in cylindrical glass tubes of equal diameter or in flat beds (slab gel) and referred to as disc or slab gel electrophoresis respectively. Slab gel is polymerized in gel casting unit. Monomers and catalyst are first mixed in buffer and then distributed into tubes 5 to 6 mm diameter and 6 to 7 cm length in case of tube gel electrophoresis. The tubes are sealed at their bases with rubber caps. They are kept vertically. The gel formation is inhibited by oxygen. To overcome this, the gel must be carefully poured by a hypodermic syringe (without needle). The gel should be filling up to the top leaving a short space for loading the sample. Polymerization of the gel may be done either by ultraviolet photo- activation with riboflavin or preferably using ammonium persulphate as a catalyst. It is necessary to include an initiator for the reaction. TEMED (tetra methyl ethylene diamine) is commonly used.

Equipment and Chemicals for Gel Electrophoresis

Equipment:

Gel Apparatus - Glass plate, alumina plate, clamp, spacers, comb, leveler and electrophoresis unit.

Power supply - Power pack

Chemicals:

Gel materials - Acrylamide, and bis acrylamide (cross linker)

Catalysts – Ammonium per sulphate, riboflavin

Initiator - TEMED

SDS-PAGE

- Separation based on molecular size
- Use of ionic detergents (SDS)
- Charge masking occurs

Merit and Demerits of PAGE

Merit

- High resolving power for small and moderate size proteins and nucleic acids
- Acceptance of relatively large sample size
- Minimal interaction of the migrating molecule with the matrix
- Physical stability of matrix

Demerit

- Chemicals are highly toxic

Care must be taken to keep the current low to prevent any significant heating effect. The principle of molecular sieving in agar, starch and polyacrylamide gels dectates that decreasing the pore size as a result of increasing cross-linkages in the gel increasingly hinders the movement of large molecules. The situation is reversed if a sephadex type gel is used as its particular nature hinders the migration of small molecules more than that of large molecules. Polyacrylamide gels have a number of features, which make them particularly useful for the separation of macromolecules because:

1. They may be produced with a wide variety of pore size.
2. They can be formed in precision bored tubes or slab and hence give reproducible results.
3. They may be used in a wide range of buffers.
4. They give rapid separation.
5. Their degree of adsorption and electroosmosis is low.
6. They do not absorb ultraviolet light at 270 nm and hence proteins separated on them may be estimated by their absorption at this wavelength.
7. They enable the macromolecules to be stained and the amount of material present determined by the use of densitometer.

(iii) Agar gel: Agar is a mixture of two-galactose polymers agrose and agaroprotein. In 1% (w/v) gel in buffer, agar has high water content and movement of ions in the gel is very rapid during electrophoresis. When electrophoresis is stopped, diffusion of the ions is rapid. This property makes agar a very suitable medium for separation and detection of antigens in immuno-electrophoresis.

Agarose gel electrophoresis

- Firstly extracted from the wall of certain marine algae
- High molecular weight linear polysaccharide of repeat units of agarobiose
- To characterize DNA and RNA

- Dissolves in aqueous buffers and forms gel
- Commercial agarose is purified form of agar

Equipment: Gel electrophoresis system (horizontal), power pack gel casting plate, comb, microwave oven or hot plate, and UV transilluminator

Reagents: Buffer, ethidium bromide, loading buffer and agarose

Advantage - Cheap nontoxic powdery mixture, suitable for the chemical staining, low diffusion resistance make it suitable for immunoelectrophoresis, good fiber structure, hence, has anti convectional properties

Disadvantage - Severe electroosmosis occurs

Purified agar is used in a buffer electrolyte and the samples are introduced in a small well or through cut out of gel. Agrose is a linear polysaccharide (molecular mass about 12000) made up of the basic repeat unit agrobiose which is made up of galactose and 3, 6 anhydrogalactose. Agrose is generally used at concentrations of 1 to 3%. Agrose gels are formed by suspending dry agrose in aqueous buffer then boiling the mixture until a clear solution is formed. This is poured and allowed to cool at room temperature to form a rigid gel. Agrose gels are used for the electrophoresis of proteins and nucleic acids.

OPERATIONAL PROCEDURE

1. Medium saturation: If the supporting medium is not a gel, it must be saturated with buffer before electrophoresis is started so that it can conduct the current. Saturation of the medium is best done before the sample is applied, otherwise spreading of the sample at the origin occurs. Chromatography paper dipped in buffer and excess moisture is removed by blotting. Cellulose acetate strips are impregnated with buffer by floating them onto the surface of buffer in a shallow tank since rapid immersion traps air bubbles, which are difficult to remove. Thin layer media are best saturated by means of capillary action.

2. Sample application: The samples are applied with micropipette or hypodermic syringe or by a glass capillary as a small spot, narrow streak or along a suitable origin. If the individual components of a mixture have opposite charges, they would be expected to migrate towards both electrodes and the origin in this case should be near the middle of the supporting medium. If movement is expected towards only one of the electrodes i.e., all components are either positively or negatively charged, the origin should be placed near the end of the medium. If a horizontal slab is used samples are introduced usually on a strip of filter paper into a slot or hole cut into the surface of gel. If the gel is a vertical cylinder (e.g., polyacrylamide gel) the sample is loaded at the upper end of cylinder by being layered in a dense solution. A 10% (w/v) sucrose, urea or sodium dodecylsulphate may be added to samples to facilitate solubilization. A tracker dye e.g., bromophenol blue is often added in samples (in gel electrophoresis) or in buffer tanks (in paper electrophoresis) which makes possible to visualize the migration of the components of the samples towards the opposite pole. The dye was fairly tightly bound to serum albumin or very low molecular weight fractions during the course of electrophoresis and did not alter the mobility of other fractions. The dye is not interfering with subsequent staining of gel or paper strips.

3. Running of samples: After the sample has been applied, the power is switched on at the required voltage for the period necessary for separation. Even with stabilized current and voltage, the equipment must be carefully watched as overheating can occur if the supporting medium has not been properly prepared. Low voltage separation is completed within one to two hours although separation of protein on paper requires a longer time period. High voltage separations are completed normally within an hour. When electrophoresis is completed the power must be switched off before the supporting medium is removed.

4. Removal of supporting media: Paper, cellulose acetate strips and thin layer plates can be removed and air dried directly, normally in an oven at 110°C. Cylindrical polyacrylamide gels are removed by forcing water from hypodermic syringe around the walls of running tubes to free the gel column to be extruded. The gel may if necessary be immersed in a fixative e.g., 7% acetic acid. Slab gel is carefully removed from glass plate.

5. Staining and recovery of compound: Biological compounds are not colored and therefore, it is necessary to visualize them in some way in order to determine their position on the supporting medium after separation is completed. The most widely applied method for detecting compounds after paper, cellulose acetate and gel electrophoresis is to treat the media with a dye which will selectively stain the compounds or fractions onto the media. For the quantitative analysis the amount of stained material may be determined by one of the following three ways:

(a) The relevant area of supporting medium is cut out and the component is eluted from the medium with a suitable solvent. The amount of stained material may then be determined by spectrophotometer.

(b) If cellulose acetate paper is used as the supporting medium, it can be dissolved in appropriate organic solvent e.g., acetone, leaving the material in solution and then detecting the amount of component by spectrophotometer.

(c) A densitometer can scan the supporting medium. The densitometer is an instrument in which the amount of light passing through narrow areas of the supporting medium is absorbed by band as the medium moved across a light beam. There is an inverse relationship between the amount of light reaching the densitometer detector and the amount of stained material present in the medium and a series of peaks is recorded which correspond to the position of the stained areas in the medium.

It is essential to reduce the background light while immersing the strips several times in washing organic solvents e.g., 1% acetic acid (for protein detection) or 40% ethanol (for the detection of lipid on the electrophoresed strips). Cellulose acetate strips have an advantage over paper strips in this respect since they can be made transparent after immersion in liquid paraffin, and colored zones may be distinguished more rapidly. Cellulose acetate strips and gel also lend themselves to destaining of excess dye by repeated immersion in a solvent e. g., dilute acetic acid. The method for visualizing enzymes separated in gel is to place the gel in a solution of substrate of the enzyme. If the product of the reaction is colored, the colored zone indicates the position of enzyme. If the compounds separated are radioactive then supporting medium should be subjected to autoradiography or scanned by a radiochromatogram scanner to detect the radioactive areas. The supporting medium may be cut into small pieces and immersed in a suitable scintillator solution and the radioactivity shall be determined by liquid scintillation counting.

Visualization

Protein is stained with coomassie brilliant blue.
Enzyme is stained with substrate.
Phenazine methansulphate (pms) and nitrobluetetrazolium (NBT)
DNA is stained with ethidium bromide.

Silver Staining

Detection of protein and nucleic acid in the nanogram range.
Based on specific reduction of silver ions.
Equipment: Platform rocker or orbital shaker, staining boxes
Solutions: Fixing solution, staining solution, developing solution

Reagents used for the detection of compounds or their fractions on electrophoresis strips are as follows:.

Components	Reagents	Remark
Protein	Nigrosine in acetic acid or TCA	Very sensitive
	Bromophenol blue or Coomasic brilliant blue	Quantitative
	Liassmine green in acetic acid	Quantitative
Glycoprotein	Periodic acid oxidation followed by treatment with Schiff's reagent	Quantitative
Lipoprotein	Sudan black in 60% ethanol	Quantitative
Peptides	ClO_2 or NaCl chlorination followed by KI, starch or benzidine acetic acid	React with all.
Nucleic acid	Methyl green pyronin	RNA red,DNAblue
Polysaccharides and acid mucopolysaccharides	Iodine Toulidine blue in methanol	 Blue

Gel electrophoresis may be either non-denaturing or denaturing type. Non-denaturing gel is used for detection of protons. Sodium dodecyl sulphate (SDS) is used for gel for breaking polypeptides.

Take desired tissue and homogenized in Tris buffer and TCA (trichloro-acetic acid) and then centrifuge at 3000 rpm for 20 minutes. The supernatant containing acid soluble substances is discarded. The residue is washed with alcohol and centrifuged again at 3000 rpm for 15 minutes. The supernatant containing alcohol soluble substances is discarded and the residue is washed with ether (thrice) and centrifuged at 3000 rpm for 15 minutes. The supernatant containing lipid or fats could be kept separately if want to study the lipid fractions in desired tissue, while the residue containing nucleic acid and protein is heated in hot TCA at 80 to 90°C for 15 minutes and centrifuged at 3000 rpm for 15 minutes. The supernatant can be stored for nucleic acid estimation, while residue is dissolved in 0.1 N NaOH and stored as protein sample for electrophoresis.

ELECTROPHORESIS OF DNA

Gel electrophoresis is also a useful method to separate nucleic acids. DNA molecule carries negative charge. When DNA is placed in an electric field they migrate towards the positive pole. The rate of migration of a molecule depends on two factors, its shape and its electric charge. Unfortunately, the most DNA molecules have the same shape and all have very similar electric charges. Fragments of different sizes cannot, therefore, be separated by standard electrophoresis. However, the size of the DNA molecule does become a factor if the electrophoresis is performed in a gel. A gel is usually made of agarose, polyacrylamide or a mixture of the two comprising a complex network of pores through which the DNA molecules travel to reach the positive electrode. The smaller the DNA molecule, the faster it can migrate through the gel. So the gel electrophoresis will separate DNA molecules according to their size. In practice the composition of the gel determines the sizes of the DNA molecules that can be separated. At the other end of the scale, a very thin (0.3 mm) 40% polyacrylamide gel (with extremely small pores) may be used to separate much smaller DNA.

APPLICATIONS OF ELECTROPHORESIS

1. The technique may be applied for separation of charged molecules e.g., protein, nucleotides etc.
2. Low voltage paper electrophoresis is widely used for the separation of amino acids, peptides, proteins, nucleotides and charged lipid and carbohydrate derivatives.
3. Better resolution of small molecules is achieved by using high voltage electrophoresis.
4. The high purity of supporting medium results in less adsorption of the samples and thus a better resolution obtained.
5. Cellulose acetate paper is advantageous for visualization of bands due to low background staining of the supporting medium.
6. Cellulose acetate paper electrophoresis has clinical application in the separation of serum proteins (glycoproteins and lipoproteins) and haemoglobins. It is also used for immunoelectrophoresis.
7. Both paper and thin layer high voltage electrophoresis are used in the separation of amino acids, which may be present in high concentration in plasma and urine due to various metabolic disorders.
8. The suitability of thin layer electrophoresis for separating relatively small molecules e.g. amino acids and nucleotides is particularly useful because of its rapidity and good resolution.
9. The use of gel as supporting media for electrophoretic separations has considerably extended the scope of the technique. The fact that the degree of porosity of a gel i.e., polyacrylamide may be selected to enhance the separation of molecules of similar charges but different sizes and/or shapes is a big advantage.
10. Gel electrophoresis is very useful for separation of mixture of nucleic acid specially RNA molecules and proteins including enzymes, isoenzymes and structural proteins.
11. The starch and agar gels have been found useful in the separation of mixture of various high molecular weight compounds.

ELECTROFOCUSING OR ISOELECTRICAL FOCUSING

Electrophoresis depends primarily upon the charge carried by a molecule but at its isoelectric pH a molecule will not migrate in an electric field. If electrophoresis is performed in an electrolyte having a pH gradient, molecule will migrate to the point where the pH of solution is the same as their isoelectric pH and remain there as long as the gradient and potential differences are maintained. It forms the basis of isoelectricfocusing technique.

In simplest term electrofocusing is electrophoresis in a pH gradient. The proteins are subjected to electrophoresis in a pH gradient. The proteins are subjected to electrophoresis in a medium which has a gradient of pH value(Andrews, 1983). The proteins migrate in an electric field but as they reach a point where the pH is the same as its isoionic point, their further movement is prevented. In other words the protein has been focused to its isoionic point (Fig. 5).

Initially a sucrose density gradient is made in the electrophoresis tube. This forms the stable medium. To this special buffering agents, called 'ampholytes' are added. One end of the tube is kept at an alkaline pH and the other end at an acidic pH· Electric current is passed through the tube. The ampholytes migrate under the influence of the field and settle down at specific points, thus forming a stable pH gradient. This is how a pH gradient is formed. When protein mixture is added and current passed again, the proteins migrate and settle down at their isoionic points. Because of the stability of sucrose gradient, there is minimal diffusion and sharp bands are obtained.

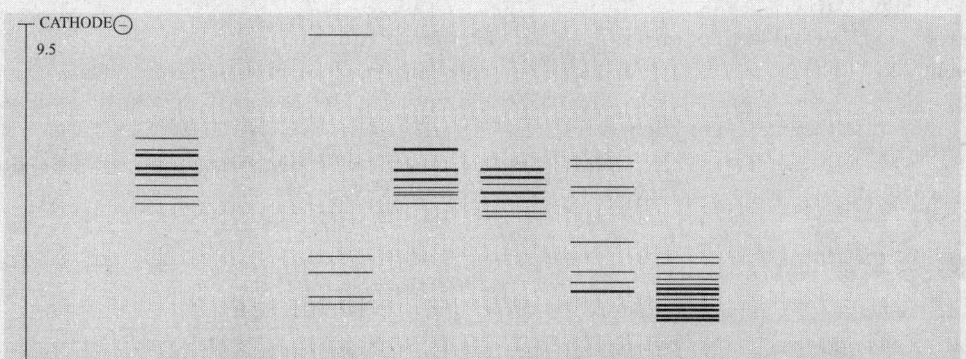

Derived from a gel-electrofocusing experiment with the Multiphor 3.5 to 9.5 pH gradient. Polyacrylamide gel composition: T = 5% C = 3%. 10 μl protein solution (conc. 1 to 2 mg/ml) was applied in filter paper (10 ′5 mm). Experiment time time: 1 hour 50 min. Final voltage = 108 V. Anode at bottom. Temp. 10°C.

Fig. 5. Showing gel electrofocusing.

SUMMARY

Electrophoresis is a technique for the separation of charged molecules. If a mixture of substances (applied as sample on an adsorbent supporting media) placed in buffer solution is introduced to electric current, the charged particles (substances or molecules) move towards the opposite pole i.e. Many biological molecules e.g., amino acids, peptides, proteins, nucleotides and nucleic acids possess ionisable groups and, therefore, at any given pH they exist in solution as electrically charged molecules either as cation (+) or as anions (–). When an electrolyte is subjected to an electric field the differential migration of ions takes place. Basically for any type of electrophoresis two items are needed i.e. an electrophoretic tank and a power supply. Electrophoresis depends primary upon the charge carried by a molecule but at its iso-electric pH a molecule will not migrate in an electric field. If electrophoresis is performed in an electrolyte having a pH gradient, molecule will migrate to the point where the pH of solution is the same as their isoelectric pH and remain there as long as the gradient and potential differences are maintained. It forms the basis of iso-electric focusing.

REFERENCES

Andrews, A. T. (1983) Electrophoresis: Theory, Techniques and Biochemical and Clinical Applications. Oxford University Press, Oxford.

Bhide, M. (2000) Detection of protein fractions in the blood serum of albino rats and other animals by paper electrophoresis In: Work Book, Academic Staff College, Dr. H. S. Gour University, Sagar, India, pp. 3-6.

Dunn, M. (1993) Gel electrophoresis, Bios. Scientific, Oxford.

Hawcroft, D.M. (1996) Electrophoresis. The Basics IRI press, Oxford.

Rodwell, V.M. (1990) Structures and function of proteins and enzymes; Protein : Structure and properties. In : Harper's Biochemistry (Eds. Murry, R.K., Granner, D.K., Mayes, P.A. and Rodwell,V.W.S.) Appleton and Lange, Connecticut, USA.

Schultz, R.M. and Liebman, M.N. (2003) Protein : Composition and Structure. In : Text Book of Biochemistry with clinical applications. (Ed. Devlin, T.M.) Wiley-Liss, New York, USA.

Suggested Readings

1. Gene VI (1997) B. Lewin, Oxford University Press. London, England.
2. Cell and Molecular Biology (1980). E. D. P. DeRobertis, and E. M. F DeRobertis, Hoit. Saunders, Tokyo, Japan.
3. Cell and Molecular Biology (1999) P. K. Gupta, Rastogi Publication, Meerut, India.
4. Laboratory Manual in Biochemistry (1985) J. Jairaman, Wiley Eastern Ltd. New Delhi.
5. Biochemistry (1975) A. L. Lehninger, IInd Ed. Kalyani, New Delhi.
6. Biotechniques, Theory and Practice (2004- 2005) S. V. S. Rana, Rastogi Publication.Meerut, India.
7. Biochemistry (1975) L. Stryer, W. H. Freeman, San Francisco.

REVIEW QUESTIONS

Long Questions

1. What is electrophoresis? Explain in brief about different types of electrophoreses.
2. Describe polyacrylamide gel electrophoresis in detail.
3. Who described electrophoresis for the first time? What are the benefits of electrophoretic separation of proteins?
4. Give a detail account of different supporting media used in the process of electrophoresis.

Multiple Choice Questions

1. Electrophoresis is a separation technique for
   (a) Protein (b) Lipid
   (c) DNA (d) All.
2. Factors effecting the migration of charged ions are
   (a) pH and buffer (b) Voltage
   (c) Net charge (d) All.
3. Agrose gel is used to separate
   (a) DNA (b) Proteins
   (c) Lipid (d) Carbohydrates
4. Paper electrophoresis is used for separation of
   (a) Negatively charged molecules (b) positively charged molecules
   (c) a and b both (d) None
5. PAGE is employed for separation of
   (a) Negatively charged molecules (b) positively charged molecules
   (c) a and b both (d) None.
6. In PAGE the gel is made up of
   (a) Agrose (b) Starch
   (c) Polyacrylamide (d) All.
7. Electrophoresis results are emphasized by
   (a) Electrophoretograms (b) Radiogram
   (c) a and b (d) None.
8. Polyacrylamide gel is formed by the
   (a) Polymerization of acrylamide and N, N′-methylene–bis–acrylamide.
   (b) Acrylamide and agar
   (c) Acrylamide and cellulose
   (d) Acrylamide and Starch.

9. Electrophoretic apparatus needs
   (a) Electrophoretic tank
   (b) Buffer solution of known pH
   (c) Power pack and electrodes
   (d) All the above.

Answers

1. (d)	2. (d)	3. (a)	4. (c)	5. (a)	6. (c)
7. (a)	8. (a)	9. (d)			

CHAPTER 53

Cytometry and Flow Cytometry

Subodh Kumar Jain

Dept. of Biotechnology, Dr. H. S. Gour University, Sagar-470 003

E-mail: Subjain@gmail.com

Cytometry refers to the measurement of the physical and chemical characteristics of a cell. Flow cytometry is a technique for counting, examining and sorting microscopic particles suspended in a stream of fluid. It allows simultaneous multiparametric analysis of the physical and or chemical characteristics of a single cell flowing through an optical or electronic detection apparatus by measuring the visible and fluorescent light concision. Flow cytometry is a technology that allows single cell to be measured for a variety of characteristics, determined by looking at how they flow in liquid. Instrument used for this purpose can gather information about cells by measuring visible and fluorescent light concision, allowing all sorting based on physical, biochemical and antigenic traits. A flow cytometer may be laser or arc lamp based, depending on the source of light used, so measures multiple cellular parameters based on scattering and fluorescence and purify physically, subpopulations of cells, leading to increasing applications of this instrument in biology and medicine. Flow cytometer sometimes called a Fluorescence Activated Cell Sorter (FACS) has several key components:

1. A light or excitation source, a laser that emits light at a particular wavelength.
2. A liquid flow that moves the suspended cells through the instrument and pass the laser.
3. A detector, in this case a photomultiplier tube, which is able to measure the brief flashes of light emitted as cells flow past the laser beam.

In a Flow cytometer, a beam of light (usually laser light) of a single frequency (colour) is directed onto a hydrodynamically focused stream of fluid. A number of detectors are aimed at the point where the stream passes through the light beam; one in line with the light beam (Forward Scatter or FSC) and several perpendicular to it, Side Scatter (SSC) and one or more fluorescent detectors. Each suspended particle passing through the beam scatters the light in same way, and fluorescent chemicals in the particle may be excited into emitting light at a lower frequency than the light source. This combination of scattered and fluorescent light is picked up by the detectors, and by analyzing fluctuations in brightness at each detector (one for each fluorescent emission peak), it is possible to deduce various facts about the physical and chemical structure of each individual particle. FSC correlates with the cell volume and SSC depends on the inner complexity of the particle (i.e., shape of the nucleus, the amount and type of cytoplasmic granules or the membrane roughness). Measurable parameters are:

- Volume and morphological complexity of cells
- Cell pigments
- DNA (cell cycle analysis, cell kinetics, proliferation, etc.) Ross, 1996.
- RNA
- Chromosome analysis and sorting (library construction, chromosome paint)
- Proteins
- Cell surface antigens (CD markers)
- Intracellular antigens (various cytokines, secondary mediators, etc.) nuclear antigens
- Enzymatic activity
- pH, intracellular ionized calcium, magnesium, membrane potential
- Membrane fluidity
- Apoptosis (quantification, measurement of DNA degradation, mitochondrial membrane potential, permeability changes)
- Cell viability
- Monitoring electropermeabilization of cells
- Oxidative burst
- Characterizing multi-drug resistance (MDR) in cancer cells
- Glutathione
- Various combinations (DNA/surface antigens, etc.).

This list is very long and constantly expanding.

INSTRUMENTATION

Unfortunately, Flow cytometers are not simple instruments (Fig. 1). As with all sophisticated measuring devices, it is important to possess a basic knowledge of the underlying principles of operation and component of the instruments, so that the significance and accuracy of the result can be assessed.

The major components of the flow cytometer can be discussed under the following headings:

1. **Fluidics:**
   (a) Fluid lines
   (b) Flow chamber

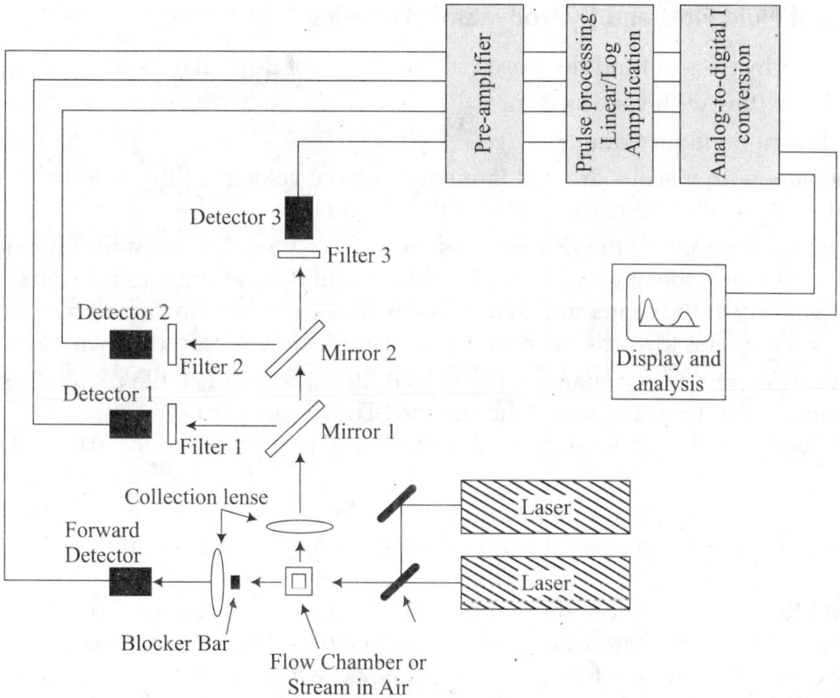

Fig. 1. A generalized flow cytometer system.

2. **Light Source:**
   (a) Lamp arc
   (b) Laser
3. **Detection:**
   (a) Optical filters:
   Colour glass filter
   Interference filter
   (b) Optical configuration
   (c) Detection devices:
   PIN diodes
   Photo Multiplier Tubes (PMTs)

FLUIDICS

"The fluidic system of a flow cytometer is used to deliver particles of a random three dimensional suspension singly to a specific point in space intersected by the illuminating beam". Fluid flow is generally achieved by injecting the cell suspension into the center of an enclosed channel through which liquid is flowing. Typically, flow chambers can transport particles upto the detection point with an accuracy of better than ±1 micron.

I. Principles of Fluid Flow and Hydrodynamic Focusing

There are two principles of fluid flow used in the design of flow chambers:

(a) Laminar flow with viscous drag

(b) Turbulent Boundary drag.

(a) Laminar flow with viscous drag: In laminar flow, the velocity of the fluid front at the inlet is same both at the walls and in the center of the channel.

Hydrodynamic focusing: As the fluid moves away from the inlet down the channel, viscous drag at the walls slows the outer layers of liquid. The fluid front changes into a parabola with the greatest velocity at the center and zero velocity at the walls of flow chamber. This velocity gradient draws particles towards the center in a process known as hydrodynamic focusing.

Inlet Length: Down in a flow channel, the velocity front forms a stable parabola, so that the particles continue to flow at a centre of the channel. The distance from the inlet to the formation of a stable velocity parabola is known as Inlet length and is approximately 50 times the channel diameter.

(b) Turbulent Boundary Drag: In turbulent boundary flow, the sample is injected into a chamber rapidly tapering to a small exit orifice. The speed of entry of the sample combined with the backpressure generated by the exit orifice results in sample turbulence. The turbulence at the interface below the sample sheath produces a viscous boundary. The drag generated at this boundary forms a velocity parabola within a shorter distance than for laminar flow. Particles are again drawn towards and contained within the center of the flow.

FLUIDIC SYSTEM

The fluidic systems comprises of:

1. Fluid lines
2. Flow chamber

1. Fluid lines/Channels: The fluidic system consist of two fluid lines for feeding the flow fluid lines for feeding the flow chambers: (1) The sheath fluid line and (2) The sample line.

In normal operations, the sheath fluid flows continuously and controlled by regulated positive air pressure acting on the sheath reservoir. Sample flow rates are controlled by a second pressure regulator acting on the sample chamber. Differential or individual pressure gauges are used to set optimum flow condition.

An alternative approach particularly for the accurate control of sample flow is the use of a finely controlled syringe pump.

2. Flow chamber: Flow chambers are designed with a wide inlet for sheath fluid, which tapers to either an exit orifice or a cylindrical or square sided channel. The small bars tube used for sample injection is located at the center of the inlet. The general design of a flow chamber can be divided into — Analytical chambers and sorting chambers.

- In laminar flow design, the tapering flow reduces the diameter of the core of sample before entry into the flow channel where the final hydrodynamic focusing of a sample has taken place.
- In turbulent boundary designs, the tapering design itself brings about hydrodynamic focusing within a short distance of the inlet.

1. Analytical Chamber

The first group of chambers is used in laser-based flow systems where fluorescence is measured at right angles to the illuminating beam. In analytical systems, the flow of fluid is usually upwards to allow air bubbles to be easily removed from the chamber. Scatter measurements can be made both at right angles to the excitation beam and in the forward direction.

The second group of chambers is used in microscope-based flow systems where fluorescence is measured in line with the optical path. The top surface of these chambers is usually a glass coverslip, so that immersion objectives of high numerical aperture can be used. Scatter measurements are restricted to within the direct optical path and are generally difficult to obtain.

Analytical chambers typically use a minimum channel bore of 250 mm to help prevent blockages and unwanted reflection from the walls.

2. Flow Chamber for Cell Sorters

The most common principle used for cell sorting is the electrostatic charging of droplets as applied to laser-based flow systems. The flow chamber design incorporates an exit orifice. The exit orifice also provides the required resistance to control sheath flow rates.

There are two basic designs of droplet sorting chambers in general use. The first type (Fig. 2A) uses laminar flow conditions in a square channel cuvette where cells intersect the laser beam above the exit orifice. The second type (Fig. 2B) uses turbulent boundary conditions, and cells intersect the laser beam immediately below the exit orifice in the exhaust jet. This second type of analysis and sorting design is known as 'stream in air' or 'jet in air'.

Scatter and fluorescence measurements are made in the same way as analytical, laser-based, flow systems with the exception that a second blocker bar is required for 'stream in air' analysis and sorting. This blocker bar is placed in front of the right angle collection lens to block laser light reflected by the cylindrical surfaces of the jet.

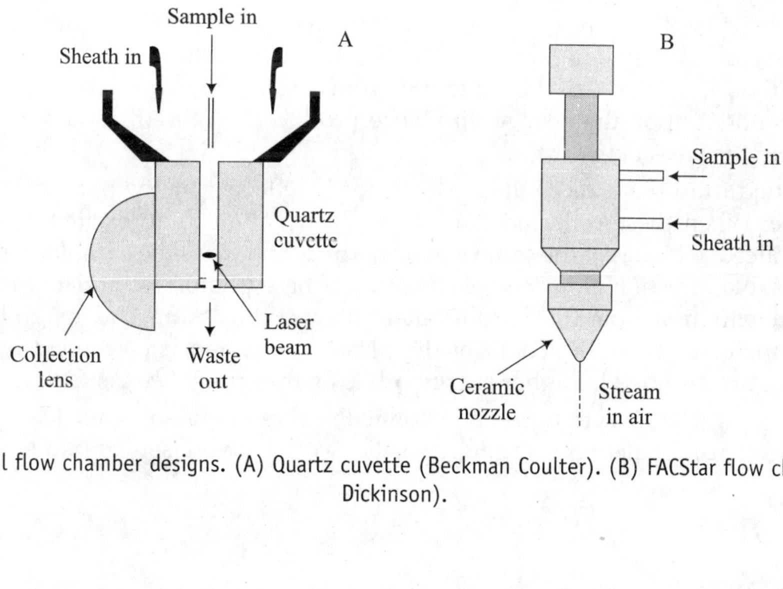

Fig. 2. Typical flow chamber designs. (A) Quartz cuvette (Beckman Coulter). (B) FACStar flow chamber (Becton Dickinson).

Microscope-based flow cytometers are less well suited for cell sorting.

Light Source

The two types of light source commonly used in flow cytometers are arc lamps and lasers.

(a) **Arc lamp:** An arc lamp is a glass envelope containing a gas or vapor at high pressure. An initial, high-voltage spark between two electrodes within the en-velope forms a luminous plasma arc, which is maintained by the application of a high current at low voltage. The arc is intrinsically unstable and is prone to flicker and wander. The average life of arc lamps is short and is, at best only a few hundred hours.

(b) **Laser lamp:** The name laser is an acronym for 'light amplification by stimulated emission of radiation'. The laser produces a coherent, plane-polarized, intense, narrow beam of light at specific selectable wavelengths. A typical gas laser is shown schematically in (Fig. 3).

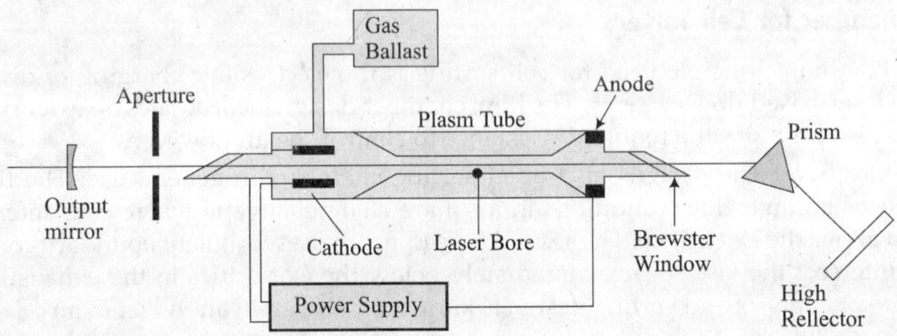

Fig. 3. A typical gas laser.

Mirrors, positioned at each end of the resonator, form an optical cavity within which the plasma tube is located. The plasma tube contains a gas at a critical pressure, which fluoresces under the application of a current, emitting light in all directions. The applied current raises electrons of the gas atoms into higher energy orbits. Light is produced when these electrons in high-energy orbits spontaneously decay to the ground state. The wavelength of the released photon is dependent upon the energy difference produced by the transition of the electron from the high to the low-energy orbit.

Light emitted from the ends of the plasma tube is reflected by the mirrors back into and along the tube. When these reflected photons strike an atom in an excited state, a second photon is produced, which is of the same wavelength and phase as the stimulating photon and travels in the same direction along the plasma tube. These photons stimulate further photon release in a chain re-action that produces light amplification. The magnitude of the amplification increases with the length of the plasma tube, and is controlled by the rate at which electrons are pumped to higher energy orbits by the applied current.

The lasers used in flow cytometers are atomic (e.g. helium-neon), ionic (e.g. argon-ion or krypton), molecular (e.g. helium-cadmium), or liquid (e.g. dye lasers). Solid-state lasers are becoming available in a form suitable for flow cytometry and have the potential to be the laser of choice in the future.

DETECTION DEVICE

(a) Optical Filtration

In cytometers employing arc lamps, filters are used to select the correct wavelength of exciting light. However, most instruments use lasers, which give monochromatic light so that further filtration is unnecessary. On the output side, filters are needed to separate the mixture of scattered and fluorescent light collected from stained particles so that specific independent but correlated measurements can be made. The separation of different wavelengths is achieved by the use of dichroic mirrors and interference and absorption filters. There are two types of glass filters in use: coloured glass and interference filters.

(i) Coloured glass filters

The most useful of these are the longpass filters which transmit light above a given wavelength and absorb light of lower wavelengths. They can be obtained with a cut-off between 300 and 700 nm.

(ii) Interference filters

These consist of dielectric layers deposited on a glass substrate. Depend-ing on the thicknesses of the layers and the wavelength of the light, the internally reflected beams interfere with one another either destructively or constructively allowing some wavelengths to pass through the filter while others are reflected.

Interference filters can deteriorate with time and their properties should occasionally be checked in a spectrophotometer to ensure that they perform to their original specification.

Interference filters come in two forms—bandpass and edge filters. Bandpass filters transmit light of the desired wavelength, over a narrow band (Fig. 4).

The major parameters of bandpass filters are the peak wavelength of transmission, the percentage of light transmitted at the peak wavelength (peak transmission), the bandwidth

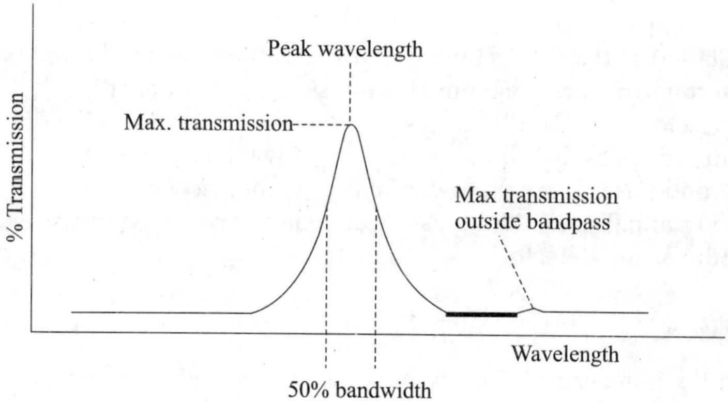

Fig. 4. Properties of a bandpass filter.

(usually measured as the separation in nanometers of the 50% transmission points), and the maximum transmission outside, the bandpass.

Edge filters are either short wavelength pass (SWP) filters, which transmit light below a given wavelength, and reflect light of a longer wavelength, or long wavelength pass (LWP) filters, which-work in the reverse way (Fig. 5).

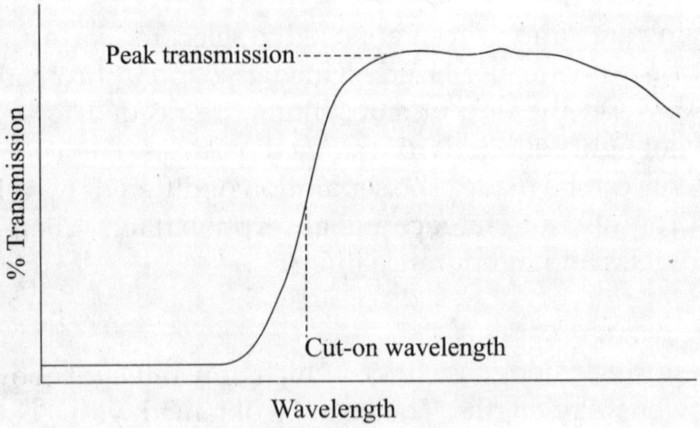

Fig. 5. Properties of an edge filter.

They are generally used in the flow cytometer as dichroic mirror (often called beam splitters).

The major parameters of edge filters are the cut-off (for LWP) or the cut-on (for SWP) wavelength (the wavelength for 50°C transmission), the peak trans-mission, and the slope.

(b) Optical Configuration

A typical optical configuration that could be used for the measurement of scattered light. Fluorescein (FITC-green) and Phycoerythrin (PE-orange) immunofluorescence is shown in Fig. 6.

Dichroic mirrors are placed at 45° to the incident beam while the absorption and interference filters are orthogonal. The first mirror in the optical path at right angle to the laser beam is a 500 nm long-pass dichroic filter, which reflects wavelengths shorter than 500 nm (the 488 nm scattered laser light) towards the right-angle scatter detector. Longer wavelengths pass on towards the second mirror, a 560 nm short-pass dichroic filter. Wavelengths greater than 560 nm are reflected towards the PE fluorescence detector and through a filter centered at 578 nm with a 28 nm half-peak bandpass. The shorter wavelengths between 500 and 560 nm incident at the second mirror pass on towards the mc fluorescence detector and through a 530 nm filter with a 30 nm half-peak bandpass. More colours could be measured by introducing further mirrors, filters, and detectors.

(c) Detection Devices

The two types of devices used in flow cytometers to detect scatter and fluorescence are PIN diodes and photomultiplier tubes (PMTs).

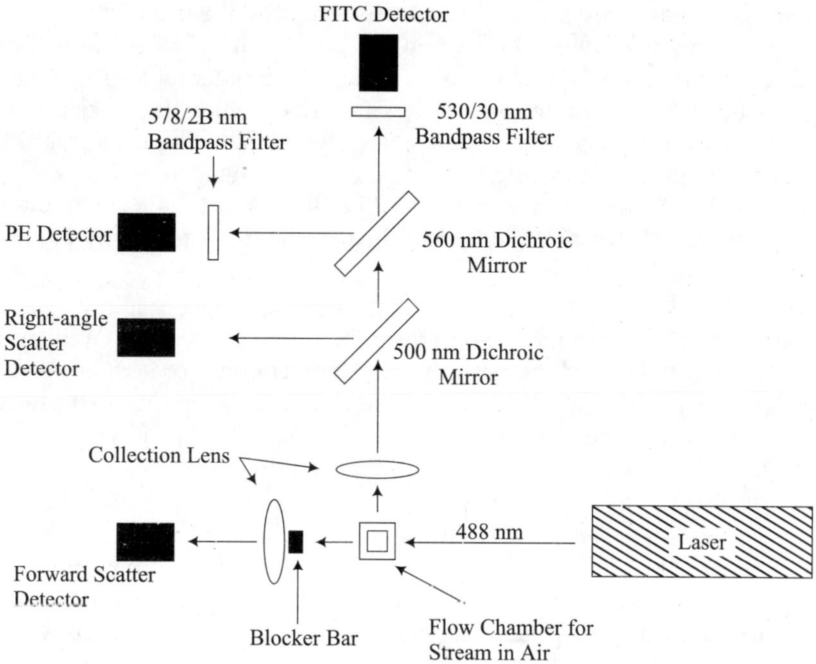

Fig. 6. An optical configuration for the simultaneous measurement of forward scatter, right-angle scatter, fluorescein 9FITC) and phycoerythrin (PE) fluorescence.

(i) PIN diodes: PIN diodes are cheap solid-state detectors of relatively low sensitivity but wide spectral characteristics and fast response. The lack of sensitivity of PIN diodes limits their usefulness and they are generally restricted to the measurement of forward scattered light.

(ii) PMT: Photomultipliers are photosensitive electron tubes with a more restricted spectral response, but with high gain and good signal-to-noise characteristics suitable, for the detection of weak fluorescence. The spectral sensitivity of PMTs is determined by the composition of the light-sensitive photocathode and this is an important factor in determining the sensitivity of a flow cytometer.

SIGNAL PROCESSING

Light falling on the photodetector surface generates a current that is fed into a filtering pre-amplifier. The output of the amplifiers is a smooth voltage pulse usually of between 0 and 10 volts. The amplitude of this pulse is proportional to the number of photons reaching the photodetector. Pulse shape is determined by the size and speed of the particle, the width of the illuminating beam, and in the case of fluorescence, and the distribution of the fluorochrome within the particle.

The pre-amplifiers are designed to follow both these rapid pulses as well as the slower pulses from larger particles without distortion.

The output from the pre-amplifier will inevitably contain some background noise. It is not desirable for further signal-processing circuits to process this noise, which may be at a high frequency and could mask true pulses. This is prevented by the use of a system threshold so that further processing only takes place when the input voltage rises above a pre-set value. Usually the threshold is set on a single signal that triggers all other measuring circuits. Forward scatter is often used as the system trigger pulse, particularly when fluorescence measurements are being made. If fluorescence was used as the trigger, negative cells might not produce signals above threshold and so would not be detected or measured. An exception is in the measurement of DNA, in which case the DNA-dye fluorescence is used as the trigger since only particles containing DNA are of interest. The threshold parameter is often referred to as the discriminator.

The pulses output by the pre-amplifiers are generally too fast to allow measurement and display circuits to function properly. Therefore, the pulse processing circuitry provides an analog voltage memory of the signal, which is held long enough for further processing to take place. The sustained voltage output from this 'sample and hold' circuit can be derived from the input signal in three ways, the output voltage can be made proportional to the height (peak height), the area (integral), or the width (time of flight) of the pulse.

Sample-and-hold circuits maintain their voltage output for a fixed period, usually between 15 and 120 m sec, and are triggered when the signal voltage rises above the pre-set threshold. During the period while the sample-and-hold circuit is operating, the electronics are unavailable for the processing of other pulses. This period is known as the 'system dead time' and events occurring during this processing period pass undetected. A large system dead time has less consequence for analysis than for sorting, the cells undetected during the system dead time will be taken randomly from the population of cells being analyzed and no bias of the final results will take place. However, the time taken to acquire a specific number of events will be increased. A large system dead time during sorting will not only reduce the maximum sorting rate but will also increase the chance of undetected cells being sorted coincidentally, thus compromising sorting purity.

The various pulse-processing modes enable different measurements to be made from the same signal.

Spectral overlap of the emission from two fluorochromes can also be corrected during signal processing. The spectral overlap produces a small signal from one fluorochrome at the detector intended to measure the fluorescence from the second fluorochrome and vice versa. Simple circuitry can be used to electronically subtract the proportion of the fluorescence due to spectral over-lap from each pulse.

WORKING (Principle of Operation)

1. Monodisperse Cells: The cells must be in single cell suspension with minimal aggregation (a monodisperse suspension). Producing a suspension of individual particles from biological samples can range from being straightforward to frustratingly difficult. Body fluids in particular blood generally contains individual cells that can be stained and processed directly on the flow cytometer e.g., a sample of blood can be stained with Thiazole orange and analyzed on the flow cytometer to obtain a count of reticulocytes.

But the solid tissues present a much more difficult and varied problem for flow cytometry. A technique which successfully releases cells from one tissue can fail totally on another for e.g. lymphoid tissues can usually be prepared by simple chopping and teasing of organ followed by scanning and density gradient centrifugation to provide a sample of separated lymphocytes. Other organs and solid tumor require enzymatic digestion before a good yield of individual cells is obtained. Detergents are often required in preparation of organelles such as isolated nuclei and chromosomes.Therefore, cells from solid tissues can be used if they can be dissociated and dispersed without breaking the cells.

2. Tagging: The monodisperse cells are labeled with a fluorescent tag (Rocederer, 1997). Typically, a monoclonal antibody to a cell surface glycoprotein receptor is used. The fluorescence can be direct or indirect. For direct fluorescence, fluorescein (or another fluorochrome) must be covalently attached to the primary monoclonal antibody. For indirect fluorescence, the primary receptor-specific antibody need not be fluorescent. It is bound to the cells first, and the excess rinsed away. Next, a secondary fluorescein-labeled anti-antibody is bound to the cell-bound primary antibody, and the excess rinsed away. The anti-antibody is purified from the serum of a rabbit or goat immunized with mouse antibody. If the primary antibody is a monoclonal mouse antibody, the secondary anti-antibody could be a rabbit anti-mouse antibody.

Data can be acquired in the flow cytofluorometer with the cells either alive, or fixed with paraformaldehyde. With fixed cells, the fluorescence and light scatter properties are stable, and data acquisition can be postponed for days or weeks.

In addition to surface receptors, intracellular components can be stained in fixed cells, such as DNA with propidium iodide. Quantitation of the total DNA per cell is very useful in analyzing the progress of cells through the cell cycle (G1 $\rightarrow$ S $\rightarrow$ G2 $\rightarrow$ M). Fluorescent sequence-specific polynucleotide probes can be hybridized *In situ* to quantitate e.g. the amount of mRNA for a particular gene. Fluorescent reporters are available for a wide variety of intracellular components, such as cytoplasmic [Ca^{++}], membrane potential, cytoplasmic pH, or actin. Changes in Ca^{++} induced by a stimulus can be measured second-by-second in living cells.

In advanced procedures, it is possible to label cells with two, three, or even four different fluorochromes simultaneously (each fluorochrome is a different color) and quantitate distinct fluorescence intensity for each.

3. Sheath Fluid and Fluidic Focusing: A suspension of tagged single cells (no clumps) is fed into the instrument. While flowing rapidly through channels in the instrument, a small amount of cell suspension joins a larger amount of cell-free buffer (called "sheath fluid"). The fluids come together under laminar flow conditions so that they flow together evenly without mixing. The "sheath" fluid surrounds a thin core thread of sample. This spaces the cells out so that only one passes the laser beam at a time, and keeps the cells centered in the flowing stream so that they pass the laser beam optimally centered.

4. Detection. The cells flow single file past a laser beam (or in some more complicated instruments, two laser beams). The momentary pulse of fluorescence emitted as the cell crosses the beam is measured by photomultipliers at a 90 degree angle from the beam. Typically, 2-3 detectors are used with different wavelength bandpass filters, allowing the simultaneous detection of emissions at different wavelengths from different fluorochromes in a single cell.

In addition to fluorescence, two types of light scatter are measured. Low-angle forward scatter (often called simply "forward scatter") is roughly proportional to the diameter of the cell. Orthogonal, 90° or "side scatter" is proportional to the granularity: neutrophil granulocytes have higher side scatter than do lymphocytes, which are agranular. Thus, in the FACScan, each cell can provide up to five numbers: size, granularity, plus green, red, and far-red fluorescence intensities.

5. Gating: Usually we want to see data only from single, viable cells. Typically one wishes to eliminate data from cell debris (particles smaller than cells), dead cells, and clumps of two or more cells. Subcellular debris and clumps can be distinguished from single cells by size (estimated by the intensity of low angle forward scatter). Dead cells have lower forward-scatter and higher side scatter than living cells. These differences are accurately preserved following paraformaldehyde fixation (despite the fact that after fixation, all the cells are dead).

The computer can be configured to display the fluorescence signals only from those particles with a specified set of scatter properties, namely, living single cells. This is called a scatter-gated fluorescence analysis. Actually, it is possible to "gate" on any set of signals. In some cases, it may be desirable to gate on a combination of fluorescence and scatter values. For example, the DNA of dead cells can be fluorescently stained under conditions, which prevent staining of living cells. Then data from dead cells can be eliminated by gating out brightly fluorescent cells.

6. Histograms and dot plots: FCM data are most often represented as dot plots or histograms (Recktenwald, 1993). Basically, histograms quantitate intensities (of scatter or fluorescence) one parameter per histogram. In contrast dot plots quantitate percentages of cells with various properties (In fact, dot plots also show intensities). Both histograms and dot plots reveal whether there are discrete subpopulations of cells with different intensities. A two-peak histogram means there are dim cells and bright cells. Histograms reveal subpopulations for a single parameter (scatter or fluorescence), while dot plots show the relations of subpopulations for two parameters.

In a dot plot, each cell is represented by a dot, positioned on the X and Y scales according to the intensities detected for that cell. Scatter dot plots (X = forward scatter intensity; Y = side scatter intensity) are often informative. Scatter scales are usually linear (Fig. 7).

Fluorescence dot plots typically plot X = green fluorescence intensity, Y = red fluorescence intensity. These two-color dot plots are often divided into four quadrants, the double negative cells, the green-only, red-only, and double positive cells. These are quantitated by giving the percentage of cells in each quadrant. Since fluorescence intensity often varies several orders of magnitude between cells, the scales are usually the logarithm of fluorescence intensity spanning four decades (a 10,000-fold range).

In a histogram, the X-axis is intensity (of scatter or fluorescence), and the Y-axis shows how many cells had each intensity (Fig. 8). Thus, histograms show the distribution of intensities for a single parameter, while dot plots show the correlated distribution for two parameters. The density of dots in a region of a dot plot shows the "number of cells", equivalent to the Y-axis of a histogram. Indeed, dot plots are sometimes represented as pseudo-3D graphs where the Z-axis is "number of cells".

The "middle" of a peak on a histogram quantitates the intensity (scatter or fluorescence) of the population of cells in that peak. However, an arithmetic average is a poor way to estimate the "middle" because it is too sensitive to the tail of the distribution. A few cells with extremely high intensities can pull the average well above the "middle" of a peak. The highest point on the histogram is the "mode". Often the mode is at the "middle" of a peak, but sometimes the

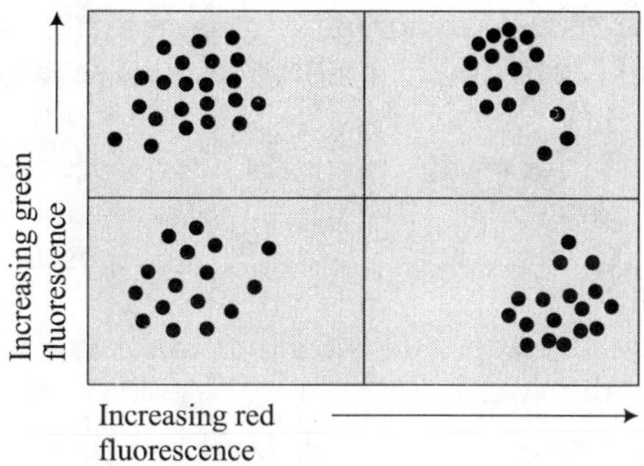

Fig. 7. A Dot plot of Flow Cytometry data. A dot plot representation of flow cytometry data. From the T-cell experiment, the CD4 positive cells would fluoresce green (the points in the upper left-hand quadrant), the CD8 positive cells would fluoresce red (the points in lower right-hand quadrant), and cells that were both CD4 and CD8 positive would fluoresce red and green (the points in the upper right-hand quadrant). The events shown in the lower left-hand quadrant are unlabelled cells.

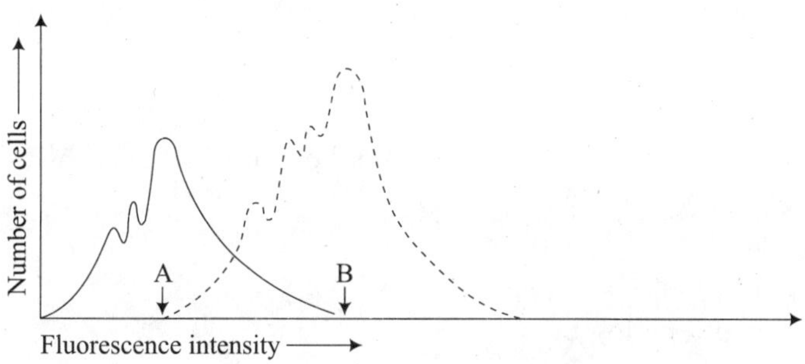

Fig. 8. A histogram of flow cytometry data. This is a histogram charting the number of cells counted and the fluorescence intensity. "A" and "B" mark the peak fluorescence intensenties . the shift shows that there is a relative increase in the average cell fluorescence for treatment "B" compared to treatment "A".

highest point is sensitive to sampling error. More seriously, sometimes the mode can be away from the peak entirely, when there is a pileup of off-scale cells at either end of the histogram. Therefore, the median is preferred as the most robust estimate of intensity characterizing a population of cells.

7. Autofluorescence: Even when no fluorescent tag or stain is added to cells, they fluoresce. This autofluorescence comes from normal cell components, which fluoresce, such as riboflavin and flavoproteins. To estimate the fluorescence intensity due to a tag, one must subtract the autofluorescence from the total fluorescence intensity measured. In order to quantitate autofluorescence, its value must be "on-scale", since if it is allowed to fall below the minimum scale value, it cannot be quantitated.

8. Analysis: Operation of the Becton-Dickinson FACSCalibur or FACScan very simple when only analysis is desired (without sorting). Calibration (setting of gains and cross-channel corrections) may be completely automated with a computer-controlled process that uses a mixture of three types of cell-sized plastic beads: non-fluorescent, green, and red. These instruments are popular both for research and clinical diagnosis because they can inexpensively and quickly analyze a large number of samples (30-50 samples/hour). Typically, data are collected for 10,000 cells/sample, and the results are available as various computer-generated graphs.

9. Sorting: Instruments capable of sorting are much more complicated than are analysis-only instruments. The FACSCalibur uses a fluidic valve mechanism to sort cells. This mechanism is limited to a throughput of 300 cells/second. Hence, it is most useful when a few parent cells are needed to start clones. In order to process cells ten times faster (about 3,000 cells/second), the high-speed FACStar vibrates the sapphire nozzle at ultrasonic frequency, which causes the stream (in air, after passing the laser beam) to break up into droplets with constant size and spacing. Typically, cells are present in less than 1/3 of the drops. The circuitry analyzes the signals from each cell, and then charges the stream with the correct time delay so that as the droplet containing that cell breaks off the stream, it is properly charged. Charging is positive, neutral or negative. The droplets then pass between plates, which set up an electrostatic field. This directs each charged droplet to the right or the left; neutral droplets (typically rejected droplets such as dead cells, cell debris, double cells etc.) fall into the middle waste collector. The operation and maintenance of the high-speed sorter are much more complicated than for the closed-flow instruments (Marty, 2001 and Weaver, 2000).

APPLICATIONS

Flow cytometry is a vast subject with long list of applications some of which could be discussed by categorizing as:

1. Applications in Research/UMass
2. Clinical Application
3. Application to cell biology.

I. APPLICATION IN RESEARCH

Research/UMass. FCM analysis is widely used in research on basic cellular and molecular mechanisms. Projects at UMass/Amherst have concerned the mechanisms by which white blood cells adhere to and kill infected cells, identification of the receptors used by viruses to infect cells, and the evolution of the genes which allow the immune system to distinguish self from foreign microorganisms. Virus projects have used fluorescent virus to measure the binding of virions to the cell surface. Our sorter has been used to sort slime mold cells expressing green fluorescent protein, bovine gamma-delta T cells, and human cell lines transfected with genes for integrin adhesion molecules.

2. CLINICAL APPLICATION

Clinical Applications. FCM analysis is widely used in hospitals in clinical research and diagnosis. A hospital or medical center typically has several FCM analyzers, which are in constant use. FCM can distinguish subtle changes in the types of cells in the blood, and discriminate many different types of blood cell cancers (leukemias and lymphomas). FCM can quantitate the depletion of T lymphocytes during immunosuppression to prevent tissue graft rejection, or the depletion of the CD4 T-helper cells in AIDS. In cancer cells taken in surgery or biopsy, the distribution of DNA/cell reveals the stage of progression. FCM results are important for diagnosis, prognosis, and optimization of therapy.

Some clinical applications of FCM that have proved useful in the clinical environments of Hematology blood transfusion, transplantation and immunology are discussed briefly under the following headings:

I. The Enumeration of Reticulocytes in Peripheral Blood

The introduction of Thiazole Orange in 1986 solved many of the problems associated with measuring reticulocytes by flow cytometry and it rapidly became the fluorochrome of choice. Its fluorescence increases by a factor of 3000 on binding to RNA, which implies a high signal-to-noise ratio giving a reliable reticulocyte count and maturity index.

2. The Enumeration of CD4+ Cells in Peripheral Blood Stem-cell Harvests and Bone-Marrow Harvests

Peripheral blood stem-cell transplants are now a treatment option for a number of malignancies. The patient's haematopoietic system is stimulated to pro-duce pluripotent cell precursors (stem cells), which exit the bone marrow and circulate in the peripheral blood. These cells can be harvested and frozen and the procedure may be repeated a number of times. The patient is subsequently treated with chemotherapy to destroy the diseased bone marrow, after which the harvested stem cells are re-infused to repopulate the bone marrow allowing haematopoietic recovery to take place.

The number of stem cells re-infused must be above a certain threshold to en-able a successful marrow recovery. Accurate enumeration of the percentage of stem cells in each harvest is therefore important and can be measured by flow cytometry using a combination of CD34 and CD45 monoclonal antibodies.

CD34+ cells in harvested bone marrow can also be estimated if these cells are to be used for bone-marrow transplantation.

3. Simultaneous Detection of Granulocyte- and Lymphocyte- Ractive Antibodies

Detection and characterization of granulocyte-reactive antibodies in blood donors and patients is of clinical importance because of their ability to cause febrile transfusion reactions, transfusion-related lung injury, and immune neutropenia. Detection of such antibodies has traditionally involved immunofluorescence techniques but the granulocyte isolation procedures involved are often complex and lengthy, resulting in considerable cell loss. Furthermore, when an antibody reacting with granulocytes is detected, its true specificity may be unclear since granulocytes express both HLA and granulocyte-specific antigens.

The problems of cell loss and discrimination between HLA and granulocyte-specific antibodies can be overcome by using a mixed cell population in an indirect immunofluorescence assay and then using flow cytometry to identify and assess the individual population.

4. Detection of Platelet-reactive Antibodies

Platelet alloantibodies recognizing the human platelet antigens (HPA) cause a variety of clinical conditions such as neonatal alloimmune thrombocytopenia, post-transfusion purpura and refractoriness to platelet transfusion. While auto-antibodies are the causative agents of autoimmune thrombocytopenias. Platelet-reactive antibodies are often detected using immunofluorescence techniques and the use of flow cytometric techniques appears to have improved the sensitivity of antibody detection methods.

5. Measurement of Cell-mediated Cytotoxicity by Flow Cytometry

Cell-mediated cytotoxicity can be measured by a dye-exclusion assay. The advantages of this arc, the lack of requirement for radioisotopes and the fact that killing can be measured on a 'per cell basis' irrespective of whether the mechanism is necrotic or apoptotic. This means that one can calculate the precise percentage of cells killed by the effector cells directly rather than in-direct extrapolation as is the case in ^{51}Cr-release assay.

6. Chromosome Analysis and Sorting by Flow Cytometry

Many types of cancer and genetic diseases are characterized by chromosomal abnormalities. These can be detected by conventional cytogenetics, which involves photographic banded chromosomes on metaphase spreads. Although this is a widely used technique in haematology, oncology, and prenatal diagnosis, it is a time-consuming process that relies heavily on the skill and experience of the cytogeneticist. An alternative approach is to prepare a monodispersed suspension of chromosomes stain with one or two fluorescent DNA dyes and pass them through a flow cytometer. The intensity of the fluorescence signal from each chromosome is recorded, the values being dependent on their DNA content.

3. Applications to Cell Biology

I. Estimating cell viability

Traditionally, the number of live cells in a suspension is estimated by counting, microscopically, the cells, which exclude an acidic dye, such as Trypan-Blue. The method can be adapted for the flow cytometer by using propidium iodide (PI) which is excluded by viable cells and, when taken up by dead or dying cells, binds to nucleic acids and fluoresces red. The use of flow cytometry has the advantage that large numbers of cells can be counted quickly and that the determination of negative/positive cells is objective.

2. Monitoring electropermeabilization of cells

Electroporation is an efficient way of introducing foreign molecules, such as DNA, into cells. The cells are permeabilized at a low temperature by a high voltage pulse and are then warmed to allow their membranes to reseal. A variety of parameters, such as the medium used, the pulse voltage, and the length and number of pulses, require optimization.

3. Measurement of oxidative burst

Phagocytes produce a variety of oxidative species when they are activated or when they ingest foreign particles. There are several probe that when oxidized, produce fluorescent compounds and can be used to explore oxidative burst.

Advantages and Disadvantages

- **Analysis.** If one is studying a *uniform (homogeneous)* population of cells (such as a cloned cell line grown in culture) and needs only population median values for receptor densities, then FCM is not necessary. ELISA or RIA methods are equally quick and easy, less expensive, and allow processing of even larger numbers of samples/day than does FCM (which can nevertheless handle hundreds/day if necessary). Also, these non-FCM methods give the single value, the average density. Nevertheless, FCM is often used even for uniform cell populations, since it is guaranteed to reveal non-uniformity, should it occur, and is precise about eliminating dead cells, debris, and clumps from the final median values.
- When studying *heterogeneous* populations of cells the FCM really shines. In a few minutes, the FCM can acquire data on all subpopulations of a sample. Not only one can tell the percentages of red cells vs. green cells, but the percentages of red-and-green doubly-labeled cells, and of cells which are neither red nor green. Moreover, there may be subpopulations of bright-green and dim-green, etc. No other method allows such rapid, quantitative, and detailed analysis of subpopulations.
- **Sorting.** FCM sorters can purify an extremely small subpopulation (1 cell in 105), or a subpopulation defined by complex criteria of size and fluorescence properties. They have a very low error rate. However, FCM sorting has several disadvantages. The major disadvantage is the rate of throughput, and losses in the yield. For reliable sorting, even on a high-speed sorter, the flow rate cannot exceed a few thousand cells/second. When pairs of cells go by too close together to distinguish, they must both be discarded. For a subpopulation, which is 20% of the starting population, the high-speed sorter will yield less than 106 cells/hour. Many experiments require far more cells than this. Running the high-speed sorter all day gets expensive, and may pose problems in keeping the cells in best condition during such a long sort. For a frame of reference, a single mouse can provide more than 108 lymphocytes, or 109 tumor cells.
- A second problem arises when the sorted cells need to be kept genobiotic ("sterile"). The high speed sorter can be used in a "sterile" mode, but this increases the complexity of the operation. Keeping the sample clean is much easier in the slow sorter, but it produces one tenth as many cells per hour. Finally there is the cost, which is hundreds of dollars per high-speed sort.
- If only a few carefully selected cells are needed to grow up large clones, the FACSCalibur can sort the desired cells quickly and inexpensively while maintaining sterility. However, when large numbers of cells are needed, particularly if sterile, one is well advised to consider whether non-FCM methods would be more satisfactory than a high-speed sort. Most non-FCM methods can handle arbitrarily large numbers of cells, and often with simultaneous parallel purification of multiple samples. Examples are deletion with antibody + complement, panning with monoclonal antibody-coated dishes, affinity columns (which can be used with living cells quite successfully), and density (isopycnic) or sedimentation (isokinetic) gradient separations.

REFERENCES

Molecular Probes Handbook.

Marti, G.E., Stetler-Stevenson, M., Bleesing, J.J.and Fleisher, T.A.. (2001). Introduction to Flow Cytometry. Sem Hematology 38(2): 93-9.

Ross, J.S. (1996). DNA Ploidy and Cell Cycle Analysis in Pathology. New York: Igaku-Shoin.

Roederer, M. (1997). Conjugation of monoclonal antibodies.

Recktenwald, D.J. (1993). Introduction to Flow Cytometry: Principles, Fluorochromes, Instrument Set-Up, Calibration. J Hematotherapy 2(3): 387-94.

Weaver, J.L. (2000). Introduction to Flow Cytometry. Methods 21(3): 199-201.

Suggested Readings

1. Melamed, M.R., Lindmo, T. and Mendelsohn, M.L. (ed.) (1990). Flow cytometry and sorting (2nd edition). Wiley, Newyork.
2. Shapiro, H.M. (1988). Practical Flow Cytometry (2nd edition). Alan R. Liss. New York.
3. Van Dilla. M.A., Dean, P.N., Laerum, O.D.; and Melamed, M.R. (ed.) (1985). Flow Cytometry instrumentation and data analysts. Academic Press, London.
4. Watson. J. V. (1991). Introduction to flow cytometry. Cambridge University Press, Cambridge.
5. Haughland, R. P. (1996). Handbook of fluorescent probes and research chemicals (6th edn.) Molecular Probes, inc., Oregon.

REVIEW QUESTIONS

Long Questions

1. What properties of a cell or particle can be measured by a flow cytometer?
2. What are three main systems in the flow cytometer?
3. How many different ways can you list to measure cellular proliferation by flow cytometry?
4. What are important concerns for a clinical diagnostic test using flow cytometry?
5. What are two flow cytometric methods for the detection of fetomaternal hemorrhage (fetal red cells in maternal blood)?

Multiple Choice Questions

1. Which of the following key cell components contribute to light scatter?
   (a) Cell membrane (b) Nucleus
   (c) Internal granular components (d) All of the above.
2. How many cells or particles should pass through the laser beam at a given time?
   (a) 1 (b) 2
   (c) 3 (d) 4
3. The wavelength of the laser used in the stream–in-airflow cytometer can be selected by adjusting one of the following optic positions at the rear of the laser.

(a) High reflector (b) low reflector
(c) Both a & b (d) none of the above.

4. What is the wavelength of the red-diode laser in the bench top analyzer?
(a) 635 nm (b) 695 nm
(c) 545 nm (d) 595 nm

5. What is the significance of the fluidics system in the flow cytometer to position the sample core in the passage of the laser beam?
(a) In the center (b) at the base
(c) at the neck (d) none of the above.

6. What is the name given to the portion of the fluidic stream where the cells are located?
(a) Sample core (b) sheath fluid
(c) both a & b (d) none of the above

7. What type of biological sample is best suited for flow cytometric analysis
(a) Suspension of 100 cells (b) suspension of 50 cells
(c) suspension of 10 cells (d) single cell suspension

8. What flow rate will be required for good data resolution for DNA studies?
(a) Low (b) high
(c) medium (d) Two of the above.

9. Which factor can affect illumination of the particles within the laser beam?
(a) Width of the stream (b) positioning of the stream
(c) Both a & b (d) none of the above.

Answers

1. (d) 2. (a) 3. (a) 4. (a) 5. (a) 6. (a)
7. (d) 8. (d) 9. (c)

PART–17

Gel Filtration and Chromatography

CHAPTER 54

Gel Filtration and Ion Exchange Chromatography

Y.C. Tripathi
Rain Forest Research Institute, P.O. Box 136, Deovan, Sotai-Ali, A.T. Road, Jorhat-785 001
E-mail: yctripathi@yahoo.com

GEL FILTRATION

Gel filtration is a chromatographic technique in which fractionation is based upon the molecular size and shape of the species in the sample. Gel filtration has also been called by different names like gel permeation, exclusion chromatography and molecular sieve chromatography. In gel permeation, separation takes place according to size of the solute molecules. In exclusion chromatography, separation of the sample components takes place according to molecular size. In molecular sieve chromatography, separations are carried out on natural and synthetic zeolites (metal-alumino silicates), called molecular sieve. A typical zeolite or molecular sieve has a general formula $M_{2/n}.Al_2O_3.xSiO_2.yH_2O$; Where M is a cation of valency n. The molecular sieve contains permanent cavities and channels in their structure. The sieve properties of the zeolite is determined by the size of these cavities, while adsorptive properties of the zeolite are determinated by the large surface area and presence of M, Al, Si and O in the zeolite structure.

Gel filtration separates molecules according to differences in size as they pass through a gel filtration medium packed in a column. Unlike ion exchange or affinity chromatography, molecules do not bind to the chromatography medium so buffer composition does not directly affect resolution. Consequently, a significant advantage of gel filtration is that conditions can be varied to suit the type of sample or the requirements for further purification, analysis or storage without altering the separation. Gel filtration differs from the other types of partition chromatography in that the particles which come from the stationary phase in the column bed is an uncharged gel.

GEL FILTRATION CHROMATOGRAPHIC TECHNIQUE

Ordinarily, gel filtration chromatography is performed on a column by the elution method. Here, the degree of retardation depends upon the extent to which the solute molecules or ions can penetrate that part of the solution phase which is held within the pores or the highly porous gel-like packing material. The solid support or gels that are used are three dimensional networks of cross-linked polymer chains. One of the most widely used polymers is prepared by cross linking the polysaccharide dextran with epichlorohydrin. A series of resins with different pore sizes can be obtained by changing the amount of epichlorohydrin. The resulting gel is called as Sephadex (Table 1). Polyacrylamide cross-linked with methylene bis-acrylamide has also been used as resins in gel chromatography. These gels have a tendency to swell in a suitable solvent and as a result of swelling of gel, the space between the polymer chains increases in size. For a gel there will be critical size of a molecule that can just penetrate the interior. Molecules or ions larger than the pores of the gel are completely excluded from the interior, while smaller species are more or less free to enter these regions. Thus larger molecules pass quickly unhindered through the column, while the molecules smaller than the critical size are retarded to a greater or lesser extent, depending upon their size and shape and sometimes upon the tendency to be absorbed by the gel. The sieve effect is now of such a magnitude that it is more convenient to specify it in terms of the molecular weight of solute which is just excluded. Exclusion limits of sephadex are given in Table 1.

Table 1. Exclusion limits of sephadex

Types of sephadex	Limiting molecular weight excluded
G-25	3500-4500
G-50	8000-10000
G-75	40000-50000
G-100	100,000
G-200	200,000

In gel chromatography, the granulated or bedded gel material is called as packing material. The solutes which are distributed through the entire gel phase is called stationary phase and the liquid flowing through the bed is called as the mobile phase. Modern stationary phases consist of micro particles of styrene-divinyl benzene copolymers (ultrastyragel), silica or porous glass.

COLUMN

The column consists of a straight tube with a bed support at the bottom. The bed support allows only the liquid to pass through without disturbing the bed material. The simplest column consists of a straight tube containing some glass wool into the bottom. The glass wool is then covered with thin layer of quartz or glass beds. In order to avoid clogging of the glass plate, a filter paper or other protective device is placed over the glass plate before performing experiment with the column.

GEL PREPARATION

There are two main methods for gel preparation:

1. A weighed amount of dry powder of gels is mixed with excess of solvent which have to be used as an eluent and then allowed to swell. The mixture is left as such till the equilibrium condition is maintained.
2. In this method, gel slurry is warmed at about 100°C in the boiling water bath. As a result the gel swells in few days. The slurry is then cooled and packed in the column. The gels can be stored in the wet state and there is no need to dry them. However, for storage purpose, the sephadex and acrylamide gels can be dried without any damage and can be brought back to wet state very easily

COLUMN PACKING

It is not desirable to pack the column with dry packing because the column may undergo due to swelling of the gels when a liquid is allowed to flow through the column. The column is first filled with eluent, and thin slurry of gel is poured into a funnel attached to the top of the column. As a result, the gel sediments into the column slowly and packing is done. The method of packing the column, however, depends on the type of gel to be used. A very careful packing is required in case of soft gel, like Dextran sephadex.

SAMPLE APPLICATION

The sample is applied at the top of the gel by the use of a plunger and then the sample solution is allowed to percolate through the bed. A small volume of element is also added and last trace of it is washed in the column bed. This is usually done with the help of a pipette having bent tip.

GEL FILTRATION THEORY

Defining the Process

Results from gel filtration are usually expressed as an elution profile or chromatogram that shows the variation in concentration (typically in terms of UV absorbance at A280 nm) of sample components as they elute from the column in order of their molecular size. Molecules that do not enter the matrix are eluted in the void volume, V_o as they pass directly through the column at the same speed as the flow of buffer. Molecules with partial access to the pores of the matrix elute from the column in order of decreasing size. Molecules with full access to the pores move down the column, but do not separate from each other. These molecules usually elute just before one total column volume, V_t of buffer has passed through the column. The behavior of each component can be expressed in terms of its elution volume, V_e determined by direct measurement from the chromatogram. As shown in Fig. 1, there are three different ways of measuring V_e dependent on the volume of sample applied to the column.

A. When very small samples are applied (small enough to be neglected compared to the elution volume), take the position of the peak maximum in the elution diagram as V_e.
B. If the sample volume cannot be neglected compared with the elution volume, measure the elution volume from half the sample volume to the position of the peak maximum.

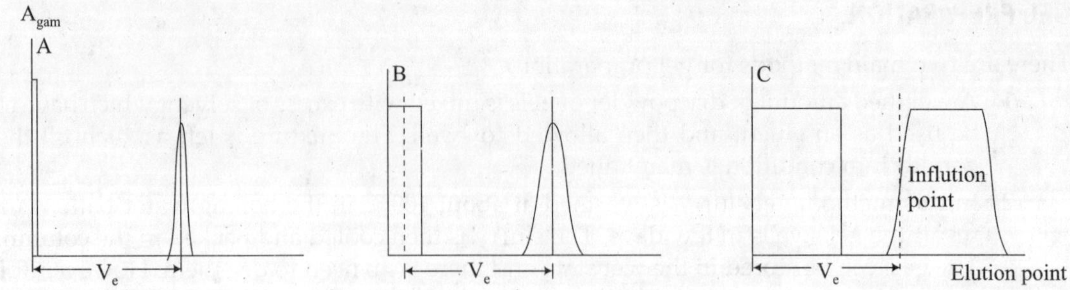

A. Sample are negligible compared with volume of packed bed.
B. Sample are not negligible compared with volume of packed bed.
C. Sample giving plateau elution curve.

Fig. 1. Measurement of elution volume V_e

C. When very large sample volumes are used (giving a plateau region in the elution curve), take the volume eluted from the start of sample application to the inflexion point of the rising part of the elution peak as V_e.

Since symmetrical peaks are common in gel filtration, elution volumes are easily determined. However, V_e does not completely define the behavior of the sample since V_e will vary with the total volume of the packed bed (V_t) and the way in which the column has been packed. The elution of a sample is best characterized by a distribution coefficient (K_d) derived as follows:

The volume of the mobile phase (buffer) is equal to the void volume, V_o, i.e. the elution volume of molecules that remain in the buffer because they are larger than the largest pores in the matrix and pass straight through the packed bed. In a well packed column the void volume is approximately 30% of the total column.

The volume of the stationary phase, V_s is equal to V_i, the volume of buffer inside the matrix which is available to very small molecules, i.e. the elution volume of molecules that distribute freely between the mobile and stationary phases minus the void volume.

Since, in practice, V_s or V_i are difficult to determine, it is more convenient to substitute the term ($V_t - V_o$). The estimated volume of the stationary phase will therefore include the volume of solid material which forms the matrix.

K_d represents the fraction of the stationary phase that is available for diffusion of a given molecular species. The stationary phase Vs can be can be substituted by the term ($V_t - V_o$) in order to obtain a value K_{av}.

$$K_{av} = (V_e - V_o)/(V_t - V_o)$$

Since ($V_t - V_o$) includes the volume of the matrix that is inaccessible to all solute molecules, K_{av} is not a true partition coefficient. However, for a given medium there is a constant ratio of K_{av}: K_d which is independent of the nature of the molecule or its concentration. K_{av} is easily determined and, like K_d defines sample behavior independently of the column dimensions and packing. Other methods of normalizing data give values which vary depending upon how well the column is packed. The approximate relationships between some of these terms are shown in Fig. 2.

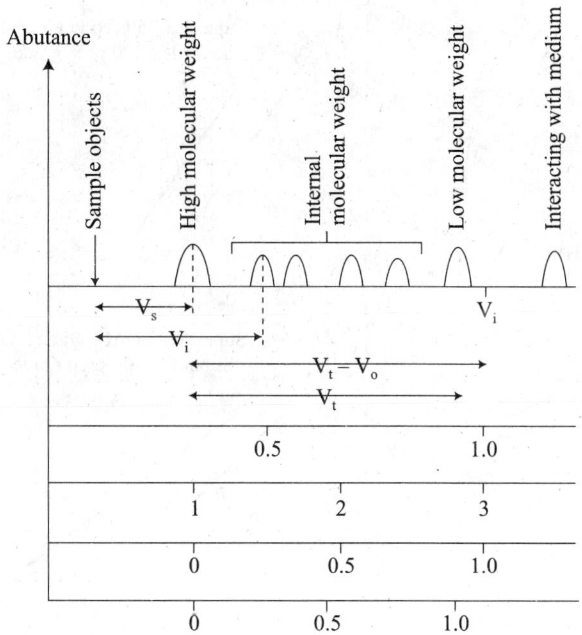

$$K_{av} = \frac{V_e - V_o}{V_t - V_o}$$

Fig. 2. Relationship between several expressions used for normalizing elution behavior.

SELECTIVITY CURVES AND MEDIA SELECTION

The partition coefficient K_{av} is related to the size of a molecule. Molecules of similar shape and density demonstrate a sigmoidal relationship between their K_{av} values and the logarithms of their molecular weights (M_r). Over a considerable range there is a linear relationship between K_{av} and log M_r. The selectivity of a gel filtration medium depends solely on its pore size distribution and is described by a selectivity curve. By plotting K_{av} against the log of the molecular weight for a set of standard proteins, selectivity curves are created for each gel filtration medium, as shown in Fig. 3.

Gel filtration media should be selected so that the high molecular weight molecules are eluted at the void volume ($K_{av} = 0$) with minimum peak broadening or dilution and minimum time on the column. The lowest molecular weight substances should be eluted near V_t ($K_{av} = 1$).

- Under ideal conditions, no molecules can be eluted with a K_{av} greater than 1 or less than 0.
- If the K_{av} is greater than 1, molecules have bound non-specifically to the medium.
- If K_{av} is less than 0 after calibration then there is channeling in the chromatography bed and the column must be repacked.

RESOLUTION

Final resolution, the degree of separation between peaks, is influenced by many factors: the ratio of sample volume to column volume, flow rate, column dimensions, particle size, particle

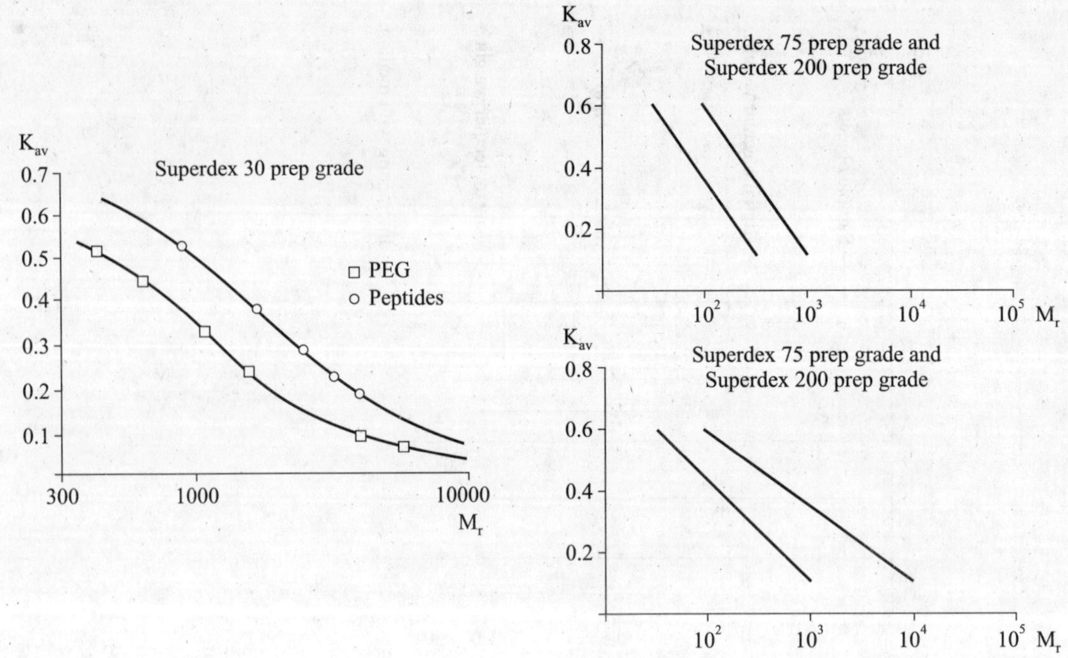

Fig. 3. Selectivity curves for Superdex 30 prep grade, Superdex 75 prep grade and Superdex 200 prep grade.

size distribution, packing density, porosity of the particle and viscosity of the mobile phase. The success of gel filtration depends primarily on choosing conditions that give sufficient selectivity and counteract peak broadening effects during the separation.

Resolution (R_s) is defined by the following expression:

$$R_s = \frac{V_{r2} - V_{r1}}{\frac{[W_1 + W_2]}{2}}$$

V_{r1} and V_{r2} are the elution volumes for two adjacent peaks measured at the center of the peak. W_1 and W_2 are the respective peak widths.

$(V_{r2} - V_{r1})$ represents the distance between the peaks and $1/2\ (W_1 + W_2)$ the mean peak width of the two peaks as shown in Fig. 4. Resolution is a function of the selectivity of the medium and the efficiency of that medium to produce narrow peaks (minimal peak broadening) as illustrated in Fig. 5. The efficiency of a packed column defines its ability to produce narrow symmetrical peaks during elution. Efficiency is particularly important in gel filtration in which separation takes place as only a single column volume of buffer passes through the column. Efficiency is defined in terms of theoretical plates per meter (N):

$$N = 5.54\ (V_e/W_1/2)2 \times 1000/L$$

where V_e = peak elution (retention) volume; $W_1/2$ = peak width at half peak height; L = bed height (mm); V_e and $W_1/2$ are in same units

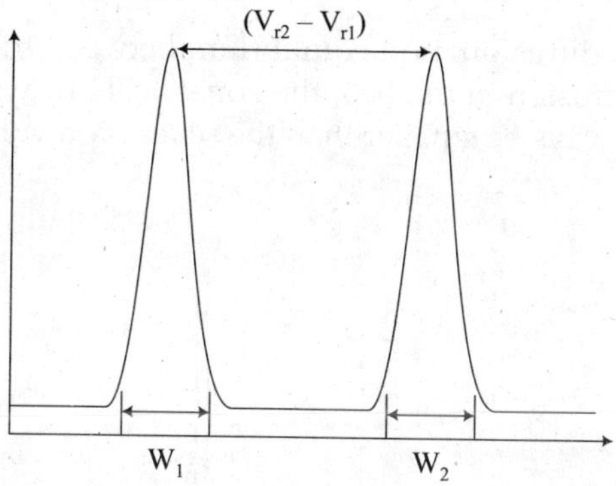

Fig. 4. Parameters used to define resolution (Rs).

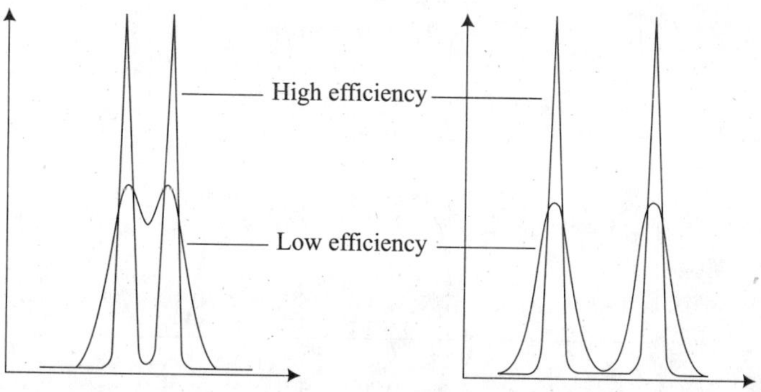

Fig. 5. Dependence of resolution on selectivity and the counteraction of peak broadening.

Efficiency can be improved by decreasing the particle size of the medium. However, using a smaller particle size may create an increase in back pressure so that flow rate needs to be decreased, lengthening the run time. The uniformity of the packed bed and the particles influences the uniformity of the flow through the column and hence affects the shape and eventual peak width. Gel filtration media with high uniformity (lower particle size distribution) facilitate the elution of molecules in narrow peaks. Gel filtration media with smaller particle sizes facilitate diffusion of sample molecules in and out of the particles by reducing the time to achieve equilibrium between mobile and stationary phases and so improve resolution by reducing peak width. Sample dilution is inevitable because sample passes through the column and diffusion occurs. In order to minimize sample dilution a maximum sample volume is used within the limits set by the separation distance i.e. the resolution required between the peaks of interest.

The sample can be regarded as a zone passing down the column. Fig. 6 shows how, if no zone broadening occurs, the maximum sample volume could be as great as the separation volume (V_{Sep}):

$$V_{Sep} = V_{eB} - V_{eA}$$

However, due to eddy diffusion, non-equilibrium between the stationary phase and the buffer, and longitudinal diffusion in the bed, the zones will always be broadened. Therefore, the sample volume must always be smaller than the separation volume.

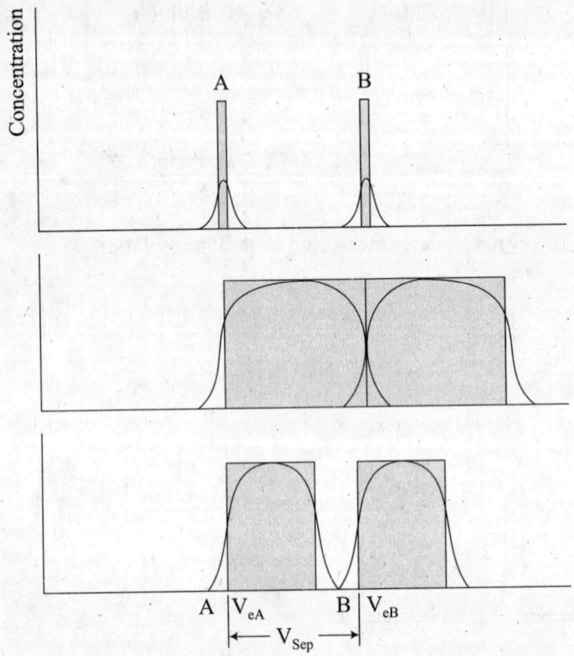

Fig. 6. Elution curves for different sample sizes. The top diagram corresponds to the application of a small sample. The center diagram corresponds to the maximum sample volume that gives complete separation if there is no zone broadening. The bottom diagram corresponds to the maximum sample volume to obtain complete separation in the conditions of the experiment. The shaded areas correspond to the elution profiles that would beobtained if there was no zone broadening.

SEPARATION BY GEL FILTRATION

For more than forty years since the introduction of Sephadex™, gel filtration has played a key role in the purification of enzymes, polysaccharides, nucleic acids, proteins and other biological macromolecules. Gel filtration is the simplest and mildest of all the chromatography techniques and separates molecules on the basis of differences in size. The technique can be applied in two distinct ways:

1. Group separations: the components of a sample are separated into two major groups according to size range. A group separation can be used to remove high or low molecular weight contaminants (such as phenol red from culture fluids) or to desalt and exchange buffers.
2. High resolution fractionation of biomolecules: the components of a sample are separated according to differences in their molecular size. High resolution fractionation can be used to isolate one or more components, to separate monomers

from aggregates, to determine molecular weight or to perform a molecular weight distribution analysis.

To perform a separation, gel filtration medium is packed into a column to form a packed bed. The medium is a porous matrix in the form of spherical particles that have been chosen for their chemical and physical stability, and inertness (lack of reactivity and adsorptive properties). The packed bed is equilibrated with buffer which fills the pores of the matrix and the space in between the particles. The liquid inside the pores is sometimes referred to as the stationary phase and this liquid is in equilibrium with the liquid outside the particles, referred to as the mobile phase. It should be noted that samples are eluted isocratically, i.e. there is no need to use different buffers during the separation.

However, a wash step using the running buffer is usually included at the end of a separation to facilitate the removal of any molecules that may have been retained on the column and to prepare the column for a new run. Fig. 7 shows the most common terms used to describe the separation and Fig. 8 illustrates the separation process of gel filtration.

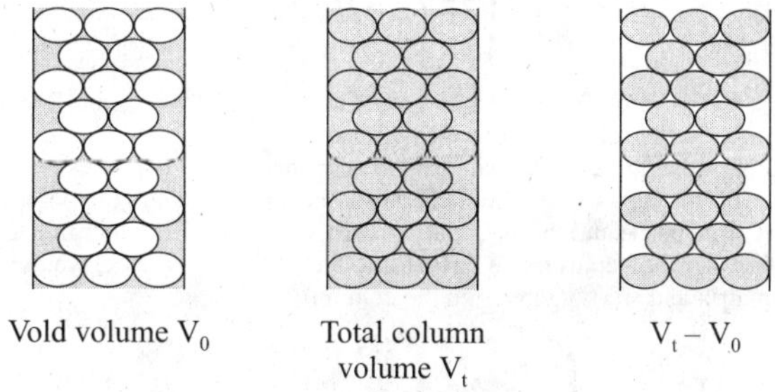

Fig. 7. Common terms of Gel filtration.

GROUP SEPARATION

Gel filtration is used in group separation mode to remove small molecules from a group of larger molecules and as a fast, simple solution for buffer exchange. Small molecules such as excess salt (desalting) or free labels are easily separated. Samples can be prepared for storage or for other chromatography techniques and assays. Gel filtration in group separation mode is often used in protein purification schemes for desalting and buffer exchange. Sephadex G-10, G-25 and G-50 are used for group separations. Fig. 9 shows the elution profile (chromatogram) of a typical group separation. Large sample volumes up to 30% of the total column volume (packed bed) can be applied at high flow rates using broad, short columns.

Large molecules are eluted in or just after the void volume (V_o) as they pass through the column at the same speed as the flow of buffer. For a well packed column the void volume is equivalent to approximately 30% of the total column volume. Small molecules such as salts that have full access to the pores move down the column, but do not separate from each other. These molecules usually elute just before one total column volume (V_t) of buffer has passed through the column. In this case the proteins are detected by monitoring their UV absorbance, usually at A 280 nm and the salts are detected by monitoring the conductivity of the buffer.

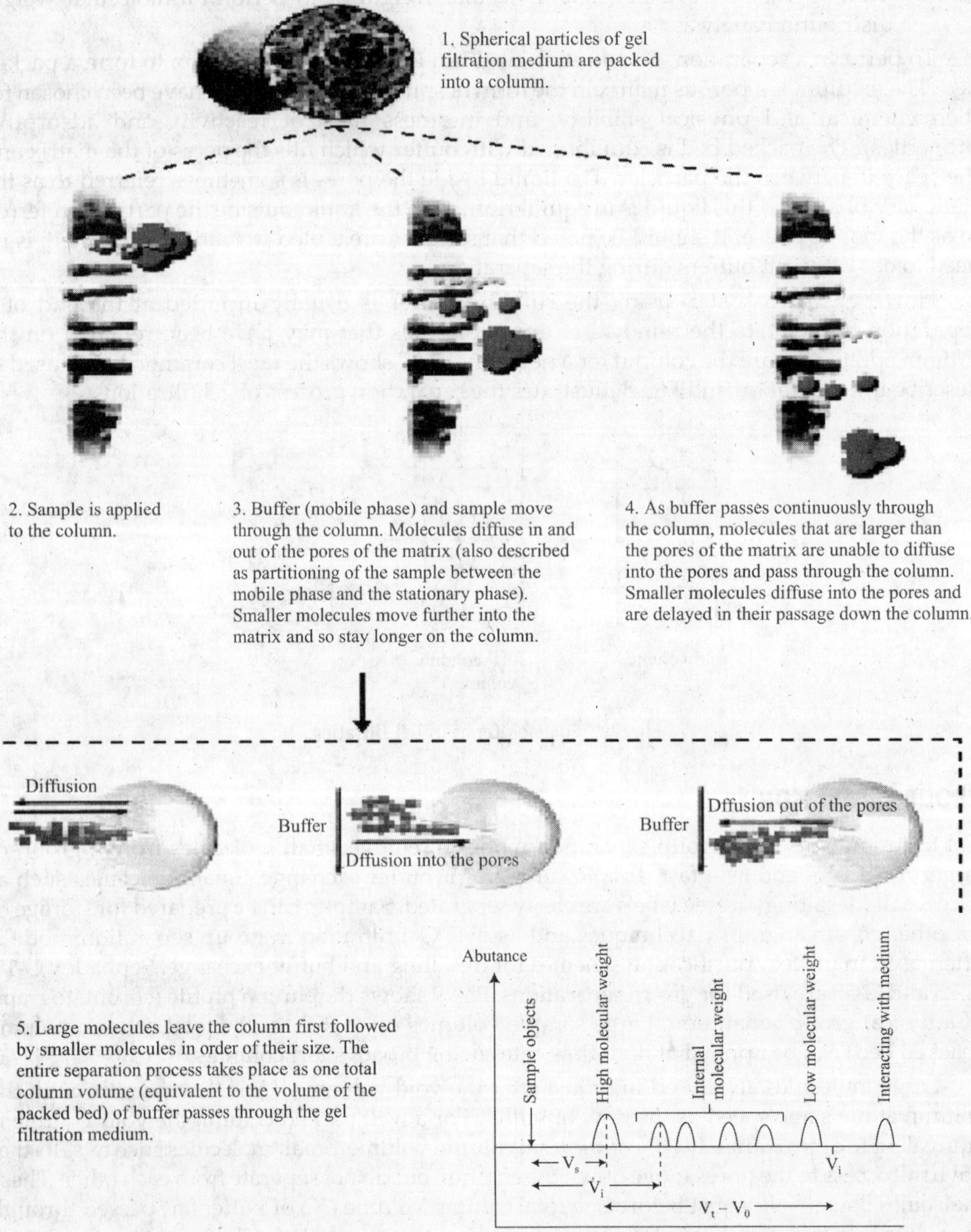

Fig. 8. Separation process of gel filtration.

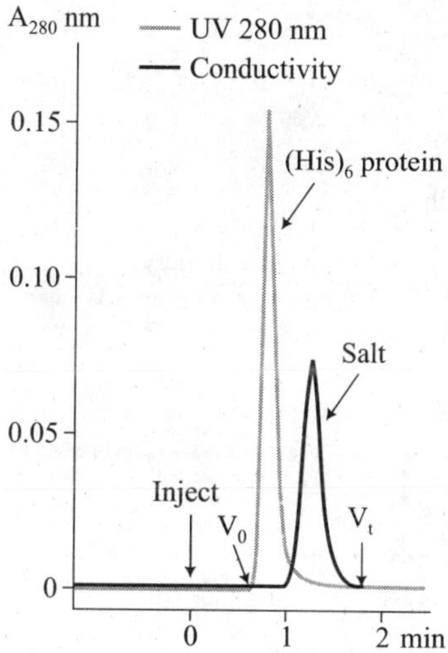

Sample: $(His)_6$ protein eluted from HiTrap™ Chelating HP with sodium phosphate 20 mM, sodium chloride 0.5 M, imidazole 0.5 M, pH 7.4

Column: HiTrap Desalting 5 ml

Buffer: Sodium phosphate 20 nM, sodium chloride 0.15 M, pH7.0

void volume V_0, total column volume V_t

Fig. 9. Typical Chromatogram of Group Separation.

HIGH RESOLUTION FRACTIONATION

Gel filtration is used in fractionation mode to separate multiple components in a sample on the basis of differences in their size. The goal may be to isolate one or more of the components, to determine molecular weight, or to analyze the molecular weight distribution in the sample. The best results for high resolution fractionation will be achieved with samples that originally contain few components or with samples that have been partially purified by other chromatography techniques.

High resolution fractionation by gel filtration (Fig. 10) is well suited for the final polishing step in a purification scheme. Monomers can be separated from aggregates (difficult to achieve by any other technique) and samples can be transferred to a suitable buffer for assay or storage. Gel filtration can be used directly after any of the chromatography techniques such as ion exchange, chromatofocusing, hydrophobic interaction or affinity since the components from any elution buffer will not affect the final separation.

Molecules that do not enter the matrix are eluted in the void volume (V_o) as they pass directly through the column at the same speed as the flow of buffer. For a well packed column the void volume is equivalent to approximately 30% of the total column volume (packed bed). Molecules with partial access to the pores of the matrix elute from the column in order of decreasing size. Small molecules such as salts that have full access to the pores move down the column, but do not separate from each other. These molecules usually elute just before one total column volume (V_t) of buffer has passed through the column.

RESOLUTION IN GEL FILTRATION

Many factors influence the final resolution viz., sample volume, the ratio of sample volume to column volume, column dimensions, particle size, particle size distribution, packing density, pore size of the particles, flow rate, and viscosity of the sample and buffer. The molecular

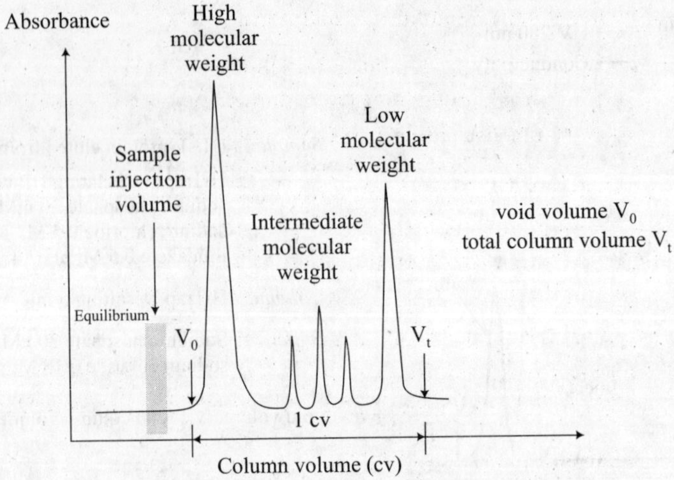

Fig. 10. High resolution Gel Filtration.

weight range over which a gel filtration medium can separate molecules is referred to as the selectivity of the medium. Resolution is a function of the selectivity of the medium and the efficiency of that medium to produce narrow peaks as illustrated in Fig. 11. The success of gel filtration depends primarily on choosing conditions that give sufficient selectivity and counteract peak broadening effects during the separation.

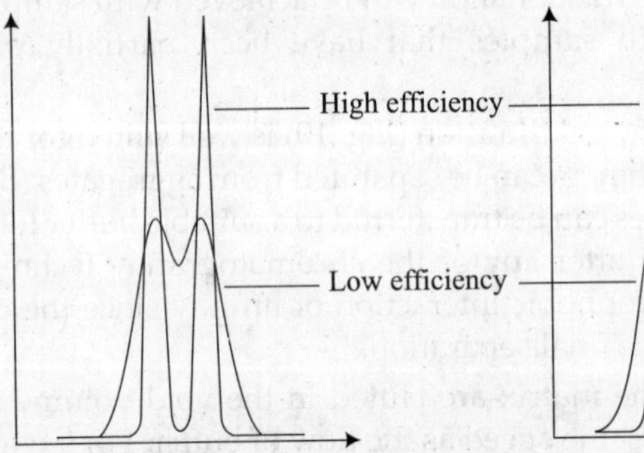

Fig. 11. Dependence of resolution on selectivity and concentration of peak broadening.

SAMPLE VOLUME AND COLUMN DIMENSIONS

Sample volumes are expressed as a percentage of the total column volume (packed bed). Using smaller sample volumes helps to avoid overlap if closely spaced peaks are eluted.

- For group separations sample volumes up to 30% of the total column volume can be applied.
- For high resolution fractionation a sample volume from 0.5–4% of the total column volume is recommended, depending on the type of medium used. For most applications the sample volume should not exceed 2% to achieve maximum resolution.

Depending on the nature of the specific sample, it may be possible to load larger sample volumes, particularly if the peaks of interest are well resolved.

- For analytical separations and separations of complex samples, start with a sample volume of 0.5% of the total column volume. Sample volumes less than 0.5% do not normally improve resolution.
- To increase the capacity of a gel filtration separation samples can be concentrated. Avoid concentrations above 70 mg/ml as viscosity may interfere with the separation.
- Sample dilution is inevitable because diffusion occurs as sample passes through the column. In order to minimize sample dilution use a maximum sample volume that gives the resolution required between the peaks of interest.

The ratio of sample volume to column volume influences resolution, where higher sample volume to column volume ratios gives lower resolution. Column volumes are normally selected according to the sample volumes to be processed. However, since larger sample volumes may require significantly larger column volumes, there may be occasions when it is more beneficial to repeat the separation several times on a smaller column and pool the fractions of interest or to concentrate the sample.

The height of the packed bed affects both resolution and the time taken for elution. The resolution in gel filtration increases as the square root of bed height. Doubling the bed height gives an increase in resolution equivalent to È2 = 1.4 (40%). For high resolution fractionation long columns will give the best results and a bed height 30–60 cm should be satisfactory. Sufficient bed height together with a low flow rate allows time for all 'intermediate' molecules to diffuse in and out of the matrix pores and give sufficient resolution. If a very long column is judged to be necessary, the effective bed height can be increased by using columns, containing the same media, coupled in series.

MEDIA SELECTION

Chromatography media for gel filtration are made from porous matrices chosen for their inertness and chemical and physical stability. The size of the pores within a particle and the particle size distribution are carefully controlled to produce a variety of media with different selectivities. Today's gel filtration media cover a molecular weight range from 100 to 80,000,000, from peptides to very large proteins and protein complexes. The selectivity of a gel filtration medium depends solely on its pore size distribution and is described by a selectivity curve, which are usually quite linear and it is this part of the curve that is used to determine the fractionation range of a gel filtration medium.

The fractionation range defines the range of molecular weights that have partial access to the pores of the matrix, i.e. molecules within this range should be separable by high resolution fractionation. The exclusion limit for a gel filtration medium, also determined from the selectivity curve, indicates the size of the molecules that are excluded from the pores of the matrix and therefore elute in the void volume.

The steeper the selectivity curve, the higher the resolution that can be achieved. While choosing an appropriate medium two main factors are considered:

- The aim of the experiment (high resolution fractionation or group separation).
- The molecular weights of the target proteins and contaminants to be separated.

The final scale of purification should also be considered. Superdex is the first choice for high resolution, short run times and high recovery. Sephacryl is suitable for fast, high recovery separations at laboratory and industrial scale. Superose offers a broad fractionation range, but is not suitable for large scale or industrial scale separations.

After deciding upon Superdex, Sephacryl or Superose, the medium is selected with the fractionation range that covers the molecular weight values of interest in the sample. In cases where two media have a similar fractionation range, the medium with the steepest selectivity curve for best resolution of all components in the sample is selected. When some one is interested in a specific component, the medium where the molecular weight of the target component falls in the middle of the selectivity curve is selected.

- Sephadex is ideal for rapid group separations such as desalting and buffer exchange and is used at laboratory and production scale, before, between or after other chromatography purification steps.
- Sephadex G-25 is recommended for the majority of group separations involving globular proteins. This medium is excellent for removing salt and other small contaminants away from molecules that are greater than M_r 5000.
- Sephadex G-10 is well suited for the separation of biomolecules such as peptides (M_r >700) from smaller molecules (M_r >100).
- Sephadex G-50 is suitable for the separation of molecules M_r >30000 from molecules M_r <1 500 such as labeled protein or DNA from free label.

For group separations gel filtration media is selected so that high molecular weight molecules are eluted at the void volume with minimum peak broadening or dilution and minimum time on the column. The lowest molecular weight substances should appear by the time, one column volume of buffer has passed through the column.

SAMPLE PREPARATION

Correct sample preparation is extremely important for gel filtration. Simple steps to clarify a sample before applying it to a column will avoid the risk of blockage, reduce the need for stringent washing procedures and extend the life of the packed chromatography medium. Samples must be clear and free from particulate matter, particularly when working with bead sizes of 34 μm or less. For small sample volumes a syringe-tip filter of cellulose acetate or PVDF can be sufficient.

SAMPLE BUFFER COMPOSITION

The pH, ionic strength and composition of the sample buffer will not significantly affect resolution as long as these parameters do not alter the size or stability of the proteins to be separated and are not outside the stability range of the gel filtration medium. The sample does not have to be in exactly the same buffer as that used to equilibrate and run through the column. Sample is exchanged into the running buffer during the separation, an added benefit of gel filtration.

SAMPLE CONCENTRATION AND VISCOSITY

Gel filtration is independent of sample mass and hence sample concentration. Hence high resolution can be maintained despite high sample concentration and with the appropriate medium, high flow rates. However, the solubility or the viscosity of the sample may limit the concentration that can be used. A critical variable is the viscosity of the sample relative to the running buffer. Too high sample viscosity causes instability of the separation and an irregular flow pattern. This leads to very broad and skewed peaks and back pressure can increase. Samples should generally not exceed 70 mg/ml protein. Sample volume is also one of the most important parameters in gel filtration.

BUFFER COMPOSITION AND PREPARATION

Buffer composition does not directly influence the resolution obtained in gel filtration since the separation should depend only on the sizes of the different molecules. The most important consideration is the effect of buffer composition on the shape or biological activity of the molecules of interest. For example, the pH and ionic strength of the buffer and the presence of denaturing agents or detergents can cause conformational changes, dissociation of proteins into subunits, dissociation of enzymes and cofactors, or dissociation of hormones and carrier proteins. Hence buffer and pH should be compatible with protein stability and activity and in which the product of interest should be collected. Buffer concentration should be sufficient to maintain buffering capacity and constant pH. To avoid non-specific ionic interactions with the matrix (shown by delays in peak elution), NaCl up to 0.15 M is used. Some proteins may precipitate in low ionic strength solutions. Volatile buffers such as ammonium acetate, ammonium bicarbonate or ethylenediamine acetate are used if the separated product is to be lyophilized.

DENATURING (CHAOTROPIC) AGENTS AND DETERGENTS

Denaturing agents such as guanidine hydrochloride or urea can be used for initial solubilization of a sample and in gel filtration buffers in order to maintain solubility of the sample. However, because they will denature the protein, they should be avoided unless denaturation is required. Modern media such as Superdex, Sephacryl and Superose are in general more suitable than classical media such as Sepharose™ or Sephadex for working under dissociating or denaturing conditions or at extreme pH values. Detergents are useful as solubilizing agents for proteins with low aqueous solubility such as membrane components and will not affect the separation. If denaturing agents or detergents are necessary to maintain the solubility of the sample, they should be present in both the running buffer and the sample buffer. If proteins that have been solubilized in a denaturant or detergent are seen to precipitate, elution should be done later than expected or add a suitable concentration of the denaturing agent or detergent to the running buffer if poorly resolved during gel filtration.

Urea or guanidine hydrochloride is very useful for molecular weight determination. The presence of these denaturing agents in the running buffer maintains proteins and polypeptides in an extended configuration. For accurate molecular weight determination the calibration standards must also be run in the same buffer. Selectivity curves are usually determined using globular proteins so they do not reflect the behavior of denatured samples. Gel filtration can be used to exchange a protein solubilized initially in, for example SDS, into a more gentle detergent such as Triton™ X-100 without losing solubility.

COLUMN AND MEDIA PREPARATION

To perform a separation, gel filtration medium is packed into a column 30–60 cm in height for high resolution fractionation and up to 10 cm in height for group separations. The volume of the packed bed is determined by the sample volumes that will be applied.

Efficient column packing is essential, particularly for high resolution fractionation. The efficiency of a packed column defines its ability to produce narrow symmetrical peaks during elution. Column efficiency is particularly important in gel filtration in which separation takes place as only a single column volume of buffer passes through the column. The uniformity of the packed bed and the particles influences the uniformity of the flow through the column and hence affects the shape and eventual peak width. Gel filtration media with high uniformity (lower particle size distribution) facilitate the elution of molecules in narrow peaks.

Efficiency can be improved by decreasing the particle size of a medium. However, using a smaller particle size may increase back pressure so that flow rate needs to be decreased, lengthening the run time. Using pre-packed columns is highly recommended to ensure the best performance and reproducible results. An evenly packed column ensures that, as the sample passes down the column, the component peaks are not unnecessarily broadened. Uneven packing causes peak broadening and high resolution results become impossible.

Buffers, media or pre-packed columns are allowed to reach the same temperature before beginning preparation. Rapid changes in temperature, for example removing packed columns from a cold room and applying buffer at room temperature, can cause air bubbles in the packing and affect the separation. Storage solutions and preservatives should be washed away thoroughly before using any gel filtration medium. Equilibrate the column with 1–2 column volumes of buffer before starting a separation.

SAMPLE APPLICATION

The choice of sample application method depends largely on the volume to be applied and on the equipment available. Ensure that the sample is not diluted on the way to the column and that the top of the column bed is not disturbed during sample application. Samples can be applied automatically or manually.

Samples are applied directly to the column via a chromatography system, a peristaltic pump or a syringe. The choice of equipment depends largely on the size of column, the type of gel filtration medium and the sample volume. For example, a chromatography system will be required for a Superdex column whereas a syringe can be used with small prepacked columns such as HiTrap Desalting. Samples are applied by gravity feed to prepacked columns such as PD-10 Desalting.

ELUTION AND FLOW RATES

Samples are eluted isocratically from a gel filtration column, using a single buffer system. After sample application the entire separation takes place as one column volume of buffer (equivalent to the volume of the packed bed) passes through the column.

Flow rates that allow time for molecules to diffuse in and out of the matrix (partitioning between the mobile phase and the stationary phase) in order to achieve a separation are maintained.

The goal for any separation is to achieve the highest possible resolution in the shortest possible time. Resolution decreases as flow rate increases, so each separation must be optimized to provide the best balance between these two parameters. Put simply, maximum resolution is obtained with a long column and a low flow rate whereas the fastest run is obtained with a short column and a high flow rate. The advantage of a higher flow rate (and consequently a faster separation) may outweigh the loss of resolution in the separation.

If peaks are well separated at a low flow rate, the flow rate is increased or the column is shorten to save time. Alternatively, the sample volume can be increased for the benefit from a higher capacity without significant loss of resolution.

For group separations such as desalting, monitor the elution of protein at A280 and follow the elution of the salt peak using a conductivity monitor. Adjust flow rate and sample volume to balance speed of separation against an acceptable level of salt in the final sample.

Flow rate is measured in simple volume terms, e.g. ml/min, but when comparing results between columns of different sizes it is useful to use the linear flow, cm/hour. Results obtained at the same linear flow on different size columns will be comparable as far as the effects of flow rate are concerned.

Selecting a smaller particle size of the same medium (if available) can also help to achieve the correct balance between flow rate and resolution. Smaller particles of the same medium can increase column efficiency, improve resolution and may allow the use of higher flow rates. However, smaller particles can also result in increased back pressure and this factor may become restrictive if the intention is to scale up the separation.

A wash step is included at the end of a run to facilitate the removal of any molecules that may have been retained on the column, to prevent cross-contamination and to prepare the column for a new separation.

CONTROLLING FLOW

Accurate, reproducible control of the flow rate is not only essential for good resolution, but also for reliability in routine preparative work and repeated experiments. A pump is used to control liquid flow for most gel filtration separations, although gravity feed has been used in the past.

A pump is used within a chromatography system (rather than a peristaltic pump or gravity feed) to fully utilize the high rigidity and excellent flow properties of Sephacryl, Superose or Superdex for high resolution fractionation. The buffer is always pumped onto a column (rather than drawing the buffer through the column with the pump below). This reduces the risk of bubble formation as a result of suction.

Gel filtration columns must not run dry. Presence of sufficient buffer for long, unattended runs should be ensured or pump should be programmed to stop the flow after a suitable time. Columns that run dry must be repacked since the packed bed has been destroyed.

Reversing flow through a gel filtration column should only be considered under cases of severe contamination. There is a risk that reversing the flow may cause channeling through the packed bed leading to poor resolution, loss of efficiency and the need to repack the column. Professionally packed columns are less likely to be affected, but extreme care must be taken.

METHOD DEVELOPMENT FOR HIGH RESOLUTION FRACTIONATION

Steps in order of priority are as follows:

- Medium that will give the best resolution of the target protein(s) should be selected.
- To ensure reproducibility and high resolution, a prepacked column is taken that is best suited to the volume of sample that needs to be processed.
- Highest flow rate is selected that maintains resolution and minimizes separation time.
- Maximum sample volume is determined that can be loaded without reducing resolution.

Higher flow rates and viscous buffers yield higher operating pressures. Hence maximum operating pressure of the packed column is checked and the upper pressure limit on the chromatography system is set accordingly.

If greater resolution is required, the bed height is increased by connecting two columns containing the same medium in series. Alternatively, a medium with the same or similar fractionation range, but with a smaller particle size is tried.

CARE OF GEL FILTRATION MEDIA

When a gel filtration medium has been in use for some time, it may be necessary to remove precipitated proteins or other contaminants that can build up. The need for cleaning may show itself as the appearance of a colored band at top of the column, as a space between the upper adaptor and the bed surface, as a loss in resolution or as a significant increase in back pressure. In all cases, prevention is better than cure and routine cleaning is recommended.

If an increase in back pressure is observed, either on the pressure monitor or by seeing the surface of the medium move downwards, check that the problem is actually caused by the column before starting the cleaning procedure. One piece of equipment is disconnected at a time (starting at the fraction collector), the pump is started and pressure is checked the after each piece is disconnected. A dirty on-line filter is a common cause of increased back pressure. Back pressure is checked at the same stage during each run, since the value can vary within a run during sample injection or when changing to a different buffer.

- Filtered buffers and samples are always used to reduce the need for additional column maintenance.
- Well degassed buffers are always used to avoid the formation of air bubbles in the packed column during a run.
- Buffers, prepacked columns and samples should be kept at the same temperature to prevent air bubbles forming in the column.
- Cleaning solutions should be filtered before use and the column should always be re-equilibrated with 2–3 column volumes of buffer before the next separation.

SCALING UP

After establishing a high resolution or group separation on a small column, it may be preferred to pack a larger column in order to process larger sample volumes in a single step. For scaling up bed height, linear flow rate and sample composition should be maintained and column diameter, volumetric flow rate and sample volume should be increased.

Following points are followed for scaling up a gel filtration column:

- Separation at small scale is optimized.
- Sample volume, column volume ratio and the sample concentration are maintained.
- Column volume is increased the by increasing the cross sectional area of the column.
- Bed height is maintained.
- Separation is run the at the same linear flow rate as used on the smaller column

Different equipment factors may affect performance after scale-up. If the large scale column has a less efficient flow distribution system, or the large scale system introduces large dead volumes, peak broadening may occur. This will cause extra dilution of the product fraction or even loss of resolution if the application is sensitive to variations in efficiency.

For certain media, e.g. Superdex, Superose or Sephadex, it is usually recommended to select a larger particle size. For high resolution fractionation, pack a small column containing the larger particles and repeat the separation to facilitate any optimization that may be needed to achieve the same resolution on the larger column.

Scaling up on Sephadex G-25, even to production scale, is a straightforward and well established process. Well known examples of commercial applications include buffer exchange in processes for removing endotoxins from albumin, and preparative steps during the production of vaccines.

APPLICATIONS

Gel Chromatography has been used with great success in the separation of the sugars, polypeptides, proteins, liquids, asphalts, butyl rubbers, polyethylenes, polystyrenes, silicon polymers and others. The extensive analytical applications of gel permeation chromatography (GPC) cover both organic and inorganic materials. GPC has been mainly applied to studies of complex, biochemical or highly polymerised molecules. GPC has also been used for the estimation of molecular weights, fractionation and purification of large molecules.

MOLECULAR WEIGHT DETERMINATION AND MOLECULAR WEIGHT DISTRIBUTION ANALYSIS

Unlike electrophoretic techniques, gel filtration provides a means of determining the molecular weight or size (Stokes radius) of native or denatured proteins under a wide variety of conditions of pH, ionic strength and temperature, free from the constraints imposed by the charge state of the molecules.

For molecular weight determination, several theoretical models have been proposed to describe the behavior of solutes during gel filtration. Most models assume that the partition of solute molecules between the particles and surrounding liquid is an entirely steric effect.

However, in practice a homologous series of compounds demonstrate a sigmodial relationship between their various elution volume parameters and the logarithm of their molecular weights. Thus molecular weight determination by gel filtration can be made by comparing an elution volume parameter, such as K_{av} of the substance of interest, with the values obtained for several known calibration standards.

A calibration curve is prepared by measuring the elution volumes of several standards, calculating their corresponding K_{av} values (or similar parameter), and plotting their K_{av} values versus the logarithm of their molecular weight. The molecular weight of an unknown substance can be determined from the calibration curve once its K_{av} value is calculated from its measured elution volume. Various elution parameters, such as V_e, V_e/V_o, K_d and K_{av} have been used in the literature for the preparation of calibration curves but the use of K_{av} is recommended since: 1) it is less sensitive to errors which may be introduced as a result of variations in column preparation and column dimensions, 2) it does not require the unreliable determination of the internal volume (V_i) as is required with K_d.

For accurate determination of molecular weight, the calibration standards must have the same relationship between molecular weight and molecular size as the substance of interest. Many of the parameters important for a successful molecular weight determination are the same as for any high resolution fractionation:

- A medium with the correct fractionation range for the molecules of interest should be used. The expected molecular weight values should fall in the linear part of the selectivity curve.
- Whenever possible a pre-packed column should be used. Homemade columns must be packed very carefully.
- Freshly prepared calibration standards be used and selected so that the expected molecular weight values are covered by the entire calibration range. Blue Dextran is always filtered before use.
- Samples may be applied in a volume less than 2% of the total column volume.
- For the separation of calibrants and sample, use of the same buffer is recommended. Recommended flow rate for the prepacked column or medium selected be used.
- If the molecular weight is unknown, a medium with a wide fractionation range such as Sephacryl HR should be used. It is also suitable for molecular weight distribution analysis and for polymeric materials such as dextrans and polyethylene glycols.
- Performing a molecular weight determination in the presence of urea, guanidine hydrochloride or SDS transforms polypeptides and proteins to a random coil configuration and so reduces structural differences. Differences will be seen in the resulting molecular weight values when compared to values acquired under non-denaturing conditions.
- Deviation from a K_{av}: log M_r calibration curve may occur if the molecule of interest does not have the same molecular shape as the standards.
- A calibrated column can be used for extended periods as long as the column is kept in good condition and not allowed to dry out, eliminating the need to set up a separate experiment for each determination.

GEL FILTRATION IN A PURIFICATION STRATEGY (CIPP)

For a high degree of purity, or when a suitable ligand is unavailable for single step affinity purification, an efficient multi-step process must be developed using the purification strategy of Capture, Intermediate Purification and Polishing (CIPP).

CIPP is used in both the pharmaceutical industry and in the research laboratory to ensure faster method development, a shorter time to pure product and good economy. As shown in Fig. 12, an important first step for any purification is correct sample preparation. Gel filtration is often used for desalting and buffer exchange during sample preparation (using Sephadex G-25), when samples volumes up to 30% of the total column volume can be applied.

In high resolution mode, gel filtration is ideal for the final polishing steps in a purification when sample volumes have been reduced (sample volume significantly influences speed and resolution in gel filtration). Samples are eluted isocratically (single buffer, no gradient) and buffer conditions can be varied to suit the sample type or the requirements for subsequent purification, analysis or storage, since buffer composition does not directly affect resolution.

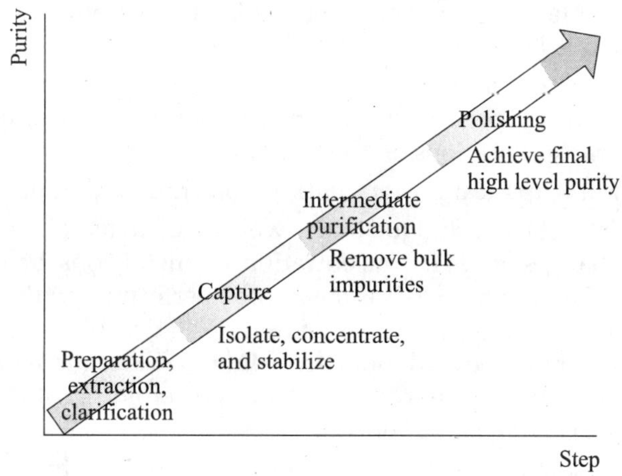

Fig. 12. Preparation and CIPP.

APPLYING CIPP

Capture, Intermediate Purification and Polishing are the three phases of purification: The purification problem associated with a particular step will depend greatly upon the properties of the starting material. Thus, the objective of a purification step will vary according to its position in the process.

(i) In the capture phase, the objectives are to isolate, concentrate and stabilize the target product. The product should be concentrated and transferred to an environment that will conserve potency/activity.

(ii) During the intermediate purification phase the objectives are to remove most of the bulk impurities, such as other proteins and nucleic acids, endotoxins and viruses.

(iii) In the polishing phase most impurities have already been removed except for trace amounts or closely related substances. The objective is to achieve final purity by removing any remaining trace impurities or closely related substances.

The optimal selection and combination of purification techniques for Capture, Intermediate Purification and Polishing is crucial for an efficient purification.

SELECTION AND COMBINATION OF PURIFICATION TECHNIQUES

Proteins are purified using purification techniques that separate according to differences in specific properties i.e., Size (Gel filtration GF), Charge (Ion exchange IEX), Hydrophobicity/ Hydrophobic interaction (HIC), Reversed phase (RPC), Biorecognition (ligand specificity) (Affinity AC). Every technique offers a balance between resolution, capacity, speed and recovery.

Capacity refers to the amount of target protein loaded during purification. In some cases the amount of sample that can be loaded will be limited by volume (as in gel filtration) or by large amounts of contaminants rather than the amount of the target protein.

Speed is most important at the beginning of purification where contaminants, such as proteases, must be removed as quickly as possible.

Recovery becomes increasingly important as the purification proceeds because of the increased value of the purified product. Recovery is influenced by destructive processes in the sample and by unfavorable conditions on the column.

A few more important applications of Gel filtration chromatography are as given below:

1. Gel filtration is a robust technique that is well suited to handling biomolecules that are sensitive to changes in pH, concentration of metal ions or co-factors and harsh environmental conditions. Separations can be performed in the presence of essential ions or cofactors, detergents, urea, guanidine hydrochloride, at high or low ionic strength, at 37 °C or in the cold room according to the requirements of the experiment. Gel filtration can also be used to facilitate the refolding of denatured proteins by careful control of changing buffer conditions.
2. Gel filtration chromatography is often the first step in sorting the proteins in a large mixture. Such is the case when cells are genetically-engineered to make a particular protein. The cloned cells are exploded and hundreds of proteins are released, including the desired protein. The mixture of proteins is initially separated into fractions using a sizing column. The fractions may be further treated and identified by other chromatographic techniques and by polyacrylamide gel electrophoresis.
3. Tightly crosslinked gels with small pore size, such as sephadex G-25 and G-50 have been used in the removal of low molecular weight molecules from high molecular weight natural product molecules.
4. Gel filtration chromatography is also applied for monitoring cellulase production from Trichoderma viride.
5. Separation of ribonuclease, trypsin, pepsin and cytochromes has been successfully carried out from higher molecular weight proteins, such as serum, albumin, haemoglobin and fibrinogen by using Sephadex-75 gel.

ION EXCHANGE CHROMATOGRAPHY

INTRODUCTION

Ion-exchange chromatography resembles liquid-solid (adsorption) chromatography with the additional characteristic dealing with the adsorption-desorption phenomena associated with ionic, or at least potentially ionic, substances. It is another type of column chromatography that separates molecules based on their overall charge at a given pH. Many proteins have a net charge on them, either positive or negative due to the kinds of amino acids present on the polypeptide. Amino acids can have no charge also. If there are more positively-charged amino acids than negative ones then the protein will carry a net positive charge.

While dealing with charged species, i.e., ionic compounds, it is not surprising that the technique is most often performed in aqueous media, as, apart from acids, bases and certain specialty solvents like liquid ammonia, such solvent systems have the highest dielectric constant and thus the greatest tendency for compounds to dissociate into ions. Ion-exchange chromatography is ideally suited to the separation of those highly polar substances which, without recourse to derivative formation, cannot be handled by GC. Into this class come amino acids, peptides, sugars, nucleic acids and salts, all of which are of the greatest importance to those working in the life sciences. Although chemically very dissimilar, ion exchange finds considerable use in the analysis of inorganic species, being perhaps in greatest demand for the separation of lanthanide, actinide and noble metals. Using what can be regarded as "classical", i.e., low-pressure ion-exchange chromatography, many of these separations have been possible for several decades albeit in most instances many hours were required to complete a separation. The use of recently developed instrumentation and particles of ion exchanger, which are much smaller in diameter than the earlier materials, can result in a very considerable improvement in the chromatographic performance and speed of analysis.

HISTORY OF ION EXCHANGE CHROMATOGRAPHY

Ion Chromatography (IC) was introduced in 1975 by Small, Stevens, and Bauman as a new analytical method. Within a short period of time, ion chromatography evolved from a new detection scheme for a few selected inorganic anions and cations to a versatile analytical technique for ionic species in general. For a sensitive detection of ions via their electrical conductance, the separator column effluent was passed through a "suppressor" column. This suppressor column chemically reduces the eluent background conductance, while at the same time increasing the electrical conductance of the analyte ions. In 1979, Fritz *et al.* described an alternative separation and detection scheme for inorganic anions, in which the separator column is directly coupled to the conductivity cell. As a prerequisite for this chromatographic setup, low capacity ion-exchange resins must be employed, so that low ionic strength eluents can be used. In addition, the eluant ions should exhibit low equivalent conductance, thus enabling sensitive detection of the sample components.

At the end of the 1970s, ion chromatographic techniques were used to analyze organic ions for the first time. The requirement for a quantitative analysis of organic acids brought about an ion chromatographic method based on the ion-exclusion process that was first described by Wheaton and Bauman in 1953. The 1980s witnessed the development of high efficiency separator columns with particle diameters between 5 μm and 8 μm, which resulted in a significant reduction of analysis time. In addition, separation methods based on the ion-pair process were introduced as an alternative to ion-exchange chromatography, because they allow the separation and determination of both anions and cations. Since the beginning of the 1990s column development has aimed to provide stationary phases with special selectivities.

The scope of ion chromatography was considerably enlarged by newly designed electrochemical and spectrophotometric detectors. A milestone of this development was the introduction of a pulsed amperometric detector in 1983, allowing a very sensitive detection of carbohydrates, amino acids, and divalent sulfur compounds. A growing number of applications utilizing post-column derivatization in combination with photometric detection opened the field of polyphosphate, poly-phosphonate, and transition metal analysis for ion chromatography, thus providing a powerful extension to conventional titrimetric and atomic spectrometry methods.

TYPES OF ION CHROMATOGRAPHY

Recent developments have made ion chromatography an integral part of both modern inorganic and organic analysis. Even though ion chromatography is still the preferred analytical method for inorganic and organic ions, meanwhile, ion analyses are also carried out with capillary electrophoresis (CE) which offers certain advantages when analyzing samples with extremely complex matrices. In terms of detection, only spectrometric methods such as UV/VIS and fluorescence detection are commercially available. Because inorganic anions and cations as well as aliphatic carboxylic acids cannot be detected very sensitively or cannot be detected at all, applications of CE are rather limited as compared to IC, with the universal conductivity detection being employed in most cases. Since the sensitivity of conductivity detection does not suffer from miniaturization, detection limits achieved for totally dissociated anions and low molecular weight organics compete well with those of ion chromatography techniques. Thus, capillary electrophoresis with suppressed conductivity detection can be regarded as a complementary technique for analyzing small ions in simple and complex matrices.

HIGH PERFORMANCE ION CHROMATOGRAPHY (HPIC)

This separation method is based on ion-exchange processes occurring between the mobile phase and ion-exchange groups bonded to the support material. In highly polarizable ions, additional non-ionic adsorption processes contribute to the separation mechanism. The stationary phase consists of polystyrene, ethylvinylbenzene, or methacrylate resins co-polymerized with divinylbenzene and modified with ion-exchange groups. Ion-exchange chromatography is used forthe separation of both inorganic and organic anions and cations. Separation of anions is accomplished with quaternary ammonium groups attached to the polymer, whereas sulfonate-, carboxyl-, or phosphonate groups are used as ion-exchange sites for the separation of cations.

HIGH PERFORMANCE ION CHROMATOGRAPHY EXCLUSION (HPICE)

The separation mechanism in ion-exclusion chromatography is governed by Donnan exclusion, steric exclusion, sorption processes and, depending on the type of separator column, by hydrogen bonding. A high-capacity, totally sulfonated cation exchange material based on polystyrene/divinylbenzene is employed as the stationary phase. In case hydrogen bonding should determine selectivity, significant amounts of methacrylate are added to the styrene polymer. Ion-exclusion chromatography is particularly useful for the separation of weak inorganic and organic acids from completely dissociated acids which elute as one peak within the void volume of the column. In combination with suitable detection systems, this separation method is also useful for determining amino acids, aldehydes, and alcohols.

ION-PAIR CHROMATOGRAPHY/MOBILE PHASE ION CHROMATOGRAPHY (MPIC)

The dominating separation mechanism in ion-pair chromatography is adsorption. The stationary phase consists of a neutral porous divinylbenzene resin of low polarity and high specific surface area. Alternatively, chemically bonded octa-decyl silica phases with even lower polarity can be used. The selectivity of the separator column is determined by the mobile phase. Besides an organic modifier, an ion-pair reagent is added to the eluant (water, aqueous buffer solution, etc.) depending on the chemical nature of the analytes. Ion-pair chromatography is particularly suited for the separation of surface-active anions and cations, sulfur compounds, amines, and transition metal complexes.

ALTERNATIVE METHODS

In addition to the three classical separation methods mentioned above, reversed-phase liquid chromatography (RPLC) can also be used for the separation of highly polar and ionic species. Long-chain fatty acids, for example, are separated on a chemically bonded octadecyl phase after protonation in the mobile phase with a suitable aqueous buffer solution. This separation mode is known as ion suppression. Chemically bonded aminopropyl phases have also been successfully employed for the separation of inorganic ions. The separation of nitrate and bromide in foods on such a phase using a phosphate buffer solution as the eluant has also been described. Separations of this kind are limited in terms of their applicability, because they can only be applied to UV-absorbing species. Moreover, applications of multidimensional ion chromatography utilizing multimode phases are very interesting, too. In those separations, ion-exchange and reversed-phase interactions equally contribute to the retention mechanism of ionic and polar species.

ION-EXCHANGE COLUMNS

Ion-exchange columns are packed with resin beads carrying a particular charge. The purpose of the ion-exchange resin beads is to separate positively-charged proteins from negatively-charged ones. An ion exchange column looks like a gel filtration column but it runs differently. The primary difference is in the type of resin and the kinds of buffers used. In ion-exchange chromatography, the resin bed is equilibrated with a buffer of a certain pH. This charges the resin beads. The sample is loaded and pushed through the column with more of the equilibration buffer. Molecules of charge opposite that of the beads attach to the beads. All other molecules wash through the column. These are collected in "wash" fractions. The bound molecules will stay attached to the beads until they are knocked off. This is called eluting the sample. The elution buffer either contains a high salt concentration or has a high or low pH. As the elution comes off, the sample is collected. All fractions are then run on a PAGE gel to visualize what was separated and collected.

In the very simplest picture, one may consider an ion-exchange material having a positive charge (an anion exchanger) which will attract or retain negatively charged species (anions) present in the mobile phase. Similarly, a cation exchanger, having negatively charged sites on its surface, will interact with ions carrying a positive charge. Perhaps of even greater importance is the development of the new generation of column packings for ion-exchange work that has been specifically designed for modern, high-speed chromatographic methodology. These materials differ considerably from the classical resins in both their physical properties and handling characteristics to such an extent that it is sometimes better to consider them separately. The use of ion-exchange methods to separate samples of biological origin is interesting in that it is often possible to examine the sample in the same form as it occurs *in vivo,* minimizing the risk of rearrangements or isomerism occurring which could complicate and possibly lead to misinterpretation of the end result.

More recently developed column packings utilize an inorganic support to impart rigidity to the ion exchanger. In many instances interactions of samples with this support or the coating on its surface contribute to the overall chromatographic selectivity obtained. The behaviour of weakly ionic organic substances on ion-exchange systems is often quite difficult to interpret in view of the various factors which can influence retention. Frequently an empirical approach yields more satisfactory results than one based on the classical concept of ion-exchange behaviour. The ion-exchange functionality of column packings is most often obtained by in-corporating ionic groups, such as sulphonate for cation exchange and quaternary ammonium for anion exchange.

MECHANISM OF ION-EXCHANGE SEPARATIONS

ION DETECTION

Ions in solution can be detected by measuring the conductivity of the solution. In ion chromatography, the mobile phase contains ions that create background conductivity, making it difficult to measure the conductivity due only to the analyte ions as they exit the column. This problem can be greatly reduced by selectively removing the mobile phase ions after the analytical column and before the detector. This is done by converting the mobile phase ions to a neutral form or removing them with an eluent suppressor, which consists of an ion-exchange column or membrane. For cation analysis, the mobile phase is often HCl or HNO_3, which can be neutralized by an eluent suppressor that supplies OH^-. The Cl^- or NO_3^- is either retained or removed by the suppressor column or membrane. The same principle holds for anion analysis. The mobile phase is often NaOH or $NaHCO_3$, and the eluent suppressor supplies H^+ to neutralize the anion and retain or remove the Na^+. Molecular ions, such as proteins, that have absorption bands in the ultraviolet or visible spectral region can be detected by absorption spectroscopy.

CLASSICAL IDEAS ON ION-EXCHANGE EQUILIBRIA

In a chromatographic column filled with a cation exchanger, the column packing, if it were in isolation, would possess a net negative charge. In practice, electro neutrality is maintained by the resin being in the "salt form", i.e., each negative site on the resin has a cation held as an ion pair. These cations are commonly hydrogen or sodium ions, thus a resin is said to be in the "hydrogen form" or "sodium form", respectively. Conversion of a resin from one form to another is accomplished by passing an ionic solution through the column containing an excess of the desired cation. Thus to convert a resin from the hydrogen form into the sodium form one passes a solution of a sodium salt through the column. The following exchange reaction takes place at the surface of the resin

$$Na^+ + [(\text{resin})^- H^+] \underset{H^+}{\overset{Na^+}{\rightleftharpoons}} H^+ + [(\text{resin})^- Na^+]$$

It is important to note that this reaction is in reality a dynamic equilibrium following the laws of mass action. If an excess of hydrogen ions were present, i. e., the solution was acidic, then the reverse reaction would take place.

The extent to which the exchange reaction occurs depends on the concentrations (strictly the activities) of the species present. The equilibrium constant defines this condition quantitatively.

$$K = \frac{[(\text{resin})^- Na^+][H^+]}{[(\text{resin})^- H^+][Na^+]}$$

A greater value of K indicates that more of the resin will be converted into the sodium form. In practice, ion-exchange chromatography is performed with the mobile phase "buffered" at a definite pH so the situation may be considering as ions in solution (counter ions) in dynamic equilibrium with similar ions forming ion pairs with the charged sites on the resin. This equilibrium condition is dependent on operational variables such as pH, temperature, and the ionic strength of the mobile phase.

If in place of the sodium ions a sample is applied which also forms cations on being introduced into the column, then a similar equilibrium distribution of this compound will be established, in competition with the distribution of the counter ions. In an analogous manner to the conversion of a resin from one ionic form to another, as described above, if no other selectivity factors were involved, the ionic species in greatest concentration would interact most strongly with the resin and thus be preferentially retained. Thus, if it were the counter ion which was in the highest concentration, one would anticipate that the sample would be eluted rapidly. Similarly, if one decreases the concentration of the counter ions in the mobile phase, increased retention of the sample would be obtained. This effect is observed in practice; thus a complex mixture of components which is retained may be eluted progressively by the application of gradient elution, where the concentration of the counter ions is increased during the course of the separation.

Although the ion-exchange resin is commonly a strongly acidic or basic material, *i.e.,* is fully ionized at all pH values, the same comment does not apply to samples, as these may be only weakly or moderately dissociated. The degree of interaction of sample with an ion-exchange resin is a function of the concentration of the ions it produces in solution, strictly the ratio of ionized to neutral molecules and not the total concentration of "ionized + non-ionized" sample. It follows that a change in the pH of the mobile phase which will cause a partially dissociated molecule to ionize further will lead to increased retention, whereas a change in pH which suppresses the dissociation will cause the sample to elute more rapidly from the column. The same feature also applies to weak ion-exchange resins, where the surface groups only ionize under favourable pH conditions, a typical example being weak cation exchangers having carboxylic acid functionality.

SECONDARY INTERACTIONS CONTRIBUTING TO THE SEPARATION

Points on the mechanism of separation outlined so far have assumed that separations in ion-exchange chromatography are due solely to ionic interactions between the sample and the column packings. However, in many applications, what might appear to be separations based on ion exchange, are, in fact, complicated by secondary interactions which are essentially non-ionic in nature. These interactions arise from adsorption or hydrogen bonding of the ample to the non-ionic part of the matrix of the column packing, or simply the limited solubility of the sample in the mobile phase. These effects are particularly noticeable when studying organic substances as distinct from the more completely ionized inorganic anions and cations. The mechanism of ion exchange is further complicated by the selectivity introduced by the charge and radius of the ions competing for the charged sites on the resin, where generally the larger the radius of the ion, the stronger will be its affinity towards the resin. This effect is described in greater detail in the section of this chapter dealing with the selection of the mobile phase.

The overall mechanism by which separations are accomplished in "ion-exchange" chromatography is thus a very complex subject owing to the many complementary processes which can affect the retention of samples. It is difficult to predict what operating conditions will provide the best separation, since the relative magnitude of the processes contributing to an overall separation varies with sample type. In the case of organic substances, it is sometimes possible to separate anions on a cation-exchange column and *vice versa,* completely contradicting the simple ideas of ion exchange. Not surprisingly, many chromato-graphers tend to adopt an empirical rather than scientific approach when examining samples by ion-exchange methods.

ION CHROMATOGRAPHIC SYSTEM

The basic components of an ion chromatograph are shown schematically in Fig. 13. It resembles the setup of conventional HPLC systems. A pump delivers the mobile phase through the chromatographic system. In general, either single-piston or dual-piston pumps are employed. A pulse-free flow of the eluant is necessary for employing sensitive UV/VIS and amperometric detectors. Therefore, pulse dampers are used with single-piston pumps and a sophisticated electronic circuitry with dual-piston pumps. The sample is injected into the system via a loop injector.

A three-way valve is required, with two ports being connected to the sample loop. The sample loading is carried out at atmospheric pressure. After switching the injection valve, the sample is transported to the separator column by the mobile phase. Typical injection volumes are between 5 µl and 100 µl. The most important part of the chromatographic system is the separator column. The choice of a suitable stationary phase and the chromatographic conditions determine the quality of the analysis. The column tubes are manufactured from inert material such as Tefzec, epoxy resins, or PEEK (polyether ether ketone). In general,

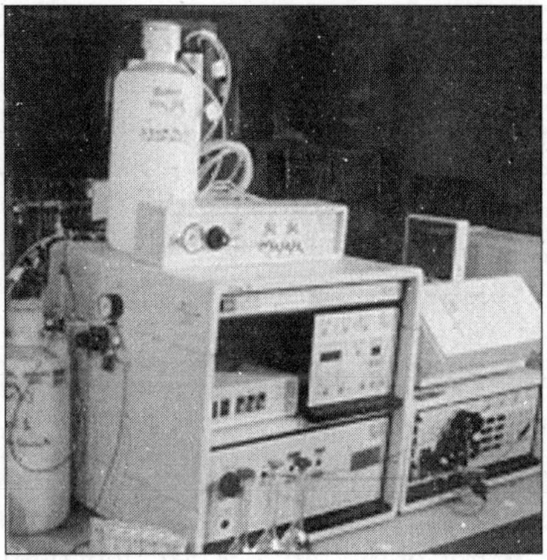

Fig. 13. An ion-exchange Chromatographic system

separation is achieved at room temperature. Only in very few cases for example for the analysis of long-chain fatty acids an elevated temperature is required to improve analyte solubility. An elevated column temperature is also recommended for the analysis of polyamines in order to improve peak efficiencies.

The analytes are detected and quantified by a detection system. The performance of any detector is evaluated according to the following criteria:

- Sensitivity
- Linearity
- Resolution (detector cell volume)
- Noise (detection limit).

The most commonly employed detector in ion chromatography is the conductivity detector, which is used with or without a suppressor system. The main function of the suppressor system as part of the detection unit is to chemically reduce the high background conductivity of the electrolytes in the eluant, and to convert the sample ions into a more conductive form. In addition to conductivity detectors, UV/VIS amperometric and fluorescence detectors are used.

The chromatographic signals can be displayed on a recorder. Quantitative results are obtained by evaluating peak areas or peak heights, both of which are proportional to the analyte concentration over a wide range. This was traditionally performed using digital integrators which are connected directly to the analog signal output of the detector. Due to low computer prices and lack of GLP/GLAP conformity, digital integrators are hardly used anymore. Modern detectors feature an additional parallel interface (e.g., RS-232-C), that enables the connection to a personal computer or a host computer with suitable chromatography software. Computers also take over control functions, thus allowing a fully automated operation of the chromatographic system.

Because corrosive eluants such as diluted acids and bases are often used in ion chromatography, all parts of the chromatographic system being exposed to these liquids should be made of inert, metal-free materials. Conventional HPLC systems with tubings and pump heads made of stainless steel are only partially suited for ion chromatography, because even stainless steel is eventually corroded by aggressive eluants. Considerable contamination problems would result, because metal ions exhibit a high affinity towards the stationary phase of ion exchangers, leading to a significant loss of separation efficiency. Moreover, metal parts in the chromatographic fluid path would make the analysis of orthophosphate, complexing agents and transition metals more difficult.

COLUMN PACKING STRUCTURE FOR ION-EXCHANGE CHROMATOGRAPHY

The most important naturally occurring materials to show ion-exchange properties are the zeolite class of alumino-silicates. These materials possess a characteristic open frame-work structure having the general composition of $M_{x/n} (AlO_{2})_{x} . (SiO_2)_y . z\,H_2O$, where the charge on the cation, M, is *n* and z is the degree of hydration. The extent of the hydration and the relative proportions of Al to Si vary from one example to another. Other types of inorganic ion exchangers have been developed commercially. Examples are those relying on a micro crystallized gel structure of zirconium oxide (anion or cation exchanger), zirconium phosphate (cation exchanger) and ammonium phosphomolybdate (cation exchanger especially selective for alkali metal ions). Inorganic ion exchangers do not enjoy particularly wide popularity as there are organic gel based materials available which offer superior performance in respect of column efficiency and the number of different types of selectivity.

Most conventional ion exchangers utilize the styrene-divinylbenzene type of copolymer as the supporting matrix, where the divinylbenzene is incorporated to cross-link the resin to give it rigidity. For use in modern LC the physical strength of the matrix of a packing material must be sufficient to withstand high pressures without compression and for this reason alone hardly any of the resins widely used in low-pressure systems can be used. Most conventional ion exchangers also swell or shrink if the pH or ionic strength of the mobile phase is varied, thus creating serious dimensional changes. These features, which are intolerable in a high-resolution system, can be reduced to some extent by heavily cross-linking the copolymer (usually with more than 8% divinylbenzene) in an attempt to make the structure more rigid. This action, however, often decreases the permeability of the column and leads to exclusion of large molecules from the resin. As an alternative, one can produce a packing material where the principal structure of the particle is inorganic, e.g., a silica or glass bead, and the ion-exchange functionality is incorporated as a thin surface coating of resin or by the ionic groups being chemically bonded directly to the surface of the support in a manner analogous to the stationary phases for partition chromatography.

The capacity of ion-exchange materials is expressed in terms of the number of equivalents of exchangeable ions available per gram of packing material. Values for the various types of ion exchangers differ considerably. For instance, a conventional resin for open-column chromatography typically has a capacity in the range of 3-7 mequiv./gm of dry resin. High-resolution copolymer resin-type beads with particle diameters in the region of 10 μm have similar exchange capacities of approximately 3-5 mequiv./gm, depending on the particular material. A particular merit of these resins is that large samples may be introduced when necessary. A general guide to sample size in ion-exchange work is to limit the maximum size of the sample to less than 5% of the total exchange capacity of the column packing. Clearly, the design of the ion-exchange chromatographic packing material has an important influence on its resultant performance characteristics. Some Column Packings for Modern Ion-Exchange Chromatography are given in Table 2.

Table 2. Some Column Packings for Modern Ion-Exchange Chromatography

Type	Name	Approx. exchange capacity	Particle size (μm)	Supplier
Pellicular strong anion	Ion-X-SA	n.d.**	30-40	Perkin-Elmer
	"Pellicular Anion"	10	-40	Varian
	AE-Pellionex SAX	10	44-53	Reeve Angel
	AS-Pellionex SAX	10	. 44-53	Reeve Angel
	Perisorb AN	30	30-40	Merck***
	Permaphase AAX*	100	25-37	DuPont
	Permaphase ABX*	60	25-37	DuPont
	Vydac Anion Exchange*	100	30-44	Separations Grou,
	Zipax SAX	12	25-37	DuPont
Pellicular weak anion	AL-Pellionex WAX	n.d.	44-53	Reeve Angel
	Zipax WAX	n.d.	25-37	DuPont
Pellicular strong cation	Ion-X-SC	n.d.	30-40	Perkin-Elmer
	"Pellicular Cation"	10	-40	Varian
	HC-Pellionex SCX	60	44-53	Reeve Angel
	HS-Pellionex SCX	8-10	44-53	Reeve Angel
	Perisorb KAT	50	30-40	Merck***
	Vydac Cation Exchanger*	100	30-44	Separations Group
	Zipax SCX	3.2	25-37	DuPont
Porous anion (polymer support)	Aminex A-14	3400	17-23	Bio-Rad
	Aminex A-25	3200	15.5-19.5	Bio-Rad
	Aminex A-27	3200	12-15	Bio-Rad
	Aminex A-28	3200	7-11	Bio-Rad
	DA-X8A	4000	6-10	Durrum
	DA-X4	2000	15-25	Durrum
	DA-X2	2000	15-25	Durrum
Porous cation (polymer support)	Aminex A-4	5000	16-24	Bio-Rad
	Aminex A-5	5000	11-15	Bio-Rad
	Aminex A-6	5000	15.5-19.5	Bio-Rad
	Aminex A-7	5000	7-11	Bio-Rad
	AA-15	5000	16-28	Beckman
	PA-28	5000	16	Beckman
	PA-35	5000	13	Beckman

Table 2 Contd.

Table 2 Contd.

	DC-1A	5000	15-21	Durrum
	DC-2A	5000	9-15	Durrum
	DC-4A	5000	6-10	Durrum
	AN-90	5200	16-28	Hamilton
	B-80	5200	10-20	Hamilton
	H-70	5200	18-30	Hamilton
Porous anion & cation (inorganic support)	Partisil-10-SAX*	n.d	10	Reeve Angel
	Vydac TP Anion Exchange	n.d.	10	Separations Group
	Partisil-10-SCX	n.d.	10	Reeve Angel
	Vydac TP Cation Exchange	n.d.	10	Separations Group

*Chemically bonded phases; ** n.d. indicates no data available; ***E.M. Labs. in the U.S.A.

FACTORS INFLUENCING SELECTION OF MOBILE PHASE

In practice, the buffer solution used in an ion-exchange separation should be selected on the basis of three factors, *viz.*

(a) **pH:** With samples or weak ion exchangers that are only partially ionized, the pH of the mobile phase will regulate the degree of ionization, hence the concentration of ions in solution. An increase in the number of ionic sites on a resin and/or the ratio of sample ions to neutral molecule in solution will lead to stronger retention of sample.

(b) **Concentration (ionic strength):** An increase in the concentration of counter ions in solution relative to ions from the sample will result in the counterions, i.e., ions from the buffer solution, being preferentially held on the resin. A decrease in ionic strength (concentration of the buffer) leads to stronger retention of the sample. Many of the separations achieved with superficially porous ion exchangers have required mobile phases containing only very dilute buffers, i.e., in the range of 1-100 *mM.* In contrast, the high-capacity porous polymer resins most often require a buffer concentration of approximately 0.1 – 10 M.

(c) **Selectivity of the counter ion:** This effect results from the ability of ion-exchange resins to discriminate between ions of similar charge but differing in their geometry. Factors influencing the selectivity include the magnitude of the ionic charge, the radius, the degree of salvation of the ion and interactions with the support matrix.

ADVANTAGES OF ION CHROMATOGRAPHY

The determination of ionic species in solution is a classical analytical problem with a variety of solutions. Whereas in the field of cation analysis both fast and sensitive analytical methods (AAS, ICP, polarography, and others) have been available for a long time, the lack of corresponding, highly sensitive methods for anion analysis is noteworthy. Conventional wet-chemical methods such as titration, photometry, gravimetry, turbidimetry, and colorimetry are all labor-intensive, time-consuming and occasionally troublesome. In contrast, ion chromatography offers the following advantages:

Speed

The time necessary to perform an analysis becomes an increasingly important aspect, because enhanced manufacturing costs for high quality products and additional environmental efforts have lead to a significant increase in the number of samples to be analyzed. With the introduction of high efficiency separator columns for ion-exchange, ion-exclusion, and ion-pair chromatography in recent years, the average analysis time could be reduced to about 10 minutes. Today, a baseline-resolved separation of the seven most important inorganic anions requires only three minutes. Therefore, quantitative results are obtained in a fraction of the time previously required for traditional wet-chemical methods, thus increasing the sample throughput.

Sensitivity

The introduction of microprocessor technology in combination with modern high efficiency stationary phases, makes it a routine task to detect ions in the medium and lower μg/L concentration range without pre-concentration. The detection limit for simple inorganic anions and cations is about 10 μg/L based on an injection volume of 50 μL. The total amount of injected sample lies in the lower ng range. Even ultra pure water, required for the operation of power plants or for the production of semiconductors may be analyzed for its anion and cation content after pre-concentration with respective concentrator columns.

With these pre-concentration techniques, the detection limit could be lowered to the ng/L range. However, it should be emphasized that the instrumentation for measuring such incredibly low amounts is rather sophisticated. In addition, high demands have to be met in the creation of suitable environmental conditions. The limiting factor for further lowering the detection limits is the contamination by ubiquitous chloride and sodium ions. High sensitivities down to the pmol range are also achieved in carbohydrate and amino acid analysis by using integrated pulsed amperometric detection.

Selectivity

The selectivity of ion chromatographic methods for analyzing inorganic and organic anions and cations is ensured by the selection of suitable separation and detection systems. Regarding conductivity detection, the suppression technique is of vital importance, because the respective counter ions of the analyte ions as a potential source of interferences are exchanged against hydronium and hydroxide ions respectively. A high degree of selectivity is achieved by using solute-specific detectors such as a UV/VIS detector to analyze nitrite in the presence of high amounts of chloride. New developments in the field of post-column derivatization show that specific compound classes such as transition metals, alkaline-earth metals, polyvalent anions, silicate, etc. can be detected with high selectivity. Such examples explain why sample preparation for ion chromatographic analyses usually involves only a simple dilution and filtration of the sample. This high degree of selectivity facilitates the identification of unknown sample components.

Simultaneous Detection

A major advantage of ion chromatography especially in contrast to other instrumental techniques such as photometry and AAS is its ability to simultaneously detect multiple sample components. Anion and cation profiles may be obtained within a short time; such profiles provide information about the sample composition and help to avoid time-consuming tests. However, the ability of ion chromatographic techniques for simultaneous quantitation is limited by extreme concentration differences between various sample components. For example, the major and minor components in a wastewater matrix may only be detected simultaneously if the concentration ratio is <1000:1. Otherwise, the sample must be diluted and analyzed in a separate chromatographic run.

Stability of the Separator Columns

The stability of separator columns very much depends on the type of packing material being used. In contrast to silica-based separator columns commonly used in conventional HPLC, resin materials such as polystyrene/divinylbenzene copolymers prevail as support material in ion chromatography. The high pH stability of these resins allows the use of strong acids and bases as eluants, which is a prerequisite for the wide-spread applicability of this method. Strong acids and bases, on the other hand, can also be used for rinsing procedures. Meanwhile, most organic polymers are compatible with organic solvents such as methanol and acetonitrile, which can be used for the removal of organic contaminants. Hence, polymer-based stationary phases exhibit a low sensitivity towards complex matrices such as wastewater, foods, or body fluids, so that a simple dilution of the sample with de-ionized water prior to filtration is often the only sample preparation procedure.

Selection of Separation and Detection Systems

The analyst usually has some information regarding the nature of the ion to be analyzed (inorganic or organic), its surface activity, its valency, and its acidity or basicity, respectively. With this information and on the basis of the selection criteria outlined schematically in Table 3, it should not be difficult for the analytical chemist to select a suitable stationary phase and detection mode. In many cases, several procedures are feasible for solving a specific separation problem. In these cases, the choice of the analytical procedure is determined by the type of matrix, the simplicity of the procedure and increasingly by financial aspects. Two examples illustrate this:

Various sulfur-containing species in the scrubber solution of a flue-gas desulfurization plant are to be analyzed. According to Table-1, non-polarizable ions such as sulfite, sulfate, and amidosulfonic acid with p*K* values below 7 are separated isocratically by HPIC using a conventional anion exchanger and are detected via electrical conductivity. Often, scrubber solutions also contain thiocyanate and thiosulfate in small concentrations. However, due to their polarizability, these anions exhibit a high affinity towards the stationary phase of conventional anion exchangers. Concentration gradients on anion exchangers reach the limit when extremely polarizable anions such as nitrilotrisulfonic acid have to be analyzed. In this case, ion-pair chromatography (MPIC) is the better separation mode, because organic solvents added to the mobile phase determine analyte retention.

According to Table 3, aliphatic carboxylic acids are separated by HPICE on a totally sulfonated cation exchange resin with subsequent conductivity detection. While this procedure is characterized by a high selectivity for aliphatic monocarboxylic acids with a small number of carbon atoms, sufficient separation cannot be obtained for the aliphatic open-chain and cyclic hydroxy acids. Only after introducing a new stationary phase with specific selectivity for hydroxycarboxylic acids did it become possible to separate the most important representatives of this class of compounds in such a matrix. Ion-exclusion chromatography is not suited for the separation of aromatic carboxylic acids. Due to π-π-interactions with the aromatic rings of the organic polymers used as support material for the stationary phase, aromatic acids are strongly retained and thus, cannot be analyzed. A good separation is achieved by reversed-phase chromatography using chemically bonded octadecyl phases with high chromatographic efficiencies. These compounds are then detected by measuring their light absorption at 254 nm.

RANGE OF APPLICABILITY

From the introductory section it should be apparent that ion exchange is the method of choice for the separation of ionic species. In general most compounds which are soluble only in water are amenable to analysis by ion-exchange methods either directly, because of inherent polarization of the molecule, or by complex formation in the aqueous phase. Into this latter category comes the separation of carbohydrates, which form anionic complexes in borate solutions. A possible exception to the general applicability of ion exchange is perhaps the separation of high-molecular-weight substances which are known to be adsorbed strongly on chromatographic packings and column walls.

APPLICATIONS IN BIOLOGICAL SCIENCES

Some of the most important applications of ion exchange are related to the biological sciences in general and separation of amino acids and nucleotides obtained by the hydrolysis of biological samples in particular.

Amino acids have been separated by ion exchange for many years and probably represent one of the earliest applications of LC where the mobile phase was forced through the column under pressure. The separation and detection of amino acids is a rather complex procedure in that gradient elution must be employed to obtain optimum resolution of components while sensitive, selective detection is obtained by using a post-column chemical reaction detector. Most widely used has been the ninhydrin reaction, whereby amino acids are reacted with ninhydrin reagent to yield an intensely blue colour. The greatest limitation to this approach is that the reaction takes approximately 15 min to go to completion and for quantitative results the column effluent, combined with the reagents, must be held in a reaction coil for this period of time. In more recent work fluorescent derivatives have become more popular as fluorigenic reagents are now available which are capable of reacting with an eluting amino acid in a matter of a few seconds. For amino acid separations, refined versions of the classical polystyrene-based ion exchangers continue to be used successfully.

Many other important species of biological origin are more amenable to LC by virtue of the greater ease of detection, e.g., separations are capable of being monitored using UV detectors.

Table 3. Schematic representation of selection criteria for separation and detection modes.

Anions	Non-polarizable ions Inorganic anions Carboxylic acids Sulfonic acids Fatty acids > C_6	pKa < 7	Inorganic ions F^-, Cl^-, SO_4^{2-} etc.	HPIC	Conductivity detection or Amperometric detection
			Chelates Polyphosphates Polyphosphonates	HPIC	Conductivity detectionUV/VIS detection withpost-column reaction
			Organic ions, > C_6 Formic acid, Acetic acid, Citric acid	HPIC HPIC	Conductivity detection Or UV/VIS detection
			Metal complexes	HPIC MPIC	Conductivity detection Or UV/VIS detection
		pKa > 7	Electroactive ions Organic ions Mercaptanes	HPIC MPIC	Amperometric detection
			Weak inorganic acids HS^-, CN^-, CO_3^{2-}	HPIC	Conductivity detection or UV detection Amperometric detection
			Alcohols, Aldehydes	HPICE	Pulsed amperometry
			Carbohydrates	HPIC	Pulsed amperometry
	Polarizable ions Inorganic anions Carboxylic acids Sulfonic acids Fatty acids< C_{20}	pKb < 7	Inorganic and Palmitic acid Stearic acid Anionic surfactants I^-, SCN –, ClO_4^-	MPIC HPIC	Conductivity detection for or UV/VIS detection
		pKa > 7	Organic ions Phenols etc	MPIC	Amperometric detection or UV/VIS detection

Table 3 Contd.

Table 3 Contd.

Cations	Non-polarizableIons Na^+, NH_4^+, K^+, Mg^{2+}, Ca^{2+} Transition metals Amines < C_6		Alkali metals Alkaline-earth metals	HPIC	Conductivity detection
			Amines < C_6	MPIC HPIC	Conductivity detection for pKb < 7 orUV/VIS detection
			Transition metals	HPIC	UV/VIS detection With post-column raction
			Amino acids	HPIC	UV/VIS detection or Fluorescence detection With post-column reaction
	Polarizable ions Aryl amines Alkyl amines < C20 Amino acids		Amines Pyrimidines Purines	MPIC HPIC	Conductivity detection for pKb < 7 or UV/VIS detection Pulsed amperometry

Areas of application related to nucleotide and purine/pyrimidine bases have been improved considerably in recent years with the advent of more advanced column technology. Cation-exchange chromatography has been employed in a comparable manner to separate organic bases such as purines and pyrimidines of biological importance.

Much interest has been shown in the possible clinical uses of ion-exchange chromatography as a means of screening biological fluids in order to highlight abnormalities. This area of application poses a major challenge in that most body fluids contain literally hundreds of different compounds that should be separated. The significance of variations in the analysis of body fluids from individual subjects is a separate and equally complex subject. Recent work in this area has dealt with the detection of the metabolites of biogenic amines and the screening of urine for possible abnormalities.

OTHER APPLICATIONS OF ION EXCHANGE

The more established, polystyrene-based, resins possess a high exchange capacity, in the order of 5 mequiv. per gram of packing. A high exchange capacity can be a distinct advantage in applications where large sample sizes are required, however in many instances the high capacity leads to very strong retention of the sample on the column. Elution of the sample components under these conditions is only possible by the action of relatively concentrated buffer or salt solutions, typically in the order of 1-5 M. High concentrations of salts may be quite acceptable in glass columns, but such solutions can be particularly aggressive towards apparatus made of stainless steel and also cause problems if salts are allowed to crystallize in fine-bore tubing.

The more recently developed ion-exchange packings, in general, possess low exchange capacities (approximately two orders lower than the micro particulate polystyrene-based materials) and are correspondingly much less retentive. As a consequence, eluting solvents require considerably lower concentrations of buffers and salts, 10-100 mM being typical, to elute the sample components. Separations using modern packing materials can often be achieved quite rapidly, usually in less than 20 min.

Many polar compounds of pharmaceutical interest may be easily chromatographed in the manner described. Some of t! well documented applications include: barbiturates, sulphonamides, analgesics, and antibiotics such as tetracyclines and cephalosporins. In addition to the more obvious separation of acids and bases, ion exchange has found widespread application in the food industry for the analysis of food-colouring agents and artificial sweeteners such as saccharin.

The methodology used in developing separations based on ion exchange can, in principle, form a very logical sequence in all separations where the mechanism of separation is solely due to ionic interactions. In practice, complications occur as secondary effects due to matrix-sample interactions often affect the elution characteristics. This situation, although frustrating when wishing to deduce a separation mechanism, should not be ignored as many highly successful separations have been reported which would appear to be possible only by the combined mechanism.

REFERENCES AND SUGGESTED READINGS

Clearfield, A. 1982) Inorganic Ion Exchange Material, CRC Press, Boca-Ratan, FL.

Eksborg, S. (1975) *Acta Pharm. Suecica*, 12: 19.

Fischer, L. (1971) An Introduction to Gel Chromatography, North-Holland, Amsterdam.

Park, J.W., Lee, S.M., Kajiuchi, Y.T.T. and Akehata, T. (1990) *Journal of Chemical Engineering of Japan*, 23: 497-499.

Persson, B.A. and Karger, B.L. (1974) J. *Chromatogr. Sci.*, 12: 521

Peterson, R. (1970) An Introduction to Ion Exchange Chromatography, Heydon-Sadtler, London.

Reiemann, W. and Walton, H.F. (1970) Ion Exchange In Analytical Chemistry, Pergamon Press, Oxford.

Samuelson, O. (1962) Ion Exchanger Separation in Analytical Chemistry, John-Willey, New York.

Scott, C.D. and Lee, N.E. (1973) *J. Chromatogr.* 83: 383.

Sharma, B.K. (2000) Instrumental Methods of Chemical Analysis 919th Edn.) Goel Publishing House, Meerut (India) pp 41-55.

Smith, F.C. Jr. and Chang, R.C. (1983) The Practice of Ion Chromatography, John-Willey, New York.

Udenfriend, S., Stein, S., Bohlen, P., Dairman, W., Leimgruber, W. and Weigele, M. (1972) *Science*, 178: 871.

REVIEW QUESTIONS

Long Questions

1. What is gel filtration chromatography? Discuss the technique involved in gel chromatography.
2. What is gel permeation? How gels are prepared for gel chromatography?
3. Discuss the methods used for packing columns in gel filtration chromatography.
4. Why is it not desirable to pack the columns with dry packing? Discuss briefly the technique used in gel filtration chromatography.
5. What is gel filtration? Give an account of the theory and applications of gel filtration chromatography.
6. What is ion exchange chromatography? Discuss its type.
7. What is ion exchange chromatography? Discuss the various factors influencing selection of mobile phase.
8. Discuss briefly the mechanism of ion exchange separations?
9. Give an account of the applications of ion exchange chromatography with special reference to biological fields.
10. What is ion exchange chromatography? Discuss its advantages.

Multiple Choice Questions

1. The stationary phase of gel filtration chromatography
   (a) A liquid
   (b) Solutes distributed through the entire gel phase
   (c) A gas
   (d) Liquid flowing through the bed

2. Size of the columns for better resolution in gel filtration chromatography
   (a) Medium in length (b) Shorter in length
   (c) Longer in length (d) No correlation
3. The solid support, or gels, that used in gel chromatography
   (a) One dimensional (b) Two dimensional
   (c) Three dimensional (d) Four dimensional
4. Examples of natural inorganic cations
   (a) Zeolites and clay (b) Clay only
   (c) Zeolites only (d) Leaf extracts
5. Amphoteric electrolytes are sorbents, capable of
   (a) Cation exchange (b) Anion exchange
   (c) Both cation and anion exchange (d) Exchange of none of charges
6. The exchange capacity ratio of organic ion exchange material than that of inorganic
   (a) Lower (b) Similar
   (c) Higher (d) Always lower
7. As the cross linking in the resin decreases, the resin swelling
   (a) Increases (b) Decreases
   (c) No correlation (d) Remains similar
8. The columns of Ion-Exchange chromatography may be operated by
   (a) Elution analysis (b) Frontal analysis
   (c) Displacement analysis (d) All above.
9. Which of the following laws, exchange of ions obeys?
   (a) Thermodynamics (b) Mass action
   (c) Electricity (d) Newton
10. Sorbability of ions increases with the increase in
   (a) Charges (b) Atomic mass only
   (c) Ionic radii and charges (d) Charges, atomic masses and ionic radii

Answers

1. (b)	2. (b)	3. (c)	4. (a)	5. (c)	6. (c)
7. (a)	8. (d)	9. (b)	10. (d)		

CHAPTER **55**

High Pressure Liquid Chromatography

Y.C. Tripathi
Rain Forest Research Institute, P. Box 136, Deovan, Sotai-Ali, A.T. Road, Jorhat–785 001
E-mail: yctripathi@yahoo.com

In today's science there is an ever growing need to visualize and characterize an ever increasing number of substances as we investigate our environment and isolate purify or create molecules and substances by numerous methods. High pressure liquid chromotography (HPLC) technology has evolved to offer numerous methods and analytical capabilities. HPLC as compared with the classical technique is characterized by:

- Small diameter (2-5 mm), reusable stainless steel columns
- Column packings with very small (3, 5 and 10 mm) particles and the continual development of new substances to be used as stationary phases
- Relatively high inlet pressures and controlled flow of the mobile phase
- Precise sample introduction without the need for large samples
- Special continuous flow detectors capable of handling small flow rates and detecting very small amounts
- Automated standardized instruments
- Rapid analysis
- High resolution.

Initially, pressure was selected as the principal criterion of modern liquid chromatography and thus the name was "high pressure liquid chromatography" or HPLC. This was, however, an unfortunate term because it seems to indicate that the improved performance is primarily due to the high pressure. This is, however, not true. In fact high performance is the result of many factors: very small particles of narrow distribution range and uniform pore size and distribution, high pressure column slurry packing techniques, accurate low volume sample injectors, sensitive low volume detectors and of course, good pumping systems. Naturally, pressure is needed to permit a given flow rate of the mobile phase; otherwise, pressure is a negative factor not contributing to the improvement in separation. Recognizing this, most experienced chromatographers today refer to the technique as high performance liquid chromatography still permitting the use of the acronym HPLC.

TYPES OF HPLC

There are many ways to classify liquid column chromatography. If this classification is based on the nature of the stationary phase and the separation process, three modes can be specified.

In adsorption chromatography the stationary phase is an adsorbent (like silica gel or any other silica based packings) and the separation is based on repeated adsorption-desorption steps.

In ion-exchange chromatography the stationary bed has an ionically charged surface of opposite charge to the sample ions. This technique is used almost exclusively with ionic or ionizable samples. The stronger the charge on the sample, the stronger it will be attracted to the ionic surface and thus the longer time it will take to elute. The mobile phase is an aqueous buffer, where both pH and ionic strength are used to control elution time.

In size exclusion chromatography the column is filled with material having precisely controlled pore sizes, and the sample is simply screened or filtered according to its solvated molecular size. Larger molecules are rapidly washed through the column; smaller molecules penetrate inside the porous of the packing particles and elute later. Mainly for historical reasons, this technique is also called gel filtration or gel permeation chromatography although, today, the stationary phase is not restricted to a "gel".

Concerning the first type, two modes are defined depending on the relative polarity of the two phases: normal and reversed-phase chromatography.

- In normal phase chromatography, the stationary bed is strongly polar in nature (e.g., silica gel), and the mobile phase is non-polar (such as n-hexane or tetrahydrofuran). Polar samples are thus retained on the polar surface of the column packing longer than less polar materials.
- Reversed-phase chromatography is the inverse of this. The stationary bed is non-polar (hydrophobic) in nature, while the mobile phase is a polar liquid, such as mixtures of water and methanol or acetonitrile. Here the more non-polar the material is, the longer it will be retained.

Above mentioned types cover almost 90% of all chromatographic applications. Eluent polarity plays the highest role in all types of HPLC. There are two elution types: isocratic and gradient. In the first type constant eluent composition is pumped through the column during the whole analysis. In the second type, eluent composition (and strength) is steadily changed during the run.

Mobile Phases

In HPLC, type and composition of the mobile phase (eluent) are one of the variables influencing the separation. Despite of the large variety of solvents used in HPLC, there are several common properties:

- Purity
- Detector compatibility

- Solubility of the sample
- Low viscosity
- Chemical inertness
- Reasonable price.

Each mode of HPLC has its own requirements. For normal phase mode solvents are mainly nonpolar, for reversed-phase eluents are usually a mixture of water with some polar organic solvents such as acetonitrile. Size-exclusion HPLC has a special requirement, SEC eluents has to dissolve polymers, but the most important is that SEC eluent has to suppress all possible interactions of the sample molecule with the surface of the packing material.

STATIONARY PHASES (ADSORBENTS)

HPLC separations are based on the surface interactions and depend on the types of the adsorption sites (surface chemistry). Modern HPLC adsorbents are the small rigid porous particles with high surface area. The main adsorbent parameters are:

- Particle size: 3 to 10 μm
- Particle size distribution: as narrow as possible, usually within 10% of the mean;
- Pore size: 70 to 300 Å
- Surface area: 50 to 250 m^2/g
- Bonding phase density (number of adsorption sites per surface unit): 1 to 5 per 1 nm^2.

The last parameter in represents an adsorbent surface chemistry. Depending on the type of the ligand attached to the surface, the adsorbent could be normal phase (–OH, $-NH_2$), or reversed-phase (C8, C18, Phenyl), and even anion (NH_4^+), or cation (–COO–) exchangers.

DETECTORS

Detectors equipped with the flow-through cell were a major breakthrough in the development of modern liquid chromatography. Such detection was first applied by the group of Tiselius, in Sweden in 1940, by continuously measuring the refractive index of the column effluent. Current LC detectors have wide dynamic range normally allowing both analytical and preparative scale runs on the same instrument. They have high sensitivities often allowing the detection of nanograms of material, and the better models are very flexible, allowing rapid conversion from one mobile phase to another and from one mode to another.

Almost all LC detectors are the on-stream monitors. The only relatively successful off-line detector is FTIR spiral disk monitor, which require sample transfer on the germanium disk and scanning in FTIR instrument. HPLC detectors always used under continuous flow conditions and the sample is always dissolved in the eluent during detection. Actual sample is only present in a ng quantity in the detector, but in trace analysis, this quantity could be fg and even the single molecule. The mobile phase is a factor which must always be considered.

In the last decade there is a significant progress in the development of LC/MS interfacing systems. MS as an on-line HPLC detector is said to be the most sensitive, selective and in the same time the most universal detector. But it is still the most expensive one. Following are some of the important detectors used in HPLC systems:

Most common Refractive index	**Less common, but important**
• UV/VIS	• Conductivity
• Fixed wavelength	• Mass-spectrometric (LC/MS)
• Variable wavelength	• Evaporative light scattering
• Diode array	
• Fluorescence	

BASIC DETECTOR REQUIREMENTS

Regardless of the principle of operation, an ideal LC detector should have the following properties:

- Low drift and noise level (particularly crucial in trace analysis).
- High sensitivity.
- Fast response.
- Wide linear dynamic range (this simplifies quantitation).
- Low dead volume (minimal peak broadening).
- Cell design which eliminates remixing of the separated bands.
- Insensitivity to changes in type of solvent, flow rate, and temperature.
- Operational simplicity and reliability.
- It should be tuneable so that detection can be optimized for different compounds.
- It should be non-destructive.

Noise and Drift

HPLC is a time-dependent process. The appearance of the component from the column in the detector is represented by the deflection of the recorder pen from the baseline. It is a problem to distinguish between the actual component and artifact caused by the pressure fluctuation, bubble, compositional fluctuation, etc. If the peaks are fairly large, one has no problem in distinguishing them. However, the smaller the peak, the more important is that the baseline be smooth, free of noise, and drift. Fig. 1 illustrates noise and drift levels of detector.

Baseline noise is the short time variation of the baseline from a straight line caused by electric signal fluctuations, lamp instability, temperature fluctuations and other factors. Noise usually has much higher frequency than actual chromatographic peak. Noise is normally measured "peak-to-peak" i.e., the distance from the top of one such small peak to the bottom of the next. Sometimes, noise is averaged over a specified period of time. Noise is the factor which limits detector sensitivity.

In trace analysis, the operator must be able to distinguish between noise spikes and component peaks. A practical limit for this is a 3 x signal-to-noise ratio, but only for qualitative purposes. Practical quantitative detection limit better be chosen as 10x signal-to-noise ratio. This ensures correct quantification of the trace amounts with less than 2% variance. Another parameter related to the detector signal fluctuation is drift. Noise is a short-time characteristic of a detector, an additional requirement is that the baseline should deviate as little as possible

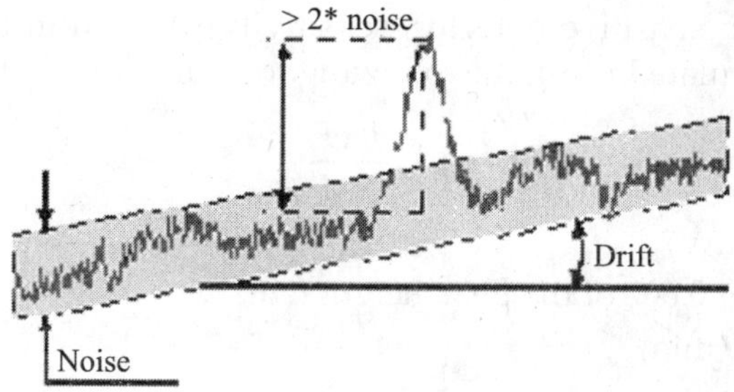

Fig. 1. Definition of noise, drift and smallest detectable peak.

from a horizontal line. It is usually measured for a specified time, e.g., 1/2 hour or one hour. Drift usually associated to the detector heat-up in the first hour after power-on.

SENSITIVITY

Detector sensitivity is one of the most important properties of a LC detector. Sensitivity of the detector is a measure of its ability to discriminate between small differences in analyte concentration. So, it is actually the slope of the calibration curve. It is also dependent on the standard deviation of the measurements. The higher the slope of your calibration curve the higher the sensitivity of your detector for that particular component, but high fluctuations of your measurements will decrease the sensitivity.

Sensitivity of a detector is not the minimum amount that can be detected. This value is influenced by the chromatographic conditions. Early eluting peaks are usually sharp, whereas the ones with long retention times are broad and sometimes difficult to discern from the noise.

SELECTIVITY

Selectivity is another highly desirable property of HPLC detectors. A selective detector allows one to see only components of interest despite of their co-elution with any others. Refractive index is an example of almost nonselective detector. Any component could make a response, but in case of poorly resolved mixture analyst will not be able to distinguish components. Fluorescence and electrochemical detectors are the most selective among the common detectors. Only about 10% of organic compounds are able to fluorescence, and by choosing excitation and emission wavelength specific for the particular compound one can detect only this compound. Usually, the more selective the detection, the lower is the signal noise and the higher the sensitivity.

RESPONSE

The definition of detector response depends on whether it is mass-sensitive or concentration-sensitive. For mass-sensitive detectors, the response **R** (mV/mass/unit time) is given by the following relationship:

$$R = \frac{hw}{sM}$$

For the concentration sensitive detector, sensitivity is given in units of mV/mass/unit volume and it can be calculated using the following formula:

$$R = \frac{hwF}{sM}$$

where, h = peak height (mV)

w = peak width at 0.607 of the peak height (cm)

F = flow rate (ml/min)

M = mass of solute injected

s = chart speed (cm/min)

In addition to factors such as the nature of the mobile phase, cell geometry, and so on, the detector response is also a function of the type of solute.

Linear Dynamic Range

The detector response is said to be linear if the difference in response for two concentrations of a given compound is proportional to the difference in concentration of the two samples. Such response appears as a straight line in the calibration curve (Fig-2). The linear dynamic range of a detector is the maximum linear response divided by the detector noise. Most detectors eventually become non-linear as sample size is increased and this upper point is usually well defined. The chromatographer should know where this occurs to avoid errors in quantification.

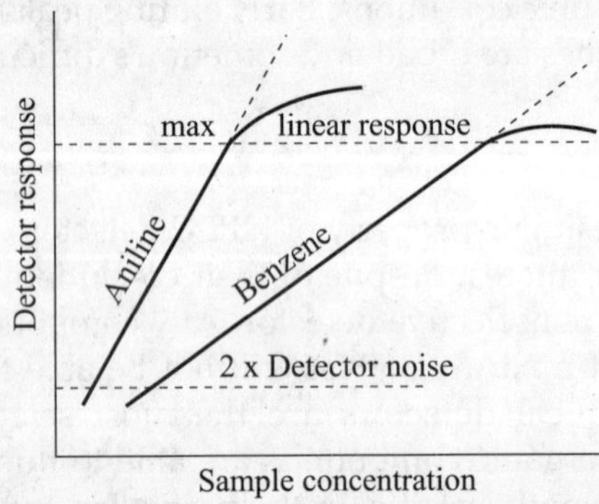

Fig. 2. Detector response.

Cell Efficiency

The heart of an efficient HPLC detector is the cell. Fig. 3 shows the schematic of a modern follow-through cell. The optics provides focusing of the light beam in the center of the cell where it is virtually unaffected by the entry-exit-window-interface disturbance, or drift induced by flow, temperature, or refractive index changes. The short, wide cell assures that maximum energy is transmitted, and a post-cell collecting lens focuses all of the existing light from the cell onto the photo-detector. Modern LC systems provide high resolution in a short

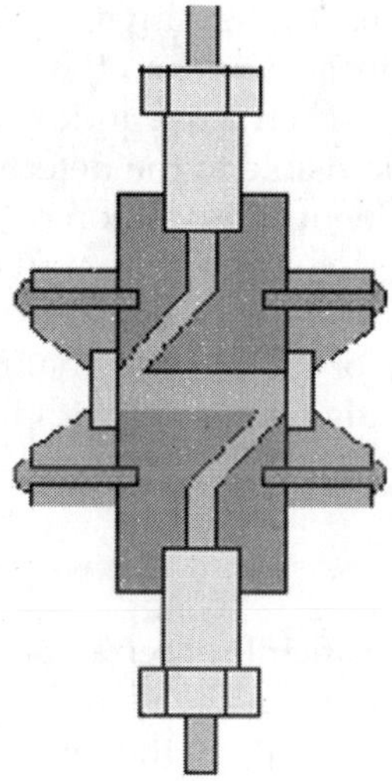

Fig. 3. Schematic of a modern follow-through cell.

time. For example, if we are using a column which provides 10,000 theoretical plates and has 15 cm length, then for the component eluted in 2 min (2 ml at 1 ml/min flow rate) we will ideally have peak width of 80 μl. Thus, having the flow-cell volume of 20 μl we will have only 4 independent measurements on this peak. This is definitely not enough to correctly describe peak shape and will introduce apparent peak broadening.

Another important feature of the flow-cell is to compensate on the refraction effect. When the components of your mixture pass though the cell they actually change (slightly) the eluent composition which will change the refraction coefficient. If the light beam is not parallel the change in the refraction will change light scattering and will contribute into the apparent adsorption readings. Unfortunately very few cell designs have this compensation.

- Apparent Peak broadening happens if the flow cell volume is more then 1/10 of the peak volume, or if the cell geometry is not optimized.
- Since the peak area is proportional to the amount of the injected analyte, the peak broadening will cause the reduction of the peak height. In case of apparent peak broadening due to the big cell volume but if the cell geometry is optimized we may not see a significant height reduction.
- The resolution may be decreased as a result of the peak broadening and peak overlapping occurs.
- If the cell geometry is not optimized one can notice an appearance of peak tailing. This indicates the presence of flow stagnant zones.

REFRACTIVE INDEX DETECTOR

The refractive index (RI) detector is the only universal detector in HPLC. The detection principle involves measuring of the change in refractive index of the column effluent passing through the flow-cell. The greater the RI difference between sample and mobile phase, the larger the imbalance will become. Thus, the sensitivity will be higher for the higher difference in RI between sample and mobile phase. On the other hand, in complex mixtures, sample components may cover a wide range of refractive index values and some may closely match that of the mobile phase, becoming invisible to the detector. RI detector is a pure differential instrument, and any changes in the eluent composition require the rebalancing of the detector. This factor is severely limiting RI detector application in the analyses requiring the gradient elution, where mobile phase composition is changed during the analysis to effect the separation. There are two basic types of RI detectors that require use of a two-path cell where the sample-containing side is constantly compared with the non-sample-containing reference side.

Deflection Detectors

This detector based on the deflection principle of refractometery, where the deflection of a light beam is changed when the composition in the sample flow-cell changes in relation to the reference side (as eluting sample moves through the system). When no sample is present in the cell, the light passing through both sides is focused on the photodetector (usually photoresistor). The optical schematic of the deflection detector is shown in Fig. 4.

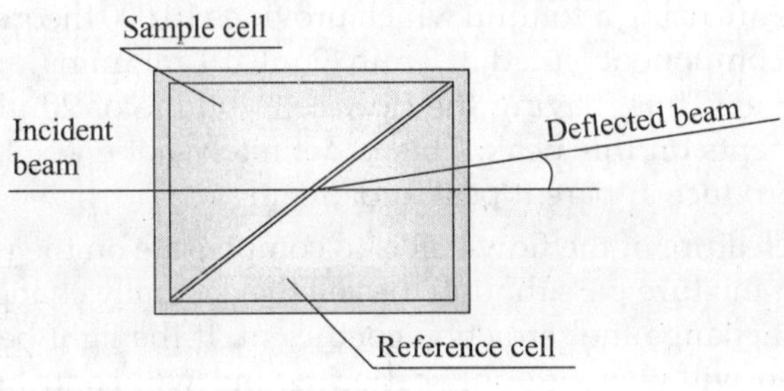

Fig. 4. Optical schematic of a deflection RI detector.

As sample elutes through one side, the changing angle of refraction moves the beam. This results in a change in the photon current falling on the detector, which unbalances it. The extent of unbalance (which can be related to the sample concentration) is recorded on a strip chart recorder.

The advantages of this type of detector are: (1) universal response; (2) low sensitivity to dirt and air bubbles in the cells; and (3) the ability to cover the entire refractive index range from 1.000 to 1.750 RI with a single, easily balanced cell. The disadvantages are the relatively low sensitivity and a general disability to easily remove and clean or replace the cell when filming or clogging occurs.

Reflective Detectors

The refractive index detector based on the Fresnel principle is relatively rear. There was only one or two commercial models and they are not in the production any more. Fig. 5 shows the optical schematic of this type of detector. Here, the light beam is reflected from the liquid-

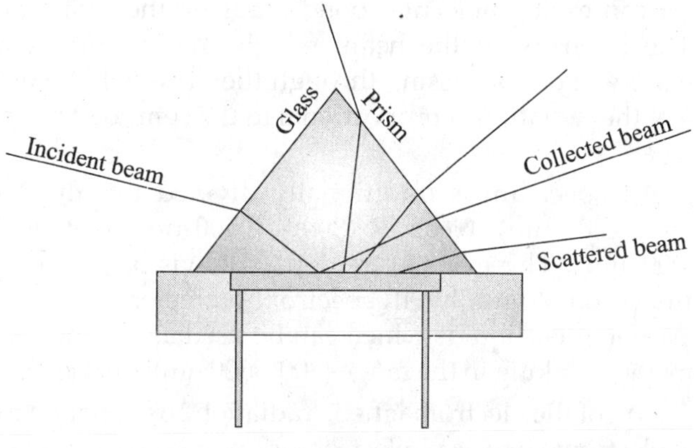

Fig. 5. Fresnel-type RI detector.

glass interface in the detecting photocell. The introduction of sample into one cell causes light to be refracted at a different angle. The deflection of the light beam from the photoresistor causes the appearance of the electrical signal. Here, too, this difference between sample cell signal and reference-cell signal is output to a recorder or data handling system as peak.

The major advantage of this type of detector is a very high sensitivity since the optics allows a higher concentration of signal in a particular RI range than is possible in other wide-range detectors. Other advantages include the ability to operate at extremely low flow rates with very low-volume cells, easy cell accessibility, and low cost. Its disadvantages are the incredible sensitivity to the flow and pressure fluctuations and the need for changing prisms to accommodate either high or low RI solvents and to manually adjust the optical path when making solvent changes.

The refractive index of an analyte is a function of its concentration. Change in concentration is reflected as a change in the RI. A refractive index detector for liquid chromatography should be sensitive to changes as small as 10^{-7} RI units (corresponding to a concentration change of 1 ppm). Presence of dissolved air, changes in solvent composition, improper mixing and column bleed will contribute to baseline drift. Eluent pressure change of 15 psi will cause the change of 1×10^{-6} RI unit and 1°C temperature variation will be equivalent to the change of 600×10^{-6} RI units. Thus it is obvious that both of these parameters must be closely controlled, especially temperature. To operate at high sensitivities, a RI detector must usually be thermostated (± 0.01°C), actually the using of the water bath connected to the detector head does not give required temperature stability and alternately passive thermo-stabilization with massive metallic block usually gives much better results.

ULTRAVIOLET/VISIBLE SPECTROSCOPIC DETECTORS

Any chemical compound could interact with the electromagnetic field. Beam of the electromagnetic radiation passed through the detector flow-cell will experience some change in its intensity due to this interaction. Measurement of this change is the basis of the most optical HPLC detectors. Radiation absorbance depends on the radiation wavelength and the functional groups of the chemical compound. Electromagnetic field depending on its energy (frequency) can interact with electrons causing their excitation and transfer onto the higher energetical level, or it can excite molecular bonds causing their vibration or rotation of the functional group. The intensity of the beam which energy corresponds to the possible transitions will decrease while it is passing through the flow-cell. According to the Lambert-Bear law absorbance of the radiation is proportional to the compound concentration in the cell and the length of the cell.

The electromagnetic spectrum is traditionally divided mainly into four regions viz., Infrared (IR) (2,500 – 50,000 nm); Near IR (800 – 2,500 nm); Visible (400 – 800 nm) and Ultraviolet (190 – 400 nm). Three major regions (IR, visible, and UV) are used in the spectroscopy. In liquid chromatography, IR spectrophotometers have found only limited use. There are few transparent polar liquids which can be used as the mobile phase. On the other hand, spectrophotometers working in the range (200 – 600 nm) are used widely as LC detectors.

UV and visible region of the electromagnetic radiation corresponds to the excitation of the relatively low energy electrons such as pi-electrons, or non-paired electrons of some functional groups. For example, n-alkanes could absorb in the UV region below 180 nm. σ-electrons require high energy radiation to get excited and to show absorption of the radiation. But any compounds which have benzene ring will show absorbance at 205–225 and 245–265 nm. The last corresponds to the excitation of conjugated p-electrons of the benzene ring.

The majority of organic compounds can be analyzed by UV/VIS detectors. Almost 70% of published HPLC analyses were performed with UV/VIS detectors. This fact, plus the relative ease of its operation, makes the UV detector the most useful and the most widely used LC detector.

UV ABSORBANCE

Absorbance is the logarithm of the ratio of the intensities of the incident light (I_o) and the transmitted light (I). It is related according to the Beer-Lambert Law to the molar absorptivity (molar extinction coefficient, ε), the thickness of the substance (i.e., the path length of the cell, b) and the molar concentration of the substance (c):

$$A = \log\left(\frac{I_o}{I}\right) = \varepsilon bc$$

In HPLC, the photodetector measures transmitted light I, but the electronics converts this signal to a logarithmic relationship (A) which is proportional to concentration.

The ordinate of the chromatogram represents the detector signal, which in general, is proportional to the analyte concentration in the cell. Since chromatographic systems permit the quantitative analysis of sample components representing many orders of magnitude—from ppm to percent concentrations, one may select various amplification ranges for visual display of components (both small and large). In UV detection, one expresses the detector range in absorbance units (A). One absorbance unit corresponds to the depreciation of the light intensity by 90% of the incident light.

Molar Absorptivity: This term is also called the molar extinction coefficient—corresponds to the absorbance for a molar concentration of the substance with a path length of 1 cm. Molar adsorptivity is dependent on the wavelength and chromatographic conditions (solvent, pH and temperature). It is a constant at a specified wavelength. Table 1 lists the molar absorptivities of a number of compounds at specified wavelengths.

Table 1. Molar Absorptivity (*e*) values of various compound types at specified wavelengths

Name	Chromophore	Wavelength [nm]	Molar extinction, e
Acetylide	–C=C	175-180	6,000
Aldehyde	–CHO	210	1,500
Amine	$–NH_2$	195	2,800
Azo	–N=N-	285-400	3-25
Bromide	–Br	208	300
Carboxyl	–COOH	200-210	50 - 70
Disulphide	-S-S-	194	5,500
Ester	–COOR	205	50
Ether	–O-	185	1,000
Ketone	>C=O	195	1,000
Nitrate	$–ONO_2$	270	12
Nitrile	–C=N	160	-
Nitrite	–ONO	220 - 230	1000-2000
Nitro	$–NO_2$	210	strong

FIXED WAVELENGTH DETECTORS

HPLC detectors which do not allow changing the wavelength of the radiation called fixed-wavelength detectors. Low-pressure mercury vapor lamps emit very intense light at 253.7 nm. By filtering out all other emitted wavelengths, manufacturers have been able to utilize this 254 nm line to provide stable, highly sensitive detectors capable of measuring subnanogram quantities of any components which contains aromatic ring. The 254 nm was chosen since the most intense line of mercury lamp is 254 nm, and most of UV absorbing compounds have some absorbance at 254 nm.

VARIABLE WAVELENGTH DETECTORS

Detectors which allow the selection of the operating wavelength called variable wavelength detectors (Fig. 6) and they are particularly useful in three cases:

- Offer best sensitivity for any absorptive component by selecting an appropriate wavelength;
- Individual sample components have high absorptivity at different wavelengths and thus, operation at a single wavelength would reduce the system's sensitivity;
- Depending on the sophistication of the detector, wavelength change is done manually or programmed on a time basis into the memory of the system.

DIODE-ARRAY DETECTORS

Diode array adds a new dimension of analytical capability to liquid chromatography because it permits qualitative information to be obtained beyond simple identification by retention time (Fig. 7).

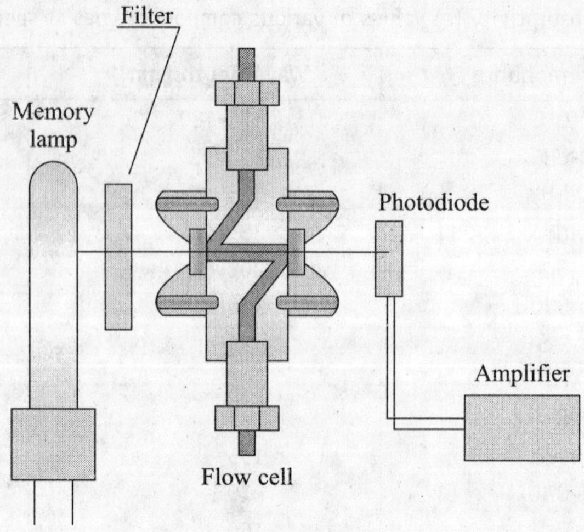

Fig. 6A. Fixed wavelength detector.

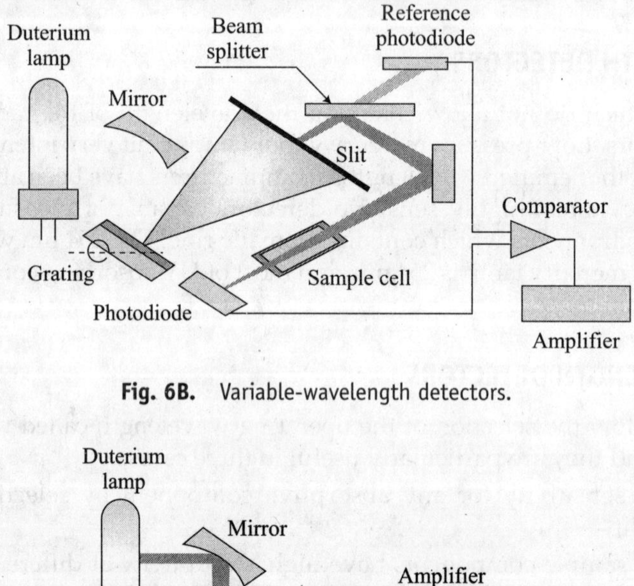

Fig. 6B. Variable-wavelength detectors.

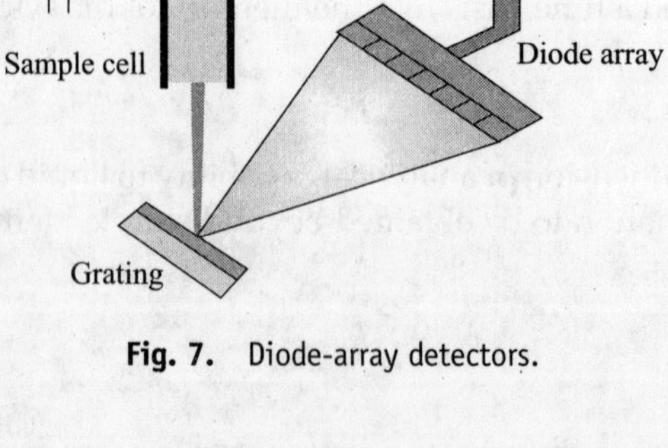

Fig. 7. Diode-array detectors.

There are two major advantages of diode array detection. In the first, it allows for the best wavelength(s) to be selected for actual analysis, which is particularly important when molar absorptivities at different wavelengths are not known. The second major advantage is related to the problem of peak purity. Often, the peak shape in itself does not reveal that it actually corresponds to two (or even more) components. In such a case, absorbance rationing at several wavelengths is particularly helpful in deciding whether the peak represents a single compound or, is in fact, a composite peak.

In absorbance rationing, the absorbance is measured at two or more wavelengths and ratios are calculated for two selected wavelengths. Simultaneous measurement at several wavelengths allows one to calculate the absorbance ratio. The ratios at chosen wavelength are continuously monitored during the analysis; if the compound under the peak is pure, the response will be a square wave function (rectangle). If the response is not rectangle, the peak is not pure.

FLUORESCENCE DETECTORS

Fluorescence detectors are probably the most sensitive among the existing modern HPLC detectors. It is possible to detect even a presence of a single analyte molecule in the flow cell. Typically, fluorescence sensitivity is 10–1000 times higher than that of the UV detector for strong UV absorbing materials. Fluorescence detectors (Fig. 8) are very specific and selective among the others optical detectors normally used as an advantage in the measurement of specific fluorescent species in samples. When compounds having specific functional groups are excited by shorter wavelength energy and emit higher wavelength radiation which called fluorescence. Usually, the emission is measured at right angles to the excitation.

Roughly about 15% of all compounds have a natural fluorescence. The presence of conjugated pi-electrons especially in the aromatic components gives the most intense fluorescent activity. Also, aliphatic and alicyclic compounds with carbonyl groups and compounds with highly conjugated double bonds fluoresce, but usually to a lesser degree.

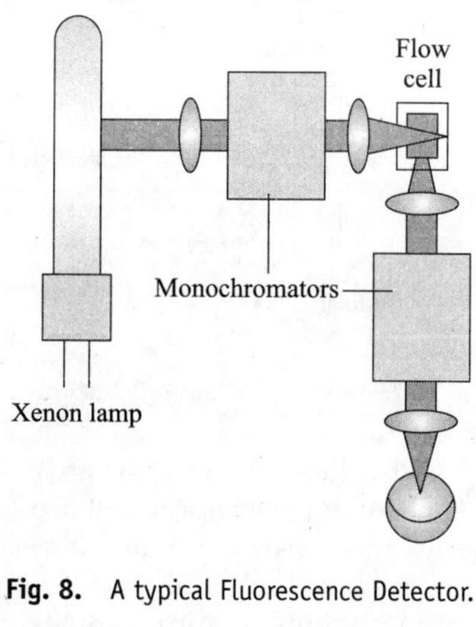

Fig. 8. A typical Fluorescence Detector.

Most unsubstituted aromatic hydrocarbons fluoresce with quantum yield increasing with the number of rings, their degree of condensation and their structural rigidity. The detectors differ in the method in which the wavelengths are controlled. Less expensive instruments utilize filters; medium priced units offer monochromator control of at least emission wavelength, and full capability research-grade instruments provide monochromator control of both excitation and emission wavelengths.

Fluorescence intensity depends on both the excitation and emission wavelength, allowing selectively detect some components while suppressing the emission of others. The detection of any component significantly depends on the chosen wavelength and if one component could be detected at 280 ex and 340 em, another could be missed. Most of the modern detectors allow fast switch of the excitation and emission wavelength, which offer the possibility to detect all component in the mixture. For example, in the very important polynuclear aromatic chromatogram the excitation and emission wavelengths were 280 and 340 nm, respectively, for the first 6 components, and then changed to the respective values of 305 and 430 nm; the latter values represent the best compromise to allow sensitive detection of compounds.

ELECTROCHEMICAL DETECTORS

The electrochemical detector (Fig. 9) is also a popular liquid chromatographic detector having additional selectivity and sensitivity for some compounds. It is based on the measurements of the current resulting from oxidation/reduction reaction of the analyte at a suitable electrode. Since the level of the current is directly proportional to the analyte concentration, this detector could be used for quantification. The eluent should contain electrolyte and be electrically conductive. The purity of the eluent is very important, because the presence of oxygen, metal contamination and halides may cause significant background current and therefore, noise and drift in the base line. Most of the analytes to be successfully detected require the pH adjustments.

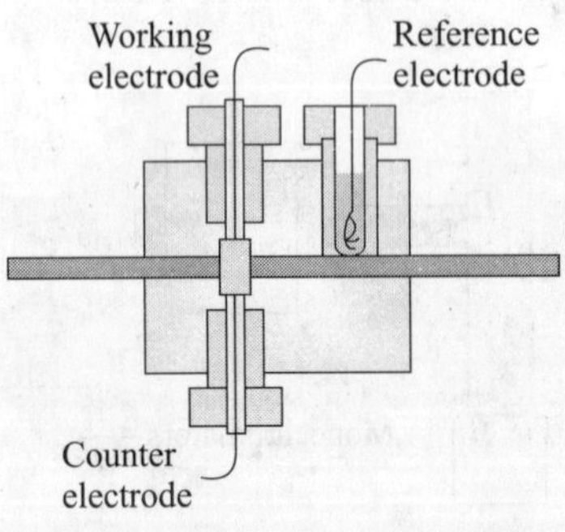

Fig. 9. Electrochemical Detectors.

The areas of application of electrochemical detection are not large, but the compounds for which it does apply, represent some of the most important drug, pollutant and natural product classes. For these, the specificity, and sensitivity make it very useful for monitoring these compounds in complex matrices such as body fluids and natural products. Sensitivities for compounds such as phenol, catecholamines, nitrosamines, and organic acids are in the picomole (nanogram) range.

ELECTROLYTIC CONDUCTIVITY DETECTOR

The conductivity of the column effluent is continuously measured and the appearance of the analyte in the cell is indicated by a change in conductivity.

Usually this is a very low volume flow-through capillary equipped with two electrodes and variations in conductivity of the mobile phase due to the eluted sample components are continuously recorded (Fig. 10). Response is linear with concentration over a wide range; quantitation of the output signal is possible with suitable preliminary calibration. Best use is made of this detector in isocratic analysis since solvent gradients will cause a proportional shift in the baseline.

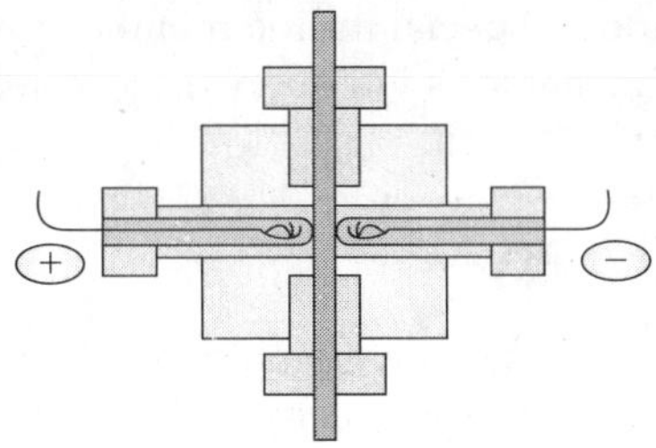

Fig. 10. Electrochemical Detectors.

Such detectors have been used most successfully in ion-exchange chromatography of anions and cation but generally, they have found only limited popular acceptance.

EVAPORATIVE LIGHT SCATTERING DETECTORS (ELSD)

Commonly described as a poor man's Mass Spectrometer, this unique type of detector offers amazing capabilities to measure a wide variety of substances. Many substance types do not produce any UV-VIS absorbance's and cannot be observed on typical UV based HPLC detection systems. The ELSD detector is basically quite similar to an electrospray part of an electrospray ESI Mass Spectrometer. In that there is a heated chamber and a flow of a carrier dispersing gas such as high grade Argon or ultra pure Nitrogen gas that served to atomize the HPLC eluent liquid mobile phase flow upon arrival and the atomized spray is then passed through a detector with a laser energy pulse that excites the molecules which in turn scatters the light. The light scattering is measured by a high sensitivity or series of detectors and compared to a reference beam measurement.

Unlike ESI Mass Spectrometers however, there is absence of a high vacuum system and time of flight tube that facilitate mass measurement. There are companies that have developed instruments that can convert ELSD data to calculate molecular mass. Yet the mass accuracy is reported as good for rough estimations of molecular weights, there are many variables and the mass accuracy is significantly lower than that achievable via true mass spectrometry. For example such a system may assign a mass of 56Kd to a protein but the actual mass of the protein may be 52 Kd so it certainly good for screening and basic evaluations of HPLC peak isolates.

The advantages of ELSD are that it allows visualization of certain types of molecules that for example do not have UV Absorbance's these include types of Glycosides. Further advantage is that it can provide ballpark mass determinations of peak isolates by comparison to results obtained running known molecular weight standards producing more accurate results than Electrophoresis running molecular weight standards.

CHEMICAL ION DETECTORS

These detectors utilize ion selective electrodes that are implanted into the flow cell to measure specific ions from substances that elute from the HPLC column. This measurement is greatly influenced by presence of various metal ions and requires either Teflon HPLC pump heads and use of "Peak" plastic tubing's and or special in-line metal trapping filters as the metal ions will interfere with measurement of target ions via quenching of signals or via poor signal to noise ratios and suboptimal detection.

HPLC-MASS SPECTROMETRY (LC-MS)

The most common type of LC-MS is via conversion of an electro-spray system to allow sample introduction via micro-flow HPLC to the electro-spray atomization chamber versus direct sample introduction via a syringe pump system. LC-MS allows that an optimized HPLC separation be effected to separate different components allowing their discrete analysis by mass ionization. LC-MS systems typically utilize either a photodiode array UV-Scanning detector or a UV detector between the column outflow and the atomizer chamber. Thus, both the absorbance trace and the mass spectra data are observed and recorded. Through various softwares most often from system manufacturers, the HPLC chromatogram peak isolates shown from the HPLC detector can be mass labeled as well as labeled with the elution time and the text report provides peak area.

In the case of systems with a photodiode array scanning UV-VIS detector the absorbance's spectra for all significant wavelength absorbance's is plotted either via color plotting using different colors or via solid, dashed or dotted etc. lines in the case of monochrome. Thus, the absorbance maxima for unknown UV-VIS light absorbing compounds, the elution time, the peak area's at various wavelengths and the mass of the eluted peak are all determined and present in the data. Examination of both peak symmetry and various absorbances determine the relative purity of eluted compounds. Peak area compared to that of standards allows analytical quantitation and mass analysis determines the molecular weight of each eluted compound isolate.

HPLC SEPARATION TECHNOLOGY

The separation of components, molecules, substances and analytes is the goal of a chromatographic separation process. This goal is achieved by careful analytical work in the choice and preparation of the mobile phase liquid solvent buffers, choice of the type of solid phase column, and choice of the gradient conditions between buffer solvents. The choice of flow rate, depth of gradient, slope of gradient, column particle size and hydrophobic interaction solid phase type are combined with sample measurement and introduction choices, column heater temperature control and the type of detector and its settings and type of data acquisition utilized to produce analytical results.

METHOD DEVELOPMENT

While performing HPLC on any type or class of materials, the first aspect is thorough examination of scientific literature about these materials and specifically evaluating published accounts of their separations characteristics. In many cases various methods advised to be used as comparing the actual chromatograms may provide great insight for choosing the best method. For unknown or novel compounds it may require that numerous runs be made with different mobile phase solvents, different gradient types, different column types, and even different detector types. The literature search for the type of most similar compounds results will act as a guide to narrow down the number of different methods to run and compare.

METHOD OPTIMIZATION

The separation can be optimized by a complete methods development with particular attention to adjusting the gradient slope and time making gradients more shallow or steep will effectively zoom-in on the region of buffer solvent mobile phase concentrations and allow the degree of gradient change (% organic solvent concentration) to change more slowly effecting better separation of individual constituents.

BASIS OF SEPARATION PROCESS

Chromatographic separation process based on the difference in the surface interactions of the analyte and eluent molecules. Let us consider a separation of a two component mixture dissolved in the eluent. Assume that component **A** has the same interaction with the adsorbent surface as an eluent, and component **B** has strong excessive interaction. Being injected into the column, these components will be forced through by eluent flow. Molecules of the component **A** will interact with the adsorbent surface and retard on it by the same way as an eluent molecules.

Thus, as an average result, component **A** will move through the column with the same speed as an eluent. Molecules of the component **B** being adsorbed on the surface (due to their strong excessive interactions) will sit on it much longer. Thus, it will move through the column slower than the eluent flow. A general shape of the chromatogram for this mixture is presented in Fig. 11.

Usually a relatively narrow band is injected (5 - 20 μl injection volume). During the run, the original chromatographic band will be spread due to the noneven flows around and inside the porous particles, slow adsorption kinetics, longitudinal diffusion and other factors. These

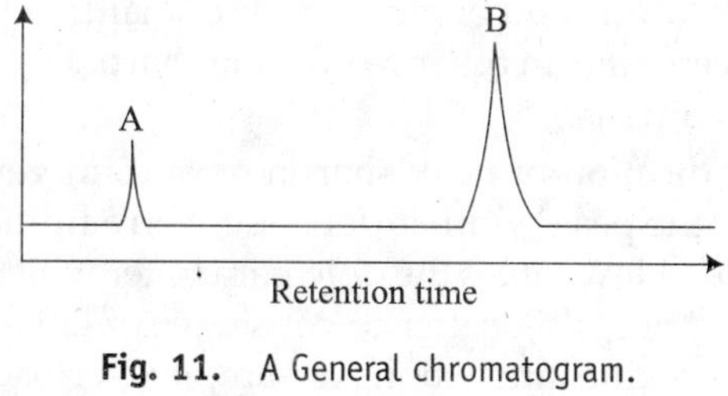

Fig. 11. A General chromatogram.

processes together produce so called band broadening of the chromatographic zone. In general, the longer the component retained on the column, the broader its zone.

Separation performance depends on both component retention and band broadening. Band broadening is, in general, a kinetic parameter, dependent on the adsorbent particle size, porosity, pore size, column size, shape, and packing performance. On the other hand, retention does not depend on the above mentioned parameters, but it reflects molecular surface interactions and depends on the total adsorbent surface.

RETENTION PARAMETERS

The easiest way to find the chromatographic retention is to measure the time between the injection point and maximum of the detector response for correspondent compound. This parameter usually called "retention time". Retention time, t_R is inversely proportional to the eluent flow rate. The product of retention time and eluent flow rate, so called "retention volume", is more of a global retention parameter. Retention volume, V_R represents the volume of the eluent passed through the column while eluting a particular component. Component retention volume V_R could be split into two parts:

1. ***Reduced retention volume*** is the volume of the eluent that passed through the column while the component was sitting on the surface.
2. ***Dead volume*** is the volume of the eluent that passed through the column while the component was moving with the liquid phase.

The second part is equal to the volume of the liquid phase in the column (dead volume, V_o), and it will be the same for any component eluted on this column. Retention volume is independent of the flow parameters for the particular run, but it depend on the geometrical parameters of the column. V_R will be different for the same compound eluted on the different columns packed with the same type of adsorbent. The more universal and fundamental retention parameter (k) is the ratio of the retention volume and dead volume ($k = V_R / V_O$).

Historically, a slightly different retention parameter, called "capacity factor" (k') was introduced by the analogy with the liquid partitioning theory and widely accepted in chromatographic practice ($k' = (V_R - V_O) / V_O$). Capacity factor is dimensionless and independent on any geometrical parameters of the column or HPLC system. It could be considered to be a thermodynamic characteristic of the adsorbent-compound-eluent system.

BAND BROADENING (COLUMN EFFICIENCY)

After injection, a narrow chromatographic band is broadened during its movement through the column. The higher the column band broadening, the smaller is the number of components that can be separated in a given time. In other words, the sharpness of the peak is an indication of how good or efficient a column is.

The peak width is an indication of peak sharpness and, in general, an indication of the column efficiency. However, the peak width is dependent on a number of parameters (column length, flow rate, particle size). Flow rate is the only parameter which can be changed from run to run on the same column. Thus, it is better to consider a relative value to express column efficiency. In absence of the specific interactions or sample overloading, the chromatographic peak can be represented by a Gaussian curve with the standard deviation(**s**) (Fig. 12). The ratio

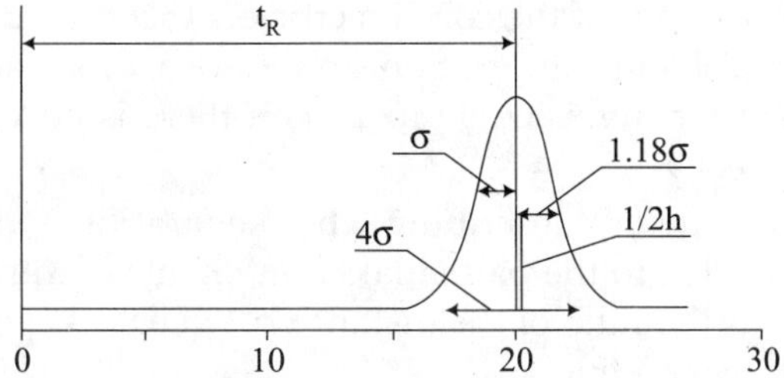

Fig. 12. Gaussian curve.

of standard deviation to the peak retention time (s/t_R) is called the relative standard deviation, which is independent on the flow rate.

In practice, the square of the reciprocal value is normally used $N - (t_R/\sigma)^2$. This has become the accepted expression of column efficiency. The reason for using the second power has come from statistics, and it is related to the fact that not the standard deviation (σ), but its square, the variance $(\sigma)^2$ is the basic measure of normal distribution. The value N is called the plate number or the number of theoretical plates. The term "theoretical plate" came form the analogy with the distillation theory. In practice, it is more convenient to measure peak width either at the base line, or at the half eight, and not at 0.609 of the peak height, which actually correspond to 2σ.

$$N = 16\ (t_R/W_b)^2;\ N = 5.545\ (t_R/W_b)^2$$

The plate number depends on column length: the longer the column, the larger the plate number. Therefore, the plate height term has been introduced to measure how efficiently column has been packed, $h = L/N$. The lower the plate height and the higher the plate number, the more efficient will be the chromatographic column.

It is well recognized now that column band broadening originates from three main sources viz., multiple path of an analyte through the column packing; molecular diffusion and effect of mass transfer between phases. The velocity of mobile phase in the column may vary significantly across the column diameter, depending on the particle shape, porosity, and the whole bed structure. Band broadening is caused by differing flow velocities through the column. Molecules disperse or mix due to the diffusion. The longitudinal diffusion (along the column long axis) leads to the band broadening of the chromatographic zone. The higher the eluent velocity, the lower the diffusion effect on the band broadening. Molecular diffusion in the liquid phase is about five orders of magnitude lower than that in the gas phase, thus this effect is almost negligible at the standard HPLC flow rates.

SELECTIVITY AND RESOLUTION

Selectivity is the ratio of the capacity factors of both peaks, or the ratio of its adjusted retention times. Selectivity represents the separation power of particular adsorbent to the mixture of these particular components.

$$\alpha = V_{R,1} - V_O\ /\ V_{R,2} - V_O = K'_1\ /\ K'_2$$

This parameter is independent of the column efficiency; it only depends on the nature of the components, eluent type, eluent composition, and adsorbent surface chemistry. In general, if the selectivity of two components is equal to 1, then there is no way to separate them by improving the column efficiency.

Resolution is the parameter describing the separation power of the complete chromatographic system relative to the particular components of the mixture. By convention, resolution (*R*) is expressed as the ratio of the distance between two peak maxima to the mean value of the peak width at the baseline:

$$R = 2 \times V_{R,2} - V_{R,1} / W_1 + W_2$$

If peaks are approximated by symmetric triangles, then, if *R* is equal to or more than 1 then component are completely separated. If *R* is less than 1, then components are overlapped.

KINETICS OF MASS TRANSFER

Mass transfer is the most questionable parameter. For the modern types of packing materials it may combine two effects: adsorption kinetics and mass transfer (mainly due to diffusion) inside the particles. 95% of all modern packing materials are the spherical, totally porous rigid particles with average diameter ~5 μm and pore diameter ~100Å. Ratio of the particle to the pore diameter is 500/1. There is no pressure propelled flow inside the particle, and molecules can move there only by diffusion. It can be shown by analogy: if consider a tunnel in the mountain which has diameter of 2 meters and one kilometer length (same ratio 500/1), and if there is a storm outside with 200 km/h wind, there will be almost no wind in 50 meters from tunnel entrance.

THERMODYNAMICS OF HPLC RETENTION

There are two basic approaches for thermodynamic description of the HPLC retention phenomena, one is based on the partitioning theory and another is based on adsorption. Partition is a concentrational change in the system due to the distribution of the components between two (or more) phases. Adsorption is the concentrational changes in the system in presence of interface with another phase and due to the surface forces.

Phase is a form of matter that is uniform throughout in chemical composition and physical state. In HPLC we have two main phases viz., the mobile phase (eluent) which is a liquid solvent or mixture of solvents which is moving through the chromatographic column and carrying analytes and the stationary phase (adsorbent), solid porous media consisting of the rigid porous particles, usually silica based, with the specific surface properties.

Adsorbent particles are considered to be nonpermeable and nonsoluble for the eluent and analyte molecules. It only introduces surface forces in the system. Consideration of the HPLC process based on the partitioning theory was transferred from GC (gas chromatography) theory, where we usually have a mobile gas phase and stationary liquid phase, and where a true partitioning occurs.

Usual description of liquid chromatography on the partitioning basis considers the assumption of the existence of the separate liquid phase which is close to the adsorbent surface. Chemically bonded phases are usually considered by this manner. The most popular bonded phase is octadecylsilica, where relatively long (21 Å) alkyl chains are chemically bonded to the silica surface. The main partitioning concept is that analyte molecules can penetrate between these alkyl chains. This process thermodynamically considered as dissolving of the analyte molecules in the surface alkyl phase.

Two restrictions exist for application of this approach in thermodynamic description i.e., the bonded chains have a certain thickness (about 3.8 Å in diameter) and depending on the bonding density the layer may not behave as a liquid and monomolecular layer would not be considered a phase in classical thermodynamics.

Adsorption from Solutions

A classic description of the adsorption process is based on the Gibbs excess adsorption theory, which basically considers two similar hypothetical adsorption systems with the same volume, temperature, pressure, and adsorbent surface area. The only difference is that the first system does not show any adsorption on the surface and the second does.

EFFECT OF TEMPERATURE

Temperature effects in HPLC are not as significant as in gas chromatography as temperature range is not always the same. Volatile solvents are not allowed to rise to higher temperatures too much, and the stability of the attached bonded ligands on the adsorbent surface may be influenced by the high temperature. So, the main temperature range is from ambient temperature to 60 or 70 C. Increasing the temperature will decrease the value of K or k', thus the actual retention time will decrease. For most of the systems these decrease will not exceed 50% of the component reduced retention time at ambient temperature. There are two other significant effects of separation under the elevated temperature.

(i) Stabilization of the column under the elevated temperature usually leads to the stabilization of the retention times.

(ii) Another effect is the increase of the column efficiency. At the elevated temperature viscosity of liquids decrease and the diffusion coefficient increase.

MOBILE PHASE EFFECT

In most of the HPLC separations binary eluents are employed. One of the solvents in the eluent is usually inert relative to the surface interactions. In Reversed-phase HPLC (RP HPLC) one of the eluent components is water, which does not interact with the hydrophobic adsorbent surface. And it does not compete with the analyte for the adsorption sutes. In normal-phase HPLC (NP HPLC) one of the eluent components is usually hexane, which also does not interact with the very polar silica surface. Another component of any binary eluent is an active one. It usually called a "modifier" because it can interact with the adsorbent surface and compete with analyte molecules for the adsorption sites. Increasing of the concentration of the "modifier" in the eluent leads to the decreasing of the analytes retention.

Capacity factor is proportional to the thermodynamic equilibrium constant, and the last is an exponent of the free Gibbs energy of the system For the binary eluent system, only the "modifier" can interact with the surface. For RP HPLC it will be an organic component of the eluent, and water is assumed not to interact with the surface.

EFFECT OF pH

A general approach to the separation of the mixtures containing an ionisible components is to suppress their ionization. Suppression of the ionization decreases a power of the molecular solvation and exposes the hydrophobic (organic) part of the molecule to the surface interaction. Ionization suppression is usually made by the adding a buffer into the solvent, which shift a pH to the certain value.

In the absence of buffer, easy ionizible components are eluted from the column as very broad peaks. According to the Le Chatelier principle, dissolved ionizible component is present in the solution as a mixture of ions and nonionized molecules

$$[AB] = [A^+] + [B^-]$$

According to the above equilibrium, about 50% of all molecules are ionized in the solution. But, the chromatographic behavior of ions and neutral molecules are different. Let us assume that neutral molecules will be retained, so during the run ions will move faster, and at the first moment they will be separated from the neutral molecules. But, according to the above equilibrium, in the absence of the neutral molecules ions will tend to form them, and this new neutral molecule will also be absorbed, and so on. This process will lead to the spreading of the component along the column and causes the appearance of the broad peak. It does not occur if the equilibrium is shifted due to the presence of the buffer with the pH at least two units apart of **pK** of the component.

RETENTION MECHANISM

In general, HPLC is a dynamic adsorption process. Analyte molecules, while moving through the porous packing bead, tend to interact with the surface adsorption sites. Depending on the HPLC mode, the different types of the adsorption forces may be included in the retention process:

- Hydrophobic interactions are the main ones in reversed-phase separations.
- Dipole-dipole (polar) interactions are dominated in normal phase mode.
- Ionic interactions are responsible for the retention in ion-exchange chromatography.

All these interactions are competitive. Analyte molecules are competing with the eluent molecules for the adsorption sites. So, the stronger analyte molecules interact with the surface, and the weaker the eluent interaction, the longer analyte will be retained on the surface.

Size-exclusion chromatography (SEC) is a special case. It is the separation of the mixture by the molecular size of its components. In this mode any positive surface interactions should be avoided (eluent molecules should have much stronger interaction with the surface than analyte molecules). Basic principle of SEC separation is that the bigger the molecule, the less possibility for her to penetrate into the adsorbent pore space, so, the bigger the molecule the less it will be retained.

INSTRUMENTATION

HPLC instrumentation includes a pump, injector, column, detector and recorder or data system, connected as shown in the Fig. 13 below. The heart of the system is the column where separation occurs. Since the stationary phase is composed of micrometer size porous particles, a high pressure pump is required to move the mobile phase through the column. The chromatographic process begins by injecting the solute onto the top of the column. Separation of components occurs as the analytes and mobile phase are pumped through the column. Eventually, each component elutes from the column as a narrow band (or peak) on the recorder.

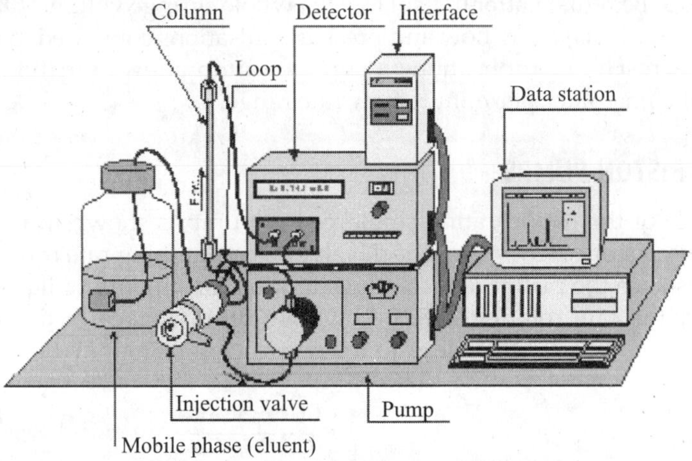

Fig. 13. Functional schematic of a modern HPLC instrument.

Detection of the eluting components is important, and this can be either selective or universal, depending upon the detector used. The response of the detector to each component is displayed on a chart recorder or computer screen and is known as a chromatogram. To collect, store and analyze the data, computers, integrators and other data processing equipment are frequently used.

CONSTANT PRESSURE PUMPS

The primary advantages of constant pressure pumps are simplicity and freedom from pulsations, resulting in smooth baselines. The most simple of these are usually inexpensive, easy to operate, and easy to maintain. However, they suffer from several disadvantages. Flow rate must be monitored carefully and constantly, especially when performing either qualitative or quantitative analysis. Flow rates can change if the solvent viscosity changes due either to a temperature or composition change. Changes in flow rate can influence both qualitative and quantitative analysis. In qualitative analysis, component identification is primarily based on matching retention volumes. If these change as a result of flow rate changes, especially during the analysis of complex mixtures, component location and identification becomes difficult. In quantitative analysis, the most common detectors (UV and RI), are concentration sensitive. Changes in flow rate affect the time that a chromatographic band passes through the detector cell, which in turn show up as changes in peak area, which is used for quantitation.

CONSTANT FLOW PUMPS

Constant-flow systems are generally of two basic types: reciprocating piston and positive displacement (syringe) pumps. The basic advantages of both systems are their ability to repeat elution volume and peak area, regardless of viscosity changes or column blockage, up to the pressure limit of the pump.

Reciprocating piston pump can maintain a liquid flow for indefinitely long time, while a syringe pump has to be refilled after it displaces the whole syringe volume. On the other hand, a syringe pump does not have any flow and pressure pulsations compared to the reciprocating pump. For the micro-HPLC applications a syringe pump allows for the maintaining of a constant flow at the microliter per minute flow rate range.

RECIPROCATING PISTON PUMPS

The basic principle of the reciprocating single piston pump is shown in Fig. 14. The piston expels liquid through a one-way valve (check valve). The pumping rate is usually adjusted by controlling the distance the piston retracts, thus limiting the amount of liquid pushed out by each stroke, or by the cam rotating speed. The main disadvantage of this type of pump is sinusoidal pressure pulsations which lead to the necessity of using pulse dampers.

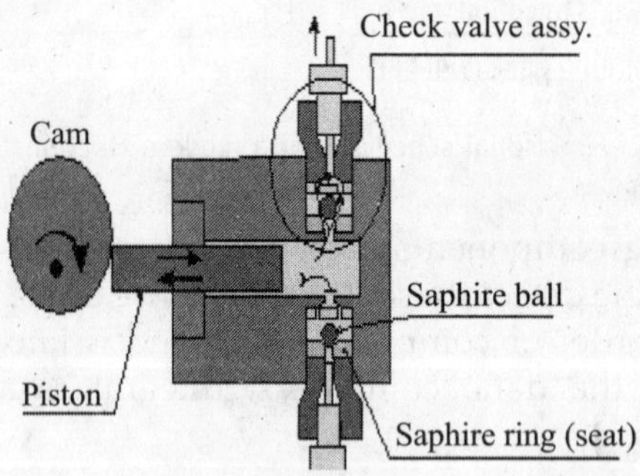

Fig. 14. Schematic of the reciprocating single piston pump.

DUAL PISTON PUMPS

A more efficient way to provide a constant and almost pulse free flow is the use of dual-headed reciprocating pumps (Fig. 15). Both pump chambers are driven by the same motor through a common eccentric cam; this common drive allows one piston to pump while the other is refilling.

As a result, the two flow-profiles overlap each other significantly reducing the pulsation downstream of the pump; this is visualized below. Since the acceleration/deceleration profile is somewhat non-linear, the more efficient types of these pumps use eccentricity-shaped cams to obtain the best overlapping of the pressure curves and to obtain smooth flow.

The advantages of this pump are the unlimited solvent reservoir allowing long-term unattended use and quick changeover and clean out capability. However, unless special care has been exercised in manufacture, these pumps may have several disadvantages. There is a

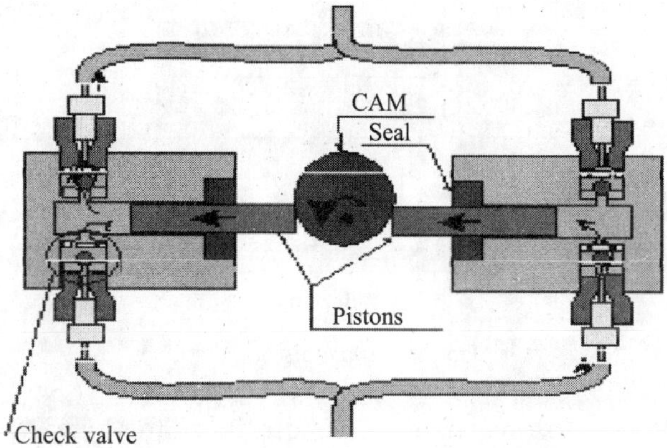

Fig. 15. A dual-head reciprocating pumps.

tendency for the incompletely compensated pulsations to be observable at high refractive index detector sensitivities, especially at low flow rates where piston cycles are widely spread. Furthermore, since each head has two check valves, pump reliability depends on the cleanliness of the mobile phase and continued sealing capability of four check valves on each cycle, with cycles normally occurring several times per minute. Recent improvements to this popular pumping system include:

- A computer-designed camshaft is used to achieve maximum overlap of pump strokes, resulting in virtually undetectable pulsation or ripple.
- Staggered inlet/outlet lines are employed to allow complete flushing when liquids are changed or if air is inadvertently drawn through the pump.
- Small-volume check valves are used to allow the pumps to function reliably at flow rates as low as 0.001 mL/min. This has the added benefit of providing excellent gradient reproducibility especially when programs start from extremely low concentrations.
- There are fewer moving parts, with all maintenance-requiring components (pump seals, check valves) readily accessible from the front of the instrument.
- A wide flow rate range (0.01 to 10 ml/min) is provided without gear change.

Check valves on the reciprocating pump are the weakest part. It may be easily contaminated or clogged which leads to the pump malfunction. Most of the recent HPLC instruments use improved dual piston pumps which have three or even two check valves.

The first piston, called low pressure, is sucks the liquid from the reservoir while the second (high pressure piston) is supplying the eluent to the system. Then the first piston refills the second piston very fast, during 1/100 of the whole pump cycle. This scheme allows the use of only 3 check valve, one of which is working under low pressure. There are no cavitations effects. Because the piston volumes are small (~100 μl), pressure pulsations are small and sharp and easy to damp. The schematic of this pump is shown in the Fig. 16.

Another type of dual piston pump uses only two check valves, but piston volumes are different. While the smaller piston dispenses an eluent in the HPLC system, the bigger piston is

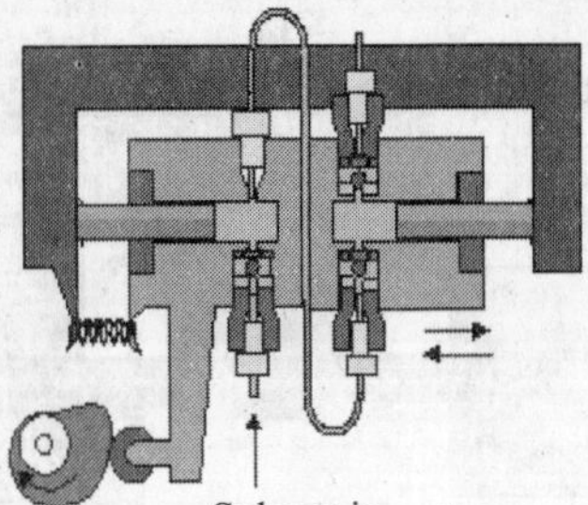

Fig. 16. Dual piston pump.

sucking an eluent. When pistons change their direction, the bigger piston simultaneously refills the smaller chamber and dispenses eluent into the system. This set-up allows only two check valves for the dual piston pump.

GRADIENT ELUTION

Blending and Mixing Mobile Phase Components

In many of the modes of modern liquid chromatography, it becomes readily apparent that the most practical system for the delivery of the mobile phase is that which can combine several liquids in different proportions at the command of the operator. This blending capability greatly speeds the process of selecting the optimum eluent mixture required for isocratic analysis and becomes absolutely necessary when programming the mobile phase composition ("gradient elution"). These techniques have been discussed earlier from the point of view of chromatography; now we shall treat them from the point of view of instrumentation. The two primary methods of blending the mobile phase components are known as high pressure mixing and low pressure mixing.

HIGH-PRESSURE MIXING

In high-pressure mixing systems, individual high pressure pumps are used to provide each liquid. The outlet of each pump is either connected to a mixing connector (usually referred to as a "T" since there are normally two inlet lines and one outlet line) or to a mixing chamber. In these, the two flows are blended en route to the injector and column. In other words, mixing is accomplished on the high-pressure side of the pumps. Most modern liquid chromatography systems utilizing this principle include pump-drive electronics which control the pumping speed. Blending is achieved through adjustment of a master controller. The flow rate which always represents the combined (total) output of both pumps is selected and then the dividing control adjusted to provide the desired ratio of the two components. These controls are commonly calibrated to read out the percentage of strong component in the mixture. The stronger component is considered Component A; thus, the mixing control is labeled "%A in A+B". When a given %A is dialed in, the mixing electronics automatically divides the output signal between the two pumps.

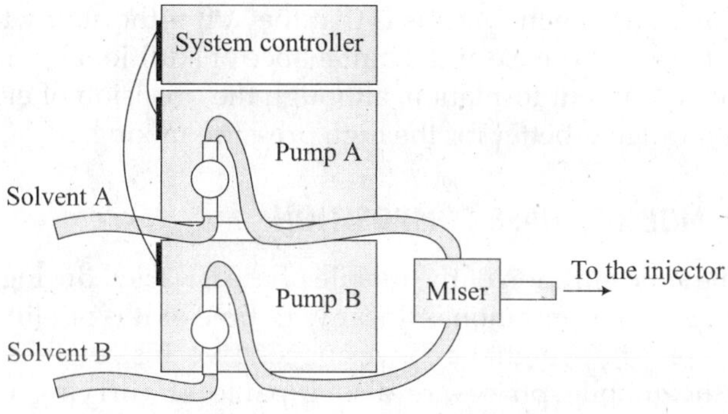

Fig. 17. Mixing of two mobile-phase components.

Pump A (strong component) receives the percentage of the signal called for in the control indicator, and Pump B (weaker component) receives the remainder of the signal, so that the total output of both pumps always equals the pre-selected flow rate. Fig. 17 shows the mixing of two mobile-phase components. It is also possible to prepare ternary mobile phases by utilizing three separate pumps. Each of these pumps will then deliver a separate solvent into the mixing chamber.

Among the advantages of high-pressure mixing is the ability to achieve precise control and repeatability of mobile phase mixtures to within 0.1% levels (only within the medium eluent composition range), rapid response to changes in concentration and the ability to utilize each pump separately if separate isocratic pumping systems are required. However, there are some disadvantages for this type of gradient formation:

- Even special mixing chambers do not provide complete mixing;
- Many HPLC solvents being mixed will change the total volume, sometimes up to 20%;
- Mixing accuracy within the 0-10% range for component A or B is usually poor;
- High pressure mixing needs two or sometimes three high pressure pumps, which increase the cost of the whole system.

Low-Pressure Mixing

In low-pressure systems, mixing is accomplished prior to the pump, at its low-pressure side and the overall flow rate is controlled by a single pump. Blending of the mobile-phase components is accomplished with controls which are calibrated in somewhat the same manner as those used for high-pressure mixers — that is, controlling the percentage of the strong component — though some systems allow the mixing of three liquids. In general, two types of blending systems are used.

In the first case, proportioning valves, normally solenoid operated, are used to deliver the individual liquids. The controller simply divides the signal according to the percentage of each component and each valve is opened for the proper period of time. Usually the valves deliver the individual liquids into a mixing chamber which then feeds the blend to the pump. In some systems, the mixing chamber may be missing and the valves feed the mobile-phase components through a mixing connector directly to the high-pressure pump.

The primary advantage of such systems is that they offer the ability to blend two or more liquids without the increased expense and maintenance of additional pump(s).This is now the most popular scheme of gradient formation, although the precision of eluent composition for medium concentration range is better for the high pressure mixing.

PROGRAMMING THE MOBILE PHASE COMPOSITION

In gradient elution, the concentration of the mobile phase is varied during a run. There are two possibilities: the change in concentration is linear with time or it is not linear. In the latter case, the plot describing the program might be convex or concave. Instruments which are capable of mixing liquids for the mobile phase are also capable of carrying out gradient elution automatically.

Each step of the linear program is described by four variables the initial and final concentration (%) of Component A in the mixed mobile phase, the rate of program (%/min) and the length (time) of the program. Three of these four fully describe the program; in practice, the instruments permit the setting of the two concentrations and either the time or the rate of the program for each step. The electronic control system will then automatically set either the rate or the time of the program. The situation with curved gradient profiles depends on the particular instrument. Some less convenient programmers require step-by-step entry of multiple percentage vs. time plateaus to approach curved gradient profiles while in more sophisticated systems, one can select from a number of programmed profiles.

In the actual instruments, one has to set the initial and final concentrations, the time of the program and select which program profile is desire. This is done by certain keys corresponding to the various exponent values. Naturally, in sophisticated instruments where various programs can be combined, curved programs can also be included together with linear programs and isocratic periods. The most important consideration in gradient elution is that the whole program – a simple as well as a complex multistage program – can be exactly reproduced upon command since inaccurate gradient repetition can contribute to significant qualitative and quantitative errors.

PROBLEMS IN PUMPING LIQUIDS

The common problems encountered in pumping mobile phases are solvent degassing, corrosion, and compressibility. Compressibility has already been discussed and since its influence is minor, we shall not deal with it further. The two other problems, however, need further analysis.

DEGASSING

As previously noted, when the mobile phase contains excessive gas which remains dissolved at the pressure produced by the column, the gas may come out of the solution at the column exit or in the detector, resulting in sharp spikes. Spikes are created by microscopic bubbles which change the nature of the flowing stream making it heterogeneous while drift may occur as these microscopic bubbles gradually collect and combine in the detector cell. The main problem is oxygen (from the air) dissolving in polar solvents, particularly water. If needed, degassing may be accomplished by one of the following methods or their combination:

- Subjecting the liquid to vacuum
- Heating the liquid until boiling occurs.
- Placing the container of liquid in an ultrasonic bath or inserting an ultrasonic probe in it.
- Bubbling a fine stream of helium through the liquid; helium has the unique ability to spurge other gases out of solutions.

The best results were found by applying vacuum to each solvent for about 5 min, with subsequent helium purging and storing under a helium atmosphere.

CORROSION

Most commercially available instruments are constructed of Type 316 stainless steel. This material represents the best compromise of corrosion resistance, workability, strength, and cost. It is inert to all bases, all organic liquids and most non-halogenated acids at pH values above 2.0. Type 316 stainless steel is, however, extremely vulnerable to contact with halogenated acids, like HCl, even at concentrations as low as 0.01 normal. When attempting to convert a thin-layer or (glass) column chromatographic method to HPLC, substitute another acid such as nitric, boric, or acetic for halogens. Ion-exchange systems should be monitored carefully for discolored eluent especially when small particle packings that exhibit high back pressure (2500+ psi) are used, since several researchers have observed highly accelerated corrosive reactions at these pressures with some buffer systems.

INJECTORS

Injectors for liquid chromatographic systems should provide the possibility of injecting the liquid sample within the range of 0.1-100ml of volume with high reproducibility and under high pressure (up to the 4000 psi). They should also produce minimum band broadening and minimize possible flow disturbances. Generally, the most useful and widely used sampling device for modern LC is the micro-sampling injector valve (Fig. 18).

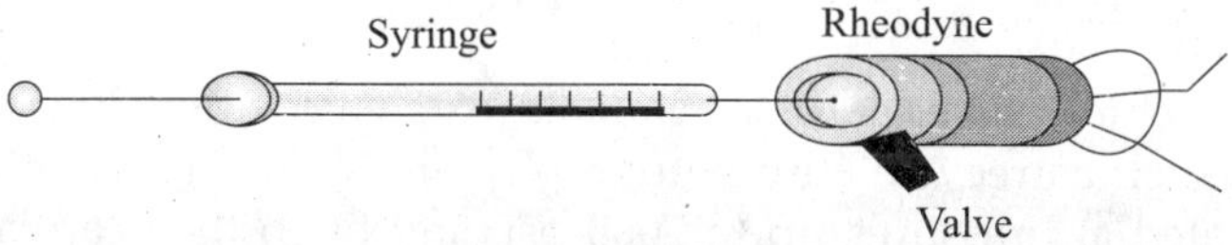

Fig. 18. Rheodyne injector.

Because of their superior characteristics valves are now used almost to the exclusion of syringe injection. With these sampling valves, samples can be introduced reproducibly into pressurized columns without significant interruption of flow, even at elevated temperatures. Fig. 19 shows schematic drawings of a six-port Rheodyne valve in which the sample fills an external loop.

Compared to shorter, wider i.d. sample loops, long, narrow loops are preferred when large sample volumes are required, because of lesser band-broadening effects. Alternatively, a specially designed syringe may be used to inject a small volume (e.g., <10, ul) into the loop

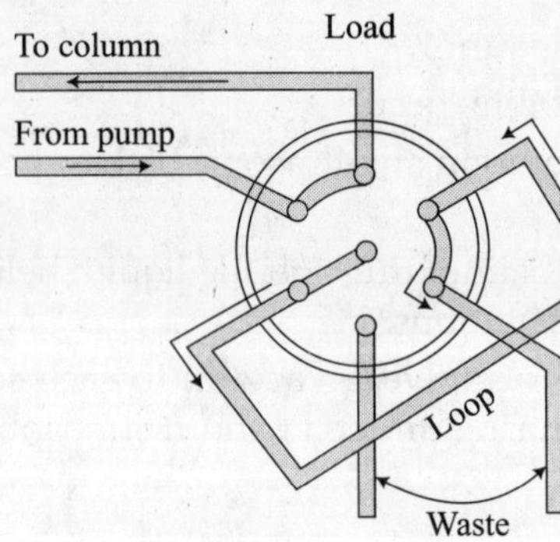

Fig. 19. Six-port Rheodyne valve.

when required, although in this case the precision in the sample introduction is dependent on the precision of syringe delivery.

A clockwise rotation of the valve rotor places the sample-filled loop into the mobile-phase stream with subsequent injection of the sample onto the top of the column through a low-volume, cleanly swept channel. Other valve types (e.g., Siemans and Valco) use an internal sample cavity consisting of an annular groove on a sliding rod that is thrust into the flowing stream. The minimum injection volume which can be made with the valve-type injectors is 60 nl.

Valve injection allows the rapid, reproducible, and essentially operator-independent delivery of a wide range of sample volumes (e.g., from 60 nl up to several milliliters) at pressures up to 7000 psi with less than 0.2% error. High-performance valves provide extra column band-broadening characteristics comparable or superior to that of syringe injection. Manually operated valves are only moderately expensive, and automated versions can be obtained at somewhat higher cost. A minor disadvantage of most sample valves is that the sample loop must be changed to obtain various sample volumes, but this can often be achieved in a few minutes. Another advantage of sampling valves is that they can be located within a temperature-controlled oven for use with samples that require handling at elevated temperatures (0° – 150°C).

Low-volume switching valves are also available (e.g., Valco, Rheodyne, Siemans) for use in special techniques such as recycle chromatography and column switching. Some of these valves can be operated at pressures up to 7000 psi, and often they can be used at elevated temperatures. The more common valves can be obtained in 3-, 4-, 6-, 8-, or 10-port configurations, for use in either the manual or automated mode.

AUTOMATIC INJECTORS

With commercially available automatic sampling devices, large numbers of samples can be routinely analyzed by LC without operator intervention. Such equipment is popular for the analysis of routine samples (e.g., quality control of drugs), particularly when coupled with automatic data-handling systems. Automatic injectors are indispensable in unattended searching (e.g., overnight) for chromatographic parameters such as solvent selectivity, flow-rate, and temperature optimization.

Most of the auto-samplers have a piston metering syringe type pump to suck the pre-established sample volume into a line and than transfer it to the relatively large loop (~100 ml) in a standard six-port valve. The simplest auto-samplers utilize the special vials with presuarization caps. A special plunger with a needle, push the cap down to the vial and displace the sample through the needle into the valve loop. Most of the auto-samplers are microprocessor controlled and can serve as a master controller for the whole instrument (Fig. 20).

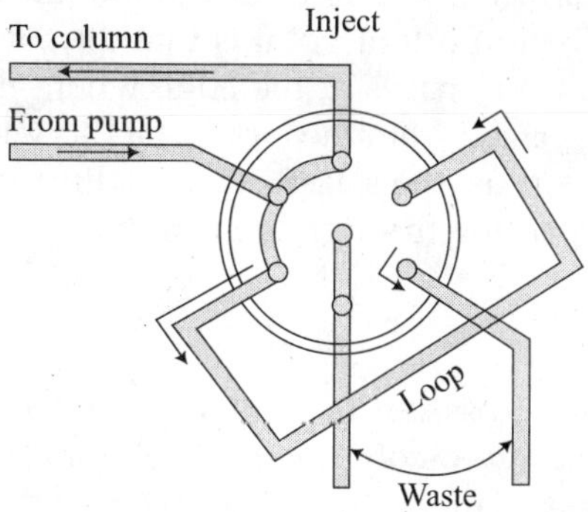

Fig. 20. Injection of sample.

COLUMN HARDWARE

Today, column material is normally Type 316 stainless steel, once again chosen because it offers the best compromise of cost, workability, and corrosion resistance. Most commercial columns available today have internal diameters of either 2.6-3 mm or 4.6-5 mm (Fig. 21). Most leading manufacturers obtain a great amount of interchangeability between column sizes without excessive fitting replacement by supplying columns with a variety of internal diameters, but a uniform ¼ in. outside diameter. Naturally, preparative columns have larger diameters: usual O.D. values are 3/8 - 5/8 in., corresponding to internal diameters of 6.4-12.7 mm, respectively.

Most modern liquid chromatography plumbing is performed with compression fittings such as Swagelok, Parker-Hannifin, Gyrolok, etc. These offer a great deal of flexibility, can be connected and disconnected frequently without loss of sealing power, are commonly available,

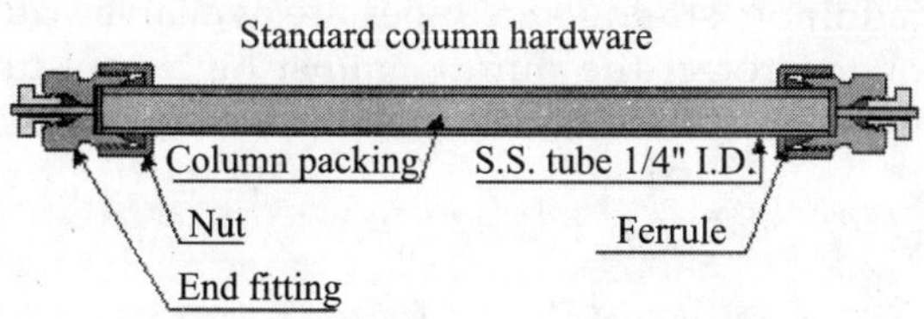

Fig. 21. Standard HPLC column.

and may be used without special seating and flanging tools allowing users to install, maintain, modify, and improvise equipment. Great care should be taken in the choice of end fittings, connectors, and unions however, since inclusion of rapid diameter changes corners and other unswept areas in a system can destroy a good separation. It is best to obtain the above parts from a reputable liquid chromatograph manufacturer since most of these companies offer fittings which are specifically made for LC.

Many reducing unions and column end fittings incorporate special devices designed to hold packings in place and/or prevent their flowing into the detector or detector tubing. The most frequently used of these consist of fine porous stainless steel fritted filter discs, usually with an average opening of 1 mm. These may be forced into the ends of columns or may be in the connector or end fitting. The important point is that the frit disc must be tightly fitted to avoid occurrence of large-diameter spaces at the edge where the disc contacts the wall. An additional benefit of the frit is derived from the fact that it fills what otherwise might be a large empty area in conversion zones thus; the sample cannot diffuse, cross mix, or undergo other efficiency-killing effects when passing through these areas.

CONNECTOR TUBING

Connector tubing must be of low cross section to hold separated components in narrow bands as they are transferred from column-to-column, from column-to-detector, or from detector-to-detector when mounted in series. Tubing having an internal diameter in excess of 0.634 mm (0.025 in) can allow peak mixing and band broadening and it is good practice to use tubing no larger than 0.12 mm (0.005 in) in areas where sample is to be transported.

LINE FILTERS

A line filter should be used between the pump and the sample injector to prevent particulates from clogging the column inlet. Porous stainless steel filters having a porosity of about 2 μm are typically used in commercial instruments; however, 0.5-μm porosity filters (e.g., Alltech Associates) are desirable with columns of less than 10-μm particles. To facilitate solvent changeover, the volume of the line filters should be small.

PRESSURE MEASUREMENT

Pressure monitors are used in the LC equipment as diagnostic tools to optimize separation and to indicate system problems (e.g., plugging or leaks). Diaphragm or Bourdon-type gauges are simple, inexpensive, and generally robust. On the other hand, strain-gauge pressure transducers are more precise and have a smaller internal volume, which facilitates the changing of solvents. In addition, straingauge types are available with high- or low-pressure alarms or cut-off circuits, to protect the pump against high-pressure overload by column plugging or the instrument against fire due to solvent leaks (low-pressure alarm).

COLUMN THERMOSTATS

It often is advantageous to run ion-exchange, size-exclusion and reverse-phase columns at higher temperatures, and to precisely control the temperature of liquid-liquid columns.

Therefore, column thermostats are a desirable feature in modern LC instruments. Generally, temperature variation within the LC column should be held within +0.2°C. Maintaining a constant temperature is especially important in quantitative analysis, since changes in temperature can seriously affect peak-size measurement. It is sometimes important to be able to work at higher temperatures for the size-exclusion chromatography of some synthetic polymers because of solubility problems, but precise temperature control is not so important in this case. High-velocity circulating air baths are most convenient in LC (as in GC). These devices usually consist of high-velocity air blowers plus electronically controlled thermostats, with configurations similar to those used in gas chromatographs. Alternatively, LC columns can be jacketed and the temperature controlled by contact heaters or by circulating fluid from a constant-temperature bath. This latter approach is practical for routine analyses, but is less convenient when columns must be changed frequently.

FRACTION COLLECTORS

Since separations are accomplished in minutes, fraction collectors are usually not needed in modern LC. Manual collection is normally used, and many commercial instruments have convenient sample-collection ports on the outlet of the detector. Fraction collectors are sometimes used in preparative chromatography and in conventional size-exclusion chromatography with larger-diameter columns, because separations are usually much slower.

PRESENTATION OF RESULTS

Chromatographic data are presented almost universally in the form of a chromatogram using chart recorder. For quantitative analysis and greater precision in retention time measurements, digital integrators, computing integrators, and computing systems may be employed. Their specification is essentially fast response time, wide linear dynamic range, and capable of accepting both narrow (fast eluting) and wide (slow-eluting) peaks. For maximum convenience, chart recorders should be provided with a wide range of chart speeds, as some chromatographic methods take but a few minutes to complete whereas others take hours.

In quantitative work, computing integrators and dedicated computers are becoming popular, for once the detector response data and other basic information have been fed into the system, and the analytical results are calculated and printed in report form by the computer. In laboratories where numerous repetitive samples are analyzed these systems can offer a significant reduction in time and effort made by the operator.

APPLICATIONS

It is quite remarkable to what extent the technique has been successfully applied. In fact, with laboratories such as those associated with the pharmaceutical industry, modern instrumental HPLC has brought about a complete change in the methods of routine quality control analysis, in most instances with a distinct reduction in analysis time and an increase in the analytical precision.

New and experienced chromatographers alike will be aware how often, after expending considerable efforts to develop a separation method, it is found that a similar procedure has already been achieved elsewhere. Scientific communications have proliferated in recent years, making, perhaps, more difficult the task of effectively covering all possible sources of information. In present scenario, HPLC has wide applications. Some of important ones are as following:

IN INORGANIC CHEMISTRY

Although the currently available commercial equipment is rather expensive, the HPLC and its various techniques have already shown to have wide applications in organic chemistry and in fact have been the doamin of organic chemistry for a long time. In the last few years it has actually been recognized that HPLC can equally be utilized for the separation of inorganic compounds. The limitation of HPLC in inorganic chemistry is that only those compounds are amenable to HPLC investigations which are soluble in solvents being compatible with HPLC columns.

There are two classes, of soluble inorganic compounds. First, there are ions (anions and cations) which are usually soluble in aqueous solutions. Second, there are molecular compounds which are more readily soluble in organic solvents. Thus, two different branches of HPLC have come into use in inorganic chemistry. Ion chromatography (IC) is used for the separation of ions in aqueous solution, while non dissociated molecular compounds are separated in reversed phase systems using bonded alkyl silica phases and organic eluting agents such as methanol or acetonitrile (Reversed phase liquid chromatog-raphy).

ANALYSIS OF NATURAL AND SYNTHETIC PHARMACEUTICAL DRUGS

HPLC is applied at several different levels in pharmacy and pharmacology. The important uses of HPLC in these areas are:

(a) In production development and product control.

(b) In isolation of pharmaceuticals.

(c) In raw material control.

(d) In clinical analysis, this closely linked to the study of pharmaceuticals.

The retention properties of 140 synthetic drugs or pharmaceuticals on a 10 μm C-18 reversed phase with various methanol water mixtures under isocratic conditions have been described. Some of the pharmaceuticals are acetyls, alicyclic acids, ascorbic acid, barbital, saccharin, caffeine, chlorothiozole, diazepam, sulphacarbamide, sulphacetamide, sulphadiazine, sulphaguanidinde, sulphathiazole, sulphanilamie, phenyl butazone, tolazamide and vanillin.

There is no small group of solvents that is suitable for the HPLC separation of all antibiotics. A number of antibiotics have actually been analyzed and separated by HPLC. These include amoxicillin, ampicillin, anisomysin, chloramphenicols, cycloheximide, erythromycin, levorin, neomycin, penicillin, polymyxin, streptomycin, trichloromycin and vancomycin.

Some other examples of HPLC analysis of various pharmaceuticals are the separation of antipyrine and benzocaine in ear drops, separation of benzoic acid and salicylic acid in ointments, separation of epinephrine in pharmaceutical preparations on a cation exchange column and analysis of insulin, glucagons and somatosatatin etc.

Other pharmaceuticals of importance separated and identified by HPLC are lignans, anthraquinone derivatives, saponins and sapogenins, arbutin drugs, biogenic amines, deterpene esters, curcubitacins, gossypol, γ-benzopyrone derivatives, khellin and visnagin, terpenes, etc. A selective treatment of all the pharmaceuticals is not possible here, because of large number of pharmaceuticals commercially available today.

PSYCHOTROPIC DRUGS IN BODY FLUIDS

The HPLC techni-que has also been used for the separation and identification of psychotropic drugs such as benzodiazepine, antidepressants, neuroleptics, phenothiazine, buty-rophenones etc. in patient's body fluids (such as urine, plasma, cere-brospinal fluids, mother's milk and saliva). Since most of the psychotropic drugs are almost com-pletely metabolized, a urinary assay allows no conclusion regarding the amount of the effective mother substance in the whole body. Hence es-timations in urine are rarely made in connection with psychotropic drugs. For correlation with therapeutic outcome, however, plasma concentrations are es-timated.

The various benzodiazepines have weakly acidic but markedly basic properties. In order to extract them with an organic solvent, the pH value of the plasma is altered to about 8 and so a concentrated buffer is generally added to the plasma before extraction. All common organic solvents, such as benzene, diethyl ether and chloroform can be used for extraction of benzodiazepines.

Antidepressants are always extracted under basic conditions. Silica column or C-18 reversed phase columns are used for their separation. Extraction has been carried out with a number of solvents such as light petroleum, hexane, diethyl ether, 1% isoamyl alcohol, 2% amyl alcohol in hexane, etc.

Phenothiazine and butyrophenones are usually called neuroleptic drugs. A number of neuroleptics have been extracted on silica and C-18 reversed phase column and many of the phenothiazines can be well identified by UV detection.

ANALYSIS AMINO ACIDS AND PROTEIN

HPLC has widely been used in the analysis and separation of a number of amino acids as well as proteins. Before HPLC, the separation was carried out on a cation exchanger and the detection was performed in the elute after separation by means of ninhydrin. Refractometric detection or UV detection at 205 nm is usually used in case of HPLC.

Proteins differ from each other in their molecular size, shape, charge and their hydrophobic properties. Chromatographic separation techniques have been developed which separate according to size, charge and hydrophobicity and hence these methods are used with success for the separation and purification of proteins. The chromatography is possible on C-18 reversed phase. The best separation is obtained at pH 2.

The nutritional need of proteins and amino acids essential to the human diet is well established. The food industry has developed food products which have amino acid compositions adequate for nutritional requirements. This involves blending proteins from different sources or fortifying foods directly with essential amino acids. The conventional amino acid analyzer, based on ion-exchange chromatography followed by derivatization with ninydrin, has been used for the determination of amino acids in a variety of proteins hydrolysates. In addition, a number of HPLC procedures have been developed for the determination of amino acids. The separation of amino acid derivatives with HPLC is generally carried out on reversed phase columns with acetonitrite aqueous buffer solvent systems. The aqueous buffer is used to minimize sample ionization and thus improving the peak shape. Both variable wavelength UV detectors and fluorescence detectors have been used for amino acid analysis, depending upon the amino acid derivative used.

SEPARATION OF COAL AND OIL PRODUCTS

The time consuming LC methods have now been replaced almost completely by modern HPLC method. The complexity of distillation residues and similar products is so large that it is almost impossible to expect the resolution of individual substances even by making use of high efficiency HPLC techniques. As a result, detailed discussion of such separations is not possible in the present Chapter.

CARBOHYDRATES

Carbohydrates are present in foods at nature or supplemented levels. Since carbohydrates do not absorb strongly in the UV spectral region, the refractive index detector is usually employed in their separation and analysis. A number of variable wavelength UV detectors at 192 nm have been employed but with no significant increase in sensitivity. The polar bonded phases, e.g., LiChrosorb -NH_2 have been found to be very effective for the separation of carbohydrates with acetonitrile–water mixtures as mobile phase. This system has actually been used in the determination of sugar content of Soyabeen extracts, dairy products and molasses. Increasing the water content of the mobile phase increases the rate of elution and this system has been used for the separation of higher car-bohydrates found in corn syrups. Elution has been found in order of increasing glucose chain length. Simple carbohydrates have also been analyzed by using ion exchange columns loaded with Ca^{2+} or other metal ions and using water as the eluting solvent. The components elute in order of decreasing molecular size.

DETERMINATION OF PRESERVATIVES AND ANTIOXIDANTS

Preservatives and antioxidants are the additives which prevent microbial or chemical deterioration of food products. Most preservatives can be determined directly using HPLC. For example, sorbic acid in beverages can be determined using a reversed phase column with an isocratic solvent system. Gradient elution on a reversed phase column is required for resolving food preservatives including p-hydroxy benzoic acid esters (PHB-esters). Most widely used food preservatives are sorbic acid, benzoic acid, PHB-methyl ester, PHB-ethyl ester, PHB-propyl ester, biphenylol, biphenyl etc.

A variety of chromatographic procedures have also been developed for the separation of antioxidants in foods, but HPLC has only been used in the quantitative determination of nine most commonly used antioxidants simultaneously without derivatization. A method for analysis of antioxidants in oils, lards and shortenings requires a liquid-liquid extraction from hexane oil into acetonitrite, followed by sample extract concentration before separation on a reversed phase column.

ANALYSIS OF VITAMINS

The HPLC has widely been used for analyzing both water soluble as well as fat soluble vitamins in a variety of food products, fortified foods and animal feeds. Advantages of HPLC include direct analysis without derivatization, multiple component determinations and resolution of isomers. Separation of fat soluble vitamins has been carried out on both silica and bonded phase columns using gradient elution. The reversed phase column has the advantage of faster equilibrium and better long term retention behavior.

SEPARATION OF STEROIDS AND LIPIDS

HPLC has successfully been applied to the separation of all classes of steroidal hormones. Thermally labile steroids, such as corticosteroids and non volatile ones such as steroid conjugates can however, be analyzed relatively easily.

The development of improved techniques for the analysis of lipids in food products has become increasingly important with the current emphasis on dietary requirements and nutritional labeling. Column chromatography was formerly used for the separation of various non polar lipids, glycolipids and phospholipids. These individual fractions were then spotted on TLC plates for analysis. HPLC techniques have been developed in order to give greater selectivity and improved quantitation. For HPLC this means that interaction of the non-polar chains with a polar stationary phase (e.g. silica) can be exploited to get the separation.

FORENSIC CHEMISTRY

The forensic chemistry in concerned with the processing of disputed questions of fact in criminal, civil and insurance law, which can be clarified by the application of chemical methods and knowledge. HPLC is widely used for the detection of a large number of poisons, intoxicants and addictive drugs etc. HPLC is also of great value in the field of chemical criminology, which is concerned mainly with residues and traced of the non-living milieu.

REFERENCES AND SUGGESTED READINGS

Brown, P.R. (1972) *High Pressure Liquid Chromatography: Biochemical and Biomedical Applications*, Academic Press, New York.

Dunges, W., Naundorf, G. and Seiler, N. (1974) *J. Chromatogr. Sci.*, 12: 655.

Engelhardt, H. (1986) *Practices to High Performance Liquid Chromatography*, Springer-Verlag, Berlin.

Hamilton, R.J. and Sewell (1982) *Introduction to High Performance Liquid Chromatography* (2nd edn.) Chapman Hall, London.

Huber, J.K.F. (1978) *Instrumentation for High Performance Liquid Chromatography*, Elsevier, Amsterdam.

Junk, G.A., Richard, J.J., Grieser, M.D., Witiak, D., Arguello, M.D., Vick, R., Svec, H.J., Fritz, J.S. and Calder, G.V. (1974) *J. Chromatogr.*, 99: 745.

Kirkland, J.J. (1973) *J. Chromatogr.*, 83: 149.

Kirkland, J.J. and Synder, L.R. (1979) *Introduction to Modern Liquid Chromatography* (2nd edn.), John Wiley, New York.

Knox, J.H. (1982) *High Performance Liquid Chromatography*, Edinburgh University Press, Edinburgh.

Liteanu,C. and Gocan, S. (1974) *Gradient Liquid Chromatography*, Ellis Harwood, Chichester.

Parris, N.A. (1980) *Instrumental Liquid Chromatography* (3rd edn.), Elsevier Scientific Publisher Company, Amsterdam, The Netherlands.

Rhys Williams, A.T. (1980) *Fluorescence Detector in Liquid Chromatography*, Perkin Elmer, Beaconsfield.

Schmidt, J.A. and Kirkland, J.J. (1971) *Modern Practice in Liquid Chromatography*, Wiley-Interscience, New York.

Scott, R.P.W. (1986) *Liquid Chromatography Detectors*, Elsevier, Amsterdam.

Scott, R.P.W. (1976) *Contemporary Liquid Chromatography*, Wiley, New York.

Simpson, C.F. (1982) *Techniques of Liquid Chromatography*, Wiley, Chichester.

Synder, L.R. (1970) *J. Chromatogr. Sci.*, 8: 692.

Synder, L.R. (1974) *J. Chromatogr.*, 92: 223.

Synder, L.R. and Kirkland, J.J. (1974) *Introduction to Modern Liquid Chromatography*, Wiley-Interscience, New York.

Synder, L.R. (1978) *Principals of Adsorption Chromatography*, Marcel-Dekker, New York.

REVIEW QUESTIONS

Long Questions

1. What is HPLC? How is it superior to other techniques?
2. What types of detectors used in HPLC? Discuss briefly the UV detector and RI detector.
3. Discuss briefly the sample injection system used in HPLC.
4. What are the detectors and their basic requirements?
5. Describe various types of pumps used in HPLC?
6. Discuss the method used in HPLC.
7. What is the effect of temperature and pH in HPLC? What are the disadvantages of an increase in temperature?
8. What is HPLC? Discuss briefly some important applications of HPLC.
9. Write short notes on:
   (a) Mobile phases (b) Stationary phases
   (c) Noise and drift

Multiple Choice Questions

1. In chromatography, the process of development is called
   (a) Elution (b) Illusion
   (c) Gradient elution (d) All the above
2. The fixed phase and moving phase of chromatography is alternatively known as
   (a) Mobile phase and stable phase
   (b) Moving phase and stable phase
   (c) Stationary phase and mobile phase
   (d) Stationary phase and moving phase, respectively

3. An in crease in the number of theoretical plates improves
   (a) Efficiency (b) Resolution
   (c) Both efficiency and resolution (d) None
4. Chromatographic development may be carried out by the following techniques
   (a) Elution development and gradient elution
   (b) Frontal analysis and elution development
   (c) Gradient elution and Displacement development
   (d) Elution development, Frontal analysis, Gradient elution and Displacement development
5. As R_f increases, the retention of solutes
   (a) Increases
   (b) Decreases
   (c) R_f and retention of solutes are not correlative
   (d) Depends on type of phase
6. The basis of adsorption column chromatography is
   (a) Distribution of a solute between two liquids in a column
   (b) Adsorption or partition on thin sheets
   (c) Distribution of a solute between a solid and liquid phase on a column
   (d) Size of solute
7. The solubility is greater affected by Hydrogen bonds
   (a) Dipole interactions
   (b) Charge transfer process between solute and solvent molecules
   (c) All the above
8. What is the general adsorption sequence of functional groups according to increasing polarity?
   (a) Acids and base, aldehydes, amino groups, saturated hydrocarbons
   (b) Amino groups, aldehydes, acids and base, saturated hydrocarbons
   (c) Saturated hydrocarbons, amino groups, aldehydes, acids and base
   (d) Aldehydes, saturated hydrocarbons, amino groups, acids and base
9. How to degas chromatographic solvents?
   (a) Subject the solvent to vacuum for 5-10 mins. to remove the gases
   (b) Subject the solvent to ultrasonics for 10-15 mins. to remove the gases
   (c) Sparse the solvent with a gas that has a very low solubility compared to the oxygen and nitrogen from the atmosphere
   (d) All the above.

Answers

1. (a)	2. (c)	3. (c)	4. (d)	5. (b)	6. (c)
7. (d)	8. (a)	9. (d)	10. (a)		

PART–18

Biophysical Methods in Biology

CHAPTER 56

Hydrodynamic Methods in Biology

D.K. Sharma
Department of Physics, J. N. Vyas University, Jodhpur-342 001
E-mail: dksjnvu@rediffmail.com

The study of physical properties of biopolymers and macromolecules is useful in many Biophysical and Biochemical applications. In many biomolecular applications the physical quantities such as molecular mass, size, shape, electric dipole moment, elasticity, hydration, and spectral analysis are commonly used. These useful parameters can be extracted from the study of the hydrodynamic properties such as viscosity, sedimentation, diffusion, rotational and translational motion of macromolecules in solution, etc. Most of the Biochemical studies of the macromolecules are conducted in water or dilute aqueous buffer. Thus, the hydrodynamic properties of water and solutions with water as solvent may provide useful information. In this chapter, we present a brief description of the hydrodynamic properties of fluids or solutions containing biopolymers and macromolecules along with the methods to measure these properties.

VISCOSITY

Viscosity is a hydrodynamic property of fluids by virtue of which a liquid opposes the relative motion between its different layers. Under steady state condition the fluid flow may be considered as motion of planes in contact (Fig. 1), each plane moving (or sliding) with constant velocity. The variation of velocity from one plane to other plane depends on the applied force. The rate of variation of velocity with respect to the displacement perpendicular to the planes is termed as velocity gradient. Consider the fluid moving in X direction . The fluid may be considered as a stack of X-Y planes, each plane sliding over the other with constant velocity in X direction. The velocity gradient is then expressed by $\Delta v/\Delta z$, where Δv is change in velocity for a displacement Δz in Z direction. The frictional force between the successive planes depends on area of the plane and velocity gradient through the relation.

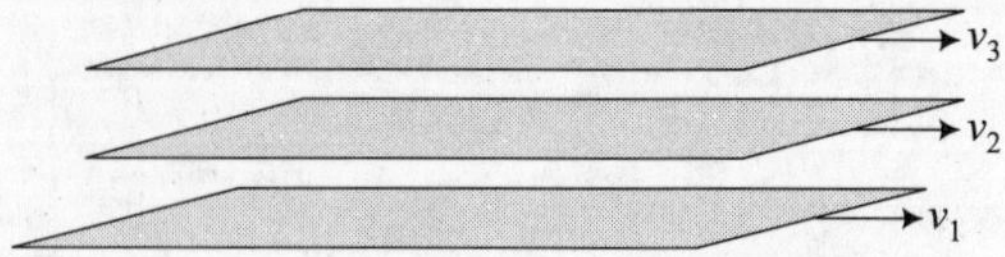

Fig. 1. Viscous flow in liquids.

$$F = \eta A \frac{dv}{dz} \quad ...(1)$$

where A is area of the plane.

The constant η (neeta) depends on the properties of the fluid and temperature, and is called the viscosity or the coefficient of viscosity. Using the defining relation (Eq. 1) η can be expressed as

$$\eta = \frac{F}{A(dv/dz)} \quad ...(2)$$

The viscosity η is defined as the force per unit area needed to maintain unit velocity gradient. The *cgs* (centimeter, gram, second) unit of η is *poise* in honour of the French scientist (Poiseuille, 1846), who investigated the viscous properties of fluids. One *poise* is defined as viscosity of the fluid for which one dyne force per unit area is required to produce a unit velocity gradient (*1/sec*). The alternative *cgs* unit for viscosity is *dynes sec/cm*^{2} which is equivalent to *poise.* The practical unit of η is *centipoise* (one hundredth of a *poise).*

The viscosity of the liquid depends on the characteristics of the liquid and its temperature. The viscosity of water is a strongly decreasing function of temperature. In biological applications the viscosity of water is adjusted to its value at 20°C. The viscosity of water at this temperature is 1.002 *centipoise* (0.01002 *poise).* The value of η for different liquids can be obtained from the standard data tables (Ed. lied, 2006). However for practical purpose the following approximate relation can be used to calculate viscosity of water at a temperature *t*°C.

$$\log_{10} \eta = \frac{1301}{998.333 + 8.1855\,(t - 20) + .00585\,(t - 20)} - 3.30233 \quad ...(3)$$

in the range $0 \le t \le 20°\text{C}$, and

$$\log_{10} \frac{\eta}{\eta_s} = \frac{1.3272(20 - t) - 0.001053(t - 20)^2}{t + 105} \quad ...(4)$$

in the range $20° \le t \le 100°\text{C}$, where η_s is the viscosity of water at 20°C.

The resistive force of viscosity is due to intermolecular interaction between molecules. The bio-molecules have a long range interaction due to their complex structure. Thus, addition of biopolymers increases the viscosity of the solvent. This increase in viscosity is very sensitive to the concentration of solutes or impurities and depends on the size and structure of the polymer. The bigger polymers when dissolved in a solvent will increase the viscosity more than the smaller molecules since their molecular domain has a greater range. The measurement of viscosity is very useful in biopolymer studies.

Assuming solute molecules to be spherical in shape, the viscosity η of the solution is related to the viscosity η_0 of the solvent through the relation

$$\eta = \eta_0\left(1 + \frac{5}{2}\phi\right) \qquad ...(5)$$

where ϕ (phy) is the ratio of the volumes of the polymer solute and the solvent in the solution and is called the volume fraction. The volume fraction can be expressed as the volume occupied by one gram of hydrated polymer v_h multiplied by weight concentration C_w (*gm/cc*) of the polymer. Therefore, using $\phi = v_h\, C_w$ in Eq. 5 we get

$$\frac{\eta}{\eta_0} - 1 = \frac{5}{2}\, v_h\, C_w \qquad ...(6)$$

or

$$\frac{\eta - \eta_0}{\eta_0\, C_w} = \frac{5}{2}\, v_h \qquad ...(7)$$

The ratio of viscosities of the solution and solvent is called the relative viscosity η_r

$$\eta_r = \frac{\eta}{\eta_0} \qquad ....(8)$$

Another useful term is specific viscosity, which is defined as relative difference between viscosities of solution and solvent, i.e.,

$$\eta_{sp} = \frac{\eta - \eta_0}{\eta_0} = \eta_r - 1 \qquad ...(9)$$

The intrinsic viscosity $[\eta]$ of the solution is defined as specific viscosity per unit concentration in the limit weight concentration approaching to zero.

$$[\eta] = \lim_{C_w \to 0}\left(\frac{\eta_{sp}}{C_w}\right) \qquad ...(10)$$

For evaluating $[\eta]$ the viscosity of the solution η is measured at different concentrations. Extrapolation method is used to obtain viscosity at $C_w = 0$ (Fig. 2).

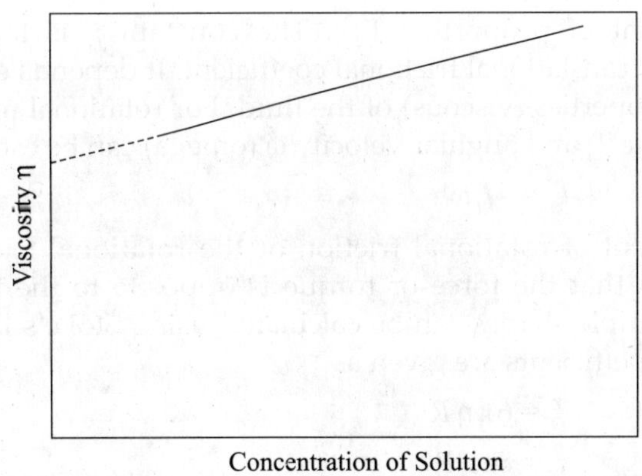

Fig. 2. Variation of viscosity η with concentration of the solution.

From Eq. 7, we can obtain the value of intrinsic viscosity for spherical molecules.

$$[\eta] = \lim_{C_w \to 0}\left(\frac{\eta - \eta_0}{\eta_0 C_w}\right) = \frac{5}{2} v_h \quad ...(11)$$

The intrinsic viscosity [η] of the solution is related to molecular mass M through the relation-

$$[\eta] = KM^{\alpha} \quad ...(12)$$

where K and α are solvent dependent constants.

When a body moves through a fluid, the layers of the fluid in contact with the body also move with the velocity of the body. This causes a relative motion between layers of the fluid. The viscous force of the fluid opposes this motion. The resistive force increases with the velocity of the body. The external force responsible for the motion of the body through the fluid is ultimately balanced by the viscous and other opposing (buoyancy) forces. Under such a condition, the body moves with a constant velocity called the terminal velocity.

The viscous force exerted on a spherical body of radius a, moving with terminal velocity v is given by

$$F = 6\pi\eta a v \quad ...(13)$$

where η is the coefficient of viscosity of the fluid. The above relation is known as Stoke's formula. The viscous force will depend on the shape and size of the macromolecules. We will discuss the opposing or frictional force on moving bodies in fluids in the next section.

FRICTIONAL COEFFICIENT

When a particle moves through a fluid, a resistive force acts on it. This resistive force is called the frictional force. The motion of the particles may be translational or rotational, depending on the force applied. For translational motion the frictional force is proportional to the velocity of the particle .

$$F_t \propto v$$

or

$$F_t = -f_t v \quad ...(14)$$

where f_t is the constant of proportionality. The constant f_t is termed as coefficient of translational friction or translational frictional coefficient. It depends on the shape and size of the particles and the properties (viscous) of the fluid. For rotational motion a similar relation between frictional torque T_r and angular velocity ω (omega) can be established.

$$T_r = -f_r \omega \quad ...(15)$$

where f_r is the coefficient of rotational friction or the rotational frictional coefficient. The negative sign indicates that the force or torque is opposite to the velocity. The frictional coefficients for some simple shapes can be calculated using Stoke's law. For solid sphere of radius R the frictional coefficients are given as

$$f_t = 6\pi\eta R \quad ...(16)$$

and

$$f_r = 8\pi\eta R^3 \quad ...(17)$$

The hydrodynamic radius R of the sphere is called the Stoke's radius. It can be observed that the translational friction depends on R, whereas the rotational friction depends on volume or R^3.

Biopolymers have complex structure but for hydrodynamic studies the shape of biopolymers can be closely approximated by some standard geometrical shapes. Apart from the spherical, other common models for biopolymers are ellipsoids, cylinders, random coils and array of spheres. The coefficient of friction for these complex structures can be obtained by using suitable geometrical relations. For non-spherical macromolecules, the commonly used geometric model is ellipsoid of revolution (Perrin, 1934, 1936). This is the geometry obtained by rotating the ellipsoid about one of its axes. The ellipsoids of revolution are often used for cylindrical rods [5]. Random coils can be modeled by a sphere of suitable radius. The effective hydrodynamic radius is related to its radius of gyration. For a random coil with N bonds, each of length b, the coefficient of translational friction is given by

$$f_t = 6\pi\eta\frac{3\sqrt{\pi}}{8}\left(\frac{b^2N}{6}\right) \qquad ...(18)$$

VISCOELASTICITY

The biopolymers have complex structure such as an array of spheres or rod (molecular chain type structure). In solution, due to various forces, deformations are produced in the structure of the biopolymers. Due to these deformations a strain is produced in the molecules. The molecular binding provides the restoring force and thus, the molecules show elastic properties. In a moving fluid, the intermolecular force opposes the movement of molecules in contact. Thus, the biopolymers exhibit simultaneously the viscous properties of a fluid and the elastic properties of a solid. Such biopolymers are called viscoelastic and the property by virtue of which the molecules posses elasticity as well as viscosity is called the viscoelasticity.

Consider a force F applied to a sample of viscoelastic material at time t = 0. There will be three types of deformations in the sample: (i) The elastic deformation (δD_b) due to bond stretching. This deformation depends on applied force only and remains constant with time. (ii) The elastic deformation (δD_s) due to strain in the structure of the molecules. This deformation increases with time and becomes constant under equilibrium conditions. (iii) The deformation (δD_f) due to motion (sliding) of molecular chains on one another. This deformation increases with time. The total deformation under a constant applied force is called the creep. The total deformation δD is given by

$$\delta D - \frac{FL_0}{A_0}\left(\frac{1}{K_i} + \frac{1}{K_b}(1 - e^{-t/\tau}) + \frac{t}{\eta}\right) \qquad ...(19)$$

where L_0 and A_0 are the undeformed length and area of cross section of the specimen, K_i is the instantaneous elastic modulus, K_b is the bulk modulus of elasticity. τ (tao) is a characteristic time depending on the viscoelastic properties of the material and is known as retardation time and η is the viscosity of the material in the solution. The creep curve (Fig. 3) provides information about the type and extent of deformation in the biopolymers.

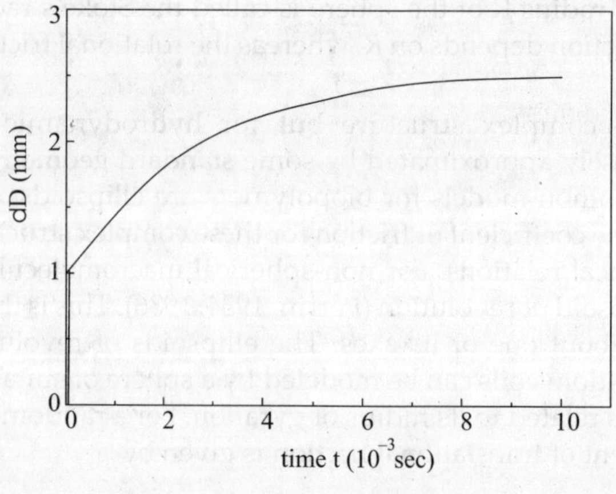

Fig. 3. Graph showing total deformation of a viscoelastic specimen with time.

SEDIMENTATION

Sedimentation may be defined as the accumulation of particles in a small volume of a solution under the action of a force. The external force commonly used for sedimentation is the centrifugal force which is generated by circular motion. Thus, another term used for sedimentation is centrifugation. Consider a particle of mass m. If this particle is rotated by angular velocity ω about an axis in a circle of radius r, then the centrifugal force acting on this particle is

$$F_{cf} = m\omega^2 r \qquad ...(20)$$

In the hydrodynamic experiments, the particles which are biopolymers or macromolecules are subjected to centrifugation using a solution. If the volume of macromolecule is v' and density of the solvent is ρ (roh) then due to force of buoyancy the effective mass of the molecule will be

$$m' = (m - \rho v') \qquad ...(21)$$

In terms of specific volume $v = v'/m$ (volume per unit mass) the above equation reduces to

$$m' = (m - m\rho v) = m(1 - \rho v) \qquad ...(22)$$

The centrifugal force is therefore given by

$$F_{cf} = m(1 - \rho v)\,\omega^2 r \qquad ...(23)$$

The mass of the molecules is usually represented in term of molecular weight M. Therefore,

$$F_{cf} = \frac{M}{N_A}(1 - \rho v)\,\omega^2 r \qquad ...(24)$$

where N_A is Avogadro's number.

The frictional force of the medium in which the particle attains a velocity v is given by

$$F_{fr} = -f_t v \qquad ...(25)$$

where f_t is the coefficient of translational friction. The velocity v is called the drift velocity of the molecules and is related to the angular speed through the relation

$$v = \frac{dr}{dt} = S\omega^2 r. \qquad \text{...(26)}$$

Where the constant S, known as sedimentation coefficient, depends on the properties of the medium (solution) and the moving particle. From above relations we get

$$F_{fr} = -f_t S\omega^2 r. \qquad \text{...(27)}$$

Under equilibrium condition the two forces will be equal, therefore

$$F_{fr} = -F_{cf}$$

$$f_t S\omega^2 r = \frac{M}{N_A}(1 - v\rho)\omega^2 r$$

$$S = \frac{M(1 - v\rho)}{N_A f_t} \qquad \text{...(28)}$$

From Eq. (26), we can express the sedimentation coefficient as

$$S = \frac{v}{\omega^2 r} = \frac{dr/dt}{\omega^2 r} \qquad \text{...(29)}$$

This equation allows us to determine S from the measured quantities. The unit of sedimentation coefficient is *seconds.* The practical unit of S is *Svedberg* (S) . One *Svedberg* (after the name of the inventor of ultracentrifuge Koining, 1975) is 10^{-13} *seconds*. The unit of angular velocity in above relations is *rad/sec*. However, the speed of the centrifuge is usually provided in the units of rotations per minute *(rpm)* as a substitute quantity. The *rpm* speed can be converted to *rad/sec* using the conversion relation

$$1\ rpm = 2\pi/60\ rad/sec.$$

For experimental determination of S, a band of molecules is marked in the solution. The position of this band is measured to obtain its radius of rotation at different time of centrifuge. Let, at start of centrifuge the radius of rotation is r_0 and after time t the radius increases to r. Using (Eq. 26), we have

$$\frac{dr}{dt} = S\omega^2 r$$

or

$$\frac{dr}{r} = S\omega^2 dt. \qquad \text{...(30)}$$

Integrating this equation, we get

$$\log_e \frac{r}{r_0} = S\omega^2 t. \qquad \text{...(31)}$$

A graph between $\log_e r/r_0$ and time t can be plotted from the experimental data (Fig. 4). The graph should be a straight line as is evident from Eq. 31. The slope of this graph can be measured, which will give us numerical value of $S\omega^2$. This value when divided by ω^2 gives the sedimentation coefficient S. However, the major problem in these experiments is to mark the band of molecules for position measurement. Various methods are available to label the molecules in a band. The most commonly used methods are radioactive and fluorescent

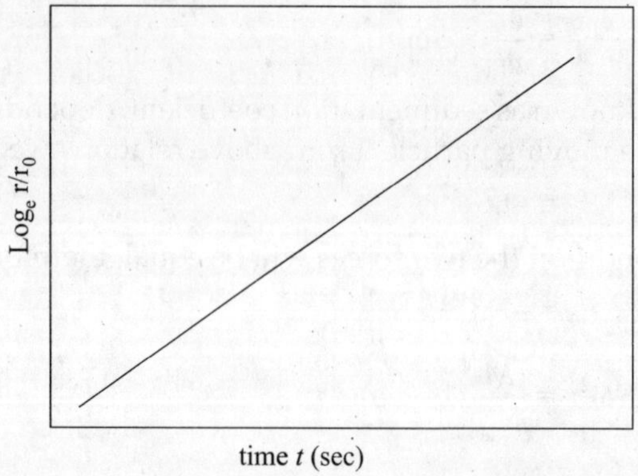

Fig. 4. Graph showing variation of $\log_e r/r_0$ with time of centrifugation.

labeling. Biological activity of the polymers can also be used to mark them using suitable reagent. If the molecules are evenly distributed in a large volume with observable boundaries, the mid point of the volume is used for position measurement during centrifuge.

DIFFUSION

The motion of macromolecules in solution is influenced by different forces. The thermal energy of the molecules plays an important role in hydrodynamic studies. In dilute solutions the molecules may be assumed to undergo Brownian motion (Perring, 1936; Koening, 1975) due to thermal agitations. The difference in concentration in different parts of a solution results in transport of molecules from higher concentration to lower concentration. This transport of molecules is called diffusion. Diffusion may also be observed when two solutions of different concentrations are mixed or separating barrier between different concentrations is removed. The motion of the molecules in diffusion is Brownian or random motion.

The diffusion process can be described by Fick's laws (Fick, 1885). The first law states that at any time t the mass flux i.e., mass of the molecules crossing unit area per unit time, in specified direction is proportional to concentration gradient (Fig. 5). If $J(x, t)$ is mass flux in X direction, then at any time t

$$J(x,t) = -D_t\left[\frac{\partial C(x,t)}{\partial x}\right]_t \qquad \text{...(32)}$$

where D_t is a constant depending on the properties of the molecules, medium and temperature. The constant D_t is termed as diffusion coefficient for translational diffusion. The Fick's second law of diffusion is based on conservation of mass. It states that the divergence of mass flux (change in mass flux per unit displacement) through a volume is equal to the rate of change of concentration in that volume. Considering only one dimensional motion we can express the second law as

$$\left[\frac{\partial J(x,t)}{\partial x}\right]_t = -\left[\frac{\partial C(x,t)}{\partial t}\right]_x \qquad \text{...(33)}$$

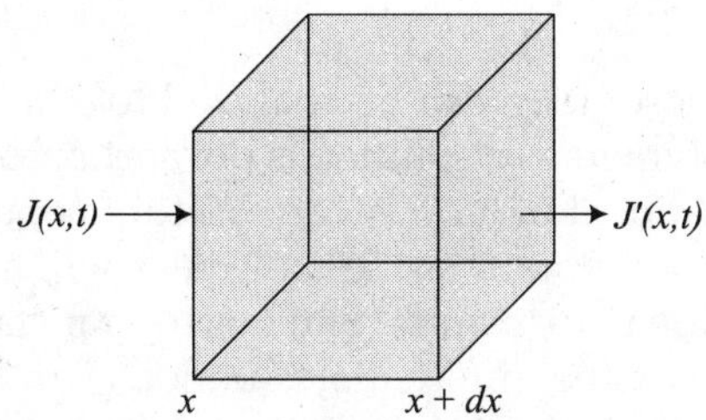

Fig. 5. Change in mass flux through a volume, in X direction. $J(x,t)$ is mass flux at the surface x = x and $J'(x,t)$ is mass flux at surface x = $x+dx$.

The negative sign indicates that if the divergence of mass flux is positive, i.e. more mass is leaving the volume than the mass entering, the concentration will decrease with time. Substituting $J(x, t)$ from Eq. 32, we get

$$\frac{\partial C}{\partial t} = D_t \frac{\partial^2 C}{\partial x^2}. \qquad ...(34)$$

Equation (34) is the differential form of the diffusion equation. The solution of this equation gives us concentration as a function of distance x and time t. The rate of diffusion increases with temperature and reduces with increasing frictional force of the medium (solution). Thus, the diffusion coefficient may be expressed as

$$D_t = \frac{k_B T}{f_t} \qquad ...(35)$$

where T is absolute temperature, f_t is the coefficient of friction and k_B is Boltzmann's constant.

Assuming the motion of the molecules similar to ideal gas, the diffusion coefficient may be related to mean free path, i.e., mean distance traveled between two successive collisions, through the relation

$$D_t = \frac{1}{3}\bar{v}\lambda \qquad ...(36)$$

where $\bar{v}$ is mean velocity and λ is mean free path. It can be shown that the mean square distance depends on time and is expressed as

$$\langle x^2 \rangle = 2D_t t. \qquad ...(37)$$

Substituting D_t from Eq. 35, we get

$$\langle x^2 \rangle = 2D_t t = \frac{2k_B T}{f_t} t. \qquad ...(38)$$

For spherical molecules $f_t = 6\pi\eta r$, therefore

$$D_t = \frac{k_B T}{f_t} = \frac{RT}{6\pi\eta N_A r} \qquad ...(39)$$

where we have used $k_B = R/N_A$ for Boltzmann constant, R is gas constant and N_A is Avogadro's number. Eq. 39 can be used to evaluate the diffusion coefficient for known values of η, temperature T and hydrodynamic radius of the molecules r.

ROTATIONAL DIFFUSION

Motion of macromolecules in a solution can be analyzed to extract many useful information. The viscous force on the biopolymers in a solution is proportional to R^3 (volume) for rotational motion. Thus, the study of rotational motion is very useful to obtain structural information of the macromolecules. The angular motion of the molecules will also result in diffusion . Such diffusion is called the rotational diffusion. Similar to translational diffusion, rotational diffusion can also be described by the differential equation (Tirado and Torre, 1980)

$$\frac{\partial C}{\partial t} = D_r \frac{\partial^2 C}{\partial \theta^2}. \qquad ...(40)$$

The rotational diffusion coefficient D_r will depend on temperature and rotational friction. Therefore, a relation similar to Eq. 35 can be used to define the rotational diffusion coefficient

$$D_r = \frac{k_B T}{f_r}. \qquad ...(41)$$

The rotational displacement can be measured by estimating the rotation of the axis of macromolecule with time. The mean square rotation can be defined as

$$\langle \phi^2 \rangle = 2D_r t. \qquad ...(42)$$

The orientation of the molecular axis will change with time. A function describing the decay of the orientation amplitude can be defined as

$$f(\theta, t) = f(\theta, 0)e^{-t/\tau_r}. \qquad ...(43)$$

It is evident from the above relations that the orientation in a specified direction will decay to $1/e$ times of the initial value in time τ_r. The characteristic time τ_r is called the relaxation time. For direct measurement of orientation magnitude, the relaxation time is related to diffusion coefficient through the relation

$$\tau_r = \frac{1}{2D_r}.$$

For spherical biopolymers

$$\tau_r = \frac{f_r}{2k_B T}.$$

Substituting $f_r = 8\pi\eta R^3$ and $k_B = R/N_A$, we get

$$\tau_r = \frac{4\pi N_A \eta R^3}{RT}.$$

For spherical shape of the macromolecules the volume is given by $V = \frac{4}{3}\pi R^3$, therefore

$$\tau_r = \frac{3V\eta}{k_B T}. \qquad ...(44)$$

The evaluation of the relaxation time τ_r gives the estimate of the rotational diffusion coefficient. There are various methods to determine the rotational diffusion coefficient. We shall outline the common methods used to study rotational diffusion.

DETERMINATION OF ROTATIONAL DIFFUSION COEFFICIENT

To determine the rotational diffusion coefficient, the optical or electric properties of the biopolymers are used. The biomolecules with their molecular axis aligned in the direction of the polarization of incident light will have maximum tendency of absorption through dipole transitions. The subsequent emission of light can be observed by spectroscopic methods (Torre and Bloomfield, 1981). The sample is first subjected to a field to rotate the molecular axis in a specified direction. Then a polarized light of specific frequency is passed through the sample. The absorption and emission profile of the polarized light is used to obtain the relaxation time τ_r.

The dielectric properties of the sample can also be used to measure the relaxation time for the polar molecules. The average dipole moment of the biomolecules in a solution will be zero due to random motion. However, if an external electric field is applied, the polar molecules will align along the direction of the field and due to polarization, the dielectric properties of the sample will change. This change in dielectric properties can be measured by different methods. If an alternating electric field of frequency ω is applied, the change in permittivity is given by

$$\delta\varepsilon = \frac{\varepsilon_0 - \varepsilon}{1 + \omega^2 \tau_r^2}. \qquad ...(45)$$

Where ε is the permittivity with maximum polarization just after the removal of polarizing field and τ_r is relaxation time. The measurement of permittivity at different frequencies of the applied field can be used to determine the relaxation time. The relaxation time can also be measured by using time domain spectroscopy (TDS).

METHODS OF MEASUREMENT

(I) Viscosity

The viscosity of a solution can be measured by using Poiseuille's method or Stoke's method. In Poiseuille's method, the solution is passed through a capillary tube of length l and radius r. A constant pressure difference P is maintained across the two ends of the capillary. If the pressure difference is not very large the fluid flow through the capillary may be assumed as streamline. Under steady state conditions the volume of the solution flowing through the capillary per second is given by

$$V = \frac{\pi P r^4}{8\eta l}. \qquad ...(46)$$

Thus, by measuring the volume of the fluid flowing through the capillary per second, the coefficient of viscosity can be evaluated using the relation

$$\eta = \frac{\pi P r^4}{8 l V}. \qquad ...(47)$$

The pressure difference P between the ends of the capillary can be measured by attaching monometer tubes at the two ends. The difference between the liquid levels in the two columns can be used to calculate pressure difference.

The other method of measuring viscosity is the Stoke's method. In this method, the solution under study is taken in the form of a long vertical column. Let the density of the solution is σ (sigma). A spherical ball of density ρ is then dropped in the solution with zero initial velocity. Due to gravitational force the ball moves downwards and its velocity increases with time. On the other hand, the viscous force acts opposite to the velocity and increases with velocity. At some value of velocity v_c the two forces will become equal and opposite (Fig. 6). Under this equilibrium condition the ball will continue to move with constant velocity v_c. The constant velocity v_c, usually called the terminal velocity, can be measured by using a scale and a telescope. If a is the radius of the spherical ball, the coefficient of viscosity is given by

$$\eta = \frac{2}{9}\frac{a^2 g(\rho - \sigma)}{v_c} \quad ...(48)$$

where g is acceleration due to gravity ($g = 980\ cm/sec^2$).

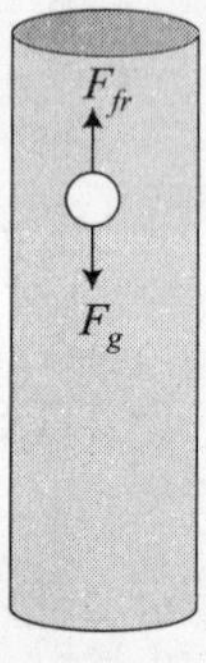

Fig. 6. Stoke's method of measuring viscosity of liquid. F_{fr} represents frictional and other resistive forces and F_g represents the gravitational force.

(2) Sedimentation

To measure sedimentation, optical properties of macromolecules, such as refractive index, are used to determine the concentration of different layers of the solution. If the polymer has been radioactively labeled, it is possible to detect very small concentration changes. In the conventional method, the optical properties of the solution are measured after a run of the centrifuge. In modern ultracentrifuge, the concentration can be monitored optically during the course of the run. For this purpose, the spectroscopic methods are used to measure the concentration at different radial distances. Light of wavelength falling in the absorption band of the macromolecules is passed through the sample (narrow band of solution). The absorption profile of the incident light is used to determine the concentration of the solution at specified position. If the concentration is initially uniform, the mid point of the boundaries formed by redistribution of solute molecules, is used for position determination. The concentration of solution at specified position may become constant during the centrifuge. This state of constant concentration is called the equilibrium state. Under equilibrium state the rate of diffusion, due to difference in concentration between neighbouring layers of the solution, and the rate of sedimentation due to centrifugal motion are balanced. The measurement of concentration in equilibrium state at different positions can also be used to obtain the sedimentation coefficient. Equating the rate of sedimentation and rate of diffusion for unit volume of solution we get:

$$C\frac{dr}{dt} = D\frac{dC}{dr}. \qquad ...(49)$$

or

$$\frac{dr}{dt} = \frac{D}{C}\frac{dC}{dr}.$$

Substituting the sedimentation drift velocity dr/dt from Eq. (26), we get

$$\frac{dC}{C} = \frac{S\omega^2}{D} r\,dr.$$

Substituting diffusion coefficient D from Eq. (35)

$$\frac{dC}{C} = \frac{S\omega^2 f_t}{k_B T} r\,dr. \qquad ...(50)$$

If C_1 is the concentration of solution at a radial distance r_1 and C_2 at radial distance r_2, under equilibrium condition, then equation (50) can be integrated to yield

$$\int_{C_1}^{C_2} \frac{dC}{C} = \frac{S\omega^2 f_t}{k_B T} \int_{r_1}^{r_2} r\,dr$$

or

$$\log\frac{C_2}{C_1} = \frac{S\omega^2 f_t}{k_B T}(r_2^2 - r_1^2). \qquad ...(51)$$

From the above equation, the sedimentation coefficient can be extracted in terms of known parameters T, ω, and f_t along with the measured concentrations C_1 and C_2, at respective position r_1 and r_2.

(3) Diffusion

There are different methods to measure the diffusion coefficient. Here we will briefly discuss a few of them. The first method is a classical method. In this method, diffusion across a porous barrier is used to estimate diffusion coefficient. The experiment is performed by taking a porous barrier of thickness Δd and area A. Solutions of different concentrations C_1 and C_2 are placed on the two sides of the barrier. The molecules should have some optical properties or be radioactively labeled so that their concentration on either side of the barrier can be measured with time. The solution is stirred continuously to keep the solution homogeneous (Bloomfield, 2000).

The net transport of molecules in time t can be expressed in terms of change in mass Δm from higher concentration C_1 to lower concentration C_2. Using Fick's first law, we can write

$$\Delta m = -AD_t\frac{\Delta C}{\Delta d} \qquad ...(52)$$

where ΔC is the difference between the concentrations ($\Delta C = C_2 - C_1$).

The experiment may be repeated for different concentrations to determine D_t. Accurate measurement of A and Δd is not possible. Thus, a sample of known D_t is used to determine $A/\Delta d$. The measurement of Δm thus gives the diffusion coefficient.

$$D_t = \frac{\Delta m}{\Delta C}\left(\frac{\Delta d}{A}\right) \qquad ...(53)$$

Dynamic laser light scattering is another method to determine the diffusion coefficient. These experiments are based on the analysis of the light scattered from the molecules. Highly monochromatic laser light is used to illuminate a portion of the solution. The light scattered from the molecules is then analyzed for intensity. Due to diffusion there are fluctuations in the scattered light intensity. The fluctuations are related to mean free path that in turn provides us the relation to measure the diffusion coefficient.

Fluorescence Photobleaching Recovery (FPR)is one of the modern methods for measuring the diffusion coefficient. This modern method is used to study the diffusion properties of the macromolecules which have fluorescence properties or to which a fluorescent label can be bound. The specified cross section (small area) of the solution containing fluorescent molecules is exposed to an intense narrow beam of laser for a short time. If the laser wave length falls in the absorption band of the fluorophore, some of the molecules exposed to the laser light will be bleached. In the bleaching process, which is a photochemical reaction, the fluorescence of the molecules is permanently quenched. The exposed area is studied under a microscope for the fluorescence intensity. Another laser beam may be used to monitor the fluorescence intensity of the cross section. Let, before the photobleaching, the fluorescent intensity of the area under observation is *F*, then just after bleaching it will be less than *F*. The unbleached molecules from the surroundings will diffuse in (or the bleached molecules will diffuse out of) the specified volume with time. Thus, the intensity will gradually recover to the initial value *F*. The recovery time τ_D is measured in this experiment. The diffusion coefficient is then given by

$$D_t = \frac{\rho^2}{4\tau_D} \qquad ...(54)$$

where ρ^2 is the mean square radius. The mean square radius can be obtained from the characteristics of the laser used. Assuming a Gaussian profile for the laser used, ρ^2 is the distance at which the intensity reduces to e^{-2} of the maximum (Fig. 7). Other advanced methods of measurement of the hydrodynamic properties are also available and interested readers may consult the literature.

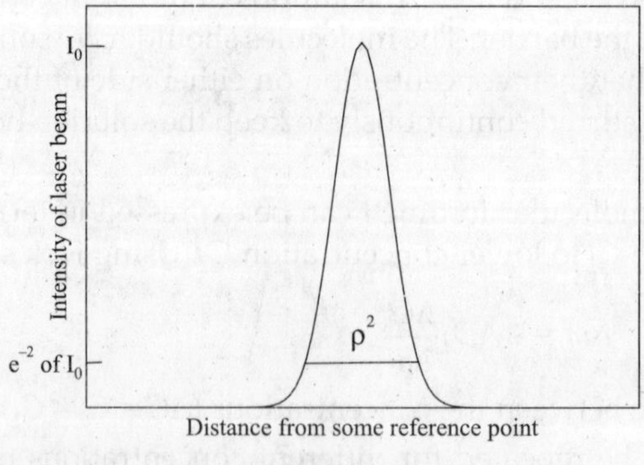

Fig. 7. Intensity distribution of a typical laser, I_0 is maximum intensity. The laser profile is used to determine the mean square radius r^2.

REFERENCES

Bloomfield, V. A. (2000) Survey of Biomolecular Hydrodynamics. In: Separations and Hydrodynamics (Schuster T. M. Edit.). On-line Biophysics Textbook., Biophysics Society, www.biophysics.org.

CRC Handbook of Chemistry and Physics, Edit. David R. Lide (2006) CRC Press, Landon.

Fick, A. (1885) Ann. Physik 23:59.

Fick, A. (1885) Ann. Physik Leipzig 170:59.

Garcia de la Torre, J. and Bloomfield, V. A. (1981) Hydrodynamic properties of complex, rigid, biological macromolecules. Q. Rev Biophys. 14:81.

Koening, S. (1975) Brownian motion of an ellipsoid. A correction to Perrin's result. Biopolymers 14:2421.

Poiseuille, J. L. M. (1846) Mémorirés présentés par divers savants á l'academie royale des sciences de l'institute de France 9:433.

Perrin, F. (1934) Mouvement brouinien d'um ellipsoide I , J. Phys. Radium 5:497.

Perrin, F. (1936) Mouvement brouinien d'um ellipsoide II , J. Phys. Radium 7:1.

Tirado, M. M. and Garcia de la Torre, J. (1980) Rotational diffusion of rigid, symmetric top macromolecules. J. Chem. Phy. 73:1986.

Suggested Reading

1. Schuster T. M. (Edit.) (2000) Separations and Hydrodynamics , On-line Biophysics Textbook, Biophysics Society, www.biophysics.org.
2. Brene, B.J. and Pecora, R. (1976) Dynamic light scattering with applications to Chemistry, Biology and Physics, Wiley-International, New York.
3. Narayanan P. (2000) Essentials of Biophysics, New Age International Publishers (India).
4. Hilyard N. C. and Biggin H. C. (1977) Physics for Applied Biologists, Edward Arnold (Publishers) Limited , Landon.

REVIEW QUESTIONS

Long Questions

1. Explain the Poseuille's method to determine viscosity of a fluid.
2. Explain the porous barrier method to obtain the diffusion coefficient.
3. What is rotational diffusion? Explain the method of estimating rotational diffusion coefficient.
4. Discuss the methods of measuring sedimentation coefficient.
5. Explain the Fluorescence Photobleaching Recovery (FPR) method of estimating diffusion coefficient.

Short Questions

1. Define viscosity, relative viscosity and specific viscosity.
2. Explain the effect of concentration of solution on the viscosity of solution.
3. Explain the difference between translational and rotational friction.
4. What do you mean by sedimentation?
5. Obtain expression for sedimentation coefficient.

6. What is diffusion? Define diffusion coefficient.
7. What do you mean by viscoelasticity?
8. What is relaxation time? How is it related to rotational diffusion coefficient?

Multiple Choice Questions

1. Coefficient of viscosity depends on
   (a) Volume of the fluid (b) Velocity gradient
   (c) Force applied (d) Temperature
2. *cgs* unit of viscosity is
   (a) *poise* (b) *dynes*
   (c) *gm/cc* (d) *pascal*
3. Speed of centrifuge is expressed in the units
   (a) cm/sec (b) meter/sec
   (c) mm/sec (d) rotation per minute
4. Diffusion depends on
   (a) difference in concentration (b) temperature
   (c) properties of the fluid (d) all of the above
5. The motion of molecules in diffusion process is
   (a) random (b) unidirectional
   (c) stream line (d) none of the above
6. The diffusion coefficient is proportional to
   (a) Temperature (b) Volume
   (c) Coefficient of friction (d) none of the above
7. Fluorescence Photobleaching Recovery (FPR) method is used to estimate
   (a) viscosity (b) diffusion
   (c) sedimentation (d) coefficient of friction
8. The practical unit of sedimentation coefficient is
   (a) dynes (b) poise
   (c) Svedberg (d) radian/sec.
9. The frictional force for rotational motion of a spherical molecule is proportional to its
   (a) radius (b) diameter
   (c) volume (d) mass
10. The total deformation in a viscoelastic material due to an external force is called
    (a) elasticity (b) creep
    (c) depression (d) elongation

Answers

1. (d)	2. (a)	3. (d)	4. (d)	5. (a)	6. (a)
7. (b)	8. (c)	9. (c)	10. (b)		

CHAPTER 57

Atomic Absorption Spectroscopy

D. K. Sharma
Dept. of Physics, J. N. Vyas University, Jodhpur–342 001
E-mail: dksjnvu@rediffmail.com

The developments in laser technology and optical instrumentation have added new dimensions to the spectroscopic studies by providing very accurate and sophisticated methods of measurement. Atoms and molecules have some characteristic spectroscopic properties such as discrete electronic states, quantized angular momentum etc. The study of spectroscopic properties can be performed through the measurement of some physical parameters characteristically representing these properties. The signature of these characteristic parameters can be used to detect the presence of a particular atom or molecule in a given sample (Kirkbright,1979). The electrons in an atom can occupy energy states which have some fixed discrete energy values, i.e.; the energy states of electrons in an atom are quantized. The electron can make transition from one energy state to other energy state by gaining or losing energy. The spectroscopic methods are based on the measurement of such transitions. When an electron makes a transition from lower energy state to higher energy state the transition is termed as absorption transition. The absorption transition can take place if the electron, initially in lower energy state, gains energy exactly equal to the difference in energy between initial and final states and jumps to the final state. The required energy can be supplied to the electron by different means. If the source of energy is electromagnetic radiations the electron will make transition by absorption of photon of appropriate frequency (energy). Atomic absorption spectroscopy deals with the methods used to detect and analyze the absorption of photons (electromagnetic radiations) by material medium (Gracía-vargas *et al.,* ,1983). Similarly, an electron from a higher energy state can make a transition to lower energy state by emitting a photon of characteristic frequency. The frequency of the emitted photon corresponds to the difference between the energy of two electronic states. The atomic emission spectroscopy covers the methods employed to detect these emitted photons.

Apart from electronic transitions in atoms,there are rotational and vibrational transitions in molecules. Free molecules rotate about their own axis with a characteristic energy of rotation. Similar to electronic states, there are discrete states of molecular rotational energy. A molecule can make a transition from lower rotational energy state to a higher state by absorbing a photon (electromagnetic radiation) of appropriate energy (frequency usually in infrared range). The study of such transitions is categorized as molecular absorption spectroscopy. The atoms bound together in a molecule are not stationary. They vibrate about their mean position with some characteristic frequency. The molecules thus possess discrete (quantized) energy values due to vibrational motion. A molecule can jump from a lower vibrational energy state to a higher state through absorption of radiations. The spectroscopic methods used to detect and analyze such transitions are studied in molecular vibrational spectroscopy.

THE ELECTROMAGNETIC SPECTRUM

Before we proceed with the atomic absorption spectroscopy it will be useful to introduce the electromagnetic spectrum. The electromagnetic waves have constant velocity in vacuum, however the wave may travel with a wide range of frequency (wavelength). The frequency ν (*new*) of electromagnetic waves decides the energy associated with the wave (energy of photon). The energy E of a photon is related to its frequency ν through Planck's relation $E = h\nu$, here h is Planck's constant ($h = 6.63 \times 10^{-34}$ *Joules second*). The wavelength λ (*lemda*) is related to frequency ν through the relation $\lambda = c/\nu$,where c is the velocity of electromagnetic waves ($c \approx 3 \times 10^8$ *m/sec* in vacuum i.e.; speed of light).

The electromagnetic spectrum is divided in several parts according to wavelength (frequency/energy) range. Each part represents a range of electromagnetic waves with some characteristic properties. Different categories of electromagnetic waves are presented in Table 1. There is no sharp boundary between two neighboring categories, rather the wavelength ranges are overlapping (Fig. 1). In the study of atomic absorption spectroscopy one is usually interested in the infrared to ultraviolet region of electromagnetic spectrum. The energy of electromagnetic waves listed in the table below is presented in electron volts (unit of energy). One electron volt (eV) is the energy gained by an electron when it is accelerated by a potential difference of one Volt ($1\,eV = 1.6 \times 10^{-19}$ *joules*).

THE EXPERIMENTAL METHOD

The experimental setup for atomic absorption spectroscopy consists of four basic units, source, sample holder, monochromator and detector (Welz, 1976). The source is a lamp or a device

Table 1. Electromagnetic Spectrum.

Wavelength Range (in meters)	Category (Nomenclature)	Energy (in electron volts)
∞ to 0.1	Radio Waves	$\leq 10^{-5}$
0.1 to 10^{-4}	Microwaves	10^{-5} to 10^{-2}
10^{-4} to 8×10^{-7}	Infrared (IR)	10^{-2} to 1
8×10^{-7} to 4×10^{-7}	Visible	1 to 3
4×10^{-7} to 10^{-9}	Ultraviolet (UV)	3 to 10^{3}
10^{-9} to 10^{-11}	X - Rays	10^3 to 10^5
Shorter then 10^{-11}	Gamma Rays	$> 10^5$

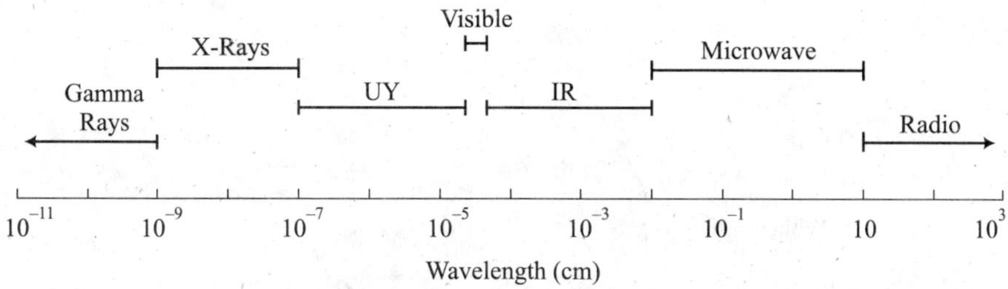

Fig. 1. The electromagnetic spectrum.

emitting electromagnetic radiations. The source may be a continuous source of radiations emitting electromagnetic radiations in a given wavelength range or it may be a discrete line source emitting wavelength of a specific frequency. There is a wide variety of radiation sources. The source is selected according to the spectroscopic properties of the sample. The wavelength/wavelength range of the radiations emitted by source should correspond to the energy required to excite the atoms of the sample from lower energy state to higher energy states. Some of the sources used in atomic absorption spectroscopy are Argon-Xenon lamps for ultraviolet and visible range, emitting radiations in the wavelength range 150 nm to 600 nm (1 nm = 10^{-9} meters). The tungsten filament halogen lamps are useful for visible range of wavelengths. Whereas nichrome wire and globar are other commonly used lamps, emitting radiations in the infrared region (wavelength range 750 nm to 40000 nm). Semiconductor gas lasers can be used as discrete wavelength sources in visible and ultraviolet range.

The next functional unit of atomic absorption spectroscopic experiment is the sample holder. The main purpose of the sample holder is to retain the sample, which may be in liquid or gas form, and insert it in the spectrometer for spectroscopic studies. Sample holders are made from materials which do not interact with the incident radiations. For visible light glass is commonly used to make sample holders. Lithium Fluoride (LiF) or quartz is used to make sample holders in the ultraviolet range of radiations. Similarly, NaCl or KBr cells can be used for ultraviolet to infrared range of radiations.

Radiations (electromagnetic) emitted by non monochromatic sources consist of a range of wavelengths. As the electromagnetic radiations pass through the sample, some of the wavelengths may get absorbed. Therefore, it is essential to analyze the individual constituent wavelengths of the incident and emergent radiations. For this purpose we need an optical instrument which can separate the electromagnetic radiations in constituent wavelengths. The optical device used for this purpose is called a monochromator. Monochromators are based on physical phenomenon such as dispersion, diffraction, interference and absorption of electromagnetic radiations (light). Prism is an example of monochromator based on dispersion of light. The incident light when passes through a prism is dispersed into its constituent colors (wavelength) due to variation in deviation. The deviation for different colors caused by refraction through prism is different. Therefore, we get spectrum of the light incident on prism (Fig. 2).

Diffraction grating is another monochromator used for incident light in visible range. The diffraction grating is made by ruling closely spaced lines on a glass plate. The ruled lines effectively convert the glass plate to an optical device with large number of parallel slits. Light

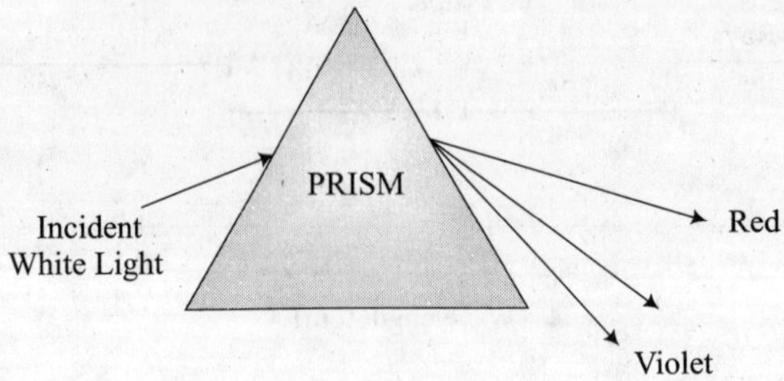

Fig. 2. Prism as a monochromator.

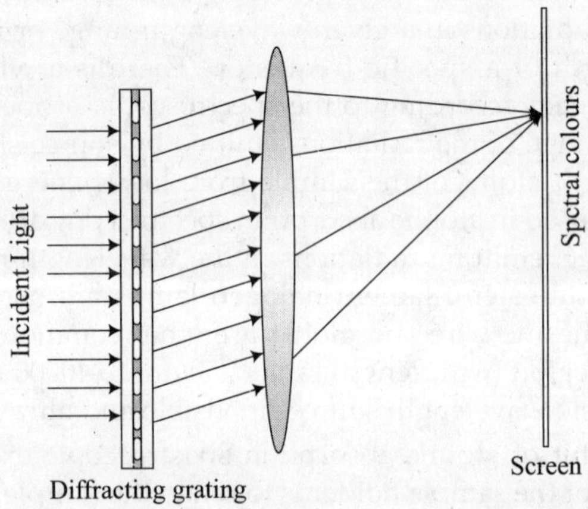

Fig. 3. Diffraction grating as a monochromator.

diffracted from these slits splits into its constituent colors due to interference (Fig. 3). Diffraction grating is a very sensitive monochromator and finds application in many spectroscopic measurements. Reflection grating can also be used for better resolution.

The third type of monochromator is a filter. The filter allows transmission of a narrow range of wavelengths through it and absorbs the other wavelengths. Different filters can be used to analyze the absorption of different wavelengths in the spectrum. The filters are categorized on the basis of physical properties of the optical device, such as transmittance and range width. Transmittance is defined as the ratio of the intensity of transmitted light of specific wavelength to the intensity of the incident light. The range width of the filter is the range of wavelengths which are allowed to pass through the filter. The transmittance for absorption filters in visible range is low (<20%) and the range width is wide (>50 nm). Another class of filters with better transmittance (>50%) and low range width (<10 nm), is interference filters. The interference filters are based on interference in thin films. A thin transparent dielectric sheet (Magnesium Fluoride or Calcium Fluoride) is used for this purpose. The thin sheet is half silvered on both sides and sandwiched between two glass plates to protect it from damage. The range width for interference filter depends on the thickness of dielectric plate (Fig. 4).

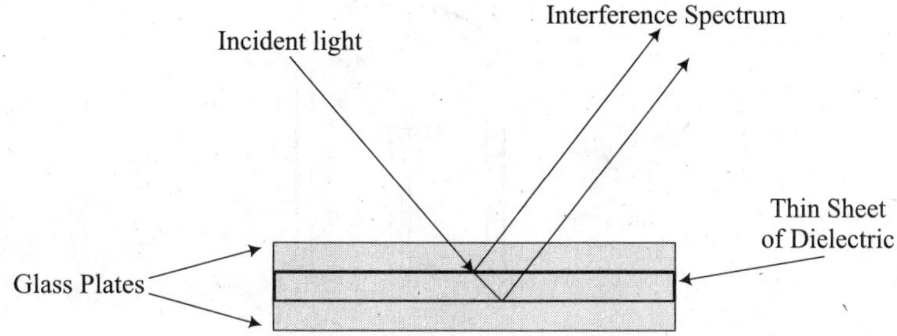

Fig. 4. Interference filter as a monochromator.

The light emergent from a monochromator is to be analyzed for wavelength and intensity. For this purpose photoelectric/thermoelectric devices called detectors are used. The radiation detectors are devices which absorb incident radiations and produce electric signals. There are mainly two types of radiation detectors (i) photodetectors, (ii) thermaldetectors. The photodetectors are based on photoelectric emission, whereas thermal detectors are based on thermoelectric effect. When photons (radiations) fall on certain substances (photo-emissive substances) the incident radiations are absorbed and electrons are emitted from the surface of the material. These electrons are collected at a plate and an equivalent signal is generated. The signal is then amplified and analyzed using electronic devices. The amplitude of the signal will be proportional to the number of electrons collected at the plate, which alternatively gives a measure of the intensity of the radiations incident on the detector.

The photo detectors are mainly phototubes (Fig. 5), photo multiplier tubes (Fig. 6) and photo diodes. The phototubes consist of a cathode plate coated with photo-emissive material and an anode. The tube is sealed with high vacuum inside. The radiations are focused on the cathode which emits electrons due to photo electric effect. The electrons are collected at anode and an equivalent current flows in the external circuit. The magnitude of the current signal is proportional to the intensity of the incident radiations. Photo multiplier tubes are basically photo tubes with an assembly of intermediate plates (dynodes), used to magnify (multiply) the number of electrons emitted from the cathode. This arrangement magnifies the signal in external circuit and thus, photo multiplier tubes are very sensitive detectors. Now-a-days semiconductor diodes are also used as detectors.

The spectroscopic measurements are carried out using two different optical arrangements,: single beam and double beam. In single beam apparatus the radiations collimated in form of a fine beam are passed through the sample. The emerging beam is then subjected to dispersion and detection. For analysis purpose the test sample is replaced by a standard sample and absorption spectrum is recorded. The relative absorption is then analyzed to find contents and concentration of the test sample. Single beam spectroscopy is simple and inexpensive, however it has some disadvantages. As the beam intensity changes with time, the incident intensity is different for test and standard samples. This reduces the precision of measurement. Further, the sensitivity of the detector varies with time and therefore introduces error in measurement. The results can be improved by using double beam arrangement. In this arrangement the beam from source is made to split into two beams using beam splitter arrangement. One part of the

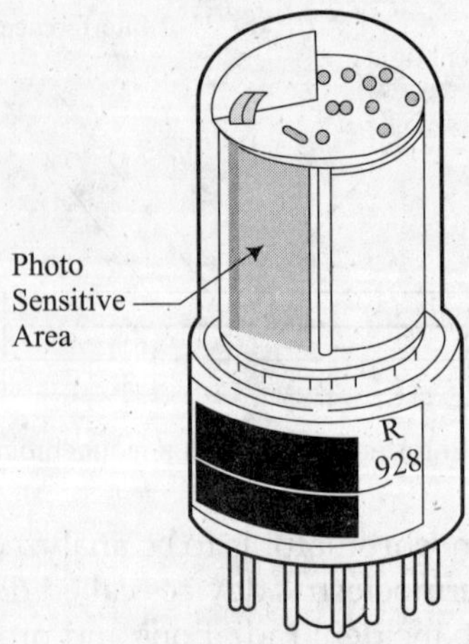

Fig. 5. Block Diagram of a Photodetector.

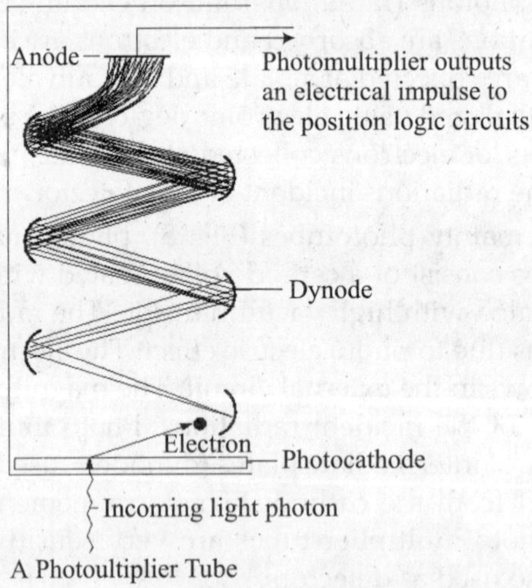

Fig. 6. Functional diagram of a Photomultiplier Tube.

splited beam is passed through standard and other through test samples. The two emerging beams are then passed through monochromators and are detected by the detector for different wavelength ranges. Since the same beam (in splited form) is used for spectroscopic analysis of the standard and the test samples, the error due to variation in the incident beam intensity is minimized. The simultaneous detection increases the precision of analysis (Fig. 7). The intensity of the emerging beam is related to the concentration of the absorptant material (atoms) in the sample. The relation between intensity and concentration is established by Beer-Lambert law.

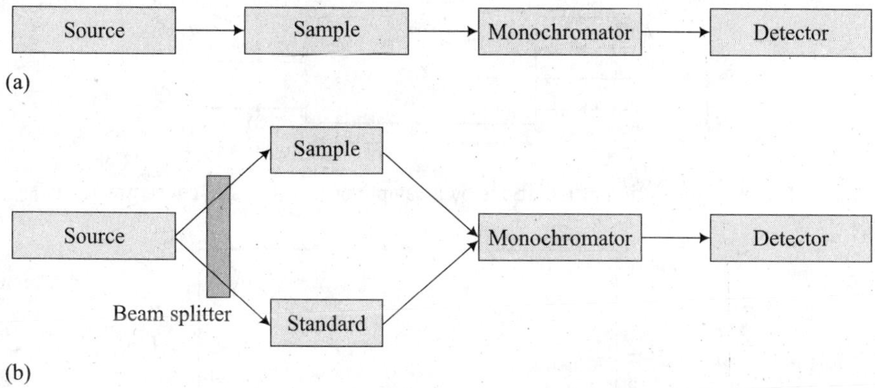

Fig. 7. Atomic Absorption Spectroscopic measurement (a) Single beam arrangement (b) Double beam arrangement.

BEER-LAMBERT LAW

The atomic absorption spectroscopy provides two types of information. The wavelength absorbed by sample determines the atoms present in the sample, as the atoms absorb characteristic (discrete) wavelengths. On the other hand, the intensity of absorbed wavelength determines the concentration of atoms in the sample. Let a beam of monochromatic light of wavelength λ and intensity I_0 (number of photons incident per unit area per second) is incident on the sample. If I is the intensity of the transmitted beam then we define a physical parameter absorbance A (or optical density) as

$$A = \log_{10} (I_0/I) \qquad ...(1)$$

The absorbance A depends on the number of atoms/molecules present in the sample (Fig. 8). If molecular concentration is c and length of the sample is L, then the absorbance A is defined as

$$A \propto cL$$

$$A = \alpha cL \qquad ...(2)$$

where α is a constant called the absorptivity of the sample. α depends on the nature of sample and wavelength of the incident radiations. The SI unit of α is $m^2\ mol^{-1}$. The absorbance A is independent of the cross-sectional area of the sample as the definition of intensity takes account of it. From Eq. (1) and (2), we can establish

$$\log_{10}\left(\frac{I_0}{I}\right) = \alpha cL \qquad ...(3)$$

Thus, for the samples of fixed length the absorbance is proportional to concentration. The length of the sample is usually of the size of the sample holder for liquid or gas samples. It may be noted that the Beer–Lambert law holds good at low concentrations (Fig. 9). At high concentrations the relation becomes non-linear. For such high concentrations a calibration curve may be obtained using the samples of known concentration. The comparison of measured absorbance for test sample with the calibration curve gives an estimate of the concentration of the test sample.

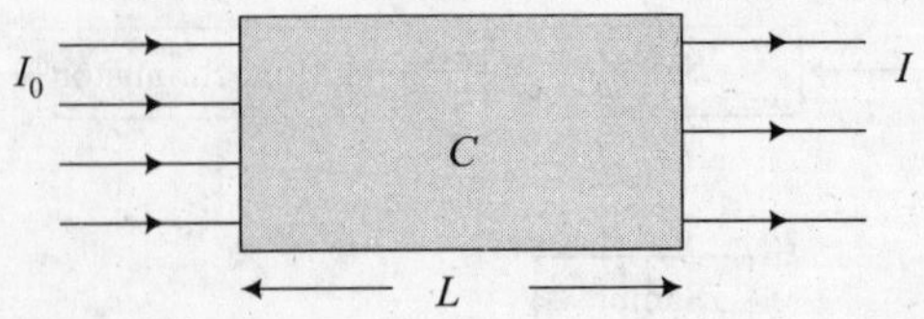

Fig. 8. Absorption of radiations by a sample of length ***L*** and concentration ***c***.

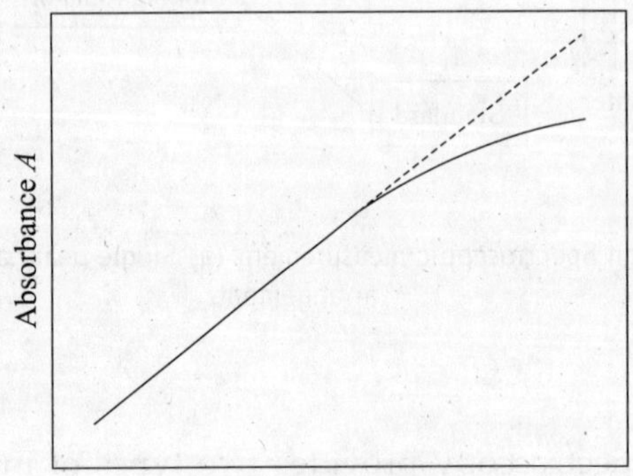

Fig. 9. Graph showing absorbance ***A*** of a sample as a function of concentration ***c***.

REFERENCES

Gracía-vargas, M.,Milla, M. and Pérez-Bustamante, J. A. (1983) Atomic absorption spectroscopy as a tool for determination if inorganic anions and organic compounds. Analyst 108, 1417.

Kirkbright, G.R. (1979) Elemental analysis of biological materials by atomic absorption spectroscopy. IAEA symp. Nuclear Activation Techniques in Life Sciences, International Atomic Energy Agency, Vienna, 71.

Welz, B. (1976) Atomic Absorption Spectroscopy.Verlag Chemic, Weinheim.

Suggested Readings

1. Narayanan P. (2000) Essentials of Biophysics, New Age International Publishers (India).
2. Hilyard N. C. and Biggin H. C. (1977) Physics for Applied Biologists, Edward Arnold (Publishers) Limited,London.
3. Welz, B.,Sperling, M. (1999) Atomic Absorption Spectroscopy 3rd ed. Wiley-VCH, Weinheim.

REVIEW QUESTIONS

Long Questions

1. What do you mean by Atomic Absorption Spectroscopy?
2. Explain the radiation sources used in spectroscopic measurements.
3. Explain the use of sample holder.
4. What is monochromator? Discuss the use of prism as a monochromator.

5. Explain the use of grating and filters in atomic absorption spectroscopic experiments.
6. Discuss the use of detectors.
7. Explain the Beer-Lambert law and discuss its utilities in spectroscopic measurements.

Multiple Choice Questions

1. An electron can make a transition from lower energy state to higher energy state through the absorption of photon of
   (a) Specific energy (b) Any energy
   (c) Specific range of energy (d) None of the above
2. Planck's law relates energy of the electromagnetic radiation (photons) to
   (a) Velocity (b) Frequency
   (c) Potential difference (d) Amplitude
3. Infrared waves have wavelength
   (a) Less than red colour (b) Less than violet colour
   (c) More than red colour (d) More than microwaves.
4. The sample holder can be constructed using
   (a) LiF (b) NaCl
   (c) KBr (d) All of the above.
5. The use of prism as a monochromator is based on
   (a) Reflection of light (b) Dispersion of light
   (c) Interference of light (d) Diffraction of light
6. Which of the following phenomenon is used for grating as monochromator?
   (a) Diffraction (b) Reflection
   (c) Refraction (d) All of the above
7. The range width of an interference filter depends on
   (a) Thickness of the dielectric plate (b) Distance from the source
   (c) Intensity of incident light (d) None of the above
8. Photo-detectors are based on
   (a) Thompson effect (b) Photo electric effect
   (c) Ohm's law (D) Coulomb's law
9. The absorbance does not depend on
   (a) Concentration of the sample.
   (b) Length of the sample
   (c) Area of cross-section of the sample
   (D) All of the above

Answers

1. (a)	2. (b)	3. (c)	4. (d)	5. (b)	6. (a)
7. (a)	8. (b)	9. (c)			

Multiple Choice Questions

(b) Any energy

(d) None of the above

(b) Frequency

(a) Less than red colour

(d) All of the above

(b) Dispersion of light

(c) Interference of light (d) Diffraction of light

(b) Reflection

(d) All of the above

(b) Distance from the source

(d) None of the above

(b) Photoelectric effect

(d) Coulomb's law

(a) Concentration of the solution

(b) Length of the sample

(c) Area of cross-section of the sample

(d) All of the above

Answer

PART–19

Bioinformatics

CHAPTER 58

Computational Biology: Concept and Applications

J. Shweta, N. Kavita and S. Trivedi*
Dept. of Zoology, J. N. Vyas University, Jodhpur–342 001
E-mail: trivedi_seema@rediffmail.com

WHAT IS BIOINFORMATICS?

There is no clear definition for computational biology/bioinformatics. Some of the views that express and explain the term suggest that the term bioinformatics was first published in 1991. One definition suggests that Bioinformatics is the application of computational techniques to the management and analysis of biological information.

Another suggests that application of information technology to the storage, management and analysis of biological information facilitated by use of computers.

The closest one can come to define the term would say that Bioinformatics (Fig. 1) is discipline that is formed by the combination of biology, computer science, and information technology. This is the filed where there is an attempt to solve biological problems by using DNA, RNA or amino acid sequences and related information with the applications of mathematical, statistical and computing methods. Therefore generally speaking to help analyze nucleic acid and protein sequences, bioinformatics key areas involve "Informatics" techniques like databases building, querying, string comparison, text search, alignment, finding patterns, data mining, graphics (surfaces, volumes), comparison and 3d matching, simulation and modelling, network modelling, significance statistics etc. (NCBI www.ncbi.nlm.nih.gov/About/primer/bioinformatics.html, Wikipedia http://en.wikipedia.org/wiki/Bioinformatics).

*Corresponding author

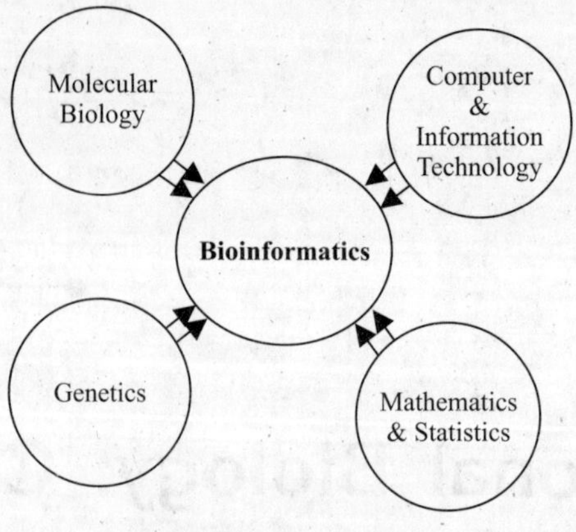

Fig.1. Bioinformatics

A comparison of molecular biology and bioinformatics can be represented in the following manner:

Central Dogma of Molecular Biology	Central Paradigm for Bioinformatics
Sequence DNA -> RNA -> Protein -> Phenotype Structure Function Processes Mechanism, Specificity, Regulation	Sequence DNA Sequence -> RNA -> Protein -> Phenotype Structure Function Processes Mechanism, Specificity, Regulation • Large Amounts of Information – Standardized – Statistical

Therefore, one can see that both involve nucleic acids or protein sequences but the main difference lies in the fact that molecular biology traditional approach involves bench work and lot of time whereas bioinformatics approach does not involve bench work at the secondary and tertiary levels and is less time consuming.

Some useful websites where one can look for more information are:

- MIT Biology Hypertextbook http://www.mit.edu
- The International Society for Computational Biology http://www.iscb.org
- National Center for Biotechnology Information (NCBI, NIH) http://www.ncbi.nlm.nih.gov
- European Bioinformatics Institute (EBI) http://www.ebi.ac.uk
- DNA Data Bank of Japan (DDBJ) http://www.ddbj.nig.ac.jp

In this article, information in field of bioinformatics is complied in sub layers where the first layer is sequencing and moves on to analysis of both nucleic acid and protein sequences.

Why Bioinformatics?

In the present times there is enormous explosion of information due to research in "Omes" and "Omics" (Genome, Transcriptome and Proteome). To be precise, information explosion is coming from different sources which are:

MOLECULAR BIOLOGY INFORMATION

Molecular biology information stems from whole genome sequencing projects, other DNA, RNA and protein sequences, functions of these sequences including transcription, translation as well as catalytic and regulatory activities. The irony with these sequences is that their alphabets are just 4 in case of nucleic acids and 20 in case of protein; but the permutation combinations of these letters form complex molecules with complex functions. For example, DNA nucleotides, AGCT may from ~ 1,000 base pairs in a small gene and ~ 3 × 109 bp in a genome (human). Similarly for proteins 20 letter alphabet (amino acids) ACDEFGHIKLMNPQRSTVWY but not BJOUXZ there are ~ 300 aa in an average protein in bacteria, there are ~ 3 × 106 known protein sequences. Further, this information explosion increases due to seeking and obtaining more information like whether these molecule are biocatalysts, whether they act as cofactor or are important for transport/storage, mechanical motion/support, immune protection, regulation of growth and differentiation etc.

Where and How is Information Generated and How Bioinformatics Helps in these Areas?

Sequencing of Nucleic Acids

Whether it is whole genome project or seeking information about any segment of nucleic acid, any wet lab would need primer to amplify sequence of interest. This is where bioinformatics in fact steps in even before any lab begins sequencing or any sequence is made available to scientists.

Primers

Primers are sequences generally about 10-20 base pairs of DNA that are complementary to sequences of either DNA sequence of interest or messenger RNA and allow either replication with help of DNA polymerases or reverse transcriptases to start copying the adjacent sequences of mRNA. In nature, DNA replication starts with help of this short sequence synthesised by RNA polymerase or "Primase" and is a short RNA molecule. However, for in vitro amplification of DNA, a complementary short molecule of DNA acts as this primer (Lodish et al. 2000) (Fig. 2).

Therefore, in laboratories primers need to be designed. Primer designing can be done manually or it can be computer-aided. Manual designing can be not only time consuming but

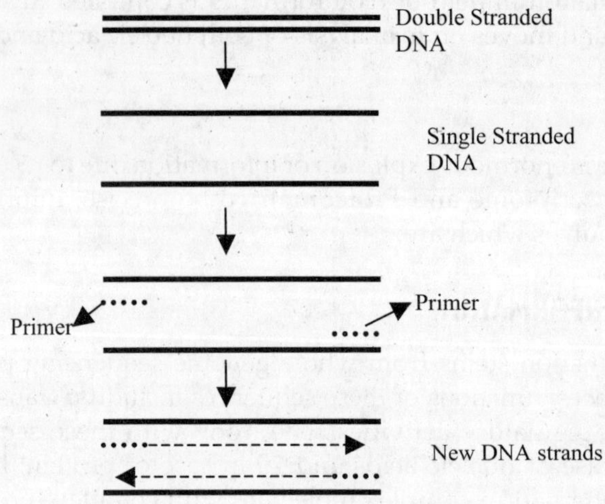

Fig. 2. Requirement for primer

also less accurate. The software tools available for primer designing help faster and efficient primer designing. Some of these tools are:

1. **Primer3** (v. 0.4.0): http://frodo.wi.mit.edu/) which picks primers from a DNA sequence.
2. **Oligo** http://www.oligo.net/
3. **Primer Premier** (http://www.premierbiosoft.com/primerdesign/index.html) is a comprehensive primer design tool for PCR, hybridization, cross species, sequencing and allele specific polymerase chain reaction.

Things to keep in mind while designing primer

When designing primers for PCR or sequencing it is necessary to make predictions about primer melting temperature (Tm) and susceptibility to form dimers with itself or other primers in the reaction. Most of the primer design softwares calculate both the Tm of the primers, as well as any undesirable pairings of primers. While designing primers, other things that should be considered are:

1. length of primer which should be around 17-28 bases
2. base composition should be 50-60% (G+C)
3. primers should end (3') in a G or C, or CG or GC
4. Tms between 55-80°C are preferred
5. 3'-ends of primers should not be complementary (i.e., base pair), as otherwise primer dimers will be synthesised preferentially to any other product.

Shotgun Sequencing

Although many viruses had already been sequenced, the first published genome sequence was of *Hemophilus influenzae* (Fleischmann *et al.*, 1995) by The Institute of Genetics Research (TIGR, at http://www.tigr.org/) started by Craig Venter and assisted by microbiologist Hamilton Smith. They used automated reading of DNA sequencing gels of random fragments of chromosome without knowledge of their location in the whole chromosome which was finally done by computer analysis. In this approach an organism's genome is shred into smaller fragments of DNA. Sequences are then entered in a computer database and compared with each other to find overlaps and produce ordered contigs of two or more sequences. The 'whole genome shotgun' method is applied to the entire genome all at once, while the 'hierarchical shotgun' method is applied to large, overlapping DNA fragments of known location in the genome (Fig. 3).

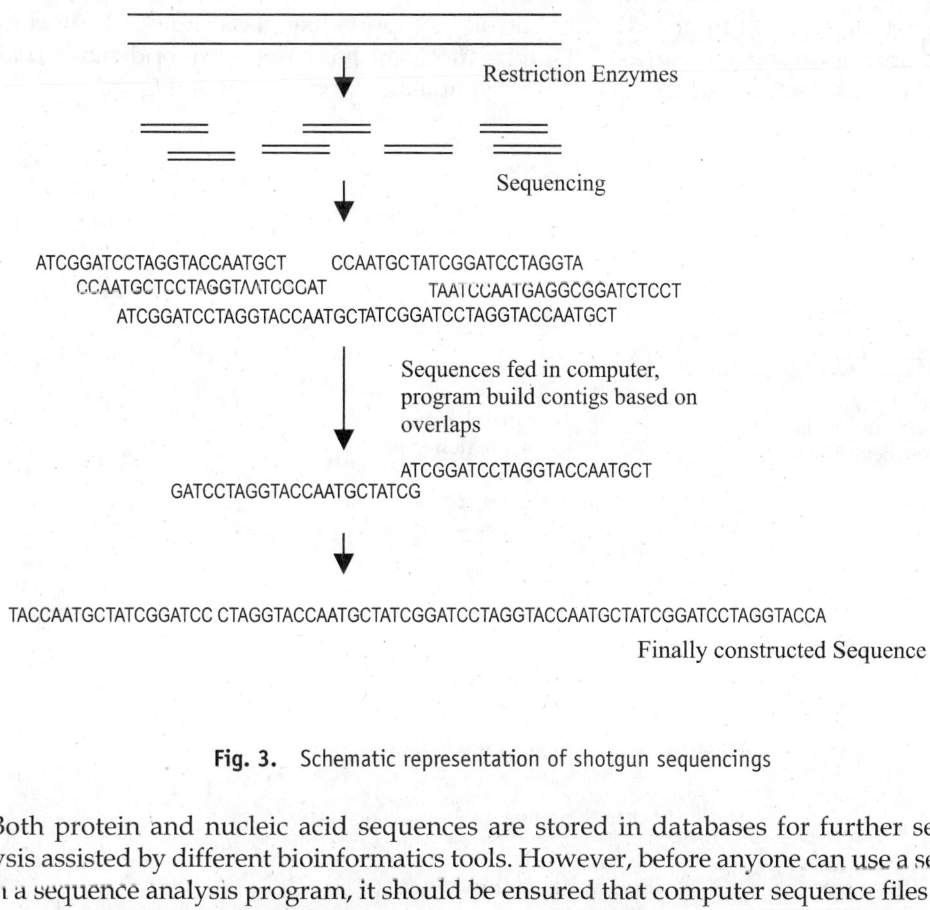

Fig. 3. Schematic representation of shotgun sequencings

Both protein and nucleic acid sequences are stored in databases for further sequence analysis assisted by different bioinformatics tools. However, before anyone can use a sequence file in a sequence analysis program, it should be ensured that computer sequence files contain only sequence characters and not special characters used by text editors. This is because a word processor can introduce undesirable changes if one is not careful to work with text or so-called ASCII files. In these files computers store sequence information as simple rows of sequence characters called strings, which are similar to the sequences seen on the computer screen. Each character is stored in binary code in the smallest unit of memory, called a byte. Each byte comprises 8 bits, with each bit having a possible value of 0 or 1, producing 255 possible combinations. These combinations have a specific definition, called their ASCII equivalent. A file with only ASCII characters is called an ASCII file (Mount 2004).

Another Application of Primers/probes

Primer design or probe designing tools can help one select the best possible probe for microarray studies.

What is DNA Microarray?

DNA microarray is a technology that helps measure level of mRNA. For microarrays, sequences (DNA, cDNA, or oligonucleotides) from thousands of different genes are immobilized, or attached in orderly way on solid supports like glass slides, silicon chips or nylon membranes. A scientist can extract information which is based on hybridization probing (fluorescently labeled nucleic acid) to identify complementary sequences (Fig. 4).

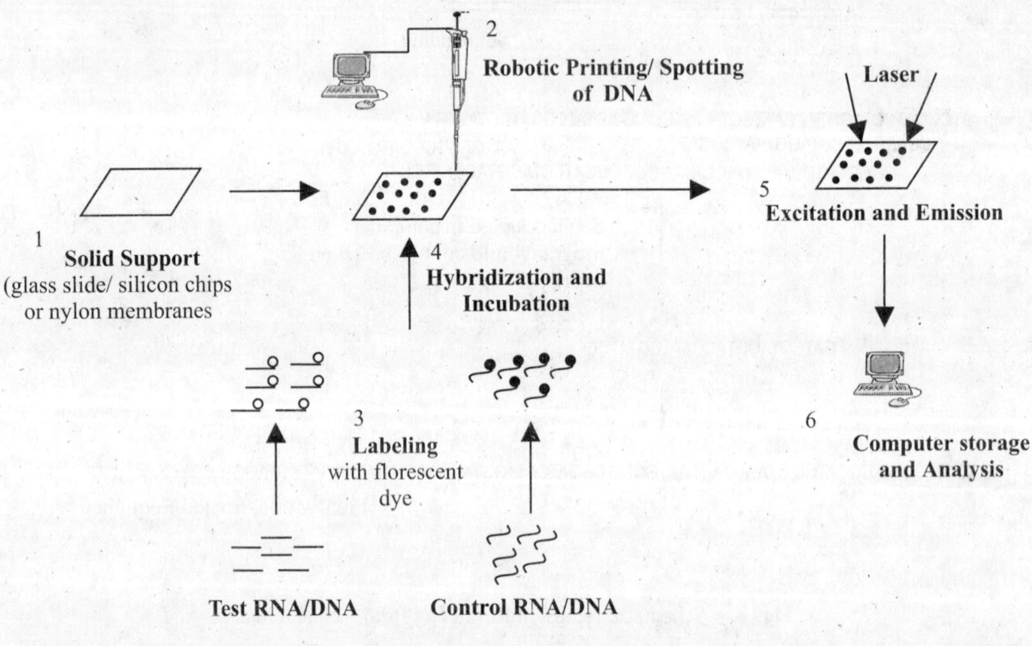

Fig. 4. Schematic representation of microarray study

The process can be briefly described in the following steps:

1. Preparation of DNA chip using chosen target DNAs.
2. Hybridization mixture containing fluorescently labelled probed is incubated with DNA chip.
3. Microarray is placed in a "reader" or "scanner".
4. Detection of bound cDNA is done by using laser technology.
5. Data in form of digital image of the array is stored in a computer.
6. Data analysis is done by using computational methods.

There are three main ways for performing microarray analyses.

- cDNA arrays (generic, multiple manufacturers)
- Oligonucleotide arrays (genechips) (Affymetrix)
- cDNA membranes (radioactive detection).

Among these methods Affimix is expensive because it uses propriety chips but it has advantage of having more than 700 thousand probes. The three options basically use the same methodology for analysis which involve designing of a 25 -70 base probes to represent gene(s). These probes are then hybridized with the nucleic acids or genome fragments of interest. Depending on the method being used either the probe may be labelled or the nucleic acid or genome fragment may be labelled (in this case this would be called a probe). Similarly, again depending on the method, either the probe may be fixed or the sequences of interest would be fixed.

Microarray data analysis is then done with help of bioinformatics tools. With help of microarry studies one can detect patterns, find linkage between gene expression data and gene sequence, their function and metabolic pathways. One can also discover common sequences in co-regulated genes. Microarray has many applications in both basic and clinical research where it helps in determining the role a gene plays in a pathway, disease, diagnostics, pharmacology, etc. (Lashkari et al., 1997 and NCBI).

After Sequencing and Deposition in Database what next?

Once the whole genome is sequenced or the draft genome is ready, there is a need for annotation which involves gene predictions, positions of exons and introns, protein coding and other RNA, etc. The following sections touch upon some of the bioinformatics approaches to solve these issues.

Bioinformatics and Gene Discovery/ Gene prediction

Among gene prediction software some are:

- GENSCAN (Burge & Karlin, 1997)
- MZEF (Zhang, 1997)

Most softwares take care of certain criteria while predicting any sequences as a gene (Fig. 5).

These criteria are:

- Detection of Start codons
- Detection of potential promotor elements
- Detection of repetitive sequences (mostly untranslated)
- Homology to known genes of related organisms.

All programs use dynamic programming for detection of the optimal solution and the methods of gene prediction can be *Ab initio*, use general knowledge of gene structure. However current best methods are all based on hidden Markov models.

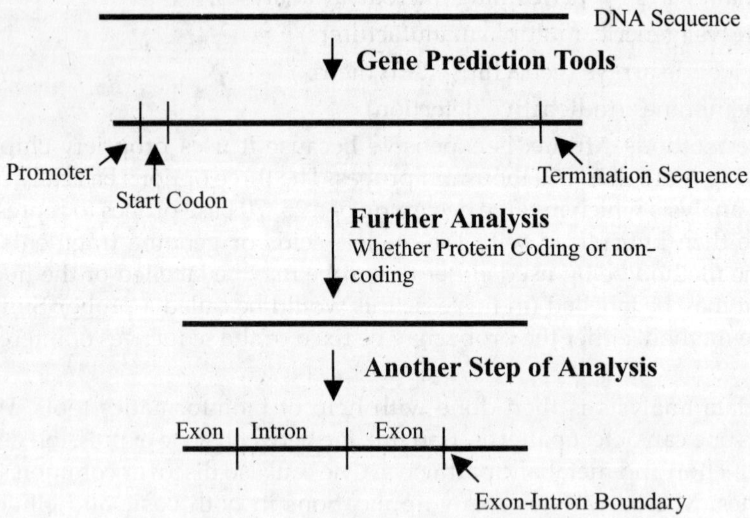

Fig. 5. Schematic diagram for gene search criteria

Similarity Search and Sequence Alignment

As described by Mount 2004, dot matrix method for comparing amino acid and nucleotide sequences was developed by Gibbs A.J. and G.A. McIntyre in 1970 described. Needleman and Wunsch (1970) developed sequence optimal global alignment algorithms called "dynamic programming" that took care of matched pairs, mismatched pairs, or extra amino acids (insertion or deletion). This was because there was a need to know whether a particular sequence of interest (which might be DNA or RNA or protein) has any similarity with any other known/unknown (possibly predicted) sequence or set of sequences. Similarly, one may want to know whether a set of sequences are similar and whether there is any conservation of sequence. In order to seek answers to these queries one can use similarity search and sequence alignment tools which can be divided into (1) Pairwise alignments and (2) Multiple sequence alignments.

Some of the sequence alignment tools are:

1. Local alignment or Smith-Waterman (Smith and Waterman, 1981; Waterman and Eggert, 1987)

   This is a modification of the Needleman-Wunsch algorithm to locate regions in DNA and protein sequences that align well. This algorithm also adjusts insertions or deletions of any size that may reflect evolutionary changes.
2. FASTA (Wilbur and Lipman, 1983; Lipman and Pearson, 1985)
3. BLAST (Basic Local Alignment Search Tool) (Altschul *et al.*, 1990)

The central idea of the BLAST algorithm is that a statistically significant alignment is likely to contain a high scoring pair of aligned words. The BLAST program has been written to allow rapid comparison of a new gene sequence with the 100s of 1000s of gene sequences in data bases.

BLAST has several options like:

- **blastp** compares an amino acid query sequence against a protein sequence database.
- **blastn** compares a nucleotide query sequence against a nucleotide sequence database.
- **blastx** compares a nucleotide query sequence translated in all reading frames against a protein sequence database.
- **tblastn** compares a protein query sequence against a nucleotide sequence database dynamically translated in all reading frames.
- **tblastx** compares the six-frame translations of a nucleotide query sequence against the six-frame translations of a nucleotide sequence database. Please note that tblastx program cannot be used with the nr database on the BLAST Web page.

BLAST has the option of filtering out low complex regions (called masking). This is because repetitive sequences increase the chance of a match during a database search.

Example of BLAST search:

>gb | EU031464.1 | Apple scar skin viroid variant ASSVd_AM10, complete genome

Length=330

Score = 111 bits (60), Expect = 1e-22

Identities = 60/60 (100%), Gaps = 0/60 (0%)

Strand=Plus/Plus

```
Query  1    GGTAAACACCGTGCGGTTCCTGTGGTTCGCCCCGCCAACGCAGATAGATAAAGAAAACGA  60
            ||||||||||||||||||||||||||||||||||||||||||||||||||||||||||||
Sbjct  1    GGTAAACACCGTGCGGTTCCTGTGGTTCGCCCCGCCAACGCAGATAGATAAAGAAAACGA  60
```

Meaning of E-Value (Expectation value) in BLAST Database Searches

Expect value of 10 for a match means that in a database of current size, one might expect to see 10 matches with a similar or better score, simply by chance alone. E-value is the most commonly used threshold which ranges between 0 to any value. Only those hits with E-values smaller than the set threshold will be reported in the output. Increasing the E-value enables one to see biologically significant but statistically insignificant matches.

Meaning of Bit Score in BLAST Database Searches

Bit score represents sequence comparisons that are independent of the size of the database. Score is opposite to E-value i.e. higher the score lower would be the E-value and would be considered a good match. More details of scoring can be found at NCBI.

PSI-BLAST

A position specific scoring matrix (PSSM) is constructed automatically from initial BLAST search. Normal E value is used. The PSSM is created as the new scoring matrix for a second BLAST search. Low E value is used E =.001. This helps one obtain distantly related sequences and could find important residues that provide function or structure.

Multiple sequence alignment (MSA)

Multiple sequence alignment matrix representations are:

(a) Profile and

(b) Position-Specific Scoring Matrix (PSSM).

Statistical and non-statistical methods like neural networks, hidden Markov models, and the expectation maximization and Gibbs sampling methods in MSA algorithms can help one find conserved regions of a structural or functional domain for protein sequences. Similarly these algorithms may help one find promoter regions or other binding sites for protein in a DNA sequences.

CLUSTAL W (Thompson et al., 1994) is one of the most commonly used programmes for Multiple Sequence Alignment.

Example:

If one has following four sequences and wishes to know the extent of their similarity then the following sequences are arranged in the following manner and submitted to the software.

>human
AGGTCCGTATCGGTCAACGTCCTCCGTATCGGTGTCAACGTCCTCCGTA
>rat
TCAACGTCCTCCTGAGGTCCGTATCGGTCAACGTCCTCCGTAGTATCGG
>chimp
GTATCGGTCCTGAGGTCCGTATCGGTCAACGTCCTCTATCGGGAGGTCCGTA
>mouse
CAACGTCCTCCGTAGTATCGTCAACGTCCTCCTGAGGTCCGTATCGGTGCAAC

Once submitted to multiple sequence alignment programme, the result may look like this:

```
rat           TCAACGTCCTCCTGAGGTCCGTATCGGTCAACGTCCTCCG-----TAGTATCGG------ 49
mouse         -CAACGTCCTCCG-----TAGTATCG-TCAACGTCCTCCTGAGGTCCGTATCGGTGCAAC 53
human         --------------AGGTCCGTATCGGTCAACGTCCTCCGTAT--CGGTGTCAACGTCCT 44
chimp         --GTATCGGTCCTGAGGTCCGTATCGGTCAACGTCCTCTATCG--GGAGGTCCGTA---- 52
                                  ****** ***********             **
```

These alignments can hep in providing information on conservation of sequences. Conservation of sequences can also be helpful in drawing phylogeny inferences.

Alignment scoring and the alignment can be read as follows:

- the 1st alignment: highly significant
- the 2nd: plausible
- the 3rd: spurious
- distinguish by alignment score
- similarities increase score
- mismatches decrease score
- gaps decrease score

Caution!!! Beware of aligning apples and oranges and grapefruit! Or in other words one should be careful about not aligning parologous versus orthologous; genomic versus cDNA; mature versus precursor sequences.

Phylogenetic Analysis and Bioinformatics

Basis of evolutionary biology in present times is the variations within a family of related nucleic acid or protein sequences. Bioinformatics tools help one find orthologes or paraloges (gene duplication events) even if each sequence that is being compared has a slightly altered or new function. The first step in phylogenetic analysis is sequence alignment and alignment scores. Most closely related sequence pairs have highest score and may be placed in outer branches of an evolutionary tree and next most-alike sequence will be represented by the next branch down on the tree. Since different methods of analysis may result in several trees with almost the same results, statistically significant sequence variation should be considered.

Most commonly used methods are:

(1) Neighbor-joining (Saitou and Nei 1987) or

(2) Maximum parsimony tree (Felsenstein 1988).

These analysis have helped one answer questions like who are our closest living relatives? Phylogenetic trees have also indicated consistently so far a recent origin of human beings in Africa.

A phylogeny tree may look like the following Fig. 6:

Fig. 6

Structural Bioinformatics

RNA Structure

As described by Mount (2004), prediction of RNA began as early as 1971 (Tinoco et al.) based on estimates of the free energy associated with stacked base pairs and destabilizing effects of loops. Nussinov and Jacobson (1980) devised a fast computer method improved by Zuker and Stiegler (1981) by considering molecular constraints and thermodynamic information.

There are several bioinformatics tools which can help one visualise RNA sequence fold into predictive secondary (Fig. 7) or active 3-dimensional structure and possibly function from the structure. The basic principle that these tools adapt are based on Watson Crick Base Pairs where RNA bases A, C, G, U can form canonical base pairs i.e. A:U, G:C and there can be "wobble" pairing like G:U. Further, the algorithms also take care of the fact that bases can only pair with one other base.

Besides base pairing, other facts that are considered are regarding backbone conformation. The algorithms take care of physical and chemical properties of the molecules involved in back bone formation. Consideration of facts like the ribose ring is never flat, the glycosidic bond connects ribose sugar to the base, and rotations of the sugar-phosphate bond etc. are integral

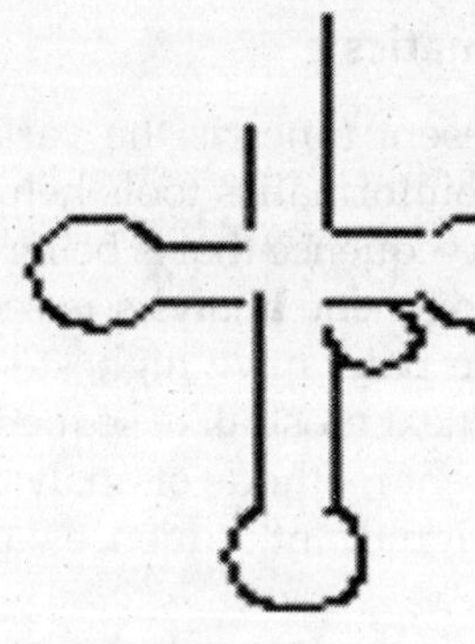

Fig. 7. Schematic representation of secondary structure of tRNA

parts of algorithms. For prediction of RNA secondary structure dynamic programming algorithm uses base pair maximization, energy minimization, thermodynamic stability, no pseudknots due to algorithm limitations, attempts to maximize the score taking into account thermodynamics, estimated structure using experimental techniques etc. The hypothesis behind these algorithms is that "Most Stable is the Most likely". Some of the software that use these algorithms are: MFOLD and ViennaRNA.

PROTEINS AND BIOINFORMATICS

Most predicted proteins in any complete genome annotation are due to applications of bioinformatics. For example, 74% of human proteins were predicted because they had matches to other known proteins or that have EST evidence. Some of these hit 8100 fly proteins (61%), 7850 worm proteins (43%), 2800 yeast (46%). Bioinformatics approach helped formation of 1308 orthologous groups. The following sections briefly touch upon how bioinformatics has been useful in analysis of protein structure and function.

Protein Structure Prediction

X-ray crystallography and nuclear magnetic resonance (NMR) methods used for solving protein structures are time-consuming and highly specialized procedures. To predict secondary structure, amino acids associated with different structure types were determined followed by scanning sequences of proteins whose structures were not known to determine whether the amino acids in each region were those often associated with one type of structure (Chou and Fasman 1978 and Garnier *et al.*, 1978). This method was improvised later by development of computational methods that can predict levels of protein structure organization (secondary, tertiary and quaternary structure). Protein with different sequences can have same three dimensional structures. Therefore, three dimensional structures based on properties of amino acid and not just the similarity of sequence has been devised (Bowie et al. 1991). Bairoch *et al.*, 1997 developed another method for predicting activity of an unknown protein based on its sequence. Proteins that had a common activity were compared. This comparison helped in finding the pattern of amino acids responsible for the said activity. Collections of these compared patterns were deposited into a database called PROSITE (http://www.expasy.ch/prosite). Using these databases one can scan unknown sequences that may have same patterns. Henikoff and Henikoff 1992 discovered additional patterns in the aligned sequences called BLOCKS (see http://www.blocks.fhcrc.org/). This was done by examining alignments of the protein sequences that make up each MOTIF. The algorithms used for these findings are called amino acid substitution matrices or BLOSUM matrices. This is similar to Margaret Dayhoff's PAM matrices. BLOSUM62 is most commonly used for aligning and searching databases for similar protein sequences.

Protein Databases that provide either online help or downloadable tools for protein study are:

- ExPASy
- Swiss-Prot
- TrEMBL

These databases provide annotations such as the description of the function of a protein, its domains structure, variants, etc. (Fig. 8).

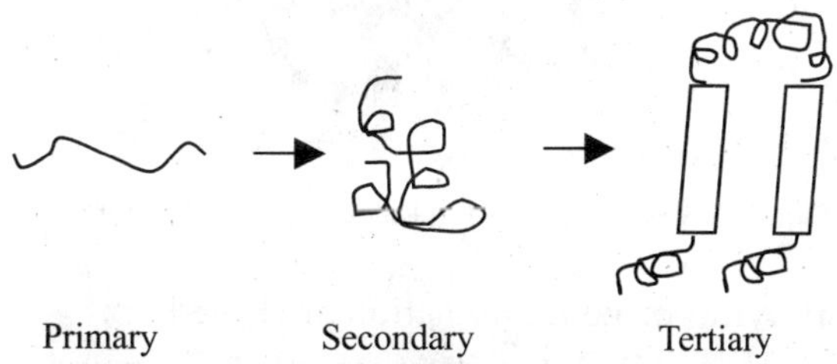

Fig. 8. Schematic representation of different levels of protein structure

Further, there is a minimal level of redundancy and high level of integration of databases with each other. Protein families can also be found by using databases like Pfam. The basic assumptions these databases adopt are that large number of different proteins can be grouped on the basis of similarities into a limited number of families. Proteins belonging to a particular family generally share functional attributes and are derived from a common ancestor. Many of these databases also provide information about protein domains. Besides the aforementioned databases some other databases that can help search protein domains are BLOCKS and PredictProtein (http://cubic.bioc.columbia.edu/predictprotein/predictprotein.html).

The standard IUPAC one-letter codes used by these databases are as follows:

- 'x' : any amino acid
- '[]' : residues allowed at the position
- '{}' : residues forbidden at the position.
- '()' : repetition of a pattern element are indicated in parenthesis. X(n) or X(n,m) to indicate the number or range of repetition
- '-' : separates each pattern element
- '<' : indicated a N-terminal restriction of the pattern
- '>' : indicated a C-terminal restriction of the pattern
- '.' : the period ends the pattern

Prosite Syntax – Example-1 [AC]-x-v-x(4)-{ED} would mean [Ala or Cys]-any-val-any-any-any-any-any but Glu or Asp.

Example-2

<A-x-[ST](2)-x(0,1)-v would mean N-terminus-Ala-any-[Ser or Thr]-[Ser or Thr]-(any or none)-val

Protein-protein interaction

Protein-protein interactions can (Fig. 9) be recognized from genome sequences by purely computational means.

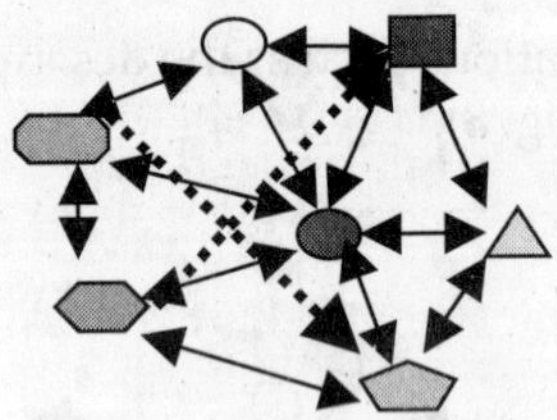

Fig. 9. Schematic representation of protein interaction

Databases that help in studying protein interaction are based on:

(1) Collection of protein-protein interaction information from research reports or

(2) Prediction of protein-protein interaction.

Pathway Studies

Pathways can be defined as a modular unit of interacting molecules to fulfil a cellular function. It is usually represented by a 2-D diagram with characteristic symbols linking the protein and non-protein entities. Generally pathway relationship is manually constructed by several scientists and static data is maintained. Type of molecular network are:

- biochemical pathways
- metabolic pathways
- signal transduction cascades
- genetic networks
- drug metabolism pathways
- Kyoto Encyclopedia of Genes and Genomes (KEGG) http://www.genome.ad.jp/kegg/
- EcoCyc http://ecocyc.pangeasystems.com/ecocyc/

For Example a reference pathway for Oxidative Phosphorylation in Homo sapiens at KEGG (http://www.genome.ad.jp/kegg/pathway/map/map00190.html) may look like this:

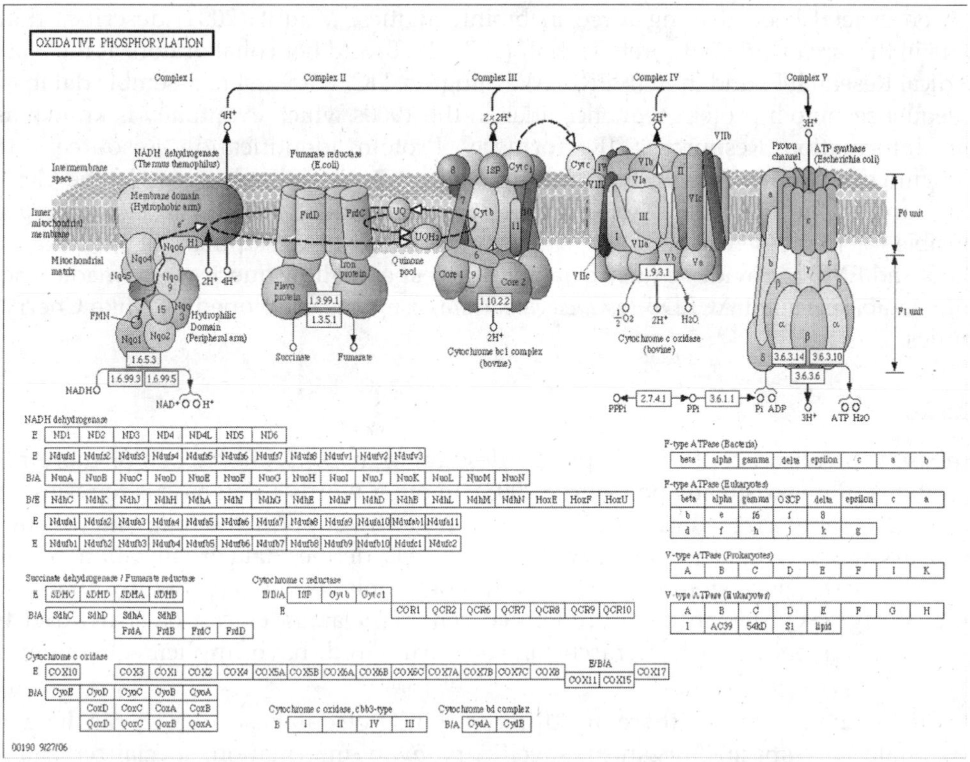

NEED TO ORGANISE EXPLOSION OF INFORMATION

The concepts and applications of some of computational biology or bioinformatics as listed above give an indication of the fact that there is tremendous explosion of information not only due to deposition of data in the databanks but also due to further analysis of these data with help of bioinformatics tools. All the data need to be organised in a manner that would not only be easily accessible to users but also comprehensible for development of knowledge databases which can further be used for understanding a research problem. However, there are difficulties in organising the information explosion which are:

- Redundancy and Multiplicity
- Different sequences can have the same structure
- Organism has many similar genes
- Single gene may have multiple functions
- Genes and proteins function in genetic and regulatory pathways.

Therefore, there are ways in which organization of all this information is done so that one can make sense of it. Organising data involves the following steps:

Storing and Analyzing Large-Scale Information

Exponential growth of data matched by development of computer technology CPU vs disk & net. Both the increase in computer speed and the ability to store large amounts of information have been crucial been driving force in bioinformatics. Mount (2004) describes that the pioneers in this area were Margaret Dayhoff (1972, 1978) and her collaborators at the National Biomedical Research Foundation (NBRF), Washington, DC, who first to assemble databases of these sequences into a protein sequence atlas in the 1960s which eventually is known as the Protein Information Resource (PIR, formerly Protein Identification Resource; http://watson.gmu.edu:8080/pirwww/index.html). Walter Goad and colleagues in the GenBank database and at the European Molecular Biology Laboratory (EMBL) in Heidelberg were responsible for first assembly of DNA sequence databases. Development of Web pages like GenBank and EMBL provide sequence database access and allow queries to be made. The first genome database, called ACEDB (a *C. elegans* database) was developed by Mike Cherry and colleagues.

Databases

A database is a collection of data, typically describing the activities of one or more related organizations (Ramakrishnan and Gehrke 2003). In other words a database is repository for a collection of computerized data files. Databases typically support the following operations: retrieval, insertion, updating, deletion, database models, defines data organization, relational, entities and relationships stored in tables. Some commonly used databases are: **Oracle, DB2, MySQL, PostgreSQL** which have predefined schema. Databases can be structured types (arrays, lists, sequences, etc.), hierarchical/semistructured, have implicit schema and are flexible.

In filed of bioinformatic there are different kinds of databases which can be general purpose, data type specific, organism specific, pathway information, specialized data, etc. Some bioinformatics databases are described as follows:

DNA and Protein Sequence Data Bases

Three databanks exchange data on a daily basis where data can be submitted and accessed at any of the following locations.

NCBI (Genebank) www.ncbi.nlm.nih.gov/Genbank/ Largest repository of bioinformatics information related to DNA, proteins, publications, data mining tools, etc.

EMBL www.ebi.ac.uk/embl/

DNA DataBank of Japan (DDBJ) www.nig.ac.jp/home.html

Organism Databases

Rat Genome Database http://rgd.mcw.edu/ Rat specific information

Pathway Databases

Kyoto Encyclopedia of Genes and Genomes (KEGG) http://www.genome.ad.jp/kegg/ Genes, genomes, metabolic pathways

Sequence Region Databases

UTRdb http://bighost.area.ba.cnr.it/BIG/UTRHome/ Collection of 3′ and 5′ UTR
While developing a database one has to take care of certain things which are:

Databases should provide

- Computer assisted analysis and visualization of integrated data
- Multi-tool in nature
- Geographically distributed, collection of cooperating tools
- Visualization and analysis should integrate both simulation and experimental data, to facilitate cross validation and comparisons
- Dissemination via web services
- Integrating data from molecular biology information because understanding the function of genomes requires integration of many diverse and complex types of information which can be:
  - (a) Metabolic pathways
  - (b) Regulatory networks
  - (c) Whole organism physiology
  - (d) Evolution, phylogeny
  - (e) Environment, ecology
  - (f) Literature (MEDLINE)
- Algorithmic developments which are helpful in:
  - (a) data storage
  - (b) data retrieval
  - (c) data analysis

Biological databanks and databases

There are a diversity of biological data which provide information regarding:

- primary sequences
- 3D structures
- functional data
- Database entry usually required for publication
- Sequences
- Structures
- Database entry may replace primary publication
- genomic approaches

Formats

Data is stored/presented in a variety of formats some of which are:

- FASTA

- GenBank
- SwissProt
- ASN.1
- XML (eXtensible Markup Language)

Each format is highly specific and example can be obtained from the provided links. However, to give an example of FASTA (or Pearson) format; it always begins with comment line followed by sequence data which does not have annotation but only the sequence.

>Example name (sequence or organism)
AATTCGATCGGTAAGGCTACGTTAAAATTCGATCGGTAAGGCTACGTTAAAATTCGATCGGTAAGGCTAC
CGGTAAGGCTACAATTCTACGTTAAAATTCGATCGATCGGTAAGGCGGTAAGGCTACGTTAAAATTCGAT

This example shows that FASTA format comment line always begins with '>' sign followed by either the sequence name or accession number or organism name. The comment line is followed by the actual sequence which can be either DNA, RNA or protein sequence.

Since there are many databases and many tools that handle different formats, it might be difficult to convert formats of large files. Fortunately, software exists to transform between many of the formats. Two tools are ReadSeq http://searchlauncher.bcm.tmc.edu/seq-util/Options/readseq.html and SEQIO http://www.cs.ucdavis.edu/~gusfield/seqio.html

Applications of Bioinformatics

- Large scale genome projects like sequencing of the Human Genome
- Identifying the location of clones and markers on the chromosome by genetic linkage analysis and physical mapping
- Assembling clone sequence reads into large (eventually complete) genome sequences
- Annotation of the genome
- Gene discovery
- Identifying coding regions in genomic DNA by database searching and other methods
- Finding homologe.
- Function assignment
- Protein interactions
- Understanding how proteins bind other molecules
- Docking and structure modelling
- Designing inhibitors
- Find proteins that are better suited for treating with drugs
- Searching for relationships and correlations in the information
- Genome comparison is done by comparing different complete genomes to infer evolutionary history and genome rearrangements
- Designing drugs
- Optimization of known therapies on individuals.

SUMMARY

Form the manuscript it is evident that bioinformatics or computational biology tools have several applications to understand molecular biology. Though there is no point which can be said to be the start of bioinformatics but this manuscript has tried ordering the information on use of bioinformatics. The start point of application of bioinformatics in this manuscript is primer designing. However, it depends on the aim and hypothesis of research problem. The start point for a person working on protein sequences could be determining the secondary, tertiary or 3D structure of protein or any other thing related to it. Therefore, the start point for this researcher may be just looking at the structure and function which have nothing to do with primer designing. The aforementioned list encompasses some of the applications of bioinformatics. In brief, computational biology or bioinformatics can be applied for both basic studies as well as for the industrial, agricultural or medical uses.

REFERENCES

Altschul, S.F., Gish W., Miller, W., Myers, E.W., and Lipman, D.J. (1990) Basic local alignment search tool. *J. Mol. Biol.* 215: 403–410.

Bairoch, A. (1991) PROSITE: A dictionary of sites and patterns in proteins. *Nucleic Acids Res.* (suppl.) 19: 2241–2245.

Bowie, J.U., Lüthy, R. and Eisenberg, D. (1991) A method to identify protein sequences that fold into a known three-dimensional structure. *Science.* 253: 164–170.

Burge, C. and Karlin, S. (1997) Prediction of complete gene structures in human genomic DNA. *J. Mol. Biol.* 268: 78-94.

Chou, P.Y. and Fasman, G.D. (1978) Prediction of the secondary structure of proteins from their amino acid sequence. *Adv. Enzymol. Relat. Areas Mol. Biol.* 47: 45–147.

Felsenstein, J. (1988) Phylogenies from molecular sequences: Inference and reliability. *Annual Review of Genetics.* 22: 521-565.

Fleischmann, R.D., Adams, M.D., White, O., Clayton, R.A., Kirkness, E.F., Kerlavage, A.R., Bult, C.J., Tomb, J.-F., Dougherty, B.A., Merrick, J.M., et al. (1995) Whole-genome random sequencing and assembly of *Haemophilus influenzae. Rd. Science.* 269: 496–512.

Garnier, J., Osguthorpe, D.J. and Robson, B. (1978) Analysis of the accuracy and implications of simple methods for predicting the secondary structure of globular proteins. *J. Mol. Biol.* 120: 97–120.

Henikoff, S. and Henikoff, J.G. (1992) Amino acid substitution matrices from protein blocks. *Proc. Natl. Acad. Sci. USA* 89: 10915-10919.

Lashkari, D.A., DeRisi, J.L., McCusker, J.H., Namath, A.F., Gentile, C., Hwang, S.Y., Brown, P.O. and Davis, R.W. (1997) Yeast microarrays for genome wide parallel genetic and gene expression analysis. *Proc. Natl. Acad. Sci. USA* 94: 13057-13062.

Lipman, D.J. and Pearson, W.R. (1985) Rapid and sensitive protein similarity searches. *Science* 227: 1435–1441.

Lodish, H., Berk, A., Zipursky, S.L., Matsudaira, P., Baltimore, D. and Darnell, J. (2000) Molecular Cell Biology. W. H. Freeman & Co

Mount, D.W. (2004) Bioinformatics: Sequence and Genome Analysis. Second Edition. Cold Spring Harbor Laboratory Press.

Ramakrishnan, R. and Gehrke, J. (2003) Database Management Systems. McGraw-Hill Professional.

Saitou, N. and Nei, M. (1987) The neighbor-joining method: a new method for reconstructing phylogenetic trees. *Mol. Biol. Evol.* 4(4): 406-425.

Smith, T.F. and Waterman, M.S. (1981). Identification of Common Molecular Subsequences. *Journal of Molecular Biology* 147: 195-197.

Thompson, J.D., Higgins, D.G. and Gibson, T.J. (1994) CLUSTAL W: Improving the sensitivity of progressive multiple sequence alignment through sequence weighting, position-specific gap penalties and weight matrix choice. *Nucleic Acids Res.* 22(22): 4673-4680.

Waterman, M.S. and Eggert, M. (1987). A new algorithm for best subsequence alignments with application to trna-rrna comparisons. *J. Mol. Biol.* 197: 723-728.

Wilbur, W.J. and Lipman, D.J. (1983) Rapid similarity searches of nucleic acid and protein data banks. *Proc. Natl. Acad. Sci.* 80: 726–730.

Zhang, M.Q. (1997) Identification of protein coding regions in the human genome by quadratic discriminant analysis. *Proc. Natl. Acad. Sci. USA.* 94: 565-568.

Suggested Readings

1. Adams M.D., Fields C., Venter J.C. (Eds.), Automated DNA sequencing and analysis. Academic Press, London (1994).
2. Baxevanis A., Ouellette F.B.F. (Eds.), Bioinformatics: a practical guide to the analysis of genes and proteins, John Wiley and Sons, New York (1998).
3. Bishop M.J., Rawlings C.J. (Eds.), Nucleic acid and protein sequence analysis. A practical approach. IRL Press, Oxford (1987).
4. Bishop M.J., Rawlings C.J. (Eds.), DNA and protein sequence analysis. A Practical approach. IRL Press, Oxford (1997).
5. Bishop M.J. (Ed.), Guide to human genome computing. Second edition. Academic Press, London (1998). Colwell R.R., Swartz D.G., McDonald M.T. (Eds.)
6. Lesk A.M. (Ed.), Computational molecular biology. Sources and methods for sequence analysis. Oxford University Press, Oxford (1988).

REVIEW QUESTIONS

Long Questions

1. How is bioinformatics or computational biology helpful in whole genome sequencing?
2. What is microarray? How are bioinformatics tools applied for the study?
3. What is a database and different types of important biological databases?
4. What is BLAST and how is it done?
5. What is multiple sequence alignement?

Multiple Choice Questions

1. MSA is full form of:

   (a) Multicoloured Sequence Analysis (b) Multiple Sequence Alignment
   (c) Mutational Sequence Analysis (d) Multilayered Sequence Alignment

2. BLAST is full form of
   (a) Basic Local Alignment Search Tool
   (b) Best Local Alignment Search Tool
   (c) Basis of Local Attachment Search Tool
   (d) Basic Level Alignment Search Tool
3. Shotgun sequencing is
   (a) Sequencing of a gun (b) Whole genome sequencing
   (c) Sequencing with help of gun (d) Sequencing of oligonucleotide
4. ORF is:
   (a) Open reading frame (b) Opening of the frame
   (c) Orange red florescence (d) Orange red fixing
5. Biological database is a place where
   (a) only DNA sequences are stored
   (b) only RNA sequences are stored
   (c) Both DNA and RNA sequences are stored
   (d) Information about sequence, function, structure etc. of DNA, RNA and proteins is stored
6. Biochemical pathways means where
   (a) biochemicals take a walk (b) biochemicals make a road
   (c) path that is made of biochemicals (d) biochemical reactions can be represented
7. What is full form of DDBJ
   (a) DNA DataBank of Japan (b) DNA DataBase of Japan
   (c) DNA DataBank of Jakarta (d) DNA DataBase of Jakarta
8. What is the full form of PSSM
   (a) Position Specific Scoring Matrix (b) Profile Specific Scoring Matrix
   (c) Position Scoring Specific Matrix (d) Profile Scoring Specific Matrix
9. tBLASTn compares protein query sequence with:
   (a) Nucleotide sequence database (b) Protein sequence database
   (c) both a and b (d) None of the above
10. Watson-Crick pairing means
    (a) purine never pairs with pyrimidine
    (b) purine pairs with purine
    (c) pyrimidine pairs with pyrimidine
    (d) C pairs with G and A pairs with T or U

Answers:

1. (b), 2. (a), 3. (b), 4. (a), 5. (d), 6. (d),
7. (a), 8. (a), 9. (a), 10. (a).

Index